Contents

ASM Specialty Handbook®

Heat-Resistant Materials

Edited by
J.R. Davis
Davis & Associates

Prepared under the direction of the
ASM International Handbook Committee

Scott D. Henry, Assistant Director of Reference Publications
Bonnie R. Sanders, Chief Copy Editor
Grace M. Davidson, Manager of Book Production
Randall L. Boring and Kathleen S. Dragolich, Production Coordinators

William W. Scott, Jr., Director of Technical Publications

First printing, October 1997

Library of Congress Cataloging-in-Publication Data

Heat-Resistant Materials/edited by J.R. Davis.

Prepared under the direction of the ASM International Handbook Committee
p. cm.—(ASM specialty handbook)

Includes bibliographical references and index.
1. Heat resistant materials—Handbooks, manuals, etc.
I. Davis, J.R. (Joseph R.)
II. Series
TA418.26.H42 1997 97-30681
620.1'1217—dc21
ISBN: 0-87170-596-6
SAN: 204-7586

ASM International®
Materials Park, OH 44073-0002

Printed in the United States of America

Preface

The evolution of heat-resistant materials closely parallels advances in superalloy technology. The development of superalloys has been motivated almost entirely by requirements to improve the efficiency, reliability, and operating life of gas turbines used for aerospace, electricity generation, gas/oil pumping, and marine propulsion applications. The differing requirements in specific parts of the engine and the different operating conditions of the various types of gas turbines have led to the development of a wide range of nickel-base superalloys with individual balances of high-temperature creep resistance, corrosion resistance, yield strength, fatigue strength, and fracture toughness. However, all of these materials have evolved from the Ni-20Cr-2.25Ti-1Al precipitation-strengthened alloy, Nimonic 80A, developed about 1940 in response to the need for a suitable turbine blade material for the first gas turbine for aircraft propulsion.

To provide improved fuel efficiency and power-to-weight performance of gas turbine engines, materials had to be developed that would withstand increased operating temperatures and higher rotational speeds. This led to the introduction of improved nickel-base superalloys made possible by a wide range of novel processing techniques such as directional solidification, single-crystal technology, powder processing, mechanical alloying, and hot isostatic pressing. Many of these alloys, which can be used at temperatures up to 85% of their melting point, contained higher refractory metal contents at the expense of chromium. As a result, the high-temperature corrosion resistance of these materials suffered. This led to the development of advanced coating systems for superalloys, including aluminide and platinum aluminide diffusion coatings, MCrAlY overlay coatings deposited by physical vapor deposition, and ceramic thermal barrier coatings applied by plasma spraying and/or electron-beam physical vapor deposition.

Despite this outstanding combination of high-temperature strength and hot-corrosion resistance, nickel-base superalloys are technologically limited by their incipient melting points (most superalloys melt in the temperature range of 1260 to 1370 °C, or 2300 to 2500 °F). In addition, many of the alloying elements that make up superalloys are either expensive or considered strategic materials (that is, they are predominantly or wholly imported metals). Subsequently, much work has been devoted to developing alternative materials. Two material classes have been explored as possible replacements for superalloys in high-temperature applications:

1. Materials that offer higher strength-to-weight ratios and/or improved corrosion resistance at similar maximum operating temperatures. Examples in this class include intermetallic compounds such as nickel aluminides and titanium aluminides.
2. Materials that, by virtue of higher melting points, offer the capability of operating at higher temperatures (in the 1905 to 2205 °C, or 2000 to 4000 °F, temperature range). Included in this class are refractory metals, ceramics, and carbon-carbon composites.

Recognizing the importance of materials for high-temperature materials and in response to the technical need for a state-of-the-art review of this subject, ASM International offers the *ASM Specialty Handbook: Heat-Resistant Materials*. The sixth in the *ASM Specialty Handbook* series, the present volume is divided into five major sections. The introductory section describes the principles associated with elevated-temperature mechanical and corrosion behavior and compares and contrasts the elevated-temperature characteristics of a wide range of engineering materials. The second section examines the properties of ferrous alloys, including high-alloy cast irons, plain carbon and low-alloy steels, and stainless steels. Alloy development and elevated-temperature performance of the metallurgically complex superalloys are reviewed in the third section. Wrought, powder metallurgy, and cast nickel-, cobalt-, and iron-base superalloys are discussed in detail, and newly developed, directionally solidified and single-crystal alloys and protective coatings are also described. The fourth section reviews nonferrous materials, including titanium alloys, refractory metal alloys, structural intermetallics, ceramics, and carbon-carbon composites. The final section, Special Topics, explores the complex interaction of various high-temperature phenomena and discusses design considerations for high-temperature applications.

I wish to extend my thanks to Gary R. Halford, NASA Lewis Research Center for his invaluable assistance on this project. Dr. Halford updated the article "Creep-Fatigue Interaction." The original version of this article was published in 1985 in *Mechanical Testing*, Volume 8 of *ASM Handbook* (formerly 9th Edition *Metals Handbook*). Thanks are also due to the following organizations for their generous support in supplying property data: Carpenter Technology Corporation; Haynes International, Inc.; Titanium Metals Corporation; Inco Alloys International, Inc.; and the Nickel Development Institute. Last, I gratefully acknowledge the efforts of Heather F. Lampman, for handling all aspects of word processing, and Scott D. Henry from the ASM staff, for his considerable patience and support throughout the project. Only through their combined efforts was the successful completion of the volume made possible.

Joseph R. Davis
Davis & Associates
Chagrin Falls, Ohio

Contents

Introduction

Elevated-Temperature Characteristics of Engineering Materials

HIGH-TEMPERATURE ALLOYS broadly refer to materials that provide strength, environmental resistance, and stability within the 260 to 1200 °C (500 to 2200 °F) temperature range. They have generally been used in the presence of combustion from heat sources such as turbine engines, reciprocating engines, power plants, furnaces, and pollution control equipment. In order to retain strength under these conditions, it is imperative that their microstructures remain stable at high operating temperatures.

During the last few decades, a better understanding of alloying effects, advances in melting technology, and the development of controlled thermomechanical processing have led to new and improved high-temperature alloys. Most such alloys have sufficient amounts of chromium (with or without additions of aluminum or silicon) to form chromia (Cr_2O_3), alumina (Al_2O_3), and/or silica (SiO_2) protective oxide scales, which provide resistance to environmental degradation. However, oxides cannot protect against failure by creep, mechanical or thermal fatigue, thermal shock, or embrittlement. In actual service, failure of a component/material is typically caused by a combination of two or more attack modes, which synergistically accelerate degradation.

This article briefly reviews the elevated-temperature characteristics of various engineered materials. Although emphasis has been placed on the most commonly employed—most notably chromium-alloyed steels and superalloys—low-density metals (titanium- aluminum-, and magnesium-base alloys), refractory metal alloys, intermetallics, ceramics, and carbon-carbon composites are also discussed. The high-temperature property data presented are comparative, the intent being to provide the reader with an understanding of alloy group rankings (see, for example, Fig. 1 to 5). More detailed information/data on specific alloys can be found in the many cited articles found throughout this Volume.

Cast Irons and Steels

Heat-resistant cast irons are basically alloys of iron, carbon, and silicon having high-temperature properties markedly improved by the addition of certain alloying elements, singly or in combination, principally chromium, nickel, molybdenum, aluminum, and silicon in excess of 3%. Silicon and chromium increase resistance to heavy scaling by forming a light surface oxide that is impervious to oxidizing atmospheres. Both elements reduce the toughness and thermal shock resistance of the metal. Although nickel does not appreciably affect oxidation resistance, it increases strength and toughness at elevated temperatures by promoting an austenitic structure that is significantly stronger than ferritic structures above 540 °C (1000 °F). Molybdenum increases high-temperature strength in both ferritic and austenitic iron alloys. Aluminum additions are very potent in raising the equilibrium temperature (A_1) and in reducing both growth and scaling, but they adversely affect mechanical properties at room temperature.

Alloy cast irons that have successfully been used for *low-stress* elevated-temperature applications include:

- High-silicon irons (4 to 6% Si), with or without molybdenum additions (0.2 to 2.5%), used at temperatures up to 900 °C (1650 °F)
- Austenitic nickel-alloyed irons (18 to 34% Ni with 0.5 to 5.5% Cr), used at temperatures up to 815 °C (1500 °F)
- High-chromium white irons (12 to 39%), used at temperatures up to 1040 °C (1900 °F)

Applications for these alloys include cylinder liners, exhaust manifolds, valve guides, gas turbine housings, turbocharger housings, nozzle rings, water pump bodies, and piston rings in aluminum pistons. More detailed information on heat-resistant irons can be found in the article "High-Alloy Cast Irons" in this Volume.

Carbon steel, the most widely used steel, is suitable where corrosion or oxidation is relatively mild. It is used for applications in condensers, heat exchangers, boilers, superheaters, and stills. The widespread usage reflects its relatively low cost, generally good service performance, and good weldability. The basic low-carbon grade contains nominally 0.15% C and is used in various tubing applications. Medium-carbon grades contain 0.35% C (max) with manganese contents ranging from 0.30 to 1.06%. These grades are used for tubing, pipe, forgings, and castings.

For low-stress applications, plain carbon steels can be used at temperatures ≤425 °C (800 °F). Temperatures up to about 540 °C (1000 °F) can be withstood for only short periods. Figures 1 to 5 compare the elevated-temperature properties of carbon steels with those of other alloy systems.

Carbon-molybdenum steels contain 0.50% Mo with a carbon content of about 0.20%. These steels are used in the same kind of equipment as carbon steel, but they can be more highly stressed because the molybdenum addition increases short-time tensile strength and reduces the creep rate for a given stress and temperature. If graphitization under service conditions is probable, the maximum service temperature for carbon-molybdenum steels is about 450 °C (850 °F).

Chromium-Molybdenum Steels. Creep-resistant low-alloy steels usually contain 0.5 to 1% Mo for enhanced creep strength, along with chromium contents between 0.5 and 9% for improved corrosion resistance, rupture ductility, and resistance against graphitization. Small additions of carbide formers such as vanadium, niobium, and titanium may also be added for precipitation strengthening and/or grain refinement. The effects of alloy elements on transformation hardening and weldability are, of course, additional factors.

The three general types of creep-resistant low-alloy steels are:

- *Plain chromium-molybdenum steels* include the 1Cr-0.5Mo and 1.25Cr-0.5Mo alloys used at temperatures up to 510 °C (950 °F); 2.25Cr-1Mo steel, the most widely employed grade, used at temperatures up to 580 °C (1075 °F); 5Cr-0.5Mo steel used at temperatures up to 620 °C (1150 °F); and the 7Cr-0.5Mo and 9Cr-1Mo alloys, used at temperatures up to 650 °C (1200 °F).
- *The chromium-molybdenum-vanadium steels* provide higher creep strengths and are used in applications where allowable design stresses may require deformations less than 1% over the life of components operating at temperatures up to 540 °C (1000 °F). The most com-

mon composition contains 1% Cr, 1% Mo, and 0.25% V.

- *Modified chromium-molybdenum steels* contain various microalloying elements such as vanadium, niobium, titanium, and boron and are used for thick-section components in hydrogen-containing environments. Depending on the grade, these modified grades can be used at temperatures up to 455 to 600 °C (850 to 1110 °F).

Chromium-molybdenum steels are widely used for pressure vessels and piping in the oil and gas industries and in fossil fuel and nuclear power plants. Product forms include forgings, tubing, pipe, castings, and plate. Figures 3 and 5 compare the properties of chromium-molybdenum steels with those of other alloy systems. Detailed information on the compositions and properties of these steels, which are covered by various ASTM specifications, can be found in the article "Elevated-Temperature Mechanical Properties of Carbon and Alloy Steels" in this Volume.

Chromium hot-work die steels (types H10 to H19) have good resistance to thermal softening (high hot hardness) up to 540 °C (1000 °F) because of their medium chromium content (5%) and the addition of carbide-forming elements such as molybdenum, tungsten, and vanadium. An increase in silicon content (up to 1.20%) improves oxidation resistance at temperatures up to 800 °C (1475 °F).

Stainless Steels and Superalloys

When the severity of the service environments precludes the use of cast irons, carbon steels, or low-alloy steels, more highly alloyed materials such as stainless steels and superalloys must be considered. As shown in Fig. 6, some wrought superalloys—particularly those based on the austenitic Fe-Ni-Cr system—evolved from stainless steel technology. In fact, some stainless steels are also considered iron-base superalloys. Examples include A-286 (UNS 566286), a precipitation-hardening austenitic stainless steel, and 19-9-DL (UNS S63198), a solid-solution-strengthened austenitic stainless steel.

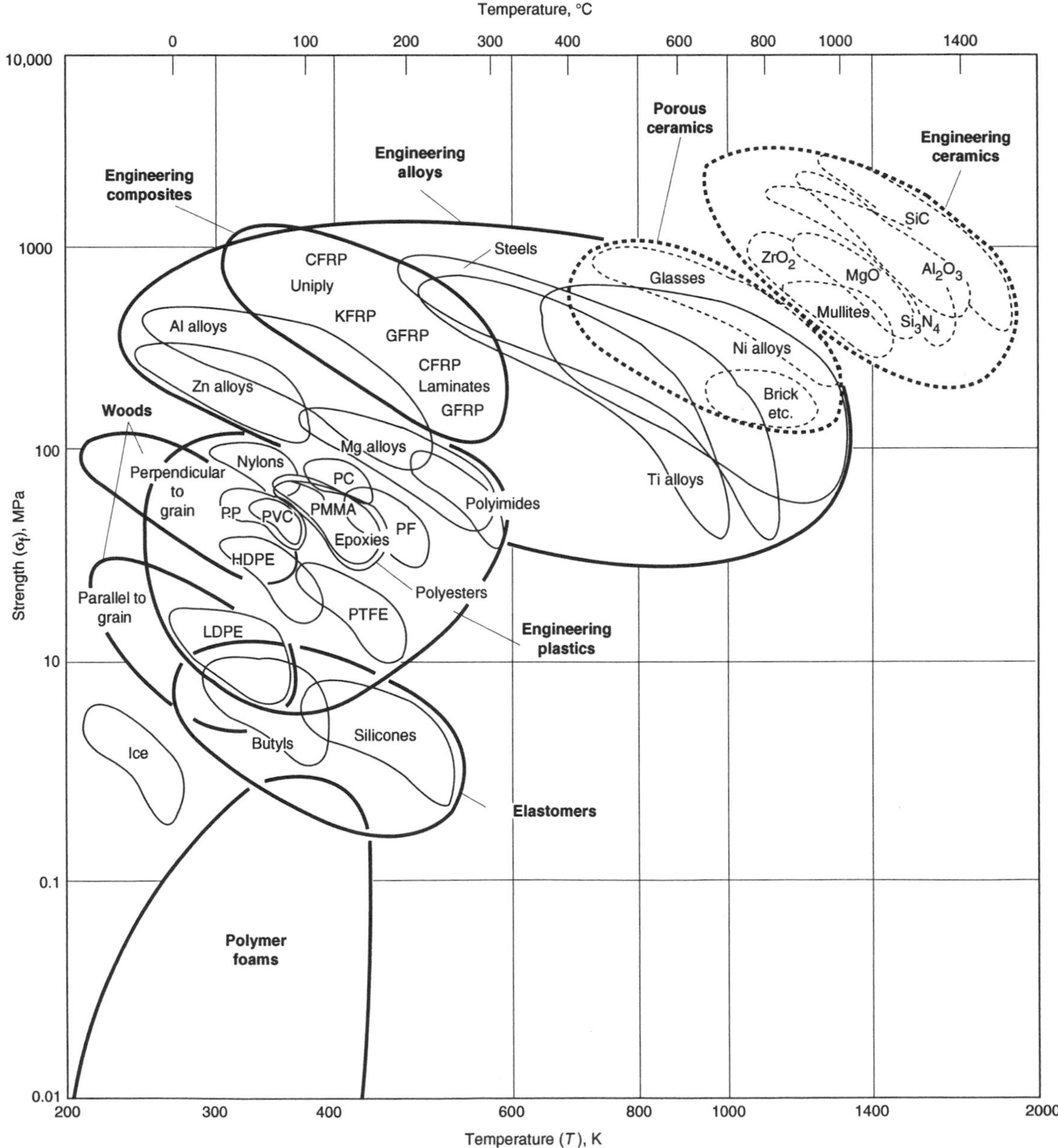

Fig. 1 Short-term high-temperature yield strength for various materials. The broken property envelope lines indicate that ceramic strength values are compressive, not tensile or yield. These data should not be used in material selection and design for long-term creep and stress rupture. Source: Ref 1, 2

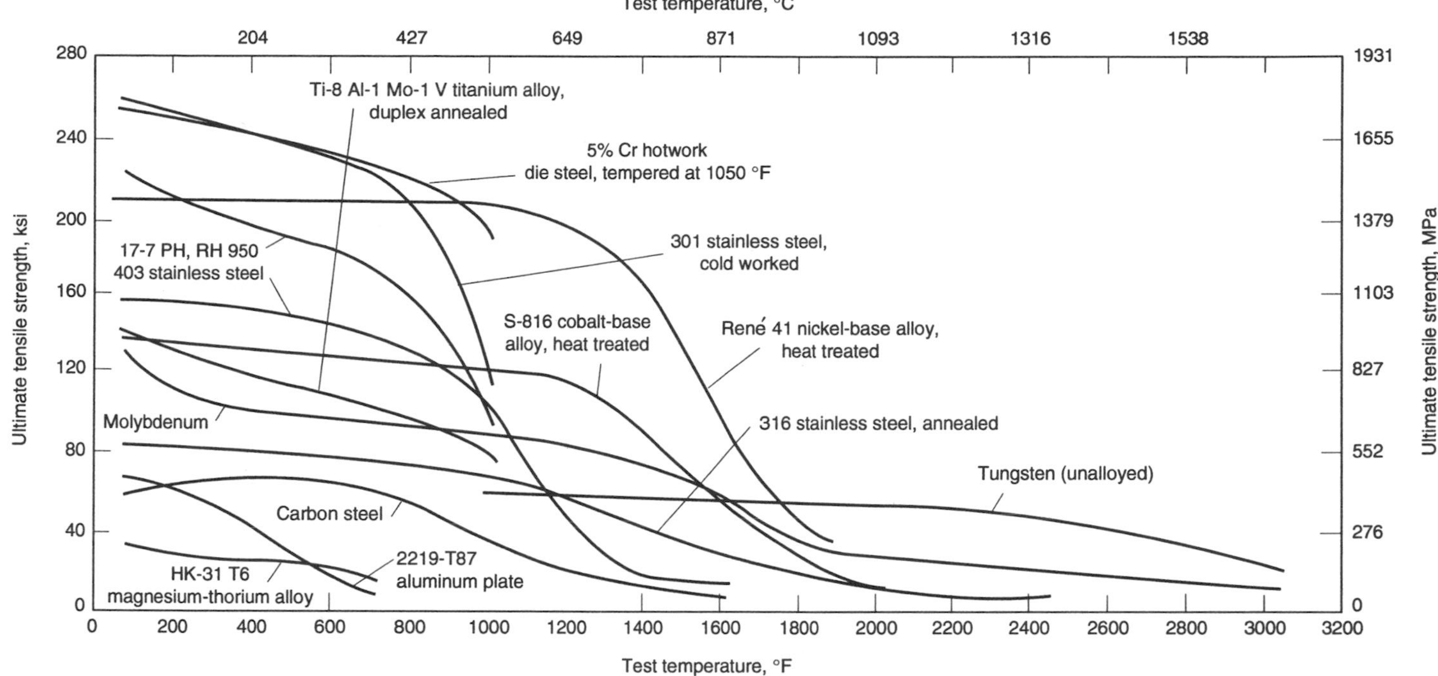

Fig. 2 Effect of temperature on the ultimate tensile strength of various metals and alloys. Source: Ref 3

Generally, however, stainless steels can be differentiated by their lower alloying content. For example, compare the analysis of type 304 stainless with Waspaloy alloy (UNS N07001), a precipitation-hardening nickel-base high-temperature alloy. Type 304 includes carbon, manganese, silicon, chromium, and nickel. Waspaloy alloy, in addition to these elements, includes molybdenum, cobalt, titanium, aluminum, zirconium, and boron. These additional elements in the right combination give Waspaloy alloy its considerable strength and corrosion resistance at temperatures of up to 870 °C (1600 °F).

Stainless Steels (Ref 7, 8)

Stainless steels for elevated-temperature applications include ferritic, martensitic, precipitation-hardening, and austenitic grades. Next to the superalloys, the stainless steels provide the best combination of high-temperature corrosion resistance and high-temperature mechanical strength of any alloy group. More detailed information on stainless steels can be found in the articles "Elevated-Temperature Mechanical Properties of Stainless Steels" and "Elevated-Temperature Corrosion Properties of Stainless Steels" in this Volume.

The ferritic grades, which have good resistance to oxidation at elevated temperature, are not known for their mechanical strength at high temperature. Instead, these alloys are primarily used in corrosion-resistant applications. Another limiting factor for high-chromium ferritic stainless steels is sigma-phase embrittlement due to high-temperature exposure. Two examples of ferritic stainless steels are given below.

The 17% Cr stainless steel, type 430, is used in applications that require oxidation and corrosion resistance up to 815 °C (1500 °F). Where elevated-temperature strength is a requirement, the use of this composition is limited because of its relatively low creep strength. Although it is ductile between about 400 and 590 °C (750 and 1100 °F), this steel will be brittle when it is cooled to ambient temperature after prolonged heating in this range (sigma-phase embrittlement). The brittleness may be eliminated by reheating to about 760 °C (1400 °F). The 27% Cr stainless steel, type 446, which has relatively low elevated-temperature strength, is used between 870 and 1095 °C (1600 and 2000 °F) in applications where the most severe oxidation is encountered. It is also subject to the same embrittling phenomena as type 430 steel. The major application of type 446 steel is in such items as furnace parts, soot blowers, and thermocouple protection tubes, where stresses are relatively low.

The martensitic stainless steels most commonly used for elevated-temperature applications are the so-called "Super 12 Chrome" steels that contain molybdenum (up to 3%) and/or tungsten (up to 3.5%) for greater strength at elevated temperatures. Other elements, such as vanadium, niobium, and nitrogen, may also be added in small amounts for additional strengthening. The 12% Cr martensitics with the aforementioned alloying additions can be used at temperatures up to 650 °C (1200 °F), but they provide only mod-

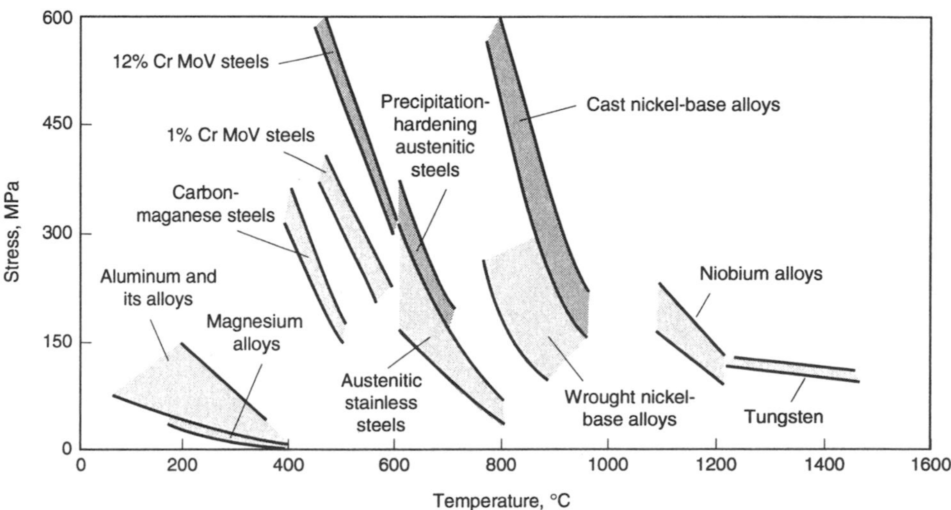

Fig. 3 Stress to produce rupture in100 h for various alloys. Source: Ref 4

erate strength above 540 °C (1000 °F). Straight (unalloyed) martensitic grades can only be used at temperatures up to 400 °C (750 °F). Property data for martensitic stainless steels are given in Fig. 3 to 5.

The austenitic stainless steels are essentially alloys of iron, chromium, and nickel. These steels as a class are the strongest steels for service above about 540 °C (1000 °F). Property data for the austenitic grades are given in Fig. 2 to 5. Some of the more commonly employed alloys are discussed below.

Type 304 is the most common grade of austenitic chromium-nickel steels, which as a group are used for handling many corrosive materials or resisting severe oxidation. Type 304 steel has excellent resistance to corrosion and oxidation, has high creep strength, and is frequently used at temperatures up to 815 °C (1500 °F). Type 304 steel is being used successfully and economically in high-temperature service in such applications as high-pressure steam pipes and boiler tubes, radiant superheaters, and oil-refinery and process industry equipment.

Types 321 and 347 stainless steels are similar to type 304 except that titanium and niobium, respectively, have been added to these steels. The titanium and niobium additions combine with carbon and minimize intergranular corrosion that may occur in certain media after welding. However, the use of niobium (or titanium) does not ensure complete immunity to sensitization and subsequent intergranular attack when the steel is exposed for long times in the sensitization range of 425 to 815 °C (800 to 1500 °F). However, types 321 and 347 stainless steels are widely used for service in this temperature range.

Type 316 stainless steel, which contains molybdenum, is used for high-strength service up to about 815 °C (1500 °F), and it will resist oxidation up to about 900 °C (1650 °F). However, above this temperature, in still air, the molybdenum will form an oxide that will volatilize and result in rapid oxidation of the steel.

For service above about 870 °C (1600 °F), types 309 and 310 stainless steels, which contain about 23 to 25% Cr, are used. These steels have the best high-temperature strength of the austenitic stainless steels at these temperatures, and because of their chromium contents, they can be used in applications where extreme corrosion or oxidation is encountered.

The highest service limits for oxidation resistance of stainless steels is achieved by the highly alloyed type 330 grade. This alloy, which contains 19% Cr, 25% Ni, and 1.0% Si, is suitable for continuous service at temperatures as high as 1150 °C (2100 °F).

Precipitation-hardening stainless steels have the highest room-temperature strengths of all the stainless steels. They fill an important gap between the chromium-free 18% Ni maraging steels and the 12% Cr quenched and tempered martensitic grades. One grade, the austenitic A-286, has moderate strength and long-term service capability up to 620 °C (1150 °F).

Valve steels are austenitic nitrogen-strengthened steels that have been used extensively in automotive/internal combustion engine valve applications. Examples of such alloys include 21-2N (21Cr, 8Mn, 2Ni + N), 21-4N (21Cr, 9Mn, 4Ni + N), 21-12N (21Cr, 12Ni, 1.25 Mn + N), and 23-8N (21Cr, 8Ni, 3.5Mn + N). The nitrogen contents in these alloys range from 0.20 to 0.50%. These engine valve grades are used at temperatures up to 760 °C (1400 °F), but they provide fairly low strength at the upper end of their temperature capability.

Cast heat-resistant alloys are primarily used in applications where service temperatures exceed 650 °C (1200 °F) and may reach extremes as high as 1315 °C (2400 °F). Many of the cast heat-resistant alloys are compositionally related to the wrought stainless steels and to the cast corrosion-resistant alloys. The major difference between these materials is their carbon content. With only a few exceptions, carbon in the cast heat-resistant alloys falls in a range from 0.3 to 0.6%, compared with the 0.01 to 0.25% C that is normally associated with the wrought and cast corrosion-resistant grades.

The standard cast heat-resistant grades have high creep strength and generally good oxidation resistance, show better carburization behavior than the corresponding wrought alloy, and are available in product forms such as tubes, retorts, and hangers at relatively low cost for the alloy content. Detailed information on the compositions and properties of cast heat-resistant alloys can be found in the article "High-Alloy Cast Steels" in this Volume.

Cast duplex nickel-chromium alloys near the eutectic composition (about 50% Cr) develop compact chromia scales rapidly and show exceptional resistance to oxidation and fuel ash corrosion. Small niobium additions, as in IN-657, improve creep strength. Such duplex alloys are described in the article "Nickel-Chromium and Nickel-Thoria Alloys" in this Volume.

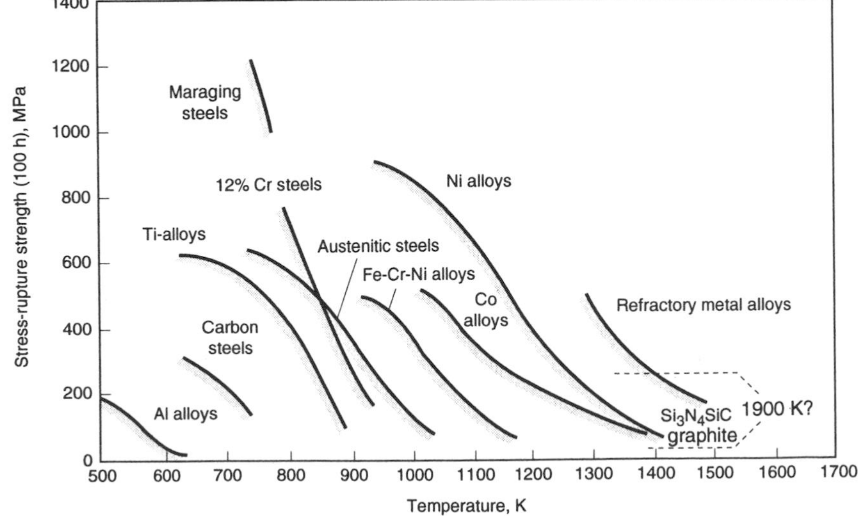

Fig. 4 Maximum service temperatures of various creep-resistant materials. Source: Ref 5

Superalloys

Superalloys are nickel-, iron-nickel-, and cobalt-base alloys generally used at temperatures above about 540 °C (1000 °F). As shown in Fig. 6, the iron-nickel-base superalloys are an extension of stainless steel technology and generally are wrought, whereas cobalt- and nickel-base superalloys may be wrought or cast, depending on the application/composition involved. Appropriate compositions of all superalloy base metals can be forged, rolled to sheet, or otherwise formed into a variety of shapes. The more highly alloyed compositions normally are processed as castings. Properties can be controlled by adjustments in composition and by processing (including heat treatment), and excellent elevated-temperature strengths are available in finished products. Figures 1 to 5 illustrate the excellent elevated-temperature characteristics of superalloys. As indicated in Fig. 5, no other alloy system has a better combination of high-temperature corrosion resistance and stress-rupture strength. Some superalloys, particularly nickel-base casting alloys, can be used at temperatures that are approximately 85% of their incipient melting point. Such alloys exhibit outstanding creep and stress-rupture properties at temperatures in excess of 1040 °C (1900 °F). The oxidation resistance of most superalloys is excellent at moderate temperatures—about 870 °C (1600 °F) and below. Some alloys can be used at temperatures up to 1200 °C (2200 °F). Coatings can further enhance high-temperature corrosion resistance. More detailed information regarding the compositions, properties, and processing of superalloys can be found in the Sections "Properties of Superalloys" and "Special Topics" in this Volume.

Superalloy Development. Nickel-base superalloys were created at approximately the turn of the century with the addition of 20 wt% Cr in an 80 wt% Ni alloy for electrical heating elements. In the late 1920s, small amounts of aluminum and titanium were added to the 80Ni-20Cr alloy, with a significant gain in creep strength at elevated temperatures. It soon became apparent that iron and cobalt alloys could be more effectively strengthened by solid-solution additions, while nickel alloys could be strengthened by a coherent phase, γ′. Concurrent with these additions, the carbon present in the alloys was identified as having a strengthening effect when combined with other alloying elements to form M_6C and $M_{23}C_6$ carbides. Other grain-boundary formers, such as boron and zirconium, were added to polycrystalline materials to hold the material together.

In the early development time period (about 1926), Heraeus Vacuumschmelze A.G. received a patent for a nickel-chromium alloy that contained up to 15 wt% W and 12 wt% Mo, thereby introducing the refractory metals into superalloy compositions. The purpose of adding refractory metals was to increase the high-temperature strength of the nickel-base alloy. By the 1930s there were two iron-base "heat-resisting alloys" containing either tungsten or molybdenum additions, and the use of these two metals was widely accepted in cobalt-base alloys. In the early 1950s, alloys containing about 5 wt% Mo were introduced in the United States. The commercial exploitation of molybdenum additions took place in 1955. A cast alloy containing 2 wt% Nb was available in the late 1950s In the early 1960s, tungsten and tantalum were widely accepted for alloying in nickel-base alloys. The demonstration of the effectiveness of rhenium additions in nickel-base alloys occurred in the late 1960s. Some highly alloyed nickel-base castings have total refractory metal contents exceeding 20 wt%. Finally, also in the late 1960s, hafnium additions were found to stabilize and strengthen grain-boundary structures. Figure 7 charts the development of superalloys from 1940 to the 1990s.

Superalloy Systems. As stated above, superalloys can be divided into three types: iron-nickel-, nickel-, and cobalt-base. Each of these is briefly reviewed below.

The most important class of iron-nickel-base superalloys includes alloys that are strengthened by intermetallic compound precipitation in a face-centered cubic (fcc) matrix. The most common precipitate is γ′, typified by A-286, V-57, or Incoloy 901, but some alloys precipitate gamma double prime (γ″), typified by Inconel 718. Other iron-nickel-base superalloys consist of modified stainless steels primarily strengthened by solid-solution hardening. Alloys in this last category vary from 19-9DL (18-8 stainless with slight chromium and nickel adjustments, additional solution hardeners and higher carbon) to Incoloy 800H (21% Cr, high nickel, and small additions of titanium and aluminum). The iron-nickel-base superalloys are used in the wrought condition.

The most important class of nickel-base superalloys is that strengthened by intermetallic-compound precipitation in an fcc matrix. The strengthening precipitate is γ′, typified by Waspaloy or Udimet 700. Another class of nickel-base superalloys is represented by Hastelloy X, which is essentially solid-solution strengthened but probably also derives some strengthening from carbide precipitation produced through a working-plus-aging schedule. A third class includes oxide-dispersion-strengthened (ODS) alloys such as MA 754 and MA 6000, which are strengthened by dispersions of inert particles such as yttria coupled in some cases with γ′ precipitation (MA 6000).

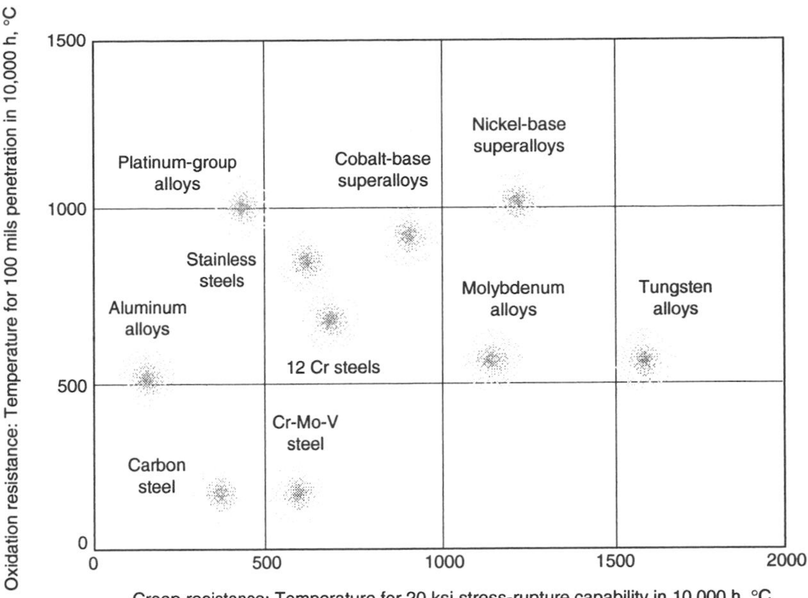

Fig. 5 Relative oxidation/stress-rupture capabilities of various alloy systems. Source: Ref 6

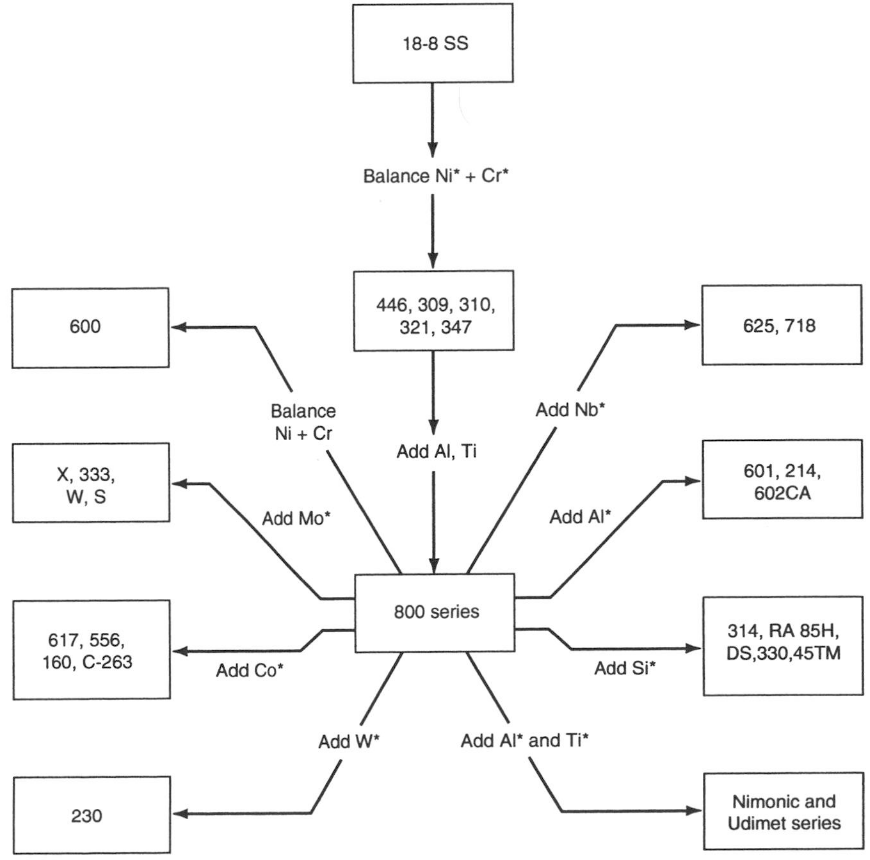

Fig. 6 Relationship of high-temperature alloy formulations for stainless steels and superalloys based on the Fe-Ni-Cr system. Asterisk indicates addition of other elements. Source: VDM Technologies Corp.

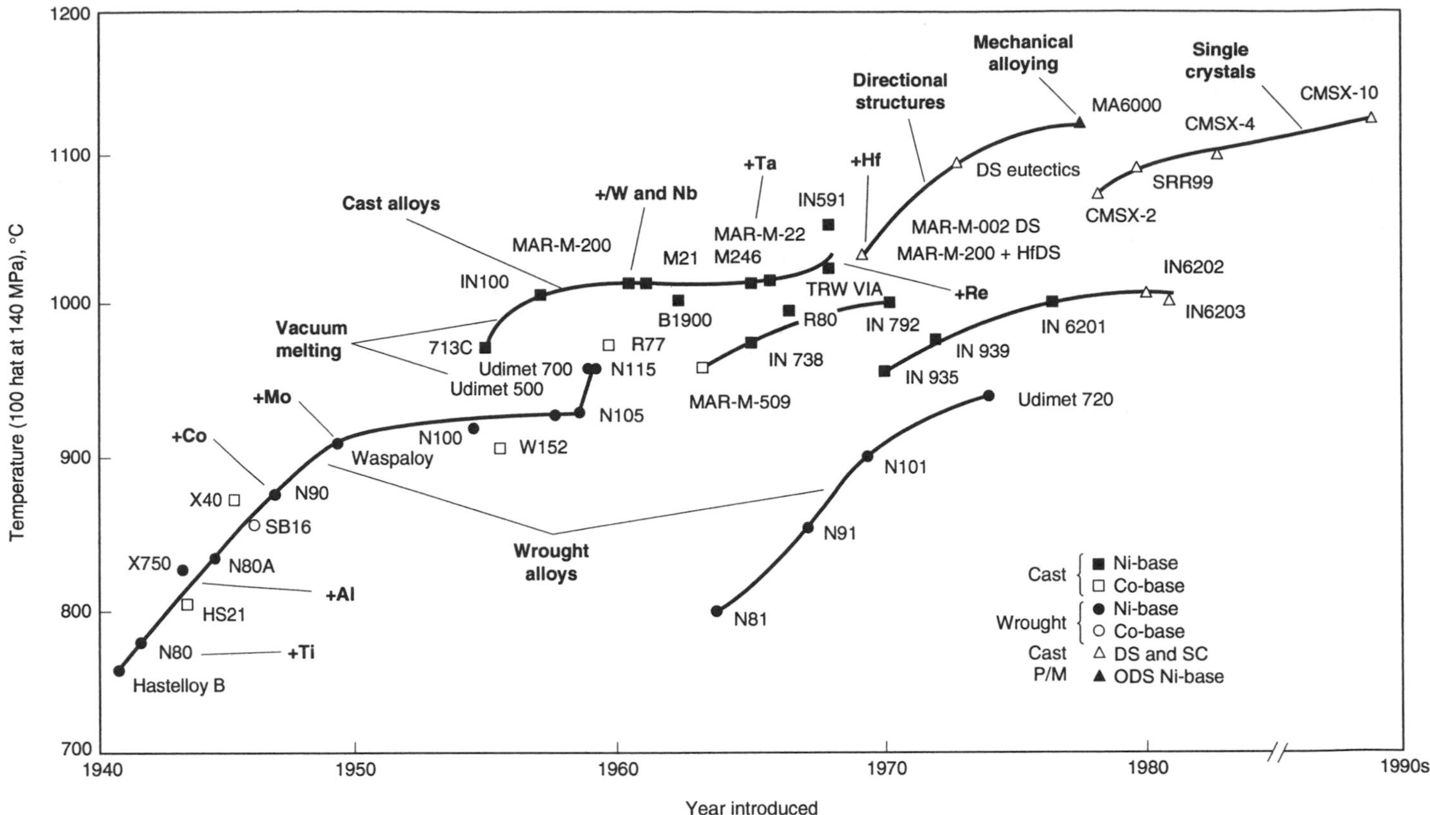

Fig. 7 Temperature capability of superalloys with approximate year of introduction. DS, directionally solidified; SC, single-crystal; P/M, powder metallurgy; ODS, oxide-dispersion-strengthened. Source: adapted from Ref 9

Nickel-base superalloys are used in both cast and wrought forms, although special processing (powder metallurgy/isothermal forging) frequently is used to produce wrought versions of the more highly alloyed compositions (Udimet 700/Astroloy, IN-100).

An additional aspect of nickel-base superalloys has been the introduction of directional-solidification and single-crystal casting technology. As shown in Fig. 7, these alloys exhibit better high-temperature properties than polycrystalline wrought or cast alloys.

Low-Density Metals

Density is very important in the design of aircraft because excess weight in engine and structural components decreases load-carrying capacity. As a result, materials with a high strength-to-weight ratio are in ever-increasing demand. Low-density alloys, which are commonly referred to as light metals, include magnesium-, aluminum-, and titanium-base alloys.

Magnesium alloys have the lowest densities (~1.8 g/cm^3) of any structural alloy. Magnesium alloys for elevated-temperature applications usually contain thorium (up to 3 wt%). The magnesium-thorium-zirconium cast alloys HK31A (Mg-3.3Th-0.7Zr) and HZ32A (Mg-3.3Th-2.1Zn-0.7Zr) are intended primarily for use at temperatures of 200 °C (400 °F) and higher. At 260 °C (500 °F) and slightly higher, HZ32A is equal to or better than HK31A in short-time and long-time creep strength at all extensions. The HK31A alloy has higher tensile, yield, and short-time creep strengths up to 370 °C (700 °F).

Another cast alloy, QH21A (Mg-2.5Ag-1.0Th-0.7Zr), retains a high yield strength up to 300 °C (570 °F). This alloy is relatively expensive due to its silver content.

Rare earth (RE) additions also contribute to improved elevated-temperature performance. Alloys WE54 (Mg-5.2Y-3.0RE-0.7Zr) and WE43 (Mg-4.0Y-3.4RE-0.7Zr) have high tensile strengths and yield strengths, and they exhibit good properties at temperatures up to 300 °C (570 °F) and 250 °C (480 °F), respectively. The WE54 alloy retains its properties at high temperature for up to 1000 h, whereas WE43 retains its properties at high temperature in excess of 5000 h.

Thorium-alloyed wrought alloys are also used at elevated temperatures. Alloy HM31A (Mg-3.0Th-1.2Mn), produced in extruded forms, is of moderate strength. It is suitable for use in applications requiring good strength and creep resistance at temperatures in the range of 150 to 425 °C (300 to 800 °F).

Alloy HM21A (Mg-2.0Th-0.6Mn), produced as forgings, is useful at temperatures up to 370 to 425 °C (700 to 800 °F) for applications in which good creep resistance is needed.

HK31A and HM21A alloys, produced in sheet and plate forms, are suitable for use at temperatures up to 315 and 345 °C (600 and 650 °F), respectively. However, HM21A has superior strength and creep resistance, as shown in the following table:

Alloy	Stress for 0.1% creep in 100 h MPa	ksi
At 205 °C (400 °F)		
HM21A	86.2	12.5
HK31A	41	6.0
At 260 °C (500 °F)		
HM21A	72.4	10.5
HK31A	28	4.0
At 315 °C (600 °F)		
HM21A	52	7.5
HK31A	14	2.0

Additional data on magnesium alloys are given in Fig. 1 to 3. As these data indicate, the elevated-temperature performance of magnesium alloys is comparable to that of some aluminum alloys. Most other alloy systems, however, have superior high-temperature properties.

Aluminum alloys have higher densities (~2.8 g/cm^3) and higher room-temperature strengths than magnesium alloys. Some cast aluminum al-

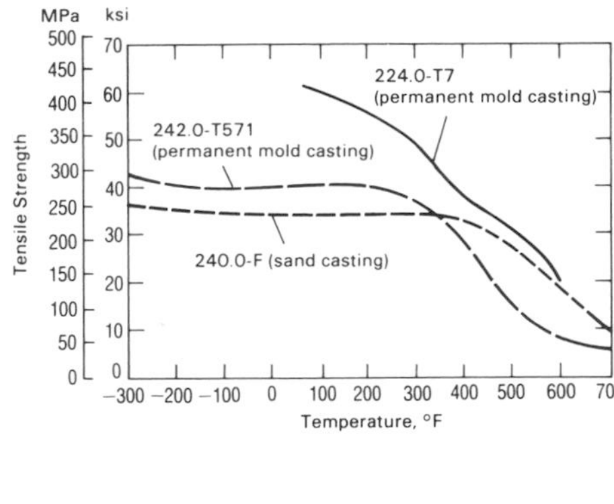

Fig. 8 Tensile strengths of aluminum alloys 240.0-F, 224.0-T7, and 242.0-T571 as functions of temperature

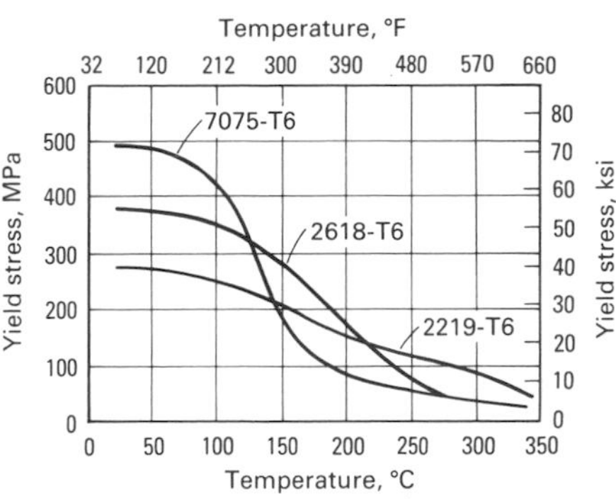

Fig. 9 Values of 0.2% yield stress of aluminum alloys after exposure for 1000 h at temperatures between 0 and 350 °C

loys are used in applications at moderately elevated temperatures (e.g., pistons in internal combustion engines). One commonly employed alloy is permanent mold cast alloy 242 (Al-4Cu-2Ni-2.5Mg). As shown in Fig. 8, this alloy retains its strength at temperatures as high as ~150 °C (300 °F). Some other cast aluminum alloys can be used at slightly higher temperatures (175 °C, or 350 °F).

The 7*xxx* series of wrought age-hardenable alloys that are based on the Al-Zn-Mg-Cu system develop the highest room-temperature tensile properties of any aluminum alloys produced from conventionally cast ingots. However, the strength of these alloys declines rapidly if they are exposed to elevated temperatures (Fig. 9), due mainly to coarsening of the fine precipitates on which the alloys depend for their strength. Alloys of the 2*xxx* series, such as 2014 and 2024, perform better above these temperatures but are not normally used for elevated-temperature applications.

Strength at temperatures above about 100 to 200 °C (200 to 400 °F) is improved mainly by solid-solution strengthening or second-phase hardening. Another approach to improve the elevated-temperature performance of aluminum alloys has been the use of rapid solidification technology to produce powders containing high supersaturations of elements such as iron or chromium that diffuse slowly in solid aluminum. In this regard, several experimental materials based on the Al-Fe-Ce, Al-Fe-V-Si, and Al-Cr-Zr systems are now available that have promising creep properties up to 350 °C (650 °F) (Ref 10). Additional data on aluminum alloys are given in Fig. 1 to 5.

Titanium alloys provide an outstanding combination of low density (~4.5 g/cm^3) and high strength (up to 1100 MPa, or 160 ksi, yield strength). Alloys have been developed that have useful strength and resist oxidation at temperatures as high as 595 °C (1100 °F). The improved elevated-temperature characteristics of these alloys, combined with their high strength-to-weight ratios, make them an attractive alternative to nickel-base superalloys for certain gas turbine components.

Most of the titanium alloys for elevated-temperature applications are near-alpha alloys based on the Ti-Al-Sn-Zr system. Important alloying elements are molybdenum, silicon, and niobium. Molybdenum enhances hardenability and enhances short-time high-temperature strength or improves strength at lower temperatures. Minor silicon additions improve creep strength, while niobium is added primarily for oxidation resistance at elevated temperature. Examples of these near-alpha alloys are Ti-1100 (Ti-6Al-2.75Sn-4Zr-0.4Mo-0.45Si) and IMI-834 (Ti-5.5Al-4Sn-4Zr-0.3Mo-1Nb-0.5Si). Specific yield strengths (density corrected) of titanium- and nickel-base alloys are compared in Fig. 10. Additional data are shown in Fig. 1, 2, and 4. As these data clearly show, titanium alloys have far greater elevated-temperature strength than do plain carbon steels and low-density aluminum- and magnesium alloys. More detailed information can be found in the article "Titanium and Titanium Alloys" in this Volume.

Refractory Metal Alloys

Refractory metals include tungsten, molybdenum, niobium, tantalum, and rhenium. These metals and their alloys have melting points in excess of 2200 °C (4000 °F), which is substantially higher than those of stainless steels or superalloys. As indicated in Fig. 1 to 5, the creep strength of some refractory metals (tungsten and niobium) exceeds that of superalloys. There are, however, a number of deficiencies of refractory metals and alloys that have precluded their being viable alternatives to superalloys in gas turbine engine applications: the open body-centered cubic structure (precluding high creep resistance relative to the melting point), lack of low-temperature ductility in the VIa metals (tungsten and molybdenum), severe lack of oxidation resistance for all, and significantly higher density than superalloys for all except niobium. Detailed information on the processing and properties of these materials, as well as efforts made to overcome some of the aforementioned deficiencies (e.g., coatings to prevent catastrophic oxidation) can be found in the article "Refractory Metals and Alloys" in this Volume.

Structural Intermetallics

The search for new high-temperature structural materials has stimulated much interest in ordered intermetallics. Recent interest has been focused on nickel aluminides based on Ni_3Al and NiAl, iron aluminides based on Fe_3Al and FeAl, and titanium aluminides based on Ti_3Al and TiAl. These aluminides possess many attributes that make them attractive for high-temperature structural applications. They contain enough aluminum to form, in oxidizing environments, thin films of aluminide oxides that often are compact and protective. They have low densities, relatively high melting points, and good high-temperature strength properties.

Nickel, iron, and titanium aluminides, like other ordered intermetallics, exhibit brittle fracture and low ductility at ambient temperatures. Poor fracture resistance and limited fabricability restrict the use of aluminides as engineering materials in most cases. Nevertheless, these materials appear the most likely to replace superalloys in high-performance applications. A brief review of these materials is given below. More detailed information can be found in the article "Structural Intermetallics" in this Volume.

Nickel Aluminides. The nickel aluminide based on NiAl has a melting point of 1638 °C (2980 °F), compared with a solidus temperature of about 1300 °C (2370 °F) for most superalloys. NiAl has excellent cyclic-oxidation resistance to 1300 °C (2370 °F), low density, and, through minor alloy additions, can provide creep strength superior to that of superalloys. While it has good ductility in single-crystal form, its polycrystalline ductility must be improved significantly, which provides a challenge to alloy developers.

Considerable research also has been conducted on the Ni_3Al compound, which has a lower melting point than NiAl but still offers strength and density advantages over current superalloys. Boron additions significantly enhance ductility over a wide temperature range up to its melting point. Ni_3Al ingots produced using conventional electroslag remelt (ESR) and vacuum induction melting (VIM) techniques be-have superplastically when thermomechanically worked at low strain rates, such as those associated with isothermal forging.

Advantages claimed for "ductilized" nickel aluminides over conventional nickel-base superalloys include:

- Lower densities due to the higher aluminum content of the aluminides
- Much simpler chemical compositions than many superalloys
- Single-phase structure
- Strength derived from their ordered structure, not from precipitates of second phases; thus, no special heat treatments, such as aging, are required

- Yield strengths that increase with increasing temperatures (as high as 650 to 750 °C, or 1200 to 1380 °F)
- Very good oxidation resistance to 1100 °C (2010 °F) due to their high aluminum content
- Potential lower cost than many superalloys when full-scale production is achieved

While as-cast properties of nickel aluminides suggest possible use in applications such as hot-forming dies, turbochargers, permanent molds, and advanced pistons, relatively little research has been performed to date on the suitability of producing nickel aluminides by investment casting.

Iron and titanium aluminides, unlike the nickel-aluminum compounds, do not offer the same creep strength at very high temperatures. They do, however, have unique specific (density-corrected) properties that should ensure their use of some rotating components. FeAl, for example, has good strength to 700 °C (1290 °F), while its high melting point (1340 °C, or 2444 °F) and good oxidation resistance may well lead to its use as a matrix material in metal-matrix composites.

Both Ti_3Al and TiAl have good specific strength at temperatures to 1100 °C (2010 °F). However, compared with superalloys, they each have limitations, such as inferior oxidation resistance (Ti_3Al) and ductility (TiAl)—areas requiring further development.

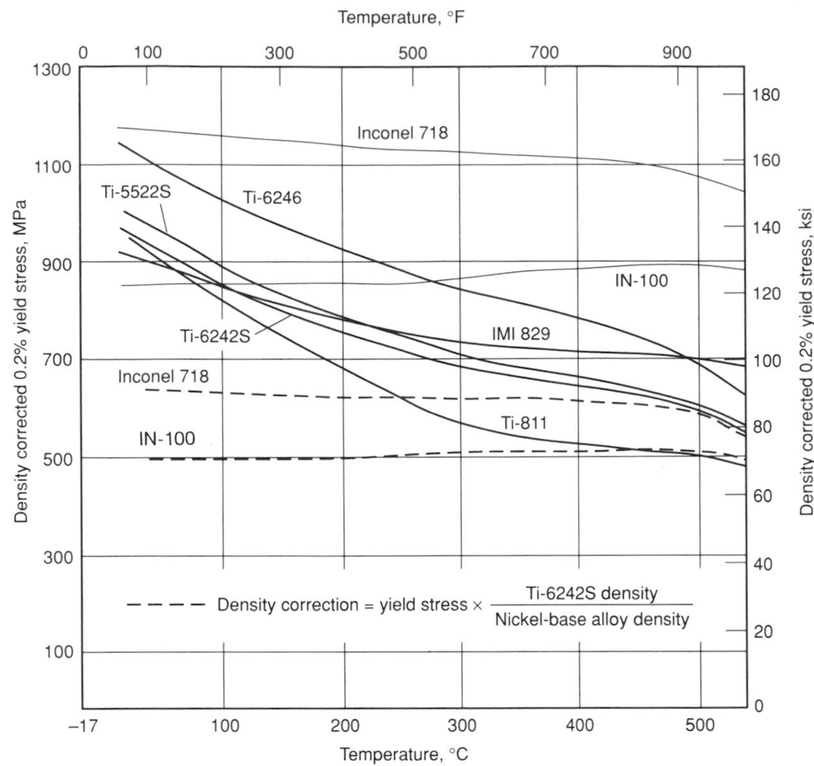

Fig. 10 Specific yield strengths for various titanium-base and nickel-base alloys

Nonmetallic Materials

Polymer-Matrix Composites. Generally, polymers or polymer-matrix composites are not considered heat-resistant materials. Most organic polymers soften or melt below 205 °C (400 °F). As a result, most polymeric materials are used at or just above ambient temperature (less than 100 °C, or 212 °F). Polymer-matrix composites are attractive, however, because they provide major weight and durability advantages. They have specific stiffness and strength values of not less than twice that of metallic structures. They also have similar advantages in fatigue strength.

The most successful high-temperature polymeric material developed to date is a polyimide resin reinforced with graphite fibers. Graphite-reinforced polyimide composites have been reported to be suitable for use in air at 288 °C (550 °F) for at least 5000 h. At 316 °C (600 °F), the useful life of these composites is in the range of 1200 to 1400 h. More detailed information on polymer-matrix composites can be found in Volume 1, *Composites,* of the *Engineered Materials Handbook* published by ASM International.

Ceramics, both in monolithic and composite forms, offer the prospect of useful heat resistance, possibly to temperatures near 1650 °C (3000 °F), coupled with low density and in some cases excellent oxidation and corrosion resistance. Unfortunately, these ceramics also are brittle, prone to thermal shock, and less thermally conductive than heat-resistant metals, leading to severe deficiencies under tensile loading. These are inherent properties determined by the nature of the interatomic bonds. Mechanical properties also are highly variable, depending sensitively on preparation technique, impurities, and surface finish; in ceramics the process basically determines the properties. Processing of ceramics is also quite costly. Nevertheless, the toughness and thermal shock resistance of silicon nitride (Si_3N_4) and its ability to form protective SiO_2 layers makes it a candidate for turbine or diesel applications. Silicon carbide (SiC) has similar properties. Other applications include heat exchangers, recuperators, and furnace components.

Oxides, such as alumina (Al_2O_3) and zirconia (ZrO_2), are also used for high-temperature applications. Zirconia has been in service as a thermal barrier coating in aircraft combustors on superalloys for many years. However, the oxide-type ceramics tend to be less desirable mechanically than are carbide-nitride ceramics, although they are very stable in oxidizing atmospheres.

Ceramic-matrix composites also show great potential. A wide variety of reinforcing materials, matrices, and corresponding processing methods have been studied. The most successful fiber-reinforced composites have been produced by hot pressing, chemical vapor infiltration, or directed metal oxidation, a process that uses accelerated oxidation reactions of molten metals to grow ceramic matrices around preplaced filler or reinforcement material preforms. Much of the work has been on glass and glass-ceramic matrices reinforced with carbon fibers. Other fiber-reinforced ceramic composites include SiC fibers in SiC, produced by chemical vapor infiltration and deposition, SiC-fiber-reinforced Al_2O_3, and zirconium carbide composites reinforced with zirconium diboride, produced by directed metal oxidation. Multidirectionally reinforced ceramics have also been produced, such as silica reinforced with Al_2O_3 or fused quartz.

The excellent high-temperature strength, oxidation resistance, and thermal shock resistance of Si_3N_4 has led to the development of SiC_w-reinforced Si_3N_4. The major phase, Si_3N_4, offers many favorable properties, and the SiC whiskers provide significant improvement in fracture toughness. Whisker-reinforced Si_3N_4 is a leading candidate material for hot-section ceramic-engine components. More detailed information on nonreinforced and reinforced ceramic materials can be found in the article "Structural Ceramics" in this Volume.

Carbon-Carbon Composites. The highest temperature capability of any material considered for high-temperature use is exhibited by carbon-carbon composites, graphite fibers in a carbon-graphite matrix. Carbon-carbon composites are now used for one-time service in rocket-nozzle and missile exit core structures and in turbine aircraft brake shoes; SiC-coated carbon-carbon parts are being used as the nose cap and heating edges of the space shuttle.

Because carbon fibers tend to increase in strength with increasing temperature, carbon-carbon composites retain their tensile strength at extreme temperatures. Carbon-carbon compos-

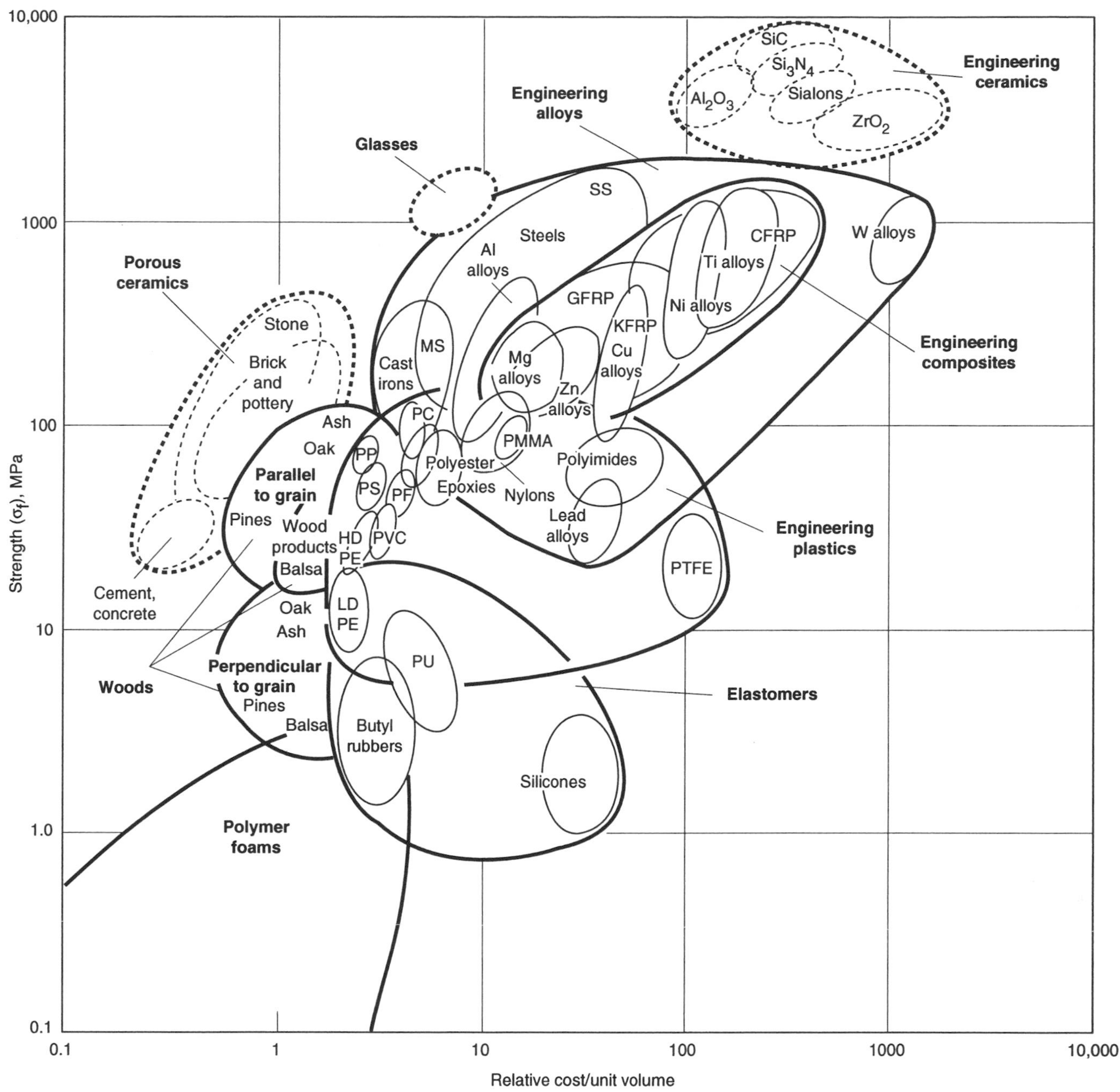

Fig. 11 Strength plotted against relative cost per unit volume for various engineered materials. Broken property envelope lines indicate that the strength values are compressive, not tensile, strengths. See text for details. Source: Ref 1, 2

ites provide unmatched specific stiffness and strength at temperatures from 1200 to 2200 °C (2192 to 3992 °F). At temperatures below 1000 °C (1832 °F), carbon-carbon composites exhibit specific strength equivalent to that of the most advanced superalloys. As a result of their carbon constituents and weakly bonded reinforced matrix, carbon-carbon composites also have superior thermal shock, toughness, ablation, and high-speed friction properties. Another outstanding feature is the low density (~1.6 g/cm^3) of carbon-carbon composites.

Because carbon-carbon is not stable in oxidizing environments at temperatures above about 425 °C (800 °F), coatings are essential. Refractory carbides such as SiC are applied by various processes. The selection and performance of coatings is the focus of the article "Carbon-Carbon Composites" in this Volume.

Cost Considerations

From a competitive standpoint, alloy selection must be based on expected cost-effectiveness. The best choice is usually the lowest-cost material able to meet design criteria. However, a higher-cost material offering greater reliability may be justified for certain components in a system that is critical and/or expensive to shut down for maintenance. Knowledge of alloy capabilities can be helpful in making a wise decision.

Figure 11 provides an overview of the costs of various materials. The relative cost is calculated by dividing the cost per kilogram of the material in question by the cost per kilogram of low-carbon (mild) steel reinforcing rod (rebar). Materials offering the greatest strength per unit cost lie toward the upper left corner. Included here would be steels and aluminum alloys. As alloying increases in steels, costs go up. Hence, stainless steels containing chromium (an expensive alloying element) are much more costly. Further alloying to produce superalloys results in even higher costs. Superalloys with high cobalt or refractory metal contents are very expensive.

Some materials are expensive because of high processing costs. An example here would be the engineering ceramics shown in Fig. 11. The costs of the principal starting materials (silicon, carbon, and nitrogen) are relatively low, but the cost of engineering ceramics is significantly increased by raw material preparation (mixing, milling, etc.), forming and fabrication (hot pressing or hot isostatic pressing), thermal processing (sintering), and finishing (grinding, polishing, and inspection).

REFERENCES

1. M.F. Ashby, *Materials Selection in Mechanical Design,* Pergamon Press, 1992, p 24–55
2. M.F. Ashby, *Materials and Process Selection Charts,* Pergamon Press, 1992, p 1–57
3. F.J. Clauss, *Engineer's Guide to High-Temperature Materials,* Addison-Wesley Publishing Co., 1969, p 2
4. G.A. Webster and R.A. Ainsworth, *High Temperature Component Life Assessment,* Chapman & Hall, 1994, p 13
5. A. Weronski and T. Hejwowski, *Thermal Fatigue of Metals,* Marcel Dekker, Inc., 1991, p 83
6. *Superalloys II,* C.T. Sims, N.S. Stoloff, and W.C. Hagel, Ed., John Wiley & Sons, 1987, p 15
7. "Steels for Elevated Temperature Service," United States Steel Corp., 1974
8. K.P. Rohrbach, Trends in High-Temperature Alloys, *Adv. Mater. Proc.,* Vol 148 (No. 4), Oct 1995, p 37–40
9. W.J. Molloy, Investment-Cast Superalloys—A Good Investment, *Adv. Mater. Proc.,* Vol 138 (No. 4), Oct 1990, p 23–30
10. J.R. Pickens, High-Strength Aluminum P/M Alloys, *Properties and Selection: Nonferrous Alloys and Special-Purpose Materials,* Vol 2, *ASM Handbook,* ASM International, 1990, p 200–215

Mechanical Properties at Elevated Temperatures

MECHANICAL PROPERTIES of interest for elevated-temperature applications include short-time elevated-temperature tensile properties, creep and stress-rupture, low-cycle and high-cycle fatigue, thermal and thermomechanical fatigue, combinations of creep and fatigue (creep-fatigue interaction), and hot hardness. Each of these will be described in this article with emphasis placed on the underlying principles associated with creep and stress-rupture properties. More detailed information on the elevated-temperature characteristics of engineering alloys and nonmetallic materials can be found throughout this Volume.

Elevated-Temperature Tensile Properties

The influence of temperature on the strength of materials can be demonstrated by running standard, short-time tensile tests at a series of increasing temperatures. Such tests are described in ASTM standards E 8 and E 21.

Material Behavior. As shown in Fig. 1, materials generally become weaker with increasing temperature. Although simple, stable alloys exhibit increased ductility behavior for most engineering materials (Fig. 1b) varies greatly. Such discontinuities in ductility with increasing temperature usually can be traced to metallurgical instabilities—carbide precipitation, for example—that affect the failure mode.

Due to the relatively high strain rates—usually 8.33×10^{-5} s^{-1} (0.5%/min) and 8.33×10^{-4} s^{-1} (5%/min)—involved in tensile testing, deformation occurs by slip (glide of dislocations along definite crystallographic planes). Thus, changes in strength and ductility with temperature generally can be related to the effect of temperature on slip. At low temperatures (less than 0.3 homologous temperature, which is the ratio of the test temperature, T, to the melting point, T_M, of the material being tested), the number of slip systems is restricted, and recovery processes are not possible. Therefore, strain-hardening mechanisms, such as dislocation intersections and pileups, lead to the increasingly higher forces required for continued deformation. This continues until the local stresses at pileups exceed the fracture stress, and failure occurs.

At higher temperatures (between 0.3 and 0.5 homologous temperature), thermally activated processes such as multiple slip and cross slip allow the high local stresses to be relaxed, and strength is decreased. For sufficiently high temperatures in excess of half the homologous temperature, diffusion processes become important, and mechanisms such as recovery, dislocation climb, recrystallization, and grain growth can reduce the dislocation density, prevent pileups, and further reduce strength.

Deformation under tensile conditions is governed to some extent by crystal structure. Face-centered cubic materials generally exhibit a gradual change in strength and ductility as temperature decreases. Such changes for type 304 stainless steel are illustrated in Fig. 1. Some body-centered cubic alloys, however, exhibit an abrupt change at the ductile-to-brittle transition temperature (approximately 200 °C, or 390 °F, for tungsten in Fig. 1), below which there is little

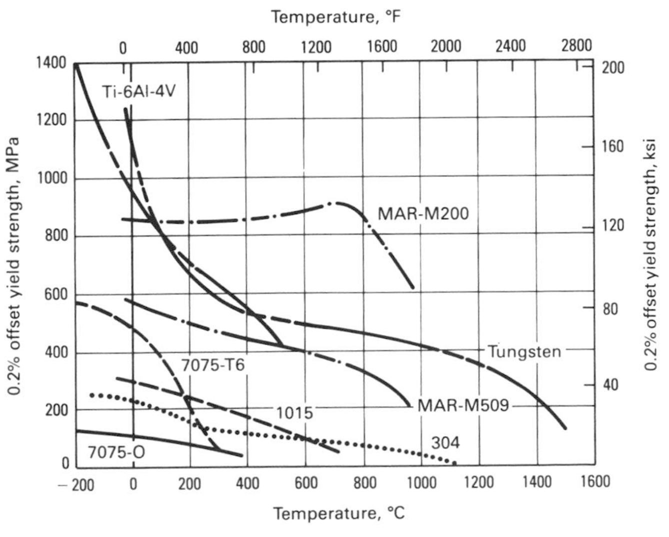

(a)

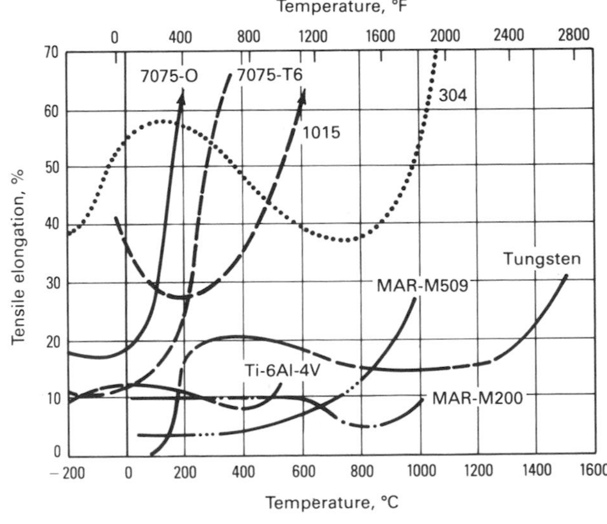

(b)

Fig. 1 Effect of temperature on strength and ductility of various materials. (a) 0.2% offset yield strength. (b) Tensile elongation. Materials tested include aluminum alloy 7075 in two heat-treated conditions (Ref 1); Ti-6Al-4V (Ref 1); AISI 1015 low-carbon steel (Ref 2); type 304 stainless steel (Ref 1); cobalt-base alloy MAR-M509 (Ref 3); directionally solidified nickel-base alloy MAR-M200 (Ref 3); and pure tungsten (Ref 4).

plastic flow. In close-packed hexagonal and body-centered cubic materials, mechanical twinning also can occur during testing. However, twinning by itself contributes little to the overall elongation; its primary role is to reorient previously unfavorable slip systems to positions in which they can be activated.

Other factors can affect tensile behavior; however, the specific effects cannot be predicted easily. For example, re-solutioning, precipitation, and aging (diffusion-controlled particle growth) can occur in two-phase alloys during heating prior to testing and during the actual testing. These processes can produce a wide variety of responses in mechanical behavior depending on the material. Diffusion processes also are involved in yield point and strain-aging phenomena. Under certain combinations of strain rate and temperature, interstitial atoms can be dragged along with dislocations, or dislocations can alternately break away and be re-pinned, producing serrations in the stress-strain curves.

There are exceptions to the above generalizations, particularly at elevated temperatures. For example, at sufficiently high temperatures, the grain boundaries in polycrystalline materials are weaker than the grain interiors, and intergranular fracture occurs at relatively low elongation. In complex alloys, hot shortness, in which a liquid phase forms at grain boundaries, or grain boundary precipitation can lead to low strength and/or ductility.

Because alloys undergoing elevated tensile testing will, in effect, be subject to annealing prior to loading, changes in microstructure can occur and produce a material that is not characteristic of the original stock. Thus, very slow heating or prolonged holds at temperature should be avoided. Figure 2 illustrates the influence of hold time on the yield strength and ductility of a precipitation-strengthened aluminum alloy. Holding at 150 °C (300 °F) changes the amount and distribution of the reinforcing phases in such a manner that strengthening initially occurs. This is subsequently followed by weakening. Clearly, the exposed material is not the same as one that is tested rapidly.

Environmental Effects. Test environment can also affect the measured properties. Generally, the atmosphere should reflect the intended or proposed use of the material. Although the environment can rarely be a complete simulation of operating conditions, it should produce the same basic effects and should not introduce foreign attack mechanisms. For example, it would be appropriate to test oxidation-resistant alloys at elevated temperature in air; however, such conditions cannot be used for refractory metals that undergo catastrophic oxidation.

Creep and Stress-Rupture

Creep is the slow deformation of a material under a stress that results in a permanent change in shape. Generally, creep pertains to rates of deformation less than 1.0%/min; faster rates are usually associated with mechanical working (processes such as forging and rolling). Shape changes arising from creep generally are undesirable and can be the limiting factor in the life of a part. For example, blades on the spinning rotors in turbine engines slowly grow in length during operation and must be replaced before they touch the housing.

Although creep can occur at any temperature, only at temperatures exceeding about 0.4 of the melting point of the material are the full range of effects visible ($T \geq 0.4\ T_M$,). At lower temperatures, creep is generally characterized by an ever-decreasing strain rate, while at elevated temperature, creep usually proceeds through three distinct stages and ultimately results in failure.

A schematic representation of creep in both temperature regimes is shown in Fig. 3. At time = 0, the load is applied, which produces an immediate elastic extension that is greater for high-temperature tests due to the lower modulus. Once loaded, the material initially deforms at a very rapid rate, but as time proceeds, the rate of deformation progressively decreases. For low temperatures, this type of behavior can continue indefinitely. At high temperatures, however, the regime of constantly decreasing strain rate (primary or first-stage creep) leads to conditions where the rate of deformation becomes independent of time and strain. When this occurs, creep is in its second-stage or steady-state regime.

Although considerable deformation can occur under these steady-state conditions, eventually the strain rate begins to accelerate with time, and the material enters tertiary or third-stage creep. Deformation then proceeds at an ever-faster rate until the material can no longer support the applied stress and fracture occurs. With ε, t, and $\dot{\varepsilon}$ representing strain, time, and strain rate, respectively, general behavior during creep can be characterized as follows:

Stage	Temperature	Characteristic
First (primary)	$T > 0.4\ T_M$ or $T \leq 0.4\ T_M$	$\dot{\varepsilon}$ decreases as t and ε increase
Secondary (steady state)	$T \geq 0.4\ T_M$	$\dot{\varepsilon}$ is constant ($\dot{\varepsilon}_{ss}$)
Third (tertiary)	$T \geq 0.4\ T_M$	$\dot{\varepsilon}$ increases as t and ε increase

In addition to affecting temperature, stress affects creep, as shown in Fig. 4. In both temperature regimes, the elastic strain on loading increases with increasing applied stress. At low temperatures (Fig. 4a), very high stresses (σ_4) near or above the ultimate tensile stress result in rapid deformation and fracture at time t_4. A somewhat lesser stress (σ_3) can result in a long period of constantly decreasing strain rate, followed by a short transition to an accelerating rate and failure at t_3. Finally, lowered stresses (σ_2 and σ_1) exhibit ever-decreasing creep rates, where σ_2 produces more elastic and plastic strain than σ_1 in the same period. The stress range over which behavior changes from that of σ_4 to that of σ_2 is small, and fracture under stress σ_3 is likely to be

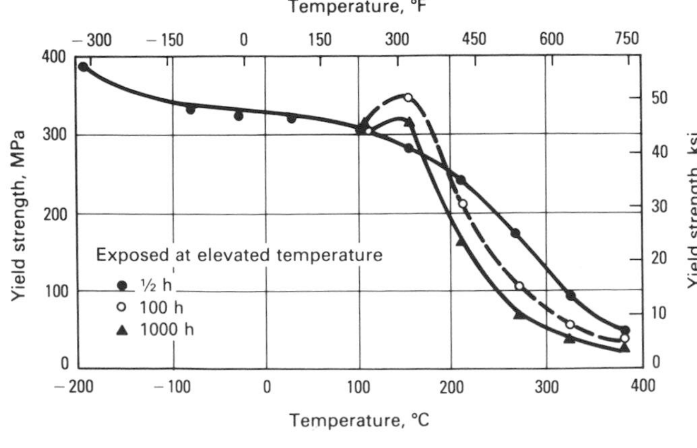

(a)

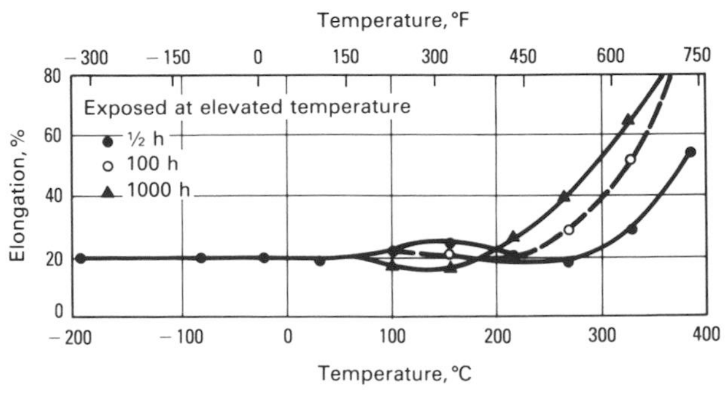

(b)

Fig. 2 Effect of exposure time and temperature on the tensile properties of naturally aged aluminum alloy 2024-T4. (a) Yield strength. (b) Percent elongation. Source: Ref 1

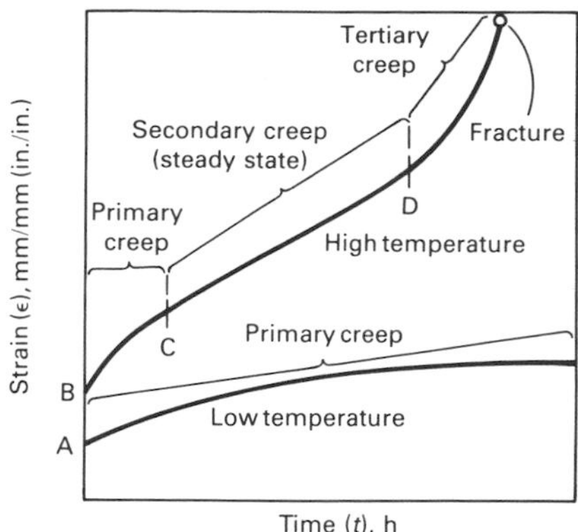

Fig. 3 Low-temperature and high-temperature creep of a material under a constant engineering stress. A and B denote the elastic strain on loading; C denotes transition from primary (first-stage) to steady-state (second-stage) creep; D denotes transition from steady-state to tertiary (third-stage) creep.

the result of microstructural and/or mechanical instabilities.

At elevated temperatures (Fig. 4b), increasing the initial stress usually shortens the period of time spent in each stage of creep. Hence, the time-to-rupture (t_6, t_7, and t_8) decreases as stress is increased. Additionally, the steady-state creep rate decreases as the applied stress is decreased. The stress range over which behavior changes from that exhibited by stresses σ_8 and σ_5 (Fig. 4b) is much broader than the range necessary to yield similar behavior at low temperatures (Fig. 4a).

Most of the behavior shown in Fig. 3 and 4 can be understood in terms of the Bailey-Orowan model (Ref 5), which views creep as the result of competition between recovery and work-hardening processes. Recovery is the mechanism(s) through which a material becomes softer and regains its ability to undergo additional deformation. In general, exposure to high temperature (stress relieving after cold working, for example) is necessary for recovery processes to be activated. Work-hardening processes make a material increasingly more difficult to deform as it is strained. The increasing load required to continue deformation between the yield stress and the ultimate tensile stress during a short-term tensile test is an example of work hardening.

After the load is applied, fast deformation begins, but this is not maintained as the material work hardens and becomes increasingly more resistant to further deformation. At low temperatures, recovery cannot occur; hence, the creep rate is always decreasing. However, at elevated temperatures, softening can occur, which leads to the steady state, in which recovery and hardening processes balance one another. As the temperature increases, recovery becomes easier to activate and overcomes hardening. Thus, the transition from primary to secondary creep generally occurs at lower strains as temperature increases.

Third-stage creep cannot be rationalized in terms of the Bailey-Orowan model. Instead, tertiary creep is the result of microstructural and/or mechanical instabilities. For instance, defects in the microstructure, such as cavities, grain-boundary separations, and cracks develop. These result in a local decrease in cross-sectional area that corresponds to a slightly higher stress in this region.

Because creep rate is dependent on stress, the strain and strain rate in the vicinity of a defect will increase. This then leads to an increase in the number and size of microstructural faults, which in turn further decreases the local cross-sectional area and increases the strain rate. Additionally the microstructural defects, as well as other heterogeneities, can act as sites for necking. Once formed, deformation tends to increase in this region, because the local stress is higher than in other parts of the specimen. The neck continues to grow, because more local deformation yields higher stresses.

Creep Experiments

The creep behavior of a material is generally determined by uniaxial loading of test specimens heated to temperature in some environment. Creep-rupture experiments measure the deformation as a function of time to failure. If strain-timed behavior is measured, but the test is stopped before failure, this is termed an interrupted creep experiment. Finally, if an inadequate strain-measuring system or no attempt to determine length is employed, and the test is run to fracture, a stress-rupture experiment results.

In terms of data that characterize creep, the stress-rupture test provides the least amount, because only the time-to-rupture and strain-at-rupture data are available for correlation with temperature and stress. These data and other information, however, can be obtained from creep-rupture experiments. Such additional measurements can include elastic strain on loading, amount of primary creep strain, time to onset of secondary creep, steady-state creep rate, amount of secondary creep, time to onset of tertiary creep, time to 0.5% strain, time to 1.0% strain, and so on. All of these data can be fitted to equations, involving temperature and stress. An interrupted creep test provides much the same data as a creep-rupture experiment within the imposed strain-time limitations.

Direction of Loading. Most creep-rupture tests of metallic materials are conducted in uniaxial tension. Although this method is suitable for ductile metals, compressive testing is more appropriate for brittle, flaw-sensitive materials. In compression, cracks perpendicular to the applied stress do not propagate as they would in tension; thus, a better measure of the inherent plastic properties of a brittle material can be obtained.

In general, loading direction has little influence on many creep properties—for example, steady-state creep rate in ductile materials (Ref 6). However, even in these materials, the onset of third-stage creep and fracture is usually delayed in compression compared to tension. This delay is due to the minimized effect of microstructural flaws and the inability to form a "neck-like" mechanical instability. For brittle materials, the difference in behavior between tension and compression can be extreme, primarily due to the response to flaws. Consequently, care must be exercised when using compressive creep properties of a brittle material to estimate tensile behavior.

Test specimens for uniaxial tensile creep-rupture tests are the same as those used in short-term tensile tests. Solid round bars with threaded or tapered grip ends or thin sheet specimens with pin and clevis grip ends are typical (refer to ASTM standard E 8). However, many other types and sizes of specimens have been used successfully where the choice of geometry was dictated by the available materials. For example, small threaded round bars with a 12 mm (0.47 in.) overall length and a 1.52 mm (0.06 in.) diameter by 5 mm (0.2 in.) long reduced section have been used to measure transverse stress-rupture properties of a 13 mm (0.51 in.) diameter directionally solidified eutectic alloy bar (Ref 7).

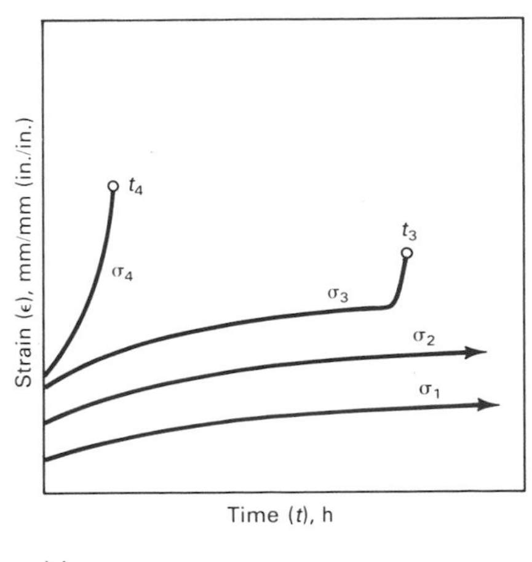

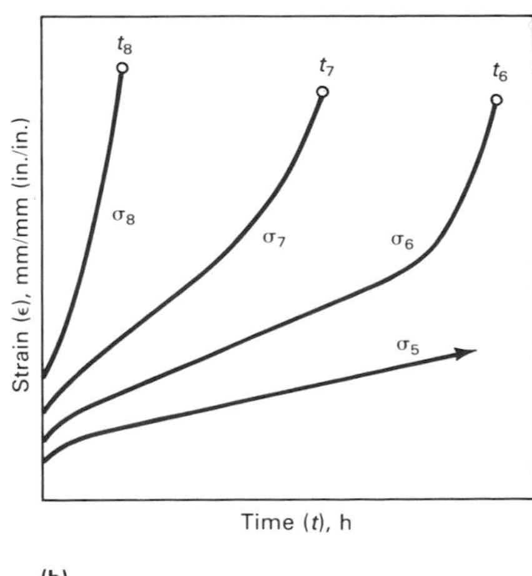

Fig. 4 Elevated-temperature creep in a material as a function of stress where the time-to-rupture is t_i for stress σ_i. (a) Low-temperature creep, where $\sigma_4 > \sigma_3 > \sigma_2 > \sigma_1$. (b) High-temperature creep, where $\sigma_8 > \sigma_7 > \sigma_6 > \sigma_5$

In the case of uniaxial compression testing, specimen design can be simple small-diameter right cylinders or parallelepipeds with length-to-diameter ratios ranging from approximately 2 to 4. Larger ratios tend to enhance elastic buckling, and smaller ratios magnify the effects of friction between the test specimen and the load-transmitting member. These specimen geometries are well suited for creep testing when only a small amount of material is available, or when the material is difficult to machine.

Environment. The optimum conditions for a creep-rupture test are those in which the specimen is influenced only by the applied stress and temperature. This rarely occurs, particularly at elevated temperatures, and these conditions do not exist for real structures and equipment operating under creep conditions. For example, turbine blades are continuously exposed to hot, reactive gases that cause corrosion and oxidation.

Reactions between the test environment and material vary greatly, ranging from no visible effect to large-scale attack. For example, creep-rupture testing of aluminum, iron-chromium-aluminum, nickel-chromium, and nickel-base superalloys at elevated temperatures in air can generally be accomplished without problems, because these materials form thin, stable, protective oxide films. This is not the case for refractory metals (molybdenum, niobium, tantalum, and tungsten) and their alloys, due to their strong reaction with oxygen, which leads to the formation of porous, nonprotective, and in some cases volatile oxides. Environmental effects such as oxidation and corrosion reduce the load-bearing cross-sectional area and can also facilitate the formation and growth of cracks.

Reactions are also possible in inert atmospheres (such as vacuum) and in reducing gas environments. Elevated-temperature testing in vacuum can result in the loss of volatile alloying elements and subsequent loss of strength. Exposure to reducing gases can result in the absorption of interstitial atoms (carbon, hydrogen, and nitrogen), which may increase strength, but also induce brittleness.

A "perfect" environment does not exist for all creep-rupture testing. The appropriate choice depends on the material, its intended use, and the available environmental protection methods. If creep mechanisms are being determined, then the atmosphere should be as inert, or nonreactive, as possible. However, if the material is to be used in an unprotected state in a reactive atmosphere, then creep-rupture testing should reflect these conditions.

Creep-rupture data from inert atmosphere tests cannot be used for design purposes when the material will be exposed to conditions of severe oxidation. However, if environmental protection methods, such as oxidation- or corrosion-resistant coatings, are available, then testing in inert gas is acceptable, and the resulting data can be used for design.

If reactions occur between the test environment and the specimen, the resultant creep-rupture data will not reflect the true creep properties of the material. Rather, the measured data are indicative of a complex interaction between creep and environmental attack, where the effects of environmental attack become more important in long-term exposure.

Strain Measurement. Care must be taken to ensure that the measured deformation occurs only in the gage section. Thus, measurements based on the relative motion of parts of the gripping system above and below the test specimen are generally inaccurate, because the site of deformation is unknown. Extensometry systems are currently available that attach directly to the specimen (shoulders, special ridges machined on the reduced section, or the gage section itself) and transmit the relative motion of the top and bottom of the gage section via tubes and rods to a sensing device such as a linear variable differential transformer. Figure 5 illustrates such a system. These systems are quite accurate and stable over long periods of time.

Other methods of direct strain measurement exist and, under certain circumstances, are suitable. At low temperatures, strain gages can be directly bonded to the gage section and can be used to follow deformation over the range of extension for which the strain gage is valid. For specimens that will undergo reasonable deformations ($\varepsilon > 1.0\%$), the distance between two gage marks can be optically tracked with a cathetometer as a function of time. While the location of strain is known, use of this technique is operator dependent and is generally limited to tests of less than 8 h or greater than 100 h in duration in order

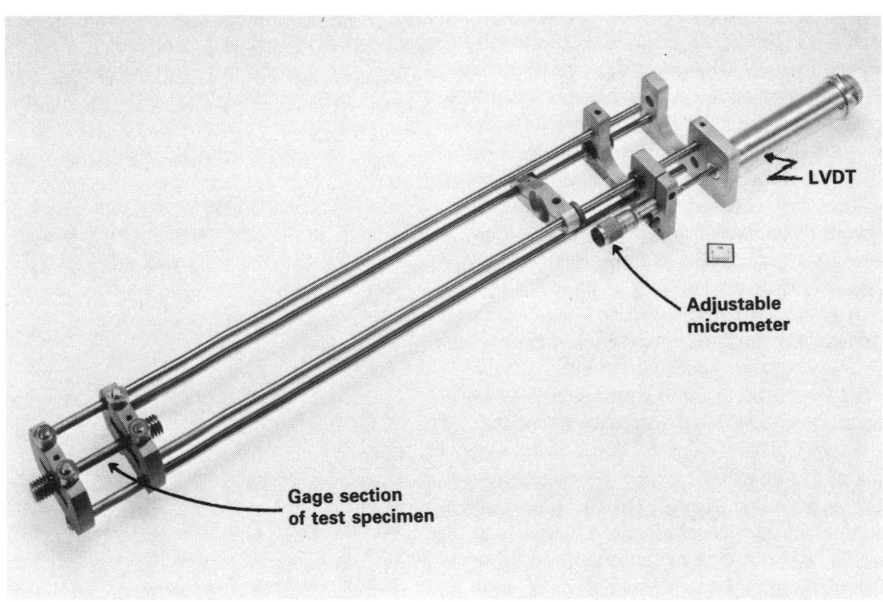

Fig. 5 Typical rod-and-tube-type extensometer for elevated-temperature creep testing. Extensometer is clamped to grooves machined in the shoulders of the test specimen. LVDT, linear variable differential transformer

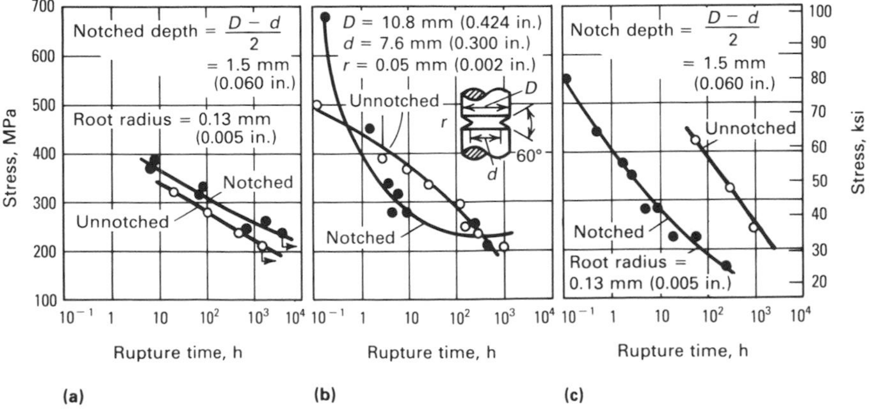

Fig. 6 Three general types of notch effects in stress-rupture tests. (a) Notch strengthening in 19-9 DL heat treated 50 h at 650 °C (1200 °F) and air cooled. (b) Mixed behavior in Haynes 88 heat treated 1 h at 1150 °C (2100 °F), air cooled, and worked 40% at 760 °C (1400 °F). (c) Notch weakening in K-42-B heat treated 1 h at 955 °C (1750 °F), water quenched, reheated 24 h at 650 °C, and air cooled

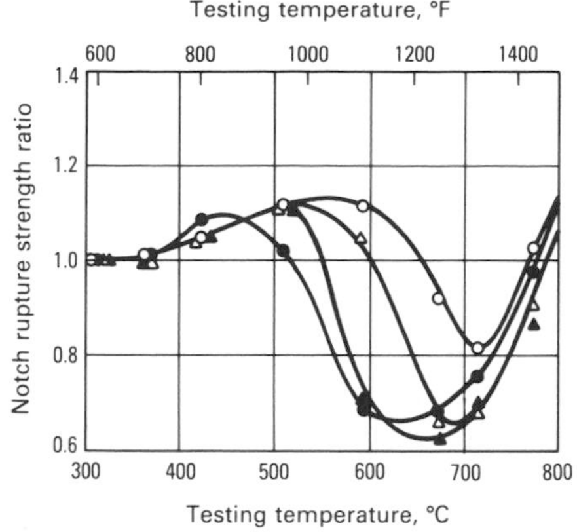

Fig. 7 Notch-rupture strength ratio vs. temperature at four different rupture times for Inconel X-750

to permit sufficient readings to properly define the creep curve.

Notched-Specimen Testing

Notched specimens are used principally as a qualitative alloy selection tool for comparing the suitability of materials for components that may contain deliberate or accidental stress concentrations. The rupture life of notched specimens is an indication of the ability of a material to deform locally without cracking under multiaxial stresses. Because this behavior is typical of superalloys, the majority of notched-specimen testing is performed on superalloys.

The most common practice is to use a circumferential 60° V-notch in round specimens, with a cross-sectional area at the base of the notch one-half that of the unnotched section. However, size and shape of test specimens should be based on requirements necessary for obtaining representative samples of the material being investigated.

In a notch test, the material being tested most severely is the small volume at the root of the notch. Therefore, surface effects and residual stresses can be very influential. The notch radius must be carefully machined or ground, because it can have a pronounced effect on test results. The root radius is generally 0.13 mm (0.005 in.) or less and should be measured using an optical comparator or other equally accurate means. Size effects, stress-concentration factors introduced by notches, notch preparation, grain size, and hardness are all known to affect notch-rupture life.

Notch-rupture properties can be obtained by using individual notched and unnotched specimens, or by using a specimen with a combined notched and unnotched test section. The ratio of rupture strength of notched specimens to that of unnotched specimens varies with notch shape and acuity, specimen size, rupture life (and therefore stress level), testing temperature, and heat treatment and processing history.

To avoid introducing large experimental errors, notched and unnotched specimens must be machined from adjacent sections of the same piece of material, and the gage sections must be machined to very accurate dimensions. For the combination specimen, the diameter of the unnotched section and the diameter at the root of the notch should be the same within ±0.025 mm (±0.001 in.).

Notch sensitivity in creep rupture is influenced by various factors, including material and test conditions. The presence of a notch may increase life, decrease life, or have no effect. When the presence of a notch increases life over the entire range of rupture time, as shown in Fig. 6(a), the alloy is said to be notch strengthened; that is, the notched specimen can withstand higher nominal stresses than the unnotched specimen. Conversely, when the notch-rupture strength is consistently below the unnotched-rupture strength, as in Fig. 6(c), the alloy is said to be

Table 1 Effect of a notch on rupture time of Waspaloy

Condition	Hours to failure at 730 °C (1350 °F) and 360 MPa (52 ksi): Unnotched bar	Notched bar
Solution heat treated 4 h at 1080 °C (1975 °F), air cooled, aged 16 h at 760 °C (1400 °F), air cooled	76.0	1.5
Solution heat treated 4 h at 1080 °C (1975 °F), air cooled, stabilized 4 h at 845 °C (1550 °F), air cooled, aged 16 h at 760 °C (1400 °F), air cooled	82.8	150(a)
Solution heat treated 4 h at 1080 °C (1975 °F), air cooled, stabilized 4 h at 870 °C (1600 °F), air cooled, aged 16 h at 760 °C (1400 °F), air cooled	87.4	150(a)
Solution heat treated 4 h at 1080 °C (1975 °F), air cooled, stabilized 1 h at 980 °C (1800 °F), air cooled, aged 16 h at 760 °C (1400 °F)	1.9	46.6

(a) No failure; test discontinued after 150 h

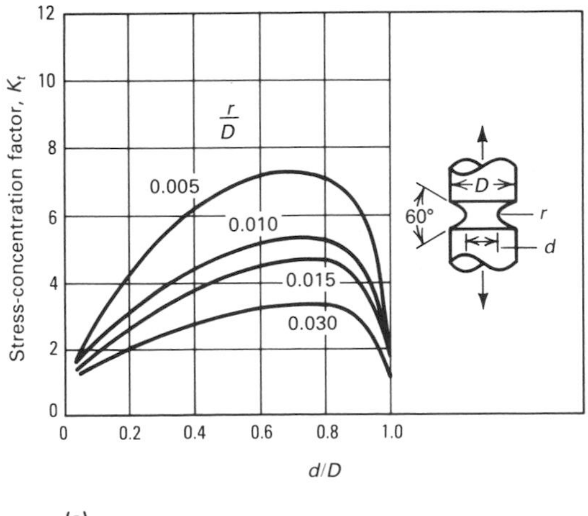

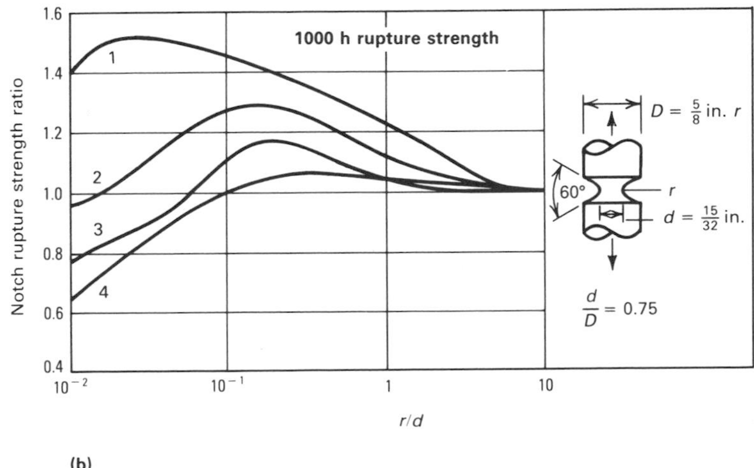

Fig. 8 Effect of notch dimensions on stress concentration and notch-rupture strength ratio. (a) Variation of stress-concentration factor with ratio of minor to major diameter and with ratio of root radius to major diameter for notched bar stressed in tension within the elastic range. (b) Variation of notch-rupture strength ratio for 1000 h life with ratio of root radius to minor diameter. Curve 1 is for 12Cr-3W steel heated 3 h at 900 °C (1650 °F) and air cooled. Grain size, ASTM No. 12; hardness, 215 HV; unnotched rupture ductility, 40%; test temperature, 540 °C (1000 °F). Curve 2 is for Refractaloy 26 oil quenched from 1010 °C (1850 °F); reheated 20 h at 815 °C (1500 °F) and air cooled; reheated 20 h at 650 °C (1200 °F) and air cooled; reheated 20 h at 815 °C and air cooled; and finally reheated 20 h at 650 °C and air cooled. Grain size, ASTM No. 7 to 8; hardness, 330 HV; unnotched rupture ductility, 7%; test temperature 650 °C. Curve 3 is for Refractaloy 26 oil quenched from 1175 °C (2150 °F); reheated 20 h at 815 °C and air cooled; reheated 20 h at 730 °C (1350 °F) and air cooled; and finally reheated 20 h at 650 °C and air cooled. Grain size, ASTM No. 2 to 3; hardness, 325 HV; unnotched rupture ductility, 10%; test temperature, 650 °C. Curve 4 is for Refractaloy 26 oil quenched from 980 °C (1800 °F); reheated 44 h at 730 °C and air cooled; and finally reheated 20 h at 650 °C and air cooled. Grain size, ASTM 7 to 8; hardness, 375 HV; unnotched rupture ductility, 3%; test temperature, 650 °C

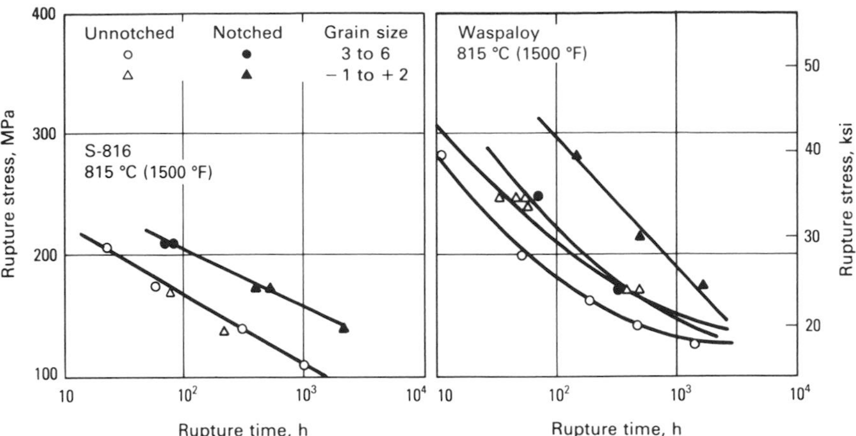

Fig. 9 Rupture strength as a function of time for notched and unnotched bars of different grain size. S-816 was heated to 1175 °C (2150 °F) and water quenched, reheated to 760 °C (1400 °F), held 12 h, and air cooled. Waspaloy was heated to 1080 °C (1975 °F), held 4 h, and air cooled; reheated to 840 °C (1550 °F), held 4 h, and air cooled; and finally reheated to 760 °C (1400 °F), held 16 h, and air cooled. Smaller grain sizes were produced by cold reducing the S-816 1%, and the Waspaloy 1.25%, by cold rolling at 24 °C (75 °F), and then heat treating. Diameter of specimens was 12.7 mm (0.5 in.), diameter at base of notch was 8.9 mm (0.35 in.), root radius was 0.1 mm (0.004 in.), and notch angle was 60°. Data are a composite of results from two laboratories.

notch sensitive, or notch weakened. Many investigators have defined a notch-sensitive condition as one for which the notch strength ratio is below unity. However, this ratio is unreliable and can vary according to class of alloy and rupture time.

Certain alloys and test conditions show notch strengthening at high nominal stresses (short rupture times) and notch weakening at lower nominal stresses (longer rupture times), with the result that the stress-rupture curve for notched specimens crosses the curve for unnotched specimens as nominal stress is reduced. Figure 6(b) shows that Haynes 88 becomes notch sensitive under high nominal stresses in a rupture time of about 2 h and that the material becomes notch strengthened again at lower nominal stresses at a rupture time of approximately 400 h. This same phenomenon has been observed in many superalloys and is illustrated in a different manner in Fig. 7. The "notch ductility trough" varies with alloy composition. For example, A-286 is notch sensitive at 540 °C (1000 °F), whereas Inconel X-750 is notch sensitive at 650 °C (1200 °F). A given alloy may show notch weakening at some temperatures and notch strengthening at others. Generally, notch sensitivity appears to increase as temperature is reduced.

Changes in heat treatment of some alloys may alter notch sensitivity significantly. For example, single low-temperature aging of some alloys may produce very low rupture ductilities, because the structure is not sufficiently stabilized. Consequently, exposure of such materials for prolonged rupture times will further reduce rupture ductility because of continued precipitation of particles that enhance notch sensitivity. On the other hand, multiple aging usually stabilizes the structure and thus reduces notch sensitivity

Notch configuration can have a profound effect on test results, particularly in notch-sensitive alloys. Most studies on notch configuration present results in terms of the elastic stress-concentration factor. The design criterion for the weakening effect of notches at normal and low temperatures is that of complete elasticity. The design stress is the yield stress divided by the elastic stress-concentration factor K_t (Fig. 8a). The value of the peak axial (design) stress depends on the configuration of the notch.

There is no simple relationship for the effect of notches at elevated temperatures. The metallurgical effects that influence the behavior of notched material are complex and include composition, fabrication history, and heat treatment. Effects of several heat treatments on rupture time of Waspaloy are shown in Table 1.

For ductile metals, the ratio of rupture strength of notched specimens to that of unnotched specimens usually increases to some maximum as the stress-concentration factor is increased. For very insensitive alloys, there may be little further change. Metals that are more notch sensitive may undergo a reduction in ratio as the notch sharpness (stress-concentration factor) is increased beyond the maximum and may show notch weakening for even sharper notches. Very notch-sensitive alloys may undergo little or no notch strengthening, even for very blunt notches (low stress-concentration factor) and may undergo progressive weakening as notch sharpness increases.

Relationships between notch configuration and the ratio of rupture strengths of notched and unnotched specimens are shown in Fig. 8(b). In curve 1, for an alloy with an unnotched rupture ductility of 40%, the notch-strengthening factor decreases as the notch is decreased in sharpness (increase in ratio r/d). In curve 2, for an alloy with unnotched rupture ductility of 7%, the notch-strength factor increases with increasing notch sharpness, reaches a peak, and then drops to a notch-strength reduction factor of less than unity. For an alloy with a still lower unnotched rupture ductility of 3% (curve 4), the notch-strength factor is only slightly greater than unity for large radii of curvature and becomes less than unity. It continues to decrease as the notches become sharper.

Effect of Grain Size and Other Variables. The effects of grain size on notched and unnotched rupture strength are shown in Fig. 9. The coarse grain sizes (ASTM –1 to +2) were obtained by reheating bars in which small strains had been introduced by cold reducing them 1 to 1.25%. Notches had a strengthening effect on both S-816 and Waspaloy when tested at 815 °C (1500 °F). There was no measured effect of grain size on either the notched or unnotched specimens of S-816. On the other hand, the coarse-grained Waspaloy specimens showed a longer rupture time at the same rupture stress for both notched and unnotched specimens.

The rupture time for Discaloy at 650 °C (1200 °F) increases with increasing hardness up to about 290 HV for notched specimens ($K_t = 3.9$) and up to 330 HV for unnotched specimens, as shown in Fig. 10. Ductility, as measured by elongation values for unnotched bars, decreases with increasing hardness.

The peak in rupture time at 650 °C (1200 °F) corresponds to a rupture elongation of 1.5%. The continual reduction in rupture elongation with increasing hardness indicates that the alloy exhib-

Table 2 Comparison of creep-rupture strengths for notched and unnotched specimens of S-816

Temperature		Static rupture strength			
		Unnotched		Notched(a)	
°C	°F	MPa	ksi	MPa	ksi
24	75	1010	147(b)	...	...
595	1100	625	91(c)	...	...
650	1200	450	65(c)	...	...
730	1350	290	42	405	59
815	1500	170	25	255	37
900	1650	97	14	...	...

(a) Circular, 60° V-notch, D = 9.5 mm (0.375 in.), d = 6.4 mm (0.25 in.), r = 0.25 mm (0.010 in.), K_t = 3.4. (b) Tensile strength. (c) Typical values; all other test specimens from same heat its time-dependent notch sensitivity. Notched bars exhibit a strengthening effect at lower hardnesses and higher ductilities; for specimens of higher hardness and lower ductility, rapid notch weakening is apparent.

For this particular alloy at this temperature, 5% rupture elongation indicates the point at which no notch strengthening or weakening occurs; this point is also indicated by the crossover of the two curves in Fig. 10 at about 318 HV. For other alloys, this crossover may occur at rupture ductilities as low as 3% or as high as 25%. Alloys with lower rupture ductilities are more notch sensitive.

The effects of notches on rupture life of three superalloys are shown in Fig. 11. Nimonic 80A, S-816, Inconel 751 (formerly Inconel X-550), and Waspaloy show various degrees of notch sensitivity. Nimonic 80A shows a notch-strengthening effect at 650 °C (1200 °F) but is notch weakened in about 100 h at 705 °C (1300 °F) and in about 40 h at 760 °C (1400 °F).

Nimonic 80A becomes notch strengthened at 815 °C (1500 °F), which illustrates that the alloy exhibits a notch ductility trough between 705 and 760 °C (1300 and 1400 °F). Using 0.13 mm (0.005 in.) as the standard notch radius, Waspaloy and Inconel 751 both exhibited notch weakening at the lower test temperatures 650 and 730 °C (1200 and 1350 °F) and notch strengthening at the higher test temperatures 815 and 870 °C (1500 and 1600 °F).

Similar tests on Inconel X-750 with rupture ductilities of 10% reduction in area at 730 °C (1350 °F) and 24% at 815 °C (1500 °F) indicated notch strengthening at both temperatures. These results support the observation that materials with high rupture ductilities under the initial test conditions will be less notch sensitive in long exposure times than materials with low initial rupture ductilities. Alloy S-816 further illustrates this theory in that the alloy has very high rupture ductilities at each test temperature and does not show signs of notch weakening at any test temperature (Table 2).

The curves for the alloys with various notch radii show that, in general, notch sensitivity increases with increased notch severity at the lower test temperatures. This effect is particularly evidenced in Waspaloy. At 650 and 730 °C (1200 and 1350 °F) and for a notch radius of 0.13 mm (0.005 in.), the alloy is highly notch sensitive, but shows notch strengthening at these same test temperatures at a blunter notch radius of 2.5 mm (0.10 in.).

The effect of the radius at the root of the notch is minimal for Inconel 751 at 730 °C (1350 °F), in that the material is notch sensitive under all conditions of notch severity at prolonged rupture times. The material does show slight notch strengthening, but only with the larger radii and very short rupture times. Thus, the larger radii compensate for notch sensitivity in Waspaloy, but not in Inconel 751 at 730 °C (1350 °F).

At the highest test temperature—815 °C (1500 °F)—Waspaloy was notch strengthened under all conditions of notch severity. The notch radius did not have any effect on alloy S-816, as evidenced by notch strengthening at all notch severities and

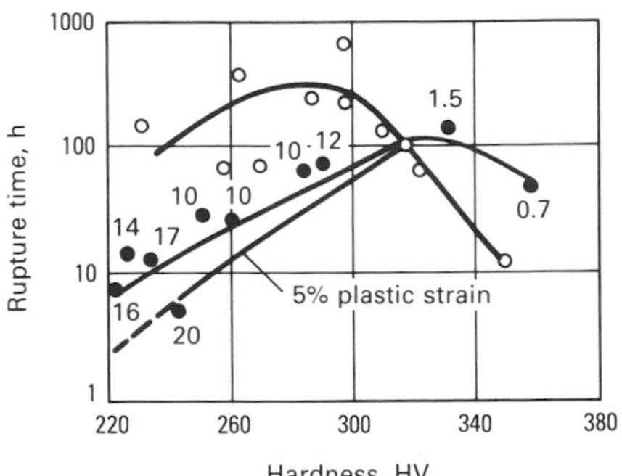

Fig. 10 Variation of rupture time at 650 °C (1200 °F) with initial hardness for Discaloy. Open symbols indicate notched-bar tests (K_t = 3.9); solid symbols indicate smooth-bar tests. Numbers adjacent to points are total elongations for these tests.

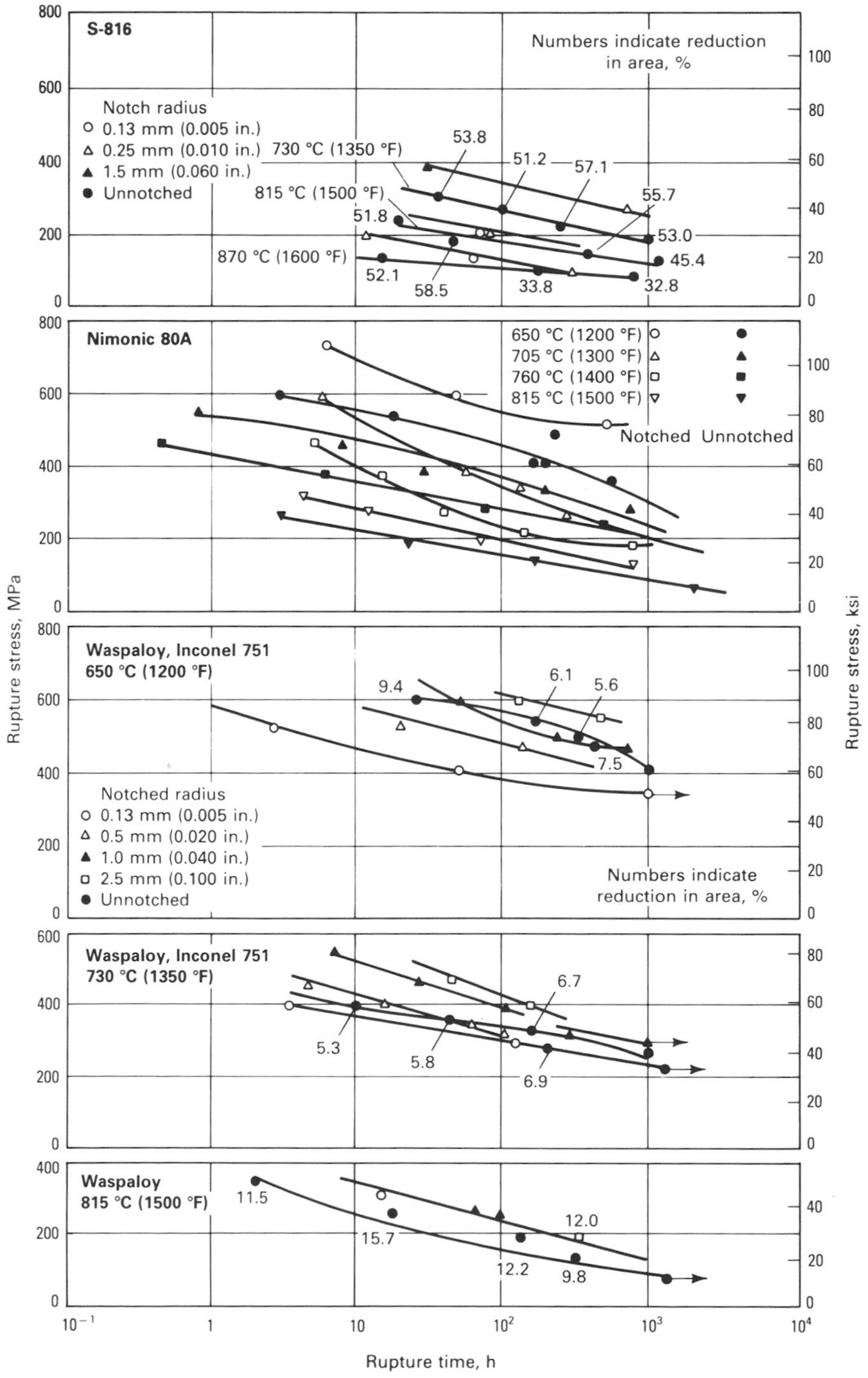

Fig. 11 Effects of notches on rupture life of three superalloys

test temperatures. However, high rupture ductilities enhanced notch strengthening.

Data Presentations

Generally, all creep and stress-rupture data are analyzed in terms of three variables: time, stress, and temperature. Other factors are also important, particularly when an understanding of the process(es) in control of deformation is desired. However, for a straightforward presentation and representation of most experimental data, these three variables are sufficient.

The time-to-rupture (t_r) from either isothermal stress-rupture or creep-rupture testing is presented as a function of stress σ as:

$$t_r = K_1 \exp(a\sigma) \quad \text{(Eq 1)}$$

and

$$t_r = K_2 \sigma^m \quad \text{(Eq 2)}$$

where K_1, K_2, and a are constants, and m is the stress exponent for rupture. An example of time-to-rupture results and the use of these equations to describe the data at several temperatures are shown in Fig. 12(a) and (b). Generally, there is little difference between the exponential (Eq 1, Fig. 12a), or power law (Eq 2, Fig. 12b) descriptions of the times-to-rupture. In both cases, the data lie on straight lines, and coefficients of determination (R^2) for the linear regression fits of the data have high values.

When temperature effects as well as stress effects on time-to-rupture are to be considered, one common presentation is:

$$t_r = K_3 \sigma^m \exp\left(\frac{Q_r}{RT}\right) \quad \text{(Eq 3)}$$

where K_3 is a constant; Q_r is the activation energy for rupture; R is the universal gas constant (8.314 kJ/mol · K); and T is the absolute temperature in degrees Kelvin. Figure 12(c) describes time-to-rupture data for several temperatures as a single line. The Larson-Miller parameter (LMP) (Ref 12) represents another approach using a single curve to represent data gathered under a variety of conditions, where:

$$\text{LMP} = T(C + \log t_r) \quad \text{(Eq 4)}$$

and

$$\text{LMP} = K_4 \log \sigma + K_5 \quad \text{(Eq 5)}$$

where C, K_4, and K_5 are constants. Originally, C was set equal to 20; however, C is currently permitted to assume a value that best describes the data. Figure 12(d) illustrates the use of the Larson-Miller parameter to combine time-to-rupture data from several temperatures and stresses into one curve. For more information on time-temperature parameters, see the article "Assessment and Use of Creep-Rupture Properties" in this Volume.

Creep curves generated by either creep-rupture or interrupted creep testing are usually presented as strain versus time or log strain versus log time. Examples of both forms are given in Fig. 13. The linear presentation (Fig. 13a) gives an accurate representation of the three stages of creep. However, the strain incurred in first- and/or second-stage creep is de-emphasized if considerable strain occurs in third-stage creep. Also, the time spent in primary and/or tertiary creep is de-emphasized if steady-state creep exists over a large fraction of the time-to-rupture.

The logarithm format, on the other hand, emphasizes the time and strain during first-stage creep and, to some extent, the strain during second-stage creep at the expense of time during second-stage creep and the strain and time of third-stage creep. De-emphasis of tertiary creep

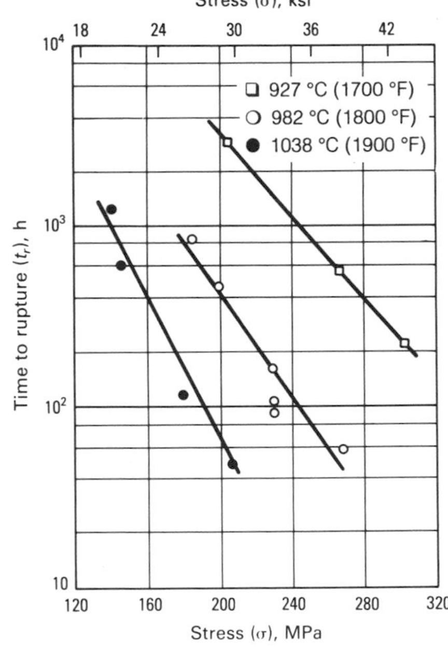

	Constants for exponential function K_1	a	Coefficient of determination, R^2
□	7.12 × 10^5	−0.0267	1.00
○	3.49 × 10^5	−0.0334	0.93
●	7.29 × 10^5	−0.0472	0.97

(a)

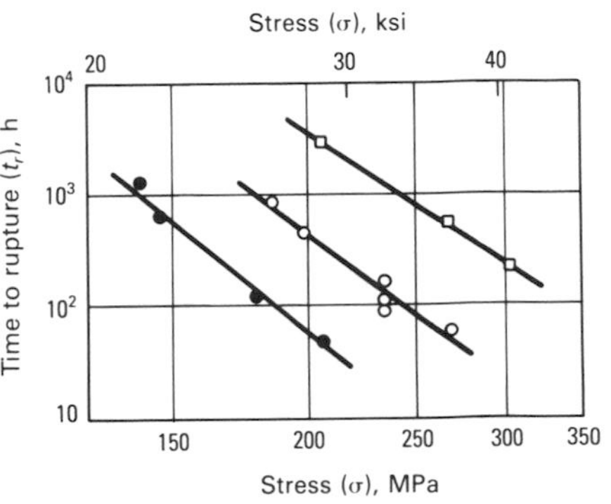

	Constants for power law function K_2	m	Coefficient of determination, R^2
□	7.96 × 10^{18}	−6.67	1.00
○	9.61 × 10^{19}	−7.54	0.95
●	2.79 × 10^{20}	−8.12	0.98

(b)

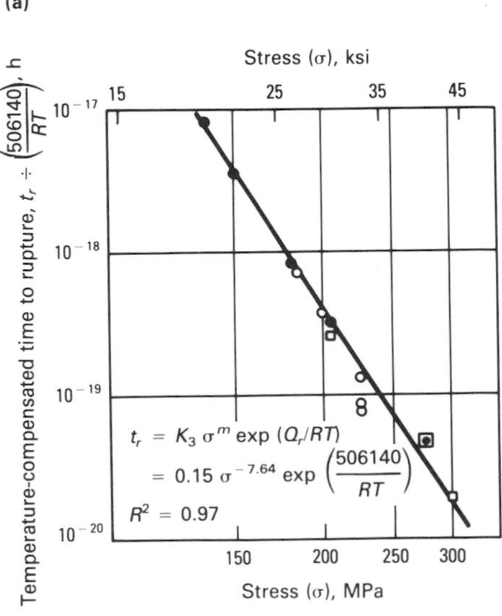

(c)

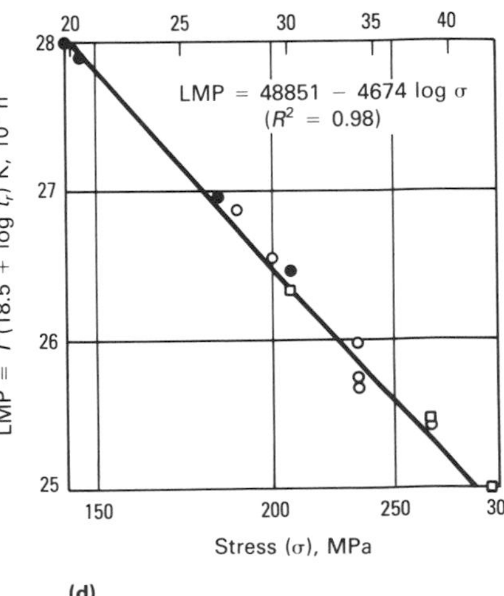

(d)

Fig. 12 Time to rupture (t_r) as a function of stress (σ) for [100] oriented Ni-5.8Al-14.6Mo-6.2Ta (wt%) single crystals tested in tension at several temperatures in air. (a) Exponential form (Eq 1). (b) Power law form (Eq 2). (c) Temperature-compensated power law form (Eq 3). (d) Larson-Miller parameter (LMP) form (Eq 5). Source: Ref 8 in the logarithm presentation allows the low strain behavior to be highlighted, which is perhaps of most interest for design purposes.

Steady-State Creep Rate. The most important creep parameter in terms of theoretical analysis is the steady-state creep rate $\dot{\varepsilon}_{ss}$. Its dependence on stress is generally expressed as:

$$\dot{\varepsilon}_{ss} = K_6\sigma^n \quad \text{(Eq 6)}$$

and for temperature and stress:

$$\dot{\varepsilon}_{ss} = K_7\sigma^n \exp\left(\frac{Q_c}{RT}\right) \quad \text{(Eq 7)}$$

where K_6 and K_7 are constants, n is the stress exponent for creep, and Q_c is the activation energy for creep. Because of theoretical development, certain values of the stress exponent for creep have been correlated with deformation mechanisms, and the activation energy for creep has been correlated to the activation energy for diffusion.

Examples of steady-state creep rates as functions of stress and temperature are shown in Fig. 14, along with the results of linear regression fits to power law creep (Eq 6) and temperature-compensated power law creep (Eq 7). Note the general agreement of the compressive test results with the tensile data in Fig. 14 and the reproducibility of $\dot{\varepsilon}_{ss}$ versus σ behavior from two castings of the same material (Fig. 14).

Deformation during Steady-State Creep. The values of n and Q_c given in Eq 7 are sensitive to the processes controlling creep (Ref 11). The region in which an individual process dominates can be obtained from a deformation mechanism map (Ref 12). A deformation mechanism map is a plot of normalized stress, σ/E, where E is Young's modulus (or shear stress divided by shear modulus) against normalized temperature, T/T_M. Figure 15 shows an example of such a plot for pure nickel. It includes contours of constant strain rate and can be used to establish the process controlling creep at a given stress and temperature.

Deformation mechanism maps show two main creep fields. In one, creep rate is governed by the glide and climb of dislocations and has a power law stress dependence. In the other, creep is controlled by the stress-directed diffusional flow of atoms. Several specific models have been proposed for these processes and the fields can be further subdivided, as indicated by the dashed lines in Fig. 15. When diffusional flow dominates, the models give $n = 1$ and the appropriate value of activation energy depends on whether grain boundary diffusion or lattice diffusion controls. When power law creep dominates, n is typically predicted to be in the range 3 to 5 (although much larger values are often measured in practice), with dislocation motion being limited at the lower temperatures by core diffusion and at the higher temperatures by lattice diffusion.

For a given mechanism, actual creep rates are dependent on material composition, microstructure, and grain size. The largest grain size dependence is observed in the diffusional flow region, with an increase in grain size resulting in a decrease in creep rate. Solid-solution and precipitation-hardening alloying additions can impede dislocation motion and influence diffusion rates. There is a general tendency for alloying additions to move mechanism boundaries to higher σ/E and T/T_M ratios (Ref 12). An illustration of this effect can be seen in Fig. 16 for the nickel-base superalloy MAR-M200 when comparison is made with pure nickel in Fig. 15.

Stress to Produce 1.0% Strain. In many cases, the objective of testing is to determine the total amount of creep strain that can be expected during stress/temperature exposures. Of particular interest are the stresses required to produce 0.5, 1.0, and 2.0% strain in a certain period of time as a function of temperature. A typical example of such a presentation is shown in Fig. 17 for several refractory alloys. Also illustrated is a case in which a metallurgical variable (grain size) was factored into the analysis to account for the difference in behavior between the two lots of the tantalum-base alloy Astar 811C.

Monkman-Grant Relationship. For elevated-temperature tensile creep-rupture experiments, the product of the time-to-rupture and steady-state creep rate raised to the power M is approximately a constant for many materials (Ref 14):

$$t_r\,\dot{\varepsilon}_{ss}^{M} = K_8 \quad \text{(Eq 8)}$$

where M and K_8 are constants with values roughly equal to 1. An example of this relationship is shown in Fig. 18 using a rearranged form of Eq 8:

$$\log t_r = \log K_8 - M\,(\log \dot{\varepsilon}_{ss}) \quad \text{(Eq 9)}$$

Once M and K_8 are known, reasonable predictions of either quantity can be made from knowledge of the other.

Other Testing Considerations

Constant Load versus Constant Stress Testing. Most uniaxial creep and stress-rupture tests are conducted under constant-load conditions. Although the method is simple, the stress in the gage section varies with strain (time). This can be seen by considering a bar of length L_0 and cross-sectional area A_0 subjected to a tensile load P. At time $t = 0$, the initial engineering stress on the bar is:

$$s_0 = \frac{P}{A_0} \quad \text{(Eq 10)}$$

With the assumption of uniform deformation during creep, the bar lengthens to L and the cross-sectional

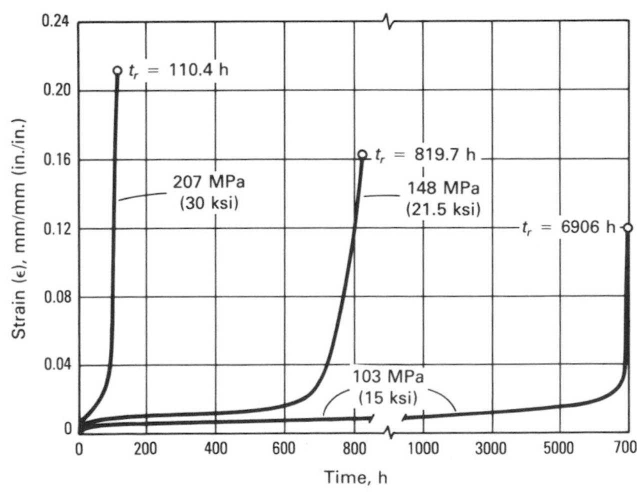

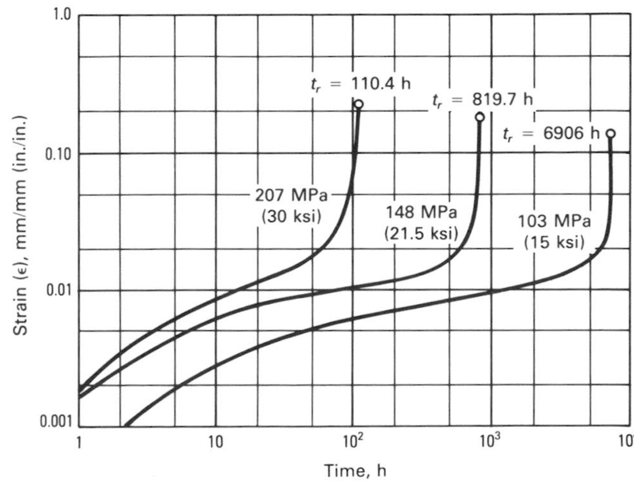

Fig. 13 Creep curves for [100] oriented NASAIR 100 [Ni-5.5Al-8.5Cr-0.7Mo-3Ta-1Ti-10W (wt%)] single crystals tested in tension to rupture at 1000 °C (1830 °F) in air. (a) Strain vs. time. (b) Log strain vs. log time. Source: Ref 10

Table 3 Life-assessment techniques and their limitations for creep-damage evaluation for crack initiation and crack propagation

Crack initiation		Crack propagation
Technique	**Issues**	
Calculation	Inaccurate	NDE Stress analysis Crack growth End-of-life criterion Remaining life
Extrapolation of past experience	Inaccurate	
Conventional NDE	Inadequate resolution	
High-resolution NDE: Acoustic emission Positron annihilation Barkhausen noise analysis	Not sufficiently developed at this time	
Strain (dimension) measurement	Uncertainty regarding original dimensions Lack of clear-cut failure criteria Difficulty in detecting localized damage	
Rupture testing	Difficulty in sample removal Difficulty in using as a monitoring technique Validity of life-fraction rule Effects of oxidation and specimen size Uniaxial-to-multiaxial correlations	**Issues:**
Microstructural evaluation: Cavitation measurement Carbide-coarsening measurements Lattice parameter Ferrite chemistry analysis Hardness monitoring	Quantitative relationships with remaining life are lacking	Uncertainties in interpretation of NDE results Lack of adequate crack growth data in creep and creep-fatigue Lack of methods for characterizing crack growth rates specific to the degraded components Lack of a clear-cut end-of-life criterion under creep conditions
Oxide scale measurements for tubes	Need data on oxide scale growth in stream Kinetics of hot-corrosion and constant-damage curves	Difficulty in assessing toughness of in-service components

Source: Ref 20

area decreases to A, because volume must be conserved:

$$LA = L_0A_0 \qquad \text{(Eq 11)}$$

Therefore,

$$A = A_0\left(\frac{L_0}{L}\right) \qquad \text{(Eq 12)}$$

and the true stress on the bar is:

$$\sigma = \left(\frac{P}{A}\right) = s_0\left(\frac{L}{L_0}\right) \qquad \text{(Eq 13)}$$

Methods have been devised to account for the change in cross-sectional area during creep. These are based on a rearranged form of Eq 13, where:

$$\sigma = \frac{P}{A} = \frac{PL}{L_0A_0} \qquad \text{(Eq 14)}$$

By maintaining PL at a fixed value, a constant stress test can be conducted. In general, the form of strain-time behavior under constant stress conditions is the same as those shown in Fig. 3 and 4. However, the period of time spent in primary and secondary creep under constant stress can be much longer than under an identical engineering stress (constant load). Hence, rupture life is longer under constant stress conditions (Fig. 19).

Typical constant-stress creep curves obtained for type 316 austenitic stainless steel at 705 °C (1300 °F) are shown in Fig. 20. In each specimen, the test was ended soon after tertiary creep began. The maximum strain reached in these tests was slightly less than 60%, although the apparatus was designed to accommodate uniform strains of at least 100%

Failure under constant stress conditions eventually occurs due to some microstructural and/or mechanical instability in the same manner as in a constant load experiment. Once a local variation in cross-sectional area is formed, the actual stress is higher than the imposed constant stress, and further deformation concentrates at this location.

In reality, the basic assumptions of conservation of volume and/or uniform deformation have been violated; therefore, Eq 14 is no longer valid. When nonuniform deformation starts in either a constant load or constant stress test, the local strain and strain rate vary along the gage section in an unknown manner.

Engineering Strain versus True Strain. In creep experiments, there is little difference between strains calculated by the engineering strain or true strain definitions when the length change is approximately 10% or less. For greater length changes, the calculated values of strain deviate greatly. Although this is of no consequence for tension, the limit of a maximum engineering strain of −1.00 in compression places an artificial barrier on the description of compressive creep. Hence, true strain is a much better indicator of

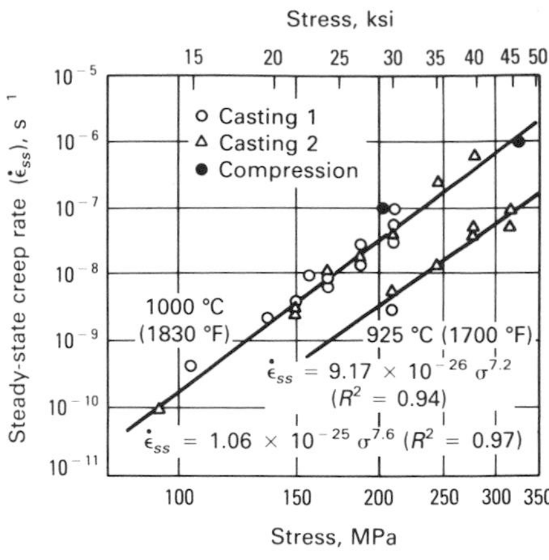

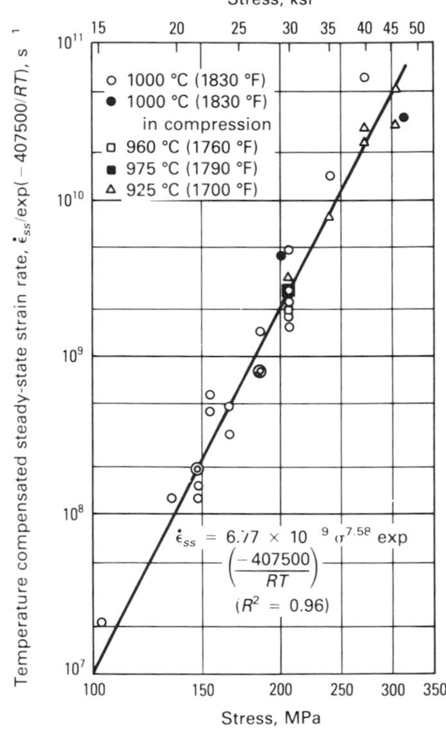

Fig. 14 Steady-state creep rate ($\dot{\varepsilon}_{ss}$) as a function of applied stress (σ) and temperature (T) for two heats of [100] oriented NASAIR 100 single crystals tested in tension at several temperatures in air. (a) Power law form (Eq 6). (b) Temperature-compensated power law form (Eq 7). Source: Ref 10

compressive ductility and creep characteristics. In particular, creep behavior measured in tension and compression can be compared only when both are expressed as true strain due to the limiting engineering strain in compression.

Effects of Creep on Microstructure and Fracture Morphology

Microstructure. During creep, significant microstructural changes occur on all levels. On the atomic scale, dislocations are created and forced to move through the material. This leads to work

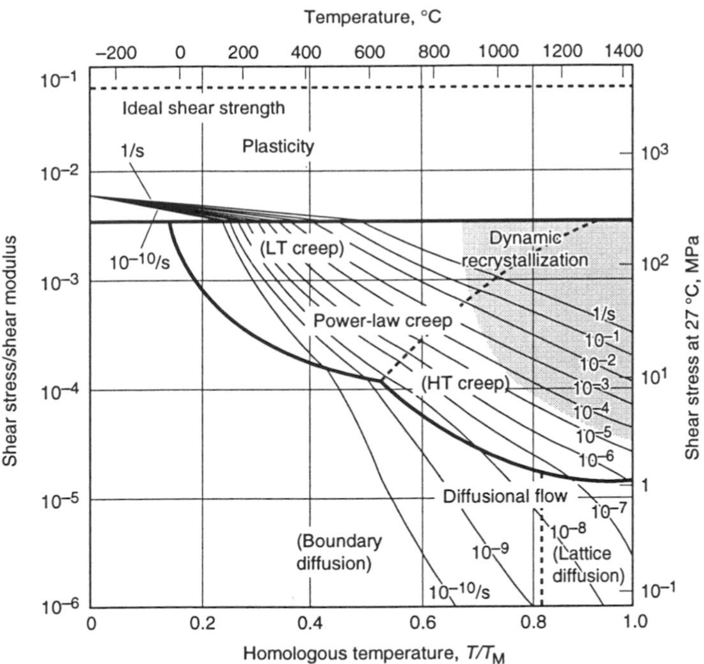

Fig. 15 Deformation mechanism map for pure nickel with a grain size of 100 μm. LT, low-temperature, HT, high-temperature. Source: Ref 12

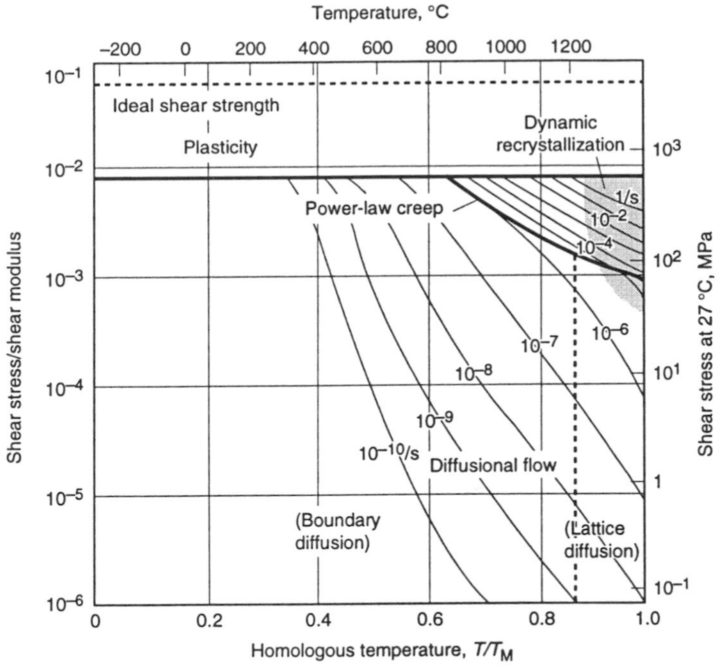

Fig. 16 Deformation mechanism map for the cast nickel-base superalloy MAR-M200 with a grain size of 100 μm. Source: Ref 12

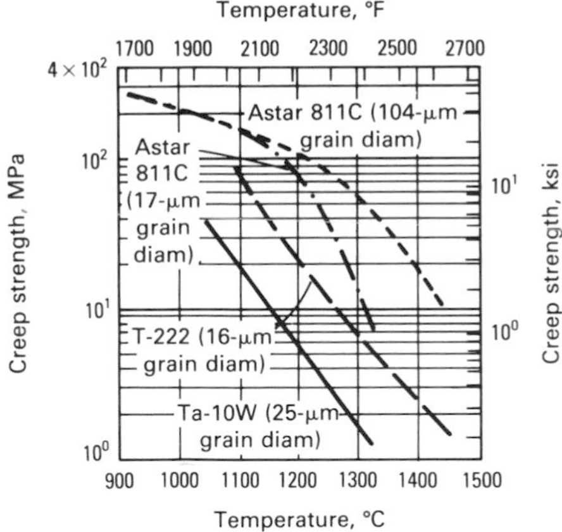

Fig.17 Stress necessary for 1.0% strain in 10,000 h as a function of temperature for several tantalum alloys tested in vacuum. Materials include Ta-10W, T-222 (Ta-2.4Hf-9.6W-0.01C), and Astar 811C (Ta-1Hf-1Re-7.5W-0.02C) (all materials in wt%). Source: Ref 13

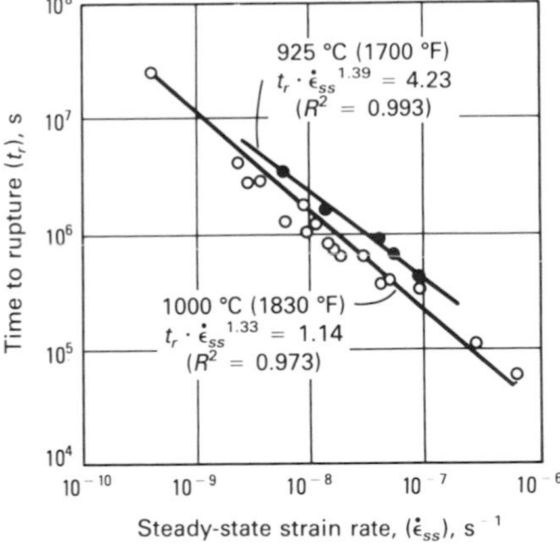

Fig. 18 Time-to-rupture (t_r) as a function of steady-state creep rate ($\dot{\epsilon}_{ss}$) for [100] oriented NASAIR 100 single crystals tested in tension at several temperatures in air. Source: Ref 10

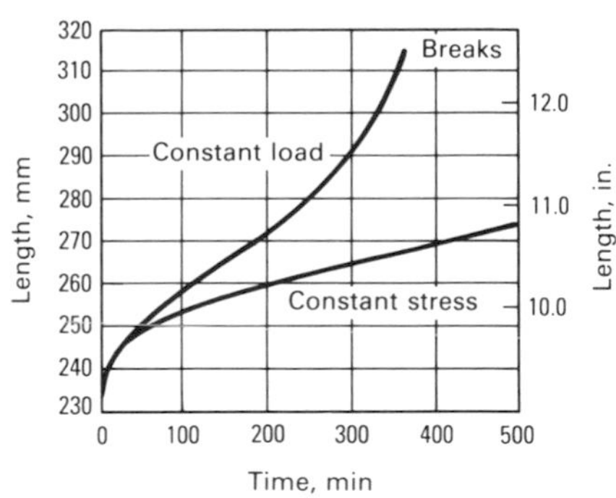

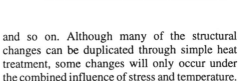

Fig. 19 Results of tests on lead wire under constant-load and constant-stress conditions. Source: Ref 15

hardening as the dislocation density increases and the dislocations encounter barriers to their motion. At low temperatures, an ever-diminishing creep rate results; however, if the temperature is sufficiently high, dislocations rearrange and annihilate through recovery events.

The combined action of hardening and recovery processes during primary creep can lead to the formation of a stable distribution of subgrains or loose three-dimensional dislocation networks in some materials, or an approximately uniform dislocation distribution without subgrains in other materials. These stable dislocation configurations are maintained and are characteristic of second-stage creep.

Creep deformation also produces change in the light optical macro- and microstructures. Such changes include slip bands, grain-boundary sliding, cavity formation and growth, and cracking (grain-boundary, interphase boundary, and transgranular). The microstructural changes that lead to cracking are described below in the section "Creep Fracture."

The microstructure of an elevated-temperature creep or stress-rupture test specimen rarely resembles the initial microstructure. Most materials are not thermodynamically stable; hence, prolonged exposure under creep conditions can result in the precipitation of new phases, dissolution or growth of desired phases, grain growth, and so on. Although many of the structural changes can be duplicated through simple heat treatment, some changes will only occur under the combined influence of stress and temperature.

Figure 21 is an example of a stress/temperature-dependent microstructure. Under normal isothermal annealing, the cube-shaped γ′ (Ni_3Al) strengthening phase (Fig. 21a) in the nickel-base superalloy NASAIR 100 undergoes Ostwald ripening (Fig. 21b), where ripening is characterized by an increase in particle size without any shape change. However, during creep testing, the individual precipitates grow together rapidly and form thin γ′ plates where the long dimensions of each plate are perpendicular to the stress in tensile creep (Fig. 21c) and parallel to the applied stress in compressive creep (Fig. 21d).

The changes in microstructure that occur during testing affect creep properties. Although such changes may be unavoidable, in many cases thermomechanical processing schedules can be es-

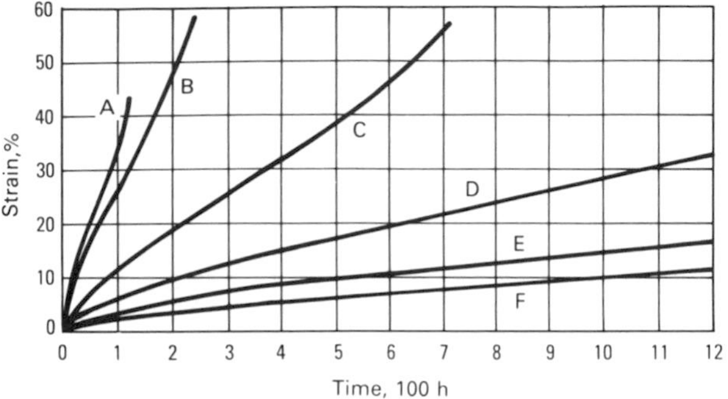

Specimen	Stress MPa	Stress ksi
A	160	23.2
B	147.5	21.4
C	128.2	18.6
D	106.9	15.5
E	91.0 (continued to 2100 h)	13.2
F	77.2 (continued to 2900 h)	11.2

Fig. 20 Typical constant-stress creep curves obtained at 705 °C (1300 °F) for type 316 austenitic stainless steel. Source: Ref 16

tablished to influence the changes so that they tend to strengthen the material or minimize the overall effect. For example, if a heterogeneous precipitate is formed during creep, a simple heat treatment or cold work followed by annealing prior to testing should give a homogeneous distribution of precipitates.

Microstructural changes due to the combined influence of temperature and stress are the most difficult to control. These changes enhance creep and therefore contribute to the observed strain. Even if the changes are essentially complete after primary creep and the resultant microstructure is more creep resistant than the original structure, the creep strain from such changes may be so great that the material cannot be used. To circumvent these changes, simulation of creep exposure prior to actual use may be necessary.

Complete microstructural examination of tested and untested materials should be an essential part of any creep experiment. As a minimum, the as-received microstructure should be compared to those at and away from the fracture site for the shortest-lived and longest-lived test specimens at each temperature. This comparison aids identification of the relevant deformation mechanism, indicates whether environment is affecting creep, and reveals any significant microstructural changes. Such information is vital for interpreting and understanding creep behavior.

Creep Fracture (Ref 18). As shown schematically in Fig. 22, intergranular creep ruptures occur by either of two fracture processes: triple-point cracking or grain-boundary cavitation. The strain rate and temperature determine which fracture process dominates. Relatively high strain rates and intermediate temperatures promote the formation of wedge cracks, or triple-point cracks (Fig. 22a). Grain-boundary sliding as a result of an applied tensile stress can produce sufficient stress concentration at grain-boundary triple points to initiate and propagate wedge cracks. Cracks can also nucleate in the grain boundary at locations other than the triple point by the interaction of primary and secondary slip steps with a sliding grain boundary. Any environment that lowers grain-boundary cohesion also promotes cracking. As sliding proceeds, grain-boundary cracks propagate and join to form intergranular decohesive fracture (Fig. 23).

At high temperatures and low strain rates, grain-boundary sliding favors cavity formation (Fig. 22b). The grain-boundary cavities resultign from creep should not be confused with microvoids formed in dimple rupture. The two are fundamentally different; the cavities are principally the result of a diffusion-controlled process, while microvoids are the result of complex slip. Even at low strain rates, a sliding grain boundary can nucleate cavities at irregularities, such as second-phase inclusion particles. The nucleation

Table 4 Neubauer classification of creep damage

Damage level (Neubauer)	Description	Recommended action
1	Undamaged	No creep damage detected
2	Isolated	Observe
3	Oriented	Observe, fix inspection intervals
4	Microcracked	Limited service until repair
5	Macrocracked	Immediate repair

Source: Ref 27

Table 5 Correlation of damage level and life fraction consumed

Damage level	Consumed life fraction range, X	Remaining life factor $(1/X - 1)$ Minimum	Maximum
1	0.00–0.12	7.33	Unknown
2	0.04–0.46	1.17	24.00
3	0.3–0.5	1.0	2.33
4	0.3–0.84	0.19	2.33
5	0.72–1.00	0 = failed	0.39

Source: Ref 28

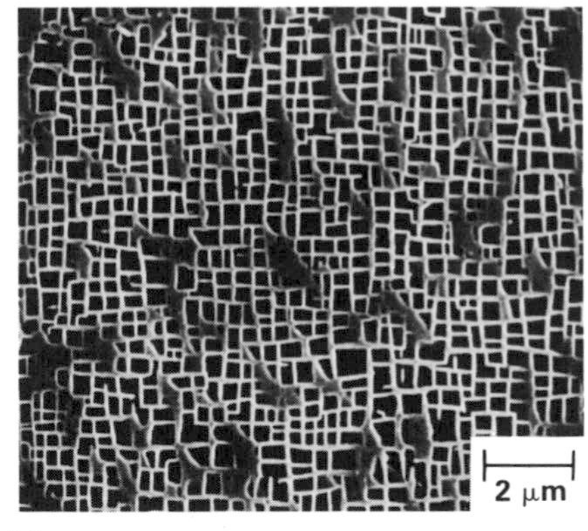

(a)

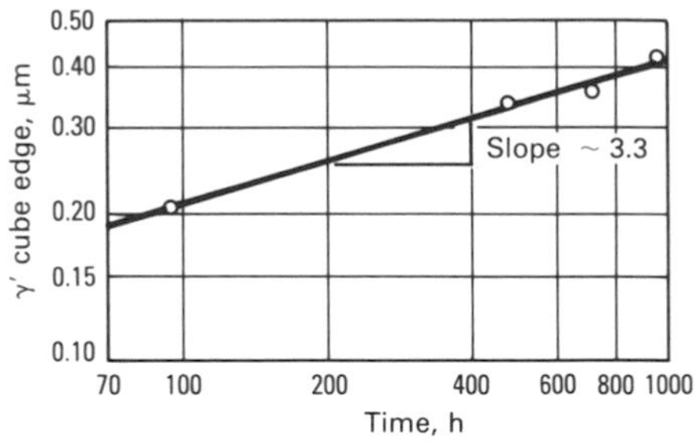

(b)

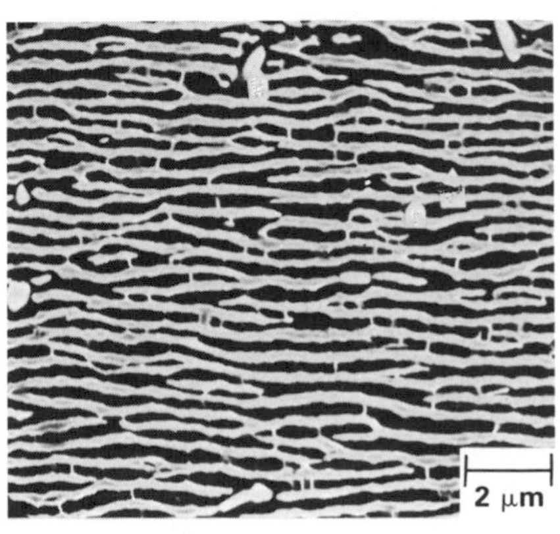

(c)

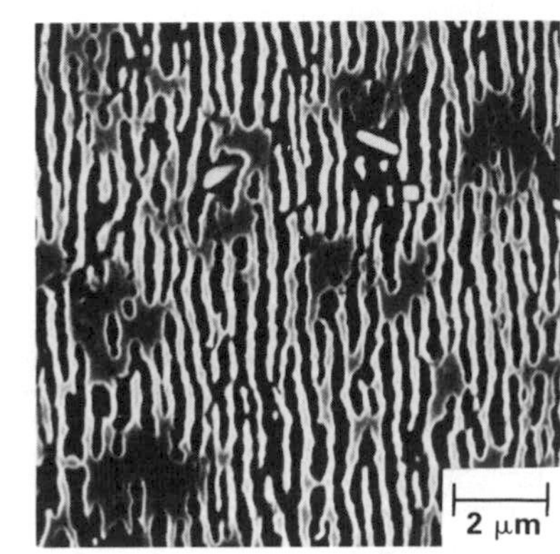

(d)

Fig. 21 Comparison of microstructural changes in a γ′-strengthened nickel-base superalloy. (a) Cube-shaped γ′ strengthening phase resulting from isothermal annealing (Ref 17). (b) Ostwald ripening of strengthening phase due to isothermal annealing (Ref 10). (c) and (d) Microstructural changes due to tensile creep and compressive creep, respectively (Ref 17). See test for details.

is believed to be a strain-controlled process (Ref 19, 20), while the growth of the cavities can be described by a diffusion growth model (Ref 21–23), and by a power-law growth relationship (Ref 24, 25). Irrespective of the growth model, as deformation continues, the cavities join to form an intergranular fracture. Figure 24 shows the microstructure and intergranular fracture morphology of a type 316 stainless steel that failed due to grain-boundary cavitation.

Instead of propagating by a cracking or a cavity-forming process, a creep rupture can occur by a combination of both. There may be no clear distinction between wedge cracks and cavities. The wedge cracks can be the result of the linkage of cavities at triple points.

There is a general trend toward transgranular failures at short rupture lifetimes (high strain rates) and relatively low temperatures. Fracture mechanism maps, similar to the steady-state creep deformation maps described above, have also been developed to differentiate between transgranular and intergranular fracture. An example is shown in Fig. 25 for an 80Ni-20Cr alloy. This figure confirms the tendency for transgranular fractures to be favored by high strains/stresses and for intergranular fractures to be favored by low strains/stresses (long rupture lifetimes).

Stress Relaxation

Traditional creep testing to develop descriptions of strain rate, stress, and temperature behavior can be time-intensive and expensive, involving many creep test stands, many specimens, and thousands of hours of testing. Stress relaxation offers the potential to eliminate this difficulty by producing strain rate/stress data over a wide range of rates from a single specimen. This information is developed when the elastic strain of a specimen extended (or compressed) to a certain, constant length is converted to plastic strain. As

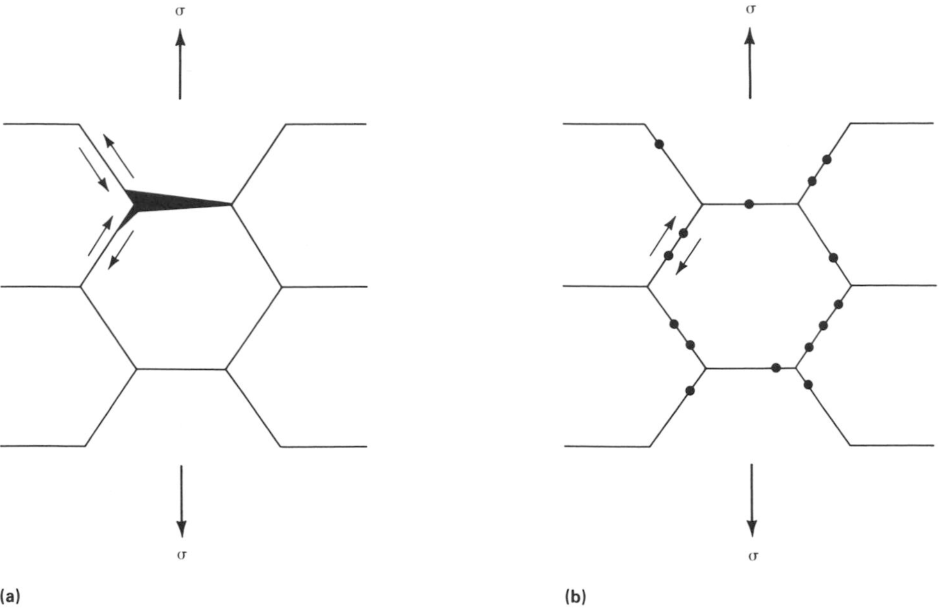

Fig. 22 Triple-point cracking (a) and cavitation (b) in intergranular creep rupture. Small arrows indicate grain-boundary sliding.

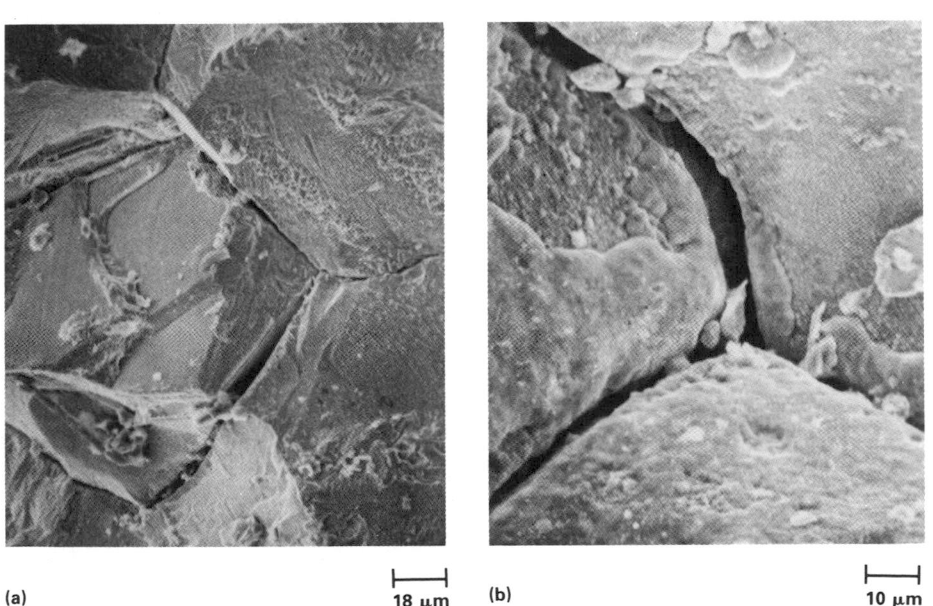

Fig. 23 Examples of intergranular creep fractures. (a) Wedge cracking in Inconel 625. (b) Wedge cracking in Incoloy 800. Source: Ref 18

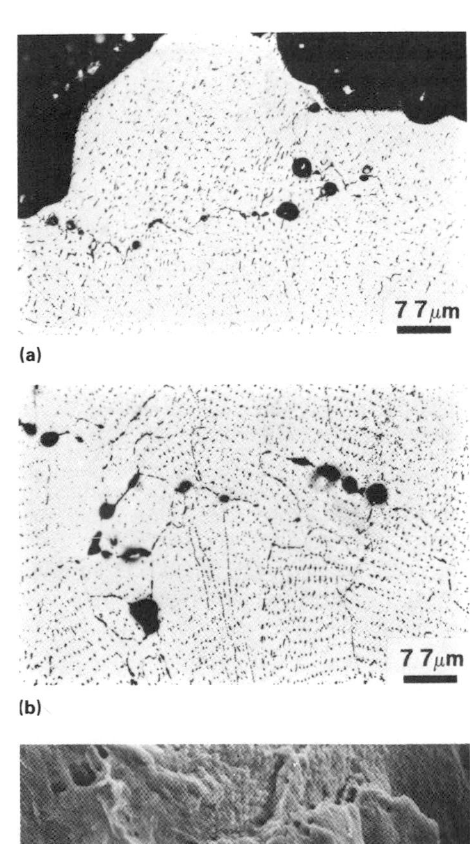

Fig. 24 Microstructure and fracture appearance of type 316 stainless steel tested in creep to fracture in air at 800 °C (1470 °F) at a load of 103 MPa (15 ksi). Time to rupture: 808 h. Light micrographs (a and b) illustrate spheroidal grain-boundary cavities that linked up to cause intergranular creep fracture. Both at 90×. The SEM fractograph (c) illustrates the intergranular fracture morphology. 1260×

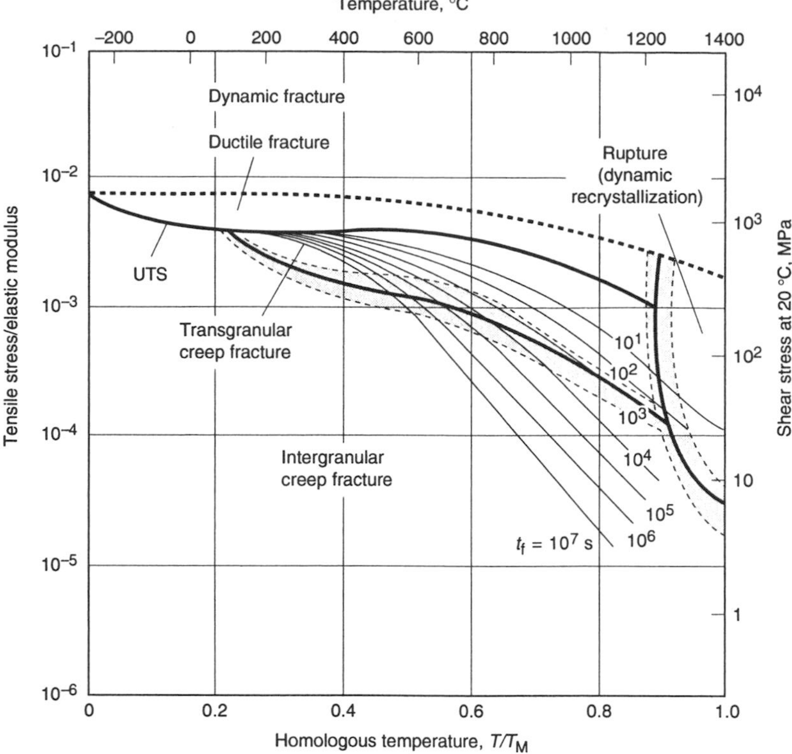

Fig. 25 Fracture mechanism map for an 80% Ni–20% Cr solid solution showing the dependence of fracture life and morphology on stress and temperature. UTS, ultimate tensile strength; t_f = time to failure. Source: Ref 19

the specimen slowly deforms, the load required to maintain the constant length is reduced, hence the term "stress relaxation" test. In addition, this type of experiment simulates the real engineering problem of the long-term loosening of tightened bolts and other fasteners.

In its simplest form, a stress relaxation test involves loading (straining) a specimen to some predetermined load (strain), fixing the position of the specimen (halting the crosshead motion in a universal test machine, for example), and measuring the load as a function of time. With knowledge of the elastic modulus of the specimen and the stiffness of the testing machine, the load-time data can be converted to stress/strain-rate data. This information can then be used to determine the stress exponents and activation energies for deformation.

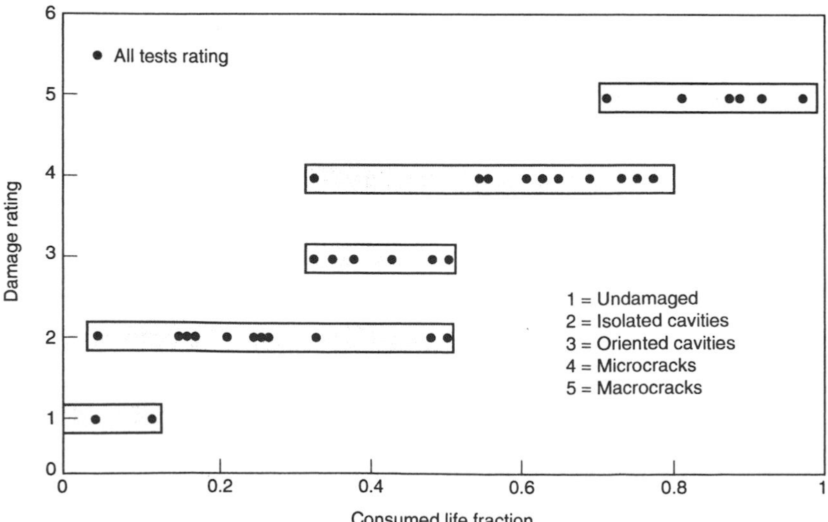

Fig. 26 Relation between Neubauer damage rating (Table 4) and consumed life fraction. Source: Ref 27, 28

One major drawback to stress relaxation testing is the demands placed on the experimental equipment. The load-measuring system must be capable of making accurate measurements of very small changes as a function of time. The effect of the loading rate on the relaxation rate should be evident. In addition, room temperature, as well as specimen temperature, must be precisely controlled throughout the experiment. This is critical, because even small fluctuations in temperature will produce thermal expansion effects that can mask changes due to relaxation. Also, the calculated stress exponents and activation energies may not be the same as those determined from creep testing. The processes that produce plastic flow could be different for these two situations. Only comparison of the results form both types of testing can detect equivalent behavior.

Remaining-Life-Assessment Methodologies (Ref 20)

Failure due to creep can be classified as resulting either from widespread bulk damage or from localized damage. The structural components that are vulnerable to bulk damage (e.g., boiler tubes) are subjected to uniform loading and uniform temperature distribution during service. If a sample of material from such a component is examined, it will truly represent the state of damage in the material surrounding it. The life of such a component is related to the creep-rupture properties. On the other hand, components that are subjected to stress (strain) and temperature gradients (typical of thick-section components) may not fail by bulk creep rupture. It is likely that at the end of the predicted creep-rupture life, a crack will develop at the critical location and propagate to cause failure. A similar situation exists where failure originates at a stress concentration or at pre-existing defects in the component. In this case, most of the life of the component is spent in crack propagation, and creep-rupture-based criteria are of little value.

Assessment of Bulk Creep Damage. The current approaches to creep damage assessment of components can be classified into two broad categories: (a) history-based methods, in which plant operating history in conjunction with standard material property data are employed to calculate the fractional creep life that has been expended, using the life-fraction rule or other damage rules; and (b) methods based on post-service evaluation of the actual component.

In history-based methods, plant records and the time-temperature history of the component are reviewed. The creep-life fraction consumed for each time-temperature segment of the history can then be calculated and summed up using the lower-bound ISO data and the life-fraction rule, or other damage rules.

The most common approach to calculation of cumulative creep damage is to compute the amount of life expended by using time or strain fractions as measures of damage. When the fractional damages add up to unity, then failure is postulated to occur. The most prominent rules are as follows:

- *Life-fraction rule (LFR)* (Ref 21):

$$\Sigma \frac{t_i}{t_{ri}} = 1 \quad \text{(Eq 15)}$$

- *Strain-fraction rule* (Ref 22):

$$\Sigma \frac{\varepsilon_i}{\varepsilon_{ri}} = 1 \quad \text{(Eq 16)}$$

- *Mixed rule* (Ref 23):

$$\Sigma \left(\frac{t_i}{t_{ri}}\right)^{1/2} \left(\frac{\varepsilon_i}{\varepsilon_{ri}}\right)^{1/2} = 1 \quad \text{(Eq 17)}$$

- *Mixed rule* (Ref 24):

$$k\Sigma \left(\frac{t_i}{t_{ri}}\right) + (1-k)\,\Sigma \left(\frac{\varepsilon_i}{\varepsilon_{ri}}\right) = 1 \quad \text{(Eq 18)}$$

where k is a constant, t_i and ε_i are the time spent and strain accrued at condition i, and t_{ri} and ε_{ri} are the rupture life and rupture strain, respectively, under the same conditions.

Goldhoff and Woodford (Ref 25) studied the Robinson life-fraction rule and determined that for a Cr-Mo-V rotor steel it worked well for small changes in stress and temperature. Goldhoff (Ref 26) assessed strain-hardening, life-fraction, and strain-fraction rules under unsteady conditions for this steel. While all gave similar results, the strain-fraction rule was found to be the most accurate.

From careful and critical examination of the available results, the following overall observations can be stated (Ref 20):

- Although several damage rules have been proposed, none has been demonstrated to have a clearcut superiority over any of the others.
- The LFR is clearly not valid for stress-change experiments. Under service conditions where stress may be steadily increasing due to corrosion-related wastage (e.g., in boiler tubes), applications of the LFR will yield nonconservative life estimates; that is, the actual life will be less than the predicted life. On the other hand, residual-life predictions using postexposure tests at high stresses will yield unduly pessimistic and conservative results.
- The LFR is generally valid for variable-temperature conditions as long as changing creep mechanisms and environmental interactions do not interfere with test results. Hence, service life under fluctuating temperatures and residual life based on accelerated-temperature tests can be predicted reasonably accurately by use of the LFR.
- The possible effects of material ductility (if any) on the applicability of the LFR need to be investigated. A major limitation in applying the LFR is that the properties of the virgin material must be known or assumed. Postexposure tests using multiple specimens often can obviate the need for assuming any damage rule.

Direct postservice evaluations represent an improvement over history-based methods, because no assumptions regarding material properties and past history are made. Unfortunately, direct examinations are expensive and time-consuming. The best strategy is to combine the two approaches. A history-based method is used to determine whether more detailed evaluations are justified and to identify the critical locations, and this is followed by judicious postservice evaluation. Table 3 summarizes the techniques that are in use for life assessment and some of the issues pertaining to each technique.

Current postservice evaluation procedures include conventional nondestructive evaluation (NDE) methods (e.g., ultrasonics, dye-penetrant inspection, etc.), dimensional (strain) measurements, and creep-life evaluations by means of accelerated creep damage and microstructural damage, which can be precursors of rapid, unanticipated failures. Due to unknown variations in the original dimensions, changes in dimensions cannot be determined with confidence. Dimensional measurements fail to provide indications of local creep damage caused by localized strains, such as those in heat-affected zones of welds and regions of stress concentrations in the base metal. Cracking can frequently occur without manifest overall strain. Furthermore, the critical strain accumulation preceding fracture can vary widely with a variety of operational material parameters and with stress state.

Surface replication is a well-known sample preparation technique that can be used to assess the condition of high-temperature power plant and petrochemical components from creep damage. The usual method of metallographic investigation involves cutting large pieces from components, which thus renders the component unfit for service. In contrast, surface replication allows examination of microstructural damage without cutting sections from the component (see the article "Replication Microscopy Techniques for NDE" in *Nondestructive Evaluation and Quality Control,* Volume 17 of *ASM Handbook*).

Replication techniques are sufficiently sophisticated to allow classifications of microstructural damage (such as in Table 4, for example) that can be directly correlated to life fractions (Fig. 26). A distinct correlation exists for these data, such that a minimum and maximum remaining life fraction can be specified (such as in Table 5, for example) (Ref 27, 28). For assessed consumed life fraction, X, after exposure time, T_{exp}, the remaining life (T_{rem}) is:

$$T_{rem} = T_{exp}(1/X - 1) \quad \text{(Eq 19)}$$

The qualitative-quantitative relation is advantageous because data from surface replication can be predictive in terms of generating a conservative minimum- and maximum-life estimate. The maximum life is useful in predictive maintenance environment, as it would dictate the planning of future repairs or replacement.

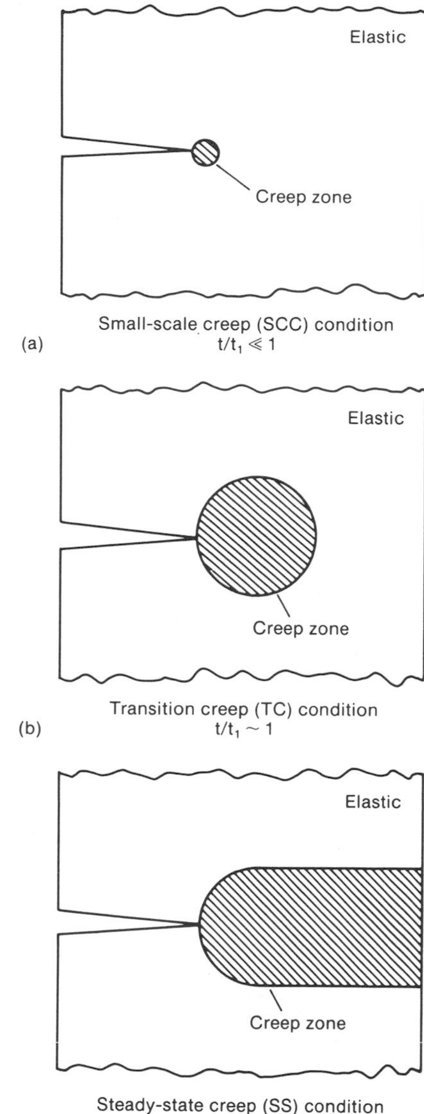

Fig. 27 Schematic representation of the levels of creep deformation under which creep-crack growth can occur. Source: Ref 20

A common method of estimating the remaining creep life is to conduct accelerated rupture tests at temperatures well above the service temperature. The stress is kept as close as possible to the service stress value, because only isostress-varied temperature tests are believed to be in compliance with the life-fraction rule. The time-to-rupture results are then plotted versus test temperature. By extrapolating the test results to the service temperature, the remaining life under service conditions is estimated.

Implementation of the above procedure requires a reasonably accurate knowledge of the stresses involved. For cyclic stressing conditions, and in situations involving large stress gradients, selection of the appropriate stress for the isostress tests is uncertain. Furthermore, the procedure in-

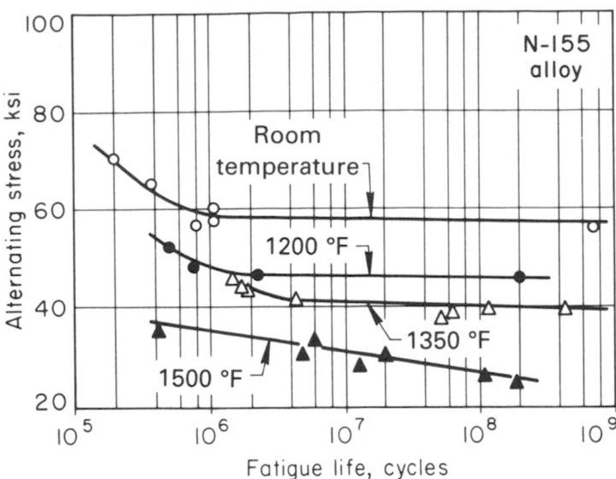

Fig. 28 Fatigue life of a specimen of N-155 alloy subjected to various temperatures and reversed-bending stress

volves destructive tests requiring removal of large samples from operating components. There are limitations on the number of available samples and the locations from which they can be taken. Periodic assessment of the remaining life is not possible. The costs of cutting out material, machining specimens, and conducting creep tests can add up to a significant expenditure. These costs are further compounded by the plant outage during this extended period of evaluation and decision making. Development of nondestructive techniques, particularly those based on metallographic and miniature-specimen approaches, has therefore been a major focus of the programs aimed at predicting crack initiation.

Assessment of Creep Crack Growth. As described in the article "Elevated-Temperature Crack Growth of Structural Alloys" in this Volume, a number of fracture mechanics parameters have been developed to describe the creep crack-growth behavior of materials. These include the integral C^* and the crack-tip driving force parameters C_t and $C(t)$. The idea of a crack-tip parameter is that identical values of the appropriate parameter in differently shaped specimens or structures generate identical conditions of stress and/or deformation near the crack tip, so that the crack-growth rate must be the same provided that the material, the environment, and the temperature at the crack tip are also the same. Thus, such a parameter may be thought of as a transfer function from specimen to structural behavior. It then will be sufficient to measure the crack-growth rate as a function of the load parameter in the laboratory, and to calculate the value of the crack-tip parameter for the crack in the structure. The expected crack-growth rate in the structure then can be estimated.

In the subcreep-temperature regime involving crack growth under elastic or elastic-plastic conditions, the fracture-mechanics approach (involving the use of the stress-intensity factor K and the J-integral) for predicting crack-growth behavior is well established. In the creep-temperature regime, the crack-tip parameter must take into account time-dependent creep deformation. Depending on the material and on the extent of creep deformation, various parameters mentioned above have been successfully correlated with rates of creep-crack growth.

Three regimes of crack growth—namely, small-scale, transient, and steady-state—can be distinguished for materials exhibiting elastic, power-law creep behavior, depending on the size of the crack-tip creep zone relative to the specimen dimensions, as shown in Fig. 27. In the early stages of crack growth, the creep zone may be very small and localized near the crack tip. This regime is defined as the small-scale creep regime. At the other extreme, cracking may occur under widespread creep conditions where the entire uncracked ligament is subjected to creep deformation, as shown in Fig. 27(c). This regime is termed the large-scale or steady-state creep condition. Even in the latter case, creep-crack growth usually begins under small-scale conditions and, as the creep proceeds, the steady-state creep conditions develop. In between, the specimen passes through the transition creep conditions shown in Fig. 27(b). The transition time, t_1, from small-scale creep to steady-state creep conditions depends on several factors, including specimen geometry and size, load level, loading rate, temperature, and the kinetics of the creep. During the small-scale and transition creep conditions, the size of the creep zone and the stress at the crack tip change continuously with time. Under large-scale creep conditions, the crack-tip stress no longer changes with time. Hence, this regime is known as the steady-state regime. The nature (plasticity or creep) and size of the crack-tip deformation zone relative to the size of the specimen determine which of the parameters K, J, C^*, $C(t)$, and C_t might be applicable to a given situation. For creeping materials, description of the phenomenology surrounding C^*, C_t, and $C(t)$ is adequate. The parameters K and J, which do not account for time-dependent strain that occurs in the creep regime, are not applicable here.

Fatigue at Elevated Temperatures

Creep-Fatigue Interaction. Failure by fatigue can usually occur at any temperature below the melting point of a metal and still maintain the characteristic features of fatigue fractures, usually with little deformation, over the whole temperature range. At high temperatures, however, both the fatigue strength and the static strength of metal generally decrease as the operational temperature increases. Figure 28 shows typical S-N curves for reversed-bending fatigue tests conducted on a structural metal alloy at various temperatures. The fatigue limit is clearly lower at the higher temperatures. Mechanical-property data on most alloys at high temperatures also show that, just as at room temperature, the fatigue strength is closely related to the tensile strength, unless the temperature is high enough for the fatigue strength to be affected by creep phenomena.

At high temperatures, application of a constant load to a metal component produces continuous deformation or creep, which will eventually lead to fracture if the load is maintained for a sufficient length of time. With increases in temperature, stress-rupture strength decreases rapidly to values that may be considerably lower than fatigue strength. Therefore, the primary requirement of a metal that will be subjected to high temperatures is that it have adequate stress-rupture strength. Many alloys that possess good creep resistance are also resistant to fatigue; however, the condition of an alloy that will provide maximum stress-rupture strength is not necessarily the condition that provides maximum fatigue strength. In practice, it is necessary to design against failure by fatigue and against excessive distortion or fracture by creep, just as it is necessary to consider combined tensile and fatigue loads at room temperature.

At room temperature, and except at very high frequencies, the frequency at which cyclic loads are applied has little effect on the fatigue strength of most metals. The effect, however, becomes much greater as the temperature increases and creep becomes more of a factor. At high temperatures, the fatigue strength often depends on the total time the stress is applied rather than solely on the number of cycles. This behavior occurs

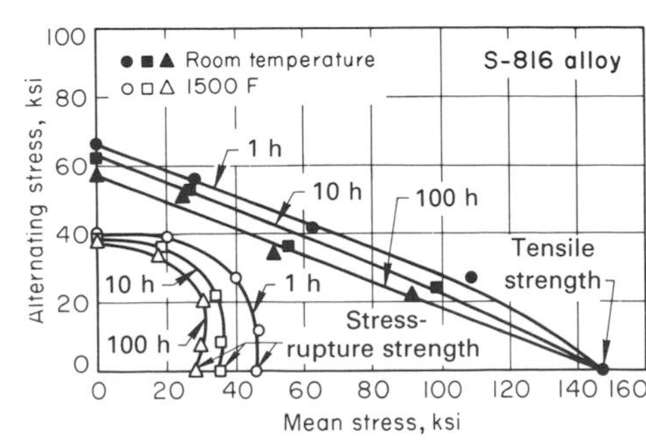

Fig. 29 Effect of temperature on the fatigue life of S-816 alloy tested under a fluctuating axial load at a frequency of 216,000 cycles per hour

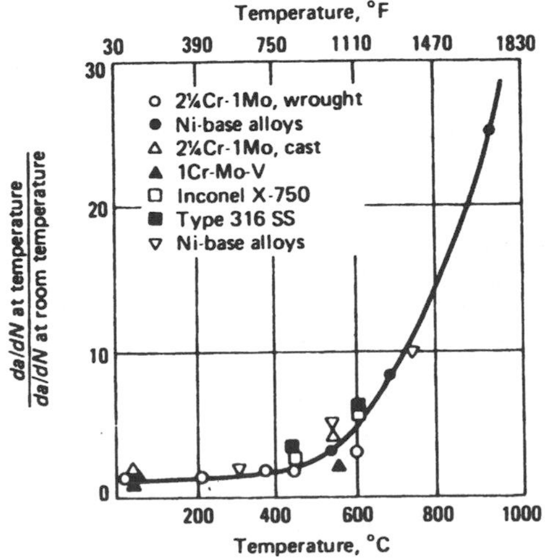

Fig. 30 Variation of fatigue crack growth rates as a function of temperature at $\Delta K = 30$ MPa$\sqrt{m}$ (27 ksi$\sqrt{in.}$). Source: Ref 20

because of continuous deformation under load at high temperatures. Under fluctuating stress, the cyclic frequency affects both the fatigue life and the amount of creep. This is shown in Fig. 29, a typical constant-life diagram that illustrates the temperature behavior of S-816 alloy tested under a fluctuating axial load. At room temperature, the curves converge at the tensile strength, plotted along the mean-stress axis. At high temperature, the curves terminate at the stress-rupture strength, which, being a time-dependent property, results in termination at a series of end points along the mean-stress axis.

The principal method of studying creep-fatigue interactions has been to conduct strain-controlled fatigue tests with variable frequencies with and without a holding period (hold time) during some portion of the test. The lower frequencies ($\leq 10^4$ cycles) and the hold times can allow creep to take place. In pure fatigue tests, at higher frequencies and short hold times, the fatigue mode dominates and failures start near the surface and propagate transgranularly. As the hold time is increased, or the frequency decreases, the creep component begins to play a role with increasing creep-fatigue interaction. In this region, fractures are of a mixed mode involving both fatigue cracking and creep cavitation. With prolonged hold times with occasional interspersed cycles, creep processes completely dominate and can be treated almost as pure cases of creep. In instances where oxidation effects contribute significantly to the creep-fatigue interaction, the situation is more complex than described above. A more detailed description of creep-fatigue effects and methodologies to predict creep-fatigue behavior can be found in the article "Creep-Fatigue Interaction" in this Volume.

Thermal and Thermomechanical Fatigue. Thermal-fatigue failure is the result of temperature cycling (without external loading), as opposed to fatigue at high temperatures caused by strain cycling. Two conditions necessary for thermal fatigue are some form of mechanical constraint and a temperature change. Thermal expansion or contraction caused by a temperature change acting against a constraint causes thermal stress. Constraint may be external—for example, constraint imposed by rigid mountings for pipes—or it may be internal, in which case it is set up by a temperature gradient within the part. In thick sections, temperature gradients are likely to occur both along and through the material, causing highly triaxial stresses and reducing material ductility, even though the uniaxial ductility often increases with increasing temperature. Reduction in the ductility of the material gives rise to fractures that have a brittle appearance, often with many cleavagelike facets in evidence.

Thermomechanical fatigue involves simultaneous changes in temperature and mechanical strain. It differs from creep-fatigue in that the latter is carried out at constant nominal temperature (isothermal) conditions. As such, the deformation and fatigue damage due to thermomechanical fatigue cannot be predicted based on isothermal creep-fatigue data.

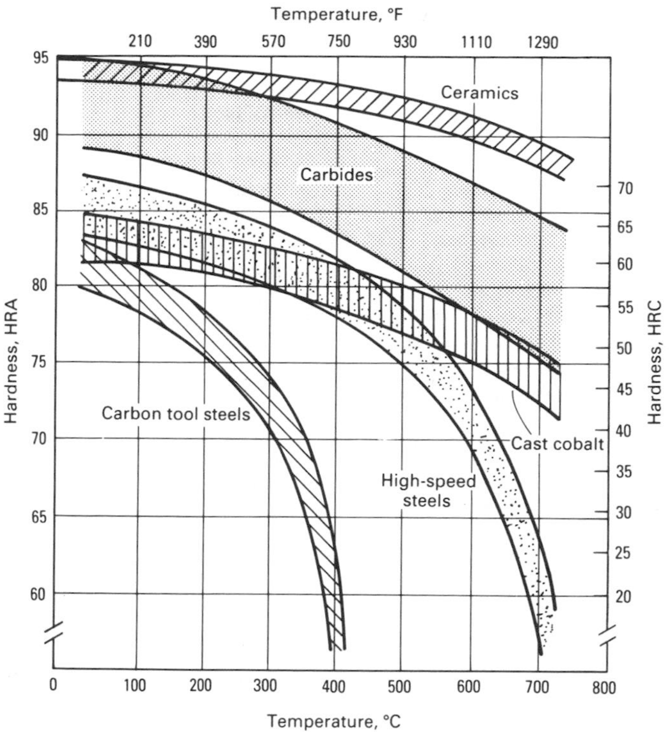

Fig. 31 Comparison of hot hardness values of various cutting tool materials

A summary of thermal and thermomechanical test methods, property data, and life prediction methods can be found in the article "Thermal and Thermomechanical Fatigue of Structural Alloys" in this Volume. This article also describes failures due to other forms of thermal cycling such as thermal shock and thermal ratcheting.

Elevated-Temperature Fatigue Crack Growth. The stress-intensity factor, K, is used for characterizing fatigue crack propagation. Most fatigue crack growth data are plotted in terms of crack growth per cycle, da/dN, versus the stress-intensity factor range, ΔK. Methods for determining K for various load/crack configurations have been derived and are listed in various publications, including Volume 8, *Mechanical Testing,* and Volume 19, *Fatigue and Fracture,* of *ASM Handbook.*

In the power-law (Paris equation) crack growth regime, the effects of temperature, stress ratio (R), and hold times have been investigated for many high-temperature alloys. Typical behavior and crack growth results for specific alloys are covered elsewhere in this Volume. However, a general comparison of temperature effects on fatigue crack growth of several different high-temperature alloys is shown in Fig. 30. Because the reported data are obtained at various ΔK ranges and temperature ranges, the general comparison is based on a constant ΔK (arbitrarily chosen as 30 MPa $\sqrt{m}$, or 27 ksi$\sqrt{in.}$). A clear trend of crack growth rate increase with increasing temperature can be seen as shown in Fig. 30. At temperatures up to about 50% of the melting point (550 to 600 °C, or 1020 to 1110 °F), the growth rates are relatively insensitive to temperature, but the sensitivity increases rapidly at higher temperatures. The crack growth rates for all the materials at temperatures up to 600 °C, relative to the room-temperature rates, can be estimated by a maximum correlation factor of 5 (2 for ferritic steels).

Besides temperature, cyclic frequency, or duration of a stress cycle (e.g., with hold time), is a key variable in high-temperature crack growth. At high frequency—that is, fast loading rate with short hold time (or no hold time)—the crack growth rate is cycle dependent and can be expressed in terms of da/dN. At low frequency (or with long hold time), however, the crack growth rate is time dependent; that is, da/dN is in proportion to the total time span of a given cycle. For tests of different cycle times, all crack growth rate data points are collapsed into a single curve of which da/dt is the dependent variable. A mixed region exists in between the two extremes. The transition from one type of behavior to another depends on material, temperature, frequency, and R (Ref 29). For a given material and temperature combination, the transition frequency is a function of R. The frequency range at which the crack growth rates remain time dependent increases as R increases (Ref 29). The limiting case is R approaching unity. It is equivalent to crack growth under sustained load, for which the crack growth rates at any frequency will be totally time dependent.

To further understand the complex interaction mechanisms of stress, temperature, time, and environmental exposure, a vast amount of experimental and analytical data was compiled (from a bibliography of 42 references) and reviewed. Crack growth behavior for 36 types of loading profiles, which were in excess of 60 combinations in material, temperature, frequency, and time variations, were examined. A compilation of the results is presented in Ref 30.

Hot Hardness

The ability to retain hardness at elevated temperatures (hot hardness) is a critical property for materials used for cutting tools/inserts and metalworking dies (e.g., hot forging dies). Tests for determining hot hardness values are carried out in standard Rockwell or Vickers hardness testers with high-temperature capability (up to 1200 °C, or 2100 °F). Figure 31 compares the hot hardness properties of a variety of cutting tool materials.

For high-speed tool steels, cobalt additions increase the hot hardness. For cemented carbides, lowering the cobalt binder content increases hot hardness. For hot-work tool steels used for hot die forging applications, high-tungsten-content grades containing 9 to 19% W have hot hardness values that are superior to those of hot-work steels. Of the ceramic tool materials, silicon-nitride-base materials exhibit higher hot hardness values than do alumina-base materials. Additional hot hardness property data for these materials can be found in the *ASM Specialty Handbook: Tool Materials.*

ACKNOWLEDGMENTS

The information in this article is largely taken from:

- J.D. Whittenberger, Introduction to Creep, Stress-Rupture, and Stress-Relaxation Testing, *Mechanical Testing,* Vol 8, *ASM Handbook,* (formerly Vol 8, 9th ed., *Metals Handbook*), American Society for Metals, 1985, p 301–307
- F.R. Morral, Testing of Superalloys, *Properties and Selection: Stainless Steels, Tool Materials and Special-Purpose Metals,* Vol 3, 9th ed., *Metals Handbook,* American Society for Metals, 1980, p 229–237
- P.F. Timmons, Failure Control in Process Operations, *Fatigue and Fracture,* Vol 19, *ASM Handbook,* ASM International, 1996, p 468–482
- A.F. Liu, High-Temperature Life Assessment, *Fatigue and Fracture,* Vol 19, *ASM Handbook,* ASM International, 1996, p 520–526
- S.D. Antolovich and A. Saxena, Fatigue Failures, *Failure Analysis and Prevention,* Vol 11, *ASM Handbook,* (formerly Vol 11, 9th ed., *Metals Handbook*), American Society for Metals, 1986, 102–135

REFERENCES

1. W.F. Brown, Jr., Ed., *Aerospace Structural Metals Handbook,* Metals and Ceramic Information Center, Columbus, OH, 1982
2. T.D. Moore, Ed., *Structural Alloys Handbook,* Metals and Ceramic Information Center, Columbus, OH, 1982
3. "High Temperature, High Strength Nickel Base Alloys," International Nickel Co., New York, 1977
4. T.E. Tietz and J.W. Wilson, *Behavior and Properties of Refractory Metals,* Stanford University Press, Stanford, CA, 1965
5. S.K. Mitra and D. McLean, Work Hardening and Recovery in Creep, *Proc. Roy. Soc.,* Vol 295, 1966, p 288–299
6. G.P. Tilly and G.F. Harrison, Interpretation of Tensile and Compressive Creep Behavior of Two Nickel Alloys, *J. Strain Anal.,* Vol 8, 1973, p 124–131
7. H.H. Gray, "Transverse Tensile and Stress Rupture Properties of γ/γ'–δ Directionally Solidified Eutectic," NASA TMX-73451, 1979
8. R.A. MacKay, "Morphological Changes of Gamma Prime Precipitates in Nickel-Base Superalloy Single Crystals," NASA TM-83698, 1984
9. F.R. Larson and J. Miller, A Time-Temperature Relationship for Rupture and Creep Stresses, *Trans. ASME,* Vol 74, 1952, p 765–775
10. M.V. Nathal, "Influence of Cobalt, Tantalum, and Tungsten on the High Temperature Mechanical Properties of Single Crystal Nickel-Base Superalloys," NASA TM-83479, 1984
11. G.A. Webster and R.A. Ainsworth, *High Temperature Component Life Assessment,* Chapman & Hall, 1994, p 10–49
12. H.J. Frost and M.F. Ashby, *Deformation-Mechanism Maps,* Pergamon Press, 1982
13. W.D. Klopp, R.H. Titran, and K.D. Sheffler, "Long-Time Creep Behavior of the Tantalum Alloy Astar 811C," NASA TP-1691, 1980
14. F.C. Monkman and N.J. Grant, "An Empirical Relationship between Rupture Life and Minimum Creep Rate," *Deformation and Fracture at Elevated Temperatures,* N.J.Grant and A.W. Mullendore, Ed., MIT Press, Cambridge, MA, 1965, p 91–103
15. E.N. Da C. Andrade, On the Viscous Flow in Metals and Allied Phenomena, *Proc. Royal Soc. London,* Vol 84, Series A, 1910–1911, p 1
16. F. Garafalo, O. Richmond, and W.F. Domis, Design of Apparatus for Constant-Stress or Constant-Load, *J. Basic Eng.,* Vol 84, 1962, p 287–293
17. M.V. Nathal and L.J. Ebert, Gamma Prime Shape Changes during Creep of a Nickel-Base Superalloy, *Scripta Met.,* Vol 17, 1983, p 1151–1154
18. V. Kerlins and A. Phillips, Modes of Fracture, *Fractography,* Vol 12, *ASM Handbook,* ASM International, 1987, p 12–71
19. M.F. Ashby, C. Gandhi, and D.M.R. Taplin, Fracture-Mechanism Maps and Their Construction for FCC Metals and Alloys, *Acta Metall.,* Vol 27, 1979, p 699–729
20. R. Viswanathan, *Damage Mechanisms and Life Assessment of High-Temperature Components,* ASM International, 1989, p 59–110, 164–167
21. E.L. Robinson, Effect of Temperature Variation on the Creep Strength of Steels, *Trans. ASME,* Vol 160, 1938, p 253–259
22. Y. Lieberman, Relaxation, Tensile Strength and Failure of E1 512 and Kh1 F-L Steels, *Metalloved Term Obrabodke Metal,* Vol 4, 1962, p 6–13
23. H.R. Voorhees and F.W. Freeman, "Notch Sensitivity of Aircraft Structural and Engine Alloys," Wright Air Development Center Technical Report, Part 11, Jan 1959, p 23
24. M.M. Abo El Ata and I. Finnie, "A Study of Creep Damage Rules," ASME Paper 71-WA/Met-1, American Society of Mechanical Engineers, Dec 1971
25. R.M. Goldhoff and D.A. Woodford, *The Evaluation of Creep Damage in a CrMoV Steel,* STP 515, American Society for Testing and Materials, 1982, p 89
26. R.M. Goldhoff, Stress Concentration and Size Effects in a CrMoV Steel at Elevated Temperatures, *Joint International Conference on Creep,* Institute of Mechanical Engineers, London, 1963
27. B. Neubauer and U. Wedel, NDT: Replication Avoids Unnecessary Replacement of Power Plant Components, *Power Eng.,* May 1984, p 44
28. J.M. Brear, et al., "Possibilistic and Probabilistic Assessment of Creep Cavitation," ICM 6, Pergamon, 1991
29. T. Nicholas and N.E. Ashbaugh, Fatigue Crack Growth at High Load Ratios in the Time-Dependent Regime, *Fracture Mechanics—19,* STP 969, ASTM, 1988, p 800–817
30. A.F. Liu, "Element of Fracture Mechanics in Elevated Temperature Crack Growth," AIAA Paper 90-0928, Collection of Technical Papers, part 2, AIAA/ASME/ASCE/AHS/ASC 31st Structures, Structural Dynamics and Materials Conference, 2–4 April 1990 (Long Beach, CA), p 981–994

Corrosion at Elevated Temperatures

HIGH-TEMPERATURE CORROSION plays an important role in the selection of materials for construction of industrial equipment ranging from gas turbines to heat treating retorts. The principal modes of high-temperature corrosion frequently responsible for equipment problems are (Ref 1):

- Oxidation
- Carburization
- Sulfidation
- Nitridation
- Halogen gas corrosion
- Ash/salt deposit corrosion
- Molten salt corrosion
- Liquid metal corrosion

Each of these modes of corrosion is described in this article, with emphasis placed in the thermodynamic considerations, the kinetics of corrosion in gases, and the principles associates with scale formation. In addition, laboratory and field tests designed for characterizing material performance for specific applications/environments are also described. Data compilations related to elevated-temperature corrosion resistance/susceptibility can be found in the articles in this Volume that deal with specific alloys or ceramics.

General Background (Ref 1)

The important environments and principal modes of high-temperature corrosion encountered in various industrial processes are briefly described below. These are summarized schematically in Fig. 1, which illustrates that in each corrosion mode there will be interactions between oxygen activity and a principal corrodent activity. Table 1 shows the various types of corrosion that can be anticipated in different industrial processes.

Oxidation is the most important high-temperature corrosion reaction. Metals or alloys are oxidized when heated to elevated temperatures in air or highly oxidizing environments, such as a combustion atmosphere with excess air or oxygen. Oxidation can also take place in reducing environments (i.e., those characterized by low oxygen activity). Most industrial environments have sufficient oxygen activities to allow oxidation to participate in the high-temperature corrosion reaction regardless of the predominant mode of corrosion. In fact, the alloy often relies on the oxidation reaction to develop a protective oxide scale to resist corrosion attack, such as sulfidation, carburization, ash/salt deposit corrosion, and so on.

Oxidation in air occurs in many industrial processes. Heat treating furnaces and chemical reaction vessels are often heated by electrical resistance in air. Under these conditions, the alloy is oxidized by oxygen.

For many other industrial processes, heat is generated by combustion, accomplished in many cases by using air and relatively "clean" fuels such as natural gas or No. 1 or No. 2 fuel oil. These fuels generally have low concentrations of contaminants, such as sulfur, chlorine, alkali metals, and vanadium. Many high-temperature processes use excess air to ensure complete combustion of the fuel. The combustion products thus consist primarily of O_2, N_2, CO_2, and H_2O. Although alloys in these environments are oxidized by oxygen, other combustion products, such as H_2O and CO_2, may play an important role in affecting oxidation behavior.

When combustion takes place under stoichiometric or substoichiometric conditions, the resultant environment becomes "reducing". This type of environment is generally charac-

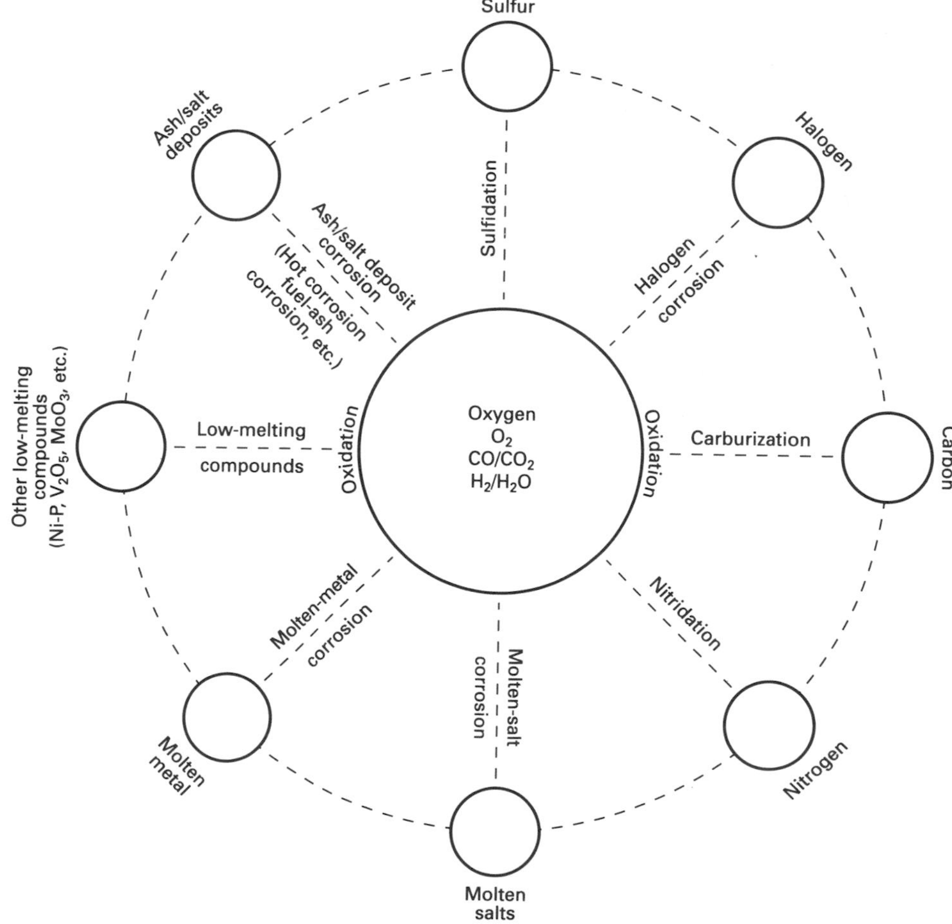

Fig. 1 Schematic showing the principal modes of high-temperature corrosion in industrial environments, as well as the interaction between oxygen activity and a principal "corrodent" activity. Source: Ref 1

Table 1 Modes of corrosion encountered in various process industries

Process/components	Temperature, °C (°F)	Type of corrosion
Chemical/petrochemical		
Ethylene steam cracking furnace tubes	to 1000 (1830)	Carburization; oxidation
Steam reforming tubes	to 1000 (1830)	Oxidation; carburization
Vinyl chloride crackers	to 650 (1200)	Halide gas
Hydrocracking heaters, reactors	to 550 (1020)	H_2S and H_2
Petroleum coke calcining recuperators	816 (1500)	Oxidation; sulfidation
Cat cracking regenerators	to 800 (1470)	Oxidation
Flare stack tips	950–1090 (1740–1995)	Oxidation; thermal fatigue; sulfidation; chlorination
Carbon disulfide furnace tubes	850 (1560)	Sulfidation; carburization; deposits
Melamine production (urea)-reactors	450–500 (840–930)	Nitriding
Other processes		
Titanium production reactor vessels	900 (1650)	Oxidation; chlorination
Nitric acid/catalyst guide	930 (1705)	Oxidation; nitriding; sulfidation
Nuclear reprocessing reactors	750–800 (1380–1470)	Oxidation (steam); fluorination (HF)
Oil-fired boiler superheaters	850–900 (1560–1650)	Fuel ash corrosion
Gas turbine blades corrosion	to 950 (1740)	Sulfates; chlorides; oxidation; ash
Waste incinerators/superheaters	480 (895)	Chlorination; sulfidation; oxidation; molten salts
Fiberglass manufacturing recuperators	1090 (1995)	Oxidation; sulfidation; molten salts

Source: Ref 2

terized by low oxygen activity. Under these conditions, the oxygen activity is typically controlled by the CO/CO_2 or H_2/H_2O ratios, and the oxidation kinetics are generally slow. The development of a protective oxide scale can be sluggish for most alloys. As a result, the effects of corrosive contaminants can become more pronounced, resulting in other modes of high-temperature corrosion. For example, if the sulfur level in the environment is high, sulfidation then becomes the predominant mode of corrosion, even though oxidation also takes part in the corrosion reaction. Thus, a majority of high-temperature corrosion problems in reducing environments are due to modes of corrosion attack other than oxidation.

The rate of oxidation for metals or alloys increases with increasing temperature. There is a large spectrum of engineering alloys available for applications in different temperature ranges. Many oxidation problems result from the use of an alloy in a temperature region exceeding the capability of that alloy. Materials of construction for high-temperature corrosive environments must withstand excessive metal loss by scale formation from oxidation and from penetration by internal oxidation products that could reduce the remaining cross-sectional area to a level that cannot sustain the load-bearing requirements. The component will then yield and may swell or distort. In some cases the internal fluid pressures can be sufficient to burst the component, releasing hot, possibly toxic or flammable fluids. Heating and cooling rates can also be significant because of possible thermal stress (fatigue) effects.

Sulfidation. When an environment has a high sulfur activity, the corrosion reaction will more likely be dominated by sulfidation. The reaction will also be influenced by oxygen activity. Lowering the oxygen activity tends to make the environment more sulfidizing, resulting in increased domination by sulfidation. Conversely, increasing the oxygen activity generally results in a less sulfidizing environment. The reaction is then increasingly dominated by oxidation. Thus, sulfidation is controlled by both sulfur and oxygen activities.

Carburization and Nitridation. Carburization behaves in a similar fashion. The reaction is controlled by both carbon and oxygen activities. Lowering the oxygen activity tends to make the environment more carburizing, and vice versa. Nitridation is the same in that the reaction becomes more severe when the environment is reducing.

In halogen corrosion, oxygen activity influences the reaction differently. For example, high-temperature corrosion in chlorine-bearing environments is generally attributed to the formation of volatile metallic chlorides. Oxidizing environments cause some alloys with high levels of molybdenum and tungsten to suffer significantly high corrosion rates, presumably by forming very volatile oxychlorides. Reducing environments often are less corrosive. Nevertheless, the reaction is controlled by both halogen and oxygen activities.

Ash/Salt Deposits. Many industrial environments may contain several corrosive contaminants that tend to form ash/salt deposits on metal surfaces during high-temperature exposure. These ash/salt deposits can play a significant role in the corrosion reaction. "Hot corrosion" of gas-turbine components is a good example. Sulfur from the fuel and NaCl from the ingested air may react during combustion to form salt vapors, such as Na_2SO_4. These salt vapors may then deposit at lower temperatures on metal surfaces, resulting in accelerated corrosion attack. In fossil-fuel-fired power generation, ash/salt deposits are also very common because of sulfur and vanadium in the fuel oil, particularly low-grade fuels, and alkali metals, chlorine, and sulfur in the coal. The accelerated corrosion due to ash/salt deposits in this case is frequently referred to as *fuel ash corrosion*. Both hot corrosion and fuel ash corrosion are generally believed to be related to liquid salt deposits, which destroy the protective oxide scale on the metal surface. Waste incineration generates very complex ash/salt deposits, which often contain sulfur, sodium, potassium, chlorine, zinc, lead, phosphorus, and other elements. Ash/salt deposits are common in waste-heat recovery systems for industrial processes, such as aluminum remelting operations and pulp and paper recovery boilers, and are also common in calcining operations for various chemical products. This mode of corrosion is referred to as *ash/salt deposit corrosion*.

Effect of Low-Melting Compounds. There are other types of low-melting compounds that can form on the metal surface during the high-temperature reaction. The most common ones include V_2O_5, MoO_3, and nickel-phosphorus compounds. These liquid phases can easily destroy the protective oxide scale and result in accelerated corrosion attack. When V_2O_5 or MoO_3 is involved, the attack is referred to as *catastrophic oxidation*. When the environment contains phosphorus, high-nickel alloys may react with it to form low-melting nickel-phosphorus eutectics, which then destroy the protective oxide scale. The subsequent oxidation or other mode of attack is thus accelerated.

Molten salt and liquid metal corrosion are two other important high-temperature corrosion modes. Oxygen activity may still play an important role in the corrosion reaction for both environments. For example, in a molten salt pot the worst attack frequently occurs at the air-salt interface, presumably because that is where oxygen activity is highest.

Fundamental Data. Essential to an understanding of the gaseous corrosion of a metal are the crystal structure and the molar volume of the metal on which the oxide builds, both of which may affect growth stresses in the oxide. For high-temperature service it is necessary to know the melting point of the metal, which indicates the practical temperature limits, and the structural changes that take place during heating and cooling, which affect oxide adherence. These data are presented in Table 2 for pure metals. For the oxides, their structures, melting and boiling points, molar volume, and oxide/metal volume ratio (Pilling-Bedworth ratio) are shown in Table 3. The structure data were taken from many sources.

Thermodynamics of High-Temperature Corrosion in Gases

Free Energy of Reaction. The driving force for reaction of a metal with a gas is the Gibbs energy change, ΔG. For the usual conditions of constant temperature and pressure, ΔG is described by the Second Law of Thermodynamics as:

$$\Delta G = \Delta H - T\Delta S \quad \text{(Eq 1)}$$

where ΔH is the enthalpy of reaction, T is the absolute temperature, and ΔS is the entropy change. No reaction will proceed spontaneously unless ΔG is negative. If $\Delta G = 0$, the system is at equilibrium, and if ΔG is positive, the reaction is thermodynamically unfavorable; that is, the reverse reaction will proceed spontaneously.

The driving force ΔG for a reaction such as aA + bB = cC + dD can be expressed in terms of the standard Gibbs energy change, ΔG^o, by:

$$\Delta G = \Delta G^o + RT \ln \frac{a_C^c\, a_D^d}{a_A^a\, a_B^b} \qquad \text{(Eq 2)}$$

where the chemical activity, a, of each reactant or product is raised to the powder of its stoichiometric coefficient, and R is the gas constant. For example, in the oxidation of a metal by the reaction:

$$x\text{M} + \frac{y}{2}\text{O}_2 = \text{M}_x\text{O}_y$$

where M is the reacting metal, M_xO_y is its oxide, and x and y are the moles of metal and oxygen, respectively, in 1 mol of the oxide.

The Gibbs energy change for the reaction is:

$$\Delta G = \Delta G^o + RT \ln \left[\frac{a_{M_xO_y}}{(a_M)^x \cdot (a_{O_2})^{y/2}} \right] \qquad \text{(Eq 3)}$$

In most cases, the activities of the solids (metal and oxide) are invariant; that is, their activities = 1 for pure solids, and for the relatively high temperatures and moderate pressures encountered in oxidation reactions, a_{O_2} can be approximated by its pressure. Therefore, at equilibrium where $\Delta G = 0$:

$$\Delta G^o = -RT \ln \left[\frac{a_{prod}}{a_{react}} \right] \approx + \frac{y}{2} RT \ln p_{O_2} \qquad \text{(Eq 4)}$$

where p_{O_2} is the partial pressure of oxygen.

In solid solutions, such as an alloy, the partial molar Gibbs energy of a substance is usually called its chemical potential μ. If 1 mol of pure A is dissolved in an amount of solution so large that the solution concentration remains virtually unchanged, the Gibbs energy change for the mole of A is:

$$\overline{\Delta G}_A = \mu_A - \mu_A^o = RT \ln a_A \qquad \text{(Eq 5)}$$

where μ_A^o is the chemical potential of 1mol of pure A, the chemical potential μ_A is the value in the solution, and a_A is the activity of A in the solution.

Metastable Oxides. Thermodynamically unstable oxides are often formed in corrosion by gases. The Gibbs energy of formation of the oxide, ΔG, is less negative than for a stable oxide, but in fact an unstable oxide can often exist indefinitely with no measurable transformation.

A common example is wustite (FeO), which is formed during the hot rolling of steel. Thermodynamically, it is unstable below 570 °C (1060 °F), but it remains the major component of mill scale at room temperature because the decomposition kinetics is extremely slow.

As another example, rapid kinetics can favor the formation of less stable oxide on an alloy. An alloy AB could oxide to form oxides AO and BO, but if BO is more stable than AO, then any AO formed in contact with B should in theory convert to BO by the reaction:

$$\text{B} + \text{AO} \rightarrow \text{BO} + \text{A}$$

Nevertheless, if AO grows rapidly compared with BO and the conversion reaction is slow, AO can be the main oxide found on the alloy.

Thermodynamically unstable crystal structures of oxides are also sometimes found. A growing oxide film tends to try to align its crystal structure in some way with that of the substrate from which it is growing. This epitaxy can cause the formation of an unstable structure that fits the substrate best. For example, cubic aluminum oxide (Al_2O_3) may form on aluminum alloys instead of the stable rhombohedral Al_2O_3.

Free Energy-Temperature Diagrams. Metal oxides become less stable as temperature increases. The relative stabilities are usually shown on a Gibbs energy-temperature diagram, sometimes called an Ellingham diagram (Fig. 2), for common metals in equilibrium with their oxides. Similar diagrams are available for sulfides, nitrides, and other gas-metal reactions. In Fig. 2, the reaction plotted in every case is:

$$\frac{2x}{y}\text{M} + \text{O}_2 = \frac{2}{y}\text{M}_x\text{O}_y$$

That is, 1 mol of O_2 gas is always the reactant, so that:

$$\Delta G^o = RT \ln p_{O_2} \qquad \text{(Eq 6)}$$

For example, the Gibbs energy of formation of Al_2O_3 at 1000 °C (1830 °F), as read from Fig. 2, is approximately –840 kJ (–200 kcal) for 2/3 mol of Al_2O_3.

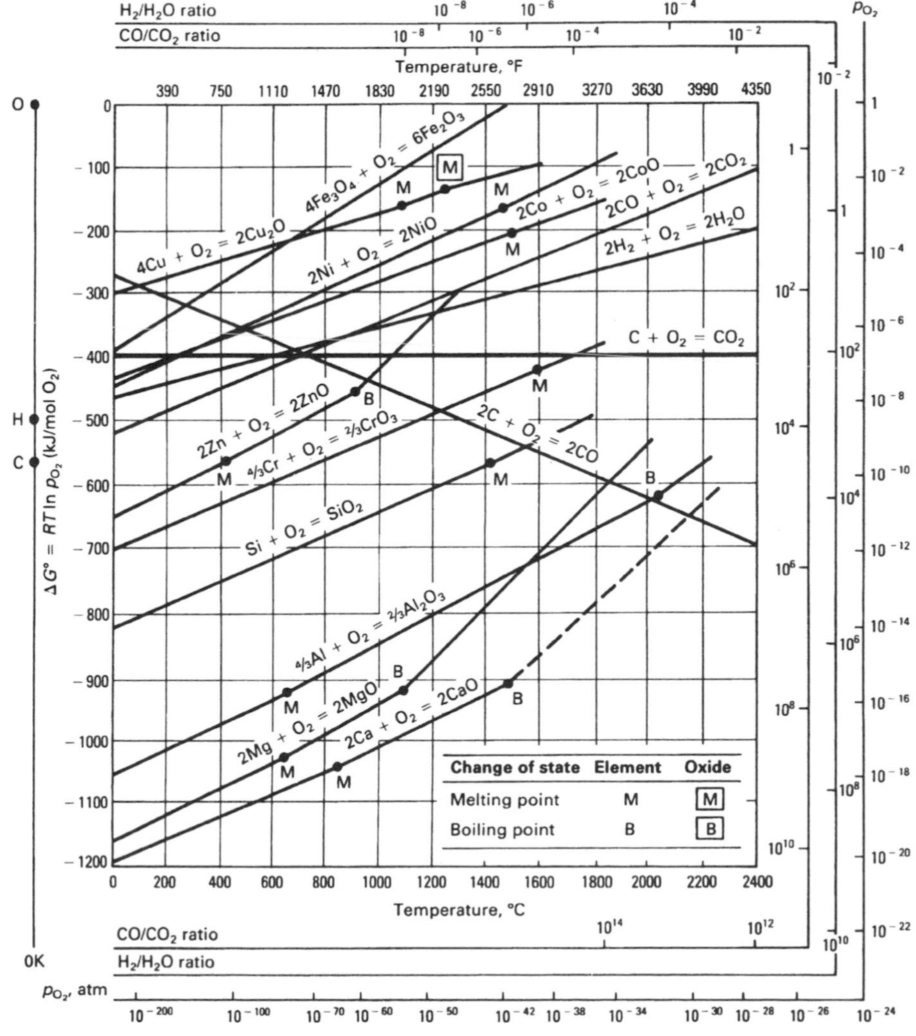

Fig. 2 Standard Gibbs energies of formation of selected oxides as a function of temperature. Source: Ref 5

Table 2 Structures and thermal properties of pure metals

Metal	Structure(a)	Transformation temperature		Volume change upon cooling(b), %	Melting point		Molar volume (c)	
		°C	°F		°C	°F	cm^3	$in.^3$
Aluminum	fcc	...	...	...	660.4	1220.7	10.0	0.610
Antimony	rhom	...	...	...	630.7	1167.3	18.18	1.109
Arsenic	rhom	...	...	...	Sublimation 615	1139	12.97	0.791
Barium	bcc	...	...	...	729	1344	39	2.380
Beryllium	(α) hcp	1250	2282	...	...	...	4.88	0.298
	(β) bcc	...	...	−2.2	1290	2354	4.99	0.304
Bismuth	rhom	...	...	...	271.4	520.5	21.31	1.300
Cadmium	hcp	...	...	...	321.1	610	13.01	0.793
Calcium	(α) fcc	448	838	...	...	...	25.9	1.581
	(β) bcc	...	...	−0.4	839	1542	...	...
Cerium	(γ) fcc	726	1339	...	...	...	20.70	1.263
	(δ) bcc	...	...	...	798	1468	...	...
Cesium	bcc	...	...	...	28.64	83.55	70.25	4.287
Chromium	bcc	...	...	...	1875	3407	7.23	0.441
Cobalt	(α) hcp	417	783	...	...	...	6.67	0.407
	(β) fcc	...	...	−0.3	1495	2723	6.70	0.408
Copper	fcc	...	...	...	1084.88	1984.78	7.12	0.434
Dysprosium	(α) hcp	1381	2518	...	...	...	19.00	1.159
	(β) bcc	...	...	0.1	1412	2573	18.98	1.158
Erbium	hcp	...	...	...	1529	2784	18.45	1.126
Europium	bcc	...	...	...	822	1512	28.98	1.768
Gadolinium	(α) hcp	1235	2255	...	...	...	19.90	1.214
	(β) bcc	...	...	−1.3	1312	2394	20.16	1.230
Gallium	ortho	...	...	...	29.78	85.60	11.80	0.720
Germanium	diamond fcc	...	...	...	937.4	1719.3	13.63	0.832
Gold	fcc	...	...	...	1064.43	1947.97	10.20	0.622
Hafnium	(α) hcp	1742	3168	...	...	...	13.41	0.818
	(β) bcc	...	...	...	2231	4048	...	...
Holmium	hcp	...	...	...	1474	2685	18.75	1.144
Indium	tetr	...	...	...	156.63	313.93	15.76	0.962
Iridium	fcc	...	...	...	2447	4437	8.57	0.523
Iron	(α) bcc	912	1674	...	...	...	7.10	0.433
	(γ) fcc	1394	2541	1.0	...	...	7.26	0.443
	(δ) bcc	...	...	−0.52	1538	2800	7.54	0.460
Lanthanum	(α) hex	330	626	...	...	...	22.60	1.379
	(β) fcc	865	...	0.5	...	...	22.44	1.369
	(γ) bcc	...	...	−1.3	918	1684	23.27	1.420
Lead	fcc	...	...	...	327.4	621.3	18.35	1.119
Lithium	(β) bcc	−193	−315	...	180.7	357.3	12.99	0.793
Lutetium	hcp	...	...	...	1663	3025	17.78	1.085
Magnesium	hcp	...	...	...	650	1202	13.99	0.854
Manganese	(α) cubic	710	1310	...	...	...	7.35	0.449
	(β) cubic	1079	1974	−3.0	...	...	7.63	0.466
	(γ) tetr	...	...	−0.0	1244	2271	7.62	0.465
Mercury	rhom	...	...	...	−38.87	−37.97	14.81	0.904
Molybdenum	bcc	...	...	...	2610	4730	9.39	0.573
Neodymium	(α) hex	863	1585	...	...	...	20.58	1.256
	(β) bcc	...	...	−0.1	1021	1870	21.21	1.294
Nickel	fcc	...	...	...	1453	2647	6.59	0.402
Niobium	bcc	...	...	...	2648	4474	10.84	0.661
Osmium	hcp	...	...	...	~2700	~4890	8.42	0.514
Palladium	fcc	...	...	...	1552	2826	8.85	0.540
Platinum	fcc	...	...	...	1769	3216	9.10	0.555
Plutonium	α, β, γ	120, 210, 315	248, 410, 599	...	...	...	α 12.04	α 0.735
	δ, δ′, ε	452, 480	846, 896	...	640	1184	ε 14.48	ε 0.884
Potassium	bcc	...	...	...	63.2	145.8	45.72	2.790
Praseodymium	(α) hex	795	1463	...	...	...	20.80	1.269
	(β) bcc	...	...	−0.5	931	1708	21.22	1.295
Rhenium	hcp	...	...	...	3180	5756	8.85	0.540
Rhodium	fcc	...	...	...	1963	3565	8.29	0.506
Rubidium	bcc	...	...	...	38.89	102	55.79	3.405
Ruthenium	hcp	...	...	...	2310	4190	8.17	0.499
Samarium	(α) rhom	734	1353	...	...	...	20.00	1.220
	(β) hcp	922	1692	...	...	...	20.46	1.249
	(γ) bcc	...	...	...	1074	1965	20.32	1.240
Scandium	(α) hcp	1337	2439	...	...	...	15.04	0.918
	(β) bcc	...	...	...	1541	2806	...	...
Selenium	(γ) hex	209	408	...	217	423	16.42	1.002

(continued)

(a) fcc, face-centered cubic; rhom, rhombohedral; bcc, body-centered cubic; hcp, hexagonal close-packed; ortho, orthorhombic; tetr, tetragonal; hex, hexagonal; bct, body-centered tetragonal. (b) Volume change upon cooling through crystallographic transformation. (c) Molar volume at 25 °C (77 °F) or at transition temperature for structures not stable at 25 °C (77 °F). Source: Ref 3

The equilibrium partial pressure of O_2 is:

$$p_{O_2} = \exp\left(\frac{\Delta G^\circ}{RT}\right) \quad \text{(Eq 7)}$$

and can also be read directly from Fig. 2 without calculation by use of the p_{O_2} scale along the bottom and right side of the diagram. A straight line drawn from the index point labeled O at the upper left of the diagram through the 1000 °C (1830 °F) point on the Al/Al_2O_3 line, intersects the p_{O_2} scale at approximately 10^{-35} atm, which is the O_2 partial pressure in equilibrium with aluminum and Al_2O_3 at 1000 °C (1830 °F). This means that any O_2 pressure greater than 10^{-35} atm tends to oxidize more aluminum, while Al_2O_3 would tend to decompose to Al + O_2 only if the pressure could be reduced to below 10^{-35} atm. Obviously, Al_2O_3 is an extremely stable oxide.

The oxidation of a metal by water vapor can be determined in the same way. The reaction is:

$$xM + yH_2O = M_xO_y + yH_2$$

The equilibrium p_{H_2}/p_{H_2O} ratio for any oxide at any temperature can be found by constructing a line from the H index point on the left side of Fig. 2. For example, for the reaction:

$$2Al(l) + 3\,H_2O(g) = Al_2O_3(s) + 3H_2(g)$$

at 1000 °C (1830 °F), the equilibrium H_2/H_2O ratio is 10^{10}. A ratio greater than this will tend to drive the reaction to the left, reducing Al_2O_3 to the metal. A ratio less than 10^{10} produces more oxide.

Similarly, the oxidation of metals by carbon dioxide (CO_2) is also shown in Fig. 2. For the reaction:

$$xM + y\,CO_2 = M_xO_y + y\,CO$$

the equilibrium carbon monoxide (CO)/CO_2 ratio is found from the index point marked C on the left side of the diagram. Oxidation of aluminum by CO_2 has an equilibrium CO/CO_2 ratio approximately 10^{10} at 1000 °C (1830 °F).

Isothermal Stability Diagrams. For situations that are more complicated than a single metal in a single oxidizing gas, it is common to fix the temperature at some practical value and plot the other variables of gas pressures or alloy composition against each other. This produces isothermal stability diagrams, or predominance area diagrams, which show the species that will be most stable in any set of circumstances.

One Metal and Two Gases. These diagrams, often called Kellogg diagrams, are constructed from the standard Gibbs energies of formation, ΔG°, of all elements and compounds likely to be present in the system. For example, for the Ni-O-S system, the ΔG° values of nickel monoxide (NiO) (s), nickel monosulfide (NiS) (l), nickel sulfate ($NiSO_4$) (s), sulfur dioxide (SO_2) (g), sulfur trioxide (SO_3) (g), and S (l) are needed.

In Fig. 3, the boundary between the Ni (s) and NiO (s) regions represents the equilibrium Ni (s) + ½O_2 (g) = NiO (s); therefore, the diagram shows that at 1250 K any O_2 pressure above

Table 2 (continued)

Metal	Structure(a)	Transformation temperature °C	Transformation temperature °F	Volume change upon cooling(b), %	Melting point °C	Melting point °F	Molar volume (c) cm^3	Molar volume (c) $in.^3$
Silicon	diamond fcc	...	...	...	1410	2570	12.05	0.735
Silver	fcc	...	...	...	961.9	1763.4	10.28	0.627
Sodium	(β) bcc	–237	–395	...	97.82	208.08	23.76	1.450
Strontium	(α) fcc	557	1035	...	...	...	34	2.075
	(β) bcc	...	...	...	768	1414	34.4	2.099
Tantalum	bcc	...	...	...	2996	5425	10.9	0.665
Tellurium	hex	...	...	...	449.5	841.1	20.46	1.249
Terbium	(α) hcp	1289	2352	...	...	...	19.31	1.178
	(β) bcc	...	...	...	1356	2472.8	19.57	1.194
Thallium	(α) hcp	230	446	...	...	...	17.21	1.050
	(β) bcc	...	...	...	303	577	...	...
Thorium	(α) fcc	1345	2453	...	...	...	19.80	1.208
	(β) bcc	...	...	...	1755	3191	21.31	1.300
Thulium	hcp	...	...	...	1545	2813	18.12	1.106
Tin	(β) bct	13.2	55.8	27	231.9	449.4	16.56	1.011
Titanium	(α) hcp	882.5	1621	...	...	...	10.63	0.649
	(β) bcc	...	...	...	1668	3034	11.01	0.672
Tungsten	bcc	...	...	...	3410	6170	9.55	0.583
Uranium	(α) ortho	661	1222	...	...	...	12.50	0.763
	(β) complex tetr	769	1416	–1.0	...	...	13.00	0.793
	(γ) bcc	...	...	–0.6	1900	3452	8.34	0.509
Ytterbium	(β) fcc	7	45	0.1	819	1506	24.84	1.516
Yttrium	(α) hcp	1478	2692	...	...	...	19.89	1.214
	(β) bcc	...	...	...	1522	2772	20.76	1.267
Zinc	hcp	...	...	...	420	788	9.17	0.559
Zirconium	(α) hcp	862	1584	...	...	...	14.02	0.856
	(β) bcc	...	...	...	1852	3366	15.09	0.921

(a) fcc, face-centered cubic; rhom, rhombohedral; bcc, body-centered cubic; hcp, hexagonal close-packed; ortho, orthorhombic; tetr, tetragonal; hex, hexagonal; bct, body-centered tetragonal. (b) Volume change upon cooling through crystallographic transformation. (c) Molar volume at 25 °C (77 °F) or at transition temperature for structures not stable at 25 °C (77 °F). Source: Ref 3

about 10^{-11} atm will tend to form NiO from metallic nickel if p_{S_2} is low. Similarly, S_2 gas pressure greater than about 10^{-7} atm will form NiS from nickel at low p_{O_2}. Also a mixed gas of 10^{-5} atm each of S_2 and O_2 should form nearly the equilibrium ratio of NiO (s) and $NiSO_4$ (s).

If the principal gases of interest were SO_2 and O_2, the same $\Delta G°$ data could be used to construct a diagram of log p_{O_2} versus p_{SO_2}, or as in Fig. 3, p_{SO_2} isobars can be added to the figure (the dotted lines). Thus, a mixed gas of 10^{-5} atm each of SO_2 and O_2 will form only NiO at 1250 K, with neither the sulfide nor sulfate being as stable.

When nickel metal is heated to 1250 K in the open air with sulfur-containing gases, $p_{SO_2} + p_{S_2} + p_{O_2} \approx 0.2$ atm. The situation is shown by the dashed line in Fig. 3 labeled $p = 0.2$ atm.

An Alloy System and a Gas. Isothermal stability diagrams for oxidation of many important alloy systems have been worked out, such as that for the Fe-Cr-O system shown in Fig. 4. In this diagram, the mole fraction of chromium in the alloy is plotted against log p_{O_2} so that for any alloy composition the most stable oxide or mixture of oxides is shown at any gas pressure.

For an alloy system in gases containing more than one reactive component, the pressures of all but one of the gases must be fixed at reasonable values to be able to draw an isothermal stability diagram in two dimensions. Figure 5 shows an example of such a situation: the Fe-Zn system in equilibrium with sulfur and oxygen-containing gases with SO_2 pressure set at 1 atm and temperature set at 1164 K.

Limitations of Predominance Area Diagrams. Isothermal stability diagrams, like all predominance area diagrams, including Pourbaix potential-pH diagrams, must be read with an understanding of their rules:

- Each area on the diagram is labeled with the predominant phase that is stable under the specified conditions of pressure or temperature. Other phases may also be stable in that area, but in smaller amounts.
- The boundary line separating two predominance areas shows the conditions of equilibrium between the two phases.

Also, the limitations of the diagrams must be understood to be able to use them intelligently:

- The diagrams are for the equilibrium situation. Equilibrium may be reached quickly in high-temperature oxidation, but if the metal is then cooled, equilibrium is often not reestablished.
- Microenvironments, such as gases in voids or cracks, can create situations that differ from

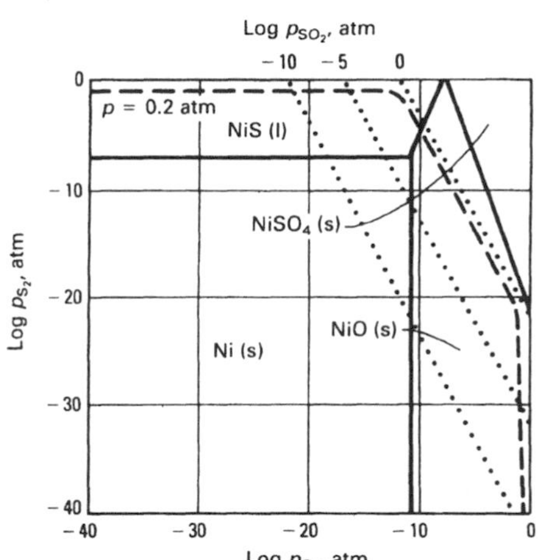

Fig. 3 The Ni-O-S system at 1205 K. Source: Ref 6

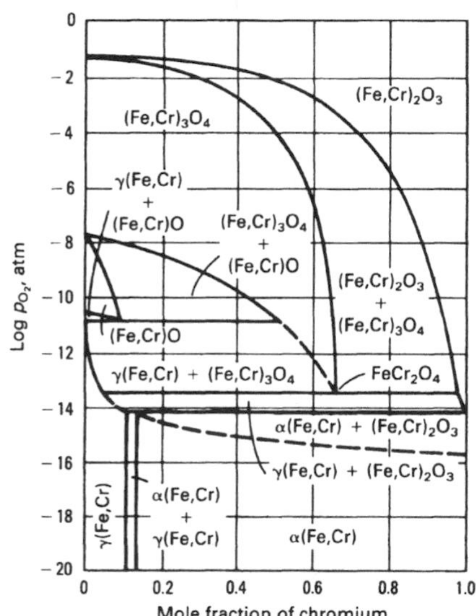

Fig. 4 Stability diagram for the Fe-Cr-O system at 1300 °C (2370 °F)

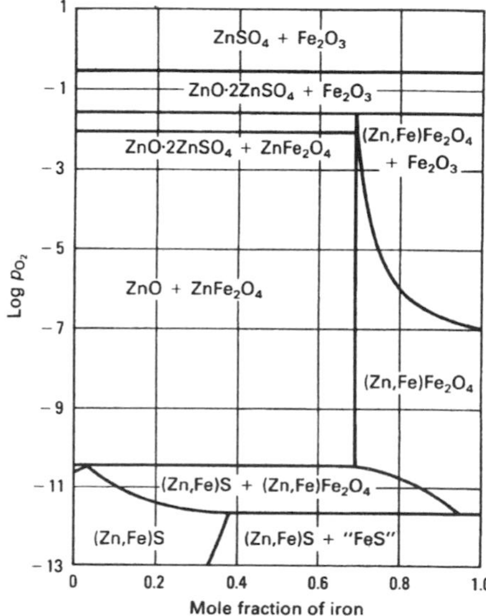

Fig. 5 The Fe-Zn-S-O system for p_{SO_2} = 1 atm at 1164 K. Source: Ref 7

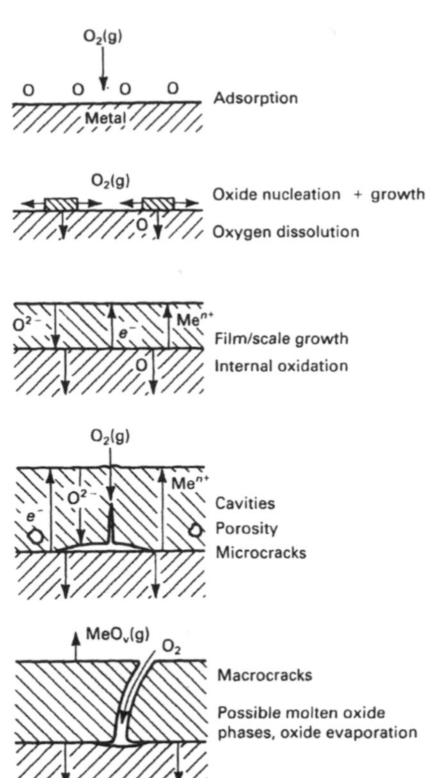

Fig. 6 Schematic illustration of the principal phenomena taking place during the reaction of metals with oxygen. Source: Ref 8

the situations expected for the bulk reactant phases.

- The diagrams often show only the major components, omitting impurities that are usually present in industrial situations and may be important.
- The diagrams are based on thermodynamic data and do not show rates of reaction.

Kinetics of Oxidation

In this section, the commonly observed kinetics of oxidation will be described and related to the corrosion mechanisms. These mechanisms are shown schematically in Fig. 6. The gas first absorbs on the metal surface as atomic oxygen. Oxide nucleates at favorable sites and most commonly grows laterally to form a complete thin film. As the layer thickens, it provides a protective scale barrier to shield the metal from the gas. For scale growth, electrons must move through the oxide to reach the oxygen atoms absorbed on the surface, and oxygen ions, metal ions, or both must move through the oxide barrier. Oxygen may also diffuse into the metal.

Growth stresses in the scale may create cavities and microcracks in the scale, modifying the oxidation mechanism or even causing the oxide to fail to protect the metal from the gas. Improved oxidation resistance can be achieved by developing better alloys and by applying protective coatings. These subjects are addressed in the articles "Design for Oxidation Resistance" and "Protective Coatings for Superalloys," respectively, in this Volume.

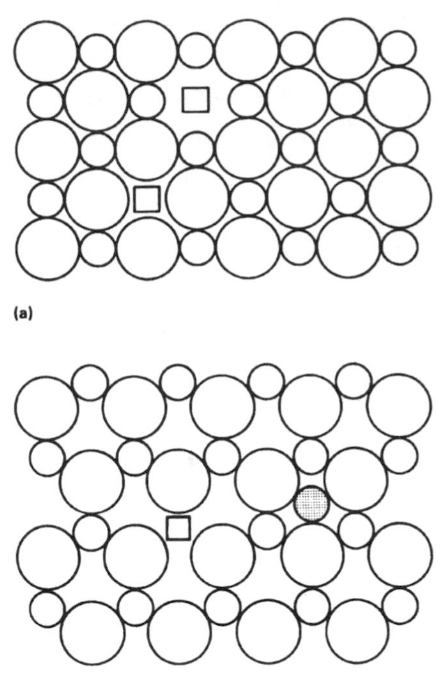

Fig. 7 Defects in ionic crystals. (a) Schottky defect. (b) Frenkel defect. Vacancies are indicated by open squares. Interstitial ion is shown as shaded circle.

Mechanisms of Oxidation

Pilling-Bedworth Theory. In 1923, N.B. Pilling and R.E. Bedworth classified oxidizable metals into two groups: those that formed protective oxide scales and those that did not (Ref 9). They suggested that unprotective scales formed if the volume of the oxide layer was less than the volume of metal reacted. For example, the oxidation of aluminum:

$$2Al + {}^3\!/_2O_2 \rightarrow Al_2O_3$$

the Pilling-Bedworth ratio is:

$$\frac{\text{Volume of 1 mol of } Al_2O_3}{\text{Volume of 2 mol of Al}}$$

where the volumes can be calculated from molecular and atomic weights and the densities of the phases.

If the ratio is less than 1, as is the case for alkali and alkaline earth metals, the oxide scales are usually unprotective, with the scales being porous or cracked due to tensile stresses and providing no efficient barrier to penetration of gas to the metal surface. If the ratio is more than 1, the protective scale shields the metal from the gas so that oxidation can proceed only by solid-state diffusion, which is slow even at high temperatures. If the ratio is much over 2 and the scale is growing at the metal/oxide interface, the large compressive stresses that develop in the oxide as it grows thicker may eventually cause the scale to spall off, leaving the metal unprotected.

Exceptions to the Pilling-Bedworth theory are numerous, and it has been roundly criticized and rejected by many. Its main flaw is the assumption that metal oxides grow by diffusion of oxygen inward through the oxide layer to the metal. In fact, it is much more common for metal ions to diffuse outward through the oxide to the gas. Also, the possibility of plastic flow by the oxide or metal was not considered. Nevertheless, historically, Pilling and Bedworth made the first step in achieving understanding of the processes by which metals react with gases. And although there may be exceptions, the volume ratio, as a rough rule-of-thumb, is usually correct. The Pilling-Bedworth volume ratios for many common oxides are listed in Table 3.

Oxide Structure and Texture

Defect Structure of Ionic Oxides. Ionic compounds can have appreciable ionic conductivity due to Schottky defects and/or Frenkel defects. Schottky defects are combinations of cation vacancies and anion vacancies in the proper ratio necessary to maintain electrical neutrality. Figure 7(a) illustrates a Schottky defect in a stoichiometric ionic crystal. With Schottky defects, the ions must diffuse into the appropriate adjacent vacancies to allow mass transfer and ionic electrical conductivity.

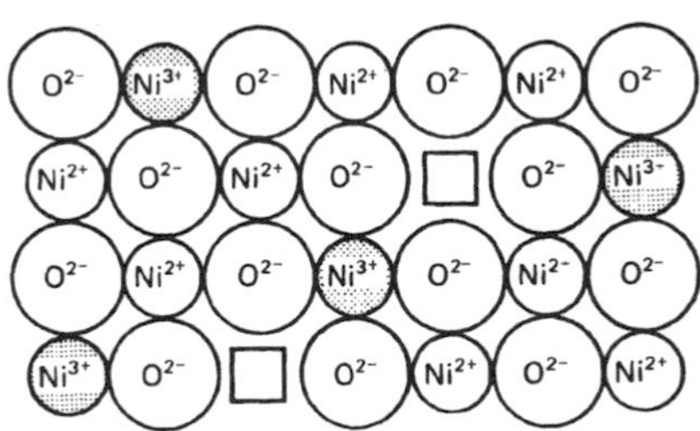

Fig. 8 Illustration of the ionic arrangement in *p*-type NiO scale. Cation vacancies are indicated as open squares. The Ni^{3+} cations are shaded.

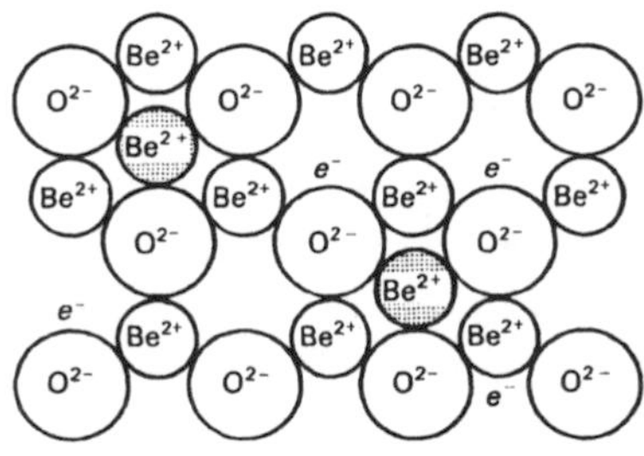

Fig. 9 Illustration of the ionic arrangement in *n*-type cation-excess BeO. Interstitial cations are shaded; free electrons are indicated as e^-.

Frenkel defects are also present in ionic crystals in such a way that electrical neutrality and stoichiometry are maintained (Fig. 7b). This type of defect is a combination of a cation vacancy and an interstitial cation. Metal cations are generally much smaller than the oxygen anions. Limited ionic electrical conductivity is possible in such crystals by diffusion of cations interstitially and by diffusion of cations into the cation vacancies.

Metallic oxides are seldom, if ever, stoichiometric and cannot grow by mere diffusion by Schottky and Frenkel defects. For oxidation to continue when a metal is protected by a layer of oxide, electrons must be able to migrate from the metal, through the oxide, to adsorbed oxygen at the oxide/gas interface. Nevertheless, Schottky and Frenkel defects may provide the mechanism for ionic diffusion necessary for oxide growth.

Defect Structure of Semiconductor Oxides. Oxides growing to provide protective scales are electronic semiconductors that also allow mass transport of ions through the scale layer. They may be conveniently categorized as *p*-type (or positive carrier), *n*-type (or negative carrier), and amphoteric semiconductors. Examples of the three types are listed in Table 4.

The p-type metal-deficit oxides are nonstoichiometric with cation vacancies present. They will also have some Schottky and Frenkel defects that add to the ionic conductivity. A typical example is NiO, a cation-deficient oxide that provides the additional electrons needed for ionic bonding and electrical neutrality by donating electrons from the 3*d* subshells of a fraction of the nickel ions. In this way, for every cation vacancy present in the oxide, two nickel ions (Ni^{3+}) will be present (Fig. 8). Each Ni^{3+} has a low-energy positively charged electron hole that electrons from other nickelous ions (Ni^{2+}) can easily move into. The positive or *p*-type semiconductors carry most of their current by means of these positive holes.

Cations can diffuse through the scale from the Ni/NiO interface by cation vacancies, to the NiO/gas interface where they react with adsorbed oxygen. Electrons migrate from the metal surface, by electron holes, to the adsorbed oxygen atoms, which then become oxygen anions. In this way, while Ni^{2+} cations and electrons move outward through the scale toward the gas, cation vacancies and electron holes move inward toward the metal. Consequently, as the scale thickens, the cation vacancies tend to accumulate to form voids at the Ni/NiO interface.

The n-type semiconductor oxides have negatively-charged free electrons as the major charge carriers. They may be either cation excess or anion deficient. Beryllium oxide (BeO) typifies the cation-excess oxides because the beryllium ion (Be^{2+}) is small enough to move interstitially through the BeO scale. Its structure is shown in Fig. 9.

Oxygen in the gas adsorbs on the BeO surface and picks up free electrons from the BeO to become adsorbed O^{2-} ions, which then react with excess Be^{2+} ions that are diffusing interstitially

Table 3 Structures and thermal properties of selected oxides

Oxide	Structure	Melting point, °C	Melting point, °F	Boiling or decomposition, d., °C	Boiling or decomposition, d., °F	Molar volume(a), cm³	Molar volume(a), in.³	Volume ratio
α-Al_2O_3	$D5_1$ (corundum)	2015	3659	2980	5396	25.7	1.568	1.28
γ-Al_2O_3	(defect-spinel)	γ→α	...	...	...	26.1	1.593	1.31
BaO	*B*1 (NaCl)	1923	3493	~2000	~3632	26.8	1.635	0.69
BaO_2	Tetragonal (CaC_2)	450	842	d.800	d.1472	34.1	2.081	0.87
BeO	*B*4 (ZnS)	2530	4586	~3900	~7052	8.3	0.506	1.70
CaO	*B*1 (NaCl)	2580	4676	2850	5162	16.6	1.013	0.64
CaO_2	*C*11 (CaC_2)	...	...	d.275	d.527	24.7	1.507	0.95
CdO	*B*1 (NaCl)	~1400	~2552	d.900	d.1652	18.5	1.129	1.42
Ce_2O_3	$D5_2$ (La_2O_3)	1692	3078	...	...	47.8	2.917	1.15
CeO_2	*C*l (CaF_2)	~2600	~4712	...	...	24.1	1.471	1.17
CoO	*B*1 (NaCl)	1935	3515	...	...	11.6	0.708	1.74
Co_2O_3	Hexagonal	...	...	d.895	d.1643	32.0	1.953	2.40
Co_3O_4	Hl_1(spinel)	→CoO	...	...	...	39.7	2.423	1.98
Cr_2O_3	*D*5 (αAl_2O_3)	2435	4415	4000	7232	29.2	1.782	2.02
Cs_2O	Hexagonal ($CdCl_2$)	...	...	d.400	d.752	66.3	4.046	0.47
Cs_2O_3	Cubic (Th_3P_4)	400	752	650	1202	70.1	4.278	0.50
CuO	*B*26 monoclinic	1326	2419	...	...	12.3	0.751	1.72
Cu_2O	*C*3 cubic	1235	2255	d.1800	d.3272	23.8	1.452	1.67
Dy_2O_3	Cubic (Tl_2O_3)	2340	4244	...	...	47.8	2.917	1.26
Er_2O_3	Cubic (Tl_2O_3)	...	...	...	...	44.3	2.703	1.20
FeO	*B*1 (NaCl)	1420	2588	...	...	12.6	0.769	1.78 on α-iron
α-Fe_2O_3	$D5_1$ (hematite)	1565	2849	...	...	30.5	1.861	2.15 on α-iron
	...	...	...	...	...	...	...	1.02 on Fe_3O_4
γ-Fe_2O_3	$D5_7$ cubic	1457	2655	...	...	31.5	1.922	2.22 on α-iron
Fe_3O_4	Hl_1 (spinel)	...	...	d.1538	d.2800	44.7	2.728	2.10 on α-iron
	...	...	...	...	...	...	...	~1.2 on FeO
Ga_2O_3	Monoclinic	1900	3452	...	...	31.9	1.947	1.35
HfO_2	Cubic	2812	5095	~5400	~9752	21.7	1.324	1.62
HgO	Defect *B*10(SnO)	...	...	d.500	d.932	19.5	1.190	1.32
In_2O_3	$D5_3$(Sc_2O_3)	...	...	d.850	d.1562	38.7	2.362	1.23
IrO_2	*C*4(TiO_2)	...	...	d.1100	d.2012	19.1	1.166	2.23
K_2O	*C*l(CaF_2)	...	...	d.350	d.662	40.6	2.478	0.45
La_2O_3	$D5_2$ hexagonal	2315	4199	4200	7592	50.0	3.051	1.10
Li_2O	*C*l (CaF_2)	~1700	~3092	1200	2192	14.8	0.903	0.57
MgO	*B*l (NaCl)	2800	5072	3600	6512	11.3	0.690	0.80
MnO	*B*l (NaCl)	...	...	...	...	13.0	0.793	1.77
MnO_2	*C*4 (TiO_2)	...	...	d.535	d.995	17.3	1.056	2.37
Mn_2O_3	$D5_3$ (Sc_2O_3)	...	...	d.1080	d.1976	35.1	2.142	2.40
α-Mn_3O_4	Hl_1 (spinel)	1705	3101	...	...	47.1	2.874	2.14
MoO_3	Orthorhombic	795	1463	...	...	30.7	1.873	3.27
Na_2O	*C*l (CaF_2)	Sublimation, 1275	2327	...	...	27.3	1.666	0.57
Nb_2O_5	Monoclinic	1460	2600	...	...	59.5	3.631	2.74
Nd_2O_3	Hexagonal	~1900	~3452	...	...	46.5	2.838	1.13
NiO	*B*l (NaCl)	1990	3614	...	...	11.2	0.683	1.70
OsO_2	*C*4 (TiO_2)	...	...	d.350	d.662	28.8	1.757	3.42
PbO	*B*10 tetragonal	888	1630	...	...	23.4	1.428	1.28
Pb_3O_4	Tetragonal	...	...	d.500	d.932	75.3	4.595	1.37
PdO	*B*17 tetragonal	870	1598	...	...	14.1	0.860	1.59
PtO	*B*17 (PdO)	...	...	d.550	d.1022	14.2	0.867	1.56
Rb_2O_3	(Th_3P_4)	489	912	...	...	62.0	3.783	0.56
ReO_2	Monoclinic	...	...	d.1000	d.1832	19.1	1.166	2.16
Rh_2O_3	$D5_1$ (α-Al_2O_3)	...	...	d.1100	d.2012	31.0	1.892	1.87
SiO	Cubic	~1700	~3092	1880	3416	20.7	1.263	1.72
SiO_2	β christobalite *C*9	1713	3115	2230	4046	25.9	1.581	2.15
SnO	*B*10 (PbO)	...	...	d.1080	d.1976	20.9	1.275	1.26
SnO_2	*C*4 (TiO_2)	1127	2061	...	...	21.7	1.324	1.31
SrO	*B*1 (NaCl)	2430	4406	~3000	~5432	22.0	1.343	0.65
Ta_2O_5	Triclinic	1800	3272	...	...	53.9	3.289	2.47
TeO_2	*C*4 (TiO_2)	733	1351	1245	2273	28.1	1.715	1.38
ThO_2	*C*l (CaF_2)	3050	5522	4400	7952	26.8	1.635	1.35
TiO	*B*l (NaCl)	1750	3182	~3000	~5432	13.0	0.793	1.22
TiO_2	*C*4 (rutile)	1830	3326	~2700	~4892	18.8	1.147	1.76
Ti_2O_3	$D5_1$ (α-Al_2O_3)	...	...	d.2130	d.3866	31.3	1.910	1.47
Tl_2O_3	$D5_3$ (Sc_2O_3)	717	1323	d.875	d.1607	44.8	2.734	1.30
UO_2	*C*1 (CaF_2)	2500	4532	...	...	24.6	1.501	1.97
U_3O_8	Hexagonal	...	...	d.1300	...	101.5	6.194	2.71
VO_2	*C*4 (TiO_2)	1967	3573	...	...	19.1	1.166	2.29
V_2O_3	$D5_1$ (α-Al_2O_3)	1970	3578	...	...	30.8	1.879	1.85
V_2O_5	$D8_7$ orthorhombic	690	1274	d.1750	d.3182	54.2	3.307	3.25
WO_2	*C*4 (TiO_2)	~1550	~2822	~1430	~2606	17.8	1.086	1.87
β-WO_3	Orthorhombic	1473	...	...	...	32.4	1.977	3.39
W_2O_5	Triclinic	Sublimation, ~850	~1562	~1530	~2786	29.8	1.819	3.12
Y_2O_3	$D5_3$ (Sc_2O_3)	2410	4370	...	...	45.1	2.752	1.13
ZnO	*B*4 (wurtzite)	1975	3587	...	...	14.5	0.885	1.58
ZrO_2	*C*43 monoclinic	2715	4919	...	...	22.0	1.343	1.57

(a) Molar volume at 25 °C (77 °F) or at transition temperature for structures not stable at 25 °C (77 °F). Source: Ref 4

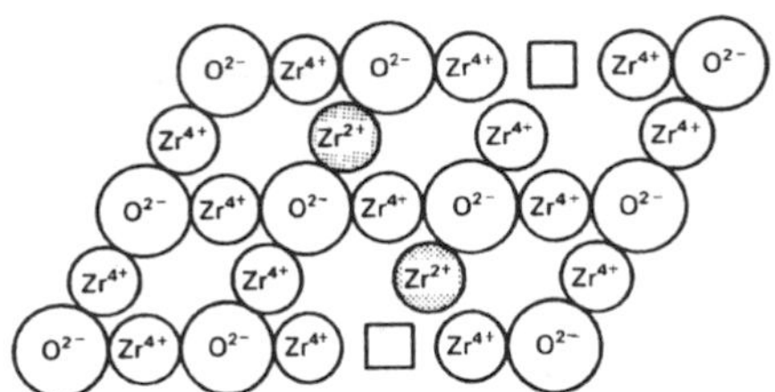

Fig. 10 Illustration of the ionic arrangement in *n*-type anion-deficient ZrO_2. Anion vacancies are indicated as open squares; Zr^{2+} ions are shaded.

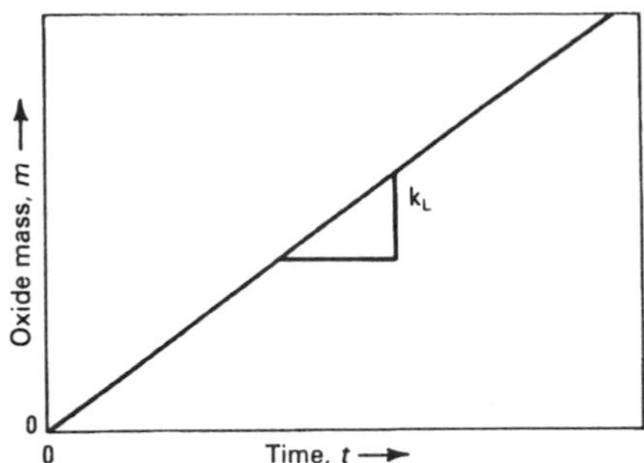

Fig. 11 Linear oxidation kinetics

Table 4 Classification of oxide and sulfide electrical conductors

Type	Compounds
Metal-excess semiconductors (*n*-type)	BeO, MgO, CaO, SrO, BaO, BaS, ScN, CeO_2, ThO_2, UO_3, U_3O_8, TiO_2, TiS_2, (Ti_2S_3), TiN, ZrO_2, V_2O_5, (V_2S_3), VN, Nb_2O_5, Ta_2O_5, (Cr_2S_3), MoO_3, WO_3, WS_2, MnO_2, Fe_2O_3, $MgFe_2O_4$, $ZnFe_2O_4$, $ZnCo_2O_4$, ($CuFeS_2$), ZnO, CdO, CdS, HgS(red), Al_2O_3, $MgAl_2O_4$, $ZnAl_2O_4$, Tl_2O_3, (In_2O_3), SiO_2, SnO_2, PbO_2
Metal-deficit semiconductors (*p*-type)	UO_2, (VS), (CrS), Cr_2O_3 (<1250 °C, or 2280 °F), $MgCr_2O_4$, $FeCr_2O_4$, $CoCr_2O_4$, $ZnCr_2O_4$, (WO_2), MoS_2, MnO, Mn_3O_4, Mn_2O_3, ReS_2, FeO, FeS, NiO, NiS, CoO, (Co_3O_4), PdO, Cu_2O, Cu_2S, Ag_2O, $CoAl_2O_4$, $NiAl_2O_4$, (Tl_2O), Tl_2S, (GeO), SnS, (PbO), (Sb_2S_3), (Bi_2S_3)
Amphoteric conductors	TiO(a), Ti_2O_3(a), VO(a), Cr_2O_3 (>1250 °C, or 2280 °F), MoO_2, FeS_2, (OsS_2), (IrO_2), RuO_2, PbS

(a) Metallic conductors. Source: Ref 10

from the beryllium metal. The free electrons coming from the metal surface as the beryllium ionizes travel rapidly through vacant high-energy levels. As with *p*-type oxides, the cation-excess *n*-type oxides grow at the oxide/gas interface as cations diffuse outward through the scale.

Another group of *n*-type semiconducting oxides is anion deficient, as exemplified by zirconium dioxide (ZrO_2). In this case, although most of the cations are contributing four electrons to the ionic bonding, a small fraction of the zirconium cations only contributes two electrons to become the zirconium ion Zr^{2+}. Therefore, to maintain electrical neutrality, an equal number of anion vacancies must be present in the oxide. This arrangement is shown in Fig. 10. The oxide grows at the metal/oxide interface by inward diffusion of O^{2-} through the anion vacancies in the oxide.

Amphoteric Oxides. A number of compounds can be nonstoichiometric with either a deficiency of cations or a deficiency of anions. An example is lead sulfide (PbS), which has a minimum in electrical conductivity at the stoichiometric composition. Thus, if the composition is $Pb_{<1}S$, it is *p*-type and if it is $PbS_{<1}$, it is *n*-type.

Similarly, intrinsic semiconductors, such as cupric oxide (CuO), have a few electron holes in their valence band and an equal number of free electrons in their nearly vacant conduction band. Current is carried both by migration of electron holes in the low-energy bonding levels and by free electrons in the high-energy conduction levels.

Oxide Texture: Amorphous Oxides. In the very early stages of oxidation and especially at low temperatures, some oxides appear to grow with an amorphous structure. In general, these oxide glasses contain more oxygen than metal in their formulas so that oxygen triangles or tetrahedra form around each of the metal ions. The random network ring structures that result allow large anions or molecular oxygen to move through them more readily than the smaller cations do. Amorphous oxides tend to crystallize as they age. Examples are silicon dioxide (SiO_2), Al_2O_3, tantalum pentoxide (Ta_2O_5), and niobium pentoxide (Nb_2O_5).

In contrast, oxides with M_2O and MO formulas have structures in which the small cations can move readily. They are apparently always crystalline. Examples are NiO, cuprous oxide (Cu_2O), and zinc oxide (ZnO).

Oxide Texture: Epitaxy. As a crystalline oxide grows on a metal surface, it often aligns its crystal structure to be compatible with the structure of the metal substrate. This epitaxy finds the best fit, not a perfect fit, between the two crystal structures. For example, either (111) or (001) planes of Cu_2O grow parallel to the Cu (001) plane with the $\langle 110 \rangle$ directions of Cu_2O parallel to the $\langle 110 \rangle$ of copper (Ref 11).

Stress develops in an epitaxial oxide layer as it grows because of the slight misfit between the oxide and metal crystals. The stress is likely to produce dislocation arrays in the oxide that would be paths of easy diffusion for mass transport through the film. A mosaic structure may develop in the oxide because of the growth stresses. The mosaic structure consists of small crystallites with orientations very slightly tilted or twisted with respect to each other. The boundaries between the crystallites are dislocation arrays that serve as easy diffusion paths.

Stresses in epitaxial layers increase as the films grow thicker until at some point the bulk scale tends to become polycrystalline and epitaxy is gradually lost. Epitaxy may last up to about 50 nm in many cases, but it is seldom strong much over 100 nm.

Oxide Texture: Preferred Orientation. As oxidation produces thicker layers, the oxide grain size increases. Crystals that are favorably oriented for growth will grow at the expense of their neighboring grains until the oxide surface consists of a few large grains with similar orientation. The variation in growth rate of different oxide grains produces the roughening of the scaled surface that is commonly observed.

Oxidation Reaction Rates

Linear Oxidation Reaction Rates. If the metal surface is not protected by a barrier of oxide, the oxidation rate usually remains constant with time, and one of the steps in the oxidation reaction is rate controlling rather than a transport process being rate controlling. This situation is to be expected if the Pilling-Bedworth ratio is less than 1, if the oxide is volatile or molten, if the scale spalls off or cracks due to internal stresses, or if a porous, unprotective oxide forms on the metals.

The linear oxidation rate is:

$$\frac{dx}{dt} = k_L \qquad \text{(Eq 8)}$$

where x is the mass or thickness of oxide formed, t is the time of oxidation, and k_L is the linear rate constant. The rate constant is a function of the metal, the gas composition and pressure, and the temperature.

Integrated, the linear oxidation equation is:

$$x = k_L t \qquad \text{(Eq 9)}$$

The oxidation never slows down; after long times at high temperatures, the metal will be completely destroyed. Figure 11 shows the relationship between oxide mass and time for linear oxidation.

Logarithmic and Inverse Logarithmic Reaction Rates. At low temperatures when only a thin film of oxide has formed (e.g., under 100 nm), the oxidation is usually observed to follow either logarithmic or inverse logarithmic kinetics. Transport processes across the film are rate controlling, with the driving force being electric fields across the film. The logarithmic equation is:

$$x = k_e \log (at + 1) \qquad \text{(Eq 10)}$$

where k_e and a are constants.

The inverse logarithmic equation is:

$$\frac{1}{x} = b - k_i \log t \qquad \text{(Eq 11)}$$

where b and k_i are constants. Under the difficult experimental conditions involved in making measurements in the thin film range, it is nearly impossible to distinguish between logarithmic and inverse logarithmic oxidation. Both equations have two constants that can be adjusted to fit the data quite well. Metals oxidizing with logarithmic or inverse log kinetics reach a limiting film thickness at which oxidation apparently stops. Figure 12 shows the curves for both logarithmic and inverse logarithmic kinetics.

Parabolic Kinetics. When the rate-controlling step in the oxidation process is the diffusion of ions through a compact barrier layer of oxide with the chemical potential gradient as the driving force, the parabolic rate law is usually observed. As the oxide grows thicker, the diffusion distance increases, and the oxidation rate slows down. The rate is inversely proportional to the oxide thickness, or:

$$\frac{dx}{dt} = \frac{k_p}{x} \quad \text{(Eq 12)}$$

Upon integration, the parabolic equation is obtained:

$$x^2 = \frac{k_p}{2} t \quad \text{(Eq 13)}$$

where k_p is the parabolic rate constant. Figure 13 shows the parabolic oxidation curve.

Other Reaction Rate Equations. A number of other kinetics equations have been fitted to the experimental data, but it is believed that they describe a combination of the new mechanisms described above, rather than any new basic process. A cubic relationship:

$$x^3 = k't \quad \text{(Eq 14)}$$

has often been reported. It can be shown mathematically to be an intermediate stage between logarithmic and parabolic kinetics (Fig. 14).

Initial Oxidation Processes: Adsorption and Nucleation

To begin oxidation, oxygen gas is chemisorbed on the metal surface until a complete two-dimensional invisible oxide layer forms. Some atomic oxygen also dissolves into the metal at the same time. After the monolayer forms, discrete nuclei of three-dimensional oxide appear on the surface and begin expanding laterally at an ever-increasing rate (Fig. 6). The nuclei may originate at structural defects, such as grain boundaries, impurity particles, and dislocations. The concentration of nuclei depends primarily on the crystal orientation of the metal, with more nuclei forming at high pressures and low temperatures.

These oxide islands grow outward rapidly by surface diffusion of adsorbed oxygen until a complete film three or four monolayers thick covers the metal. The oxidation rate then drops abruptly. If chemisorption were still the rate-controlling (slow) step in oxidation after the thin film is completed, a logarithmic rate law should be observed. A logarithmic rate law is found, but it is more likely the result of the strong electric field across the film that affects the oxidation.

Thin-Film Mechanisms

A large number of theories have been proposed to explain the oxidation mechanism at low temperatures or in the early stages of high-temperature oxidation where logarithmic kinetics is commonly observed. None of the theories is completed accepted yet, and perhaps none is completely correct, but they have common threads of agreement that indicate reasonably well what is happening. Some of the most important theories will be briefly described.

The Cabrera-Mott theory, probably the best established theory of thin film oxidation, applies to films up to about 10 nm thick (Ref 12). It proposes that electrons from the metal easily pass through the thin film by tunneling to reach adsorbed oxygen at the oxide/gas surface and form oxygen anions. A potential of approximately 1 V is set up between the external oxide surface and the metal. For a film 1 nm thick, the field strength would be 10^7 V/cm, powerful enough to pull cations from the metal and through the film. The rate-controlling step is the transfer of cations (or anions) into the oxide or the movement of the ions through the oxide. The electric field reduces this barrier. The structure of the oxide determines whether cations or anions migrate through the oxide. As the film grows thicker, the field strength decreases until it has so little effect on the ions that the rate-controlling mechanism changes.

N. Cabrera and N.F. Mott developed an inverse logarithmic kinetic equation to describe the mechanism. A logarithmic equation is more commonly observed, but it can be derived from the Cabrera-Mott mechanism if the activation energy for ionic migration is a function of film thickness. Such a situation would exist if the oxide film were initially amorphous and become more crystalline with aging, giving a constant field strength through the film instead of a constant voltage.

The Hauffe-Ilschner theory, a modification of Mott's original concept of a space charge developed across the oxide film, proposes that quantum-mechanical tunneling of electrons is the rate-controlling step (Ref 13). After the film thickness reaches about 10 nm, tunneling becomes increasingly difficult, and the observed reaction rate decreases greatly. For film thicknesses up to perhaps 20 nm, a logarithmic equation results. For films from 20 to 200 nm thick, the inverse logarithmic relationship holds. Potentials across thin films have been measured; a change in sign of the potential is interpreted as a change from electronic transport control to ionic control.

The Grimley-Trapnell Theory (Ref 14). T.B. Grimley and B.M.W. Trapnell used the Cabrera-Mott model, but assumed a constant electric field instead of a constant potential, They assumed that the adsorbed oxygen layer would always be complete, even at high temperatures and low pressure. The adsorbed oxygen would take electrons from cations in the oxide, not from the metal, so that a space charge would develop at the MO/O^-_{ads} interface and be independent of the oxide thickness. If the rate-controlling step is diffusion of cations through vacancies, logarithmic kinetics should be observed. If some other process is rate controlling, linear kinetics is most likely.

The Uhlig theory, developed by H.H. Uhlig and amplified by Fromhold, also predicts logarithmic kinetics at temperatures up to 600 K (Ref 15). The rate-controlling step is the thermal emission of electrons from the metal into the oxide (or electron holes from the adsorbed O^- to the oxide) under the combined effects of induced potential and applied field. The field is created by the diffusing ions. Because growth of the film depends on the electronic work function of the metal, the theory explains oxidation rate changes at crystal and magnetic transformations; most other theories do not.

Solid-State Diffusion

Diffusion processes in solids play a key role in the oxidation of metals. Mass transfer may be the result of diffusion of metal ions from the metal surface through the oxide layer to the adsorbed oxygen anions at the oxide/gas interface, or the result of the diffusion of anions inward through

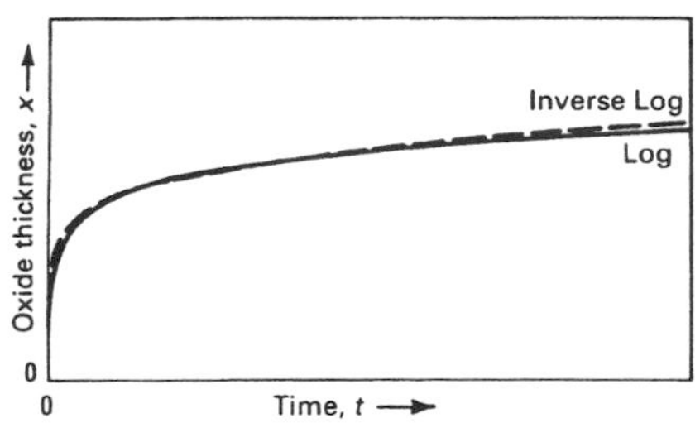

Fig. 12 Logarithmic and inverse logarithmic oxidation kinetics

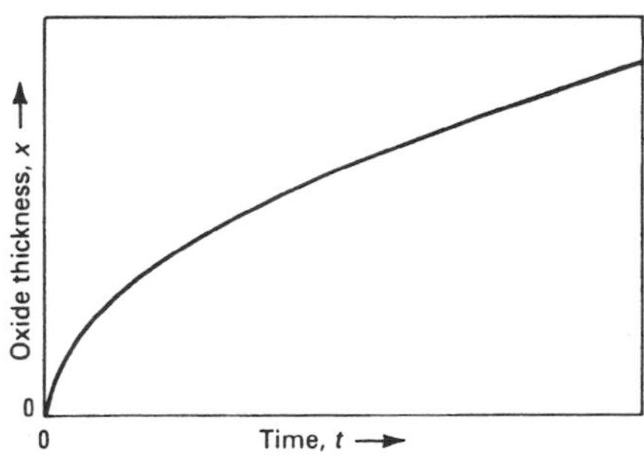

Fig. 13 Parabolic oxidation kinetics

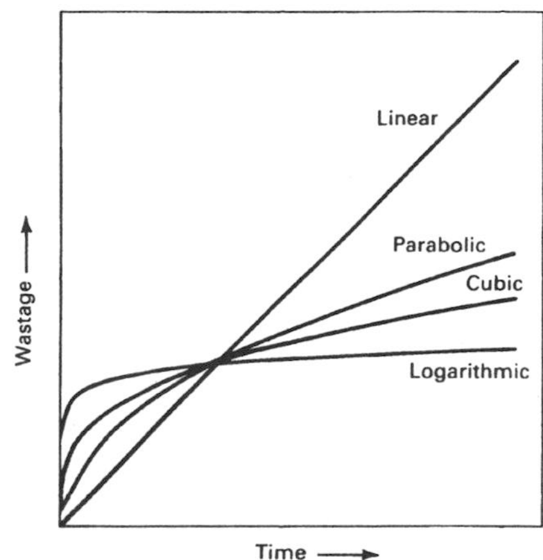

Fig. 14 A comparison of the forms of kinetic curves that represent various thermal degradation processes

the oxide to the metal. The diffusion of atomic oxygen into the metal from the oxide or the gas can also be involved. Within an alloy, the diffusion of oxidizable metal atoms toward the surface and the back diffusion of unreactive atoms from the metal surface inward to the unaltered alloy can occur.

Diffusion Mechanisms. Atoms or ions diffuse through solids by any of several mechanisms. The most common is vacancy diffusion. A metal crystal always contains large numbers of vacancies, while ionic oxides contain Schottky and Frenkel defects that also involve vacancies. An atom or ion sitting on a regular lattice site can diffuse by jumping to a vacant identical site nearby (Fig. 15a). For metal atoms, this is relatively easy because the jump distances are short. For ionic crystals, the jump distances are much longer because cation sites are surrounded by anion sites, and vice versa.

Small interstitial atoms diffuse readily from one interstitial position to another. In ionic oxides, the cations may diffuse interstitially, but the anions are usually not small enough to do so. Interstitial diffusion is shown in Fig. 15(b).

In ionic crystals, an interstitial ion may crowd into a regular lattice site, displacing an ion, which is forced to move into an interstitial position or to the next lattice site. This "crowdion" effect may extend for several atomic spacings along a line or equivalent direction. Figure 15(c) shows "crowdion" diffusion.

Fick's Law. In 1855, A. Fick formulated his two laws of diffusion for the simplest sort of diffusion system: a binary system at constant temperature and pressure, with net movement of atoms in only one direction. This is the usual situation for diffusion through an oxide growing on a pure metal.

Fick's first law states that the rate of mass transfer is proportional to the concentration gradient. Mathematically:

$$J = -D\left(\frac{\partial c}{\partial x}\right) \quad \text{(Eq 15)}$$

where J is the flux or mass diffusing per second through a unit cross section in the concentration gradient $(\partial c/\partial x)$ and D is the diffusion coefficient, or diffusivity in square centimeters per second, which is a function of the diffusing atoms, the structure through which they are diffusing, and the temperature. For diffusion of cations through a protective oxide, the entry of cations into the oxide at the metal/oxide interface will very nearly equal the flux of cations delivered to the oxygen at the oxide/gas interface.

For diffusion of oxygen into the metal, the concentration of oxygen changes with time inside the metal. Fick's second law describes this change as:

$$\frac{\partial c}{\partial t} = \frac{\partial}{\partial x}\left(D\frac{\partial c}{\partial x}\right) \quad \text{(Eq 16)}$$

Equation 16 must be solved for the particular geometry and boundary conditions involved (flat or round specimens, and so on). For oxygen atoms diffusing inward from a flat surface, with a constant diffusivity and a constant interfacial concentration, the solution to Eq 16 for the concentration of oxygen at any distance x from the metal surface is:

$$\frac{C_M - C_X}{C_M - C_0} = \text{erf}\left(\frac{x}{2\sqrt{Dt}}\right) \quad \text{(Eq 17)}$$

(assuming D is independent of composition) where C_M is the oxygen concentration at the metal/oxide interface, C_X is the concentration at time t and distance x from the surface, and C_0 is the initial constant concentration at any distance x when $t = 0$. The error function, erf, is tabulated in many books on probability. Figure 16 shows the variation of C_X with x.

As oxidation proceeds, C_M and C_0 remain constant; therefore, for any fixed value of C_X, Eq 17 reduces to:

$$X \propto \sqrt{t} \quad \text{(Eq 18)}$$

The Diffusion Coefficient. The diffusion coefficient D in rather stoichiometric compounds can usually be assumed to be proportional to the defect concentration. The diffusion coefficient can also vary with crystal orientation in noncubic crystals. For oxides that are epitaxial or have a preferred orientation, the diffusivity can be several times more or less than it would be for a random polycrystalline oxide. The diffusion coefficient D is independent of orientation in cubic crystals. Temperature has a major effect on the diffusion coefficient; D increases exponentially with temperature according to the Arrhenius equation:

$$D = D_0 \exp\left(\frac{-Q}{RT}\right) \quad \text{(Eq 19)}$$

where D_0 is a constant called the frequency factor that is a function of the diffusing species and the diffusion medium, Q is the activation energy for diffusion, R is the gas constant, and T is the absolute temperature. Activation energies for interstitial diffusion are much lower than those for vacancy diffusion.

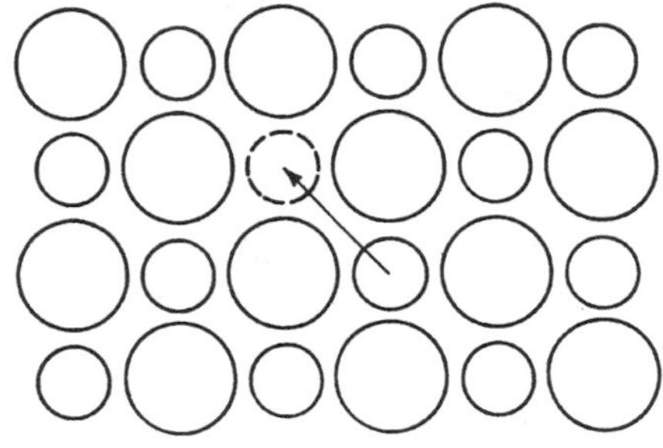
(a)

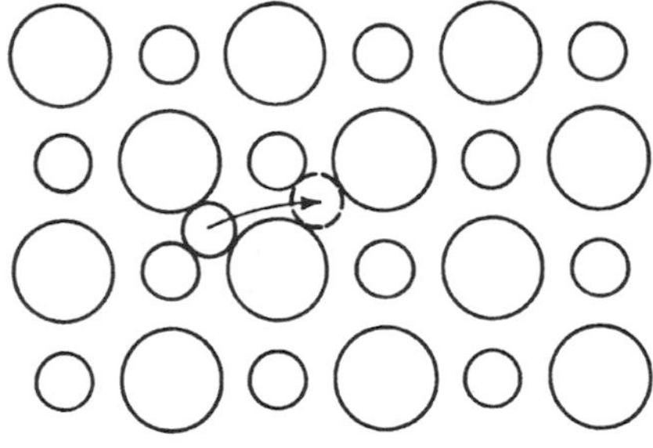
(b)

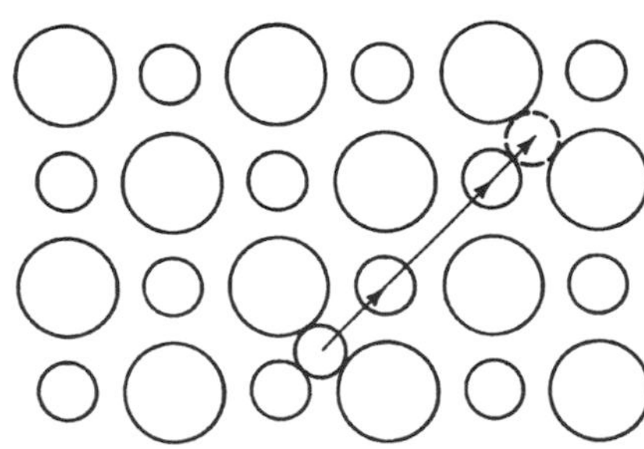
(c)

Fig. 15 Diffusion mechanisms. (a) Vacancy diffusion. (b) Interstitial diffusion. (c) Crowdion diffusion

Table 5 Selected diffusion data in metal oxides

Metal oxide	Temperature °C	Temperature °F	Frequency factor(a), cm²/s	Activation energy for diffusion J/mol	Activation energy for diffusion Btu/mol
Copper in Cu_2O	800–1050	1470–1920	0.12	151.0	143
Nickel in NiO	740–1400	1365–2550	0.017	234	222
Oxygen in Fe_2O_3	1150–1250	2100–2280	10^{11}	610	578
Iron in Fe_3O_4	750–1000	1380–1830	5.2	230	218
Iron in $Fe_{0.92}O$	690–1010	1275–1850	0.014	126.4	120
Chromium in Cr_2O_3	1000–1350	1830–2460	4000	420	398
Oxygen in UO_2	450–600	840–1110	2.6×10^{-5}	124	118
Magnesium in MgO	1400–1600	2550–2910	0.25	330	313

(a) The frequency factor, D_0, is a function of the diffusing species and the diffusion medium. See Eq 19 and corresponding text. Source: Ref 7

The activation energy is a measure of the temperature dependence of a diffusion process. A high value of Q means that the diffusion proceeds much more rapidly at high temperatures, but very much slower at low temperatures. Typical values of D_0 and Q for diffusion in oxides are listed in Table 5.

If the diffusivity D is plotted on a natural logarithm (base e) scale as a function of $1/T$, the slope of the resulting straight line is $-Q/R$. If the graph shows two intersecting lines, it indicates that one diffusion mechanism is operative at low temperatures, such as grain-boundary diffusion, and that another mechanism with a higher activation energy, such as volume diffusion, has gained control at high temperatures.

Effect of Impurities. All oxides contain some substitutional impurity cations from the alloy before oxidation or from the gas phase during oxidation. Although the solubility limit for foreign ions is low, they can have a great effect on diffusivity in the oxide, and consequently on oxidation rate.

In p-type oxides, such as NiO, any substitutional cation with a valence greater than the Ni^{2+} ion it replaces tends to increase the concentration of cation vacancies. Two aluminum ions (Al^{3+}) replacing two Ni^{2+} ions in the structure also supply two extra valence electrons to the oxygen, so that an additional cation vacancy in the oxide structure will be necessary to maintain the balance of charge between anions and cations. Additional cation vacancies in the NiO increase the diffusivity of Ni^{2+} cations through the oxide. Conversely, substitutional cations with a valence lower than +2 should reduce diffusion in NiO, reducing the number of cation vacancies. Cations with a +2 valence should have little effect on diffusion if substituting for Ni^{2+} ions.

For n-type oxides, the effect is reversed. If Al^{3+} is substituted for some titanium ions (Ti^{4+}) in titanium dioxide (TiO_2), more anion vacancies will be present in the oxide. For every two Al^{3+} ions substituted, one additional O^{2-} vacancy must be present in the oxide to maintain the electronic charge balance. Diffusion of oxygen ions inward then increases because of the increase in anion vacancy concentration. Higher valent impurity ions would decrease oxygen diffusion in n-type oxides.

The impurity effect is especially important for diffusion at low temperatures at which the native defect concentration is low; therefore, the activation energy is associated with only the movement of ions. At high temperatures, the activation energy increases because it involves formation of defects as well as the motion of the ions.

Short-Circuit Diffusion. The activation energies for diffusion along line and surface defects in solids are much less than those for volume diffusion. Dislocations, grain boundaries, porosity networks, and external surfaces offer rapid diffusion paths at low temperatures at which volume diffusion has virtually stopped. In metals, diffusion along dislocations is more important than volume diffusion below about one-half of the absolute melting point. At high temperatures, of course, volume diffusion predominates in both metals and oxides.

Wagner Theory of Oxidation (Ref 16)

C. Wagner derived the parabolic rate equation for scale growth on a metal in which diffusion of ions or electrons is rate controlling (Ref 17). It was assumed that the oxidation of a metal (M) by a single oxidant (X) from the gas phase will result in the formation of one or more compounds of M and X. If a compound forms a continuous layer on the metal, diffusion of M and/or X through this layer will be required for further reaction to occur. Beyond some thickness of the layer this diffusion will control the overall rate of reaction and the rate of thickening of a layer of thickness, x, is observed to follow the equation

$$\frac{dx}{dt} = \frac{k'}{x} \quad \text{(Eq 20)}$$

where k′ is called the parabolic rate constant, which generally has the units cm^2/s. The integrated form of this equation is:

$$X^2(t) - X^2(t_0) = 2k'(t - t_0) \quad \text{(Eq 21)}$$

where t_0 is the time at which diffusion control begins. The extent of reaction may also be expressed in terms of mass change per unit area, $\Delta M/A$, according to:

$$\left(\frac{\Delta M}{A}\right)^2 - \left(\frac{\Delta M}{A}\right)^2_{t_0} = 2k''(t - t_0) \quad \text{(Eq 22)}$$

where

$$k'' = \frac{M_x}{\overline{V}Z_x} 2_{k'} \quad \text{(Eq 23)}$$

for which M_x is the atomic weight of the nonmetal X, $\overline{V}$ is the equivalent volume of the reaction product layer, and Z_x is the valence of X. The units of k″are $g^2 \cdot cm^{-4} \cdot s^{-1}$ (Ref 18).

Wagner (Ref 17) has provided a theoretical treatment for k′ based on the idealized model for oxide formation in Fig. 17 and subject to the following assumptions:

- The product layer is a compact, perfectly adherent scale.
- Migration of ions or electrons across the scale is the rate-controlling process.
- Thermodynamic equilibrium is established at both the metal-scale and scale-gas interfaces.
- The scale shows only small deviations from stoichiometry.
- Thermodynamic equilibrium is established locally throughout the scale.
- The scale is thick compared with the distances over which space charge effects (electrical double layer) occur.
- Nonmetal solubility in the metal may be neglected.

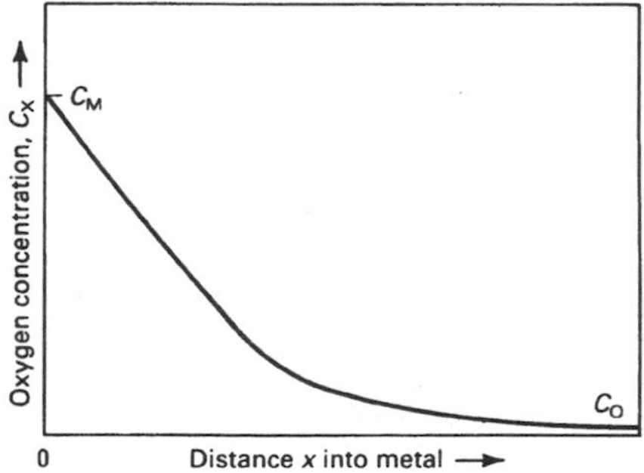

Fig. 16 Nonsteady-state diffusion. Fick's second law

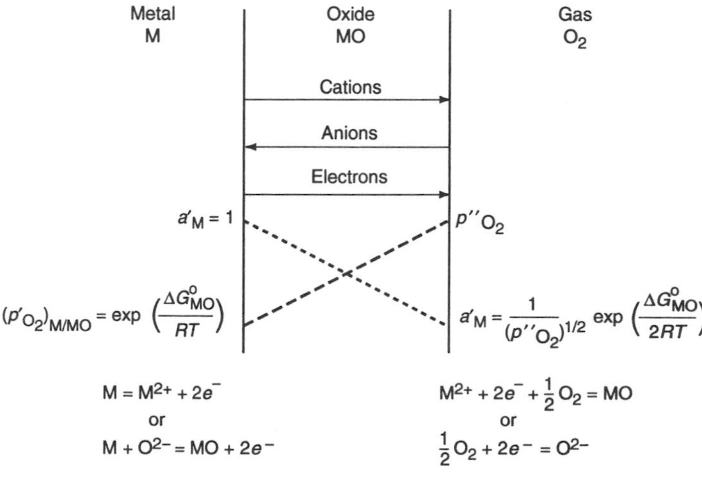

Fig. 17 Diagram of scale formation according to Wagner's model. Source: Ref 16

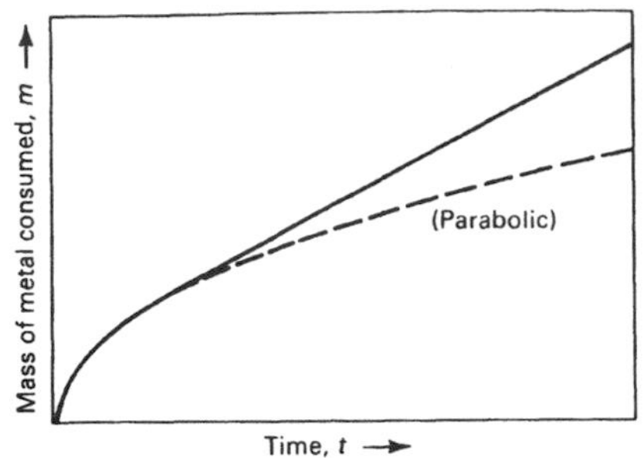

Fig. 18 Paralinear oxidation. Linear region is tangential to initial parabolic curve.

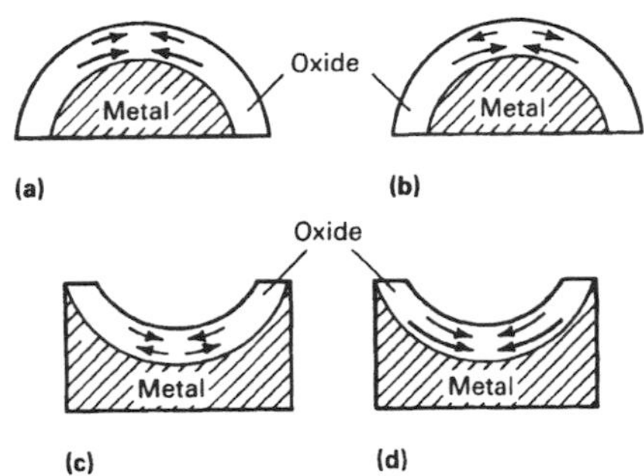

Fig. 19 Stress development on curved surfaces. (a) Oxide grows by cation diffusion on a convex surface. (b) Oxide grows by anion diffusion on a convex surface. (c) Oxide grows by cation diffusion on a concave surface. (d) Oxide grows by anion diffusion on a concave surface.

In terms of diffusivities, the results of this treatment may be expressed as:

$$k' = \frac{1}{RT}\int_{\mu'_O}^{\mu''_O}\left(D_O + \frac{Z_M}{Z_O}D_M\right)d\mu_O \qquad \text{(Eq 24)}$$

where D_M and D_O are the diffusivities of M and O, respectively, in the scale and μ'_O and μ''_O are the chemical potentials of O at the metal-scale and scale-gas interfaces, respectively. Good agreement has been found between experimentally determined rate constants and those calculated from the Wagner theory for a number of metals reacting with oxygen, sulfur, and halogens (Ref 19). The majority of data, however, deviate from the Wagner model because one or more of Wagner's assumptions are not satisfied in the real system. This is particularly true when the assumption that migration of ions or electrons across the scale is the rate-controlling process is not satisfied, in which case the rate of scale growth will not follow a simple parabolic expression such as given in Eq 21 and 22. Nevertheless, the use of parabolic rate constants is a convenient way to compare the relative rates of oxidation for the many cases that approximate parabolic behavior.

The above discussion was limited to oxidation of a pure metal by a single oxidant. Obviously the oxidation of alloys is more complex than that of pure metals, as a result of some or all of the following (for the case of reaction with oxygen):

- The metals in the alloy will have different affinities for oxygen, reflected by the different free energies of formation of the oxides.
- Ternary and higher oxides may be formed.
- A degree of solid solubility between the oxides may exist.
- The various metal ions will have different mobilities in the oxide phases.
- The various metals will have different diffusivities in the alloy.
- Dissolution of oxygen into the alloy may result in subsurface precipitation of oxides of one or more alloying elements (internal oxidation).

Following Wagner (Ref 20), alloys may be grouped as (a) noble parent with base alloying element, and (b) base parent with base alloying element, where both elements are reactive. Typical oxidation morphologies within these two groups are discussed below in the section "Selective Oxidation" in this article.

Properties of Scales

The desired characteristics for a protective oxide scale are:

- High thermodynamic stability (highly negative Gibbs free energy of formation) so that it forms preferentially to other possible reaction products
- Low vapor pressure so that the oxide forms as a solid and does not evaporate into the atmosphere
- Pilling-Bedworth ratio greater than 1.0 so that the oxide completely covers the metal surface
- Low coefficient of diffusion of reactant species (metal cations and corrodent anions) so that the scale has a slow growth rate
- High melting temperature
- Good adherence to the metal substrate, which usually involves a coefficient of thermal expansion close to that of the metal, and sufficient high-temperature plasticity to resist fracture from differential thermal expansion stresses

Multiple Scale Layers

A pure metal that can oxidize to more than one valence will form a series of oxides, usually in separate layers. For example, iron has valences of +2 and +3 at high temperatures and forms wustite (FeO), magnetite ($FeO{\cdot}Fe_2O_3$, commonly written as Fe_3O_4), and hematite (Fe_2O_3) scale layers. The layers will be arranged with the most metal-rich oxide next to the metal, progressively less metal in each succeeding layer, and finally the most oxygen-rich layer on the outside. Within each layer a concentration gradient exists with higher metal ion concentration closest to the metal.

If the oxygen partial pressure in the gas is so low that it is below the dissociation pressures of the outer oxygen-rich oxides, then only the thermodynamically stable, inner oxides will form. In general, the lowest-valence, inner oxide will usually be *p*-type because of the ease with which electron holes and cation vacancies can form. The outermost oxide is often *n*-type because of its anion vacancies. A scale consisting of an inner layer with cations diffusing outward and an outer layer with anions diffusing inward will grow at the oxide/oxide interface.

Relative Thickness. When diffusion is rate controlling, the relative thickness of the layers is proportional to the relative diffusion rates if no porosity develops in the layers. For compact layers growing by a single diffusion mechanism, the ratio of thicknesses should be related to the ratio of the parabolic rate constants by:

$$\frac{x_1}{x_2} = \left(\frac{k_{p1}}{k_{p2}}\right)^{1/2} \qquad \text{(Eq 25)}$$

where subscripts 1 and 2 refer to layers 1 and 2. The thickness ratio, consequently, is a constant and does not change with time. Because the ions diffusing through the various layers are likely to be different and because the crystal structures of the layers are certainly different, the thickness ratio is commonly found to be quite far from unity. One layer is usually much thicker than the others.

When diffusion controls the growth of each layer, the entire scale will appear to follow the parabolic oxidation equation with an effective parabolic rate constant k_p. However, this effective parabolic rate constant does not follow the Arrhenius equation (Eq 26 given below) unless the thickness ratios remain far from unity throughout the temperature range (i.e., unless one layer predominates at all temperatures). If an inner scale predominates, its growth is independent of oxygen pressure, but if the outermost scale is the major part of the scale, the rate constant will vary with oxygen pressure.

Effects of Temperature. The parabolic oxidation rate increases exponentially with temperature, following the Arrhenius equation:

$$k_p = k_o \exp\frac{-Q}{RT} \qquad \text{(Eq 26)}$$

where k_o is a constant that is a function of the oxide composition and the gas pressure. For cation-deficient or cation-excess oxides where the diffusion of cations is much greater than the diffusion of anions, the activation energy Q for oxide growth has been found to be the same as the activation energy for diffusion of cations in the oxide. For anion-deficient oxides, such as ZrO_2, where the diffusion of anions is much greater than the diffusion of cations, the activation energy for oxide growth is the same as that for anion diffusion, verifying that ionic diffusion is the rate-controlling process.

Paralinear Oxidation. With some metals, the oxidation begins as parabolic, but the protective scale gradually changes to a nonprotective outer layer. If the inner protective layer remains at a constant thickness, then the diffusion through the layer of constant thickness results in a linear rate of oxidation. The outer layer may become unprotective by sublimation, transformation to a porous layer, fracture, and so on. This type of oxidation behavior that is initially parabolic and gradually transforms to linear is termed paralinear oxidation (Fig. 18). The mass of metal m consumed at time t can be calculated from:

$$m = \frac{k_p}{k_l} \ln \frac{k_p}{k_p - k_l(m - k_l t)} \quad \text{(Eq 27)}$$

where k_p is the parabolic rate constant for formation of the inner layer, and k_l is the linear rate constant for formation of the outer layer (Ref 21).

Oxide Evaporation

At very high temperatures, the evaporation of a protective oxide may limit its protective qualities or remove the oxide entirely, because evaporation rates increase exponentially with temperature. The platinum metals and refractory metals, in particular, tend to have volatile oxides. Suboxides and unusual valences are also often found at high temperatures. Aluminum, for example, forms only one stable solid oxide, Al_2O_3, but it vaporizes as Al_2O and AlO. Evaporation may be much worse in gases containing water or halide vapor if volatile hydroxides (hydrated oxides) or oxyhalides form.

Theoretically, the evaporation rate should be directly proportional to the sublimation vapor pressure of the oxide if no evaporating molecules return to the surface, but at gas pressure above 10^{-3} to 10^{-4} atm, a gaseous stagnant boundary layer slows the escape of evaporated oxide molecules. The boundary layer becomes thinner at higher gas velocities, leading to higher evaporation losses.

As evaporation removes material from the oxide layer, diffusion through the oxide increases until the two rates finally become equal. The oxide thickness then remains constant, and the metal has oxidized paralinearly. If more than one oxide layer protects the metal, the higher-valent outermost oxide always has the higher vapor pressure and is more volatile.

Stresses in Scales

As oxide scales grow, stresses develop that influence the protective properties of the scale. During growth, recrystallization may occur in either the metal or oxide to alter the stress situation radically. Changes in temperature usually have a great effect on the stress state for metals in engineering service. Stress relief, if it damages the scale, will result in partial or complete loss of oxidation protection.

Growth Stresses. The magnitude of growth stresses is not well defined by the Pilling-Bedworth ratio, indicating that other factors also play a major role. Crystalline oxides will attempt to grow on a metal substrate with an epitaxial relationship. However, because the fit of the oxide crystal on the metal crystal is never perfect, the stresses that develop tend to limit the epitaxy to about the first 50 nm of oxide. Therefore, the stresses in thick scales are not greatly affected by the original epitaxy.

Polycrystalline oxides develop stresses along their grain boundaries because of more rapid growth of grains oriented in preferred directions. Short-circuit diffusion along oxide grain boundaries may lead to oxide formation at the boundaries that increase compressive stresses. Furthermore, any second phases or foreign inclusions in a metal may oxidize at a rate different from that of the parent metal and create high stresses within the oxide.

The compositional variation across a scale due to deviations from stoichiometry also create stresses within the oxide. Wustite is an important example, varying from $Fe_{0.95}O$ in equilibrium with the metal to as little as $Fe_{0.84}O$ in equilibrium with Fe_3O_4 at 1370 °C (2500 °F).

Composition will also tend to vary in a metal alloy in which one component is preferentially oxidized. If diffusion in the alloy is too slow to maintain a constant composition at the metal/oxide interface, stress develops in the metal. Similarly, any diffusion of oxygen into the metal from the oxide creates compressive stresses in the metal.

Surface geometry will contribute an additional effect to the growth stresses in an adherent oxide. Whether the contribution is compressive or tensile depends on whether the surface profile is convex or concave and whether the scale grows at the oxide/gas interface or at the metal/oxide interface. Figure 19 shows the four possibilities for changes in stress state in an oxide scale as it grows, assuming that the original growth stresses are compressive (Ref 22).

In Fig. 19(a), for a convex surface on which oxide grows at the oxide/gas interface by cation diffusion outward through the scale, the metal surface will gradually recede, increasing the compressive stresses at the metal/oxide interface as long as adhesion is maintained. Figure 19(b) shows oxidation of a convex surface on which the oxide grows at the metal/oxide interface by anion diffusion inward. The compressive stresses that develop at the metal/oxide interface are due only to the volume change of the reaction. Oxide that is pushed away from the growth area will gradually reduce its compressive stress until the outer surface may even be in tension.

For concave surfaces, Fig. 19(c) illustrates oxide growth at the oxide/gas interface by outward diffusion of cations. As the metal surface recedes, the compressive growth stresses are reduced and may eventually even become tensile if oxidation continues long enough. Figure 19(d) shows growth on a concave surface by anion diffusion inward for reaction at the metal/oxide interface. Very high compressive stresses develop during growth until they exceed the cohesive strength of the oxide.

Transformation Stresses. Preferential oxidation of one component in an alloy may alter the alloy composition to the point that a crystallographic phase transformation occurs. A change in temperature could also cause crystallographic transformation of either metal or oxide. The volume change accompanying a transformation creates severe stresses in both the metal and the oxide. Some oxides form initially in an amorphous structure and gradually crystallize as the film grows thicker. The tensile stress created by volume contraction may partially counteract the compressive growth stresses usually present.

Thermal Stresses. A common cause of failure of oxide protective scales is the stress created by cooling from the reaction temperature. The stress generated in the oxide is directly proportional to the difference in coefficients of linear expansion between the oxide and the metal. Examples of the coefficients are listed in Table 6 for a few important metal/oxide systems. In most cases, the thermal expansion of the oxide is less than that of the metal; therefore, compressive stress develops in the oxide during cooling. Multilayered scales will develop additional stresses at the oxide/oxide interface.

Table 6 Coefficients of linear thermal expansion of metals and oxides

System	Oxide coefficient, m/m · K	Metal coefficient, m/m · K	Temperature range, °C	Temperature range, °F
Fe/FeO	12.2×10^{-6}	15.3×10^{-6}	100–900	212–1650
Fe/Fe_2O_3	14.9×10^{-6}	15.3×10^{-6}	20–900	70–1650
Ni/NiO	17.1×10^{-6}	17.6×10^{-6}	20–1000	70–1830
Co/CoO	15.0×10^{-6}	14.0×10^{-6}	25–350	75–660
Cr/Cr_2O_3	7.3×10^{-6}	9.5×10^{-6}	100–1000	212–1830
Cu/Cu_2O	4.3×10^{-6}	18.6×10^{-6}	20–750	70–1380
Cu/CuO	9.3×10^{-6}	18.6×10^{-6}	20–600	70–1110

Source: Ref 22

Stress Relief

Stresses develop in oxide scale during growth or temperature change. The oxide may develop porosity as it grows. If temperatures are high, the stress may reach the yield strength of either the metal or the oxide so that plastic deformation relieves the stress. A brittle oxide may crack. A strong oxide may remain intact until the internal stresses exceed the adhesive forces between metal and oxide so that the oxide pulls loose from the metal. The ways in which stresses can be relieved are discussed below.

Porosity. For oxides that grow by cation diffusion outward through cation vacancies (p-type oxides, such as NiO), the vacancies are created at the oxide/gas interface and diffuse inward through the scale as they exchange places with an equal number of outward-diffusing cations. The

Table 7 Low-melting oxides

Oxide	Melting point °C	°F
V_2O_5	674	1245
MoO_3	795	1463
MoO_2-MoO_3 eutectic	778	1432
Bi_2O_3	817	1503
PbO	885	1625
WO_3	1470	2678

vacancies are annihilated within the oxide, at the metal/oxide interface, or within the metal, depending on the system.

In some oxides, the vacancies collect together within the oxide to form approximately spherical cavities. The preferred sites for cavity formation are along paths of rapid diffusion, such as grain boundaries and dislocation lines.

Vacancies that are annihilated at the metal/oxide interface may cause detachment of the scale, because voids that form there reduce adhesion of oxide to metal. If detachment is only partial, the oxidation rate slows down, because the cross-sectional area available for diffusion decreases.

Plastic Flow of Oxide. Dislocations form in the oxide as it grows epitaxially on the metal because of the crystallographic lattice mismatch between oxide and metal. As the growth continues, these glissile slip dislocations move out into the oxide by the process of glide. Once out in the oxide, they become sessile growth dislocations.

Although dislocations are present in the oxide, slip is not an important process in relieving growth stress (Ref 23). Plastic deformation of the oxide occurs only at high temperatures at which creep mechanisms become operative. The three important creep mechanisms in oxides are grain-boundary sliding, Herring-Nabarro creep, and climb. Grain-boundary sliding allows relative motion along the inherently weak boundaries. Herring-Nabarro creep allows grain elongation by diffusion of ions away from grain-boundary areas in compression over to boundaries in tension. Within the grains, dislocation climb is controlled by diffusion of the slower-moving ions. The creep rate increases with the amount of porosity in the oxide.

Cracking of Oxide. Tensile cracks readily relieve growth stresses in the oxide scale. As shown in Fig. 19(b) and (c), tensile stresses may eventually develop in oxide growing on curved surfaces and cause fracture. In the case of anion diffusion inward on convex surfaces, if oxygen diffuses on into the metal, the tensile stresses near the oxide/gas interface develop much more quickly.

Shear cracks can form in oxide having high compressive stresses near the metal surface, as shown in Fig. 19(a) and (d), if the scale cohesion is weak and adhesion to the metal is strong. If the metal/oxide interface is planar, a shear crack initiating at the interface can extend rapidly across the surface. However, if the interface is rough because oxidation has concentrated at the grain boundaries of the metal and keyed the oxide into the metal, rapid crack extension may be prevented and scale adherence may be improved.

Periodic cracking of a protective oxide results in interruption of the parabolic oxidation by a sudden increase in rate when the gas can react directly with the bare metal surface. As oxide begins to cover the metal surface again, parabolic oxidation is resumed. A typical oxidation curve for this repeated process in shown in Fig. 20. The time periods between successive parabolic steps are sometimes fairly uniform, because a critical scale thickness is reached that causes cracks to be initiated. The overall oxidation of the metal becomes approximately a slow linear process.

Occasionally, a metal oxidizes parabolically until the scale cracks or spalls off, and from that time on, the oxidation is linear. The oxide, originally protective, completely loses its protective properties. The breakaway oxidation (Fig. 21) commonly occurs if many cracks form continuously and extend quickly through the oxide. It can also occur for alloys that have had one component selectively oxidized. When the protective scale spalls off, the metal surface is so depleted in the component that the same protective scale cannot re-form. Breakaway oxidation leaves bare metal continually exposed, unlike paralinear oxidation (Fig. 18), in which an inner protective scale always remains.

Decohesion and Double-Layer Formation. For scale growth by cation diffusion, a protective scale may eventually reach a thickness at which it can no longer deform plastically to conform to the receding metal surface. At this point, decohesion begins at some places along the metal/oxide interface. However, oxidation continues at the oxide/gas interface, because cations continue to diffuse outward through the detached scale, driven by the chemical potential gradient across the scale.

At the inner scale surface, the cation concentration decreases, thus increasing the chemical potential of oxygen there. The increased chemical potential of oxygen has associated with it an increased pressure of O_2 gas that will form in the space between the scale and the metal. The O_2 then migrates to the metal surface, and because its pressure is greater than the dissociation pressure of oxide in equilibrium with metal, an inner layer of oxide begins to form on the metal.

The inner oxide layer forms a porous fine-grain structure, although it has essentially the same composition as the compact outer layer. Initially, it started forming as mounds from the nucleation sites. If the growth of these mounds were controlled by cation diffusion, the inner oxide would quickly thicken in the areas where it was thinnest and where diffusion distances were shortest. However, in this situation, the diffusion of O_2 from the outer layer to the inner layer is rate controlling, so that the inner layer grows mainly at its high points and forms a porous scale (Ref 24). Figure 22 shows the mechanism of the double-layer scale.

Deformation of Metal. Foils and thin-wall tubes are often observed to deform during oxidation, thus relieving the oxide growth stresses. At high temperatures, slip and creep mechanisms can both be operative in metals while the temperature is still too low for plastic deformation of the oxide. The deformation processes are facilitated by accumulation of porosity in the metal, which is caused by cation diffusion outward from the metal and by selective oxidation of one component of an alloy.

Logarithmic Oxidation of Scales

A logarithmic rate law for thick oxide scales has been developed for situations in which cavities or precipitates hinder the diffusion so that only part of the oxide is available for diffusion (Ref 25). The equation developed for logarithmic thick scale growth should not be confused with thin-film logarithmic behavior, which has an entirely different mechanism.

Even where the whole oxide scale originally serves as a diffusion path for ions or vacancies, the area available for diffusion may reduce with time. This can occur when vacancies collect at the

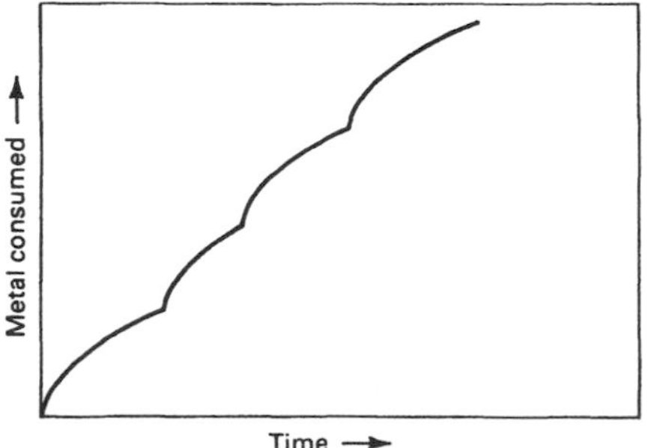

Fig. 20 Periodic cracking of scale

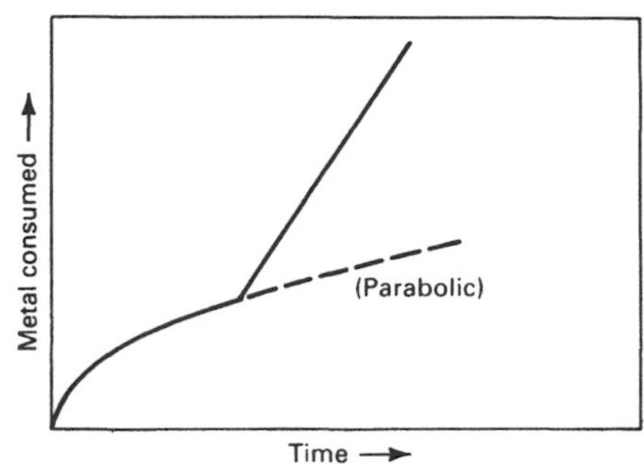

Fig. 21 Breakaway oxidation

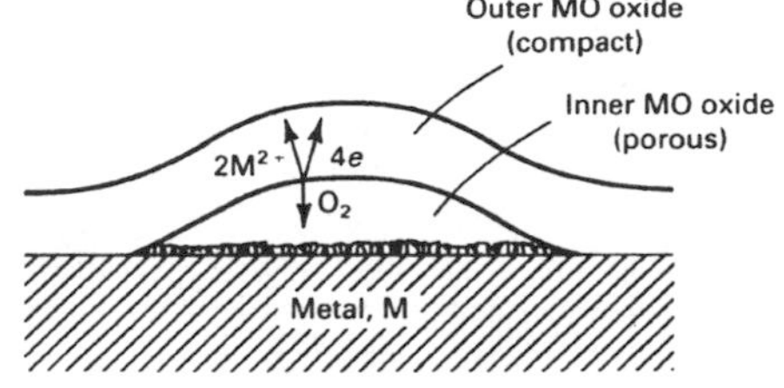

Fig. 22 Structure of double-layer single-phase scale

metal/oxide interface to cause partial detachment or cavity formation, or in the case of an A-B alloy, it can occur where particles of a very stable but slow-growing BO oxide form and restrict the growth of a much faster-growing AO oxide layer. It has even been suggested that the mechanism might hold when growth stresses are so high that they cause the scale to crack in short cracks that run parallel to the metal surface.

The equation developed for these situations will approximate the standard logarithmic equation:

$$m = \mathrm{k} \log (at + 1) \qquad \text{(Eq 28)}$$

where m is the mass increase per unit area, and k and a are constants. The oxidation begins rapidly but then almost comes to a stop as the cross-sectional area available for diffusion decreases.

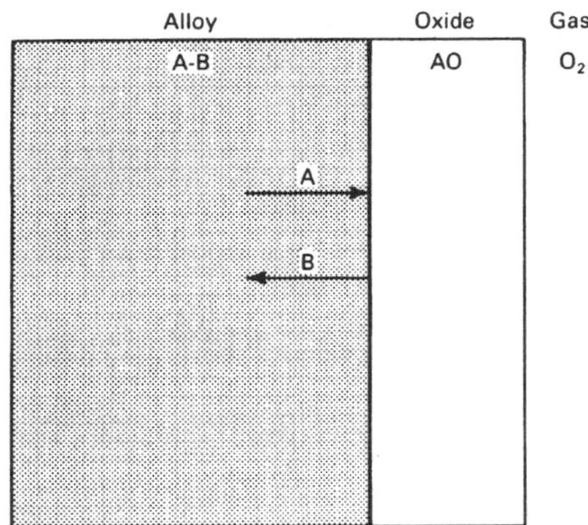

Fig. 23 Selective oxidation of alloy A-B, with oxide AO forming. Directions for diffusion of A and B are indicated.

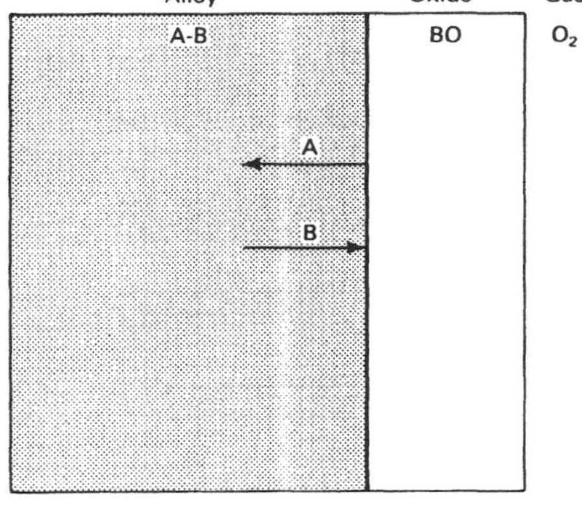

Fig. 24 Selective oxidation of alloy A-B for $N_B > N_B^*$

Catastrophic Oxidation

Although many oxidation failures can be described as catastrophies, the term catastrophic oxidation is reserved for the special situation in which a liquid phase is formed in the oxidation process. This can occur either when the metal is exposed to the vapors of a low-melting oxide or during oxidation of an alloy having a component that forms a low-melting oxide. The low-melting oxides that have commonly caused catastrophic oxidation are listed in Table 7.

The exact mechanism of catastrophic oxidation is disputed, but the evidence shows that it occurs at the metal/oxide interface. A liquid phase seems to be essential. The liquid usually forms at the oxide/gas surface and penetrates the scale along grain boundaries or pores to reach the metal. The penetration paths can also serve as paths for rapid diffusion of reacting ions. Once at the metal/oxide interface, the liquid spreads out by capillary action, destroying adherence of the solid scale.

Oxidation in the absence of any protective scale proceeds linearly or even at an ever increasing rate if the metal heats up from the exothermic oxidation reaction. The Wagner mechanism clearly does not apply to catastrophic oxidation.

Internal Oxidation

Internal oxidation is the term used to describe the formation of fine oxide precipitates found within an alloy. It is sometimes called subscale formation. Oxygen dissolves in the alloy at the metal/oxide interface or at the bare metal surface if the gas pressure is below the dissociation pressure of the metal oxides. The oxygen diffuses into the metal and forms the most stable oxide that it can. This is usually the oxide of the most reactive component of the alloy. Internal oxides can form only if the reactive element diffuses outward more slowly than the oxygen diffuses inward; otherwise, only surface scale would form.

Because diffusion of oxygen is usually the rate-controlling process in internal oxidation, parabolic behavior is observed. Wagner developed an equation that, with simplifications, is approximately (Ref 26):

$$x = \left(\frac{2N_O^s D_O t}{\nu N_B^i} \right)^{1/2} \qquad \text{(Eq 29)}$$

where N_O^s is the mole fraction of oxygen in the alloy at its surface, N_B^i is the mole fraction of reactive metal B initially in the alloy, D_O is the diffusion coefficient of oxygen in the alloy, ν is the ratio of oxygen atoms to metal atoms in the oxide compound that forms, and x is the subscale thickness at the time t. It is assumed that counter-diffusion of B atoms is negligible and that oxygen has a very low solubility limit in the alloy.

Because the diffusion coefficients of B atoms and oxygen both vary exponentially with temperature and because their activation energies are different, it is possible that internal oxidation will form in an alloy only in a certain temperature range. Because internal oxide precipitates reduce the cross section available for oxygen diffusion, only surface oxide forms above some critical solute concentration in the alloy.

In an A-B alloy, in which BO is a more stable oxide than AO, a mixed (A,B)O oxide may form as the internal oxide. This will occur if AO and BO have considerable mutual solubility so that the free energy of the system is lowered by precipitation of the mixed oxide. For example, the internal oxide that forms in unalloyed steels is (Fe, Mn)O (Ref 27).

Alloy Oxidation: The Doping Principle (Ref 28)

For oxides that form according to the Wagner mechanism and contain wrong-valent impurity cations that are soluble in the oxide, the impurities alter the defect concentration of the scale. Consequently, the oxide growth rate may also be altered by the alloy impurities. Whether oxidation increases or decreases depends on the relative valences of the cations and on the type of oxide.

***p*-type Oxides.** In a p-type semiconducting oxide (typified by NiO), the oxidation rate is controlled by cation diffusion through cation vacancies. If the number of cation vacancies can be decreased, the oxidation is slowed. The cation vacancies are present in the first place because a small fraction of nickel cations have a higher-

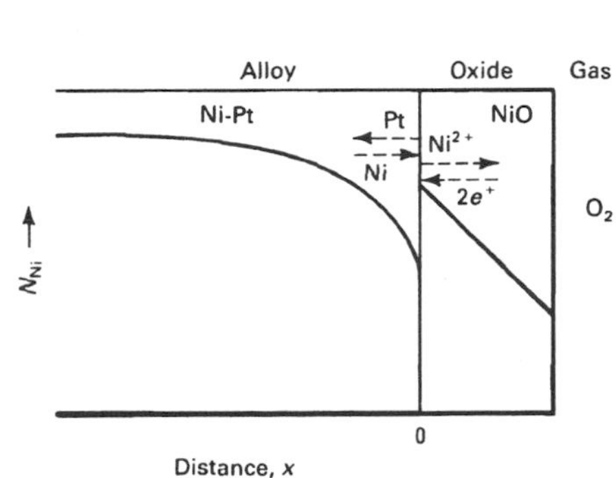

Fig. 25 Schematic diagram of concentration gradients of nickel in nickel-platinum alloy and NiO oxide. Electron holes are indicated as e^+.

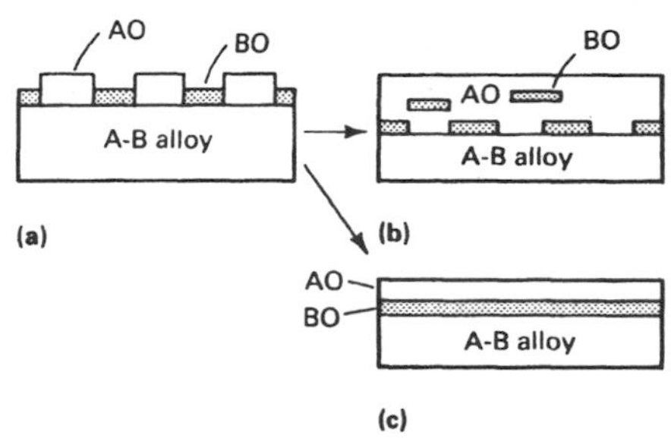

Fig. 26 Simultaneous growth of competing oxides. BO is more stable, but AO grows faster. (a) Early stage with nucleation of both oxides. (b) Later stage if diffusion in alloy is rapid. (c) Final stage if diffusion in alloy is slow.

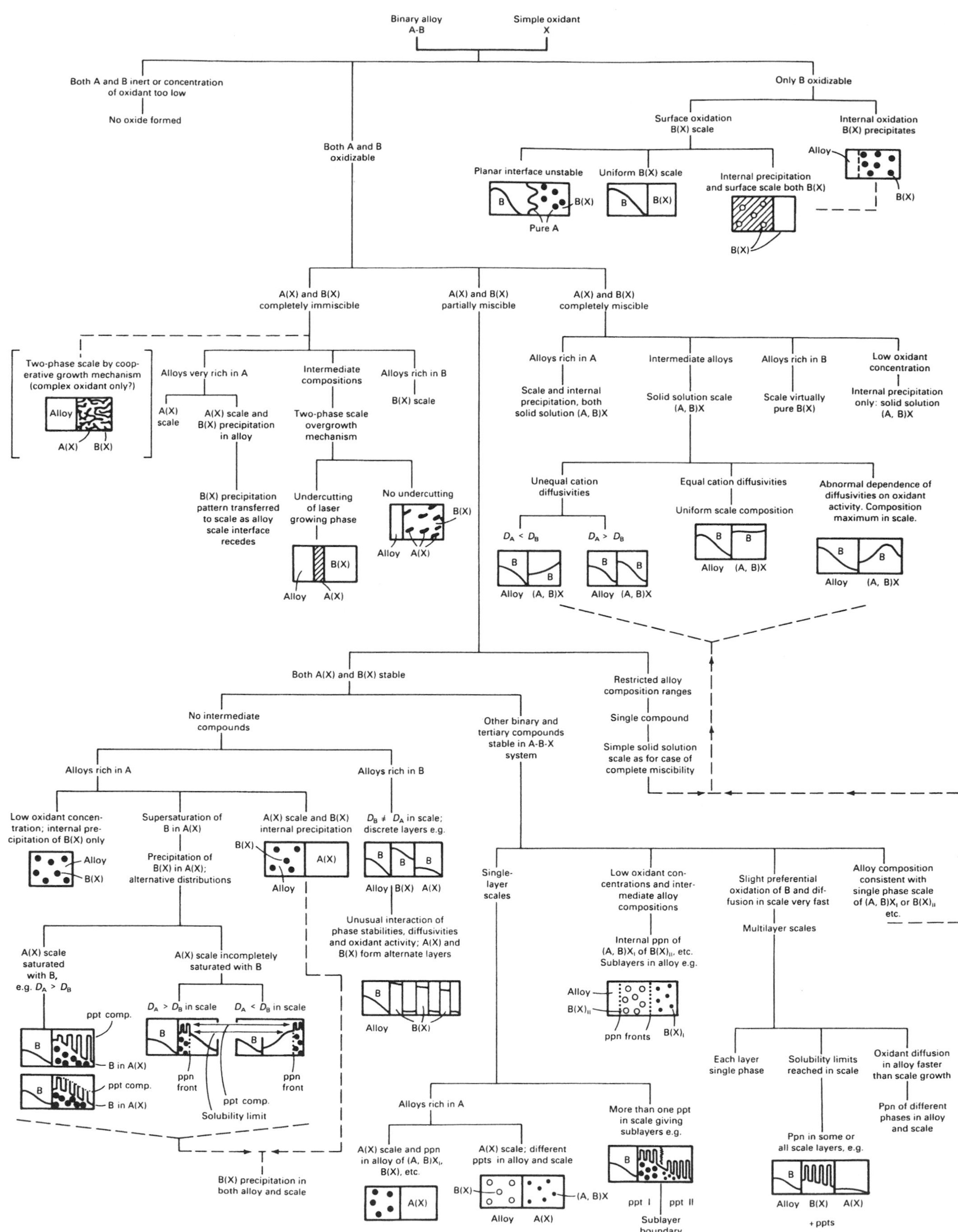

Fig. 27 Schematic showing the relationships between scale morphologies on binary alloys. ppn, precipitation; ppt, precipitate. Source: Ref 29

than-normal valence, contributing more than their share of valence electrons to the oxygen and thus allowing a small fraction of cations to be absent from the structure.

If a few cations with a higher valence are substituted for the regular cations in an oxide, the vacancy concentration is increased. For example, if a small amount of aluminum is alloyed with nickel and then oxidized to form a scale of NiO with a few substititional Al^{3+} cations, the oxidation rate will be faster than for pure NiO because of the increased cation vacancy concentrations. For every two Al^{3+} cations in the oxide, contributing one additional electron each, another cation vacancy V''_{Ni} must be present. For the same reason, adding lower-valent cations, such as lithium ions (Li^+), to the NiO will reduce the cation vacancy concentration. Substituting ions with the same valence as the rest of the cations in the oxide should have little effect.

***n*-type Oxides.** The *n*-type semiconducting oxides behave exactly opposite to the *p*-type oxides. For the oxides that grow by anion diffusion through anion vacancies (typified by ZrO_2), the anion vacancies exist because some cations have a lower-than-normal valence, contributing fewer electrons to the oxygen than required by the structural arrangement. Consequently, anion vacancies are present to facilitate anion diffusion. If additional low-valent cations replace the regular high-valent cations, the oxidation rate increases. Conversely, substituting higher-valent cations reduces the anion vacancy concentration and reduces the oxidation rate.

For *n*-type oxides that grow by interstitial diffusion of cations, as with BeO, substitution of higher-valent cations, for example, Al^{3+}, for a few of the Be^{2+} ions in the oxide structure leaves the oxide negatively charged with the excess electrons. The free electrons tend to prevent the ionization of the metal:

$$Be \rightarrow Be_i^{2+} + 2e^-$$

at the metal/oxide interface. The concentration of interstitial cations Be_i^{2+} is consequently reduced; therefore, the oxidation rate is reduced. However, if lower-valent cations are substituted in the *n*-type semiconducting oxide, (e.g., a small amount of Li^+ in BeO), the Li^+ supplies only one valence electron to the oxygen in place of the two electrons that would have been supplied by the replaced Be^{2+}. The oxygen must obtain more electrons from the surface of the metal, increasing the ionization and formation of interstitial Be_i^{2+} cations. Alloying with a metal that forms cations of the same valence as the rest of the cations in the oxide will have little effect.

Alloy Development. As it turns out, the doping principle has been useful in verifying the Wagner mechanism, but it is not very helpful in developing oxidation-resistant alloys. First, the concentration of foreign cations that can be put into solid solution in the oxide is severely restricted by low solubility limits. Second, the choices of foreign ions that could be used is extremely limited by their valence. For *p*-type oxides, the alloying element should have a lower valence than the metal being oxidized in order to slow corrosion, but *p*-type oxides are commonly formed by metals with +1 and +2 valences. The *n*-type oxides form on metals with high valences, but very few alloying elements have even higher valences to slow the oxidation.

For any stoichiometric oxides, nitrides, or sulfides that are ionic conductors instead of semiconducting electronic conductors, the doping principle can be applied in theory to reduce oxidation at room temperature. However, at high temperatures, perhaps all these protective scales carry current principally by free electrons (*n*-type) or electron holes (*p*-type).

Selective Oxidation

An alloy is selectively oxidized if one component, usually the most reactive one, is preferentially oxidized. The simplest case would be a binary alloy with a uniform scale composed entirely of the only oxide that one of the components can form. This situation will be described to illustrate the principles involved. The alloy is formed from metals A and B where B is more reactive than A is; that is, oxide BO is thermodynamically more stable than AO, and AO and BO are immiscible. Which scale, AO or BO, will form on the A-B alloy depends on the relative nobility of A and B, their concentrations in the alloy, the oxygen pressure, and the temperature.

Alloying with a Noble Metal. An obvious example of selective oxidation would be scale formation on alloy A-B where A is so noble that AO is not thermodynamically stable at the environmental pressure and temperature. That is, the oxygen partial pressure in the gas is less than the

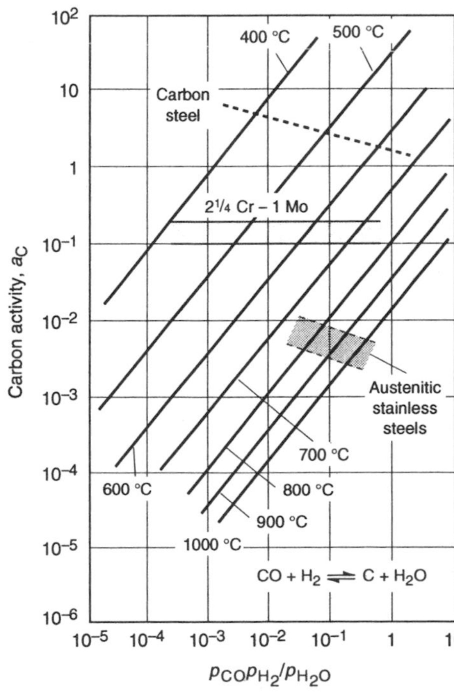

Fig. 28 Carbon activity of an environment based on the reaction $CO + H_2 = C + H_2O$ compared to carbon steel (a_c in equilibrium with Fe_3C) and 2.25 Cr-1Mo and austenitic stainless steels (both measured a_c). Source: Ref 30

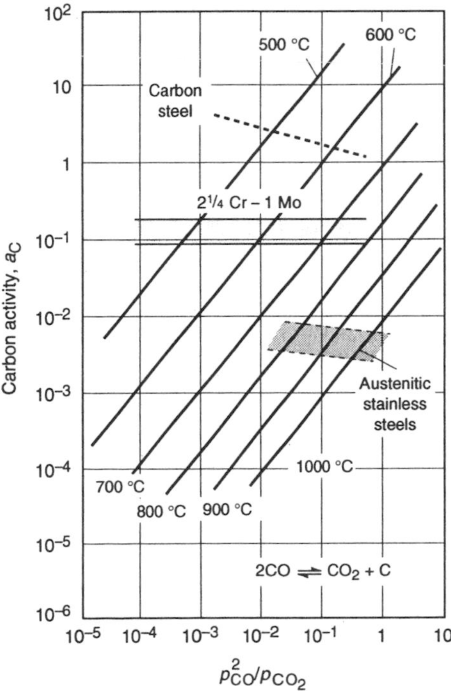

Fig. 29 Carbon activity of an environment based on the reaction $2CO = CO_2 + C$ compared to carbon steel (a_c in equilibrium with Fe_3C) and 2.25Cr-1Mo and austenitic stainless steels (both measured a_c). Source: Ref 30

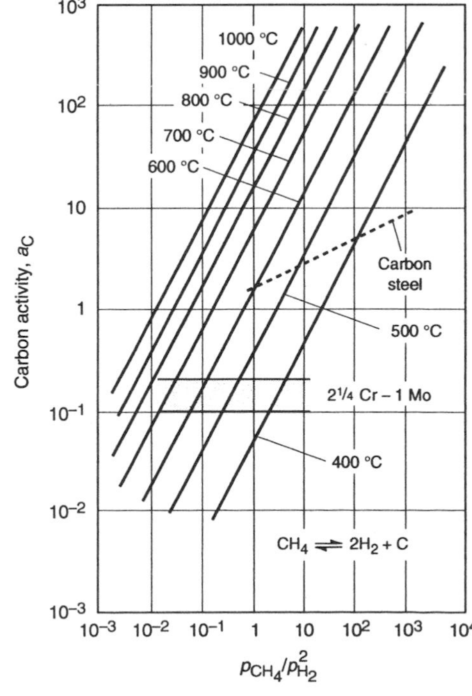

Fig. 30 Carbon activity of an environment based on the reaction $CH_4 = 2H_2 + C$ compared to carbon steel (a_c equilibrium with Fe_3C) and 2.25 Cr-1Mo (measured a_c). Carbon activities of austenitic stainless steels are below 10^{-2} at 800 to 1000 °C (1470 to 1830 °F). Source: Ref 30

equilibrium (dissociation) pressure of AO oxide. Then, only BO scale can form if it is stable.

Both Elements Reactive. For situations in which both A and B are reactive with oxygen at the temperature and gas pressure involved by A is somewhat more noble than B, the alloy composition determines which oxide forms. If the alloy contains some very low concentration of B, the activity of B could be so low that the free energy for formation of BO could actually be positive. That is, for the reaction:

$$B_{(alloy)} + \tfrac{1}{2}O_2 \rightarrow BO$$

the driving force is:

$$\Delta G \approx \Delta G^{o} + RT \ln \frac{1}{N'_B \cdot p_{O_2}^{1/2}} \quad \text{(Eq 30)}$$

where ΔG is the Gibbs energy change; ΔG^o is the Gibbs energy change under standard conditions; the activity of the pure solid oxide is approximately 1 (invariant); the small mole fraction of B in the alloy is N'_B, which approximates its activity; and the pressure of $O_2(p_{O_2})$ closely approximates its fugacity. It can be seen that if N'_B is smaller than some critical value N_B^* described below, then the Gibbs energy for formation of BO could be positive even though ΔG^o is negative. Because it is stable, AO will grow on the alloy, while BO cannot.

The situation for selective oxidation of A is illustrated in Fig. 23. Alloying element A becomes depleted as it reacts at the metal/oxide interface. This creates a concentration gradient of A, causing A to diffuse from the interior of the alloy toward the surface. At the same time, depleting A near the metal surface increases the concentration of BO so that B should diffuse inward. If the diffusion rates of A and B in the alloy are similar to the rate-controlling diffusion through the oxide, then element A never becomes seriously depleted at the metal/oxide interface and AO continues to form. However, if diffusion through the alloy is much slower than through the oxide, the concentration of B will increase at the alloy surface until it reaches N_B^*, the critical concentration at which formation of BO is thermodynamically favorable. When that time comes, BO will form along with O.

At a high concentration of B in an alloy, BO forms while A diffuses back into the alloy (Fig. 24). In this case, the mole fraction N_A is less than the minimum or critical concentration N_A^* at which AO oxide could be stable. However, if diffusion into the metal is slow, the concentration of A may eventually build up to N_A^*, and the AO oxide would then begin to form.

Reducing Oxidation Rates. It may at first seem that the oxidation rate of a reactive metal could be reduced by alloying it with a noble metal, but this is not always the case. If the concentration of the reactive metal is far greater than the critical concentration needed to form an external scale, that is, if $N_B >> N_B^*$, alloying with a noble metal will have very little effect on the oxidation rate.

For example, Wagner considered alloying nickel with a noble metal, such as platinum. For oxidation of pure nickel, the rate is proportional to the concentration gradient across the scale (Fick's first law); therefore, the rate constant k is given by:

$$k = \text{Constant } [(p_{O_2}^{o})^{1/n} - (p_d)^{1/n}] \quad \text{(Eq 31)}$$

where $p_{O_2}^o$ is the oxygen pressure in the gas at the outer surface of the oxide, p_d is the dissociation pressure for NiO in equilibrium with the pure nickel, and n is a constant theoretically equal to 6 for NiO. Figure 25 shows the concentration gradients of nickel in the alloy and in the scale.

Oxidation of nickel-platinum alloy would have a rate constant of:

$$k' = \text{Constant} \cdot [\,(p_{O_2}^{o})^{1/n} - (p_{O_2}^{i})\,]^{1/n} \quad \text{(Eq 32)}$$

where $p_{O_2}^i$ is the equilibrium oxygen pressure for NiO in equilibrium with the nickel-platinum alloy at the inner surface of the oxide. Comparing the oxidation rate of the alloy with the rate for pure nickel yields:

$$\frac{k'}{k} = \frac{(p_{O_2}^{o})^{1/n} - (p_{O_2}^{i})^{1/n}}{(p_{O_2}^{o})^{1/n} - (p_d)^{1/n}} \quad \text{(Eq 33)}$$

If activity of nickel in the alloy can be approximated by its concentration, Eq 33 can be written as:

$$\frac{k'}{k} = \frac{1 - (N_{Ni}^{*}/N_{Ni}^{i})^{2/n}}{1 - (N_{Ni}^{*})^{2/n}} \quad \text{(Eq 34)}$$

where N_{Ni}^* is the critical concentration of nickel in an alloy in equilibrium with NiO, and N_{Ni}^i is extremely small for a reactive metal, such as nickel. Therefore, alloying nickel with a small amount of platinum, but still keeping $N_{Ni}^i >> N_{Ni}^*$, will have very little effect on the oxidation. Oxide growth will slow to a stop only when enough platinum is added to make N_{Ni}^i approach N_{Ni}^*. The oxidation rate is then no longer controlled by diffusion of nickel through the oxide but by interdiffusion of platinum and nickel in the alloy.

Composite External Scales

Wagner has shown that for an A-B alloy that is forming both AO and BO oxides, the mole fraction of B at the alloy surface must not exceed:

$$N_B^i \leq (1 - N_A^*) = \frac{V}{z_B M_O}\left(\frac{\pi k_p}{D}\right)^{1/2} \quad \text{(Eq 35)}$$

where N_A^* is the minimum concentration of A necessary to form AO, V is the molar volume of the alloy, z_B is the valence of B, M_O is the atomic weight of oxygen, k_p is the parabolic rate constant for growth of BO, and D is the diffusion coefficient of B in the alloy. Equation 35 does not take into account any complications, such as porosity and internal oxidation. If the concentration of B lies anywhere between N_B^* and $(1 - N_A^*)$, thermodynamics predicts that both AO and BO form.

Competing Oxides. When oxides of both metals form, their relative positions and distribution depend on the thermodynamic properties of the oxides and the alloy, the diffusion processes, and reaction mechanisms. This section will discuss the two common situations in which both metals in a binary alloy oxidize to form two separate oxide phases.

The first situation involves immiscible oxides with the more stable oxide growing slowly. With both AO and BO stable but with rapid growth of AO and slow growth of BO, the more stable BO may nucleate first, but gradually becomes overwhelmed and surrounded by fast-growing AO. Figures 26(a) and (b) illustrate the situation. If diffusion in the alloy is rapid, the oxidation proceeds to form an AO scale with BO islands scattered through it. However, if diffusion in the alloy is slow, the metal becomes depleted of A near the metal/oxide interface, while the growth of BO continues until it forms a complete layer, undercutting the AO (Fig. 26c). Pockets of AO at the metal/oxide interface will gradually be eliminated by the displacement reaction:

$$AO + B_{(alloy)} \rightarrow BO + A_{(alloy)}$$

because BO is thermodynamically more stable than AO. This reaction continues even if the oxygen supply is cut off.

The second situation involves two oxides that are partially miscible. For alloys rich in A, an AO scale will form with some B ions dissolved substitutionally in the AO structure. If the solubility limit is exceeded when B ions continue to diffuse into the scale, BO precipitates as small islands throughout the AO layer. Even if the solubility limit is not reached, the more stable BO may nucleate within the AO scale and precipitate.

For alloys rich in B, a BO layer first forms. If B ions diffuse through the scale faster than A ions do, the concentration of A ultimately builds up in the scale close to the metal/oxide interface. An AO layer then forms underneath the BO. On the other hand, if A ions diffuse through the original BO scale more quickly than the B ions, the AO layer eventually forms on top of the BO layer. In addition, if A ions diffuse rapidly through BO and B ions diffuse rapidly through AO, alternate layers of BO/AO/BO/AO may even form.

Double Oxides. A great deal of research has been directed toward developing alloys that form slow-growing complex oxides. The silicates are particularly important, because they can form glassy structures that severely limit diffusion of ions. Therefore, silicide coatings on metals have been very successful. In addition to forming a protective SiO_2 outer layer, much of the silicon diffuses into the underlying alloy, where it gradually oxidizes to ternary silicates.

Spinels often have extremely low diffusion rates. Spinels are double oxides of a metal with +2 valence and a metal with +3 valence, having the general formula $MO \cdot Me_2O_3$ and also having the crystal structure of the mineral spinel ($MgO \cdot Al_2O_3$). The iron oxide Fe_3O_4 has an inverse spinel structure. On iron-chromium alloys, the spinel phase can be either stoichiometric $FeO{\cdot}Cr_2O_3$ or the solid solution $Fe_{3-x}Cr_xO_4$. Although many ternary oxides tend to be brittle, much research has been devoted to minor alloy addition to improve the high-temperature mechanical properties of those ternary scales that are extremely protective.

Summary Outline of Alloy Oxidation. A schematic diagram has been constructed to show the morphologies of scale growth on binary alloys (Ref 29). It is shown in Fig. 27. The diagram illustrates those types of structures that are known to form, not all those that would be theoretically possible.

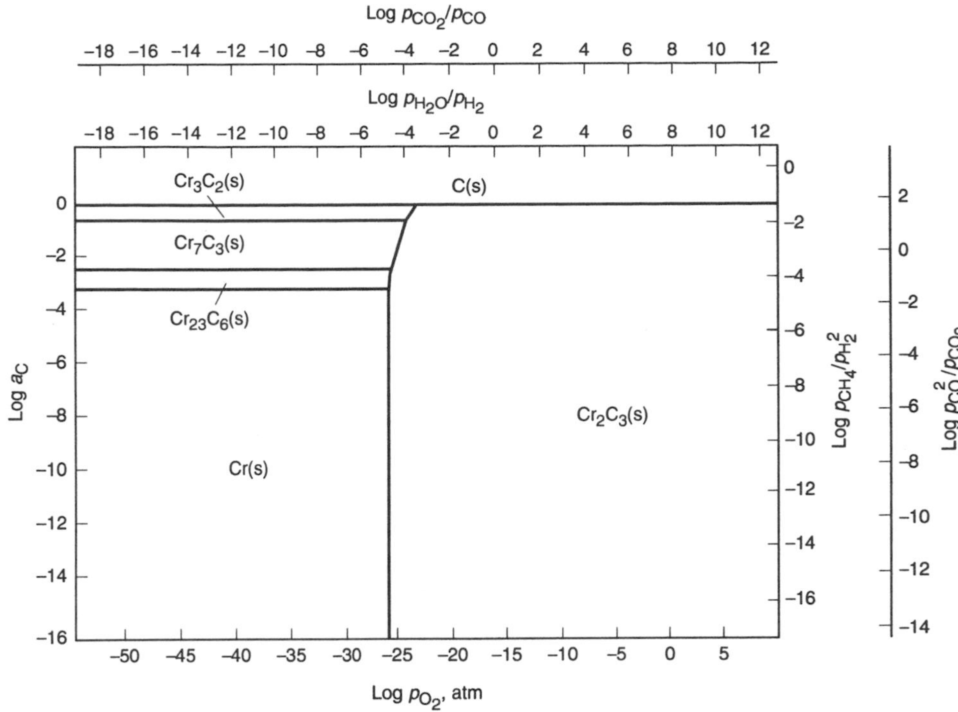

Fig. 31 Stability diagram of the Cr-C-O system at 870 °C (1600 °F). Source: Ref 35

Carburization (Ref 1)

Metals or alloys are generally susceptible to carburization when exposed to an environment containing CO, methane (CH_4), or other hydrocarbon gases, such as propane (C_3H_8), at elevated temperatures. Carburization attack generally results in the formation of internal carbides, which often cause the alloy to suffer embrittlement as well as other mechanical property degradation.

Carburization problems are quite common to heat treating equipment, particularly furnace retorts, baskets, fans, and other components used for case hardening of steels by gas carburizing. A common commercial practice for gas carburizing is to use an endothermic gas as a carrier enriched with one of the hydrocarbon gases, such as CH_4 or C_3H_8. An endothermic gas enriched with about 10% CH_4 is a commonly used atmosphere. The typical endothermic gas consists of 39.8% N_2, 20.7% CO, 38.7% H_2, and 0.8% CH_4, with a dew point of –20 to –4 °C (–5 to +25 °F). Gas carburizing is typically performed at temperatures of 840 to 930 °C (1550 to 1700 °F). Furnace equipment and components repeatedly subjected to these service conditions frequently suffer brittle failures as a result of carburization attack.

In the petrochemical industry, carburization is one of the major modes of high-temperature corrosion for processing equipment. The pyrolysis furnace tubes for the production of ethylene and olefins are a good example. Ethylene is formed by cracking petroleum feedstocks, such as ethane and naphtha, at temperatures up to 1150 °C (2100 °F). This generates strong carburizing environments inside the tubes.

Production of carbon fibers also generates carburizing atmospheres in a furnace. As a result, the furnace's retorts, fixtures, and other components require frequent replacement because of carburization attack.

There is another form of carburization attack that generally results in metal wastage, such as pitting and/or thinning. It is commonly referred to as "metal dusting." The environment is highly carburizing, with enrichment of H_2, CO, or hydrocarbons. A stagnant atmosphere is particularly conducive to metal dusting. Metal dusting occurs at between 430 and 900 °C (800 and 1650 °F) and has been encountered in heat treating, refining and petrochemical processing, and other operations.

Thermodynamic Considerations

Whether an alloy is likely to be carburized or decarburized depends on the carbon activity, a_C, in the environment and that of the alloy. The thermodynamic condition that dictates either carburization or decarburization can be simply described as follows:

$$(a_C)_{environment} > (a_C)_{metal} = \text{carburization}$$

$$(a_C)_{environment} < (a_C)_{metal} = \text{decarburization}$$

Thus, in order to predict whether an alloy will be carburized or not, one needs to know the carbon activities of both the environment and the alloy.

Carbon Activity for the Environment. Carburization can proceed by one of the following reactions when the environment contains CH_4, CO, or H_2 and CO:

$$CO + H_2 = C + H_2O \quad \text{(Eq 36)}$$

$$2CO = C + CO_2 \quad \text{(Eq 37)}$$

$$CH_4 = C + 2H_2 \quad \text{(Eq 38)}$$

Assuming that carburization follows Eq 36, the carbon activity in the environment can be calculated by:

$$\Delta G^o = -RT \ln\left(\frac{a_C \cdot p_{H_2O}}{p_{CO} \cdot p_{H_2}}\right) \quad \text{(Eq 39)}$$

Rearranging the equation, it becomes:

$$a_C = e^{-\Delta G^o/RT}\left(\frac{p_{CO} \cdot p_{H_2}}{p_{H_2O}}\right) \quad \text{(Eq 40)}$$

From Eq 40, one can construct graphs of carbon activity as a function of gaseous composition for various temperatures in terms of $(p_{CO} \cdot p_{H_2})/p_{H_2O}$ (see Fig. 28).

Similarly, if carburization follows Eq 37, the carbon activity of the environment can be calculated as follows:

$$\Delta G^o = -RT \ln\left(\frac{a_C \cdot p_{CO_2}}{p^2_{CO}}\right) \quad \text{(Eq 41)}$$

$$a_C = e^{-\Delta G^o/RT}\left(\frac{p^2_{CO}}{p_{CO_2}}\right) \quad \text{(Eq 42)}$$

Plots of carbon activities as a function of gas composition in terms of p^2_{CO}/p_{CO_2} for various temperatures are shown in Fig. 29.

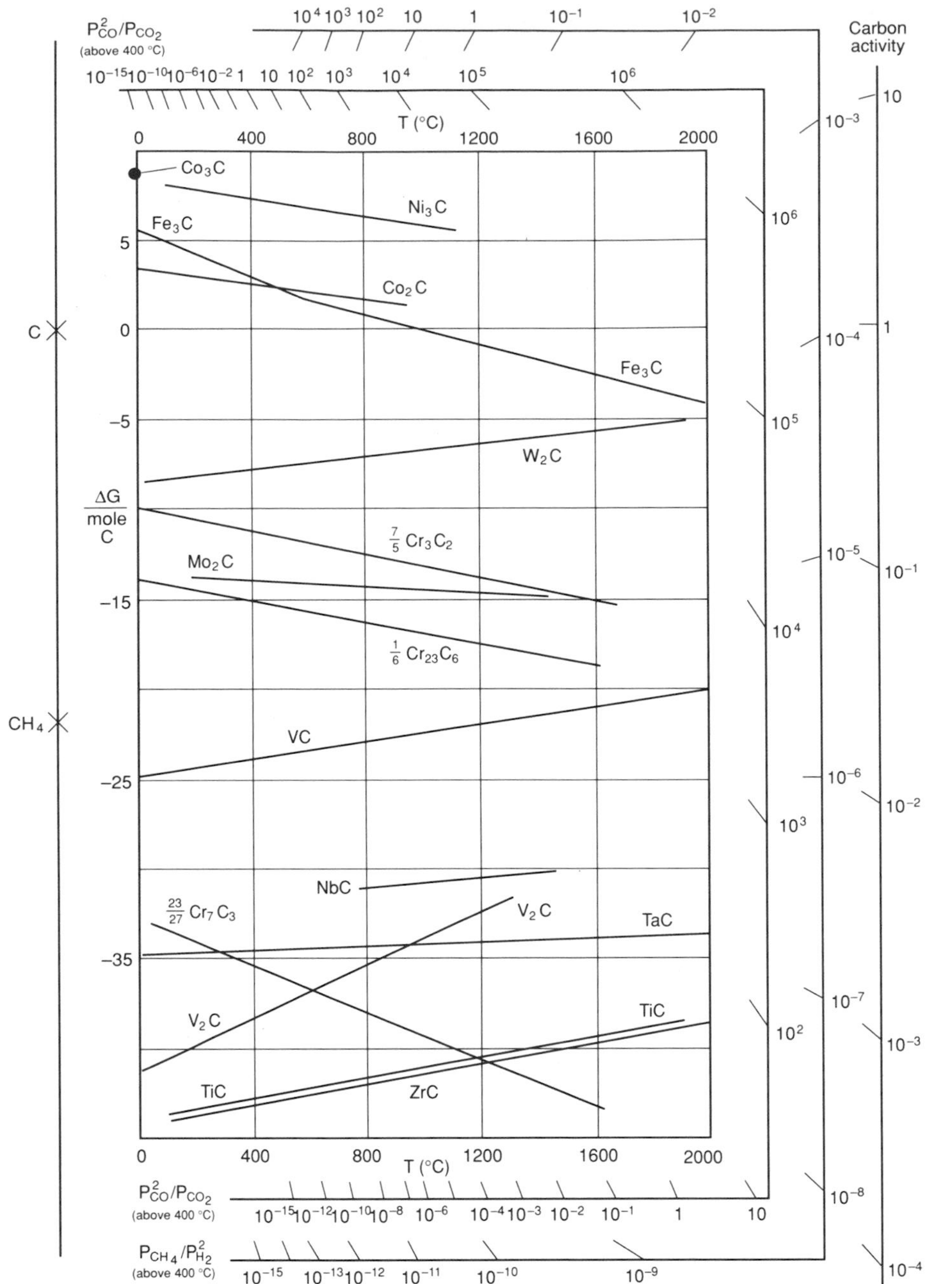

Fig. 32 Standard free energies of formation for carbides. Source: Ref 37

When carburization follows Eq 38, the carbon activity in the environment is:

$$a_c = e^{-DELTAG^o/RT}\left(\frac{p_{CH_4}}{p^2_{H_2}}\right) \quad \text{(Eq 43)}$$

Carbon activities as a function of $(p_{CH_4}/p^2_{H_2})$ are plotted in Fig. 30.

The reactions described by Eq 36 and 37 have a similar characteristic, showing lower carbon activities with increasing temperature (Fig. 28, 29). On the other hand, the reaction described by Eq 38 shows increased carbon activities with increasing temperature (Fig. 30). If the environment contains CH_4, the carbon activity of the environment will be dominated by the reaction $CH_4 = C + 2H_2$. When no CH_4 is present in the environment, the reactions described by Eq 36 and 37 will control carbon activity.

It is generally assumed that a gas mixture will reach an equilibrium condition when reacted with metals or alloys at elevated temperatures. The inlet gas mixture is in a nonequilibrium condition. One may use this nonequilibrium gas composition to obtain a "ballpark" estimate of the carbon activity in the environment. Computer programs are available for more precise determination of an equilibrium gas composition, as well as the thermodynamic potentials, such as carbon activity and oxygen potential (Ref 31, 32).

Carbon Activity of the Alloy. Data on carbon activities of commercial alloys at temperatures below 1200 °C (2190 °F) are very limited. Natesan (Ref 33) reported that a_C for 2.25Cr-1Mo steel is in the range of 1×10^{-1} to 2×10^{-1} from 550 to 750 °C (1020 to 1380 °F). Natesan and Kassner (Ref 34) reported the carbon activities of Fe-18Cr-8Ni alloys. These values are superimposed in Fig. 28 to 30.

For carbon steels, carbon activity can be estimated by assuming that it is in equilibrium with cementite (Fe_3C). This is illustrated below:

$$3Fe + C \leftrightarrow Fe_3C \quad \text{(Eq 44)}$$

$$\Delta G^o = -RT \ln\left(\frac{a_{Fe_3C}}{a_C \cdot a^3_{Fe}}\right) \quad \text{(Eq 45)}$$

$$\Delta G^o = -RT \ln\left(\frac{1}{a_C}\right) \quad \text{(Eq 46)}$$

where a_{Fe_3C} and a_{Fe} are assumed to be unity and

$$a_C = e^{\Delta G^o/RT} \quad \text{(Eq 47)}$$

The a_C values for carbon steel based on Eq 47 are plotted in Fig. 28 to 30. Using such graphs one can make a quick determination as to whether an environment has a carbon potential high enough to carburize the alloy of interest.

The environment can also be characterized in terms of a_C and p_{O_2} to determine the relative severity of its carburization potential. The environment can then be presented in a stability diagram of a metal-carbon-oxygen system (Fig. 31). From the stability diagram, the possible phases that the alloy may form at the gas-metal interface can be predicted. As the activities of both carbon and oxygen are decreasing from the gas-metal interface to the metal interior, the possible phases that the alloy may form beneath the gas-metal interface can also be predicted.

For carbon and alloy steels with relatively low levels of chromium, ingress of carbon into the metal or alloy may result in the formation of iron carbides. Several forms of iron carbides have been reported with composites ranging from Fe_4C to Fe_2C (Ref 36). Cementite (Fe_3C) is the most stable carbide.

In ferritic and austenitic stainless steels and nickel- and cobalt-base alloys, ingress of carbon into the alloy results in the formation of mainly chromium carbides. There are three forms of chromium carbides: $Cr_{23}C_6$, Cr_7C_3, and Cr_3C_2. During carburization the relative stabilities of these carbides can be best described by a stability diagram, such as the one shown in Fig. 31. If the carbon and oxygen activities of the environment are in the Cr_3C_2 region, conditions will favor formation of Cr_3C_2 on the surface and/or in the underlying metal. As carbon diffuses farther into the interior of the alloy, carbon activity will be lowered, thus favoring Cr_7C_3. Moving even farther into the interior, carbon activities will be still lower, favoring the formation of $Cr_{23}C_6$.

For many high-temperature alloys, particularly superalloys, there are other alloying elements, such as titanium, tantalum, niobium, molybdenum, and tungsten, that form carbides. The carbides of these alloying elements are important to the physical metallurgy of high-temperature alloys in that they provide an important strengthening mechanism (see the article "Metallurgy, Processing, and Properties of Superalloys" in this Volume for details). The relative stabilities of various binary metallic carbides are plotted in Fig. 32.

Sulfidation (Ref 1)

Sulfur is one of the most common corrosive contaminants in high-temperature industrial environments. Sulfur is generally present as an impurity in fuels or feedstocks. Typically, fuel oils are contaminated with sulfur varying from fractions of 1% (No. 1 or No. 2 fuel oil) to about 3% (No. 6 fuel oil). U.S. coal may contain from 0.5 to 5% S, depending on where it is mined. Feedstocks for calcining operations in mineral and chemical processing are frequently contaminated with various amounts of sulfur.

When combustion takes place with excess air to ensure complete combustion of fuel for generating heat in many industrial processes, such as coal- and oil-fired power generation, sulfur in the fuel reacts with oxygen to form SO_2 and SO_3. An atmosphere of this type is generally oxidizing. Oxidizing environments are usually much less corrosive than reducing environments, where sulfur is in the form of H_2S. However, sulfidation in oxidizing environments (as well as in reducing environments) is frequently accelerated by other fuel impurities, such as sodium, potassium, and chlorine, which may react among themselves and/or with sulfur during combustion to form salt vapors. These salt vapors may then deposit at lower temperatures on metal surfaces, resulting in accelerated sulfidation attack. Corrosion of this type (e.g., hot corrosion in gas turbines, fuel ash corrosion in fossil-fired boilers) is discussed below in the section "Ash/Salt Deposit Corrosion" in this article.

Oxidizing environments that contain high concentrations of SO_2 can be produced by the chemical process used to manufacture sulfuric acid. Sulfur, in this case, is used as a feedstock. Combustion of sulfur with excess air takes place in a sulfur furnace at about 1150 to 1200 °C (2100 to 2200 °F). The product gas typically contains about 10 to 15% SO_2 along with 5 to 10% O_2 (balance N_2), which is then converted to SO_3 for sulfuric acid.

In many industrial processes, combustion is carried out under stoichiometric of substoichiometric conditions in order to convert feedstocks to process gases consisting of H_2, CO, CH_4, and other hydrocarbons. Sulfur is converted to H_2S. The environment, in this case, is reducing and is characterized by low oxygen potentials. Coal gasification, which converts coal to substitute natural gas or medium- and low-Btu fuel gases, is a common example of a process that generates this type of atmosphere. Reducing conditions may also prevail in localized areas, in some cases even when combustion is taking place with excess air. Furthermore, ash deposits on the metal surface can sometimes turn an oxidizing condition in the gaseous environment into a reducing condition beneath the deposits.

In most cases, metals and alloys rely on oxide scales to resist sulfidation attack; most high-temperature alloys relay on chromium oxide scales. In oxidizing environments, oxide scales form much more readily because of high oxygen activities. Thus, oxidation is likely to dominate the corrosion reaction.

When the environment is reducing (i.e., characterized by low oxygen potentials), the corrosion reaction becomes a competition between oxidation and sulfidation. Thus, lowering the oxygen activity tends to make the environment more sulfidizing, resulting in increased domination by sulfidation. Conversely, increasing the oxygen activity generally results in a less sul-

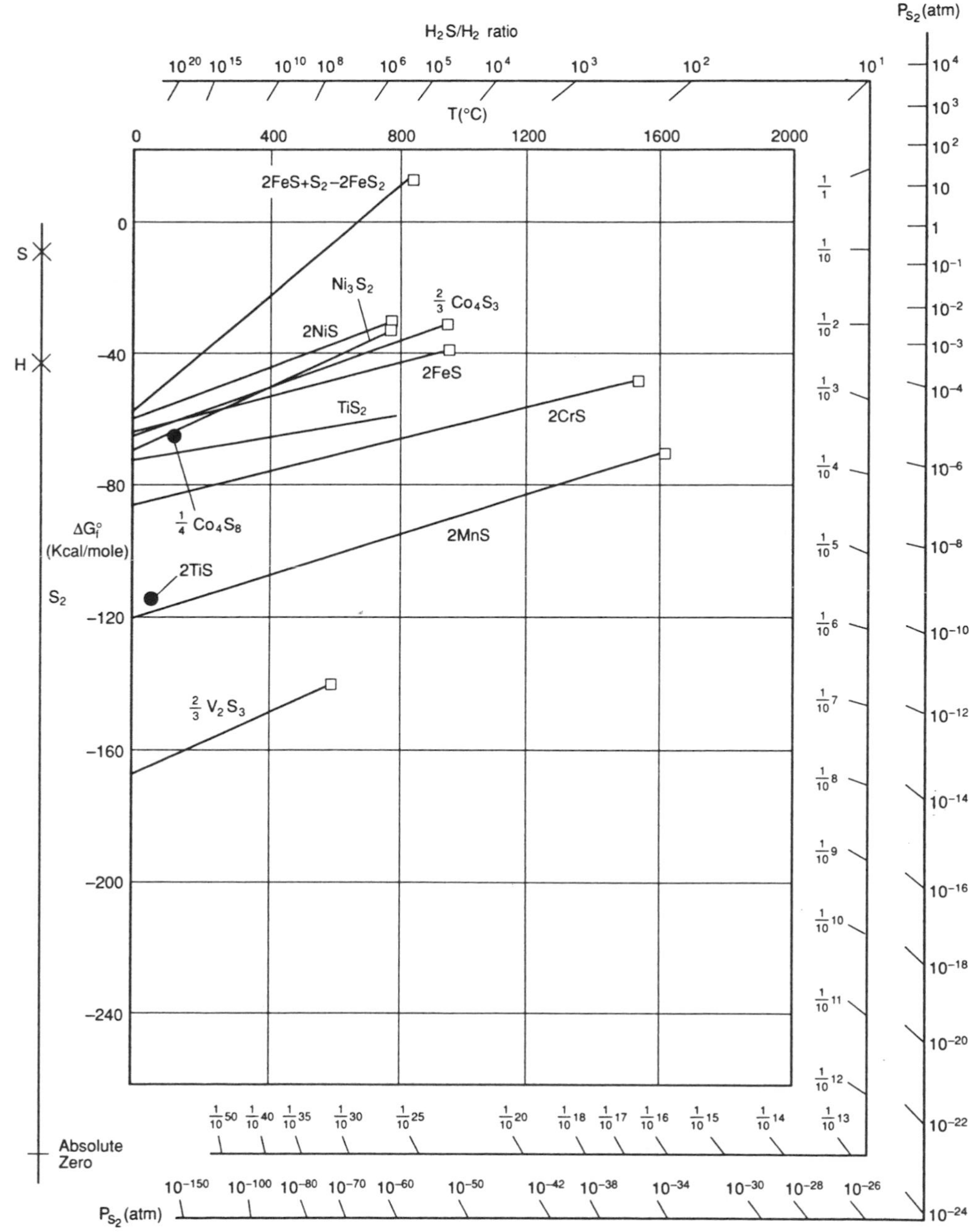

Fig. 33 Standard free energies of formation for selected sulfides. Source: Ref 38

fidizing environment with increased domination by oxidation. Sulfidation is thus controlled by both sulfur and oxygen activities. When corrosion involves more than one mode, including sulfidation, sulfidation generally dictates materials selection.

Thermodynamic Considerations. An Ellingham diagram such as the one shown in Fig. 33 can help determine whether an environment has a sulfur potential high enough to form sulfides. The sulfur potential is presented by either p_{S_2} or p_{H_2S}/p_{H_2}. The sulfur partial pressure (p_{S_2}) in equilibrium with a sulfide can be read from Fig. 33 by drawing a straight line from point S through the free energy line of the sulfide at the temperature of interest. The line intersects with the p_{S_2} scale, giving the sulfur partial pressure in equilibrium with the sulfide. The p_{H_2S}/p_{H_2} value in equilibrium with a sulfide can be obtained similarly by using the starting point H for the straight line; the intersection with the H_2S/H_2 scale gives the equilibrium H_2S/H_2 value. It is clear from Fig. 33 that the sulfide of chromium is more stable than that of iron, nickel, or cobalt.

Most sulfidizing environments exhibit both sulfur and oxygen activities. It is thus better to characterize the environment in terms of p_{S_2} and p_{O_2}. The environment can then be presented in a stability diagram of a metal-sulfur-oxygen system, as shown in Fig. 3 and 34. Various metal-sulfur-oxygen stability diagrams at different temperatures can be found in Ref 39.

Perkins (Ref 40) proposed that in an environment within the upper-right region of the Cr-S-O diagram (Fig. 34), both CrS and Cr_2O_3 will form initially on the metal surface of chromium and high-chromium alloys. The Cr_2O_3 will grow and overtake the sulfide because it is a stable phase in this phase stability region. The alloying elements that form stable sulfides can diffuse through the oxide scale and eventually form sulfides on the surface of the oxide scale, leading to breakaway corrosion. It is particularly serious when the sulfides become liquid. Some metal-metal sulfide eutectics have low melting points: 635 °C (1175 °F) for Ni-Ni_3S_2, 880 °C (1616 °F) for Co-Co_4S_3, and 985 °C (1805 °F) for Fe-FeS (Ref 41). Most high-temperature alloys are chromia formers, which rely on chromium oxide scale for protection against sulfidation. Most industrial environments have sufficient oxygen activities to alloy these alloys to form a chromium oxide scale. This scale may eventually break down, leading to breakaway corrosion. Perkins (Ref 40) referred to this region as "limited life"; see the upper-right region of the Cr-S-O diagram shown in Fig. 34. The environments within this region typically contain H_2, H_2O, CO, CO_2, and H_2S and are referred to as reducing mixed-gas environments.

Another type of sulfidizing environment with common characteristics for the sulfidation of various alloy is the upper-left region in Fig. 24. In this region, sulfides are stable phases. Alloys in these environments form sulfide scales. The environments include sulfur vapor and H_2/H_2S mixtures with extremely low oxygen activities, such that Cr_2O_3 is not thermodynamically stable.

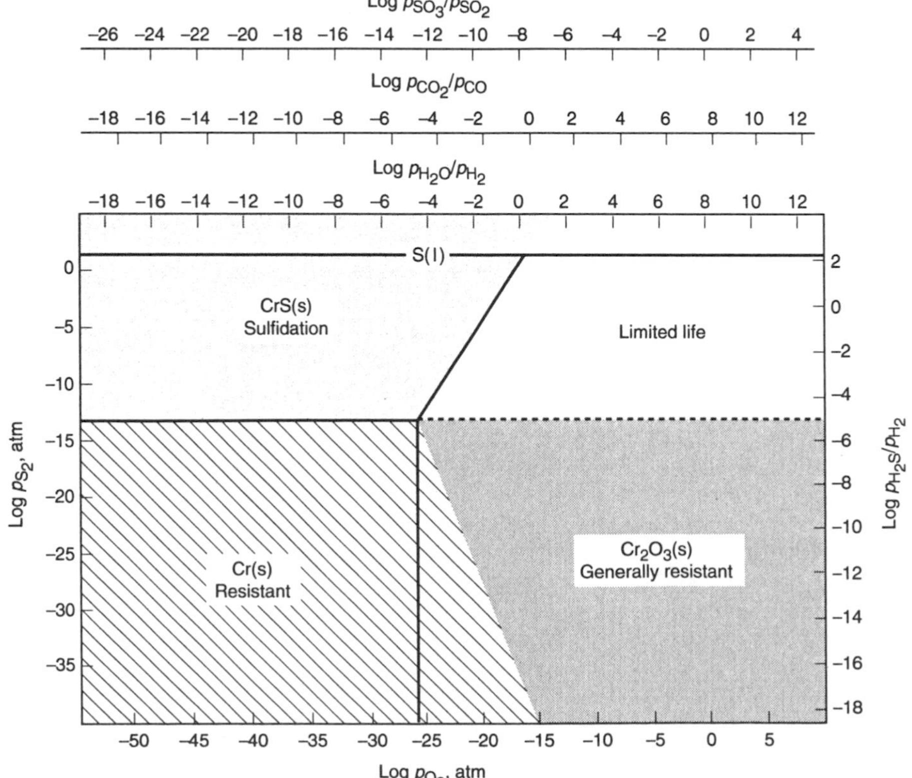

Fig. 34 Stability diagram of the Cr-S-O system at 870 °C (1600 °F). Source: Ref 39, 40

Nitridation (Ref 1)

Metals or alloys are generally susceptible to nitridation attack when exposed to ammonia-bearing environments at elevated temperatures. Nitrogen-base atmospheres can also be nitriding, particularly when the environments are reducing (i.e., characterized by low oxygen potentials). As the environment becomes more oxidizing, nitridation attack generally becomes less severe. Highly oxidizing combustion atmospheres or air normally does not result in nitridation attack. During nitridation, the alloy absorbs nitrogen from the environment. When nitrogen in the alloy exceeds its solubility limit, nitrides then precipitate out in the matrix as well as grain boundaries. As a result, the alloy can become embrittled.

Ammonia is a commonly used nitriding gas for case hardening of steel, typically performed at temperatures from 500 to 590 °C (925 to 1100 °F). Furnace equipment and components repeatedly subjected to these service conditions frequently suffer brittle failures as a result of nitridation attack.

Carbonitriding is another important method of core hardening. It produces a surface layer of both carbides and nitrides. The process is typically carried out at 700 to 900 °C (1300 to 1650 °F) in ammonia, with additions of carbonaceous gases, such as CH_4. Thus, the heat treat retort, fixtures, and other furnace equipment are subject to both nitridation and carburization.

Cracked ammonia (i.e., ammonia that is completely dissociated into H_2 and N_2) provides an economical protective atmosphere for processing metals and alloys. Many bright annealing operations for stainless steels utilize a protective atmosphere consisting of N_2 and H_2, generated by dissociation of ammonia. Once nitrogen molecules are formed from the dissociation of ammonia, the potential for nitriding the metal is significantly reduced. With three parts H_2 and one part N_2 produced in cracked ammonia, heat treating equipment is generally less susceptible to nitridation attack.

Nitrogen atmospheres are becoming increasingly popular as protective atmospheres in heating treating and sintering operations. Although molecular nitrogen is less reactive than ammonia, it can be severely nitriding when the temperature is sufficiently high. Long-term exposure to a nitrogen atmosphere at high temperatures may cause metallic equipment to suffer premature failure.

In the chemical processing industry, nitriding environments are generated by processes employed for production of ammonia, nitric acid, melamine, and nylon 6-6. Ammonia is produced by reacting nitrogen with hydrogen over a catalyst at temperatures of typically 500 to 550 °C (930 to 1020 °F) and pressures of 200 to 400 atm. The converter, where the ammonia synthesis reaction takes place, may suffer nitridation attack. Brittle failure of the welds for the waste heat boiler of an ammonia plant has also been reported (Ref 42).

Table 8 Nitrides of important alloying elements for engineering alloys

Element	Nitrides
Iron	Fe_4N, Fe_2N
Chromium	CrN, Cr_2N
Molybdenum	MoN, Mo_2N
Tungsten	WN, W_2N
Aluminum	AlN
Titanium	TiN, Ti_2N
Niobium	NbN, Nb_2N, Nb_4N_3
Tantalum	TaN, Ta_2N, Ta_3N_5
Zirconium	ZrN
Hafnium	HfN, Nf_3N_2, Hf_4N_3
Silicon	Si_3N_4
Manganese	Mn_4N, Mn_2N, Mn_3N
Vanadium	VN, V_2N
Boron	BN
Magnesium	Mg_3N

Source: Ref 45

Production of nitric acid involves the oxidation of ammonia over a platinum gauze catalyst at temperatures of about 900 °C (1650 °F). The catalyst grid support structure and other processing components in contact with ammonia may also be susceptible to nitridation attack.

Thermodynamic Considerations. When metal is exposed to nitrogen gas at elevated temperatures, nitridation proceeds according to:

$$1/2\, N_2\,(g) \leftrightarrow [N]\ \text{(dissolved in metal)} \qquad \text{(Eq 48)}$$

$$[\%N] = k(p_{N_2})^{1/2} \qquad \text{(Eq 49)}$$

where k is the equilibrium constant and p_{N_2} is the partial pressure of N_2 in the atmosphere. In nitrogen-base atmospheres, the nitriding potential is proportional to $(p_{N_2})^{1/2}$. Increasing the nitrogen partial pressure (or nitrogen concentration) increases the thermodynamic potential for nitridation. Molecular nitrogen is relatively inert in terms of nitridation of metals. However, when metals or alloys are heated to excessively high temperatures (e.g., 1000 °C or higher), nitridation may become significant.

When the environment is ammonia (NH_3) or contains NH_3, metals or alloys may undergo rapid nitridation reactions. It is precisely for this reason that NH_3 is frequently used for case hardening. Ammonia is metastable and dissociates into molecular N_2 and molecular H_2 when heated to elevated temperatures. Once NH_3 is completely dissociated into N_2 and H_2, the nitriding potential is greatly reduced and is then defined by Eq 49. To increase nitrogen absorption by steel, molecular NH_3 should be allowed to dissociate on the steel surface, thus allowing dissociated atomic nitrogen to be dissolved at the surface (Ref 43, 44). Thus, to increase nitridation reactions, it is necessary to bring as much fresh, uncracked, NH_3 as possible in contact with the surface to minimize the production of molecular nitrogen. At temperatures below 600 °C (1110 °F) and at high gas flow rates, the production of molecular nitrogen is minimized and the nitrogen solubility at the surface of iron is determined by:

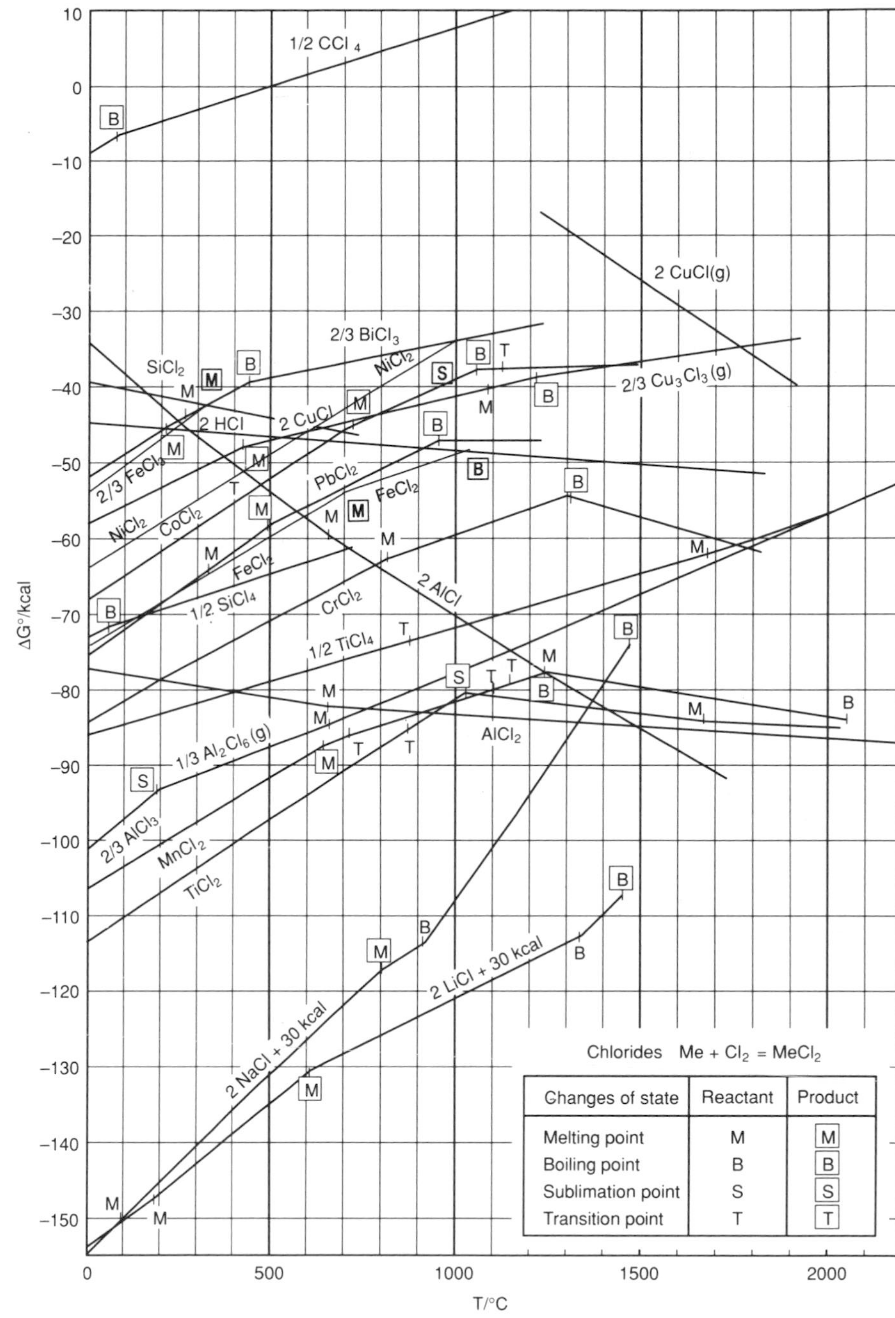

Fig. 35 Standard free energies of formation for chlorides. Source: Ref 48

$$NH_3 \leftrightarrow 3/2\, H_2 + [N]\ \text{(dissolved in Fe)} \qquad \text{(Eq 50)}$$

$$[\%\,N] = k\,\frac{p_{NH_3}}{(p_{H_2})^{3/2}} \qquad \text{(Eq 51)}$$

where p_{NH_3} and p_{H_2} are partial pressures of NH_3 and H_2, respectively. The nitriding potential is proportional to $p_{NH_3}/(p_{H_2})^{3/2}$. Increasing ammonia partial pressure (or concentration) in the atmosphere increases the thermodynamic potential for nitridation.

When nitrogen in the metal exceeds its solubility limit, nitrides will then precipitate out. The nitrides of important alloying elements for engineering alloys are tabulated in Table 8. For iron, nickel, and cobalt, three important alloy bases for high-temperature alloys, only iron forms nitrides. No nitrides of nickel or cobalt have been reported.

Halogen Gas Corrosion (Ref 1)

Many metals react readily with halogen gases at elevated temperatures to form volatile metal halides. Many metal halides also exhibit low melting points (Ref 46, 47). Some metal halides even sublime at relatively low temperatures. As a result, alloys containing elements that form volatile halides can suffer severe high-temperature corrosion.

Industrial environments often contain halogen gases. Because of high vapor pressures of many metal chlorides, the chlorination process is an important step in processing metallurgical ores for production of titanium, zirconium, tantalum, niobium, and tungsten. Chlorination is also used for extraction of nickel from iron laterites and for detinning of tin plate. Production of TiO_2 and SiO_2 involves processing environments containing Cl_2 and/or HCl, along with O_2 and other combustion products. Calcining operations for production of lanthanum, cerium, and neodymium for electronic and magnetic materials, as well as for production of ceramic ferrites for permanent magnets, frequently generate environments contaminated with chlorine. In the chemical processing industry, many processing streams also contain chlorine. Manufacturing of ethylene dichloride, which is an intermediate for the production of vinyl chloride monomer, generates chlorine-bearing environments. The reactor vessels, calciners, and other processing equipment for the above operations require alloys resistant to high-temperature chlorination attack.

In the manufacture of fluorine-containing compounds, such as fluorocarbon plastics, refrigerants, and fire-extinguishing agents, the processing equipment requires alloys with good resistance to corrosion by fluorine and hydrogen fluoride at elevated temperatures. During the refining operation in the production of uranium, UO_2 is fluorinated at elevated temperatures (e.g., 500 to 600 °C, or 930 to 1110 °F) with HF to produce UF_4 or UF_6 for separation of U_{235}. The materials of construction for this processing equipment must resist corrosion by HF at both high and low temperatures.

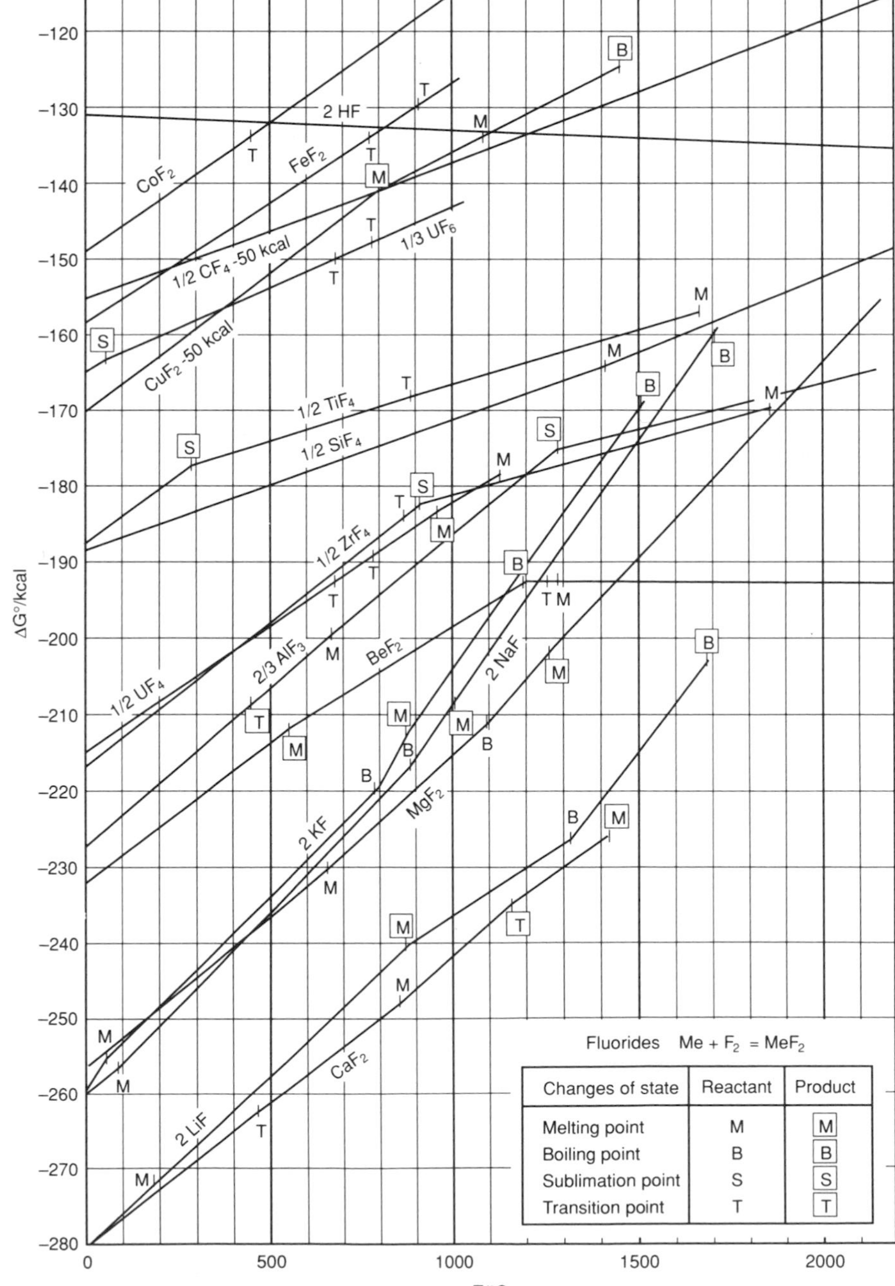

Fig. 36 Standard free energies of formation for fluorides. Source: Ref 48

Thermodynamic Considerations. The relative stabilities of various chlorides and fluorides are presented in Fig. 35 and 36 in terms of standard free energies of formation versus temperature. Because most industrial environments also exhibit oxygen activities, a stability diagram, such as the one shown in Fig. 37 for the Fe-O-Cl system, can be used to describe the possible corrosion products. Similar diagrams can be constructed for other major alloying elements of high-temperature alloys. With stability diagrams, one can predict the possible phases that the alloy is likely to form at the gas-metal interface. The phases that the alloy may form depend on the reaction path, which in turn may depend on alloy composition, among other factors. For example, the reaction between the test environment and iron can follow any one of numerous possible reaction paths, as shown in Fig. 37. Reaction paths A and B could lead to the formation of liquid $FeCl_2$ deposits, which would be expected to flux the oxide and cause accelerated corrosion, while reaction path C would produce only oxides and be expected to produce much less corrosion. The reaction paths can be used to rationalize the observation of particular corrosion products and to predict the effects of changes in the environment on corrosion kinetics.

The metal-halogen reaction (halogenation) differs from other reactions, such as oxidation, in that most reaction products are characteristic of high vapor pressures and, sometimes, low melting points. The volatile halides (the reaction products) formed on the metal surface can no longer provide protection against further corrosion. This is in contrast to most oxides, which generally exhibit very low vapor pressures. Even under conditions where the metal forms both oxides and halides, the corrosion reaction may still proceed at a rapid rate.

Ash/Salt Deposit Corrosion

Deposition of ashes and salts on surfaces of process components is quite common in some industrial environments. The corrosion process under these conditions involves both the deposit and corrosive gases. The deposit may alter the thermodynamic potentials of the environment on the metal surface beneath the deposit. In an environment with both oxygen and sulfur activities, for example, the deposit tends to lower the oxygen activity and raise the sulfur activity beneath the deposit. As a result, formation of a protective oxide scale on the metal surface may become more difficult for most alloys. In most cases, the deposit involves some type of salt. This may lead to chemical reactions between the protective oxide scale and the salt, resulting in the breakdown of the scale. It is particularly damaging when the salt deposit becomes liquid.

Hot Corrosion. High-temperature corrosion of hot-section gas turbine components, such as nozzle guide vanes and rotor blades, due to salt deposits (principally sodium sulfate) is known as "hot corrosion." Hot corrosion generally proceeds in two stages: an incubation period exhibiting low corrosion rates, followed by accelerated corrosion attack. The incubation period is related to the formation of a protective oxide scale. Accelerated corrosion attack is believed to be initiated by breakdown of the protective oxide scale due to the presence of molten sodium sulfate-rich deposits. More detailed information on hot corrosion of gas turbine components can be found in the article "Elevated-Temperature Corrosion Properties of Superalloys" in this Volume.

Fireside corrosion of components such as superheaters and reheaters in fossil-fired boilers is commonly referred to as fuel ash corrosion. The corrosion process in fuel ash corrosion is related to alkali-iron trisulfates for coal-fired boilers and vanadium salts (a mixture of vanadium pentoxide and sodium oxide or sodium sulfate) for oil-fired boilers. These low-melting-point-salts flux the protective scale from the metal surface. More detailed information on ash/salt deposit corrosion of boiler components can be found in the articles "Elevated-Temperature Corrosion Properties of Carbon and Alloy Steels" and "Elevated-Temperature Corrosion Properties of Stainless Steels" in this Volume.

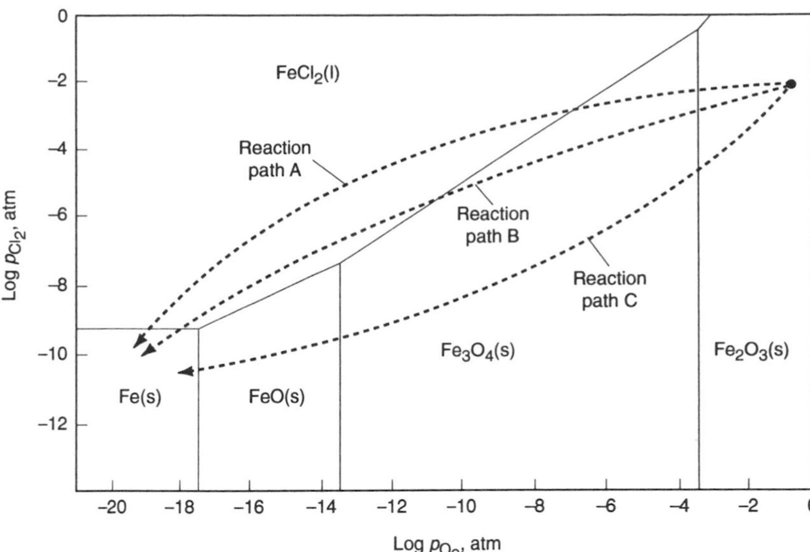

Fig. 37 Stability diagram of the Fe-Cl-O system at 900 °C (1650 °F). Also shown are an arbitrary test environment and several hypothetical reaction paths indicating the phases that may form at the gas-metal interface. See text for details. Source: Ref 1

Methods of Testing and Evaluation

Laboratory testing and field testing are aimed at characterizing material performance for specific applications, or at evaluating the effect of variations in test conditions on candidate materials (Ref 49). The test environment should duplicate as closely as possible, the environment of the actual application. This includes simulating the chemical composition of the atmosphere, temperature and thermal cycle profiles, stress states, fatigue conditions, and design. Test racks placed in actual use are frequently employed. In other cases, prototype or scale-downed apparatus are designed and constructed to duplicate the end-use application. Laboratory service tests are commonly engineered specifically to evaluate the effect of one or more critical aspects of the exposure conditions. The data on material performance are then used to determine failure mechanisms, components, or material life or to screen materials for the application.

This section focuses on high-temperature gaseous environment testing used to evaluate materials where the applications simultaneously involve high-temperature and corrosive conditions. Basic principles will be described that an investigator must consider in conducting and evaluating these tests. Additional information on testing in high-temperature gases can be found in Ref 49 and 50.

Laboratory Testing

Laboratory testing is important in understanding the behavior of metals and alloys in various corrosive environments. Most of the scientific theories on high-temperature corrosion have been developed through laboratory study. Laboratory testing has also contributed significantly to the wealth of corrosion data that allows engineers to make informed materials selections for various processing equipment.

There are, however, several drawbacks. A laboratory test cannot simulate exactly the operating environment and conditions of a processing system. Another drawback is the relatively short test duration compared to the design life of the equipment. Extrapolation becomes necessary to make materials behavior predictions. One major problem with extrapolation is the unpredictability of breakaway corrosion. In many cases, metals and alloys cannot resist high-temperature corrosion attack unless they form protective scales (mostly oxide scales). Although it is generally understood that the protective scale may eventually break down, leading to breakaway corrosion, it is not currently possible to predict the onset of breakaway corrosion. Thus, laboratory tests are often conducted under accelerated conditions (e.g., higher temperatures and/or more corrosive environments) in order to increase the confidence level for the selected alloy. The accelerated test is also frequently used for initial alloy screening to narrow down the viable candidates for long-term tests and/or field trials. Extreme care should be taken if the results of short-term tests and/or accelerated tests are used for life extrapolation.

Test Methods. The gravimetric method is widely used to study oxidation and other forms of high-temperature corrosion. This test method involves measuring the weight of a specimen as a function of time. Test apparatus that continuously monitors the weight of a specimen during testing with a recording balance is very popular in academia for conducting oxidation studies. One such test apparatus is shown schematically in Fig. 38. This particular experimental arrangement allows the inlet test gas to be preheated to the test temperature in a small tube prior to reaction with the test sample in the reaction chamber. Preheating the inlet gas is particularly desirable when the test involves a mixture of several gaseous components, because preheating allows the gas mixture to reach thermodynamic equilibrium prior to reaction with the test sample.

The major advantage of a gravimetric apparatus with an automatic recording balance is the continuous record of the reaction kinetics. It is an excellent way of studying kinetics and mechanisms of a high-temperature corrosion reaction. The disadvantage is that only one sample can be tested each time, so generating comparative data for a number of candidate alloys is a lengthy process. Therefore, the method is not very suitable for generating an engineering database.

Industrial laboratories prefer to test multiple samples simultaneously. By conducting tests for various lengths of time, the kinetic curves for several alloys can be obtained. It is possible to generate corrosion data for a large number of alloys within a reasonable time frame using this method. It is common practice to include a control sample in each test run to monitor the reproducibility of the tests from run to run. This also allows the investigator to check whether any test abnormalities have occurred in any test runs. The alloy used for the control sample should be sensitive to variations in test conditions. For example, an alloy very resistant to the specific test

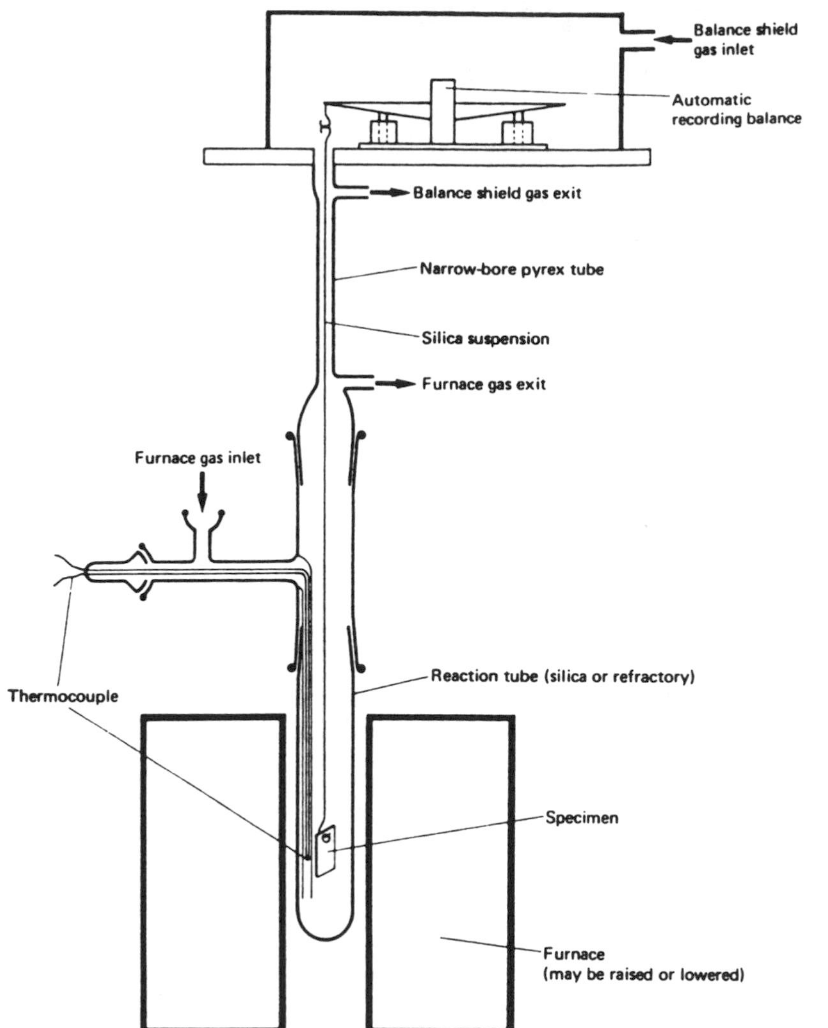

Fig. 38 Schematic of a thermogravimetric apparatus with an automatic recording balance for oxidation and other high-temperature gas-metal reactions. Source: Ref 51

environment is not a good choice because variations in test conditions (e.g., gas composition) from run to run may not be detected.

Most industrial process gas streams contain many gaseous components. When combustion is stoichiometric or substoichiometric, combustion products are generally comprised of H_2, CO, CO_2, H_2O, and products of impurities coming from fuel and/or feedstock, such as H_2S. Laboratory testing in a multicomponent gas mixture to simulate the industrial process environment has been widely used, generating a significant amount of sulfidation and carburization data. Test systems, although varying from laboratory to laboratory, typically consist of a multicomponent gas mixing system with leaktight test retorts where test samples are exposed to a flowing gas mixture. Gaseous components are metered by flowmeters or mass flow controllers to make up the test gas mixture.

Most oxidation data have been generated in air. These data are easy to generate and are also quite relevant to many industrial processes where metallic components are oxidized through reaction with air or highly oxidizing combustion atmospheres containing few or no contaminants.

Oxidation of alloys has also been studied in combustion environments using a burner rig test system. The burner rig test has been widely used in the gas turbine industry to evaluate the oxidation behavior of gas turbine alloys. The test environment is produced by combustion of fuel with air. An example of such a system is shown schematically in Fig. 39. This particular burner rig system produces a high-velocity (0.3 mach) gas stream and is capable of imposing severe thermal cycling on the test samples (1.0 mach = 750 mph). Samples are loaded in a carousel, rotating at a constant speed (e.g., 30 rpm) to ensure that all samples are subjected to the same test conditions, and then, are automatically withdrawn from the hot zone, for example, once every 30 min, and fan-cooled to less than 260 °C (500 °F) for 2 min before reinsertion.

The burner rig test system can also be used to investigate the performance of alloys in combustion environments contaminated with salts. Hot corrosion of superalloys has been widely studied using a burner rig with injection of sea salts into the combustion test environment. A burner rig test system used to study hot corrosion of gas turbine alloys is shown schematically in Fig. 40. Details of the design and operation of such a rig are described by Doering and Bergman (Ref 52). The rig can also be used for injecting other salts or impurities to study ash/salt deposit corrosion related to non-gas turbine industries.

Other test methods for studying hot corrosion include the modified Dean rig (Ref 53) and the salt-coated test (Ref 54). The modified Dean rig uses a two-zone furnace with independent temperature controls for each zone. A crucible containing a suitable salt is placed in the first zone, while the sample is placed in the second zone. A test gas is passed over the salt crucible to carry the salt vapor prior to the reaction with the sample located downstream in the second zone. The salt vapor will condense on the sample, if the second zone is cooler. In the salt-coated test, the sample is first coated with a suitable salt (or salt mixture) and then exposed to a test gas. Coatings of salt are normally applied by spraying with aqueous solutions while the sample is being heated by, for example, a hot plate.

Specimen Evaluation and Data Presentation. After testing, specimens should always first be visually examined, which may reveal some important features of corrosion, such as localized corrosion. Visual examination is also necessary to determine the best location for subsequent metallographic examination.

The most commonly used method for determining alloy performance is to measure specimen weight change (or mass change) as a result of corrosion. This is determined by the difference in specimen weight before and after testing, divided by specimen surface area (before testing).

This method of data presentation has several drawbacks. Most high-temperature corrosion data for commercial alloys are generated by industrial laboratories using multiple specimens per test run. Frequently, corrosion products may spall from the specimens. With several specimens in a test retort, it is difficult to account for all the spalled corrosion products. As a result, weighing specimens with partially spalled scales (or corrosion products) may obscure the overall results. Furthermore, oxide scales or corrosion products formed on different alloy systems have different compositions, and thus different densities and vapor pressures. This can further complicate the weight change data. The most important drawback of the weight change method is that it fails to reveal the depth of the corrosion attack.

High-temperature corrosion generally causes two forms of damage to metals or alloys; metal loss (or metal wastage) and internal penetration (or subscale attack). Different alloys sometimes exhibit different forms of attack under the same test conditions. Figure 41 shows alloy 188 (Co-Cr-Ni-W-La) and alloy C-276 (Ni-Cr-Mo-W) after exposure to Ar-20O_2-2Cl_2 at 900 °C (1650 °F) for 8 h (Ref 55). Alloy 188 suffered mainly internal penetration with very little scale (or metal loss). On the other hand, alloy C-276 suffered mainly heat loss with no internal penetration. Thus, the best method for presenting high-temperature corrosion data is to use the total depth of attack, which is the sum of metal loss and internal penetration. ASTM G 54, "Practice for Simple Static Oxidation Testing," recommends the following formula:

$$\text{Total depth of attack (mm)} = \frac{t_o - t_m}{2}$$

where t_o is the original specimen thickness of diameter (mm) and t_m is the remaining good metal (mm), as shown in Fig. 42. This method is especially suitable for specimens suffering severe localized attack such as pitting or intergranular attack.

When a specimen exhibits general corrosion attack, the metal loss or wastage is generally determined from weight change measurements using the following formula:

$$\text{Metal loss (mm)} = \frac{\Delta W}{(A)(d)}$$

where ΔW is the difference in weight between the original specimen and the specimen tested and descaled (mg), A is surface area of the specimen (mm^2), and d is the density of the alloy (mg/mm^3).

Following descaling (per ASTM G 1, "Practice for Preparing, Cleaning, and Evaluating Corrosion Test Specimens") and weighing of the specimen for determination of metal loss, the specimen is evaluated metallographically for internal penetration. Prior to descaling, the scales and/or corrosion products formed on the specimen are frequently examined by scanning electron microscopy and x-ray diffraction analysis to determine their morphology, composition, and phases.

Certain high-temperature corrosion reactions, such as carburization and nitridation, result in the formation of internal phases or precipitates. Test results are generally presented in terms of weight gain, mass absorption (e.g., carbon), concentration profile (e.g., carbon), or penetration depth. However, these data may not reflect the real damage for an alloy. Coupon testing is preferably used for initial alloy screening. Final materials selection should include data that reflect the effect of carburization (or nitridation) on the mechanical properties of the candidate alloys.

Analyses of Scales and Corrosion Products. It is always worthwhile to examine the scale and/or corrosion products formed on specimens before descaling and/or metallographic examination. Valuable information regarding the corrosion reaction can be revealed through such analysis.

Many modern analytical techniques are available for analyzing scales and corrosion products. The scanning electron microscope (SEM) is probably the most widely used equipment for examining corrosion samples. The modern SEM is frequently equipped with an energy-dispersive x-ray spectrometer (EDX) or a wavelength x-ray spectrometer (WDX) for performing qualitative (or quantitative) chemical analysis on a selected area. The major advantage of the EDX system is that it can measure a wide range of x-ray energies simultaneously (i.e., almost all the elements). In a typical SEM/EDX system, sodium is the lightest element that can be detected. The WDX system, however, is able to detect light elements down to boron. The major disadvantage of the WDX system is that it typically requires a much longer analysis time. A general description of various electron optical systems for analysis of microconstituents can be found in Volume 10, *Materials Characterization,* of the *ASM Handbook.*

With a modern SEM, chemical constituents as well as morphology of protective oxide scales and/or corrosion products can be determined. In-

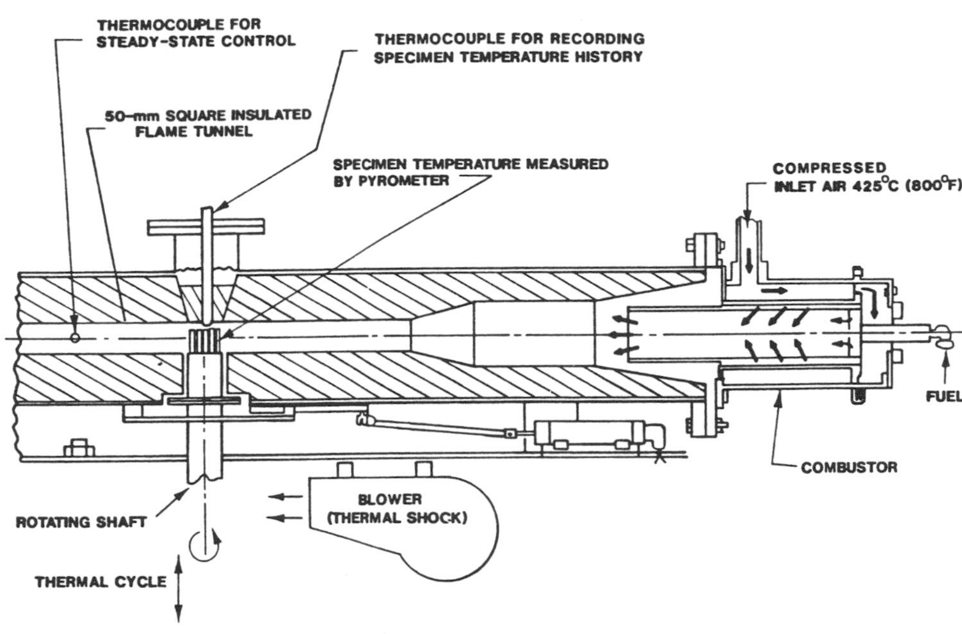

Fig. 39 Schematic of a burner rig dynamic oxidation test system.

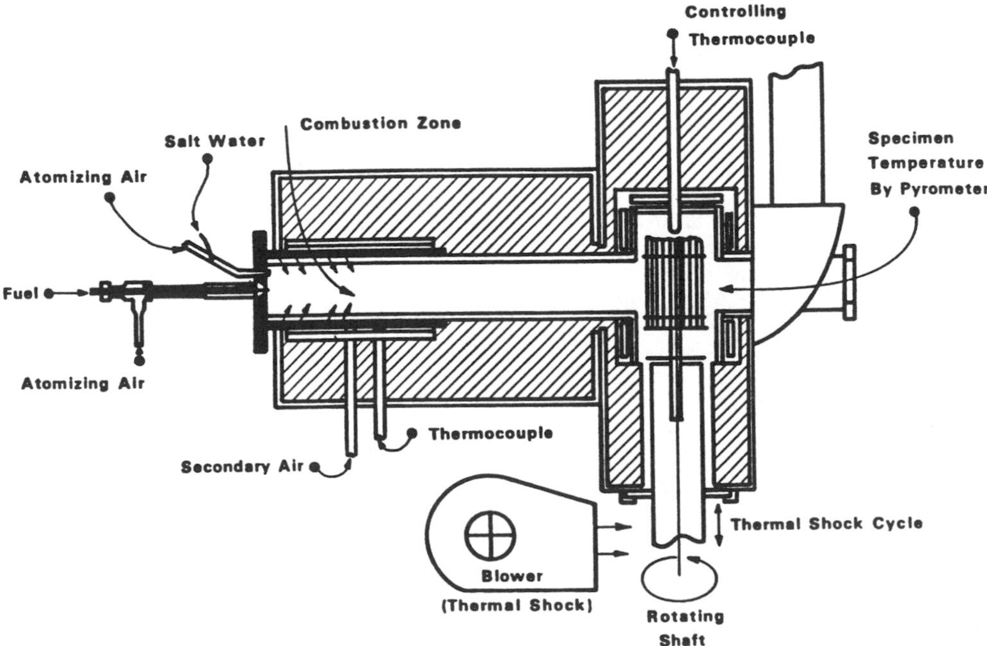

Fig. 40 Schematic of a hot corrosion test system

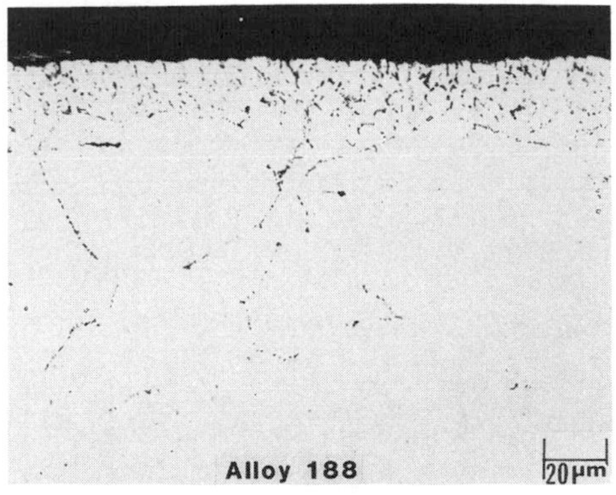

(a)

(b)

Fig. 41 Examples of different alloy systems suffering different forms of damage after exposure to the same test conditions. After exposure to Ar-20O_2-2Cl_2 (vol.%) at 900 °C (1650 °F) for 8 h, the Co-Cr-Ni-W-La alloy (alloy 188) suffered primarily internal attack, with very little metal loss (or wastage), while the Ni-Cr-Mo-W alloy (alloy C-276) suffered primarily metal loss, with no internal attack. Source: Ref 55

formation about elemental distribution in a selected area (e.g., a cross section of the corrosion scale or internal penetration) can also be easily generated. Sometimes, a detailed analysis of chemical compositions for a few atom layers of the protective oxide scale may be needed. This can be accomplished using the scanning Auger microprobe.

Once the chemical constituents of the oxide scales or corrosion products are known, it is useful to determine their phases or chemical compounds using x-ray diffraction techniques.

Field Testing

Field testing is an effective means of generating performance data in the operating environment. It allows a group of candidate alloys to be ranked according to performance before a selection is made. An industrial environment may contain corrosive impurities, such as sulfur, sodium, potassium, chlorine, vanadium, phosphorus, zinc, and lead. In many cases, the primary corrosive contaminants are not known to the operators, so field testing is also useful in identifying the principal corrosive contaminants as well as the principal corrosion mode. Combining alloy ranking data with the principal mode of corrosion makes informed materials selection possible.

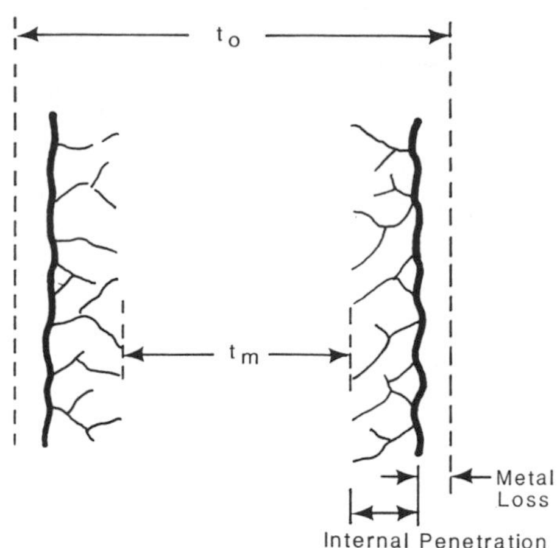

Fig. 42 Schematic showing the method for determining the total depth of attack for static oxidation testing as recommended by ASTM G54

Test Methods. Rack tests are the most commonly used method for generating field data. Coupons of various candidate alloys are grouped together in a rack. It is important to include a coupon of the material currently being used, as a reference, if the test is intended to help select an alternate alloy. The test coupons are normally separated by alumina spacers. A typical test rack is shown in Fig. 43. The rack is installed within the operating system (e.g., inside a reaction vessel or process line) to expose the samples to the actual environment and conditions for a predetermined duration, from several weeks to a year or longer. Actual exposure time depends on the corrosivity of the environment as well as the operation schedule. It is very important to expose the test rack for a suitable duration in order to obtain meaningful results. Prior materials experience can generally provide some guidance.

Sometimes it is not feasible to install a rack inside the operating system, for example, when access to the reaction vessel is a problem. Under these circumstances, other test methods have been successfully used to obtain data. One method involves installing individual samples instead of a test rack. Thermowells provide an excellent location for generating field data. The alloy of interest can be made into a thermowell for testing. If no tubing is available for that alloy, the alloy can be coated as a weld overlay on an off-the-shelf thermowell. The advantage of using a thermowell is that almost every high-temperature industrial processing system has thermowells or sample probes. The disadvantage is that the data generated may not be applicable to remote areas, where conditions (e.g., temperature) can be very different.

Generating field data for heat exchanger alloys can be challenging, primarily because of the difficulty in simulating the tube temperatures and ash or salt deposits. Special methods are required to simulate heat exchanger conditions unless the candidate alloys are tested as part of the actual heat exchanger. Battelle Columbus Laboratories (Ref 56) has developed a special probe for generating field data relevant to boiler tubes in waste incinerators. The probe has a specially tapered wall, which will produce different temperatures at different locations along the wall. The probe is internally cooled by high-velocity air, and the temperature is monitored by thermocouples inserted into the probe. It is capable of generating corrosion rates as a function of temperature in each exposure. However, this method is costly. Furthermore, it is suitable only for large vessels, such as power plant boilers, where constant injection of air into the system will not affect the process.

Another method developed by Russell et al. (Ref 57) has been used successfully for generating heat exchanger corrosion data in industrial furnaces. In their field test, a specially designed test chamber was installed to evaluate the corrosion behavior of candidate recuperator alloys for an aluminum remelting furnace. The flue gas entered the test chamber at the bottom and reentered the furnace stack at the top of the chamber. Test tubes made of selected alloys were installed in the test chamber and exposed to the furnace flue gas at the outside diameter while air was allowed to flow through the internal diameter of the tubes to simulate recuperator conditions.

Specimen Evaluation and Data Presentation. When the test rack or test samples are returned for evaluation, they should first be examined visually for possible localized attack, pitting, perforation, or other unusual features. It is important to collect corrosion products and deposits prior to metallurgical evaluation. Analysis of the deposits can yield valuable information about the environment, particularly the corrosive contaminants. Evaluation of the test samples should include (a) analysis of the deposits and corrosion products that form on the sample surface, as well as the internal penetration, and (b) determination of the total depth of attack, including metal loss (or metal wastage) and internal penetration. Total depth of attack can be determined using established laboratory testing procedures, discussed earlier in this section.

Analyses of Deposits and Corrosion Products. Unlike laboratory test conditions, many industrial environments are not very well defined or understood. As a result, deposits and corrosion products are a major source of information about the flue gas environment, particularly the principal corrosive contaminants that influence the corrosion reaction. Therefore, it is always good policy to analyze these deposits using the analytical techniques described above in the section "Laboratory Testing" in this article.

ASTM Test Standards (Ref 49)

ASTM G 54, "Practice for Simple Static Oxidation Testing." This service test practice covers determination of preliminary information on the relative growth, scaling, and microstructural characteristics of an oxide on the surface of a pure metal or alloy under isothermal conditions in still air.

ASTM B 76, "Method for Accelerated Life Test of Nickel-Chromium and Nickel-Chromium-Iron Alloys for Electrical Heating." This service test method covers the determination of the resistance to oxidation of nickel-chromium-iron electrical heating alloys at elevated temperatures under intermittent heating. Procedures for a constant temperature test are provided. This test method is used for comparative purposes only.

ASTM B 78, "Method for Accelerated Life Test of Iron-Chromium-Aluminum Alloys for Electrical Heating." This service test method covers the determination of the resistance to oxidation of iron-chromium-aluminum alloys for electrical heating alloys at elevated temperatures under intermittent heating using a constant-temperature-cycle test. This test is used for comparative purposes only.

ASTM G 79, "Practice for Evaluation of Metals Exposed to Carburization Environments." This service test method covers procedures for the identification and measurement of the extent of carburization in a metal sample and for the interpretation and evaluation of the effects of carburization. It applies mainly to iron- and nickel-base alloys for high-temperature applications. Four methods are described:

- Method A, Total Mass Gain
- Method B, Metallographic Evaluation
- Method C, Carbon Diffusion Profile
- Method D, Change in Mechanical Properties

These methods are intended, with the interferences as noted, to evaluate either laboratory specimens or commercial product samples that have been exposed in either laboratory or commercially produced environments.

Molten Salt Corrosion

The corrosion of metal containers by molten, or fused, salts has been observed for an extended period of time, but over the last several decades, more effort has been directed toward understanding corrosion phenomena at higher temperatures. Annotated bibliographies of molten-salt corrosion for different media and metal combinations have been published and are cited in Ref 58 to 61.

Thermodynamics and Kinetics of Molten Salt Corrosion. The chemistry of molten salts can be as complicated as one wishes to make it, based on the definition of a molten salt and whether or not the media may be wholly ionic. For simplicity, most of the processes considered in this article involve electrode processes.

According to Inman and Lovering (Ref 58), except in rare cases in which hydrogen is a part of the molten salt or the melts are exposed to hydrogen atmosphere, the hydrogen ion plays a very small role. The oxygen ions are generally quite important in matters of corrosion. The function pO^{2-} (equivalent to pH in aqueous environments) defines the oxide ion activity. The higher the value of pO^{2-}, the more corrosion of metal will occur. Also, the concentration of oxide ions can influence the corrosive effects of certain nonoxygen-containing melts that have been subject to hydrolysis through contact with atmospheric moisture. In molten salt systems, corrosion is rarely inhibited because of the reactivity of the molten salts and the high temperatures. Molten salts often act as fluxes, thus removing oxide layers on container materials that generally might prove to be protective. Molten salts are generally good solvents for precipitates; therefore, passivation, because of deposits, generally does not occur.

One of the most familiar mechanisms of corrosion arises from ions of metals more noble than the container material, that is, the metal being corroded. In some cases, the more noble metal can be a constituent part of the molten salt, and in others, it can occur as an impurity in the system.

Another mechanism is best described by the example of silver in molten sodium chloride (NaCl). Thermodynamics would not predict a corrosion product. However, the reaction occurs because sodium, as a result of the formation of silver chloride (AgCl), can dissolve in molten NaCl and distill out of the system. Thus, the reaction proceeds.

If a molten salt contains oxyanion constituents that can be reduced, oxide ions are released. Corrosion will occur on a metal in contact with the salt.

Lastly, oxygen itself can be reduced to oxide ions. However, uncombined oxygen is rarely found in molten salts because of limited solubility.

The potential, *E*, versus pO^{2-} diagram is often used as the equivalent to the *E* versus pH (Pourbiax) diagrams for aqueous corrosion (Ref 62). Both of these diagrams are used to establish the stability characteristics of a metal in the respective media. A typical *E* versus pO^{2-} diagram for iron in a molten salt at an elevated temperature is shown in Fig. 44. Areas of corrosion, immunity, passivation, and passivity breakdown are evident.

Actually, the *E* versus pO^{2-} diagram is probably more useful than the Pourbaix diagram because of the absence of kinetic limitations at elevated temperatures. The following problems, however, do exist:

- Molten salt electrode reactions and the concomitant thermodynamic data are not readily available.
- Products from the reactions are often lost by vaporization.
- Diagrams based on pure component thermodynamic data are unrealistic because of departure from ideality.
- There may be lack of passivity even where predictions would show passive behavior.
- Stable oxides other than the O^{2-} species may exist.

Mechanisms of Molten-Salt Corrosion. Two general mechanisms of corrosion can exist in molten salts. One is the metal dissolution caused by the solubility of the metal in the melt. This dissolution is similar to that in molten metals but is not common. The second and more common mechanism is the oxidation of the metal to ions, which is similar to aqueous corrosion. For this reason, molten-salt corrosion has been identified as an intermediate form of corrosion between molten metal and aqueous corrosion.

General, or uniform, metal oxidation and dissolution is a common form of molten-salt corrosion but is not the only form of corrosion seen. Selective leaching is very common at higher temperatures, as are pitting and crevice corrosion at lower temperatures. All the forms of corrosion observed in aqueous systems, including stress-assisted corrosion, galvanic corrosion, erosion-corrosion, and fretting corrosion, have been seen in fused salts.

Electrochemically, the molten salt/metal surface interface is very similar to the aqueous solution/metal surface interface. Many of the principles that apply to aqueous corrosion also apply to molten-salt corrosion, such as anodic reactions leading to metal dissolution and cathodic reduction of an oxidant.

The concept of acid and base behavior of the melt is very similar to its aqueous counterpart. The corrosion process is mainly electrochemical

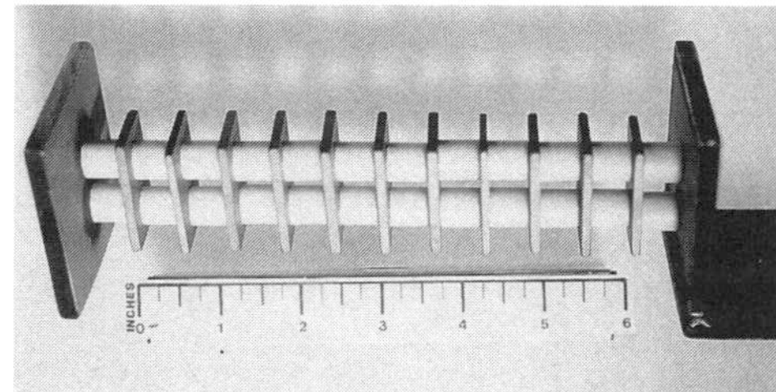

Fig. 43 Test rack containing coupons of candidate alloys for field testing. Alumina spacers are used to separate the coupons.

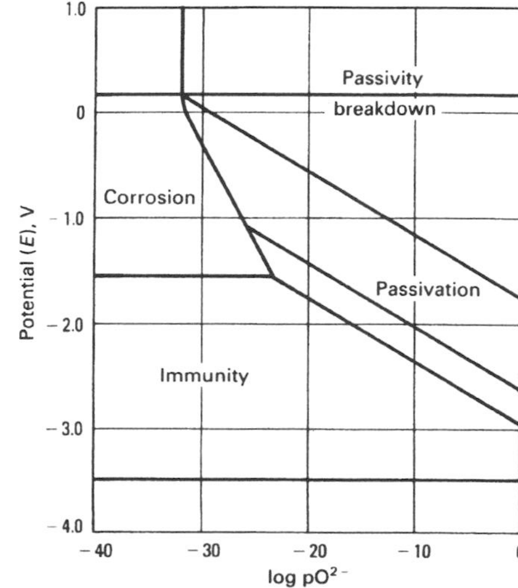

Fig. 44 Typical *E* versus pO^{2-} diagram for iron in a molten salt of an elevated temperature

in nature because of the excellent ionic conductivity of most molten salts. Some investigators feel that dissolved water enhances the electrochemical corrosion nature of the molten salts.

Even though the corrosion mechanism is similar, there are major differences between molten-salt and aqueous corrosion. The differences arise mainly from the fact that molten salts are partially electronic conductors as well as ionic conductors. This fact allows for reduction reactions to take place in the melt as well as the metal/melt interface. This behavior also allows increase in frequency of cathodic reactions and can therefore lead to a substantial increase in corrosion rate over a similar electrochemically controlled aqueous system, especially if the corrosion media contain very few oxidants. Because of property differences between water and molten salt, the rate-controlling step in most molten-metal systems is ion diffusion into the bulk solution, not the charge transfer reaction that is typical of aqueous systems. Molten-salt systems operate at higher temperatures than aqueous systems, which leads to different forms of corrosion attack.

In aqueous systems in which specific elements are removed from alloys, such as the dezincification of zinc from brass, it is not clear whether the nobler element is oxidized at some stage in the process and is subsequently plated out or whether the element agglomerates by a surface diffusion mechanism. In both cases, bulk diffusion is unlikely to play a major role. In high-temperature molten-salt systems, dissolution of the less noble element is the most probable mechanism and will even occur when the element is present at low concentration. Unlike aqueous systems, the rate of dissolution is therefore related to the bulk diffusion of the selectively leached element.

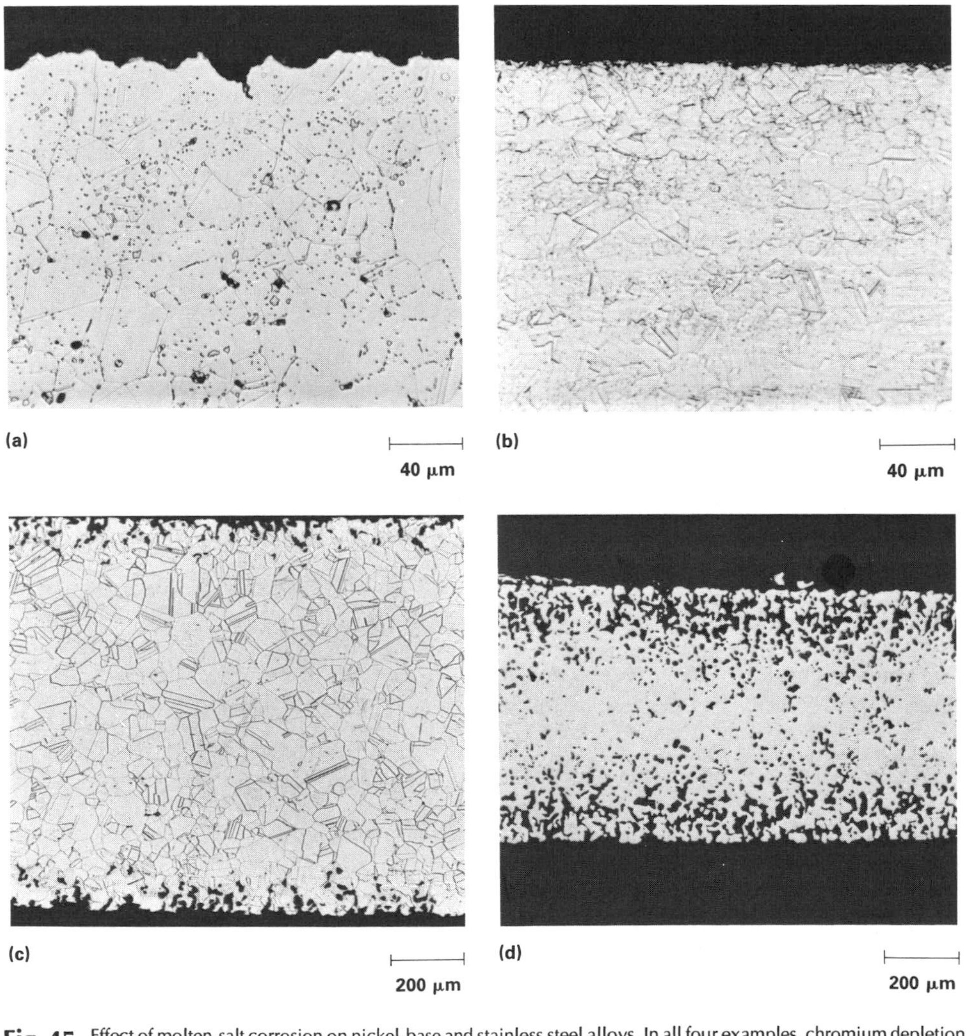

Because molten-salt corrosion reactions are reduction/oxidation controlled, the relative nobilities of the salt melts and the metals are important. The corrosion potential of the melt is often controlled by impurities in the melt or gas phase, which increases the cathodic reaction rate or changes the acidic or basic nature of the melt. The aggressiveness of the salt melts is typically governed by its redox equilibria.

Thermal gradients in the melt can cause dissolution of metal at hot spots and metal deposition at cooler spots. The result is very similar to aqueous galvanic corrosion, and like aqueous galvanic corrosion, a continuous electrical path is necessary between the hot and cold areas. Crevice corrosion has been observed, and wash-line attack caused by oxygen concentration corrosion is not uncommon at the metal/molten salt/air interface.

High-temperature corrosion in molten salts often exhibits selective attack and internal oxidation. Chromium depletion in iron-chromium-nickel alloy systems can occur by the formation of a chromium compound at the surface and by the subsequent removal of chromium from the matrix, leaving a depleted zone (Fig. 45). Thus, the selectively removed species move out, while vacancies move inward and eventually form voids. The voids tend to form at grain boundaries in most chromium-containing metals, but in some high-nickel alloys, the voids form in the grains. Specific examples of the types of corrosion expected for the different metal-fused salt systems will follow.

Molten Fluorides. Fluoride melts are important because of their consideration for nuclear reactor cooling systems. Corrosion in many fluoride molten-salt melts is accelerated because protective surface films are not formed. In fact, the fluoride salts act as excellent fluxes and dissolve the various corrosion products.

Typically, nickel-base alloys show better corrosion resistance than iron-base alloys. Studies have also shown that most nickel-iron-base alloys that contain varying amounts of chromium show void formation to varying depths. In nonisothermal flow systems, the metallic material shows void formation in the hot section and deposits in the cold section. In most alloys that contain chromium, depletion of the chromium accompanies the void formation (Fig. 45). Analysis of the attacked metal and salt clearly shows that selective removal and outward diffusion of chromium result from oxidation by the fluoride mixture.

Inconel alloy 600 and certain stainless steels become magnetic after exposure to molten fluoride salt. This magnetism is caused by the selective removal of the chromium and the formation of a magnetic iron-nickel alloy covering the surface. The conditions of the melt need to be controlled to minimize the selective removal of the alloying elements. This control can be done by

(a) 40 μm

(b) 40 μm

(c) 200 μm

(d) 200 μm

Fig. 45 Effect of molten-salt corrosion on nickel-base and stainless steel alloys. In all four examples, chromium depletion (dealloying) was the result of prolonged exposure. Accompanying chromium depletion was the formation of subsurface voids, which did not connect with the surface or with each other. As chromium was leached from the surface of the metal, a concentration (activity) gradient resulted and caused chromium atoms from the underlying region to diffuse toward the surface, leaving behind a zone enriched with vacancies. These vacancies would then agglomerate at suitable sites—primarily at grain boundaries and impurities. The vacancies become visible as voids, which tended to agglomerate and grow with increasing time and temperature. (a) Microstructure of Hastelloy alloy N exposed to $LiF-BeF_2-ThF_4$ (68, 20, 11.7, 0.3 mol%, respectively) for 4741 h at 700 °C (1290 °F). (b) Microstructure of Hastelloy alloy N after 2000 h exposure to $LiF-BeF_2-ThF_4$ (73, 2, 25 mol%, respectively) at 676 °C (1249 °F). (c) Microstructure of type 304L stainless steel exposed to $LiF-BeF_2-ZrF_4-ThF_4-UF_4$ (70, 23, 5, 1, 1 mol%, respectively) for 5700 h at 688 °C (1270 °F). (d) Microstructure of type 304L stainless steel exposed to $LiF-BeF_2-ZrF_4-ThF_4-UF_4$ (70, 23, 5, 1, 1, mol%, respectively) for 5724 h at 685 °C (1265 °F)

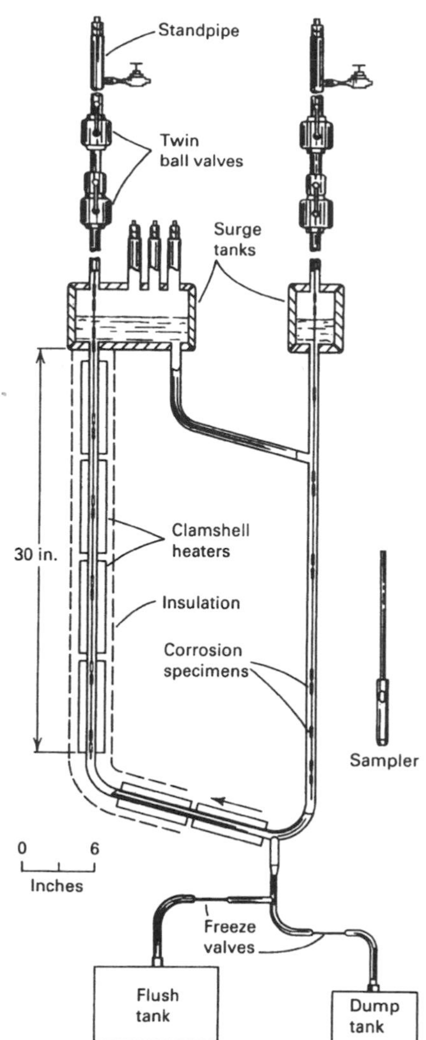

Fig. 46 Thermal convection loop for testing of materials in molten salts

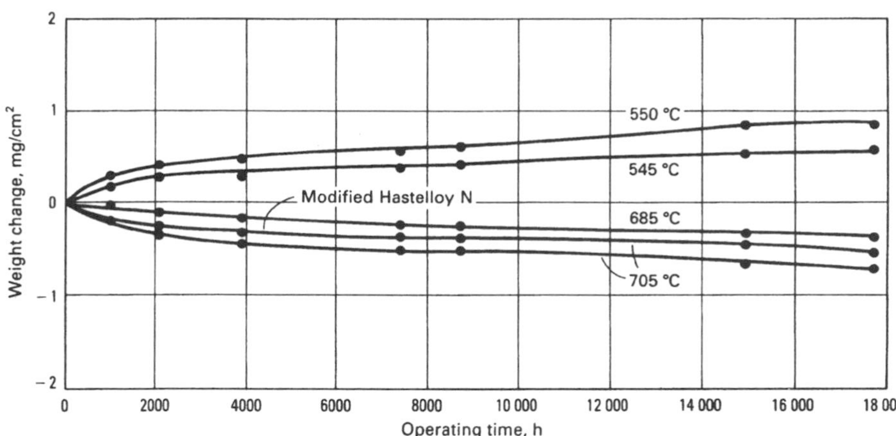

Fig. 47 Weight changes of Hastelloy N exposed to $LiF-BeF_2-ThF_4-UF_4$ (68, 20, 11.7, and 0.3 mol%, respectively), tested for various times in a thermal convection loop as shown in Fig. 46

maintaining the melt in a reducing condition. Addition of beryllium metal to the melt is one way to slow the corrosion rate.

Chloride Salts. Molten salts consisting of chlorides are important, but because they have limited use, they have been studied less than fluoride systems. In general, chloride salts attack steels very rapidly with preferential attack of the carbides. Aluminum coatings on steels are not effective, while the addition of nickel to steel is beneficial. Nickel-base alloys decrease in resistance with increased oxygen partial pressure. In the chloride salts, no protective oxide scale is formed on the nickel-base alloys. The attack of metal surfaces in pure sodium chloride has been observed at temperatures above 600 °C (1110 °F).

In most cases with iron-nickel-chromium alloys, the corrosion takes the form of intergranular attack. An increase of chromium in the alloy from 10 to 30% increases the corrosion rate by a factor of seven, while changes in nickel content have no effect. Thus, the intergranular attack is most likely selective with respect to chromium. The chromium removal begins at the grain boundary and continues with diffusion of the chromium from within the grain to the boundary layer, gradually enlarging the cavity in the metal. The gross corrosive attack is probably caused by the free chlorine, which is a highly oxidizing material, attacking the highly active structure-sensitive sites, such as dislocations and grain boundaries.

Selective attack of Iconel alloy 600 is observed in molten chloride melts. The resulting attacked region is a layer of porous spongelike material. The pores are not interconnected and are typically located at the grain boundaries.

Intergranular attack and rapid scale growth occur with zirconium alloys over 300 °C (570 °F). Platinum can be protected by the addition of oxide ions to the melt; this addition assists in the formation of a passive film.

Molten nitrates are commonly used for heat treatment baths; therefore, a great deal of material compatibility information exists. Plain carbon and low-alloy steels form protective iron oxide films that effectively protect the metal surface to approximately 500 °C (930 °F). Chromium additions increase the corrosion resistance of the steel, and hydroxide additions to the melt further increase the resistance of chromium-containing steels. Aluminum and aluminum alloys should never be used to contain nitrate melts, because of the danger of explosion.

Molten Sulfates. High-temperature alloys containing chromium perform well in sulfate salts because they form a protective scale. If the chromium content is not sufficient, the alloy will suffer severe external corrosion and internal sulfidation. Non-film-forming metals, such as copper and silver, corrode very rapidly.

Hydroxide Melts. Stainless steels perform poorly in hydroxide melts because of selective oxidation of the chromium, which leads to pore formation in the metal. Nickel is more resistant than stainless steels or unalloyed steels.

The peroxide content of the melt controls the corrosion rate. An increase in water vapor and a decrease in oxygen pressure reduce the peroxide content and subsequently reduce the corrosion rate. An exception to this rule occurs with silver, which shows an increase in corrosion rate with increased water vapor. Most glasses and silica are rapidly attacked by hydroxide melts, while alumina is more resistant.

Carbonate Melts. Austenitic stainless steels perform well in carbonate melts up to 500 °C (930 °F). If temperatures to 600 °C (1110 °F) are required, nickel-base alloys containing chromium are needed. For temperatures to 700 °C (1290 °F), high-chromium alloys containing at least 50% Cr are required. Above 700 °C (1290 °F), the passive films that form at lower temperatures will break down and preclude the use of metals. Aluminum coatings on steel structures perform well to 700 °C (1290 °F). For higher temperatures, alumina is required. Nickel does not provide adequate protection because of intergranular attack caused by the formation of nickel oxides.

Test Methods. A number of kinetic and thermodynamic studies have been carried out in capsule-type containers. These studies can determine the nature of the corroding species and the corrosion products under static isothermal conditions and do provide some much-needed information. However, to provide the information needed for an actual flowing system, corrosion studies must be conducted in thermal convection loops (Fig. 46) or forced convection loops, which will include the effects of thermal gradients, flow, chemistry changes, and surface area effects. These loops can also include electrochemical probes and gas monitors. Examples of the types of information gained from thermal convection loops during an intensive study of the corrosion of various alloys by molten salts are shown in Fig. 47 and 48.

Corrosion in Liquid Metals

Concern about corrosion of solids exposed to liquid-metal environments, that is, liquid-melt corrosion, dates from the earliest days of metals processing, when it became necessary to handle and contain molten metals. Corrosion considerations also arise when liquid metals are used in applications that exploit their chemical or physical properties. Liquid metals serve as high-temperature reducing agents in the production of metals (such as the use of molten magnesium to produce titanium) and because of their excellent heat transfer properties, they have been used or considered as coolants in a variety of powder-producing systems. Examples of such applications include molten sodium for liquid-metal fast breeder reactors and central receiver solar stations as well as liquid lithium for fusion and space nuclear reactors. In addition, tritium breeding in deuterium-tritium fusion reactors necessitates the exposure of lithium atoms to fusion neutrons. Breeding fluids of lithium or lead-lithium are attractive for this purpose. Molten lead or bismuth can serve as neutron multipliers to raise the tritium breeding yield if other types of lithium-containing breeding materials are used. Liquid metals can also be used as two-phase working fluids in Rankine cycle power conversion devices (molten cesium or potassium) and in heat pipes (potassium, lithium, sodium, sodium-potassium). Because of their high thermal conductivities, sodium-potassium alloys which can be any of a wide range of sodium-potassium combinations that are molten at or near room temperature, have also been used as static heat sinks in automotive and aircraft valves.

Whenever the handling of liquid metals is required, whether in specific uses as discussed above or as melts during processing, a compatible containment material must be selected. At low temperatures, liquid-metal corrosion is often insignificant, but in more demanding applications, corrosion considerations can be important in selecting the appropriate containment material and/or operating parameters.

Corrosion Reactions. The forms in which liquid-metal corrosion are manifested can be divided into the following categories.

- Dissolution for a surface by (a) direct dissolution; (b) surface reaction, involving solid-metal atom(s), the liquid metal, and an impurity element present in the liquid metal; or (c) intergranular attack
- Impurity and interstitial reactions
- Alloying
- Compound reduction

Definitions and descriptions of these types of reactions are given below. However, it is important to note that this classification is somewhat arbitrary, and as will become clear during the following discussion, the individual categories are not necessarily independent of one another. Additional information on liquid-metal corrosion can be found in Ref 63 to 68.

The simplest corrosion reaction that can occur in a liquid-metal environment is direct dissolution. Direct dissolution is the release of atoms of the containment material into the melt in the absence of any impurity effects. Such a reaction is a simple solution process and therefore is governed by the elemental solubilities in the liquid metal and the kinetics of the rate-controlling step of the dissolution reaction. The net rate, J, at which an elemental species enters solution, can be described as:

$$J = k(C - c) \qquad \text{(Eq 52)}$$

where k is the solution rate constant for the rate-controlling step, C is the solubility of the particular element in the liquid metal, and c is the actual instantaneous concentration of this element in the melt. Under isothermal conditions, the rate of this dissolution reaction would decrease with time as c increases. After a period of time the actual elemental concentration becomes equal to the solubility, and the dissolution rate is then 0. Therefore, in view of Eq 52, corrosion by the direct dissolution process can be minimized by selecting a containment material whose elements have low solubilities in the liquid metal of interest and/or by saturating the melt before actual exposure. However, if the dissolution kinetics are relatively slow, that is, for low values of k, corrosion may be acceptable for short-term exposures. The functional dependence and magnitude of the solution rate constant, k, depend on the rate-controlling step, which in the simplest cases can be a transport across the liquid-phase boundary layer, diffusion in the solid, or a reaction at the phase boundary. Measurements of weight changes as a function of time for a fixed $C - c$ (see discussion below) yield the kinetic information necessary for determination of the rate-controlling mechanism.

Corrosion resulting from dissolution in a non-isothermal liquid-metal system is more complicated than the isothermal case. Although Eq 52 can be used to describe the net rate at any particular temperature, the movement of liquid—for example, due to thermal gradients or forced circulation—tends to make c the same around the liquid-metal system. Therefore, at temperatures where the solubility (C) is greater than the bulk concentration (c), dissolution of an element into the liquid metal will occur, but at lower temperatures in the circuit where $C < c$, a particular element will tend to come out of solution and be deposited on the containment material (or it may remain as a suspended particulate). A schematic of such a mass transfer process is shown in Fig. 49.

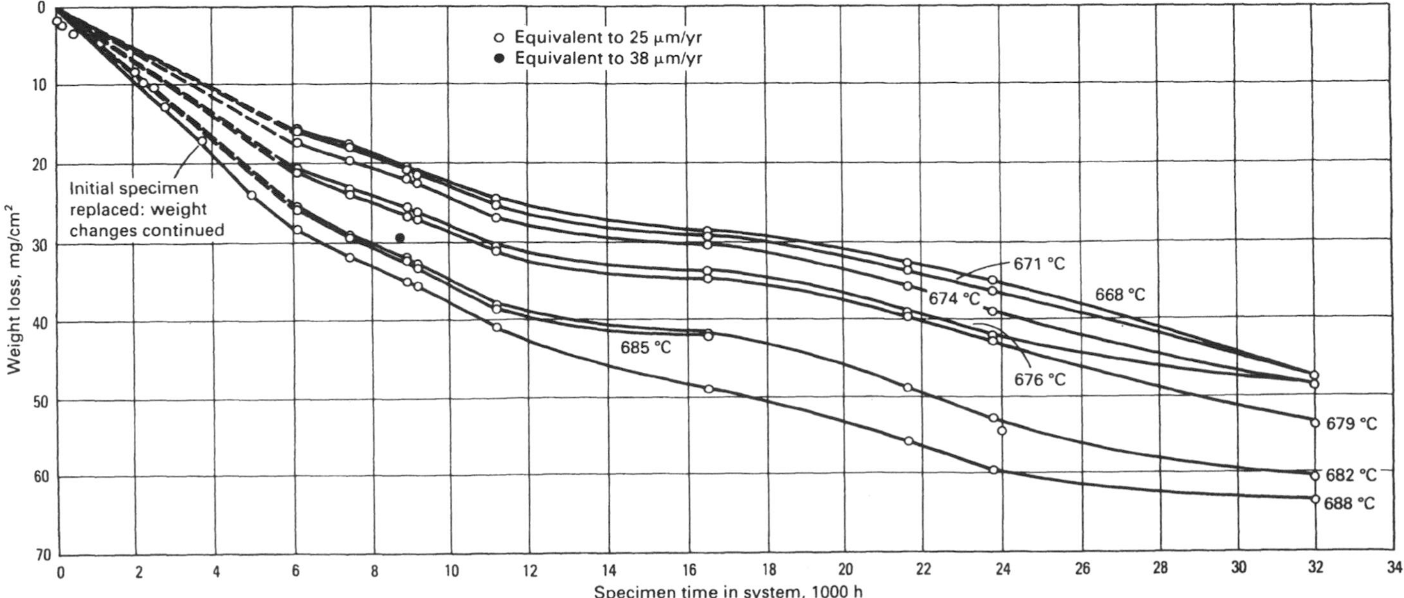

Fig. 48 Weight changes of type 304L stainless steel specimens exposed to $LiF-BeF_2-ZrF_4-ThF_4-UF_4$ (70, 23, 5, 1, and 1 mol%, respectively), tested for various times and temperatures in a thermal convection loop as shown in Fig. 46

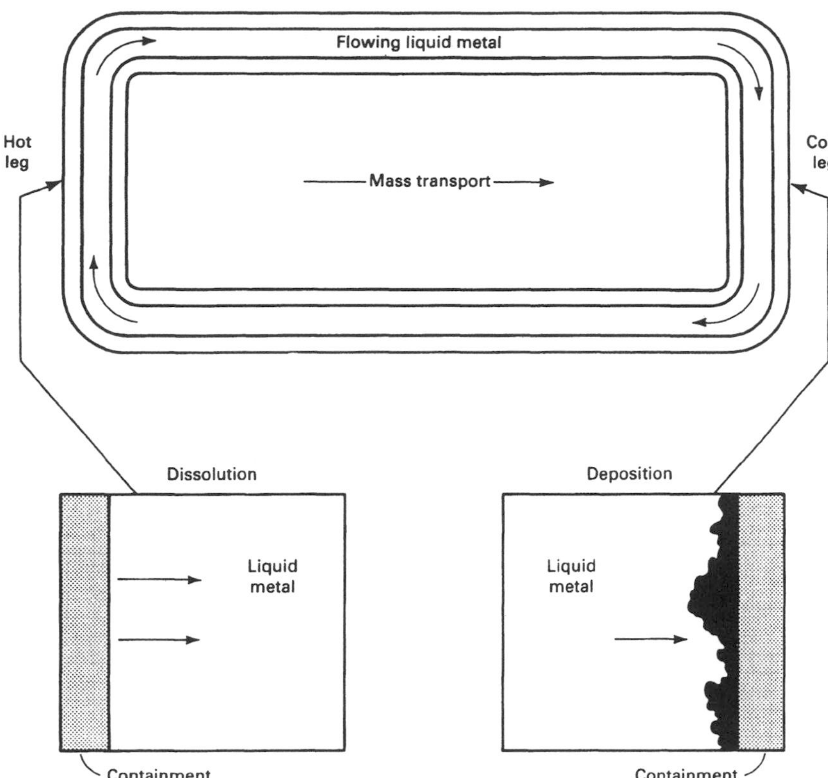

Fig. 49 Schematic of thermal gradient mass transfer in a liquid-metal circuit. Source: Ref 69

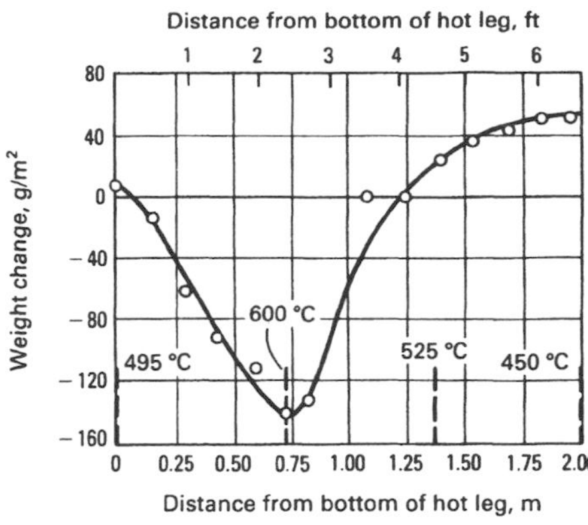

Fig. 50 Mass transfer as characterized by the weight changes of type 316 stainless steel coupons exposed around a nonisothermal liquid lithium type 316 stainless steel circuit for 9000 h. Source: Ref 70

If net dissolution or deposition is measured by weight changes, a mass transfer profile such as the one shown in Fig. 50 can be established. Such mass transfer processes under nonisothermal conditions can be of prime importance when, in the absence of dissimilar-metal effects (see below), forced circulation (pumping) of liquid metals used as heat transfer media exacerbates the transport of materials from hotter to cooler parts of the liquid-metal circuit. Normally, the concentration in the bulk liquid, c, rapidly becomes constant with time such that, at given temperature, the concentration driving force ($C - c$, Eq 52) is then also constant. However, much more elaborate analyses based on Eq 52 are required to describe nonisothermal mass transfer precisely. Such treatments must take into account the differences in k around the circuit as well as the possibility that the rate constant for dissolution (or deposition) may not vary monotonically with temperature because of changes in the rate-controlling step within the temperature range of dissolution (deposition). The presence of more than one elemental species in the containment material further complicates the analysis; the transfer of each element typically has to be handled with its own set of thermodynamic and kinetic parameters. Although a thermal gradient increases the amount of dissolution, plugging of coolant pipes by nonuniform deposition of dissolved species in cold zones often represents a more serious design problem than metal loss from dissolution (which sometimes may be handled by corrosion allowances). The most direct way to control deposition, however, is usually to minimize dissolution in the hot zone by use of more corrosion-resistant materials and/or inhibition techniques. An example of a mass transfer deposit is shown in Fig. 51.

Mass transfer may even occur under isothermal conditions if an activity gradient exists in the system. Under the appropriate conditions, dissolution and deposition will act to equilibrate the activities of the various elements in contact with the liquid metal. Normally, such a process is chiefly limited to interstitial element transfer between dissimilar metals, but transport of substitutional elements can also occur. Elimination (or avoidance) of concentration (activity) gradients across a liquid-metal system is the obvious and, most often, the simplest solution to any problems rising from this type of mass transport process.

Corrosion can also occur because of the interaction of light elements present in the containment material (interstitials) or the liquid metal (impurities). Examples of such reactions include the decarburization of steel in lithium and the oxidation of steel in sodium or lead of high oxygen activity. In many cases, when the principal elements of the containment material have low solubilities in liquid metals (e.g., refractory metals in sodium, lithium, and lead), reactions in-

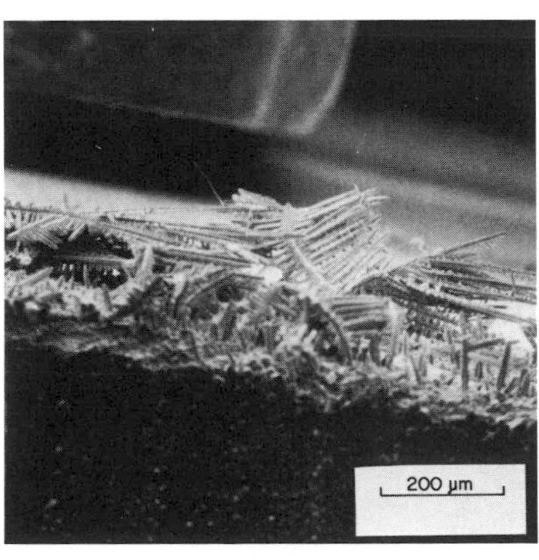

(a)

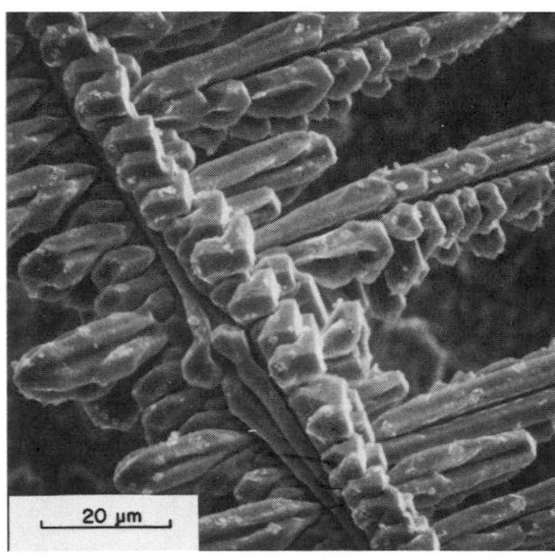

(b)

Fig. 51 Scanning electron micrographs of chromium mass transfer deposits found at the 460 °C (860 °F) position in the cold leg of a lithium/type 316 stainless steel thermal convection loop after 1700 h. (a) Cross section of specimen on which chromium was deposited. (b) Top view of surface

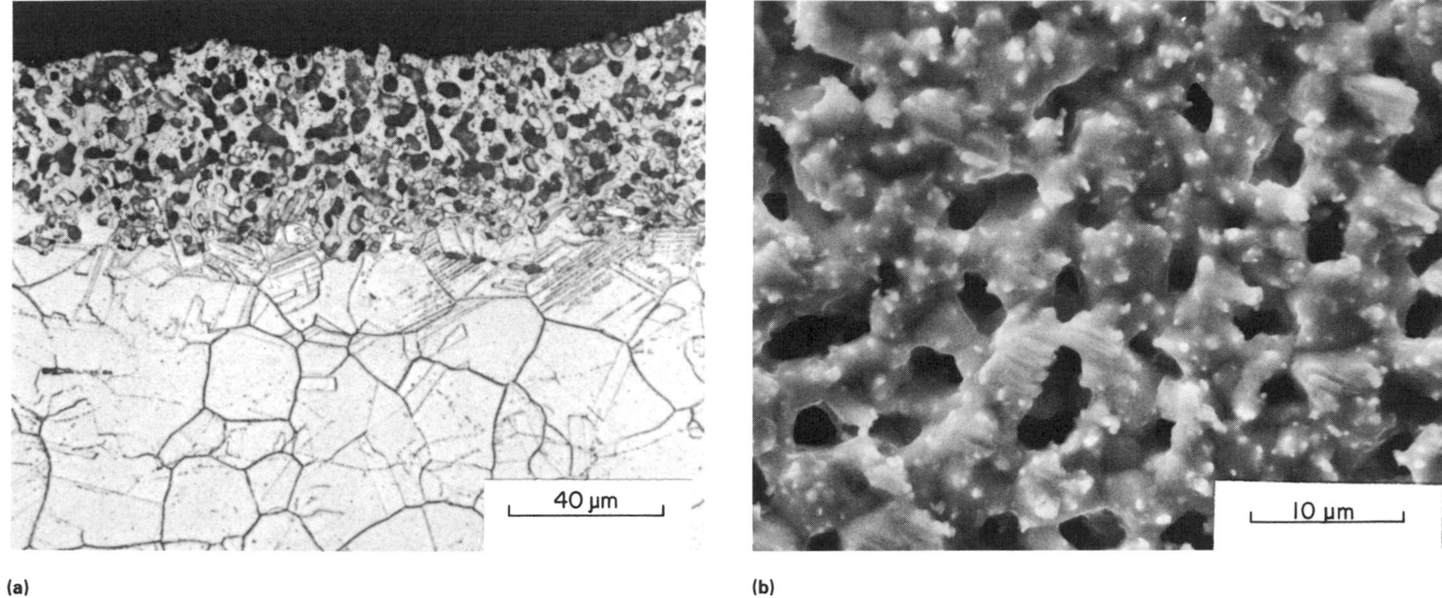

Fig. 52 Corrosion of type 316 stainless steel exposed to thermally convective lithium for 7488 h at the maximum loop temperature of 600 °C (1110 °F). (a) Light micrograph of polished and etched cross section, showing porous ferritic surface layer. (b) Scanning electron micrograph showing the top view of the porous surface

volving light elements such as oxygen, carbon, and nitrogen dominate the corrosion process.

Reactions between atoms of the liquid metal and those of the constituents of the containment material may lead to the formation of a stable product on the solid without the participation of impurity or interstitial elements:

$$xM + yL = M_x L_y \qquad \text{(Eq 53)}$$

Alloying is not a common form of liquid-metal corrosion, particularly with the molten alkali metals, but it can lead to detrimental consequences if it is not understood or expected. Alloying reactions, however, can be used to inhibit corrosion by adding an element to the liquid metal to form a corrosion-resistant layer by reaction of this species with the contaminant material. An example is the addition of aluminum to a lithium melt contained by steel. A more dissolution-resistant aluminide surface layer forms, and corrosion is reduced.

Attack of ceramics exposed to liquid metals can occur because of reduction of the solid by the melt (compound reduction). In very aggressive situations, such as when most oxides are exposed to molten lithium, the effective result of such exposure is the loss of structural integrity by reduction-induced removal of the nonmetallic element from the solid. The tendency for reaction under such conditions can be qualitatively evaluated by consideration of the free energy of formation of the solid oxide relative to the oxygen/oxide stability in the liquid metal.

Factors influencing corrosion of materials by liquid metals include:

- Composition, impurity content, and stress condition of the metal or alloy
- Exposure temperature and temperature range
- Impurity content of the liquid metal
- Circulating or static inventory
- Heating/cooling conditions
- Single or two-phase coolant
- Liquid-metal velocity
- Presence/control of corrosion inhibition elements
- Exposure time
- Monometallic or multialloy system components

These factors have a varied influence, depending on the combination of containment material and liquid metal or liquid-metal alloy. In most cases, the initial period of exposure (of the order of 100 to 1000 h, depending on temperature and liquid metal involved), is a time of rapid corrosion that eventually reaches a much slower steady-state condition as factors related to solubility and activity differences in the system approach a dynamic equilibrium. In some systems, this eventually leads to the development of a similar composition on all exposed corroding surfaces. High-nickel alloys and stainless steel exposed together in the high-temperature region of a sodium system will, for example, all move toward a composition that is more than 95% Fe.

Compatibility of a liquid metal and its containment varies widely. For a pure metal, surface attrition may proceed in an orderly, planar fashion, being controlled by either dissolution or a surface reaction. For a multicomponent alloy,

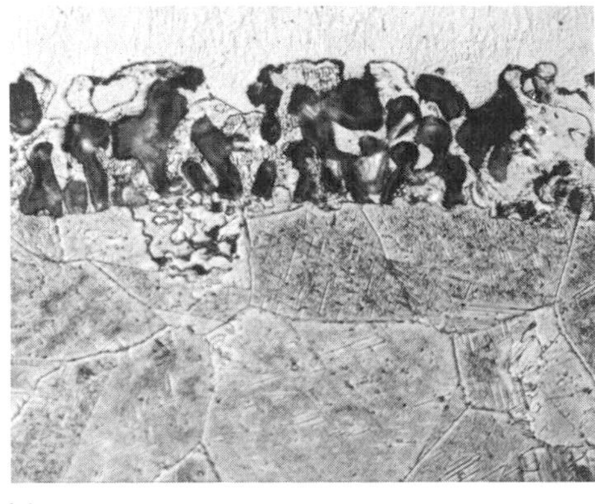

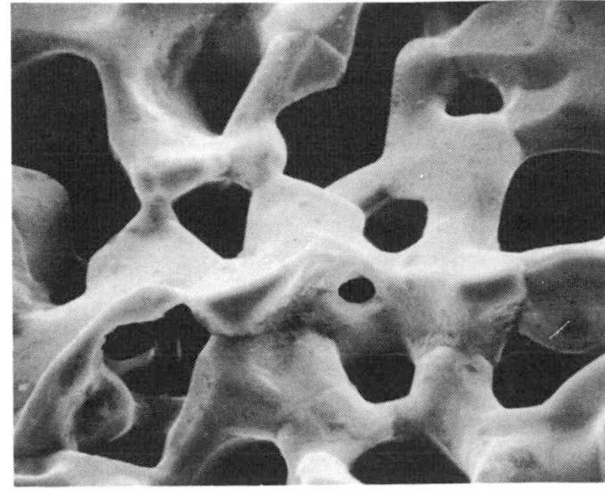

Fig. 53 Corrosion of Inconel alloy 706 exposed to liquid sodium for 8000 h at 700 °C (1290 °F); hot leg of circulating system. A porous surface layer has formed with a composition of ~95% Fe, 2% Cr, and <1% Ni. The majority of the weight loss encountered can be accounted for by this subsurface degradation. Total damage depth: 45 μm. (a) Light micrograph. (b) Scanning electron micrograph of the surface of the porous layer

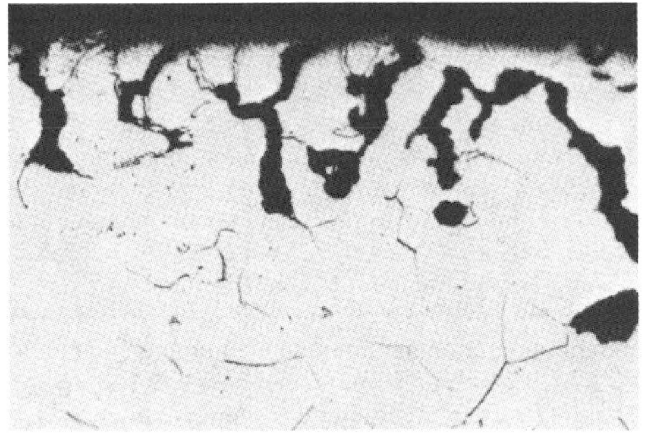

Fig. 54 Intergranular attack of unalloyed niobium exposed to lithium at 1000 °C (1830 °F) for 2 h. Light micrograph. Etched with 25% HF, 12.5 HNO_3, 12.5% H_2SO_4 in water

selective loss of certain elements may lead to a phase transformation. For example, loss of nickel from austenitic stainless steel exposed to sodium may result in the formation of a ferritic surface layer (Fig. 52). In high-nickel alloys, the planar nature of the corroding surface may be lost altogether, and a porous, spongelike layer may develop (Fig. 53). A more insidious situation can produce intergranular attack; liquid lithium, for example, will penetrate deep into refractory metals if precautions are not taken to ensure that the impurity element oxygen is in an oxide form more stable than Li_2O or LiO solutions, and is not left free in solid solution. Figure 54 illustrates intergranular attack in niobium.

As a general rule, assuming impurity and dissimilar metal effects are controlled and not overriding, the lower the nickel content of an alloy, the better its corrosion resistance in liquid metals (Ref 68). The materials shown in Table 9 have proven to be corrosion resistant to the specified liquid metals up to the temperature limit indicated. For additional information on materials compatibility, the reader is referred to Ref 66 and 67.

Testing and Evaluation. Historically, testing routines have generally progressed from an evaluation of specimens encapsulated in static liquid metals through natural convection loops, forced convection loops, loops simulating reactor circuits, and large engineering experiments. As the tools have become more sophisticated, researchers have sought to identify the corrosion effects of system parameters such as temperature, geometry, flow velocity, turbulence, heat transfer rates, and system impurities.

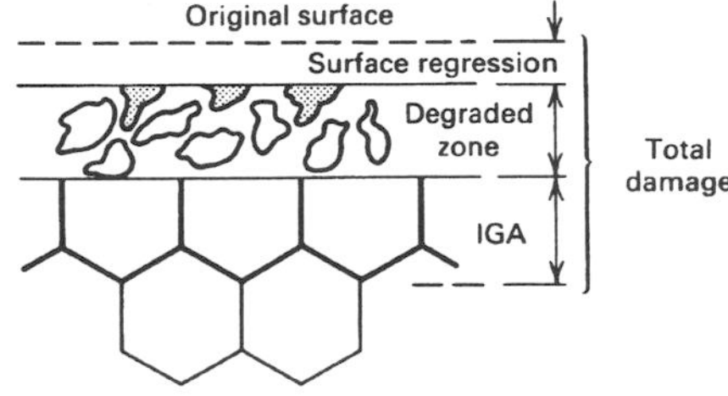

Fig. 55 Representative modes of surface damage in liquid-metal environments. IGA, intergranular attack. Source: Ref 71

Table 9 Range of applicability for various metals and alloys with liquid metals

Material	Maximum temperature of operation for nominal corrosion (°C)			
	Na	Li	K	Pb, Pb-Bi
Stainless steels	600	450	600	400
Cr-Mo steels	600	500	...	425
Ti and alloys	...	550	...	...
Nb-1Zr	900	1300	750	...

Source: Ref 68

In the case of the alkali metals, impurities such as oxygen, nitrogen, and carbon can have a significant effect on the corrosion of steels and refractory metals. For this reason, it is imperative that impurities initially present in the alkali metal and those entering during testing be carefully monitored and controlled. In effect, any corrosion testing program must be designed around the sampling procedures and analytical techniques used to measure impurities in the alkali metals. An excellent review of laboratory test procedures is given in Ref 68.

Three factors—surface attrition, depth of depleted zone (for an alloy), and the presence of intergranular attack—should be evaluated collectively in any liquid-metal system. This evaluation will lead to an assessment of total damage, which may be presented either as a rate or as a cumulative allowance that must be made for the exposure of a given material over a given time. In sodium studies, a total damage function was developed to express the corrosion process, as shown in Fig. 55.

ACKNOWLEDGMENTS

The information in this article is largely taken from:

- G.Y. Lai, *High-Temperature Corrosion of Engineering Alloys*, ASM International, 1990
- S.A. Bradford, Fundamentals of Corrosion in Gases, *Corrosion*, Vol 13, *ASM Handbook* (formerly Vol 13, 9th ed., *Metals Handbook*), ASM International, 1987, p 61-76
- P.F. Tortorelli, Fundamentals of High-Temperature Corrosion in Liquid Metals, *Corrosion*, Vol 13, *ASM Handbook* (formerly Vol 13, 9th ed., *Metals Handbook*), ASM International, 1987, p 56-60
- J.W. Kroger, Fundamentals of High-Temperature Corrosion in Molten Salts, *Corrosion*, Vol 13, *ASM Handbook* (formerly Vol 13, 9th ed., *Metals Handbook*), ASM International, 1987, p 50-55
- J.W. Kroger, Molten-Salt Corrosion, *Corrosion*, Vol 13, *ASM Handbook* (formerly Vol 13, 9th ed., *Metals Handbook*), ASM International, 1987, p 88-91
- C. Bagnall, Corrosion in Liquid Metals, *Corrosion*, Vol 13, *ASM Handbook* (formerly Vol 13, 9th ed., *Metals Handbook*), ASM International, 1987, p 91-96

REFERENCES

1. G.Y. Lai, *High-Temperature Corrosion of Engineering Alloys*, ASM International, 1990
2. P. Elliot, "Practical Guide To High-Temperature Alloys," NIDI Technical Series 10, 056, Nickel Development Institute
3. Properties of Pure Metals, *Properties and Selection: Nonferrous Alloys and Pure Metals,* Vol 2, 9th ed., *Metals Handbook,* American Society for Metals, 1979, p 714–831
4. R.C. Weast, E., Physical Constants of Inorganic Compounds, *Handbook of Chemistry and Physics,* 6th ed., The Chemical Rubber Company, 1984, p B68–B161
5. N. Birks and G.H. Meier, *Introduction to High Temperature Oxidation of Metals,* Edward Arnold, 1983
6. C.S. Giggins and F.S. Pettit, Corrosion of Metals and Alloys in Mixed Gas Environments at Elevated Temperatures, *Oxid. Met.,* Vol 14 (Nov. 5), 1980, p 363–413
7. T. Rosenqvist, Phase Equilibria in the Pyrometallurgy of Sulfide Ores, *Metall. Trans. B,* Vol 9, 1978, p 337–351
8. P. Kofstad, Oxidation Mechanisms for Pure Metals in Single Oxidant Gases, *High Temperature Corrosion,* R.A. Rapp, Ed., National Association of Corrosion Engineers, 1983, p 123–138
9. N.B. Pilling and R.E. Bedworth, The Oxidation of Metals at High Temperatures, *J. Inst. Met.,* Vol 29, 1923, p 529–582
10. O. Kubashewski and B.E. Hopkins, *Oxidation of Metals and Alloys,* 2nd ed., Butterworths, 1962
11. K.R. Lawless and A.T. Gwathmey, The Structure of Oxide Films on Different Faces of a Single Crystal of Copper, *Acta Mettall.,* Vol 4., 1956, p 153–163
12. N. Cabrera and N.F. Mott, Theory of Oxidation of Metals, *Rep. Prog. Phys.,* Vol 12, 1948-1949, p 163–184
13. K. Hauffe and B. Ilschner, Defective-Array States and Transport Processes in Ionic Crystals, *Z. Elektrochem,* Vol 58, 1954, p 467–477
14. T.B. Gimley and B.M.W. Trapnell, The Gas/Oxide Interface and the Oxidation of Metals, *Proc. R. Soc.* (*London*) A, Vol 234, 1956, p 405–418
15. H.H. Uhlig, Initial Oxidation Rate of Metals and the Logarithmic Equation, *Acta Metall.,* Vol 4, 1956, p 541–554
16. N. Birks, G.H. Meier, and F.S. Pettit, High Temperature Corrosion, *Superalloys, Supercomposites, and Superceramics,* Academic Press, 1989, p 439–489
17. C. Wagner, Contributions to the Theory of the Tarnishing Process, *Z. Phys. Chem.,* Vol B21, 1933, p 25–41

18. N. Birks and G.H. Meier, Chapter 2, *Introduction to High Temperature Oxidation of Metals,* Edward Arnold, London, 1983
19. R.A. Rapp, *Metall. Trans. A,* Vol 15, 1984, p 765
20. C. Wagner, *J. Electrochem. Soc.,* Vol 63, 1959, p 772
21. E.W. Haycock, Transitions from Parabolic to Linear Kinetics in Scaling of Metals, *J. Electrochem. Soc.,* Vol 106, 1959, p 771–775
22. P. Hancock and R.C. Hurst, The Mechanical Properties and Breakdown of Surface Films at Elevated Temperatures, *Advances in Corrosion Science and Technology,* Vol 4, R.W. Staehle and M.G. Fontana, Ed., Plenum Press, 1974, p 1–84
23. D.L. Douglass, Exfoliation and the Mechanical Behavior of Scales, *Oxidation of Metals and Alloys,* American Society for Metals, 1971, p 137–156
24. S. Mrowec and T. Werber, *Gas Corrosion of Metals*, National Center for Scientific, Technical and Economic Information, 1978
25. U.R. Evans, *The Corrosion and Oxidation of Metals,* Edward Arnold, 1960, p 836–837
26. C. Wagner, Types of Reaction in the Oxidation of Alloys, *Z. Elektrochem.,* Vol 63, 1959, p 772–782
27. S.A. Bradford, Formation and Composition of Internal Oxides in Dilute Iron Alloys, *Trans. AIME,* Vol 230, 1964, p 1400–1405
28. K. Hauffe, *Oxidation of Metals,* Plenum Press, 1965
29. B.D. Bastow, G.C. Wood, and D.P. Whittle, Morphologies of Uniform Adherent Scales on Binary Alloys, *Oxid. Met.,* Vol 16, 1981, p 1–28
30. F.N. Mazandarany and G.Y. Lai, *Nucl. Technol.,* Vol 43, 1979, p 349
31. T.M. Besmann, "SOLGASMIX-PV, A Computer Program to Calculate Equilibrium Relationships in Complex Chemical Systems," ORNL/TM-5755, Oak Ridge National Laboratory, April 1977
32. C. Bresseleers, R. Gavison, J. Harrison, G. Kemeny, J. Norton, H. Rother, M. Vande Voorde, and D. Whittle, Report EUR 6203EN, Joint Research Center, Petten Establishment, The Netherlands, 1978
33. K. Natesan, *Nucl. Technol.,* Vol. 28, 1976, p 441
34. K. Natesan and T.F. Kassner, *Metall. Trans.,* Vol 4, 1973, p 2557
35. P.L. Hemmings and R.A. Perkins, "Thermodynamic Phase Stability Diagrams for the Analysis of Corrosion Reactions in Coal Gasification/Combustion Atmospheres," Report FP-539, Electric Powder Research Institute, Palo Alto, CA, Dec 1977
36. R. Hultgren, P. Desai, D. Hawkins, M. Gleiser, and K.K. Kelley, *Selected Values of the Thermodynamic Properties of Elements and Binary Alloys,* American Society for Metals, 1973
37. S.R. Shatynski, *Oxid. Met.,* Vol 13 (No. 2), 1979, p 105
38. S.R. Shatynski, *Oxid. Met.,* Vol. 11 (No. 6), 1977, p 307
39. P.L. Hemmings and R.A. Perkins, "Thermodynamic Phase Stability Diagrams for the Analysis of Corrosion Reactions in Coal Gasification/Combustion Atmospheres," EPRI Report FP-539, Lockheed Palo Alto Research Laboratories, 1977
40. R.A. Perkins, in *Environmental Degradation of High Temperature Materials,* Series 3, No. 13, Vol 2, 1980, p 5/1
41. M. Hansen and K. Anderko, *Constitution of Binary Alloys,* McGraw- Hill, 1958
42. J.M.A. Van der Horst, in *Corrosion Problems in Energy Conversion and Generation,* C.S. Tedmon, Jr., Ed., The Electrochemical Society, 1974
43. M.B. Bever and C.F. Floe, in *Source Book on Nitriding,* American Society for Metals, 1977, p 125
44. B.J. Lightfoot and D.H. Jack, in *Source Book on Nitriding,* American Society for Metals, Metals Park, OH, 1977, p 248
45. K.N. Strafford, *Corros. Sci.,* Vol 19, 1979, p 49
46. P.L. Daniel and R.A. Rapp, *Advances in Corrosion Science and Technology*, Vol 5, M.G. Fontana and R.W. Staehle, Ed., Plenum Press, 1970
47. O. Kubaschewski and E. Evans, *Metallurgical Thermochemistry,* Pergamon Press, 1958
48. T. Rosenquist, *Principles of Extractive Metallurgy,* McGraw-Hill, 1974
49. G.D. Smith, P. Ganesan, C.S. Tassen, and C. Conder, High-Temperature Gases, *Corrosion Tests and Standards: Application and Interpretation,* R. Baboian, Ed., ASTM, 1995, p 359–371
50. L.R. Scharfstein and M. Henthorne, "Testing at High Temperatures, *Handbook on Corrosion Testing and Evaluation*, W.H. Ailor, Ed., John Wiley and Sons, Inc., 1971, p 291–366
51. N. Birks and G.H. Meier, *Introduction to High Temperature Oxidation of Metals,* Edward Arnold, London, 1983
52. H. Von E. Doering and P. Bergman, *Mat. Res. Stand.,* Vol 9 (No. 9), Sept 1969, p 35
53. A.V. Dean, "Investigation into the Resistance of Various Nickel and Cobalt Base Alloys to Sea Salt Corrosion at Elevated Temperatures," NGTE Report, 1964
54. J. Stringer, in *Proc. Symp. Properties of High Temperature Alloys with Emphasis on Environmental Effects,* Z.A. Foroulis and F.S. Pettit, Ed., The Electrochemical Society, 1976, p 513
55. J. Oh, M.J. McNallen, G.Y. Lai, and M.F. Rothman, *Metall. Trans. A.,* Vol 17, June 1986, p 108
56. H.H. Krause, Paper 401, presented at Corrosion/87, National Association of Corrosion Engineers, Houston
57. A.D. Russell, C.E. Smeltzer, and M.E. Ward, "Waste Heat Recuperation for Aluminum Furnaces," Final Report, Report GRI 8119160, Gas Research Institute, Chicago, Dec 1982
58. D. Inman and D.G. Lovering, *Comprehensive Treatise of Electrochemistry,* Vol 7, Plenum Publishing, 1983
59. G.J. Janz and R.P.T. Tompkins, *Corrosion,* Vol 35, 1979, p 485
60. C.B. Allen and G.T. Janz, *J. Hazard. Mater.,* Vol 4, 1980, p 145
61. R.J. Gale and D.G. Lovering, *Molten Salt Techniques*, Vol 2, Plenum Press, 1984, p 1
62. R. Littlewood, *J. Electrochem. Soc.,* Vol 109, 1962, p 525
63. W.E. Berry, *Corrosion in Nuclear Applications,* John Wiley & Sons, 1971
64. H.U. Borgstedt, Ed., *Material Behavior and Physical Chemistry in Liquid Metal Systems,* Plenum Press, 1982
65. J.E. Draley and J.R. Weeks, Ed., *Corrosion by Liquid Metals,* Plenum Press, 1970
66. *Liquid-Metals Handbook*, 2nd ed., R.N. Lyon, Ed., NAVEXOS P-733(Rev), Atomic Energy Commission, Dept. Of the Navy, Washington, DC, June 1952
67. *Liquid-Metals Handbook, Sodium-NaK Supplement,* C.B. Jackson, Ed., TID 5277, Atomic Energy Commission, Dept. of the Navy, Washington, DC, 1 July 1995
68. C. Bagnall, P.F. Tortorelli, J.H. DeVan, and S.L. Schrock, Liquid Metals, *Corrosion Tests and Standards: Application and Interpretation,* R. Baboian, Ed., ASTM, 1995, p 387–402
69. J.E. Selle and D.L. Olson, in *Materials Considerations in Liquid Metal Systems in Powder Generation*, National Association of Corrosion Engineers, 1978, p 15–22
70. P.F. Tortorelli and J.H. Devan, *J. Nucl. Mater.,* Vol 85 and 86, 1979, p 289–293
71. J.H. DeVan and C. Bagnall, in *Proc. International Conference on Liquid Metal Engineering and Technology,* Vol 3, The British Nuclear Energy Society, 1985, p 65–72

Industrial Applications of Heat-Resistant Materials

FOR HIGH-TEMPERATURE APPLICATIONS, proper alloy selection is important for safety and economic reasons. Since all high-temperature materials have certain limitations, the optimum choice is often a compromise between the mechanical property constraints (creep and stress-rupture strength), environmental constraints (resistance to various high-temperature degradation phenomena), fabricability characteristics, and cost.

This article presents material selection recommendations for a wide variety of heat-resistant applications, including furnace parts and fixtures, resistance heating elements, thermocouples, hot-working tools, high-temperature bearings, components used in petroleum refining units and petrochemical plants, power generation equipment, emission-control equipment, and aerospace components. Additional application areas for heat-resistant materials can be found in other articles throughout this Volume.

Heat-Treating Furnace Parts and Fixtures

The many parts used in industrial heat-treating furnaces can be divided into two categories. The first consists of parts that go through the furnaces and are therefore subjected to thermal and/or mechanical shock: trays, fixtures, conveyor chains and belts, and quenching fixtures. The second comprises parts that remain in the furnace with less thermal or mechanical shock: support beams, hearth plates, combustion tubes, radiant tubes, burners, thermowells, roller and skid rails, conveyor rolls, walking beams, rotary retorts, pit-type retorts, muffles, recuperators, fans, and drive and idler drums.

These heat-resistant alloys are supplied in either wrought or cast forms. In some situations, they may be a combination of the two. The properties and costs of the two forms vary, even though their chemical compositions are similar. There are many foundries and fabricators experienced in the design and application of these products, and it is important to seek their advice when purchasing high-alloy parts.

The great majority of heat-treating furnaces use iron-chromium-nickel or iron-nickel-chromium alloys because the straight iron-chromium alloys do not have sufficient high-temperature strength to be useful. Some iron-chromium alloys (more than 13% Cr) are susceptible to so-called 475 °C (885 °F) embrittlement. Because of increasing temperatures (e.g., >980 °C, or 1800 °F), more and more applications use nickel-base alloys for their improved creep-rupture strengths and oxidation resistance. Cobalt-base alloys are generally too expensive except for very special applications. Therefore, this discussion is limited to the use and properties of the iron-chromium-nickel, iron-nickel-chromium, and nickel-base alloys.

Room-temperature mechanical properties have limited value when selecting materials or designing for high-temperature use, but they may be useful in checking the quality of the alloys. The useful high-temperature properties of these alloys are summarized in Table 1 for castings and Table 2 for wrought products. The tables include nominal composition of the alloys and the stress required to produce 1% creep in 10,000 h and rupture in 10,000 h and 100,000 h, at temperatures of 650, 760, 870, and 980 °C (1200, 1400, 1600, and 1800 °F). A design stress figure commonly used for uniformly heated parts not subjected to thermal or mechanical shock is 50% of the stress to produce 1% creep in 10,000 h, but this should be used carefully and should be verified with the supplier.

In general, these materials contain iron, nickel, and chromium as the major alloying elements. Carbon, silicon, and manganese also are present and affect the foundry pouring and rolling characteristics of these alloys, as well as their properties at elevated temperature. Nickel influences primarily high-temperature strength and toughness. Chromium increases oxidation resistance by the formation of a protective scale of chromium oxide on the surface. An increase in carbon content increases strength.

Since the mid to late 1970s, a number of heat-resistant wrought alloys have been used in the heat-treating industry. Some of these alloys, such as Haynes alloys 230 (UNS N06230) and 556 (UNS R30556) and Inconel alloy 617 (UNS N06617), were originally developed for gas turbines, which require alloys with high creep-rupture strengths, good oxidation resistance, good fabricability, and good thermal stability. These alloys, commonly referred to as solid-solution-strengthened alloys, use molybdenum and/or tungsten for strengthening. The alloys are also strengthened by carbides. Another alloy with high creep strength, originally developed for gas turbine combustors, is Incoloy alloy MA 956, which is strengthened by oxide dispersion. This alloy is produced by a mechanical alloying process, using the high-energy milling of metal powders. These wrought heat-resistant alloys, along with chemical compositions and major characteristics, are tabulated in Table 3.

All of the alloys commonly used in castings for furnace parts have essentially an austenitic structure. The iron-chromium-nickel alloys (HF, HH, HI, HK, and HL) may contain some ferrite, depending on composition balance (see the article "High-Alloy Cast Steels" in this Volume). If exposed to a temperature in the range of 540 to 900 °C (1000 to 1650 °F), these compositions may convert to the embrittling σ phase. This can be avoided by using the proper proportions of nickel, chromium, carbon, and associated minor elements. Chromium and silicon promote ferrite, whereas nickel, carbon, and manganese favor austenite. Use of the iron-chromium-nickel types should be limited to applications in which temperatures are steady and are not within the σ-forming temperature range. Transformation from ferrite to σ phase at elevated temperature is accompanied by a change from ferromagnetic material and from a soft to a very hard, brittle material. All heat-resistant alloys of the iron-nickel-chromium group are wholly austenitic and are not as sensitive to composition balance as is the iron-chromium-nickel group. Also, the iron-nickel-chromium alloys contain large primary chromium carbides in the austenitic matrix and, after exposure to service temperature, show fine, precipitated carbides. The iron-nickel-chromium alloys are considerably stronger than the iron-chromium-nickel alloys and may be less expensive per part if the increased strength is considered when designing for a known load.

The life expectancy of trays and fixtures is best measured in cycles rather than hours, particularly if the parts are quenched. It may be cheaper to replace all trays after a certain number of cycles to avoid expensive shutdowns caused by wrecks in the furnace. Chains or belts that cycle from room temperature to operating temperature several times a shift will not last as long as stationary parts that do not fluctuate in temperature. Parts for carburizing furnaces will not last as long as those used for straight annealing.

Finally, alloy parts represent a sizable portion of the total cost of a heat-treating operation. Alloys should be selected carefully, designed properly, and operated with good controls throughout to keep costs at a minimum.

Material Comparison for Heat-Resistant Cast and Wrought Components

The selection of a cast or fabricated component for furnace parts and fixtures depends primarily on the operating conditions associated with heat-treating equipment in the specific processes, and secondarily on the stresses that may be involved. The factors of temperature, loading conditions, work volume, rate of heating, and furnace cooling or quenching need to be examined for the operating and economic tradeoffs. Other factors that enter into the selection include furnace and fixture design, type of furnace atmosphere, length of service life, and pattern availability or justification.

Some of the factors affecting the service life of alloy furnace parts, not necessarily in order of importance, are alloy selection, design, maintenance procedures, furnace and temperature control, atmosphere, contamination of atmosphere or work load, accidents, number of shifts operated, thermal cycle, and overloading. High-alloy parts may last from a few months to many years, depending on operating conditions. In the selection of a heat-resistant alloy for a given application, all properties should be considered in relation to the operating requirements to obtain the most economical life.

If either cast or wrought alloy fabrications can be used practically, both should be considered. Similar alloy compositions in cast or wrought form may have varying mechanical properties, different initial costs, and inherent advantages and disadvantages. Castings are more adaptable to complicated shapes, and fabrications to similar parts, but a careful comparison should be made to determine the overall costs of cast and fabricated parts. Initial costs, including pattern or tooling costs, maintenance expenses, and estimated life are among the factors to be included in such a comparison. Lighter-weight trays and fixtures will use less fuel in heating. Cast forms are stronger than wrought forms of similar chemical composition. They will deform less rapidly than wrought products, but they may crack more rapidly under conditions of fluctuating temperatures. Selection should be based on the practical advantages, with all facts considered.

General Considerations. Both cast and wrought alloys are well accepted by the designers and users of furnaces requiring high-temperature furnace load-carrying components. There are certain advantages for each type of manufactured component; often, the compositions are similar, if the carbon and silicon levels in the castings versus the wrought material are ignored. In general, the specifications of the wrought grades have carbon contents below 0.25%, and many are nominally near 0.05% C. In contrast, the cast alloys have from 0.25 to 0.50% C. This difference has an effect on hot strength. The difficulty in hot working the higher-carbon alloys accounts for their scarcity in the wrought series. Castings and fabricated parts are not always competitive; each product has advantages:

Advantages of cast alloys

- *Initial cost:* A casting is essentially a finished product as-cast; its cost per pound is frequently less than that of a fabricated item.
- *Strength:* Similar alloy compositions are inherently stronger at elevated temperatures than are wrought alloys.
- *Shape:* Some designs can be cast that may not be available in wrought form; also, even if wrought material is available, it may not be possible to fabricate it economically.
- *Composition:* Some alloy compositions are available only in castings; they may lack suffi-

Table 1 Composition and elevated-temperature properties of selected cast heat-resistant alloys used for furnace parts and fixtures

		Approximate composition, %			Temperature		Creep stress to produce 1% creep in 10,000 h		Stress to rupture in 10,000 h		Stress to rupture in 100,000 h	
Grade	UNS number	C	Cr	Ni	°C	°F	MPa	ksi	MPa	ksi	MPa	ksi
Iron-chromium-nickel alloys												
HF	J92603	0.20–0.40	19–23	9–12	650	1200	124	18.0	114	16.5	76	11.0
					760	1400	47	6.8	42	6.1	28	4.0
					870	1600	27	3.9	19	2.7	12	1.7
					980	1800	...	...	...	...	...	...
HH	J93503	0.20–0.50	24–28	11–14	650	1200	124	18.0	97	14.0	62	9.0
					760	1400	43	6.3	33	4.8	19	2.8
					870	1600	27	3.9	15	2.2	8	1.2
					980	1800	14	2.1	6	0.9	3	0.4
HK	J94224	0.20–0.60	24–28	18–22	650	1200	...	...	...	...	...	...
					760	1400	70	10.2	61	8.8	43	6.2
					870	1600	41	6.0	26	3.8	17	2.5
					980	1800	17	2.5	12	1.7	7	1.0
Iron-nickel-chromium alloys												
HN	J94213	0.20–0.50	19–23	23–27	650	1200	...	...	...	...	...	...
					760	1400	...	...	...	...	...	...
					870	1600	43	6.3	33	4.8	22	3.2
					980	1800	16	2.4	14	2.1	9	1.3
HT	J94605	0.35–0.75	15–19	33–37	650	1200	...	...	...	...	...	...
					760	1400	55	8.0	58	8.4	39	5.6
					870	1600	31	4.5	26	3.7	16	2.4
					980	1800	14	2.0	12	1.7	8	1.1
HU	...	0.35–0.75	17–21	37–41	650	1200	...	...	...	...	...	...
					760	1400	59	8.5	...	...	...	...
					870	1600	34	5.0	23	3.3	...	...
					980	1800	15	2.2	12	1.8	...	...
HX	...	0.35–0.75	15–19	64–68	650	1200	...	...	...	...	...	...
					760	1400	44	6.4	...	...	...	...
					870	1600	22	3.2	...	...	...	...
					980	1800	11	1.6	...	...	...	...

Note: Some stress values are extrapolated.

Table 2 Composition and elevated-temperature properties of selected wrought heat-resistant alloys used for furnace parts and fixtures

Grade	UNS number	Approximate composition, % C	Cr	Ni	Other	Temperature °C	°F	Creep stress to produce 1% creep in 10,000 h MPa	ksi	Stress to rupture in 10,000 h MPa	ksi
Iron-chromium-nickel alloys											
309S	S30908	0.08 max	22–24	12–15	...	650	1200	48	7.0	...	...
						760	1400	14	2.0	...	...
						870	1600	3	0.5	10	1.45
						980	1800	...	...	3	0.5
310S	S31008	0.08 max	24–26	19–22	...	650	1200	63	9.2	...	...
						760	1400	17	2.5	...	...
						870	1600	9	1.3	13.5	1.95
						980	1800	...	...	4	0.6
Iron-nickel-chromium alloys											
RA 330	N08330	0.08 max	17–20	34–37	...	760	1400	25	3.6	30	4.4
						870	1600	13	1.9	12	1.8
						980	1800	3.5	0.52	4.5	0.65
RA 330 HC	...	0.4 max	17–22	34–37	...	760	1400	47	6.8	54	7.8
						870	1600	18	2.6	18	2.6
						980	1800	5	0.7	5	0.7
RA 333	N06333	0.08 max	24–27	44–47	3 Mo, 3 Co, 3 W	760	1400	43	6.2	65	9.4
						870	1600	21	3.1	21	3.1
						980	1800	6	0.9	7	1.05
Incoloy 800	N08800	0.1 max	19–23	30–35	0.15–0.60 Al, 0.15–0.60 Ti	760	1400	19	2.8	23	3.3
						870	1600	4	0.61	12	1.7
						980	1800	1	0.23	6	0.8
Incoloy 802	N08802	0.2–0.5	19–23	30–35	...	760	1400	83	12.0	79	11.5
						870	1600	30	4.4	33	4.8
						980	1800	8	1.1	11.5	1.65
Nickel-base alloys											
Inconel 600	N06600	0.15 max	14–17	72 min	...	760	1400	28	4.1	41	6.0
						870	1600	14	2.0	16	2.3
						980	1800	4	0.56	8	1.15
Inconel 601	N06601	0.10 max	21–25	58–63	1.0–1.7 Al	760	1400	28	4.0	42	6.1
						870	1600	14	2.0	19	2.7
						980	1800	5.5	0.79	8	1.2

cient ductility to be worked into wrought material configurations.

Advantages of wrought alloys

- *Section size:* There is practically no limit to section sizes available in wrought form.
- *Thermal-fatigue resistance:* The ductility of the fine-grain microstructure of wrought alloys may promote better thermal fatigue resistance.
- *Soundness:* Wrought alloys are normally free of internal or external defects; they have smoother surfaces that may be beneficial for avoiding local hot spots.
- *Availability:* Wrought alloys are frequently available in many forms from stock.

Shape, complexity, and number of duplicate parts (eventually affecting cost) usually determine the choice between casting or wrought part. Where section thickness and configuration permit, castings are usually cheaper. The cost per pound of the casting metal is comparable to that of a fabricated part. The total projected cost of the fabrication is usually higher because the cost of forming, joining, and/or assembling must be added to the cost of the material. However, when only one or two types of parts are to be made, the pattern cost precludes the use of a casting.

In energy-intensive heat-treating industries, the use of wrought fabrications allows fuel savings through reduced heat-treating time cycles. At the present level of energy costs, wrought fabrications may be economically preferable because of improvements in thermal efficiency.

Fabrications are preferred for thin sections and for parts where less weight or greater heat transfer may be required. Where thick walls are necessary

Table 3 More recently developed wrought heat-resistant alloys used for furnace parts and fixtures

Alloy	UNS number	Composition, wt% Fe	Ni	Co	Cr	Mo	W	C	Other	Major characteristics
253 MA(a)	S30815	Bal	11	...	21	...	...	0.08	1.7 Si, 0.17 N, 0.04 Ce	Oxidation resistance
RA85H(b)	S30615	Bal	14.5	...	18.5	...	...	0.2	3.6 Si, 1.0 Al	Carburization resistance
Fecralloy A(c)	...	Bal	...	...	15.8	...	...	0.03	4.8 Al, 0.3 Y	Oxidation resistance
HR-120(d)	...	Bal	37	...	25	...	...	0.05	0.7 Nb, 0.2 N	Creep-rupture strength
556(d)	R30556	Bal	20	18	22	3	2.5	0.1	0.6 Ta, 0.2 N, 0.02 La	Creep-rupture strength
HR-160(d)	...	2	Bal	29	28	...	...	0.05	2.75 Si	Sulfidation resistance
214(d)	...	3	Bal	...	16	...	...	0.05	4.5 Al, Y (present)	Oxidation resistance
230(d)	N06230	...	Bal	...	22	2	14	0.1	0.005 B, 0.02 La	Creep-rupture strength/oxidation resistance
Inconel 617(e)	N06617	1.5	Bal	12.5	22	9	...	0.07	1.2 Al	Creep-rupture strength/oxidation resistance
Incoloy MA 956(e)	...	Bal	...	...	20	...	...	...	0.5 Y_2O_3, 4.5 Al, 0.5 Ti	Creep-rupture strength/oxidation resistance

(a) 253 MA is a registered trademark of Avesta Jernverks Aktiebolag. (b) RA85H is a registered trademark of Rolled Alloys, Inc. (c) Fecralloy A is a trademark of UK Atomic Energy. (d) HR-120, HR-160, 556, 214, and 230 are trademarks of Haynes International, Inc. (e) Inconel and Incoloy are registered trademarks of Inco family of companies.

for strength or where heavy loads are transported or pushed, the cost of fabricated sections may be prohibitive. Wrought materials have a greater degree of acceptance in fabricated baskets used under carburizing or carbonitriding conditions.

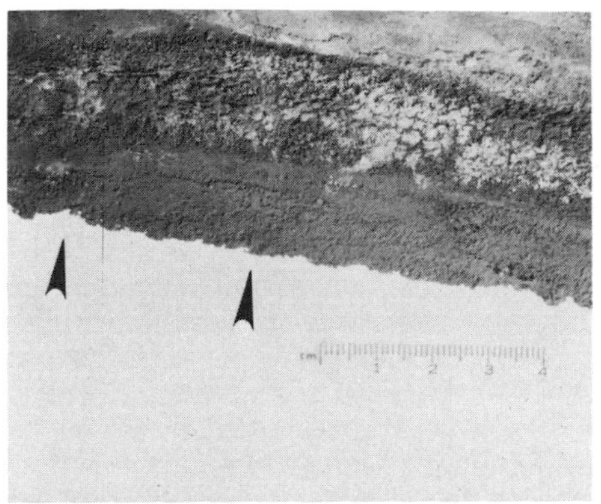

(a)

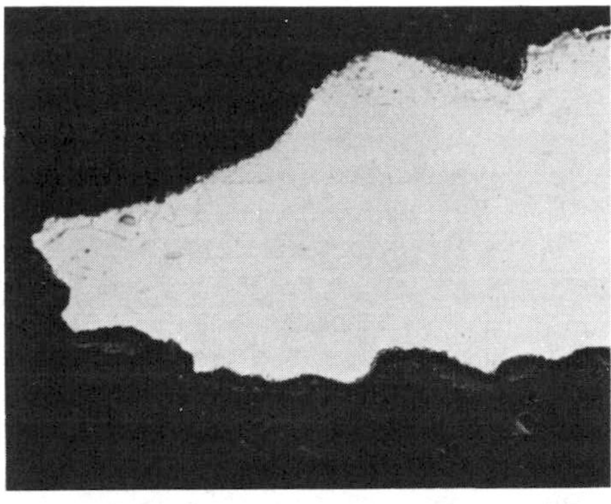

(b)

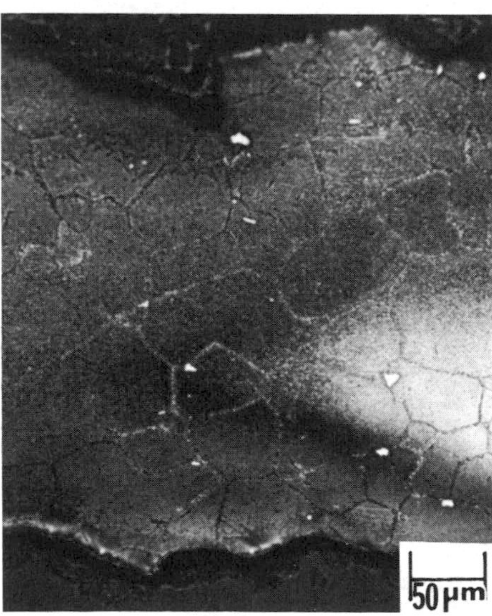

(c)

Fig. 1 Metal dusting of a Multimet alloy component at the refractory interface in a carburizing furnace. (a) Perforation of the component (arrows). (b) Cross section of the sample showing severe pitting. (c) Severe carburization beneath the pitted area

A factor that must be considered in evaluating castings and fabrications is the importance of good welding techniques, particularly for parts that are used in case hardening atmospheres. Castings have replaced fabricated products because of weld failures in multiwelded fabrications.

Although cast alloys exhibit greater high-temperature strength, it is possible to place too much emphasis on this characteristic in materials selection. Strength is rarely the only requirement and frequently is not the major one. More failures are due to brittle fracture from thermal fatigue than from stress rupture or creep. However, high-temperature strength is important where severe thermal cycling is required.

Specific Applications. Recommended alloy applications for parts and fixtures of various types of heat-treating furnaces, based on atmosphere and temperature, are summarized in Tables 4, 5, and 6. Where more than one alloy is recommended, each has proved adequate, although service life varies in different installations because of differences in exposure conditions.

Corrosion of Heat-Treating Furnace Accessories

The medium or environment used for heat treating varies from process to process. The high-temperature corrosion of furnace components depends heavily on the environment (or atmosphere) involved in the operation. Typical environments are air, combustion atmospheres, carburizing and nitriding atmospheres, molten salts, and protective atmospheres (such as endothermic atmospheres, nitrogen, argon, hydrogen, and vacuum). Protective atmospheres are used to prevent metallic parts from forming heavy oxide scales during heat treatment. The environment can often be contaminated by impurities, which can greatly accelerate corrosion. These contaminants (such as sulfur, vanadium, and sodium) generally come from fuels used for combustion, from fluxes used for specific operations, and from drawing compounds, lubricants, and other substances that are left on the parts to be heat treated.

The modes of high-temperature corrosion that are most frequently responsible for the degradation of furnace accessories are oxidation, carburization, sulfidation, molten-salt corrosion, and molten-metal corrosion. Each mode of corrosion, along with the corrosion behavior of important engineering alloys, is discussed in detail in this section.

Oxidation is probably the predominant mode of high-temperature corrosion encountered in the heat-treating industry. The oxidation discussed in this section involves air or combustion atmospheres with little or no contaminants, such as sulfur, chlorine, alkali metals, and salt.

Carbon steel and alloy steels generally have adequate oxidation resistance for reasonable service lives for furnace accessories at temperatures to 540 °C (1000 °F) (Ref 1). At intermediate temperatures of 540 to 870 °C (1000 to 1600 °F), heat-resistant stainless steels, such as AISI types 304, 316, 309, and 446, generally exhibit good oxidation resistance (Ref 1). Very few oxidation data have been reported in this temperature range. As the temperature increases above 870 °C (1600 °F), many stainless steels begin to suffer rapid oxidation. Better heat-resistant materials, such as the nickel-base high-performance alloys, are needed for furnace components in order to combat oxidation at these high temperatures.

Numerous oxidation tests on commercial alloys have been performed at 980 °C (1800 °F) or higher. For example, in one investigation, cyclic oxidation tests were conducted in air, with each cycle consisting of exposing the samples at 980 °C (1800 °F) for 15 min, followed by a 5-min air cooling (Ref 2). The performance ranking, in order of decreasing performance, was found to be as follows: Inconel alloy 600, Incoloy alloy 800, type 310 stainless steel, type 309 stainless steel, type 347 stainless steel, and type 304 stainless steel. Similar cyclic oxidation tests performed in air at 1150, 1205, and 1260 °C (2100, 2200, and 2300 °F), cycling to room temperature by air cooling after every 50 h at temperature, showed Inconel alloy 601 to be the best performer, followed by Inconel alloy 600 and Incoloy alloy 800 (Ref 3).

In another study, ferritic stainless steels such as E-Brite and type 446 were shown to be significantly better than type 310 and Incoloy alloy 800H in terms of cyclic oxidation resistance in air (Ref 4). These test results showed weight change data of 2.2 mg/cm^2 for E-Brite, 10.0 mg/cm^2 for type 446 stainless steel, –83.2 mg/cm^2 for alloy 800, and –90.3 mg/cm^2 for type 310 stainless steel after exposure of the samples for a total of 1000 h with 15 min at 980 °C (1800 °F) and 5 min at room temperature. A separate test was also conducted. This test involved exposure of the samples at 980 °C (1800 °F) for 1000 h in air with interruptions after 1, 20, 40, 60, 80, 100, 220, 364, and 512 h for cooling to room temperature. The weight change results of these four alloys were –12.9, 9.2, 1.7, and 3.0 mg/cm^2 for E-Brite, type 446 stainless steel, type 310 stainless steel, and alloy 800, respectively (Ref 4).

An oxidation database for a wide variety of commercial alloys, including stainless steels, iron-nickel-chromium alloys, nickel-chromium-iron alloys, and high-performance alloys, was recently generated (Ref 5). Tests were conducted in air at 980, 1095, 1150, and 1205 °C (1800, 2000, 2100, and 2200 °F) for 1008 h. The samples were cooled to room temperature once a week (each 168 h) for visual inspection. The results are summarized in Table 7.

Type 304 stainless steel and type 316 stainless steel both exhibited severe oxidation attack at 980 °C (1800 °F), while type 446 showed relatively mild attack. Many higher alloys, such as Incoloy alloy 800 and the nickel- and cobalt-base alloys, showed little attack. At 1095 °C (2000 °F), type 446 stainless steel suffered severe oxidation. Iron-nickel-chromium alloys, such as Incoloy alloy 800H and RA330, also suffered significant oxidation. Many nickel-base alloys, however, still exhibited little oxidation. At 1150

Table 4 Recommended materials for furnace parts and fixtures for hardening, annealing, normalizing, brazing, and stress relieving

Retorts, muffles, radiant tubes			Chain link		Sprockets, rolls, guides, trays	
Wrought	Cast	Mesh belts, wrought	Wrought	Cast	Wrought	Cast
595–675 °C (1100–1250 °F)						
430	HF	430	430	HF	430	HF
304		304	304		304	
675–760 °C (1250–1400 °F)						
304	HF	309	309	HF	304	HF
347	HH			HH	316	HH
309					309	
760–925 °C (1400–1700 °F)						
309	HH	309	314	HH	310	HH
310	HK	314	RA 330 HC	HL	RA 330	HK
253 MA	HT	253 MA	800H/800HT	HT	800H/800HT	HL
RA 330	HL	RA 330	HR-120		HR-120	HT
800H/800HT	HW					
HR-120						
600						
925–1010 °C (1700–1850 °F)						
RA 330	HK	314	314	HL	310	HL
800H/800HT	HL	RA 330	RA 330 HC	HT	RA 330	HT
HR-120	HW	600	802	HX	601	HX
600	HX	601	601		617	
601		214	617		X	
617			X		556	
X			556		230	
214			230			
556						
230						
1010–1095 °C (1850–2000 °F)						
601	HK	80–20	80–20	HL	601	HL
617	HL	600	617	HT	617	HX
X	HW	601	X	HX	X	
	HX	214	556		214	
556	NA22H		230		556	
230					230	
1095–1205 °C (2000–2200 °F)						
601	HL	601	601	HX	601	HL
617	HU	214	617		617	HX
230	HX		230		230	

Table 5 Recommended materials for parts and fixtures for carburizing and carbonitriding furnaces

	815–1010 °C (1500–1850 °F)	
Part	Wrought	Cast
Retorts, muffles, radiant tubes, structural parts	RA 330	HK
	800H/800HT	HT
	HR-120	HU
	600	HX
	601	
	617	
	X	
	214	
	556	
	230	
Pier caps, rails	RA 330	HT
	800H/800HT	
	HR-120	
	600	
	601	
Trays, baskets, fixtures	RA85H	HT
	RA 330	HT (Nb)
	800H/800HT	HU
	HR-120	HU (Nb)
	600	HX
	601	
	617	
	X	
	556	
	214	
	230	

°C (2100 °F), most alloys suffered unacceptable oxidation, with the exception of only a few nickel-base alloys. At 1205 °C (2220 °F), all alloys except Haynes alloy 214 suffered severe attack. Alloy 214 showed negligible oxidation at all the test temperatures. This alloy is different from all of the other alloys tested in that it forms an aluminum oxide (Al_2O_3) scale when heated to elevated temperatures. Other alloys tested form chromium oxide (Cr_2O_3) scales when heated to elevated temperatures.

The alloy performance rankings (Ref 6) generated from the field in the furnace atmosphere produced by the combustion of natural gas were found to correspond closely to the air oxidation data presented in Table 7. The alumina-forming alloy 214 was found to be the best performer (Ref 6).

Carburization. Materials problems due to carburization are quite common in heat-treating components associated with carburizing furnaces. The environment in the carburizing furnace typically has a carbon activity that is significantly higher than that in the alloy of the furnace component. Therefore, carbon is transferred from the environment to the alloy. This results in the carburization of the alloy, and the carburized alloy becomes embrittled.

Nickel-base alloys are generally considered to be more resistant to carburization than stainless steels. The results of 25 h carburization tests performed at 1095 °C (2000 °F) in a gas mixture consisting of 2% methane (CH_4) and 98% hydrogen revealed weight gain data of 2.78, 5.33, 18.35, and 18.91 mg/cm^2 for Inconel alloy 600, Incoloy alloy 800, type 310 stainless steel, and type 309 stainless steel, respectively (Ref 7). Extensive carburization tests were recently performed to investigate 22 commercial alloys, including stainless steels, iron-chromium-nickel alloys, nickel-chromium-iron alloys, and nickel- and cobalt-base alloys (Ref 8). Tests were performed for 215 h at 870 and 925 °C (1600 and 1700 °F) and for 55 h at 980 °C (1800 °F) in a gas mixture consisting of 5 vol% hydrogen, 5 vol% CH_4, 5 vol% carbon monoxide (CO), and the balance argon. The results failed to reveal any correlation between carburization resistance and the alloy base. Nevertheless, it was found that the alumina-forming Haynes alloy 214 was the most resistant to carburization among all of the alloys tested.

These findings were confirmed in 24 h tests performed at 1095 °C (2000 °F) in the same gas mixture. The carburization data are summarized in Table 8. In this study, alloy 214 (an alumina former) was found to be significantly better than the chromia formers tested. Among the chromia formers, however, there is some question regarding the significance of the differences within the carbon absorption range of 9.9 to 14.4 mg/cm^2. Perhaps less severe environments are required to separate the capabilities of these alloys. Field testing will be an excellent way of determining alloy performance ranking. However, few data are available. Field tests were recently conducted in a heat-treating furnace used for carburizing, carbonitriding, and neutral hardening operations (Ref 9). Both RA333 and Inconel alloy 601 were found to exhibit better carburization resistance than any of the alloys tested, which included RA330, Incoloy alloy 800, and alloy DS.

Metal dusting is another frequently encountered mode of corrosion that is associated with carburizing furnaces. Metal dusting tends to occur in a region where the carbonaceous gas atmosphere becomes stagnant. The alloy normally suffers rapid metal wastage. The corrosion products (or wastage) generally consist of carbon soots, metal, metal carbides, and metal oxides. The attack is normally initiated from the metal surface that is in contact with the furnace refractory. The furnace components that suffer metal dusting include thermowells, probes, and anchors. Figure 1 illustrates the metal dusting attack on Multimet alloy. The component was perforated as a result. Metal dusting problems have also been reported in petrochemical processing (Ref 10).

Table 6 Recommended materials for parts and fixtures for salt baths

Process and temperature range	Electrodes	Pots	Thermocouple protection tubes
Salt quenching at 205–400 °C (400–750 °F)	Low-carbon steel	Low-carbon steel	Low-carbon steel, 446
Tempering at 400–675 °C (750–1250 °F)	Low-carbon steel, 446, 35-18(a)	Aluminized low-carbon steel, 309	Aluminized low-carbon steel, 446
Neutral hardening at 675–870 °C (1250–1600 °F)	446, 35-18(a)	35-18(a), HT, HU, ceramic, 600, 556	446, 35-18(a)
Carburizing at 870–940 °C (1600–1720 °F)	446, 35-18(a)	Low-carbon steel(b), 35-18(a), HT	446, 35-18(a)
Tool steel hardening at 1010–1315 °C (1850–2400 °F)	Low-carbon steel(c), 446	Ceramic	446, 35-18(a), ceramic

Note: Where more than one material is recommended for a specific part and operating temperature, each has proved satisfactory in service. Multiple choices are listed in order of increasing alloy content (except ceramic parts). (a) A series of alloys generally of the 35Ni-15Cr type or modifications that contain from 30 to 40% Ni and 15 to 23% Cr and include RA 330, 35-19, Incoloy, and other proprietary alloys. (b) Immersed electrode furnaces only. (c) Low-carbon steel is recommended for completely submerged electrodes only.

Metal dusting has been encountered with straight chromium steels, austenitic stainless steels, and nickel- and cobalt-base alloys. All of these alloys are chromia formers; that is, they form Cr_2O_3 scales when heated to elevated temperatures. No work has been reported on the alloy systems that form a much more stable oxide scale, such as Al_2O_3. The Al_2O_3 scale was found to be much more resistant to carburization attack than the Cr_2O_3 scale (Ref 8). Because metal dusting is a form of carburization, it would appear that alumina formers, such as Haynes alloy 214, would also be more resistant to metal dusting.

Sulfidation. Furnace environments can sometimes be contaminated with sulfur. Sulfur can come from fuels, fluxes used for specific operations, and cutting oil left on the parts to be heat treated, among other sources. Sulfur in the furnace environment could greatly reduce the service lives of components through sulfidation attack.

The sulfidation of metals and alloys has been the subject of numerous investigations. However, the investigations involving commercial alloys examined only a limited number of alloys in each case. This frequently does not provide designers or engineers with a sufficient number of alloys to make an informed materials selection. A comprehensive sulfidation study was recently undertaken to determine the relative alloy rankings of base alloys (Ref 11). Tests were performed at 760, 870, and 980 °C (1400, 1600, and 1800 °F) for 215 h in a gas mixture consisting of 5% hydrogen, 5% CO, 1% carbon dioxide (CO_2), 0.15% hydrogen sulfide (H_2S), 0.1% H_2O, and the balance argon. The cobalt-base alloys were found to be the best performers, followed by iron-base alloys and then nickel-base alloys, which, as a group, were generally the worst performers. It is well known that nickel-base alloys are highly susceptible to catastrophic sulfidation due to the formation of nickel-rich sulfides, which melt at about 650 °C (1200 °F). Among iron-base alloys, the iron-nickel-cobalt-chromium alloy 556 was better than iron-nickel-chromium alloys such as Incoloy alloy 800H and type 310 stainless steel. The test results of representative alloys from each alloy base group are summarized in Table 9.

Molten-Salt Corrosion. Molten salts are widely used in the heat-treating industry for tempering, annealing, hardening, reheating, carburizing, and other operations. The salts that are commonly used include nitrates, carbonates, cyanides, chlorides, and caustics, depending on the operation. For example, a mixture of nitrates and nitrites is normally used for tempering and quenching. An alkali chloride-carbonate mixture is used for annealing ferrous and nonferrous metals. Neutral salt baths containing mixed chlorides are used for hardening steel parts.

Carbon steels, alloy steels, stainless steels, and iron-nickel-chromium alloys have been used for various furnace parts, such as electrodes, thermocouple protection tubes, and pots for salt baths. However, few corrosion data have been reported involving heat-treating salts.

Recent investigations have provided corrosion data in molten sodium-potassium nitrate ($NaNO_3 \cdot KNO_3$) salts (Ref 11, 12). The results of one study are given in Table 10. The nickel-chromium-iron-aluminum-yttrium alloy (alloy 214), nickel-chromium-iron alloys (Inconel alloys 600 and 601), and nickel-chromium-molybdenum alloys (Hastelloy alloys N and S) performed significantly better than stainless steels and iron-

Table 7 Results of 1008 h cyclic oxidation test in flowing air at temperatures indicated

Specimens were cycled to room temperature once a week.

	Oxidation rate at temperature															
	980 °C (1800 °F)				1095 °C (2000 °F)				1150 °C (2100 °F)				1205 °C (2200 °F)			
	Metal loss		Average metal affected(a)		Metal loss		Average metal affected		Metal loss		Average metal affected		Metal loss		Average metal affected	
Alloy	mm	mils	mm	mils	mm	mils	mm	mils	mm	mils	mm	mils	mm	mils	mm	mils
Haynes alloy 214	0.0025	0.1	0.005	0.2	0.0025	0.1	0.0025	0.1	0.005	0.2	0.0075	0.3	0.005	0.2	0.018	0.7
Haynes alloy 230	0.0075	0.3	0.018	0.7	0.013	0.5	0.033	1.3	0.058	2.3	0.086	3.4	0.11	4.5	0.2	7.9
Hastelloy alloy S	0.005	0.2	0.013	0.5	0.01	0.4	0.033	1.3	0.025	1.0	0.043	1.7	>0.81	>31.7(b)	>0.81	>31.7
Haynes alloy 188	0.005	0.2	0.015	0.6	0.01	0.4	0.033	1.3	0.18	7.2	0.2	8.0	>0.55	>21.7	>0.55	>21.7
Inconel alloy 600	0.0075	0.3	0.023	0.9	0.028	1.1	0.041	1.6	0.043	1.7	0.074	2.9	0.13	5.1	0.21	8.4
Inconel alloy 617	0.0075	0.3	0.033	1.3	0.015	0.6	0.046	1.8	0.028	1.1	0.086	3.4	0.27	10.6	0.32	12.5
AISI type 310	0.01	0.4	0.028	1.1	0.025	1.0	0.058	2.3	0.075	3.0	0.11	4.4	0.2	8.0	0.26	10.3
RA333	0.0075	0.3	0.025	1.0	0.025	1.0	0.058	2.3	0.05	2.0	0.1	4.0	0.18	7.1	0.45	17.7
Haynes alloy 556	0.01	0.4	0.028	1.1	0.025	1.0	0.067	2.6	0.24	9.3	0.29	11.6	>3.8	>150.0	>3.8	>150.0
Inconel alloy 601	0.013	0.5	0.033	1.3	0.03	1.2	0.067	2.6	0.061	2.4	0.135	5.3	0.11	4.4	0.19	7.5
Hastelloy alloy X	0.0075	0.3	0.023	0.9	0.038	1.5	0.069	2.7	0.11	4.5	0.147	5.8	>0.9	>35.4	>0.9	>35.4
Inconel alloy 625	0.0075	0.3	0.018	0.7	0.084	3.3	0.12	4.8	0.41	16.0	0.46	18.2	>1.21	>47.6	>1.21	>47.6
RA330	0.01	0.4	0.11	4.3	0.02	0.8	0.17	6.7	0.041	1.6	0.22	8.7	0.096	3.8	0.21	8.3
Incoloy alloy 800H	0.023	0.9	0.046	1.8	0.14	5.4	0.19	7.4	0.19	7.5	0.23	8.9	0.29	11.3	0.35	13.6
Haynes alloy 25	0.01	0.4	0.018	0.7	0.23	9.2	0.26	10.2	0.43	16.8	0.49	19.2	>0.96	>37.9	>0.96	>37.9
Multimet	0.01	0.4	0.033	1.3	0.226	8.9	0.29	11.6	>1.2	>47.2	>1.2	>47.2	>3.72	>146.4	>3.72	>146.4
AISI type 446	0.033	1.3	0.058	2.3	0.33	13.1	0.37	14.5	>0.55	>21.7	>0.55	>21.7	>0.59	>23.3	>0.59	>23.3
AISI type 304	0.14	5.5	0.21	8.1	>0.69	>27.1	>0.69	>27.1	>0.6	>23.6	>0.6	>23.6	>1.7	>68.0	>1.73	>68.0
AISI type 316	0.315	12.4	0.36	14.3	>1.7	>68.4	>1.7	>68.4	>2.7	>105.0	>2.7	>105.0	>3.57	>140.4	>3.57	>140.4

(a) Average metal affected = metal loss = metal loss + internal penetration. (b) All figures shown as greater than stated value represent extrapolation of tests in which samples were consumed in less than 1008 h. Source: Ref 5

Table 8 Results of 24 h carburization tests performed at 1095 °C (2000 °F) in Ar-5H_2-5CO-5CH_4

Alloy	Carbon absorption, mg/cm^2
Haynes alloy 214	3.4
Inconel alloy 600	9.9
Inconel alloy 625	9.9
Haynes alloy 230	10.3
Hastelloy alloy X	10.6
Hastelloy alloy S	10.6
AISI type 304	10.6
Inconel alloy 617	11.5
AISI type 316	12.0
RA333	12.4
Incoloy alloy 800H	12.6
RA330	12.7
Haynes alloy 25	14.4

Table 9 Results of 215 h sulfidation tests conducted in Ar-5H_2-5CO-1CO_2-0.15H_2S-0.1H_2O

Alloy	Average metal affected at temperature(a)					
	760 °C (1400 °F)		870 °C (1600 °F)		980 °C (1800 °F)	
	mm	mils	mm	mils	mm	mils
Alloy 6B	0.038	1.5	0.064	2.5	0.11	4.2
Haynes alloy 25	0.046	1.8	0.036	1.4	0.046	1.8
Haynes alloy 188	0.084	3.3	0.074	2.9	0.048	1.9
Haynes alloy 556	0.097	3.8	0.297	11.7	0.05	2.0
AISI type 310	0.23	9.1	0.34	13.5	0.19	7.4
Incoloy alloy 800H	0.28	11.2	0.49	19.2	0.59	23.2
Haynes alloy 214	0.42	16.7	>0.45	>17.7	>0.45	>17.7
Inconel alloy 600	0.55	>21.7	>0.55	>21.7	>0.55	>21.7
Hastelloy alloy X	0.749	>29.5	>0.55	>21.7	>0.55	>21.7
Inconel alloy 601	0.749	>29.5	>0.55	>21.7	>0.55	>21.7

(a) Average metal affected = metal loss + internal penetration. Source: Ref 10

nickel-chromium alloys such as Incoloy alloy 800H and RA330. The data generated from 1-month tests in a neutral salt bath containing a mixture of barium, potassium, and sodium chlorides ($BaCl_2$, KCl, and NaCl) at 845 °C (1550 °F) were recently reported (Ref 6). The results are summarized in Table 11. The cobalt-base Haynes alloy 188 was found to be the best performer; the nickel-chromium-iron Inconel alloy 600 was the worst. Somewhat similar results were obtained in laboratory tests conducted for 100 h at 845 °C (1550 °F) in NaCl. These data are given in Table 12. Haynes alloy 188 was found to be the best, and Inconel alloy 600 the worst.

In some cases, the salt vapors could cause high-temperature corrosion attack that is significantly worse than that caused by contact with the molten salt. For example, it was found that the corrosion of a nickel-base alloy salt pot containing molten $BaCl_2$-KCl-NaCl mixture at 1010 °C (1850 °F) was significantly different between the air side (outside of the pot) and the molten salt side (inside) (Ref 14). The outside of the pot (i.e., the air side contaminated with salt vapors) suffered three times as much attack as the inside of the pot, which was in contact with molten salt. No corrosion data were reported.

Molten-Metal Corrosion. Some heat-treating operations involve molten metals. Lead is used as a heat-treating medium. Cast iron and carbon steels have been used for components in contact with molten lead at temperatures to 480 °C (900 °F). In a 1242 h test in molten lead in an open crucible at 600 °C (1110 °F), Inconel alloy 600 was not appreciably attacked either at or below the liquid-metal surface (Ref 15). At 675 °C (1250 °F) for 1281 h in molten lead, Inconel alloy 600 suffered severe corrosion attack (Ref 15). Few corrosion data in molten lead have been reported in the literature.

The molten-zinc bath is used for galvanizing processes. Few corrosion data are available in the literature to allow engineers to make an informed materials selection for furnace components in contact with molten zinc. Carbon steel is generally used for the furnace components. Iron-nickel-cobalt-chromium alloys such as Haynes alloy 556 also have been reported for use as baskets.

Resistance Heating Elements

Resistance heating alloys are used in many varied applications, from small household appliances to large industrial process heating systems and furnaces. In appliances or industrial process heating, the heating elements are usually either open helical coils of resistance wire mounted with ceramic bushings in a suitable metal frame, or enclosed metal-sheathed elements consisting of a smaller-diameter helical coil of resistance wire electrically insulated from the metal sheath by compacted refractory insulation. In industrial furnaces, elements often must operate continuously at temperatures as high as 1300 °C (2350 °F) for furnaces used in metal-treating industries, 1700 °C (3100 °F) for kilns used for firing ceramics, and occasionally 2000 °C (3600 °F) or higher for special applications.

The primary requirements of materials used for heating elements are high melting point, high electrical resistivity, reproducible temperature coefficient of resistance, good oxidation resistance, absence of volatile components, and resistance to contamination. Other desirable properties are good elevated-temperature creep strength, high emissivity, low thermal expansion, and low modulus (both of which help minimize

Table 10 Corrosion rates in molten $NaNO_3$-KNO_3 at 675 and 705 °C (1250 and 1300 °F)

Alloy	Corrosion rate					
	675 °C (1250 °F), 14-day test		675 °C (1250 °F), 80-day test		705 °C (1300 °F), 30-day test	
	mm/yr	mils/yr	mm/yr	mils/yr	mm/yr	mils/yr
Haynes alloy 214	...	...	0.4	16	0.53	21
Inconel alloy 600	0.3	12	0.25	10	0.99	39
Hastelloy alloy N	0.33	13	0.23	9	1.22	48
Inconel alloy 601	0.48	19	0.48	19	1.24	49
Inconel alloy 617	0.36	14	...	...	...	...
Hastelloy alloy S	0.4	16	...	...	...	...
Inconel alloy 690	0.56	22	...	...	...	...
RA333	0.69	27	...	...	...	...
Inconel alloy 625	0.74	29	...	...	...	...
Hastelloy alloy X	1.04	41	...	...	...	...
Incoloy alloy 800	1.07	42	1.85	73	6.58	259
Haynes alloy 556	1.75	69	...	...	...	...
AISI type 310	2.0	79	...	...	...	...
AISI type 316	2.1	81	...	...	...	...
AISI type 317	2.11	83	...	...	...	...
AISI type 446	2.38	94	...	...	...	...
AISI type 304	2.67	105	...	...	...	...
RA330	2.77	109	...	...	...	...
253MA	2.97	117	...	...	...	...
Nickel 200	8.18	322	...	...	...	...

Source: Ref 11

Table 11 Results of 30-day field tests performed in a neutral salt bath containing $BaCl_2$, KCl, and NaCl at 845 °C (1550 °F)

Alloy	Average metal affected(a)	
	mm	mils
Haynes alloy 188	0.69	27
Multimet	0.75	30
Hastelloy alloy X	0.97	38
Hastelloy alloy S	0.1	40
Haynes alloy 556	0.11	44
Haynes alloy 214	1.8	71
AISI type 304	1.9	75
AISI type 310	2.0	79
Inconel alloy 600	2.4	96

(a) Average metal affected = metal loss + internal penetration. Source: Ref 6 thermal fatigue), good resistance to thermal shock, and good strength and ductility at fabrication temperatures.

Properties of Resistance Heating Alloys

Table 13 gives physical and mechanical properties of, and Table 14 presents recommended maximum operating temperatures for, resistance heating materials for furnace applications. Of the four groups of materials listed in these tables, the first group (Ni-Cr and Ni-Cr-Fe alloys) serves by far the greatest number of applications. Table 15 provides elevated-temperature tensile strengths for selected resistance heating materials.

The ductile wrought alloys in the first group have properties that enable them to be used at both low and high temperatures in a wide variety of environments. The Fe-Cr-Al compositions (second group) are also ductile alloys. They play an important role in heaters for the higher temperature ranges, which are constructed to provide more effective mechanical support for the element. The pure metals that comprise the third group have much higher melting points. All of them except platinum are readily oxidized and are restricted to use in nonoxidizing environments. They are valuable for a limited range of application, primarily for service above 1370 °C (2500 °F). The cost of platinum prohibits its use except in small, special furnaces.

The fourth group, nonmetallic heating element materials, are used at still higher temperatures. Silicon carbide can be used in oxidizing atmospheres at temperatures up to 1650 °C (3000 °F); three varieties of molybdenum disilicide are effective up to maximum temperatures of 1700, 1800, and 1900 °C (3100, 3270, and 3450 °F) in air. Molybdenum disilicide heating elements are gaining increased acceptance for use in industrial and laboratory furnaces. Among the desirable properties of molybdenum disilicide elements are excellent oxidation resistance, long life, constant electrical resistance, self-healing ability, and resistance to thermal shock. The nonmetallic heating elements described are considerably more fragile than metal heating alloys.

Effect of Atmosphere on Heating-Element Performance

Based on element temperature, Table 16 rates serviceabilities of various heating element materials as good, fair, or not recommended for the temperatures and atmospheres indicated. With the exception of molybdenum, tantalum, tungsten, and graphite, commonly used resistor materials have satisfactory life in air and in most other oxidizing atmospheres.

Oxidizing Atmospheres. Nickel-chromium and nickel-chromium-iron alloys are the most widely used heating materials in electric heat-treating furnaces. The 80Ni-20Cr alloys are more commonly used than the 60Ni-16Cr-20Fe or the 35Ni-20Cr-45Fe types. In fact, most electric-furnace manufacturers provide 80Ni-20Cr elements as standard, both because they permit a wider range of furnace temperatures and because it is usually more economical to stock only a limited number of heater materials. The 80Ni-20Cr alloys permit a wider range of operating temperatures because they have the greatest resistance to oxidation, and therefore can be used at higher temperatures than other Ni-Cr and Ni-Cr-Fe alloys. Heating elements of lower nickel content are required for certain special applications, such as where an oxidizing atmosphere contaminated with sulfur, lead, or zinc is present.

The iron-chromium-aluminum alloys are widely used in furnaces operating at 800 to 1300 °C (1500 to 2350 °F). In general, Ni-Cr heating elements are unsuitable above 1150 °C (2100 °F) because the oxidation rate in air is too great and the operating temperature is too close to the melting point of the alloy, although some Ni-Cr ele-

Table 12 Results of 100 h tests performed in NaCl at 845 °C (1550 °F)

Alloy	Average metal affected(a)	
	mm	mils
Haynes alloy 188	0.05	2.0
Haynes alloy 556	0.066	2.6
Haynes alloy 214	0.079	3.1
AISI type 304	0.081	3.2
AISI type 446	0.081	3.2
AISI type 316	0.081	3.2
Hastelloy alloy X	0.097	3.8
AISI type 310	0.107	4.2
Inconel alloy 800H	0.110	4.3
Inconel alloy 625	0.112	4.4
RA330	0.117	4.6
Inconel alloy 617	0.122	4.8
Haynes alloy 230	0.14	5.5
Hastelloy alloy S	0.168	6.6
RA333	0.19	7.5
Inconel alloy 600	0.196	7.7

(a) Average metal affected = metal loss + internal penetration. Source: Ref 13

Table 13 Typical properties of resistance heating materials

Basic composition	Resistivity(a), $\Omega \cdot mm^2/m$(b)	Average change in resistance(c), %, from 20 °C to:				Thermal expansion, μm/m · °C, from 20 °C to:			Tensile strength		Density	
		260 °C	540 °C	815 °C	1095 °C	100 °C	540 °C	815 °C	MPa	ksi	g/cm^3	$lb/in.^3$
Nickel-chromium and nickel-chromium-iron alloys												
78.5Ni-20Cr-1.5Si (80-20)	1.080	4.5	7.0	6.3	7.6	13.5	15.1	17.6	655–1380	95–200	8.41	0.30
77.5Ni-20Cr-1.5Si-1Nb	1.080	4.6	7.0	6.4	7.8	13.5	15.1	17.6	655–1380	95–200	8.41	0.30
68.5Ni-30Cr-1.5Si (70-30)	1.180	2.1	4.8	7.6	9.8	12.2	...	...	825–1380	120–200	8.12	0.29
68Ni-20Cr-8.5Fe-2Si	1.165	3.9	6.7	6.0	7.1	...	12.6	...	895–1240	130–180	8.33	0.30
60Ni-16Cr-22Fe-1.5Si	1.120	3.6	6.5	7.6	10.2	13.5	15.1	17.6	655–1205	95–175	8.25	0.30
37Ni-21Cr-40Fe-2Si	1.08	7.0	15.0	20.0	23.0	14.4	16.5	18.6	585–1135	85–165	7.96	0.288
35Ni-20Cr-43Fe-1.5Si	1.00	8.0	15.4	20.6	23.5	15.7	15.7	...	550–1205	80–175	7.95	0.287
35Ni-20Cr-42.5Fe-1.5Si-1Nb	1.00	8.0	15.4	20.6	23.5	15.7	15.7	...	550–1205	80–175	7.95	0.287
Iron-chromium-aluminum alloys												
83.5Fe-13Cr-3.25Al	1.120	7.0	15.5	...	...	10.6	...	...	620–1035	90–150	7.30	0.26
81Fe-14.5Cr-4.25Al	1.25	3.0	9.7	16.5	...	10.8	11.5	12.2	620–1170	90–170	7.28	0.26
73.5Fe-22Cr-4.5Al	1.35	0.3	2.9	4.3	4.9	10.8	12.6	13.1	620–1035	90–150	7.15	0.26
72.5Fe-22Cr-5.5Al	1.45	0.2	1.0	2.8	4.0	11.3	12.8	14.0	620–1035	90–150	7.10	0.26
Pure metals												
Molybdenum	0.052	110	238	366	508	4.8	5.8	...	690–2160	100–313	10.2	0.369
Platinum	0.105	85	175	257	305	9.0	9.7	10.1	345	50	21.5	0.775
Tantalum	0.125	82	169	243	317	6.5	6.6	...	345–1240	50–180	16.6	0.600
Tungsten	0.055	91	244	396	550	4.3	4.6	4.6	3380–6480	490–940	19.3	0.697
Nonmetallic heating-element materials												
Silicon carbide	0.995–1.995	−33	−33	−28	−13	4.7	...	...	28	4	3.2	0.114
Molybdenum disilicide	0.370	105	222	375	523	9.2	...	...	185	27	6.24	0.225
$MoSi_2$ + 10% ceramic additives	0.270	167	370	597	853	13.1	14.2	14.8	...	...	5.6	0.202
Graphite	9.100	−16	−18	−13	−8	1.3	...	...	1.8	0.26	1.6	0.057

(a) At 20 °C (68 °F). (b) To convert to Ω · circ mil/ft, multiply by 601.53. (c) Changes in resistance may vary somewhat, depending on cooling rate.

ments have been used at element temperatures up to 1200 °C (2200 °F). The Fe-Cr-Al elements historically have been recommended for operation in air. Recent use in many reducing atmospheres indicates that the performance of Fe-Cr-Al elements is comparable to Ni-Cr resistance elements in most environments. In addition, Fe-Cr-Al elements can generally be used at higher temperatures than Ni-Cr elements.

For temperatures above 1300 °C (2350 °F), silicon carbide or molybdenum disilicide elements are employed in industrial furnaces. Here again, maximum life of heating elements is obtained in air. These nonmetallic materials give fair service life in slightly reducing atmospheres at temperatures up to 1300 °C (2350 °F) for SiC and 1500 °C (2750 °F) for $MoSi_2$. They can be used in both oxidizing and slightly reducing atmospheres more commonly than Fe-Cr-Al elements, which are recommended only for service in air or inert atmospheres.

Platinum has been used in some small laboratory furnaces up to 1480 °C (2700 °F) in air. Because of the high cost of platinum, it is used only in special applications where silicon carbide cannot be worked into the furnace design. Platinum is restricted to service in air and cannot be used in reducing atmospheres. Although the initial cost of platinum is high, it has a high salvage value.

Elements made of 90% molybdenum disilicide and 10% refractory oxide mixtures perform well at continuous temperatures of 1700 and 1800 °C (3100 and 3270 °F) (depending on type) in air and in other oxidizing or inert atmospheres.

Carburizing Atmospheres. Unpurified exothermic (type 102) and purified exothermic (type 202) atmospheres are less harmful than endothermic (type 301) or charcoal (type 402) atmospheres, which are higher in carbon potential. The atmospheres that have high carbon potential have a tendency to carburize Ni-Cr alloys, especially at higher temperatures. Chromium is a strong carbide former and may pick up enough carbon to lower the melting point of the alloy, causing localized fusion in the heating element. For this reason, in reducing atmospheres of high-carbon potential, it is safer to limit the operating temperature of 80Ni-20Cr to about 1000 °C (1850 °F).

Unprotected heating elements made of Ni-Cr alloys are not recommended for use at more than 30 V in enriched endothermic carburizing or carbonitriding (type 309) atmospheres. Short life of heating elements in these types of atmospheres may be caused by carbon deposits on the element or refractory and by carburization of the element alloy. Carbon deposits may also short out extension wires and terminals, and cause them to melt. In recent years, a coated Ni-Cr heating element has been developed for use in carburizing atmospheres; in this element, the alloy is protected with a high-temperature ceramic coating that resists carburization. The element is designed to operate at low voltage (8 to 10 V) to prevent arcing at the terminals in carbon-impregnated brickwork.

Table 14 Recommended maximum furnace operating temperatures for resistance heating materials

Basic composition, %	Approximate melting point °C	Approximate melting point °F	Maximum furnace operating temperature in air °C	Maximum furnace operating temperature in air °F
Nickel-chromium and nickel-chromium-iron alloys				
78.5Ni-20Cr-1.5Si (80-20)	1400	2550	1150	2100
77.5Ni-20Cr-1.5Si-1Nb	1390	2540	...	...
68.5Ni-30Cr-1.5Si (70-30)	1380	2520	1200	2200
68Ni-20Cr-8.5Fe-2Si	1390	2540	1150	2100
60Ni-16Cr-22Fe-1.5Si	1350	2460	1000	1850
35Ni-30Cr-33.5Fe-1.5Si	1400	2550	...	...
35Ni-20Cr-43Fe-1.5Si	1380	2515	925	1700
35Ni-20Cr-42.5Fe-1.5Si-1Nb	1380	2515	...	...
Iron-chromium-aluminum alloys				
83.5Fe-13Cr-3.25Al	1510	2750	1050	1920
81Fe-14.5Cr-4.25Al	1510	2750	...	...
79.5Fe-15Cr-5.2Al	1510	2750	1260	2300
73.5Fe-22Cr-4.5Al	1510	2750	1280	2335
72.5Fe-22Cr-5.5Al	1510	2750	1375	2505
Pure metals				
Molybdenum	2610	4730	400(a)	750(a)
Platinum	1770	3216	1500	2750
Tantalum	3000	5400	500(a)	930(a)
Tungsten	3400	6150	300(a)	570(a)
Nonmetallic heating-element materials				
Silicon carbide	2410	4370	1600	2900
Molybdenum disilicide	(b)	(b)	1700–1800	3100–3270
$MoSi_2$ + 10% ceramic additives	(b)	(b)	1900	3450
Graphite	3650–3700(b)	6610–6690(c)	400 (d)	750(d)

(a) Recommended atmospheres for these metals are a vacuum of 10^{-4} to 10^{-5} mm Hg, pure hydrogen, and partly combusted city gas dried to a dew point of 4 °C (40 °F). In these atmospheres the recommended temperatures would be:

Element	Vacuum	Pure H_2	City gas
Molybdenum	1650 °C (3000 °F)	1760 °C (3200 °F)	1700 °C (3100 °F)
Tantalum	2480 °C (4500 °F)	Not recommended	Not recommended
Tungsten	1650 °C (3000 °F)	2480 °C (4500 °F)	1700 °C (3100 °F)

(b) See the property data on molybdenum disilicide at the end of this article. (c) Graphite volatilizes without melting at 3650 to 3700 °C (6610 to 6690 °F). (d) At approximately 400 °C (750 °F) (threshold oxidation temperature), graphite undergoes a weight loss of 1% in 24 h in air. Graphite elements can be operated at surface temperatures up to 2205 °C (4000 °F) in inert atmospheres.

As more carburizing furnaces have been converted from fossil fuels to electricity, it has been found that molybdenum disilicide can operate safely in both carburizing and reducing atmospheres at element temperatures up to 1500 °C (2700 °F). This makes it possible to eliminate the radiant tubes often used to protect metallic heating elements, thereby greatly increasing the efficiency of these furnaces by allowing faster recovery when a cold charge is placed in the furnace. Radiant tubes form a thermal barrier that slows heat transfer from the heating elements to the charge.

Reducing Atmospheres. In conventional heat-treating terminology, a reducing atmosphere is one that will reduce iron oxide on steel. Reduc-

Table 15 Elevated-temperature tensile strength of selected resistance heating materials

Heating material	Tensile strength at: 425 °C (800 °F) MPa	425 °C (800 °F) ksi	650 °C (1200 °F) MPa	650 °C (1200 °F) ksi	870 °C (1600 °F) MPa	870 °C (1600 °F) ksi	1100 °C (2000 °F) MPa	1100 °C (2000 °F) ksi
Nickel-chromium and nickel-chromium-iron alloys								
68.5Ni-30Cr-1.5Si	735	107	675	98	205	30	75	11
78.5Ni-20Cr-1.5Si	715	104	620	90	170	25	75	11
68Ni-20Cr-8.5Fe-1.5Si	760	110	655	95	195	28	75	11
Iron-chromium-aluminum alloys								
79.5Fe-15Cr-5.2Al	480	70	205	30	48	7	...	...
73.5Fe-22Cr-4.5Al	525	76	165	24	14	2	...	...
72.5Fe-22Cr-5.5Al	550	80	345	50	52	7.5	26	3.8
Pure metals								
Tungsten	560	81	525	76	395	57	295	43
Molybdenum	620	90	585	85	365	53	235	34
Tantalum	315	46	315	46	280	41	195	28

Table 16 Comparative life of heating-element materials in various furnace atmospheres

	Relative life and maximum operating temperature in							
Element material	Oxidizing (air)	Reducing: dry H_2 or type 501	Reducing: type 102 or 202	Reducing: type 301 or 402	Carburizing: type 307 or 309	Reducing or oxidizing, with sulfur	Reducing, with lead or zinc	Vacuum
Nickel-chromium and nickel-chromium-iron alloys								
80Ni-20Cr	Good to 1150 °C	Good to 1175 °C	Fair to 1150 °C	Fair to 1000 °C	NR(a)	NR	NR	Good to 1150 °C
60Ni-16Cr-22Fe	Good to 1000 °C	Good to 1000 °C	Good to fair to 1000 °C	Fair to poor to 925 °C	NR	NR	NR	...
35Ni-20Cr-43Fe	Good to 925 °C	Good to 925 °C	Good to fair to 925 °C	Fair to poor to 870 °C	NR	Fair to 925 °C	Fair to 925 °C	...
Iron-chromium-aluminum alloys								
Fe, 22Cr, 5.8Al, 1Co	Good to 1400 °C	Fair to poor to 1150 °C(b)	Good to 1150 °C(b)	Fair to 1050 °C(b)	NR	Fair	NR	Good to 1150 °C(b)
22Cr, 5.3Al, bal Fe	Good to 1400 °C	Fair to poor to 1050 °C(b)	Good to 1050 °C(b)	Fair to 950 °C(b)	NR	Fair	NR	Good to 1050 °C(b)
Pure metals								
Molybdenum	NR(c)	Good to 1650 °C	NR	NR	NR	NR	NR	Good to 1650 °C
Platinum	Good to 1400 °C	NR	NR	NR	NR	NR	NR	...
Tantalum	NR	NR	NR	NR	NR	NR	NR	Good to 2500 °C
Tungsten	NR	Good to 2500 °C(d)	NR	NR	NR	NR	NR	Good to 1650 °C
Nonmetallic heating element materials								
Silicon carbide	Good to 1600 °C	Fair to poor to 1200 °C	Fair to 1375 °C	Fair to 1375 °C	NR	Good to 1375 °C	Good to 1375 °C	NR
Graphite	NR	Fair to 2500 °C	NR	Fair to 2500 °C	Fair to poor to 2500 °C	Fair to 2500 °C in reducing	Fair to 2500 °C	...
Molybdenum disilicide	Good to 1850 °C	1350 °C	1600 °C	1400 °C	1350 °C	...	...	...

NR, not recommended. Note: Inert atmosphere of argon or helium can be used with all materials. Nitrogen is recommended only for the nickel-chromium group. Temperatures listed are element temperatures, not furnace temperatures. (a) Special 80Ni-20Cr elements with ceramic protective coatings designated for low voltage (8 to 16 V) can be used. (b) Must be oxidized first. (c) Special molybdenum heating elements with $MoSi_2$ coating can be used in oxidizing atmospheres. (d) Good with pure H_2 only

Table 17 Types and compositions of standard furnace atmospheres

See Table 16 for comparative life of heating elements in these atmospheres.

		Composition, vol%					Typical dew point	
Type	Description	N_2	CO	CO_2	H_2	CH_4	°C	°F
Reducing atmospheres								
102(a)	Exothermic unpurified	71.5	10.5	5.0	12.5	0.5	27	80
202	Exothermic purified	75.3	11.0	...	13.0	0.5	−40	−40
301	Endothermic	45.1	19.6	0.4	34.6	0.3	10	50
502	Charcoal	64.1	34.7	...	1.2	...	−29	−20
501	Dissociated ammonia	25	...	...	75	...	−51	−60
Carburizing atmospheres								
307	Endothermic + hydrocarbon	No standard composition					...	...
309	Endothermic + hydrocarbon + ammonia	No standard composition					...	...

(a) This atmosphere, refrigerated to obtain a dew point of 4 °C (40 °F), is widely used.

ing atmospheres are of several types, as shown in Table 17.

With the exception of dry hydrogen and dissociated ammonia (type 501), all atmospheres listed as reducing in Table 16 are oxidizing to Ni-Cr and Fe-Cr-Al alloys. Even hydrogen or dissociated ammonia will selectively oxidize chromium in a Ni-Cr alloy unless the gas is extremely dry. The type of oxide produced in "reducing" atmospheres is entirely different from that produced in air. The oxide produced in air is a green-to-black, impervious type that retards further oxidation of the underlying metal. It is usually a combination of Cr_2O_3 and $NiO{\cdot}Cr_2O_3$. The oxide produced on Ni-Cr elements in reducing atmospheres is green and porous and allows the atmosphere to internally oxidize the base metal. This type of attack, frequently referred to as green rot, takes place over a limited temperature range—870 to 1040 °C (1600 to 1900 °F)—in any atmosphere that is oxidizing to chromium and reducing to nickel, and occurs as particles or stringers of Cr_2O_3 surrounding metallic nickel.

Among the reducing atmospheres listed in the tables, type 501 has the smallest effect on Ni-Cr heating elements. At temperatures above 1100 °C (2000 °F), a Ni-Cr element will have better life in dry hydrogen than in air, because oxidation in air occurs more rapidly at elevated temperatures. Wet hydrogen, on the other hand, will cause preferential oxidation, and around 950 °C (1750 °F), green rot will occur.

Graphite heating elements have been used for laboratory applications at temperatures near 1370 °C (2500 °F) in atmospheres free from O_2, CO_2, and H_2O. Silicon carbide elements give fair life in some reducing atmospheres; however, the maximum operating temperature is lower than for operation in air.

The poorest life for silicon carbide elements is obtained in hydrogen or dissociated ammonia. All atmospheres, including air, must be relatively dry; wet atmospheres shorten the life of silicon carbide elements. Silicon carbide is not recommended for use in carburizing atmospheres because it absorbs carbon, thus reducing electrical resistance and overloading the power supply.

Molybdenum disilicide heating elements can be safely used in carbon monoxide environments at 1500 °C (2730 °F), in dry hydrogen at 1350 °C (2460 °F), and in moist hydrogen at 1460 °C (2660 °F). As a rule of thumb, any combination of temperature and atmosphere that does not attack silica glass is compatible with molybdenum disilicide.

Atmosphere Contamination. Sulfur, if present, will appear as hydrogen sulfide in reducing atmospheres and as sulfur dioxide in oxidizing atmospheres. Sulfur contamination usually comes from one or more of the following sources: high-sulfur fuel gas used to generate the protective atmosphere; residues of sulfur-base cutting oil on the metal being processed; high-sulfur refractories, clays, or cements sued for sealing carburizing boxes; and the metal being processed in the furnace. Sulfur is destructive to Ni-Cr and Ni-Cr-Fe heating elements. Pitting and blistering of the alloy occur in oxidizing atmospheres, and a Ni-S eutectic that melts at 645 °C (1190 °F) may form in any type of atmosphere. The higher the nickel content, the greater the attack. Therefore, if sulfur is present and cannot be eliminated, Fe-Cr-Al elements are preferred over those made of nickel-base alloys.

Lead and zinc contamination of a furnace atmosphere may come from the work being processed. This is a common occurrence in sintering furnaces for processing powder metallurgy parts. In the presence of a reducing atmosphere, lead will vaporize from leaded bronze bushings) and attack the heating elements, forming lead chromate. Metallic lead vapors are even more harmful than sulfur to Ni-Cr alloys, and they will cause severe damage to a heating element in a matter of hours if unfavorable conditions of concentration and temperature exist. Higher-nickel alloys are affected more than lower-nickel alloys. Elements made of 35Ni-20Cr-43.5Fe-1.5Si give satisfactory life for sintering lead-bearing bronze powders at 845 °C (1550 °F) in reducing atmospheres; 80Ni-20Cr elements give poor life in this application.

Zinc contamination results from zinc stearate used as a lubricant and binder when powder metallurgy compacts are pressed. The zinc stearate volatilizes when the compacts are heated and may carburize the heating element. (Brazing of nickel silvers, which contain at least 18% Zn, also results in a high concentration of zinc vapors in the furnace atmosphere.) Zinc vapors, which alloy with Ni-Cr heating elements and result in poor life, may be eliminated at the higher sintering temperatures by using a separate burn-off furnace at 650 °C (1200 °F), with heating elements protected by full muffle, by sheathing, or by a high-temperature ceramic protective coating. If these precautions are not feasible, silicon carbide elements (which are not affected by sulfur, lead, or zinc contamination) should be used at both low and high temperatures when contamination is anticipated.

The Ni-Cr, Ni-Cr-Fe, Fe-Cr-Al, and molybdenum disilicide heating elements should not be used in the presence of uncombined chlorine or other halogens.

Vacuum Service. For vacuum heating, 80Ni-20Cr elements have been used at temperatures up to 1150 °C (2100 °F). The 80Ni-20Cr alloys generally are not satisfactory much above 1150 °C (2100 °F), because the vapor pressure of chromium is high enough for chromium to vaporize from the elements, resulting in poor life, contamination of the material being processed, and loss of vacuum. Because of this, watt density must be kept low, especially at higher temperatures.

In vacuum heating, the estimated maximum operating temperature at which weight loss by evaporation from refractory-metal heating elements will not exceed 1% in 100 h is:

Metal	Temperature °C	Temperature °F
Tungsten	2550	4620
Tantalum	2400	4350
Molybdenum	1900	3470
Platinum	1600	2910

Molybdenum disilicide heating elements are not suitable for use in high vacuum.

Thermocouples

Thermocouples are the most important contact-type electrical temperature sensors used in the metals industry. It is estimated that well over 90% of the temperature-sensing devices used in this industry are thermocouples.

Table 18 Properties of standard thermocouples

Type	Thermoelements	Base composition	Melting point, °C	Resistivity, nΩ · m	Recommended service	Max temperature °C	Max temperature °F
J	JP	Fe	1450	100	Oxidizing or reducing	760	1400
	JN	44Ni-55Cu	1210	500			
K	KP	90Ni-9Cr	1350	700	Oxidizing	1260	2300
	KN	94Ni-Al, Mn, Fe, Si, Co	1400	320			
N	NP	84Ni-14Cr-1.4Si	1410	930	Oxidizing 1260	2300	
	NN	95Ni-4.4Si-0.15Mg	1400	370			
T	TP	OFHC Cu	1083	17	Oxidizing or reducing	370	700
	TN	44Ni-55Cu	1210	500			
E	EP	90Ni-9Cr	1350	700	Oxidizing	870	1600
	EN	44Ni-55Cu	1210	500			
R	RP	87Pt-13Rh	1860	196	Oxidizing or inert	1480	2700
	RN	Pt	1769	104			
S	SP	90Pt-10Rh	1850	189	Oxidizing or inert	1480	2700
	SN	Pt	1769	104			
B	BP	70Pt-30Rh	1927	190	Oxidizing, vacuum or inert	1700	3100
	BN	94Pt-6Rh	1826	175			

Thermocouples consist of two dissimilar wires joined at one end, forming a measuring, or hot, junction. The other end, which is connected to the copper wire of the measuring instrument circuitry, is called the reference, or cold, junction. The electrical signal output in millivolts is proportional to the difference in temperature between the measuring junction (hot) and the reference junction (cold). The principles on which thermocouples depend, as well as their design considerations, are described in the article "Thermocouple Materials" in *Properties and Selection: Nonferrous and Special-Purpose Materials*, Volume 2 of *ASM Handbook*.

Thermocouple Materials

Commercially available thermocouples are grouped according to material characteristics (base metal or noble metal) and standardization. At present, five base-metal thermocouples and three noble-metal thermocouples have been standardized and given letter designations by ANSI (American National Standards Institute), ASTM (American Society for Testing and Materials), and ISA (Instrument Society of America). Among the remaining thermocouples in use, some have not been assigned letter designations because of limited usage, and some are being considered for standardization.

Standard Thermocouples. The base compositions, melting points, and electrical resistivities of the individual thermoelements of the seven standard thermocouples are presented in Table 18. Maximum operating temperatures and limiting factors in environmental conditions are listed also.

The type J thermocouple is widely used, primarily because of its versatility and low cost. In this couple, the positive thermoelement is iron and the negative thermoelement is constantan, a 44Ni-55Cu alloy. Type J couples can be used in both oxidizing and reducing atmospheres at temperatures up to about 760 °C (1400 °F). They find extensive use in heat-treating applications in which they are exposed directly to the furnace atmosphere.

Type K thermocouples, like type J couples, are also widely used in industrial applications. The positive thermoelement is a 90Ni-9Cr alloy; the negative thermoelement is a 94% Ni alloy containing silicon, manganese, aluminum, iron, and cobalt as alloying constituents. Type K thermocouples can be used at temperatures up to 1250 °C (2280 °F) in oxidizing atmospheres.

Type K couples should not be used in elevated-temperature service in reducing atmospheres or in environments containing sulfur, hydrogen, or carbon monoxide. At elevated temperatures in oxidizing atmospheres, uniform oxidation takes place, and the oxide formed on the surface of the positive (90Ni-10Cr) thermoelement is a spinel, $NiO\text{-}Cr_2O_3$. However, in reducing atmospheres, preferential oxidation of chromium takes place, forming only Cr_2O_3. The presence of this greenish oxide (commonly known as "green rot") depletes the chromium content, causing a very large negative shift of the thermal electromotive force (emf) and rapid deterioration of the thermoelement.

The Nicrosil/Nisil thermocouple (type N in Table 18) was developed for oxidation resistance and emf stability superior to those of type K thermocouples at elevated temperatures. The positive thermoelement is Nicrosil (nominal composition 14 Cr, 1.4 Si, 0.1 Mg, bal Ni), and the negative thermoelement is Nisil (nominal composition 4.4 Si, 0.1 Mg, bal Ni). These couples have been shown to have longer life and better emf stability than type K thermocouples at elevated temperatures in air, both in the laboratory and in several industrial applications.

The positive thermoelement of the type E thermoelement is 90Ni-9Cr, the same as that of the type K thermocouple; the negative element is 44Ni-55Cu, the same as that of the type T couple described below. The recommended maximum operating temperature for type E thermocouples is 870 °C (1600 °F). Like type K thermocouples, type E couples should be used only in oxidizing atmospheres, because their use in reducing atmospheres results in preferential oxidation of chromium (green rot).

The type S thermocouple served as the interpolating instrument for defining the International Practical Temperature Scale of 1968 (amended in 1975) from the freezing point of antimony

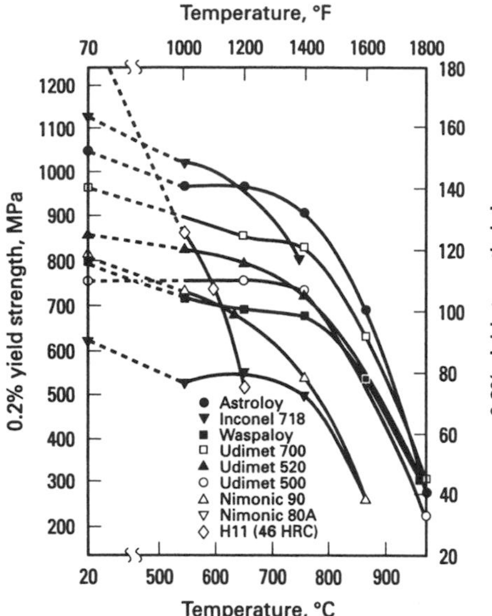

Fig. 2 Elevated-temperature yield strengths of some wrought nickel-base superalloys. A hot-work die steel (H11) is included for comparison.

(630.74 °C, or 1167.33 °F) to the freezing point of gold (1064.43 °C, or 1947.97 °F). It is characterized by a high degree of chemical inertness and stability at high temperatures in oxidizing atmospheres. The materials used in the legs of this thermocouple, Pt-10Rh and platinum, both are ductile and can be drawn into fine wire (as small as 0.025 mm, or 0.001 in., in diameter for special applications).

The type S couple is widely used in industrial laboratories as a standard for calibration of base-metal thermocouples and other temperature-sensing instruments. It is commonly used for controlling processing of steel, glass, and many refractory materials. It should be used in air or in oxidizing or inert atmospheres. It should not be used unprotected in reducing atmospheres in the presence of easily reduced oxides, atmospheres containing metallic vapors such as lead or zinc, or atmospheres containing nonmetallic vapors such as arsenic, phosphorus, or sulfur. It should not be inserted directly into metallic protection tubes and is not recommended for service in vacuum at high temperatures except for short periods of time. Because the negative leg of this couple is fabricated from high-purity platinum (approximately 99.99% for commercial couples and 99.995%+ for special grades), special care should be taken to protect the couple from contamination by the insulators used as well as by the operating environment.

The type R thermocouple (Pt-13Rh/Pt) has characteristics and end-use applications similar to those of the type S couple.

Type B thermocouples (Pt-30Rh/Pt-6Rh) may be used in still air or inert atmospheres for extended periods at temperatures up to 1700 °C (3100 °F) and intermittently up to 1760 °C (3200 °F) (Pt-6Rh leg melts at approximately 1826 °C, or 3319 °F). Because both of its legs are platinum-rhodium alloys, the type B couple is less sensitive than type R or type S to pickup of trace impurities from insulators or from the operating environment. Under corresponding conditions of temperature and environment, type B thermocouples exhibit less grain growth and less drift in calibration than type R or type S thermocouples.

The type B couple also is suitable for short-term use in vacuum at temperatures up to about 1700 °C (3100 °F). It should not be used in reducing atmospheres, or in those containing metallic or nonmetallic vapors, unless suitably protected with ceramic protection tubes. It should never be inserted directly into a metallic primary protection tube.

Nonstandard thermocouples include:

- *The 19 alloy/20 alloy thermocouple,* developed for use in hydrogen or in reducing atmospheres from 0 to 1260 °C (32 to 2300 °F). The positive thermoelement is the 20 alloy, which has a nominal composition of 82Ni-18Mo. The negative thermoelement is the 19 alloy, the nominal composition of which is 99Ni-1Co.
- *Iridium-rhodium thermocouples,* including 60Ir-40Rh/Ir, 50Ir-50Rh/Ir, and 40Ir-60Rh/Ir, which are suitable for use for limited periods of time in air or other oxygen-carrying atmospheres at temperatures up to about 2000 °C (3600 °F).
- *The Pt-5Mo/Pt-0.1Mo thermocouple,* used for measuring temperatures from 1100 to about 1500 °C (2000 to about 2700 °F) under neutron irradiation.
- *Platinel thermocouples,* an all-noble-metal combination. Actually, two combinations have been produced: Platinel I and Platinel II. Both have negative legs of 65Au-35Pd. The positive leg of the Platinel I couple is 83Pd-14Pt-3Au, and the positive leg of the Platinel II couple consists of 55Pd-31Pt-14Au. Platinel couples can be used unprotected (insulators only) in air to 1200 °C (2190 °F) for extended periods of time and to 1300 °C (2370 °F) for shorter periods.
- *Tungsten-rhenium thermocouples,* including W/W-26Re, doped W/W-25Re (dopants include potassium, silicon, and aluminum compounds), and W-5Re/W-26Re. All three thermocouples have been used at temperatures up to 2760 °C (5000 °F).

Table 19 Maximum service temperatures for thermocouple protection tubes

Materials	Maximum service temperature °C	°F
Carbon steel	540	1000
Wrought iron	700	1300
Cast iron	700	1300
304 stainless steel	870	1600
316 stainless steel	870	1600
Chrome iron (446)	980	1800
Nickel	980	1800
Inconel	1150	2100
Porcelain	1650(a)	3000(a)
Silicon carbide	1650	3000
Sillimanite	1650(a)	3000(a)
Aluminum oxide	1760(a)	3200(a)

(a) Horizontal tubes should receive additional support above 1480 °C (2700 °F). Source: Adapted from ANSI MC96.1

Insulation and Protection. To operate properly, thermocouple wires must be electrically insulated from on another at all points other than the measuring junction and must be protected from the operating environment. At temperatures above approximately 300 °C (570 °F), ceramic insulators are used on most bare thermocouple elements. Common examples include:

- *For base-metal thermocouples:* mullite, aluminum oxide, and steatite
- *For Pt-Rh thermocouples:* quartz, mullite, sillimanite, porcelain, and aluminum oxide (99.5% min Al_2O_3)
- *For Ir-Rh and W-Re thermocouples:* beryllia, thoria, and hafnia

Closed-end tubes made of metal, porcelain, mullite, sillimanite, quartz, or pyrex-glass may be used to prevent contamination of thermocouple sensing elements by the environment and to provide mechanical protection and support. Such tubes are called protection tubes. Table 19 lists maximum service temperatures for commonly used protection tube materials.

Hot-Working Tools

Materials for hot-working tools (e.g., forging anvils and dies, extrusion dies, hot shear blades) require high hot hardness, ability to withstand impact stresses and thermal shock, and adequate abrasion resistance. The most commonly used materials are the hot-work tool steels with about 0.4% C and various additions of tungsten, chromium, vanadium, molybdenum, and cobalt. These steels maintain a high resistance to deformation up to about 550 to 600 °C (1020 to 1110 °F). The important factors are the temperature of the tool, the time at temperature, and the strength of the work material. Even in cases where tool steels give satisfactory performance, advantage can often be gained by using superalloy tools whose longer life offsets the increased cost. Figure 2 compares the hot yield strength of H11 tool steel and certain nickel-base superalloys. The latter are employed as die materials for forging of titanium alloys and nickel-base superalloys. Die temperatures in the range of 760 to 1200 °C (1400 to 2200 °F) are encountered during forging for these applications. More detailed information on the selection of die materials for hot/isothermal forging can be found in the *ASM Specialty Handbook: Tool Materials.*

High-Temperature Service Bearings

Rolling-element bearings are fabricated from a wide variety of steels. In a broad sense, bearing steels can be divided into two classes: standard bearing steels are intended for normal service conditions (maximum temperatures are of the order of 120 to 150 °C, or 250 to 300 °F); whereas special-purpose bearing steels are used for either extended fatigue life or excessive operating conditions of temperature and corrosion.

When bearing service temperatures exceed about 150 °C (300 °F), common low-alloy steels cannot maintain the necessary surface hardness to provide satisfactory fatigue life. Table 20 lists the compositions of certain bearing steels suited for high-temperature service. These steels are typically alloyed with carbide-stabilizing elements such as chromium, molybdenum, vanadium, and silicon to improve their hot hardness and temper resistance. The listed maximum operating temperatures are those at which the hardness at temperature falls below a minimum of 58 HRC. Table 21 indicates the effect of extended exposure to elevated temperatures on the recovered (room-temperature) hardness of various steels, both carburized and through-hardened.

An important application of the high-temperature bearing steels is aircraft and stationary turbine engines. Bearings made from M50 steel have been used in engine applications for many years. Jet engine speeds are being continually increased in order to improve performance and efficiency; therefore, the bearing materials used in these engines must have increased section toughness to withstand the stresses that result from higher centrifugal forces. For this reason, the carburizing high-temperature bearing steels, such as M50-NiL and CBS-1000M, are receiving much attention. The core toughness of these steels is more than twice that of the through-hardening steels.

Components for Petroleum Refining and Petrochemical Operations

Most petroleum refining and petrochemical plant operations involve flammable hydrocarbon streams, highly toxic or explosive gases, and strong acids or caustics that are often at elevated temperatures and pressures. Among the many metals and alloys that are available, relatively few can be used for the construction of process equipment and piping (Ref 16). These include carbon steel, some cast irons, certain low-alloy steels and stainless steels, and, to a much lesser degree, aluminum, copper, nickel, titanium, and their alloys.

Materials Selection Criteria

The selection of materials of construction has a significant impact on the operability, economics, and reliability of refining units and petrochemical plants. For this reason, materials selection should be a cooperative effort between the materials engineer and plant operations and maintenance personnel. Reliability can often be equated to predictable materials performance under a wide range of exposure conditions. Ideally, a material should provide some type of warning before it fails; materials that fracture spontaneously and without bulging as a result of brittle fracture or stress-corrosion cracking (SCC) should be avoided. Uniform corrosion of equipment can be readily detected by various inspection techniques. In contrast, isolated pitting is potentially much more serious, because leakage can occur at highly localized areas that are difficult to detect. The effect of environment on the mechanical properties of a material can also be significant. Certain exposure conditions can convert a normally ductile material into a very brittle material that may fail without warning. A material must not only be suitable for normal process conditions but must also be able to handle transient conditions encountered during start-up, shutdown, emergencies, or extended standby. It is often during these time periods that equipment suffers serious deterioration or that failure occurs.

Of particular concern is what will happen to equipment during a fire. Unexpected exposure to elevated temperatures cannot only affect mechanical properties but can also produce detrimental side effects. Although all possible precautions should be taken to minimize the probability of a fire, the engineer responsible for materials selection must recognize that a fire may occur and that the equipment is expected to retain its integrity in order to avoid fueling the fire. This limits the application of materials with low melting points or those that may become subject to damage by thermal shock when fire-fighting water is applied, particularly in the case of refinery piping and equipment used to handle highly flammable hydrocarbon streams. On the other hand, fire resistance need not be considered for cooling-water or instrument-air systems. Although petrochemical plants may include some processes that involve nonflammable or nonhazardous streams, most equipment must be resistant to fires. Lack of fire resistance rules out the use of plastic components in refineries and petrochemical plants despite their excellent resistance to many types of corrosives. In addition, plastic components tend to be damaged by steam-out during a shutdown; this is required in order to free components of hydrocarbon residues and vapor before inspection or maintenance operations. The final step in the materials selection process is a reliability review of the materials and the corrosion control techniques that were selected. There must be total assurance that a plant will provide reliable service under all conditions, including those that occur during start-ups, shutdowns, downtime, standby, and emergencies.

Carbon and Low-Alloy Steels. Carbon steel is probably used for at least 80% of all components in refineries and petrochemical plants because it is inexpensive, readily available, and easily fabricated. Every effort is made to use

Table 20 Nominal compositions of high-temperature bearing steels

Steel	Composition, %								Maximum operating temperature(a)	
	C	Mn	Si	Cr	Ni	Mo	V	Other	°C	°F
M50	0.85	...	...	4.10	...	4.25	1.00	...	315	600
M50-NiL	0.13	0.25	0.20	4.20	3.40	4.25	1.20	...	315	600
Pyrowear 53	0.10	0.35	1.00	1.00	2.00	3.25	0.10	2.00 Cu	205	400
CBS-600	0.19	0.60	1.10	1.45	...	1.00	...	0.06 Al	230	450
Vasco X2-M	0.15	0.29	0.88	5.00	...	1.50	0.5	1.50 W	230	450
CBS-1000M	0.13	0.55	0.50	1.05	3.00	4.50	0.40	0.06 Al	315	600
BG42	1.15	0.50	0.30	14.5	...	4.00	1.20	...	370	700

(a) Maximum service temperature, based on a minimum hot hardness of 58 HRC

Table 21 Effect of elevated-temperature exposure on the recovered (room-temperature) hardness of various bearing steels

Steel type	Exposure time, 10^3 h	Hardness as heat treated, HRC	Minimum HRC after exposure for indicated time at °C (°F)						
			205 (400)	260 (500)	315 (600)	370 (700)	425 (800)	480 (900)	540 (1000)
Hardness of case layers (0.70–1.0% C)									
CBS-600	3	62	60	60	60	57	...	...	...
CBS-1000M	1	60	60	60	60	60	60	60	51
CBS-1000M(a)	1	60	59	59	59	58	58	...	...
9310	1	60	58	55	53	...	...	...	...
8620	1	60	58	56	53	47	...	...	...
52100	1	61	58	56	53	47	...	...	...
M50	1	62	62	62	62	62	62	60	52
Hardness of core regions(b)									
CBS-600	3	41	41	41	41	41	...	...	...
CBS-1000M	1	46	46	46	46	46	46	42	28
CBS-1000M(a)	1	44	44	44	44	44	44	40	27

(a) Oil quenched from 955 °C (1750 °F) rather than from the standard temperature of 1095 °C (2000 °F). (b) Core carbon 0.20% for CBS-600; 0.15% for CBS-1000M

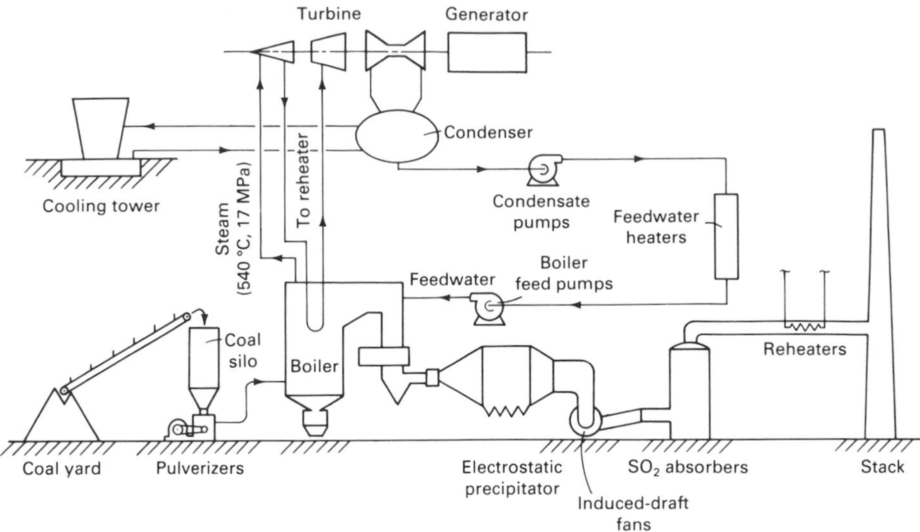

Fig. 3 Schematic of a coal-fired power plant

carbon steel, even if process changes are required to obtain satisfactory service from carbon steel (Ref 17). For example, process temperatures can be decreased, hydrocarbon streams dried up, or additives injected in order to reduce potential corrosion problems with carbon steel (Ref 18). In refineries, fractionation towers, separator drums, heat-exchanger shells, storage tanks, most piping, and all structures are generally fabricated from carbon steel. Carbon molybdenum steels, primarily the C-0.5 Mo grade, can offer substantial savings over carbon steels at temperatures between 425 and 540 °C (800 and 1000 °F). Because C-0.5 Mo steel has better resistance than carbon steel to high-temperature hydrogen attack, it has been extensively used for reactor vessels, heat-exchanger shells, separator drums, and piping for processes involving hydrogen at temperatures above 260 °C (500 °F). However, questions have been raised regarding the effect of long-term hydrogen exposure on C-0.5 Mo steel. As a result, low-alloy steels are preferred for new construction.

Low-alloy steels for refinery service are the chromium-molybdenum steels containing less than 10% Cr. These steels have excellent resistance to certain types of high-temperature sulfidic corrosion as well as to high-temperature hydrogen attack. To improve resistance to hydrogen stress cracking, low-alloy steels normally require postweld heat treatment. For refinery reactor vessels, which operate at high temperatures and pressures, 2.25Cr-1 Mo steel is widely used. For improved corrosion resistance, these are often overlayed with stainless steel. Other applications for low-alloy steels are furnace tubes, heat-exchanger shells, and piping and separator drums. Additional information is provided in the articles "Elevated-Temperature Mechanical Properties of Carbon and Alloy Steels" and "Elevated-Temperature Corrosion Properties of Carbon and Alloy Steels" in this Volume.

Stainless steels are extensively used in petrochemical plants because of the highly corrosive nature of the catalysts and solvents that are often used. In refineries, stainless steels have been primarily limited to applications involving high-temperature sulfidic corrosion and other forms of high-temperature attack (Ref 19). Most stainless steels will pit in the presence of chlorides (Ref 20).

Martensitic stainless steels, such as type 410 (S41000), must be postweld heat treated after welding to avoid hydrogen stress cracking problems as a result of exposure to hydrogen-sulfide-containing environments. Typical applications include pump components, fasteners, valve trim, turbine blades, and tray valves and other tray components in fractionation towers. Low-carbon varieties of type 410 stainless steel (S41008) are preferred for furnace tubes and piping, often in combination with aluminizing. Ferritic stainless steels, such as type 405 (S40500), are not subject to hydrogen stress cracking and are therefore a better choice than type 410 (S41000) stainless steel for vessel linings that are attached by welding (Ref 21). Austenitic stainless steels, such as type 304 (S30400) or type 316 (S31600), have excellent corrosion resistance, but are subject to SCC by chlorides. If sensitized, they are also subject to SCC by polythionic acids (Ref 22, 23). Typical applications include linings and tray components in fractionation towers: piping; heat-exchanger tubes; reactor cladding; tubes and tube hanger in furnaces; various components for compressors, turbines, pumps, and valves; and reboiler tubes. Additional information is available in the articles "Elevated-Temperature Mechanical Properties of Stainless Steels" and Elevated-Temperature Corrosion Properties of Stainless Steels" in this Volume.

Cast irons, because of their brittleness and low strength, are normally not used for pressure-retaining components for handling flammable hydrocarbons. The main exceptions are pump and

Table 22 Property requirements and materials of construction for coal-fired steam power plant components

Component	Major property requirements	Typical materials
Boiler		
Waterwall tubes	Tensile strength, corrosion resistance, weldability	C and C-Mo steels
Drum	Tensile strength, corrosion resistance, weldability, corrosion-fatigue strength	C, C-Mo, and C-Mn steels
Headers	Tensile strength, weldability, creep strength	C, C-Mo, C-Mn, and Cr-Mo steels
Superheater/reheater tubes	Weldability, creep strength, oxidation resistance, low coefficient of thermal expansion	Cr-Mo steels; austenitic stainless steels
Steam pipe	Same as above	Same as above
Turbine		
HP-IP rotors/disks	Creep strength, corrosion resistance, thermal-fatigue strength, toughness	Cr-Mo-V steels
LP rotors/disks	Toughness, stress-corrosion resistance, fatigue strength	Ni-Cr-Mo-V steels
HP-IP blading	Creep strength, fatigue strength, corrosion and oxidation resistance	12% Cr steels
LP blading	Fatigue strength, corrosion-fatigue pitting resistance	12% Cr steels, 17-4 PH stainless steel, Ti-6Al-4V
Inner casings, steam sheets, valves	Creep strength, thermal-fatigue strength, toughness, yield strength	Cr-Mo steels
Bolts	Proof stress, creep strength, stress-relaxation resistance, toughness, notch ductility	Cr-Mo-V and 12Cr-Mo-V steels
Generators		
Rotor	Yield strength, toughness, fatigue strength, magnetic permeability	Ni-Cr-Mo-V steels
Retaining rings	High yield strength, hydrogen- and stress-corrosion resistance, nonmagnetic	18Mn-5Cr and 18Mn-18Cr steels
Condensers		
Condensers	Corrosion and erosion resistance	Cupronickel, titanium, brass, stainless steels

HP, high pressure; IP, intermediate pressure; LP, low pressure valve components, ejectors, jets, strainers, and fittings in which the high hardness of cast iron reduces the velocity effects of corrosion, such as impingement, erosion, and cavitation. High-silicon cast irons (with 14% Si) are extremely corrosion resistant because of a passive surface layer of silicon oxide that forms during exposure to many chemical environments (except hydrofluoric acid). Typical refinery and petrochemical plant applications include valve and pump components for corrosive service. High-nickel cast irons (with 13 to 36% Ni and up to 6% Cr) have excellent corrosion, wear, and high-temperature resistance because of the relatively high alloy content (Ref 24). Typical uses are valve components, pump components, dampers, diffusers, tray components, and compressor parts. Additional information is provided in the article "High-Alloy Cast Irons" in this Volume.

Nickel alloys are especially resistant to sulfuric acid, hydrochloric acid, hydrofluoric acid, and caustic solutions, all of which can cause corrosion problems in certain refinery and petrochemical operations (Ref 25). As the nickel content is increased above 30%, austenitic alloys become, for all practical purposes, immune to chloride SCC. Nickel also forms the basis for many high-temperature alloys, but nickel alloys can be attacked and embrittled by sulfur-bearing gases at elevated temperatures. High-nickel alloys, including alloy 625 (N06625) and alloy 825 (N08825), are used to reduce the polythionic acid corrosion of flare-stack tips. Alloy B-2 (N10665) is particularly well suited to handling hydrochloric acid at all concentrations and temperatures (including the boiling point), but is attacked if oxidizing salts are present (Ref 26, 27). Alloy B-2 (N10665), alloy C-4 (N10002), and alloy C-276 (N10276) have excellent resistance to all concentrations of sulfuric acid up to at least 95 °C (200 °F). Although expensive, these alloys are used for specific applications to overcome unusually severe corrosion problems.

Components in Coal-Fired Steam Power Plants

Figure 3 shows an illustration of the various components of a coal-fired steam power plant. Water is first preheated to a relatively low temperature in feedwater heaters and pumped into tubes contained in a boiler. The water is heated to steam by the heat of combustion of pulverized coal in the boiler and then superheated. Superheated and pressurized steam is then allowed to expand in a high-pressure (HP) steam turbine and to cause rotation of the turbine shaft. The outlet steam from the HP turbine may once again be reheated and made to expand through an intermediate-pressure (IP) turbine and then through a low-pressure (LP) turbine. The turbine shafts are all connected to one or more generator shafts, which in turn rotate and convert the mechanical energy of rotation into electrical energy in the generator. The exit steam from the LP turbine is condensed in the condenser and is once again fed back to the boiler through the feedwater heaters and pumps. A closed loop of the water and steam is thus maintained. A second water loop through a cooling tower provides the cooling water needed to condense the steam exiting from the LP turbine. Combusted gases from the boiler are passed over more heat exchangers to preheat the incoming air to the boiler, are cleaned in scrubbers, and then are allowed to escape into the environment through the stacks. The pressure, temperature, and specific volume of the steam at various stages are illustrated in Fig. 4. It is thus clear that a variety of materials of construction are needed to withstand a wide range of these conditions in the plant, depending on the local conditions of pressure, temperature, and chemical environment. The capacity, reliability, efficiency, availability, and safety of plants depend critically on the integrity of the components and materials employed. A number of damage phenomena, such as embrittlement, creep, thermal fatigue, hot corrosion, oxidation, and erosion, can impair plant integrity at elevated temperatures.

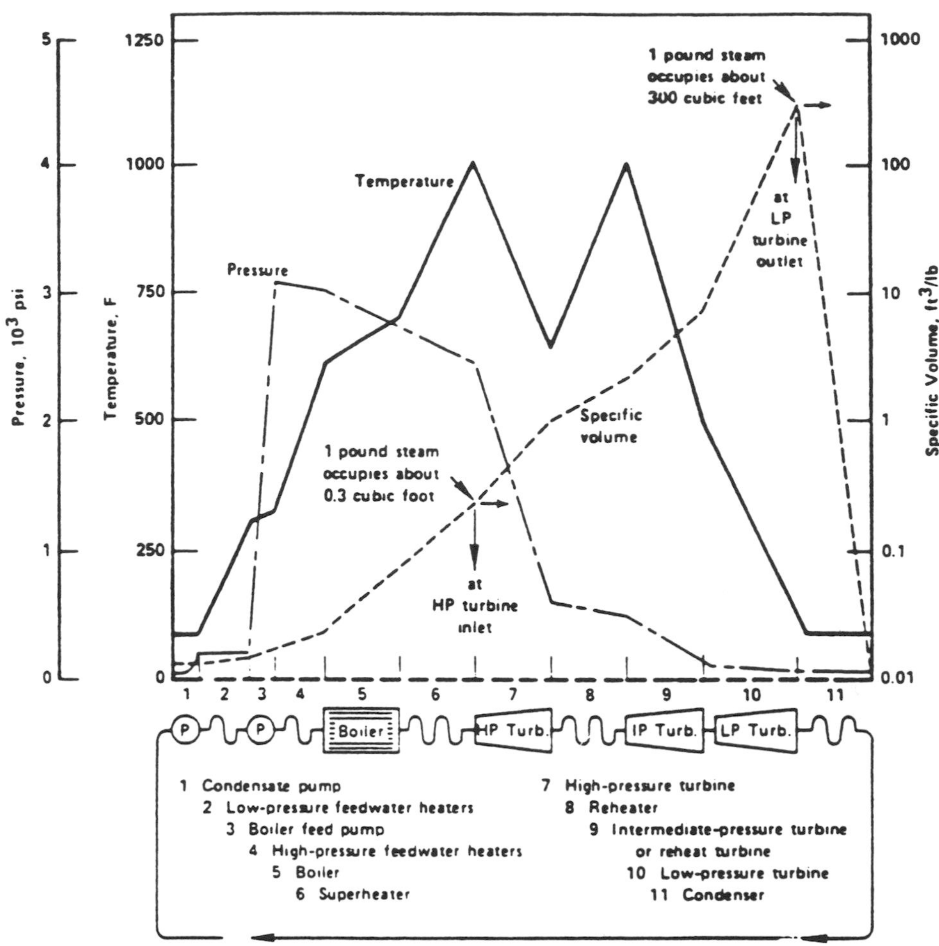

Fig. 4 Relationships between steam pressure, temperature, and specific volume in the various components of a large steam power plant

A list of key components, property requirements, and materials of construction for steam power plants is presented in Table 22. As can be seen from this table, carbon steels and low-alloy ferritic steels containing chromium, molybdenum, and/or vanadium constitute the bulk of the materials used in steam power plants. Additional information on ferritic steels used for high-temperature applications can be found in the article "Elevated-Temperature Mechanical Properties of Carbon and Alloy Steels" in this Volume.

Emission-Control Equipment

Material selection for emission-control equipment can be difficult because of the varied corrosive compounds present and the severe environments encountered. Therefore, a number of the more common emission-control applications are discussed in this article. More detailed information on these applications is available in the references cited at the end of the article.

Flue Gas Desulfurization Systems

The control of sulfur dioxide (SO_2) emissions is a primary concern in coal-fired power plants. This can be done either by removing sulfur from the coal before combustion or by flue gas desulfurization (FGD) in scrubbers. A variety of methods of both types are being developed, and some are already in commercial operation.

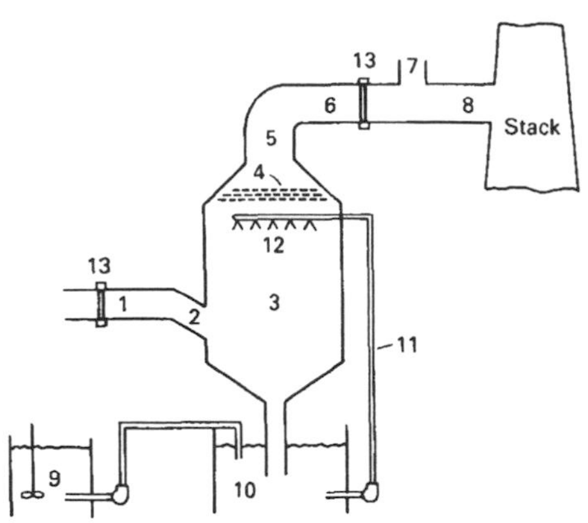

Fig. 5 Schematic of a wet flue gas desulfurization system. 1, inlet ductwork; 2, absorber inlet; 3, absorber body; 4, mist eliminators; 5, absorber outlet; 6, outlet ductwork; 7, reheaters and associated ductwork; 8, stack breeching; 9, alkali tanks; 10, recycle tanks; 11, slurry piping; 12, slurry nozzles; 13, dampers

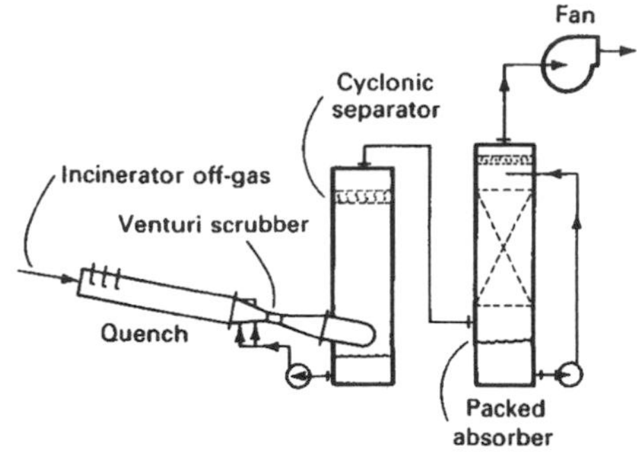

Fig. 6 Schematic of a general scrubber system arrangement

Among the FGD processes, lime and limestone wet scrubbing are the most developed. A schematic of a typical system of this type is shown in Fig. 5. Hot, dry flue gas at about 150 °C (300 °F) contacts a spray of an aqueous lime or limestone slurry in the absorber. The SO_2 in the flue gas dissolves in the slurry to form sulfites, which may be oxidized to sulfates by oxygen in the flue gas. The pH of the absorber slurry typically ranges from 5 to 6.5. Scrubbed and cooled gas at 50 to 55 °C (120 to 130 °F) passes through mist eliminators and outlet ducting to the stack. The scrubber gas is sometimes reheated to 60 to 95 °C (140 to 200 °F) to increase its buoyancy by mixing with unscrubbed gas (bypass reheat) or by using special reheaters. When bypass reheat is used, the SO_2 and sulfur trioxide (SO_3) in the gas readily react with the water in the scrubbed gas to form sulfuric acid (H_2SO_4), which can create an extremely oxidizing condensate on the duct wall.

Materials Selection. A variety of construction materials have been used in operating FGD systems. These include metals, organic linings and plastics, and ceramic and inorganic materials. Metals ranging from carbon steel to nickel-base alloys have been used in most components of FGD systems.

Where pH is neutral or higher, austenitic stainless steels (AISI types 304, 316, and 317, L grades preferred) perform well even at elevated temperatures. If pH is as low as 4 and chloride content is low (less than 100 ppm) but temperatures are above approximately 65 °C (150 °F), then Incoloy 825, Inconel 625, Hastelloy G-3, and alloy 904L (UNS N08904) or their equivalents are usually acceptable. Table 23 lists compositions of alloys commonly used in FGD systems.

When chloride content is up to 0.1% and pH approaches 2, only Hastelloys C-276, G, and G-3, and Inconel 625 can be successfully used. The other alloys mentioned above would be subjected to pitting and crevice corrosion. If a region is encountered with pH as low as 1 and chloride content above 0.1%, one of the only successful alloys acceptable is reported to be Hastelloy C-276 or its equivalent. In terms of metals selection, the higher the molybdenum content in an alloy, the more severe the corrosive environment it can withstand in the FGD system (Ref 28).

Waste Incineration

In a number of ways, the problems associated with materials for incinerator off-gas treatment equipment are similar to those used for FGD systems. Depending on the wastes being burned, however, significantly higher gas temperatures as well as more varied and more highly corrosive compounds may be encountered. Materials selection for waste incineration parallels that for FGD systems to some extent, but can often be more demanding.

The importance of incineration for the treatment of domestic and industrial wastes has increased as the availability of sanitary landfills has lessened and their costs have escalated. At the same time, environmental safety regulations have limited the use of deep below-ground and sea-disposal sites for untreated wastes.

Incineration provides a viable, although not inexpensive, alternative that produces scrubbable gaseous and particulate contaminants from a myriad of waste products. Incinerators are used to burn municipal solid wastes, industrial chemical wastes, and sewage sludge. In general, the off-gases can be classified according to their corrosiveness in descending order as follows: industrial chemical, municipal solid, and sewage sludge.

Industrial chemical gases are characterized by extremely high temperatures (1000 °C, or 1830 °F, is not uncommon) and the presence of halogenated compounds. In many cases, chlorinated hydrocarbons and plastics are burned, producing HCl, chlorine, hydrogen fluoride (HF), and possibly hydrogen bromide. Some sulfur and phosphorus compounds may also be produced.

The typical treatment system uses a gas quench to saturate and cool the gases, a wet venturi scrubber (if particulates pose a problem), a packed tower absorber, exhaust fan, ducting, liquid piping, and liquid recirculation pumps. Figure 6 shows a standard system arrangement.

Because of high temperatures, the presence of chlorides, and the fact that the gas becomes saturated with water vapor within the quench, very few materials can be successfully used for the quench construction. The major problem is not uniform attack but local pitting and crevice corrosion of many metals. In particular, chloride SCC severely affects austenitic stainless steels.

The materials that have been found to perform very well are such high-nickel alloys as Hastelloy C-276, Inconel 625, and titanium for the highest-

Table 23 Compositions of some alloys used in flue gas desulfurization systems

Alloy	Composition, % (a)						
	C	Fe	Ni	Cr	Mo	Mn	Others
Type 304L	0.03 max	Bal(b)	10.0	19.0	...	2.0 max	0.045 max P, 0.03 max S, and 1.00 max Si
Type 316L	0.03 max	Bal	12.0	17.0	2.5	2.0 max	1.00 max Si, 0.045 max P, and 0.03 max S
Type 317L	0.03 max	Bal	13.0	19.0	3.5	2.0 max	1.00 max Si, 0.045 max P, and 0.03 max S
Inconel alloy 625	0.10 max	5.0 max	Bal	21.5	9.0	0.50 max	0.40 max Al, 0.40 max Ti, 3.65 Nb, 0.015 max P, 0.015 max S, and 0.50 max Si
Inconel alloy 825	0.05 max	Bal	42.0	21.5	3.0	1.0 max	0.8 Ti, 0.5 max Si, 0.2 max Al, 2.25 Cu, and 0.03 max S
INCO alloy G	0.05 max	19.5	Bal	22.25	6.5	1.5	1.0 max Si, 2.125 Nb, 2.5 max Co, 2.0 Cu, 1.0 max W, and 0.04 max P
INCO alloy G-3	0.15 max	19.5	Bal	22.25	7.0	1.0 max	5.0 Co, 2.0 Cu, 0.04 max P, 1.0 max Si, 0.03 max S, 1.5 max w, and 0.50 max Nb + Ta
INCO alloy C-276	0.02 max	5.5	Bal	15.5	16.0	1.0 max	2.5 max Co, 0.03 max P, 0.03 max S, 0.08 max Si, and 0.35 max V
INCO alloy 904L	0.02 max	Bal	25.5	21.0	4.5	2.0 max	1.5 Cu, 1.0 max Si, 0.045 max P, and 0.035 max S

(a) Nominal composition unless otherwise specified. (b) bal, balance temperature cases and Hastelloy G and G-3 for slightly less severe cases. These materials have been used in other critical areas of the treatment system, such as fan wheels, dampers, liquid spray nozzles, and piping. Multiple-year service life histories have been reported with these alloys (Ref 29).

Refractory linings for the quench have also been used with some success. This can sometimes prove to be a more economical alternative to the use of high-nickel alloys. Problems do occur, however, because of attack on the binding substances employed and on the carbon steel base material, if exposed.

Following the quench, where temperatures are typically less than 95 °C (205 °F), the major equipment (venturis, tower shells, sump tanks, fan housings, and pump bodies) can be constructed of fiberglass-reinforced plastics (FRP).

Minicipal Solid Waste. The byproducts of solid municipal wastes can be similar to those found in chemical incineration. The levels of the worst contaminants—chlorides, for example—are usually lower. The nature of the requirements for burning these wastes, which contain large portions of cellulose, result in lower off-gas temperatures than those for chemical incineration.

Nevertheless, corrosion problems are severe, and materials selection is not very different from that for industrial chemicals incineration. Reference 30 provides a ranking of metals with respect to corrosion resistance on the basis of corrosion tests in this service. Reference 31 shows the results of corrosion tests for a very wide range of alloys in six distinct system zones.

Sewage Sludge. The burning of sewage sludge presents the least corrosive discharge of the three types under discussion. This can be attributed to limited halogen compounds in the gas and somewhat lower temperatures (typically 315 to 650 °C, or 600 to 1200 °F).

Type 304 and 316 stainless steels are suitable for construction in most areas of the system, including the quenching area, whether as a separate quench or part of the wet scrubber. Thermoplastics, FRP, and lined carbon steel can be used in the cooler regions.

The predominant contaminants in the environment are odorous sulfur compounds, both organic (mercaptans) and inorganic (hydrogen sulfide, H_2S), and particulate. Chlorides can exist, but they normally originate from the water used for makeup. Their presence sometimes requires the use of high-nickel alloys for such components as fan wheels and pump impellers.

Aerospace Components

Some of the most dramatic improvements in materials performance have originated from the aerospace community. The evolution of the aircraft gas turbine engine depended on the development of materials that could withstand the high operating temperatures and stresses encountered and exhibit outstanding oxidation resistance. In fact, the development of superalloys—particularly nickel-base superalloys—has almost entirely been motivated by the need to improve the efficiency, reliability, and operating life of gas turbines. There have been other peripheral applications, but at present, about 90% of superalloys produced are used in gas turbines for a range of applications, including aerospace, electricity generation, gas/oil pumping, and marine propulsion. The differing requirements in specific parts of the engine and the different operating conditions of the various types of gas turbine have led to the development of a wide range of nickel-base superalloys with individual balances of high-temperature creep resistance, corrosion resistance, yield strength, and fracture toughness.

Although metals have been the primary material of choice for high-temperature components for manned aircraft and aerospace vehicles, nonmetallic materials are being used increasingly often. One outstanding example, described below, is the use of ceramics and carbon-carbon composites for the thermal protection system of the U.S. Space Shuttle.

Gas Turbine Engines

Jet gas turbine engine designs push components to their limits for mechanical strength and temperature capability. New engine designs for commercial airplanes and military fighter jets introduce new requirements for thrust and performance. The fact is that the hotter the combustion in the engine, the more powerful and efficient the engine becomes. Advances such as new superalloys, improved cooling flow designs, and thermal barrier coatings were developed to stretch the capabilities of these components and to take advantage of the thrust and efficiency gains associated with increased combustion temperatures. These advances are described in detail in the articles contained in the Section "Properties of Superalloys" in this Volume. A chronology of superalloy developments can also be found in Fig. 7 in the article "Elevated-Temperature Characteristics of Engineering Materials."

The major engine subsystems consist of the fan, the high-pressure compressor (HPC), the combustor, the high- and low-pressure turbines (HPT and LPT), and the exhaust nozzle. The engine design contains one nonrotating system and two concentric rotating systems. The nonrotating (stator) system is made up of structural frames and casings. The low-pressure rotating system consists of the fan disk(s) and fan blades, the LPT disks and turbine blades, and a connecting shaft. The high-pressure rotating system consists of the HPC disks/spools and compressor blades, the HPT disks and turbine blades, and a connecting shaft.

Operating environments vary widely between different sections of the engine and depend on where the engine is in its mission. Temperatures may vary from subzero to above 1095 °C (2000 °F) and rotational speeds may climb to more than 15,000 rev/min. Components may also be subjected to ingested particle impacts (see the discussion on the effects on ingested sand particles in the article "Design for Oxidation Resistance" in this Volume).

The wide variety of operating conditions means that a wide variety of materials must be used to meet the design needs of the engine. Aluminum and titanium alloys, plastics, and resin-matrix graphite composites are frequently used in the fan and the engine nacelle. The HPC uses titanium alloys, nickel-base superalloys, such as Inconel 718, and steels, such as M152, 17-4PH, and A286. The combustor requires heat-

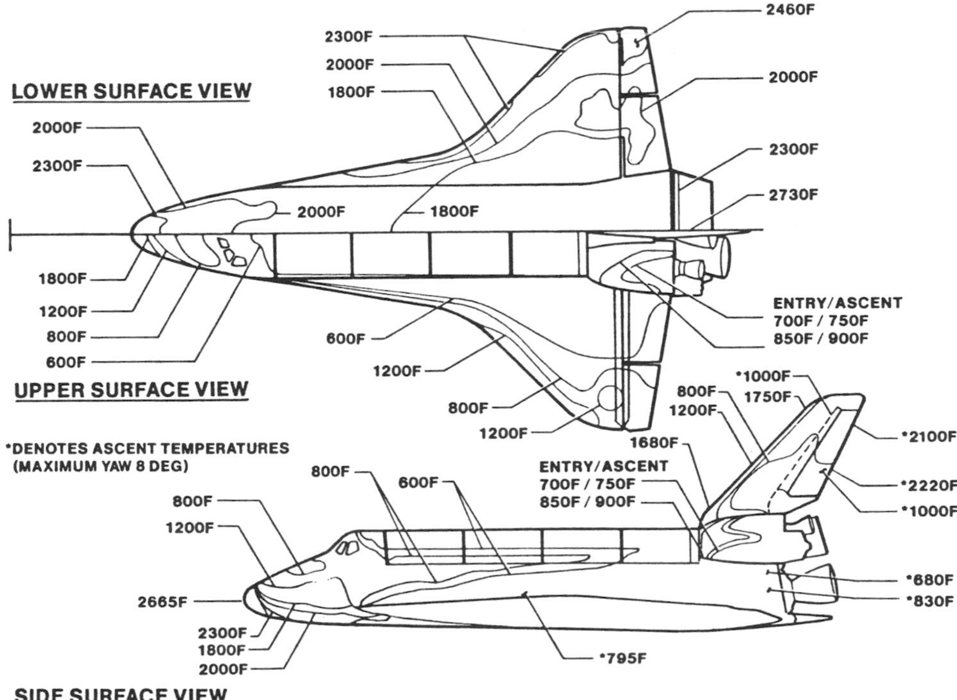

Fig. 7 Orbiter isotherms for a typical trajectory. Source: Ref 32

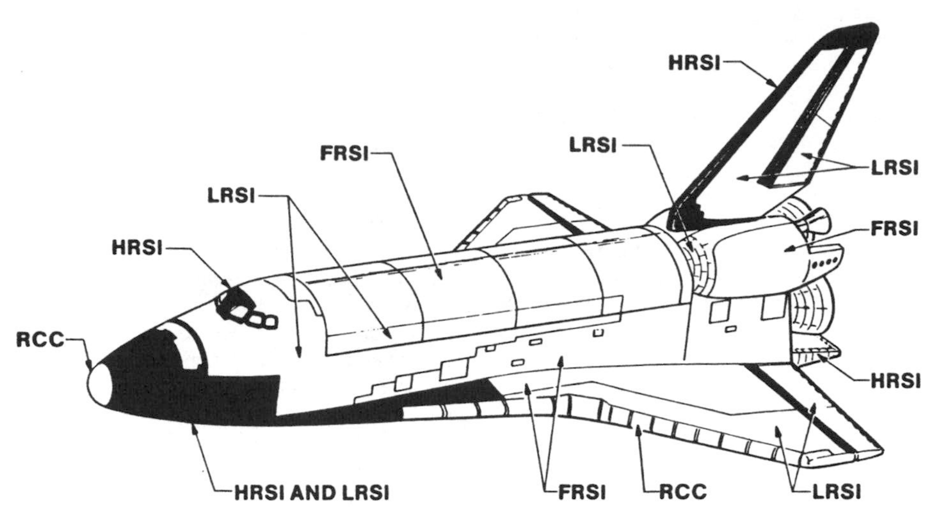

Material generic name	Material temperature capability, °C (°F)(a)	Material composition	Areas of orbiter
Reinforced carbon-carbon (RCC)	to 1650 (3000)	Pyrolized carbon-carbon, coated with SiC	Nose cone, wing leading edges, forward external tank separation panel
High-temperature reusable surface insulation (HRSI)	650–1260 (1200–2300)	SiO_2 tiles, borosilicate glass coating with SiB_4 added	Lower surfaces and sides, tail leading and trailing edges, tiles behind RCC
Low-temperature reusable surface insulation (LRSI)	400–650 (750–1200)	SiO_2 tiles, borosilicate glass coating	Upper wing surfaces, tail surfaces, upper vehicle sides, OMS(b) pods
Felt reusable surface insulation (FRSI)	to 400 (750)	Nylon felt, silicone rubber coating	Wing upper surface, upper sides, cargo bay doors, sides of OMS(b) pods

(a) 100 missions; higher temperatures are acceptable for a single mission. (b) Orbital maneuvering system (OMS) engines

Fig. 8 Thermal protection system materials for the U.S. Space Shuttle. More than 30,000 ceramic tiles are included in the system. Other materials making up the system are reinforced carbon-carbon composites (44 panels and the nose cap) and 333 m^2 (3581 ft^2) of felt reusable surface insulation. Source: Ref 32

resistant nickel or cobalt alloys, such as Hastelloy X or Haynes 188, and stainless steels for fuel tubing. The turbine sections rely on cobalt and nickel superalloys, such as Inconel X750, MAR-M-509, René 77, René 80, René 125, and advanced directionally solidified and single-crystal alloys.

Space Shuttle Thermal Protection System

A key to the success of the Space Shuttle orbiter is the development of a fully reusable thermal protection system (TPS) capable of being used for up to 100 missions. The key element of the TPS is the thousands of ceramic tiles that protect the Shuttle during reentry. Figure 7 shows the orbiter and the temperatures reached during reentry in a typical trajectory. During reentry of the Shuttle into the earth's atmosphere, its surface reaches 1260 °C (2300 °F) where the ceramic tiles are used. Even hotter regions (up to 1650 °C, or 3000 °F) occur at the nose tip and the wing leading edges, where reinforced carbon-carbon (RCC) composites must be employed. Figure 8 indicates the materials chosen for various areas of the TPS.

The basic tile system is composed of four key elements: a ceramic tile, a nylon felt mounting pad, a filler bar, and a room-temperature vulcanizing (RTV) silicone adhesive. The tile, coated with a high emittance layer of glass, functions as both radiator (to dissipate heat) and insulator (to block heating to the structure). The felt mounting pad, called a strain isolator pad (SIP), isolates the tile from the thermal and mechanical strains of the substructure. The filler bar, also a nylon felt material coated with silicone rubber, protects the structure under the tile-to-tile gap from overheating. The RTV adhesive bonds the tile to the SIP and the SIP and filler bar to the substructure (Fig. 9). The substructure (tile substrate) consists of aluminum alloys or graphite-epoxy composites. More detailed information on the tile system can be found in Ref 32.

The ceramic tiles are made from very-high-purity amorphous silica fibers ~1.2 to 4 μm in diameter and 0.32 cm (0.125 in.) long, which are felted from a slurry and pressed and sintered at ~1370 °C (2500 °F) into blocks. Two tile densities are used: 144 kg/m^3 (9 lb/ft^3) (LI-900) and 352 kg/m^3(22 lb/ft^3) (LI-2200). The LI-900 depends on a colloidal silica binder to achieve a

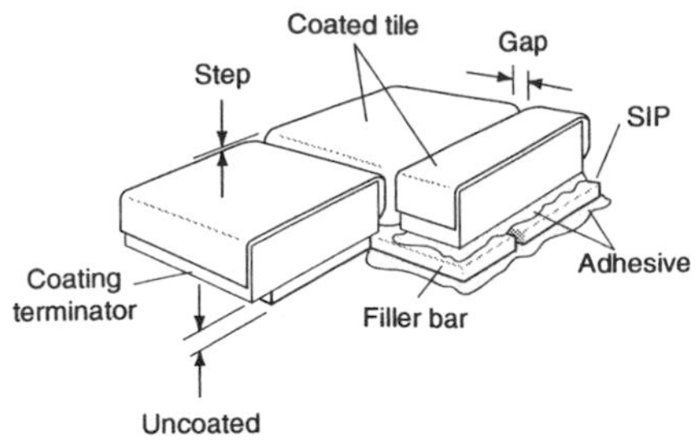

Fig. 9 The silica tile thermal protection system configuration. SIP, strain isolator pad. Source: Ref 32

fiber-to-fiber bond, whereas LI-2200 depends entirely on fiber-to-fiber sintering.

Most of the tiles are made from LI-900; however, in areas where higher strength is required, LI-2200 is used. All tiles have a borosilicate glass coating on five sides to provide the proper thermal properties. Those on the underside of the vehicle appear black because of the addition of silicon tetraboride for high emittance at high temperatures, whereas those on upper surfaces generally are white to limit on-orbit system temperature. The black tiles are usually a 15.2 by 15.2 cm (6 by 6 in.) square planform and typically 1.3 to 8.9 cm (0.5 to 3.5 in.) thick, as required. White tiles are generally a 20.3 by 20.3 cm (8 by 8 in.) square planform, 0.5 (0.2) to ~2.5 cm (1 in.) thick. There are special shapes or sizes—as small as ~4.4 cm (1.75 in.) square—as vehicle geometry dictates in some areas.

The silica tiles offer many advantages. The tile is 93% void; thus, it is an excellent insulator having conductivities (through the thickness) as low as 0.017 to 0.052 W/m · K (0.01 to 0.03 Btu · ft/h · ft^2 · °F). The low coefficient of expansion of amorphous silica, as well as the low modulus of the tile, eliminates thermal-stress and thermal-shock problems. Because of the very high purity (99.62%), devitrification is limited, avoiding the high stresses associated with the expansion or contraction of the crystalline phase (cristobalite) Silica has high temperature resistance; it is capable of exposure above 1480 °C (2700 °F) for a limited time. Since it is an oxide, further protection is unnecessary, whereas RCC and refractory metals must have oxidation-protective coatings. Oxidation of the silicon tetraboride in the coating results in boria and silica, the basic ingredients of the glass itself.

ACKNOWLEDGMENTS

The information in this article is largely taken from:

- G.Y. Lai, Heat-Resistant Materials for Furnace Parts, Trays, and Fixtures, *Heat Treating,* Vol 4, *ASM Handbook,* ASM International, 1991, p 510–518
- G.Y. Lai, Corrosion of Heat-Treating Furnace Accessories, *Corrosion,* Vol 13, *ASM Handbook* (formerly Vol 13, 9th ed., *Metals Handbook*), ASM International, 1987, p 1311–1314

- J. Gutzeit, R.D. Merrick, and L.R. Scharfstein, Corrosion in Petroleum Refining and Petrochemical Operations, *Corrosion,* Vol 13, *ASM Handbook* (formerly Vol 13, 9th ed., *Metals Handbook*), ASM International, 1987, p 1262–1287
- W.J. Gilbert and R.J. Chironna, Corrosion of Emission-Control Equipment, *Corrosion,* Vol 13, *ASM Handbook* (formerly Vol 13, 9th ed., *Metals Handbook*) ASM International, 1987, p 1367–1370
- R.A. Watson et al., Electrical Resistance Alloys, *Properties and Selection: Nonferrous Alloys and Special-Purpose Materials,* Vol 2, *ASM Handbook* (formerly Vol 2, 10th ed., *Metals Handbook*), ASM International, 1990, p 822–830
- T.P. Wang, Thermocouple Materials, *Properties and Selection: Nonferrous Alloys and Special Purpose Materials,* Vol 2, *ASM Handbook* (formerly Vol 2, 10th ed., *Metals Handbook*), ASM International, 1990, p 869–888
- Steels for Rolling-Element Bearings, *ASM Specialty Handbook: Carbon and Alloy Steels,* J.R. Davis, Ed., ASM International, 1996, p 654–658

REFERENCES

1. *Selection of Stainless Steels,* American Society for Metals, 1968
2. E.N. Skinner, J.F. Mason, and J.J. Moran, *Corrosion,* Vol 16, p 593
3. INCONEL alloy 601 brochure, INCO Alloys International, Inc.
4. F.K. Kies and C.D. Schwartz, *J. Test. Eval.,* Vol 2 (No. 2), March 1974, p 118
5. M.F. Rothman, Cabot Corp., private communication, 1985
6. D.E. Fluck, R.B. Herchenroeder, G.Y. Lai, and M.F. Rothman, *Met. Prog.,* Sept 1985, p 35
7. INCO Alloys International, Inc. unpublished research
8. G.Y. Lai, in *High Temperature Corrosion in Energy Systems,* M.F. Rothman, Ed., Symp. Proc., The Metallurgical Society, 1985, p 551
9. G.R. Rundell, Paper 377, presented at Corrosion/86, Houston, TX, National Association of Corrosion Engineers, March 1986
10. G.L. Swales, in *Behavior of High Temperature Alloys in Aggressive Environments,* I. Kirnan et al., Ed., Proc. Petten International Conf., The Metals Society, 1980, p 45
11. G.Y. Lai, in *High Temperature Corrosion in Energy Systems,* M.F. Rothman, Ed., Symp. Proc., The Metallurgical Society, 1985, p 227
12. J.W. Slusser, J.B. Titcomb, M.T. Heffelfinger, and D.R. Dunbobbin, *J. Met.,* July 1985, p 24
13. M.F. Rothman and G.Y. Lai, *Ind. Heat.,* Aug 1986, p 29
14. D.E. Fluck, Cabot Corp., private communication, 1985
15. INCO Alloys International, Inc., unpublished research
16. B.B. Morton, Metallurgical Methods for Combatting Corrosion and Abrasion in the Petroleum Industry, *J. Inst. Petrol.,* Vol 34 (No. 289), 1948, p 1–68
17. E.L. Hildebrand, Materials Selection for Petroleum Refineries and Petrochemical Plants, *Mater. Prot. Perform.,* Vol 11 (No. 7), 1972, p 19–22
18. A.J. Freedman, G.F. Tisinai, and E.S. Troscinski, Selection of Alloys for Refinery Processing Equipment, *Corrosion,* Vol 16 (No. 1), 1960, p 19t–25t
19. *The Role of Stainless Steels in Petroleum Refining,* American Iron and Steel Institute, 1977
20. Selection of Steel for High-Temperature Service in Petroleum Refinery Applications, in *Properties and Selection of Metals,* Vol 1, 8th ed., *Metals Handbook,* American Society for Metals, 1961, p 585–603
21. G.E. Moller, I.A. Franson, and T.J. Nichol, Experience with Ferritic Stainless Steel in Petroleum Refinery Heat Exchangers, *Mater. Perform.,* Vol 20 (No. 4), 1981, p 41–50
22. A.J. Brophy, Stress Corrosion Cracking of Austenitic Stainless Steels in Refinery Environments, *Mater. Perform.,* Vol 13 (No. 5), 1974, p 9–15
23. A.S. Couper and H.F. McConomy, Stress Corrosion Cracking of Austenitic Stainless Steels in Refineries, *Proc. API,* Vol 46 (III), 1966, p 321–326
24. T.P. May, J.F. Mason, Jr., and W.K. Abbott, Austenitic Nickel Cast Irons in the Petroleum Industry, *Mater. Prot.,* Vol 1 (No. 8), 1962, p 40–55
25. J. Kolts, J.B.C. Wu, and A.I. Asphahani, Highly Alloyed Austenitic Materials for Corrosion Service, *Met. Prog.,* Vol 125 (No. 10), 1983, p 25–36
26. *Corrosion Resistance of Hastelloy Alloys,* The Cabot Corp., 1978
27. A.I. Asphahani, Corrosion Resistance of High Performance Alloys, *Mater. Perform.,* Vol 19 (No. 12), 1980, p 33-43
28. R.W. Kirchner, Materials of Construction for Flue-Gas-Desulfurization Systems, *Chem. Eng.,* 19 Sept 1983, p 81–86
29. D.C. Agarwal and F.G. Hodge, "Material Selection Processes and Case Histories Associated with the Hazardous Industrial and Municipal Waste Treatment Industries," Cabot Corp.
30. R.W. Kirchner, Corrosion of Pollution Control Equipment, *Chem. Eng. Prog.,* Vol 71 (No. 3), 1975, p 58–63
31. H.D. Rice, Jr. and R.A. Burford, "Corrosion of Gas-Scrubbing Equipment in Municipal Refuse Incinerators," paper presented at the International Corrosion Forum, National Association of Corrosion Engineers, 19–23 March 1973
32. L.J. Korb, C.A. Morant, R.M. Calland, and C.S. Thatcher, The Shuttle Orbiter Thermal Protection System, *Ceram. Bull.,* Vol 60 (No. 11), 1981, p 1188–1193

SELECTED REFERENCES

- *Heat-Resistant Materials,* K. Natesan and D.J. Tillack, Ed., *Proc. First International Conf. on Heat-Resistant Materials,* 23–26 Sept 1991, Fontana, WI, ASM International, 1991
- *Heat-Resistant Materials II,* K. Natesan, P. Ganesan, and G. Lai, Ed., *Proc. Second International Conf. on Heat-Resistant Materials,* 11–14 Sept 1995, Gatlinburg, TN, ASM International, 1995

Properties of Ferrous Heat-Resistant Alloys

Elevated-Temperature Mechanical Properties of Carbon and Alloy Steels

CARBON AND LOW-ALLOY STEELS are used extensively at elevated temperatures in fossil-fired power generating plants, aircraft power plants, chemical processing plants, and petroleum processing plants. Carbon steels are often used at up to about 370 °C (700 °F) under continuous loading, but allowable stresses up to 540 °C (1000 °F) are defined in Section VIII of the ASME Boiler and Pressure Vessel Code. Carbon-molybdenum steels with 0.5% Mo are used up to 540 °C (1000 °F), while steels alloyed with 0.5 to 1% Mo in combination with 0.5 to 9.0% Cr (and sometimes with other carbide formers, such as vanadium, tungsten, niobium, and titanium) are often used at up to about 650 °C (1200 °F). For temperatures above 650 °C (1200 °F), stainless steels are generally used. However, these general maximum-use temperature limits do not apply in all applications. Tables 1 to 3 list maximum-use temperatures in three sample applications with different design criteria.

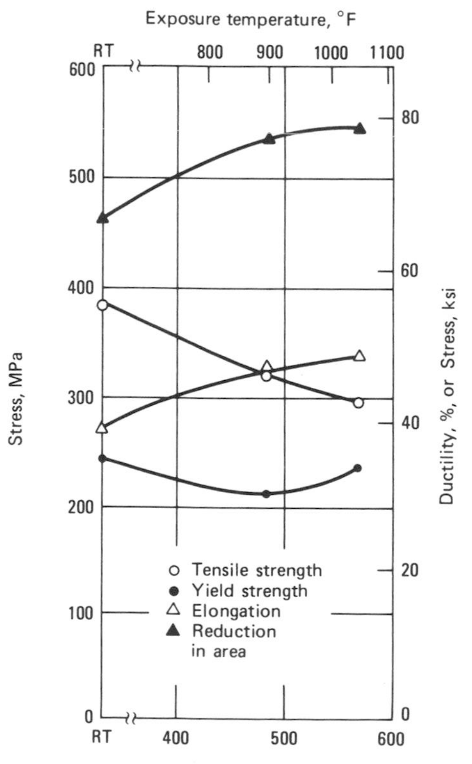

Fig. 1 Effect of elevated-temperature exposure on the room-temperature tensile properties of normalized 0.17% C steel after exposure (without stress) to indicated temperature for 83,000 h

This article emphasizes the elevated-temperature mechanical properties of carbon steels and low-alloy steels with ferrite-pearlite and ferrite-bainite microstructures for use in boiler tubes, pressure vessels, and steam turbines. In these applications, the selection of steels to be used at elevated temperatures generally involves compromise between the higher efficiencies obtained at higher operating temperatures and the cost of equipment, including materials, fabrication, replacement, and downtime costs. The more highly alloyed stainless steels, which depend on an austenitic matrix for their high-temperature properties, generally have higher resistance to mechanical and chemical degradation at elevated temperatures than the low-alloy ferritic steels. However, a higher alloy content generally means higher cost. Therefore, carbon and low-alloy ferritic steels are extensively used in various forms (e.g., piping, pressure vessel plates, bolts, structural parts) in a number of applications that involve exposure to elevated temperatures. In addition, interest in ferritic steels has increased recently because their relatively lower thermal expansion coefficient and higher thermal conductivity make them more attractive than austenitic steels in applications where thermal cycling is present.

Table 1 Temperature limits of superheater tube materials covered in ASME Boiler Codes

Material	Maximum-use temperature: Oxidation/graphitization criteria, metal surface(a) °C	°F	Strength criteria, metal midsection °C	°F
SA-106 carbon steel	400–500	750–930	425	795
Ferritic alloy steels				
0.5Cr-0.5Mo	550	1020	510	950
1.2Cr-0.5Mo	565	1050	560	1040
2.25Cr-1Mo	580	1075	595	1105
9Cr-1Mo	650	1200	650	1200
Austenitic stainless steel				
Type 304H	760	1400	815	1500

(a) In the fired section, tube surface temperatures are typically 20–30 °C (35–55 °F) higher than the tube midwall temperature. In a typical U.S. utility boiler, the maximum metal surface temperature is approximately 625 °C (1155 °F).

As an example of the tonnage requirements for carbon and low-alloy steels in industrial construction, 1360 metric tonnes (1500 tons) of pressure tubing were required for the construction of a single 500 MW coal-fired generating plant. The quantities of the various carbon and low-alloy steels used in the pressure tubing were as follows:

Steel type	Tons	% of total tonnage
Carbon	540	36
C-0.5Mo	150	10
1.25Cr-1Mo	495	33
2.25Cr-1Mo	150	10
9Cr-1Mo	165	11

This table of carbon and low-alloy steels is for pressure tube applications and does not include the chromium-molybdenum-vanadium steels that are used for turbine rotors, high-temperature bolts, and pressure tubing. Some of the major property requirements for such applications are listed in Table 4.

Information pertaining to chemical degradation in elevated-temperature environments can be found in the article "Elevated-Temperature Corrosion Properties of Carbon and Alloy Steels" in this Volume.

Alloy Designations and Specifications

Carbon and alloy steels used for elevated-temperature service are usually identified by American Iron and Steel Institute (AISI) designation; Aerospace Material Specification (AMS), American Society of Mechanical Engineers (ASME), or ASTM specification number; nominal composition; or trade name. These steels have also been assigned numbers in the Unified Numbering System. In addition, military and U.S. government specifications cover many of these steels.

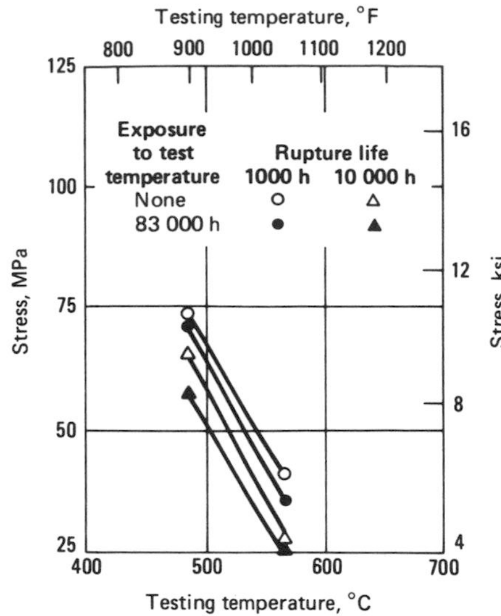

Fig. 2 Effect of exposure to elevated temperature on stress-to-rupture of carbon steel. Stress-to-rupture in 1000 and 10,000 h at the indicated temperature for specimens of normalized 0.17% C steel exposed to the test temperature (without stress) for 83,000 h and for similar specimens not exposed to elevated temperature prior to testing

Table 2 Suggested maximum temperatures in petrochemical operations for continuous service based on creep or rupture data

Material	Maximum temperature based on creep rate		Maximum temperature based on rupture	
	°C	°F	°C	°F
Carbon steel	450	850	540	1000
C-0.5Mo	510	950	595	1100
2.5Cr-1Mo	540	1000	650	1200
Type 304 stainless steel	595	1100	815	1500
Alloy C-276 nickel-base alloy	650	1200	1040	1900

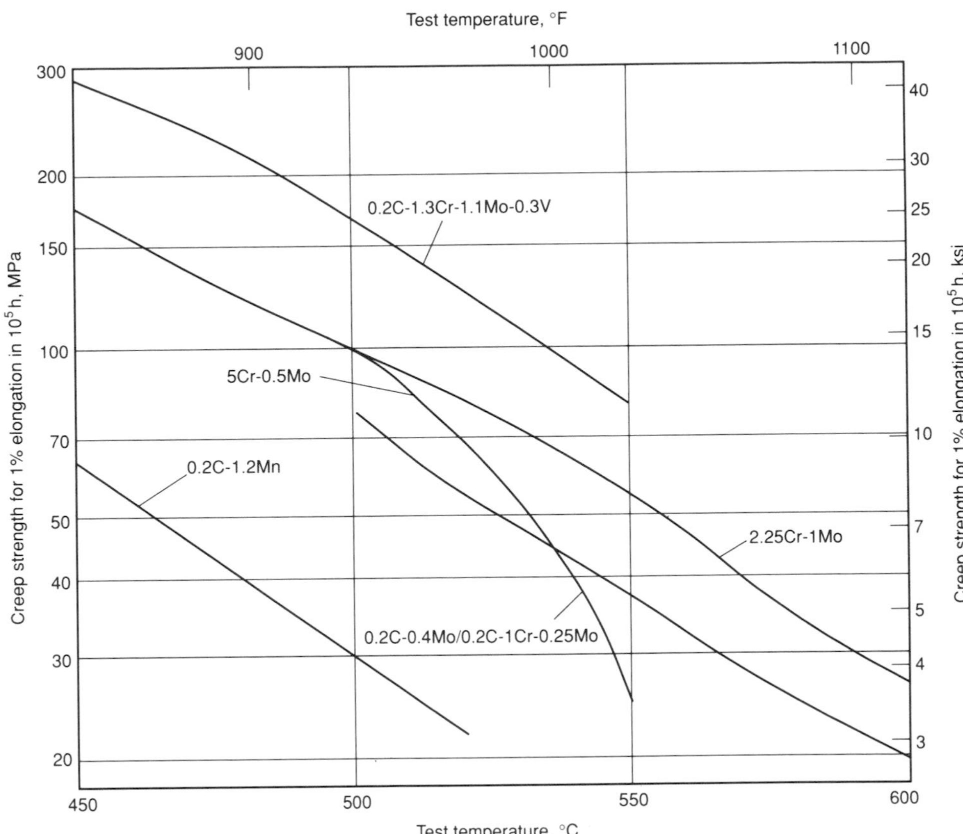

Fig. 3 General comparison of creep strengths of various creep-resistant low-alloy steels

Steel products manufactured for use under the ASME Boiler and Pressure Vessel Code must comply with the appropriate ASME specification. Each specification includes information on ranges and limits of composition, dimensions and tolerances, minimum mechanical properties, and other functional requirements. The designations applied to these products include the letters "SA," the number of the specification, and possibly other letters or numbers to distinguish among various types, grades, and classes within a single specification. Most ASTM specifications are identical to the ASME specification of the same number, except that the ASTM designations begin with the letter "A." Tables 5 and 6 give some examples of ASME specifications for elevated-temperature steels, as well as their compositions and typical room-temperature mechanical properties.

Aerospace material specifications, as the name suggests, are specifications for products intended for the aerospace industry. Table 7 gives the nominal compositions, typical applications, and typical mechanical properties of steels that are often identified by AMS numbers.

The AISI designation for steels intended for elevated-temperature service is a three-digit number beginning with a 6, such as 601. The AISI designations are included in Table 7. It should be noted that the AISI 6*xx* designations for steels are used far less often today than they were previously.

Carbon Steels

Carbon steels are the predominant materials in pressure vessel fabrication because of their low cost, versatile mechanical properties, and availability in fabricated forms. They are the most common materials used in noncorrosive environments in the temperature range of –29 to 425 °C (–20 to 800 °F) in oil refineries and chemical plants. Although the ASME code gives allowable stresses for temperatures greater than 425 °C (800 °F), it also notes that prolonged exposure at these temperatures may result in conversion of the carbide phase of the carbon steel to graphite. This phenomenon, known as graphitization, is a cumulative process that depends on the amount of time the material is kept at or above 425 °C (800 °F). The result is a weakening of the steel after high-temperature exposure (Fig. 1). Carbon steels are also increasingly affected by creep at temperatures above 370 °C (700 °F). Figure 2 shows the effect of temperature on the stress-to-rupture life of a carbon steel.

Creep-Resistant Low-Alloy Steels

Creep-resistant low-alloy steels usually contain 0.5 to 1.0% Mo for enhanced creep strength, along with 0.5 to 9% Cr for improved corrosion resistance, rupture ductility, and resistance to graphitization. Small additions of carbide formers, such as vanadium, niobium, and titanium, may also be added for precipitation strengthening and/or grain refinement. Of course, the effects of alloy elements on transformation hardening and weldability are additional factors.

The three general types of creep-resistant low-alloy steels are chromium-molybdenum steels, chromium-molybdenum-vanadium steels, and modified chromium-molybdenum steels. Chromium-molybdenum steels are used primarily for tube, pipe, and pressure vessels, where the allow-

Table 3 Maximum temperature limits for steels used in coal-fired steam power plants

Specification	Nominal composition	Product form	Minimum tensile strength, ksi	Minimum yield strength, ksi	High heat input furnace walls	Other furnace walls and enclosures	SH RH econ.	Unheated connector pipe (<10.75 in. OD)	Headers and pipe (>10.75 in. OD)	Drums	Recommended maximum use temperature, °F	Footnotes
SA-178A	C-steel	ERW tube	(47.0)	(26.0)	X	X	X				950	(a)(b)
SA-192	C-steel	Seamless tube	(47.0)	(26.0)	X	X	X	X			950	(a)
SA-178C	C-steel	ERW tube	60.0	37.0		X	X				950	(b)
SA-210A1	C-steel	Seamless tube	60.0	37.0	X	X	X	X			950	...
SA-106B	C-steel	Seamless pipe	60.0	35.0				X	X		950	(c)
SA-178D	C-steel	ERW tube	70.0	40.0	X	X	X				950	(b)
SA-210C	C-steel	Seamless tube	70.0	40.0		X	X	X			950	...
SA-106C	C-steel	Seamless pipe	70.0	40.0				X	X		950	(c)
SA-216WCB	C-steel	Casting	70.0	36.0		X	X	X	X		950	...
SA-105	C-steel	Forging	70.0	36.0		X	X	X	X		950	(c)
SA-181-70	C-steel	Forging	70.0	36.0		X	X	X	X		950	(c)
SA-266C12	C-steel	Forging	70.0	36.0					X		800	...
SA-516-70	C-steel	Plate	70.0	38.0					X	X	800	...
SA-266C13	C-steel	Forging	75.0	37.5					X		800	...
SA-299	C-steel	Plate	75.0	40.0						X	800	...
SA-250T1A	C-Mo	ERW tube	60.0	32.0		X	X				975	(d)(e)
SA-209T1A	C-Mo	Seamless tube	60.0	32.0		X	X	X			975	(d)
SA-335P1	C-Mo	Seamless pipe	55.0	30.0				X			875	...
SA-250T2	0.5Cr-0.5Mo	ERW tube	60.0	30.0	X		X		X		1025	(f)(g)
SA-213T2	0.5Cr-0.5Mo	Seamless tube	60.0	30.0	X		X				1025	(f)
SA-250T12	1Cr-0.5Mo	ERW tube	60.0	32.0			X				1050	(e)(g)
SA-213T12	1Cr-0.5Mo	Seamless tube	60.0	32.0			X				1050	(h)
SA-335P12	0.5Cr-0.5Mo	Seamless pipe	60.0	32.0					X		1050	(h)
SA-250T11	1.25Cr-0.5Mo-Si	ERW tube	60.0	30.0			X				1050	(e)
SA-213T11	1.25Cr-0.5Mo-Si	Seamless tube	60.0	30.0			X				1050	...
SA-335P11	1.25Cr-0.5Mo-Si	Seamless pipe	60.0	30.0				X	X		1050	...
SA-217WC6	1.25Cr-0.5Mo	Casting	70.0	40.0		X	X	X	X		1100	...
SA-250T22	2.25Cr-1Mo	ERW tube	60.0	30.0			X				1115	(e)(g)
SA-213T22	2.25Cr-1Mo	Seamless tube	60.0	30.0			X				1115	
SA-335P22	2.25Cr-1Mo	Seamless pipe	60.0	30.0				X	X		1100	...
SA-217WC9	2.25Cr-1Mo	Casting	70.0	40.0			X	X	X		1115	...
SA-182F22A	2.25Cr-1Mo	Forging	60.0	30.0			X		X		1115	...
SA-336F22A	2.25Cr-1Mo	Forging	60.0	30.0					X		1100	...
SA-213T91	9Cr-1Mo-V	Seamless tube	85.0	60.0			X				1200	...
SA-335P91	9Cr-1Mo-V	Seamless pipe	85.0	60.0				X	X		1200	...
SA-182F01	9Cr-1Mo-V	Forging	85.0	60.0			X				1200	...
SA-336F91	9Cr-1Mo-V	Forging	85.0	60.0					X		1200	...
SA-213TP304H	18Cr-8Ni	Seamless tube	75.0	30.0			X				1400	...
SA-213TP347H	18Cr-10Ni-Cb	Seamless tube	75.0	30.0			X				1400	...
SA-213TP310H	25Cr-20Ni	Seamless tube	75.0	30.0			X				1500	...
SB-407-800H	Ni-Cr-Fe	Seamless tube	65.0	25.0			X				1500	...
SB-423-825	Ni-Fe-Cr-Mo-Cu	Seamless tube	85.0	35.0			X				1000	...

ERW, electric resistance welded; SH, superheaters; RH, reheaters; econ., economizers. (a) Values in parentheses are not required minimums, but are expected minimums. (b) Requires special inspection if used at 100% efficiency above 850 °F. (c) Limited to 800 °F maximum for piping 10.75 in. OD and larger and outside the boiler setting. (d) Limited to 875 °F maximum for applications outside the boiler setting. (e) Requires special inspection if used at 100% efficiency. (f) Maximum OD temperature is 1025 °F. Maximum mean metal temperature for Code calculations is 1000 °F. (g) Requires use of a Code Case now. Will not later. (h) 32 ksi minimum yield strength requires use of Code Case 2070, which is being incorporated into the Code. Source: ASME Boiler and Pressure Vessel Code

able stresses may permit creep deformation up to about 5% over the life of the component. Typical creep strengths of various chromium-molybdenum steels are shown in Fig. 3. Figure 3 also shows the creep strength of a chromium-molybdenum steel with vanadium additions. Chromium-molybdenum-vanadium steels provide higher creep strengths and are used for high-temperature bolts, compressor wheels, or steam turbine rotors, where allowable stresses may require deformations less than 1% over the life of the component.

Chromium-Molybdenum Steels

Chromium-molybdenum steels are widely used in oil refineries, chemical industries, and electrical power generating stations for piping, heat exchangers, superheater tubes, and pressure vessels. The main advantage of these steels is the improved creep strength from molybdenum and chromium additions and the enhanced corrosion resistance from chromium. The creep strength of chromium-molybdenum steels is derived mainly from two sources: solid-solution strengthening of the ferrite matrix by carbon, molybdenum, and chromium; and precipitation hardening by carbides. Creep strength generally, but not always, increases with higher amounts of molybdenum and chromium. The effects of chromium and molybdenum on creep strength are complex (see the section "Effects of Composition" in this article). In Fig. 3, for example, 2.25Cr-1Mo steel has a higher creep strength than 5Cr-0.5Mo steel.

As indicated in Table 8, chromium-molybdenum steels are available in several product forms. In actual applications, boiler tubes are used mostly in the annealed condition, whereas piping is used mostly in the normalized and tempered condition. Bend sections used in piping, however, are closer to an annealed condition than to a normalized condition. As a result of the cooling rates employed in these treatments, the microstructures of chromium-molybdenum steels may vary from ferrite-pearlite aggregates to ferrite-bainite aggregates. Bainite microstructures have better creep resistance under high-stress, short-time conditions, but they degrade more rapidly at high temperatures than pearlitic structures. As a result, ferrite-pearlite material has better intermediate-term, low-stress creep resistance. Because both microstructures will eventually spheroidize, it is expected that over long service lives the two microstructures will converge to similar creep strengths.

The 0.5Mo steel with 0.15% C is used for piping and superheater tubes operating at metal temperatures up to 455 °C (850 °F). Above this temperature, spheroidization and graphitization may increase the possibility of failure in service (Fig. 4). Use of carbon-molybdenum steel at higher temperatures has been largely discontinued because of graphitization. Chromium steels are highly resistant to graphitization and are

Table 4 Property requirements and materials of construction for fossil/steam power plant components

Component	Major property requirements	Typical materials
Boiler		
Waterwall tubes	Tensile strength, corrosion resistance, weldability	C and C-Mo steels
Drum	Tensile strength, corrosion resistance, weldability, corrosion-fatigue strength	C, C-Mo, and C-Mn steels
Headers	Tensile strength, weldability, creep strength	C, C-Mo, C-Mn, and Cr-Mo steels
Superheater/reheater tubes	Weldability, creep strength, oxidation resistance, low coefficient of thermal expansion	Cr-Mo steels and austenitic stainless steels
Steam pipe	Same as above	Same as above
Turbine		
HP-IP rotors/disks	Creep strength, corrosion resistance, thermal-fatigue strength, toughness	Cr-Mo-V steels
LP rotors/disks	Toughness, stress-corrosion resistance, fatigue strength	Ni-Cr-Mo-V steels
HP-IP blading	Creep strength, fatigue strength, corrosion and oxidation resistance	12% Cr steels
LP blading	Fatigue strength, corrosion-fatigue pitting resistance	12% Cr steels, 17-4 PH stainless steel, Ti-6Al-4V
Inner casings, steam chests, valves	Creep strength, thermal-fatigue strength, toughness, yield strength	Cr-Mo steels
Bolts	Proof stress, creep strength, stress-relaxation resistance, toughness, notch ductility	Cr-Mo-V and 12Cr-Mo-V steels
Generators		
Rotor	Yield strength, toughness, fatigue strength, magnetic permeability	Ni-Cr-Mo-V steels
Retaining rings	High yield strength, hydrogen- and stress-corrosion resistance, nonmagnetic	18Mn-5Cr and 18Mn-18Cr steels
Condensers		
Condensers	Corrosion and erosion resistance	Cupronickel, titanium, brass, stainless steels

HP, high pressure; IP, intermediate pressure; LP, low pressure

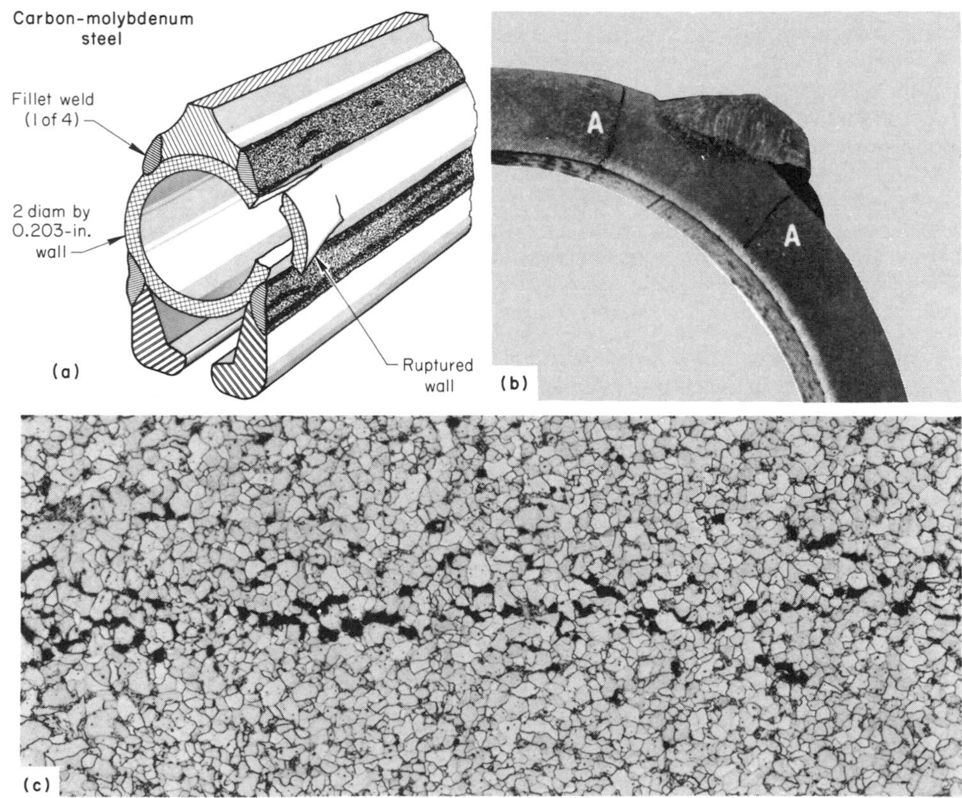

Fig. 4 Carbon-molybdenum steel tube that ruptured in a brittle manner after 13 years in service because of graphitization at weld heat-affected zones (HAZs). (a) View of tube showing dimensions (in inches), locations of welds, and rupture. (b) Macrograph showing graphitization along edges of a weld HAZ (letter A); this was typical of all four welds. 2×. (c) Micrograph of a specimen etched in 2% nital, showing chainlike array of embrittling graphite nodules (black) at the edge of a HAZ. 100×

therefore preferred for service above 455 °C (850 °F).

The 1Cr-0.5Mo steel is used for piping, cracking-still tubes, and boiler tubes at service temperatures up to 510 or 540 °C (950 or 1000 °F). The similar 1.25Cr-0.5Mo steel is used at up to 590 °C (1100 °F) and has stress-rupture and creep properties comparable to those of the 1Cr-0.5Mo alloy (Fig. 5).

The 2.25Cr-1Mo steel has better oxidation resistance and creep strength than the steels mentioned above. The 2.25Cr-1Mo steel is a highly favored alloy for service up to 650 °C (1200 °F) without the presence of hydrogen or 480 °C (900 °F) in a hydrogen environment. This steel, whose elevated-temperature properties have been thoroughly documented (Ref 2-6), is discussed in more detail in the section "Elevated-Temperature Behavior of 2.25Cr-1Mo Steel" in this article.

The 5, 7, and 9% Cr steels are generally lower in stress rupture and creep strength than the lower-chromium steels, because the strength at elevated temperatures typically drops off with an increase in chromium. However, this may not always be the case, depending on the service temperature (Fig. 6) and the exposure (Fig. 7, 8). Heat treatment is also an important factor. The main advantage of these steels is improved oxidation resistance from the increased chromium content.

Chromium-Molybdenum-Vanadium Steels

The chromium-molybdenum-vanadium steels are manufactured with higher carbon ranges (such as 0.28 to 0.33% and 0.40 to 0.50%) and are used in the normalized and tempered or quenched and tempered conditions. Because of the relatively high yield strengths (Fig. 9) and creep strengths (Fig. 3), these steels are suitable for bolts, compressor wheels in gas turbines and steam turbine rotors, and other parts operating at temperatures up to 540 °C (1000 °F). The most common low-alloy composition contains 1% Cr, 1% Mo, and 0.25% V.

Bolt Applications. The basic composition of low-alloy, high-temperature bolt steels has evolved from chromium-molybdenum to chromium-molybdenum-vanadium. The chromium-molybdenum steels used until the late 1940s had creep strengths adequate for service at temperatures up to about 480 °C (895 °F). With the increasing need for a higher-strength steel, a 1Cr-1Mo-0.25V steel strengthened by stable V_4C_3 precipitates was developed. This alloy was found to be adequate for steam temperatures up to 540 °C (1000 °F). When steam temperatures reached about 565 °C (1050 °F) in the mid-1950s, a 1Cr-1Mo-0.75V steel was developed in which vanadium and carbon had been stoichiometrically optimized to get the largest volume fraction of V_4C_3 and hence the highest creep strength. Unfortunately, this development overlooked the importance of rupture ductility, and many creep-rupture failures of bolts occurred due to notch sensitivity. The loss in rupture ductility was subsequently

Table 5 Compositions of steels for elevated-temperature service

ASME specification	UNS designation	Nominal composition	Product form	C	Mn	Si	P	S	Cr	Ni	Mo	Other
				Composition, %								
SA-106A	K02501	C	Seamless carbon steel pipe	0.25(a)	0.27–0.93	0.10(b)	0.048(a)	0.058(a)	...	...	...	...
SA-106B	K01700	C-Si	Seamless carbon steel pipe	0.30(a)	0.29–1.06	0.10(b)	0.048(a)	0.058(a)	...	...	...	...
SA-285A	K03006	C	Carbon steel PV plate	0.17(a)	0.90(a)	...	0.035(a)	0.045(a)	...	...	...	0.25 Cu(a)
SA-299	K02803	C-Mn-Si	C-Mn-Si steel PV plate	0.28(a)	0.90–1.40	0.15–0.30	0.035(a)	0.040(a)	...	...	...	...
SA-204A	K11820	C-0.5Mo	Mo alloy steel PV plate	0.18(a)	0.90(a)	0.15–0.30	0.035(a)	0.040(a)	...	...	0.45–0.60	...
SA-302A	K12021	Mn-Mo	Mn-Mo-Mn and Mo-Ni alloy PV plate	0.20(a)	0.95–1.30	0.15–0.30	0.035(a)	0.040(a)	...	...	0.45–0.60	...
SA-533B2	K12539	Mn-Mo-Ni	Mn-Mo-Mn and Mo-Ni alloy steel PV plate	0.25(a)	1.15–1.50	0.15–0.30	0.035(a)	0.040(a)	...	0.40–0.70	0.45–0.60	0.10 Cu(a)
SA-517F	K11576	...	High-strength alloy steel PV plate	0.10–0.20	0.60–1.00	0.15–0.35	0.035(a)	0.040(a)	0.40–0.65	0.70–1.00	0.40–0.60	0.002–0.006 B, 0.15–0.050 Cu, 0.03–0.08 V
SA-335P12	K11562	1Cr-0.5Mo	Seamless ferritic alloy steel pipe for high-temperature service	0.15(a)	0.30–0.61	0.50(a)	0.045(a)	0.045(a)	0.50–1.25	...	0.44–0.65	...
SA-217WC6	J12072	1.25Cr-0.5Mo	Alloy steel castings	0.20(a)	0.50–0.80	0.60(a)	0.04(a)	0.045(a)	1.00–1.50	...	0.45–0.65	...
SA-387Gr22	K21590	2.25Cr-1Mo	Cr-Mo alloy steel PV plate	0.15(a)	0.30–0.60	0.50(a)	0.035(a)	0.035(a)	2.0–2.5	...	0.90–1.10	...
SA-387Gr5	S50100	5Cr-0.5Mo	Cr-Mo alloy steel PV plate	0.15(a)	0.30–0.60	0.50(a)	0.040(a)	0.030(a)	4.0–6.0	...	0.45–0.65	...
SA-217C12	J82090	9Cr-1Mo	Alloy steel castings	0.02(a)	0.35–0.65	1.00(a)	0.04(a)	0.045(a)	8.0–1.0	...	0.90–1.20	...

PV, pressure vessel. (a) Maximum. (b) Minimum

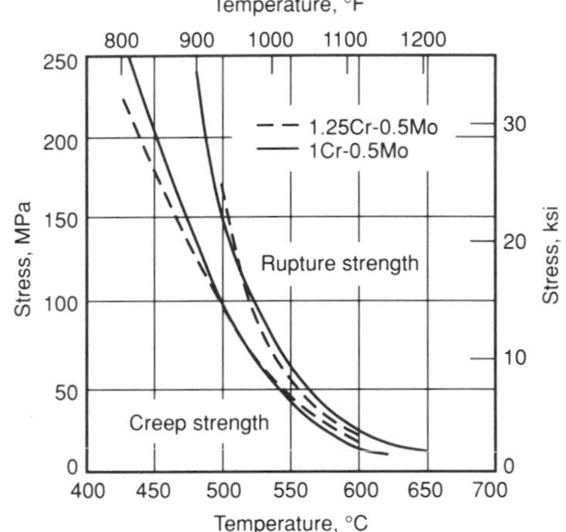

Fig. 5 Creep strength (0.01%/1000 h) and rupture strength (100,000 h) of 1Cr-0.5Mo and 1.25Cr-0.5Mo steel. Source: Ref 1

Table 6 Room-temperature mechanical properties of steels for elevated-temperature service listed in Table 5

ASME specification	Tensile strength, MPa	Tensile strength, ksi	Yield strength, minimum, MPa	Yield strength, minimum, ksi	Minimum elongation in 50 mm (2 in.), %	Minimum reduction in area, %
	Mechanical properties					
SA-106A	330	48(a)	207	30	35(b), 25(c)	...
SA-106B	415	60(a)	241	35	30(b), 16.5(c)	...
SA-285A	310–380	45–55	165	24	27(d), 30	...
SA-299	515–620	75–90	290	42	16(d)	
SA-204A	445–530	65–77	255	37	19(d), 23	...
SA302A	515–655	75–95	310	45	15(d), 19	...
SA-533B2	620–790	90–115	475	70	16	...
SA-517F	795–930	115–135	690	100	16	35–45
SA-335P12	415	60(a)	207	30	30(b), 20(c)	...
SA-217WC6	485–620	70–90	275	40	20	35
SA-387Gr22-1	415–585	60–85	207	30	18(d), 45	40
SA-387Gr5-2	515–690	75–100	310	45	18(d), 22	45
SA-217C12	620–795	90–115	415	60	18	35

(a) Minimum. (b) Longitudinal. (c) Transverse. (d) Elongation in 200 mm (8 in.)

Table 7 Compositions and mechanical properties of AISI and AMS steels for elevated-temperature service

AISI designation	AMS designation	Commercial designation	UNS designation	Typical applications	C	Mn	Si	Cr	Mo	V
					Nominal composition, %					
601	6304	...	K14675	Bolting and structural parts	0.46	0.60	0.26	1.00	0.50	0.30
602	6302, 6385, 6458	17-22 AS	K23015	Bolting and structural parts	0.30	0.55	0.65	1.25	0.50	0.25
603	6303, 6436	17-22 AV	K22770	Turbine rotors and aircraft parts	0.27	0.75	0.65	1.25	0.50	0.85
610	6437, 6485	H11 mod	T20811	Ultrahigh-strength components	0.40	0.30	0.90	5.00	1.30	0.50

AISI designation	Room-temperature tensile properties: Yield strength, MPa	Yield strength, ksi	Tensile strength, MPa	Tensile strength, ksi	Elongation in 50 mm (2 in.), %	Reduction in area, %	Temperature at which 70 MPa (10 ksi) will cause rupture in: 1000 h, °C	1000 h, °F	10,000 h, °C	10,000 h, °F	Temperature to produce min creep rate at 70 MPa (10 ksi): 1 μm/m · h, °C	1 μm/m · h, °F	0.1 μm/m · h, °C	0.1 μm/m · h, °F
601	710	103	855	124	29	61	620	1150	595	1100	...	...	...	...
602	745–930	108–135	880–1060	128–154	16–21	53–63	625	1160	590	1090	...	...	555	1030
603	1000	145	1100	160	17	52	650	1200	613	1135	...	...	565	1050
610	1480	215	1805	262	10	36	630	1170	595	1100	560	1040	540	1000

countered by grain refinement and by compositional modifications involving titanium and boron. Melting practice is another factor in improving rupture ductility.

High-Temperature Rotor Applications. Since its introduction in the 1950s, 1Cr-1Mo-0.25V steel has remained the industry standard in turbine rotor applications, although a few higher-alloy martensitic rotor steels (e.g., 12Cr-1Mo-0.3V) have been developed. It is well recognized that 1Cr-1Mo-0.25V rotor steels are limited by their creep strength for service up to about 540 °C (1000 °F).

The desired properties in chromium-molybdenum-vanadium steel rotors is made possible by careful control of heat treatment and composition. In the United States, the usual practice has been to air cool the rotors from the austenitizing temperature in order to achieve a highly creep-resistant, but somewhat less tough, upper bainitic microstructure. In Europe, however, manufacturers have resorted to oil quenching of rotors from the austenitizing temperature to achieve a better compromise between creep strength and toughness. Oil quenching of 1Cr-1Mo-0.25V rotors may shift the transformation product increasingly toward lower bainite, but it is unlikely that the cooling rates needed for formation of martensite (that is, 10,000 °C/h or 20,000 °F/h) are ever encountered.

Numerous investigators have compared the creep properties of chromium-molybdenum-vanadium steels with different microstructures (Ref 8). There is consensus that upper-bainitic structures provide the best creep resistance, coupled with adequate ductility. Toughness properties are discussed in Ref 9.

Turbine Casing Applications. Chromium-molybdenum-vanadium steels are also used for turbine casings. The table below compares the maximum application temperatures of various low-alloy steels used for turbine casings (Ref 10):

Casing material	Maximum application temperature °C	°F
C-0.5Mo (0.25 C max, 0.20–0.50 Si, 0.5–1.0 Mn, 0.50–0.70 Mo)	480	895
Cr-0.5Mo (0.15 C max, 0.60 Si max, 0.5–0.8 Mn, 1.0–1.5 Cr, 0.45–0.65 Mo)	525	975
2.25Cr-1Mo (0.15 C max, 0.45 Si max, 0.4–0.8 Mn)	540	1000
Cr-Mo-V (0.15 C max, 0.15–0.30 Si max, 0.4–0.6 Mn, 0.7–1.2 Cr, 0.7–1.2 Mo, 0.25–0.35 V)	565	1050
0.5Cr-Mo-V (0.1–0.15 C, 0.45 Si max, 0.4–0.7 Mn, 0.4–0.6 Cr, 0.4–0.6 Mo, 0.22–0.28 V)	565	1050

Modified Chromium-Molybdenum Steels

To achieve higher process efficiencies in future coal conversion plants, chemical processing plants, and petrochemical refining plants, several modified versions of chromium-molybdenum pressure vessel steels have been investigated for operation at higher temperatures and pressures than those currently encountered. The higher temperatures affect the elevated-temperature strength, the dimensional deformation, and the metallurgical stability of an alloy, while higher operating pressures require either higher-strength alloys or thicker sections.

Of the unmodified ferritic steels, SA-387 grade 22, class 2 (normalized and tempered 2.25Cr-1Mo unmodified steel) meets the requirements for the fabrication of large pressure vessels per Section VIII, Division 2 of the ASME Boiler and Pressure Vessel Code. Unfortunately, the thick-section hardenability is insufficient to prevent the formation of cementite, even with accelerated cooling procedures and lower tempering conditions (Ref 11). This is a concern because of the possibility of hydrogen attack (Ref 12). Other unmodified chromium-molybdenum steels resist

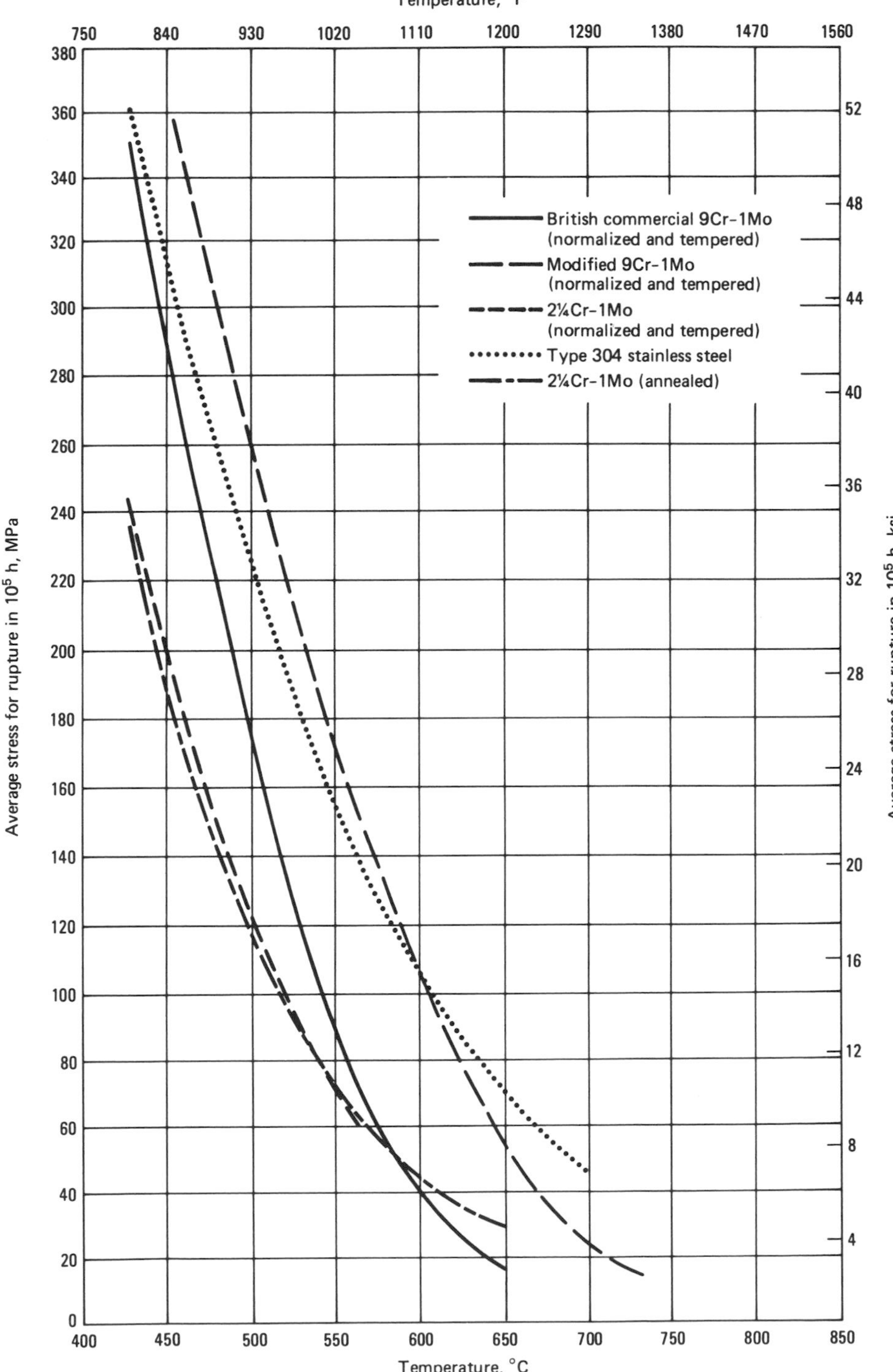

Fig. 6 Variation of 10^5 h creep-rupture strength as a function of temperature for 2.25Cr-1Mo steel, standard 9Cr-1Mo, modified 9Cr-1Mo, and 304 stainless steel. Source: Ref 7

hydrogen attack, such as 3Cr-1Mo and 5Cr-0.5Mo (SA-387 grades 21 and 5), but the design allowables are below those of 2.25Cr-1Mo steel at some temperatures of interest. Higher-chromium alloys, such as 7Cr-0.5Mo and 9Cr-1Mo, also have strengths below normalized and tempered 2.25Cr-1Mo and so have not been considered in the United States for heavy-wall vessels.

Therefore, several modified chromium-molybdenum alloys have been investigated for thick-section vessels in a hydrogen environment. These modified chromium-molybdenum alloys contain various microalloying elements, such as vanadium, niobium, titanium, and boron. Three categories (Ref 11) of modified chromium-molybdenum steels investigated for thick-section applications in a hydrogen environment are:

- *3Cr-1Mo modified with vanadium, titanium, and boron* (Ref 13): This steel is approved for service up to 455 °C (850 °F), is fully hardenable, resists hydrogen attack, and has strengths capable of meeting the design allowables of normalized and tempered 2.25Cr-1Mo steel.
- *9Cr-1Mo steel modified with vanadium and niobium* (Ref 14, 15): This steel has strengths exceeding those of 2.25Cr-1Mo and is approved for use at temperatures above 600 °C (1110 °F) for steam and hydrogen service.
- *2.25Cr-1Mo steel modified with vanadium, titanium, and boron* (Ref 4, 16): Vanadium-modified 2.25Cr-1Mo steel is fully hardenable, resists hydrogen attack, and exceeds the strength of normalized and tempered 2.25Cr-1Mo steel.

Other modified chromium-molybdenum alloys, such as 3Cr-1.5Mo-0.1V-0.1C, have also been investigated (Ref 13, 17). The modified alloys have improved hardenability over unmodified 2.25Cr-1Mo steel, but those with bainitic microstructures undergo a strain softening (Ref 18–21), which may be a limitation in applications with cyclic stresses.

The modified 9Cr-1Mo steel is an attractive alloy because its strengths (Fig. 6) meet or exceed the allowable stresses of stainless steel (Fig. 10). Microstructural studies have indicated that the improved strength of the modified alloy derives from two factors. First, fine $M_{23}C_6$ precipitate particles nucleate on Nb(C,N), which first appears during the heat treatment. Second, the vanadium enters $M_{23}C_6$ and retards its growth at the service temperature. The finer distribution of $M_{23}C_6$ adds to the strength, and its retarded grain-size growth holds the strength for long periods of time at the service temperature. Figure 11 shows the grain-coarsening behavior of the modified 9Cr-1Mo steel as a function of normalizing temperature and time-temperature exposure.

Hot-Work Tool and Die Steels

Many manufacturing operations involve punching, shearing, or forming of metals at high temperatures. Hot-work steels (group H) have been developed to withstand the combinations of heat, pressure, and abrasion associated with such operations. Table 9 gives composition limits for hot-work steel used for tool and die applications.

Classification of Hot-Work Steels

Group H tool steels usually have medium carbon contents (0.35 to 0.45%) and chromium,

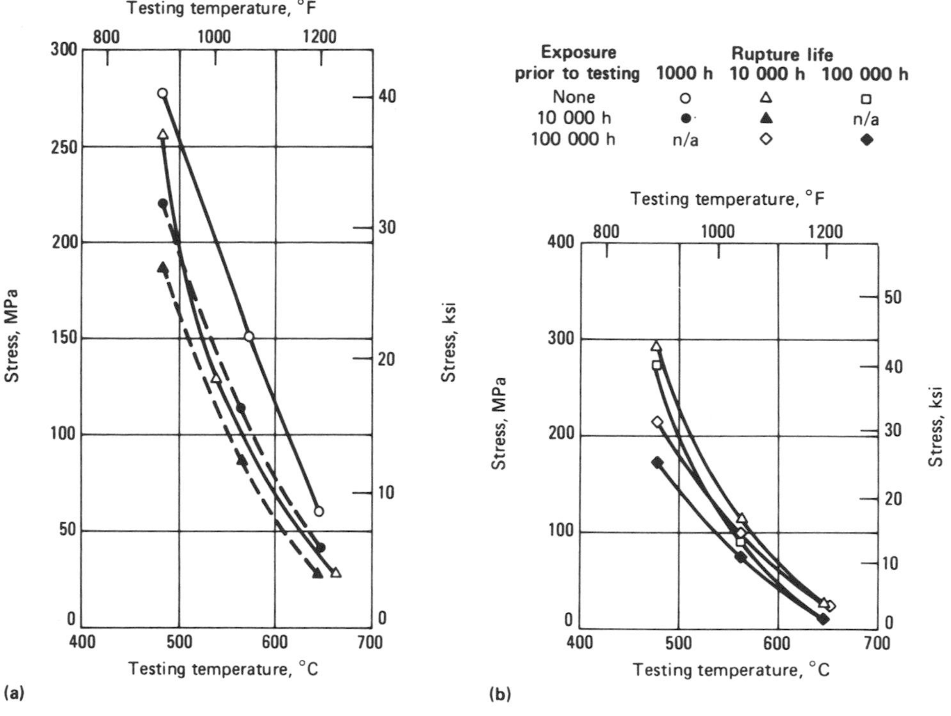

Fig. 7 Effect of elevated-temperature exposure on stress-rupture behavior of (a) normalized and tempered 2.25Cr-1Mo steel and (b) annealed 9Cr-1Mo steel. Exposure to stress-rupture testing was at the indicated test temperatures (without stress) and was 10,000 h long for the 2.25Cr-1Mo steel and 100,000 h long for the 9Cr-1Mo steel. n/a, data not available at indicated exposure and rupture life

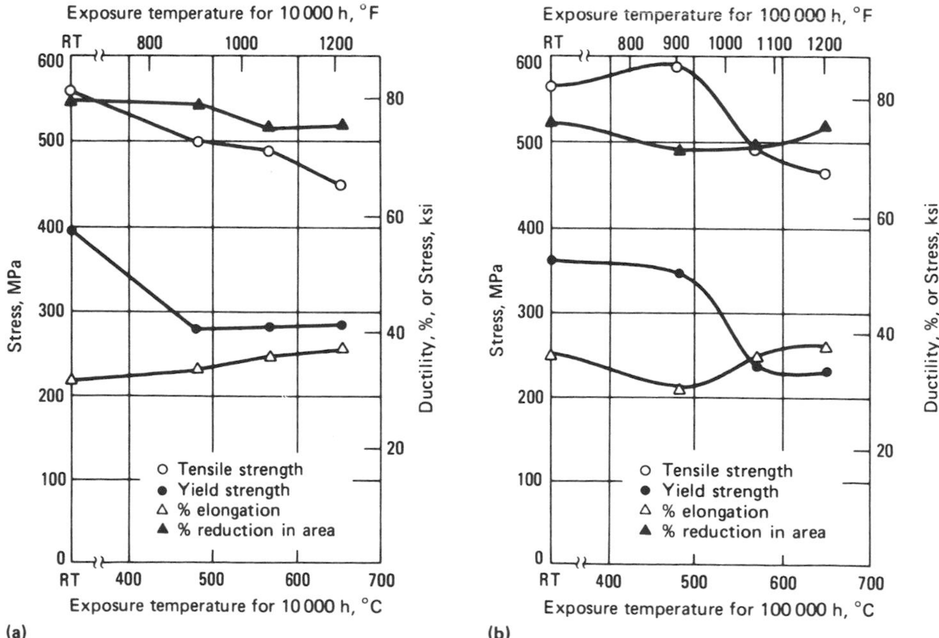

Fig. 8 Effect of temperature exposure on the room-temperature properties of (a) normalized (900 °C, or 1650 °F) and tempered (705 °C, or 1300 °F) 2.25Cr-1Mo steel after exposure (without stress) to indicated temperature for 10,000 h and (b) annealed 9Cr-1Mo steel after exposure (without stress) to indicated temperatures for 100,000 h

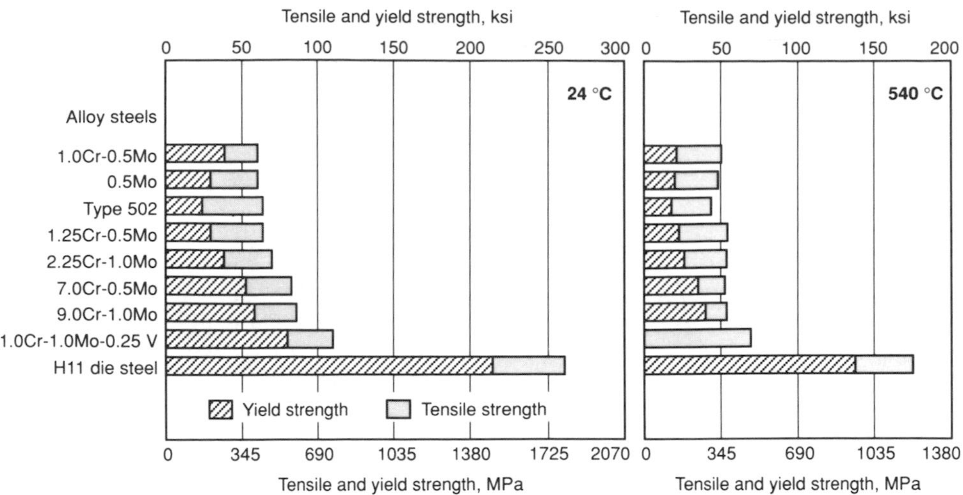

Fig. 9 Room-temperature and short-term elevated-temperature tensile strengths and yield strengths of selected steels containing less than 10% alloy content. The 1.0Cr-0.5Mo steel, 0.5Mo steel, type 502 (5.0Cr-1.0Mn-0.5Mo), and 2.25Cr-1.0Mo steel were annealed at 845 °C (1550 °F). The 1.25Cr-0.5Mo steel was annealed at 815 °C (1500 °F). The 7.0Cr-0.5Mo and 9.0Cr-1.0Mo steels were annealed at 900 °C (1650 °F). The 1.0Cr-1.0Mo-0.25V steel was normalized at 955 °C (1750 °F) and tempered at 650 °C (1200 °F). H11, hardened 1010 °C (1850 °F), tempered 565 °C (1050 °F)

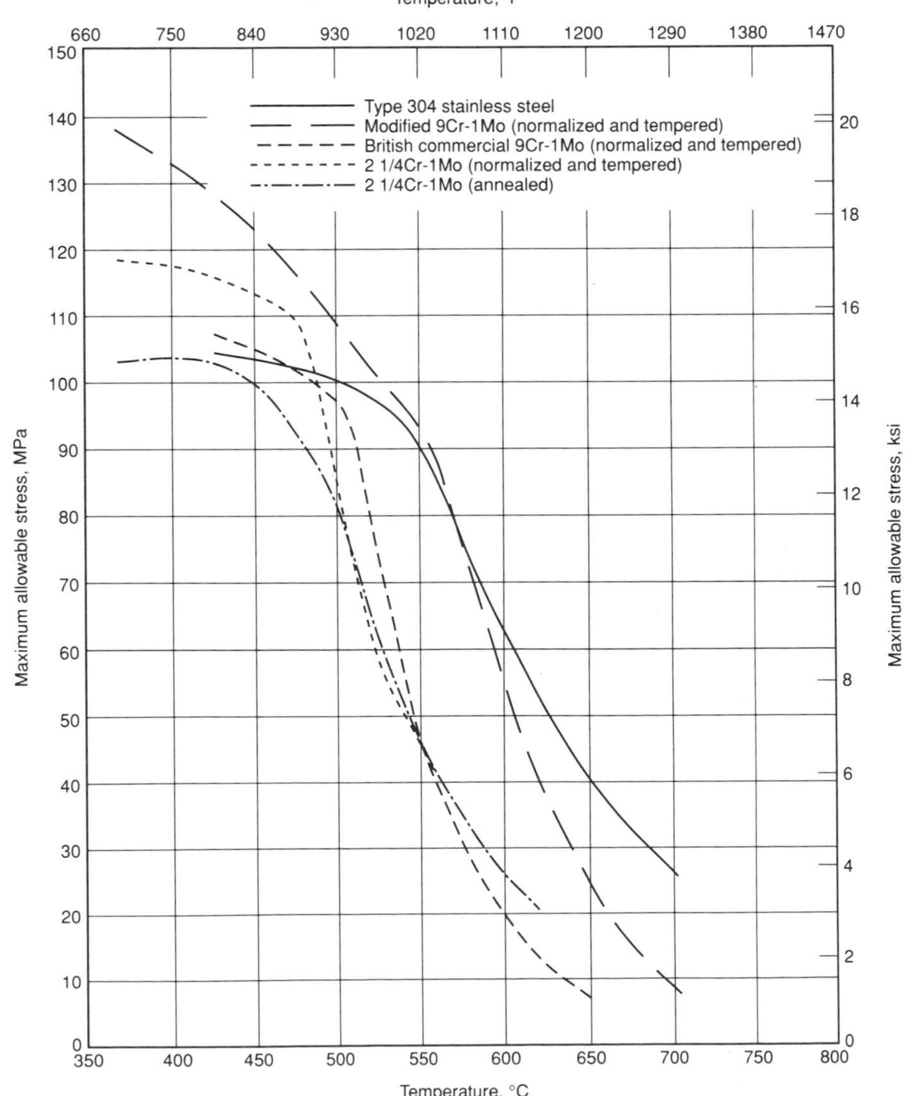

Fig. 10 Estimated design allowable stresses (Section VIII of ASME Boiler and Pressure Vessel Code) as a function of temperature for modified 9Cr-1Mo steel, standard 9Cr-1Mo, 2.25Cr-1Mo steel, and 304 stainless steel. Source: Ref 7

tungsten, molybdenum, and vanadium contents of 6 to 25%. These steels are divided into three subgroups: chromium hot-work steels (types H10 to H19), tungsten hot-work steels (types H21 to H26), and molybdenum hot-work steels (types H42 and H 43).

Chromium hot-work steels (types H10 to H19) have good resistance to heat softening because of their medium chromium content and the addition of carbide-forming elements such as molybdenum, tungsten, and vanadium. The low carbon and low total alloy contents promote toughness at the normal working hardnesses of 40 to 55 HRC. Higher tungsten and molybdenum contents increase hot strength but slightly reduce toughness. Vanadium is added to increase resistance to washing (erosive wear) at high temperatures. An increase in silicon content improves oxidation resistance at temperatures up to 800 °C (1475 °F). The most widely used types in this group are H11, H12, H13, and, to a lesser extent, H19.

Tungsten Hot-Work Steels. The principal alloying elements of tungsten hot-work steels (types H21 to H26) are carbon, tungsten, chromium, and vanadium. The higher alloy contents of these steels make them more resistant to high-temperature softening and washing than H11 and H13 hot-work steels. However, high alloy content also makes them more prone to brittleness at normal working hardnesses (45 to 55 HRC) and makes it difficult for them to be safely water cooled in service.

Molybdenum Hot-Work Steels. There are only two active molybdenum hot-work steels: type H42 and type H43. These alloys contain molybdenum, chromium, vanadium, carbon, and varying amounts of tungsten. They are similar to tungsten hot-work steels, having almost identical characteristics and uses. Although their compositions resemble those of various molybdenum high-speed steels, they have a low carbon content and greater toughness. The principal advantage of types H42 and H43 over tungsten hot-work steels is their lower initial cost. They are also more resistant to heat checking.

Properties of Hot-Work Steels

Hot Hardness. In all hot forging applications, the die steel should have a high hot hardness and should retain this hardness over extended periods of exposure to elevated temperatures. Figure 12 shows hot hardnesses of five AISI hot-work die steels at various temperatures. All of these steels were heat treated to about the same initial hardness. Hardness measurements were made after holding the specimens at testing temperature for 30 min. Except for H12, all the die steels considered have about the same hot hardness at temperatures below about 315 °C (600 °F). The differences in hot hardness become apparent only at temperatures above 480 °C (900 °F).

Thermal Softening. Figure 13 shows the resistance of some hot-work die steels to softening at elevated temperatures after 10 h of exposure. All of these steels have about the same initial hardness after heat treatment. For the die steels shown, there is not much variation in resistance to softening at temperatures below 540 °C (1000 °F). However, for longer periods of exposure at higher temperatures, high-alloy hot-work steels, such as H19, H21, and H10 modified, retain hardness better than medium-alloy steels, such as H11.

Resistance to Plastic Deformation. Failure of forging dies to perform properly because of plastic deformation can be measured by hot hardness or yield strength. In general, the yield strength of a steel decreases with increasing temperature. However, yield strength is also dependent on the prior heat treatment, composition, and hardness. The higher the initial hardness, the greater the yield strength at various temperatures. In addition, the yield strengths of different die steels increase with alloy content: tungsten hot-work die steels are harder than chromium hot-work die steels, which are themselves harder than the low-alloy steels (Fig. 14). Not shown in Fig. 14 is the fact that the molybdenum hot-work die steels are even harder than those made of tungsten and thus manifest the greatest resistance to plastic deformation of all die steels.

Low-Strain-Rate Toughness. Figure 15 shows the ductility of various hot-work steels at elevated temperatures, as measured by the percent reduction in area of a specimen before fracture in a standard tensile test. As the curves show, high-alloy hot-work steels, such as H19 and H21, have less ductility than medium-alloy hot-work steels, such as H11.

High-Strain-Rate Toughness. Figure 16 shows the results of elevated-temperature Charpy V-notch tests on various die steels. The data show that toughness decreases as the content of the steel increases. Medium-alloy steels, such as H11, H12, and H13, have better resistance to brittle fracture than H14, H19, and H21, which have higher alloy contents. Increasing the hardness of a steel lowers its impact strength. On the other hand, wear resistance and hot strength decrease with decreasing hardness. Thus, a compromise is made in actual practice, and the dies are tempered to near-maximum hardness levels at which they have sufficient toughness to withstand loading.

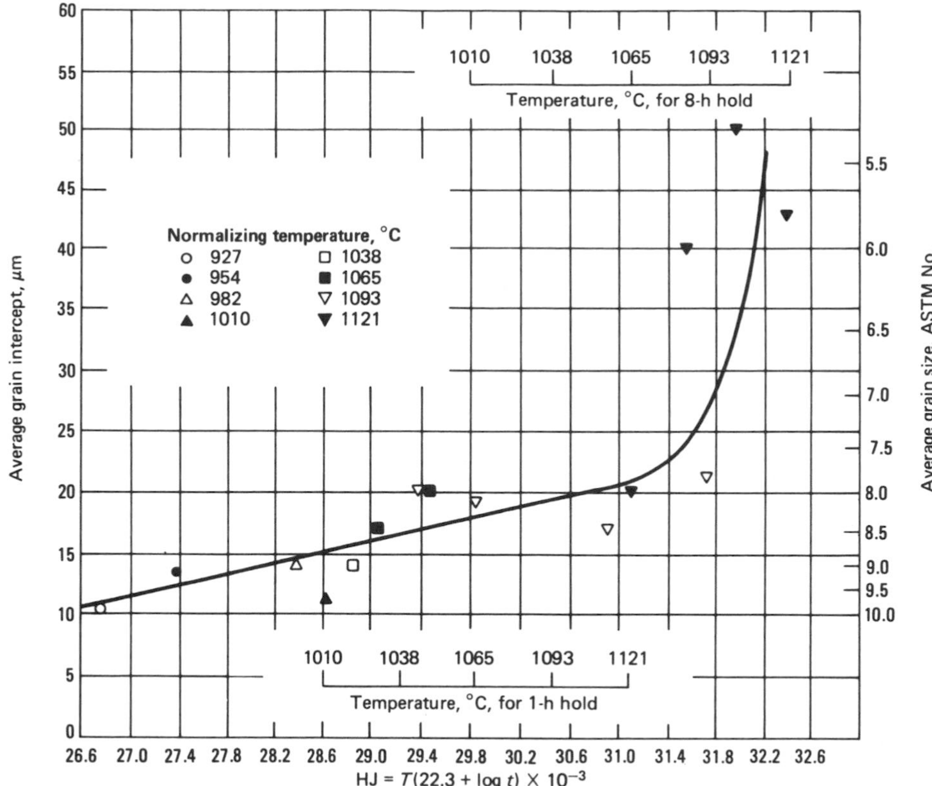

Fig. 11 Grain-coarsening behavior of a modified 9Cr-1Mo steel (9Cr-1Mo steel with 0.06 to 0.10% Nb and 0.18 to 0.25% V). Source: Ref 7

Thermal fatigue is the gradual deterioration and eventual cracking of a material from cyclic thermal transients. Conventional hot forging is intermittent by nature, so there may be thermal fatigue of die surfaces. This results in the second most common reason for rejecting dies: heat checking.

Heat checking is the development of minute surface cracks, mainly in corners or on projections within the die cavity. It results from high surface temperatures that cause greater thermal expansion at working surfaces than in the interior of the die block. The difference in expansion causes plastic deformation at the surface, which results in tensile stresses and cracking on cooling. Once surface cracks are started, working pressures (especially impact) will cause them to grow, and if crack growth is allowed to continue, the die

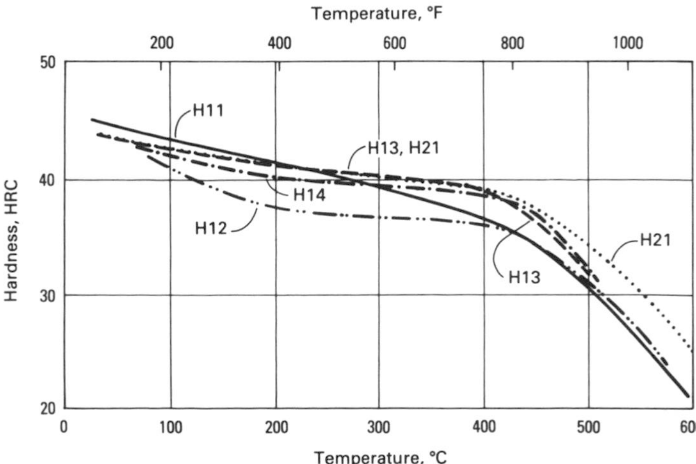

Fig. 12 Hot hardnesses of AISI hot-work tool steels. Measurements were made after holding at the test temperature for 30 min.

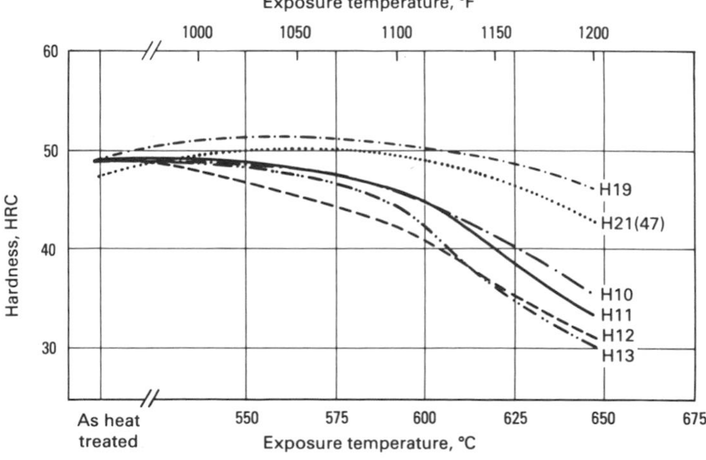

Fig. 13 Resistance of hot-work AISI tool steels to softening during 10 h elevated-temperature exposure, as measured by room-temperature hardness. Unless otherwise specified by values in parentheses, initial hardness of all specimens was 49 HRC.

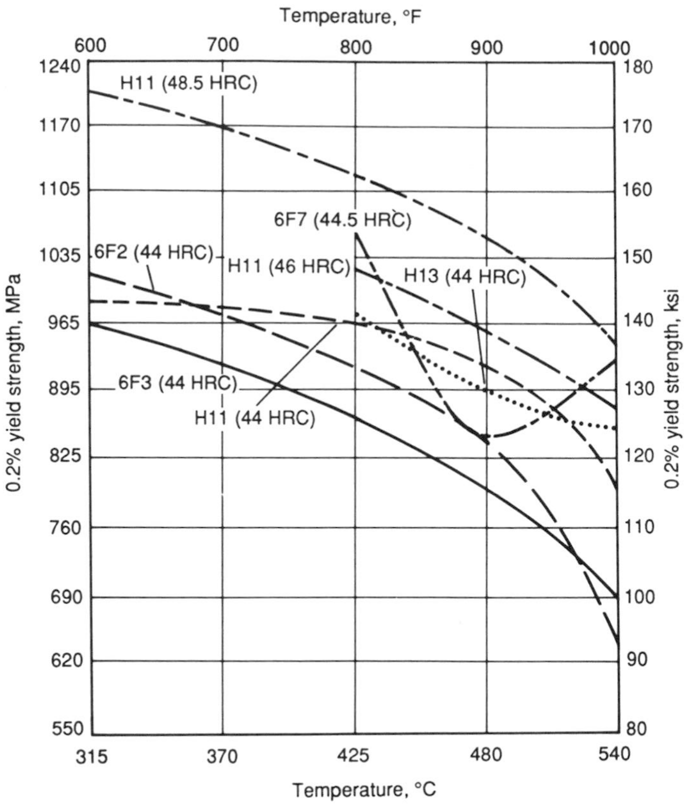

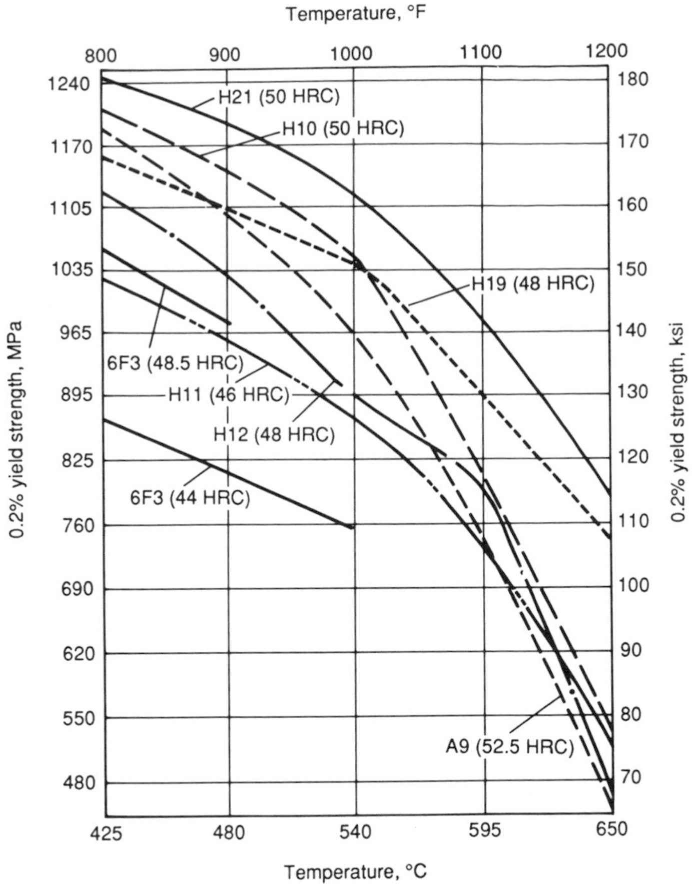

Fig. 14 Resistance of selected die steels to plastic deformation at elevated temperatures. Values in parentheses indicate room-temperature hardness.

block will break. An example of heat checking is shown in Fig. 17. The resistance to thermal fatigue can be improved by using a steel with higher yield strength, lowering the maximum die temperature variations, or treating the surface.

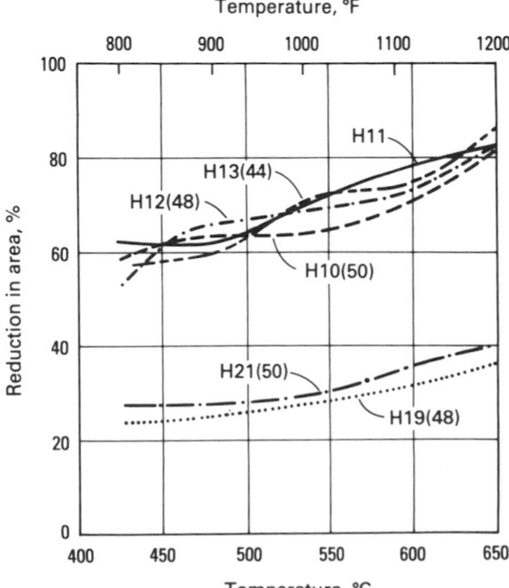

Fig. 15 Elevated-temperature ductilities of various hot-work die steels. Values in parentheses indicate room-temperature Rockwell C hardness.

Microstructural Effects on Thermal Fatigue. Ductility has a large effect on the number of thermal cycles that a forging die can undergo prior to forming cracks. Therefore, microstructure can have a significant impact on the frequency of heat checking. The most important microstructural variables are cleanliness, grain size, and microstructural uniformity. Inclusions act as nuclei for crack initiation, so a slightly more expensive steel that has been refined to remove inclusions may be a wise investment. Grain size can also affect thermal fatigue resistance, because grain size has a large influence on crack initiation. Fine-grained material tends to perform better in this respect. Lastly, steels whose chemistry and microstructure are uniform (i.e., free of segregation) tend to have uniform thermal expansion coefficients. They are thus able to resist thermal stresses and strains that may develop due to variations in the temperature field.

Maraging Steels

Maraging steels comprise a special class of high-strength steels. Instead of being hardened by a metallurgical reaction involving carbon, they are strengthened by the precipitation of intermetallic compounds at temperatures of about 480 °C (900 °F). The term *maraging* is derived from the term *martensite age hardening* and denotes the age hardening of a low-carbon, iron-nickel lath martensite matrix.

Commercial maraging steels are designed to provide specific levels of yield strength, from 1030 to 2420 MPa (150 to 350 ksi). Some experimental maraging steels have yield strengths as high as 3450 MPa (500 ksi). These steels typically have very high nickel, cobalt, and molybdenum contents and very low carbon contents. Carbon, in fact, is an impurity in these steels and is kept as low as commercially feasible in order to minimize the formation of titanium carbide (TiC), which can adversely affect strength, ductility, and toughness. Other varieties of maraging steel have been developed for special applications. Composition ranges for maraging steels are listed in Table 10.

Elevated-Temperature Characteristics. The effects of temperature on the mechanical properties of maraging steels are illustrated in Fig. 18. Maraging steels can be used for prolonged service at temperatures up to approximately 400 °C (750 °F). Yield and tensile strengths at 400 °C (750 °F) are about 80% of the room-temperature values. Long-term stress-rupture failures can occur at 400 °C (750 °F), but the rupture stresses are fairly high. At temperatures above 400 °C (750 °F), reversion of the martensite matrix to austenite becomes dominant, and long-term load-carrying capacities decay fairly quickly.

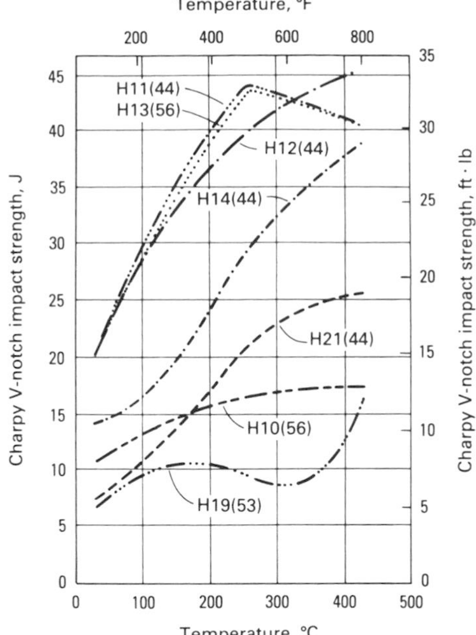

Fig. 16 Effect of hardness, composition, and testing temperature on Charpy V-notch impact strength of hot-work die steels. Values in parentheses indicate Rockwell C hardness at room temperature.

Table 8 Product forms and ASTM specifications for chromium-molybdenum steels

Type	Forgings	Tubes	Pipe	Castings	Plate
0.5Cr-0.5Mo	A 182-F2	...	A 335-P2 A 369-FP2 A 426-CP2	...	A 387-Gr2
1Cr-0.5Mo	A 182-F12 A 336-F12	...	A 335-P12 A 369-FP12 A 426-CP12	...	A 387-Gr12
1.25Cr-0.5Mo	A 182-F11 A 336-F11/F11A A 541-C11C	A 199-T11 A 200-T11 A 213-T11	A 335-P11 A 369-FP11 A 426-CP11	A 217-WC6 A 356-Gr6 A 389-C23	A 387-Gr11 ...
2.25Cr-1Mo	A 182-F22/F22a A 336-F22/F22A A 541-C22C/22D	A 199-T22 A 200-T22 A 213-T22	A 335-P22 A 369-FP22 A 426-CP22	A 217-WC9 A 356-Gr10	A 387-Gr22 A 542
3Cr-1Mo	A 182-F21 A 336-F21/F21A	A 199-T21 A 200-T21 A 213-T21	A 335-P21 A 369-FP21 A 426-CP21	...	A 387-Gr21
3Cr-1MoV	A 182-F21b	...	...	...	...
5Cr-0.5Mo	A 182-F5/F5a A 336-F5/F5A A 473-501/502	A 199-T5 A 200-T5 A 213-T5	A 335-P5 A 369-FP5 A 426-CP5	A 216-C5	A 387-Gr5
5Cr-0.5MoSi	...	A 213-T5b	A 335-P5b A 4260CP5b	...	...
5Cr-0.5MoTi	...	A 213-T5c	A 335-P5c	...	...
7Cr-0.5Mo	A 182-F7 A 473-501A	A 199-T7 A 200-T7 A 213-T7	A 335-P7 A 369-FP7 A 426-CP7	...	A 387-Gr7
9Cr-1Mo	A 182-F9 A 336-F9 A 473-501B	A 199-T9 A 200-T9 A 213-T9	A 335-P9 A 369-FP9 A 426-CP9	A 217-C12	A 387-Gr9

Mechanical Properties at Elevated Temperatures

The allowable design stresses for steels at elevated temperatures may be controlled by different mechanical properties, depending on the application and temperature exposure. For applications with temperatures below the creep-temperature range, the tensile strength or the yield strength at the expected service temperature generally controls allowable stresses. For temperatures in the creep range, allowable stresses are determined from either creep-rupture properties or the degree of deformation from creep. In recent years, the worldwide interest in life extension of high-temperature components has also promoted considerably more interest in elevated-temperature fatigue. This effort has led to tests and methods for evaluating the effects of creep-fatigue interaction on the life of elevated-temperature components.

Ductility and toughness are not commonly considered when allowable stresses are set, but these properties can be important. In elevated-temperature applications, ductility and toughness often change with temperature and with time at temperature. Sometimes the changes are beneficial, but more often they are deleterious. The changes are of interest both at service temperature and, because of shutdowns, at ambient temperatures. Ductility also influences notch sensitivity and creep-fatigue interaction.

The types of tests used to evaluate the mechanical properties of steels at elevated temperatures include:

- Short-term elevated-temperature tests
- Long-term elevated-temperature tests
- Fatigue tests (including thermal fatigue and thermal shock tests)
- Time-dependent fatigue tests
- Ductility and toughness tests
- Short-term and long-term tests following long-term exposure to elevated temperatures

Several methods are used to interpret, interpolate, and extrapolate the data from some of these tests, as described in the section "Methods for Correlating, Interpolating, and Extrapolating Elevated-Temperature Mechanical Property Data" in this article.

Short-Term Elevated-Temperature Tests

Short-term elevated-temperature tests include the elevated-temperature tensile tests (described in ASTM E 21), a test for elastic modulus (ASTM E 231), compression tests, pin bearing load tests, and the hot hardness test.

The mechanical properties determined by means of the tensile tests include ultimate tensile strength, yield strength, percent elongation, and percent reduction in area. Because elevated-temperature tensile properties are sensitive to strain rate, these tests are conducted at carefully controlled strain rates. Figure 19 shows tensile

Fig. 17 Gross heat checking in a low-alloy tool steel forging die due to excessive temperature. Heat checking occurred after an undetermined number of 225 kg (500 lb) nickel-base alloy preforms had been forged from an average temperature of 1100 °C (2000 °F).

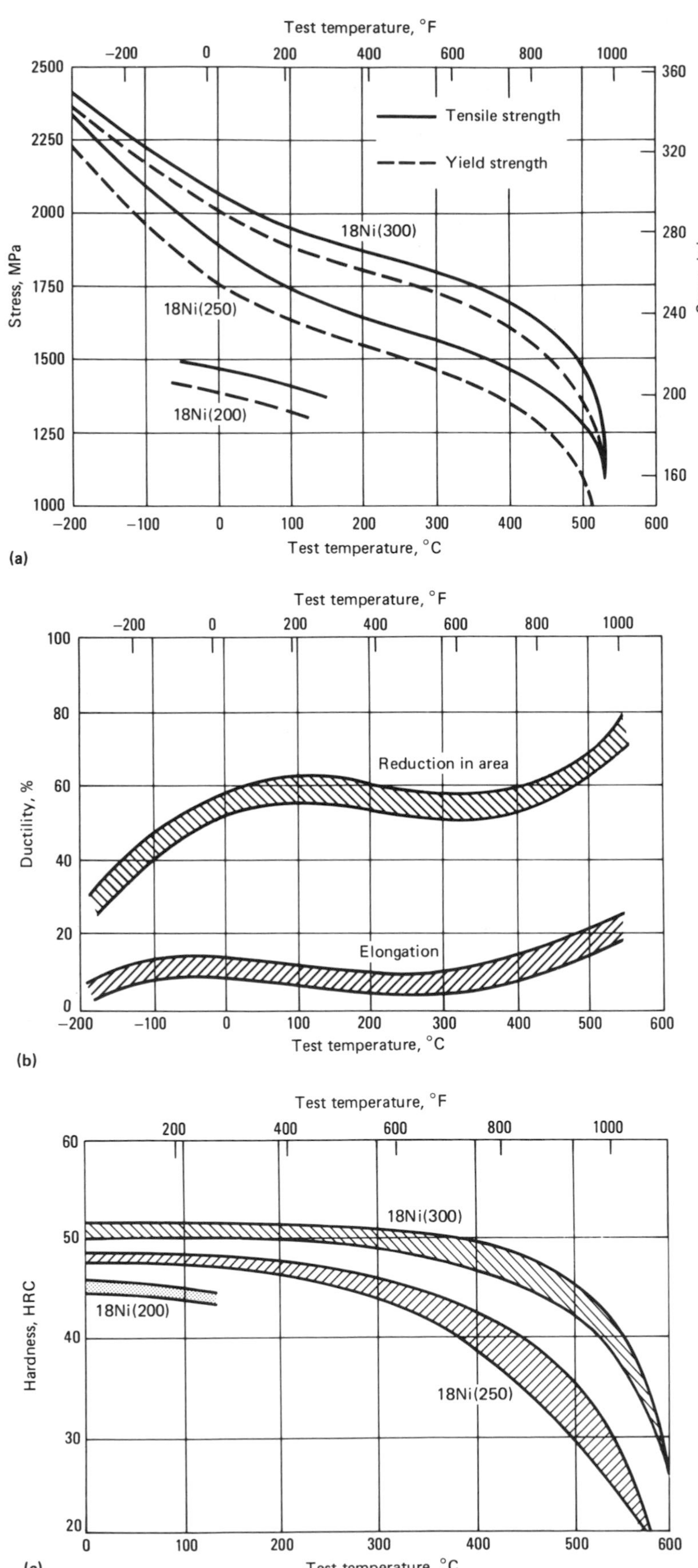

Fig. 18 Effect of temperature on the mechanical properties of 18Ni maraging steels. (a) Strength. (b) Ductility. (c) Hardness

strength data obtained on specimens of annealed 2.25Cr-1Mo steel at various temperatures and at strain rates ranging from 2.7×10^{-6} s^{-1} to 144 s^{-1}.

In designing components that will be produced from low-alloy steels and will be exposed to temperatures up to 370 °C (700 °F), the yield and ultimate strengths at the maximum service temperature can be used much as they would be used in the design of components for service at room temperature. Figure 9 compares the short-term elevated-temperature yield and tensile strengths of selected alloys. Certain codes require that appropriate factors be applied in calculating allowable stresses.

Elevated-temperature values of elastic modulus can be determined during tensile testing or dynamic testing by measuring the natural frequency of a test bar at the designated test temperature. Figure 20 shows values of elastic modulus at temperatures between room temperature and 650 °C (1200 °F) for several low-alloy steels, determined during static tensile loading and dynamic loading.

Compression tests and pin bearing load tests (ASTM E 209 and E 238) can be used to evaluate materials for applications in which the components will be subjected to these types of loading at elevated temperatures. Hot hardness tests can be used to evaluate materials for elevated-temperature service and can be applied to the qualification of materials in the same way in which room-temperature hardness tests are applied.

Components for many elevated-temperature applications are joined by welding. Elevated-temperature properties of both the weld metal and the heat-affected zones can be determined by the same methods used to evaluate the properties of the base metal.

Long-Term Elevated-Temperature Tests

Long-term elevated-temperature tests are used to evaluate the effects of creep, which is defined as the time-dependent strain that occurs under constant load at elevated temperatures. Creep is observed in steels at temperatures above about 370 °C (700 °F). In general, creep occurs at a temperature slightly above the recrystallization temperature of a metal or alloy. At such a temperature, atoms become sufficiently mobile to allow time-dependent rearrangement of the structure. In time, creep may lead to excessive deformation, and even fracture, at stresses considerably below those determined in room-temperature and elevated-temperature short-term tension tests.

Typical creep behavior consists of three distinct stages, as shown in Fig. 21. Following initial elastic-plastic strain resulting from the immediate effects of the applied load, there is a region of increasing plastic strain at a decreasing strain rate (first-stage creep, also called primary creep). Next, there is a region where the creep strain increases at a minimum, and almost constant rate of plastic strain (second-stage creep, also called secondary creep). This nominally constant creep rate is generally known as the minimum creep rate and is widely employed in research and engineering studies. Finally, there is a region of drastically increased strain rate with rapid extension to fracture (third-stage creep, also called tertiary creep). Tertiary creep has no distinct beginning, but this term refers to the region with an increasing rate of extension that is followed by fracture.

Of all the parameters pertaining to the creep curve, the most important for engineering applications are the creep rate and the time to rupture. These parameters are determined from long-term elevated-temperature tests that include creep, creep-rupture, and stress-rupture tests (ASTM E 139) and notched-bar rupture tests (ASTM E 292). In addition, relaxation tests (ASTM E 328) are used to evaluate the effect of creep behavior on the performance of high-temperature bolt steels.

Creep Strength. When the rate or degree of deformation is the limiting factor, the design stress is based on the minimum (secondary) creep rate and design life after allowing for initial transient creep. The stress that produces a specified minimum creep rate of an alloy or a specified amount of creep deformation in a given time (for example, 1% total creep in 100,000 h) is called the limiting creep strength or limiting stress. Typical creep strengths of various low-alloy steels are shown in Fig. 3. Table 2 also lists some suggested maximum service temperatures of various low-alloy steels based on creep rate. Figure 22 shows the 0.01%/1000 h creep strength of carbon steel as a function of room-temperature tensile strength.

Stress Rupture. When fracture is a limiting factor, stress-rupture values are used in design. Stress-rupture values of various low-alloy chromium-molybdenum steels are shown in Fig. 5 to 7. Figures 23 and 24 show typical creep-rupture values of carbon and 1Cr-1Mo-0.25V steel, respectively.

Long-term creep and stress-rupture values (e.g., 100,000 h) are often extrapolated from

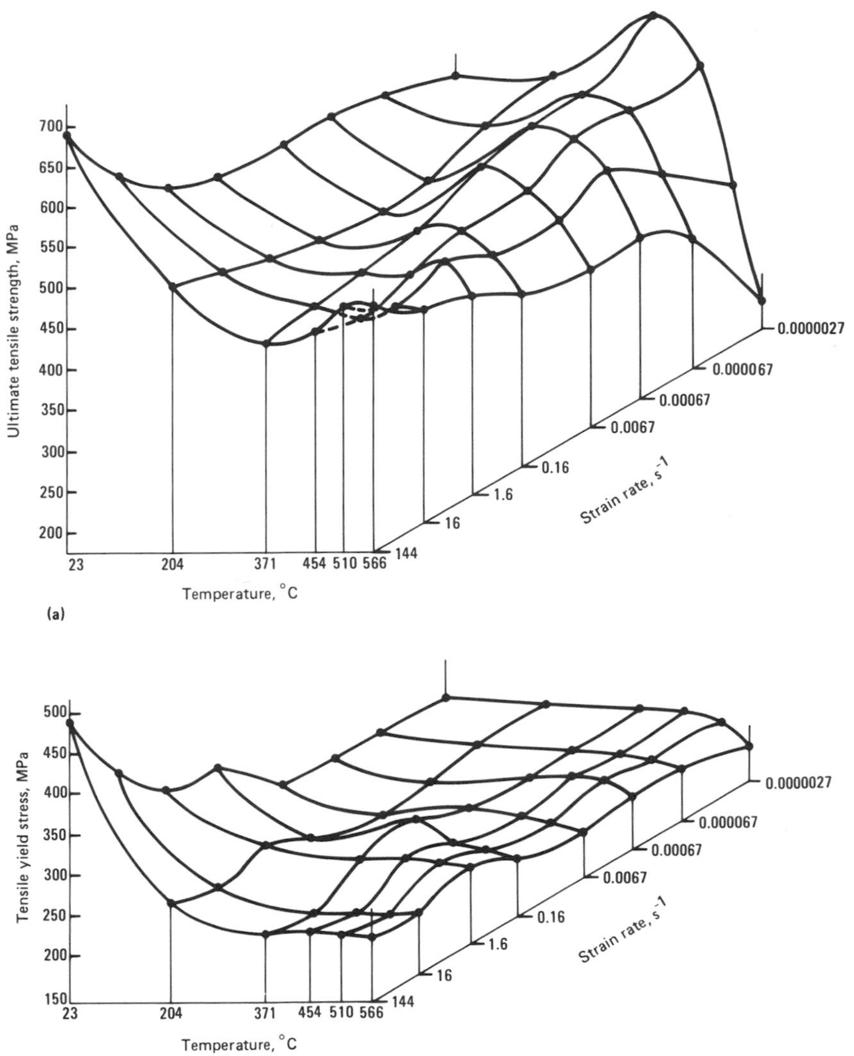

Fig. 19 Effect of test temperature and strain rate on the strength of annealed 2.25Cr-1Mo steel. Tensile strength (a) and yield strength (b) of 2.25Cr-1Mo steel tested at various temperatures and strain rates. Source: Ref 22

Table 9 Composition limits of hot-work tool steels

Designation		Composition, %								
AISI	UNS	C	Mn	Si	Cr	Ni	Mo	W	V	Co
Chromium hot-work steels										
H10	T20810	0.35–0.45	0.25–0.70	0.80–1.20	3.00–3.75	0.30 max	2.00–3.00	...	0.25–0.75	...
H11	T20811	0.33–0.43	0.20–0.50	0.80–1.20	4.75–5.50	0.30 max	1.10–1.60	...	0.30–0.60	...
H12	T20812	0.30–0.40	0.20–0.50	0.80–1.20	4.75–5.50	0.30 max	1.25–1.75	1.00–1.70	0.50 max	...
H13	T20813	0.32–0.45	0.20–0.50	0.80–1.20	4.75–5.50	0.30 max	1.10–1.75	...	0.80–1.20	...
H14	T20814	0.35–0.45	0.20–0.50	0.80–1.20	4.75–5.50	0.30 max	...	4.00–5.25	...	...
H19	T20819	0.32–0.45	0.20–0.50	0.20–0.50	4.00–4.75	0.30 max	0.30–0.55	3.75–4.50	1.75–2.20	4.00–4.50
Tungsten hot-work steels										
H21	T20821	0.26–0.36	0.15–0.40	0.15–0.50	3.00–3.75	0.30 max	...	8.50–10.00	0.30–0.60	...
H22	T20822	0.30–0.40	0.15–0.40	0.15–0.40	1.75–3.75	0.30 max	...	10.00–11.75	0.25–0.50	...
H23	T20823	0.25–0.35	0.15–0.40	0.15–0.60	11.00–12.75	0.30 max	...	11.00–12.75	0.75–1.25	...
H24	T20824	0.42–0.53	0.15–0.40	0.15–0.40	2.50–3.50	0.30 max	...	14.00–16.00	0.40–0.60	...
H25	T20825	0.22–0.32	0.15–0.40	0.15–0.40	3.75–4.50	0.30 max	...	14.00–16.00	0.40–0.60	...
H26	T20826	0.45–0.55(a)	0.15–0.40	0.15–0.40	3.75–4.50	0.30 max	...	17.25–19.00	0.75–1.25	...
Molybdenum hot-work steels										
H42	T20842	0.55–0.70(a)	0.15–0.40	...	3.75–4.50	0.30 max	4.50–5.50	5.50–6.75	1.75–2.20	...

(a) Available in several carbon ranges

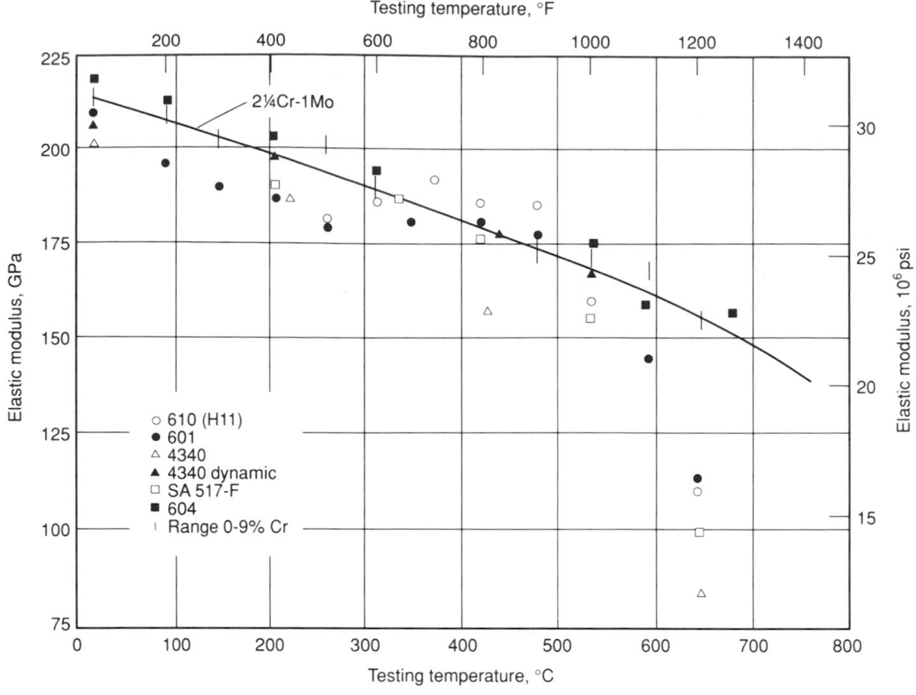

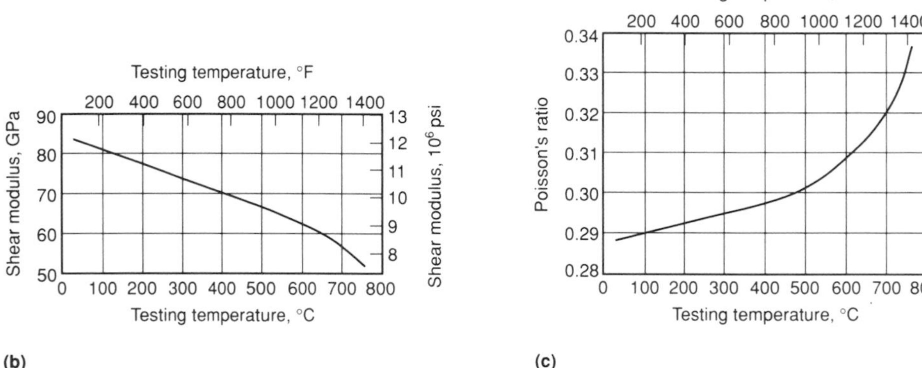

Fig. 20 Effect of test temperature on elastic modulus, shear modulus, and Poisson's ratio. (a) Effect of test temperature on elastic modulus for several steels commonly used at elevated temperatures. Dynamic measurements of elastic modulus were made by determining the natural frequencies of test specimens; static measurements were made during tensile testing. (b) Effect of test temperature on shear modulus of 2.25Cr-1Mo steel. (c) Effect of test temperature on Poisson's ratio of 2.25Cr-1Mo steel. Source: Ref 23

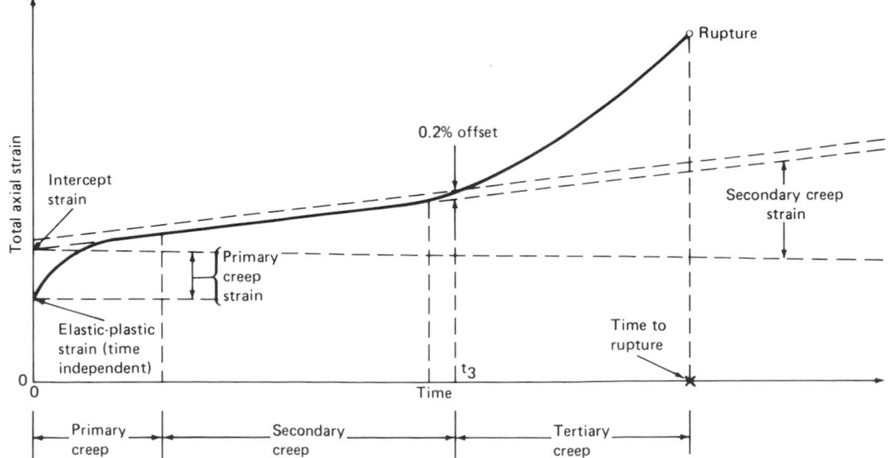

Fig. 21 Schematic representation of classical creep behavior

shorter-term tests. Whether these property values are extrapolated or determined directly often has little bearing on the operating life of high-temperature parts. The actual material behavior is often difficult to predict accurately because of the complexity of the service stresses relative to the idealized, uniaxial loading conditions in the standardized tests and because of attenuating factors, such as cyclic loading, temperature fluctuations, or metal loss from corrosion.

For alloys in which failure occurs before a well-defined start of tertiary creep, it is useful to use notched specimens or specimens with both smooth and notched test sections (with the cross-sectional area of the notch equal to that of the smooth test section). If the material is notch sensitive, the specimen will fail in the notch before failure occurs in the smooth section. It has been well recognized for many years that notch sensitivity is related to creep ductility. It has been suggested that a minimum smooth-bar creep ductility of about 10% in terms of reduction in area may be desirable for avoidance of notch sensitivity (Ref 25, 26). Limited published data on notched stress-rupture properties of low-alloy ferritic steels for elevated temperatures indicate that these steels generally are not notch sensitive. Representative stress-rupture data for notched and unnotched specimens of AISI 603 steel are presented in Fig. 25.

Relaxation Tests. Creep tests on metals are usually carried out by keeping constant either the applied load or the stress, and by noting the specimen strain as a function of time. In another type of test, known as the stress relaxation test, a sample is first deformed to a given strain and then the stress is measured as a function of time, such that the total strain remains constant. Stress relaxation tests are more difficult to carry out than ordinary creep tests and are more difficult to interpret. However, stress relaxation is an important elevated-temperature property in the design of bolts or other devices intended to hold components in contact under pressure. If the service temperature is high enough, the extended-time stress on the bolt causes a minute amount of creep, which results in a reduction in the restraining force.

Because of their low relaxed stresses, carbon steels are usually used only at temperatures below 370 °C (700 °F). Various low-alloy steels have been widely used up to metal temperatures of about 540 °C (1000 °F). Modified 12% Cr martensitic steels can be used for slightly higher temperatures. The common austenitic stainless steels are seldom used because of their low yield strength in the annealed condition, but they are used in the cold-worked condition. Nickel-base superalloys are usually employed only at the highest temperatures. The comparative 1000 h relaxation strengths of this class of alloys are shown in Fig. 26(a). Additional data are provided in Ref 29.

Carbon steel is not recognized as a high-temperature bolting material under ASTM standards or by the ASME Boiler Code. One of the most widely used low-alloy steels for moderately high-

Table 10 Nominal compositions of commercial maraging steels

Grade	Composition(a), %					
	Ni	Mo	Co	Ti	Al	Nb
Standard grades						
18Ni(200)	18	3.3	8.5	0.2	0.1	...
18Ni(250)	18	5.0	8.5	0.4	0.1	...
18Ni(300)	18	5.0	9.0	0.7	0.1	...
18Ni(350)	18	4.2(b)	12.5	1.6	0.1	...
18Ni(Cast)	17	4.6	10.0	0.3	0.1	...
12-5-3(180)(c)	12	3	...	0.2	0.3	...
Cobalt-free and low-cobalt-bearing grades						
Cobalt-free 18Ni(200)	18.5	3.0	...	0.7	0.1	...
Cobalt-free 18Ni(250)	18.5	3.0	...	1.4	0.1	...
Low-cobalt 18Ni(250)	18.5	2.6	2.0	1.2	0.1	0.1
Cobalt-free 18Ni(300)	18.5	4.0	...	1.85	0.1	...

(a) All grades contain no more than 0.03% C. (b) Some producers use a combination of 4.8% Mo and 1.4% Ti, nominal. (c) Contains 5% Cr

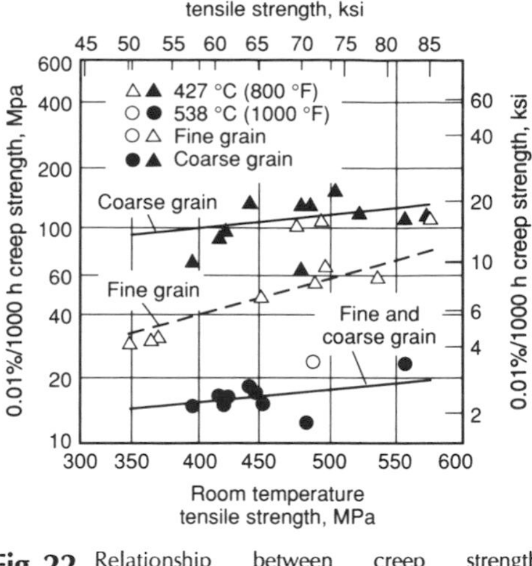

Fig. 22 Relationship between creep strength (0.01%/1000 h) and ultimate tensile strength of a carbon steel. Creep strength estimates made using isothermal lot constants. Source: Ref 24

temperature bolting applications is quenched and tempered 4140, in accordance with ASTM A 193, grade B7. Its relaxation behavior is approximately indicated by the solid lines for steels with 0.65 to 1.10% Cr and 0.10 to 0.30% Mo in Fig. 26(b) and (c). The relaxation strength of 4140 is greater after normalizing and tempering than in the quenched and tempered condition, in order to obtain more consistent mechanical properties. Chromium-molybdenum steels that contain approximately 0.50% Mo (A 193, grade B7A) have also been widely used. They are similar to 4140 but have slightly higher relaxation strength (and are less readily available).

The strongest low-alloy steels are those with approximately 1% Cr, 0.5% Mo, and 0.25% V, in the normalized and tempered condition (A 193, grade B14) or the quenched and tempered condition (A 193, grade B16). Some of these grades are produced with rather high silicon contents (~0.75%), which seems to increase resistance to tempering. These grades have been satisfactory in service up to 540 °C (1000 °F) in the absence of excessive follow-up or retightening. However, they are somewhat notch sensitive in creep rupture and in impact at room temperature, especially in the normalized and tempered condition.

Fatigue

At room temperature and in nonaggressive environments (and except at very high frequencies), the frequency at which loads are applied has little effect on the fatigue strength of most metals. The effects of frequency, however, become much greater as the temperature increases or as the presence of corrosion becomes more significant. At high temperatures, creep becomes more of a factor, and the fatigue strength seems to depend on the total time for which stress is applied rather than solely on the number of cycles. The behavior occurs because the continuous deformation (creep) under load at high temperatures affects the propagation of fatigue cracks. This effect is referred to as creep-fatigue interaction. The primary objectives in time-dependent fatigue tests are quantification of creep-fatigue interaction effects and application of this information to life prediction procedures. Time-dependent fatigue tests are also used to assess the effect of load frequency on corrosion fatigue.

Effect of Load Frequency on Corrosion Fatigue. In aggressive environments, fatigue strength is strongly dependent on frequency. Corrosion fatigue strength (endurance limit at a prescribed number of cycles) generally decreases as the cyclic frequency is decreased. This effect is most important at frequencies of less than 10 Hz.

The frequency dependence of corrosion fatigue is thought to result from the fact that the interaction of a material and its environment is essentially a rate-controlled process. Low frequencies, especially at low strain amplitudes or when there is substantial elapsed time between changes in stress levels, allow time for interaction between material and environment. High frequencies do not, particularly when high strain amplitude is also involved. At very high frequencies or in the plastic-strain range, localized heating may seriously affect the properties of the part. Such effects normally are not considered to be related to a corrosion fatigue phenomenon.

When environments have a deleterious effect on fatigue behavior, a critical range of frequencies of loading may exist in which the mechanical/environmental interaction is significant. Above this range the effect usually disappears, while below this range the effect may diminish. Additional information on corrosion fatigue characteristics of steels can be found in the article "Mechanisms of Corrosion Fatigue" in Volume 19 of the *ASM Handbook.*

Creep-fatigue interaction is an elevated-temperature phenomenon that can seriously reduce fatigue life and creep-rupture strength. Figure 27 illustrates the effect of time-dependent fatigue when the elevated temperature is within the creep range of a material. It shows a continuous strain cycling waveform (Fig. 27a) and a hold time cycling waveform (Fig. 27b) for fatigue strength testing. Figure 27(c) shows the fatigue life from a continuous strain cycle and from cycling with two different hold times. This decrease in fatigue life with increasing hold time or decreasing fre-

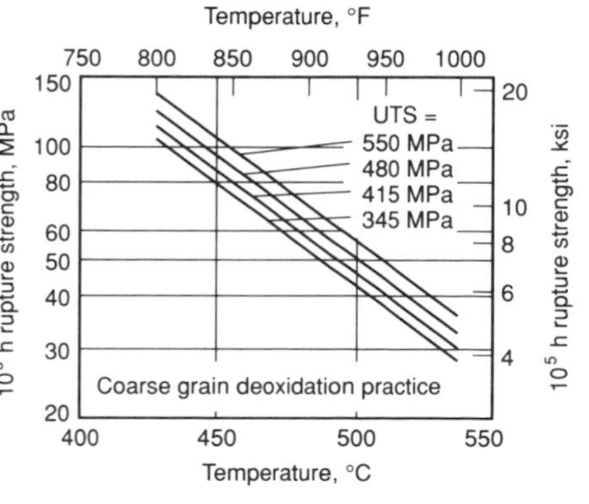

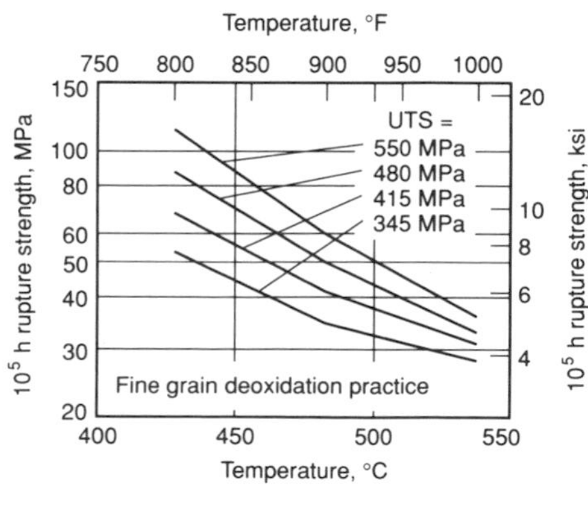

Fig. 23 Predicted 10^5 h creep-rupture strengths of carbon steel with (a) coarse-grain deoxidation practice and (b) fine-grain deoxidation practice. Source: Ref 24

quency, which occurs at temperatures within the creep range, is referred to as time-dependent fatigue or creep-fatigue interaction. It has been attributed to a number of factors, including the formation of intergranular voids or classical creep damage (which permits intergranular crack propagation under cyclic loading conditions), environmental interaction (corrosion fatigue), mean stress effects, and microstructural instabilities of defects produced as a result of stress and/or thermal aging, irradiation damage, and fabrication processing.

Most of these changes can occur at elevated temperatures and depend on time and possibly waveform. There is also ample evidence to show that rupture ductility has a major influence on creep-fatigue interaction. Because this effect is believed to be caused by the influence of rupture ductility on the creep-fracture component, endurance in continuous-cycle and in high-frequency or short-hold-time fatigue tests (where fracture is fatigue-dominated) will be relatively unaffected. As the frequency is decreased or as the hold time is increased, the effect of rupture ductility becomes more pronounced. Endurance data for several ferritic steels, in relation to the range of rupture ductility exhibited by them, are illustrated in Fig. 28. The lower the ductility, the lower the creep-fatigue endurance. In addition, long hold periods, small strain ranges, and low ductility favor creep-dominated failures. Short hold periods, intermediate strain ranges, and high creep ductility favor creep-fatigue-interaction failures.

To determine the effect of cyclic loading superimposed on a constant load at elevated temperatures, several types of fatigue testing can be employed: continuous alternating stress, continuous alternating strain, tension-tension loading with the stress ratio greater than 0, and special waveforms that provide specific holding times at maximum load. Results of these tests show which factors are most contributory to deformation and fracture of the specimens for the testing conditions employed. Further information on time-dependent fatigue is available in the article "Creep-Fatigue Interaction" in this Volume.

Fatigue-Crack Growth. Plots of stress against the number of cycles (*S-N* curves) are a basic tool for design against fatigue. However, their limitations have become increasingly obvious. One of the more serious limitations is that they do not distinguish between crack initiation and crack propagation. Particularly in the low-stress regions, a large fraction of the life of a component may be spent in crack propagation, thus allowing for crack tolerance over a large portion of the life. Engineering structures often contain flaws or cracklike defects that may altogether eliminate the crack-initiation step. A methodology that quantitatively describes crack growth as a function of the loading variables is therefore of great value in design and in assessing the remaining lives of components. Such a methodology, which is based on fracture mechanics, is described in various articles in Volume 19 of the *ASM Handbook.*

Fatigue crack growth rates are obtained at various stress-intensity factor ranges (ΔK) and temperature ranges, so it is difficult to compare the various types of materials directly. At a constant ΔK (arbitrarily chosen as 30 MPa$\sqrt{m}$, or 27 ksi$\sqrt{in.}$), there is a clear trend of increasing crack growth rate with increasing temperature, as shown in Fig. 29. In this figure, the growth rates are relatively insensitive to temperature at temperatures up to about 50% of the melting point (550 to 600 °C, or 1020 to 1110 °F), but the sensitivity increases rapidly at higher temperatures. The crack growth rates for all the materials at temperatures up to 600 °C (1110 °F), relative to the room-temperature rates, can be estimated by a maximum correlation factor of 5 (2 for ferritic steels). More detailed information on the effects of temperature on fatigue crack growth mechanisms can be found in the article "Elevated-Temperature Crack Growth of Structural Alloys" in this Volume.

Thermal Fatigue. In the past, thermal fatigue traditionally has been treated as synonymous with isothermal low-cycle fatigue (LCF) at the maximum temperature of the thermal cycle. Life-prediction techniques also have evolved from the LCF literature. More recently, advances in finite-element analysis and in servohydraulic test systems have made it possible to analyze complex thermal cycles and to conduct thermomechanical fatigue (TMF) tests under controlled conditions. The assumed equivalence of isothermal LCF tests and TMF tests has been brought into question as a result of a number of studies. It has been shown that for the same total strain range, the TMF test can be more damaging under certain conditions than the pure LCF test. Information on the thermal fatigue of materials is provided in Ref

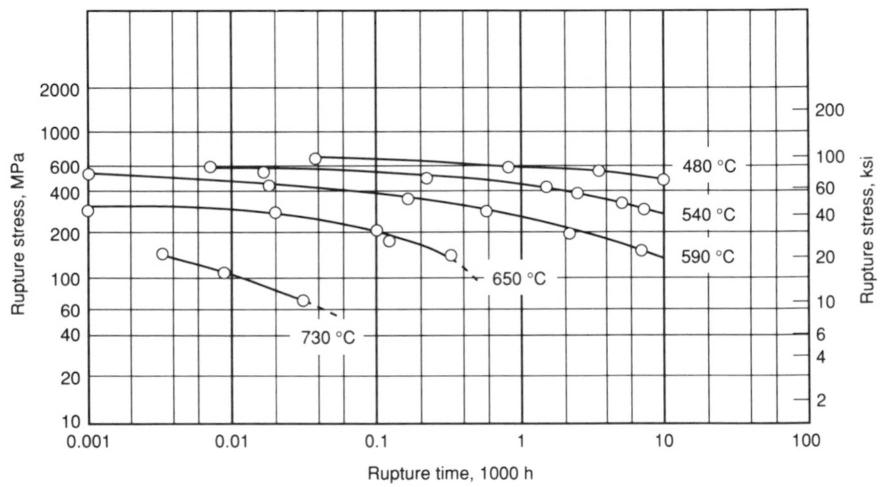

Fig. 24 Time-temperature-rupture data of a 1Cr-1Mo-0.25V steel

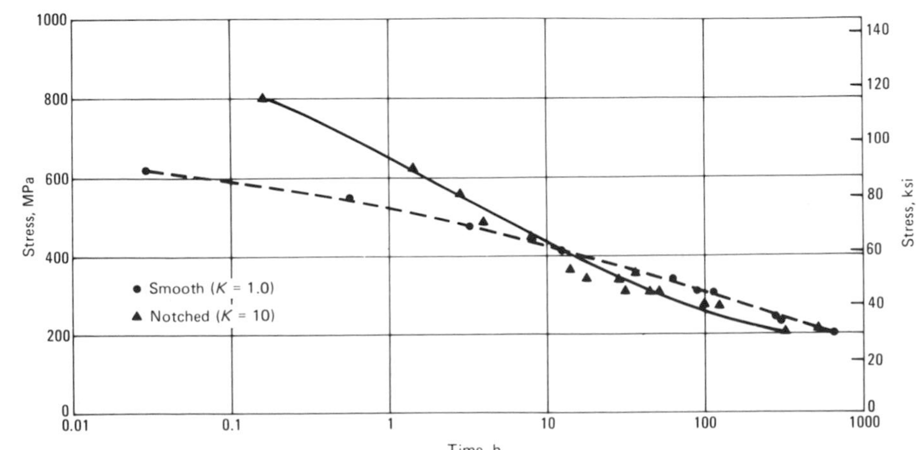

Fig. 25 Effect of notch on stress-rupture behavior. Stress-rupture behavior of smooth (K = 1.0) and notched specimens of AISI 603 steel (see Table 7 for composition) tested at 595 °C (1100 °F). All specimens were normalized at 980 °C (1800 °F) and tempered 6 h at 675 °C (1250 °F). Source: Ref 27

33 and in the article "Thermal and Thermomechanical Fatigue of Structural Alloys" in this Volume.

High-cycle thermal fatigue frequently results from intermittent wetting of a hot surface by a coolant having a considerably lower temperature. In this case, thermal fatigue cracks may initiate at the surface after a sufficient number of cycles. In other cases, the thermal cycling or ratcheting may result in plastic deformation. Thermal ratcheting is progressive, cyclic inelastic deformation that occurs as a result of cyclic strains caused by thermal or secondary mechanical stresses. Sustained primary loading often contributes to thermal ratcheting. Salt pots used to contain heat

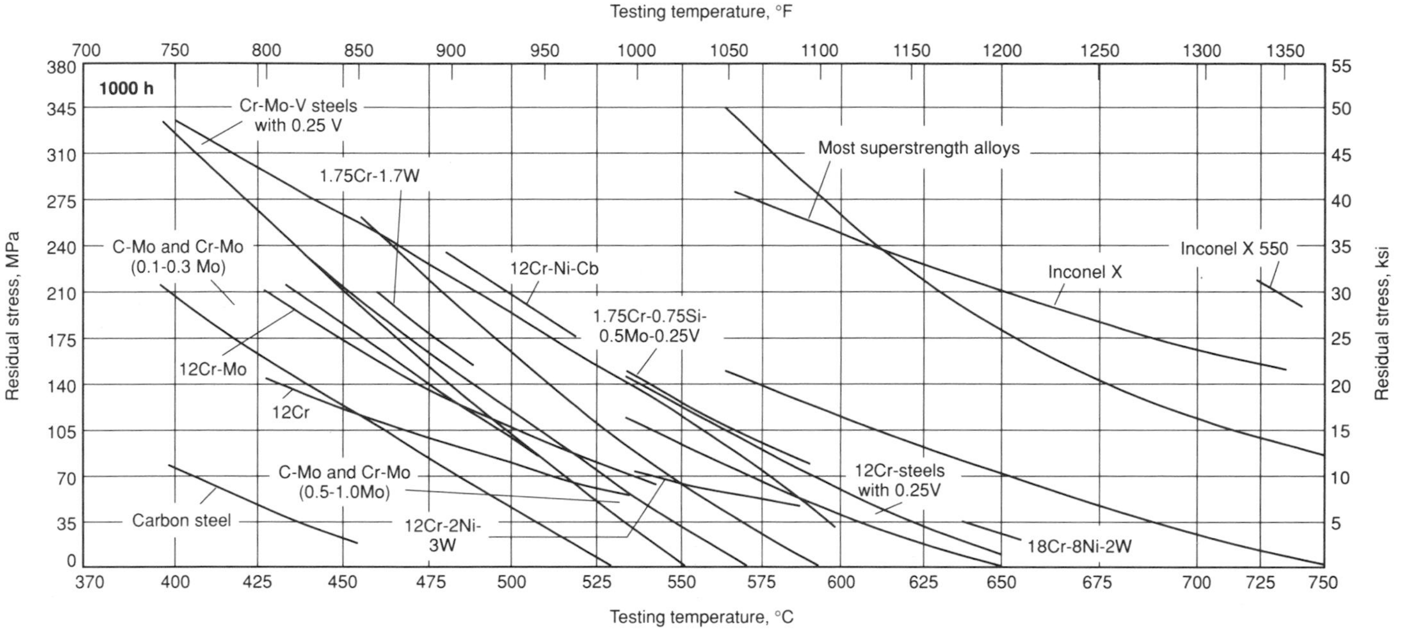

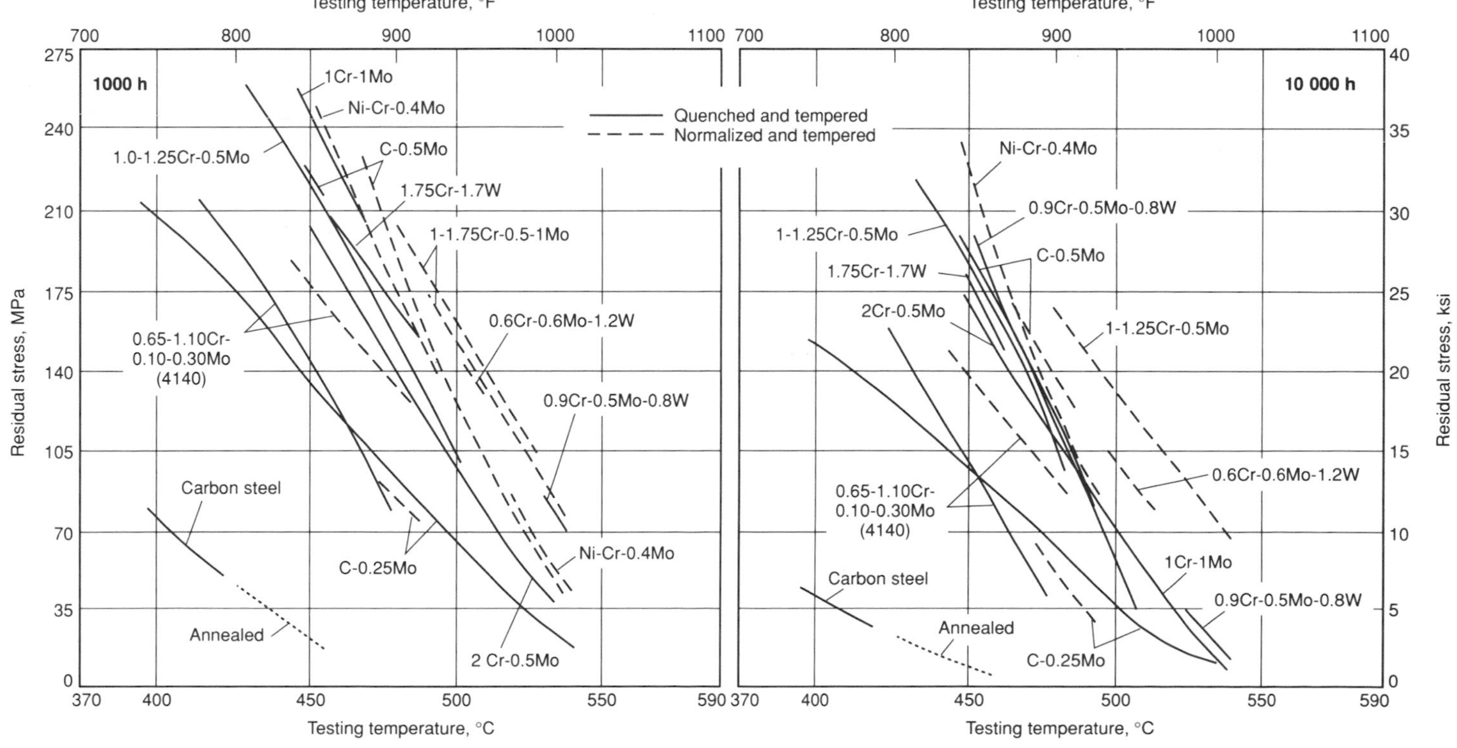

Fig. 26 Comparison of relaxation strengths (residual stress) of various steels. (a) Comparison of low-alloy steels with stainless steels and superalloys. (b) Low-alloy steels at 1000 h. (c) Low-alloy steels at 10,000 h. Source: Ref 28

Table 11 Room-temperature mechanical properties of 2.25Cr-1Mo steel in various product forms

ASME specification	Grade	Product form	Mechanical properties: Yield strength, MPa	Yield strength, ksi	Ultimate tensile strength, MPa	Ultimate tensile strength, ksi	Minimum elongation in 50 mm (2 in.), %	Minimum reduction in area, %
SA-182	F22	Pipe flanges, fillings, and valves	275	40	485	70	...	...
SA-199	T22	Seamless cold-drawn tubes	170	25	415	60	30	...
SA-213	T22	Seamless ferritic alloy steel tubes	170	25	415	60	30	...
SA-217	WC9	Alloy steel castings	275	40	480	70	20	33
SA-333	P22	Welded and seamless pipe	205	30	415	60	30	20
SA-336	F22	Alloy steel forgings	310	45	515–755	75–110	18	25
	F22a	Alloy steel forgings	205	30	415–585	60–85	20	35
SA-369	FP22	Ferritic alloy steel forged and bored pipe	205	30	415	60	20–30	...
SA-387	GR22, class 1	Chromium-molybdenum PV plate	205	30	415–585	60–85	18(a), 45	40
	GR22, class 2	Chromium-molybdenum PV plate	310	45	515–690	75–100	18(a), 45	40
SA-426	CP22	Centrifugally cast ferritic alloy steel pipe	275	40	480	70	20	35
SA-542	Class 1	Chromium-molybdenum alloy steel plate	585	85	725–860	105–125	14	...
	Class 2	Chromium-molybdenum alloy steel plate	690	100	790–930	115–135	15	...

Elongation in 200 mm (8 in.)

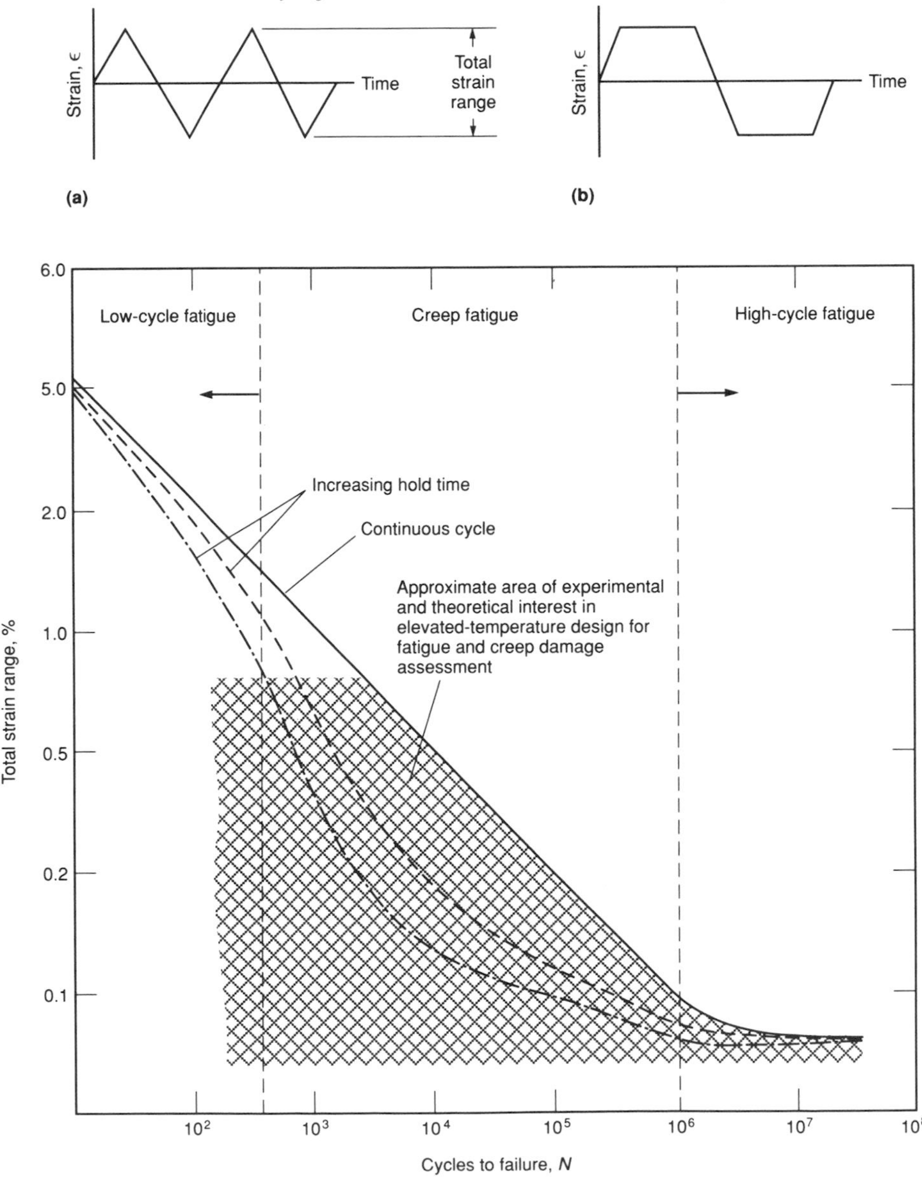

Fig. 27 Range of conditions to be considered in studies of elevated-temperature fatigue and the effect of (a) continuous cycling and (b) strain hold on (c) elevated-temperature fatigue. Source: Ref 30

treating salt are subject to thermal ratcheting whenever the salt goes through a freeze-melt cycle.

Ductility and Toughness

Steels typically have adequate ambient temperature toughness and excellent elevated-temperature ductility, but several embrittling mechanisms can occur during elevated-temperature exposure. Consequently, ductility and toughness tests are useful in assessing embrittling mechanisms.

Temper embrittlement (also known as temper brittleness, two-step temper embrittlement, or reversible temper embrittlement) is associated with tempered alloy steels that are heated within, or slowly cooled through, a critical temperature range, generally 300 to 600 °C (570 to 1110 °F) for low-alloy steels. This treatment causes a decrease in toughness as determined with Charpy V-notch impact specimens. It is a particular problem for heavy-section components, such as pressure vessels and turbine rotors, that are slowly cooled through the embrittling range after tempering and also experience service at temperatures within the critical range.

Temper-embrittled steels exhibit an increase in their ductile-to-brittle transition temperature (DBTT) and a range in fracture mode in the brittle test temperature range from cleavage to intergranular. The DBTT can be assessed in several ways. The most common is the temperature for 50% ductile and 50% brittle fracture (50% fracture appearance transition temperature, or FATT, or the lowest temperature at which the fracture is 100% ductile fibrous criterion). Transition temperatures based on absorbed energy values are not normally employed.

Temper embrittlement is reversible; that is, the toughness of embrittled steels can be restored by tempering them above the critical region, followed by cooling (e.g., water quenching). This decreases the DBTT and changes the low-temperature (that is, below the 50% FATT) intergranular brittle appearance back to the cleavage mode.

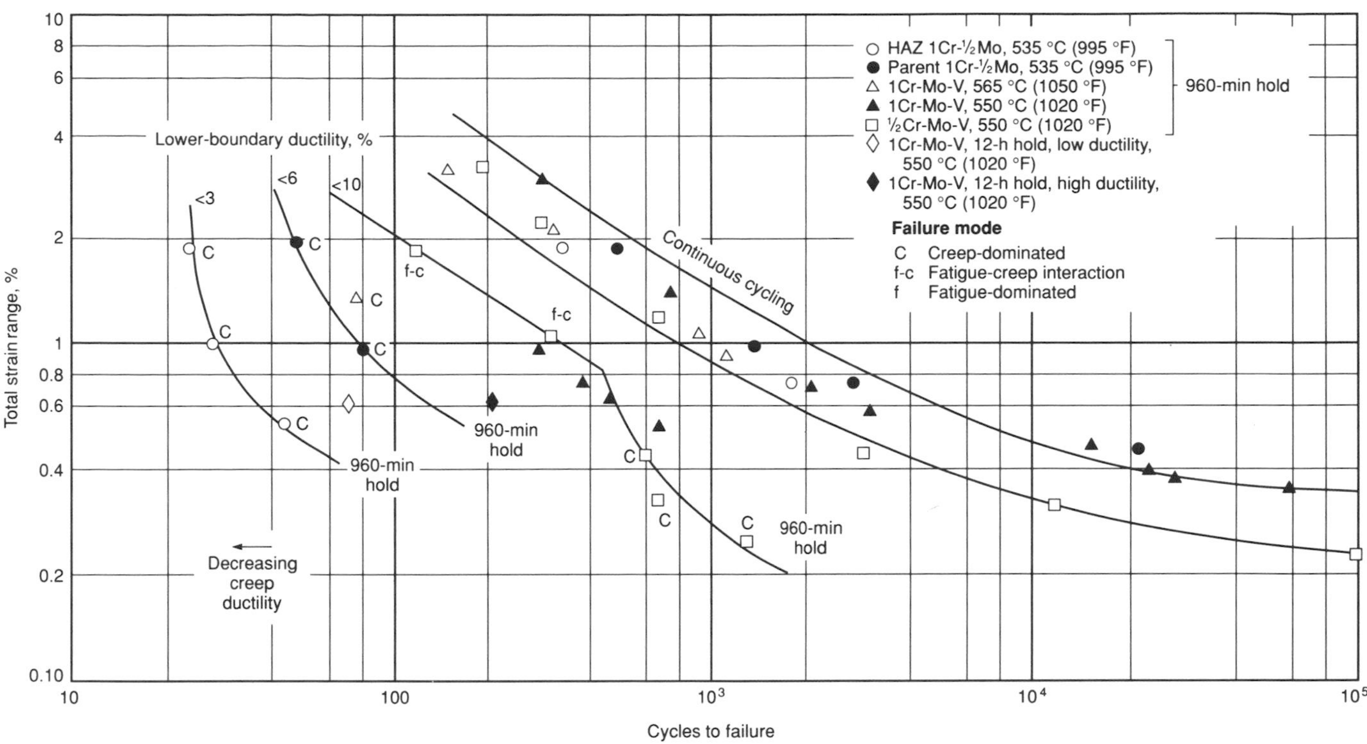

Fig. 28 Effect of ductility on endurance limit of ferritic steels. Source: Ref 31

Temper embrittlement occurs only in alloy steels, not in plain carbon steels, and the degree of embrittlement varies with alloy steel composition. Therefore, it is important to know what alloying elements are present, and their levels. Impurities, in decreasing order of influence in terms of weight percent, are antimony, phosphorus, tin, and arsenic. Of these elements, phosphorus is most commonly present in alloy steels, and it has captured the most attention in research studies. Manganese and silicon also increase the susceptibility to embrittlement.

Figure 30 shows the increase in the DBTT of a 3140 steel as the result of temper embrittlement. This low-alloy steel, which contained nominally 1.15% Ni and 0.65% Cr, was embrittled by isothermal tempering and slow furnace cooling through the critical range of about 375 to 575 °C (705 to 1070 °F). More detailed information on temper embrittlement can be found in the article "Thermally Induced Embrittlement" in the *ASM Specialty Handbook: Carbon and Alloy Steels.*

Tempered martensite embrittlement (TME), also known as one-step temper embrittlement, of high-strength alloy steels occurs upon tempering in the range of 205 to 370 °C (400 to 700 °F). It differs from temper embrittlement in terms of the strength of the material and the temperature exposure range. In temper embrittlement, the steel is usually tempered at a relatively high temperature, producing lower strength and hardness, and embrittlement occurs upon slow cooling after tempering and during service at temperatures within the embrittlement range. In TME, the steel is tempered within the embrittlement range, and service exposure is usually at room temperature. This is why temper embrittlement is often called two-step temper embrittlement, while TME is often called one-step temper embrittlement.

It is well established that lower bainite is also embrittled when tempered in the range of 205 to 370 °C (400 to 700 °F). Other structures, such as upper bainite and pearlite/ferrite, are not embrittled by tempering in this range.

While temper embrittlement is evaluated by the change in the DBTT, most studies of TME have evaluated only the change in room-temperature impact energy (Fig. 31). In general, when an as-quenched alloy steel is tempered, the toughness at room temperature increases with tempering temperature, up to about 200 °C (390 °F). With further increases in tempering temperature, the toughness decreases. Then, with increasing tempering temperatures above about 400 °C (750 °F), the toughness increases again. This change in toughness with tempering temperature is not apparent when examining hardness or tensile strength data, which generally decrease with increasing tempering temperatures. More detailed information on TME can be found in the article "Thermally Induced Embrittlement" in the *ASM Specialty Handbook: Carbon and Alloy Steels.*

Creep Embrittlement. A third method of assessing the effects of embrittlement mechanisms is by measuring ductility (reduction of area). Creep embrittlement effects, for example, are usually reported in terms of a ductility minimum in stress-rupture tests, while temper embrittlement is usually recorded as an upward shift in Charpy V-notch transition temperature.

Creep embrittlement occurs in roughly the same temperature range as temper embrittlement, but is not reversible with heat treatment. Creep embrittlement also seems to depend on tempering reactions inside grains and on the presence of a carbide-denuded zone at prior-austenite grain boundaries. In contrast, segregation effects producing temper embrittlement occur at distances only a few atomic diameters from the grain boundary. Some investigators maintain that im-

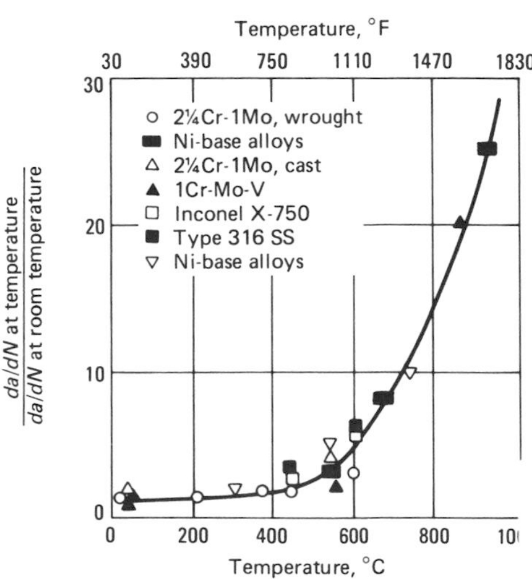

Fig. 29 Variation of fatigue-crack growth rates as function of temperature at ΔK = 3 MPa$\sqrt{m}$ (27 ksi$\sqrt{in.}$). Source: Ref 32

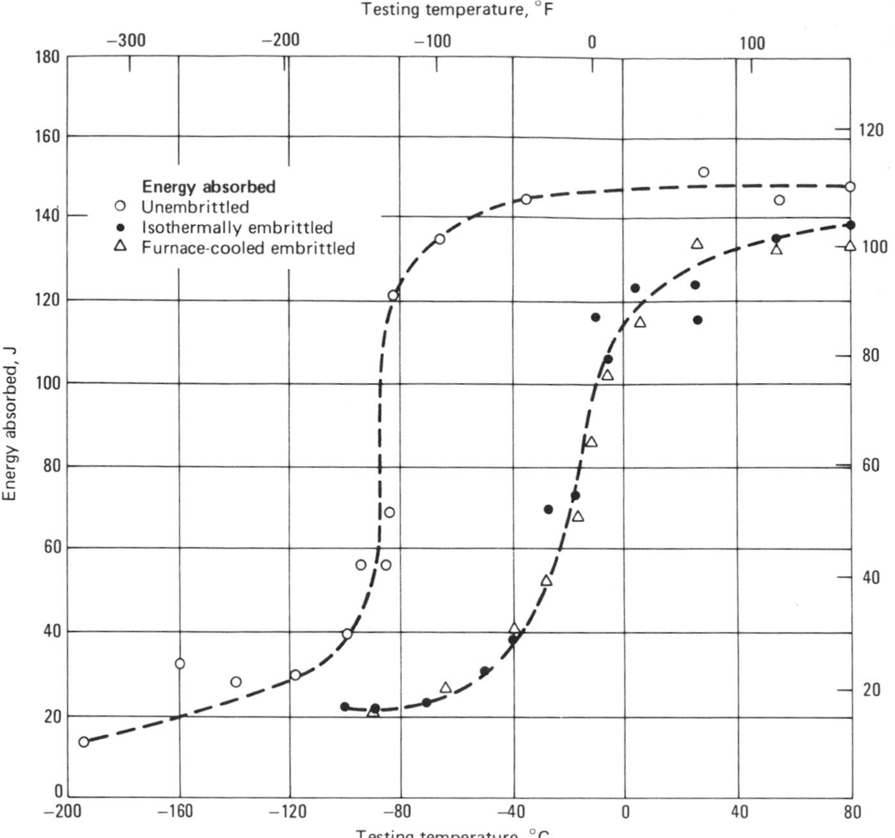

Fig. 30 Shift in impact transition curve to higher temperature as a result of temper embrittlement produced in SAE 3140 steel by isothermal holding and furnace cooling through the critical range. Source: Ref 34

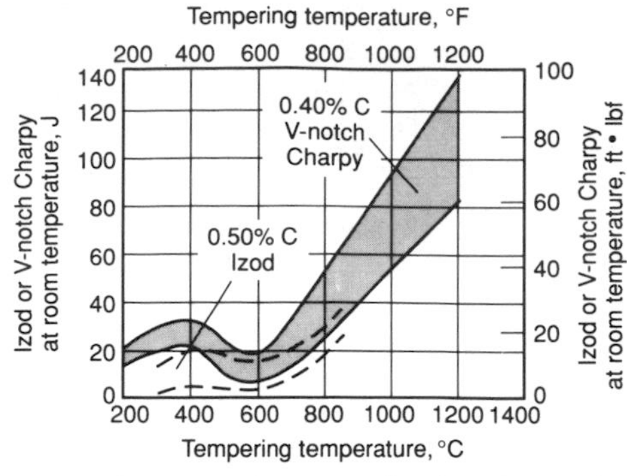

Fig. 31 Impact toughness as a function of tempering temperature of hardened, low-alloy, medium-carbon steels. Source: Ref 34

purities known to produce temper embrittlement also contribute to the development of creep embrittlement. Some general characteristics of creep embrittlement are:

- Creep embrittlement has been shown to occur in the temperature range 425 to 595 °C (800 to 1100 °F) for alloy steels having ferrite-plus-carbide microstructures.
- The lower the embrittling temperature, the more time required for creep embrittlement to appear and the greater its severity.
- Creep embrittlement is manifested by a loss and then partial recovery of stress-rupture ductility with decreasing stress.
- The development of embrittlement is invariably associated with a transition from transgranular to intergranular fracture. Voids and microcracks are found throughout a creep-embrittled microstructure. These voids form along prior-austenite grain boundaries transverse to the tensile direction.
- The mechanism for creep embrittlement appears to be closely associated with tempering reactions inside grains and at the grain boundaries during the creep process. The formation of fine, needlelike precipitates in grain interiors, accompanied by the development of a denuded zone and elongated alloy carbides at grain boundaries, seems to contribute significantly to the embrittlement process.
- Loss in toughness produced by creep embrittlement is largely unaffected by subsequent heat treatments.
- Void formation caused by creep is irreversible.

Long-Term Exposure and Microstructure

Long-term exposure to elevated temperature may affect either short-term or long-term properties. For example, the initial microstructure of creep-resistant chromium-molybdenum steels consists of bainite and ferrite containing Fe_3C carbides, ε carbides, and fine M_2C carbides. Although a number of different carbides may be present, the principal carbide phase responsible for strengthening is a fine dispersion of M_2C carbides, where M is essentially molybdenum. With increasing aging in service or tempering in the laboratory (Fig. 32), a series of transformations of the carbide phases take place that eventually transform M_2C into M_6C and $M_{23}C$ (where the M in the latter two metal carbides is mostly chromium). Such an evolution of the carbide structure results in coarsening of the carbides, changes in the matrix composition, and an overall decrease in creep strength. The effect of exposure on the stress-rupture strength of two chromium-molybdenum steels is shown in Fig. 7.

Other metallurgical changes (such as spheroidization and graphitization) and corrosion effects may also occur during long-term exposure at elevated temperature. Therefore, tests after long-term exposure may be useful in determining the effect of these metallurgical changes on short-term or long-term properties.

Data Presentation and Analysis

Presentation of Tensile and Yield Strength. One method for comparing steels of different strengths is to report elevated-temperature strength as a percentage of room-temperature strength. This method is illustrated in Fig. 33. The strength levels of the steels represented in Fig. 33 varied from 480 to 1100 MPa (70 to 160 ksi).

Presentation of Creep Data. Four different presentations of the same creep data for 2.25Cr-1Mo steel are given in Fig. 34. In Fig. 34(a) to (c), only the creep strain is plotted. In the isochronous stress-strain diagram (Fig. 34d), total strain is used. The overall format of Fig. 34(d) is particularly useful in design problems in which total strain is a major consideration.

Methods for Correlating, Interpolating, and Extrapolating Elevated-Temperature Mechanical Property Data. The behavior of steels at elevated temperatures can be affected by many variables, including time, temperature, stress, and environment. A variety of methods have been devised for correlating, interpolating, and extrapolating elevated-temperature mechanical property data. Further information on the analysis of elevated-temperature data is contained in Ref 32 and in the articles contained in the section "Special Topics" in this Volume.

Larson-Miller Parameter. Several parameters have been used for comparison of, and interpolation between, stress-rupture data. The most widely used is the Larson-Miller parameter, *P*, defined by the equation:

$$P = T(\mathrm{C} + \log t) \times 10^{-3} \quad \text{(Eq 1a)}$$

where *T* is the test temperature in degrees Rankine, *t* is the rupture time in hours, and C is a constant whose value is approximately 20 for low-alloy steels. If *T* is given in degrees Kelvin, the equation is:

$$P = 1.8\,T(\mathrm{C} + \log t) \times 10^{-3} \quad \text{(Eq 1b)}$$

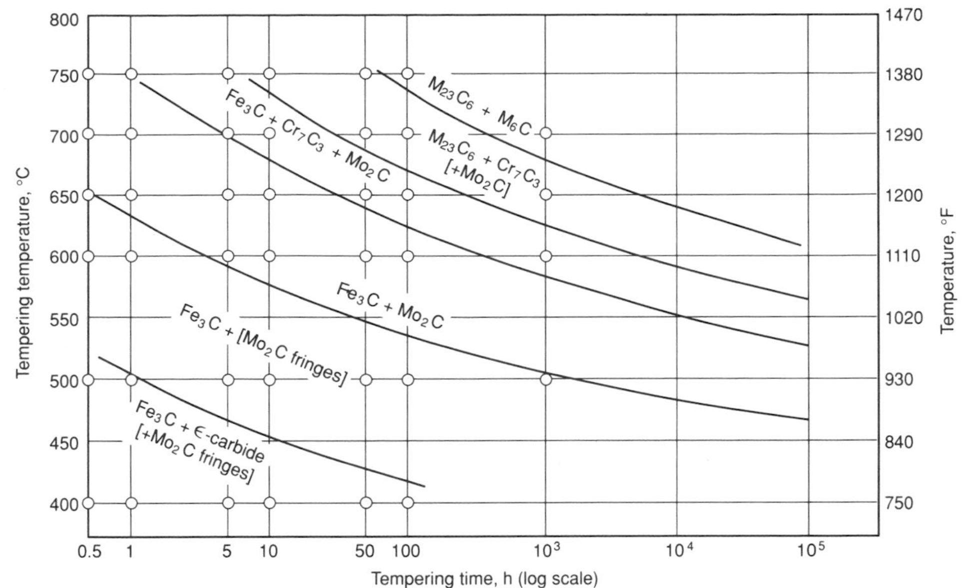

Fig. 32 Isothermal diagram showing the sequence of carbide formation on tempering of normalized 2.25Cr-1Mo steel. Source: Ref 12

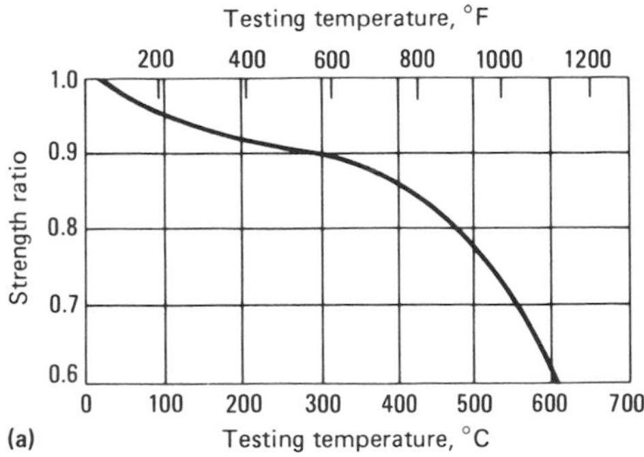

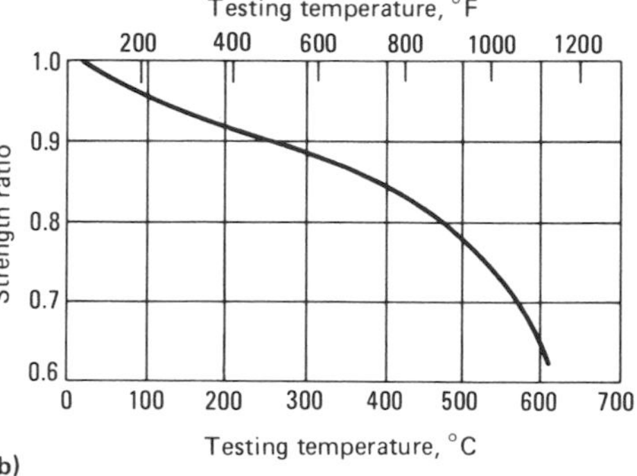

Fig. 33 Ratios of elevated-temperature strength to room-temperature strength for hardened and tempered 2.25Cr-1Mo steel tempered to room-temperature tensile strengths ranging from 480 to 1100 MPa (70 to 160 ksi). (a) Tensile strength. (b) Yield strength. Source: Ref 2

The Larson-Miller parameter is used with an experimentally determined graph, such as that shown in Fig. 35, to correlate stress, temperature, and rupture time. Each graph should include the ranges of time and temperature for which the data apply. Extrapolation beyond these ranges is generally not appropriate.

A similar parameter, P', was used by Smith (Ref 36) to describe the creep behavior of 9Cr-1Mo steel:

$$P' = T(20 - \log r) \times 10^{-3} \quad \text{(Eq 2a)}$$

where r is the minimum creep rate in percent per hour. If T is given in degrees Kelvin, the equation is:

$$P' = 1.8\,T(20 - \log r) \times 10^{-3} \quad \text{(Eq 2b)}$$

The creep rate parameter is used with an experimentally determined graph, such as the one shown in Fig. 36 for 2.25Cr-1Mo steel.

Extrapolation of Creep and Rupture Data. Long-term creep and stress-rupture values (for example, 100,000 h) are often extrapolated from shorter-term tests. Whether these property values are extrapolated or determined directly often has little bearing on the operating life of high-temperature parts. The actual material behavior is often difficult to predict accurately because of the complexity of the service stresses relative to the idealized, uniaxial loading conditions in the standardized tests and because of attenuating factors such as cyclic loading, temperature fluctuations, or metal loss from corrosion.

Marked changes in the slope of stress-rupture curves (see, e.g., the lower plot in Fig. 23b near 480 °C, or 900 °F) must also be considered in data extrapolation. These changes often indicate microstructural changes. Marked differences in slope between curves representing temperatures separated by less than 100 °C (180 °F) are evidence that the slope of the lower-temperature curve will change over the time period of extrapolation, indicating the need for longer tests or careful approximations of the probable influence of the change in slope. Such changes in slope are almost always in the direction of lower stress-rupture strength than would be predicted by straight-line extrapolation.

Because of microstructural instabilities, deviations from the ideal creep may be virtually absent or may be excessive and extend over long periods of time. Secondary creep may persist only for very short time periods or may exhibit nonclassical behavior. The creep behavior of annealed 2.25Cr-1Mo steel, for example, exhibits creep curves that differ from a classical three-stage creep curve in that two steady-state stages occur (Ref 37). During the first steady-state stage, the creep rate is controlled by the motion of dislocations that contain atmospheres of carbon and molybdenum atom clusters, a process that is called interaction solid-solution hardening (see the section "Strengthening Mechanisms" in this article). Eventually, the precipitation of Mo_2C removes molybdenum and carbon from solution, and the creep rate increases to a new steady state where the creep rate is controlled by atmosphere-free dislocations moving through a precipitate field. These nonclassical curves occur at intermediate stresses. As the stress decreases, the first steady-state stage disappears because the dislocation velocity decreases and the molybdenum-carbon atmosphere will be able to diffuse with the dislocations. At high stresses, a classical curve occurs when the creep rate is controlled by a combination of processes that operate in the two steady-state stages of the nonclassical curves (Ref 37). Such factors indicate the need to experimentally check values of deformation predicted by extrapolation of secondary creep data.

The extrapolation of stress-rupture ductility with parametric techniques has been considered a potential method for predicting long-term ductility from short-term tests (Ref 38). The stress-rupture ductility of many alloys used at elevated temperatures varies with temperature and stress, so the objective is to develop a combined (stress, temperature) parameter that can be correlated to rupture ductility over a wide range of stresses and temperatures. Reference 38 compares the correlation between some parametric models and rupture ductility data for a 1.25Cr-0.5Mo steel in the temperature range of 510 to 620 °C (950 to 1150 °F).

Methods for Predicting Time-Dependent Fatigue (Creep-Fatigue Interaction) Behavior. Many methods have been employed to extrapolate available data to estimate the time-dependent fatigue life of materials. Development of a mathematical formulation for life prediction is one of the most challenging aspects of creep-fatigue interaction. It is complicated by the fact that any proposed formulation must account for strain rate, relaxation at constant strain, creep at constant load, the difference between tension and compression creep and/or relaxation, or combinations of all of these.

Linear damage summation is perhaps the most widely known and simplest of the many life prediction methods. It has been used extensively in the evaluation of creep-fatigue interaction. In this method, fatigue damage is expressed as a cycle fraction of damage and creep damage is ex-

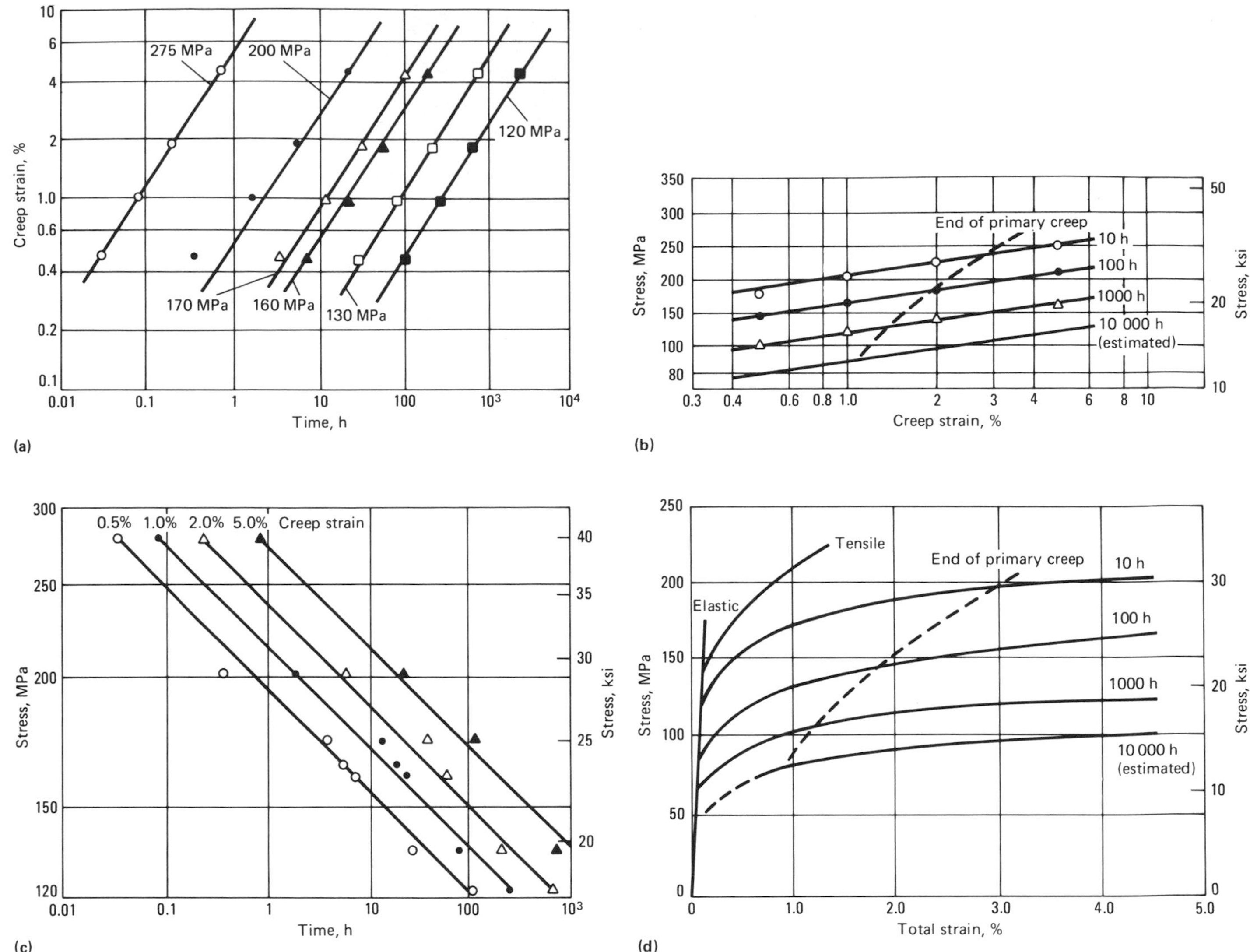

Fig. 34 Analysis of creep data. Creep behavior of 2.25Cr-1Mo steel tested at 540 °C (1000 °F). (a) Creep strain-time plot; constant-stress lines have been drawn parallel. (b) Stress-creep strain plot. (c) Stress-time plot; constant-strain lines have been drawn parallel. (d) Isochronous stress-strain curves. Source: Ref 35

pressed as a time fraction of damage. It is also assumed that these quantities can be added linearly to represent damage accumulation. Failure occurs when this summation reaches a certain value.

Other methods include the ductility exhaustion approach, the frequency modified approach, and strain range partitioning. These methods are reviewed in Ref 30 and 32 and in the article "Creep-Fatigue Interaction" in this Volume.

Factors Affecting Mechanical Properties

The factors affecting the mechanical properties of steels include the nature of the strengthening mechanisms, the microstructure, the heat treatment, and the alloy composition. This section describes these factors, with emphasis on chromium-molybdenum steels (especially 2.25Cr-1Mo) used for elevated-temperature service.

In addition, various service factors, such as thermal exposure and environmental conditions, can induce metallurgical changes that may affect the mechanical properties of steels used at elevated temperatures. These metallurgical changes include spheroidization, graphitization, decarburization, and carburization. Depending on the temperature and exposure environment, ferritic steels used at elevated temperatures may also be susceptible to various types of thermally induced embrittlement (see the earlier section "Ductility and Toughness" in this article) or environmentally induced embrittlement. The latter category includes such embrittlement mechanisms as hydrogen embrittlement, stress-corrosion cracking, and liquid-metal and solid-metal embrittlement. Each of these are reviewed in the *ASM Specialty Handbook: Carbon and Alloy Steels.*

Strengthening Mechanisms

The creep strength of a steel is affected by the typical strengthening mechanisms: grain refinement, solid-solution hardening, and precipitation hardening. Of these various strengthening mechanisms, the refinement of grain size is perhaps the most unique because it is the only strengthening mechanism that also increases toughness. Figure 22 shows the effect of grain size on the creep strength of a carbon steel.

The creep strength of chromium-molybdenum steels is mainly derived from a complex combination of solid-solution (primarily interaction solid-solution strengthening) and precipitation effects, as illustrated in Fig. 37. In the early stages of creep, solid-solution effects are the largest contributor to creep resistance. As time progresses, the precipitation of carbides (primarily Mo_2C in the case of molybdenum steels) contributes more to the creep resistance. As time pro-

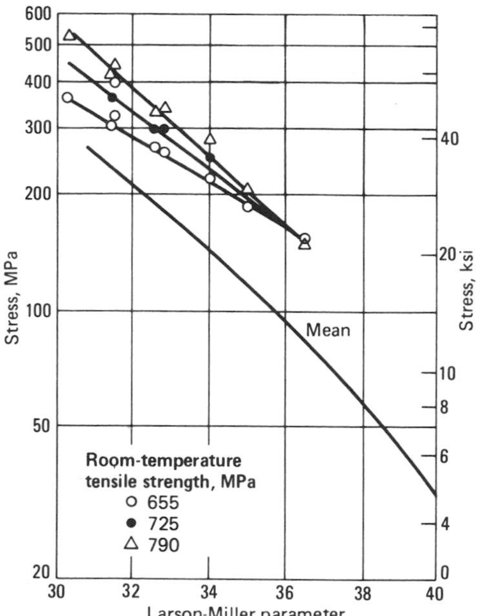

Fig. 35 Larson-Miller plot of stress-rupture behavior of 2.25Cr-1Mo steel. Variation in Larson-Miller parameter with stress to rupture for normalized and tempered and hardened and tempered specimens of 2.25Cr-1Mo steel tested between 425 and 650 °C (800 and 1200 °F) for rupture life to 10,000 h; the data are grouped according to the room-temperature tensile strength of the steel. Larson-Miller plot for annealed steel (mean curve) included for comparison. Source: Ref 2

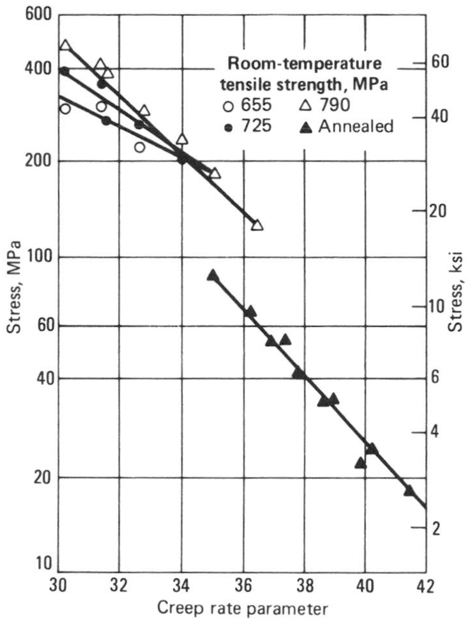

Fig. 36 Modified Larson-Miller plot of creep behavior of 2.25Cr-1Mo steel. Variation in creep rate parameter with creep stress for normalized and tempered and hardened and tempered specimens of 2.25Cr-1Mo steel tested between 425 and 650 °C (800 and 1200 °F) for test duration to 10,000 h. The data are grouped according to the room-temperature tensile strength of the steel. Creep rate data for annealed steel included for comparison. Source: Ref 2, 36

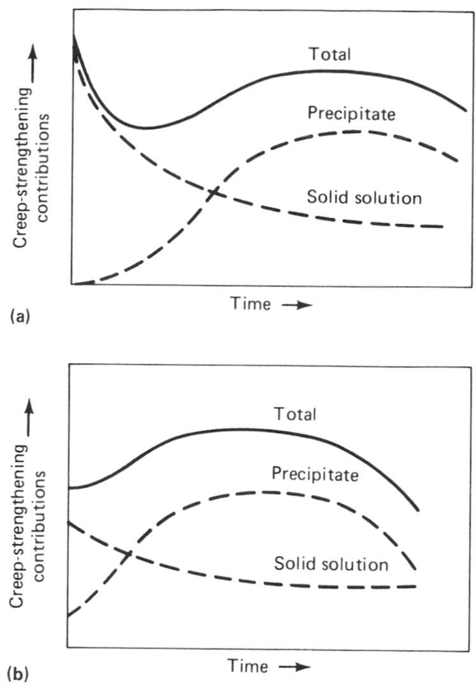

Fig. 37 Schematic of changes in creep strengthening contributions at 550 °C (1020 °F) in (a) normalized molybdenum steel and (b) normalized and tempered molybdenum steel. Source: Ref 39

gresses still further, the strengthening effect of the carbides is reduced as the carbides coarsen (Ostwald ripening) and diffuse into stabler but weaker structures.

Both of these strengthening mechanisms become unstable at high temperatures. In solid-solution hardening, an increase in temperature increases the diffusion rates of solute atoms in the dislocation atmospheres, while at the same time dispersing the atoms of the atmospheres. Both effects make it easier for dislocations to move. In precipitation hardening, heating of the alloy to an excessively high temperature can cause solutionizing of the precipitates. At intermediate temperatures, the precipitates can coarsen and become less effective impediments to dislocation motion. High stresses and high-strain cyclic loading also can lead to accelerated softening.

The solid-solution strengthening effect illustrated in Fig. 37 occurs primarily from interaction solid-solution strengthening (or hardening), a mechanism that involves the interaction of substitutional and interstitial solutes (Ref 39–41). This process occurs in ferritic alloys that in solid solution contain interstitial and substitutional elements that have an affinity for each other. As a result of this strong attraction, atom pairs or clusters could form dislocation atmospheres that hinder dislocation motion and therefore strengthen the steel. Other solid-solution effects from either a pure substitutional solute or a pure interstitial solute do not alone provide significant creep strengthening in carbon-manganese steels and

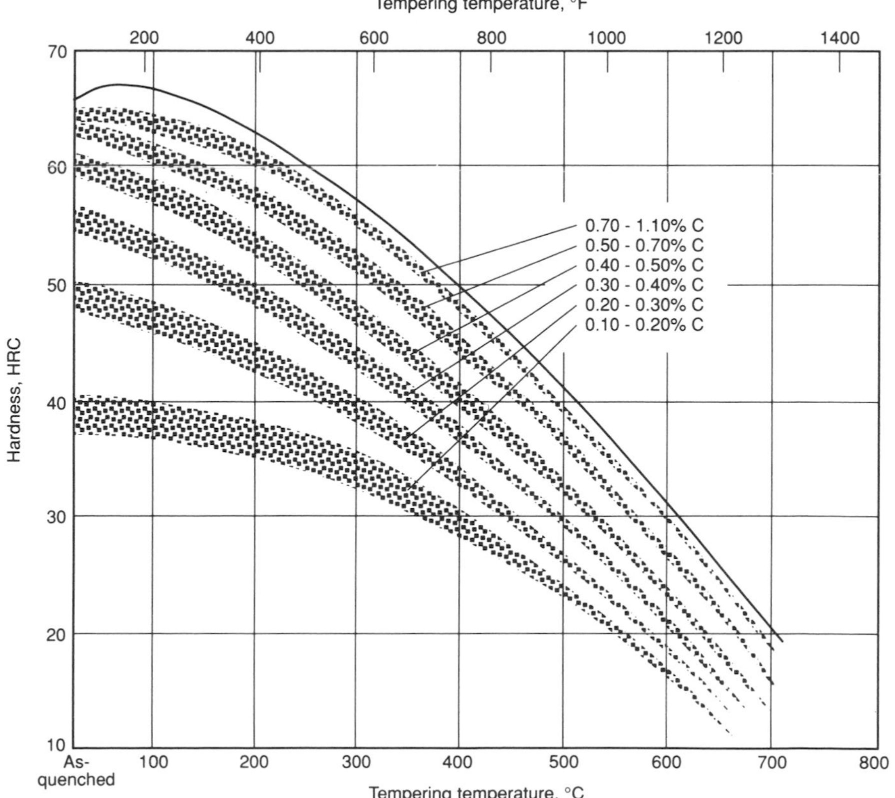

Fig. 38 Decrease in hardness with increasing tempering temperature for steels of various carbon contents. Source: Ref 43

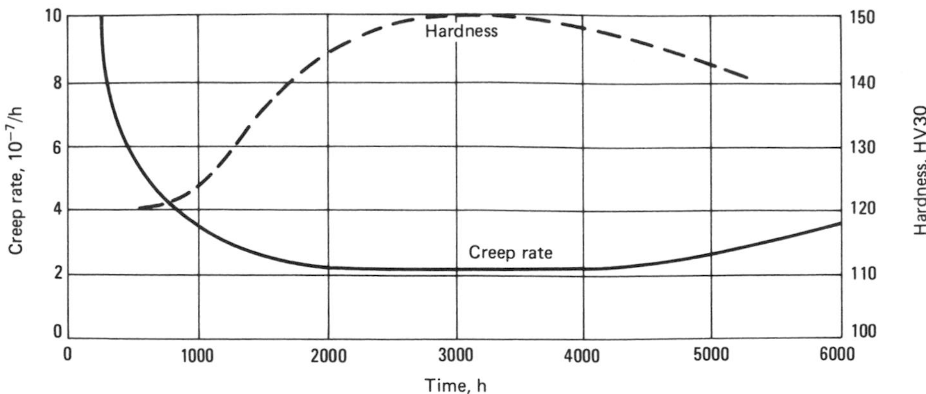

Fig. 39 Relationship between change in creep rate and change in room-temperature hardness during creep of normalized 1% Mo steel tested at 123 MPa (17.8 ksi) at 550 °C (1020 °F). Under these test conditions, secondary creep coincided with maximum precipitation hardening. Source: Ref 39

carbon-molybdenum steels. The addition of interstitial solutes to iron has no significant creep-strengthening effect above 450 °C (840 °F), while the substitutional solutes manganese, chromium, and molybdenum give rise to only modest increases in strength in the absence of interstitial solutes (Ref 41). However, when certain combinations of substitutional and interstitial solutes are present together (for example, manganese-nitrogen, molybdenum-carbon, and molybdenum-nitrogen), there is a substantial increase in creep strength (Ref 41).

Precipitation-strengthening effects are probably negligible in the carbon-manganese steels typically used for elevated-temperature applications (Ref 39), although strengthening by fine NbC particles has been observed (Ref 42). Precipitation strengthening is more significant in molybdenum steels, for which the strengthening precipitates are mainly Mo_2C and Mo_2N. Further increases in precipitation strengthening can be achieved with additions of niobium or vanadium to chromium-molybdenum steels. The stability of the carbides increases in the following order of alloying elements: chromium, molybdenum, vanadium, and niobium. Fine and closely dispersed precipitates of NbC and VC are thus desirable, followed by the other carbides. This precipitation strengthening effect in creep-resistant chromium-molybdenum steels is related to secondary hardening, as discussed below.

Secondary Hardening

If the mechanical properties of tempered steels need to be maintained at elevated service temperatures, the problem is to reduce the amount of softening during tempering so that higher strength (hardness) can be achieved at higher temperatures. One way to reduce softening is with strong carbide formers, such as chromium, molybdenum, and vanadium. These carbide formers induce an effect known as secondary hardening. Without these elements, iron-carbon alloys and low-carbon steels soften rapidly with increasing tempering temperature, as shown in Fig. 38. This softening is largely due to the rapid coarsening of cementite with increasing tempering temperature, a process dependent on the diffusion of carbon and iron. If present in a steel in sufficient quantity, however, the carbide-forming elements not only retard softening but also form fine alloy carbides that produce a hardness increase at higher tempering temperatures. This hardness increase is frequently referred to as secondary hardening. The hardening can also occur during elevated-temperature service and is related to creep strength, as shown in Fig. 39.

Secondary hardening allows higher tempering temperatures, and this increases the range of serv-

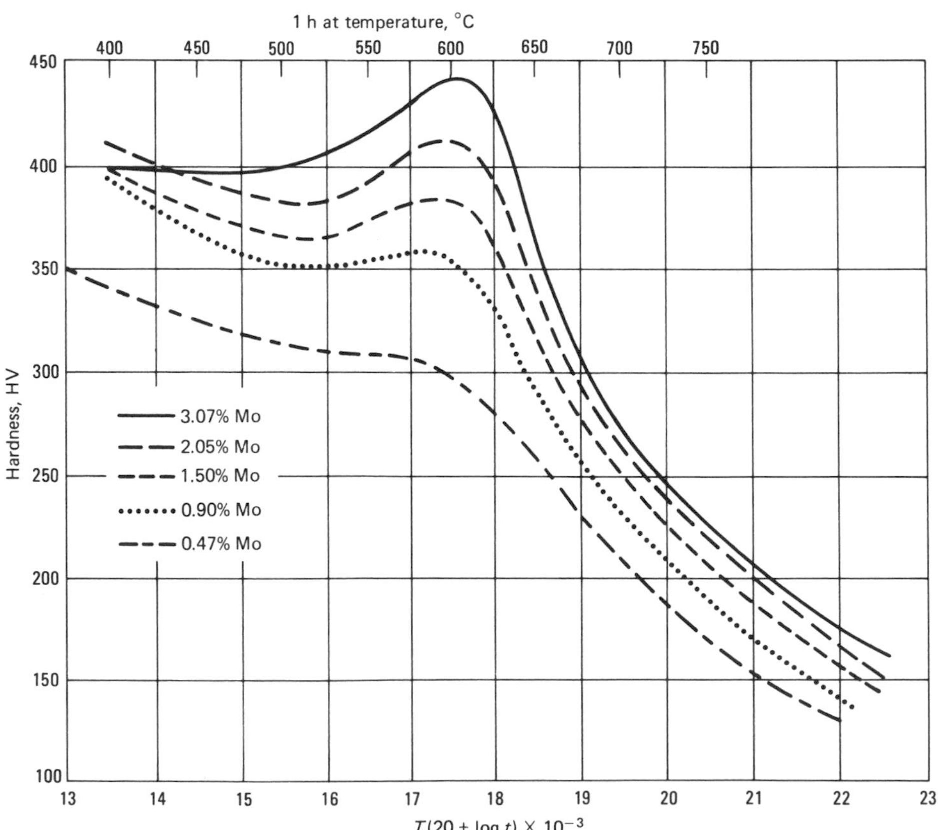

Fig. 40 Retardation of softening and secondary hardening during tempering of steels with various molybdenum contents. Source: Ref 43

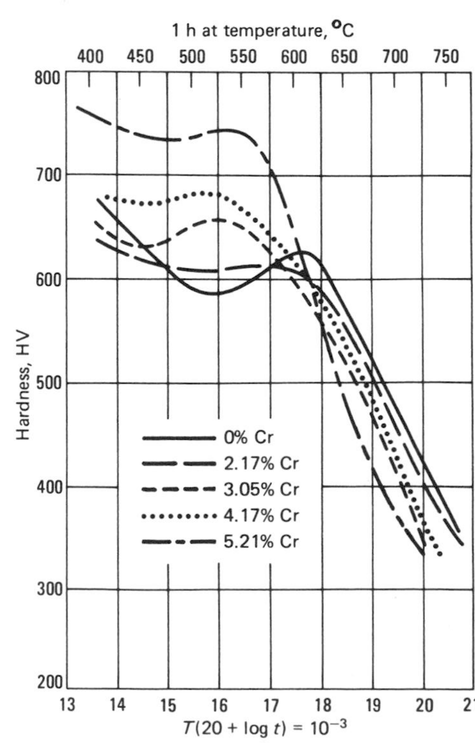

Fig. 41 Effect of chromium on the tempering characteristics of 0.45C-1.75Mo-0.75V steels. Source: Ref 44

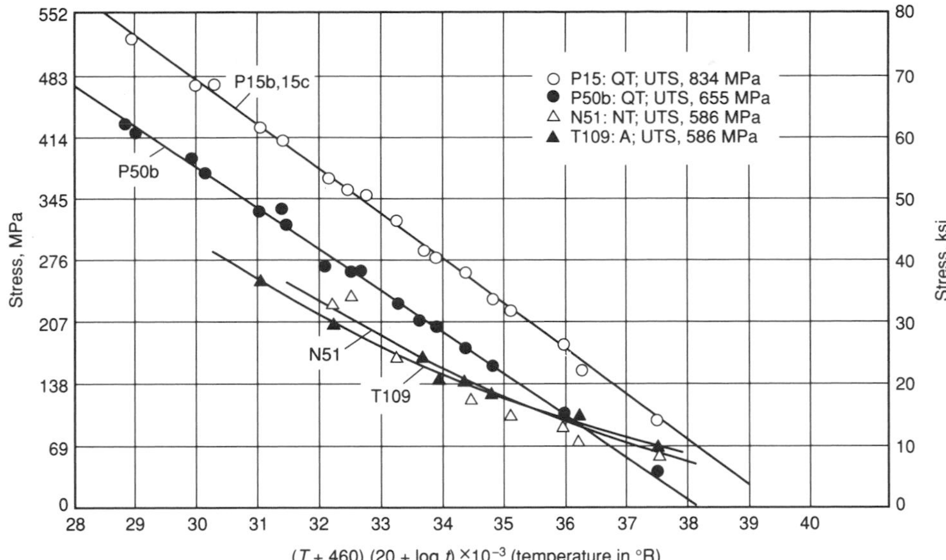

Fig. 42 Variation in stress-rupture strength of 2.25Cr-1Mo steel under different heat treatments. QT, quenched and tempered; NT, normalized and tempered; A, annealed; UTS, ultimate tensile strength. Source: Ref 45

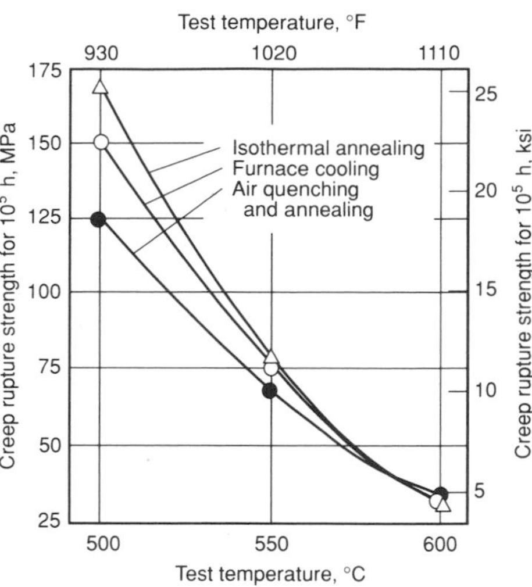

Fig. 43 Influence of heat treatment on 10^5 h creep-rupture strength of 2.25Cr-1Mo steel. Source: Ref 48

ice temperatures. Figure 40 shows secondary hardening in a series of steels containing molybdenum. The secondary hardening peaks develop only at high tempering temperatures, because alloy carbide formation depends on the diffusion of the carbide-forming elements, a more sluggish process than that of carbon and iron diffusion. As a result, a finer dispersion of particles is produced, and the alloy carbides, once formed, are quite resistant to coarsening. The latter characteristic of the fine-alloy carbides is used to advantage in tool steels that must not soften, even though high temperatures are generated by their use in hot-working dies or high-speed machining. Also, ferritic low-carbon steels containing chromium and molybdenum are used in pressure vessels and reactors operated at temperatures around 540 °C (1000 °F), because the alloy carbides are slow to coarsen at those temperatures.

The beneficial property changes from secondary hardening can be improved by increasing the intensity of secondary hardening, decreasing the rate of overaging of the secondary-hardening carbide, or increasing the temperature of secondary hardening. The intensity of secondary hardening can be increased by increasing the mismatch between the carbide precipitate and the matrix (Ref 44). Although this tends to cause more rapid overaging, the net effect can be beneficial, so that a higher strength after tempering is achieved. Increased mismatch is produced by:

- Increasing the lattice parameter of the carbide precipitate
- Decreasing the lattice parameter of the matrix

The carbide Mo_2C can dissolve both chromium and vanadium. Chromium, being a smaller atom than molybdenum, reduces the lattice parameter of Mo_2C, but vanadium increases it. Chromium therefore tends to decrease the intensity of secondary hardening. Moreover, chromium causes the Mo_2C carbide to be less stable (i.e., it gives maximum secondary hardening at lower tempering temperatures) and accelerates overaging (Fig. 41). On the other hand, vanadium increases the latter parameter of Mo_2C and stabilizes the carbide. The result is a greater intensity of secondary hardening.

Effects of Microstructure

It is widely accepted that the strength and impact toughness of carbon and chromium-molybdenum steels with fully bainitic microstructures are better than those with a ferritic-bainitic microstructure. Bainitic microstructures also have better creep resistance under high-stress, short-time conditions, but they degrade more rapidly at high temperatures than pearlitic structures. As a result, ferrite-pearlite material has better intermediate-term, low-stress creep resistance. Because both microstructures will eventually spheroidize, the creep strengths of the two microstructures converge over long service lives. This convergence can be estimated to occur in about 50,000 h at 540 °C (1000 °F) for 2.25Cr-1Mo steel, based on the limited data presented in Fig. 42. Investigations of chromium-molybdenum steels for one application concluded that tempered bainite is the optimum microstructure for creep resistance (Ref 46). However, the carbide precipitates are also an

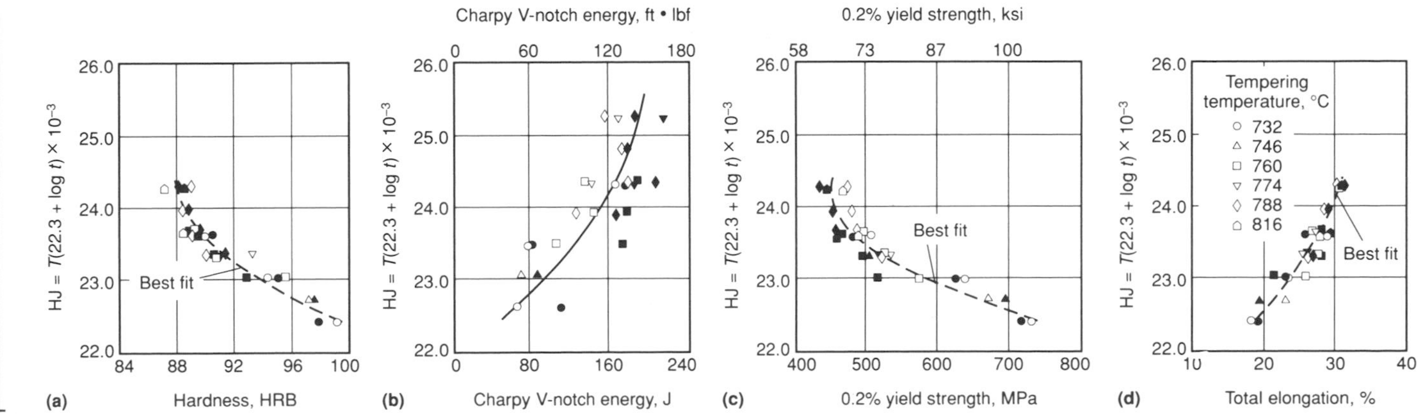

Fig. 44 Room-temperature properties of two heats (open or closed symbols) of a modified 9Cr-1Mo steel correlated with the Holloman-Jaffe (HJ) tempering parameter. (a) Hardness. (b) Charpy energy. (c) 0.2% yield strength. (d) Total elongation at room temperature. Source: Ref 7

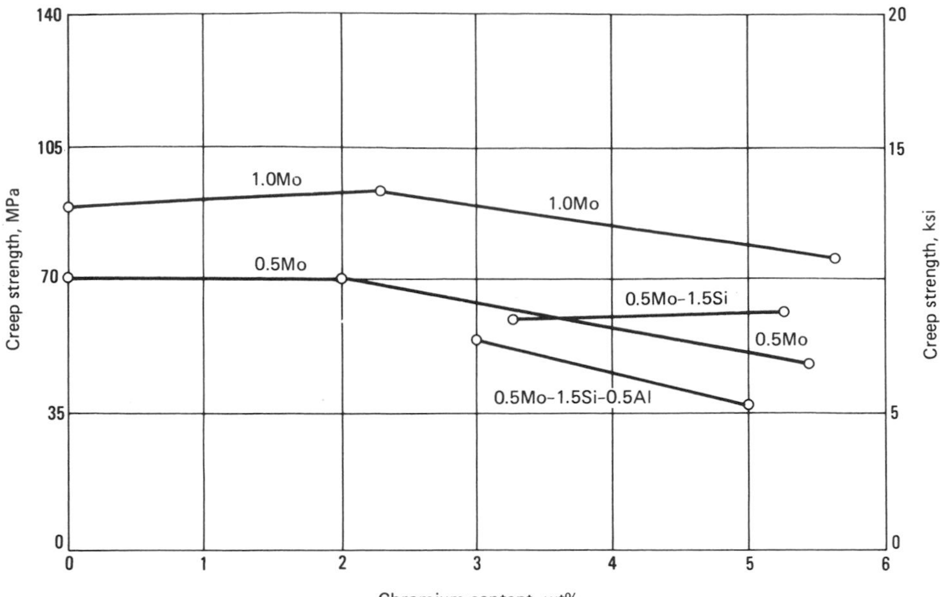

Fig. 45 Effect of chromium on the creep strength (stress to produce a minimum creep rate of 0.0001% per hour) of several steels containing small amounts of molybdenum, silicon, and aluminum at 540 °C (1000 °F). Source: Ref 50

important microstructural factor in achieving optimum creep behavior, and for some microstructures an untempered condition may be desirable (see the following section, "Effects of Heat Treatment," in this article). Moreover, even though bainitic microstructures improve strength, toughness, and creep resistance, chromium-molybdenum steels with bainitic and tempered martensitic microstructures also undergo strain softening during mechanical cycling. This effect of strain softening of bainitic chromium-molybdenum steels has undergone several investigations (Ref 18–21).

Microstructure may also influence the carbide precipitation and strengthening mechanism of chromium-molybdenum steels. In 2.25Cr-1Mo steel, for example, precipitation reactions are known to occur much more rapidly in bainite than in proeutectoid ferrite (Ref 12). In addition, the interaction solid-solution strengthening of 2.25Cr-1Mo steel is influenced by microstructure. In tensile studies, it was concluded that interaction solid-solution hardening in bainitic (normalized and tempered) 2.25Cr-1Mo steel is due to chromium-carbon interactions, while it is due to molybdenum-carbon interactions in the proeutectoid ferritic of annealed steel (Ref 47).

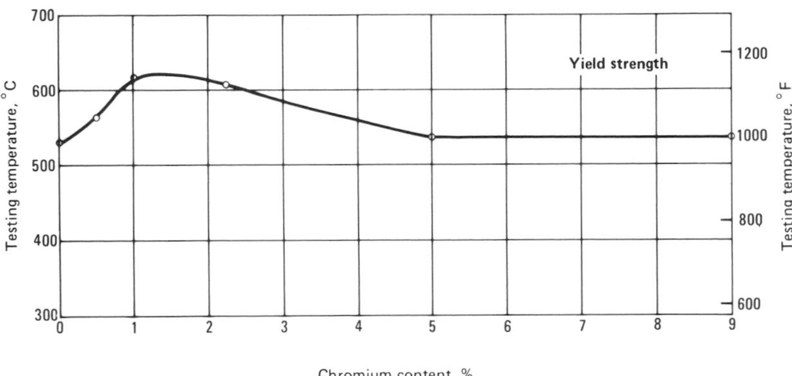

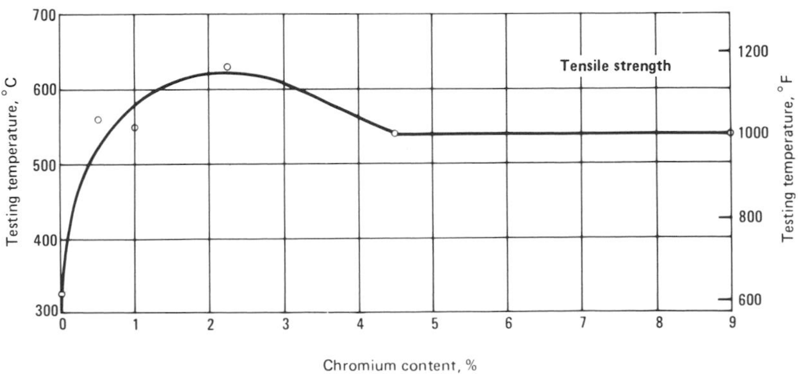

Fig. 46 Effect of chromium content on strength. Test temperature required to reduce tensile strength and yield strength to 60% of their room-temperature values for chromium-molybdenum steels containing 0.5 to 1.0% Mo and the indicated amount of chromium

Effects of Heat Treatment

Figure 43 shows the general effect of three heat treatments on the creep-rupture strength of 2.25Cr-1Mo steel. As with the long-term exposure data in Fig. 42, the creep-rupture strengths converge.

Tempering has an important influence on the level of precipitation strengthening and solid-solution strengthening in chromium-molybdenum steel (Fig. 37). In a normalized molybdenum steel (Fig. 37a), the initial contribution from solid-solution strengthening is greater than that of the normalized and tempered steel. In the normalized and tempered molybdenum steels (Fig. 37b), the initial contribution from precipitation strengthening will be larger than that from the normalized steel. In addition, the precipitation-strengthening effect in the normalized and tempered steel will reach a maximum and begin to decline at an earlier stage, due to the earlier incidence of overaging in tempered material. This is a potential consideration in applications requiring creep resistance over long times and at high temperatures.

As noted in the previous section "Effects of Microstructure," an investigation for one application concluded that tempered bainite resulted in the optimum creep resistance. In ferrite-pearlite or ferrite-bainite structures, however, it has been suggested that the best creep resistance at relatively high stresses is obtained in the untempered condition, because the dislocations introduced on loading are then able to nucleate a finer dispersion of particles in ferrite grains than is obtained by tempering in the absence of strain (Ref 39). This concept does not apply to bainitic structures. In bainitic steels, where the dislocation density is already higher than that introduced upon straining of a ferrite/pearlite steel, the use of untempered structures is unlikely to prove beneficial to short-term creep strength (Ref 39). Ultimately, it is the balance of hardness (or strength) and toughness required in service that determines the conditions of tempering for a given application. Figure 44 shows the variation of properties from the tempering of a modified 9Cr-1Mo alloy (9Cr-

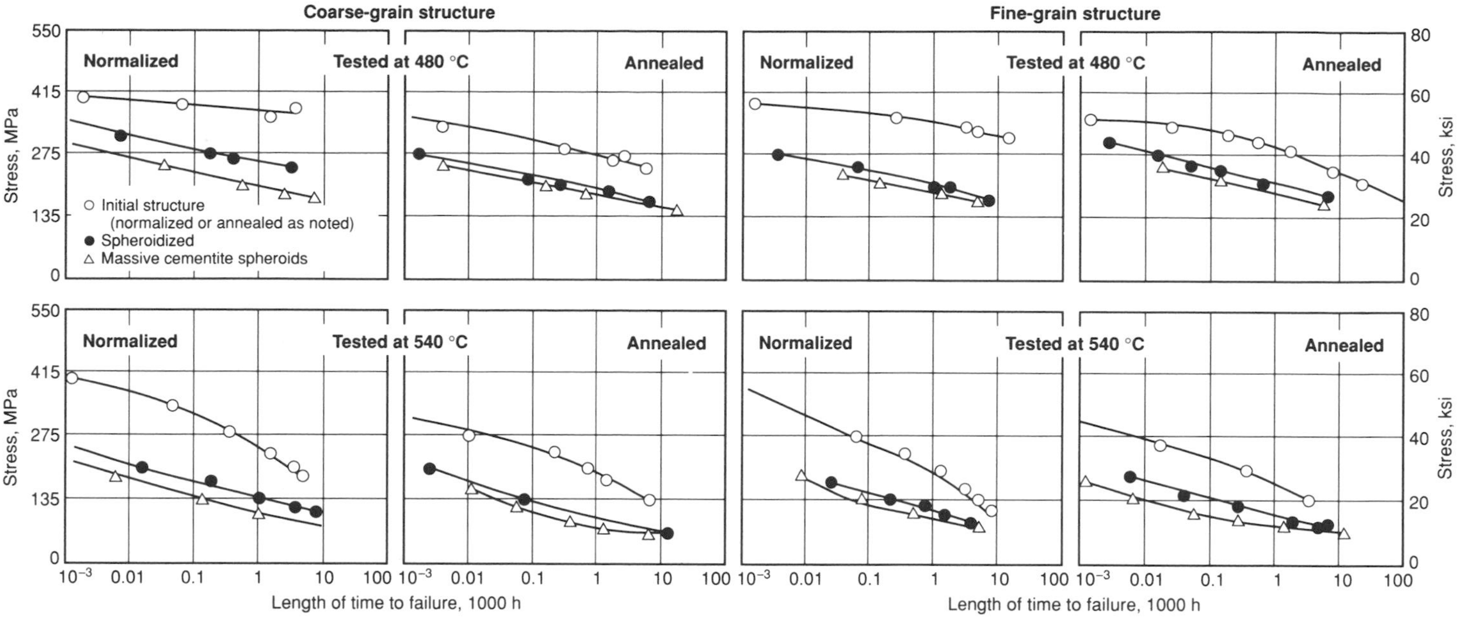

Fig. 47 Effect of spheroidization on the rupture strength of carbon-molybdenum steel (0.17C-0.88Mn-0.20Si-0.42Mo). Source: Ref 55

1Mo with 0.06 to 0.10 wt% Nb and 0.18 to 0.25 wt% V).

Effects of Composition

The mechanical properties of carbon and low-alloy steels are determined primarily by composition and heat treatment. The effects of alloying elements in annealed, normalized and tempered, and quenched and tempered steels are discussed below.

Carbon increases both the strength and hardenability of steel at room temperature but decreases the weldability and impact toughness. In plain carbon and carbon-molybdenum steels intended for elevated-temperature service, carbon content is usually limited to about 0.20%. In some classes of tubing for boilers, however, carbon may be as high as 0.35%. For chromium-containing steels, carbon content is usually limited to 0.15%. Carbon increases short-term tensile strength but does not add appreciably to creep resistance at temperatures above 540 °C (1000 °F), because carbides eventually become spheroidized at such temperatures.

Manganese, in addition to its normal function of preventing hot shortness by forming dispersed manganese sulfide inclusions, also appears to enhance the effectiveness of nitrogen in increasing the strength of plain carbon steels at elevated temperatures. Manganese significantly improves hardenability but contributes to temper embrittlement.

Phosphorus and sulfur are considered undesirable because they reduce the elevated-temperature ductility of steel. This reduction in ductility is demonstrated by reductions in stress-rupture life and thermal fatigue life. Phosphorus contributes to temper embrittlement.

Silicon increases the elevated-temperature strength of steel. It also increases the resistance to scaling of the low-chromium steels in air at elevated temperatures. Silicon is one factor in temper embrittlement.

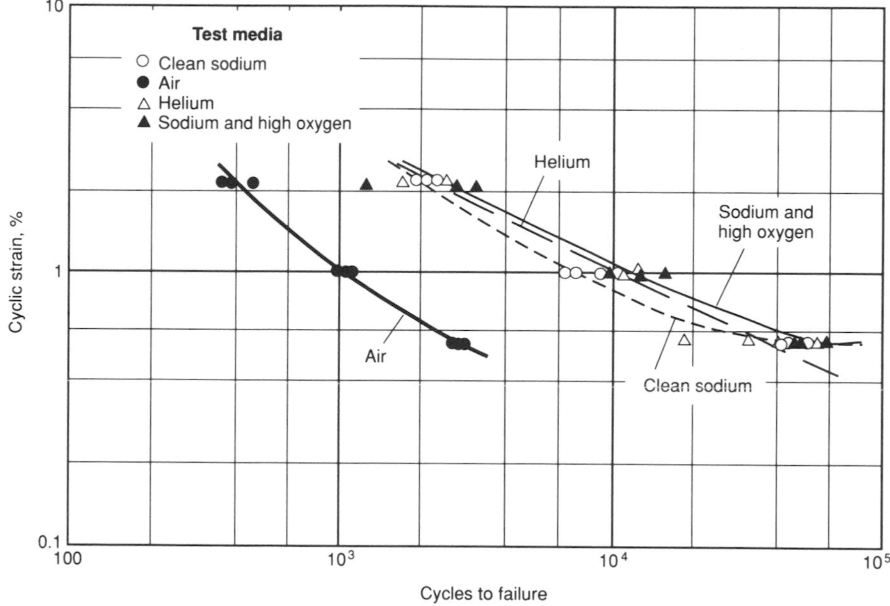

Fig. 48 Fatigue test results of 2.25Cr-1Mo steel in sodium, air, and helium at 593 °C (1100 °F). Source: Ref 63

Chromium in small amounts (~0.5%) is a carbide former and stabilizer. In larger amounts (up to 9% or more), it increases the resistance of steels to corrosion. Chromium also influences hardenability.

The effect of chromium in ferritic creep-resistant steels is complex. By itself, chromium enhances creep strength, although increasing the chromium content in lower-carbon grades does not increase resistance to deformation at elevated temperatures (Ref 41). When added to molybdenum steel, chromium generally leads to some reduction in creep strength (Ref 49), such as that shown in Fig. 45. For the 1Mo steel in Fig. 45, the optimum creep strength occurs with about 2.5% Cr. Chromium is most effective in strengthening molybdenum steels (0.5 to 1% Mo) when it is used in amounts of 1 to 2.5%.

Figure 46 summarizes the effects of chromium content on the tensile and yield strengths of chromium-molybdenum steels containing 0.5 to 1.0% Mo and various amounts of chromium. The effect of temperature is reported as the test temperature

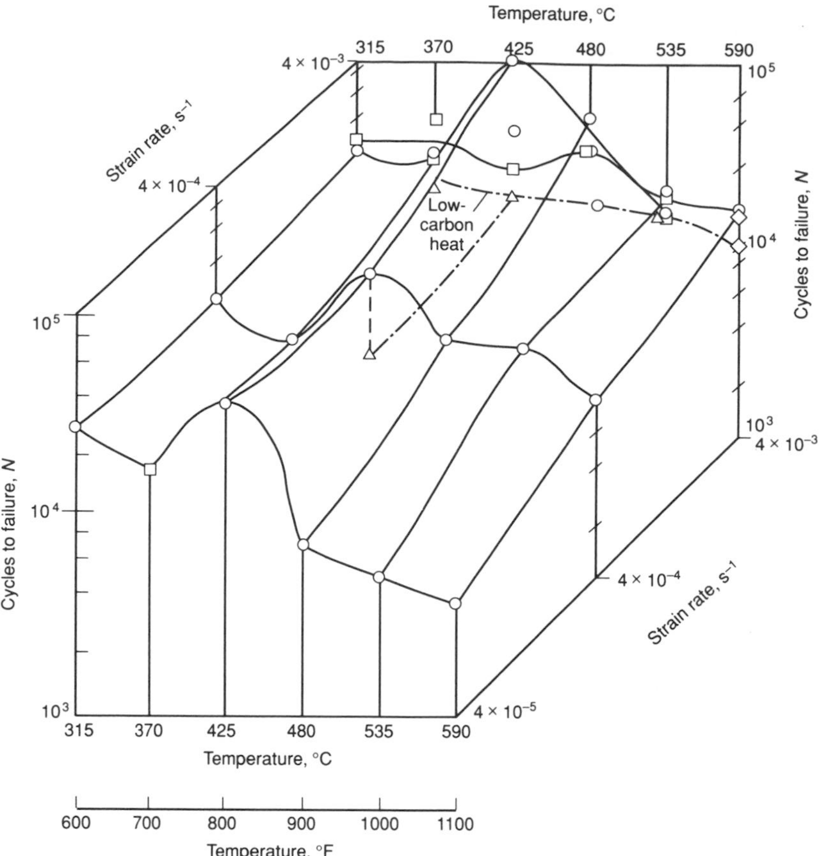

Fig. 49 Cycles to failure as a function of temperature and strain rate (continuous cycling) for various heats of isothermally annealed 2.25Cr-1Mo steel. Source: Ref 63

at which strength is reduced to 60% of its room-temperature value. Chromium is most effective in strengthening these chromium-molybdenum steels when it is used in amounts of 1 to 2.5%.

Molybdenum is an essential alloying element in ferritic steels where good creep resistance above 450 °C (840 °F) is required. Even in small amounts (0.1 to 0.5%), molybdenum increases the resistance of these steels to deformation at elevated temperatures. Much greater creep strength can be obtained by increasing the molybdenum level to about 1%, but at the expense of greatly reduced rupture ductility (Ref 51). Additions of chromium can improve rupture ductility.

Molybdenum is a carbide stabilizer and prevents graphitization. For certain ranges of stress and temperature, the dissolving of iron carbide and the concurrent precipitation of molybdenum carbide cause strain hardening in these steels. Molybdenum in amounts of 0.5% or less also minimizes temper embrittlement.

Niobium and vanadium are added to improve elevated-strength properties. Vanadium is also added to some of the higher-carbon steels to provide additional resistance to tempering and to retard the growth of carbides at service temperatures. Niobium is sometimes added to these steels to increase their strength through the formation of carbides. Niobium and vanadium improve resistance to hydrogen attack, but they may promote hot (reheat) cracking.

Boron is added to increase hardenability. Boron can cause hot shortness and impair toughness.

Tungsten behaves like molybdenum in simple steels and has been proposed for replacing molybdenum in nuclear applications. This idea is described in Ref 52 to 54 and the article "Neutron Irradiation Damage of Steels" in the *ASM Specialty Handbook: Carbon and Alloy Steels.*

Thermal Exposure and Aging

Thermal exposure over time is one of the main service conditions affecting mechanical properties, because the metallurgical structure of steel changes with time at temperature. For example, a ferritic matrix may be either fine- or coarse-grained initially, and the carbides may vary from lamellar to completely spheroidized. With increasing time at service temperatures, the metallurgical structure slowly approaches a more stable state. For example, there may be some increase in ferrite grain size, the carbides may spheroidize, and the structure of carbon and carbon-molybdenum steels may approach the graphitized condition, with large irregular nodules of graphite in a ferrite matrix and few, if any, remaining carbides.

The thermal exposure of molybdenum and molybdenum-chromium ferritic steels also contributes to complex aging phenomena, which are governed by the complicated carbide precipitation processes that occur in the steel. Figure 32, for example, shows the sequence of carbide formations in 2.25Cr-1Mo steel. The M_2C carbide (where M is primarily molybdenum) is the principal carbide for strengthening in this steel. The Mo_2C first precipitates during heat treatment and/or elevated-temperature exposure. The Mo_2C forms a high density of fine needles or platelets and thus contributes to strengthening by dispersion hardening. During thermal exposure, however, the unstable Mo_2C carbide eventually transforms into large globular particles of $M_{23}C$ and η carbide. These particles are thought to have little strengthening effect, although there are some indications that $M_{23}C$ and η carbide that are present after long aging times in 2.25Cr-1Mo steel can enhance rupture strength (Ref 37). Precipitation kinetics also depend on microstructure. The strengthening carbide Mo_2C precipitates more rapidly in bainite than proeutectoid ferrite. Similarly, the Mo_2C is replaced more quickly by more stable carbides in bainite than in proeutectoid ferrite. In either case, these precipitation reactions influence the strength in a similar way, regardless of whether the microstructure is bainite or proeutectoid ferrite.

Spheroidization and Graphitization

Spheroidization of the carbides in a steel occurs over time, because spheroidized microstructures are the most stable microstructure found in steels. This phenomenon reduces strength and increases ductility.

The effect of spheroidization on the rupture strength of a typical carbon-molybdenum steel containing 0.17% C and 0.42% Mo, at 480 and 540 °C (900 and 1000 °F), is shown in Fig. 47 for several initial metallurgical structures (normalized or annealed, fine- or coarse-grained). In these tests, the structure of the steel affected the rupture strength. For example, the stress for failure of a spheroidized structure in a given time was sometimes only half that of a normalized structure.

At 480 °C (900 °F), a coarse-grain normalized structure was the strongest for both short-time and long-time tests. The spheroidized structures were weaker than the normalized or annealed structures for short-time tests at both 480 and 540 °C (900 and 1000 °F). As the test time increased, the rupture values for all the structures tended to approach a common value.

The rate of spheroidization depends on the initial microstructure. The slowest spheroidizing is associated with pearlitic microstructures, especially those with coarse interlamellar spacings. Spheroidizing is more rapid if the carbides are initially in the form of discrete particles, as in bainite, and even more rapid if the initial structure is martensite.

Graphitization is a microstructural change that sometimes occurs in carbon or low-alloy steels subjected to moderate temperatures for long periods of time. The microstructure of carbon and carbon-molybdenum steels used for high-temperature applications is normally composed of pearlite, which is a mixture of ferrite with some iron carbide (cementite). However, the stable form of carbon is graphite rather than cementite. Therefore, the pearlite can decompose into ferrite and randomly dispersed graphite, while the cementite will tend to disappear in these materials if they are in service long enough at metal temperatures higher than 455 °C (850 °F). This phenomenon can embrittle steel parts, especially when the graphite particles form along a continuous zone through a load-carrying member. Graphite particles that are randomly distributed throughout the microstructure cause only moderate loss of strength. Graphitization can be resisted by steels containing more than 0.7% Cr because such steels always contain at least 0.5% Mo as well, largely to impart elevated-temperature strength and resistance to temper embrittlement. Figure 4 shows the failure of a carbon-molybdenum steel due to graphitization.

Decarburization and Carburization

Decarburization is a loss of carbon from the surface of a ferrous alloy as a result of heating in a medium (e.g., hydrogen) that reacts with carbon. Unless special precautions are taken, the risk of losing carbon from the surface of steel is always present during heating to high temperatures in an oxidizing atmosphere. A marked reduction in fatigue strength is noted in steels with decarburized surfaces. The effect of decarburization is much greater on steels with high tensile strength than on steels with low tensile strength.

Carburization. As in the case of sulfide penetration, carburization of high-temperature alloys is thermodynamically unlikely except at very low oxygen partial pressures. This is true because the protective oxides of chromium and aluminum are generally more likely to form than the carbides. However, carburization can occur kinetically in many carbon-containing environments. Carbon transport across continuous nonporous scales of alumina or chromia is very slow, and the alloy pretreatments that are likely to promote such scales (e.g., initially smooth surfaces or preoxidation) are generally effective in decreasing carburization attack.

The suitability of a carburized metal for further service can be determined by evaluating its properties and condition. The mechanical properties of the carburized layer vary markedly from those of the unaffected metal. Room-temperature ductility and toughness are decreased, and hardness is increased greatly. This deterioration is important if the carburized layer is stressed in tension, because cracking is quite likely to occur. Weldability is adversely affected. Welds in carburized materials frequently show cracks because of thermal tensile stresses, even with preheating and postheating. Ductility at temperatures above 400 °C (750 °F) is usually adequate. The corrosion resistance of the low-alloy chromium-molybdenum steels commonly used for elevated-temperature applications is reduced because of the reduction in effective chromium content. For the same reason, carburized stainless steels may have a relatively low resistance to general corrosion and to intergranular corrosion, particularly while the equipment is shut down. Minor amounts of carburization do not affect creep and rupture strengths significantly.

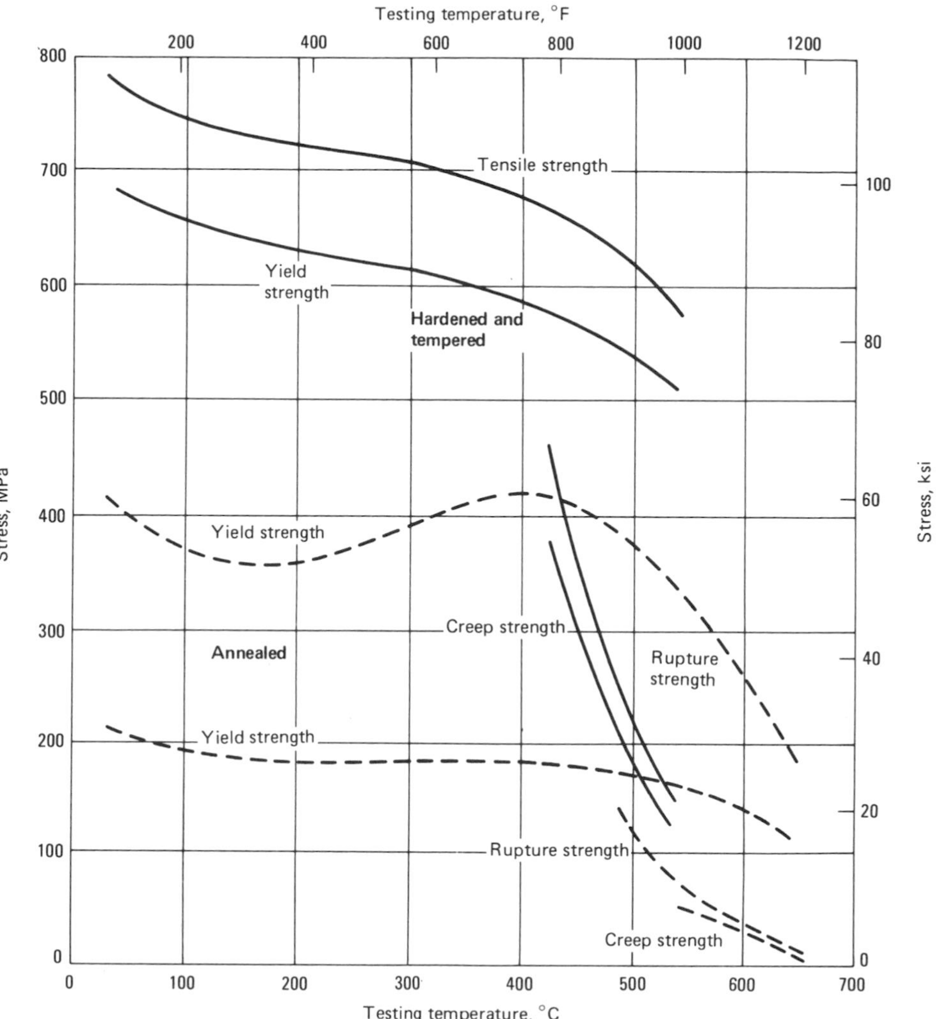

Fig. 50 Effect of test temperature on strength of 2.25Cr-1Mo steel. Effect of test temperature on tensile strength, yield strength, creep strength (for creep rate of 0.1 µm/m · h), and stress to rupture (for life of 100,000 h) of annealed specimens (dashed lines) and hardened and tempered specimens (solid lines) of 2.25Cr-1Mo steel. Source: Ref 2

Factors Affecting Fatigue Strength

As described in the section "Creep-Fatigue Interaction" in this article, the hold times (dwell periods) and the waveform of cyclic strains influence the fatigue strength of metals at elevated temperatures. These factors affect the assessment of low-cycle fatigue and creep fatigue (Fig. 27). In addition, environmental effects and strain aging influence fatigue strength.

Environmental Effects. It has long been recognized that oxidation at elevated temperatures can have a marked effect (usually an acceleration) on fatigue crack initiation and growth (Ref 56–62). In many alloys, intergranular oxidation initiates intergranular cracks at a temperature that is approximately one-half the melting point temperature (the exact temperature depends on waveform or frequency) (Ref 59). Another possibility is penetration of oxygen along slip bands with subsequent localized embrittlement and cracking. An example of the effects of environment on the fatigue strength of 2.25Cr-1Mo steel is shown in Fig. 48. These tests were conducted in bending at a frequency of 0.05 Hz.

Dynamic Strain Aging. In addition to the effects of environment and temperature, ferritic low-carbon and alloy steels undergo dynamic strain aging when subjected to inelastic deformation in certain ranges of temperature, strain, and strain rate. Dynamic strain aging (which involves the interaction of interstitials and/or carbide or nitride formers such as chromium, molybdenum, and manganese with strain-induced dislocations) has been shown to markedly influence the hardening characteristics that depend on cyclic strain rate, thus affecting both the initiation and growth of fatigue cracks in ferritic materials (Ref 63). Figure 49 shows the deleterious effect of decreasing strain rate on the fatigue strength of an annealed 2.25Cr-1Mo steel at various temperatures. Typically, strength is increased and ductility is decreased over the temperature ranges where aging occurs, and both low- and high-cycle fatigue properties can be influenced accordingly. Thus, it

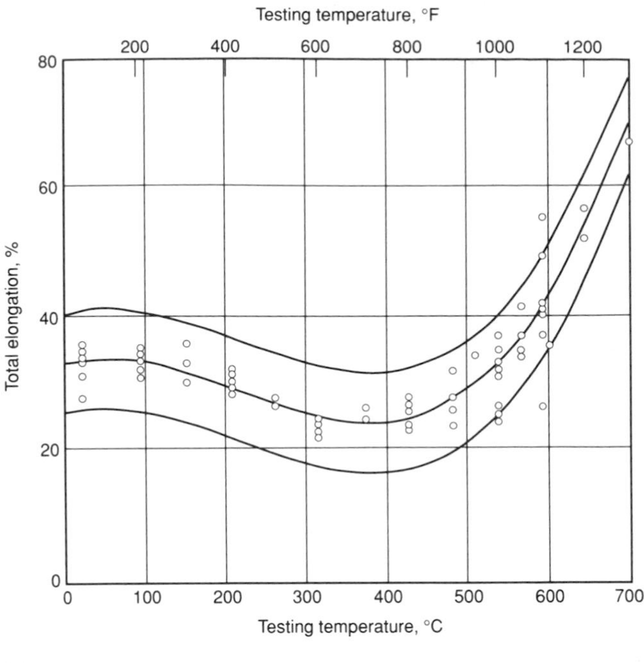

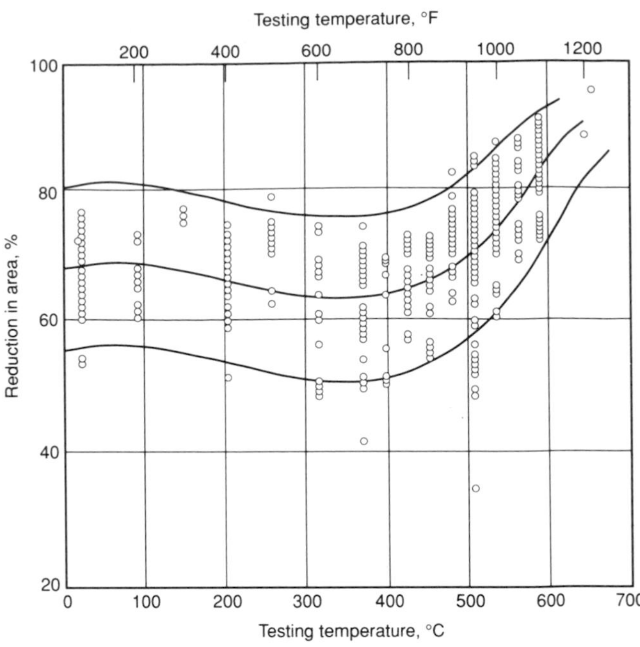

Fig. 51 Effect of test temperature on ductility. (a) Elongation in 50 mm (2 in.) and (b) reduction in area for annealed specimens of 2.25Cr-1Mo steel tested at the indicated temperatures. Source: Ref 66

is important to understand fatigue, creep fatigue, environment, and strain aging interactions in the intermediate- to high-cycle life region.

Elevated-Temperature Behavior of 2.25Cr-1Mo Steel

The elevated-temperature behavior of 2.25Cr-1Mo steel has been studied more thoroughly than that of any other creep-resistant steel. The available data on annealed and normalized and tempered 2.25Cr-1Mo steel are summarized in Ref 64 and 65. The rupture strength and creep ductility of 2.25Cr-1Mo steel in various heat-treated conditions are reviewed in Ref 64. The following conclusions were reached:

- The stress-rupture strength generally increases linearly with room-temperature tensile strength up to about 565 °C (1050 °F) for times up to 10,000 h.
- At a given strength level, tempered bainite results in higher creep strength than tempered martensite or ferrite-pearlite aggregates for temperatures up to 565 °C (1050 °F) and times up to 100,000 h. For higher temperatures and times, the ferrite-pearlite structure is the strongest.
- Rupture ductility generally decreases with rupture time, reaches a minimum, and then increases again. Test temperature, room-temperature tensile strength, austenitizing temperature, and impurity content increase the rate of decrease of ductility with time and cause the ductility minimum to occur at shorter times.

In terms of application, this steel has an excellent service record in both fossil fuel and nuclear fuel plants for generating electricity. The severe operating conditions in these plants have justified extensive studies of the behavior of 2.25Cr-1Mo steel under complex loading conditions and in unusual environments. This steel has become a reference against which the performance of other steels can be measured.

Specifications, Steelmaking Practices, and Heat Treatments. Some of the specifications for 2.25Cr-1Mo steel in the ASME Boiler and Pressure Vessel Code are listed in Table 11, which also includes product forms and room-temperature mechanical property requirements. For some of these specifications, composition ranges and limits differ slightly from those given in Tables 5 and 6.

In the United States, 2.25Cr-1Mo steel is normally manufactured in an electric furnace. In Japan, basic oxygen processes are used. For certain critical applications, vacuum arc remelting or electroslag remelting is appropriate.

The austenitizing temperature for 2.25Cr-1Mo steel is about 900 °C (1650 °F). Heat treatments commonly employed with 2.25Cr-1Mo steel include:

- *Normalize and temper:* Austenitize at 910 to 940 °C (1650 to 1725 °F), cool in air, temper at 580 to 720 °C (1075 to 1325 °F).
- *Oil quench and temper:* Austenitize at 940 to 980 °C (1725 to 1800 °F), quench in oil, temper at 570 to 705 °C (1065 to 1300 °F).

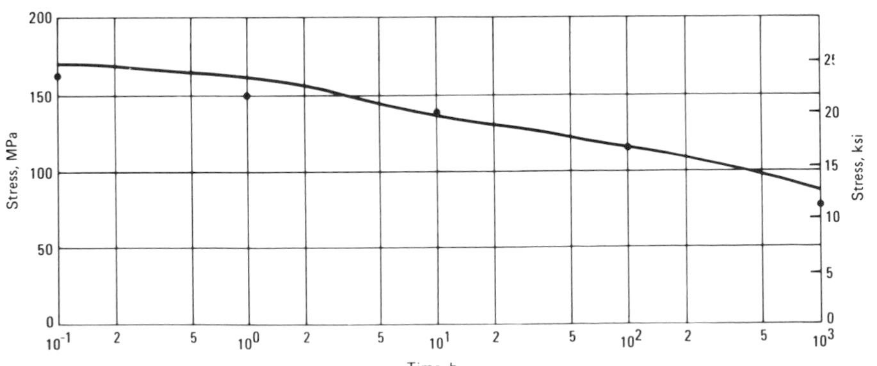

Fig. 52 Relaxation behavior of 2.25Cr-1Mo steel. Specimens were stressed to level indicated on ordinate of graph and exposed to elevated temperature for indicated duration; remaining stress indicated on graph. Source: Ref 66

Short-Term Elevated-Temperature Mechanical Properties of 2.25Cr-1Mo Steel. The effects of test temperature on the tensile and yield strengths of 2.25Cr-1Mo steel are illustrated in

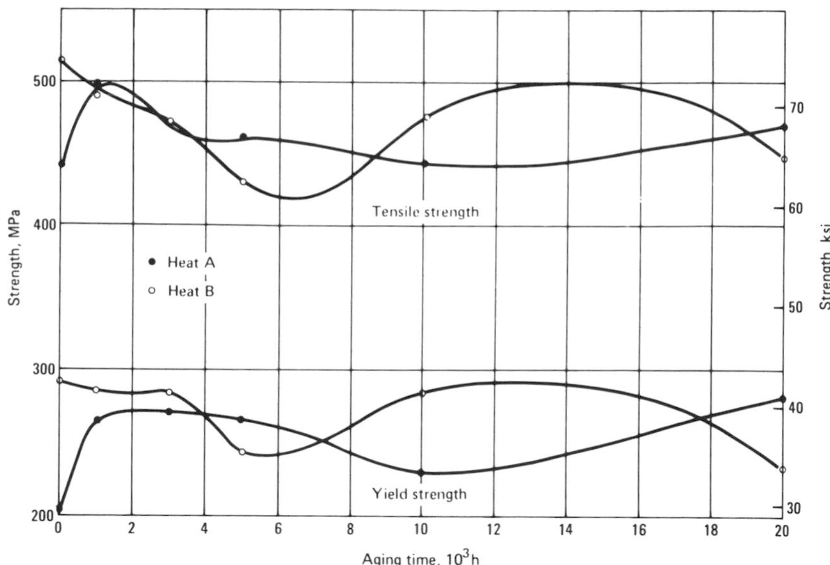

Fig. 53 Effect of exposure to elevated temperature on the strength of 2.25Cr-1Mo steel. Variation in tensile and yield strengths of two different heats of 2.25Cr-1Mo steel after exposure (without stress) to test temperature of 455 °C (850 °F)

Fig. 19, 33, and 50. Data for annealed specimens and for hardened and tempered specimens are also included. The large variations in both tensile strength and yield strength with temperature and strain rate (Fig. 19) are caused by strain rate, temperature, and microstructure.

The effects of elevated temperatures, elongation, and reduction in area for annealed specimens tested at standard strain rates are illustrated in Fig. 51. Specimens tested at about 400 °C (750 °F) showed an increase in strength and a reduction in ductility, both of which were caused by strain aging. However, the reduction in ductility was relatively small.

The effects of temperature on modulus of elasticity, shear modulus, and Poisson's ratio are shown in Fig. 20. The modulus of elasticity diminishes from 215 GPa (31×10^6 psi) at room temperature to 140 GPa (20.3×10^6 psi) at 760 °C (1400 °F); similarly, the shear modulus diminishes from 83 GPa (12.05×10^6 psi) at room temperature to 52.4 GPa (7.6×10^6 psi) at 760 °C (1400 °F). Poisson's ratio increases from 0.288 at room temperature to 0.336 at 760 °C (1400 °F).

Long-Term Elevated-Temperature Mechanical Properties of 2.25Cr-1Mo Steel. The creep and stress-rupture behavior of annealed specimens and hardened and tempered specimens of 2.25Cr-1Mo steel are illustrated in Fig. 34, 35, 36, and 50. With regard to rupture life and creep rate, the hardened and tempered specimens were able to withstand higher stresses than the annealed specimens.

The ductility exhibited by stress-rupture specimens can be roughly correlated with stress level or rupture life. In general, specimens tested at high stress levels have short rupture lives, and such specimens exhibit greater reduction in area than similar specimens tested at lower stress levels. These data show considerable scatter but no evidence of brittle behavior by this steel. The relaxation behavior of 2.25Cr-1Mo steel is illustrated in Fig. 52.

Long-term exposure to elevated temperature can reduce the room-temperature and elevated-temperature properties of 2.25Cr-1Mo steel. Some of these effects are illustrated in Fig. 7(a), 8(a), 53, and 54.

Figure 8(a) shows the changes in room-temperature tensile properties caused by exposure (without stress) to elevated temperatures.

Figure 53 shows the effect of variations in aging time (without stress) at 455 °C (850 °F) on the ultimate tensile and yield strengths of two heats of 2.25Cr-1Mo steel tested at the same temperature. The differences in strength between these two heats was observed even before the tests; they were probably caused by variations in composition and microstructure. The same factors account for strength changes during aging, because they affect both the size and distribution of carbides in the steel.

Figure 54 shows that prolonged aging without stress at 565 °C (1050 °F) can reduce time to rupture for annealed 2.25Cr-1Mo steel. Similarly, the data in Fig. 7(a) show that prolonged exposure to high temperatures without stress substantially reduces stress to rupture in a fixed time. The amount of reduction in stress to rupture is greatest for exposure at 480 °C (900 °F).

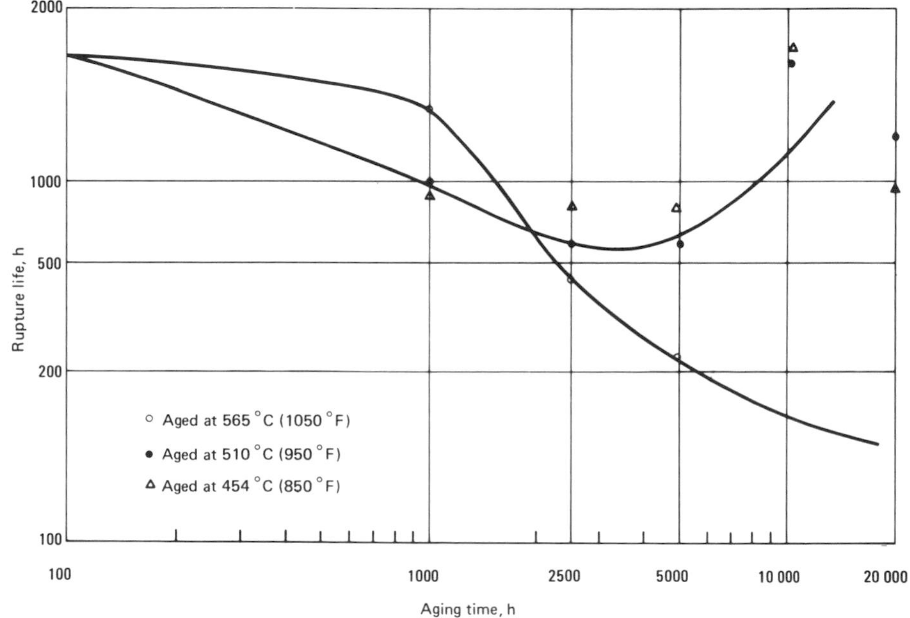

Fig. 54 Effect of exposure to elevated temperature on the stress-rupture behavior of 2.25Cr-1Mo steel. Variation in rupture life for specimens of annealed 2.25Cr-1Mo steel exposed to various elevated temperatures for the durations indicated. After aging, all specimens were stressed to 140 MPa (20 ksi) and tested at 565 °C (1050 °F).

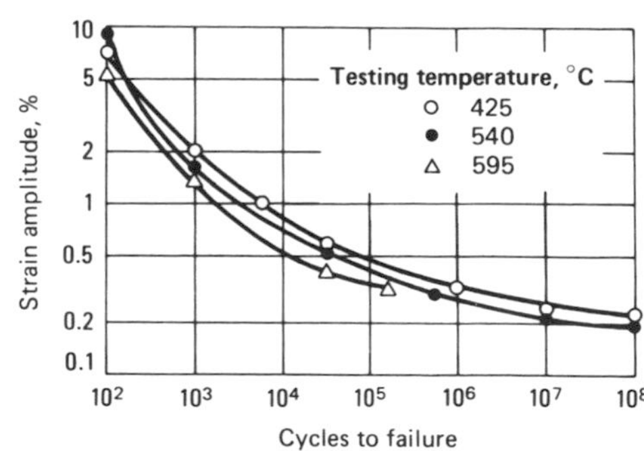

Fig. 55 Effect of elevated temperature on strain-controlled fatigue behavior of annealed 2.25Cr-1Mo steel. Strain rate was greater than 4 mm/m s. Source: Ref 66

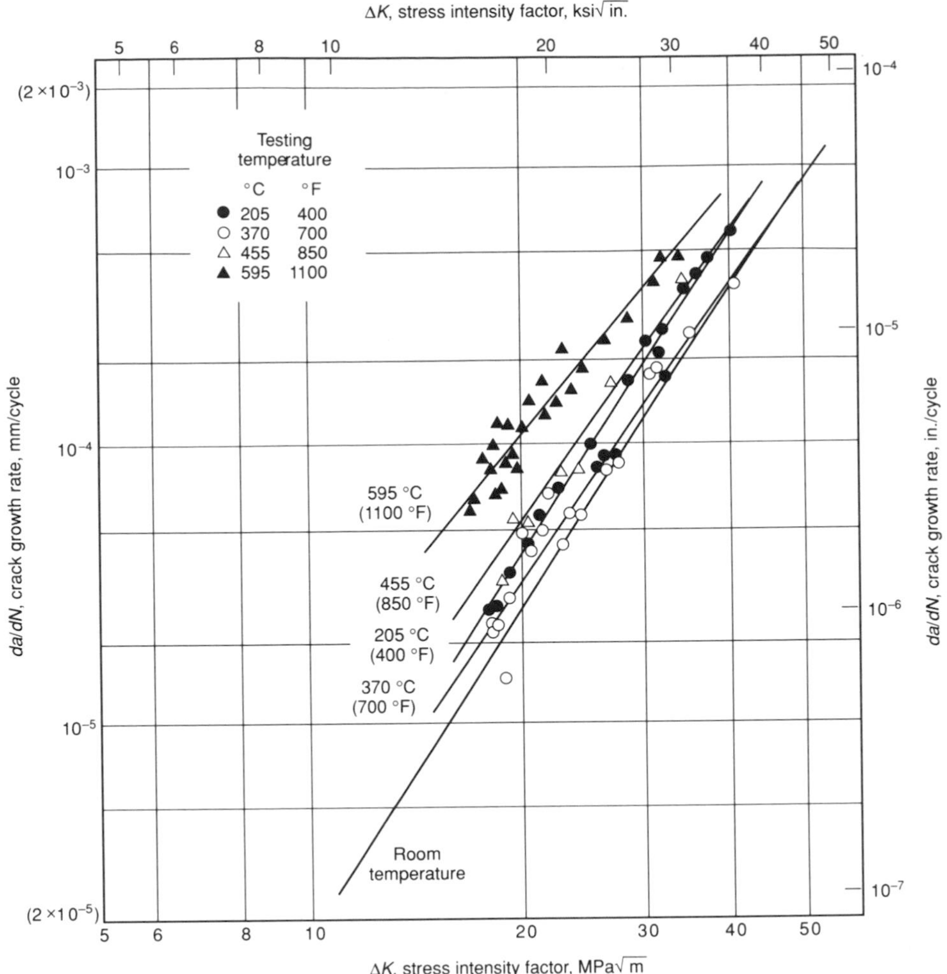

Fig. 56 Effect of temperature on fatigue crack growth rate. Variations in fatigue crack growth rate with test temperature for specimens of 2.25Cr-1Mo steel tested in air. Stress ratio was 0.05; cyclic frequency was 400/min. Source: Ref 67

because the values shown do not trend toward a unique value, the linear damage summation method is highly questionable for data extrapolation. The primary reason that the damage sums are less than 1 is that significant environmental interaction or corrosion fatigue occurs in air. This oxidation can substantially reduce the time for crack initiation in smooth bar tests, depending on waveform, and it is not adequately accounted for by the simple linear damage summation of fatigue and creep damage fractions. Environmental interaction is discussed in the section "Factors Affecting Fatigue Strength" in this article.

Properties of Welds in 2.25Cr-1Mo Steel. Welding is often required in the fabrication of pressure vessels, boilers, heat exchangers, and similar structures for use at elevated temperatures in power plants, refineries, chemical processing plants, and similar applications. Therefore, in evaluating materials for these structures, it is important to consider the mechanical properties of welded joints.

In one investigation, the elevated-temperature tensile and creep-rupture properties of weldments in 2.25Cr-1Mo steel were measured (Ref 69, 70). Specimens were cut from the weld metal and the base metal; other specimens had transverse welds. All specimens were tempered at 705 °C (1300 °F) before testing. In all these tests, the weld metal was stronger than either the base metal or the specimens containing transverse welds. Specimens with transverse welds invariably fractured in the base metal. The high strength of the weld metal relative to that of the base metal was attributed to differences in microstructure. The base metal, which had been normalized and tempered, contained more ferrite and less bainite than the weld metal. In these tests, the base metal was the weakest part of the welded structure.

Elevated-Temperature Fatigue Behavior of 2.25Cr-1Mo Steel. The results of strain-controlled fatigue tests at 425, 540, and 595 °C (800, 1000, and 1100 °F) on specimens of annealed 2.25Cr-1Mo steel are shown in Fig. 55. Within this range, the test temperature had relatively little effect on the number of cycles to failure. Other strain-controlled fatigue tests (Fig. 49) have shown that reducing the carbon content to 0.03% decreases the fatigue strength. Furthermore, because of variations in strain-aging effect, specimens from one heat with a higher carbon content ran longer at 425 °C (800 °F) than at 315 °C (600 °F).

The crack growth rate data shown in Fig. 56 and 57 were obtained from precracked specimens subjected to cyclic loading at a constant maximum load. Crack extension was measured at intervals during testing. The stress-intensity factor range increased as crack length was increased. Figure 56 illustrates the increase in crack growth rate with increasing test temperature. The data in Fig. 57 indicate that in elevated-temperature tests at a given stress-intensity factor range, crack growth rate increases as cyclic frequency is decreased. These fracture mechanics data can be applied to the design of structural components that may contain undetected discontinuities or that may develop cracks in service.

The introduction of a holding period at the peak strain of each fatigue cycle reduces fatigue life, as described in the section "Creep-Fatigue Interaction" in this article. From studies conducted on annealed 2.25Cr-1Mo steel in air (Ref 6, 63, 68), investigators have concluded that (Ref 30):

- Compressive hold periods are more damaging than tensile hold periods. Hold periods imposed on the tension-going side of the hysteresis loop are more damaging (in terms of reduced cycle life) than hold periods on the compression-going side.
- Linear damage summation of fatigue and creep damage does not sum to a unique value.

The fact that the damage sums are less than 1 indicates apparent creep-fatigue interaction, but

ACKNOWLEDGMENTS

The information in this article is largely taken from:

- S. Lampman, Elevated-Temperature Properties of Ferritic Steels, *Properties and Selection: Irons, Steels, and High-Performance Alloys*, Vol 1, *ASM Handbook* (formerly 10th ed., *Metals Handbook*), ASM International, 1990, p 617–652
- G.F. Vander Voort, Embrittlement of Steels, *Properties and Selection: Irons, Steels, and High-Performance Alloys,* Vol 1, *ASM Handbook* (formerly 10th ed., *Metals Handbook*), ASM International, 1990, p 689–736
- Steels for Coal-Fired Steam Power Plants, *ASM Specialty Handbook: Carbon and Alloy Steels,* J.R. Davis, Ed., ASM International, 1996, p 654–658
- Classification and Properties of Tool and Die Steels, *ASM Specialty Handbook: Tool Materials,* J.R. Davis, Ed., ASM International, 1995, p 119–153
- Selection of Materials for Hot Forging Tools, *ASM Specialty Handbook: Tool Materials,* J.R.

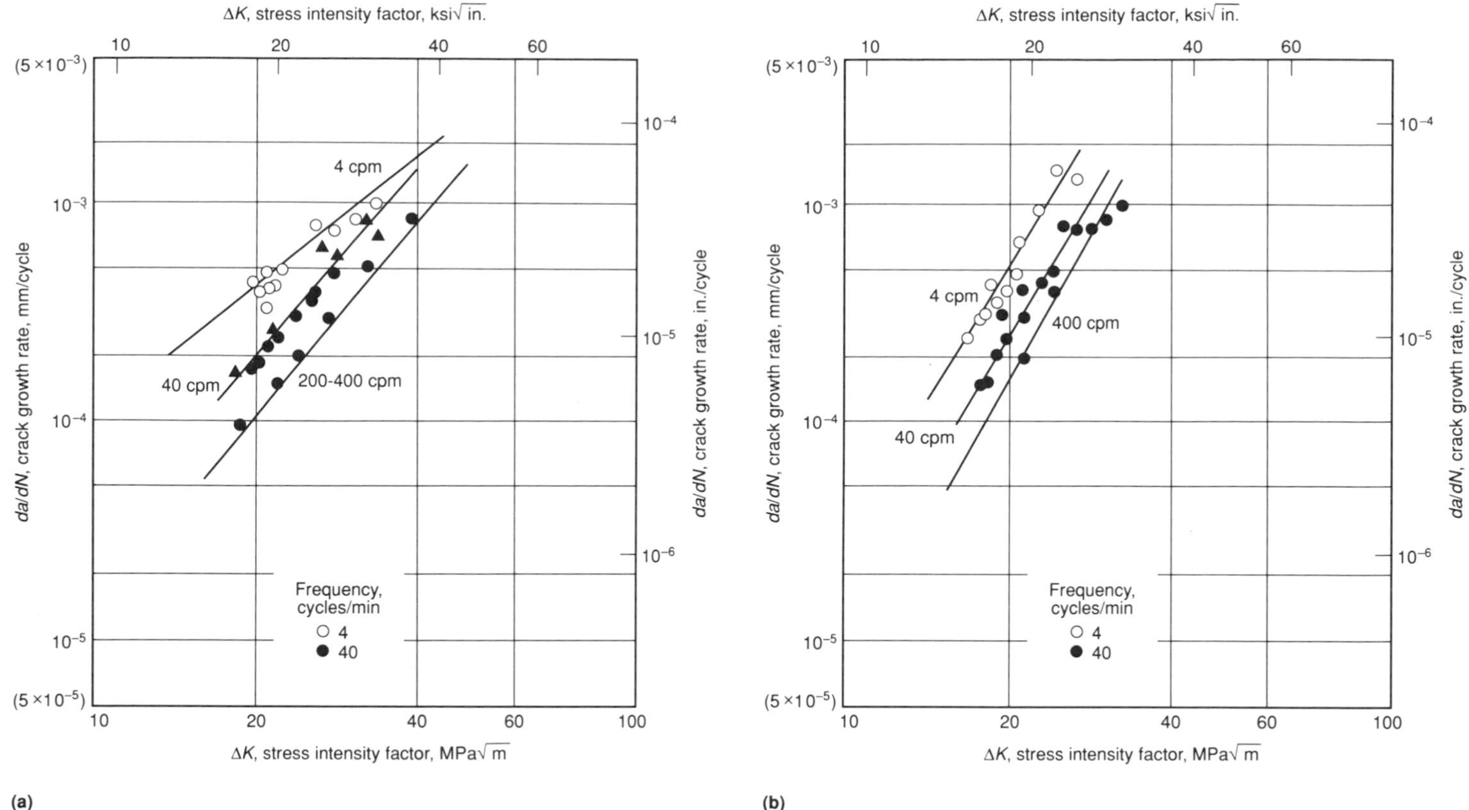

Fig. 57 Effect of cyclic frequency on fatigue crack growth rate. Variations in fatigue crack growth rate with cyclic frequency for specimens of 2.25Cr-1Mo steel tested in air. Stress ratio was 0.05. (a) Tested at 510 °C (950 °F). (b) Tested at 595 °C (1100 °F). Source: Ref 67

Davis, Ed., ASM International, 1995, p 218–235

REFERENCES

1. G.V. Smith, *Evaluation of the Elevated Temperature Tensile and Creep-Rupture Properties of ½Cr–½Mo, 1Cr-½Mo, and 1¼Cr-Mo-Si Steels,* DS 50, American Society for Testing and Materials, 1973
2. G.V. Smith, *Supplemental Report on the Elevated-Temperature Properties of Chromium-Molybdenum Steels (An Evaluation of 2¼Cr-1Mo Steel),* DS 6 S2, American Society for Testing and Materials, March 1971
3. G.S. Sangdahl and H.R. Voorhees, Quenched-and-Tempered 2¼Cr-1Mo Steel at Elevated Temperatures—Tests and Evaluation, in *2¼ Chrome-1 Molybdenum Steel in Pressure Vessels and Piping,* American Society of Mechanical Engineers, 1972
4. G.S. Sangdahl and M. Semchyshen, Ed., *Application of 2¼Cr-1Mo for Thick-Wall Pressure Vessels,* STP 755, American Society for Testing and Materials, 1987
5. *Low Carbon and Stabilized 2¼% Chromium 1% Molybdenum Steels,* American Society for Metals, 1973
6. C.R. Brinkman et al., Time-Dependent Strain-Controlled Fatigue Behavior of Annealed 2¼Cr-1Mo Steel for Use in Nuclear Steam Generator Design, *J. Nucl. Mater.,* Vol 62, 1976, p 181–204
7. V.K. Sikka, "Development of a Modified 9Cr-1Mo Steel for Elevated Temperature Service," *Proc. Topical Conf. on Ferritic Alloys for Use in Nuclear Energy Technologies,* The Metallurgical Society of AIME, 1984, p 317–327
8. R. Viswanathan, Strength and Ductility of CrMoV Steels in Creep at Elevated Temperatures, *ASTM J. Test. Eval.,* Vol 3 (No. 2), 1975, p 93–106
9. R. Viswanathan and R.I. Jaffee, Toughness of Cr-Mo-V Steels for Steam Turbine Rotors, *ASME J. Eng. Mater. Tech.,* Vol 105, Oct 1983, p 286–294
10. R. Crombie, High Integrity Ferrous Castings for Steam Turbines—Aspects of Steel Development and Manufacture, *Mater. Sci. Tech.,* Vol 1, Nov 1985, p 986–993
11. J.A. Todd et al., New Low Chromium Ferritic Pressure Vessel Steels, *Mi-Con 86: Optimization of Processing, Properties, and Service Performance through Microstructural Control,* STP 979, American Society for Testing and Materials, 1986, p 83–115
12. R.G. Baker and J. Nutting, *J. Iron Steel Inst.,* Vol 192, 1959, p 257–268
13. T. Ishiguro et al., *Research on Chrome Moly Steels,* R.A. Swift, Ed., MPC-21, American Society of Mechanical Engineers, 1984, p 43–51
14. V.K. Sikka, M.G. Cowgill, and B.W. Roberts, Creep Properties of Modified 9Cr-1Mo Steel, *Conf. on Ferritic Alloys for Use in Nuclear Energy Technologies,* American Institute of Mining, Metallurgical and Petroleum Engineers, 1984, p 413–423
15. V.K. Sikka, G.T. Ward, and K.C. Thomas, in *Ferritic Steels for High Temperature Applications,* American Society for Metals, 1982, p 65–84
16. R.L. Klueh and R.W. Swindeman, The Microstructure and Mechanical Properties of a Modified 2.25Cr-1Mo Steel, *Metall. Trans. A,* Vol 17A, 1986, p 1027–1034
17. R.L. Klueh and A.M. Nasreldin, *Metall. Trans. A,* Vol 18A, 1987, p 1279–1290
18. W.B. Jones, Effects of Mechanical Cycling on the Substructure of Modified 9Cr-1Mo Ferritic Steel, *Ferritic Steels for High-Temperature Applications,* A.K. Khare, Ed., American Society for Metals, 1983, p 221–235
19. J.L. Handrock and D.L. Marriot, Cyclic Softening Effects on Creep Resistance of Bainitic Low Alloy Steel Plain and Notched Bars, *Properties of High Strength Steels for High-Pressure Containments,* E.G. Nisbett, Ed., MPC-27, American Society of Mechanical Engineers, 1986
20. R.W. Swindeman, Cyclic Stress-Strain-Time Response of a 9Cr-1Mo-V-Nb Pressure Vessel Steel at High Temperature, *Low Cycle Fatigue,* STP 942, American Society for Testing and Materials, 1987, p 107–122
21. S. Kim and J.R. Weertman, Investigation of Microstructural Changes in a Ferritic Steel Caused by High Temperature Fatigue, *Metall. Trans. A,* Vol 19A, 1988, p 999–1007

22. R.L. Klueh and R.E. Oakes, Jr., High Strain Rate Tensile Properties of Annealed 2¼Cr-1Mo Steel, *J. Eng. Mater. Technol.*, Vol 98, Oct 1976, p 361–367
23. *Digest of Steels for High Temperature Service*, 6th ed., The Timken Roller Bearing Co., 1957
24. M. Prager, *Factors Influencing the Time-Dependent Properties of Carbon Steels for Elevated Temperature Pressure Vessels*, MPC 19, American Society of Mechanical Engineers, 1983, p 12, 13
25. R.M. Goldhoff, Stress Concentration and Size Effects in a CrMoV Steel at Elevated Temperatures, *Joint International Conference on Creep*, Institute of Mechanical Engineers, London, 1963
26. R. Viswanathan and C.G. Beck, Effect of Aluminum on the Stress Rupture Properties of CrMoV Steels, *Metall. Trans. A*, Vol 6A, Nov 1975, p 1997–2003
27. *Aerospace Structural Metals Handbook*, AFML-TR-68-115, Army Materials and Mechanics Research Center, 1977
28. J.W. Freeman and H. Voorhees, *Relaxation Properties of Steels and Superstrength Alloys at Elevated Temperatures*, STP 187, American Society for Testing and Materials
29. H.R. Voorhees and M.J. Manjoine, *Compilation of Stress-Relaxation Data for Engineering Alloys*, DS 60, American Society for Testing and Materials, 1982
30. C.R. Brinkman, High-Temperature Time-Dependent Fatigue Behavior of Several Engineering Structural Alloys, *Int. Met. Rev.*, Vol 30 (No. 5), 1985, p 235–258
31. D.A. Miller, R.H. Priest, and E.G. Ellison, A Review of Material Response and Life Prediction Techniques under Fatigue-Creep Loading Conditions, *High Temp. Mater. Proc.*, Vol 6 (No. 3 and 4), 1984, p 115–194
32. R. Viswanathan, *Damage Mechanisms and Life Assessment of High-Temperature Components*, ASM International, 1989
33. *Thermal Fatigue of Materials and Components*, STP 612, American Society for Testing and Materials, 1976
34. M.A. Grossman and E.C. Bain, *Principles of Heat Treatment*, 5th ed., American Society for Metals, 1964
35. J.B. Conway, Parametric Considerations Applied to Isochronous Stress-Strain Plots, *The Generation of Isochronous Stress-Strain Curves*, A.O. Schaefer, Ed., American Society of Mechanical Engineers, 1972
36. G.V. Smith, *Evaluation of the Elevated Temperature Tensile and Creep-Rupture Properties of 3–9% Chromium-Molybdenum Steels*, DS 58, American Society for Testing and Materials, 1975
37. R.L. Klueh, Interaction Solid Solution Hardening in 2.25Cr-1Mo Steel, *Mater. Sci. Eng.*, Vol 35, 1978, p 239–253
38. R. Viswanathan and R.D. Fardo, Parametric Techniques for Extrapolating Rupture Ductility, *Ductility and Toughness Considerations in Elevated Temperature Service*, G.V. Smith, Ed., MPC-8, American Society of Mechanical Engineers, 1978
39. J.D. Baird et al., Strengthening Mechanisms in Ferritic Creep Resistant Steels, *Creep Strength in Steel and High Temperature Alloys*, The Metals Society, 1974, p 207–216
40. J.D. Baird and A. Jamieson, *J. Iron Steel Inst.*, Vol 210, 1972, p 841
41. J.D. Baird and A. Jamieson, *J. Iron Steel Inst.*, Vol 210, 1972, p 847
42. B.B. Argent et al., *J. Iron Steel Inst.*, Vol 208, 1970, p 830–843
43. G. Krauss, *Principles of Heat Treatment of Steel*, American Society for Metals, 1980
44. F.B. Pickering, *Physical Metallurgy and the Design of Steels*, Applied Science, 1978
45. R. Viswanathan, Strength and Ductility of 2¼Cr-1Mo Steels in Creep, *Met. Tech.*, June 1974, p 284–293
46. J. Orr, F.R. Beckitt, and G.D. Fawkes, The Physical Metallurgy of Chromium-Molybdenum Steels for Fast Reactor Boilers, *Ferritic Steels for Fast Reactor Steam Generators*, S.F. Pugh and E.A. Little, Ed., British Nuclear Energy Society, 1978, p 91
47. R.L. Klueh, *J. Nucl. Mater.*, Vol 68, 1977, p 294
48. J. Ewald et al., Over 30 Years Joint Long-Term Research on Creep Resistant Materials in Germany, *Advances in Material Technology for Fossil Power Plants*, R. Viswanathan and R.I. Jaffee, Ed., ASM International, 1987, p 33–39
49. A. Krisch, *Jernkontorets Ann.*, Vol 155, 1971, p 323–331
50. G.V. Smith, *Properties of Metals at Elevated Temperatures*, McGraw-Hill, 1950, p 231
51. J.D. Baird, *Jernkontorets Ann.*, Vol 151, 1971, p 311–321
52. R.L. Klueh and P.J. Maziasz, Reduced-Activation Ferritic Steels: A Comparison with Cr-Mo Steels, *J. Nucl. Mater.*, Vol 155–157, 1988, p 602–607
53. R.L. Klueh and E.E. Bloom, The Development of Ferritic Steels for Fast Induced-Radioactive Decay for Fusion Reactor Applications, *Nuclear Engineering and Design/Fusion 2*, North-Holland, 1985, p 383–389
54. R.L. Klueh and P.J. Maziasz, Low-Chromium Reduced-Activation Ferritic Steels, *Reduced-Activation Materials for Fusion Reactors*, STP 1046, American Society for Testing and Materials
55. S.H. Weaver, The Effect of Carbide Spheroidization upon the Rupture Strength and Ductility of Carbon Molybdenum Steel, *Proc. ASTM*, Vol 46, 1946, p 856–866
56. L.F. Coffin, *Metall. Trans.*, Vol 3, 1972, p 1777–1788
57. L.A. James, *J. Eng. Mater. Technol.*, Vol 98, July 1976, p 235–243
58. M. Gell and G.R. Leverant, *Fatigue at Elevated Temperatures*, STP 520, American Society for Testing and Materials, 1973, p 37–66
59. J.C. Runkle and R.M. Pelloux, *Fatigue Mechanisms*, STP 675, J.T. Fong, Ed., American Society for Testing and Materials, 1979, p 501–527
60. D.J. Duquette, Environmental Effects I: General Fatigue Resistance and Crack Nucleation in Metals and Alloys, *Fatigue and Microstructure*, American Society for Metals, 1979, p 335–363
61. H.L. Marcus, Environmental Effects II: Fatigue-Crack Growth in Metals and Alloys, *Fatigue and Microstructure*, American Society for Metals, 1979, p 365–383
62. P. Marshall, in *Fatigue at High Temperature*, R.P. Skelton, Ed., Applied Science, 1983, p 259–303
63. C.R. Brinkman et al., Time-Dependent Strain-Controlled Fatigue Behavior of Annealed 2¼Cr-1Mo Steel for Use in Nuclear Steam Generator Design, *J. Nucl. Mater.*, Vol 62, 1976, p 181–204
64. R. Viswanathan, Strength and Ductility of 2Cr-1Mo Steels in Creep at Elevated Temperatures, *Met. Technol.*, June 1974, p 284–294
65. G.V. Smith, Elevated Temperature Strength and Ductility of Q&T 2Cr-1Mo Steel, *Current Evaluation of 2¼Cr-1Mo Steel in Pressure Vessels and Piping*, American Society of Mechanical Engineers, 1972
66. M.K. Booker, T.L. Hebble, D.O. Hobson, and C.R. Brinkman, Mechanical Property Correlations for 2¼Cr-1Mo Steel in Support of Nuclear Reactor Systems Design, *Int. J. Pressure Vessels Piping*, Vol 5, 1977
67. C.R. Brinkman, W.R. Corwin, M.K. Booker, T.L. Hebble, and R.L. Klueh, "Time Dependent Mechanical Properties of 2¼Cr-1Mo Steel for Use in Steam Generator Design," ORNL-5125, Oak Ridge National Laboratory, 1976
68. J.J. Burke and V. Weiss, in *Fatigue Environment and Temperature Effects*, Plenum Press, 1983, p 241–261
69. R.L. Klueh and D.A. Canonico, Microstructure and Tensile Properties of 2¼Cr-1Mo Steel Weldments with Varying Carbon Contents, *Weld J.* (Research Supplement), Sept 1976
70. R.L. Klueh and D.A. Canonico, Creep-Rupture Properties of 2Cr-1Mo Steel Weldments with Varying Carbon Content, *Weld. J.* (Research Supplement), Dec 1976

Elevated-Temperature Mechanical Properties of Stainless Steels

STAINLESS STEELS are iron-base alloys that contain a minimum of approximately 1.1% Cr, the amount needed to prevent the formation of rust in unpolluted atmospheres (hence the designation *stainless*). Few stainless steels contain more than 30% Cr or less than 50% Fe. They achieve their stainless characteristics through the formation of an invisible and adherent chromium-rich oxide surface film. This oxide forms and heals itself in the presence of oxygen. Other elements added to improve particular characteristics include nickel, molybdenum, copper, titanium, aluminum, silicon, niobium, nitrogen, sulfur, and selenium. Carbon is normally present in amounts ranging from less than 0.03% to over 1.0% in certain martensitic grades. Figure 1 provides a useful summary of some of the compositional and property linkages in the stainless steel family.

This article deals with the wrought stainless steels used for high-temperature applications (see the article "High-Alloy Cast Steels" in this Volume for the elevated-temperature properties of cast stainless steels). Corrosion resistance is often the first criterion used to select stainless steel for a particular application. However, strength is also a significant factor in a majority of elevated-temperature applications and may even be the key factor governing the choice of a stainless steel. The stainless steels used in applications in which high-temperature strength is important are sometimes referred to as *heat-resistant steels.* Many stainless steels used for elevated-temperature applications are designed for service at temperatures up to 650 °C (1200 °F).

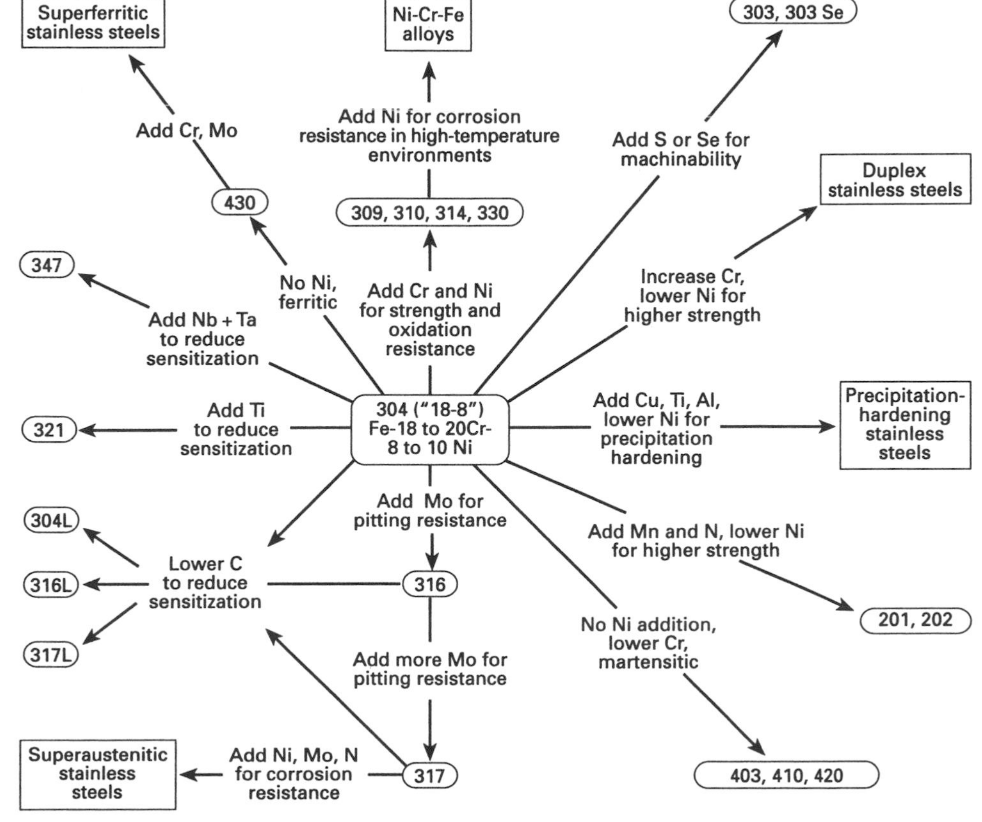

Fig. 1 Compositional and property linkages in the stainless steel family of alloys

Classification of Stainless Steels

Stainless steels can be divided into five families. Four are based on the characteristic crystallographic structure/microstructure of the alloys in the family: ferritic, martensitic, austenitic, or duplex (austenitic plus ferritic). The fifth family, the precipitation-hardening (PH) alloys, is based on the type of heat treatment used, rather than microstructure. Table 1 lists the compositions of stainless steels considered for use in elevated-temperature applications.

Martensitic stainless steels are essentially alloys that possess a body-centered tetragonal (bct) crystal structure (martensitic) in the hardened condition. They are ferromagnetic, hardenable by heat treatments, and generally resistant to corrosion only in relatively mild environments. Chromium content is generally in the range of 10.5 to 18%, and carbon content may exceed 1.2%. The chromium and carbon contents are balanced to ensure a martensitic structure after hardening. Elements such as niobium, silicon, tungsten, and vanadium may be added to modify the tempering response after hardening. Small amounts of nickel may be added to improve corrosion resistance in some media and to improve toughness. Sulfur or selenium is added to some grades to improve machinability.

In the annealed condition, martensitic stainless steels have a tensile yield strength of approximately 275 MPa (40 ksi) and can be moderately hardened by cold working. However, martensitic alloys are typically heat treated by both hardening and tempering in order to produce strength levels up to 1900 MPa (275 ksi), depending on carbon level, primarily. These alloys have good ductility and toughness properties, which decrease as strength increases. Depending on the heat treatment, hardness values range from approximately 150 HB (80 HRB) for materials in

Table 1 Nominal compositions of wrought heat-resistant stainless steels

Designation	UNS number	Composition, % C	Cr	Ni	Mo	N	Nb	Ti	Other
Ferritic stainless steels									
405	S40500	0.15 max	13.0	...	...	...	...	...	0.2 Al
406	...	0.15 max	13.0	...	...	...	...	...	4.0 Al
409	S40900	0.08 max	11.0	0.5 max	...	...	...	6 × C min	...
429	S42900	0.12 max	15	...	...	...	...	...	...
430	S43000	0.12 max	16.0	...	...	...	...	...	...
434	S43400	0.12 max	17.0	...	1.0	...	...	...	...
439	S43035	0.07 max	18.25	...	...	...	...	12 × C min	1.10 Ti max
18 SR	...	0.05	18.0	...	...	...	...	0.40 max	2.0 Al max
18Cr-2Mo	S44400	...	18.5	...	2.0	...	(a)	(a)	0.8 (Ti + Nb) max
446	S44600	0.20 max	25.0	...	...	0.25	...	...	...
E-Brite 26-1	S44627	0.01 max	26.0	...	1.0	0.015 max	0.1	...	...
26-1Ti	S44626	0.04	26.0	...	1.0	...	...	10 × C min	...
29Cr-4Mo	S44700	0.01 max	29.0	...	4.0	0.02 max	...	...	...
Quenched and tempered martensitic stainless steels									
403	S40300	0.15 max	12.0	...	...	...	...	...	...
410	S41000	0.15 max	12.5	...	...	...	...	...	...
410Cb	S41040	0.15 max	12.5	...	...	...	0.12	...	...
416	S41600	0.15 max	13.0	...	0.6(b)	...	...	...	0.15 min S
422	S42200	0.20	12.5	0.75	1.0	...	...	...	1.0 W, 0.22 V
H-46	...	0.12	10.75	0.50	0.85	0.07	0.30	...	0.20 V
Moly Ascoloy	...	0.14	12.0	2.4	1.80	0.05	...	...	0.34 V
Greek Ascoloy	S41800	0.15	13.0	2.0	...	...	...	...	3.0 W
Jethete M-152	...	0.12	12.0	2.5	1.7	...	...	...	0.30 V
Almar 363	...	0.05	11.5	4.5	...	...	...	10 × C min	...
431	S43100	0.20 max	16.0	2.0	...	...	...	...	...
Lapelloy	S42300	0.30	11.5	...	2.75	...	...	...	0.25 V
Precipitation-hardening martensitic stainless steels									
Custom 450	...	0.05 max	15.5	6.0	0.75	...	8 × C min	...	1.5 Cu
Custom 455	...	0.03	11.75	8.5	...	...	0.30	1.2	2.25 Cu
15-5 PH	S15500	0.07	15.0	4.5	...	...	0.30	...	3.5 Cu
17-4 PH	S17400	0.04	16.5	4.25	...	...	0.25	...	3.6 Cu
PH 13-8 Mo	S13800	0.05	12.5	8.0	2.25	...	...	...	1.1 Al
Precipitation-hardening semiaustenitic stainless steels									
AM-350	S35000	0.10	16.5	4.25	2.75	0.10	...	...	...
AM-355	S35500	0.13	15.5	4.25	2.75	0.10	...	...	...
17-7 PH	S17700	0.07	17.0	7.0	...	...	...	...	1.15 Al
PH 15-7 Mo	S15700	0.07	15.0	7.0	2.25	...	...	...	1.15 Al
Austenitic stainless steels									
304	S30400	0.08 max	19.0	10.0	...	...	...	...	...
304H	S30409	0.04–0.10	19.0	10.0	...	...	...	...	...
304L	S30403	0.03 max	19.0	10.0	...	...	...	...	...
304N	S30451	0.08 max	19.0	9.25	...	0.13	...	...	...
309	S30900	0.2 max	23.0	13.0	...	...	...	...	...
309H	S30909	0.04–0.10	23.0	13.0	...	...	...	...	...
310	S31000	0.25 max	25.0	20.0	...	...	...	...	...
310H	S31009	0.04–0.10	25.0	20.0	...	...	...	...	...
316	S31600	0.08 max	17.0	12.0	2.5	...	...	...	...
316L	S31603	0.03 max	17.0	12.0	2.5	...	...	...	...
316N	S31651	0.08 max	17.0	12.0	2.5	0.13	...	...	...
316H	S31609	0.04–0.10	17.0	12.0	2.5	...	...	...	...
316LN	S31653	0.035 max	17.0	12.0	2.5	0.13	...	...	...
317	S31700	0.08 max	19.0	13.0	3.5	...	...	...	...
317L	S31703	0.035 max	19.0	13.0	3.5	...	...	...	...
321	S32100	0.08 max	18.0	10.0	...	...	...	5 × C min, 0.70 max	...
321H	S32109	0.04–0.10	18.0	10.0	...	...	...	4 × C min, 0.60 max	...
347	S34700	0.08 max	18.0	11.0	...	...	10 × C min(c)	...	1.0 (Nb + Ta) max
347H	S34709	0.04–0.10	18.0	11.0	...	...	8 × C min(c)	...	1.0 (Nb + Ta) max
348	S34800	0.08 max	18.0	11.0	...	...	8 × C min(c)	...	0.10 Ta max, 1.0 (Nb + Ta) max
348H	S34809	0.04–0.10	18.0	11.0	...	...	8 × C min(c)	...	0.10 Ta max, 1.0 (Nb + Ta) max
19-9 DL	K63198	0.30	19.0	9.0	1.25	...	0.4	0.3	1.25 W
19-9 DX	K63199	0.30	19.2	9.0	1.5	...	...	0.55	1.2 W
17-14-CuMo	...	0.12	16.0	14.0	2.5	...	0.4	0.3	3.0 Cu
201	S20100	0.15 max	17	4.2	...	0.25 max	...	...	...
202	S20200	0.09	18.0	5.0	...	0.10	...	...	8.0 Mn
204 (Nitronic 30)	S20400	0.03 max	16.0	2.25	...	0.30	...	...	16.0 Mn
205	S20500	0.18	17.2	1.4	...	0.36	...	...	...
216	S21600	0.05	10.0	6.0	2.5	0.35	...	...	8.5 Mn
21-6-9	S21900	0.04 max	20.25	6.5	...	0.30	...	...	9.0 Mn
Nitronic 32	S24100	0.10	18.0	1.6	...	0.34	...	...	12.0 Mn
Nitronic 33	S24000	0.08 max	18.0	3.0	...	0.30	...	...	13.0 Mn
Nitronic 50	...	0.06 max	21.0	12.0	2.0	0.30	0.20	...	5.0 Mn
Nitronic 60	S21800	0.10 max	17.0	8.5	2.0	...	...	...	8.0 Mn, 0.20 V, 4.0 Si
Carpenter 18-18 Plus	S28200	0.10	18.0	<0.50	1.0	0.50	...	...	16.0 Mn, 0.40 Si, 1.0 Cu
Tenelon	S21400	0.12 max	18.0	0.75 max	...	0.35	...	...	15.0 Mn

(a) Ti + Nb = (0.20 + 4C + 4N) min. (b) Optional. (c) Minimum for Nb + Ta

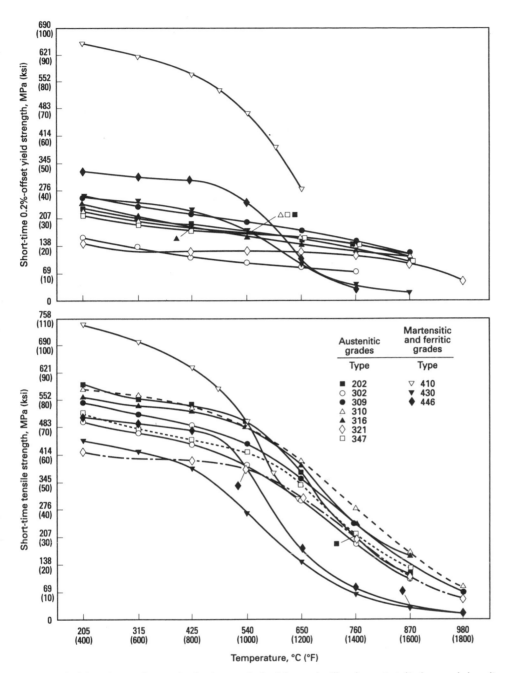

Fig. 2 Typical short-time tensile strengths of various standard stainless steels. All steels were tested in the annealed condition except for the martensitic type 410, which was heat treated by oil quenching from 980 °C (1800 °F) and tempering at 650 °C (1200 °F).

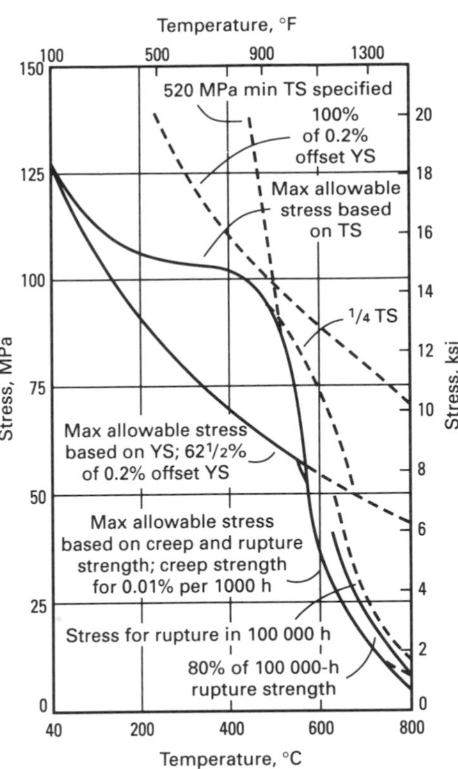

Fig. 3 Example of relationships between temperature and high-temperature strengths previously used in the ASME Boiler Code to establish maximum allowable stresses in tension for type 18-8 austenitic stainless steel. The current code uses two-thirds of the yield strength instead of 62.5% of the yield strength. YS, yield strength; TS, tensile strength

the annealed condition to levels greater than 600 HB (58 HRC) for fully hardened materials.

Martensitic stainless steels are specified when the application requires good tensile strength, creep, and fatigue strength properties, in combination with moderate corrosion resistance and heat resistance up to approximately 650 °C (1200 °F). In the United States, low- and medium-carbon martensitic steels (e.g., type 410 and modified versions of this alloy) have been used primarily in steam turbines, jet engines, and gas turbines.

Ferritic stainless steels are essentially chromium-containing alloys with body-centered cubic (bcc) crystal structures. Chromium content is usually in the range of 10.5 to 30%. Some grades may contain molybdenum, silicon, aluminum, titanium, and niobium to confer particular characteristics. Sulfur or selenium may be added to improve machinability. The ferritic alloys are ferromagnetic, and unlike martensitic grades, they cannot be hardened by heat treatment.

In general, ferritic stainless steels do not have particularly high strength. Their annealed yield strengths range from 275 to 350 MPa (40 to 50 ksi), and their poor toughness and susceptibility to sensitization limit their fabricability and the usable section size. Their chief advantages are their resistance to chloride stress-corrosion cracking, atmospheric corrosion, and oxidation at a relatively low cost.

The high-temperature strengths of ferritic stainless steels are relatively poor compared to those of martensitic or austenitic grades. They are used in elevated-temperature applications because of their good oxidation resistance, and some ferritic grades have been used extensively in automotive exhaust systems.

Austenitic stainless steels have a face-centered cubic (fcc) structure. This structure is attained through the liberal use of austenitizing elements such as nickel, manganese, and nitrogen. These steels are essentially nonmagnetic in the annealed condition and can be hardened only by cold working. They usually possess excellent cryogenic properties and good high-temperature strength. Chromium content generally varies from 16 to 26%; nickel content, up to about 35%; and manganese content, up to 15%. The 200-series steels contain nitrogen, 4 to 15.5% Mn, and up to 7% Ni. The 300-series steels contain larger amounts of nickel and up to 2% Mn. Molybdenum, copper, silicon, aluminum, titanium, and niobium may be added to confer certain characteristics such as halide pitting resistance or oxidation resistance. Sulfur or selenium may be added to certain grades to improve machinability.

Typical 300-series chromium-nickel stainless steels have room-temperature tensile yield

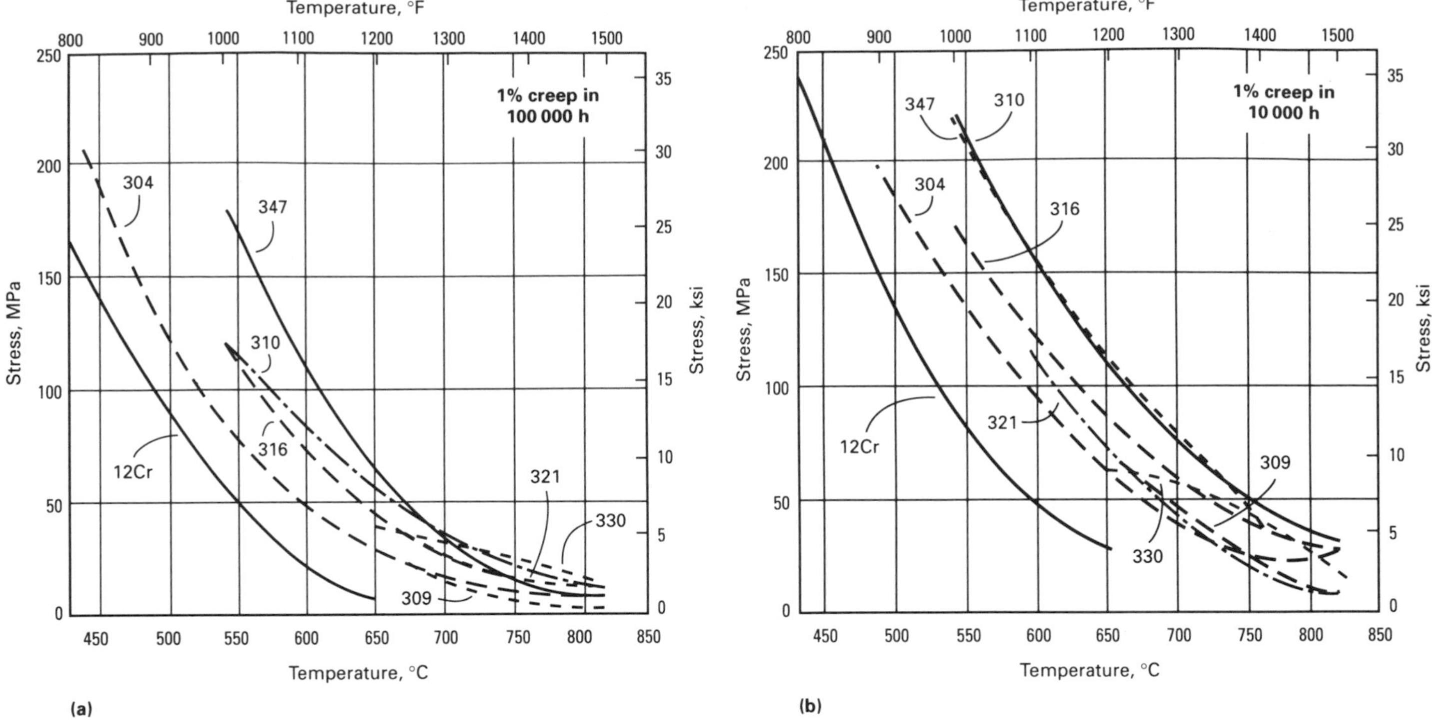

Fig. 4 Creep rate curves for several annealed H-grade austenitic stainless steels. (a) 1% creep in 100,000 h. (b) 1% creep in 10,000 h. Source: Ref 1

strengths from 200 to 275 MPa (30 to 40 ksi) in the annealed condition, whereas the high-nitrogen alloys have yield strengths up to 500 MPa (70 ksi). Of all of the stainless steel types, austenitic grades have the highest strengths at high-temperatures.

Duplex stainless steels have a mixed structure of bcc ferrite and austenite. The exact amount of each phase is a function of composition and heat treatment. Most alloys are designed to contain about equal amounts of each phase in the annealed condition. The principal alloying elements are chromium and nickel, but nitrogen, molybdenum, copper, silicon, and tungsten may be added to control structural balance and to impart certain corrosion-resistance characteristics.

Duplex stainless steels are capable of tensile yield strengths ranging from 550 to 690 MPa (80 to 100 ksi) in the annealed condition, significantly higher than those of their austenitic or ferritic counterparts. Despite these relatively high room-temperature properties, duplex grades are subject to embrittlement and loss of mechanical properties, particularly toughness, through prolonged exposure to high temperatures. As a result, duplex stainless steels are generally not recommended for elevated-temperature applications.

Precipitation-hardening (PH) stainless steels are chromium-nickel grades that can be hardened by an aging treatment. These grades are classified as austenitic, semiaustenitic, or martensitic. The classification is determined by their solution-annealed microstructure. The semi-austenitic alloys are subsequently heat treated so that the austenite transforms to martensite. Cold work is sometimes used to facilitate the aging reaction. Various alloying elements, such as aluminum, titanium, niobium, or copper, are used to achieve aging.

Like the martensitic stainless steels, PH alloys can attain high tensile yield strengths, up to 1700 MPa (250 ksi). Cold working prior to aging can result in even higher strengths. However, the PH martensitic and semiaustenitic grades lose strength rapidly at temperatures above about 425 °C (800 °F).

Thermally Induced Embrittlement

Stainless steels are susceptible to embrittlement during thermal treatment or elevated-temperature service. Thermally induced forms of embrittlement of stainless steels include sensitization, 475 °C (885 °F) embrittlement, and σ-phase embrittlement, as briefly described below. Additional information on these embrittlement mechanisms can be found in the *ASM Specialty Handbook: Stainless Steels.*

Sensitization. Stainless steels become susceptible to localized intergranular corrosion when chromium carbides form at the grain boundaries during high-temperature exposure. This depletion of chromium at the grain boundaries is termed "sensitization" because the alloys become more sensitive to localized attack in corrosive environments. Sensitization and intergranular corrosion can occur in austenitic, duplex, and ferritic stainless steels, depending on the alloy content and the time-temperature exposures required for carbide precipitation.

Austenitic stainless steels become susceptible to intergranular corrosion when subjected to temperatures in the range of 480 to 815 °C (900 to 1500 °F). Several approaches have been taken to minimize or prevent the sensitization of austenitic stainless steels. If sensitization results from welding heat and the component is small enough, solution annealing will dissolve the precipitates and restore immunity. However, in many cases this cannot be done because of distortion problems or the size of the component. In these cases, a low-carbon version of the grade or a stabilized composition should be used. Complete immunity requires a carbon content below about 0.015 to 0.02%. Additions of niobium or titanium to tie up the carbon are also effective in preventing sensitization, as long as the ratio of these elements to the carbon content is high enough. Stabilizing heat treatments are not very effective.

Ferritic stainless steels can be susceptible to sensitization, but ferritic grades with less than 15% Cr can be immune from sensitization. Reducing the carbon and nitrogen interstitial levels improves the intergranular corrosion resistance of ferritic stainless steels.

Sensitization can occur in titanium-stabilized ferritic stainless steels. The thermal treatment that causes sensitization, however, is altered by the addition of titanium.

Duplex stainless steels are resistant to intergranular corrosion when aged in the region of 480 to 700 °C (895 to 1290 °F). It has been recognized for some time that duplex grades with 20 to

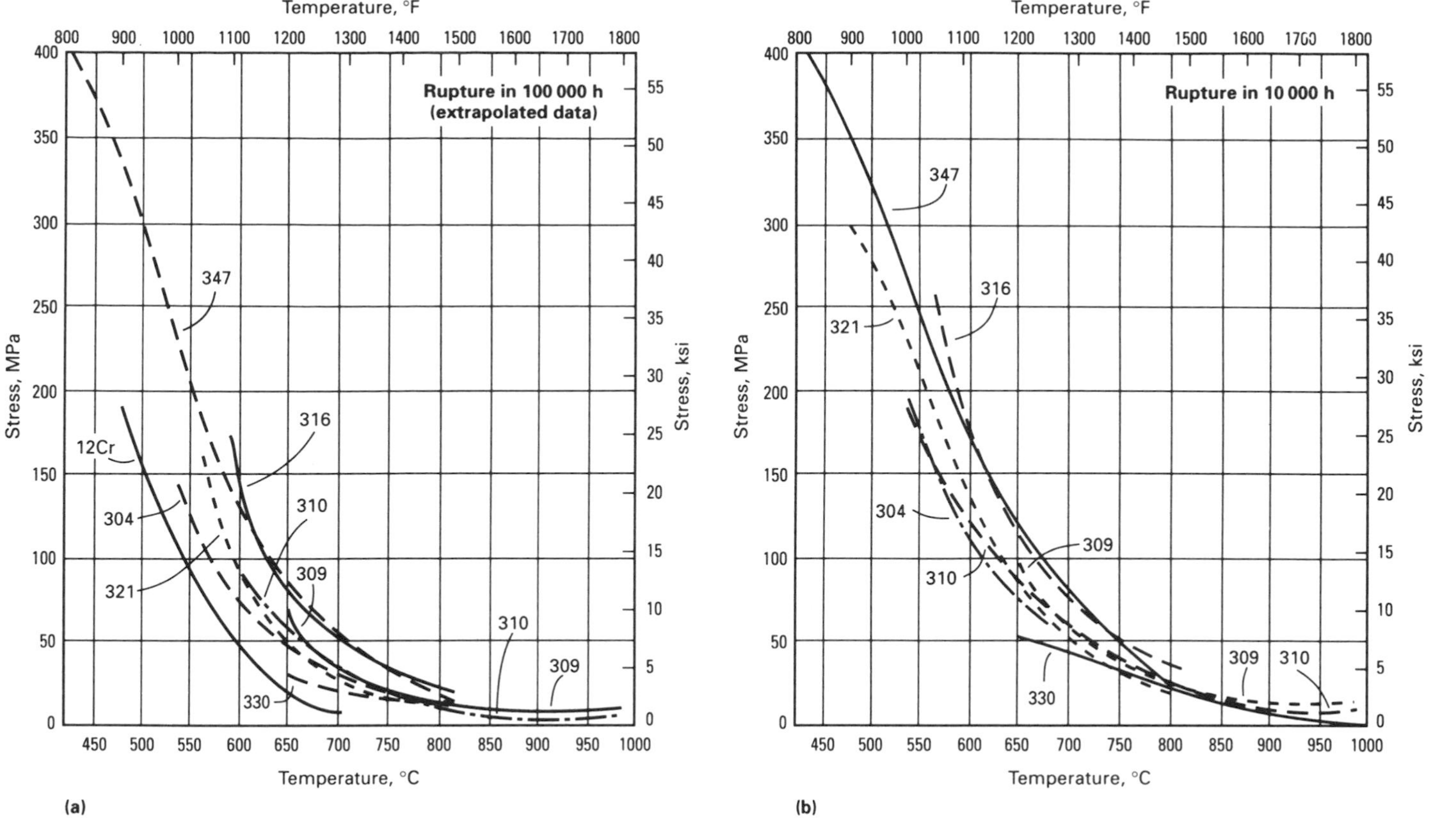

Fig. 5 Stress-rupture curves for several annealed H-grade austenitic stainless steels. (a) Extrapolated data for rupture in 100,000 h. (b) Rupture in 10,000 h. Source: Ref 1

Table 2 Effect of hold-period length in tension-hold-only testing on fatigue resistance of type 304 stainless steel

Tested in air at 650 °C (1200 °F) and a strain rate of 4×10^{-3} s^{-1} at a strain range of about 2.0%

Hold period, min	Cycles to failure, N_f	
	Test 1	Test 2
0	592	546
0.1	570	545
1.0	329	331
10.0	193	201
30.0	146	165
60.0	144	158
180.0	150	120

Table 3 Test results of type 304 stainless steel obtained using a 30 min hold period in tension plus a short hold period in compression

Tested in air at 650 °C (1200 °F) and a strain rate of $4 \times 10^{-3} s^{-1}$

Hold period tension, min	Hold period compression, min	Total strain range, %	Cycles to failure, N_f	
			Test 1	Test 2
0	0	1.98	592	546
30	30	1.98	380	416
30	0	2.08	146	...
30	0	2.02	...	165
30	3	1.98	308	...
30	3	2.00	...	336

40 vol% ferrite exhibit excellent resistance to sensitization and intergranular corrosion.

475 °C Embrittlement. Iron-chromium alloys containing 13 to 90% Cr are susceptible to embrittlement when held within or cooled slowly through the temperature range of 550 to 400 °C (1020 to 750 °F). This phenomenon, called 475 °C (885 °F) embrittlement, increases tensile strength and hardness and decreases tensile ductility, impact strength, electrical resistivity, and corrosion resistance.

475 °C embrittlement occurs with iron-chromium ferritic and duplex ferritic-austenitic stainless steels, but not with austenitic grades. Aging at 475 °C (885 °F) can cause a rapid rate of hardening with aging between about 20 and 120 h because of homogeneous precipitation. The rate of hardening is much slower with continued aging from 120 to 1000 h. During this aging period, precipitation increases. Aging beyond 1000 h produces little increase in hardness because of the stability of the precipitates.

Even for a severely embrittled alloy, 475 °C embrittlement is reversible. Properties can be restored within minutes by reheating the alloy to 675 °C (1250 °F) or above. The degree of embrittlement increases with chromium content, but embrittlement is negligible below 13% Cr. Carbide-forming alloying additions, such as molybdenum, vanadium, titanium, and niobium, appear to increase embrittlement, particularly with higher chromium levels. Increased levels of carbon and nitrogen also enhance embrittlement and, of course, are detrimental to nonembrittled properties as well. Cold work prior to 475 °C (885 °F) exposure accelerates embrittlement, particularly for higher-chromium alloys.

Sigma-Phase Embrittlement. The existence of σ-phase in iron-chromium alloys has been identified in over fifty binary systems and in other commercial alloys. Sigma phase has a tetragonal crystal structure and a hardness equivalent to approximately 68 HRC (940 HV). Because of its brittleness, σ-phase often fractures during indentation.

In general, σ-phase forms with long-time exposure in the range of 565 to 980 °C (1050 to 1800 °F), although this range varies somewhat with composition and processing. Sigma formation exhibits C-curve behavior, with the shortest time for formation (nose of the curve) generally occurring between about 700 and 810 °C (1290 and 1490 °F). The temperature that produces the greatest amount of σ with time is usually somewhat lower.

In commercial austenitic and ferritic stainless steels, even small amounts of silicon markedly accelerate the formation of σ-phase. In general, all of the elements that stabilize ferrite promote σ-phase formation. Molybdenum has an effect similar to that of silicon; aluminum has a lesser influence. Increasing the chromium content also favors σ-phase formation. Small amounts of nickel and manganese increase the rate of σ-

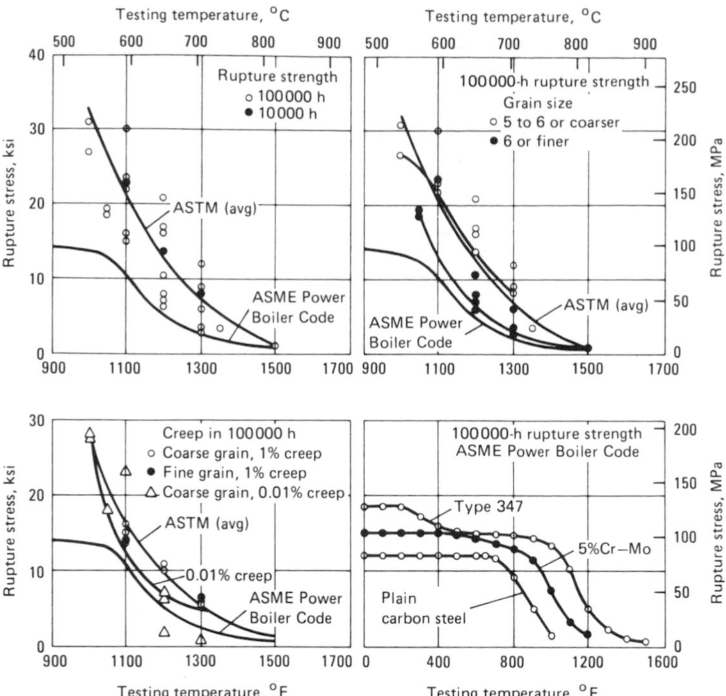

Fig. 6 Rupture strength and creep properties of type 347 stainless steel compared with stresses permitted by the ASME Boiler Code

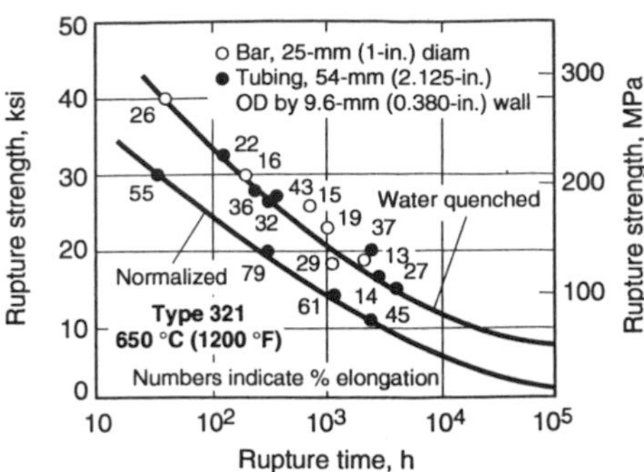

Fig. 7 Stress-rupture characteristics of type 321 stainless steel, cold worked and solution treated. Specimens indicated by lower curve were normalized at 955 °C (1750 °F), which resulted in grain sizes of ASTM No. 8 or finer. Specimens water quenched (top curve) from between 1040 and 1120 °C (1900 and 2050 °F) exhibited grain sizes of ASTM No. 5 to 8.

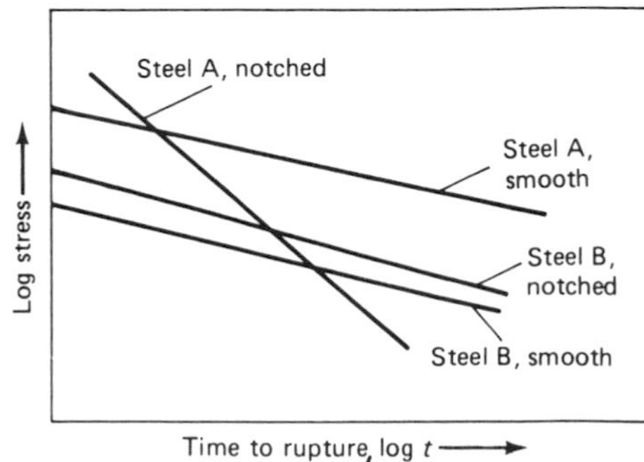

Fig. 8 Notched-bar rupture behavior for a creep-brittle steel (A) and a creep-ductile steel (B)

phase formation, but large amounts, which stabilize austenite, retard σ-phase formation. Carbon additions decrease σ-phase formation by forming chromium carbides, thereby reducing the amount of chromium in solid solution. Additions of tungsten, vanadium, titanium, and niobium also promote σ-phase formation.

As might be expected, σ-phase forms more readily in ferritic stainless steels than in austenitic stainless steels. Coarse grain sizes from high solution-annealing temperatures retard σ-phase formation, and prior cold working enhances it. The influence of cold work on σ-phase formation depends on the amount of cold work and its effect on recrystallization. If the amount of cold work is sufficient to produce recrystallization at the service temperature, σ-phase formation is enhanced. If recrystallization does not occur, the rate of σ-phase formation may not be affected. Small amounts of cold work that do not promote recrystallization may actually retard σ-phase formation.

The most sensitive room-temperature property for assessing the influence of σ-phase is the impact strength. High-temperature exposure can produce a variety of phases, and embrittlement is not always due solely to σ-phase formation. Therefore, each situation must be carefully evaluated to determine the true cause of the degradation of properties.

Mechanical Property Considerations

For service at elevated temperatures, the first property considered is the tensile strength during short-term exposure at elevated temperatures.

For applications involving short-term exposure to temperatures below about 480 °C (900 °F), the short-time tensile properties are usually sufficient in the mechanical design of steel components. Typical short-time tensile strengths of various standard stainless steel grades are shown in Fig. 2.

For temperatures above 480 °C (900 °F), the design process must include other properties, such as creep rate, creep-rupture strength, creep-rupture ductility, and creep-fatigue interaction.

Various methods, depending on the application, are used to establish the design criteria for using materials at elevated temperatures. One method, for example, develops allowable stresses by multiplying tensile strengths, yield strengths, creep strength, and/or rupture strength with safety factors. This method is illustrated in Fig. 3, where various safety factors are used to establish reliable stresses for 18-8 austenitic stainless steel at various temperatures. This method does not take into account environmental interactions, aging effects from long-term temperature exposure, or the possibility of creep-fatigue interaction.

Creep and Stress Rupture. Creep is defined as the time-dependent strain that occurs under load at elevated temperatures. Creep is operative in most applications when metal temperatures exceed 480 °C (900 °F). In time, creep may lead to excessive deformation and even fracture at stresses considerably below those determined in room-temperature and elevated-temperature short-term tension tests. The designer must usually determine whether the serviceability of the component in question is limited by the rate or the degree of deformation.

When the rate or degree of deformation is the limiting factor, the design stress is based on the minimum creep rate and design life after allowing for initial transient creep. The stress that produces a specified minimum creep rate of an alloy or a specified amount of creep deformation in a given time (e.g., 1% total creep in 100,000 h) is referred to as the *limiting creep strength* or *limiting stress*. Of the various types of stainless steels, the austenitic types provide the highest limiting creep strength. Figure 4 plots typical creep rates of various austenitic stainless steels. The original data for Fig. 4 were generated on steels with carbon contents greater than 0.04%; the steels were solution annealed at sufficiently high temperatures to meet H-grade requirements (see the section "H-grades" in this article). Today, the 300-series stainless steels are usually low carbon, unless an H-grade is specified.

When fracture is the limiting factor, stress-to-rupture values can be used in design. Typical stress-to-rupture values of various austenitic stainless steels are shown in Fig. 5. The values were generated from steels meeting H-grade requirements.

It should be recognized that long-term creep and stress-rupture values (e.g., 100,000 h) are often extrapolated from shorter-term tests, conducted at high stresses, in which creep is dislocation controlled. Whether these property values are extrapolated or determined directly often has little bearing on the operating life of high-temperature parts, where operating stresses are lower and where the mechanisms of creep are diffusion controlled. The actual material behavior can also be difficult to predict accurately because of (a) the complexity of the service stresses relative to the idealized, uniaxial loading conditions in the standardized tests and (b) attenuating factors such as cyclic loading, temperature fluctuations, and metal loss from corrosion.

Creep and Stress Rupture Data Spread. Many of the values for stress-rupture and creep given in the data compilations in this article are typical or average values. There is a spread above and below these values caused by differences among heats of metal, methods of processing, and variations in conducting the standard test procedures.

A typical spread in 100,000 h rupture strength is shown at upper left in Fig. 6 for type 347 stainless steel. At 590 °C (1100 °F), the test values range from 105 to 205 MPa (15 to 30 ksi), and at 650 °C (1200 °F) from 41 to 145 MPa (6 to 21 ksi). The average curve is well above the stress levels allowed by the ASME Power Boiler Code. The graph at lower left in Fig. 6 shows the total range in stresses for a creep rate of 1% per 100,000 h. On a percentage basis, the spread is about the same as in the upper left-hand chart. The stresses permitted by the ASME Power Boiler Code are below the average curve, except for low values for fine-grain material at 650 and 700 °C (1200 and 1300 °F). The results of an analysis of these tests are plotted in the upper right-hand chart in Fig. 6, classified according to grain size. Above 590 °C (1100 °F) the results fall into two distinct groups on the basis of grain size, with coarse-grain materials having higher rupture strengths.

The total range of rupture-strength values for fine- and coarse-grain type 347 stainless steel is shown at upper left in Fig. 6. However, when the total spread for coarse-grain material is considered alone (upper right), the range in values at 650 °C (1200 °F) is 95 to 145 MPa (14 to 21 ksi) instead of 41 to 145 MPa (6 to 21 ksi).

Stresses allowed by the ASME Boiler Code are determined by, and vary considerably with, the type of steel. Allowable stresses for two heat-resistant steels are compared with those for carbon steel at lower right in Fig. 6.

Effect of Grain Size on Rupture Strength. In steam-generating equipment, many premature failures of type 321 superheater tubes have occurred at a metal temperature of 650 °C (1200 °F) and a maximum fiber stress of 34 MPa (5 ksi) in the tube wall. This stress is in accordance with the maximum allowable stress of the Boiler Code and is supposed to result in a tube life of at least 100,000 h. The tubes ranged approximately from 50 to 75 mm (2 to 3 in.) in outside diameter and had a maximum wall thickness of 13 mm (0.500

Table 4 Cyclic oxidation resistance of ferritic stainless steels compared with other alloys

Specimens were exposed for 100 h in air containing 10% water vapor, cooled to room temperature every 2 h, then reheated to test temperature.

Alloy	Weight change (scale not removed), g/m², at: 705 °C (1300 °F)	815 °C (1500 °F)	980 °C (1800 °F)	1090 °C (2000 °F)
409	+0.1	+0.8	+1430(a)	−10,000(b)
430	+0.4	+1.3	−1660(c)	−10,000(b)
18 SR	+0.1	+0.3	+2.5	+7.4
304	+0.2	+1.7	−3400	−10,000(d)
309	+0.2	+2.7	−120	−910
201	+0.8	+3.1	+10	−150
Incoloy 800	+0.3	+3.2	+8.6	−560
Inconel 601	+0.1	+1.2	+10	−2.1

(a) Removed after 36 h. (b) Removed after 12 h. (c) Removed after 30 h. (d) Removed after 24 h

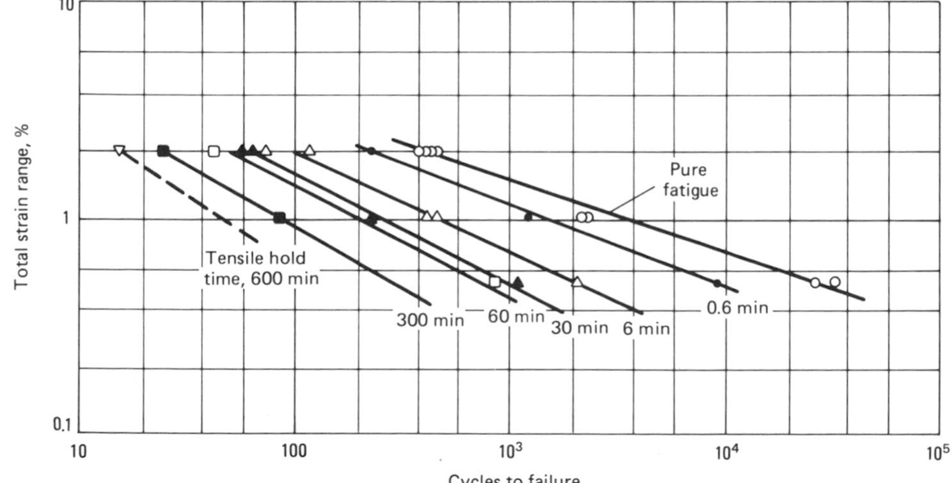

Fig. 9 Effect of tensile hold time in the temperature range of 550 to 625 °C (1020 to 1160 °F) on fatigue endurance of type 316 stainless steel. Source: Ref 2

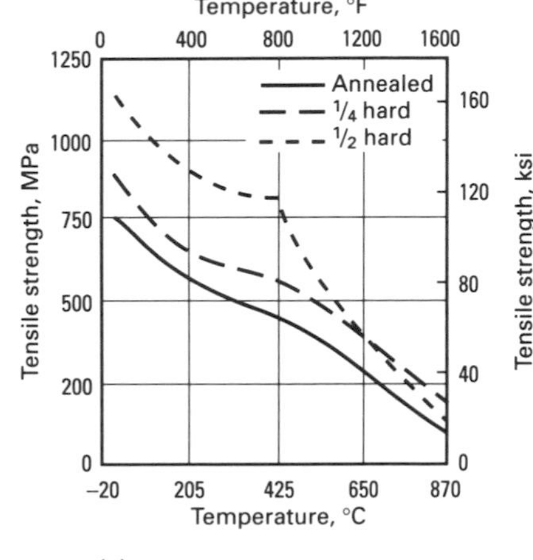

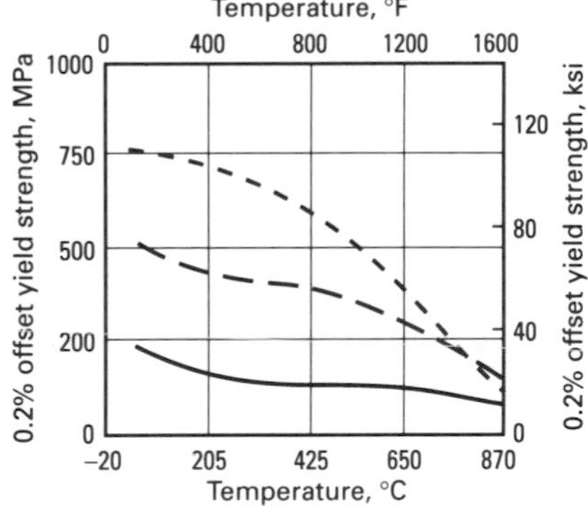

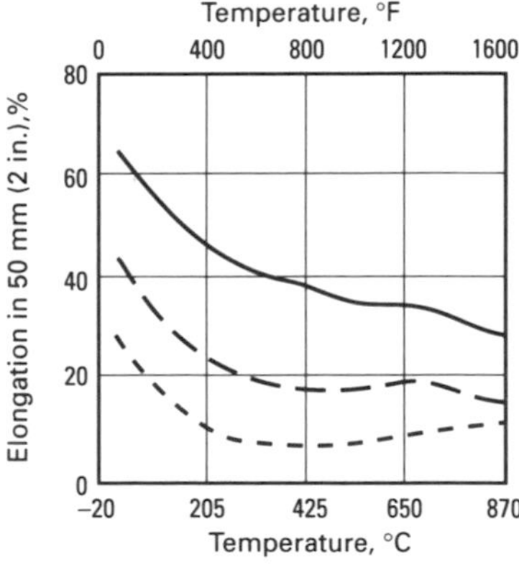

Fig. 10 Effect of short-term elevated temperature on tensile properties of cold-worked type 301 stainless steel. (a) Tensile strength. (b) Yield strength. (c) Elongation

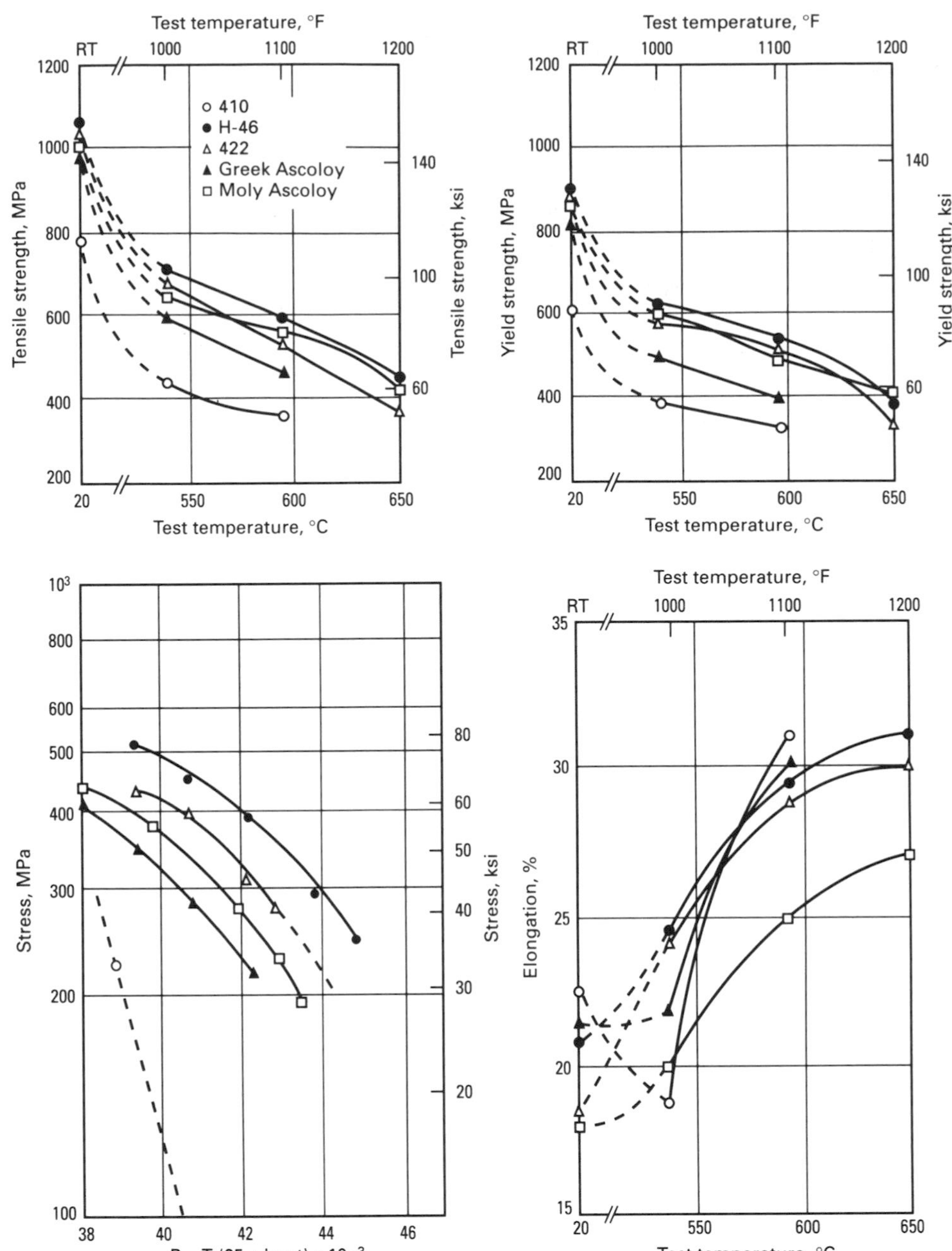

Fig. 11 Comparison of mechanical properties of martensitic stainless steels. Heat-treating schedules were as follows. Type 410: 1 h at 980 °C (1800 °F), oil quench; then 2 h at 650 °C (1200 °F), air cool. H-46: 1 h at 1150 °C (2100 °F); air cool; then 2 h at 650 °C (1200 °F), air cool. Type 422: 1 h at 1040 °C (1900 °F), oil quench; then 2 h at 650 °C (1200 °F), air cool. Greek Ascoloy: 1 h at 955 °C (1750 °F), oil quench; then 2 h at 650 °C (1200 °F), air cool. Moly Ascoloy: 30 min at 1050 °C (1925 °F), oil quench; then 2 h at 650 °C (1200 °F), air cool.

Fig. 12 Approximate effects of time and stress on tempering of types 422 and H-46. Circles indicate specimens heated to 1150 °C (2100 °F) and rapidly cooled, tempered 2 h at 705 °C (1300 °F), and tested (to fracture) at a temperature of 540 °C (1000 °F) and a stress of 380 MPa (55 ksi). Triangles indicate specimens heated to 980 °C (1800 °F) and rapidly cooled, tempered 2 h at 705 °C (1300 °F), and tested (to fracture) at 540 °C (1000 °F) and 275 MPa (40 ksi). Open symbols represent data taken at the unstressed specimen shoulder; solid symbols represent data taken within the stressed gage length.

Table 5 Creep and stress-rupture properties of types 304N and 316N stainless steels

Type	Temperature °C	Temperature °F	Stress, MPa (ksi) for rupture in: 10,000 h	Stress, MPa (ksi) for rupture in: 100,000 h	Stress, MPa (ksi) for a minimum creep rate of: 0.0001 %/h	Stress, MPa (ksi) for a minimum creep rate of: 0.00001 %/h
304N	565	1050	234 (34)	186 (27)	214 (31)	172 (25)
	650	1200	124 (18)	86 (12.5)	103 (15)	70 (10.2)
	730	1350	61 (8.8)	41 (6)	52 (7.5)	38 (5.5)
	815	1500	33 (4.8)	23 (3.3)	28 (4)	...
316N	565	1050	286 (41.5)	228 (33)	255 (37)	200 (29)
	650	1200	179 (26)	131 (19)	117 (17)	86 (12.5)
	730	1350	97 (14)	64 (9.3)	60 (8.7)	42 (6.1)
	815	1500	45 (6.5)	28 (4)	28 (4)	18 (2.6)

Source: Ref 18

in.). They were produced by cold working (cold drawing or tube reducing), followed by solution heat treating at elevated temperature.

Although the tubes were designed for a life of 100,000 h, many exhibited considerable bulging (creep) within a few thousand hours, and some actually ruptured in 30,000 h or less. All tubes that bulged or ruptured had an ASTM grain size of 8 or finer, whereas those that performed satisfactorily had a grain size of 8 or coarser. The contribution of grain size to performance is considered minor, however. It is overshadowed by the effects of the solution treating temperature and the rate of cooling from the solution treating temperature.

The fine-grain tubes were solution treated at 950 °C (1750 °F) and cooled in air, whereas the coarser-grain tubes were solution treated at 1040 °C (1900 °F) or higher and water quenched. The rupture characteristics of both fine-grain and coarse-grain tubes are shown in Fig. 7.

Both solution-treating temperature and quenching rate affect performance. Solution treating at a temperature of 950 °C (1750 °F) does not place sufficient titanium carbide in solution to exert the strengthening effect at 650 °C (1200 °F) that is produced by solution treating at a temperature of 1040 °C (1900 °F). Also, depending on section size, water quenching provides sufficiently rapid cooling to retain titanium carbide in solution. In larger sections, cooling in air may be too slow to produce this desirable result.

Rupture ductility is an important mechanical property when stress concentrations and localized defects such as notches are a factor in design. It varies inversely with creep and rupture strength. It influences the growth of cracks or defects and thus affects notch sensitivity. This general effect of rupture ductility on rupture strength is shown conceptually in Fig. 8. When smooth parts are tested, the rupture strength is higher for the steel with lower ductility (steel A). When a notch is introduced, the rupture strength of steel A plummets, but the rupture strength of steel B is less notch sensitive because of its higher ductility. It is clear from Fig. 8 that for low-stress, long-term applications, steel B would be preferred to steel A. This is true even though steel B is weaker than steel A, as shown by results of smooth-bar rupture tests.

In many service conditions, the amount of deformation is not critical, and relatively high rupture ductility can be used in design. Under such conditions, with the combined uncertainties of actual stress, temperature, and strength, it may be important that failure not occur without warning and that the metal retain high elongation and reduction in area throughout its service life. In the oil and chemical industries, for instance, many applications of tubing under high pressure require high long-time ductility, and impending rupture will be evident from the bulging of the tubes.

Values of elongation and reduction in area obtained in rupture tests are used in judging the ability of metal to adjust to stress concentration. The requirements are not well defined and are controversial. Most engineers are reluctant to use alloys with elongations of less than 5%, and this limit is sometimes considerably higher. Low ductility in a rupture test almost always indicates high resistance to the relaxation of stress by creep and possible sensitivity to stress concentrations. There is also ample evidence that rupture ductility has a major influence on creep-fatigue interaction (see the section below). Large changes in elongation with increasing fracture time usually indicate extensive changes in metallurgical structure or surface corrosion.

Creep-fatigue interaction can have a detrimental effect on the performance of metal parts or components operating at elevated temperatures. When temperatures are high enough to produce creep strains, and when cyclic (i.e., fatigue) strains are present, the two can interact. For example, it has been found that creep strains can seriously reduce fatigue life and that fatigue strains can seriously reduce creep life. This effect occurs in both stainless steels and low-alloy or carbon steels when temperatures are in the creep range.

Creep-fatigue interaction causes a reduction in fatigue life when either the frequency of the cycling stress is reduced or the cycling waveform has a tensile (and sometimes compressive) hold time (Fig. 9). Early studies (Ref 3–8) on type 316 stainless steel showed that tensile hold periods in the temperature region from 550 to 625 °C (1020 to 1160 °F) were very damaging, as shown in Fig. 9. Because the strain ranges were fairly high and the hold periods were short, failures were dominated by fatigue. More recent results at lower strain ranges and longer hold periods have revealed that creep-dominated failures also occur in stainless steels (Ref 9–11). Creep-dominated failures have been observed for tensile hold times up to 16 h at 600 °C (1110 °F) (Ref 9) and in tests at 625 °C (1160 °F) with tensile hold times up to 48 h (Ref 10).

Some investigators have observed a saturation in the detrimental effects of tensile hold periods, that is, a recovery of the endurance occurring at longer hold periods. Table 2 summarizes the data obtained in the evaluation of this effect at 2.0%

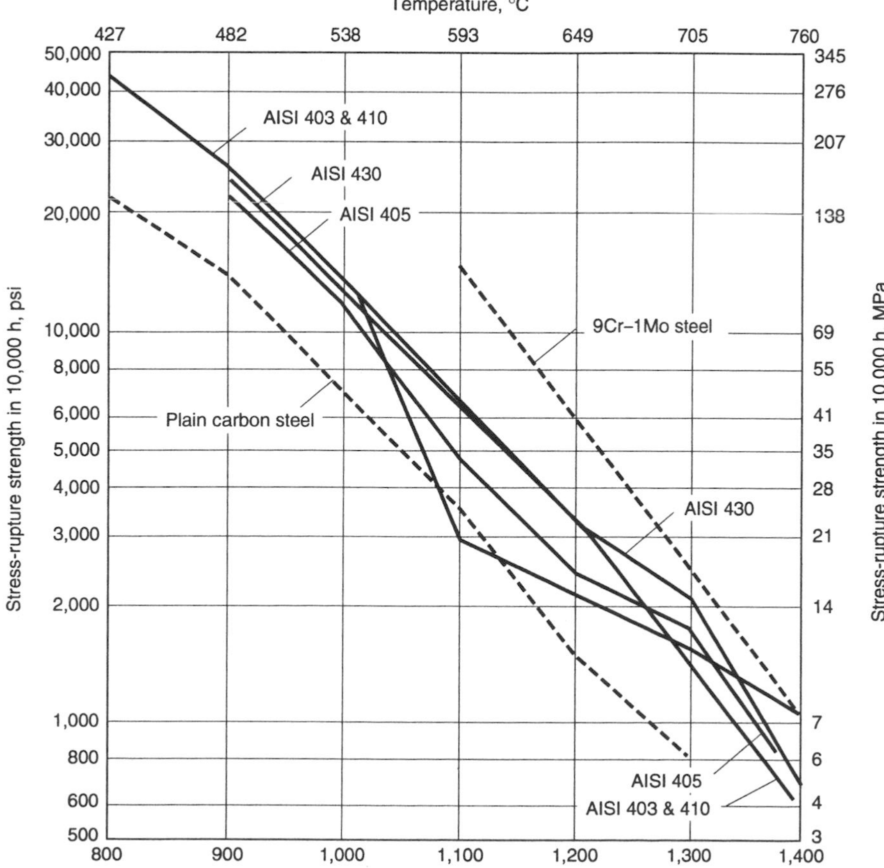

Fig. 13 10,000 h rupture strengths of 400-series stainless steels. Source: Ref 17

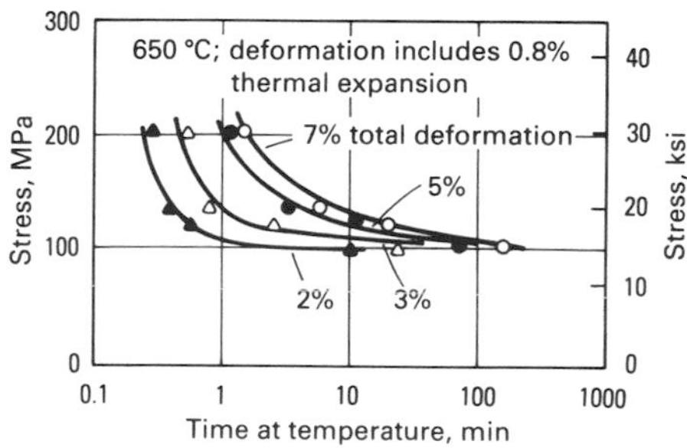

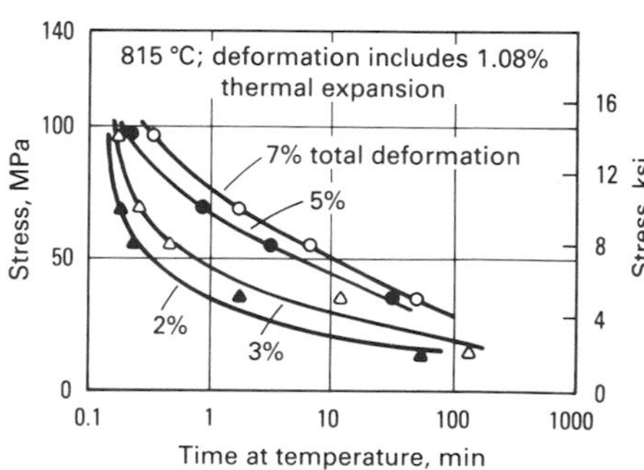

Fig. 14 Stress-time deformation curves for type 410 stainless steel sheet, showing effect of time at temperature on total deformation at specific stress levels. Design curves in the chart at the top represent a heating rate of 90 °C/s (160 °F/s) to 650 °C (1200 °F). Those at the bottom represent a heating rate of 105 °C/s (190 °F/s) to 815 °C (1500 °F). Room-temperature properties of the sheet used in these tests were: tensile strength, 650 to 695 MPa (94.5 to 101 ksi); yield strength at 0.2% offset, 555 to 565 MPa (80.7 to 82.3 ksi); and elongation in 50 mm (2 in.), 9.6 to 16% after air cooling from the normalizing temperature of 955 °C (1750 °F).

Table 6 Heat treating schedules for precipitation-hardening semiaustenitic stainless steels

Alloy	Mill heat treatment (solution anneal)	Conditioning and hardening treatment	Aging or tempering treatment
17-7 PH	1065 °C (1950 °F), air cool	10 min at 955 °C (1750 °F), air cool, 8 h at −75 °C (−100 °F)	1 h at 510, 565, or 620 °C (950, 1050, or 1150 °F)
		1 ½ h at 760 °C (1400 °F), air cool to 15 °C (60 °F), hold ½ h	1 h at 510, 565, or 620 °C (950, 1050, or 1150 °F)
15-7 Mo	1065 °C (1950 °F), air cool	10 min at 955 °C (1750 °F), air cool, 8 h at −75 °C (−100 °F)	1 h at 510, 565, or 620 °C (950,1050, or 1150 °F)
		1½ h at 790 °C (1450 °F), air cool to 15 °C (60 °F), hold ½ h	1 h at 510, 565, or 620 °C (950, 1050, or 1150 °F)
AM-350	1040–1080 °C (1900–1975 °F), air cool	930 °C (1710 °F), air cool, 3 h at −75 °C (−100 °F), 3 h at 745 °C (1375 °F), air cool to 27 °C (80 °F) max	3 h at 455 or 540 °C (850 or 1000 °F), 3 h at 455 °C (850 °F)
AM-355	3 h at 775 °C (1425 °F), oil or water quench to 27 °C (80 °F) max, 3 h at 580 °C (1075 °F), air cool	1040 °C (1900 °F), water quench, 3 h at −75 °C (−100 °F), reheat to 955 °C (1750 °F), air cool, 3 h at −75 °C (−100 °F)	3 h at 455 or 540 °C (850 or 1000 °F)

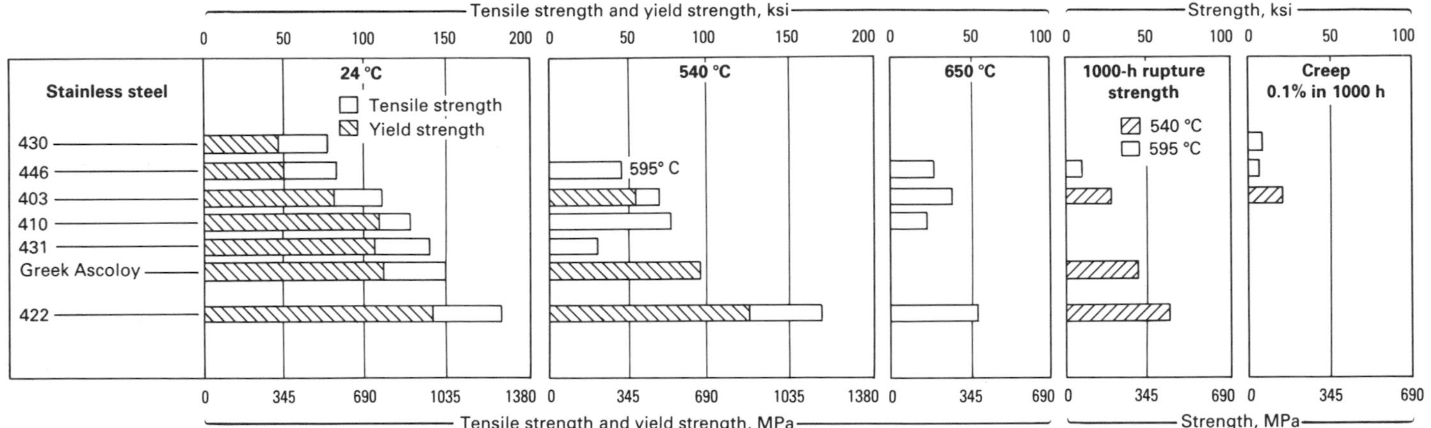

Fig. 15 Tensile, yield, rupture, and creep strengths for seven ferritic and martensitic stainless steels. Types 430 and 446 were annealed. Type 403 was quenched from 870 °C (1600 °F) and tempered at 620 °C (1150 °F). Type 410 was quenched from 955 °C (1750 °F) and tempered at 590 °C (1100 °F). Type 431 was quenched from 1025 °C (1875 °F) at 590 °C (1100 °F). Greek Ascoloy was quenched from 955 °C (1750 °F) and tempered at 590 °C (1100 °F). Type 422 was quenched from 1040 °C (1900 °F) and tempered at 590 °C (1100 °F).

strain range. A saturation effect is observed when the hold period approaches 30 min. This has been attributed to microstructural changes that lead to increases in ductility. Aging and the concomitant precipitation and growth of large carbides prior to testing have been shown to eliminate creep-fatigue effects altogether in type 316 stainless steel at 650 °C (1200 °F) (Ref 12).

Tests involving a 30 min hold period in tension, plus a shorter hold period in compression (asymmetrical holding), have shown that the detrimental effect of a hold period in tension can be significantly reduced by a short hold period in the compression portion of the cycle (Table 3). When the tension period is 30 min and a 3 min compression hold period is introduced, the fatigue life is within 80% of the fatigue life observed in the 30 min symmetrical-holding tests. Without this short hold period in compression, the fatigue life is reduced to about 40% of the 30 min symmetrical-holding fatigue life. In this type of testing, the hold period in compression exerts a "heating" effect (reduces the tendency for internal void formation).

The effect of slow/fast cycles (in which the strain increases slowly during the tension cycle but rapidly during the compression-going cycle) on the endurance of type 304, type 316, and other stainless steels has been investigated (Ref 13–16). In other tests, lower strain rates in the tension cycle were found to reduce endurance. Additional data on creep-fatigue interaction for various stainless steels are given in the article "Creep-Fatigue Interaction" in this Volume.

Effect of Product Form on Properties. The elevated-temperature properties of any of the stainless steel grades are influenced to some extent by the form of the product. These properties depend largely on the specific alloy characteristics, such as oxidation resistance, type of

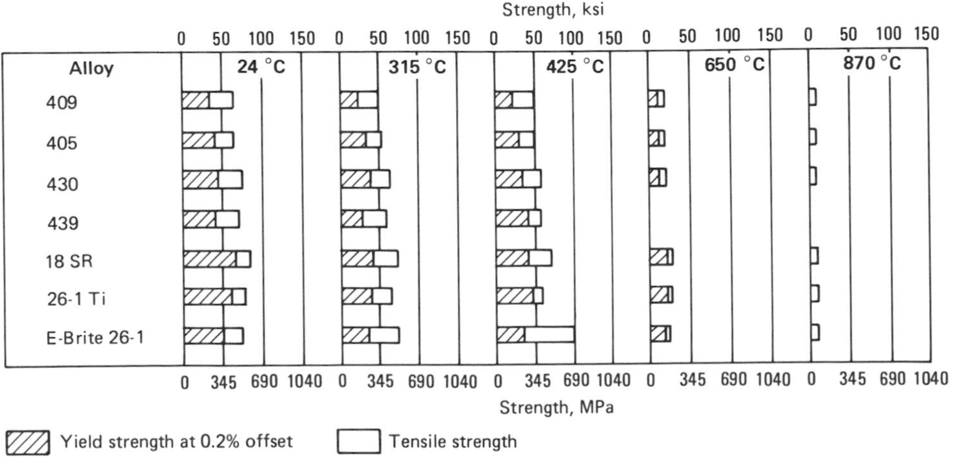

Fig. 16 Room-temperature and high-temperature tensile properties of selected ferritic stainless steels from Table 1. All alloys were in the annealed condition, fast cooled from 815 to 925 °C (1500 to 1700 °F).

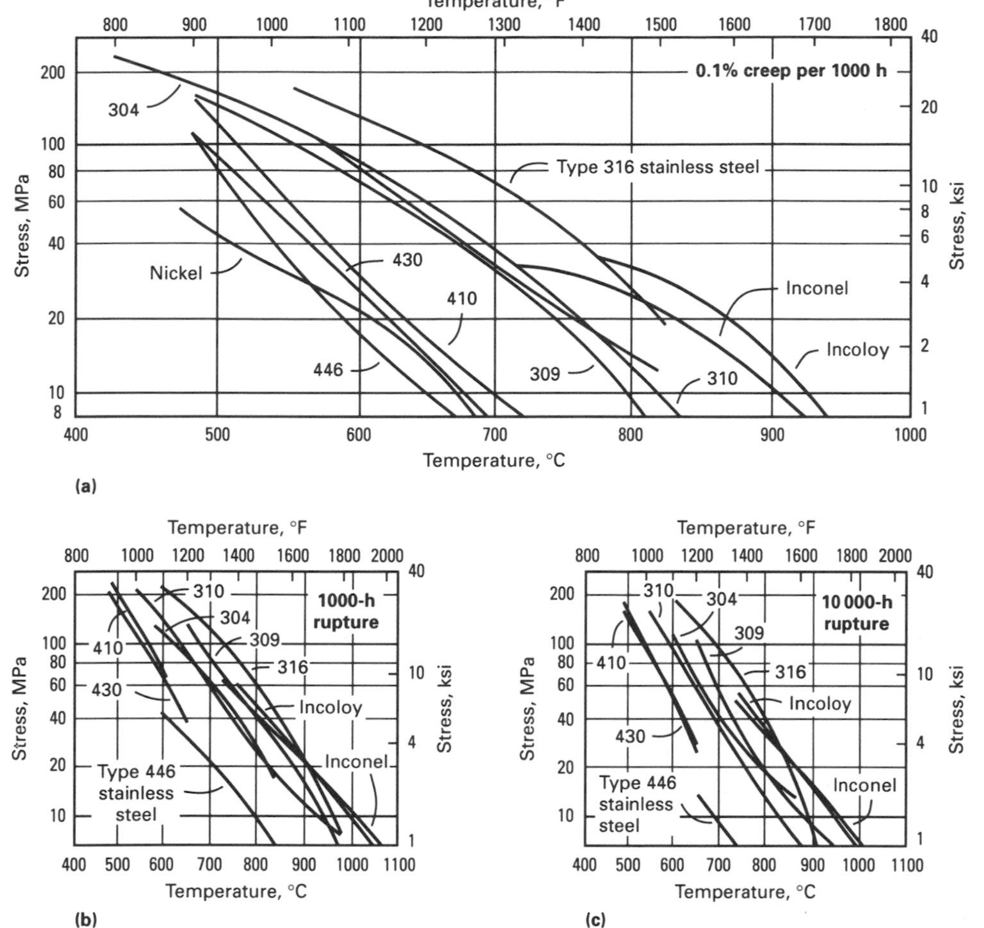

Fig. 17 Creep and rupture behavior of selected heat-resistant alloys as a function of temperature. (a) Stresses for a creep rate of 0.1% in 1000 h. (b) Stresses for rupture in 1000 h. (c) Stresses for rupture in 10,000 h

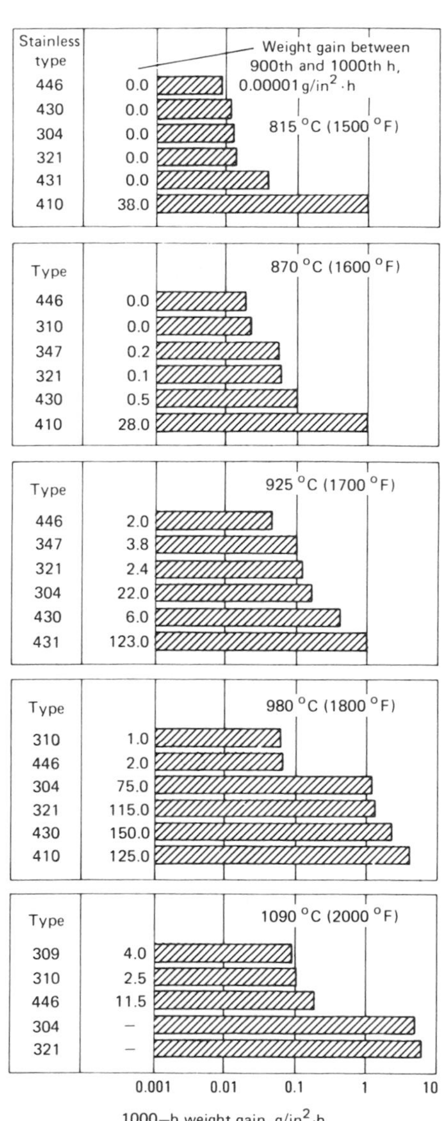

Fig. 18 The 1000 h oxidation resistance of selected stainless steels

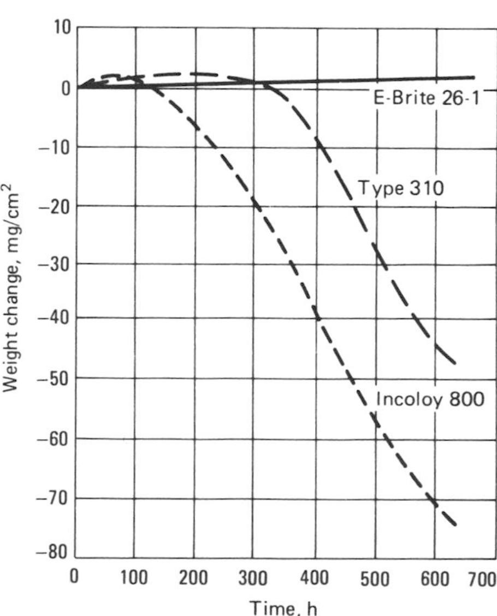

Fig. 19 Cyclic oxidation behavior of three iron-base heat-resistant alloys at 980 °C (1800 °F)

oxide scale, thermal conductivity, and thermal expansion. Time-temperature exposure and the duration and type of loading are also significant factors, as are differences in properties among different product forms.

For alloys that form thin, tenacious scales at elevated temperatures, the stress-rupture properties of bar and sheet of the same alloy will be about the same. For alloys that are less resistant to oxidation, rupture values are likely to be significantly lower for sheet than for the same alloy in bar form. This is true because of the greater ratio of surface area to volume, which allows greater interaction between the environment and the substrate metal. In the case of oxidation, a fixed depth of oxidation (such as 75 to 125 μm, or 3 to 5 mils) will more drastically affect properties in 1.3 mm (50 mil) sheet than in 6.5 mm (250 mil) bar stock.

Effect of Cold Working on Properties. The high-temperature strength of the heat-resistant austenitic alloys can be increased by such cold-working processes as rolling, swaging, or hammering. The increased strength is retained, however, only up to the recrystallization temperature. Figure 10, for example, shows the effect of temperature on the tensile properties of cold-worked 301 stainless steel. In particular, cold-worked products have poor resistance to creep, which generally occurs at temperatures slightly above the recrystallization temperature of the metal. During long-term high-temperature exposure, the benefit of cold working is lost, and stress-rupture strength may even fall below annealed strength.

Martensitic Stainless Steels

Quenched and tempered martensitic stainless steels are essentially martensitic and harden when air cooled from the austenitizing temperature. These alloys offer good combinations of mechanical properties, with usable short-time strength up to 590 °C (1100 °F) and relatively good corrosion resistance. The strength levels at temperatures up to 590 °C (1100 °F) that can be attained in these alloys through heat treatment are considerably higher than those attainable in ferritic stainless steels, but the martensitic alloys have inferior corrosion resistance. Also, the martensitic stainless steels are not very tough.

These alloys are normally purchased in the annealed or fully treated (hardened and tempered) condition. They are used in the hardened and tempered condition. For best long-time thermal stability, these alloys should be tempered at

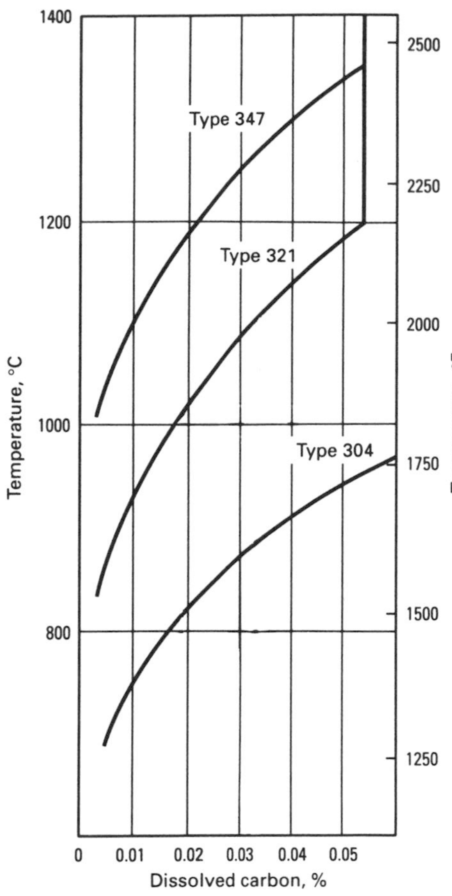

Fig. 20 Dissolved carbon under equilibrium conditions calculated from the solubility products for a type 347 stainless steel (0.054% C, 0.76% Nb), a type 321 stainless steel (0.054% C, 0.42% Ti), and an unstabilized 18Cr-8Ni stainless steel

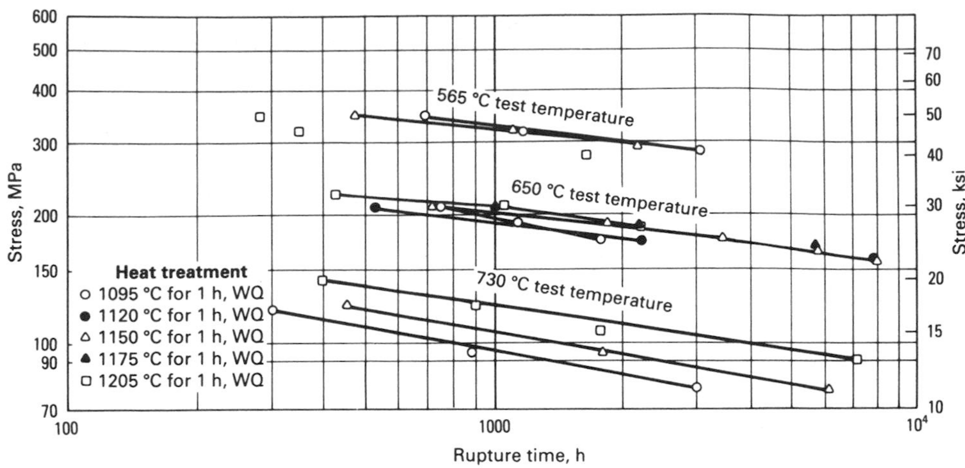

Fig. 21 Stress-rupture strength of type 347H stainless steel treated at different solution-annealing temperatures. WQ, water quench

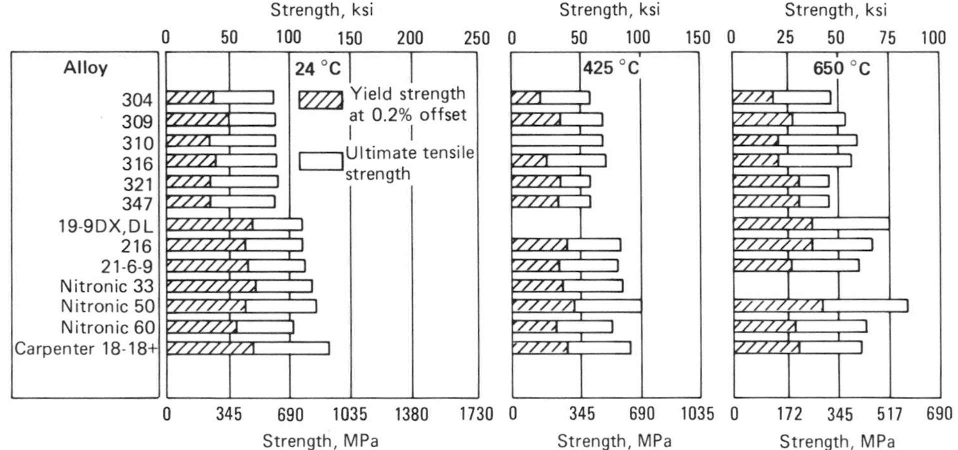

Fig. 22 Effect of testing temperature on tensile properties of austenitic stainless steels. Heat-treating schedules were as follows. Type 304: 1065 °C (1950 °F), water quench. Type 309: 1090 °C (2000 °F), water quench. Type 310: 1120 °C (2050 °F), water quench. Type 316: 1090 °C (2000 °F), water quench. Type 321: 1010 °C (1850 °F), water quench. Type 347: 1065 °C (1950 °F), water quench. 19-9 DX and 19-9 DL: 705 °C (1300 °F), air cool. Type 216: 1050 °C (1950 °F), water quench. 21-6-9: 1065 °C (1950 °F), water quench. Nitronic 33: 1056 °C (1950 °F), water quench. Nitronic 50: 1090 °C (2000 °F), water quench. Nitronic 60: 1065 °C (1950 °F), water quench. Carpenter 18-18 Plus: 1065 °C (1950 °F), water quench

a temperature that is 110 to 165 °C (200 to 300 °F) above the expected service temperature.

Properties. Quenched-and-tempered martensitic stainless steels can be grouped according to increasing strength and heat resistance:

- *Group 1:* Types 403, 410, and 416 (lowest strength and heat resistance)
- *Group 2:* Greek Ascoloy and type 431
- *Group 3:* Moly Ascoloy and Jethete M-152
- *Group 4:* H-46 and type 422 (highest strength and heat resistance)

A general comparison of mechanical property data is presented in Fig. 11 for some of these alloys. Data for type 410 are typical of group 1 alloys. Data for Greek Ascoloy are typical of type 431 (group 2). Data for Moly Ascoloy are typical of group 3 alloys (the composition of Jethete M-152 is very similar to that of Moly Ascoloy). Although H-46 and type 422 are similar in strength, their compositions are somewhat different; therefore, data are shown for both alloys.

The short-time tensile and rupture data shown in Fig. 11 were generated in tests of material that had been given austenitizing treatments typical for the specific alloys tested. These alloys are normally used at service temperatures near 540 °C (1000 °F) (although they may be used up to 590 °C, or 1100 °F). Therefore, data are shown for a relatively high tempering temperature of 650 °C (1200 °F), which results in good thermal stability in these alloys at 540 °C (1000 °F).

It should be noted that the group 1 alloys, of which type 410 is typical, show the lowest values of strength capability as a function of test temperature. Greek Ascoloy is considerably stronger than type 410, with a yield strength (0.2% offset) of 480 MPa (70 ksi) and a tensile strength of 585 MPa (85 ksi) at 540 °C (1000 °F). The tensile strength capabilities of H-46, Moly Ascoloy, and type 422 are fairly similar and are the highest in this group of alloys. Tensile elongation data for all these alloys are similar: from about 20% elongation at 21 °C (70 °F) to about 30% at 650 °C (1200 °F).

Stress-rupture data for alloys typical of each subgroup are compared in Fig. 11 by means of a Larson-Miller stress-rupture plot. The niobium-containing H-46 alloy has the highest stress-rupture capability, and type 422, Moly Ascoloy, and Greek Ascoloy have increasingly lower rupture capabilities, in that order. Type 410 has a very low stress-rupture capability and is the weakest of all the martensitic stainless alloys being considered. The niobium-containing alloys such as H-46 usually show an advantage in stress-rupture capability (creep resistance) for short testing times (100 to 1000 h) but lose their strength advantage when tested for periods of about 10,000 h or more. The favorable effects of niobium additions on short-time stress-rupture properties are attributed to a finely dispersed precipitation of NbC. The favorable effects tend to diminish as tempering temperature is increased, and a coarsely dispersed precipitate is formed.

The effect of tempering in service is shown by the hardness data in Fig. 12 for types 422 and

H-46. Compared with type 422, the H-46 alloy shows a larger hardness drop for extended thermal exposure at a testing temperature of 540 °C (1000 °F).

Applications. Quenched-and-tempered martensitic stainless steels find their greatest application in steam and gas turbines, where they are used in blading at temperatures up to 540 °C (1000 °F). Other uses include steam valves, bolts, and miscellaneous parts requiring corrosion resistance and good strength up to 540 °C (1000 °F).

Types 410 and 403. Type 410 is the basic, general-purpose stainless steel used for steam valves, pump shafts, bolts, and miscellaneous parts requiring corrosion resistance and moderate strength up to 540 °C (1000 °F). Type 403 is similar to 410, but the chemical composition is adjusted to prevent formation of δ-ferrite in heavy sections. It is used extensively for steam turbine rotor blades and gas turbine compressor blades operating at temperatures up to 480 °C (900 °F). For this type of application the steel is tempered at 590 °C (1100 °F) or above, after which embrittlement is negligible in the service temperature range of 370 to 480 °C (700 to 900 °F).

A satisfactory heat treatment for types 410 and 403 is to austenitize at 950 to 980 °C (1750 to 1800 °F), cool rapidly in air or oil, and temper. Cooling from the hot-rolling temperature and tempering without intermediate austenitizing is sometimes practical but may result in a structure that contains free ferrite, which is detrimental to transverse properties. Warm or cold work after tempering sets up residual stresses that can be relieved by heating to approximately 620 °C (1150 °F).

Stress-rupture strengths of types 403 and 410 martensitic steels are shown in Fig. 13. As this figure indicates, their stress-rupture properties lie approximately midway between plain carbon steels and 9Cr-1Mo alloy steels.

Stress-time deformation curves for type 410 sheet are given in Fig. 14. These values are useful for special applications where heating rates are high.

Greek Ascoloy, type 431, and type 422 are variants of type 410, modified by the addition of such elements as nickel, tungsten, aluminum, molybdenum, and vanadium. Nickel serves a useful purpose by causing the steel to be entirely austenitic at conventional heating temperatures when the carbon and chromium contents are such that a two-phase structure would exist if nickel were absent. The tempering temperature for Greek Ascoloy may be 55 °C (100 °F) higher than that for type 410 of equivalent strength and hardness, or even higher. Type 422 develops the highest mechanical properties and at 650 °C (1200 °F) has a tensile strength equivalent to that of type 403 at 590 °C (1100 °F). The rupture strength of type 422 at 540 °C (1000 °F) is considerably higher than those of the other steels in this series (Fig. 15).

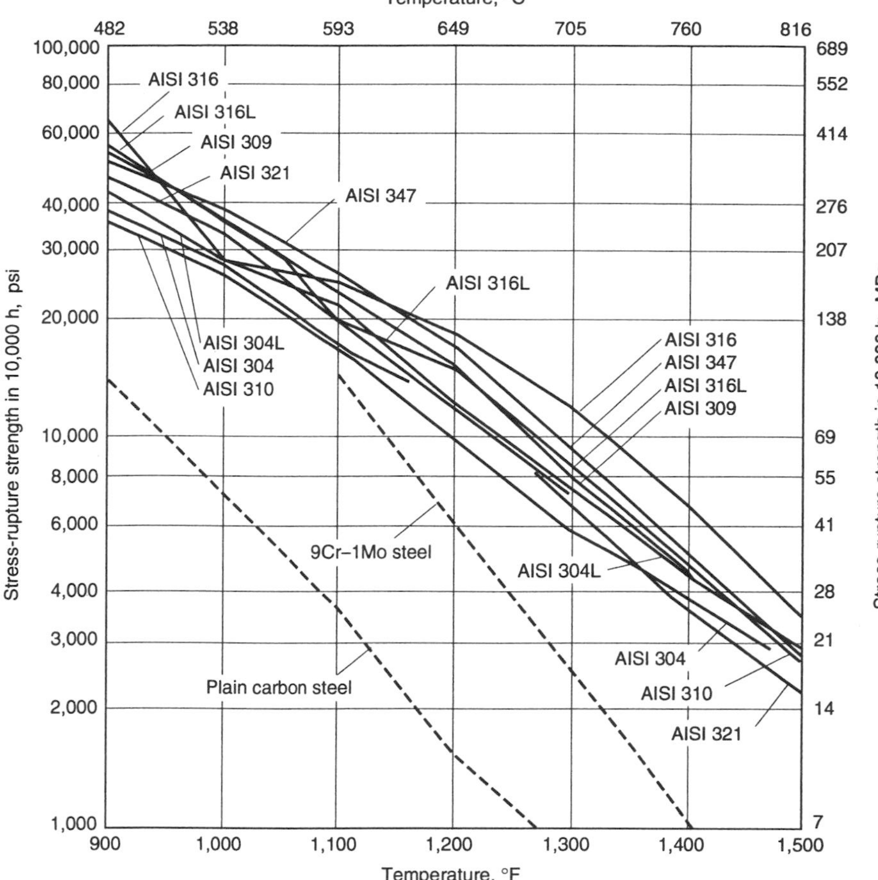

Fig. 23 10,000 h rupture strengths of 300-series stainless steels. Source: Ref 17

Ferritic Stainless Steels

Embrittlement. An important high-temperature characteristic of all ferritic stainless steels is precipitation of α′, a chromium-rich ferrite, when the steel is exposed to temperatures from 370 to

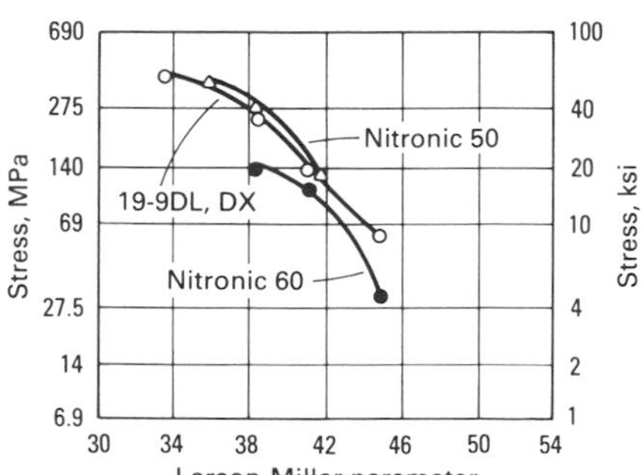

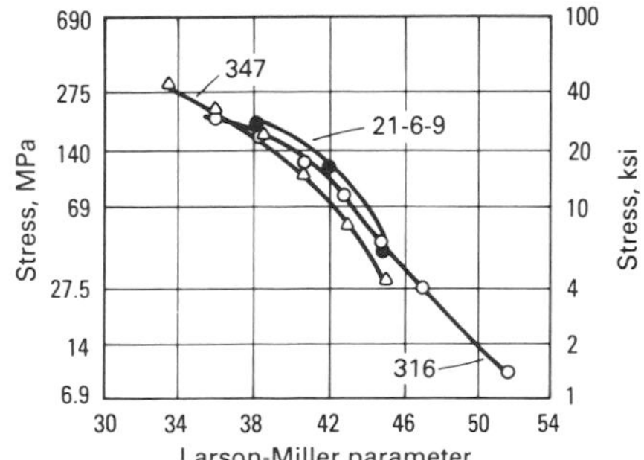

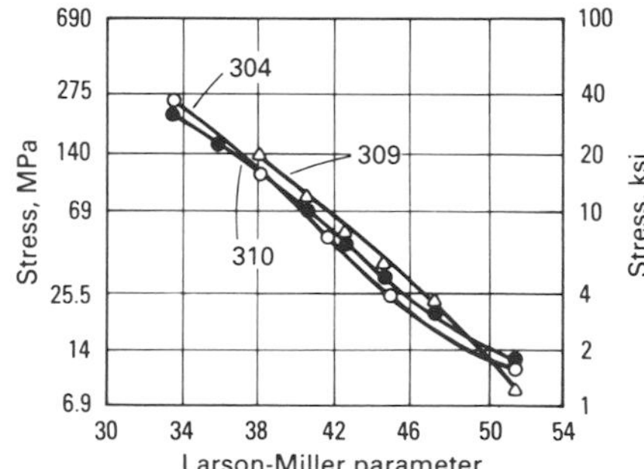

Fig. 24 Stress-rupture plots for various austenitic stainless steels, Heat-treating schedules were as follows. Type 304: 1065 °C (1950 °F), water quench. Type 309: 1090 °C (2000 °F), water quench. Type 310: 1120 °C (2050 °F), water quench. Type 316: 1090 °C (2000 °F), water quench. Type 347: 1065 °C (1950 °F), water quench. 21-6-9: 1065 °C (1950 °F), water quench. 19-9 DX and 19-9 DL: for tests above 705 °C (1300 °F), 1065 °C (1950 °F) and water quench, then 705 °C (1300 °F) and air cool; for tests below 705 °C (1300 °F), 705 °C (1300 °F) and air cool. Nitronic 50: 1090 °C (2000 °F), water quench. Nitronic 60: 1065 °C (1950 °F), water quench. Larson-Miller parameter = $T/1000\,(20 + \log t)$ where T is temperature in °R and t is time in h. All data taken from 1000 h tests

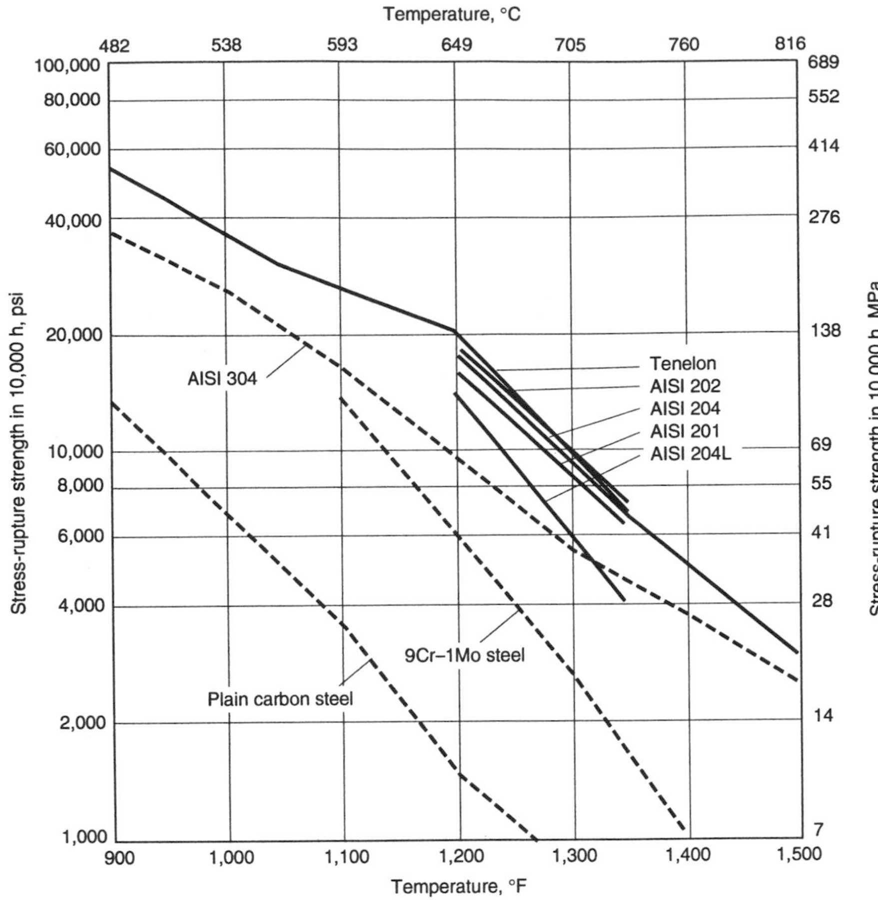

Fig. 25 10,000 h rupture strengths of 200-series stainless steels. Source: Ref 17

540 °C (700 to 1000 °F). This precipitation results in an increase in hardness and a drastic reduction in room-temperature toughness, which is known as 475 °C (885 °F) embrittlement (see the section "Thermally Induced Embrittlement" in this article). This embrittlement occurs in ferritic grades that have chromium contents above approximately 13%, and its severity increases at higher chromium levels. (Low-chromium grades, such as type 409, are not nearly as susceptible to embrittlement). This characteristic has to be considered for applications involving exposure to temperatures in the range from 370 to 540 °C (700 to 1000 °F), because subsequent room-temperature ductility will be severely impaired. In the higher-chromium alloys, such as 18Cr-2Mo, type 446, 26-1, and 29-4 (Table 1), σ-phase is encountered at temperatures above 565 °C (1050 °F). The χ-phase will also form in 26-1 Ti and 29Cr-4Mo. The high-molybdenum steels such as 29Cr-4Mo will also form a χ-phase, which has an embrittling effect similar to that of the σ-phase. The χ-phase is formed only in high-molybdenum steels. Titanium has little, if any, effect on the formation of χ-phase.

Properties and Applications. Tensile and yield strengths of ferritic stainless steels in the annealed condition are shown in Fig. 16. At room temperature, these properties are nearly equivalent to those of austenitic stainless steels. At higher temperatures, however, ferritic steels are much lower in strength. The rupture strength and creep strength of types 430 and 446 are illustrated in Fig. 17. Long-time and short-time high-temperature strengths of ferritic steels are relatively low compared with those of austenitic steels. As shown in Fig. 13, the 10,000 h stress-rupture values for types 405, 430, and 446 lie between those of carbon steels and chromium-molybdenum creep-resistant steels.

The main advantage of ferritic stainless steels for high-temperature use is their good oxidation resistance, which is comparable to that of austenitic grades. In view of their lower alloy content and lower cost, ferritic steels should be used in preference to austenitic steels, stress conditions permitting. Oxidation resistance of stainless steel is affected by many factors, including temperature, time, type of service (cyclic or continuous), and atmosphere. For this reason, selection of a material for a specific application should be based on tests that duplicate anticipated conditions as closely as possible.

Figure 18 compares the oxidation resistance of type 430, type 446, and several martensitic and austenitic grades in 1000 h continuous exposure to water-saturated air at temperatures from 815 to 1095 °C (1500 to 2000 °F). Table 4 presents data on cyclic oxidation resistance of ferritic stainless steels in air containing 10% water vapor. At 705 and 815 °C (1300 and 1500 °F), all the alloys listed are resistant to oxidation. At 980 °C (1800 °F), the lower-alloy types 409, 430, and 304 exhibit high corrosion. At 1090 °C (2000 °F), only 18 SR and Inconel 601 have adequate oxidation resistance. The cycling oxidation resistances of E-Brite 26-1, type 310, and Incoloy 800 are compared in Fig. 19.

Type 409, the lowest-alloy stainless steel with a nominal chromium content of 11.0%, is used extensively because of its good fabricating characteristics, including weldability. Its best-known high-temperature applications are in automotive exhaust systems, where metal temperature in catalytic converters exceeds 540 °C (1000 °F). Type 409 is also used for exhaust ducting and silencers in gas turbines. Type 405 is used in stationary vanes and spacers in steam turbines and in various furnace components. Types 430 and 439 are used for heat exchangers, hot-water tanks, condensers, and furnace parts. Type 18 SR,

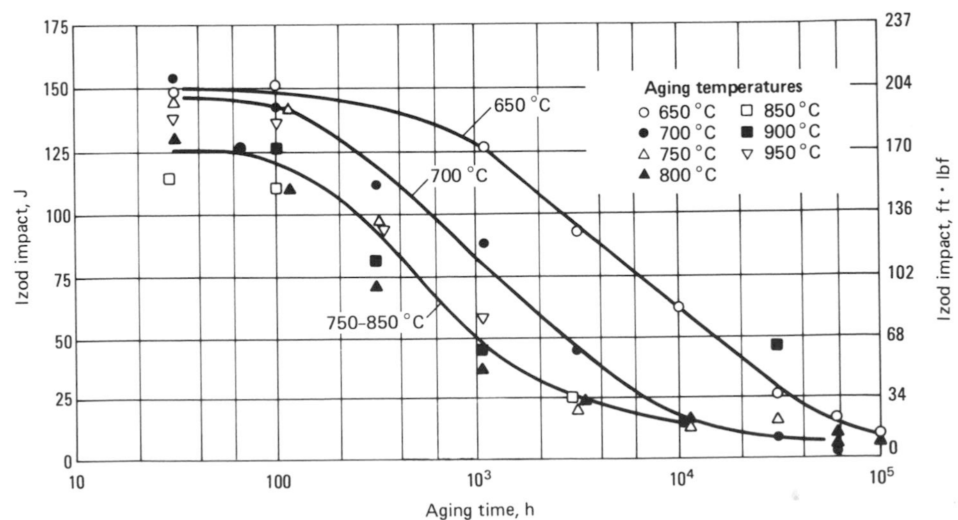

Fig. 26 Room-temperature impact toughness of type 316 stainless steel after aging at indicated temperatures. Source: Ref 19

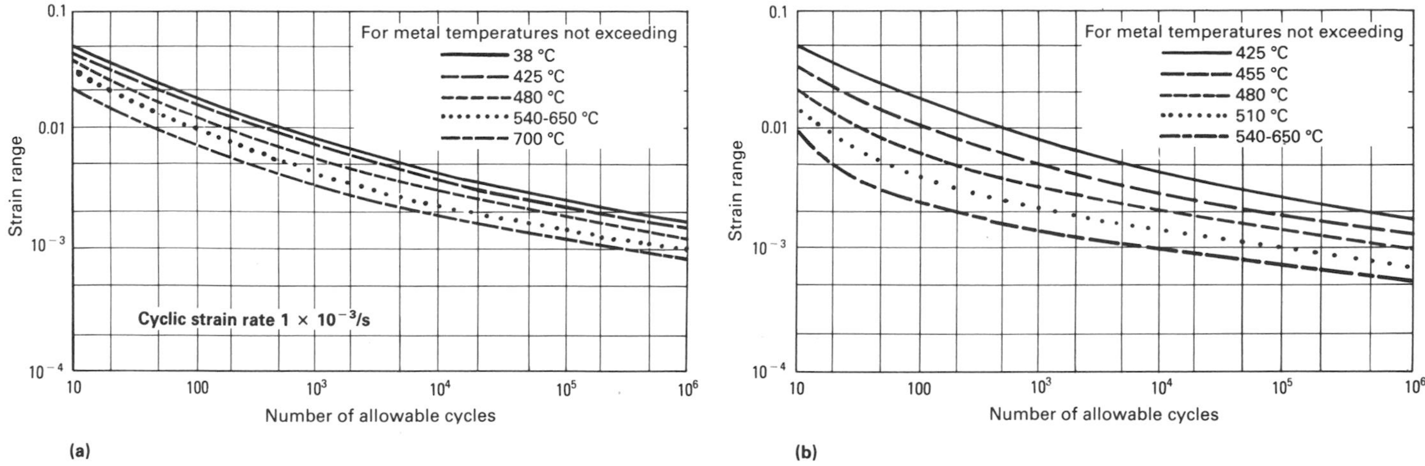

Fig. 27 Design fatigue-strain range curves for types 304 and 316 stainless steel. (a) Design curves with continuous cycling (pure fatigue). (b) Design curves with hold times (creep-fatigue interaction)

like type 446, is used in industrial ovens, blowers, exhaust systems, furnace equipment, annealing boxes, kiln liners, and pyrometer tubes. E-Brite 26-1 and 26-1Ti were developed primarily for corrosion service, but they also can be used in applications similar to those of types 18 SR and 446.

Austenitic Stainless Steels

The austenitic stainless steels listed in Table 1 can be grouped into three categories, based primarily on composition:

- *300-series alloys* are essentially chromium-nickel and chromium-nickel-molybdenum austenitic stainless steels to which small amounts of other elements have been added.
- *19-9 DL, 19-9 DX,* and *17-14-CuMo* contain 1.25 to 2.5% Mo and 0.3 to 0.55% Ti. Other elements used include 1.25% W and 3% Cu in 17-14-CuMo.
- *Chromium-nickel-manganese alloys* include types 201 and 202; 21-6-9; Nitronics 32, 33, 50, and 60; and Carpenter 18-18 Plus. These alloys contain 5 to 18% Mn and 0.10 to 0.50% N. The manganese in these steels saves nickel and increases the solubility for nitrogen, which is used for structure control, strengthening, and improving corrosion resistance.

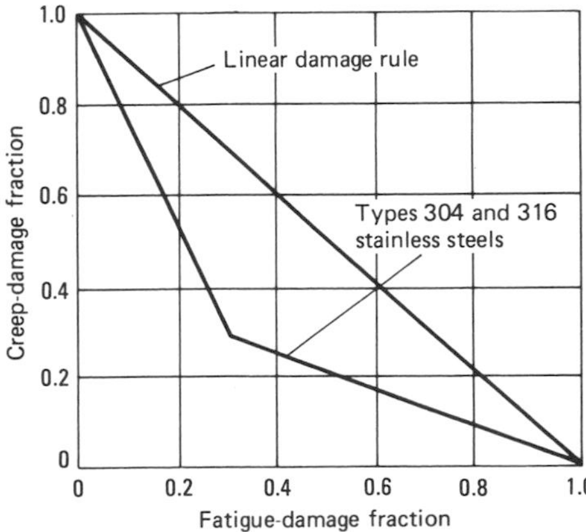

Fig. 28 Comparison of linear damage rule of creep-fatigue interaction with design envelopes in ASME Code Case N-47 for types 304 and 316 stainless steel. The creep-damage fraction is time/time-to-rupture (multiplied by a safety factor). The fatigue-damage fraction is number of cycles/cycles to failure (multiplied by a safety factor).

Austenitic stainless steels are noted for high strength and for exceptional toughness, ductility, and formability. As a class, they exhibit considerably better corrosion resistance than martensitic or ferritic stainless steels, and they also have excellent strength and oxidation resistance at elevated temperatures.

Solution heat treatment of these alloys is done by heating to about 1095 °C (2000 °F), followed by rapid cooling. Carbides that are dissolved at these temperatures may precipitate at grain boundaries upon exposure to temperatures from 425 to 870 °C (800 to 1600 °F), causing chromium depletion in grain-boundary regions (sensitization). In this condition, the metal is sensitive to intergranular corrosion in oxidizing acids. The precipitation of chromium carbides can be controlled by reducing carbon content, as in types 304L and 316L, or by adding the stronger carbide formers titanium and niobium, as in types 321 and 347. These alloys are normally purchased and used in the annealed condition. The reduced carbon in solution in the low-carbon (304L, 316L) and stabilized grades (321, 347) results in reduced creep strength and creep-rupture strength. Additional information on sensitization of austenitic stainless steels can be found in the section "Thermally Induced Embrittlement" in this article.

H-Grades. For the best creep strength and creep-rupture strength, the H-grades of austenitic stainless steels are specified. These steels have carbon contents of 0.04 to 0.10% (Table 1) and are solution annealed at temperatures high enough to produce improved creep properties. The minimum annealing temperatures specified in ASTM A 312 and A 240 for H-grades in general corrosion service are:

- *1040 °C (1900 °F)* for hot-finished or cold-worked 304H and 316H steels
- *1095 °C (2000 °F)* for cold-worked 321H, 347H, and 348H steels
- *1050 °C (1095 °F)* for hot-finished 321H, 347H, and 348H steels

The stabilized grades (such as 321H, 347H, and 348H) have additions of strong carbide-forming elements, which lower the amount of dissolved carbon at a given annealing temperature (Fig. 20). The carbide formers in the stabilized grades, such as niobium in 347H, increase the resistance to intergranular corrosion by making less dissolved carbon available for chromium carbide formation, thereby preventing the depletion of chromium in grain-boundary regions.

When intergranular corrosion is of concern, annealing temperatures must be low enough to keep dissolved carbon at low levels. For example, at the minimum annealing temperature of 1095 °C (2000 °F) (specified in ASME SA213), type 347H tube would have only about 0.01% soluble carbon (Fig. 20), and the alloy should be stabilized against chromium depletion in the grain boundaries. However, annealing temperatures above 1065 °C (1950 °F) may still impair the intergranular corrosion resistance of stabilized grades such as 321, 321H, 347, 347H, 348, and 348H. When types 321H and 347H are used in applications where intergranular corrosion may be a problem, it is possible to apply an additional stabilizing treatment at a temperature near 900 °C (1650 °F) to reduce free carbon content by carbide precipitation.

Lower annealing temperatures improve resistance to intergranular corrosion but also reduce creep strength and creep-rupture strength. Figure 21 shows the creep-rupture strength of 347H tube treated at different annealing temperatures. In applications where intergranular corrosion is not a concern, better creep properties can be obtained

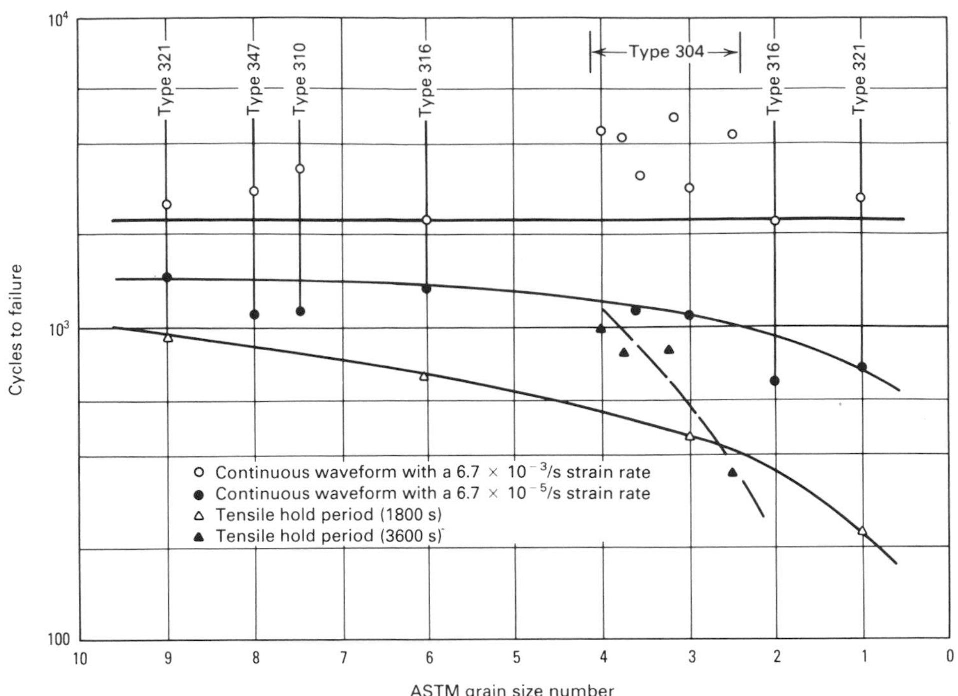

Fig. 29 Effect of strain rate and grain size on the fatigue life of various stainless steels at elevated temperatures. Grain size has the greatest influence on fatigue life when hold times are increased. Test conditions: total strain range, 1.0%; test temperature, 593 to 600 °C (1100 to 1110 °F). Source: Ref 20

with higher annealing temperatures. Types 321H and 347H, for example, are solution annealed at temperatures above 1120 °C (2050 °F) and 1150 °C (2100 °F), respectively, to put carbides in solution and to coarsen the grain structure, thereby ensuring the best creep strength and creep-rupture strength.

Nitrogen Additions. During the course of studies on the H-grades, it was found that controlled additions of nitrogen improved the high-temperature strength of types 304 and 316. Creep and stress-rupture properties of the N-grades, which contain from 0.10 to 0.13% N, are given in Table 5.

Tensile Properties. Typical mechanical property data for austenitic stainless steels are given in Fig. 22. Room-temperature tensile properties of annealed 300-series alloys are similar. At higher testing temperatures (425 and 650 °C, or 800 and 1200 °F), types 321, 347, and 309 appear to have yield strengths somewhat higher than those of types 304, 310, and 316. Types 309, 310, and 316 have the highest tensile strengths at 650 °C (1200 °F). Similar data on H-grades are shown in Fig. 2.

Tensile and yield strengths of 19-9 DL and 19-9 DX are higher than those of any 300-series alloy. However, 19-9 DL and 19-9 DX are heat treated at 705 °C (1300 °F), compared with an average of 1065 °C (1950 °F) for 300-series alloys. Also, 19-9 DL and 19-9 DX are normally strengthened by controlled amounts of hot and cold work.

Tensile and yield strengths of chromium-nickel-manganese alloys are higher than those of 300-series alloys at both room and elevated temperatures. Carpenter 18-18 Plus exhibits the highest room-temperature tensile strength, whereas Nitronic 50 has the highest tensile strength at 650 °C (1200 °F).

Stress-rupture properties of various 300-series stainless steels are shown in Fig. 5. The H-grades of types 347 and 316 appear to be the two strongest alloys over a range of temperatures.

Similar data on 300-series stainless steels are shown in Fig. 23. This figure also illustrates the superior high-temperature strength of stainless steels when compared with plain carbon steel or 9Cr-1Mo alloy steel.

The 19-9 DL and 19-9 DX alloys have rupture strengths superior to those of all 300-series alloys over the limited temperature and time range for which rupture data are available (1000 h rupture strength at 540 to 815 °C, or 1000 to 1500 °F) (Fig. 24). At longer times or higher temperatures, the 300 series may be superior. For the time-temperature range in Fig. 24, the chromium-nickel-manganese alloys have higher stress-rupture capabilities than the 300-series alloys, except that type 316 is superior to 21-6-9, and types 316 and 347 are stronger than Nitronic 60. The spread in rupture strength capability among these alloys is greater at the lower testing temperatures (540 to 700 °C, or 1000 to 1300 °F) and becomes progressively smaller as temperature is increased to

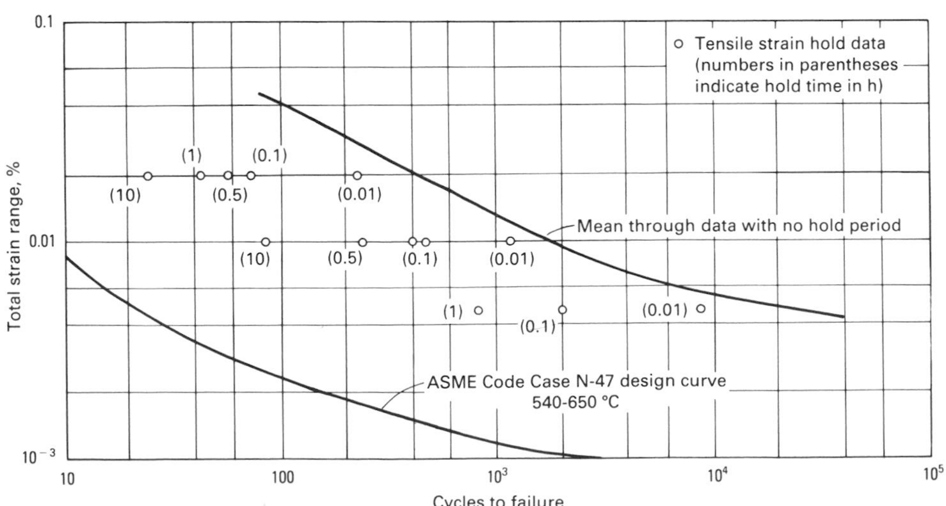

Fig. 30 Influence of tensile hold times at peak strain on failure life of a single heat of type 316 stainless steel tested at 593 °C (1100 °F). Source: Ref 20

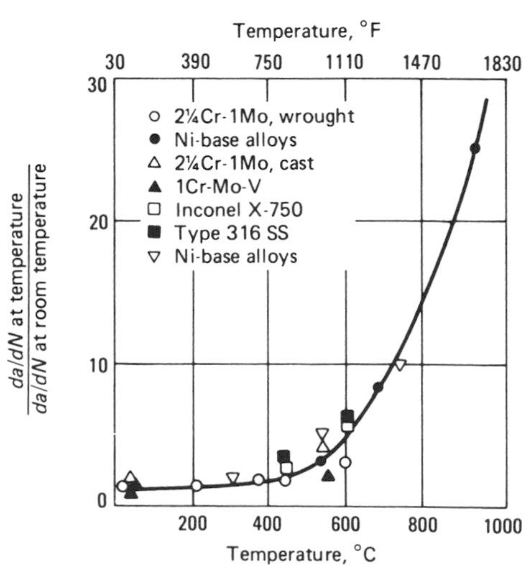

Fig. 31 Variation of fatigue crack growth rates as a function of temperature at $\Delta K = 30$ MPa$\sqrt{m}$ (27 ksi$\sqrt{in.}$). Source: Ref 2

approximately 980 °C (1800 °F), where all the alloys exhibit 1000 h rupture stresses of about 7 to 10 MPa (1 to 1.5 ksi). Types 304 and 310 have the lowest stress-rupture strengths.

A comparison of 200- and 300-series stainless steels is also shown in Fig. 25. Once again it is evident that both austenitic types perform better at elevated temperatures than carbon or alloy steels.

Aging and the degradation of mechanical properties occur in austenitic steel because of two principal factors: precipitation reactions that occur during prolonged exposure at elevated temperatures, and environmental effects, such as corrosion or nuclear irradiation.

The precipitation reactions in austenitic stainless steels that occur during prolonged exposure at elevated temperatures are complex, but some general guidance to the precipitates formed is provided by the constitutive diagrams in Ref 19. The precipitates formed at temperatures in the range of about 500 to 600 °C (930 to 1110 °F) are predominantly carbides, whereas at higher temperatures they are in the form of intermetallic phases. About 30 phases have been identified in stainless steels, but in plant-serviced alloys and weld metals exposed to about 550 °C (1020 °F), the predominant precipitates found are $M_{23}C_6$ and MC carbides in unstabilized and stabilized steels, respectively. At temperatures above 600 °C (1110 °F), σ-phase and Fe_2Mo are also formed. The aging process also tends to occur more rapidly in weld metals that contain δ-ferrite (Ref 19). Phases found in wrought stainless steels are also discussed in the article "Microstructures of Wrought Stainless Steels" in the *ASM Specialty Handbook: Stainless Steels.*

Impact Toughness. Solution-annealed austenitic stainless steels are very tough, and their impact energies are very high. However, following elevated-temperature thermal aging, the impact energy decreases with increasing time at temperature. The impact energy of type 316 stainless steel (Fig. 26) shows a continuous fall with increasing exposure time and temperature in the range of 650 to 850 °C (1200 to 1560 °F). Service-exposed type 316 stainless steel exhibited impact energies of 80 and 300 J (60 and 220 ft · lbf) in the serviced and re-solution heat-treated conditions, respectively, when tested at room temperature (Ref 19).

Fatigue properties at elevated temperatures are dependent on several variables, including strain range, temperature, cyclic frequency, hold times, and the environment. The fatigue design curves in Fig. 27(a) show the simple case of pure fatigue (continuous cycles without hold times) for types 304 and 316 stainless steel. These design curves (from Code Case N-47 in the ASME Boiler Code) have a built-in factor of safety: they are established by applying a safety factor of 2 with respect to strain range or a factor of 20 with respect to the number of cycles, whichever gives the lower value. The creep-life fraction is determined by the time-life fraction per cycle, using assumed stresses 1.1 times the applied stress and using the minimum stress-rupture curves incorporated in the code. The total damage must not exceed the envelope defined by the bilinear damage curve shown in Fig. 28.

The design curves in Fig. 27(a) are based on a strain rate of 1×10^{-3}/s. If the strain rate decreases, fatigue life also decreases. In Fig. 29, for example, the fatigue lives of several stainless steels are shown for continuous cycling at two different strain rates. Fatigue life is reduced with a lower strain rate, while grain size has little effect on fatigue life when life is determined from pure fatigue (or continuous cycling).

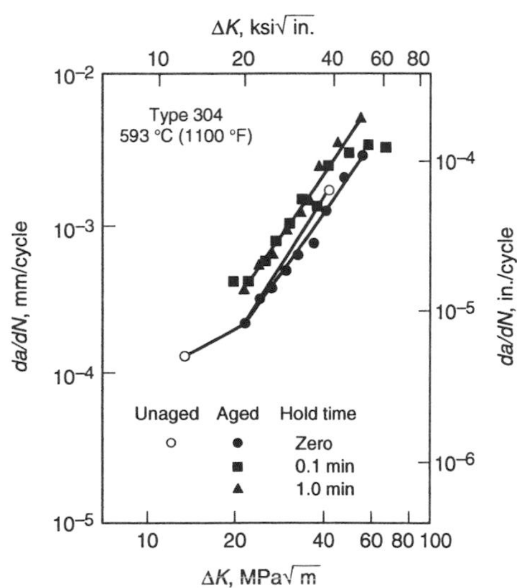

(a)

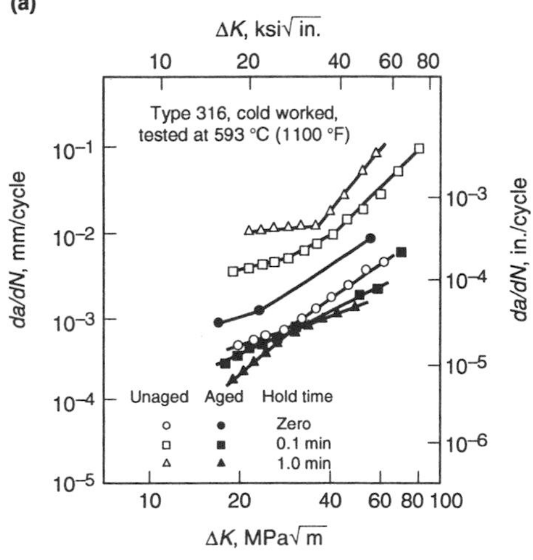

(b)

Fig. 32 Effect of aging and hold times on fatigue crack growth rates. (a) Effect of aging at 593 °C (1100 °F) for 5000 h, and hold times of 0.1 and 1.0 min for each cycle, on fatigue crack growth rates of longitudinal-transverse oriented specimens of type 304 stainless steel tested in air at 0.17 Hz and an *R* ratio of 0. (b) Effect of exposure at 593 °C (1100 °F) for 5000 h, and hold times during cycling, on fatigue crack growth rate of 20% cold-worked type 316 stainless steel at 593 °C in air. Source: Ref 21

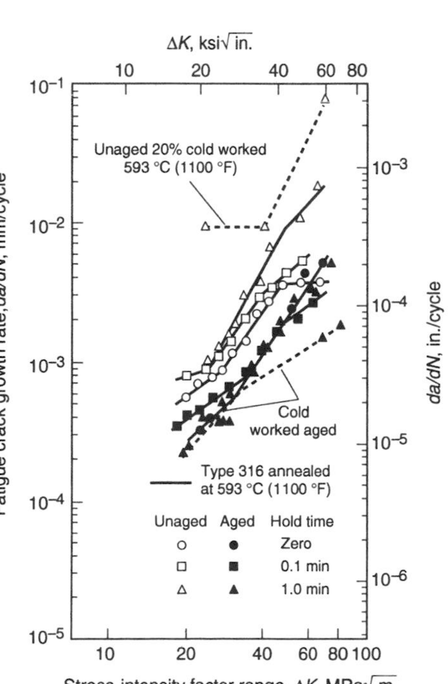

Fig. 33 Effect of exposure in air at 593 °C (1100 °F) for 5000 h, and hold times, on fatigue crack growth rates for annealed type 316 stainless steel at 593 °C in air

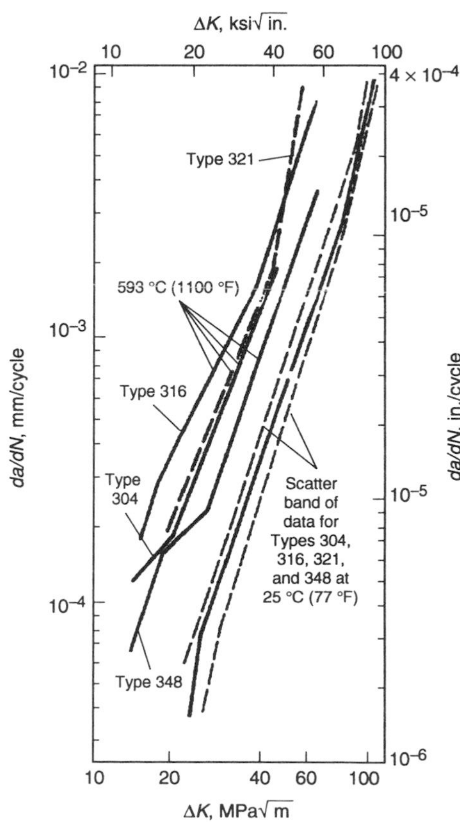

Fig. 34 Fatigue crack growth rates for annealed types 304, 316, 321, and 348 stainless steels in air at room temperature and 593 °C (1100 °F), longitudinal-transverse (L-T) orientation, 0.17 Hz, and an *R* ratio of 0. Tests were made on single-edge-notch cantilever specimens of types 321 and 348 stainless steels from the L-T orientation at 0.17 Hz with an *R* ratio of zero at room temperature and at elevated temperatures to 593 °C (1100 °F).

When hold times are introduced, a different set of design curves is used (Fig. 27b) to determine the allowable fatigue-life fraction (creep-life fraction is determined the same way as for continuous cycling). These allowable fatigue-life curves are a more conservative set of curves than those of Fig. 27(a). They incorporate the effect of creep damage by applying a fatigue life reduction factor, which includes hold time effects in addition to the factor of safety (2 in strength and 20 in cycles, whichever gives the lower value). Figure 30 compares the 540 to 650 °C (1000 to 1200 °F) design curve in Fig. 27(b) with actual fatigue life results from testing type 316 stainless at 593 °C (1100 °F) and various hold times. When hold times are introduced, the influence of grain size may also be more pronounced (Fig. 29).

Fatigue Crack Growth. Although *S-N* curves (stress/number of cycles) have been used in the past as the basic design tool against fatigue, their limitations have become increasingly obvious. One of the more serious limitations is that they do not distinguish between crack initiation and crack propagation. Particularly in the low-stress regions, a large fraction of the life of a component may be spent in crack propagation, allowing crack tolerance over a large portion of the life. Engineering structures often contain flaws or cracklike imperfections that may altogether eliminate the crack initiation step. A methodology that quantitatively describes crack growth as a function of the loading variables is of great value in design and in assessing the remaining lives of components.

Fatigue crack growth rates are obtained at various ΔK and temperature ranges (ΔK is the stress-intensity factor). Therefore it is difficult to compare the various types of materials directly. At a constant ΔK (arbitrarily chosen as 30 MPa$\sqrt{m}$, or 27 ksi$\sqrt{in.}$), a clear trend of increasing crack growth rate with increasing temperature can be seen (Fig. 31). At temperatures up to about 50% of the melting point (550 to 600 °C, or 1020 to 1110 °F), the growth rates are relatively insensitive to temperature, but sensitivity increases rapidly at higher temperatures.

Effect of Aging on Fatigue Crack Growth. Because the expected service lives of most components of austenitic stainless steels are many years, an evaluation of the effect of long-time aging at service temperatures is important. Figure 32 (Ref 21) shows results of fatigue crack growth rate tests on specimens that were tested in the unaged and aged conditions (5000 h at 593 °C, or 1100 °F. After aging for 5000 h at this temperature, precipitation of $M_{23}C_6$ carbides is essentially complete. These results indicate that at 593 °C (1100 °F) there are no deleterious effects of aging on the crack growth rates of specimens that are continuously cycled. When a holding time of 0.1 or 1.0 min is included in each loading cycle, there tends to be a slight increase in the fatigue crack growth rate at a given ΔK level.

As shown in Fig. 32(a), aging of cold work specimens type 304 at 593 °C (1100 °F) for 5000 h tends to increase slightly the fatigue crack growth rates of specimens that are continuously cycled at 593 °C (1100 °F) (Ref 21). With holding times of 0.1 to 1.0 min, the fatigue crack growth rates also increases (Ref 21). However, an opposite effect is observed in aged 316 with hold times. Effects of holding times on cyclic loading of unaged and aged specimens of 20% cold worked type 316 are shown in Fig. 32(b) for tests at 593 °C (1100 °F). The frequency for specimens cycled with zero holding time was 0.17 Hz, and the *R* ratio was zero. Aging was done for 5000 h at 593 °C (1100 °F), and testing was done in air. For the unaged specimens, increasing the holding time significantly increased the fatigue crack growth rates, as shown. For the aged specimens, holding at maximum load for 0.1 or 1.0 min for each loading cycle reduced the fatigue crack growth rates over those obtained with no holding time. These data indicate that cold working and aging at 593 °C (1100 °F) before or during service exposure can lead to improved fatigue crack growth resistance and that short hold times at maximum load reduce fatigue crack growth rates.

Similar results have been obtained for annealed type 316 after aging (Fig. 33). Long-time expo-

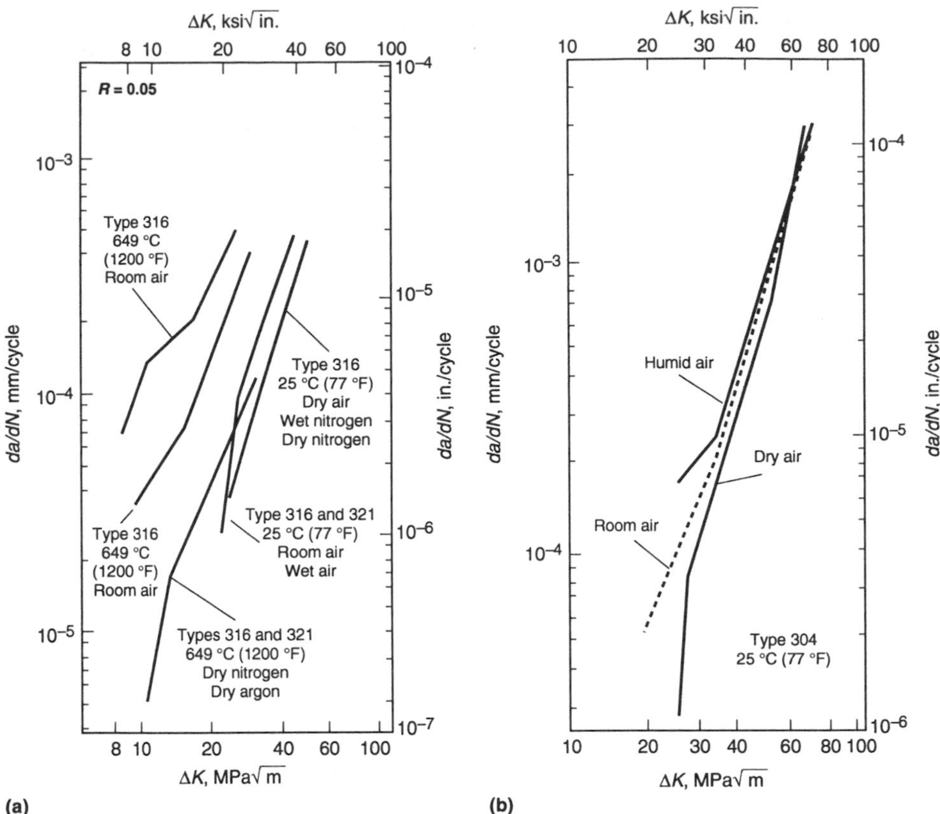

Fig. 35 Effect of environments on fatigue crack growth rates. (a) Types 316 and 321 stainless steels at 25 and 649 °C (77 and 1200 °F). Compact specimens were tested in fatigue loading according to a sine wave loading pattern at 5 Hz with an *R* ratio of 0.05 in room air, dry air, humid air, dry nitrogen, wet nitrogen, and dry argon. (b) Effect of humidity on fatigue crack growth rates for type 304 stainless steel tested at room temperature, 0.17 Hz, and an *R* ratio of 0. Source: Ref 24

Table 7 Elevated-temperature tensile properties of A-286 PH austenitic stainless steel

Tests were carried out on 22 mm (7/8 in.) diam bar stock solution treated to 980 °C (1800 °F) for 1 h, oil quenched, aged to 720 °C (1325 °F) for 16 h, air cooled

Test temperature		0.02% offset yield strength		0.2% offset yield strength		Tensile strength		Elongation in 2 in. (50.8 mm),	Reduction of area,
°C	°F	MPa	ksi	MPa	ksi	MPa	ksi	%	%
21	70	621	90	655	95	1000	145	24.0	45.0
204	400	524	76	645	93.5	986	143	21.5	52.0
427	800	496	72	641	93	951	138	18.5	35.0
538	1000	427	62	603	87.5	903	131	18.5	31.0
593	1100	445	64.5	621	90	841	122	21.0	23.0
649	1200	431	62.5	607	88	714	103.5	13.0	14.5
704	1300	472	68.5	593	86	596	86.5	11.0	10.0
760	1400	307	44.5	427	62	441	64	18.5	23.0
816	1500	214	31	228	33	252	36.5	68.5	37.5

Source: Carpenter Technology Corp.

sure (5000 h of aging) at 593 °C (1100 °F) in air substantially reduced the fatigue crack growth rates at ΔK levels from 18 to 55 MPa$\sqrt{m}$ (16 to 50 ksi$\sqrt{in.}$) for the continuous cycling tests and over the whole testing range for specimens cycled with 0.1 and 1.0 min holding times for each cycle. Fatigue crack growth rates for specimens tested without prior exposure and with holding times of 0.1 and 1.0 min for each cycle were higher than those for specimens cycled continuously under the same conditions. However, the effect of hold time was less significant for specimens that had been aged at 593 °C (1100 °F) before testing at the same temperature. In other work (Ref 22), holding times of up to 8 min did not cause significant increases in fatigue crack growth rates at 593 °C (1100 °F) in aged specimens. However, holding for 16 min caused marked increases in fatigue crack growth rates.

In annealed type 321 stainless steel aged at 593 °C (1100 °F) for 5000 h and then tested at 593 °C, long-time exposure at the service temperature did not reduce the fatigue crack propagation resistance in air (Ref 23). Aged specimens tested with zero holding time had lower crack growth rates than corresponding specimens that were not aged.

Environmental Effects on Fatigue Crack Growth. The curves in Fig. 34 show that, at room temperature, the fatigue rates for types 304, 316, 321, and 348 all fall in a narrow band. For tests at 593 °C (1100 °F), however, specimens of type 316 had the least fatigue crack propagation resistance, whereas specimens of type 348 had the highest fatigue crack propagation resistance, over the ΔK range studied. Results of tests on specimens of type 304 and 321 were nearly the same at 593 °C (1100 °F) in air.

The presence of moisture and oxygen causes an increase in crack growth rates, as shown in Fig. 35(a) for types 316 and 321. Fatigue crack growth rate data at 25 °C (77 °F) show that crack growth rates increased slightly with increased humidity when oxygen was present but that high humidity in an inert gas had no significant effect. Fatigue crack growth rates in room air at room temperature were the same for types 316 and 321 stainless steels. Furthermore, in tests at 649 °C (1200 °F) in dry nitrogen, fatigue crack growth rates for types 316 and 321 also were the same. In air, however, fatigue crack growth rates in type 316 specimens increased by a factor of about 22 over rates in an inert environment at the same temperature. For specimens of type 321, the corresponding increase in fatigue crack growth rates was about 5 times that for the inert environment at 649 °C (1200 °F). If components of these stainless steels are exposed to inert environments instead of to air or oxygen-containing environments, fatigue crack growth rates will be substantially lower than those expected on the basis of tests in air.

Figure 35(b) shows the effects of humid air environments on the room-temperature fatigue crack growth rates of specimens of annealed type 304 stainless steel, cycled at 0.17 Hz with an R ratio of zero. At the lower end of the ΔK range, fatigue crack growth rates in humid air are substantially greater than crack growth rates in dry air.

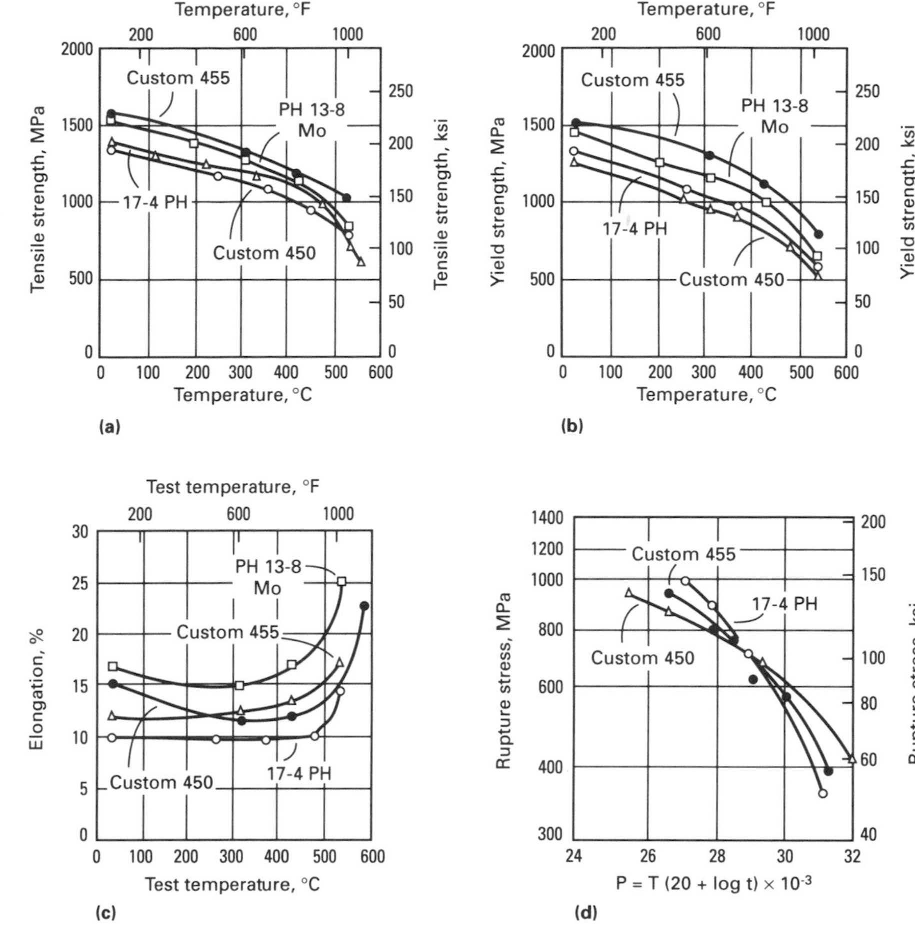

Fig. 36 Comparison of mechanical properties of precipitation-hardening martensitic stainless steels. (a) Tensile strength. (b) Yield strength. (c) Elongation. (d) Rupture strength. Heat-treating schedules were as follows. Custom 450: 1 h at 1040 °C (1900 °F), water quench; then 4 h at 480 °C (900 °F), air cool. 17-4 PH: 30 min at 1040 °C (1900 °F), oil quench; then 4 h at 480 °C (900 °F), air cool. Custom 455: 30 min at 815 °C (1500 °F), water quench; then 4 h at 510 °C (950 °F), air cool. PH 13-8Mo: oil quenched from 925 °C (1700 °F); then 4 h at 540 °C (1000 °F), air cool

Applications that use the heat-resisting capabilities of austenitic stainless steels to advantage include furnace parts, heat exchanger tubing, steam lines, exhaust systems in reciprocating engines and gas turbines, afterburner parts, and similar parts that require strength and oxidation resistance.

Type 304 has good resistance to atmospheric corrosion and oxidation. Types 309 and 310 rank higher in these properties because of their higher nickel and chromium contents. Type 310 is useful where intermittent heating and cooling are encountered, because it forms a more adherent scale than type 309. Types 309 and 310 are used for parts such as firebox sheets, furnace linings, boiler baffles, thermocouple wells, aircraft cabin heaters, and jet engine burner liners.

Table 8 Stress-rupture properties of A-286 PH austenitic stainless steel

See Table 7 for heat treatment conditions.

Test temperature		Stress for rupture					
		100 h		Elongation in $4D$,	1000 h		Elongation in $4D$,
°C	°F	MPa	ksi	%	MPa	ksi	%
538	1000	683	99	3.0	607	88	3.0
593	1100	562	81.5	3.0	493	71.5	3.0
649	1200	421	61	5.0	317	46	8.5
704	1300	307	44.5	12.0	200	29	30.0
732	1350	241	35	29.0	145	21	35.0
816	1500	90	13	55.0	53	7.7	...

Source: Carpenter Technology Corp.

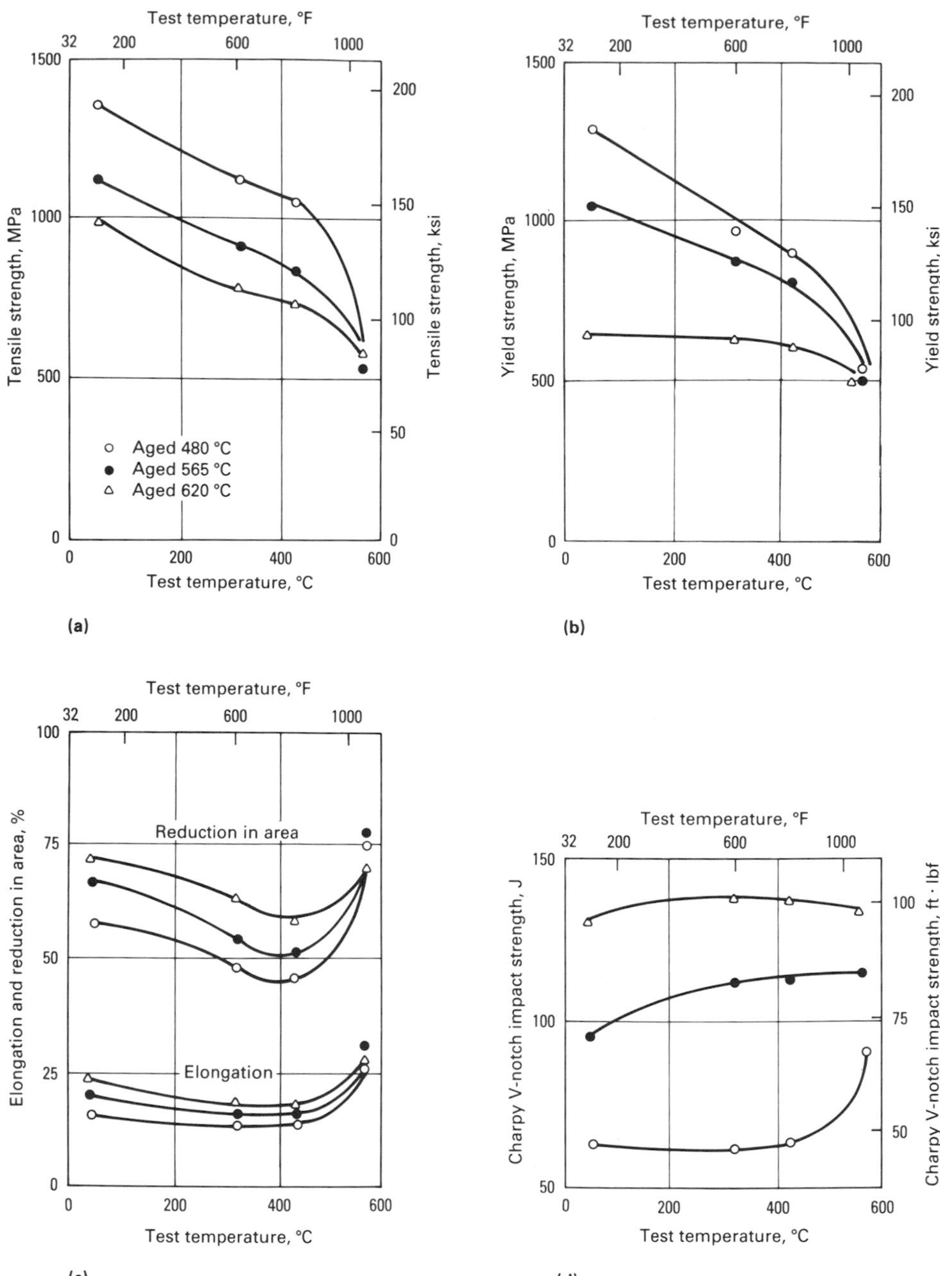

Fig. 37 Effect of temperature on (a) tensile strength, (b) yield strength, (c) elongation, and (d) impact toughness of Custom 450. Material used for testing was round bar stock, 25 mm (1 in.) in diameter, that had been solution treated by heating 1 h at 1040 °C (1900 °F) and water quenching.

The austenitic stainless steels may become susceptible to intergranular corrosion in some environments, usually limited to aqueous environments. This occurs when these steels are exposed to temperatures between 500 and 900 °C (950 and 1650 °F) and carbon diffuses to the grain boundaries to form chromium carbides. The formation of these carbides can significantly reduce the amount of free chromium in the alloy, thus impairing the corrosion resistance of the grain boundaries. The intergranular corrosion that occurs in the heat-affected zones of welds is often called weld knife-edge attack. Some types are resistant to intergranular corrosion, including types 321 and 347 (when correctly heat treated to form titanium or niobium carbides, respectively) and the naturally low-carbon grades 304L and 316L.

When types 321H and 347H are used in applications where intergranular corrosion may be a problem, it is possible to apply an additional stabilizing heat treatment at approximately 900 °C (1650 °F) to produce stable titanium and niobium carbides and thus reduce the free carbon content.

Types 321 and 347 can be used where solution treatment after welding is not feasible, such as in steam lines, superheater tubes, and exhaust systems in reciprocating engines and gas turbines that operate at 425 to 870 °C (800 to 1600 °F). The low-carbon types 304L and 316L are used for similar applications, but they are more susceptible to intergranular attack during long exposure to high temperatures.

Type 316 has better mechanical properties than types 304 or 321 and is more resistant to corrosion in some media, such as fatty acids at elevated temperatures and mild sulfuric acid solutions. The tensile and yield strengths of types 304, 304L, 316, and 316L can be increased by alloying these grades with nitrogen. These modifications are designated as 304N (UNS S30451), 304LN (UNS S30454), 316N (UNS S31651), and 316LN (UNS S31653).

Duplex Stainless Steels

The duplex steels are susceptible to 475 °C (885 °F) embrittlement at 370 to 480 °C (700 to 900 °F), and they are susceptible to σ-phase formation at 650 to 815 °C (1200 to 1500 °F). For this reason, they are not very suitable for high-temperature applications and are generally used in applications up to 300 °C (570 °F).

The thermal cycling properties of a duplex stainless steel, 23Cr-5Ni-1.5Mo, was compared with those of a standard austenitic of type 316 (Ref 25). Different thermal cycling tests at temperatures up to 1125 °C (2060 °F) showed remarkable acceleration of failures in the duplex steel as compared with the austenitic steel. This is related to the large differences in internal stresses between the two phases in the duplex steel and to an extensive grain growth during thermal cycling.

Thermal cycling of a 24Cr-4Ni-1.3Si duplex steel between room temperature and 900 °C (1650 °F) raised microstresses varying from grain to grain (Ref 26). The plastification caused an accumulating change of the shape of the specimen, which gave high residual stresses and internal cracks and damage.

The creep-fatigue behavior of the duplex steel 2205 (22Cr-5.5Ni-3Si) was studied in a sulfur-containing environment of Ar + 3% SO_2 at 700 °C (Ref 27). Severe sulfidation attack occurred at the external surface of both 2205 duplex and 316 austenitic stainless steel under a combination of creep-fatigue loading and atmosphere. However, the attack on the duplex steel was less severe than the attack on the austenitic stainless steel.

Precipitation-Hardening Martensitic Stainless Steels

The PH martensitic stainless steels fill an important position between the chromium-free 18% Ni maraging steels and the 12% Cr, low-nickel, quenched and tempered martensitic stainless alloys. These PH alloys contain 12 to 16% Cr for

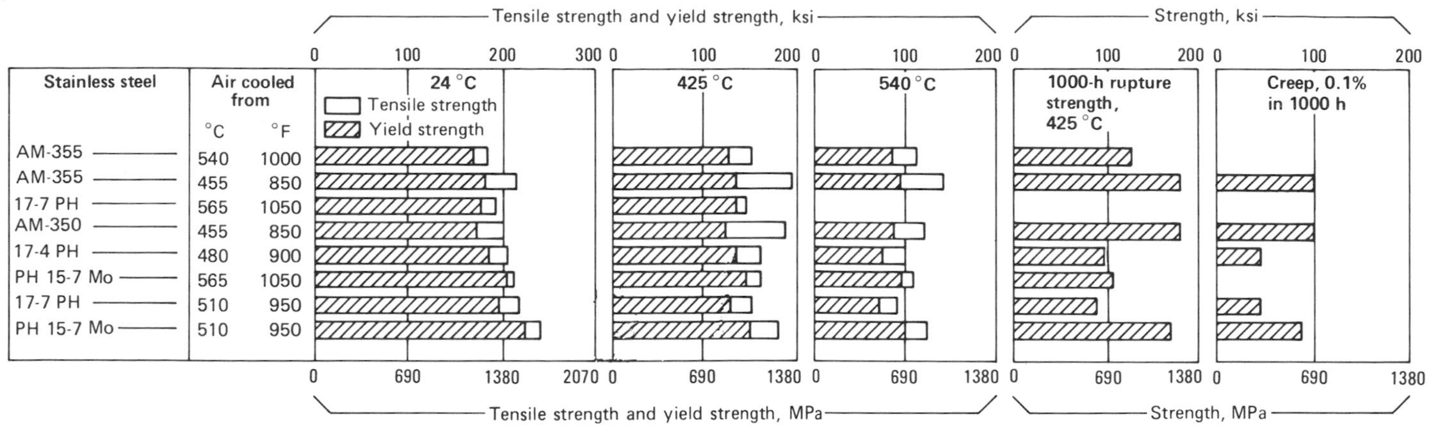

Fig. 38 Short-time tensile, rupture, and creep properties of PH stainless steels. AM-355 was finish hot worked from a maximum temperature of 980 °C (1800 °F), reheated to 930 to 955 °C (1710 to 1750 °F), water quenched, treated at –75 °C (–100 °F), and aged at 540 and 455 °C (1000 and 850 °F). 17-7 PH and PH 15-7 Mo were solution treated at 1040 to 1065 °C (1900 to 1950 °F). 17-7PH (TH1050) and PH 15-7 Mo (TH1050) were reheated to 760 °C (1400 °F), air cooled to 15 °C (60 °F) within 1 h, and aged 90 min at 565 °C (1050 °F). 17-7 PH (RH950) and PH 15-7 Mo (RH950) were reheated to 955 °C (1750 °F) after solution annealing, cold treated at –75 °C (–100 °F), and aged at 510 °C (950 °F). 17-4 PH was aged at 480 °C (900 °F) after solution annealing. AM-350 was solution annealed at 1040 to 1065 °C (1900 to 1950 °F), reheated to 930 °C (1710 °F), air cooled, treated at –75 °C (–100 °F), and aged at 455 °C (850 °F).

corrosion resistance and scaling resistance at elevated temperatures.

Heat Treatment. The PH martensitic alloys are normally purchased in the solution-annealed condition. Depending on the application, they may be used in the annealed condition or in the annealed plus age hardened condition. In some cases, material will be supplied in an overaged condition to facilitate the forming of parts. The formed parts are then solution annealed following fabrication.

Properties. The PH martensitic alloys listed in Table 1 include Custom 450, 17-4 PH, 15-5 PH, Custom 455, and PH 13-8 Mo. Property data for these alloys are shown in Fig. 36. Data for 15-5 PH are not shown separately because the properties of this alloy are very similar to those of 17-4 PH.

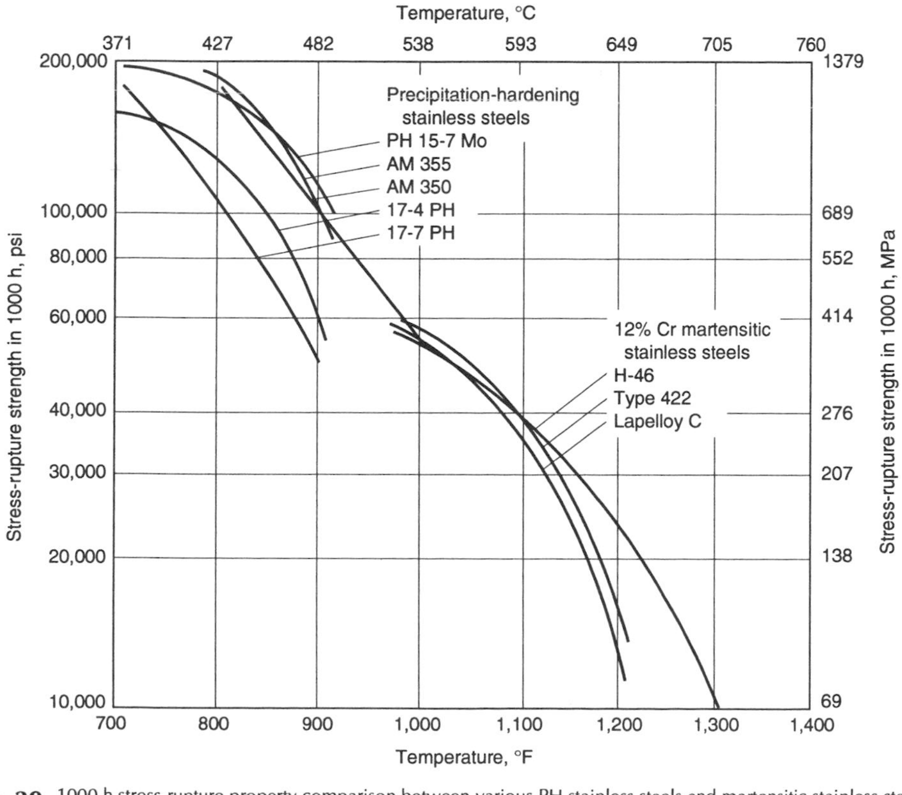

Short-time tensile data indicate that Custom 455 and PH 13-8 Mo have higher strengths than Custom 450, 17-4 PFC, or 15-5 PH. For all of these alloys, tensile and yield strengths drop rapidly at temperatures above 425 °C (800 °F), and tensile elongation is greater than 10% over the temperature range from ambient to 540 °C (1000 °F).

Stress-rupture data are compared in Fig. 36 by means of a Larson-Miller plot. Data were developed at testing temperatures of 425 and 480 °C (800 and 900 °F) over time periods of 100 and 1000 h. It should be noted that 17-4 PH appears to have better stress-rupture strength at 425 °C (800 °F), whereas Custom 450 is superior in this respect at 480 °C (900 °F). The stress-rupture data available for Custom 455 appear to indicate that the alloy is intermediate in rupture strength between 17-4 PH and Custom 450.

It is possible to produce a wide variety of useful properties in a given alloy by varying the aging temperature. An example of this can be seen in Fig. 37, where tensile strength, yield strength (0.2% offset), ductility, and impact data are shown for Custom 450 at three different aging temperatures. Aging at 480 °C (900 °F) can produce significant strengthening at testing temperatures as high as 450 °C (850 °F), but it also results in lower toughness values (as measured by Charpy V-notch testing) than aging at either of the two higher temperatures.

Applications. Precipitation-hardening martensitic stainless steels are used for short-time elevated temperature exposures in industrial and military applications when it is necessary to have resistance to corrosion and high mechanical properties at temperatures up to 425 °C (800 °F). Typical uses include valve parts, ball bearings, forgings, turbine blades, mandrels, conveyor chain, miscellaneous hardware, and mechanical and structural components for aircraft.

Fig. 39 1000 h stress-rupture property comparison between various PH stainless steels and martensitic stainless steels. Source: Ref 17

Precipitation-Hardening Semiaustenitic Stainless Steels

The PH semiaustenitic heat-resistant stainless steels are modifications of standard 18-8 austenitic stainless steels. Nickel contents are lower, and such elements as aluminum, copper, molybdenum, and niobium are added. These steels are used at temperatures up to 480 °C (900 °F).

Heat Treatment. Typical schedules for heat treating PH semiaustenitic alloys are given in Table 6. These alloys are solution annealed above 1046 °C (1900 °F). In this condition they can be formed, stamped, stretched, and otherwise cold worked to about the same extent as 18-8 alloys, although they are less ductile and may require intermediate annealing.

All the semiaustenitic stainless steels can also be used in the cold-worked condition in either sheet or wire form. Cold working causes partial transformation of the rather unstable austenite to martensite because of plastic deformation. Aging or tempering is performed after cold working.

Properties. Typical short-time tensile, rupture, and creep properties of several PH semiaustenitic alloys are compared in Fig. 38 and 39. Different hardening heat treatments may produce a wide variety of useful properties for the same alloy. For example, the PH 15-7 Mo alloy treated to condition RH950 has a higher rupture strength than the same alloy treated to TH1050 (Fig. 40). Strengths also degrade after long-term exposure at elevated temperatures because of averaging (coarsening) of precipitates.

Compressive and tensile yield strengths are approximately equal for all PH semiaustenitic stainless steels. For sheet, the yield strengths of specimens taken transverse and parallel to the direction of rolling may vary appreciably. The magnitude of this effect varies from grade to grade and with the heat treatment for a given grade.

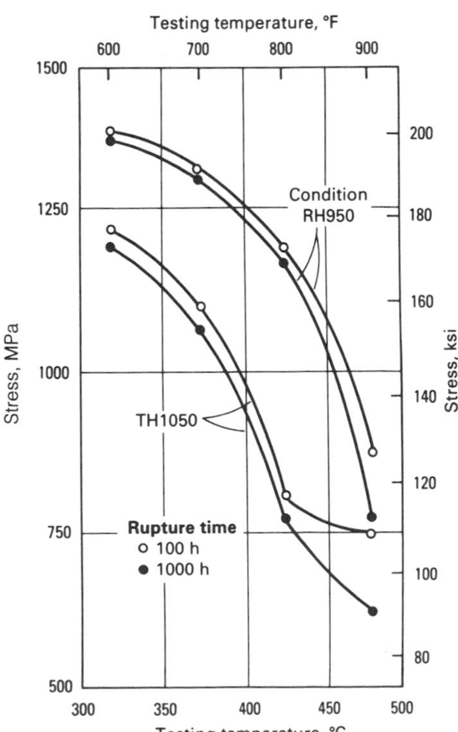

Fig. 40 Stress-rupture curves of PH 15-7 Mo stainless steel in the TH1050 and RH950 conditions

Applications. The PH semiaustenitic stainless steels are used for industrial and military applications that require resistance to corrosion as well as high mechanical properties at temperatures up to 425 °C (800 °F). Typical uses of these steels include landing-gear hooks, poppet valves, fuel tanks, hydraulic lines, hydraulic fittings, compressor casings, miscellaneous hardware, and structural components for aircraft. The higher-carbon grade (0.13% C) is used for compressor blades, spacers, frames and casings for gas turbines, oil well drill rods, and rocket casings.

Precipitation-hardening semiaustenitic steels can be cold rolled and tempered. Work-hardening rates are higher than for type 301 stainless steel and can be varied by regulating the annealing temperature. Compared with cold-rolled type 301, cold-rolled PH semiaustenitic stainless steels have higher ductility at a given strength level, higher modulus of elasticity in compression in the rolling direction, and less reduction in strength with increasing temperature.

Precipitation-Hardening Austenitic Stainless Steels

Alloy A-286 is the prototype PH austenitic stainless steel. Its high alloy content (15Cr-25Ni-1Mo-2Ti) provides higher corrosion resistance than that of the martensitic or semiaustenitic types. Furthermore, it contains no magnetic phases, which makes it suitable for use in high magnetic fields, such as those associated with superconducting magnets used for fusion energy research. Aging at about 730 °C (1350 °F) causes nickel-titanium intermetallic compound particles to precipitate, hardening the alloy.

Properties. The fully hardened yield strength of A-286 is approximately 590 MPa (86 ksi), which is considerably lower than that available from the martensitic or semiaustenitic grades. Alloy A-286 can, however, be used at temperatures up to 700 °C (1300 °F). Table 7 provides elevated-temperature tensile properties for A-286. Stress-rupture values are given in Table 8.

Applications. Alloy A-286 has been used in jet engines, superchargers, and various high-temperature applications, such as turbine wheels and blades, frames, casings, afterburner parts, and fasteners.

ACKNOWLEDGMENTS

The information in this article is largely taken from:

- Elevated-Temperature Properties, *ASM Specialty Handbook: Stainless Steels,* J.R. Davis, Ed., ASM International, 1994
- S. Lampman, Fatigue and Fracture Properties of Stainless Steels, *Fatigue and Fracture,* Vol 19, *ASM Handbook,* ASM International, 1996

REFERENCES

1. W.F. Simmons and J.A. Van Echo, *Report on the Elevated-Temperature Properties of Stainless Steels,* ASTM Data Series, Publication DS-5-S1 (formerly STP 124), ASTM
2. R. Viswanathan, *Damage Mechanism and Life Assessment of High-Temperature Components,* ASM International, 1989
3. J. Wareing, *Met. Trans. A,* Vol 8, 1977, p 711–721
4. C.R. Brinkman, G.E. Korth, and R.R. Hobbins, *Nucl. Tech.,* Vol 16, 1972, p 299–307
5. Y. Asada and S. Mitsuhaski, *Fourth International Conference on Pressure Vessel Technology,* Vol 1, 1980, p 321
6. C.R. Brinkman and G.E. Korth, *Met. Trans.,* Vol 5, 1974, p 792
7. J. Wareing, *Met. Trans. A,* Vol 6, 1975, p 1367
8. J. Wareing, H.G. Vaughan, and B. Tomkins, Report NDR-447S, United Kingdom Atomic Energy Agency, 1980
9. I.W. Goodall, R. Hales, and D.J. Walters, *Proceedings of IUTAM 103,* International Union of Theoretical and Applied Mechanics, 1980
10. D.S. Wood, J. Wynn, A.B. Baldwin, and P. O'Riordan, *Fatigue Eng. Mater. Struct.,* Vol 3, 1980, p 89
11. J. Wareing, *Fatigue Eng. Mater. Struct.,* Vol 4, 1981, p 131
12. C.E. Jaske, M. Mindlin, and J.S. Perrin, Development of Elevated Temperature Fatigue Design Information for Type 316 Stainless Steel, *International Conference on Creep and Fatigue,* Conference Publication 13, Institute of Mechanical Engineers, 1973, p 163.1–163.7
13. S. Majumdar and P.S. Maiya, *J. Eng. Mater. Technol., (Trans. ASME),* Vol 102 (No. 1), 1980, p 159
14. V.B. Livesey and J. Wareing, *Met. Sci.,* Vol 17, 1983, p 297
15. D. Gladwin and D.A. Miller, *Fatigue Eng. Mater. Struct.,* Vol 5, 1982, p 275–286
16. D.A. Miller, R.E. Priest, and E.G. Ellison, A Review of Material Response and Life Prediction Techniques under Fatigue-Creep Loading Conditions, *High Temp. Mater. Process.,* Vol 6 (No. 3 and 4), 1984, p 115–194
17. F.J. Claus, Stainless Steels, *Engineer's Guide to High-Temperature Materials,* Addison-Wesley, 1969, p 86–128
18. *Elevated-Temperature Properties as Influenced by Nitrogen Additions to Types 304 and 316 Austenitic Stainless Steels,* STP 522, ASTM, 1973
19. P. Marshall, *Austenitic Stainless Steels Microstructure and Mechanical Properties,* Elsevier, 1984
20. C.R. Brinkman, High-Temperature Time-Dependent Fatigue Behavior of Several Engineering Structural Alloys, *Int. Met. Rev.,* Vol 30 (No. 5), 1985, p 235–258

21. D.J. Michel and H.H. Smith, NRL-MR-3627, Naval Research Laboratory, Oct 1977
22. D. Michel and H. Smith, *Acta Met.,* Vol 28, 1980, p 999
23. D. Michel and H. Smith, "Effect of Hold Time and Thermal Aging on Elevated Temperature Fatigue Crack Propagation in Austenitic Stainless Steels," Report NRL-MR-3627, Naval Research Laboratory, Washington, Oct 1977
24. *Application of Fracture Mechanics,* American Society for Metals, 1982, p 105–169
25. S.K. Kamachie et al., Thermal Fatigue by Impact Heating and Stresses of Two Phase Stainless Steels at Elevated Temperature, *Progress in Science and Engineering of Composites, ICCM-IV,* Tokyo, 1982, p 1383–1389
26. F.D. Fischer, F.G. Rammerstorfer, and F.J. Bauer, Fatigue and Fracture of High-Alloyed Steel Specimens Subjected to Purely Thermal Cycling, *Met. Trans. A,* Vol 21, April 1990, p 935–948
27. E. Aghion and C.A. Molaba, Creep-Fatigue Failure of SAF 2205 and 316 Stainless Steels in Ar + 3% SO_2 Environment at 700 °C, *J. Mater. Sci.,* Vol 29, 1994, p 1758–1764

SELECTED REFERENCES

- *Austenitic Chromium-Nickel Stainless Steels: Engineering Properties at Elevated Temperatures,* INCO Europe Limited, 1963
- *High Temperature Characteristics of Stainless Steels,* American Iron and Steel Institute, 1979
- R.A. Lula, *Source Book on the Ferritic Stainless Steels,* American Society for Metals, 1982
- P. Marshall, *Austenitic Stainless Steels Microstructure and Mechanical Properties,* Elsevier, 1984
- T.D. Parker, Strength of Stainless Steels at Elevated Temperature, *Selection of Stainless Steels,* American Society for Metals, 1968, p 47–66
- G.V. Smith, *Evaluations of the Elevated Temperature Tensile and Creep Rupture Properties of 12 to 27 Percent Chromium Steels,* DS 59, ASTM, 1980
- R. Viswanathan, *Damage Mechanisms and Life Assessment of High-Temperature Components,* ASM International, 1989

Elevated-Temperature Corrosion Properties of Carbon and Alloy Steels

HIGH-TEMPERATURE CORROSION plays an important role in the selection of steels for such key industries as petroleum refining, petrochemical processing, and fossil fuel power generation. This article reviews the more common forms of elevated-temperature corrosion that directly influence the maximum allowable service temperature of carbon and alloy steels including:

- Oxidation
- Sulfidation (sulfidic corrosion)
- Hydrogen attack
- Ash/salt deposit corrosion
- Molten salt corrosion
- Molten metal corrosion

Emphasis has been placed on how these modes of corrosion influence the properties of steels. The thermodynamics and reaction kinetics associated with elevated-temperature corrosive reactions, as well as the fundamental mechanisms of scale formation, are not addressed in detail. Such information, however, can be found in the article "Corrosion at Elevated Temperatures" in this Volume.

Oxidation

Oxidation from steam or air is a serious problem that can occur at elevated temperatures. When metal is exposed to an oxidizing gas at elevated temperature, corrosion can occur by direct reaction with the gas. This type of corrosion is referred to as tarnishing, high-temperature oxidation, or scaling. The rate of attack increases substantially with temperature. The surface film typically thickens as a result of reaction at the scale/gas or metal/scale interface due to cation or anion transport through the scale, which behaves as a solid electrolyte.

Steels intended for high-temperature applications are designed to have the capability of forming protective oxide scales. Alloying requirements for the production of specific oxide scales have been translated into minimum levels of the scale-forming elements, or combinations of elements, depending on the base alloy composition and the intended service temperature. Figure 1 schematically represents the oxidation rate of iron-chromium alloys (1000 °C, or 1830 °F, in 0.13 atm oxygen) and depicts the types of oxide scale associated with various alloy types.

Carbon versus Chromium-Molybdenum Steels. Carbon steel is probably the most widely used engineering material. It is extensively used for high-temperature applications in power generation, chemical and petrochemical processing, oil refining, industrial heating, metallurgical processing, and other industries. Components made of carbon steel include boiler tubes in power plants and pulp and paper recovery boilers, reactor vessels in the chemical process industry, hot-dip galvanizing tanks, vessels or tanks for holding molten metals, heat treat fixtures, and automotive exhaust train piping, just to name a few.

At temperatures below 570 °C (1060 °F), iron oxidizes to form Fe_3O_4 and Fe_2O_3. Above 570 °C (1060 °F), it oxidizes to form FeO, Fe_3O_4, and Fe_2O_3, as shown in Fig. 1. The oxidation behavior of carbon steel in air at 430, 540, 650, and 760 °C (800, 1000, 1200, and 1400 °F) is summarized in Fig. 2. After 700 h of exposure at 430 and 540 °C (800 and 1000 °F), carbon steel suffered negligible oxidation attack, with less than 20 mg/cm^2 of weight loss. As the temperature was increased to 650 °C (1200 °F), the oxidation rate was sig-

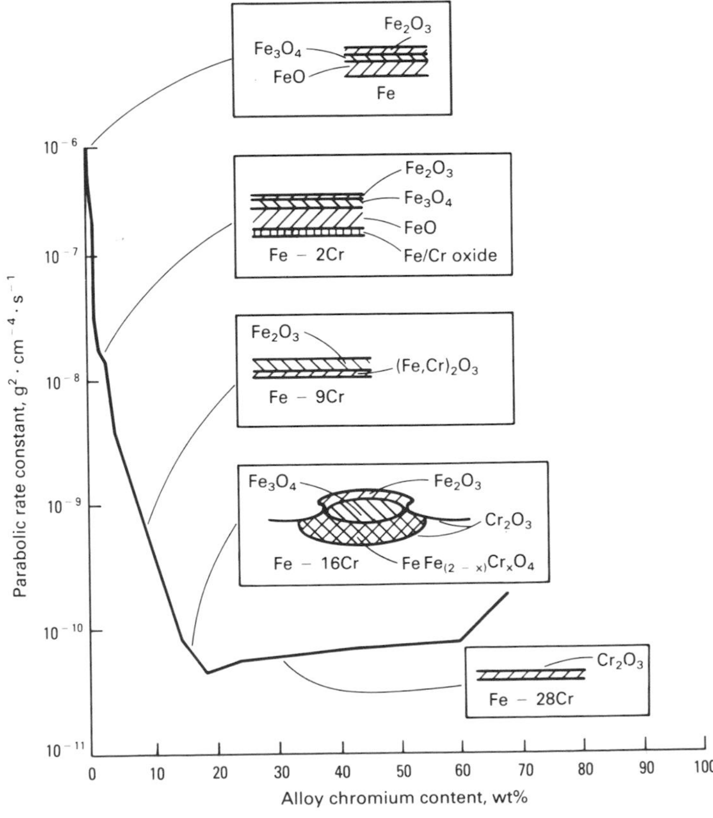

Fig. 1 Schematic of the variation of the oxidation rate and oxide scale structure (based on isothermal studies at 1000 °C, or 1830 °F, in 0.13 atm oxygen) of steels with varying chromium contents

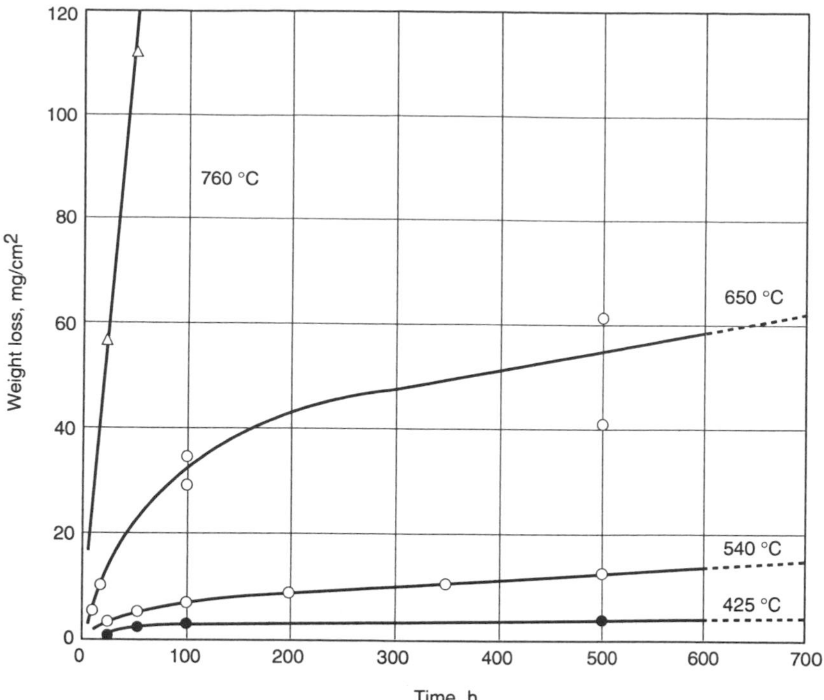

Fig. 2 Oxidation behavior of plain low-carbon steel in air at 430, 540, 650, and 760 °C (800, 1000, 1200, and 1400 °F). Source: Ref 1

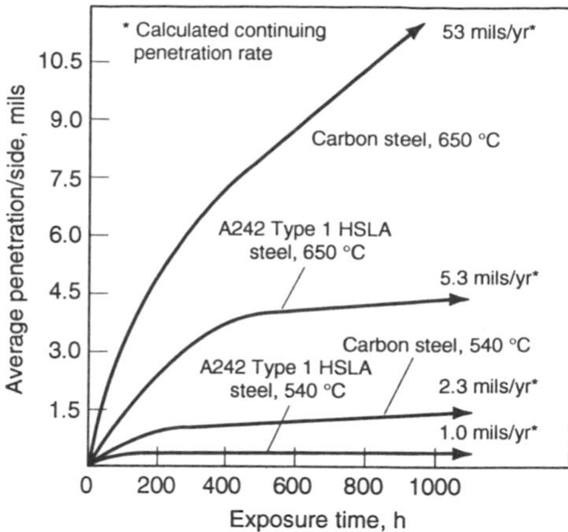

Fig. 3 Oxidation of carbon steel and high-strength low-alloy (HSLA) steel in air. Source: Ref 2, 3

nificantly increased. At 760 °C (1400 °F), carbon steel suffered rapid oxidation, exhibiting essentially a linear rate of attack. These results suggest that carbon steel may not be suitable in air or highly oxidizing atmospheres at temperatures in excess of 650 °C (1200 °F). This was substantiated by the results of Vrable et al. (Ref 2), as illustrated in Fig. 3. At 650 °C (1200 °F), carbon steel suffered an oxidation rate of about 1.3 mm/year, or 53 mils/year. The oxidation rate is expected to be much higher when carbon steel is exposed to temperatures higher than 650 °C (1200 °F). Figure 3 also illustrates that high-strength low-alloy (HSLA) steel is significantly better than carbon steel in oxidation resistance, presumably due to minor alloying elements such as manganese, silicon, chromium, and nickel.

Chromium-molybdenum steels are used at higher temperatures than carbon steel because of higher tensile and creep-rupture strengths. Molybdenum and chromium provide not only solid solution strengthening but also carbide strengthening. Some of the major applications for chromium-molybdenum steels include superheater tubes and steam pipes.

Low-alloy steels with chromium and silicon additions exhibit better oxidation resistance than carbon steel. The beneficial effects of chromium and silicon additions to carbon steel are summarized in Fig. 4. Silicon is very effective in improving the oxidation resistance of chromium-molybdenum steels. Addition of 1.5% Si to 5Cr-0.5Mo steel significantly improved its oxidation resistance, although silicon additions also reduce creep strength and may promote temper embrittlement when other impurities are present. The most important alloying element for improving oxidation resistance is chromium. As shown in Fig. 4, for 0.5% Mo-containing steels, increasing chromium from 1 to 9% significantly increases oxidation resistance. The 7Cr-0.5Mo and 9Cr-1Mo steels showed negligible oxidation rates at temperatures up to 680 °C (1250 °F) and 700 °C (1300 °F), respectively. Further increases in chromium improve oxidation resistance even more (see the article "Elevated-Temperature Corrosion Properties of Stainless Steels" in this Volume).

Factors Influencing Oxidation Rates. The scaling data given in Fig. 4 were obtained in the presence of air. If other variables affecting oxidation are changed, such as gas composition, heating method, temperature, pressure, or velocity, different rates of scaling can be expected. Elements such as sulfur, vanadium, and sodium can change the nature of metal oxidation, sometimes increasing it to a catastrophic level of several inches per year.

At elevated temperatures, steam decomposes at metal surfaces to hydrogen and oxygen and may cause steam oxidation of steel, which is somewhat more severe than air oxidation at the same temperature. Fluctuating steam temperatures tend to increase the rate of oxidation by causing scale to spall, which exposes fresh metal to further attack. Table 1 gives the maximum-use temperatures for several boiler alloys for which code standards exist. The strength criteria are based on the wall midsection temperatures, which are typically 25 °C (45 °F) lower than the outer surface temperature.

In a water environment, corrosion is significantly influenced by the concentrations of dissolved species, pH, temperature, suspended particles, and bacteria. Temperature plays a dual role with respect to oxygen corrosion. Increasing the temperature will reduce oxygen solubility. In open systems, in which oxygen can be released from the system, corrosion will increase up to a maximum at 80 °C (175 °F), where the oxygen solubility is 3 mg/L. Beyond this temperature, the reduced oxygen content limits the oxygen reduction reaction, preventing occurrence of the iron dissolution process. Thus, the corrosion rate of carbon steel decreases, and at boiling water conditions, the temperature effect is similar to room temperature with a high oxygen content. For closed systems, in which oxygen cannot escape, corrosion continues to increase linearly with temperature. The other factors affected by temperature are the diffusion of oxygen to the metal surface, the viscosity of water, and solution con-

Table 1 Temperature limits of superheater tube materials covered in ASME Boiler Codes

	Maximum-use temperature			
	Oxidation/graphitization criteria, metal surface(a)		Strength criteria, metal midsection	
Material	°C	°F	°C	°F
SA-106 carbon steel	400–500	750–930	425	795
Ferritic alloy steels				
0.5Cr-0.5Mo	550	1020	510	950
1.2Cr-0.5Mo	565	1050	560	1040
2.25Cr-1Mo	580	1075	595	1105
9Cr-1Mo	650	1200	650	1200
Austenitic stainless steel, type 304H	760	1400	815	1500

(a) In the fired section, tube surface temperatures are typically 20-30 °C (35-55 °F) higher than the tube midwall temperature. In a typical U.S. utility boiler, the maximum metal surface temperature is approximately 625 °C (1155 °F).

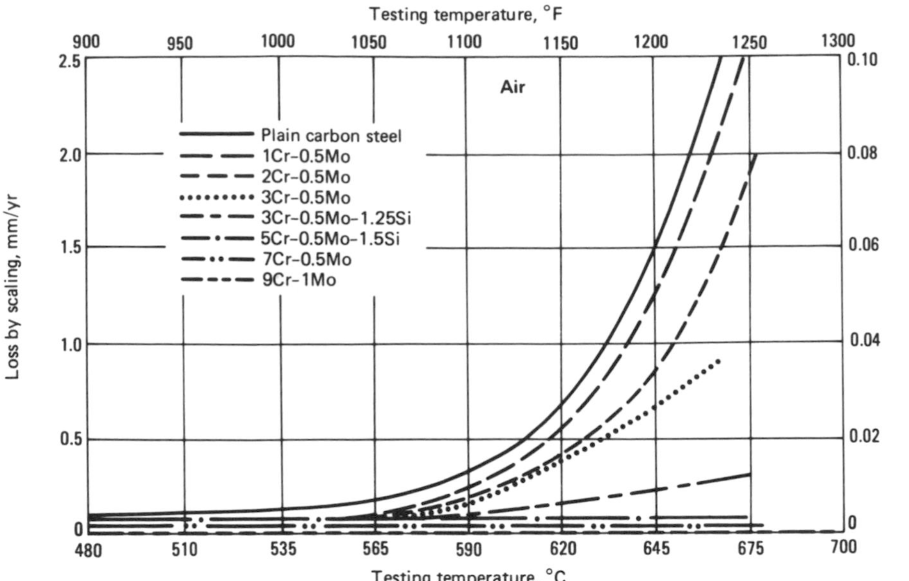

Fig. 4 Effect of chromium and/or silicon on the oxidation resistance of steels in air. Source: Ref 4

ductivity. Increasing the temperature will increase the rate of oxygen diffusion to the metal surface, thus increasing corrosion rate because more oxygen is available for the cathodic reduction process. The viscosity will decrease with increasing temperature, which will aid oxygen diffusion.

Sulfidation

Corrosion by various sulfur compounds at temperatures between 260 and 540 °C (500 and 1000 °F) is a common problem in many petroleum-refining processes and, occasionally, in petrochemical processes. Sulfur compounds originate with crude oils and include polysulfides, hydrogen sulfide, mercaptans, aliphatic sulfides, disulfides, and thiophenes (Ref 5). With the exception of thiophenes, sulfur compounds react with metal surfaces at elevated temperatures, forming metal sulfides, certain organic molecules, and hydrogen sulfide (Ref 6, 7). The relative corrosivity of sulfur compounds generally increases with temperature. Depending on the process particulars, corrosion is in the form of uniform thinning, localized attack, or erosion-corrosion. Corrosion control depends almost entirely on the formation of protective metal sulfide scales that exhibit parabolic growth behavior (Ref 8). The combination of hydrogen sulfide and hydrogen can be particularly corrosive. As described below, carbon and alloy steels at times do not provide adequate sulfidation resistance, and austenitic stainless steels are required for effective corrosion control.

Sulfidation Without Hydrogen Present. This type of corrosion occurs primarily in various components of crude distillation units, catalytic cracking units, and hydrotreating and hydrocracking units upstream of the hydrogen injection line. Crude distillation units that process mostly sweet crude oils (less than 0.6% total sulfur, with essentially no hydrogen sulfide) experience relatively few corrosion problems. Preheat-exchanger tubes, furnace tubes, and transfer lines are generally made from carbon steel, as is corresponding equipment in the vacuum distillation section. The lower shell of distillation towers, where temperatures are above 230 °C (450 °F), is usually lined with stainless steel containing 12% Cr, such as type 405 (S40500). This prevents impingement attack under the highly turbulent flow conditions encountered, for example, near downcomers. For the same reason, trays are made of stainless steel containing 12% Cr. Even with low corrosion rates of carbon steel, certain tray components, such as tray valves, may fail in a short time because attack occurs from both sides of a relatively thin piece of metal.

Crude distillation units that process mostly sour crude oils require additional alloy protection against high-temperature sulfidic corrosion. The extent of alloying needed also depends on the design and the operating practices of a given unit. Typically, such units require low-alloy steels containing a minimum of 5% Cr for furnace tubes, headers and U-bends, and elbows and tees in transfer lines. In vacuum furnaces, tubes made from chromium steels containing 9% Cr are often used. Distillation towers are similar to those of units that process mostly sweet crude oils. Where corrosion problems persist, upgrading with steels containing a greater amount of chromium is indicated.

The high processing temperatures encountered in the reaction and catalyst regeneration section of catalytic cracking units require extensive use of refractory linings to protect all carbon steel components from oxidation and sulfidation. Refractory linings also provide protection against erosion by catalyst particles, particularly in cyclones, risers, standpipes, and slide valves. Cobalt-base hard-facing is used on some components to protect against erosion. When there are no erosion problems and when protective linings are impractical, austenitic stainless steels, such as type 304 (S30400), can be used. Cyclone dip legs, air rings, and other internals in the catalyst regenerator are usually made from type 304 (S30400) stainless steel, as is piping for regenerator flue gas. Reactor feed piping is made from low-alloy steel, such as 5Cr-0.5Mo or 9Cr-1Mo, to control high-temperature sulfide corrosion.

The main fractionation tower is usually made of carbon steel, with the lower part lined with stainless steel containing 12% Cr, such as type 405 (S40500) (Ref 9). Slurry piping between the bottom of the main fractionation tower and the reactor may receive an additional corrosion allowance as protection against excessive erosion. As a rule, there are few corrosion problems in the reaction, catalyst regeneration, and fractionation sections (Ref 10).

Hydrocracking and hydrotreating units usually require alloy protection against both high-temperature sulfidic corrosion and high-temperature hydrogen attack (Ref 11, 12). Low-alloy steels may be required for corrosion control ahead of the hydrogen injection line.

The so-called McConomy curves can be used to predict the relative corrosivity of crude oils and their various fractions (Ref 13). Although this method relates corrosivity to total sulfur content and thus does not take into account the variable effects of different sulfur compounds, it can provide reliable corrosion trends if certain corrections are applied. Plant experience has shown that the McConomy curves, as originally published, tend to predict excessively high corrosion rates. The curves apply only to liquid hydrocarbon streams containing 0.6 wt% S (unless a correction factor for sulfur content is applied) and do not take into account the effects of vaporization and flow regime. The curves can be particularly useful, however, for predicting the effect of operational changes on known corrosion rates.

Over the years, it has been found that corrosion rates predicted by the original McConomy curves should be decreased by a factor of roughly 2.5, resulting in the modified curves shown in Fig. 5. The curves demonstrate the beneficial effects of alloying steel with chromium in order to reduce corrosion rates. Corrosion rates are roughly halved when the next higher grade of low-alloy steel (for example, 2.25Cr-1Mo, 5Cr-0.5Mo, 7Cr-0.5Mo, or 9Cr-1Mo steel) is selected. Essentially, no corrosion occurs with stainless steels containing 12% or more chromium. Although few data are available, plant experience has shown that corrosion rates start to decrease as temperatures exceed 455 °C (850 °F). Two explanations frequently offered for this phenomenon are the possible decomposition of reactive sulfur compounds and the formation of a protective coke layer.

Metal skin temperatures, rather than stream temperatures, should be used to predict corrosion rates when significant differences between the

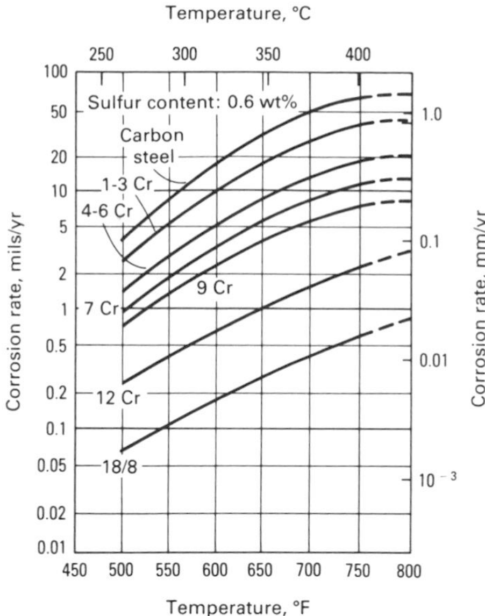

Fig. 5 Modified McConomy curves showing the effect of temperature on high-temperature sulfidic corrosion of various steels. Source: Ref 14

Fig. 6 High-temperature sulfidic corrosion of 150-mm (6-in.) diam carbon steel tube from radiant section of crude preheat furnace at crude distillation unit. Note accelerated attack on fire side.

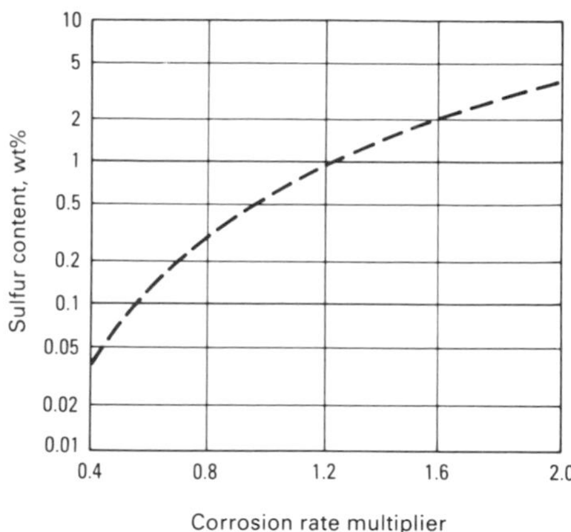

Fig. 7 Effect of sulfur content on corrosion rates predicted by modified McConomy curves in the temperature range of 290 to 400 °C (550 to 750 °F). Source: Ref 14

two arise. For example, metal temperatures of furnace tubes are typically 85 to 110 °C (150 to 200 °F) higher than the temperature of the hydrocarbon stream passing through the tubes. Furnace tubes normally corrode at a higher rate on the hot side (fire side) than on the cool side (wall side), as shown in Fig. 6. Convective-section tubes often show accelerated corrosion at contact areas with tube hangers because of locally increased temperatures. Similarly, replacement of bare convective-section tubes with finned or studded tubes can further increase tube metal temperatures by 85 to 110 °C (150 to 200 °F).

Correction factors for process streams with various total sulfur contents, average of those proposed originally by McConomy, are shown in Fig. 7. As can be seen, doubling the sulfur content can increase corrosion rates by approximately 30%. To allow for the fact that the proportion of noncorrosive thiophenes is greater in high-boiling cuts (and residuum) than in the original crude charge, a corrosion factor ranging from 0.5 to 1 may have to be applied to the total sulfur content so that realistic corrosion rates can be obtained for such cuts. The degree of vaporization and the resultant two-phase flow regimes can have a significant effect on high-temperature sulfidation.

Sulfidation With the Presence of Hydrogen. The presence of hydrogen increases the severity of high-temperature sulfidic corrosion. Hydrogen converts organic sulfur compounds to hydrogen sulfide; corrosion becomes a function of hydrogen sulfide concentration (or partial pressure).

A number of researchers have proposed various corrosion rate correlations for high-temperature sulfidic corrosion in the presence of hydrogen (Ref 15–21), but the most practical correlations seem to be the so-called Couper-Gorman curves. The Couper-Gorman curves are based on a survey conducted by National Association of Corrosion Engineers (NACE) Committee T-8 on Refining Industry Corrosion (Ref 22).

The Couper-Gorman curves differ from those previously published in that they reflect the influence of temperature on corrosion rates throughout a whole range of hydrogen sulfide concentrations. Total pressure was found not to be a significant variable between 1 and 18 MPa (150 and 2650 psig). It was also found that essentially no corrosion occurs at low hydrogen sulfide concentrations and temperatures above 315 °C (600 °F) because the formation of iron sulfide becomes thermodynamically impossible. Curves are available for carbon steel, Cr-0.5Mo steel, 9Cr-1Mo steel, 12% Cr stainless steel, and 18Cr-8Ni austenitic stainless steel. For the low-alloy steels, two sets of curves apply, depending on whether the hydrocarbon stream is naphtha or gas oil. The curves again demonstrate the beneficial effects of alloying steel with chromium to reduce the corrosion rate.

Modified Couper-Gorman curves are shown in Fig. 8. To facilitate the use of these curves, original segments of the curves were extended (dashed lines). In contrast to sulfidic corrosion in the absence of hydrogen, there is often no real improvement in corrosion resistance unless chromium content exceeds 5%. Therefore, the curves for 5Cr-0.5Mo steel also apply to carbon steel and low-alloy steels containing less than 5% Cr. Stainless steels containing at least 18% Cr are often required for essentially complete immunity to corrosion. Because the Couper-Gorman curves are primarily based on corrosion rate data for an all-vapor system, partial condensation can be expected to increase corrosion rates because of droplet impingement.

When selecting steels for resistance to high-temperature sulfidic corrosion in the presence of hydrogen, the possibility of high-temperature hydrogen attack should be considered. Conceivably, this problem arises when carbon steel and low-alloy steels containing less than 1% Cr are chosen for temperatures exceeding 260 °C (500 °F) and hydrogen partial pressures above 700 kPa (100 psia) and when corrosion rates are expected to be relatively low (see the discussion below on "Hydrogen Attack").

Hydrogen Attack

The term hydrogen attack (or, more specifically, high-temperature hydrogen attack) refers to the deterioration of the mechanical properties of steels in the presence of hydrogen gas at elevated temperatures and pressures. Hydrogen attack is potentially a very serious problem with regard to the design and operation of refinery equipment in hydrogen service (Ref 23, 24). It is of particular concern in hydrotreating, reforming, and hydrocracking units at above roughly 260 °C (500 °F) and hydrogen partial pressures above 689 kPa (100 psia) (Ref 25). Under these conditions, molecular hydrogen (H_2) dissociates at the steel surface to atomic hydrogen (H), which readily diffuses into the steel. At grain boundaries, dislocations, inclusions, gross discontinuities, laminations, and other internal voids, atomic hydrogen will react with dissolved carbon and with metal carbides to form methane. The large size of its molecule precludes methane diffusion. As a result, internal methane pressures become high enough to blister the steel or to cause intergranular fissuring (Ref 26). If temperatures are high enough, dissolved carbon diffuses to the steel surface and combines with atomic hydrogen to evolve methane. Hydrogen attack now takes the form of decarburization rather than blistering or cracking.

The overall effect of hydrogen attack is the partial depletion of carbon in pearlite (decarburization) and the formation of fissures in the metal, as shown in Fig. 9. As attack proceeds, these effects become more pronounced.

Damage to steels by high-temperature, high-pressure hydrogen is preceded by a period of time when no noticeable change in properties can be detected by the usual test methods. The length of time before hydrogen attack is detected is referred to as the incubation time. Consequently,

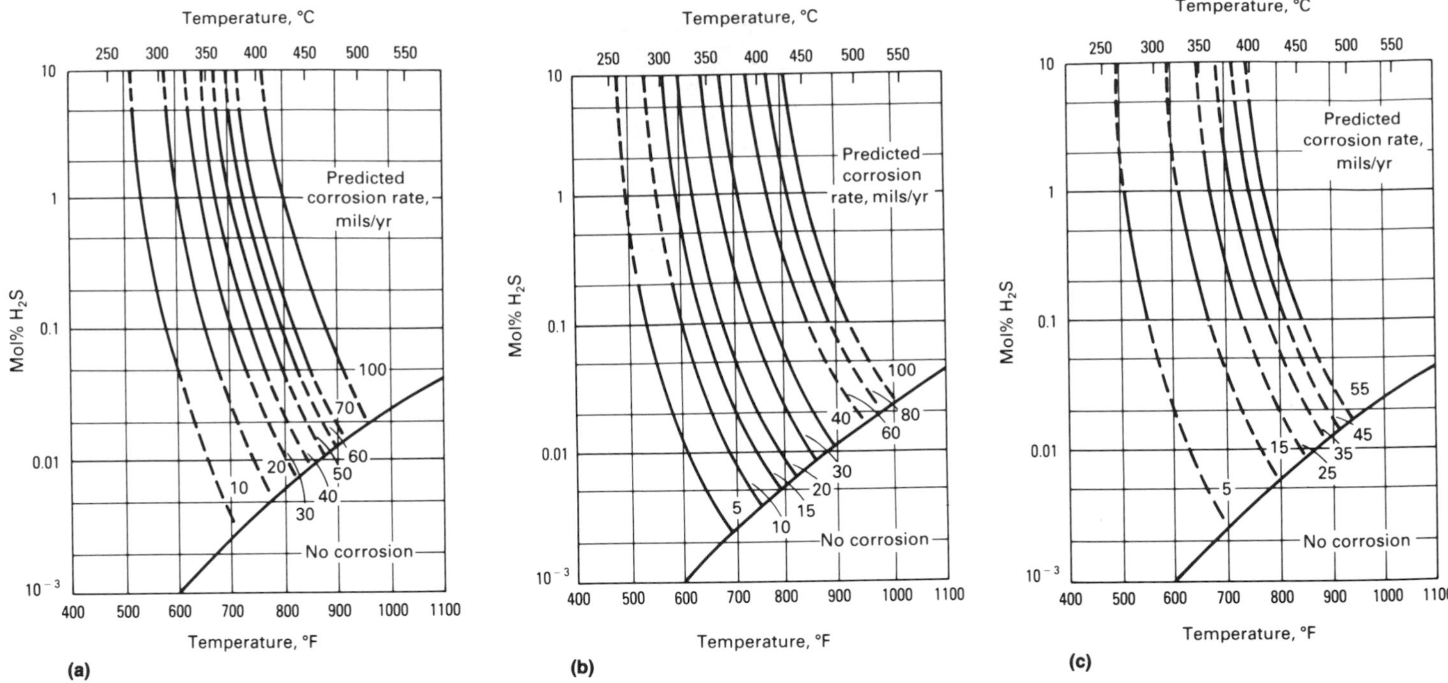

Fig. 8 Effect of temperature and hydrogen sulfide content on high-temperature H_2S/H_2 corrosion of (a) carbon steel, (b) 5Cr-0.5Mo steel, and (c) 9Cr-1Mo steel. These corrosion rates are based on the use of gas oil desulfurizers; corrosion rates with naphtha desulfurizers may be slightly less severe. Source: Ref 14

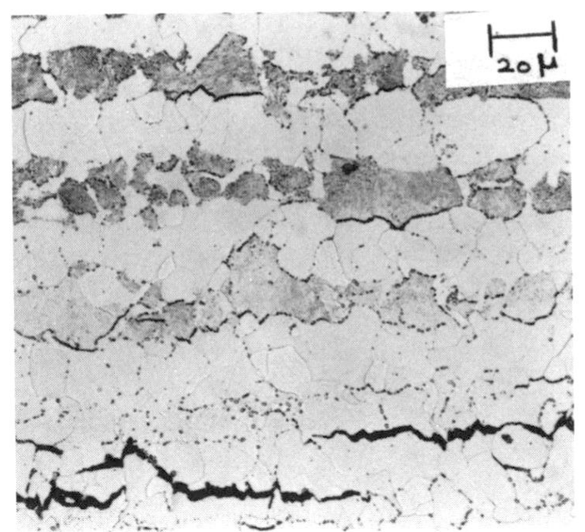

Fig. 9 Decarburization and fissuring in a pressure-vessel steel due to hydrogen attack

unexpected failure of equipment without prior warning signs is the primary cause for concern.

Hydrogen attack often initiates at areas of high stress or stress concentration in the steel because atomic hydrogen preferentially diffuses to these areas. Isolated fingers of decarburized and fissured material are often found adjacent to weldments and are associated with the initial stages of hydrogen attack. It is also evident that the fissures tend to be parallel to the edge of the weld rather than the surface. This orientation of fissures is probably the result of residual stress adjacent to the weldment. Fissures in this direction can form through-thickness cracks.

Effects of Hydrogen Attack on Mechanical Properties. Hydrogen attack is manifested by losses in room-temperature tensile strength, ductility, impact energy, and density, as shown in Fig. 10. In all cases, these effects start to occur only when a critical temperature of prior exposure to hydrogen is exceeded. With increasing alloy content of the steel, the critical exposure temperature for susceptibility is shifted to increasingly higher temperatures. The effect of hydrogen attack on creep-rupture life at 540 °C (1000 °F) for a steel containing 0.5% Mo is shown in Fig. 11. A pronounced reduction in rupture life is observed. The effect of hydrogen pressure on rupture time for Cr-0.5Mo steel at 540 °C (1000 °F) has been investigated by Holmes et al. (Ref 30). Figure 12, based on their results, shows that rupture time is appreciably reduced by increasing the pressure of hydrogen.

The effect of temperature on the rupture strength of quenched-and-tempered 1.25Cr-0.5Mo steels in hydrogen has been investigated by Watanabe et al. (Ref 31). Their results (see Fig. 13) show a significant drop in the rupture strength at all temperatures down to 500 °C (930 °F).

Prevention of Hydrogen Attack. The only practical way to prevent hydrogen attack is to use only steels that, based on plant experience, have been found to be resistant to this type of deterioration. The following general rules are applicable to hydrogen attack:

- Carbide-forming alloying elements, such as chromium and molybdenum, increase the resistance of steel to hydrogen attack.
- Increased carbon content decreases the resistance of steel to hydrogen attack.
- Heat-affected zones are more susceptible to hydrogen attack than the base or weld metal.

For most refinery and petrochemical plant applications, low-alloy chromium- and molybdenum-containing steels are used to prevent hydrogen attack. However, questions have recently been raised regarding the effect of long-term hydrogen exposure on C-0.5Mo steel (Ref 32). As a result, chromium-containing low-alloy steels are preferred over C-0.5Mo steel for new construction.

The conditions under which different steels can be used in high-temperature hydrogen service are listed in American Petroleum Institute (API) 941 (Ref 33). The principal data are presented in the form of Nelson curves, as shown in Fig. 14 and 15. The curves are based on long-term refinery experience, rather than on laboratory studies. The curves are periodically revised by the API Subcommittee on Materials Engineering and Inspection, and the latest edition of API 941 should be consulted to ensure that the proper steel is selected for the operating conditions encountered.

Ash/Salt Deposit Corrosion (Ref 34)

Deposition of ashes and salts on the surfaces of process components is quite common in some industrial environments, such as fossil-fuel-fired power plants and waste incineration plants. The corrosion process under these conditions involves both the deposit and the corrosive gases. The deposit may alter the thermodynamic potentials of the environment on the metal surface beneath the deposit. In an environment with both oxygen and sulfur activities, for example, the deposit tends to lower the oxygen activity and raise the sulfur activity beneath the deposit. As a result, formation of a protective oxide scale on the metal surface may become more difficult for most alloys. In most cases, the deposit involves some type of salt. This may lead to chemical

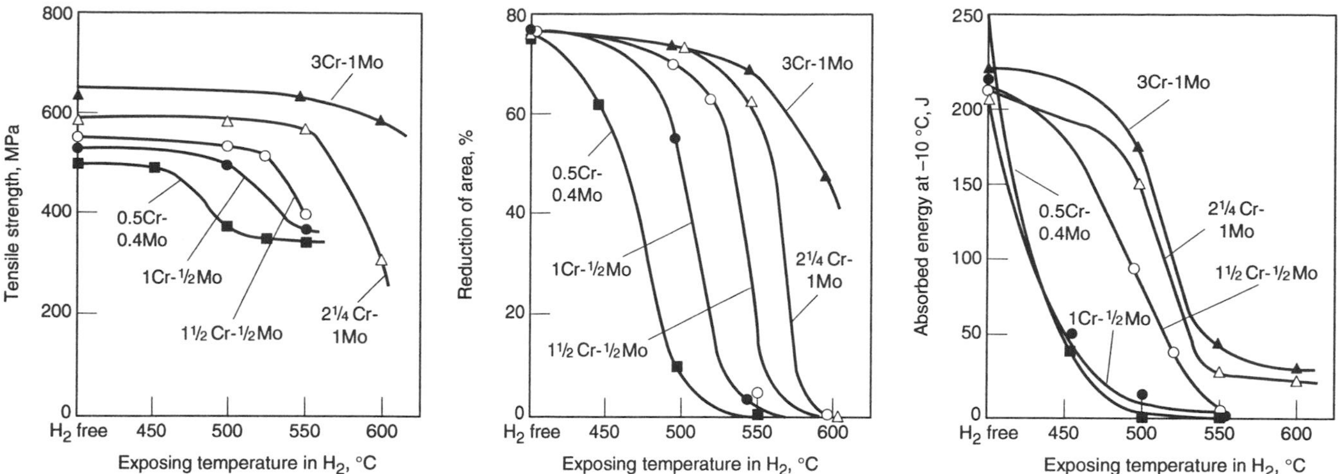

Fig. 10 Effects of hydrogen-exposure temperature on the mechanical properties of normalized-and-tempered chromium-molybdenum steels with carbon contents of 0.12% tested in 30 MPa (4350 psi) hydrogen. Source: Ref 27, 28

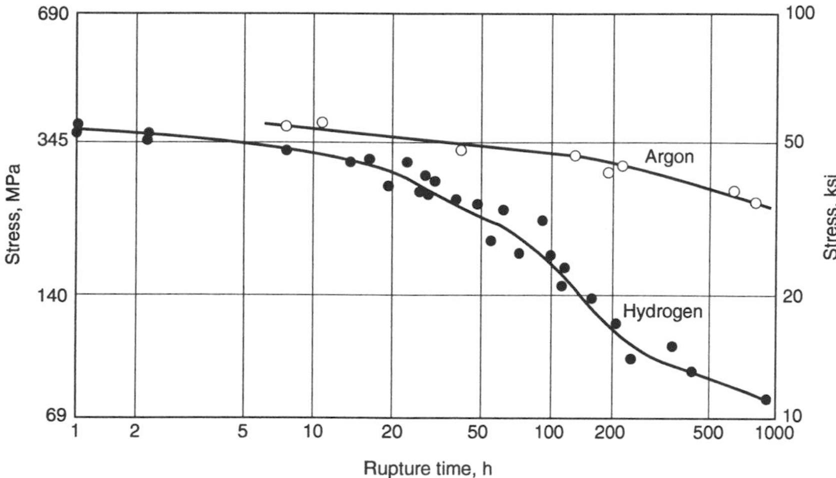

Fig. 11 Loss in creep-rupture life of 0.5Mo steel at 540 °C (1000 °F) in 5-MPa (725-psi) hydrogen. Source: Ref 29

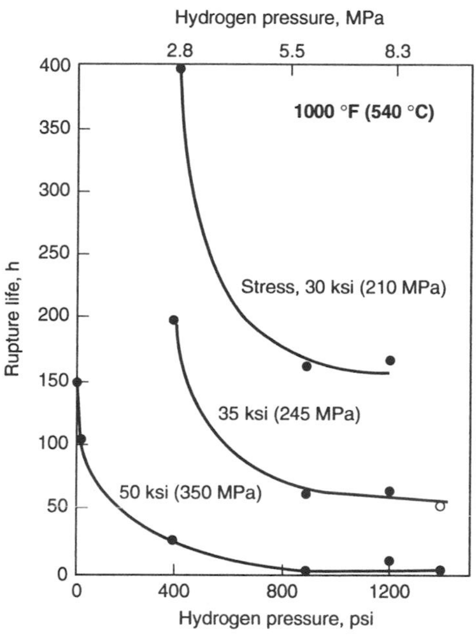

Fig. 12 Relationships between rupture life and hydrogen pressure for a 1Cr-0.5Mo steel at various applied stresses. Source: Ref 30

reactions between the protective oxide scale and the salt, resulting in the breakdown of the scale. A salt deposit that becomes liquid is particularly damaging.

Fireside corrosion of components such as superheaters and reheaters in fossil-fired boilers is commonly referred to as "fuel ash corrosion." The corrosion process in fuel ash corrosion is related to alkali-iron trisulfates for coal-fired boilers and vanadium salts (a mixture of vanadium pentoxide and sodium oxide or sodium sulfate) for oil-fired boilers. Hot corrosion in oil-fired boilers is referred to as "oil ash corrosion."

Hot Corrosion Mechanisms in Coal- and Oil-Fired Boilers. Corrosion from the firing of coal or oil is essentially related to specific impurities in the fuels, which can lead to the formation of nonprotective scales or can disrupt normally protective oxide scales. The relevant impurities in coal are sulfur (about 0.5 to 5.2% in United States coals), sodium (0.01 to 0.7%), potassium (0.2 to 0.7%), and chlorine (0.01 to 0.28%). In oil, the important impurities are sodium, which from United States refineries may range up to 300 ppm by weight; vanadium (up to 150 ppm by weight); and sulfur (0.6 to 3.6%). During combustion, these impurities can be melted or vaporized and will deposit upon contact with surfaces at temperatures lower than the condensation temperatures of the specific species. This provides a mechanism for the accumulation of deposits of fly ash on surfaces downstream of the burners. Figure 16 shows an illustration of the deposit formed on superheater tubes in a coal-fired boiler. The deposit may be several inches thick, with the outer layers formed of loosely sintered fly ash. The outer part of this deposit will be essentially at the gas temperature. The white inner layer is rich in alkali sulfate. The black layer is largely Fe_3O_4. Sulfur prints indicate that sulfide is present at the metal surface.

Water Wall Corrosion. Water walls (or furnace walls) are made of carbon steel tubes connected by carbon steel webs to form an enclosure where combustion of coal takes place. The temperature in the combustion zone can be higher than 1500 °C (2730 °F). The outer metal skin temperature of the furnace walls is usually kept at temperatures below 450 °C (840 °F) by water/steam in the furnace tubes. Water is converted to steam in the evaporator section. The fireside corrosion of furnace tubes generally occurs in the outer tube metal temperature range of 260 to 450 °C (500 to 840 °F).

Water wall fireside corrosion in coal- and oil-fired boilers is generally found in regions around the burners. The thick, hard external scales formed may be quite smooth, but often exhibit cracks or grooves that can resemble an alligator hide. Cracking or grooving is usually circumferential, and it is more common in supercritical than subcritical boilers. It occurs in the areas of

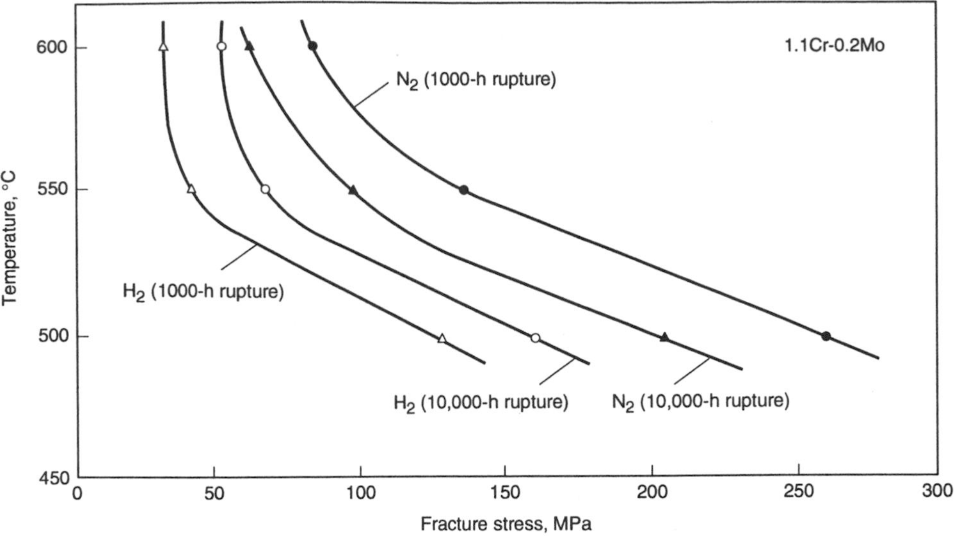

Fig. 13 Relationships between temperature and fracture stress in internal-pressure rupture tests of 1.1Cr-0.2Mo steel. Source: Ref 31

the water walls that receive the highest heat flux and is apparently a result of superimposed thermal stress. Figure 17 illustrates the appearance of grooving. Figure 18 shows the grooves to be sharp-pointed cracks filled with corrosion product (mostly iron oxide) but with a central core of iron sulfide. Figure 19 shows a cross section of an ASME SA-213, grade T-11 water wall tube that has formed a thick, smooth scale. The outer scale is a mixture of iron sulfide and iron oxide, with the chromium from the alloy dispersed in the inner layer. The major causes of corrosion of the water wall tubes are, first, the reducing (substoichiometric) conditions caused by impingement of incompletely combusted coal particles and flames and, second, molten salt or slag-related attack.

Superheater/Reheater Corrosion. Steam from the evaporator section of furnace walls is further heated in a superheater before going to a high-pressure steam turbine. After expanding through the high-pressure steam turbine, the steam is returned to the boiler to be reheated in a reheater. The metal temperatures of superheaters and reheaters may reach 650 °C (1200 °F) in coal-fired boilers and up to 815 °C (1500 °F) in oil-fired boilers.

Alloys that have been used for superheaters and reheaters are chromium-molybdenum steels (e.g., 1.25Cr-0.25Mo, 2.25Cr-1Mo, 5Cr-0.5Mo, and 9Cr-1Mo), 300-series austenitic stainless steels, and superalloys. Figure 20 shows the results of a 10,000 h field test in a boiler fired with fuel oil containing 2.65% S, 49 ppm V, and 44 ppm Na. At 500 to 650 °C (930 to 1200 °F), the low-alloy ferritic steels were significantly better than the austenitic grades.

Hot Corrosion in Boilers Burning Municipal Solid Waste. The corrosion problems experienced in boilers fueled with municipal refuse are different from those encountered with fossil fuels in that chlorine rather than sulfur is primarily responsible for the attack. The average chlorine content of municipal solid waste is 0.5%, of which about one-half is present as polyvinyl

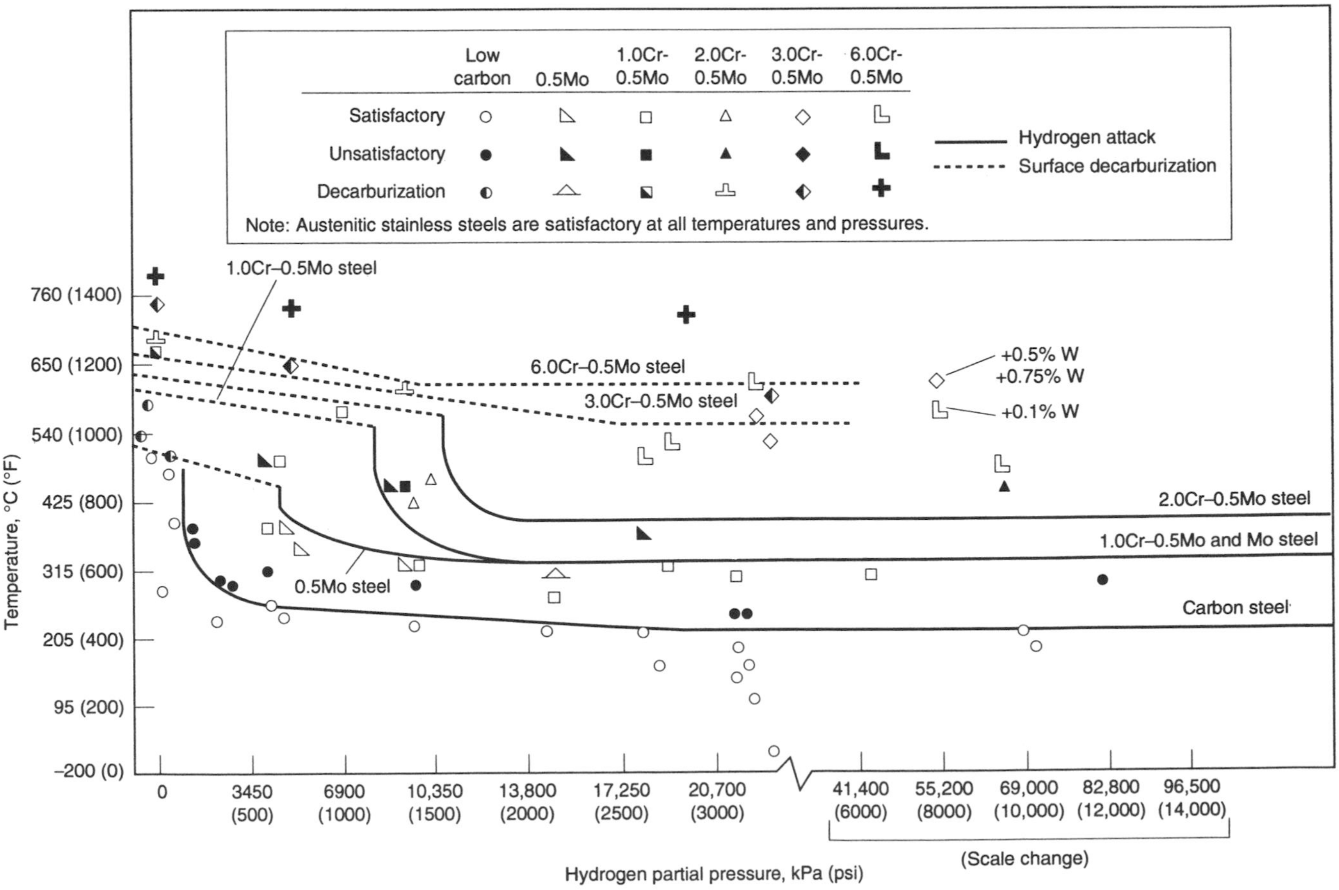

Fig. 14 Nelson curves for operating limits for carbon and alloy steels in contact with hydrogen at high temperatures and pressures

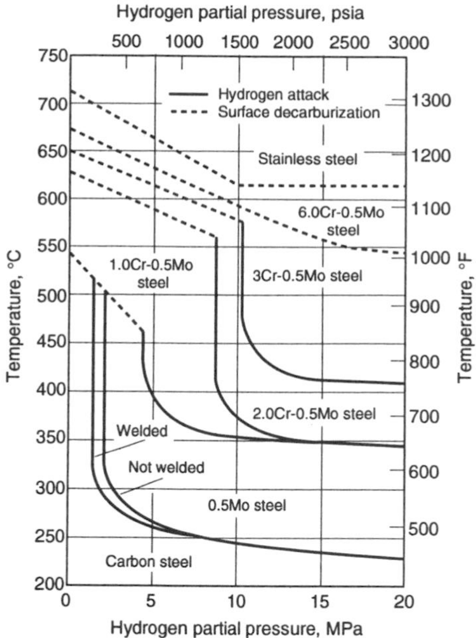

Fig. 15 Safe operating zones for steels in hydrogen service adapted from the Nelson curves, shown in Fig. 14

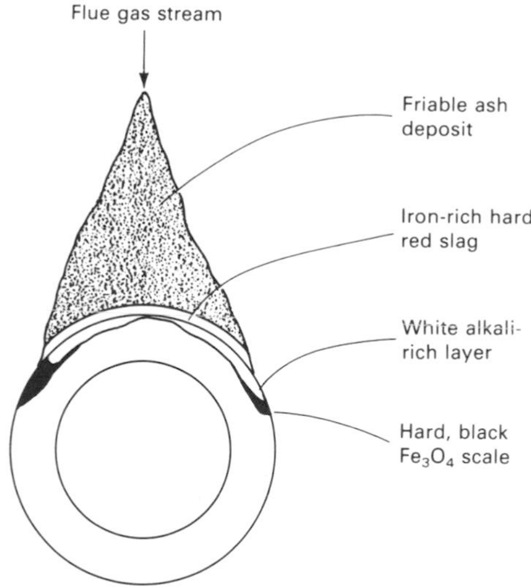

Fig. 16 Deposit layers on a corroding superheater or reheater tube

chloride (PVC) plastic. The other half is inorganic, principally NaCl. The chlorine in the plastic is converted to hydrochloric acid (HCl) in the combustion process. The inorganic chlorides are vaporized in the flame and ultimately condense in the boiler deposits or pass through the boiler with the flue gases. Zinc, lead, and tin in the refuse also play a role in the corrosion process by reacting with the HCl to form metal chlorides and/or eutectic mixtures with melting points low enough to cause molten salt attack at wall tube metal temperatures.

Hydrochloric acid alone has little effect on carbon steel at temperatures below 260 °C (500

Fig. 17 Example of grooving in water wall tubes. Note the thick, adherent scale remaining in some areas.

°F) (Ref 36). However, in the presence of excess air in the boiler, the reaction to form ferrous chloride ($FeCl_2$) on the steel surface occurs more readily. This corrosion product is stable at water wall tube temperatures, and the curve for metal wastage versus time is parabolic. However, at a metal temperature of about 400 °C (750 °F), which may occur in the superheater, the $FeCl_2$ is further chlorinated to the readily volatile ferric chloride ($FeCl_3$) (Ref 37). If the gas temperature in the area exceeds 815 °C (1500 °F), the $FeCl_3$ will evaporate rapidly, and breakaway corrosion can then occur. These effects are shown in Fig. 21, which presents data from corrosion probe exposures conducted in a municipal incinerator.

Krause et al. (Ref 38) conducted field tests on carbon steels (A106 and T-11), stainless steels, and high-nickel alloys at temperatures varying from 145 to 650 °C (290 to 1200 °F) in a municipal waste incinerator. Test results obtained from an 828-h exposure for carbon steels and stainless steels are shown in Fig. 22. Corrosion rates depended greatly on temperature. The rates for carbon steels were approximately 0.13 mm (5 mils) per month at about 200 to 240 °C (390 to 470 °F), around 0.25 mm (10 mils) per month at 360 to 430 °C (680 to 810 °F), and around 0.5 mm (20 mils) per month at 500 to 540 °C (940 to 1000 °F). Stainless steels were generally better than carbon steels.

Prevention of Corrosion. Solutions to hot corrosion due to ash/salt deposits include changes in operating procedures and changes in materials. Because changes in operating procedures can exact some penalty in boiler efficiency, upgrading the material to a more corrosion-resistant alloy such as a stainless steel is often preferred. Cladding carbon steel with stainless steel is another option.

Fig. 18 Cross section of circumferential cracks on the fireside surface of a water wall tube

Molten Salt Corrosion

Molten salt technology plays an important role in various industries. In the heat treating industry, molten salts are commonly used as a medium for heat treatment of metals and alloys (e.g., annealing, tempering, hardening, quenching, and cleaning) as well as for surface treatment (e.g., case hardening). In nuclear and solar energy systems, they have been used as a medium for heat transfer and energy storage. Other applications include extraction of aluminum, magnesium, sodium, and other reactive metals, refining of refractory metals, and high-temperature batteries and fuel cells.

The Corrosion Process. Molten salts generally are a good fluxing agent, effectively removing oxide scale from a metal surface. The corrosion reaction proceeds primarily by oxidation, which is then followed by dissolution of metal oxides in the melt. Oxygen and water vapor in the salt thus often accelerate molten salt corrosion.

Corrosion can also take place through mass transfer due to thermal gradient in the melt. This mode of corrosion involves dissolution of an alloying element at hot spots and deposition of that element at cooler spots. This can result in severe fouling and plugging in a circulating system. Corrosion is also strongly dependent on temperature and velocity of the salt.

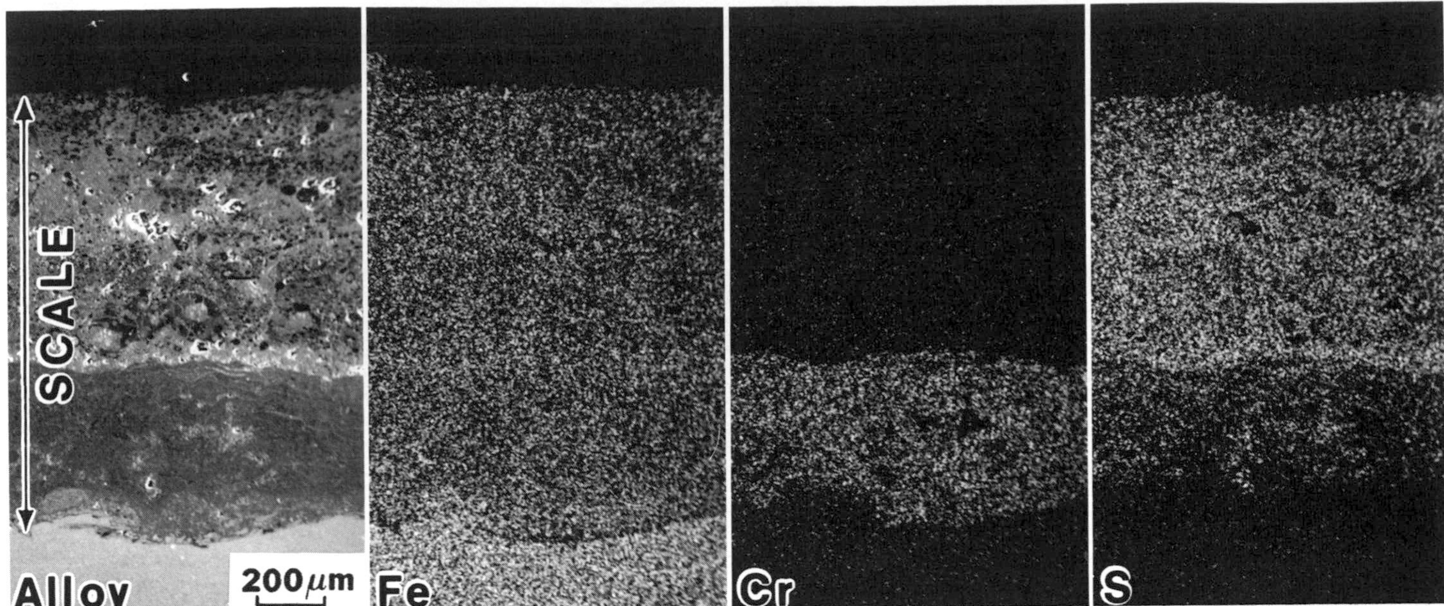

Fig. 19 Cross section of the fireside face of a water wall tube subjected to reducing (substoichiometric) combustion conditions. From left to right: optical micrograph of deposit and x-ray elemental dot maps for iron, chromium, and sulfur, respectively

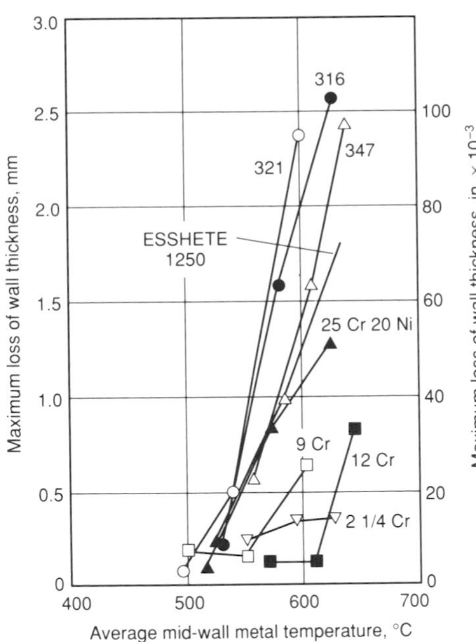

Fig. 20 Results of a 10,000 h field test in the superheater inlet zone (horizontal position) in an oil-fired boiler. Source: Ref 35

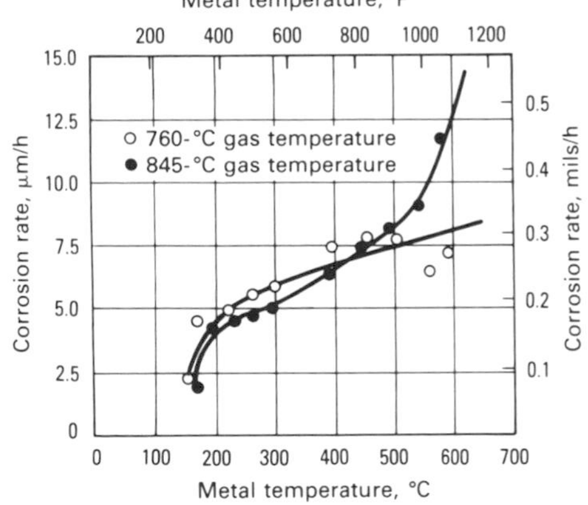

Fig. 21 Corrosion rates at two temperatures of carbon steel in a boiler burning municipal refuse

Corrosion can take the form of uniform thinning, pitting, or internal or intergranular attack. In general, molten salt corrosion is quite similar to aqueous corrosion. More detailed information can be found in the article "Corrosion at Elevated Temperatures" in this Volume.

Suitable Steels. Generally, carbon and alloy steels are not sufficiently corrosion resistant to handle or contain molten salts. Molten fluorides, chloride salts, sulfates, hydroxides, and carbonates must be contained using austenitic stainless steels or nickel-base alloys. Figure 23 shows the severe corrosion of carbon and alloy steels when exposed to chloride salts. One exception, however, is the use of carbon and alloy steels for handling the molten nitrates commonly used for heat treatment baths. These steels form protective oxide films that effectively protect the metal surface to approximately 500 °C (930 °F). Table 2 compares the corrosion rates of various materials, including plain carbon and chromium-molybdenum steels, in molten $NaNO_3$-KNO_3 salt.

Molten Metal Corrosion (Ref 41)

Molten metals can attack steels in various ways. One form of attack that can occur is dissolution and/or intergranular penetration by molten metal as illustrated in Fig. 24. The unfortunate aspect of this mode of attack is that it can result in a loss of strength without any large weight loss or change in appearance. In this respect, it resembles the more familiar aqueous intergranular corrosion. Molten or liquid metal corrosion differs from liquid metal embrittlement (LME). The latter requires that both stress and a liquid metal be present. Liquid metal embrittlement is not a corrosion-, dissolution-, or diffusion-controlled intergranular penetration process, but rather a special case of brittle fracture that occurs in the absence of an inert environment and at low temperatures. Time- and temperature-dependent processes are not responsible for the occurrence of LME. Also in most cases of LME, there is little or no penetration of liquid metal into the solid metal.

A brief overview of the compatibility of steels in various molten-metal environments is given below. More detailed information can be found in the article "Corrosion at Elevated Temperatures" in this Volume.

Sodium and Sodium-Potassium Alloys. Plain carbon and low-alloy steels are generally suitable for long-term use in these media at temperatures to 450 °C (840 °F). Beyond these temperatures, stainless steels are required.

The principal disadvantage of ferritic chromium-molybdenum steels in sodium systems is the decarburization potential and its possible deleterious effect on mechanical properties.

Lithium is somewhat more aggressive to steels than sodium or sodium-potassium. As a result, alloy steels should not be considered for long-term use above 300 °C (570 °F). At higher temperatures, the ferritic stainless steels show better results.

Cadmium. Low-alloy steels exhibit good serviceability to 700 °C (1290 °F).

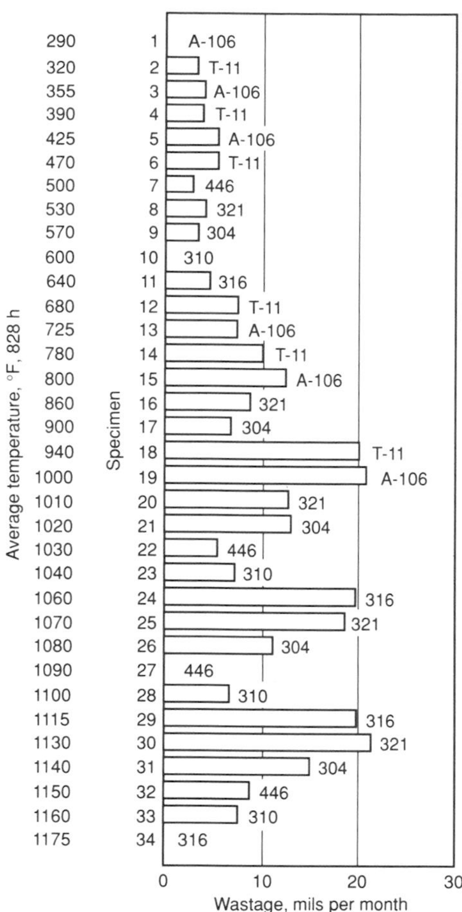

Fig. 22 Corrosion rates of carbon steels (A106 and T-11) and stainless steels obtained from field tests in a municipal waste incinerator. Source: Ref 38

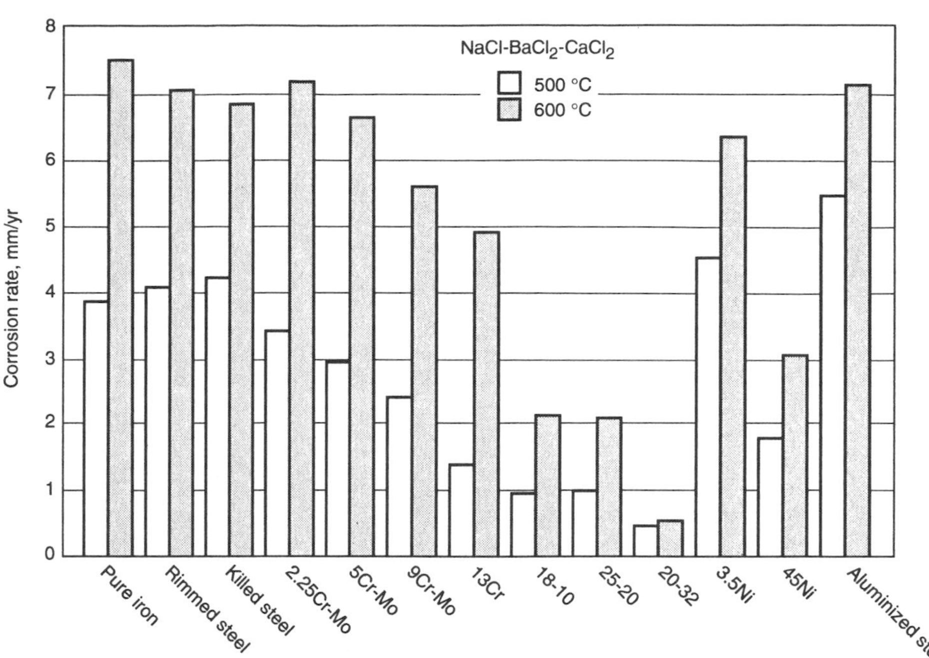

Fig. 23 Corrosion rates of steel, Fe-Cr, Fe-Ni, and Fe-Cr-Ni alloys in molten $20NaCl\text{-}30BaCl_2\text{-}50CaCl_2$ at 500 and 600 °C (930 and 1110 °F). Source: Ref 39

Zinc. Most engineering metals and alloys show poor resistance to molten zinc, and carbon steels are no exception.

Antimony. Low-carbon steels have poor resistance to attack by antimony.

Mercury. Although plain carbon steels are virtually unattacked by mercury under nonflowing or isothermal conditions, the presence of either a temperature gradient or liquid flow can lead to drastic attack. The corrosion mechanism seems to be one of dissolution, with the rate of attack increasing rapidly with temperature above 500 °C (930 °F). Alloy additions of chromium, titanium, silicon, and molybdenum, alone or in combination, show resistance to 600 °C (1110 °F). Where applicable, the attack of ferrous alloys by mercury can be reduced to negligible amounts by the addition of 10 ppm Ti to the mercury; this raises the useful range of operating temperatures to 650 °C (1200 °F). Additions of metal with a higher affinity for oxygen than titanium, such as sodium or magnesium, may be required to prevent oxidation of the titanium and loss of the inhibitive action.

Aluminum. Steels are not satisfactory for the long-term containment of molten aluminum.

Gallium is one of the most aggressive of all liquid metals and cannot be contained by carbon or alloy steels at elevated temperatures.

Table 2 Corrosion rates of selected metals and alloys in molten $NaNO_3\text{-}KNO_3$

Alloy	Temperature		Corrosion rate	
	°C	°F	mm/yr	mils/yr
Carbon steel	460	860	0.120	4.7
2.25Cr-1Mo	460	860	0.101	4.0
	500	932	0.026	1.0
9Cr-1Mo	550	1020	0.006	0.2
	600	1110	0.023	0.9
Aluminized Cr-Mo steel	600	1110	<0.004	<0.2
12Cr steel	600	1110	0.022	0.9
Type 304	600	1110	0.012	0.5
Type 316	600	1110	0.007–0.010	0.3–0.4
	630	1170	0.106	4.2
Alloy 800	565	1050	0.005	0.2
	600	1110	0.006–0.01	0.2–0.4
	630	1170	0.075	3.0
Alloy 600	600	1110	0.007–0.01	0.3–0.4
	630	1170	0.106	4.2
Nickel	565	1050	>0.5	>20
Titanium	565	1050	0.04	1.6
Aluminum	565	1050	<0.004	<0.2

Source: Ref 40

Indium. Carbon and alloy steels have poor resistance to molten indium.

Lead, Bismuth, Tin, and Their Alloys. Alloy steels have good resistance to lead up to 600 °C (1110 °F), to bismuth up to 700 °C (1290 °F), and to tin only up to 150 °C (300 °F). The various alloys of lead, bismuth, and tin are more aggressive.

Coatings for High-Temperature Corrosion Resistance

The service life of carbon and alloy steel components used in aggressive environments can be extended by the use of various surface treatments. These include hot-dip aluminum coating, pack aluminizing, thermal spraying, and cladding.

Hot-Dip Aluminum Coatings. Aluminum-coated steel products are used successfully in corrosive and oxidizing environments in which the temperature ranges from that of outdoor exposure to 1150 °C (2100 °F). Aluminum coatings protect steel from attack by forming a very resistant barrier between the corrosive atmosphere and the steel. The aluminum oxide that forms on the aluminum surface is highly resistant to a wide range of environments.

Successful application of aluminum-coated steel for resistance to oxidation and corrosion at

Table 3 Applications of diffused aluminum hot-dip coatings for resistance to oxidation and corrosion at 455 to 980 °C (850 to 1800 °F)

Product and base metal	Type of service
Heat-treating equipment	
Burner pipes, 5Cr-0.5Mo	Oxidation, 870 °C (1600 °F)
Fixtures, low-carbon and medium-alloy steels	Carburizing, carbonitriding
Flue stacks, low-carbon steel	Oxidation, sulfur corrosion
Furnace insulation supports, low-carbon and medium-alloy steels	Oxidation, 540-650 °C (1000-1200 °F)
Pyrometer protection tubes, 310 and 316 stainless steel, low-carbon steel	Oxidation, 980 °C (1800 °F)
Heat exchanger components	
Boiler soot blowers, 1Cr-0.5Mo	Oxidation, sulfur attack
Boiler tubing, 2Cr-0.5Mo	Oxidation, 540-595 °C (1000-1100 °F)
Cylinder barrel, air-cooled engine, Nitralloy	Oxidation to 480 °C (900 °F)
Preheater tubing, 1Cr-0.5Mo	Oxidation, 650 °C (1200 °F)
Tubing, low-carbon steel, 1.5Cr-0.5Mo	Hydrogen sulfide gases
Fasteners	
Steel fasteners for chemical piping and boilers	Oxidation to 480 °C (900 °F)
High-temperature fasteners	Oxidation to 760 °C (1400 °F)
Studs, 4140, for chemical and oil refineries	Oxidation and ease of removal after service at 480 °C (900 °F)
Miscellaneous equipment	
Chemical reactor tubing, low-carbon steel	Carbonization, iron contamination
Chimney caps, low-carbon steel	Oxidation and corrosion
Recuperator tubing, 2.5Cr-0.5Mo	Oxidation and sulfidation
Refinery tubing, 304 stainless steel, 2.25Cr-1Mo	Oxidation and sulfidation
Sulfuric acid converters, 5Cr-0.5Mo	Sulfur dioxide corrosion, 705 °C (1300 °F)

Table 4 Partial list of commercial applications of pack cementation aluminizing of carbon and low-alloy steels

Industry	Component	Typical materials aluminized
Hydrocarbon processing	Refinery heater tubes	2¼% Cr-1% Mo steel
	Hydrodesulfurizer furnace tubes	2¼% Cr-1% Mo steel
	Delayed coker furnace tubes	9% Cr-1% Mo steel
	Catalyst reactor grating	Carbon steel
Sulfuric acid	Gas-to-gas heat exchanger tubes	Carbon steel
Industrial furnace components	Aluminum plant furnace parts	Carbon steel
	Heat treating pots	Carbon steel
	Structural members	High-nickel alloy steel
	Thermowells	Carbon and stainless steels
Steam power and cogeneration	Waterwall tubes	2¼% Cr-1% Mo steel
	Fluidized bed combustor tubes	2¼% Cr-1% Mo steel
	Waste heat boiler tubes	Carbon steel
	Economizer and air preheater tubes	2¼% Cr-1% Mo steel
	Superheater tubes	2¼% Cr-1% Mo steel

elevated temperatures depends on the physical and mechanical properties of the alloy chemical bond between the aluminum and the steel. It is important that the hot strength of the steel be suitable for the stress and temperatures encountered. Low-carbon HSLA steels alloyed with titanium or niobium offer improved high-temperature creep resistance when used as substrates for aluminum coatings.

Aluminum coatings that contain from 5 to 11% Si minimize the thickness of the iron-aluminum alloy bond and improve formability. Undiffused, such coatings retain excellent heat reflectivity at temperatures to 480 °C (900 °F).

Above 480 °C (900 °F), further alloying occurs between the aluminum coating and steel base. Because the rate of alloying is dependent on time and temperature, all coating converts to aluminum-iron-silicon alloy with sufficient time at temperature. The refractory alloy formed is extremely heat resistant and resistant to spalling up to 680 °C (1250 °F). Spalling at service temperatures above 680 °C (1250 °F) can be overcome by the use of heat-resistant aluminized steel that contains sufficient titanium to stabilize carbon and nitrogen as well as maintain excess titanium in solution (Ref 42).

Table 3 lists applications for steels that have been prepared by batch hot dipping in aluminum and then heat treated to diffuse the aluminum into the steel. This treatment eliminates spalling and provides an impervious protective coating during high-temperature service.

The use of aluminum-coated plain carbon steel for complicated heat-treating fixtures subjected to temperatures lower than 870 °C (1600 °F) may decrease overall fixture cost in comparison to fixtures made of the highly alloyed austenitic steels normally used for this application. Figure 25 shows the effects of coating thickness and operating temperature on oxidation resistance for coated and uncoated heat-treating fixtures made of 1020 steel.

Compared with solid stainless steel, aluminum-coated carbon steel offers greater resistance to attack by hydrogen sulfide. One set of test data

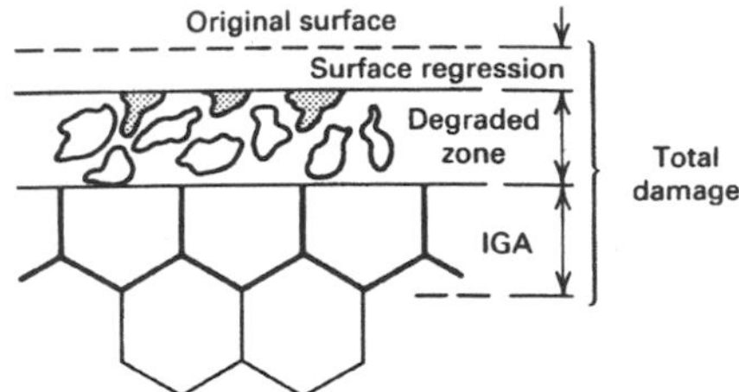

Fig. 24 Schematic of the dissolution and intergranular attack (IGA) that take place during the molten-metal corrosion process

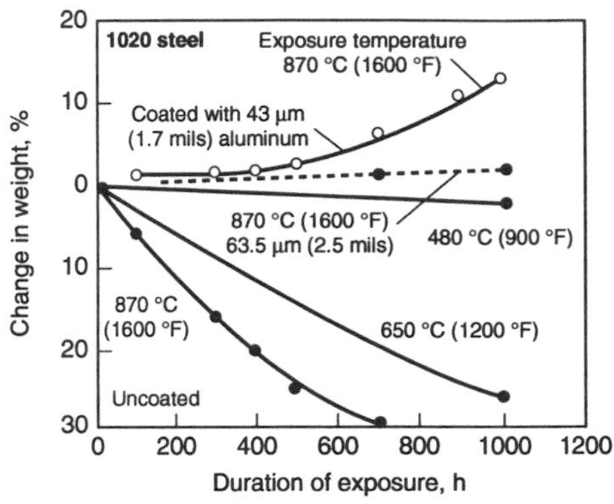

Fig. 25 Effects of coating thickness and exposure temperature on oxidation of coated and uncoated steel. Oxidation at 480 to 870 °C (900 to 1600 °F). Steel 6.4 mm (¼ in.) thick was completely oxidized after 700 h at 870 °C (1600 °F).

indicated that, at 595 °C (1110 °F), aluminized carbon steels offer more than 100 times the resistance of 18-8 stainless steel to pure hydrogen sulfide. Other data have shown them to be 25 times more resistant than straight chromium steel.

The pack cementation aluminizing process is also used to improve the performance of steels in high-temperature corrosive environments. The complex aluminide intermetallic coatings formed during the process exhibit superior resistance to oxidation, carburization, and sulfidation. Table 4 provides a partial listing of commercial applications for the pack aluminizing process. The structure, including the diffusion zone, of pack aluminized low-carbon steel, is shown in Fig. 26.

The action of pack aluminized steels under heat is similar to that of unprocessed stainless steels. Stainless steels owe their oxidation resistance to the formation of a thin chromium-rich oxide (Cr_2O_3) film that protects the underlying steel. A similar reaction occurs with aluminized steels, which form an even more protective oxide, Al_2O_3, from the aluminum in the coating. This oxide grows at a lower rate than Cr_2O_3 and does not exhibit volatility in the presence of oxygen above about 927 °C (1700 °F) as does Cr_2O_3. While a type 304 stainless steel scales excessively at temperatures of ≥870 °C (≥1600 °F) in an oxidizing environment, a pack aluminized carbon steel exhibits only slight discoloration of the surface under these conditions. Figure 27 compares aluminized carbon steel with bare car-

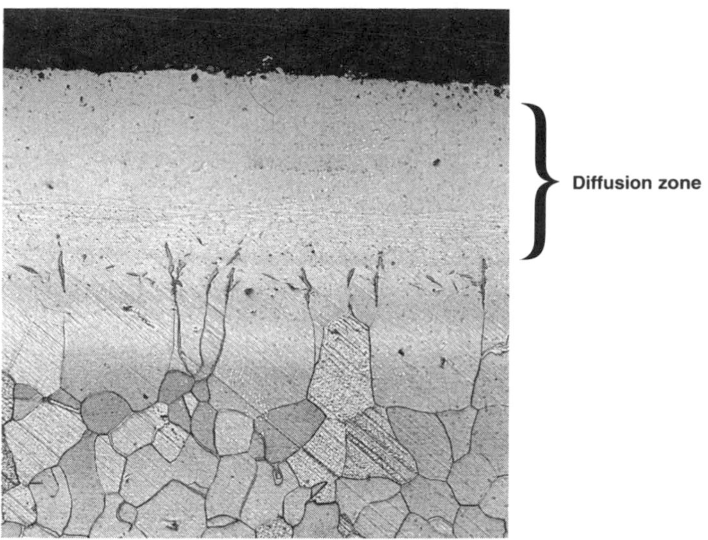

Fig. 26 The structure of a pack aluminized low-carbon steel

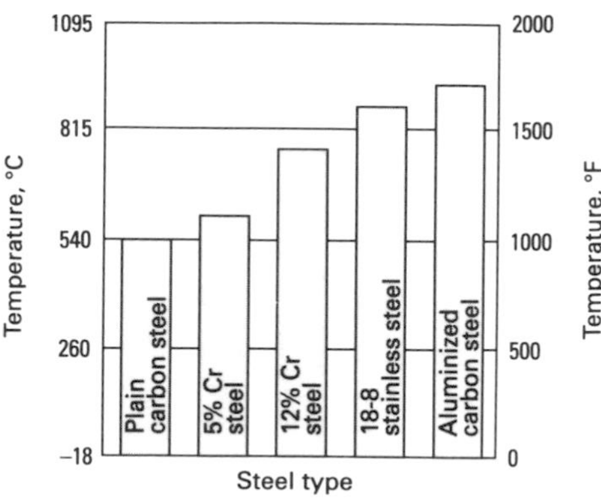

Fig. 27 Oxidation of steels in air at the temperature at which scaling is less than 10 mg/cm^2. Source: Ref 43

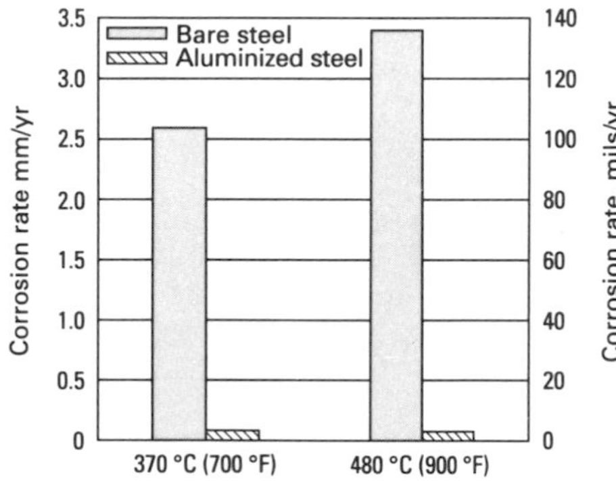

Fig. 28 Relative corrosion rates of 9Cr-1Mo alloy steel in 5 mol% H_2S at 3550 kPa (515 psi) for 300 h. Source: Ref 44

Table 5 Thermal spray coatings used for elevated-temperature service

Service temperature	Coating metal or alloy	Coating thickness µm	Coating thickness mils
Up to 550 °C (1020 °F)	Aluminum	175	7
Up to 550 °C (1020 °F) in the presence of sulfurous gases	Ni-43Cr-2Ti	375	15
550–900 °C (1020–1650 °F)	Aluminum or aluminum-iron	175	7
900–1000 °C (1650–1830 °F)	Nickel-chromium or MCrAlY	375	15
900–1000 °C (1650–1830 °F) in the presence of sulfurous gases	Nickel-chromium, followed by aluminum	375/100	15/4

bon steel, 5% Cr steel, 12% Cr steel, and 18Cr-8Ni steel with respect to the temperature at which scaling is less than 10 mg/cm^2 for oxidation in air. High-strength, highly alloyed stainless steels are also frequently aluminized to improve their oxidation resistance in elevated-temperature applications.

Pack aluminized steels also exhibit resistance to the corrosive attack of gases such as hydrogen sulfide (H_2S), sulfur dioxide (SO_2), and sulfur trioxide (SO_3) as well as many other sulfur-bearing atmospheres at temperatures exceeding 230 °C (450 °F). The diffusion zone typically contains a minimum of 20% Al while the alloy surface is about 50 at.% Al and is far more resistant to high-temperature sulfide corrosion than stainless steels containing as much as 28% Cr. The maximum temperature at which these materials can be used in sulfur-bearing environments is limited by the high-temperature mechanical properties of the base steel. Figure 28 provides experimental data on the relative corrosion rates of bare and aluminized 9Cr-1Mo steel in H_2S environments.

When exposed at high temperatures to carbon-rich atmospheres, alloy steels will carburize, become extremely brittle, and lose their heat- and corrosion-resistant properties. In certain environments, metal dusting associated with carburization will rapidly destroy the steel. Pack aluminized steels, however, suffer negligible deterioration from either carburization or metal dusting in high-temperature, carbon-rich atmospheres, thus greatly lengthening the service life of the steels in these environments (Ref 45).

Thermal spray coatings are extensively used by industry to protect steel components and structures from heat oxidation at surface temperatures to 1095 °C (2000 °F). By ensuring long-term protection, thermal spray coatings show real economic advantages during the service lives of such items. Coatings such as pure aluminum, aluminum-iron, nickel-chromium, and MCrAlY are particularly effective in protecting low-alloy and carbon steels. Table 5 lists various thermal spray coatings used to protect steels from high-temperature corrosive environments.

Metal Cladding. Carbon steels can be bonded to more corrosion- and/or heat-resistant materials by hot-roll bonding, cold-roll bonding, extrusion, or weld cladding. The resultant lamellar composite has specific properties not obtainable in a single material. Commonly used cladding alloys include austenitic stainless steels (e.g., types 304 and 310), 50Ni-50Cr alloys, and nickel-base alloys (e.g., Inconel 625). Clad carbon steels find various applications in the marine, chemical-processing, power, and pollution control industries. Specific uses include heat exchangers, reaction and pressure vessels, furnace tubes, tubes and tube elements for boilers, scrubbers, and other systems involved in the production of chemicals.

ACKNOWLEDGMENTS

The information in this article is largely taken from:

- J. Gutzeit, R.D. Merrick, and L.R. Scharfstein, Corrosion in Petroleum Refining and Petrochemical Operations, *Corrosion,* Vol 13, *ASM Handbook* (formerly 9th ed., *Metals Handbook*), ASM International, 1987, p 1262–1287
- B.C. Syrett, et al., Corrosion in Fossil Fuel Power Plants, *Corrosion,* Vol 13, *ASM Handbook* (formerly 9th ed., *Metals Handbook*), ASM International, 1987, p 985–1010
- S. Lampman, Elevated-Temperature Properties of Ferritic Steels, *Properties and Selection: Irons, Steels, and High-Performance Alloys,* Vol 1, *ASM Handbook* (formerly 10th ed., *Metals Handbook*), ASM International, 1990, p 617–652
- G.Y. Lai, *High-Temperature Corrosion of Engineering Alloys,* ASM International, 1990
- R. Viswanathan, *Damage Mechanisms and Life Assessment of High-Temperature Components,* ASM International, 1989
- Protection of Steel from Corrosion, *ASM Specialty Handbook: Carbon and Alloy Steels,* J.R. Davis, Ed., ASM International, 1996, p 520–572

REFERENCES

1. W.R. Patterson, in *Designing for Automotive Corrosion Prevention,* Proceedings P-78, 8–10 Nov 1978 (Troy, MI), Society of Automotive Engineers, p 71
2. J.B. Vrable, R.T. Jones, and E.H. Phelps, "The Application of Corrosion-Resistant High-Strength Low-Alloy Steels in the Chemical Industry," presented at Fall Meeting (Chicago), American Society for Metals, 1977
3. R.T. Jones, in *Process Industry Corrosion,* National Association of Corrosion Engineers, 1986, p 373
4. A.W. Zeuthen, *Heating, Piping and Air Conditioning,* Vol 42 (No. 1), 1970, p 152
5. Z.A. Foroulis, High Temperature Degradation of Structural Materials in Environments Encountered in the Petroleum and Petrochemical Industries: Some Mechanistic Observations, *Anti-Corros.,* Vol 32 (No. 11), 1985, p 4–9
6. A.S. Couper and A. Dravnieks, High Temperature Corrosion by Catalytically Formed Hydrogen Sulfide, *Corrosion,* Vol 18 (No. 8), 1962, p 291t–298t
7. A.S. Couper, High Temperature Mercaptan Corrosion of Steels, *Corrosion,* Vol 19 (No. 11), 1963, p 396t–401t
8. K.N. Strafford, The Sulfidation of Metals and Alloys, *Metall. Rev.,* Vol 138, 1969
9. F.A. Hendershot and H.L. Valentine, Materials for Catalytic Cracking Equipment (Survey), *Mater. Prot.,* Vol 6 (No. 10), 1967, p 43–47
10. N. Schofer, Corrosion Problems in a Fluid Catalytic Cracking and Fractionating Unit, *Corrosion,* Vol 5 (No. 6), 1949, p 182–188
11. S.L. Estefan, Design Guide to Metallurgy and Corrosion in Hydrogen Processes, *Hydrocarbon Process,* Vol 49 (No. 12), 1970, p 85–92
12. L.T. Overstreet and R.A. White, Materials Specifications and Fabrication for Hydrocracking Process Equipment, *Mater. Prot.,* Vol 4 (No. 6), 1965, p 64–71
13. H.F. McConomy, High-Temperature Sulfide Corrosion in Hydrogen-Free Environment, *Proc. API,* Vol 43 (III), 1963, p 78–96
14. J. Gutzeit, High Temperature Sulfide Corrosion of Steels, in *Process Industries Corrosion—The Theory and Practice,* National Association of Corrosion Engineers, 1986
15. E.B. Backensto, R.D. Drew, and C.C. Stapleford, High Temperature Hydrogen Sulfide Corrosion, *Corrosion,* Vol 12 (No. 1), 1956, p 6t–16t
16. G. Sorell and W.B. Hoyt, Collection and Correlation of High Temperature Hydrogen Sulfide Corrosion Data, *Corrosion,* Vol 12 (No. 5), 1956, p 213t–234t
17. C. Phillips, Jr., High Temperature Sulfide Corrosion in Catalytic Reforming of Light Naphthas, *Corrosion,* Vol 13 (No. 1), 1957, p 37t–42t
18. G. Sorell, Compilation and Correlation of High Temperature Catalytic Reformer Corrosion Data, *Corrosion,* Vol 14 (No. 1), 1958, p 15t–26t
19. W.H. Sharp and E.W. Haycock, Sulfide Scaling Under Hydrorefining Conditions, *Proc. API,* Vol 39 (III), 1959, p 74–91
20. J.D. McCoy and F.B. Hamel, New Corrosion Data for Hydrosulfurizing Units, *Hydrocarbon Process.,* Vol 49 (No. 6), 1970, p 116–120
21. J.D. McCoy and F.B. Hamel, Effect of Hydrosulfurizing Process Variables on Corrosion Rates, *Mater. Prot. Perform.,* Vol 10 (No. 4), 1971, p 17–22
22. A.S. Couper and J.W. Gorman, Computer Correlations to Estimate High Temperature H_2S Corrosion in Refinery Streams, *Mater. Prot. Perform.,* Vol 10 (No. 1), 1971, p 31–37
23. G. Sorell and M.J. Humphries, High Temperature Hydrogen Damage in Petroleum Refinery Equipment, *Mater. Perform.,* Vol 17 (No. 8), 1978, p 33–41
24. A.R. Ciuffreda and W.R. Rowland, Hydrogen Attack of Steel in Reformer Service, *Proc. API,* Vol 37 (III), 1957, p 116–128
25. R.D. Merrick and A.R. Ciuffreda, Hydrogen Attack of Carbon-0.5 Molybdenum Steels, *Proc. API,* Vol 61 (III), 1982, p 101–114
26. T. Skei, A. Wachter, W.A. Bonner, and H.D. Burnham, Hydrogen Blistering of Steel in Hydrogen Sulfide Solutions, *Corrosion,* Vol 9 (No. 5), 1953, p 163–172
27. H. Ishizuka and R. Chiba, Japan Steel Works Research Laboratories Report, 1967, p 43–66
28. Japan Pressure Vessel Research Council, Subcommittee on Hydrogen Embrittlement, Temper Embrittlement and Hydrogen Attack in Pressure Vessel Steels, Report No. 2, May 1979
29. F.H. Vitovec, *Proc. Amer. Petroleum Inst.,* Vol 44 (No. III), 1964, p 179
30. E. Holmes, P.C. Rosenthal, P. Thoma, and F.H. Vitovec, "Progress Report to Subcommittee on Corrosion," No. 3, American Petroleum Institute, 1964
31. J. Watanabe, H. Ishizuka, K. Onishi, and R. Chiba, *Proc. 20th National Symp. Strength, Fatigue, and Fracture,* (Japan), 1975, p 101
32. R. Chiba, K. Ohnishi, K. Ishii, and K. Maeda, Effect of Heat Treatment on Hydrogen Attack Resistance of C-0.5Mo Steels for Pressure Vessels, Heat Exchangers, and Piping, *Corrosion,* Vol 41 (No. 7), 1985, p 415–426
33. "Steels for Hydrogen Service at Elevated Temperatures and Pressures in Petroleum Refineries and Petrochemical Plants," Publication 941, 3rd ed., American Petroleum Institute, 1983
34. B.C. Syrett, et al., Corrosion in Fossil Fuel Power Plants, *Corrosion,* Vol 13, *ASM Handbook* (formerly 9th ed., *Metals Handbook*), ASM International, 1987, p 985–1010
35. J.C. Parker and D.F. Rosborough, *J. Inst. Fuel,* Feb 1972, p 95
36. M.H. Brown, W.B. DeLong, and J.R. Auld, Corrosion by Chlorine and by Hydrogen Chloride at High Temperatures, *Ind. Eng. Chem.,* Vol 39, 1947, p 839–844
37. Y. Ihara, H. Ohgame, and K. Sakiyama, The Corrosion Behavior of Iron in Hydrogen Chloride Gas and Gas Mixtures of Hydrogen Chloride and Oxygen at High Temperatures, *Corros. Sci.,* Vol 21 (No. 12), 1981, p 805–817
38. H.H. Krause, P.W. Cover, and W.E. Berry, in *Proc. First Conf. Advanced Materials for Alternative Fuel Capable Directly Fired Heat Engines,* 31 July–3 Aug 1979, Maine Maritime Academy, Castine, ME
39. K. Takehara and T. Ueshiba, *J. Soc. Mater. Sci. Jpn.,* Vol 179, 1968, p 755
40. R.W. Bradshaw and R.W. Carling, "A Review of the Chemical and Physical Properties of Molten Alkali Nitrate Salts and Their Effect on Materials Used for Solar Central Receivers," SAND 87-8005, Sandia National Laboratories, Livermore, CA, April 1987
41. S. Lampman, Elevated-Temperature Properties of Ferritic Steels, *Properties and Selection: Irons, Steels, and High-Performance Alloys,* Vol 1, *ASM Handbook* (formerly 10th ed., *Metals Handbook*), ASM International, 1990, p 617–652
42. Y.-W. Kim and R.A. Nickola, "A Heat Resistant Aluminized Steel for High Temperature Applications," Technical Paper, Series 300316, Society of Automotive Engineers, 1980
43. W. Beck, Comparison of Carbon Steel, Alonized Type 304 for Use as Dummy Slabs in Reheat Furnace Operations," Alon Processing, Inc., Tarentum, PA
44. T. Perng, "A Fundamental Study of the Noxso NO_x/SO_2 Flue Gas Treatment," Noxso, 1984
45. "Alonized Steels for High-Temperature Corrosion Resistance," Alon Processing, Inc., Tarentum, PA, 1990

Elevated-Temperature Corrosion Properties of Stainless Steels

STAINLESS STEELS are among the most popular construction materials for applications requiring resistance to corrosion at elevated temperatures (≥540 °C, or 1000 °F). Much attention has been given to the compatibility of stainless steels with air or oxygen. However, recent trends in nuclear reactor design, and in chemical process and steam-generation equipment, have resulted in renewed interest in stainless steel oxidation in CO, CO_2, and water vapor. Other gases, such as sulfur dioxide, hydrogen sulfide, hydrocarbons, ammonia, hydrogen, and the halogens, may be present in variable proportions and may strongly affect corrosive conditions. Information concerning the practical limits of operation of stainless steels is available for the simpler gaseous environments. However, it is exceedingly more difficult to predict service lives for complex, multicomponent environments without the aid of controlled field testing.

This article summarizes the practical information available about the compatibility of stainless steels with the gaseous environments mentioned above, as well as with molten salts and molten metals. Emphasis is on the wrought stainless steels. Information on the elevated-temperature corrosion behavior of cast stainless steels can be found in the article "High-Alloy Cast Steels" in this Volume.

Oxidation

Oxidation reactions are the most frequent causes of high-temperature corrosion of stainless steels. This is not surprising, because most environments are either air, oxygen, CO_2, or steam, or complex atmospheres containing one or more of these gases. (Sulfur dioxide may also be included, but it has been deferred to the section "Sulfidation" in this article.)

At elevated temperatures, stainless steels resist oxidation primarily because of their chromium content and their ability to form protective chromium oxide scales (see discussion below on "Oxidation Mechanisms"). Increased nickel minimizes scale spalling when temperature cycling occurs. Table 1 lists generally accepted maximum safe service temperatures for wrought stainless steels. Maximum temperatures for intermittent service are lower for the austenitic stainless steels and higher for most of the martensitic and ferritic stainless steels listed.

Oxidation Mechanisms. When iron is heated in highly oxidizing gases at temperatures exceeding 540 °C (1000 °F), a multilayered scale is formed consisting of wustite (FeO) at the metal-oxide interface, an intermediate layer of magnetite (Fe_3O_4), and hematite (Fe_2O_3) at the oxide-gas interface. As shown in Fig. 1 at 0 wt% Cr, the layers are arranged with the most metal-rich oxide next to the metal, progressively less metal in each succeeding layer, and finally the most oxygen-rich layer on the outside. Within each layer a concentration gradient exists with higher metal ion concentration closest to the metal. Unfortunately, iron oxide scales do not provide adequate protection to the base metal, and oxidation proceeds at a rapid rate.

Alloys intended for high-temperature applications are designed to have the capability of forming protective oxide scales. Alternatively, where the alloy has ultrahigh-temperature strength capabilities (which is usually synonymous with reduced levels of protective scale-forming elements), it must be protected by a specially designed coating. The only oxides that effectively meet the criteria for protective scales listed above and can be formed on practical alloys are chromium oxide (Cr_2O_3), alumina (Al_2O_3), and possibly silicon dioxide (SiO_2). The Cr_2O_3 scale is applicable to stainless steels.

Alloying requirements for the production of specific oxide scales have been translated into minimum levels of the scale-forming elements,

Table 1 Generally accepted maximum service temperatures in air for AISI stainless steels

AISI type	Maximum service temperature: Intermittent service, °C	Intermittent service, °F	Continuous service, °C	Continuous service, °F
Austenitic grades				
201	815	1500	845	1550
202	815	1500	845	1550
301	840	1545	900	1650
302	870	1600	925	1700
304	870	1600	925	1700
308	925	1700	980	1795
309	980	1795	1095	2000
310	1035	1895	1150	2100
316	870	1600	925	1700
317	870	1600	925	1700
321	870	1600	925	1700
330	1035	1895	1150	2100
347	870	1600	925	1700
Ferritic grades				
405	815	1500	705	1300
406	815	1500	1035	1895
430	870	1600	815	1500
442	1035	1895	980	1795
446	1175	2145	1095	2000
Martensitic grades				
410	815	1500	705	1300
416	760	1400	675	1250
420	735	1355	620	1150
440	815	1500	760	1400

Source: Ref 1

or combinations of elements, depending on the base alloy composition and the intended service temperature. Figure 1 schematically represents the oxidation rate of Fe-Cr alloys (1000 °C, or 1830 °F, in 0.13 atm O_2) and depicts the types of oxide scale associated with various alloy types. As Fig. 1 indicates, a minimum chromium content of approximately 20 wt% is needed to develop a continuous Cr_2O_3 scale against further oxidation in this environment.

In assessing the potential high-temperature oxidation behavior of an alloy, a useful guide is the reservoir of scale-forming element contained by the alloy in excess of the minimum level (around 20 wt% for Fe-Cr alloys at 1000 °C, or 1830 °F, according to Fig. 1). The greater the reservoir of scale-forming element required in the alloy for continued protection, the more likely it is that service conditions will cause repeated loss of the protective oxide scale. Extreme cases require chromizing or aluminizing to enrich the surface regions of the alloy, or an external coating that is rich in the scale-forming elements.

In the majority of cases, the breakdown of protective scales based on Cr_2O_3 appears to originate through mechanical means. The most common is spallation as a result of thermal cycling, or loss through impact or abrasion. Typical scale structures on an Fe-18Cr alloy after thermal cycling are shown in Fig. 2. Cases in which the scales have been destroyed chemically are usually related to reactions occurring beneath deposits, especially where these consist of molten species. An additional mode of degradation of protective Cr_2O_3 scale is through oxidation to the volatile chromium trioxide (CrO_3), which becomes prevalent above about 1010 °C (1850 °F) and is greatly accelerated by high gas-flow rates.

These protective oxide scales form wherever the alloy surface is exposed to the ambient environment, so they form at all surface discontinuities. Therefore, the possibility exists that notches of oxide will form at occluded angles in the surface and may eventually initiate or propagate cracks under thermal cycling conditions. The ramifications of stress-assisted oxidation (and of oxidation assisting the applied stress) are not well understood, but stress-assisted oxidation is an important consideration in practical failure analysis.

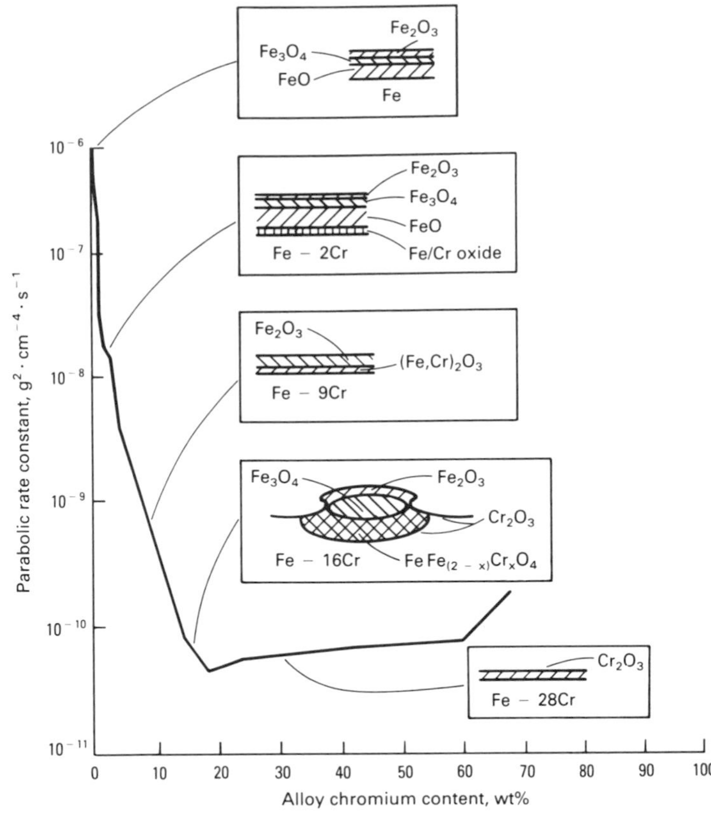

Fig. 1 Schematic of the variation of the oxidation rate and oxide scale structure with alloy chromium content (based on isothermal studies at 1000 °C, or 1830 °F, in 0.13 atm oxygen)

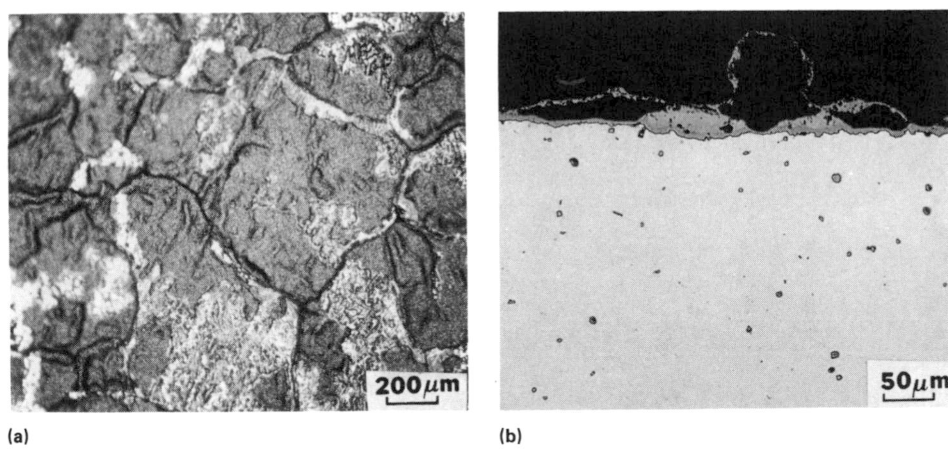

Fig. 2 Topography (a) and cross section (b) of oxide scale formed on Fe-18Cr alloy at 1100 °C (2012 °F). The bright areas on the alloy surface (a) are areas from which scale has spalled. The buckled scale and locally thickened areas (b) are iron-rich oxide. The thin scale layer adjacent to the alloy is Cr_2O_3, which controls the oxidation rate.

Oxidation Resistance of Martensitic and Ferritic Grades. Figure 3 illustrates the superior oxidation resistance of martensitic and ferritic stainless steels to that of carbon and low-alloy Cr-Mo steels. As chromium content in the straight chromium steels increases from 9 to 27%, resistance to oxidation improves significantly. The ferritic 27% Cr steel (type 446) is the most oxidation resistant among the 400-series stainless steels, due to the development of a continuous Cr_2O_3 scale on the metal surface.

Cyclic oxidation studies conducted by Grodner (Ref 3) also revealed that type 446 was the best performer in the 400-series stainless steels, followed by types 430, 416, and 410 (Fig. 4). Another ferritic stainless steel, 18SR (about 18% Cr), was found to be as good as, and sometimes better than, type 446 (27% Cr), as shown in Table 2. This was attributed to the addition of 2% Al and 1% Si to the alloy. Furthermore, 18SR and type 446 showed better cyclic oxidation resistance than some austenitic stainless steels, such as types 309 and 310, when cycled to 980 to 1040 °C (1800 to 1900 °F), as shown in Tables 2 and 3.

Oxidation Resistance of Austenitic Grades. When the service temperature is above 640 °C (1200 °F), ferritic stainless steels, which have a body-centered cubic (bcc) crystal structure, drastically lose their strengths. At these temperatures, alloys with a face-centered cubic (fcc) crystal structure are preferred because of their higher creep strengths. Nickel is added to Fe-Cr steels to stabilize the fcc austenitic structure. The austenitic structure is inherently stronger and more creep resistant than ferrite (Ref 5). The 300-series austenitic stainless steels have been widely used for high-temperature applications in various industries. These alloys exhibit higher high-temperature strengths than ferritic stainless steels. Furthermore, they do not suffer 475 °C (885 °F)

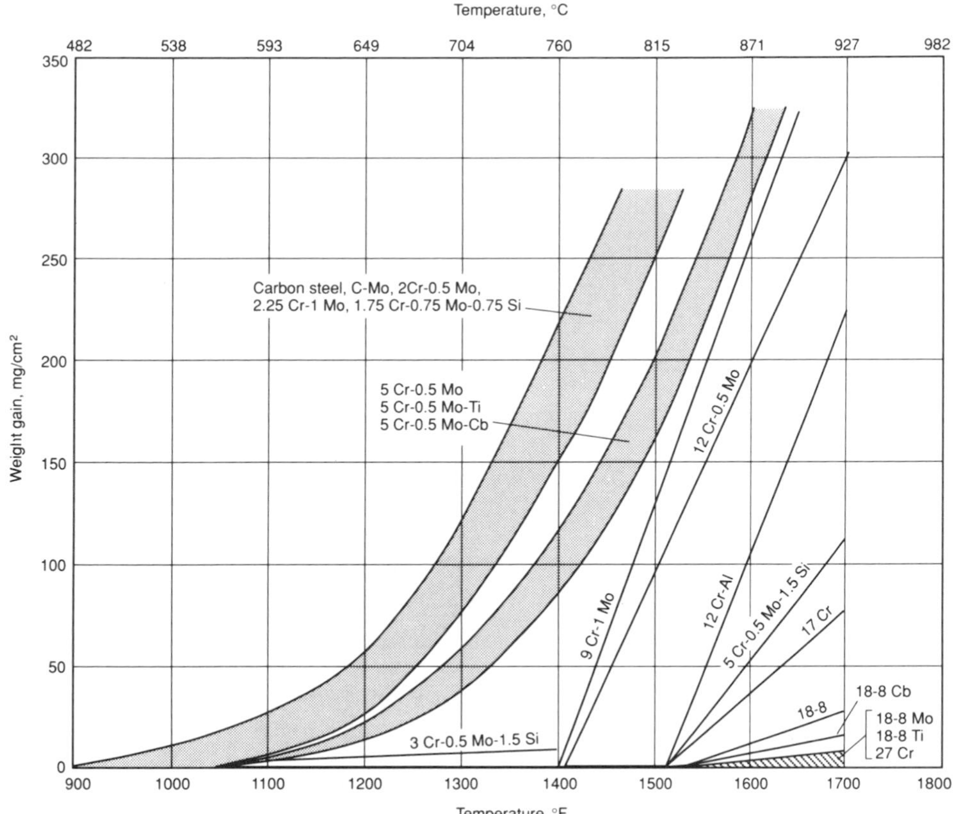

Fig. 3 Oxidation resistance of carbon, low-alloy, and stainless steels in air after 1000 h at temperatures from 590 to 930 °C (1100 to 1700 °F). Source: Ref 2

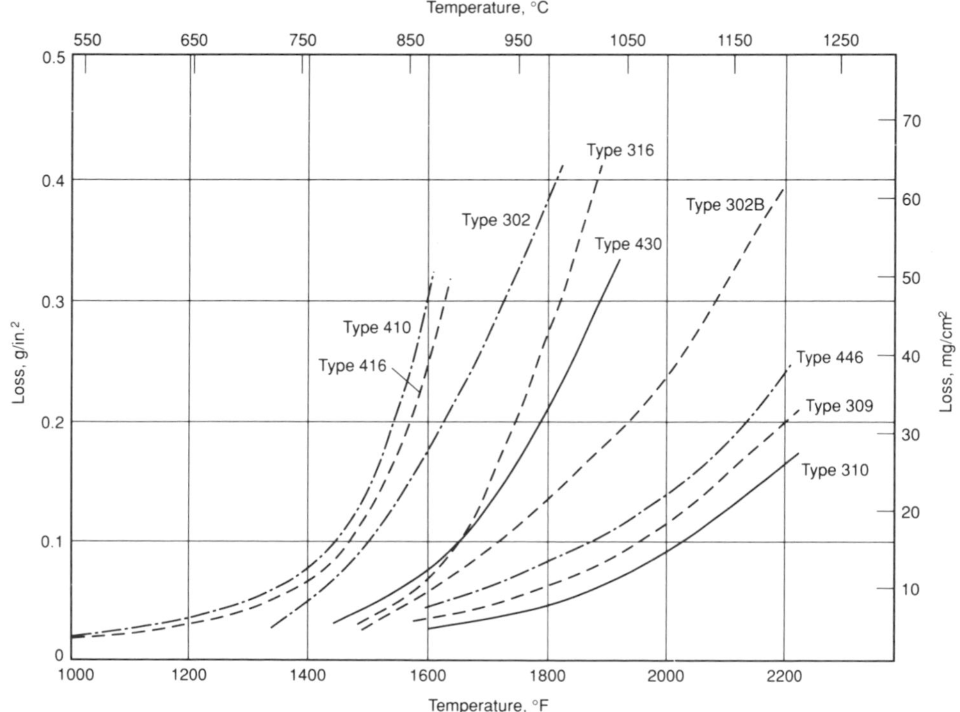

Fig. 4 Oxidation resistance of several stainless steels as a function of temperature. Source: Ref 3

embrittlement or ductility problems in thick sections and in heat-affected zones, as do ferritic stainless steels. Nevertheless, some austenitic stainless steels can suffer significant ductility loss or embrittlement upon long-term exposure to intermediate temperatures (540 to 800 °C, or 1000 to 1500 °F) due to σ-phase formation. Embrittlement of stainless steels is described in the article "Metallurgy and Properties of Wrought Stainless Steels" in *ASM Specialty Handbook: Stainless Steels.*

Table 2 Cyclic oxidation resistance of several stainless steels in air cycling to 870 to 930 °C (1600 to 1700 °F) temperature range

15 min in furnace and 15 min out of furnace

Alloy	Specimen weight changes after indicated cycles, mg/cm² 288 cycles	480 cycles	750 cycles	958 cycles
409 + Al	Destroyed	...	...	...
430	9.9	Destroyed	...	...
22-13-5	0.5	−3.0	−18.8	−35.7
442	0.7	1.2	1.5	1.5
446	0.3	0.4	0.2	0.1
309	0.3	−4.6	−23.7	−32.6
18SR	0.3	0.4	0.5	0.6

Source: Ref 7

Table 3 Cyclic oxidation resistance of several stainless steels in air cycling to 980 to 1040 °C (1800 to 1900 °F) temperature range

15 min in furnace and 15 min out of furnace

Alloy	Specimen weight changes after indicated cycles, mg/cm² 130 cycles	368 cycles	561 cycles	753 cycles	1029 cycles
446	0.4	0.5	−0.2	7.0	−19.4
18SR	0.7	1.1	1.5	2.2	3.0
309	−24.2	−77.5	−178.3	−242	−358
310	1.5	−11.3	−29.3	−62.8	−107

Source: Ref 4

The oxidation resistance of several austenitic stainless steels is illustrated in Fig. 5. Nickel improves the resistance of alloys to cyclic oxidation. Similar results have also been observed by Grodner (Fig. 4) and by Moccari and Ali (Ref 7). Brasunas et al. (Ref 8) studied the oxidation behavior of about 80 experimental Fe-Cr-Ni alloys exposed to an air-H_2O mixture at 870 to 1200 °C (1600 to 2190 °F) for 100 and 1000 h. They observed that increases in nickel in excess of 10% in alloys containing 11 to 36% Cr improved the oxidation resistance of the alloys. Figure 5 also shows several high-nickel alloys that exhibited better oxidation resistance than austenitic stainless steels.

Comparative Data from Engine Atmospheres. Oxidation studies have been carried out by Kado et al. (Ref 9) and Michels (Ref 10) to evaluate materials for automobile emission-control devices, such as thermal reactors and catalytic converters. In cyclic oxidation tests performed by Kado et al. (Ref 9) in still air at 1000 °C (1830 °F) for 400 cycles (30 min in the furnace and 30 min out of the furnace), types 409 (12% Cr), 420 (13% Cr), and 304 (18Cr-8Ni) suffered severe attack. Type 420 was completely

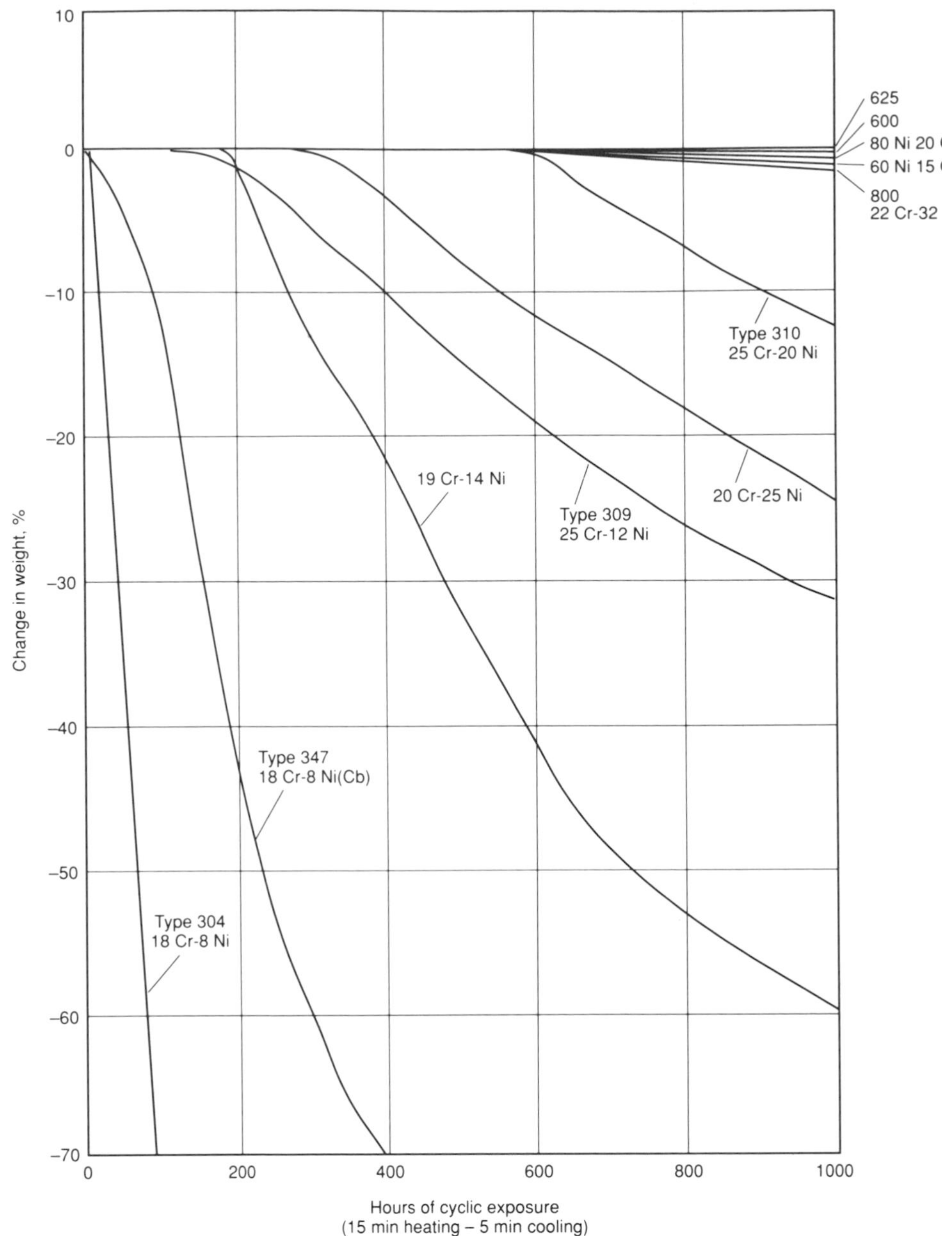

Fig. 5 Cyclic oxidation resistance of several stainless steels and nickel-base alloys in air at 980 °C (1800 °F). Source: Ref 6

oxidized after only 100 cycles, although the sample did not show any weight changes. Alloys that performed well under these conditions were types 405 (14% Cr), 430 (17% Cr), 446 (27% Cr), 310 (25Cr-20Ni), and DIN 4828 (19Cr-12Ni-2Si steel), as illustrated in Fig. 6. When cycled to 1200 °C (2190 °F) for 400 cycles (30 min in the furnace and 30 min out of the furnace), all the alloys tested except F-1 alloy (Fe-15Cr-4Al) suffered severe oxidation attack (Fig. 7). This illustrates the superior oxidation resistance of alumina formers (i.e., alloys that form Al_2O_3 scales when oxidized at elevated temperatures). Two familiar alumina-forming, ferritic alloys are Kanthal A (5% Al) and Fecralloy (4.7% Al and 0.3% Y).

Kado et al. (Ref 9) also investigated oxidation behavior in a combustion environment that simulated a gasoline engine. Their test involved air-to-fuel ratios of 9 to 1 and 14.5 to 1, using regular gasoline that contained 0.01 wt% S and 0.04 L/kL Pb. Exhaust gas taken from the exhaust manifold was mixed with air before being piped into the furnace retort where the tests were performed. Test specimens were exposed to the mixture of exhaust gas and air. The gas mixture coming from the combustion environment with an air-to-fuel ratio of 14.5 to 1 consisted of 72.4% N_2, 9.7% H_2O, 9.93% O_2, 8% CO_2, and 507 ppm NO_x, while that coming from the combustion environment with an air-to-fuel ratio of 9 to 1 consisted of 70.6% N_2, 13.7% H_2O, 3.21% O_2, 12.5% CO_2, and 34 ppm NO_x. Test results and air oxidation data are summarized in Fig. 8.

There were no significant differences between air and exhaust gas test environments when tested at 800 °C (1470 °F). All the alloys tested showed negligible attack. When the test temperature was increased to 1000 °C (1830 °F), all the 400-series stainless steels with less than 17% Cr (i.e., types 405, 409, 410, and 430) and type 304 exhibited significant oxidation attack. Type 310, type 446, DIN 4828 (Fe-19Cr-12Ni-2Si), F-1 alloy (Fe-15Cr-4Al), A-1 alloy (Fe-16Cr-13Ni-3.5Si), and A-2 alloy (Fe-20Cr-13Ni-3.5Si) performed well. At 1200 °C (2190 °F), all the alloys tested suffered severe oxidation attack. At both 1000 and 1200 °C (1830 and 2190 °F), the exhaust gas test environment was found to be more aggressive than air. The authors attributed the enhanced attack to the presence of sulfur in the exhaust gas environment, although low-sulfur (0.01%) gasoline was used for testing. Sulfur segregation to the scale-metal interface was detected.

In another study (Ref 10), an engine combustion atmosphere was also found to be significantly more corrosive than an air-10% H_2O environment. The engine combustion exhaust gas contained about 10% H_2O, along with 2% CO, 0.33 to 0.55% O_2, 0.05 to 0.24% hydrocarbon, and 0.085% NO_x. The balance was presumably N_2 (not reported in the paper). The engine exhaust gas was piped into the furnace retort where the tests were performed. The results generated in both the air-10H_2O environment and the engine exhaust environment are shown in Fig. 9. After exposure to the air-10H_2O environment at 980 °C (1800 °F) for 102 h, type 309, type 310, 18SR, alloy OR-1 (Fe-13Cr-3Al), alloy 800, and alloy 601 were all relatively unaffected. On the other hand, only 18SR and alloy 601 were relatively unaffected by the engine exhaust gas environment. Alloy OR-1, type 309, type 310, and alloy 800 suffered severe oxidation attack. The sulfur content in the gasoline used in this test was not reported. The relatively high gas velocity, about 6 to 9 m/s (20 to 30 ft/s), was considered one of the possible factors responsible for accelerated oxidation attack (Ref 10).

Oxidation data generated in combustion atmospheres is relatively limited. No systematic studies have been reported in which combustion conditions, such as air-to-fuel ratios, were varied. In combustion atmospheres, the oxidation behavior of metals or alloys is not controlled by just oxygen. The combustion products, such as H_2O, CO, CO_2, hydrocarbon, and others, are expected to influence oxidation behavior. Water vapor, for example, has been found to be detrimental to the oxidation resistance of austenitic stainless steels.

Figure 10 illustrates the effect of moist air on the oxidation of types 302 and 330. Type 302 undergoes rapid corrosion in wet air at 1095 °C (2000 °F), whereas a protective film is formed in dry air. The higher-nickel type 330 is less sensitive to the effects of moisture, so it is assumed that increased chromium and nickel permit higher operating temperatures in moist air (Ref 11).

Catastrophic Oxidation (Ref 12). As temperature increases, metals and alloys generally suffer increasingly higher rates of oxidation. When the temperature is excessively high, stainless steels can suffer *catastrophic oxidation.* An

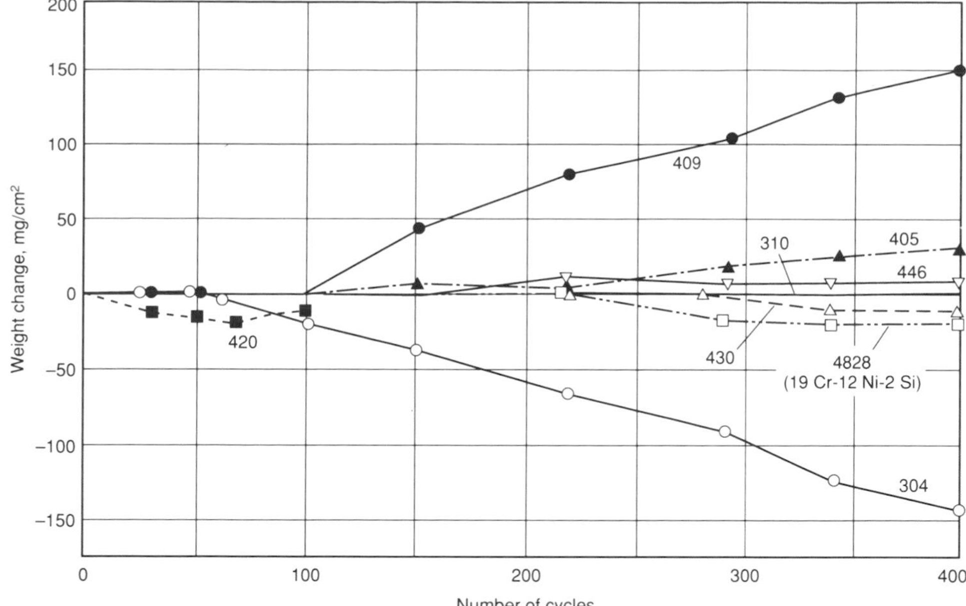

Fig. 6 Cyclic oxidation resistance of several ferritic and austenitic stainless steels in still air at 1000 °C (1830 °F) for up to 400 cycles (30 min in furnace and 30 min out of furnace). Source: Ref 9

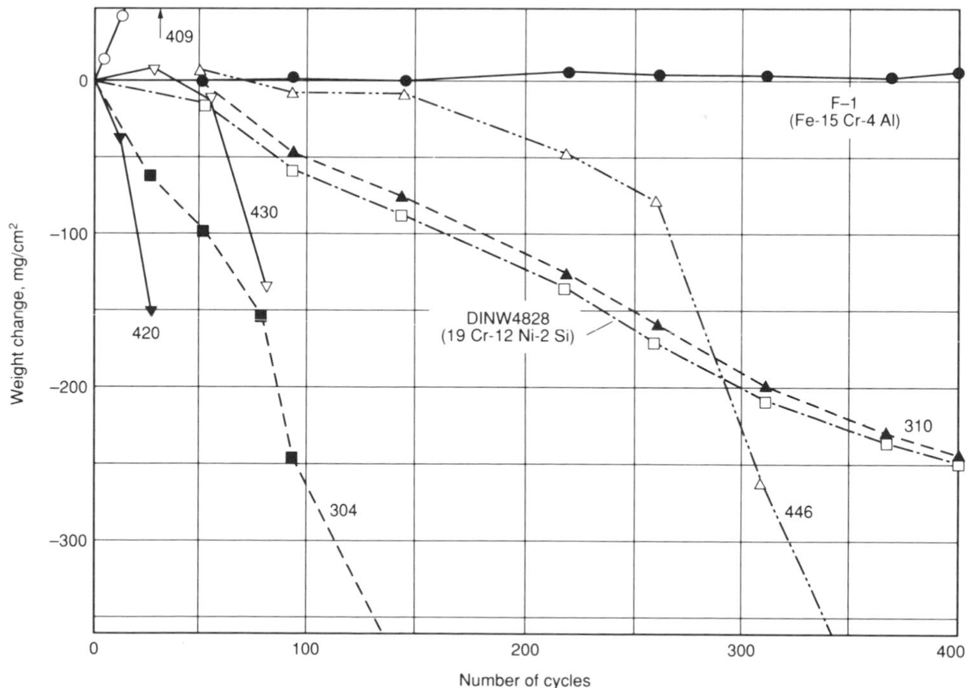

Fig. 7 Cyclic oxidation resistance of several ferritic and austenitic stainless steels in still air at 1200 °C (2190 °F) for up to 400 cycles (30 min in furnace and 30 min out of furnace). Source: Ref 9

example of this is shown in Fig. 11 for austenitic Fe-25Cr-10Ni steel.

Another mode of catastrophic oxidation takes place at relatively lower temperatures. This mode is generally associated with the formation of a liquid oxide that disrupts and dissolves the protective oxide scale, causing the alloy to suffer catastrophic oxidation. Leslie and Fontana (Ref 13) observed an unusually rapid oxidation rate for Fe-25Ni-16Cr alloy containing 6% Mo when heated to 900 °C (1650 °F) in static air. The same alloy exhibited good oxidation resistance when heated to the same temperature in flowing air. The authors postulated that the rapid oxidation was due to the accumulation of gaseous MoO_3 on the metal surface and the thermal dissociation of MoO_3 into MoO_2 and O. However, Meijering and Rathenau (Ref 14), Brasunas and Grant (Ref 15), and Brennor (Ref 16) attributed the rapid oxidation to the presence of a liquid oxide phase. The MoO_3 oxide melts at about 795 °C (1463 °F), and the 19Cr-9Ni steel suffered catastrophic oxidation attack in the presence of MoO_3 at 770 °C (1420 °F) in air (Ref 17). This temperature is very close to the eutectic temperature of MoO_2-MoO_3-Cr_2O_3.

Other oxides, such as PbO and vanadium pentoxide (V_2O_5), can also cause metals or alloys to suffer catastrophic oxidation in air at intermediate temperatures of 640 to 930 °C (1200 to 1700 °F) (Ref 12). PbO and V_2O_5 melt at 888 °C (1630 °F) and 690 °C (1270 °F), respectively. The deleterious effect of lead oxide is believed to be related to exhaust-valve failures in gasoline engines, because gasoline additives are a primary source for lead compounds. Vanadium is an important contaminant in residual or heavy fuel oils, and V_2O_5 plays a significant role in oil-ash corrosion.

Sawyer (Ref 18) indicated that accelerated oxidation of type 446 stainless steel in the presence of lead oxide can proceed at temperatures where the liquid phase does not exist. Experiments carried out by Brasunas and Grant (Ref 19) showed that 16-25-6 alloy specimens placed adjacent to, but not in contact with, 0.5 g samples of WO_3 oxides suffered accelerated oxidation attack when tested in air at 868 °C (1585 °F), which is well below the melting point of WO_3 (1473 °C, or 2683 °F). However, there are no reported data on mixed oxides involving WO_3.

Sulfidation

General Principles. When the sulfur activity (partial pressure, concentration) of the gaseous environment is sufficiently high, sulfide phases, instead of oxide phases, can be formed. The mechanisms of sulfide formation in gaseous environments and beneath molten-salt deposits have been determined in recent years. In the majority of environments encountered in practice by oxidation-resistant stainless steel alloys, Cr_2O_3 should form in preference to any sulfides, and destructive sulfidation attack occurs mainly at sites where the protective oxide has broken down. Sulfur, once it has entered the alloy, appears to tie up the chromium as sulfides, effectively redistributing the protective scale-forming elements near the alloy surface and thus interfering with the process of formation or re-formation of the protective scale. If sufficient sulfur enters the alloy so that all immediately available chromium is converted to sulfides, then the less stable sulfides of the base metal may form because of morphological and kinetic factors. These base metal sulfides are often responsible for the observed accelerated attack, because they grow much faster than the oxides or sulfides of chromium. In addition, they have relatively low melting points, so that molten slag phases often form.

Sulfur can transport across continuous protective scales of Cr_2O_3 under certain conditions, with the result that discrete sulfide precipitates can be observed immediately beneath the scales on alloys that are behaving in a protective manner. For reasons indicated above, as long as the amount of sulfur present as sulfides is small, there is little danger of accelerated attack. However, once sulfides have formed in the alloy, there is a tendency for the sulfide phases to be preferentially oxidized by the encroaching reaction front and for the sulfur to be displaced inward. New sulfides then form deeper in the alloy, often in grain boundaries or at the sites of other chromium-rich phases, such as carbides. In this way, fingerlike protrusions of oxide/sulfide can be formed from the alloy surface inward, which may act to localize stress or otherwise reduce the load-bearing section. Figure 12 shows such attack, experienced by an austenitic stainless steel in a coal gasifier product gas.

Sulfidation Resistance of Stainless Steels. As with oxidation, resistance to sulfidation relates to chromium content. Unalloyed iron will be converted rather rapidly to iron sulfide scale, but when iron is alloyed with chromium, sulfidation resistance is enhanced, as illustrated in Fig. 13. Other alloying elements that provide some protection against sulfidation include silicon, aluminum, and titanium.

However, the low-melting-point nickel/nickel sulfide eutectic may be formed on austenitic stainless steels containing more than 25% Ni, even in the presence of high chromium. The occurrence of molten phases during high-temperature service can lead to catastrophic destruction of the alloy.

In addition to the usual factors of time, temperature, and concentration, sulfidation depends on the form in which the sulfur exists. Of particular interest are the effects of sulfur dioxide, sulfur vapor, hydrogen sulfide, and flue gases.

Reaction with Sulfur Dioxide. Stainless steels containing more than 18 to 20% Cr are resistant to dry sulfur dioxide (Ref 21). In 24 h tests over the temperature range 590 to 870 °C (1100 to 1600 °F), only a heavy tarnish was formed on type 316 in atmospheres varying from 100% O_2 to 100% SO_2. The corrosion rate of type 316 in SO_2-O_2-N_2 atmospheres was 0.12 mm/yr (4.9 mils/yr) at 640 to 655 °C (1185 to 1210 °F). Fe-15Cr (types 430 and 440), Fe-30Cr (type 446), and Fe-18Cr-8Ni (type 304) show increasing resistance to sulfur dioxide, in the order indicated.

Reaction with Sulfur Vapor. Sulfur vapor readily attacks the austenitic grades. In tests, the following relatively high corrosion rates were encountered in flowing sulfur vapor at 570 °C (1060 °F) (Ref 21):

Type	Corrosion rate(a)	
	mm/yr	mils/yr
314	0.43	16.9
310	0.48	18.9
309	0.57	22.3
304	0.69	27.0
302B	0.76	29.8
316	0.79	31.1
321	1.39	54.8

(a) Corrosion rate based on 1295 h tests

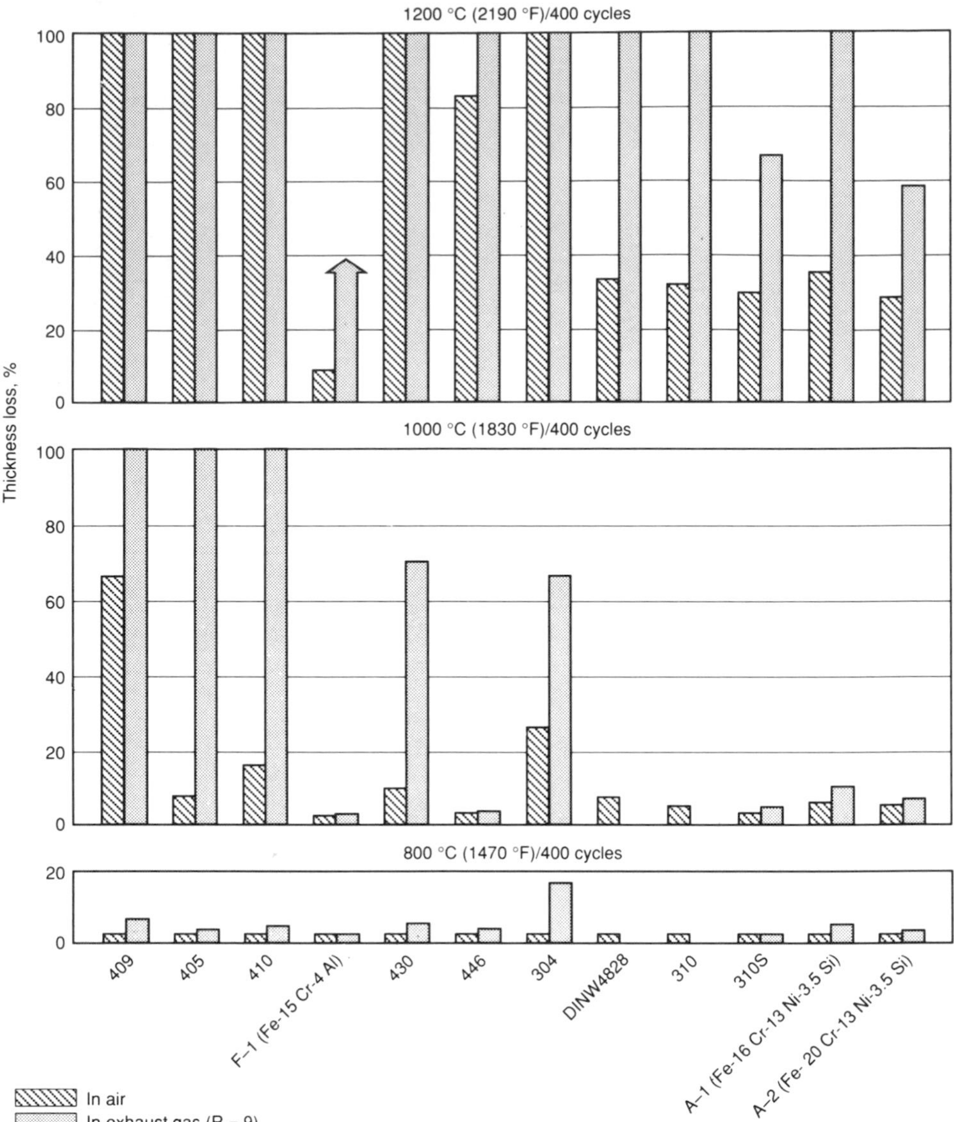

Fig. 8 Comparison of cyclic oxidation resistance between air and gasoline engine exhaust gas environments at 800, 1000, and 1200 °C (1470, 1830, and 2190 °F) for 400 cycles (30 min in hot zone and 30 min out of hot zone). Alloy F-1 suffered localized attack at 1200 °C in engine exhaust gas. Source: Ref 9

In liquid sulfur, most austenitic grades are resistant up to 200 °C (400 °F), with the stabilized types 321 and 347 showing satisfactory service up to 445 °C (830 °F) (Ref 20).

Reaction with Hydrogen Sulfide. The rate of corrosion in hydrogen sulfide depends on concentration, temperature, pressure, and the permeability of the sulfide scale. The presence of chromium in the steel helps to stabilize the scale and slow the diffusion process. However, at high pressure and temperature when hydrogen is present, the attack is more aggressive, to the extent that low-chromium steels are not adequate.

Hydrogen-hydrogen sulfide (H_2-H_2S) mixtures are characteristic of the gas stream in catalytic reforming units (Ref 22). Catalytic reforming is used in petroleum refineries to upgrade the octane number of gasoline (Ref 23). A large amount of hydrogen is present in catalytic reforming, and sulfur in the naphtha charge reacts with hydrogen to form hydrogen sulfide (Ref 23). Severe corrosion attack on processing equipment by hydrogen sulfide has been encountered in several catalytic reforming and desulfurizing units (Ref 22–24).

Sorell (Ref 22) reported hydrogen sulfide corrosion data that was generated by various refinery operators in laboratory tests, pilot plant testing, field testing in commercial units, and inspection of commercial operating equipment. Results are summarized in Table 4. Austenitic stainless steels (18Cr-8Ni) were most resistant, followed by straight chromium stainless steels (12 to 16% Cr). Low-chromium steels (0 to 9% Cr) were worst. It is interesting to note that adding a moderate amount of nickel to Fe-Cr alloys significantly improves the alloy sulfidation resistance of the alloy in H_2-H_2S environments (Fig. 14).

Sulfidation of alloys in H_2-H_2S mixtures has been described by isocorrosion rate curves that

Table 4 Summary of corrosion data in H_2-H_2S mixtures typical of catalytic reforming generated by various petroleum refining companies

Source	Type of corrosion data(a)	Temperature, °C (°F)	Test duration, days	Hydrogen pressure, atm (psi)	H_2S concentration, vol%	Materials	Corrosion rate, mm/yr (mils/yr)
American Oil	I	510 (950)	36–89	18–27 (265–400)	0.03–0.07	1¼Cr, 2¼Cr	1.5 (59)
	C	510-550 (950–1025)	89	18–27 (265–400)	0.03–0.07	0–9Cr	1.5 (59)
						12Cr	0.76 (30)
						18Cr–8Ni	0.33 (13)
Atlantic	L, P, C, I	480 (900)	1–365	34 (500)	0.013–0.074	0–5Cr	0.1–2.5 (5–100)
					0.011–0.13	11½-13½Cr	0.03–1.5 (1–60)
					0.015–0.27	18Cr–8Ni	0.03–0.2 (1–8)
	I	480 (900)	180	34 (500)	0.036	2¼Cr	1.0 (38)
					0.016	12Cr	0.18 (7)
D-X Sunray	I	465 (870)	180	27 (400)	0.04	5Cr	1.0 (40)
Canadian Petrofina	C	480–490 (890–920)	90	21–22 (310–330)	0.008	0-9Cr	0.4 (16)
						12Cr	0.2 (8)
						18Cr–8Ni	0.06 (2.5)
Humble	C	480–510 (900–956)	1–4	20 (300)	0.013–0.14	0–12Cr	1.9–10 (73–398)
						18Cr–8Ni	0.1–1.1 (5–42)
	P	540 (1000)	4	20 (300)	0.007–0.15	0–12Cr	0.03–10 (1–400)
	I	450–520 (850–975)	127	20 (300)	0.035	1¼Cr	2.3 (90)
		480–550 (900–1025)				CS, 2¼Cr	5.1 (200)
Major U.S. Ref. Comp.	C	470–520 (875–960)	139–577	34–36 (500–535)	0.035–0.09	2¼Cr, 5Cr	0.3–1.0 (13.5–41.5)
						12Cr	0.28–0.7 (11–27)
						18Cr–8Ni	0.1 (3.7)
Major West Coast Ref. Comp.	C	480–490 (900–920)	72	37–38 (540–565)	0.015	0–5Cr	0.5 (18.9)
						12Cr	0.2 (8.3)
						18Cr–8Ni	0.1 (4.5)
Pure Oil	I	480–540 (900–1000)	240	34 (500)	0.026	2Cr	0.8 (33)
Richfield	L, P, C, I	510 (950)	16.7	27 (400)	0.011–1.0	0–9Cr	0.4–7.9 (15–310)
					0.014–1.0	12Cr	0.3–6.1 (10–240)
					0.025–1.0	18Cr–8Ni	0.1–0.5 (3.5–20)
Shell Oil	C	510 (950)	71	43 (625)	0.015	0–9Cr	0.5 (18)
						12Cr	0.2 (8)
	I	510 (950)	148–758	46 (680)	0.05–0.08	5Cr	1.4–3.4 (56–135)
		510–610 (955–1135)	758	46 (670)	0.08	9Cr	1.7 (68)
	C	500 (925)	148–251	46 (680)	0.05–0.08	0–9Cr	0.9–1.2 (35–47)
						12Cr	0.4–0.6 (15–22)
						18Cr–8Ni	0.1–0.2 (4.3–7.2)
Sinclair	L	510 (950)	4.2	34 (500)	0.08–1.0	0–7Cr	2.0–11 (80–450)
					0.1–1.0	13Cr	0.5–5.3 (20–210)
					0.1–1.0	18Cr–8Ni	0.1–1.2 (3–48)
	C	490–500 (910–940)	212–382	34 (500)	0.001–0.0089	0–7Cr	0.1–0.3 (2.4–10)
						13Cr	0.05–0.2 (1.9–6.5)
						18Cr–8Ni	0.01-0.02 (0.55-0.6)
	I	480–550 (900–1025)	120–270	34 (500)	0.005	5Cr	0.2 (9)
		480–510 (900–950)				12Cr	0.2 (6.5)
Socony Mobil	L	530 (985)	6.3–20.8	33 (485)	0.009–0.2	0–5Cr	0.03–7.2 (1–285)
					0.017–0.2	7–16Cr	0.03–2.6 (1–102)
					0.06–0.2	Cr–Ni	0.03-0.2 (1-8.7)
	L	490 (905)	21	31 (460)	0.085	5Cr	1.1 (44)
	C	470–500 (885–935)	55	12 (175)	0.5	0–9Cr	2.7 (108)
						12–16Cr	0.5 (21)
						Cr–Ni	0.1 (5)
	C	470–490 (885–920)	50.5–213	37–39 (550–575)	0.018–0.04	0–9Cr	0.5–1.6 (18–62)
						12Cr	0.3–0.6 (13–23)
						Cr–Ni	0.02–0.1 (0.7–4.3)
Standard Oil (Ind.)	L	480 (900)	0.7–11.7	19 (285)	0.013–0.44	0–9Cr	0.3–13 (10–500)
						12Cr	0.1–5 (4–200)
						18Cr–8Ni	0.03–1.6 (1.3–62.5)
	C	510 (950)	16.5–46	20 (300)	0.023	CS	1.8 (70)
		510–540 (950–1000)			0.032		1.1 (45)
	C	550 (1030)	47	23 (340)	0.029	0–9Cr	1.0 (37.5)
						18Cr–8Ni	0.1 (3.3)
Sun Oil	C	500–520 (925–960)	188–395	20–41 (300–600)	0.017–0.04	0–5Cr	0.2–0.8 (6.1–29.7)
						12Cr	0.1–0.3 (5.8–11.7)
						18Cr–8Ni	0.02-0.05 (0.7-1.9)
Texas	C	500 (940)	133–668	41 (600)	0.0016	0–5Cr	0.07 (2.9)
						12Cr	0.05 (1.9)
						18Cr–8Ni	0.01 (0.3)

(a) L, laboratory corrosion tests; P, pilot plant corrosion tests; C, commercial unit corrosion tests; I, inspection of commercial operating equipment. Source: Ref 22

show corrosion rate as a function of hydrogen sulfide concentration and temperature (Ref 25). The isocorrosion curves in Fig. 15 show the effects of hydrogen sulfide and temperature on the sulfidation resistance of austenitic stainless steels.

Effect of Sulfur in Flue and Process Gases. It is extremely difficult to generalize corrosion rates in flue and process gases, because gas com-

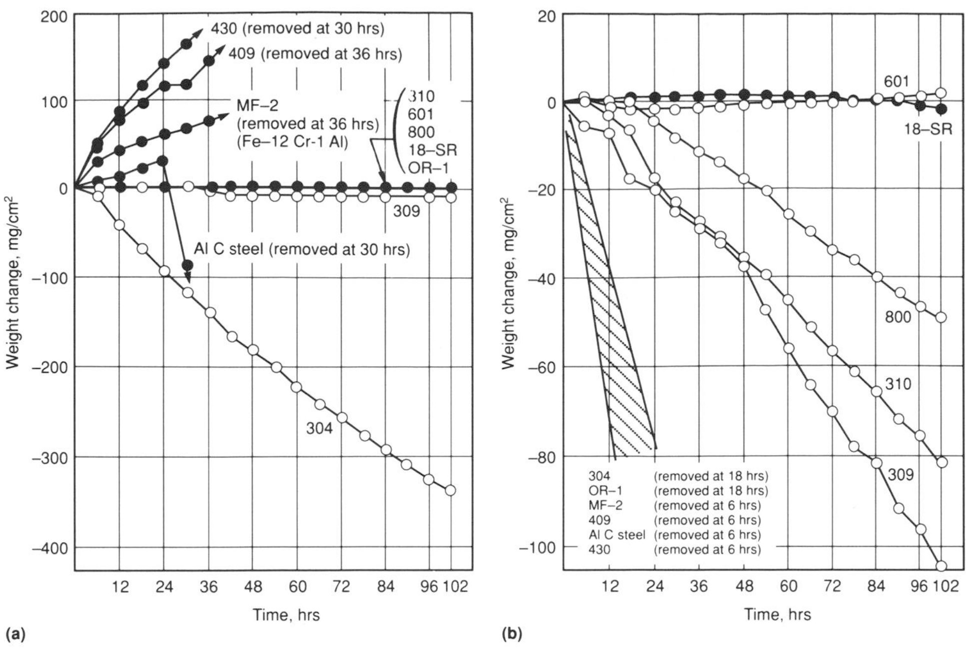

Fig. 9 Cyclic oxidation resistance of several ferritic and austenitic stainless steels in (a) air/10% H_2O at 980 °C (1800 °F), cycled every 2 h, and (b) gasoline engine exhaust gas at 980 °C (1800 °F), cycled every 6 h. Source: Ref 10

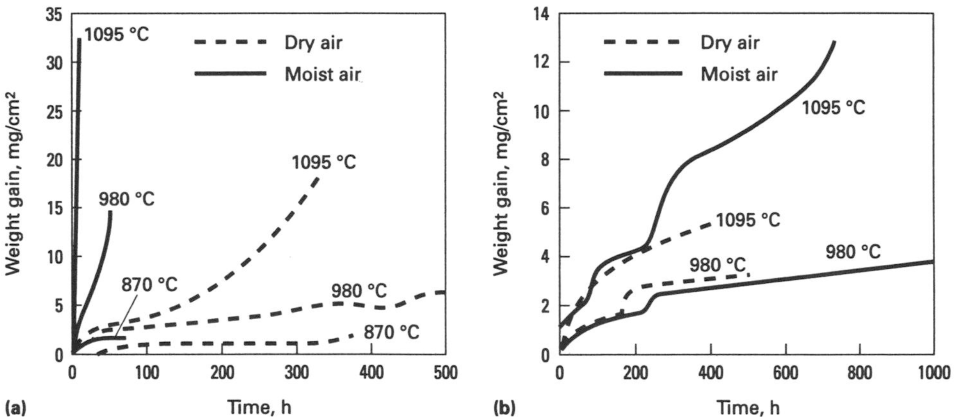

Fig. 10 Oxidation rates of type 302 (a) and type 330 (b) in moist and dry air. Source: Ref 11

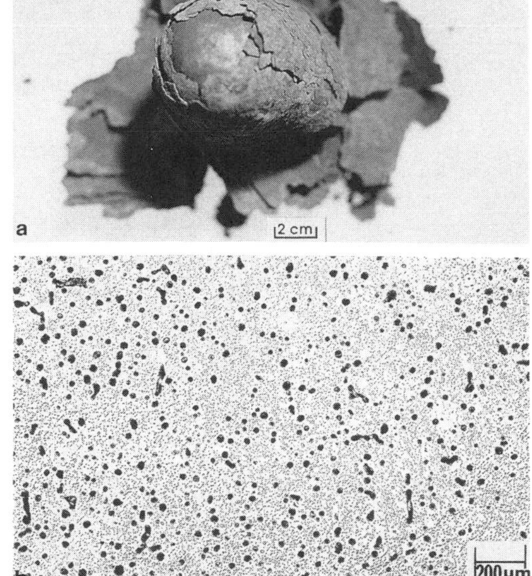

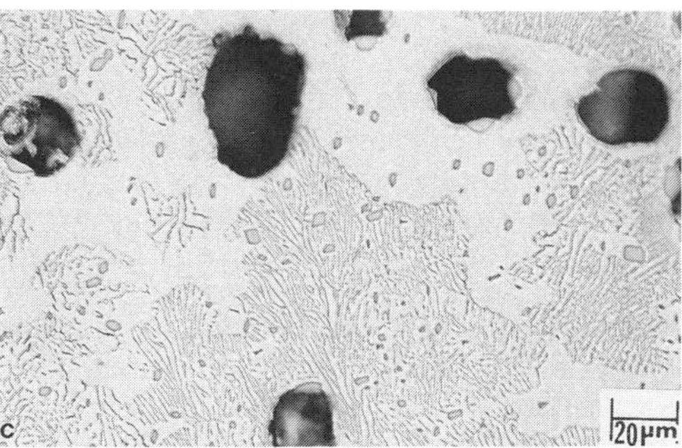

Fig. 11 25Cr-10Ni steel that suffered catastrophic oxidation in a natural gas-fired furnace operating at temperatures up to 1230 °C (2250 °F), above the upper temperature limit of the alloy for long-term applications. The alloy suffered not only extensive scaling (a), but also extensive internal void formation (b and c). Source: Ref 12

position and temperature may vary considerably within the same process unit. Combustion gases normally contain sulfur compounds; sulfur dioxide is present as an oxidizing gas along with CO, CO_2, nitrogen, and excess oxygen. Protective oxides are generally formed, and depending on the exact conditions, the corrosion rate may be approximately the same as in air, or slightly greater. The resistance of stainless steels to normal combustion gases is increased by successive increments in chromium content. Table 5 indicates the beneficial effect of chromium and the influence of fuel source.

Reducing flue gases contain varying amounts of CO, CO_2, hydrogen, hydrogen sulfide, and nitrogen. The corrosion rates encountered in these environments are sensitive to hydrogen sulfide content and temperature, and satisfactory material selection often necessitates service tests. Table 6 illustrates the effect of sulfur content on the corrosion of types 309, 310, and 330 in oxidizing and reducing flue gases. The deleterious effect of high nickel content is apparent (type 330).

Carburization and Metal Dusting

Carburization of stainless steels can occur in CO, methane (CH_4), and other hydrocarbon gases, such as propane (C_3H_8), at elevated temperatures. Carburization can also occur when stainless steels that are contaminated with oil or grease are annealed without sufficient oxygen to burn off the carbon. This can occur during vacuum or inert gas annealing, as well as during open air annealing of oily parts whose shapes restrict air access.

Carburization problems are quite common to heat treating equipment, particularly furnace retorts, baskets, fans, and other components used for case hardening of steels by gas carburizing. In the petrochemical industry, carburization is one of the major modes of high-temperature corrosion for processing equipment. Pyrolysis tubes for the production of ethylene and olefins are a good example. The majority of the stainless steels

Table 5 Corrosion rates of stainless steels after 3-month exposure to flue gases

Type	Corrosion rate, mm/yr (mils/yr)		
	Coke oven gas at 815 °C (1500 °F)	Coke oven gas at 980 °C (1800 °F)	Natural gas at 815 °C (1500 °F)
430	2.31 (91)	6.0 (236)(a)	0.3 (12)
446 (26Cr)	0.76 (30)	1.0 (40)	0.1 (4)
446 (28Cr)	0.68 (27)	0.36 (14)	0.07 (3)
302B	2.6 (104)	5.7 (225)(a)	...
309S	0.94 (37)(b)	1.14 (45)	0.07 (3)
310S	0.96 (38)(b)	0.64 (25)	0.07 (3)
314	0.58 (23)(b)	2.39 (94)	0.07 (3)

(a) Specimen destroyed. (b) Pitted specimens—average pit depth. Source: Ref 21 used for these two application areas are cast alloys. The carburization behavior of cast stainless steels is described in the article "High-Alloy Cast Steels" in this Volume. The remainder of this section is devoted to wrought stainless steels.

General Principles. As in the case of sulfide penetration, carburization of high-temperature alloys is thermodynamically unlikely, except at very low oxygen partial pressures, because the protective oxides of chromium are generally more likely to form than the carbides. While carburization can occur kinetically in many carbon-containing environments, carbon transport across continuous nonporous scales of Cr_2O_3 is very slow. Furthermore, the alloy pretreatments that are likely to promote such scales, such as initially smooth surfaces or preoxidation, have generally been effective in decreasing carburization attack. In practice, the scales formed on high-temperature alloys often consist of multiple layers of oxides resulting from localized bursts of oxide formation in areas where the original scale was broken or lost. The protection is derived from the innermost layer, which is usually richest in chromium. Concentration of gaseous species, such as CO, in the outer porous oxide layers appears to be one means by which sufficiently high carbon activities can be generated at the alloy surface for carburization to occur in otherwise oxidizing environments. The creation of localized microenvironments is also possible under deposits that create stagnant conditions not permeable by the ambient gas.

Once carbon is inside the alloy, its detrimental effects depend on the location, composition, and morphology of the carbides formed. Austenitic steels carburize more readily than ferritic steels because of the high solubility of carbon in austenite. Iron-chromium alloys containing less than about 13% Cr contain various amounts of austenite, depending on temperature, and are susceptible to carburization, while alloys with 13 to 20% Cr form austenite as a result of absorption of small amounts of carbon. Iron-chromium alloys containing more than ~20% Cr can absorb considerable amounts of carbon before austenite forms, becoming principally $(CrFe)_{23}C_6$ and ferrite. An example of rapid high-temperature carburization attack of an austenitic stainless steel is shown in Fig. 16.

Effect of Carburization on Stainless Steels. Diffusion of carbon in stainless steel results in the formation of additional carbides, which may take the form of M_7C_3 $M_{23}C_7$, or M_3C_2. Figure 17(a) shows a typical example of a microstructure developed in stainless steel as a result of carburization: islands of massive carbides have formed in the austenite matrix. In the same steel before carburization, carbides were small and were restricted to the grain boundaries (Fig. 17b). In the carburized steel, chromium has migrated to the carbides, depleting the matrix. As chromium depletion progresses, the relative percentages of nickel and iron have increased.

The primary significance of carburization is its effect on properties. A carburized alloy has slightly increased creep strength, and a change in volume results from the increased amount of carbide, which has a lower density than that of the original alloy. A density gradient, predicated on carbon content, develops across the carburized zones and into those zones that have not been carburized. In stainless steel heater tubes used in petroleum refineries, carburization occurs on the interior walls of the tubes. This leads to the development of an inner layer of carburized metal and an outer layer of uncarburized metal, each with a different density and a different coefficient of thermal expansion. During thermal cycling, these

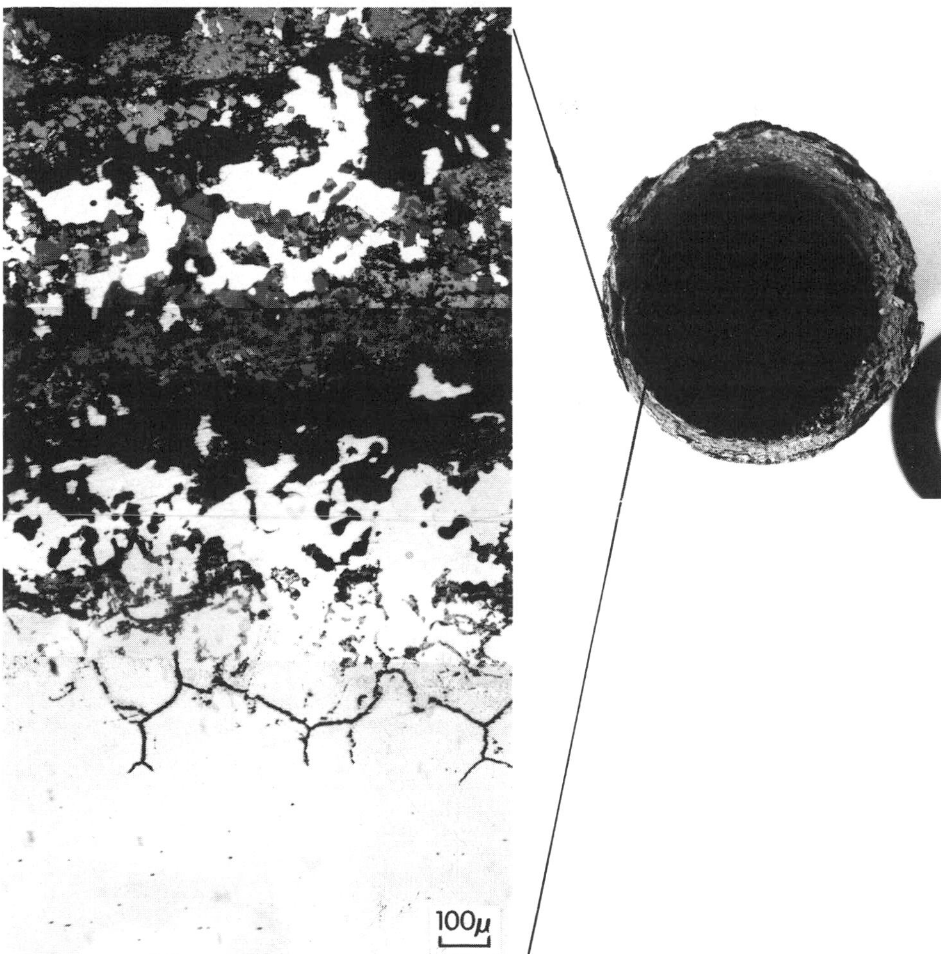

Fig. 12 Example of high-temperature sulfidation attack in a type 310 heat-exchanger tube after ~100 h at 705 °C (1300 °F) in coal gasifier product gas

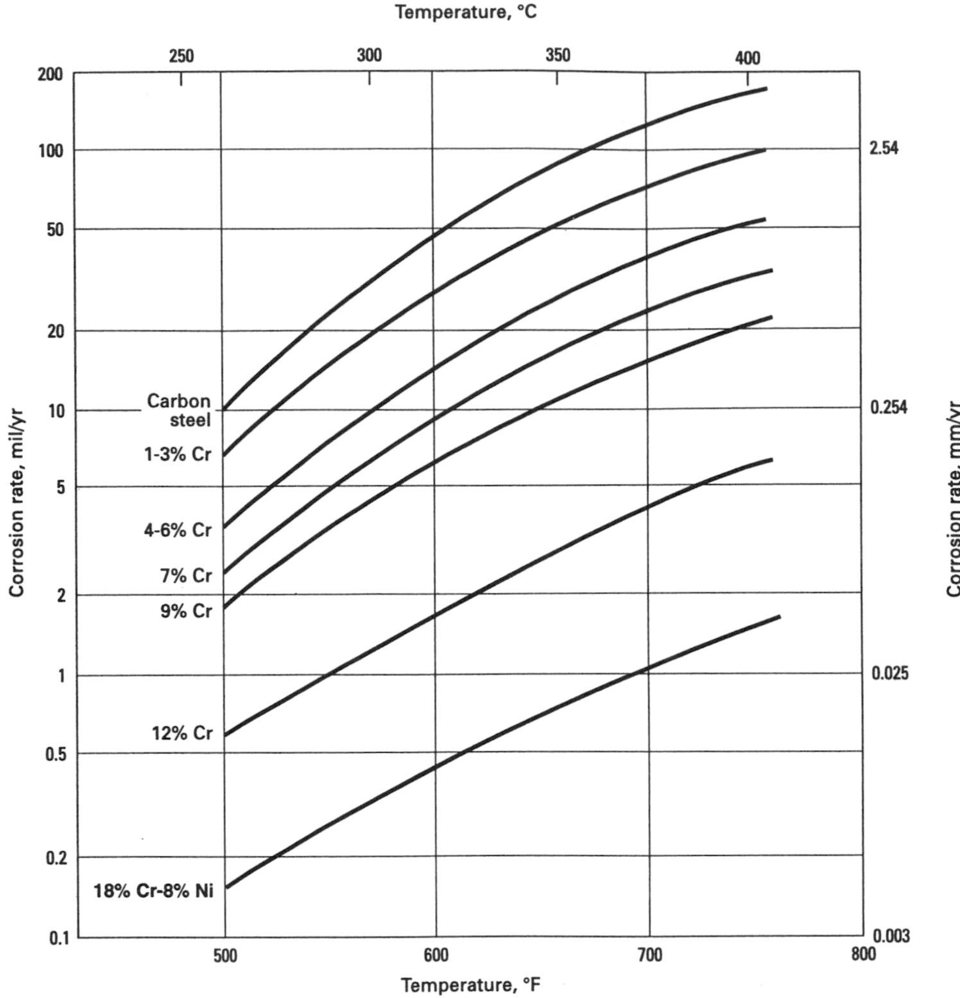

Fig. 13 Effect of chromium content on the sulfidation resistance of iron-base alloys in a hydrogen-free environment. Source: Ref 20

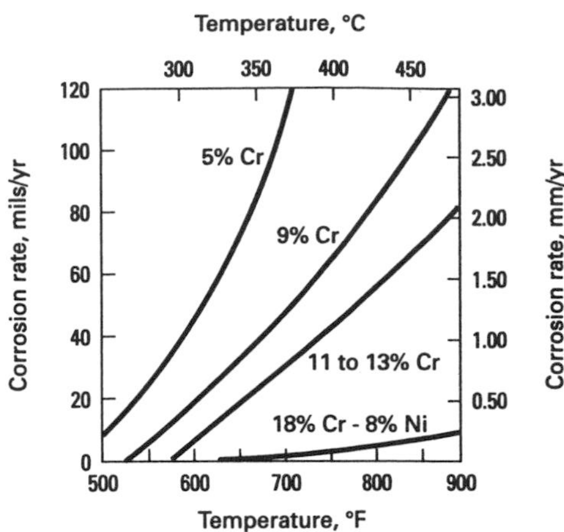

Fig. 14 Corrosion rates of Fe-Cr and Fe-Cr-Ni alloys in H_2-H_2S environments (H_2S concentrations above 1 mol%). Source: Ref 20

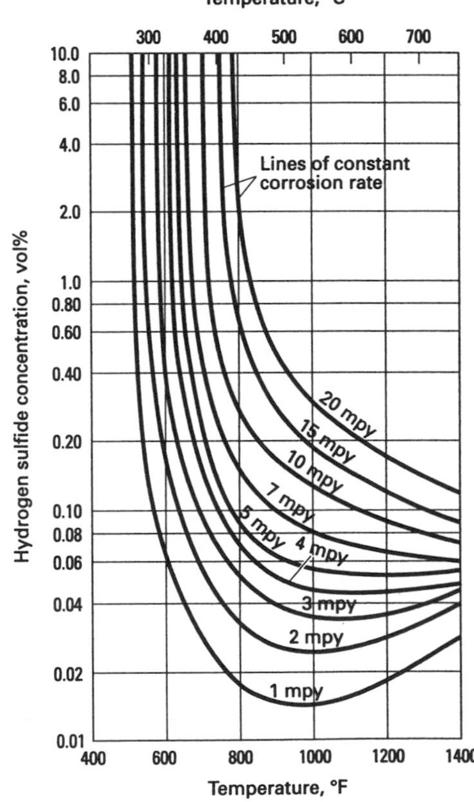

Fig. 15 Isocorrosion diagram showing the effect of hydrogen sulfide concentration on corrosion rate (in mils per year, or mpy) of austenitic stainless steels in hydrogen atmospheres at 1200 to 3450 kPa (175 to 500 psig). Exposure time was longer than 150 h. Source: Ref 25

Table 6 Corrosion rates of cast stainless steels in air and oxidizing and reducing flue gases

Type	Corrosion rate, mm/yr (mils/yr)					
	Air		Oxidizing flue gas at 1095 °C (2000 °F)		Reducing flue gas at 1095 °C (2000 °F)	
	1095 °C (2000 °F)	1200 °C (2200 °F)	5 gr S(a)	100 gr S(a)	5 gr S(a)	100 gr S(a)
309	1-2.3 (40-90)	1.5 (60)	1.3-1.8 (50-70)	1-2.5 (40-100)	0.5-1.3 (20-50)	0.76 (30)
310	1 (40)	1.3-2 (50-80)	1.3 (50)	1 (40)	0.5-1.3 (20-50)	0.76 (30)
330	1.3 (50)	2.5-25 (100-1000)	1.5-7.6 (60-300)	2.5-12.7 (100-500)	1.3-5 (50-200)	7.6-20 (300-800)

(a) Grains of sulfur per 100 ft^3. Source: Ref 21

differences promote the generation of high thermal stresses that can result in tube failure at elevated temperatures.

In addition to its adverse effects on density and thermal expansion, carburization of stainless steels contributes to embrittlement by producing a high volume percentage of carbide and an increased susceptibility to attack by oxidation. Oxidation attack is promoted by the depletion of chromium from the matrix by preferential formation of chromium carbides. The attack occurs when metal surfaces are simultaneously or intermittently exposed to heavily carburizing and oxidizing environments. Under these conditions, it is thermodynamically possible for the dual action of carburization and oxidation to weaken the metal. Specifically, in carburized metal, oxidation attack occurs at grain boundaries. Loss of grain-boundary strength is followed by detachment of grains and subsequent erosion.

Carburization Resistance. Mason et al. (Ref 26) investigated various wrought stainless steels by performing pack carburization tests. The results are shown in Table 7. Silicon is again noted for its beneficial alloying effect, as illustrated by type 330 (0.47% Si) versus type 330 (1.0% Si) and by type 304 (0.39% Si) versus type 302B (2.54% Si). The improved carburization of silicon-alloyed stainless steels is attributed to the formation of SiO_2 scale, which is more impervious to carbon ingress than Cr_2O_3 scale. Chromium was found to be beneficial in Fe-Cr alloys, as shown by type 446 (27% Cr) versus 430 (16% Cr). Small additions of titanium or niobium appeared to be beneficial when comparing type 321 and type 347 to type 304. Increasing nickel in Fe-Cr-Ni alloys also improves carburization re-

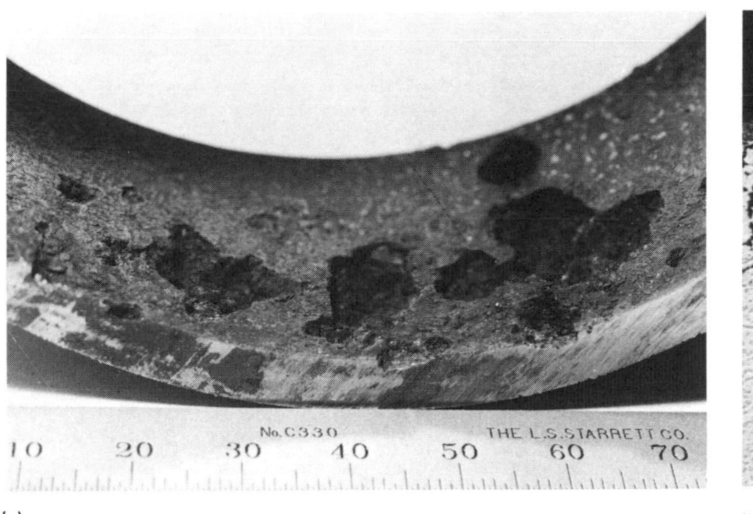

(a)

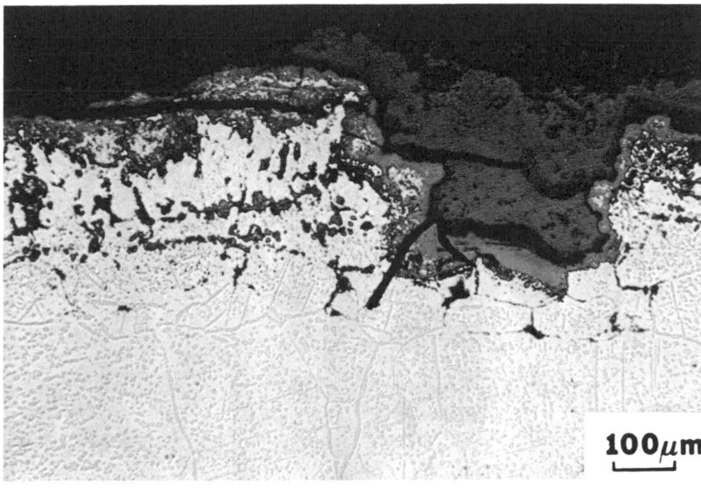

(b)

Fig. 16 Example of high-temperature carburization attack pitting in a type 310 reactor wall after ~4000 h exposure to coal gasification product gas. The pits were formed during operation under conditions of high carbon activity in the gas. (a) Overall view of pitting. (b) Section through a pit

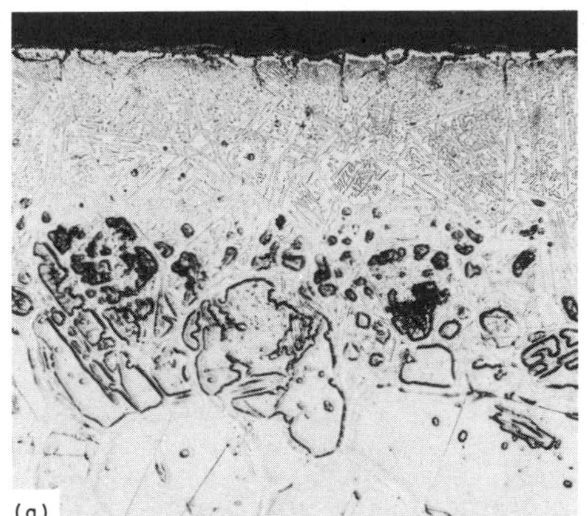

(a)

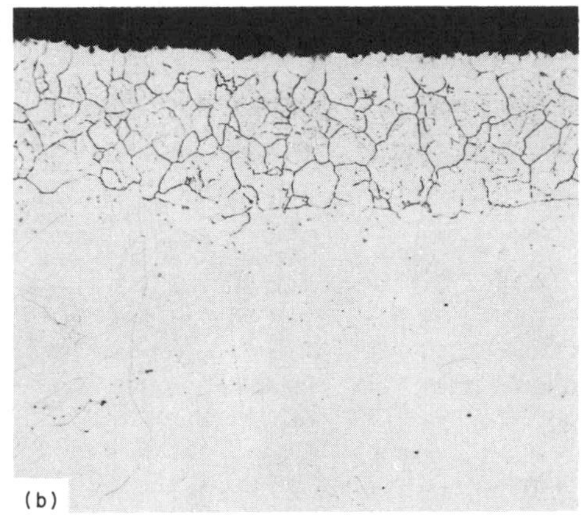

(b)

Fig. 17 Effect of carburization on the microstructure of stainless steel. (a) Typical microstructure after carburization, showing massive carbides in the austenite matrix. Massive carbides are formed by the reaction of carbon with chromium, which depletes the matrix of chromium in regions adjacent to the carbides. (b) Microstructure of the same steel before carburization. Both etched with a mixture of 10 mL HNO_3, 20 mL HCl, 20 mL glycerol, 10 mL H_2O_2, and 40 mL water. 50×

Table 7 Results of pack carburization tests for various stainless steels

Alloy	Nominal composition	Silicon content, %	Increase in carbon content(a), %
800	34Ni-21Cr	0.34	0.04
330	35Ni-15Cr	0.47	0.23
330	35Ni-15Cr + Si	1.00	0.08
310	20Ni-25Cr	0.38	0.02
314	20Ni-25Cr + Si	2.25	0.03
309	12Ni-25Cr	0.25	0.12
347	8Ni-18Cr + Nb	0.74	0.57
321	8Ni-18Cr + Ti	0.49	0.59
304	8Ni-18Cr	0.39	1.40
302B	8Ni-18Cr + Si	2.54	0.22
446	28Cr	0.34	0.07
430	16Cr	0.36	1.03

Note: 40 cycles of 25 h at 980 °C. Carburizer renewed after each cycle. (a) By bulk analysis. Source: Ref 26

sistance, because nickel reduces the diffusivity of carbon.

Metal dusting is a form of carburization attack that generally results in metal wastage, such as pitting and/or thinning. Metal dusting occurs between 430 and 900 °C (800 and 1650 °F) and has been encountered in heat treating, refining and petrochemical processing, and other operations. The environment must be highly carburizing, with enrichment of H_2, CO, or hydrocarbons. A stagnant atmosphere is particularly conducive to metal dusting. Figure 18 shows a stainless steel component that was subjected to metal dusting in a furnace used to manufacture carbon fibers for reinforced composites. These carbon fibers were manufactured from polyacrylonitrile (see the article "Carbon-Carbon Composites" in this Volume for manufacturing details). The last step of manufacturing, "carbonization," was carried out at about 900 °C (1650 °F) in an environment enriched with CO, CO_2, CH_4, N_2, HCN, H_2O, and so forth. As indicated in Fig. 18, furnace components made of iron-base alloys were found to suffer general metal thinning and pitting attack.

Nitridation

Stainless steels are generally susceptible to nitridation attack when exposed to ammonia-bearing environments at elevated temperatures. Ammonia is a commonly used nitriding gas for case hardening of steel, which is typically performed at temperatures from 500 to 590 °C (925 to 1100 °F). Furnace equipment and components that are repeatedly subjected to these service conditions frequently suffer brittle failures as a result of nitridation attack.

Carbonitriding, another important method of case hardening, produces a surface layer of both carbides and nitrides. The process is typically carried out at 700 to 900 °C (1300 to 1650 °F) in ammonia, with additions of carbonaceous gases, such as CH_4. Thus, the heat treat retort, fixtures, and other furnace equipment are subject to both nitridation and carburization.

The production of ammonia is still another source of nitridation corrosion. Ammonia is produced by reacting nitrogen with hydrogen over a catalyst, typically at temperatures of 500 to 550 °C (930 to 1020 °F) and pressures of 200 to 400 atm.

Nitridation Resistance. Nitridation attack is different from other modes of high-temperature corrosion in that metals or alloys do not suffer metal loss or metal wastage. Nitrogen from the environment is absorbed on the metal surface and then diffuses into the interior. Once nitrogen exceeds its solubility limit, nitrides precipitate out.

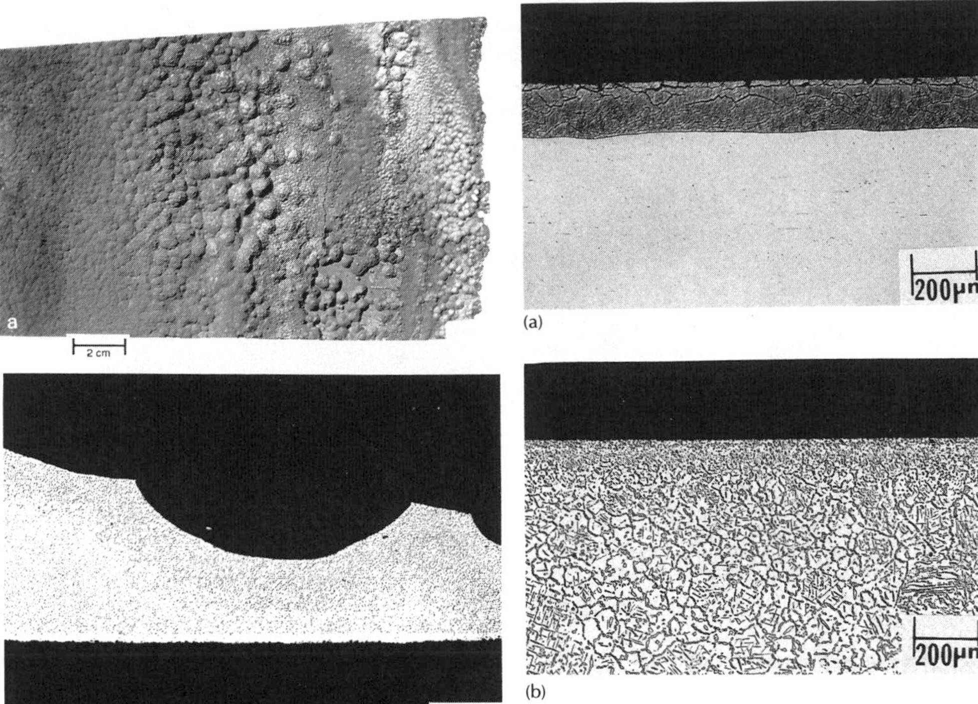

Fig. 18 Type 310 stainless steel component suffering metal dusting in a furnace used in manufacturing carbon fibers. (a) General view. (b) Cross section of the sample, showing pitting attack

Fig. 19 Typical morphology of nitrides for type 310 stainless steel formed in ammonia at (a) 650 °C (1200 °F) for 168 h and (b) 1090 °C (2000 °F) for 168 h

When the temperature is low, such as 500 °C (930 °F), the diffusion of nitrogen is slow.

Nitridation at these temperatures generally results in the formation of a surface nitride layer (Fig. 19a). At higher temperatures (e.g., 1000 °C, or 1830 °F), the diffusion of nitrogen is rapid. In this case nitridation leads to the formation of internal nitrides in the matrix and at grain boundaries (Fig. 19b). With either a surface nitride or internal nitrides, the metal or alloy can become brittle.

Verma et al. (Ref 27) reported that an ammonia cracker unit, used to produce nitrogen and hydrogen, failed after only about 1000 h of operation. The preheater tubes (operating at 350 to 400 °C, or 660 to 750 °F) were made of type 304 stainless steel, while the furnace tubes (operating at about 600 °C, or 1100 °F) were made of type 310. Both suffered severe nitridation attack. To select alternate alloys, nitriding tests were performed on various alloy samples at 600 °C (1100 °F) in an environment consisting of 6 to 8% NH_3, 75.77 to 77.5% N_2, and 16.25 to 16.5% H_2 (by wt). Test results are summarized in Table 8. The alloys that performed well include types 347, 316, 321, SLX-254, and HV-9A. Type 347 was the best performer, having a linearly extrapolated penetration rate of about 0.13 mm/yr (5 mils/yr). Alloy 800, which contains more nickel than any of the above stainless steels, did not perform as well. Furthermore, type 304 was found to suffer attack two orders of magnitudes higher than that of type 316L. The results also showed that titanium suffered severe nitridation attack, which resulted in severe sample cracking. The temperature of 600 °C (1100 °F) may be too high for both carbon steel and 1Cr-0.5Mo steel in ammonia environments, since both alloys suffered decarburization after only 50 h of exposure.

The nitridation resistance of various alloys was studied by Moran et al. (Ref 28) in an ammonia converter and preheater line. The results are summarized in Table 9. Corrosion rates were found to depend strongly on the concentration of ammonia. Type 304, for example, suffered corrosion rates that increased from about 0.02 to 2.5 mm/yr (0.6 to 99 mils/yr) as the concentration of NH_3 was increased from 5 to 6% (in the ammonia converter) to 99% (in the ammonia preheater line) at about 500 °C (930 °F). In an ammonia converter with about 5 to 6% NH_3 and 490 to 550 °C (910 to 1020 °F), all stainless steels tested (i.e., 430, 446, 302B, 304, 316, 321, 309, 314, 310, and 330) showed negligible nitridation attack, with corrosion rates of about 0.03 mm/yr (1 mil/yr) or less. For the plant ammonia line (preheater exit), which was exposed to 99% NH_3, stainless steels (e.g., 446, 304, 316, and 309) suffered severe nitridation attack, with corrosion rates of about 2.54 mm/yr (100 mils/yr) or more. Moran et al. also found that type 304 was better than type 316, contrary to the observations of Verma et al. (Ref 27).

Rorbo (Ref 29) reported the performance experience of several alloys in a Topsoe-type ammonia converter. Most of the components made of type 304, exposed to temperatures up to 500 °C (930 °F) with ammonia concentrations up to 20%, exhibited negligible nitridation rates of attack (about 0.01 to 0.1 mm/yr, or 0.4 to 4 mils/yr). One type 304 sample showed a slightly higher corrosion rate (0.25 mm/yr, or 10 mils/yr), presumably due to a higher temperature. Alloy 600 (Ni-Cr-Fe alloy) had significantly better nitridation resistance than stainless steels, with corro-

Table 8 Nitridation attack of various alloys in an ammonia-bearing environment at 600 °C (1110 °F) for indicated exposure times

Alloy	Penetration depth of nitridation attack, mm (mils), after: 50 h	100 h	300 h	600 h	1000 h	1500 h
Carbon steel	Decarb.	Decarb.	Decarb.	Decarb.	Decarb.	Decarb.
1Cr-0.5Mo steel	Decarb.	Decarb.	0.033 (1.3)	0.033 (1.3)	0.033 (1.3)	0.3 (11.8)
Titanium	0.0066 (0.3)	0.0133 (0.5)	0.233 (9.2)	0.266 (10.5)	Cracked(a)	Cracked(a)
304	...	0.013 (0.5)	0.013 (0.5)	0.03 (1.2)	0.06 (2.4)	4.2 (165)
316L	0.02 (0.8)	0.02 (0.8)	0.02 (0.8)	0.03 (1.2)	0.04 (1.6)	0.04 (1.6)
329	...	0.066 (2.6)	0.10 (3.9)	0.10 (3.9)	0.20 (7.9)	0.40 (15.7)
310	...	0.03 (1.2)	0.13 (5.1)	0.16 (6.3)	0.33 (13.0)	0.40 (15.7)
321	0.013 (0.5)	0.013 (0.5)	0.013 (0.5)	0.016 (0.6)	0.06 (2.4)	0.06 (2.4)
347	...	0.013 (0.5)	0.013 (0.5)	0.013 (0.5)	0.02 (0.8)	0.02 (0.8)
SLX-254(b)	0.013 (0.5)	0.013 (0.5)	0.026 (1.0)	0.026 (1.0)	0.06 (2.4)	0.06 (2.4)
HV-9A(c)	0.01 (0.4)	0.10 (3.9)	0.10 (3.9)	0.10 (3.9)	0.10 (3.9)	0.10 (3.9)
800	0.02 (0.8)	0.10 (3.9)	0.20 (7.9)	...	0.20 (7.9)	0.20 (7.9)

Decarb., decarburized. (a) Nitridation through thickness. (b) SLX-254: Fe-19.7Cr-24.5Ni-4.35Mo-1.43Cu. (c) HV-9A: Fe-21.2Cr-24.6Ni-3.8Mo-1.5Cu. Source: Ref 27

Table 9 Corrosion behavior in an ammonia converter and plant ammonia line

Alloy	Corrosion rate, mm/yr (mils/yr): Ammonia converter(a)	Plant ammonia line(b)
430	0.022 (0.90)	...
446	0.028 (1.12)	4.18 (164.5)
302B	0.019 (0.73)	...
304	0.015 (0.59)	2.53 (99.5)
316	0.012 (0.47)	>13.21 (520)
321	0.012 (0.47)	...
309	0.006 (0.23)	2.41 (95)
314	0.003 (0.10)	...
310	0.004 (0.14)	...
330 (0.47Si)	0.002 (0.06)	...
330 (1.00Si)	0.001 (0.02)	0.43 (17.1)
600	...	0.16 (6.3)
80Ni-20Cr	...	0.19 (7.4)
Ni	...	2.01 (79.0)

(a) 5 to 6% NH_3, 29164 h at 490 to 550 °C (910 to 1020 °F) and 354 atm (5200 psi) (Haber-Bosch converter). (b) 99.1% NH_3, 1540 h at 500 °C (930 °F). Source: Ref 28

Table 10 Corrosion behavior of various alloys in an ammonia converter

Converter No.	Component	Material	Gas temperature, °C	NH_3, %	Time of operation, yr	Thickness of nitride, μm (mils)	Average nitriding, μm/yr (mils/yr)
1	Lining	304	525	15–20	4	1000 (39.4)	250 (9.8)
	Plate, 2nd bed	304	475	15–20	7	100 (3.9)	14 (0.6)
2	Bolt	302	...	...	7	375 (14.8)	54 (2.1)
	Wire mesh, 2nd bed	Alloy 600	520	...	7	8 (0.3)	1 (0.04)
3	Perforated plate, 1st bed	304	500	13	5	270 (10.6)	54 (2.1)
	Inner shell, 2nd bed	304	440	8–10	5	45 (1.8)	9 (0.4)
	Perforated plate, 2nd bed	304	440	8–10	5	60 (2.4)	12 (0.5)
	Center tube, 2nd bed	304	485	16	5	440 (17.3)	88 (3.5)
	Nut, bottom	304	480	16	5	260 (10.2)	52 (2.0)
	Bolt, bottom	403	480	16	5	540 (21.3)	108 (4.3)
	Wire mesh	Alloy 600	500		4	6 (0.2)	1.5 (0.06)
4	Thermowell	304	500	3.5	8	200 (7.9)	25 (1.0)

Note: Topsoe-type ammonia converter operated at 22 MPa (3.2 ksi)

Table 11 Depth of nitridation for various alloys after exposure for 1 and 3 yr in a Casale ammonia converter

Alloy	Nitridation depth, mm (mils) 1 yr	3 yr
502 (5Cr steel)	2.88 (113.2)	Completely nitrided
446	1.06 (41.7)	1.15 (45.3)
304	1.08 (42.7)	1.12 (44.0)
316	0.46 (18.2)	0.48 (18.7)
321	0.46 (18.3)	0.60 (23.6)
347	0.49 (19.2)	0.45 (17.6)
309	0.24 (9.5)	0.24 (9.6)
310	0.22 (8.8)	0.23 (9.2)
800	0.14 (5.4)	0.13 (5.3)
804 (30Cr-42Ni)	0.03 (1.2)	0.03 (1.2)
600	0.16 (6.4)	0.16 (6.4)
Nickel 200	None	None

Note: Operated at 540 °C (1000 °F) and 76 MPa (11 ksi). Source: Ref 30

sion rates one or two orders of magnitude lower. These results are summarized in Table 10.

McDowell (Ref 30) reported field test results performed in a Casale converter (540 °C, or 1000 °F, and 76 MPa, or 11,000 psi) for 1 and 3 years. The results are summarized in Table 11. AISI 502 (5Cr steel) was extremely susceptible to nitridation attack, with over 2.54 mm (100 mils) of nitridation depth in a year. Results showed a general trend of increased resistance to nitridation as alloy nickel content increased. One striking observation was that after 3 years of exposure, the alloys showed essentially the same depths of nitridation attack as they did after 1 year.

Table 12 Corrosion of stainless steels in chlorine

Chlorine pressure was approximately 1.0 atm.

Alloy	Temperature °C	°F	Flow rate, L/min	Linear rate constant(a), μm/min	Corrosion rate(b) mm/yr	mils/yr
Ferritic stainless (Fe-17Cr)	300	572	15	4×10^{-4}	0.2	7.9
	360	680	15	3.8×10^{-3}	2	79
	440	824	15	6.7×10^{-2}	40	1.6 in.
	540	1004	15	1.35	700	28 in.
Austenitic stainless (Fe-18Cr-9Ni-Ti)	418	784	15	1.1×10^{-3}	0.6	24
	450	842	15	4.3×10^{-2}	20	787
	480	896	15	0.13	70	2.8 in.
	535	995	15	0.47	200	7.9 in.
	640	1184	15	46	20,000	787 in.
Austenitic stainless (Fe-18Cr-8Ni-Mo)	315	599	28	1.4×10^{-3}	0.8	31
	340	644	28	2.9×10^{-3}	1.5	59
	400	752	28	5.9×10^{-3}	3	118
	450	842	28	2.9×10^{-2}	15	590
	480	896	28	5.9×10^{-2}	30	1.2 in.
Austenitic stainless (Fe-18Cr-8Ni)	290	554	28	1.5×10^{-3}	0.8	31
	315	599	28	2.9×10^{-3}	1.5	59
	340	644	28	5.9×10^{-3}	3	118
	400	752	28	2.9×10^{-2}	15	590
	450	842	28	5.9×10^{-2}	30	1.2 in.

(a) Duration of these tests was 60 to 360 min for the first two alloys and 120 to 1200 min for the last two alloys. (b) Estimated metal loss after one year of exposure. Source: Ref 31

Halogen Gas Corrosion

Halogen gas corrosion involves reactions between metals and chlorides, fluorides, and hydrogen halides such as hydrogen chloride (HCl) and hydrogen fluoride (HF). Halogen and halogen compounds generally attack via the gaseous phase or molten salt compounds. Salts cause slagging and disintegration of the oxide layer; the gas-phase halogens penetrate deeply into the material without destroying the oxide layer. Therefore, preoxidation is of no benefit. This section deals primarily with gas-phase halogenation. Corrosion due to molten salt compounds is discussed in the section "Ash/Salt Deposit Corrosion" in this article.

Resistance to Halogen Gas Corrosion. In Cl_2- and HCl-bearing environments, the corrosion behavior of various alloy systems is strongly dependent on whether the environment is oxidizing or reducing. For Cl_2-bearing environments with no measurable oxygen, iron and steel are very susceptible to chlorination attack. Adding chromium and/or nickel to iron improves the corrosion resistance of the alloy. Thus, ferritic and austenitic stainless steels can resist chlorination attack at higher temperatures than cast iron and carbon steels. Nickel and nickel-base alloys, including Ni-Cr-Fe, Ni-Cr-Mo, and Ni-Mo alloys, are significantly more resistant to chlorination attack than stainless steels and Fe-Ni-Cr alloys such as 800-type alloys. Table 12 lists the corrosion resistance of stainless steels in chlorine.

In oxidizing environments containing both chlorine and oxygen, molybdenum and tungsten are detrimental to the resistance of the alloy to chlorination attack, presumably due to the formation of highly volatile oxychlorides such as WO_2Cl_2 and MoO_2Cl_2. Thus, nickel-base alloys containing high levels of tungsten or molybdenum, such as alloy 188 (14% W) and alloy C-276 (16% Mo, 4% W), suffer higher rates of corrosion attack than Fe-Ni-Cr and Ni-Cr-Fe alloys, such as type 310 stainless steel and alloy 600. The addition of aluminum improves the chlorination resistance of nickel-base alloys such as alloy 214 (Ni-Cr-Al). Table 13 compares the corrosion rates of nickel-base alloys and stainless steels in an oxygen- and chlorine-containing environment.

In reducing environments containing HCl, nickel and nickel-base alloys are generally more resistant than iron-base alloys, such as austenitic stainless steels. Table 14 compares the corrosion rates of various materials, including stainless steels, in dry HCl.

Stainless steels are very susceptible to corrosion in fluorine gas at temperatures as low as 300 °C (570 °F) or even lower (Table 15). Commercially pure nickel probably has the best resistance to fluorine corrosion at elevated temperatures, attributable to the formation of an adherent nickel fluoride scale. Many nickel-base alloys are significantly worse than pure nickel.

Table 13 Corrosion of selected alloys in Ar-20O_2-2Cl_2 at 900 °C (1650 °F) for 8 h

Alloy	Metal loss		Average metal affected(a)	
	mm	mils	mm	mils
214	0	0	0.012	0.48
R-41	0.004	0.16	0.028	1.12
600	0.012	0.48	0.035	1.36
310SS	0.012	0.48	0.041	1.60
S	0.053	2.08	0.063	2.48
X	0.020	0.80	0.071	2.80
C-276	0.079	3.12	0.079	3.12
6B	0.014	0.56	0.098	3.84
188	0.014	0.56	0.116	4.56

(a) Metal loss + average internal penetration. Source: Ref 32

Stainless steels also exhibit poor resistance to fluorination attack in HF environments. As shown in Table 16, pure nickel and nickel-base alloys such as alloy 400 and alloy 600 are far more resistant to HF attack than stainless steels.

Table 14 Corrosion of alloys in dry HCl

Based on short-term laboratory tests

Alloy	Approximate temperature, °C (°F), at which given corrosion rate is exceeded			
	0.8 mm/yr (30 mils/yr)	1.5 mm/yr (60 mils/yr)	3.0 mm/yr (120 mils/yr)	15 mm/yr (600 mils/yr)
Nickel	455 (850)	510 (950)	565 (1050)	675 (1250)
600	425 (800)	480 (900)	538 (1000)	675 (1250)
B	370 (700)	425 (800)	480 (900)	650 (1200)
C	370 (700)	425 (800)	480 (900)	620 (1150)
D	288 (550)	370 (700)	455 (850)	650 (1200)
18-8Mo	370 (700)	370 (700)	480 (900)	593 (1100)
25-12Cb	345 (650)	400 (750)	455 (850)	565 (1050)
18-8	345 (650)	400 (750)	455 (850)	593 (1100)
Carbon steel	260 (500)	315 (600)	400 (750)	565 (1050)
Ni-resist	260 (500)	315 (600)	370 (700)	538 (1000)
400	230 (450)	260 (500)	345 (650)	480 (900)
Cast iron	205 (400)	260 (500)	315 (600)	455 (850)
Copper	93 (200)	148 (300)	205 (400)	315 (600)

Source: Ref 33

Table 15 Corrosion of several alloys in fluorine

Tests were conducted in flowing fluorine.

Alloy	Exposure time, h	Corrosion rate, mm/yr (mils/yr)		
		200 °C (400 °F)	370 °C (700 °F)	540 °C (1000 °F)
400	5	0.013 (0.5)	0.048 (1.9)	0.76 (29.8)
	24	0.013 (0.5)	0.043 (1.7)	0.29 (11.3)
	120	0.003 (0.1)	0.031 (1.2)	0.18 (7.2)
Ni-200	5	0.084 (3.3)	0.043 (1.7)	0.62 (24.5)
	24	0.013 (0.5)	0.031 (1.2)	0.41 (16.1)
	120	0.003 (0.1)	0.010 (0.4)	0.35 (13.8)
304	5	0.155 (6.1)	40 (1565)	...
304L	24	0.191 (7.5)	153 (6018)	...
	120	0.65 (25.4)	...	...
347	5	0.102 (4.0)	108 (4248)	...
Illium R	5	0.152 (6.0)	0.32 (12.7)	103 (4038)
600	5	0.015 (0.6)	2.0 (78.0)	88 (3451)

Source: Ref 34

Ash/Salt Deposit Corrosion

Corrosion due to fuel ash or salt deposits is a serious problem in many installations, including incinerators, boilers, heat exchangers, gas turbines, calciners, and recuperators. Except for most gaseous fuels, combustion of fossil fuels produces solid, liquid, and gaseous compounds that can be corrosive to structural components (e.g., superheater tubes) and heat-transfer surfaces. In addition, deposits of solid and liquid residues in gas passages can alter the heat-transfer characteristics of the system, with potentially severe effects on system efficiency and tube-wall temperatures.

Residues from the combustion process, called ash, normally constitute 6 to 20% of bituminous coals, but they may run as high as 30%. The composition of coal ash varies widely, but it is composed chiefly of silicon, aluminum, iron, and calcium compounds, with smaller amounts of magnesium, titanium, sodium, and potassium compounds.

Wood, bagasse (crushed juiceless remains of sugar cane), and other vegetable wastes used as fuel in some industrial plants contain lower amounts of ash than coal does. In many respects, however, the compositions of these vegetable ashes resemble that of coal ash.

Fuel oils have ash contents that seldom exceed 0.2%. Even so, corrosion and fouling of oil-fired boilers can be particularly troublesome because of the nature of oil-ash deposits. The main contaminants in fuel oil are vanadium, sodium, and sulfur—elements that form a wide variety of compounds, many of which are extremely corrosive.

Regardless of the fuel, all rapid fireside corrosion circumstances are similar in certain respects. A liquid phase forms in the ash deposit adjacent to the tube surface. Once a liquid phase forms, the protective oxide scale on the tube surface is dissolved, and rapid wastage follows. In coal-fired boilers, the liquid phase is a mixture of sodium and potassium iron trisulfate: ($Na_3Fe(SO_4)_3$ and $K_3Fe(SO_4)_3$, respectively. Mixtures of these substances melt at temperatures as low as 555 °C (1030 °F). In oil-fueled boilers the liquid phase that forms is a mixture of V_2O_5 with either sodium oxide (Na_2O) or sodium sulfate (Na_2SO_4). Mixtures of these compounds have melting points below 540 °C (1000 °F). The formation of liquid salts is responsible for accelerated corrosion attack by oil-ash corrosion.

Coal-Ash Corrosion. During combustion of coal, the minerals in the burning coal are exposed to high temperatures and to the strongly reducing effects of generated gases, such as CO and hydrogen. Aluminum, iron, potassium, sodium, and sulfur compounds are partly decomposed, releasing volatile alkali compounds and sulfur oxides (predominantly SO_2, plus small amounts of SO_3). The remaining portion of the mineral matter reacts to form glassy particles known as fly ash.

Coal-ash corrosion starts with the deposition of fly ash on surfaces that operate predominantly at temperatures from 540 to 705 °C (1000 to 1300 °F), primarily surfaces of superheater and reheater tubes. These deposits may be loose and powdery, or they may be sintered or slag-type masses that are more adherent. Over an extended period of time, volatile alkali and sulfur compounds condense on the fly ash and react with it to form complex alkali sulfates, such as $KAl(SO_4)_2$ and $Na_3Fe(SO_4)_3$, at the boundary

Table 16 Corrosion of various metals and alloys in anhydrous HF

Material	Corrosion rate, mm/yr (mils/yr)		
	500 °C (930 °F)	550 °C (1020 °F)	600 °C (1110 °F)
Nickel	0.9 (36)	...	0.9 (36)
400	1.2 (48)	1.2 (48)	1.8 (72)
600	1.5 (60)	...	1.5 (60)
Copper	1.5 (60)	...	1.2 (48)
Aluminum	4.9 (192)	...	14.6 (576)
Magnesium	12.8 (504)	...	...
Carbon steel (1020)	15.5 (612)	14.6 (576)	7.6 (300)
304	...	...	13.4 (528)
347	183 (7200)	457 (18,000)	177 (6960)
309Cb	5.8 (228)	43 (1680)	168 (6600)
310	12.2 (480)	100 (3960)	305 (12,000)
430	1.5 (60)	9.1 (360)	11.6 (456)

Source: Ref 35

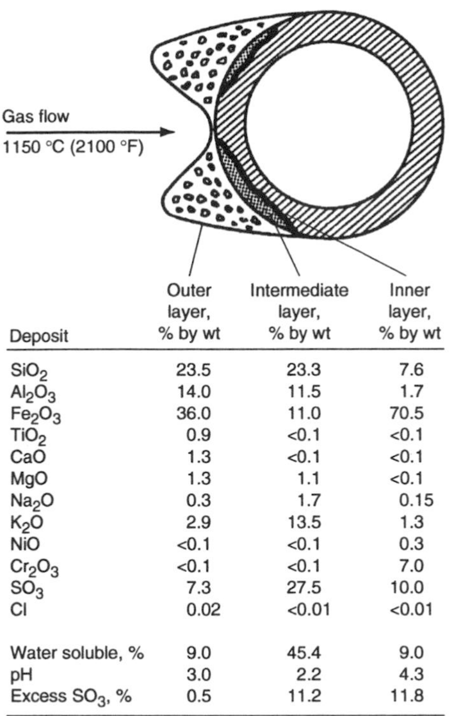

Deposit	Outer layer, % by wt	Intermediate layer, % by wt	Inner layer, % by wt
SiO_2	23.5	23.3	7.6
Al_2O_3	14.0	11.5	1.7
Fe_2O_3	36.0	11.0	70.5
TiO_2	0.9	<0.1	<0.1
CaO	1.3	<0.1	<0.1
MgO	1.3	1.1	<0.1
Na_2O	0.3	1.7	0.15
K_2O	2.9	13.5	1.3
NiO	<0.1	<0.1	0.3
Cr_2O_3	<0.1	<0.1	7.0
SO_3	7.3	27.5	10.0
Cl	0.02	<0.01	<0.01
Water soluble, %	9.0	45.4	9.0
pH	3.0	2.2	4.3
Excess SO_3, %	0.5	11.2	11.8

Fig. 20 Analyses of typical ash deposit from 18Cr-8Ni superheater tube. Source: Ref 36

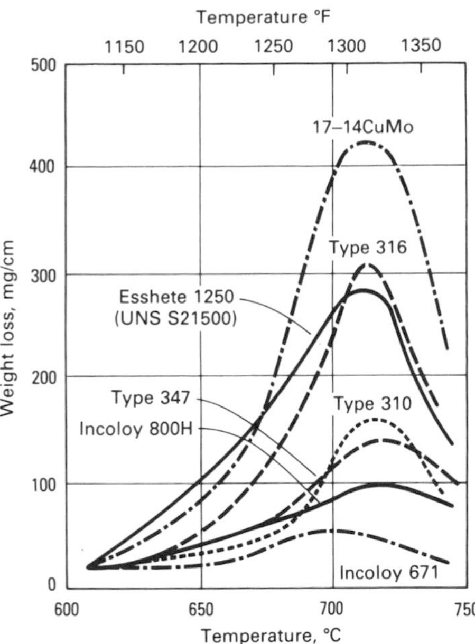

Fig. 21 Corrosion rates of alloys in a laboratory test using synthetic ash (37.5 mol% Na_2SO_4, 37.5 mol% K_2SO_4, and 25 mol% Fe_2O_3) in a synthetic flue gas (80% nitrogen, 15% CO_2, 4% oxygen, and 1% SO_2 saturated with water). Exposure time 50 h. Source: Ref 37

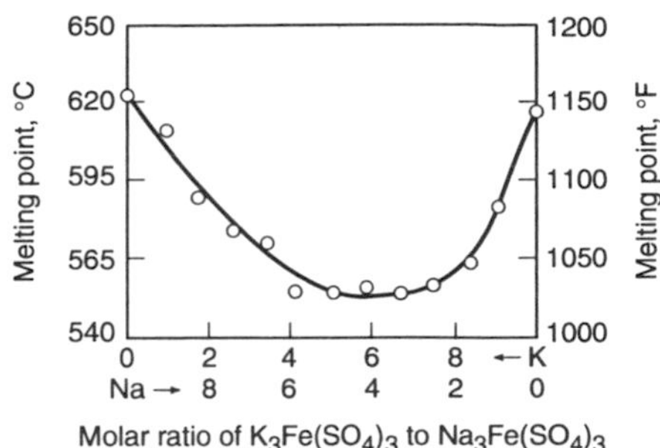

Fig. 22 Melting point curve for the $K_3Fe(SO_4)_3$-$Na_3Fe(SO_4)_3$ system. Source: Ref 39

between the metal and the deposit. The reactions that produce alkali sulfates are believed to depend in part on the catalytic oxidation of SO_2 to SO_3 in the outer layers of the fly-ash deposit. The exact chemical reactions between the tube metal and the ash deposit are not well defined. However, in every example studied, the ash and corrosion products contain both carbon and sulfides. The sulfides cannot exist in a strongly oxidizing environment, so, at least on the tube surface, the conditions are reducing. The following coal-ash corrosion characteristics are common:

- Rapid attack occurs at temperatures between the melting temperature of the sulfate mixture and the limit of thermal stability for the mixture.
- Corrosion rate is a nonlinear function of metal temperature, being highest at temperatures from 675 to 730 °C (1250 to 1350 °F).
- Corrosion is almost always associated with sintered or glassy slag-type deposits.
- The deposit consists of three distinct layers (Fig. 20). The porous, outermost layer comprises the bulk of the deposit and is composed essentially of the same compounds as those found in fly ash. The innermost layer (heavy black lines, Fig. 20) is a thin, glassy substance composed primarily of corrosion products of iron. The middle layer, called the white layer, is whitish or yellowish in color, is often fused, and is largely water-soluble, producing an acid solution. The fused layers contain sulfides as part of the corrosion product and can be easily detected by means of a sulfur print.
- Coal-ash corrosion can occur with any bituminous coal, but it is more likely when the coal contains more than 3.5% S and 0.25% Cl.
- None of the common tube materials is immune to attack, although the 18-8 austenitic stainless steels corrode at slower rates than lower-alloy grades. Corrosion rates vary widely depending on environment and operating conditions. 2.25Cr-1Mo steel corrodes 0.4 to 1 mm (15 to 40 mils, or 0.015 to 0.040 in.) per year, while the 18-8 stainless steels similar to 304 corrode 0.08 to 0.1 mm (3 to 5 mils, or 0.003 to 0.005 in.) per year.
- The overall appearance of the wasted tube surface is rough, with an "alligator hide" appearance.

Because the layer next to the tube contains sulfides, the likely net corrosion reaction is:

$$2C + SO_2 + Fe \leftrightarrow FeS + 2CO$$

The corrosion occurs in the presence of carbon, so it seems probable that coal-ash corrosion of superheaters and reheaters can be reduced by more complete combustion of the coal.

Particles of fly-ash are deposited on superheater and reheater tubes in a characteristic pattern, depending on the direction of flue-gas flow. The tube surfaces are corroded most heavily beneath the thickest portions of the deposit (Fig. 20). When deposits are removed, shallow macropitting can be seen. Eventually, the tube wall becomes thinned to the point at which the material can no longer withstand the pressure within the tube, and the tube ruptures.

In coal-ash corrosion, the corrosion rate exhibits a "bell-shaped" curve with respect to temperature for austenitic stainless steels (Fig. 21). The rate increases with temperature to a maximum, then decreases with increasing temperature. Maximum corrosion rates occur between 675 and 730 °C (1250 and 1350 °F). The accelerated corrosion associated with this bell-shaped curve is related to the formation of molten alkali metal-iron-trisulfate [$(Na,K)_3Fe(SO_4)_3$] (Ref 38, 39). Variation of the sodium-to-potassium ratio greatly affects the melting point of the complex sulfate in an ash deposit (Ref 39). Figure 22 illustrates a wide variation of melting points of mixtures of sodium iron trisulfate and potassium iron trisulfate (Ref 22). Corrosion rate was also affected by the sodium-to-potassium ratio in the complex sulfate (Ref 39).

Nelson and Cain (Ref 40) conducted a series of laboratory tests with flowing synthetic flue gas (N_2-$15CO_2$-$3.6O_2$-$0.25SO_2$) over a mixture of potassium sulfate, sodium sulfate, and iron oxide (molecular ratio 1.5:1.5:1.0) that covered the test coupons. Samples were exposed at different temperatures for 5 days. Results are summarized in Fig. 23. The corrosion rate was greatly enhanced when the sulfate was molten. Corrosion rate also depends on the SO_2 concentration in the flue gas and on the $Na_2SO_4 + K_2SO_4$ concentration in the ash deposit, as illustrated in Fig. 24 and 25. For alkali sulfates, only the acid-soluble alkalis are of concern in coal-ash corrosion (Ref 42).

Oil-Ash Corrosion. During combustion of fuel oils, organic compounds (including those containing vanadium or sulfur) decompose and react with oxygen. The resulting volatile oxides are carried along in the flue gases. Sodium, which is usually present in the oil as a chloride, reacts with the sulfur oxides to form sulfates. Initially, V_2O_5 condenses as a semifluid slag on furnace walls, boiler tubes, and/or superheater tubes, in fact, virtually anywhere in the high-temperature region of the boiler. Sodium oxide reacts with V_2O_5 to form complex compounds, especially vanadates ($nNa_2O \cdot V_2O_5$) and vanadylvanadates ($nNa_2O \cdot V_2O_4 \cdot mV_2O_5$). These complex compounds, some of which have melting temperatures as low as 249 °C (480 °F), foul and actively corrode the tube surfaces. Table 17 lists the melting points of various oil-ash constituents.

Slag of equilibrium thickness (3.0 to 6.4 mm, or 0.12 to 0.25 in.) has developed in experimental furnaces within periods of time as short as 100 h. Slag insulates the tubes, resulting in an increase

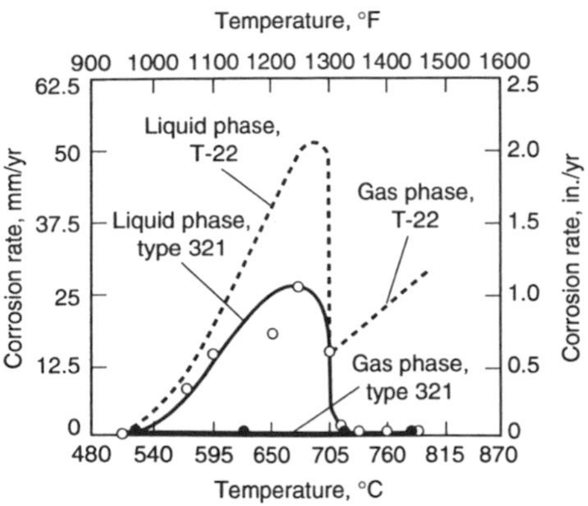

Fig. 23 Results of laboratory tests with flowing synthetic flue gas (N_2-15CO_2-3.6O_2-0.25SO_2) over a synthetic coal ash (K_2SO_4, Na_2SO_4, and Fe_2O_3 with a molecular ratio of 1.5:1.5:1.0) that covered the test coupons. Source: Ref 40

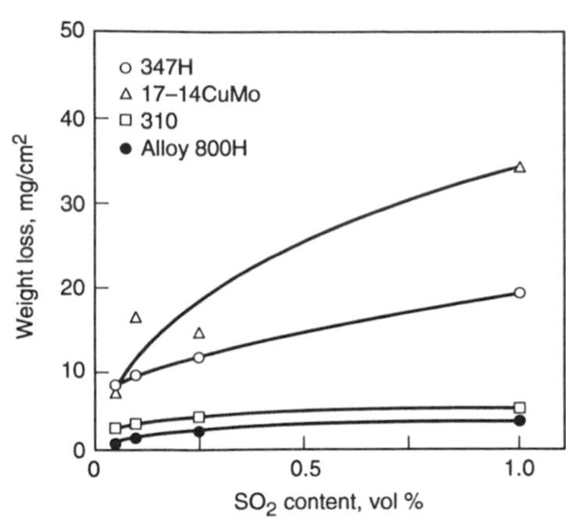

Fig. 24 Effect of SO_2 content in flue gas on the corrosion of several superheater/reheater materials exposed to synthetic ash containing 5 wt% (Na_2SO_4 + K_2SO_4) at 650 °C (1200 °F). Source: Ref 41

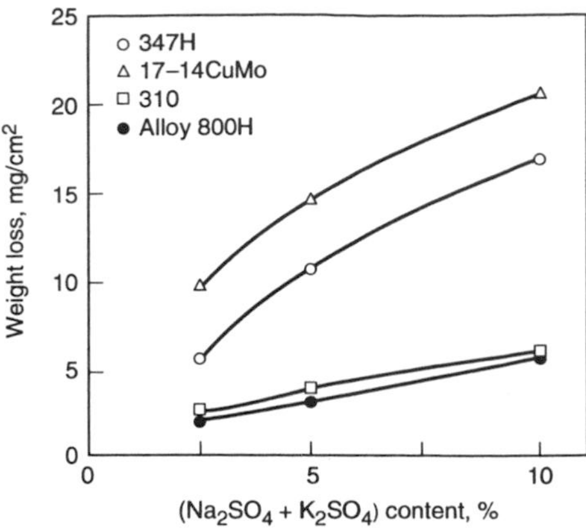

Fig. 25 Effect of Na_2SO_4 + K_2SO_4 content in synthetic ash on the corrosion of several superheater/reheater materials at 650 °C (1200 °F) in flue gas containing 0.25% SO_2. Source: Ref 41

in the temperature of the slag, which in turn increases the rate of corrosion and promotes further deposition of ash. Thicker slag deposits generally lead to greater corrosion, because slag temperatures are higher and more of the corrodent is present to react with tube materials. However, higher slag temperatures also make the slag more fluid, so that it flows more readily on vertical surfaces. Consequently, slag generally builds up in corners and on horizontal surfaces, such as at the bases of water walls and around tube supports in the superheater.

Resistance to Oil-Ash Corrosion. McDowell and Mihalisin (Ref 43) conducted extensive field rack tests in boilers fired with Bunker "C" oils containing high concentrations of vanadium (150 to 450 ppm). Test racks were exposed in the superheater section. A variety of alloys, ranging from low-alloy steels to iron- and nickel-base alloys, suffered severe corrosion attack. The results of one test rack are shown in Table 18. Specimens included 5Cr steel, stainless steels (both 400 and 300 series), Fe-Ni-Cr alloys, Ni-Cr-Fe alloy, and 50Ni-50Cr alloy, along with two cast stainless steels (HE and HH alloys). All tested alloys exhibited unacceptable corrosion rates. Even the best performer (50Ni-50Cr alloy) suffered a corrosion rate of 3.1 mm/yr (121 mils/yr). Figure 26 shows that as the concentration of vanadium in fuel oil increases, the rate of corrosion increases accordingly.

Superheaters and reheaters, which operate at much lower temperatures than tube supports, are also susceptible to oil-ash corrosion. Bolt (Ref 44) evaluated various superheater and reheater materials in an experimental boiler firing with heavy oil containing 2.2% S, 200 ppm V, and 50 ppm Na. The test was conducted on type 347H tubes and several coextruded tubes, including type 310 over ESSHETE 1250 (UNS 21500), type 446 over alloy 800H, CR35A over alloy 17-14CuMo, and alloy 671 over alloy 800H. Type 347H suffered the worst corrosion attack, followed by type 310, with maximum corrosion rates occurring at about 670 °C (1240 °F). Both alloys showed unacceptably high corrosion rates (>1 mm/yr, or 39 mils/yr) at 630 to 675 °C (1170 to 1250 °F). Three high-chromium cladding materials performed significantly better than types 347H and 310: type 446 (27Cr), Cr35A (35Cr-45Ni-Fe), and alloy 671 (47Cr).

Besides cladding, another effective method of combatting oil-ash corrosion problems is to inject additives (high-melting-point compounds) into the fuel to raise the melting point of the oil-ash deposit. Useful additives include (Ref 45):

- *Magnesium compounds:* MgO, $Mg(OH)_2$, $MgCO_3$, $MgSO_4$, $MgCO_3 \cdot CaCO_3$, organic magnesium compounds
- *Calcium compounds:* CaO, $Ca(OH)_2$, $CaCO_3$
- *Barium compounds:* BaO, $Ba(OH)_2$
- *Aluminum and silicon compounds:* Al_2O_3, SiO_2, $3Al_2O_3 \cdot 2SiO_2$

The additive reacts with vanadium compounds to form reaction products with higher melting points. When magnesium compounds are used, some of the reaction products and their melting points are:

- $MgO \cdot V_2O_5$: 671 °C (1240 °F) melting point
- $2MgO \cdot V_2O_5$: 835 °C (1535 °F) melting point
- $3MgO \cdot V_2O_5$: 1191 °C (2176 °F) melting point

When the injection involves magnesium compounds, increasing the Mg/V ratio increases the melting point of the oil-ash deposits (Ref 46), which results in lower corrosion rates. Figure 27 shows the effectiveness of the $Mg(OH)_2$ additive injection in reducing the corrosion rate of type 321 superheater tubes. Disadvantages of the additive injection approach include additional operating costs and a substantial increase in ash volume, which may require

Table 17 Melting points of various oil-ash constituents

Compound	Melting point	
	°C	°F
Aluminum oxide, Al_2O_3	1799	3270
Aluminum sulfate, $Al_2(SO_4)_3$	771(a)	1420(a)
Calcium oxide, CaO	2572	4662
Calcium sulfate, $CaSO_4$	1449	2640
Ferric oxide, Fe_2O_3	1566	2850
Ferric sulfate, $Fe_2(SO_4)_3$	480(a)	896(a)
Nickel oxide, NiO	2091	3796
Nickel sulfate, $NiSO_4$	841(a)	1546(a)
Silicon dioxide, SiO_2	1721	3130
Sodium sulfate, Ni_2SO_4	885	1625
Sodium bisulfate, $NaHSO_4$	249(a)	480(a)
Sodium pyrosulfate, $Na_2S_2O_7$	399(a)	750(a)
Sodium ferric sulfate, $Na_3Fe(SO_4)_3$	538	1000
Vanadium trioxide, V_2O_3	1971	3580
Vanadium tetroxide, V_2O_4	1971	3580
Vanadium pentoxide, V_2O_5	691	1276
Sodium metavanadate, $Na_2O \cdot V_2O_5(NaVO_3)$	630	1166
Sodium pyrovanadate, $2Na_2O \cdot V_2O_5$	641	1186
Sodium orthovanadate, $3Na_2O \cdot V_2O_5$	849	1560
Sodium vanadylvanadates		
$Na_2O \cdot V_2O_4 \cdot V_2O_5$	627	1161
$5Na_2O \cdot V_2O_4 \cdot 11V_2O_5$	535	995

(a) Decomposes at a temperature around the melting point. Source: Ref 36

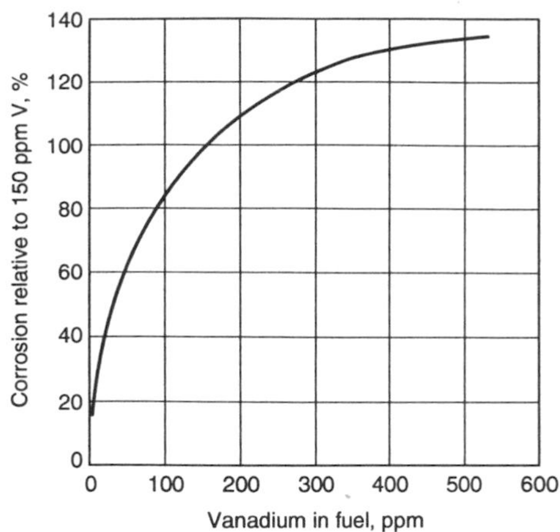

Fig. 26 Effect of vanadium concentration on oil-ash corrosion. Source: Ref 36

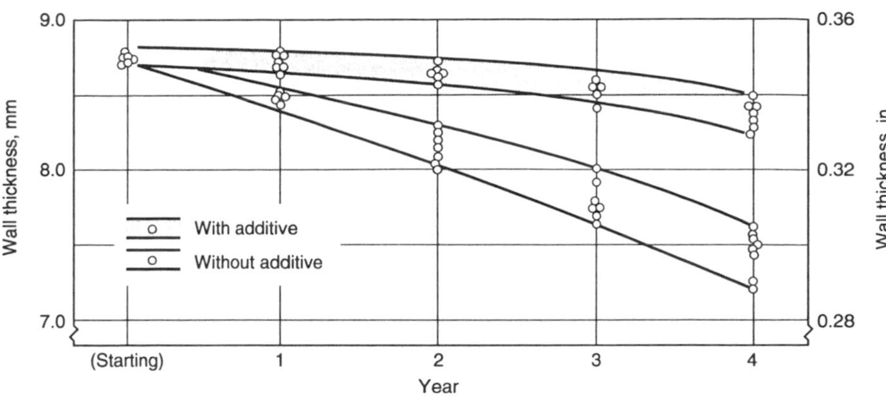

Fig. 27 Corrosion of type 321 superheater tubes with and without $Mg(OH)_2$ injection in an oil-fired boiler. Source: Ref 45

Table 18 Results of a field test (uncooled specimens) exposed in the superheater section at 815 °C (1500 °F) in a boiler fired with high-vanadium-content (150–450 ppm) Bunker "C" fuel

Alloy	Corrosion rate, mm/yr (mils/yr)
5Cr steel	32–45 (1270–1775)
406	5.7 (224)
431	17–23 (655–925)
446	3.8 (149)
302B	14–16 (533–644)
309	5.0 (196)
321	9–13 (346–505)
347	6.2 (243)
310	4.7 (187)
Incoloy 800	9–12 (364–458)
Incoloy 804	13–18 (495–710)
Inconel 600	5.0 (196)
50Cr-50Ni	3.1 (121)
HE	4.8 (187)
HH	12–16 (467–645)

Source: Ref 43

Table 19 Various low-melting-point salt deposits formed by burning municipal wastes

Eutectic mixture, mole %	Melting point °C	Melting point °F
$25NaCl-75FeCl_3$	156	313
$37PbCl_2-63FeCl_3$	175	347
$60SnCl_2-40KCl$	176	349
$70SnCl_2-30NaCl$	183	361
$70ZnCl_2-30FeCl_3$	200	392
$20ZnCl_2-80SnCl_2$	204	400
$55ZnCl_2-45KCl$	230	446
$70ZnCl_2-30NaCl$	262	504
$60KCl-40FeCl_2$	355	671
$58NaCl-42FeCl_2$	370	698
$70PbCl_2-30NaCl$	410	770
$52PbCl_2-48KCl$	411	772
$72PbCl_2-28FeCl_2$	421	790
$90PbCl_2-10MgCl_2$	460	860
$80PbCl_2-20CaCl_2$	475	887
$49NaCl-50CaCl_2$	500	932

Source: Ref 59

additional furnace downtime for tube cleaning (Ref 47).

Fireside Corrosion in Waste Incinerators. Incineration has become a viable technology for disposing of various types of wastes, including municipal, hospital, chemical, and hazardous wastes. An increasing number of incineration plants have been constructed and put into operation. However, equipment problems resulting from fireside corrosion have been frequently encountered in various incinerators. The major problem is the complex nature of the feed (wastes) as well as its corrosive impurities, which are in greater number and higher concentrations than those of regular fuels such as coal and oil.

Sulfur is almost always present in the environment generated by incineration of wastes. Sulfidation has frequently been observed to occur in fireside corrosion in various waste incinerators. High-nickel alloys, such as alloys 690, X, and C-276, have suffered severe sulfidation attack in municipal waste incinerators when exposed in a temperature range of 750 to 980 °C (1400 to 1800 °F) (Ref 48, 49). "Erosion" shields made of stainless steels (a common design intended to protect superheater tubing from corrosion attack) frequently suffer sulfidation attack in municipal waste incinerators. Figure 28 illustrates a failed type 310 stainless steel erosion shield that shows significant sulfidation attack.

Chlorine is another contaminant that is almost always present in incineration environments. Chlorine has been found in flue gases produced in incinerators burning hospital wastes (Ref 50), low-level radioactive wastes (Ref 51, 52), and municipal wastes (Ref 53–58). Krause (Ref 55) indicated that the average chlorine content for municipal waste is about 0.5%, present as polyvinylchloride plastic as well as inorganic substances such as NaCl. Corrosion of furnace wall and superheater tubes in municipal waste incinerators is attributed to attack by HCl and chlorine, resulting from interaction between sulfur oxides and chlorides in the deposits (Ref 54). Corrosion may be enhanced when localized reducing conditions exist in the vicinity of the furnace walls. Formation of low-melting-point salts may also be responsible for fireside corrosion of furnace tubes and superheater tubes (Ref 54, 55, 58).

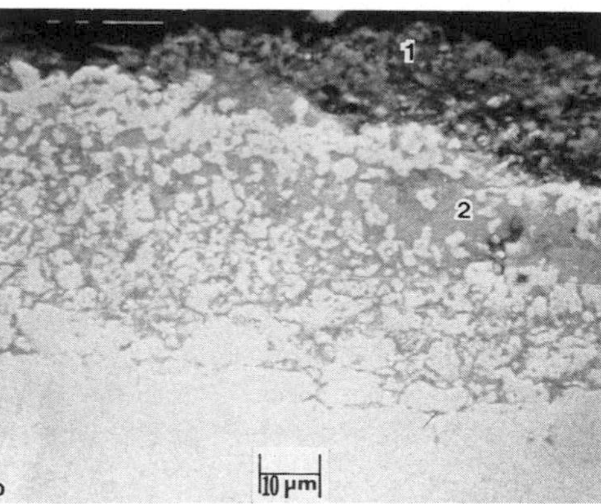

Fig. 28 Corroded erosion shield (type 310 stainless steel) of a superheater in a municipal waste incinerator, showing significant sulfidation attack. (a) General view. (b) SEM micrograph with electron dispersive x-ray analysis of the specimen cross section. Area 1 composition: 38Cr-27Fe-18Ni-3Si-10S-4Zn; area 2 composition: 32Cr-31Fe-4Ni-33S

In addition to sulfur and chlorine, other constituents frequently detected in significant amounts in deposits include potassium, sodium, zinc, and lead. These elements may contribute to the formation of low-melting-point salts. Various low-melting-point eutectics are shown in Table 19. Many salt mixtures become molten in the temperature range of the furnace wall tubes and superheater tubes, so molten salt deposit corrosion may also be a likely corrosion mechanism.

Extensive corrosion studies in municipal incinerators have shown that in the temperature range

Table 20 Corrosion of carbon steel, chromium-molybdenum steels, and stainless steels in liquid sodium under isothermal conditions

Materials	Test temperature, °C	Test temperature, °F	Exposure time, h	Test system	Weight change rate, mg/cm² per month
1010 steel	593	1100	1000	Flowing	–0.49
	593	1100	1000	Static	–0.37
2.25Cr-1Mo	552	1026	943	Flowing	–0.12
	556	1033	902	Static	–0.12
	593	1100	1000	Flowing	–0.14
	593	1100	1000	Static	–0.09
5Cr-0.5Mo	552	1026	943	Flowing	+0.22
	566	1033	1913	Static	–0.06
	593	1100	500	Flowing	+0.23
	593	1100	500	Static	–0.08
9Cr-1Mo	552	1026	943	Flowing	+0.35
	566	1033	902	Static	–0.05
	593	1100	500	Flowing	+0.70
	593	1100	500	Static	+0.29
410	593	1100	1000	Flowing	+0.38
	593	1100	1000	Static	+0.35
420	593	1100	1000	Flowing	+0.33
	593	1100	1000	Static	+0.31
304	593	1100	1000	Flowing	+0.17
	593	1100	1000	Static	+0.15
310	593	1100	500	Flowing	+0.75
	593	1100	500	Static	+0.27
316	593	1100	1000	Flowing	+0.10
	593	1100	1000	Static	+0.13
347	593	1100	500	Flowing	+1.46
	593	1100	500	Static	+0.22
410	1000	1830	400	Static	+29.8
430	1000	1830	400	Static	+46.8
446	1000	1830	400	Static	+28.2
304	1000	1830	400	Static	+25.5
316	1000	1830	400	Static	+29.6
310	1000	1830	400	Static	+28.2
347	1000	1830	400	Static	+44.2
600	1000	1830	400	Static	+18.7

Note: The liquid sodium contained a maximum of 100 ppm oxygen. Source: Ref 62

of 150 to 315 °C (300 to 600 °F), a number of alloys provide good performance in resisting high-temperature corrosion (Ref 60). In decreasing order, the better alloys are Incoloy 825; AISI types 446, 310, 316L, 304, and 321 stainless steels; and Inconel alloys 600 and 601. However, when subjected to moist deposits, simulating boiler downtime conditions, all of the austenitic stainless steels undergo chloride stress-corrosion cracking. Type 446 stainless steel, Inconel 600, and Inconel 601 suffer pitting. Consequently, unless the boilers are to be maintained at a temperature above the HCl dew point during downtime, only Incoloy 825 is recommended. Additional data on fireside corrosion behavior of stainless steels and other engineering alloys can be found in the articles "Elevated-Temperature Corrosion Properties of Carbon and Alloy Steels" and "Elevated-Temperature Corrosion Properties of Superalloys" in this Volume.

Corrosion in Molten Salts and Liquid Metals

Molten salts, often called fused salts, can cause corrosion by the solution of constituents of the container material, selective attack, pitting, electrochemical reactions, mass transport due to thermal gradients, reaction of constituents of the molten salt with the container material, reaction of impurities in the molten salt with the container material, and reaction of impurities in the molten salt with the alloy.

The comparative corrosion behavior of various alloys (including stainless steels) in molten chlorides, nitrates/nitrites, sodium hydroxide, fluorides, and carbonates is reviewed in the article "Corrosion at Elevated Temperatures" in this Volume. In general, corrosion rates of metals and alloys are strongly dependent on temperature, and they can generally be reduced by decreasing the temperature. For example, the corrosion behavior of stainless steels in molten heat treating salts, such as $NaNO_3$-KNO_3, is reasonably well understood. For applications at temperatures up to 630 °C (1170 °F), many austenitic stainless steels are capable of handling the molten salt. At higher temperatures, nickel-base alloys are preferred because of the increased salt corrosivity.

Reducing oxidizing impurities, such as oxygen and water vapor, in the melt can also significantly reduce the corrosiveness of the molten salt. Thermal gradients in the melt, in the case of circulating systems, may cause dissolution of an alloying element at the hot leg and deposition of that element at the cold leg, leading to potential tube plugging problems.

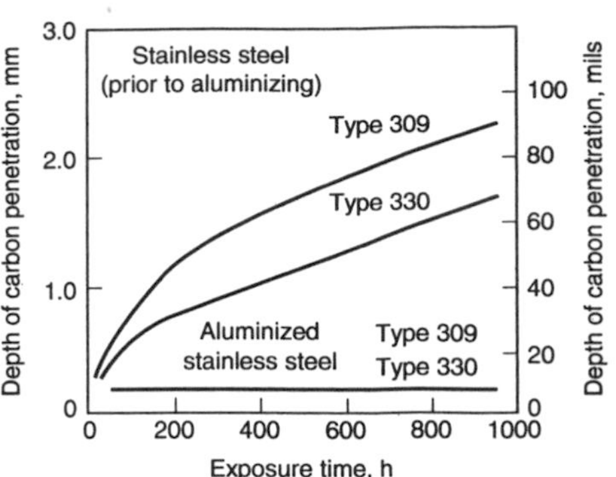

Fig. 29 Carburization resistance of bare and aluminized stainless steels at 925 °C (1700 °F). Source: Ref 63

Corrosion in Liquid Metals. The liquid metal of most interest to stainless steel users is sodium. This interest stems from the development of liquid-metal (sodium) fast-breeder reactors. Stainless steels are a very important part of these reactor systems, so extensive studies of the compatibility of stainless steel with sodium have been made. The 18-8 stainless steels are highly resistant to liquid sodium or sodium-potassium alloys. Mass transfer is not expected up to 540 °C (1000 °F), and it remains at moderately low levels up to 870 °C (1600 °F). Accelerated attack of stainless steels in liquid sodium occurs with oxygen contamination, with a noticeable effect occurring at about 0.02% O_2 by weight (Ref 61). Table 20 summarizes the corrosion rates of stainless steels in liquid sodium.

Exposure to molten lead under dynamic conditions often results in mass transfer in common stainless alloy systems. Particularly severe corrosion can occur in strongly oxidizing conditions. Stainless steels are generally attacked by molten aluminum, zinc, antimony, bismuth, cadmium, and tin. Additional information on the corrosion behavior of various metals and alloys, including stainless steels, can be found in the article "Corrosion at Elevated Temperatures" in this Volume.

Coatings for High-Temperature Corrosion Resistance

Although stainless steels do not undergo coating treatments as frequently as carbon and low-alloy steels, coating treatments are available to extend the service lives of stainless steel components in aggressive corrosive environments. The coatings most commonly used for this purpose are diffusion coatings (via pack cementation) and overlay coatings (via physical vapor deposition).

Stainless steel components that have been aluminized by the pack cementation process include:

- Catalyst reactor screens made from type 347 stainless steel, used in the hydrocarbon processing industry
- Flue gas scrubber NO_x/SO_x removal units made from type 304 stainless steel

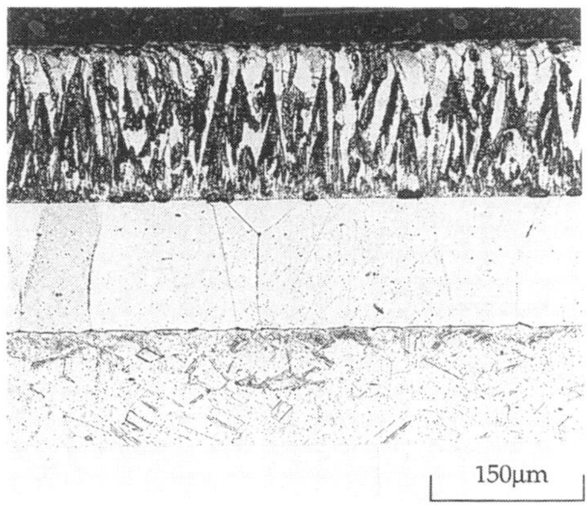

Fig. 30 Polished cross section of an Fe-Cr-Al-Y coating on austenitic stainless steel after 100 h at 1000 °C (1830 °F), electrolytically etched in CrO_3/H_2O to reveal the grain structure of the coating and substrate

- Reactor vessels and tubing made from type 304 or 316 stainless steels, used in the chemical processing industry

Figure 29 illustrates the improved carburization resistance afforded to stainless steels by the pack cementation aluminizing process.

Figure 30 shows the structure of an oxidation-resistant FeCrAlY overlay coating, applied by physical vapor deposition on stainless steel after 100 h of oxidation at 1000 °C (1830 °F). The substrate is virtually immune from oxidation attack.

ACKNOWLEDGMENTS

The information in this article is largely taken from:

- High-Temperature Corrosion, *ASM Specialty Handbook: Stainless Steels,* J.R. Davis, Ed., ASM International, 1994, p 205–228
- G.Y. Lai, *High-Temperature Corrosion of Engineering Alloys,* ASM International, 1990

REFERENCES

1. L.A. Morris, Resistance to Corrosion in Gaseous Atmospheres, *Handbook of Stainless Steels,* D. Peckner and I.M. Bernstein, Ed., McGraw-Hill, 1977, p 17-1 to 17-33
2. *The Making, Shaping and Treating of Steel,* H.E. McGarrow, Ed., United States Steel Corp., 1971, p 1136
3. A. Grodner, *Weld. Res. Counc. Bull.,* No. 31, 1956
4. S.B. Lasday, *Ind. Heat.,* March 1979, p 12
5. O.D. Sherby, *Acta Metall.,* Vol 10, 1962, p 135
6. H.E. Eiselstein and E.N. Skinner, in STP No. 165, ASTM, 1954, p 162
7. A. Moccari and S.I. Ali, *Br. Corros. J.,* Vol 14 (No. 2), 1979, p 91
8. A. de S. Brasunas, J.T. Gow, and O.E. Harder, *Proc. ASTM,* Vol 46, 1946, p 870
9. S. Kado, T. Yamazaki, M. Yamazaki, K. Yoshida, K. Yabe, and H. Kobayashi, *Trans. Iron Steel Inst. Jpn.,* Vol 18 (No. 7), 1978, p 387
10. H.T. Michels, *Met. Eng. Quart.,* Aug 1974, p 23
11. "Design Guidelines for the Selection and Use of Stainless Steel," Document 9014, Nickel Development Institute
12. G.Y. Lai, *High-Temperature Corrosion of Engineering Alloys,* ASM International, 1990, p 15–46
13. W.C. Leslie and M.C. Fontana, Paper 26, presented at the 30th Annual Convention of the American Society for Metals (Philadelphia), 25–29 Oct 1948
14. J.K. Meijering and G.W. Rathenau, *Nature,* Vol 165, 11 Feb 1950, p 240
15. A.D. Brasunas and N.J. Grant, *Iron Age,* 17 Aug 1950, p 85
16. S.S. Brennor, *J. Electrochem. Soc.,* Vol 102 (No. 1), Jan 1955, p 16
17. G.W. Rathenau and J.L. Meijering, *Metallurgia,* Vol 42, 1950, p 167
18. J.C. Sawyer, *Trans. TMS-AIME,* Vol 221, 1961, p 63
19. A. de S. Brasunas and N.J. Grant, *Trans. ASM,* Vol 44, 1950, p 1133
20. "High-Temperature Characteristics of Stainless Steels," Document 9004, Nickel Development Institute
21. L.A. Morris, Corrosion Resistance of Stainless Steels at Elevated Temperatures, *Selection of Stainless Steels,* American Society for Metals, 1968, p 30–47
22. G. Sorell, "Compilation and Correlation of High Temperature Catalytic Reformer Corrosion Data," Tech. Comm. Rep., Publication 58-2, National Association of Corrosion Engineers, 1957
23. E.B. Backensto, R.E. Drew, J.E. Prior, and J.W. Sjoberg, "High-Temperature Hydrogen Sulfide Corrosion of Stainless Steels," Tech. Comm. Rep., Publication 58-3, National Association of Corrosion Engineers, 1957
24. E.B. Backensto, "Corrosion in Catalytic Reforming and Associated Processes," Summary Report of the Panel on Reformer Corrosion to the Subcommittee on Corrosion, Division of Refining, American Petroleum Institute, 22nd Midyear Meeting (Philadelphia), 13 May 1957
25. E.B. Backensto and J.W. Sjoberg, "Iso-Corrosion Rate Curves for High Temperature Hydrogen-Hydrogen Sulfide," Tech. Comm. Rep., Publication 59-10, National Association of Corrosion Engineers, 1958
26. J.F. Mason, J.J. Moran, and E.N. Skinner, *Corrosion,* Vol 16, 1960, p 593t
27. K.M. Verma, H. Ghosh, and J.S. Rai, *Br. Corros. J.,* Vol 13 (No. 4), 1978, p 173
28. J.J. Moran, J.R. Mihalisin, and E.N. Skinner, *Corrosion,* Vol 17 (No. 4), 1961, p 191t
29. K. Rorbo, in *Environmental Degradation of High Temperature Materials,* Series 3, No. 13, Vol 2, The Institution of Metallurgists, London, 1980, p 147
30. D.W. McDowell, Jr., *Mat. Protect.,* Vol 1 (No. 7), July 1962, p 18
31. P.L. Daniel and R.A. Rapp, *Advances in Corrosion Science and Technology,* Vol 5, M.G. Fontana and R.W. Staehle, Ed., Plenum Press, 1970
32. S. Baranow, G.Y. Lai, M.F. Rothman, J.M. Oh, M.J. McNallan, and M.H. Rhee, Paper 16, Corrosion/84, National Association of Corrosion Engineers, 1984
33. M.H. Brown, W.B. DeLong, and J.R. Auld, *Ind. Eng. Chem.,* Vol 39 (No. 7), 1947, p 839
34. R.B. Jackson, "Corrosion of Metals and Alloys by Fluorine," NP-8845, Allied Chemical Corp., 1960
35. W.R. Myers and W.B. DeLong, *Chem. Eng. Prog.,* Vol 44 (No. 5), 1948, p 359
36. Fuel Ash Effect in Boiler Design and Operation, Chap 20, *Steam: Its Generation and Use,* S.C. Stultz and J.B. Kitto, Ed., Babcock & Wilcox Co., 1992, p 20-1 to 20-7
37. J. Stringer, Corrosion of Superheaters and High-Temperature Air Heaters, *Corrosion,* Vol 13, *ASM Handbook* (formerly 9th ed., *Metals Handbook*), ASM International, 1987, p 998–999
38. R.W. Borio, A.L. Plumley, and W.R. Sylvester, in *Ash Deposits and Corrosion Due to Impurities in Combustion Gases,* R.W. Bryers, Ed., Hemisphere Publishing/McGraw-Hill, 1978, p 163
39. C. Cain, Jr. and W. Nelson, *J. Eng. Power (Trans. ASME),* Oct 1961, p 468
40. W. Nelson and C. Cain Jr., *J. Eng. Power (Trans. ASME),* July 1960, p 194
41. J.L. Blough and S. Kihara, Paper 129, presented at Corrosion/88, National Association of Corrosion Engineers, 1988
42. R.W. Borio and R.P. Hensel, *J. Eng. Power (Trans. ASME),* Vol 94, 1972, p 142
43. D.W. McDowell, Jr. and J.R. Mihalisin, Paper 60-WA-260, presented at ASME Winter Annual Meeting (New York), 27 Nov to 2 Dec, 1960
44. N. Bolt, in *Proc. Tenth Int. Cong. Metallic Corrosion,* Oxford and IBH Publishing, New Delhi, 1988, p 3593
45. T. Kawamura and Y. Harada, "Control of Gas Side Corrosion in Oil Fired Boilers," Mitsubishi Tech. Bull. 139, Mitsubishi Heavy Industries, Tokyo, May 1980
46. M. Fichera, R. Leonardi, and C.A. Farina, *Electrochim. Acta,* Vol 32 (No. 6), 1987, p 955
47. J.R. Wilson, Paper 12, presented at Corrosion/76, National Association of Corrosion Engineers, 1976
48. D.E. Fluck, G.Y. Lai, and M.F. Rothman, Paper 333, presented at Corrosion/85, National Association of Corrosion Engineers, 1985
49. S.K. Srivastava, G.Y. Lai, and D.E. Fluck, Paper 398, presented at Corrosion/87, National Association of Corrosion Engineers, 1987
50. G.Y. Lai, Paper 209, presented at Corrosion/89, National Association of Corrosion Engineers, 1989
51. R.L. Tapping, E.G. McVey, and D.J. Disney, "Corrosion of Metallic Materials in the CRNL Radwaste Incinerator," presented at Chemical

Waste Incineration Conference (Manchester, UK), 12–13 March 1990
52. G.R. Smolik and J.D. Dalton, Paper 207, presented at Corrosion/89, National Association of Corrosion Engineers, 1989
53. D.A. Vaughan, H.H. Krause, and W.K. Boyd, *Ash Deposits and Corrosion Due to Impurities in Combustion Gases,* R.W. Bryers, Ed., Hemisphere Publishing, 1978, p 473
54. H.H. Krause, *High Temperature Corrosion in Energy Systems,* M.F. Rothman, Ed., The Metallurgical Society of AIME, 1985, p 83
55. H.H. Krause, *Corrosion,* Vol 13, *Metals Handbook,* 9th ed., 1987, p 997
56. H.H. Reichel and U. Schirmer, *Werkst. Korros.,* Vol 40, 1989, p 135
57. W. Steinkusch, *Werkst. Korros.,* Vol 40, 1989, p 160
58. P.L. Daniel, J.L. Barna, and J.D. Blue, in *Proc. National Waste Processing Conf.,* American Society of Mechanical Engineers, 1986, p 221
59. H.H. Krause, Paper 401, presented at Corrosion/87, National Association of Corrosion Engineers, 1987
60. H.H. Krause, D.A. Vaughan, and P.D. Miller, Corrosion and Deposits from Combustion of Solid Waste, *J. Eng. Power (Trans. ASME),* Vol 95 (No. 1), 1973, p 45–52
61. F.L. LaQue and H.R. Copson, Ed., *Corrosion Resistance of Metals and Alloys,* Reinhold, 1963, p 375–445
62. W.E. Berry, *Corrosion in Nuclear Applications,* John Wiley & Sons, 1971
63. "Alonized Steels for High-Temperature Corrosion Resistance," Alon Processing. Inc., Tarentum, PA, 1990

Alloy Cast Irons

THE SELECTION of cast irons for elevated-temperature applications is one of the most demanding tasks for designers and engineers. Properties of importance include:

- Dimensional stability (growth)
- Resistance to scaling (oxidation)
- Short-time tensile properties at elevated temperatures
- Retention of hardness at elevated temperatures (hot hardness)
- Creep resistance
- Stress rupture at temperature
- Thermal fatigue resistance

Further complicating the selection process is the fact that many of these properties are dependent on many factors. For example, the thermal fatigue resistance of a cast iron is dependent on its thermal conductivity, elastic modulus, strength, ductility, and resistance to stress relaxation. The relative importance of these properties varies with the application.

The role of alloying elements is another key consideration. Usually the greatest benefits are achieved when various alloying elements are used in combination. For example, as a single alloy addition in gray irons, chromium produces the greatest increase in resistance to microstructural decomposition, growth, and oxidation; however, it has a minor effect on elevated-temperature strength and creep resistance. Molybdenum produces the greatest increase in strength and in creep-rupture properties at elevated temperatures, but it has little or no beneficial effect on growth, structural stability, or thermal conductivity. When molybdenum and chromium are combined, the effects appear to be synergistic: structural stability is greatly increased and, as a result, both growth and creep-rupture properties are greatly increased.

Classification of Cast Irons

The five types of commercial cast iron are gray, ductile, malleable, compacted graphite, and white iron. With the exception of a white cast iron, all cast irons have in common a microstructure that consists of a graphite phase in a matrix that may be ferritic, pearlitic, bainitic, tempered martensitic, or combinations thereof. The four types of graphitic cast irons are roughly classified according to the morphology of the graphite phase. Gray iron has flake-shaped graphite, ductile iron has nodular or spherically shaped graphite, compacted graphite iron (also called vermicular graphite iron) is intermediate between these two, and malleable iron has irregularly shaped globular or "popcorn"-shaped graphite that is formed during tempering of white cast iron.

White cast irons, so named because of their characteristically white fracture surfaces, do not have any graphite in the microstructures. Instead, the carbon is present in the form of carbides, chiefly of the types Fe_3C and Cr_7C_3. Often, complex carbides are also present, such as $(Fe,Cr)_3C$ from additions of 3 to 5% Ni and 1.5 to 2.5% Cr, $(Cr,Fe)_7C_3$ from additions of 11 to 35% Cr, or those containing other carbide-forming elements.

Cast irons may also be classified as either unalloyed cast irons or alloy cast irons. Unalloyed cast irons are essentially iron-carbon-silicon alloys containing small amounts of manganese, phosphorus, and sulfur. The range of composition for typical unalloyed cast irons is given in Table 1.

Alloy cast irons are considered to be those casting alloys based on the Fe-C-Si system that contain one or more alloying elements intentionally added to enhance one or more useful properties. The addition of a small amount of a substance (e.g., ferrosilicon, cerium, or magnesium) that is used to control the size, shape, and/or distribution of graphite particles is termed inoculation rather than alloying. The quantities of material used for inoculation neither change the basic composition of the solidified iron nor alter the properties of individual constituents. Alloying elements, including silicon when it exceeds about 3%, are usually added to increase the strength, hardness, hardenability, or corrosion resistance of the basic iron. They are often added in quantities sufficient to affect the occurrence, properties, or distribution of constituents in the microstructure.

In gray and ductile irons, small amounts of alloying elements such as chromium, molybdenum, or nickel are used primarily to achieve high strength or to ensure the attainment of a specified minimum strength in heavy sections. For example, in moderately alloyed gray iron (≤3 wt% of total alloying elements), the typical ranges for alloying elements are as follows:

Element	Composition, %
Chromium	0.2–0.6
Molybdenum	0.2–1
Vanadium	0.1–0.2
Nickel	0.6–1
Copper	0.5–1.5
Tin	0.04–0.08

Such castings are commonly referred to as low-alloy cast irons.

High-alloy cast irons differ from unalloyed and low-alloy cast irons mainly in the higher content of alloying elements (>3%), which promote microstructures having special properties for elevated-temperature applications, corrosion resistance, and wear resistance. A classification of the main types of high-alloy cast irons is shown in Fig. 1. As this figure indicates, high-alloy cast irons include graphite-free white irons and graphite-bearing (graphitic) irons in both flake and spheroidal graphite form. The remainder of this section is devoted to the heat-resistant grades listed in Fig. 1. More detailed information on the metallurgy and properties of cast irons, including the abrasion-resistant and corrosion-resistant grades listed in Fig. 1, can be found in the *ASM Specialty Handbook: Cast Irons.*

High-Silicon Irons. Graphitic irons alloyed with 4 to 6% Si provide good service and low cost in many elevated-temperature applications. These irons, whether gray or ductile, provide

Table 1 Range of compositions for typical unalloyed cast irons

Type of iron	Composition, % C	Si	Mn	P	S
Gray	2.5–4.0	1.0–3.0	0.2–1.0	0.002–1.0	0.02–0.25
Compacted graphite	2.5–4.0	1.0–3.0	0.2–1.0	0.01–0.1	0.01–0.03
Ductile	3.0–4.0	1.8–2.8	0.1–1.0	0.01–0.1	0.01–0.03
Malleable	2.2–2.9	0.9–1.9	0.15–1.2	0.02–0.2	0.02–0.2

Source: Ref 2

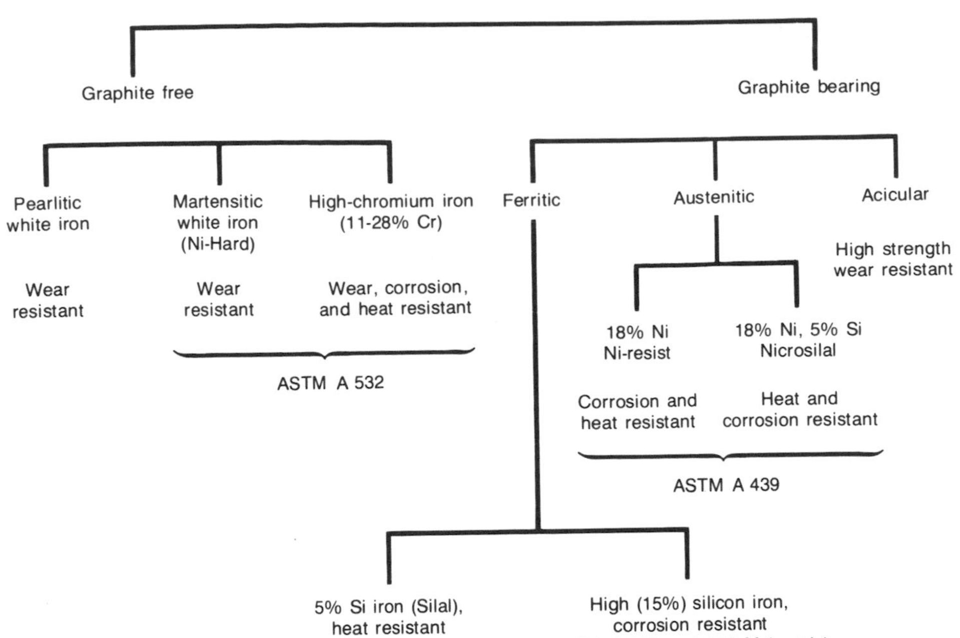

Fig. 1 Classification of special high-alloy cast irons. Source: Ref 1

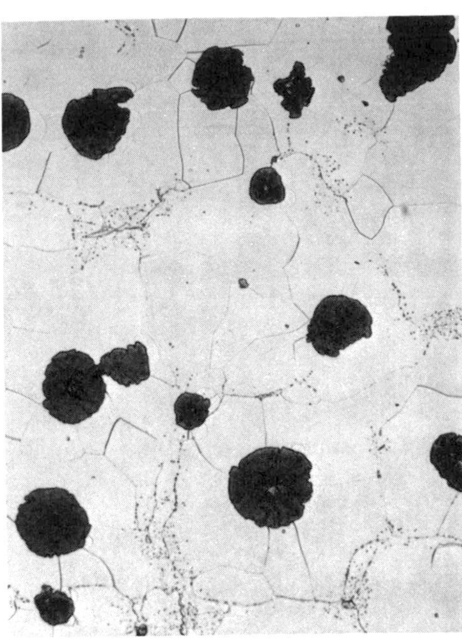

Fig. 2 Photomicrograph of a 4Si-Mo ductile iron showing nodular graphite structure. 400×

good oxidation resistance and stable ferritic matrix structures that will not go through a phase change at temperatures up to 900 °C (1650 °F). The elevated silicon content of these alloys reduces the rate of oxidation at elevated temperatures, because it promotes the formation of a dense, adherent oxide at the surface, which consists of iron silicate rather than iron oxide. This oxide layer is highly resistant to oxygen penetration, and its effectiveness improves with increasing silicon content. More detailed information on the oxidation resistance of high-silicon cast irons can be found in the section "Growth and Scaling of High-Alloy Irons" in this article.

The high-silicon gray irons were developed in the 1930s by the British Cast Iron Research Association (BCIRA) and are commonly called Silal. In Silal, the advantages of a high critical temperature (A_1), a stable ferritic matrix, and a fine, undercooled type D graphite structure are combined to provide good growth and oxidation resistance. Oxidation resistance is further improved with additions of chromium, which in these grades can approach levels of 2% Cr. An austenitic grade called Nicrosilal was also developed, but the nickel-alloyed austenitic irons described below have replaced this alloy.

Although quite brittle at room temperature, the high-silicon gray irons are reasonably tough at temperatures above 260 °C (500 °F). They have been successfully used for furnace and stoker parts, burner nozzles, and heat treatment trays.

The advent of ductile iron led to the development of high-silicon ductile irons, which currently constitute the greatest tonnage of these types of iron being produced. Converting the eutectic flake graphite network into isolated graphite nodules (Fig. 2) further improves oxidation resistance and growth. The higher strength and ductility of the ductile iron version of these alloys qualify it for more rigorous service.

The high-silicon ductile iron alloys are designed to extend the upper end of the range of service temperatures for ferritic ductile irons. These irons are used to temperatures of 900 °C (1650 °F). Raising the silicon content to 4% raises the A_1 to 815 °C (1500 °F), and at 5% Si the A_1 is above 870 °C (1600 °F). The mechanical properties of these alloyed irons at the lower end of the range (4 to 4.5% Si) are similar to those of standard ferritic ductile irons. At 5 to 6% Si, oxidation resistance is improved and A_1 is increased, but the iron can be very brittle at room temperature. At higher silicon levels, the impact transition temperature rises well above room temperature, and upper-shelf energy is dramatically reduced. Ductility is restored when temperatures exceed 425 °C (800 °F).

For most applications, alloying with 0.5 to 1% Mo provides adequate elevated-temperature strength and creep resistance. Higher molybdenum additions are used when maximum elevated-temperature strength is needed. Figure 3 illustrates the improvement in yield and tensile strengths at 700 °C (1300 °F) that can be achieved with molybdenum contents up to 2.5%. Table 2 gives the tensile properties for six 4% silicon irons with increasing molybdenum contents, at temperatures from 20 to 700 °C (68 to 1300 °F). High molybdenum additions (>1%) tend to generate interdendritic carbides of the Mo_2C type, which persist even through annealing, and tend to reduce toughness and ductility at room temperature. Additional information about the effects of molybdenum on the elevated-temperature strength of ductile irons can be found in subsequent sections of this article.

Silicon lowers the eutectic carbon content, which must be controlled in order to avoid graph-

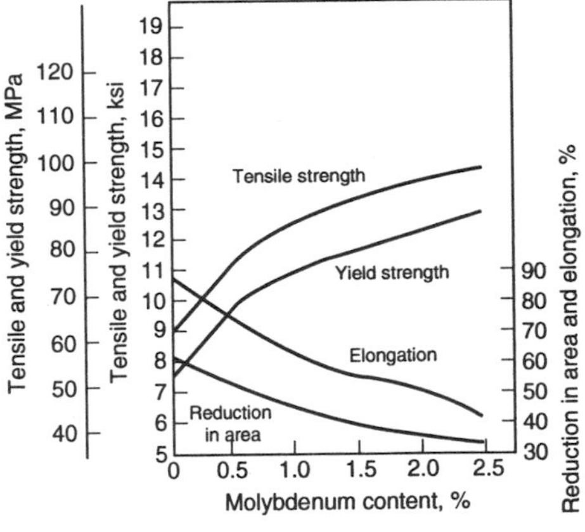

Fig. 3 Effect of molybdenum content on the elevated-temperature (705 °C, or 1300 °F) tensile properties of 4% Si ductile irons that were annealed at 790 °C (1450 °F). Source: Ref 2

ite flotation. For 4% Si irons the carbon content should range from 3.2 to 3.5%, depending on section size, and at 5% Si it should be about 2.9%.

The high-silicon and silicon-molybdenum ductile irons are currently produced as manifolds and turbocharger housings for trucks and some automobiles. They are also used in heat treating racks.

Austenitic Nickel-Alloyed Gray and Ductile Irons. The nickel-alloyed austenitic irons (Ni-Resists) are produced in both gray and ductile cast iron versions for elevated-temperature service. Austenitic gray irons date back to the 1930s, when they were specialized materials of minor importance. After the invention of ductile iron, austenitic grades of ductile iron were also developed. These nickel-alloyed austenitic irons have been used in applications requiring corrosion resistance, wear resistance, and elevated-temperature stability and strength. Additional ad-

Table 2 Effect of varying molybdenum content on the elevated-temperature tensile properties of 4% Si ductile irons

Iron(a)	0.2% offset yield strength, MPa	0.2% offset yield strength, psi	Tensile strength, MPa	Tensile strength, psi	Elongation, %	Reduction in area, %
20 °C (68 °F)						
B	443	64,300	565	81,900	19.5	26.5
C	470	68,200	596	86,400	17.0	23.0
D	474	68,800	610	88,400	14.5	13.5
E	485	70,300	626	90,700	12.0	10.0
F	478	69,300	635	92,100	10.0	9.0
G	490	71,000	654	94,800	10.5	10.0
320 °C (600 °F)						
B	388	56,300	490	71,100	6.5	7.5
C	407	59,000	524	76,000	6.2	6.4
D	414	60,000	546	79,200	4.6	5.8
E	426	61,700	557	80,800	4.6	4.6
F	446	64,600	579	84,000	3.8	5.5
G	439	63,600	594	86,100	3.8	5.2
425 °C (800 °F)						
B	345	50,000	383	55,500	1.9	1.8
C	365	52,900	415	60,200	2.3	3.4
D	387	56,100	422	61,200	1.5	3.4
E	396	57,400	434	63,000	1.5	3.0
F	403	58,400	446	64,600	1.9	3.0
G	411	59,600	488	70,700	1.5	4.2
540 °C (1000 °F)						
B	224	32,500	246	35,600	45.0	43.0
C	250	36,300	277	40,100	35.5	37.5
D	273	39,600	302	43,800	24.5	29.5
E	262	38,000	299	43,300	23.0	23.0
F	281	40,700	320	46,400	21.5	19.5
G	273	39,600	321	46,600	20.5	23.5
650 °C (1200 °F)						
B	67	9,700	83	12,100	59.0	56.0
C	112	16,300	130	18,800	71.5	47.5
D	111	16,100	130	18,800	73.0	39.5
E	119	17,300	138	20,000	48.5	36.0
F	120	17,400	145	21,000	42.5	34.5
G	123	17,800	148	21,500	31.0	24.5
700 °C (1300 °F)						
B	50	7,300	61	8,800	87.5	62.0
C	68	9,800	77	11,100	75.5	52.5
D	76	11,000	89	12,900	59.0	43.5
E	79	11,500	90	13,000	55.5	39.0
F	85	12,300	97	14,100	51.5	36.0
G	88	12,800	99	14,400	40.5	32.5

(a) All irons subcritically annealed at 790 °C (1450 °F). Iron B, 0.2% Mo; Iron C, 0.49% Mo; Iron D, 0.98% Mo; Iron E, 1.45% Mo; Iron F, 1.93% Mo; Iron G, 2.50% Mo. Source: Ref 2

Table 3 Compositions of flake-graphite (gray) austenitic cast irons per ASTM A 436

Type	UNS number	Composition, % TC(a)	Si	Mn	Ni	Cu	Cr
1(b)	F41000	3.00 max	1.00–2.80	0.50–1.50	13.50–17.50	5.50–7.50	1.50–2.50
1b	F41001	3.00 max	1.00–2.80	0.50–1.50	13.50–17.50	5.50–7.50	2.50–3.50
2(c)	F41002	3.00 max	1.00–2.80	0.50–1.50	18.00–22.00	0.50 max	1.50–2.50
2b	F41003	3.00 max	1.00–2.80	0.50–1.50	18.00–22.00	0.50 max	3.00–6.00(d)
3	F41004	2.60 max	1.00–2.00	0.50–1.50	28.00–32.00	0.50 max	2.50–3.50
4	F41005	2.60 max	5.00–6.00	0.50–1.50	29.00–32.00	0.50 max	4.50–5.50
5	F41006	2.40 max	1.00–2.00	0.50–1.50	34.00–36.00	0.50 max	0.10 max(e)
6(f)	F41007	3.00 max	1.50–2.50	0.50–1.50	18.00–22.00	3.50–5.50	1.00–2.00

(a) Total carbon. (b) Type 1 is recommended for applications in which the presence of copper offers corrosion-resistance advantages. (c) Type 2 is recommended for applications in which copper contamination cannot be tolerated, such as handling of foods or caustics. (d) Where some machining is required, 3.0 to 4.0% Cr is recommended. (e) Where increased hardness, strength, and heat resistance are desired, and where increased expansivity can be tolerated, Cr may be increased to 2.5 to 3.0%. (f) Type 6 also contains 1.0% Mo.

vantages include low thermal expansion coefficients, nonmagnetic properties, and good toughness at low temperatures. Compared with corrosion and heat-resistant steels, nickel-alloyed irons have excellent castability and machinability.

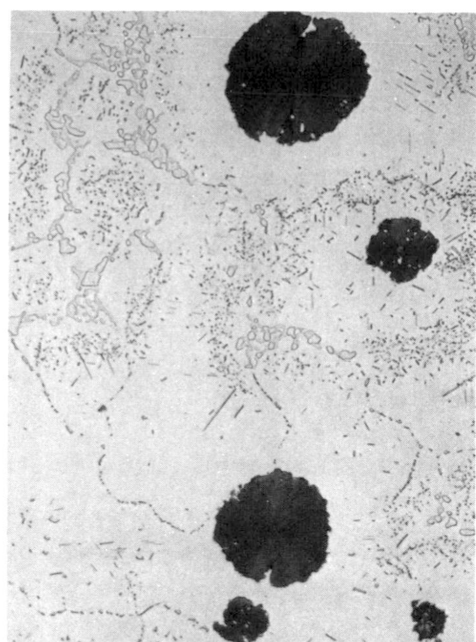

Fig. 4 Photomicrograph of a D5S Ni-Resist ductile iron casting showing nodular graphite structure. 400×

ASTM specification A 436 defines eight grades of austenitic gray iron alloys, four designed for elevated-temperature applications (2, 2b, 3, and 5 in Table 3) and four designed for corrosion resistance (1, 1b, 4, and 6 in Table 3). The nickel produces a stable austenitic microstructure with good corrosion resistance and strength at elevated temperatures. The nickel-alloyed irons are additionally alloyed with chromium and silicon for wear resistance and oxidation resistance at elevated temperatures. Types 1 and 1b, which are designed exclusively for corrosion-resistant applications, are alloyed with 13.5 to 17.5% Ni and 6.5% Cu. Types 2b, 3, and 5, which are principally used for elevated-temperature service, contain 18 to 36% Ni, 1 to 2.8% Si, and 0 to 6% Cr. Type 4 is alloyed with 29 to 32% Ni, 5 to 6% Si, and 4.5 to 5.5% Cr and is recommended for stain resistance.

ASTM A 439 defines the group of austenitic ductile irons (Table 4). The nine austenitic ductile iron alloys have compositions similar to those of the austenitic gray iron alloys but have been treated with magnesium to produce nodular graphite. The ductile family of alloys is available in every type but Type 1; its high copper content is not compatible with production of spheroidal graphite. The ductile iron alloys have high strength and ductility combined with the desirable properties of the gray iron alloys. They provide resistance to frictional wear, corrosion resistance, strength and oxidation resistance at elevated temperatures, nonmagnetic characteristics and, in some alloys, low thermal expansivity at ambient temperatures. Figure 4 illustrates the microstructure typical of austenitic ductile iron.

High-nickel ductile irons are considerably stronger and tougher than the comparable gray

Table 4 Compositions of nodular-graphite (ductile) austenitic cast irons per ASTM A 439

Type	UNS number	Composition, %					
		TC(a)	Si	Mn	P	Ni	Cr
D-2	F43000	3.00 max	1.50–3.00	0.70–1.25	0.08 max	18.0–22.0	1.75–2.75
D-2b	F43001	3.00 max	1.50–3.00	0.70–1.25	0.08 max	18.0–22.0	2.75–4.00
D-2c	F43002	2.90 max	1.00–3.00	1.80–2.40	0.08 max	21.0–24.0	0.50 max
D-3	F43003	2.60 max	1.00–2.80	1.00 max	0.08 max	28.0–32.0	2.50–3.50
D-3a	F43004	2.60 max	1.00–2.80	1.00 max	0.08 max	28.0–32.0	1.00–1.50
D-4	F43005	2.60 max	5.00–6.00	1.00 max	0.08 max	28.0–32.0	4.50–5.50
D-5	F43006	2.60 max	1.00–2.80	1.00 max	0.08 max	34.0–36.0	0.10 max
D-5b	F43007	2.40 max	1.00–2.80	1.00 max	0.08 max	34.0–36.0	2.00–3.00
D-5S	...	2.30 max	4.9–5.5	1.00 max	0.08 max	34.0–37.0	1.75–2.25

(a) Total carbon

Table 5 Tensile strength at temperature for aluminum-alloyed gray irons

Alloy	Tensile strength, MPa (ksi), at:			Room-temperature hardness, HB
	27 °C (80 °F)	427 °C (800 °F)	538 °C (1000 °F)	
2.4% Al iron	348 (50.5)	200 (29.0)	175 (25.4)	260
4.3% Al iron	324 (47.0)	213 (31.0)	193 (28.0)	270
6% Al iron	279 (40.5)	192 (27.8)	168 (24.4)	281
24% Al iron	124 (18.0)	123 (17.8)	102 (14.8)	179

Source: Ref 3

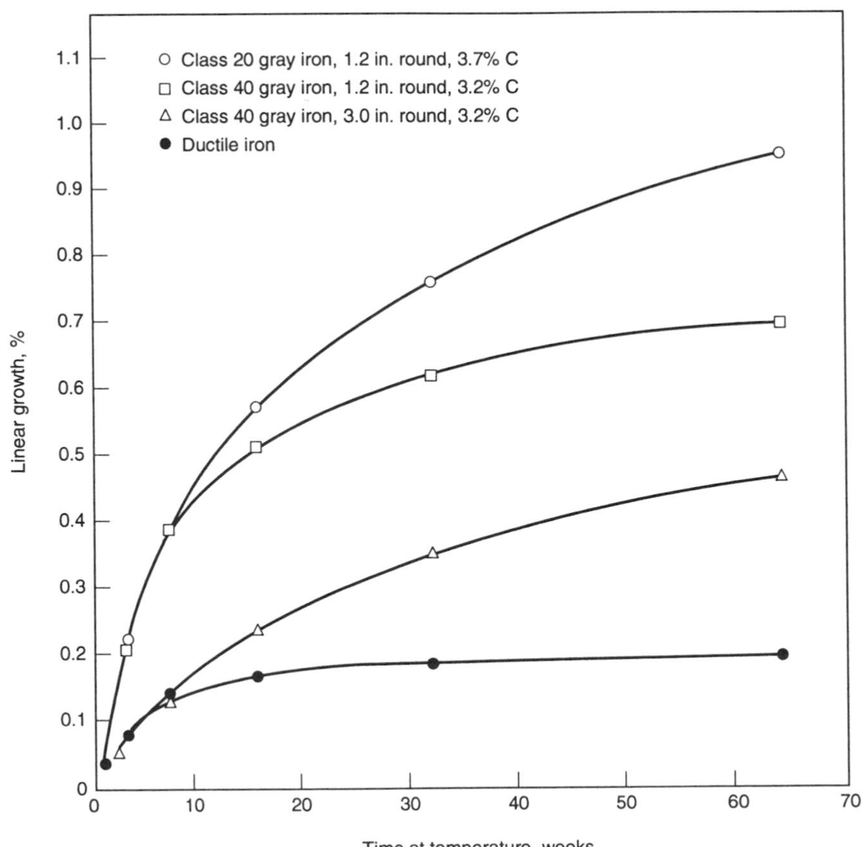

Fig. 5 Results of growth tests at 500 °C (930 °F) in air for gray and ductile irons. Source: Ref 6

irons. Tensile strengths of 400 to 470 MPa (58 to 68 ksi), yield strengths of 205 to 275 MPa (30 to 40 ksi), and elongations of 10 to 40% can be realized.

For applications requiring higher strength at elevated temperatures, the ductile Ni-Resists should be selected. The chromium-bearing types D-2, D-2B, D-3, D-4, and D-5B and the high-silicon type D5S provide useful mechanical properties at temperatures as high as 800 °C (1475 °F). Elevated-temperature mechanical properties of Ni-Resist castings are described below.

Ni-Resists exhibit resistance to elevated-temperature scaling and growth up to 815 °C (1500 °F) in most oxidizing atmospheres and good performance in steam service up to 530 °C (990 °F). They can withstand sour gases and liquids up to 400 °C (750 °F). The maximum temperature of use is 540 °C (1000 °F) if appreciable sulfur is present in the atmosphere.

In elevated-temperature applications, nickel-alloyed irons are used as cylinder liners, exhaust manifolds, valve guides, gas turbine housings, turbocharger housings, nozzle rings, water pump bodies, and piston rings in aluminum pistons.

The aluminum-alloyed irons consist of two groups of gray and ductile irons. The low-alloy group contains 1 to 7% Al, and the aluminum essentially replaces the silicon as the graphitizing element in these alloys. The high-alloy group contains 18 to 25% Al. Irons alloyed with aluminum in amounts between these two ranges will be white irons as-cast and will have no commercial importance.

Table 5 lists tensile strengths and hardness values for gray irons containing various levels of aluminum. As these data indicate, the alloy irons containing 1 to 7% Al have both higher strengths and hardnesses. Elevated-temperature tensile strengths are also higher.

The aluminum greatly enhances oxidation resistance at elevated temperatures and also strongly stabilizes the ferrite phase to very high temperatures, up to and beyond 980 °C (1800 °F). Like the silicon-alloyed irons, the aluminum irons form a tight, adherent oxide on the surface of the casting that is very resistant to further oxygen penetration. The oxidation resistance of aluminum-alloyed irons is described in the section "Growth and Scaling of High-Alloy Irons" in this article.

Unfortunately, the aluminum-alloyed irons are difficult to cast without dross inclusions and laps (cold shuts). The aluminum in the iron is very reactive at the temperatures of the molten iron, and contact with air and moisture must be negligible. Care must be taken not to draw the oxide skin, which forms during pouring, into the mold in order to avoid dross inclusions. Methods for overcoming these problems in commercial practice are under development.

At present, there is no ASTM standard covering the chemistry and expected properties of these alloys, and commercial production is very limited. In the past, the 1.5 to 2.0% irons have been used in the production of truck exhaust manifolds.

White Irons for Elevated-Temperature Service. Because of castability and cost, high-chromium white iron castings can often be used for complex and intricate parts in elevated-temperature applications at considerable savings compared to stainless steel. These cast iron grades are alloyed with 12 to 35% Cr at temperatures up to 1040 °C (1900 °F) for scaling resis-

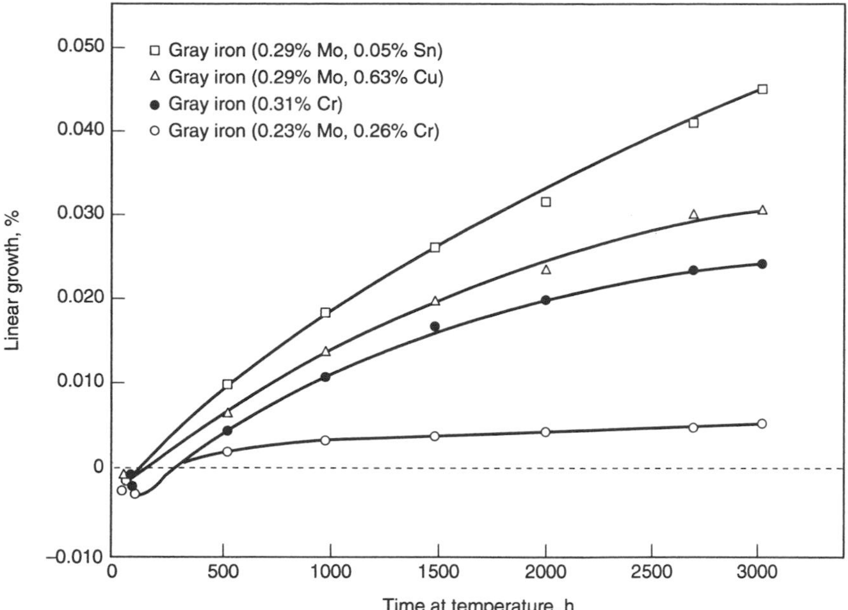

Fig. 6 Growth of four gray irons produced from the same base iron (3.3% C, 2.2% Si) and tested at 455 °C (850 °F) in air. Source: Ref 7

tance. Chromium causes the formation of an adherent, complex, chromium-rich oxide film at elevated temperatures. The high-chromium irons designated for use at elevated temperatures fall into one of three categories, depending on the matrix structure:

- Martensitic irons alloyed with 12 to 28% Cr
- Ferritic irons alloyed with 30 to 35% Cr
- Austenitic irons that contain 15 to 30% Cr as well as 10 to 15% Ni to stabilize the austenite phase

The carbon content of these alloys ranges from 1 to 2%. The choice of an exact composition is critical to preventing σ-phase formation at intermediate temperatures and to preventing the ferrite-to-austenite transformation during thermal cycling, which leads to distortion and cracking. Typical applications include recuperator tubes; breaker bars and trays in sinter furnaces; grates, burner nozzles, and other furnace parts; glass bottle molds; and valve seats for combustion engines.

Growth and Scaling of Gray Irons

When gray cast irons are exposed for long times to elevated temperatures below the critical temperature, or when they are subject to prolonged periods of cyclic heating or cooling, they have a tendency to grow in size and exhibit oxidation at the surface. Growth may occur from one or a combination of the following causes (Ref 4, 5):

- Decomposition of carbides
- The structural breakdown of the pearlite to ferrite, which is accompanied by the formation of the bulkier graphite (graphitization)
- Internal cracking (due to differential expansions and contractions during cyclic heating), which accelerates oxidation
- Carbon deposition on graphite flakes in atmospheres containing carbon monoxide at 350 to 550 °C (660 to 1025 °F)

Deterioration of mechanical properties occurs concurrently with growth as a result of structural decomposition. Knowledge of the growth characteristics of cast iron is also important to the proper interpretation of creep data. In order to determine the true contribution of mechanical stress to creep, dimensional changes caused by growth must be subtracted from the elongations in creep specimens.

Growth occurs in gray iron more rapidly than in ductile, compacted graphite, or malleable irons because of the graphite structure. Growth is most rapid in irons with higher carbon content, as shown in Fig. 5. Growth can be reduced either by producing a ferritic matrix with no pearlite to decompose at elevated temperatures or by stabilizing the carbides to prevent their breakdown into ferrite and graphite. Growth decreases with an increase in the coarseness of the graphite flakes, probably because of a lower combined carbon content and a decrease in the rate of pearlite decomposition resulting from a greater mean-free path between graphite flakes (Ref 5).

The most effective growth reduction in gray iron has been achieved with strong carbide stabi-

Table 6 Summary of growth and scaling tests carried out by the British Cast Iron Research Association

Iron No.	Description	64 weeks(a): Growth, % increase in length, 450 °C	64 weeks(a): Growth, % increase in length, 500 °C	64 weeks(a): Scaling, $g/cm^2 \times 10^{-3}$, 400 °C	64 weeks(a): Scaling, $g/cm^2 \times 10^{-3}$, 450 °C	64 weeks(a): Scaling, $g/cm^2 \times 10^{-3}$, 500 °C	6 years: Growth, % increase in length, 400 °C	6 years: Scaling, $g/cm^2 \times 10^{-3}$, 350 °C	6 years: Scaling, $g/cm^2 \times 10^{-3}$, 400 °C	11½ years: Growth, % increase in length, 350 °C	11½ years: Growth, % increase in length, 400 °C	11½ years: Scaling, $g/cm^2 \times 10^{-3}$, 350 °C	11½ years: Scaling, $g/cm^2 \times 10^{-3}$, 400 °C	Alloy variations
1	Base iron(b)	0.31	0.72	1.86	3.24	9.93	0.035	1.70	3.24	0.005	0.075	2.03	3.90	...
2	Silicon series	0.25	0.75	1.92	4.06	11.95	0.058	1.70	3.56	0.002	0.085	2.08	4.40	2.0Si
3		0.20	0.76	1.86	3.51	8.39	0.042	1.75	3.62	0.007	0.071	2.14	4.40	1.2Si
4		0.16	0.63	2.14	3.84	7.57	0.020	1.75	3.40	0.003	0.038	2.14	4.23	1.3Si-1.2Ni
5	Carbon series	0.33	0.98	2.69	4.55	15.85	0.047	3.45	5.21	0.013	0.070	3.90	6.10	3.7C
7		0.20	0.50	2.63	4.11	10.10	0.025	2.41	4.50	0.002	0.035	2.69	5.44	3.15C
9	Mn, S, P series	0.12	0.62	1.54	2.80	10.36	0.028	1.75	2.96	0.008	0.048	1.98	3.63	1.5Mn
10		0.41	0.78	1.98	3.56	9.49	0.040	1.64	3.51	0.002	0.065	2.20	4.34	0.25S
11		0.22	0.67	1.86	3.84	8.94	0.033	1.48	3.56	0.003	0.071	1.87	4.34	0.64P
12	Cr and Cr + Ni series	0.12	0.43	2.03	4.06	8.44	0.025	1.75	3.51	0.003	0.040	2.20	4.23	0.3Cr
13		0.05	0.30	1.97	4.22	7.68	0.020	1.75	3.62	0	0.028	2.08	4.40	0.6Cr
14		0.11	0.49	1.64	3.73	7.24	0.035	1.75	2.85	0.003	0.053	2.08	3.52	0.3Cr + 0.6Ni
15		0.05	0.27	1.92	3.40	6.86	0.015	1.64	3.35	0	0.033	1.92	4.11	0.6Cr + 1.2Ni
16	Ni, Cr, and Mo series	0.02	0.32	2.03	3.84	8.50	0.012	1.75	3.40	0.002	0.012	2.08	4.28	0.3Cr + 0.6Mo
17		0.05	0.56	1.65	3.28	8.44	0.015	1.70	3.40	0.010	0.018	2.03	3.96	0.6Mo + 0.6Ni
18		0.02	0.27	1.81	3.56	8.12	0.012	1.70	3.35	0.008	0.008	2.14	4.07	0.4Cr + 0.6Mo + 0.6Ni
19	As and Sn series	0.21	0.73	2.14	3.29	8.78	0.030	1.64	3.29	0.018	0.058	2.03	4.07	0.07As
20		0.22	0.67	1.98	3.13	5.87	0.020	1.59	3.29	0.002	0.025	1.92	4.01	0.06Sn
21	Nodular iron	0.09	0.22	1.26	2.41	5.27	0.020	0.77	2.03	0.005	0.028	1.04	2.14	3.4C + 1.7Si + 0.4Mn

(a) Growth occurring in 64 weeks was insignificant at temperatures of 400 °C (750 °F) and below. (b) Base iron composition: 3.25% C, 1.58% Si, 0.65% Mn, 0.11% S, 0.25% P (produced from refined iron having low residual alloy levels). Source: Ref 6, 8-11

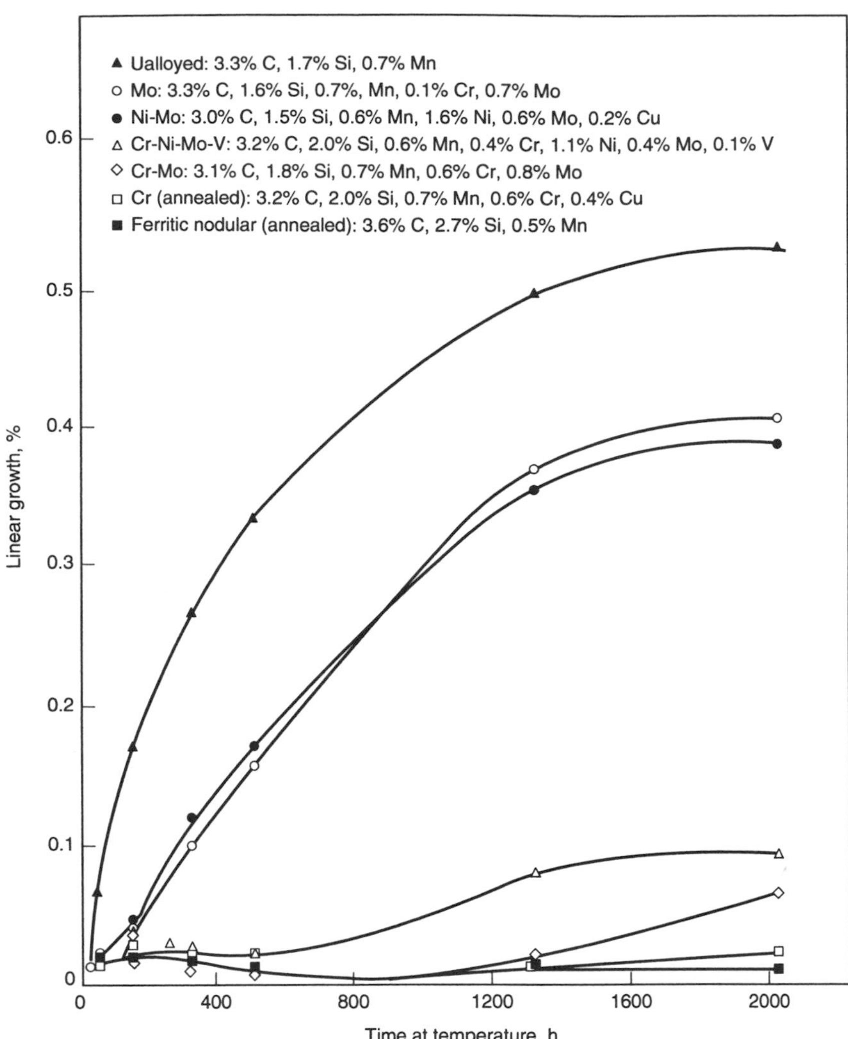

Fig. 7 Growth of six gray irons and one ductile iron tested at 540 °C (1000 °F) in air. Source: Ref 12

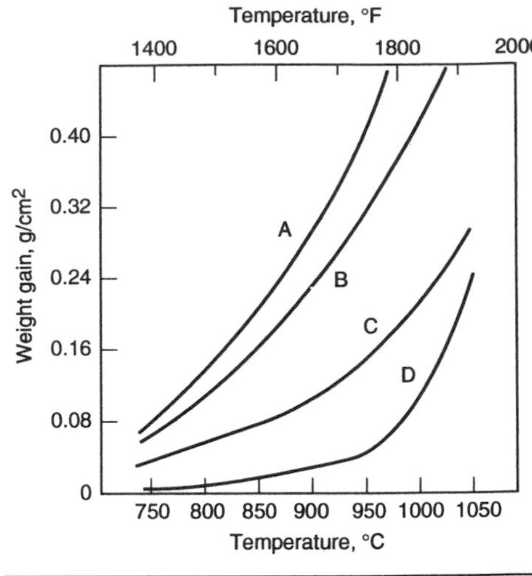

Iron	Type	Chemical analysis, % TC	Si	Mn	P	Ni	Cr	Cu
A	Low-Si gray iron	2.98	1.14	1.07	0.18	...	...	...
B	Ni-Cr gray iron	3.45	1.64	1.03	0.15	1.18	0.92	...
C	Ni-Resist	2.74	1.93	0.89	0.04	14.62	2.11	7.16
D	Nicrosilal	1.78	4.63	0.91	0.10	22.46	2.52	...

TC, total carbon

Fig. 8 Effect of temperature and alloying on the scaling behavior of gray irons after 200 h at temperature in air. Source: Ref 14

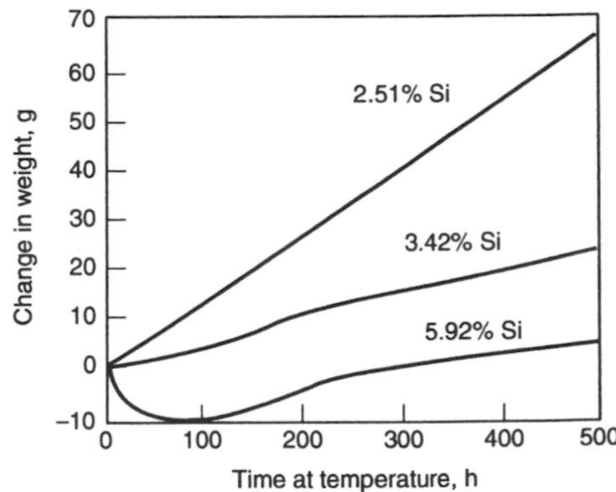

Fig. 9 Effect of silicon content on the oxidation behavior of gray irons tested at 800 °C (1470 °F). Source: Ref 15

lizing elements such as chromium and molybdenum. Bevan (Ref 7) demonstrated the beneficial effects of adding Cr and Cr + Mo on the dimensional stability of gray iron at 455 °C (850 °F). Growth tests were conducted for 3000 h on specimens that had been stress relieved at 565 °C (1050 °F), in order to evaluate gray irons for potential use as housings for passenger car turbines. The results presented in Fig. 6 show that an addition of 0.23% Mo to a chromium-bearing iron reduced growth by a factor of four and limited growth to an insignificant level (<0.01%).

The most extensive growth and scaling tests carried out on gray irons have been performed by BCIRA (Ref 6, 8–11). These tests, which are summarized in Table 6, were conducted at 350 to 500 °C (660 to 930 °F). The results show that no growth occurs in flake graphite irons exposed at 350 °C (660 °F) for up to six years. Slight growth (<0.1%) occurred in unalloyed gray irons exposed at 400 °C (750 °F) for 11.5 years, as shown in Table 6. Additions of Cr + Mo and Cr + Ni + Mo limited growth at 400 °C (750 °F) to insignificant levels (<0.02%).

Table 6 also presents the growth test results for cast irons after 64 weeks (10,000+ h) of exposure in air at 500 and 450 °C (930 and 840 °F). Moderate alloy additions, particularly Cr + Mo and Cr + Ni + Mo, reduced growth up to 500 °C (930 °F). Similar results were demonstrated by Kattus and McPherson (Ref 12) in growth tests at 540 °C (1000 °F) for 2000 h of exposure (Fig. 7).

Scaling. Prolonged exposure of gray iron in air at elevated temperatures produces a surface scale, primarily by oxidation. The scale formed on gray iron when heating in air consists of a mixture of three iron oxides: an outer layer of Fe_2O_3, an intermediate layer of Fe_3O, and an inner layer of FeO. This scale is usually continuous and adherent up to 800 °C (1470 °F) (Ref 13). In matching a material and application, it is important to consider whether the scale is an adherent, protective type or a cracking type that will permit continued oxidation. Generally, the relative amount of scaling is measured by the change in weight of the specimen. The absorption of oxygen during scale formation usually increases the specimen weight, whereas removal of the scale decreases the weight, because the oxidized iron is removed. An adherent scale is associated with a gradually diminishing rate of increase in weight, while a loose or cracked scale permits a continually rapid gain in weight (Ref 4).

The amount of scaling or gain in weight for a particular gray iron increases with the service temperature and exposure time. The relative changes depend on atmosphere as well as the iron composition (Ref 4). Figure 8 illustrates the effect of temperature on four different gray irons. The surface oxidation of gray iron is considered to be greatest at temperatures above 760 °C (1400 °F); the specific effect is related to the carbon, silicon, and alloy contents of the iron. Steam is typically a more aggressive atmosphere than air, as are atmospheres produced by hydrogen-containing fuels such as oil or natural gas (Ref 14). Scaling can be minimized through the use of hydrogen or carbon monoxide atmospheres. Sulfur-bearing atmospheres are deleterious, but higher-chromium-content irons are resistant.

The quantity and shape of the graphite have an appreciable influence on gray iron scaling, particularly up to 500 °C (900 °F) (Ref 4). In general, the degree of scaling increases markedly as the amount of graphite increases. Matrix structure apparently has little influence on the scaling of gray iron. Test results indicate that heat treatment to obtain ferritic conditions has a relatively minor effect on scaling propensity (Ref 4).

Fig. 10 Effect of silicon content on growth of unalloyed and 2% Cr gray irons tested at 800 °C (1470 °F). Source: Ref 15

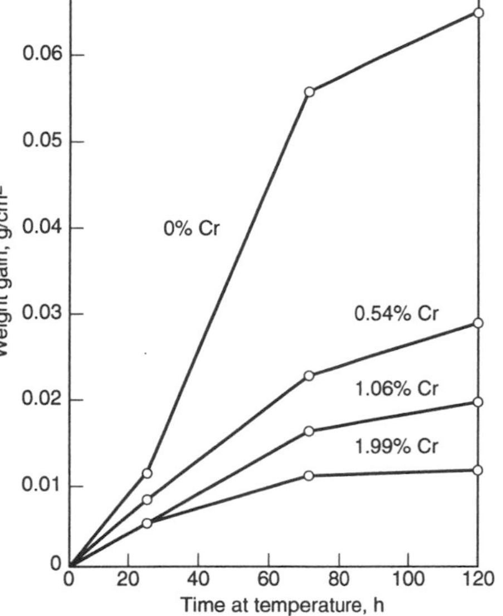

Fig. 11 Effect of chromium content on the oxidation of gray irons tested at 800 °C (1470 °F). Source: Ref 16

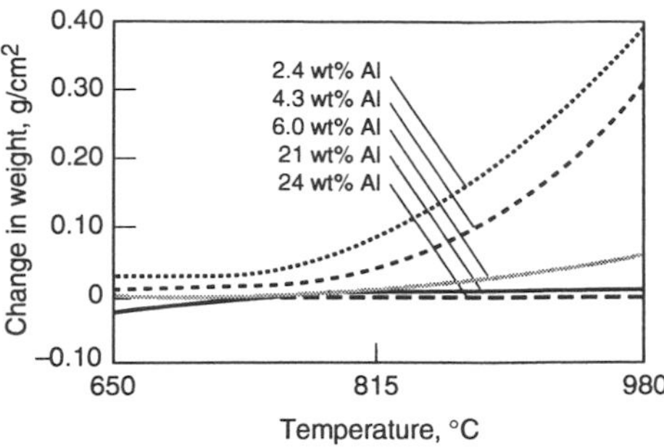

Fig. 12 Effect of aluminum content on the oxidation behavior of gray irons tested in air for 200 h. Source: Ref 13

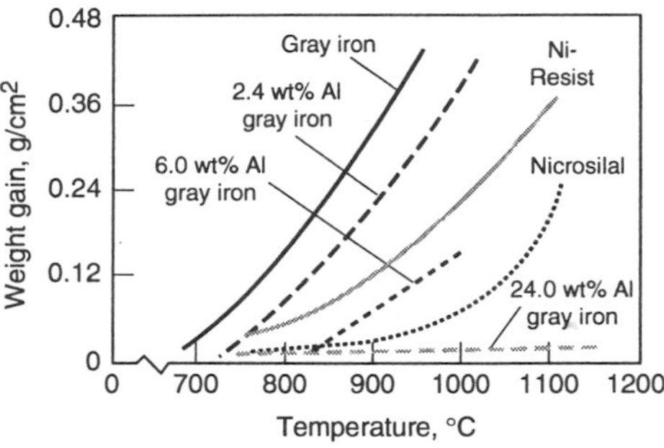

Fig. 13 Comparison of oxidation resistance of various gray irons tested in air for 200 h. Source: Ref 3

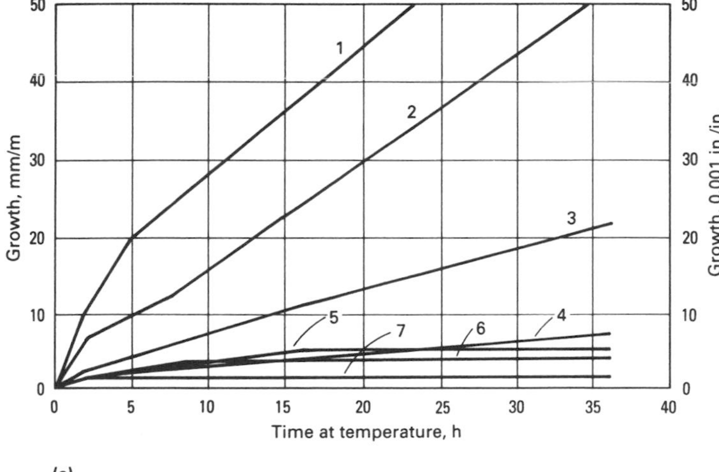

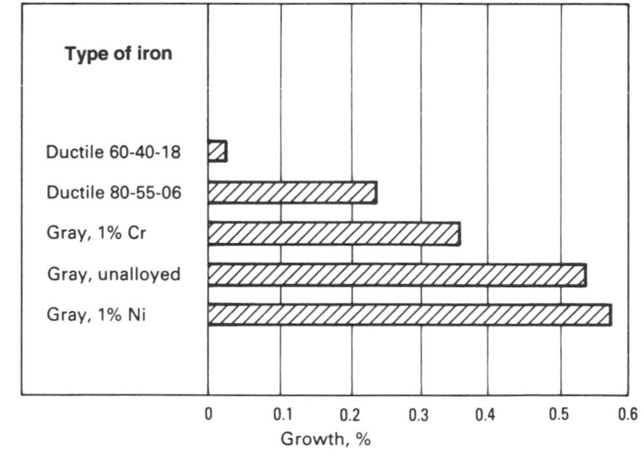

Type of iron	Composition, %							
	TC	Mn	Si	P	S	Cr	Ni	Mg
1 Class 25 gray iron	3.27	0.68	2.27	0.20	0.15	…	…	…
2 Class 35 alloy gray iron	3.19	0.93	2.10	0.16	0.092	0.37	1.12	…
3 Class 40 gray iron	3.15	0.90	1.28	0.092	0.10	…	…	…
4 Class 40 alloy gray iron	3.06	1.01	1.55	0.078	0.069	0.341	0.98	…
5 Ductile iron, as-cast(a)	3.43	0.47	2.19	0.13	0.009	…	1.95	0.077
6 Ductile iron, annealed(b)	3.50	0.54	2.58	0.12	0.010	…	1.50	0.055
7 Ductile iron, annealed(a)	3.43	0.47	2.19	0.13	0.009	…	1.95	0.07

TC, total carbon. (a) Cut from 115 mm (4.5 in.) wide keel block. (b) Cut from experimental piston after heat treating

Fig. 14 Growth of ductile and gray irons at elevated temperature. (a) Plot of growth vs. time at 900 °C (1650 °F). (b) Growth of selected ductile and gray irons after 6696 h at 540 °C (1000 °F)

Table 7 Growth and scaling of ductile irons in air

Grade	Exposure temperature °C	Exposure temperature °F	Exposure time, years	Growth mm/mm	Growth in./in.	Scaling weight gain, g/m^2
700/2 (pearlite)	350	660	0	0.000	0.000	0.00
			4.9	0.005	0.005	8.77
			10.4	–0.003	–0.003	12.61
			21.3	–0.003	–0.003	13.71
	400	750	0	0.000	0.000	0.00
			4.9	0.020	0.020	21.93
			10.4	0.018	0.018	29.60
			21.3	0.015	0.015	33.44
500/7 (pearlite + ferrite)	350	660	0	0.000	0.000	0.00
			4.9	0.003	0.003	5.48
			10.4	0.000	0.000	9.87
			21.3	–0.005	–0.005	9.87
	400	750	0	0.000	0.000	0.00
			4.9	0.005	0.005	16.45
			10.4	0.075	0.075	26.31
			21.3	0.047	0.047	32.34
400/12 (ferrite)	350	660	0	0.000	0.000	0.00
			4.9	0.005	0.005	6.03
			10.4	0.000	0.000	10.42
			21.3	–0.017	–0.017	9.87
	400	750	0	0.000	0.000	0.00
			4.9	0.008	0.008	17.54
			10.4	0.007	0.007	24.12
			21.3	0.000	0.000	27.96

Table 8 Oxide penetration of ductile iron and other materials at 705 °C (1300 °F)

Material	Oxide penetration 2000 h mm/year	2000 h mils/year	3000 h mm/year	3000 h mils/year	4000 h mm/year	4000 h mils/year
Ductile iron						
80-55-06, 2.5% Si	1.35	53	...	...	1.05	41
60-40-18, 2.5% Si	1.4	56	...	...	0.7	28
Ferritic ductile						
4.0% Si	1.05	41	...	...	0.85	34
5.5% Si	0	0	...	...	0.08	3.3
Cast steel	5.1	201	5.05	199	...	...
Gray iron						
1.5% Si	3.3	131	2.8	110	...	...
1.65% Si	2.95	116	3.3	131	...	...
2.0% Si	4.6	181	3.3	131	...	...
2.5% Si	4.8	188	3.6	143	...	...
Pearlitic malleable	3.3	131	...	...	...	...

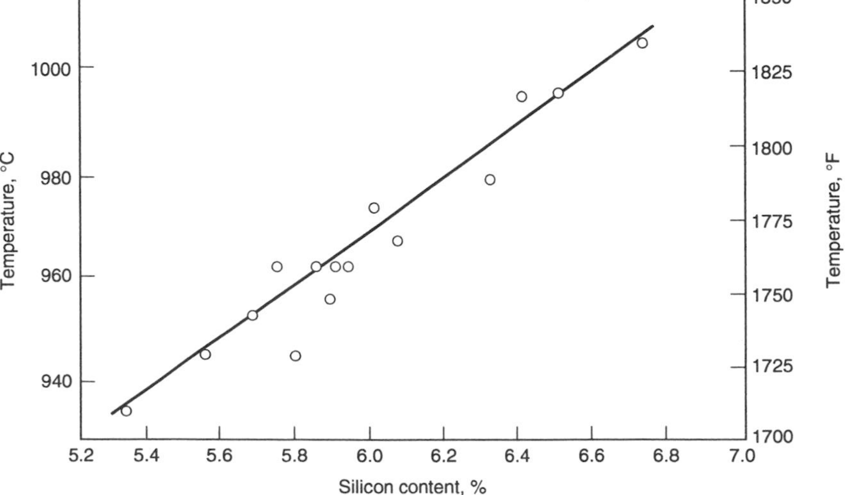

Fig. 15 Effect of silicon content on the critical temperature in cast irons. Source: Ref 18

The percentage of carbon in a gray iron will affect its ability to resist scaling. Even at temperatures as low as 250 °C (500 °F), higher-carbon irons will exhibit considerably more scaling than irons having low carbon contents (Ref 6).

The most extensive study on scaling in unalloyed and low-alloy gray irons was performed at BCIRA by Gilbert (Ref 6) and Palmer (Ref 10), who monitored weight gain in specimens exposed to air for up to 11.5 years at 350 and 400 °C (660 and 750 °F) and for up to 64 weeks at 450 and 500 °C (840 and 930 °F). Their results, which are presented in Table 6, can be summarized as follows (Ref 5):

Temperature	Extent of scaling
350 °C (566 °F)	4 mg/cm^2 in 11.5 years
400 °C (750 °F)	4 mg/cm^2 in 11.5 years
450 °C (840 °F)	3-4.5 mg/cm^2 in 64 weeks
500 °C (930 °F)	6-16 mg/cm^2 in 64 weeks

There was little distinction in resistance to scaling among the 18 low-alloyed irons tested at or below 450 °C (840 °F), with the exception of a high-carbon-equivalent iron (3.7% C) that exhibited somewhat greater weight gains. At 500 °C (930 °F) the range in scaling resistance was greater, again largely because one iron had a significantly higher graphitic carbon content. Differences in alloy content had minor effects on scaling resistance, although all alloyed irons exhibited slightly lower weight gains (6 to 9 mg/cm^2) over the unalloyed base iron (10 mg/cm^2).

Table 9 Growth of high-nickel irons in superheated steam at 480 °C (900 °F)

Type of iron	Growth, mm/m or 0.001 in./in., after: 500 h	1000 h	2500 h
Gray iron (unalloyed)	2.3	5.2	14
High-nickel gray iron (20% Ni)	0.5	1.0	1.5
High-nickel gray iron (30% Ni)	0.3	0.45	0.48
High-nickel ductile iron (20% Ni)	0.3	0.5	0.5
High-nickel ductile iron (30% Ni)	0.3	nil	nil

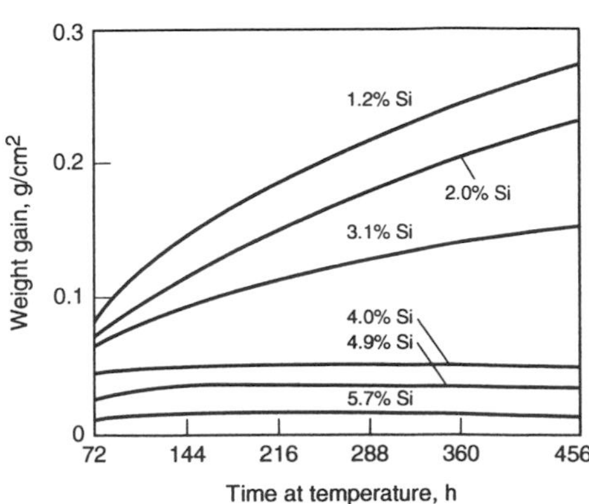

Fig. 16 Effect of silicon on the oxidation of ferritic ductile iron in air at 650 °C (1200 °F). Source: Ref 18

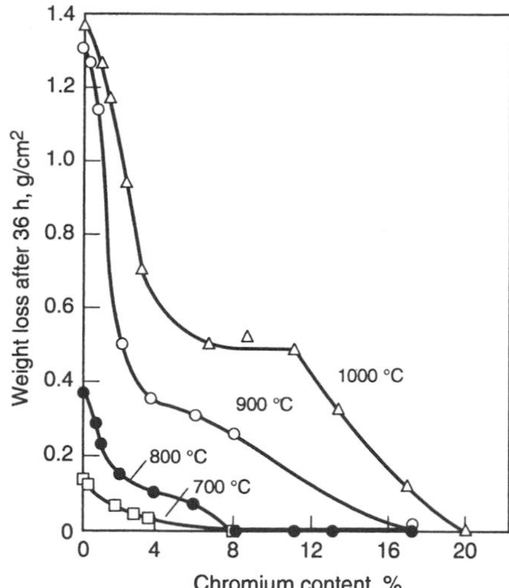

Fig. 17 Effect of chromium content on the oxidation behavior of alloy cast irons. Source: Ref 19

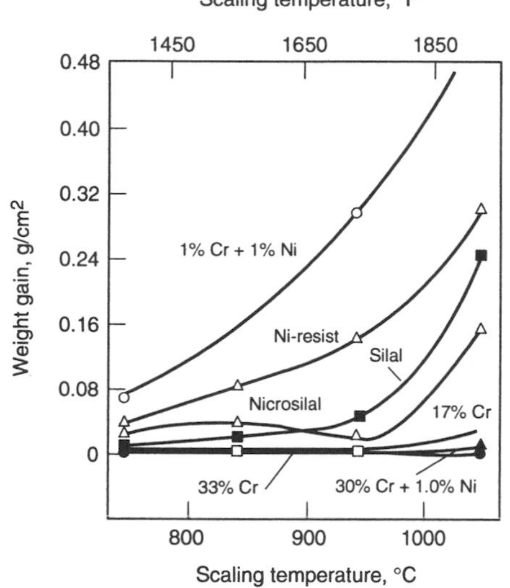

Fig. 18 Oxidation behavior of alloyed cast irons held at temperature for 200 h. Source: Ref 14

High-Alloy Gray Irons. At temperatures above 650 °C (1200 °F), severe growth will occur in normal cast irons due to oxidation. In flake graphite irons, growth may result in a 40% increase in volume, and continued growth at sufficiently elevated temperatures generally leads to complete disintegration of the component. Raising the silicon content or alloying with chromium makes it possible to increase the practical operating temperatures.

High-silicon-content gray irons such as Silal (6% Si) are suitable for use at temperatures up to about 800 °C (1470 °F). In Silal, the advantages of a high critical temperature, a ferritic matrix (no combined carbon), and a fine, undercooled type D graphite structure provide good growth and scaling resistance (Ref 5). Figure 9 shows the effect of silicon additions on the oxidation resistance of gray irons. For limiting growth in plain or 2% Cr gray irons for use at temperatures up to

Table 10 Oxidation of plain and alloy cast irons and one stainless steel

	Composition, % (a)				Oxide penetration				Growth at 815 °C (1500 °F)(c)	
					At 760 °C (1400 °F)(b)		At 815 °C (1500 °F)(c)			
Iron	TC	Si	Cr	Ni	mm/yr	mils/yr	mm/yr	mils/yr	mm/yr	mils/yr
Austenitic(d)	2.69	1.96	2.05	13.96	4.7	184	9.5	374	0.4	15
Austenitic(e)	2.97	2.38	4.87	(14.0)	2.4	96	5.9	232	0.2	8
Austenitic	2.40	1.57	2.98	30.28	2.1	83	6.3	249	0.2	8
Austenitic	(1.8)	(6.0)	(5.0)	(30.0)	0.05	2	1.3	53	0.2	8
Austenitic	(2.8)	(1.7)	(2.0)	(20.0)	4.2	166	7.9	312	0.2	8
Austenitic	(2.7)	(2.5)	(5.0)	(20.0)	1.9	74	3.6	143	0.4	15
Plain ferritic	(3.2)	(2.2)	...	...	>20(f)	>800(f)	>85(f)	>3300(f)	2.0	78
Low-alloy ferritic	(3.3)	(1.5)	(0.6)	(1.5)	>20(f)	>800(f)	>90(f)	>3500(f)	1.4	54
Low-alloy ferritic	(3.3)	(2.2)	(1.0)	(1.0)	5.8	228	25.9	1020	1.2	47
Low-alloy ferritic	(3.1)	(2.2)	(0.9)	(1.5)	7.2	284	29.0	1140	1.6	62
Type 309 stainless	...	...	(25.0)	(12.0)	nil	nil	nil	nil	nil	nil

(a) Parenthetical values are estimates. Phosphorus and sulfur contents in all iron samples were about 0.10%. TC, total carbon. (b) Exposure of 2000 h in electric furnace at 760 °C (1400 °F) with air atmosphere containing 17-19% O. (c) Exposed for 492 h in gas-fired heat treating furnace at 815 °C (1500 °F). (d) 6.05% Cu. (e) 6.0% Cu. (f) Specimen completely burned. Source: International Nickel Co., Inc.

Table 11 Oxidation of ferritic and austenitic cast irons and one stainless steel

	Composition, % (a)				Growth		Oxide penetration	
Iron	TC	Si	Cr	Ni	mm/yr	mils/yr	mm/yr	mils/yr
After 3723 h at 745-760 °C (1375-1400 °F) in electric furnace, air atmosphere								
Ferritic	3.05	2.67	0.90	1.55	2.0	78	(b)	(b)
Austenitic	2.97	1.63	1.89	20.02	0.8	31	6.9	270
Austenitic	2.52	2.67	5.16	20.03	nil	nil	0.2	6
Austenitic	2.32	1.86	2.86	30.93	nil	nil	2.0	78
Austenitic	1.86	5.84	5.00	29.63	nil	nil	<0.1	<3
309 stainless	...	...	(25.0)	(12.0)	nil	nil	<0.1	<3
After 1677 h at 815-915 °C (1500-1700 °F) in gas-fired furnace, slightly reducing atmosphere								
Ferritic	(3.2)	(2.2)	...	...	3.2	125	(b)	(b)
Austenitic(c)	(3.0)	(2.4)	(5.0)	(14.0)	0.4	15	8.4	330
Austenitic	(2.7)	(2.5)	(5.0)	(20.0)	0.4	15	5.6	220
Austenitic	(2.4)	(1.6)	(3.0)	(30.0)	0.4	15	6.9	270
Austenitic	(1.8)	(6.0)	(5.0)	(30.0)	0.4	15	0.1	5
309 stainless	...	...	(25.0)	(12.0)	nil	nil	0.1	5

(a) Parenthetical values are estimates. Phosphorus and sulfur contents in all iron samples were about 0.10%. TC, total carbon. (b) Sample was completely burned. (c) 6.0% Cu (est). Source: International Nickel Co., Inc.

Table 12 Oxide penetration in ductile irons and one stainless steel

	Estimated composition, %					Oxide penetration			
						Test 1		Test 2	
Iron	TC	Si	Cr	Ni	Cu	mm/yr	mils/yr	mm/yr	mils/yr
After 15 months at 870 °C (1600 °F)(a)									
Austenitic	1.80	6.0	5.0	30.0	...	1.1	44	...	...
18-8 stainless	...	...	18.0	8.0	...	0.2	9	...	...
25-12 stainless	...	...	25.0	12.0	...	0.1	3	...	...
Austenitic	2.90	2.0	2.0	14.0	6.0	(b)	(b)	...	...
Austenitic	2.8	1.7	2.0	20.0	...	(b)	(b)	...	...
Austenitic	2.7	2.5	5.0	20.0	...	(b)	(b)	...	...
Air atmosphere, 400 h at 705 °C (1300 °F) (test 1) air atmosphere(c) (test 2)									
2.5 Si ductile	3.40	2.50	...	...	...	1.1	42	12.7	500
5.5 Si ductile	2.6	5.50	...	...	...	0.1	4	1.3	51
Austenitic ductile	2.3	2.5	1.7	20.0	...	1.1	42	4.4	175
Austenitic ductile	2.3	2.0	...	22.0	...	1.8	70	...	...
Austenitic ductile	2.1	5.5	5.0	30.0	...	1.1	4	nil	nil
Austenitic	3.0	1.6	1.9	20.0	...	2.5	98	7.6	300
309 stainless	...	...	25.0	12.0	...	nil	nil	nil	nil

TC, total carbon. (a) Exposed to flue gases from powdered coal containing 1.25-2.00% S. (b) Completely oxidized. (c) Exposed to a heat cycle, 600 h at 870-925 °C (1600-1700 °F), 600 h at 870-925 °C (1600-1700 °F) and 425-480 °C (800-900 °F), 600 h at 425-480 °C (800-900 °F). Source: International Nickel Co., Inc.

800 °C (1470 °F), silicon contents ranging from 3.0 to 4.5% provide optimum results (Fig. 10).

High-alloy austenitic cast irons, such as Nicrosilal and Ni-Resist, also have good growth resistance in the temperature range of 650 to 900 °C (1200 to 1650 °F). In these irons, the resistance to growth is partly due to the absence of phase transformations and partly to improved oxidation resistance. Figure 8 shows the superior scaling behavior of austenitic gray irons.

Moderate additions of chromium have a significant growth-retarding effect in gray irons, as shown in Fig. 11. Growth and oxidation at 800 °C (1470 °F) were shown to be minimal with a 2% Cr addition.

As shown in Fig. 12, additions of aluminum up to 24% have been shown to markedly reduce scaling in gray irons at temperatures up to 980 °C (1800 °F). Petitbon and Wallace (Ref 17) demonstrated that oxidation resistance increased continuously with aluminum contents from 2.4 to 24.4%. The values of weight gain or weight loss compare favorably with those of high-alloy gray irons (Fig. 13).

The resistance to scaling may be explained by the presence of an aluminum oxide (Al_2O_3) film on the surface of these aluminum-alloyed irons (Ref 13). The adherent oxide film retards the diffusion of oxygen in air to the graphite in the cast iron. This then limits the rate of which oxygen can combine with graphite to decompose into carbon monoxide and carbon dioxide. Thus the rate of cast iron scaling and graphite decomposition may be retarded even more by the addition of increased amounts of aluminum to form a more dense adherent oxide film.

Growth and Scaling of Ductile Irons

Growth. Ductile irons are much more resistant to growth than gray irons. Unlike gray irons, which experience growth by both graphitization and subsurface oxidation, the growth of ductile irons results primarily from graphitization.

Annealed ferritic ductile iron is essentially free from growth up to 815 °C (1500 °F), while pearlitic irons begin to grow due to decomposition of the pearlite into ferrite and graphite at temperatures above 538 °C (1000 °F). Pearlite will completely break down at the transformation temperature (approximately 760 °C, or 1400 °F). Both types will show larger amounts of growth above 815 °C (1500 °F), with pearlitic structures growing more than the ferritic grades.

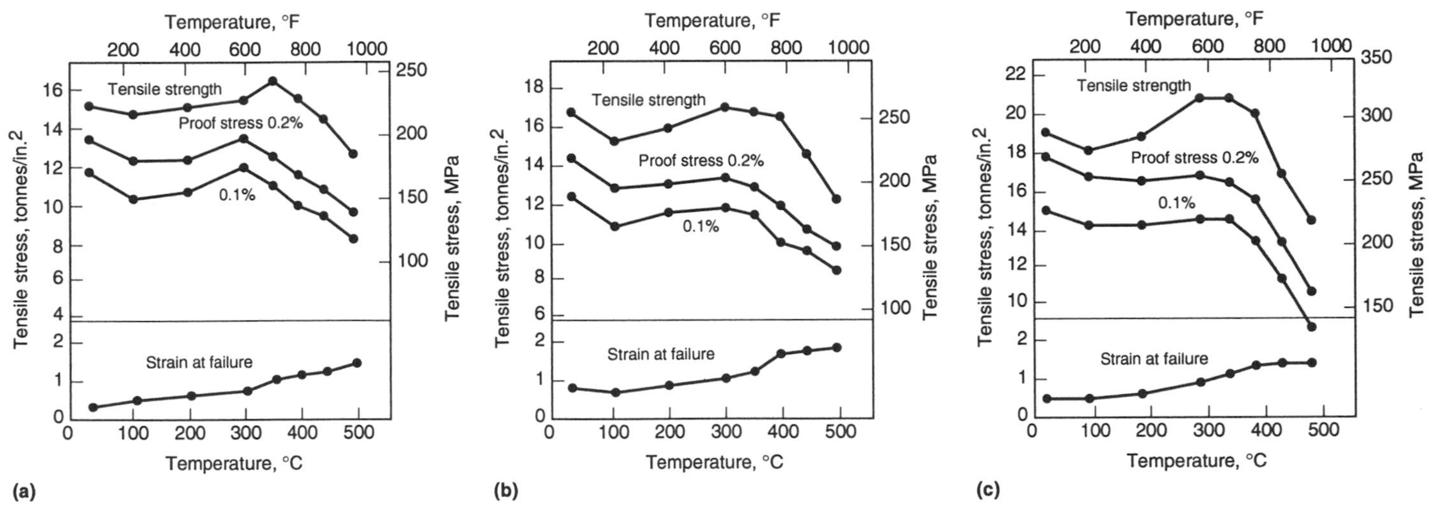

Fig. 19 Variation in tensile properties resulting from increasing temperature for three British gray iron grades. (a) Grade 14 (216 MPa, or 31 ksi, minimum tensile strength). (b) Grade 17 (263 MPa, or 38 ksi, minimum tensile strength). (c) Grade 20 (309 MPa, or 45 ksi, minimum tensile strength). Source: Ref 5

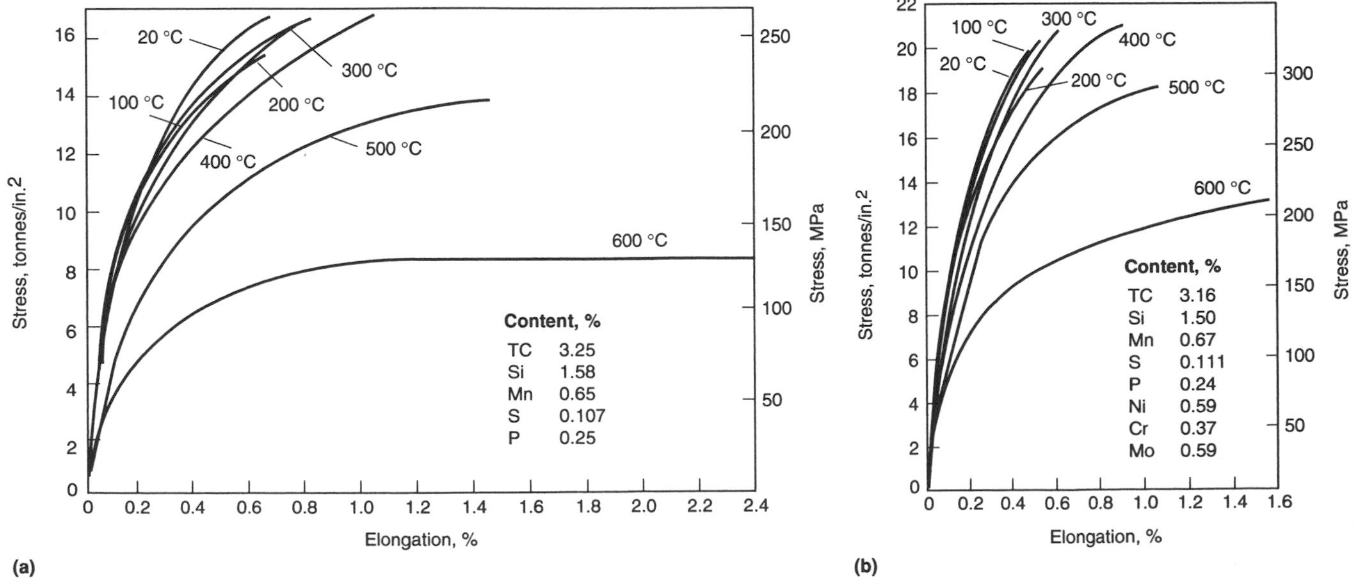

Fig. 20 Effect of temperature on the stress-strain curves of (a) unalloyed gray iron and (b) low-alloy gray iron. TC, total carbon. Source: Ref 6

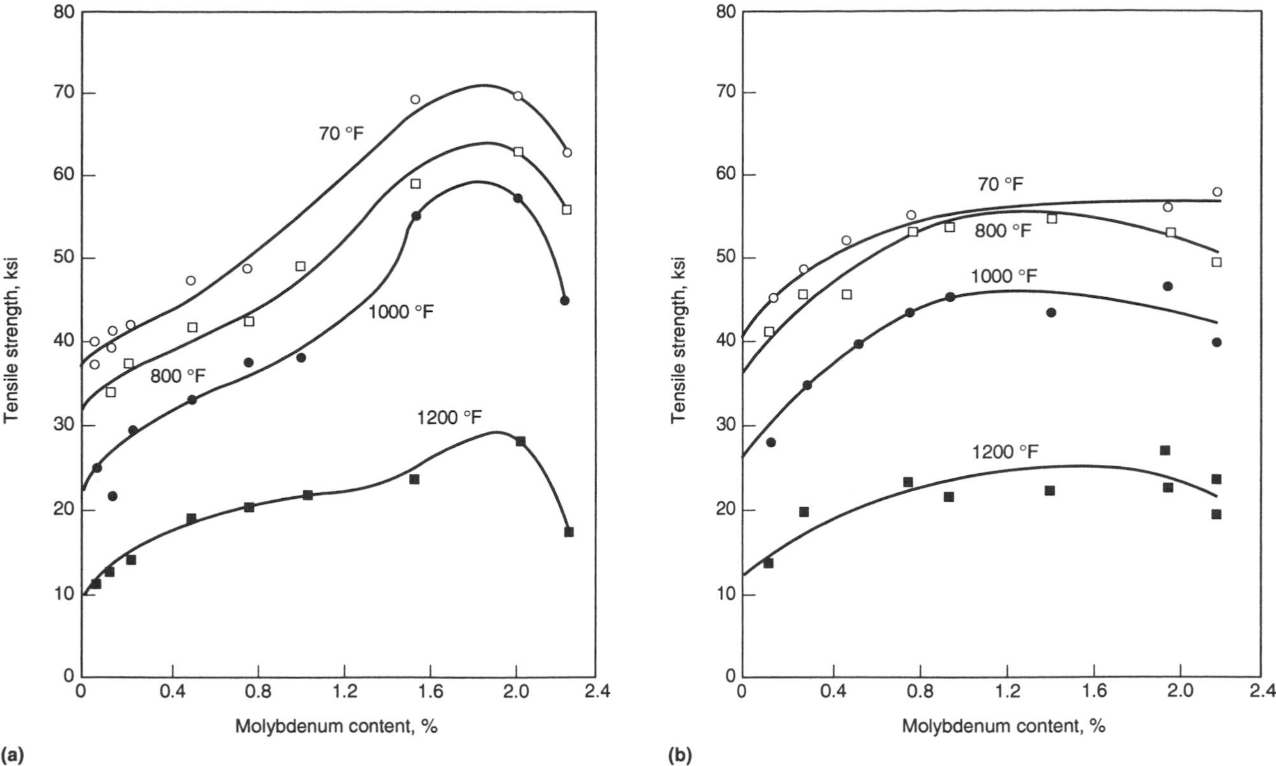

Fig. 21 Tensile strength at room and elevated temperatures as a function of molybdenum content in (a) unalloyed and (b) 0.6% Cr-alloyed gray irons. Source: Ref 20

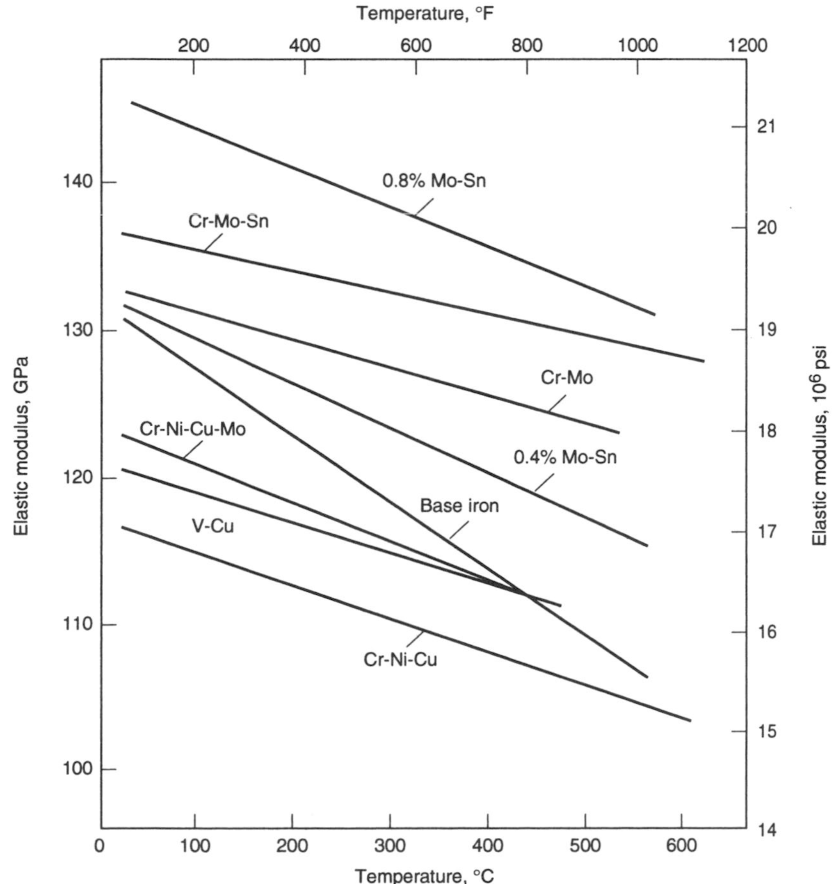

Fig. 22 Elastic modulus as a function of temperature and alloying content. Source: Ref 5

Growth decreases with increasing section size and can be retarded by increasing the silicon content and by alloying with chromium and molybdenum. The effects of silicon contents up to 4% or more are described below in the section "Growth and Scaling of High-Alloy Irons" in this article. Annealed ductile irons also exhibit lower growth rates.

Figure 14(a) compares the growth of ductile and gray irons at 900 °C (1650 °F). The compositions of the irons are shown in the table below the graph. Data on the growth of ductile and gray irons at 540 °C (1000 °F) for 6696 h are plotted in Fig. 14(b), which emphasizes the excellent performance of ASTM grade 60-40-18 ferritic ductile iron. Additional data are given in Table 7.

Scaling. Unalloyed grades of ductile iron provide excellent oxidation resistance at temperatures as high as 650 °C (1200 °F). Resistance to oxidation increases with increasing silicon content. Table 8 compares the oxidation behavior of ductile irons with that of other ferrous alloys at 705 °C (1300 °F). The data indicate that ductile iron of normal composition has much better oxidation resistance than cast steel, gray iron, or pearlitic malleable iron. Additional data are given in Table 7.

Growth and Scaling of High-Alloy Irons

High-silicon ductile irons containing 4 to 6% Si, either alone or combined with up to 2% Mo, were developed to meet the increasing demands

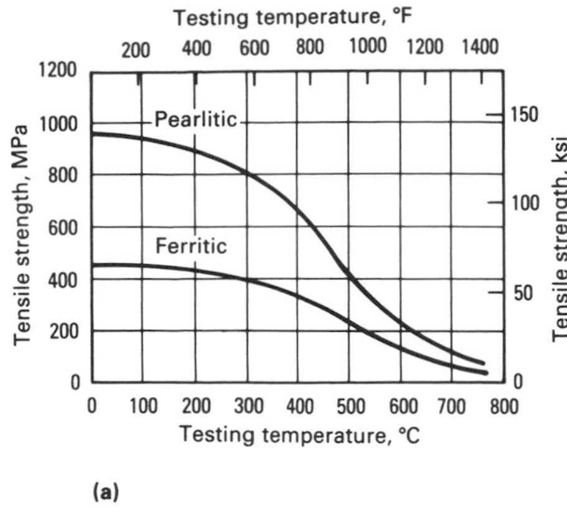

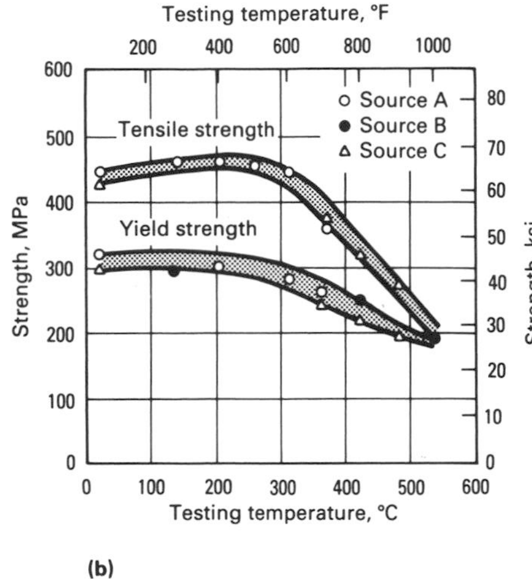

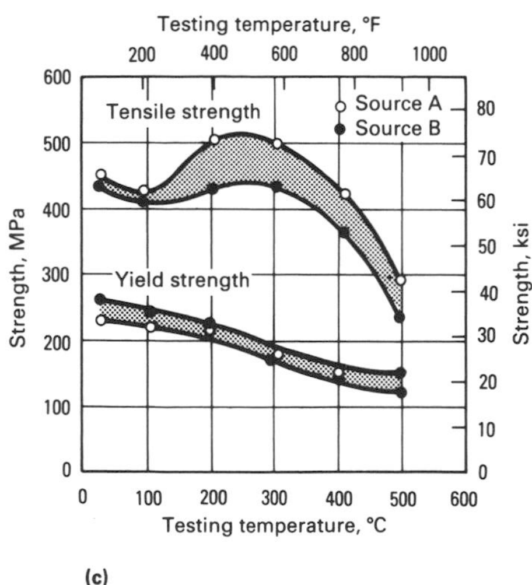

Fig. 23 Comparison of short-time tensile properties of cast products at elevated temperatures. (a) Pearlitic and ferritic ductile irons. (b) Ferritic ductile iron from three sources. (c) Low-carbon cast steel from two sources

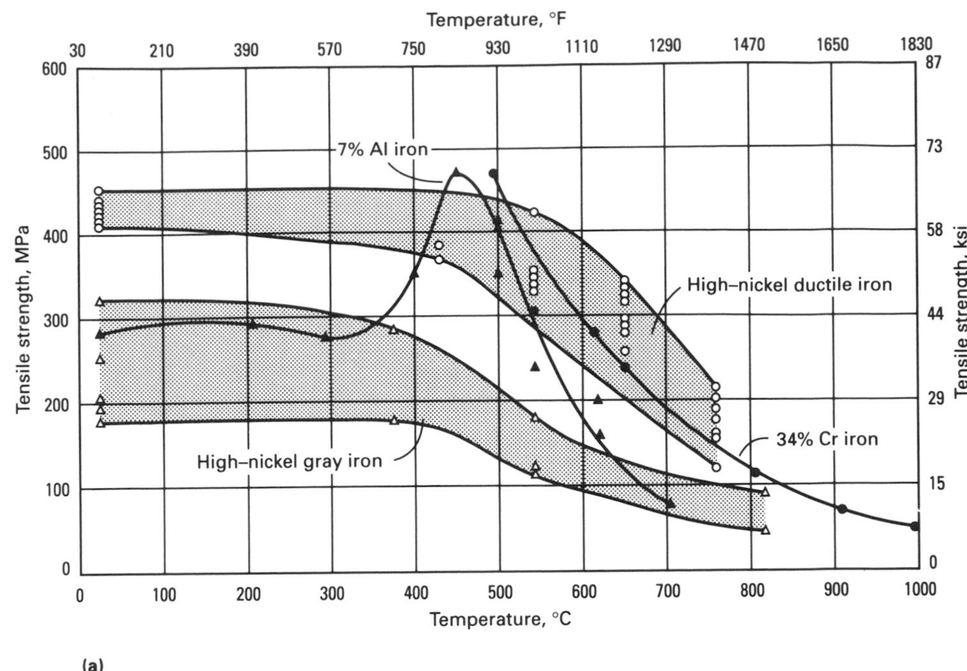

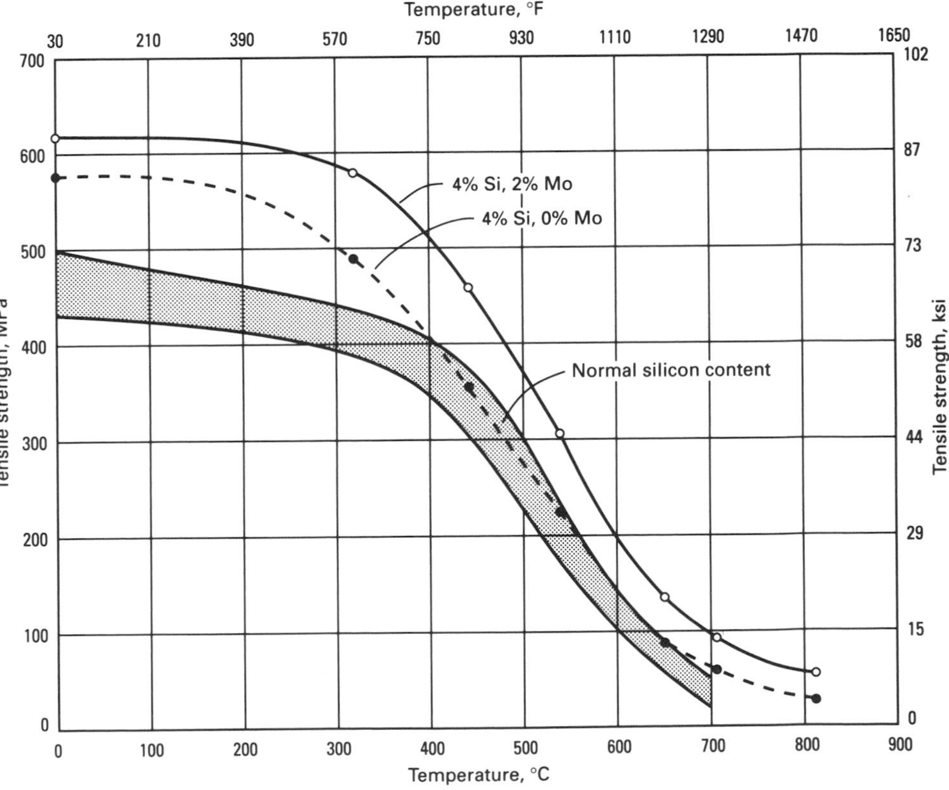

Fig. 24 Short-time elevated-temperature tensile strengths of several (a) alloy irons and (b) ferritic nodular irons

for high-strength ductile irons capable of operating at elevated temperatures in applications such as exhaust manifolds or turbocharger casings. The primary properties required for such applications are:

- Oxidation resistance
- Structural stability
- Elevated-temperature strength
- Resistance to thermal cycling

Silicon enhances the performance of ductile iron at elevated temperatures by stabilizing the ferritic matrix and forming a silicon-rich surface layer that inhibits further oxidation. Stabilization of the ferrite phase reduces elevated-temperature growth in two ways. First, silicon raises the critical temperature at which ferrite transforms to austenite (Fig. 15). The critical temperature is considered to be the upper limit of the useful temperature range for ferritic ductile irons. Above this temperature, the expansion and con-

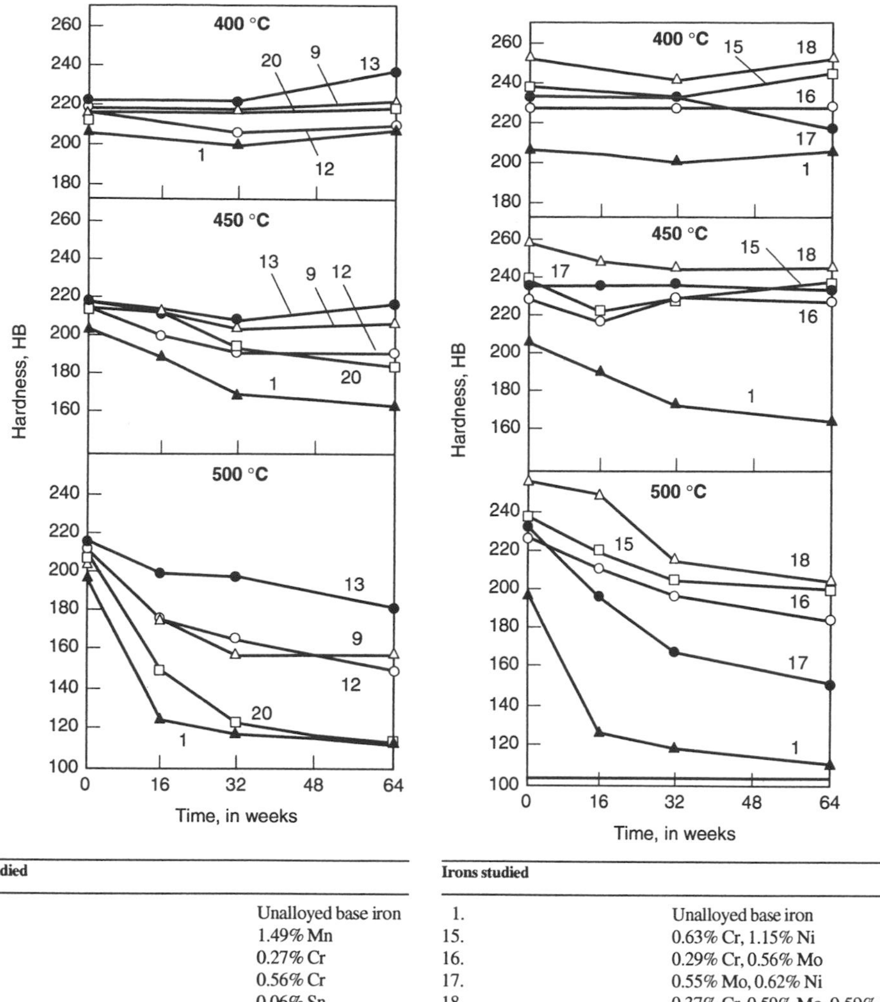

Irons studied	
1.	Unalloyed base iron
9.	1.49% Mn
12.	0.27% Cr
13.	0.56% Cr
20.	0.06% Sn

Irons studied	
1.	Unalloyed base iron
15.	0.63% Cr, 1.15% Ni
16.	0.29% Cr, 0.56% Mo
17.	0.55% Mo, 0.62% Ni
18.	0.37% Cr, 0.59% Mo, 0.59% Ni

Fig. 25 Effect of temperature and alloying content on the hardness retention of gray irons. The base iron contained 3.06-3.25% C, 1.5-1.75% Si, and 0.5-0.8% Mn. Source: Ref 6

Table 13 Elevated-temperature tensile strengths of various ductile iron grades

Grade	Strength, MPa (ksi), at: 95 °C (200 °F)	205 °C (400 °F)	315 °C (600 °F)	425 °C (800 °F)	540 °C (1000 °F)	650 °C (1200 °F)
120-90-02	841 (122)	827 (120)	745 (108)	565 (82)	310 (45)	117 (17)
100-70-03	703 (102)	690 (100)	607 (88)	469 (68)	283 (41)	103 (15)
80-55-06	565 (82)	565 (82)	517 (75)	414 (60)	248 (38)	100 (14.5)
60-45-12	469 (68)	469 (68)	441 (64)	345 (50)	200 (29)	93 (13.5)
60-40-18	407 (59)	407 (59)	365 (53)	276 (40)	172 (25)	86 (12.5)
4% Si + 1% Mo	607 (88)	...	...	421 (61)	303 (44)	131 (19)

Source: QIT-Fer et Titane, Inc., Montreal, Quebec, Canada

Table 14 Tensile design stresses for three ductile irons at elevated temperatures

Grade	Structure	Temperature °C	°F	Maximum tensile design stress MPa	ksi	Basis for maximum design stress
420/12	Ferrite	20	68	136	20	0.52 × (0.1% offset yield strength)
		100	212	130	19	
		200	390	124	18	
		300	570	124	18	
600/3	Pearlite	20	68	148	21	0.45 × (0.1% offset yield strength)
		100	212	131	19	
		200	390	130	19	
		300	570	130	19	
700/2	Pearlite	20	68	173	25	0.45 × (0.1% offset yield strength)
		100	212	154	22	
		200	390	150	22	
		300	570	147	21	

Source: Ref 22

traction associated with the transformation of ferrite to austenite can cause distortion of the casting and cracking of the surface oxide layer, reducing oxidation resistance. Second, the strong ferritizing tendency of silicon stabilizes the matrix against the formation of carbides and pearlite, thus reducing the growth associated with the decomposition of these phases at elevated temperature.

Silicon is highly effective in protecting ductile iron from oxidation through the formation of a silicon-rich oxide layer that inhibits further oxidation. The oxidation protection offered by silicon increases with increasing silicon content (Fig. 16). Silicon levels above 4% are sufficient to prevent any significant weight gain after the formation of an initial oxide layer.

High-Nickel Ductile Irons. Tables 9 to 12 provide growth and oxidation resistance data for nickel-alloyed austenitic irons.

High-Chromium White Irons (Ref 2). Chromium inhibits graphitization in iron and is particularly effective in retarding growth as well as oxidation at elevated temperatures. White irons containing 15 to 35% Cr show little or no growth at temperatures up to 950 °C (1470 °F) and resist oxidation at temperatures up to 1040 °C (1900 °F).

Figure 17 demonstrates the beneficial effect of chromium on the scaling resistance of cast iron at 700 to 1000 °C (1290 to 1830 °F). The oxidation resistance of the high-chromium irons is compared with that of other alloy cast irons in Fig. 18. All the irons plotted in Fig. 18 were simultaneously exposed to air and sulfurous atmospheres, simulating conditions in electric furnaces and in oil-fired or coal-fired furnaces. The results of the laboratory tests were supported by testing the same irons as furnace parts under industrial conditions.

Short-Time Tensile Properties at Elevated Temperatures

Gray Irons (Ref 5). The tensile properties of unalloyed gray irons exhibit small changes at temperatures up to 400 °C (750 °F), as demonstrated in Fig. 19. In general, there is a slight decrease in strength as the temperature increases to about 100 °C (210 °F) and then a gradual increase as the temperature increases to about 350 °C (660 °F). Above 400 °C (750 °F) the tensile strength decreases quite rapidly. The proof (yield) stress tends to follow the same trend.

Gilbert (Ref 6) conducted an extensive investigation of the tensile properties of several plain and alloyed cast irons at temperatures up to 600 °C (1110 °F). Figure 20 illustrates a series of stress-strain curves obtained at the various temperatures for two gray irons. At temperatures up to 400 °C (750 °F), the amount of curvature in the stress-strain curve increased slightly with temperature, but that above 400 °C (750 °F) the ability of gray iron to resist plastic deformation decreased rapidly. The molybdenum-alloyed irons appeared to be more resistant to plastic

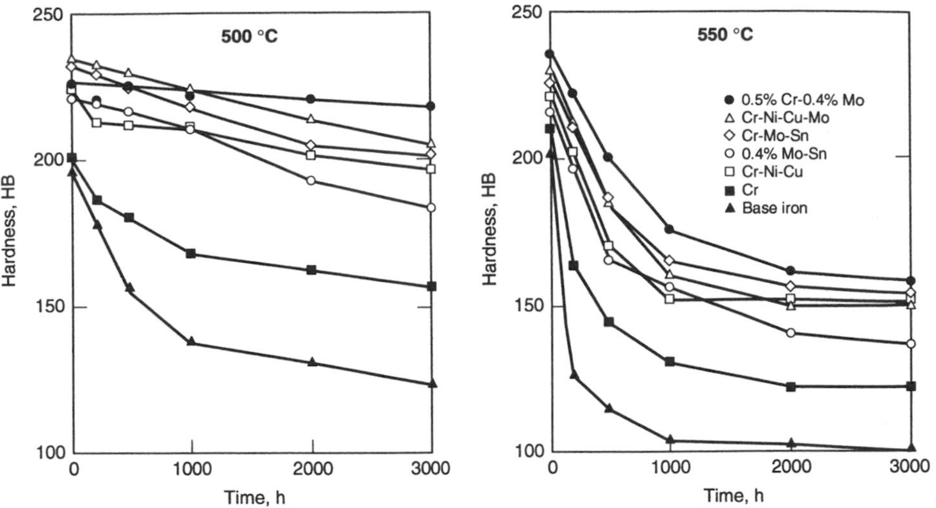

Fig. 26 Effect of temperature and alloying content on the hardness retention of gray irons. Source: Ref 21

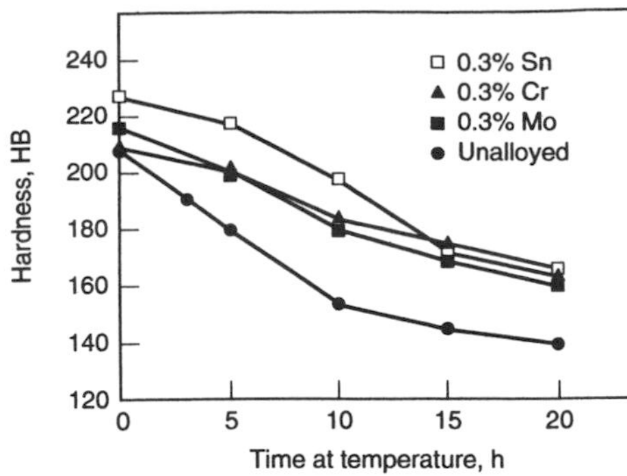

Fig. 27 Effect of tin, chromium, and molybdenum on the hardness of 50 mm (2 in.) diam gray iron bars heated to 650 °C (1200 °F). Source: Ref 23

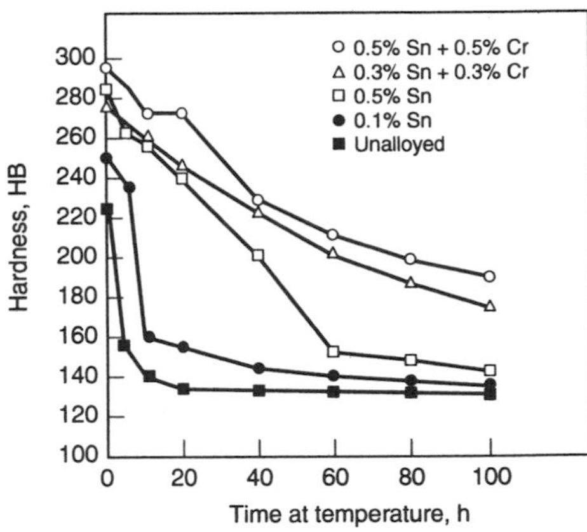

Fig. 28 The combined effects of tin and chromium on the hardness of 22 mm (0.875 in.) diam gray iron bars heated to 650 °C (1200 °F). Source: Ref 23

Table 15 Maximum tensile design stresses for unalloyed gray irons at temperatures up to 450 °C (840 °F) recommended by the British Cast Iron Research Association

Gray iron grades(a)	Maximum tensile design stress at various temperature, MPa (tonnes/in.^{2})(b)								
	Room temperature	100 °C (210 °F)	200 °C (340 °F)	300 °C (570 °F)	350 °C (660 °F)	375 °C (710 °F)	400 °C (750 °F)	425 °C (800 °F)	450 °C (840 °F)
154 (10)	39 (2.5)	36 (2.3)	32 (2.1)	29 (1.9)	...	...	...	...	...
185 (12)	46 (3.0)	43 (2.8)	40 (2.6)	37 (2.4)	...	...	...	...	...
216 (14)	54 (3.5)	51 (3.3)	48 (3.1)	45 (2.9)	49 (3.2)	39 (2.5)	29 (1.9)	22 (1.4)	17 (1.1)
263 (17)	66 (4.3)	63 (4.1)	60 (3.9)	57 (3.7)	51 (3.3)	42 (2.7)	31 (2.0)	23 (1.5)	19 (1.2)
309 (20)	77 (5.0)	74 (4.8)	71 (4.6)	68 (4.4)	54 (3.5)	43 (2.8)	32 (2.1)	25 (1.6)	19 (1.2)
355 (23)	89 (5.8)	86 (5.6)	83 (5.4)	80 (5.2)	74 (4.8)	66 (4.3)	51 (3.3)	42 (2.7)	31 (2.0)
402 (26)	100 (6.5)	97 (6.3)	94 (6.1)	91 (5.9)	74 (4.8)	66 (4.3)	51 (3.3)	42 (2.7)	31 (2.0)

(a) Grade irons grades are classified according to their room-temperature tensile strengths, given in MPa (tonnes/in.^{2}). (b) Room-temperature design stresses are based on ¼ × tensile strength. Design stresses from 100 °C (210 °F) to 300 °C (570 °F) are based on 0.384 × 0.1% proof stress value; design stresses from 350 °C (660 °F) to 450 °C (840 °F) are based on ⅓ stress to produce rupture in 100,000 h. Source: Ref 5

deformation at temperatures above 400 °C (750 °F).

Turnbull and Wallace (Ref 20) determined the elevated-temperature strength of two series of gray irons alloyed with Mo and Cr + Mo at 427, 538, and 650 °C (800, 1000, and 1200 °F). Increasing the molybdenum content increased the strength of both plain and 0.6% Cr-alloyed irons to maximum values with 1.4 to 1.9% Mo (Fig. 21).

The effects of various alloy additions on the retention of strength at 540 °C (1000 °F) are readily seen in the ratio of tensile strength at 540 °C to that at room temperature. The strength ratios varies from 56 to 75%, and among the irons studied in Ref 5 and 21 the Cr-Mo alloyed irons exhibited the highest strengths and strength ratios at 540 °C (1000 °F). The presence of either molybdenum or chromium appeared to contribute to retention of strength at elevated temperatures, while additions of nickel and copper did not.

Gundlach (Ref 21) has demonstrated that the elastic modulus decreases linearly with increasing temperature. As shown in Fig. 22, most alloying additions reduce the rate at which modulus changes with temperature.

Ductile Irons. The short-time tensile strength of unalloyed pearlitic ductile irons decreases more or less continuously with increasing temperature, and at 400 °C (750 °F) it is about two-thirds the room-temperature strength (Fig. 23a). For ferritic irons, the decrease is less pronounced: at 400 °C (750 °F) the strength is about three-fourths the room-temperature value (Fig. 23a). Proof stress, however, for both ferritic and pearlitic irons is more or less maintained up to 350 to 400 °C (660 to 750 °F), above which it falls rapidly. Figures 23(b) and (c) summarize the elevated-temperature tensile properties of annealed ductile iron compared to those of cast steel. Table 13 lists the elevated-temperature tensile strengths of specific grades.

For temperatures up to 300 °C (570 °F), static design stress can be based on proof stress values obtained at room temperature. Design stresses are given in Table 14. At temperatures above 350 °C (660 °F), design stresses should be based on creep data.

High-Alloy Irons. Figure 24 shows short-time elevated-temperature tensile strengths for gray, ductile, and white alloy irons. The ductile Ni-Resists and ductile irons containing 4% Si and 2% Mo provide the best short-time property values. Table 2 and Figure 3 show the influence of molybdenum alloying additions on elevated-temperature tensile data of 4% Si ductile irons.

Hot Hardness

Gray Irons. Depending on alloy content, gray irons experience a loss of hardness retention at temperatures ≥450 °C (≥840 °F). Loss of hardness corresponds to the microstructural decomposition of the casting. For example, pearlitic irons exposed for 64 weeks at 500 °C (930 °F) exhibited structures consisting of decomposed (spheroidized) pearlite and ferrite. The carbide in the pearlite was fine and widely dispersed, indicating that decomposition was almost complete (Ref 5). Concurrent with this microstructural decomposition is a rapid loss in hardness.

Several investigators have studied the hardness of gray irons after exposure to elevated temperatures. Bevan (Ref 7) indicates no significant change in the hardness of growth test specimens of irons alloyed with Cr, Cr-Mo, or Mo-Sn following 3000 h exposure at 455 °C (850 °F). Gilbert (Ref 6) reports a significant change in the hardness of unalloyed and tin-alloyed (0.06% Sn)

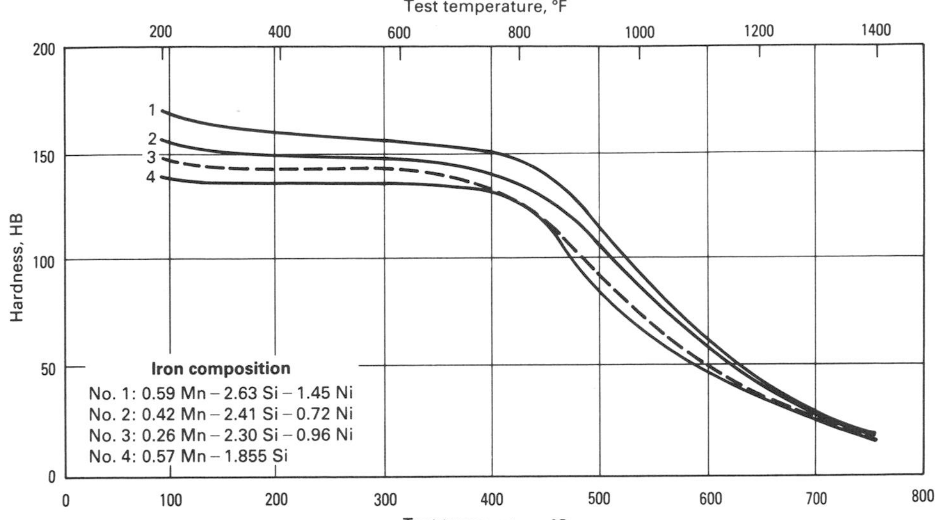

Fig. 29 Hot hardness of four annealed (ferritic) ductile irons. Source: Ref 24

irons at 450 °C (840 °F) after 64 weeks (10,000 h), but little change in the hardness of irons alloyed with Mn, Mo-Ni, Cr, Cr-Mo, or Cr-Ni-Mo (Fig. 25). After exposure for 64 weeks at 500 °C (930 °F), all alloys exhibited significant reduction in hardness; however, irons containing chromium exhibited the greatest resistance to softening.

Gundlach (Ref 21) studied the influence of various alloy additions to pearlitic irons on microstructural stability at 500 and 550 °C (930 and 1020 °F) for times up to 3000 h. Figure 26 illustrates the reduction in hardness with time and demonstrates the benefits of various alloying combinations. At 500 °C (930 °F) an iron alloyed with 0.5% Cr + 0.4% Mo was remarkably stable, while at 550 °C (1020 °F) none of the irons exhibited good microstructural stability.

Thwaites and Pryterch (Ref 23) studied the influence of adding tin, in larger than normal amounts (>0.1%), on the structural stability of chromium-alloyed irons between 500 and 700 °C

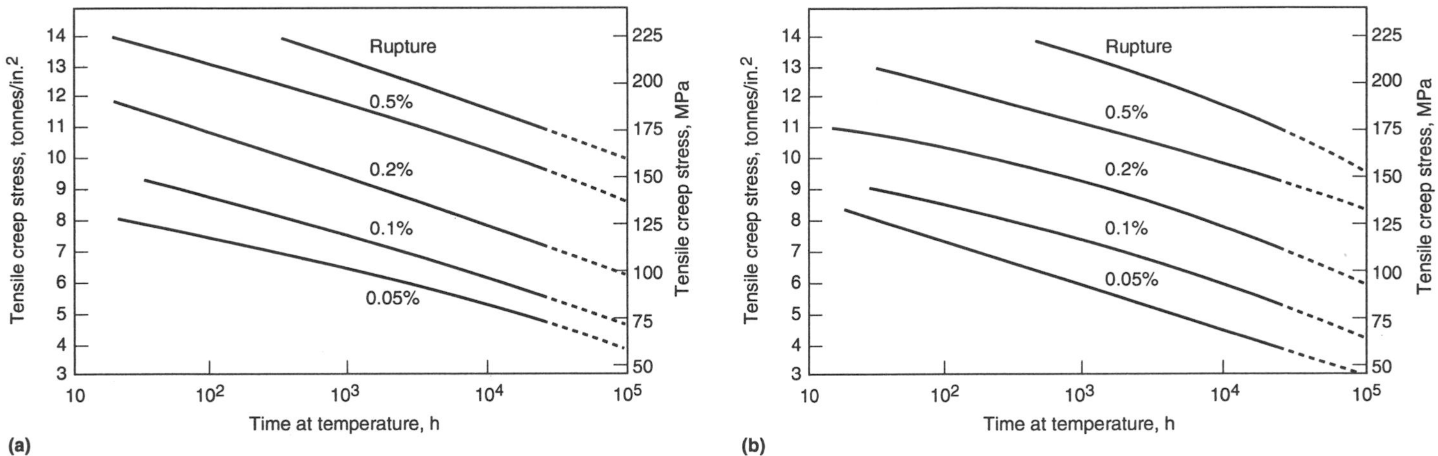

Fig. 30 Typical creep strain and stress-rupture curves for an unalloyed gray iron tested at 350 °C (660 °F). (a) 30 mm (1.2 in.) cast section. (b) 75 mm (3.0 in.) cast section. As these data indicate, section size has little effect on creep and stress-rupture properties. Source: Ref 25

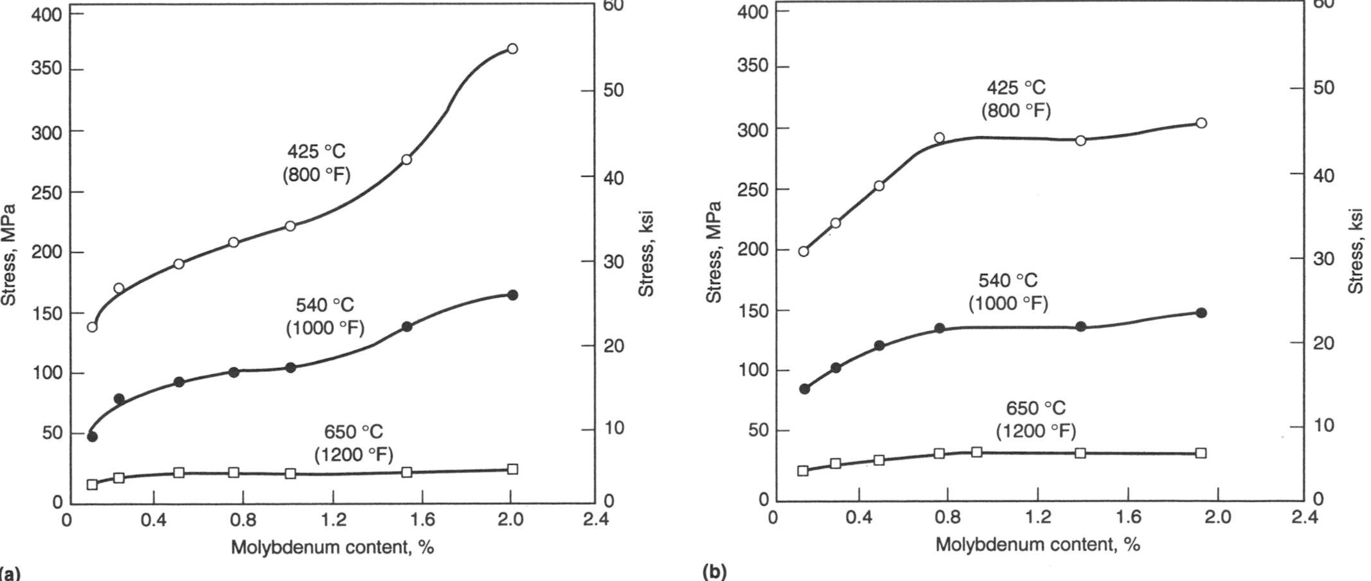

Fig. 31 Effect of molybdenum on the stress to produce rupture in 100 h in gray irons at various temperatures. (a) Unalloyed base iron. (b) 0.6% Cr-alloyed base iron. Source: Ref 20

Table 16 Creep strength of several ductile irons at 425 °C (800 °F)

Chemical composition, %										Stress for minimum creep rate of:			
										0.0001 %/h		0.00001 %/h	
TC	Mn	P	Si	Ni	Mo	Cu	Mg	Condition(a)	Pearlite, %	MPa	ksi	MPa	ksi
3.54	0.40	0.017	2.26	0.56	...	0.15	0.05	Ann, 950 °C (1740 °F)	1	48	7	...	...
3.54	0.40	0.017	2.26	0.56	...	0.15	0.05	Ann, 870 °C (1600 °F)	10	58.5	8.5	...	...
3.49	0.78	0.086	2.46	1.08	...	0.56	0.072	Ann, 870 °C (1600 °F)	20	141	20.5	...	...
3.49	0.78	0.086	2.46	1.08	...	0.56	0.072	Ann, 950 °C (1740 °F)	10	172	25	103	15
3.59	0.40	0.02	2.43	1.08	0.24	0.1	0.05	Ann, 870 °C (1600 °F)	2	124	18	86	12.5
3.61	0.47	0.02	2.43	1.19	0.81	0.1	0.05	Ann, 950 °C (1740 °F)	3	186	27	152	22
3.54	0.40	0.017	2.26	0.56	...	0.15	0.05	As-cast, stress relieved	30	79	11.5	...	...
3.49	0.37	0.085	2.50	1.22	...	0.1	0.064	As-cast, stress relieved	94	145	21	...	...
3.49	0.78	0.086	2.46	1.08	...	0.56	0.072	As-cast, stress relieved	100	172	25	114	16.5
3.61	0.47	0.02	2.43	1.19	0.81	0.1	0.05	As-cast, stress relieved	50	...	...	145	21

TC, total carbon. (a) Ann, annealed. Source: Ref 28

(930 and 1290 °F). Figure 27 compares the influence of individual additions (0.3%) of tin, chromium, and molybdenum on hardness after heating at 650 °C (1200 °F) and demonstrates the potency of a 0.3% Sn addition on resistance to softening. The hardness data in Fig. 28 show that Sn + Cr is more potent than tin alone in stabilizing the structure. Thwaites and Prytrech also showed that, even at 700 °C (1290 °F), a 0.1% Sn addition significantly reduced the rate of decomposition of pearlite.

Ductile Irons. The hardnesses of all standard grades of ductile iron are relatively constant up to about 425 °C (800 °F). Hot hardness data for four annealed (ferritic) ductile irons are shown in Fig. 29. The hot hardness is lower with a lower silicon content.

Creep and Stress-Rupture Properties

Creep and stress-rupture information are considerably more important than short-time tensile tests when applications involving extended elevated-temperature exposure are considered. Much of the creep data associated with cast irons have been expressed in the form of minimum creep rates (percent deformation/elongation), which were usually measured in tests lasting 1000 h or less. The usefulness of this kind of data is limited because it ignores the elongation that occurs in first-stage creep, and because the design life of a component is frequently many years.

More useful data are obtained from long-time creep tests (≥10^4 h) that measure the total creep strain produced over a given time. These tests provide information regarding the stresses that can be applied without causing excessive deformation. Stress-rupture data are presented in graphs that plot the stress versus the time necessary to produce fracture at a constant temperature. Extrapolation methods, such as the well-known Larson-Miller parameter, allow for the prediction of long-time (100,000 h) stress-rupture properties from relatively short-time (10,000 h) results.

Gray Irons. For designs requiring high dimensional stability of unalloyed gray irons in long-term applications, design stress should be based on one-third of the stress needed to produce rupture in 100,000 h, which has been shown to be about 80% of the stress needed to produce 0.1% creep strain in 100,000 h for unalloyed gray irons (Ref 11). BCIRA has presented recommended design stresses for various classes of gray cast irons for use up to 450 °C (840 °F), as listed in Table 15. For 300 °C (570 °F) and below, design stresses are based on the 0.1% proof stress; for 350 °C (660 °F) and above, design stresses are based on one-third of the stress needed to produce rupture in 100,000 h. Above 450 °C (840 °F), unalloyed gray irons are not recommended where good dimensional stability is required because scaling and growth become significant. Creep strain and stress-rupture curves for unalloyed gray iron specimens are given in Fig. 30.

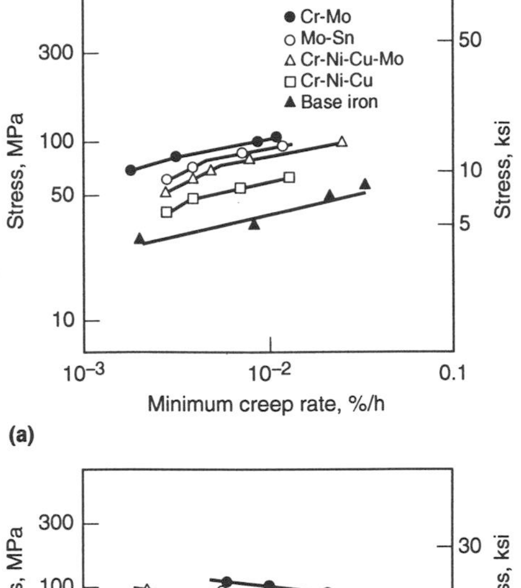

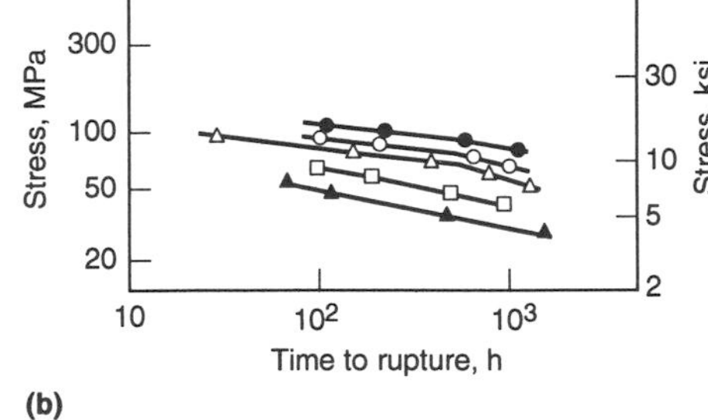

Fig. 32 Effect of alloying elements on the elevated-temperature behavior of ASTM Class 40 gray irons tested at 540 °C (1000 °F) for 1000 h. (a) Stress vs. minimum creep rate. (b) Stress-rupture characteristics. Source: Ref 26

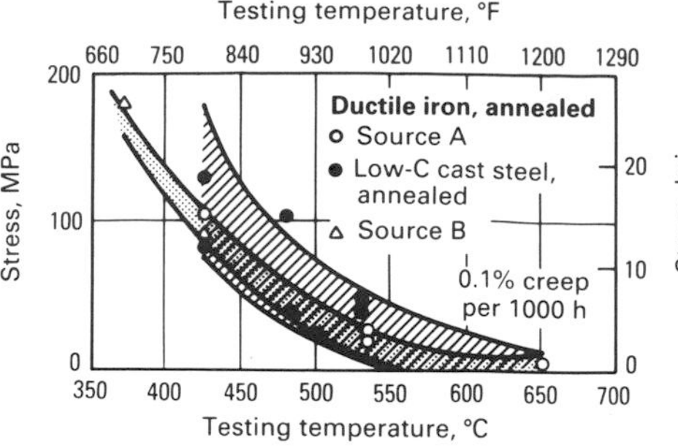

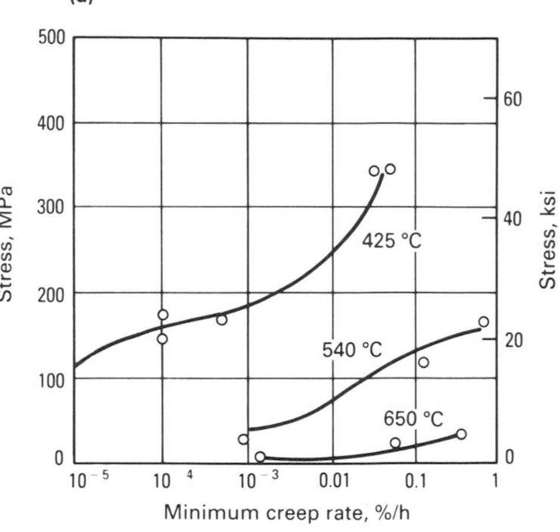

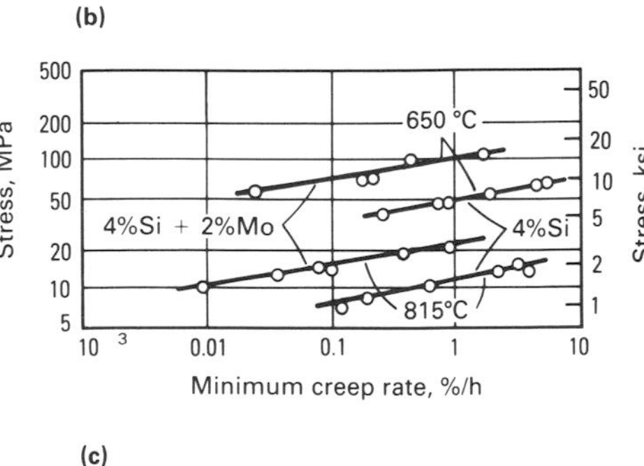

Fig. 33 Creep characteristics of ductile iron. (a) Stress vs. testing temperature for annealed ferritic ductile irons. (b) Stress vs. minimum creep rate for pearlitic ductile iron. (c) Stress vs. minimum creep rate for 4% Si ductile iron and 4Si-2Mo ductile iron at 650 °C (1200 °F) and 815 °C (1500 °F). Source: Ref 27

A number of investigators have studied the influence of alloying on creep and the stress-rupture properties of gray irons. Turnbull and Wallace (Ref 20) demonstrated the beneficial effects of chromium and molybdenum additions on the stress-rupture properties of gray irons at 425, 540, and 650 °C (800, 1000, and 1200 °F) (Fig. 31). Their test results show that molybdenum additions up to 2% to an unalloyed iron continuously raise the stress-to-rupture in 100 h at all three temperatures. In a 0.6% Cr-alloyed base iron, the stress-to-rupture in 100 h increased with increasing molybdenum content up to 0.8% Mo at the same temperatures.

Gundlach (Ref 26) also demonstrated the benefits of various alloy additions on the stress-to-rupture and creep properties at 540 °C (1000 °F) in

Table 17 Stress required to produce specific creep strain or rupture in ductile iron

Type of ductile iron	Creep strain or rupture, %	Stress required at indicated time and temperature, MPa (ksi) 350 °C (660 °F) 1000 h	10,000 h	30,000 h	100,000 h(a)	400 °C (750 °F) 1000 h	10,000 h	30,000 h	100,000 h(a)
Pearlitic grade 700/2	0.1	239 (34.5)	178 (26)	145 (21)	124 (18)	120 (17.5)	70 (10)	50 (7.5)	28 (4)
	0.2	276 (40)	219 (32)	199 (29)	151 (22)	147 (21.5)	93 (13.5)	77 (11)	40 (6)
	0.5	312 (45)	270 (39)	246 (35.5)	222 (32)	199 (29)	140 (20.5)	114 (16.5)	80 (11.5)
	1.0	355 (51.5)	297 (43)	278 (40.5)	256 (37)	239 (34.5)	184 (26.5)	150 (22)	128 (18.5)
	Rupture	430 (62.5)	370 (53.5)	352 (51)	317 (46)	309 (45)	255 (37)	195 (28.5)	160 (23)
Ferritic grade 400/12	0.1	185 (27)	159 (23)	142 (20.5)	120 (17.5)	96 (14)	60 (8.5)	43 (6.5)	26 (4)
	0.2	204 (29.5)	171 (25)	158 (23)	137 (20)	111 (16)	75 (11)	59 (8.5)	35 (5)
	0.5	222 (32)	195 (28.5)	176 (25.5)	167 (24)	130 (19)	94 (13.5)	77 (11)	59 (8.5)
	1.0	241 (39.5)	210 (30.5)	192 (28)	175 (25.5)	142 (20.5)	106 (15.5)	88 (13)	71 (10.5)
	Rupture	298 (43)	264 (35.5)	246 (35.5)	225 (32.5)	195 (28)	154 (22.5)	136 (20)	114 (16.5)

(a) Extrapolated from stress/log time curve

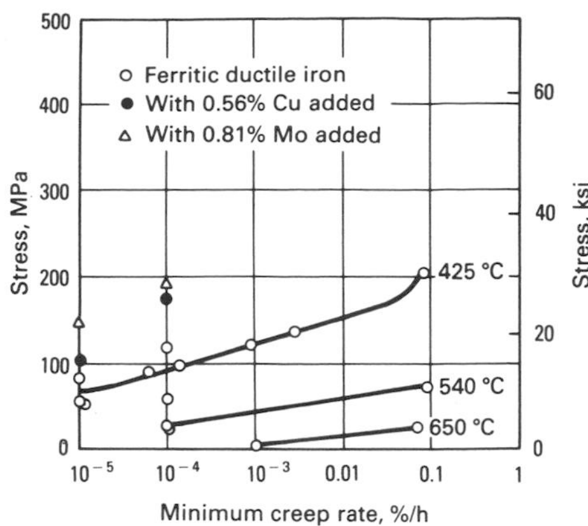

Fig. 34 Effect of molybdenum and copper additions on the creep characteristics of ferritic ductile iron. Source: Ref 27

one unalloyed and four alloyed ASTM Class 40 gray irons. As shown in Fig. 32, alloy additions increased 1000 h rupture strength more than 140% and dramatically reduced creep rates at a given stress level. An iron alloyed with 0.5% Cr + 0.4% Mo produced the highest rupture strength and resistance to creep, followed by an iron alloyed with 0.4% Mo + 0.08% Sn.

Ductile Irons. The creep strength of ductile iron depends on composition and microstructure. As shown in Fig. 33(a), ferritic ductile iron has a creep resistance comparable to that of annealed low-carbon cast steel up to 650 °C (1200 °F). Table 16 shows the creep strength of ductile iron of various compositions. Representative minimum creep rates versus applied stress are shown in Fig. 33(b) and (c). Table 16 and Fig. 34 show the improvement in creep strength that can be achieved by adding molybdenum or copper. Figure 35 shows the stress necessary at a given temperature to produce a minimum creep rate of 0.0001%/h in ferritic and pearlitic ductile irons.

Table 17 lists the stresses needed to produce 0.1, 0.2, or 1% strain or rupture in 1000, 10,000, 30,000, and 100,000 h (extrapolation) at 350 and 400 °C (660 and 750 °F).

High-Alloy Irons. As stated above, alloying high-silicon irons with molybdenum enhances their elevated-temperature performance. There is a continuous increase in the stress-rupture strength and a reduction in creep rates as molybdenum contents are increased from 0 to 2.5%, but the greatest response is realized from the additions of 0.5 and 1.0% Mo (Ref 2). Figure 36 shows creep data for 4% Si irons alloyed with 0.96 to 1.86% Mo. Figure 37 gives stress-rupture data at 815 °C (1500 °F) for various molybdenum-alloyed silicon irons.

Typical stress-rupture data for the austenitic 20Ni, 20Ni-1Mo, 30Ni, and 30Ni-1Mo ductile irons are given in Fig. 38, along with data for ferritic 4Si and 4Si-1Mo. For comparison, stress-rupture data are included for a cast 19Cr-9Ni stainless steel (ASTM A 297, grade HF). The addition of 1% Mo to types D-2, D-3, and D-5B increases the creep and stress-rupture strengths to levels equal or superior to those of cast stainless steels. The ductile Ni-Resists offer creep properties that are superior to those of conventional (flake graphite) Ni-Resists.

Thermal Fatigue

When castings are used in an environment where frequent changes in temperature occur, or where temperature differences are imposed on a part, thermal stresses occur. These may result in elastic and plastic strains and finally in crack formation, so that a casting can be destroyed as a result of thermal fatigue. Changes in microstructure, associated with stress-inducing volume

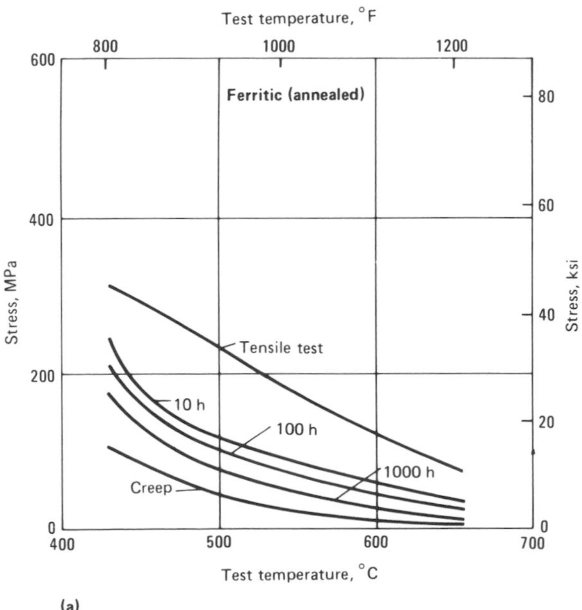

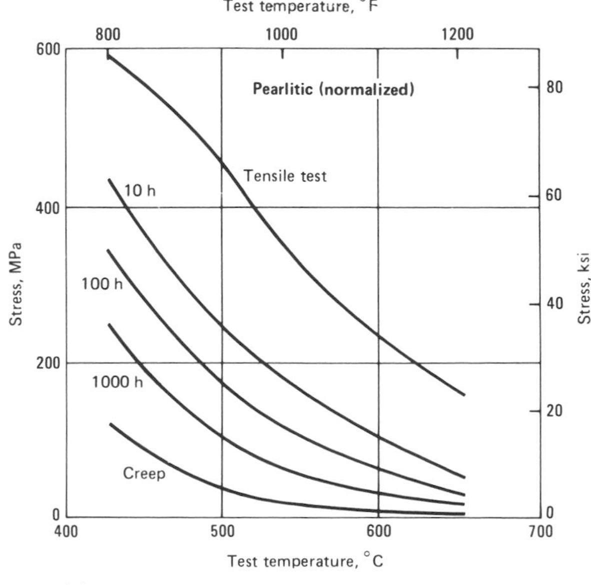

Fig. 35 Stress-rupture properties of ductile iron. (a) Ferritic (annealed). (b) Pearlitic (normalized). The curve labeled "creep" shows the stress-temperature combination that will result in a creep rate of 0.0001%/h. Source: Ref 29, 30

changes, and surface and internal oxidation may also be associated with stresses induced by temperature differences.

Test Methods. The interpretation of thermal fatigue tests is complicated by the many test methods employed by various investigators. The two widely accepted methods are constrained thermal fatigue and finned-disk thermal shock tests (Ref 32, 33).

In the constrained thermal fatigue test, a specimen (see Fig. 39a for dimensions) is mounted between two stationary plates that are held rigid by two columns, heated by high frequency (450 kHz) induction current, and cooled by conduction of heat to water-cooled grips (Fig. 39b). The thermal stress that develops in the test specimen is monitored by a load cell installed in one of the grips holding the specimen.

Initially the specimen develops compressive stress upon heating due to constrained thermal expansion (Fig. 40). Some yielding and stress relaxation occur during holding at 540 °C (1000 F), and upon subsequent cooling the specimen develops residual tensile stress. During subsequent thermal cycling, the maximum compressive stress that has developed upon heating decreases continuously, and the maximum tensile stress upon cooling increases, as shown for six different irons in Fig. 41. The specimen accumulates fatigue damage in a fashion similar to that in mechanical fatigue testing; ultimately, the specimen fails by fatigue.

Experimental results (Fig. 42) point to higher thermal fatigue resistance for compacted graphite iron than for gray iron and also indicate the beneficial effect of molybdenum. In fact, regression analysis of experimental results indicates that the main factors influencing thermal fatigue are tensile strength (TS) and molybdenum content:

$$\log N = 0.934 + 0.026 - \mathrm{TS} + 0.861 \times \mathrm{Mo} \qquad \text{(Eq 1)}$$

where N is the number of thermal cycles to failure, tensile strength is in ksi, and molybdenum is in percent.

In the finned-disk thermal shock test, the specimen (see Fig. 43(a)for dimensions) is cycled between a moderate-temperature environment and an elevated-temperature environment, which causes thermal expansion and contraction. The test apparatus is shown in Fig. 43(b). Because thermal conductivity plays a significant role in this type of test, gray iron has shown much greater resistance to cracking than compacted graphite iron (Ref 33). Major cracking occurred in less than 200 cycles in all compacted graphite iron specimens, while the unalloyed gray iron developed minor cracking after 500 cycles and major cracking after 775 cycles. The alloyed gray iron, because of its higher elevated-temperature strength, did not show any sign of cracking even after 2000 cycles (Ref 33). The compacted graphite iron containing more ferrite had slightly better thermal fatigue resistance than the compacted graphite iron with less ferrite.

Selection Guidelines. In general, for good resistance to thermal fatigue, cast irons must have high thermal conductivity, low modulus of elasticity, high strength at room and elevated temperatures, and, for use above 500 to 550 °C (930 to 1020 °F), resistance to oxidation and structural change. The relative ranking of irons varies with test conditions. When high cooling rates are encountered, experimental data and commercial experience show that thermal conductivity and a low modulus of elasticity are most important. Consequently, gray irons of high carbon content (3.6 to 4%) are superior (Ref 32, 33). When cooling rates are intermediate, ferritic ductile and compacted graphite irons have the highest resistance to cracking, but they are subject to distor-

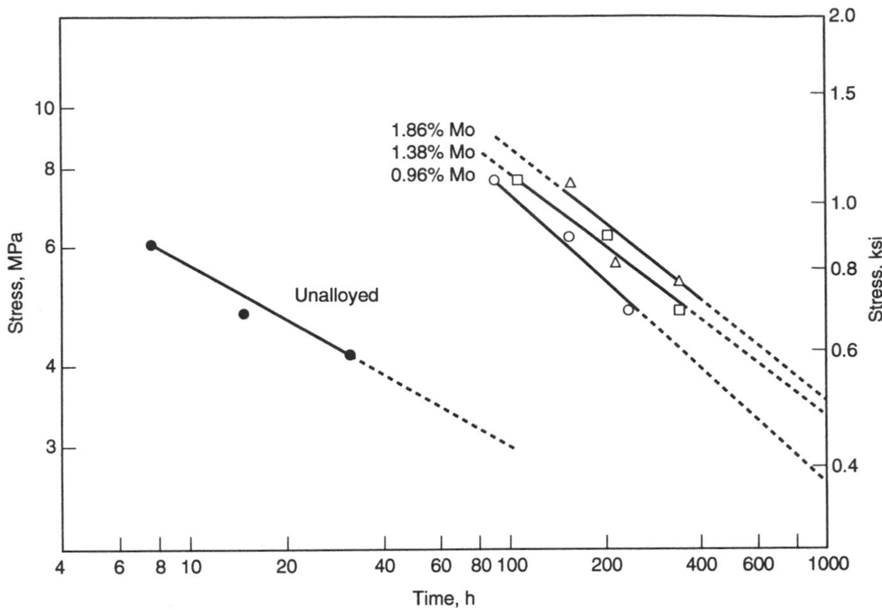

Fig. 36 Effect of molybdenum content on the time and stress to induce 1% creep at 815 °C (1500 °F) for 4% Si ductile irons. Source: Ref 2

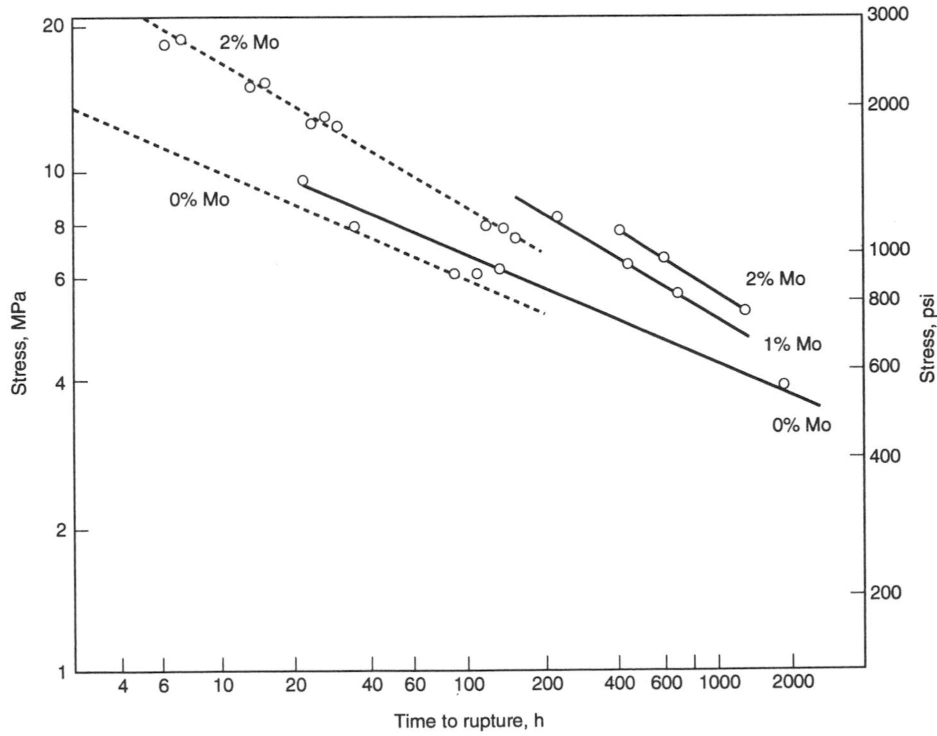

Fig. 37 Effect of molybdenum content on the 815 °C (1500 °F) stress-rupture data for 4% Si ductile irons subcritically annealed at 800 °C (1475 °F). The dashed lines represent samples graphitized at 900 °C (1650 °F) and control cooled. Source: Ref 2

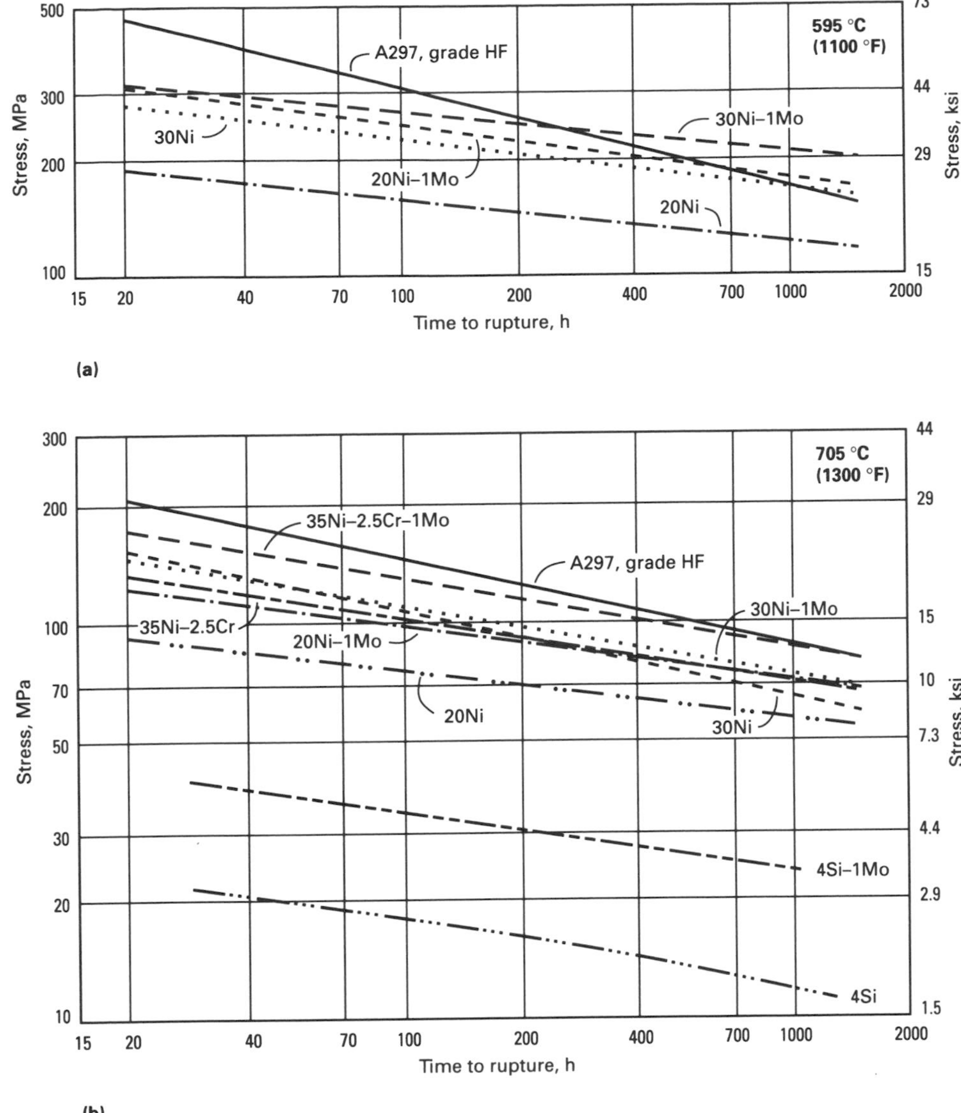

Fig. 38 Typical stress-rupture properties of high-nickel heat-resistant ductile irons. (a) At 595 °C (1100 °F). (b) At 705 °C (1300 °F). Source: Ref 31

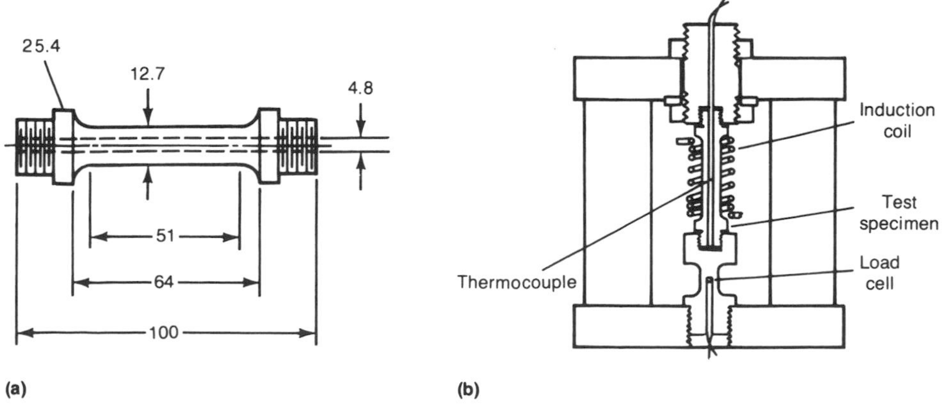

Fig. 39 (a) Dimensions of constrained fatigue test specimen. Dimensions are given in millimeters.(b) Schematic of apparatus for constrained fatigue tests. Source: Ref 33

tion. When cooling rates are low, high-strength pearlitic ductile irons or ductile irons alloyed with silicon and molybdenum are most resistant to cracking and distortion (Fig. 44).

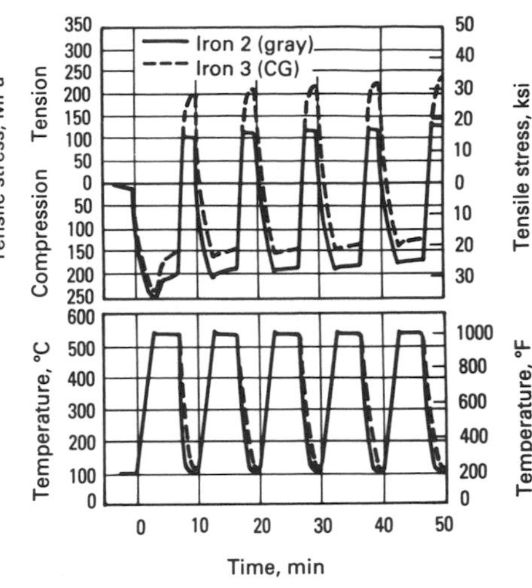

Fig. 40 Typical thermal stress cycles at the beginning of the test for gray and compacted graphite irons. Source: Ref 33

ACKNOWLEDGMENTS

The information in this article is largely taken from:

- Elevated-Temperature Properties, *ASM Specialty Handbook: Cast Irons,* J.R. Davis, Ed., ASM International, 1996, p 409–427
- Metallurgy and Properties of High-Alloy Graphitic Irons, *ASM Specialty Handbook: Cast Irons,* J.R. Davis, Ed., ASM International, 1996, p 123–130

REFERENCES

1. R. Elliott, *Cast Iron Technology,* Butterworths, 1988
2. White and High Alloy Irons, *Iron Castings Handbook,* C.F. Walton and T.J. Opar, Ed., Iron Castings Society, 1981, p 399–461
3. R.P. Walson, Aluminum Alloyed Cast Iron Properties Used in Design, *Trans. AFS,* Vol 85, 1977, p 51–58
4. Mechanical Properties of Gray Iron, Chapter 6, Section A, *Iron Castings Handbook,* C.F. Walton and T.J. Opar, Ed., Iron Castings Society, Inc., 1981, p 203–295
5. R.B. Gundlach, The Effects of Alloying Elements on the Elevated Temperature Properties of Gray Irons, *Trans. AFS,* Vol 91, 1983, p 389-422
6. G.N.J. Gilbert, The Growth and Scaling Characteristics of Cast Irons in Air and Steam, *BCIRA J.,* Vol 7, 1959, p 478–566
7. E. Bevan, "Effect of Molybdenum on Dimensional Stability and Tensile Properties of Pearlitic Gray Irons at 600 to 850 °F (315 to 455 °C)," Internal Report, Climax Molybdenum Co.
8. D.G. White, Growth and Scaling Characteristics of Cast Irons with Undercooled and Normal Flake Graphite, *BCIRA J.,* Vol 11, 1963, p 223–230

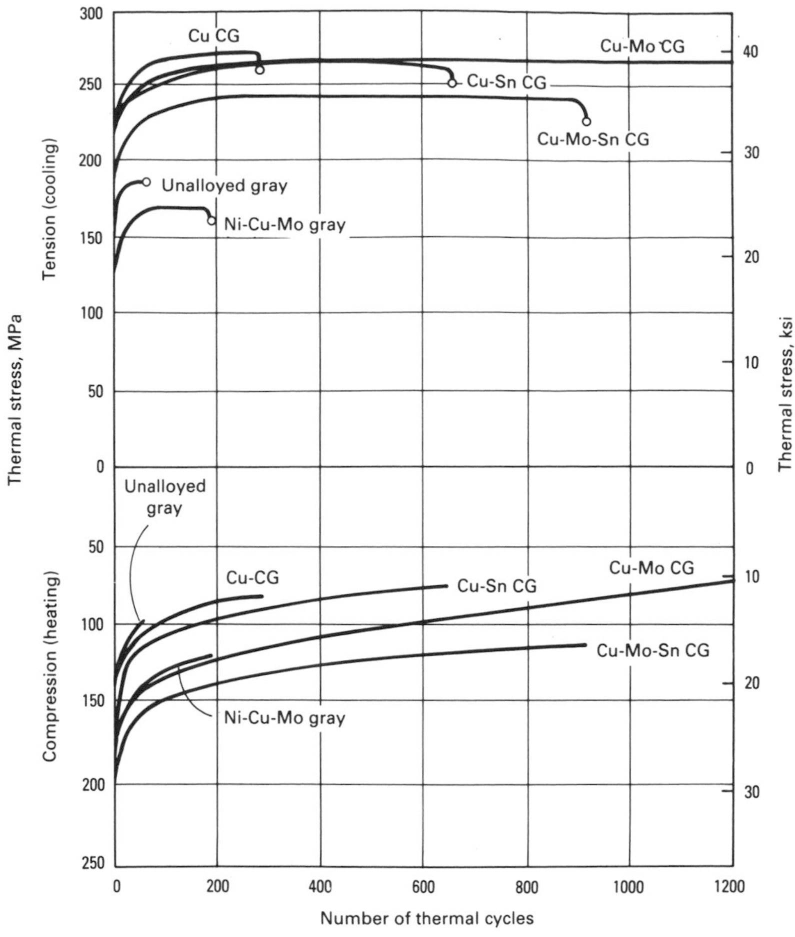

Fig. 41 The shift in thermal stress vs. the number of cycles for six irons cycled between 100 and 540 °C (212 and 1000 °F). CG, compacted graphite (iron). Source: Ref 33

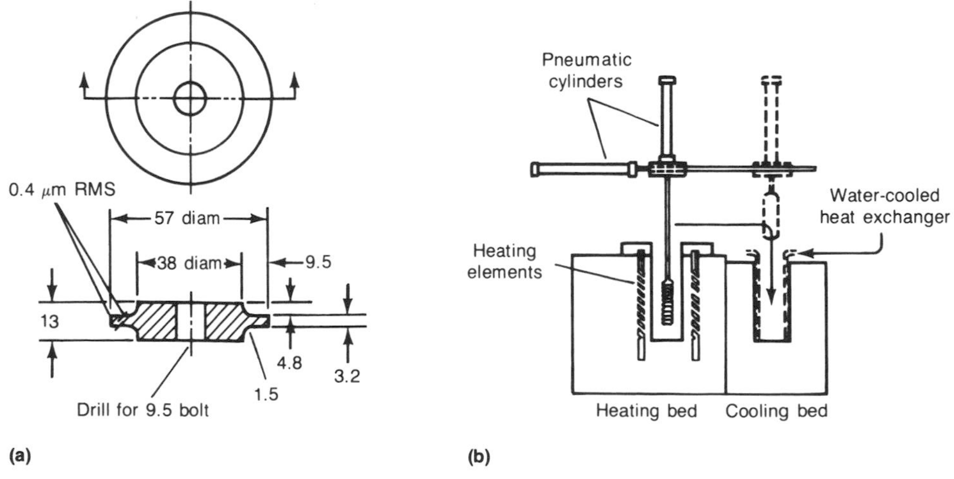

Fig. 43 (a) Dimensions (in millimeters) of finned-disk specimens. RMS, root mean square. (b) Schematic of apparatus for finned-disk thermal shock test. Source: Ref 33

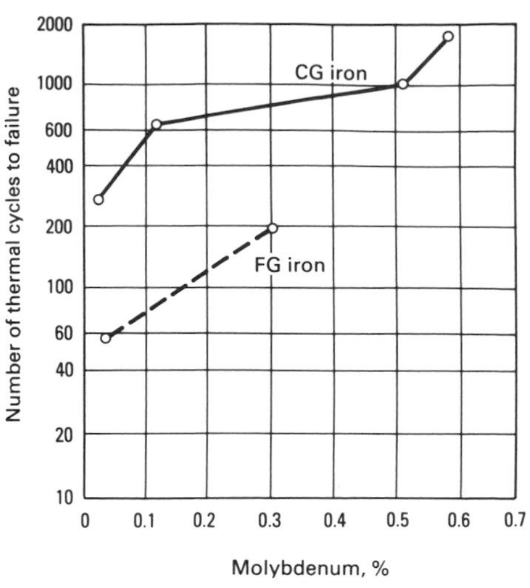

Fig. 42 Results of constrained thermal fatigue tests conducted between 100 and 540 °C (212 and 1000 °F). CG, compacted graphite; FG, flake graphite. Source: Ref 33

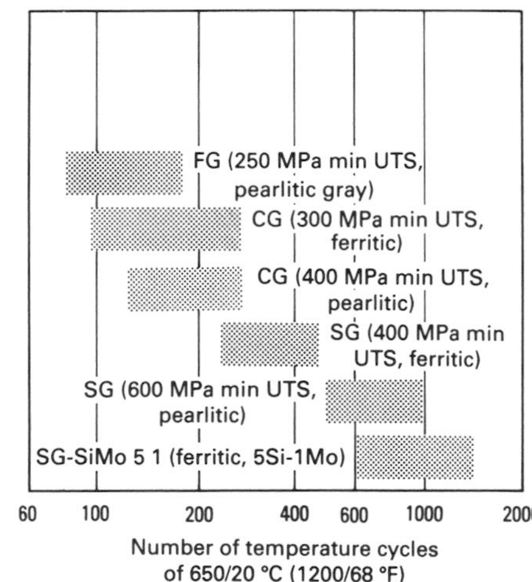

Composition(a)					
C	Si	Mn	P	Mg	Other
2.96	2.90	0.78	0.66	...	0.12 Cr
3.52	2.61	0.25	0.051	0.015	...
3.52	2.25	0.40	0.054	0.015	1.47 Cu
3.67	2.55	0.13	0.060	0.030	...
3.60	2.34	0.50	0.053	0.030	0.54 Cu
3.48	4.84	0.31	0.067	0.030	1.02 Mo

(a) Compositions of the irons shown in figure from top to bottom

Fig. 44 Results of thermal fatigue tests on various cast irons; specimens cycled between 650 and 20 °C (1200 and 70 °F). FG, flake graphite; UTS, ultimate tensile strength; CG, compacted graphite; SG, spheroidal graphite. Source: Ref 32

9. G.N.J. Gilbert and D.G. White, Growth and Scaling Characteristics of Flake Graphite and Nodular Graphite Cast Irons Containing Tin, *BCIRA J.*, Vol 11, 1963, p 295–318
10. K.B. Palmer, "Design with Cast Irons at High Temperatures, Part 1: Growth and Scaling," Report 1248, British Cast Iron Research Association
11. K.B. Palmer, "High Temperature Properties of Cast Irons," paper presented at Conference on Engineering Properties and Performance of Modern Iron Castings, Loughborough, 1970 (partially reproduced in *Iron and Steel*, 1971, p 39–46)
12. J.R. Kattus and B. McPherson, Properties of Cast Iron at Elevated Temperatures, STP 248, ASTM, 1959
13. H.T. Angus, *Cast Irons: Physical and Engineering Properties*, Butterworths, 1976, p 253–278

14. M.M. Hallett, Tests on Heat Resisting Cast Irons, *J. Iron Steel Inst.*, Vol 70, April 1952, p 321
15. C.O. Burgess and R.W. Bishop, An Optimum Silicon Range in Plain and 2.0 Percent Chromium Cast Irons Exposed to Elevated Temperatures, *Trans. ASM,* Vol 33, 1944, p 455–476
16. C.O. Burgess, Influence of Chromium on the Oxidation Resistance of Cast Irons, *Proc. ASTM,* Vol 39, 1939, p 604
17. E.U. Petitbon and J.F. Wallace, Aluminum Alloyed Gray Iron: Properties at Room and Elevated Temperature, *AFS Cast Metals Research J.*, Vol 9 (No. 3), 1973, p 127–134
18. Ductile Iron Data for Design Engineers, QIT-Fer et Titane Inc., 1990, p 5-1 to 5-3
19. Cast Metals Handbook, Chapter 7, American Foundrymen's Society, 1944, p 540–566
20. G.K. Turnbull and J.F. Wallace, Molybdenum Effect on Gray Iron Elevated Temperature Properties, *AFS Trans.,* Vol 67, 1959, p 35–46
21. R.B. Gundlach, Thermal Fatigue Resistance of Alloyed Gray Irons for Diesel Engine Components, *AFS Trans.,* Vol 87, 1979, p 551–450
22. G.N.J. Gilbert, Engineering Data on Nodular Cast Irons: SI Units, British Cast Iron Research Association, 1986
23. C.J. Thwaites and J.C. Pryterch, Structural Stability of Flake Graphite Iron Alloyed with Tin and Chromium, *Foundry Trade J.,* Jan 1969, p 115–121
24. H.D. Merchant and M.H. Moulton, Hot Hardness and Structure of Cast Irons, *Br. Foundryman,* Vol 57 (Part 2), Feb 1964, p 62–73
25. K.B. Palmer, "Design with Cast Irons at High Temperatures, Part 2: Tensile, Creep and Rupture Properties," Report 1251, British Cast Iron Research Association
26. R.B. Gundlach, Elevated Temperature Properties of Alloyed Gray Irons for Diesel Engine Components, *AFS Trans.,* Vol 86, 1978, p 55–64
27. C.F. Walton, Ed., *Gray and Ductile Iron Castings Handbook,* Gray and Ductile Founders' Society, 1971
28. D.L. Sponseller, W.G. Scholz, and D.F. Rundle, Development of Low-Alloy Ductile Irons for Service at 1200-1500 F, *AFS Trans.,* Vol 76, 1968, p 353–368
29. C.R. Wilks, N.A. Matthews, and R.W. Kraft, Jr., Elevated Temperature Properties of Ductile Cast Irons, *Trans. ASM,* Vol 47, 1954
30. F.B. Foley, Mechanical Properties at Elevated Temperatures of Ductile Cast Iron, *Trans. ASME,* Vol 78, 1956, p 1435–1438
31. "Engineering Properties and Applications of the Ni-Resists and Ductile Ni-Resists," International Nickel Co., 1978
32. K. Roehrig, Thermal Fatigue of Gray and Ductile Irons, *Trans. AFS,* Vol 86, 1978, p 75
33. Y.J. Park, R.B. Gundlach, R.G. Thomas, and J.F. Janowak, Thermal Fatigue Resistance of Gray and Compacted Graphite Irons, *Trans. AFS,* Vol 93, 1985, p 415

High-Alloy Cast Steels

CAST HEAT-RESISTANT STEELS are primarily used in applications where service temperatures exceed 650 °C (1200 °F) and may reach temperatures as high as 1315 °C (2400 °F). Materials selection considerations in such applications include resistance to corrosion at elevated temperatures, stability (resistance to warping, cracking, or thermal fatigue), creep strength (resistance to plastic flow), and stress-rupture strength.

As the title implies, emphasis in this article has been placed on high-alloy (stainless) cast steels, which are compositionally related to wrought stainless steels and to cast corrosion-resistant stainless steels. The major difference between these materials is their carbon content. With only a few exceptions, carbon in the cast heat-resistant alloys falls in a range from 0.3 to 0.6%, compared with 0.01 to 0.25% C normally associated with the wrought and cast corrosion-resistant grades. This difference in carbon results in significant changes in properties, for example, the higher rupture strength of the cast heat-resistant steels.

Also briefly described in this article are some alloys that are technical extensions of the more highly alloyed cast stainless steels. These castings include some nickel-chromium-iron and chromium-nickel alloys. These materials lie between what are considered stainless steels and superalloys.

Standard Grade Designations and Compositions

Cast heat-resistant alloys are most often specified on the basis of composition using the designation system of the High Alloy Product Group of the Steel Founders' Society of America. (The High Alloy Product Group has replaced the Alloy Casting Institute, or ACI, which formerly administered these designations.) The first letter of the designation indicates whether the alloy is intended primarily for high-temperature service (H) or liquid corrosion service (the latter steels will not be further described in this article). The second letter denotes the nominal chromium-nickel type of the alloy (Fig. 1). As nickel content increases, the second letter of the designation is changed. The numeral or numerals following the first two letters indicate maximum carbon content; for example, HK-40 is an alloy containing approximately 26% Cr and 20% Ni with 0.40% C. Table 1 lists the compositions of standard cast heat-resistant grades. These materials, which are also recognized in ASTM specifications, fall in a range of 0 to 68% Ni and 8 to 32% Cr with the balance consisting of iron plus up to 2.5% Si and 2.0% Mn.

Standard heat-resistant alloys may also be simply classified on the basis of structure alone. Grades HA and HC with 8 to 30% Cr and up to 4% Ni are ferritic. Grades HD, HE, HF, and HH may exhibit duplex structures of ferrite and austenite, while the remaining alloys HK to HX are fully austenitic.

A third alternative classification is based on the order of increasing quantity of major elements, which breaks down into the following four groups:

- Iron-chromium
- Iron-chromium-nickel
- Iron-nickel-chromium
- Nickel-iron-chromium

Iron-Chromium Alloys. The three alloys normally considered in this group are HA, HC, and HD, although only the first of these is technically an iron-chromium alloy. The other two grades contain 26 to 30% Cr and up to 7% Ni. These grades are mainly used in environments containing sulfur-bearing gases, where high-temperature strength is not an important consideration.

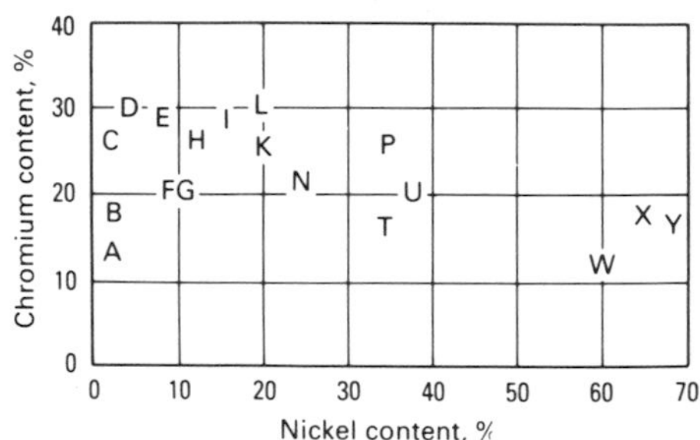

Fig. 1 Chromium and nickel contents in ACI standard grades of heat- and corrosion-resistant steel castings. See text for details.

Table 1 Compositions of standard heat-resistant casting alloys

ACI designation	UNS number	ASTM specifications(a)	Composition(b), % C	Cr	Ni	Si (max)
HA	...	A 217	0.20 max	8–10	...	1.00
HC	J92605	A 297, A 608	0.50 max	26–30	4 max	2.00
HD	J93005	A 297, A 608	0.50 max	26–30	4–7	2.00
HE	J93403	A 297, A 608	0.20–0.50	26–30	8–11	2.00
HF	J92603	A 297, A 608	0.20–0.40	19–23	9–2	2.00
HH	J93503	A 297, A 608, A 447	0.20–0.50	24–28	11–14	2.00
HI	J94003	A 297, A 567, A 608	0.20–0.50	26–30	14–18	2.00
HK	J94224	A 297, A 351, A 567, A 608	0.20–0.60	24–38	18–22	2.00
HK-30	J94203	A 351	0.25–0.35	23.0–27.0	19.0–22.0	1.75
HK-40	J94204	A 351	0.35–0.45	23.0–27.0	19.0–22.0	1.75
HL	N08604	A 297, A 608	0.20–0.60	28–32	18–22	2.00
HN	J94213	A 297, A 608	0.20–0.50	19–23	23–27	2.00
HP	N08705	A 297	0.35–0.75	24–28	33–37	2.00
HP-50WZ(c)	...	...	0.45–0.55	24–28	33–37	2.50
HT	N08605	A 297, A 351, A 567, A 608	0.35–0.75	13–17	33–37	2.50
HT-30	N08603	A 351	0.25–0.35	13.0–17.0	33.0–37.0	2.50
HU	N08005	A 297, A 608	0.35–0.75	17–21	37–41	2.50
HW	N08006	A 297, A 608	0.35–0.75	10–14	58–62	2.50
HX	N06050	A 297, A 608	0.35–0.75	15–19	64–68	2.50

(a) ASTM designations are the same as ACI designations. (b) Rem Fe in all compositions. Manganese content: 0.35 to 0.65% for HA, 1% for HC, 1.5% for HD, and 2% for the other alloys. Phosphorus and sulfur contents: 0.04% (max) for all but HP-50WZ. Molybdenum is intentionally addd only to HA, which has 0.90 to 1.20% Mo; maximum for other alloys is set at 0.5% Mo. HH also contains 0.2% N (max). (c) Also contains 4 to 6% W, 0.1 to 1.0% Zr, and 0.035% S (max) and P (max)

Table 2 Nominal compositions of selected nonstandard heat-resistant alloys

Generic alloy base	Composition, wt % C	Mn	Si	Cr	Ni	Mo	W	Nb	Co	Ti
20Ni-25Cr										
HK-Nb-Ti	0.4	1	1	25	22	...	...	0.3	...	0.1
HK-Nb	0.3	1	1	24	25	...	...	1.5	...	...
HK-Co-W	0.4	1	1	23	25	...	2	...	15	...
35Ni-25Cr										
HP-Nb	0.45	1	1	25	35	...	...	1.25	...	...
HP-Nb-Si	0.45	1	1.7	25	35	...	...	1.25	...	...
HP-Nb-L.C	0.15	1	1	25	35	...	...	1.3	...	...
HP-Nb-Ti	0.5	1	1	26	35	...	...	0.5	...	0.1-0.3
HP-Nb-W	0.4	1	1.5	25	37	...	1.5	1.5	...	...
HP-Nb-W-Mo	0.45	1	1.6	25	35	0.5	1.3	1.3	...	...
HP-Mo	0.45	1.5	1.5	2	36	1.3	...	...	...	...
HP-W	0.45	1	1.5	26	36	...	4	...	...	...
HP-W-Co	0.5	0.5	1.3	26	35	...	5	...	15	...
45Ni-30Cr										
30-45-Nb-W	0.4	1	1	34	44	...	0.5	0.5	...	...
30-45-Nb-Ti	0.42	1	1	34	45	...	...	1	...	0.1-0.3
30-45-W	0.5	1	1	28	47	...	5	...	...	...
30-45-W-Co	0.45	1	1	26	47	...	5	...	3	...
30-45-W-Al	0.2	0.4	0.2	33	50	...	16	...	...	...

Note: Sulfur and phosphorus typically specified at less than 0.03% of these alloys. Some alloys may also contain microalloying additions of aluminum.

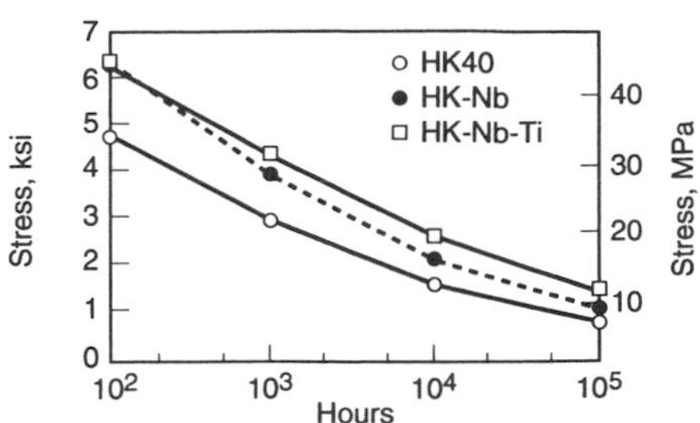

Fig. 2 The effects of niobium and niobium plus titanium on the rupture stresses of HK-40 base alloy tested at 980 °C (1800 °F). Source: Ref 1

Iron-Chromium-Nickel Alloys. Alloys in this group contain 18 to 32% Cr and 8 to 22% Ni, with chromium always exceeding nickel, and include the grades HE, HF, HH, HI, HK, and HL. While these alloys are considered to be austenitic, the lower nickel compositions will contain some ferrite. Transformation of the ferrite to brittle σ phase is a concern with this group, even in the higher nickel grades, particularly if their compositional balance leans to ferrite. The high-temperature strength of this group is greater than that of the iron-chromium alloys, and their creep and rupture strengths increase as nickel is raised.

Iron-Nickel-Chromium Alloys. The four standard grades in this group, HN, HP, HT, and HU, contain 15 to 28% Cr and 23 to 41% Ni. Nickel always exceeds the chromium content. These alloys have stable austenitic structures, good high-temperature strength, and enhanced resistance to thermal cycling and thermal stresses, combined with high resistance to oxidizing and reducing environments.

Nickel-Iron-Chromium Alloys. Two standard grades, HX and HW, fall into this group, which contains 58 to 68% Ni and 10 to 19% Cr. Usually referred to as high-alloy steels, these materials are more correctly described as nickel-base alloys. While possessing moderate high-temperature rupture strength, their creep strength is low. These grades have the highest carburization resistance of the standard alloys.

Nonstandard Grade Compositions

Nonstandard (proprietary) grades of heat-resistant alloys are generally more highly alloyed. Single or multiple additions of the elements aluminum, cobalt, molybdenum, niobium, the rare earth metals (cerium, lanthanum, and yttrium), titanium, tungsten, and zirconium are added to enhance specific properties, such as high-temperature strength, carburization resistance, and resistance to thermal cycling. Some of the more common nonstandard grades have the following base compositions:

- HK-20Ni-25Cr
- HP-35Ni-25Cr
- 45Ni-30Cr
- 50Cr-50Ni (with or without niobium)
- 60Cr-40Ni

Tables 2 and 3 list the nominal compositions of these nonstandard grades. A number of other proprietary alloys with compositions outside these ranges have also been developed for specific applications (see, for example, the alloys listed in Table 2) as described below.

20Ni-25Cr. The HK composition has served as a base for further strengthening, without overly increasing the base metal cost. One of the first of these alloys, Inco's IN519, made use of small additions of niobium, while increasing the nominal nickel level from 20 to 24% in order to offset the development of σ phase. Other manufacturers have developed even higher strength materials with tungsten or niobium combined with microalloying additions of titanium, as shown in Fig. 2.

35Ni-25Cr. Research sponsored by the Steel Founders' Society of America in the 1960s was an important factor in the development of a range of alloys in which the HP alloy is used as a compositional base (Ref 2). These alloys make use of the carbide-forming elements niobium, molybdenum, titanium, tungsten, and zirconium, together with noncarbide forming additions of aluminum, copper, and cobalt to increase strength and/or carburization resistance. Silicon also must be considered as an alloying addition because it is used at low levels to maximize creep and rupture strength and at high levels to enhance carburization. Carbon, which is used as a strengthener, is reduced in some modifications to increase the resistance to thermal cycling and shock.

One of the most successful and widely used modifications of HP alloy contains 0.5 to 1.5% Nb. This lead to the development of other variants with optimized levels of carbon and silicon, as well as the use of titanium as a microalloy strengthening addition.

45Ni-30Cr. These alloys are primarily intended for high-temperature service where strength and/or carburization resistance are the prime considerations. Because of high-temperature applications, good oxidation resistance is also a prerequisite. Like the preceding group, use is made of single or multiple additions of tung-

Table 3 Chemical composition requirements for chromium-nickel alloys described in ASTM A 560

Grade	Composition, wt % (a) C	Mn	Si	S	P	N	N + C	Fe	Ti	Al	Nb	Cr	Ni
50Cr-50Ni	0.10	0.30	1.00	0.02	0.02	0.30	...	1.00	0.50	0.25	...	48.0–52.0	Balance
60Cr-40Ni	0.10	0.30	1.00	0.02	0.02	0.30	...	1.00	0.50	0.25	...	58.0–62.0	Balance
50Cr-50Ni-Nb	0.10	0.30	0.50	0.02	0.02	0.16	0.20	1.00	0.50	0.25	1.4–1.7	47.0–52.0	Balance

(a) The total of the nickel, chromium, and niobium contents must exceed 97.5%.

sten, niobium, cobalt, aluminum, and microalloying with titanium.

Chromium-Nickel Alloys. ASTM A 560 describes three grades of chromium-nickel casting alloys: 50Cr-50Ni, 60Cr-40Ni, and 50Cr-50Ni-Nb. Table 3 lists the composition ranges for these alloys. These alloys (not strictly considered nickel-base alloys) are cast into tube supports and other firebox fittings for certain stationary and marine boilers. The chief attribute of the chromium-nickel alloys is their resistance to hot-slag corrosion in boilers that fire oil high in vanadium content. Hot slag high in vanadium pentoxide (V_2O_5) content is extremely destructive to most other heat-resistant alloys. More detailed information on the resistance of Cr-Ni alloys to molten V_2O_5 can be found below in the section "Ash/Salt Deposit Corrosion" in this article.

Metallurgical Structures

The structures of chromium-nickel and nickel-chromium cast steels must be wholly austenitic, or mostly austenitic with some ferrite, if these alloys are to be used for heat-resistant service. Depending on the chromium and nickel content, the structures of these iron-base alloys can be austenitic (stable), ferritic (stable, but also soft, weak, and ductile), or martensitic (unstable). Therefore, chromium and nickel levels should be selected to achieve good strength at elevated temperatures combined with resistance to carburization and hot-gas corrosion.

A fine dispersion of carbides or intermetallic compounds in an austenitic matrix increases high-temperature strength considerably. For this reason, heat-resistant cast steels are higher in carbon content than are corrosion-resistant alloys of comparable chromium and nickel content. By holding at temperatures where carbon diffusion is rapid (such as above 1200 °C) and then rapidly cooling, a high and uniform carbon content is established, and up to about 0.20% C is retained in the austenite. Some chromium carbides are present in the structures of alloys with carbon contents greater than 0.20%, regardless of solution treatment.

Castings develop considerable segregation as they freeze. In standard grades, either in the as-cast condition or after rapid cooling from a temperature near the melting point, much of the carbon is in supersaturated solid solution. Subsequent reheating precipitates excess carbides. The lower the reheating temperature, the slower the reaction and the finer the precipitated carbides. Fine carbides increase creep strength and decrease ductility. Intermetallic compounds, such as Ni_3Al, have a similar effect if present. Reheating material containing precipitated carbides in the range between 980 and 1200 °C (1800 and 2200 °F) will agglomerate and spheroidize the carbides, which reduces creep strength and increases ductility. Above 1100 °C (2000 °F), so many of the fine carbides are dissolved or spheroidized that this strengthening mechanism loses its importance. For service above 1100 °C (2000 °F), certain proprietary alloys of the iron-nickel-chromium type have been developed. Alloys for this service contain tungsten to form tungsten carbides, which are more stable than chromium carbides at these temperatures.

Aging at a low temperature, such as 760 °C (1400 °F), where a fine, uniformly dispersed carbide precipitate will form, confers a high level of strength that is retained at temperatures up to those at which agglomeration changes the character of the carbide dispersion (overaging temperatures). Solution heat treatment or quench annealing, followed by aging, is the treatment generally employed to attain maximum creep strength.

Ductility is usually reduced when strengthening occurs. In some alloys the strengthening treatment corrects an unfavorable grain-boundary network of brittle carbides, and both properties benefit. However, such treatment is costly and may warp castings excessively. Hence, this treatment is applied to heat-resistant castings only for the small percentage of applications for which the need for premium performance justifies the high cost.

Carbide networks at grain boundaries are generally undesirable in iron-base heat-resistant alloys. Grain-boundary networks usually occur in very high carbon alloys or in alloys that have cooled slowly through the high-temperature ranges in which excess carbon in the austenite is rejected as grain-boundary networks rather than as dispersed particles. These networks confer brittleness in proportion to their continuity.

Carbide networks also provide paths for selective attack in some atmospheres and in certain molten salts. Therefore, it is advisable in some salt bath applications to sacrifice the high-temperature strength imparted by high carbon content and gain resistance to intergranular corrosion by specifying that carbon content be no greater than 0.08%.

Prolonged exposure of the iron-chromium alloys to temperatures in the range from 1200 to

(a)

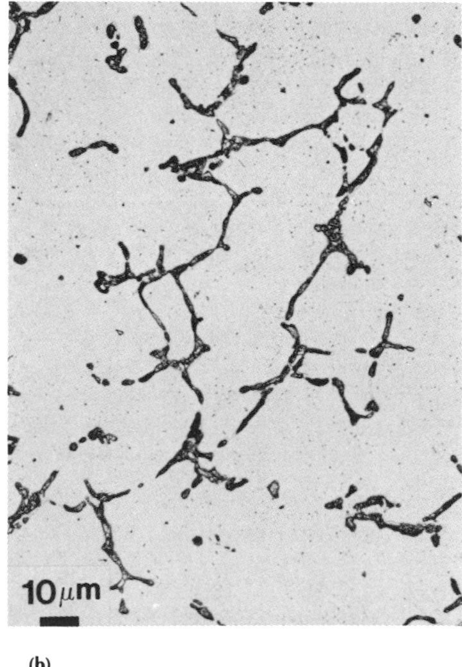

(b)

Fig. 3 Cracked 25Cr-12Ni cast stainless steel quenching fixture. (a) Macrograph of part of the fixture. (b) Microstructure showing substantial σ-phase. Electrolytically etched with 10 *N* KOH. 500×

1600 °F (650 to 870 °C) leads to the transformation of the ferrite to sigma (σ) phase. Sigma is extremely brittle, and if it forms a continuous phase with carbides at the grain boundaries, it can cause dramatic brittle fractures, even if subjected to light impact at room temperature.

Some of the iron-chromium-nickel alloys may also form σ if the compositional balance leans toward ferrite stabilization rather than austenite. Sigma forms much more slowly in these alloys than in the iron-chromium compositions. However, cracks caused by the formation of very small amounts of several years of σ after several years of exposure to critical temperatures are common.

Figure 3 shows part of a broken hook used to hold a heat treatment basket during austenitization and quenching. The hook was made from cast 25Cr-12Ni heat-resisting steel. However, the composition was not properly balanced, and a higher-than-normal δ-ferrite content was present in the hook. The δ-ferrite transformed to σ during the periods that the hook was in the austenitizing furnace (temperatures from 815 to 900 °C, or 1500 to 1650 °F, generally). The micrograph in Fig. 3(b) shows a very heavy, nearly continuous grain-boundary σ network.

Table 4 Typical room-temperature properties of heat-resistant casting alloys

Alloy	Condition	Tensile strength, MPa	Tensile strength, ksi	Yield strength, MPa	Yield strength, ksi	Elongation, %	Hardness, HB
Standard grades							
HA	N + T(a)	738	107	558	81	21	220
HC	As-cast	760	110	515	75	19	223
	Aged(b)	790	115	550	80	18	...
HD	As-cast	585	85	330	48	16	90
HE	As-cast	655	95	310	45	20	200
	Aged(b)	620	90	380	55	10	270
HF	As-cast	635	92	310	45	38	165
	Aged(b)	690	100	345	50	25	190
HH, type 1	As-cast	585	85	345	50	25	185
	Aged(b)	595	86	380	55	11	200
HH, type 2	As-cast	550	80	275	40	15	180
	Aged(b)	635	92	310	45	8	200
HI	As-cast	550	80	310	45	12	180
	Aged(b)	620	90	450	65	6	200
HK	As-cast	515	75	345	50	17	170
	Aged(c)	585	85	345	50	10	190
HL	As-cast	565	82	360	52	19	192
HN	As-cast	470	68	260	38	13	160
HP	As-cast	490	71	275	40	11	170
HT	As-cast	485	70	275	40	10	180
	Aged(c)	515	75	310	45	5	200
HU	As-cast	485	70	275	40	9	170
	Aged(d)	505	73	295	43	5	190
HW	As-cast	470	68	250	36	4	185
	Aged(e)	580	84	360	52	4	205
HX	As-cast	450	65	250	36	9	176
	Aged(d)	505	73	305	44	9	185
Nonstandard grades							
HK-Nb-Ti	As-cast	510	74	...	...	18	176
HK-Nb	As-cast	524	76	255	37	20	176
HK-Co-W	As-cast	579	84	344	50	15	195
HP-Nb	As-cast	517	75	276	40	12	181
HP-Nb-LC	As-cast	606	88	283	41	44	...
HP-Nb-Ti	As-cast	531	77	255	37	12	185
HP-Nb-W	As-cast	538	78	296	43	13	...
HP-Nb-W-Mo	As-cast	525	76	250	36	11	181
HP-Mo	As-cast	592	86	317	46	13	181
HP-W	As-cast	524	76	310	45	13	185
HP-W-Co	As-cast	510	74	303	44	8	185
30-45-Nb-Ti	As-cast	586	85	290	42	10	195
30-45-W	As-cast	517	75	290	42	10	171
30-45-W-Co	As-cast	531	77	303	44	10	...
30-45-W-Al	As-cast	710	103	...	...	...	256
50Cr-50Ni	As-cast	550	80	340	49	5	...
60Cr-40Ni	As-cast	760	110	590	80	...	...
50Cr-50Ni-Nb	As-cast	550	80	345	50	5	...

(a) Normalized and tempered at 675 °C (1250 °F). (b) Aging treatment: 24 h at 760 °C (1400 °F), furnace cool. (c) Aging treatment: 24 h at 760 °C (1400 °F), air cool. (d) Aging treatment: 48 h at 980 °C (1800 °F), air cool. (e) Aging treatment: 48 h at 980 °C (1800 °F), furnace cool

Mechanical Properties

Room-Temperature Tensile Properties. ASTM and many industrial specifications cite minimum tensile properties that must be met by cast heat-resistant alloys at room temperature. While such requirements can be a useful tool for quality control purposes, they have little relevance to the performance of a casting in service at elevated temperatures.

Representative room-temperature tensile properties and hardness values are shown in Table 4. These values should be used only as a guide because the properties within different sections of the same casting may show a wide range due to differences in section thickness, pouring temperatures, and solidification characteristics.

Elevated-Temperature Tensile Properties. Short-term tensile properties undergo major changes with increasing temperature. Both yield and tensile strength decrease, while elongation and reduction of area increase. The short-term elevated-temperature test, in which a standard tension test bar is heated to a designated uniform temperature and then strained to fracture at a standardized rate, identifies the stress due to a short-term overload that will cause fracture in uniaxial loading. The manner in which the values of tensile strength and ductility change with increasing temperature during testing is shown in Fig. 4 for selected alloys. Representative tensile properties at temperatures between 760 and 980 °C (1400 and 1800 °F) are given in Table 5 for several heat-resistant cast steel standard grades. Similar data for nonstandard grades are given in Table 6.

Creep and Stress-Rupture Properties. Creep is defined as the time-dependent strain that occurs under load at elevated temperature and is operative in most applications of heat-resistant, high-alloy castings at the normal service temperatures. In time, creep may lead to excessive deformation and even fracture at stresses considerably below those determined in room-temperature and elevated-temperature short-term tension tests.

When the rate or degree of deformation is the limiting factor, the design stress is based on the minimum creep rate and design life after allowing for initial transient creep. The stress that produces a specified minimum creep rate of an alloy or a specified amount of creep deformation in a given time (for example, 1% total creep in 100,000 h) is referred to as the *limiting creep strength,* or *limiting stress.* Tables 7 to 9 list the creep strengths of various H-type castings at specific temperatures. Figure 5 shows creep rates as a function of temperature.

When rupture stresses are used in design, the most widely used value is the minimum stress required to cause rupture in 100,000 h (11.4 years). Such long-term data are invariably extrapolated from tests of shorter duration. Graphical extrapolation is feasible if some 10,000 h data, or longer, are available. Most extrapolations rely on parametric equations involving rupture time and temperature. While several of these equations have been developed, the most widely used by producers and users of heat-resistant alloys is the Larson-Miller expression (see the article "Assessment and Use of Creep-Rupture Data" in this Volume for details). A typical stress versus Larson-Miller parameter plot is shown in Fig. 6.

The number assigned to the constant in the Larson-Miller expression is typically in a range

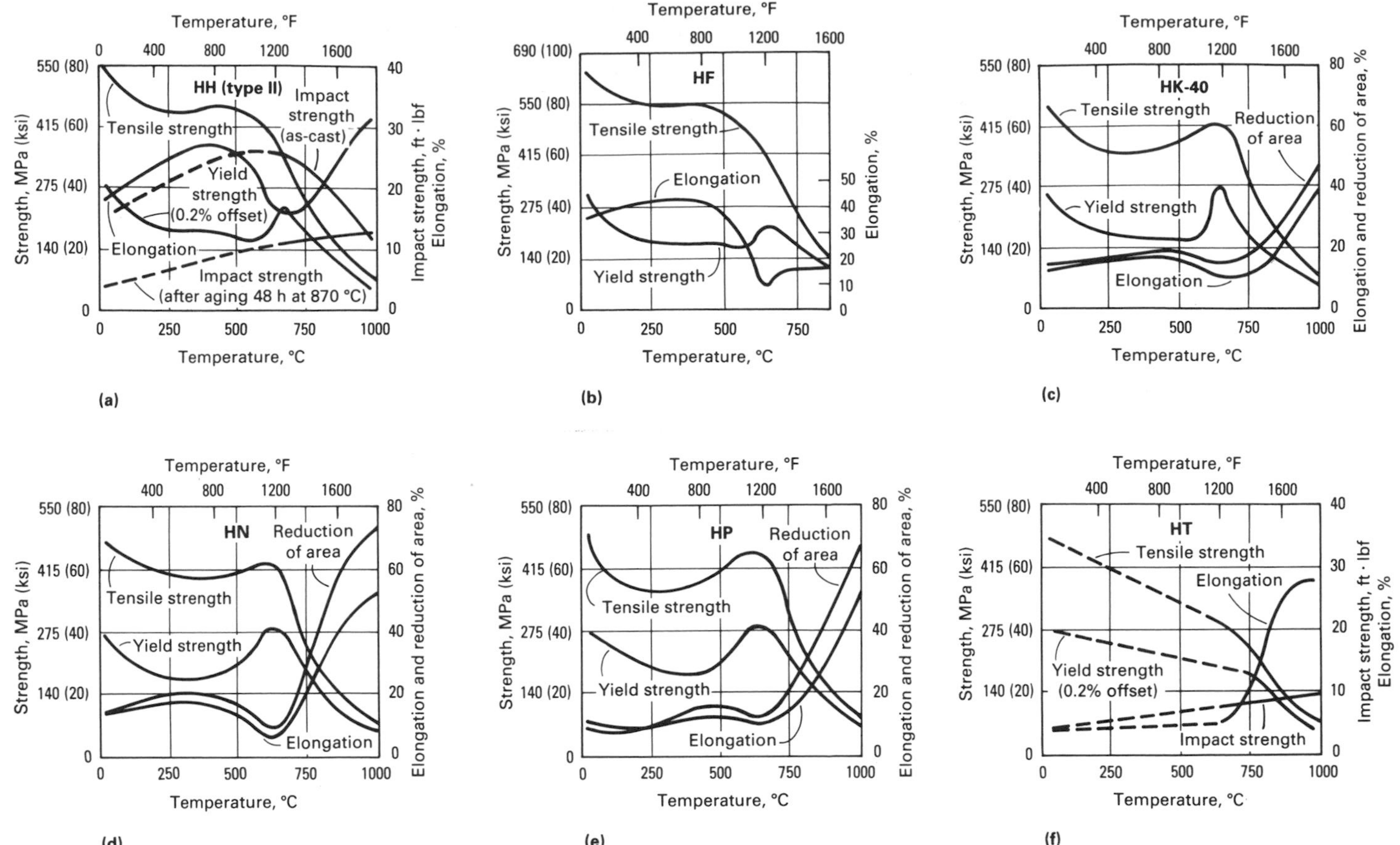

Fig. 4 Effect of short-term elevated-temperature exposure on the tensile properties of wholly austenitic (type II) HH cast steel (a) and of five other heat-resistant cast steels: (b) HF cast steel, (c) HK-40 cast steel, (d) HN cast steel, (e) HP cast steel, and (f) HT cast steel. Long-term elevated-temperature exposure reduces the strengthening effects between 500 to 750 °C (900 to 1400 °F) in (c), (d), and (e). Tensile properties of alloy HT in (f) include extrapolated data (dashed lines) below 750 °C but should be similar to alloy HN in terms of yield and tensile strengths. Source: Ref 3

Table 5 Representative short-term tensile properties of standard cast heat-resistant alloys at elevated temperatures

Alloy	Property at indicated temperature: 760 °C (1400 °F), Ultimate tensile strength, MPa	760 °C, Ultimate tensile strength, ksi	760 °C, Yield strength at 0.2% offset, MPa	760 °C, Yield strength at 0.2% offset, ksi	760 °C, Elongation, %	870 °C (1600 °F), Ultimate tensile strength, MPa	870 °C, Ultimate tensile strength, ksi	870 °C, Yield strength at 0.2% offset, MPa	870 °C, Yield strength at 0.2% offset, ksi	870 °C, Elongation, %	980 °C (1800 °F), Ultimate tensile strength, MPa	980 °C, Ultimate tensile strength, ksi	980 °C, Yield strength at 0.2% offset, MPa	980 °C, Yield strength at 0.2% offset, ksi	980 °C, Elongation, %
HA	462(a)	67(a)	220(b)	32(b)	...	...	...	...	...	...	...	...	...	...	...
HD	248	36	...	...	14	159	23	...	...	18	103	15	...	...	40
HF	262	38	172	25	16	145	21	107	15.5	16	...	...	...	...	...
HH (type I)(c)	228	33	117	17	18	127	18.5	93	13.5	30	62	9	43	6.3	45
HH (type II)(c)	258	37.4	136	19.8	16	148	21.5	110	16	18	75	10.9	50	7.3	31
HI	262	38	...	...	6	179	26	...	...	12	...	...	...	...	...
HK	258	37.5	168	24.4	12	161	23	101	15	16	85.5	12.4	60	8.7	42
HL	345	50	...	...	...	210	30.5	...	...	...	129	18.7	...	...	...
HN	...	...	...	...	...	140	20	100	14.5	37	83	12	66	9.6	51
HP	296	43	200	29	15	179	26	121	17.5	27	100	14.5	76	11	46
HT	240	35	180	26	10	130	19	103	15	24	76	11	55	8	28
HU	275	40	...	...	...	135	19.5	...	...	20	69	10	43	6.2	28
HW	220	32	158	23	...	131	19	103	15	...	69	10	55	8	40
HX	310(d)	45(d)	138(d)	20(d)	8(d)	141	20.5	121	17.5	48	74	10.7	47	6.9	40

(a) In this instance, test temperature was 540 °C (1000 °F). (b) Test temperature was 590 °C (1100 °F). (c) Type I and II per ASTM A 447. (d) Test temperature was 650 °C (1200 °F).

from 12 to 25 (it is 16 in Fig. 6). The constant is obtained from curve fitting equations or by experimentation, but unfortunately, it is common for different suppliers of the same alloy to derive different constants from their data banks. This can lead to conflicting claims for long-term stresses. In sensitive design applications, the users may need to critically examine the actual data banks and compositional variance in order to resolve this problem to their satisfaction.

Figure 7 compares the creep-rupture strengths of various H type steel castings at 10,000 and 100,000 h. Whether these property values are extrapolated or determined directly often has little bearing on the operating life of high-temperature parts. The actual material behavior is often difficult to predict accurately because of the complexity of the service stresses relative to the idealized, uniaxial loading conditions in the standardized tests, and because of the attenuating factors such as cyclic loading, temperature fluc-

Table 6 Representative short-term tensile properties of nonstandard heat-resistant alloys at elevated temperatures

	Property at indicated temperature														
	760 °C (1400 °F)					870 °C (1600 °F)					980 °C (1800 °F)				
	Ultimate tensile strength		Yield strength		Elongation,	Ultimate tensile strength		Yield strength		Elongation,	Ultimate tensile strength		Yield strength		Elongation,
Alloy	MPa	ksi	MPa	ksi	%	MPa	ksi	MPa	ksi	%	MPa	ksi	MPa	ksi	%
HK-Nb-Ti	324	47	145	21	21	186	27	103	15	27	107	15.5	69	10	33
HK-Nb	317	46	138	20	27	179	26	103	15	32	103	15	90	13	35
HK-Co-W	...	...	...	...	...	234	34	...	...	35	159	23	...	...	43
HP-Nb	310	45	138	20	22	193	28	97	14	36	110	16	66	9.5	48
HP-Nb-Si	303	44	145	21	20	193	28	97	14	29	117	17	62	9	34
HP-Nb-Ti	317	46	138	20	26	193	28	110	16	41	97	14	62	9	49
HP-Nb-W-Mo	317	46	159	23	26	200	29	110	16	40	110	16	62	9	50
HP-Mo	...	...	...	...	...	214	31	145	21	32	152	22	103	15	60
HP-W	...	...	...	...	...	228	33	...	...	21	159	23	...	...	37
HP-W-Co	...	...	...	...	...	248	36	193	28	18	165	24	124	18	21
30-45-Nb-Ti	352	51	165	24	23	207	30	117	17	32	124	18	72	10.5	39
30-45-W	...	...	...	...	...	221	32	...	...	27	131	19	...	...	34
30-45-W-Co	...	...	...	...	...	207	30	...	...	26	124	18	...	...	32
30-45-W-Al	...	...	...	...	...	372	54	...	...	...	262	38	...	...	...

Source: Ref 1

Table 7 Creep properties of selected standard cast heat-resistant alloys

	Temperature		Creep stress to produce 1% creep in 10,000 h		Stress to rupture in 10,000 h		Stress to rupture in 100,000 h	
Grade	°C	°F	MPa	ksi	MPa	ksi	MPa	ksi
Iron-chromium-nickel alloys								
HF	650	1200	124	18.0	114	16.5	76	11.0
	760	1400	47	6.8	42	6.1	28	4.0
	870	1600	27	3.9	19	2.7	12	1.7
	980	1800	...	...	...	...	...	...
HH	650	1200	124	18.0	97	14.0	62	9.0
	760	1400	43	6.3	33	4.8	19	2.8
	870	1600	27	3.9	15	2.2	8	1.2
	980	1800	14	2.1	6	0.9	3	0.4
HK	650	1200	...	...	...	...	...	...
	760	1400	70	10.2	61	8.8	43	6.2
	870	1600	41	6.0	26	3.8	17	2.5
	980	1800	17	2.5	12	1.7	7	1.0
Iron-nickel-chromium alloys								
HN	650	1200	...	...	...	...	...	...
	760	1400	...	...	...	...	...	...
	870	1600	43	6.3	33	4.8	22	3.2
	980	1800	16	2.4	14	2.1	9	1.3
HT	650	1200	...	...	...	...	...	...
	760	1400	55	8.0	58	8.4	39	5.6
	870	1600	31	4.5	26	3.7	16	2.4
	980	1800	14	2.0	12	1.7	8	1.1
HU	650	1200	...	...	...	...	...	...
	760	1400	59	8.5	...	...	...	...
	870	1600	34	5.0	23	3.3	...	...
	980	1800	15	2.2	12	1.8	...	...
HX	650	1200	...	...	...	...	...	...
	760	1400	44	6.4	...	...	...	...
	870	1600	22	3.2	...	...	...	...
	980	1800	11	1.6	...	...	...	...

Note: Some stress values are extrapolated. Source: Ref 6

tuations, and metal loss from corrosion. The designer should anticipate the synergistic effects of these variables.

Table 7 lists the stress required to produce 1% creep in 10,000 h and rupture in 10,000 h and 100,000 h at temperatures of 650, 760, 870, and 980 °C (1200, 1400, 1600, and 1800 °F). Similar data are given in Tables 9 and 10. A design stress figure commonly used for uniformly heated parts not subjected to thermal or mechanical shock is 50% of the stress to produce 1% creep in 10,000 h. This should be used carefully and should be verified with the supplier (Ref 6).

Thermal fatigue failure involves cracking caused by heating and cooling cycles. Very little experimental thermal fatigue information is available on which to base a comparison of the various alloys, and no standard test as yet has been adopted. Field experience indicates that resistance to thermal fatigue is usually improved with an increase in nickel content. Niobium-modified alloys have been employed successfully when a high degree of thermal fatigue resistance is desired, such as in reformer outlet headers.

Thermal Shock Resistance. Thermal shock failure may occur as a result of a single, rapid temperature change or as a result of rapid cyclic temperature changes that induce stresses high enough to cause failure. Thermal shock resistance is influenced by the coefficient of thermal expansion and the thermal conductivity of materials. Increases in the thermal expansion coefficient or decreases in thermal conductivity reduce the resistance against thermal shock. Table 11 lists the thermal conductivities and expansion coefficients for heat-resistant castings at various temperatures. The HA, HC, and HD alloys, because of their predominantly ferritic microstructure, have the lowest thermal expansion coefficients and the highest thermal conductivities.

High-Temperature Corrosion

An important factor pertaining to the corrosion behavior of heat-resistant alloys is chromium content. Chromium imparts resistance to oxidation and sulfidation at high temperatures by forming a passive oxide film. Heat-resistant casting alloys must also have good resistance to carburization.

The atmospheres most commonly encountered by heat-resistant cast steels are air, flue gases, and process gases. Such gases may be either oxidizing or reducing, and they may be sulfidizing or carburizing if sulfur and carbon are present. The corrosion of heat-resistant alloys by the environment at elevated temperatures varies significantly with alloy type, temperature, velocity, and the nature of the specific environment to which the part is exposed. Table 8 reviews the general corrosion characteristics of various heat-resistant castings. Table 12 presents a general ranking of the standard cast heat-resistant grades in various environments at 980 °C (1800 °F).

Table 8 General corrosion characteristics of heat-resistant cast steels and typical limiting creep stress values at indicated temperatures

Alloy	Corrosion characteristics	Creep test temperature °C	Creep test temperature °F	Limiting creep stress (0.0001 %/h) MPa	Limiting creep stress (0.0001 %/h) ksi
HA	Good oxidation resistance to 650 °C (1200 °F); widely used in oil refining industry	650	1200	21.5	3.1
HC	Good sulfur and oxidation resistance up to 1095 °C (2000 °F); minimal mechanical properties; used in applications where strength is not a consideration or for moderate load bearing up to 650 °C (1200 °F)	870	1600	5.15	0.75
HD	Excellent oxidation and sulfur resistance plus weldability	980	1800	6.2	0.9
HE	Higher temperature and sulfur resistant capabilities than HD	980	1800	9.5	1.4
HF	Excellent general corrosion resistance to 815 °C (1500 °F) with moderate mechanical properties	870	1600	27	3.9
HH(a)	High strength; oxidation resistant to 1090 °C (2000 °F); most widely used	980	1800	7.5 (type I) 14.5 (type II)	1.1 (type I) 2.1 (type II)
HI	Improved oxidation resistance compared to HH	980	1800	13	1.9
HK	Because of its high-temperature strength, widely used for stressed parts in structural applications up to 1150 °C (2100 °F); offers good resistance to corrosion by hot gases, including sulfur-bearing gases, in both oxidizing and reducing conditions (although HC, HE, and HI are more resistant in oxidizing gases); used in air, ammonia, hydrogen, and molten neutral salts; widely used for tubes and furnace parts	1040	1900	9.5	1.4
HL	Improved sulfur resistance compared to HK; especially useful where excessive scaling must be avoided	980	1800	15	2.2
HN	Very high strength at high temperatures; resistant to oxidizing and reducing flue gases	1040	1900	11	1.6
HP	Resistant to both oxidizing and carburizing atmospheres at high temperatures	980	1800	19	2.8
HP-50WZ	Improved creep rupture strength at 1090 °C (2000 °F) and above compared to HP	1090	2000	4.8	0.7
HT	Widely used in thermal shock applications; corrosion resistant in air, oxidizing and reducing flue gases, carburizing gases, salts, and molten metals; performs satisfactorily up to 1150 °C (2100 °F) in oxidizing atmospheres and up to 1095 °C (2000 °F) in reducing atmospheres, provided that limiting creep stress values are not exceeded	980	1800	14	2.0
HU	Higher hot strength than HT and often selected for its superior corrosion resistance	980	1800	15	2.2
HW	High hot strength and electrical resistivity; performs satisfactorily to 1120 °C (2050 °F) in strongly oxidizing atmospheres and up to 1040 °C (1900 °F) in oxidizing or reducing products of combustion that do not contain sulfur; resistant to some salts and molten metals	980	1800	9.5	1.4
HX	Resistant to hot-gas corrosion under cycling conditions without cracking or warping; corrosion resistant in air, carburizing gases, combustion gases, flue gases, hydrogen, molten cyanide, molten lead, and molten neutral salts at temperatures up to 1150 °C (2100 °F)	980	1800	11	1.6

(a) Two grades: type I (ferrite in austenite) and type II (wholly austenitic), per ASTM A 447

Oxidation. Resistance to oxidation increases directly with chromium content (Fig. 8). For the most severe service at temperatures above 1095 °C (2000 °F), 25% or more chromium is required. Additions of nickel, silicon, manganese, and aluminum promote the formation of relatively impermeable oxide films that retard further scaling. Thermal cycling is extremely damaging to oxidation resistance because it leads to breaking, cracking, or spalling of the protective oxide film. The best performance is obtained with austenitic alloys containing 40 to 50% combined nickel and chromium. Figure 9 illustrates the corrosion behavior of H-type grades in air and oxidizing flue gases.

Sulfidation environments are becoming increasingly important. Petroleum processing, coal conversion, utility and chemical applications, and waste incineration have heightened the need for alloys resistant to sulfidation attack in relatively weak oxidizing or reducing environments. Fortunately, high chromium and silicon contents increase resistance to sulfur-bearing environments. On the other hand, nickel has been found to be detrimental in the presence of the most aggressive gases. The problem is attributable to the formation of low-melting nickel-sulfur eutectics. These produce highly destructive liquid phases at temperatures even below 815 °C (1500 °F). Once formed, the liquid may run onto adjacent surfaces and rapidly corrode other metals. The behavior of H-type grades in sulfidation environments is represented in Fig. 10.

Carburization. High alloys are often used in nonoxidizing atmospheres in which carbon diffusion into metal surfaces is possible. Depending on chromium content, temperature, and carburizing potential, the surface may become extremely rich in chromium carbides, rendering it hard and possibly susceptible to cracking. As described below, silicon and nickel enhance resistance to carburization.

Effects of Alloying Additions. Carburization is a major controlling factor for the performance of pyrolysis furnace tubes for ethylene and olefin plants. Furnace tubes are typically constructed of Fe-Ni-Cr cast alloys, such as HK-40 (Fe-25Cr-20Ni-0.4C-1.25Si). In order to meet the increasing demand for better furnace alloys to handle higher temperatures and pressures in ethylene cracking and steam reforming operations, a wide variety of modified HK and HP alloys have been produced and made available to industry. These modifications involved additions of niobium, tungsten, molybdenum, and silicon. Some also involved increases in nickel and/or chromium. It has been found that these additions and increases improve carburization resistance. Figures 11 to 14 illustrate the carburization resistance of some of these modified alloys compared to HK-40.

In an extensive study undertaken by Steel and Engel (Ref 11) on the relative influence of nickel and chromium on the carburization resistance of Fe-Ni-Cr alloys, standard heats of ASTM grades varying from HC to HX were investigated, along with many experimental cast alloys. They found nickel to be beneficial, as illustrated in Fig. 15. The role of chromium, however, appeared to be different for different levels of nickel, as shown in Fig. 16. For iron-base alloys with 25% or less nickel, increasing chromium significantly reduced carbon pickup. A slight decrease in carbon pickup with increasing chromium was noted for alloys containing 26 to 45% Ni. For alloys containing 46 to 70% Ni, increasing chromium resulted in an increase in carbon pickup.

Small additions of some minor elements, such as titanium, niobium, tungsten, and rare earth elements, also can improve resistance of an alloy to carburization; see Table 13. The superior carburization resistance of TMA 4750 alloy to HK-40 + 2% Si can be attributed to small additions of titanium, niobium, tungsten, and rare earth elements.

Silicon also has been found to be very effective in improving carburization resistance. Figure 17 shows how increasing the silicon content in modified HK-40 alloys significantly decreased carbon pickup.

Effects of Surface Finish. Other factors such as surface finish have been found to be very important in affecting carburization reactions. Machining the metal surface to improve the surface finish can significantly increase carburization resistance of an alloy. It is common practice to bore or hone the internal diameter of a centrifugally cast tube to remove surface shrinkage pores. Figure 18 illustrates the significant improvement in carburization resistance as a result of surface machining. A cast metal surface with shrinkage pores can generate stagnant conditions in crevices, which are very conducive to carburization attack. In addition, a machined surface exhibits a cold work layer, which tends to accelerate the diffusion process and results in rapid formation of oxide scale or film, thus slowing subsequent carbon ingress.

Effects of Sulfur Injections. Injecting sulfur compounds into the processing stream is another approach that has been used in the petrochemical industry to mitigate carburization problems. The

Table 9 Creep properties of selected nonstandard cast heat-resistant alloys

Alloy	Temperature		Creep stress to produce 0.0001 %/h creep in 100,000 h		Stress to produce rupture in 100,000 h	
	°C	°F	MPa	ksi	MPa	ksi
25Cr-20Ni						
HK-Nb-Ti	760	1400	...	...	59.3	8.60
	870	1600	...	...	31.7	4.60
	980	1800	...	...	11.0	1.60
	1095	2000	...	...	...	...
HK-Nb	760	1400	...	...	51.0	7.40
	870	1600	...	...	23.0	3.34
	980	1800	...	...	8.3	1.21
	1095	2000	...	...	...	...
HK-Co-W	760	1400	...	...	...	...
	870	1000	...	...	...	...
	980	1800	24.8	3.6	13.8	2.00
	1095	2000	11.4	1.66	2.4	0.35
25Cr-35Ni						
HP-Nb	760	1400	...	...	63.4	9.20
	870	1600	...	...	33.1	4.80
	980	1800	19.3	2.8	13.8	2.00
	1095	2000	...	...	4.3	0.63
HP-Nb-Si	760	1400	...	...	...	...
	870	1600	...	...	29.4	4.27
	980	1800	17.7	2.57	10.8	1.56
	1095	2000	...	...	3.2	0.46
HP-Nb-Ti	760	1400	...	...	...	...
	870	1600	...	...	40.7	5.90
	980	1800	19.3	2.8	16.5	2.40
	1095	2000	...	...	5.3	0.77
HP-Nb-W-Mo	760	1400	...	...	56.5	8.20
	870	1600	...	...	27.6	4.00
	980	1800	13.8	2.0	10.8	1.56
	1095	2000	5.4	0.78	3.0	0.43
HP-W	760	1400	6.4	9.3	62.7	9.10
	870	1600	38.6	5.6	32.1	4.65
	980	1800	17.9	2.6	13.2	1.92
	1095	2000	5.1	0.74	4.0	0.58
HP-W-Co	760	1400	...	...	...	...
	870	1600	34.5	5.0	31.0	4.50
	980	1800	15.2	2.2	12.1	1.75
	1095	2000	4.8	0.70	5.2	0.75
30Cr-45Ni						
30-45-Nb-Ti	760	1400	...	...	57.8	8.39
	870	1600	...	...	31.4	4.55
	980	1800	21.4	3.1	12.3	1.79
	1095	2000	...	...	3.1	0.45
30-45-W	760	1400	...	...	...	...
	870	1600	...	...	...	...
	980	1800	21.4	3.1	14.5	2.10
	1095	2000	8.8	1.27	4.8	0.69
30-45-W-Co	760	1400	...	...	...	...
	870	1600	...	...	31.7	4.6
	980	1800	22.1	3.2	13.1	1.90
	1095	2000	11.0	1.6	5.2	0.76
30-45-W-Al	760	1400	...	...	...	...
	870	1600	...	...	41.4	6.0
	980	1800	...	...	15.7	2.27
	1095	2000	8.6	1.25	5.2	0.75

Source: Ref 1

beneficial effect of sulfur in reducing carburization attack was demonstrated by Norton and Barnes (Ref 14) in their laboratory tests. Figure 19 shows that carburization of HK-40 alloy was significantly reduced when 100 ppm H_2S was injected into the test environment.

Ash/Salt Deposit Corrosion. Fireside corrosion can be significant in oil-fired boilers when low-grade fuels with high concentrations of vanadium, sulfur, and sodium are used for firing. During combustion, vapors of vanadium pentoxide (V_2O_5) and alkali metal sulfates are formed. These vapors, combined with other ash constituents, then deposit onto cooler component surfaces. Vanadium pentoxide and alkali metal sulfates in the ash deposits react to form low-melting-point salts, which flux the protective oxide scale from the metal surface and result in accelerated corrosion attack. Thus fireside corrosion in oil-fired boilers or furnaces is frequently referred to as "oil ash corrosion."

Vanadium pentoxide and sodium sulfate are the principal constituents responsible for oil ash corrosion. Reactions between vanadium and sodium compounds result in formation of complex vanadates of low melting points. Table 14 lists various oil ash constituents and their melting points. It is clear from the table that the melting point of the ash salt deposit may vary widely, depending on composition. The formation of liquid salts is responsible for accelerated corrosion attack by oil ash corrosion.

McDowell and Mihalisin (Ref 16) conducted extensive field rack tests in boilers fired with Bunker "C" oils containing high concentrations of vanadium (150 to 450 ppm). Test racks were exposed in the superheater section. Alloys ranging from low-alloy steels to iron- and nickel-base alloys suffered severe corrosion attack. The results of one test rack are shown in Table 15. Specimens included 5Cr steel, stainless steels (both 400 and 300 series), Fe-Ni-Cr alloys, Ni-Cr-Fe alloy, and 50Ni-50Cr alloy, along with two cast stainless steels (HE and HH alloys). All the alloys tested exhibited unacceptable corrosion rates. Even the best performer (50Ni-50Cr alloy) suffered a corrosion rate of 3.1 mm/year (121 mils/year).

Spafford (Ref 17) reported good performance of the cast 50Ni-50Cr alloy (see Table 3) in refinery heaters for coking and catalytic reformer units. The heaters were fired with heavy fuel oil containing 2.5 to 4% S and 50 to 70 ppm V (occasionally up to 150 ppm). The hangers and tube supports made of cast HH alloy (25Cr-12Ni steel) suffered severe corrosion attack. Metal temperatures were in the range of 730 to 890 °C (1350 to 1630 °F). The highest corrosion rates were 6.4 to 9.5 mm/year (250 to 375 mils/year). Replacements of IN-657 (a 50Ni-50Cr alloy) were reported to perform very well, with minimal maintenance and repair (Ref 17). In a field rack test in a crude oil heater at 700 °C (1290 °F), alloy 657 performed 10 times better than HH and HK alloys, as illustrated in Fig. 20. Swales and Ward (Ref 18) reported numerous field experiences for IN-657 as tube supports in refinery heaters. They concluded that the alloy provided satisfactory service at temperatures up to 900 °C (1650 °F). At temperatures higher than 900 °C (1650 °F), IN-657 often suffered severe corrosion attack.

Molten Salt Corrosion. Chloride salts are widely used in the heat treating industry for annealing and normalizing of steels. These salts are commonly referred to as neutral salt baths. The most common neutral salt baths are barium, sodium, and potassium chlorides, used separately or in combination in the temperature range of 760 to 980 °C (1400 to 1800 °F). Compositions of some common neutral salt baths are (Ref 19):

- 50NaCl-50KCl
- $50KCl-50Na_2CO_3$
- $20NaCl-25KCl-55BaCl_2$
- $25NaCl-75BaCl_2$
- $21NaCl-31BaCl_2-48CaCl_2$

Jackson and LaChance (Ref 19) performed an extensive study on the corrosion of cast Fe-Ni-Cr alloys in the $NaCl-KCl-BaCl_2$ salt bath. They found that the alloys suffered intergranular attack significantly more than metal loss. Corrosion data in terms of metal loss and intergranular attack are shown in Fig. 21 and 22, respectively. These figures also

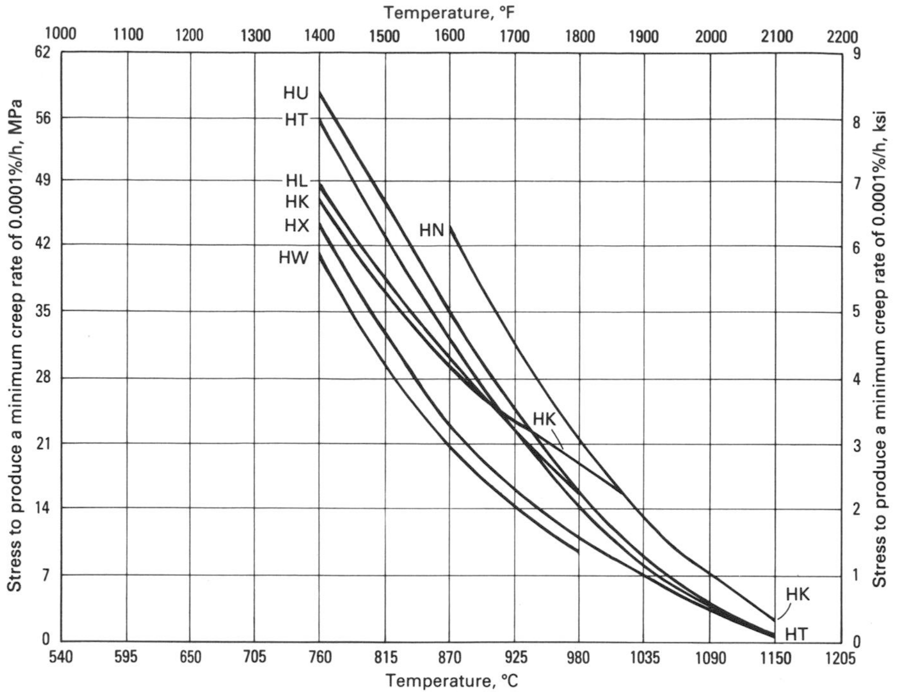

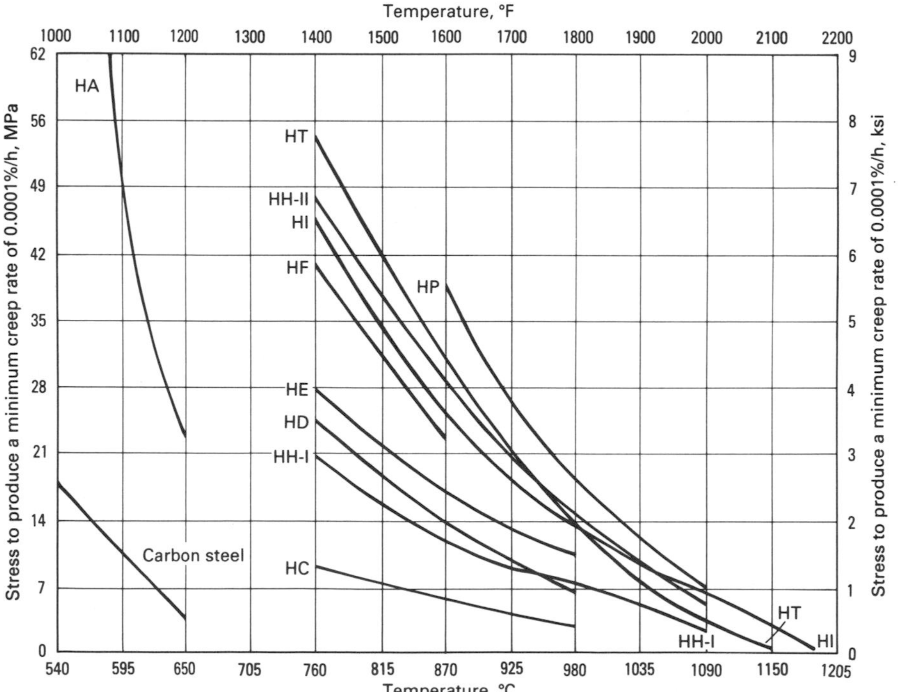

Fig. 5 Creep strength of heat-resistant alloy castings (HT curve is included in both graphs for ease of comparison). Source: Ref 5

indicate that resistance to the molten salt (NaCl-KCl-$BaCl_2$) increases with decreasing chromium and increasing nickel in Fe-Ni-Cr alloys. HW alloy (12Cr-60Ni) was consistently the best performer among the four commercial cast alloys (HW, HT, HK, and HH alloys) studied. These authors further noted that intergranular attack generally followed grain boundary carbides. Thus, lowering carbon content from 0.4% to about 0.07% resulted in a threefold improvement. Decreasing grain size also improved alloy resistance to intergranular attack. Five different neutral salt baths were compared for HW, HT, and three Fe-Cr alloys, as shown in Fig. 23. In general, the four chloride salt baths were quite similar. The KCl-Na_2CO_3 salt bath was significantly less aggressive than the chloride baths. It is also interesting to note that Fe-17Cr alloy was better than HW (Fe-12Cr-60Ni) and HT (Fe-15Cr-35Ni) alloys in NaCl-KCl, NaCl-KCl-$BaCl_2$, and NaCl-$BaCl_2$-$CaCl_2$ salt baths.

Alloy Selection

Factors that must be considered when selecting heat-resistant alloy castings include:

- The expected normal service temperature and the minimum and maximum temperatures likely to be encountered. In extreme cases the liquidus and solidus temperatures may be required.
- Frequency and rate of thermal cycling
- Severity of thermal gradients in the component
- Thermal expansion properties of the alloy
- Applied loads and manner of loading
- Design life
- Short-term high-temperature tensile properties and long-term creep and rupture strength
- Service environment. Is it oxidizing reducing, sulfidizing, carburizing, or others? Does it contain corrodents, such as vanadates, sodium, and chlorides., or does it contain liquid metals that could flux protective oxides?
- Weldability in the as-cast condition and after service exposure
- Machinability

Comprehensive compilations of the physical and mechanical properties of the standard ACI alloys, together with recommended welding and machining procedures needed to assist in alloy selection, are contained in Handbook Supplements available from the Steel Founders' Society of America. Alloy data sheets and technical brochures are also available from the manufacturers or licensees of proprietary alloys.

Material cost should not be a deterrent to upgrading to a more highly alloyed grade. The upgraded material may have greater strength, allowing redesign of the component to a lighter section. Table 16 compares a number of different materials suitable for reformer tubes, which were originally made with HK-40 alloy. The alternative materials cost from 50 to 140% more than the HK-40. Upgrading to HP-Nb alloy reduces the tube wall thickness from 18.3 to 7.1 mm (0.721 to 0.280 in.) and lowers the tube weight from 59 to 23 kg/m (40 to 15.5 lb/ft). Consequently, despite the higher unit cost of the alloy, the relative tube assembly cost may be as much as 25% lower than that of the original HK-40.

Experienced metallurgists and foundry engineers should be consulted to assist in alloy selection and component design. These two simple steps can result in reduced capital cost and enhanced service performance.

Commercial applications of heat-resistant castings include metal treatment furnaces, gas turbines, aircraft engines, military equipment, oil refinery furnaces, cement mill equipment, petrochemical furnaces, chemical process equipment, power plant equipment, steel mill equipment, turbochargers, and equipment used in manufacturing glass and synthetic rubber. Oftentimes a variety of heat-resistant alloys are utilized for a

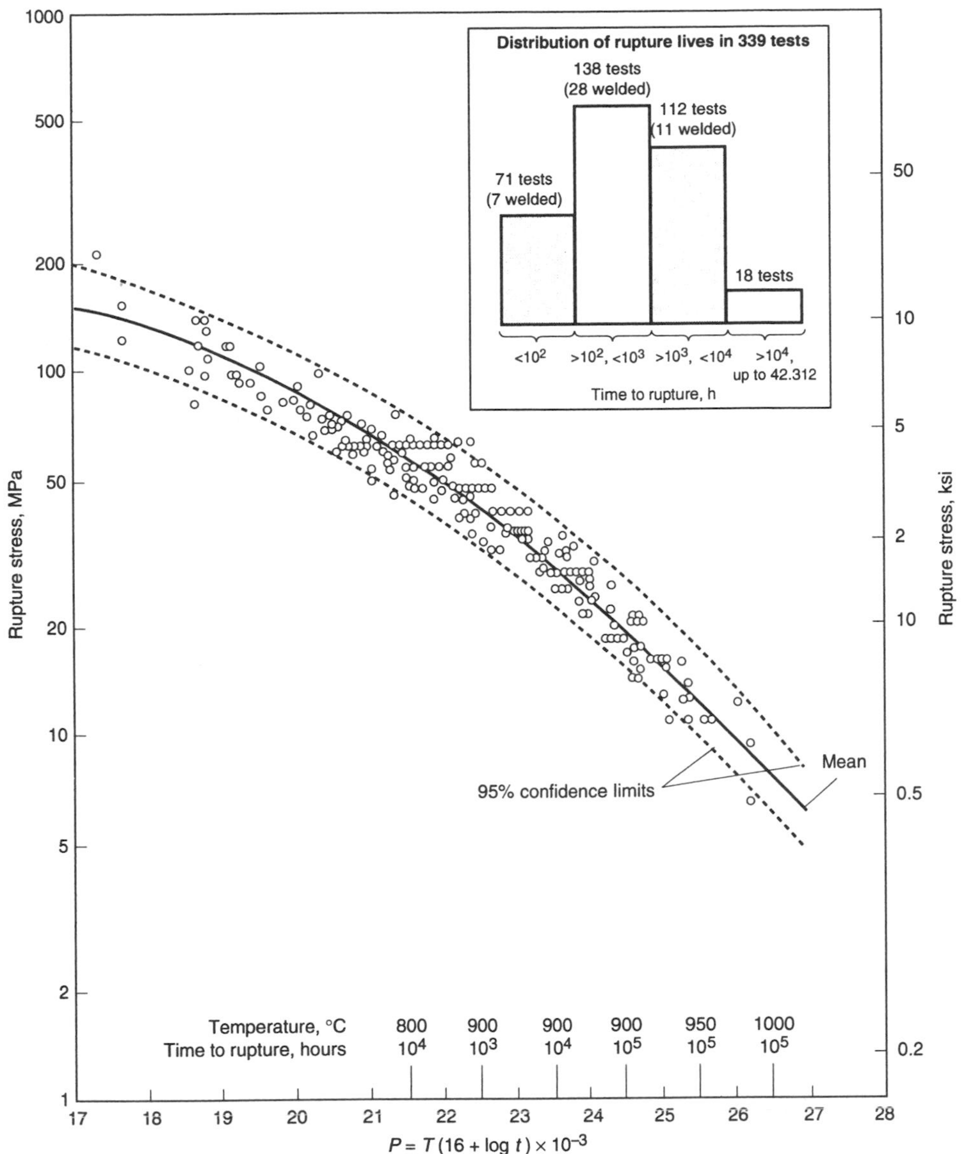

Fig. 6 A typical Larson-Miller parameter versus stress plot for IN-519 (modified HK-Nb in Table 2). Temperature (T) is in degrees Kelvin, and time (t) is in hours. Source: Ref 4

specific application area. For example, Fig. 24 shows the cast heat-resistant alloys used for various components of a cement mill. Environmental conditions that provide the basis for selection are listed for each component. Additional selection guidelines for other applications are described in the following sections.

Selection of Straight Chromium Heat-Resistant Castings

Iron-chromium alloys, also known as straight chromium alloys, contain either 9 or 28% Cr. HC and HD alloys are included among the straight chromium alloys, although they contain low levels of nickel.

HA alloy (9Cr-1Mo), a heat treatable material, contains enough chromium to provide good resistance to oxidation at temperatures up to about 650 °C (1200 °F). The 1% Mo is present to provide increased strength. HA alloy castings are widely used in oil refinery service. A higher-chromium modification of this alloy (12 to 14% Cr) is widely used in the glass industry.

HA alloy has a structure that is essentially ferritic; carbides are present in pearlitic areas or as agglomerated particles, depending on prior heat treatment. Hardening of the alloy occurs upon cooling in air from temperatures above 815 °C (1500 °F). In the normalized-and-tempered condition, the alloy exhibits satisfactory toughness throughout its useful temperature range.

HC alloy (28% Cr) resists oxidation and the effects of high-sulfur flue gases at temperatures up to 1100 °C (2000 °F). It is used for applications in which strength is not a consideration, or in which only moderate loads are involved, at temperatures of about 650 °C (1200 °F). It is also used where appreciable nickel cannot be tolerated, as in very-high-sulfur atmospheres, or where nickel may act as an undesirable catalyst and destroy hydrocarbons by causing them to crack.

HC alloy is ferritic at all temperatures. Its ductility and impact strength are very low at room temperature, and its creep strength is very low at elevated temperatures unless some nickel is present. In a variation of HC alloy that contains more than 2% Ni, substantial improvement in all three of these properties is obtained by increasing the nitrogen content to 0.15% or more.

HC alloy becomes embrittled when heated for prolonged periods at temperatures between 400 and 550 °C (750 and 1025 °F), and it shows low resistance to impact. The alloy is magnetic and has a low coefficient of thermal expansion, comparable to that of carbon steel. It has about eight times the electrical resistivity and about half the thermal conductivity of carbon steel. Its thermal conductivity, however, is roughly double the value for austenitic iron-chromium-nickel alloys.

HD alloy (28Cr-5Ni) is very similar in general properties to HC, except that its nickel content gives it somewhat greater strength at high temperatures. The high-chromium content of this alloy makes it suitable for use in high-sulfur atmospheres.

HD alloy has a two-phase, ferrite-plus-austenite structure that is not hardenable by conventional heat treatment. Long exposure at 700 to 900 °C (1300 to 1650 °F), however, may result in considerable hardening and severe loss of room-temperature ductility through the formation of a phase. Ductility may be restored by heating uniformly to 980 °C (1800 °F) or higher and then cooling rapidly to below 650 °C (1200 °F).

Selection of Iron-Chromium-Nickel Heat-Resistant Castings

Heat-resistant ferrous alloys in which the chromium content exceeds the nickel content are made in compositions ranging from 20Cr-10Ni to 30Cr-20Ni.

HE alloy (28Cr-10Ni) has excellent resistance to corrosion at elevated temperatures. Because of its higher chromium content, it can be used at higher temperatures than HF alloy and is suitable for applications up to 1100 °C (2000 °F). This alloy is stronger and more ductile at room temperature than the straight chromium alloys.

In the as-cast condition, HE alloy has a two-phase, austenite-plus-ferrite structure containing carbides. HE castings cannot be hardened by heat treatment; however, as with HD castings, long exposure to temperatures near 815 °C (1500 °F) will promote formation of σ-phase and consequent embrittlement of the alloy at room temperature. The ductility of this alloy can be improved somewhat by quenching from about 1100 °C (2000 °F).

Castings of HE alloy have good machining and welding properties. Thermal expansion is about 50% greater than that of either carbon steel or the iron-chromium alloy HC. Thermal conductivity is much lower than for HD or HC, but electrical

Table 10 Nominal compositions and comparative stress-rupture data for various nonstandard cast heat-resistant alloys

Group	Composition, %							Mean stress to rupture in 10,000 h, psi			Mean stress to rupture in 10,000 h, psi		
	C	Si	Mn	Cr	Ni	Nb	Other	1600 °F	1800 °F	2000 °F	1600 °F	1800 °F	2000 °F
HK-40	0.2–0.6	<2.0	<2.0	24–28	18–22	...	...	2700	1200	450	4200	1900	830
						...	0.3Ti, 0.5W	3831	1479	440	5500	2600	1000
						0.25	0.1Ti, 0.5W	4925	2183	546	6700	3200	1200
	0.4	1.3	1.0	25	20	...	<0.5Mo	2562	877	232	4379	1595	400
	0.35–0.45	<1.5	<1.75	23–27	19–22	...	...	3100	1250	...	4600	2250	870
	0.39–0.41	0.98	1.05	25.4	20.0	...	...	2600	950	...	4260	1800	670
	0.4–0.45	<1.0	<1.0	23–25	21–23	...	...	2701	1137	...	4120	1849	782
	0.4	<1.5	<1.5	25	20	...	...	2830	1170	...	4060	1950	...
HP-45 and HP-Mod	0.45	1.0	0.5	25	35	...	...	2700	1500	580	3900	2250	900
	0.45	1.0	0.5	25	36	...	5W	4000	1750	...	5820	2650	...
	0.50	1.6	0.7	26	35	...	5W, 15Co	5200	2800	970	7400	4000	1550
	0.45	1.0	0.5	26	36	...	0.3Ti, 0.5W	4610	2105	624	6300	3200	1400
						0.25	0.1Ti, 0.5W	5500	2730	675	7300	3900	1500
	0.4	0.75–2.0	1.0	25	35	0.7–1.5	...	4490	1895	...	6500	2586	...
	0.1	1.3	1.0	25	35	1.2	...	3200	1075	280	4300	1750	520
	0.5	1.7	0.5	25	35(+Co)	...	5W	3300	1225	375	5050	2480	770
	0.35–0.45	<1.5	<1.5	23–27	32–35	<1.5	...	4300	1750	460	6200	4500	2900
	0.35–0.45	<2.0	<2.0	24–27	32–35	1–2	1–2W	4200	1450	420	6000	4500	2500
	0.37–0.47	<1.5	<1.5	24–28	34–37	...	...	3128	1166	...	4550	2100	670
	0.4–0.5	<2.0	<0.5	24–28	34–37	0.6–1.5	...	4806	2000	640	6542	3128	1100
						0.6–1.5	3.5–4.5W	4400	1850	500	5200	2400	824
						0.6–1.5	0.5–1.5W						
						1.1–1.7	0.3–0.8Mo	4123	1500	426	5545	2560	780
							W + Mo + Nb ≤3.0						
	0.45–0.55	1.6–2.0	<1.5	23–27	34–37	...	...	3128	1137	...	4692	2130	853
	0.45–0.55	<1.5	<1.5	24–29	36–37	...	...	3200	890	...	4612	2275	...
	0.4–0.5	<1.5	1-2	24–26	34–36	1–1.5	...	3555	1136	...	4977	2417	853
	0.4	1.5	1.5	25	35	...	...	2900	1200	360	4350	2100	800
	0.4	1.5	1.5	25	35	1.5	...	4400	1950	510	6090	3210	750
	0.40–0.50	<1.5	<1.25	25–28	35–38	...	1.6W	...	1500	550	6000	2800	1100
	0.40–0.50	<2.0	<2.0	24–28	33–37	...	1.0Mo	3400	1150	500	5200	2210	...
	0.35–0.45	<2.0	<2.0	24–28	33–37	1.0	...	4742	1840	550	5250	2300	900
	0.40–0.50	<2.0	<2.0	24–28	33–37	1.5	1.5W	...	1860	575	...	2700	1075
	0.40–0.50	<2.0	<2.0	24–28	33–37	1.5	3.5W	5500	1940	610	...	3000	1280
	0.10M	<2.0	<2.0	24–28	33–37	1.5	...	3100	900	260	4300	1650	380
28/48W	0.25–0.50	<1.5	<1.25	24–27	44–47	...	36W, 3.6Mo, 3.6Co	...	...	...	...	2500	940
	0.15–0.25	<0.3	<0.3	32–34	48–52	...	16.0W	...	...	...	...	2700	1310
	0.40–0.60	<1.75	<1.5	26–30	46–50	...	5.0W, 3.6Co	5100	2400	...	...	2900	1200
	0.5	1.5	0.8	28	48	...	5W	...	1235	375	...	2260	780
	0.45–0.55	<1.5	<1.5	26–30	46–50	...	4–6W	4550	1635	398	7110	2700	995
	0.45	1.5	1.5	28	48	...	5W	...	1834	550	...	2780	1160

Source: Ref 7

resistivity is about the same. HE alloy is weakly magnetic.

HF alloy (20Cr-10Ni) is the cast version of 18-8 stainless steel, which is widely used for its outstanding resistance to corrosion. HF alloy is suitable for use at temperatures up to 870 °C (1600 °F). When this alloy is used for applications requiring resistance to oxidation at elevated temperatures, it is not necessary to keep the carbon content at the low level specified for corrosion-resistant castings. Molybdenum, tungsten, niobium, and titanium are sometimes added to the basic HF composition to improve elevated-temperature strength.

In the as-cast condition, HF alloy has an austenitic matrix that contains interdendritic eutectic carbides and, occasionally, a lamellar constituent presumed to consist of alternating platelets of austenite and carbide or carbonitride. Exposure at service temperatures usually promotes precipitation of finely dispersed carbides, which increases room-temperature strength and causes some loss of ductility. If improperly balanced, as-cast HF may be partly ferritic. HF is susceptible to embrittlement due to σ-phase formation after long exposure at 760 to 815 °C (1400 to 1500 °F).

Table 11 Thermal conductivity and mean coefficient of linear thermal expansion of heat-resistant cast steels at various temperatures

Alloy	Mean coefficient of linear thermal expansion for a temperature change				Thermal conductivity, W/m · K, at:		
	From 21 to 540 °C (700 to 1000 °F)		From 21 to 1090 °C (70 to 2000 °F)				
	mm/mm/°C × 10^{-6}	in./in./°F × 10^{-6}	mm/mm/°C × 10^{-6}	in./in./°F × 10^{-6}	100 °C (212 °F)	540 °C (1000 °F)	1090 °C (2000 °F)
HA	12.8	7.1	...	...	26.0	27.2	...
HC	11.3	6.3	13.9	7.7	21.8	31.0	41.9
HD	13.9	7.7	16.6	9.2	21.8	31.0	41.9
HE	17.3	9.6	20.0	11.1	14.7	21.5	31.5
HF	17.8	9.9	19.3	10.7	14.4	21.3	...
HH (type I)(a)	17.1	9.5	19.3	10.7	14.2	20.8	30.3
HH (type II)(a)	17.1	9.5	19.3	10.7	14.2	20.8	30.3
HI	17.8	9.9	19.4	10.8	14.2	20.8	30.3
HK	16.9	9.4	18.7	10.4	13.7	20.4	32.2
HL	16.6	9.2	18.2	10.1	14.2	21.1	33.4
HN	16.7	9.3	18.4	10.2	13.0	19.0	29.4
HP	16.6	9.2	19.1	10.6	13.0	19.0	29.4
HT	15.8	8.8	18.0	10.0	12.1	18.7	28.2
HU	15.8	8.8	17.5	9.7	12.1	18.7	28.2
HW	14.2	7.9	16.7	9.3	12.5	19.2	29.4
HX	14.0	7.8	17.1	9.5	12.5	19.2	29.4

(a) Type I and II specified per ASTM A 447

HH alloy (26Cr-12Ni) is basically austenitic and holds considerable carbon in solid solution, but carbides, ferrite (soft, ductile, and magnetic), and σ-phase (hard, brittle, and nonmagnetic) may also be present in the microstructure. The amounts of the various structural constituents present depend on composition and thermal history. In fact, two distinct grades of material can

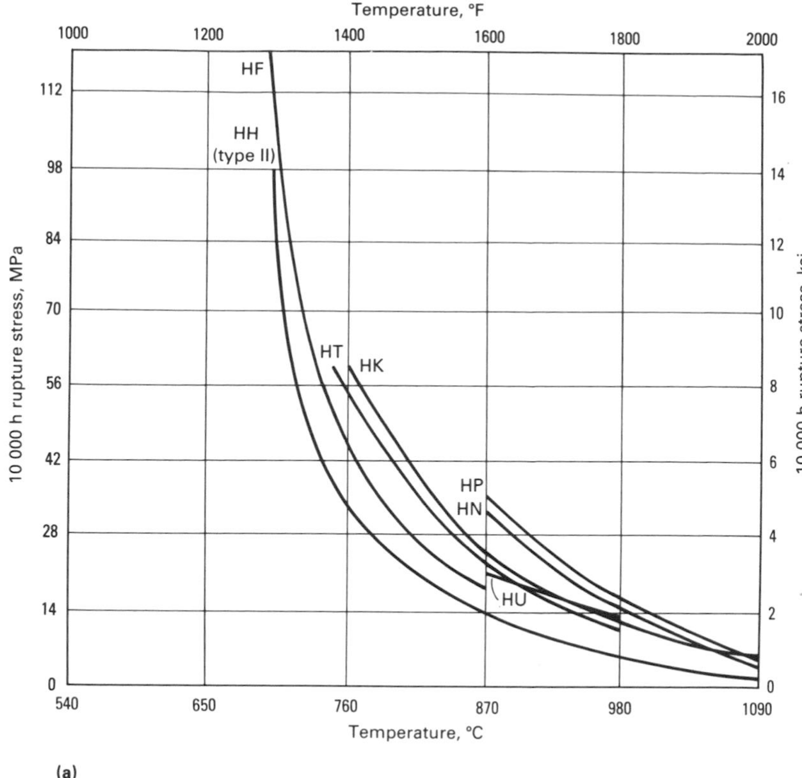

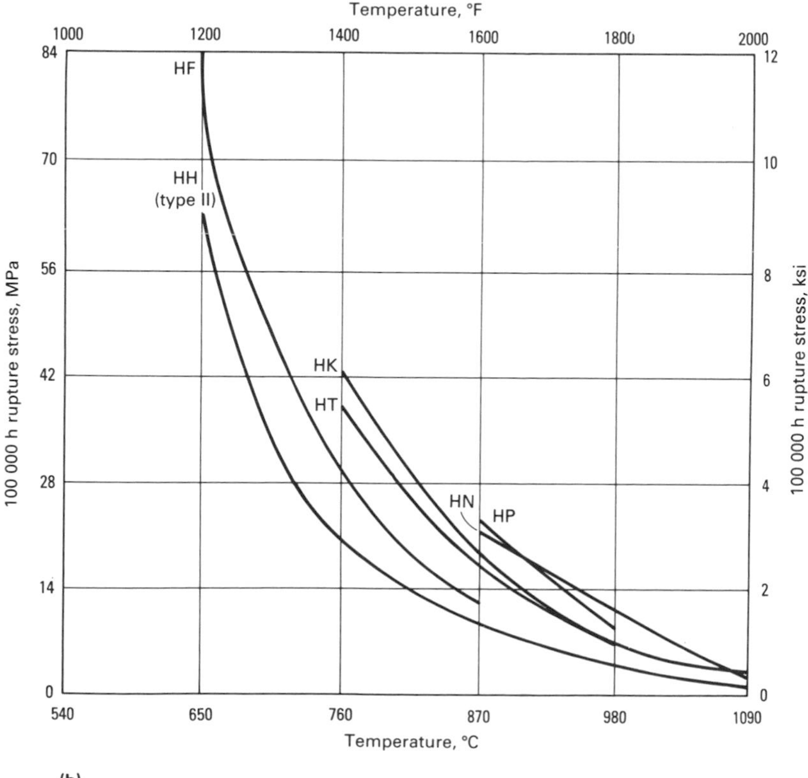

Fig. 7 Stress-rupture properties of several heat-resistant alloy castings. (a) 10,000 h rupture stress. (b) 100,000 h rupture stress. Source: Ref 5

be obtained within the stated chemical compositional range of the type alloy HH. These grades are defined as type I (partially ferritic) and type II (wholly austenitic) in ASTM A 447.

The partially ferritic (type I) alloy HH is adapted to operating conditions that are subject to changes in temperature level and applied stress. A plastic extension in the weaker, ductile ferrite

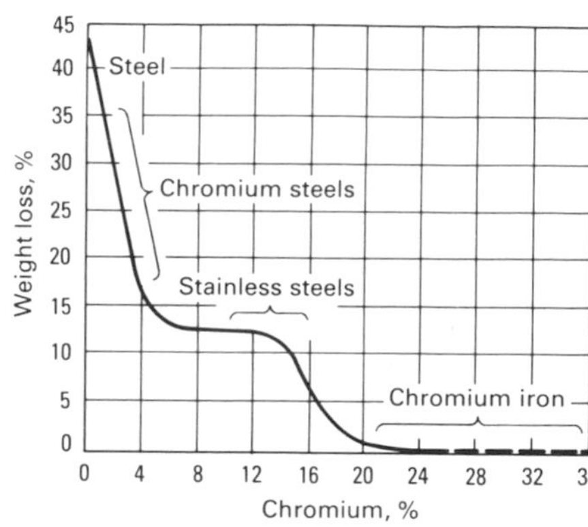

Fig. 8 Effect of chromium on oxidation resistance of cast steels. Specimens (13 mm, or 0.5 in., cubes) were exposed for 48 h at 1000 °C (1830 °F). Source: Ref 8

under changing load tends to occur more readily than in the stronger austenitic phase, thereby reducing unit stresses and stress concentrations and permitting rapid adjustment to suddenly applied overloads without cracking. Near 870 °C (1600 °F), the partially ferritic alloys tend to embrittle from the development of σ-phase, while close to 760 °C (1400 °F), carbide precipitation may cause comparable loss of ductility. Such possible embrittlement suggests that 930 to 1100 °C (1700 to 2000 °F) is the best service temperature range, but this is not critical for steady temperature conditions in the absence of unusual thermal or mechanical stresses.

To achieve maximum strength at elevated temperatures, the HH alloy must be wholly austenitic. Where load and temperature conditions are comparatively constant, the wholly austenitic (type II) alloy HH provides the highest creep strength and permits the use of maximum design stress. The stable austenitic alloy is also favored for cyclic temperature service that might induce σ-phase formation in the partially ferritic type. When HH alloy is heated to between 650 and 870 °C (1200 and 1600 °F), a loss in ductility may be produced by either of two changes within the alloy: precipitation of carbides or transformation of ferrite to σ-phase. When the composition is balanced so that the structure is wholly austenitic, only carbide precipitation normally occurs. In partly ferritic alloys, both carbides and σ-phase may form.

The wholly austenitic (type II) HH alloy is used extensively in high-temperature applications because of its combination of relatively high strength and oxidation resistance at temperatures up to 1100 °C (2000 °F). The HH alloy (type I or II) is seldom used for carburizing applications because of embrittlement from carbon absorption. High silicon content (over 1.5%) will fortify the alloy against carburization under mild conditions but will promote ferrite formation and possible σ-embrittlement.

For the wholly austenitic (type II) HH alloy, composition balance is critical in achieving the desired austenitic microstructure. An imbalance of higher levels of ferrite-promoting elements compared to levels of austenite-promoting ele-

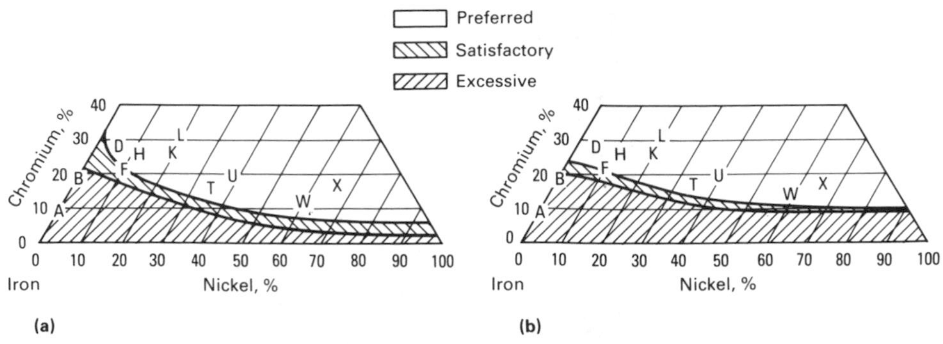

Fig. 9 Corrosion behavior of H-type (heat-resistant) alloy castings in air (a) and in oxidizing flue gases containing 5 grains of sulfur per 2.8 m^3 (100 ft^3) of gas (b). Letters represent the nickel content of the alloy with A denoting the lowest nickel content and X the highest. Source: Ref 8

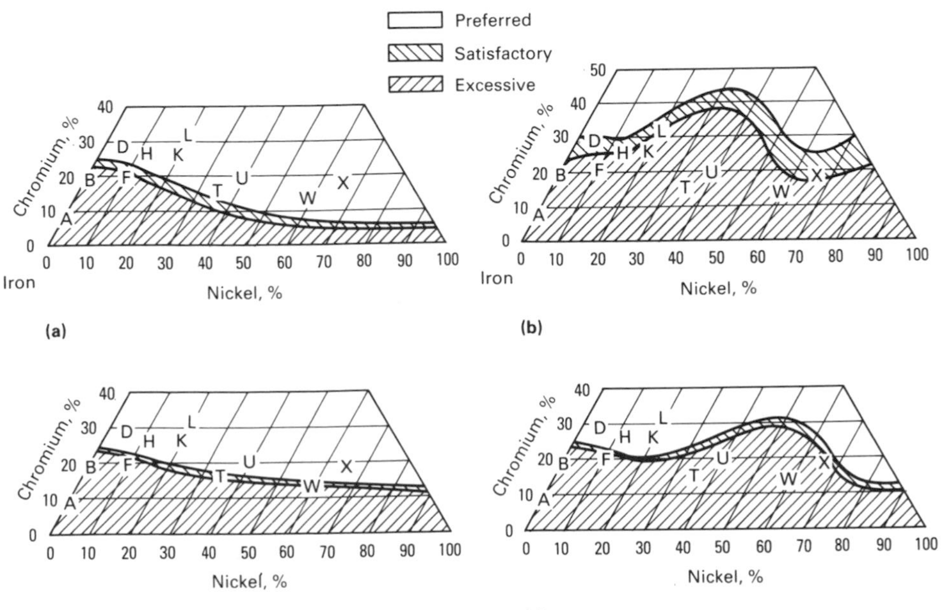

Fig. 10 Corrosion behavior of H-type alloys in 100 h tests at 980 °C (1800 °F) in reducing sulfur-bearing gases. (a) Gas contained 5 grains of sulfur per 2.8 m^3 (100 ft^3) of gas. (b) Gas contained 300 grains of sulfur per 2.8 m^3 (100 ft^3) of gas. (c) Gas contained 100 grains of sulfur per 2.8 m^3 (100 ft^3) of gas; test at constant temperature. (d) Same sulfur content as gas in (c), but cooled to 150 °C (300 °F) each 12 h. Source: Ref 8

Table 12 Corrosion resistance of heat-resistant cast steels at 980 °C (1800 °F) in 100 h tests in various atmospheres

	Corrosion rating(a) in indicated atmosphere					
Alloy	Air	Oxidizing flue gas(b)	Reducing flue gas(b)	Reducing flue gas(c)	Reducing flue gas (constant temperature)(d)	Reducing flue gas cooled to 150 °C (300 °F) every 12 h(d)
HA	U	U	U	U	U	U
HC	G	G	G	S	G	G
HD	G	G	G	S	G	G
HE	G	G	G	...	G	...
HF	S	G	S	U	S	S
HH	G	G	G	S	G	G
HI	G	G	G	S	G	G
HK	G	G	G	U	G	G
HL	G	G	G	S	G	G
HN	G	G	G	U	S	S
HP	G	G	G	G	G	...
HT	G	G	G	U	S	U
HU	G	G	G	U	S	U
HW	G	G	G	U	U	U
HX	G	G	G	S	G	U

(a) G, good (corrosion rate r <1.27 mm/yr, or 50 mils/yr); S, satisfactory (r <2.54 mm/yr, or 100 mils/yr); U, unsatisfactory (r >2.54 mm/yr, or 100 mils/yr). (b) Contained 2 g of sulfur/m^3 (5 grains S/100 ft^3). (c) Contained 120 g/S/m^3 (300 grains S/100 ft^3). (d) Contained 40 g/S/m^3 (100 grains S/100 ft^3)

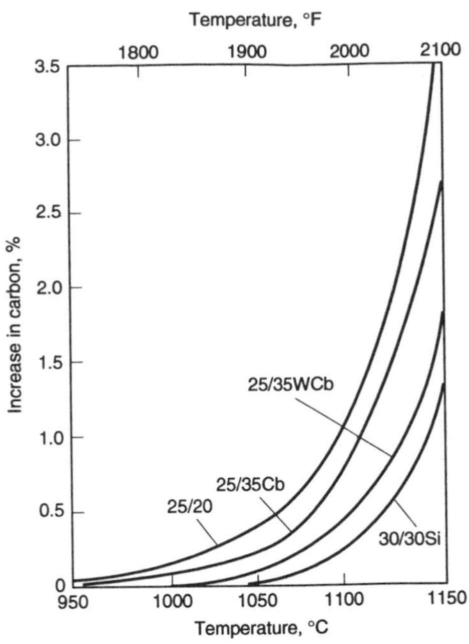

Fig. 11 Carburization resistance of HK (25Cr-20Ni) and several HP alloys (Cr-Ni) as a function of temperature. Carburized at temperature for 260 h in granular carbon. Source: Ref 8

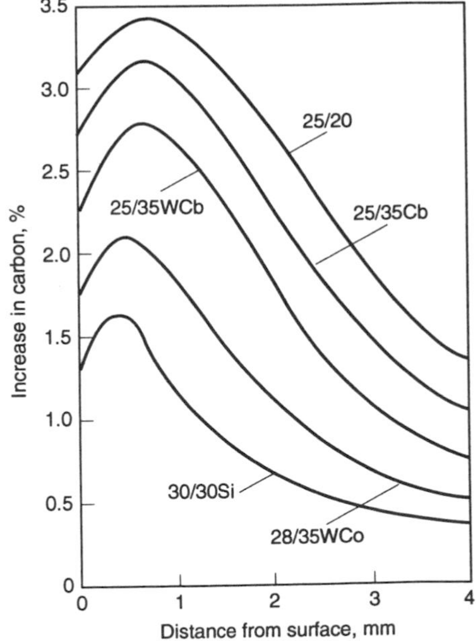

Fig. 12 Carbon concentration profiles for HK (25Cr-20Ni) and several HP alloys (Cr-Ni) carburized at 1100 °C (2010 °F) for 520 h in granular carbon. Source: Ref 8

ments may result in substantial amounts of ferrite, which improves ductility but decreases strength at high temperatures. If a balance is maintained between ferrite-promoting elements (such as chromium and silicon) and austenite-promoting elements (such as nickel, carbon, and nitrogen), the desired austenitic structure can be obtained. In commercial HH alloy castings with the usual carbon, nitrogen, manganese, and silicon contents, the ratio of chromium to nickel

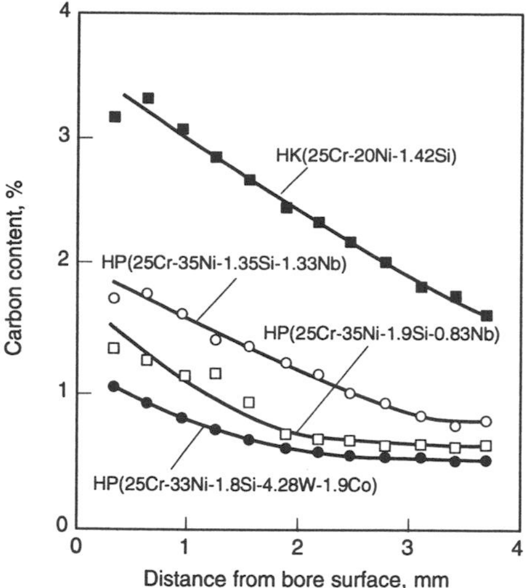

Fig. 13 Carbon concentration profiles of HK and HP alloys exposed at 1050 °C (1920 °F) for 1200 h in $3CH_4$-$20CO$-$40H_2$-$37N_2$. Source: Ref 9

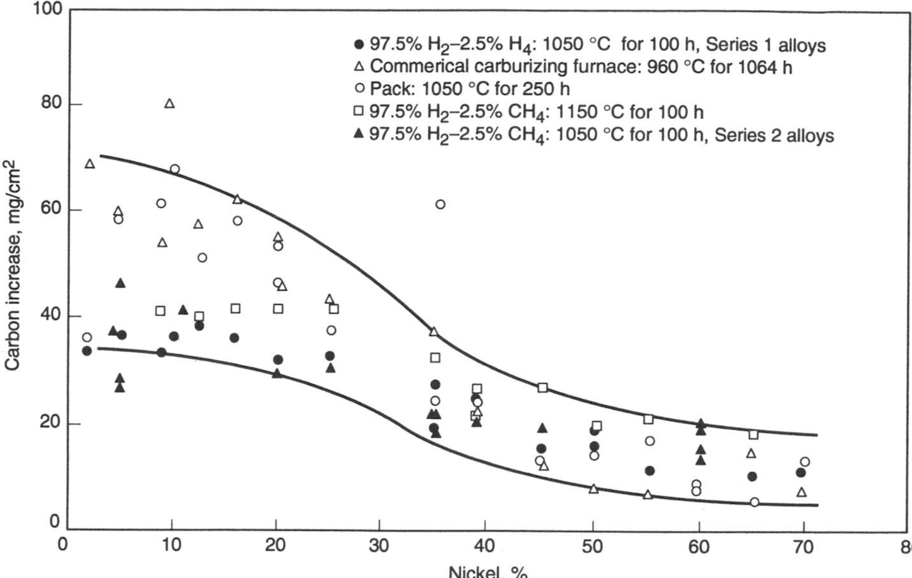

Fig. 15 Effect of nickel on the carburization resistance of Fe-Ni-Cr alloys. Source: Ref 11

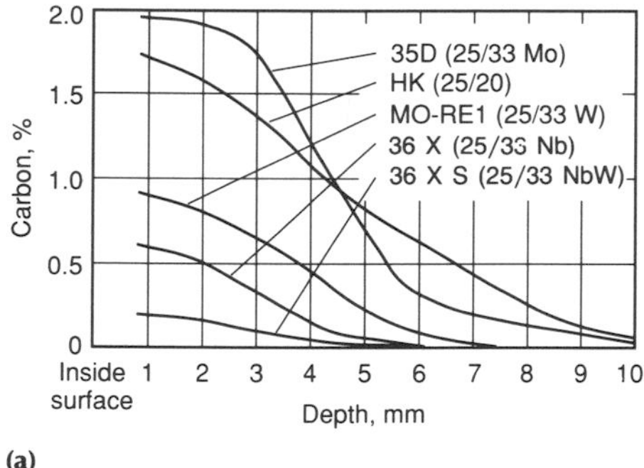

(a)

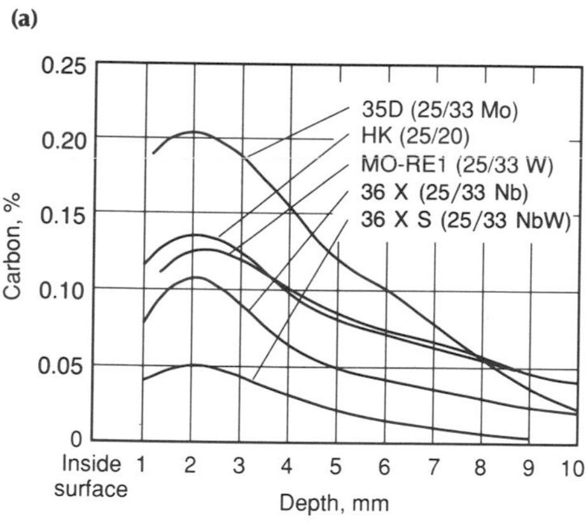

(b)

Fig. 14 Carbon concentration profiles of several centrifugally cast alloys in (a) the as-cast surface condition and (b) the machined surface condition after one year of field testing in an ethylene furnace. Source: Ref 10

necessary for a stable austenitic structure is expressed by:

$$\frac{\%Cr - 16\ (\%C)}{\%Ni} < 1.7 \qquad \text{(Eq 1)}$$

Silicon and molybdenum have definite effects on the formation of σ-phase. A silicon content in

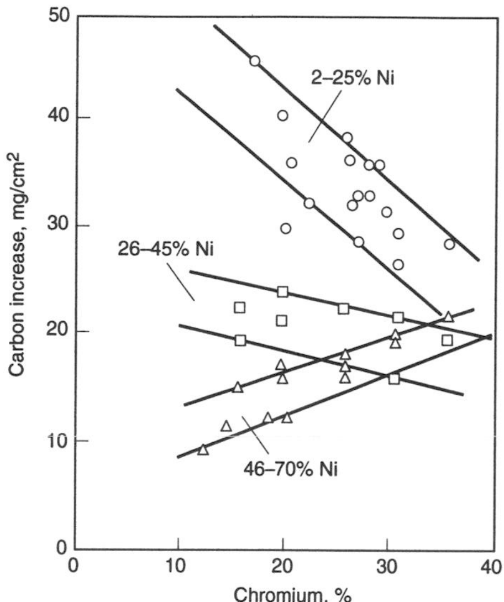

Fig. 16 Effect of chromium on the carburization resistance of Fe-Ni-Cr alloys after gas carburizing ($97.5H_2$-$2.5CH_4$) at 1050 °C (1920 °F) for 100 h. Source: Ref 11

excess of 1% is equivalent to a chromium content three times as great, and any molybdenum content is equivalent to a chromium content four times as great.

Before HH alloy is selected as a material for heat-resistant castings, it is advisable to consider the relationship between chemical composition and operating-temperature range. For castings that are to be exposed continuously at temperatures appreciably above 870 °C (1600 °F), there is little danger of severe embrittlement from either the precipitation of carbide or the formation of σ-phase, and composition should be 0.50% C (max) (0.35 to 0.40% preferred), 10 to 12% Ni, and 24 to 27% Cr. On the other hand, castings to be used at temperatures from 650 to 870 °C (1200 to 1600 °F) should have compositions of 0.40% C (max), 11 to 14% Ni, and 23 to 27% Cr. For applications involving either of these temperature ranges, that is, 650 to 870 °C (1200 to 1600 °F) or appreciably above 870 °C (1600 °F), composition should be balanced to provide an austenitic structure. For service from 650 to 870 °C (1200 to 1600 °F), for example, a combination of 11% Ni and 27% Cr is likely to produce σ-phase and its associated embrittlement, which occurs most rapidly around 870 °C (1600 °F). It is preferable, therefore, to avoid using the maximum chromium content with the minimum nickel content.

Short-time tensile testing of fully austenitic HH alloys shows that tensile strength and elongation depend on carbon and nitrogen contents. For maximum creep strength, HH alloy should be fully austenitic in structure. In the design of load-carrying castings, data concerning creep stresses should be used with an understanding of the limitations of such data. An extrapolated limiting creep stress for 1% elongation in 10,000 h cannot necessarily be sustained for that length of time without structural damage. Stress-rupture testing is a valuable adjunct to creep testing and a useful aid in selecting section sizes to obtain appropriate levels of design stress.

Because HH alloys of wholly austenitic structure have greater strength at high temperatures than partly ferritic alloys of similar composition, measurement of ferrite content is recommended. Although a ratio calculated from Eq 1 that is less than 1.7 indicates wholly austenitic material, ratios greater than 1.7 do not constitute quantitative indications of ferrite content. It is possible, however, to measure ferrite content by magnetic analysis after quenching from about 1100 °C (2000 °F). The magnetic permeability of HH alloys increases with ferrite content. This measurement of magnetic permeability, preferably after

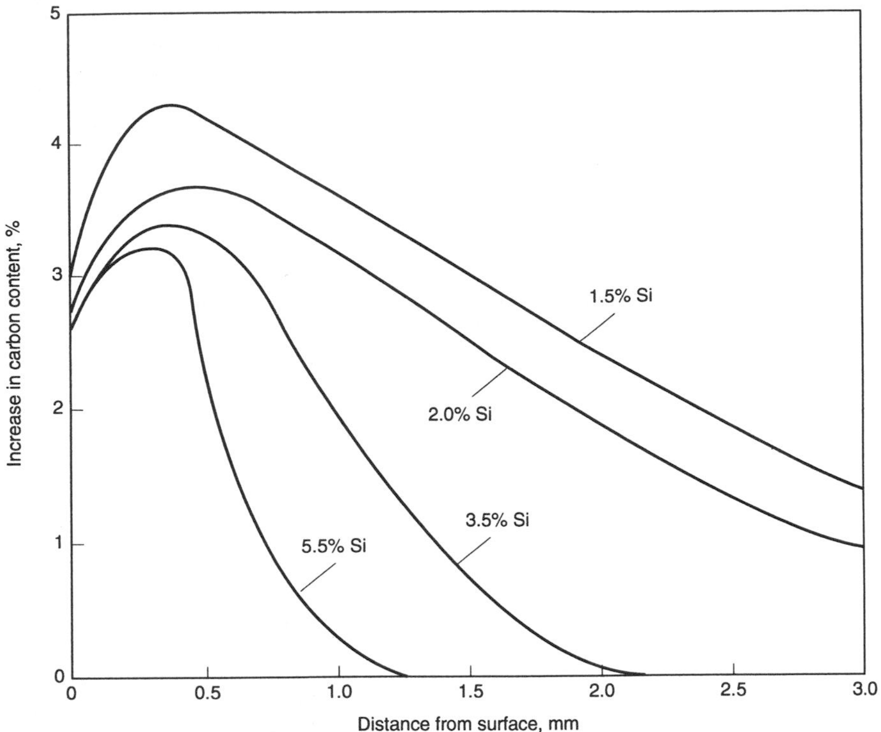

Fig. 17 Effect of silicon on the carburization resistance of HK-40 steel tested at 1100 °C (2010 °F) for 520 h in carbon granulate. Source: Ref 13

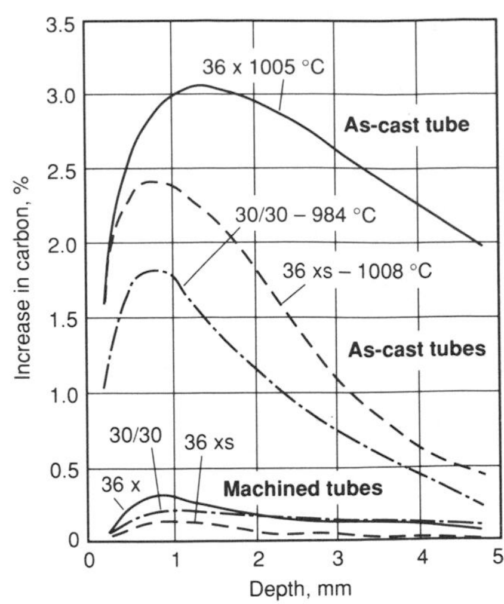

Fig. 18 Comparative carburization resistance of as-cast and machined tubes of alloys 30Cr-30Ni, 36X (25Cr-35Ni-1.5Nb), and 36XS (36X + W) after 3 years of field testing in an ethylene pyrolysis furnace. Source: Ref 10

Table 13 Carburization behavior of cast heat-resistant alloys after 100 h at 1090 °C (2000 °F) in H_2-CH_4-H_2O mixtures

Alloy	Nominal composition, wt %	Weight gain, mg/cm²	
		H_2-8.6CH_4-7H_2O	H_2-12CH_4-10H_2O
HK-40 (1% Si)	Fe-0.43C-0.60Mn-0.96Si-25.4Cr-20.7Ni	25.0	21.8
HK-40 (2% Si)	Fe-0.41C-0.60Mn-1.98Si-25.0Cr-20.7Ni	16.8	10.2
TMA-4750	Fe-0.44C-0.69Mn-1.99Si-24.9Cr-20.8Ni-0.11Ti-0.29Nb-0.3W + REM(a)	2.0	1.0
HP-45	Fe-0.51C-0.54Mn-1.65Si-25.5Cr-36.1Ni	19.0	4.3
TMA-6350	Fe-0.50C-0.70Mn-1.84Si-25.1Cr-38.4Ni-0.13Ti-0.28Nb-0.27W + REM(a)	3.8	2.3

(a) REM denotes rare earth metal additions. Source: Ref 12

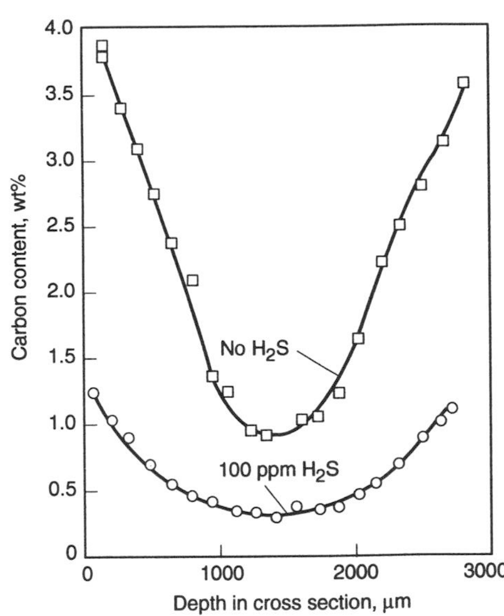

Fig. 19 Carbon concentration profiles for HK-40 alloy after testing at 1000 °C (1830 °F) for 100 h in a carburizing environment with and without injection of 100 ppm H_2S. Source: Ref 14

holding 24 h at 1100 °C (2000 °F) and then quenching in water, can be related to creep strength, which also depends on structure.

HH alloys are often evaluated by measuring percentage elongation in room-temperature tension testing of specimens that have been held 24 h at 760 °C (1400 °F). Such a test may be misleading because there is a natural tendency for engineers to favor compositions that exhibit the greatest elongation after this particular heat treatment. High ductility values are often measured for alloys that have low creep resistance, but conversely, low ductility values do not necessarily connote high creep resistance.

HI alloy (28Cr-15Ni) is similar to HH but contains more nickel and chromium. The higher chromium content makes HI more resistant to oxidation than HH, and the additional nickel serves to maintain good strength at high temperatures. Exhibiting adequate strength, ductility, and corrosion resistance, this alloy has been used extensively for retorts operating with an internal vacuum at a continuous temperature of 1175 °C (2150 °F). It has an essentially austenitic structure that contains carbides and that, depending on the exact composition balance, may or may not contain small amounts of ferrite. Service at 760 to 870 °C (1400 to 1600 °F) results in precipitation of finely dispersed carbides that increase strength and decrease ductility at room temperature. At service temperatures above 1100 °C (2000 °F), however, carbides remain in solution, and room-temperature ductility is not impaired.

HK alloy (26Cr-20Ni) is somewhat similar to wholly austenitic HH alloy in general characteristics and mechanical properties. Although less resistant to oxidizing gases than HC, HE, or HI, HK alloy contains enough chromium to ensure good resistance to corrosion by hot gases, including sulfur-bearing gases, under both oxidizing and reducing conditions. The high nickel content of this alloy helps make it one of the strongest heat-resistant casting alloys at temperatures above 1040 °C (1900 °F). Accordingly, HK alloy castings are widely used for stressed parts in structural applications at temperatures up to 1150 °C (2100 °F). As normally produced, HK alloy is a stable austenitic alloy over its entire range of service temperatures. The as-cast microstructure consists of an austenitic matrix containing relatively large carbides in the form of either scattered islands or networks. After the alloy has been exposed to service temperatures, fine, granular carbides precipitate within the grains of austenite and, if the temperature is high enough, undergo subsequent agglomeration. These fine,

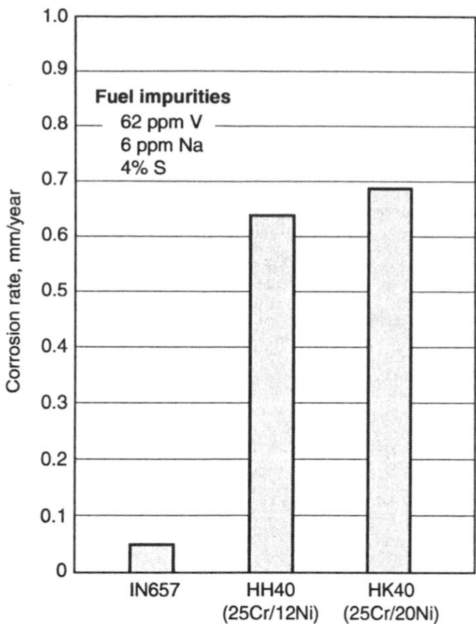

Fig. 20 Results of a field test at 700 °C (1290 °F) for 6 months in a refinery (crude oil) heater comparing IN-657 to HH and HK alloys. Source: Ref 18

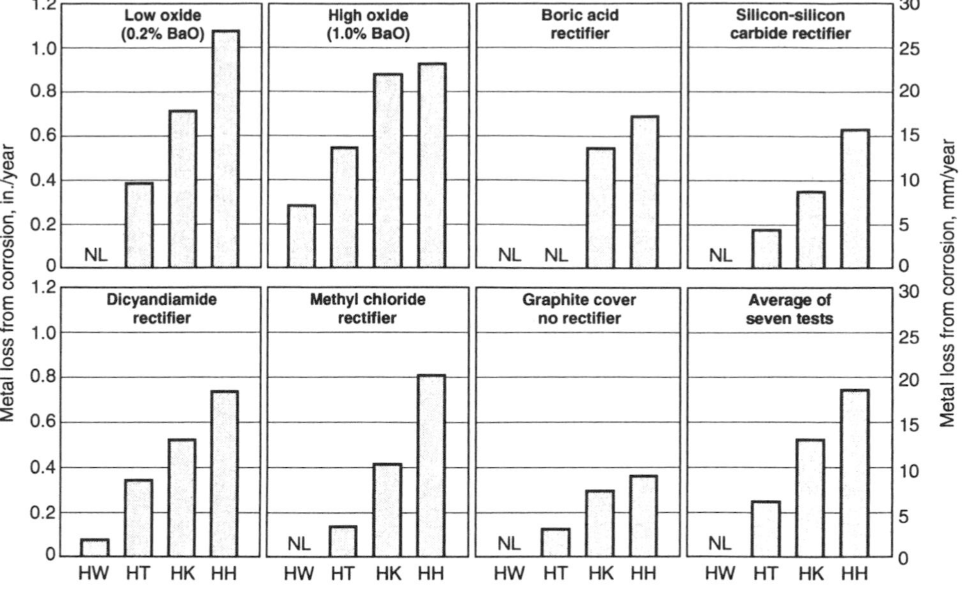

Fig. 21 Corrosion rates in terms of metal loss for four commercial cast Fe-Ni-Cr alloys in a 20NaCl-25KCl-55BaCl$_2$ salt bath under different conditions of rectification at 870 °C (1600 °F) for 60 h. NL represents virtually no weight loss from corrosion. Alloy composition: HW = 12% Cr-60% Ni-0.50% C; HT = 15% Cr-35% Ni-0.45% C; HK = 25% Cr-20% Ni-0.45% C; H-H = 25% Cr-12% Ni-0.45% C. Source: Ref 19

Table 14 Melting points of various oil ash constituents

Compound	Melting point °C	°F
Aluminum oxide, Al_2O_3	1799	3270
Aluminum sulfate, $Al_2(SO_4)_3$	771	1420(a)
Calcium oxide, CaO	2572	4662
Calcium sulfate, $CaSO_4$	1449	2640
Ferric oxide, Fe_2O_3	1566	2850
Ferric sulfate, $Fe_2(SO_4)_3$	480	896(a)
Nickel oxide, NiO	2091	3796
Nickel sulfate, $NiSO_4$	841	1546(a)
Silicon dioxide, SiO_2	1721	3130
Sodium sulfate, Na_2SO_4	885	1625
Sodium bisulfate, $NaHSO_4$	249	480(a)
Sodium pyrosulfate, $Na_2S_2O_7$	399	750(a)
Sodium ferric sulfate, $Na_3Fe(SO_4)_3$	538	1000
Vanadium trioxide, V_2O_3	1971	3580
Vanadium tetroxide, V_2O_4	1971	3580
Vanadium pentoxide, V_2O_5	691	1276
Sodium metavanadate, $Na_2O \cdot V_2O_5(NaVO_3)$	630	1166
Sodium pyrovanadate, $2Na_2O \cdot V_2O_5$	641	1186
Sodium orthovanadate, $3Na_2O \cdot V_2O_5$	849	1560
Sodium vanadylvanadates,		
$Na_2O \cdot V_2O_4 \cdot V_2O_5$	627	1161
$5Na_2O \cdot V_2O_4 \cdot 11V_2O_5$	535	995

(a) Decomposes at a temperature around the melting point. Source: Ref 15

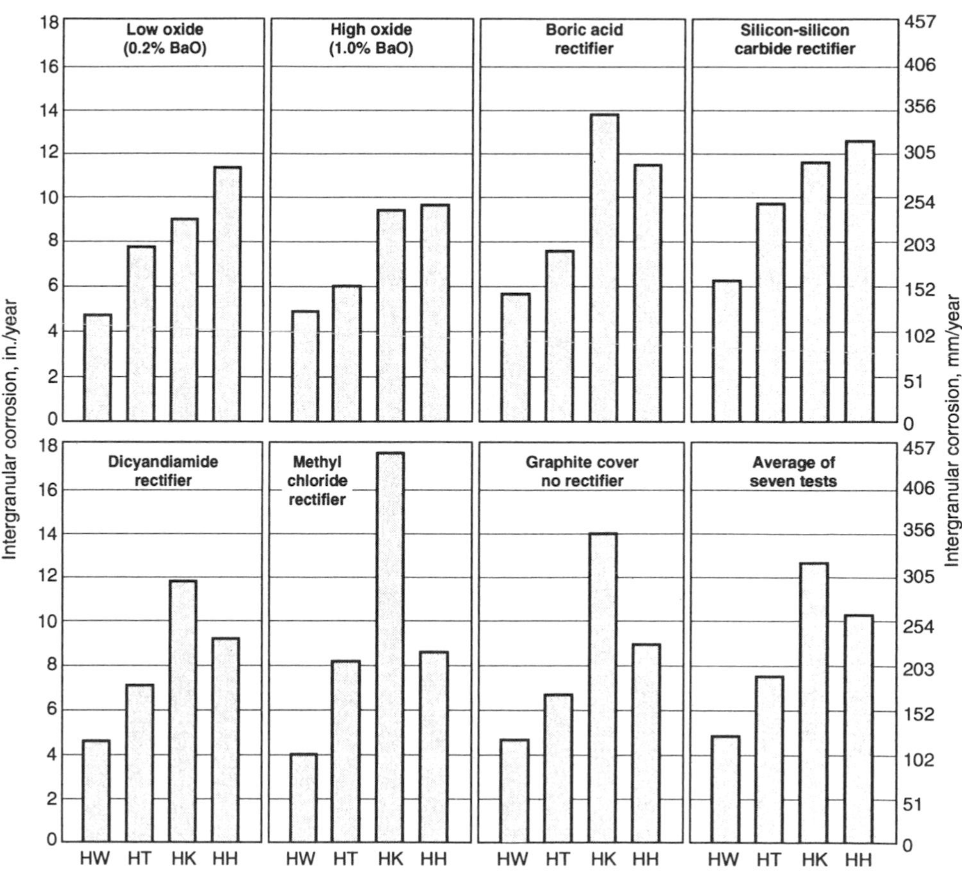

Fig. 22 Corrosion rates in terms of intergranular attack for four commercial cast Fe-Ni-Cr alloys in a 20NaCl-25KCl-55BaCl$_2$ salt bath under different conditions of rectification at 870 °C (1600 °F) for 60 h. Alloy composition: HW = 12% Cr-60% Ni-0.50% C; HT = 15% Cr-35% Ni-0.45% C; HK = 25% Cr-20% Ni-0.45% C; H-H = 25% Cr-12% Ni-0.45% C. Source: Ref 19

dispersed carbides contribute to creep strength. A lamellar constituent that resembles pearlite, but is presumed to be carbide or carbonitride platelets in austenite, is also frequently observed in HK alloy.

Unbalanced compositions are possible within the standard composition range for HK alloy; hence some ferrite may be present in the austenitic matrix. Ferrite will transform to brittle σ-phase if the alloy is held for more than a short time at about 815 °C (1500 °F) with consequent embrittlement upon cooling to room temperature. Direct transformation of austenite to σ-phase can occur in HK alloy in the range of 760 to 870 °C (1400 to 1600 °F), particularly at lower carbon levels (0.20 to 0.30%). The presence of σ-phase

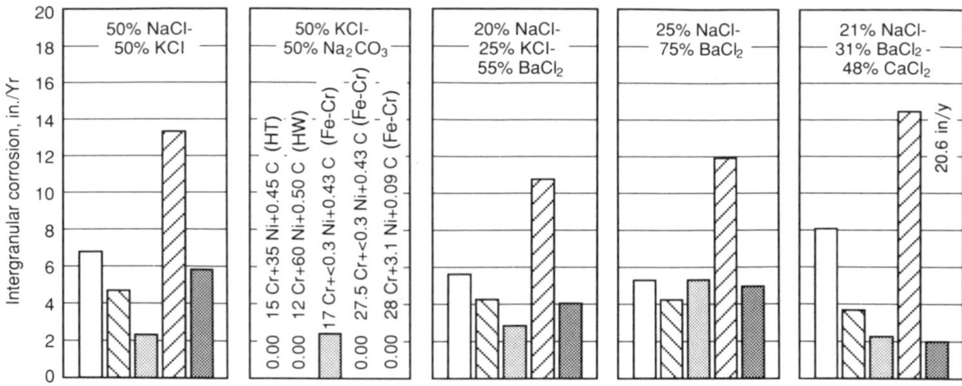

Fig. 23 Comparison of different neutral salt baths for HT, HW, and Fe-Cr alloys at 870 °C (1600 °F) for 60 h. Source: Ref 19

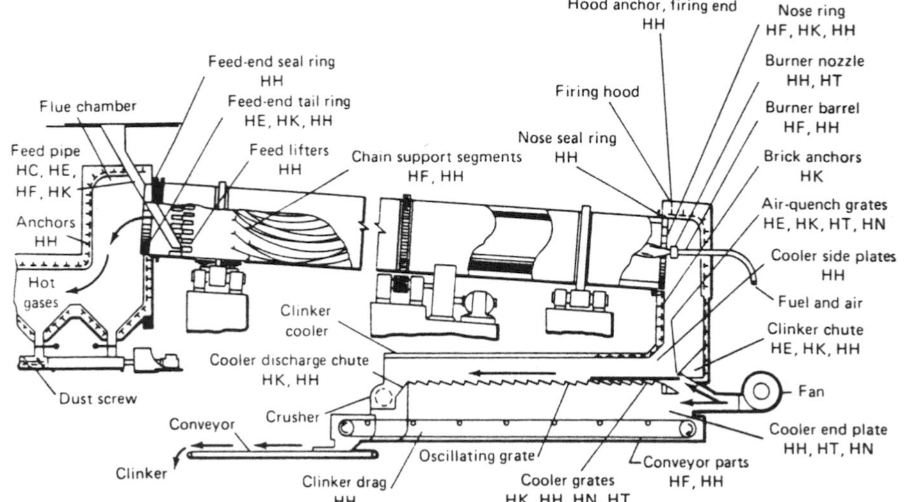

Maximum operating temperature, °C	Maximum operating temperature, °F	Part name	Environmental conditions	Alloys used	Service life, yr
650	1200	Conveyor parts	Severe abrasion and oxidation	HF, HH-II	Indefinite
650	1200	Cooler discharge chute	Severe abrasion and oxidation	HH_2, HK	3 to 5
650	1200	Clinker drag	Severe abrasion and oxidation	HH-II	5 to 10
760	1400	Feed-end seal ring	Some abrasion and oxidation	HH-II	Indefinite
815	1500	Brick anchors	Even temperature	HK	Indefinite
815	1500	Burner barrel	Slight abrasion and oxidation	HF, HH-II	5 to 10
815	1500	Hood, anchor firing end	Even temperature, oxidation	HH-II	Indefinite
815	1500	Clinker chute	Severe abrasion, impact, oxidation	HE, HH-II, HK	Indefinite
815	1500	Air-quench grates	Severe abrasion and oxidation	HE, HK, HN, HT	3 to 7
925	1700	Anchors	Even temperature	HH-II	Indefinite
980	1800	Feed pipe	Moderate abrasion inside feed and dust particles outside, thermal shock, oxidation and sulfur gases	HC, HE, HF, HK	2 to 7
980	1800	Feed-end tail ring	Abrasive dust particles, thermal shock and oxidation	HE, HH-II, HK	10 to 15
980	1800	Feed lifters	Some abrasion, thermal shock, oxidation and sulfur gases	HH-II	5 to 10
980	1800	Chain support segments	Intermittent temperature surges, light abrasion, sulfur gases	HF, HH-II	Indefinite
980	1800	Cooler end plates	Severe abrasion and oxidation	HH-II, HN, HT	1 to 5
980	1800	Cooler grates	Severe abrasion and oxidation	HH-II, HK, HN, HT	1 to 5
980	1800	Cooler side plate	Severe abrasion and oxidation	HH-II	1 to 5
1100	2000	Nose seal ring	Some abrasion, oxidation and sulfur gases	HH-II	3 to 10
1100	2000	Burner nozzle	Some abrasion, oxidation and sulfur gases	HH-II, HT	1 to 3
1200	2200	Nose ring	Extreme abrasion, oxidation and sulfur gases	HF, HH-II, HK	3 to 5

Fig. 24 Heat-resistant alloys used for various components of a cement mill (kilns and related components)

can cause considerable scatter in property values at intermediate temperatures.

The minimum creep rate and average rupture life of HK alloy are strongly influenced by variations in carbon content. Under the same conditions of temperature and load, alloys with higher carbon content have lower creep rates and longer lives than lower-carbon compositions. Room-temperature properties after aging at elevated temperatures are affected also. The higher the carbon, the lower will be the residual ductility. For these reasons, three grades of HK alloys with carbon ranges narrower than the standard HK alloy in Table 1 are recognized: HK-30, HK-40, and HK-50. In these designations, the number indicates the midpoint of a 0.10% C range. HK-40 is widely used for high-temperature processing equipment in the petroleum and petrochemical industries.

HL alloy (30Cr-20Ni) is similar to HK; its higher chromium content gives it greater resistance to corrosion by hot gases, particularly those containing appreciable amounts of sulfur. Because essentially equivalent high-temperature strength can be obtained with either HK or HL, the superior corrosion resistance of HL makes it especially useful for service in which excessive scaling must be avoided. The as-cast and aged microstructures of HL alloy, as well as its physical properties and fabricating characteristics, are similar to those of HK.

Selection of Iron-Nickel-Chromium Heat-Resistant Castings

Iron-nickel-chromium alloys generally have more stable structures than iron-base alloys in which chromium is the predominant alloying element. There is no evidence of an embrittling phase change in iron-nickel-chromium alloys that would impair their ability to withstand prolonged service at elevated temperatures. Experimental data indicate that composition limits are not critical; therefore, the production of castings from these alloys does not require the close composition control necessary for making castings from iron-chromium-nickel alloys.

The following general observations should be considered in the selection of iron-nickel-chromium alloys:

- As nickel content is increased, the ability of the alloy to absorb carbon from a carburizing atmosphere decreases.
- As nickel content is increased, tensile strength at elevated temperatures decreases somewhat, but resistance to thermal shock and thermal fatigue increases.
- As chromium content is increased, resistance to oxidation and to corrosion in chemical environments increases.
- As carbon content is increased, tensile strength at elevated temperatures increases.
- As silicon content is increased, tensile strength at elevated temperatures decreases, but resistance to carburization increases somewhat.

When used for fixtures and trays for heat-treating furnaces, which are subjected to rapid heating and cooling, high-nickel alloys (HP, HT, HU, HW, and HX) have exhibited excellent service life. Because these compositions are not as readily carburized as iron-chromium-nickel alloys, they are used extensively for parts of carburizing furnaces. Because they form an adherent scale

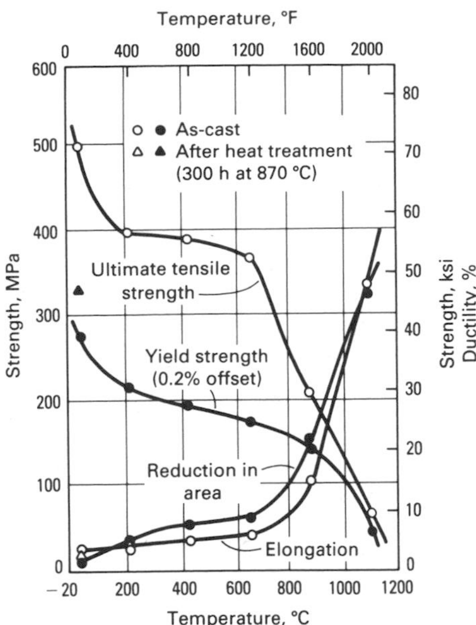

Fig. 25 Tensile properties versus temperature for heat-resistant alloy HP-50WZ

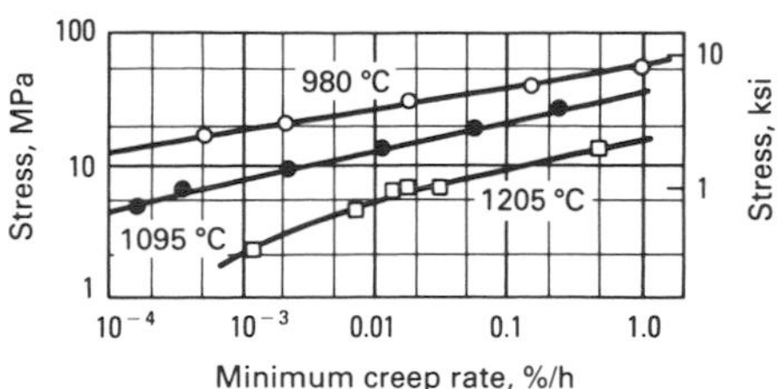

Fig. 26 Minimum creep rate versus stress and temperature for alloy HP-50WZ

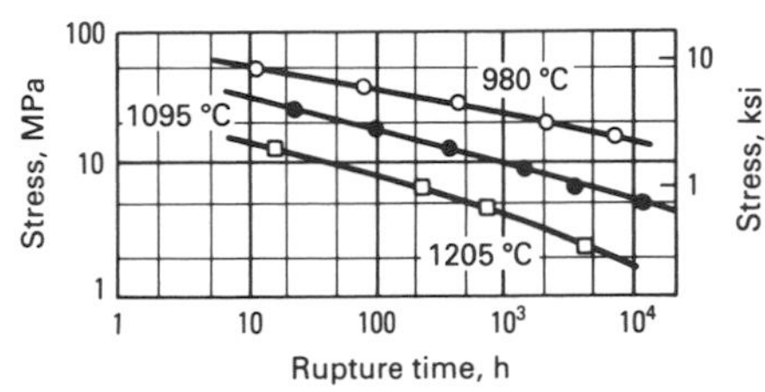

Fig. 27 Rupture time versus stress and temperature for alloy HP-50WZ

Table 15 Results of a field test (uncooled specimens) exposed in the superheater section at 815 °C (1500 °F) in a boiler fired with high vanadium (150-450 ppm) Bunker "C" fuel

Alloy	Corrosion rate mm/yr	Corrosion rate mils/yr
5Cr steel	32–45	1270–1775
406	5.7	224
431	17–23	655–925
446	3.8	149
302B	14–16	533–644
309	5.0	196
321	9–13	346–505
347	6.2	243
310	4.7	187
Incoloy 800	9–12	364–458
Incoloy 804	13–18	495–710
Inconel 600	5.0	196
50Cr-50Ni	3.1	121
HE	4.8	187
HH	12–16	467–645

Source: Ref 16

Table 16 The relative decrease in tube assembly costs resulting from the weight reductions achieved by the use of more highly alloyed, higher strength materials

Minimum sound wall thickness, in.	Minimum sound wall thickness, mm	Alloy	Relative metal cost/lb	Relative tube assembly cost/lb
0.721	18.31	25Cr-20Ni (HK-40)	100	100
0.330	8.38	25Cr-35Ni (HP)	149	78
0.280	7.11	25Cr-35Ni (HP-Nb)	157	75
0.232	5.89	30Cr-45Ni-5W	235	91

Note: Design parameters: temperature, 1650 °F (899 °C); tube inside diameter, 4 in. (101.6 mm); internal pressure, 450 psi (310 MPa). Source: Ref 1

that does not flake off, castings of these alloys are also used in enameling applications in which loose scale would be detrimental.

Four of these high-nickel alloys (HT, HU, HW, and HX) also exhibit good corrosion resistance with molten salts and metal. They have excellent corrosion resistance to tempering and to cyaniding salts and fair resistance to neutral salts with proper control. These alloys exhibit excellent resistance to molten lead, good resistance to molten tin to 345 °C (650 °F), and good resistance to molten cadmium to 410 °C (775 °F). The alloys have poor resistance to antimony, babbitt, soft solder, and similar metals.

HN alloy (25Ni-20Cr) contains enough chromium for good high-temperature corrosion resistance. HN has mechanical properties somewhat similar to those of the much more widely used HT alloy, but it has better ductility (see Fig. 4d and 4f for a comparison of HN and HT tensile properties above 750 °C, or 1400 °F). It is used for highly stressed components in the temperature range of 980 to 1100 °C (1800 to 2000 °F). In several specialized applications (notably, brazing fixtures), it has given satisfactory service at temperatures from 1100 to 1150 °C (2000 to 2100 °F). HN alloy is austenitic at all temperatures: Its composition limits lie well within the stable austenite field. In the as-cast condition, it contains carbide areas, and additional fine carbides precipitate with aging. HN alloy is not susceptible to σ-phase formation, and increases in its carbon content are not especially detrimental to ductility.

HP alloy (35Ni-26Cr) is related to HN and HT alloys but is higher in alloy content. It contains the same amount of chromium but more nickel than HK, and the same amount of nickel but more chromium than HT. This combination of elements makes HP resistant to both oxidizing and carburizing atmospheres at high temperatures. It has stress-rupture properties that are comparable to, or better than, those of HK-40 and HN alloys (Fig. 7).

HP alloy is austenitic at all temperatures and is not susceptible to σ-phase formation. Its microstructure consists of massive primary carbides in an austenitic matrix; in addition, fine secondary carbides are precipitated within the austenite grains upon exposure to elevated temperatures. This precipitation of carbides is responsible for the strengthening between 500 and 750 °C (900 and 1400 °F) shown in Fig. 4(e). This strengthening, which is reduced after long-term exposure at high temperatures, also occurs for the cast stainless steels shown in Fig. 4(c) and (d).

HP-50WZ alloy (Table 1) is a modified version of alloy HP with a narrower carbon content range that also contains tungsten for enhanced elevated-temperature performance. Figures 25 to 27 show elevated-temperature properties for alloy HP-50WZ.

HT alloy (35Ni-17Cr) contains nearly equal amounts of iron and alloying elements. Its high nickel content enables it to resist the thermal shock of rapid heating and cooling. In addition, HT is resistant to high-temperature oxidation and carburization and has good strength at the temperatures ordinarily used for heat treating steel. Except in high-sulfur gases, and provided that limiting creep-stress values are not exceeded, it performs satisfactorily in oxidizing atmospheres at temperatures up to 1150 °C (2100 °F) and in reducing atmospheres at temperatures up to 1100 °C (2000 °F).

HT alloy is widely used for highly stressed parts in general heat-resistant applications. It has an austenitic structure containing carbides in amounts that vary with carbon content and thermal history. In the as-cast condition, it has large carbide areas at interdendritic boundaries, but fine carbides precipitate within the grains after exposure to service temperatures, causing a decrease in room-temperature ductility. Increases in carbon content may decrease the high-temperature ductility of the alloy. A silicon content above about 1.6% provides additional protection against carburization but at some sacrifice in elevated-temperature strength. HT can be made still more resistant to thermal shock by the addition of up to 2% Nb.

HU alloy (39Ni-18Cr) is similar to HT, but its higher chromium and nickel contents give it greater resistance to corrosion by either oxidizing or reducing hot gases, including those that contain sulfur. Its high-temperature strength and resistance to carburization are essentially the same as those of HT alloy; thus its superior corrosion resistance makes it especially well suited for severe service involving high stress and/or rapid thermal cycling, in combination with an aggressive environment.

HW alloy (60Ni-12Cr) is especially well suited for applications in which wide and/or rapid fluctuations in temperature are encountered. In addition, HW exhibits excellent resistance to carburization and high-temperature oxidation. HW alloy has good strength at steel-treating temperatures, although it is not as strong as HT. HW performs satisfactorily at temperatures up to about 1120 °C (2050 °F) in strongly oxidizing atmospheres and up to 1040 °C (1900 °F) in oxidizing or reducing products of combustion provided that sulfur is not present in the gas. The generally adherent nature of its oxide scale makes HW suitable for enameling furnace service, where even small flakes of dislodged scale could ruin the work in process.

HW alloy is widely used for intricate heat-treating fixtures that are quenched with the load and for many other applications (such as furnace retorts and muffles) that involve thermal shock, steep temperature gradients, and high stresses. Its structure is austenitic and contains carbides in amounts that vary with carbon content and thermal history. In the as-cast condition, the microstructure consists of a continuous interdendritic network of elongated eutectic carbides. Upon prolonged exposure at service temperatures, the austenitic matrix becomes uniformly peppered with small carbide particles except in the immediate vicinity of eutectic carbides. This change in structure is accompanied by an increase in room-temperature strength, but there is no change in ductility.

HX alloy (66Ni-17Cr) is similar to HW but contains more nickel and chromium. Its higher chromium content gives it substantially better resistance to corrosion by hot gases (even sulfur-bearing gases), which permits it to be used in severe service applications at temperatures up to 1150 °C (2100 °F). However, it has been reported that HX alloy decarburizes rapidly at temperatures from 1100 to 1150 °C (2000 to 2100 °F). High-temperature strength, resistance to thermal fatigue, and resistance to carburization are essentially the same as for HW. Hence HX is suitable for the same general applications in which its corrosion microstructures, as well as its mechanical properties and fabricating characteristics, are similar to those of HW.

ACKNOWLEDGMENTS

The information in this article is largely taken from:

- Metallurgy and Properties of Cast Stainless Steels, *ASM Specialty Handbook: Stainless Steels,* J.R. Davis, Ed., ASM International, 1994, p 66–88
- Heat-Resistant High Alloy Steels, *Steel Casting Handbook,* 6th ed., M. Blair and T.L. Stevens, Ed., Steel Founders' Society of America and ASM International, 1995, p 22-1 to 22-13

REFERENCES

1. Heat-Resistant High Alloy Steels, *Steel Castings Handbook,* 6th ed., M. Blair and T.L. Stevens, Ed., Steel Founders' Society of America and ASM International, 1995, p 22-1 to 22-13
2. "Investigation of Strengthening Mechanisms and Surface Protection of Cast Alloys Above 2000 °F," Project No. 49, Alloy Casting Institute, 1966
3. High Alloy Data Sheet, Heat Series, *Steel Castings Handbook Supplement 9,* Steel Founders' Society of America
4. "IN-519 Cast Chromium-Nickel-Niobium Heat-Resisting Steel," INCO Databook, 1976
5. "Heat and Corrosion-Resistant Castings," The International Nickel Company, 1978
6. G.Y. Lai, Heat-Resistant Materials for Furnace Parts, Trays, and Fixtures, *Heat Treating,* Vol 4, *ASM Handbook,* ASM International, 1991, p 510–518
7. C.M. Schillmoller, "HP-Modified Furnace Tubes for Steam Reformers and Steam Crackers," NiDI Technical Series No. 10,058, Nickel Development Institute, April 1992
8. C.M. Schillmoller, *Chem. Eng., 6,* Jan 1986, p 87
9. D.J. Hall, M.K. Hossain, and J.J. Jones, *Mater. Perform.,* Jan 1985, p 25
10. J.A. Thuillier, *Mater. Perform.,* Nov 1976, p 9
11. C. Steel and W. Engel, *AFS Int. Cast Metals J.,* Sept 1981, p 28
12. I.Y. Khandros, R.G. Bayer, and C.A. Smith, Paper No. 10, presented at Corrosion/84, National Association of Corrosion Engineers, Houston
13. U. Van den Bruck and C.M. Schillmoller, Paper No. 23, presented at Corrosion/85, National Association of Corrosion Engineers, Houston
14. J.F. Norton and J. Barnes, in *Corrosion in Fossil Fuel Systems,* I.G. Wright, Ed., The Electrochemical Society, 1983, p 277
15. Fuel-Ash Effects and Boiler Design and Operation, *Steam—Its Generation and Use,* Babcock and Wilcox, 1972, p 15–21
16. D.W. McDowell, Jr. and J.R. Mihalisin, Paper No. 60-WA-260, presented at ASME Winter Annual Meeting, 27 Nov to 2 Dec 1960, New York
17. B.F. Spafford, in *UK Corrosion '83,* Conference Proceedings, Birmingham, UK, 15–17 Nov 1982, Institution of Corrosion Science & Technology, 1982, p 67
18. G.L. Swales and D.M. Ward, Paper No. 126, presented at Corrosion/79, National Association of Corrosion Engineers, Houston
19. J.H. Jackson and M.H. LaChance, *Trans. ASM,* Vol 46, 1954, p 157

Properties of Superalloys

Metallurgy, Processing, and Properties of Superalloys

SUPERALLOYS are a group of nickel-, iron-nickel-, and cobalt-base materials that are used at temperatures of 540 °C (1000 °F) and above. A noteworthy feature of nickel-base alloys is their use in load-bearing applications at temperatures in excess of 80% of their incipient melting temperatures (0.85 T_m), a fraction that is higher than for any other class of engineering alloys. Superalloys exhibit some combination of high strength at temperature; resistance to environmental attack (including nitridation, carbonization, oxidation, and sulfidation); excellent creep resistance, stress-rupture strength, toughness, and metallurgical stability; useful thermal expansion characteristics; and resistance to thermal fatigue and corrosion.

Superalloys were initially developed for use in aircraft piston engine turbosuperchargers, and their development over the last 60 years has been paced by the demands of advancing gas turbine engine technology. The vast majority of use by tonnage of nickel-base superalloys is found in turbines, both for aerospace applications and for land-based power generation. These applications require a material with high strength, good creep and fatigue resistance, good corrosion resistance, and the ability to be operated continuously at elevated temperatures.

Nickel-base superalloys are used primarily in turbine blades (called "buckets" in land-based power turbines), turbine disks, burner cans, and vanes. The operating temperatures of these components range from the relatively mild temperature of 150 °C (300 °F) up to almost 1500 °C (2730 °F). Additionally, several components experience large temperature gradients; for example, turbine disks range from 150 °C (300 °F) at the center to 550 °C (1020 °F) at the rim where the blades are attached. In addition to the high temperatures they must endure, the blades are also subject to an extremely corrosive environment—namely, the products of combustion. The primary loading, which results mainly from centripetal acceleration of the rotating blades and disk, in conjunction with the high temperature, leads to creep deformation. Finally, fatigue cycles result from each engine startup and shutdown as the load changes from zero to maximum and back to zero. For some military engines, thrust settings are varied so greatly that they can also be considered as a fatigue cycle.

Turbine components thus experience thermomechanical loading and fatigue as well as creep-fatigue interactions. The good combination of strength and toughness, as well as an unusual yield behavior (in which the yield strength increases with increased temperature up to about 700 °C, or 1290 °F), continues to make nickel-base superalloys the material of choice for high-performance, high-temperature applications.

Other uses for nickel-, iron-nickel-, and cobalt-base alloys include:

- Heat treatment fixtures and furnace parts
- Heating elements used for domestic or industrial applications
- Nuclear and fossil fuel power plant components (both rotating parts and structural components such as piping and pump hardware)
- Pyrochemical and pyrometallurgical processing equipment, particularly where high temperature is combined with high stress and/or an aggressive environment
- Environmental remediation equipment such as that used to process stack gases, recuperators used to recover waste heat, and incinerators or other hardware used for high-temperature destruction of waste

Many technical considerations, such as formability, strength, creep resistance, fatigue strength, and surface stability, must be evaluated when selecting a superalloy for any of the applications identified above. Unfortunately, those compositional and microstructural variables that benefit one property may result in undesirable performance in another area. For example, fine grain size is desirable for low-temperature tensile strength, fatigue strength, and high-temperature formabil-

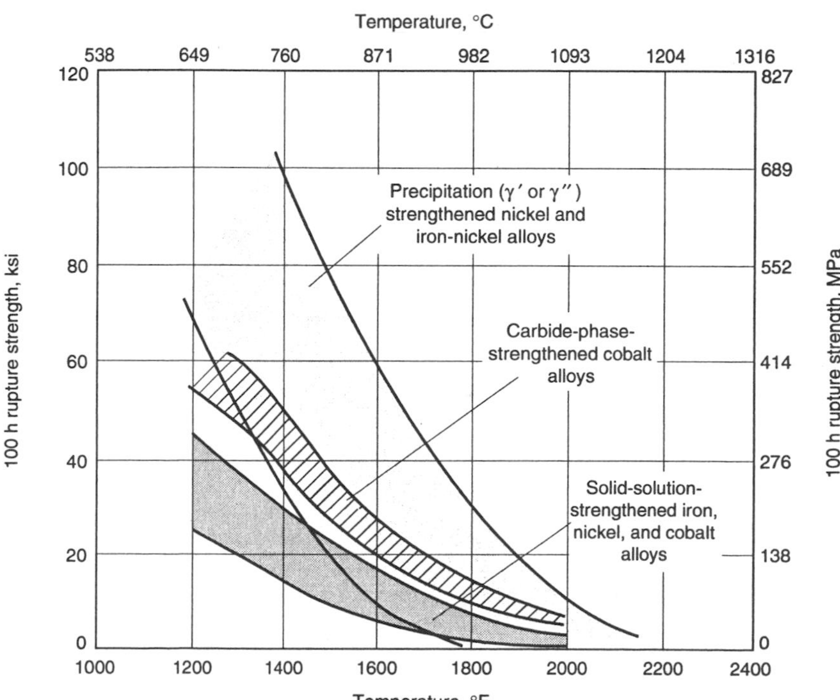

Fig. 1 Stress-rupture characteristics of wrought superalloys

Table 1 Nominal compositions of wrought superalloys

Alloy	UNS Number	Composition, %										
		Cr	Ni	Co	Mo	W	Nb	Ti	Al	Fe	C	Other
Solid-solution alloys												
Iron-nickel-base												
Alloy N-155 (Multimet)	R30155	21.0	20.0	20.0	3.00	2.5	1.0	...	...	32.2	0.15	0.15 N, 0.2 La, 0.02 Zr
Haynes 556	R30556	22.0	21.0	20.0	3.0	2.5	0.1	...	0.3	29.0	0.10	0.50 Ta, 0.02 La, 0.002 Zr
19-9 DL	S63198	19.0	9.0	...	1.25	1.25	0.4	0.3	...	66.8	0.30	1.10 Mn, 0.60 Si
Incoloy 800	N08800	21.0	32.5	...	...	...	...	0.38	0.38	45.7	0.05	...
Incoloy 800H	N08810	21.0	33.0	...	...	...	...	...	...	45.8	0.08	...
Incoloy 800HT	N08811	21.0	32.5	...	...	...	...	0.4	0.4	46.0	0.08	0.8 Mn, 0.5 Si, 0.4 Cu
Incoloy 801	N08801	20.5	32.0	...	...	...	...	1.13	...	46.3	0.05	...
Incoloy 802	...	21.0	32.5	...	...	...	...	0.75	0.58	44.8	0.35	...
Nickel-base												
Haynes 214	...	16.0	76.5	...	...	...	...	...	4.5	3.0	0.03	
Haynes 230	N06230	22.0	55.0	5.0 max	2.0	14.0	...	...	0.35	3.0 max	0.10	0.015 max B, 0.02 La
Inconel 600	N06600	15.5	76.0	...	...	...	...	...	...	8.0	0.08	0.25 Cu
Inconel 601	N06601	23.0	60.5	...	...	...	...	...	1.35	14.1	0.05	0.5 Cu
Inconel 617	N06617	22.0	55.0	12.5	9.0	...	...	...	1.0	...	0.07	...
Inconel 625	N06625	21.5	61.0	...	9.0	...	3.6	0.2	0.2	2.5	0.05	...
RA 333	N06333	25.0	45.0	3.0	3.0	3.0	...	...	...	18.0	0.05	...
Hastelloy B	N10001	1.0 max	63.0	2.5 max	28.0	...	...	...	...	5.0	0.05 max	0.03 V
Hastelloy N	N10003	7.0	72.0	...	16.0	...	...	0.5 max	...	5.0 max	0.06	
Hastelloy S	N06635	15.5	67.0	...	15.5	...	...	...	0.2	1.0	0.02 max	0.02 La
Hastelloy W	N10004	5.0	61.0	2.5 max	24.5	...	...	...	...	5.5	0.12 max	0.6 V
Hastelloy X	N06002	22.0	49.0	1.5 max	9.0	0.6	...	...	2.0	15.8	0.15	...
Hastelloy C-276	N10276	15.5	59.0	...	16.0	3.7	...	...	...	5.0	0.02 max	...
Haynes HR-120	N08120	25.0	37.0	3.0	2.5	2.5	0.7	...	0.1	33.0	0.05	0.7 Mn, 0.6 Si, 0.2 N, 0.004 B
Haynes HR-160	N12160	28.0	37.0	29.0	...	...	...	...	...	2.0	0.05	2.75 Si, 0.5 Mn
Nimonic 75	N06075	19.5	75.0	...	...	...	...	0.4	0.15	2.5	0.12	0.25 max Cu
Nimonic 86	...	25.0	65.0	...	10.0	...	...	...	...	...	0.05	0.03 Ce, 0.015 Mg
Cobalt-base												
Haynes 25 (L605)	R30605	20.0	10.0	50.0	...	15.0	...	...	...	3.0	0.10	1.5 Mn
Haynes 188	R30188	22.0	22.0	37.0	...	14.5	...	...	...	3.0 max	0.10	0.90 La
Alloy S-816	R30816	20.0	20.0	42.0	4.0	4.0	4.0	...	...	4.0	0.38	...
MP35-N	R30035	20.0	35.0	35.0	10.0	...	...	...	...	...	...	...
MP159	R30159	19.0	25.0	36.0	7.0	...	0.6	3.0	0.2	9.0	...	...
Stellite B	N07718	30.0	1.0	61.5	...	4.5	...	...	...	1.0	1.0	...
UMCo-50	...	28.0	...	49.0	...	...	...	...	...	21.0	0.12	...
Precipitation-hardening alloys												
Iron-nickel-base												
A-286	S66286	15.0	26.0	...	1.25	...	...	2.0	0.2	55.2	0.04	0.005 B, 0.3 V
Discaloy	S66220	14.0	26.0	...	3.0	...	...	1.7	0.25	55.0	0.06	...
Incoloy 903	N19903	0.1 max	38.0	15.0	0.1	...	3.0	1.4	0.7	41.0	0.04	...
Pyromet CTX-1	...	0.1 max	37.7	16.0	0.1	...	3.0	1.7	1.0	39.0	0.03	...
Incoloy 907	N19907	...	38.4	13.0	...	...	4.7	1.5	0.03	42.0	0.01	0.15 Si
Incoloy 909	N19909	...	38.0	13.0	...	...	4.7	1.5	0.03	42.0	0.01	0.4 Si
Incoloy 925	N09925	20.5	44.0	...	2.8	...	...	2.1	0.2	29	0.01	1.8 Cu
V-57	...	14.8	27.0	...	1.25	...	...	3.0	0.25	48.6	0.08 max	0.01 B, 0.5 max V
W-545	S66545	13.5	26.0	...	1.5	...	...	2.85	0.2	55.8	0.08 max	0.05 B
Nickel-base												
Astroloy	N13017	15.0	56.5	15.0	5.25	...	...	3.5	4.4	<0.3	0.06	0.03 B, 0.06 Zr
Custom Age 625 PLUS	N07716	21.0	61.0	...	8.0	...	3.4	1.3	0.2	5.0	0.01	...
Haynes 242	...	8.0	62.5	2.5 max	25.0	...	...	...	0.5 max	2.0 max	0.10 max	0.006 max B
Haynes 263	N07263	20.0	52.0	...	6.0	...	...	2.4	0.6	0.7	0.06	0.6 Mn, 0.4 Si, 0.2 Cu
Haynes R-41	N07041	19.0	52.0	11.0	10.0	...	...	3.1	1.5	5.0	0.09	0.5 Si, 0.1 Mn, 0.006 B
Inconel 100	N13100	10.0	60.0	15.0	3.0	...	...	4.7	5.5	<0.6	0.15	1.0 V, 0.06 Zr, 0.015 B
Inconel 102	N06102	15.0	67.0	...	2.9	3.0	2.9	0.5	0.5	7.0	0.06	0.005 B, 0.02 Mg, 0.03 Zr
Incoloy 901	N09901	12.5	42.5	...	6.0	...	...	2.7	...	36.2	0.10 max	...
Inconel 702	N07702	15.5	79.5	...	...	...	...	0.6	3.2	1.0	0.05	0.5 Mn, 0.2 Cu, 0.4 Si
Inconel 706	N09706	16.0	41.5	...	...	...	...	1.75	0.2	37.5	0.03	2.9 (Nb + Ta), 0.15 max Cu
Inconel 718	N07718	19.0	52.5	...	3.0	...	5.1	0.9	0.5	18.5	0.08 max	0.15 max Cu
Inconel 721	N07721	16.0	71.0	...	...	...	...	3.0	...	6.5	0.04	2.2 Mn, 0.1 Cu
Inconel 722	N07722	15.5	75.0	...	...	...	...	2.4	0.7	7.0	0.04	0.5 Mn, 0.2 Cu, 0.4 Si
Inconel 725	N07725	21.0	57.0	...	8.0	...	3.5	1.5	0.35 max	9.0	0.03 max	
Inconel 751	N07751	15.5	72.5	...	...	...	1.0	2.3	1.2	7.0	0.05	0.25 max Cu
Inconel X-750	N07750	15.5	73.0	...	...	...	1.0	2.5	0.7	7.0	0.04	0.25 max Cu
M-252	N07252	19.0	56.5	10.0	10.0	...	...	2.6	1.0	<0.75	0.15	0.005 B
Nimonic 80A	N07080	19.5	73.0	1.0	...	...	...	2.25	1.4	1.5	0.05	0.10 max Cu
Nimonic 90	N07090	19.5	55.5	18.0	...	...	...	2.4	1.4	1.5	0.06	...
Nimonic 95	...	19.5	53.5	18.0	...	...	...	2.9	2.0	5.0 max	0.15 max	+B, +Zr
Nimonic 100	...	11.0	56.0	20.0	5.0	...	...	1.5	5.0	2.0 max	0.30 max	+B, +Zr
Nimonic 105	...	15.0	54.0	20.0	5.0	...	...	1.2	4.7	...	0.08	0.005 B

(continued)

Table 1 (continued)

Alloy	UNS Number	Composition, %										
		Cr	Ni	Co	Mo	W	Nb	Ti	Al	Fe	C	Other
Nickel-base (continued)												+
Nimonic 115	...	15.0	55.0	15.0	4.0	...	...	4.0	5.0	1.0	0.20	0.04 Zr
C-263	N07263	20.0	51.0	20.0	5.9	...	...	2.1	0.45	0.7 max	0.06	...
Pyromet 860	...	13.0	44.0	4.0	6.0	...	...	3.0	1.0	28.9	0.05	0.01 B
Pyromet 31	N07031	22.7	55.5	...	2.0	...	1.1	2.5	1.5	14.5	0.04	0.005 B
Refractaloy 26	...	18.0	38.0	20.0	3.2	...	...	2.6	0.2	16.0	0.03	0.015 B
René 41	N07041	19.0	55.0	11.0	10.0	...	...	3.1	1.5	<0.3	0.09	0.01 B
René 95	...	14.0	61.0	8.0	3.5	3.5	3.5	2.5	3.5	<0.3	0.16	0.01 B, 0.05 Zr
René 100	...	9.5	61.0	15.0	3.0	...	...	4.2	5.5	1.0 max	0.16	0.015 B, 0.06 Zr, 1.0 V
Udimet 500	N07500	19.0	48.0	19.0	4.0	...	...	3.0	3.0	4.0 max	0.08	0.005 B
Udimet 520	...	19.0	57.0	12.0	6.0	1.0	...	3.0	2.0	...	0.08	0.005 B
Udimet 630	...	17.0	50.0	...	3.0	3.0	6.5	1.0	0.7	18.0	0.04	0.004 B
Udimet 700	...	15.0	53.0	18.5	5.0	...	...	3.4	4.3	<1.0	0.07	0.03 B
Udimet 710	...	18.0	55.0	14.8	3.0	1.5	...	5.0	2.5	...	0.07	0.01 B
Unitemp AF2-1DA	N07012	12.0	59.0	10.0	3.0	6.0	...	3.0	4.6	<0.5	0.35	1.5 Ta, 0.015 B, 0.1 Zr
Waspaloy	N07001	19.5	57.0	13.5	4.3	...	...	3.0	1.4	2.0 max	0.07	0.006 B, 0.09 Zr

ity, but creep resistance is usually adversely affected. Similarly, high chromium contents in nickel alloys improve the resistance to oxidation and hot corrosion, but result in lower tensile and creep strengths and promote the formation of σ-phase. Further, the more temperature resistant the alloy, the more likely it is to be segregation prone and, perhaps, brittle, and thus formable only by casting to shape or by using powder processing. For these and other compelling reasons, the interplay between composition, microstructure, consolidation method, mechanical properties, and surface stability is emphasized in this article.

General Characteristics of Superalloys

The high-temperature strength of all superalloys is based on the principle of a stable face-centered cubic (fcc) matrix combined with either precipitation strengthening and/or solid-solution hardening. In general, superalloys have an austenitic (γ-phase) matrix and contain a wide variety of secondary phases. The most common second phases are metal carbides (MC, $M_{23}C_6$, M_6C, and M_7C_3) and γ′, the ordered fcc strengthening phase [$Ni_3(Al,Ti)$] found in iron-nickel- and nickel-base superalloys. In alloys containing niobium or niobium and tantalum, the primary strengthening phase is γ″, a body-centered tetragonal (bct) phase. Cobalt-base superalloys may develop some precipitation strengthening from carbides, but no intermetallic-phase strengthening equal to the γ′ strengthening in nickel-base alloys has been discovered in cobalt-base superalloys.

Other phases, generally undesirable, may be observed due to variations in composition or processing or due to high-temperature exposure. Included in this group are orthorhombic δ-phase (Ni_3Nb), σ-phase, Laves, and the hexagonal close-packed (hcp) η–phase (Ni_3Ti). Nitrides are also commonly observed, and borides may be present in some alloys.

The physical metallurgy of these systems is extremely complex, perhaps more challenging than that of any other alloy system. In addition, as demonstrated in Tables 1 and 2, the compositions of these alloys are complex as well. Table 3 summarizes the functions of elements in superalloys.

Nickel-Base Alloys. Nickel-base high-temperature alloys are basically of three types: solid-solution strengthened, precipitation hardenable, and oxide-dispersion strengthened (ODS). The solid-solution alloys contain little or no aluminum, titanium, or niobium. The precipitation-hardenable alloys contain several percent aluminum and titanium, and a few contain substantial niobium. The ODS alloys contain a small amount of fine oxide particles (0.5 to 1% Y_2O_3) and are produced by powder metallurgy techniques. These alloys are described in the article "Powder Metallurgy Superalloys" in this Volume.

The age-hardenable alloys are strengthened by γ′ precipitation by the addition of aluminum and titanium, by carbide, and by solid-solution alloying. The nature of the γ′ is of primary importance in obtaining optimum high-temperature properties. Compositionally, the aluminum and titanium contents and the aluminum/titanium ratio are very important, as is heat treatment. Increasing the aluminum/titanium ratio improves high-temperature properties. The volume fraction, size, and spacing of γ′ are important parameters to control.

As illustrated in Fig. 1, precipitation-hardening alloys have considerably higher strength values when compared to solid-solution strengthened alloys. For the most demanding of elevated-temperature applications, γ′-strengthened alloys are preferred. Solid-solution strengthened alloys are preferred, when service conditions allow their use, because of their ease of fabrication, especially weldability.

Iron-Nickel-Base Alloys. Several types of iron-nickel-base alloys have been developed. These alloys contain at least 10% Fe, but generally 18% to approximately 55%. The most important iron-nickel-base alloys are those with an austenitic matrix that are strengthened by γ′, such as A-286. Some of these alloys are quite similar to wrought austenitic stainless steels with the addition of the γ′ strengthening agent. Other iron-nickel-base alloys, such as Inconel 718, contain less iron plus additions of niobium and tantalum to obtain strengthening from γ″. Another group of iron-nickel-base alloys contains rather high carbon contents and is strengthened by carbides, nitrides, carbonitrides, and solid-solution strengthening.

Cobalt-Base Alloys. Cobalt-base superalloys are strengthened by solid-solution alloying and carbide precipitation. The grain-boundary carbides inhibit grain-boundary sliding. Unlike the iron-nickel- and nickel-base alloys, no intermetallic phase has been found that will strengthen cobalt-base alloys to the same degree that γ′ or γ″ strengthens the other superalloys. Gamma is not stable at high temperatures in cobalt-base alloys. The carbides in cobalt-base superalloys are the same as those in the other systems and include Cr_7C_3 and $M_{23}C_6$.

Nickel-Base Superalloys

Nickel-base superalloys are the most complex, the most widely used for the hottest parts, and, to many metallurgists, the most interesting of all superalloys. They currently constitute over 50% of the weight of advanced aircraft engines. Their use in cast form extends to the highest homologous temperature of any common alloy system.

The principal characteristics of nickel as an alloy base are the high phase stability of the fcc nickel matrix and the capability to be strengthened by a variety of direct and indirect means. Further, the surface stability of nickel is readily improved by alloying with chromium and/or aluminum. In order to adequately describe mechanical behavior, however, it is first necessary to consider the composition and microstructure of the various classes of nickel alloys.

Chemical Composition

The compositions of many representative wrought and cast nickel-base superalloys are listed in Tables 1 and 2. The nickel-base superalloys discussed below are considered to be complex because they incorporate as many as a dozen elements. In addition, deleterious elements such

Table 2 Nominal compositions of cast superalloys

Alloy designation	Nominal composition, %											
	C	Ni	Cr	Co	Mo	Fe	Al	B	Ti	Ta	W	Zr	Other
Nickel-base													
B-1900	0.1	64	8	10	6	...	6	0.015	1	4(a)	...	0.10	...
CMSX-2	...	66.2	8	4.6	0.6	...	56	...	1	6	8	6	...
Hastelloy X	0.1	50	21	1	9	18	...	...	...	...	1	...	...
Inconel 100	0.18	60.5	10	15	3	...	5.5	0.01	5	...	...	0.06	1 V
Inconel 713C	0.12	74	12.5	...	4.2	...	6	0.012	0.8	1.75	...	0.1	0.9 Nb
Inconel 713LC	0.05	75	12	...	4.5	...	6	0.01	0.6	4	...	0.1	...
Inconel 738	0.17	61.5	16	8.5	1.75	...	3.4	0.01	3.4	...	2.6	0.1	2 Nb
Inconel 792	0.2	60	13	9	2.0	...	3.2	0.02	4.2	...	4	0.1	2 Nb
Inconel 718	0.04	53	19	...	3	18	0.5	...	0.9	...	...	...	0.1 Cu, 5 Nb
X-750	0.04	73	15	...	...	7	0.7	...	2.5	...	...	...	0.25 Cu, 0.9 Nb
M-252	0.15	56	20	10	10	...	1	0.005	2.6	...	...	...	...
MAR-M 200	0.15	59	9	10	...	1	5	0.015	2	...	12.5	0.05	1 Nb(b)
MAR-M 246	0.15	60	9	10	2.5	...	5.5	0.015	1.5	1.5	10	0.05	...
MAR-M 247	0.15	59	8.25	10	0.7	0.5	5.5	0.015	1	3	10	0.05	1.5 Hf
PWA 1480	...	bal	10	5.0	...	...	5.0	...	1.5	12	4.0	...	...
René 41	0.09	55	19	11.0	10.0	...	1.5	0.01	3.1	...	...	...	...
René 77	0.07	58	15	15	4.2	...	4.3	0.015	3.3	...	...	0.04	...
René 80	0.17	60	14	9.5	4	...	3	0.015	5	...	4	0.03	...
René 80 Hf	0.08	60	14	9.5	4	...	3	0.015	4.8	...	4	0.02	0.75 Hf
René 100	0.18	61	9.5	15	3	...	5.5	0.015	4.2	...	...	0.06	1V
René N4	0.06	62	9.8	7.5	1.5	...	4.2	0.004	3.5	4.8	6	...	0.5 Nb, 0.15 Hf
Udimet 500	0.1	53	18	17	4	2	3	...	3	...	...	...	...
Udimet 700	0.1	53.5	15	18.5	5.25	...	4.25	0.03	3.5	...	...	...	...
Udimet 710	0.13	55	18	15	3	...	2.5	...	5	...	1.5	0.08	...
Waspaloy	0.07	57.5	19.5	13.5	4.2	1	1.2	0.005	3	...	...	0.09	...
WAX-20 (DS)	0.20	72	...	...	...	...	6.5	...	...	...	20	1.5	...
Cobalt-base													
AiResist 13	0.45	...	21	62	...	...	3.4	...	...	2	11	...	0.1 Y
AiResist 213	0.20	0.5	20	64	...	0.5	3.5	...	...	6.5	4.5	0.1	0.1 Y
AiResist 215	0.35	0.5	19	63	...	0.5	4.3	...	...	7.5	4.5	0.1	0.1 Y
FSX-414	0.25	10	29	52.5	...	1	...	0.010	...	...	7.5	...	...
Haynes 21	0.25	3	27	64	...	1	...	...	...	...	...	...	5 Mo
Haynes 25; L-605	0.1	10	20	54	...	1	...	...	...	...	15	...	...
J-1650	0.20	27	19	36	...	...	...	0.02	3.8	2	12	...	...
MAR-M 302	0.85	...	21.5	58	...	0.5	...	0.005	...	9	10	0.2	...
MAR-M 322	1.0	...	21.5	60.5	...	0.5	...	...	0.75	4.5	9	2	...
MAR-M 509	0.6	10	23.5	54.5	...	...	...	...	0.2	3.5	7	0.5	...
MAR-M 918	0.05	20	20	52	...	...	...	...	...	7.5	...	0.1	...
NASA Co-W-Re	0.40	...	3	67.5	...	...	...	...	1	...	25	1	2 Re
S-816	0.4	20	20	42	...	4	...	...	...	...	4	...	4 Mo, 4 Nb, 1.2 Mn, 0.4 Si
V-36	0.27	20	25	42	...	3	...	...	...	...	2	...	4 Mo, 2 Nb, 1 Mn, 0.4 Si
WI-52	0.45	...	21	63.5	...	2	...	...	...	...	11	...	2 Nb + Ta
X-40 (Stellite alloy 31)	0.50	10	22	57.5	...	1.5	...	...	...	...	7.5	...	0.5 Mn, 0.5 Si

(a) B-1900 + Hf also contains 1.5% Hf. (b) MAR-M 200 + Hf also contains 1.5% Hf.

as silicon, phosphorus, sulfur, oxygen, and nitrogen must be controlled through appropriate melting practices. Other trace elements, such as selenium, bismuth, and lead, must be held to very small (parts per million) levels in critical parts.

Many nickel-base superalloys contain 10 to 20% Cr, up to about 8% Al and Ti combined, 5 to 15% Co, and small amounts of boron, zirconium, hafnium, and carbon. Other common additions are molybdenum, niobium, tantalum, rhenium, and tungsten, all of which play dual roles as strengthening solutes and carbide formers. Chromium and aluminum are also necessary to improve surface stability through the formation of Cr_2O_3 and Al_2O_3, respectively. The functions of the various elements in nickel alloys are summarized in Table 3. Other alloys that have been developed primarily for low-temperature service, often in corrosive environments (refer to the Hastelloy series and Inconel 600 shown in Table 1), are likely to contain chromium, molybdenum, iron, or tungsten in solution, with little or no second phase present.

Microstructure

The major phases that may be present in nickel-base alloys are:

- *Gamma matrix,* γ, in which the continuous matrix is an fcc nickel-base nonmagnetic phase that usually contains a high percentage of solid-solution elements such as cobalt, iron, chromium, molybdenum, and tungsten. All nickel-base alloys contain this phase as the matrix.
- *Gamma prime,* γ′, in which aluminum and titanium are added in amounts required to precipitate fcc γ′ (Ni_3Al,Ti), which precipitates coherently with the austenitic gamma matrix. Other elements, notably niobium, tantalum, and chromium, also enter γ′. This phase is required for high-temperature strength and creep resistance.
- *Gamma double prime,* γ″, in which nickel and niobium combine in the presence of iron to form bct Ni_3Nb, which is coherent with the gamma matrix, while inducing large mismatch strains of the order of 2.9%. This phase provides very high strength at low to intermediate temperatures, but it is unstable at temperatures above about 650 °C (1200 °F). This precipitate is found in nickel-iron alloys.
- *Grain-boundary* γ′, a film of γ′ along the grain boundaries in the stronger alloys, produced by heat treatments and service exposure. This film is believed to improve rupture properties.
- *Carbides,* in which carbon that is added in amounts of about 0.02 to 0.2 wt% combines with reactive elements, such as titanium, tantalum, hafnium, and niobium, to form metal carbides. During heat treatment and service, MC

carbides tend to decompose and generate other carbides, such as $M_{23}C_6$ and/or M_6C, which tend to form at grain boundaries. Carbides in nominally solid-solution alloys may form after extended service exposures.

- *Borides,* a relatively low density of boride particles formed when boron segregates to grain boundaries
- *Topologically close-packed (TCP)-type phases,* which are platelike or needle-like phases such as σ, μ, and Laves that may form for some compositions and under certain conditions. These cause lowered rupture strength and ductility. The likelihood of their presence increases as the solute segregation of the ingot increases.

Table 4 summarizes data on the commonly encountered constituents in nickel-base superalloys. The major alloying elements that may be present in nickel-base superalloys are illustrated in Fig. 2. The height of the element blocks indicates the amounts that may be present. Additional information on the alloying of superalloys and the phases present in these materials can be found in Ref 2 and 3.

The Gamma Matrix (γ). Pure nickel does not display an unusually high elastic modulus or low diffusivity (two factors that promote creep rupture resistance), but the gamma matrix is readily strengthened for the most severe temperature and time conditions. Some superalloys can be used at >0.85 T_m and, for times up to 100,000 h, at somewhat lower temperatures. These conditions can be tolerated because of three factors (Ref 3):

- The high tolerance of nickel for solutes without phase instability, because of its nearly filled third electron (*d*–) shell
- The tendency, with chromium additions, to form Cr_2O_3 having few cation vacancies, thereby restricting the diffusion rate of metallic elements outward and the rate of oxygen, nitrogen, and sulfur inward (Ref 4)
- The additional tendency, at high temperatures, to form Al_2O_3 barriers, which display exceptional resistance to further oxidation

Gamma prime (γ′) is an intermetallic compound of nominal composition Ni_3Al, which is stable over a relatively narrow range of compositions (Fig. 3). It precipitated as spheroidal particles in early nickel-base alloys, which tended to have a low volume fraction of particles (see Fig. 4a). Later, cuboidal precipitates were noted in alloys with higher aluminum and titanium contents (Fig. 4b). The change in morphology is related to a matrix-precipitate mismatch. It is observed that γ′ occurs as spheres for 0 to 0.2% mismatches, becomes cuboidal for mismatches of 0.5 to 1%, and is platelike at mismatches above about 1.25%.

To understand fully the vital role of γ′ in the nickel-base superalloys, it is necessary to consider the structure and properties of this phase in some detail. Gamma prime is a superlattice that possesses the Cu_3Au ($L1_2$)-type structure, which exhibits long-range order to its melting point of 1385 °C (2525 °F). It exists over a fairly restricted range of composition, but alloying elements may substitute to a considerable degree for either of its constituents. In particular, most nickel-base alloys are strengthened by a precipitate in which up to 60% of the aluminum can be substituted by titanium and/or niobium. Also, nickel sites in the superlattice may be occupied by iron or cobalt atoms.

The volume fraction of γ′ clearly influences alloy strength. Early superalloys contained less than 25 vol% of γ′. Most wrought nickel-base superalloys contain between 20 and 45 vol% of γ′. Gamma prime contents above approximately 45% render the alloy difficult to deform by hot or cold rolling. Nickel-base cast superalloys contain approximately 60 vol% γ′. This increased level of γ′ results in greater alloy creep strength (Fig. 5).

Both single crystals and polycrystals of unalloyed γ′ exhibit a startling, reversible increase in flow stress between –196 and about 800 °C (–320 and 1470 °F), which is highly dependent on solute content (Ref 5), as shown in Fig. 6. Several other superlattices, such as Ni_3Si, Co_3Ti, Ni_3Ge, and Ni_3Ga, all of $L1_2$ structure, display increasing strength over a temperature range comparable to that of Ni_3Al.

The magnitude and temperature position of the peak in flow stress of γ′ may be shifted by alloying elements such as titanium, chromium, and niobium (Fig. 6). There is no simple relation

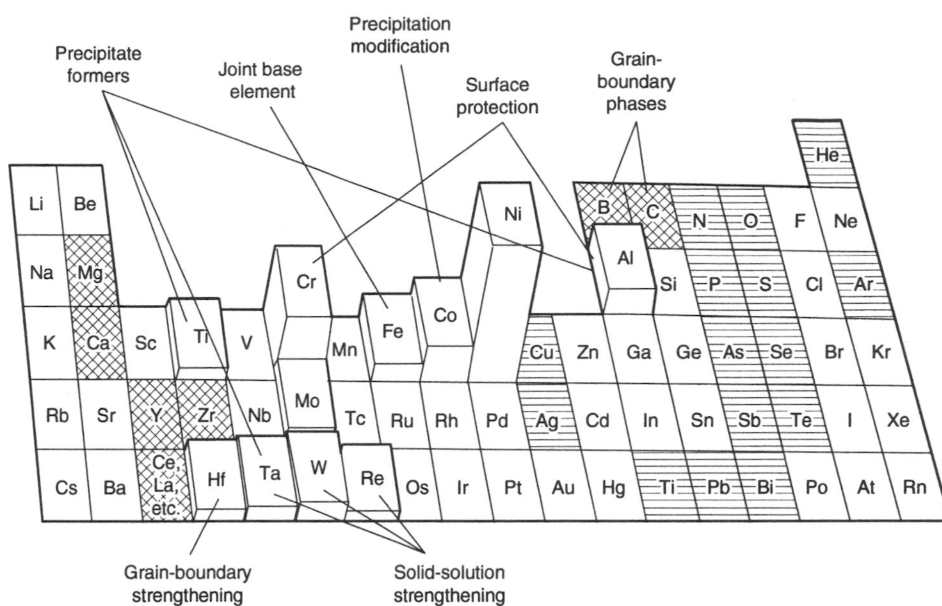

Fig. 2 Alloying elements used in nickel-base superalloys. Beneficial trace elements are marked with cross hatching and harmful trace elements are marked with horizontal line hatching.

Table 3 Role of elements in superalloys

Effect(a)	Iron-base	Cobalt-base	Nickel-base
Solid-solution strengtheners	Cr, Mo	Nb, Cr, Mo, Ni, W, Ta	Co, Cr, Fe, Mo, W, Ta, Re
fcc matrix stabilizers	C, W, Ni	Ni	...
Carbide form:			
MC	Ti	Ti	W, Ta, Ti, Mo, Nb, Hf
M_7C_3	...	Cr	Cr
$M_{23}C_6$	Cr	Cr	Cr, Mo, W
M_6C	Mo	Mo, W	Mo, W, Nb
Carbonitrides: M(CN)	C, N	C, N	C, N
Promotes general precipitation of carbides	P	...	...
Forms γ′ $Ni_3(Al,Ti)$	Al, Ni, Ti	...	Al, Ti
Retards formation of hexagonal η (Ni_3Ti)	Al, Zr	...	...
Raises solvus temperature of γ′	...	...	Co
Hardening precipitates and/or intermetallics	Al, Ti, Nb	Al, Mo, Ti(b), W, Ta	Al, Ti, Nb
Oxidation resistance	Cr	Al, Cr	Al, Cr, Y, La, Ce
Improve hot corrosion resistance	La, Y	La, Y, Th	La, Th
Sulfidation resistance	Cr	Cr	Cr, Co, Si
Improves creep properties	B	...	B, Ta
Increases rupture strength	B	B, Zr	B(c)
Grain-boundary refiners	...	...	B, C, Zr, Hf
Facilitates working	...	Ni_3Ti	...

(a) Not all these effects necessarily occur in a given alloy. (b) Hardening by precipitation of Ni_3Ti also occurs if sufficient Ni is present. (c) If present in large amounts, borides are formed. Source: Adapted from Ref 1 between the magnitude of flow stress increase and the change in the temperature of the peak. Tantalum, niobium, and titanium are effective solid-solution hardeners of γ′ at room temperature. Tungsten and molybdenum are strengtheners at both room and elevated temperatures, while cobalt does not solid-solution strengthen γ′.

Gamma double prime **(γ″),** a bct coherent precipitate of composition Ni_3Nb, precipitates in nickel-iron-base alloys such as Inconel 706 and Inconel 718. In the absence of iron, or at temperatures and times shown in the transformation diagram of an iron-containing alloy (Fig. 7), an orthorhombic precipitate of the same Ni_3Nb composition (delta phase) forms instead. The latter is invariably incoherent and does not confer strength when present in large quantities. However, small amounts of delta phase can be used to control and refine grain size, resulting in improved tensile properties, fatigue resistance, and creep rupture ductility. Careful heat treatment is required to ensure precipitation of γ″ instead of δ.

Gamma double prime often precipitates together with γ′ in Inconel 718, but γ″ is the principal strengthening phase under such circumstances. Unlike γ′, which causes strengthening through the necessity to disorder the particles as they are sheared, γ″ strengthens by virtue of high coherency strains in the lattice. A more detailed description of the physical metallurgy of γ′/γ″ alloys appears in the section "Iron-Base Superalloys" in this article.

Eta phase **(η)** has a hexagonal $D0_{24}$ crystal structure with a Ni_3Ti composition. Eta can form in iron-nickel-, nickel-, and cobalt-base superalloys, especially in grades with high titanium/aluminum ratios that have had extended high-temperature exposure. Eta phase has no solubility for other elements and will grow more rapidly and form larger particles than γ′, although it precipitates slowly.

Two forms of η may be encountered. The first develops at grain boundaries as a cellular constituent similar to pearlite, with alternate lamellae

Table 4 Constituents observed in superalloys

Phase	Crystal structure	Lattice parameter, nm	Formula	Comments
γ′	fcc (ordered $L1_2$)	0.3561 for pure Ni_3Al to 0.3568 for $Ni_3(Al_{0.5}Ti_{0.5})$	Ni_3Al $Ni_3(Al,Ti)$	Principal strengthening phase in many nickel- and nickel-iron-base superalloys; crystal lattice varies slightly in size (0 to 0.5%) from that of austenite matrix; shape varies from spherical to cubic; size varies with exposure time and temperature. Gamma prime is spherical in iron-nickel-base and in some of the older nickel-base alloys, such as Nimonic 80A and Waspaloy. In the more recently developed nickel-base alloys, γ′ is generally cuboidal. Experiments have shown that variations in molybdenum content and in the aluminum/titanium ratio can change the morphology of γ. With increasing γ/γ′ mismatch, the shape changes in the following order: spherical, globular, blocky, cuboidal. When the γ/γ′ lattice mismatch is high, extended exposure above 700 °C (1290 °F) causes undesirable η (Ni_3Ti) or δ (Ni_3Nb) phases to form.
η	hcp ($D0_{24}$)	$a_0 = 0.5093$ $c_0 = 0.8276$	Ni_3Ti (no solubility for other elements)	Found in iron-, cobalt-, and nickel-base superalloys with high titanium/aluminum ratios after extended exposure; may form intergranularly in a cellular form or intragranularly as acicular platelets in a Widmanstätten pattern
γ″	bct (ordered $D0_{22}$)	$a_0 = 0.3624$ $c_0 = 0.7406$	Ni_3Nb	Principal strengthening phase in Inconel 718; γ″ precipitates are coherent disk-shaped particles that form on the {100} planes (avg diam approximately 600 Å, thickness approximately 50 to 90 Å). Bright-field TEM examination is unsatisfactory for resolving γ″ due to the high density of the precipitates and the strong contrast from the coherency strain field around the precipitates. However, dark-field TEM examination provides excellent imaging of the γ″ by selective imaging of precipitates that produce specific superlattice reflections. In addition, γ″ can be separated from γ′ using the dark-field mode, because the γ″ dark-field image is substantially brighter than that of γ′
Ni_3Nb (δ)	Orthorhombic (ordered Cu_3Ti)	$a_0 = 0.5106–0.511$ $b_0 = 0.421–0.4251$ $c_0 = 0.452–0.4556$	Ni_3Nb	Observed in overaged Inconel 718; has an acicular shape when formed between 815 and 980 °C (1500 and 1800 °F); forms by cellular reaction at low aging temperatures and by intragranular precipitation at high aging temperatures
MC	Cubic	$a_0 = 0.430–0.470$	TiC NbC HfC	Titanium carbide has some solubility for nitrogen, zirconium, and molybdenum; composition is variable; appears as globular, irregularly shaped particles that are gray to lavender; "M" elements can be titanium, tantalum, niobium, hafnium, thorium, or zirconium
$M_{23}C_6$	fcc	$a_0 = 1.050–1.070$ (varies with composition)	$Cr_{23}C_6$ $(Cr,Fe,W,Mo)_{23}C_6$	Form of precipitation is important; it can precipitate as films, globules, platelets, lamellae, and cells; usually forms at grain boundaries; "M" element is usually chromium, but nickel-cobalt, iron, molybdenum, and tungsten can substitute
M_6C	fcc	$a_0 = 1.085–1.175$	Fe_3Mo_3C Fe_3W_3C-Fe_4W_2C Fe_3Nb_3C Nb_3Co_3C Ta_3Co_3C	Randomly distributed carbide; may appear pinkish; "M" elements are generally molybdenum or tungsten; there is some solubility for chromium, nickel-niobium, tantalum, and cobalt
M_7C_3	Hexagonal	$a_0 = 1.398$ $c_0 = 0.4523$	Cr_7C_3	Generally observed as a blocky intergranular shape; observed only in alloys such as Nimonic 80A after exposure above 1000 °C (1830 °F), and in some cobalt-base alloys
M_3B_2	Tetragonal	$a_0 = 0.560–0.620$ $c_0 = 0.300–0.330$	Ta_3B_2 V_3B_2 Nb_3B_2 $(Mo,Ti,Cr,Ni,Fe)_3B_2$ Mo_2FeB_2	Observed in iron-nickel- and nickel-base alloys with about 0.03% B or greater; borides appear similar to carbides, but are not attacked by preferential carbide etchants; "M" elements can be molybdenum, tantalum, niobium, nickel, iron, or vanadium
MN	Cubic	$a_0 = 0.4240$	TiN (Ti,Nb,Zr)N (Ti,Nb,Zr) (C,N) ZrN NbN	Nitrides are observed in alloys containing titanium, niobium, or zirconium; they are insoluble at temperatures below the melting point; easily recognized as-polished, having square to rectangular shapes and ranging from yellow to orange
μ	Rhombohedral	$a_0 = 0.475$ $c_0 = 2.577$	Co_2W_6 $(Fe,Co)_7(Mo,W)_6$	Generally observed in alloys with high levels of molybdenum or tungsten; appears as coarse, irregular Widmanstätten platelets; forms at high temperatures
Laves	Hexagonal	$a_0 = 0.475–0.495$ $c_0 = 0.770–0.815$	Fe_2Nb Fe_2Ti Fe_2Mo Co_2Ta Co_2Ti	Most common in iron-base and cobalt-base superalloys; usually appears as irregularly shaped globules, often elongated, or as platelets after extended high-temperature exposure
σ	Tetragonal	$a_0 = 0.880–0.910$ $c_0 = 0.450–0.480$	FeCr FeCrMo CrFeMoNi CrCo CrNiMo	Most often observed in iron- and cobalt-base superalloys, less commonly in nickel-base alloys; appears as irregularly shaped globules, often elongated; forms after extended exposure between 540 and 980 °C (1005 to 1795 °F)

of γ and η; the second develops intragranularly as platelets with a Widmanstätten pattern. The cellular form is detrimental to notched stress-rupture strength and creep ductility, and the Widmanstätten pattern impairs stress-rupture strength but not ductility.

Carbides encountered in superalloys include MC, $M_{23}C_6$, M_6C, and M_7C_3. Carbides in these alloys serve three principal functions. First, grain-boundary carbides, when properly formed, strengthen the grain boundary, prevent or retard grain-boundary sliding, and permit stress relaxation. Second, if fine carbides are precipitated in the matrix, strengthening results. This is important in cobalt-base alloys that cannot be strengthened by γ'. Third, carbides can tie up certain elements that would otherwise promote phase instability during service.

The MC carbide usually exhibits a coarse, random, cubic (Fig. 8), or script morphology. The carbide $M_{23}C_6$ is found primarily at grain boundaries (Fig. 8) and usually occurs as irregular, discontinuous, blocky particles, although plates and regular geometric forms have been observed. The M_6C carbide also can precipitate in blocky form in grain boundaries and less often in a Widmanstätten intragranular morphology. Although data are sparse, it appears that continuous grain-boundary $M_{23}C_6$ and Widmanstätten M_6C, caused by an improper choice of processing or heat treatment temperatures, are to be avoided for best ductility and rupture life.

MC carbides, fcc in structure, usually form in superalloys during freezing. They are distributed heterogeneously through the alloy, both in intergranular and transgranular positions, often interdendritically. Little or no orientation relation with the alloy matrix has been noted. MC carbides are a major source of carbon for subsequent phase reactions during heat treatment and service (Ref 4). In some alloys, such as Incoloy 901 and A-286, MC films may form along grain boundaries and reduce ductility.

These carbides, for example, TiC and HfC, are among the most stable compounds in nature. The preferred order of formation (in order of decreasing stability) in superalloys for these carbides is HfC, TaC, NbC, and TiC. This order is not the same as that of thermodynamic stability, which is HfC, TiC, TaC, and NbC. In these carbides, M atoms can readily substitute for each other, as in (Ti, Nb)C. However, the less reactive elements, principally molybdenum and tungsten, can also substitute in these carbides. For example, (Ti,Mo)C is found in Udimet 500, M-252, and René 77. It appears that the change in stability order cited above is due to the molybdenum or tungsten substitution, which weakens the binding forces in MC carbides to such an extent that degeneration reactions, discussed later, can occur. This typically leads to the formation of the more stable compounds $M_{23}C_6$- and M_6C-type carbides in the alloys during processing or after heat treatment and/or service. Additions of niobium and tantalum tend to counteract this effect. Recent alloys with high niobium and tantalum contents contain MC carbides that do not break down easily during processing or solution treatment in the range of 1200 to 1260 °C (2190 to 2300 °F).

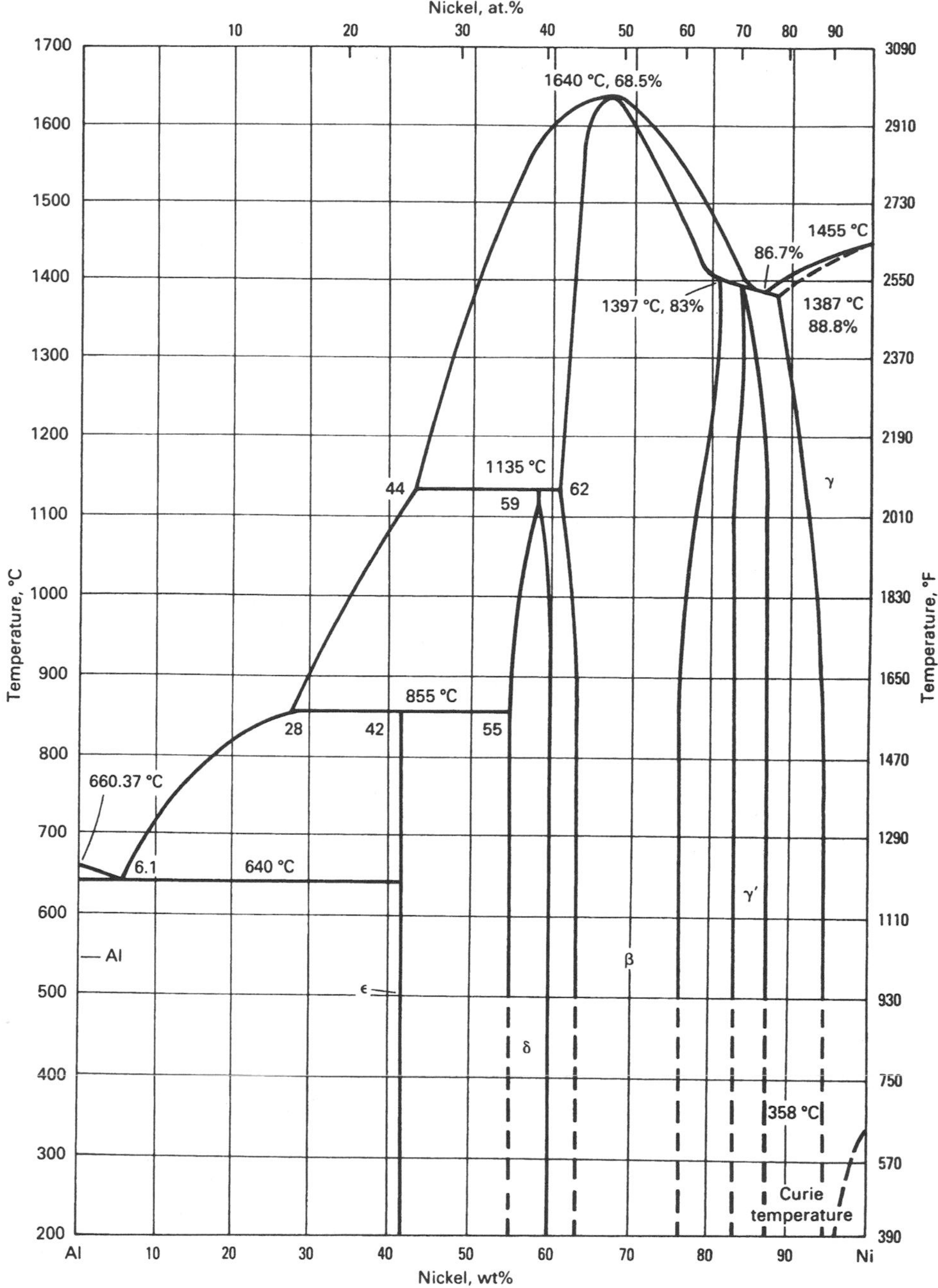

Fig. 3 Nickel-aluminum phase diagram showing the narrow composition range over which the γ' (Ni_3Al) phase is stable

The $M_{23}C_6$ carbides readily form in alloys with moderate to high chromium content. They form during lower-temperature heat treatment and service (that is, 760 to 980 °C, or 1400 to 1800 °F), both from the degeneration of MC carbide and from soluble residual carbon in the alloy matrix. Although usually seen at grain boundaries (Fig. 8), they occasionally occur along twin bands, stacking faults, and at twin ends. The carbides $M_{23}C_6$ have a complex cubic structure, which, if the carbon atoms were removed, would closely approximate the structure of the TCP σ-phase. In fact, σ-phase plates often nucleate on $M_{23}C_6$ particles.

When tungsten or molybdenum is present, the approximate composition of $M_{23}C_6$ is $Cr_{21}(Mo,W)_2C_6$, although it also has been shown that appreciable nickel can substitute in the carbide. It is also possible for small amounts of cobalt or iron to substitute for chromium.

The $M_{23}C_6$ particles strongly influence the properties of nickel alloys. Rupture strength is improved by the presence of discrete particles, apparently through the inhibition of grain-boundary sliding. Eventually, however, failure can in-

Table 5 Effect of temperature on the ultimate tensile strengths of wrought nickel-, iron-, and cobalt-base superalloys

		Ultimate tensile strength at:										
		21 °C (70 °F)		540 °C (1000 °F)		650 °C (1200 °F)		760 °C (1400 °F)		870 °C (1600 °F)		
Alloy	Form	MPa	ksi	MPa	ksi	MPa	ksi	MPa	ksi	MPa	ksi	Condition of test material(a)
Nickel-base												
Astroloy	Bar	1415	205	1240	180	1310	190	1160	168	775	112	1095 °C (2000 °F)/4 h/OQ + 870 °C (1600 °F)/8 h/AC + 980 °C (1800 °F)/4 h/AC + 650 °C (1200 °F) 24 h/AC + 760 °C (1400 °F)/8 h/AC
Cabot 214	...	915	133	715	104	675	98	560	84	440	64	1120 °C (2050 °F)
D-979	Bar	1410	204	1295	188	1105	160	720	104	345	50	1040 °C (1900 °F)/1 h/OQ + 845 °C (1550 °F)/6 h/ AC + 705 °C (1300 °F)/16 h/AC
Hastelloy C-22	Sheet	800	116	625	91	585	85	525	76	...	...	1120 °C (2050 °F)/RQ
Hastelloy G-30	Sheet	690	100	490	71	...	...	...	...	...	...	1175 °C (2150 °F)/RAC-WQ
Hastelloy S	Bar	845	130	775	112	720	105	575	84	340	50	1065 °C (1950 °F)/AC
Hastelloy X	Sheet	785	114	650	94	570	83	435	63	255	37	1175 °C (2150 °F)/1 h/RAC
Haynes 230	...	870	126	720	105	675	98	575	84	385	56	1230 °C (2250 °F)/AC
Inconel 587(b)	Bar	1180	171	1035	150	1005	146	830	120	525	76	...
Inconel 597(b)	Bar	1220	177	1140	165	1060	154	930	135	...	...	...
Inconel 600	Bar	660	96	560	81	450	65	260	38	140	20	1120 °C (2050 °F)/2 h/AC
Inconel 601	Sheet	740	107	725	105	525	76	290	42	160	23	1150 °C (2100 °F)/2 h/AC
Inconel 617	Bar	740	107	580	84	565	82	440	64	275	40	1175 °C (2150 °F)/AC
Inconel 617	Sheet	770	112	590	86	590	86	470	68	310	45	1175 °C (2150 °F)/0.2 h/AC
Inconel 625	Bar	965	140	910	132	835	121	550	80	275	40	1150 °C (2100 °F)/1 h/WQ
Inconel 706	Bar	1310	190	1145	166	1035	150	725	105	...	...	980 °C (1800 °F)/1 h/AC + 845 °C (1550 °F)/3 h/AC + 720 °C (1325 °F)/8 h/FC + 620 °C (1150 °F)/8 h/AC
Inconel 718	Bar	1435	208	1275	185	1228	178	950	138	340	49	980 °C (1800 °F)/1 h/AC + 720 °C (1325 °F)/8 h/FC + 620 °C (1150 °F)/18 h/AC
Inconel 718 Direct Age	Bar	1530	222	1350	196	1235	179	...	...	...	...	735 °C (1325 °F)/8 h/SC + 620 °C (1150 °F)/8 h/AC
Inconel 718 Super	Bar	1350	196	1200	174	1130	164	...	...	...	...	925 °C (1700 °F)/1 h/AC + 735 °C (1325 °F)/8 h/SC + 620 °C (1150 °F)/8 h/AC
Inconel X750	Bar	1200	174	1050	152	940	136	...	...	...	...	1150 °C (2100 °F)/2 h/AC + 845 °C (1550 °F)/24 h/AC + 705 °C (1300 °F)/20 h/AC
M-252	Bar	1240	180	1230	178	1160	168	945	137	510	74	1040 °C (1900 °F)/4 h/AC + 760 °C (1400 °F)/16 h/AC
Nimonic 75	Bar	745	108	675	98	540	78	310	45	150	22	1050 °C (1925 °F)/1 h/AC
Nimonic 80A	Bar	1000	145	875	127	795	115	600	87	310	45	1080 °C (1975 °F)/8 h/AC + 705 °C (1300 °F)/16 h/AC
Nimonic 90	Bar	1235	179	1075	156	940	136	655	95	330	48	1080 °C (1975 °F)/8 h/AC + 705 °C (1300 °F)/16 h/AC
Nimonic 105	Bar	1180	171	1130	164	1095	159	930	135	660	96	1150 °C (2100 °F)/4 h/AC + 1060 °C (1940 °F)/16 h/AC + 850 °C (1560 °F)/16 h/AC
Nimonic 115	Bar	1240	180	1090	158	1125	163	1085	157	830	120	1190 °C (2175 °F)/1.5 h/AC + 1100 °C (2010 °F)/6 h/AC
Nimonic 263	Sheet	970	141	800	116	770	112	650	94	280	40	1150 °C (2100 °F)/0.2 h/WQ + 800 °C (1470 °F)/8 h/AC
Nimonic 942(b)	Bar	1405	204	1300	189	1240	180	900	131	...	...	...
Nimonic PE.11(b)	Bar	1080	157	1000	145	940	136	760	110	...	...	...
Nimonic PE.16	Bar	885	128	740	107	660	96	510	74	215	31	1040 °C (1900 °F)/4 h/AC + 800 °C (1470 °F)/2 h/AC + 1100–1115 °C (2010–2040 °F)/0.25 h/AC + 850 °C (1500 °F)/4 h/AC
Nimonic PK.33	Sheet	1180	171	1000	145	1000	145	885	128	510	74	1100–1115 °C (2010–2040 °F)/0.25 h/AC + 850 °C (1500 °F)/4 h/AC
Pyromet 860(b)	Bar	1295	188	1255	182	1110	161	910	132	...	...	1095 °C (2000 °F)/2 h/WQ + 830 °C (1525 °F)/2 h/AC + 760 °C (1400 °F)/24 h/AC
René 41	Bar	1420	206	1400	203	1340	194	1105	160	620	90	1065 °C (1950 °F)/4 h/AC + 760 °C (1400 °F)/16 h/AC
René 95	Bar	1620	235	1550	224	1460	212	1170	170	...	...	900 °C (1650 °F)/24 h + 1105 °C (2025 °F)/1 h/OQ + 730 °C (1350 °F)/64 h/AC
Udimet 400(b)	Bar	1310	190	1185	172	...	...	...	...	...	...	...
Udimet 500	Bar	1310	190	1240	180	1215	176	1040	151	640	93	1080 °C (1975 °F)/4 h/AC + 845 °C (1550 °F)/24 h/AC + 760 °C (1400 °F)/16 h/AC
Udimet 520	Bar	1310	190	1240	180	1175	170	725	105	515	75	1105 °C (2025 °F)/4 h/AC + 845 °C (1550 °F)/24 h/AC + 760 °C (1400 °F)/16 h/AC
Udimet 630(b)	Bar	1520	220	1380	200	1275	185	965	140	...	...	...
Udimet 700	Bar	1410	204	1275	185	1240	180	1035	150	690	100	1175 °C (2150 °F)/4 h/AC + 1080 °C (1975 °F)/4 h/AC + 845 °C (1550 °F)/24 h/AC + 760 °C (1400 °F)/16 h/AC
Udimet 710	Bar	1185	172	1150	167	1290	187	1020	148	705	102	1175 °C (2150 °F)/4 h/AC + 1080 °C (1975 °F)/4 h/AC + 845 °C (1550 °F)/24 h/AC + 760 °C (1400 °F)/16 h/AC
Udimet 720	Bar	1570	228	...	...	1455	211	1455	211	1150	167	1115 °C (2035 °F)/2 h/AC + 1080 °C (1975 °F)/4 h/ OQ + 650 °C (1200 °F)/24 h/AC + 760 °C (1400 °F)/8 h/AC
Unitemp AF2-1DA6	Bar	1560	226	1480	215	1400	203	1290	187	...	...	1150 °C (2100 °F)/4 h/AC + 760 °C (1400 °F)/16 h/AC
Waspaloy	Bar	1275	185	1170	170	1115	162	650	94	275	40	1080 °C (1975 °F)/4 h/AC + 845 °C (1550 °F)/24 h/AC + 760 °C (1400 °F)/16 h/AC
Iron-base												
A-286	Bar	1005	146	905	131	720	104	440	64	...	...	980 °C (1800 °F)/1 h/OQ + 720 °C (1325 °F)/16 h/AC
Alloy 901	Bar	1205	175	1030	149	960	139	725	105	...	...	1095 °C (2000 °F)/2 h/WQ + 790 °C (1450 °F)/2 h/AC + 720 °C (1325 °F)/24 h/AC
Discaloy	Bar	1000	145	865	125	720	104	485	70	...	...	1010 °C (1850 °F)/2 h/OQ + 730 °C (1350 °F)/20 h/AC + 650 °C (1200 °F)/20 h/AC
Haynes 556	Sheet	815	118	645	93	590	85	470	69	330	48	1175 °C (2150 °F)/AC
Incoloy 800(b)	Bar	595	86	510	74	405	59	235	34	...	...	...

(continued)

(a) OQ, oil quench; AC, air cool; RQ, rapid quench; RAC-WQ, rapid air cool-water quench; FC, furnace cool; SC, slow cool; CW, cold worked. (b) Ref 13. (c) Ref 14. (d) Annealed. (e) Precipitation hardened. (f) Ref 15. (g) Ref 1. (h) Work strengthened and aged. (i) At 700 °C (1290 °F). (j) At 900 °C (1650 °F). Source: Ref 12, except as noted

Table 5 (continued)

Alloy	Form	Ultimate tensile strength at: 21 °C (70 °F) MPa	ksi	540 °C (1000 °F) MPa	ksi	650 °C (1200 °F) MPa	ksi	760 °C (1400 °F) MPa	ksi	870 °C (1600 °F) MPa	ksi	Condition of test material(a)
Iron-base (continued)												
Incoloy 801(b)	Bar	785	114	660	96	540	78	325	47	...	...	...
Incoloy 802(b)	Bar	690	100	600	87	525	76	400	58	195	28	...
Incoloy 807(b)	Bar	655	95	470	68	440	64	350	51	220	32	...
Incoloy 825(c)(d)	...	690	100	~590	~86	~470	~68	~275	~40	~140	~20	...
Incoloy 903	Bar	1310	190	...	...	1000	145	...	...	...	...	845 °C (1550 °F)/1 h/WQ + 720 °C (1325 °F)/8 h/FC + 620 °C (1150 °F)/8 h/AC
Incoloy 907(c)(e)	...	~1365	~198	~1205	~175	~1035	~150	~655	~95	...	...	...
Incoloy 909	Bar	1310	190	1160	168	1025	149	615	89	...	...	980 °C (1800 °F)/1 h/WQ + 720 °C (1325 °F)/8 h/FC + 620 °C (1150 °F)/8 h/AC
N-155	Bar	815	118	650	94	545	79	428	62	260	38	1175 °C (2150 °F)/1 h/WQ + 815 °C (1500 °F)/4 h/AC
V-57	Bar	1170	170	1000	145	895	130	620	90	...	...	980 °C (1800 °F)/2–4 h/OQ + 730 °C (1350 °F)/16 h/AC
19-9 DL(f)	...	815	118	615	89	517	75	...	...	...	...	...
16-25-6(f)	...	980	142	...	...	620	90	415	60	...	...	...
Cobalt-base												
AirResist 213(g)	...	1120	162	...	...	960	139	485	70	315	46	...
Elgiloy(g)	...	690(d)–2480(h)	100(d)–360(h)	...	...	...	...	...	...	...	...	...
Haynes 188	Sheet	960	139	740	107	710	103	635	92	420	61	1175 °C (2150 °F)/1 h/RAC
L-605	Sheet	1005	146	800	116	710	103	455	66	325	47	1230 °C (2250 °F)/1 h/RAC
MAR-M 918	Sheet	895	130	...	...	...	...	...	...	...	...	1190 °C (2175 °F)/4 h/AC
MP35N	Bar	2025	294	...	...	...	...	...	...	...	...	53% CW + 565 °C (1050 °F)/4 h/AC
MP159	Bar	1895	275	1565	227	1540	223	...	...	...	...	48% CW + 665 °C (1225 °F)/4 h/AC
Stellite 6B(g)	Sheet	1010	146	...	...	...	...	...	...	385	56	2 mm (0.063 in.) sheet heat treated at 1232 °C (2250 °F) and RAC
Haynes 150(g)	...	925	134	...	...	325(i)	47	...	...	155(j)	22.8	...

(a) OQ, oil quench; AC, air cool; RQ, rapid quench; RAC-WQ, rapid air cool-water quench; FC, furnace cool; SC, slow cool; CW, cold worked. (b) Ref 13. (c) Ref 14. (d) Annealed. (e) Precipitation hardened. (f) Ref 15. (g) Ref 1. (h) Work strengthened and aged. (i) At 700 °C (1290 °F). (j) At 900 °C (1650 °F). Source: Ref 12, except as noted

itiate either by fracture of particles or by decohesion of the carbide/matrix interface. In some alloys, cellular structures of $M_{23}C_6$ have been noted. These can cause premature failures, but they can be avoided by proper processing and/or heat treatment.

The M_6C carbides have a complex cubic structure. They form when the molybdenum and/or tungsten content is more than 6 to 8 at.%, typically in the range of 815 to 980 °C (1500 to 1800 °F). The M_6C forms with $M_{23}C_6$ in René 80, René 41, and AF 1753. Typical formulas for M_6C are $(Ni,Co)_3Mo_3C$ and $(Ni,Co)_2W_4C$, although a wider range of compositions has been reported for Hastelloy X. Therefore, M_6C carbides are formed when molybdenum or tungsten acts to replace chromium in other carbides; unlike the more rigid $M_{23}C_6$, the compositions can vary widely. Because M_6C carbides are stable at higher levels than are $M_{23}C_6$ carbides, M_6C is more commercially important as a grain-boundary precipitate for controlling grain size during the processing of wrought alloys.

The M_7C_3 Carbides. Although M_7C_3 is not widely observed in superalloys, it is present in some cobalt-base alloys and in Nimonic 80A, a nickel-chromium-titanium-aluminum superalloy, when heated above 1000 °C (1830 °F). Additions of such elements as cobalt, molybdenum, tungsten, or niobium to nickel-base alloys prevents formation of M_7C_3. Massive Cr_7C_3 is formed in Nimonic 80A in the grain boundaries after heating to 1080 °C (1975 °F). Subsequent aging at 700 °C (1290 °F) to precipitate γ' impedes precipitation of $M_{23}C_6$ due to the previously formed Cr_7C_3, which generally exhibits a blocky shape when present at grain boundaries.

Carbide Reactions. MC carbides are a major source of carbon in most nickel-base superalloys below 980 °C (1800 °F). However, MC decomposes slowly during heat treatment and service, releasing carbon for several important reactions.

The principal carbide reaction in many alloys is believed to be the formation of $M_{23}C_6$ (Ref 4):

$$MC + \gamma \rightarrow M_{23}C_6 + \gamma' \text{ or}$$

$$(Ti,Mo)C + (Ni,Cr,Al,Ti) \rightarrow Cr_{21}Mo_2C_6 + Ni_3(Al,Ti) \qquad \text{(Eq 1)}$$

The carbide M_6C can form in a similar manner.

Also, M_6C and $M_{23}C_6$ interact, forming one from the other (Ref 3):

$$M_6C + M' \rightarrow M_{23}C_6 + M'' \qquad \text{(Eq 2)}$$

or

$$Mo_3(Ni,Co)_3C + Cr \leftrightarrow Cr_{21}Mo_2C_6 + (Ni,Co,Mo) \qquad \text{(Eq 3)}$$

depending on the alloy. For example, René 41 and M-252 can be heat treated to generate MC and M_6C initially, with long-time exposure causing the conversion of M_6C to $M_{23}C_6$ (Ref 3).

These reactions lead to carbide precipitation in various locations, but typically at grain boundaries. Perhaps the most beneficial reaction (for high creep resistance applications), one that is controlled in many heat treatments, is that shown in Eq 1. Both the blocky carbides and the γ' produced are important in that they may inhibit grain-boundary sliding. In many cases, the γ' generated by this reaction coats the carbides, and the grain boundary becomes a relatively ductile, creep-resistant region.

Borides. Small additions of boron are essential to improved creep-rupture resistance of superalloys. Borides are hard particles, blocky to half moon in appearance, that are observed at grain boundaries. The boride found in superalloys is of the form M_3B_2, with a tetragonal unit cell. At least two types of borides have been observed in Udimet 700; the type observed depends on the thermal history of the alloy. These boride types are $(Mo_{0.48}Ti_{0.07}Cr_{0.39}Ni_{0.3}Co_{0.3})_3B_2$ and $(Mo_{0.31}Ti_{0.07}Cr_{0.49}Ni_{0.08}Co_{0.07})_3B_2$. However, when Phacomp procedures are used to estimate long-term alloy stability, the composition $(Mo_{0.5}Ti_{0.15}Cr_{0.25}Ni_{0.10})_3B_2$ is usually assumed. Because the level of boron added rarely exceeds 0.03 wt%, and because it is often substantially less than the solubility limit of 0.01%, the volume fraction of borides tends to be quite small. In fact, direct observation has been made of boron segregation to grain boundaries in Udimet 700 containing 0.03 wt% B, but grain boundaries decorated with fine borides were difficult to find in these observations (Ref 7).

TCP Phases. In some alloys, if composition has not been carefully controlled, undesirable phases can form either during heat treatment or,

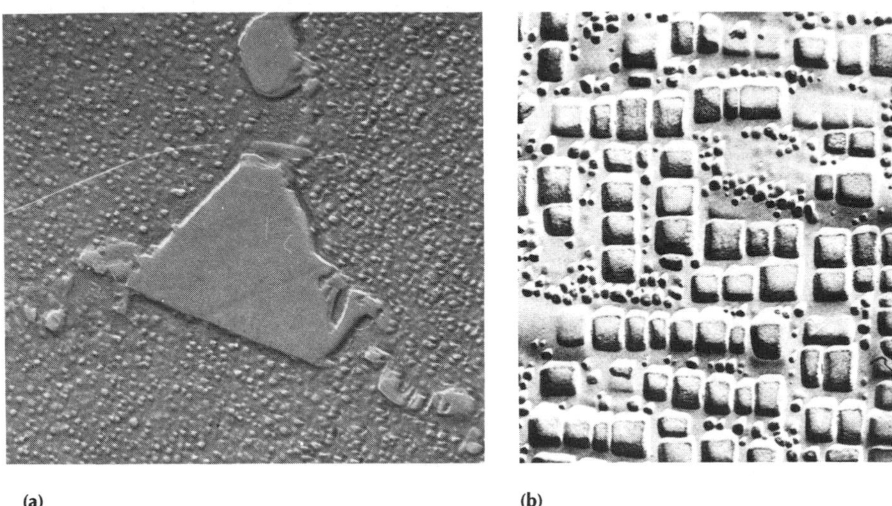

(a) (b)

Fig. 4 Microstructure of (a) fully heat-treated Nimonic 80, showing a grain-boundary carbide ($M_{23}C_6$) and uniformly dispersed spheroidal γ′ in a γ matrix. 15,000×. (b) Fully heat-treated Udimet 700 showing cubical γ′. 6000×

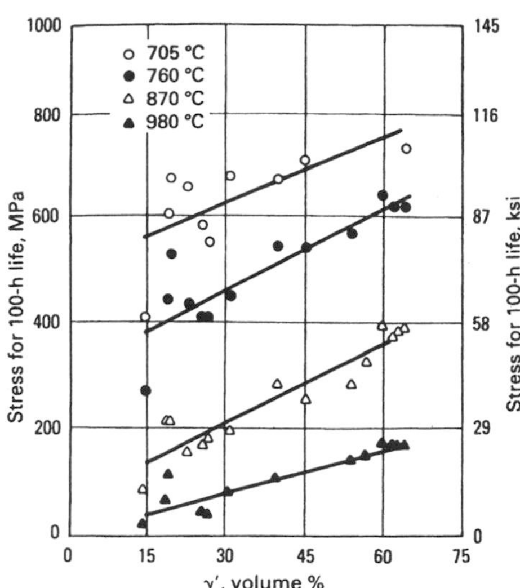

Fig. 5 The relationship between γ′ volume percent and stress-rupture strength for nickel-base superalloys. Source: Ref 4

more commonly, during service. These precipitates, known as TCP phases, are composed of close-packed layers of atoms parallel to {111} planes of the γ matrix. Usually harmful, they may appear as long plates or needles, often nucleating on grain-boundary carbides. Nickel alloys are especially prone to the formation of σ and μ. The formula for σ is $(Fe,Mo)_x$ $(Ni,Co)_y$, where x and y can vary from 1 to 7. Alloys containing a high level of body-centered cubic (bcc) transition metals (tantalum, niobium, chromium, tungsten, and molybdenum) are most susceptible to TCP formation.

The σ hardness and its platelike morphology cause premature cracking, leading to low-temperature brittle failure, although yield strength is unaffected. However, the major effect is on elevated-temperature rupture strength, as shown for Udimet 700 in Fig. 9. Sigma formation must deplete refractory metals in the γ matrix, causing loss of strength of the matrix. Also, high-temperature fracture can occur along σ plates rather than along the normal intergranular path, resulting in sharply reduced rupture life.

Platelike μ can form also, but little is known about its detrimental effects. A general formula for μ is $(Fe,Co)_7$ $(Mo,W)_6$. Nickel can substitute for part of the iron or the cobalt.

Laves phase has a $MgZn_2$ hexagonal crystal structure with a composition of the AB_2 type. Typical examples include Fe_2Ti, Fe_2Nb, and Fe_2Mo, but a more general formula is $(Fe,Cr,Mn,Si)_2(Mo,Ti,Nb)$. They are most commonly observed in the iron-nickel-base alloys as coarse intergranular particles; intragranular precipitation may also occur. Silicon and niobium promote formation of Laves phase in Inconel 718. Excessive amounts will impair room-temperature tensile ductility and creep properties.

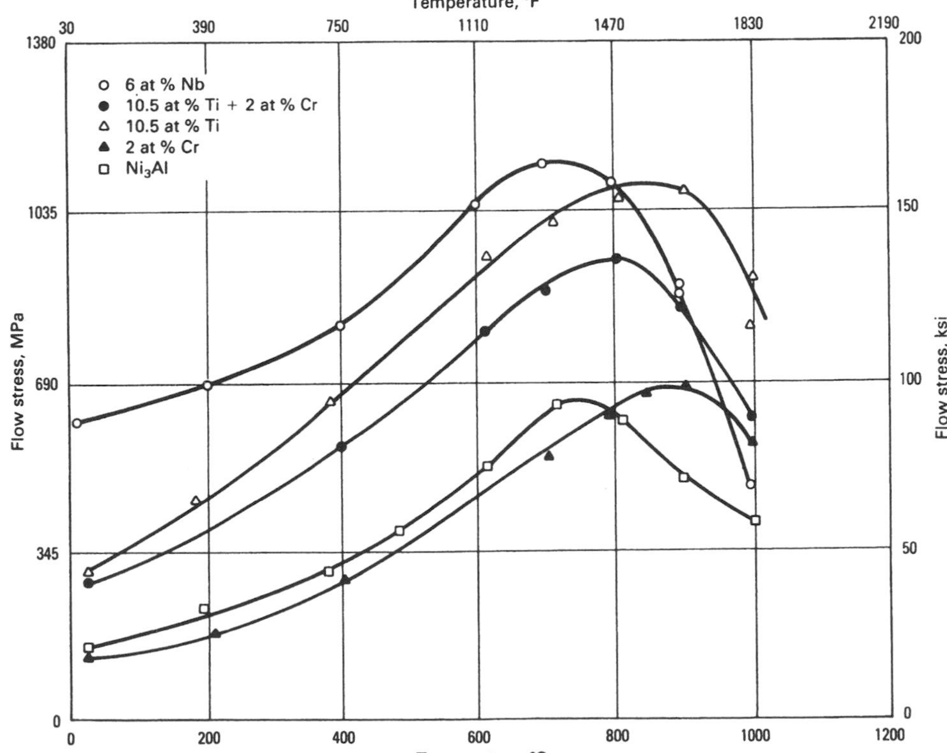

Fig. 6 Flow stress peak in γ′ and influence of several solutes. Source: Ref 5

High-Temperature Behavior of Nickel-Base Superalloys

The principal microstructural variables of superalloys are:

- The precipitate amount and its morphology
- Grain size and shape
- Carbide distribution

Nickel and iron-nickel superalloy properties are controlled by all three variables; the first variable is essentially absent in cobalt-base superalloys. Structure control is achieved through composition selection/modification and by processing. For a given nominal composition, there are property advantages and disadvantages of the structures produced by deformation processing and by casting. Cast superalloys generally have coarser grain sizes, more alloy segregation, and improved creep and rupture characteristics. Wrought superalloys generally have more uniform, and usually finer, grain sizes and improved tensile and fatigue properties.

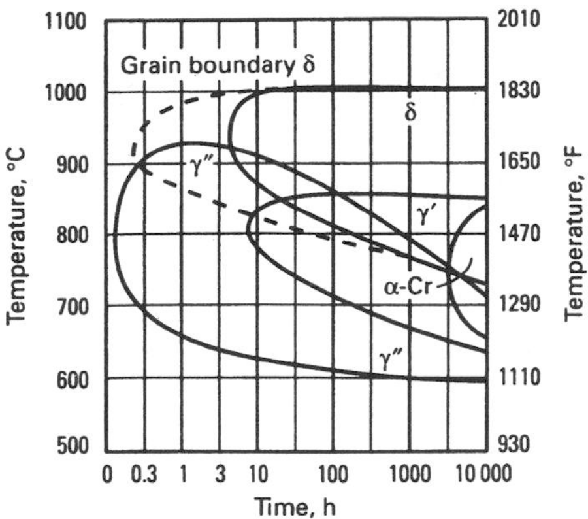

Fig. 7 Transformation diagram for vacuum-melted and hot-forged Inconel 718 bar. Source: Ref 6

Nickel- and iron-nickel-base superalloys typically consist of γ′ dispersed in a γ matrix, and the strength increases with increasing γ′ volume fraction. The lowest volume fractions of γ′ are found in iron-nickel-base and first-generation nickel-base superalloys, where the γ′ volume fracture is generally less than about 0.25 (25 vol%). The γ′ is commonly spheroidal in lower-volume fraction γ′ alloys but often cuboidal in higher-volume-fraction (*L symbol*> }0.35) nickel-base superalloys. The inherent strength capability of such superalloys is controlled by the intragranular distribution; however, the usable strength in polycrystalline alloys is determined by the condition of the grain boundaries, particularly as affected by the carbide-phase morphology and distribution. Satisfactory properties are achieved by optimizing the γ′ volume fraction and morphology (not necessarily independent characteristics) in conjunction with securing a dispersion of discrete globular carbides along the grain boundaries. Discontinuous (cellular) carbide or γ′ at grain boundaries increases surface area and drastically reduces rupture life, even though tensile and creep strength may be relatively unaffected.

Wrought nickel- and iron-nickel-base superalloys generally are processed to have optimum tensile and fatigue properties. At one time, when wrought alloys were used for creep-limited applications such as gas-turbine high-pressure turbine blades, heat treatments different from those used for tensile-limited uses were applied to the same nominal alloy composition in order to maximize creep-rupture life. Occasionally, the nominal composition of an alloy such as IN-100 or Udimet 700 varies according to whether it is in the cast or wrought condition.

Grain Size. The strength of superalloys is very dependent on grain size and its relation to component thickness. Richards (Ref 9) found that rupture life and creep resistance increased as the ratio of component thickness to grain size increased. With a wrought superalloy, provided that the ratio was kept constant, life and creep resistance increased with grain size. Cast superalloys show the same dependence of life and creep resistance on thickness-to-grain-size ratio.

These conditions can be serious when large grains occur in thin section. Thin sections usually exhibit reduced creep rupture resistance: the thinner the section, the lower the rupture strength compared to thick sections.

In modern cast superalloys, control of grain size is vital. A balance must be struck to avoid excessively fine grains, which decrease creep and rupture strength, and excessively large grains, which lower tensile strength (but conversely have good rupture strength).

Grain-Boundary Chemistry. The improvement of creep properties by very small additions of boron and zirconium is a notable feature of nickel-base superalloys. Improved forgeability and better properties have also resulted from magnesium additions of 0.01 to 0.05% (Ref 3). It is believed that this is due primarily to the tying up of sulfur, a grain-boundary embrittler, by the magnesium.

Mechanisms for these property effects are unclear. However, it is believed that boron and zirconium segregate to grain boundaries because of their large size misfit with nickel. Because, at higher temperatures, cracks in superalloys usually propagate along grain boundaries, the importance of grain-boundary chemistry is apparent. Although early work on coarse-grain materials suggested that boron and zirconium influence rupture properties because of their effects on carbide and γ′ distribution, recent work on powder metallurgy superalloys has revealed no such effects with ultrafine grain size.

Boron and zirconium also improve the rupture life of γ′-free alloys, cobalt alloys, and stainless steels, so that microstructural alterations cannot, in any case, apply to all systems. Boron may also reduce carbide precipitation at grain boundaries by releasing carbon into the grains. Magnesium may have a similar effect in a nickel-chromium-titanium-aluminum alloy in which intergranular MC has been noted (Ref 3). Finally, the segregation of misfitting atoms to grain boundaries may reduce grain-boundary diffusion rates. Direct evidence of a lowering of grain-boundary diffusivity by 0.11% Zr in Ni-20Cr alloys over the range of 800 to 1200 K has recently been reported (Ref 10). This effect was accompanied by the precipitation of several precipitates containing zirconium at grain boundaries. The creep strength increased significantly, and the fracture mode changed from intergranular to transgranular ductile rupture, with a corresponding substantial increase in ductility.

Metallurgical Instabilities. Another characteristic of high-temperature service involves metallurgical instability. Stress, time, temperature, and environment may change the metallurgical structure during operation and, thereby, contribute to failure by reducing strength and/or ductility. In a few cases, strength may be enhanced. The structural changes or metallurgical instabilities are best described in terms of their influence on stress-rupture properties. A sharp change downward in the slope of the stress-rupture curve indicates that failures will occur in shorter times and at lower stresses than originally predicted. Instabilities usually are associated with aging (phase precipitation), overaging (phase coalescence and coarsening), phase decomposition (generally involving carbides, borides, and nitrides), intermet-

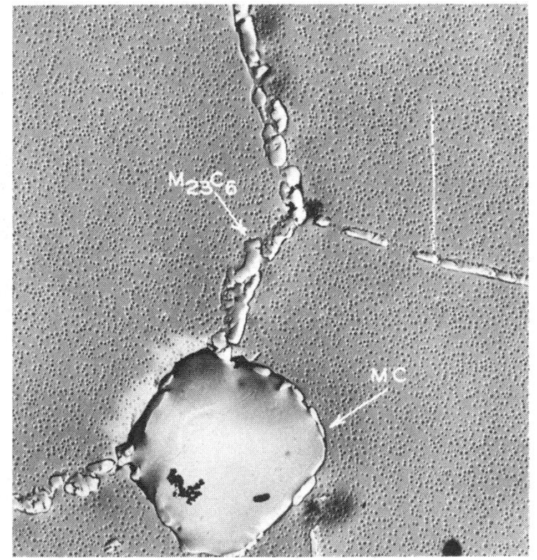

Fig. 8 Microstructure of fully heat-treated Waspaloy showing $M_{23}C_6$ and MC carbides. 3400×

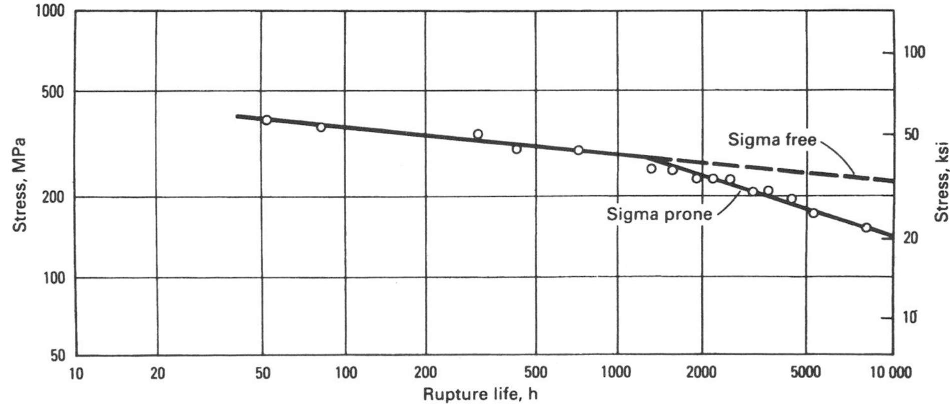

Fig. 9 Log stress vs. log rupture life at 815 °C (1500 °F) for Udimet 700. Source: Ref 8

Table 6 Effect of temperature on the yield strength, elongation, and modulus of elasticity of wrought nickel-, iron-, and cobalt-base superalloys

		Yield strength at 0.2% offset at:										Tensile elongation, % at:				
		21 °C (70 °F)		540 °C (1000 °F)		650 °C (1200 °F)		760 °C (1400 °F)		870 °C (1600 °F)						
Alloy	Form	MPa	ksi	MPa	ksi	MPa	ksi	MPa	ksi	MPa	ksi	21 °C (70 °F)	540 °C (1000 °F)	650 °C (1200 °F)	760 °C (1400 °F)	870 °C (1600 °F)
Nickel-base																
Astroloy	Bar	1050	152	965	140	965	140	910	132	690	100	16	16	18	21	25
Cabot 214	...	560	81	510	74	505	73	495	72	310	45	38	19	14	9	11
D-979	Bar	1005	146	925	134	980	129	655	95	305	44	15	15	21	17	18
Hastelloy C-22	Sheet	405	59	275	40	250	36	240	35	...	...	57	61	65	63	...
Hastelloy G-30	Sheet	315	46	170	25	...	...	...	...	...	...	64	75	...	...	...
Hastelloy S	Bar	455	65	340	49	320	47	310	45	220	32	49	50	56	70	47
Hastelloy X	Sheet	360	52	290	42	275	40	260	38	180	26	43	45	37	37	50
Haynes 230	(b)(c)	390	57	275	40	270	39	285	41	225	32	48	56	55	46	59
Inconel 587(a)	Bar	705	102	620	90	615	89	605	88	400	58	28	22	21	20	16
Inconel 597(a)	Bar	760	110	720	104	675	98	665	96	...	...	15	15	15	16	...
Inconel 600	Bar	285	41	220	32	205	30	180	26	40	6	45	41	49	70	80
Inconel 601	Sheet	455	66	350	51	310	45	220	32	55	8	40	34	33	78	128
Inconel 617	Bar	295	43	200	29	170	25	180	26	195	28	70	68	75	84	118
Inconel 617	Sheet	345	50	230	33	220	32	230	33	205	30	55	62	61	59	73
Inconel 625	Bar	490	71	415	60	420	61	415	60	275	40	50	50	34	45	125
Inconel 706	Bar	1005	146	910	132	860	125	660	96	...	...	20	19	24	32	...
Inconel 718	Bar	1185	172	1065	154	1020	148	740	107	330	48	21	18	19	25	88
Inconel 718 Direct Age	Bar	1365	198	1180	171	1090	158	...	...	...	...	16	15	23	...	...
Inconel 718 Super	Bar	1105	160	1020	148	960	139	...	...	...	...	16	18	14	...	...
Inconel X750	Bar	815	118	725	105	710	103	...	...	...	...	27	26	10	...	...
M-252	Bar	840	122	765	111	745	108	720	104	485	70	16	15	11	10	18
Nimonic 75	Bar	285	41	200	29	200	29	160	23	90	13	40	40	46	67	68
Nimonic 80A	Bar	620	90	530	77	550	80	505	73	260	38	39	37	21	17	30
Nimonic 90	Bar	810	117	725	105	685	99	540	78	260	38	33	28	14	12	23
Nimonic 105	Bar	830	120	775	112	765	111	740	107	490	71	16	22	24	25	27
Nimonic 115	Bar	865	125	795	115	815	118	800	116	550	80	27	18	23	24	16
Nimonic 263	Sheet	580	84	485	70	485	70	460	67	180	26	39	42	27	21	25
Nimonic 942(a)	Bar	1060	154	970	141	1000	145	860	125	...	...					
Nimonic PE.11(a)	Bar	720	105	690	100	670	97	560	81	...	...					
Nimonic PE.16(a)	Bar	530	77	485	70	485	70	370	54	140	20	37	26	30	42	80
Nimonic PK.33(a)	Sheet	780	113	725	105	725	105	670	97	420	61	30	30	26	18	24
Pyromet 860(a)	Bar	835	121	840	122	850	123	835	121	...	...	22	15	17	18	...
René 41	Bar	1060	154	1020	147	1000	145	940	136	550	80	14	14	14	11	19
René 95	Bar	1310	190	1255	182	1220	177	1100	160	...	...	15	12	14	15	...
Udimet 400(a)	Bar	930	135	830	120	...	...	...	...	...	...	30	26	...	...	...
Udimet 500	Bar	840	122	795	115	760	110	730	106	495	72	32	28	28	39	20
Udimet 520	Bar	860	125	825	120	795	115	725	105	520	75	21	20	17	15	20
Udimet 630(a)	Bar	1310	190	1170	170	1105	160	860	125	...	...	15	15	7	5	...
Udimet 700	Bar	965	140	895	130	855	124	825	120	635	92	17	16	16	20	27
Udimet 710	Bar	910	132	850	123	860	125	815	118	635	92	7	10	15	25	29
Udimt 720	Bar	1195	173	...	...	1130	164	1050	152	...	...	13	...	17	9	...
Unitemp AF2-1DA6	Bar	1015	147	1040	151	1020	148	995	144	...	...	20	19	18	16	...
Waspaloy	Bar	795	115	725	105	690	100	675	98	520	75	25	23	34	28	35
Iron-base																
A-286	Bar	725	105	605	88	605	88	430	62	...	...	25	19	13	19	...
Alloy 901(a)	Bar	895	130	780	113	760	110	635	92	...	...	14	14	13	19	...
Discaloy	Bar	730	106	650	94	630	91	430	62	...	...	19	16	19	...	...
Haynes 556	Sheet	410	60	240	35	225	33	220	32	195	29	48	54	52	49	53
Incoloy 800(a)	Bar	250	36	180	26	180	26	150	22	...	...	44	38	51	83	...
Incoloy 801(a)	Bar	385	56	310	45	305	44	290	42	...	...	30	28	26	55	...
Incoloy 802(a)	Bar	290	42	195	28	200	29	200	29	150	22	44	39	25	15	38
Incoloy 807(a)	Bar	380	55	255	37	240	35	225	32.5	185	26.5	48	40	35	34	71
Incoloy 825(d)(e)	...	310	45	~234	~34	~220	~32	180	~26	~105	~15	45	~44	~35	~86	~100
Incoloy 903	Bar	1105	160	...	...	895	130	...	...	...	...	14	...	18	...	...
Incoloy 907(d)(f)	...	~1110	~161	~960	~139	~895	~130	~565	~82	...	...	~12	~11	~10	~20	...
Incoloy 909	Bar	1020	148	945	137	870	126	540	78	...	...	16	14	24	34	...
N-155	Bar	400	58	340	49	295	43	250	36	175	25	40	33	32	32	33
V-57	Bar	830	120	760	110	745	108	485	70	...	...	26	19	22	34	...
19-9 DL(g)	...	570	83	395	57	360	52	...	...	...	...	43	30	30	...	...
16-25-6(g)	...	770	112	...	...	517	75	345	50	255	37	23	...	12	11	9
Cobalt-base																
AirResist 213(h)	...	625	91	...	...	425	66	385	56	220	32	14	...	28	47	55
Elgiloy(i)	Sheet	480(e)–2000 (i)	70–290	...	...	...	...	...	...	...	...	34	...	...	...	...
Haynes 188	Sheet	485	70	305	44	305	44	290	42	260	38	56	70	61	43	73
L-605	Sheet	460	67	250	36	240	35	260	38	240	35	64	59	35	12	35
MAR-M 918	Sheet	895	130	...	...	...	...	...	...	...	...	48	...	...	...	...
MP35N	Bar	1620	235	...	...	...	...	...	...	...	...	10	...	...	...	...
MP159	Bar	1825	265	1495	217	1415	205	...	...	...	...	8	8	7	...	...
Stellite 6B(h)	Sheet	635	92	...	...	...	...	...	...	270	39	11	...	...	...	18
Haynes 150(g)	...	317	46	...	...	160	23	...	...	...	...	8	...	...	...	...

(continued)

(a) Ref 13. (b) Cold-rolled and solution-annealed sheet, 1.2 to 1.6 mm (0.048 to 0.063 in.) thick. (c) Ref 16. (d) Ref 14. (e) Annealed. (f) Precipitation hardened. (g) Ref 15. (h) Ref 1. (i) Work strengthened and aged. (j) Data for bar, rather than sheet. Source: Ref 12, except as noted

Table 6 (continued)

		Dynamic modulus of elasticity(a) at:									
		21 °C (70 °F)		540 °C (1000 °F)		650 °C (1200 °F)		760 °C (1400 F)		870 °C (1600 °F)	
Alloy	Form	GPa	10^6 psi	GPa	10^6 psi	GPa	10^6 psi	GPa	10^6 psi	GPa	10^6 psi
Nickel base											
D-979	Bar	207	30.0	178	25.8	167	24.2	156	22.6	146	21.2
Hastelloy S	Bar	212	30.8	182	26.4	174	25.2	166	24.1	...	...
Hastelloy X	Sheet	197	28.6	161	23.4	154	22.3	146	21.1	137	19.9
Haynes 230	(b)(c)	211	30.6	184	26.4	177	25.3	171	24.1	164	23.1
Inconel 587	Bar	222	32.1	...	...	...	...	...	...	...	...
Inconel 596	Bar	186	27.0	...	...	...	...	...	...	...	...
Inconel 600	Bar	214	31.1	184	26.7	176	25.5	168	24.3	157	22.8
Inconel 601(j)	Sheet	207	30.0	175	25.4	166	24.1	155	22.5	141	20.5
Inconel 617	Bar	210	30.4	176	25.6	168	24.4	160	23.2	150	21.8
Inconel 625	Bar	208	30.1	179	25.9	170	24.7	161	23.3	148	21.4
Inconel 706	Bar	210	30.4	179	25.9	170	24.7	...	...	...	...
Inconel 718	Bar	200	29.0	171	24.8	163	23.7	154	22.3	139	20.2
Inconel X750	Bar	214	31.0	184	26.7	176	25.5	166	24.0	153	22.1
M-252	Bar	206	29.8	177	25.7	168	24.4	156	22.6	145	21.0
Nimonic 75	Bar	221	32.0	186	27.0	176	25.5	170	24.6	156	22.6
Nimonic 80A	Bar	219	31.8	188	27.2	179	26.0	170	24.6	157	22.7
Nimonic 90	Bar	226	32.7	190	27.6	181	26.3	170	24.7	158	22.9
Nimonic 105	Bar	223	32.3	186	27.0	178	25.8	168	24.4	155	22.5
Nimonic 115	Bar	224	32.4	188	27.2	181	26.3	173	25.1	164	23.8
Nimonic 263	Sheet	222	32.1	190	27.5	181	26.2	171	24.8	158	22.9
Nimonic 942	Bar	196	28.4	166	24.1	158	22.9	150	21.8	138	20.0
Nimonic PE.11	Bar	198	28.7	166	24.0	157	22.8	...	...	...	...
Nimonic PE.16	Bar	199	28.8	165	23.9	157	22.7	147	21.3	137	19.9
Nimonic PK.33	Sheet	222	32.1	191	27.6	183	26.5	173	25.1	162	23.5
Pyromet 860	Bar	200	29.0	...	...	...	...	...	...	...	...
René 95	Bar	209	30.3	183	26.5	176	25.5	168	24.3	...	...
Udimet 500	Bar	222	32.1	191	27.7	183	26.5	173	25.1	161	23.4
Udimet 700	Bar	224	32.4	194	28.1	186	27.0	177	25.7	167	24.2
Udimet 710	Bar	222	32.1	...	...	...	...	...	...	...	...
Waspaloy	Bar	213	30.9	184	26.7	177	25.6	168	24.3	158	22.9
Iron-base											
A-286	Bar	201	29.1	162	23.5	153	22.2	142	20.6	130	18.9
Alloy 901(a)	Bar	206	29.9	167	24.2	153	22.1	...	...	...	...
Discaloy	Bar	196	28.4	154	22.3	145	21.0	...	...	...	...
Haynes 556	Sheet	203	29.5	165	23.9	156	22.6	146	21.1	137	19.9
Incoloy 800	Bar	196	28.4	161	23.4	154	22.3	146	21.1	138	20.0
Incoloy 801	Bar	208	30.1	170	24.7	162	23.5	154	22.3	144	20.9
Incoloy 802	Bar	205	29.7	169	24.5	161	23.4	156	22.6	152	22.0
Incoloy 807	Bar	184	26.6	155	22.4	146	21.2	137	19.9	128	18.5
Incoloy 903	Bar	147(e)	21.3	152(f)	22.1	...	...	...	...	...	...
Incoloy 907(d)(f)	...	165(e)	23.9	165(f)	23.9	159(e)	23	...	...	...	...
N-155	Bar	202	29.3	167	24.2	159	23.0	149	21.6	138	20.0
V-57	Bar	199	28.8	163	23.6	153	22.2	144	20.8	130	18.9
19-9 DL(g)	...	203	29.5	...	...	152	22.1	...	...	...	...
16-25-6(g)	...	195	28.2	...	...	123	17.9	...	...	...	...
Cobalt-base											
Haynes 188	Sheet	207	30	...	...	...	...	...	...	...	...
L-605	Sheet	216	31.4	...	...	185	26.8	...	...	166	24.0
MAR-M 918	Sheet	225	32.6	186	27.0	176	25.5	168	24.3	159	23.0
MP35N	Bar	231(e)	33.6	...	...	...	...	...	...	...	...
Haynes 150(g)	...	217(e)	31.5	...	...	...	...	...	...	...	...

(a) Ref 13. (b) Cold-rolled and solution-annealed sheet, 1.2 to 1.6 mm (0.048 to 0.063 in.) thick. (c) Ref 16. (d) Ref 14. (e) Annealed. (f) Precipitation hardened. (g) Ref 15. (h) Ref 1. (i) Work strengthened and aged. (j) Data for bar, rather than sheet. Source: Ref 12, except as noted

allic phase precipitation, order-disorder transition, internal oxidation, and stress corrosion. Typical instability problems with the γ' nickel-base superalloys involve intermetallic phase precipitation. Close-packed σ, μ, and Laves phases form at elevated temperatures, generally with deteriorating effects on stress-rupture properties. More detailed information on how metallurgical instabilities influence properties can be found in the article "Microstructural Degradation of Superalloys" in this Volume.

Effect of Carbide Precipitation. Carbides exert a profound influence on properties by their precipitation on grain boundaries. In most superalloys, $M_{23}C_6$ forms at the grain boundaries after a postcasting or postsolution treatment thermal cycle such as aging. A chain of discrete globular $M_{23}C_6$ carbides were found to optimize creep-rupture life by preventing grain-boundary sliding in creep rupture while concurrently providing sufficient ductility in the surrounding grain for stress relaxation to occur without premature failure.

In contrast, if carbides precipitate as a continuous grain-boundary film, properties can be severely degraded. $M_{23}C_6$ films were reported to reduce impact resistance of M-252, and MC films were blamed for lowered rupture lives and ductility in forged Waspaloy (Ref 11). At the other extreme, when no grain-boundary carbide precipitate is present, premature failure also will occur because grain-boundary movement essentially is unrestricted, leading to subsequent cracking at grain-boundary triple points.

Property Data for Wrought Alloys (Ref 12). Mechanical and physical properties for nickel-, iron-, and cobalt-base wrought superalloys are given in the tables and graphs briefly described here. Although much data are presented comprehensively in this section, descriptions of iron- and cobalt-base superalloys are provided independently in respective sections, along with representative data. Ultimate tensile strength values at a variety of temperatures are given in Table 5, while tensile yield strengths and elongations are provided in Table 6. Rupture strengths measured at 1000 h are given in Table 7, while stress-rupture curves for selected superalloys are shown in Fig. 10 and 11. Physical properties of wrought alloys are given in Table 8.

Property Data for Cast Alloys. The introduction of commercial vacuum induction melting and vacuum investment casting in the early 1950s provided further potential for γ' exploitation. Many nickel-base alloy developments resulted, continuing through the 1960s (Fig. 12). The development of new polycrystalline alloys continued through the 1970s, however, at a more moderate rate. Attention was concentrated instead on process development, with specific interest directed toward grain orientation and directional-solidification (DS) turbine blade and vane casting technology.

Applied to turbine blades and vanes, the DS casting process results in the alignment of all component grain boundaries such that they are parallel to the blade/vane stacking fault axis, essentially eliminating transverse grain boundaries. Turbine blades/vanes encounter major operating stress in the direction that is near normal to the stacking fault axis, so transverse grain boundaries provide relatively easy fracture paths. The elimination of these paths provides increased strain elasticity by virtue of the lower <001> elastic modulus, thereby creating opportunities for further exploitation of the nickel-base alloy potential.

The logical progression to grain-boundary reduction is the total elimination thereof. Thus, single-crystal turbine blade/vane casting technology soon developed, providing further opportunity for nickel-base alloy design innovation.

Tables 9 to 12 list mechanical and physical properties of polycrystalline nickel- and cobalt-base cast superalloys. Stress-rupture curves for selected cast superalloys are shown in Fig. 13. The advantages of DS or single-crystal alloys over conventionally cast superalloys are clearly evident in Fig. 14. These new process technologies, which are more fully discussed in the article "Directionally Solidified and Single-Crystal Superalloys" in this Volume, have contributed dramatically to improvements in gas turbine engine operating efficiency.

Alloying for Surface Stability

Low-Temperature Corrosion. High chromium contents are required in nickel-base alloys for good resistance to corrosive media such as aqueous solutions and acids at low temperatures. A series of Hastelloy and Inconel alloys has been developed for such applications. Hastelloy C and Inconel 600 are typical: the former contains 16.5% Cr and 17% Mo as principal components for corrosion resistance, while Inconel 600 contains 15.5% Cr and 8% Fe. Other Hastelloy alloys contain up to 28% Mo, sometimes with small additions of tungsten.

Oxidation and Hot Corrosion. At elevated temperatures, oxidation resistance is provided by Al_2O_3 or Cr_2O_3 protective films. Accordingly, nickel-base alloys must contain one or both of these elements, even where strength is not a principal factor. For example, Hastelloy X, one of the most oxidation and (hot) corrosion resistant of all nickel-base alloys, contains 22% Cr, 9% Mo, and 18.5% Fe as principal solutes (Table 1). Because Hastelloy X is essentially a solid-solution alloy when placed into service (carbides precipitate after long-term exposure), the alloy is much weaker than superalloys containing γ' or γ'' as strengthening precipitates.

Because chromium is known to degrade the high-temperature strength of γ' (see Fig. 6), there has been a strong incentive to lower chromium content in modern superalloys. Thus, the level of chromium decreased from 20% in earlier wrought alloys to as little as 9% in modern cast alloys. Unfortunately, this compositional change degraded hot corrosion resistance to the point that superalloys used in gas turbines had to be coated.

Further, as turbine blade temperatures exceed 1000 °C (1830 °F), Cr_2O_3 tends to decompose to CrO_3, which is more volatile and therefore less

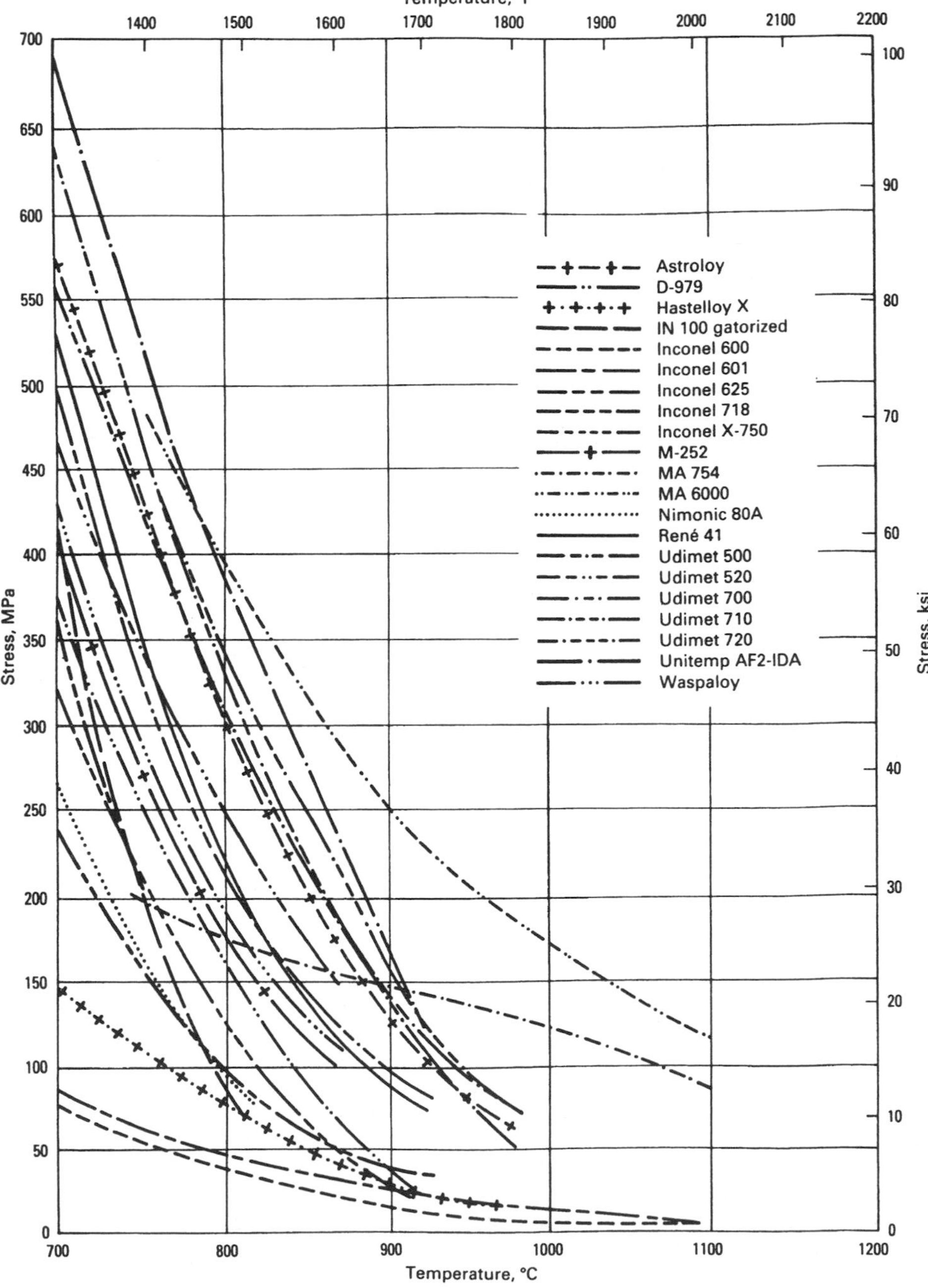

Fig. 10(a) 1000 h rupture strengths of selected wrought nickel-base superalloys

protective. To some extent, the loss of oxidation resistance has been compensated for by raising aluminum contents, although aluminum resides primarily in γ′. (Aluminum in small quantities promotes the formation of Cr_2O_3.) However, Al_2O_3 is less protective than Cr_2O_3 under sulfidizing conditions, making coatings indispensable in aircraft turbines and, more recently, in industrial turbines.

Other elements that contribute to oxidation and hot corrosion resistance are tantalum, yttrium, and lanthanum. The rare earths appear to improve oxidation resistance by preventing spalling of the oxide, while the mechanism for improvement with tantalum is not known. Yttrium is now widely used in overlay coatings of the NiCrAlY type.

Molybdenum and tungsten are considered to be the most deleterious solutes in terms of hot corrosion resistance. Nevertheless, one or both of these elements are required for strength (e.g., in most of the γ′-strengthened alloys), so that alloying for improved surface stability is often in conflict with alloying for strength. The two most prominent solutes that provide both strength and surface stability are aluminum and tantalum.

Protection Against Oxidation and Corrosion. Superalloys are often diffusion or overlay coated in order to improve their corrosion resistance. The overlay coatings used are based on iron-chromium-aluminum-yttrium, cobalt-chromium-aluminum-yttrium, and nickel-chromium-aluminum-yttrium alloys. The earlier, less protective diffusion coatings are based on the reaction of aluminum with the substrate to form one or more aluminum-rich intermetallics (NiAl and/or Ni_2Al_3).

An important consideration for coated superalloys is the reduction in incipient melting temperature of the system (coating/base metal) that may result from the change in composition caused by the diffusion of coating components inward from the surface. Incipient melting (Table 13) reduces grain-boundary strength and ductility and thus reduces stress-rupture capabilities. Once an alloy has been heated above its incipient melting point, the alloy properties cannot be restored by heat treatment. Generally, a cast alloy should not be used for a structural application at any temperature higher than the point about 125 °C (225 °F) below its incipient melting temperature. Oxidation behavior and strength will determine how closely actual metal temperatures may approach this suggested upper limit. Wrought alloys are used for applications, such as turbine disks and rotating seals, that typically operate at much lower temperatures. Therefore, coatings are much less likely to be used for such applications.

An important difference between nickel- and cobalt-base superalloys is related to the superior hot corrosion resistance claimed for cobalt-base alloys in atmospheres containing sulfur, sodium salts, halides, vanadium oxides, and lead oxide, all of which can be found in fuel-burning systems. In part, this apparent superiority may arise from the higher chromium content that is characteristic of cobalt-base alloys. Nickel forms low-melting-point eutectics with nickel sulfide; in sulfur-bearing gases, the attack on nickel alloys may be devastating. Generalizations about the comparative oxidation and hot corrosion resistance of nickel- and cobalt-base alloys must be treated with some caution, because resistance to corrosion varies widely within each alloy group. Small variations in service conditions (temperature, fuel impurity content, gas flow rates) can sharply alter results. More detailed information on high-temperature corrosion of superalloys can be found in the articles "Elevated-Temperature Corrosion Properties of Superalloys" and "Protective Coatings for Superalloys" in this Volume.

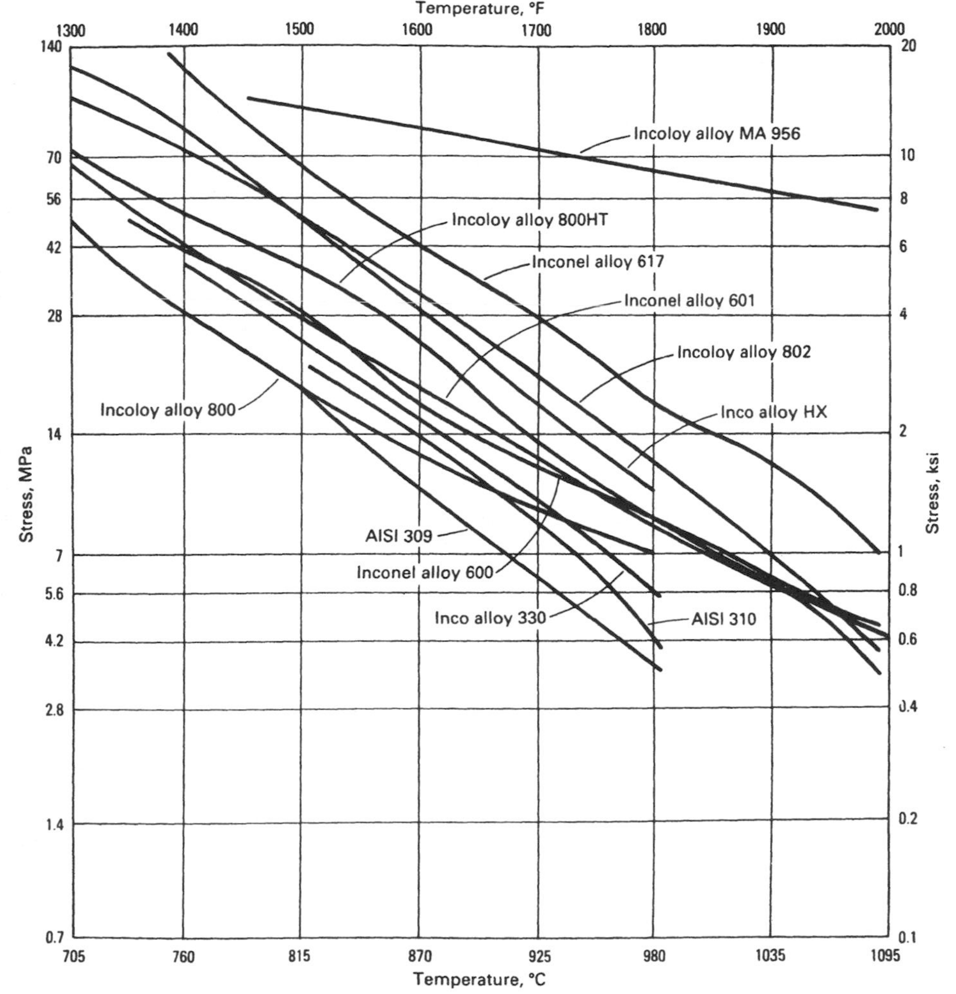

Fig. 10(b) 10,000 h rupture strengths of nickel-base alloys and stainless steels

Iron-Base Superalloys

Microstructure. Iron-base superalloys evolved from austenitic stainless steels and are based on the principle of combining a close-packed fcc matrix with (in most cases) both solid-solution hardening and precipitate-forming elements. The austenitic matrix is based on nickel and iron, with at least 25% Ni needed to stabilize the fcc phase. Other alloying elements, such as chromium, partition primarily to the austenite for solid-solution hardening. The strengthening precipitates are primarily ordered intermetallics, such as γ′ Ni_3Al, η Ni_3Ti, and γ″ Ni_3Nb, although carbides and carbonitrides may also be present. Elements that partition to grain boundaries, such as boron and zirconium, perform a function similar to that which occurs in nickel-base alloys; that is, grain-boundary fracture is suppressed under creep rupture conditions, resulting in significant increases in rupture life. Compositions of wrought iron-nickel alloys are listed in Table 1. A summary of the function of various alloying elements is found in Table 3.

Several groupings of iron-nickel alloys based on composition and strengthening mechanisms have been established. Alloys that are strengthened by ordered fcc γ′, such as V-57 and A-286, and contain 25 to 35 wt% Ni, represent one subgroup. The γ′-phase is titanium-rich in these alloys, and care must be taken to avoid an excessively high titanium-to-aluminum ratio, resulting in the replacement of fcc γ′ by hcp η(Ni_3Ti), a less effective strengthener.

A second iron-rich subgroup, of which Pyromet 860 and Incoloy 901 are examples, contains at least 40% Ni, as well as higher levels of

Table 7 1000 h rupture strengths of wrought nickel-, cobalt-, and iron-base superalloys

		Rupture strength at:							
		650 °C (1200 °F)		760 °C (1400 °F)		870 °C (1600 °F)		980 °C (1800 °F)	
Alloy	Form	MPa	ksi	MPa	ksi	MPa	ksi	MPa	ksi
Nickel-base									
Astroloy	Bar	770	112	425	62	170	25	55	8
Cabot 214	...	...	...	...	...	30	4	15	2
D-979	Bar	515	75	250	36	70	10	...	...
Hastelloy S	Bar	...	...	90	13	25	4	...	...
Hastelloy X	Sheet	215	31	105	15	40	6	15	2
Haynes 230	...	...	...	125	18	55	8	15	2
Inconel 587(a)	Bar	...	...	285	41	...	...	...	...
Inconel 597(a)	Bar	...	...	340	49	...	...	...	...
Inconel 600	Bar	...	...	...	...	30	4	15	2
Inconel 601	Sheet	195	28	60	9	30	4	15	2
Inconel 617	Bar	360	52	165	24	60	9	30	4
Inconel 617	Sheet	...	...	160	23	60	9	30	4
Inconel 625	Bar	370	54	160	23	50	7	20	3
Inconel 706	Bar	580	84	...	...	...	...	...	...
Inconel 718	Bar	595	86	195	28	...	...	...	...
Inconel 718 Direct Age	Bar	405	59	...	...	...	...	...	...
Inconel 718 Super	Bar	600	87	...	...	...	...	...	...
Inconel X750	Bar	470	68	...	...	50	7	...	...
M-252	Bar	565	82	270	39	95	14	...	...
Nimonic 75	Bar	170	25	50	7	5	1	...	...
Nimonic 80A	Bar	420	61	160	23	...	...	...	...
Nimonic 90	Bar	455	66	205	30	60	9	...	...
Nimonic 105	Bar	...	...	330	48	130	19	30	4
Nimonic 115	Bar	...	...	420	61	185	27	70	10
Nimonic 942(a)	Bar	520	75	270	39	...	...	...	...
Nimonic PE.11(a)	Bar	335	49	145	21	...	...	...	...
Nimonic PE.16	Bar	345	50	150	22	...	...	...	...
Nimonic PK.33	Sheet	655	95	310	45	90	13	...	...
Pyromet 860(a)	Bar	545	79	250	36	...	...	...	...
René 41	Bar	705	102	345	50	115	17	...	...
René 95	Bar	860	125	...	...	...	...	...	...
Udimet 400(a)	Bar	600	87	305	44	110	16	...	...
Udimet 500	Bar	760	110	325	47	125	18	...	...
Udimet 520	Bar	585	85	345	50	150	22	...	...
Udimet 700	Bar	705	102	425	62	200	29	55	8
Udimet 710	Bar	870	126	460	67	200	29	70	10
Udimet 720	Bar	670	97	...	...	...	...	...	...
Unitemp AF2-1DA6	Bar	885	128	360	52	...	...	...	...
Waspaloy	Bar	615	89	290	42	110	16	...	...
Iron-base									
A-286	Bar	315	46	105	15	...	...	...	...
Alloy 901	Sheet	525	76	205	30	...	...	...	...
Discaloy	Bar	275	40	60	9	...	...	...	...
Haynes 556	Sheet	275	40	125	18	55	8	20	3
Incoloy 800(a)	Bar	165	24	66	9.5	30	4.4	13	1.9
Incoloy 801(a)	Bar	...	...	...	...	...	...	...	...
Incoloy 802(a)	Bar	170	25	110	16	69	10	24	3.5
Incoloy 807(a)	Bar	...	...	105	15	43	6.2	19	2.7
Incoloy 903	Bar	510	74	...	...	...	...	...	...
Incoloy 909	Bar	345	50	...	...	...	...	...	...
N-155	Bar	295	43	140	20	70	10	20	3
V-57	Bar	485	70	...	...	...	...	...	...
Cobalt-base									
Haynes 188	Sheet	...	...	165	24	70	10	30	4
L-605	Sheet	270	39	165	24	75	11	30	4
MAR-M 918	Sheet	...	...	60	9	20	3	5	1
Haynes 150(b)	...	...	...	40(c)	5.8	...	...	...	...

(a) Ref 13. (b) Ref 15. (c) At 815 °C (1500 °F). Source: Ref 12, except as noted

solid-solution strengthening and precipitate-forming elements.

A third iron-rich subgroup, of which N-155 (Multimet) and Haynes 556 are examples, are iron-nickel-chromium-cobalt alloys that display little or no precipitation strengthening. These alloys exhibit excellent resistance to high-temperature corrosive environments.

Another iron-rich group, based on the iron-nickel-cobalt system strengthened by fcc γ', combines low thermal expansion coefficients and relatively high strength to a temperature of 650 °C (1200 °F). These alloys are typified by Incoloy 903, 907, and 909, Pyromet CTX-1, and Pyromet CTX-3, which do not have any chromium and oxidize and spall readily. The unusually low thermal expansion coefficients result from the elimination of ferrite-stabilizing elements. In each of these alloys, a small quantity of titanium is present to allow the formation of metastable $\gamma'(Ni_3Ti,Al)$ in the austenite matrix during aging. Another phase that may occur after extended aging of the Pyromet alloys is $\eta(Ni_3Ti)$, a platelike, stable hcp phase that contributes less to strengthening than does γ'. The alloys are used for shafts, rings, and casings to permit the reduction of clearances between rotating and static components.

Alloys With High Nickel Content. One group of iron-nickel alloys actually contain more nickel than iron. Examples include Incoloy 706 and Inconel 718 (Table 1), both of which are strengthened by a coherent bct phase known as γ'' (Ni_3Nb). These alloys contain 3 and 5 wt% Nb, respectively. The iron acts principally as a catalyst for the formation of γ'', which is metastable. These alloys also contain smaller quantities of aluminum and titanium, thereby leading to the formation of γ' Ni_3Al,Ti. Improper heat treatment can lead to the formation of a stable orthorhombic δ-phase with the composition Ni_3Nb. Because proper heat treatment procedures are essential for these alloys, time-temperature-transformation diagrams such as that shown in Fig. 7 are used to establish appropriate heat treatment schedules.

The maximum-use temperature of Inconel 718 is about 650 °C (1200 °F) because of instability of the γ'' precipitate. Other problems that have arisen with this alloy are notch sensitivity from about 525 to 750 °C (975 to 1380 °F) when tested in air (generally caused by improper processing) and a significantly higher rate of crack growth in air than in vacuum. Recent work has shown that lowering the carbon content can result in equal or better properties than those of standard Inconel 718 (Ref 21). Further, increasing the aluminum and niobium contents and the aluminum-to-titanium ratio in Inconel 718 enhances its mechanical properties by producing more of the stable γ' phase and less of the undesirable δ-phase (Ref 22).

Boron in quantities of 0.003 to 0.03 wt% and, less frequently, small additions of zirconium are added to improve stress-rupture properties and hot workability. Zirconium also forms the MC carbide ZrC. Another MC carbide, NbC, is found in alloys that contain niobium, such as Inconel 706 and Inconel 718. Vanadium also is added in small quantities to iron-nickel alloys to improve both notch ductility at service temperatures and hot workability. Manganese and rare earth elements may also be present as deoxidizers.

Inconel 718 is one of the strongest (at low temperatures) and most widely used of all superalloys, but it rapidly loses strength in the range of 650 to 815 °C (1200 to 1500 °F). Figures 15 and 16 show the effect of temperature on the yield strength and stress-rupture strength, respectively, of Inconel 718, A-286, and various high-nickel-content superalloys. The rapid loss of high-temperature strength in Inconel 718 is probably due to the high lattice misfit associated with the precipitation of γ'' in the austenitic matrix. Additional data on iron-nickel-base superalloys can be found in Tables 5 to 8 and Fig. 11.

Cobalt-Base Superalloys

Cobalt-base alloys, unlike other superalloys, are not strengthened by a coherent, ordered precipitate. Rather, they are characterized by a solid-solution strengthened austenitic fcc matrix in which a small quantity of carbides is distributed. (Cast cobalt alloys rely upon carbide strengthening to a much greater extent.) Cobalt crystallizes in the hcp structure below 417 °C (780 °F). At higher temperatures, it transforms to fcc. To avoid this transformation during service, virtually all cobalt-base alloys are alloyed with nickel in order to stabilize the fcc structure between room temperature and the melting point.

Melting point is not a reliable guide to the temperature capability of iron-, nickel-, and cobalt-base superalloys. Although nickel possesses the lowest melting point of the three alloy bases, it has by far the highest temperature capability under moderate-to-high stresses. Further, the incipient melting temperatures of nickel- and cobalt-base alloys are very similar (see Table 13). Cobalt-base alloys, with a flatter rupture stress-temperature relationship, may actually display better creep rupture properties above about 1000 °C (1830 °F) than some of the other superalloys. Also, cobalt-base alloys display superior hot corrosion resistance at high temperatures, probably a consequence of the considerably higher chromium contents that are characteristic of these alloys. Lists of wrought and cast cobalt-base alloys and their compositions appear in Tables 1 and 2, respectively. The function of various solutes is summarized in Table 3.

Cobalt-base alloys generally exhibit better weldability and thermal-fatigue resistance than do nickel-base alloys. Another advantage of cobalt-base alloys is the capability to be melted in air or argon, in contrast to the vacuum melting required for nickel-base and iron-nickel-base alloys containing the reactive metals aluminum and titanium. However, unlike nickel-base alloys, which have a high tolerance for alloying elements in solid solution, cobalt-base alloys are more likely to precipitate undesirable platelike σ, Laves, and similar TCP phases.

Microstructure. Virtually all wrought cobalt alloys are based on an fcc matrix obtained by alloying with 10% or more nickel (some cast alloys contain no nickel). Iron, manganese, and carbon additions also stabilize the fcc phase, while nickel and iron additions improve workability. Exerting the opposite hcp stabilizing tendency are other common alloying elements, such as tungsten, added primarily for solid-solution strengthening, as in L-605, and chromium, added primarily for oxidation and hot corrosion resistance. Tungsten is favored over molybdenum as a strengthener, even though the latter is more effective per atomic percent because tungsten,

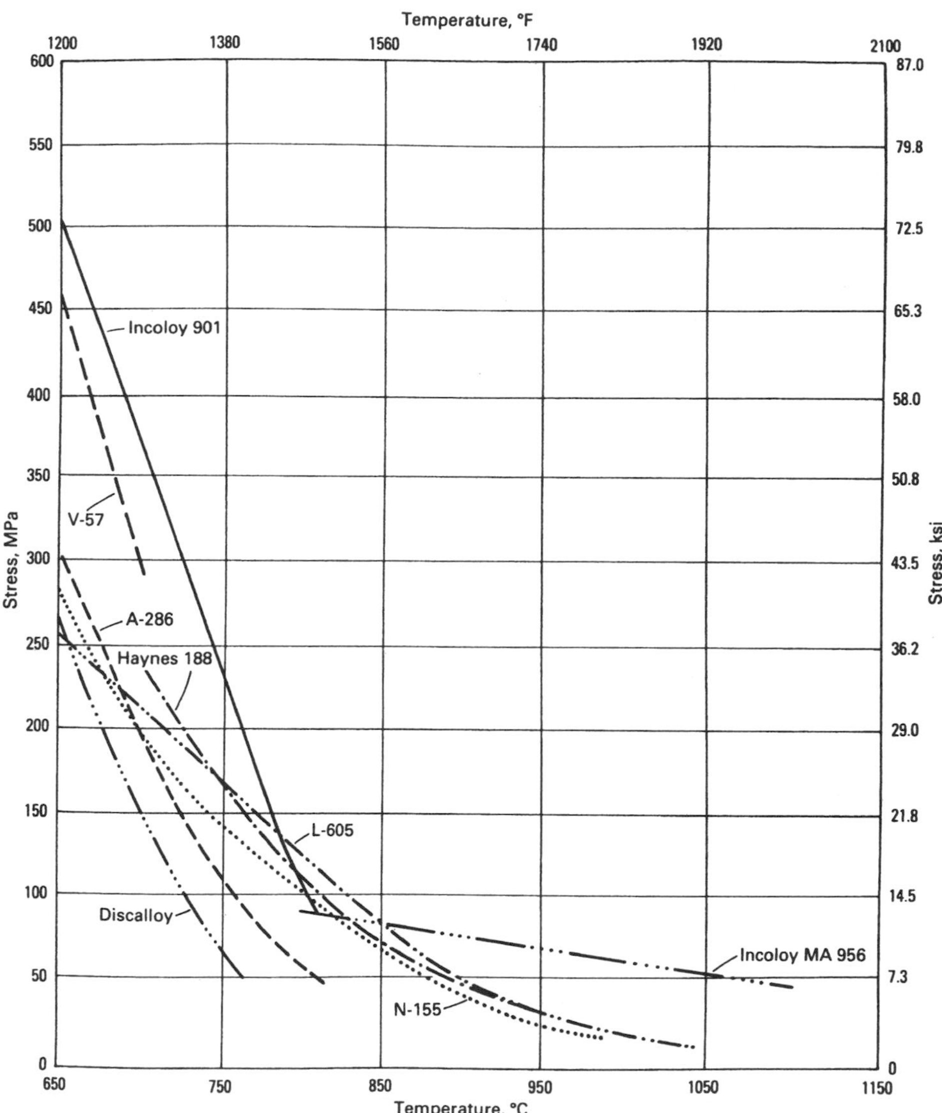

Fig. 11 1000 h stress-rupture curves of wrought cobalt-base (Haynes 188 and L-605) and wrought iron-base superalloys

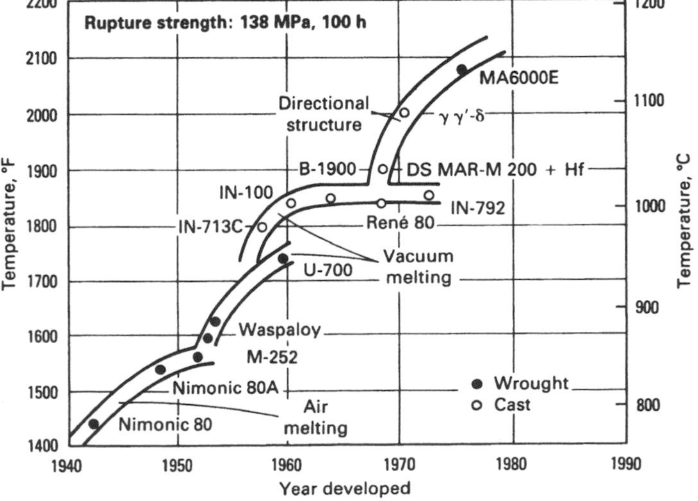

Fig. 12 Progress in the high-temperature capabilities of superalloys since the 1940s

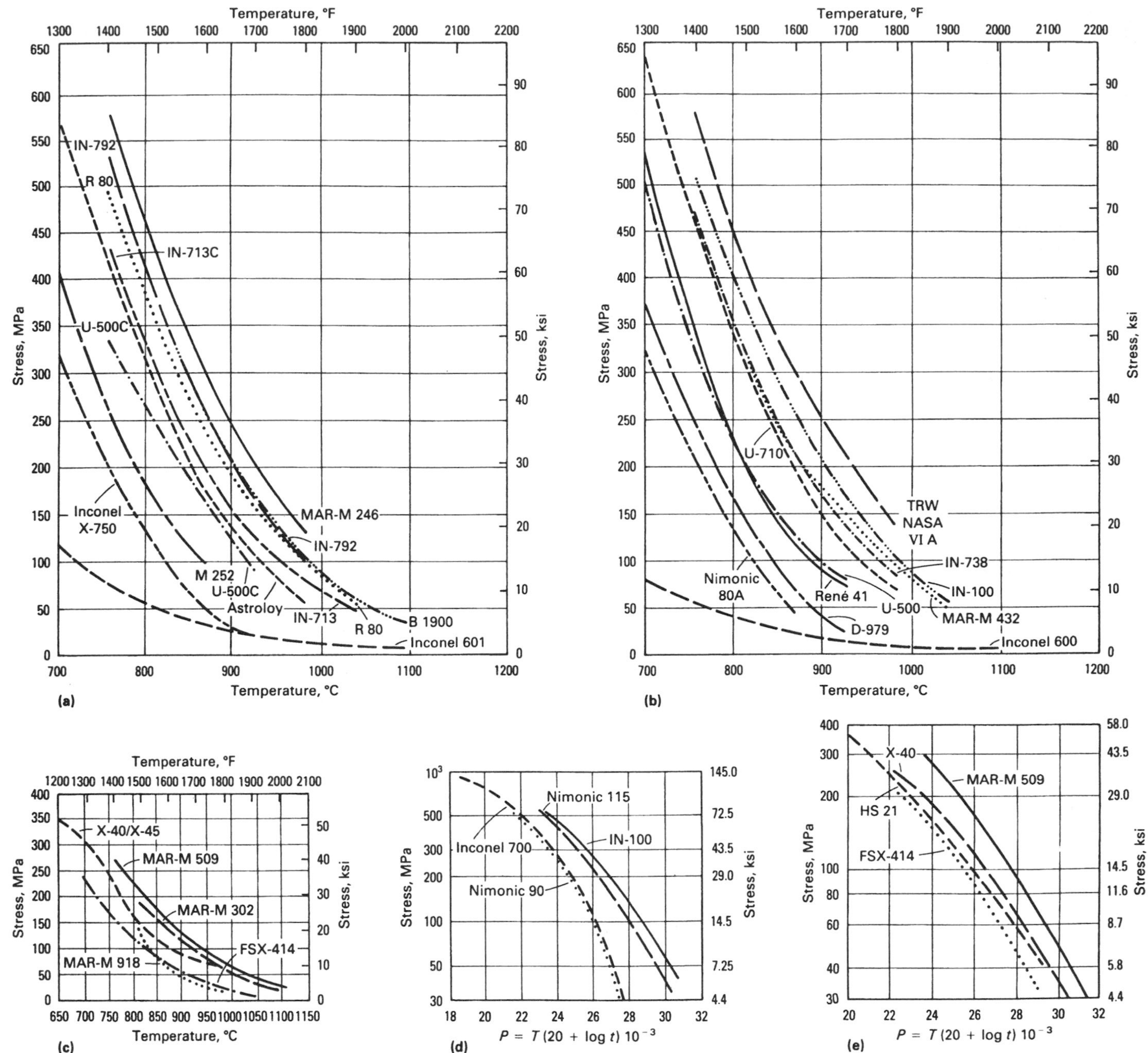

Fig. 13 Stress-rupture curves for selected superalloys. (a) and (b) Nickel-base superalloys. 1000 h. (c) Cobalt-base superalloys. 1000 h. Source: Ref 12. (d) Larson-Miller stress-rupture curves for selected nickel-base superalloys. (e) Larson-Miller stress-rupture curves for selected cobalt-base superalloys. Source: Ref 19

alone among potential solutes in cobalt, raises the melting temperature of cobalt. Tantalum has been used as a replacement for tungsten in sheet alloys MAR-M 918 and S-57, while molybdenum contents of up to 10 wt% are found in the work-hardened Multiphase (MP) alloys.

Improved oxidation and corrosion resistance with 5 wt% Al have been noted in a few cobalt-base alloys (Ref 23). Titanium additions also have been made in order to precipitate coherent, ordered Co_3Ti as a strengthening phase. Unfortunately, this phase is stable only to about 700 °C (1290 °F), which is much lower than for γ' Ni_3Al,Ti in nickel-base alloys.

Increased susceptibility to Laves-phase precipitation has been noted in the presence of silicon in tungsten-containing alloys such as L-605. Haynes 188 was developed with reduced tungsten, increased nickel, and controlled silicon contents in order to avoid this problem.

As in the case of nickel-base alloys, a variety of carbides have been found in cobalt alloys. These include $M_{23}C_6$, M_6C, and MC carbides. In both L-605 and Haynes 188, M_6C transforms in $M_{23}C_6$ during exposure to temperatures in the range of 816 to 927 °C (1500 to 1700 °F) for 3000 h. MC carbides are found only in alloys containing tantalum, niobium, zirconium, titanium, or hafnium.

In addition to carbides, small quantities of intermetallic phases such as Co_3W, Co_2W, and Co_7W_6 have been found in L-605. Other alloys display the compounds CoAl, Co_3Ti, and $Co_2(Ta,Nb,Ti)$ are TCP phases that are likely to cause the deterioration of mechanical properties. A uniform dispersion of $Co_3(Ti,Al)$ has been

Table 8 Physical properties of selected wrought superalloys

Designation	Form	Density, g/cm³	Melting range		Specific heat capacity at: 21 °C (70 °F)		538 °C (1000 °F)		871 °C (1600 °F)	
			°C	°F	J/kg · K	Btu/lb · °F	J/kg · K	Btu/lb · °F	J/kg · K	Btu/lb · °F
Nickel-base										
Astroloy	Bar	7.91	...	...	...	...	...	...	...	...
D-979	Bar	8.19	1200–1390	2225–2530	...	...	...	...	...	...
Hastelloy X	Sheet	8.21	1260–1355	2300–2470	485	0.116	...	...	700	0.167
Hastelloy S	Bar	8.76	1335–1380	2435–2515	405	0.097	495	0.118	595	0.142
Inconel 597	Bar	8.04	...	...	...	...	...	...	...	...
Inconel 600	Bar	8.41	1355–1415	2470–2575	445	0.106	555	0.132	625	0.149
Inconel 601	Bar	8.05	1300–1370	2375–2495	450	0.107	590	0.140	680	0.162
Inconel 617	Bar	8.36	1330–1375	2430–2510	420	0.100	550	0.131	630	0.150
Inconel 625	Bar	8.44	1290–1350	2350–2460	410	0.098	535	0.128	620	0.148
Inconel 690	Bar	8.14	1345–1375	2450–2510	...	...	...	...	...	...
Inconel 706	Bar	8.08	1335–1370	2435–2500	445	0.106	580	0.138	670	0.159
Inconel 718	Bar	8.22	1260–1335	2300–2435	430	0.102	560	0.133	645	0.153
Inconel X750	Bar	8.25	1395–1425	2540–2600	430	0.103	545	0.130	715	0.171
Haynes 230	...	8.8	1300–1370	2375–2500	397	0.095	473	0.112	595	0.145
M-252	Bar	8.25	1315–1370	2400–2500	...	...	...	...	...	...
Nimonic 75	Bar	8.37	...	...	460	0.11	...	...	...	...
Nimonic 80A	Bar	8.16	1360–1390	2480–2535	460	0.11	...	...	...	...
Nimonic 81	Bar	8.06	...	...	460	0.11	585	0.14	670	0.16
Nimonic 90	Bar	8.19	1335–1360	2435–2480	460	0.11	585	0.14	670	0.16
Nimonic 105	Bar	8.00	...	...	420	0.10	545	0.13	670	0.16
Nimonic 115	Bar	7.85	...	...	460	0.11	...	...	...	...
Nimonic 263	Sheet	8.36	...	...	460	0.11	...	...	...	...
Nimonic 942	Bar	8.19	1240–1300	2265–2370	420	0.10	...	...	...	...
Nimonic PE.11	Bar	8.02	1280–1350	2335–2460	420	0.10	585	0.14	...	...
Nimonic PE.16	Bar	8.02	...	...	545	0.13	...	...	...	...
Nimonic PK.33	Sheet	8.21	...	...	420	0.10	545	0.13	670	0.16
Pyromet 860	Bar	8.21	...	...	...	...	...	...	...	...
René 41	Bar	8.25	1315–1370	2400–2500	...	...	545	0.13	725	0.173
René 95	Bar	...	...	...	...	...	...	...	...	...
Udimet 500	Bar	8.02	1300–1395	2375–2540	...	...	...	...	...	...
Udimet 520	Bar	8.21	1260–1405	2300–2560	...	...	...	...	...	...
Udimet 700	Bar	7.91	1205–1400	2200–2550	...	...	575	0.137	590	0.141
Unitemp AF2-1DA	Bar	8.26	...	...	420	0.100	...	...	...	...
Waspaloy	Bar	8.19	1330–1355	2425–2475	...	...	...	...	...	...
Iron-base										
Alloy 901	Bar	8.21	1230–1400	2250–2550	...	...	...	...	...	...
A-286	Bar	7.91	1370–1400	2500–2550	460	0.11	...	...	...	...
Discaloy	Bar	7.97	1380–1465	2515–2665	475	0.113	...	...	...	...
Haynes 556	Sheet	8.23	...	...	450	0.107	...	...	...	...
Incoloy 800	Bar	7.95	1355–1385	2475–2525	455	0.108	...	...	...	...
Incoloy 801	Bar	7.95	1355–1385	2475–2525	455	0.108	...	...	...	...
Incoloy 802	Bar	7.83	1345–1370	2450–2500	445	0.106	...	...	...	...
Incoloy 807	Bar	8.32	1275–1355	2325–2475	...	...	...	...	...	...
Incoloy 825(a)	...	8.14	1370–1400	2500–2550	440	0.105	...	...	...	...
Incoloy 903	Bar	8.14	1320–1395	2405–2540	435	0.104	...	...	...	...
Incoloy 904	Bar	8.12	...	...	460	0.11	...	...	...	...
Incoloy 907(a)	...	8.33	1335–1400	2440–2550	431	0.103	...	...	...	...
Incoloy 909(a)	...	8.30	1395–1430	2540–2610	427	0.102	...	...	...	...
N-155	Bar	8.19	1275–1355	2325–2475	430	0.103	...	...	...	...
19-9 DL(b)	...	7.9	1425–1430	2600–2610	...	...	...	...	...	...
16-25-6(b)	...	8.0	...	...	...	...	...	...	...	...
Cobalt-base										
Haynes 188	Sheet	9.13	1300–1330	2375–2425	405	0.097	510	0.122	565	0.135
L-605	Sheet	9.13	1330–1410	2425–2570	385	0.092	...	...	...	...
Stellite 6B(a)	...	8.38	1265–1354	2310–2470	421	0.101	...	...	...	...
Haynes 150(b)	...	8.05	1395	2540	...	...	...	...	...	...
MP35N(b)	...	8.41	1315–1425	2400–2600	...	...	...	...	...	...
Elgiloy(b)	...	8.3	1495	2720	...	...	...	...	...	...

Designation	Form	Thermal conductivity at: 21 °C (70 °F)		538 °C (1000 °F)		871 °C (1600 °F)		Mean coefficient of thermal expansion, 10^{-6}/K		Electrical resistivity, nΩ · m
		W/m · K	Btu/ft² · in. · h · °F	W/m · K	Btu/ft² · in. · h · °F	W/m · K	Btu/ft² · in. · h · °F	At 538 °C (1000 °F)	At 871 °C (1600 °F)	
Nickel-base										
Astroloy	Bar	...	...	...	...	...	...	13.9	16.2	...
D-979	Bar	12.6	87	18.5	128	...	...	14.9	17.7	...
Hastelloy X	Sheet	9.1	63	19.6	136	26.0	180	15.1	16.2	1180(c)
Hastelloy S	Bar	...	...	20.0	139	26.1	181	13.3	14.9	...
Inconel 597	Bar	...	...	18.2	126	...	...	...	...	...
Inconel 600	Bar	14.8	103	22.8	158	28.8	200	15.1	16.4	1030(c)

(continued)

(a) Ref 14. (b) Ref 8. (c) Ref 17. (d) At 21 to 93 °C (70 to 200 °F). (e) At 25 to 427 °C (77 to 800 °F). (f) At 705 °C (1300 °F). (g) At 980 °C (1800 °F). Source: Ref 13, unless otherwise noted

Table 8 (continued)

Designation	Form	Thermal conductivity at: 21 °C (70 °F) W/m · K	Btu/ft² · in. · h · °F	538 °C (1000 °F) W/m · K	Btu/ft² · in. · h · °F	871 °C (1600 °F) W/m · K	Btu/lb · °F	Mean coefficient of thermal expansion, 10^{-6}/K At 538 °C (1000 °F)	At 871 °C (1600 °F)	Electrical resistivity, nΩ · m
Nickel-base (continued)										
Inconel 601	Bar	11.3	78	20.0	139	25.7	178	15.3	17.1	1190(d)
Inconel 617	Bar	13.6	94	21.5	149	26.7	185	13.9	15.7	1220(d)
Inconel 625	Bar	9.8	68	17.5	121	22.8	158	14.0	15.8	1290(c)
Inconel 690	Bar	13.3	95	22.8	158	27.8	193	...	...	148(c)
Inconel 706	Bar	12.6	87	21.2	147	...	...	15.7	...	...
Inconel 718	Bar	11.4	79	19.6	136	24.9	173	14.4	...	1250(d)
Inconel X750	Bar	12.0	83	18.9	131	23.6	164	14.6	16.8	1220(d)
Haynes 230	...	8.9	62	18.4	133	24.4	179	14.0	15.2	1250
M-252	Bar	11.8	82	...	...	...	...	13.0	15.3	...
Nimonic 75	Bar	...	...	...	...	...	...	14.7	17.0	1090(d)
Nimonic 80A	Bar	8.7	60	15.9	110	22.5	156	13.9	15.5	1240(d)
Nimonic 81	Bar	10.8	75	19.2	133	25.1	174	14.2	17.5	1270(d)
Nimonic 90	Bar	9.8	68	17.0	118	...	...	13.9	16.2	1180(d)
Nimonic 105	Bar	10.8	75	18.6	129	24.0	166	13.9	16.0	1310(d)
Nimonic 115	Bar	10.7	74	17.6	124	22.6	154	13.3	16.4	1390(d)
Nimonic 263	Sheet	11.7	81	20.4	141	26.2	182	13.7	16.2	1150(d)
Nimonic 942	Bar	...	...	...	...	...	...	14.7	16.5	...
Nimonic PE.11	Bar	...	...	...	...	...	...	15.2	...	...
Nimonic PE.16	Bar	11.7	81	20.2	140	26.4	183	15.3	18.5	1100(d)
Nimonic PK.33	Sheet	10.7	74	19.2	133	24.7	171	13.1	16.2	1260(d)
Pyromet 860	Bar	...	...	...	...	...	...	15.4	16.4	...
René 41	Bar	9.0	62	18.0	125	23.1	160	13.5	15.6	1308(c)
René 95	Bar	8.7	60	17.4	120	...	...	...	...	...
Udimet 500	Bar	11.1	77	18.3	127	24.5	170	14.0	16.1	1203(c)
Udimet 700	Bar	19.6	136	20.6	143	27.7	192	13.9	16.1	...
Unitemp AF2-1DA	Bar	10.8	75	16.5	114	19.5	135	12.4	14.1	...
Waspaloy	Bar	10.7	74	18.1	125	24.1	167	14.0	16.0	1240
Iron-base										
Alloy 901	Bar	13.3	92	...	...	...	...	15.3	...	...
A-286	Bar	12.7	88	22.5	156	...	...	17.6	...	...
Discaloy	Bar	13.3	92	21.1	146	...	...	17.1	...	...
Haynes 556	Sheet	11.6	80	17.5	121	...	...	16.2	17.5	...
Incoloy 800	Bar	11.6	80	20.1	139	...	...	16.4	18.4	989
Incoloy 801	Bar	12.4	86	20.7	143	25.6	177	17.3	18.7	...
Incoloy 802	Bar	11.9	82	19.8	137	24.2	168	16.7	18.2	...
Incoloy 807	Bar	...	...	...	...	...	...	15.2	17.6	...
Incoloy 825(a)	...	11.1	76.8	...	...	...	...	14.0(d)	...	1130
Incoloy 903	Bar	16.8	116	20.9	145	...	...	8.6	...	610
Incoloy 904	Bar	16.8	116	22.4	155	...	...	...	...	...
Incoloy 907(a)	...	14.8	103	...	...	...	...	7.7(e)	...	697
Incoloy 909(a)	...	14.8	103	...	...	...	...	7.7(e)	...	728
N-155	Bar	12.3	85	19.2	133	...	...	16.4	17.8	...
19-9 DL(b)	...	...	...	...	...	...	...	17.8	...	...
16-25-6(b)	...	...	...	15	104	...	...	16.9	...	...
Cobalt-base										
Haynes 188	Sheet	...	...	19.9	138	25.1	174	14.8	17.0	922
L-605	Sheet	9.4	65	19.5	135	26.1	181	14.4	16.3	890
Stellite 6B(c)	...	14.7	101	...	...	...	...	15.0	16.9	910
Haynes 150(b)	...	...	...	0.75(f)	5.2	...	...	...	16.8(g)	810
MP35N(b)	...	...	...	...	...	...	...	...	15.7(g)	1010
Elgiloy(b)	...	1.0	7.2	1.4	10	...	...	...	15.8(g)	995

(a) Ref 14. (b) Ref 8. (c) Ref 17. (d) At 21 to 93 °C (70 to 200 °F). (e) At 25 to 427 °C (77 to 800 °F). (f) At 705 °C (1300 °F). (g) At 980 °C (1800 °F). Source: Ref 13, unless otherwise noted

achieved in CM-7 (modified L-605 containing aluminum and titanium) by solution treatment at 1200 °C (2190 °F) and aging at 800 °C (1470 °F). However, this microstructure is unstable at temperatures of 815 °C (1500 °F) and above for times over 1000 h. Figure 17 shows the distribution of phases after various heat treatments for two age-hardened cobalt-base alloys, Haynes 25 and Haynes 188.

An additional source of strengthening, in the Multiphase alloys, arises from the strain-induced transformation of the γ matrix to an hcp structure. This transformation is closely linked to the occurrence of stacking faults in cobalt-base alloys. The lower the stacking fault energy of the γ matrix (as in alloys with low nickel and high refractory element contents), the more readily faulting will occur.

Applications and Properties. Various wrought cobalt-base alloys can be grouped according to use:

- Alloys for use primarily at high temperatures from 650 to 1150 °C (1200 to 2100 °F), including S-816, Haynes 25, Haynes 188, Haynes 556, and UMCo-50
- Corrosion-resistant alloys MP35N and MP159, for use to about 650 °C (1200 °F)
- Wear-resistant Stellite 6B and similar alloys

All alloys in the heat-treated and softened condition have fcc crystal structures. However, alloys MP35N and MP159 develop controlled amounts of hcp structure during the thermomechanical processing recommended before service applications. They are typically used as fasteners for service to 650 °C (1200 °F). Stellite 6B, when heat treated between 650 and 1060 °C (1200 and 1900 °F), and Haynes 25, when exposed for 1000 h or more at temperatures near 650 °C (1200 °F), may partly transform to an hcp structure.

Table 9 Effect of temperature on the mechanical properties of cast nickel-base and cobalt-base alloys

Alloy	Ultimate tensile strength at:						0.2% yield strength at:						Tensile elongation, % at:			Dynamic modulus of elasticity at:					
	21 °C (70 °F)		538 °C (1000 °F)		1093 °C (2000 °F)		21 °C (70 °F)		538 °C (1000 °F)		1093 °C (2000 °F)		21 °C (70 °F)	538 °C (1000 °F)	1093 °C (2000 °F)	21 °C (70 °F)		538 °C (1000 °F)		1093 °C (2000 °F)	
	MPa	ksi	MPa	ksi	MPa	ksi	MPa	ksi	MPa	ksi	MPa	ksi				GPa	10^6 psi	GPa	10^6 psi	GPa	10^6 psi
Nickel-base																					
IN-713 C	850	123	860	125	...	...	740	107	705	102	...	...	8	10	...	206	29.9	179	26.2	...	...
IN-713 LC	895	130	895	130	...	...	750	109	760	110	...	...	15	11	...	197	28.6	172	25.0	...	...
B-1900	970	141	1005	146	270	38	825	120	870	126	195	28	8	7	11	214	31.0	183	27.0	...	...
IN-625	710	103	510	74	...	...	350	51	235	34	...	...	48	50	...	...	...	...	...	...	...
IN-718	1090	158	...	...	...	...	915	133	...	...	...	...	11	...	...	...	...	...	...	...	...
IN-100	1018	147	1090	150	(380)	(55)	850	123	885	128	(240)	(35)	9	9	...	215	31.2	187	27.1	...	...
IN-162	1005	146	1020	148	...	...	815	118	795	115	...	...	7	6.5	...	197	28.5	172	24.9	...	...
IN-731	835	121	...	...	275	40	725	105	...	...	170	25	6.5	...	...	...	...	...	...	...	...
IN-738	1095	159	...	...	...	...	950	138	...	...	...	...	...	...	...	201	29.2	175	25.4	...	...
IN-792	1170	170	...	...	...	...	1060	154	...	...	...	...	4	...	...	...	...	...	...	...	...
M-22	730	106	780	113	...	...	685	99	730	106	...	...	5.5	4.5	...	...	...	...	...	...	...
MAR-M 200	930	135	945	137	325	47	840	122	880	123	...	...	7	5	...	218	31.6	184	26.7	...	...
MAR-M 246	965	140	1000	145	345	50	860	125	860	125	...	...	5	5	...	205	29.8	178	25.8	145	21.1
MAR-M 247	965	140	1035	150	...	...	815	118	825	120	...	...	7	...	...	...	...	...	...	...	...
MAR-M 421	1085	157	995	147	...	...	930	135	815	118	...	...	4.5	3	...	203	29.4	...	...	141	20.4
MAR-M 432	1240	180	1105	160	...	...	1070	155	910	132	...	...	6	...	...	...	...	...	...	...	...
MC-102	675	98	655	95	...	...	605	88	540	78	...	...	5	9	...	...	...	...	...	...	...
Nimocast 75	500	72	...	...	...	...	179	26	...	...	...	...	39	...	...	...	...	...	...	...	...
Nimocast 80	730	106	...	...	...	...	520	75	...	...	...	...	15	...	...	...	...	...	...	...	...
Nimocast 90	700	102	595	86	...	...	520	75	420	61	...	...	14	15	...	...	...	...	...	...	...
Nimocast 242	460	67	...	...	...	...	300	44	...	...	...	...	8	...	...	...	...	...	...	...	...
Nimocast 263	730	106	...	...	...	...	510	74	...	...	...	...	18	...	...	...	...	...	...	...	...
René 77	...	...	...	...	...	...	...	...	...	...	...	...	...	...	...	...	...	...	...	...	...
René 80	...	...	...	...	...	...	...	...	...	...	...	...	...	...	...	208	30.2	...	...	...	...
Udimet 500	930	135	895	130	...	...	815	118	725	105	...	...	13	13	...	...	...	...	...	...	...
Udimet 710	1075	156	...	...	240	35	895	130	...	...	170	25	8	...	...	...	...	...	...	...	...
CMSX-2(a)(b)	1185	172	1295(c)	188(c)	...	...	1135	165	1245(c)	181(c)	...	...	10	17(c)	...	...	...	...	...	...	...
GMR-235(b)	710	103	...	...	...	...	640	93	...	...	...	...	3	...	18(d)	...	...	...	...	...	...
IN-939(b)	1050	152	915(c)	133(c)	325(d)	47(d)	800	116	635(c)	92(c)	205(d)	30(d)	5	7(c)	25(d)	...	...	...	...	...	...
MM 002(b)(e)	1035	150	1035(c)	150(c)	550(d)	80(d)	825	120	860(c)	125(c)	345(d)	50(d)	7	5(c)	12(d)	...	...	...	...	...	...
IN-713 Hf(b)(f)	1000	145	895(c)	130(c)	380(d)	55(d)	760	110	620(c)	90(c)	240(d)	35(d)	11	6(c)	20(d)	...	...	...	...	...	...
René 125 Hf(b)(g)	1070	155	1070(c)	155(c)	550(d)	80(d)	825	120	860(c)	125(c)	345(d)	50(d)	5	5(c)	12(d)	...	...	...	...	...	...
MAR-M 246 Hf(b)(h)	1105	160	1070(c)	155(c)	565(d)	82(d)	860	125	860(c)	125(c)	345(d)	50(d)	6	7(c)	14(d)	...	...	...	...	...	...
MAR-M 200 Hf(b)(i)	1035	150	1035(c)	150(c)	540(d)	78(d)	825	120	860(c)	125(c)	345(d)	50(d)	5	5(c)	10(d)	...	...	...	...	...	...
PWA-1480(a)(b)	...	...	1130(c)	164(c)	685(d)	99(d)	895	130	905(c)	131(c)	495(d)	72(d)	4	8(c)	20(d)	...	...	...	...	...	...
SEL(b)	1020	148	875(c)	127(c)	...	...	905	131	795(c)	115(c)	...	...	6	7(c)	...	...	...	...	...	...	...
UDM 56(b)	945	137	945(c)	137(c)	...	...	850	123	725(c)	105(c)	...	...	3	5(c)	...	...	...	...	...	...	...
SEL-15(b)	1060	154	1090(c)	158(c)	...	...	895	130	815(c)	118(c)	...	...	9	5(c)	...	...	...	...	...	...	...
Cobalt-base																					
AiResist 13(j)	600	87	420(c)	61(c)	...	...	530	77	330(c)	48(c)	...	...	1.5	4.5(c)	...	...	...	...	...	...	...
AiResist 215(j)	690	100	570(k)	83(k)	...	...	485	70	315(k)	46(k)	...	...	4	12(k)	...	...	...	...	...	...	...
FSX-414	...	...	...	...	...	...	...	...	...	...	...	...	...	...	...	...	...	...	...	...	...
Haynes 1002	770	112	560	81	115	17	470	68	345	50	95	14	6	8	28	210	30.4	173	25.1	...	...
MAR-M 302	930	135	795	115	150	22	690	100	505	73	150	22	2	...	21	...	...	...	...	...	...
MAR-M 322(j)	830	120	595(c)	86(c)	...	...	630	91	345(c)	50(c)	...	...	4	6.5(c)	...	...	...	...	...	...	...
MAR-M 509	785	114	570	83	...	...	570	83	400	58	...	...	4	6	...	225	32.7	...	...	...	...
WI-52	750	109	745	108	160	23	585	85	440	64	105	15	5	7	35	...	...	...	...	...	...
X-40	745	108	550	80	...	...	525	76	275	40	...	...	9	17	...	...	...	...	...	...	...

(a) Single crystal [001]. (b) Data from Ref 12. (c) At 760 °C (1400 °F). (d) At 980 °C (1800 °F). (e) RR-7080. (f) MM 004. (g) M 005. (h) MM 006. (i) MM 009. (j) Data from Volume 3, 9th Edition, Metals Handbook, 1980. (k) At 650 °C (1200 °F). Source: Nickel Development Institute, except as noted

None of the cobalt-base superalloys is a complete solid-solution alloy because all contain secondary phases in the form of carbides (M_6C, $M_{23}C_3$, or MC) or intermetallic compounds. Aging causes additional second-phase precipitation, which generally results in some loss of room-temperature ductility. Of the high-temperature group, alloy S-816 originally was used extensively in turbochargers, as well as in gas turbine wheels, blades, and vanes, but has been largely replaced by higher-strength, lower-density nickel-base alloys with improved resistance to adverse environments.

Haynes 25 is perhaps the best-known wrought cobalt-base alloy and has been widely used for hot sections of gas turbines, components of nuclear reactors, devices for surgical implants, and, in the cold-worked condition, for fasteners and wear pads.

Haynes 188 was specially designed for sheet metal components, such as combustors and transition ducts, in gas turbines. The basic composition, provided that lanthanum, silicon, aluminum, and manganese contents are judiciously controlled, provides excellent qualities, such as oxidation resistance at temperatures up to 1100 °C (2000 °F), hot corrosion resistance, creep resistance, room-temperature formability, and ductility after long-term aging at service temperatures. As shown in Fig. 18, Haynes 188 displays creep resistance comparable to that of solid-solution strengthened nickel-base alloys.

Figure 19 compares the static oxidation resistance of Haynes 188 with that of several other alloys. It should be noted that Haynes 188 has excellent oxidation resistance in dry air (Fig. 19a) but displays inferior properties in moist air (Fig. 19b). Haynes 188, like Haynes 25, MP35N, and MP159, can be work hardened to relatively high hardness and tensile strength; after 50% cold reduction, Haynes 188 has a tensile strength of 1690 MPa (245 ksi) at room temperature and 1585 MPa (230 ksi) at 540 °C (1000 °F). Additional property data for wrought cobalt alloys can be found in Tables 5 to 8.

UMCo-50, which contains about 21% Fe, is not as strong as Haynes 25 or 188. It is not used extensively in the United States, and especially not in gas turbine applications. In Europe, on the other hand, it is used extensively for furnace parts and fixtures.

The last group of wrought cobalt alloys consists of Stellite 6B, and similar wear-resistant alloys, which are characterized by high hot hardness and relatively good resistance to oxidation. The latter property is derived chiefly from the high chromium content of these alloys (about 30%), whereas the hot hardness property is obtained through the formation of complex carbides of the Cr_7Co_3 and $M_{23}C_6$ types. Stellite 6B is widely used for erosion shields in steam turbines, for wear pads in gas turbines, and for bends in tube systems carrying particulate matter at high temperatures and high velocities.

Cast cobalt-base superalloys, like their wrought counterparts, are also used in heat-resistant, wear-resistant, and corrosion-resistant applications. Table 2 lists compositions of cast cobalt alloys.

Cobalt-base castings used for gas turbine engine components (blades, vanes, combustion chambers) include FSX-414, MAR-M 302, MAR-M 509, and the AiResist alloys. Some of these alloys contain higher carbon (up to 1.0%) and higher tantalum (up to 9%) than wrought alloys. Stress-rupture curves for cobalt-base aerospace alloys are shown in Fig. 13(c), (d), and (e). Additional property data are given in Tables 9, 11, and 12.

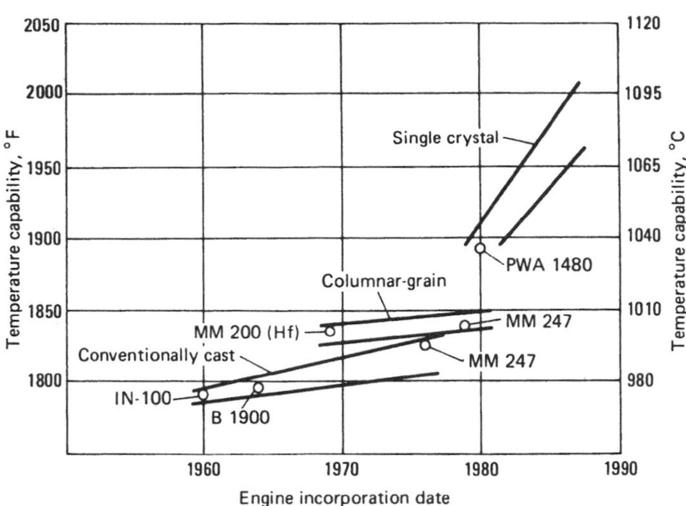

Fig. 14 Advances in turbine blade materials and processes since 1960. Source: Ref 20

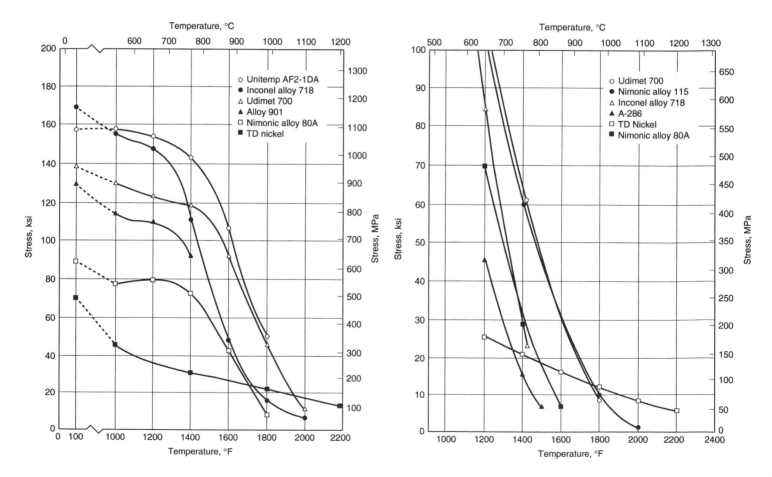

Fig. 15 Effect of temperature on the 0.2% offset yield strength of Inconel 718 and competing alloys. Product form: bar stock. Source: Ref 13

Fig. 16 1000 h stress-rupture strength of Inconel 718 and competing alloys. Product form: bar stock. Source: Ref 13

One of the most important application areas for corrosion-resistant cast cobalt alloys is that of orthopedic implants. One commonly used investment cast alloy is ASTM F 75, which has the following chemical composition:

Element	wt %
C	0.35
Mn	1.00
Si	0.40
Cr	27.0–30.0
Ni	1.00
Mo	6.0
Fe	1.50
Co	bal

Table 10 Stress-rupture strengths for selected cast nickel-base superalloys

	Rupture stress at:					
	815 °C (1500 °F)		870 °C (1600 °F)		980 °C (1800 °F)	
Alloy	100 h MPa (ksi)	1000 h MPa (ksi)	100 h MPa (ksi)	1000 h MPa (ksi)	100 h MPa (ksi)	1000 h MPa (ksi)
IN-713 LC	425 (62)	325 (47)	295 (43)	240 (35)	140 (20)	105 (15)
IN-713 C	370 (54)	305 (44)	305 (44)	215 (31)	130 (19)	70 (10)
IN-738 C	470 (68)	345 (50)	330 (48)	235 (34)	130 (19)	90 (13)
IN-738 LC	430 (62)(a)	315 (46)	295 (43)(a)	215 (31)	140 (20)(a)	90 (13)
IN-100	455 (66)	365 (53)	360 (52)	260 (38)	160 (23)	90 (13)
MAR-M 247 (MM 0011)	585 (85)	415 (60)	455 (66)	290 (42)	185 (27)	125 (18)
MAR-M 246(a)	525 (76)	435 (62)	440 (63)	290 (42)	195 (28)	125 (18)
MAR-M 246 Hf (MM 006)	530 (77)	425 (62)	425 (62)	285 (41)	205 (30)	130 (19)
MAR-M 200	495 (72)(a)	415 (60)(a)	385 (56)(a)	295 (43)(a)	170 (25)	125 (18)
MAR-M 200 Hf (MM 009)(b)	…	…	…	305 (44)	…	125 (18)
B-1900	510 (74)	380 (55)	385 (56)	250 (36)	180 (26)	110 (16)
René 77(a)	…	…	310 (45)	215 (31.5)	130 (19)	62 (9.0)
René 80	…	…	350 (51)	240 (35)	160 (23)	105 (15)
IN-625(a)	130 (19)	110 (16)	97 (14)	76 (11)	34 (5)	28 (4)
IN-162(a)	505 (73)	370 (54)	340 (49)	255 (37)	165 (24)	110 (16)
IN-731(a)	505 (73)	365 (53)	…	…	165 (24)	105 (15)
IN-792(a)	515 (75)	380 (55)	365 (53)	260 (38)	165 (24)	105 (15)
M-22(a)	515 (75)	385 (56)	395 (57)	285 (41)	200 (29)	130 (19)
MAR-M 421(a)	450 (65)	305 (44)	310 (46)	215 (31)	125 (18)	83 (12)
MAR-M 432(a)	435 (63)	330 (48)	295 (40)	215 (31)	140 (20)	97 (14)
MC-102(a)	195 (28)	145 (21)	145 (21)	105 (15)	…	…
Nimocast 90(a)	160 (23)	110 (17)	125 (18)	83 (12)	…	…
Nimocast 242(a)	110 (16)	83 (12)	90 (13)	59 (8.6)	45 (6.5)	…
Udimet 500(a)	330 (48)	240 (35)	230 (33)	165 (24)	90 (13)	…
Udimet 710(a)	420 (61)	325 (47)	305 (44)	215 (31)	150 (22)	76 (11)
CMSX-2(b)	…	…	…	345 (50)	…	170 (25)
GMR-235(b)	…	…	…	180 (26)	…	75 (11)
IN-939(b)	…	…	…	195 (28)	…	60 (9)
MM 002(b)	…	…	…	305 (44)	…	125 (18)
IN-713 Hf (MM 004)(b)	…	…	…	205 (30)	…	90 (13)
René 125 Hf (MM 005)(b)	…	…	…	305 (44)	…	115 (17)
SEL-15(b)	…	…	…	295 (43)	…	75 (11)
UDM 56(b)	…	…	…	270 (39)	…	125 (18)

(a) Ref 12. (b) Ref 18

Table 11 Stress-rupture strengths for selected cast cobalt-base superalloys

	Rupture stress at:							
	815 °C (1500 °F)		870 °C (1600 °F)		980 °C (1800 °F)		1095 °C (2000 °F)	
Alloy	100 h MPa (ksi)	1000 h MPa (ksi)	100 h MPa (ksi)	1000 h MPa (ksi)	100 h MPa (ksi)	1000 h MPa (ksi)	100 h MPa (ksi)	1000 h MPa (ksi)
HS-21	150 (22)	95 (14)	115 (17)	90 (13)	60 (9)	50 (7)	…	…
X-40 (HS-31)	180 (26)	140 (20)	130 (19)	105 (15)	75 (11)	55 (8)	…	…
MAR-M 509	270 (39)	225 (33)	200 (29)	140 (20)	115 (17)	90 (13)	55 (8)	41 (6)
FSX-414	150 (22)	115 (17)	110 (16)	85 (12)	55 (8)	35 (5)	21 (3)	…
WI-52	…	195 (28)	175 (25)	150 (22)	90 (13)	70 (10)	…	…

Cast cobalt-chromium-molybdenum alloys such as ASTM F 75 have been used for medical implants since the 1950s. An endoprosthesis consists of a large ball that replaces the femoral head attached to a shaft that extends down into the intermedullary canal of the femur. Endoprostheses remain in wide use today.

More recently, total hip replacement has become common. In this surgery, a plastic socket is placed into the acetabulum (the natural socket of the pelvis), while a metallic hip stem consisting of a shaft and ball is placed in the femur. The shaft is often held in place by a polymethyl methacrylate cement that acts like a grout between the bone and the stem. This ball and socket arrangement replaces a diseased joint to provide mobility and pain relief. These types of devices are shown in Fig. 20.

Cobalt-base alloys are also widely used in total knee replacements and, to a lesser extent, as implants that fix bone fractures, that is, bone screws, staples, and plates. The support structures for heart valves are also often fabricated from cobalt-base alloys. A variety of dental implants have also been produced from cobalt alloys.

Oxidation and Hot Corrosion. The influence of solute elements on the oxidation of cobalt-base alloys has been summarized by Sims (Ref 23) and Beltran (Ref 24) (see Table 14). Among refractory metals, only tantalum is considered beneficial, but tantalum is not added to wrought alloys. Tungsten, molybdenum, vanadium, and niobium are decidedly harmful. Caution must be exercised because carbide distribution may also play a role in oxidation resistance. Beltran (Ref 24) has suggested that the same refractory metals that degrade oxidation resistance play a similar role in hot corrosion. Niobium is cited as being harmful to both the oxidation and corrosion resistance of Wi-52, a cast alloy, but less so for wrought S-816, even though the latter contains twice the niobium content.

Pettit and Giggins (Ref 25) note that cobalt and cobalt-aluminum alloys with chromium and aluminum contents below the level needed to form external scales of Cr_2O_3 or Al_2O_3 are susceptible to a form of catastrophic hot corrosion known as basic fluxing. This type of attack is inhibited by higher levels of chromium or aluminum. In general, cobalt-chromium alloys are less susceptible than nickel-aluminum and nickel-aluminum-X alloys, where X = Mo, W, or V, to two other forms of hot corrosion: alloy-induced acidic degradation and sulfur-induced degradation.

Processing of Superalloys

As indicated in Fig. 21, superalloys are produced by three distinct processing routes: casting; powder metallurgy processing; and conventional wrought (deformation) processing. Regardless of the processing route chosen, the original feedstock (ingots) must be melted and cast with due regard for the volatility and reactivity of the elements present. Vacuum melting processes are a necessity for many nickel- and iron-nickel-base alloys because of the presence of aluminum and titanium as solutes. Most wrought cobalt-base alloys, on the other hand, do not usually contain these elements and therefore may be melted in air.

In addition to discussing melting processes, this section also briefly reviews deformation (wrought) processing of wrought alloys, thermomechanical processing, investment casting, hot isostatic pressing of castings, and joining. Heat treating, which is another critical processing step for superalloys, is described in the article "Effect of Heat Treating on Superalloy Properties" in this Volume.

Melt Processes

A number of superalloys, particularly cobalt- and iron-base alloys, are air melted by various methods applicable to stainless steels. However, for most nickel- or iron-nickel-base superalloys,

Table 12 Physical properties of cast nickel-base and cobalt-base alloys

Alloy	Density, g/cm³	Melting range °C	Melting range °F	Specific heat at 21 °C (70 °F) J/kg · K	Btu/lb · °F	Specific heat at 538 °C (1000 °F) J/kg · K	Btu/lb · °F	Specific heat at 1093 °C (2000 °F) J/kg · K	Btu/lb · °F	Thermal conductivity at 93 °C (200 °F) W/m · K	Btu · in./h · ft² · °F	Thermal conductivity at 538 °C (1000 °F) W/m · K	Btu · in./h · ft² · °F	Thermal conductivity at 1093 °C (2000 °F) W/m · K	Btu · in./h · ft² · °F	Mean coefficient of thermal expansion, 10^{-6}/K(a) at: 93 °C (200 °F)	538 °F (1000 °C)	1093 °C (2000 °F)
Nickel-base																		
IN-713 C	7.91	1260–1290	2300–2350	420	0.10	565	0.135	710	0.17	10.9	76	17.0	118	26.4	183	10.6	13.5	17.1
IN-713 LC	8.00	1290–1320	2350–2410	440	0.105	565	0.135	710	0.17	10.7	74	16.7	116	25.3	176	10.1	15.8	18.9
B-1900	8.22	1275–1300	2325–2375	...	...	...	...	...	...	(10.2)	(71)	16.3	113	...	...	11.7	13.3	16.2
Cast alloy 625	8.44	...	...	...	...	...	...	...	...	...	...	...	...	...	...	...	...	...
Cast alloy 718	8.22	1205–1345	2200–2450	...	...	...	...	...	...	...	...	...	...	...	...	...	...	...
IN-100	7.75	1265–1335	2305–2435	...	...	480	0.115	605	0.145	...	...	17.3	120	...	...	13.0	13.9	18.1
IN-162	8.08	1275–1305	2330–2380	...	...	...	...	...	...	...	...	...	...	...	...	12.2	14.1	...
IN-731	7.75	...	...	...	...	...	...	...	...	...	...	...	...	...	...	...	...	...
IN-738	8.11	1230–1315	2250–2400	420	0.10	565	0.135	710	0.17	...	...	17.7	123	27.2	189	11.6	14.0	...
IN-792	8.25	...	...	...	...	...	...	...	...	...	...	...	...	...	...	...	...	...
M-22	8.63	...	...	...	...	...	...	...	...	...	...	...	...	...	...	12.4	13.3	...
MAR-M 200	8.53	1315–1370	2400–2500	400	0.095	420	0.10	565	0.135	13.0	90	15.2	110	29.7	206	...	13.1	17.0
MAR-M 246	8.44	1315–1345	2400–2450	...	...	...	...	...	...	...	...	18.9	131	30.0	208	11.3	14.8	18.6
MAR-M 247	8.53	...	...	...	...	...	...	...	...	...	...	...	...	...	...	...	...	...
MAR-M 421	8.08	...	...	...	...	...	...	...	...	...	...	19.1	137	32.0	229	...	14.9	19.8
MAR-M 432	8.16	...	...	...	...	...	...	...	...	...	...	...	...	...	...	...	14.9	19.3
MC-102	8.84	...	...	...	...	...	...	...	...	...	...	...	...	...	...	12.8	14.9	...
Nimocast 75	8.44	1410(b)	2570(b)	...	...	...	...	...	...	...	...	...	...	...	...	12.8	14.9	...
Nimocast 80	8.17	1310–1380	2390–2515	...	...	...	...	...	...	...	...	...	...	...	...	12.8	14.9	...
Nimocast 90	8.18	1310–1380	2390–2515	...	...	...	...	...	...	...	...	...	...	...	...	12.3	14.8	...
Nimocast 242	8.40	1225–1340	2235–2445	...	...	...	...	...	...	...	...	...	...	...	...	12.5	14.4	...
Nimocast 263	8.36	1300–1355	2370–2470	...	...	...	...	...	...	...	...	...	...	...	...	11.0	13.6	...
René 77	7.91	...	...	...	...	...	...	...	...	...	...	...	...	...	...	...	...	...
René 80	8.16	...	...	...	...	...	...	...	...	...	...	...	...	...	...	...	...	...
Udimet 500	8.02	1300–1395	2375–2540	...	...	...	...	...	...	...	...	...	...	...	...	13.3	...	...
Udimet 710	8.08	...	...	...	...	...	...	...	...	12.1	84	18.1	126	...	...	...	...	...
Cobalt-base																		
FSX-414	8.3	...	...	...	...	...	...	...	...	...	...	...	...	...	...	...	...	...
Haynes 1002	8.75	1305–1420	2380–2590	420	0.10	530	0.126	645	0.154	11.0	76	21.8	151	32.1	222	12.2	14.4	...
MAR-M 302	9.21	1315–1370	2400–2500	...	...	...	...	...	...	18.7	130	22.2	154	...	...	...	13.7	16.6
MAR-M 322	8.91	1315–1360	2400–2475	...	...	...	...	...	...	...	...	...	...	...	...	...	...	...
MAR-M 509	8.85	...	...	...	...	...	...	...	...	...	...	27.9	194	44.6	310	9.8	15.9	18.2
WI-52	8.88	1300–1355	2425–2475	420	0.10	...	...	...	...	24.8	172	27.4	190	40.3	280	...	14.4	17.5
X-40	8.60	...	...	...	...	...	...	...	...	11.8	82	21.6	150	...	...	...	15.1	...

(a) From room temperature to indicated temperature. (b) Liquidus temperature. Source: Nickel Development Institute

vacuum induction melting (VIM) is required as the primary melting process. The use of VIM reduces interstitial gases (O_2, N_2) to low levels, enables higher and more controllable levels of aluminum and titanium (along with other relatively reactive elements) to be achieved, and results in less contamination from slag or dross formation than air melting. The benefits of reduced gas content and ability to control aluminum plus titanium are shown in Fig. 22 and 23. Vacuum arc remelting (VAR) and electroslag remelting (ESR) are the commonly employed secondary melting techniques.

The VIM process generally is used as the initial melting process for superalloys and may be the only melting process used when material for investment castings is being produced. However, for material that will be subjected to conventional wrought processing, particularly when the material is one of the higher-strength superalloys that must be hot worked to produce larger gas turbine parts, a secondary remelting operation is required. VIM ingots generally have coarse and nonuniform grain sizes, shrinkage, and alloying element segregation. Although these factors cause no problem in producing primary material that will be remelted for casting, such factors restrict the hot workability of forging alloys such as Incoloy 901, Waspaloy, Inconel 718, and Astroloy. The above problems are resolved by the use of VIM followed by VAR or ESR. In addition to refining the alloy composition, VAR and ESR refine the solidification structure of the resulting ingot. In some advanced nickel-base superalloys having high volume fractions of γ', even VIM-VAR or VIM-ESR does not provide a satisfactory ingot structure for subsequent hot working. Such superalloys have been processed by powder metallurgy techniques (see the section "Powder Processing" in this article). Recent improvements in melting/ingot casting technology, such as vacuum arc double electrode remelting (VADER), appear to offer the promise of sufficient structural refinement to permit such high-strength-alloy ingots to be processed in ingot form by appropriate extrusion and forging routes.

Electron-beam remelting/refining (EBR) has been evaluated as an alternative process for improving nickel- and iron-nickel-base superalloy properties and processability through a further lowering of impurity levels and drastic reductions in dross/inclusion content. EBR can help to produce improved feedstock for casting operations or to provide more workable starting ingot for wrought processing. The expanded use of secondary melt processes such as EBR, argon-oxygen degassing (AOD), and VADER will be governed by the extent to which they each provide an economical means for superalloy processing.

Melting of cobalt-base superalloys generally does not require the sophistication of vacuum processing. An air induction melt is commonly used, but VIM and ESR also have found application, the latter being used to produce stock for subsequent deformation processing. Alloys containing aluminum or titanium (J-1570) and tantalum or zirconium (MAR-M 302, MAR-M 509) must be melted by VIM. Vacuum melting of other cobalt-base superalloys may enhance properties such as strength and ductility because of the improved cleanliness and compositional control associated with this process.

Deformation Processing

Wrought heat-resistant alloys are manufactured in all mill forms common to the metal industry. Iron-base, cobalt-base, and nickel-base superalloys are produced conventionally as bar, billet, extrusions, plate, sheet, strip, wire, and forgings by primary mills. Inconel and Hastelloy alloys also are available as rod, bar, plate, sheet,

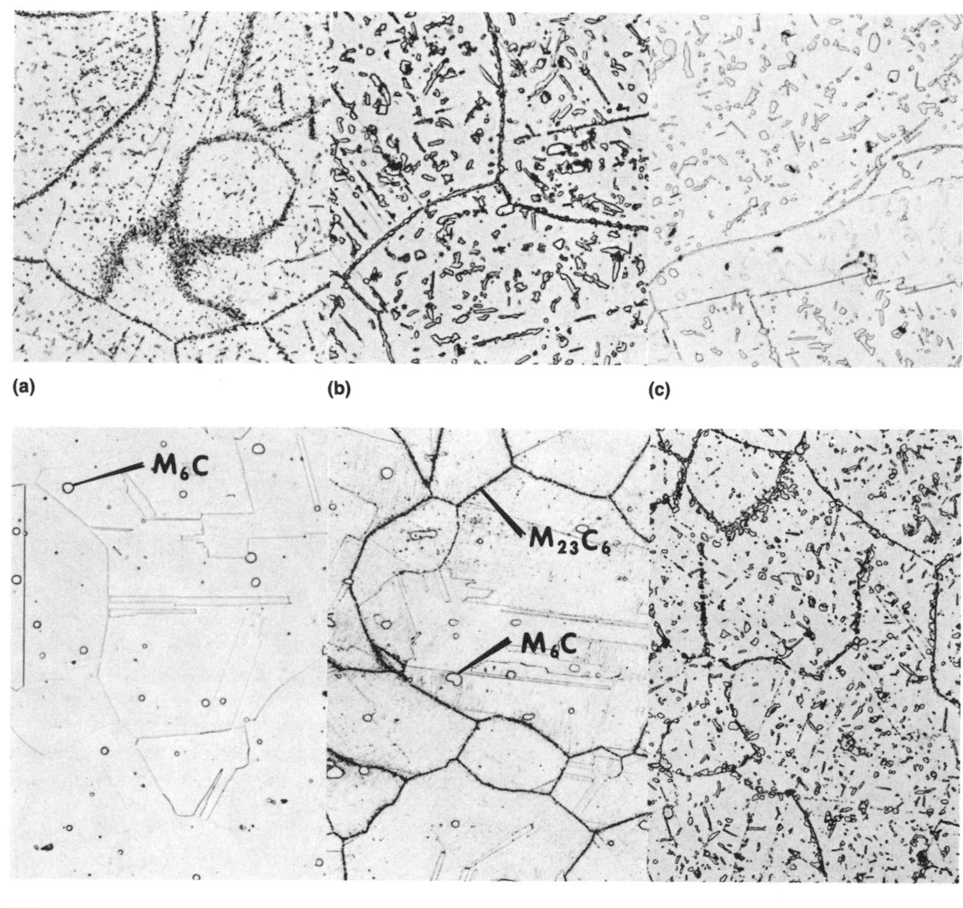

Fig. 17 Microstructures of Haynes 25 and 188 after various heat treatments. (a) Haynes 25, solution annealed at 1204 °C (2200 °F) and aged for 3400 h at 816 °C (1500 °F). Structure is made up of precipitates of M_6C and Co_2W intermetallic compound in an fcc matrix. (b) Haynes 25, solution annealed at 1204 °C (2200 °F) and aged for 3400 h at 871 °C (1600 °F). Structure is same as (a). (c) Haynes 25, solution annealed at 1204 °C (2200 °F) and aged for 3400 h at 927 °C (1700 °F). Structure is same as (a), but M_6C is considered primary, while Co_2W is considered secondary. (d) Haynes 188, cold rolled, 20% solution annealed at 1177 °C (2150 °F) for 10 min before water quenching. Fully annealed structure is M_6C particles in an fcc matrix. (e) Haynes 188, solution annealed at 1177 °C (2150 °F) and aged at 649 °C (1200 °F) for 3400 h. Microstructure is particles of M_6C and $M_{23}C_6$ in an fcc matrix. (f) Haynes 188, solution annealed at 1177 °C (2150 °F) and aged at 871 °C (1600 °F) for 6244 h. Structure is $M_{23}C_6$, Laves phase, and probably M_6C in an fcc matrix. All 500×

strip, tube, pipe, shapes, wire, forging stock, and specialty items from secondary converters.

Cast structure is refined by a working operation known as cogging (Ref 27). In this process, hydraulic presses are used with open dies. Uniform billet structure and improved surface finish are additional objectives of this process. Initial ingot breakdown is followed by methods such as rolling, forging, or extrusion.

Sheet, bar, and ring rolling are commonly employed for secondary hot working of superalloys. Working temperature ranges for many nickel-base alloys are listed in Fig. 24. Edge cracking is minimized by rapid handling. Forgings are used in both the turbine disk and compressor sections of gas turbine engines. Rates of die closure vary from 0.3 mm/s (0.012 in./s) for hydraulic presses to 7.5 m/s (295 in./s) for hammers. While most forging is carried out with steel tooling heated in the range of 200 to 430 °C (390 to 805 °F), isothermal forging is now widely used for near-net shape processing. Superalloy or molybdenum alloy dies are heated to the same temperature as the forging, in the range of 650 to 980 °C (1200 to 1800 °F). Isothermal forging produces a uniform microstructure while requiring less material, thereby lowering machining costs. For temperatures at which superplastic behavior of superalloys occurs, less powerful presses or hammers and slower die closer rates can be used. The benefits outweigh the increased costs of hot die tooling for expensive input materials.

Extrusion is used for the conversion of ingot to billet, especially for the stronger, crack-prone alloys, as well as for powder-processed alloys. Canning in mild steel or stainless steel is required to avoid chilling and surface cracking. Glass is the most common lubricant (Ref 28). Seamless tubing is also produced by this method.

Sheet and other semifinished products are produced by rolling, often with many reheats, frequent conditioning, and possibly encasing of the alloy in a can. Open-die forging is carried out in flat or swaging dies. Closed-die forging is used to produce shapes that match the impressions of dies attached to the ram and anvil. Both hammers and presses are used; for the hammers, stresses required for deformation are higher because of the higher strain rates employed.

A few superalloys can be cold formed on high-capacity equipment by drawing, extrusion, pressing and deep drawing, spinning, or rolling. For all processes except rolling, a lubricant is usually required, and speeds are relatively low. The uniform, fine-grain microstructures resulting from cold working and reannealing lead to improved mechanical properties.

Thermal-Mechanical Processing

Thermal-mechanical processing (TMP) refers to the control of temperature and deformation during processing to enhance specific properties. Special TMP sequences have been developed for a number of nickel- and iron-nickel-base alloys.

The design of TMP sequences relies on a knowledge of the melting and precipitation temperatures for the alloy of interest. Table 15 lists these temperatures for several nickel-base alloys. Although nickel-base (as well as iron- and cobalt-base) alloys form various carbides, the primary precipitate of concern in the processing of such materials is the γ'-strengthening precipitate. In iron-nickel alloys such as alloy 718, titanium, niobium, and, to a lesser extent, aluminum combine with nickel to form γ' or γ''. Iron-nickel base alloys are also prone to the formation of other phases, such as Ni_3Ti (η) in titanium-rich alloy 901, or orthorhombic Ni_3Nb (δ) in niobium-rich alloy 718.

Early forging practice of nickel- and iron-nickel-base alloys consisted of forging from, and solution heat treating at, temperatures well in excess of the γ' solvus temperature. High-temperature solution treatment dissolved all of the γ', annealed the matrix, and promoted grain growth (typical grain size ≈ASTM 3 or coarser). This was followed by one or more aging treatments that promoted controlled precipitation of γ' and carbide phases. Optimal creep and stress rupture properties above 760 °C (1400 °F) were thus achieved. Later in the development of forging practice, it was found that using preheat furnace temperatures slightly above the recrystallization temperature led to the development of finer grain sizes (ASTM 5 to 6). Coupling this with modified heat treating practices resulted in excellent combinations of tensile, fatigue, and creep properties.

TMP of Nickel-Base Superalloys. State-of-the-art forging practices for nickel-base alloys rely on the following microstructural effects:

- Dynamic recrystallization is the most important softening mechanism during hot working.
- Grain boundaries are preferred nucleation sites for recrystallization.
- The rate of recrystallization decreases with the temperature and/or the extent of deformation.
- Precipitation that may occur during the recrystallization can inhibit the softening process.

Recrystallization cannot be completed until the precipitate coarsens to a relatively ineffective morphology.

Forging temperature is carefully controlled during TMP of nickel- and nickel-iron base alloys to make use of the structure-control effects of second phases such as γ'. Above the optimal forging temperature range (Table 16), the structure-control phase goes into solution and loses its effect. Below this range, extensive fine precipitates are formed, and the alloy becomes too stiff to process. Two examples of specific TMP sequences are given below.

Waspaloy. A typical TMP treatment of nickel-base alloys is that used for Waspaloy (UNS N07001) to obtain good tensile and creep properties. This consists of initial forging at 1120 °C (2050 °F) and finish forging below approximately 1010 °C (1850 °F) to produce a fine, equiaxed grain size of ASTM 5 to 6. Solution treatment is then done at 1010 °C (1850 °F), and aging is conducted at 845 °C (1550 °F) for 4 h, followed by air cooling plus 760 °C (1400 °F) for 16 h and then air cooling.

René 95. Initial forging of René 95 is done at a temperature between 1095 and 1140 °C (2000 and 2080 °F). Following an in-process recrystallization anneal at 1175 °C (2150 °F), finish forging (reduction ~40 to 50%) is then imposed below the γ' solvus, typically at temperatures between 1080 and 1105 °C (1975 and 2025 °F). The large grains formed during high-temperature recrystallization are elongated and surrounded by small recrystallized grains that form during finish forging.

TMP of Iron-Nickel-Base Alloys. Grain structure may be controlled by TMP in several iron-nickel-base alloys that have two precipitates present, the primary strengthening precipitate (γ'' Ni_3Nb in Inconel 718 and γ' Ni_3Ti in Inconel 901) and a secondary precipitate (δ in Inconel 718 and η Ni_3Ti in Inconel 901). The secondary precipitate is produced first, by an appropriate heat treatment (8 h at 900 °C, or 1650 °F, for 901), followed by working at about 950 °C (1740 °F), below the η solvus. Finally, the alloy is aged by standard procedures. The result is a fine-grain alloy with higher tensile strength and improved fatigue resistance.

As shown in Table 16, the critical warm-working temperature range for Inconel 718 is 915 to 995 °C (1675 to 1825 °F). The upper limit avoids grain coarsening at higher temperatures due to re-solution of δ, while the lower limit is established to avoid an excessively high flow stress during working. Delta and γ'' precipitates compete for the available niobium. Therefore, any factor suppressing δ tends to favor γ'' formation, and vice versa. Delta does not strengthen Inconel 718, but it reduces the room-temperature ductility. However, when some δ-phase is precipitated prior to or during working, the grain size can be reduced substantially, leading to increased tensile and fatigue strength.

Powder Processing

Powder techniques are being used extensively in superalloy production. Principally, high-strength gas-turbine disk alloy compositions such as IN-100, which are difficult or impractical to forge by conventional methods, have been powder processed. Inert atmospheres are used in the production of powders, often by gas atomization, and the powders are consolidated by extrusion or hot isostatic pressing (HIP). HIP has been used either to produce shapes directly for final machining or to consolidate billets for subsequent forging. Extruded or hot isostatically pressed (HIPed) billets often are isothermally forged to configurations for final machining. Minimal segregation, reduced inclusion sizes, ability to use very high γ'-content compositions, and ease of grain size control are significant advantages of the powder process.

Powder techniques also have been used to produce turbine blade/vane ODS alloys. Mechanical alloying is the principal technique for introducing the requisite oxide/strain energy combination to achieve maximum properties. Rapid-solidification-rate (RSR) technology has been applied to produce highly alloyed superalloys. RSR and ODS alloys can benefit from aligned crystal growth in the same manner as can directionally

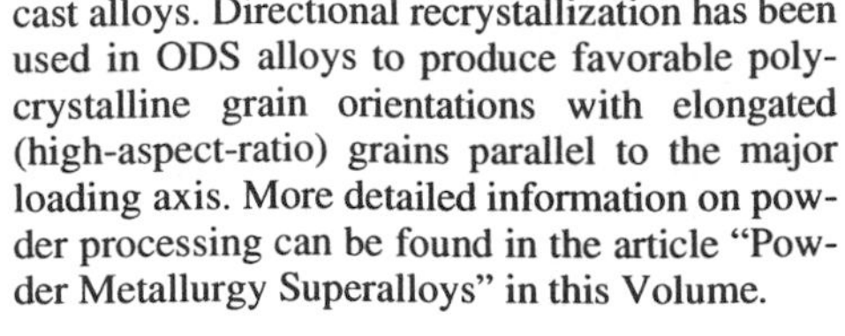

cast alloys. Directional recrystallization has been used in ODS alloys to produce favorable polycrystalline grain orientations with elongated (high-aspect-ratio) grains parallel to the major loading axis. More detailed information on powder processing can be found in the article "Powder Metallurgy Superalloys" in this Volume.

Investment Casting (Ref 30)

A number of casting processes can provide near-net shape superalloy cast parts, but essentially all components are produced by investment casting. The characteristic physical and mechanical properties and complex, hollow shape-making capabilities of investment casting have made it ideal for amplifying the unusual high-temperature properties of superalloys.

Cast superalloys are made in a wider range of compositions than are wrought alloys. Creep and rupture properties of a given superalloy composition are maximized by the casting and heat treatment processes. Ductility and fatigue properties of polycrystalline materials are generally lower in castings than in their wrought counterparts of similar composition. The gap, however, is being reduced by new technological developments to eliminate casting defects and refine grain size.

Patterns, Cores, and Molds. The first step in the investment casting process is to produce an exact replica or pattern of the part in wax, plastic, or a combination thereof. Pattern dimensions must compensate for wax, mold, and metal shrinkage during processing. If the product con-

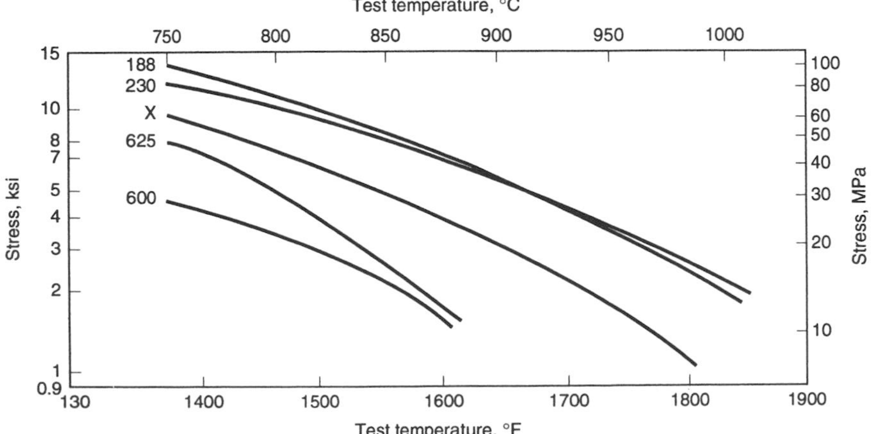

Fig. 18 Comparison of the stress to produce 1% creep in 1000 h for a cobalt-base alloy (188) and four solid-solution strengthened nickel alloys. Source: Haynes International, Inc.

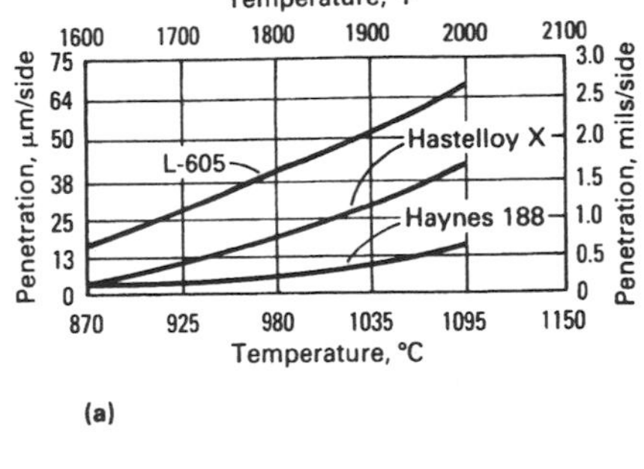

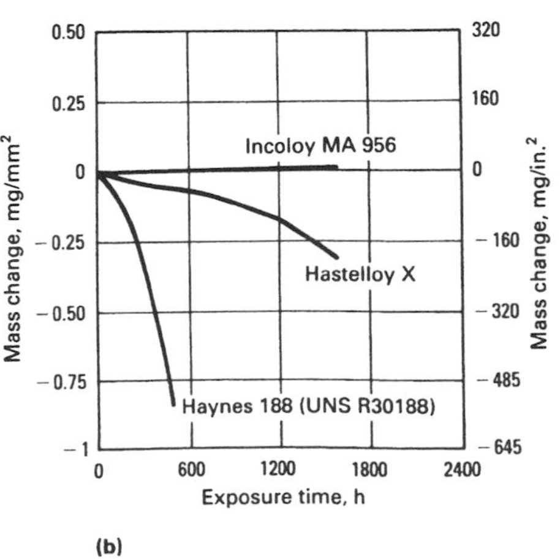

Fig. 19 Oxidation resistance. (a) In dry air for Haynes 188 vs. Hastelloy X and L-605 alloys, showing continuous penetration from original thickness. (b) Static values at 1100 °C (2010 °F) in air with 5% water vapor

tains internal passages, a preformed ceramic core is inserted in the die cavity, around which the pattern material is injected. Except for large or complex castings, a number of patterns may be assembled in a cluster and held in position in order to channel the molten metal into the various mold cavities. Design and positioning of the runners and gating is critical to achieving sound, metallurgically acceptable castings. Today the molds are produced by first immersing the pattern assembly in an aqueous ceramic slurry. A dry, granular ceramic stucco is applied immediately after dipping to strengthen the shell. These steps are repeated several times to develop a rigid shell. After slow, thorough drying, the wax is melted out of the shell, and the mold is fired to increase substantially its strength for handling and storage. An insulating blanket is tailored to the mold configuration to minimize heat loss during the casting operation and to control solidification. More information on the production of patterns, cores, and shells for investment casting is available in the article "Investment Casting" in Volume 15 of the *ASM Handbook.*

To make equiaxed-grain castings, the mold is preheated to enhance mold filling, control solidification, and develop the proper microstructure. For vacuum casting, the alloy charge is melted in an isolated chamber before the preheated mold is inserted, and the pressure is maintained at about 1 μm for pouring. After casting, exothermic material is applied as a hot top for feeding purposes, and the mold is allowed to cool. A different procedure is followed in the production of directionally solidified (DS) and single-crystal (SC) superalloy castings; this is described in the article "Directionally Solidified and Single-Crystal Superalloys" in this Volume.

Table 13 Incipient melting temperatures of selected wrought superalloys

Alloy	Incipient melting temperature °C	°F
Hastelloy X (Ni)	1250	2280
L-605 (Co)	1329	2425
Haynes 188 (Co)	1302	2375
Incoloy 800 (Ni)	1357	2475
Incoloy 825 (Fe)	1370	2500
Incoloy 617 (Ni)	1333	2430
Inconel 625 (Ni)	1288	2350
Inconel X750 (Ni)	1393	2540
Nimonic 80A (Ni)	1360	2480
Nimonic 90 (Ni)	1310	2390
Nimonic 105 (Ni)	1290	2354
René 41 (Ni)	1232	2250
Udimet 500 (Ni)	1260	2300
Udimet 700 (Ni)	1216	2220
Waspaloy (Ni)	1329	2425

Source: Ref 1

Table 14 Oxidation role of alloying additions to cobalt-base materials

Alloying element	Probable effect of addition on the oxidation behavior of a Co(20–30)Cr base
Titanium	Innocuous at low levels
Zirconium	Innocuous at low levels
Carbon	Slightly deleterious; ties up chromium
Vanadium	Harmful, even at 0.5%
Niobium	Harmful, even at 0.5%
Tantalum	Beneficial to moderate levels (<5%)
Molybdenum	Harmful; forms volatile oxides
Tungsten	Innocuous below ~1000 °C (1800 °F); harmful >1000 °C (1800 °F); forms volatile oxides
Yttrium	Beneficial; improves scale adherence
Nickel	May be slightly deleterious
Manganese	Beneficial; induces the formation of spinels
Iron	Tends to induce spinel formation

Source: Ref 23

The Casting Process. Most superalloys are cast in vacuum to avoid the oxidation of reactive elements in their compositions. Some cobalt-base superalloys are cast in air using induction or indirect arc rollover furnaces. The vacuum casting of equiaxed-grain products is usually done in a furnace divided into two major chambers, each held under vacuum and separated by a large door or valve. The upper chamber contains an induction-heated reusable ceramic crucible in which the alloy is melted. Zirconia crucibles are commonly employed; single-use silica liners may be specified when alloy cleanliness is especially critical.

The preweighed charge is introduced through a lock device and is melted rapidly to a predetermined temperature, usually 85 to 165 °C (150 to 300 °F) above the liquidus temperature. Precise optical measurement of this temperature is crucial. Metal temperature during casting is much more critical than mold temperature in controlling grain size and orientation; it also strongly affects the presence and location of microshrinkage. When the superheat condition is satisfied, the preheated mold is rapidly transferred from the preheat furnace to the lower chamber, which is then evacuated. The mold is raised to the casting position, and the molten superalloy is quickly poured into the cavity; speed and reproducibility are essential in order to achieve good fill without cold shuts and other related imperfections. Pre-

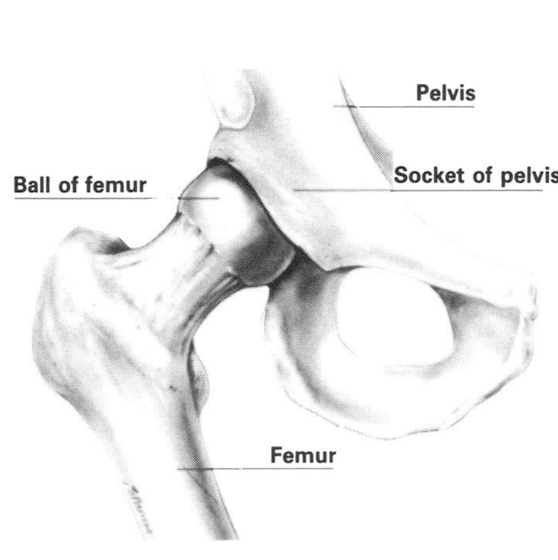

(a)

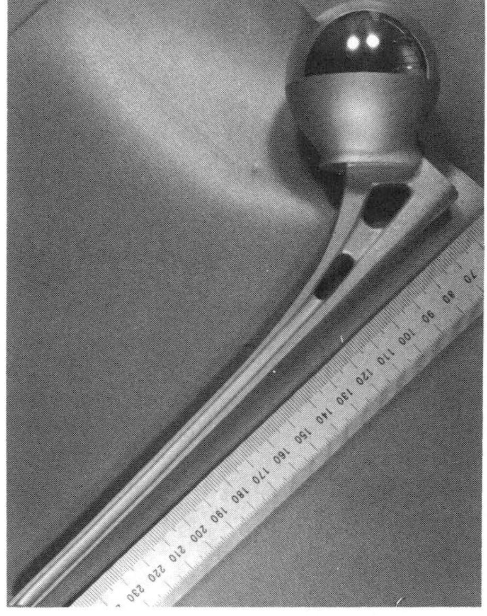

(b)

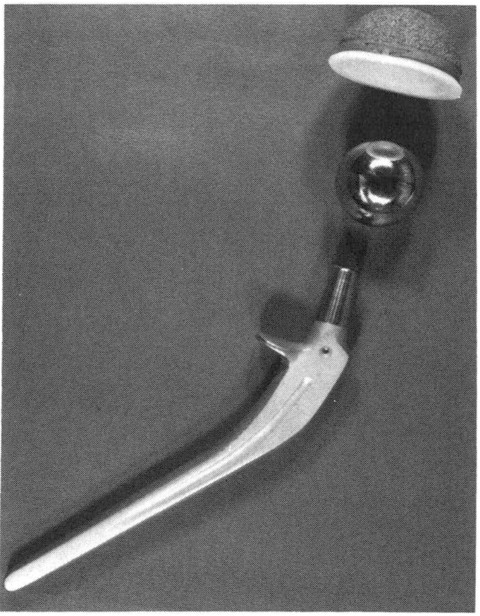

(c)

Fig. 20 The natural hip joint (a) and two types of artificial joints (b and c). The cast cobalt-chromium-molybdenum endoprosthesis (b) has a large diameter head that articulates in the acetabulum (the natural socket of the pelvis). The total hip replacement (c) replaces both the femoral head and the socket of the pelvis. The prosthesis shown in (c) uses a modular cobalt-chromium-molybdenum head that fits on a forged cobalt-chromium-molybdenum stem; the socket component is ultrahigh-molecular-weight polyethylene with a metal backing.

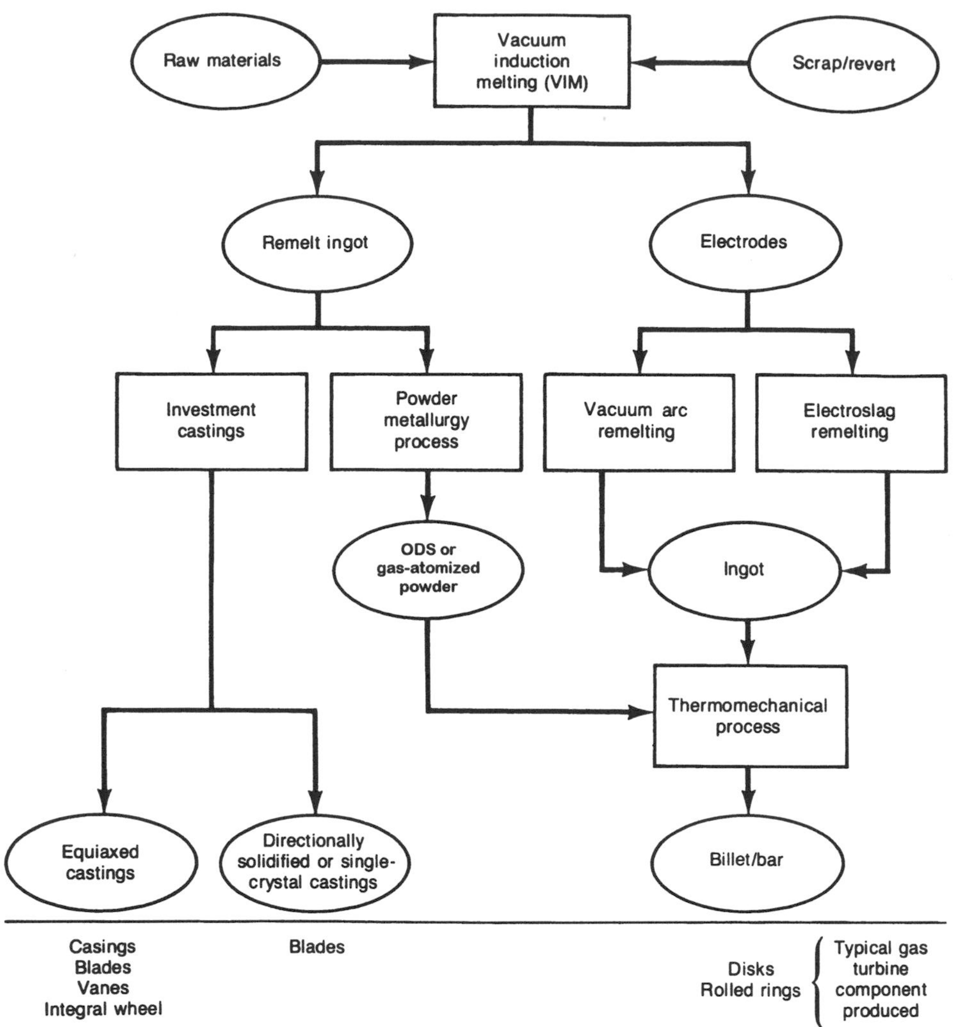

Fig. 21 Flow diagram of processes widely used to produce superalloy components. ODS, oxide-dispersion-strengthened. Source: Ref 26

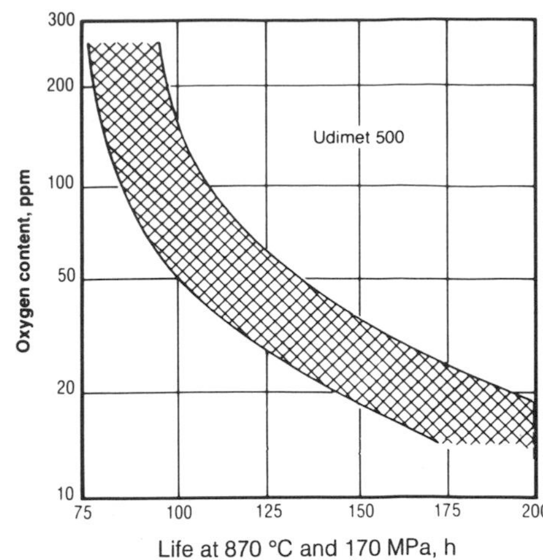

Fig. 22 Improvement of rupture life at 870 °C (1600 °F) and 170 MPa (25 ksi) by reduced oxygen content produced by vacuum melting

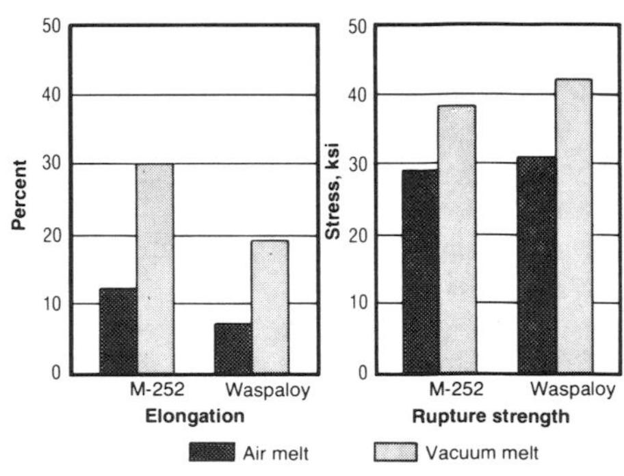

Fig. 23 Effects of vacuum melting, incorporating beneficial modifications in composition, on properties of two nickel-base superalloys

cise mold positioning and pour rates also are imperative. For maximum consistency, melting and casting are automated with programmed closed-loop furnace control. The filled mold is lowered and removed from the furnace.

Shrinkage during solidification is minimized in part by maintaining a head of molten metal to feed the casting; this is achieved by adding an exothermic material immediately after mold removal from the furnace.

Because of thermal expansion differences, the shell mold usually fractures upon cooling, facilitating its removal by mechanical or hydraulic means. Before grit- and sand-blasting operations, the individual castings are separated from the cluster by abrasive cutoff. After shell removal, the cluster is checked by one of several commercially available emission or x-ray fluorescence instruments to verify the alloy identification.

A major portion of the casting cost is in the finishing operations, which remain labor intensive. Superficial surface defects are blended out abrasively within specified limits, and the castings may require mechanical straightening operations before and after heat treatment to satisfy dimensional requirements.

Control of Casting Microstructure (Ref 31)

The solidification of investment cast superalloy components is precisely controlled so that the microstructure, which ultimately determines mechanical properties, remains consistent. For example, once the process for a particular component has been defined, the production of these components does not deviate from the agreed-upon steps for the entire production run, which may last many years. If steps are changed, it must be shown that the new steps do not cause a degradation in the properties of the component.

To control the solidification of equiaxed-grain castings, the investment caster has several tools: facecoats that encourage grain nucleation, pour temperature of the metal, preheat temperature of the shell, shell thickness, part orientation, part spacing, gating locations, insulation to wrap the shells, pouring speed, and shell agitation. However, the investment caster must first fill the shell cavity, prevent hot tears or other cracks, and minimize porosity. If the first two objectives can be met, the investment caster has some freedom to produce the desired structure. If the desired structure still cannot be made, more complex techniques may be employed, including changing the thermal conductivity of the shell.

Dendrites are probably the most visible microstructural feature in superalloy castings. Primary and secondary dendrite arm spacings are controlled by the cooling rate. As the dendrite arm spacing is reduced, segregation in the dendrite core and interdendritic regions is also reduced, thereby benefiting mechanical properties.

Carbides. Conventional equiaxed-grain nickel-base superalloys typically have 0.05 to 0.20 wt% C, while cobalt-base alloys contain up to about 1.0% C. Both alloy systems may use carbon to increase grain-boundary strength. Cobalt-base alloys require more because internal carbides are one of the primary strengthening mechanisms.

Carbide morphology is controlled by solidification or composition. For example, by increasing the cooling rate, more discrete, blocky-type MC carbides are formed in IN-713 C, and this often results in an improvement of at least two

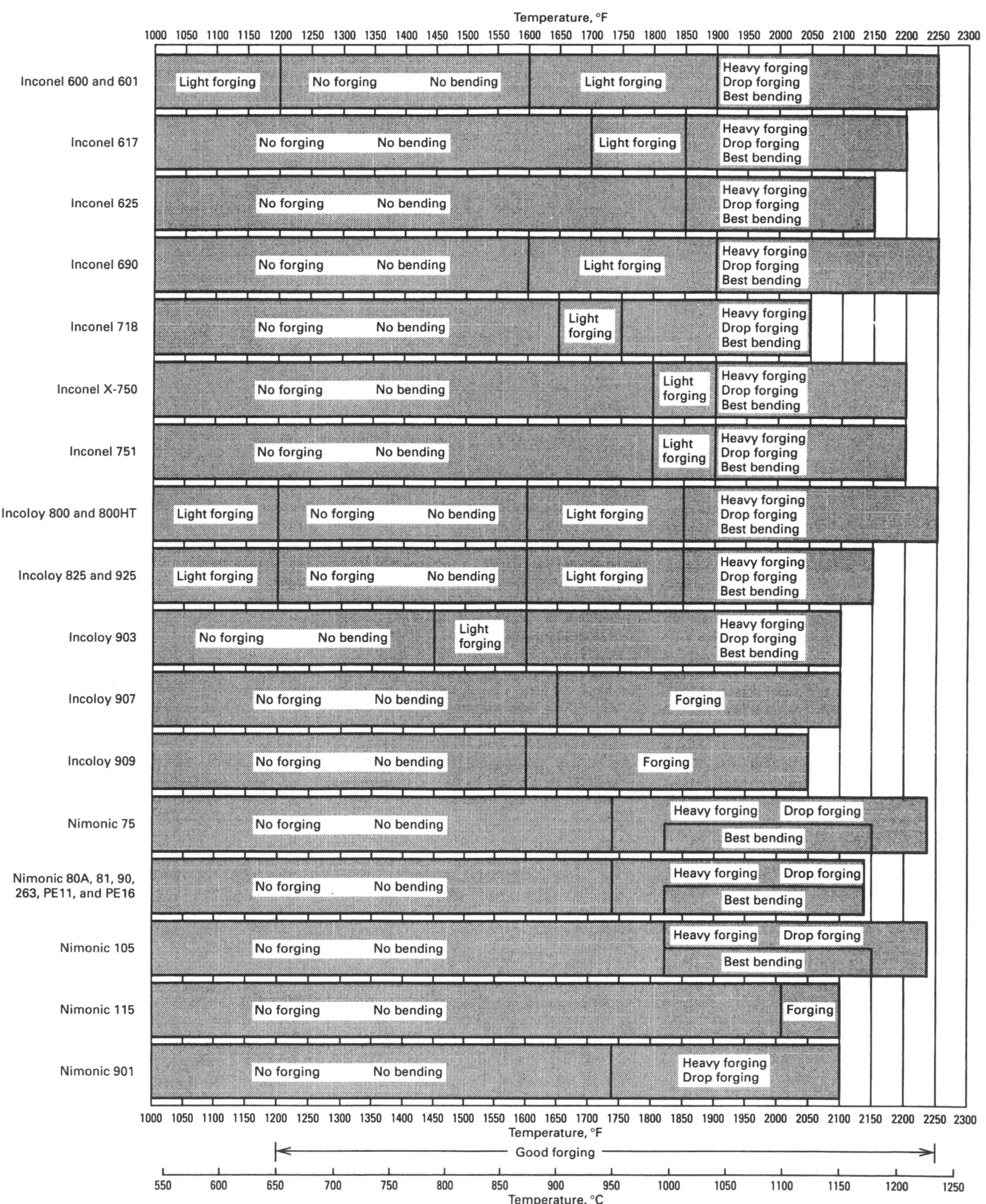

Fig. 24 Hot working temperature ranges for various superalloys

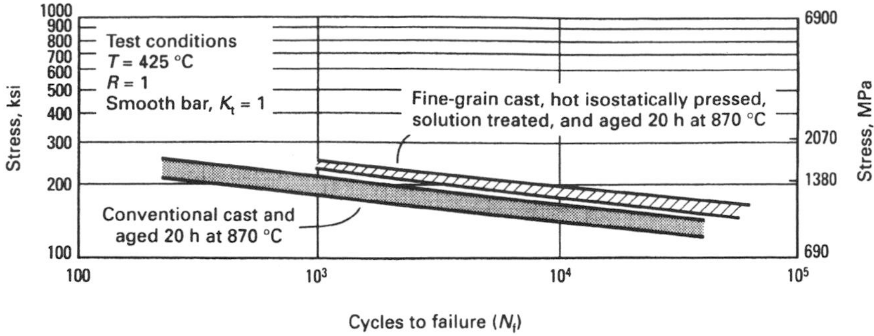

Fig. 25 Effect of grain size control and hot isostatic pressing on the strain-controlled (axial) low-cycle fatigue of CM 247

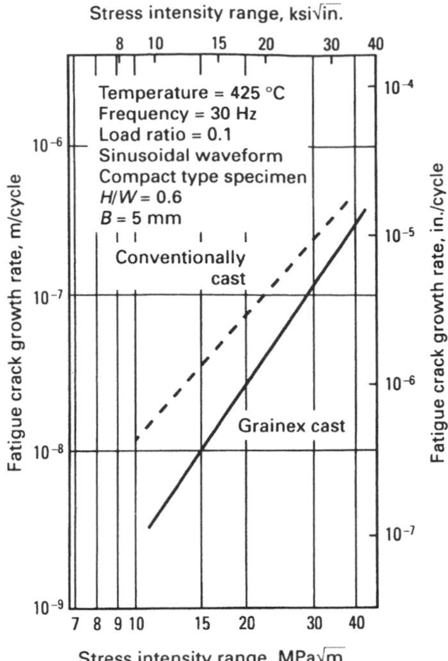

Fig. 26 Influence of grain refinement on fatigue crack growth rate. Source: Ref 30

times to low-cycle fatigue properties (Ref 32, 33). If it is not possible to influence the cooling rate of a casting significantly, adding small amounts of magnesium, calcium, cerium, or other rare earth metals acting as nucleating agents will assist in carbide shape control.

Eutectic Segregation. By the very nature of solidification, segregation is introduced into the component. Important segregants of interest in cast superalloys are eutectics, which often are found in interdendritic or intergranular regions. In nickel-base alloys, eutectic pools are the last constituents to solidify and have a cellular appearance. The composition of the eutectic pools varies, but they typically contain excess γ′, carbides, borides, and low-melting-point phases. Control of the eutectic pool is done primarily through composition. However, it has been shown that while the volume fraction of eutectic remained constant near 0.10 vol% in IN-713, the size of the eutectic pool increased from 11 to 19 μm as the cooling rate decreased from 0.56 to 0.036 °C/s (1 to 0.065 °F/s) (Ref 31).

In cobalt-base alloys, the eutectics typically form lamellar γ and $M_{23}C_6$ pools or colonies. Heat treatment between 1150 and 1230 °C (2100 and 2250 °F) for 4 h resolutions these eutectic colonies, redistributing much of the carbon.

Porosity. It is important to minimize the porosity in castings because the pores serve as initiation sites for fracture, especially fatigue cracks. There are three primary sources of porosity in superalloy investment castings: undissolved gas, microshrinkage caused by poor feeding between dendrites, and macroshrinkage caused by inadequate gating. Undissolved gas is gas that has come out of solution, but with today's vacuum technology, it is seldom experienced. Usually made up of oxygen, nitrogen, or hydrogen, this gas can form spherical voids up to two or more times the diameter of the dendrite arm spacing. Gas porosity can be essentially eliminated by maintaining a vacuum during remelting and casting.

Microshrinkage (microporosity) is inherent to castings that experience dendritic solidification. The pores are spherical, but they typically have a diameter less than the dendrite spacing. Microshrinkage forms just ahead of the advancing solidus interface because liquid metal feeding is impeded by the tortuous path through and around the secondary dendrite arms (a fluid flow problem).

The 2 to 6% shrinkage experienced upon solidification by superalloy castings makes macroshrinkage (solidification shrinkage) a problem. This type of porosity tends to be confined within the thickest section of the casting, where the last solidification takes place.

The investment caster can control solidification shrinkage to a great extent by gating or feeding those areas that are the last to solidify. With complex geometries and the desire to produce net-shape castings, the size and placement of gating is based on experience, which necessitates experimentation before the process can be defined. Modeling with computers, however, is changing this practice. It has become possible to model the solidification process of simple castings by taking into account the thermal properties of the metal and the shell. Thus, the areas of solidification shrinkage can be predicted. Once the shrinkage areas are located, various gating schemes can be evaluated until the shrink within the part is pulled into the gate. At this point, the model is verified by an experiment, significantly reducing the overall time it takes to design gating configurations. More information on the use of modeling to predict solidification shrinkage and other casting variables is available in the Section "Computer Applications in Metal Casting" in Volume 15 of the *ASM Handbook*.

Grain Size. The control of grain size is an important means for developing and maintaining

Table 15 Critical melting and precipitation temperatures for several nickel-base superalloys

Alloy	UNS No.	Melting temperature °C	Melting temperature °F	Precipitation temperature °C	Precipitation temperature °F
Alloy X	N06002	1260	2300	760	1400
Alloy 718	N07718	1260	2300	845	1550
Waspaloy	N07001	1230	2250	980	1800
Alloy 901	N09901	1200	2200	980	1800
Alloy X-750	N07750	1290	2350	955	1750
M-252	N07252	1200	2200	1010	1850
Alloy R-235	...	1260	2300	1040	1900
René 41	N07041	1230	2250	1065	1950
U500	N07500	1230	2250	1095	2000
U700	...	1230	2250	1120	2050
Astroloy	N13017	1230	2250	1120	2050

Table 16 Structure-control phases and working temperature ranges for various heat-resistant alloys

Alloy	UNS No.	Phase for structure control	Working temperature range °C	Working temperature range °F
Nickel-base alloys				
Waspaloy	N07001	γ′ ($Ni_3(Al,Ti)$)	955–1025	1750–1875
Astroloy	N13017	γ′ ($Ni_3(Al,Ti)$)	1010–1120	1850–2050
IN-100	...	γ′ ($Ni_3(Al,Ti)$)	1040–1175	1900–2150
René 95	...	γ′ ($Ni_3(Al,Ti)$)	1025–1135	1875–2075
Nickel-iron-base alloys				
901	N09901	η (Ni_3Ti)	940–995	1725–1825
718	N07718	δ (Ni_3Nb)	915–995	1675–1825
Pyromet CTX-1	...	η (Ni_3Ti), δ (Ni_3Nb), or both	855–915	1575–1675

Source: Ref 29

both physical and mechanical properties. Generally, a number of randomly oriented equiaxed grains in a given cross section is preferred to provide consistent properties, but often this is difficult to achieve in thin sections. To meet this objective, mold facecoat nucleants, mold and metal temperature, and other parameters are chosen to accelerate grain nucleation and solidification.

Finer grain size generally improves tensile, fatigue, and creep properties at low to intermediate temperatures (Fig. 25 and 26). The finer grain size produced by relatively rapid solidification is accompanied by a finer distribution of γ' particles and a tendency to form blocky carbide particles. The latter morphology is preferred to the script-type carbides produced by slow solidification rates, particularly for a fatigue-sensitive environment. Under these conditions, the carbide particles do not contribute to superalloy properties. As the service temperature increases, they impart important grain-boundary strengthening, provided that continuous films or necklaces are avoided.

For high-temperature rupture performance, slower solidification and cooling rates are preferred to coarsen both the grain size during solidification and the γ' precipitated during cooling. While this benefits high-temperature strength through a reduction in grain-boundary content, more property scatter can be expected due to (random) crystallographic orientation effects. For turbine blades, the desired microstructure is difficult to achieve because the thin airfoils operating at the highest temperatures should have coarse grains, and the heavier-section root attachment area, being less rupture dependent, should have a fine-grain microstructure. Where conventional practice fails, a gate, or gutter, along the airfoil edges may be employed, through which metal is caused to flow, thereby creating deliberate hot spots to retard the local solidification rate.

A significant foundry advancement has been the development of processes to produce fine-grain superalloy castings. In the late 1960s, experiments were conducted on a grain refinement technique for integral turbine wheels. The technique used the mechanical motion of a mold to shear dendrites from the solidifying metal. These dendrites then acted as nucleation sites for additional grains. However, the process was not commercially introduced because it produced castings with unacceptable levels of porosity.

In the mid-1970s, developmental work on this process resumed when it was realized that HIP could be used to eliminate residual casting porosity. The Howmet Corporation process that developed from this work, known as Grainex, results in ASTM grain sizes as fine as No. 2. A further Howmet Corporation development, the Microcast-X process, has led to a greater refinement in grain size (ASTM No. 3 to 5). Figure 27 compares the microstructures of grain-refined rotors with those of a conventionally cast part. References 32 to 36 provide additional information on grain size control/property relationships and on fine-grain casting process development work performed by others in the precision investment casting industry.

Hot Isostatic Pressing of Castings

Hot isostatic pressing subjects a cast component to both elevated-temperature and isostatic gas pressure in an autoclave. The most widely used pressurizing gas is argon. For the processing of castings, argon is applied at pressures between 103 and 206 MPa (15 and 30 ksi), with 103 MPa (15 ksi) being the most common. Process temperatures of 1200 to 1220 °C (2200 to 2225 °F) are common for polycrystalline superalloy castings.

When castings are HIPed, the simultaneous application of heat and pressure virtually eliminates internal voids and microporosity through a combination of plastic deformation, creep, and diffusion. The elimination of internal defects leads to improved nondestructive testing ratings, increased mechanical properties, and reduced data scatter (Fig. 28, 29).

In the past 25 years, HIP has become an integral part of the manufacturing process for high-integrity aerospace castings. The growth of HIP has paralleled the introduction of advanced nickel-base superalloys and increasingly complex casting designs, both of which tend to increase levels of microporosity. In addition, to optimize mechanical properties, turbine engine manufacturers have become more stringent in allowances for microporosity. The requirement for reduced porosity levels and increased mechanical properties has been achieved in many cases through the use of HIP.

In selecting HIP process parameters for a particular alloy, the primary objective is to use a combination of time, temperature, and pressure that is sufficient to achieve closure of internal voids and microporosity in the casting. There are also material considerations for avoiding such deleterious effects as incipient melting, grain growth, and the degradation of constituent phases such as carbides.

If encountered, incipient melting can be avoided by pre-HIP homogenization heat treatments or by lowering HIP temperatures. If the temperature is lowered, an increase in processing pressure may be required to obtain closure in certain alloys. For example, hafnium-bearing nickel-base superalloys such as C 101 and MAR-M 247, when cast with heavy sections (e.g., integral wheels), have been found to undergo incipient melting when HIPed at 1205 °C (2200 °F) and

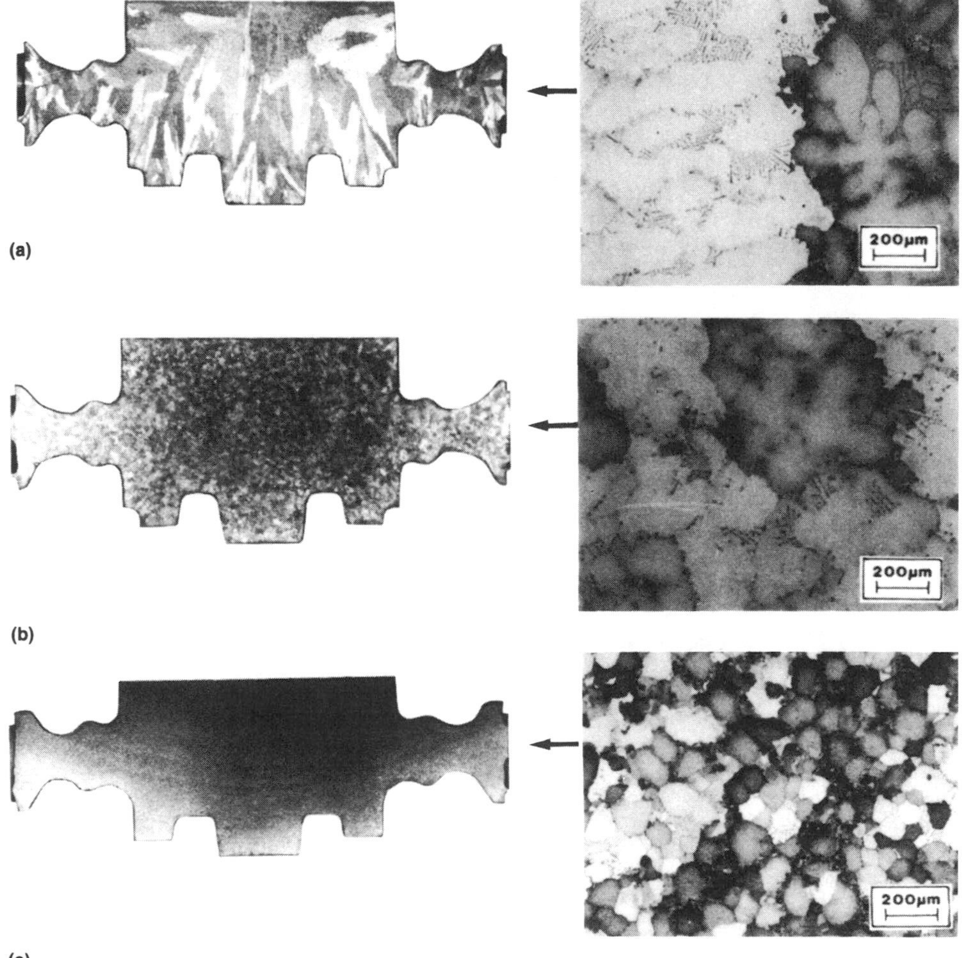

Fig. 27 Structure of conventionally cast turbine wheel (a) compared to wheels cast using the Grainex (b) and Microcast-X (c) processes

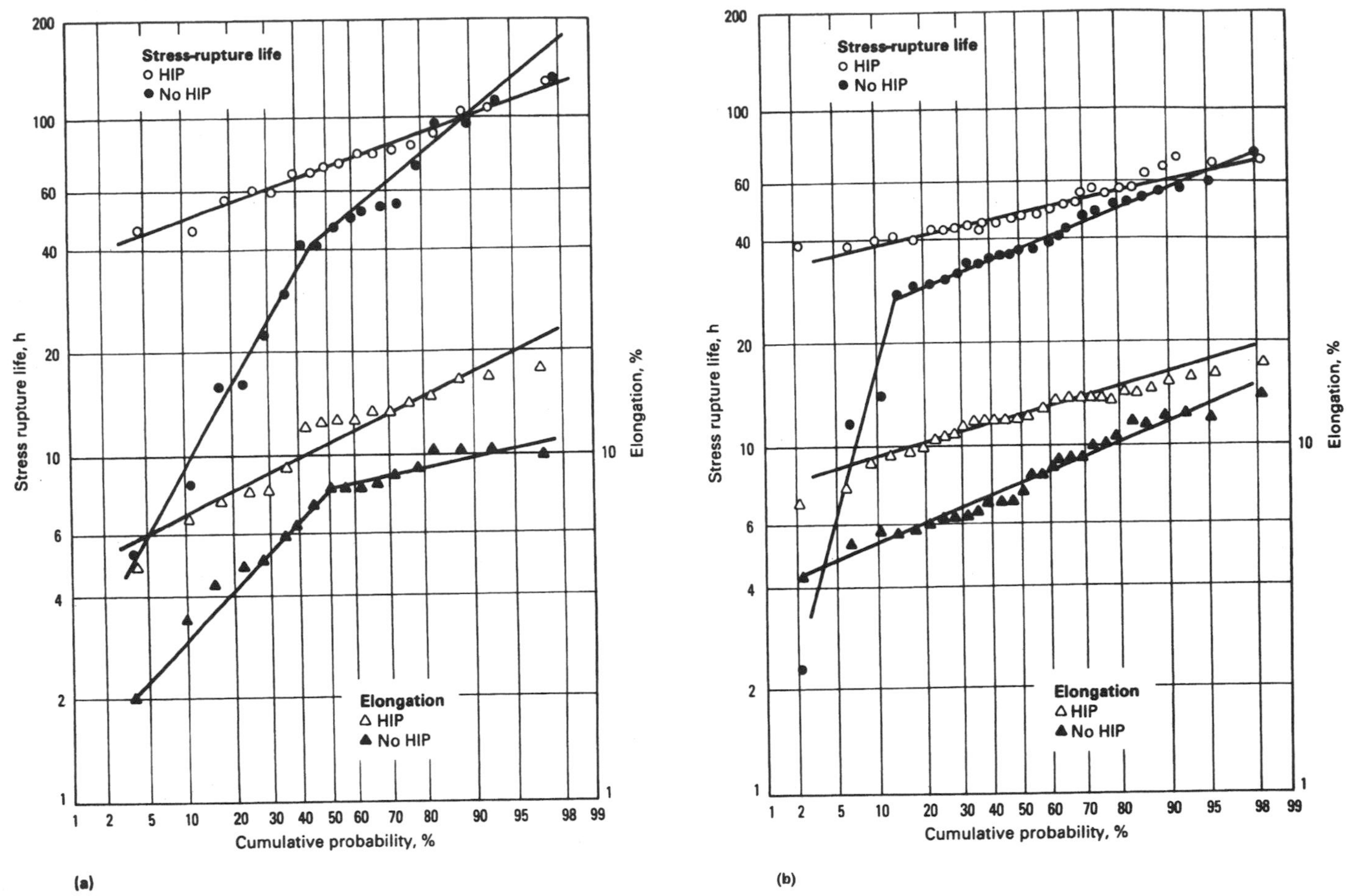

Fig. 28 Effect of hot isostatic pressing on stress-rupture properties of cast IN-738. Test material was hot isostatically pressed at 1205 °C (2200 °F) and 103 MPa (15 ksi) for 4 h. (a) Test conditions: 760 °C (1400 °F) and 586 MPa (85 ksi). (b) Test conditions: 980 °C (1800 °F) and 152 MPa (22 ksi). Source: Howmet Corp.

103 MPa (15 ksi) for 4 h. To prevent incipient melting and still obtain closure, the HIP parameters were changed to 1185 °C (2165 °F) and 172 MPa (25 ksi) for 4 h. This tradeoff between temperature and pressure can also be used to prevent grain growth and to prevent or limit carbide degradation while obtaining closure of microporosity.

Time at temperature and pressure will obviously affect processing cost. For most alloys, 2 to 4 h is sufficient. Exceptions are massive section sizes, which require additional thermal soaking time.

Joining

Fusion Welding. Cobalt-base superalloys are readily welded by gas-metal arc (GMA) or gas-tungsten arc (GTA) techniques. Cast alloys such as WI-52 and wrought alloys such as Haynes 188 have been extensively welded. Filler metals generally have been less highly alloyed cobalt-base alloy wire, although parent rod or wire (the same composition as the alloy being welded) have been used. Cobalt-base superalloy sheet also is successfully welded by resistance techniques. Gas turbine vanes that crack in service have been repair welded using the above techniques (e.g., WI-52 vanes using Haynes 25 filler rod and 540 °C, or 1000 °F, preheat). Appropriate preheat techniques are needed in GMA and GTA welding to eliminate tendencies for hot cracking. Electron-beam (EB) and plasma-arc (PA) welding can be used on cobalt-base superalloys but usually are not required because alloys of this class are so readily weldable.

Solid-solution strengthened nickel- and iron-nickel-base alloys are also readily weldable by various fusion welding techniques. The most widely employed processes for welding solid-solution strengthened alloys are GTA, GMA, and shielded metal arc (SMA) welding. Submerged arc welding and electroslag welding have limited applicability, as does the PA process.

Precipitation-strengthened nickel- and iron-nickel-base alloys are considerably less weldable than solid-solution strengthened superalloys. Because of the presence of the γ' strengthening

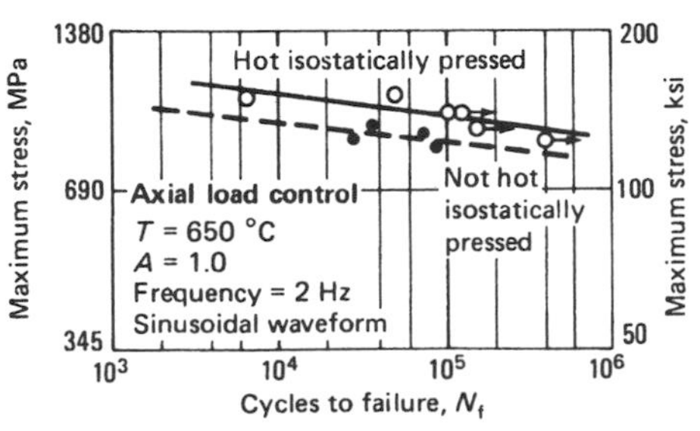

Fig. 29 Improvement of fatigue properties by the elimination of microporosity through hot isostatic pressing. Source: Ref 30

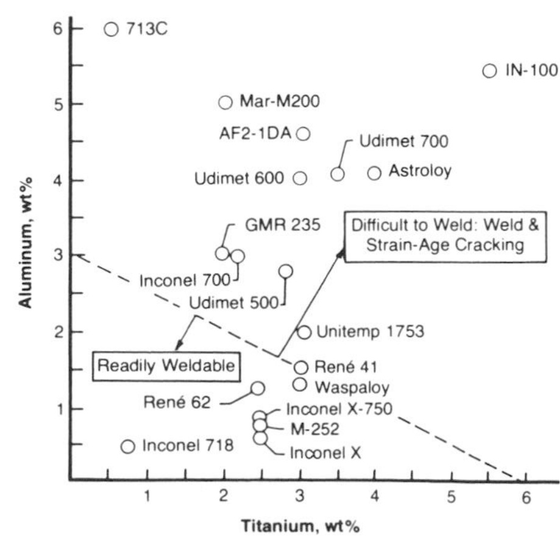

Fig. 30 Weldability diagram for some γ'-strengthened nickel- and iron-nickel-base superalloys, showing influence of total aluminum + titanium hardeners. Source: Ref 37

(a)

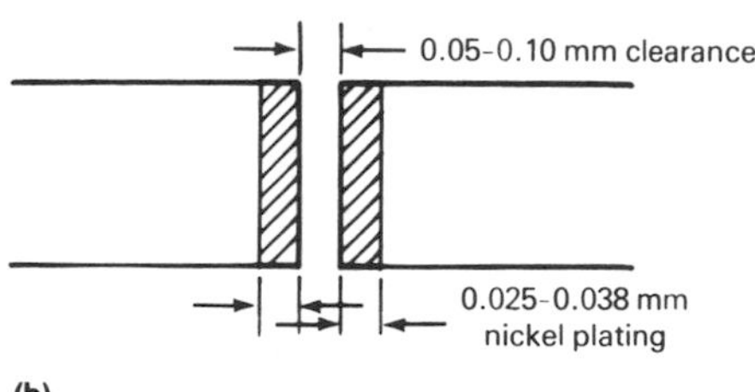

(b)

Fig. 31 Diffusion brazed joint illustrating how the brazing filler metal and nickel plate diffused into the base metal, almost obliterating the joint. Specimen: nickel-plated Nimonic 80A 6.4 mm (0.252 in.) tensile test bar machined from rectangular brazed blocks. Brazing procedure: 30 min at 1175 °C (2150 °F); Nicrobraz 125 brazing filler metal; aged after brazing. (a) Macrostructure of joint. 8×. (b) Schematic of joint

phase, these alloys tend to be susceptible to strain-age cracking. The susceptibility of an alloy to strain-age cracking tends to increase with increasing aluminum and titanium contents (γ' formers), as shown in Fig. 30. Alloys that contain niobium as a hardening element exhibit a sluggish aging response and are therefore less sensitive to strain-age cracking. Strain-age cracking is more prevalent in highly restrained weldments or in the presence of high residual stresses. Cracking is caused by the combination of stress, precipitation strengthening, and the volumetric contraction associated with γ' precipitation. Guidelines to follow to avoid strain-age cracking include:

- Heat as rapidly as possible through the region of aging temperatures when stress-relief annealing.
- Minimize heat input during welding to avoid partial melting of grain boundaries adjacent to the fusion line.
- Heat treat in an inert atmosphere if possible; the presence of oxygen tends to increase intergranular embrittlement, which can result in cracking.
- Always stress-relief anneal before aging. For heavy-section weldments in aluminum-titanium aged materials, multiple stress-relief anneals may be necessary during welding.
- Avoid welding on material with rough or poor surface conditions.

Inconel 718 does not undergo strain-age cracking. The age hardening develops around a Ni_3Nb, γ'', precipitate. The γ'' precipitates at a much slower rate than the γ'. This allows alloys to be heated into the solution temperature range without suffering aging and the resultant strain-age cracking.

Precipitation-hardenable superalloys have been welded by GMA, GTA, EB, laser, and PA techniques. Filler metals, when used, usually are weaker, more ductile austenitic alloys so as to minimize hot cracking. Occasionally base metal compositions are employed as fillers. Welding is generally restricted to wrought alloys with lower γ' contents ($\lesssim$40 vol%).

Some precipitation-hardenable alloys often require welding in the aged condition. IN-713C, for example, is a nickel-chromium alloy with very high levels of aluminum and titanium that is used for gas turbine vanes. It is normally produced as a casting that precipitation hardens during cooldown. It is considered nonweldable by fusion processes. However, experience has shown that blade and vane tips can be restored by fusion welding with a nonhardenable filler metal (ERNiCrMo-3) without cracking (Ref 38). Thus, crack-free welds can sometimes be accomplished in very-difficult-to-weld materials where residual shrinkage stresses can be kept low and cooling rates are slow. Both conditions occur naturally when welding on an edge (blade and vane tips).

Other welding and brazing processes that have been used to join nickel- and iron-nickel-base alloys include friction welding, ultrasonic welding, resistance welding (for sheet applications), diffusion (solid-state) bonding, and brazing. As with welding, furnace brazing of precipitation-hardening presents several difficulties not normally encountered with solid-solution strengthened alloys. As stated above, the precipitation-hardenable contain appreciable amounts of aluminum and titanium (see Fig. 30). The oxides of these elements are almost impossible to reduce in a controlled atmosphere (vacuum, hydrogen). Therefore, nickel plating or the use of a flux is necessary to obtain a surface that allows wetting by the filler metal.

Generally, nickel-base superalloys are brazed with nickel alloys containing boron and/or silicon, which serve as melting-point depressants. In many commercial brazing filler metals, the levels are 2 to 3.5% B and 3 to 10% Si. Phosphorus is another effective melting-point depressant for nickel and is used in filler metals from 0.02 to 10%. It is also used where good flow is important in applications of low stress, where service temperatures do not exceed 760 °C (1400 °F).

In addition to boron, silicon, and phosphorus, chromium is often present to provide oxidation and corrosion resistance. The amount may be as high as 20%, depending on the service conditions. Higher amounts, however, tend to lower brazement strength. Other brazing filler metals used for joining nickel-base superalloys include gold-base alloys and nickel-palladium alloys.

Cobalt-base filler metals are used mainly for brazing cobalt-base components, such as first-stage turbine vanes for jet engines. Most cobalt-base filler metals are proprietary. In addition to containing boron and silicon, these alloys usually contain chromium, nickel, and tungsten to provide corrosion and oxidation resistance and to improve strength.

Another method for joining nickel-base superalloys that is widely used by the aerospace community is diffusion brazing, which is also referred to as transient liquid-phase bonding. Diffusion brazing has been found to be very useful, principally in turbine parts of aircraft gas-turbine engines. The distinguishing characteristic is that, although a lower-temperature bond is made as in brazing, subsequent diffusion occurs at the bonding temperature, leading to a fully solidified joint which has a composition similar to that of the base metal and a microstructure indistinguishable from it (Fig. 31). Consequently, the resultant joint can have a melting temperature and properties very similar to those of the base metal.

ACKNOWLEDGMENTS

The information in this article is largely taken from:

- N.S. Stoloff, Wrought and P/M Superalloys, *Properties and Selection: Irons, Steels, and High-Performance Alloys,* Vol 1, *ASM Handbook* (formerly Vol 1, 10th ed., *Metals Handbook*), ASM International, 1990, p 950–980
- G.L. Erickson, Polycrystalline Cast Superalloys, *Properties and Selection: Irons, Steels, and High-Performance Alloys,* Vol 1, *ASM Handbook* (formerly Vol 1, 10th ed., *Metals Handbook*), ASM International, 1990, p 981–994
- M.J. Donachie, Jr., Superalloys, *Metals Handbook Desk Edition,* American Society for Metals, 1984, p 16-5 to 16-18
- S.L. Semiatin and H.H. Ruble, Forging of Nickel-Base Alloys, *Forming and Forging,* Vol 14, *ASM Handbook* (formerly Vol 14, 9th ed., *Metals Handbook*), ASM International, 1988, p 261–266

REFERENCES

1. F.R. Morral, Ed., Wrought Superalloys, *Properties and Selection: Stainless Steels, Tool Materials and Special-Purpose Metals,* Vol 3, 9th ed., *Metals Handbook,* American Society for Metals, 1980
2. C.T. Sims, Superalloys: Genesis and Character, *Superalloys II,* C.T. Sims, N.S. Stoloff, and W.C. Hagel, Ed., John Wiley & Sons, 1987, p 3–26
3. E.W. Ross and C.T. Sims, Nickel-Base Alloys, *Superalloys II,* C.T. Sims, N.S. Stoloff, and W.C. Hagel, Ed., John Wiley & Sons, 1987, p 97–135

4. R.F. Decker, "Strengthening Mechanisms in Nickel-Base Superalloys," paper presented at the Climax Molybdenum Co. Symp., Zurich, May 1969
5. P.H. Thornton et al., *Metall. Trans.,* Vol 1, 1970, p 207
6. J.W. Brook and P.J. Bridges, in *Superalloys 1988,* The Metallurgical Society, 1988, p 33–42
7. J.M. Walsh and B.H. Kear, *Metall. Trans. A,* Vol 6, 1975, p 226–229
8. D. Moon and F. Wall, in *Proc. Symp. on Structural Stability in Superalloys,* American Institute of Mining, Metallurgical, and Petroleum Engineers, 1968, p 115
9. E.G. Richards, *J. Inst. Met.,* Vol 96, 1968, p 365
10. J.H. Schneibel, C.L. White, and M.H. Yoo, *Metall. Trans. A.,* Vol 6, 1985, p 651
11. M.J. Donachie, *Superalloy Source Book,* American Society for Metals, 1984, p 105
12. Appendix B, compiled by T.P. Gabb and R.L. Dreshfield, *Superalloys II,* C.T. Sims, N.S. Stoloff, and W.C. Hagel, Ed., John Wiley & Sons, 1987, p 575–596
13. "High-Temperature High-Strength Nickel Base Alloys," Inco Alloys International Ltd., distributed by Nickel Development Institute
14. "Product Handbook" Publication 1A1-38, Inco Alloys International, Inc., 1988
15. *Materials Selector 1988,* Penton, 1987
16. Alloy 230 Product Literature, Haynes International
17. B.A. Cowles, D.L. Sims, and J.R. Warren, NASA-CR-159409, National Aeronautics and Space Administration, 1978
18. R.W. Fawley, Superalloy Progress, *The Superalloys,* C.T. Sims and W.C. Hagel, Ed., John Wiley & Sons, 1972, p 12
19. W. Betteridge, *Cobalt and Its Alloys,* Ellis Horwood, 1982
20. M. Gell and D.N. Duhl, The Development of Single-Crystal Superalloy Turbine Blades, *Advanced High-Temperature Alloys: Processing and Properties,* American Society for Metals, 1986, p 41–49
21. J.M. Moyer, in *Proceedings of Superalloys 1984 Conference,* The Metallurgical Society, 1984, p 445
22. J.P. Collier, A.O. Selius, and J.K. Tien, in *Proceedings of Superalloys 1988 Conference,* The Metallurgical Society, 1988, p 43
23. C.T. Sims, *J. Met.,* Vol 21 (No. 12), 1969, p 27
24. A.M. Beltran, *Cobalt,* No. 46, 1970, p 3
25. F.S. Pettit and C.S. Giggins, *Superalloys II,* C.T. Sims, N.S. Stoloff, and W.C. Hagel, Ed., John Wiley & Sons, 1987, p 327
26. G.E. Maurer, Primary and Secondary Melt Processing—Superalloys, *Superalloys, Supercomposites, and Superceramics,* Academic Press, 1989, p 64–96
27. W.H. Couts, Jr. and T.E. Howson, in *Superalloys II,* C.T. Sims, N.S. Stoloff, and W.C. Hagel, Ed., John Wiley & Sons, 1987, p 441
28. L.A. Jackman, in *Proc. Symp. Properties of High Temperature Alloys,* Electrochemical Society, 1976, p 42
29. D.R. Muzyka, in *MiCon 78: Optimization of Processing, Properties, and Service Performance through Microstructural Control,* H. Abrams et al., Ed., American Society for Testing and Materials, 1979, p 526
30. W.R. Freeman, Jr., Chapter 15, *Superalloys II,* C.T. Sims, N.S. Stoloff, and W.C. Hagel, Ed., John Wiley & Sons, 1987, p 411–439
31. G.K. Bouse and J.R. Mihalisin, Metallurgy of Investment Cast Superalloy Components, *Superalloys, Supercomposites and Superceramics,* Academic Press, 1989, p 99–148
32. G.L. Erickson, K. Harris, and R.E. Schwer, "Optimized Superalloy Manufacturing Process for Critical Investment Cast Components," Cannon-Muskegon Corp., 1982, p 6–9
33. M. Lamberigts, S. Ballarati, and J.M. Drapier, Optimization of the High Temperature, Low Cycle Fatigue Strength of Precision Cast Turbine Wheels, *Proc. Fifth Internat. Symp. on Superalloys,* American Institute of Mining, Metallurgical, and Petroleum Engineers, 1984, p 13–22
34. B.A. Ewing and K.A. Green, Polycrystalline Grain Controlled Castings for Rotating Compressor and Turbine Components, *Proc. Fifth Internat. Symp. on Superalloys,* American Institute of Mining, Metallurgical, and Petroleum Engineers, 1984, p 33–42
35. M.J. Woulds and H. Benson, Development of a Conventional Fine Grain Casting Process, *Proc. Fifth Internat. Symp. on Superalloys,* American Institute of Mining, Metallurgical, and Petroleum Engineers, 1984, p 3–12
36. S.J. Veeck, L.E. Dardi, and J.A. Butzer, "High Fatigue Strength, Investment Cast Integral Rotors for Gas Turbine Applications," paper presented at the TMS-AIME Annual Meeting, The Metallurgical Society, Dallas, Feb 1982
37. M. Prager and C.S. Shira, Welding of Precipitation-Hardening Nickel-Base Alloys, *Weld. Res. Counc. Bull.,* No. 128, 1968
38. D.S. Duvall and J.R. Doyle, *Repair of Turbine Blades and Vanes,* Publication 73-GT-44, American Society of Mechanical Engineers, 1973

Directionally Solidified and Single-Crystal Superalloys

DIRECTIONALLY SOLIDIFIED (DS) columnar-grained superalloys and single-crystal (SX) nickel-base superalloys have found widespread use as blades and vanes in advanced gas turbine engines due to their outstanding elevated-temperature properties. There are two primary reasons why DS and SX superalloys are superior to conventionally cast (equiaxed-grain) superalloys. Alignment (or elimination, in the case of SX superalloys) of the grain boundaries normal to the stress axis enhances elevated-temperature ductility by eliminating the grain boundary as the failure initiation site. This permits the γ' microstructure to be refined with a solution heat treatment that increases alloy creep strength. The second reason is that the DS process provides a preferred low-modulus ⟨001⟩ texture or orientation parallel to the solidification direction. This results in a significant enhancement in thermal fatigue resistance, which is important in elevated-temperature components. The three solidification modes (equiaxed grain, columnar grain, and single crystal) are compared in Fig. 1.

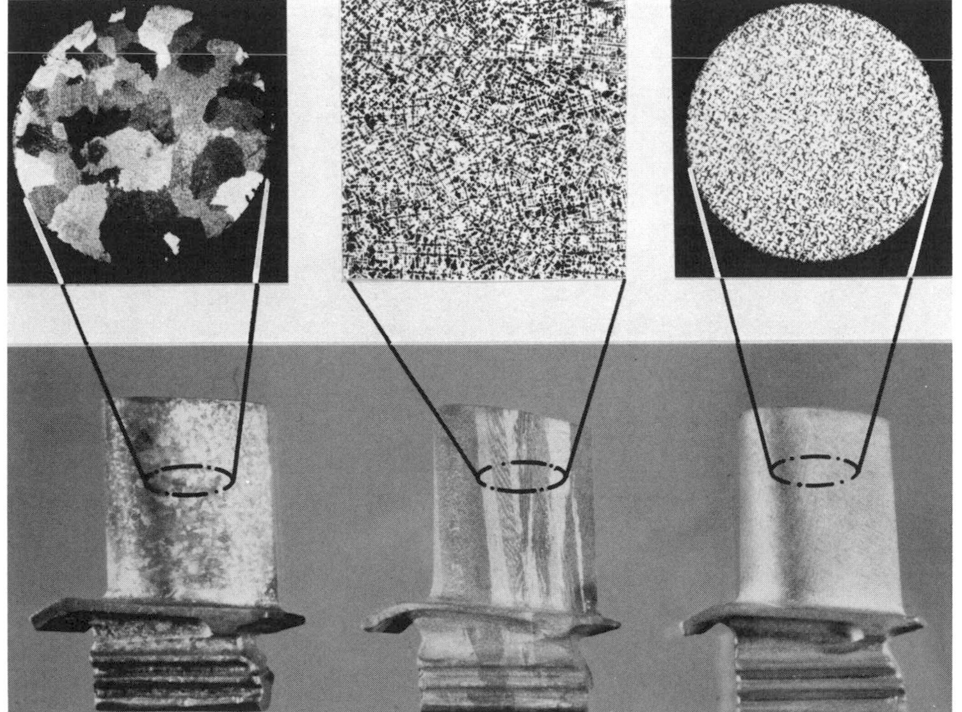

Fig. 1 Comparison of macro- and microstructures in (from left) equiaxed, directionally solidified, and single-crystal turbine blades

Alloy Development

The primary goals in the continuing development of aircraft gas turbines are increased operating temperatures and improved efficiencies. A more efficient turbine is required to achieve lower fuel consumption. Higher turbine inlet temperature and increased stage loading result in fewer parts, shorter engine lengths, and reduced weight. Engine operating costs can be reduced if higher temperatures are possible without increasing part life-cycle costs.

Critical turbine components include high-pressure turbine blades, vanes, and disks. During the last 15 years, turbine inlet temperatures have increased by 278 °C (500 °F). About half of this increase is due to a more efficient design for the air cooling of turbine blades and vanes, while the other half is due to improved superalloys and casting processes (Ref 1). The cooling that is now possible with serpentine cores and multiple shaped-hole film cooling (Fig. 2) enables high-pressure turbine blades and vanes to operate with turbine inlet temperatures of typically 1343 °C (2450 °F), which is above the melting point of the superalloy materials. Turbine inlet temperatures as high as 1600 °C (2910 °F) and a metal temperature in excess of 1100 °C (2010 °F) are current parameters for several advanced fighter engines (Ref 2). It is forecast that by the late 1990s, fabricated single-crystal airfoils with ultraefficient transpiration cooling schemes will be capable of operating in gas temperatures greater than 1650 °C (3000 °F) with sufficient durability and reliability for man-rated flight turbine engines.

For the past 35 years, high-pressure turbine blades and vanes have been made from cast nickel-base superalloys. The higher-strength al-

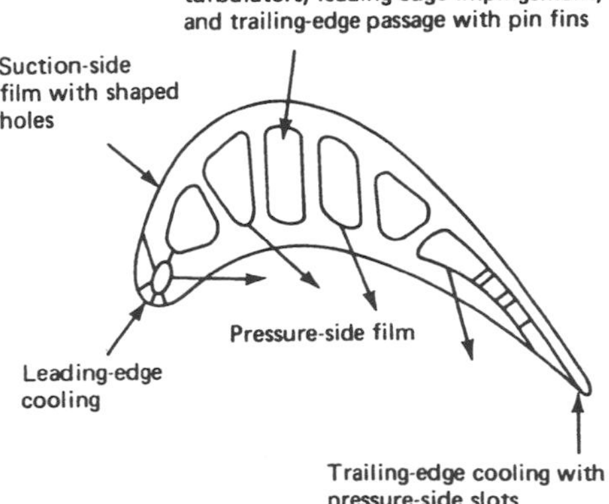

Fig. 2 Shaped holes, turbulators, pin fins, and other techniques used in turbine rotor blade cooling

Table 1 Chemical compositions of nickel-base directionally solidified castings

Alloy	Nominal composition, wt %													
	C	Cr	Co	Mo	W	Nb	Re	Ta	Al	Ti	B	Zr	Hf	Ni
First generation														
MAR-M 200 Hf	0.13	8.0	9.0	...	12.0	1.0	...	...	5.0	1.9	0.015	0.03	2.0	bal
René 80H	0.16	14.0	9.0	4.0	4.0	...	...	...	3.0	4.7	0.015	0.01	0.8	bal
MAR-M 002	0.15	8.0	10.0	...	10.0	...	...	2.6	5.5	1.5	0.015	0.03	1.5	bal
MAR-M 247	0.15	8.0	10.0	0.6	10.0	...	...	3.0	5.5	1.0	0.015	0.03	1.5	bal
PWA 1422	0.14	9.0	10.0	...	12.0	1.0	...	...	5.0	2.0	0.015	0.10	1.5	bal
Second generation														
CM 247 LC	0.07	8.0	9.0	0.5	10.0	...	...	3.2	5.6	0.7	0.015	0.010	1.4	bal
CM 186 LC	0.07	6.0	9.0	0.5	8.4	...	3.0	3.4	5.7	0.7	0.015	0.005	1.4	bal
PWA 1426	0.10	6.5	10.0	1.7	6.5	...	3.0	4.0	6.0	...	0.015	0.10	1.5	bal
René 142	0.12	6.8	12.0	1.5	4.9	...	2.8	6.35	6.15	...	0.015	0.02	1.5	bal

loys are hardened by a combination at approximately 60 vol% γ′ [$Ni_3(Al,Ti)$] precipitated in a γ matrix, with solid-solution strengthening provided by the powerful strengtheners tantalum, tungsten, and molybdenum. The γ phase, which has an ordered face-centered cubic structure, is coherent with the γ matrix, their lattice parameters being almost identical (<1% mismatch). This allows homogeneous nucleation of the precipitate with low surface energy and long-time stability at temperature, ensuring the potential usefulness of the alloys at elevated temperatures up to 0.85 T_m (melting point) for extended periods of time. Tantalum, tungsten, and hafnium substitute for some of the aluminum and titanium in the γ′ phase, thus stiffening this phase because of their relatively large atomic size. Initially, the blades were made as isotropic polycrystal or equiaxed casting. Under aerospace turbine engine operating conditions, failure of these equiaxed-grain components usually occurred at the grain boundaries from a combination of creep, thermal fatigue, and oxidation.

Development of DS casting technology to produce blades and vanes with low-modulus ⟨100⟩-oriented columnar grains, aligned parallel to the longitudinal, or principal-stress, axis (Fig. 3), resulted in significant improvements in creep strength and ductility as well as in thermal fatigue resistance (5× improvement). VerSnyder and coworkers at Pratt and Whitney Aircraft (PWA) pioneered this process (Ref 3, 4). The alloy they originally used was MAR M-200, a nickel-base alloy containing 12.5% W used in turbine components. The solidified structure consisted of tungsten-rich dendrites with high strength and creep resistance that grew to the length of the casting. The grain-boundary material, which was parallel to the dendrites, was strong enough to withstand the transverse stresses on the components.

In columnar structures, the primary dendrites are aligned, as are the grain boundaries. The primary dendrites form around spines of the highest-melting constituent to freeze. As freezing continues, the solid rejects solute into the residual liquid (segregation occurs) until the final low-melting eutectic has frozen at the grain boundaries.

Fig. 3 Directionally solidified turbine blade CM 247 LC

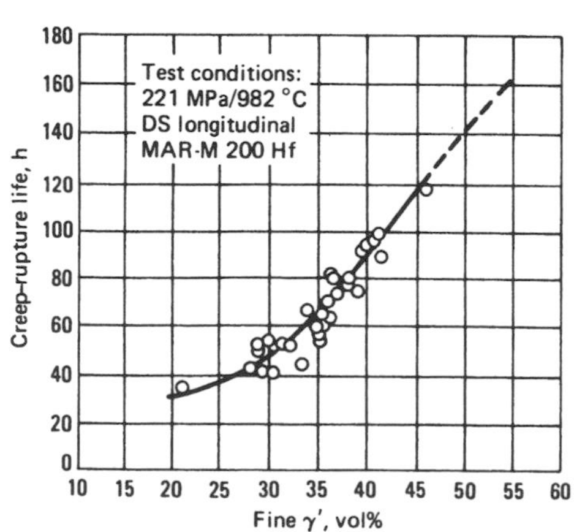

Fig. 4 Rupture life vs. volume fraction of fine γ′ at a fixed total amount of fine and coarse γ′ for DS MAR-M 200 Hf alloy

There has been recent interest in DS blades, not only for small to medium-size airfoils for industrial turbines that burn natural gas, but also for large base-load electricity-generating machines. Improved fuel efficiency requirements, along with the desire for high-temperature exhaust gases from the gas turbine (to produce steam suitable for co-generation electricity production), have resulted in the development and application engineering of DS blades with component

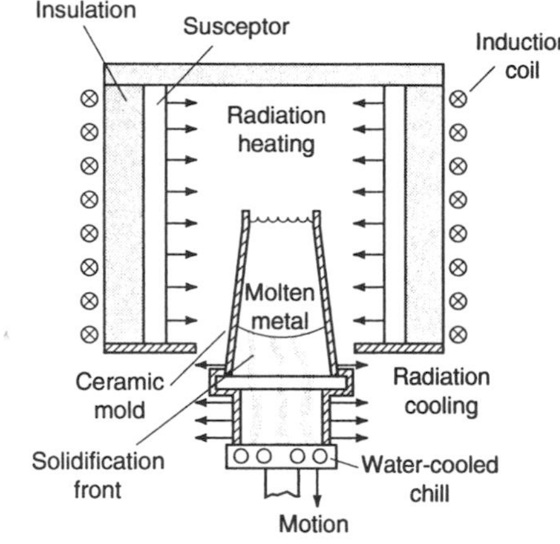

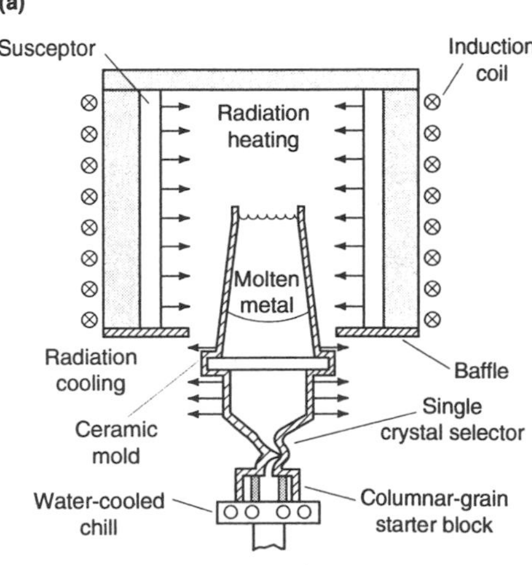

Fig. 5 Schematics showing (a) the directional solidification process and (b) the single-crystal solidification process. Source: Ref 6

lengths in the range of 305 to 635 mm (12 to 25 in.).

Development of SX Casting Technology. Pratt and Whitney Aircraft pioneered SX casting processes in the mid-1960s (Ref 5, 6). Initially, there was limited interest in the development of single-crystal blades because the conventional heat treatments being applied to MAR-M 200-type single-crystal components did not produce improvements in creep strength, thermal fatigue strength, and oxidation resistance that significantly exceeded the results achieved with the DS columnar-grain MAR-M 200 Hf. Only ductility and transverse creep resistance were improved. Around 1975, the beneficial role of γ' solutioning heat treatment applied to DS MAR-M 200 Hf was shown by PWA (Ref 7). It was found that creep strength was a direct function of the volume fraction of solutioned and reprecipitated fine γ' (Fig. 4). Experimental work by PWA showed that elimination of grain-boundary strengthening elements (boron, hafnium, zirconium, and carbon) resulted in a substantial increase in the incipient melting temperature of the alloy (Ref 6). Consequently, the complete solutioning of the γ' phase, with appreciable solutioning of the γ/γ' eutectic phase, became possible without provoking incipient melting of the alloy.

Single-crystal alloy PWA 1480 offered a 25 to 50 °C (45 to 90 °F) temperature capability improvement in terms of time-to-1% creep, compared to the extensively used DS MAR-M 200 Hf alloy (Ref 5). The creep property improvement, which increased with temperature, depended on optimized single-crystal microstructures with full solutioning of the as-cast coarse γ'. The PWA 1480 alloy was developed to utilize the relatively low thermal gradient, single-crystal casting facilities already available as DS production units, without the freckling problems of alloy 444 (single-crystal MAR-M 200 with no carbon, boron, hafnium, zirconium, or cobalt) (Ref 5). Alloy PWA 1480, with its high tantalum (12%) and low tungsten (4%) contents, proved to be unique with this castability feature. Multistep homogenization/solutioning treatments with tight temperature control were developed to completely solution the γ' PWA 1480 without inducing incipient melting. Since 1982, PWA has had more than 5 million flight hours of successful experience using turbine blade and vane parts of single-crystal alloy PWA 1480 in commercial and military engines (Ref 8).

Alloy Design. The basic differences in composition between DS and SX superalloys are:

- The hafnium added to DS superalloys to prevent grain-boundary cracking
- The balancing of alloying elements in SC superalloys to provide incipient melting temperature above the γ' solvus
- The balancing of refractory elements to minimize defects such as freckles. In SX alloys, this

Table 2 Chemical compositions of nickel-base single-crystal castings

Alloy	Compsition, wt %												Density, g/cm^3
	Cr	Co	Mo	W	Ta	Re	V	Nb	Al	Ti	Hf	Ni	
First generation													
PWA 1480	10	5	...	4	12	...	...	...	5.0	1.5	...	bal	8.70
PWA1483	12.8	9	1.9	3.8	4	...	...	...	3.6	4.0	...	bal	...
René N4	9	8	2	6	4	...	...	0.5	3.7	4.2	...	bal	8.56
SRR 99	8	5	...	10	3	...	...	...	5.5	2.2	...	bal	8.56
RR 2000	10	15	3	...	...	...	1	...	5.5	4.0	...	bal	7.87
AM1	8	6	2	6	9	...	...	...	5.2	1.2	...	bal	8.59
AM3	8	6	2	5	4	...	...	...	6.0	2.0	...	bal	8.25
CMSX-2	8	5	0.6	8	6	...	...	...	5.6	1.0	...	bal	8.56
CMSX-3	8	5	0.6	8	6	...	...	...	5.6	1.0	0.1	bal	8.56
CMSX-6	10	5	3	...	2	...	...	...	4.8	4.7	0.1	bal	7.98
CMSX-11B	12.5	7	0.5	5	5	...	...	0.1	3.6	4.2	0.04	bal	8.44
CMSX-11C	14.9	3	0.4	4.5	5	...	...	0.1	3.4	4.2	0.04	bal	8.36
AF 56 (SX 792)	12	8	2	4	5	...	...	...	3.4	4.2	...	bal	8.25
SC 16	16	...	3	...	3.5	...	...	...	3.5	3.5	...	bal	8.21
Second generation													
CMSX-4	6.5	9	0.6	6	6.5	3	...	...	5.6	1.0	0.1	bal	8.70
PWA 1484	5	10	2	6	9	3	...	...	5.6	...	0.1	bal	8.95
SC 180	5	10	2	5	8.5	3	...	...	5.2	1.0	0.1	bal	8.84
MC2	8	5	2	8	6	...	...	...	5.0	1.5	...	bal	8.63
René N5	7	8	2	5	7	3	...	...	6.2	...	0.2	bal	...
Third generation													
CMSX-10	2	3	0.4	5	8	6	...	0.1	5.7	0.2	0.03	bal	9.05
René N6	4.2	12.5	1.4	6	7.2	5.4	...	...	5.75	...	0.15	bal	8.98

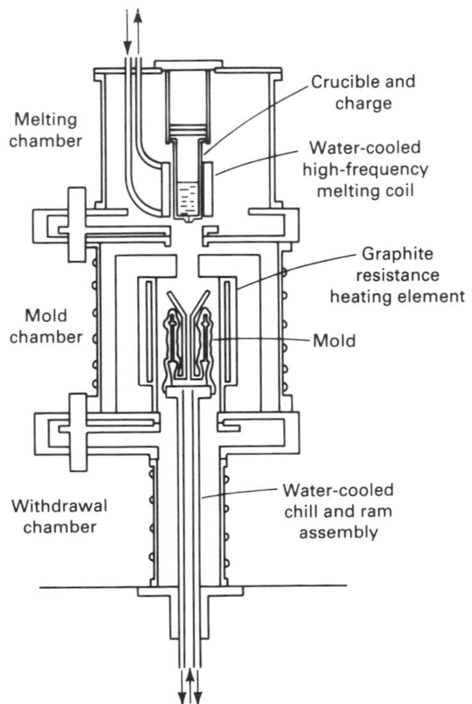

Fig. 6 Configuration of one type of directional solidification furnace. Source: Ref 14

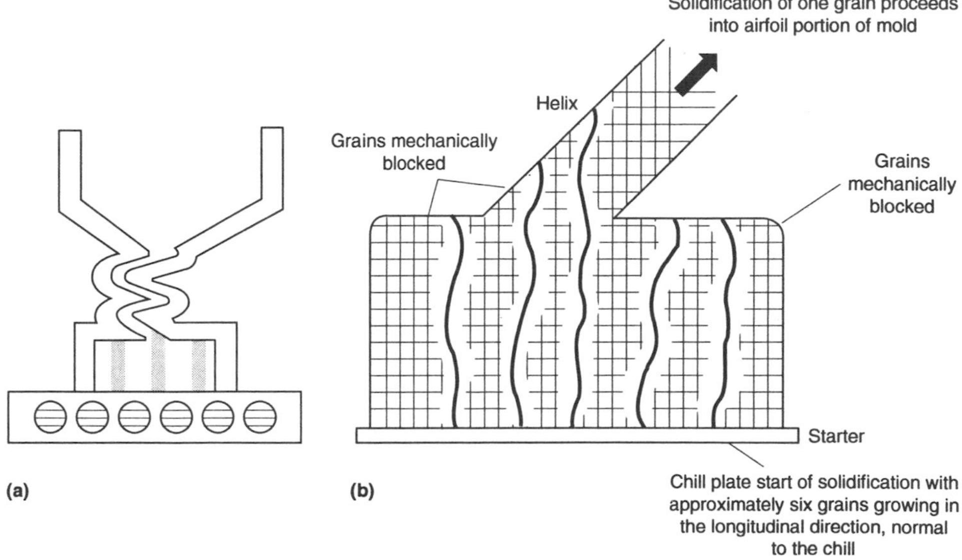

Fig. 7 Single-crystal processing via the use of a multiple-turn constriction device. (a) Schematic of the helical mold section. (b) Schematic of grains entering the helix single-crystal system. Although two grains may initially enter the helix, after one or two turns of the spiral, only one grain survives and fills the mold as a single grain. Source: Ref 9

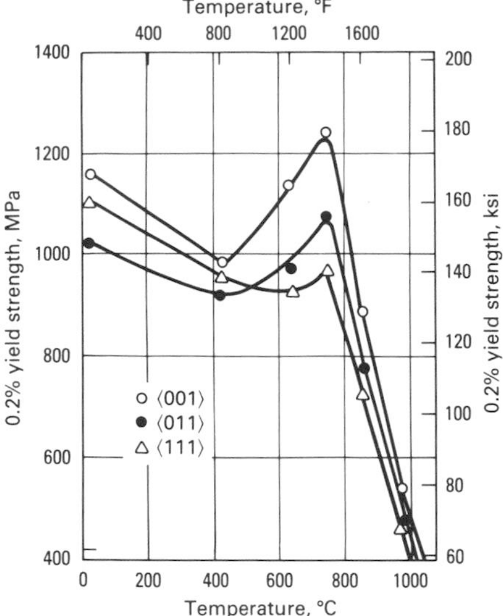

Fig. 8 Yield strength of single-crystal PWA 1480 alloy as a function of temperature and orientation. Source: Ref 15

is accomplished by a closer tantalum-to-tungsten ratio.

- The absence (or low level) of the grain-boundary strengthening elements boron, carbon, hafnium, and zirconium in SX alloys

Additional information on the effects of various alloying elements can be found below and in Ref 9. Tables 1 and 2 list the chemical compositions of DS and SX alloys, respectively.

Directional Solidification Processing

Heat Flow Control. To obtain a directionally solidified structure, it is necessary to cause the dendrites to grow from one end of the casting to the other. This is accomplished by removing the bulk of the heat from one end of the casting. To this end, a strong thermal gradient is established in the temperature zone between the liquidus and solidus temperatures of the alloy and is passed from one end of the casting to the other at a rate that maintains the steady growth of the dendrite, as shown in Fig. 5(a). If the thermal gradient is moved through the casting too rapidly, nucleation of grains ahead of the solid/liquid interface will result; if the gradient is passed too slowly, excessive macrosegregation will result, along with the formation of freckles (equiaxed grains of interdendritic composition) (Ref 10). Therefore, the production of directionally solidified castings requires that both the thermal gradient and its rate of travel be controlled. For the case of nickel-base alloys, thermal gradients of 36 to 72 °C/cm (165 to 330 °F/in.) have been found to be effective (Ref 11), and rates of travel of 30 cm/h (12 in./h) can be used. There is, however, no upper limit on the allowable gradient, and higher gradients usually produce better castings than lower gradients.

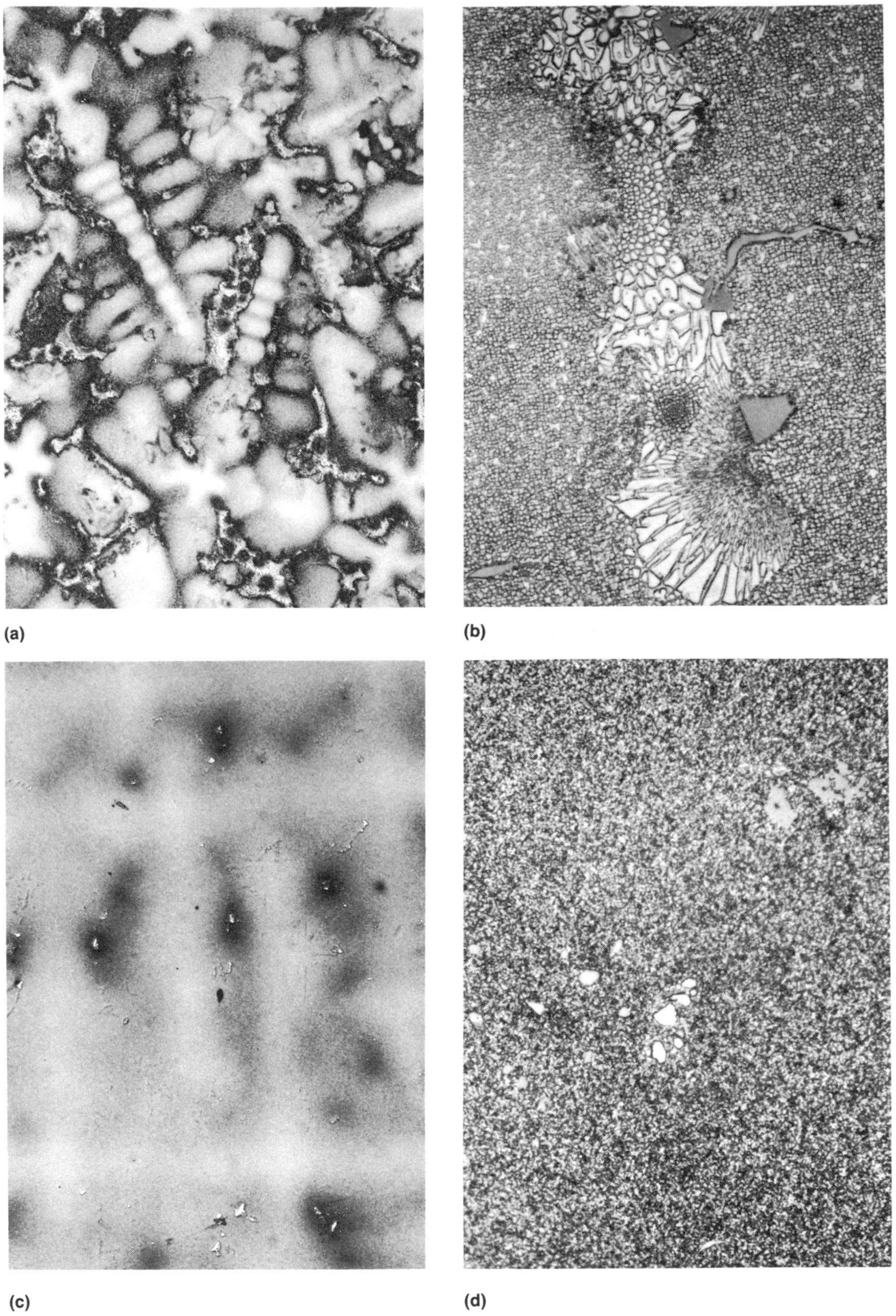

Fig. 9 CM 247 LC directionally solidified turbine blade, as-cast, and supersolutioned microstructures. Micrographs taken from airfoil, transverse orientation. (a) As-cast. 90×. (b) As-cast. 905×. (c) Supersolutioned. 90×. (d) Supersolutioned. 905×

The lower limit on the thermal gradient is a function of alloy composition and casting geometry.

The most effective way to control heat flow is to use a thin-wall mold, such as an investment casting mold, that is open at the bottom. The mold is placed on a chill (which is usually water cooled) and heated above the liquidus temperature of the alloy. Molten metal is poured into the mold, and the mold is cooled from the chilled end by withdrawing the mold from the mold-heating device.

The chill is used to ensure that there is good nucleation of grains to start the process. Because of the low thermal conductivity of nickel-base alloys, the thermal effect of the chill extends only about 50 to 60 mm (2 to 2.4 in.) (Ref 12, 13). Although the grains originally nucleate with random orientations, those with the preferred growth

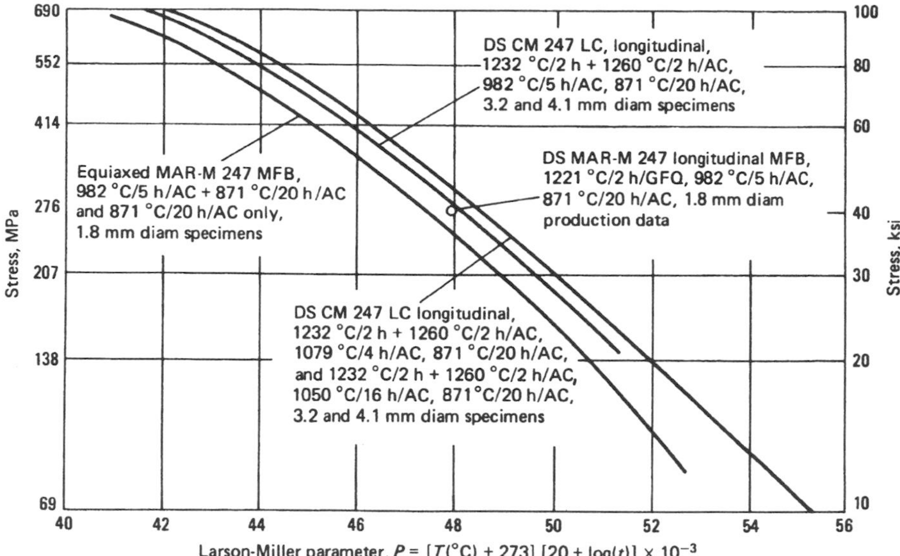

Fig. 10 Larson-Miller stress-rupture strength of DS CM 247 LC vs. that of DS and equiaxed MAR-M 247. MFB, machined from blade; GFQ, gas furnace quenched; AC, air cooled

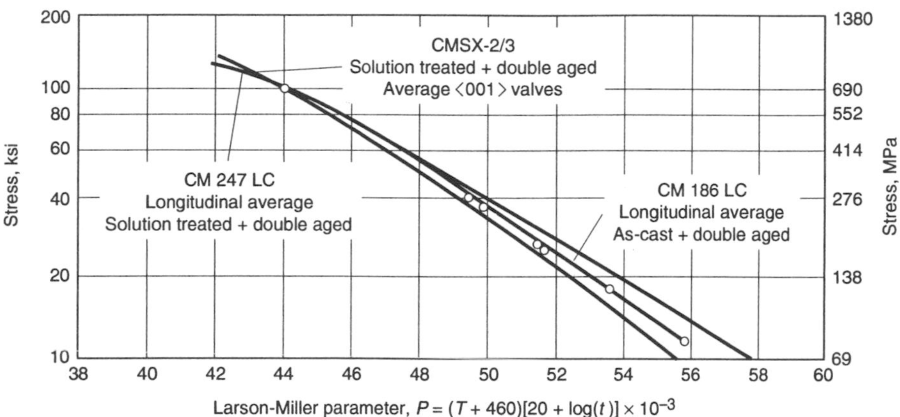

Fig. 11 Stress-rupture strength comparison of DS alloy CM 247 LC (no rhenium), CM 186 LC (3% Re), and first-generation single-crystal alloys CMSX-2 and CMSX-3. Source: Ref 19

direction normal to the chill surface grow and crowd out the other grains. Therefore, those grains that grow through the casting are all aligned in the direction of easiest growth. For nickel-base alloys, the preferred growth direction is ⟨001⟩; therefore, in castings made of these alloys, the grains are aligned in the ⟨001⟩ direction. Passing the thermal gradient through the casting at a uniform rate ensures that the secondary dendrite arm spacing is uniform throughout the casting (Ref 12).

Equipment. In the directional solidification process, an investment casting mold, open at the bottom as well as the top, is placed on a water-cooled copper chill and raised into the hot zone of the furnace (Fig. 6). The mold is heated to a temperature above the liquidus temperature of the alloy to be poured. Meanwhile, the alloy is melted under vacuum in an upper chamber of the furnace. When the mold is at the proper temperature and the charge is molten, the alloy is poured into the mold. After a pause of a few minutes to allow the grains to nucleate and begin to grow on the chill, during which the most favorably oriented grains are established, the mold is withdrawn from the hot zone and moved to the cold zone.

The furnace shown in Fig. 6 has a relatively small chill diameter (140 mm, or 5.5 in.) to enhance the thermal gradient, a resistance-heated hot zone, and an unconventional melting method in which the charge melts through a plate in a bottom-pour crucible instead of being poured. However, other furnace designs use larger chill plates (up to 500 mm, or 20 in.), induction-heated graphite susceptors in their hot zones, and conventional pouring to produce these castings.

Gating. Castings can be gated either into the top of the mold cavity or the bottom. Bottom gating heats the mold just above the chill and sets up a very high gradient that encourages well-aligned dendrites. Particular care is taken to keep the transition between the hot and cold zones as sharp as possible through the use of radiation baffles made of refractory materials; these baffles are placed at the chill level between the hot and cold zones.

Mold Design. In designing molds for the process, consideration must be given to the orientation of the part on the cluster. Because heat transfer is by radiation, parts must be placed to minimize shadowing. Internal radiation baffles are sometimes added to the mold, particularly around the center downsprue, to distribute radiation energy to those parts of the mold that would otherwise be shadowed. Some furnace designs use a heating source or cooling baffle around the center downpole (the chill is designed with a circular cutout at its center) to increase the gradient. Because castings solidify directionally, it is possible to stack them on top of each other to increase the number of castings that can be made in each heat.

Process Control. A very high degree of control must be exercised over the process; therefore, the furnaces are highly automated. Completely automated furnaces (which charge, melt, heat the mold, pour, hold, and withdraw according to a programmed cycle) are commonly used, and even in those furnaces in which melting is done manually the solidification (withdrawal) cycle is automated. Thermocouples are placed within the mold cavity on large clusters to ensure that the molds are at the proper temperature before pouring.

Withdrawal rates during solidification are not necessarily constant. Large differences in section size in specific castings change the solidification rate, and the withdrawal rate can be changed to compensate for this. In selecting a solidification cycle for a hollow part, the effect of the core must be included. Cores lengthen the time required to preheat the mold and slow the withdrawal rate, because the heat they contain must also be removed in the process.

Casting Defects. Directionally solidified castings are routinely inspected by etching their surfaces and examining the surface visually for defects. Most of these defects can be avoided by careful control of the solidification parameters of the casting.

Equiaxed grains are most often freckles, which are caused by segregation of eutectic liquid that is less dense than the bulk liquid in many alloys. This liquid forms jets within the mushy zone, and as these jets freeze they form equiaxed grains. Freckles are usually cured by increasing the thermal gradient and solidification rate in the casting (Ref 10).

Misoriented grains occur when the temperature ahead of the interface falls below the liquidus temperature and new grains nucleate. These grains will have a random orientation, but because they are growing in gradient, they will be columnar. They can be eliminated by increasing the gradient.

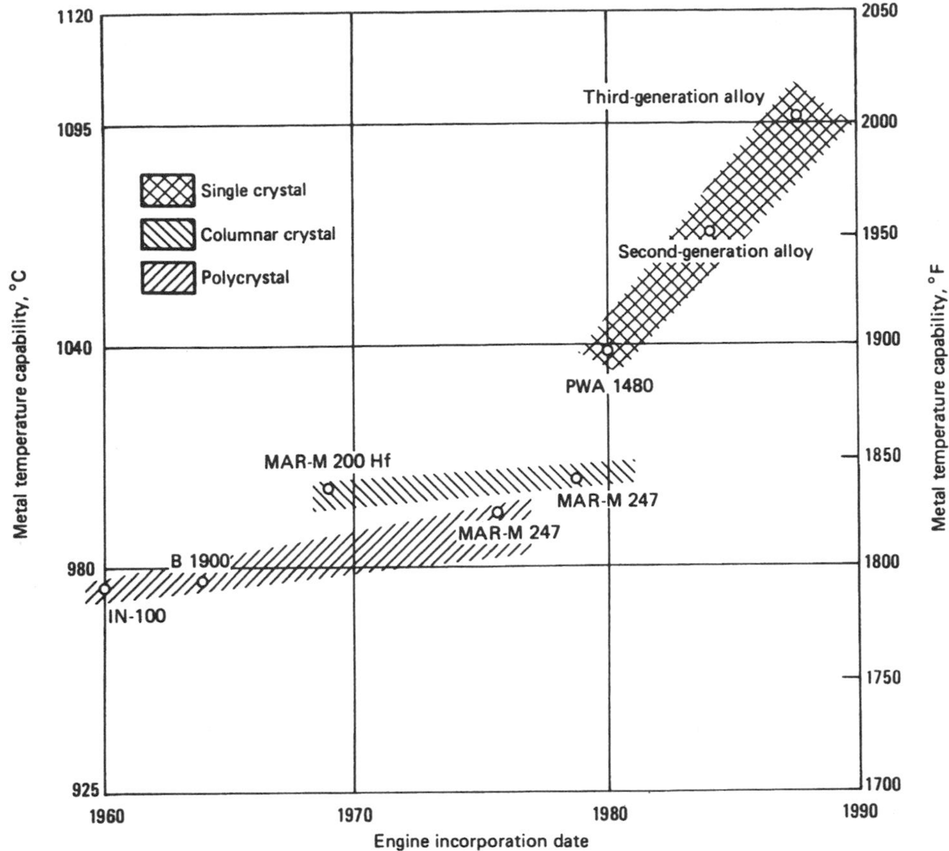

Fig. 12 Progress in turbine airfoil metal temperature capability. Source: Ref 20

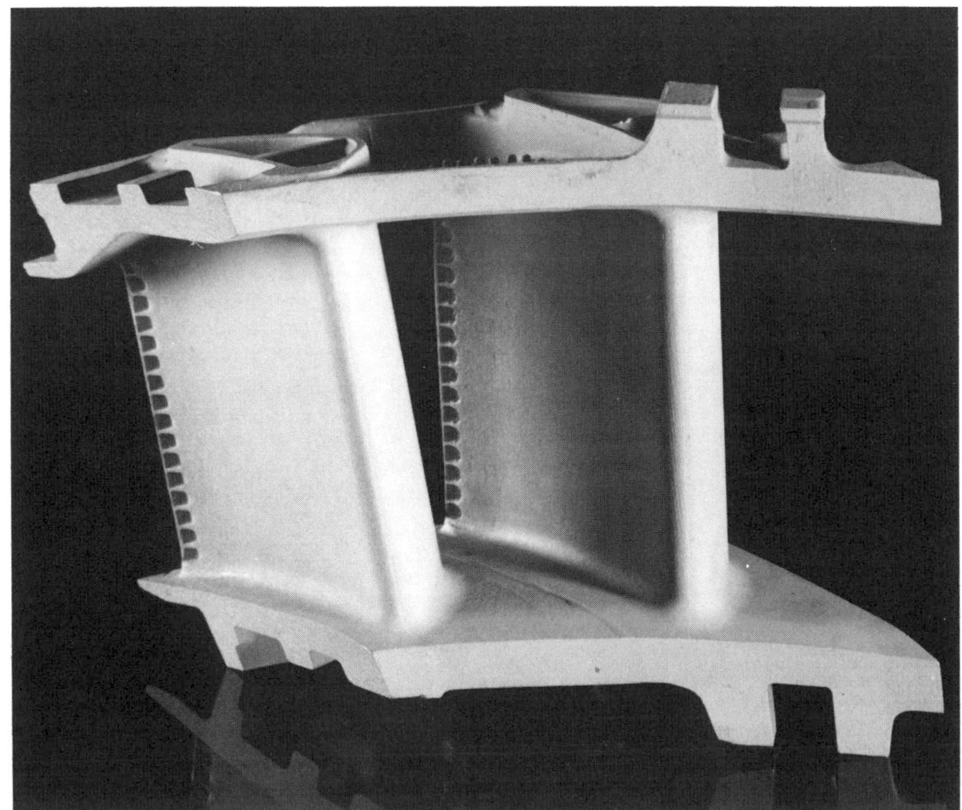

Fig. 13 F-100 paired single-crystal vane in PWA 1480

Grain-Boundary Cracking. When hollow air-cooled turbine airfoil castings are required, a ceramic core, which is subsequently leached out, is used to provide the hollow cavity of required dimensions in the casting. The ceramic core has a lower coefficient of expansion than the DS superalloy, so hoop stresses are set up in the alloy during cooling as the superalloy shrinks around the ceramic core. In the case of DS castings these stresses can result in cracks forming at the grain boundaries. To avoid this, 0.8 to 2.0% Hf is added to DS superalloys (see Table 1).

Shrinkage is sometimes encountered on the upper surfaces of directionally solidified castings. There is no way to feed these surfaces; the addition of risers to these surfaces usually interferes with radiation heat transfer from another part of the casting. The most common solution is to invert the casting in order to minimize the surface area that is susceptible to shrinkage.

Microporosity may occur in directionally solidified castings if the length of the mushy zone (length of the casting that is between the liquidus and solidus temperatures during solidification) becomes too great for feed metal to reach into the areas where solidification is taking place. Increasing the thermal gradient (which shortens the length of the mushy zone) usually solves this problem.

Mold or Core Distortion. A frequent cause of scrap in directionally solidified castings results from mold or core distortion. Because the mold and core are held at high temperatures for long times while the casting solidifies, it is possible for the mold or core to sag or to undergo local allotropic transformations of the refractory materials from which it is made. The resulting changes in mold or core dimensions are reflected in the casting dimensions. Careful control of the core and mold composition, their uniformity, and the firing conditions under which they are made is required in order to avoid these dimensional problems.

Single-Crystal Processing

It was early recognized that if columnar-grain castings could be produced, the production of castings that contained only a single crystal (more accurately, a single grain or primary dendrite) could be produced by suppressing all but one of the columnar grains. Single-crystal castings are produced using techniques similar to those used for DS castings, with one important difference: A method of selecting a single, properly oriented grain is required.

Helical Mold Sections. In one method, a helical section of mold is placed between the chill and the casting (Fig. 5b). This helix, or spiral grain selector, acts as a filter and permits only a single grain to pass through. This is because superalloys solidify by dendritic growth. Each dendrite can grow only in the three mutually orthogonal ⟨001⟩ directions. The continually changing direction of the helix, combined with the orthogonal nature of dendritic growth, gradually constricts all but one favored grain, resulting in a single crystal emitting from the top of the helix, as shown in Fig. 7. The selected grain then fills the mold cavity, as for the case with the DS casting, and a ⟨001⟩-oriented SX foil casting results.

Crystal Growth from Seeds. Seeding is the second method for producing single-crystal castings. This method is particularly useful for growing crystals with an orientation other than the ⟨001⟩ growth direction. With the use of seed crystals, ⟨001⟩-, ⟨011⟩-, and ⟨111⟩-oriented crystals can be grown. Using this method, the helix and DS grain starter block shown in Fig. 5(b) are replaced by a seed crystal. The seed crystal should be of the desired alloy or of an alloy that has an equivalent or higher melting temperature. It is positioned so that its orientation will be repeated in the alloy that fills the mold cavity. The seed sits on the chill, and the temperature at the top of the seed is controlled so that the seed crystal does not melt completely, thereby allowing the molten alloy in the mold cavity to solidify with the same orientation as the seed. Figure 8 shows how crystal orientation influences properties of SX superalloys.

Molds. Because single-crystal alloys have higher incipient melting temperatures than conventional alloys, mold preheat temperatures will normally be higher for their manufacture than for columnar-grain castings. Therefore, mold composition control is of particular importance in the production of these castings.

Casting Defects. With the exception of grain-boundary cracking, SX castings are subject to the same solidification defects as DS castings. However, SX castings are also subject to the formation of low-angle boundaries, boundaries that have a small lattice misorientation. Although the misorientation is usually less than 15°, these boundaries can act as crack initiation sites. The permissible amount of misorientation is a function of alloy and application, but boundaries above 10° are usually not permitted (a tolerance of ±5° of the required orientation is often specified). Single-crystal castings are inspected by back reflection Laué techniques to determine crystallographic orientation.

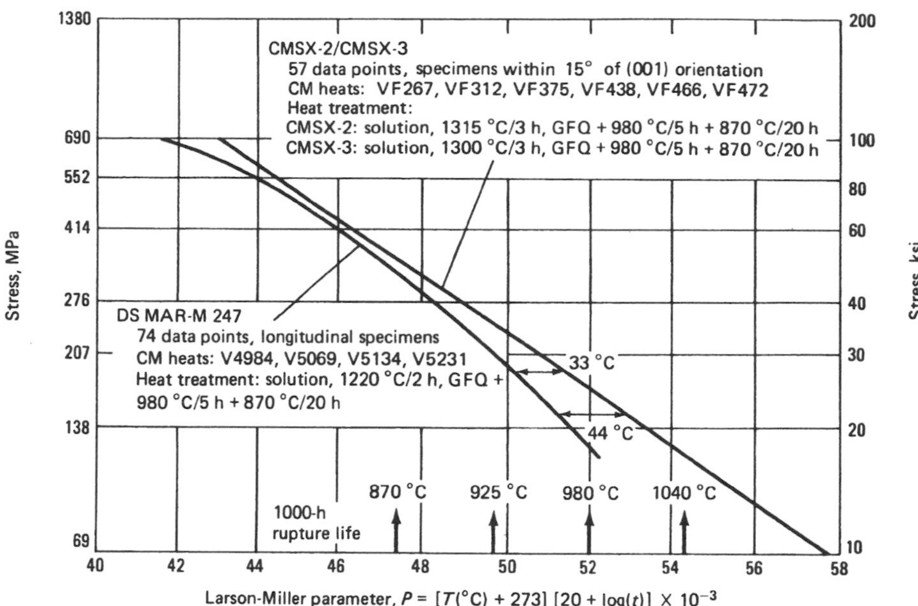

Fig. 14 Larson-Miller stress-rupture strength of CMSX-2/CMSX-3 vs. that of DS MAR-M 247, using 1.8 mm (0.070 in.) specimens machined from blades. GFQ, gas furnace quenched

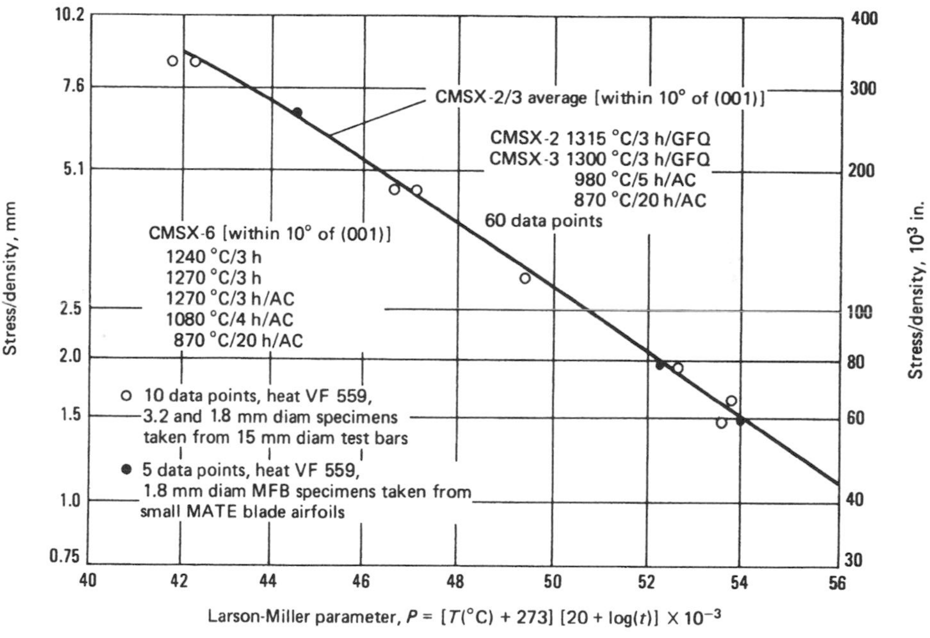

Fig. 15 Larson-Miller specific stress-rupture strength of CMSX-6 vs. that of CMSX-2/3. MFB, machined from blade; GFQ, gas furnace quenched; AC, air cooled; MATE, Materials for Advanced Technology Engines (NASA program)

Directionally Solidified Superalloys

Chemistry and DS Castability. Early work with directionally solidified columnar-grain turbine blades in the 1960s involved the superalloys used for conventionally cast, equiaxed blades containing approximately 60 vol% γ' such as IN-100 and MAR-M 200. The problems encountered ranged from little longitudinal stress-rupture improvement with IN-100 to the lack of transverse ductility and DS grain-boundary cracking with MAR-M 200.

Pioneering work by Martin Metals resulted in the addition of hafnium to conventionally cast equiaxed superalloys to improve 760 °C (1400 °F) stress-rupture ductility and castability. For directional solidification, PWA added hafnium to MAR-M 200, which reduced DS grain-boundary cracking and increased transverse ductility. Although hafnium levels of up to 2% and greater in MAR-M 200 Hf combatted DS grain-boundary cracking, increasing levels of hafnium also increased the DS airfoil component rejection rate and the number of quality assurance problems. This was due to the occurrence of HfO inclusions that usually resulted from hafnium-ceramic reactions (core, shell-mold). Other first-generation DS alloys that were successfully and extensively adopted by turbine engine companies included René 80H (René 80 + Hf) by GE, MAR-M 002 by Rolls-Royce, and MAR-M 247 by Garrett. Both MAR-M 002 and MAR-M 247 were originally developed by Martin Metals to contain hafnium for optimized equiaxed turbine blade mechanical properties and castabilities. The nominal compositions of these first-generation DS superalloys are listed in Table 1. Directionally solidified superalloy turbine blades employed in large commercial turbofan engines for long-distance flights have been used for up to 15,000 h with high reliability.

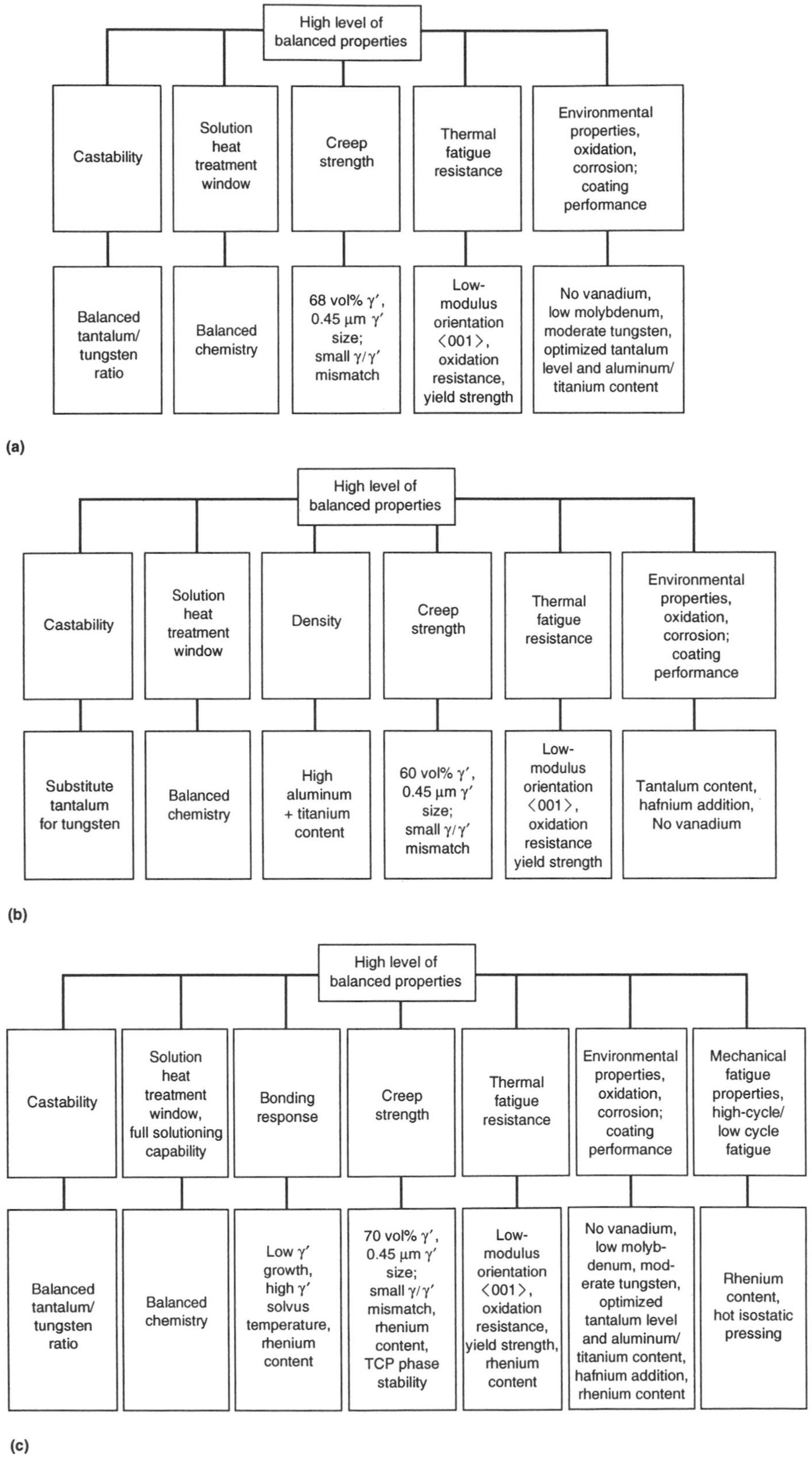

Continuing improvements in airfoil cooling techniques have usually led to significant gains in gas turbine operating efficiencies. However, these cooling techniques often result in very complex cored, thin-wall (0.5 to 1 mm, or 0.02 to 0.04 in.) airfoil designs, which can be susceptible to grain-boundary cracking during the DS casting of high-creep-strength alloys, particularly with modern high-thermal-gradient casting processes. Thus, the need for improved DS castability resulted in CM 247 LC, a second-generation alloy from the MAR-M 247 composition that was developed by the Cannon-Muskegon Corporation (Ref 16). The nominal composition of this superalloy, which is also known as René 108, is given in Table 1. The CM 247 LC alloy has particularly excellent resistance to DS grain-boundary cracking. In addition, it is capable of essentially 100% γ' solutioning to maximize creep strength without incipient melting or deleterious M_6C platelet formation upon subsequent high-temperature stress exposure, but with adequate transverse ductility retention.

With respect to DS grain-boundary cracking, zirconium and silicon are generally considered detrimental. Small amounts of a brittle, hafnium-rich eutectic phase containing high concentrations of zirconium and silicon have been found in DS crack-prone tests (Ref 16). It has also been observed that very small reductions in zirconium and titanium contents, combined with a very tight control of silicon and sulfur, dramatically reduces the DS grain-boundary cracking tendency of a high-creep strength superalloy such as MAR-M 247 (Ref 17). The major microstructural effect of the lower titanium content in CM 247 LC, compared to MAR-M 247, is to significantly reduce the size of the γ/γ' eutectic nodules and lower the volume fraction of the eutectic from approximately 4 vol% in MAR-M 247 to 3 vol% in CM 247 LC DS components. This factor is also believed to be significant in reducing the DS grain-boundary cracking tendency of CM 247 LC.

Heat Treatment and Mechanical Properties. Multistep solutioning techniques, based on a slow temperature increase between steps and temperatures up to 1254 °C (2290 °F), are used to supersolution heat treat CM 247 LC DS airfoil components to attain microstructures such as those shown in Fig. 9. Resultant stress-rupture property improvements are illustrated in Fig. 10.

The advent of single-crystal technology is not likely to preempt the need for DS airfoil components in the intermediate term. Directionally solidified airfoils will continue to be used for vane segments and low-pressure blades in advanced turbine engines because of the producibility of the components, which makes them cost effective.

Second-generation DS superalloys containing rhenium have been developed that have stress-rupture strength values close to those of the first-generation single-crystal alloys (Ref 18). These new alloys are particularly useful for DS vanes where load-bearing capability, such as to support

Fig. 16 Alloy development goals for (a) CMSX-2 first-generation SX superalloy, (b) CMSX-6 first-generation low-density SX superalloy, and (c) CMSX-4 second-generation SX superalloy. TCP, topologically close-packed

(a)
High level of balanced properties
Castability — Balanced tantalum/tungsten ratio
Solution heat treatment window — Balanced chemistry
Creep strength — 68 vol% γ', 0.45 µm γ' size; small γ/γ' mismatch
Thermal fatigue resistance — Low-modulus orientation <001>, oxidation resistance, yield strength
Environmental properties, oxidation, corrosion; coating performance — No vanadium, low molybdenum, moderate tungsten, optimized tantalum level and aluminum/titanium content

(b)
High level of balanced properties
Castability — Substitute tantalum for tungsten
Solution heat treatment window — Balanced chemistry
Density — High aluminum + titanium content
Creep strength — 60 vol% γ', 0.45 µm γ' size; small γ/γ' mismatch
Thermal fatigue resistance — Low-modulus orientation <001>, oxidation resistance yield strength
Environmental properties, oxidation, corrosion; coating performance — Tantalum content, hafnium addition, No vanadium

(c)
High level of balanced properties
Castability — Balanced tantalum/tungsten ratio
Solution heat treatment window, full solutioning capability — Balanced chemistry
Bonding response — Low γ' growth, high γ' solvus temperature, rhenium content
Creep strength — 70 vol% γ', 0.45 µm γ' size; small γ/γ' mismatch, rhenium content, TCP phase stability
Thermal fatigue resistance — Low-modulus orientation <001>, oxidation resistance, yield strength, rhenium content
Environmental properties, oxidation, corrosion; coating performance — No vanadium, low molybdenum, moderate tungsten, optimized tantalum level and aluminum/titanium content, hafnium addition, rhenium content
Mechanical fatigue properties, high-cycle/low cycle fatigue — Rhenium content, hot isostatic pressing

a bearing, is an important design consideration. Examples of these rhenium-bearing alloys include CM 186 LC and PWA 1426 (see Table 1).

One of the second-generation rhenium-containing DS alloys listed in Table 1 is CM 186 LC (3% Re). It is primarily intended for DS columnar, complex-cooled vents and relatively large low-pressure turbine blade components. The alloy exhibits excellent resistance to grain-boundary cracking in casting complex-cored, thin-wall turbine airfoils. Additionally, it is particularly attractive for use in components that are prone to recrystallization during solution heat treatment (resulting from residual casting stresses), because the alloy is used in the as-cast plus double-aged condition.

In the as-cast plus double-aged condition, CM 186 LC exhibits a 180 °C (320 °F) metal temperature advantage relative to fully solutioned DS CM 247 LC at the 982 °C/248 MPa (1800 °F/36.0 ksi) test condition, and 24 °C (43 °F) greater capability based on time to 1.0% plastic strain. This essentially equates to first-generation single-crystal alloy strength. At higher temperatures, the alloy strength is about midway between that of DS CM 247 LC and CMSX-2/3 (Fig. 11).

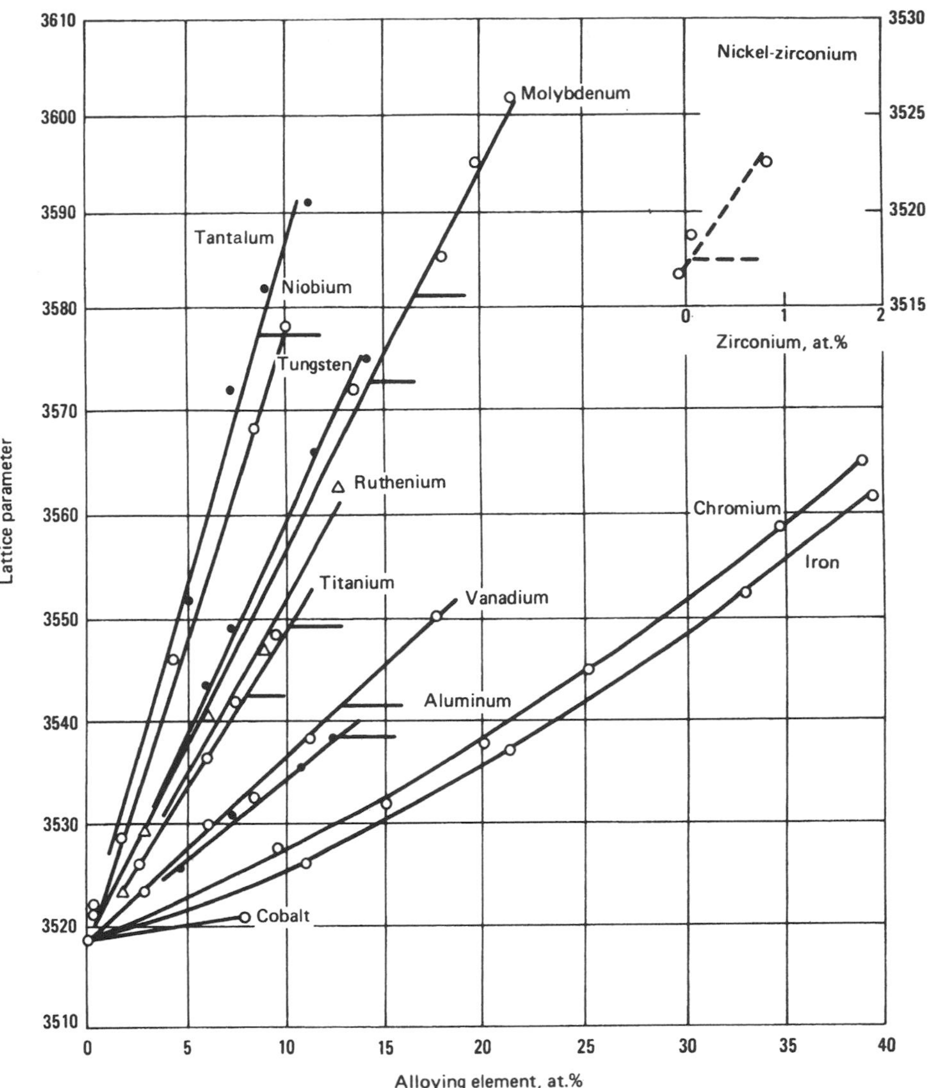

Fig. 17 Influence of alloying elements on the lattice parameter of binary nickel alloys. Source: Ref 30

Single-Crystal Superalloys

The greatest advance in the metal temperature capability of turbine blades in the last 25 years has been the single-crystal superalloy and process technology pioneered by PWA (Fig. 12). The dramatic improvement in the durability of the F-100 fighter engine turbine, as evidenced by the service performance of the F-220 version, is largely due to the PWA 1480 superalloy single-crystal first- and second-stage blades and vanes (Fig. 13).

Other pioneering single-crystal alloy development work resulted in the derivation of several single-crystal compositions from MAR-M 247 during the Garrett/NASA Materials for Advanced Technology Engines (MATE) Program, which began in 1977 (Ref 21, 22). The two alloys studied extensively were NASAIR 100 and NASAIR Alloy 3; the latter contained a minor hafnium addition.

The compositions of the first-generation single-crystal superalloys, many of which are being used in turbine engine applications, are shown in Table 2: René N-4 developed by General Electric (Ref 23); SRR 99 and RR 2000 by Rolls-Royce (Ref 24); AM1 and AM3 by the Office National d'Etudes et de Recherches Aerospatiales (ONERA) and Societe Nationale d'Etude et de Construction de Moteurs (SNECMA) (Ref 25); CMSX-2, CMSX-3 (Ref 26), and CMSX-6 (Ref 27) by Cannon-Muskegon Corporation; and SC 16 by ONERA (Ref 28). These alloys are characterized by approximately the same creep-rupture strength (density corrected) but have differing SX castabilities, grain qualities, solution heat treatment windows, propensities for recrystallization upon solution treatment, environmental oxidation and hot corrosion properties, and densities. Typical stress-rupture properties for first-generation SX alloys are shown in Fig. 14 and 15.

As indicated in Table 2, the compositions of second-generation generally include rhenium. Alloys CMSX-4, PWA 1484, SC 180, and René N5 all contain 3% Re. As described below, the rhenium-containing second-generation alloys have higher elevated-temperature strengths than either DS or first-generation SX superalloys. MC2 second-generation SX superalloy, also developed at ONERA, is the strongest non-rhenium-containing superalloy developed to date.

Rhenium content also defines third-generation SX superalloys. As shown in Table 2, SMSX-10 and René N6 contain 6.0% and 5.4% Re, respectively. These alloys offer the highest elevated-temperature strength of all superalloys.

Chemistry and SX Castability

First-Generation Alloys. Alloy CMSX-2 was developed in 1979 from the MAR-M 247 composition, using some of the experience of the Garret/NASA MATE program (Ref 21). A multidimensional development approach was used to achieve a high level of balanced properties (Fig. 16a). The chemistry modifications applied to MAR-M 247 to develop CMSX-2 (Table 2) are summarized below with respect to function and objectives:

- Grain-boundary strengthening elements (boron, hafnium, zirconium, and carbon) were removed to achieve a very high incipient melting temperature (1335 °C, or 2435 °F).
- Tantalum was partially substituted for tungsten (CMSX-2 has 6% Ta) for good single-crystal castability, high γ' volume fraction (68%), improved γ' precipitate strength, microstructural stability (freedom from α-tungsten and tungsten, molybdenum-rich μ phases), good oxidation resistance, and coating stability.
- Cobalt content was maintained to increase solid solubility and microstructural stability.

- The chemistry balance was designed to ensure a wide and practical solution heat treatment temperature range (difference between the γ′ solvus and the incipient melting temperature), that is, a range of at least 22 °C (40 °F).
- Electron hole calculations, or phase computation (PHACOMP) calculations (Ref 29), were performed to control the chemistry of the alloy and prevent deleterious topologically close-packed phases.

Figure 17 shows the relative potency of tantalum, tungsten, and molybdenum as solid-solution strengtheners in binary nickel alloys, where tantalum is the most powerful strengthener on an atomic percent basis. An increase in the lattice parameter of the γ′ phase due to alloy additions increases the solid-solution strengthening. Tantalum also partitions strongly to the γ′ phase, increasing the volume fraction and stiffening the γ′ phase due to its relatively large atomic size. The strength of the γ′ phase is important in superalloys with a high volume fraction of γ′ (>50%) because γ′ shearing is the primary strengthening mechanism. Because the mean free edge-to-edge distance in the γ matrix between the precipitates is smaller than the average precipitate size itself, dislocation shearing of the γ′ particle is favored over Orowan dislocation looping around the γ′ particles.

Detailed transmission electron microscopy studies of dislocation movement in cast high-strength superalloys, such as MAR-M 002 and its single-crystal derivative SRR 99, have shown the importance of ensuring that the antiphase boundary energy is high, so that the stacking fault mode of creep deformation occurs at temperatures up to 850 °C (1562 °F), thus ensuring high creep strength (Ref 24). Tantalum additions raise the antiphase boundary energy relative to the stacking fault energy (Ref 24), leading to the increased tendency for stacking faults to be formed at lower temperatures.

The CMSX-2 alloy is designed to provide good SX foundry performance, because castability is a crucial alloy performance criterion for any complex, thin-wall turbine blade or vane component. This characteristic is sometimes given limited attention in alloy design, but it affects not only the yield and cost of components but also the defect level and therefore component performance. Single-crystal casting defects of concern are:

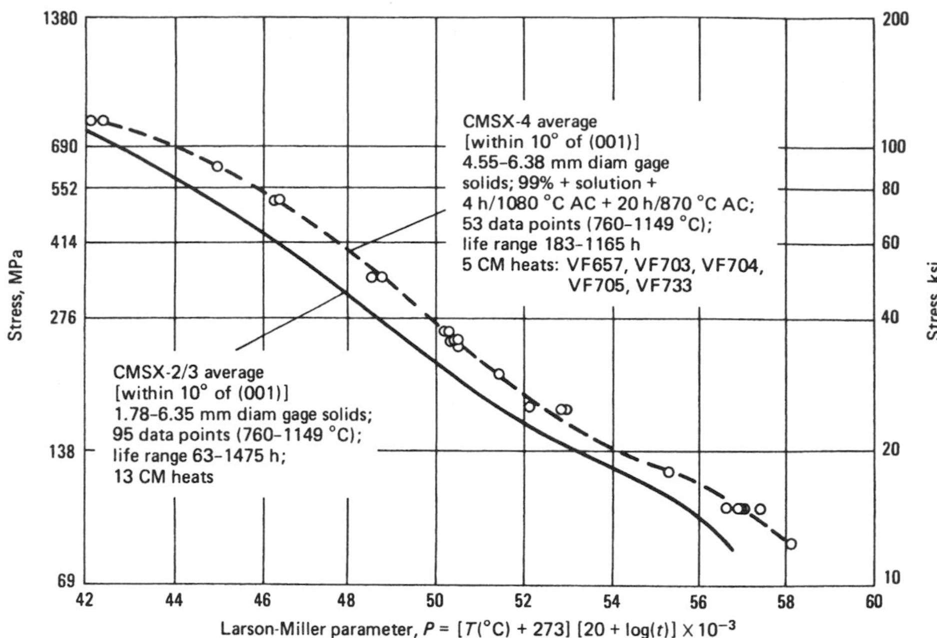

Fig. 18 Larson-Miller stress-rupture strength of CMSX-4 vs. that of CMSX-2/3

- *Freckling:* A spiral of equiaxed grains caused by elemental segregation in the liquid state
- *Slivers:* Moderate-angle grain defects
- *Microporosity:* A uniform distribution of interdendritic micropores
- *Spurious grains:* High-angle grain boundaries
- *Stable oxide inclusions:* e.g., Al_2O_3
- *Carbides:* e.g., TiC

The partial substitution of tantalum for tungsten in the CMSX-2 alloy helps overcome the freckling problems inherent in the low-tantalum, high-tungsten single-crystal alloys. The strong γ′-forming elements, aluminum and titanium, which are also low density, tend to segregate to the last liquid to solidify in the interdendritic spaces during the SX solidification process. This can create density changes and consequential flow in the liquid metal close to the solidification front, which can nucleate freckle trails of equiaxed grains. This can occur particularly under conditions of low or changing thermal gradients. Tantalum, which is a strong γ′-forming element of high density, also tends to segregate to the last liquid to solidify in the interdendritic spaces, so it evens out density changes in the liquid, or mushy, zone and reduces freckling tendencies.

Several studies undertaken in the United States, Europe, and Japan confirm that high nitrogen and oxygen levels in single-crystal superalloy ingot adversely affect SX casting grain yield, supporting the importance of low nitrogen and oxygen levels in the master alloy. Carbon, sulfur, and oxygen master alloy impurities are shown to

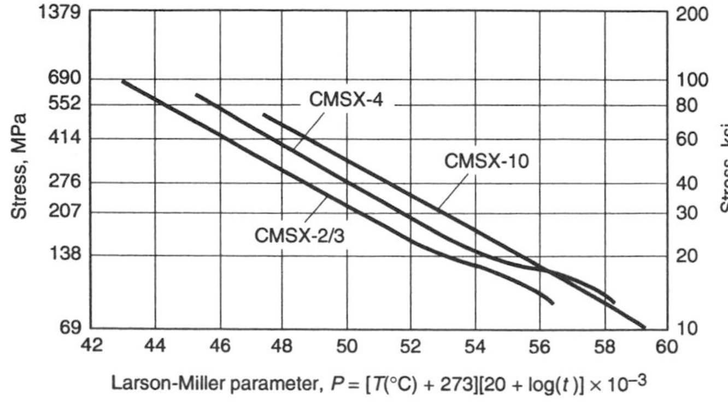

Fig. 19 Average Larson-Miller stress rupture of CMSX-10 vs. that of CMSX-4 and CMSX-2/3. Source: Ref 34

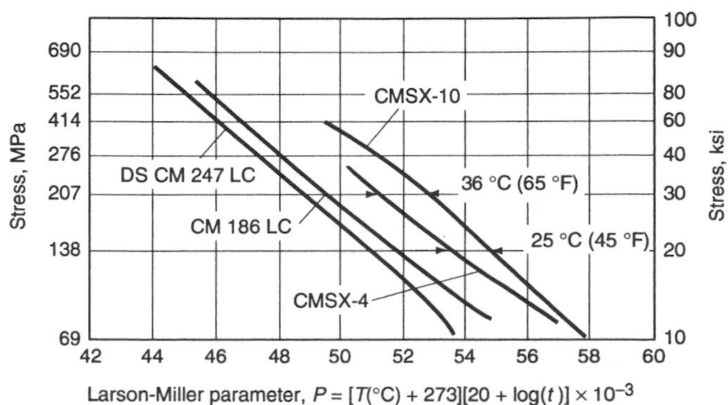

Fig. 20 Average Larson-Miller 1.0% creep strength of CMSX-10 vs. that of CSMX-4, DS CM 186 LC, and DS CM 247 LC. Source: Ref 34

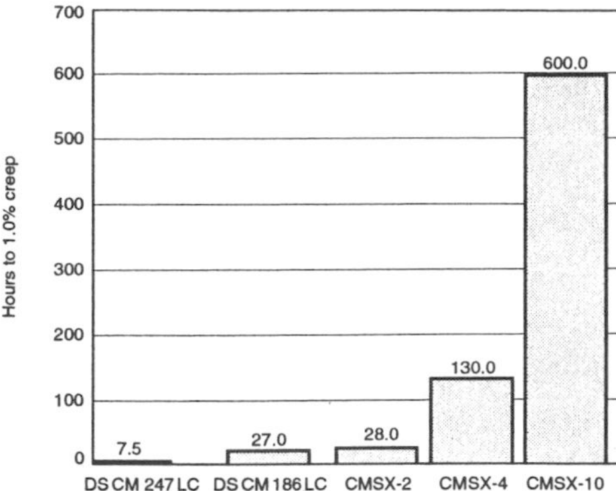

Fig. 21 Average 1.0% creep strengths at 982 °C/248 MPa (1800 °F/36 ksi) test condition for CMSX-10, CMSX-4, CMSX-2, DS CM 186 LC, and DS CM 247 LC. Source: Ref 34

Table 3 Chemical composition of the γ' phase in CMSX-2

Element	Composition, wt %
Nickel	69.25
Cobalt	3.15
Chromium	2.05
Molybdenum	0.30
Tungsten	7.25
Aluminum	7.55
Titanium	1.30
Tantalum	9.15

Source: Ref 35

(a)

(b)

(c)

(d)

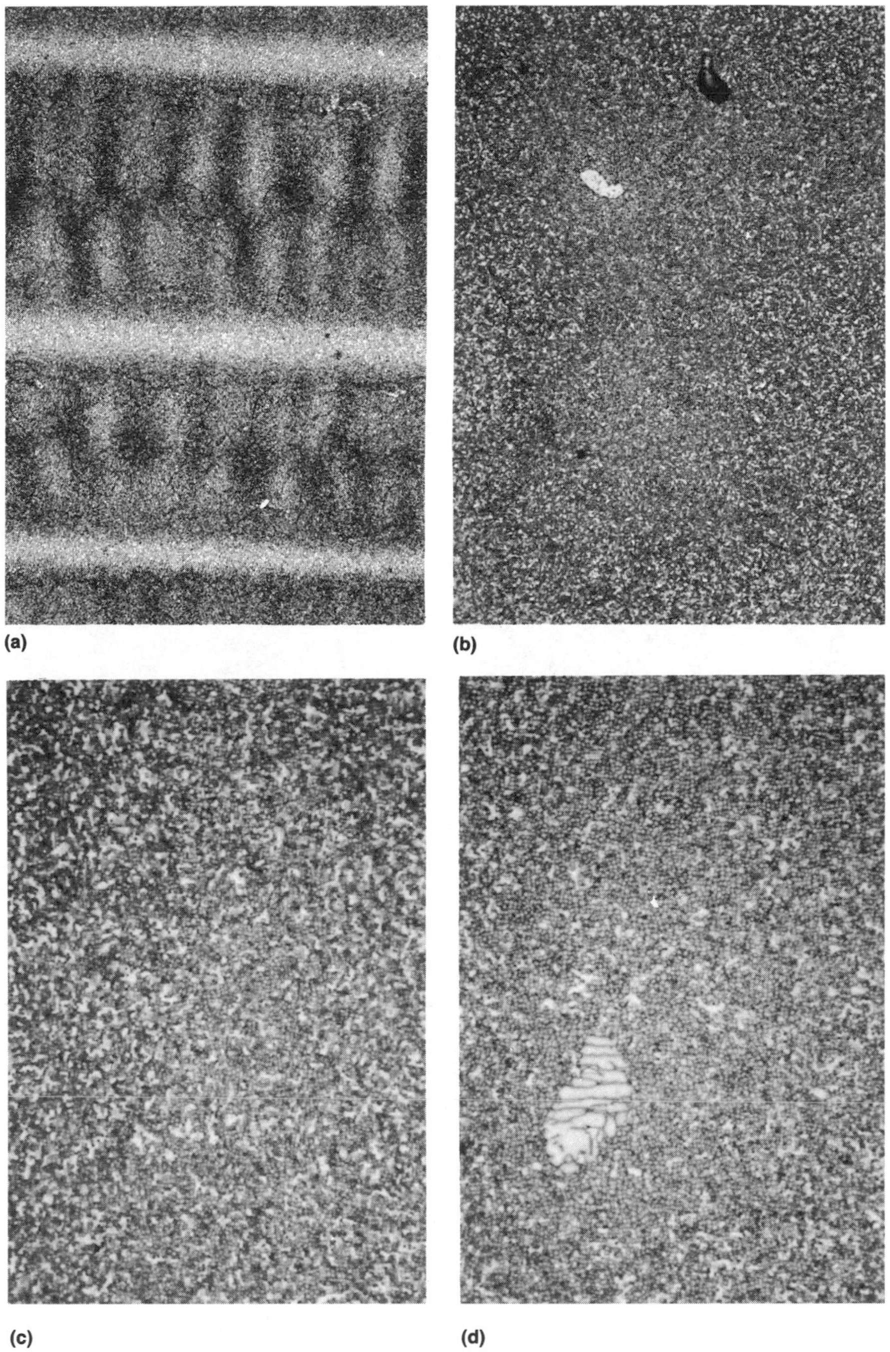

Fig. 22 CMSX-4 flat specimen, 25 mm wide by 1.25 mm thick by 100 mm long (1 in. by 0.05 in. by 4 in.). Specimen was cast, 99% solutioned, and double aged. Micrographs taken from longitudinal orientation. (a) 90×. (b) 365×. (c) 905×. (d) 905×

transfer nonmetallic inclusions, such as Al_2O_3, (Ti,Ta) C/N, and $(Ti,Ta)_xS$, to SX parts (Ref 31). Grain defects can nucleate on these inclusions.

Alloy CMSX-6 is a low-density (7.98 g/cm^3) superalloy developed to serve as a replacement for low-density equiaxed alloys IN-100 (7.76 g/cm^3) and IN-6212, which has density/mechanical properties similar to IN-100. In turbine blade applications, the advantages of a material of low relative density are:

- Reduction in the steady stress in the blade due to centrifugal forces, effecting an increase in the margin of safety with respect to dynamic loading (high-cycle fatigue and low-cycle fatigue)
- Reduction in load on the blade-disk "firtree" connection
- Lower disk rim loading, leading to smaller-section, lighter turbine disks

The prime objective in the development of the single-crystal alloy CMSX-6 was to limit its density to 8.0 g/cm^3 while maintaining density-corrected, specific stress-rupture strength comparable to that of existing first-generation single-crystal alloys of higher relative density (e.g., CMSX-2 with a density of 8.56 g/cm^3). As shown in Fig. 16(b) and Table 2, lower density was made possible by the high aluminum plus titanium content (9.5 wt%) and the elimination of tungsten. As shown in Fig. 15, the density-corrected stress-rupture strength of CMSX-6 is equivalent to that of CMSX-2 and CMSX-3.

Second-Generation Alloys. A number of rhenium-containing SX superalloys have been developed for turbine engine applications. Four typical compositions are given in Table 2. Rhenium partitions mainly to the γ matrix; this retards coarsening of the γ'-strengthening phase and increases γ/γ' misfit (Ref 32). Atom-probe microanalysis of rhenium-containing modifications of the PWA 1480 and CMSX-2 alloys reveals the occurrence of short-range order in the matrix, with small rhenium clusters (~1.0 nm) detected in the γ in the alloys (Ref 33). The rhenium clusters can act as more efficient obstacles against dislocation movement compared to isolated solute atoms in the γ solid solution; therefore, they play a significant role in improving the creep strength.

The major chemistry differences between the second-generation CMSX-4 alloy and the precursor first-generation single crystal alloys CMSX-2 and CMSX-3 are:

- Addition of 3% Re and reduction of tungsten content
- Increase of total Ta + W + Re + Mo refractory element content from 14.6 to 16.5%
- Reduction in chromium content (from 8 wt% to 6.5%) and increase in cobalt content (from 5 wt% to 9%) to assist solid solubility and to ensure the essential absence of topologically close-packed phases following high-temperature exposure

The CMSX-4 alloy composition was designed using a multidimensional approach to achieve a high level of balanced properties (Fig. 16c), building on the experience gained with CMSX-2 and CMSX-3.

The Larson-Miller stress-rupture comparison of CMSX-4 and CMSX-2/3 is shown in Fig. 18. The stress-rupture temperature capability advantage of CMSX-4 over CMSX-2/3 is 27 °C (48 °F) (density corrected) in the region of 248 MPa/982 °C (36 ksi/1800 °F). In the region of 103 MPa/1121 °C (15 ksi/2050 °F), the stress-rupture temperature capability advantage is 30 °C (54 °F) (density corrected). The data also indicate that CMSX-4 has a potential peak-use temperature under stress of at least 1150 °C (2100 °F).

Third-generation SX alloys include CMSX-10 (6 wt% Re) and René N6 (5.4 wt% Re). Both of these alloys are characterized by their high refractory metal level (W + Ta + Re + Mo) and their relatively low chromium content (2 to 4 wt%). Despite this low chromium content, these alloys perform well in oxidative environments because of their high aluminum levels (~5.7 wt% Al). Compositions of third-generation alloys are listed in Table 2.

The CMSX-10 alloy is thought to be the strongest nickel-base superalloy currently available for gas turbine industry use (Ref 34). It provides an approximate 30 °C (54 °F) creep-rupture advantage relative to the CMSX-4 material. As illustrated in Fig. 19, the 30 °C advantage prevails to about 1110 °C (2030 °F), with significant advantage occurring again at temperatures beyond about 1163 °C (2125 °F). Whereas the CMSX-4 maximum-use temperature is generally thought to be about 1163 °C (2125 °F), γ′ particle observations following elevated-temperature creep testing suggest that CMSX-10 may extend superalloy functionality (usefulness) to about 1204 °C (2200 °F).

The creep strength advantage of the alloy is more dramatic when compared on the basis of time to 1.0% creep deformation. For this case, as shown in Fig. 20, CMSX-10 exhibits about 36 °C (65 °F) greater strength than CMSX-4 at 207 MPa (30 ksi) stress, it exhibits about 25 °C (45 °F) greater strength at 138 MPa (20 ksi) stress, and it appears to maintain an advantage at stresses as low as 83 MPa (12 ksi). An additional illustration of the creep strength is provided in Fig. 21, where time to 1.0% creep data at 982 °C/248 MPa (1800 °F/36 ksi) is compared for CMSX-10, CMSX-4, CMSX-2, DS CM 186 LC, and DS CM 247 LC alloys. The CMSX-10 alloy exhibits advantage in time to 1% creep of 4.6 to 80 times that of the comparative materials.

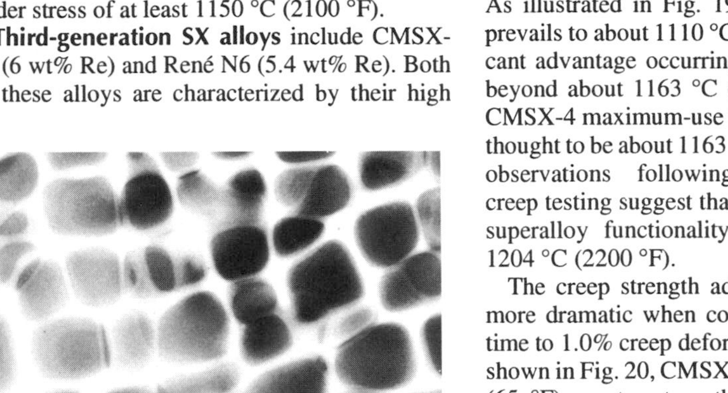

(a)

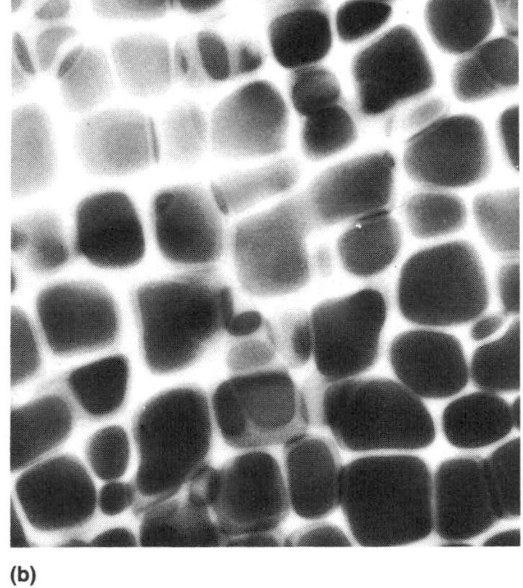

(b)

Fig. 23 Morphology of γ′ precipitates in CMSX-2 alloy. (a) After T_2 heat treatment. Source: Ref 37

Single-Crystal Heat Treatments and Microstructures

Solutioning. With regard to solutioning, the latest multistep ramped cycles developed for single-crystal components are designed to completely solution the γ′ and most of the γ/γ′ eutectic without incipient melting. Alloy CMSX-4, which is solutioned at a maximum temperature of 1320 °C (2410 °F) in commercial vacuum heat treatment furnaces, readily attains the 99%+ (<1% remnant γ/γ′ eutectic) solutioned microstructure, as illustrated in Fig. 22.

Aging and Creep Response. With regard to aging, the volume fraction of γ′ in CMSX-2 is approximately 68% (Fig. 16c). The resulting chemistry of the γ′ precipitates is given in Table 3. Both γ′ volume fraction and chemistry are independent of the aging treatments used. The measured lattice parameter of the γ′ is 0.35865 nm with a γ/γ′ misfit at room temperature of 0.14% (Ref 35) and –0.33% at 1050 °C (1922 °F) (Ref 36).

It has been reported by ONERA that a high-temperature aging heat treatment (T_2) [16 h/1050 °C (1922 °F) air cooled (AC)] following solution treatment, with subsequent intermediate temperature aging [20 h/871 °C (1600 °F) or 48 h/850 °C (1562 °F)], gives CMSX-2 cuboidal γ′ with a

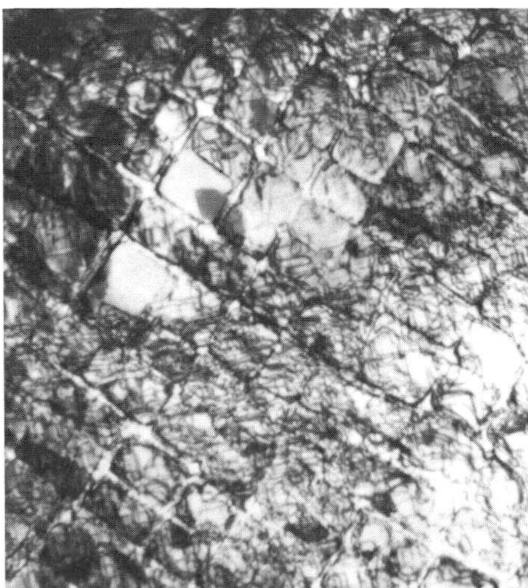

Fig. 24 Homogeneous deformation in CMSX-2 (T_2 heat treatment) after 0.16% creep strain at 760 °C (1400 °F). Source: Ref 37

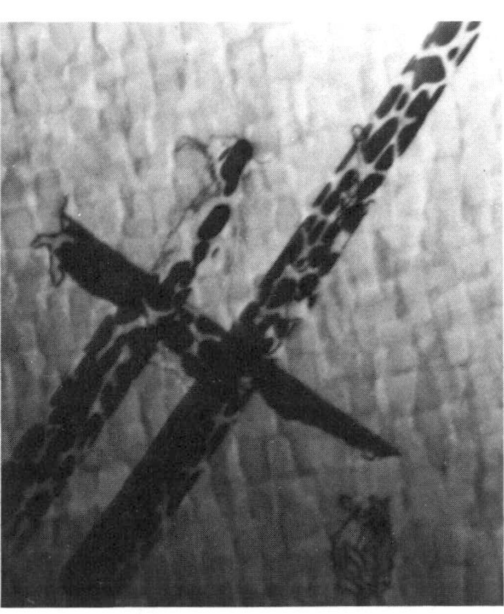

Fig. 25 Inhomogeneous deformation in CMSX-2 (T_1 heat treatment) during primary creep at 760 °C (1400 °F). Source: Ref 37

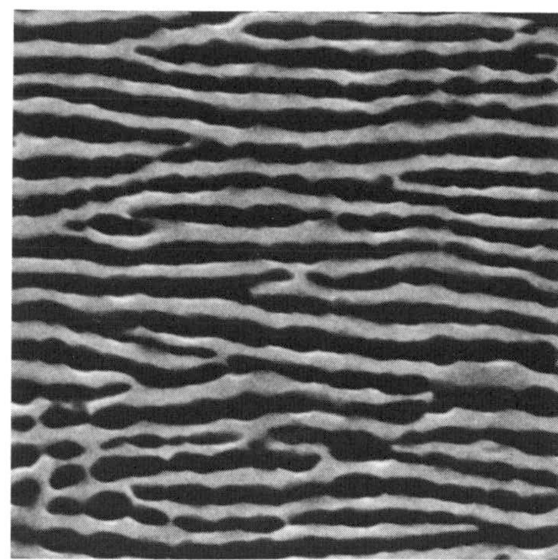

Fig. 26 Oriented coalescence (rafting) of the γ′ phase in CMSX-2 after 20 h of creep at 1050 °C (1920 °F) under 120 MPa (17.4 ksi). Tensile stress axis is [001]. Source: Ref 37

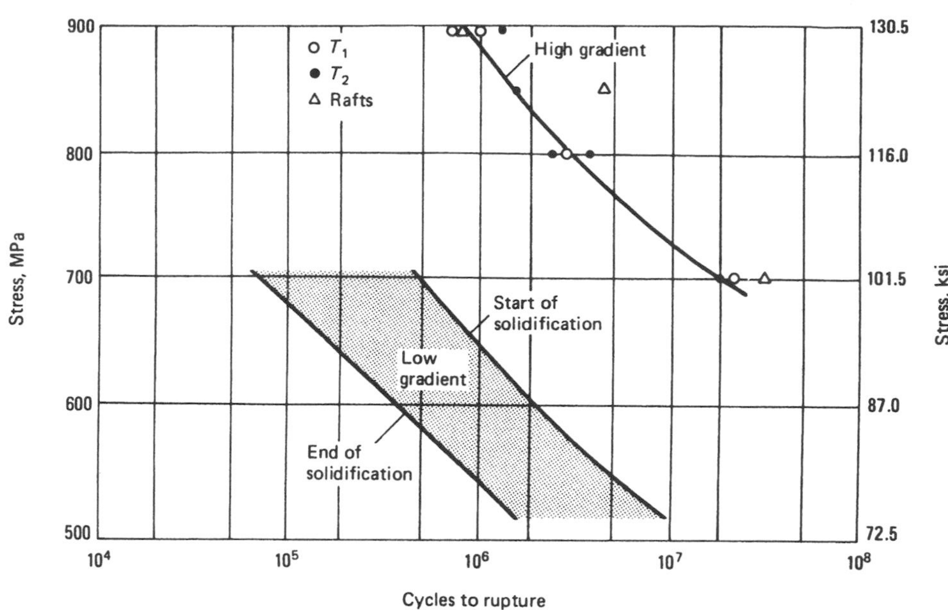

Fig. 27 Effect of thermal gradient and heat treatments on the high-cycle fatigue behavior of CMSX-2 at 870 °C (1598 °F) with frequency of 50 Hz. Source: Ref 37

mean size of 0.5 μm, which optimizes creep response (Ref 35, 37). Similar γ′ morphology and size are obtained with a 4 h/1079 °C (1975 °F) AC postsolution treatment. The morphology of the γ′ in CMSX-2 with this ONERA-type aging treatment is shown in Fig. 23(a), which should be compared to the conventional irregularly shaped γ′ particles with a mean size of 0.3 μm shown in Fig. 23(b). The particles shown in Fig. 23(b) result from a 5 h/982 °C (1800 °F) AC + 48 h/850 °C (1562 °F) aging (T_1). Specimens of the T_2-type at 760 °C (1400 °F) deform in a homogeneous manner in the early stage of creep (Fig. 24). The homogeneous nature of the deformation leads to a rapid strain hardening of the material, causing a decrease in the creep rate. The T_1 heat treatment, which produces smaller, irregularly shaped particles, favors inhomogeneous deformation within the specimen due to the precipitate shearing that occurs during the early stages of creep (Fig. 25). In this case, the amplitude of primary creep is high, and the strain hardening of the material is achieved at a much later stage, compared with that of the T_2-type heat-treated specimens.

Rafting. During creep at high temperature, the γ′ precipitates in the form of rafts perpendicular to the stress axis. The kinetics of raft formation depend on the testing temperature, among other factors. At 1050 °C (1922 °F) under a stress of 120 MPa (17.4 ksi), the rafts form within a few hours (Fig. 26). The rafts have a high aspect ratio in the T_2-type heat-treated specimens in which the cuboidal γ′ precipitates are already aligned. The lateral extension of the γ′ phase in the form of rafts causes the specimen to creep at a much lower rate, compared with the creep rate of the material in which the γ′ phase coalesces irregularly. The CMSX-2 and CMSX-3 alloys that show this type of rafted γ′ morphology possess very long rupture lives at high temperatures. In these alloys, the misfit between the γ and γ′ phases is found to be negative at high temperatures (Ref 38).

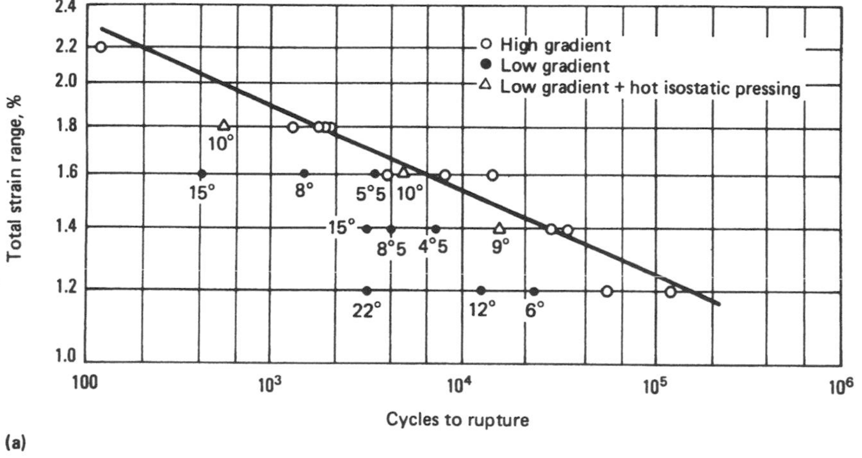

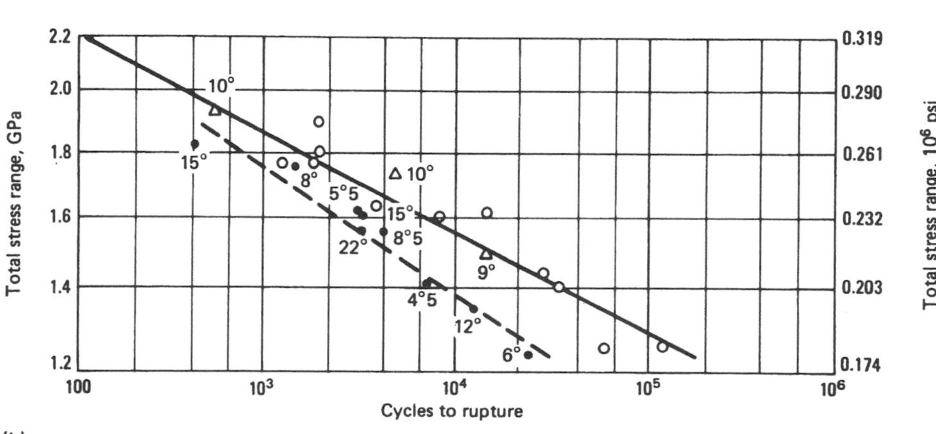

Fig. 28 Effect of thermal gradient, orientation, and hot isostatic pressing on the strain-controlled low-cycle fatigue behavior of CMSX-2 (fully reversed, with frequency of 0.33 Hz) at 760 °C (1400 °F). Numbers represent the deviation, in degrees, from the [001] orientation. (a) Strain vs. cycles to failure. (b) Stress vs. cycles to failure. Source: Ref 27

Effect of Orientation on Creep Properties. Work by ONERA and Ishikawajima-Harima Heavy Industries Company, Ltd. shows some interesting effects of the crystal orientation and heat treatments on the creep behavior and strength of several single-crystal superalloys (Ref 39).

At intermediate temperatures (760 to 849 °C, or 1400 to 1560 °F), the creep behavior of nickel-base single-crystal superalloys is extremely sensitive to crystal orientation and γ′ precipitate size. For a γ′ size in the range of 0.35 to 0.5 μm, the highest creep strength is obtained near [001], while orientations near the [111]-[011] boundary of the standard stereographic triangle exhibit very short creep lives. When the γ′ size decreases to 0.2 μm, the longest creep lives are exhibited, in decreasing order, by the crystals oriented near [111], [001], and [110]. The anisotropy in creep between the [001] and [111] orientations can therefore be reduced by appropriate precipitation heat treatments. The creep strengths, however, remain poor near the [011] orientation.

At high temperatures (982 to 1049 °C, or 1800 to 1920 °F), the creep behavior of the single-crystal superalloys is much less sensitive to crystal orientation and γ′ size than it is at intermediate temperatures. The [001]-oriented single crystals develop a rafted γ′ structure normal to the tensile stress axis, while the γ′ precipitates coarsen irregularly in the [111] specimens.

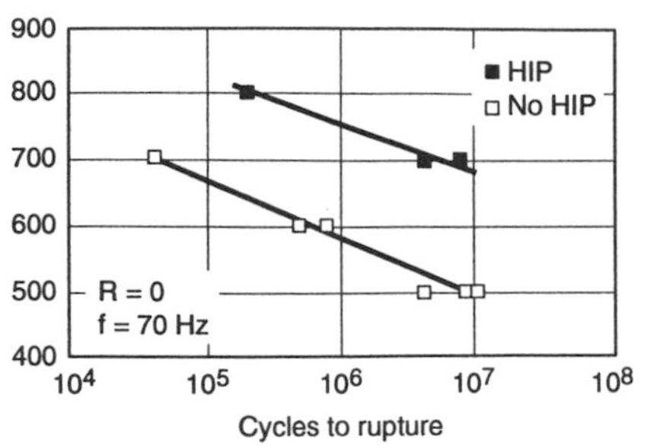

Fig. 29 Effect of hot isostatic pressing (HIP) on the high-cycle fatigue behavior at 870 °C (1600 °F) of [001] AM3 single crystals. Source: Ref 40

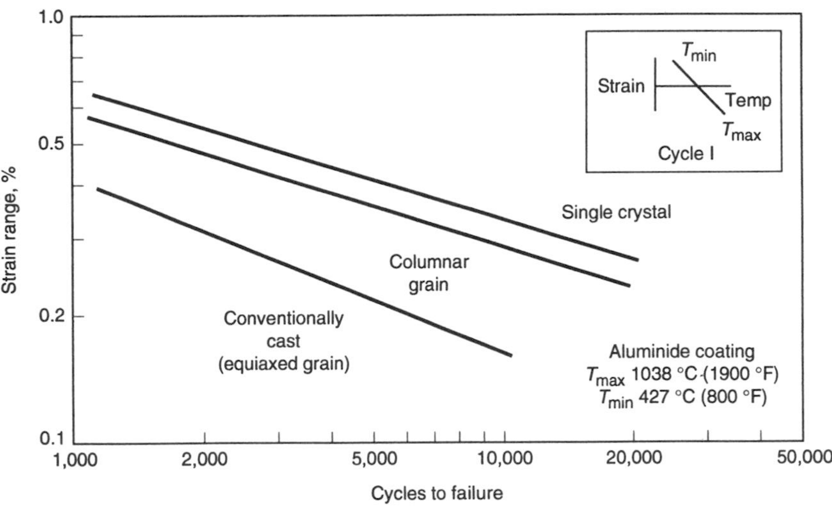

Fig. 30 Comparison of the average thermal fatigue lives of conventionally cast, DS cast, and SX cast nickel-base superalloys. Source: Ref 41

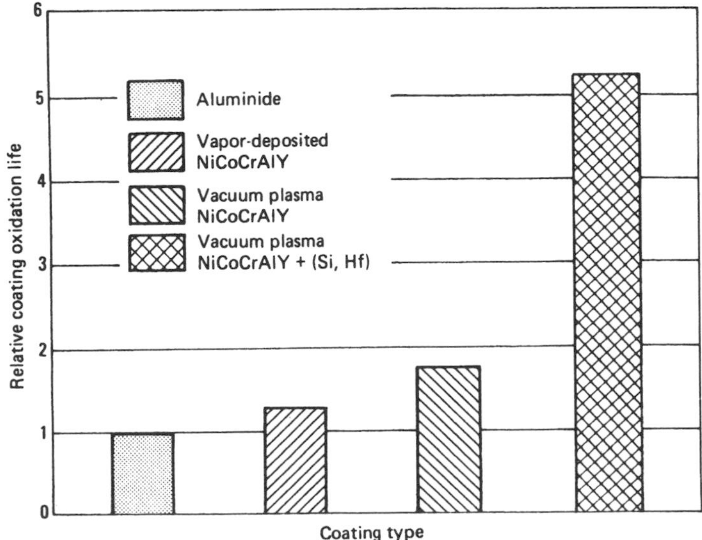

Fig. 31 Coating performance of PWA 1480 at 1149 °C (2100 °F). Burner rig oxidation test. Source: Ref 20

Fatigue Properties of SX Alloys

High-Cycle Fatigue. An important property that must be considered when selecting single-crystal superalloys for turbine blade applications is fatigue strength. Single crystals of CMSX-2 have been cast under both low- and high-gradient conditions and then subjected to high-cycle fatigue tests in the repeated tension mode at 870 °C (1598 °F) (Ref 37); the results are reported in Fig. 27. The fatigue resistance of specimens cast under a very high temperature gradient (laboratory conditions) is much superior to that of material cast under industrial conditions, primarily because of the very small pore size (<10 μm) inherent in the high-gradient specimens. The single crystals cast under industrial conditions have a more heterogeneous structure where the interdendritic spacing and the level of porosity vary along the length of the bar. Specimens corresponding to the beginning of solidification exhibit better fatigue resistance than those corresponding to the end of solidification. Fatigue tests have also been performed on specimens in which a rafted γ′ morphology was developed prior to testing. It is interesting to note that the fatigue behavior was not significantly affected by the rafted γ′ morphology (Ref 37).

Low-Cycle Fatigue. Strain-controlled, fully reversed low-cycle fatigue tests performed at 760 °C (1400 °F) confirm the much better fatigue behavior of single crystals cast under a high thermal gradient (Fig. 28). In this type of test, the higher the deviation from the [001] orientation, the shorter the fatigue life. It can be seen in Fig. 28(a) that for a total strain range of 1.2%, the fatigue life is decreased by an order of magnitude when the crystal orientation, relative to the [001] orientation, moves away from 6 to 22°. The decrease in fatigue life is a consequence of the increase in stress level through the increase of elastic modulus. Because the plastic strain component at 760 °C (1400 °F) is small, the results can be plotted as total stress versus the number of cycles to failure (Fig. 28b). In Fig. 28, the effect of crystalline orientation on the fatigue life of the industrially processed single crystals is not apparent, and all the results of low-gradient single crystals can be represented by a single curve.

An examination of fracture surfaces shows that the cracks are initiated at microporosity, which indicates that these defects (microporosity) are of primary importance in determining the fatigue life of CMSX-2. The microporosity in industrial single crystals can be as large as 50 to 80 μm, but it is rarely more than 10 μm in single crystals cast under laboratory conditions with very high temperature gradients.

Effects of Hot Isostatic Pressing. In DS and SX superalloys, interdendritic casting microporosity invariably acts as the crack initiation site during service (Ref 40), although columnar grain boundaries are also known to initiate thermal fatigue cracks in DS components. Reduction in the density and, more importantly, the size of the inherent porosity can be expected to improve fatigue lives substantially. The application of hot isostatic pressing (HIP) for closing up interdendritic microporosity has been recognized as a valuable tool for improving the creep and fatigue lives of DS and SX superalloys. However, early attempts at applying the technology to DS and SX superalloys met with only limited success, because typical HIP temperatures and pressures produced excessive plastic flow around the pores, which led to localized recrystallization. Creep and fatigue cracks then initiated at the isolated recrystallized grains, thus negating the benefits of eliminating the porosity. In this respect, HIP maps and flow equations are powerful tools for selecting the appropriate pressurization ratios for avoiding recrystallization.

Improvement in high-cycle fatigue properties of ONERA-processed SX alloy AM3 is shown in Fig. 29. Beneficial effects of HIP have also been observed in low-cycle fatigue tests. Figure 28 shows that the low-cycle fatigue behavior of CMSX-2 cast under low thermal gradients after HIP can be improved to that of high-gradient-processed specimens.

Thermal fatigue commonly is measured using a thin-wall tube specimen driven through an independently controlled thermomechanical fatigue cycle (Ref 41). A wide variety of cycles can be imposed, but that with the independently controlled temperature and strain 180° out-of-phase (maximum temperature at minimum strain), referred to as cycle 1, is considered to represent realistic conditions for the leading edge of an air-cooled gas turbine airfoil. The strain range (tension plus compression) through which the specimen is cycled is plotted versus the number of cycles to 50% load drop, which is taken as the failure point or thermal fatigue life. In this test, single crystals exhibit a much better thermal fatigue life than polycrystalline superalloys do, as shown in Fig. 30. These tests generally are conducted with coated specimens, because single-crystal turbine airfoils are used in the coated condition for most applications. Thermal fatigue life is related to alloy creep strength at the maximum cycle temperature, with stronger alloys exhibiting longer thermal fatigue lives. This is why single-crystal superalloys, with their superior high-temperature creep strength, have better thermal fatigue lives than columnar-grained DS superalloys (Fig. 30). More detailed information can be found in the article "Thermal and Thermomechanical Fatigue of Structural Alloys" in this Volume.

Oxidation and Hot Corrosion of SX Alloys

Oxidation/hot corrosion resistance is primarily a function of alloy composition, not solidification mode (Ref 41). Grain-boundary oxidation is eliminated with SX superalloys, but this is not normally a problem at the high temperatures where polycrystalline superalloys are used. Little, if any, effect of crystallographic orientation has been observed on oxidation/hot corrosion rate.

Active element additions can be made to all superalloys to enhance oxidation resistance, but the magnitude of the enhancement in oxidation resistance is much greater with SX superalloys that are free of the detrimental grain-boundary strengthening elements, such as boron.

There has recently been increased interest in the use of SX alloys for land-based turbine applications involving the burning of fossil fuels. Higher-chromium alloys such as CMSX-11B (12.5% Cr), CMSX-11C (14.5% Cr), and SC 16 (16% Cr) have been developed for these applications (see Table 2).

Coatings perform well with single-crystal alloys, particularly when they are optimized to the base alloy system, as shown with PWA 1480 alloy in Fig. 31. The absence of grain boundaries, rosette clusters of carbides, and elemental segregation in heat-treated single-crystal alloys contributes to improved coating performance.

Figure 32 shows how coating performance varies from alloy to alloy. In this study (Ref 42), a NiCoCrAlHfSiY overlay coating was deposited on PWA 1480 and PWA 1484. Burner rig oxidation tests were carried out on the coated specimens at temperatures between 1095 and 1175 °C (2000 and 2150 °F). As Fig. 32 shows, the service life of the overlay coating on PWA 1484 was 70% greater than that on PWA. This was most likely due to the 3% Re content in PWA 1484, which slowed down element diffusion at the coating/substrate interface.

Test Results. Cyclic oxidation testing at 900 and 1050 °C (1652 and 1922 °F) has shown that uncoated CMSX-4 and CMSX-2 have excellent and similar performance (Ref 43). This work also

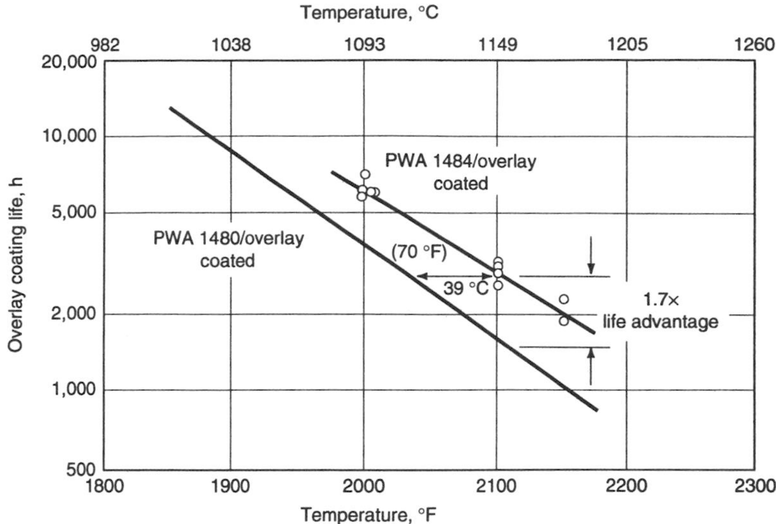

Fig. 32 Burner rig oxidation test results on overlay-coated SX alloys PWA 1480 and PWA 1484. Source: Ref 42

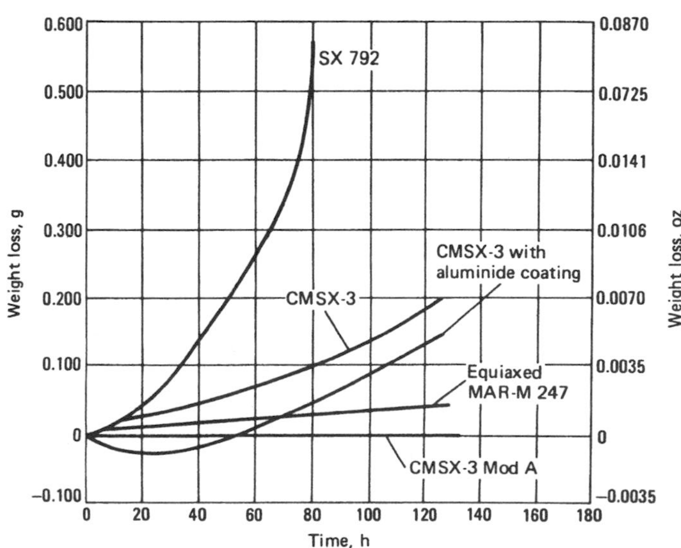

Fig. 33 Dynamic cyclic oxidation, 1177 °C (2150 °F). Source: Allison Gas Turbine Division, General Motors Corp.

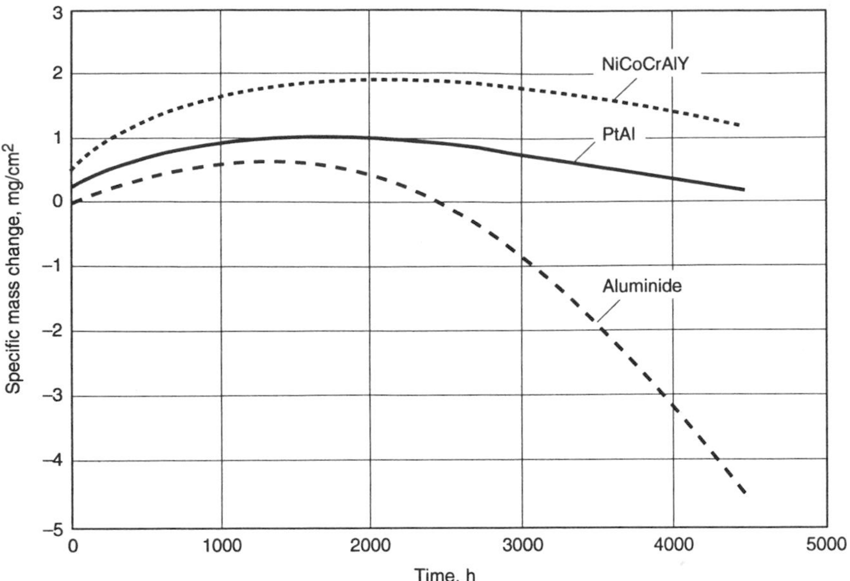

Fig. 34 Dynamic cyclic oxidation, 1040 °C (1900 °F), test results of coated CMSX-4. Source: Ref 42

showed uncoated CMSX-4 has better molten-salt hot corrosion characteristics than CMSX-2, probably because of its lower tungsten content. Ultra-high-temperature cyclic burner rig oxidation studies undertaken at 1177 °C (2150 °F) showed that uncoated, modified CMSX-3 (designated CMSX-3 Mod A) has excellent performance, similar to that of uncoated MAR-M 247 (which contains 1.4% Hf), with little or no attack after 140 h of testing (Fig. 33). Alloy CMSX-4 Mod A showed similar results. Figure 34 illustrates the advantages of PtAl diffusion and NiCoCrAlY overlay coatings over simple aluminide diffusion coatings.

ACKNOWLEDGMENTS

The information in this article is largely taken from:

- K. Harris, G.L. Erickson, and R.E. Schwer, Directionally Solidified and Single-Crystal Superalloys, *Properties and Selection: Irons, Steels, and High-Performance Alloys,* Vol 1, *ASM Handbook* (formerly Vol 1, 10th ed., *Metals Handbook*), ASM International, 1990, p 995–1006
- T.S. Piwonka, Directional and Monocrystal Solidification, *Casting,* Vol 15, *ASM Handbook* (formerly Vol 15, 9th ed., *Metals Handbook*), ASM International, 1988, p 319–323

REFERENCES

1. F.E. Pickering, Advances in Turbomachinery, Cliff Garrett Award Lecture, *Aerosp. Eng.,* Jan 1986, p 30–35
2. T. Khan and P. Caron, Advanced Single Crystal Ni-Base Superalloys, *Advances in High Temperature Structural Materials and Protective Coatings,* A.K. Koul, Ed., National Research Council of Canada, 1994, p 11–31
3. B.J. Piearcey and F.L. VerSnyder, *J. Aircr.,* Vol 3 (No. 5), 1960, p 390
4. F.L. VerSnyder and M.E. Shank, *Mater. Sci. Eng.,* Vol 6 (No. 4), 1970, p 321
5. M. Gell, The Science and Technology of Single Crystal Superalloys, *Proc. Japan-U.S. Seminar on Superalloys,* International Iron and Steel Institute, 1984
6. M. Gell, D.N. Duhl, and A.F. Giamei, The Development of Single Crystal Superalloy Turbine Blades, *Proc. Fourth International Symp. on Superalloys,* American Society for Metals, 1980, p 205–214
7. J.J. Jackson, M.J. Donachie, R.J. Henricks, and M. Gell, The Effects of Volume % of Fine γ' on Creep in DS MAR M 200 Hf, *Metall. Trans. A,* Vol 8 (No. 10), 1977, p 1615
8. A.D. Cetel and D.N. Duhl, Second-Generation Nickel-Base Single Crystal Superalloy, *Proc. Sixth International Symp. on Superalloys,* The Metallurgical Society, 1988, p 235–244
9. D.N. Duhl, Directionally Solidified Superalloys, *Superalloys II,* C.T. Sims, N.S. Stoloff, and W.C. Hagel, Ed., John Wiley & Sons, 1987, p 189–214
10. S.M. Copley, A.F. Giamei, S.M. Johnson, and M.F. Hornbecker, *Metall. Trans.,* Vol 1, Aug 1970, p 2193
11. F.L. VerSnyder, *High Temperature Alloys for Gas Turbines 1982,* Reidel, 1982, p 1
12. T.S. Piwonka and P.N. Atanmo, *Proc. 1977 Vacuum Metallurgy Conf.,* Science Press, 1977, p 507
13. S. Morimoto, A. Yoshinari, and E. Niyama, in *Superalloys 1984,* The Metallurgical Society, 1984, p 177
14. M.J. Goulette, P.D. Spilling, and R.P. Anthony, in *Superalloys 1984,* The Metallurgical Society, 1984, p 167
15. D.M. Shah and D.N. Duhl, in *Superalloys 1984,* The Metallurgical Society, 1984, p 105
16. J.J. Burke, H.L. Wheaton, and J.R. Feller, paper presented at the Annual TMS-AIME Meeting (Denver, CO), 1978
17. K. Harris, G.L. Erickson, and R.E. Schwer, MAR M 247 Derivations: CM 247 LC DS Alloy, CMSX Single Crystal Alloys, Properties and Performance, *Proc. Fifth International Symp. on Superalloys,* The Metallurgical Society, 1984, 221–230
18. K. Harris, G.L. Erickson, and R.E. Schwer, CMSX Single Crystal, CM DS and Integral Wheel Alloys: Properties and Performances, *Cost 50/501 Conf. on High Temperature Alloys for Gas Turbines and Other Applications,* Reidel, 1986
19. G.L. Erickson, Superalloy Developments for Aero and Industrial Gas Turbines, *Advanced Materials and Coatings for Combustion Turbines,* V.P. Swaminathan and N.S. Cheruvu, Ed., ASM International, 1994, p 29–41
20. M. Gell, D.N. Duhl, D.K. Gupta, and K.D. Sheffler, Advanced Superalloy Airfoils, *J. Met.,* July 1987, p 11–15
21. T.E. Strangman, G.S. Hoppin III, et al., Development of Exothermically Cast Single Crystal Mar M 247 and Derivative Alloys, *Proc. Fourth International Symp. on Superalloys,* American Society for Metals, 1980, 215–224
22. G.S. Hoppin III and W.P. Danesi, Manufacturing Processes for Long Life Gas Turbines, *J. Met.,* July 1986
23. C.S. Wukusick, Final Report, Contract N62269-78-C-0315, Naval Air Systems Command, 25 Aug 1980
24. D.A. Ford and R.P. Arthey, Development of Single Crystal Alloys for Specific Engine Applications, *Proc. Fifth International Symp. on Superalloys,* The Metallurgical Society, 1984, p 115–124
25. E. Bachelet and G. Lamanthe, "AM1, High Temperature Superalloy for Turbine Blades," paper presented at the National Symp. on Single Crystal Superalloys, Villard-de-Lans, France, Feb 1986
26. K. Harris, G.L. Erickson, and R.E. Schwer, "Development of the Single Crystal Alloys CMSX-2 and CMSX-3 for Advanced Technology Turbines," Paper 83-GT-244, American Society of Mechanical Engineers
27. J. Wortmann, R. Wege, K. Harris, G.L. Erickson, and R.E. Schwer, Low Density Single Crystal Superalloy CMSX-6, *Proc. Seventh World Conf. on Investment Casting* (Munich, West Germany), European Investment Casters' Federation, 1988
28. T. Khan and P. Caron, Development of a New Single Crystal Superalloy for Industrial Gas Turbine Blades, *High Temperature Materials for Power Engineering, Part II,* E. Bachelet et al., Ed., Kluwer Academic Publishers, 1990, p 1261–1270

29. C.T. Sims, Prediction of Phase Composition, *Superalloys II,* C.T. Sims, N.S. Stoloff, and W.C. Hagel, Ed., John Wiley & Sons, 1987, p 217–240
30. I.I. Kornilov and A.Y. Snetkov, Lattice Parameter Limitations of Some Solid Solution Elements in Nickel, *Izv. Akad. Nauk.,* 1960, p 106–111
31. S. Isobe et al., "The Effects of Impurities on Defects in Single Crystals of NASAIR 100," Paper presented at the International Gas Turbine Congress, Tokyo, International Iron and Steel Institute, 1983
32. A.F. Giamei and D.L. Anton, *Metall. Trans. A,* Vol 16, 1985, p 1997
33. D. Blavette, P. Caron, and T. Khan, *Scr. Metall.,* Vol 20 (No. 10), Oct 1986
34. G.L. Erickson, "The Development of CMSX-10, A Third-Generation SX Casting Superalloy," paper presented at the *Second Pacific Rim International Conf. on Advanced Materials and Processing* (Kyongju, Korea), 18–22 June 1995
35. T. Khan and P. Caron, The Effect of Processing Conditions and Heat Treatments on the Mechanical Properties of a Single Crystal Superalloy, The Institute of Metals, 1985
36. A. Fredholm and J.L. Strudel, On the Creep Resistance of Some Nickel Base Single Crystals, *Proc. Fifth International Symp. on Superalloys,* The Metallurgical Society, 1984, p 220–221
37. T. Khan, P. Caron, D. Fournier, and K. Harris, "Single Crystal Superalloys for Turbine Blades: Characterization and Optimization of CMSX-2 Alloy," paper presented at the 11th Symp. on Steels & Special Alloys for Aerospace (Paris), L'Association Aéronautique et Astronautique de France, June 1985
38. T. Khan and P. Caron, Effect of Heat Treatment on the Creep Behavior of a Ni-Base Single Crystal Superalloy, *Fourth RISO International Symp. on Metallurgy and Material Sciences,* Office National d'Etudes et de Recherches Aerospatiales, 1983, p 173
39. T. Khan, P. Caron, Y.U. Nakagawa, and Y. Ohta, Creep Deformation Anisotropy in Single Crystal Superalloy, *Proc. Sixth International Symp. on Superalloys,* The Metallurgical Society, 1988, p 215–224
40. R. Castillo, A.K. Khoul, J.P. Immarigeon, and P. Lowden, Processing of Superalloy Investment Castings through HIP, *Advances in High Temperature Structural Materials and Protective Coatings,* A.K. Koul, Ed., National Research Council of Canada, 1994, p 147–167
41. D.N. Duhl, Single Crystal Superalloys, *Superalloys, Supercomposites, and Superceramics,* Academic Press, 1989, p 149–181
42. P.S. Burkholder et al., "Allison Engine Testing CMSX-4 Single Crystal Turbine Blades and Vanes," paper presented at Third International Charles Parsons Turbine Conf. Materials Engineering in Turbines and Compressors (Newcastle-upon-Tyne, U.K.), 25–27 April 1995
43. M. Matsubara, A. Nitta, and K. Kuwabara, paper presented at the International Gas Turbine Congress (Tokyo), International Iron and Steel Institute, Oct 1987

Powder Metallurgy Superalloys

SUPERALLOY PARTS made by powder metallurgy (P/M) techniques are being used in advanced turbine engines. The advantages of the P/M process, as applied to turbine-engine hardware, are:

- Ability to produce near-net shapes, which results in reduced materials usage and reduced machining
- Improved property uniformity and alloy-development flexibility, due to the elimination of macrosegregation
- Reduced energy requirements and shorter delivery time, because the P/M process requires fewer steps than conventional ingot technology

The majority of P/M superalloys fall into one of two categories. The first involves production of spherical prealloyed powder, screening to remove oversize particles, blending the powders to homogenize the powder size distribution, loading the powder into containers, vacuum outgassing and sealing the containers, and then consolidating the powder to full density by hot isostatic pressing (HIP) or another consolidation process. Category two includes mechanically alloyed P/M alloys strengthened by an ultrafine dispersed oxide such as yttria (Y_2O_3). Mechanically alloyed superalloy powders are consolidated by hot extrusion and hot rolling.

P/M Disk Alloys

Aircraft gas turbine disks, designed to operate at about 650 °C (1200 °F) in current high-performance engines, require forgeable alloys with high yield strength (to tolerate overspeed without burst), high creep resistance, and good damage tolerance. The crack growth rate must be kept low even under conditions of environmental attack and hold times under stress.

Several P/M superalloys have replaced forged alloys as turbine disks, including MERL 76, LC Astroloy, IN-100, and René 95. Table 1 gives compositions. In general, the strength of these alloys is a direct function of the γ' content. Powder processing permits the attainment of a fine grain size, which lends the alloys superplastic forming capability, as in the Pratt and Whitney Gatorizing process. The alloys are characterized by a high homogeneous concentration of both solid-solution strengthening elements and the γ'-forming elements aluminum and titanium. These factors would limit forgeability of conventionally cast and wrought alloys.

The advantages of powder-processed disk alloys are:

- More uniform composition and phase distribution
- Finer grain size
- Reduced carbide segregation
- Higher material yields
- Increased flexibility in alloy design

The superalloy compressor disk shown schematically in Fig. 1 is a good example of the material and fabrication savings attainable with P/M technology. It is noteworthy that the P/M process required markedly fewer processing steps and reduced the material input weight from 95 to 18 kg (210 to 40 lb).

However, several problems arise directly from powder techniques:

- Increased residual gas content
- Carbon contamination
- Ceramic inclusions
- Formation of prior particle boundary oxide and/or carbide films

These problems can lead directly to inferior mechanical properties. For example, oxide and other inclusions more readily cause crack nucleation, and cause greater variability in fatigue life, than they do in wrought alloys. For these reasons major efforts have been made to produce a cleaner master melt and to reduce contamination during the preparation

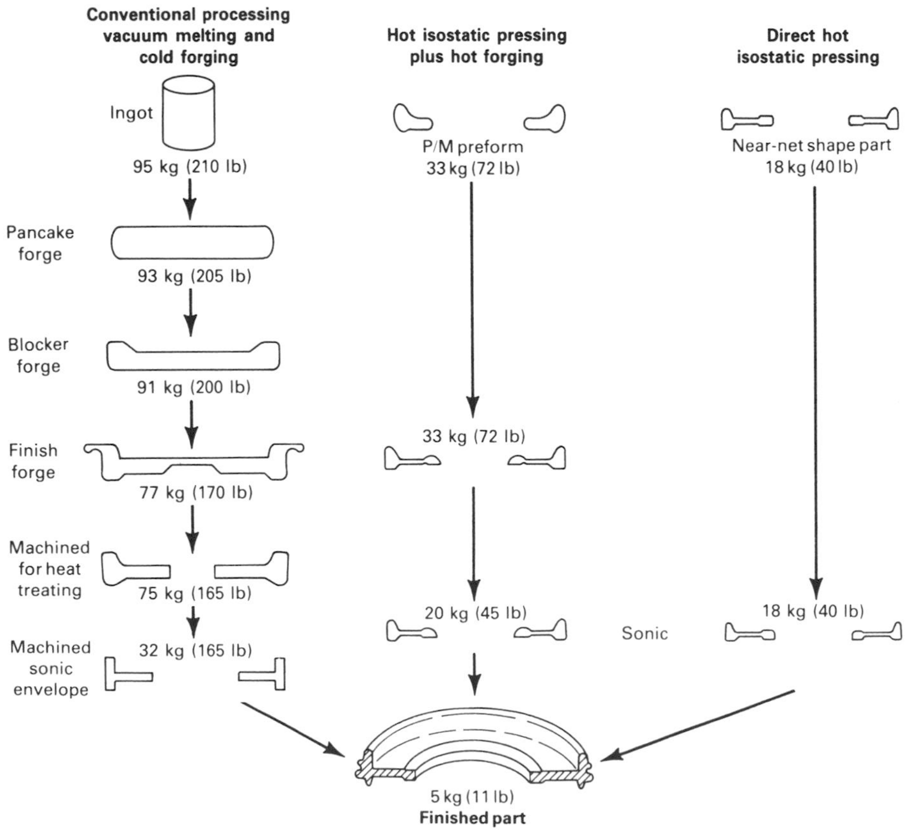

Fig. 1 Material and fabrication savings attainable with powder metallurgy (P/M) processing of superalloys

Table 1 Chemical compositions of some nickel-base superalloys produced by powder metallurgy

Alloy	Composition, %													
	C	Ni	Cr	Co	Mo	W	Ta	Nb	Hf	Al	Ti	V	B	Zr
IN-100	0.07	bal	12.4	18.5	3.2	...	...	...	...	5.0	4.3	0.8	0.02	0.06
LC Astroloy	0.023	bal	15.1	17.0	5.2	...	...	...	...	4.0	3.5	...	0.024	<0.01
Nimonic AP1	...	bal	15.0	17.0	5.0	...	...	...	...	4.0	3.5	...	...	...
Waspaloy	0.04	bal	19.3	13.6	4.2	...	...	...	...	1.3	3.6	...	0.005	0.048
NASA II B-7	0.12	bal	8.9	9.1	2.0	7.6	10.1	...	1.0	3.4	0.7	0.5	0.023	0.080
René 80	0.20	bal	14.5	10.0	3.8	3.8	...	...	...	3.1	5.1	...	0.014	0.05
Unitemp AF2-1DA	0.35	bal	12.2	10.0	3.0	6.2	1.7	...	...	4.6	3.0	...	0.014	0.12
MAR-M 200	0.15	bal	9.0	10.0	...	12.0	...	1.0	...	5.0	2.0	...	0.015	0.05
IN-713 LC	0.05	bal	12.0	0.08	4.7	...	...	(2.0)	...	6.2	0.8	...	0.005	0.1
IN-738	0.17	bal	16.0	8.5	1.7	2.6	1.7	0.9	...	3.4	3.4	...	0.01	0.1
IN-792	0.12	bal	12.4	9.0	1.9	3.8	3.9	...	...	3.1	4.5	...	0.02	0.10
AF-115	0.045	bal	10.9	15.0	2.8	5.7	...	1.7	0.7	3.8	3.7	...	0.016	0.05
MERL 76	0.025	bal	12.2	18.2	3.2	...	...	1.3	0.3	5.0	4.3	...	0.02	0.06
René 95	0.08	bal	12.8	8.1	3.6	3.6	...	3.6	...	3.6	2.6	...	0.01	0.053
Modified MAR-M 432	0.14	bal	15.4	19.6	...	2.9	0.7	1.9	0.7	3.1	3.5	...	0.02	0.05
New alloys														
RSR 103	...	bal	...	...	15.0	...	...	...	...	8.4	...	...	...	...
RSR 104	...	bal	...	...	18.0	...	...	...	...	8.0	...	...	...	...
RSR 143	...	bal	...	...	14.0	...	6.0	...	...	6.0	...	...	...	...
RSR 185	0.04	bal	...	...	14.4	6.1	...	...	...	6.8	...	...	...	...

Source: Ref 1

of powders. The semicontinuous carbide films on prior particle boundaries cannot be altered by simple heat treatment, but they can be affected by proper hot-working schedules. These boundaries may also be minimized by the proper selection and balance of solute elements.

Powder Production

Various means of producing superalloy powders are summarized in Table 2. All involve gas atomization processes that produce spherical powders. More detailed information on these processes can be found in Volume 7, *Powder Metallurgy,* of the *ASM Handbook,* and in Ref 1 to 3.

Inert gas atomization is the most common technique of producing superalloy powders (Fig. 2). An ingot is first cast, typically by induction melting, in order to minimize the oxygen and nitrogen contents. In some cases remelting may be carried out by electron beam heating, arc melting under argon, or plasma heating. Atomization is carried out by pouring the master melt through a refractory orifice. A high-pressure inert gas stream (typically argon) breaks up the alloy into liquid droplets, which are solidified at a rate of about 10^2 K/s.

The spherical powder is collected at the outlet of the atomization chamber. The maximum particle diameter resulting from this process depends on the surface tension (γ), viscosity (η), and density (ρ) of the melt, as well as the velocity (v) of the atomizing gas. The principal factor is gas velocity. Oxygen contents are of the order of 100 ppm, depending on particle size. Finer particle sizes are obtained by screening. Generally, spherical fine particles are desired for further processing (Ref 1).

Soluble Gas Atomization. Another important powder production method, the soluble gas process, is based on the rapid expansion of gas-saturated molten metal. A fine spray of molten droplets forms as the dissolved gas, usually hydrogen, is suddenly released (Ref 3) (Fig. 3). The droplets solidify at a rate of about 10^3 K/s, and the cooled powder is collected under vacuum in another chamber, which is sealed and backfilled with an inert gas. This method is capable of atomizing up to 1000 kg (2200 lb) of superalloy in one heat and produces spherical powder that can be made very fine (Fig. 4). This method has been successfully employed for LC Astroloy, MERL 76, and IN-100.

Centrifugal Atomization. The third method of powder preparation is based on centrifugal atomization. The melt is accelerated and disintegrated by rotating it under vacuum or in a protective atmosphere. One example of this method is the rotating electrode process (REP) used in the early production of IN-100 and René 95 powder. In this process, a bar of the desired composition, 15 to 75 mm (0.6 to 3 in.) in diameter, serves as a consumable electrode. The face of this positive electrode, which is rotated at high speed, is melted by a direct current electric arc between the consumable electrode and a stationary tungsten negative electrode (Fig. 5). Centrifugal force causes spherical molten droplets to fly off the

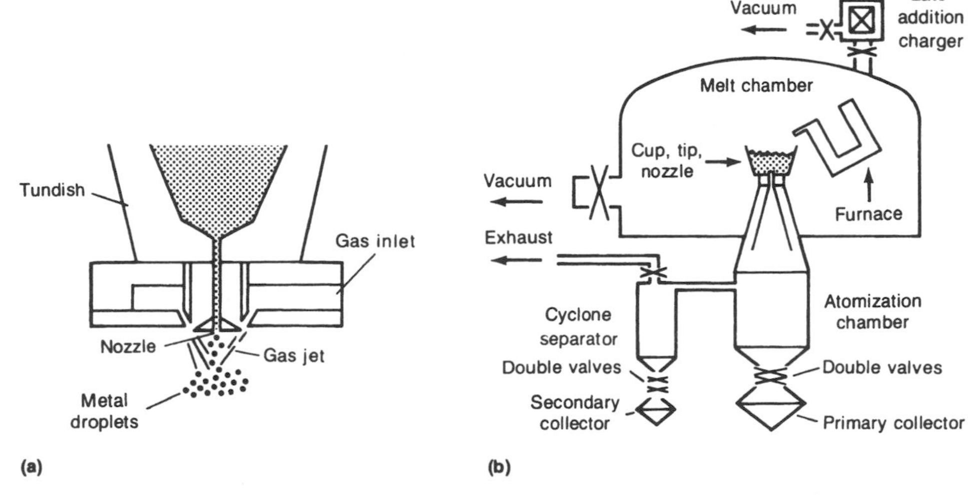

Fig. 2 Gas atomization system for superalloy powder production. (a) Atomization nozzle. (b) Typical system. Source: Ref 3

Table 2 Powder production methods

Step	Inert gas atomization(a)	Soluble gas process	Rotating electrode process(b)	Plasma rotating electrode process	Centrifugal atomization with forced convective cooling (RSR)(c)
Melting 1	VIM; ceramic crucible	VIM; ceramic crucible	VIM, VAR, ESR	VIM, VAR, ESR	VIM; ceramic crucible
Melting 2	...	...	Argon arc	Plasma	...
Melt disintegration system/environment	Nozzle; argon stream	Expansion of dissolved hydrogen against vacuum and Ar + H_2 mixture	Rotating consumable electrode; argon or helium	Rotating consumable electrode; argon	Rotating disk; forced helium convective cooling

(a) VIM, vacuum induction melting. (b) VAR, vacuum arc remelting; ESR, electroslag remelting. (c) RSR, rapid solidification rate. Source: Ref 1

rotating electrode. These droplets freeze and are collected at the bottom of the tank, which is filled with helium or argon. A major advantage of this process is the elimination of ceramic inclusions and the lack of any increase in the gas content of the powder relative to that of the alloy electrode.

A variant on the REP process is the plasma rotating electrode process (PREP). Instead of a tungsten electrode, a plasma arc is used to melt the superalloy electrode surface. Cooling rates are higher, up to 10^5 K/s for IN-100 powder. On average, particle sizes are nearly twice as large in these processes as in gas atomization. Neither REP nor PREP processes are currently in active production for superalloys (Ref 3).

Rapid Solidification Processing. When solidification rates of powder exceed 10^5 K/s, the process is referred to as rapid solidification rate (RSR). For superalloys, the objective of the high rates is to obtain a microcrystalline alloy rather than an amorphous material. Apart from extremely fine grain size, such powders display nonequilibrium solubilities and very uniform compositions, because very fine dendritic arm spacings result from rapid solidification. Both conventional superalloys and new alloys based on Ni-Al-Mo-*X* alloys, where *X* is tantalum or tungsten, have been prepared by RSR (see Table 1).

Rapid solidification processing can be done by centrifugal atomization with forced convective cooling, as in the method shown in Fig. 6. In this method, the alloy is vacuum induction melted in the upper part of a chamber. The chamber is then backfilled with helium, and the alloy is poured into a preheated tundish. The liquid is poured through the tundish nozzle onto a rotor that is spun at 24,000 rev/min. The melt is accelerated to rim speed and then ejected longitudinally as droplets. Further atomization and cooling of the droplets is accomplished by the injection of helium gas through annular nozzles. Spherical powder in the range of 10 to 100 μm (400 to 4000 μin.) in diameter is produced. The cooling rate typically varies between 10^5 and 10^7 K/s, the higher rates being achieved with the smaller particles.

As shown in Fig. 7, the powder produced by this process is typically free of attached satellites. The microstructure depends on both particle size and alloy composition. Structures typical of nickel-base superalloys are shown in the particle cross sections of Fig. 8. They tend to be either dendritic (Fig. 8) or crystalline. The presence of the latter form varies with alloy composition and particle size.

Rapidly solidified powders may also be prepared from melt-spun ribbon that is pulverized after solidification. The ribbon is produced by pouring the melt through an orifice under pressure and impinging it on a rotating wheel of, for example, copper, which acts as a heat sink. Typical cooling rates are approximately 10^6 K/s, and ribbon thicknesses are less than 25 μm (1000 μin.). The ribbon must be mechanically pulverized, and this method is generally limited to small quantities of experimental alloys.

Powder Consolidation

Although virtually every P/M consolidation technique has been applied to superalloys, production of superalloy disks is usually accomplished by HIP and related processes or by extrusion plus isothermal forging. Full density can be achieved by these processes.

A key feature of any consolidation process is the necessity to minimize contamination, especially from absorbed surface gases and organic material mixed in with the powder. Therefore, powders are packed into sheet metal containers under dynamic vacuum (either warm or cold). The evacuated container is then sealed, heated to the desired temperature, and compacted, either isostatically under gas pressure or in a closed die.

Hot Isostatic Pressing. For HIP, powder-filled stainless steel containers are placed in an autoclave that is subsequently heated and pressurized. Superalloys are normally hot isostatically pressed to full density at temperatures ranging from 1095 to 1205 °C (2000 to 2200 °F) under a pressure of 103 MPa (15 ksi).

The advantage of HIP is that a fully dense product can be obtained without retained prior particle boundaries. The densification mode is either plastic deformation or creep, depending on the HIP cycle time, temperature, and pressure. Grain size control during HIP is achieved by

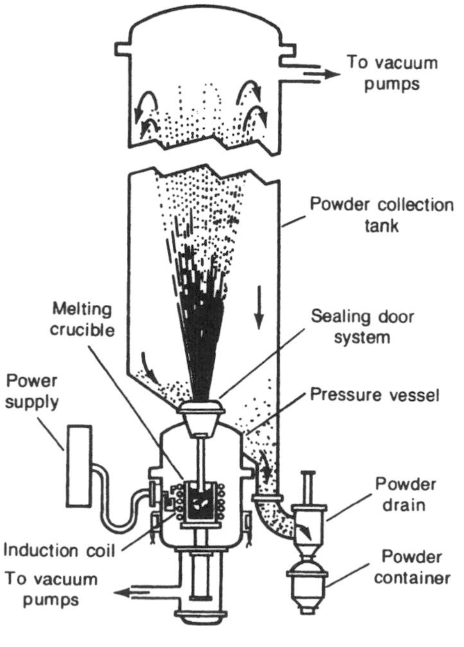

Fig. 3 Soluble gas atomization system for producing superalloy powder. Source: Ref 3

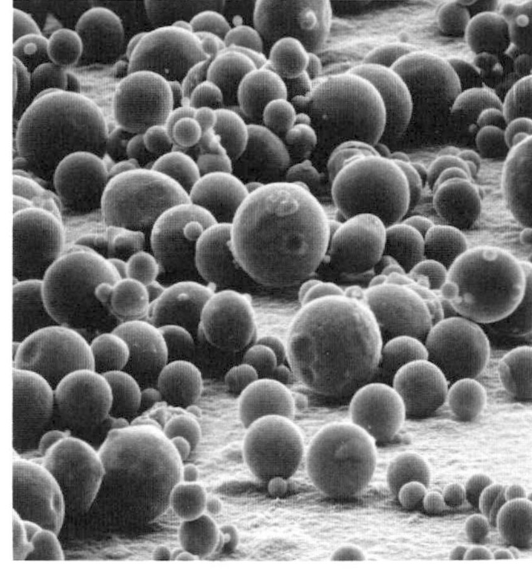

Fig. 4 Scanning electron micrograph of soluble-gas atomized nickel-base powder (–100 mesh). 650×

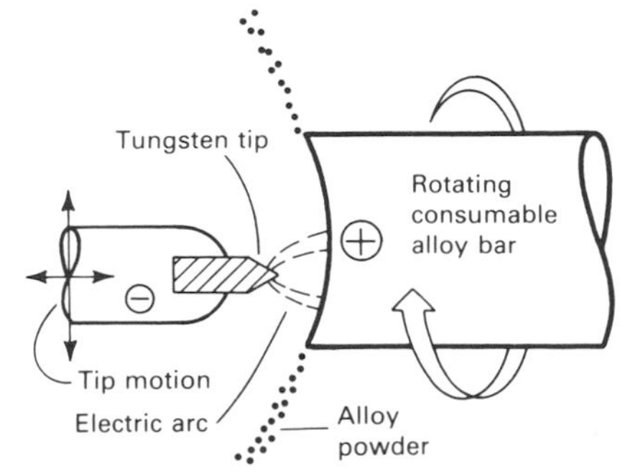

Fig. 5 Schematic of rotating electrode process

choosing a temperature either above or below the γ' solvus.

Recently, both glassy and ceramic containers have been used for HIP (Ref 1). Glassy containers are weak and cannot support the weight of a large part. However, a ceramic container inserted into a large steel container allows the production of intricate shapes. During the HIP cycle, the outer metal jacket can be subjected to the autoclave gas pressure. The pressure is transmitted to the ceramic mold through a bed of fine alumina powder.

Other powder consolidation techniques related to HIP are the consolidation by atmospheric pressure (CAP) process and the fluid die process (Ref 2). Neither technique requires an expensive HIP unit. The CAP process resembles a vacuum sintering process, assisted by low pressure exerted on the surface of a shaped glass container. The fluid die process incorporates a cavity surrounded by a dense, incompressible mass of material. The higher mass of these containers, compared to the mass of sheet containers, allows greater vibration during filling and sealing, ensuring more complete filling and uniform tap density. The outer material softens appreciably at the compaction temperature, allowing pressure to be transmitted to the powder. Convectional die forging equipment is capable of much higher ram pressures than those possible with HIP autoclaves; full consolidation occurs in less than 1 s.

HIP Plus Isothermal and Hot Die Forging. Hot isostatically pressed (HIPed) billets can be further thermodynamically processed by isothermal forging or hot die forging. Isothermal forging involves heating the workpiece and tooling to the same temperature and forging at low strain rates. The hot die forging process is characterized by die temperature higher than those of conventional forging, but lower than those in isothermal forging.

Hot die and isothermal forging processes have greater shapemaking capabilities than conventional press forging for these alloys, due to the use of low strain rates and hot tooling, which eliminates die chill problems. Advantages of these processes over conventional press forging include:

- Reduced flow stresses in the workpiece
- Enhanced workability
- Greater control over final microstructure
- Greater dimensional precision

The low strain rates capitalize on the fine grain size of P/M preforms for improved workability. These advantages allow production of near-net shapes, which results in significant cost savings due to reduced materials usage and reduced machining.

Disadvantages of HIP plus isothermal and hot die forging include:

- Long forging cycles (typically 15 min or longer dwell times)
- High cost of forging equipment (atmosphere chambers and integral heating stations)

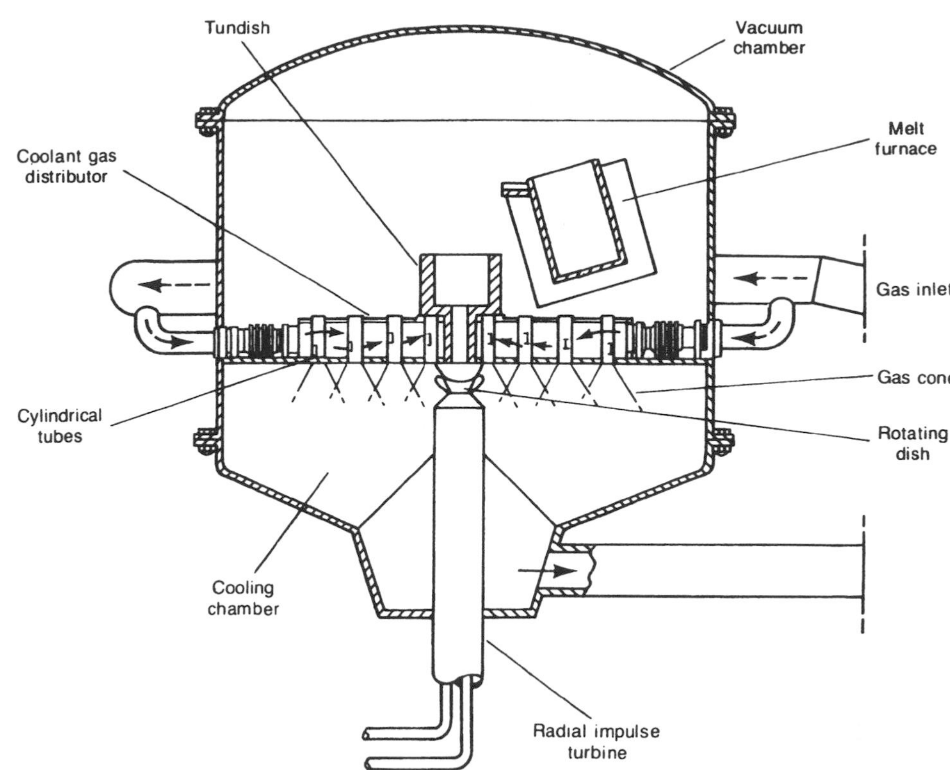

Fig. 6 Schematic of Pratt and Whitney powder rig for producing rapidly solidified powders. Source: Ref 1

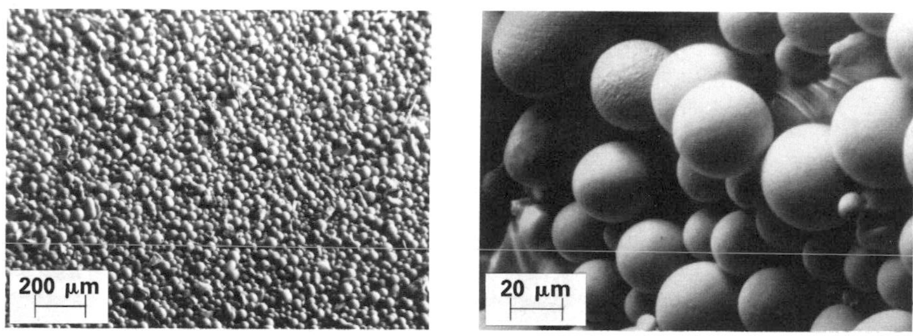

Fig. 7 Scanning electron micrograph of typical rapid solidification rate powder. Note the lack of attached satellite droplets and the generally spherical shape.

Table 3 Heat treatments, grain size, and tensile properties of various René 95 product forms

Heat treatment/property	Extruded and forged(a)	Hot isostatic pressing(b)	Cast and wrought(c)
Heat treatment	1120 °C (2050 °F)/1 h AC + 760 °C (1400 °F)/8 h AC	1120 ° C (2050 °F)/1 h AC + 760 °C (1400 °F)/8 h AC	1220 °C (2230 °F)/1 h AC + 1120 °C (2050 °F)/1 h AC + 760 °C (1400 °F)/8 h AC
Grain size, μm (mils)	5 (0.2) (ASTM No. 11)	8 (0.3) (ASTM No. 8)	150 (6) (ASTM No. 3-6)
40 °C (100 °F) tensile properties			
0.2% yield strength, MPa (ksi)	1140 (165.4)	1120 (162.4)	940 (136.4)
Ultimate tensile strength, MPa (ksi)	1560 (226.3)	1560 (226.3)	1210 (175.5)
Elongation, %	8.6	16.6	8.6
Reduction in area, %	19.6	19.1	14.3
650 °C (1200 °F) tensile properties			
0.2% yield strength, MPa (ksi)	1140 (165.4)	1100 (159.5)	930 (134.7)
Ultimate tensile strength, MPa (ksi)	1500 (217.6)	1500 (217.6)	1250 (181.3)
Elongation, %	12.4	13.8	9.0
Reduction in area, %	16.2	13.4	13.0

(a) AC, air cooled. Processing: –150 mesh powder, extruded at 1070 °C (1900 °F) to a reduction of 7 to 1 in area, isothermally forged at 1100 °C (2012 °F) to 80% height reduction. (b) Processing: –150 mesh powder, HIP processed at 1120 °C (2050 °C) at 100 MPa (15 ksi) for 3 h. (c) Processing: cross-rolled plate, heat treated at 1218 °C (2225 °F) for 1 h. Source: Ref 3, 4

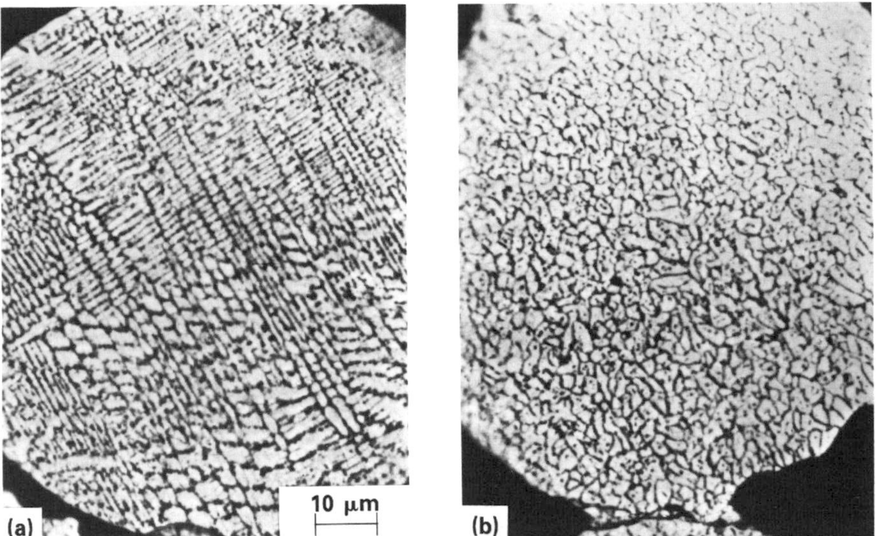

Fig. 8 Representative cross-sectional internal microstructure observed in rapid solidification rate nickel-base superalloy powders. (a) Dendritic. (b) Microcrystalline. The relative occurrence of the two forms has been found to depend on both particle size and alloy compositions.

- Cost of forging dies (expensive alloys such as TZM molybdenum alloy, rather than lower-cost tool steels)
- Need for an inert atmosphere or vacuum to protect tooling

Despite these costs, this process combination is economical for producing parts from expensive alloys, where near-net shape production minimizes expensive machining and material losses.

Hot Extrusion Plus Forging. Superalloy powders also can be consolidated into billet shapes by extrusion. Powder, usually containerized, is hot extruded at a reduction ratio of at least 9 to 1 to achieve a fully consolidated billet. The extruded length is then sectioned into workpiece billets for subsequent hot working by conventional press forging, isothermal forging, or hot die forging. For press forging, two steps are required. First, the billet is forged into a preform shape. Then a forged blank is produced from the preform. Preform shaping operations also may be added to the isothermal and hot die forging steps to reduce cycle time in the low-strain-rate deformation step.

Mechanical Properties

Properties of P/M superalloys used for turbine disk applications are directly related to the composition and structure of the alloy. In turn, the structure is related to powder particle size, consolidation process, and heat treatment. For example, the effects of grain size and heat treatment on the tensile properties of René 95 are obvious in Table 3, which shows that tensile strength and ductility increase as grain size becomes finer (Ref 3, 4). However, when René 95 is heat treated above the γ' solvus temperature, both grain size and stress-rupture life increase, while tensile yield decreases, as shown in Table 4. Table 5 summarizes the effect of heat treatment on the tensile properties and low-cycle fatigue (LCF) resistance of HIPed and forged Nimonic AP1.

Figure 9 compares the mechanical properties of cast and wrought products with those of HIPed and forged P/M products. Tensile yield and ultimate tensile strength levels of the P/M materials are superior to those of the ingot metallurgy material. Creep properties are similar for the two materials over the temperature range shown. In addition to producing material with improved microstructural homogeneity and improved mechanical properties, HIP plus forging is more economical than conventional cast or wrought practices. Material utilization is improved, and fewer forging steps are required (refer to Fig. 1).

Although HIP plus forging yields suitable mechanical properties, increased material costs prompted development of a less expensive, more reliable as-HIPed powder metallurgy production process for engine applications. The as-HIPed production process was established after 3 years of extensive development and testing. This method was considerably less expensive than the original cast and wrought process, because it used 40% less input material. It was also less expensive than HIP plus forging, because fewer processing steps were involved. Currently, the successful production of as-HIPed engine components continues. Mechanical property data in Fig. 10 show that parts produced by direct HIP, followed by heat treatment, inspection, and machining, have property levels in excess of design requirements. Additional comparative data are given in Table 6.

Controlling the cyclic mechanical properties of superalloys, particularly LCF, is important to gas turbine engine performance and reliability improvements. Low-cycle fatigue is primarily controlled by defects in the material that result in the initiation of fatigue cracks by any mechanism

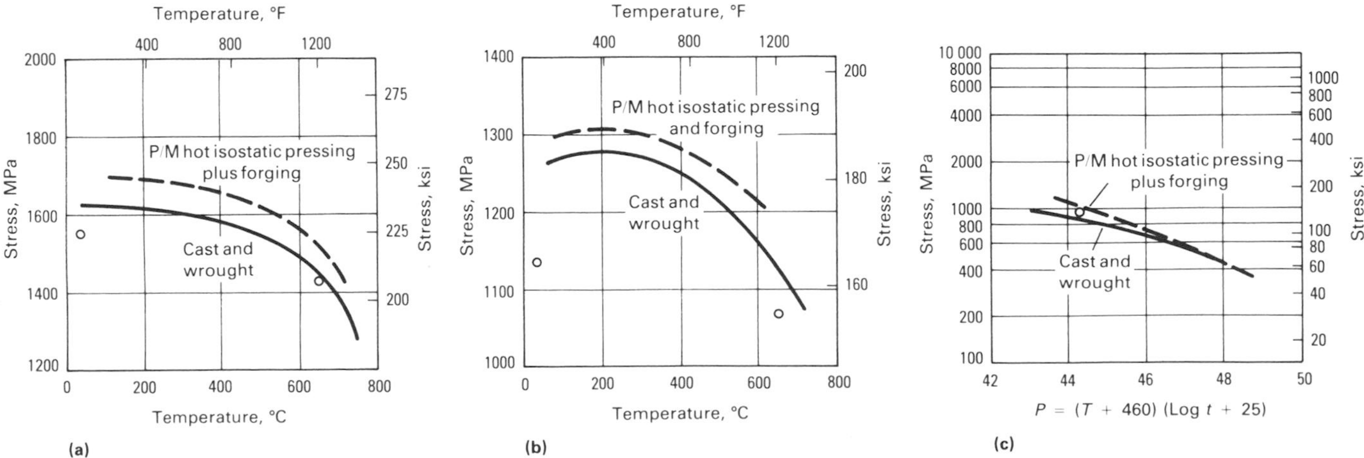

Fig. 9 Comparison of mechanical properties of T-700 engine hardware produced from René 95 by conventional cast and wrought processing and by hot isostatic pressing plus forging of powder metallurgy (P/M) materials. (a) Ultimate tensile strength. (b) 0.2% yield strength. (c) 0.2% creep strength. *P*, Larson-Miller parameter; *T*, temperature (°F); *t*, time (h). Open circles indicate T-700 design requirements. Source: Ref 6

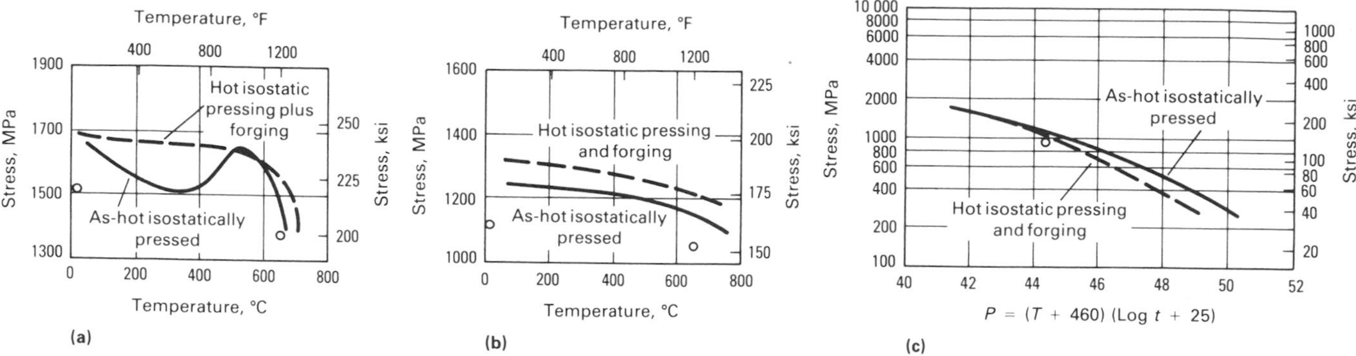

Fig. 10 Mechanical property comparisons between hot isostatically pressed plus forged hardware and direct hot isostatically pressed hardware of René 95 for T-700 engine application. (a) Ultimate tensile strength. (b) 0.2% yield strength. (c) 0.2% creep strength. *P*, Larson-Miller parameter; *T*, temperature (°F); *t*, time (h). Open circles indicate T-700 design requirements. Source: Ref 6

that is noncrystallographic. The effects of defects on LCF depend on the size and location of the defects in the specimen. (The defects themselves, in terms of size and distribution, depend on powder particle size and consolidation process.) A defect of a given size is less detrimental when located internally rather than externally, primarily due to the absence of adverse environmental interactions (Ref 3).

The effects of thermomechanical working on LCF also are important. As shown in Fig. 11, extruded and forged material shows longer life than the average HIPed or HIPed and forged materials (Ref 3, 8). The effects of forging are that defects are dispersed during processing, the size of defects may be reduced, and the grain size is further refined. Controlling the thermomechanical process can reduce or even eliminate defects in prior particle boundaries. The LCF life of a thermomechanically processed P/M superalloy does appear to be limited by small ceramic defects (Ref 3).

Test conditions and alloy chemistry also are factors in fatigue resistance. This may best be seen in Fig. 12, which compares the strain-controlled fatigue lives of seven superalloys. At low strain amplitudes, the strongest alloys (e.g., René 95 and MERL 76) are the most fatigue resistant, while at high strain amplitudes, the weakest alloys (Waspaloy and HIP Astroloy) display the longest lives. The crossover in ranking of the various alloys occurs because strength is the most important variable at low strain amplitudes, while ductility dominates at high strain amplitudes.

For alloys that are highly oxidation resistant, the stress-rupture properties of bar and sheet of the same alloy differ little. However, less oxidation-resistant alloys are likely to exhibit lower properties for sheet than for bar of the same composition. Similarly, elevated-temperature fatigue properties are very susceptible to changes in

Table 4 Powder metallurgy René 95 tensile properties at 650 °C (1200 °F)

Alloy	ASTM grain size	Ultimate tensile strength, MPa	Ultimate tensile strength, ksi	0.2% offset yield strength, MPa	0.2% offset yield strength, ksi	Elongation, %	Reduction in area, %	Stress-rupture life at 965 MPa (140 ksi), h
HIP René 95(a)	8	1505	218	1140	165	13.8	15	87,150
HIP René 95 with supersolvus heat treatment(b)	6	1415	205	1000	145	16	19	244,298

(a) Processing: –150 mesh powder, HIP processed at 1120 °C (2050 °F), 103 MPa (15 ksi)/3 h + 1120 °C (2050 °F)/1 h + 870 °C (1600 °F)/1 h + 650 °C (1200 °F)/24 h. (b) Processing: –150 mesh powder, HIP processed at 1120 °C (2050 °F), 103 MPa (15 ksi)/3 h + 1200 °C (2200 °F)/1 h + 1120 °C (2050 °F)/1 h + 870 °C (1600 °F)/1 h + 650 °C (1200 °F)/24 h. Source: Ref 3

Table 5 Effect of heat treatment on the properties of hot isostatically pressed plus conventionally forged Nimonic alloy AP 1

Processing temperature, °C	Processing temperature, °F	Size of sample disk, mm	Size of sample disk, in.	Solution treatment	Tensile properties(a): Yield point, 0.2% offset, MPa	Yield point, 0.2% offset, ksi	Ultimate tensile strength, MPa	Ultimate tensile strength, ksi	Elongation, %	Reduction in area, %	Notched tensile strength, MPa	Notched tensile strength, ksi	Stress rupture(b): Plain life, h	Elongation, %	Notch life, h	Low-cycle fatigue(c), cycles
1150	2100	150	6	4 h at 1110 °C (2030 °F), air cool	971	141	1307	190	30.4	31.6	1869	271	42	30.1	195	>276,000
1150	2100	150	6	4 h at 1080 °C (1980 °F), oil quench	1120	162	1513	219	23.2	24.2	1992	289	64	15.3	159	>307,000
1150	2100	150	6	4 h at 1110 °C (2030 °F), quenched and aged(d)	1037	150	1381	200	30.4	46.7	1776	258	88	20.4	163	>214,000
1220	2230	150	6	4 h at 1110 °C (2030 °F), air cool	999	145	1328	193	28.6	32.7	1868	270	45	20.5	188	>155,000
1220	2230	150	6	4 h at 1080 °C (1980 °F), oil quench	1085	157	1463	212	23.2	23.4	1941	281	66	17.2	247	>228,000
1150	2100	150	6	4 h at 1110 °C (2030 °F), quenched and aged(d)	1052	153	1383	201	25.0	25.8	1844	267	74	16.9	315	>242,000
1150	2100	475	19	4 h at 1110 °C (2030 °F), air cool	952	138	1320	191	29.5	31.4	1521	221	85	22.9	>500	>35,000
1150	2100	475	19	4 h at 1080 °C (1980 °F), oil quench	993	144	1356	197	26.1	28.0	1785	259	113	20.3	>450	>100,000

(a) At 650 °C (1200 °F). (b) 750 MPa (110 ksi) at 705 °C (1300 °F). (c) 1080 MPa (157 ksi) at 600 °C (1110 °F). (d) 50% water-soluble polymeric compound, 50% water. All material aged 24 h; 650 °C (1200 °F); air cooled, 8 h; 760 °C (1400 °F); air cooled. Source: Ref 5

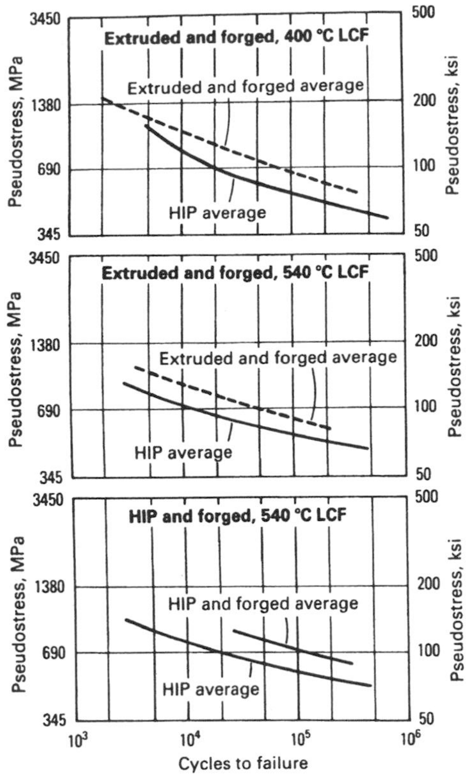

Fig. 11 Comparison of average low-cycle fatigue (LCF) lives of hot isostatically pressed (HIP) René 95 vs. extruded and forged René 95 and vs. HIP and forged René 95. Source: Ref 3, 8

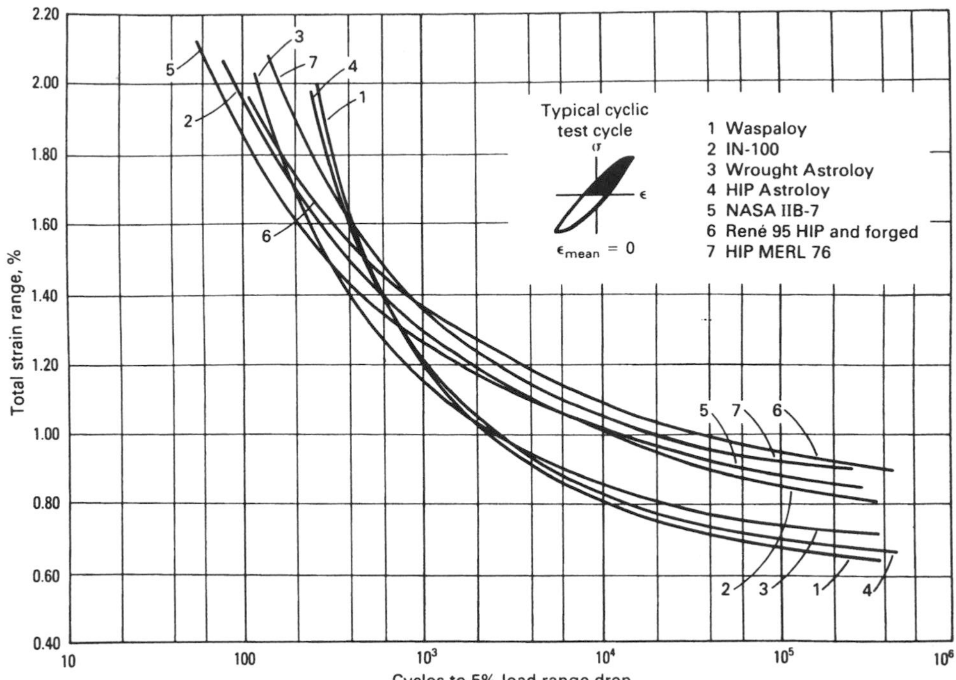

Fig. 12 Comparison of cyclic strain control low-cycle fatigue properties of seven nickel-base alloys tested at 650 °C (1200 °F). HIP, hot isostatically pressed. Source: Ref 9

environment. Oxygen and sulfur are particularly harmful with respect to crack growth at elevated temperatures. The application of a coating may actually decrease fatigue resistance at high stress or strain levels, because the coating treatment can alter the microstructure. Long-time exposures of coated materials, however, should yield superior fatigue properties.

Selected P/M superalloy mechanical properties are provided in Table 7. As stated previously, mechanical properties can change significantly with changes in processing conditions and chemical compositions. In particular, changes in grain size and precipitate distribution that result from processing can have a substantial influence on strength.

Dual-Alloy/Dual-Property Engine Components

Powder metallurgy technology is currently under development to produce dual-alloy, dual-property turbine disks. By use of this technology, the high tensile and LCF property requirements of the bore can be attained with a fine-grained P/M product. Creep and stress rupture requirements of the rim and/or blades can be attained by special processing techniques, such as single-crystal casting, directionally solidified (DS) casting, or mechanical alloying.

One concept uses a blade ring made by tack welding of individual cast single-crystal blades. The ring is sandwiched between two halves of a ceramic mold. The mold is then filled with prealloyed superalloy powder, and the assembly is hot isostatically pressed to consolidate the powder and bond it to the blades. Figure 13 shows turbine wheels made by this technique. Figure 14 shows the microstructure of the bond line between the powder and cast materials.

Another dual-property wheel technique involves the use of a preconsolidated HIPed powder metallurgy hub and a directionally solidified cast blade ring. The two components are joined by HIP. Figure 15 shows a completed wheel manufactured by this process. The coarse-grained directionally solidified rim and the fine-grain P/M hub are easily visible.

A third method for fabricating dual-alloy integral turbine wheel ("blisks") involves vacuum plasma structural deposition. Using this process, the disk material is sprayed onto a preassembled, or integrally cast, ring of airfoils. Conventional preparation of the cast surface provides good, but inconsistent, joint properties. However, by modifying the surface, consistent properties can be obtained and the interface cannot be detected

Table 6 Comparison of mechanical properties of several powder metallurgy superalloys

Condition	Test temperature °C	Test temperature °F	0.2% offset yield strength MPa	0.2% offset yield strength ksi	Ultimate tensile strength MPa	Ultimate tensile strength ksi	Reduction in area, %	Total elongation, %
René 95								
Hot isostatically pressed(a)	23	74	1214	176	1636	237	15	16
Hot isostatically pressed and forged	23	74	1179	171	1629	236	23	18
Cast and wrought	23	74	1144	166	1434	208	12	10
Minimum hot isostatically pressed	650	1202	1120	162	1514	220	17	16
Hot isostatically pressed and forged	650	1202	1122	163	1480	215	14	13
Cast and wrought	650	1202	1055	153	1282	186	10	8
Astroloy								
Hot isostatically pressed	23	74	936	136	1379	200	31	27
Hot isostatically pressed and forged	23	74	1055	153	1517	220	23	27
Hot isostatically pressed	650	1202	881	128	1234	179	36	31
Hot isostatically pressed and forged	650	1202	975	141	1261	183	25	38
IN-100								
Hot isostatically pressed	650	1202	1286	187	942	137	...	21
Hot isostatically pressed and forged	650	1202	1200	174	1000	145	...	8
Hot isostatically pressed and extruded	650	1202	1350	196	1000	145	...	18

(a) 1120 °C (2050 °F) at 103 MPa (15 ksi) for 3 h, solution treated at 1150 °C (2100 °F) 1 h, hot salt quench to 535 °C (1000 °F), aged at 870 °C (1600 °F) 1 h, then 650 °C (1200 °F) 24 h; air cooled. Source: Ref 7

Table 7 Selected powder metallurgy superalloy properties

Property	MERL 76(a)	RSR 185(b)	IN-100(c)	LC Astroloy(d)	Astroloy(c)	René 95(e)	IN-718(f)
Condition	Gatorized	As-extruded	...	HIPed(g)	...	HIPed at 1121 °C (2050 °F)(g)	RS(h)
Grain size, μm	20	16–20	5	...	5	...	...
Properties at 25 °C (77 °F)							
0.2% yield strength, MPa (ksi)	1035 (150)	1380 (200)	940 (136)	932 (135)	936 (135)	1215 (176)	1240 (180)
Ultimate tensile strength, MPa (ksi)	1505 (218)	1860 (270)	1130 (164)	1380 (200)	1393 (202)	1636 (237)	1450 (210)
Elongation, %	38	8	8	26	...	...	18
Reduction in area, %	30	...	...	...	...	15	33.5
Properties at 649 °C (1200 °F)							
0.2% yield strength, MPa (ksi)	1050 (152)	...	1080 (157)	863 (125)	1025 (149)	1120 (162)	1035 (150)
Ultimate tensile strength, MPa (ksi)	1276 (185)	...	1290 (187)	1290 (187)	1300 (189)	1514 (220)	1173 (170)
Elongation, %	20	...	16	25	25	17	20
Reduction in area, %	20	...	...	...	...	...	32
Properties at 704 °C (1300 °F)							
0.2% yield strength, MPa (ksi)	1050 (152)	1104 (160)	1065 (154)	...	1030 (149)	...	...
Ultimate tensile strength, MPa (ksi)	1320 (191)	1310 (190)	1270 (184)	...	1160 (168)	...	...
Elongation, %	16	...	20	...	24	...	...
Reduction in area, %	23	...	...	...	...	...	...

(a) Source: Ref 10. (b) Source: Ref 11. (c) Source: Ref 12. (d) Source: Ref 13. (e) Source: Ref 14. (f) Source: Ref 15. (g) HIPed, hot isostatically pressed. (h) RS, rapidly solidified

metallographically. Blisks for small gas-turbine engines have the most current potential.

Mechanically Alloyed Superalloys

Mechanical alloying was originally developed for the manufacture of nickel-base superalloys strengthened by both an oxide dispersion and γ' precipitate. It provides a means for producing P/M dispersion-strengthened alloys of varying compositions with a unique set of properties. Commercial alloy compositions, which are listed in Table 8, are based on either nickel-chromium or iron-chromium matrices. The most common mechanically alloyed (MA) oxide-dispersion-strengthened (ODS) alloys are MA 754, MA 758, MA 956, MA 6000, and MA 760.

It has been known for some time that the strength of metals at high temperature could be increased by the addition of a dispersion of fine refractory oxides. While many methods can produce such dispersions in simple metal systems, these techniques are not applicable to the production of the more highly alloyed materials required for gas turbines and critical industrial applications. Conventional P/M techniques, for example, either do not produce an adequate dispersion or do not permit the use of reactive elements such as aluminum and chromium. These elements confer beneficial characteristics, including corrosion resistance and intermediate temperature strength. In contrast, the mechanical alloying process was developed to introduce a fine inert oxide dispersion into superalloy matrices that contain reactive alloying elements.

Applications for Mechanically Alloyed Alloys

Mechanically alloyed ODS alloys were used first in aircraft gas turbine engines and later in industrial turbines. Components include vane airfoils and platforms, blades, nozzles, and combustor/augmentor assemblies. As experience was gained with production, fabrication, and use of the alloys, this knowledge was applied to the manufacture of component parts in numerous industries. These include diesel-engine glow plugs, heat treatment fixtures (including shields, baskets, trays, mesh belts, and skid rails for steel plate and billet heating furnaces), burner hardware for coal- and oil-fired power stations, gas sampling tubes, thermocouple tubes, and a wide variety of components used in the production or handling of molten glass.

Mechanical Alloying Process

The mechanical alloying process may be defined as a method for producing composite metal powders with a controlled microstructure. The process involves repeated fracturing and rewelding of a mixture of powder particles in a highly energetic ball mill charge. On a commercial scale, the process is carried out in vertical attritors or horizontal ball mills.

During each collision of the grinding balls, many powder particles are trapped and plastically deformed. The process is illustrated schematically in Fig. 16. Sufficient deformation occurs to rupture any absorbed surface-contaminant film and expose clean metal surfaces. Cold welds are formed where metal particles overlay, producing composite metal particles. At the same time, other powder particles are fractured. Figure 16

Fig. 13 Dual-property turbine wheel made from cast blades and a powder metallurgy hub. Source: Ref 16

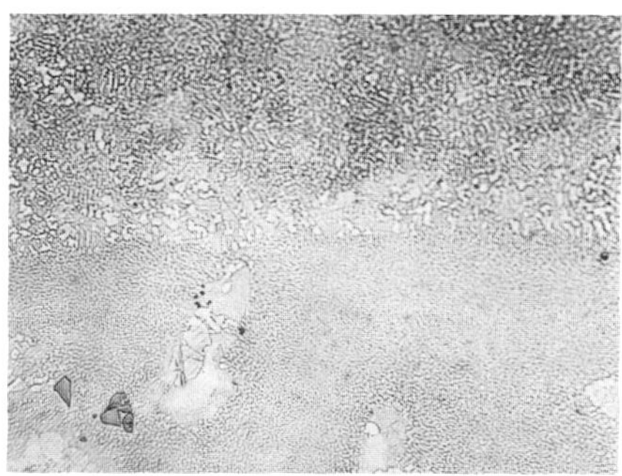

Fig. 14 Microstructure at hot isostatically pressed bond line between cast alloy C103 (top) and powder metallurgy alloy PA-101 (bottom). Source: Ref 16

Fig. 15 Dual-property turbine wheel made by bonding (hot isostatically pressing) a fine-grain hub to a coarse-grain directionally solidified cast blade ring

shows two metallic constituents, indicated by light and dark particles, although in a commercial alloy there may be several constituents (Fig. 17).

As the process progresses, most of the particles become microcomposites, similar to the one produced in the collision illustrated in Fig. 16. The cold welding (which increases the size of the particles involved) and the fracturing of the particles (which reduces particle size) reach a steady-state balance, resulting in a relatively coarse and stable overall particle size. The internal structure of the particles, however, is continually refined by the repeated plastic deformation.

Production of ODS Superalloy Powders. The production of ODS alloys requires the development of a uniform distribution of submicron refractory oxide particles in a highly alloyed matrix. The powder mixture must be more varied in composition and particle size than indicated schematically in Fig. 16. A typical powder mixture may consist of fine (4 to 7 μm) nickel powder, −150 μm chromium powder, and −150 μm master alloy (nickel-titanium-aluminum). The master alloy may contain a wide range of elements selected for their roles as alloying constituents or for gettering of contaminants. About 2 vol% of very fine yttria (25 nm, or 250 Å) is added to form the dispersoid. The yttria becomes entrapped along the weld interfaces between fragments in the composite metal powders. After completion of the powder milling, a uniform interparticle spacing of about 0.5 μm (20 μin.) is achieved.

Consolidation and Property Development. The production of powder containing a uniform dispersion of fine refractory oxide particles in a superalloy matrix is only the first step in achieving the full potential of ODS alloys. The powder must be consolidated and worked under conditions that develop coarse grains during a secondary recrystallization heat treatment. Figure 18 shows the key operations in the production sequence of a selected product. After mechanical alloying, the powder is put into low-carbon steel tubes and extruded to full density. Extrusion temperature and reduction are critical because the final microstructure of the product is affected, and the ability to control this microstructure is important. For nickel-base alloys, reduction ratios of 13 to 1 and temperatures of 1065 to 1120 °C (1950 to 2050 °F) are typical.

Table 8 Nominal composition of selected mechanically alloyed materials

Alloy designation	Ni	Fe	Cr	Al	Ti	W	Mo	Ta	Y_2O_3	C	B	Zr
MA 754	bal	...	20	0.3	0.5	...	...	...	0.6	0.05	...	...
MA 758	bal	...	30	0.3	0.5	...	...	...	0.6	0.05	...	...
MA 760	bal	...	20	6.0	...	3.5	2.0	...	0.95	0.05	0.01	0.15
MA 6000	bal	...	15	4.5	2.5	4.0	2.0	2.0	1.1	0.05	0.01	0.15
MA 956	...	bal	20	4.5	0.5	...	...	...	0.5	0.05	...	...

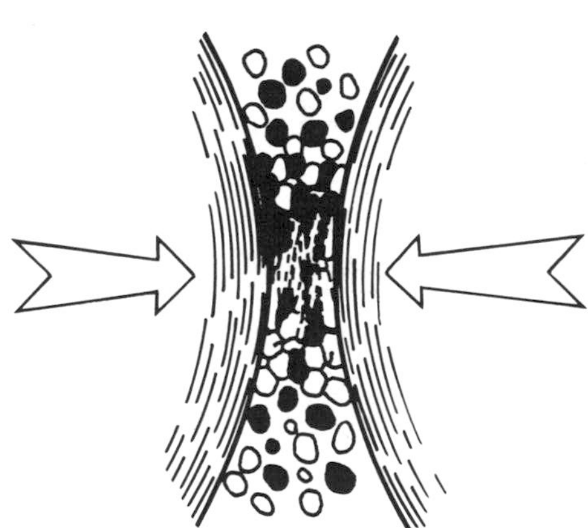

Fig. 16 Schematic depicting the formation of composite powder particles at an early stage in the mechanical alloying process

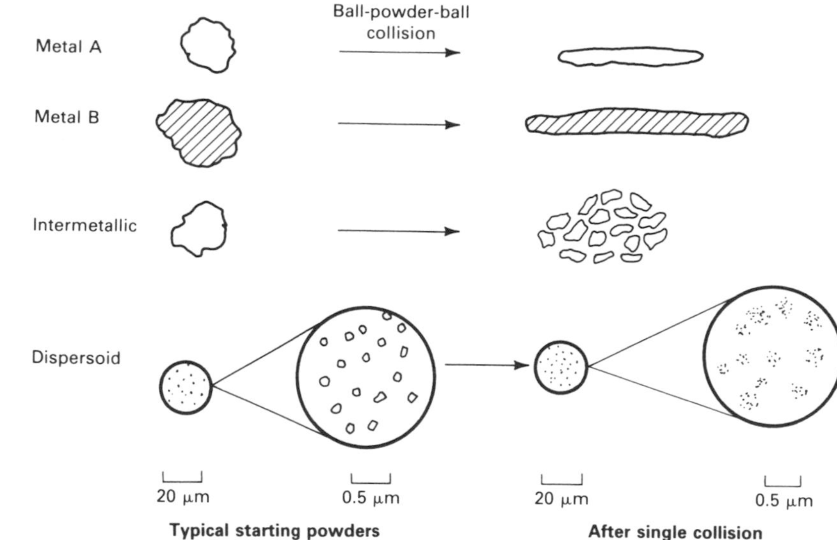

Fig. 17 Representative constituents of starting powders used in mechanical alloying, showing deformation characteristics during ball milling

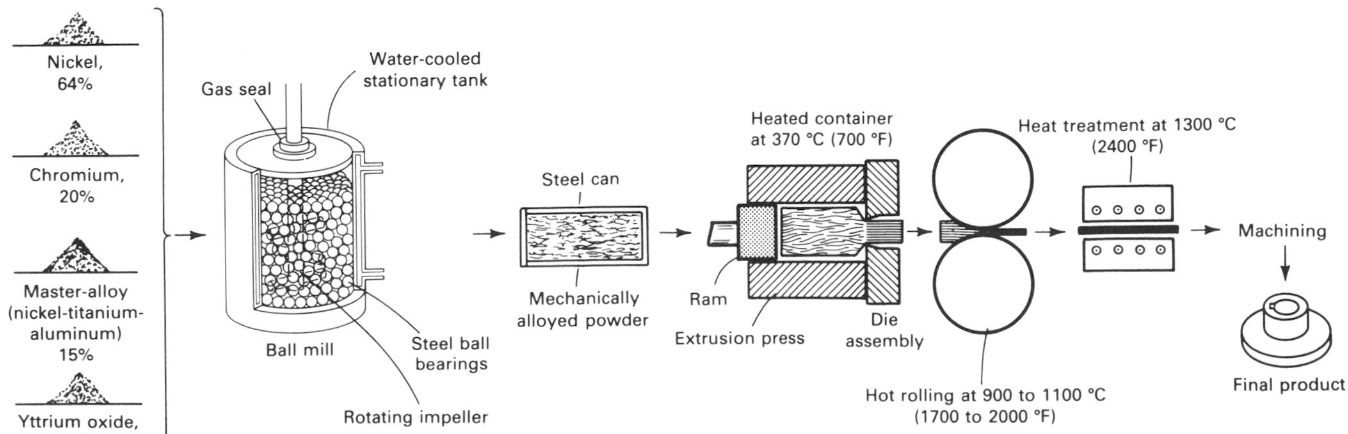

Fig. 18 Schematic showing typical process operations used in the production of mechanically alloyed oxide-dispersion-strengthened products

The extruded stock, now fully consolidated, is then hot rolled into mill shapes. As with extrusion, the thermomechanical treatment followed during rolling is critical to the final microstructure and properties of the product. For iron- and nickel-base superalloys, rolling temperatures are normally between 950 and 1065 °C (1750 and 1950 °F), with rolling resulting in highly directional working of the product.

Mechanically alloyed ODS alloys are now commercially available as bar, plate, sheet, tube, wire, shapes, and forgings. However, not all alloys are available in all product forms. Table 9 lists available product forms for various alloys. Property data for various bar products are listed in Table 10.

Effect of Grain Structure. As noted previously, the final properties of mechanically alloyed ODS alloys are dependent on the grain structure as well as the presence of the fine dispersoid. In most products, a grain-coarsening anneal at about 1315 °C (2400 °F) is provided after fabrication. In bar stock, for example, relatively coarse grains are formed that are elongated in the direction of extrusion and working. This elongated structure is necessary for achievement of maximum elevated-temperature properties. Grain aspect ratio (the average grain dimension parallel to the applied stress, divided by the average grain dimension perpendicular to the applied stress) has a strong effect on elevated-temperature stress-rupture properties (Fig. 19). For many bar products, a high grain aspect ratio is desirable. Plate and sheet products tend to exhibit pancake-shape grains. Through careful control of processing conditions, these grains can be made to be equiaxed in the plane of the sheet, thus providing nearly isotropic properties in the plane of the sheet. Structures for tubing and forgings may be more complex.

Alloy MA 754

Alloy MA 754 was the first mechanically alloyed ODS superalloy to be produced on a large scale. This material is basically a Ni-20Cr alloy strengthened by about 1 vol% Y_2O_3 (see Table 8). It is comparable to thoria-dispersed (TD) Ni-Cr (an earlier ODS alloy strengthened by thoria, ThO_2), but it has a nonradioactive dispersoid.

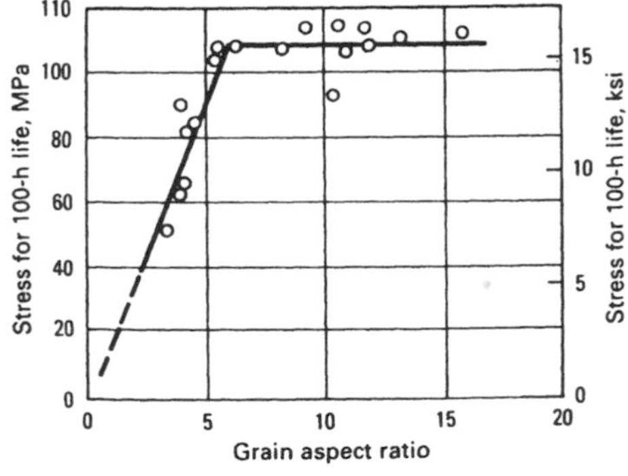

Fig. 19 Influence of grain aspect ratio on stress for 100 h life for MA 753 at 1040 °C (1900 °F). Source: Inco Alloys Internat., Inc.

Table 9 Available product forms for mechanically alloyed oxide-dispersion-strengthened alloys

Product form	Alloy MA 956	Alloy MA 754	Alloy MA 758	Alloy MA 6000
Hot finished rounds	X	X	X	...
Hot finished flats	X	X	X	X
Extruded rounds	X	X	X	X
Extruded section	X	X	...	X
Extruded tube	X	...	...	...
Hot rolled plate	X	X	X	...
Hot rolled sheet	X	...	...	...
Cold rolled sheet	X	...	...	...
Cold drawn round	X	...	...	...
Cold drawn tube	X	...	...	...
Hot rolled wire	X	X	...	...
Cold drawn wire	X	...	...	...

Source: Inco Alloys International, Inc.

Table 10 Material properties of oxide-dispersion-strengthened superalloys

Property	Inconel MA 754(a)	Incoloy MA 956(b)	Inconel MA 6000(c)
Mechanical			
Ultimate tensile strength, MPa (ksi)			
At 21 °C (70 °F), longitudinal	965 (140)	645 (94)	1295 (188)
At 540 °C (1000 °F), longitudinal	760 (110)	370 (54)	1155 (168)
At 1095 °C (2000 °F)			
Longitudinal(d)	148 (21.5)	91 (13.2)	222 (32.2)
Transverse(d)	131 (19)	90 (13.0)	177 (25.7)
Yield strength, 0.2% offset			
At 21 °C (70 °F), longitudinal	585 (85)	555 (80)	1285 (186)
At 540 °C (1000 °F), longitudinal	515 (75)	285 (41)	1010 (147)
At 1095 °C (2000 °F)			
Longitudinal(d)	134 (19.5)	84.8 (12.3)	192 (27.8)
Transverse(d)	121 (17.5)	82.7 (12.0)	170 (24.7)
Elongation, %			
At 21 °C (70 °F), longitudinal	21	10	4
At 540 °C (1000 °F), longitudinal	19	20	6
At 1095 °C (2000 °F)			
Longitudinal(d)	12.5	3.5	9.0
Transverse(d)	3.5	4.0	2.0
Reduction in area at 1095 °C (2000 °F), %			
Longitudinal(d)	24	...	31.0
Transverse(d)	1.5	...	1.0
1000 h rupture strength, MPa (ksi)			
At 650 °C (1200 °F)	255 (37)	110(e) (16)	...
At 980 °C (1800 °F)	130 (19)	65 (10)	185 (27)
Physical			
Melting range, °C (°F)	...	1480(f)(g) (2700)	...
Specific heat capacity at 21 °C (70 °F), J/kg · K (Btu/lb · °F)	...	469(g) (0.112)	...
Thermal conductivity at 21 °C (70 °F), W/m · K (Btu/ft^2 · in. · h · °F)	...	10.9(g) (76)	...
Mean coefficient of thermal expansion at 538 °C (1000 °F), 10^{-6}/K	...	11.3(g)	...
Electrical resistivity, nΩ · m	...	1310(g)	...

(a) Data for bar form. Condition of test material was 1315 °C (2400 °F)/1 h/air cooled (AC). (b) Data for sheet. Condition of test material was 1300 °C (2375 °F)/1 h/AC. (c) Data for bar. Condition of test material was 1230 °C (2250 °F)/0.5 h/AC + 955 °C (1750 °F)/2 h/AC + 845 °C (1550 °F)/24 h/AC. (d) Ref 3. (e) At 760 °C (1400 °F). (f) Approximate solidus temperature. (g) Ref 17

Table 11 Stress-rupture properties of alloy MA 754 bars

Temperature		Longitudinal, stress to produce rupture in:				Long transverse, stress to produce rupture in:			
		100 h		1000 h		100 h		1000 h	
°C	°F	MPa	ksi	MPa	ksi	MPa	ksi	MPa	ksi
650	1200	284	41.2	256	37.2	241	35.0	208	30.2
760	1400	214	31.1	199	28.8	172	25.0	149	21.6
870	1600	170	24.7	158	22.9	108	15.6	91	13.2
980	1800	136	19.7	129	18.7	63	9.1	46	6.6
1095	2000	102	14.8	94	13.6	38	5.5	24	3.5
1150	2100	90	13.1	78	11.3	23	3.4	17	2.4

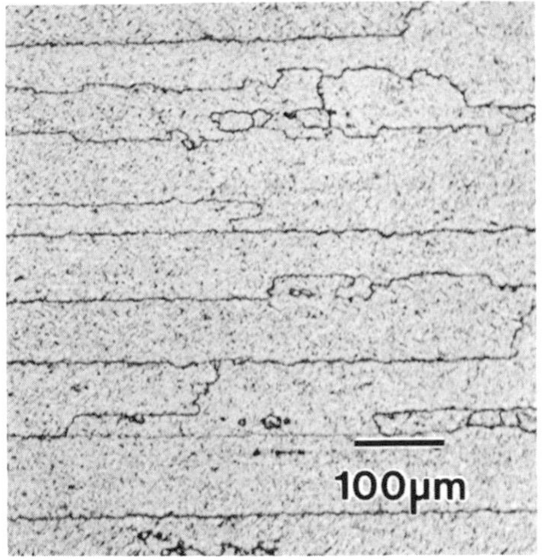

(a)

(b)

Fig. 20 Alloy MA 754 microstructure shown from two different views. (a) Longitudinal. (b) Transverse. Note high grain aspect ratio shown in the longitudinal section. Grain aspect ratios may be as high as 10 to 1.

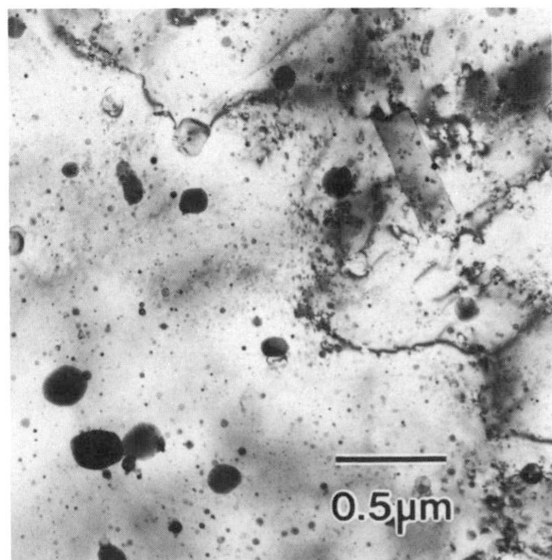

Fig. 21 Transmission electron micrograph of alloy MA 754, showing uniform distribution of fine oxides and scattered coarser carbonitrides

Because of its higher strength, it has been extensively used for aircraft gas turbine vanes and high-temperature test fixtures.

Microstructure. The microstructure of a commercially produced rectangular bar shows the elongation of the grains along the direction of working. Grain width in the long transverse direction is somewhat greater than the grain thickness. The details of the grain structure in the longitudinal and transverse sections are shown in Fig. 20. The longitudinal view shows the maximum and minimum grain dimensions, whereas the transverse view shows the extreme irregularity of grain boundaries typical of ODS materials. Although it is not obvious from the photomicrograph, this alloy possesses a strong (100) crystallographic texture in the longitudinal direction. This texture has been associated with optimum thermal fatigue resistance.

The oxide dispersoid distribution in MA 754 is shown in Fig. 21. The very fine, dark particles are the uniform dispersion of stable yttrium aluminates formed by the reaction between the added yttria, excess oxygen in the powder, and aluminum added to the getter oxygen. The larger dark particles are titanium carbonitrides.

Elevated-Temperature Strength. The tensile properties of MA 754 bar are shown in Fig. 22(a),

Table 12 Physical properties of selected mechanically alloyed oxide-dispersion-strengthened materials

Alloy	Melting point		Modulus of elasticity		Mass density		Coefficient of expansion, at 20 to 980 °C (70 to 1800 °F)	
	°C	°F	GPa	psi × 10^6	g/cm^3	lb/in.^{3}	μm/m · K	μin./in. · °F
MA 754	1400	2550	151	22	8.3	0.30	16.9	9.41
MA 956	1480	2700	269	39.0	7.2	0.26	14.8	8.22
MA 6000	1296–1375	2365–2507	203	29.4	8.11	0.29	16.7	9.3

Table 13 Stress-rupture properties of alloy MA 956 sheet

Temperature		Longitudinal, Stress to produce rupture in: 10 h		100 h		1000 h		Transverse, Stress to produce rupture in: 10 h		100 h		1000 h	
°C	°F	MPa	ksi	MPa	ksi	MPa	ksi	MPa	ksi	MPa	ksi	MPa	ksi
980	1800	84	12.2	75	10.9	67	9.7	72	10.4	70	10.2	63	9.1
1100	2010	64	9.3	57	8.3	51	7.4	64	9.3	57	8.3	...	...
1150	2100	50	7.3	39	5.7	...	...	31	4.5	24	3.5	...	...

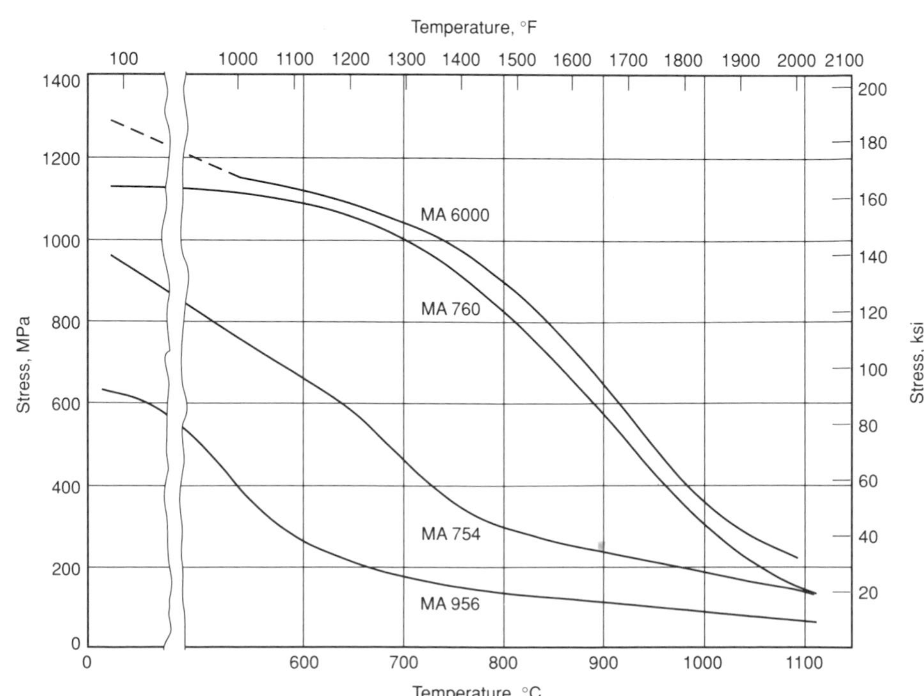

Fig. 22(a) Effect of temperature on the tensile strength of selected mechanically alloyed oxide-dispersion-strengthened alloys. Data are for the longitudinal direction.

22(b), and 22(c). The properties shown are for the longitudinal direction. Long transverse strength is similar, but ductility is considerably lower.

In Fig. 23, the 1050 °C (2000 °F) longitudinal stress-rupture properties of MA 754 bar are compared to those of a TD Ni-Cr alloy, TD nickel bar, alloy MAR-M 509 (a cast cobalt-base alloy), and alloy 80A, a conventional nickel-base alloy having a composition similar to the matrix of MA 754. MA 754, like other ODS materials, has a very flat log stress-log rupture life slope compared to conventional alloys.

The elevated-temperature stress-rupture properties of MA 754 bar are dependent on testing direction, as indicated in Table 11. The rupture-stress capability in the longitudinal direction is consistently higher than that in the long transverse direction, reflecting the differences in grain aspect ratio in the two directions. When MA 754 is produced as cross-rolled plate with coarse equiaxed pancake grains, equal longitudinal and transverse stress-rupture properties are observed. In this form, the rupture strength is about 80% that of the longitudinal bar.

Physical Properties. Important physical properties of alloy MA 754 are given in Table 12. The relatively high melting point, 1400 °C (2550 °F), and low room-temperature modulus of elasticity in the longitudinal direction, 151 GPa (22 × 10^6 psi), are especially important. The low modulus, indicating a strong (100) crystallographic texture in the direction of the long grain dimension, has been shown to give superior thermal fatigue resistance.

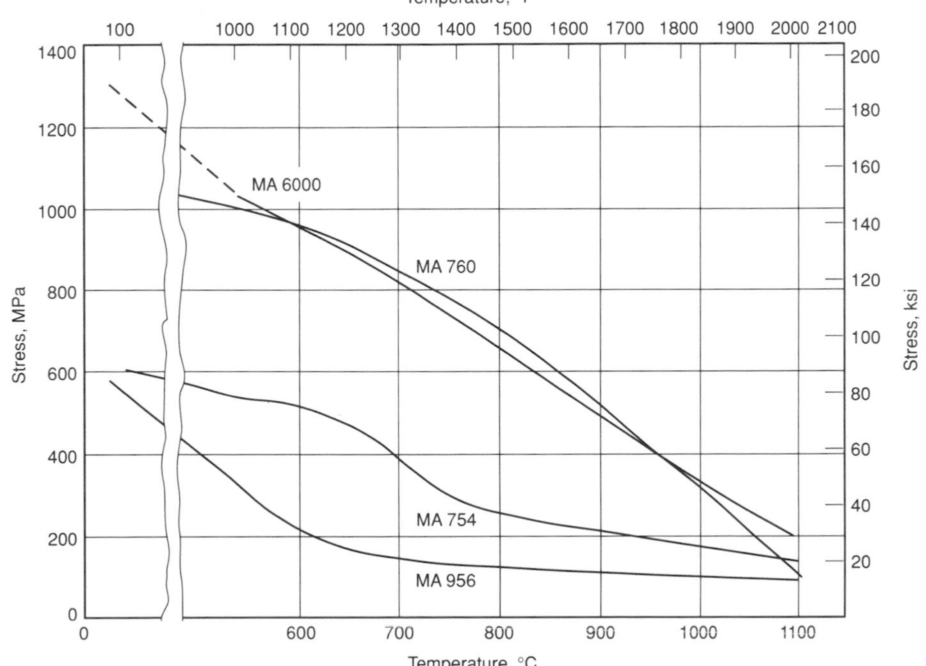

Fig. 22(b) Effect of temperature on the yield strength (0.2% offset) of selected mechanically alloyed oxide-dispersion-strengthened alloys. Data are for the longitudinal direction.

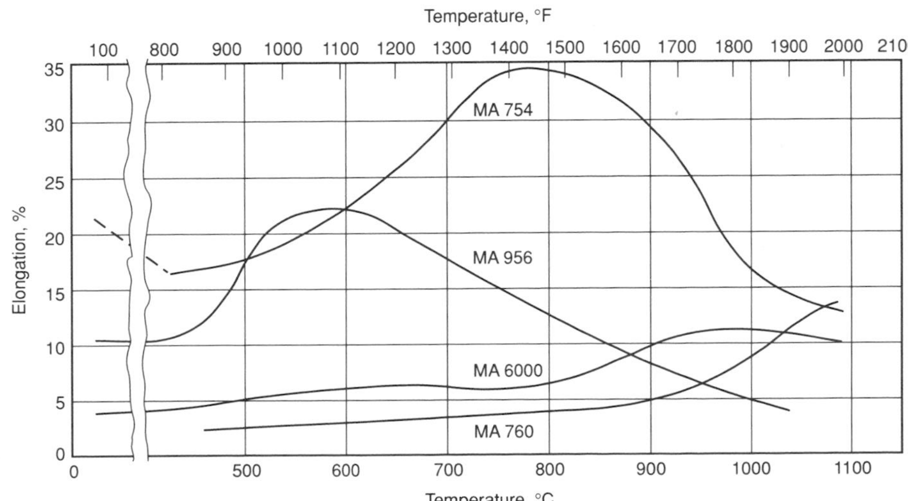

Fig. 22(c) Effect of temperature on the elongation of selected mechanically alloyed oxide-dispersion-strengthened alloys. Data are for the longitudinal direction.

Alloy MA 758

Alloy MA 758 is a higher-chromium version of MA 754 (see Table 8). This alloy was developed for applications in which the higher chromium content is needed for greater oxidation resistance. The mechanical properties of this alloy are similar to those of MA 754 when identical product forms and grain structures are compared. This alloy has found applications in the thermal processing industry and the glass processing industry.

Alloy MA 956

The production of alloy MA 956 demonstrates the ability to add large amounts of metallic aluminum by mechanical alloying (see Table 8). This material is a ferritic iron-chromium-aluminum alloy, dispersion-strengthened with yttrium aluminates formed by the addition of about 1 vol% of yttria. Because of its generally good hot and cold fabricability, MA 956 has been produced in the widest range of product forms of any mechanically alloyed ODS alloy (see Table 9). In sheet form, this alloy is produced by a sequence of hot and cold working, which yields large pancake-shape grains following heat treatment. This grain structure ensures excellent isotropic properties in the plane of the sheet. MA 956 is used in the heat treatment industry for furnace fixturing, racks, baskets, and burner nozzles. It also is used in advanced aerospace sheet and bar components, where good oxidation and sulfidation resistance are required in addition to high-temperature strength properties.

Mechanical Properties. The tensile properties of MA 956 are shown in Fig. 22(a), 22(b), and 22(c). The tensile strength of this alloy is quite a bit lower than that of the other MA materials at low temperatures. However, the strength-versus-temperature curve is extremely flat; the strength of this alloy exceeds that of all non-ODS sheet materials at approximately 1095 °C (2000 °F).

Table 13 gives the stress-rupture properties of MA 956 at elevated temperatures in both the longitudinal and transverse directions.

Physical Properties. Alloy MA 956 has a very high melting point (1480 °C, or 2700 °F), a relatively low density (7.2 g/cm^3, or 0.26 lb/in.^{3}) compared to competitive materials, and a relatively low thermal expansion coefficient (see Table 12). This combination of properties makes the alloy well suited for sheet applications such as gas turbine combustion chambers.

Alloy MA 6000

The composition of alloy MA 6000 is similar to that of more sophisticated cast and wrought γ'-strengthened superalloys. It contains a critical balance of elements to produce strength at intermediate and elevated temperatures, along with

Table 14 Weld transverse properties of selected oxide-dispersion-strengthened alloys

ODS alloy	Condition(a)	Welding process(b)	Joint type	Filler metal	Postweld heat treatment(c) °C	Postweld heat treatment(c) °F	Test temperature °C	Test temperature °F	Stress MPa	Stress ksi	Life(e), strength, h	Longitudinal base-metal approximate %	Comment
MA 956	FG	FRW	Butt	...	1315	2400	982	1800	77.9	11.3	HTT	75	...
	CG	FRW	Butt	...	1315	2400	982	1800	6.9	1.0	24.0	8	...
	FG	FRW	Butt	...	1315	2400	982	1800	6.9/13.8(d)	1.0/2.0	13.0	17	...
MA 758	CG	GMAW	Butt	52	...	...	1093	2000	13.8	2.0	40.4	11	...
	CG	GMAW	Butt	52	...	...	1093	2000	6.9	1.0	1601	8	...
	CG	GMAW	Butt	52	1315	2400	1093	2000	13.8	2.0	16.5	10	...
	CG	GMAW	Butt	52	1315	2400	1093	2000	6.9	1.0	259.9	7	...
MA 956	CG	EBW	Butt	...	...	...	982	1800	13.8	2.0	5.4	15	90° butt
	CG	EBW	Butt	...	...	...	982	1800	27.6	4.0	38	33	5° butt
	CG	EBW	Single lap	...	...	...	1093	2000	13.8	2.0	710	25	Double weld
	CG	EXW	Single lap	...	1100(f)	2012(f)	1100	2012	18/28/38(d)	2.6/4.1/5.5	14.0	58	Stress is base metal
	FG	EXW	Single lap	...	1315	2400	1100	2012	18/28/38(d)	2.6/4.1/5.5	8.0	56	Stress is base metal
	CG	EXW	Single lap	...	...	...	1100	2012	18	2.6	2.0	25	Stress is base metal
	FG	RSW	Double lap	...	(g)	(g)	1093	2000	37.9	5.5	HTT	42	Single-spot stress
	FG	RSW	Double lap	...	(g)	(g)	982	1800	44.1(h)	6.4(h)	HTT	43	Single-spot stress

(a) FG, fine grain; CG, coarse grain. (b) FRW, friction welding; GMAW, gas-metal arc welding; EBW, electron-beam welding; EXW, explosion welding; RSW, resistance spot welding. (c) 1 h, except where noted. (d) Step-loaded test. Stress was increased every 24 h until failure. (e) HTT, high-temperature tensile test. (f) 10 h. (g) 1204 °C (2200 °F) for 24 h plus 1315 °C (2400 °F) for 1 h. (h) Attachment pin sheared through sheet

oxidation and hot-corrosion resistance. Alloy MA 6000 combines γ′ hardening from its aluminum, titanium, and tantalum content (for intermediate strength) with oxide-dispersion-strengthening from a yttria addition (for strength and stability at very high temperatures). Oxidation resistance comes from its aluminum and chromium contents, while titanium, tantalum, chromium, and tungsten act in concert to provide sulfidation resistance. The tungsten and molybdenum also act as solid-solution strengtheners in this alloy. MA 6000 is an ideal alloy for gas turbine vanes and blades where exceptional high-temperature strength is required.

Microstructure. Alloy MA 6000 has a highly elongated coarse-grain structure that results from thermomechanical processing followed by high-temperature annealing. Zone annealing is useful in achieving the optimum grain aspect ratio for this alloy. As a result, the only product forms presently available are bar or small forgings.

Zone annealing is performed by slowly passing a heating element down the axis of the bar. In practice, either resistance or induction heating can be used. A typical zone annealing speed is 100 mm/h (4 in./h).

The microstructure of MA 6000 is shown in Fig. 24. Note the high volume fraction of γ′ (45 to 50 vol%), and the very fine dispersoid particles present in both the γ′ (dark irregular particles) and lighter matrix.

Mechanical Properties. Figure 25 compares the elevated-temperature properties of alloy MA 6000 (the specific rupture strength for 1000 h life as a function of temperature) with those of a TD nickel alloy bar and directionally solidified alloy DS MAR-M 200 containing hafnium. This diagram clearly shows the effect of the two strengthening mechanisms in alloy MA 6000. At intermediate temperatures, around 815 °C (1500 °F), the strength of MA 6000 approaches that of the com-

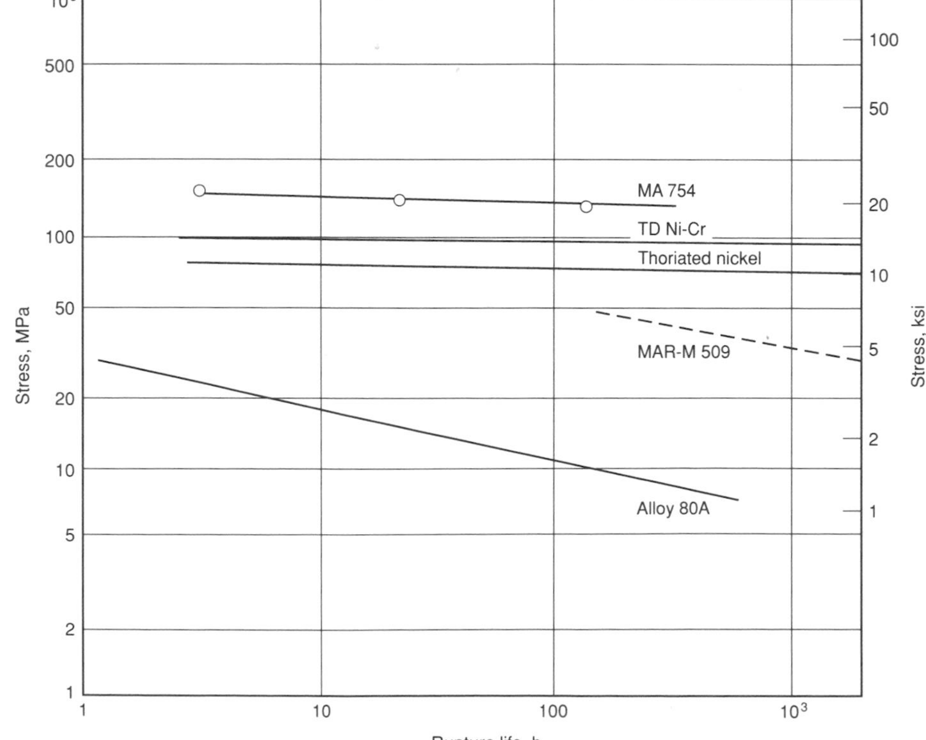

Fig. 23 Comparison of stress-rupture properties of alloy MA 754 bar to other alloy material bars at 1095 °C (2000 °F)

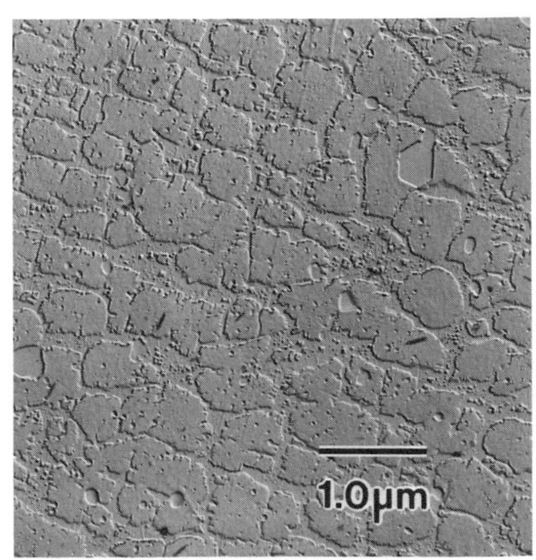

Fig. 24 Microstructure of heat-treated alloy MA 6000, showing high-volume fraction γ′ and dispersoid phases

plex, highly alloyed DS MAR-M 200 containing hafnium, and it is almost four times that of an unalloyed ODS metal such as TD nickel. At high temperatures (~1095 °C, or 2000 °F), where the alloy DS MAR-M 200 containing hafnium has lost most of its strength due to growth and dissolution of its γ' precipitate, MA 6000 has useful strength due to the presence of the oxide dispersion. At temperatures between these extremes, the strength of MA 6000 is superior to both the cast nickel-base superalloy and the TD nickel, because the two strengthening mechanisms supplement one another.

Physical properties of MA 6000 are listed in Table 9.

Alloy MA 760

Alloy MA 760 is an age-hardened nickel-base alloy with a composition designed to provide a balance of high-temperature strength, long-term structural stability, and oxidation resistance. Its primary use is expected to be for industrial gas turbines (it is currently in limited production). The composition of this alloy is shown in Table 8. It is similar to MA 6000 in that its strength is supplemented by γ' age hardening. Its properties also benefit from zone annealing to give coarse elongated grains. The stress-rupture properties of alloy MA 760 exceed those of MA 754 but are exceeded by those of MA 6000 (see Fig. 22 and 25).

Oxidation and Hot-Corrosion Properties

The mechanically alloyed ODS alloys are normally used uncoated at very high temperatures in hostile environments, so their resistance to oxidation, carburization, sulfidation, and oxide fluxing is important. While all of the alloys are resistant to the effects of these deleterious chemical processes, the relative resistance varies with the specific alloy and environment.

Figure 26 shows the resistance of the mechanically alloyed ODS alloys to cyclic oxidation at various temperatures. All of the alloys form protective scales at 1000 °C (1830 °F) and thus are highly resistant to oxidation. At higher temperatures, the relatively low-chromium MA 6000 shows increased weight loss, whereas the other alloys are highly stable. Under the most severe conditions, MA 6000 would require coating for long-time exposure. The oxidation resistance of MA 956 is unsurpassed by any existing commercial ODS sheet alloy. However, accelerated oxidation occurs in this alloy during long-time exposure at temperatures above 1200 °C (2190 °F), depending on the environment.

Evaluation of oxidation-sulfidation resistance of gas turbine alloys is frequently done in a burner rig test. Representative data for selected alloys tested at 927 °C (1700 °F) is shown in Fig. 27. MA 956 exhibits extremely high resistance to this form of attack. MA 966, although less resistant than MA 956, was comparable to the cast alloy IN-738 in this test.

The mechanically alloyed ODS materials have also been evaluated in a wide range of specialized environments. MA 956 has proved to be especially resistant to carburization, as illustrated in Fig. 28.

These alloys also show excellent resistance to attack by molten C glass and lime glass, which have the following compositions:

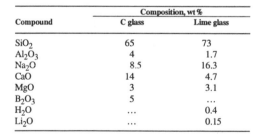

Compound	Composition, wt%	
	C glass	Lime glass
SiO_2	65	73
Al_2O_3	4	1.7
Na_2O	8.5	16.3
CaO	14	4.7
MgO	3	3.1
B_2O_3	5	...
H_2O	...	0.4
Li_2O	...	0.15

Based on a 5-day immersion test, MA 754 and MA 758 demonstrate high corrosion resistance to molten C glass:

Alloy	Metal loss	
	mm	mil
MA 754	0.04	1.6
MA 758	0.03	1.2

Based on a 240 h immersion test in lime glass at 1150 °C (2100 °F), MA 754 has corrosion-resistance properties intermediate between those of MA 956 and MA 758:

Alloy	Mass change	
	$Mg \cdot cm^{-2}$	$lb \cdot in.^{-2} \times 10^{-4}$
MA 956	4	0.57
MA 754	28	4.0
MA758	42	6.0

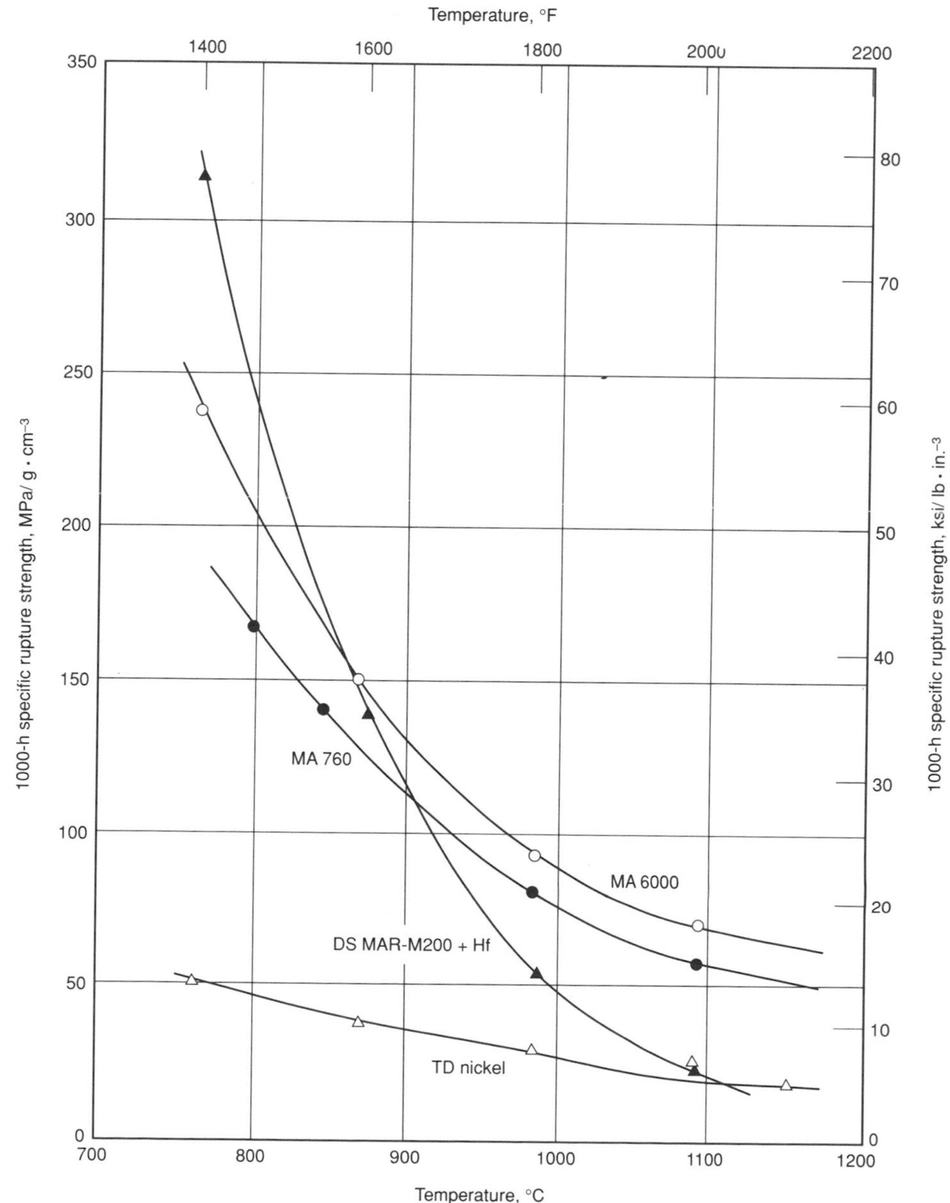

Fig. 25 Effect of temperature on the 100 h specific rupture strength of MA 760, MA 6000, DS MAR-M 200, and TD nickel

It is well known that the relative performance of alloys in glass depends on glass composition, temperature, impurity level, velocity, and other factors. Consequently, the tables above should be considered only illustrative of the generally high resistance of mechanically alloyed ODS alloys in molten glass environments.

Fabrication of Mechanically Alloyed ODS Alloys

The mill product forms of mechanically alloyed ODS alloys vary (Table 9), depending on factors such as ease of fabrication and applicable forming methods.

Bars. All of the alloys are available as bars, and much of the data reported in the literature refer to bar properties. All of the bar products can be precision forged, and MA 754 forged airfoils have been in commercial use for years. The high-temperature properties of forgings can be equivalent to those of annealed bar, provided that care is taken in the design of the part and the thermomechanical processing is controlled to produce the desired grain structure and orientation. Forgings of MA 6000 and MA 760 with optimal properties in the airfoil axis can be obtained by zone annealing after forging. Both seamless and flat-butt-welded rings with desired properties in the hoop direction have been made from MA 754 and MA 758.

Plate products are available for MA 754, MA 758, and MA 956. Equiaxed properties can be obtained through control of rolling conditions. Plate is readily amenable to a variety of hot-forming operations, including hot shear spinning. Optimal formability and minimum flow stress is obtained when the plate is in the fine-grain (unrecrystallized) condition. The standard grain-coarsening anneal is then applied to the formed component.

Sheet. The only alloy currently available in sheet form is MA 956. This material, which is readily cold rolled to standard sheet tolerance, is commercially available in gages down to thicknesses of 0.25 mm (0.010 in.) and widths up to 610 mm (24 in.). A wide variety of components have been cold formed from MA sheet by standard metal-forming operations. Experience has shown that warming to about 95 °C (200 °F) is necessary to prevent cracking, because this alloy undergoes a ductile-to-brittle transition in the vicinity of room temperature.

Additional Product Forms. MA 956 has also been produced in a number of other forms for special applications, including pipe, thin-wall tube, and fine wire. MA 754 has also been produced as hot-rolled wire.

Joining of Mechanically Alloyed ODS Alloys

Many applications for mechanically alloyed ODS alloys require some method of joining. Procedures that involve fusion of the base metal destroy the unique microstructure that is responsible for the high-temperature strength of these alloys. Accordingly, fusion welds that are needed for attachment or positioning for brazing should be located in areas of relatively low stress.

Factors affecting the elevated-temperature strength of welded ODS alloys are discussed below. More detailed information can be found in Volume 6, *Welding, Brazing, and Soldering*, of the *ASM Handbook*.

Weld Design. The high-temperature creep strength obtained from a cast weld structure, using ODS alloys, is generally comparable to that obtained with conventional wrought alloys and does not exhibit the same exceptional high-temperature creep strength of unwelded ODS materials. The stable oxides, or dispersoids, that are used in the ODS alloy systems have essentially no equilibrium solubility in the liquid phase. If melting occurs, then the dispersoids rapidly agglomerate, and buoyancy floats the oxide to an upper surface, if possible. In fusion welds, this results in the loss of the primary means of high-temperature strengthening. Agglomerated semicontinuous dispersoids in the partially melted zones of fusion welds are also suspected of contributing to early weld-service failures.

A number of successful strategies have been employed in joining ODS alloys. Conventional fusion welds or brazes are often placed in cooler locations in order to retain acceptable strength levels. A design is often altered so that conventional fusion welds are not the principal load-carrying members, but are attachments instead. Solid-state joining techniques can often be used, but the resulting grain structure must also be considered.

Grain Structure and Weldment Properties. At very high temperatures, grain boundaries become the weak link, so to speak, and fracture becomes intergranular. In order to minimize such a grain-boundary strength limitation, ODS alloys are produced with a macro grain size, where the grain-boundary area normal to the principal stresses are minimized without incurring the difficulties inherent in producing single crystals. As described above, this elongated grain structure is often referred to as a high grain aspect ratio. Welds or joints that significantly interrupt this large interlocked grain structure will fail prematurely at extreme temperatures, even if they are produced by solid-state means. A friction weld in a rod, where there is a straight grain-boundary interface through the thickness and normal to the

Table 15 Properties of furnace-brazed oxide-dispersion-strengthened alloys

ODS alloy	Joint type	Filler metal	Test temperature °C	Test temperature °F	Stress(a) MPa	Stress(a) ksi	Life, h	Longitudinal base-metal strength, approximate %	Comment
MA 754	Double lap	B93(b)	982	1800	110	16	392	85	Sheared braze
	Double lap	AM788(c)	982	1800	110	16	168	84	Sheared braze and base metal
	Double lap	AM788	1093	2000	41.4	6	178	40	Sheared braze
	Double lap	TD6(d)	1093	2000	56.2	8	105.5	53	Sheared braze
MA 956	Single lap (8t)	B93	982	1800	41.4	6	89.4	30	Sheared braze
	Single lap (8t)	AM788	982	1800	48.3	7	23.5	58	Sheared braze

(a) Stresses are believed to be based on base metal. (b) General Electric alloy B50TF108, available in powder form from Alloy Metals, Inc. Composition: 14% Cr, 9.5% Co, 4.9% Ti, 4% W, 4% Mo, 3% Al, 4.5% Si, 0.7% B, bal Ni. (c) Available in powder form from Alloy Metals, Inc. Composition: 22% Cr, 21% Ni, 14% W, 2% B, 0.03% La, bal Co. (d) General Electric alloy B5DTF83, available in powder form. Composition: 16% Cr, 17% Mo, 5% W, 5% Si, 1% Al, bal Ni

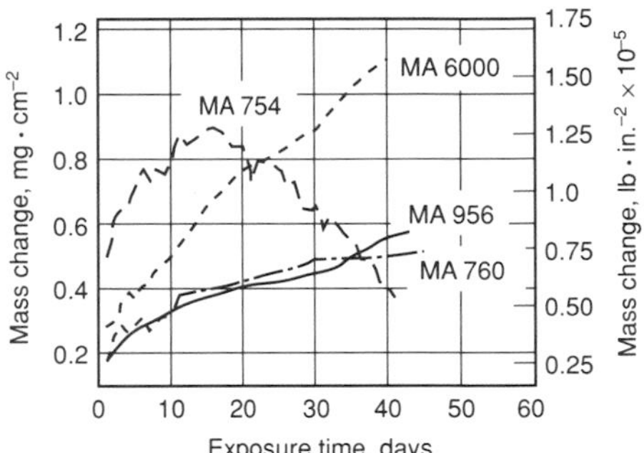

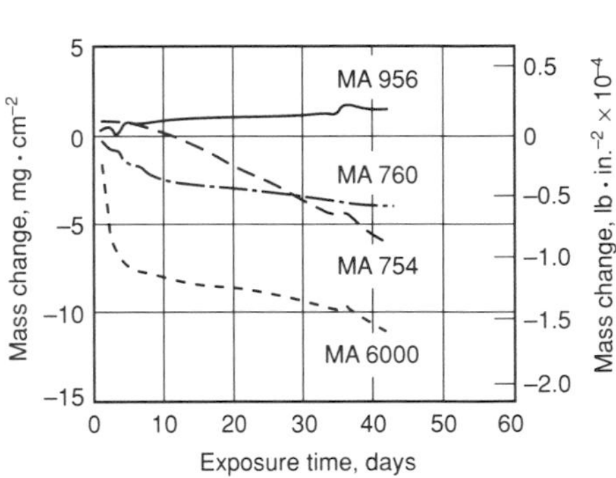

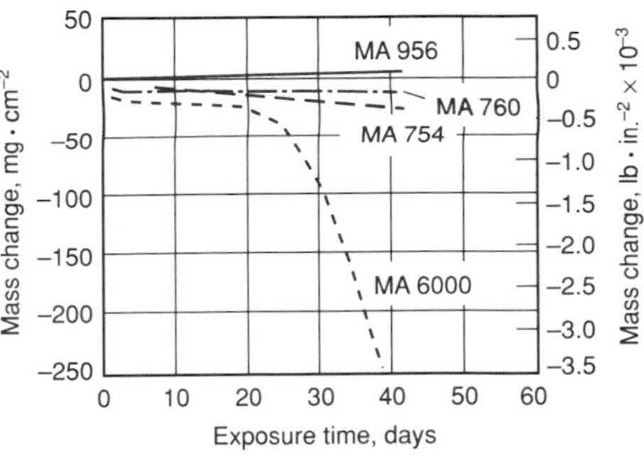

Fig. 26 Effect of temperature on mass change for four mechanically alloyed materials exposed to air containing 5% H_2O vapor. (a) 1000 °C (1830 °F). (b) 1100 °C (2010 °F). (c) 1200 °C (2190 °F)

principal stresses, typically will not perform significantly better than a conventional fusion weld in which a similar structure is produced. The large, interlocking grain structure in such materials is a key attribute for good performance at extreme temperatures.

Because of grain structure considerations, it is often preferable to use resistance welds, braze welds, or both in a lap joint, rather than a butt joint, where there is invariably a through-thickness noninterlocked grain structure normal to the principal stresses across the joint. When lap welds are used, double-lap joints outperform single-lap joints, which fail by peeling at lower stresses.

Postweld recrystallization anneals have been used with all types of welds to grow grains across the welds to the greatest extent possible. The best structural and strength results are generally obtained by joining base metals in a fine, unrecrystallized condition and by performing a grain-coarsening recrystallization anneal after welding. However, a postweld recrystallization anneal typically improves properties only in narrow joints, such as those commonly produced by electron-beam, diffusion, resistance, and friction welding techniques. Creep-rupture properties from wide-gap joints, such as those produced in gas-metal or gas-tungsten arc welds, can be reduced by such an anneal. This anneal is typically 1316 °C (2400 °F) for 1 h, when MA 754 and MA 956 alloys are used. The fine-grain approach, if adopted, requires coordination with materials suppliers, because most ODS alloys are typically supplied in the coarse-grain, recrystallized condition. However, some measure of improved properties may be obtained in narrow joints through postweld recrystallization, even where the base metal was joined in the coarse-grain condition. Table 14 provides transverse property data on selected alloys joined by various fusion and nonfusion (solid-state) welding processes.

Applicable Fusion Welding Methods. Procedures such as gas-tungsten arc welding (GTAW), gas-metal arc welding (GMAW), electron-beam welding (EBW), and pulsed laser-beam welding (LBW) have all been used successfully with ODS alloys on a limited scale. MA 956 sheet assemblies have also been made using resistance spot welding (RSW). Table 14 provides elevated-temperature properties of ODS weldments produced by GMAW, EBW, and RSW.

Applicable Solid-State Welding Methods. As might be expected, nonfusion solid-state processes are required in order to obtain tensile and stress-rupture properties approaching those of the parent metal. Riveting operations using similar alloy rivets have also been applied for nonaircraft applications.

Friction welding (FRW) is capable of creating sound solid-state joints in ODS alloys. However, the plastic deformation and flow at the bond interface typically produce a relatively straight, flat grain-boundary interface through-thickness after annealing, which is undesirable because it is normal to the principal stresses. It produces properties that are no better than those of most fusion processes. The best results have been obtained by welding fine-grain unrecrystallized rod and then performing a postweld recrystallization anneal. Properties of ODS weldments produced by FRW are listed in Table 14.

Explosion welding (EXW) generally creates a favorable grain structure by virtue of the lap joints used. In a limited study in which MA 956 was joined, the best results were obtained by bonding coarse-grain material and giving it a postweld diffusion treatment of 1100 °C (2012 °F) for 10 h, as indicated in Table 14.

Furnace brazing of ODS alloys has been extensively used by the aircraft industry. Iron- and nickel-base ODS alloys containing significant aluminum or titanium are typically plated with a very thin barrier of nickel to prevent oxidation of these elements during the brazing cycle. If oxidation occurs, then the flow of the brazing filler metals is inhibited.

Mechanically alloyed ODS alloys that are not strengthened by γ' precipitates (e.g., MA 754 and MA 956) are the easiest to braze. Vacuum, hydrogen, or inert atmospheres can be used for brazing. Prebraze cleaning consists of grinding or machining the faying surfaces and washing with a solvent that evaporates without leaving a residue. Generally, brazing temperatures should not exceed 1315 °C (2400 °F) unless demanded by a specific application that has been well examined and tested. The brazing filler metals for use with these ODS alloys usually are not classified by the American Welding Society (AWS). In most cases, the filler metals used with these alloys have brazing temperatures in excess of 1230 °C (2250 °F). These include proprietary alloys that are based on nickel, cobalt, gold, or palladium.

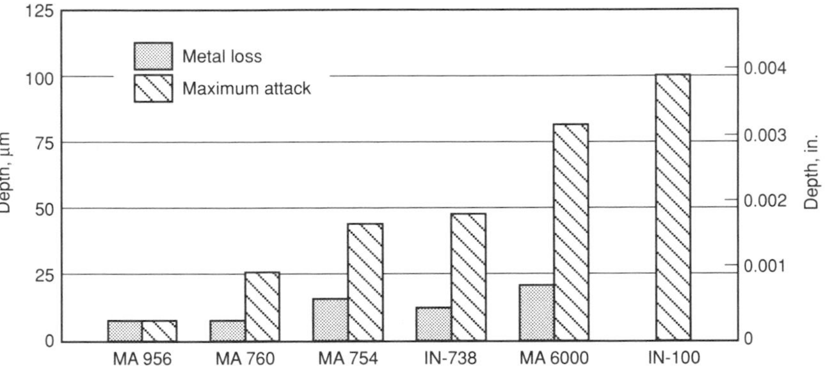

Fig. 27 Comparison of the oxidation-sulfidation resistance of mechanically alloyed oxide-dispersion-strengthened alloys with that of superalloys IN-738 and IN-100. Tested in a burner rig for 500 h at 925 °C (1700 °F) using an air-to-fuel ratio that varied from 27:1 to 21:1. JP-5 fuel contained 0.3% S. Temperature test cycle consisted of the alloy held at temperature for 1 h and then cooled for 3 min. There are no metal loss data for IN-100 because the sample was destroyed in 50 h. Metal loss is defined as loss of diameter due to oxide and sulfide scale formation. Maximum attack is defined as loss of diameter due to internal oxidation and sulfidation.

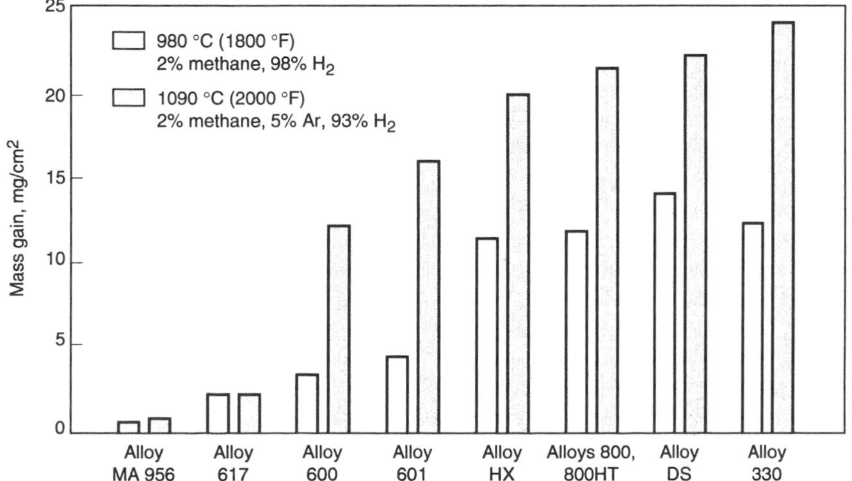

Fig. 28 Comparison of the carburization resistance of MA 956 with that of conventionally processed wrought nickel-base and iron-nickel-base alloys. Source: Inco Alloys Internat., Inc.

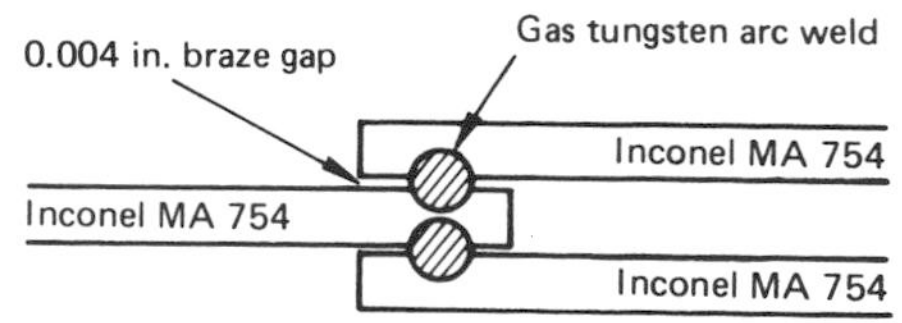

Fig. 29 Brazing specimen assembly used to evaluate stress-rupture properties of brazed MA 754

Brazements made of ODS alloys to be used at elevated temperature must be tested at elevated temperature to prove fitness for purpose. In the case of stress-rupture testing, AWS specification C3.2 may be used as a guide because it gives the actual joint configuration. The configuration shown in Fig. 29 is preferred by some as a test model, although any test configuration without stress raisers is adequate. The elevated-temperature brazement properties for MA 754 should meet the following requirements:

Shear stress		Temperature		Service
MPa	ksi	°C	°F	Life, h
26	3.8	980	1900	>1000
9.0	1.3	1095	2000	>1000

Actual properties of furnace-brazed ODS alloys are shown in Table 15. Attempts to improve properties have made use of special techniques, such as diffusion brazing, transient liquid-phase bonding, and pressure brazing.

The nickel-base ODS alloy that is most difficult to braze is MA 6000. The amount of alloying elements in the base metal matrix plus the γ' precipitations cause a formidable problem. MA 6000 has a solidus temperature of 1300 °C (2372 °F); therefore, the brazing temperature should be no higher than 1250 °C (2282 °F). Additionally, because 1230 °C (2250 °F) is the γ' solution treatment temperature, it becomes important to select the brazing filler metal carefully and to heat treat the assembly after brazing. The AWS-specified BNi and BCo filler metals, as well as specially formulated filler metals, have been used for this alloy.

MA 6000 is used for its high-temperature strength and corrosion resistance; unfortunately, the passive oxide scale that provides good corrosion resistance also prevents wetting and flow of brazing filler metal. Therefore, correct cleaning procedures are very important. Surfaces to be brazed should be mechanically cleaned with a water-cooled, low-speed belt or wheel of approximately 320 grit and stored in a solvent, such as methanol, until immediately before the beginning of the brazing cycle.

Cobalt P/M Superalloys

Cobalt-base alloys produced by P/M processing are generally not used in heat-resistant applications. They are more commonly used in corrosion-resistant applications (e.g., Co-Cr-Mo alloys used for orthopedic implants) and wear-resistant applications (e.g., cutting tools and hardfacing alloys), although cobalt hardfacing alloys also provide high-temperature corrosion resistance. The powder processing of cobalt alloys is similar to that of nickel-base P/M alloys (i.e., spherical powders produced by gas atomization are consolidated by HIP).

ACKNOWLEDGMENTS

The information in this article is largely taken from:

- N.S. Stoloff, Wrought and P/M Superalloys, *Properties and Selection: Irons, Steels, and High-Performance Alloys,* Vol 1, *ASM Handbook* (formerly Vol 1, 10th ed., *Metals Handbook*), ASM International, 1990, p 950–980
- J.J. deBarbadillo and J.J. Fischer, Dispersion-Strengthened Nickel-Base and Iron-Base Alloys, *Properties and Selection: Nonferrous Alloys and Special-Purpose Materials,* Vol 2, *ASM Handbook* (formerly Vol 2, 10th ed., *Metals Handbook*), ASM International, 1990, p 943-949
- P.E. Price, Hot Isostatic Pressing of Metal Powders, *Powder Metallurgy,* Vol 7, *ASM Handbook* (formerly Vol 7, 9th ed., *Metals Handbook*), American Society for Metals, 1984, p 419-450
- B.L. Ferguson, Forging and Rolling of P/M Billets, *Powder Metallurgy,* Vol 7, *ASM Handbook* (formerly Vol 7, 9th ed., *Metals Handbook*), American Society for Metals, 1984, p 522-529
- B.L. Ferguson, Aerospace Applications, *Powder Metallurgy,* Vol 7, *ASM Handbook* (formerly Vol 7, 9th ed., *Metals Handbook*), American Society for Metals, 1984, p 646-656
- D. O'Donnell, Joining of Oxide-Dispersion-Strengthened Materials, *Welding, Brazing, and Soldering,* Vol 6, *ASM Handbook,* ASM International, 1993, p 1037-1040
- D. Manente, Brazing of Heat-Resistant Alloys, Low-Alloy Steels, and Tool Steels, *Welding, Brazing, and Soldering,* Vol 6, *ASM Handbook,* ASM International, 1993, p 924-930

REFERENCES

1. G.H. Gessinger, *Powder Metallurgy of Superalloys,* Butterworths, 1984
2. G.H. Gessinger, *Powder Metall. Int.,* Vol 13, 1981, p 93
3. S. Reichman and D.S. Chang, in *Superalloys II,* C.T. Sims, N.S. Stoloff, and W.C. Hagel, Ed., John Wiley & Sons, 1987, p 459
4. R.V. Miner and J. Gayda, *Int. J. Fatigue,* Vol 6 (No. 3), 1984, p 189
5. C. Symonds et al., Properties and Structures of Hot Isostatic Pressed and Hot Isostatic Pressed plus Forged Superalloys, *Powder Metall. Int.,* Vol 15 (No. 1), 1983, p 35
6. J.L. Bartos, P/M Superalloys for Military Gas Turbine Applications, *Powder Metallurgy in Defense Technology,* Vol 5, Metal Powder Industries Federation, Princeton, NJ, 1980, p 81–113
7. R. Evans, "Review of European Powder Metallurgy of Superalloys," Report No. EOARD-TR-80-10, prepared for the U.S. Air Force by the University of Swansea, 1979
8. G.I. Friedman and G.S. Ansell, in *The Superalloys,* C.T. Sims and W.C. Hagel, Ed., John Wiley & Sons, 1972, p 427
9. B.A. Cowles, D.L. Sims, and J.R. Warren, NASA-CR-159409, National Aeronautics and Space Administration, 1978
10. H. Caless and D.F. Paulonis, in *Superalloys 1988,* The Metallurgical Society, 1988, p 101–110
11. D.B. Miracle, K.A. Williams, and H.A. Lipsitt, in *Rapid Solidification Processing: Principles and Technologies III,* Claitor's Publishing Division, Baton Rouge, LA, 1982, p 234–239
12. C. Ducrocq, A. Lasalmonie, and Y. Honnorat, in *Superalloys 1988,* The Metallurgical Society, 1988, p 63–72
13. R.M. Pelloux and J.S. Huang, in *Creep-Fatigue Environment Interactions,* The Metallurgical Society of AIME, 1980, p 151–164
14. S.J. Choe, S.V. Golwalker, D.J. Duquette, and N.S. Stoloff, in *Superalloys 1984,* The Metallurgical Society of AIME, 1984, p 309–318
15. J.F. Radovich and D.J. Myers, in *Superalloys 1984,* The Metallurgical Society of AIME, 1984, p 347-356
16. J.H. Moll, J.H. Schwertz, and V.K. Chandhok, P/M Dual-Property Wheels for Small Engines, *Progress in Powder Metallurgy,* Vol 37, Metal Powder Industries Federation, Princeton, NJ, 1982
17. "Product Handbook," Publication 1A1-38, Inco Alloys International, Inc., 1988

SELECTED REFERENCES

- E. Arzt and L. Schultz, Ed., *New Materials by Mechanical Alloying Techniques,* Deutsche Gesellschaft fur Metallkunde, Oberursel, West Germany, 1989
- J.S. Benjamin, Dispersion Strengthened Superalloys by Mechanical Alloying, *Met. Trans.,* Vol 1, 1970, p 2943–2951
- R.C. Benn and G.M. McColvin, The Development of ODS Superalloys for Industrial Gas Turbines, in *Superalloys 1988,* S. Reichman, D.N. Duhl, G. Maurer, S. Antolovich, and C. Lund, Ed., The Metallurgical Society, 1988, p 75–80
- L.R. Curwick, The Mechanical Alloying Process: Powder to Mill Product, *Frontiers of High Temperature Alloys I: Proc. First Int. Conf. on Oxide Dispersion Strengthened Alloys by Mechanical Alloying,* J.S. Benjamin, Ed., Inco Alloys International, Inc., 1981, p 310
- J.J. Fischer, I. Astley, and J.P. Morse, *Proc. Third Internat. Symp. on Superalloys: Metallurgy and Manufacture* (Seven Springs, PA), Sept 1976, p 361–371
- J.J. Fischer and R.M. Haeberle, Commercial Status of Mechanically Alloyed Materials, *Mod. Devel. Powder Metall.,* Vol 18–21, Metal Powder Industries Federation, 1988, p 461–477
- Frontiers of High Temperature Materials, in *Proc. Internat. Conf. on Oxide Dispersion Strengthened Superalloys by Mechanical Alloying,* Inco Alloys International, Inc., 1981
- Frontiers of High Temperature Materials II, in *Proc. Second Internat. Conf. on Oxide Dispersion Strengthened Superalloys by*

Mechanical Alloying, J.S. Benjamin and R.C. Benn, Ed., Inco Alloys International, Inc., 1983

- G.A. Hack, Fundamentals of Mechanical Alloying, *Frontiers of High Temperature Alloys II: Proc. Second Internat. Conf. on Oxide Dispersion Strengthened Alloys by Mechanical Alloying,* J.S. Benjamin and R.C. Benn, Ed., Inco Alloys International, Inc., 1983, p 3–18
- H.D. Hedrich, H.C. Mayer, G. Haufler, M. Koph, and N. Reheis, "Joining of ODS-Superalloys," Paper 91-GT-411, The Internat. Gas Turbine and Aeroengine Congress and Exposition (Orlando, FL), American Society of Mechanical Engineers, June 1991
- T.J. Kelly, Joining Mechanical Alloys for Fabrication, *Frontiers of High Temperature Alloys II: Proc. Second Internat. Conf. on Oxide Dispersion Strengthened Alloys by Mechanical Alloying,* I.S. Benjamin and R.C. Benn, Ed., Inco Alloys International, Inc., 1983, p 129–148
- T.J. Kelly, Welding of Mechanically Alloyed ODS Materials, *Trends in Welding Research in the United States,* S.A. David, Ed., American Society for Metals, 1982, p 471–488
- R. Petkovic-Luton, D.-J. Sroiovitz, and M.J. Luton, Microstructural Aspects of Creep in Oxide Dispersion Strengthened Alloys, *Frontiers of High Temperature Alloys II: Proc. Second Internat. Conf. on Oxide Dispersion Strengthened Alloys by Mechanical Alloying,* I.S. Benjamin and R.C. Benn, Ed., Inco Alloys International, Inc., 1983, p 73–97
- G.D. Smith and J.J. Fischer, High Temperature Corrosion Resistance of Mechanically Alloyed Products in Gas Turbine Environments, *Proc. 1990 ASAFE Turbo Exposition* (Brussels, Belgium), American Society of Mechanical Engineers
- J.H. Weber, High-Temperature Oxide Dispersion Strengthened Alloys, The 1980s: Pay Off Decade for Advanced Materials, *SAMPE J.,* Vol 25, 1980

Effect of Heat Treating on Superalloy Properties

THE HIGH-TEMPERATURE STRENGTH of all superalloys is based on the principle of a stable face-centered cubic (fcc) matrix combined with either precipitation strengthening and/or solid-solution hardening. In age-hardenable nickel-base alloys, the γ' intermetallic (Ni_3Al,Ti) is generally present for strengthening, while the nonhardenable nickel-, cobalt-, and iron-base alloys rely on solid-solution strengthening of the fcc (γ) matrix. Iron-base and nickel-iron superalloys may also develop, in addition to γ', second-phase strengthening from the γ'' (Ni_3Nb) intermetallic and perhaps η (Ni_3Ti). Cobalt-base superalloys may develop some precipitation strengthening from carbides (Cr_7C_3, $M_{23}C_6$), but no intermetallic-phase strengthening equal to γ' strengthening in nickel-base alloys has been discovered in cobalt-base superalloys.

This article describes the procedures for heat treatment of superalloys and the resulting properties. The first part briefly reviews the heat-treating processes and the special considerations required by superalloys. The later sections describe these processes for both the wrought and cast versions of the age-hardenable and solid-solution-strengthening superalloys. Solid-solution-strengthened iron-, nickel-, and cobalt-base superalloys are generally distinguishable from the precipitation-strengthened (age-hardenable) superalloys by their relatively low content of precipitate-forming elements such as aluminum, titanium, or niobium. There are, of course, some exceptions to this, particularly as regards niobium content. Typical compositions for precipitation-strengthened and solid-solution-strengthened superalloys are given in the article "Metallurgy and Processing of Superalloys" in this Volume.

Overview of Heat-Treating Operations

Commonly employed heat treatment procedures for superalloys include stress relieving, annealing, solution treating and quenching, and aging. Each of these will be briefly described below. Additional information on these operations can be found in the sections of this article that describe heat treating of:

- Precipitation-strengthened nickel-iron-base superalloys
- Precipitation-strengthened nickel-base superalloys
- Solution-strengthened iron-, nickel-, and cobalt-base superalloys

Stress Relieving

Stress relieving of superalloys frequently entails compromise between maximum relief of residual stress and effects deleterious to high-temperature properties and/or corrosion resistance. True stress relieving of wrought material is usually, but not always, confined to non-age-hardenable alloys because a full-solution treatment is useful prior to age hardening. In addition, the stress-relieving temperature would typically fall within the upper temperature range for age hardening.

Time and temperature cycles vary considerably depending on the metallurgical characteristics of the alloy and on the type and magnitude of the residual stresses developed in the prior fabrication processes. Stress-relieving temperatures are usually below the annealing or recrystallization temperatures.

Annealing

When applied to superalloys, annealing implies full annealing, that is, complete recrystallization and attaining maximum softness. The practice is usually applied to nonhardening wrought alloys. For most of the age-hardenable alloys, the annealing cycles are the same as for solution treating. However, the two treatments serve different purposes. Annealing is used mainly to reduce hardness and increase ductility to facilitate forming or machining, prepare for welding, relieve stresses after welding, produce specific microstructures, or soften age-hardened structures by resolution of second phases. Solution treating is intended to dissolve second phases to produce maximum corrosion resistance or to prepare for age hardening.

Most wrought superalloys can be cold formed but are more difficult to form than austenitic stainless steels. Severe cold-forming operations may require several intermediate annealing operations. Full annealing must be followed by fast cooling.

Annealing should be conducted after welding the age-hardenable alloys if highly restrained joints are involved. For alloys sensitive to strain age cracking, heating rates must be rapid. If the configuration of the weldment does not permit annealing, aging may be used for stress relieving in alloys not prone to strain age cracking.

Reheating for hot working is similar to annealing in that the aim is to promote adequate formability of the metal being deformed. Control of temperature can be critical to resultant properties as varying degrees of recrystallization and control of grain growth may be desired. In most standard operations, heating or reheating for hot working is a full annealing step with recrystallization and dissolution of all or most secondary phases. Additional discussion of controlled heating for hot working can be found in the section "Thermomechanical Processing" in this article.

Solution Treating and Quenching

Solution Treating. The first step in heat treating superalloys is usually solution treatment. In some wrought alloys, the solution-treating temperature will depend on the properties desired. A higher temperature is specified for optimum creep-rupture properties; a lower temperature is used for optimum short-time tensile properties at elevated temperature, improved fatigue resistance (via finer grain size), or improved resistance to notch rupture sensitivity.

The higher solution-treating temperature will result in some grain growth and more extensive dissolving of carbides. The principal objective is to put hardening phases into solution and dissolve some carbides. After aging, the resulting microstructure of these wrought alloys consists of large grains that contain the principal aging phases (γ', γ'', η) and a heavy concentration of carbides in the grain boundaries. The lower solution-treating temperature dissolves the principal aging phases without grain growth or significant carbide solution.

For some wrought superalloys (such as Nimonic 80A and Nimonic 90), an intermediate

Table 1 Typical solution-treating and aging cycles for wrought superalloys

Alloy	Solution treating: Temperature, °C	Solution treating: Temperature, °F	Solution treating: Time, h	Solution treating: Cooling procedure	Aging: Temperature, °C	Aging: Temperature, °F	Aging: Time, h	Aging: Cooling procedure
Iron-base alloys								
A-286	980	1800	1	Oil quench	720	1325	16	Air cool
Discaloy	1010	1850	2	Oil quench	730	1350	20	Air cool
					650	1200	20	Air cool
N-155	1165–1190	2125–2175	1	Water quench	815	1500	4	Air cool
Incoloy 903	845	1550	1	Water quench	720	1325	8	Furnace cool
					620	1150	8	Air cool
Incoloy 907	980	1800	1	Air cool	775	1425	12	Furnace cool
					620	1150	8	Air cool
Incoloy 909	980	1800	1	Air cool	720	1325	8	Furnace cool
					620	1150	8	Air cool
Incoloy 925	1010	1850	1	Air cool	730(a)	1350(a)	8	Furnace cool
					620	1150	8	Air cool
Nickel-base alloys								
Astroloy	1175	2150	4	Air cool	845	1550	24	Air cool
	1080	1975	4	Air cool	760	1400	16	Air cool
Custom Age 625 PLUS	1038	1900	1	Air cool	720	1325	8	Furnace cool
					620	1150	8	Air cool
Inconel 901	1095	2000	2	Water quench	790	1450	2	Air cool
					720	1325	24	Air cool
Inconel 625	1150	2100	2	(b)	...	...	...	...
Inconel 706	925–1010	1700–1850	...	...	845	1550	3	Air cool
					720	1325	8	Furnace cool
					620	1150	8	Air cool
Inconel 706(c)	980	1800	1	Air cool	730	1350	8	Furnace cool
					620	1150	8	Air cool
Inconel 718	980	1800	1	Air cool	720	1325	8	Furnace cool
					620	1150	8	Air cool
Inconel 725	1040	1900	1	Air cool	730(a)	1350	8	Furnace cool
					620	1150	8	Air cool
Inconel X-750	1150	2100	2	Air cool	845	1550	24	Air cool
					705	1300	20	Air cool
Nimonic 80A	1080	1975	8	Air cool	705	1300	16	Air cool
Nimonic 90	1080	1975	8	Air cool	705	1300	16	Air cool
René 41	1065	1950	1/2	Air cool	760	1400	16	Air cool
Udimet 500	1080	1975	4	Air cool	845	1550	24	Air cool
					760	1400	16	Air cool
Udimet 700	1175	2150	4	Air cool	845	1550	24	Air cool
	1080	1975	4	Air cool	760	1400	16	Air cool
Waspaloy	1080	1975	4	Air cool	845	1550	24	Air cool
Cobalt-base alloys								
S 816	1175	2150	1	(b)	760	1400	12	Air cool

Note: Alternate treatments may be used to improve specific properties. (a) If furnace size/load prohibits fast heat up to initial age temperature, a controlled ramp up from 590 to 730 °C (1100 to 1350 °F) is recommended. (b) To provide adequate quenching after solution treating, it is necessary to cool below about 540 °C (1000 °F) rapidly enough to prevent precipitation in the intermediate temperature range. For sheet metal parts of most alloys, rapid air cooling will suffice. Oil or water quenching is frequently required for heavier sections that are not subject to cracking. (c) Heat treatment of Inconel 706 to enhance tensile properties instead of creep resistance for tensile-limited applications

solution-treating temperature is selected to produce a compromise of the properties. For other alloys (such as Udimet 500 and Udimet 700), the intermediate-temperature aging treatment is used to tailor the grain boundaries for improved creep-rupture properties.

Quenching. The purpose of quenching after solution treating is to maintain at room temperature the supersaturated solid solution obtained during solution treating. Quenching permits a finer age-hardening precipitate size. Cooling methods commonly used include oil and water quenching as well as various forms of air or inert gas cooling. Internal stresses resulting from quenching can also accelerate overaging in some age-hardenable alloys.

Aging Treatments

Aging treatments strengthen age-hardenable alloys by causing the precipitation of additional quantities of one or more phases from the supersaturated matrix that is developed by solution treating. Factors that influence the selection and number of aging steps and aging temperature include:

- Type and number of precipitating phases available
- Anticipated service temperature
- Precipitate size
- The combination of strength and ductility desired and heat treatment of similar alloys

Typical heat treatments of wrought age-hardenable superalloys are given in Table 1. Because dimensional changes can occur during aging, it is recommended that finish machining be done after aging.

Aging Precipitates. Principal aging phases in the superalloys usually include one or more of the following: γ' (Ni_3Al or N_3Al,Ti), η (Ni_3Ti), or γ'' (body-centered tetragonal Ni_3Nb). Secondary phases that may be present include: carbides ($M_{23}C_6$, M_7C_3, M_6C, and MC), nitrides (MN), carbonitrides (MCN), and borides (M_3B_2), as well as Laves phase (M_2Ti) and δ phase (orthorhombic Ni_3Nb). The above phases occur principally in nickel-base alloys. The primary phases in cobalt-base alloys are $M_{23}C_6$, M_7C_3, M_6C, and MC. The primary phases in iron-base superalloys will be similar to those in nickel alloys, although η is more apt to be found because the Ti-to-Al ratio is generally higher in iron alloys than in nickel alloys.

When more than one phase is capable of precipitating from the alloy matrix, judicious selection of a single aging temperature (or a double aging treatment that produces different sizes and types of precipitates at different temperatures) is important. The principal strengthening precipitates are γ' or γ'', while other precipitates such as η, δ, or Laves phase can provide grain size control in nickel-iron superalloys (see, for example, the next section "Thermomechanical Processing"). In some alloys, if composition has not been carefully controlled, undesirable phases, such as σ and μ, can also form either during heat treatment or, more commonly, during service. These precipitates, known as topologically close-packed (TCP) phases, are composed of close-packed layers of atoms parallel to {111} planes of the γ matrix. Usually harmful, they may appear as long plates or needles, often nucleating on grain-boundary carbides. Alloys containing a high level of body-centered cubic (bcc) transition metals (tantalum, niobium, chromium, tungsten, and molybdenum) are most susceptible to TCP phase formation. More detailed information on the constituents observed in superalloys can be found in the article "Metallurgy and Processing of Superalloys" in this Volume.

The size distribution of precipitates is affected by aging temperature. Exposure to temperatures higher than the optimum aging temperature results in a decrease in strength through the process of averaging (coarsening of precipitates); at still higher temperatures, resolution may occur. High aging temperatures will produce coarser γ′ particles than lower temperatures and result in higher creep-rupture properties. For optimum short-time elevated-temperature properties, small, finely dispersed particles of γ′ precipitate are desired. Therefore, final aging temperatures are lower than those used to obtain high creep-rupture properties. For all γ′ dispersions, care must be taken to ensure the correct carbide distribution.

Double-Aging Treatment. A two-step aging treatment is commonly used to control the size distribution of γ′ and γ″ precipitates. A principal reason for two-step aging sequences, in addition to γ′ or γ″ control, is to precipitate or control grain-boundary carbide morphology. In some alloys, such as Incoloy 901 and A-286, MC films may form along grain boundaries and reduce ductility.

Double or multistep aging treatments vary according to the alloy type and design objectives. In some alloys, a second aging step up to about 850 °C (1560 °F) is added (see Examples 1 and 2 below).

In other alloys, however, aging may involve an initial treatment in the range of 850 to 1100 °C (1560 to 2010 °F) over a period of up to 24 h. Aging at one or more lower temperatures, for example at 760 °C (1400 °F) for 16 h, completes the precipitation of γ′. The finer γ′ produced in the second aging treatment is advantageous for tensile strength as well as for rupture life. This type of aging treatment is used for nickel-base superalloys such as Udimet 700, Astroloy, and Udimet 710.

Example 1: Double-Aging Treatment of A-286. Adding a second aging treatment can improve properties to meet a requirement. For example, borderline values of yield strength, 615 and 630 MPa (89 and 91 ksi), were obtained in two heats of material with a heat treatment of:

- 900 °C (1650 °F) solution treatment for 2 h and oil quench
- Aging treatment of 705 °C (1300 °F) for 16 h with air cooling

By aging a second time at 650 °C (1200 °F) for 16 h (air cooled), the yield strengths were improved to 635 and 698 MPa (92 and 101 ksi).

Example 2: Double-Aging Sequence of Udimet 500 for Stabilization of Grain Boundary Carbides. Udimet 500 is typical of wrought precipitation-hardened superalloys that contain MC and $M_{23}C_6$ carbides and are strengthened by γ′. For a good balance of tensile strength and stress-rupture life, the alloy is:

- Solution heat treated at 1080 °C (1975 °F) for 4 h (air cooled)
- Stabilized at 845 °C (1550 °F) for 24 h (air cooled)
- Aged at 760 °C (1400 °F) for 16 h (air cooled)

The solution exposure dissolves all phases except MC carbides, and γ′ precipitates nucleate during cooling from the solution temperature. The stabilization at 845 °C (1550 °F) precipitates discontinuous $M_{23}C_6$ at grain boundaries as well as γ′. Final aging increases the volume fraction of γ′. The grain boundary $M_{23}C_6$ increases stress-rupture life as long as it is not a continuous carbide film, which markedly decreases rupture ductility.

Thermomechanical Processing

In recent years, there has been more interest in the interdependence of hot working and heat-treating operations. In many critical applications, the desired final properties are not attainable via heat treatment if the hot working operation has not been conducted under controlled temperature and deformation parameters. This requires a study of hot working and heat treating, known as thermomechanical processing. One application of thermomechanical processing is the development of direct age 718 for turbine disk applications (Ref 1). Proper heating temperatures and forging operations also influence the microstructure and distribution of phases in alloys such as 718.

Grain Size Control. An important objective of thermomechanical processing is grain size control (Ref 2-4). For example, grain structure may be controlled by thermomechanical processing in several iron-nickel-base alloys that have two precipitates present, such as the primary strengthening precipitate (γ″Ni_3Nb in Inconel 718 and γ′Ni_3Ti in Inconel 901) and a secondary precipitate (δ in Inconel 718 and γ′Ni_3Ti in Inconel 901) (Ref 3, 4). The secondary precipitate is produced first by an appropriate heat treatment (8 h at 900 °C, or 1650 °F, for Inconel 901) followed by working at about 950 °C (1740 °F) below the δ solvus. Final working is carried out below the recrystallization temperature, and the alloy is subsequently recrystallized below the δ solvus. Finally, the alloy is aged by standard procedures. The result is a fine-grain alloy with higher tensile strength and improved fatigue resistance.

The critical warm-working temperature range for Inconel 718 is 955 to 995 °C (1750 to 1820 °F). The upper limit avoids grain coarsening at higher temperatures due to re-solution of δ, while the lower limit is established to avoid an excessively high flow stress during working. Delta and γ″ precipitates compete for the available niobium. Therefore, any factor suppressing δ tends to favor γ″ formation and vice versa. Delta does not strengthen Inconel 718, but it reduces the room-temperature ductility. However, when some δ phase is precipitated prior to or during working, the grain size can be reduced substantially, leading to increased tensile and fatigue strength.

Other thermomechanical working schedules are used to produce a double necklace structure of fine grains surrounding the large grains formed during high-temperature recrystallization. Reductions of 25 to 50% are needed in the final working operations at 1080 to 1110 °C (1975 to 2030 °F) to produce the small recrystallized grains in cast/wrought René 95.

Surface-Related Problems Due to Heat Treatment

Although superalloys offer resistance to surface degradation during elevated-temperature service, heat treatment temperatures (particularly solution treatment) can degrade surface characteristics. The potential forms of surface degradation include oxidation, carbon pickup, alloy depletion, and contamination.

Oxidation. In general, superalloys have good oxidation resistance within their normal range of service temperature and above their aging temperatures (typical range 760 to 980 °C, or 1400 to 1800 °F) depending on the alloy. Others may require coatings to provide suitable resistance in the service environment, such as blade and vane alloys for turbine engines and compressors. However, at higher temperatures, such as those used for solution treating, these alloys are susceptible to intergranular oxidation, a defect that adversely affects thermal fatigue.

Oxidation resistance is enhanced by adding chromium, aluminum, and certain other elements. Chromium allows formation of a protective chromium oxide and is most beneficial at intermediate temperatures and below (≲870 °C, or 1600 °F), or in the presence of adequate aluminum, it permits the formation of aluminum oxide, which is more protective at higher temperatures (>870 °C, or 1600 °F). The principal mode of intergranular attack involves preferential oxidation of chromium, aluminum, titanium, zirconium, and boron. Molybdenum increases susceptibility to intergranular attack in age-hardenable alloys. In relation to intergranular oxidation, aluminum is preferable to titanium as a hardening element because:

- Increased aluminum reduces formation of the η phase from γ′

- Aluminum oxide provides a denser and less permeable barrier to the diffusion of oxygen
- Lower density

However, a smooth tight oxide layer is not possible in chromium-free low-expansion (Invar-type) superalloys such as Incoloy alloys 903, 907, and 909 and Pyromet alloys CTX-1, CTX-3, and CTX-909.

Carbon pickup can occur if the solution-treating atmosphere has a carburizing potential. For instance, the carbon content of the surface of A-286 has been observed to increase from 0.05 to 0.30%. The added carbon forms a stable carbide (TiC), thus removing titanium from solid solution and preventing normal precipitation hardening in the surface layers. TiN can be formed in the same manner as a result of nitrogen contamination.

Alloy Depletion. In addition to oxidation, exposure to high-temperature environments can cause changes in the composition of the alloy near the surface. As certain elements are preferentially consumed by the scale layer, the bulk composition can become depleted. In addition, some alloys can be very susceptible to deboronization. This can affect the properties of the surface layers and can be of considerable concern, for instance in sheet products.

Miscellaneous Contaminants. All exposed surfaces of superalloy parts should be kept free of dirt, fingerprints, oil, grease, forming compounds, lubricants, and scale. Lubricants or fuel oils that contain sulfur compounds are particularly active in corroding and embrittling the metal surface by first forming Cr_2S_3 and then, as the attack progresses, also forming a Ni-Ni_3S_2 eutectic that melts at 645 °C (1190 °F), particularly at low pressures of less than 10^{-2} Pa (10^{-4} torr) in vacuum.

Scale and slag from furnace hearths are another source of contamination. Contact with steel scale, slag, and furnace spallings should be avoided; low-melting constituents can form on the metal surface and promote corrosion.

Selection of Protective Atmospheres to Prevent Surface Problems

Protective atmospheres are used in annealing or solution treating if heavy oxidation cannot be tolerated. If oxidation can be tolerated, because of subsequent stock removal, superalloys can be solution treated in air or in the normal mixture of air and combustion products found in gas-fired furnaces. A vacuum environment is desirable.

Vacuum atmosphere generally below 1/4 Pa (2×10^{-3} torr) is commonly used above 815 °C (1500 °F). It is particularly desirable when parts are at or close to final dimensions.

Inert Gas. Dry argon with a dew point of –50 °C (–60 °F) or lower should be used if no oxidation can be tolerated. It is mandatory that this type of atmosphere be used in a sealed retort or sealed furnace chamber. A purge of at least ten times the volume of the retort is recommended before the retort is placed in the furnace. The argon must he kept flowing continually during and after the treatment until the workpieces have cooled nearly to room temperature to prevent the formation of an oxide film.

Alloys containing stable-oxidic formers, such as aluminum and titanium, with or without boron, must be bright annealed in a vacuum or in a chemically inert gas such as argon. If used, argon must be pure and dry with a dew point of –50 °C (–60 °F) or lower. If the argon has a slightly higher dew point, but not more than –40 °C (–40 °F), oxidation will be limited to a thin surface film that can usually be tolerated.

Dry hydrogen with a dew point of –50 °C (–60 °F) or lower is used in reference to dissociated ammonia for bright annealing. If the hydrogen is prepared by catalytic gas reactions instead of by electrolysis, residual hydrocarbons, such as methane, should be limited to about 50 parts per million, to prevent carburizing. Hydrogen is not recommended for bright annealing of alloys containing significant amounts of elements (such as aluminum or titanium) that form stable oxides not reducible at normal heat-treating temperatures and dew points. Hydrogen is not recommended for annealing or solution treating alloys that contain boron because of the danger of deboronization through formation of boron hydrides. Titanium hydrides also can form.

Exothermic Atmosphere. A lean and dilute exothermic atmosphere is relatively safe and economical. The surface scale formed in such an atmosphere can be removed by pickling or by salt bath descaling and pickling. Such an atmosphere, formed by burning fuel gas with air, contains about 85% nitrogen, 10% carbon dioxide, 1.5% carbon monoxide, 1.5% hydrogen, and 2% water vapor. This atmosphere will produce a scale rich in chromium oxides.

Endothermic atmospheres prepared by reacting fuel gas with air in the presence of a catalyst are not recommended because of their carburizing potential. Similarly, the endothermic mixture of nitrogen and hydrogen formed by dissociating ammonia is not used because of the probability of nitriding.

Atmosphere for Aging. Air is the most common aging atmosphere. The smooth, tight oxide layer that is formed is usually unobjectionable on the finished product (except when heat treating chromium-free low-expansion superalloys). However, if this oxide layer must be minimized, a lean exothermic gas (air-to-gas ratio of about 10 to 1) or vacuum can be employed. It will not entirely prevent oxidation, but the oxide layer will be very light. The use of gases containing hydrogen and carbon monoxide for aging cycles is dangerous because of the explosion hazard at temperatures below 760 °C (1400 °F).

Precipitation-Strengthened Nickel-Iron-Base Superalloys

There is also an important distinction between the nickel-base and nickel-iron precipitation-strengthened superalloys. Many of the nickel-iron alloys allow precipitation of additional intermetallics (such as η, δ, and Laves phases) in addition to the principal strengthening precipitates (γ' and/or γ''). These additional precipitates, with appropriate heat treatment, can provide some grain size control in nickel-iron superalloys. For example:

- The η phase in Inconel 901 and A-286 reduces grain growth when temperatures are below the η solvus (see Example 3 described below).
- The δ phase can control grain size in γ''-strengthened superalloys such as Inconel 718.
- The Laves phase (solvus at about 1040 °C, or 1900 °F) in Incoloy 909 and Pyromet CTX3 controls grain size when annealing in a temperature range of 980 to 1010 °C (1800 to 1850 °F).

Alloy Types

Precipitation-strengthened nickel-iron-base superalloys can be grouped into two classes according to the main strengthening phase. These classes include alloys strengthened by γ' and γ'' phases, respectively.

Alloys strengthened by the γ' phase include such materials as A-286, which contain relatively low levels of nickel (25 to 35 wt%), and alloys such as Incoloy 901 and Inconel X-750, which contain higher levels of nickel (more than 40 wt%) and higher volume fractions of γ' for increased strength. Iron-rich Incoloy 903 and 909, which are iron-nickel-cobalt alloys developed to exhibit low coefficient of thermal expansion properties along with high strength, are also strengthened by γ' precipitates.

Alloys strengthened primarily by the γ'' phase are nickel rich, contain niobium for formation of γ'', and are represented by Inconel alloys 706 and 718. Inconel 718 is particularly important. In fact, Inconel 718 constitutes approximately 45% of all wrought nickel-base and nickel-iron-base superalloys produced.

Effects of Cold Working on Heat Treatment Response

Cold working of age-hardenable nickel-iron-base superalloys affects the response of the alloy during heat treatment. Cold work affects the recrystallization and grain growth behavior during subsequent solution treatment and the reaction kinetics of aging. The cold working itself is usually performed on solution-treated alloys because of the markedly lower strength and increased ductility of the material before aging (see Table 2).

Effect of Cold Work on Grain Growth during Solution Treatment. Larger amounts of cold work refine the grain size during solution treatment, but smaller amounts of cold work can lead to critical grain growth. The effect of varying amounts of cold work on grain growth in A-286 during solution treating is illustrated in Fig. 1. The initial material was solution treated with a maximum grain size of ASTM 5. Cold working in the range of 1 to 5% caused excessive grain growth during subsequent solution treating at 900 °C (1650 °F). Above about 5% of cold work,

Table 2 Typical effects of aging on room-temperature mechanical properties of solution-treated heat-resisting alloys

Alloy	Yield strength (0.2% offset) Not aged MPa	ksi	Aged MPa	ksi	Elongation in 50 mm (2 in.), % Not aged	Aged
A-286	240	35	760	110	52	33
René 41	620	90	1100	160	45	15
X-750	410	60	650	92	45	24
Haynes alloy 25	480	69	480	70	55	45

Table 3 Effect of heat treatment on the properties of A-286

See also text accompanying Example 3.

Heat treatment	Tensile properties at 21 °C (70 °F): 0.2% yield strength, MPa	ksi	Ultimate tensile strength, MPa	ksi	Elongation, %	Reduction in area, %	Stress rupture at 650 °C (1200 °F) with 450 MPa (65 ksi): Life, h	Elongation, %	Reduction in area, %
980 °C (1800 °F) for 1 h, oil quench (OQ) + 720 °C (1325 °F) for 16 h, air cool	690	100	1070	156	24	46	85	10	15
900 °C (1650 °F) for 2 h, OQ + 720 °C (1325 °F) for 16 h, air cool	740	108	1100	160	25	46	64	15	20

Source: Ref 5

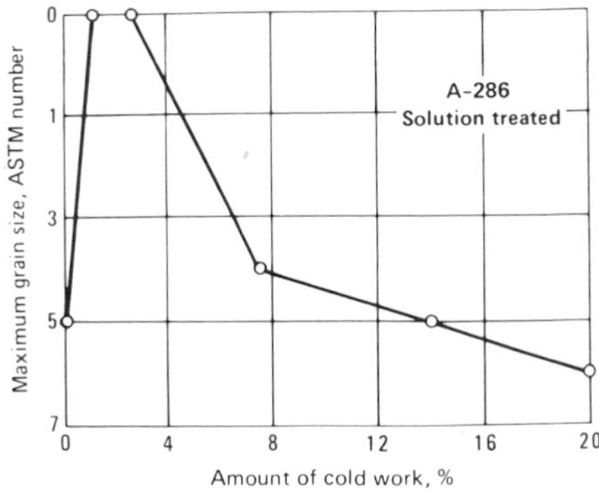

Fig. 1 Effect of cold work on grain size of A-286 alloy solution treated at 900 °C (1650 °F) for 1 h and oil quenched

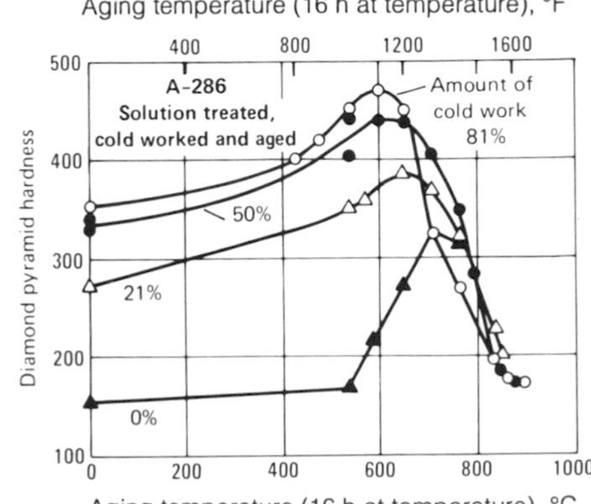

Fig. 2 Effect of cold work and aging on diamond pyramid hardness of A-286

critical grain growth did not occur, and the recrystallized grain size decreased with increasing cold work. Excessive grain growth, especially excessive localized grain growth, would likely degrade tensile properties. Therefore, on parts subjected to cold or hot work prior to solution treating, the critical amount of work (about 1 to 6% cold work, depending on the alloy, and about 10% hot work) must be exceeded in all areas to avoid the growth of abnormally large grains. This rule applies to items such as cold-headed bolts, spun or stretch-formed sheets, and parts formed by simple bending.

Effect of Cold Work on Aging Response. Cold working accelerates the aging reaction, causing the early appearance of precipitates at normal aging temperature and the appearance of a precipitate at temperatures below the normal aging temperature if the alloy is cold worked sufficiently and held at temperature long enough. In other words, cold working can make a material more prone to overaging at the normal aging temperature. Even relatively small amounts of residual strain can lead to strain-induced overaging of the low-expansion (Invar-type) superalloys such as Incoloy 907 and 909 and Pyromet CTX-3 and CTX-909. This effect can result in an extreme loss in tensile properties. Thus care must be taken to provide strain (dislocation)-free structures prior to aging. Properly charted information correlating cold work with aging response, however, can be used to shorten the required aging period or to lower the normal aging temperature. The effect of cold work and aging on hardness in A-286 is shown in Fig. 2.

The hardness after cold working and before aging, as well as the peak hardness after aging, increases greatly with increasing amounts of cold rolling. The temperature at which peak hardness is attained decreases with increasing amounts of reduction. The temperature at which softening occurs also decreases with increasing amounts of deformation. In Fig. 2, for example, the material cold worked 81% and then aged at 760 °C (1400 °F) for 16 h is softer than material that was not cold worked before being aged.

Double Aging of Cold-Worked Parts to Obtain Uniformity of Properties. In parts that have been cold worked nonuniformly, higher-than-normal aging temperatures are sometimes used, followed by a second aging cycle at lower-than-normal aging temperature. For example, in A-286 parts with varying amounts of cold work, double aging at 760 °C (1400 °F) and 720 °C (1300 °F), respectively, provides more uniform hardness, short-time tensile properties, and creep-rupture properties than the normal 720 °C (1325 °F) aging treatment. The higher aging temperature also improves the structural stability of the part in service.

Effect of η Phase on Grain Growth

As stated above, the η phase reduces grain growth when temperatures are below the η transus. A specific example of grain size control with the η phase is described below.

Example 3: Control of Grain Growth in A-286 with η. In A-286, the γ′ solvus is approximately 855 °C (1575 °F), and the η solvus is approximately 910 °C (1675 °F). The alloy can be worked and heat treated above the η solvus, or the γ′ can be fully dissolved with some η phase retained for grain size control. A-286 has been processed to fine grain sizes by finish forging at temperatures in the vicinity of the η solvus, solution treating below the η solvus but above the γ′ solvus (for example, hold at 900 °C, or 1650 °F, for 2 h and oil quench), then aging below the γ′ solvus (for example, 720 °C, or 1325 °F, for 16 h and air cool). Recrystallization occurs above the γ′ solvus either during forging or solution heat treatment, but the η phase controls grain growth. If the alloy is solution heat treated above the η solvus (for example, at 980 °C, or 1800 °F, for 1 h with oil quenching), a coarser grain size results. Table 3 shows typical tensile and stress-rupture data for A-286 obtained by each approach. The finer-grain-size material has better room-temperature tensile strength, better stress rupture ductility (including notched bar rupture ductility), but lower stress-rupture life.

Heat Treatment of Nickel-Iron-Base Alloys Strengthened by γ′

For the nickel-iron-base superalloys strengthened by γ′, heat treatments are designed to control grain size and develop the desired morphology of possible phases, primarily γ′, η, and MC carbides. Precipitation of the η phase occurs at temperatures higher than the γ′ solvus and can provide control of grain growth during thermomechanical processing or during heat treatment (see, for instance, Example 3 in this article). Adding a second aging treatment can also improve properties (see Example 1).

Acceptable mechanical properties do not always result from the solution treating and aging procedure first tried for a given alloy or applica-

Table 4 Effects of alternative aging treatments on rupture properties of A-286

Treatment	Creep rupture(a) Life, h	Elongation in 50 mm (2 in.), %	Location of failure
Original heat treatment(b)			
900 °C (1650 °F), 2 h, oil quench; 720 °C (1325 °F), 16 h, air cool	7–69	...	Notch
Revised heat treatments(c)			
Original plus additional age at 650 °C (1200 °F), 12 h, air cool	74–142	5.6–7.7	Smooth bar
900 °C (1650 °F), 2 h, oil quench; 730 °C (1350 °F), 16 h, air cool	24–82	4.9–7.7	Smooth bar

(a) Specification requirements for creep-rupture at 650 °C (1200 °F) and 450 MPa (65 ksi); life, 23 h min; no failure in notch permitted. (b) Ten specimens tested. (c) Five specimens for each treatment

tion. To develop specified properties, changes of the following kinds are often required:

- Adjust the solution temperature or time
- Adjust the aging temperature
- Add an intermediate (stabilizing) aging treatment at a temperature higher than that of the final aging
- Add a second (final) aging treatment at a lower temperature
- Adjust one or both aging temperatures in a double aging cycle
- Add a third aging treatment
- Increase the aging time

An example for A-286 is shown in Table 4, which shows the effects of alternative approaches to the problem of insufficient notch ductility in heat treated A-286 alloy (namely, an additional aging treatment and the substitution of a higher aging temperature for single aging). Although both approaches provided required creep-rupture properties, the higher aging temperature (with no increase in the solution-treating temperature) is recommended over the additional aging treatment.

Such procedures can be especially important when high-temperature annealing (such as during brazing) results in a coarse-grain microstructure. This is definitely the case for low-expansion superalloys because initial aging temperature and time must be increased to provide averaging necessary for improved notched bar rupture strength if grain size is coarse.

Heat Treatment of Incoloy 901. Incoloy 901 has a γ' solvus of approximately 940 °C (1725 °F) and an η solvus of approximately 995 °C (1825 °F). Just as for A-286 (Example 3), solution heat treating above the γ' solvus but below the η solvus is used to obtain a finer grain structure (for improved tensile properties), and solution heat treating above the η solvus is used to yield a coarser grain structure (for improved creep characteristics). In addition, for Incoloy 901, a stabilization heat treatment is carried out before aging to achieve a desirable grain-boundary carbide morphology. The stabilization heat treatment promotes formation of globular MC (primarily TiC) carbides as opposed to a continuous grain-boundary film. Low stress-rupture ductility is associated with a continuous intergranular MC film, while the globular MC morphology promotes good rupture life and ductility.

Heat treatment to achieve a coarser grain size of ASTM 2 to 4 with discrete intergranular carbides and a γ' strengthened matrix might be:

- Solution heat treating at 1080 to 1105 °C (1975 to 2025 °F) for 2 h (air cooling or faster)
- Stabilizing at 775 to 800 °C (1425 to 1475 °F) for 2 to 4 h (air cooling)
- Aging at 705 to 745 °C (1300 to 1375 °F) for 24 h (air cooling)

To preserve a finer grain size achieved by forging, lower solution, stabilizing, and aging temperatures are chosen. Heat treatment for finer grain size would be:

- Solution heat treating at 980 to 1040 °C (1800 to 1900 °F) for 1 to 2 h (air cooling or faster)
- Stabilizing at 705 to 730 °C (1300 to 1350 °F) for 6 to 20 h (air cooling or faster)
- Aging at 635 to 665 °C (1175 to 1225 °F) for 12 to 20 h (air cooling)

This approach typically yields a dynamically recrystallized grain size of ASTM 5 to 7, with grain boundaries strengthened by discrete carbides and possibly containing some η phase, and a matrix strengthened by γ' precipitates. Tensile properties for an Incoloy 901 forged and heat-treated turbine disk that are representative of the above two heat treatment approaches are presented for room temperature and 540 °C (1000 °F) in Tables 5 and 6, respectively. The lower solution temperature yields higher tensile strengths and ductility.

It is important to note, however, that not only heat treatment influences tensile properties; the control of structure during thermomechanical processing is also important. A method for achieving fine grain structures in Incoloy 901 (less than ASTM 10) by controlling structure during thermomechanical processing with η phase (Ref 2) is briefly described in the section

Table 5 Tensile properties at room temperature at various locations in disk forgings of Incoloy 901 in two heat-treated conditions

Condition	Test location	Yield strength MPa	Yield strength ksi	Ultimate tensile strength MPa	Ultimate tensile strength ksi	Elongation in 50 mm (2 in.), %	Reduction in area, %
1095 °C (2000 °F) for 2 h, water quench + 790 °C (1450 °F) for 2 h, water quench + 730 °C (1350 °F) for 24 h, air cool	Rim-radial-top	859	124.6	1178	170.8	15	16
	Rim-radial-bottom	907	131.6	1168	169.4	13	14
	Rim-radial-middle	880	127.6	1179	171.0	15	17
	Rim-axial-middle	858	124.4	1054	152.9	...	...
	Rim-tangent-middle	883	128.0	1175	170.4	13	17
	Bore-radial-top	874	126.8	1200	174.0	14	17
	Bore-radial-bottom	889	129.0	1131	164.0	...	...
	Bore-radial-middle	869	126.0	1172	170.0	16	20
	Bore-axial-middle	840	121.8	1154	167.4	...	...
	Bore-tangent-middle	859	124.6	1167	169.2	15	17
1010 °C (1850 °F) for 2 h, water quench + 730 °C (1350 °F) for 20 h, water quench + 650 °C (1200 °F) for 20 h, air cool	Rim-radial-top	924	134.0	1234	179.0	17	20
	Rim-radial-bottom	952	138.0	1240	179.8	17	21
	Rim-radial-middle	980	142.0	1258	182.4	19	29
	Rim-axial-middle	972	141.0	1255	182.0	21	31
	Rim-tangent-middle	986	143.0	1274	184.8	18	25
	Bore-radial-top	978	141.9	1248	181.0	18	24
	Bore-radial-bottom	976	141.6	1255	182.0	20	31
	Bore-radial-middle	968	140.4	1252	181.6	21	34
	Bore-axial-middle	940	136.4	1081	156.8	5	9
	Bore-tangent-middle	965	140.0	1253	181.8	20	31

Source: Ref 6

Table 6 Tensile properties at 1000 °F at various locations in disk forgings of Incoloy 901 in two heat-treated conditions

Condition	Test location	Yield strength		Ultimate tensile strength		Elongation in 50 mm (2 in.), %	Reduction in area, %
		MPa	ksi	MPa	ksi		
1095 °C (2000 °F) for 2 h, water quench + 790 °C (1450 °F) for 2 h, water quench + 730 °C (1350 °F) for 24 h, air cool	Rim-radial-top	772	112.0	1037	150.4	13	18
	Rim-tangent-middle	772	112.0	1048	152.0	12	18
	Web-radial-top	781	113.2	1049	152.1	13	21
	Web-tangent-middle	772	112.0	1041	151.0	14	21
	Bore-radial-top	782	113.4	1045	151.6	13	20
	Bore-tangent-middle	772	112.0	1027	149.0	14	22
1010 °C (1850 °F) for 2 h, water quench + 730 °C (1350 °F) for 20 h, water quench + 650 °C (1200 °F) for 20 h, air cool	Rim-radial-top	832	120.6	1066	154.6	14	27
	Rim-tangent-middle	910	132.0	1117	162.0	17	38
	Web-radial-top	853	123.7	1091	158.2	20	39
	Web-tangent-middle	876	127.0	1089	158.0	19	39
	Bore-radial-top	855	124.0	1069	155.0	17	30
	Bore-tangent-middle	876	127.0	1105	160.2	17	38

Source: Ref 6

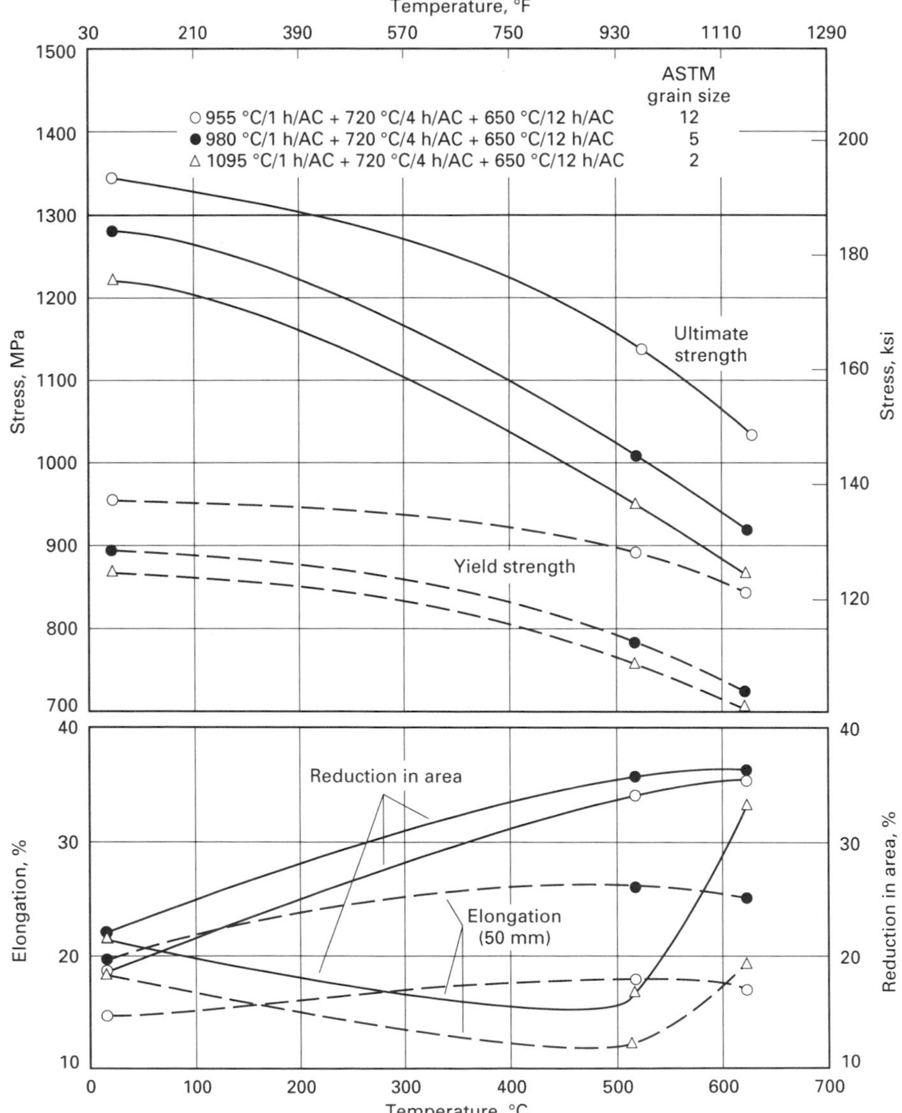

Fig. 3 Effects of temperature and grain size on tensile properties of Incoloy 901 forgings in the solution-treated, stabilized, and aged condition. AC, air cooled. Source: Ref 6

"Thermomechanical Processing" in this article. Figure 3 shows the effect of solution temperatures from 955 to 1905 °C (1750 to 2000 °F) and grain sizes from ASTM 2 to 12 on tensile properties of Incoloy 901 processed using this method. The high-cycle fatigue strength of Incoloy 901

Table 7 Effect of grain size on the high-cycle fatigue properties of Incoloy 901 at 455 °C (850 °F)

Incoloy 901 grain size	455 °C (850 °F) Fatigue strength (10^7 cycles)		Fatigue ratio (FS/UTS)(a)
	MPa	ksi	
ASTM 2	315	46	0.32
ASTM 5	439	64	0.42
ASTM 12	624	91	0.55

(a) FS/UTS, fatigue strength to ultimate tensile strength. Source: Ref 2

doubled as a result of the reduction in grain size from ASTM 2 to 12 (Table 7). The low-cycle fatigue also improved greatly with decreasing grain size (Table 8).

An example of the effect of the stabilization heat treatment for Incoloy 901 is given in Table 9. Adding the stabilization treatment greatly improved the rupture ductility, although at some sacrifice of yield strength and rupture life.

Adding a second aging treatment can improve selected properties as needed. In one application, single aging at 720 °C (1325 °F) for 24 h (air cooled) after a 2-h 1085 °C (1985 °F) solution and stabilization (water quenched) failed to provide sufficient yield strength to meet specifications for room-temperature tensile strength. A second aging treatment at 650 °C (1200 °F) for 12 h (air cooled) provided additional strength to meet requirements. Table 10 lists the properties obtained by the two treatments.

In another Incoloy 901 application, a problem of low room-temperature yield strength and low rupture ductility was solved by increasing the temperature of the 2-h stabilization (and air cooling) treatment from 775 to 790 °C (1425 to 1450 °F) and by adding a second aging treatment. Table 11 lists the mechanical properties obtained using the original and the revised heat treatments.

Another example showing the effect of a stabilization treatment on rupture ductility is presented in Table 12. Initially, the heat treatment included a solution treatment at 1085 °C (1985 °F) for 2 h (air cooling) and aging at 720 °C (1325

Table 8 Effect of grain size on the low-cycle fatigue properties of Incoloy 901 at 455 °C (850 °F)

Incoloy 901 grain size	Stress MPa	Stress ksi	Temperature °C	Temperature °F	Cycles to failure(a)
ASTM 2	205 ± 448	30 ± 65	455	850	9 000
ASTM 5	205 ± 448	30 ± 65	455	850	26 000
ASTM 12	205 ± 448	30 ± 65	455	850	200 000+
ASTM 2	205 ± 530	30 ± 77	455	850	5 000
ASTM 5	205 ± 530	30 ± 77	455	850	16 000
ASTM 12	205 ± 530	30 ± 77	455	850	137 000

(a) Average of 8 tests at 455 °C (850 °F). Source: Ref 2

Table 9 Effect of stabilization on typical properties of Incoloy 901

Condition	Ultimate tensile strength MPa	Ultimate tensile strength ksi	Yield strength MPa	Yield strength ksi	Elongation in 50 mm (2 in.), %	Reduction in area, %	Creep-rupture life, h
Tested at 20 °C (70 °F)							
No intermediate aging(a):							
Heat A	1050	152	790	115	12	13	...
Heat B	1080	157	790	114	17	16	...
With intermediate aging(b):							
Heat A	1040	151	730	106	12	15	...
Heat B	1040	151	710	103	12	13	...
Tested at 650 °C (1200 °F)							
No intermediate aging(a):							
Heat A	...	...	...	...	1.0	...	76
Heat B	...	...	...	...	1.5	...	118
With intermediate aging(b):							
Heat A	...	...	...	...	11	...	45
Heat B	...	...	...	...	7	...	54

(a) Heat treatment: 1120 °C (2050 °F) for 2 h, water quench; 745 °C (1375 °F) for 24 h, air cool. (b) Heat treatment: 1120 °C (2050 °F) for 24 h, water quench; 815 °C (1500 °F), 4 h; air cool; 745 °C (1375 °F), 24 h; air cool

Table 10 Effect of single and double aging on room-temperature properties of Incoloy 901

Condition	Ultimate tensile strength MPa	Ultimate tensile strength ksi	Yield strength MPa	Yield strength ksi	Elongation in 50 mm (2 in.), %	Reduction in area, %
Specification	1140	165	827	120	12	15
Single aging(a)	1150–1160	167–169	800–810	116–118	20–23	24–29
Double aging(b)	1190–1210	173–175	830–890	121–129	18–22	24–29

Single and double aging data reflect the results of four tests. (a) Solution treated at 1085 °C (1985 °F) for 2 h, water quenched; aged at 770 °C (1450 °F) for 2 h, air cooled; aged at 720 °C (1325 °F) for 24 h, air cooled. (b) Re-aged at 650 °C (1200 °F) for 12 h and air cooled

Table 11 Effects of adding a third aging treatment on properties of Incoloy 901

Ultimate tensile strength MPa	Ultimate tensile strength ksi	Yield strength MPa	Yield strength ksi	Elongation in 50 mm (2 in.), %	Reduction in area, %	Creep-rupture elongation(a)
Specification						
1140	165	830	120	12	15	4% in 23 h(b)
Double aging(c)						
1160-1210(d)	169–175(d)	810–900(d)	118–131(d)	22–23(d)	25–30(d)	4.9% in 31 h to 2.8% in 85 h(e)
Triple aging(f)						
1200-1240(d)	174–180(d)	850–930(d)	123–135(d)	18–20(d)	23–29(d)	7% in 64 h to 6.3% in 74 h(g)

(a) At 650 °C (1200 °F) and 620 MPa (90 ksi). (b) Minimum values. (c) Solution treated at 1085 °C (1985 °F) for 2 h, water quenched; aged at 775 °C (1425 °F) for 2 h, air cooled; aged at 720 °C (1325 °F) for 24 h, air cooled. (d) Seven tests. (e) Three tests. (f) Solution treated at 1085 °C (1985 °F) for 2 h; water quenched; aged at 790 °C (1450 °F) for 2 h, air cooled; aged at 720 °C (1325 °F) for 24 h, air cooled; aged at 650 °C (1200 °F) for 12 h, air cooled. (g) Two tests

Table 12 Effect of revision in aging treatment on creep-rupture properties of Incoloy 901

Test No.	Original treatment(b) Life, h	Original treatment(b) Elongation, %	Revised treatment(c) Life, h	Revised treatment(c) Elongation in 50 mm (2 in.), %
1	72	4	74	13
2	126	4	115	12
3	161	4	160	13
4	111	4	110	9
5	127	4	84	9
6	76	4	84	8
7	127	4	98	9

Creep rupture(a). (a) At 650 °C (1200 °F) and 552 MPa (80 ksi) specified minimum: life, 23h; elongation, 5%. (b) Solution treated at 1085 °C (1985 °F) for 2 h, cooled; aged at 720 °C (1325 °F) for 24 h, air cooled. (c) Same conditions as in (b) except that temperature of first aging was 810 °C (1490 °F)

°F) for 24 h (air cooling). The rupture ductility was significantly improved with a small penalty in rupture life by adding an 810 °C (1490 °F) stabilization treatment.

Inconel X-750 heat treatment in general depends on the application and the properties desired. For service above 595 °C (1100 °F), optimum properties in rods, bars, and forgings are achieved by:

- Solution treating at 1150 °C (2100 °F) for 2 to 4 h (air cooling)
- Stabilization treating at 845 °C (1550 °F) for 24 h (air cooling)
- Aging at 705 °C (1300 °F) for 20 h (air cooling)

This treatment maximizes creep and rupture strength.

For service below 595 °C (1100 °F), the optimum heat treatment is:

- Solution heat treating at 980 °C (1800 °F) for 1 h (air cooling)
- Aging at 730 °C (1350 °F) for 8 h (furnace cooling) to 620 °C (1150 °F)
- Hold for total time of 18 h and then air cool

This treatment maximizes tensile properties. Heat treatments for various product forms of Inconel X-750 are summarized in Table 13.

Heat Treatment of Nickel-Iron-Base Alloys Strengthened by γ''

The heat treatment of nickel-iron-base superalloys strengthened by the γ'' phase are fashioned using many of the same principles that guide the heat treatments of A-286 and Incoloy 901. In these alloys, δ phase, which is present at temperatures above the γ'' solvus, can be used for grain size control, just as η phase can be utilized in A-286 or Incoloy 901. However, careful heat treatment is required to ensure adequate precipitation of γ'' instead of the δ (orthorhombic) phase of the same Ni_3Nb composition. The latter is invariably incoherent and does not confer

Table 13 Typical thermal treatments for precipitation hardening of Inconel X-750 products

Form	Desired property	Thermal treatment
Rods, bars, and forgings	Strength and optimum ductility up to 595 °C (1100 °F)	Equalize: 885 °C (1625 °F), 24 h, air cool Precipitation: 705 °C (1300 °F), 20 h, air cool
	Optimum tensile strength up to 595 °C (1100 °F)	Solution: 980 °C (1800 °F), air cool Furnace-cool, precipitation: 730 °C (1350 °F), 8 h, furnace cool to 620 °C (1150 °F), hold 8 h, air cool
	Maximum creep strength above 595 °C (1100 °F)	Full solution: 1150 °C (2100 °F), 2-4 h, air cool Stabilize: 845 °C (1550 °F), 24 h, air cool Precipitation: 750 °C (1300 °F), 20 h, air cool
Sheet, strip, and plate	High strength at high temperatures	Annealed + Precipitation: 705 °C (1300 °F), 20 h, air cool
	High strength and higher tensile properties to 705 °C (1300 °F)	Annealed + Furnace-cool, precipitation: 730 °C (1350 °F), 8 h, furnace cool to 620 °C (1150 °F), hold 8 h, air cool(a)
Tubing	High strength at high temperatures	Annealed + Precipitation: 705 °C (1300 °F), 20 h, air cool
No. 1 temper wire	Service up to 540 °C (1000 °F)	Solution treated + cold drawn (15–20%) + 730 °C (1350 °F), 16 h, air cool
Spring temper wire	Service up to 370 °C (700 °F)	Solution treated + cold drawn (30–65%) + 650 °C (1200 °F), 4 h, air cool
	Service at 480-650 °C (900-1200 °F)	Cold drawn (30–65%) + 1150 °C(2100 °F), 2 h, air cool + 845 °C (1550 °F), 24 h, air cool + 705 °C (1300 °F), 20 h, air cool

(a) Equivalent properties in a shorter time can be developed by the following precipitation treatment: 760 °C(1400 °F) for 1 h, furnace cool (FC) to 620 °C (1150 °F), hold 3 h, air cool (AC). Source: Ref 7

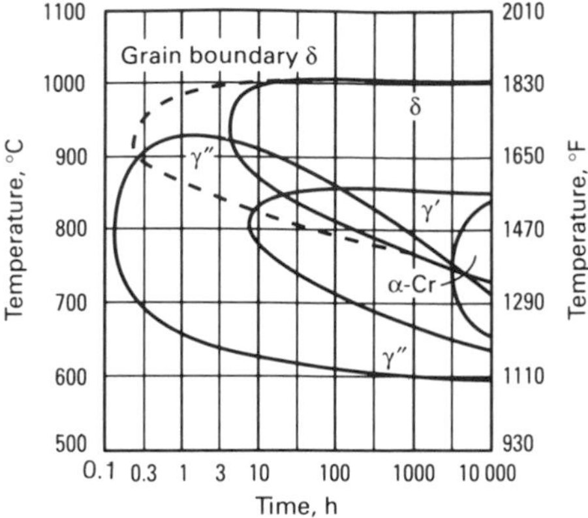

Fig. 4 Transformation diagram for vacuum-melted and hot-forged Inconel 718 bar. Source: Ref 8

strength when present in large quantities. In the absence of iron, or at temperatures and times shown in the transformation diagram of an iron-containing alloy (Fig. 4), the δ phase forms instead of the γ″ phase.

The γ″ phase forms in alloys such as Inconel 718 and Inconel 706. In Inconel alloy 718, γ″ forms in the range of 705 to 900 °C (1300 to 1650 °F) and has a solvus temperature of about 910 °C (1675 °F). The δ phase, depending on exposure time (Fig. 4), precipitates in the approximate temperature range of 870 to 1010 °C (1600 to 1850 °F) and has a solvus temperature of about 1010 °C (1850 °F). The alloy can be worked and solution heat treated above the δ solvus, or at a temperature below the δ solvus but above the γ″ solvus for grain size control. Gamma double prime often precipitates together with γ′ in Inconel 718 (Fig. 4), but γ″ is the principal strengthening phase under such circumstances. Unlike γ′, which causes strengthening through the necessity to disorder the particles as they are sheared, γ″ strengthens by virtue of high coherency strains in the lattice. More detailed descriptions of the physical metallurgy of γ′/γ″ alloys appear in the article "Metallurgy and Processing of Superalloys" in this Volume.

Heat Treatment of Inconel 718. Inconel 718 is usually used in the solution and aged condition; the exact conditions of the temperatures, times, and cooling rates depend on the application and mechanical property needs. Many aerospace applications requiring high tensile and fatigue strength, as well as good stress-rupture properties, use a solution treatment below the δ solvus and a two-step aging treatment as follows:

- Solution heat treat at 925 to 1010 °C (1700 to 1850 °F) for 1 to 2 h (air cooling or faster)
- Age at 720 °C (1325 °F) for 8 h followed by furnace cooling to 620 °C (1150 °F)
- Hold at 620 °C (1150 °F) for a total aging time of 18 h (air cooled)

Higher strength levels are achieved by forging below the δ solvus, quenching after forging, and performing the two-step aging process directly without a solution treatment. This approach, termed direct aging, requires high quality, uniform billet material, and careful control of the forging process to achieve the high strengths and the uniform ASTM 10 grain size throughout the forging.

The effect of the solution temperature, or direct aging (no solution treatment), on tensile and stress-rupture properties of Inconel 718 is shown in Table 14. The highest tensile strength is achieved with the direct-aged material at some sacrifice of stress-rupture life. The lower solution temperatures produce better strength, while the higher solution temperatures (up to 1010 °C, or 1850 °F) result in better stress-rupture strength.

Typical properties for Inconel 718 in three forms are plotted as a function of temperature in Fig. 5. The standard solution-treated and aged condition of Inconel 718 with an average grain size of ASTM 4 to 6 is labeled STD 718. This material condition is used for noncritical or difficult-to-make shapes. High-strength 718, labeled HS 718, with ASTM 8 average grain size, is used for more highly stressed components with less complex configurations. This material is also solution-treated and aged, but tighter controls are placed on input billet material and the forging practice. The highest strengths and fatigue properties are achieved by Inconel 718 in the direct-aged condition, termed DA 718, with ASTM 10 average grain size. Figure 5(b) again illustrates that these strengths are obtained at some reduction of stress-rupture life at low stress and high temperature. To obtain the DA 718 properties in Fig. 5, very high-quality and uniform input material is required along with tightly controlled forging temperatures and processing.

A slightly different solution-and-age heat treatment is sometimes used for optimum ductility, impact properties, and low-temperature toughness in heavy sections, although it can also pro-

Table 14 Properties of Inconel 718 as function of heat treatment

Solution treatment(a)	Room-temperature tensile properties: 0.2% yield strength, MPa	ksi	Ultimate tensile strength, MPa	ksi	Elongation, %	Reduction in area, %	Stress rupture at 650 °C (1200 °F) with 690 MPa (100 ksi): Life, h	Elongation, %	Reduction in area, %
None (direct aged)	1330	193	1525	221	19	34	95	24	31
940 °C (1725 °F), 1 h, air cooled	1240	180	1460	212	18	34	194	11	16
955 °C (1750 °F), 1 h, air cooled	1180	171	1420	206	20	38	122	14	19
970 °C (1775 °F), 1 h, air cooled	1145	166	1405	204	23	41	218	13	15
980 °C (1800 °F), 1 h, air cooled	1172	170	1405	204	24	43	200	6	10
1010 °C (1850 °F), 1 h, air cooled	1185	172	1390	202	22	46	270	6	12
1040 °C (1900 °F), 1 h, air cooled	1165	169	1365	198	25	48	225	2	8

(a) All aged 720 °C (1325 °F) for 8 h, cool 55 °C/h (100 °F/h) to 620 °C (1150 °F), hold for 8 h, air cool (AC). (b) Notch sensitive at K_t = 3.8. Source: Ref 9

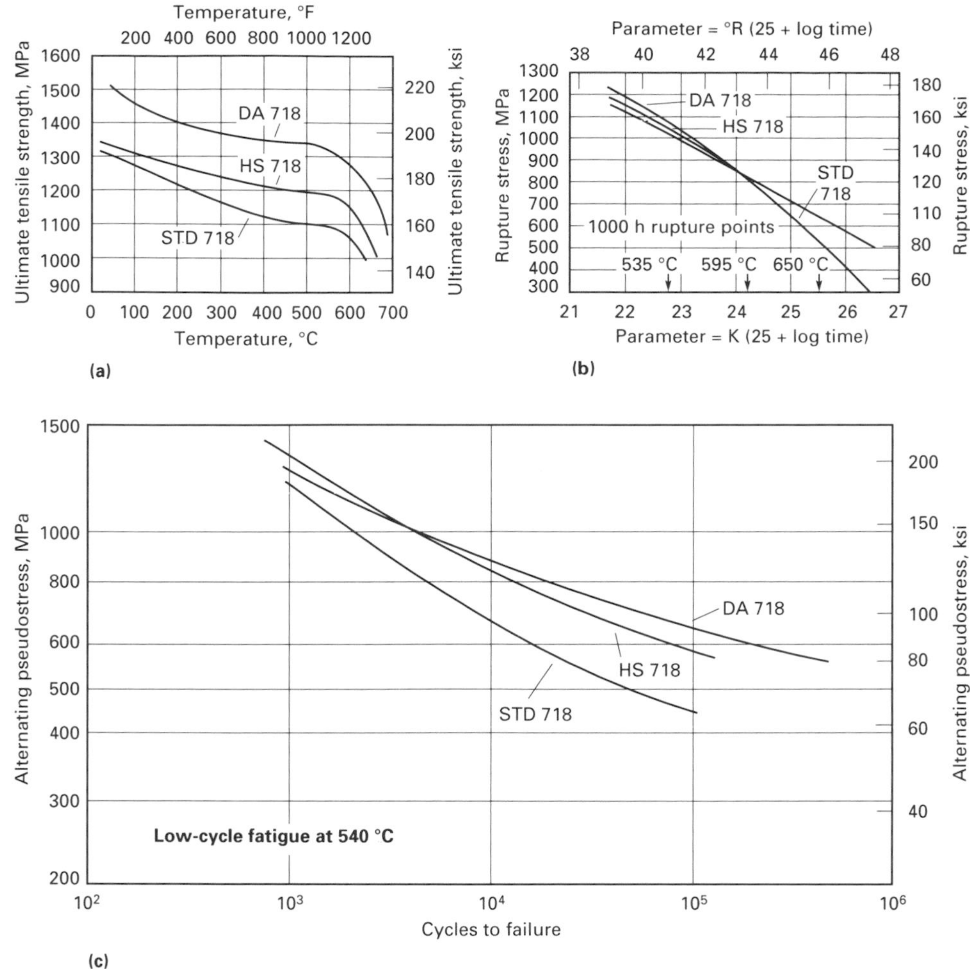

Fig. 5 Typical properties of Inconel 718 as a function of processing. (a) Ultimate tensile strength. (b) Rupture stress. (c) Fatigue at 540 °C (1000 °F). Source: Ref 1

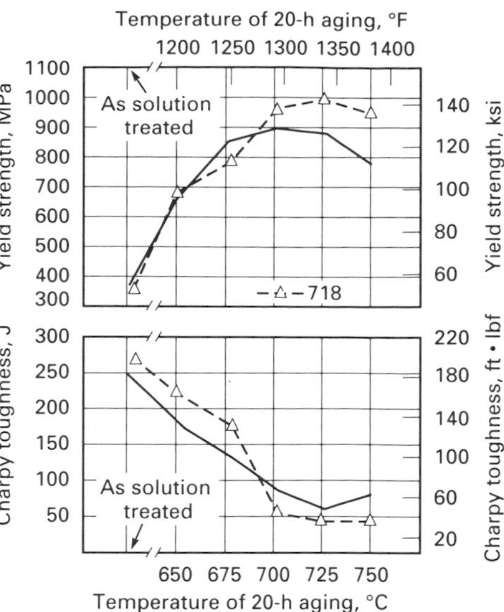

Fig. 6 Effect of aging temperature on the yield strength and toughness of Inconel 718. Source: Ref 12

duce notch brittleness in stress rupture. This heat treatment involves:

- Solution heat treatment at 1040 to 1065 °C (1900 to 1950 °F) for ½ to 1 h, air cooled (or faster)
- Aging at 760 °C (1400 °F) for 10 h with furnace cool to 650 °C (1200 °F)
- Hold at 650 °C (1200 °F) for a total aging time of 20 h (air cool)

Tensile properties as a function of the two heat treatment approaches are listed in Table 15 for hot-rolled rounds of various diameters and in Table 16 for pancake forgings.

For applications of Inconel 718 in the oil and gas industry, the very high strength levels needed in aerospace applications are unnecessary. Instead, heat treatments are designed to address other needs, including good toughness with adequate strength and resistance to hydrogen embrittlement and stress-corrosion cracking. Single-step aging treatments are used to develop mechanical properties in oil and gas applications. Solution treatment in the range of 1010 to 1065 °C (1850 to 1950 °F) is followed by a single-step aging at 650 to 815 °C (1200 to 1500 °F). Room-temperature properties achieved by single-step and two-step aging treatments are compared in Table 17. The single-step aging treatments result in lower strengths but higher fracture toughness. Both general and pitting corrosion resistance can be reduced at higher aging temperatures. This behavior has been attributed to carbide formation and chromium depletion in localized stress. The trade-off between strength and toughness is illustrated in Fig. 6, which shows yield strength and toughness as a function of aging temperature for material after solution-and-age treatment. The exact choice of heat treatment parameters depends on the balance of properties required.

Precipitation-Strengthened Nickel-Base Superalloys

To achieve optimal microstructures for the intended applications, wrought precipitation-strengthened nickel-base superalloys usually receive a solution heat treatment stage followed by one or more precipitation aging heat treatments. The solution heat treatment may be above or below the γ′ solvus temperature depending on a number of factors, including intended grain size, whether strengthening from warm working is to be retained, the desired γ′ morphology, and considerations of carbide precipitation. These microstructural goals are determined by the property requirements. Aging heat treatments are designed to precipitate γ′ and control its volume fraction, size, morphology, and distribution. The aging heat treatments may also precipitate and stabilize various carbide phases and control their morphologies. The temperatures and times of aging heat treatments depend on the alloy composition and on the specific property goals.

The simplest heat treatment, for wrought alloys with relatively low alloying contents and low γ′ volume fractions, is a high-temperature solution treatment followed by air or more rapid cooling, and then low-temperature aging. The solution heat treatment temperature is above the γ′ solvus (thus dissolving γ′ and possibly carbides, depending on the alloy) and causes recrystallization and grain growth to a desired size. The aging precipitates γ′ homogeneously and $M_{23}C_6$ typically at grain boundaries. An example of this basic method for Nimonic 80A or Nimonic 90 is as follows:

- Solution heat treatment at 1080 °C (1975 °F) for 8 h and air cool
- Aging at 700 °C (1290 °F) for 16 h and air cool

The hardness changes as a result of aging Nimonic 80A at various temperatures after a solution exposure of 1080 °C (1975 °F) for 8 h and water quenching are shown in Fig. 7. The curves exhibit normal precipitation hardening characteristics; the lower temperatures and longer aging times provide the highest hardnesses.

Heat treatment affects the types and morphologies of carbides and the γ′ morphology in more highly alloyed superalloys with higher γ′ volume fractions. Consideration of these effects has led to modification of the above two-step (solution treat-and-age) treatments. In Nimonic 80A, the

Table 15 Tensile properties in longitudinal orientations of Inconel 718 hot-rolled rounds of several sizes and in various heat-treated conditions

Diameter		Condition	Yield strength		Ultimate tensile strength		Elongation in 50 mm (2 in.), %	Reduction in area, %
mm	in.		MPa	ksi	MPa	ksi		
16	5/8	As-rolled	566	82.1	962	139.5	46	60
		955 °C (1750 °F) 1 h, air cool	546	79.2	958	139.0	50	49
		1065 °C (1950 °F) 1 h, air cool	332	48.2	803	116.5	61	66
		955 °C (1750 °F) 1 h, air cool + age(a)	1239	179.8	1435	208.2	21	39
		1065 °C (1950 °F) 1 h, air cool + age(b)	1086	157.5	1339	194.2	22	30
25	1	As-rolled	448	65.0	896	130.0	54	67
		955 °C (1750 °F) 1 h, air cool	445	64.5	889	129.0	55	61
		1065 °C (1950 °F) 1 h, air cool	359	52.0	776	112.5	64	68
		955 °C (1750 °F) 1 h, air cool + age(a)	1206	175.0	1389	201.5	20	36
		1065 °C (1950 °F) 1 h, air cool + age(b)	1048	152.0	1296	188.0	21	34
38	1½	As-rolled	727	105.5	1013	147.0	40	52
		955 °C (1750 °F) 1 h, air cool	500	72.5	976	141.5	46	45
		1065 °C (1950 °F) 1 h, air cool	379	55.0	827	120.0	58	60
		955 °C (1750 °F) 1 h, air cool + age(a)	1155	167.5	1413	205.0	20	28
		1065 °C (1950 °F) 1 h, air cool + age(b)	1055	153.0	1316	191.0	24	36
100	4	955 °C (1750 °F) 1 h, air cool	379	55.0	810	117.5	53	52
		1065 °C (1950 °F) 1 h, air cool	331	48.0	776	112.5	60	63
		955 °C (1750 °F) 1 h, air cool + age(a)	1138	165.0	1323	192.0	17	24
		1065 °C (1950 °F) 1 h, air cool + age(b)	1138	165.0	1348	195.5	21	34

(a) Age 720 °C (1325 °F) for 8 h, furnace cool to 620 °C (1150 °F), hold for a total age of 18 h, air cool. (b) Age 760 °C (1400 °F) for 10 h, furnace cool to 650 °C (1200 °F), hold for a total age of 20 h, air cool. Source: Ref 10

Table 16 Tensile properties of Inconel 718 pancake forgings of various sizes in different heat-treated conditions

Forging size	Condition	Orientation	Yield strength		Ultimate tensile strength		Elongation in 50 mm (2 in.), %	Reduction in area, %
			MPa	ksi	MPa	ksi		
200 mm diam (8 in.) × 63.5 mm (2½ in.)	Solutioned 925 °C (1700 °F) for 1 h and air cooled + age(a)	Radial:						
		Top edge	1096	159.0	1255	182.0	10	10.5
		Center	1103	160.0	1351	196.0	24	33.0
		Bottom edge	1100	159.5	1286	186.5	16	19.0
		Tangential:						
		Top edge	1248	181.0	1441	209.0	19	27.5
		Bottom edge	1234	179.0	1448	210.0	18	29.5
175 mm diam (7 in.) × 25 mm (1 in.)	Solutioned 1065 °C (1950 °F) for ½ h and air cooled + age(b)	Radial	1055	153.0	1307	189.5	19	29.8
		Tangential	1056	153.2	1277	185.2	19	27.2
140 mm diam (5½ in.) × 25 mm (1 in.)	Solutioned 980 °C (1800 F) for 1 h, water quenched + age(a)	Radial	1189	172.5	1398	202.8	19	25.6
	Solutioned 1065 °C (1950 °F) for 1 h, water quenched + age(b)	Radial	1048	152.0	1310	190.0	18	24.3

(a) Age 720 °C (1325 °F) for 8 h, furnace cool to 620 °C(1150 °F), hold for total age of 18 h. (b) Age 760 °C (1400 °F) for 10 h, furnace cool to 650 °C (1200 °F), hold for total age of 20 h. Source: Ref 10

Table 17 Effect of heat treatment on typical room-temperature properties of Inconel 718

Heat treatment	Yield strength (0.2% offset)		Tensile strength		Elongation, %	Reduction in area, %	Hardness, HRC	Fracture toughness J_{Ic}	
	MPa	ksi	MPa	ksi				MPa · m	psi · in.
Solution anneal: 1025 °C (1875 °F) for 1 h, water quench; Age: 790 °C (1450 °F) for 6-8 h, air cool	855	124	1200	174	28	51	35	334	1908
Solution anneal: 1050 °C (1925 °F) for 1 , water quench; Age: 760 °C (1400 °F) for 6 h, air cool	855	124	1205	175	27	42	38	286	1631
Solution anneal: 955 °C (1750 °F) for 2 h, water quench or air cool; Age: 720 °C (1325 °F) for 8 h, cool 55 °C/h (100 °F/h) to 620 °C (1150 °F), hold 8 h, air cool	1130	164	1330	193	23	48	42	100	572
Solution anneal: 1050 °C (1925 °F) for 1 h, air cool; Age: 760 °C (1400 °F) for 6 h, furnace cool 55 °C/h (100 °F/h) to 650 °C (1200 °F), hold 8 h, air cool	1255	182	1415	205	17	41	44	84	480
Solution anneal: 1065 °C (1950 °F) for 1 h, air cool	1110	161	1310	190	19	...	40	96	546

Source: Ref 11

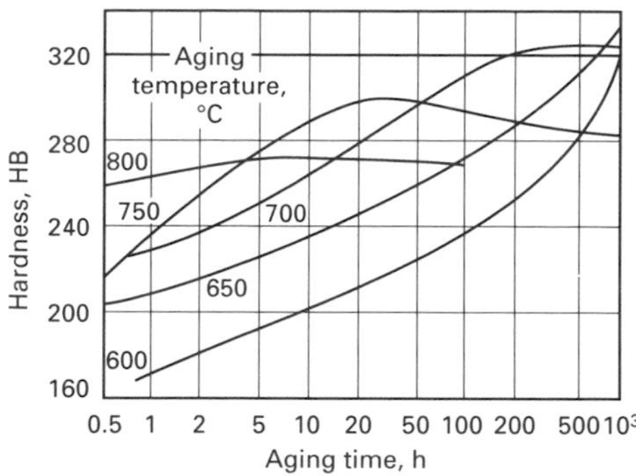

Fig. 7 Hardness changes on aging Nimonic 80A at various temperatures. Prior treatment: solution 1080 °C (1975 °F) for 8 h, water quenched. Source: Ref 13

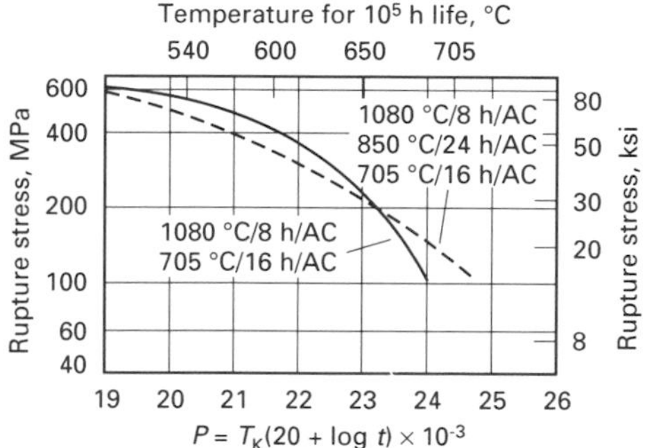

Fig. 8 Effect of increasing intermediate aging on rupture properties of wrought Nimonic 80A. AC, air cooled. Source: Ref 14

above two-step heat treatment yielded good tensile and short-time rupture properties but did not stabilize the microstructure sufficiently to produce optimized properties for long-term elevated-temperature service. For a stabilized microstructure, a higher-temperature aging cycle of 850 °C (1560 °F) for 24 h (air cooling) was added before the final 700 °C (1290 °F) aging. The purpose of the stabilization treatment is to force the development of grain boundary $M_{23}C_6$ carbides by a reaction in which MC carbides react with matrix constituents to form $M_{23}C_6$ carbides and γ'. Thus the stabilization aging heat treatment results in grain boundaries with coarse discrete $M_{23}C_6$ carbides surrounded by a layer of γ'. The stabilization of the grain boundary carbides led to better long-time (low stress, high temperature) stress-rupture properties as shown in Fig. 8.

Carbide distributions can also be controlled by modifying the temperature for solution treatment. For example, a solution treatment of René 41 at 1175 °C (2150 °F) leads to precipitation of a grain-boundary film of $M_{23}C_6$, with deleterious effects on mechanical properties. Therefore, a lower solution treatment temperature (about 1075 °C, or 1970 °F) is used to preserve the fine-grain as-worked structure with well-dispersed M_6C (Ref 15). With exposures of about 1075 to 1095 °C (1970 to 2000 °F) for 8 to 10 h, M_6C remains stable, but γ' goes into solution (Fig. 9).

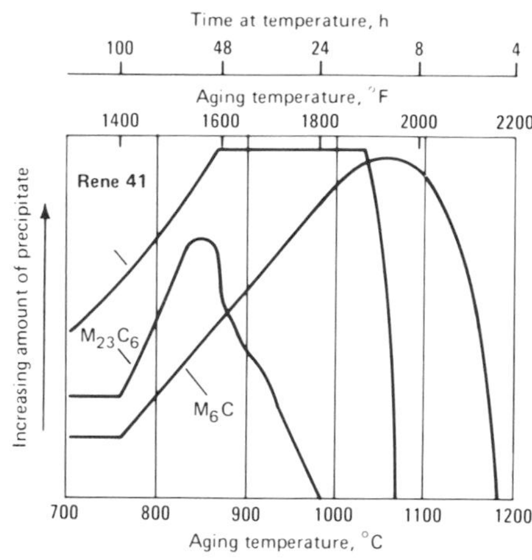

Fig. 9 Relative amounts of precipitates resulting from aging René 41 at various times and temperatures after solution treating at 1205 °C (2200 °F) and water quenching

Heat Treatment of Udimet 500. Udimet 500 is typical of wrought precipitation-hardened superalloys that contain MC and $M_{23}C_6$ carbides and are strengthened by γ'. For a good balance of tensile strength and stress-rupture life, the alloy is given a stabilization treatment before final aging (see Example 2 in this article). If high room-temperature tensile properties are desired and stress-rupture requirements are less important, a two-step heat treatment consisting of just the 1080 °C (1975 °F) solution treatment and the 760 °C (1400 °F) age treatment may be utilized. Table 18 shows how the room-temperature tensile ductility and ultimate strength in Udimet 500 were improved by eliminating the 845 °C (1550 °F) stabilization treatment.

For optimum creep strength, an initial high-temperature solution treatment may be added to develop a coarse grain size. For Udimet 500, a four-step heat treatment includes an initial solution treatment at 1175 °C (2150 °F) for 2 h (with air cooling) followed by the three-step heat treatment described in Example 2 of this article.

Heat Treatment of Waspaloy. Waspaloy is a widely used superalloy for which the heat treatment approach is similar to that for Udimet 500. The heat treatment temperatures and times vary according to working temperatures and microstructural goals. For applications (such as turbine disks) requiring fine-grain structures for optimum tensile, stress rupture, and cyclic properties, a typical heat treatment is:

- Partial solution treatment below the γ' solvus at 995 to 1040 °C (1825 to 1900 °F) for 4 h with air cooling (or faster)
- Stabilization at 845 °C (1550 °F) for 4 h with air cooling
- Aging at 760 °C (1400 °F) for 16 h with air cooling

Before requirements for finer grain sizes led to the use of lower solution temperatures, higher solution temperatures and longer stabilization treatments were used. This was because the higher solution temperature more effectively dissolved carbides, and because a longer stabilization (for example, 845 °C, or 1550 °F, for 24 h with air cooling) was required in order to restore sufficient ductility by spheroidizing the reprecipitating grain-boundary carbides. Now, the lower-temperature solution treatment starts the carbide precipitation process, and a shorter stabilization can be used.

The Waspaloy heat treatment for applications (such as turbine blades) requiring better creep resistance (and therefore a coarser grain size) involves a higher solution temperature above the γ' solvus and a longer stabilization exposure. For turbine blade applications, Waspaloy typically receives:

- 1080 °C (1975 °F) for 4 h with air cooling (or faster)
- 845 °C (1550 °F) for 24 h with air cooling
- 760 °C (1400 °F) for 16 h with air cooling

Figure 10 illustrates the effect of the different heat treatments on the tensile and stress-rupture properties of Waspaloy. A disk heat treatment yields better tensile and rupture properties at lower temperatures. The blade heat treatment results in better rupture strength at higher temperatures.

Heat Treatment of Udimet 700 and Udimet 710. Udimet 700, which is very similar to Astroloy, and Udimet 710 are examples of more highly alloyed wrought superalloys with a relatively high γ' volume fraction of about 40%. Different heat treatments are administered depending on whether a coarse-grained microstructure for turbine bucket and vane applications

Table 18 Effect of eliminating intermediate aging on typical room-temperature mechanical properties of Udimet 500

Condition	Ultimate tensile strength MPa	Ultimate tensile strength ksi	Yield strength (0.2% offset) MPa	Yield strength (0.2% offset) ksi	Elongation in 50 mm (2 in.), %	Reduction in area, %
Specified min	1030	150	690	100	10	15
Obtained with intermediate aging(a)						
Test 1	1030	149	830	120	7	11
Test 2	970	141	810	118	4	5
Obtained without intermediate aging(b)						
Test 1	1170	170	800	116	14.5	17
Test 2	1230	179	850	123	14	16

(a) Heat treatment: 4 h at 1080 °C (1975 °F), air cool; 24 h at 845 °C (1550 °F), air cool (intermediate aging); 16 h at 760 °C (1400 °F), air cool. (b) Same as (a), but without intermediate aging

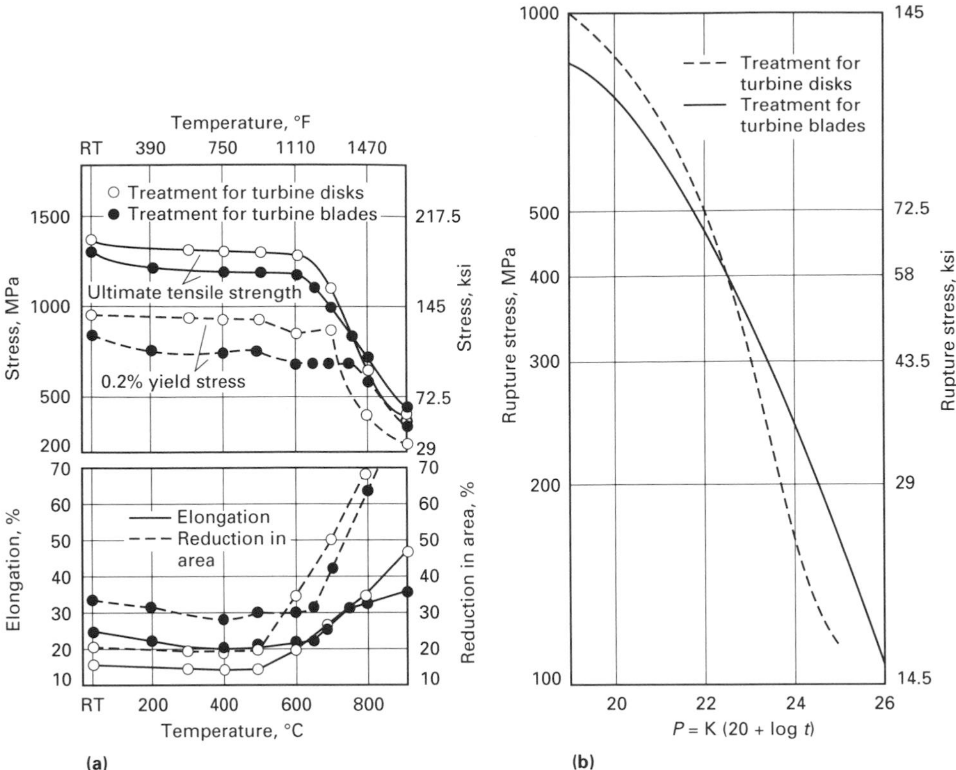

Fig. 10 Influence of different treatments on (a) the tensile properties of Waspaloy and (b) the Larson-Miller plot of Waspaloy. Source: Ref 16

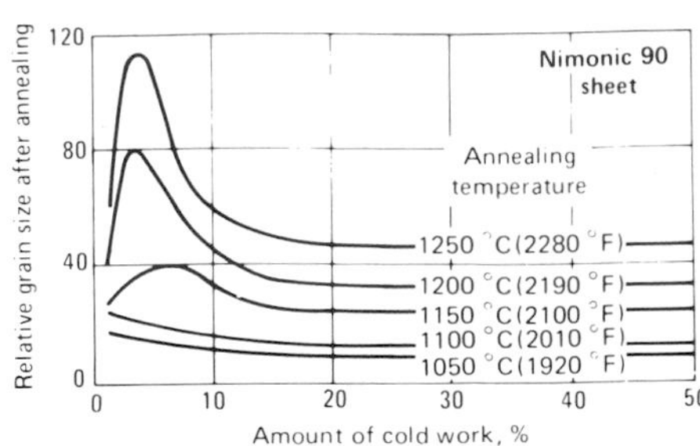

Fig. 11 Effect of cold work and annealing on grain size for Nimonic 90 sheet cold rolled in steps from 1.8 to 0.9 mm (0.072 to 0.036 in.) thick and annealed at five temperatures

is desired, or a fine-grained microstructure for disk applications. For a coarse-grained microstructure, the following heat treatment is typical:

- 1175 °C (2150 °F) for 4 h with air cooling
- 1080 °C (1975 °F) for 4 h with air cooling
- 845 °C (1550 °F) for 24 h with air cooling
- 760 °C (1400 °F) for 16 h with air cooling

The 1175 °C (2150 °F) annealing is a full-solution treatment above the γ' solvus that dissolves the precipitate and allows the grains to coarsen. During the 1080 °C (1975 °F) aging exposure, about half of the γ' that is ultimately formed precipitates and forms a coarse dispersion of particles 0.2 to 0.6 μm in diameter. The two subsequent aging heat treatments build precipitates of fine γ' between the coarser γ' particles, and also precipitate $M_{23}C_6$ at grain boundaries. This heat treatment yields an average grain size of about 225 μm with γ' volume fraction of about 45%. The coarse-grained structure is tailored for good creep-rupture strength.

For a fine-grained microstructure, the heat treatment is of the following form:

- 1105 °C (2020 °F) for 4 h with air cooling (or faster)
- 870 °C (1600 °F) for 8 h with air cooling
- 980 °C (1800 °F) for 4 h with air cooling
- 650 °C (1200 °F) for 24 h with air cooling
- 760 °C (1400 °F) for 8 h with air cooling

The 1105 °C (2020 °F) anneal is a partial solution treatment below the γ' solvus that retains some of the γ' to limit grain growth. The subsequent treatments precipitate carbides and γ'. The two-step exposures of first 870 °C (1600 °F) and then 980 °C (1800 °F) are designed to maximize first the nucleation of precipitates and then the rate of growth of the precipitates. The average grain size of the structure produced is about 11 μm with a γ' volume fraction of about 35%. The fine-grained structure has better mechanical properties at turbine disk application temperatures than that from coarse-grained heat treatment, which is designed for higher-temperature applications.

The effect of the amount of cold working on the recrystallization and grain growth during subsequent solution treating of the nickel-base superalloy Nimonic 90 is shown in Fig. 11. The effect is similar to the behavior shown for A-286 in Fig. 1. The critical amount of deformation that leads to abnormally large grains is in the range of 2 to 10% reduction in thickness, and the grain growth accelerates rapidly at temperatures above 1100 °C (2010 °F).

The precipitation-hardened superalloys that undergo extensive deformation processing, as in sheet forming, usually require in-process annealing to maintain temperatures, relieve forming stresses, and enhance microstructural changes. The annealing practice can also have a marked effect on response to solution treating and aging. This is illustrated by the following two examples for René 41. Like solution-treatment temperatures (Fig. 9), high annealing temperatures can dissolve M_6C carbides, which are useful in preventing formation of $M_{23}C_6$ grain boundary films during aging.

Example 4: Effect of Annealing Temperature on the Grain-Boundary Carbides and Ductility of René 41 Sheet. In one case, parts formed from René 41 sheet showed strain age cracking after solution treatment at 1080 °C (1975 °F) for ½ h, air cooling, and then aging at 760 °C (1400 °F) for 16 h. Cracking has been attributed to a carbide network in the grain boundaries. Cause of the carbide network was traced to in-process annealing at 1180 °C (2150 °F). At 1180 °C (2150 °F), the M_6C carbide was dissolved. Subsequent exposure to temperatures between 760 and 870 °C (1400 and 1600 °F) produced an $M_{23}C_6$ carbide network in the grain boundaries that reduced ductility to an unacceptable level. If the annealing temperature is kept below 1095 °C (2000 °F), M_6C does not dissolve (Fig. 9) and ductility can be improved. A similar effect can occur in weldments of nickel-base alloys if they are annealed at temperatures above 1095 °C (2000 °F).

Example 5: Effect of Thermomechanical Processing on the Grain-Boundary Carbides and Ductility of René 41 Bar Stock. A problem similar to that described in the preceding example occurred in René 41 bar stock. Grain-boundary carbide network reduced ductility and caused difficulty (sometimes cracking) during forming and welding. Investigation of the cause of the grain-boundary network indicated that the bar stock was produced with a final rolling temperature of 1180 °C (2150 °F). Light reductions were taken during final rolling to ensure size for the finished bar stock and to eliminate the possibility of surface tearing. This high rolling temperature, coupled with relatively light reductions (in the range of 2 to 3% produced grain-boundary network because:

- The M_6C carbides were dissolved at the rolling temperature.
- Slow cooling through the range of 870 to 760 °C (1600 to 1400 °F) produced $M_{23}C_6$ in an unfavorable morphology (grain-boundary carbide film).

Rolling temperatures of 1150 °C (2100 °F) maximum, coupled with a final reduction in rolling of at least 10 to 15%, eliminated the grain-boundary carbide film and produced bars that could be welded and formed.

Solid-Solution-Strengthened Iron-, Nickel-, and Cobalt-Base Superalloys

Solid-solution-strengthened iron-, nickel-, and cobalt-base superalloys are generally distinguishable from the precipitation-strengthened superalloys by their relatively low content of precipitate-forming elements such as aluminum, titanium, or niobium. There are, of course, some exceptions to this, particularly as regards niobium content.

As their classification implies, these alloys derive a significant proportion of their strength from solution strengthening, most typically associated with a high content of refractory metals, such as molybdenum or tungsten. Not to be overlooked, however, is the equally significant contribution of carbon, which serves both as a potent solution-strengthening element, and as a source of both primary and secondary carbide strengthening. Primary carbides, carried over from final melting operations, serve to control grain structure and thus contribute somewhat to alloy strength; however, the formation of secondary carbides, which is critical to developing the best strength, is also the key issue in formulating and performing alloy heat treatments.

Solid-solution-strengthened superalloys are usually supplied in the solution-heat-treated condition, where virtually all of the secondary carbides are dissolved, or "in solution." Microstructures generally consist of primary carbides dispersed in a single-phase matrix, the grain boundaries of which are reasonably clean. This is the optimum condition for good elevated-temperature strength and generally best room-temperature fabricability. When the carbon is mostly in solution, exposure at elevated temperatures below the solution temperature will result in secondary carbide precipitation. In service, where the alloy component is subjected to operating stresses, this carbide precipitation will occur both on grain boundaries and intragranularly on areas of high dislocation density. It is the latter that provides for increased strength in service. When exposure to temperatures below the solution temperature occurs during component heat-treating cycles, the result is usually to precipitate secondary carbides only on grain boundaries. This is not normally beneficial for subsequent fabrication, and it reduces the capability of the alloy to develop in-service strengthening by depleting carbon from solution.

Generally speaking, then, solid-solution-strengthened alloy components will exhibit highest strength when placed in service in the fully solution-heat-treated condition; however, the reality of modern complex component designs dictates what can and cannot be done in terms of final heat treatments. Quite often the compromise between component manufacturability and performance will mean something less than optimal alloy structure.

Annealing and Stress Relieving

In the case of solid-solution-strengthened superalloys, heat treatments performed at temperatures below the secondary carbide solvus or solutioning temperature range are classified as mill annealing or stress-relief treatments. Of the two treatments, mill annealing is the most commonly used.

Mill annealing treatments are generally employed for restoring formed, partially fabricated, or otherwise as-worked alloy material properties to a point where continued manufacturing operations can be performed. Such treatments may also be used in finished raw materials to produce structures that are optimum for specific forming operations, such as fine grain size structure for deep drawing applications.

Mill-annealed products may also be used in preference to solution heat treatments for final components where properties other than creep and stress-rupture strength are vital. For example, where low-cycle fatigue properties are important, mill annealing may be used to produce a finer grain size. A finer grain size from mill annealing may also be useful in applications where yield strength instead of creep strength is the limiting design criterion. Finally, mill annealing may be selected in preference to solution annealing because of external constraints, such as avoidance of component distortion at full solution annealing temperatures, or limits to temperature imposed by the melting point of component braze joints.

Because mill annealing is performed below the secondary carbide solvus temperature, some decoration of grain boundaries can be expected in the microstructure. Depending upon the annealing temperature, the particular alloy, and the nature of the secondary carbide involved, this decoration may take the form of either discrete, globular particles or a more continuous film-like morphology. Cooling rates will markedly influence the appearance of this carbide precipitation, as most alloys of this type exhibit the most significant amount of precipitation in the temperature range from about 650 to 870 °C (1200 to 1600 °F). It is always recommended that components be cooled as rapidly as is feasible through this range, within the constraints of equipment used and with due consideration to avoiding component distortion from thermal stresses.

Typical minimum mill annealing temperatures for various alloys are given in Table 19. These temperatures vary significantly from alloy to alloy. They are based principally upon the ability of the treatment to develop a recrystallized grain structure starting from a cold-worked or warm-worked condition and to produce low enough yield strength and high enough ductility for subsequent cold forming operations. Grain size would be expected to increase somewhat, although perhaps not markedly, when higher mill annealing temperatures are used.

The same basic temperatures would apply for mill annealing hot-worked material, although solution annealing is more common. Hot-worked material is usually dynamically recrystallized during the hot-working operation, and the main effect of mill annealing is to promote uniformity of the structure throughout the piece.

Table 19 Minimum mill annealing temperatures for solid-solution-strengthened alloys

Alloy	Approximate minimum temperature for mill annealing °C	°F
Hastelloy X	1010	1850
Hastelloy S	955	1750
Alloy 625	925	1700
RA 333	1035	1900
Inconel 617	1035	1900
Haynes 230	1120	2050
Haynes 188	1120	2050
Alloy L-605	1120	2050
Alloy N-155	1035	1900
Haynes 556	1035	1900

Times at temperature required for mill annealing are governed by several factors. Sufficient furnace time should be allowed to ensure that all parts of the piece are at temperature for the requisite time. The requisite time should be long enough to ensure that structure changes, such as recovery, recrystallization, and carbide dissolution (if any), are essentially complete. Generally, about 5 to 20 min at temperature is sufficient, particularly in thin sections. In continuous thin-strip annealing operations, as little as 1 to 2 min will often suffice. Excessive time at temperature for mill annealing is not necessarily deleterious, but is most often not beneficial. Use of a thermocouple on the actual piece undergoing annealing is always appropriate.

Stress Relief. Unlike mill annealing, stress-relief treatments for solid-solution-strengthened superalloys are not well defined. Dependent upon the particular circumstances, stress relief may be achieved with relatively low-temperature annealing, or may require the equivalent of mill or even solution annealing. In any case, such treatments represent a major compromise between the effectiveness of stress relief and the harm done to the structure or dimensional stability of the component.

Strictly speaking, stress-relief annealing should be considered only if the material is not recrystallized by the treatment. If the intent is to relieve stresses in a piece or component that would otherwise be mill annealed or solution treated, then the first choice is the equivalent of a solution heat treatment or mill annealing to accomplish the required stress relief. Temperatures below the mill annealing temperature range, particularly in the range of 650 to 870 °C (1200 to 1600 °F), will likely result in significant carbide precipitation or other phase formation in some alloys, which may significantly impair alloy performance. Treatments below 650 °C (1200 °F) may be less deleterious but are likely to be less effective in relieving residual stresses.

To relieve stresses in a partially cold- or warm-worked piece or component (that is, a finish-formed component that cannot be mill- or solution-annealed), then the stress-relief treatment should be restricted to a temperature less than that

Table 20 Typical solution annealing temperatures for solid-solution-strengthened alloys

Alloy	Typical solution annealing temperatures °C	°F
Hastelloy X	1165–1190	2125–2175
Hastelloy S	1050–1135	1925–2075
Alloy 625	1095–1205	2000–2200
RA 333	1175–1205	2150–2200
Inconel 617	1165–1190	2125–2175
Haynes 230	1165–1245	2125–2275
Haynes 188	1165–1190	2125–2175
Alloy L-605	1175–1230	2150–2250
Alloy N-155	1165–1190	2125–2175
Haynes 556	1165–1190	2125–2175

which will induce recrystallization. In this class of material, that temperature will vary with the particular alloy and degree of cold or warm work, but will generally be less than about 815 °C (1500 °F). In some materials (such as Inconel 625 and Haynes alloy 214), age-hardening reactions occurring at these lower temperatures must be considered in addition to the more general carbide precipitation encountered in other alloys.

Times at temperature required to effect a significant amount of stress relief are equally ill defined. For the equivalent to mill and solution annealing, similar times should be used. For lower-temperature stress-relief treatments, no specific guidelines are offered, but excessive times should be avoided for obvious reasons.

Solution Heat Treating

Solution heat treating is the most common form of finishing operation applied to solid-solution-strengthened superalloys. As mentioned earlier, a solution treatment places virtually all the secondary carbides into solution. The temperatures at which all secondary carbides are dissolved vary somewhat from alloy to alloy and can differ as a function of the type of secondary carbide involved and the carbon content.

Typical solution treatment temperatures for various alloys are given in Table 20. For some alloys, the temperature range is broader than others; in most cases, such as Haynes 230, this is related to desired flexibility in controlling the grain size in the solution-treated piece. In Haynes 230, for example, an 1175 °C (2150 °F) solution treatment might produce an ASTM grain size between 7 and 9, while a solution treatment at 1230 °C (2250 °F) could be expected to yield a grain size of ASTM 4 to 6, assuming starting material in a sufficiently cold-reduced condition.

Recrystallization and Grain Size. A major function of the solution annealing treatment is to recrystallize warm- or cold-worked structure fully and to develop the required grain size. Aspects such as heating rate and time at temperature are important considerations. Rapid heating to temperature is usually desirable to help minimize carbide precipitation and to preserve the stored energy from cold or warm work required to provide recrystallization and/or grain growth during the solution treatment itself. For much the same reason that re-solution-treating an already annealed piece often does not coarsen grain size without increasing the temperature, slow heating of a cold- or warm-worked material to the solution-treating temperature can produce a finer grain size than may be desired or required.

Table 21 Cooling rate effects on time to 0.5% creep at 870 °C (1600 °F) with 48 MPa (7 ksi) load

Solution treat at 1175 °C (2150 °F) and cool at the rate indicated	Time to 0.5% creep, h Hastelloy X	Haynes 188	Inconel 617
Water quench	8	148	302
Air cool	7	97	15
Furnace cool to 650 °C (1200 °F) and then air cool	6	48	9

Time at temperature considerations for solution heat treatments are similar to those for mill annealing, although slightly longer exposures are generally indicated to ensure full dissolution of secondary carbides. For minimum temperature solution treatments, heavier sections should generally be exposed at temperature for about 10 to 30 min, and thinner sections should be exposed for somewhat shorter times. Solution treatments at the high end of the prescribed temperature range can be shorter, similar to mill annealing. Although very massive parts, such as forgings, may benefit from somewhat longer times at temperature, in no case should any component be exposed to solution treatment temperatures for excessive periods (such as overnight). Long exposures at solution treatment temperatures can result in partial dissolution of primary carbides, with consequent grain growth or other adverse effects.

The effects of cooling rate upon alloy properties following solution heat treatment can be much more pronounced than those related to mill annealing. Because the solution treatment places the alloy in a state of greater supersaturation relative to carbon, the propensity for carbide precipitation upon cooling is significantly increased over that for mill annealing. It is therefore even more important to cool from the solution treatment temperature as fast as possible, bearing in mind the constraints of the equipment and the need to avoid component distortion due to thermal stresses. The sensitivity of individual alloys to property loss from slower cooling down to about 650 °C (1200 °F) varies, but most alloys will suffer at least some property degradation as a result of secondary carbide precipitation. This is shown by the data in Table 21, in which the effects of various cooling practices on the low-strain creep properties of three alloys are described.

Solution Treating Combined with Brazing. Unlike mill annealing, which is usually performed as a manufacturing step itself, solution treating may sometimes be combined with another operation, which imposes significant constraints upon both heating and cooling practices. A good example of this is vacuum brazing. Often performed as the final manufacturing step in the fabrication of components, such a process precludes subsequent solution treatment by virtue of the limits imposed by the melting point of the brazing compound. Therefore, the actual brazing temperatures are sometimes adjusted to allow simultaneous solution heat treating of the component. Unfortunately, the nature of vacuum brazing furnace equipment specifically, and vacuum furnace equipment in general, is such that relatively slow heating and cooling rates are a given. In these circumstances, even with the benefit of advanced forced gas cooling equipment, the structure and properties of alloy components are likely to be less optimal than those achievable with solution treatments performed in other types of equipment.

Relationship of Processing History to Heat Treatment

As for most other alloy materials, the response of solid-solution-strengthened superalloys to heat treatment is very much dependent upon the initial material condition. Generally speaking, when the material is not in the cold- or warm-worked condition, the principal response to heat treatment is a change in the amount and morphology of secondary carbide phases present. Relief of minor residual stresses or relaxation of internal strains, either of which may influence alloy properties to some degree, may also occur. Grain structure, however, may often be substantially unaltered by heat treatment when cold or warm work is absent.

Hot-worked products, in particular those produced at high finishing temperatures, undergo recovery, recrystallization, and grain growth during the working operation itself. If finish working temperatures are too high relative to the final mill-annealing or solution-treatment temperatures, a significant degree of control over the structure resides in the working operation, rather than in the heat treatment. Similarly, if the final hot-working reductions are small, the piece to be heat treated often is initially nonuniform and responds nonuniformly to heat treatment. Material finished at a very high temperature may be best heat treated at temperatures near the high end of the allowable range, and almost always at a temperature above the finish hot-working temperature. For cases with small finish reductions, temperatures at the low end of the range would probably be advisable to minimize the nonuniformity in structure. This last approach might be particularly advisable for pieces with very heavy section thickness, such as large forgings, large-size bars, and thick plates.

Fortunately, solid-solution-strengthened superalloys as a group exhibit relatively wide hot-working ranges, which allow finishing temperatures low enough to produce a warm-worked condition. They are also readily manufactured using cold working processes. In the warm-worked or cold-worked condition, grain structure control resides basically in the heat treatment, but results can be significantly influenced by the amount of work in the piece. As an example of this, the data presented in Table 22 show the

Table 22 Effect of cold reduction and annealing temperature on grain size of 556 alloy

Cold reduction, %	5-min subsequent annealing temperature °C	°F	Degree of recrystallization	ASTM grain size
0	None	None	...	5.0–6.0
10	1010	1850	Incomplete	...
20	1010	1850	Incomplete	...
30	1010	1850	Partial	...
40	1010	1850	Partial	7.5–9.5
50	1010	1850	Full	9.0–10.0
10	1065	1950	Incomplete	...
20	1065	1950	Incomplete	...
30	1065	1950	Full	7.5–9.5
40	1065	1950	Full	8.0–9.5
50	1065	1950	Full	8.5–10.0
10	1120	2050	Full	5.0–5.5
20	1120	2050	Full	7.5–8.5
30	1120	2050	Full	7.0–7.5
40	1120	2050	Full	7.5–9.0
50	1120	2050	Full	8.0–9.5
10	1175	2150	Full	5.0–5.5
20	1175	2150	Full	6.0–6.5
30	1175	2150	Full	4.5–6.5
40	1175	2150	Full	4.5–6.5
50	1175	2150	Full	5.5–6.0

Table 23 Effect of small strains on abnormal grain growth of Hastelloy X

Prior cold work, %	5-min subsequent annealing temperature °C	°F	ASTM grain size
0	None	None	4.5–6.5
1	1120	2050	4.5–6.5
2	1120	2050	4.0–6.5
3	1120	2050	4.0–6.0
4	1120	2050	3.5–6.0
5	1120	2050	3.5–6.0
8	1120	2050	3.5–6.0
1	1175	2150	5.0 + 0 at surface
2	1175	2150	5.0–5.5 + 0 at surface
3	1175	2150	00–0.5
4	1175	2150	4.5–5.0 + 1.0–1.5
5	1175	2150	3.0–3.5 + 1.0–1.5
8	1175	2150	4.5–5.0 (recrystallized)

influence of initial cold work on the grain size of final heat-treated Haynes 556 sheet.

Cold-Worked Products. The particular sequence of cold-work/annealing cycles used in multistep material manufacturing or component fabrication can also affect the structure and properties of these alloys. One general guideline is to keep the temperatures used for intermediate annealing steps at or below the final annealing temperature. Intermediate annealing at temperatures above the final annealing temperature can reduce the degree of structure control possible in the alloy.

The minimum level of cold work shown in Table 22, 10%, is an important rough dividing line between normal recrystallization behavior and possible abnormal grain growth in these alloys. Introduction of small amounts of cold or warm work prior to solution heat treating should be avoided where possible to minimize the potential for abnormal grain growth phenomena. The effects of very small amounts of cold work on the grain size response to annealing for Hastelloy X are shown in Table 23. The samples used to generate these data were carefully strained tensile test specimens, subsequently exposed to the annealing temperatures shown. Strains from 1 to 8% produced little effect for mill annealing temperatures up to 1120 °C (2050 °F); however, for solution annealing at 1175 °C (2150 °F), abnormal grain growth was observed for strains of 1 to 5%.

Unfortunately, in everyday fabrication of complex components, it is difficult if not impossible to avoid situations where such low levels of cold work or strain are present. Some alloys are more tolerant of this than others, but virtually all will exhibit abnormal grain growth under some conditions. Procedures that may be effective for minimizing the problem are:

- Solution treating at the low end of allowable temperature ranges
- Mill annealing in preference to solution annealing for intermediate heat treatments during component fabrication
- Stress-relief annealing directly prior to final solution annealing

Cast Superalloy Heat Treatment

Heat treatment of cast superalloys in the traditional sense was not employed until the mid-1960s. Before the use of shell molds, the heavy-walled investment mold dictated a slow cooling rate with its associated aging effect on the casting. As faster cooling rates with shell molds developed, the aging response varied with section size and the many possible casting variables. These factors, coupled with significant γ' alloying additions, provided the opportunity to minimize property scatter by heat treatment. The combination of hot isostatic pressing (HIP) plus heat treatment has also greatly enhanced properties.

Generally, heat treating cast superalloys involves homogenization and solution heat treatments or aging heat treatments. A stress-relief heat treatment may also be performed to reduce residual casting, welding, or machining stresses. Cobalt-base alloy heat treatments may be done in an air atmosphere unless unusually high-temperature treatments are required, in which case vacuum or inert gas environments are used. Conversely, nickel-base alloys are always heat treated in a vacuum or in an inert gas medium. Detailed information can be found in Ref 17.

Like wrought superalloys, the solution heat-treating procedures of cast superalloys must be optimized to stabilize the carbide morphology. High-temperature exposure may cause extensive carbide degeneration, resulting in grain-boundary carbide overload and compromised mechanical properties. Unlike wrought superalloys, however, many polycrystalline materials are used in the as-cast plus aged condition without any specific solution step. Cast cobalt-base superalloys, for example, are not usually solution treated (although they may be given stress-relief and/or aging treatments). When required, cast cobalt-base superalloys are generally aged at 760 °C (1400 °F) to promote formation of discrete

Table 24 Typical heat treatments for precipitation-strengthened cast superalloys

Alloy	Heat treatment (temperature/duration in h/cooling)(a)
Polycrystalline (conventional) castings	
B-1900/B-1900 + Hf	1080 °C (1975 °F)/4/AC + 900 °C (1650 °F)/10/AC
IN-100	1080 °C (1975 °F)/4/AC + 870 °C (1600 °F)/12/AC
IN-713	As-cast
IN-718	1095 °C (2000 °F)/1/AC + 955 °C (1750 °F)/1/AC + 720 °C (1325 °F)/8/FC + 620 °C (1150 °F)/8/AC
IN-718 with HIP	1150 °C (2100 °F)/4/FC + 1190 °C (2175 °F)/4/15 ksi (HIP) + 870 °C (1600 °F)/10/AC + 955 °C (1750 °F)/1/AC + 730 °C (1350 °F)/8/FC + 655 °C (1225 °F)/8/AC
IN-738	1120 °C (2050 °F)/2/AC + 845 °C (1550 °F)/24/AC
IN-792	1120 °C (2050 °F)/4/RAC + 1080 °C (1975 °F)/4/AC + 845 °C (1550 °F)/24/AC
IN-939	1160 °C (2120 °F)/4/RAC + 1000 °C (1830 °F)/6/RAC + 900 °C (1650 °F)/24/AC + 700 °C (1290 °F)/16/AC
MAR-M246+Hf	1220 °C (2230 °F)/2/AC + 870 °C (1600 °F)/24/AC
MAR-M 247	1080 °C (1975 °F)/4/AC + 870 °C (1600 °F)/20/AC
René 41	1065 °C (1950 °F)/3/AC + 1120 °C (2050 °F)/0.5/AC + 900 °C (1650 °F)/4/AC
René 77	1163 °C (2125 °F)/4/AC + 1080 °C (1975 °F)/4/AC + 925 °C (1700 °F)/24/AC + 760 °C (1400 °F)/16/AC
René 80	1220 °C (2225 °F)/2/GFQ + 1095 °C (2000 °F)/4/GFQ + 1050 °C (1925 °F)/4/AC + 845 °C (1550 °F)/16/AC
Udimet 500	1150 °C (2100 °F)/4/AC + 1080 °C (1975 °F)/4/AC + 760 °C (1400 °F)/16/AC
Udimet 700	1175 °C (2150 °F)/4/AC + 1080 °C (1975 °F)/4/AC + 845 °C (1550 °F)/24/AC + 760 °C (1400 °F)/16/AC
Waspaloy	1080 °C (1975 °F)/4/AC + 845 °C (1550 °F)/4/AC + 760 °C (1400 °F)/16/AC
Directionally-solidified (DS) castings	
DS MAR-M 247	1230 °C (2250 °F)/2/GFQ + 980 °C (1800 °F)/5/AC + 870 °C (1600 °F)/20/AC
DS MAR-M200+Hf	1230 °C (2250 °F)/4/GFQ + 1080 °C (1975 °F)/4/AC + 870 °C (1600 °F)/32/AC
DS René 80H	1190 °C (2175)/2/GFQ + 1080 °C (1975 °F)/4/AC + 870 °C (1600 °F)/16/AC
Single-crystal castings	
CMSX-2	1315 °C (2400 °F)/3/GFQ + 980 °C (1800 °F)/5/AC + 870 °C (1600 °F)/20/AC
PWA 1480	1290 °C (2350 °F)/4/GFQ + 1080 °C (1975 °F)/4/AC + 870 °C (1600 °F)/32/AC
René N4	1270 °C (2320 °F)/2/GFQ + 1080 °C (1975 °F)/4/AC + 900 °C (1650 °F)/16/AC

(a) AC, air cooling; FC, furnace cooling; GFQ, gas furnace quench; RAC, rapid air cooling

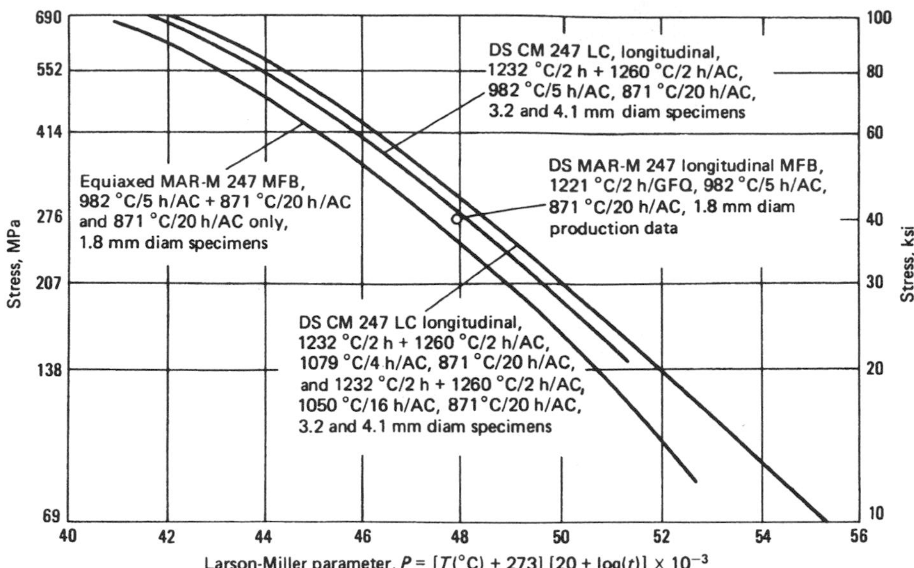

Fig. 12 Effect of various heat treatments on the Larson-Miller stress-rupture strength of directionally solidified (DS) CM 247 LC and DS and equiaxed MAR-M 247. MFB, machined from blade. GFQ, gas furnace quenched; AC, air cooled

$Cr_{21}C_6$ particles. Higher-temperature aging can result in acicular and/or lamellar precipitates.

Precipitation-Strengthened Castings

Precipitation-strengthened nickel- or iron/nickel-base superalloys are cast using the investment casting process. The resultant casting comprises a large number of grains and is referred to as a polycrystalline or conventional casting. If the casting is solidified under a thermal gradient, a columnar-grained directionally solidified casting will result. Directionally solidified (DS) airfoil castings are used in the turbine sections of gas turbine engines to enhance durability and performance. Additional benefits can be achieved using directional-solidification investment casting to cast turbine airfoils as single crystals. Precipitation-strengthened nickel-base superalloys are primarily utilized for turbine airfoils, while iron-nickel alloys are employed as large investment-cast structural castings.

Superalloys are heat treated to control the morphology of the precipitating phases (γ′, γ″, carbides, and δ) that are responsible for the mechanical properties of the alloy. Three basic heat treatment steps are used:

- Solution
- Stabilization
- Aging

Representative heat treatments for several alloys are listed in Table 24. The effects of multistep solution heat treating and aging treatments on the stress-rupture strength of DS nickel-base alloys are shown in Fig. 12.

Solution Treating. Polycrystalline cast nickel-base superalloys may or may not be given solution treatment. Because alloys respond differently to γ′ solution treatment, some are only given an aging treatment. For those that do respond to partial solution treatment, the treatment is performed at a temperature safely below the incipient melting point of the alloy for times ranging from 2 to 6 h at temperature. The solution heat treatment is employed to dissolve the phases in the as-cast microstructure, in the ideal case returning the alloy microstructure to a single-phase γ (fcc) solid solution, and to homogenize the segregated as-cast microstructure.

The solution treatment is performed at a temperature above or near the γ′ solvus temperature. A protective atmosphere, such as vacuum, argon, helium, or hydrogen, can be used to prevent oxidation of the casting. When a vacuum furnace is employed, a partial pressure of an inert gas, such as argon, is used rather than a hard vacuum to prevent surface depletion of chromium and aluminum from the castings. When the solution heat treatment temperature is very close to the incipient melting temperature of the alloy, varied heating rates are used to homogenize the castings during the time taken to reach the solution temperature. This is done to prevent incipient melting, which might occur if a segregated casting were heated very rapidly. Many conventionally cast (polycrystalline) nickel-base alloys are not solution heat treated, but all DS alloys, either columnar-grained or single-castings, are solution treated.

Gamma prime and other phases precipitate as the casting is cooled below the γ′ solvus. In nickel-base superalloys, such as those used for turbine airfoils, growth of the fine γ′ precipitate phase is very rapid at a few hundred degrees Fahrenheit below the temperature involved in solution heat treatment. Therefore, it is necessary to cool the casting as rapidly as possible to prevent coarsening of the γ′ during the cooling cycle, which can degrade the mechanical properties of the casting. Rapid cooling rates are achieved in a vacuum furnace by introducing additional cold inert gas and circulating the gas in the furnace. This is sometimes referred to as a gas furnace quench (GFQ). A retort, which contains the castings in a protective gas environment, is employed when a conventional furnace (not vacuum) is used for solution heat treatment. Removing the retort from the furnace and passing cold gas through it provides the desired rapid cooling rate. With iron-nickel-base alloys, the solution heat treatment homogenizes the casting and dissolves the δ and γ″ phases, which facilitates weld repair of the casting. For these alloys a rapid cooling rate from the solution temperature is not required.

The stabilization heat treatment is employed to enhance creep-rupture properties. A temperature between the solution and aging temperature is used. The purpose of this heat treatment is to optimize the γ′ size and morphology and to assist decomposition of the coarse, as-cast MC carbides into fine, grain-boundary carbides. With nickel-base alloys used for turbine airfoils, the stabilization heat treatment is often combined with the heat treatment used to bond or diffuse a coating onto the alloy substrate.

In iron-nickel-base alloys, a stabilization heat treatment can be used to precipitate δ phase at the grain boundaries for good notch rupture properties. Like solution heat treatment, the stabilization heat treatment is carried out in a protective atmosphere, such as argon, helium, hydrogen, or vacuum, to prevent excessive oxidation of the casting. Retorts and conventional furnaces are used to provide the stabilization heat treatment under a protective atmosphere.

Cooling rates equivalent to air cooling or faster are normally used. As the stabilization heat treatment temperature is normally several hundred degrees Fahrenheit lower than the solution temperature, coarsening of the γ′ or other strengthening precipitate phases will be much slower, and a rapid cooling rate is not as critical. For iron-nickel-base alloy castings, which are commonly weld repaired in the solution-treated condition, the stabilization heat treatment also serves as a stress relief.

The aging heat treatment is employed to precipitate additional γ′ as very fine precipitates. This is important to achieve tensile and lower-temperature creep-rupture properties. With iron-nickel-base superalloys, γ″ also precipitates during the aging heat treatment. Cooling from the aging temperature is not critical, but rates equivalent to air cooling or greater are often used. Protective atmospheres are less critical at the lower temperatures employed for aging, but they are usually used. Equipment similar to that employed for the stabilization heat treatment is used for the aging heat treatment.

Hot isostatic pressing (HIP) is a process wherein hydrostatic pressure and elevated temperature are applied concurrently. It is utilized on superalloy castings to eliminate casting porosity. HIP is usually conducted at or near the solution temperature. Use of an inert gas such as argon under high pressure as the pressure transfer medium precludes achieving a rapid cooling rate upon completion of the cycle. As a result, cast-

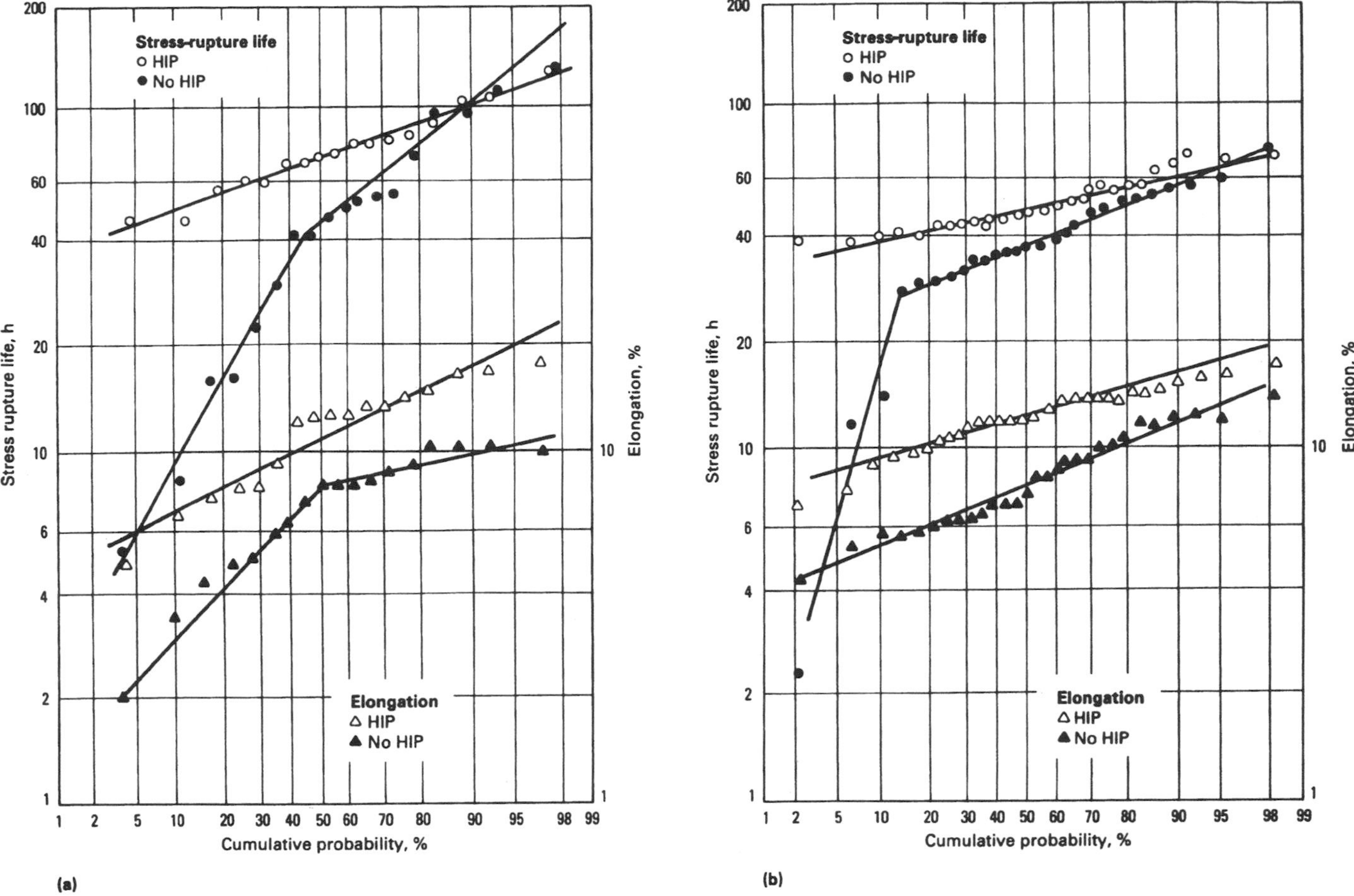

Fig. 13 Effect of hot isostatic pressing (HIP) on the stress-rupture properties of cast IN-738. Test material was hot isostatically pressed at 1205 °C (2200 °F) and 103 MPa (15 ksi) for 4 h. (a) Test conditions: 760 °C (1400 °F) and 586 MPa (85 ksi). (b) Test conditions: 980 °C (1800 °F) and 152 MPa (22 ksi)

Table 25 Typical heat treatments for solid-solution-strengthened cast superalloys

Alloy	Heat treatment
Hastelloy C	1220 °C (2225 °F)/0.5 h/air cool
Hastelloy S	1050 °C (1925 °F)/1 h, air cool
Hastelloy X	As-cast
Inconel 600	As-cast
Inconel 625	1190 °C (2175 °F)/1 h/air cool
FSX-414	1150 °C (2100 °F)/4 h/ furnace cool + 980 °C (1800 °F)/4 h/ furnace cool
MAR-M509	As-cast
WI-52	As-cast
X-40	As-cast

ings receiving a HIP cycle that require a rapid cool from the HIP temperature are given a subsequent heat treatment at atmospheric pressure so that the castings can be rapidly cooled. A pre-HIP homogenization cycle is used for some large iron-nickel alloy castings to increase the local melting temperature by homogenization of the local alloy composition. Figure 13 shows the effect of HIP on the stress-rupture properties of a polycrystalline nickel-base superalloy.

Stress-relief heat treatments are performed following welding or other processing on the casting that increases residual stress. They are usually carried out between the stabilization and aging temperatures in a protective atmosphere. The DS nickel-base alloys, with their higher γ' solvus temperatures, require higher stress-relief temperatures than the iron-nickel-base alloys. Stress-relief temperatures of 870 to 1080 °C (1600 to 1975 °F) are employed to stress relieve precipitation-strengthened superalloy castings.

Solid-Solution-Strengthened Castings

Nonprecipitation-strengthened or solid-solution-strengthened high-temperature superalloy castings are generally distinguishable from the precipitation-strengthened cast superalloys by their relatively low content of precipitate-forming elements such as aluminum, titanium, or niobium. These iron-nickel-, nickel-, and cobalt-base high-temperature alloys are heat treated to homogenize the casting and relieve any stresses in the casting as a result of either the casting process or welding. These alloys primarily derive their strength from solid solution strengthening, with carbides being the only other phases present. With no phase reactions to control enhancement of mechanical properties by heat treatment, many of these alloys do not require any heat treatment and are often used in the as-cast condition. Representative heat treatments for several alloys are listed in Table 25.

Subsequent processing following casting, such as application of a surface coating, welding, or brazing, may impose additional heat treatment requirements. Heat treatments to bond coatings to a cast superalloy substrate are usually performed at temperatures of 980 to 1090 °C (1800 to 2000 °F). Stress relief, following joining or for other purposes, can be carried out over a broad range of temperatures. The particular temperature represents a compromise between the effectiveness of stress relief and the damage to the structure or dimensional stability of the casting. Although not truly a stress relief, some stress-relief heat treatments are conducted at temperatures sufficiently high to cause recrystallization. In a cast component, local recrystallization does not normally have significant detrimental effects on the mechanical properties of the casting and is usually tolerated.

The annealing or stress-relieving heat treatments that are given to this class of nonprecipitating high-temperature alloys are normally done in a protective atmosphere to prevent oxidation or surface contamination of the casting. Heat treatment is usually conducted in a batch-type furnace; the high-temperature superalloy castings are loaded into a retort that contains a protective atmosphere such as argon or hydrogen. Vacuum can also be used. Rapid cooling rates are not usually required.

ACKNOWLEDGMENT

The information in this article is largely taken from D.A. DeAntonio, D. Duhl, T. Howson, and M.F. Rothman, Heat Treating of Superalloys, *Heat Treating*, Vol 4, *ASM Handbook*, ASM International, 1991, p 793–814.

REFERENCES

1. D.D. Krueger, The Development of Direct Age 718 for Gas Turbine Engine Disk Applications, in *Proceedings of Superalloy 718—Metallurgy and Applications,* E.A. Loria, Ed., The Metallurgical Society, 1989, p 279–296
2. E.E. Brown et al., Minigrain Processing of Nickel-Base Alloys, in *Superalloys—Processing,* American Institute of Mechanical Engineers, 1972, section L
3. L.A. Jackman, in *Proceedings of the Symposium on Properties of High-Temperature Alloys,* Electrochemical Society, 1976, p 42
4. N.A. Wilkinson, *Met. Technol.,* July 1977, p 346
5. E.E. Brown and D.R. Muzyka, in *Superalloys II,* C.T. Sims, N.S. Stoloff, and W.C. Hagel, Ed., John Wiley & Sons, 1987, p 180
6. H. Hucek, Ed., *Aerospace Structural Metals Handbook,* MPDC, Battelle Columbus, 1990, Section 4107, p 5–8
7. H. Hucek, Ed., *Aerospace Structural Metals Handbook,* MPDC, Battelle Columbus, 1990, Section 4105, p 7
8. J.W. Brook and P.J. Bridges, in *Superalloys 1988,* The Metallurgical Society, 1988, p 33–42
9. E.E. Brown and D.R. Muzyka, in *Superalloys II,* C.T. Sims, N.S. Stoloff, and W.C. Hagel, Ed., John Wiley & Sons, 1987, p 185
10. H. Hucek, Ed., *Aerospace Structural Metals Handbook,* MPDC, Battelle Columbus, 1990, Section 4103, p 16
11. O.A. Onyeiouenyi, Alloy 718—Alloy Optimization for Applications in Oil and Grease Production, in *Proceedings of Superalloy 718—Metallurgy and Applications,* E.A. Loria, Ed., The Metallurgical Society, 1989, p 350
12. J. Kolts, Alloy 718 for the Oil and Gas Industry, in *Proceedings of Superalloy 718—Metallurgy and Applications,* E.A. Loria, Ed., The Metallurgical Society, 1989, p 332
13. W. Betteridge, *The Nimonic Alloys,* Edward Arnold, Ltd., 1959, p 77
14. E.W. Ross and C.T. Sims, Jr., *in Superalloys II,* C.T. Sims, N.S. Stoloff, and W.C. Hagel, Ed., John Wiley & Sons, 1987, p 127
15. E.W. Ross and C.T. Sims, in *Superalloys II,* C.T. Sims, N.S. Stoloff, and W.C. Hagel, Ed., John Wiley & Sons, 1987, p 927
16. F. Schubert, Temperature and Time Dependent Transformation: Application to Heat Treatment of High Temperature Alloys, in *Superalloys Source Book,* M.J. Donachie, Jr., Ed., ASM International, 1989, p 88
17. G.K. Bouse and J.R. Mihilisin, Metallurgy of Investment Cast Superalloy Components, in *Superalloys, Supercomposites and Superceramics,* Academic Press, 1989, p 99–148

Elevated-Temperature Corrosion Properties of Superalloys

HIGH-TEMPERATURE DEGRADATION of superalloys is typically caused by oxidation, carburization, nitridation, sulfidation, and/or halogenation. Each type of degradation is caused by specific corrosive media and may be minimized by the addition of appropriate alloying elements as summarized in Table 1. Most high-temperature alloys have sufficient amounts of chromium (with or without additions of aluminum or silicon) to form chromia (Cr_2O_3), alumina (Al_2O_3), and/or silica (SiO_2) protective oxide scales, which provide resistance to environmental degradation.

Oxidation

Alloying Effects. Resistance to oxidizing environments at high temperatures is a primary requirement for superalloys, whether they are used coated or uncoated. Therefore, an understanding of superalloy oxidation and how it is influenced by alloy characteristics and exposure conditions is essential for effective design and application of superalloys. As the nickel content in the Fe-Ni-Cr system increases from austenitic stainless steels to 800-type alloys, and to Ni-Cr alloys, the alloys become more stable in terms of metallurgical structure and more resistant to creep deformation. The alloys also become more resistant to scaling as nickel content increases. Figure 1 illustrates that the scaling resistance of several high-nickel alloys was better than that of austenitic stainless steels when cycled to 980 °C (1800 °F) in air (Ref 1).

A majority of high-temperature alloys rely on chromia scale to resist oxidation attack. Chromium has a very high affinity for oxygen, so it can readily react with oxygen to form chromia. A sufficiently high level of chromium is needed in the alloy to develop an external chromia scale for oxidation protection. The Fe-Cr, Ni-Cr, and Co-Cr alloys exhibit the lowest oxidation rates when chromium concentration is about 15 to 30% (Ref 2). Therefore, most iron-, nickel-, and cobalt-base commercial austenitic alloys contain about 16 to 25% Cr. Because of the high solubility of chromium in austenitic alloys, these commercial alloys have been developed without melting and processing difficulties. Excessive amounts of chromium tend to promote embrittling intermetallic phases, such as σ-phase, or even α-ferrite.

Aluminum is also very effective in improving alloy oxidation resistance. Cyclic oxidation in air to 1090 °C (2000 °F) showed that alloy 601 (about 1.3% Al) was better than alloy 600 (0% Al), as illustrated in Fig. 2. As temperature increases, the amount of aluminum needed in the alloy to resist cyclic oxidation also increases. Figure 3 shows a linear weight loss for alloy 601 when cycled to 1150 °C (2100 °F) in air for 42 days (Ref 4). In contrast, alloy 214 (about 4.5% Al) showed essentially no changes in weight (Ref 4). With only about 1.3% Al, alloy 601 forms a chromia scale when heated to elevated temperatures. The chromia begins to convert to volatile CrO_3 as the temperature exceeds about 1000 °C (1830 °F), thus losing its protective capability against oxidation attack (Ref 5, 6). Alloy 214, with about 4.5% Al, on the other hand, forms an alumina scale that is significantly more protective. The excellent oxidation resistance of alloy 214 might also be due partially to yttrium in the alloy. The beneficial effects of rare earth elements (e.g., Y, La, Ce, etc.) on the oxidation resistance of alloys have been extensively studied for alumina or chromia formers (Ref 7–13).

Many mechanisms have been proposed to explain these effects. One widely accepted mechanism is the formation of oxide intrusions (or pegs), which mechanically key the scale to the substrate (Ref 8). Funkenbusch et al. (Ref 14) proposed that rare earth elements (and other reactive elements), which are strong sulfide formers, reduce the segregation of sulfur to the oxide scale-metal interface, thereby improving scale adherence. Rare earth elements have been used in the development of superalloys in recent years. Some of these commercial alloys include alloy 214 using yttrium, and alloys 188, S, and 230 using lanthanum.

Table 1 Alloying elements and their effects in nickel- and iron-base superalloys

Element	Effects on alloy
Chromium	Improves oxidation resistance provided temperature does not exceed 950 °C for long periods; decreases carbon ingress—helps carburization resistance; detrimental to fluorine-containing environment at high temperatures; detrimental to nitridation resistance; high chromium beneficial to oil ash corrosion and attack by molten glass; improves sulfidation resistance
Silicon	Improves resistance to oxidation, nitridation, sulfidation, and carburization; synergistically acts with chromium to improve scale resilience; detrimental to non-oxidizing chlorination process
Aluminum	Independently and synergistically with chromium raises oxidation resistance; helps sulfidizing and carburization resistance; detrimental to nitridation resistance
Titanium	Detrimental to nitridation resistance
Niobium	Increases short-term creep strength; may be beneficial in carburization resistance; detrimental to nitridation resistance
Molybdenum, tungsten	Improves high-temperature strength, good in reducing chlorination resistance; improves creep strength; detrimental for oxidation resistance at higher temperatures
Nickel	Improves carburization, nitridation, and chlorination resistance; detrimental to sulfidation resistance
Carbon	Improves strength; helps nitridation resistance; beneficial to carburization resistance; oxidation resistance adversely affected
Yttrium, rare earths	Improves adherence and spalling resistance of oxide layer, hence improves oxidation, sulfidation, and carburization resistance
Manganese	Slight positive effect on high-temperature strength and creep; detrimental to oxidation resistance, increases solubility of nitrogen
Cobalt	Reduces rate of sulfur diffusion, hence helps with sulfidation resistance; improves solid-solution strength

Source: VDM Technologies Corp.

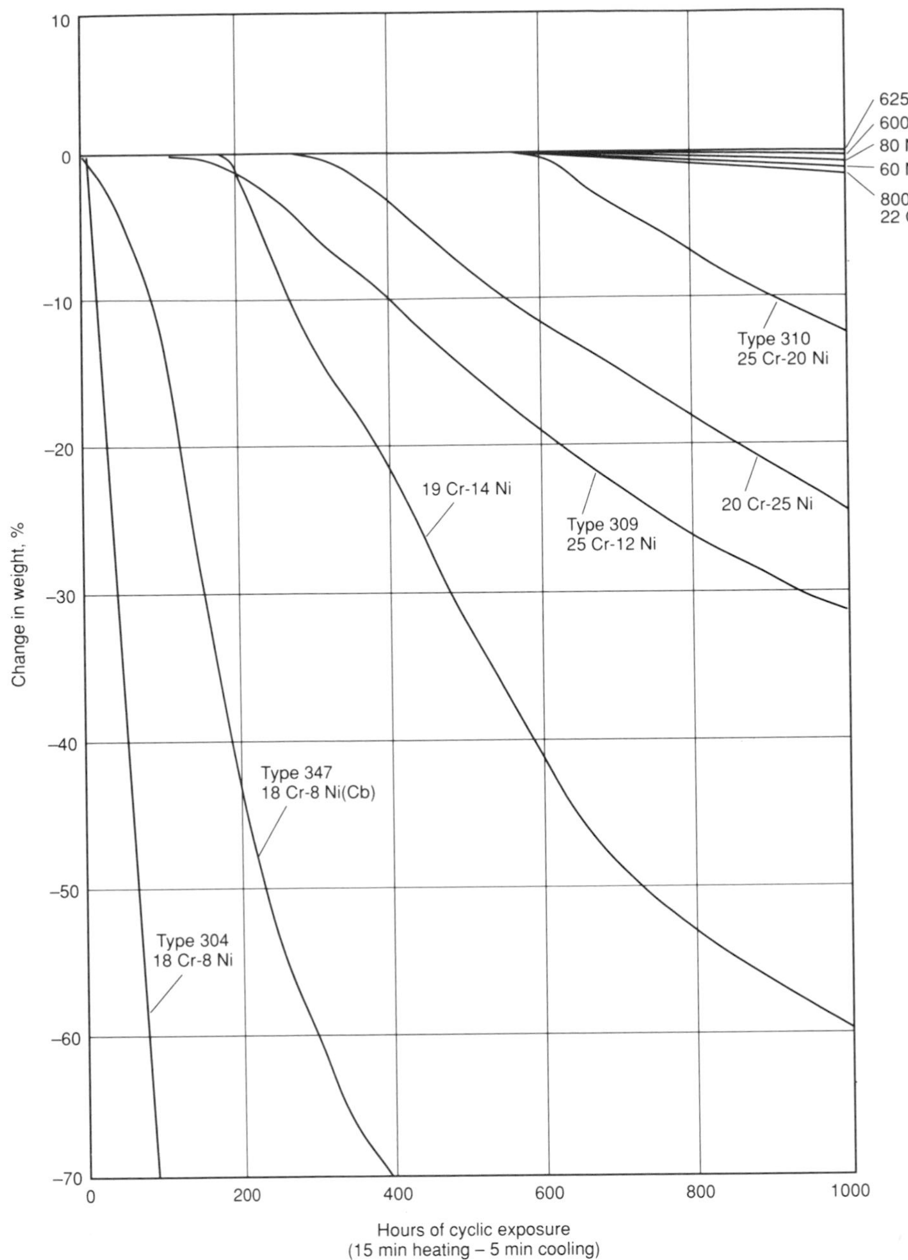

Fig. 1 Cyclic oxidation resistance of several stainless steels and nickel-base alloys in air at 980 °C (1800 °F). Source: Ref 1

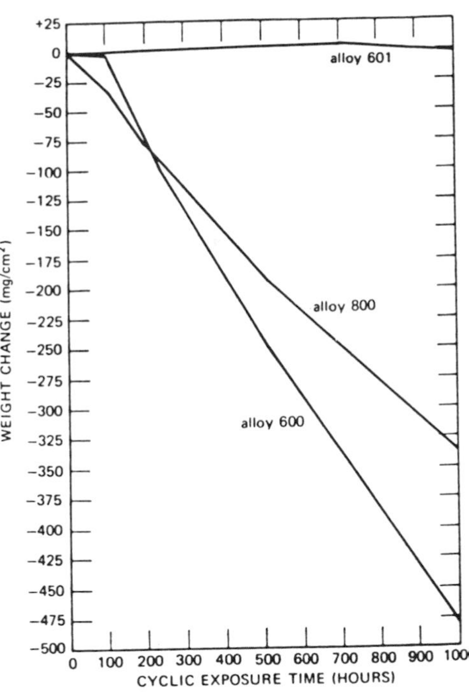

Fig. 2 Cyclic oxidation resistance of alloy 601 compared to alloys 600 and 800 in air at 1090 °C (2000 °F) (15 min in hot zone and 5 min out of hot zone). Source: Ref 3

Long-term oxidation data in air at an intermediate temperature were generated by Barrett (Ref 15) for 33 high-temperature alloys, from ferritic stainless steels to superalloys. Tests were performed at 815 °C (1500 °F) for 10,000 h with 10,000 h cycles (a total of 10 cycles). The results are shown in Fig. 4. Also included for comparison is the maximum metal loss value for chromia formers in isothermal tests (10,000 h in air at 815 °C, or 1500 °F). As expected, many alloys suffered more attack in cyclic tests than in isothermal tests. Alloys with no chromium or low chromium content, such as alloy B (Ni-Mo), alloy N (Ni-Mo-Cr), and type 409 stainless steel, suffered severe oxidation. It was surprising to find that type 321 exhibited severe metal loss, significantly more than types 304, 316, and 347. No explanation of this was offered in the study. It may be due to titanium, which is generally detrimental to alloy oxidation resistance, as discussed below.

An oxidation database for a wide variety of commercial alloys, including stainless steels, high-nickel alloys, and superalloys, has also been generated (Ref 16). Tests were conducted in flowing air (30 cm/min) at 980, 1095, 1150, and 1200 °C (1800, 2000, 2100, and 2200 °F) for 1008 h. The samples were cooled to room temperature for visual examination once a week (168 h). The results, presented in terms of metal loss and average metal affected (metal loss plus internal penetration), are summarized in Table 2.

For Fe-Cr and Fe-Ni-Cr alloys, types 304 and 316 stainless steel were found to be the least resistant to oxidation, both suffering severe attack at 980 °C (1800 °F). As the temperature was increased to 1095 °C (2000 °F) or higher, the oxidation rates of these two alloys became so high that the samples were consumed before the end of the 1008 h test. Type 304 performed slightly better than type 316, presumably due to a slightly higher chromium level. The high-chromium ferritic stainless steel type 446 (Fe-27Cr) exhibited good oxidation resistance at 980 °C (1800 °F), but suffered very severe oxidation attack at 1095 °C (2000 °F) and higher. Types 310, RA330, and alloy 800H were, in general, better than types 446, 304, and 316, particularly at higher temperatures. This may be attributed to higher nickel contents.

The oxidation resistance of alloy 556 is significantly better than that of Multimet alloy, even though both are iron-base superalloys with molybdenum and tungsten additions for solid solution strengthening. Multimet alloy begins suffering rapid oxidation at 1095 °C (2000 °F), and the rates become unacceptably high at higher temperatures. Alloy 556 shows excellent oxidation resistance up to 1095 °C (2000 °F). Oxidation attack becomes high at 1150 °C (2100 °F) and unacceptably high at 1200 °C (2200 °F).

Alloy 556 and Multimet alloy have similar chemical compositions. Alloy 556 was developed by making minute changes of minor alloying elements in Multimet alloy. The improved oxidation resistance resulted from replacement of niobium (columbium) with tantalum as well as the controlled addition of reactive elements, such

Table 2 Results of 1008 h static oxidation tests on iron-, nickel-, and cobalt-base alloys in flowing air at indicated temperatures

	980 °C (1800 °F)				1095 °C (2000 °F)				1150 °C (2100 °F)				1205 °C (2200 °F)			
	Metal loss		Average metal affected		Metal loss		Average metal affected		Metal loss		Average metal affected		Metal loss		Average metal affected	
Alloy	mm	mils	mm	mils	mm	mils	mm	mils	mm	mils	mm	mils	mm	mils	mm	mils
214	0.0025	0.1	0.005	0.2	0.0025	0.1	0.0025	0.1	0.005	0.2	0.0075	0.3	0.005	0.2	0.018	0.7
601	0.013	0.5	0.033	1.3	0.03	1.2	0.067	2.6	0.061	2.4	0.135	5.3	0.11	4.4	0.19	7.5
600	0.0075	0.3	0.023	0.9	0.028	1.1	0.041	1.6	0.043	1.7	0.074	2.9	0.13	5.1	0.21	8.9
230	0.0075	0.3	0.018	0.7	0.013	0.5	0.033	1.3	0.058	2.3	0.086	3.4	0.11	4.5	0.20	7.9
S	0.005	0.2	0.013	0.5	0.01	0.4	0.033	1.3	0.025	1.0	0.043	1.7	>0.81	31.7	>0.81	31.7
617	0.0075	0.3	0.033	1.3	0.015	0.6	0.046	1.8	0.028	1.1	0.086	3.4	0.27	10.6	0.32	12.5
333	0.0075	0.3	0.025	1.0	0.025	1.0	0.058	2.3	0.05	2.0	0.1	4.0	0.18	7.1	0.45	17.7
X	0.0075	0.3	0.023	0.9	0.038	1.5	0.069	2.7	0.11	4.5	0.147	5.8	>0.9	35.4	>0.9	35.4
671	0.0229	0.9	0.043	1.7	0.038	1.5	0.061	2.4	0.066	2.6	0.099	3.9	0.086	3.4	0.42	16.4
625	0.0075	0.3	0.018	0.7	0.084	3.3	0.12	4.8	0.41	16.0	0.46	18.2	>1.2	47.6	>1.2	47.6
Waspaloy	0.0152	0.6	0.079	3.1	0.036	1.4	0.14	5.4	0.079	3.1	0.33	13.0	>0.40	15.9	>0.40	15.9
R-41	0.0178	0.7	0.122	4.8	0.086	3.4	0.30	11.6	0.21	8.2	0.44	17.4	>0.73	28.6	>0.73	28.6
263	0.0178	0.7	0.145	5.7	0.089	3.5	0.36	14.2	0.18	6.9	0.41	16.1	>0.91	35.7	>0.91	35.7
188	0.005	0.2	0.015	0.6	0.01	0.4	0.033	1.3	0.18	7.2	0.2	8.0	>0.55	21.7	>0.55	21.7
25	0.01	0.4	0.018	0.7	0.23	9.2	0.26	10.2	0.43	16.8	0.49	19.2	>0.96	37.9	>0.96	37.9
150	0.01	0.4	0.025	1.0	0.058	2.3	0.097	3.8	>0.68	26.8	>0.68	26.8	>1.17	46.1	>1.17	46.1
6B	0.01	0.4	0.025	1.0	0.35	13.7	0.39	15.2	>0.94	36.9	>0.94	36.9	>0.94	36.8	>0.94	36.8
556	0.01	0.4	0.028	1.1	0.025	1.0	0.067	2.6	0.24	9.3	0.29	11.6	>3.8	150.0	>3.8	150.0
Multimet	0.01	0.4	0.033	1.3	0.226	8.9	0.29	11.6	>1.2	47.2	>1.2	47.2	>3.7	146.4	>3.7	146.4
800H	0.023	0.9	0.046	1.8	0.14	5.4	0.19	7.4	0.19	7.5	0.23	8.9	0.29	11.3	0.35	13.6
RA330	0.01	0.4	0.11	4.3	0.02	0.8	0.17	6.7	0.041	1.6	0.22	8.7	0.096	3.8	0.21	8.3
310	0.01	0.4	0.028	1.1	0.025	1.0	0.058	2.3	0.075	3.0	0.11	4.4	0.2	8.0	0.26	10.3
316	0.315	12.4	0.36	14.3	>1.7	68.4	>1.7	68.4	>2.7	105.0	>2.7	105.0	>3.57	140.4	>3.57	140.4
304	0.14	5.5	0.21	8.1	>0.69	27.1	>0.69	27.1	>0.6	23.6	>0.6	23.6	>1.7	68.0	>1.73	68.0
446	0.033	1.3	0.058	2.3	0.33	13.1	0.37	14.5	>0.55	21.7	>0.55	21.7	>0.59	23.3	>0.59	23.3

Source: Ref 16

as lanthanum and aluminum (Ref 17). Irving et al. (Ref 18) found that tantalum is beneficial and niobium is detrimental to oxidation resistance in Co-20Cr-base alloys.

Among the cobalt-base superalloys tested, alloy 188 was best. However, oxidation attack for this alloy became severe when the temperature was increased to 1150 °C (2100 °F). At 1200 °C (2200 °F) the attack was so severe that the alloy should not be considered for long-term service at this temperature. Alloy 188 is significantly better than alloy 25 in oxidation resistance. This improvement was attributed to the addition of lanthanum and the control of minor elements such as silicon, aluminum, titanium, and so forth. Higher nickel content may also have contributed to improved oxidation resistance. Irving et al. (Ref 19) observed a noticeable reduction in oxide scale thickness when the nickel content in Co-15Cr-Ni alloys was increased to 20% and higher. Alloy 150 (or UMCo 50) exhibited unacceptably high oxidation rates at 1150 °C (2100 °F) and higher, while alloy 6B, widely used for wear applications, suffered rapid oxidation at 1095 °C (2000 °F) and higher.

Among the three groups of nickel-base alloys, the precipitation-strengthened alloys, particularly γ'-strengthened alloys, are overall the least resistant to oxidation. These alloys contain relatively high levels of titanium. Titanium is very active in scale formation. The formation of titanium-rich oxides apparently disrupts the chromia scale. The poor oxidation resistance of type 321 stainless

Table 3 Comparative oxidation resistance of iron-, nickel-, and cobalt-base alloys in flowing air between 168 h and 25 h cycles at 1095 °C (2000 °F)

	Metal loss, mm (mils)		Extrapolated metal loss, mm/yr (mils/yr)		Average metal affected, mm (mils)		Extrapolated oxidation rate, mm/yr (mils/yr)	
Alloy	1008 h/168 h	1050 h/25 h	168 h cycle	25 h cycle	1008 h/168 h	1050 h/25 h	168 h cycle	25 h cycle
214	0.003 (0.1)	0.003 (0.1)	0.025 (1)	0.025 (1)	0.003 (1.0)	0.025 (1)	0.025 (1)	0.20 (8)
601	0.030 (1.2)	0.231 (9.1)	0.25 (10)	1.93 (76)	0.066 (2.6)	0.297 (11.7)	0.58 (23)	2.49 (98)
600	0.028 (1.1)	0.079 (3.1)	0.25 (10)	0.66 (26)	0.041 (1.6)	0.185 (7.3)	0.36 (14)	1.55 (61)
671	0.038 (1.5)	0.462 (18.2)	0.33 (13)	3.86 (152)	0.061 (2.4)	0.584 (23.0)	0.53 (21)	4.88 (192)
230	0.013 (0.5)	0.015 (0.6)	0.10 (4)	0.13 (5)	0.003 (1.3)	0.086 (3.4)	0.28 (11)	0.71 (28)
S	0.010 (0.4)	0.013 (0.5)	0.10 (4)	0.10 (4)	0.003 (1.3)	0.061 (2.4)	0.28 (11)	0.51 (20)
G-30	0.038 (1.5)	0.036 (1.4)	0.33 (13)	0.30 (12)	0.122 (4.8)	0.203 (8.0)	1.07 (42)	1.70 (67)
617	0.015 (0.6)	0.168 (6.6)	0.13 (5)	1.40 (55)	0.046 (1.8)	0.267 (10.5)	0.41 (16)	2.24 (88)
333	0.025 (1.0)	0.033 (1.3)	0.23 (9)	0.28 (11)	0.058 (2.3)	0.130 (5.1)	0.51 (20)	1.09 (43)
625	0.084 (3.3)	0.353 (13.9)	0.74 (29)	2.95 (116)	0.122 (4.8)	0.414 (16.3)	1.07 (42)	3.45 (136)
Waspaloy	0.036 (1.4)	0.279 (10.9)	0.30 (12)	2.31 (91)	0.137 (5.4)	0.414 (16.3)	1.19 (47)	3.15 (136)
263	0.089 (3.5)	0.439 (17.3)	0.76 (30)	3.66 (144)	0.361 (14.2)	0.478 (18.8)	3.12 (123)	3.99 (157)
188	0.010 (0.4)	0.028 (1.1)	0.10 (4)	0.23 (9)	0.003 (1.3)	0.058 (2.3)	0.28 (11)	0.48 (19)
25	0.234 (9.2)	0.419 (16.5)	2.03 (80)	3.51 (138)	0.259 (10.2)	0.490 (19.3)	2.26 (89)	4.09 (161)
150	0.058 (2.3)	0.223 (8.8)	0.51 (20)	1.85 (73)	0.097 (3.8)	0.353 (13.9)	0.84 (33)	2.95 (116)
6B	0.348 (13.7)	>0.800 (31.5)	3.02 (119)	>6.68 (263)	0.394 (15.5)	>0.800 (31.5)	3.43 (135)	>6.68 (263)
556	0.025 (1.0)	0.038 (1.5)	0.23 (9)	0.33 (13)	0.066 (2.6)	0.117 (4.6)	0.58 (23)	0.97 (38)
Multimet	0.223 (8.9)	0.328 (12.9)	1.96 (77)	2.74 (108)	0.295 (11.6)	0.381 (15.0)	2.57 (101)	3.18 (125)
800H	0.137 (5.4)	0.328 (12.9)	1.19 (47)	2.74 (108)	0.188 (7.4)	0.406 (16.0)	1.63 (64)	3.40 (134)
RA330	0.020 (0.8)	0.269 (10.6)	0.18 (7)	2.24 (88)	0.170 (6.7)	0.442 (17.4)	1.47 (58)	3.68 (145)
310	0.025 (1.0)	0.058 (2.3)	0.23 (9)	0.48 (19)	0.058 (2.3)	0.112 (4.4)	0.51 (20)	0.94 (37)
446	0.333 (13.1)	0.579 (22.8)	2.92 (115)	4.83 (190)	0.368 (14.5)	0.655 (25.8)	3.20 (126)	5.46 (215)

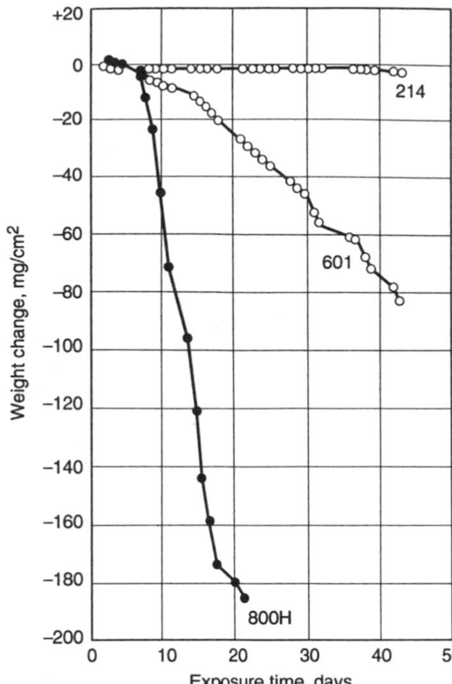

Fig. 3 Cyclic oxidation resistance of alloy 214 compared to alloys 601 and 800H in still air at 1150 °C (2100 °F) cycled once a day every day except weekends. Source: Ref 4

Table 4 Cyclic oxidation resistance of several iron-, nickel-, and cobalt-base alloys at 1100 °C (2010 °F) for 504 h in air-5H_2O(a)

Alloy	Specific weight change (descaled), mg/cm²		Metal loss		Maximum attack	
	Mean	Range	mm	mils	mm	mils
AC1 grade HK	−105.8	−98 to −124	0.25	9.9	0.35	13.8
310SS	−149.0	−92 to −235	0.31	12.2	0.38	15.0
800	−168.6	−83 to −223	0.39	15.4	0.59	23.2
601	−11.0	−6.3 to −17.2	<0.02	0.8	0.12	4.7
617	−13.5	−6.5 to −17.5	...	...	...	...
X	−20.0	−10.0 to −29.5	0.05	2.0	0.25	9.9
RA333	−30.5	...	...	...	...	...
IN-814	−3.5	−2.5 to −4.9	0	0	<0.01	0.4
188	−25.0	−13.0 to −40.5	0.03	1.2	0.15	5.9
MA-956	−1.0	−0.3 to −1.5	<0.01	0.4	0.02	0.8

(a) 15 min in furnace and 5 min out of furnace. Source: Ref 23

Table 5 Dynamic oxidation resistance of iron-, nickel-, and cobalt-base alloys in high-velocity combustion gas stream at 1090 °C (2000 °F) for 500 h(a)

Alloy	Metal loss		Maximum metal affected(b)	
	mm	mils	mm	mils
214	0.013	0.5	0.046	1.8
230	0.056	2.2	0.15	5.7
RA333	0.10	4.0	0.22	8.7
188	0.19	7.5	0.27	10.7
556	0.22	8.7	0.30	11.7
X	0.23	9.0	0.34	13.5
RA330	0.28	10.9	0.35	13.6
S	0.30	11.8	0.39	15.2
600	0.44	17.2	0.53	20.7
310	0.54	21.2	0.61	24.1
601	0.27	10.7	0.61(c)	24.0(c)
617	0.32	12.4	0.61(c)	24.0(c)
800H	0.77(d)	30.5(d)	0.86(d)	34.0(d)
625	>0.79(f)	31.0(f)	>0.79(f)	31.0(f)
Multimet	1.25(e)	49.1(e)	1.42(e)	55.8(e)

(a) Gas velocity was 0.3 mach (100 m/s, or 225 mph); samples were cycled to less than 260 °C (500 °F) once every 30 min; 50 to 1 air-to-fuel ratio; two parts No. 1 fuel oil and one part No. 2 fuel oil. (b) Metal loss plus maximum internal penetration. (c) Internal nitrides through thickness. (d) Extrapolated from 400 h; sample was about to be consumed after 400 h. (e) Extrapolated from 225 h; sample was about to be consumed after 225 h. (f) Sample was consumed in 500 h. Source: Ref 21

Table 6 Dynamic oxidation resistance of iron-, nickel-, and cobalt-base alloys in high-velocity combustion gas stream at 980 °C (1800 °F) for 1000 h(a)

Alloy	Metal loss		Maximum metal affected(b)	
	mm	mils	mm	mils
214	0.010	0.4	0.031	1.2
230	0.020	0.8	0.089	3.5
188	0.028	1.1	0.107	4.2
556	0.043	1.7	0.158	6.2
X	0.069	2.7	0.163	6.4
S	0.079	3.1	0.17	6.6
RA333	0.064	2.5	0.18	7.0
625	0.12	4.9	0.19	7.6
617	0.069	2.7	0.27	10.7
RA330	0.20	7.8	0.30	11.8
Multimet	0.30	11.8	0.38	14.8
800H	0.31	12.3	0.39	15.3
310	0.35	13.7	0.42	16.5
600	0.31(c)	12.3(c)	0.45(c)	17.8(c)
601	0.076	3.0	0.51	20.0
304	>9.0(d)	354(d)	>9.0(d)	354(d)
316	>9.0(d)	354(d)	>9.0(d)	354(d)

(a) Gas velocity was 0.3 mach (100 m/s, or 225 mph); samples were cycled to less than 260 °C (500 °F) once every 30 min; 50 to 1 air-to-fuel ratio; two parts No. 1 fuel oil and one part No. 2 fuel oil. (b) Metal loss plus maximum internal penetration. (c) Extrapolated from 917 h; sample was about to be consumed after 917 h. (d) Extrapolated from 65 h; sample was consumed in 65 h. Source: Ref 21

steel as compared to types 304, 316, and 347 (Ref 15) may also be attributed to titanium. Nagai et al. (Ref 11) found that titanium was detrimental to the oxidation resistance of Ni-20Cr alloy. In the Fe-Cr-Al system, however, the addition of 1% Ti to Fe-18Cr-6Al was found to improve resistance to cyclic oxidation at 950 °C (174 °F) in air (Ref 20). Niobium is another alloying element that is detrimental to alloy oxidation resistance. The relatively poor oxidation resistance of alloy 625 at 1095 and 1150 °C (2000 and 2100 °F) may be attributed to niobium.

For Ni-Cr-Fe alloys with various amounts of aluminum, alloy 600 (which generally contains 0.2 or 0.3% Al from melting practices) exhibits oxidation attack similar to that of alloy 601, with about 1.4% Al. The level of aluminum in alloy 601 is well below the level needed to form a continuous alumina scale. Thus, alloy 601 is still a chromia former. Because of insufficient aluminum in alloy 601 to form a continuous alumina scale, aluminum tends to segregate to form internal oxides as well as internal nitrides. This sometimes results in much deeper internal penetration. With about 4.5% Al, alloy 214 readily forms an alumina scale. It is clearly shown in Table 2 that alloy 214, exhibiting less than 0.025 mm (1 mil) of total depth of attack at temperatures up to 1200 °C (2200 °F), is significantly better than all the other alloys tested. With a continuous alumina scale, alloy 214 showed little or no internal oxidation attack.

Cyclic Oxidation. Very few industrial processes are able to operate without intermittent shutdowns. Therefore, studies on oxidation should always include the cyclic effect. The oxide scales of some alloys may be more susceptible to cracking or spalling during cooling, thus increasing subsequent oxidation attack when the system resumes operation. The oxidation data presented in Table 2 were generated with a weekly cycle (168 h) to room temperature. When the cyclic frequency was increased to 25 h cycles, oxidation rates were increased for all the alloys tested. However, some alloys are more sensitive to cyclic oxidation than others. The effect of thermal cycling on the oxidation resistance of various alloys in air at 1095 °C (2000 °F) is illustrated in Table 3, which compares once-a-week (168 h) cycle data (Ref 16) with 25 h cycle data (Ref 21). Alloy 214 (an alumina former) was found to suffer little or no attack in tests with either 168 h or 25 h cycling. Some chromia formers (230, S, 333, 188, 556, and type 310) showed good resistance to thermal cycling. It is not possible to correlate the performance of these alloys with their chemical compositions. However, most of these alloys (i.e., alloys 230, S, 188, and 556) contain many minor reactive elements as well as a rare earth element, lanthanum.

Kane et al. (Ref 22) studied the cyclic oxidation resistance of several high-temperature alloys in air-5H_2O at 1100 °C (2010 °F) for exposure times up to 2016 h with 24 h cycles. Their results in terms of weight change as a function of exposure time are shown in Fig. 5. The alumina-forming MA956 alloy (about 4.5% Al) exhibited no changes in weight, while all chromia formers (alloy 601, HK alloy, alloy 800, and type 310) suffered weight losses, with type 310 being the

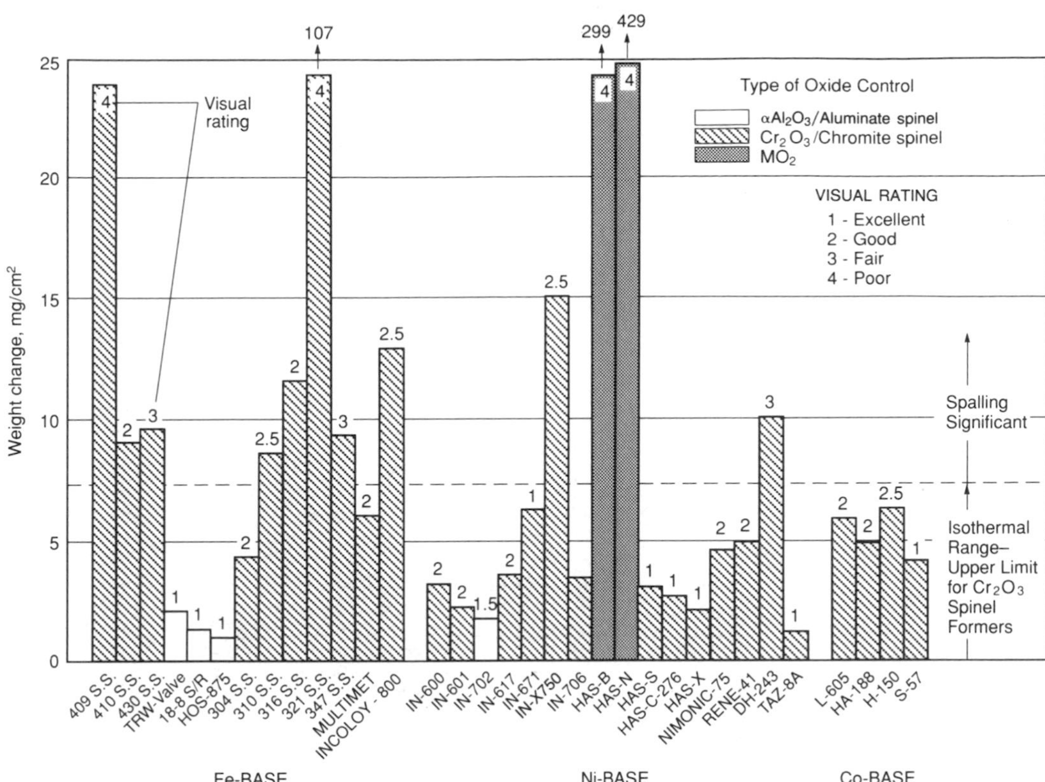

Fig. 4 Long-term oxidation tests (10,000 h) in air at 815 °C (1500 °F) with 1000 h cycles for iron-, nickel-, and cobalt-base alloys. Also included is the upper limit of the metal loss for isothermal tests (i.e., 10,000 without cycling) for same alloys. Source: Ref 15

Table 7 Results of field test in a natural gas-fired tube at 1010 °C (1850 °F) for 3000 h

Alloy	Metal loss		Maximum metal affected(a)		Oxidation rate	
	mm	mils	mm	mils	mm/yr	mils/yr
214	0.003	0.1	0.025	1	0.076	3
601	0.023	0.9	0.076	3	0.23	9
230	0.028	1.1	0.10	4	0.30	12
556	0.018	0.7	0.10	4	0.30	12
310	0.041	1.6	0.10	4	0.30	12
600	0.018	0.7	0.15	6	0.46	18
RA330	0.048	1.9	0.15	6	0.46	18
800H	0.12	4.7	0.30	12	0.89	35
309	0.50	19.7	0.51	20	1.5	58
304	>1.5(b)	60(b)	>1.5	60(b)	>4.4	175

(a) Metal loss plus maximum internal penetration. (b) Sample was consumed. Source: Ref 24

Table 8 Results of field test in a natural gas-fired furnace for reheating nickel- and cobalt-base alloy ingots and slabs for 113 days at 1090 to 1230 °C (2000 to 2250 °F) with frequent cycles to 540 °C (1000 °F)

Alloy	Metal loss		Maximum affected(a)	
	mm	mils	mm	mils
214	0.013	0.5	0.11	4.5
RA330	0.39	15.5	0.65	25.5
601	0.18	7.2	0.95	37.2
600	0.64	25.0	1.1	45.0
800H	>0.79(b)	31.0(b)	>0.79(b)	31.0(b)
310SS	>1.0(b)	41.0(b)	>1.0(b)	41.0(b)
304SS	>1.5(b)	60.0(b)	>1.5(b)	60.0(b)
316SS	>1.6(b)	63.0 (b)	>1.6(b)	63.0 (b)
446SS	>0.61(b)	24.0(b)	>0.61(b)	24.0(b)

(a) Metal loss plus internal penetration. (b) Samples were consumed. Source: Ref 25

worst. This was in contrast to the 25 h cyclic test results observed by Lai (Ref 21), where type 310 was found to be more resistant to cyclic oxidation than some nickel-base alloys, (such as alloy 601) and Fe-Ni-Cr alloys (such as alloy 800H). There was a difference between the two test environments. The air environment used by Kane et al. (Ref 22) contained 5% H_2O, while that used by Lai (Ref 21) was basically dry air (the air was passed through an air filter prior to entering the test retort). Table 4 shows the results of another cyclic oxidation study, done by Kane et al. (Ref 23) in air-5H_2O at 1100 °C (2010 °F) for 504 h with 15-min cycles. Two alumina formers, MA-956 and IN-814, were the best performers, followed by cobalt- and nickel-base alloys (alloys 188, 601, 617, X, and 333). Three iron-base alloys (HK alloy, type 310, and alloy 800) were not as good as the nickel-base alloys.

Oxidation Behavior in Combustion Atmospheres. Many industrial processes involve combustion. The burner rig test system has been quite popular in the gas turbine industry to study the oxidation of gas turbine alloys in combustion atmospheres. It has also been routinely used during the development of new alloys for gas turbines. Lai (Ref 21) has recently studied the oxidation behavior of a wide variety of alloys, from stainless steels to superalloys, in combustion atmospheres using a burner rig test system. The combustion of fuel oil with preheated, pressurized air results in a high-velocity gas stream (0.3 mach, or 225 mph). Because of this high-velocity stream, the test is also referred to as a "dynamic" oxidation test. The sample holder, which can hold as many as 24 samples, is rotated during testing to ensure that all samples are subjected to identical conditions. Furthermore, severe thermal cycling is imposed, which involves automatically lowering the samples every 30 min, blasting them with fan air to cool them to below 260 °C (500 °F) in 2 min, then returning them to the combustion test tunnel. The air-to-fuel ratio is 50 to 1,

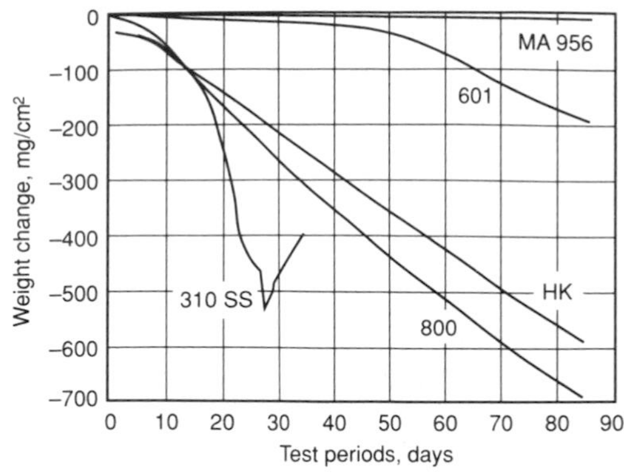

Fig. 5 Cyclic oxidation resistance of several high-temperature alloys in air-5H_2O at 1100 °C (2010 °F) for times up to 2016 h with 24 h cycles. Source: Ref 22

and the fuel oil (mixture of two parts No. 1 and one part No. 2) typically contains about 0.15% S. Table 5 tabulates the test results generated at 1090 °C (2000 °F) for 500 h with 30-min cycles for iron-, nickel-, and cobalt-base alloys.

With an alumina scale, alloy 214 suffered very little attack. The alloy showed no sign of breakaway corrosion after 500 h and the scale consisted of aluminum-rich oxides. After 1000 h (2000 cycles) of testing, the scale remained aluminum-rich. The maximum metal affected (metal loss plus maximum internal penetration) remained about the same after 1000 h compared to after 500 h. All the other alloys tested were chromia formers, with performance varying quite significantly from alloy to alloy.

In general, the Fe-Ni-Cr and Ni-Cr-Fe alloys were the poorest performers; however, RA330 was the best in the group. Alloys containing about 1% Al are generally susceptible to the formation of internal aluminum nitrides in air or highly oxidizing combustion atmospheres. For example, alloy 601, which suffered a metal loss similar to that of RA330, exhibited internal aluminum nitrides throughout the sample thickness. Alloy 617 is another alloy that frequently suffers the formation of internal aluminum nitrides.

Of all the alloys tested at 1090 °C (2000 °F), Multimet alloy was one of the worst. The specimen (about 1.4 mm, or 0.056 in., thick) was consumed in only about 225 h. Alloy 556, a modified version of Multimet alloy with some changes of minor alloying elements (Ref 17), exhibited significantly better oxidation resistance. This is a good example of how the addition and modification of minor alloying elements can improve oxidation resistance.

The results of dynamic oxidation tests at 980 °C (1800 °F) for 1000 h (2000 cycles) are presented in Table 6. The performance ranking was similar to that of 1090 °C (2000 °F) tests, with the exception of alloy 625, which was ranked higher. RA330 continued to be the best among the Fe-Ni-Cr and Ni-Cr-Fe alloys. The 18Cr-8Ni steels suffered extremely rapid oxidation attack. Types 304 and 316 stainless steel samples (1.17 mm, or 0.046 in., thick) were consumed in about 65 h.

Very limited field test data generated in "clean" combustion atmospheres have been reported. Ta-

Table 9 Carburization resistance of various alloys in Ar-5H_2-5CO-5CH_4 at 1090 °C (2000 °F) for 24 h

Alloy	Carbon absorption, mg/cm²
214	3.4
600	9.9
625	9.9
230	10.3
X	10.6
S	10.6
304	10.6
617	11.5
316	12.0
333	12.4
800H	12.6
330	12.7
25	14.4

Source: Ref 27

Table 10 Carburization in H_2-2CH_4 at 1000 °C (1830 °F) for 100 h

Alloy	Weight gain, mg/cm²
MA-956	<0.3
601	10
800	19
310SS	36
HK (1.07% Si)	34
HK (2.54%)	29

Source: Ref 28

Table 11 Nitridation resistance of various alloys at 650 °C (1200 °F) for 168 h in ammonia(a)

Alloy	Alloy base	Nitrogen absorption, mg/cm²	Depth of nitride penetration, mm	Depth of nitride penetration, mils
C-276	Nickel	0.7	0.02	0.6
230	Nickel	0.7	0.03	1.2
HR-160	Nickel	0.8	0.01	0.5
600	Nickel	0.8	0.03	1.3
625	Nickel	0.9	0.01	0.5
RA333	Nickel	1.0	0.03	1.0
601	Nickel	1.1	0.03	1.0
188	Cobalt	1.2	0.02	0.6
S	Nickel	1.3	0.03	1.1
617	Nickel	1.3	0.03	1.0
214	Nickel	1.5	0.04	1.5
X	Nickel	1.7	0.04	1.5
825	Nickel	2.5	0.06	2.2
800H	Iron	4.3	0.10	4.1
556	Iron	4.9	0.09	3.5
316	Iron	6.9	0.19	7.3
310	Iron	7.4	0.15	6.0
304	Iron	9.8	0.21	8.4

(a) 100% NH_3 in the inlet gas and 30% NH_3 in the exhaust gas. Source: Ref 29

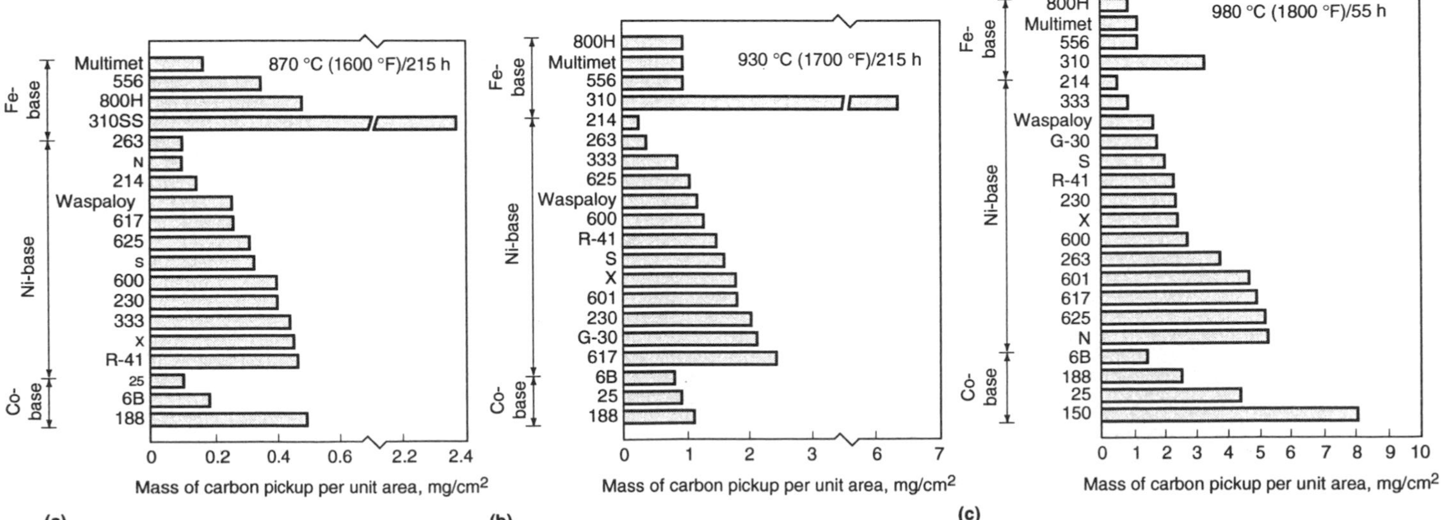

Fig. 6 Carburization resistance of various iron-, nickel-, and cobalt-base alloys tested in Ar-5H_2-5CO-5CH_4 at (a) 870 °C/215 h, (b) 930 °C/215 h, and (c) 980 °C/55 h. Source: Ref 26

ble 7 summarizes the results of a field test performed inside a radiant tube fired with natural gas with an average temperature of 1010 °C (1850 °F) (Ref 24). The rack containing coupons of various alloys was exposed for about 3000 h. Many chromia formers, such as alloys 601, 230, 556, 310SS, 600, and 330, were found to perform well, with oxidation rates of less than 0.5 mm/yr (20 mils/yr). The test also showed that the temperature of 1010 °C (1850 °F) was too high for type 304 stainless steel. Similar to the results of air oxidation tests and burner rig dynamic oxidation tests, the alumina former (alloy 214) was found to be the best performer. In another field test (Ref 25) performed in a natural gas-fired furnace used for reheating ingots and slabs of nickel- and cobalt-base alloys, samples of various alloys were exposed for about 113 days at temperatures from 1090 to 1230 °C (2000 to 2250 °F) with frequent cycles to 540 °C (1000 °F). The results are summarized in Table 8. All of the chromia formers tested suffered severe oxidation attack. On the other hand, the alumina former (alloy 214) exhibited little attack. The scales formed on alloy 214 consisted of essentially aluminum-rich oxides (Ref 25).

The alumina scale formed on alloy 214 is very protective in both laboratory and field tests at temperatures up to 1230 °C (2250 °F). This oxide scale remains protective even when the temperature is increased to 1320 °C (2400 °F). The sample showed no sign of breakaway corrosion after testing in air at 1320 °C (2400 °F) for 200 h with 25 h cycles (Ref 21). The scale consisted of mainly aluminum-rich oxides and showed no sign of cracking.

Table 12 Nitridation resistance of various alloys at 980 °C (1800 °F) for 168 h in ammonia(a)

Alloy	Alloy base	Nitrogen absorption, mg/cm^2	Depth of nitride penetration, mm	Depth of nitride penetration, mils
214	Nickel	0.3	0.04	1.4
600	Nickel	0.9	0.12	4.8
S	Nickel	0.9	0.18	7.2
601	Nickel	1.2	0.17	6.6
230	Nickel	1.4	0.12	4.9
617	Nickel	1.5	0.38	15.0
HR-160	Nickel	1.7	0.18	7.2
188	Cobalt	2.3	0.19	7.4
625	Nickel	2.5	0.17	6.9
6B	Cobalt	3.1	0.15	5.8
253MA	Iron	3.3	0.48	19.0
25	Cobalt	3.6	0.26	10.4
X	Nickel	3.2	0.19	7.4
RA333	Nickel	3.7	0.42	16.4
RA330	Iron	3.9	0.52	20.6
800H	Iron	4.0	0.28	11.1
825	Nickel	4.3	0.58	23.0
150	Cobalt	5.3	0.38	15.1
Multimet	Iron	5.6	0.35	13.6
316	Iron	6.0	0.52	20.3
556	Iron	6.7	0.37	14.7
304	Iron	7.3	>0.58	23.0
310	Iron	7.7	0.38	15.1
446	Iron	12.9	>0.58	23.0

(a) 100% NH_3 in the inlet gas and less than 5% NH_3 in the exhaust gas. Source: Ref 29

Carburization

In addition to oxygen attack, high-temperature alloys are frequently subjected to attack by carbon compounds (carburization). Materials problems due to carburization are quite common in heat-treating components associated with carburizing furnaces. The environment in the carburizing furnace typically has a carbon activity that is significantly higher than that in the alloy of the furnace component. Therefore, carbon is transferred from the environment to the alloy. This results in the carburization of the alloy, and the carburized alloy becomes embrittled.

Lai (Ref 26) studied the carburization resistance of more than 20 commercial wrought alloys, ranging from stainless steels to superalloys. The test environments were characterized by a unit carbon activity and oxygen potentials such that a chromia scale was not expected to form on the metal surface. Oxides of silicon, titanium, and aluminum were expected to be stable under the test conditions. The relative performance rankings for alloys tested at 870, 930, and 980 °C (1600, 1700, and 1800 °F) are shown in Fig. 6. No obvious correlations between performance and alloy composition were noted, perhaps because most of the alloys contained many alloying elements, including solid solution strengthening elements (e.g., molybdenum, tungsten, niobium), precipitate-forming elements (e.g., aluminum, titanium, niobium), and other minor elements for oxidation resistance, melting control, and so forth. Some of these alloying elements may play an important role in affecting the performance of an alloy.

One alloy that was found to be consistently better than the others was alloy 214, containing about 4.5% Al. The excellent resistance of this alloy was attributed to the alumina film formed on the metal surface. The existence of this film was confirmed by Auger analysis (Ref 26). Figure 7 illustrates the optical microstructure of alloy 214 compared to those of alloys X, 601, and 150

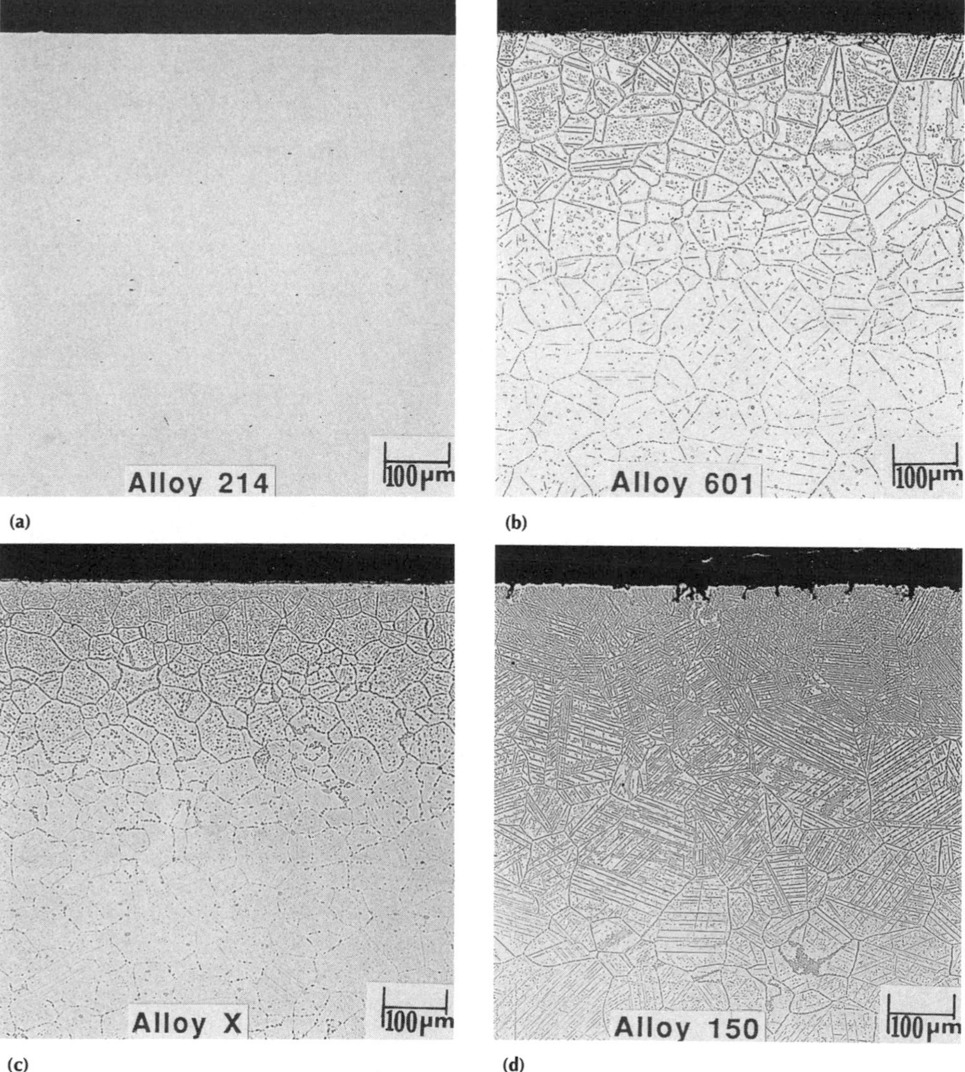

Fig. 7 Optimal microscope of several high-temperature alloys tested at 980 °C (1800 °F) for 55 h in Ar-5H_2-5CO-5CH_4.

after exposure to the test environment at 980 °C (1800 °F) for 55 h (Ref 26). Tests in the same gas mixture at 1090 °C (2000 °F) showed that the mass of carbon absorbed by alloy 214 was only about one-third of that by the best chromia former (Table 9). A similar observation was made by Kane et al. (Ref 28), who compared MA-956 (an oxide dispersion strengthened alloy with about 4.5% Al) with alloys 601, 800H, 310, and HK. MA-956 (an alumina former) was significantly better than the other alloys tested (chromia formers), as shown in Table 10.

Metal dusting, which is sometimes referred to as catastrophic carburization, is another frequently encountered mode of corrosion that is associated with carburizing furnaces. Metal dusting tends to occur in a region where the carbonaceous gas atmosphere becomes stagnant. The alloy normally suffers rapid metal wastage. The corrosion products (or wastage) generally consist of carbon soots, metal, metal carbides, and metal oxides. The attack is normally initiated from the metal surface that is in contact with the furnace refractory. The furnace components that suffer metal dusting include thermowells, probes, and anchors. Figure 8 illustrates the metal dusting attack on Multimet alloy. The component was perforated as a result of metal dusting.

Metal dusting has been encountered with straight chromium steels, austenitic stainless steels, and nickel- and cobalt-base alloys. All of these alloys are chromia formers. Because metal dusting is a form of carburization, it would appear that alumina formers, such as alloy 214, would also be more resistant to metal dusting.

(a)

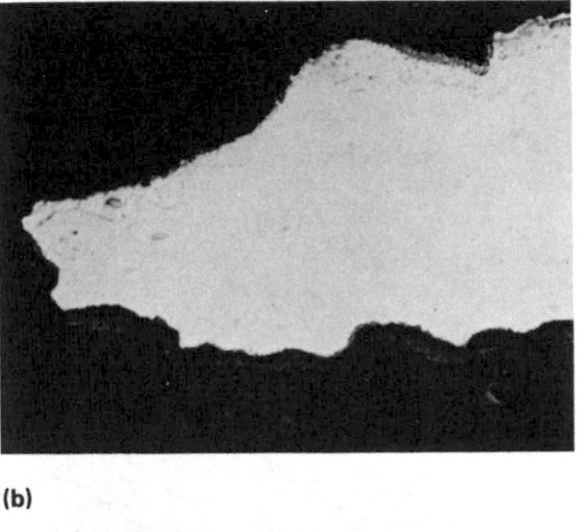

(b)

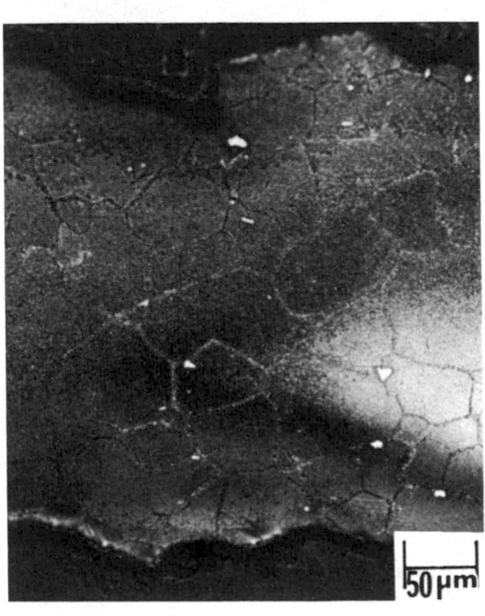

(c)

Fig. 8 Metal dusting of a Multimet alloy component at the refractory interface in a carburizing furnace. Perforation of the component (arrows). (b) Cross section of the sample showing severe pitting. (c) Severe carburization beneath the pitted area

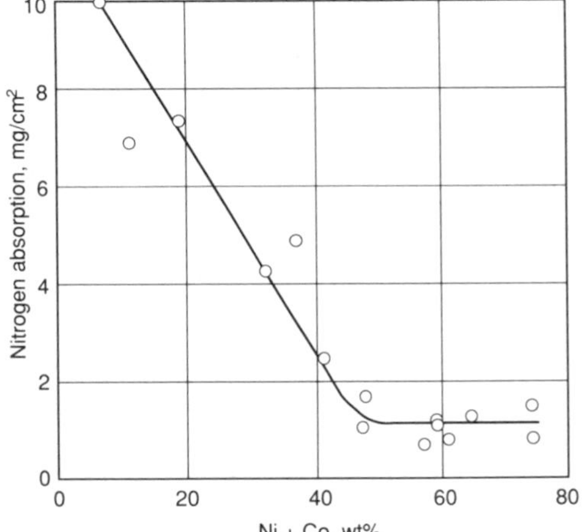

Fig. 9 Effect of Ni + Co content in iron-, nickel-, and cobalt-base alloys on nitridation resistance at 650 °C (1200 °F) for 168 h in ammonia (100% NH_3 in the inlet gas and 30% NH_3 in the exhaust). Source: Ref 29

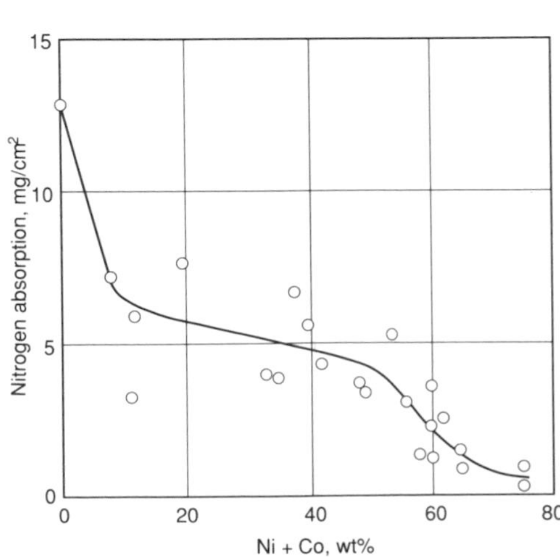

Fig. 10 Effect of Ni + Co content in iron-, nickel-, and cobalt-base alloys on nitridation resistance at 980 °C (1800 °F) for 168 h in ammonia (100% NH_3 inlet gas and $<NH_3$ in the exhaust). Source: Ref 29

Nitridation

Metals or alloys are generally susceptible to nitridation when exposed to ammonia-bearing or nitrogen-base atmospheres at elevated temperatures. Nitridation typically results in the formation of nitrides. As a result, the alloy can become embrittled.

Ammonia-Base Atmospheres. Studies by Barnes and Lai (Ref 29) have examined the nitridation behavior of a variety of commercial alloys in ammonia. The alloys studied included stainless steels, Fe-Ni-Cr alloys, and nickel- and cobalt-base superalloys. The superalloys under study contained many alloying elements, such as aluminum, titanium, zirconium, niobium, chromium, molybdenum, tungsten, and iron. These are all nitride formers. The test results generated at 650, 980, and 1090 °C (1200, 1800, and 2000 °F) are summarized in Tables 11 to 13. Again, it was found that nickel-base alloys are generally better than iron-base alloys, and that increasing nickel content generally improves the resistance of an alloy to nitridation attack. Increased cobalt content appears to have the same effect. When

Table 13 Nitridation resistance of various alloys at 1090 °C (2000 °F) for 168 h in ammonia(a)

Alloy	Alloy base	Nitrogen absorption, mg/cm²	Depth of nitride penetration, mm	Depth of nitride penetration, mils
600	Nickel	0.2	0	0
214	Nickel	0.2	0.02	0.7
S	Nickel	1.0	0.34	13.4
230	Nickel	1.5	0.39	15.3
25	Cobalt	1.7	>0.65	25.5
617	Nickel	1.9	>0.56	22
188	Cobalt	2.0	>0.53	21
HR-160	Nickel	2.5	0.46	18
601	Nickel	2.6	>0.58	23
RA330	Iron	3.1	>0.56	22
625	Nickel	3.3	>0.56	22
316	Iron	3.3	>0.91	36
304	Iron	3.5	>0.58	23
X	Nickel	3.8	>0.58	23
150	Cobalt	4.1	0.51	20
556	Iron	4.2	>0.51	20
446	Iron	4.5	>0.58	23
6B	Cobalt	4.7	>0.64	25
Multimet	Iron	5.0	>0.64	25
825	Nickel	5.2	0.58	23
RA333	Nickel	5.2	>0.71	28
800H	Iron	5.5	>0.76	30
253MA	Iron	6.3	>1.5	60
310	Iron	9.5	>0.79	31

(a) 100% NH_3 in the inlet gas and less than 5% in the exhaust gas. Source: Ref 29

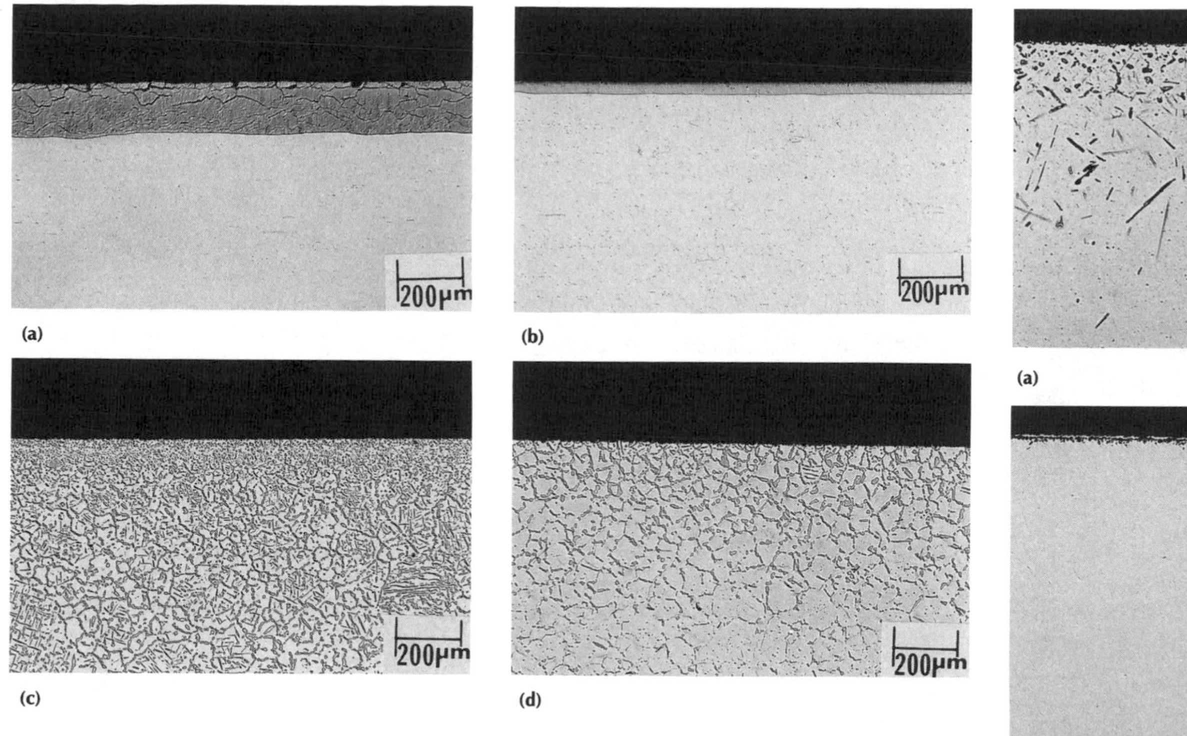

Fig. 11 Typical morphology of nitrides formed in ammonia at 650 °C (1200 °F) for 168 h for (a) type 310 stainless steel

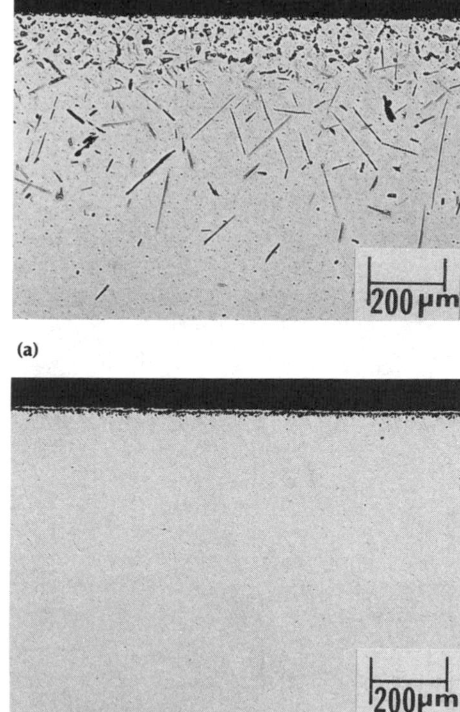

Fig. 12 Extensive internal AlN formation (needle phases) in alloy 601 (about 1.3% Al) and (b) insignificant AlN formation in alloy 214 (about 4.5 Al) after nitriding at 1090 °C (2000 °F) for 168 h in 100% NH_3 (inlet gas)

nitrogen absorption was plotted against Ni + Co content in the alloy for the 650 °C (1200 °F) test data (Fig. 9), resistance to nitridation improved with increasing Ni + Co content up to about 50 wt%. Further increases up to about 75% did not seem to affect the nitridation resistance of an alloy. For maximum resistance to nitridation attack at 650 °C (1200 °F), it appears that alloys with at least 50% Ni or 50% (Ni + Co) are most suitable.

This finding is in general agreement with the results reported by Moran et al. (Ref 30). Their results suggested that improvement in nitridation resistance begins to level off at about 40% Ni, with no improvement resulting from further increases in nickel up to 80%. Pure nickel, however, showed significantly lower nitridation resistance (Ref 30).

At higher temperatures (e.g., 980 °C, or 1800 °F), a slightly different relationship was observed, as shown in Fig. 10. Nitrogen absorption was reduced dramatically with an initial 15% Ni or 15% (Ni + Co). As Ni + Co content increased from 15 to 50%, no dramatic improvement in nitridation resistance was noted. Further increases in Ni + Co content in excess of about 50% caused a sharp improvement. Alloys with Ni (or Ni + Co) in excess of about 60% showed the most resistance to nitridation. No data are available for alloys with 80% Ni (or Ni + Co) and higher. It is generally believed that the beneficial effect of nickel or cobalt in increasing nitridation resistance is due to the reduced solubility of nitrogen in the alloy. Nickel and cobalt were found to reduce the solubility of nitrogen in iron (Ref 31, 32).

Nitridation at low temperatures (e.g., 650 °C, or 1200 °F) generally results in a surface nitride layer consisting of mostly iron nitrides (Fe_2N or Fe_4N). High temperatures, on the other hand, result in internal nitrides, mostly CrN, Cr_2N, and/or $(Fe,Cr)_2N$. Figure 11 illustrates the morphology of nitrides formed at both low and high temperatures. In addition to iron and chromium nitrides, aluminum nitrides, which are needle phases, are frequently observed in some alloys containing aluminum. Aluminum is a very strong nitride former. When alloys contain relatively low aluminum (e.g., about 1%), such as alloys 601 and 617, a significant amount of internal aluminum nitrides is generally formed, as shown in Fig. 12(a). For alloys containing relatively high aluminum (e.g., 4.5% Al in alloy 214), very few aluminum nitrides are formed, as illustrated in Fig. 12(b). This may be related to the alumina

Table 14 Nitridation resistance of several iron- and nickel-base alloys in pure nitrogen

Alloy	982 °C (1800 °F)/1008 h(a) nitrided depth		1093 °C (2000 °F)/900 h(b) nitrided depth		1204 °C (2200 °F)/100 h(c) nitrided depth	
	mm	mils	mm	mils	mm	mils
600	1.30	51	1.85	73	2.16	85
601	1.55	61	2.79	110	>3.81	150
800	1.85	73	>3.81	150	>3.81	150
520	...	...	>3.81	150	...	...
RA330	2.57	101	>3.81	150	>3.81	150
314SS	>3.81	150	>3.81	150	...	...

(a) Specimens were cycled to room temperature once every 24 h for the first three days and then weekly for the remainder of the test. (b) Specimens were cycled to room temperature once every 96 h (four days). (c) Isothermal exposure. Source: Ref 33

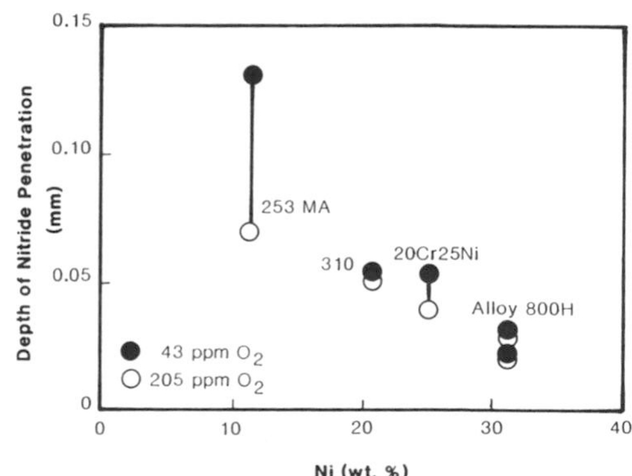

Fig. 13 Results of nitridation tests in nitrogen with two different levels of oxygen contamination at 825 °C (1500 °F) for 400 h. Source: Ref 34

film developed in alloy 214. Alloys such as alloy 601, with only about 1.3% Al, are generally not capable of developing an alumina continuous surface film. It is believed that alumina is thermodynamically stable in the test environments used in the studies by Barnes and Lai (Ref 29). It should be noted that alumina scale formation is favored kinetically at high temperatures, such as 980 °C (1800 °F) or above. Higher temperatures generally produce a more protective alumina scale.

Nitrogen-Base Atmospheres. There have been very few studies on the nitridation behavior of alloys in nitrogen-base atmospheres. This type of atmosphere is becoming more popular as a protective atmosphere in heat-treating and sintering operations. Smith and Bucklin (Ref 33) investigated the gas-metal reactions for several iron- and nickel-base alloys in several nitrogen-base atmospheres. Their test results, generated in nitrogen at 980, 1090, and 1200 °C (1800, 2000, and 2200 °F), are shown in Table 14. Both AlN and Cr_2N were found in alloys 600 and 800 after exposure to nitrogen at 1200 °C (2200 °F) for 100 h. For RA330, only Cr_2N was detected after exposure to the same test conditions.

The nitridation attack was surprisingly severe in pure nitrogen. Alloys 601, 800, and 330 suffered more than 3.8 mm (150 mils) of penetration depth after only 100 h of exposure at 1200 °C (2200 °F). Even the best performer, alloy 600, suffered a penetration depth of 2.2 mm (85 mils) in 100 h. When the temperature was lowered to 1090 °C (2000 °F), nitridation attack was still quite severe. For exposure of 960 h at 1090 °C (2000 °F), alloys 800, 520, 330, and 314 suffered more than 3.8 mm (150 mils) of penetration. Alloys 600 and 601 exhibited 1.9 mm (73 mils) and 2.8 mm (110 mils) of attack, respectively. The nitridation attack remained quite severe even at 980 °C (1800 °F). After 1008 h, alloys 600 and 601 showed 1.3 and 1.6 mm (51 and 61 mils) of attack, respectively, whereas iron-base alloys suffered much more attack.

According to the above results, generated by Smith and Bucklin (Ref 33), it appears that stainless steels, Fe-Ni-Cr alloys, and Ni-Cr-Fe alloys may not be suitable for long-term service in pure nitrogen at temperatures above 980 °C (1800 °F). The pure nitrogen environments employed in these studies had a dew point of –66 °C (–86.8 °F) and 1 ppm O_2. Oxygen potentials may play a significant role in nitridation behavior. Odelstam et al. (Ref 34) found that nitride penetration was significantly increased in pure nitrogen when the level of oxygen was reduced from 205 to 43 ppm for alloy 253MA, as illustrated in Fig. 13. Other higher-nickel alloys, such as alloy 800H, showed no significant changes. Smith and Bucklin (Ref 33) found that in nitrogen containing 1% H_2 with a dew point of 5.6 °C (42 °F), none of the alloys tested (314, 330, 800, 601, and 600) showed signs of nitridation after exposure at 930 °C (1700 °F) for 1032 h. It is not clear whether this was due to a lower temperature (100 °F lower in this case) or to a higher dew point. It is reasonable to assume that an oxide scale tends to reduce

(a)

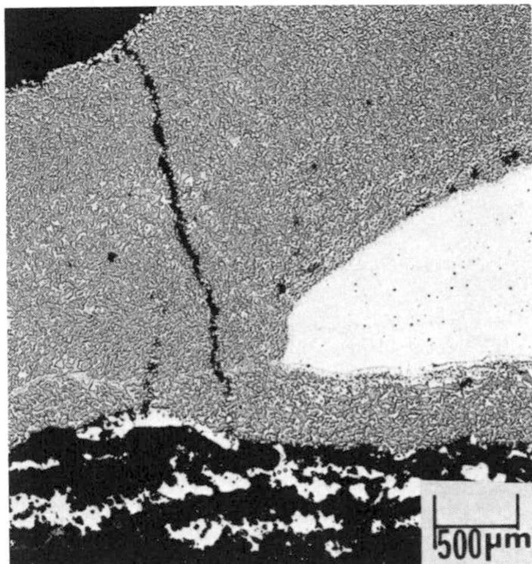

(b)

(c)

Fig. 14 Catastrophic sulfidation of an alloy 601 furnace tube. The furnace atmosphere was contaminated with sulfur; the component failed after less than 1 month at 930 °C (1700 °F). (a) General view. (b) Cross section of the perforated area showing liquid-like nickel-rich sulfides. (c) Higher-magnification view of nickel-rich sulfides

Table 15 Sulfidation resistance of alloy HR-160 compared to that of cobalt-base alloys at 870 °C (1600 °F) for 500 h(a)

Alloy	Weight change, mg/cm^2	Metal loss, mm	Metal loss, mils	Maximum metal affected(b), mm	Maximum metal affected(b), mils
HR-160	–0.5	0.01	0.2	0.13	5.2
556	183.6	0.52	20.6	0.90	35.6
188	40.6	0.19	7.6	0.60	23.6
25	33.0	0.10	4.1	0.37	14.6
150	131.7	0.26	10.3	0.72	28.3
6B	10.7	0.01	0.3	0.08	3.3

(a) $Ar-5H_2-5CO-1CO_2-0.15H_2S$; $P_{O_2} = 3 \times 10^{-19}$ atm. $P_{S_2} = 0.9 \times 10^{-6}$ atm. (b) Metal loss plus maximum internal penetration. Source: Ref 40

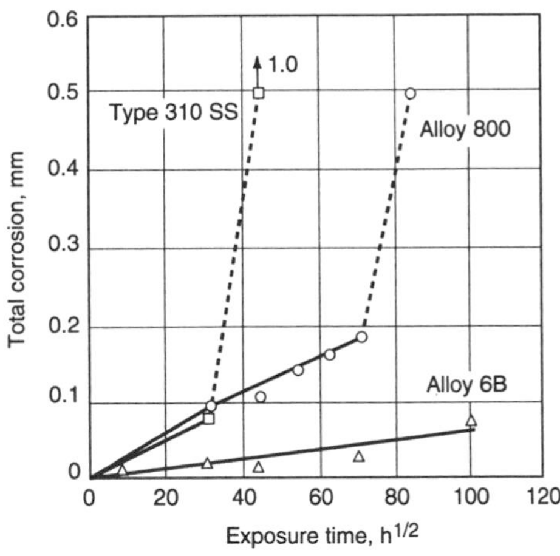

Fig. 15 Corrosion behavior of type 310 stainless steel, alloy 800, and alloy 6B at 980 °C (1800 °F) in a coal gasification atmosphere. Inlet gas: $24H_2-18CO-12CO_2-5CH_4-1NH_3-0.5H_2S$ (balance H_2O) at 6.9 MPa, or 100 psig. Source: Ref 35

Table 16 Corrosion of selected alloys in chlorine

Alloy	Approximate temperature, °C (°F), at which given corrosion rate is exceeded: 0.8 mm/yr (30 mils/yr)	1.5 mm/yr (60 mils/yr)	3.0 mm/yr (120 mils/yr)	15 mm/yr (600 mils/yr)
Nickel	510 (950)	538 (1000)	593 (1100)	650 (1200)
Alloy 600	510 (950)	538 (1000)	565 (1050)	650 (1200)
Alloy B	510 (950)	538 (1000)	593 (1100)	650 (1200)
Alloy C	480 (900)	538 (1000)	565 (1050)	650 (1200)
Chromel A	425 (800)	480 (900)	538 (1000)	650 (1150)
Alloy 400	400 (750)	455 (850)	480 (900)	538 (1000)
18-8Mo	315 (600)	345 (650)	400 (750)	455 (850)
18-8	288 (550)	315 (600)	345 (650)	400 (750)
Carbon steel	120 (250)	175 (350)	205 (400)	230 (450)
Cast iron	93 (200)	120 (250)	175 (350)	230 (450)

Source: Ref 43

nitridation kinetics. It may reduce the rate of surface absorption of nitrogen molecules and/or the diffusion of nitrogen into the metal. Nevertheless, more studies in pure nitrogen are needed, with emphasis on the effect of oxygen potentials and other alloy systems, such as alumina formers.

Sulfidation

Sulfidation involves the interaction of metal with sulfur to form sulfide scale. Because sulfur is one of the most common corrosive contaminants in high-temperature industrial environments (it is generally present in fuels or feedstocks), this mode of attack is frequently encountered. Sulfidation is influenced by both sulfur and oxygen activities, and formation of metal sulfides leads to severe damage. One example of this damage is development of porous layers offering little protection. Because the volume of metal sulfides is 2.5 to 2.9 times greater than that of the corresponding metal oxides, the resulting stresses lead to severe flaking. Another reason they are so damaging is that the metal sulfides have lower melting points than corresponding oxides or carbides. As a result, corrosive attack is catastrophic because of the increase in the diffusion rate by several orders of magnitude via the liquid phase.

Figure 14 illustrates catastrophic failure of a Ni-Cr-Fe alloy tube used in a furnace. The liquid-appearing nickel-rich sulfide phase, which melts at about 650 °C (1200 °F), is clearly visible. Figure 14 also shows that sulfidation attack is quite localized in many cases. In this particular case, the high-nickel alloy suffered sulfidation attack at about 930 °C (1700 °F) in a furnace that was firing ceramic tiles. The cross section at the corroded area showed sulfides through the section of the component. The breakdown of a protective oxide scale (i.e., the chromia scale for most high-temperature alloys) usually signifies the initiation of breakaway corrosion, which is generally followed by rapid corrosion attack.

Breakaway corrosion is illustrated in Fig. 15 for both type 310 stainless steel and alloy 800H tested in a coal gasification atmosphere. Both of these alloys followed a parabolic reaction rate prior to rapid corrosion attack. Figure 16 shows the oxide scales formed on alloy 800H during the protective stage and after breakaway corrosion in a coal gasification environment.

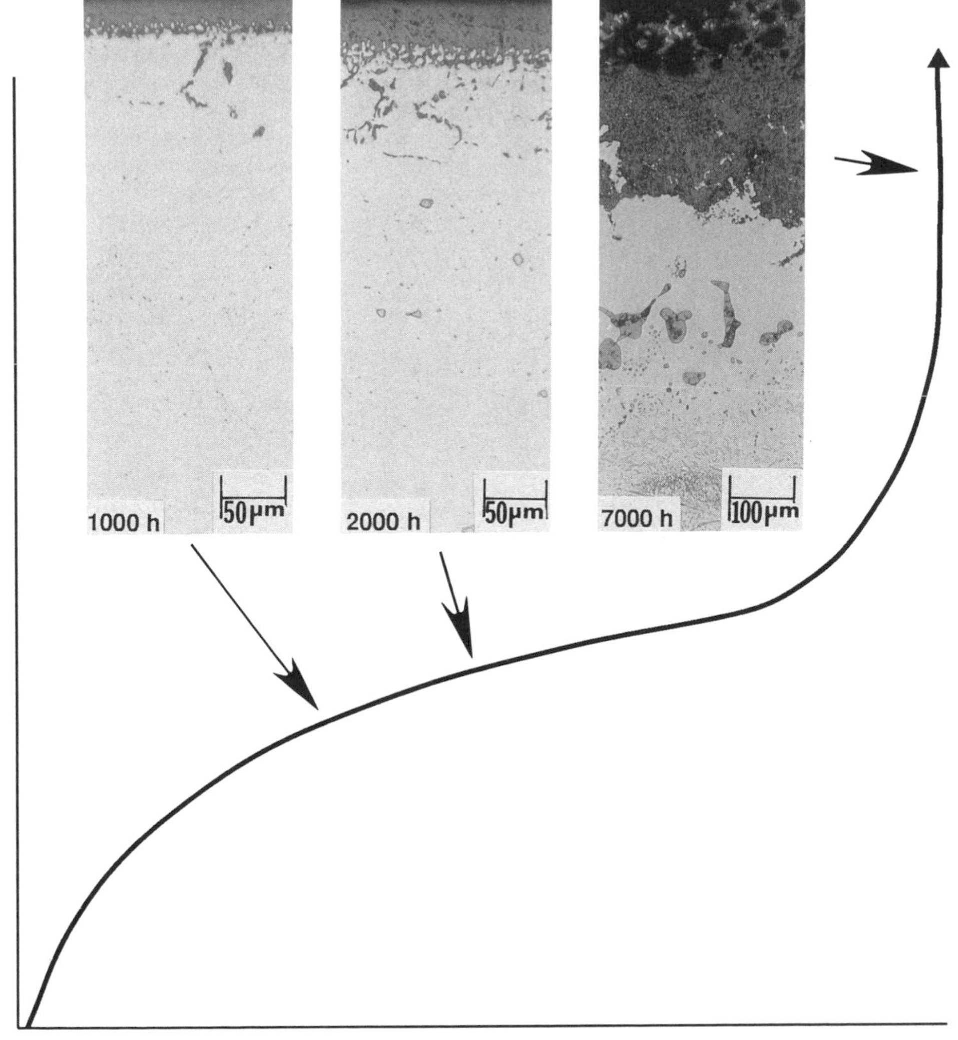

Fig. 16 Corrosion behavior of alloy 800 in a coal gasification atmosphere (0.5% H_2S at 900 °C, or 1650 °F, and 6.9 MPa, or 1000 psig), showing oxide scales during the protective stage and after breakaway corrosion. Source: Ref 36

Table 17 Corrosion of several alloys in Ar-30Cl_2 after 500 h at 400 to 705 °C (750 to 1300 °F)

Alloy	Descaled weight loss, mg/cm²			
	400 °C (750 °F)	500 °C (930 °F)	600 °C (1110 °F)	705 °C (1300 °F)(a)
Ni-201	0.2	0.3	47–101	97
600	0.02	5	127–180	160
601	0.3	3	85–200	215
625	0.7	7	...	180
617	0.6	7	...	190
800	6	13	200–270	890
310	28	370	...	820
304	108	1100	...	>1000
347	215	Total	...	Total

(a) 24 h test period. Source: Ref 44

Table 18 Corrosion of selected alloys in Ar-20O_2-2Cl_2 at 900 °C (1650 °F) for 8 h

Alloy	Metal loss		Average metal affected(a)	
	mm	mils	mm	mils
214	0	0	0.012	0.48
R-41	0.004	0.16	0.028	1.12
600	0.012	0.48	0.035	1.36
310SS	0.012	0.48	0.041	1.60
S	0.053	2.08	0.063	2.48
X	0.020	0.80	0.071	2.80
C-276	0.079	3.12	0.079	3.12
6B	0.014	0.56	0.098	3.84
188	0.014	0.56	0.116	4.56

(a) Metal loss plus average internal penetration. Source: Ref 46

Table 19 Corrosion of various alloys in Ar-20O_2-0.25Cl_2 for 400 h at 900 and 1000 °C (1650 and 1830 °F)

Alloy	Weight loss, mg/cm²	
	900 °C (1650 °F)	1000 °C (1830 °F)
214	4.28	9.05
601	20.67	124.99
600	72.08	254.96
800H	26.91	87.05
310SS	47.15	97.40
556	40.29	82.74
X	54.41	153.49
625	99.07	220.09
R-41	63.83	207.32
263	82.57	229.53
188	139.77	156.30
S	228.21	248.98
C-276	132.05	298.85

Source: Ref 49

High-nickel alloys such as 600/601 are particularly susceptible to sulfidation (Fig. 17). However, resistance to sulfidation can generally be improved by increasing chromium levels to ≥25%. It can also be improved by the presence of cobalt: cobalt-base alloys, as well as cobalt-containing alloys, provide higher resistance to sulfidation attack. In addition, high cobalt levels in nickel-base alloys reduce the rate of diffusion of sulfur in the matrix, and they reduce the risk of developing low-melting-point eutectics, because the CO-CO_4S_3 eutectic forms only above 880 °C (1620 °F). By comparison, the Ni-Ni_3S_2 eutectic occurs at 635 °C (1180 °F). Results generated by Lai (Ref 38) at 760, 870, and 980 °C (1400, 1600, and 1800 °F) also revealed that cobalt-base alloys were more sulfidation resistant than nickel-base and Fe-Ni-Cr alloys with similar chromium contents (Fig. 18).

The beneficial effect of titanium was also demonstrated in the Lai study (Ref 38). As shown in Fig. 18, alloy R-41 and Waspaloy alloy (both contain about 3% Ti) were the best of the nickel-base alloys tested. In fact, their performance approached that of some cobalt-base alloys. Alloy 263 (2.5% Ti), while performing well at 760 °C (1400 °F), suffered severe sulfidation attack at 870 and 980 °C (1600 and 1800 °F).

More recently developed alloys for resisting sulfidation attack are based on the Ni-Co system with high chromium and silicon contents. For example, alloy HR-160 combines high chromium (28%) with relatively high silicon (2.75%) to form a very protective oxide scale (Ref 39–41). Silicon is also effective in improving sulfidation resistance. Increasing cobalt in the Ni-Co-Cr-Fe-Si alloy by replacing iron significantly improved sulfidation resistance. Figure 19 shows the sulfidation resistance of alloy HR-160 compared to that of some existing commercial alloys such as 556, 800H, and 600 (Ref 41). Norton (Ref 42) conducted corrosion tests on several Fe-Ni-Cr alloys and alloy HR-160 at 700 °C (1290 °F) for up to 1000 h in H_2-7CO-1.5H_2O-0.6H_2S. After exposure for 1000 h, alloy HR-160 exhibited about 1.0 mg/cm^2 weight gain compared to about 7 mg/cm^2 for alloys 556 and HR-120, and about 200 to 300 mg/cm^2 for type 321, type 347, and alloy 800H. The gravimetric data are shown in Fig. 20. Comparison of alloy HR-160 with cobalt-base alloys is shown in Table 15. The alloy is significantly better than alloys 188, 25, and 150 (or UMCo-50) and approaches alloy 6B in performance.

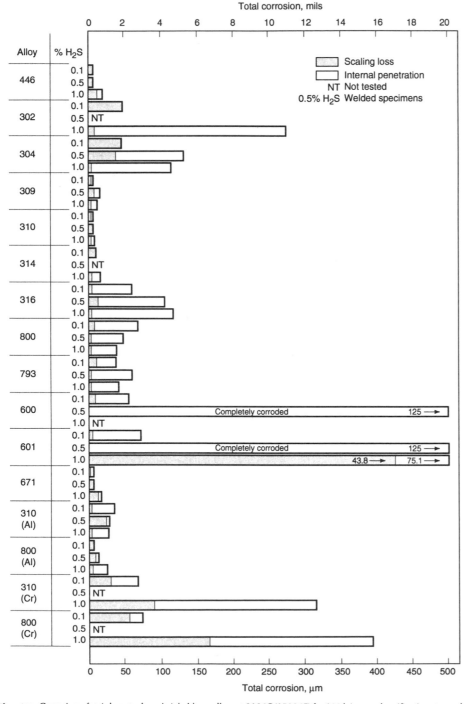

Fig. 17 Corrosion of stainless steels and nickel-base alloys at 816 °C (1500 °F) for 100 h in a coal gasification atmosphere with 0.1, 0.5, and 1.0% H_2S. Source: Ref 37

Halogenation

Halogen and halogen compounds generally attack via the gaseous phase or molten salt compounds. Salts cause slagging and disintegration of the oxide layer; the gas-phase halogens penetrate deeply into the material without destroying the oxide layer. Therefore, preoxidation is of no benefit.

Corrosion in Chlorine-Bearing Environments. Nickel and nickel-base alloys are widely used in chlorine-bearing environments. Nickel reacts with chlorine to form $NiCl_2$, which has a relatively high melting point (1030 °C, or 1886 °F) compared to $FeCl_2$ and $FeCl_3$ (676 and 303 °C, or 1249 and 577 °F, respectively). This may be an important factor, making nickel more resistant to chlorination attack than iron. Brown et al. (Ref 43) conducted short-term laboratory tests in chlorine on various commercial alloys. The results (see Table 16) suggested that, in an environment of 100% Cl_2, carbon steel and cast iron are useful at temperatures up to 150 to 200 °C (300 to 400 °F) only. The 18-8 stainless steels can be used at higher temperatures, up to 320 to 430 °C (600 to 800 °F). Nickel and nickel-base alloys (e.g., Ni-Cr-Fe, Ni-Mo, and Ni-Cr-Mo) were most resistant. The beneficial effect of nickel on the resistance of chlorination attack in chlorine environments is illustrated in Fig. 21. This trend is also reflected in long-term tests (Table 17).

Corrosion in Oxygen-Chlorine Environments. Many industrial environments may contain both chlorine and oxygen. Metals generally

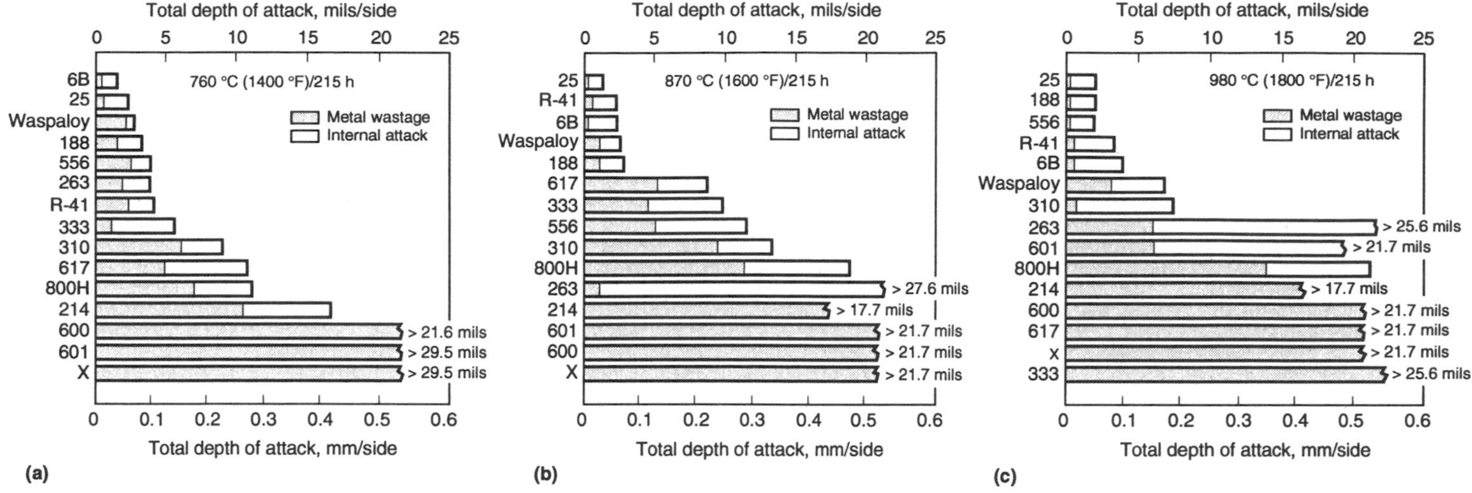

Fig. 18 Corrosion of iron-, nickel-, and cobalt-base alloys after 215 h at (a) 760 °C (1400 °F), (b) 870 °C (1600 °F), and (c) 980 °C (1800 °F) in Ar-$5H_2$-5CO-$1CO_2$-$0.15H_2S$. Source: Ref 38

follow a parabolic rate law by forming condensed phases of oxides, if the environment is free of chlorine. With the presence of both oxygen and chlorine, corrosion of metals then involves a combination of condensed and volatile chlorides. Depending on the relative amounts of oxides and chlorides formed, corrosion can follow either a paralinear rate law (a combination of weight gain due to oxidation and weight loss due to chlorination) or a linear rate law due to chlorination. This is illustrated by the results of Maloney and McNallan (Ref 45) on corrosion of cobalt in Ar-$50O_2$-Cl_2 mixtures (Fig. 22).

A number of studies have examined the corrosion behavior of commercial alloys in mixed oxygen-chlorine environments. Short-term test results generated in Ar-$20O_2$-$2Cl_2$ at 900 °C (1650 °F) for 8 h are summarized in Table 18. The results indicate several interesting trends. Two aluminum-containing nickel-base alloys (214 and R-41) performed best. The two worst alloys were cobalt-base alloys containing tungsten. Molybdenum-containing nickel-base alloys did not perform well at all. Accordingly, alloy 188 (14% W), alloy C-276 (16% Mo, 4% W), alloy 6B (4.5% W, 1.5% Mo), alloy X (9% Mo), and alloy S (14.5% Mo) suffered relatively high rates of corrosion attack. Simple iron- and nickel-base alloys, such as type 310 stainless steel (Fe-Ni-Cr) and alloy 600 (Ni-Cr-Fe), performed better than molybdenum- or tungsten-containing alloys. The thermogravimetric results for representative alloys are summarized in Fig. 23.

With about 1.5% Al and 3% Ti, alloy R-41 exhibited good resistance to chlorination attack in a short-term test despite a relatively high molybdenum content (about 10%). The results of long-term tests in Ar-$20O_2$-$0.25Cl_2$ by Rhee et al. (Ref 48) and McNallan et al. (Ref 49) showed that these nickel-base alloys with molybdenum, such as alloys R-41 and 263 (6% Mo, 2.4% Ti, and 0.6% Al), eventually suffered severe attack despite the presence of aluminum and titanium. Figure 24 shows gravimetric data for aluminum-containing nickel-base alloys with and without molybdenum, tested at 900 °C (1650 °F). Test results for all the alloys tested at 900 °C (1650 °F) and 1000 °C (1830 °F) are summarized in Tables 19 and 20.

The beneficial effect of aluminum, as well as the detrimental effect of molybdenum and tungsten, on resistance to chlorination attack in oxidizing environments is further substantiated by the results of long-term tests in another environ-

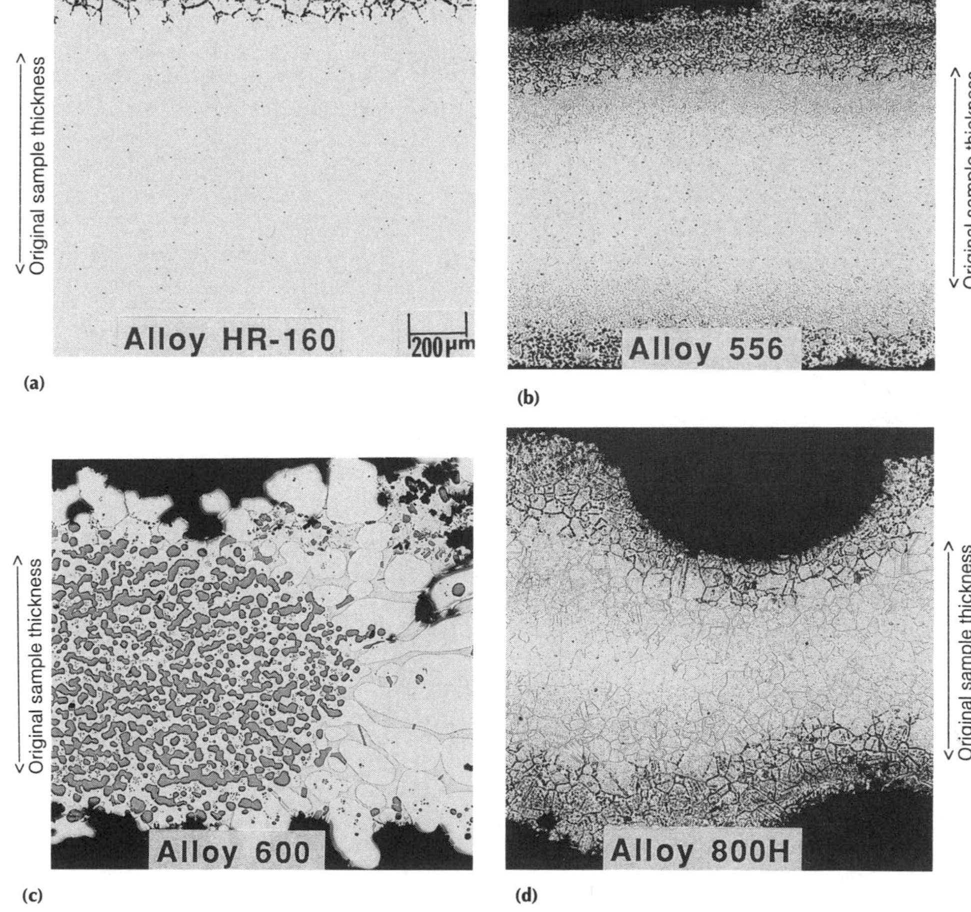

Fig. 19 Sulfidation resistance of alloy HR-160 compared to those of alloys 556, 800H, and 600 after 215 h at 870 °C (1600 °F) in Ar-$5H_2$-5CO-$1CO_2$-$0.15H_2S$. Samples were cathodically descaled before mounting for metallographic ex-

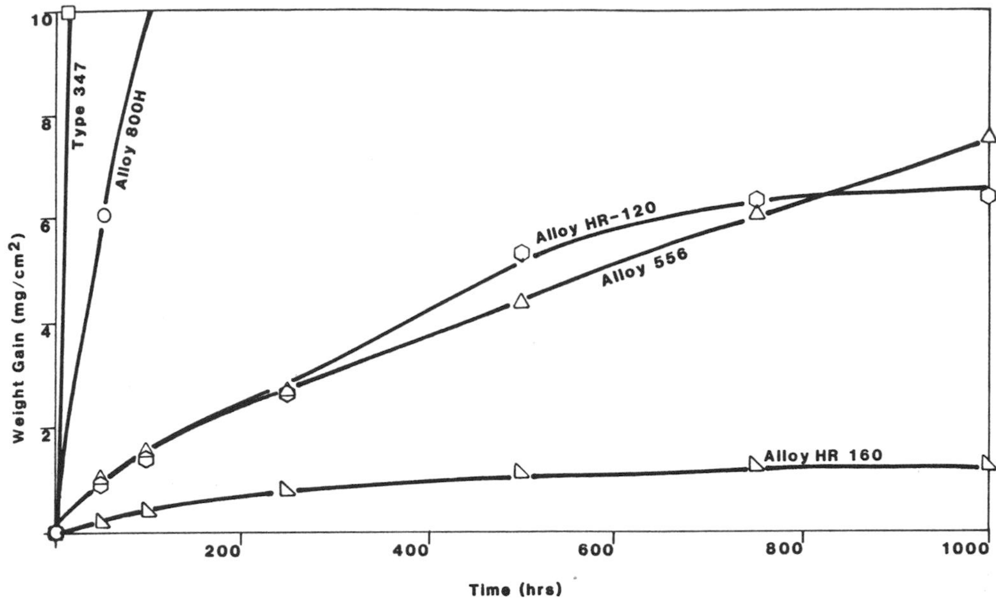

Fig. 20 Corrosion behavior of type 347 stainless steel, alloy 800H, alloy HR-120, alloy 556, and alloy HR-160 at 700 °C (1290 °F) in H_2O-7CO-1.5H_2O-0.6H_2S. Source: Ref 42

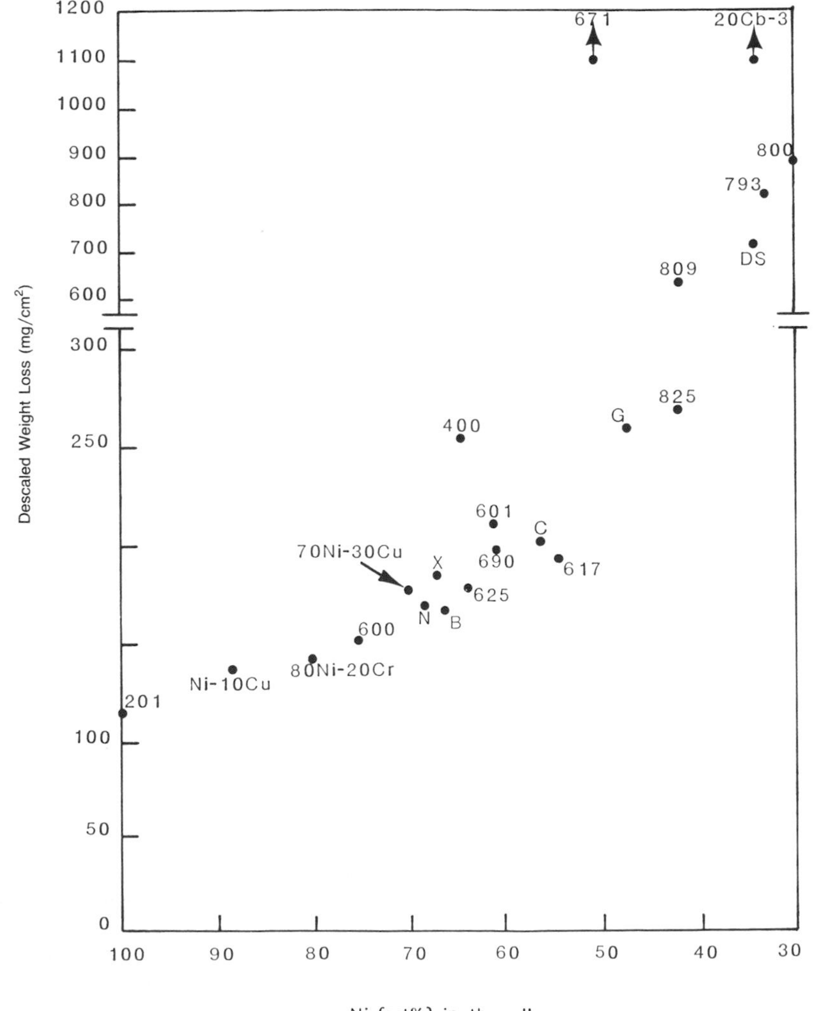

Fig. 21 Effect of nickel on the corrosion resistance of alloys in Ar-30Cl_2 at 704 °C (1300 °F) for 24 h. Source: Ref 44

ment with a higher concentration of chlorine, as shown in Fig. 25. Similar results were obtained by Elliott et al. (Ref 50) from tests conducted in air-2Cl_2 at 900 °C (1650 °F) for 50 h (Fig. 26).

McNallan et al. (Ref 49) reported corrosion behavior in Ar-20O_2-0.25Cl_2 at 900 and 1000 °C (1650 and 1830 °F). This was followed by a study (Ref 51) using the same environment to investigate the same alloys at lower temperatures: 700, 800, and 850 °C (1290, 1470, and 1560 °F) (see Table 21). The corrosion behavior of alloys at temperatures from 700 to 1000 °C (1290 to 1830 °F) are summarized in Fig. 27 using three alloy systems:

- Ni-Cr-Mo (alloy S)
- Fe-Ni-C (alloy 800H)
- Ni-Cr-Al (alloy 214)

As discussed earlier, refractory metals, such as molybdenum and tungsten, are detrimental to chlorination resistance in oxidizing environments. Alloy S was less corrosion resistant than alloy 900H. Both alloys suffered increasing corrosion attack with increasing temperatures. This represents a typical trend for most alloys in oxidizing environments. One exception is the Ni-Cr-Al system. As illustrated in Fig. 27, alloy 214 showed a sudden decrease in corrosion attack as the temperature was increased from 900 to 1000 °C (1650 to 1830 °F). This sharp reduction in corrosion attack at 1000 °C (1830 °F) was attributed to the formation of a protective alumina scale. At lower temperatures, such as ≤900 °C (≤1650 °F), the kinetics of alumina formation was not fast enough to form a protective oxide scale.

Corrosion in HCl Environments. Hossain et al. (Ref 52) performed long-term tests in HCl on several commercial alloys. Their results are summarized in Table 22 and Fig. 28. Type 310 stainless steel was the worst among the alloys tested. The molybdenum-containing Ni-Cr alloys, such as alloys 625 and C-4, were the best performers. This is in contrast to the discussion above about oxygen-chlorine environments, where molybdenum- or tungsten-containing nickel-base alloys performed poorly. Table 22 also shows that unalloyed nickel performed reasonably well in HCl until the temperature reached 700 °C (1290 °F). At 700 °C, nickel was inferior to many nickel-base alloys, such as alloys 600, 625, and C-4.

In reducing environments, such as Ar-4HCl-4H_2 investigated by Baranow et al. (Ref 46), Ni-Cr-Mo alloys, such as alloys C0276 and S, were significantly better than alloys 600, 625, 188, and X. Nickel and nickel alloys also exhibit excellent high-temperature corrosion resistance in dilute HCl environments.

Corrosion in Fluorine-Bearing Environments. Additions of alloying elements to nickel generally are detrimental to fluorine corrosion resistance. Many nickel-base alloys have been found to be significantly more susceptible than nickel to fluorine corrosion or corrosion in fluorine-bearing environments such as hydrogen fluoride (Ref 53, 54).

Hot Corrosion

Hot corrosion is generally described as a form of accelerated attack experienced by the hot gas components of gas turbine engines. As described below, two forms of hot corrosion can be distinguished.

Type I Hot Corrosion. Most of the corrosion encountered in turbines burning fuels can be described as type I hot corrosion, which occurs primarily in the metal temperature range of 850 to 950 °C (1550 to 1750 °F). This is a sulfidation-based attack on the hot gas path parts involving the formation of condensed salts, which are often molten at the turbine operating temperature. The major components of such salts are sodium sulfate (Na_2SO_4) and/or potassium sulfate (K_2SO_4), apparently formed in the combustion process from sulfur from the fuel and sodium from the fuel or the ingested air. Because potassium salts act very much like sodium salts, alkali specifications for fuel or air are usually taken to the sum total of sodium plus potassium. An example of the corrosion morphology typical of type I hot corrosion is shown in Fig. 29.

Very small amounts of sulfur and sodium or potassium in the fuel and air can produce sufficient Na_2SO_4 in the turbine to cause extensive corrosion problems because of the concentrating effect of turbine pressure ratio. For example, a threshold level has been suggested for sodium in air of 0.008 ppm by weight, below which hot corrosion will not occur. Type I hot corrosion, therefore, is possible even when premium fuels are used. Other fuel (or air) impurities, such as vanadium, phosphorus, lead, and chlorides, may combine with Na_2SO_4 to form mixed salts having reduced melting temperature, thus broadening the range of conditions over which this form of attack can occur. Also, agents such as unburned carbon can promote deleterious interactions in the salt deposits.

Hot corrosion generally proceeds in two stages: an incubation period exhibiting low corrosion rates, followed by accelerated corrosion attack. The incubation period is related to the formation of a protective oxide scale. Initiation of accelerated corrosion attack is believed to be related to the breakdown of the protective oxide scale. Many mechanisms have been proposed to explain accelerated corrosion attack; the salt fluxing model is probably the most widely accepted. Oxides may dissolve in Na_2SO_4 as anionic species (basic fluxing) or cationic species (acid fluxing), depending on the salt composition. Salt is acidic when it is high in SO_3 and basic when low in SO_3. More detailed information on the hot corrosion mechanism by salt fluxing can be found in Ref 55 to 60.

Research has led to greater definition of the relationships among temperature, pressure, salt concentration, and salt vapor-liquid equilibria so that the location and rate of salt deposition in an engine can be predicted. Additionally, it has been demonstrated that a high chromium content is required in an alloy for good resistance to type I hot corrosion (Table 23). The trend to lower chromium levels with increasing alloy strength has therefore rendered most superalloys inherently susceptible to this type of corrosion. The effects of other alloying additions, such as tungsten, molybdenum, and tantalum, have been documented, and their effects on rendering an alloy more or less susceptible to type I hot corrosion are known and mostly understood (Ref 61-66).

Although various attempts have been made to develop figures of merit to compare superalloys, these have not been universally accepted. Nonetheless, the near standardization of such alloys as

Table 20 Depth of attack after 400 h at 900 and 1000 °C (1650 and 1830 °F) in Ar-20O_2-0.25Cl_2

Alloy	900 °C (1650 °F)				1000 °C (1830 °F)			
	Metal loss		Total depth(a)		Metal loss		Total depth(a)	
	mm	mils	mm	mils	mm	mils	mm	mils
214	0.023	0.9	0.150	5.9	0.013	0.5	0.051	2.0
601	0.061	2.4	0.264	10.4	0.203	8.0	0.295	11.6
600	0.127	5.0	0.252	9.9	0.330	13.0	0.386	15.2
800H	0.043	1.7	0.191	7.5	0.203	8.0	0.424	16.7
301SS	0.086	3.4	0.152	6.0	0.191	7.5	0.246	9.7
556	0.046	1.8	0.152	6.0	0.152	6.0	0.300	11.8
X	0.099	3.9	0.218	8.6	0.318	12.5	0.434	17.1
625	0.208	8.2	0.272	10.7	0.356	14.0	0.437	17.2
R-41	0.114	4.5	0.244	9.6	0.381	15.0	0.457	18.0
263	0.130	5.1	0.193	7.6	0.368	14.5	0.424	16.7
188	0.216	8.5	>0.356	14.0	0.254	10.0	0.417	16.4
S	0.315	12.4	0.353	13.9	0.419	16.5	0.472	18.6
C-276	0.300	11.8	0.320	12.6	0.419	16.5	0.450	17.7

(a) Metal loss plus internal penetration. Source: Ref 49

Table 21 Depth of attack after 400 h at 700, 800, and 850 °C (1290, 1470, and 1560 °F) in Ar-20O_2-0.25Cl_2

Alloy	700 °C (1290 °F)				800 °C (1470 °F)				850 °C (1560 °F)			
	Metal loss		Total depth		Metal loss		Total depth		Metal loss		Total depth	
	mm	mils	mm	mils	mm	mils	mm	mils	mm	mils	mm	mils
214	0.010	0.4	0.010	0.4	0.018	0.7	0.061	2.4	0.018	0.7	0.066	2.6
600	...	...	...	...	0.020	0.8	0.086	3.4	0.038	1.5	0.132	5.2
800H	0.025	1.0	0.033	1.3	0.023	0.9	0.046	1.8	0.031	1.2	0.097	3.8
310SS	...	...	...	...	0.036	1.4	0.053	2.1	0.031	1.2	0.061	2.4
556	...	...	...	...	0.020	0.8	0.051	2.0	0.020	0.8	0.079	3.1
S	0.079	3.1	0.081	3.2	0.145	5.7	0.150	5.9	0.224	8.8	0.257	10.1
C-276	0.033	1.3	0.046	1.8	0.066	2.6	0.071	2.8	0.163	6.4	0.175	6.9
188	...	...	...	...	0.058	2.3	0.074	2.9	0.025	1.0	0.264	10.4

Source: Ref 51

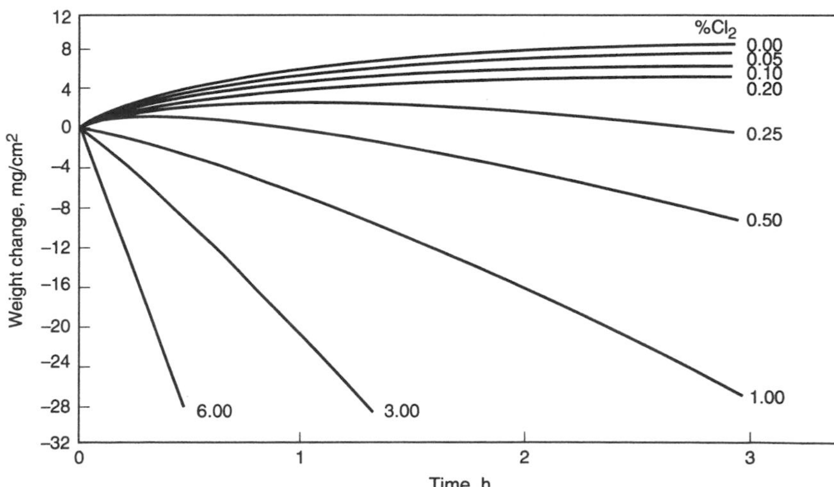

Fig. 22 Corrosion of cobalt in Ar-50O_2-Cl_2 mixtures at 927 °C (1700 °F). Source: Ref 45

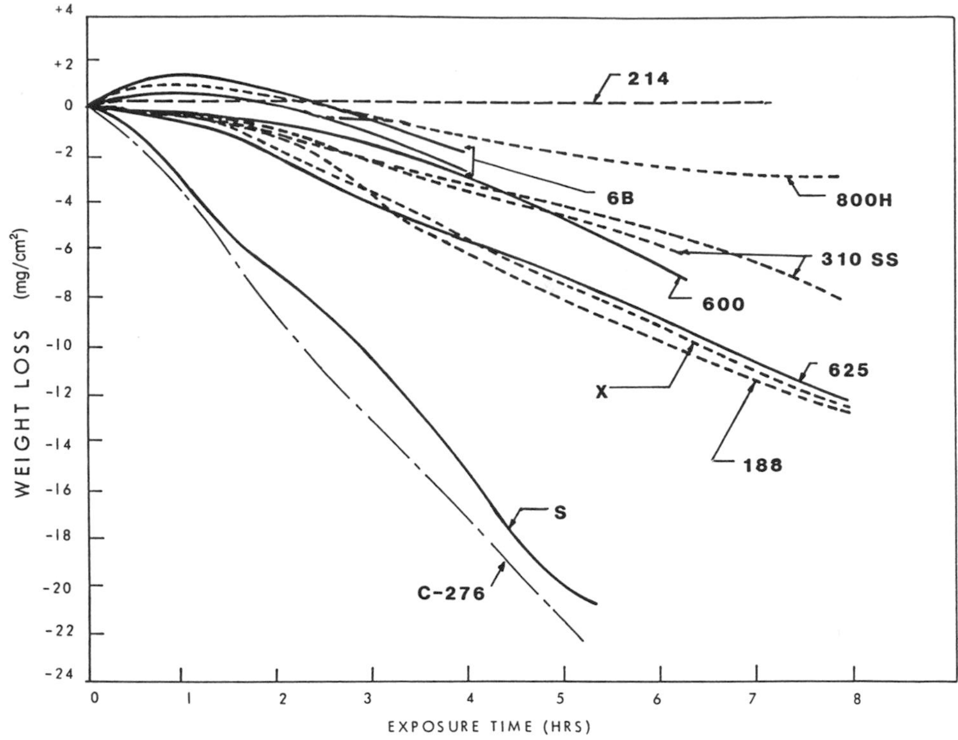

Fig. 23 Corrosion of selected commercial alloys in Ar-20O_2-2Cl_2 at 900 °C (1650 °F) in terms of weight change of specimen as a function of time. Source: Ref 47

IN-738 and IN-939 for first-stage blades/buckets, and FSX-414 for first-stage vanes/nozzles, implies that these are the accepted best compromises between high-temperature strength and hot corrosion resistance. It has also been possible to devise coatings with alloying levels adjusted to resist this form of hot corrosion. The use of such coatings is essential for the protection of most modern superalloys intended for duty as first-stage blades or buckets. Coatings to prevent hot corrosion are described in the article "Protective Coatings for Superalloys" in this Volume.

Type II, or low-temperature hot corrosion, occurs in the metal temperature range of 650 to 700 °C (1200 to 1400 °F), well below the melting temperature of Na_2SO_4, which is 884 °C (1623 °F). This form of corrosion produces characteristic pitting, which results from the formation of low-melting mixtures of essentially Na_2SO_4 and cobalt sulfate ($CoSO_4$), a corrosion product resulting from the reaction of the blade/bucket surface with sulfur trioxide (SO_3) in the combustion gas. The melting point of the Na_2SO_4-$CoSO_4$) eutectic is 540 °C (1004 °F). Unlike type I hot corrosion, a partial pressure of SO_3 in the gas is critical for the reactions to occur. Knowledge of the SO_3 partial pressure-temperature relationships inside a turbine allows some prediction of where type II hot corrosion can occur. Cobalt-free nickel-base alloys (and coatings) may be more resistant to type II hot corrosion than cobalt-base alloys; it has also been observed that resistance to type II hot corrosion increases with the chromium content of the alloy or coating.

Table 22 Corrosion of selected alloys in HCl at 400, 500, 600, and 700 °C (750, 930, 1110, and 1290 °F)

Alloy	Metal loss, mg/cm²: 400 °C (750 °F) 300 h	400 °C (750 °F) 1000 h	500 °C (930 °F) 100 h	500 °C (930 °F) 300 h	500 °C (930 °F) 1000 h	600 °C (1110 °F) 100 h	600 °C (1110 °F) 300 h	700 °C (1290 °F) 96 h
Ni-201	1.19	0.91	1.60	2.89	4.86	11.46	37.7	377
601	1.58	1.47	2.57	4.14	9.38	9.01	19.46	102.5
310	3.26	5.16	6.74	13.65	46.60	15.65	32.6	102.5
625	0.74	1.1	2.42	3.78	8.64	6.79	14.6	26.5
C-4	0.55	1.12	2.09	3.36	7.24	7.31	19.14	34.9
B-2	0.75	0.76	2.10	2.65	5.87	12.93	62.3	126.4
600	0.93	0.81	1.69	3.31	7.81	7.67	17.3	49.6

Source: Ref 52

Corrosion in Waste Incineration Environments

As shown in Table 24, combustion environments generated by incineration of municipal, hospital, chemical, and hazardous wastes contain common corrosive contaminants such as sulfur and chlorine. At temperatures higher than 650 °C (1200 °F), sulfidation and/or chloride attack are frequently responsible for the corrosion reaction. Figure 30 illustrates sulfidation and chloride attack of an Fe-Ni-Co-Cr alloy used in an industrial waste incinerator. Thus, alloys resistant to both sulfidation and chloride attack are preferred for applications at temperatures higher than 650 °C (1200 °F). In addition to sulfur and chlorine, other constituents frequently detected in significant amounts in deposits on incinerator components include potassium, sodium, zinc, and lead (Table 24). These elements may contribute to the formation of low-melting-point salts. Many salt mixtures become molten in the temperature range of the furnace wall tubes and superheater tubes. Thus, molten salt deposit corrosion may also be a likely corrosion mechanism.

Molten Salt Corrosion

Molten salts generally are a good fluxing agent, effectively removing oxide scale from a metal surface. The corrosion reaction proceeds primarily by oxidation, which is then followed by dissolution of metal oxides in the melt. Oxygen and water vapor in the salt thus often accelerate molten salt corrosion.

Corrosion can also take place through mass transfer due to thermal gradient in the melt. This

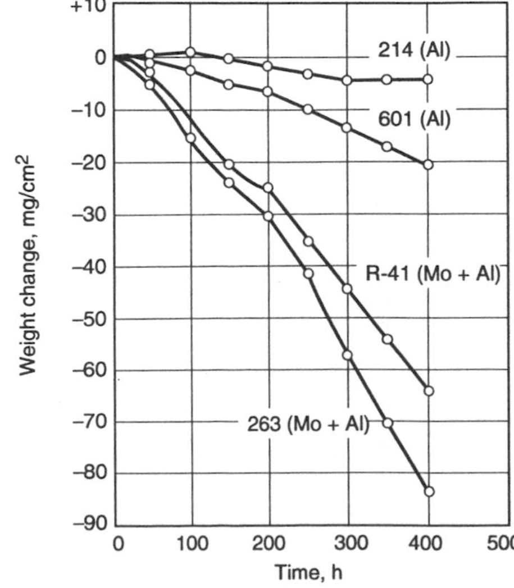

Fig. 24 Corrosion of several aluminum-containing nickel-base alloys with and without molybdenum in Ar-20O_2-0.25Cl_2 at 900 °C (1650 °F). Source: Ref 49

mode of corrosion involves dissolution of an alloying element at hot spots and deposition of that element at cooler spots. This can result in severe fouling and plugging in a circulating system. Corrosion is also strongly dependent on temperature and velocity of the salt.

Corrosion can take the form of uniform thinning, pitting, or internal or intergranular attack. In general, molten salt corrosion is quite similar to aqueous corrosion. A more complete discussion on the mechanisms of molten salt corrosion can be found in the article "Corrosion at Elevated Temperatures" in this Volume.

Corrosion in Molten Chlorides. Chloride salts are widely used in the heat treating industry for annealing and normalizing of steels, These salts are commonly referred to as neutral salt baths. The most common neutral salt baths are barium, sodium, and potassium chlorides, used separately or in combination in the temperature range of 760 to 980 °C (1400 to 1800 °F).

Lai et al. (Ref 69) evaluated various wrought iron-, nickel-, and cobalt-base alloys in a NaCl-KCl-$BaCl_2$ salt bath at 840 °C (1550 °F) for 1 month (Fig. 31). Surprisingly, two high-nickel alloys (alloys 600 and 601) suffered more corrosion attack than stainless steels such as types 304 and 310. The Co-Fe-Ni-W, Fe-Ni-Co-Cr, and Ni-Cr-Fe-Mo alloys performed best. Laboratory testing in a simple salt bath failed to reveal the correlation between alloying elements and performance. Tests were conducted at 840 °C (1550 °F) for 100 h in a NaCl salt bath with fresh salt for each test run. Results are tabulated in Table 25. Similar to the field test results, Co-Ni-Cr-W and Fe-Ni-Co-Cr alloys performed best.

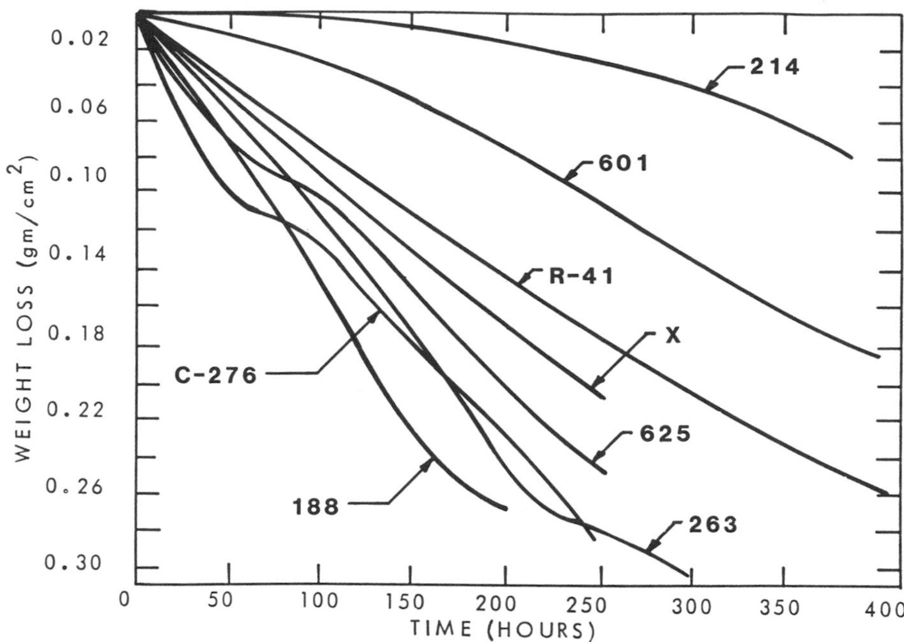

Fig. 25 Corrosion of several nickel- and cobalt-base alloys in Ar-$20O_2$-$1Cl_2$ at 900 °C (1650 °F). Source: Ref 46

Intergranular corrosion is the major corrosion morphology by molten chloride salts. Figures 32 and 33 show typical intergranular corrosion by molten chloride salt. Figure 32 shows the intergranular attack of a Ni-Cr-Fe alloy (alloy 600) coupon welded to a heat treating basket that underwent heat treat cycles involving a molten KCl salt bath at 870 °C (1600 °F) and a quenching salt bath of molten sodium nitrate and sodium nitrite at 430 °C (800 °F) for 1 month (Ref 71). Figure 33 shows the intergranular attack of a heating treat basket made of the same alloy after service for 6 months in the same heat treating cycling operation (Ref 71).

Corrosion in Molten Nitrates/Nitrites. Molten nitrates or nitrate-nitrite mixtures are widely used for heat treat salt baths, typically operating from 160 to 590 °C (325 to 1100 °F). They are also used as a medium for heat transfer or energy storage.

Numerous studies have been carried out to determine potential candidate containment materials for handling molten drawsalt ($NaNO_3$-KNO_3). Bradshaw and Carling (Ref 72) summarized these studies as well as the results of their own study in Table 26. The data suggest that, for temperatures up to 630 °C (1170 °F),

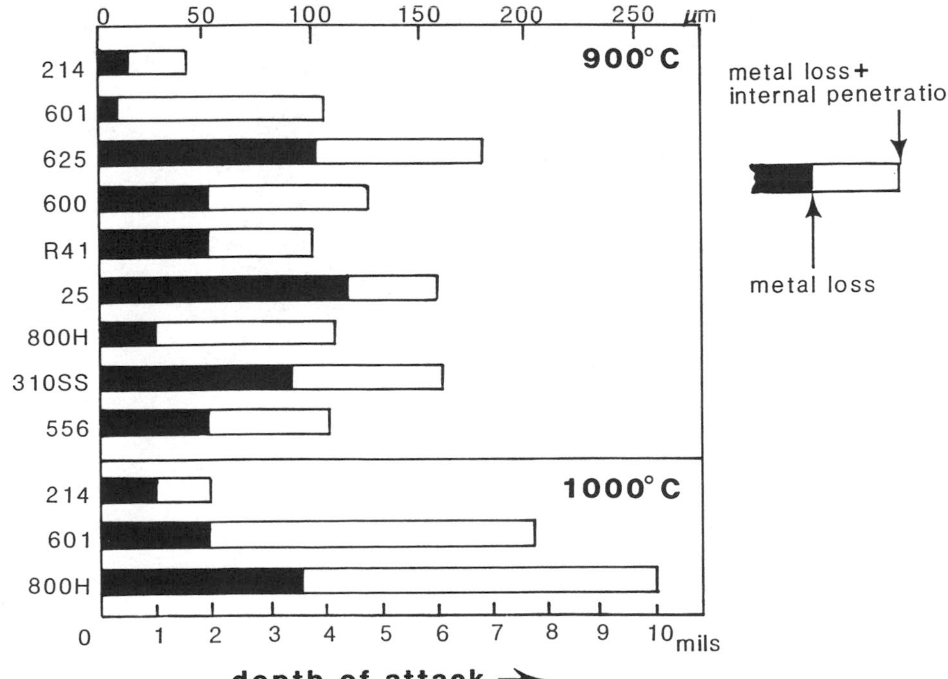

Fig. 26 Corrosion of several iron- and nickel-base alloys in air-$2Cl_2$ at 900 and 1000 °C (1650 and 1830 °F) for 50 h. Source: Ref 50

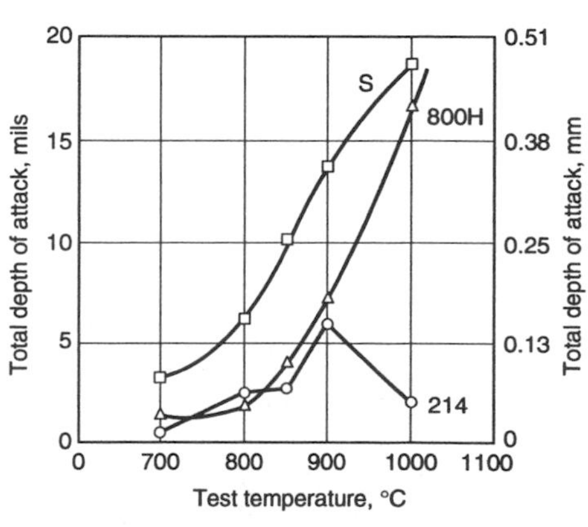

Fig. 27 Corrosion behavior of alloy 214 (Ni-Cr-Al-Y), alloy S (Ni-Cr-Mo), and alloy 800H (Fe-Ni-Cr) in Ar-$20O_2$-$0.25Cl_2$ for 400 h at 700 to 1000 °C (1290 to 1830 °F). Source: Ref 49, 51

Table 23 Effect of chromium content on the hot corrosion resistance of nickel- and cobalt-base alloys

Alloy	Chromium content in alloy, %	Loss in sample diameter (a), mm (mils) 870 °C (1600 °F) 500 h	950 °C (1750 °F) 1000 h	980 °C (1800 °F) 1000 h	1040 °C (1900 °F) 1000 h
SM-200	9.0	1.6 (64.4)	3.3+ (130+)	...	...
IN-100	10.0	3.3+ (130+)	3.3+ (130+)	...	...
SEL-15	11.0	3.3+ (130+)	3.3+ (130+)	...	...
IN-713	13.0	3.3+ (130+)	2.0+ (77+)	...	...
U-700	14.8	1.7+ (66+)	1.6 (63.9)	...	...
SEL	15.0	1.2 (45.8)	1.3 (51.8)	0.3 (11.4)	...
U-500	18.5	0.2 (7.6)	0.8 (31.7)	0.7 (29.3)	...
René 41	19.0	0.3 (10.3)	...	0.8 (30.8)	
Hastelloy alloy X	22.0	...	0.3 (12.0)	0.4 (15.2)	...
L-605 (alloy 25)	20.0	...	0.4 (15.3)	0.3 (11.3)	1.1 (41.9)
WI-52	21.0	0.5 (21.4)	0.5 (18.2)	...	1.9 (73.9)
MM-509	21.5	...	0.3 (10.9)	...	0.8 (31.8)
SM-302	21.5	0.14 (5.4)	0.3 (10.0)	...	0.6 (23.1)
X-40	25.0	0.11 (4.2)	0.3 (11.6)	...	0.5 (18.5)

(a) Results of burner rig tests with 5 ppm sea salt injection. Source: Ref 61

Table 24 Temperatures and principal contaminants encountered in various types of incinerators

Incinerator type	Temperatures and principal contaminants
Municipal waste incinerators	980–1090 °C (1800–2000 °F), S, Cl, K, Zn, etc.
	700–760 °C (1300–1400 °F), S, Cl, K, Zn, Pb
Industrial waste incinerator	870–903 °C (1600–1700 °F), S, Cl, K, etc.
Hospital waste incinerator	650–760 °C (1200–1400 °F), S, Cl, Zn, etc.
Low-level radioactive waste incinerator	590–760 °C (1100–1400 °F), S, Cl, Zn, P, Pb, etc.
Chemical waste incinerator	480 °C (900 °F), Pb, K, S, P, Zn, and Ca

Source: Ref 67, 68

many alloys are adequate for handling molten $NaNO_3$-KNO_3 salt. Carbon steel and 2.25Cr-1Mo steel exhibited low corrosion rates (mm/yr, or 5 mils/yr) at 460 °C (860 °F). At 500 °C (932 °F), 2.25Cr-1Mo steel exhibited a corrosion rate of about 0.026 mm/yr (1 mil/yr). Aluminized Cr-Mo steel showed higher resistance, with a corrosion rate of less than 0.004 mm/yr (0.2 mils/yr) at 600 °C (1110 °F). Austenitic stainless steels, alloy 800, and alloy 600 were more resistant than carbon steel and Cr-Mo steels. Nickel, however, suffered high corrosion rates.

Slusser et al. (Ref 73) evaluated the corrosion behavior of a variety of alloys in molten $NaNO_3$-KNO_3 (equimolar volume) salt with an equilibrium nitrite concentration (about 6 to 12 wt%) at 675 °C (1250 °F) for 336 h. A constant purge of air in the melt was maintained during testing. Nickel-base alloys were generally much more resistant than iron-base alloys. Increasing nickel content improved alloy corrosion resistance to molten nitrate-nitrite salt. However, pure nickel suffered rapid corrosion attack. Figure 34 shows the corrosion rates of various alloys as a function of nickel content (Ref 73). Silicon-containing alloys, such as RA330 and Nicrofer 3718, performed poorly. A long-term test (1920 h exposure) at 675 °C (1250 °F) was performed on selected alloys, showing corrosion rates similar to those obtained from 336 h exposure tests (Table 27). Alloy 800, however, exhibited a higher corrosion rate in the 1920 h test than in the 336 h test. As the temperature was increased to 700 °C (1300 °F), corrosion rates became much higher, particularly for iron-base alloy 800, which suffered an unacceptably high rate (Table 27).

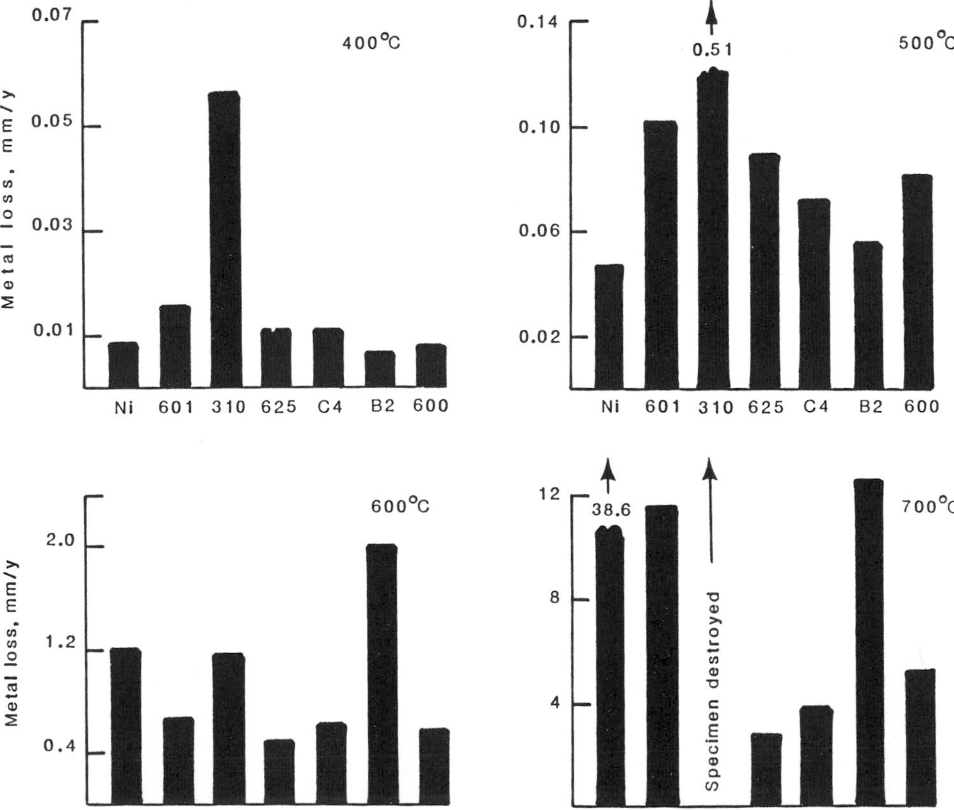

Fig. 28 Corrosion rates of several iron- and nickel-base alloys in HCl at 400 to 700 °C (750 to 1290 °F). Source: Ref 52

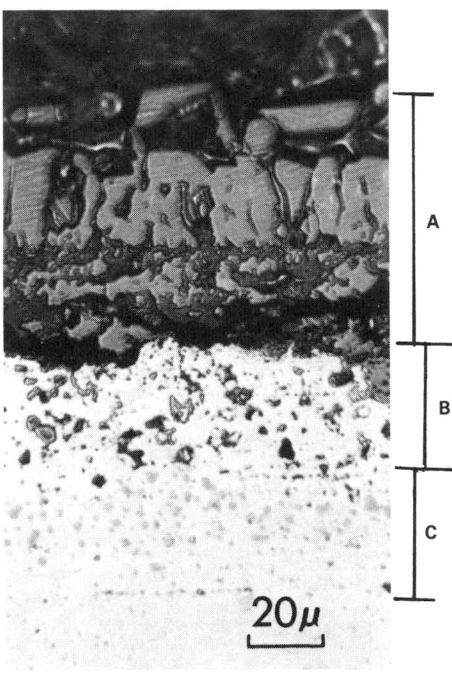

Fig. 29 Ni-20Cr-2ThO$_2$ after simulated type I hot corrosion exposure (coated with Na_2SO_4 and oxidized in air at 1000 °C, or 1832 °F). A, nickel-rich scale; B, Cr_2O_3 subscale; C, chromium sulfides

Table 25 Results of laboratory tests in a NaCl salt bath(a) at 840 °C (1550 °F) for 100 h

Alloy	Total depth of attack	
	mm	mils
188	0.051	2.0
25	0.064	2.5
556	0.066	2.6
601	0.066	2.6
Multimet	0.069	2.7
150	0.076	3.0
214	0.079	3.1
304	0.081	3.2
446	0.081	3.2
316	0.081	3.2
X	0.097	3.8
310	0.107	4.2
800H	0.109	4.3
625	0.112	4.4
RA330	0.117	4.6
617	0.122	4.8
230	0.140	5.5
S	0.168	6.6
RA330	0.191	7.5
600	0.196	7.7

(a) A fresh salt bath was used for each test run; air was used for the cover gas. (b) Mainly intergranular attack; no metal wastage. Source: Ref 69, 70

Boehme and Bradshaw (Ref 74) attributed the increased corrosion rate with increasing temperature to higher alkali oxide concentration.

Slusser et al. (Ref 73) found that adding sodium peroxide (Na_2O_2) to the salt increased the salt corrosivity.

Corrosion in Molten Sodium Hydroxide (Caustic Soda). The reaction of metals with molten sodium hydroxide (NaOH) leads to metal oxide, sodium oxide, and hydrogen. Nickel is most resistant to molten NaOH, particularly low-carbon nickel such as Ni 201 (Ref 74). Gregory et al. (Ref 75) reported several nickel-base alloys obtained from static tests at 400 to 680 °C (750 to 1256 °F) (Table 28). Molybdenum and silicon appear to be detrimental alloying elements in molten NaOH salt. Iron may also be detrimental.

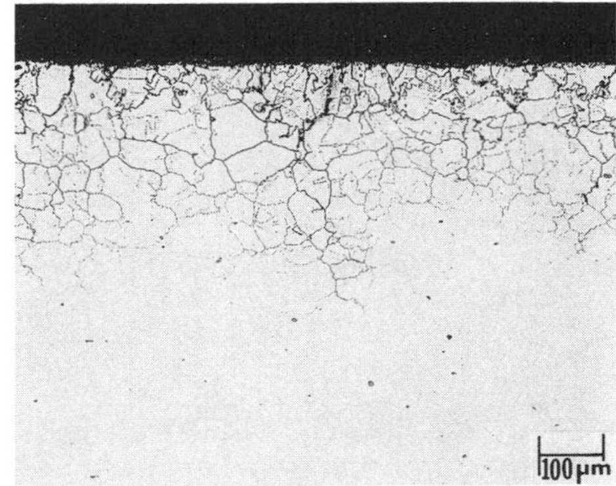

Fig. 32 Intergranular attack of an alloy 600 coupon welded to a heat treating basket after service for 1 month in a heating treat operation cycling between a molten KCl bath at 870 °C (1600 °F) and a quenching salt bath of molten sodium nitrate-nitrite at 430 °C (800 °F). Source: Ref 71

Table 26 Corrosion rates of selected metals and alloys in molten $NaNO_3$-KNO_3

Alloy	Temperature		Corrosion rate	
	°C	°F	mm/yr	(mils/yr)
Carbon steel	460	860	0.120	4.7
2.25Cr-1Mo	460	860	0.101	4.0
	500	932	0.026	1.0
9Cr-1Mo	550	1020	0.006	0.2
	600	1110	0.023	0.9
Aluminized Cr-Mo steel	600	1110	<0.004	0.2
12Cr steel	600	1110	0.022	0.9
304SS	600	1110	0.012	0.5
316SS	600	1110	0.007–0.010	0.03–0.4
	630	1170	0.106	4.2
800	565	1050	0.005	0.2
	600	1110	0.006–0.01	0.02–0.4
	630	1170	0.075	3.0
600	600	1110	0.007–0.01	0.3–0.4
	630	1170	0.106	4.2
Nickel	565	1050	>0.5	20
Titanium	565	1050	0.04	1.6
Aluminum	565	1050	<0.004	0.2

Source: Ref 72

Molybdenum and iron were found to be selectively removed from nickel-base alloys with less than 90% Ni, leading to the formation of internal voids (Ref 76).

Corrosion in Molten Fluorides. Corrosion of alloys in molten fluoride salts has been extensively studied for nuclear reactor applications. The nuclear reactor uses a LiF-BeF_2 base salt as a fuel salt, containing various amounts of UF_4, ThF_4, and ZrF_4 (Ref 77). The reactor coolant salt is a $NaBF_4$-NaF mixture. A nickel-base alloy, Hastelloy N, has proved to be the most corrosion resistant in molten fluoride salts (Ref 78). The alloy was the primary containment material for a molten salt test reactor successfully operated from 1965 to 1969 (Ref 79). Koger (Ref 77) reported a corrosion rate of less than 0.0025 mm/yr (0.1 mil/yr) at 704 °C (1300 °F) in the LiF-BeF_2 base salt (fuel salt) and about 0.015 mm/yr (0.6 mil/yr) at 607 °C (1125 °F) in the $NaBF_4$-NaF coolant salt for alloy N.

Corrosion in Molten Carbonates. Molten carbonates are generally less corrosive than molten chlorides or hydroxides. Corrosion data generated in molten carbonate salt at 900 °C (1650 °F) for 504 h are summarized in Table 29. The

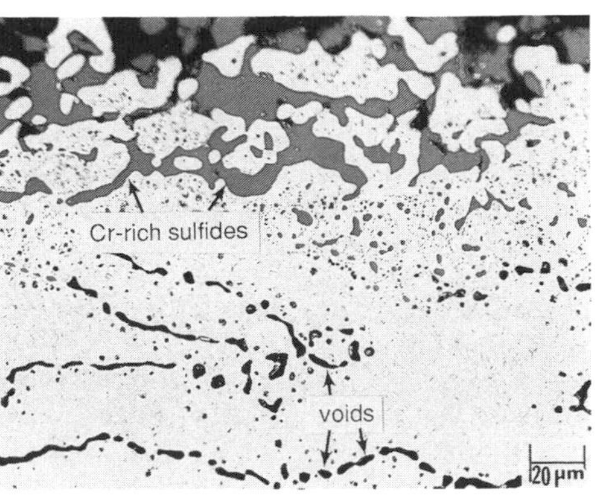

Fig. 30 Fe-Ni-Co-Cr alloy showing both sulfidation and chloride attack. Internal voids were formed by volatile metal chlorides.

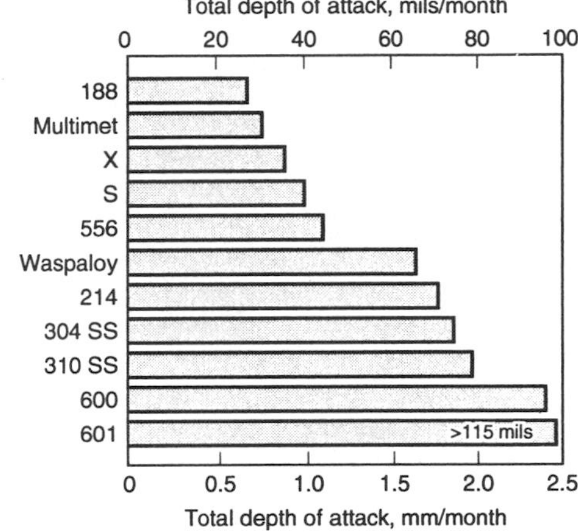

Fig. 31 Results of a field rack test in a NaCl-KCl-$BaCl_2$ salt bath at 840 °C (1550 °F) for 1 month. Source: Ref 69

Table 27 Corrosion rates of selected alloys at 675 and 700 °C (1250 and 1300 °F) in sodium-potassium nitrate-nitrite salt

Alloy	Corrosion rate, mm/yr (mils/yr)	
	675 °C (1250 °F), 1920 h	700 °C (1300 °F), 720 h
214	0.41 (16)	0.53 (21)
600	0.25 (10)	0.99 (39)
N	0.23 (9.1)	1.22 (48)
601	0.48 (19)	1.25 (49)
800	1.85 (73)	6.6 (259)

Source: Ref 73

Table 28 Corrosion rates of selected nickel-base alloys obtained from static tests in molten sodium hydroxide

Alloy	Corrosion rate, mm/yr (mils/yr)			
	400 °C (750 °F)	500 °C (930 °F)	580 °C (1080 °F)	680°C (1260 °F)
Ni-201	0.023 (0.9)	0.033 (1.3)	0.06 (2.5)	0.96 (37.8)
C	...	2.54 (100)	(a)	...
D	0.018 (0.7)	0.056 (2.2)	0.25 (9.9)	(a)
400	0.046 (1.8)	0.13 (5.1)	0.45 (17.6)	...
600	0.028 (1.1)	0.06 (2.4)	0.13 (5.1)	1.69 (66.4)
301SS	0.043 (1.7)	0.08 (3.2)	0.26 (10.4)	1.03 (40.7)
75	0.028 (1.1)	0.36 (14.3)	0.53 (20.8)	1.21 (47.6)

(a) Severe corrosion. Source: Ref 75

best performer was Ni-Cr-Fe-Mo alloy (alloy X), followed by Ni-Cr-Fe-Al alloy (214 alloy) and Co-Ni-Cr-W alloy (alloy 188). Ni-Cr-Mo alloy (alloy S) was severely corroded. There was no evidence of a systematic trend in the correlation between alloying elements and performance. There was, however, an anomaly in the test results. A significant difference in corrosion was observed between two samples of alloy 800 obtained from different suppliers. This marked difference could not be attributed to the different suppliers. Two samples of alloy 600, however, showed good agreement.

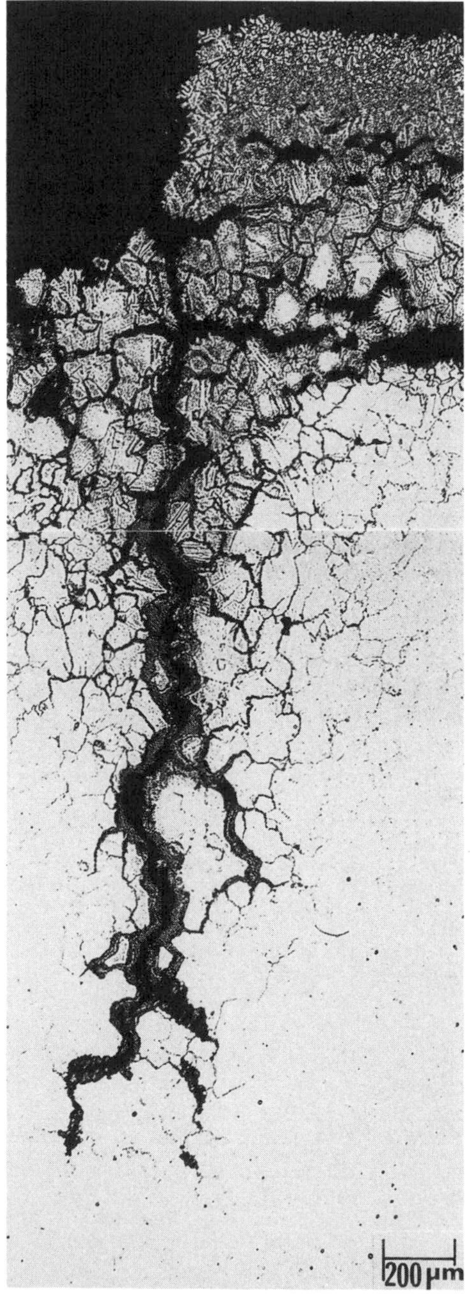

Fig. 33 Intergranular attack of an alloy 600 heat treating basket after service for 6 months in the same heat treating cycling operation described in Fig. 32. Source: Ref 71

Molten Metal Corrosion

Molten metal corrosion of containment metal is most often related to its solubility in the molten metal. This type of corrosion is simply a dissolution-type attack. A containment metal with a higher solubility in the molten metal generally exhibits a higher corrosion rate. In the case of an alloy, the solubilities of the major alloying elements could dictate the corrosion rate. The solubility of an alloying element in a molten metal typically increases with increasing temperature. As the temperature increases, the diffusion rate also increases. Thus, the alloy corrosion rate increases with increasing temperature.

Corrosion by molten metal can also proceed by alloying of the containment metal (or its major alloying elements) with the molten metal to form an intermetallic compound. This requires some solubility of the liquid metal in the containment metal.

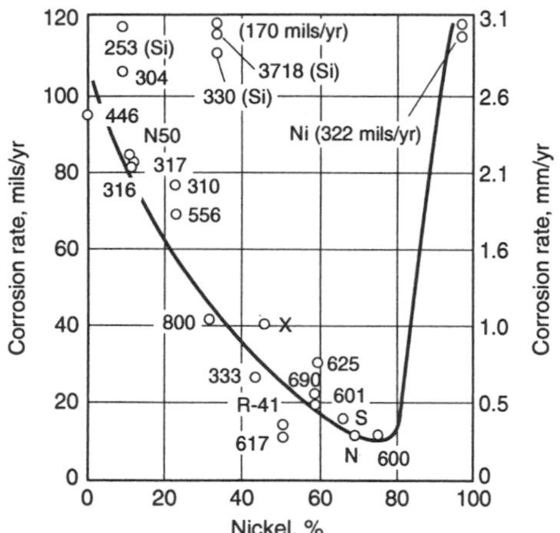

Fig. 34 Corrosion rates of various alloys as a function of nickel content in molten $NaNO_3$-KNO_3 salt at 675 °C (1250 °F). Source: Ref 73

Corrosion in Molten Aluminum. Aluminum melts at 660 °C (1220 °F). Iron, nickel, and cobalt, along with their alloys, are readily attacked by molten aluminum. Extremely high corrosion rates of iron-, nickel-, and cobalt-base alloys in molten aluminum are illustrated by the laboratory test results shown in Table 30.

Corrosion in Molten Zinc. Zinc melts at 420 °C (787 °F). Molten zinc is widely used in the hot dip galvanizing process to coat steel for corrosion protection. Galvanizing tanks, along with baskets, fixtures, and other accessories, require materials resistant to molten zinc corrosion.

Nickel and high-nickel alloys react readily with molten zinc by direct alloying. Iron- and cobalt-base alloys are generally corroded by dissolution, even those containing up to 33% Ni, such as alloy 800H. The results of static tests in molten zinc for selected iron-, nickel-, and cobalt-base alloys are summarized in Table 31. Nickel-base alloys suffered the worst attack, followed by austenitic stainless steels, Fe-Ni-Cr alloys, and Fe-Cr alloys. Cobalt-base alloys generally performed better. However, an Fe-Ni-Co-Cr alloy (556 alloy) performed as well as cobalt-base alloys.

Corrosion in Molten Lead. Lead melts at 327 °C (621 °F). Nickel and nickel-base alloys generally have poor resistance to molten lead corrosion (Ref 82, 83). The solubility of nickel in molten lead is higher than that of iron. Cast iron, steels, and stainless steels are commonly used for handling molten lead.

Corrosion in Liquid Lithium. Nickel has a very high solubility in liquid lithium. Thus, nickel and nickel-base alloys are not considered good candidates for handling liquid lithium (Ref 84). Cobalt-base alloys are only slightly more resistant than nickel-base alloys (Ref 85, 86).

Corrosion in Liquid Sodium. High-nickel alloys do not exhibit good resistance to liquid sodium. Exposure can lead to the development of a highly porous surface layer (Fig. 35) and/or intergranular corrosion.

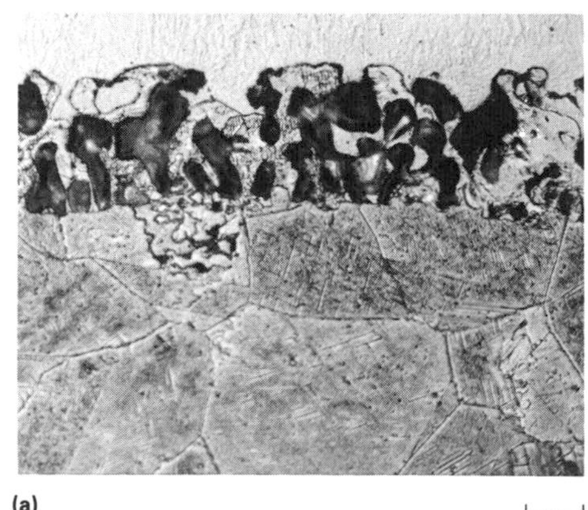

(a) 20 μm

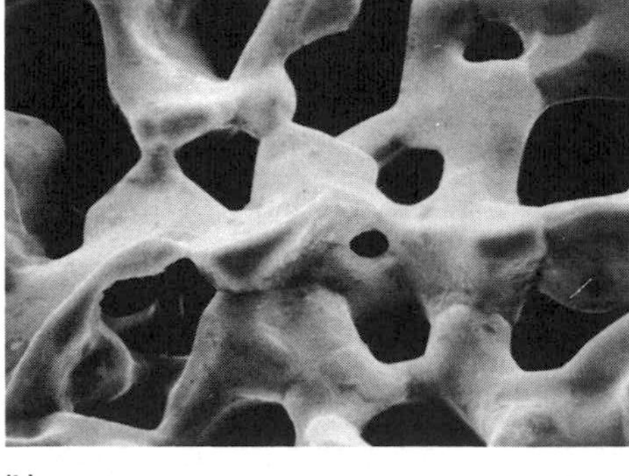

(b) 5 μm

Fig. 35 Corrosion of alloy 706 exposed to liquid sodium for 8000 h at 700 °C (1290 °F); hot leg of circulating system. A porous surface layer has formed with a composition of ~95% Fe, 2% Cr, and % Ni. The majority of the weight loss encountered can be accounted for by this subsurface degradation. Total damage depth: 45 μm. (a) Light micrograph. (b) SEM of the surface of the porous layer

Table 29 Results of corrosion tests in molten eutectic sodium-potassium carbonate at 900 °C (1650 °F) for 504 h(a)

Alloy	Total depth of corrosion attack(b)	
	mm	mils
X	0.12	4.7
214	0.19	7.5
188	0.22	8.7
556	0.26	10.2
X-750	0.27	10.6
600(c)	0.34	13.4
600(c)	0.44	17.3
R-41	0.42	16.5
N	0.51	20.1
304SS	0.54	21.3
316SS	0.63	24.8
230	0.77	30.3
Nickel	>0.30	11.8
800(c)	0.25	9.8
800(c)	>0.8	31.5
S	>1.43	56.3

(a) N_2-$1CO_2$-(1-$10O_2$) was used for the cover gas, (b) All alloys showed metal loss only except nickel, which suffered 0.2 mm (7.9 mils) metal loss and more than 0.11 mm (4.3 mils) intergranular attack. (c) Two samples from two different suppliers. Source: Ref 80

Table 30 Results of static immersion tests in molten aluminum at 760 °C (1400 °F) for 4 h

Alloy	Maximum depth of corrosion attack	
	mm	mils
Titanium	0.22	8.5
6B	0.43	16.8
188	0.51	20.2
150	0.73	28.9
556	>0.5(a)	20.6(a)
X	>0.6(a)	23.8(a)
671	>0.7(a)	26.3(a)
Carbon steel	>1.6(a)	63.1(a)

(a) Sample was consumed. Source: Ref 81

Table 31 Results of static immersion tests in molten zinc at 454 °C (850 °F) for 50 h

Alloy	Depth of corrosion attack	
	mm	mils
556	0.04	1.6
25	0.06	2.3
188	0.06	2.5
446SS	0.24	9.3
800H	0.28	11.0
304SS	0.36	14.1
625	>0.61	24.0(a)
X	>0.61	24.0(a)

(a) Complete alloying. Source: Ref 81

ACKNOWLEDGMENTS

The information in this article is largely taken from:

- G.Y. Lai, *High-Temperature Corrosion of Engineering Alloys,* ASM International, 1990
- I.G. Wright, High Temperature Corrosion, *Corrosion,* Vol 13, *ASM Handbook,* (formerly 9th ed., Metals Handbook), 1987, p 96–103

REFERENCES

1. H.E. Eiselstein and E.N. Skinner, in *STP No. 165,* ASTM, 1954, p 162
2. G.C. Wood, I.G. Wright, T. Hodgkiess, and D.P. Whittle, *Werkst. Korros.,* Vol 21, 1970, p 900
3. "Inconel Alloy 601," booklet, Huntington Alloys, Inc., 1969
4. G.Y. Lai, *J. Met.,* Vol 37 (No. 7), July 1985, p 14
5. E.A. Gulbranson and K.F. Andrew, *J. Electrochem. Soc.,* Vol 104, 1957, p 334
6. E.A. Gulbranson and S.A. Jansson, in *Heterogeneous Kinetics at Elevated Temperatures,* G.R. Belton and W.L. Worrell, Ed., Plenum Press, New York, 1970, p 181
7. K.N. Strafford, *High Temp. Technol.,* Nov 1983, p 307
8. D.P. Whittle and J. Stringer, *Philos. Trans. R. Soc.,* Vol A295, 1980, p 309
9. K.N. Strafford and J.M. Harrison, *Oxid. Met.,* Vol 10, 1976, p 347
10. C.C. Wood and J. Bonstead, *Corros. Sci.,* Vol 8, 1968, p 719
11. H. Nagai, M. Okaboyashi, and H. Mitani, *Trans. Jpn. Inst. Met.,* Vol 21, 1980, p 341
12. M.H. Lagrange, A.M. Huntz, and J.H. Davidson, *Corros. Sci.,* Vol 24 (No. 7), 1984, p 617
13. J.C. Pivin, D. Delaunay, C. Rogues-Carmes, A.M, Huntz, and P. Lacombe, *Corros. Sci.,* Vol 20, 1980, p, 351
14. A.W. Funkenbusch, J.G. Smeggil, and N.S. Borstein, *Metall. Trans. A.,* Vol 16, 1985, p 1164
15. C.A. Barrett, *Proc. Conf. Environmental Degradation of Engineering Materials,* M.R. Louthan, Jr. and R.P. McNitt, Ed., Virginia Polytechnic Institute, Blacksburg, VA, 1977, p 319
16. M.F. Rothman, internal report, Cabot Corporation, 1985
17. R.B. Herchenroeder, *Behavior of High Temperature Alloys in Aggressive Environments,* Proc. Petten Int. Conf., The Netherlands, 15–18 Oct 1979, The Metals Society, London
18. G.N. Irving, J. Stringer, and D.P. Whittle, *Corros. Sci.,* Vol 15, 1975, p 337
19. G.N. Irving, J. Stringer, and D.P. Whittle, *Oxid. Met.,* Vol 8 (No. 6), 1974, p 393
20. G.M. Kim, E.A. Gulbranson, and G.H. Meier, *Proc. Conf. Fossil Energy Materials Program,* ORNL/FMP 87/4, compiled by R.R. Judkins, Oak Ridge National Laboratory, 19–21 May 1987, p 343
21. G.Y. Lai, unpublished results, Haynes International, Inc., 1988
22. R.H. Kane, G.M. McColvin, T.J. Kelly, and J.M. Davidson, Paper 12, presented at Corrosion/84, National Association of Corrosion Engineers, 1984
23. R.H. Kane, J.W. Schultz, H.T. Michels, R.L. McCarron, and F.R. Mazzotta, Ref 30 of the paper by R.H. Kane in *Process Industries Corrosion,* B.J. Moniz and W.I. Pollock, Ed., National Association of Corrosion Engineers, 1986, p 45
24. G.Y. Lai, M.F. Rothman, and D.E. Fluck, Paper 14, presented at Corrosion/85, National Association of Corrosion Engineers, 1985
25. J.J. Barnes and S.K. Srivastava, Paper 527, presented at Corrosion/89, National Association of Corrosion Engineers, 1989
26. G.Y. Lai, *High Temperature Corrosion in Energy Systems,* Proc. TMS-AIME Symposium, M.F. Rothman, Ed., The Metallurgical Society of AIME, 1985, p 551
27. G.Y. Lai and C.R. Patriarca, *Corrosion,* Vol 13, 9th ed., *Metals Handbook,* ASM International, 1987, p 1311
28. R.H. Kane, G.M. McColvin, T.J. Kelly, and J.M. Davison, Paper 12, presented at Corrosion/84, National Association of Corrosion Engineers, 1984
29. J.J. Barnes and G.Y. Lai, paper presented at TMS Annual Meeting, Las Vegas, 1989
30. J.J. Moran, J.R. Mihalisin, and E.N. Skinner, *Corrosion,* Vol 17 (No. 4), 1961, p 191t
31. H. Schenck, M.G. Frohberg, and F. Reinders, *Stahl Eisen,* Vol 83, 1963, p 93
32. H.A. Wriedt and O.D. Gonzalez, *Trans. Met. Soc. AIME,* Vol 221, 1961, p 532
33. G.D. Smith and P.J. Bucklin, Paper 375, presented at Corrosion/86, National Association of Corrosion Engineers, 1986
34. T. Odelstam, B. Larsson, C. Martensson, and M. Tynell, Paper 367, presented at Corrosion/86, National Association of Corrosion Engineers, 1986
35. M.A.H. Howes, "High Temperature Corrosion in Coal Gasification Systems," Final Report GRI-8710152, Gas Research Institute, Chicago, Aug 1987
36. S.K. Verma, Paper 336, presented at Corrosion/85, National Association of Corrosion Engineers, 1985
37. J.L. Blough, V.L. Hill, and B.A. Humphreys, in *The Properties and Performance of Materials in the Coal Gasification Environment,* V.L. Hill and H.L. Black, Ed., American Society for Metals, 1981, p 225
38. G.Y. Lai, in *High Temperature Corrosion in Energy Systems,* M.F. Rothman, Ed., The Metallurgical Society of AIME, 1985, p 227
39. G.Y. Lai, "Sulfidation-Resistant Co-Cr-Ni Alloy with Critical Contents of Silicon and Cobalt," U.S. Patent No. 4711763, Dec 1987
40. G.Y. Lai, Paper 209, presented at Corrosion/89, National Association of Corrosion Engineers, 1989
41. G.Y. Lai, *J. Met.,* Vol 41 (No. 7), 1989, p 21
42. J.F. Norton, unpublished data, Joint Research Centre, Petten Establishment, Petten, The Netherlands, 1989
43. W.M.H. Brown, W.B. DeLong, and J.R. Auld, *Ind. Eng. Chem.,* Vol 39 (No. 7), 1947, p 839
44. R.H. Kane, in *Process Industries Corrosion,* B.J. Moritz and W.I. Pollock, Ed., National Association of Corrosion Engineers, 1986, p 45
45. M.J. Maloney and M.J. McNallan, *Met. Trans. B,* Vol 16, 1995, p 751

46. S. Baranow, G.Y. Lai, M.F. Rothman, J.M. Oh, M.J. McNallan, and M.H. Rhee, Paper 16, presented at Corrosion/84, National Association of Corrosion Engineers, 1984
47. J.M. Oh, M.J. McNallan, G.Y. Lai, and M.F. Rothman, *Metall. Trans. A,* Vol 17, June 1986, p 1087
48. M.H. Rhee, M.J. McNallan, and M.F. Rothman, in *High Temperature Corrosion in Energy Systems,* M.F. Rothman, Ed., The Metallurgical Society of AIME, 1985, p 483
49. M.J. McNallan, M.H. Rhee, S. Thongtem, and T. Hensler, Paper 11, presented at Corrosion/85, National Association of Corrosion Engineers, 1985
50. P. Elliott, A.A. Ansari, R. Prescott, and M.F. Rothman, Paper 13, presented at Corrosion/85, National Association of Corrosion Engineers, 1985
51. S. Thongtem, M.J. McNallan, and G.Y. Lai, Paper 372, presented at Corrosion/86, National Association of Corrosion Engineers, 1986
52. M.K. Hossain, J.E. Rhoades-Brown, S.R.J. Saunders, and K. Ball, in *Proc. UK Corrosion/83,* p 61
53. C.F. Hale, E.J. Barber, H.A. Bernhardt, and K.E. Rapp, "High Temperature Corrosion of Some Metals and Ceramics in Fluorinating Atmospheres," Report K-1459, Union Carbide Nuclear Co., Sept 1960
54. M.J. Steindler and R.C. Vogel, "Corrosion of Materials in the Presence of Fluorine at Elevated Temperatures," ANL-5662, Argonne National Laboratory, Jan 1957
55. F.S. Pettit and C.S. Giggons, Hot Corrosion, *Superalloys II,* C.T. Sims, N.S. Stoloff, and W.C. Hagel, Ed., John Wiley & Sons, 1987, p 327–358
56. J.A. Goebel, F.S. Pettit, and G.W. Goward, *Metall. Trans.,* Vol 4, 1973, p 261
57. J. Stringer, *Am. Rev. Mat. Sci.,* Vol 7, 1977, p 477
58. R.A. Rapp, *Corrosion,* Vol 42 (No. 10), 1986, p 568
59. J. Stringer, R.I. Jaffee, and T.F. Kearns, Ed., *High Temperature Corrosion of Aerospace Alloys,* AGARD-CP120, Advisory Group for Aerospace Research and Development, North Atlantic Treaty Organizations, 1973
60. Hot Corrosion Problems Associated *with Gas Turbines,* STP 421, ASTM, 1967
61. P.A. Bergman, A.M. Beltran, and C.T. Sims, "Development of Hot Corrosion-Resistant Alloys for Marine Gas Turbine Service," Final Summary Report to Marine Engineering Lab., Navy Ship R&D Center, Annapolis, MD, Contract N600 (61533) 65661, 1 Oct 1967
62. J. Clelland, A.F. Taylor, and L. Wortley, in *Proc. 1974 Gas Turbine Materials in the Marine Environment Conf.,* MCIC-75-27, J.W. Fairbanks and I. Machlin, Ed., Battelle Columbus Laboratories, 1974, p 397
63. M.J. Zetlmeisl, D.F. Laurence, and K.J. McCarthy, *Mater. Perform.,* June 1984, p 41
64. J. Stringer, *Proc. Symp. Properties of High Temperature Alloys with Emphasis on Environmental Defects,* Z.A. Foroulis and F.S. Pettit, Ed., The Electrochemical Society, 1976, p 513
65. A.M. Beltran, *Cobalt,* Vol 46, March 1970, p 3
66. P.A. Bergman, C.T. Sims, and A.M. Beltran, *Hot Corrosion Problems Associated with Gas Turbines,* STP 421, ASTM, 1967, p 38
67. G.R. Smolik and J.D. Dalton, Paper 207, presented at Corrosion/89, National Association of Corrosion Engineers, 1989
68. G.Y. Lai, "Alloy Performance in Incineration Plants," paper presented at Chemical Waste Incineration Conference, 12–13 March 1990, Manchester, UK
69. G.Y. Lai, M.F. Rothman, and D.E. Fluck, Paper 14, presented at Corrosion/85, National Association of Corrosion Engineers, 1985
70. G.Y. Lai, unpublished results, Haynes International, Inc., 1986
71. S.K. Srivastava, unpublished results, Haynes International, Inc., 1989
72. R.W. Bradshaw and R.W. Carling, "A Review of the Chemical and Physical Properties of Molten Alkali Nitrate Salts and Their Effect on Materials Used for Solar Central Receivers," SAND 87-8005, Sandia Laboratory, April 1987
73. J.W. Slusser, J.B. Titcomb, M.T. Heffelfinger, and B.R. Dunbobbin, *J. Met.,* July 1985, p 24
74. R.R. Miller, "Thermal Properties of Hydroxide and Lithium Metal," Quarterly Progress Report 1 May–1 Aug 1952, NRL-3230-201/52, 1952
75. J.N. Gregory, N. Hodge, and J.V.G. Iredale, "The Static Corrosion of Nickel and Other Materials in Molten Caustic Soda," AERE-C/M-272, March 1956
76. G.P. Smith and E.E. Hoffman, *Corrosion,* Vol 13, 1957, p 627t
77. J.W. Koger, *Corrosion,* Vol 29 (No. 3), 1973, p 115
78. J.W. Koger, *Corrosion,* Vol 30 (No. 4), 1974, p 125
79. J.R. Keiser, D.L. Manning, and R.E. Clausing, *Proc. Int. Symp. Molten Salts,* J.P. Pemsler, et al., Ed., Electrochemical Society, 1976, p 315
80. R.T. Coyle, T.M. Thomas, and G.Y. Lai, *High Temperature Corrosion in Energy Systems,* M.F. Rothman, Ed., The Metallurgical Society of AIME, 1985, p 627
81. G.Y. Lai, unpublished results, Haynes International, Inc., 1985
82. F.R. Morrall, *Wire and Wire Products,* Vol 23, 1948, p. 484, 571
83. L.R. Kelman, W.D. Wilkinson, and F.L. Yaggee, "Resistance of Materials to Attack by Liquid Metals," Report ANL-4417, Argonne National Laboratory, July 1950
84. R.E. Cleary, S.S. Blecherman, and J.E. Corliss, "Solubility of Refractory Metals in Lithium and Potassium," USAEC Report TIM-950, Nov 1965
85. M.S. Freed and K.J. Kelly, "Corrosion of Columbium Base and Other Structural Alloys in High Temperature Lithium," Report PWAC-355, Pratt and Whitney Aircraft-CANEL, Division of United Aircraft Corp., June 1961 (declassified in June 1965)
86. E.E. Hoffman, "Corrosion of Materials by Lithium at Elevated Temperatures," USAEC Report ORNL-2924, Oct 1960

Microstructural Degradation of Superalloys

MANY SUPERALLOYS respond to heat treatment, and thus exposure of these alloys to elevated temperatures, with or without stress, can cause microstructural changes that affect properties. Generally, the higher the exposure temperature is, the more rapid will be the structural change. As exposure temperature decreases, the type of microstructural degradation may change. At the highest exposure temperatures, an alloy may be subjected to incipient melting. In addition, oxidation and surface corrosion may occur at temperatures for which these alloys normally are specified.

This article deals principally with the effects of microstructural changes, melting, and corrosion on nickel-base and cobalt-base superalloys at temperatures above about 725 °C (1340 °F). It also touches briefly on microstructural changes affecting nickel-base, iron-base, and iron-nickel superalloys at temperatures below 700 °C (1300 °F).

When times or temperatures exceed normal test or operating levels, alloys are exposed to substantially different operating environments from those ordinarily experienced. They may behave in a manner that could not have been predicted from normal test data. Additional information on behavior at elevated temperatures may be found in failure analysis studies and in Ref 1 to 10.

Overheating

Overheating in the broad sense consists of exposing a metal to excessively high temperatures for short periods of time. Allowable metal temperatures for wrought superalloys in structural applications generally do not exceed about 950 °C (1740 °F). In applications where the component does not bear a load, allowable temperatures may exceed 1200 °C (2200 °F). In general, any temperature can be considered to be in the overheating range when it: (a) causes melting; (b) causes strengthening phases to dissolve in the matrix or to coalesce excessively; or (c) causes extensive oxidation or corrosion. Results of overheating depend on the maximum temperature reached by the metal.

Nickel-base and cobalt-base superalloys generally have incipient melting temperatures above 1200 °C (2200 °F). Table 1 gives melting temperatures for some representative wrought superalloys. Figure 1 shows the microstructures of two typical wrought superalloys before and after incipient melting. Incipient melting reduces grain-boundary strength and ductility. It thus significantly reduces alloy rupture capabilities.

Once an alloy has exceeded its incipient melting point, normal properties cannot be restored by any known heat treatment. In cast alloys, incipient melting may occur at temperatures substantially below the temperatures predicted from alloy composition. This behavior results from alloy segregation to grain boundaries and interdendritic areas during solidification. This is particularly troublesome with cast superalloys because, based on their excellent high-temperature strength, they often are used for very high-temperature applications, such as jet engine turbine airfoils. In wrought alloys, incipient melting takes place at a temperature much closer to the general alloy melting temperature (solidus), and overheating actually may cause significant portions of the structure to melt.

Effects on Oxidation Resistance. Overheating may deplete alloying elements that provide oxidation resistance. Oxidation attack is thus accelerated on return of the alloy to normal operating temperatures. Even if prolonged exposure to overtemperature does not result in mechanical failure, it frequently will cause excessive surface corrosion. This, in turn, can reduce the strength of the alloy. Even if an alloy is coated for oxidation resistance, the coating will be degraded by surface attack, and eventually mechanical properties of the alloy will be affected. The alloy under a coating may also degrade rapidly because of excessive interdiffusion of alloying elements if temperatures exceed the normal operating range. Coatings eventually fail because of diffusion in both directions—the coating composition diffusing into the substrate and the substrate alloy diffusing outward into the coating. This diffusion changes the coating composition over time so that it no longer provides effective protection. Overheating thus hastens coating failure. More detailed information on coating degradation can be found in the article "Protective Coatings for Superalloys" in this Volume.

As a rule-of-thumb, alloys used in structural applications should not be exposed to temperatures within about 125 °C (225 °F) of their incipient melting temperatures. The strength and oxidation resistance of the alloy and the operating environment will determine how close actual metal temperatures may approach the suggested upper limit.

Effects on Strength. An important factor to consider when dealing with coated superalloys is reduction in incipient melting temperature of the system (coating/base metal) that may result from the change in composition brought about by diffusion. For example, for aluminide coatings on an alloy such as U-700, which has an incipient melting temperature of about 1215 °C (2220 °F), incipient melting may occur in the inner diffusion zone at temperatures between 1175 and 1190 °C (2150 and 2175 °F). Incipient melting in coated systems leads to accelerated degradation of the coating.

At temperatures that cause neither incipient melting nor surface degradation, alloy strength may still be reduced because strengthening phases are taken into solution, or they may become less effective as strengtheners because of

Table 1 Incipient melting temperatures of select wrought superalloys

Alloy	Incipient melting temperature	
	°C	°F
Hastelloy X	1250	2280
Haynes 25 (L-605)	1329	2425
Haynes 188	1302	2375
Incoloy 800	1357	2475
Incoloy 825	1370	2500
Inconel 617	1333	2430
Inconel 625	1288	2350
Inconel X-750	1393	2540
Nimonic 80A	1360	2480
Nimonic 90	1310	2390
Nimonic 105	1290	2354
René 41	1232	2250
Udimet 500	1260	2300
Udimet 700	1216	2220
Waspaloy	1329	2425

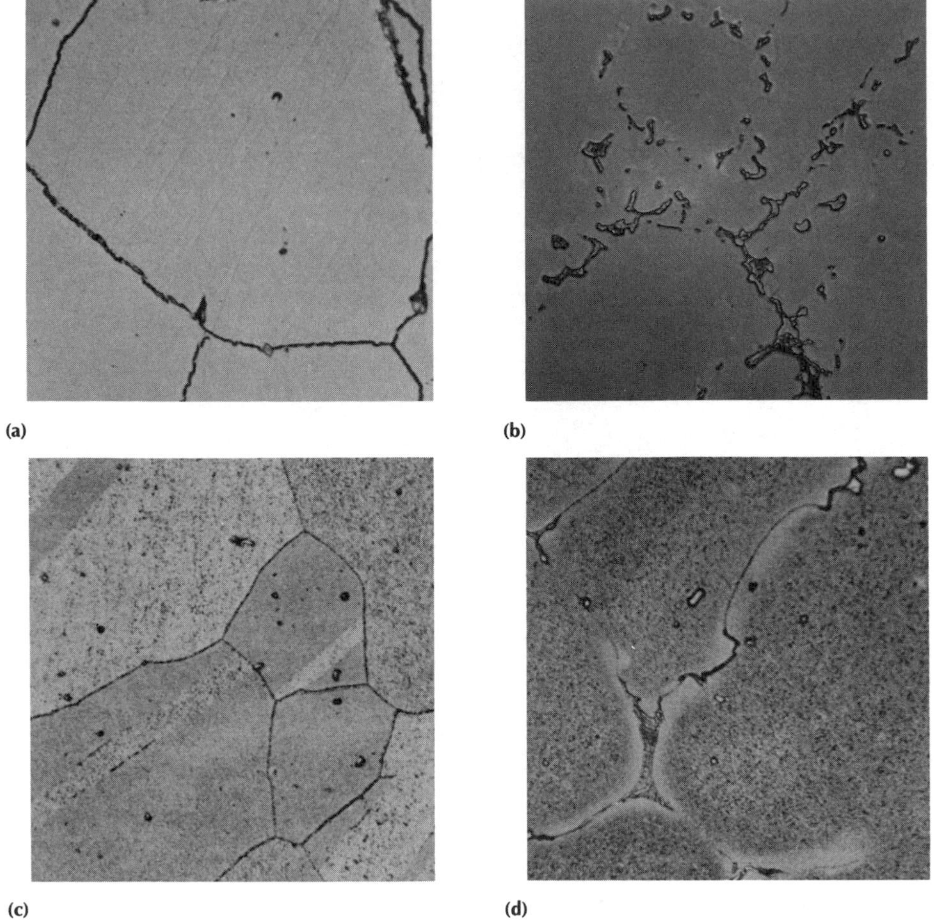

Fig. 1 Effect of incipient melting on microstructure of two nickel-base superalloys. (a) Inconel 617 (no melting). 500×, unetched. (b) Inconel 617 (incipient melting). 500×, unetched. (c) Udimet 700 (no melting). 500×, lactic acid + HCl + HNO_3 etch. (d) Udimet 700 (incipient melting). 500×, lactic acid + HCl + HNO_3 etch

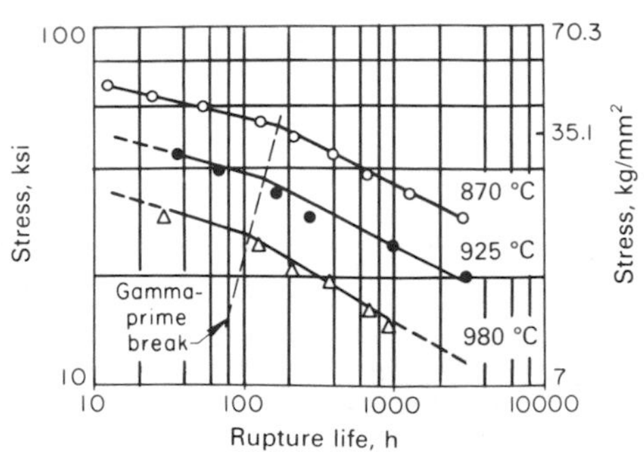

Fig. 2 Logarithmic plot of stress-rupture versus rupture life for nickel-base alloy B-1900. The increasing slope of the curves to the right of the γ′ break is believed to be caused by the γ′ coarsening due to overaging.

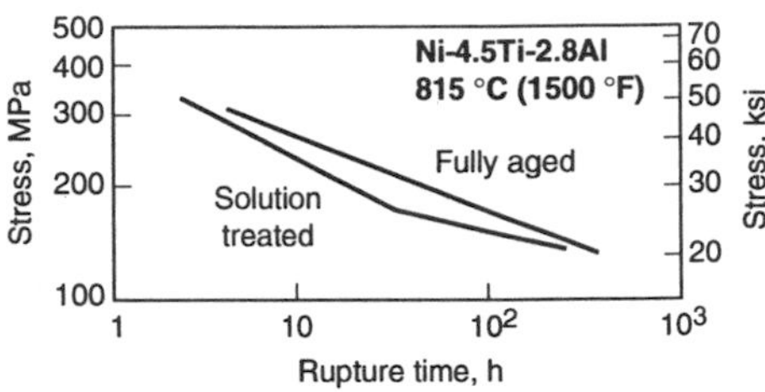

Fig. 3 Stress-rupture plot for a γ′-hardening nickel alloy. Source: Ref 9

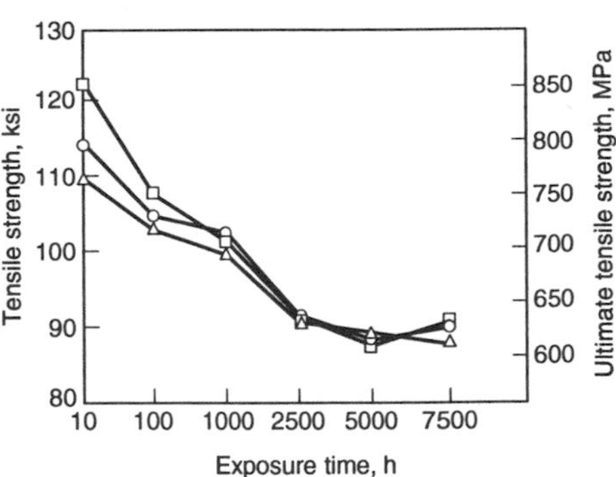

Fig. 4 Room-temperature tensile strength versus exposure time at 1040 °C (1900 °F) in air for Hastelloy X. Source: Ref 10

Table 2 Solution treatments for select wrought nickel-base superalloys

Alloy	Solution temperature(a) °C	Solution temperature(a) °F	Time, h
Inconel X-750	1150	2100	4
Nimonic 90	1080	1975	8
Nimonic 105	1125–1150	2060–2100	4
Udimet 500	1175	2150	2
Udimet 700	1175	2150	4
Waspaloy	1080	1975	4

(a) All materials air cooled after solution treatment

coalescence. Wrought nickel-base alloys frequently are strengthened by a precipitation of the phase γ′, $Ni_3(Al,Ti)$, a face-centered cubic (fcc) intermetallic compound. In wrought alloys, the γ′ can be taken into solution at temperatures of about 1175 °C (2150 °F) or less (see Table 2). Exposure at or near the solution temperature will reduce the amount of γ′ by solid solution and thereby reduce alloy strength. Exposure at temperatures below the solution temperature, but still above normal operating temperature, will also reduce strength as the precipitated γ′ strengthening particles coalesce to coarser size and become less effective in strengthening.

Prolonged operation at temperatures within the solution range is inadvisable although occasional excursions into this range may be tolerated if the part can be re-heat-treated before excessive creep occurs. If γ′ is dissolved, it can be reprecipitated as fine particles by subsequent aging; original property levels can be reasonably recovered, assuming additional damage due to stress or oxidation does not occur. However, properties will not be recovered if slow cooling is used after extensive solution has occurred or if the material is held at a high temperature so that coarse γ′ forms while fine γ′ dissolves.

Figure 2 shows the effect of γ′-coarsening on the stress-rupture life of a cast nickel-base superalloy. The dashed line in Fig. 2 designated γ′ break indicates the beginnings of slight increases in slope of the curves, which are believed to be caused by γ′-coarsening due to overaging. Actually, the coarsening was not apparent in the microstructure at the points where the breaks occurred but was observed metallographically after longer tests were completed.

Figure 3 compares the service life of a nickel-base superalloy when it is only solution treated with the service life when it is fully heat treated. Under high-stress applications, strength is significantly reduced if a γ′-strengthened alloy has been exposed to a solution treatment temperature and not re-aged. During long-time, low-stress applications within the aging temperature range, a lesser reduction in strength occurs. More detailed information on solution treatment procedures for γ′-strengthened alloys can be found in the article "Effect of Heat Treating on Superalloy Properties" in this Volume.

Carbide phases in superalloys behave somewhat like γ′, but subsequent reprecipitation is not as easily controlled. Carbides are taken into solution or agglomerated during over-temperature exposure, and there can be substantial variations in the amount, form, and distribution of the resulting carbide structure. In Hastelloy X, a

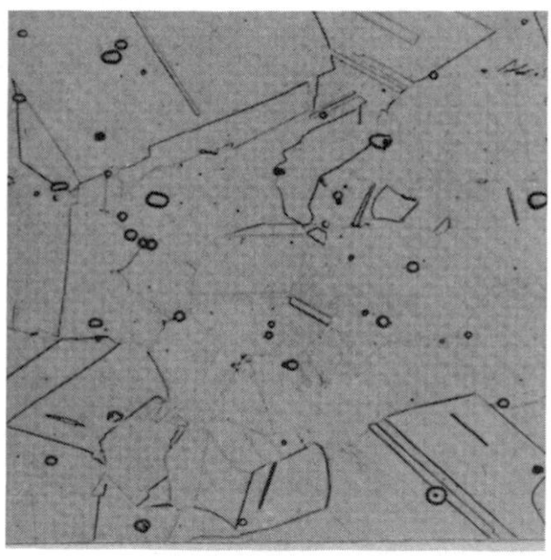

(a)

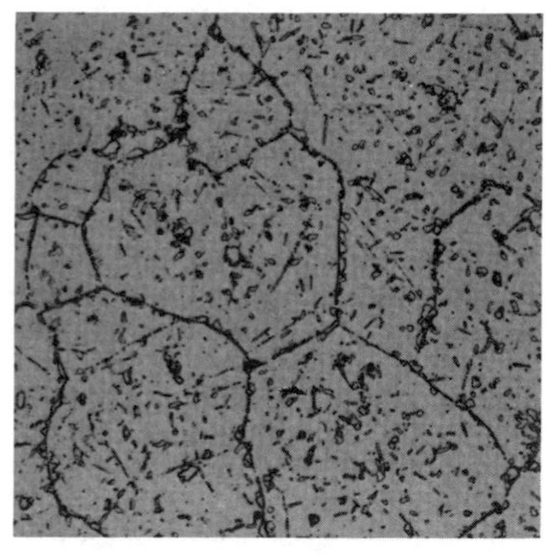

(b)

Fig. 5 Microstructure of Haynes 188. (a) Solution treated structure. (b) Structure after exposure at 870 °C (1600 °F) for 6244 h. Both shown at 500× and etched HCl + H_2O_2. Source: Ref 11

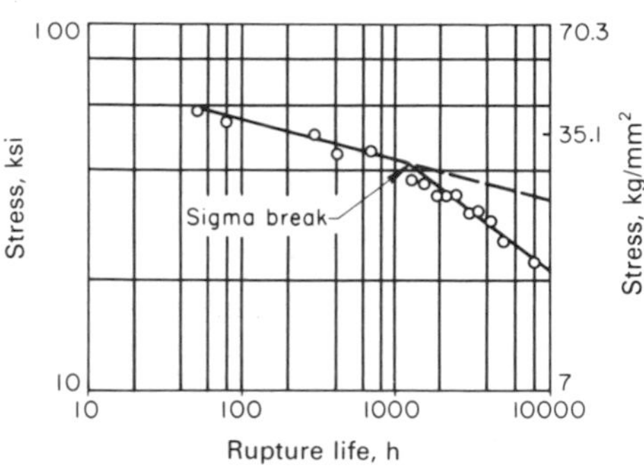

Fig. 6 Logarithmic plot of stress-rupture stress versus rupture life for nickel-base alloy U-700 at 815 °C (1500 °F). The increasing slope of the curve to the right of the σ break is caused by σ-phase formation.

solid-solution-strengthened nickel superalloy, there are large differences in the volume fractions and structure of carbides between normal-temperature and over-temperature exposures. The volume fraction of M_6C is about 12 vol% after approximately 7500 h at a temperature of 980 °C (1800 °F). During over-temperature exposure, carbide form and distribution are changed considerably due to agglomeration and rounding particles. As a result, mechanical properties are reduced, as shown in Fig. 4, for room-temperature strength after exposure at 1040 °C (1900 °F) in air. Similar results were obtained from a series of exposures at about 1100 °C (2000 °F).

Extensive carbide precipitation frequently can occur in alloys when there is a change from the original carbide that was present in the mill annealed condition. Figure 5 shows Haynes 188, a cobalt-base wrought alloy, before and after exposure for approximately 6000 h at 870 °C (1600 °F). During exposure, M_6C carbides dissolved and were replaced predominantly by $M_{23}C_6$ carbides and, to a lesser extent, by Laves phase. Significant losses in ductility resulted from the precipitation shown in Fig. 5.

In alloys where extensive carbide precipitation or agglomeration occurs, it is frequently possible to recover the original carbide distribution and a major portion of alloy properties by solution heat treatment. For example, Haynes 188 can recover original room-temperature ductility by heat treatment at 1175 °C (2150 °F) for 15 min.

Alloy Stability

Obviously, alloy stability and the ability of a material to resist damaging effects due to overheating are of particular importance to superalloys that are intended for elevated-temperature service. Clearly, the nickel-base superalloys strengthened by a secondary precipitated phase are the most complex and indeed the most remarkable of all the superalloys. These alloys are used in the most demanding applications relative to stress and temperature. They have demonstrated remarkably useful strength at the highest fraction of the base metal melting point of any alloy system ever developed.

As discussed above, a significant characteristic of high-temperature service is metallurgical instability. Stress, time, temperature, and environment may act to change the metallurgical structure during operation and, thereby, contribute to failure by reducing strength and/or ductility. It should be noted that in a few cases strength may be enhanced. These structural changes or metallurgical instabilities are best described in terms of their influence on stress-rupture properties. A sharp change downward in the slope of the stress-rupture curve indicates that failures will occur in shorter times and at lower stresses than originally predicted. Instabilities are associated with aging (phase precipitation), overaging (phase coalescence and coarsening), phase decomposition (usually carbides, borides, and nitrides), intermetallic phase precipitation, internal oxidation, and stress corrosion.

Typical instability problems with the γ' nickel-base superalloys involve intermetallic phase formation. In certain alloys where composition has not been carefully controlled, undesirable hard, brittle phases can form either during heat treatment or service. These hard compounds have been identified as close-packed σ, μ, and Laves phases, and it has been determined that they form at elevated temperatures with a deteriorating effect on stress-rupture properties. These phases are characterized by platelike structures, often nucleating on grain-boundary carbides.

Figure 6 shows the effect of σ-phase formation on the stress-rupture life of nickel-base alloy U-700 at 815 °C (1500 °F). Starting at about 1000 h, a pronounced break was found in the slope of the rupture curve. The difference between the extrapolated life and the actual life at 207 MPa (30 ksi) was about 5500 h, representing a decrease of about 50% in expected life. Sigma phase was identified in this alloy system and was clearly associated with the failure because the voids formed by creep occurred along the periphery of σ-phase particles.

Still another instability characteristic relates to carbide reactions. A variety of carbides are found in superalloys identified by this chemical composition, such as TiC, Cr_7C_3, etc. In many instances, the carbide composition is complex and is comprised of many of the superalloy alloying elements, such as chromium, molybdenum, tungsten, niobium, and tantalum. These complex metal combinations are identified by the letter "M." Thus, superalloys are characterized by carbides identified as MC, M_6C, and $M_{23}C_6$. Although temperature and stress affect both carbides within grains and at grain boundaries, the effects on grain-boundary carbides are usually a much more significant factor in altering creep and rupture behavior. Grain-boundary morphology is indeed important relative to high-temperature properties. The presence of carbides at grain boundaries as strengtheners is necessary for optimum creep and rupture life, but alteration in shape or breakdown to other carbide forms may cause property degradation. The best carbide formation at grain boundaries for optimum strength is discrete, blocky particles. Continuous carbide films at the boundaries substantially reduce stress-rupture life.

In superalloy selection, it is important to ensure that the composition is such that carbide morphology at grain boundaries is proper and that detrimental σ, μ, or Laves phases do not form at operating temperatures.

Microstructural Degradation

Effects of Service below 700 °C (1300 °F). For superalloys normally used at temperatures of 425 to 725 °C (800 to 1340 °F), melting by overtemperature is not a problem; microstructural changes are the important consideration. Hot corrosion and other types of accelerated oxidation are important for some alloys. The superalloys used at these temperatures are relatively stable. Superalloys such as A-286, Incoloy 901, V-57, Waspaloy, and Astroloy generally are considered microstructurally stable at the temperatures for which they are used as disks. Their properties are almost exclusively determined by prior heat treatment.

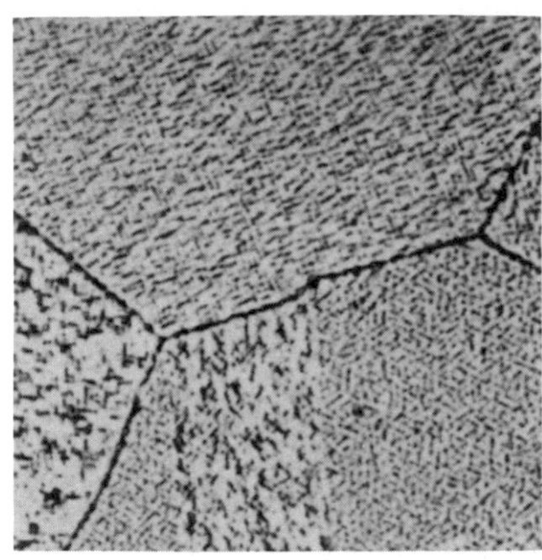
(a)

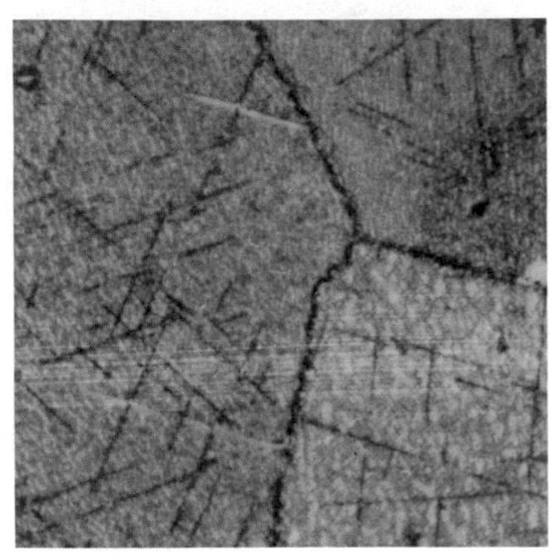
(b)

Fig. 7 Microstructure of Udimet 700. (a) Solution treated and aged structure. (b) Structure after exposure at 870 °C (1600 °F) for 500 h. Both shown at 100× and etched in Kelling's reagent. Source: Ref 12

If nickel-base alloys such as Waspaloy, Astroloy, and René 95 are exposed for prolonged times at metal temperatures that exceed about 650 °C (1200 °F), the strengthening phase may coarsen slightly. The same effect may occur above 600 °C (1100 °F) for iron-base or iron-nickel superalloys such as A-286. However, unless the alloys are operated at temperatures well into or above their aging-temperature ranges, no significant microstructural effects will be noted. Carbide precipitation at dislocations may be significant when operating times at 480 to 650 °C (900 to 1200 °F) exceed 10,000 h. However, there are no published data to support the existence of significant effects of thermal exposure on creep-rupture behavior although notch behavior may become important. At these temperatures, surface oxidation generally is not a problem although long-term exposure to certain highly corrosive environments may produce surface attack such as hot corrosion (sulfidation).

Inconel 718 deserves some mention because it often is used at temperatures above its final aging temperature of 620 °C (1150 °F). Although some minor coarsening of the γ'' phase may take place, no detrimental effects normally occur at temperatures up to 650 °C (1200 °F). If overheating to above 700 °C (1300 °F) should occur, the strengthening γ'' phase is degraded; significant σ phase precipitation can then occur with resultant losses in strength. With respect to carbide changes, superalloys are not as sensitive to corrosive attack as some austenitic stainless steels are because of the many carbide formers in superalloy compositions.

Effects of Service above 700 °C (1300 °F). Whereas some superalloys are exposed to extremely high temperatures in furnace and petrochemical applications, alloys for gas turbine applications generally are exposed to the most demanding combinations of high temperature and stress. The normal operating regime of turbine blades and vanes is about 725 to 850 °C (1340 to 1560 °F) for wrought alloys and up to 1050 °C (1920 °F) for cast alloys. Within these ranges, microstructural changes readily occur with time at temperature. Furthermore, when stress is applied, the changes may be accelerated.

The principal changes are: (a) breakdown of primary carbides and formation of secondary carbides; (b) agglomeration of primary geometrically close-packed (gcp) strengthening phases such as γ'; and (c) formation of topologically close-packed (tcp) phases, such as σ, Laves, and μ. Processes described under (a) and (b) are an extension of the normal strengthening process. These reactions are recognized during the design of components, and allowances are made for their effects on alloy strength at moderate times. Carbide transition and γ' agglomeration do reduce strength with time but generally are not as detrimental as formation of tcp phases.

Figure 7 shows the microstructure of Udimet 700 nickel-base alloy before and after exposure at 870 °C (1600 °F). During exposure, the γ' became agglomerated, and $M_{23}C_6$ carbides formed. Generally, design parameters take into account γ' agglomeration if it is experienced in the moderate times used to test alloys (most often, 20 to 1000 h). However, for significantly longer times at normal temperatures, γ' agglomeration may reduce alloy strength below predicted values.

Changes in carbide phases also adversely affect strength although initially there may be increases in strength as additional carbides precipitate. In the wrought cobalt-base solid-solution alloy Haynes 25 (L-605), carbide precipitation at 815 °C (1500 °F) is responsible for alloy hardening in both early and late stages of exposure. In the late stages, $M_{23}C_6$, M_6C, and Laves phase participate in strengthening. In Hastelloy X, extensive precipitation occurs at 705 to 790 °C (1300 to 1450 °F). As a result, σ, μ, and a dense intragranular secondary M_6C carbide can occupy as much as 27 vol% of the structure. As primary M_6C carbides coalesce, there is a continual reduction in strength; formation of small secondary carbides and tcp phases enhances strength but also reduces ductility.

ACKNOWLEDGMENTS

The information in this article is largely taken from:

- Microstructural Degradation, Overheating, Stability, *Superalloys: A Technical Guide,* E.F. Bradley, Ed., ASM International, 1988, p 89–97
- Elevated-Temperature Failures, *Failure Analysis and Prevention,* Vol 11, *ASM Handbook* (formerly 9th ed., *Metals Handbook*), 1986, p 263–297

REFERENCES

1. J.L. Johnson and M.J. Donachie, Jr., "Microstructure of Precipitation Strengthened Nickel-Base Superalloys," Report System Paper C 6-18.1, American Society for Metals, 1966
2. G.P. Sabol and R. Stickler, Microstructure of Nickel-Based Superalloys, *Physica Status Solidi,* Vol 35, 1969, p 11–52
3. P.S. Kotval, The Microstructure of Superalloys, *Metallography,* Vol 1, 1969, p 251–285
4. C.P. Sullivan et al., *Cobalt Base Superalloys-1970,* Cobalt Information Center, Brussels, 1970
5. C.T. Sims and W.C. Hagel, Ed., *The Superalloys,* John Wiley & Sons, Inc., 1972
6. R. Stickler, Phase Stability in Superalloys, *High Temperature Materials in Gas Turbines,* Elsevier Applied Science Publishers Ltd., 1974, p 115–149
7. D. Coutsouradis, et al., *High Temperature Alloys for Gas Turbines,* Applied Science Publishers Ltd., 1978
8. E.F. Bradley, *Source Book on Materials for Elevated Temperature Applications,* American Society for Metals, 1979
9. N. Rogen and N.J. Grant, Aging Characteristics of Ni-Cr Alloys Containing Appreciable Amounts of Ti and Al, *Proc. ASTM,* Vol 58, 1958, p 697
10. W.L. Clark, Jr. and G.W. Titus, Oxidation and Structural Stability Investigations, *Evaluation Study of Hastelloy X as a Nuclear Cladding,* Vol 1, Nuclear Div., Aerojet—General Corp., San Ramon, CA, June 1968
11. R.B. Herchenroeder et al., Haynes Alloy No. 188, *Cobalt,* No. 54, March 1972, p 3
12. J. Johnson, M.S. thesis, Rensselaer Polytechnic Institute, Troy, NY, 1965

Protective Coatings for Superalloys

A SUPERALLOY is a material designed for use in applications at elevated temperatures where good mechanical properties, heat resistance, and corrosion resistance (surface stability) are required. Current superalloys are divided into three classes: nickel base, cobalt base, and nickel-iron base (which have characteristics similar to those of nickel-base superalloys, but contain large amounts of iron). These materials primarily are used in aircraft gas turbine engines and industrial land-based turbines, as well as in coal-conversion plants and chemical processing plants. While great strides in superalloy performance have been made over the past 50 years, there is an ongoing demand for even higher performance. For example, primary goals in the continuing development of aircraft engines are increased operating temperatures and improved engine efficiencies. Higher operating temperatures require alloys that have higher mechanical properties and improved resistance to high-temperature environmental attack.

Superalloys are used at a higher proportion of their actual melting temperatures than any other metallurgical material class. For example, nickel-base superalloys are used in load-bearing applications at temperatures in excess of 80% of the incipient melting temperature. Table 1 shows the maximum use temperature of some metals. While incremental increases in use-temperature capabilities have been achieved via advanced alloy design and materials processing, protective coatings play a major role in extending the performance limits of superalloys. The largest use of coatings on superalloys is on components in the hot gas section of turbine engines (i.e., in critical turbine components such as high-pressure turbines, turbine blades, vanes, and disks).

Coating Requirements

Coatings are relied on to protect superalloy components from environmental attack. Protective coating development for superalloys has been driven by advances in propulsion technology for aircraft and space vehicles since about 1960. These advances have placed increasing temperature and structural demands on materials for service at temperatures up to 1010 °C (1850 °F) and higher. This, coupled with weight-associated penalties for flight systems, has driven materials and design technology to maximize the hot strength of structural components.

To achieve the mechanical properties required by modern gas-turbine technology, superalloys have had to sacrifice the oxidation and corrosion resistance that was inherent in previously used heat-resistant alloys containing higher chromium contents.

For example, early use of superalloys (stainless steels) was at a moderate service temperature of about 700 °C (1500 °F), and the nominal chromium content in the range of 16 to 20% was sufficient to impart the necessary protection against oxidation and hot corrosion. However, as strength requirements increased with increasing operating temperature, chromium was reduced in favor of higher aluminum additions, which result in the formation of a more stable aluminum oxide (Al_2O_3) protective scale (more stable than chromium oxide, or Cr_2O_3) and provide increased strength as well. Unfortunately, simultaneously increasing strength and protection from environmental attack are conflicting goals, and so protective coatings were developed.

The exclusive purpose of a coating on a superalloy is to prevent environmental attack of the substrate for the maximum possible time with the maximum degree of reliability. Figure 1 illustrates the benefits derived from the use of a protective coating at high temperatures; surface-corrosion resistance is significantly increased.

A coating is essentially a layer of material that can prevent or inhibit direct interaction between a substrate and the harmful environment. Damage to the substrate can be in the form of metal wastage due to oxidation and/or corrosion, or a reduction in mechanical properties due to diffusion at high temperatures of harmful species into the substrate.

Coatings used to protect superalloys from environmental attack do not function as inert barriers (i.e., they are not in equilibrium with the substrate). They differ from the substrate mechanically, physically, and chemically. They provide protection by forming a dense, tightly adherent oxide scale (Cr_2O_3 and Al_2O_3) via interaction of chromium and aluminum with oxygen in the environment. These scales inhibit the diffusion of harmful species (e.g., oxygen, nitrogen, and sulfur) into the substrate. Thus, coatings should contain sufficient aluminum, chromium, and silicon that can continually participate in the formation of the protective scale (Ref 1).

Coatings must be reasonably compatible with the substrates to which they are applied. The components of the coating and the method of application should be selected to prevent undesirable reaction phases between the coating and the substrate, and to prevent rapid penetration of the substrate by coating elements. The presence of such phases and/or interdiffusion leads to void formation, or cracking at the interface, and spalling of the coating. These coating imperfections compromise the mechanical performance capability of the system and the protection it provides.

At the high temperatures involved, interdiffusion between coating and substrate occurs, although at a relatively low rate in well-balanced coating/substrate systems. Coatings eventually fail due to this interdiffusion, which causes coating chemistry to change substantially so that it is no longer protective.

Table 1 Current maximum use temperature of some metals

Metal	Melting point, T_m, K	Melting point, T_m, °C	Potential operating temperature, $T = \frac{2}{3} T_m$, K	Potential operating temperature, $T = \frac{2}{3} T_m$, °C	Maximum T/T_m actually achieved
Aluminum	93	660	62	350	0.56 (RR58 at 250 °C)
Copper	135	1083	90	630	...
Nickel	172	1453	115	880	0.74 (Nimonic 115 at 980 °C)
Iron	180	1536	120	930	0.47 (ferritic steel at 575 °C) 0.57 (austenitic steel at 750 °C)
Titanium	194	1668	129	1020	~0.4
Zirconium	212	1852	142	1150	...
Chromium	217	1900	145	1180	0.6 possible if chromium could be made sufficiently ductile
Hafnium	249	2222	164	1370	...
Niobium	274	2468	183	1550	0.54 possible if a satisfactory coating could be found
Molybdenum	288	2610	192	1650	...
Tantalum	326	2996	218	1910	...
Tungsten	368	3410	243	2160	0.76 (electric light filament at 2500 °C)

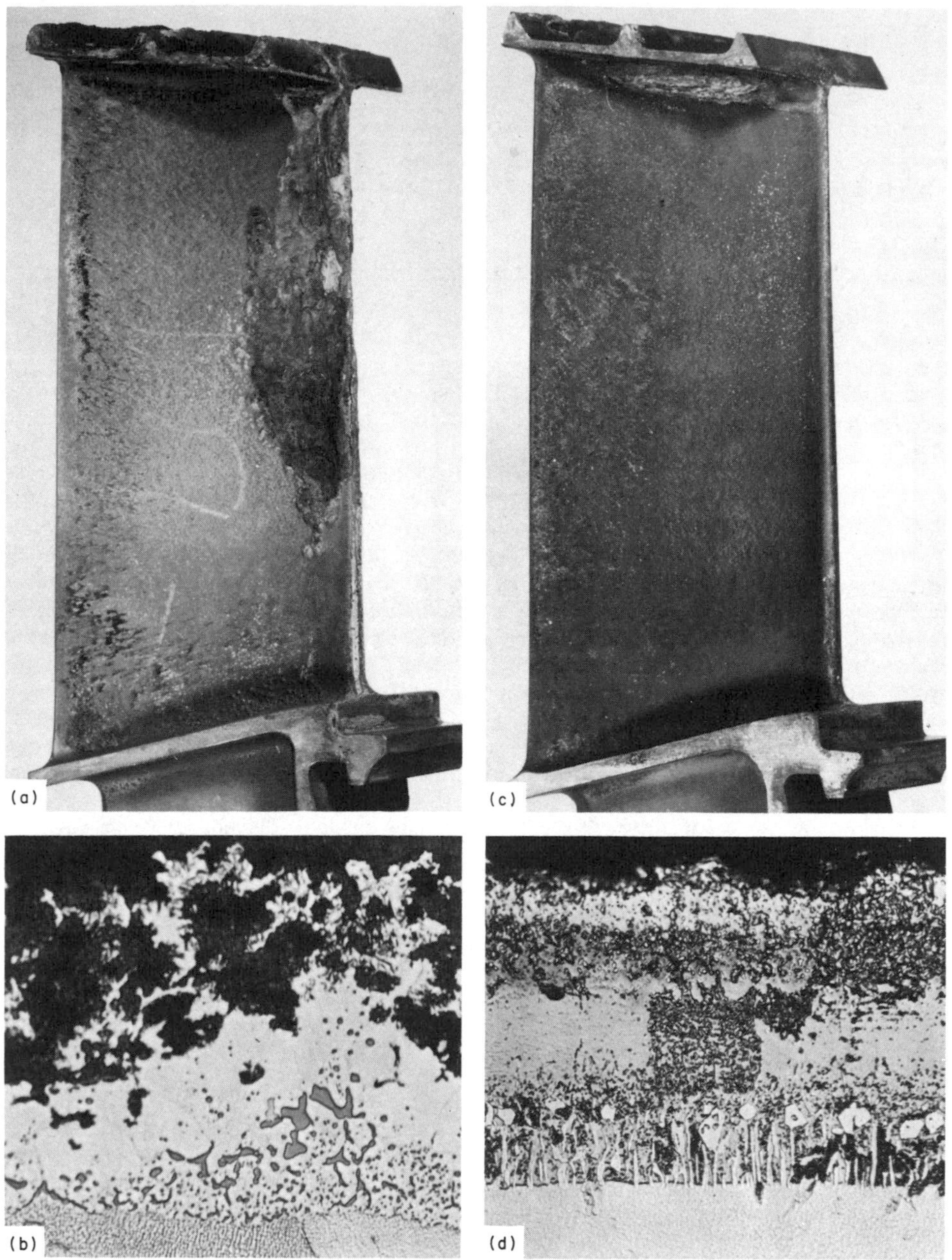

Fig. 1 Effect of an aluminide-type (nickel-aluminum) coating on the hot corrosion resistance of an IN-713 turbine blade compared with that of an uncoated blade. (a) Uncoated blade after 118 programmed cycles. 1.8×. (b) Severe degradation of the uncoated blade by hot corrosion. 450×. (c) Aluminide-coated blade after accelerated hot corrosion testing in an engine. 1.8×. (d) Slight degradation of the aluminide coating. 450×. Micrographs etched in ferric chloride

In addition to corrosion resistance, coatings for superalloys are required to resist both thermal cycling, often at rapid rates, and mechanical forces acting on hardware, without cracking. The protective Al_2O_3 film, on which superalloy coatings rely for protection, does not readily span even small gaps. Cracking of protective coatings is soon followed by a breakdown in protection, and localized substrate attack follows shortly. The thermal expansion match between coating and substrate is important, and coatings with a modicum of ductility are desirable from a corrosion standpoint. The coatings also must withstand combustion char and solid particulates ingested during turbine operations that result in erosive action.

The goals in the design, development, and application of coated systems are to apply a coating material that provides oxidation and corrosion protection, and to provide a coated system that is adequate in all respects for the intended purpose.

The development of advanced military aircraft gas turbines was the driving force behind the development of coating technology. However, the performance and fuel-economy advantages associated with advanced turbine technology led to the use of advanced materials, including coatings, in commercial aircraft, marine, and land-based turbines.

Environmental Effects

Two general types of environmental effects on superalloys are oxidation and hot corrosion. Both processes involve rapid attack of the surface of superalloy parts, resulting in a decrease in material stability. Coatings are designed to resist the specific effects of each process.

Oxidation. Advanced high-strength superalloys generally react with oxygen, which is the primary environmental condition that affects service life. At moderate temperatures of about 870 °C (1600 °F) and lower, general uniform oxidation is not a major problem. At higher temperatures, commercial nickel- and cobalt-base superalloys are attacked by oxygen. The level of oxidation resistance at temperatures below about 980 °C (1800 °F) is a function of chromium content; Cr_2O_3 forms as a protective oxide scale. At temperatures above 980 °C (1800 °F), aluminum content becomes more important in oxidation resistance; Al_2O_3 forms as a protective oxide scale. Chromium and aluminum can contribute to oxidation protection in an interactive manner: as the chromium level increases, less aluminum may be required to form a highly protective Al_2O_3 layer. However, the aluminum contents of many superalloys are insufficient to provide long-term Al_2O_3 protection. Protective coatings are used to provide satisfactory service life, prevent selective attack that occurs along grain boundaries and at surface carbides (Fig. 2), and inhibit internal oxidation or subsurface interaction of O_2/N_2 with γ' envelopes, a process believed to occur in nickel-base superalloys.

Because oxidation resistance at elevated temperatures is provided by Al_2O_3 or Cr_2O_3 protective scale, nickel-base alloys must contain either chromium or aluminum, or both, even where strength is not a principal factor. For example, Hastelloy X, one of the nickel-base alloys that is most resistant to oxidation and hot corrosion, contains 22% Cr, 9% Mo, and 18.5% Fe as principal solutes. Hastelloy X is essentially a solid-solution alloy when placed into service (carbides precipitate after long-term exposure). Therefore, the alloy is much weaker than superalloys containing γ' or γ'' as strengthening precipitates.

Because chromium is known to degrade the high-temperature strength of γ', there has been a strong incentive to lower chromium content in modern superalloys. Thus, the level of chromium decreased from 20% in earlier wrought alloys to as little as 9% in modern cast alloys. Unfortunately, this compositional change degraded hot corrosion resistance to the point that superalloys used in gas turbines had to be coated.

Further, as turbine blade temperatures exceed 1000 °C (1830 °F), Cr_2O_3 tends to decompose to CrO_3, which is more volatile and therefore less protective. To some extent, the loss of oxidation resistance has been compensated for by raising aluminum contents, although aluminum resides

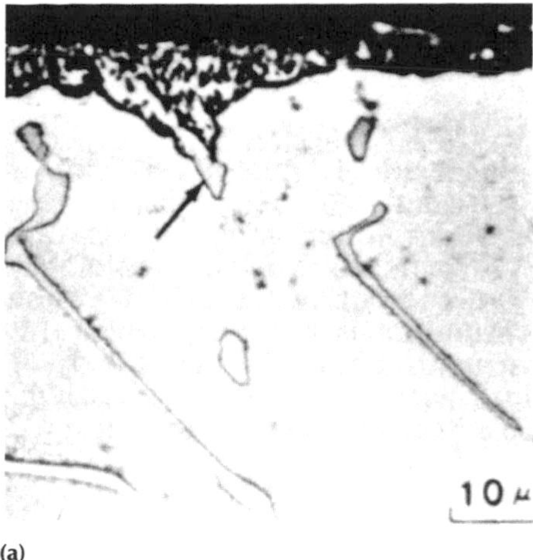

(a)

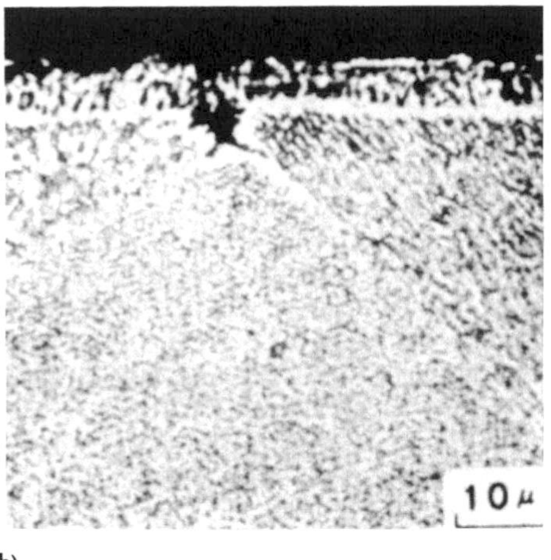

(b)

Fig. 2 Effects of oxidation on superalloys. (a) Accelerated oxidation of MC carbide (arrow) at surface of MAR-M 200 at 925 °C (1700 °F). (b) Accelerated oxidation of grain boundary in Udimet 700 at 760 °C (1400 °F). 1000×

primarily in γ′. (Aluminum in small quantities promotes the formation of Cr_2O_3.) However, Al_2O_3 is less protective than Cr_2O_3 under sulfidizing conditions, making coatings indispensable in aircraft turbines and, more recently, in industrial turbines.

Other elements that contribute to oxidation and hot corrosion resistance are tantalum, yttrium, and lanthanum. The rare earths appear to improve oxidation resistance by preventing spalling of the oxide, while the mechanism for improvement with tantalum is not known. Yttrium is now widely used in overlay coatings of the NiCrAlY type. The two most prominent solutes that provide both strength and surface stability are aluminum and tantalum.

For either Cr_2O_3 or Al_2O_3 formers to exhibit satisfactory lives under cyclic conditions, the chromium content in the original alloy should be about 20 wt%. However, this level of chromium is not compatible with the high-temperature strength and microstructural stability demanded for the most rigorous service conditions. Therefore, alloys having 10 to 15 wt% Cr, typical of many superalloys, require protective coatings.

Hot Corrosion. In-lower temperature operating conditions, such as ≤870 °C (≤1600 °F), accelerated oxidation can occur in superalloys through the operation of selective fluxing agents. One of the most well-known accelerated oxidation processes is hot corrosion (sometimes known as sulfidation). The hot corrosion process is separated into two regimes: low temperature and high temperature. High-temperature attack (type I) occurs at temperatures between 900 and 1050 °C (1650 and 1920 °F), and low-temperature attack (type II) occurs at temperatures between 680 and 750 °C (1255 and 1380 °F). Hot corrosion is triggered by the presence of sulfur in fuel and impurities, particularly salt, in the environment.

Both hot corrosion and oxidation involve rapid attack and consumption of hardware, and coatings are designed to resist the specific attack expected for particular turbine applications. Marine propulsion turbines, for example, are susceptible to both forms of hot corrosion, whereas aircraft gas turbines are more subject to oxidation, and sometimes to high-temperature hot corrosion, if the aircraft operates consistently in a coastal or marine environment. The problem of hot corrosion first became evident in large industrial and power-generated gas turbines that used low-quality fuels containing sulfur and sodium contaminants and other impurities, or that were located in areas where harmful species were ingested through air intakes, such as in marine and desert environments.

The principal method for combatting hot corrosion is the use of a high chromium content (≥20 wt%) in the base alloy. Although cobalt-base superalloys and many iron-nickel alloys have chromium levels in this range, many nickel-base alloys do not (especially those having high creep-ruptures strengths at high temperatures), because a high chromium content is not compatible with the high volume fraction of γ′ required. As chromium is increased in these alloys, the γ′ solid solubility temperature is decreased, so that there is a decreased amount of γ′ available for strengthening at high temperatures.

Higher titanium-aluminum ratios also seem to reduce attack on uncoated superalloys. Alloys have been produced with improved resistance to hot corrosion, based on slightly increased chromium contents and appropriate titanium-aluminum modifications. For maximum uncoated hot corrosion resistance, however, chromium contents in excess of 20 wt% appear to be required. Such alloys are not capable of achieving the strengths of the high-volume-fraction γ′ alloys, such as MAR-M 200 and B-1900. Consequently, coatings that protect the base metal (overlay coatings seem to provide the best surface protection), or sometimes environmental inhibitors, are used to suppress hot corrosion attack in high-strength (high-volume-fraction γ′) nickel-base superalloys.

An important consideration for coated superalloys is the reduction in incipient melting temperature of the system (coating/base metal) that may result from the change in composition that is caused by the diffusion of coating components inward from the surface. Incipient melting (Table 2) reduces grain-boundary strength and ductility and thus reduces stress-rupture capabilities. Once an alloy has been heated above its incipient melting point, the alloy properties cannot be restored by heat treatment. Generally, a cast alloy should not be used for a structural application at any temperature higher than the point about 125 °C (225 °F) below its incipient melting temperature. Oxidation behavior and strength will determine how closely the actual metal temperatures may approach this suggested upper limit. Wrought alloys are used for applications, such as turbine disks and rotating seals, that typically operate at much lower temperatures. Therefore, coatings are much less likely to be used for such applications.

Table 2 Incipient melting temperatures of selected wrought superalloys

Alloy	Incipient melting temperature, °C	Incipient melting temperature, °F
Hastelloy X (Ni)	1250	2280
L-605 (Co)	1329	2425
Haynes 188 (Co)	1302	2375
Incoloy 800 (Ni)	1357	2475
Incoloy 825 (Fe)	1370	2500
Incoloy 617 (Ni)	1333	2430
Inconel 625 (Ni)	1288	2350
Inconel X750 (Ni)	1393	2540
Nimonic 80A (Ni)	1360	2480
Nimonic 90 (Ni)	1310	2390
Nimonic 105 (Ni)	1290	2354
René 41 (Ni)	1232	2250
Udimet 500 (Ni)	1260	2300
Udimet 700 (Ni)	1216	2220
Waspaloy (Ni)	1329	2425

An important difference between nickel- and cobalt-base superalloys is related to the superior hot corrosion resistance claimed for cobalt-base alloys in atmospheres containing sulfur, sodium salts, halides, vanadium oxides, and lead oxide, all of which can be found in fuel-burning systems. In part, this apparent superiority may arise from the higher chromium content that is characteristic of cobalt-base alloys. Nickel forms low-melting-point eutectics such as nickel sulfide in sulfur-bearing gases and the subsequent attack on nickel alloys can be devastating.

Aluminide coatings developed for oxidation resistance in aircraft engines were not effective in inhibiting severe hot corrosion attack, and so coatings were developed that are specifically designed to fight hot corrosion (Ref 1).

Types of Coatings

Superalloy coatings are divided into two main categories: coatings that alter the substrate outer layer by their contact and interaction with certain metal species (*diffusion coatings*); and coatings that are formed by the deposition of protective metallic species onto a substrate surface, with some element interdiffusion providing coating adhesion (*overlay coatings*).

Diffusion Coatings

Diffusion coatings are based on the formation of intermetallic compounds such as βNiAl and βCoAl via a diffusion process. Their usefulness comes from the protective nature of an Al_2O_3 scale that forms on the coated part at the service operating temperature. Developed in the 1950s, diffusion coatings still are widely used, accounting for nearly 90% of the world market (Ref 2). The majority of these coatings are manufactured by means of slurry-fusion, pack cementation, and related gas-phase (out-of-contact) processes. Table 3 lists processing parameters for both pack cementation and gas-phase diffusion coating processes.

Diffusion aluminide coatings on superalloys are classified by microstructure as being of the "inward diffusion" or "outward diffusion" type. They differ in the conditions under which they are applied to the substrate in the pack-cementation process.

Pack cementation is a type of vapor deposition process. In the process, both the component to be coated and the reactants that combine to form a vapor are contained in an airtight retort. The reactants (designated as the "pack") consist of an aluminum-containing powder, a chemical activator, and an inert filler such as Al_2O_3. Upon heating in an inert atmosphere, the reactants form a vapor that reacts with the component surface enriching it with aluminum. The aluminum penetrates into the substrate to form a zone, the thickness and morphology of which are a function of the time and temperature of the process. Aluminide phases of interest for nickel-base alloys are Ni_3Al, NiAl, and Ni_2Al_3. Aluminides that form on cobalt-base alloys are CoAl and FeAl, respectively.

An inward coating is produced when aluminum activity is high compared with that of nickel. The aluminum diffuses inward faster than nickel diffuses outward through the nickel-aluminide intermetallics that initially form on the surface.

Inward coatings usually are formed by conducting the diffusion reaction process at a relatively low temperature, in the range of 700 to 800 °C (1290 to 1475 °F). Predominantly inward diffusion of aluminum occurs in the δNi_2Al_3 phase. Such a coating is not practical for use because it is brittle and has a low melting point. Therefore, the coating must be heat treated at a higher temperature, such as between 1040 and 1095 °C (1900 and 2000 °F), to convert the low-melting δ phase to the more refractory βNiAl monoaluminide phase for service (Ref 1).

Because the coating is initially formed by the inward diffusion of aluminum, substrate modification of the coating is maximized as substrate elements are locked in place. The outer layer of final-processed inward diffusion coatings is multiphased and, depending on the final heat treatment, can be quite rich in aluminum (hyperstoichiometric). In practice, inward diffusion coatings are applicable only to nickel-base superalloys. Cobalt-base superalloys, which do not contain aluminum, can form brittle, refractory phases at the substrate coating interface during service, and these phases compromise coating integrity.

Outward diffusion coatings are formed when aluminum activity is low compared with that of nickel (low pack aluminum/activator contents) and processing is carried out at a high reaction temperature, between 980 and 1090 °C (1800 and 2000 °F). Under these conditions, coatings are formed by the selective diffusion of nickel or cobalt outward through the monoaluminide layer. Compared to inward diffusion coatings, these coatings are modified to a lesser degree by the slower diffusing refractory elements in the substrate, and the overall aluminum contents are hypostoichiometric, providing somewhat better ductility in the coating than inward diffusion coatings provide. Outward diffusion coatings can be modified by elements supplied during the coating process, such as silicon, manganese, or chromium added to the cocoon or pack mix, whereas inward diffusion coatings usually cannot.

Both inward and outward coatings can be present on a single component, due to temperature and composition gradients in the pack or variations in surface geometry and surface condition. Between the extreme cases of inward and outward type coatings, an intermediate structure exists in which a βNiAl monoaluminide layer forms in the outer coating layer and supports inward aluminum diffusion (and, to a much lesser extent, outward nickel diffusion) (Ref 2).

Both coating types contain the high-melting point βNiAl phase. Pack mixtures often are adjusted to produce a more oxidation-resistant hyperstoichiometric NiAl in the outer layer, because NiAl is stable for aluminum contents ranging from about 45 to 60 at.%. The solubility of most other superalloy substrate elements in NiAl is small, so they are rejected from the NiAl outer layer and precipitate out as carbides ($M_{23}C_6$, M_6C, and MC), metals such as αCr, and topologically close-packed phases such as σ and η.

The classification of inward and outward diffusion coatings is derived from the work of Goward and Boone (Ref 3), who studied aluminide coating formation on Udimet 700 (a typical nickel-base superalloy having a nominal chemical composition of Ni-15Cr-17Co-5Mo-4Al-3.5Ti). They observed that for pack mixes containing aluminum (unit or "high" activity), coatings were formed by predominant inward diffusion of aluminum through Ni_2Al_3, and deeper in the coating, through aluminum-rich NiAl (for pure nickel, by inward diffusion through Ni_2Al_3 only). The diffusion rates are abnormally high: practical coating thicknesses could be achieved in a few hours at 760 °C (1400 °F).

A typical as-coated microstructure is shown in Fig. 3(a). Upon further heat treatment at, for example, 1080 °C (1975 °F) for 4 hours, the microstructure shown in Fig. 3(b) is formed. The coating matrix is now NiAl. The single-phase region in the center of the coating is nickel-rich NiAl, grown by predominant outward diffusion of nickel from the substrate alloy to react with aluminum from the top layer. The inner layer, or so-called interdiffusion zone, consists of refractory metal (tungsten, molybdenum, tantalum, etc.) carbides, and/or complex intermetallic phases in a NiAl and/or Ni_3Al matrix, formed by the removal of nickel from the underlying alloy, thereby converting its Ni-Ni_3Al structure to those phases.

Conversely, if the activity of aluminum in the source is reduced by alloying with, for example, nickel or chromium, to a level where nickel-rich NiAl is formed at the surface, the coating grows by predominant outward diffusion of nickel from the substrate to form NiAl by reaction with aluminum from the source (Fig. 3c). The lower layer of this coating is formed as previously described. Diffusion rates are relatively low, so the coating process must be carried out at higher temperatures, usually greater than 1000 °C (1830 °F). At the high-aluminum limit of NiAl, diffusion is by predominant motion of aluminum. Figure 3(d) shows a coating with a matrix of NiAl formed by this diffusion mechanism. Upon further heat treatment, this coating will stabilize with a structure similar to that shown in Fig. 3(b).

The structure and composition of an aluminide coating will vary for different superalloys for a given set of coating parameters. The structure and

Table 3 Examples of pack mixes/sources and processing parameters for various coatings on nickel and cobalt superalloys

Coating type	Source composition	Processing parameters
Pack aluminizing, inward diffusion in Ni_2Al_3 in nickel alloys	5-20% Al (Al-10Si), 0.5-3% NH_4Cl, balance Al_2O_3 powder	1 to 4 h at 650 to 680 °C (1200 to 1255 °F) in air, argon, H_2; heat treat 4 to 6 h at 1095 °C (2000 °F) in argon
Pack aluminizing, inward diffusion in NiAl in nickel alloys	44% Al, 56% $CrNH_4Cl$, balance Al_2O_3 powder	5 to 10 h at 1040 °C (1900 °F) in vacuum (argon, H_2)
Pack aluminizing, outward diffusion in NiAl in nickel alloys	2-3% Al, 20% Cr, 0.25% NH_4HF_2, balance Al_2O_3 powder	25 h at 1040 °C (1900 °F) in argon
Pack aluminizing of cobalt alloys	8% Al, 22% Cr 1% NH_4F, balance Al_2O_3 powder	4 to 20 h at 980 to 1150 °C (1800 to 2100 °F) in argon
Gas-phase aluminizing, outward diffusion in NiAl in nickel alloys	10% Co_2Al_5, 2.5% NaCl, 2.5% $AlCl_3$, balance Al_2O_3 powder	3 h at 1095 °C (2000 °F) in argon
Gas-phase aluminizing, outward diffusion in NiAl in nickel alloys	30% Al-70% Cr alloy granules, NH_4F	4 h at 1150 °C (2100 °F) in argon
Pack or gas-phase chromizing of nickel alloys	15% Cr, 4% Ni, 1% Al, 10.25% NH_4Br or NH_4Cl, balance Al_2O_3 powder	3 h at 1040 °C (1900 °F) in argon

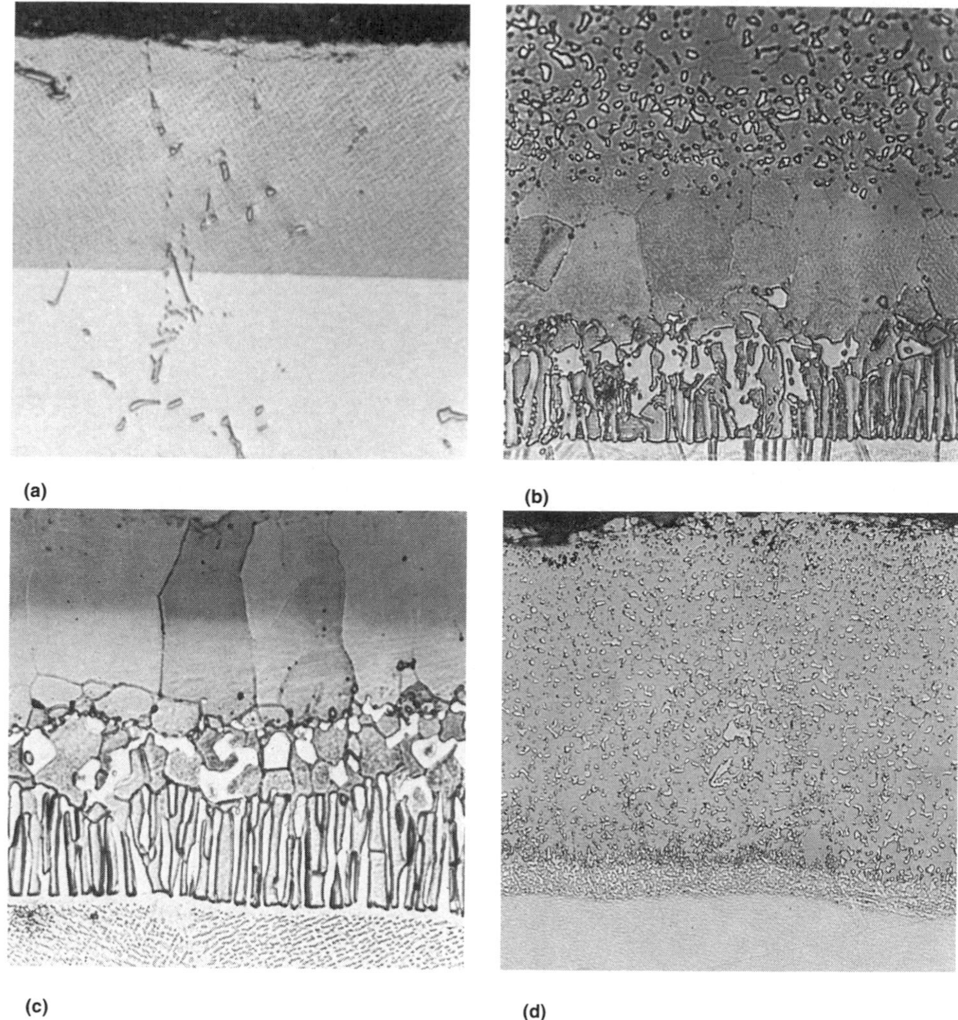

Fig. 3 Typical microstructures of aluminide coatings on a nickel-base superalloy. (a) Inward diffusion based on Ni_2Al_3 (and aluminum-rich NiAl). (b) Same as (a) but heat treated at 1080 °C (1975 °F). (c) Outward diffusion of nickel in nickel-rich NiAl. (d) Inward diffusion of aluminum in aluminum-rich NiAl

composition depend on the rate at which alloying elements such as chromium, molybdenum, and tungsten in the superalloy segregate to the coating/substrate interface. It is possible to predict the microstructure and constitution of a coating if the isothermal ternary phase diagram at diffusion temperatures, the aluminum activity in the pack, and the chemical composition of the superalloy substrate is known (Ref 2).

For aircraft gas turbine blades fabricated from moderately corrosion-resistant nickel alloys (those containing 12 to 15% Cr), simple aluminide coatings of the inward and outward diffusion types provide adequate protection for many contemporary engines. When alloys are more corrosion prone (those containing 7 to 10% Cr), current practice is to modify aluminide coatings by chromizing prior to aluminizing. Aluminizing should be by inward diffusion of aluminum to locate the higher chromium concentration in the outer layer of the coating. For more severe type I hot corrosion resulting from exposure of engines to salt spray from marine environments, modification of aluminide coatings with silicon (or better, by platinum) can prove to be the more cost-effective solution.

Major developments in diffusion coatings for superalloys over the past two decades include modifications of aluminide diffusion coatings with chromium (Ref 4), platinum (Ref 5), and to a lesser extent, silicon. The theory of codeposition of combinations of aluminum, chromium, silicon, and the so-called reactive metals (yttrium, rare earths, hafnium, etc.) has been refined and straightforward processes have been developed (Ref 6).

Platinum-modified aluminide coatings on nickel-base superalloys are based on the formation of intermetallic phases including $PtAl_2$, Pt_2Al_3, and PtAl in the outer coating layer, which depends on the type of platinum-aluminide coating. A range of coating structures may be possible within certain processing parameters, such as initial platinum thickness, diffusion of platinum into the substrate prior to aluminizing, pack-aluminum activity, and post-aluminizing treatment.

For cobalt-base superalloys, a low-activity pack cementation process is used to produce a βCoAl type diffusion coating, where cobalt is diffused in the outward direction. The reaction temperature is about 50 to 60 °C (90 to 110 °F) higher than that used for nickel-base superalloys. A typical coating comprises a thin outer layer of Co_2Al_3, a thick βCoAl intermediate layer, and an inner layer of tungsten-chromium carbides. There is no inner diffusion zone as with nickel-base superalloys. The carbide inner layer can limit the diffusion of cobalt into the coating from the substrate. The βCoAl coatings are more brittle than βNiAl coatings (Ref 2).

A given set of processing conditions produces different coatings on different substrate superalloys, due to the compositional complexities of the alloys. For example, a particular coating will be thinner on a cobalt-base alloy than on a nickel-base alloy due to lower aluminum diffusivity in cobalt. A coating can also have different characteristics when applied to nickel-base alloys having different compositions; this is especially evident with the phase structure of the interdiffusion zone.

Chemical vapor deposition (CVD) offers advantages for certain applications of diffusion coatings. In this out-of-contact process, a vapor of predetermined composition is produced independent of the coating process and introduced into a coating chamber where it reacts with the surface of the part.

The major advantage of the CVD process over pack cementation is the ability to coat internal cooling passages of film-cooled turbine airfoils. The vapor is pumped through the internal passages, resulting in a relatively uniform coating. By comparison, pack cementation has a very limited "throwing power." CVD also has compositional flexibility because the thermodynamics of vapor formation are completely separate from those of the metal-vapor reaction.

CVD processes have been used to deposit elements other than aluminum, such as chromium and silicon. In addition, noble metals such as platinum and palladium can be plated onto a substrate surface prior to aluminizing in duplex processes (Ref 1).

The inward and outward diffusion mechanisms described above for the pack cementation process apply equally to coatings formed by out-of-contact, or CVD, processes, from slurry "slip packs," and from aluminum alloy powders deposited on superalloys by slurry spraying or by slurry electrophoresis. These coatings are applied by spraying or electrophoretically depositing pure aluminum or low-melting aluminum alloys (e.g., Al-10Si) and then heat treating. The coatings form by dissolution of the superalloy into the melt until the melt solidifies, followed by diffusion of aluminum similar to that described above.

All known aluminide-base coatings on nickel-base superalloys, including those modified by chromium, platinum, and silicon, have one of the archetypical microstructures described above. For pure nickel and nickel alloys containing no aluminum (or <0.2% Al), the interdiffusion zone does not form. For pure nickel, Kirkendall voids form at the coating/substrate interface, along with

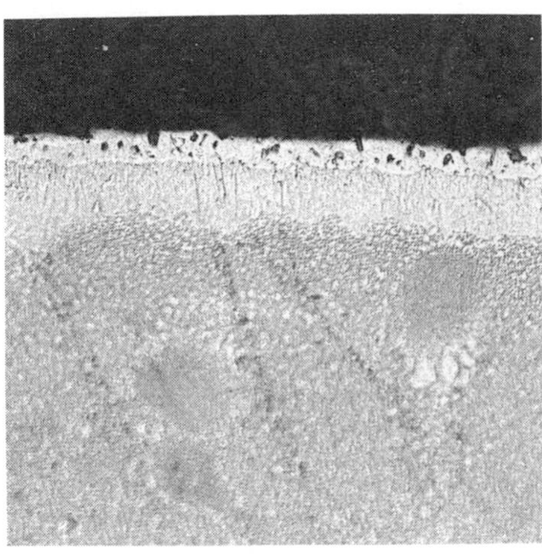

(a)

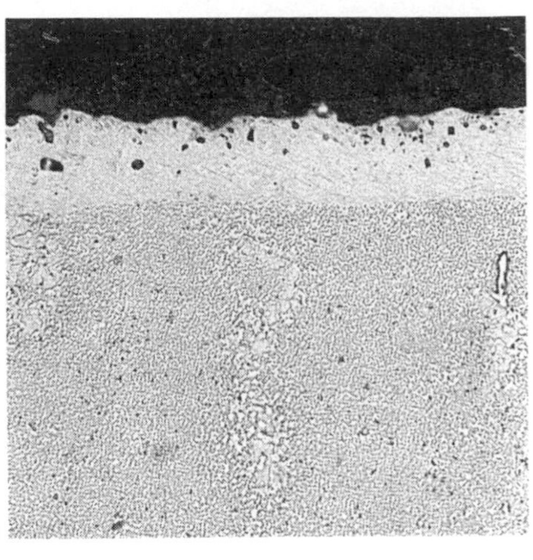

(b)

Fig. 4 Chromium diffusion coatings on a nickel superalloy by (a) pack cementation and (b) out-of-contact gas-phase (chemical vapor deposition) processing. Both at 500×

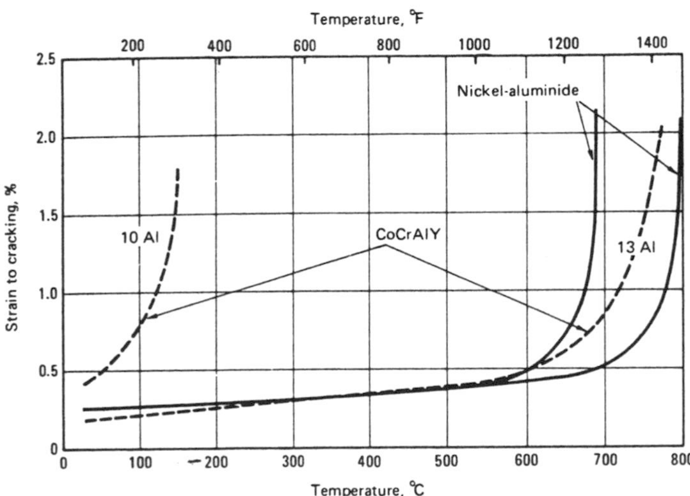

Fig. 5 Ductility of CoCrAlY overlay and diffusion aluminide coatings. CoCrAlY overlay coatings show significant ductility (1 to 3%) at lower temperature ranges (ambient temperature to 650 °C, or 1200 °F), with ductility continuously increasing as the volume fraction of CoAl is decreased.

Al_2O_3 from oxygen in the nickel. For nickel alloys containing no aluminum, voids, refractory metal layers, and Al_2O_3 form at the interface. The adherence of the resulting coatings is compromised and they may not be practically useful. It is believed that similar mechanisms apply to the coating of cobalt superalloys. Again, the absence of aluminum in many of these alloys precludes the formation of the interdiffusion zone that is common to most nickel superalloys. Rather, a refractory metal carbide (tungsten carbide, chromium carbide) forms at the juncture with the base alloy. As described for similar nickel-base alloys, a refractory metal carbide and Al_2O_3 (formed from oxygen in the aluminum-free alloys) can compromise the adherence of these coatings. Special processing conditions, involving slow coating growth at high temperatures (up to 1095 °C, or 2000 °F) from relatively low aluminum activity sources, can sometimes be used to achieve satisfactory coating adherence. Minor additions of aluminum (1 to 2%) to cobalt superalloys prevent these problems: stable interdiffusion zones then form, analogous to those on most nickel superalloys.

Figure 4 illustrates diffusion chromide coatings formed on a nickel superalloy by pack cementation and out-of-contact processes. The coating deposited by pack cementation is overlaid with a thin layer of α-chromium, as shown in Fig. 4(a). Users generally require that this phase be absent. It must then be removed chemically, or alternatively, the coating must be applied by an out-of-contact process to produce the structure shown in Fig. 4(b). These coatings usually then contain 20 to 25% Cr at the outer surface. Coating formation, from chromium-alumina-activator packs (usually ammonium chloride) or from out-of-contact sources (powders or chromium granules), involves approximately equal rates of interdiffusion of chromium and nickel. Significant depletion of titanium and aluminum from the alloy surface occurs because the sources do not contain these elements. The desired coatings are thus solid solutions of chromium in the remaining nickel-base alloy. Internal oxides of aluminum and titanium can form because the oxygen potential of the sources is normally sufficient to cause internal oxidation. This can be avoided by adding aluminum to the sources in amounts just below that which would cause aluminizing rather than chromizing.

Overlay Coatings

Overlay coatings differ from diffusion coatings in that interdiffusion of the coating with the substrate is not required to generate the desired coating structure/composition. Overlay coatings do not rely on reaction with the substrate for their formation. Instead, a prealloyed material applied over the substrate determines the coating composition and microstructure, and adhesion of the coating to the substrate is effected by some elemental interdiffusion.

Coatings of this type in current use are generally called MCrAlY overlay coatings, where M represents Ni, Co, Fe, or some combination of these metals. These coatings essentially comprise a monoaluminide (MAl) component contained in a more ductile matrix of solid solution (γ), in the case of CoCrAlY coatings, or a mixture of γ and $\gamma'Ni_3Al$, phase in the case of NiCrAlY coatings. NiCoCrAlY and FeCrAlY modifications are also available.

The matrix is nickel- or cobalt-base and contains a rather large amount of chromium and an intermediate amount of aluminum. Aluminum provides the primary protection against oxidation through the formation of a slow-growing Al_2O_3 scale. The supply of aluminum for formation of protective Al_2O_3 scales comes largely from the dispersed MAl phase during the useful life of such coatings. The contained chromium is important in combatting hot corrosion and also increases the effective aluminum chemical activity. The solid-solution matrix provides ductility in this coating class that is generally not possible with diffusion coatings (Fig. 5), and it imparts much improved resistance to thermal fatigue cracking (Fig. 6). A small amount of yttrium is usually included in overlay coatings to improve the adherence of the oxidation product.

CoCrAlY coatings are recognized as being superior in hot corrosion resistance, whereas NiCrAlY coatings possess the better oxidation resistance. The range of NiCoCrAlY coatings may be tailored for a desired compromise between oxidation and hot corrosion resistance.

The composition of overlay coatings, as well as that of the substrate, affects the extent of coating-substrate interdiffusion during service. Coating composition may also be adjusted to achieve a good thermal expansion fit with a given superalloy substrate. MCrAlY coatings may contain 15 to 25% Cr, 10 to 15% Al, and 0.2 to 0.5% Y.

The general features of a MCrAlY coating are an oxide scale on the outer surface, material immediately beneath the scale that has a modified composition depleted in aluminum, a layer of the

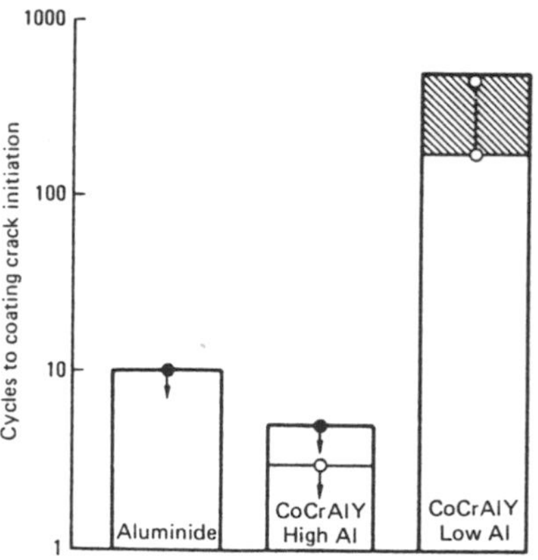

Fig. 6 Thermal mechanical fatigue behavior of brittle (high-aluminum) and ductile (low-aluminum) overlay coatings, compared with aluminide diffusion coating. When thermal strain reaches its maximum value of relatively low temperatures, substantial improvement in thermal fatigue behavior can be achieved with a ductile MCrAlY coating. All specimens were cycled between 425 °C (800 °F) and 925 °C (1700 °F). Open circles, ±3% strain; closed circles, ±0.25% strain

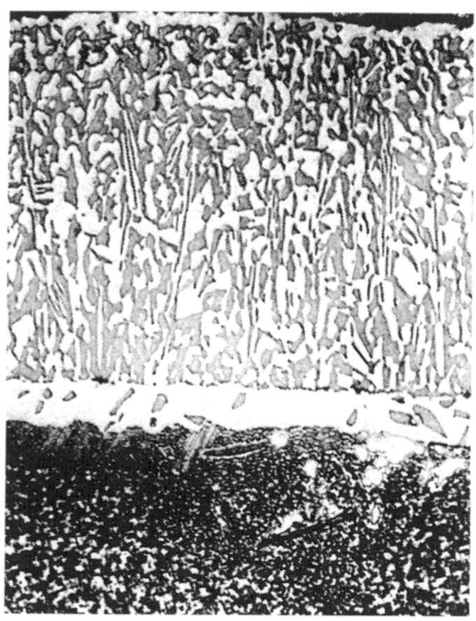

Fig. 7 Microstructure of a two-phase CoCrAlY overlay coating

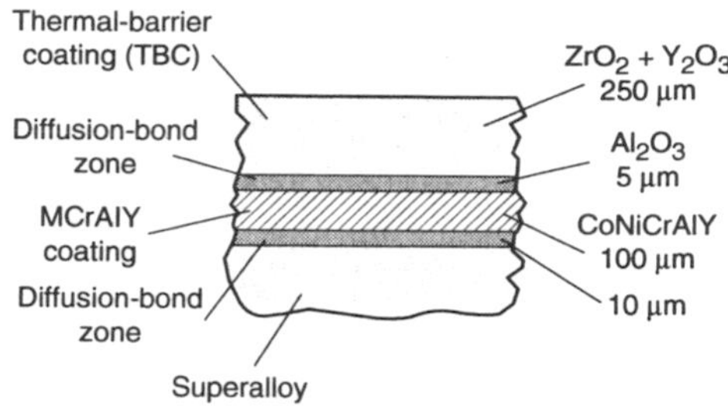

Fig. 8 Schematic of a multilayer thermal barrier coating system produced using a combination of electron-beam physical vapor deposition and plasma spray techniques

coating alloy, and an interdiffusion zone in contact with the substrate. The composition and microstructure of the coating alloy depend on post-deposition treatment and service exposure. The interdiffusion zone functions as a bonding layer between coating and underlying alloy substrate. Figure 7 shows the microstructure of a CoCrAlY overlay coating.

Overlay coating processes of primary importance today are electron-beam physical vapor deposition (EB-PVD) and plasma spraying. While CVD can be used, it is not practical from a cost standpoint because of very low deposition rates and the difficulty in controlling complex coating compositions. The EB-PVD technique generally produces higher quality coatings, while plasma spraying has an equipment-cost advantage.

Electron-beam physical vapor deposition emerged in the 1960s as the primary overlay coating production method. The process allows the deposition of metals via vapor transport in a vacuum without the need for a chemical reaction. The vapor source can be produced by several methods, but EB evaporation is the most commonly used technique for coating turbine components.

A cloud of metal atoms impinges on the preheated part surface and condenses out into equilibrium or metastable phases. Carrying out the process at elevated temperatures promotes the formation of a dense coating and coating adhesion.

The process results in an as-deposited coating structure that is typically oriented perpendicular to the substrate surface (columnar structure). This is caused by fast-growing grains of the coating alloy that propagate through the coating thickness. Separations between adjacent grains of the deposited coating, known as "leader defects," or columnar voids, often are present, especially on convex surfaces. Shot peening and laser glazing can be used to close these defects to prevent premature corrosive attack and thermal fatigue cracking. The microstructure of an EB-PVD coating can be altered by varying the substrate temperature and by bombarding the substrate with energetic particles, such as plasma or ion beam, that can break up the columnar structure and improve coating density (Ref 1, 2).

The composition and microstructure of the deposited film are the two major factors that determine its corrosion resistance. The typical composition of a MCrAlY film is 20% Cr, 10% Al, and 0.3% Y, with the balance M (Fe, Co, or Co-Ni). Acceptable composition tolerances depend on the specified nominal composition. The compositions of films obtained in practice usually fall within acceptable tolerance ranges. For example, the chromium, aluminum, and yttrium contents of a NiCoCrAlY deposit typically are 80%, 60%, and 60%, respectively, of the acceptable tolerance ranges. Similarly, the cobalt, chromium, aluminum, and yttrium contents of a CoCrAlY deposit are 60%, 45%, 60%, and 60%, respectively. Stated another way, if the chromium range specified for a NiCoCrAlY deposit is 16 to 22%, the range for a production coating will be approximately 16.5 to 21.5%.

Plasma-sprayed coatings are produced by injecting a prealloyed powder into a high-temperature plasma gas stream (via a plasma-spray gun) and depositing the melted particles on the substrate surface. The molten particles solidify on contact, forming the coating. The process generally is carried out in a low-pressure vacuum chamber (hence the term low-pressure plasma spraying, LPPS), which minimizes the formation of oxide defects within the as-deposited coating.

The high velocity at which the molten metal particles are directed at the substrate causes the molten droplets to "splat" against the substrate and spread out in a direction parallel to the surface. A typical as-coated microstructure contains splat interfaces parallel to the surface. A diffusion heat treatment eliminates the individual splat layers, and the resulting structure assumes a two-phase nature similar to that of an EB-PVD coating.

The surface finish of a plasma-sprayed coating generally is rougher than that of an EB-PVD coating. A finishing operation, such as abrasive slurry or controlled vapor blasting, can be used to achieve the required smooth surface (Ref 1).

Thermal-Barrier Coatings

The hot corrosion (and oxidation) resistance of coated blades can be further improved by applying a layer of thermal insulation. This thermal-barrier coating (TBC) must be sufficiently thick, have a low thermal conductivity and high thermal-shock resistance, and have a high concentration of internal voids to further reduce thermal conductivity to a value well below that of the bulk material. The temperature difference between the outer surface of a TBC and the outer surface of the underlying corrosion-resistant film can be as high as 150 °C (270 °F). In addition to reducing the temperature at the surface of the superalloy, these coatings also reduce thermal-shock loads on the blades: rapid changes in ambient temperature are moderated and attenuated before they reach the substrate.

A TBC system consists of an insulating ceramic outer layer (top coat) and a metallic inner layer (bond coat) between the ceramic and the substrate. Both the top coat and bond coat can be applied by plasma spraying (air and low pressure) and EB-PVD. A schematic of a typical system is shown in Fig. 8. Current state-of-the-art TBCs are zirconium oxide (ZrO_2), or zirconia, with 6 to 8% (by weight) of yttrium oxide (Y_2O_3), or yttria, to partially stabilize the tetragonal phase for good strength, fracture toughness, and resistance to thermal cycling. The coatings are relatively inert, have a high melting point, and have low thermal conductivity.

Air-plasma-sprayed coatings contain porosity and microcracks that help to redistribute thermal stresses but also provide corrosion paths through the coating. Low-pressure plasma spray coatings provide high coating purity and essentially eliminate oxides and porosity. EB-PVD coatings have a columnar grain morphology (Fig. 9) in which individual grains are strongly bonded at their base, but have a weak bond between grains. The

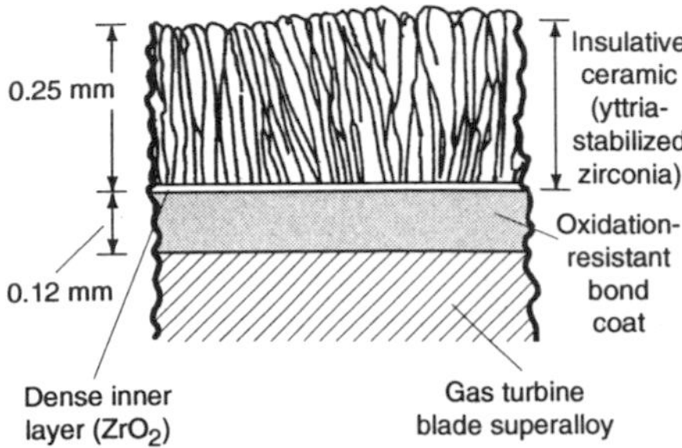

Fig. 9 Cross section illustrating the strain-tolerant columnar ZrO_2 microstructure of EB-PVD zirconia thermal barrier coatings. Source: Ref 7

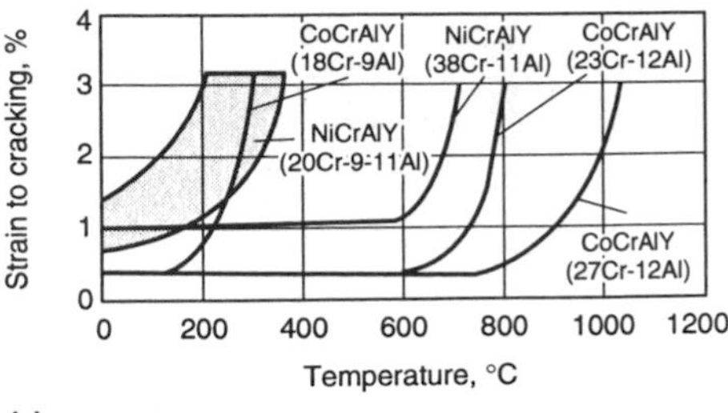

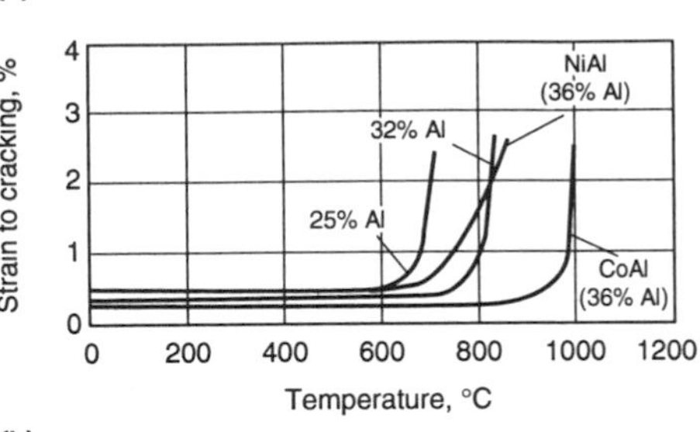

Fig. 10 Ductility/temperature characteristics of (a) MCrAlY coatings and (b) aluminide diffusion coatings

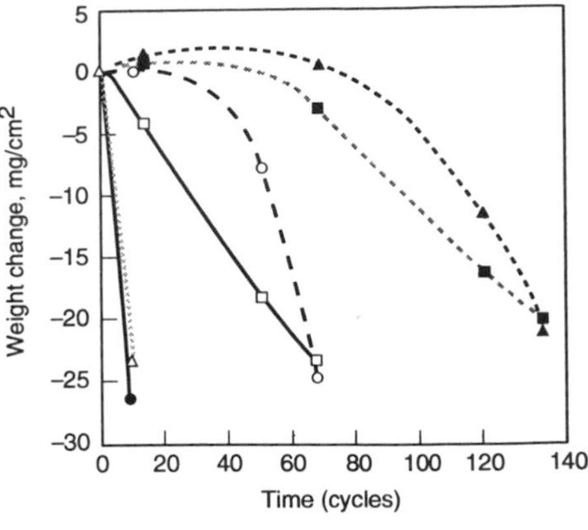

Fig. 11 Cyclic oxidation data at 1200 °C (2190 °F) for uncoated superalloys and for superalloy-diffusion coated with aluminides and aluminides containing platinum (Pt). Source: Ref 11

major advantage of this columnar outer structure lies in the fact that it reduces stress buildup within the body of the coating. Strain within the coating is accommodated by free expansion (or contraction) of the columns into the gaps, which results in negligible stress buildup (Ref 8). The columnar structure of EB-PVD zirconia TBCs has the disadvantage, however, of increasing heat conductivity by a factor of about 2 as compared to plasma-sprayed TBCs (Ref 9).

Figure 9 also shows that a thin, dense ZrO_2 layer occurs between the bond coat surface and the upper columnar zirconia structure. This phase grows under oxygen-deficient conditions just at the beginning of zirconia deposition, and its thickness is controlled by how quickly the oxygen bleed is activated in the vacuum chamber of the EB-PVD apparatus after zirconia coating commences. This dense, interfacial ZrO_2 film is critical to the life of the EB-PVD coating in that it provides for chemical bonding between the columnar zirconia and the oxidation-resistant bond coat. If, however, this interfacial film becomes too thick (>2 μm), it may sustain and transmit compressive stresses sufficient to cause cracking within the outer zirconia coating.

The metallic (MCrAlY) bond coat aids in the adhesion of the ceramic topcoat, protects the substrate from hot corrosion and oxidation, and helps in handling expansion mismatch between the ceramic and superalloy. For best adhesion of EB-PVD TBCs, the bond coat surface should be smooth, or preferably polished, in contrast to plasma-sprayed TBCs, which require a rough bond coat.

High-pressure/high-velocity oxygen fuel (HP/HVOF) spraying also is being evaluated to produce MCrAlY coatings, which could be used as bond coats for TBCs in aircraft turbine applications (Ref 10).

Coating Performance

The ability of a coating to provide satisfactory performance in high-temperature applications is measured by whether it can remain intact, resist oxidation and corrosion, and avoid cracking. Generally, diffusion aluminide coatings are limited by their oxidation behavior, overlay coatings are limited by their susceptibility to thermal-fatigue cracking in cyclic conditions, and TBCs are limited by the thermal-expansion mismatch between the ceramic and metallic layers, and by environmental attack of the bond coat.

Coating Ductility

Coatings must be capable of tolerating strain due to thermal expansion mismatch and mechanical loads in order to retain coating-substrate integrity. The ductile-to-brittle transition temperature (DBTT) is used to describe the ability of a coating to tolerate strain. Above the DBTT, the coating behaves in a ductile manner; below the DBTT the coating is brittle. The DBTT is affected by several factors, including coating composition, coating application process, coating thickness, surface finish, and strain rate. The DBTT should be as low as possible so that cracking in the coating does not occur in service, since the cracks may then propagate into the substrate.

MCrAlY overlay coatings generally have a lower DBTT than diffusion coatings, because their chemical composition can be controlled. The DBTT of aluminide diffusion coatings is a function of aluminum content, increasing with increasing aluminum. Platinum additions also raise the DBTT of aluminide coatings. Figure 10 shows the DBTT of aluminide and overlay coatings.

For overlay coatings, NiCrAlY coatings have a higher DBTT than CoCrAlY coatings (due to a lower DBTT of NiAl). The DBTT of both coatings is increased with increasing chromium and/or aluminum content (Ref 2). NiCoCrAlY coatings containing 20 to 26% Co are significantly more ductile than either NiCrAlY or CoCrAlY coatings.

Ductility of aluminide diffusion coatings can sometimes be enhanced by laser remelting of the coating. A thin surface layer of the coating is melted and rapidly quenched by the cold bulk solid substrate, which refines the grain size.

Oxidation

Aluminide coatings degrade in service through cyclic oxidation, hot corrosion, erosion, interdiffusion, and thermomechanical fatigue cracking. Aluminum in the coating combines with oxygen at the substrate surface, forming a protective Al_2O_3 scale. When the scale cracks and spalls from thermal cycling, additional aluminum from the coating diffuses to the surface to reform the scale. Aluminum from the coating also diffuses into the substrate, and as aluminum is depleted in the coating, βNiAl converts to $\gamma'Ni_3Al$ and then to γNi solid solution. When the aluminum content in the coating drops to about 4 to 5 wt%, the Al_2O_3 scale can no longer form and rapid oxidation occurs. The rate of aluminum diffusion is influenced by substrate composition.

If the coating undergoes thermal fatigue cracking, refractory elements in the diffusion zone may be exposed to the oxidizing environment and oxidize rapidly. This effect must be considered when selecting a coating for use in cyclic conditions.

Incipient melting in the diffusion zone also can result in rapid oxidation penetration. Such melting can occur at temperatures as low as 1120 °C (2050 °F), well below the melting point of NiAl (1590 °C, or 2900 °F).

It is well established that platinum in diffusion aluminides extends the lives of such coatings in oxidizing environments (Fig. 11). The mechanism by which platinum affects coating service lives is not completely understood. It may improve Al_2O_3 adherence, causing an Al_2O_3 scale to develop with slower transport properties due to higher purity, and it may inhibit interdiffusion

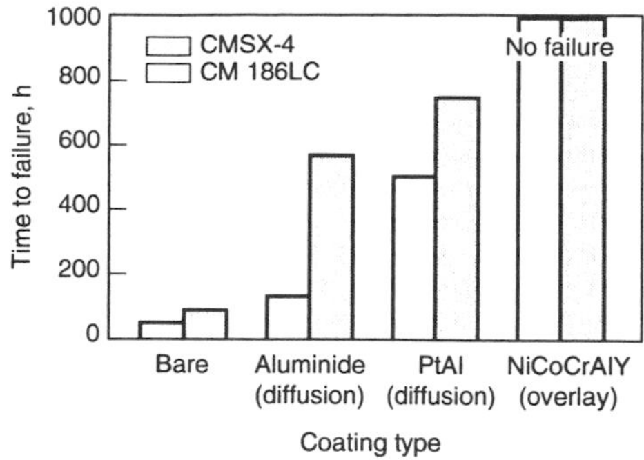

Fig. 12 Hot corrosion resistance of uncoated and coated CMSX-4 (nickel-base single-crystal superalloy) and CM186LC (nickel-base directionally solidified superalloy) test specimens. The accelerated hot corrosion tests were carried out at 900 °C (1650 °F) in fuel containing 1% sulfur and 10 ppm salt. The diffusion coatings were applied by a proprietary electrophoresis process. The NiCoCrAlY coating was applied by electron-beam physical vapor deposition. Source: Ref 12

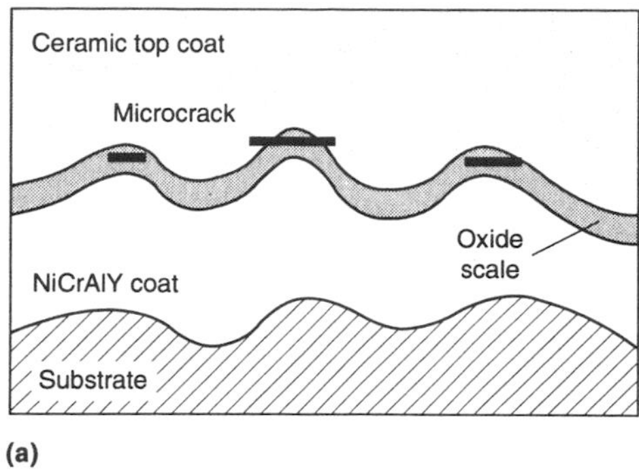

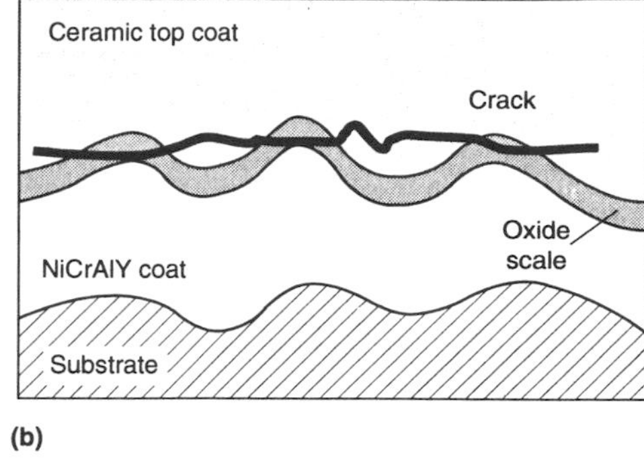

Fig. 13 Failure mechanism in a thermal barrier coating (TBC). Microcracks develop in the aluminum oxide scale on the bond coat (a) and propagate along the lamellar interfaces, leading to spalling of the ceramic TBC or complete delamination (b).

between the coating and the substrate (Ref 11). Other precious elements such as rhodium produce similar effects, and more recent results show that palladium may also produce beneficial effects (Ref 11).

Enhanced oxidation resistance may also be possible through the use of ion implantation. The addition of yttrium or hafnium by this technique improves coating adherence and reduces scale growth rate.

Overlay Coatings. The oxidation pattern of MCrAlY overlay coatings is similar to that of diffusion aluminide coatings. Elements such as chromium and yttrium improve the resistance of the Al_2O_3 scale to spallation. The most widely used overlay coating for oxidation resistance is the NiCoCrAlY type. The addition of cobalt also improves coating ductility.

In the oxidation process, grains of aluminum-rich β phase convert to islands of γ', eventually leaving only the less oxidation-resistant γ matrix phase. Substrate composition can influence oxidation resistance.

MCrAlY overlay coatings have a higher melting point than diffusion coatings, so melting does not occur in the interdiffusion zone at temperatures lower than the melting point of the bulk coating. Coatings have survived exposure temperatures as high as 1290 °C (2350 °F) without melting.

The addition of elements such as silicon, tantalum, and hafnium can improve oxidation resistance at the expense of some ductility. Compositional flexibility offers the opportunity to tailor coatings for optimum performance.

Hot Corrosion

Hot corrosion resistance is required more in marine and industrial turbine applications than in aircraft applications. Coatings are used to prevent catastrophic failure of components in conditions involving dirty fuels or contaminants in the atmosphere. Figure 12 compares the resistance to hot corrosion of diffusion and overlay coatings.

Diffusion Coatings. Conventional coatings have unsatisfactory performance in hot corrosion applications. Platinum-aluminide coatings offer improved hot corrosion resistance by increasing aluminum activity at the surface and by enhancing scale adherence and the rate of scale formation (Fig. 12). Degradation of platinum coatings results when the aluminum is depleted to a point where a protective scale cannot form. Molten alkali metal salts speed up the destruction of the scale, which accelerates aluminum consumption.

As aluminum is consumed, the β matrix phase is transformed to γ', which ends effective corrosion protection. In the advanced stages of attack, chromium-rich internal sulfides form in the substrate and in the substrate/coating interdiffusion zone.

The low ductility of platinum-aluminide coatings is of concern for use in aircraft applications involving severe cyclic conditions. However, these coatings have provided satisfactory service for land-based turbines operating in corrosive environments.

Overlay coatings have very good resistance to hot corrosion. The best in this regard are cobalt-base coatings having high chromium-aluminum ratios. As indicated in Fig. 12, NiCoCrAlY coatings also offer superior resistance to hot corrosion. Nickel- and iron-base coatings offer good oxidation resistance and good resistance to mild hot corrosion environments. Increasing the Cr-Al ratio in cobalt-base coatings increases hot corrosion resistance at the expense of some oxidation resistance.

MCrAlY coating degradation is characterized by the presence of sulfides and oxides in the coating; chromium-rich sulfides precede oxides. However, coating failure occurs due to aluminum and chromium depletion, which are needed to form the protective scale. Thicker coatings can provide the required protection in applications where thermal cycling is not severe.

Low-Temperature Corrosion

Coatings developed to resist oxidation and hot corrosion generally are ineffective against low-temperature corrosion (i.e., 680 to 750 °C, or 1255 to 1380 °F). Instead, Cr_2O_3 and silica (SiO_2) scales offer the best protection.

Diffusion Coatings. Chromide diffusion coatings rapidly form a continuous, adherent Cr_2O_3 scale for protection. Because interdiffusion is small in lower-temperature environments, chromide coatings are relatively thin (0.04 to 0.05 mm, or 1.5 to 2.0 mils). The thinner coating is more favorable with respect to mechanical properties, because chromium compositions are prone to cracking due to lower ductility.

Overlay Coatings. High CrCo-, nickel-, and iron-base coatings are effective against low-temperature hot corrosion. Cobalt-base coatings with high chromium content are preferred, because the operating temperature range in many turbine airfoil applications necessitates protection against both oxidation and corrosion (Ref 2).

TBC Performance

As discussed above, TBCs generally consist of a metallic bond coating (typically MCrAlY applied by vacuum plasma spraying or PVD) and a thick ceramic top coat (typically stabilized ZrO_2 deposited by low-pressure plasma spray or EB evaporation). Despite their columnar microstructure, thin PVD coatings can be dense and serve as a diffusion gas barrier at high temperatures, and the columnar structure guarantees an improved strain and stress tolerance. Thermal stress within the coatings occurs due to a mismatch between the thermal expansion coefficients of the metallic substrate and the coating, and due to transient thermal gradients during rapid thermal cycling. Depending on deposition conditions, the PVD technique may also induce some stress within the coating. This intrinsic compressive stress of the sputtered bond coating may act as a "prestress" that diminishes coating failure due to tensile thermal stresses at elevated temperatures.

The plasma spraying process also induces some residual stress, due to substrate heating and deposition of a stress-free coating at the deposition temperature. Thermal residual stress develops during cooling to room temperature.

Failure of TBCs in service generally is attributed to stress that develops during cooling after high-temperature exposure, and to transient thermal stress that develops during rapid thermal cycling. Failure occurs primarily due to thermal expansion mismatch between the ceramic and metallic layers and environmental attack of the bond coat.

The stress state in the ceramic layer is biaxial compressive in the plane of the coating. Strains induced by these stresses increase during repeated thermal cycling, which results in crack initiation and eventual spalling of the coating, as shown in Fig. 13. The use of Y_2O_3 to stabilize the ceramic coat, together with an MCrAlY-type bond coat, significantly improves the thermal fatigue resistance of TBCs.

REFERENCES

1. C.T. Sims, N.S. Stoloff, and W.C. Hagel, *Superalloys II,* John Wiley & Sons, 1987, p 359–382
2. A.K. Kovl et al., *Advances in High Temperature Structural Materials and Protective Coatings,* National Research Council of Canada, 1994, p 169–238
3. G.W. Goward and D.H. Boone, *Oxid. Met.,* Vol 3, 1971, p 475–495
4. K. Godlewski and E. Godlewska, *Oxid. Met.,* Vol 26, 1986, p 125–128
5. G. Lehnardt and H. Meinhardt, *Electrodeposition and Surface Treatment,* Vol 1, 1972, p 189–193
6. R. Bianco and R.A. Rapp, Chap 9, *Metallurgical and Ceramic Coatings,* K.H. Stern, Ed., Chapman and Hall, 1993
7. R.L. Jones, Thermal Barrier Coatings, *Metallurgical and Ceramic Protective Coatings,* K.H. Stern, Ed., Chapman & Hall, 1996, p 194–235
8. T.E. Strangman, *Thin Solid Films,* Vol 127, 1985, p 93–105
9. S.M. Meier, D.M. Nissley, and K.D. Sheffler, "Thermal Barrier Coating Life Prediction Model Development," NASA Contractor Report 189111, NASA Lewis Research Center, 1991
10. G. Irons and V. Zanchuk, "Comparison of MCrAlY Coatings Sprayed by HVOF and Low Pressure Processes," *Proc. 1993 National Thermal Spray Conference,* ASM International, 1993
11. N. Birks, G.H. Meier, and F.S. Pettit, Degradation of Coatings by High Temperature Atmospheric Corrosion and Molten Salt Deposits, *Metallurgical and Ceramic Protective Coatings,* K.H. Stern, Ed., Chapman & Hall, 1996, p 290–305
12. P.S. Korinko, M.J. Barber, and M. Thomas, "Coating Characterization and Evaluation of Directionally Solidified CM 186 LC and Single Crystal CMSX-4," paper presented at ASME Turbo Expo '96, Birmingham, UK, 10–13 June 1996

SELECTED REFERENCES

- *Advances in High Temperature Structural Materials and Protective Coatings,* A.K. Koul, V.R. Parameswaran, J-P. Immarigeon, and W. Wallace, Ed., National Research Council of Canada, 1994
- *High Temperature Coatings,* M. Khobaib and R.C. Krutenat, Ed., TMS-AIME, 1986
- *Elevated Temperature Coatings,* N.B. Dahotre, J.M. Hampikian, and J.J. Stiglich, Ed., TMS-AIME, 1995
- *High-Temperature Protective Coatings,* S.C. Singhal, Ed., TMS-AIME, 1983

Properties of Nonferrous Heat-Resistant Materials

Titanium and Titanium Alloys

TITANIUM ALLOYS for use in gas turbine engines, hot sections of airframes, and other applications have been developed; they have useful strength and resist oxidation at temperatures as high as 595 °C (1100 °F). The improved elevated-temperature characteristics of these alloys combined with their high strength-to-weight ratios make them an attractive alternative to nickel-base superalloys for certain gas turbine components. Titanium alloys possess a weight reduction advantage of approximately 40% over their nickel-base counterparts.

This article reviews recent advances in high-temperature alloy development with emphasis on conventional (wrought and cast) titanium alloys. Table 1 lists titanium alloys developed for high-temperature use in the chronological order of their introduction. As described below, the compositions of these alloys were intended to increase temperature capabilities and improve other mechanical properties.

Alloy Types

Titanium exists in two crystallographic forms. At room temperature, unalloyed titanium has a hexagonal close-packed (hcp) crystal structure referred to as α (alpha) phase. At 883 °C (1621 °F), this transforms to a body-centered cubic (bcc) structure known as β (beta) phase. The manipulation of these crystallographic variations through alloying additions, heat treatment, and thermomechanical processing is the basis for the development of a wide range of alloys and properties.

Table 2 lists some of the common alloying elements in titanium and classifies them as either α or β stabilizers. Titanium alloys for high-temperature aerospace applications contain both α and β stabilizing elements in various proportions depending on the application and, therefore, the required mechanical properties. The addition of alloying elements also divides the single temperature for equilibrium transformation into two temperatures: the α transus, above which the α phase begins transformation to β, and the β transus, above which the alloy is all β. Between these temperatures, both α and β are present. Transus temperatures vary with impurity levels and the range of alloying additions.

These α and β phases also provide a convenient way to categorize titanium mill products and castings. Based on the phases present, titanium alloys can be classified as either α, near-α, α-β, or β alloys. Table 3 lists the compositions, densities, and β transus temperatures of selected titanium and titanium alloys. Typical room-temperature tensile properties for these alloys are given in Table 4. More detailed information on the physical metallurgy, processing, and properties of titanium alloys can be found in Ref 1 to 3.

Commercially pure titanium grades, which have minimum titanium contents ranging from about 98.5 to 99.5 wt% (Table 3), are used primarily for corrosion resistance. Unalloyed grades are also useful in applications requiring high ductility for fabrication but relatively low strength. Yield strengths of commercially pure grades vary from 170 to 520 MPa (25 to 75 ksi) simply as a result of variations in the interstitial and impurity levels. Oxygen and iron are the primary variants in these grades; strength increases with increasing oxygen and iron contents.

Alpha and Near-α Alloys. Alpha alloys that contain aluminum, tin, and/or zirconium are preferred for high-temperature as well as cryogenic applications. Alpha-rich alloys generally are more resistant to creep at high temperature than α-β or β alloys.

Unlike α-β and β alloys, α alloys containing no β stabilizers cannot be strengthened by heat treatment. Generally, α alloys are annealed or recrystallized to remove residual stresses induced by cold working. Alpha alloys have good weldability because they are insensitive to heat treatment. They generally have poorer forgeability and narrower forging-temperature ranges than α-β or β alloys, particularly at temperatures below the β transus.

Alpha alloys that contain small additions of β stabilizers listed in Table 2 are classified as "near-α" alloys. Although they contain some retained β phase, these alloys consist primarily of α and behave more like conventional α alloys than α-β alloys. Near-α alloys can, however, be strengthened by heat treatment or thermomechanical processing. Figure 1 shows how the room-temperature tensile and yield strengths were improved in two near-α alloys as the result of solution heat treatment and a combination of solution heat treating and hot forging. As described in the section "Development of Near-α Alloys," heat treating above the β transus temperature can also

Table 1 Upper temperature limit for titanium alloys developed for elevated-temperature applications

Alloy	Alloy type	Year of introduction	Useful maximum temperature °C	°F
Ti-6Al-4V (Ti-64)	α-β	1954	300	580
Ti-4Al-2Sn-4Mo-0.5Si (IMI-550)	α-β	1956	400	750
Ti-8Al-1Mo-1V (Ti-811)	near-α	1961	425	800
Ti-2Al-11Sn-5Zr-1Mo-0.2Si (IMI-679)	near-α	1961	450	840
Ti-6Al-2Sn-4Zr-6Mo (Ti-6246)	α-β	1966	450	840
Ti-6Al-2Sn-4Zr-2Mo (Ti-6242)	near-α	1967	450	840
Ti-3Al-6Sn-4Zr-0.5Mo-0.5Si (Hylite 65)	near-α	1967	520	970
Ti-6Al-5Zr-0.5Mo-0.25Si (IMI-685)	near-α	1969	520	970
Ti-5Al-5Sn-2Zr-2Mo-0.2Si (Ti-5522S)	near-α or α-β	1972	520	970
Ti-6Al-2Sn-1.5Zr-1Mo-0.1Si-0.3Bi (Ti-11)	near-α	1972	540	1000
Ti-6Al-2Sn-4Zr-2Mo-0.1Si (Ti-6242S)	near-α	1974	520	970
Ti-5Al-5Sn-2Zr-4Mo-0.1Si (Ti-5524S)	near-α or α-β	1976	500	930
Ti-5.5Al-3.5Sn-3Zr-0.3Mo-1Nb-0.3Si (IMI-829)	near-α	1976	580	1080
Ti-5.5Al-4Sn-4Zr-0.3Mo-1Nb-0.5Si-0.06C (IMI-834)	near-α	1984	590	1100
Ti-6Al-2.75Sn-4Zr-0.4Mo-0.45Si (Ti-1100)	near-α	1987	590	1100
Ti-15Mo-3Al-2.75Nb-0.25Si (Beta-21S)	β	1988	590	1100

Table 2 Ranges and effects of some alloying elements used in titanium

Alloying element	Range (approx), wt%	Effect on structure
Aluminum	2–8	α stabilizer
Tin	2–11	α stabilizer
Vanadium	2–20	β stabilizer
Molybdenum	2–15	β stabilizer
Chromium	2–12	β stabilizer
Copper	2–6	β stabilizer
Niobium	1–2	β stabilizer
Zirconium	2–8	β stabilizer
Silicon	0.05 to 1	Improves creep resistance

Table 3 Compositions, densities, and β transus temperatures for selected commercial purity titanium grades and high-temperature titanium alloys

Alloy	Impurity limits, wt% max: N	C	H	Fe	O	Max others (each or total)	Alloying elements(a): Al	Sn	Zr	Mo	V	Others	Density, g/cm³	β transus temperature: °C	°F
Commercially pure grades															
ASTM grade 3 (UNS R50500)	0.05	0.1	0.015	0.3	0.35	0.4	...	...	...	...	...	...	4.51	920	1685
ASTM grade 4 (UNS R50700)	0.05	0.1	0.015	0.5	0.4	0.4	...	...	...	...	...	...	4.51	950	1740
ASTM grade 7 (UNS R52400)(b)	0.03	0.1	0.015	0.3	0.25	0.4	...	...	...	...	...	0.12–0.25 Pd	4.52	913	1675
ASTM grade 12 (UNS R53400)	0.03	0.08	0.015	0.3	0.25	0.4	...	...	...	...	...	0.2–0.4 Mo, 0.6–0.9 Ni	4.52	890	1635
Near-α alloys															
Ti-8Al-1Mo-1V (UNS 54810)	0.05	0.08	0.015	0.3	0.12	0.005 Y	8.0	...	...	1.0	1.0	...	4.37	~1040	~1900
IMI-679	...	...	0.125 max	0.20 max	0.20 max	...	2.25	11.0	5.0	1.0	...	0.25 Si	4.84	950 ± 10	1740 ± 20
IMI-685	0.03 max	0.08 max	0.01 max	0.05 max	0.20 max	...	6.0	...	5.0	0.50	...	0.25 Si	4.45	1020	1870
Ti-6242S (UNS 54620)	0.05	0.05	0.0125	0.25	0.12	...	6.0	2.0	4.0	2.0	...	0.08 Si	4.54	995 ± 15	1825 ± 25
IMI-829	0.03 max	...	0.0060 max	...	0.115	...	5.6	3.5	3.0	0.25	...	0.30 Si, 1.0 Nb	4.54	1015 ± 10	1860 ± 20
IMI-834	0.03 max	0.06	0.006 max	0.05 max	0.10	...	5.8	4.0	3.5	0.5	...	0.35 Si, 0.7 Nb	4.55	1045 ± 10	1915 ± 20
Ti-1100	0.03 max	0.04 max	...	0.02 max	0.07	...	6.0	2.7	4.0	0.40	...	0.45 Si	4.5	1015	1860
α–β alloys															
Ti-6Al-4V (UNS 56400)	0.05	0.10	0.015	0.3	0.2	...	6.0	...	...	...	4.0	...	4.43	995	1825
IMI-550	0.27 N+O max	...	0.0125 max	0.2 max	0.27 O+N max	...	4.0	2.0	...	4.0	...	0.5	4.60	975 ± 10	1787 ± 18
Ti-6246 (UNS 56260)	0.04	0.04	0.0125	0.15	0.15	...	6.0	2.0	4.0	6.0	...	...	4.65	935	1715
β alloys															
β-21S (UNS 58210)	0.05 max	0.05 max	0.015 max	0.3	0.13	...	3.0	...	...	15.0	...	2.8 Nb, 0.20 Si	4.94	~793 to 810	~1460 to 1490

(a) Unless a range or maximum content is specified, valves are nominal quantities. (b) ASTM grade 11, which also contains 0.12-0.25 Pd, contains lower oxygen (0.18%) and iron (0.20) contents than does grade 7.

Table 4 Room-temperature properties for selected commercial purity titanium grades and high-temperature titanium alloys

Alloy	Condition	Tensile strength: MPa	ksi	Yield strength: MPa	ksi	Elongation, %	Reduction in area, %	Hardness
Commercially pure grades								
Grade 3	Annealed	450–517	65–75	380–448	55–65	25	45	225 HB
Grade 4	Annealed	550–662	80–96	480–586	70–85	20	40	265 HB
Grade 7	Annealed	340–434	50–63	280–345	40–50	28	50	200 HB
Grade 12	Annealed	480–517	70–75	380–448	55–65	25	42	180–235 HB
Near-α alloys								
Ti-8Al-1Mo-1V	Duplex annealed	900–1000	130–145	830–951	120–138	15	28	35 HRC
IMI-679	900 °C (1650 °F) (1 h), OQ + 500 °C (930 °F) (24 h, AC	≥1110	≥160	≥970	≥140	≥8	≥25	...
IMI-685	1050 °C (1920 °F) (30 min.), OQ + 550 °C (1020 °F) (24 h), AC	≥950	≥138	≥880	≥128	≥6	≥15	...
Ti-6242S	Duplex annealed	896–993	130–144	827–903	120–131	10	25	32–36 HRC
IMI-829	1050 °C (1920 °F) (30 min.), AC + 625 °C (1155 °F) (2 h), AC	≥930	≥135	≥820	≥119	≥9	≥15	...
IMI-834	SHT + aged (α-β processed)	≥1030	≥149	≥910	≥132	≥6	≥15	...
Ti-1100	β forged	1000	145	910	132	8	15	...
α–β alloys								
Ti-6Al-4V	SHT + aged	1172	170	1103	160	10	25	41 HRC
IMI-550	SHT + aged	1100	160	940	136	9 in 5 *D*	20–25	360 HV
Ti-6246	SHT + aged	1269	189	1172	170	10	23	36–42 HRC
β alloys								
β-21S	845 °C (1550 °F) (30 min.), AC + 540 °C (1000 °F) (8 h), AC	1331–1386	193–201	1248–1289	181–187	4–6.5	...	...

Note: OQ, oil quenched; AC, air cooled; SHT, solution heat treated

improve the creep behavior of some near-α alloys.

Alpha-β alloys contain one or more α stabilizers or α-soluble elements plus one or more β stabilizers. These alloys retain more β phase after solution treatment than do near-α alloys; the specific amount depends on the quantity of β stabilizers present and on heat treatment.

Alpha-β alloys can be strengthened by solution treating and aging. Solution treating usually is done at a temperature high in the two-phase α-β field and is followed by quenching in water, oil, or other suitable quenchant. As a result of quenching, the β phase present at the solution treating temperature may be retained or may be partly transformed during cooling by either martensitic transformation or nucleation and growth. The specific response depends on alloy

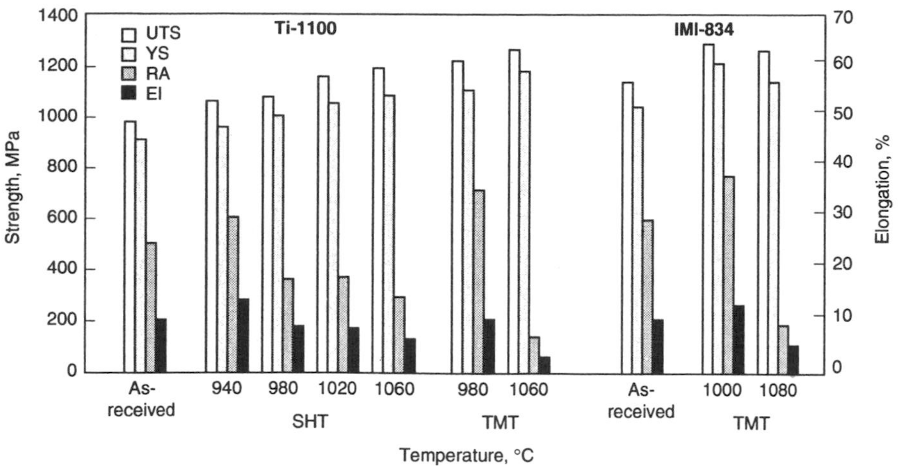

Heat-treatment conditions

Ti-1100

As-received: 206 mm (8.1 in.) bar, hot forged at 980 °C (1795 °F), air-cooled to room temperature

Solution heat treatments (SHT): specimens heated for 20 min at 940 °C (1725 °F), 980 °C (1795 °F), 1020 °C (1870 °F), and 1060 °C (1940 °F), then water quenched. Annealing followed for 4 h at 600 °C (1110 °F) plus air cool

Thermomechanical treatment (TMT): cylindrical bars were swaged in 7 steps at 950 °C (1740 °F) to a deformation of 83%, then air cooled. Afterward, one set of alloys was heat treated for 20 min at 40 °C (72 °F) below the β transus temperature (980 °C, or 1795 °F) and the other at 40 °C (72 °F) above the β transus temperature (1060 °C, or 1940 °F). All specimens were then annealed for 4 h at 600 °C (1110 °F).

IMI-834

As-received: hot-rolled to 20 mm (0.79 in.) diameter bars, solution heat treated at 1025 °C (1875 °F), then oil quenched and aged at 700 °C (1290 °F)

TMT: cylindrical bars were swaged in 7 steps at 950 °C (1740 °F) to a deformation of 74%, then air cooled. Afterward, one set of alloys was heat treated to 40 °C (72 °F) below the β transus temperature (1000 °C, or 1830 °F) and the other at 40 °C (72 °F) above the β transus temperature (1080 °C, or 1975 °F). All specimens were then annealed for 4 h at 600 °C (1110 °F).

Fig. 1 Effect of heat treatment and thermomechanical processing on room-temperature tensile properties of near-α alloys Ti-1100 and IMI-834. UTS, ultimate tensile strength; YS, yield strength; RA, reduction in area; El, elongation. Source: Ref 4

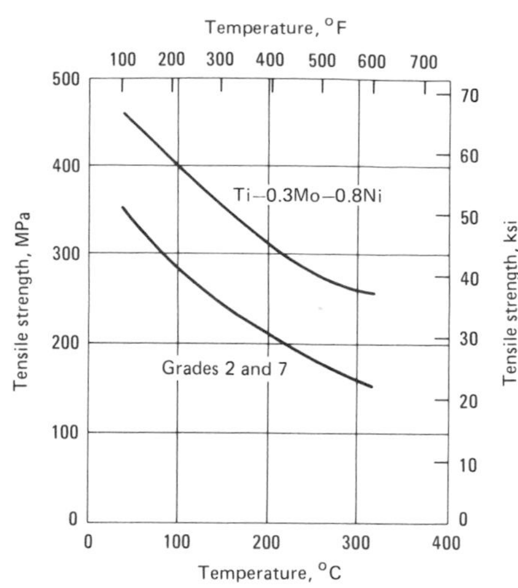

Fig. 2 Effect of temperature on minimal tensile strength of commercially pure titanium grades. Source: Ref 1

composition, solution treating temperature (β-phase composition at the solution temperature), cooling rate, and section size. Solution treatment is followed by aging, normally at 480 to 650 °C (900 to 1200 °F), to precipitate α and produce a fine mixture of α and β in the retained or transformed β phase.

Solution treating and aging can increase the strength of α-β alloys 30 to 50%, or more, over the annealed or over-aged conditions. Response to solution treating and aging depends on section size; alloys relatively low in β stabilizers (Ti-6Al-4V, for example) have poor hardenability and must be quenched rapidly to achieve significant strengthening. For Ti-6Al-4V, the cooling rate of a water quench is not rapid enough to significantly harden sections thicker than about 25 mm (1 in.). As the content of β stabilizers increases, hardenability increases; Ti-5Al-2Sn-2Zr-4Mo-4Cr, for example, can be through hardened with relatively uniform response throughout sections up to 150 mm (6 in.) thick. For some alloys of intermediate β-stabilizer content, the surface of a relatively thick section can be strengthened, but the core may be 10 to 20% lower in hardness and strength. The strength that can be achieved by heat treatment is also a function of the volume fraction of β phase present at the solution treating temperature. Alloy composition, solution temperature, and aging conditions must be carefully selected and balanced to produce the desired mechanical properties in the final product.

Beta alloys are richer in β stabilizers and leaner in α stabilizers than α-β alloys. They are characterized by high hardenability, with β phase completely retained on air cooling of thin sections or water quenching of thick sections. Beta alloys have excellent forgeability and, in sheet form, can be cold formed more readily than high-strength α-β or α alloys. After solution treating, β alloys are aged at temperatures of 450 to 650 °C (850 to 1200 °F) to partially transform the β

Table 5 Elevated-temperature properties of commercially pure titanium grades

Temperature		Ultimate tensile strength		Tensile yield strength(a)		Compressive yield strength		Shear strength		Ultimate bearing strength		Bearing yield strength(b)	
°C	°F	MPa	ksi	MPa	ksi	MPa	ksi	MPa	ksi	MPa	ksi	MPa	ksi
ASTM grade 3													
21	70	550	80	450	65	450	65	380	55	680	99	565	82
93	200	440	64	340	49	315	46	315	46	580	84	455	66
204	400	310	45	200	29	200	29	240	35	435	63	345	50
316	600	250	36	140	20	200	29	195	28	345	50	240	35
427	800	200	29	115	17	180	26	160	23	290	42	220	32
538	1000	165	24	90	13	140	20	110	16	180	26	145	21
ASTM grade 4													
21	70	630	90	550	80	550	80	450	65	830	120	695	101
93	200	495	72	415	60	385	56	380	55	700	102	560	81
204	400	345	50	250	36	250	36	390	42	530	77	390	57
316	600	275	40	165	24	240	35	220	32	415	60	295	43
427	800	220	32	145	21	220	32	185	27	345	50	270	39
538	1000	185	27	110	16	165	24	140	20	215	31	170	26

(a) 0.2% offset. (b) 2% permanent set. Source: Ref 1

Table 6 Elevated-temperature tensile properties of ASTM grade 12 commercially pure titanium

Temperature		Ultimate tensile strength		Tensile yield strength (0.2%)		Elongation,
°C	°F	MPa	ksi	MPa	ksi	%
25	77	510	74	415	60	33
205	400	345	50	250	36	37
316	600	325	47	205	30	32

Source: Ref 1

Table 7 Creep and stress-rupture data for commercial purity (CP) titanium

Alloy	Stress to 1.0% creep in 1000 h at 250 °C (480 °F)		Stress to rupture in 1000 h at 250 °C (480 °F)	
	MPa	ksi	MPa	ksi
CP grade 1	90	13	103	14
CP grade 2	103	15	117	17
CP grade 3	131	19	138	20
Ti-0.3Mo-0.8Ni, grade 12	221	32	297	43

Source: Ref 1

phase to α. The α forms as finely dispersed particles in the retained β, and strength levels comparable or superior to those of aged α-β alloys can be attained.

In the solution treated condition (100% retained β), β alloys have good ductility and toughness, relatively low strength, and excellent formability. Solution-treated β alloys begin to precipitate α phase at slightly elevated temperatures and thus are unsuitable for elevated-temperature service without prior stabilization or over-aging treatment.

Beta alloys, despite the name, actually are metastable because cold work at ambient temperature or heating to a slightly elevated temperature can cause partial transformation to α. The principal advantages of β alloys are that they have high hardenability, excellent forgeability, and good cold formability in the solution-treated condition.

Properties of Commercially Pure Titanium

The maximum service temperature of unalloyed grades of titanium for stress-free isothermal oxidation in clean air or oxygen is about 400 °C (750 °F). Modified grades containing palladium (grades 7 and 11) or molybdenum (0.3%) and nickel (0.8%) additions (grade 12) have slightly better high-temperature characteristics, including retention of strength at temperature. Tables 5 and 6 show the effects of elevated temperatures on the mechanical properties of commercially pure grades. Figure 2 compares the elevated-temperature strength of several grades. As these data indicate, unalloyed grades lose much of their strength at moderately low temperatures.

Creep can occur, not only at elevated temperatures, but at ambient temperature in commercially pure titanium as well as other metals. At ambient temperature, significant creep is encountered at relatively low percentages of 0.2% yield strength. As temperature increases, strain aging occurs and enhances creep performance. For instance, 1% creep will occur in 1000 h at room temperature for low-iron grade 2 at stresses less than 70% of the 0.2% yield strength. But at 250 °C (480 °F), the 1000 h/1% creep stress is very near the yield strength. Similar behavior exists for grades 1 and 3 commercially pure low-iron titanium. Creep data for several unalloyed grades are compared in Table 7 and Fig. 3. Creep and stress-rupture values for grades 3 and 4 are shown at temperatures ranging from 25 to 540 °C (80 to 1000 °F) in Fig. 4.

Development of Near-α Alloys

As indicated in Table 1, the majority of titanium alloys used for elevated-temperature service are the thermally stable near-α alloys. Thermal stability is the ability of alloys to retain their original mechanical properties after prolonged service at elevated temperature. An alloy is thermally unstable if it undergoes microstructural changes during use at elevated temperature that affect its properties adversely. Instability may cause either embrittlement or softening, depending on the nature of the microstructural changes. Thermal stability is measured by comparing the properties of an alloy at room temperature before or after exposure (stressed or unstressed) at elevated temperature.

The near-α alloys discussed in this section are generally stable at temperatures ranging from ~540 to 595 °C (1000 to 1100 °F) for exposure periods of 1000 h or more, except that alloys high in aluminum, such as Ti-8Al-1Mo-1V, will undergo a small, but generally tolerable, amount of hardening and some loss in ductility because of formation of $Ti_3Al(\alpha_2)$ in the microstructure.

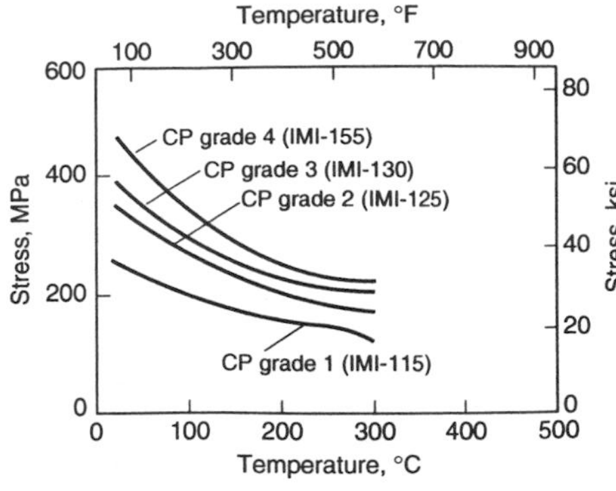

Fig. 3 Stress-rupture characteristics of various unalloyed titanium grades based on Larson-Miller interpolation on tests extending 10^4 h at 250 °C (480 °F) and 1000 h at 350 °C (660 °F). CP, commercially pure. Source: Ref 1

Elongation and reduction in area after exposure to the elevated temperature will be 10% or more.

Effects of Alloying Additions (Ref 6)

The variety in composition in near-α alloys arises in part because some alloys are designed for optimization of certain properties like short-term strength where a higher β stabilizer content is required (for example, Ti-6Al-2Sn-4Zr-6Mo [Ti-6246]) or long-term creep strength where increased α stabilizer content is required (for example, Ti-6Al-2Sn-4Zr-2Mo [Ti-6242]). Table 2 lists the common α and β stabilizers in titanium alloys.

Aluminum. The addition of aluminum, an α stabilizer, increases tensile and creep strengths and moduli while reducing alloy density. The maximum solid solution strengthening that can be achieved by aluminum is limited because about 7% Al promotes ordering and Ti_3Al formation with associated embrittlement. Thus Tables 1 and 3 show that, with one exception, the aluminum content of all alloys is below 7%.

Tin is a less potent α stabilizer and a solid-solution strengthener that is often used in conjunction with aluminum to achieve higher strength without embrittlement. It is used as the main α stabilizer in IMI-679 (2Al-11Sn-5Zr-1Mo-0.2Si). This alloy has a good combination of strength and temperature capability but higher density and lower modulus than Ti-6Al-2Sn-4Zr-2Mo-0.1Si (Ti-6242S), which uses aluminum as its main α stabilizer.

Zirconium forms a continuous solid solution with titanium and is a weak β stabilizing element. It increases strength at low and intermediate temperatures. The use of zirconium above 5 to 6% may reduce ductility and creep strength.

Oxygen content is usually kept fairly low, 0.10 to 0.15%, in this class of alloys. Although it strengthens titanium, the beneficial effects deteriorate at temperatures above 300 °C (570 °F). Increased oxygen content also tends to lower ductility, toughness, and long-term high-temperature stability.

Molybdenum is the prime β stabilizer in near-α alloys, as it markedly increases the heat treat-

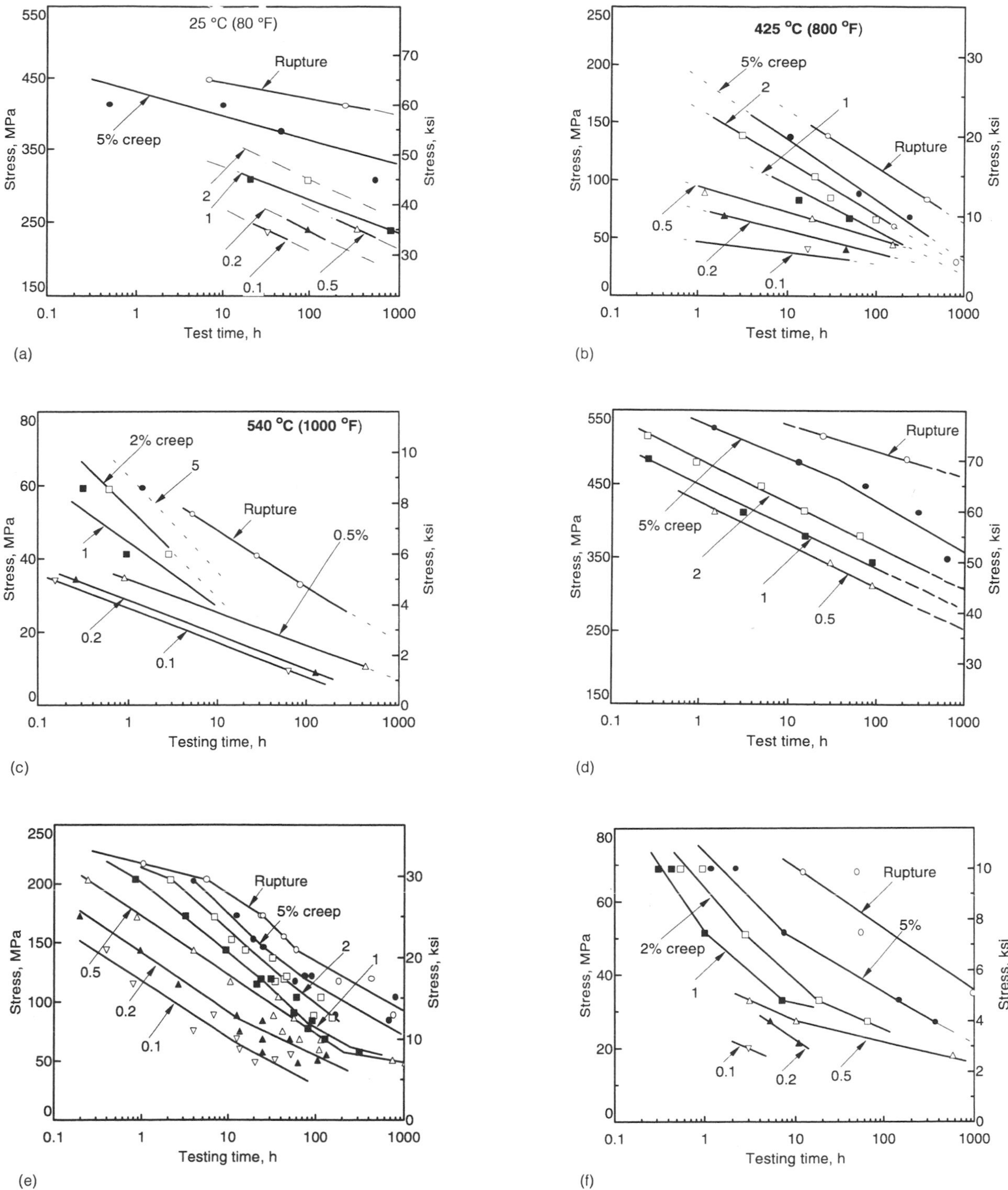

Fig. 4 Creep characteristics of commercially pure titanium grades. (a)-(c): mill-annealed grade 3 with a minimum yield strength of 380 MPa (55 ksi) at (a) 25 °C (80 °F), (b) 425 °C (800 °F), and (c) 540 °C (1000 °F). (d)-(f): Mill-annealed grade 4 with a minimum yield strength of 480 MPa (70 ksi) at (d) 25 °C (80 °F), (e) 425 °C (800 °F), and (f) 540 °C (1000 °F). Source: Ref 5

ment response of the alloys. It promotes high strength in quenched and aged material as well as increased hardenability. Alloys aimed at long-term creep strength (such as Ti-1100 and Ti-6242S in Tables 1 and 3) have lower molybdenum content. Conversely, alloys aimed at short-term high-temperature strength (such as Ti-5524S) or superior strength at lower temperature (such as Ti-6246) have greater molybdenum percentages. The addition of molybdenum makes the alloy more difficult to weld.

Niobium, which is also a β stabilizer, is added primarily to improve surface stability (oxidation resistance) during high-temperature exposure in alloys such as IMI-829.

Silicon is an important element in high-temperature titanium alloys since it increases strength at all temperatures and has a marked beneficial effect on creep resistance (Fig. 5). The alloy IMI-550 was the first commercial alloy to use silicon to improve high-temperature performance capabilities (see Table 1). More recently developed alloys (for example, IMI-834 and Ti-1100) use higher silicon contents (~0.5%). Beyond these levels, post-creep ductility (stability) is compromised with no further creep enhancement. The maximum effective concentration of

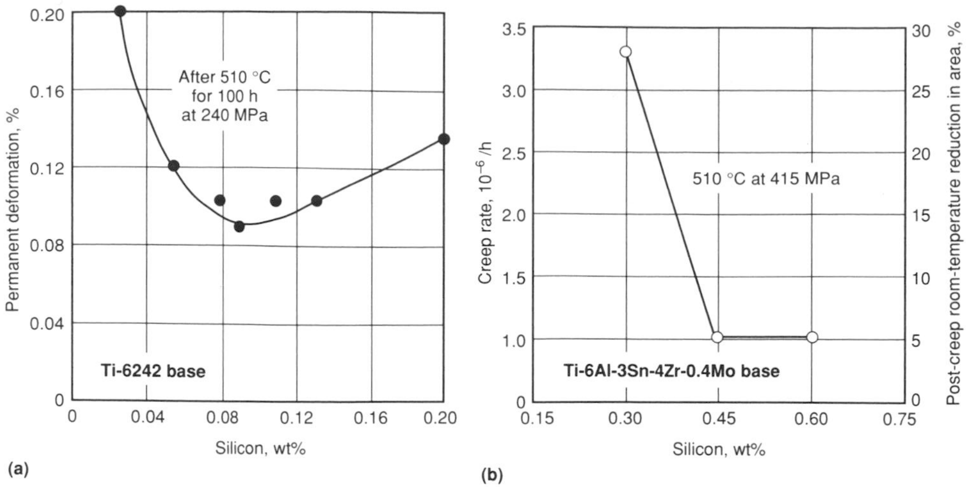

Fig. 5 Effect of silicon content on the creep behavior of (a) Ti-6Al-2Sn-4Zr-2Mo base composition and (b) Ti-6Al-3Sn-4Zr-0.4Mo base composition. Source: Ref 7 and 8

silicon in high-temperature alloys is governed by its tendency to form zirconium silicides. These are undesirable because they remove silicon from solid solution in the α phase and reduce ductility. If the local solubility is exceeded, silicides may form during solidification. More often, silicides precipitate slowly at the aging or operating temperatures for high-temperature titanium alloys, resulting in a reduction in stability.

Microstructural Effects

Because near-α alloys contain some β stabilizers, they can exhibit microstructural variations (Fig. 6) similar to that of α-β alloys. The microstructures can range from equiaxed α (Fig. 6a), when processing is performed in the α-β region, to an acicular structure (Fig. 6c) of transformed β after processing above the β transus. Near-α titanium alloys with equiaxed (Fig. 6a) or duplex (Fig. 6b) structures exhibit:

- Higher ductility and formability
- Higher threshold stress for hot-salt corrosion
- Slightly higher tensile and yield strengths (for equivalent heat treatment) at room and elevated temperatures
- Superior low-cycle fatigue properties (at room and elevated temperatures)
- Superior high-cycle fatigue properties (at room and elevated temperatures)

Near-α titanium alloys with an acicular α structure exhibit:

- Superior creep properties
- Higher fracture toughness values
- Slightly lower tensile and yield strengths (for equivalent heat treatment) at room and elevated temperatures
- Superior stress-corrosion resistance
- Lower crack-propagation rates

The relationship between microstructure and properties can be illustrated best by comparing high-temperature alloys IMI-834 and Ti-1100. As indicated in Table 1, both of these are near-α alloys that can be used at temperatures up to 595 °C (1100 °F). IMI-834, which contains equiaxed grains of primary α in a matrix of transformed β containing fine acicular α (duplex structure similar to Fig. 6b), is α-β processed as follows:

- Beta transus of 1055 °C (1930 °F)
- Alpha-β forged 98% at 1020 °C (1870 °F)

Table 8 Creep property comparison of near-α alloys Ti-1100 and IMI-834

Alloy	Minimum creep rate, 10^{-4}/h	Time to 0.2% creep strain, h	Reduction in area after 500 h creep, %
Creep exposure: 510 °C (414 MPa) or 950 °F (60 ksi)			
IMI-834 (α-β)	3.7	330	...
Ti-1100 (β)	0.5	4010	12.5
IMI-834 (β)	0.5	920	...
Creep exposure: 565 °C (275 MPa) or 1050 °F (40 ksi)			
IMI-834 (α-β)	8.4	140	11.5
Ti-1100 (β)	1.6	850	7.6
IMI-834 (β)	2.9	520	3.1
Creep exposure: 595 °C (165 MPa) or 1100 °F (24 ksi)			
IMI-834 (α-β)	2.4	610	12.2
Ti-1100 (β)	1.0	2410	6.0
IMI-834 (β)	1.1	1600	2.1

Note: α-β is alpha-beta processed (duplex structure); β is beta processed (acicular structure). See text for description of alloy conditions. Source: Ref 9

Table 9 Fracture toughness comparison of near-alpha alloys Ti-1100 and IMI-834

Alloy	Test condition	Fracture toughness, K_{Ic} MPa$\sqrt{m}$	Fracture toughness, K_{Ic} ksi$\sqrt{in.}$
IMI-834	No exposure	37.9	34.5
	300 h at 650 °C (1200 °F)	30.5	27.8(a)
Ti-1100	No exposure	62.8	57.2
	300 h at 650 °C (1200 °F)	43.4	39.5

Note: See text for description of alloy conditions. (a) K_Q value. Source: Ref 9

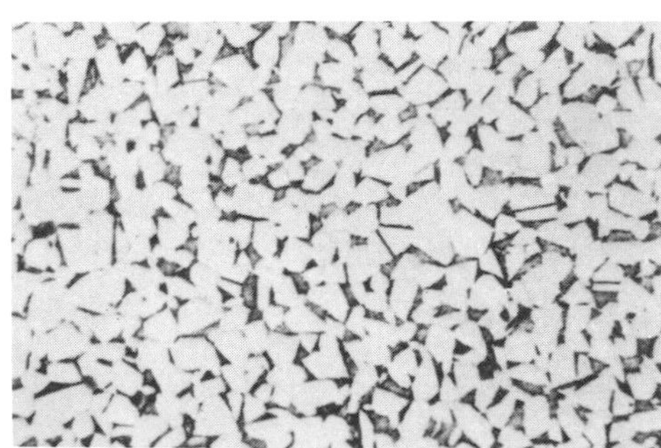

(a) Forged with a starting temperature of 900 °C (1650 °F), which is below the normal temperature range for forging Ti-8Al-1Mo-1V

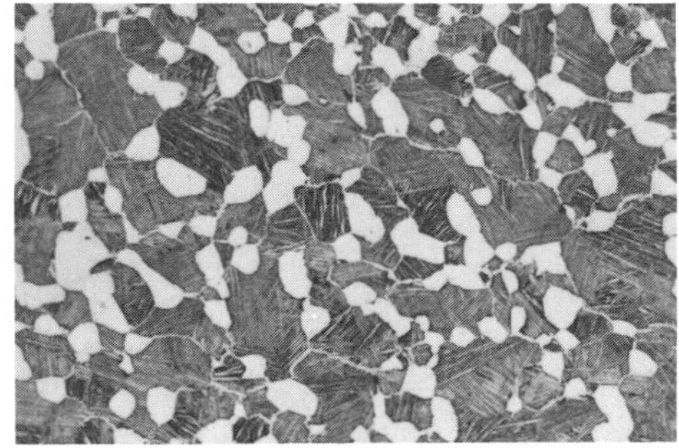

(b) Forged with a starting temperature of 1005 °C (1840 °F), which is within the normal range, and air cooled

(c) Forged with a starting temperature of 1095 °C (2000 °F), which is above the β transus temperature, and rapidly air cooled after finish forging

Fig. 6 Microstructures of near-α alloy Ti-8Al-1Mo-1V after forging with different starting temperatures. (a) Equiaxed α grains (light) in a matrix of α and β (dark). (b) Equiaxed grains of primary α (light) in a matrix of transformed β (dark) containing fine acicular α. (c) Transformed β containing coarse and fine acicular α (light). Etchant: Kroll's reagent (ASTM 192). All micrographs at 250×

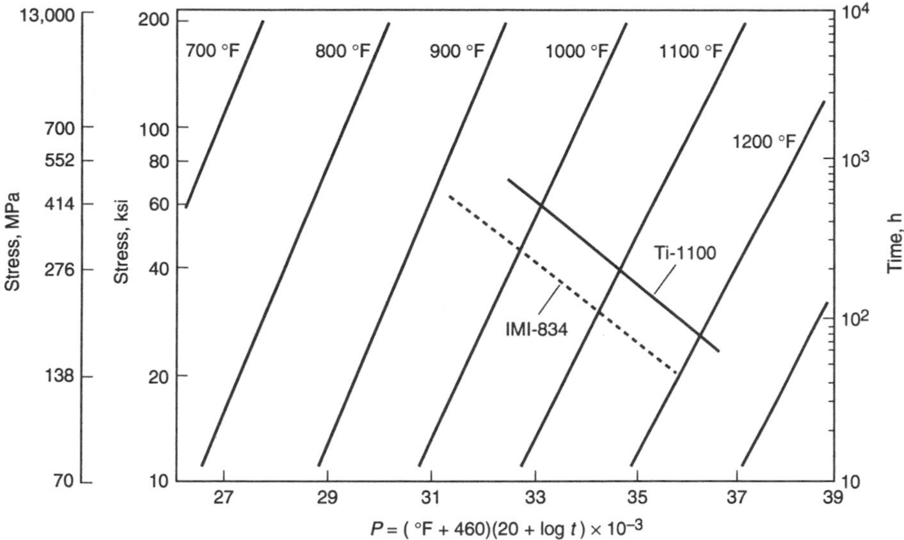

Fig. 7 Stress-rupture comparison of near-α alloys Ti-1100 and IMI-834. Source: Ref 9

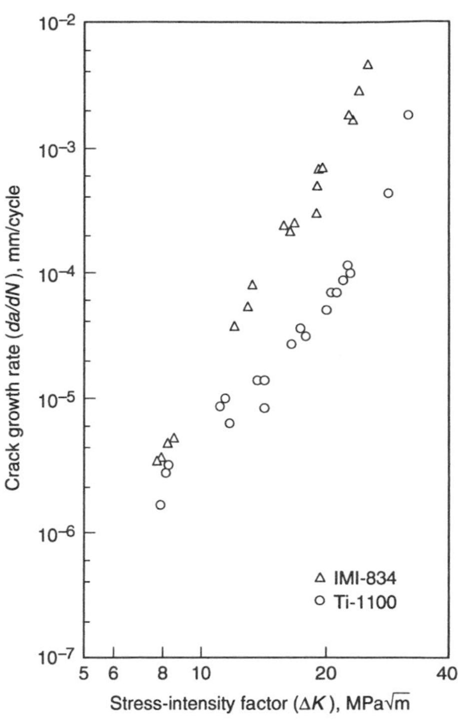

Fig. 8 Fatigue crack growth rate comparison of near-α alloys Ti-1100 and IMI-834. Exposure: 300 h at 650 °C (1200 °F). Stress ratio, R, equals 0.1. Source: Ref 9

Table 10 Tensile property comparison of near-α alloys Ti-1100 and IMI-834

Alloy	Ultimate tensile strength		Yield strength		Elongation, %	Reduction in area, %
	MPa	ksi	MPa	ksi		
Room temperature						
IMI-834	1043	151.2	936	135.8	13.0	19.8
Ti-1100	992	143.8	891	129.2	8.3	12.8
205 °C (400 °F)						
IMI-834	883	128.0	735	106.6	13.8	25.8
Ti-1100	827	119.9	683	99.1	9.3	17.1
595 °C (1100 °F)						
IMI-834	694	100.7	568	82.4	23.3	56.4
Ti-1100	652	94.6	565	82.0	6.1	17.3

Note: See text for description of alloy conditions. Source: Ref 9

- Solution treated at 1030 °C (1885 °F)/2 h/oil quenched
- Aged at 730 °C (1350 °F)/2 h/ air cooled

Ti-1100, which contains acicular α (similar to Fig. 6c), is β processed as follows:

- Beta transus of 1015 °C (1860 °F)
- Beta forged 110% at 1095 °C (2000 °F)/furnace air cooled
- Beta annealed (stabilized) at 595 °C (1100 °F)/8 h

In terms of properties, Ti-1100 has an advantage in creep (Fig. 7 and Table 8), toughness (Table 9), and crack growth (Fig. 8); IMI-834 has an advantage in strength (Table 10), low-cycle fatigue (Fig. 9), and high-cycle fatigue (Fig. 10). Microstructure does not play a key role in the oxidation behaviors of both alloys. As shown in Fig. 11, their oxidation characteristics are very similar for the temperature range of anticipated application. The superiority of IMI-834 at temperatures above 700 °C (1290 °F) is attributable to its niobium (1 wt%) content.

Properties of Near-α Alloys

As indicated in Table 1, a number of near-α high-temperature titanium alloys have been developed since the early 1960s. Some of the more commonly employed (commercially important) alloys as well as representative data are described

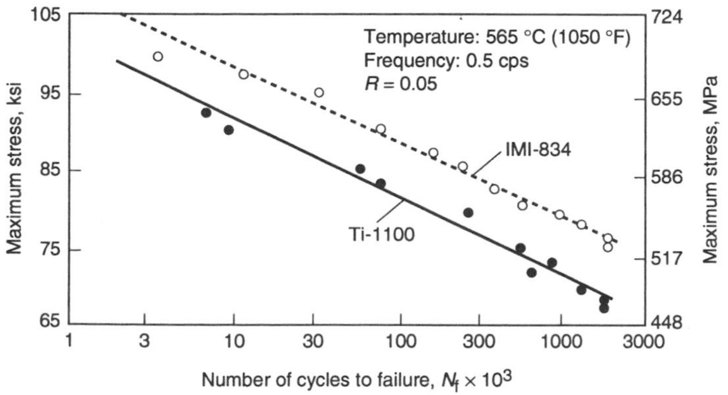

Fig. 9 Comparison of smooth-specimen low-cycle fatigue behavior of near-alpha alloys Ti-1100 and IMI-834. cps, cycles per second. Source: Ref 9

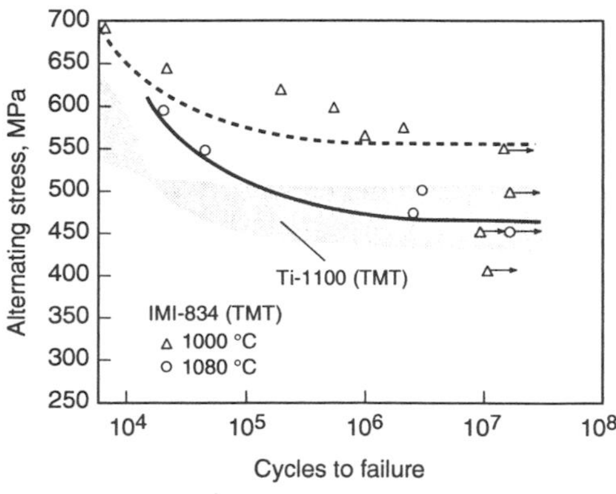

Fig. 10 High-cycle fatigue comparison of near-α alloys Ti-1100 and IMI-834. Test temperature: 400 °C (750 °F). See table accompanying Fig. 1 for processing details. TMT, thermomechanical treatment. Source: Ref 4

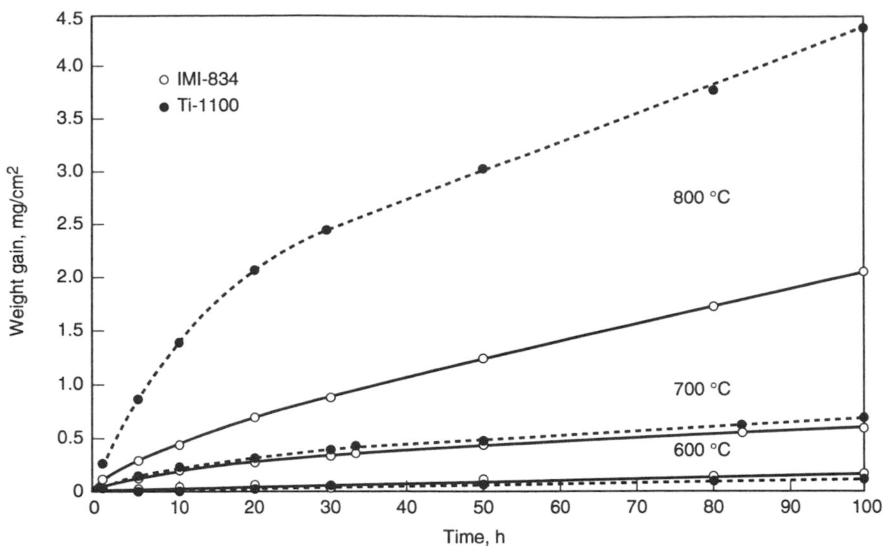

Fig. 11 Oxidation weight-gain curves of Ti-1100 and IMI-834 determined at 600, 700, and 800 °C in laboratory air. Source: Ref 9

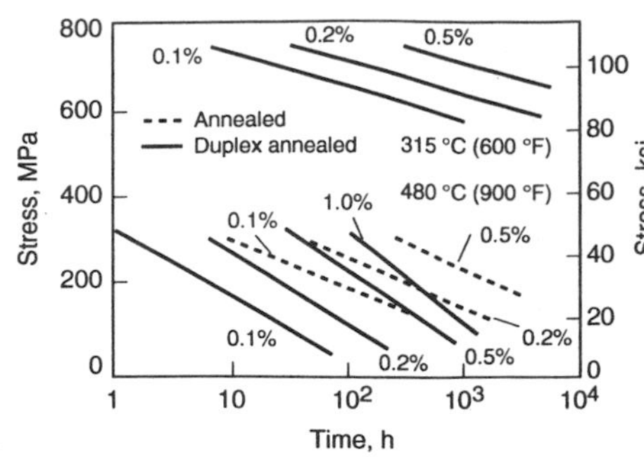

Fig. 12 Creep strength of Ti-8Al-1Mo-1V at 315 °C (600 °F) and 480 °C (900 °F). Mill annealed 1.3 mm (0.050 in.) thick sheet was heated at 790 °C (1450 °F) for 8 h; duplex annealed 1.3 mm (0.050 in.) thick sheet was heated at 1010 °C (1850 °F) for 5 min, air cooled, and held at 745 °C (1375 °F) for 15 min, air cooled. Source: Ref 1

below. More detailed information on the properties of these alloys can be found in Ref 1.

Ti-8Al-1Mo-1V (Ti-811) was one of the first near-α alloys developed for high-temperature gas turbine engine applications, specifically compressor blades and wheels. It exhibits good creep resistance at temperatures up to about 425 °C (800 °F). Elevated-temperature tensile data for Ti-811 are given in Table 11. Creep data are shown in Fig. 12.

Ti-11Sn-5Zr-2.25Al-1Mo-0.25Si (IMI-679) has a maximum use temperature of 450 °C (840 °F). Its creep strength is superior to that of Ti-811, but inferior to that of Ti-6Al-2Sn-4Zr-2Mo at temperatures above 480 °C (900 °F). This alloy has largely been superseded by Ti-6242S described below.

Ti-6Al-5Zr-0.5Mo-0.25Si (IMI-685) was the first near-α alloy developed with high-temperature stability exceeding 500 °C (930 °F). As shown in Fig. 13, its creep lies between that of Ti-6242S and IMI-834.

Ti-6Al-2Sn-4Zr-2Mo-0.08Si (Ti-6242S) has an outstanding combination of tensile strength, creep strength, toughness, and high-temperature stability. It has been reported that 1000-h exposure up to 540 °C (1000 °F) without load has no deleterious effect on subsequent room-temperature tensile properties of duplex annealed sheet. In contrast, 1000-h exposure with load produces a serious loss in ductility at 480 °C (900 °F), 137 MPa (20 ksi) for duplex annealed sheet and at 540 °C (1000 °F), 69 MPa (10 ksi) for triplex annealed sheet. At 510 °C (950 °F), 103 MPa (15 ksi), no damage is done to triplex annealed sheet for 1000-h exposure. Bar in the recommended duplex annealed condition exposed 1000 h with load at temperatures up to 540 °C (1000 °F) exhibits no significant instability (see Table 12). However, losses in ductility are observed for forgings in the recommended duplex annealed condition at exposure temperatures as low as 425 °C (800 °F). Creep data for Ti-6242S are given in Fig. 13 and 14.

Ti-5Al-3.5Sn-3.0Zr-1Nb-0.3Si (IMI-829) is regarded as having good creep performance up to around 550 °C (1020 °F) and somewhat higher for short-time applications. At 540 °C (1000 °F), a total plastic strain of less than 0.1% in 100 h is achieved under a stress of about 300 MPa (43.5 ksi). Creep data for IMI-829 are given in Fig. 13.

Table 11 Elevated-temperature tensile properties of near-α alloy Ti-8Al-1Mo-1V

Test temperature		Ultimate tensile strength		Tensile yield strength (0.2% offset)		Elongation in 50 mm (2 in.), %	Reduction of area, %
°C	°F	MPa	ksi	MPa	ksi		
RT	RT	1020	148	896	130	13	19
93	200	951	138	793	115	13	20
205	400	855	124	668	97	12	21
315	600	779	113	586	85	11	23
425	800	724	105	551	80	10	26
540	1000	655	95	503	73	14	30

Note: 13 mm (½ in.) round specimens, rolled from 1065 °C (1950 °F), annealed 1 h, at 1065 °C (1950 °F), air cooled, reheated 8 h at 595 °C (1100 °F). Source: Ref 1

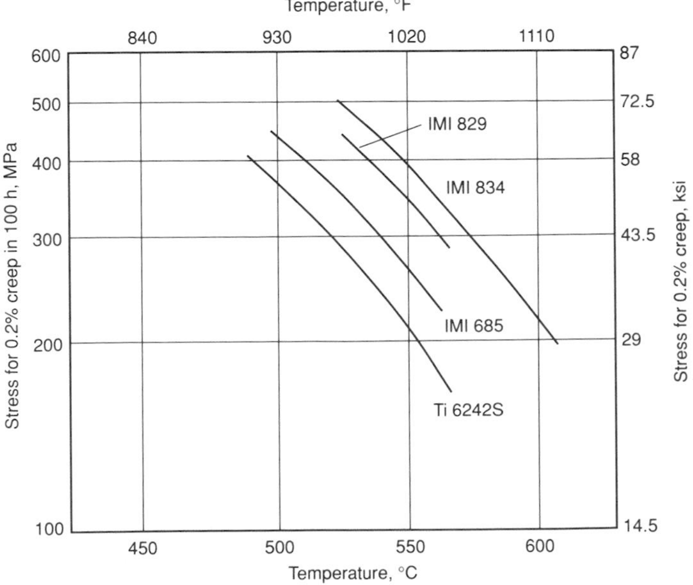

Fig. 13 Creep strength comparison of various near-α alloys. Source: Ref 3

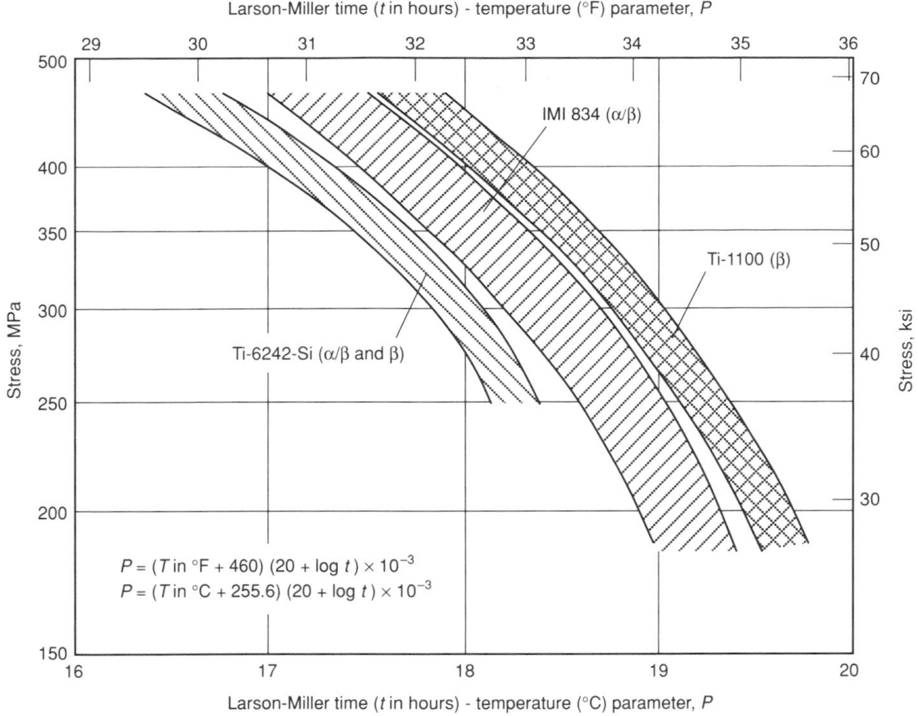

Fig. 14 Creep properties of Ti-6Al-2Sn-4Zr-2Mo(Si), IMI-834, and Ti-1100 alloys with α-β or β processing as indicated. Source: Ref 10

Ti-5.8Al-4Sn-3.5Zr-0.7Nb-0.5Mo-0.35Si (IMI-834) combines high fatigue resistance and temperature capability up to about 595 °C (1100 °F). Property data for IMI-834 are given in Tables 8 and 9 and Fig. 7 to 11, 13, and 14.

Ti-6Al-2.75Sn-4Zr-0.4Mo-0.45Si (Ti-1100) is the most creep-resistant of all titanium alloys. It was developed to offer an approximately 40 °C (75 °F) improvement in creep, as measured on a Larson-Miller time to 0.2% creep deformation plot, over Ti-6242S. Property data for Ti-1100 are given in Tables 8 and 9 and Fig. 7 to 11, 14.

Properties of α–β Alloys

The stability of commercial α-β alloys depends on composition and heat treatment. In the mill-annealed condition, the alloys may be considered stable up to 315 to 370 °C (600 to 700 °F), although measurable changes in properties will usually accompany exposure to stress and temperature for longer times. Properly fabricated and heat treated, these alloys are generally stable up to about 450 °C (840 °F) in the heat treated condition for periods of 1000 h or more.

Ti-6Al-4V (Ti-64), which was the first high-temperature titanium alloy developed (see Table 1), is the most widely used alloy in the titanium family (it accounts for more than 50% of all titanium tonnage in the world). Ti-6Al-4V is used up to about 300 to 350 °C (570 to 660 °F). As shown in Fig. 15, the creep properties of Ti-6Al-4V are inferior to the α and near-α compositions.

Ti-4Al-4Mo-2Sn-0.5Si (IMI-550) is regarded as having useful creep performance up to about 400 °C (750 °F) giving less than 0.1% total plastic strain in 100 h at 465 MPa (67 ksi). Figure 16 compares the elevated-temperature properties of IMI-550 and Ti-6Al-4V.

Ti-6Al-2Sn-4Zr-6Mo (Ti-6246) is a heat-treatable α-β alloy designed to combine the long-term, elevated-temperature strength properties of near-α Ti-6242S with much-improved short-term strength properties of a fully hardened α-β alloy. It is used for forgings in intermediate-temperature sections of gas turbine engines, particularly in compressor disks and fan blades. This alloy is used at lower temperature than Ti-6242S,

Table 12 Effects of elevated-temperature exposure on the properties of near-α alloy Ti-6242S

1000 h exposure					Subsequent RT tensile properties					
Temperature		Stress		Deformation,	Ultimate tensile strength		Tensile yield strength		Elongation in 25 mm	Reduction of area,
°C	°F	MPa	ksi	%	MPa	ksi	MPa	ksi	(1 in.), %	%
900 °C (1650 °F), 1 h, AC + 595 °C (1100 °F), 8 h, AC										
None					944	137.0	832	120.7	15.5	31.2
425	800	379	55	0.11	1016	147.4	912	132.3	15.5	34.3
455	850	344	50	0.15	990	143.6	912	132.3	18.5	34.8
480	900	275	40	0.17	982	142.4	890	129.1	17.5	28.9
540	1000	103	15	0.12	979	142.1	904	131.2	15.0	31.6
955 °C (1750 °F), 1 h, AC + 595 °C (1100 °F), 24 h, AC										
None					962	139.5	857	124.3	17.5	32.8
425	800	448	65	0.22	942	136.7	849	123.2	18.0	28.2
455	850	344	50	0.14	959	139.2	862	125.0	20.5	32.2
		379	55	0.18	941	136.5	846	122.7	18.0	30.2
480	900	275	40	0.22	942	136.6	850	123.3	19.0	30.2
		310	45	0.23	952	138.1	865	125.5	19.5	33.5
540	1000	138	20	0.17	957	138.9	888	128.9	16.5	30.8
1025 °C (1875 °F), 1 h, AC + 595 °C (1100 °F), 24 h, AC										
None					954	138.4	834	121.0	15.0	21.2
425	800	448	65	0.19	955	138.5	869	126.1	13.5	23.4
455	850	344	55	0.14	962	139.6	867	125.8	14.5	21.9
480	900	310	45	0.16	964	139.8	873	126.7	14.5	23.4
540	1000	138	20	0.10	977	141.7	894	129.7	12.5	18.3

Note: Specimens were 57 mm (2.25 in.) diameter rolled bar duplex annealed in full sections as indicated. Subsequent tensile properties were measured in the outside longitudinal direction and were tested after exposure without surface conditioning. Source: Ref 1

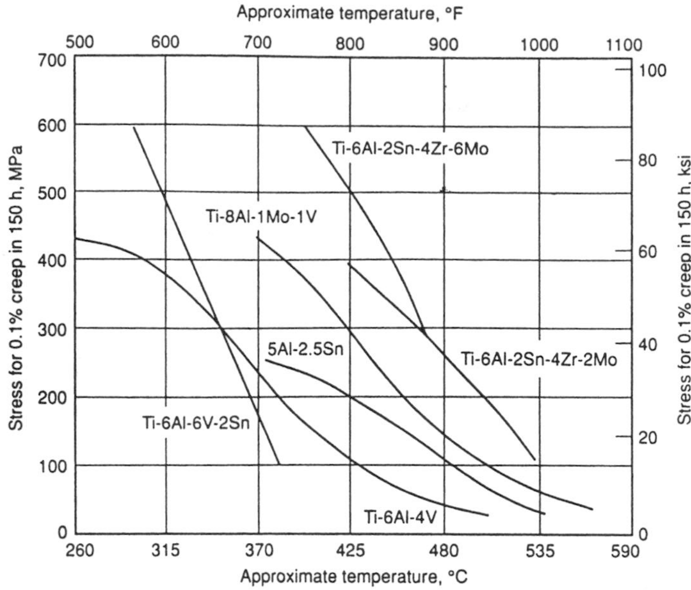

Fig. 15 Comparison of creep strengths for α (Ti-5Al-2.5Sn), near-α (Ti-8Al-1Mo-1V and Ti-6Al-2Sn-4Zr-2Mo), and α-β (Ti-6Al-4V and Ti-6Al-2Sn-4Zr-6Mo) alloys. Source: Ref 3

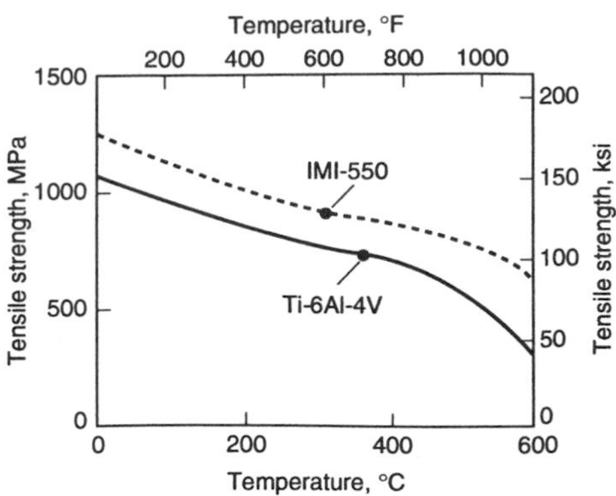

Fig. 16 Tensile strength versus temperature comparison for α-β alloys IMI-550 and Ti-6Al-4V. Source: Ref 1

but should be considered for long-term load-carrying applications at temperatures up to 400 °C (750 °F) and short-term load-carrying applications at temperatures up to 540 °C (1000 °F). Creep data for Ti-6246 are shown in Fig. 15.

Properties of β Alloys

In general, metastable β titanium alloys are cold rollable (therefore strip and foil producible) and have excellent formability, weldability, and age-hardening characteristics. However, their limitation is often temperature capability, generally due to poor creep resistance and oxidation behavior. However, a recently developed metastable β alloy, designated β-21S, circumvents these common shortcomings of metastable β alloys. The β-21S alloy (Ti-15Mo-2.7Nb-3Al-0.2Si) is so named because it is a β alloy containing 21% alloying additions, and it contains Si.

Creep Resistance. Figure 17 shows a Larson-Miller 0.2% creep strain plot for β-21S as well as several other common titanium alloys. In general, metastable β alloys fall to the left (lower creep resistance) of the Ti-6Al-4V band (see, for example, Ti-15-3 in Fig. 17). However, β-21S is to the right of the Ti-6Al-4V curve, indicating an approximately 30 °C (50 °F) advantage in creep over Ti-6Al-4V. The near-α alloys Ti-6242S or Ti-1100 are, however, more creep resistant than β-21S.

Stability. Table 13 shows that for unaged (solution annealed) material, long-term exposures as low as 260 °C (500 °F) can lead to very low ductility. If the alloy is cold worked, an even lower temperature can be expected to promote embrittlement. Thus unaged material should not be used in applications exceeding 200 °C (390 °F).

In the fully aged condition, Table 13 shows that the situation is quite different. Thermal exposure at 540 °C (1000 °F), augmented by an applied stress of 210 MPa (30 ksi), produced only moderate strengthening and virtually no change in ductility after up to 1000 h of exposure. Clearly, if the alloy is to be used in warm or hot service, it should be fully aged prior to service.

Oxidation Resistance. Table 14 provides a comparison of the oxidation resistance of β-21S versus other titanium alloys and some common nickel-base alloys. Clearly, β-21S provides superior oxidation resistance when compared to β alloy Ti-15V-3Al-3Cr-3Sn and commercially pure titanium. However, β-21S is not as resistant as α-2 or γ (gamma) titanium aluminide materials and certainly not as good as the nickel-base alloys René 41 or IN718. In fact, the β-21S weight gain at 650 °C (1200 °F) is approximately equivalent to the nickel-base materials at 815 °C (1500 °F). Thus while β-21S is exceptionally good for a β titanium alloy, it is not competitive with nickel-base alloys at the higher temperatures.

Corrosion Resistance. Beta-21S is very resistant to corrosion and hydrogen absorption in hot reducing acids. This corrosion behavior led to testing of the alloy in commercial aircraft hydraulic fluid. Prior to the development of β-21S, use of titanium and titanium alloys for certain warm (205 to 370 °C, or 400 to 700 °F) engine components on commercial aircraft has been limited because of the risk of catastrophic failure of titanium components due to corrosion and hydrogen embrittlement caused by contact with hydraulic fluid. These phosphate-based fluids (such as "Skydrol") can decompose above about 175 °C (350 °F) to form concentrated phosphoric acid solutions. Hot reducing acids such as phosphoric acid are very aggressive on most titanium alloys.

Tests were carried out on various titanium alloys and nickel-base Alloy 625 in hydraulic fluid at temperatures ranging from 230 to 315 °C (450 to 600 °F). As shown in Fig. 18, β-21S exhibited

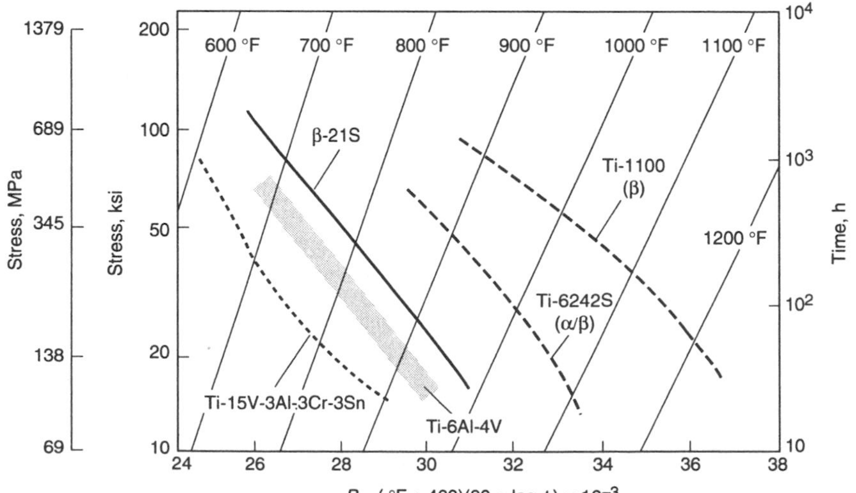

Fig. 17 Creep-rupture comparison of β-21S with other β, α-β, and near-α alloys. Source: Ref 11

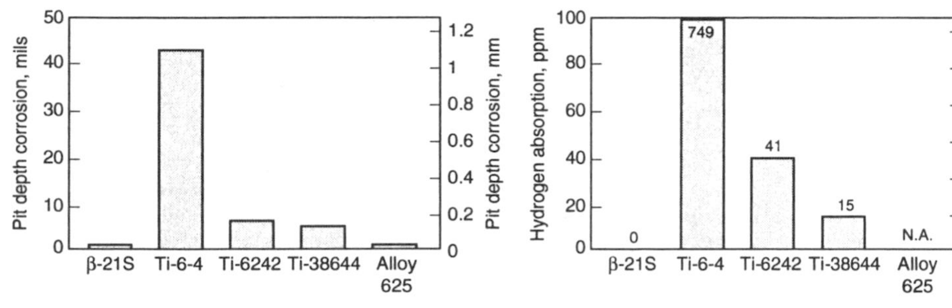

Alloy	Composition, wt% Al	Cr	Fe	Mo	Ni	V	Other	Ti
β-21S	3.0		0.15	15.0			2.7Nb,0.2Si	Balance
Ti-6-4	6.0		0.2			4.0		Balance
Ti-6242	6.0		0.2	2.0			2.0Sn,4.0Zr	Balance
Ti-38644	3.0	6.0	0.2	4.0		8.0	4.0Zr	Balance
Alloy 625		22.0	5.0	9.0	Balance		3.5Nb	...

Fig. 18 Effect of aircraft hydraulic fluid on the corrosion of various titanium alloys and nickel-base Alloy 625. Test conditions: 96 h drip test (one drip every 3 min); specimens at 45° incline; temperature range of 230 to 315 °C (450 to 600 °F); maximum attack displayed. Source: Ref 13

only minimal pitting and virtually no susceptibility toward hydrogen absorption when exposed to intermittent (dropwise) contact with aircraft hydraulic fluid. Under the test conditions described in Fig. 18, the performance of β-21S was equivalent to Alloy 625. However, Alloy 625 exhibits greater tolerance under certain conditions involving immersion.

Advanced Titanium Alloys and Coating Systems

As has been described in this article, the current temperature limit of conventionally processed wrought and cast titanium alloys is 595 °C (1100 °F). This limit is due mainly to long-term surface and bulk metallurgical stability problems, though creep strength obviously decreases continuously with temperature. In order to further improve this limit to 700 °C (1290 °F) and above, either new alloy systems or advanced coatings must be developed. In the paragraphs that follow, seven approaches for improving elevated-temperature performance are described. These include:

Advanced titanium alloys

- Titanium-matrix composites
- Titanium aluminides
- Titanium powder metallurgy alloys

Advanced coating systems

- Noble metal coatings
- Aluminized coatings
- Ion-implanted titanium alloys
- Sol-gel coatings

Titanium-matrix composites containing continuous silicon carbide (SiC) fiber reinforcement have been developed to extend the elevated-temperature performance of titanium and its alloys. One commercially available type, the SCS-6 fiber, has a 140 μm diameter, a 33 μm carbon core, and a carbon-rich surface (Fig. 19). SiC fibers generally constitute 35 to 40 vol% of this composite.

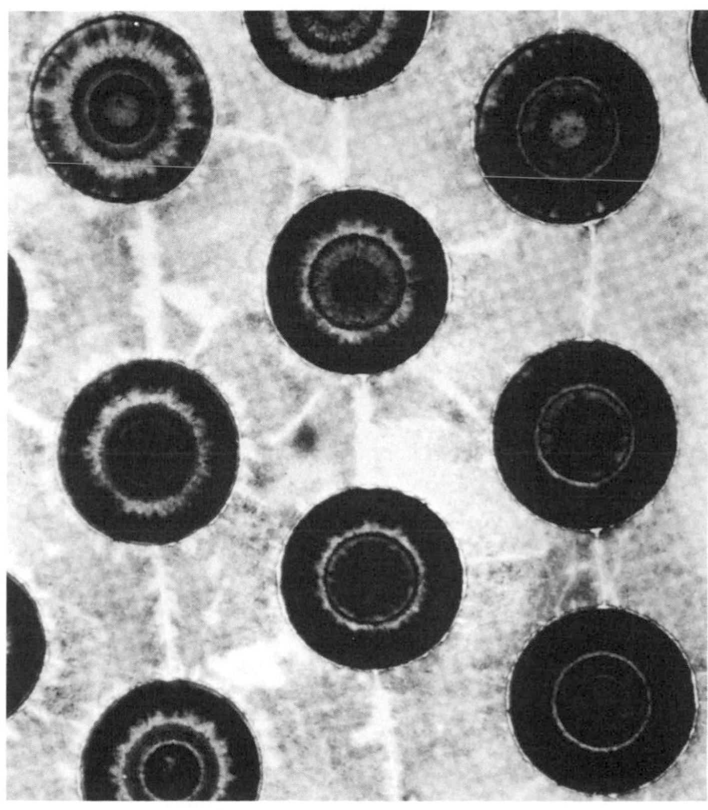

(a)

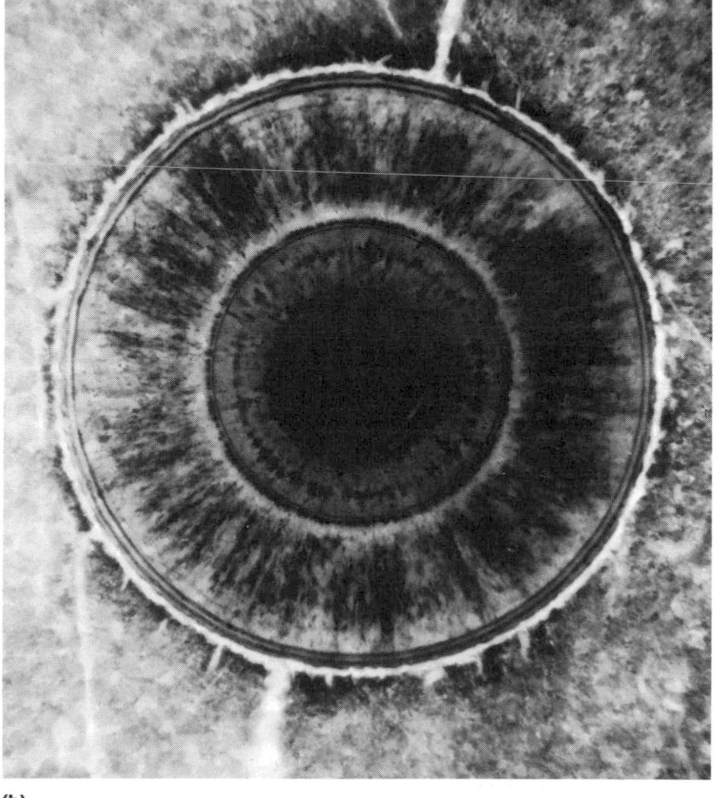

(b)

Fig. 19 Continuous-fiber-reinforced titanium-matrix MMCs. (a) Hot-pressed SiC fibers (SCS-6, 35 vol%) in a Ti-6Al-4V matrix. Fiber thickness, 140 μm; density, 3.86 g/cm^3. (b) Chemical vapor deposited SiC fiber (SCS-6) showing the central carbon monofilament substrate and the carbon-rich surface. Fiber properties: thickness, 140 μm; tensile strength, 3450 MPa (500 ksi); modulus of elasticity, 400 GPa (58 × 10^6 psi); density, 3.0 g/cm^3

Table 13 Effect of elevated-temperature exposure on the bulk stability of β-21S

Condition	Exposure	Tensile strength MPa	Tensile strength ksi	Yield strength MPa	Yield strength ksi	Elongation, %
β annealed	None	875	127	855	124	16.0
	205 °C (400 °F)/1000 h	950	138	940	136	12.0
	260 °C (500 °F)/1000 h	1105	160	1105	160	2.0
Aged 595 °C (1100 °F)/8 h	None	1060	154	965	140	12.5
	540 °C (1000 °F)/210 MPa (30 ksi)/500 h	1140	165	1050	152	11.5
	540 °C (1000 °F)/210 MPa (30 ksi)/1000 h	1180	171	1075	156	14.5

Source: Ref 12

Table 14 Oxidation resistance comparison of various titanium alloys, titanium intermetallics, and nickel-base alloys

Alloy	96 h exposure temperature °C	96 h exposure temperature °F	Weight gain, mg/cm^2
Ti-15V-3Al-3Cr-3Sn	650	1200	3.18
	815	1500	174.0
ASTM grade 2	650	1200	0.53
	815	1500	26.1
β-21S	650	1200	0.11
	815	1500	2.78
Ti-14Al-20Nb (α-2)	815	1500	0.73
Ti-33Al-1V (γ)	815	1500	0.78
René 41	815	1500	0.32
IN-718	815	1500	0.14

Note: Test specimens were 1.5-mm (0.06-in.) thick coupons. Source: Ref 12

Matrix materials used to date include Ti-6Al-4V, Ti-15V-3Sn-3Cr-3Al, and β-21S. The latter alloy, which can withstand temperatures as high as 800 °C (1500 °F) when reinforced with SiC, has been considered for use on Boeing 777 engine nacelles. Typical elevated-temperature properties of β-21S/SCS-6 fiber-reinforced four-ply unidirectional laminates are (Ref 14):

Temperature	Tensile strength
20 °C (70 °F)	1724 MPa (250 ksi)
650 °C (1000 °F)	1138 MPa (165 ksi)
815 °C (1500 °F)	517 MPa (75 ksi)

A number of processing techniques have been evaluated for titanium metal-matrix composites (MMC), but only high-temperature/short-time roll bonding, hot isostatic pressing, and vacuum hot pressing have been used to any substantial degree. Plasma spraying is used to deposit a titanium matrix onto the fibers.

Titanium Aluminides. New classes of materials based on titanium-aluminide intermetallics have recently emerged. These materials have essentially the same density as titanium but can be used at much higher temperatures. The three alloy systems that have emerged as primary candidates for structural applications are Ti_3Al-based (designated α-2), TiAl-based (designated γ), and Ti_2AlNb-based orthorhombic intermetallics. The potential service temperatures for these materials is expected to range from 600 to 760 °C (1110 to 1400 °F). The α-2 alloys typically contain 23 to 25 at.% Al and 11 to 18 at.% Nb. Other alloying elements include vanadium (up to 3 at.%) and molybdenum (~1 at.%). The γ alloys contain 48 to 54 at.% Al and 1 to 10 at.% of one of the following: vanadium, chromium, manganese, niobium, tantalum, tungsten, and molybdenum. The orthorhombic alloys typically contain 21 to 25 at.% Al and 25 to 30 at.% Nb. A property comparison of titanium aluminides with conventional titanium alloys is given in Table 15.

Aluminides, however, are difficult to process and fabricate into structural components because they have limited ductility and toughness at lower temperature ranges and therefore require very high processing temperatures. These drawbacks are significant when working aluminides into product forms that require a large amount of deformation, such as sheet for honeycomb-panel or truss-panel cores in aircraft/aerospace structures. Detailed information on titanium aluminides can be found in the article "Structural Intermetallics" in this Volume.

Powder Metallurgy Alloys. Powder metallurgy (P/M) production techniques such as rapid solidification processing, mechanical alloying, and blended elemental powder processing are being used to produce titanium with compositions that would be impossible to achieve through conventional processing.

Work on rapid solidification has focused on increasing the temperature capability of both conventional and intermetallic alloys by dispersion strengthening (Ref 15, 16). Dispersoids that are being studied include rare-earth additions, yttrium, erbium, neodymium, and gadolinium and elements such as carbon and boron. The rapid solidification technique has two major drawbacks. The first is that it has not been possible to produce more than about 6 vol% of second-phase particles. The second drawback is that rapid solidification results in a β grain size that is much smaller than is desirable and that there is a lack of elongated α phase, which would be formed on cooling into the α-β phase field after a β anneal. Unfortunately, a β anneal, which corrects the second drawback, results in unacceptable coarsening of the dispersoids.

Mechanical alloying of titanium alloys is at an early stage of development, but it exhibits the potential to increase the volume percentage of dispersoids for elevated-temperature applications (Ref 17, 18). In addition, there are indications that the normally immiscible titanium and magnesium can be combined by mechanical alloying to produce a low-density titanium alloy.

Preliminary results on the mechanical alloying of titanium-magnesium and titanium-eutectoid formers such as nickel and copper suggest a very fine nanoscale microstructure (about 10^{-9} m scale) can be obtained; such a microstructure could have novel physical and mechanical properties.

Particulate-reinforced titanium-matrix composites can also be produced by P/M processing (Ref 1). This family of materials includes a choice of several ceramic or intermetallic additions (for example, TiC, TiB_2, or TiAl) at various loading levels (for example, 10 to 20 wt%) in a select titanium alloy matrix such as Ti-6Al-4V. These materials are produced by blended elemental P/M processing to near-net shape by cold isostatic pressing and can be further consolidated/refined by forging or extrusion of the powder preform. Compared to unreinforced titanium alloys, particulate-reinforced materials offer improved tensile strength and elastic modulus, both at room and elevated temperature, thereby increasing the use temperature at approximately the same density.

Noble metal coatings are applied on titanium alloy substrates to improve both creep and oxidation resistance. High-temperature coatings for titanium alloys include gold, platinum, and platinum-tungsten coatings applied by ion plating.

Air exposure surface oxidation rates in Ti-6242 are shown in Table 16, which demonstrates the long-term oxidation protection up to 590 °C (1100 °F) by platinum ion plating. However, with the use of tungsten as a primer and platinum as a secondary coat, it is possible to extend this range up to 700 °C (1300 °F), which is the goal temperature for titanium intermetallics. Additionally, the platinum ion plating improves the creep resistance of the conventional high-temperature titanium alloys (Ref 19).

Table 15 Property comparison of titanium aluminides with conventional titanium alloys and superalloys

Property	Ti alloys	α-2	γ	Superalloys
Density, g/cm^3 ($lb/in.^3$)	4.54 (0.16)	4.84 (0.17)	4.04 (0.14)	8.3 (0.30)
Stiffness, GPa (10^6 psi)	110 (16)	145 (21)	176 (25)	207 (30)
Creep temperature (max), °C (°F)	540 (1000)	730 (1345)	900 (1650)	1090 (1995)
Oxidation temperature (max), °C (°F)	590 (1095)	705 (1300)	815 (1500)	1090 (1995)
Ductility, RT, %	15	2–4	1–3	3–10
Ductility, operating temperature, %	15	5–12	5–12	10–20

Source: Ref 1

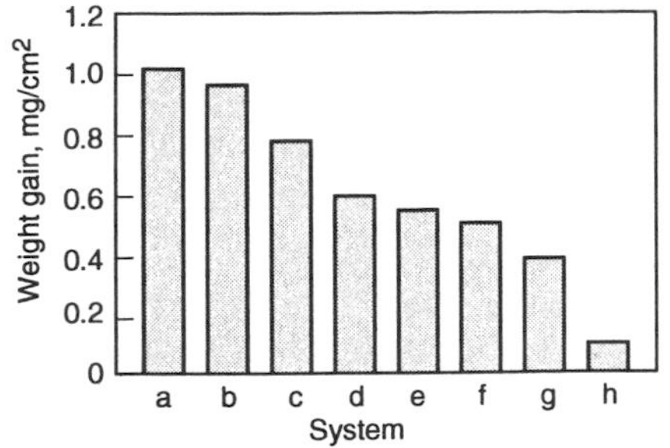

Fig. 20 Weight gain for 1 h at 980 °C (1800 °F) in air for (a) uncoated Ti_3Al-Nb-based intermetallic alloy, (b) sputtered yttria, (c) boron oxide from solution, (d) sodium aluminum borophosphate from solution, (e) calcium phosphate from solution, (f) calcium aluminate from solution, (g) calcium aluminophosphate from solution, and (h) magnesium silicophosphate from sol-gel. Source: Ref 27

Table 16 500-hour weight gain of Ti-6242 alloy in air

Ion plating materials	Exposure temperature		Weight gain rate, $mg/cm^2/h$
	°C	°F	
No coating	590	1100	6.9×10^{-2}
Gold	430	800	2.2×10^{-4}
Gold	480(a)	900	2.6×10^{-3}
Platinum	590(a)	1100	1.2×10^{-3}
Tungsten/platinum	650	1200	3.3×10^{-4}
Tungsten/platinum	700(a)	1300	1.7×10^{-3}

(a) Highest temperatures under which no spalling or loss of the coating was detected after 500 h. Source: Ref 19

Aluminized Coatings. Aluminizing has been carried out via pack processes or deposition and diffusion processes to improve the oxidation resistance of titanium alloys. In a pack process, the compound layers formed depend on the processing parameters. For example, at 760 °C (1400 °F) for 20 h, 50 μm of $TiAl_3$ is formed. $TiAl_3$ is produced at the surface at 900 °C (1650 °F) for 16 h, and thinner layers of TiAl and Ti_3Al form sequentially toward the substrate of $TiAl_2$. Cracks through the compound layers are often present, and such cracks can lessen the effectiveness of aluminized coatings. Nonetheless at 900 °C (1650 °F), the weight gain of the aluminized coatings has been found to be 50 times less than for pure titanium, 30 times less than for Ti-6Al-4V, and 5 times less than for a silicide coating. After 10 h, assuming a linear rate, the oxidation of the aluminide coating was 300 times slower than for the pure titanium (Ref 20). Studies have also shown that at lower temperatures, the beginning of rapid oxidation can be delayed (Ref 22). The $TiAl_3$ surface layers provided better oxidation resistance than the TiAl layer. Successful pack aluminizing of intermetallic alloy Ti-14Al-24Nb has also been achieved (Ref 22). Similar pack-aluminized intermetallic material has been subjected to cyclic oxidation, and excellent protection from cyclic oxidation was measured to 1000 °C (1830 °F). Thicker coatings offered limited resistance to oxidation because of cracking of the coatings after about five cycles; after more cycles, large crack growths were observed. Optimum coating thicknesses were found to be 40 to 70 μm (Ref 23). Thinner aluminized coating can be obtained by electron beam evaporation followed by a diffusion anneal (Ref 24).

Ion Implantation. Oxidation resistance of ion-implanted titanium alloys has been researched using a variety of alloying elements. Oxidation for 50 min in dry oxygen at 600 °C (1110 °F) has been reduced by implantation to a level of ~2.10^{16} ions/cm^2 of species that have either oxides with high negative free energy of formation or large ionic radii for smaller heats of formation (Ref 25). Particularly beneficial effects were obtained from implanting barium, rubidium, cesium, strontium, calcium, ytterbium, and europium (Ref 26). The reduction in oxidation was thought to be caused by implanted impurities obstructing short-circuit diffusion paths or the formation of Ti + M (implant) mixed oxides, for example, $BaTiO_3$.

Sol-gel coatings on titanium alloys are being studied for use as oxidation protection surface treatments. Surface coatings applied on a Ti_3Al-Nb-based intermetallic alloy include (Ref 27):

- Sputtered oxides of magnesium, yttrium, zirconium, and hafnium
- Sputtered fluorides of calcium and yttrium
- 1 μm thick layers of silicon, aluminum, and boron oxides applied both by sputtering and from sols
- Calcium, sodium, and phosphorus oxides applied from sols

A reaction barrier undercoat expected to be stable with the titanium alloy was combined with oxidation barrier topcoats. Weight gains for various systems after 1 h at 980 °C (1800 °F) in air are shown in Fig. 20. The measured values for weight gain are indicative of the relative film integrity after exposure; sputtered coatings were often cracked. A sputtered yttria coating was improved by dipping in boric acid, which during heating, forms boron oxide, which was believed to fill the cracks in the yttria layer, consequently improving the protection achieved.

A multilayered system involving 1 to 2 μm of reaction barrier coating is applied (either barium titanate or yttrium-stabilized zirconia), over which is applied a diffusion-barrier coating containing magnesium phosphate and silica glasses. In cyclic oxidation tests to 980 °C (1800 °F), the performance of this coating equaled or exceeded that of other thicker reaction coatings (Ref 28).

ACKNOWLEDGMENTS

The editor thanks John C. Fanning, Manager of Structural Applications Development, Titanium Metals Corporation, for his significant contributions to this article. An additional key source of information was:

- A. Bloyce, P.H. Morton, and T. Bell, Surface Engineering of Titanium and Titanium Alloys, *Surface Engineering*, Vol 5, *ASM Handbook*, ASM International, 1995, p 835–851.

REFERENCES

1. *Materials Properties Handbook: Titanium Alloys,* R. Boyer, G. Welsch, and E.W. Collings, Ed., ASM International, 1994
2. *Titanium, A Technical Guide,* M.J. Donachie, Jr., Ed., ASM International, 1988
3. S. Lampman, Wrought Titanium and Titanium Alloys, *Properties and Selection: Nonferrous Alloys and Special Purpose Materials,* Vol 2, *ASM Handbook,* ASM International, 1990, p 592–633
4. M. Peters, Y.T. Lee, K.-J. Grundhoff, H. Schurmann, and G. Welsch, Influence of Processing on Microstructure and Mechanical Properties of Ti-1100 and IMI-834, Paper presented at TMS Fall Meeting (Detroit, MI), TMS, 7–11 Oct 1990
5. *Properties and Selection of Metals,* Vol 1, *Metals Handbook,* 8th ed., American Society for Metals, 1961
6. D. Eylon, S. Fujishiro, P.J. Postans, and F.H. Froes, High-Temperature Titanium Alloys—A Review, *Titanium Technology: Present Status and Future Trends,* Titanium Development Association, 1985, p 87–93
7. S.R. Seagle, G.S. Hall, and H.B. Bomberger, "High-Temperature Properties of Ti-6Al-2Sn-4Zr-2Mo-0.09Si, *Met. Eng. Q.,* Feb 1975, p 48–54
8. P.J. Bania, Ti-1100: A New High-Temperature Titanium Alloy, *Sixth World Conference on Titanium Proceedings (Part I),* Société Française de Métallurgie, 1988, p 825–830
9. P.J. Bania, "High Temperature Alloy Comparison—IMI 834 vs TIMETAL 1100," Titanium Metals Corporation, Henderson, NV, March 1994
10. J.S. Park, et al., The Effects of Processing on the Properties of Forgings from Two New High Temperature Titanium Alloys, *Sixth World Conference on Titanium Proceedings (Part I),* Société Française de Métallurgie, 1988, p 1283–1288
11. P.J. Bania and W.M. Parris, "Beta-21S: A High Temperature Metastable Beta Titanium Alloy," Paper presented at the 7th Annual World Conference on Titanium (San Diego, CA), June 1992
12. P.J. Bania, Next Generation Titanium Alloys for Elevated Temperature Service, *ISIJ Int.,* Vol 31 (No. 8), Aug 1991
13. J.S. Grauman and J.C. Fanning, "Recent Developments Regarding the Resistance of Titanium to Hydraulic Fluid at Elevated Temperatures," Paper presented at AeroMat 1993 (Anaheim, CA), ASM International, June 9, 1993
14. "Data Sheet: TIMETAL 21S (Ti-15Mo-3Nb-3Al-0.2Si) High Strength, Oxidation Resistant Strip Alloy," Titanium Metals Corporation, Henderson, NV
15. F.H. Froes and R.G. Rowe, Rapidly Solidified Titanium, *Rapidly Solidified Alloys and Their*

Mechanical and Magnetic Properties, Vol 58, Materials Research Society, 1986, p 309–334

16. R.G. Rowe and F.H. Froes, Titanium Rapid Solidification—Alloys and Processes, *Processing of Structural Metals by Rapid Solidification*, ASM International, 1987, p 163–173
17. R. Sundaresan, A.G. Jackson, and F.H. Froes, Dispersion Strengthened Titanium Alloys Through Mechanical Alloying, *Proceedings of the Sixth World Conference on Titanium*, Société Française de Métallurgie, 1988, p 855
18. R. Sundaresan and F.H. Froes, Development of the Titanium-Magnesium Alloy System Through Mechanical Alloying, *Proceedings of the Sixth World Conference on Titanium*, Société Française de Métallurgie, 1988, p 931
19. S. Fujishiro and D. Eylon, Improved Mechanical Properties of Alpha + Beta Ti Alloys by Pt Ion Plating, *Titanium '80, Science and Technology*, TMS-AIME, 1980, p 1175–1182
20. R. Streff and S. Poize, Oxidation of Aluminide Coatings on Unalloyed Titanium, *International Conference on High Temperature Corrosion* (San Diego, CA), March 1981, p 591–597
21. J. Subrahamanyam and J. Annapurna, High Temperature Cyclic Oxidation of Aluminide Layers on Titanium, *Oxid. Met.*, Vol 26 (No. 3/4), 1986, p 275–285
22. J. Subrahamanyam, Cyclic Oxidation of Aluminized Ti-14Al-24Nb Alloy, *J. Mater. Sci.*, Vol 23, 1988, p 1906–1910
23. J.L. Smialek, M.A. Gedwell, and P.K. Brindley, Cyclic Oxidation of Aluminide Coatings on Ti_3Al+Nb, *Scr. Metall. Mater.*, Vol 24, 1990, p 1291–1296
24. R.K. Clark, J. Unnam, and K.E. Wiedemann, Effect of Coatings on Oxidation of Ti-6Al-2Sn-4Zr-2Mo Foil, *Oxid. Met.*, Vol 29 (No. 3/4), 1988, p 255–269
25. J.D. Benjamin and G. Dearnaley, Further Investigation of the Effects of Ion Implantation on the Thermal Oxidation of Titanium, *Inst. Phys. Conf. Ser.*, Vol 28, 1976, p 141–146
26. A. Galerie, High Temperature Oxidation of Ion Implanted Metals, *Ion Implantation into Metals, Proceedings 3rd International Conference on Modification of Surface Properties by Ion Implantation*, UIST 1981, Pergamon, 1982, p 190–200
27. K.E. Wiedemann, P.J. Taylor, R.K. Clark, and T.A. Wallace, "Thin Coatings for Protecting Titanium Aluminides in Hot-Corrosion Environments," Paper presented at TMS Fall Meeting (Detroit, MI), TMS, 1990
28. NASA Langley Research Center, Ultra-Thin Coatings Protect Titanium From Adverse Environments, *Adv. Mater.*, Vol 23, 1992, p 4

Refractory Metals and Alloys

THE REFRACTORY METALS include niobium (also known as columbium), tantalum, molybdenum, tungsten, and rhenium. With the exception of two of the platinum-group metals, osmium and iridium, the refractory metals have the highest melting temperatures (>2000 °C, or >3630 °F) and the lowest vapor pressures of all metals. They are readily degraded by oxidizing environments at moderately low temperatures, a property that has restricted their applicability in low-temperature or non-oxidizing high-temperature environments. Protective coating systems have been developed, mostly for niobium alloys, to permit their use in high-temperature oxidizing aerospace applications.

At one time refractory metals were limited to use in lamp filaments, electron tube grids, heating elements, and electrical contacts. However, they have since found widespread application in the aerospace, electronics, nuclear and high-energy physics, and chemical process industries. Each of the refractory metals, with the exception of rhenium, is consumed in quantities exceeding 900 metric tonnes annually worldwide.

Table 1 compares the physical, thermal, electrical, magnetic, and optical properties of pure refractory metals. As described below, these properties are sensitive to purity, processing, and other factors. Figures 1 to 3 compare the temperature-dependent ultimate tensile strengths, elastic moduli, and creep-rupture strengths of the refractory metals. The values for hexagonal close-packed (hcp) rhenium are quite different from those of the other metals, which have body-centered cubic (bcc) structures. Comparative high-temperature data for refractory metals and their alloys are given in Table 2.

Molybdenum and Molybdenum Alloys (Ref 3)

Molybdenum combines a high melting point (2610 °C, or 4730 °F) with strength retention at high temperatures. Molybdenum also boasts a high specific elastic modulus, which makes it attractive for applications that require both high stiffness and low weight. The high thermal conductivity, low coefficient of thermal expansion, and low specific heat of this metal provide resistance to thermal shock and fatigue, and these properties are also important in electronic applications. In addition, molybdenum is stable in a wide variety of chemical environments, and it has good electrical conductivity, although this physical property is seldom the critical factor in materials selection. Table 1 lists the physical properties of unalloyed molybdenum.

Molybdenum has a bcc crystal structure and displays the ductile-to-brittle transition (DBTT) behavior typical of such metals; that is, the DBTT is sensitive to test conditions (stress concentrations, testing mode, strain rate), alloy composition, and microstructure. For instance, the DBTT decreases as the amount of deformation experienced by the material increases, and stress-relieved (recovery-annealed) material has a lower DBTT than recrystallized material (Fig. 4).

Primary consolidation of molybdenum and its alloys can be done by either vacuum-arc casting (VAC) or powder metallurgy (P/M) techniques. Both mechanical pressing and cold isostatic pressing (CIP) are used to consolidate P/M billets, although most P/M mill products originate as CIPed billets. P/M billets are typically sintered in

Table 1 Property comparison of pure refractory metals

Property	Niobium	Tantalum	Molybdenum	Tungsten	Rhenium
Structure and atomic properties					
Atomic number	41	73	42	74	75
Atomic weight	92.9064	180.95	95.94	183.85	186.31
Density at 20 °C (70 °F), g/cm^3 (lb/in.^{3})	8.57 (0.310)	16.6 (0.600)	10.22 (0.369)	19.25 (0.695)	21.04 (0.760)
Crystal structure	bcc	bcc	bcc	bcc	hcp
Lattice constants, nm					
a	0.3294	0.3303	0.3147	0.3165	0.27609
c	...	...	...	...	0.45829
Slip plane at room temperature	110	110	112	...	0001–1010
Thermal properties					
Melting temperature, °C (°F)	2468 (4474)	2996 (5425)	2610 (4730)	3410 (6170)	3180 (5755)
Boiling temperature, °C (°F)	4927 (8901)	5427 (9801)	5560 (10040)	5700 (10290)	5760 (10400)
Vapor pressure at 2500 K, mPa (torr)	5.3 (4×10^{-5})	0.11 (8×10^{-7})	80 (6×10^{-4})	0.0093 (7×10^{-8})	0.17 (1.3×10^{-6})
Coefficient of expansion, near RT(a), µm/m · K (µin./in. · °F)	7.3 (4.1)	6.5 (3.6)	4.9 (2.7)	4.6 (2.6)	6.7 (3.7)
Specific heat at 20 °C (70 °F), kJ/kg · K (Btu/lb · °F)	0.268 (0.0643)	0.139 (0.0333)	0.276 (0.0662)	0.138 (0.0331)	0.138 (0.0331)
Latent heat of fusion, kJ/kg (Btu/lb)	290 (125)	145-174 (62-75)	270 (115)	220 (95)	177 (76)
Latent heat of vaporization, kJ/kg (Btu/lb)	7490 (3202)	4160–4270 (1790–1840)	5123 (2160)	4680 (2010)	3415 (1470)
Thermal conductivity, W/m · K (Btu/ft · h · °F)					
At 20 °C (70 °F)	52.7 (30.4)	54.4 (31.4)	142 (81.9)	155 (89.4)	71 (41)
At 500 °C (930 °F)	63.2 (36.5)	66.6 (38.4)	123 (71.0)	130 (75)	...
Electrical properties					
Electrical conductivity at 18 °C (64 °F), %IACS(a)	13.2	13.0	33.0	30.0	8.1
Electrical resistivity, at 20 °C (70 °F), nΩ · m	160	135	52	53	193
Electrochemical equivalent, mg/C	0.1926	0.375	0.166	0.318	0.276
Hall coefficient, nV · m/A · T	0.09	0.095	...	...	...
Magnetic properties					
Magnetic susceptibility (volume) at 25 °C (75 °F), mks system	28×10^{-6}	10.4×10^{-6}	1.17×10^{-8}	4.1×10^{-8}	0.37×10^{-6}
Optical properties					
Total emissivity at 1500 °C (2730 °F), %	0.19	0.21	0.19	0.23	...
Spectral emittance at $\lambda = 650$ nm, %	0.37	0.49	0.37	0.43	...
Additional properties					
Poisson's ratio at 25 °C (75 °F)	0.38	0.35	0.32	0.28	0.49
Elastic modulus, GPa	103	185	324	400	469

(a) RT, room temperature. (b) IACS, International Annealed Copper Standard

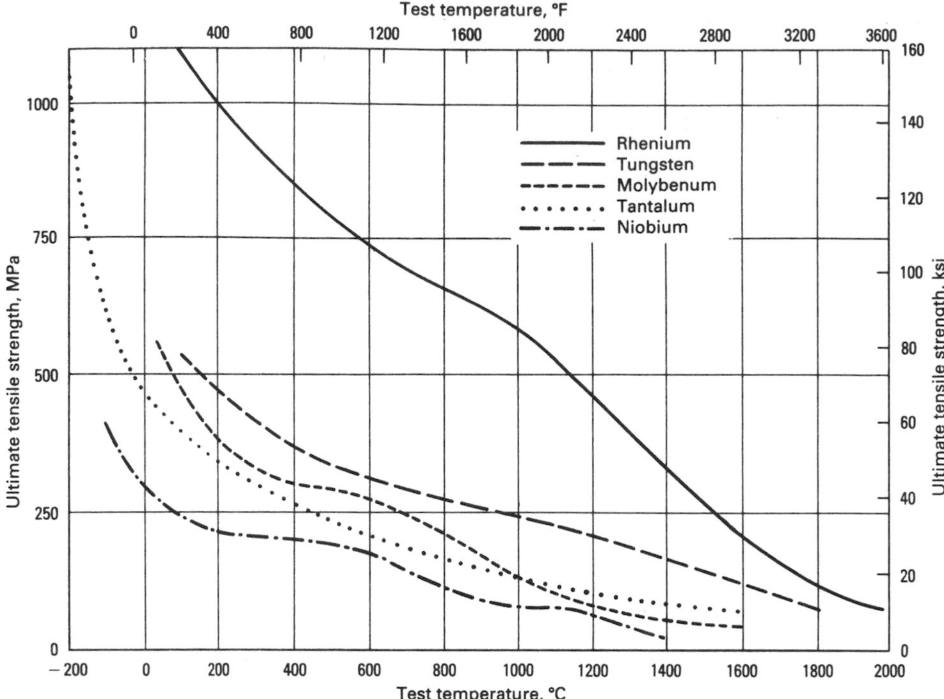

Fig. 1 Test temperature vs. ultimate tensile strength for pure refractory metals

hydrogen because hydrogen reduces molybdenum oxides and further purifies the material. Vacuum sintering is used by some manufacturers.

Extrusion, rolling, or forging can be used to work large P/M billets, but VAC ingots must first be hot extruded because of the propensity for the as-cast material to develop brittle intergranular fracture under tensile stresses. After billets and ingots have been reduced to more convenient sizes, the materials can be processed by conventional mill techniques, such as hot rolling, cold rolling, or swaging.

More molybdenum is consumed annually than any other refractory metal. Most molybdenum is used as an alloying element in irons, steels, and superalloys. Molybdenum-base mill products represent about 5% of total usage.

Classes of Molybdenum-Base Alloys

As shown in Table 3, there are several classes of commercial molybdenum-base alloys:

- *Carbide-strengthened alloys* rely on the formation of fine reactive-metal carbides to disperse strengthen the material, and to increase the recrystallization temperature above that of pure molybdenum by stabilizing the dislocation structure formed during processing. These alloys are produced in both VAC and P/M grades.
- *Solid-solution alloys* are produced in both VAC and P/M grades.
- *Combination alloys,* such as HWM-25 (Mo-W-Hf-C), contain both carbide-forming and substitutional elements to provide improved high-temperature strength.
- *Dispersion-strengthened P/M alloys* rely on second-phase particles, introduced or produced during powder processing, to increase the resistance to recrystallization and stabilize the recrystallized grain structure. These alloys have enhanced high-temperature strength and improved low-temperature ductility.

Some of the important properties of molybdenum and its alloys are compared in Table 2.

Carbide-strengthened alloys were the first molybdenum alloys to be commercialized. Mo-0.5Ti, the initial alloy, is no longer commercially available. Its high-temperature strength and recrystallization resistance were improved by adding about 0.08% zirconium, resulting in the alloy known as TZM. TZC, a higher-alloy-content modification of TZM, has improved properties and responds to an age-hardening heat treatment. However, TZC has not replaced TZM as the commercial alloy of choice, primarily due to economic considerations. More recently, alloys strengthened with hafnium carbide (MHC) and combinations of reactive metal carbides (ZHM) have been marketed. Figure 5 compares elevated-temperature properties of carbide-strengthened alloys. Additional property data are listed in Tables 2 to 6.

Both TZM and MHC are used as tooling materials in the isothermal forging of nickel-base su-

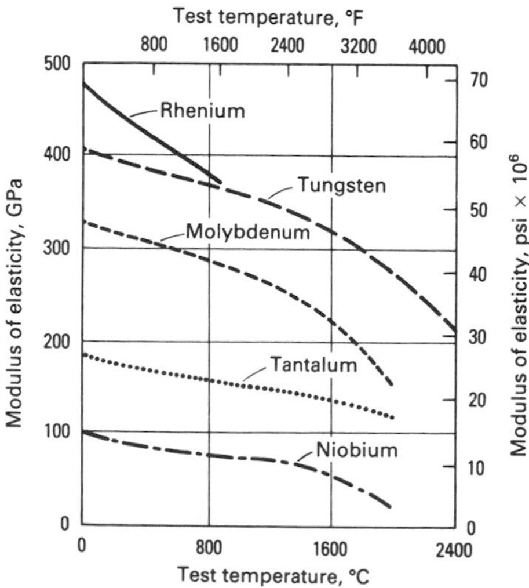

Fig. 2 Test temperature vs. modulus of elasticity for pure refractory metals

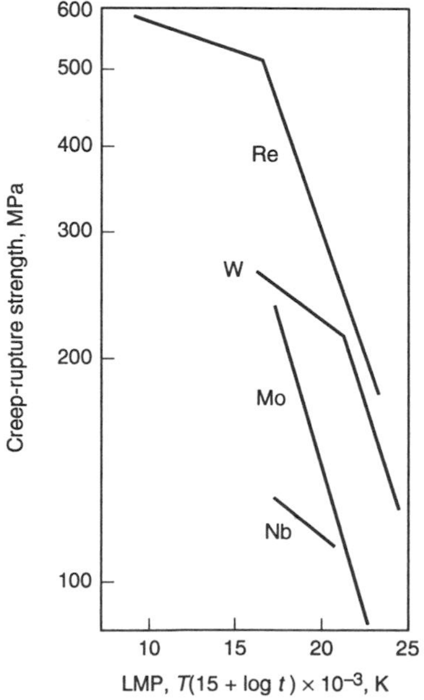

Fig. 3 Creep-rupture strength comparison for pure refractory metals. LMP, Larson-Miller parameter. Source: Ref 1

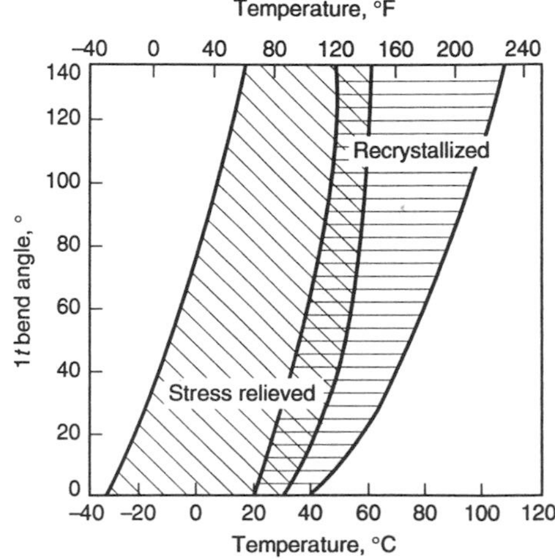

Fig. 4 Effect of heat treatment on the ductile-to-brittle transition behavior (DBTT) of vacuum-arc cast molybdenum sheet (thickness, t, = 3.2 mm, or 0.125 in.), as measured in bending over a $1t$ radius. Stress-relieved (recovery-annealed) material has a lower DBTT than recrystallization-annealed molybdenum. In addition, the DBTT decreases as the amount of deformation increases. Source: Ref 3

Table 2 Property comparison of refractory metal alloys

Nominal alloy additions, wt %	Common designation	Most common product form	Usual condition(a)	Low-temperature ductility class(b)	Typical high-temperature strength: Temperature, °C	Tensile, MPa	Temperature, °C	10 h rupture, MPa
Niobium and its alloys								
None	Unalloyed Nb	All	Rx	A	1095	69	1095	37
1 Zr	Nb-1Zr	All	Rx	A	1095	158	1095	96
1 Zr, 0.1 C	PWC-11	All	Rx	A	1095	130	...	...
10 Hf, 1 Ti, 0.7 Zr	C-103 (KBI 3)	All	Rx	A	1095	186	...	...
10 Ta, 10 W	SCb-291	Bar, sheet	Rx	A	1095	220	1095	62
10 W, 10 Hf, 0.1 Y	C-129Y	Sheet	Rx	A	1315	179	1095	103
28 Ta, 11 W, 0.8 Zr	FS-85	Sheet	Rx	A	1315	158	1315	83
Molybdenum and its alloys								
None	Unalloyed Mo(c)	All	SRA	B-C	1000	50	980	175
0.5 Ti, 0.08 Zr, 0.03 C	TZM (MT-104)(c)	All	SRA	B-C	1000	600	1315	140
1.2 Ti, 0.3 Zr, 0.1 C	TZC	All	SRA	B-C	1000	800	1315	190
1.2 Hf, 0.05 C	MHC (HCM)	All	SRA	B-C	1000	800	1315	210
0.5 Zr, 1.5 Hf, 0.2 C	ZHM	All	SRA	B-C	1000	800	1400	200
25 W	25W	All	SRA	B-C	1000	330	...	...
30 W	30W(c)	All	SRA	B-C	1000	350	1095	140
1 Hf, 0.07 C, 25 W	HWM-25 (Mo25WH)	All	SRA	C	1000	900	1300	200
5 Re	5 Re	All	SRA or Rx	B	1000	400	1650	7
41 Re	41 Re	All	SRA or Rx	B	1000	600	...	...
47.5 Re	50 Re	All	SRA or Rx	B	1000	580	1600	27
0.5 ZrO_2	Z-6	All	SRA	B-C	1000	280	...	...
150 K, 300 Si (ppm)	MH (HD)	Wire, sheet	CW, SRA	B	1000	300	...	...
200 K, 300 Si, 100 Al (ppm)	KW	Wire, sheet	CW, SRA	B	1000	300	...	...
Tantalum and its alloys								
None	Unalloyed Ta	All	Rx	A	1315	59	1315	7
7.5 W (P/M alloy)	FS-61	Wire, strip	CW	A	25	1140	...	...
2.5 W, 0.15 Nb	FS-63	All	Rx	A	95	315	...	...
25 W	KBI-6	All	Rx	A	95	315	...	...
10 W	Ta-10W	All	Rx	A	1315	345	1315	140
8 W, 2 Hf	T-111	All	Rx	A	1315	255	...	...
8 W, 1 Re, 1 Hf, 0.025 C	Astar 811C	All	Rx	A	1315	275	...	...
40 Nb	KBI-40	All	Rx	A	260	290	...	...
37.5 Nb, 2.5 W, 2 Mo	KBI-41	All	Rx	A	260	515	...	...
Tungsten and its alloys								
None	Unalloyed W	Wrought bar	SRA	D	1000	620	...	...
		Sheet, wire	SRA	D	1650	120	1650	50
None	Unalloyed W	CVD sheet	Rx	D	1000	565	...	...
Al, K, Si (ppm levels)	Doped AKS W	Wire	CW	C	1650	650	...	...
1 ThO_2	W-1%ThO_2	Bar, sheet, wire	SRA	D	1650	255	...	...
2 ThO_2	W-2%ThO_2	Bar, sheet, wire	SRA	D	1650	205	1650	125
15 Mo	W-15Mo	Bar, sheet	SRA	D	1650	250	1650	85
4 Re	W-4Re	Wire, ribbon	SRA or Rx	C	1650	150	...	...
25 Re	W-25Re	All	SRA or Rx	B	1650	275	1650	95
4 Re, HfC	W-4Re-HfC	Wire	SRA	C	1650	620	...	...
Rhenium								
None	Unalloyed Re(c)	All	Rx	A	20	1172/2324(d)	1600	56
					500	786/1196(d)	2200	21
					1000	588/855(d)	2800	4.2
					1500	262/276(d)	...	...
					2000	.../103(d)	...	...
					2300	53/...(d)	...	...

(a) CW, cold worked; Rx, recrystallized; SRA, stress-relief annealed. (b) A, excellent cryogenic ductility; B, excellent room-temperature ductility; C, may have marginal ductility at room temperature; D, normally brittle at room temperature. (c) Available in both powder metallurgy and arc-cast forms. (d) First value, annealed material; second value, wrought material cold worked 15%. Source: Ref 2

peralloy parts for aircraft gas-turbine engines. In addition, one manufacturer has capitalized on the improved high-temperature strength offered by MHC by developing a rapid-solidification-rate process for making titanium and refractory-metal alloy powders. In this process, a rapidly rotating molybdenum alloy disc is used to quench a stream of molten metal impinging on the disc surface. Upgrading the disc alloy from TZM to MHC permits an increase in disc rotation speed to 45,000 rpm, which decreases powder particle size and dramatically increases particle cooling rate. A higher cooling rate permits alloys to be processed that formerly were not candidates for rapid-solidification-rate processing.

The carbide-strengthened alloys also are of interest as higher-strength alternatives to ultralow-carbon pure molybdenum in glassmaking applications. A low carbon content is required because the element can react with molten glass to form gas bubbles, which render the product unacceptable. However, glassmaking equipment also must resist creep, because operating temperatures often exceed 1400 °C (2550 °F). The reactive-metal carbides resist decomposition at glassmaking temperatures, preventing gas bubble formation. Thus, use of carbide-strengthened alloys can extend glassmaking equipment life without harming the product.

The microstructures of carbide-strengthened alloys are similar in appearance to that of pure molybdenum, in that there is no optical evidence of the precipitate. However, there is a significant difference between the microstructures of these alloys as produced by arc casting and P/M techniques. P/M alloys containing reactive metals are

Table 3 Compositions of selected commercial molybdenum alloys

Alloy	Alloying additions, wt%	Recrystallization temperature, °C (°F)
Unalloyed Mo	...	1100 (2010)
Reactive-metal-carbide alloys		
TZM (MT-104)	0.5 Ti, 0.08 Zr, 0.03 C	1400 (2550)
TZC	1.2 Ti, 0.3 Zr, 0.1 C	1550 (2820)
MHC (HCM)	1.2 Hf, 0.05 C	1550 (2820)
ZHM	0.5 Zr, 1.5 Hf, 0.2 C	1550 (2820)
Solid-solution alloys		
25W	25 W	1200 (2190)
30W	30 W	1200 (2190)
5 Re	5 Re	1200 (2190)
41 Re	41 Re	1300 (2370)
50 Re	47.5 Re	1300 (2370)
Combination alloy		
HWM-25 (Mo25WH)	1 Hf, 0.07 C, 25 W	1650 (3000)
Dispersion-strengthened alloys		
Z-6	0.5 ZrO_2	1250 (2280)
MH (HD)	150 K, 300 Si (ppm)	1800 (3270)
KW	200 K, 300 Si, 100 Al (ppm)	1800 (3270)

Source: Ref 3

Table 4 Typical tensile properties of TZM

Temperature °C	Temperature °F	Tensile strength MPa	Tensile strength ksi	Yield strength at 0.2% offset MPa	Yield strength at 0.2% offset ksi	Elongation in 50 mm (2 in.), %
Stress-relieved condition						
20	70	965	140	860	125	10
1095	2000	490	71	435	63	...
1650	3000	83	12	62	9	...
Recrystallized material						
20	70	550	80	380	55	20
1095	2000	505	73	...	...	...
1315	2400	369	53.5	...	...	...

Table 5 Room- and elevated-temperature tensile properties of 380 μm (15 mil) molybdenum wire

Material designation	Composition	Temperature °C	Temperature °F	Ultimate tensile strength MPa	Ultimate tensile strength ksi	Elongation, %
Unalloyed molybdenum	Mo	20	70	1350	196	4.1
		1000	1830	305	44	2.4
		1100	2010	140	20	10.3
		1200	2190	115	17	12.5
MT-104	Mo-0.5Ti-0.08Zr-0.01C	20	70	1565	227	3.1
		1000	1830	1020	148	2.7
		1100	2010	795	115	3.2
		1200	2190	675	98	2.8
Mo + 45 W	Mo-45W	20	70	1980	287	3.6
		1000	1830	1095	159	2.3
		1100	2010	950	138	2.3
		1200	2190	745	108	2.2
HCM	Mo-1.1Hf-0.07C	20	70	1795	260	2.9
		1000	1830	1270	184	3.4
		1100	2010	1185	172	3.3
		1200	2190	1035	150	3.0
HWM-25	Mo-25W-1.0Hf-0.035C	20	70	1935	281	3.2
		1000	1830	1350	196	3.3
		1100	2010	1225	178	3.1
		1200	2190	1075	156	4.6
HWM-45	Mo-45W-0.9Hf-0.03C	20	70	2135	310	3.6
		1000	1830	1460	212	3.3
		1100	2010	1295	188	2.6
		1200	2190	1170	170	2.4

Table 6 Creep test results for several molybdenum-base materials

See Table 3 for chemical compositions.

Material	Time to failure, h (a)	Time to failure, h (b)	Elongation, % (a)	Elongation, % (b)	Reduction in area, % (a)	Reduction in area, % (b)	Creep rate, %/h (a)	Creep rate, %/h (b)
TZM	39	9.4	13	16.5	68	80	0.052	0.18
MHC	44	16.7	3	3.9	1	4.9	0.038	0.07
TZC	89	16.0	7	19.2	19	74.4	0.021	0.14

(a) Results at stress of 330 MPa (48 ksi) and at 1200 °C (2190 °F). (b) Results at stress of 207 MPa (30 ksi) and at 1315 °C (2400 °F). Source: Battelle Columbus Laboratories

prone to have alloy oxides in their microstructures. Even though these materials are typically sintered under reducing atmospheres, which eliminate free oxygen, a substantial amount of oxygen can be carried into the sintering furnace as either molybdenum oxide or absorbed oxygen on powder particle surfaces. During sintering, the reactive-metal alloy additions can scavenge this oxygen. ASTM specifications for TZM (ASTM B 386 and B 387) acknowledge this fact by permitting higher oxygen contents in the P/M product (0.030 vs. 0.0030 wt% max). In the extreme case, the zirconium in the alloy may be present primarily as ZrO_2, which makes the material little more than a Mo-0.5Ti alloy having ZrO_2 inclusions. Oxide inclusions will be present in even the most carefully produced P/M materials.

Solid-Solution Alloys. Tungsten and rhenium are the substitutional elements of interest in solid-solution molybdenum alloys. In addition to several "standard" alloys, other compositions can be special ordered. With the exception of Mo-30W, which is also available as a VAC product, the solid-solution molybdenum alloys are normally made via P/M.

The molybdenum-tungsten alloys were developed for their chemical resistance, and they are primarily used in equipment for handling molten zinc. They are lower-cost, lighter-weight alternatives to pure tungsten in these applications. The commercial importance of molybdenum-rhenium alloys is due to the so-called "rhenium effect" (Ref 4), which results in the material's having a significantly lower DBTT. The most common alloys contain 5, 41, and 50% (actually 47.5%) Re. The 5Re and 41Re alloys are used for thermocouple wire and for structural applications in the aerospace market. The W-50Re alloy is typically specified for high-temperature structural components.

Dispersion-Strengthened Alloys. While carbide-hardened alloys are difficult to produce due to oxidation of their reactive-metal constituents, other dispersion-strengthened molybdenum alloys rely exclusively on P/M manufacturing techniques. Using a powder precursor makes it possible to produce fine dispersions of second phases, which can stabilize a wrought structure, preventing recrystallization, or stabilize an elongated recrystallized grain structure, preventing a transition to equiaxed grains. The latter effect provides significant improvements in the low-temperature ductility of material in the recrystallized condition.

Doped molybdenum alloys were the first dispersion-strengthened materials. They are analogous to the Al-K-Si-doped tungsten alloys (AKS tungsten) developed for lamp filaments. They were initially designed to satisfy lighting industry requirements for creep-resistant molybdenum parts. Doped alloys such as MH and KW (see Table 3) do not have particularly high strength at low temperatures, but these materials are extremely resistant to recrystallization and have excellent creep resistance.

The dopant strengthening mechanism is unique. Small "bubbles" of dopant are retained in the sintered ingot and are highly elongated by subsequent deformation. Annealing causes the elongated dopant regions to "pinch off," forming arrays of extremely fine bubbles that impede grain boundary motion during recrystallization. The result is recrystallized material having an elongated, interlocked grain structure, which is much more ductile at ambient temperatures than the equiaxed recrystallized structure of pure molybdenum and other molybdenum alloys. This stable grain structure also reduces grain boundary creep, producing a one-to-two order of magni-

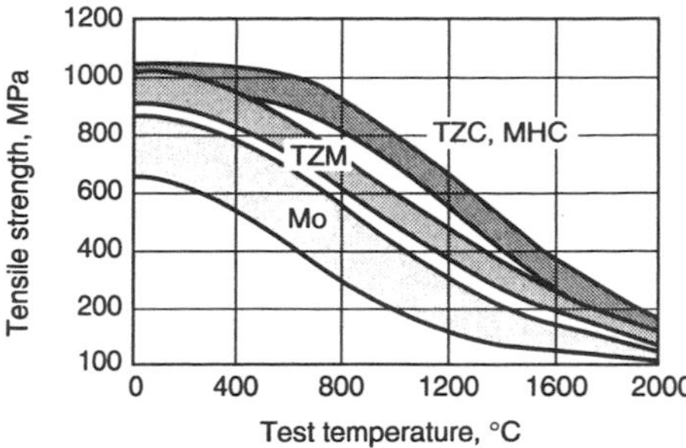

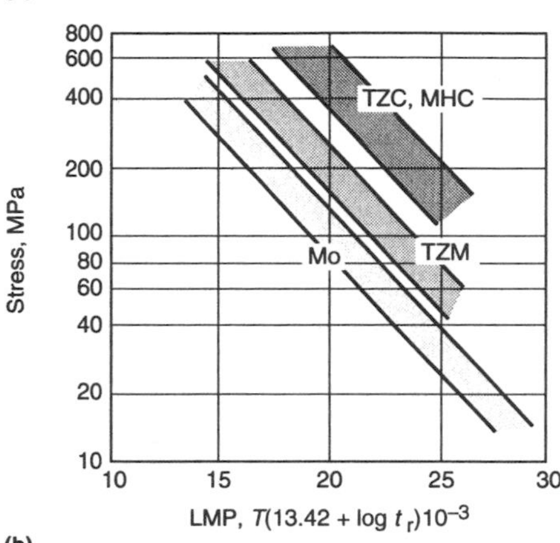

Fig. 5 Elevated-temperature properties of molybdenum and molybdenum alloys. (a) Tensile strength. (b) Larson-Miller parameter (LMP) with temperature given in degrees Kelvin and the time to rupture, t_r, given in hours. Source: Ref 3

tude decrease in steady-state creep rate, compared with that of pure molybdenum.

A disadvantage of doped alloys is that they require very large deformations (≥95%) for maximum effectiveness. Typically, this requires nearly unidirectional deformation, which means that only the longitudinal properties dramatically improve. This can be of concern for flat products that must retain ductility in both in-plane directions, but not for wire and bar.

Dispersions of reactive-metal oxides also are used to improve the elevated-temperature strength and creep resistance of molybdenum. An example is the zirconia-dispersion-strengthened alloy Z-6 (0.5% ZrO_2). A fine, initial oxide dispersion is critical to alloy performance, because the "pinching-off" mechanism of doped alloys is not operative. The improvement in steady-state creep rate provided by the zirconia-stabilized grain structure is similar to that produced by doping.

Oxidation of Molybdenum

The primary disadvantage of molybdenum that prohibits its use in many high-temperature applications is its rapid and catastrophic oxidation in air at temperatures above roughly 790 °C (1450 °F) (Ref 5). Oxidation produces molybdenum trioxide (MoO_3), which is very volatile and sublimes readily from the solid. Vaporization of the oxide prevents molybdenum from generating a protective film that would retard further oxidation, and as a result, molybdenum literally "goes up in smoke" when heated in an oxidizing atmosphere. Oxidation can be promoted by melting as well as by vaporization of the oxides. MoO_3 melts at 795 °C (1465 °F), which is well below the melting point of metallic molybdenum (2610 °C, or 4730 °F), and MoO_2, which is the oxide that is thought to form at the metal-oxide interface, forms a eutectic with MoO_3 that melts at 778 °C (1432 °F). The liquid oxide can promote oxidation, even if it is nonvolatile, by allowing easy transport of molybdenum and oxygen ions through the oxide.

Improvements in oxidation resistance depend on the development of improved coatings or new oxidation-resistant alloys. Promising alloys have been identified in the Mo-Cr-Pd and Mo-W-Cr-Pd systems (Ref 6), but they are a long way from commercialization. Coating-development efforts also continue, although the pace of progress is slow. Incompatibility of thermal expansion coefficients between the coating and molybdenum substrate is an ongoing issue. A variety of coating systems are described below.

High-Temperature Oxidation-Resistant Coatings for Molybdenum. Ductile metallic overlays were among the first materials investigated as coatings for molybdenum. The materials and processes used are summarized in Table 7. Chromium is the most promising of the metallic coatings for molybdenum. It offers excellent oxidation protection and is compatible with the substrate, but it is embrittled by nitrogen and will crack and spall upon repeated thermal cycling. A nickel or nickel alloy overlay will protect chromium from nitriding. A duplex coating will give good service for molybdenum up to 1200 °C (2195 °F).

Table 7 Metallic coatings for molybdenum

Process type	Material	Thickness range µm	Thickness range µin.
Electrodeposition	Chromium, nickel, gold, iridium, palladium, platinum, rhodium	12.7-76.2	500-3002
Flame sprayed	Nickel-chromium-boron, nickel-silicon-boron, nickel-chromium, nickel-molybdenum	127-254	5004-10,007
Clad or bonded	Platinum, nickel, nickel-chromium, platinum-rhodium	50.8-508	2002-20,015
Molten bath	Chromium	12.7-25.4	500-1001

Table 8 Aluminide coatings for molybdenum

Type	Compositions	Deposition process	Thickness range µm	Thickness range µin.
Aluminum-chromium	20% Al + 80% (55Cr-40Si-3Fe-1Al)	Thermal spray	178-254	7013-10,008
Aluminum-silicon	88% Al-12% Si	Thermal spray, hot dip	12.7-178	500-7013
Tin-aluminum	90% (Sn-25Al)-10% $MoAl_3$	Slurry dip or spray	50.8-203	2002-7998

Noble metal coatings, such as platinum and iridium, can be used on molybdenum at temperatures up to about 1425 °C (2595 °F). They provide a significantly longer useful life at all temperatures than the duplex chromium-nickel coatings.

Coatings based on compounds with aluminum were initially developed in an attempt to provide molybdenum parts for aircraft gas turbines. Coatings of aluminum-chromium-silicon, aluminum-silicon, and aluminum-tin systems are listed in Table 8. The aluminide coatings are applied as thick overlays, using a variety of spray or dip processes. These coatings provide good oxidation protection to molybdenum at temperatures up to 1540 °C (2805 °F). The life at higher temperatures is very short, less than 1 h, probably as a result of rapid interdiffusion.

Interest in coatings for molybdenum shifted from aluminides to silicides in the mid-1950s. A list of the basic silicide coatings is given in Table 9. Most of the silicide coatings are deposited by pack-cementation diffusion processes. Table 10 gives time/temperature cycles for applying oxidation-resistant coatings on refractory metal substrates. A major deficiency in the performance of silicide-based coatings appears when the system is used in low-pressure environments. As shown in Fig. 6, silicide coatings that protect TZM substrate for more than 4 h at 1650 to 1760 °C (3000 to 3200 °F) in air at 1 atm cannot be used above 1480 °C (2695 °F) in air at pressures of 0.1 to 1.0 torr (13 to 130 Pa).

Refractory oxides or ceramics are the only materials suitable for the oxidation protection of molybdenum above 1650 °C (3000 °F) for any length of time. The types of oxide coatings that have been used on molybdenum are summarized in Table 11. Ceramic coatings suffer from one common problem: they crack upon thermal cycling and tend to spall from the substrate. For only minutes, during single-cycle uses such as in rocket motors, ceramic coatings provide useful protection from oxidation to 1930 °C (3505 °F) with Al_2O_3 and to 2200 °C (3990 °F) with ZrO_2.

Tungsten and Tungsten Alloys (Ref 7)

Tungsten has the highest melting point of any metal (3410 °C, or 6170 °F) and is among the most dense, at 19.26 g/cm^3 (0.696 lb/in.^{3}). It also has an unusually high elastic modulus (414 GPa, or 60 psi × 10^6) and is the only elastically isotropic metal. Tungsten is found in group 6 of the periodic table. It has an atomic weight of 183.85

Table 9 Silicide coatings for molybdenum

Type	Deposition process
Molybdenum silicide	Fluidized bed, pack cementation, slip pack, plasma spray, electrophoresis
Molybdenum silicide and chromium	Pack cementation
Molybdenum silicide and chromium, boron	Pack cementation
Molybdenum silicide and chromium, aluminum, boron	Slip pack
Molybdenum silicide and tin-aluminum	Cementation and impregnation

Table 10 Cycles for application of silicide and other oxidation-resistant ceramic coatings by pack cementation

Processing cycles suitable for depositing coatings of silicon, chromium, boron, aluminum, titanium, zirconium, vanadium, hafnium, and iron

Substrate metal	Processing cycle Temperature(a) °C	°F	Time, h(b)
Niobium alloys	1040–1260	1900–2300	4–16
Molybdenum alloys	1040–1150	1900–2100	4–16
Tantalum alloys	1040–1150	1900–2100	4–12
Tungsten alloys	1040–1370	1900–2500	3–16

(a) Tolerances: ±6 °C (±10 °F) at 1040 °C (1900 °F); ±14 °C (±25 °F) at 1260 °C (2300 °F). (b) Tolerance, ±10 min

and forms a bcc crystal structure at all temperatures.

The high melting temperature and low vapor pressure of tungsten, along with its ability to be drawn into fine wire, were responsible for its initial commercial application in lamp filaments at the turn of the century. The dawn of the nuclear and space ages in the 1950s led to the growth of tungsten metallurgy. Today, demanding high-temperature applications in aerospace propulsion and energy production rely on tungsten and its alloys. In addition to its high-temperature capabilities, tungsten is commercially important for applications that rely on its high density, strength, and elastic modulus.

Tungsten and tungsten alloys are used in mill products, as an alloying element in tool steels and superalloys, in tungsten carbide cutting tools, and in a variety of tungsten-base chemicals. In terms of refractory metal consumption, tungsten ranks second to molybdenum, with more than 8500 metric tonnes consumed annually. Cutting tools account for 59% of the total; mill products, 26%; alloying, 9%; and chemicals and miscellaneous applications, 6%.

Extraction and Processing Methods (Ref 7)

The principal ores of tungsten are wolframite, $(Fe,Mn)WO_4$, and scheelite, $CaWO_4$. There are two mines in the United States, both in California, where tungsten is mined and concentrated. However, the United States imports roughly 75% of required tungsten concentrates, mostly from China, Bolivia, and Germany.

Tungsten concentrates are chemically processed to yield high-purity tungstic acid or ammonium paratungstate (APT). Both tungstic acid and APT can be converted to tungsten oxide, which is then reduced with hydrogen to pure metal powder. During the reduction cycle, efforts are made to control the resultant tungsten particle size by controlling the initial oxide particle size, the reduction time and temperature, and the moisture content of the reducing atmosphere. Highly pure, fine powders with diameters from 1 to 10 μm (40 to 400 μin.) are commonly available. Dilute alloy additions are sometimes made prior to hydrogen reduction. For example, a small volume fraction of thoria (ThO_2) may be added to high-strength tungsten alloys, and minute quantities of aluminum, potassium, and silicon are added to produce so-called AKS tungsten for lamp filaments.

Tungsten is consolidated to full density by three principal methods, two of which are P/M processes: solid-state sintering and mechanical working (wrought P/M tungsten), liquid-phase sintering of powders, and chemical vapor deposition (CVD). Tungsten and tungsten alloys may also be produced by arc casting or electron beam melting, but these methods are not of significant commercial interest. Although melt-processed tungsten can exhibit higher purity than P/M or CVD products, the slight improvement in mechanical and physical properties does not justify the added expense and engineering challenge of melting tungsten.

Processing Wrought P/M Tungsten. P/M processing of tungsten has its roots in the Coolidge process, which was developed in 1909 by W.D. Coolidge at General Electric Co., Schenectady, N.Y. In the process, a cold-pressed tungsten powder bar is self-resistance heated to around 2500 °C (4530 °F) and sintered to a density of about 90% of theoretical. The bar is then swaged and drawn into wire, or alternatively, the sintered tungsten bar may be forged or rolled into bar or sheet products. The as-sintered bar is brittle at low temperatures because it is fully recrystallized and not fully dense. Thus, initial hot working of tungsten is conducted at temperatures above 1500 °C (2730 °F) until full density is obtained.

The DBTT of fully dense unalloyed tungsten is approximately 300 °C (570 °F) and the recrystallization temperature is 1700 °C (3090 °F). The common thermomechanical processing scheme is to deform the metal at temperatures between the DBTT and the recrystallization temperature. As the amount of deformation increases, both the DBTT and the recrystallization temperature of tungsten decrease. Thus, at room temperature, hot-worked tungsten bar is brittle, but heavily worked tungsten wire exhibits significant ductility. Although tungsten lamp filaments had been used prior to the development of the Coolidge process, their brittleness prevented them from being successfully fabricated into an optimum filament geometry.

The processing of tungsten sheet, rod, and wire is primarily unidirectional in nature, which results in highly elongated grains having a pronounced crystallographic texture (Fig. 7). For rolled sheet, the texture consists of (001) planes in the rolling plane of the sheet with [110] directions aligned along the principal rolling direction. For swaged and drawn rod and wire, the texture consists of [110] directions along the axis of deformation. These textures are commonly seen in heavily worked bcc metals and are often referred to as the "bcc fiber texture."

Dispersion-strengthened tungsten alloys represent a unique class of materials. They are stronger than any other metal at temperatures above 2000 °C (3630 °F). These materials contain dispersions of ThO_2 or hafnium carbide (HfC) and may also include solid-solution additions of rhenium. The alloy systems that have received the most study include W-ThO_2 ("thoriated" tungsten), W-Re-HfC, and W-Re-ThO_2. In these alloys, rhenium serves as a solid-solution strengthening agent as well as a low-temperature ductilizer. The ThO_2 and HfC dispersions retard recrystallization of the alloy and pin dislocations.

Thoriated tungsten is commercially available in a number of bar, rod, and wire forms. It is manufactured by a P/M process similar to the Coolidge process. Thoriated tungsten cannot be worked at temperatures as low as those used for unalloyed tungsten because the ThO_2 particles crack or exhibit matrix decohesion. Thus, the DBTT of thoriated tungsten will be higher than that of unalloyed tungsten. W-Re-HfC alloys can be produced either by P/M processing or vacuum arc melting. These alloys are of interest as fiber reinforcements for nickel-base superalloys, and current research is directed at optimized wire-making processes.

Liquid-phase sintering is used to produce "tungsten-heavy alloys," which are microcomposites of spherical tungsten grains in a continuous ductile matrix of, typically, Ni + Fe or Ni + Cu (Fig. 8). In this method, a mixture of tungsten and

Table 11 Oxide coatings for molybdenum

Type	Deposition process	Thickness range μm	mils
Zirconium oxide/glass	Frit, enamel	130–760	5–30
Chromium-glass	Frit, enamel	130–250	5–10
Chromium/alumina oxide	Thermal spray over chromium plate	200–380	8–15
Alumina oxide	Thermal spray	25–2500	1–100
Zirconium oxide	Thermal spray	25–2500	1–100
Zirconium oxide	Troweling	2500–7600	100–300

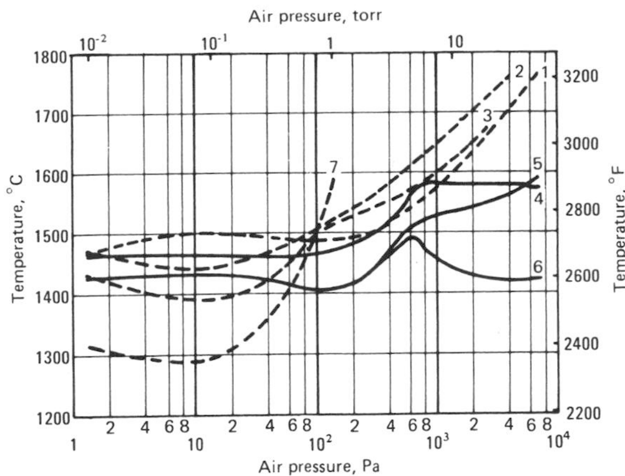

Fig. 6 The effect of air pressure on the maximum temperature for a 4 h life of silicide-coated refractory metals. Alloy/coating: 1, TZM/PRF-6; 2, TZM/Disil; 3, TAM/Durak-B; 4, Cb-752/PFR-32; 5, Cb-752/CrTiSi; 6, B-66/CrTiSi; 7, Ta-10W/Sn-Al

matrix powders is heated in a reducing atmosphere to a temperature sufficient for formation of a liquid phase. Densification is achieved mostly by tungsten particle rearrangement upon liquation and through tungsten particle growth (which is promoted by the rapid transport of tungsten through the liquid phase). With careful control of powder size, powder purity, and sintering variables (atmosphere composition, sintering temperature, and sintering time), sintered tungsten articles of near-theoretical density may be produced.

A variant of liquid-phase sintering is liquid infiltration. The process involves sintering tungsten powder to a density of 70 to 80% of theoretical and then placing the porous preform into contact with a liquid pool of a low-melting metal such as silver or copper. The liquid metal is absorbed by the preform through capillary forces, and a material of nearly full density is obtained.

The ductility of tungsten heavy alloys is directly related to the degree of tungsten-tungsten particle contact in the as-sintered product. Thus, increasing the tungsten content generally leads to a decrease in tensile ductility. The strength of a tungsten heavy alloy is related to both matrix composition and tungsten particle size. Recent research by U.S. tungsten producers has focused on methods to optimize the mechanical behavior of tungsten heavy alloys via microstructural control. In one case, a tungsten heavy alloy was consolidated to full density by rapid application of heat and pressure. The degree of tungsten particle growth and tungsten-tungsten particle contact was greatly diminished relative to the traditional liquid-phase-sintered product. Other areas of current interest include selection of alternative matrix compositions for reduced grain growth during sintering, and microstructural refinement via mechanical alloying.

Production of CVD tungsten begins by introducing tungsten hexafluoride (WF_6) and hydrogen gases into a reaction chamber. The hydrogen reduces the WF_6, and fully dense, high-purity tungsten is deposited onto a suitable substrate. Typically, the substrate is a radially symmetric body such as a cylindrical shell, conical shell, or tube. It is rotated during deposition to yield a deposit of uniform thickness. Following deposition, the substrate is chemically or mechanically removed. Tungsten produced via the classical CVD process has highly oriented columnar grains extending through the entire section thickness (Fig. 9a). Grain orientation, determined by growth kinetics, is such that (001) planes lie in the plane of the deposited sheet with [001] axes oriented perpendicular to the plane of the sheet.

The microstructure of CVD-produced sheet and tubing is fundamentally different from that of the wrought P/M product. CVD sheet, for example, is fully recrystallized and has a different crystallographic texture. In recent years, there have been steady improvements in CVD technology and the use of CVD tungsten has expanded to a variety of applications, including thin-wall tubing and small rocket thrusters. Tungsten also is deposited via CVD in microelectronic applications.

A recent development is the production of fine-structure CVD tungsten, which is expected to have better mechanical properties than the traditional columnar deposit. An example of fine-structure CVD tungsten is shown in Fig. 9(b).

Properties and Application of Tungsten and Tungsten Alloys (Ref 7)

Tungsten and tungsten alloys used for mill products can be grouped as follows:

- *Unalloyed tungsten* with chemical composition requirements conforming to ASTM B 760
- *Doped tungsten* containing minute quantities of aluminum (15 ppm), silicon (50 ppm), potassium (90 ppm), and oxygen (35 ppm)
- *Solid-solution alloys* containing various amounts of molybdenum (2 to 20%) or rhenium (1 to 25%)
- *Dispersion-strengthened alloys* containing 1 to 2% ThO_2, or tungsten-rhenium alloys with a dispersion of ThO_2 or HfC added
- *Tungsten heavy alloys,* which consist of tungsten-nickel-copper or tungsten-nickel-iron alloys

Fig. 7 Microstructure of wrought powder metallurgy tungsten sheet. Note the heavily worked anisotropic grains. X-ray analysis of this unalloyed tungsten product would reveal a pronounced crystallographic texture with [110] directions oriented along the elongated axes of the grains. 500×. Courtesy of J.P. Wittenauer, Lockheed Martin Missiles & Space Co., Inc.

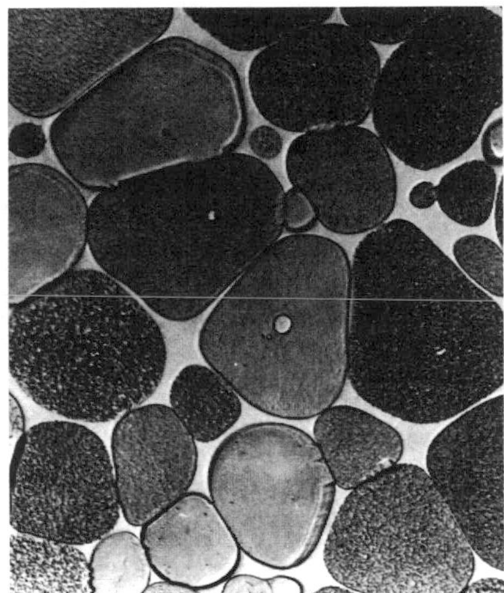

Fig. 8 Liquid-phase sintered heavy tungsten alloy (95W-3.5Ni-1.5Fe). The structure consists of spherical tungsten grains surrounded by a ductile phase rich in iron and nickel. 350×

Table 1 and Fig. 1 to 3 compare the properties of pure tungsten with those of other refractory metals. Properties of tungsten alloys are compared in Table 2.

Properties of Unalloyed Tungsten. The mechanical properties of wrought P/M are strongly dependent on deformation history, purity, and testing orientation. Generally, an increased level of deformation below the recrystallization temperature raises strength and ductility and lowers the DBTT. When wrought P/M tungsten is fully

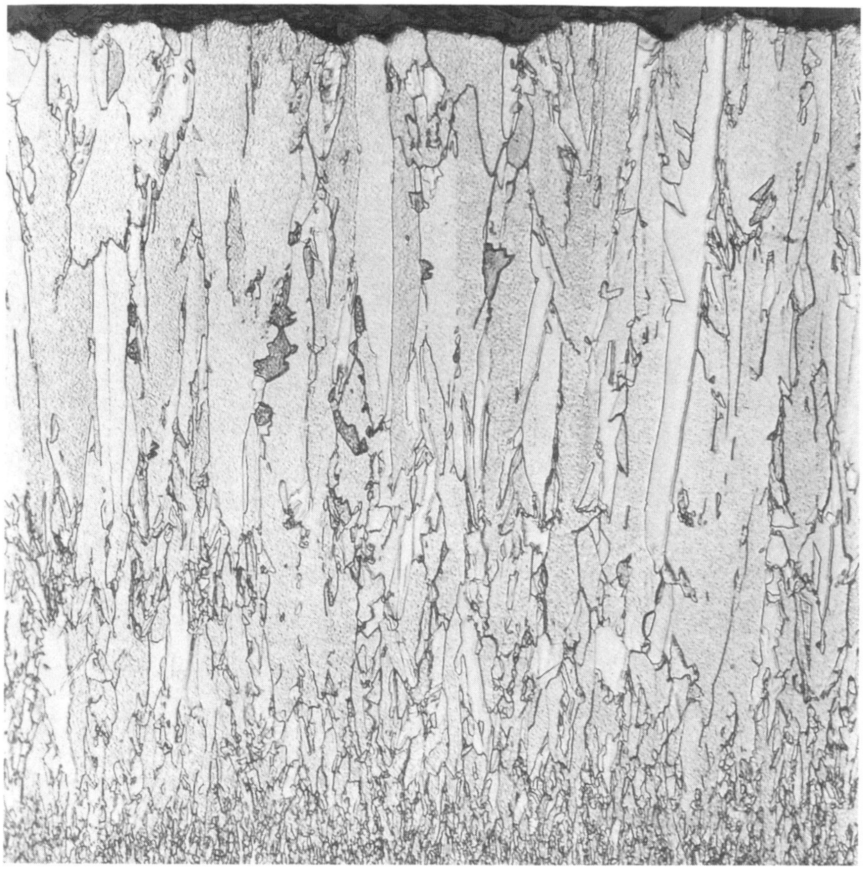

(a)

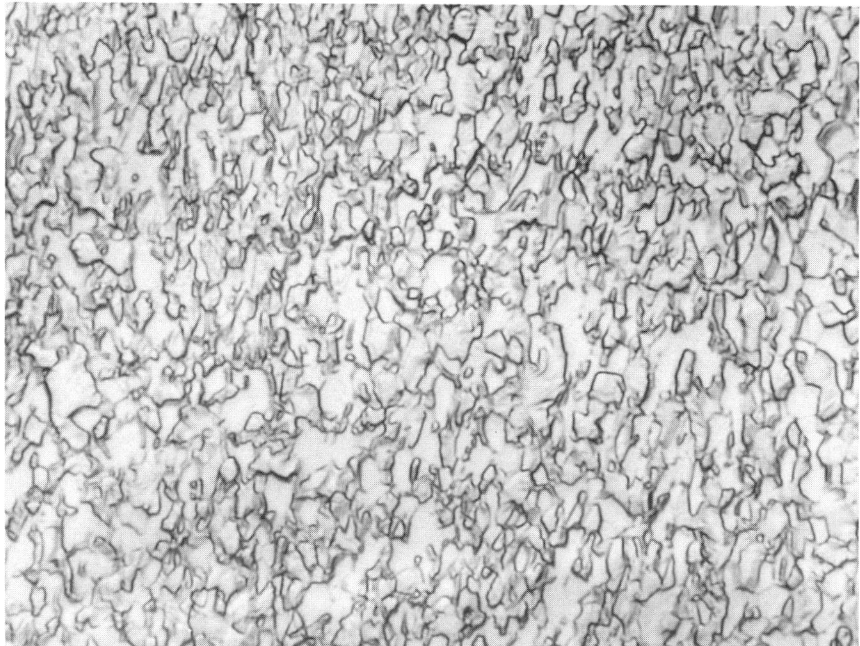

(b)

Fig. 9 Microstructure of tungsten produced by chemical vapor deposition (CVD). (a) Microstructure of tungsten sheet produced by classic CVD from gaseous WF_6 and H_2 precursors. Initially, tungsten grains of random orientation nucleate on a substrate (at bottom of photo). A columnar grain structure results, because growth kinetics favor only those grains with a [001] axis oriented perpendicular to the substrate. 200×. (b) Fine-structure CVD tungsten results when the growth of columnar grains is periodically interrupted, allowing for the nucleation of new grains. The strength and ductility of the fine-structure product are comparable to those of wrought powder metallurgy tungsten at temperatures above the transition temperature. 1000×. Courtesy of J.P. Wittenauer, Lockheed Martin Missiles & Space Co., Inc.

recrystallized, its DBTT is in the range of 200 to 300 °C (390 to 570 °F). Tensile failure of recrystallized tungsten is primarily a result of grain boundary separation, although individual grains may plastically deform a great deal prior to intergranular failure. By contrast, as-wrought (unrecrystallized) P/M tungsten can have a DBTT below room temperature for the extreme case of heavily drawn wire. Tensile failure of as-wrought tungsten is characterized by a combination of transgranular cleavage and ductile rupture across the grains with secondary longitudinal splitting among the fibrous grains.

For both wrought and fully recrystallized P/M tungsten, the ductility is higher and the DBTT is lower for finer grain sizes and increased purity. The ductility and strength of recrystallized tungsten is less than that of as-worked tungsten at all temperatures. The observed deformation-related mechanical property enhancement is associated with at least two factors:

- Deformation imparts an anisotropic grain structure that is elongated in the direction of deformation. The grain boundary is most often the "weak link" in tensile failure of tungsten. Relatively little grain boundary area is oriented perpendicular to a tensile testing axis (mechanical properties of sheet, rod, and wire are most often measured in the principal direction of deformation). Thus, greater plastic flow can occur in the material prior to failure via grain boundary separation. By contrast, tungsten microstructures characterized by the fibrous texture have virtually zero ductility when tested in an orientation transverse to the fiber axis.
- Pure tungsten has a low solubility for the interstitial elements carbon, oxygen, and nitrogen. In recrystallized material, these elements tend to segregate to grain boundaries (low-energy positions), resulting in a decrease in ductility. However, heavy working of the tungsten produces a multitude of additional low-energy sites (dislocation tangles). Segregation of impurities to these sites, rather than to grain boundaries, helps as-wrought tungsten retain high grain-boundary strength, which results in greater overall strength and ductility.

Applications of Unalloyed Tungsten. Wire made of unalloyed tungsten may be woven into a mesh and used as electrical-resistance heating elements in high-temperature vacuum furnaces (to 3000 °C, or 5430 °F). Similarly, tungsten sheet is used for heat shields in high-temperature furnaces, and sheet and foil are used for radiation shields in medical x-ray equipment. Unalloyed tungsten sheet can be fabricated using conventional metalworking operations, such as shearing, blanking, and roll forming, if the sheet is heated to a temperature above its DBTT. Since the DBTT of tungsten is lowered by increasing amounts of deformation, thinner sheet generally requires a lower preheat temperature.

A demanding application of unalloyed tungsten sheet is the wall of the plasma-containment chamber in a magnetohydrodynamic (MHD) power generation system (Ref 7). In the MHD

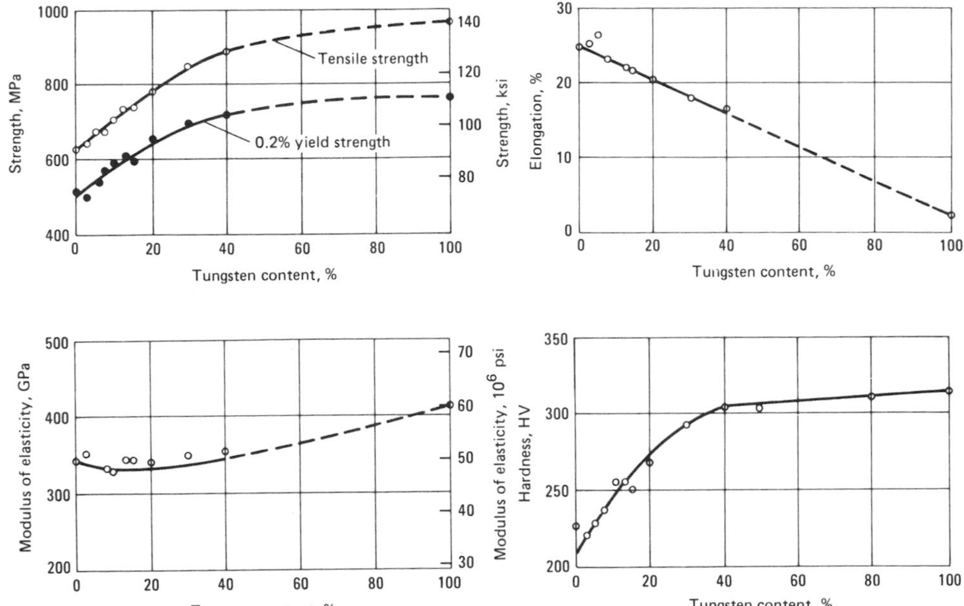

Fig. 10 Effect of tungsten content on the room-temperature mechanical properties of tungsten-molybdenum alloys

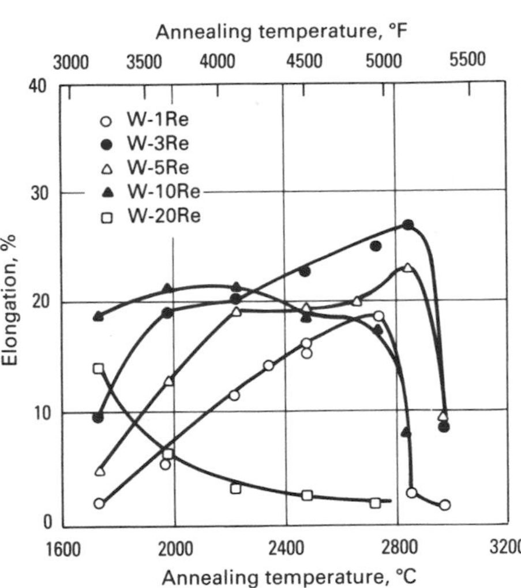

Fig. 11 Room-temperature ductility of annealed wire for five tungsten-rhenium alloys

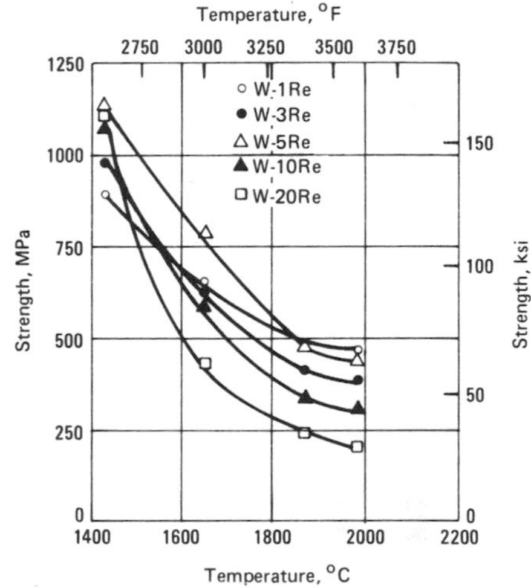

Fig. 12 Short-time tensile strengths of five tungsten-rhenium alloys

Table 12 Properties of tungsten-molybdenum alloys

Alloy content, wt%		Melting point, °C	Density, g/cm^3	Thermal coefficient, µm/m · K, at 20 to 100 °C	Brinell hardness
Molybdenum	Tungsten				
100	0	2600	10.2	4.75×10^{-3}	200
90	10	2620	11.2	4.02×10^{-3}	210
80	20	2640	12.1	3.50×10^{-3}	230
72.5	27.5	2675	12.8	3.25×10^{-3}	250
51	49	2850	14.8	2.90×10^{-3}	300
20	80	3075	17.5	3.20×10^{-3}	330
0	100	3370	19.3	4.82×10^{-3}	350

Source: Ref 8

system, which is still under development as of this writing, an advanced coal combustor is used to produce a high-temperature plasma that is accelerated to supersonic velocity within a chamber. The plasma interacts with a magnetic field to generate electrical current. The walls of the chamber used to contain the plasma are subjected to high-temperature electrochemical corrosion, electrical-arc erosion, and particle erosion. Wrought P/M tungsten was chosen for the wall material because of its high thermal conductivity and excellent resistance to both corrosion and wear. Compared with coal-fired power generation, MHD promises improved thermal efficiency and decreased SO_x and NO_x emissions.

Other uses of unalloyed tungsten include foil screens for photoetching in the electronics industry, crucibles for processing liquid metals, and sputtering targets (high-purity tungsten) in vapor deposition systems. Tungsten also has been evaluated as a hydrogen barrier coating for rocket nozzles, an application that exploits its low solubility for hydrogen.

Solid-Solution Alloys. Both tungsten-molybdenum and tungsten-rhenium alloys have been extensively studied. Tungsten-molybdenum alloys are of interest because molybdenum helps refine the grain size of arc-cast tungsten. They also are less dense than unalloyed tungsten, which is a plus for aerospace applications. In addition, molybdenum lowers the melting point of tungsten, which makes it easier to produce material by melt processing. On the other hand, tungsten-molybdenum alloys have lower maximum useful operating temperatures than unalloyed tungsten. The effects of various molybdenum additions on the properties of tungsten-molybdenum alloys are shown in Table 12 and Fig. 10. A current application of tungsten-molybdenum alloys is a rotating shaft that operates in a corrosive environment. Other materials met requirements for corrosion resistance, but the tungsten-base alloy also offered superior wear resistance.

Tungsten-rhenium alloys are of tremendous interest because of the ductilizing effect achieved by the rhenium addition. Tungsten-rhenium alloys are stronger than unalloyed tungsten, and they have a lower DBTT and higher recrystallization temperature. (Rhenium is the only alloying element known to lower the DBTT of tungsten.) The effects of various rhenium additions on the properties of tungsten-rhenium alloys are shown in Fig. 11 and 12. Applications of tungsten-rhenium alloys include high-temperature thermocouple elements, propulsion-system components, and rotating anodes for x-ray tubes (the rhenium addition provides increased resistance to thermal shock and thermal fatigue).

Dispersion-Strengthened Alloys. Thoriated tungsten was originally developed as a filament alloy for the lighting industry. The ThO_2 dispersion modifies the grain structure, resulting in extended filament life. ThO_2 also increases the high-temperature strength and lowers the work function of tungsten. (A lower work function translates into less evaporative loss of tungsten at the high filament operating temperature.)

Today, thoriated tungsten is also used as a high-temperature structural material. The ThO_2 particles retard recrystallization in wrought tungsten, which inhibits grain growth at high tempera-

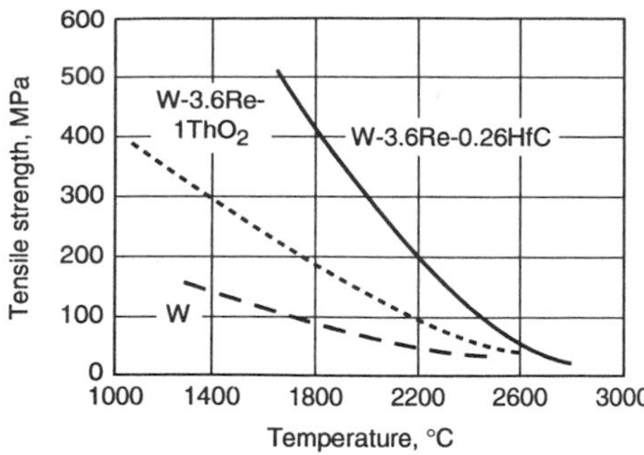

Fig. 13 Elevated-temperature tensile strength of dispersion-strengthened tungsten alloys. Source: Ref 7

tures and results in a maximum temperature capability that is several hundred degrees higher than that of unalloyed tungsten. ThO_2 particles also interact with dislocations during deformation, which strengthens the material via the classical dispersion-hardening mechanism. Thoriated tungsten is stronger than unalloyed tungsten at all temperatures.

Thoriated tungsten is commercially available in a number of wire, rod, bar, and plate products. Care must be taken when using large-diameter bars: insufficient deformation during processing can result in centerline porosity, surface-to-center variability in the DBTT, and poor machining characteristics.

A successful application of large-diameter thoriated tungsten bars is forged valve bodies for altitude-control systems aboard spacecraft, which operate at temperatures as high as 1650 °C (3000 °F). Thoriated tungsten also is being evaluated by the National Aeronautics and Space Administration (NASA) as an electron emitter for space-based thermionic power systems. This application takes advantage of the increased electron emittance provided by the ThO_2 addition, as well as the high creep strength of the material.

As noted above, the addition of rhenium to tungsten simultaneously raises high-temperature strength (via solid-solution strengthening) and low-temperature ductility. To further improve the strength of tungsten-rhenium alloys, a dispersion of ThO_2 or HfC may be added. Research also has been conducted on the strengthening effects of zirconia particles formed by internal oxidation of tungsten-zirconium alloys.

Dispersion-strengthened tungsten alloys are among the strongest materials ever produced for service at temperatures up to 2000 °C (3630 °F). Of the tungsten alloys, those in the W-Re-HfC system have the highest strength (Fig. 13). W-Re-HfC alloys currently are in an advanced stage of development and are being evaluated for a number of applications, including reinforcing fibers in cast nickel-base superalloy turbine blades (to increase the temperature capability) and as structural materials in space-based power generating systems. The use of W-Re-HfC fibers in superalloys is discussed below in the section "Refractory Metal Fiber-Reinforced Composites" in this article.

Developing Sag-Free Lamp Filaments. Drawn tungsten wire is extraordinarily strong. For example, room-temperature tensile strengths of 5800 MPa (850 ksi) have been recorded for fine tungsten filaments. The wire also has sufficient ductility to be coiled, at room temperature, into lamp filaments. The strength and ductility are results of the large amount of reduction imposed during wire drawing.

When the lamp is lighted for the first time, filament temperature may be as high as 2500 °C (4530 °F). At this temperature, the tungsten filament readily recrystallizes and the fibrous microstructure of the as-drawn wire is replaced by recrystallized, somewhat equiaxed grains. Loading of the filament (due to gravity or vibrational effects) during service can eventually cause creep deformation (sagging). The creep-deformation mechanism most often observed in lamp filaments is grain boundary sliding, which causes local thinning of the filament. In these thin areas, the local electrical resistance of the wire increases and filament temperature rises, which, in turn, causes an increase in the rate of evaporative loss of tungsten, further filament thinning, and, ultimately, lamp failure. Thus, the key to producing longer-lasting lamp filaments is to fabricate a wire that does not sag.

Early studies of tungsten filament materials demonstrated that a ThO_2 addition could impart antisagging characteristics to tungsten wire. Although still in commercial use, thoriated tungsten wire has largely been replaced by AKS tungsten for common light bulb filaments. AKS tungsten provides longer filament life for the same cost.

Recrystallized AKS tungsten has a highly anisotropic, interlocking grain boundary structure with exaggerated grain growth extending along the direction of the drawn wire (Fig. 14). No grain boundaries directly span the entire wire cross section; instead, the grain boundaries form an acute angle with the axis of the wire. This structure is due to preferential segregation of the aluminum, potassium, and silicon additives along grain boundaries during wire fabrication. Figure 15 shows the effect of doping to produce sag-free lamp filaments.

The additives in AKS tungsten form longitudinal grain boundary stringers in the fibrous as-drawn microstructure. When the wire recrystallizes upon initial heating (lamp lighting), the additive particles retard grain growth in the through-thickness of the wire. What results is the

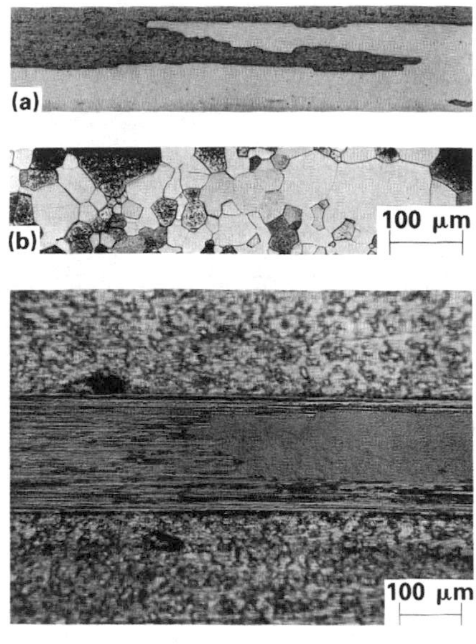

Fig. 14 Recrystallized structure of tungsten wires, 180 μm (7 mil) in diameter. (a) Doped lamp grade exhibiting interlocking grain structure. (b) Undoped grade exhibiting equiaxed grain structure. (c) Finger-type grain growth in doped tungsten wire.

characteristic exaggerated grain growth in the longitudinal direction.

Current research in the lamp filament arena is aimed at two aspects of microstructural control: optimized dopant distribution through processing modifications, and improved filament recrystallization heat treatment schedules.

Tungsten Heavy Alloys. The mechanical properties of tungsten heavy alloys are closely linked to tungsten content, sintering variables, and the amount of post-sintering mechanical working. Most applications seek to exploit the high density of the material. However, as density increases with increasing tungsten content, ductility decreases due to more tungsten-tungsten particle contact). Mechanical working greatly increases the strength of tungsten heavy alloys, but at the expense of ductility. Thus, for applications where mechanical properties are important, both composition and degree of mechanical working must be carefully chosen. Table 13 lists properties of various W-Ni-Fe alloys.

Tungsten heavy alloys are principally used in counterweights and balances, radiation shielding,

Table 13 Mechanical properties of tungsten-nickel-iron heavy alloys

W, %	Condition	Density, g/cm^3	Hardness, HRC	0.2% offset strength, MPa	(ksi)	Tensile strength, MPa	(ksi)	Ductility, %
90.7	As-sintered	17.21	28	585	85	925	134	38
93	As-sintered	17.70	30	600	87	945	137	35
95	As-sintered	18.09	30.5	600	87	950	138	30
97.3	As-sintered	18.59	31	600	87	930	135	16
97.3	Lightly worked	18.59	40	895	130	1115	162	7
97.3	Heavily worked	18.59	45.5	1180	171	1250	181	1

Source: Ref 7

heat sinks and electrical contacts, and antiarmor kinetic-energy penetrators.

Tungsten Net Shapes Produced by CVD. The tensile properties of CVD tungsten differ markedly from those of P/M wrought material. CVD material is in the fully recrystallized condition and thus does not have its mechanical properties enhanced by heavy deformation.

Generally, the tensile behavior of recrystallized tungsten is dominated by grain boundary effects. For fully recrystallized P/M tungsten, a fine grain size lowers the DBTT, while at temperatures above the DBTT, tensile failure invariably occurs by grain boundary separation. For the columnar deposits produced by traditional CVD processing, a single grain boundary may span the entire cross section, and grain boundary separation during tensile loading results in immediate specimen failure. In contrast, the grain size of wrought P/M tungsten (whether in the worked or recrystallized condition) is rather fine relative to the section thickness. This is why the tensile and creep properties of conventional CVD tungsten (having columnar grains) are generally inferior to those of P/M wrought materials. However, the strength and ductility of the recently developed fine-structure CVD tungsten are comparable to those of wrought P/M products at temperatures above the DBTT.

The primary advantage of CVD processing is its ability to yield net-shape products and products that are difficult or impossible to make from P/M tungsten. Most applications of CVD tungsten are currently in the developmental stage including thin-wall tubing; advanced hybrid materials such as compositionally graded, clad, or bimetallic structures; and CVD-coated wire mesh for advanced energy applications. CVD alloys also have been used to produce tungsten-rhenium alloys by co-reducing a mixture of tungsten- and rhenium-fluoride gases.

Fig. 15 Comparison of the effect of a large, interlocking grain structure from doped tungsten and equiaxed grain structure on the sag of a lamp filament. (a) Doped. (b) Undoped. Note filament sag.

Oxidation of Tungsten

The oxidation resistance of tungsten is intermediate to that of molybdenum and niobium. Tungsten trioxide becomes significantly volatile in the range of 800 to 1000 °C (1475 to 1830 °F) (Ref 5). Although some new alloys based on the W-Mo-Cr system exhibit improved oxidation resistance, coatings are usually recommended for high-temperature oxidizing environments.

High-Temperature Protective Coatings for Tungsten. A more restrictive situation exists with tungsten than with coated molybdenum. Silicide coatings are protective for a short time up to 1925 °C (3500 °F). In the temperature range of 1090 to 1650 °C (1995 to 3000 °F), lives of 10 to 50 h are attained.

Modifying the silicides with tungsten, zirconium, or titanium may be a more successful approach. A tungsten-modified silicide pack coating with HfB_2 additives is available commercially. A multi-element umbrella type, consisting of iridium electrodeposited from a fused salt bath, followed by a plasma-sprayed ZrO_2 layer, prevents volatilization losses during very high-temperature oxidation. This may be serviceable at 1800 °C (3270 °F) for several hours. In the ceramic approach, a coating of ThO_2 over a tungsten wire mesh can be used at 2925 °C (5300 °F), and a coating of HfO_2 can be used at 2700 °C (4890 °F).

Niobium and Niobium Alloys

Niobium, also called columbium, melts at 2468 °C (4474 °F). It is a tough, ductile, silver-gray metal having a unique combination of physical and mechanical properties. Of the five major refractory metals, it has the lowest melting temperature, density (8.57 g/cm^3, 0.31 lb/in.^{3}), and elastic modulus at room temperature (103 GPa, or 14.9 psi $\times$ 10^6), while its coefficient of linear thermal expansion (7.3 μm/m · K) is the highest. Niobium has a thermal-neutron cross section of only 1 barn/atom (1.1 $\times$ 10^{-28} m^2/atom), and it is extremely stable in a wide variety of normal, corrosive environments. Table 1 compares the properties of pure niobium with those of other refractory metals. Comparative data are also given in Fig. 1 to 3.

The crystal structure of niobium is bcc, and the ductile-to-brittle transition behavior is typical of bcc refractory metals. The DBTT of commercially pure, polycrystalline niobium sheet is about –160 °C (–255 °F), lower than that of all refractory metals except tantalum.

Discovered and identified in the early 19th century, niobium was considered little more than a laboratory curiosity for nearly 100 years. The first commercial use of any importance is believed to have occurred around 1925, when microalloy additions of niobium were used to improve the mechanical properties of carbon, stainless, and tool steels. Presently, the use of niobium as an alloying element in steels, superalloys, and nonferrous alloys accounts for about 95% of production; the consumption of niobium and niobium-base alloys accounts for the remaining 5%. Niobium ranks third, behind molybdenum and tungsten, in terms of refractory metal production.

Production of Niobium

Prior to about 1960, niobium metal was produced by P/M processes, such as high-temperature vacuum sintering and carbon reduction of the metal oxide. The resulting metal or alloys actually exhibited better mechanical properties than current higher-purity materials do, because of their higher trace-element metallic impurities.

At present, the major process for the recovery of niobium is the aluminothermic reduction of pyrochlore concentrates to ferroniobium. Niobium metal is purified by a chlorination process wherein volatile $NbCl_5$ is distilled and then hydrolyzed to pentoxide. The metal is then recovered by a second aluminothermic reduction:

$$3Nb_2O_5 + 10Al \rightarrow 6Nb + 5Al_2O_3$$

During the exothermic reaction, oxide impurities slag from the molten niobium. After aluminothermic reduction of the recovered oxide, the metal is further purified by electron beam melting.

Another niobium recovery process involves the collection of niobium oxide as a byproduct in the processing of tantalum ores. The ore is digested in mixed acids containing hydrofluoric acid and sulfuric acid, and solvent extraction is employed for purification.

A third method for producing niobium, less frequently used, is P/M processing. Powders are produced from ingot by hydriding, crushing, and dehydriding. In addition, some recent efforts have been directed toward producing complex metastable alloy powders, such as niobium-aluminum and niobium-silicon alloys, by liquid metal atomization and rapid quenching.

Niobium alloys are made by subsequent vacuum arc remelting with the appropriate elemental additions. The most common alloy additions are zirconium, titanium, hafnium, tantalum, and tungsten.

Fabrication of the common alloys is generally accomplished by high-temperature extrusion or forging near the alloy recrystallization temperature, which is typically 1095 to 1370 °C (2000 to 2500 °F). Secondary fabrication is completed by warm working and cold working to final shapes with appropriate stops for recrystallization annealing, which is performed under vacuum. Most commercial alloys are ductile enough to be processed into various mill products, such as sheet, foil, rod, wire, and tubing.

Properties and Applications of Niobium Alloys

The most common niobium alloys developed to date are listed in Table 14. The majority of these alloys were developed during the 1960s and early 1970s for a number of applications (Ref 9). These included highly fabricable and weldable low-strength alloys, such as Nb-1Zr and PWC-11 (Nb-1Zr-0.1C), for containment of liquid alkali metals in space nuclear power systems. These alloys typically contained additions of zirconium or hafnium, which removed oxygen from solution in niobium by precipitation of ZrO_2 or HfO_2, providing enhanced alkali metal corrosion resistance. Another application was for moderate-strength fabricable alloys, generally produced in the form of sheet for aerospace applications. Alloys such as FS-85 (Nb-28Ta-10W), C129Y (Nb-10W-10Hf-0.2Y), and B-66 (Nb-5Mo-5V-1Zr) were typical of this group. Higher-strength alloys for use in aircraft gas turbine blades were also developed, such as B-88 (Nb-28W-2Hf-0.067C) and Cb-1 (Nb-30W-1Zr-0.06C).

All of the niobium alloys listed in Table 14 are solid-solution-strengthened alloys. However, second-phase particles are observed in the microstructures of these alloys. The compositions of these particles vary, but they are generally associated with impurities such as oxides, nitrides, and carbides. Often the size and distribution of second phases can have a strong influence on mechanical properties and recrystallization behavior (Ref 10). A variation of the Nb-1Zr alloy, commonly known as PWC-11, contains an intentional addition of 0.1% C to form carbide precipitates that significantly improve high-temperature creep properties. More detailed information on the effects of carbides on high-temperature properties of niobium alloys can be found in the section "Niobium Alloys for Space Power Reactors" in this article.

Table 14 Compositions of selected niobium alloys

Common designation	Nominal alloy additions, %
Low strength and high ductility	
Nb-1Zr	1 Zr
D-14, B-33	5 Zr
C-103	10 Hf, 1 Ti, 0.7 Zr
Cb-753	5 V, 1.25 Zr
D-36	10 Ti, 5 Zr
Moderate strength and ductility	
PWC-11	1 Zr, 0.1 C
SCb-291	10 W, 10 Ta
C-129Y	10 W, 10 Hf, 0.1 Y
Cb-752	10 W, 2.5 Zr
D-43	10 W, 1 Zr, 0.1 C
FS-85	28 Ta, 11 W, 0.8 Zr
SU-16	10 W, 3 Mo, 2 Hf
B-66	5 Mo, 5 V, 1 Zr
AS-55	10 W, 1 Zr, 0.06 Y
High strength	
Cb-1	30 W, 1 Zr, 0.05 C
B-88	28 W, 2 Hf, 0.07 C
SU-31	17 W, 3.5 Hf, 0.12 C, 0.05 Si
Cb-132M	20 Ta, 15 W, 5 Mo, 1 Zr, 0.1 C
F-48	15 W, 5 Mo, 1 Zr, 0.05 C
F-50	15 W, 5 Mo, 5 Ti, 1 Zr, 0.05 C
WC-3009	30 Hf, 9 W

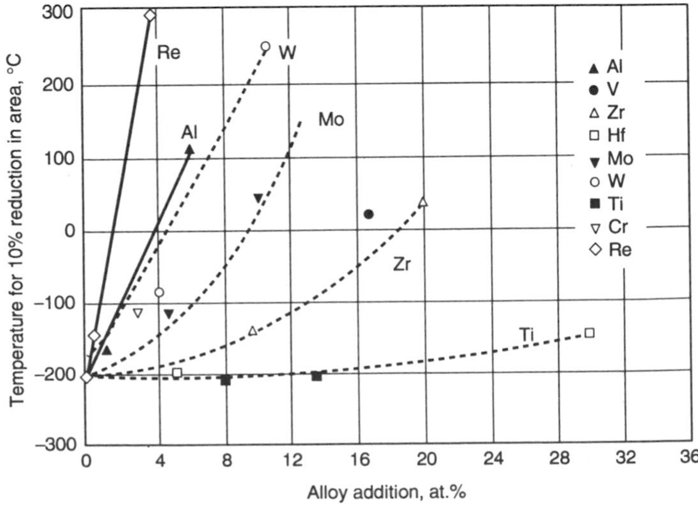

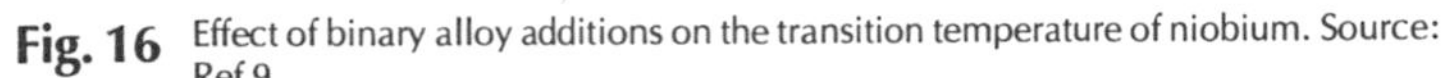
Fig. 16 Effect of binary alloy additions on the transition temperature of niobium. Source: Ref 9

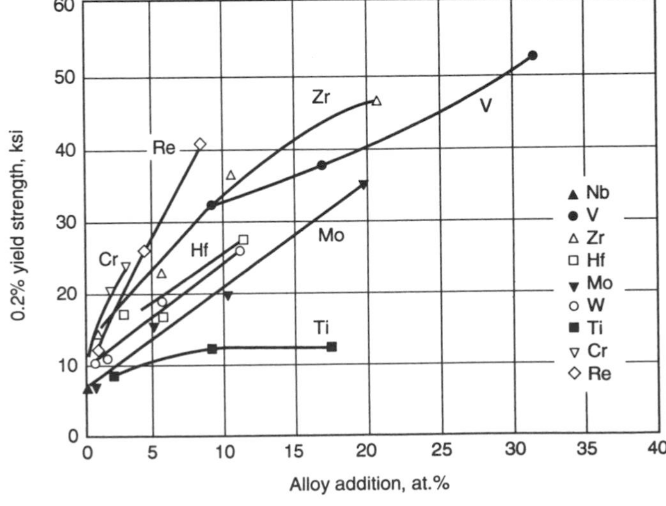

Fig. 17 Effect of binary alloy additions on the yield strength of niobium at 1095 °C (2000 °F). Source: Ref 9

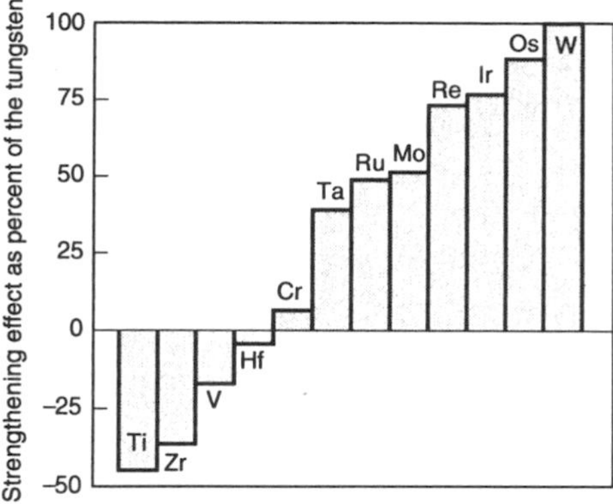

Fig. 18 Creep strengthening effect of alloying elements in niobium at 1200 °C (2190 °F). Source: Ref 9

Alloying Effects and Properties (Ref 9). The alloying additions used in commercial and advanced developmental niobium alloys include:

- Substitutional solutes (Mo, W, V, Ta)
- Reactive elements that have a higher negative free energy of formation for carbides, nitrides, and/or oxides than the matrix element (Zr, Hf, Ti)
- Interstitials (C, N)

The effects of these alloying elements on the DBTT, elevated-temperature yield strength, and creep strength of niobium alloys are reviewed below.

A key objective in developing alloys for structural applications is to achieve appropriately high strength levels while maintaining satisfactory fabricability and ductility. Niobium, in common with other metals having a bcc crystal structure, undergoes a ductile-to-brittle transition at low temperatures. In pure niobium, the tensile transition temperature is quite low, on the order of –200 °C (–325 °F), but solutes can increase the transition temperature rather dramatically. Figure 16 shows the effect of binary alloy additions on the fracture transition temperature of niobium. In early studies, the group VA elements (chromium, molybdenum, and tungsten) had a pronounced effect in raising this temperature, as did rhenium and aluminum. The reactive elements zirconium and hafnium had much less effect, and titanium and tantalum had no significant influence. These studies indicated that only a limited number and range of substitutional solutes could be used if a balanced combination of properties was to be achieved.

Similar data on elevated-temperature (1095 °C, or 2000 °F) yield strength are shown in Fig. 17. Again, significant strengthening was exhibited by alloys containing additions of tungsten, molybdenum, chromium, rhenium, and vanadium, while titanium had no effect on short-time elevated-temperature strength properties. However, in these tests appreciable strengthening was exhibited by alloys containing zirconium and hafnium. Because these alloys were produced with niobium of moderate interstitial level (250 ppm C, 150 ppm N, and 450 ppm O), part of the strengthening may be attributed to the interaction of the interstitials with the reactive solutes zirconium and hafnium.

The relative effect of alloy additions on the creep strength of niobium is quite different. McAdam (Ref 11) investigated the effects of various substitutional solutes on the creep strength of niobium. The strengthening contributions of the individual elements are summarized in Fig. 18, where the contributions are expressed as a percentage of the effect of tungsten, the most potent solute. The data are for 1200 °C (2190 °F). McAdam's results showed that the group VIA elements, particularly tungsten and molybdenum, improved the creep strength of niobium, as did rhenium and tantalum. Unfortunately these elements, with the exception of tantalum, also raised the fracture transition temperature of niobium significantly. McAdam also observed that vanadium, hafnium, zirconium, and titanium make a negative contribution to the creep strength of niobium alloys, even though, as stated above, vanadium has a potent effect in increasing the short-time elevated-temperature strength of niobium. The weakening of niobium by titanium additions has been observed by a number of investigators. McAdam showed that the precious metals osmium, iridium, and ruthenium were effective strengthening elements, but they were not considered for engineering alloys because of the obvious cost and availability problems.

In summary, only a limited number and range of alloy additions can be effectively used in niobium alloys intended for high-temperature applications (Ref 9). Tungsten and molybdenum are effective strengtheners but significantly degrade low-temperature ductility and weldability. Rhenium has the same characteristics and provides no benefit compared to tungsten as an alloy addition. Furthermore, all of these elements increase alloy density. Tantalum provides moderate strengthening without degrading low-temperature ductility or fabricability, but this benefit is achieved at the expense of alloy density. Vanadium is a very effective strengthener with respect to short-time properties, but it degrades creep strength. Titanium additions result in ductile and fabricable alloys but greatly degrade strength properties. Alloy additions of zirconium and hafnium are used primarily to form carbides and carbonitrides for dispersed-phase strengthening. Strengthening of niobium alloys arising from the precipitation of reactive metal carbides and nitrides has been shown by many investigators, and this strengthening mechanism is a key contributor to the properties of the higher-strength niobium alloys, such as B-88, Cb-1, and Cb-132M. Experience with high-strength niobium alloys indicates that hafnium additions are equally effective carbide formers as zirconium, and that they provide somewhat superior fabricability than zirconium at equivalent additions.

The high-temperature tensile and creep properties of common niobium alloys are shown and compared to those of tantalum alloys and TZM molybdenum in Fig. 19 and 20. In Fig. 21, the long-term creep behavior of Nb-1Zr is compared to that of other refractory alloys, type 316 stainless steel, a cobalt-base superalloy (HS-188), and a nickel-base superalloy (Hastelloy X). Figure 22 compares the short-term creep behavior of nickel-base alloys and high-strength niobium alloys.

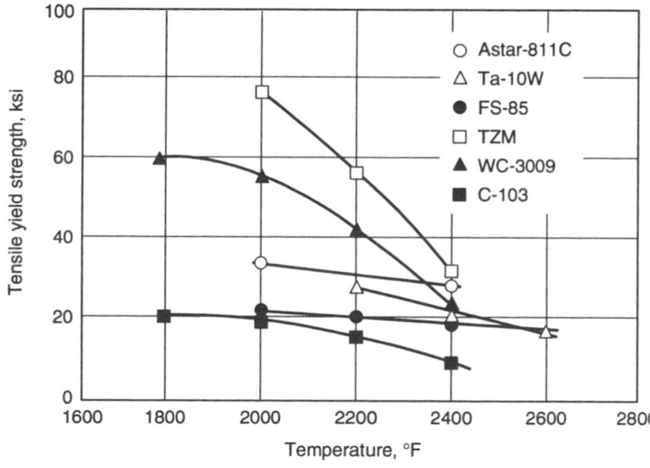

Fig. 19 Tensile yield strength vs. temperature for various refractory metal alloys. Source: Ref 10

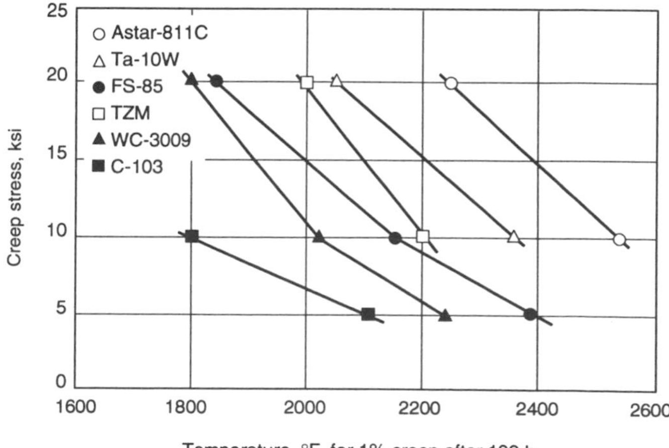

Fig. 20 Creep stress vs. temperature for various refractory metal alloys. Source: Ref 10

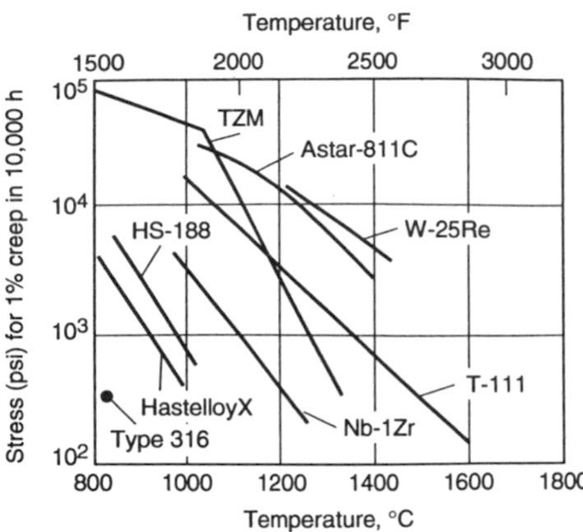

Fig. 21 Long-term creep properties of various refractory metal alloys and superalloys. Source: Ref 10

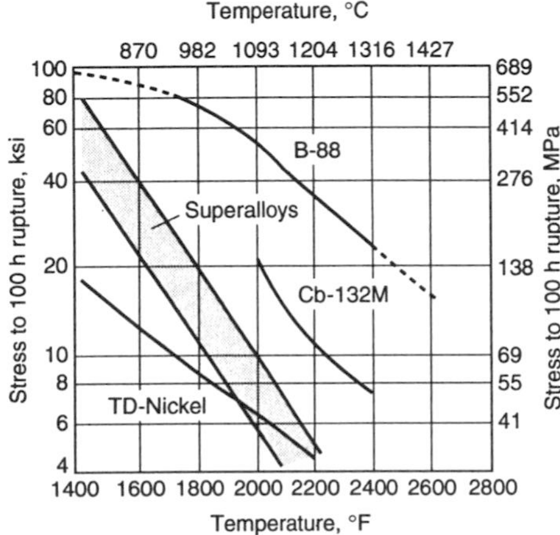

Fig. 22 100 h rupture strength vs. temperature for niobium alloys and other gas turbine bucket materials. Source: Ref 9

Applications. Niobium and niobium alloys find use in the aerospace, nuclear, and chemical processing industries. Applications of unalloyed niobium metal include heat exchangers for chromium electroplating solutions, cathodic protection systems, and electronic and nuclear components. Ultrahigh-purity niobium, having <25 ppm total interstitial elements (oxygen, carbon, nitrogen, and hydrogen), is used in superconducting microwave-cavity electron accelerators.

Alloy C-103 has been widely used for rocket components that require moderate strength at about 1095 to 1370 °C (2000 to 2500 °F). Alloy Nb-1Zr is used in nuclear applications because it has a low thermal neutron absorption cross section, good corrosion resistance, and good resistance to radiation damage. It is used extensively for liquid metal systems operating at 980 to 1205 °C (1800 to 2200 °F). The Nb-1Zr alloy combines moderate strength with excellent fabricability. As a result, it is used for parts in sodium vapor and magnesium vapor lamps.

Vapor deposition of Nb-1Zr or niobium on the inside surface of type 316 stainless steel tubing improves the performance of the tubing in many chemical process applications without degrading the mechanical properties of the stainless steel. An intermediate layer of pure niobium under Nb-1Zr improves adherence to the steel substrate.

Alloys C-129Y, FS-85, and Cb-752 have shown higher elevated-temperature tensile and creep strengths than C-103 while maintaining good fabricability, coatability, and thermal stability. They are used for leading edges, nose caps for hypersonic flight vehicles, rocket nozzles, gas turbines, and guidance structures for reentry vehicles. Alloy C-3009 is being evaluated for potential use in fasteners and gun barrels.

As described in the next section, niobium alloys are also being evaluated for use in space nuclear power systems. Alloy PWC-11 shows promise for such applications.

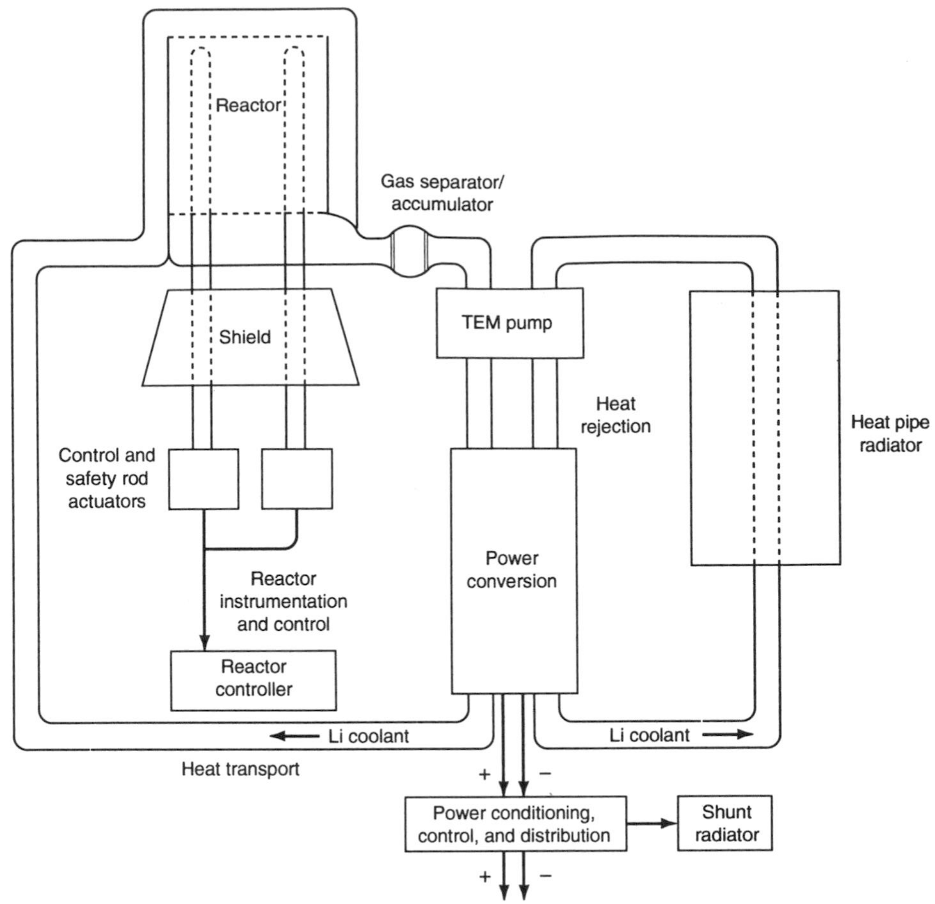

Fig. 23 Schematic of the SP-100 fast-spectrum reactor. TEM, thermoelectric electromagnetic. Source: Ref 12

Niobium Alloys for Space Power Reactors (Ref 12)

In the 1980s the United States, through its Department of Energy, Department of Defense, and NASA, initiated a three-phase joint program to develop the technology necessary for a space nuclear power system for military and civil applications. The program is called SP-100, for its goal of a 100 kW space power reactor (Ref 13). Current spacecraft require from a few hundred watts to about 75 kW in electric power. The space station is expected to initially require about 100 kW. Other future missions, such as a lunar base or a manned flight to Mars, are expected to require tens to hundreds of megawatts.

Initial studies resulted in the selection of a concept using a fast-spectrum reactor coupled by magnetically pumped liquid lithium to thermoelectric energy conversion modules, which in turn are coupled to individual heat-pipe-based waste-heat-rejection radiator panels (Fig. 23). The technical specifications for the SP-100 system include design life of 10 years, full-power (100 kW) life of 7 years, a mass goal of 30 kg/kW, and a stowed length of one-third that of the space shuttle cargo bay. It is anticipated that SP-100-type reactors will be able to support a broad spectrum of space activities, including communications, navigation, surveillance, and materials processing.

Materials-related constraints for the SP-100 dictate the use of niobium-base alloys. The material used must withstand exposure to flowing liquid lithium at a temperature higher than 1075 °C (1970 °F) for 7 years, while exhibiting no more than 2% creep strain. It also must be capable of providing a maximum system weight of 3000 kg (6600 lb).

Even more constraining was the decision to rely on off-the-shelf alloys and the existing 1974 state-of-the-art refractory metals technology. Of

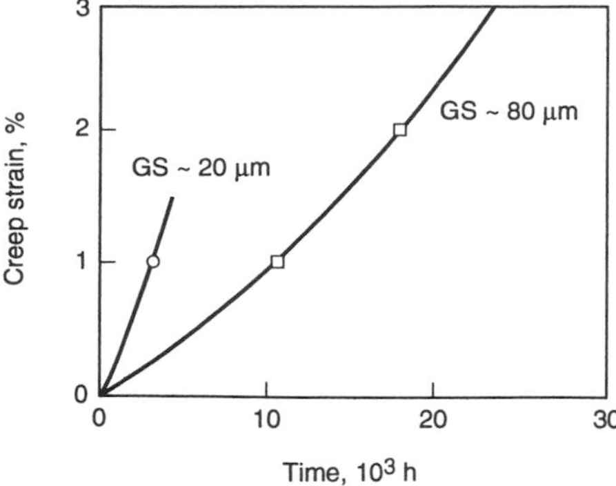

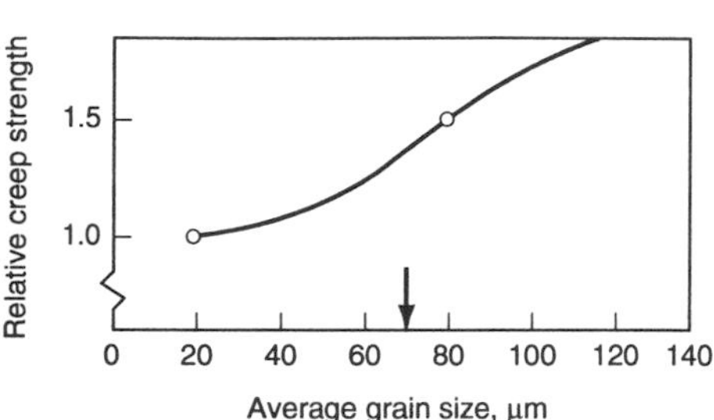

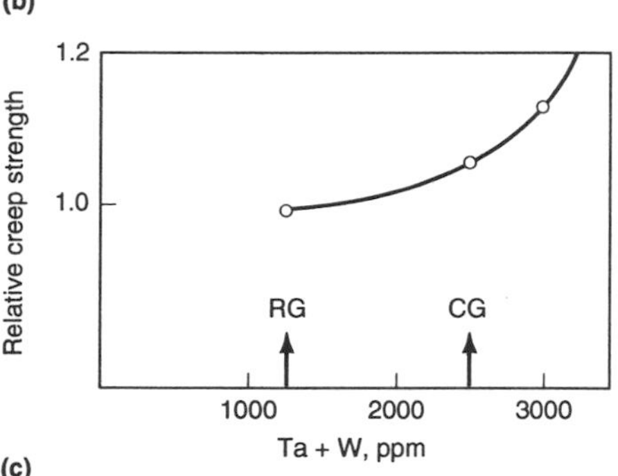

Fig. 24 Effect of grain size and trace-element strengtheners on the creep strength of Nb-1Zr alloy tested at 1075 °C (1970 °F) and 10 MPa (1450 psi). (a) Increasing the grain size (GS) from 20 to 80 µm provides a threefold increase in the time to 1% strain. (b) Nb-1Zr having the optimum 70 µm grain size is about 50% stronger than 20 µm material. (c) Raising the Ta + W content increases the relative creep strength. Noted are specification maximums for reactor grade (RG) and commercial grade (CG) material (1300 and 2500 ppm, respectively). Source: Ref 12

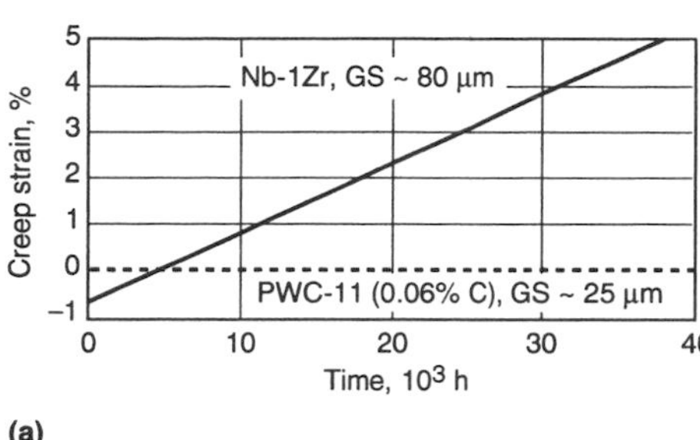

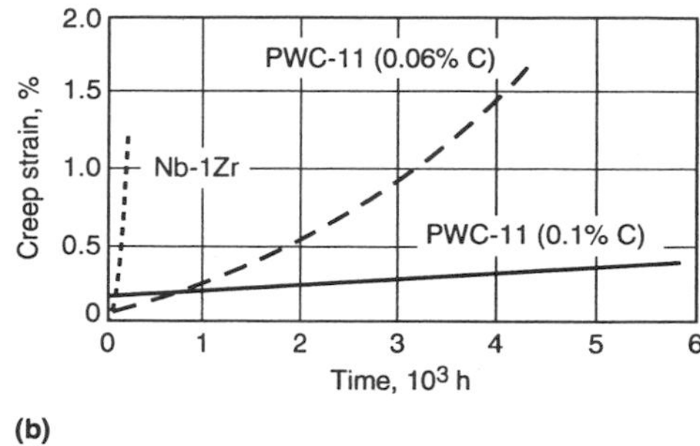

Fig. 25 Creep curves for polycrystalline Nb-1Zr and PWC-11. All tests were performed in a vacuum of 10^{-7} Pa (7.5×10^{-10} torr) and at stress levels similar to those that may be encountered by fuel pin claddings in an SP-100 space power reactor. (a) Long-time data for tests at 1075 °C (1970 °F) and 10 MPa (1450 psi). Note the substantial increase in creep strength for PWC-11, which is basically Nb-1Zr with the addition of 0.06% C. (b) Creep curves for tests at 1075 °C (1970 °F) and 34.5 MPa (5000 psi) also show the additional strengthening effect obtained by raising the carbon content of PWC-11 from 0.06% to 0.1%. Grain size of the alloys: 25 µm. Source: Ref 12

the niobium-base alloys that had been proposed for aerospace applications, only two could be considered commercially available: C-103 (Nb-10Hf-1Ti-0.7Zr) and Nb-1Zr.

Based on the preponderance of 1960s nuclear-related research data, Nb-1Zr was chosen by the SP-100 program as the reactor material that would be in contact with the liquid lithium coolant. However, studies of the creep behavior and thermal stability of more recently produced heats of the alloy (high-purity, electron-beam-processed) showed that Nb-1Zr had marginal creep strength in terms of SP-100 design criteria. The conclusion was that in light of the 30 kg/kW requirement, use of the lower-creep-strength material could seriously decrease system safety and reliability.

From among the variety of methods that could be used to increase the creep strength and thermal stability of Nb-1Zr, materials scientists chose to focus first on the obvious and proven techniques: increase the grain size and/or add trace amounts of solid-solution strengtheners. One attempt to raise the 7-year, 1075 °C (1970 °F), 2% creep strength of the higher-purity Nb-1Zr alloy to the level provided by its less pure 1960s counterpart involved increasing the grain size from 20 to 80 µm. This resulted in a threefold increase in the time to 1% strain, and apparently a similar advantage at 2% strain (Fig. 24a). In addition, Nb-1Zr having a grain size of 70 µm (the "ideal" size) proved to be about 50% stronger than 20 µm material (Fig. 24b). (Because five or more grains are required in the fuel cladding thickness, grain sizes larger then 80 µm were of little interest.) However, the increase in creep strength due to a larger grain size was considered marginal with respect to design criteria.

Heats having increased Ta + W contents also were creep tested. Creep strength rose when Ta + W passed the 1300 ppm mark (Fig. 24c). However, these Ta + W levels exceeded reactor-grade Nb-1Zr specifications, and the nucleonics and welding response of the stronger alloys were less than those required.

This lack of success with traditional strengthening methods prompted consideration of a particulate-strengthened Nb-1Zr material that would still satisfy the materials constraints for SP-100. The alloy selected for evaluation was PWC-11, a Nb-1Zr-0.1C alloy developed in the 1960s (Ref 14). It had been previously tested in lithium and had proved to be compatible under the conditions anticipated for SP-100. However, several questions that arose during these earlier studies still had to be answered. For example, the weldability of PWC-11, the effects of welding on creep properties, and the long-term stability of carbide precipitates (which are believed to be responsible for the improved strength) had not been fully characterized.

Testing and Evaluation of PWC-11. Creep tests of both Nb-1Zr and PWC-11 were performed at low stress levels, similar to those that may be encountered by SP-100 fuel pin claddings (Ref 15, 16). The test conditions were 1075 °C (1970 °F) and 10 MPa (1450 psi). After more than 35,000 h (4 years), PWC-11 had not shown any measurable creep deformation, while Nb-1Zr reached 1% creep at 11,000 h and 2% creep at 18,800 h (Fig. 25a). Results of other creep tests (Fig. 25b) showed that PWC-11 also is substantially stronger at 1075 °C (1970 °F) and 34.5 MPa (5000 psi). In addition, these data revealed that additional strengthening of PWC-11 can be obtained by increasing its carbon content from 0.06% to 0.1%.

The creep strength advantage of PWC-11 over Nb-1Zr at temperatures greater than half the melting point has been attributed to the presence of very fine precipitates of $(Nb,Zr)_2C$ and/or (Nb,Zr)C, which range in diameter from 1 to 10 µm. However, in all precipitation-strengthened alloys, the beneficial effect of the precipitate on creep strength over the long term is suspect. For PWC-11, it has been postulated that welding and/or isothermal aging could result in a significant (>50%) loss in elevated-temperature creep strength.

To verify or disprove the assumption that welding is detrimental, specimens of PWC-11 welded by electron beam (EB) were subjected to creep and short-time creep-rupture tests. The EB welds in the test specimens were perpendicular to the

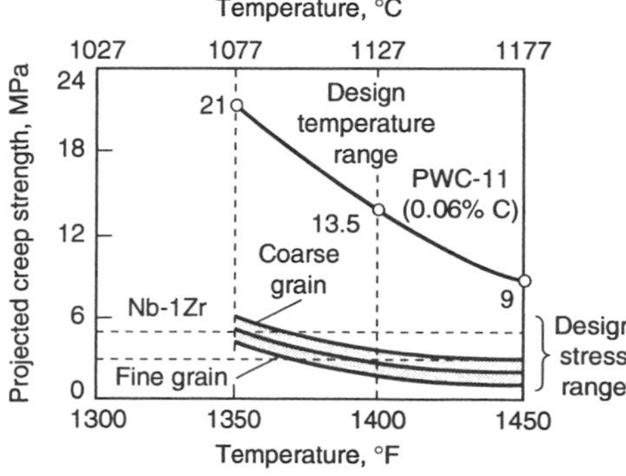

Fig. 26 Projected temperature and creep strength potentials for PWC-11 (0.06% C) and Nb-1Zr based on creep test data. The creep strength is that for 2% creep strain over 7 years (61,000 h). Also shown are current SP-100 design requirements. Source: Ref 12

test axis. Specimens were postweld heat treated for 1 h at 1200 °C (2190 °F), then aged for 1000 h at 1075 °C (1970 °F). Tests were conducted in a 10^{-5} Pa (7.5×10^{-8} torr) vacuum at 1075 °C (1970 °F). In all the creep-rupture tests, failure occurred in the unaffected base metal, proving that the weld region was stronger.

Based on the results of PWC-11 creep tests, projections were made for the stress for 1% and 2% creep in 7 years. Figure 26 compares these projections and those for the best new (high-purity) Nb-1Zr alloy with the design requirements for SP-100A. The comparison shows that at 1075 °C (1970 °F), PWC-11 (0.06% C) is four times stronger than the SP-100 design temperature range (20 vs. 5 MPa, or 2900 vs. 725 psi). This also means that PWC-11 could potentially be used to cope with future increases in the SP-100 design stress criterion.

Studies on the strength advantage of PWC-11 over Nb-1Zr are ongoing. Researchers are focusing on the effects of thermomechanical processing on thermal stability and mechanical properties. Of particular interest is the concern about overaging of PWC-11 precipitates during extended exposure at higher temperatures. Precipitate morphology and composition are being characterized for PWC-11 specimens that have been creep tested at 1075 to 1175 °C (1970 to 2150 °F) for more than 35,000 h (Ref 17, 18).

High-Temperature Corrosion of Niobium

Oxidation. Niobium is easily oxidized. It will oxidize in air above 200 °C (390 °F), but the reaction does not become rapid until above red heat (about 500 °C, or 930 °F). At 980 °C (1795 °F), the oxidation rate is 430 mm/yr (17 in./yr). At 1100 °C (2010 °F), a 3 mm (1/8 in.) thick plate of niobium would be completely converted to oxide after 24 h. In pure oxygen, the attack is catastrophic at 390 °C (735 °F). Oxygen diffuses freely through the metal, and this causes embrittlement.

As with molybdenum, extensive programs have been conducted to improve the oxidation of niobium by alloying. Significant improvements in oxidation resistance have been achieved with niobium alloys containing additions of chromium and aluminum with relatively large additions of titanium (e.g., Nb-20Ta-10Ti-5Cr-1Al) or very large additions of titanium and tungsten (Nb-20W-10Ti-3V). However, the improvement was obtained at the expense of strength and fabricability. As a result, most of the research on improving the oxidation resistance of niobium alloys has centered around coatings.

High-Temperature Oxidation-Resistant Coatings for Niobium. Niobium and its alloys are the only refractory metals for which large parts are coated to prevent oxidation in high-temperature service (>425 °C, or >800 °F). Early coating development for niobium centered on pack cementation and CVD. However, experience with these two processes showed that they increased the DBTT of the metal and caused part distortions. Later, spurred by the needs of the Apollo space program, techniques were devised for applying slurry coatings of complex aluminides and, subsequently, silicides. When properly prepared and applied, these coatings were reliable, exhibited excellent cyclic performance characteristics without drastic mechanical property deterioration, and caused minimal hardware distortion.

Today the mainstay coating is Si-20Cr-20Fe, made using elemental powder suspended in nitrocellulose lacquer with a thermotropic gelling agent. Common substrates include alloys Cb-752, (see, for example, Fig. 6), C-129Y, and FS-85. Other variants contain hafnium silicide, which gives the final coating a higher remelt temperature. Methods for applying the slurry include dipping, spraying, and touch-up painting. Following application of approximately 0.08 mm (0.003 in.) of slurry per side, the coating is heated to about 1300 to 1400 °C (2370 to 2550 °F) for reaction bonding and diffusion.

The improvements in high-temperature oxidation resistance with the use of slurry coating technology have not been achieved without side ef-

Table 15 Oxidation-resistant coating systems for niobium

System concept	Composition	Process
Silicide coatings		
Complex silicide multilayered	Cr, Ti, Si	Vacuum pack and vacuum slip-pack, fused slurry and pack, fluidized bed, electrophoresis, chemical vapor deposition, electrolytic fused salt
Modified silicide	Si + V, Cr, Ti	Fluidized bed
Modified silicide	Si	Pack cementation, iodine
Modified silicide	Si-B-Cr	Pack cementation, multicycle
Modified silicide	Si-Cr-Al	Pack cementation, multicycle
Modified silicide	Si + additives	Pack cementation, single cycle
Modified silicide	Si + Cr, Al, B	Pack cementation, single cycle
Silicide	Si	Chemical vapor deposition
Molybdenum disilicide	Si + Mo	MoO_3 reduction and chemical vapor deposition
Modified silicide	Si + additives	Pack cementation, fluidized bed, fused salt, slurry dip
Modified silicide	Si + additives	Pack cementation, single cycle
Liquid phase-solid matrix	Se, Sn, Al	Porous silicide applied by pack or CVD-impregnated with Sn-Al
Multilayered complex silicide	40Mo-40Si, 10CrB-10Al	Plasma spray-diffuse
Modified silicide	Si + additives	Pack cementation
Complex silicide	Si-20Fe-20Cr	Fused silicides
Complex silicide	Si-20Cr-5Ti	Fusion of eutectic mixtures
Complex silicide	Si + (Cr, Ti, V, Al, Mo, W, B, Fe, Mn)	...
Complex silicide	V-Cr-Ti-Si	Vacuum and high-pressure pack
Complex silicide	Mo-Cr-Ti-Si, V-Cr-Ti-Si, V-Al-Cr-Ti-Si, Mo-Cr-Si	Multicycle vacuum pack
Glass-sealed silicide	Si + glass	Silicide by pack cementation or CVD + glass slip overcoat
Multilayered	Mo, Ti + Si and glass	Slurry sinter application of Mo + Ti powder, pack silicide, glass slurry seal
Aluminide coatings		
Modified aluminide	Al + B	Pack cementation
Modified aluminide	88Al-10Cr-2Si	Fused slurry
Modified aluminide	Al-Si-Cr	Fused slurry
Modified aluminide	Al-Si-Cr, Ag-Si-Al	Hot dip
Multilayered systems	Fe, Cr, Al, Ni, Mo, Si, VSi_2, $TiCr_2$, $CrSi_2$, B + Al	Powder metallize + aluminum hot dip
Simple aluminide	Al	Pack cementation
Multilayered systems	Cr, FeB, NiB, Si, Al_2O_3, SiO_2, ThO_2, + Al	Electroplate dispersions in Ni + Al hot dip
Modified aluminide	Al + (Si, Ag, Cr)	Silver plate + Al, Cr, Si hot dip
Multilayered systems	Al_2O_3 + Ti + Al	(Al_2O_3 + TiH), spray-sinter + Al hot dip
Oxide-metal composite	Al_2O_3 + Al	Slurry fusion of Al_2O_3-Al mixture
Aluminide	Al + additives	Fused slurry
Modified aluminide	Al + Sn	Hot dip
Zinc coating		
Self-healing intermetallic	Zn and Zn + Al, Ti, Co, Cu, Cr, Fe, Zr, Cu, Si	Vacuum distillation and hot dip
Oxide coating		
Glass-sealed oxide	Al_2O_3 + glass (baria, alumina, silica)	Flame spray Al_2O_3 + glass slurry
Nickel-chromium coating		
Oxidation-resistant alloy	Ni-Cr	Flame spray, detonation gun, plasma arc
Chromium carbide coating		
Carbide	Cr-C	Plasma spray
Noble metal coatings		
Clad	Pt, Rh	Roll bonding and hermetic sealing
Barrier-layer-clad	Pt, Rh + Re, Be, Al_2O_3, W, ZrO_2, MgO, SiC, Hf	Noble metal clad over barrier layer-diffusion couple study
Pure metal	Ir	Fused-salt deposition

fects. Compared with base metal, coated metal has lower strength and ductility, higher emissivity, and increased weight. A more serious limitation imposed by the presence of coatings is a reduction in design and fabrication options. For aerospace applications, sharp edges must be eliminated because of the coating requirement. Spot welding and riveting are not recommended for coated hardware fabrication.

Other coating processes used to improve oxidation resistance include thermal spraying and various metal cladding processes (e.g., diffusion bonding, forging, extrusion, and rolling). Coating systems and processes for niobium are reviewed in Table 15.

High-Temperature Corrosion in Other Gases. Niobium reacts with nitrogen above 350 °C (660 °F), with water vapor above 300 °C (570 °F), with chlorine above 200 °C (390 °F), and with carbon dioxide, carbon monoxide, and hydrogen above 250 °C (480 °F). At 100 °C (212 °F), niobium is inert in most common gases (e.g., bromine, chlorine, nitrogen, hydrogen, oxygen, carbon dioxide, argon monoxide, and wet or dry sulfur dioxide).

Corrosion in Liquid Metals. Niobium resists attack in many liquid metals to relatively high temperatures. These include bismuth below 510 °C (950 °F); gallium below 400 °C (750 °F); lead below 850 °C (1560 °F); lithium below 1000 °C (1830 °F); mercury below 600 °C (1110 °F); sodium, potassium, and sodium-potassium alloys below 1000 °C (1830 °F); thorium-magnesium eutectic below 850 °C (1560 °F); uranium below 1400 °C (2550 °F); and zinc below 450 °C (840 °F). The presence of excessive amounts of nonmetallic impurities (e.g., gases) may reduce the resistance of niobium to these liquid metals.

Because liquid metals are excellent heat transfer media, they can be used in very compact thermal systems, such as the fast breeder reactor, reactors for space vehicles, and fusion reactors. Niobium is a serious candidate as a material for high-efficiency reactors (see the section "Niobium Alloys for Space Power Reactors" in this article).

Niobium resists attack by sodium vapor at high temperatures and pressures. The Nb-1Zr alloy is used in the end caps in high-pressure sodium vapor lamps used for highway lighting.

Tantalum and Tantalum Alloys

Tantalum provides a combination of properties not found in many refractory metals. Its high melting point (2996 °C, or 5425 °F), tolerance for interstitial elements, and reasonable modulus of elasticity make it an attractive alloy base material. Tantalum also possesses excellent room-temperature ductility (>20% tensile elongation), is readily weldable, has a very low DBTT (about 25 K, or about –250 °C) in both the welded and unwelded conditions, and exhibits relatively high-solid solubility for other refractory and reactive metals (Ref 19). Properties of unalloyed tantalum are compared with those of other pure refractory metals in Table 1 and in Fig. 1 and 2.

Tantalum has excellent resistance to corrosion by a large number of acids, by most aqueous solutions of salts, by organic chemicals, and by various combinations and mixtures of these agents. Tantalum also exhibits good resistance to many corrosive and common gases and to many liquid metals. At moderate temperatures (below about 150 °C, or 300 °F), the only media that can affect it are fluorine, hydrofluoric acid, sulfur trioxide (including fuming sulfuric acid), concentrated strong alkalis, and certain molten salts. The corrosion resistance of tantalum is about the same as that of glass, but it withstands higher temperatures and offers the intrinsic fabrication advantages of a metal.

Tantalum derives its exceptional corrosion resistance from the tough, impermeable oxide film that forms on the metal when it is exposed to normal atmospheric conditions. The oxide film also is rectifying under an applied current, which discourages the formation of corrosion-promoting electrolytic cells. Because it is relatively thin, the film also is suitable as a heat transfer surface.

Although the initial cost of parts and equipment made of tantalum may be high, extended service life can result in a reduction in total life-cycle cost. Important materials selection factors thus include the costs of process downtime and product contamination stemming from use of less durable materials.

The single largest use for tantalum is as powder and anodes for electronic capacitors, representing about 50% of total consumption. Mill products constitute about 25% of tantalum consumption. In terms of total refractory metal consumption, tantalum ranks behind molybdenum, tungsten, and niobium.

Production of Tantalum

The production of tantalum metal is accomplished by the extraction of tantalum from ores, such as $(Fe,Mn)(Ta,Nb)_2O_6$, or from certain tin slags, primarily those from Thailand and Malaysia. Further extraction is necessary to separate tantalum from other metals present. The purified extract is recovered by precipitation of $Ta(OH)_5$, which is then calcined to the pentoxide, or by crystallization with potassium fluoride to the intermediate salt, potassium fluorotantalate (K_2TaF_7).

Several methods for reducing tantalum compounds to tantalum metal have been developed, but sodium reduction of K_2TaF_7 to produce tantalum metal powder is the most commonly used. The powder product of the sodium reduction can be further refined by EB melting and converted into ingots measuring up to 305 mm (12 in.) in diameter and weighing up to 2725 kg (6000 lb).

Alloying can take place in either the EB furnace or during a subsequent vacuum arc remelting (VAR) operation. All tantalum and tantalum-alloy products are EB processed; whether EB + VAR is used depends primarily on the application.

Certain metallurgical properties of tantalum, such as deep drawability, can be enhanced by P/M processing. Relatively high-purity, chemically produced powder is consolidated in a hot-isostatic press and then vacuum sintered. The near-theoretical-density compact is mechanically converted into sheet and wire products.

Metallurgical-grade tantalum can be supplied as powder, bar, ingot, rod, plate, wire, fine wire, sheet, foil, and tubing. It also is available as expanded mesh and value-added fabrications.

Properties and Applications of Tantalum Alloys

Table 16 lists the most common tantalum alloys developed to date. As with niobium alloys, most of these alloys were developed prior to 1973.

Properties. As indicated in Fig. 19 and 20, the elevated-temperature properties of tantalum alloys are generally superior to those of niobium alloys but lower than those of either molybdenum or tungsten alloys. Alloys strengthened from a dispersed second phase by the addition of carbon exhibit superior creep and high-temperature yield properties (Table 17). Some of these alloys (e.g., Astar-811C) have creep properties similar to those of molybdenum and tungsten alloys (Fig. 21). High-temperature annealing treatments are recommended for carbide-strengthened alloys in order to increase grain size, which in turn improves creep strength (Fig. 27). Properties of various tantalum alloys are compared in Tables 2 and 18.

Applications (Ref 20). The corrosion and heat resistance of tantalum and its relatively high heat transfer coefficient make it attractive for both

Table 16 Compositions of selected tantalum alloys

Common designation	Nominal alloy additions, %
Tantaloy 63	2.5 W, 0.15 Nb
FS-61	7.5 W
Ta-10W	10 W
T-111	8 W, 2 Hf
T-222	10 W, 2.5 Hf, 0.01 C
Astar-811C	8 W, 1 Re, 1 Hf, 0.025 C
Astar-1211C	12 W, 1 Re, 1 Hf, 0.025 C
Astar-1511C	15 W, 1 Re, 1 Hf, 0.025 C
KBI 40	40 Nb
WC-640	40 Nb, 0.5 W
Ta-Mo	Various Mo contents ranging from 10 to <50%

Table 17 Mechanical behavior of tantalum alloys at 1316 °C (2400 °F)

Alloy	Carbon content	Yield strength, MPa	Yield strength, ksi	Stress for 1% creep elongation in 10,000 h
Ta-10W	(a)	120	17.4	5
T-111	(a)	190	27.6	4
T-222	0.01	260	37.7	20
Astar-811C	0.025	215	31.2	60
Astar-1211C	0.025	267	38.7	90
Astar-1511C	0.025	288	41.8	96

(a) No carbon deliberately added to alloys. Source: Ref 19

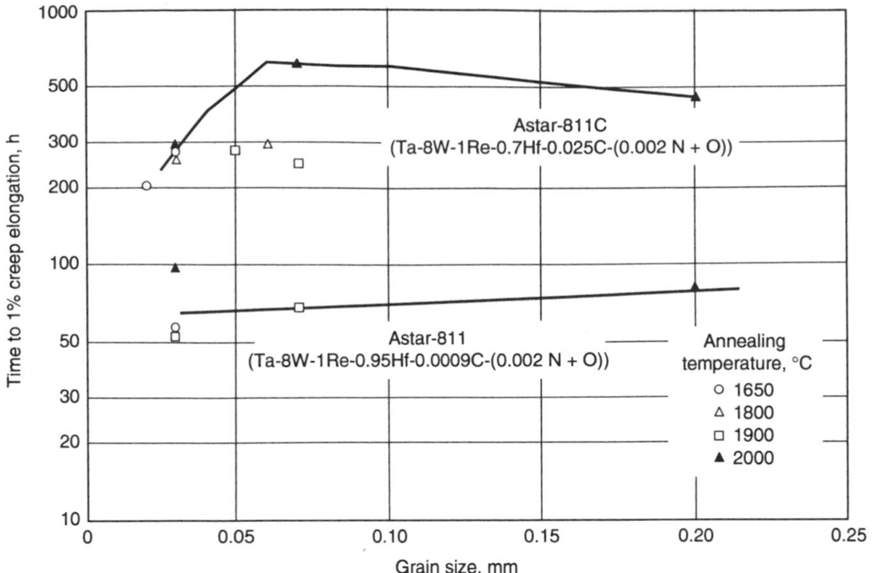

Fig. 27 Effect of grain size on the creep strength of Astar-811C and Astar-811 at 1316 °C (2400 °F) and 103.4 MPa (15 ksi). Source: Ref 19

heating and cooling applications. Its thermal conductivity is higher than that of many steels, silicon-iron, and nickel alloys. Because tantalum does not require an allowance for corrosion when specifying thickness, it can be used to reduce the size and cost of equipment such as heat exchangers. Tantalum also resists biofouling, which can produce a thermally insulating layer on other materials.

In the chemical processing industry, tantalum is a preferred material of construction for heat exchangers, coils, condensers, coolers, and bayonet heaters. Other chemical processing applications range from relatively small parts (e.g., thermocouple wells, dip pipes, orifices, valves, and diaphragms) to tubing and linings for tanks and pipe. Tantalum also can be used to repair damaged or flawed glass-lined steel equipment.

Large, thin plates of tantalum, typically measuring 1.2 by 37 by 0.9 mm (4 by 12 by 0.035 in.), can be clad to thicker plates of steel, copper, or aluminum by explosive bonding. The tantalum-clad product is then used to fabricate very large, but relatively economical, chemical process vessels.

A variety of high-temperature chemical reactions, many performed under vacuum, are frequently carried out in tantalum crucibles. Examples include high-temperature fusions, distillations, and melting of special glasses. In metallizing and other thin-film deposition applications, tantalum crucibles are often used to hold the materials to be evaporated.

Tantalum is used in high-temperature vacuum and inert-gas furnaces for susceptors, resistance heaters, trays, and structural parts. Fabricability and stability at high temperatures make tantalum particularly suitable for heat shields. Tantalum thermocouple sheaths have been used in nuclear reactors, and tantalum expanded mesh is used to reduce warping and aging of resistance-heating elements.

Tantalum also resists attack by some liquid metals. For example, it has been used in contact with liquid sodium at 1375 °C (2510 °F).

Capacitor components using tantalum include powder anodes, capacitor cans, and lead wires. The chief advantage of tantalum in capacitor manufacture is its ability to form an efficient dielectric oxide layer, unmatched by other metals. Tantalum is preferred over ceramic materials because it provides stable capacitance over a wide temperature range: –55 to 125 °C (–65 to 255 °F). A variety of capacitor-grade tantalum powders are available.

Primarily because of its unique combination of ductility and high density (16.6 g/cm^3), tantalum has become the material of choice for several advanced antiarmor weapon systems. In the past, such weapons had to be aimed and fired from a position close to the target. Today these systems have sophisticated target-location sensors that permit accurate firing from much greater distances. A saucer-shaped warhead "liner" is explosively formed into a long, rod-shaped, armor-piercing penetrator while it is being accelerated toward the target. Tantalum replaced copper as the liner material of choice (the density of tantalum is nearly twice that of copper).

Tantalum is supplied as plate, typically 4 to 8 mm (0.15 to 0.3 in.) thick, or as circular discs, which can be shaped into saucers, hemispheres, and cones. Some liner designs employ tantalum bar stock. These ballistic applications require consistent mechanical properties, a uniform grain structure, and high purity to ensure ductility at high strain rates.

The total inertness of tantalum to body fluids has resulted in its use for occluding holes in the skull and other locations in the human body.

Table 18 Typical properties of tantalum and tantalum-base alloys

Grade(a)	Hardness	Density		Melting point		Temperature		Tensile strength		Yield strength		Elongation,	Modulus of elasticity	
	HV	g/cm^3	lb/in.^{3}	°C	°F	°C	°F	MPa	ksi	MPa	ksi	%	GPa	10^6 psi
Commercially pure tantalum, EB melted	110	16.9	0.609	3000	5430	20	70	205	30	165	24	40	185	27
						200	390	190	27.5	69	10	30	…	…
						750	1380	140	20	41	6	45	160	23
						1000	1830	90	13	34	5	33	…	…
Commercially pure tantalum, P/M	120	16.6	0.600	3000	5430	20	70	310	45	220	32	30	185	27
63 metal, EB melted	130	16.7	0.602	3005(b)	5440(b)	20	70	345	50	230	33	40	195	28
						200	390	315	46	195	28	33	…	…
						750	1380	180	26	83	12	22	…	…
						1000	1830	125	18	69	10	20	…	…
Ta-10W, EB melted	245	16.8	0.608	3030	5490	20	70	550	80	460	67	25	205	30
						200	390	515	75	400	58	…	…	…
						750	1380	380	55	275	40	…	150	22
						1000	1830	305	44	205	30	…	…	…
Ta-7.5W, P/M														
Wire	325	16.8	0.606	3025(b)	5477(b)	20	70	1035	150	1005	146	6	200	29
Sheet	400	16.8	0.606	3025(b)	5477(b)	20	70	1165	169	875	127	7	200	29
Ta-40Nb, EB melted	…	12.1	0.437	2705	4900	…	…	275	40	193	28	25	…	…

(a) EB, electron beam; P/M, powder metallurgy. (b) Estimated

Tantalum surgical staples are used in open-heart surgery to close veins and arteries. There is no recorded instance of tantalum being the cause of an allergic reaction.

High-purity tantalum is an important ingredient of the superalloys used for turbine blades and other parts for aircraft and stationary gas-turbine engines. When used as an alloying element, tantalum provides solid-solution strengthening, forms carbides, and improves the thermal stability of intermetallics. Cast single-crystal alloys for airfoils may contain as much as 12% Ta.

High-Temperature Corrosion of Tantalum

Oxidation. Tantalum and its alloys cannot be used for extended service in air or other oxidizing atmospheres at temperatures above 260 °C (500 °F) unless a protective inert atmosphere or vacuum is provided. In the absence of a protective atmosphere, two approaches for imparting a degree of oxidation resistance are to:

- Form a denser, more adherent oxide film by alloying additions to the tantalum that alter and modify the oxide phase.
- Provide a protective coating to inhibit oxygen attack.

Of these two approaches, greater success has been achieved with coatings.

Oxidation-Resistant Coatings for Tantalum. Much of the work on tantalum alloys is the outgrowth of approaches taken with niobium alloys. Most coatings were developed in the 1970s, with emphasis on the diffusional growth of intermetallic layers. Following are brief descriptions of some commercial and semicommercial coatings:

- A series of Sn-Al-Mo coatings are used by a slurry process. These are limited primarily by poor resistance to reduced pressure and erosion at very high temperatures.
- Electrophoretical deposition of binary disilicides combined with a molybdenum-vanadium system shows no pest phenomenon during service. This process can be considered commercial for small parts.
- A fluidized-bed, three-step Si-V-Si process may be used for Ta-10W.
- Complex titanium-, molybdenum-, tungsten-, and vanadium-modified silicide coatings can be applied by a two-step method: a slurry plus high-temperature sinter of an alloy layer, followed by a straight silicide pack. In furnace tests, protection for hundreds of hours at 870 °C (1600 °F) and 1320 °C (2410 °F) has been obtained.
- A fused-silicide coating system, particularly Si-20Ti-10Mo, appears to be practical for coating large, complex aerospace sheet metal components.
- A duplex coating consisting of a sintered hafnium boride/molybdenum silicide layer overlaid with a hafnium-tantalum slurry is serviceable at 1820 to 1870 °C (3310 to 3400 °F).

Corrosion in Liquid Metals. Tantalum and some tantalum-base alloys exhibit good resistance to many liquid metals. Table 19 lists the effects of various liquid metals on tantalum. Tantalum materials exhibit remarkable resistance to several liquid metals, even up to high temperatures (900 to 1100 °C, or 1650 to 2010 °F) in the absence of oxygen or nitrogen. A more detailed account of the compatibility of tantalum with various liquid metals can be found in the article "Corrosion of Tantalum" in Volume 13 of the *ASM Handbook.*

Rhenium and Rhenium-Bearing Alloys

Rhenium differs from the other refractory metals in that it has an hcp structure. The scarcity and high cost of rhenium have limited its use. For example, the addition of 3% Re to tungsten wire doubles the cost of the wire. Nevertheless, the unique properties of rhenium allow it to be used in important, albeit specialized, applications.

Production of Rhenium

Most rhenium occurs in porphyry copper deposits. Rhenium is available as perrhenic acid ($HReO_4$), ammonium perrhenate (NH_4ReO_4), and metal powder.

Ammonium perrhenate is converted to metal powder by hydrogen reduction. The reduction is carried out at 380 °C (715 °F) and is followed by a purification and reduction cycle at 700 to 800 °C (1290 to 1470 °F) to remove any residual rhenium oxide. The powder is generally consolidated by cold pressing at about 205 MPa (30 ksi) to a density of 35 to 40%, using stearic acid in ether as a lubricant on the punch and the die walls. Subsequent sintering at 1200 °C (2190 °F) for 2 h in vacuum results in little densification but increases the mechanical strength of the compact and burns off volatile impurities. Finally, resistance heating in a vacuum or hydrogen atmosphere at 2700 to 2900 °C (4890 to 5250 °F) produces sintered compacts with densities of more than 90%.

Electron beam remelting is sometimes used to reduce the impurity content of rhenium compacts. CVD is an alternative fabrication method.

Properties of Rhenium (Ref 1)

Rhenium does not have a DBTT, as it retains its ductility from subzero to high temperatures. Rhenium also has a very high modulus of elasticity that, among metals, is second only to those of iridium and osmium. As temperature increases from room temperature to 725 °C (1340 °F), the modulus decreases only about 20%. This implies that structures made of rhenium will have excellent mechanical stability and rigidity, permitting the design of parts having thin sections.

A high recrystallization temperature is a prerequisite for good creep resistance. Among the refractory metals, that of rhenium is the highest: 1625 °C, or 2960 °F. Compared with other refractory metals, rhenium has superior tensile and creep-rupture strengths over a wide temperature range. With approximately 35% elongation and a tensile strength that decreases from 1172 MPa (170 ksi) at room temperature to 48 MPa (7 ksi) at 2710 °C (4910 °F), rhenium is virtually immune to thermal shock. At temperatures up to 2800 °C (5070 °F) and at high stresses, the rupture life of rhenium is longer than that of tungsten. The metal also accommodates wide swings in operating temperature (large thermal expansions and contractions) without incurring mechanical damage. For example, rhenium-metal rocket thruster nozzles have withstood more than 100,000 thermal fatigue cycles, from room temperature to above 2225 °C (4040 °F), without any evidence of failure. Properties of rhenium are given in Tables 1 and 2 and in Fig. 1 to 3.

Table 19 Effects of molten metals on tantalum

Metal	Remarks	Temperature, °C (°F)	Code(a)
Aluminum	Forms Al_3Ta	Molten	NR
Antimony	...	to 1000 (1830)	NR
Bismuth	...	to 900 (1650)	E
Cadmium	...	Molten	E
Gallium	...	to 450 (840)	E
Lead	...	to 1000 (1830)	E
Lithium	...	to 1000 (1830)	E
Magnesium	...	to 1150 (2100)	E
Mercury	...	to 600 (1110)	E
Potassium	...	to 900 (1650)	E
Sodium	...	to 900 (1650)	E
Sodium-potassium alloys	...	to 900 (1650)	E
Zinc	...	to 500 (930)	E/V
Tin	...	...	V
Uranium	...	...	V
Mg-37Th	In helium	to 800 (1470)	S
Bi-5 to 10U	In helium	to 1100 (2010)	S
Bi-5U-0.3Mn	In helium	to 1050 (1920)	S
Bi-10U-0.5Mn	In helium	to 1160 (2120)	S
Al-18Th-6U	Failed	to 1000 (1830)	NR
U-10Fe	Failed	to 900 (1650)	NR
U-Cr (eutectic)	Failed	to 900 (1650)	NR
YSb-intermetallic compound	...	1800–2000 (3270–3630)	S
YBi-intermetallic compound	...	1800–2000 (3270–3630)	S
ErSb-intermetallic compound	...	1800–2000 (3270–3630)	S
LaSb-intermetallic compound	...	1800–2000 (3270–3630)	S
Plutonium-cobalt-cerium	...	to 650 (1200)	V

(a) E, no attack; S, satisfactory; V, variable, depending on temperature and concentration; NR, not resistant

Applications for Rhenium (Ref 1)

Platinum-rhenium reforming catalysts are the major rhenium end-use products and account for about 85% of rhenium consumption. Rhenium catalysts are exceptionally resistant to poisoning from nitrogen, sulfur, and phosphorus. They are used for the hydrogenation of fine chemicals and for hydrocracking, reforming, and the disproportionation of olefins, including increasing the octane rating in the production of lead-free petroleum products.

Alloying Additions. Because of its hcp structure, rhenium has a high solubility in transition metals having bcc and face-centered cubic (fcc) structures. This characteristic was used to advantage in the development of tungsten-rhenium and molybdenum-rhenium alloys, which derive their properties from the so-called "rhenium effect": A rhenium addition simultaneously improves strength, plasticity, and weldability, lowers the DBTT of wrought products, and reduces the degree of recrystallization embrittlement. A rhenium addition changes the deformation mechanism during warm and cold working from slipping only to twinning and slipping, and it neutralizes the embrittling effects of carbon and oxygen. The greatest improvements in properties are obtained with additions of 10 to 26 wt% Re to tungsten and 11 to 50 wt% Re to molybdenum.

Data for 7 to 12 mm (0.3 to 0.5 in.) thick rolled plates of molybdenum-rhenium and tungsten-rhenium alloys show that hardness increases with increasing amounts of rhenium beyond about 5%. The hardness of W-26Re is 20% greater than that of pure tungsten, and the tensile strength is more than twice as high, with an elongation of 15 to 18%. The recrystallized grain size decreases from 62 μm (0.002 in.) for pure tungsten to 32 μm (0.001 in.) for W-26Re. The hardness of Mo-41Re is 60% greater than that of pure molybdenum, and the strength of the alloy is 75% higher, while its elongation is an acceptable 17%.

Rhenium is also a solid-solution-strengthening alloying element in superalloys. A number of directionally solidified and single-crystal nickel-base superalloys contain up to 3% Re. One recently developed single-crystal superalloy (CMSX-10) contains 6% Re as well as 0.25 to 0.20% Mo, 3.5 to 7.5% W, 7.0 to 10.0% Ta, and 0.5% Nb.

High-Temperature Coatings. Rhenium is unique among refractory metals in that it does not form a stable carbide. The solubility for carbon is rather high, resulting in a eutectic melting point at about 2500 °C (4530 °F) and 0.85 wt% C. Although carbon diffusion into rhenium causes some solid-solution hardening, catastrophic embrittlement (such as that caused by carbide formation in other refractory metals) does not occur.

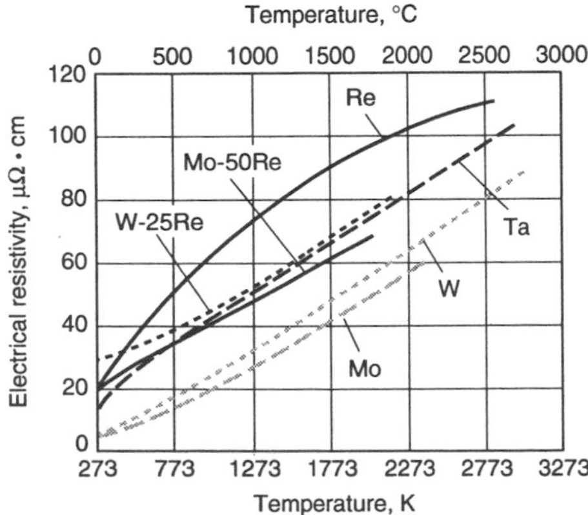

Fig. 28 Effect of rhenium additions on the electrical resistivity of refractory metals. Source: Ref 1

Consequently, rhenium can be used in contact with graphite and carbon composites.

Rhenium-coated carbon has been evaluated for rocket engine exhaust nozzles that were required to operate for 20 s bursts at about 3725 °C (6740 °F) and at gas pressures up to 17 MPa (2500 psi). Rhenium also is a potential heat-resistant coating for other graphite and carbon/carbon components that must withstand a wide range of operating temperatures in low-oxygen environments.

Heating Equipment. Because rhenium-containing alloys have relatively high electrical resistivity compared with other transition-metal alloys (Fig. 28), their use as heating elements for resistance furnaces is a potential application. They can be used in vacuum, hydrogen, or inert atmospheres, and they do not embrittle during service. Mo-50Re wire and sheet components are candidates for use at temperatures up to 2125 °C (3860 °F) in heaters, reflectors, and work stands. Moreover, these components can be fabricated by welding. Alloy W-20Re is already used in about 50 industrial applications, including heating elements for resistance furnaces and electrical bulbs, and in components of medical and industrial x-ray equipment. Alloy W-5Re can be fabricated into thin (100 to 150 mm, 0.004 to 0.006 in.) spiral heating elements.

Thermocouples based on tungsten-rhenium alloys containing 3 to 26% Re can be used to measure and control temperatures up to 2325 °C (4220 °F) with excellent reproducibility and reliability. (Thermocouples are shielded from the atmosphere by protection tubes made of a material such as alumina and filled with an inert gas such as argon.) Short-time use up to 2725 °C (4940 °F) is possible. Molybdenum-rhenium thermocouples are useful for measuring lower temperatures in non-oxidizing environments such as hydrogen. Also, because molybdenum carburizes more slowly than tungsten, molybdenum-rhenium thermocouples are potentially more useful than thermocouples in carbon-rich environments at temperatures up to 1725 °C (3140 °F). Mo-50Re, in the form of seamless tubing, also is an excellent sheathing material for high-temperature thermocouples.

High-Temperature Corrosion of Rhenium

Oxidation. Rhenium oxidizes catastrophically above 600 °C (1110 °F). Oxidation occurs as a result of the formation of rhenium heptoxide (Re_2O_7), which has a melting point of 297 °C (567 °F) and a boiling point of 363 °C (685 °F).

Oxidation-Resistant Coatings for Rhenium. Iridium is currently used as an oxidation-resistant coating for rhenium at high temperatures. Iridium-coated rhenium nozzles for small chemical rockets and resistojet thrusters are used in space for satellite orientation. NASA researchers report that iridium-coated rhenium combustion chambers outlast silicide-coated niobium parts by a factor of 10 at an operating temperature of 1400 °C (2550 °F) (Ref 1).

Corrosion in Liquid Metals. Rhenium is resistant to attack by liquid lithium, tin, zinc, silver, copper, and aluminum.

Refractory Metal Fiber-Reinforced Composites

In spite of their poor oxidation resistance and high density, refractory metal wires have received a great deal of attention as fiber reinforcement materials for use in high-temperature composites. Although the theoretical specific strength potential of refractory alloy fiber-reinforced composites is less than that of ceramic fiber-reinforced composites, the more ductile metal fiber systems are more tolerant of fiber-matrix reactions and thermal expansion mismatches. When refractory metal fibers are used to reinforce a ductile and oxidation-resistant matrix, they are protected from oxidation, and the specific strength of the composite is much higher than that of superalloys at elevated temperatures.

The majority of the studies conducted on this topic have been on refractory wire and nickel-base superalloy composites that use tungsten or molybdenum wire (available as lamp filament or thermocouple wire) as the reinforcement material. These refractory alloy wires were not designed for use in composites, nor were they developed to achieve optimum mechanical properties in the temperature range of interest for component applications (1000 to 1200 °C, or 1830 to 2190 °F). The stress-rupture properties of a tungsten lamp filament wire used in early studies were superior to those of rod and bulk forms of tungsten, and this wire showed promise for use as composite reinforcement. After the need for stronger wire was recognized, high-strength tungsten, tantalum, molybdenum, and niobium alloys that were originally used for rod and/or sheet fabrication were drawn into wire.

Excellent progress has been made in providing wires with increased strength. Dispersion-strengthened tungsten alloy wires have been fabricated that have tensile strengths 2.5 times higher than those obtained for potassium-doped tungsten lamp filament wire. The strongest wire fabricated, tungsten/rhenium/hafnium carbide, has a tensile strength of 2165 MPa (314 ksi) at 1093 °C (2000 °F), which is more than six times the strength of the strongest nickel-base or cobalt-base superalloy. Although the ultimate tensile strength values of the tungsten alloy wires were higher than those obtained for molybdenum, tantalum, or niobium wires, their advantage is lessened when the higher density of tungsten is taken into account. Nevertheless, high-strength tungsten alloy wires rank alongside molybdenum wires as offering the most promise for composite applications.

Mechanical and Thermal Properties. Refractory fiber-reinforced superalloy composites have demonstrated strengths significantly above those of the strongest superalloys. Tungsten fiber-reinforced superalloy composites, in particular, are potentially useful at high temperatures

(1000 to 1200 °C, or 1830 to 2190 °F) because of their microstructural stability and superior resistance to stress-rupture, creep deformation, thermal shock, and low- and high-cycle fatigue. Compared with conventional superalloys, refractory metal fiber-reinforced composites have better ductility, impact damage resistance, and thermal conductivity.

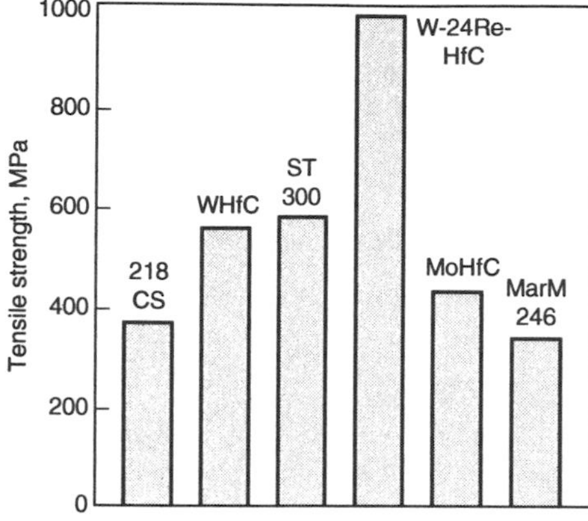

Fig. 29 Elevated-temperature (1093 °C, or 2000 °F) tensile strength of Waspaloy reinforced with 50 vol% refractory metal wire. 218 CS represents potassium-doped tungsten. ST 300 is a W-1.0ThO_2 alloy. Comparative data are included for unreinforced MarM 246, a nickel-base superalloy. Source: Ref 21

Figure 29 compares the elevated-temperature tensile strength of a nickel-base superalloy (Waspaloy) reinforced with various refractory wires. As this figure indicates, a composite consisting of 50 vol% W-24Re-HfC had the highest strength at 1093 °C (2000 °F).

Refractory fiber-reinforced niobium alloy composites have demonstrated a potential for use above 1200 °C (2190 °F). The tensile and creep strength properties of these composites have been improved by an order of magnitude by adding 50 vol% W fiber to the niobium alloys. Both potassium-doped tungsten and tungsten-thoria wires have been evaluated.

Applications. Tungsten fiber-reinforced superalloy composites are being developed for use in rocket engine turbine blades (Fig. 30). Tungsten fiber-reinforced superalloy composites have a highly attractive combination of properties at temperatures from 870 to 1100 °C (1600 to 2010 °F); these properties make them well suited for advanced rocket engine turbopump blade applications. The composites offer the potential for significantly improved operating life, higher operating temperature capability, and reduced strains induced by transient thermal conditions during engine startup and shutdown.

Tungsten fiber-reinforced niobium alloy systems are being investigated for potential long-term high-temperature applications in space power systems. In addition, molybdenum-base fibers are being proposed for use in intermetallic-matrix composites for aerospace applications.

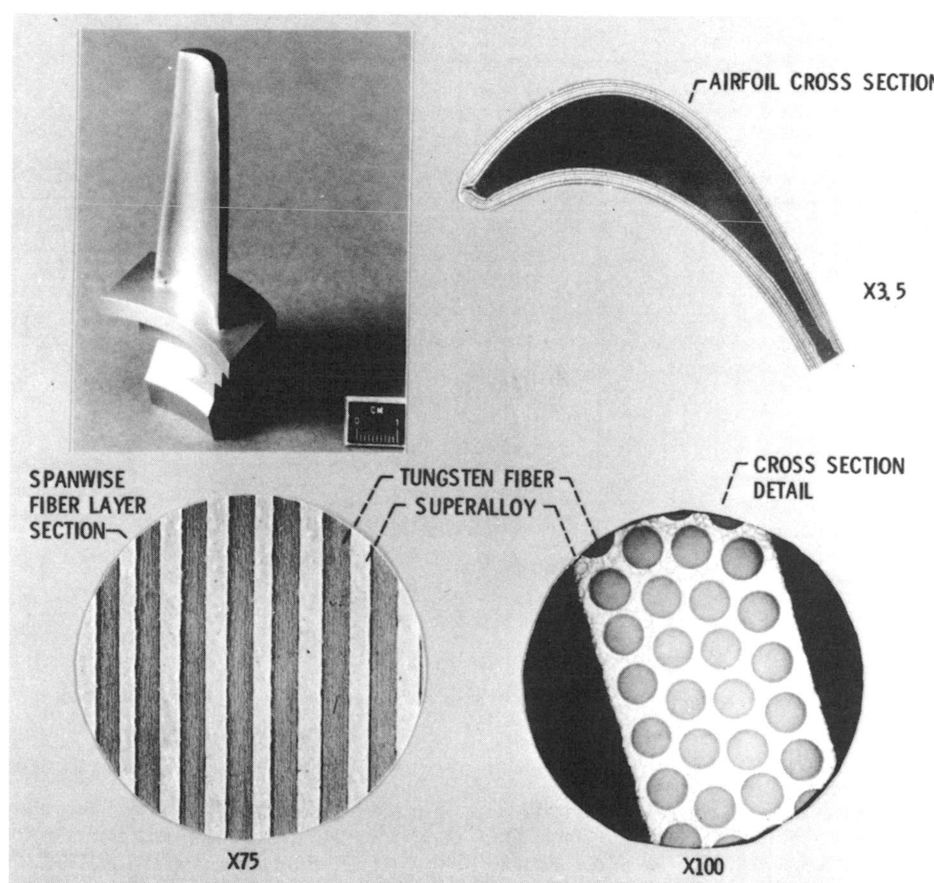

Fig. 30 Location and structure of tungsten fibers in fiber-reinforced superalloy composite turbine blades for rocket engine turbopumps

ACKNOWLEDGMENTS

The information in this article is largely taken from the "Refractory Metals Forum," published in *Advanced Materials & Processes* in September 1992 (Part I, tantalum, rhenium, and tungsten), October 1992 (Part II, molybdenum), and November 1992 (Part III, niobium). Specific articles are listed below in Ref 1, 3, 7, 12, and 20. Other important information sources include:

- J.A. Shields, Jr., Surface Engineering of Refractory Metals and Alloys, *Surface Engineering*, Vol 5, *ASM Handbook*, ASM International, 1994, p 856–863
- J.B. Lambert et al., Refractory Metals and Alloys, *Properties and Selection: Nonferrous Alloys and Special-Purpose Alloys*, Vol 2, *ASM Handbook*, ASM International, 1990, p 557–585
- T.L. Yau and R.T. Webster, Corrosion of Niobium and Niobium Alloys, *Corrosion*, Vol 13, *ASM Handbook*, ASM International, 1987, p 722–724
- M. Schussler and C. Pokross, Corrosion of Tantalum, *Corrosion*, Vol 13, *ASM Handbook*, ASM International, 1987, p 725–739

REFERENCES

1. B.D. Bryskin, Rhenium and Its Alloys, *Advanced Materials & Processes*, Vol 142 (No. 3), Sept 1992, p 22—27
2. Datasheet: Typical Properties of Refractory Metals and Their Alloys, *Advanced Materials & Processes*, Vol 142 (No. 3), Sept 1992, p 54
3. J.A. Shields, Jr., Molybdenum and Its Alloys, *Advanced Materials & Processes*, Vol 142 (No. 4), Oct 1992, p 28–36
4. R.I. Jaffee, C.T. Sims, and J.J. Harwood, The Effect of Rhenium on the Fabricability and Ductility of Molybdenum and Tungsten, *Proc. Third Plansee Seminar*, Reutte, Austria, 1958, p 380–411
5. F.J. Clauss, Refractory Metals and Their Alloys, *Engineer's Guide to High-Temperature Materials*, Addison-Wesley Publishing Co., 1969, p 1762–217
6. D.B. Lee and G. Simkovich, "Oxidation of Mo-Cr-Pd and Mo-W-Cr-Pd Alloys," paper presented at TMS-AIME Annual Meeting, San Diego, CA, 1992
7. J.P. Wittenaurer, T.G. Nieh, and J. Wadsworth, Tungsten and Its Alloys, *Advanced Materials & Processes*, Vol 142 (No. 3), Sept 1992, p 28–37
8. R.L. Heestand, A History of Tungsten- and Molybdenum-Base Alloys, *Evolution of Re-*

fractory Metals and Alloys, TMS-AIME, 1994, p 109–117

9. R.T. Begley, Columbium Alloy Development at Westinghouse, *Evolution of Refractory Metals and Alloys*, TMS-AIME, 1994, p 29–48
10. C.C. Wojcik, High Temperature Niobium Alloys, *High Temperature Niobium Alloys*, TMS-AIME, 1991, p 1–12
11. G.D. McAdam, Substitutional Niobium Alloys of High Creep Strength, *J. Inst. Met.*, Vol 93, 1964–65, p 59–564
12. R.H. Titran, Niobium and Its Alloys, *Advanced Materials & Processes*, Vol 142 (No. 5), Nov 1992, p 34–41
13. "Refractory Alloy Technology for Space Nuclear Power Applications," R.H. Cooper and E.E. Hoffman, Ed., CONF-8308130, U.S. Dept. of Energy, Washington, D.C., 1984
14. E.J. DelGrosso, E.C. Carlson, and J.J. Kaminsky, Development of Niobium-Zirconium-Carbon Alloys, *Journal of Less Common Metals*, Vol 12, 1967, p 173–201
15. R.H. Titran, "Creep Strength of Niobium Alloys," NASA TM-102390, 1989
16. R.H. Titran, T.L. Grobstein, and D.L. Ellis, "Advanced Materials for Space Nuclear Power Systems," NASA TM-105171, 1991
17. T.L. Grobstein and R.H. Titran, "Characterization of Precipitates in a Niobium-Zirconium-Carbon Alloy," NASA TM-100848, 1986
18. M. Uz and R.H. Titran, "Thermal Stability of the Microstructure of an Aged Nb-Zr-C Alloy," NASA TM-103647, 1991
19. R.W. Buckman, Jr. and R.L. Ammon, Evolution of Tantalum Alloy Development, *Evolution of Refractory Metals and Alloys*, TMS-AIME, 1994, p 49–79
20. S.M. Cardonne, P. Kumar, C.A. Michaluk, and H.D. Schwartz, Tantalum and Its Alloys, *Advanced Materials & Processes*, Vol 142 (No. 3), Sept 1992, p 16–20
21. F.J. Ritzert and R.L. Dreshfield, Progress toward a Tungsten Alloy Wire/High Temperature Alloy Composite Blade, *Tungsten and Tungsten Alloys—1992*, Metals Powder Industries Federation, 1993, p 455–462

Nickel-Chromium and Nickel-Thoria Alloys

THE NICKEL-CHROMIUM ALLOYS described in this article are essentially binary systems where the total alloy content (excluding nickel and chromium) totals less than approximately 2.5 wt%. This excludes the more complex alloys such as the solid-solution strengthened or precipitation-strengthened nickel-base superalloys.

Two distinct groups of nickel-chromium alloys are discussed in this article. The first group, which includes cast alloys containing 40 to 50% Ni and 50 to 60% Cr, are used for heat-resistant and elevated-temperature corrosion applications such as structural members, containers, supports, hangers, spacers, and the like in corrosive environments up to 1090 °C (2000 °F). The second group, which includes wrought alloys containing 70 to 80% Ni and 20 to 30% Cr, are used as resistance heating elements for oxidizing (air) atmospheres up to 1150 °C (2100 °F). The thoria-dispersed nickel alloys (TD Nickel) described in this article contain thorium oxide additions (~2 wt%) for increased elevated-temperature strength up to 1200 °C (2200 °F).

Table 1 Amounts of various constituents found in residual fuel oils from various crudes

Source of crude oil	Content, ppm V	Ni	Na
Africa			
1	5.5	5	22
2	1	5	...
Middle East			
3	7	...	1
4	173	51	...
5	47	10	8
U.S.			
6	13	...	350
7	6	2.5	120
8	11	...	84
Venezuela			
9	...	6	480
10	57	13	72
11	380	60	70
12	113	21	49
13	93	...	38

Source: Ref 2

Nickel-Chromium Alloys to Resist Fuel Oil Ash Corrosion (Ref 1)

A growing problem over the past 35 years has been corrosion of heat-resistant metals and refractories by the ash from combustion of heavy residual fuel oils used in various industrial furnaces. This has stemmed from the growing use of heavy residual fuel oil, owing to its relatively attractive price, and extension of its use in furnaces operating at increasingly higher temperatures. An additional factor is the general trend (often governed by petroleum marketing considerations) toward maximizing production of distillates with corresponding reduction in fuel oil yield. As a result, greater amounts of the ash-forming constituents responsible for high-temperature corrosion, particularly organo-metallic vanadium compounds, are concentrated in the residual fuel oil.

It has been well established for many years that vanadium pentoxide (V_2O_5) and alkali metal sulfates are the primary ash constituents responsible for oil ash corrosion. Vanadium exists in certain crude oils as an oil-soluble porphyrin complex in variable quantities. While fuel oil derived from some crude oil contains relatively low amounts of vanadium (e.g., 20 ppm), that from certain Middle Eastern crudes, particularly Venezuelan crudes, often contains several hundred parts per million (Table 1).

Various mechanisms have been advanced to explain oil ash corrosion. The simplest explanation is that low-melting constituents formed from vanadium pentoxide and alkali metal sulfates, when molten, exert a fluxing action on the protective oxide films of heat-resistant alloys, allowing corrosion to progress in an accelerated manner at temperatures ranging from 500 to 700 °C (930 to 1300 °F).

Although a number of methods have been used or proposed to mitigate oil ash corrosion, it is best controlled by proper alloy selection. As described below, the 50/50-type nickel-chromium alloys are well-established materials for resisting oil ash corrosion.

Alloy development of cast 50Ni-50Cr and 40Ni-60Cr alloys began in the mid-1950s with an extensive series of laboratory screening tests on established and experimental heat-resistant alloys. These tests were conducted by the U.S. Naval Boiler & Turbine Laboratory and the U.S. National Bureau of Standards under contract from the U.S. Navy. This work was undertaken to find materials for service as uncooled tube spacers and supports in marine boiler superheaters that would offer significantly improved resistance to oil ash from Bunker C fuel oil, compared with standard HH (25Cr-12Ni) and HK (25Cr-20Ni) heat-resistant steel castings. The results

Table 2 Chemical composition requirements for nickel-chromium alloys described in ASTM A 560

Grade	Composition, wt % (a) C	Mn	Si	S	P	N	N+C	Fe	Ti	Al	Nb	Cr	Ni
50Ni-50Cr	0.10	0.30	1.00	0.02	0.02	0.30	...	1.00	0.50	0.25	...	48.0–52.0	bal
40Ni-60Cr	0.10	0.30	1.00	0.02	0.02	0.30	...	1.00	0.50	0.25	...	58.0–62.0	bal
50Ni-50Cr-Nb	0.10	0.30	0.50	0.02	0.02	0.16	0.20	1.00	0.50	0.25	1.4–1.7	47.0–52.0	bal

(a) The total of the nickel, chromium, and niobium contents must exceed 97.5%.

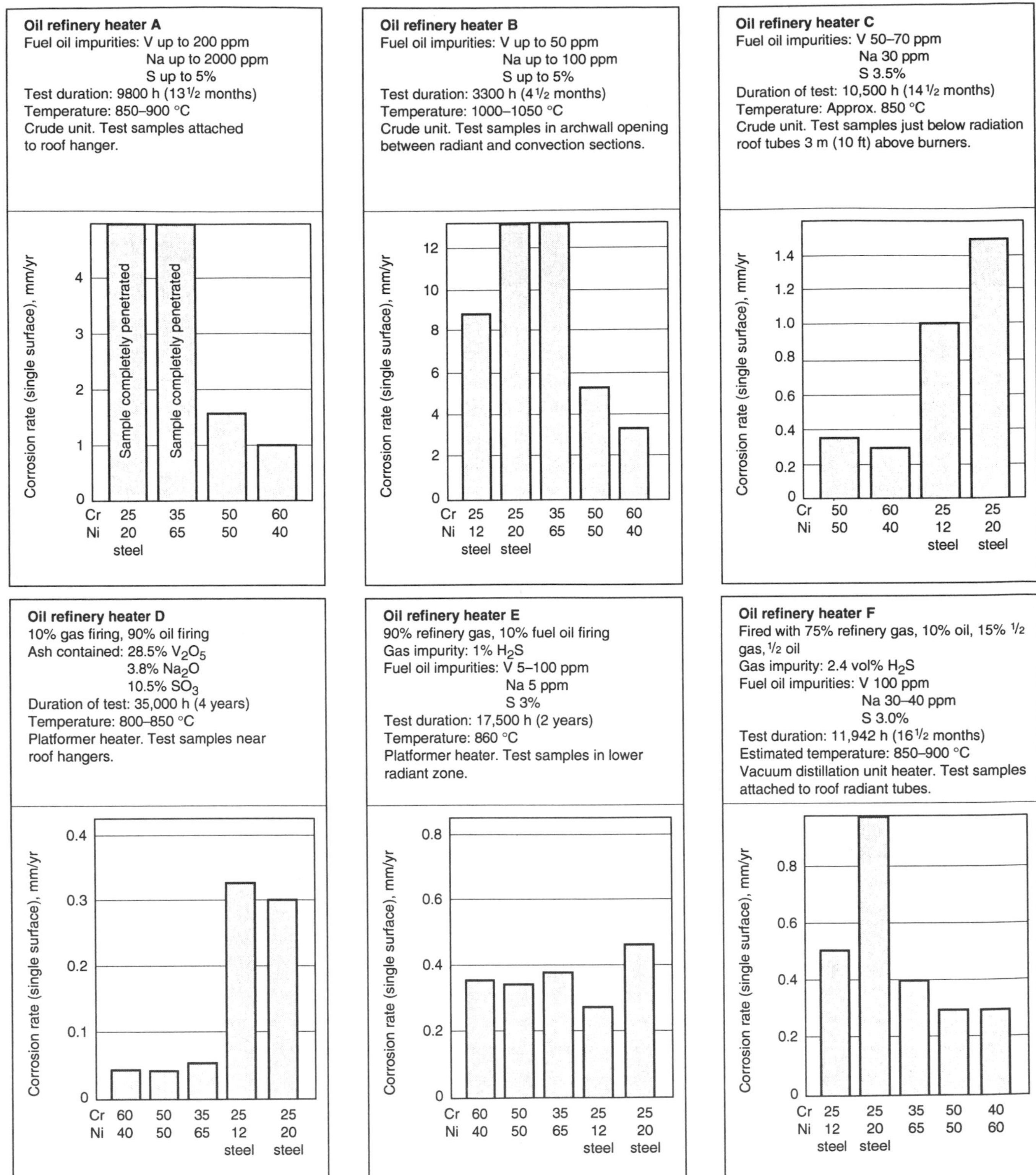

Fig. 1 Results of field tests of oil ash corrosion in oil refinery heaters. Source: Ref 1

demonstrated that the cast nickel-chromium alloys had much improved resistance to synthetic oil ash attack (Ref 1). In fact, these alloys exhibited corrosion resistance 12 to 45 times that of 25Cr-20Ni steels.

Further studies by Swales and Ward (Ref 3), McDowell and Mihalisin (Ref 4), and Spafford (Ref 5) confirmed the good performance of cast nickel-chromium in high-temperature oil ash environments. These alloys have similar resistance to oil ash attack at temperatures up to 900 °C (1650 °F). At temperatures ranging from 900 to 1090 °C (1650 to 2000 °F), the higher-chromium alloy is recommended. However, due to its superior as-cast ductility, machinability, and weld-

Table 3 Minimum room-temperature property requirements for cast nickel-chromium alloys described in ASTM A 560

Alloy	Tensile strength, MPa	Tensile strength, ksi	Yield point, MPa	Yield point, ksi	Elongation in 50 mm (2 in.), %	Charpy unnotched impact strength, J	Charpy unnotched impact strength, ft · lbf
50Ni-50Cr	550	80	340	49	5	78	50
40Ni-60Cr	760	110	590	85	...	14	10
50Ni-50Cr-Nb	550	80	345	50	5	...	...

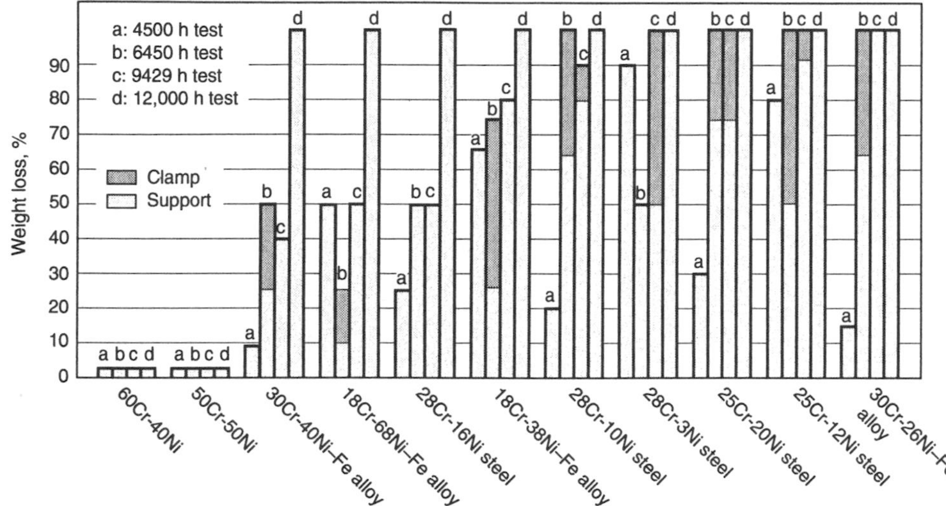

Fig. 2 Results of field tests of oil ash corrosion in power station boiler components. The tests were carried out in Bunker C oil at 815 °C (1500 °F). See text for details. Source: Ref 1

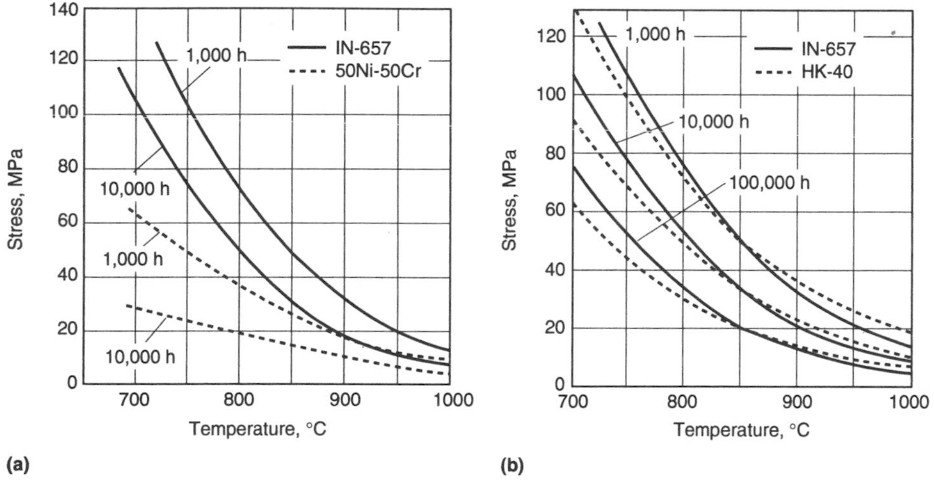

Fig. 3 Stress-rupture properties for nickel-chromium and steel castings derived from Larson-Miller curves. (a) IN-657 vs. 50Ni-50Cr. (b) IN-657 vs. HK-40. Source: Ref 1

ability, the 50Ni-50Cr alloy is used in the majority of applications in power plants, oil refinery heaters, and marine boilers involving temperatures less than about 900 °C (1650 °F). The 50/50-type alloy also has better foundry characteristics and lower cost.

Both of these alloys, as well as a modified higher-strength alloy containing approximately 1.5% Nb, are covered by ASTM A 560. The basic compositional and property requirements of A 560 are given in Tables 2 and 3. Additional information on the higher-strength nickel-chromium-niobium cast alloy is presented below.

Corrosion of Oil-Refinery Heaters. Figure 1 summarizes the results of a number of quantitative field tests in crude oil distillation, platformer, and catalytic-cracking heaters. These tests, which covered a reasonably wide practical range of fuel quality, temperature, and operating conditions, confirmed the general superiority of the nickel-chromium alloys, as far as oil ash corrosion is concerned, over the standard cast heat-resistant steels used for tube supports. In cases where residual fuel was used for only a small proportion of the test duration and refinery gas was used for the remainder, high-chromium alloys did not show any advantage over the conventional 25Cr-20Ni and 25Cr-12Ni steels in resisting attack in oxidizing sulfurous gases at high temperatures. Indeed, they may be inferior (e.g., compare the results for oil refinery heaters D and E in Fig. 1). Consequently, there is no point in specifying the more expensive high-chromium alloys for heaters unless it is probable that heavy residual fuel oils will be used for a substantial proportion of the firing operations.

Corrosion in Power Stations. Figure 2 shows the results of a comprehensive test carried out by the Consolidated Edison Company of New York at their Hudson Avenue Generating Station, following simple 500 h rack tests at their Hell Gate station. Seventy-five sets of castings representing 12 different materials were installed in the first-pass zone of the superheater. Gas temperatures were up to 980 °C (1800 °F) and the temperatures of the metal supports averaged about 820 °C (1500 °F). The fuel oil during the test period had vanadium contents of 150 to 250 ppm, sodium contents of 20 to 80 ppm, and sulfur contents of the order of 2.5%. After 4500, 6450, 9429, and 12,000 h, a set of castings, comprising one casting of each material, was removed for examination. After 12,000 h all the castings, with the exception of the nickel-chromium alloys, had to be removed because of their poor condition. Both the 50Ni-50Cr and 40Ni-60Cr alloys were reported to be still in service several years later.

A high-strength nickel-chromium alloy, commonly referred to as IN-657, has been developed that provides the same good resistance to fuel oil ash corrosion exhibited by the standard 50Ni-50Cr alloys, but with improved creep and stress-rupture properties. This was achieved by modifying the standard alloy with a controlled addition of niobium and reducing the upper permissible limits of nitrogen and silicon as specified in ASTM A 560. The basic compositional requirements for this alloy are given in Tables 2 and 3.

The elevated-temperature strength of IN-657 is better than that of the standard 50Ni-50Cr alloy, as illustrated by the mean stress-rupture relationships shown in Fig. 3(a). These are based on tests made on six commercial and six laboratory heats. Similar comparisons between IN-657 and cast 25Cr-20Ni steel (HK-40), based on published data for the latter material, shows their stress-rupture strengths to be roughly equal (Fig. 3b). The stress-rupture ductility values of IN-657, an important consideration in high-temperature service, are generally equal or superior to those of 50Ni-50Cr alloy and 25Cr-20Ni steel, depending on the test temperature.

In common with most cast heat-resistant alloys, including the standard 50Ni-50Cr type, IN-657 suffers a loss of room-temperature tensile ductility after service at high temperatures. However, the niobium-containing nickel-chromium

Alloy	Time, h	Temperature, °C	Weight loss after descaling, mg/cm² (20, 40, 60, 80, 100, 200, 400, 600, 800)
IN-657	300	800	
		900	
50Cr-50Ni	300	800	
		900	
HK-40	16	800	
		900	

Fig. 4 Comparison of fuel oil ash corrosion resistance for nickel-chromium alloys and HK-40 steel. Source: Ref 1

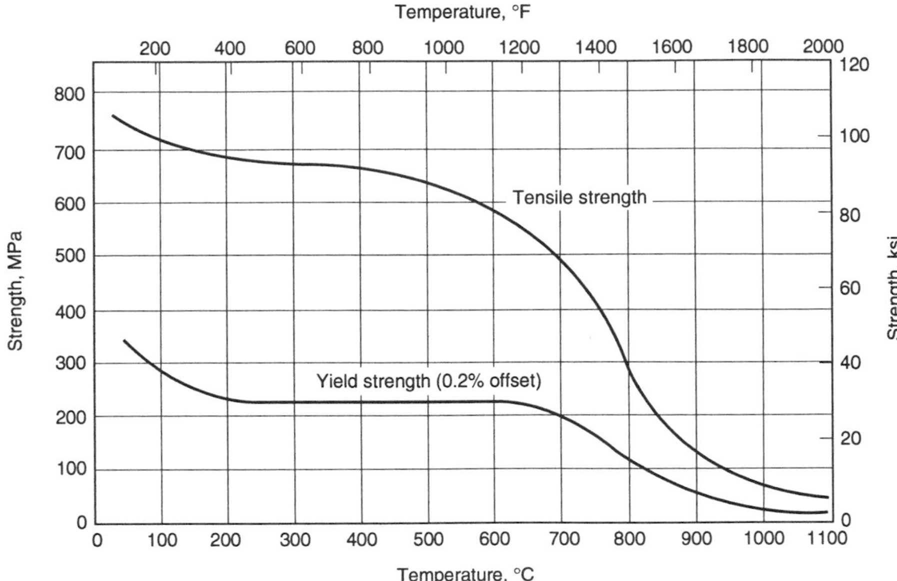

Fig. 5 Effect of temperature on the tensile properties of an annealed 80Ni-20Cr resistance heating element alloy. Source: Inco Alloys International

Table 4 Elevated-temperature tensile properties of TD Nickel ($98Ni\text{-}2ThO_2$) and TD NiCr ($78Ni\text{-}20Cr\text{-}2ThO_2$)

Temperature		Tensile strength		0.2% yield strength		Elongation in 25 mm
°C	°F	MPa	ksi	MPa	ksi	(1 in.), %
TD Nickel (bar)						
20	70	690	100	550	80	25
540	1000	310	45	295	43	14
650	1200	260	38	250	36	12
760	1400	230	33	215	31	11
870	1600	195	28	180	26	9
980	1800	165	24	140	20	7
1095	2000	140	20	110	16	6
TD Nickel (sheet)						
20	70	450	65	310	45	15
540	1000	230	33	205	30	6
650	1200	195	28	180	26	5
760	1400	170	25	160	23	4
870	1600	145	21	130	19	4
980	1800	125	18	115	17	4
1095	2000	97	14	83	12	6
TD NiCr (sheet)						
20	70	945	137	615	89	20
540	1000	685	99	450	65	14.5
650	1200	435	63	365	53	11
760	1400	270	39	260	38	5
870	1600	185	27	180	26	3
980	1800	140	20	130	19	2
1095	2000	110	16	105	15	1.5

Source: Ref 7

alloys show less impairment of ductility than the standard 50Ni-50Cr alloy (Ref 6).

Figure 4 compares IN-657, 50Ni-50Cr, and 25Cr-20Ni (HK-40) steel for resistance to oil ash corrosion. The weight losses shown were determined after half-immersion tests in 80% V_2O_5 + 20% Na_2SO_4. The results recorded for HK-40 relate to only 16 h of exposure, compared with 300 h for the other two alloys. Preliminary field service experience of up to about 3 years has supported these laboratory data, indicating that there is essentially no difference between 50Cr-50Ni alloy and IN-657 with respect to oil ash corrosion resistance.

Nickel-Chromium Alloys for Resistance Heating Elements

Resistance heating alloys are used in many varied applications, from small household appliances to large industrial process heating systems and furnaces. In appliances or industrial process heating, the heating elements are usually either open helical coils, consisting of resistance wire mounted with ceramic bushings in a suitable metal frame, or enclosed metal-sheathed elements, consisting of a smaller-diameter helical coil of resistance wire that is electrically insulated from the metal sheath by compacted refractory insulation.

The primary requirements of materials used for heating elements are high melting point, high electrical resistivity, reproducible temperature coefficient of resistance, good oxidation resistance, absence of volatile components, and resistance to contamination. Other desirable properties are good elevated-temperature creep strength, high emissivity, low thermal expansion, and low modulus (both of which help minimize thermal fatigue), good resistance to thermal shock, and good strength and ductility at fabrication temperatures.

Materials used for resistance heating applications include:

- Nickel-chromium alloys (80Ni-20Cr and 70Ni-30Cr)
- Iron-nickel-chromium alloys
- Iron-chromium-aluminum alloys
- Pure metals (e.g., platinum, molybdenum, and tungsten)
- Nonmetallic materials (e.g., silicon carbide and graphite)

Of these five material groups, nickel-chromium and iron-nickel-chromium alloys are the most widely used heating materials in electric heat treating furnaces. In fact, most manufacturers of electric furnaces provide 80Ni-20Cr elements as standard, both because they permit a wider range of furnace temperatures and because it is usually more economical to stock only a limited number of heater materials. The 80Ni-20Cr alloys permit a wider range of operating temperatures because they have the greatest resistance to oxidation, and therefore they can be used at higher temperatures than other nickel-chromium and iron-nickel-chromium alloys. Small sili-

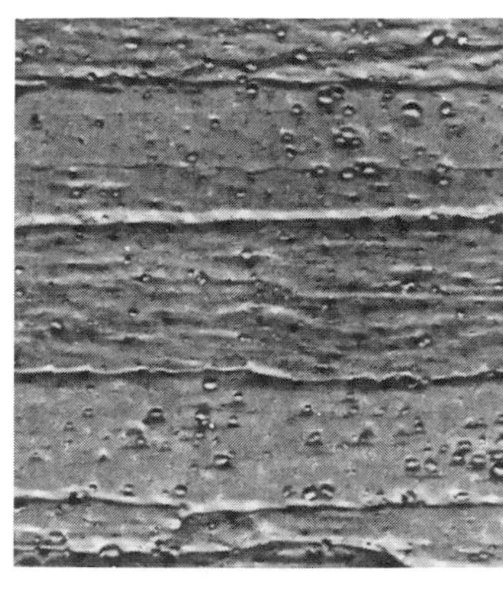

(a)

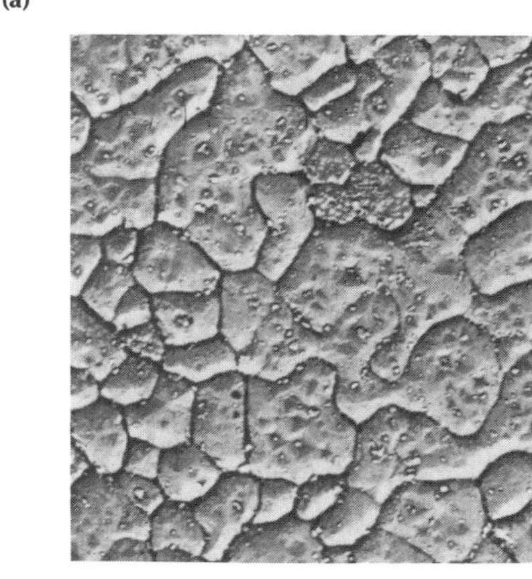

(b)

Fig. 6 Replica electron micrographs of annealed TD Nickel sheet, showing grain structure and dispersion of ThO_2. (a) Longitudinal (rolling) direction. (b) Transverse section. Both at 7000×

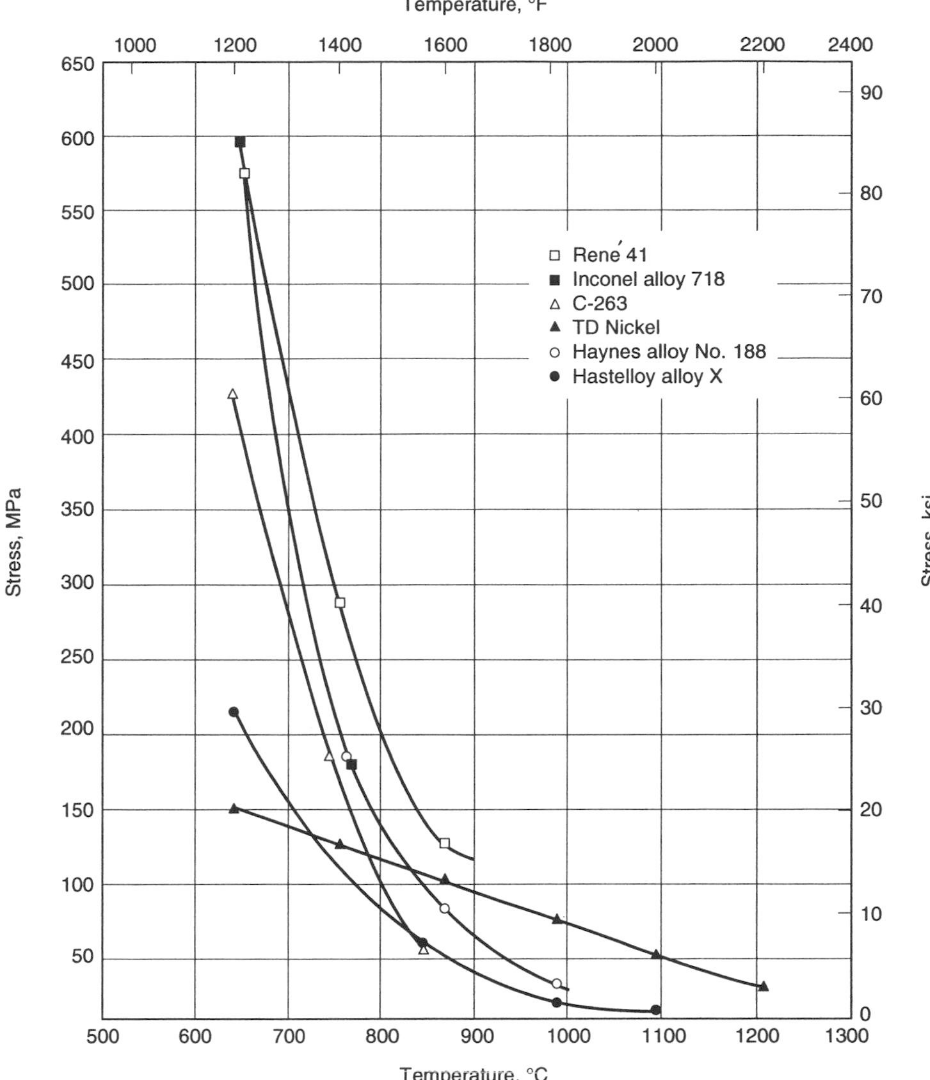

Fig. 7 The 1000 h stress-rupture strength of TD Nickel and other heat-resistant alloys. Source: Ref 7

con additions (0.75 to 1.75%) can also improve oxidation resistance. They are not recommended for use in sulfur-bearing and reducing atmospheres at high temperatures.

In general, nickel-chromium heating elements are unsuitable above 1150 °C (2100 °F) because the oxidation rate in air is too great and the operating temperature is too close to the melting point of the alloy (the approximate melting point of 80Ni-20Cr is 1400 °C, or 2550 °F). As shown in Fig. 5, the strength of 80Ni-20Cr drops significantly when temperature exceeds 650 °C (1200 °F). Additional information on the properties of 80Ni-20Cr resistance heating alloys, as well as the other types of heating element materials listed above, can be found in the article "Industrial Applications of Heat-Resistant Materials" in this Volume.

Nickel-Thoria Alloys

Dispersion-strengthened nickel was one of the first commercially available dispersion-strengthened materials. In these powder metallurgy alloys, either nickel or a 78Ni-20Cr alloy is mixed with a fine dispersoid of thoria (ThO_2). Thoria contents range from 1.80 to 2.60%, with the normal content being 2.0% ThO_2. A typical microstructure is shown in Fig. 6.

Commonly referred to as TD Nickel or TD NiCr, these alloys were developed for use in components in combustion systems of advanced gas turbine engines, fixtures for high-temperature tensile testing, and specialized furnace components and heating elements. Elevated-temperature tensile properties and stress-rupture characteristics are given in Table 4 and Fig. 7, respectively. As Fig. 7 shows, TD Nickel demonstrates superior stress-rupture properties at very high temperatures (850 to 1200 °C, or 1600 to 2200 °F).

Unfortunately, there are several disadvantages associated with TD Nickel that have limited its commercial viability (Ref 8). This material has poor oxidation resistance and should be coated for long-term high-temperature surface stability. It is difficult to process and cannot be hot worked, and it is mildly radioactive (thorium is an actinide metal). Possession of this material requires licensing by the U.S. Nuclear Regulatory Commission. As a result of the problems, production of TD Nickel has ceased. Today yttria (Y_2O_3) dispersion-strengthened nickel-base superalloys are being used for elevated-temperature applications. These alloys are described in the article "Powder Metallurgy Superalloys" in this Volume.

ACKNOWLEDGMENT

Much of the information in this article was adapted from various publications available from the Nickel Development Institute.

REFERENCES

1. "High-Chromium Cr-Ni Alloys to Resist Fuel Oil Ash Corrosion: A Review of Developments

and Experience 1955–1975," Publication 4299, Nickel Development Institute, 1975

2. Fuel Ash Effects on Boiler Design and Operation, *Steam: Its Generation and Use,* 40th ed., S.C. Stultz and J.B. Kitto, Ed., Babcock & Wilcox Co., 1992
3. G.L. Swales and D.M. Ward, Paper 126, presented at Corrosion/79, National Association of Corrosion Engineers, 1979
4. D.W. McDowell, Jr. and J.R. Mihalisin, Paper 60-WA-260, presented at ASME Winter Annual Meeting (New York), 27 Nov to 2 Dec 1960
5. B.F. Spafford, *Conf. Proc. U.K. Corrosion '83,* Institution of Corrosion Science & Technology, Birmingham, U.K., 1982, p 67
6. P.J. Penrice, A.J. Stapley, and J.A. Towers, Nickel Chromium Alloys with 30–60% Chromium in Relation to Their Resistance to Corrosion by Fuel Ash Deposits, Part II: Mechanical Properties and the Influence of Exposure at High Temperatures on Tensile and Impact Properties, *J. Inst. Fuel,* Vol 39, Jan 1966, p 14–21
7. "High-Temperature High-Strength Nickel Base Alloys," Publication 393, Nickel Development Institute, 1995
8. J. Gadbut, D.E. Wenschhof, and R.B. Herchenroeder, Properties of Nickels and Nickel Alloys, *Properties and Selection: Stainless Steels, Tool Materials, and Special-Purpose Metals,* Vol 3, *Metals Handbook,* 9th ed., American Society for Metals, 1980, p 128–170

Structural Intermetallics

ALLOYS based on ordered intermetallic compounds constitute a unique class of metallic material that form long-range ordered crystal structures (Fig. 1) below a critical temperature, generally referred to as the critical ordering temperature (T_c). These ordered intermetallics usually exist in relatively narrow compositional ranges around simple stoichiometric ratios.

The recent research for new high-temperature structural materials has simulated much interest in ordered intermetallics. These compounds generally exhibit promising high-temperature properties, because the long-range ordered superlattice lowers dislocation mobility and diffusion processes at elevated temperatures. However, because of the brittleness problem, the intermetallics have been used mainly as strengthening constituents in structural materials. For example, high-temperature nickel-base superalloys owe their outstanding strength properties to a fine dispersion of precipitation particles of the ordered γ' phase (Ni_3Al) embedded in a ductile disordered matrix. (See the article "Metallurgy, Processing, and Properties of Superalloys" in this Volume).

Recent research has focused on understanding the brittle fracture and low ductility in ordered intermetallics. In some cases, the brittleness results from strong resistance to the motion of dislocations, to the point that cleavage or intergranular fracture may be favored. In many cases, however, the dislocations are relatively mobile. Brittleness results either from low-symmetry crystal structures that do not possess enough independent slip systems to permit arbitrary deformation or from the presence of grain boundaries that are too weak to resist the propagation of cracks. It has also been found that quite a number of ordered intermetallics, such as iron aluminides (Ref 1, 2), exhibit environmental embrittlement at ambient temperatures. The embrittlement involves the reaction of water vapor in air with reactive elements (aluminum, for example) in intermetallics to form atomic hydrogen, which drives into the metal and causes premature fracture.

In recent years, alloying and processing have been employed to control the ordered crystal structure, microstructural features, and grain-boundary structure and composition to overcome the brittleness problem of ordered intermetallics. Success in this work has inspired parallel efforts aimed at improving strength properties. The results have led to the development of a number of attractive intermetallic alloys having useful ductility and strength.

Alloy design work has been centered primarily on aluminides of nickel, iron, and titanium. These materials possess a number of attributes that make them attractive for high-temperature applications. They contain sufficient amounts of aluminum to form, in oxidizing environments, thin films of alumina (Al_2O_3) that often are compact and protective (Ref 3). These materials have low densities, relatively high melting points (Table 1), and good high-temperature strength properties. Table 2 lists the major attributes and upper use temperature limits for selected intermetallic compounds.

Crystal structures showing the ordered arrangements of atoms in several of these aluminides are illustrated in Fig. 2. For most of the aluminides listed in Table 1, the critical ordering temperature is equal to the melting temperature. Others disorder at somewhat lower temperatures, and Fe_3Al passes through two ordered structures ($D0_3$ and $B2$) before becoming disordered. Deviations from stoichiometry are accommodated either by the incorporation of vacancies in the lattice (for example, NiAl) (Ref 5–8) or by the location of antisite atoms in one of the sublattices. Many of the aluminides exist over a range of compositions, but the degree of order decreases as the deviation from stoichiometry increases. Additional elements can be incorporated without losing the ordered structure. For example, in Ni_3Al, silicon atoms are located in aluminum sites, cobalt atoms on nickel sites, and iron atoms on either (Ref 9). In many instances, the so-called intermetallic compounds can be used as bases for alloy development to improve or optimize properties for specific applications.

This article summarizes research and development of nickel aluminides based on Ni_3Al and NiAl, iron aluminides based on Fe_3Al and FeAl, titanium aluminides based on Ti_3Al and TiAl, and other intermetallics such as silicides. Information on other intermetallics used for nonstructural applications (e.g., $SmCo_5$ used for permanent magnets and Nb_3Sn used in superconductors) can be found in Volume 2 of the *ASM Handbook*. Additional information on structural intermetallics can be found in Ref 10 to 20.

Nickel Aluminides

The nickel-aluminum phase diagram shows two stable intermetallic compounds, Ni_3Al and NiAl, formed on the nickel-rich end (Fig. 3). The compound Ni_3Al has an $L1_2$ crystal structure, a

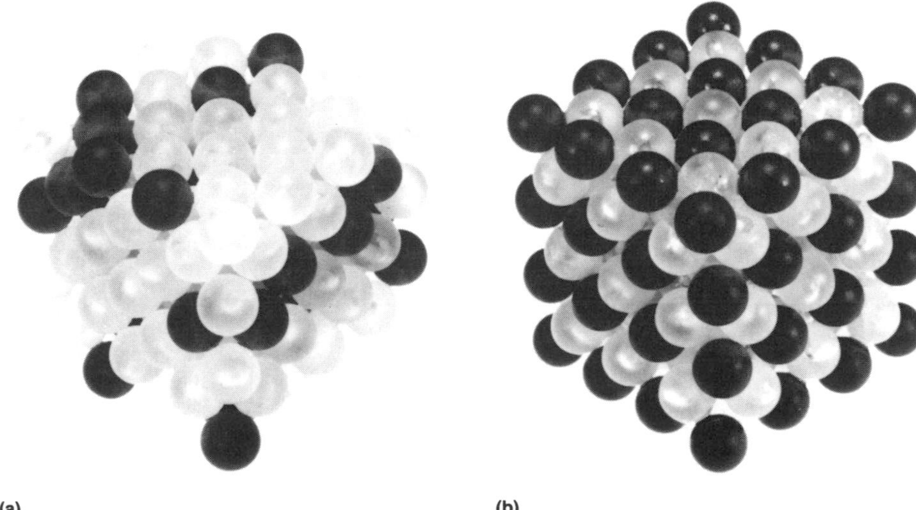

Fig. 1 Atomic arrangements of conventional alloys and ordered intermetallic compounds. (a) Disordered crystal structure of a conventional alloy. (b) Long-range ordered crystal structure of an ordered intermetallic compound

Table 1 Properties of nickel, iron, and titanium aluminides

Alloy	Crystal structure(a)	Critical ordering temperature (T_c)		Melting point (T_m)		Material density,	Young's modulus	
		°C	°F	°C	°F	g/cm^3	GPa	10^6 psi
Ni_3Al	$L1_2$ (ordered fcc)	1390	2535	1390	2535	7.50	179	25.9
NiAl	*B*2 (ordered bcc)	1640	2985	1640	2985	5.86	294	42.7
Fe_3Al	$D0_3$ (ordered bcc)	540	1000	1540	2805	6.72	141	20.4
	*B*2 (ordered bcc)	760	1400	1540	2805	...	...	...
FeAl	*B*2 (ordered bcc)	1250	2280	1250	2280	5.56	261	37.8
Ti_3Al	$D0_{19}$ (ordered hcp)	1100	2010	1600	2910	4.2	145	21.0
TiAl	$L1_0$ (ordered tetragonal)	1460	2660	1460	2660	3.91	176	25.5
$TiAl_3$	$D0_{22}$ (ordered tetragonal)	1350	2460	1350	2460	3.4	...	...

(a) fcc, face-centered cubic; bcc, body-centered cubic; hcp, hexagonal close packed

Table 2 Attributes and upper use temperature limits for nickel, iron, and titanium aluminides

Alloy	Attributes	Maximum use temperature, °C (°F)	
		Strength limit	Corrosion limit
Ni_3Al	Oxidation, carburization, and nitridation resistance; high-temperature strength	1000 (1830)	1150 (2100)
NiAl	High melting point; high thermal conductivity; oxidation, carburization, and nitridation resistance	1200 (2190)	1400 (2550)
Fe_3Al	Oxidation and sulfidation resistance	600 (1110)	1100 (2010)
FeAl	Oxidation, sulfidation, molten salt, and carburization resistance	800 (1470)	1200 (2190)
Ti_3Al	Low density; good specific strength	760 (1400)	650 (1200)
TiAl	Low density; good specific strength; wear resistance	1000 (1830)	900 (1650)

Source: Ref 4

Table 3 Nominal compositions of selected nickel aluminide alloys

Alloy (a)	Composition, wt %						
	Al	Cr	Fe	Zr	Mo	B	Ni
IC-50	11.3	...	...	0.6	...	0.02	bal
IC-74M	12.4	...	...	...	...	0.05	bal
IC-218	8.5	7.8	...	0.8	...	0.02	bal
IC-218 LZr	8.7	8.1	...	0.2	...	0.02	bal
IC-221	8.5	7.8	...	1.7	...	0.02	bal
IC-357	9.5	7.0	11.2	0.4	1.3	0.02	bal
IC-396M	8.0	7.7	...	0.8	3.0	0.01	bal

(a) Designations used by Oak Ridge National Laboratory, Oak Ridge, TN

derivative of the face-centered cubic (fcc) crystal structure; NiAl has a *B*2 structure, a derivative of the body-centered cubic (bcc) crystal structure (see Fig. 2). Because of the different crystal structures, the two nickel aluminides have quite different physical and mechanical properties.

Ni_3Al Aluminides

Intergranular Fracture and Alloying Effects. The aluminide Ni_3Al is of interest because of its excellent strength and oxidation resistance at elevated temperatures (see Table 2). As mentioned earlier, Ni_3Al is the most important strengthening constituent in nickel-base superalloys. Single crystals of Ni_3Al are ductile at ambient temperatures, but polycrystalline Ni_3Al fails by brittle grain-boundary fracture with very little plasticity (Ref 23–23). This effect persists even in very high-purity materials where no grain-boundary segregation of impurities can be detected, suggesting that the brittleness is an intrinsic feature (Ref 21, 24–26). The observation of this characteristic turned attention toward a search for segregants that might act in a beneficial way.

Studies of segregants led to the discovery that small (~0.1 wt%) boron additions not only eliminated the brittle behavior of Ni_3Al but converted the material to a highly malleable form exhibiting tensile ductility as high as 50% at room temperature (Fig. 4) (Ref 21, 24). The beneficial effect of boron is, however, dependent on stoichiometry, and boron is effective in increasing the ductility of Ni_3Al only in alloys containing less than 25 at.% aluminum (Ref 24, 27). Both Auger spectroscopy and imaging atom probe studies have demonstrated the strong segregation of boron to grain boundaries, although the extent of segregation varies along boundaries and even at different points along the same boundary. Segregation is less strong in aluminum-rich alloys. The beneficial effect of boron has been attributed to an increase in the intrinsic strength or cohesion of the grain boundary (Ref 24, 28–31) and enhancement of dislocation generation and facilitation of slip transmission along grain boundaries (Ref 32–35). Both suggested mechanisms can be rationalized from boron-induced atomic disordering in the grain-boundary region (Ref 35–39).

Since gaining the knowledge that microalloying with boron can ductilize polycrystalline Ni_3Al, a number of Ni_3Al alloy compositions have been developed. As shown in Table 3, macroalloying additions include chromium, iron, zirconium, and molybdenum. These alloying additions were made to improve strength, castability, hot workability, and corrosion resistance. The effects of these as well as other alloying additions are described in the paragraphs that follow.

Environmental Embrittlement at Elevated Temperatures. Ductility at elevated temperatures depends on the test environment (Ref 40–43). In vacuum, the ductility of boron-doped nickel-rich alloys (<23% Al) remains high at all test temperatures, although a moderate minimum appears in the vicinity of 800 °C (1470 °F) (see Fig. 5). In tests conducted in environments containing oxygen, the ductility minimum is much deeper. In fact, ductility approaches zero with full intergranular fracture in tests conducted in air at 760 °C (1400 °F). The loss in ductility is due to a dynamic effect that requires the simultaneous application of a tensile stress and the presence of oxygen (Ref 40–43). It appears that the formation of protective Al_2O_3 films is too slow to deter rapid intergranular crack propagation. Chromium additions in the range of 6 to 10% restore the intermediate temperature ductility (Ref 44–46) possibly because of the more rapid formation of protective chromia (Cr_2O_3) films. Ductility at intermediate temperatures can also be improved by the production of an elongated grain structure in Ni_3Al (Ref 47).

Anomalous Dependence of Yield Strength on Temperature. Ni_3Al is one of a number of intermetallic alloys that exhibit an engineering yield strength (0.2% offset) that increases with increasing temperature. This is shown in Fig. 6, which is a plot of yield stress as a function of test temperature. The anomalous yielding effect, which is lower at lower strains, occurs because of the extremely rapid work hardening. The work hardening is caused by the cross slip of screw dislocation segments from the primary {111} slip planes to {100} planes, where they become pinned and much less mobile (Ref 49, 50). Driving forces for cross slip include anisotropy of the energy of antiphase boundaries formed between the superdislocation pairs required for deformation in the ordered lattice (Ref 49, 50) and the torque exerted between the screw dislocation pairs arising from elastic anisotropy (Ref 51, 52). In either case, the cross slip pinning process is thermally activated, which leads to the positive temperature dependence of the yield strength shown in Fig. 6. The reduction of yield strength at high temperatures occurs because of enhanced dislocation mobility on {100} planes, which reduces the effectiveness of the pinning centers formed by the cross slip process. The anomalous yielding behavior makes Ni_3Al stronger than many commercial solid-solution alloys (such as type 316 stainless steel and Hastelloy alloy X) at elevated temperatures (Fig. 6).

As with boron-doped Ni_3Al, Ni_3Al alloys (Table 3) also exhibit rising yield strength with rising temperature. Figure 7 shows that the yield strength of four nickel aluminide alloys tends to rise to a maximum in the temperature range of ~400 to 650 °C (~750 to 1200 °F). Above this temperature range, the yield strength declines. The effect of cold working on the yield strength of IC-50 alloy is shown in Fig. 8. Up to test temperatures of ~600 °C (~1110 °F), cold working had an effect on the yield strength, but at 800 °C (1470 °F), there was no significant enhancement of yield strength.

Solid-Solution Hardening. Ni_3Al doped with boron serves as a design base for ductile and strong materials for structural uses. The aluminide is capable of being hardened by solid-solution effects because it can dissolve substantial alloying additions without losing the advantage of long-range order. One study constructed the solubility lobes of ternary Ni_3Al phase ($L1_2$) at 1000 °C (1830 °F) for various alloying elements (Ref 53). The elements that dissolve substantially in Ni_3Al can be divided into three groups (Fig. 9). The first group of elements, including silicon, germanium, titanium, vanadium, and hafnium, substitutes almost exclusively on aluminum sublattice sites. The second group, consisting of copper, cobalt, and platinum, substitutes on nickel sublattice sites. The third group, which includes elements such as iron, manganese, and chromium, substitutes on both sublattice sites. Guard and Westbrook (Ref 54) first suggested that the electronic structure (i.e. the position of elements in the periodic table) rather than the atom size factor plays a dominant role in the substitution behavior. The extent of solid solution in Ni_3Al, however, is controlled by the atomic size misfit and the difference in the heats of formation between Ni_3Al and Ni_3X.

The room-temperature solid-solution hardening of Ni_3Al depends on the substitutional behavior of alloying elements, atomic size misfit, and the degree of non-stoichiometry of the alloy. A review of the mechanical properties of Ni_3Al indicated that it could be hardened more effectively by the elements substituting on aluminum sites and, to a lesser degree, by the elements substituting on nickel or on both nickel and aluminum sites (Ref 55). In addition, the strengthening is most pronounced for stoichiometric alloys and aluminum-rich alloys; it is much less pronounced for nickel-rich alloys. The solid-solution hardening is quite complex at elevated temperatures (Ref 56, 57). The hardening effects depend strongly on crystal orientation and test temperature, which cannot be explained by the classic solid-solution models developed based on the elastic interaction between dislocations and symmetric defects (Ref 58). For example, hafnium is found to be more effective in hardening at 850 °C (1560 °F) than at room temperature (Ref 48). These observations suggest that the unusual hardening is related to how solutes affect the cross slip pinning processes occurring at elevated temperatures. Hafnium and zirconium are most effective in strengthening Ni_3Al at elevated temperatures (Ref 48, 59).

Mechanical Properties. The study of ductility and strength of Ni_3Al has led to the development of ductile nickel aluminide alloys for structural applications with the following composition range (in atomic percent):

Ni-(14–18)Al-(6–9)Cr-(1–4)Mo-(0.01–1.5)Zr/Hf-(0.01–0.20)B

In these aluminide alloys, 6 to 9 at.% Cr is added to reduce environmental embrittlement in oxidizing environments at elevated temperatures. Zirconium and hafnium additions most effectively improve high-temperature strength via solid-solution hardening effects. Molybdenum additions improve strength at ambient and elevated temperatures. Microalloying with boron reduces moisture-induced hydrogen embrittlement and enhances grain-boundary cohesive strength, resulting in sharply increased ductility at ambient temperatures. In some

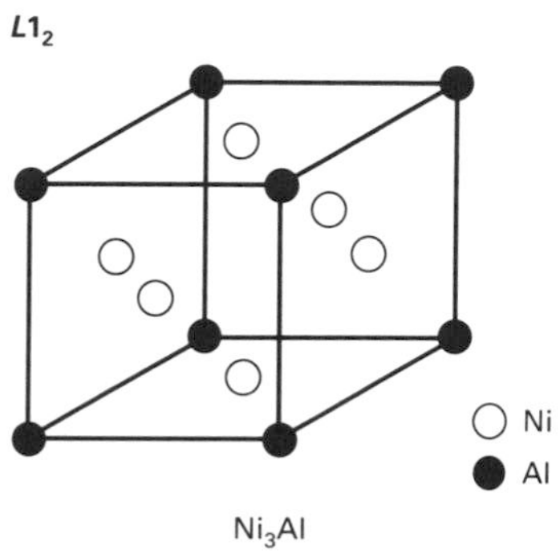

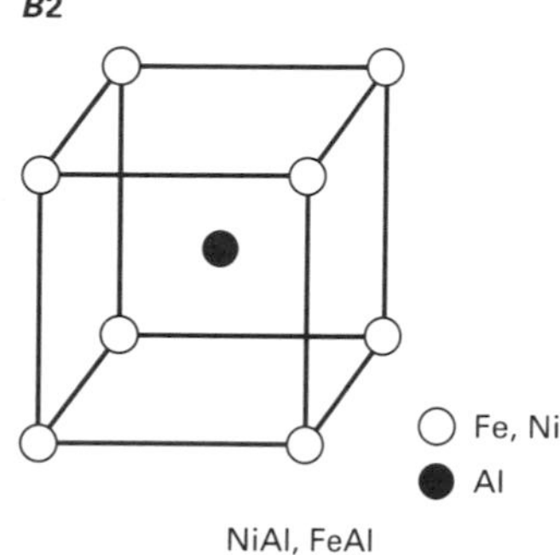

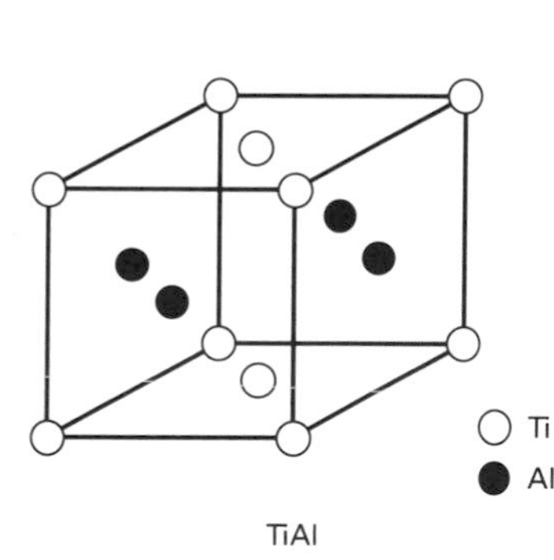

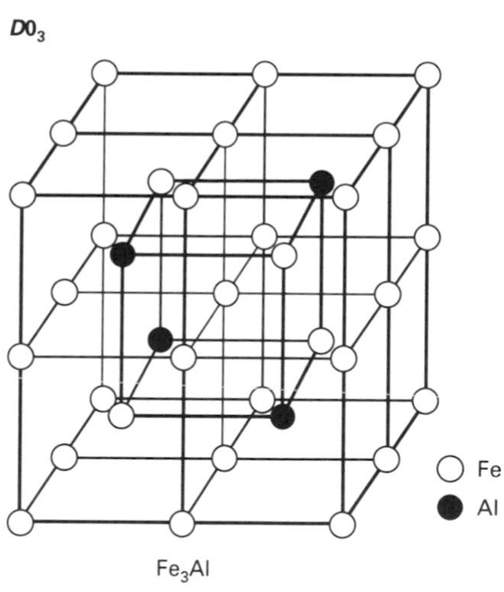

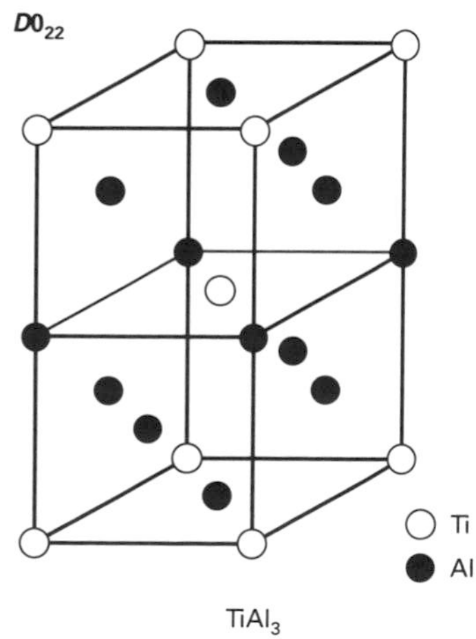

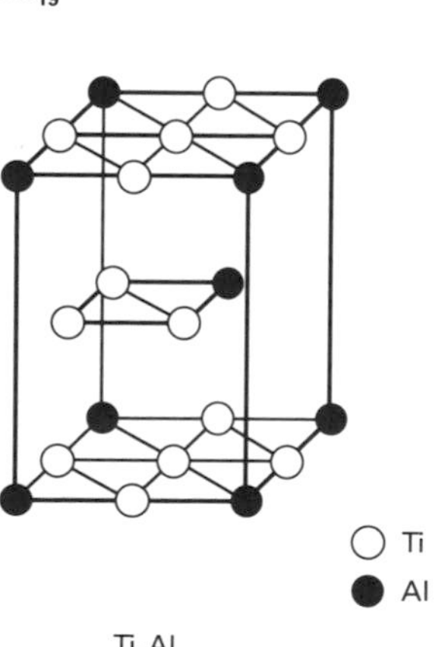

Fig. 2 Crystal structures of nickel, iron, and titanium aluminides

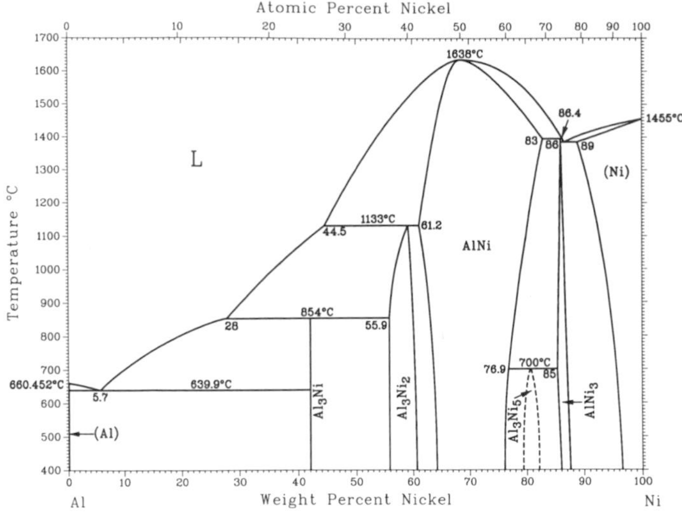

Fig. 3 The Al-Ni phase diagram showing both NiAl and Ni_3Al compounds on the nickel-rich end

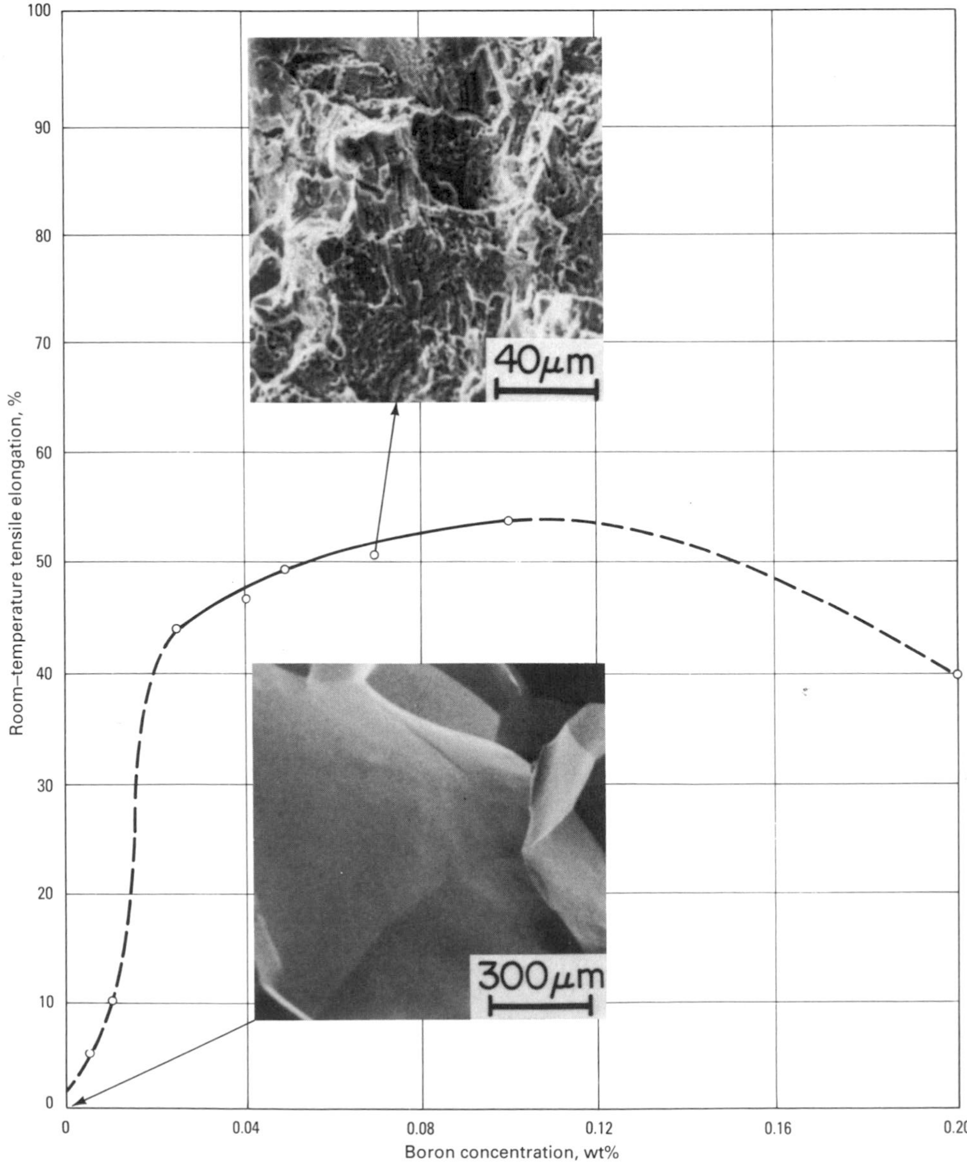

Fig. 4 Effect of boron additions on the room-temperature tensile elongation and fracture behavior of Ni_3Al (24 at.% Al)

cases, certain amounts (<20 at.%) of cobalt and iron are added to replace nickel, and aluminum and nickel, respectively, in order to further improve hardness and corrosion resistance. The alloys with optimum properties usually contain 5 to 15 vol % of the disordered γ phase, which has the beneficial effect of reducing environmental embrittlement in oxidizing atmospheres and improving creep properties at elevated temperatures.

Figure 6 shows the yield strength of a cast Ni_3Al alloy, which is stronger than nickel-base alloy IN-713C at elevated temperatures. The Ni_3Al alloys generally possess ductilities of 25 to 40% at temperatures up to 700 °C (1290 °F), and 15 to 30% at up to 100 °C (1830 °F) in air.

Figure 10 compares the tensile properties of cast Ni_3Al alloys with cast HU alloy (Fe-20Cr-39Ni-2.5Si). It is clear from this figure that the Ni_3Al alloys are nearly twice as strong at room temperature and six times as strong at 1000 °C (1830 °F). The tensile properties of high-aluminum (4.5 wt% Al) alloy 214 and cast and powder metallurgy (P/M) processed Ni_3Al alloys are compared in Fig. 11. The P/M processed IC-221 alloy in Fig. 11 is significantly stronger than alloy 214 or the cast Ni_3Al-base alloys.

Creep properties of Ni_3Al alloys have been characterized as functions of stress, temperature, and composition. Hafnium and zirconium additions are most effective in improving the creep resistance of Ni_3Al (Ref 44). Figure 12 shows creep data for the polycrystalline nickel aluminide alloy IC-221. The creep properties of Ni_3Al alloys, like those of nickel-base superalloys, are sensitive to grain size, but the Ni_3Al alloys have better creep resistance for coarse-grain materials (e.g., cast materials). For applications where creep resistance is important, coarse-grain material is more desirable at temperatures greater than 700 °C (1290 °F). Creep properties of single-crystal Ni_3Al alloys containing refractory elements such as tantalum have been studied at temperatures up to 1000 °C (1830 °F) (Ref 61). In general, the creep resistance of Ni_3Al is comparable to that of most of the nickel-base superalloys, but it is not as good as that of some advanced single-crystal nickel-base superalloys used for jet engine turbine blades. Figure 13 compares the creep-rupture strength of Ni_3Al alloys with that of a wrought nickel-base superalloy.

Fatigue and fatigue crack growth are substantially better in Ni_3Al alloys that in nickel-base superalloys in tests below the range of the ductility minimum (Ref 62–64); see Fig. 14 for room-temperature fatigue crack growth. The good fatigue resistance of Ni_3Al and other ordered intermetallic alloys has been attributed to fine planar slip and superlattice dislocation structure. Dynamic embrittlement in oxidizing environments severely reduces the fatigue resistance of Ni_3Al at temperatures above 500 °C (930 °F) (Ref 62); however, this problem has been alleviated by adding moderate amounts (for example, 8 at.%) of chromium to Ni_3Al (Ref 62–64). Fatigue/creep interactions in single crystals and directionally solidified Ni_3Al alloys have been characterized for temperatures up to 800 °C

(1470 °F) (Ref 65, 66). Limited results indicate that the performance of single-crystal Ni_3Al alloyed with hafnium and boron is superior to that of Udimet 115 at 760 °C (1400 °F). Figure 15 compares the fatigue properties at 650 °C (1200 °F) of IC-221M with that of IN-713C. The Ni_3Al alloy exhibits a fatigue life an order of magnitude better than the IN-713C, cast under the same conditions.

Corrosion Resistance. The oxidation and carburization resistance of Ni_3Al alloys are compared in Fig. 16 and 17. It is clear from Fig. 16 that the Ni_3Al alloys that form a protective Al_2O_3 scale on the surface have significantly better oxidation resistance that aluminum-free alloy 800. Carburization resistance is also high under both oxidizing or reducing environments (Fig. 17).

Physical Properties. Few data exist on the physical properties of the whole series of nickel aluminide alloys; however, the information in Table 4 summarizes the effect of small changes in stoichiometry of Ni_3Al on several physical properties of the material (Ref 68). Data showing the effect of temperature on electrical resistivity, thermal conductivity, and the mean coefficient of thermal expansion are also given in Table 4.

Table 4 Physical properties of selected nickel-aluminum binary alloys

Data are for arc-melted, homogenized, and annealed specimens produced from high-purity nickel and aluminum stock.

Temperature		Electrical resistivity		Thermal conductivity		Mean coefficient of thermal expansion	
°C	°F	$(\Omega \cdot m) \times 10^{-8}$	mΩ in.	W/m · K	Btu · in./ft² · h · °F	$(m/m \cdot K) \times 10^{-6}$	in./in. · °F
24.0 at.% aluminum content(a)							
27	81	51.3	20.1	21.4	148	...	...
427	801	72.6	28.5	30.1	208	...	...
727	1341	84.2	33.0	32.7	227	...	...
27–327	81–621	...	...	...	...	12.5	6.9
427–727	801–1341	...	...	...	...	14.8	8.2
727–1027	1341–1881	...	...	...	...	16.8	9.3
25.0 at.% aluminum content(b)							
27	81	32.6	12.8	28.9	200	...	...
427	801	55.3	21.7	37.1	257	...	...
727	1341	73.0	28.6	36.1	250	...	...
27–327	81–621	...	...	...	...	12.5	6.9
427–727	801–1341	...	...	...	...	14.6	8.1
727–1027	1341–1881	...	...	...	...	16.8	9.3

(a) Density, 7.49 g/cm³ (0.270 lb/in.³) at 22 °C (72 °F). (b) Density, 7.43 g/cm³ (0.268 lb/in.³) at 22 °C (72 °F). Source: Ref 68

Processing and Fabrication. Nickel aluminide parts can be produced by P/M, casting, and ingot metallurgy. Each of these processes is briefly described below.

The nickel aluminide powder can be produced by argon or nitrogen-gas atomization. The nickel aluminide powder can be readily consolidated to 100% density by extruding the powder in a mild steel can at 1100 °C (2010 °F) to a reduction ratio ≥9:1 (Ref 67). The powder can also be consolidated by hot isostatic pressing. The extrusion-process-consolidated powder has a grain structure that can be superplastic under proper conditions of temperature and strain rate.

Near-net-shape or net-shape casting of parts directly from liquid metal is a highly desirable method for the fabrication of nickel aluminides. Among the near-net-shape methods is the method of sheer fabrication by bringing liquid metal in direct contact with a rotating drum (Ref 67). The sheet thickness is controlled by the speed of the drum. The as-cast sheets are highly ductile, and their strength can be enhanced significantly by cold rolling.

Conventional fabrication techniques (such as hot rolling for large ingots) are ineffective because regions near the surface cool to the range of the ductility minimum, which leads to the formation of large intergranular surface cracks. Isothermal forging offers excellent possibilities for fabrication because the alloys exhibit superplastic behavior above about 1000 °C (1830 °F). Conventional hot forging is feasible for fine-grain alloys containing less than 0.3 at.% Zr or Hf. Cold fabrication is effective if the materials can be cast into sheet or rod forms that can be cold formed further without the need for repeated recrystallization treatments.

Alloys based on Ni_3Al are susceptible to weld cracking; however, if welding is done with care, sound welds can be made in most of the alloys (Ref 69–71). Welding speed should be reduced, and the boron level has to be limited to about 0.1 at.% to avoid hot cracking. Oxygen is particularly detrimental: oxide scale on alloy surfaces should be removed prior to welding, and the atmosphere must be controlled to reduce oxygen during welding. Certain alloying additions (iron, for example) have been found to promote weldability.

Structural Applications. Although the properties of alloys based on Ni_3Al approach those of established superalloys, the Ni_3Al alloys are unlikely to displace superalloys in aircraft engine applications. The opportunity exists, however, for enhancing the properties of Ni_3Al alloys further through the incorporation of second phases. In addition, alloys based on Ni_3Al could provide an attractive matrix for composite development (Ref 72–76). Monolithic aluminide alloys are likely to find near-term use in applications that take advantage of some of their unique or unusual properties. The potential applications (and the properties they would exploit include:

- Heat-treating furnace parts (superior carburization resistance, high-temperature strength, and thermal fatigue resistance)
- Gas, water, and steam turbines (the excellent cavitation, erosion, and oxidation resistance of the alloys)
- Aircraft fasteners (low density and ease of achieving the desired strength)
- Automotive turbochargers (high fatigue resistance and low density)
- Pistons and valves (wear resistance and capability of developing a thermal barrier by high-temperature oxidation treatment)
- Bellows for expansion joints to be used in corrosive environments (good aqueous corrosion resistance)
- Tooling (high-temperature strength and wear resistance developed through preoxidation)
- Permanent molds (the ability to develop a thermal barrier coating by high-temperature oxidation)

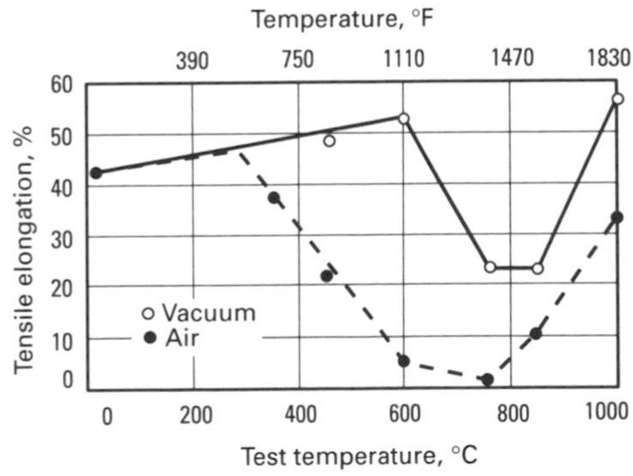

Fig. 5 Tensile elongation of alloy IC-145 (Ni-21.5Al-0.5Hf-0.1B at.%) in vacuum and air

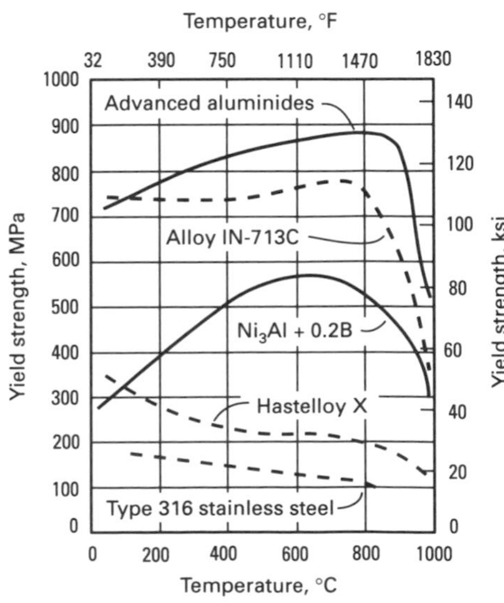

Fig. 6 Yield strength vs. test temperature for Ni_3Al alloys, two superalloys, and type 316 stainless steel. Source: Ref 48

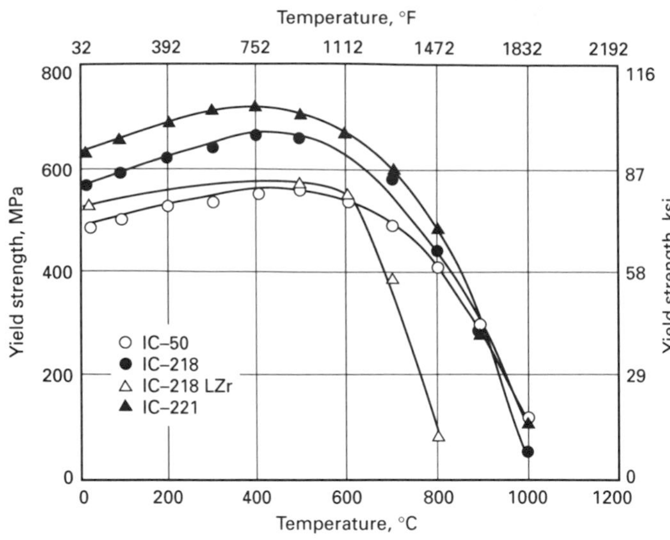

Fig. 7 Variation of yield strength with test temperature for selected nickel aluminide alloys. Strain rate, 0.5 mm/mm/min. See Table 3 for alloy compositions.

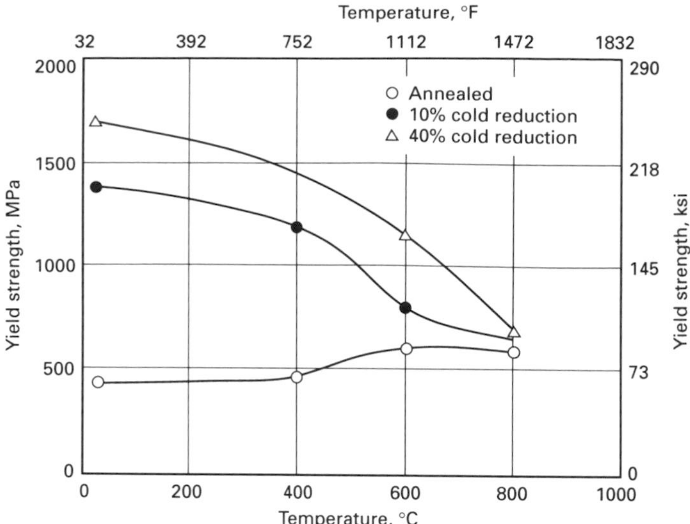

Fig. 8 Plot of yield strength vs. test temperature of IC-50 nickel aluminide alloy as a function of cold working

NiAl Aluminides

Nickel-aluminum containing more than about 40 at.% Ni starts to form a single-phase *B*2-type ordered crystal structure based on the bcc lattice (Fig. 2). In terms of physical properties, *B*2 NiAl offers more potential for high-temperature applications than *L*1$_2$ Ni_3Al. It has a higher melting point (1638 °C, or 2980 °F), a substantially lower density (5.86 g/cm^3 for NiAl versus 7.50 g/cm^3 for Ni_3Al), and a higher Young's modulus (294 GPa, or 4.27×10^6 psi, versus 179 GPa, or 25.9×10^6 psi). In addition, NiAl offers excellent oxidation resistance at high temperatures (Ref 77). In the 1950s and 1960s, NiAl alloys were employed as coating materials for hot components in corrosive environments. The oxidation resistance of NiAl can be further improved by alloying with yttrium and other refractory elements such as hafnium and zirconium (Ref 78, 79).

Structure and Property Relationships. The use of NiAl in structural members suffers two major drawbacks: poor ductility at ambient temperatures and low strength and creep resistance at elevated temperatures. Single crystals of NiAl are quite ductile in compression, but both single and polycrystalline NiAl appear to be brittle in tension at room temperature. The nickel aluminide exhibits mainly {100} slip, rather than {111} slip as commonly observed for bcc materials (Ref 80–82). The lack of sufficient slip systems has been regarded as the major cause of low ductility in NiAl. The aluminide shows a sharp increase in ductility above 400 °C (750 °F) and becomes very ductile above 600 °C (1110 °F) (Ref 83); therefore, fabrication of NiAl at high temperatures presents no major problems.

The ordered *B*2 structure exists in NiAl over a solubility range of about 15 at.%. Deviations from stoichiometry are accommodated by the incorporation of vacancies in aluminum-rich alloys and by the formation of antisite defects in nickel-rich alloys (Ref 5–8). The presence of lattice defects has a strong effect on low-temperature strength, with minimum strength occurring at the stoichiometric composition (Fig. 18). The minimum becomes less pronounced with increasing temperature and is essentially eliminated at 600 °C (1110 °F). At all compositions, the yield strength decreases with increasing temperature (Ref 83). Abrupt drops in strength in the range of 400 to 600 °C (750 to 1110 °F) are accompanied by a sharp increase in ductility. The alloys are highly ductile but extremely weak at higher temperatures. For example, Ni-50Al (at.%) showed a yield strength of 35 MPa (5 ksi) and a tensile ductility of greater than 50% at 1000 °C (1830 °F (Ref 83).

NiAl is quite weak in creep at elevated temperatures (Ref 85–88). However, its creep properties can be substantially improved by alloy additives (Ref 86). Figure 19 shows the compressive creep rate of NiAl alloyed with up to 5 at.% ternary additions. The strength at 1300 K of alloys containing tantalum, niobium, and hafnium is comparable to or even greater than that of the superalloy IN-100. These alloying elements showed very low solubility in NiAl, and the improvement apparently comes from the precipitation of fine second-phase particles that impede the motion of dislocations. It has recently been reported that alloying with 15 at.% Fe to replace nickel lowers diffusion rates and thus reduces the creep rate of NiAl (Ref 87).

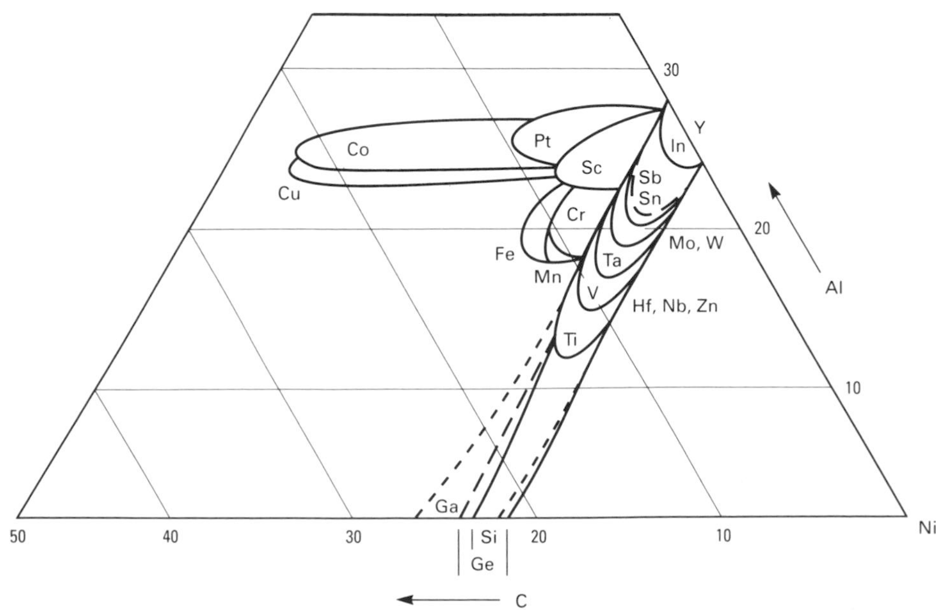

Fig. 9 Semischematic depiction of the solubility lobes of ternary Ni_3Al phase around 1000 °C (1830 °F). Source: Ref 53

The creep properties of NiAl can also be greatly improved by solid-solution hardening and particle strengthening. The best creep resistance obtained so far was produced by reinforcing NiAl with AlN dispersoids that were introduced by milling NiAl powder in liquid nitrogen (Ref 89, 90).

In development of strong polycrystalline NiAl alloys, Liu et. al. (Ref 91) have identified the following alloy compositions (at.%) for structural use at elevated temperatures in hostile environments:

Ni-(49.0±0.6)Al-(1.0±0.8)Mo-(0.7±0.5)Nb/Ta/Zr/Hf-(0–0.5)Fe-(0.01–0.03)B/C

In this case, the alloys were prepared by melting and casting, followed by hot extrusion at 900 to 1050 °C (1650 to 1920 °F). The alloys show a yield strength as high as 700 MPa (102 ksi) at room temperature and 350 MPa (31 ksi) at 1000 °C (1830 °F). The creep rate of the alloys is lower than that of binary NiAl by five to six orders of magnitude at 816 °C in air. Among the alloying elements added to NiAl, zirconium and hafnium improve creep resistance best, but reduce the tensile ductility of the alloys.

Effect of Grain Size and Alloy Stoichiometry on Ductility. Considerable efforts have been devoted to improving the ductility of NiAl at ambient temperatures by controlling microstructure and alloy additions. Schulson and Barker (Ref 92) studied ductile-to-brittle transition as a function of grain size and temperature for NiAl and other aluminides and silicides. NiAl specimens produced by hot extrusion of P/M material, even those with extremely fine grain sizes, were brittle at ambient temperatures. They found that grain refinement alone did not improve the room-temperature ductility of NiAl. On the other hand, NiAl (49 at.% Al) showed a sharp brittle-to-ductile transition at a critical grain size at elevated temperatures. For example, NiAl exhibited a sharp increase in ductility at 400 °C (750 °F) for grain sizes less than 20 μm (Fig. 20). A tensile elongation of 40% was achieved at 400 °C (750 °F) for NiAl with a grain size of 3 μm.

Hahn and Vedula (Ref 83) investigated tensile elongation and fracture behavior as functions of aluminum concentration in NiAl prepared by hot extrusion of cast ingot. They found that NiAl alloys with off-stoichiometric compositions fractured with no appreciable plastic deformation, whereas an alloy with a near-stoichiometric composition exhibited significant tensile elongation (~2%) at both room temperature and 200 °C (390 °F) (Fig. 21). Their findings essentially confirm previous results reported in Ref 93. Although the stoichiometric effect is not well understood, it is believed that yield strength plays a dominant role in the ductility observed for near-stoichiometric NiAl. As indicated in Fig. 21, the yield strength for the near-stoichiometric composition (Ni-50.3Al) is distinctly lower than that for off-stoichiometric compositions. The low yield strength apparently prevents the initiation and propagation of brittle fracture until a high stress level is reached by strain hardening through plastic deformation.

Alloying Effect and Ductility Improvement. Because NiAl with on- and off-stoichiometric composition shows mainly brittle intergranular fracture (Ref 83), it is possible to improve its ductility by the control of grain-boundary composition through microalloying. Auger analyses have revealed that grain boundaries in NiAl (50% Al) are clean and free of any segregated impurities (Fig. 22), indicating that they are intrinsically brittle (Ref 94). Boron added to NiAl has a strong tendency to segregate to the grain boundaries and suppress intergranular fracture (Fig. 22). However, there is no attendant improvement in tensile ductility because boron is an extremely potent solid-solution strengthener in NiAl. Unlike boron, both carbon and beryllium are ineffective in suppressing intergranular fracture in NiAl. Beryllium slightly improves the room-temperature tensile ductility of NiAl. In these microalloyed alloys, the nickel and aluminum contents of the grain boundaries are not significantly different from the bulk levels, and no evidence of strong boron-nickel cosegregation has been found (Ref 94).

Attempts have also been made to improve room-temperature ductility to macroalloying NiAl with alloy additions that might change its deformation behavior in bulk material. Additions of chromium, manganese, and vanadium were reported to promote ⟨111⟩ slip vectors in NiAl; however, no improvement in ductility was observed (Ref 95, 96). On the other hand, limited tensile ductility was obtained in NiAl alloyed with cobalt (Ref 97), which promotes additional deformation modes through possible martensitic transformation. Sufficient iron additions to nickel-rich NiAl, such as Ni-30Al-20Fe (at.%) with a *B*2 structure, also result in approximately 2% plastic elongation (Ref 98). The same alloy with a fine-grain structure shows 5% elongation when produced by a rapid solidification tech-

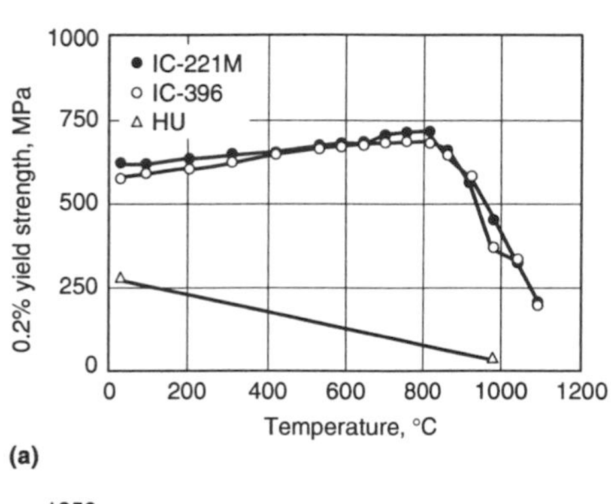

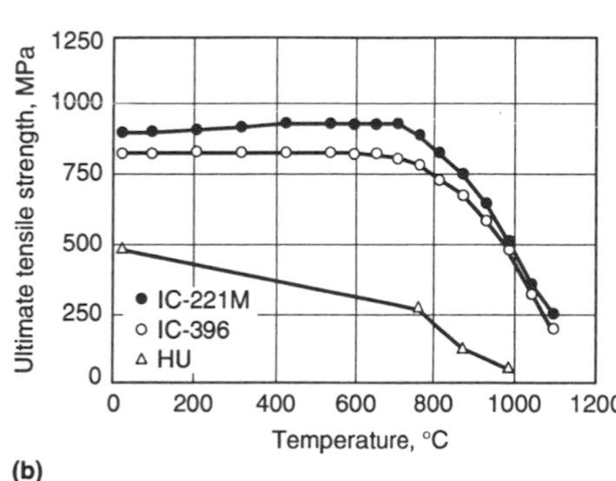

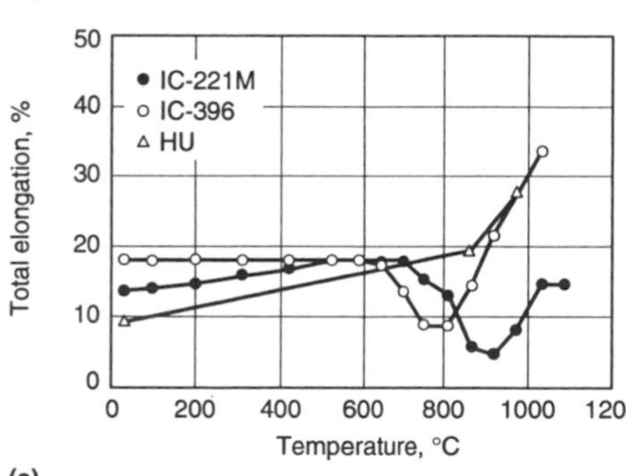

Fig. 10 Comparison of tensile properties of cast Ni_3Al alloys with cast HU alloy. (a) 0.2% yield strength. (b) Ultimate tensile strength. (c) Total elongation. Source: Ref 4

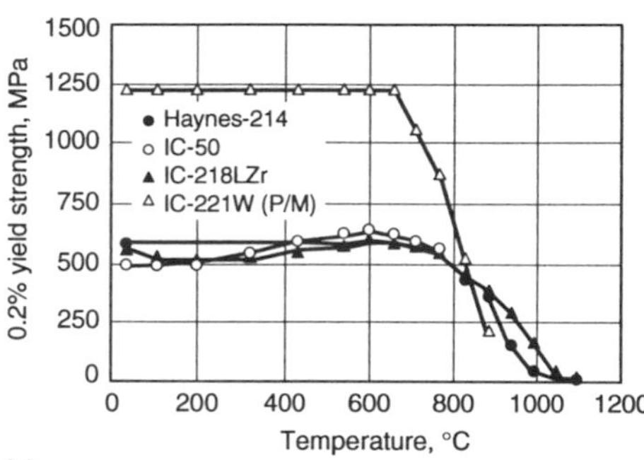

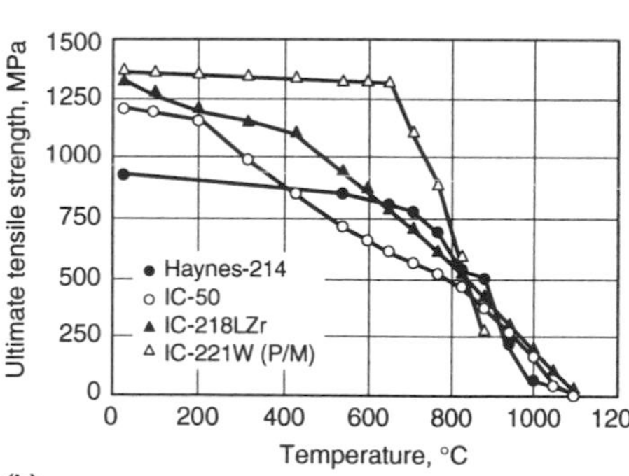

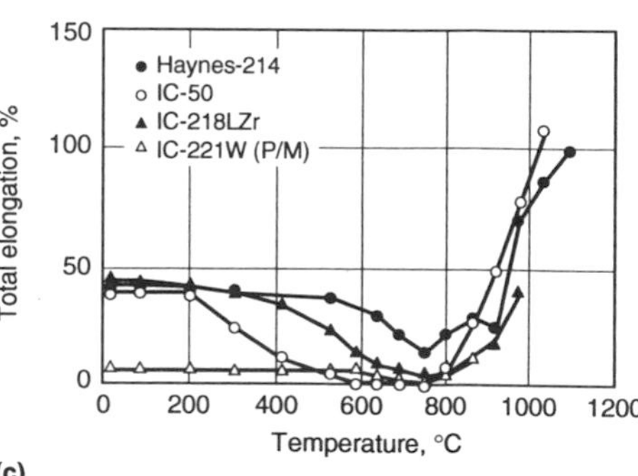

Fig. 11 Comparison of tensile properties of cast and P/M Ni_3Al alloys with Haynes 214. (a) 0.2% yield strength. (b) Ultimate tensile strength. (c) Total elongation. Source: Ref 4

nique (Ref 99). The alloy Ni-20Al-30Fe, which has a two-phase structure (NiAl + Ni_3Al), exhibited a tensile ductility of 22% when produced by hot extrusion (Ref 100) and a lower ductility (10 to 17%) when produced by rapid solidification (Ref 99, 101).

Potential and Future Work. Although significant progress has been made, NiAl has not yet developed into an engineering material for structural use. Further efforts must be devoted to improving both low-temperature ductility and high-temperature strength. The design options offered by NiAl are certainly attractive enough to motivate additional research into the development of NiAl alloys. At present, major NiAl development efforts are going on in a number of laboratories. NiAl and its alloys are also attractive as matrix materials for intermetallic composites, and a great deal of work has been conducted in this area (Ref 102).

Iron Aluminides

Phase Stability and Potential for Structural Use. Iron aluminides form bcc ordered crystal structures over the composition range of 25 to 50 at.%. The aluminide Fe_3Al exists in the ordered $D0_3$ structure up to 540 °C (1000 °F) and in the $B2$ structure between 540 and 760 °C (1000 and 1400 °F); it has a disordered structure above 760 °C (1400 °F). The $D0_3 \rightarrow B2$ transition temperature decreases and the $B2$ ordered temperature increases with an increase in aluminum concentration above 25%. Only the $B2$ structure is stable at aluminum levels above 36%, and the single-phase field extends to approximately 50 at.% Al (FeAl). The iron-aluminum phase diagram is shown in Fig. 23.

The iron aluminides based on Fe_3Al and FeAl possess unique properties and have development potential as new materials for structural use. This potential is based on the capability of the aluminides to form protective aluminum oxide scales in oxidizing and sulfidizing environments at elevated temperatures (Ref 103–106). In addition to excellent corrosion resistance, the aluminides offer low material cost, low density, and conservation of strategic elements. However, the major drawbacks of the aluminides are their low ductility and fracture toughness at ambient temperature and their poor strength at temperatures above 600 °C (1110 °F). Recently, considerable efforts have been devoted to understanding and improving their mechanical properties through control of grain structure, alloy additions, and material processing.

Mechanical Behavior of Fe_3Al and FeAl. The aluminides show low ductility and brittle fracture at room temperature, and their fracture mode depends on aluminum concentration. The aluminides containing less than 40 at.% Al exhibit mainly transgranular cleavage fracture (Ref 107–111). The fracture mode is also sensitive to other parameters, such as grain size and impurities. Fe_3Al (25 at.% Al) has been reported to fracture intergranularly when it contains excess carbon (Ref 112). Also, some FeAl alloys with less than 40% Al show grain-boundary fracture when prepared by P/M techniques and contaminated with oxygen (Ref 108).

Mechanical properties of the iron aluminides have been characterized as functions of test temperature and alloy composition in a number of studies (Ref 107–111). In general, yield strength is not sensitive to temperature below 600 to 650 °C (1110 to 1200 °F); above that temperature range, strength shows a sharp drop with temperature. For intermediate and coarse-grain materials, the aluminides with less than about 40% Al generally show a small increase in yield strength with temperature, with strength reaching a peak at temperatures of about 550 to 650 °C (1020 to 1200 °F). For the aluminides with higher levels of aluminum, the increase in yield strength is suppressed, possibly because of grain-boundary sliding at elevated temperatures (Ref 109, 108, 111).

The room-temperature yield strength of Fe_3Al that contains approximately 25 at.% Al is quite high (550 to 700 MPa, or 800 to 100 ksi) because of the low mobility of single dislocations and particle strengthening due to the presence of a disordered bcc phase (Ref 111, 113). Strength decreases with increasing aluminum concentra-

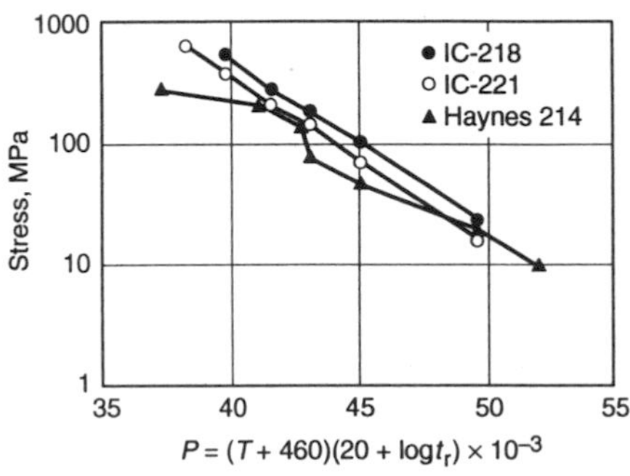

Fig. 13 Comparison of creep-rupture strength of Ni_3Al alloys with Haynes 214. Source: Ref 4

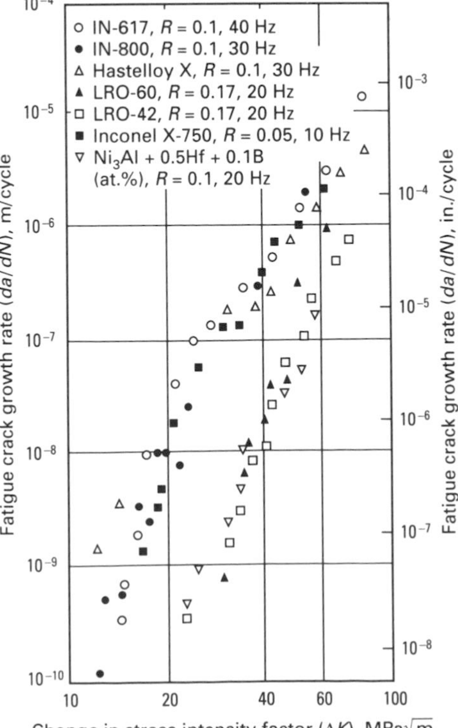

Fig. 14 Crack growth rates of nickel aluminide (Ni-23.5Al-0.5Hf-0.1B, at.%), LRO alloys [$(Fe,Ni)_3(V,Ti)$], and several high-temperature alloys tested in air at 25 °C (80 °F). Source: Ref 63

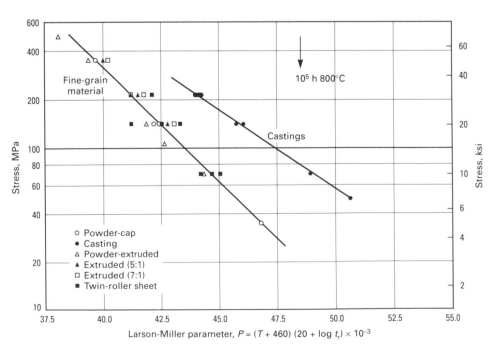

Fig. 12 Larson-Miller parameter (P) plot showing the effect of processing on the creep-rupture properties of IC-221. Tests were conducted in the temperature range of 650 to 870 °C (1200 to 1600 °F) for times ranging from 10 to 12,464 h. Source: Ref 60

tion and reaches a minimum at an aluminum content of about 30%. With further increase in aluminum concentration, strength shows a moderate increase. Unlike NiAl, the FeAl aluminide shows a substantial increase in strength and hardness when it approaches a stoichiometric composition (Fig. 18). The cause for this different behavior is not well understood.

The fatigue and crack growth behavior of Fe_3Al alloys have been studied as functions of aluminum concentration (23.7 to 28.7 at.% Al) and test temperature (20 to 600 °C, or 70 to 1110 °F) (Ref 63). At room temperature, the hyperstoichiometric alloy with 28.7% Al has better fatigue resistance than the hypostoichiometric alloy with 23.7% Al. The better fatigue properties of the higher-aluminum alloys have been attributed to the presence of superdislocations, which cause crack initiation to be more difficult than in the hypostoichiometric alloy. However, the trend is reversed at 500 °C (930 °F). Figure 24 compares the crack growth behavior of Fe_3Al with that of other materials. At room temperature and 500 °C (930 °F), Fe_3Al aluminides are generally intermediate in crack growth rates between nickel-base superalloys and Ni_3Al + B or (Fe, $Ni)_3$ V alloys at low changes in stress level (ΔK); at high ΔK, the crack growth is most rapid in Fe_3Al. At 600 °C (1110 °F), the crack growth rate of Fe_3Al is unusually high as compared with that of the other materials. The reason for this sharp increase in crack growth rate is not well known, but it may be related to the transition of $D0_3$ to $B2$ in Fe_3Al.

Ductility and Slip Behavior. In terms of ductility, the aluminides with less than 40% Al have a room-temperature tensile elongation of about 2 to 4% for coarse-grain materials (i.e., materials with grain sizes of 150 to 200 μm) (Ref 1). Elongation increases to 6 to 8% when the grain structure is refined (Ref 114, 115), indicating the effect of grain size on the ductility. A recent study found that the ductility of FeAl with 40% Al was substantially improved by mechanical alloying (Ref 116); the mechanical alloying probably reduced the grain size, thereby reducing the tendency toward brittle intergranular fracture.

Figure 25 shows the effect of temperature and aluminum concentration on the tensile ductility of FeAl. At temperatures up to 400 °C (750 °F), the ductility shows a general trend of decreasing with increasing aluminum level. At higher temperatures, the alloys exhibit a peak ductility around 35 to 38 at.% Al. The crease in tensile elongation of the alloys with higher aluminum concentrations is due to grain-boundary cavitation during deformation at elevated temperatures.

The aluminides, like iron and steels, slip by {111} dislocations at ambient temperatures. However, the slip changes to {100} type at elevated temperatures. The transition depends on the aluminum concentration (Ref 118, 119) with a general trend of decreasing transition temperature with increasing aluminum level. For example, a transition temperature of about 1000 °C (1830 °F) was reported for the 35% Al alloy; the transition temperature was below 400 °C (750 °F) for the 50% Al alloy (Ref 118). There is no sharp change in ductility around the transition temperature.

Environmental Embrittlement. For more than 45 years, the iron aluminides have been known to be brittle at ambient temperatures; however, the major cause of the brittleness has only recently been identified. Researchers have found that the aluminides are intrinsically quite ductile and that the poor ductility commonly observed in air tests is due to an extrinsic effect—environmental embrittlement (Ref 1, 2). The data in Fig. 26 and Table 5 indicate the effect of test environment on the room-temperature tensile properties of FeAl (36.5% Al) and Fe_3Al (28% Al). Yield strength is not sensitive to environment, but ultimate tensile strength is generally

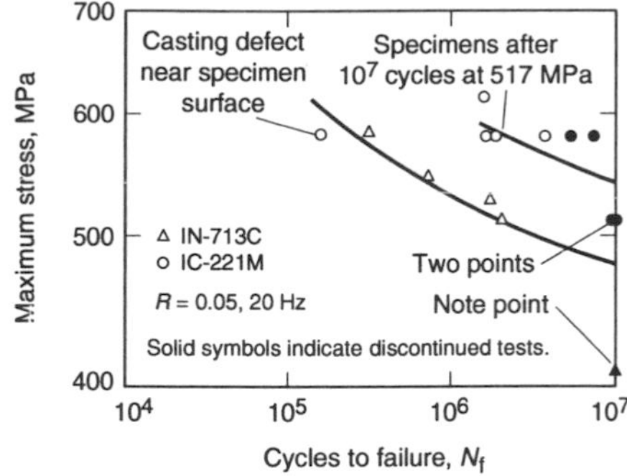

Fig. 15 Fatigue data in air at 650 °C (1200 °F) for precision-cast test bars of IN-713C and IC-221M. Source: Ref 67

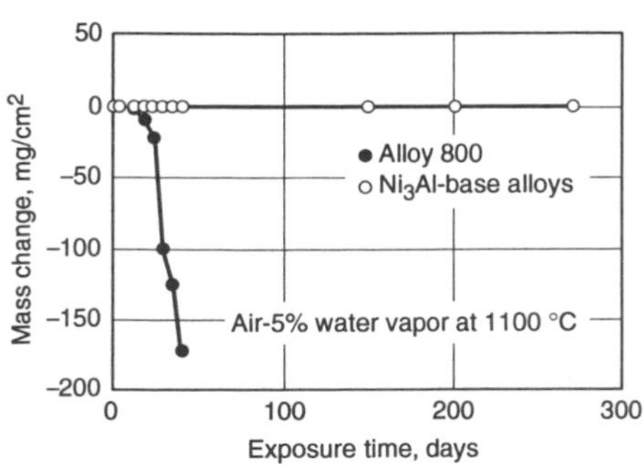

Fig. 16 Comparison of the oxidation resistance of Ni_3Al alloys with that of alloy 800 in air with 5% water vapor at 1100 °C (2010 °F). Source: Ref 4

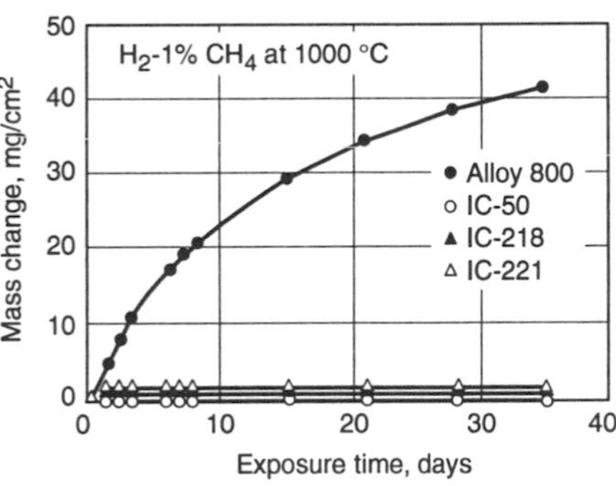

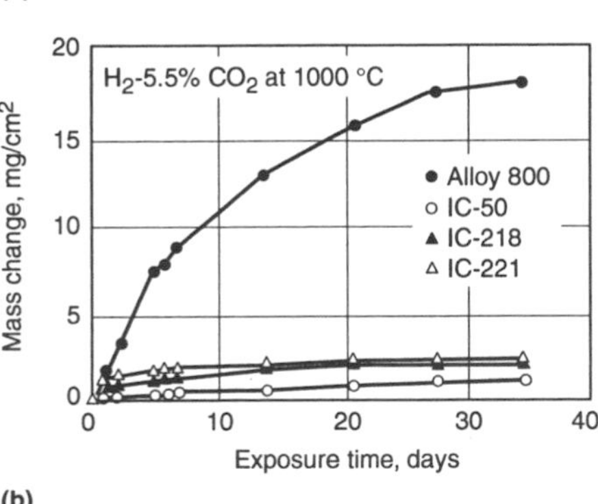

Fig. 17 Comparison of the carburization resistance of Ni_3Al alloys with that of alloy 800. (a) Oxidizing carburizing environment. (b) Reducing carburizing environment. Source: Ref 4

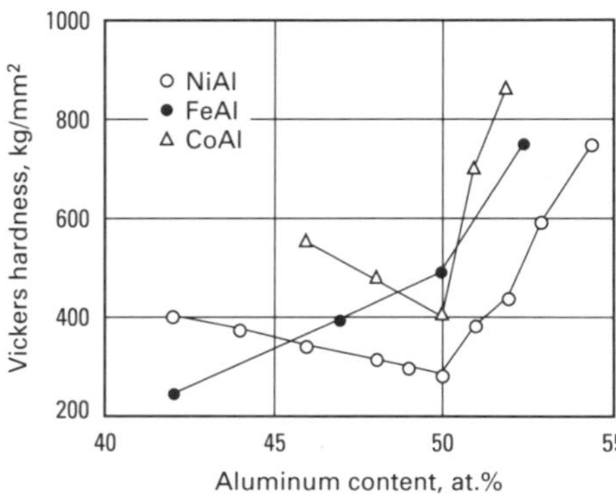

Fig. 18 Vickers hardness of CoAl, FeAl, and NiAl as a function of aluminum content. Source: Ref 84

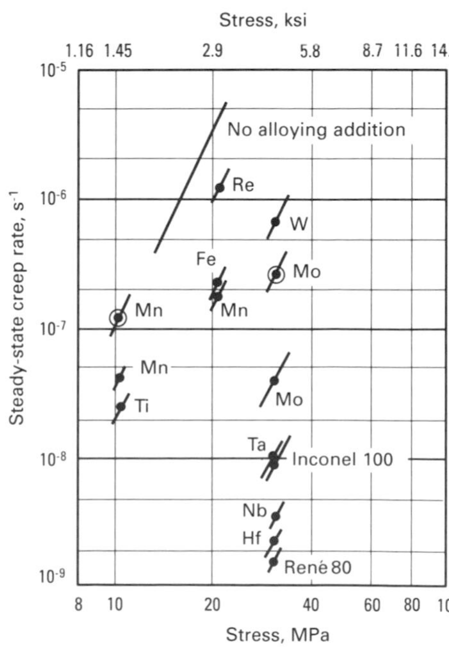

Fig. 19 Compressive creep data for NiAl alloys containing various alloying elements tested at 1300 K. Data for selected conventional superalloys are provided for comparison. Lines drawn through data points are expected slopes. Source: Ref 86

correlated with tensile ductility, which depends strongly on test environment. Aluminides tested in air had a ductility of 2 to 4%. FeAl tested in dry oxygen had a ductility of 17.6%, and Fe_3Al tested in vacuum and dry oxygen had a ductility of 12 to 13%. The water vapor test confirmed the low ductility found in the air tests, indicating that moisture in air is the embrittling agent.

Embrittlement is expected to involve the following chemical reaction at metal surfaces:

$$2Al + 3H_2O \rightarrow Al_2O_3 + 6H^+ + 6e^-$$

The reaction of water vapor with aluminum atoms at crack tips results in the formation of atomic hydrogen that drives into the metal and causes crack propagation. The fact that yield strength (Table 5) is insensitive to ductility and test environment is consistent with the mechanisms of hydrogen embrittlement observed in other ordered intermetallic alloys (Ref 120–128). Molecular hydrogen causes much less embrittlement in the aluminides, possibly because of its lower activity as compared with that of the atomic hydrogen produced from the water vapor reaction.

Alloying Effects in Fe_3Al Aluminides. Grain structure refinement by material processing and alloy additions has been shown to be useful in increasing ductility in Fe_3Al aluminides (Ref 129, 130). Additions of titanium diboride (TiB_2) to Fe_3Al powders are very effective in reducing grain size, and they increase the tensile ductility of recrystallized materials from 2% to 5 to 7%. Stress relief following hot working of the same materials results in ductilities as high as 18%. The presence of TiB_2 particles increases the recrystallization temperature from 650 to 1100 °C (1200 to 2010 °F), which means that wrought materials will retain room-temperature ductility even after exposure to temperatures as high as 100 °C (1830 °F). For these materials, ductility is very high at temperatures above 600 °C (1110 °F), and conventional hot fabrication techniques can be employed without difficulty.

Strength properties of these aluminides are also sensitive to microstructure and the level of aluminum (Ref 105, 129–131) (Fig. 27). Room-temperature yield strength drops sharply with an increase of aluminum above 25%. This drop is, as mentioned before, a result of both the increase in mobility of paired dislocations and the elimination of particle strengthening from the disordered bcc phase. Additions of TiB_2, which reduce the grain size of recrystallized material and stabilize the wrought structure, increase the strength significantly and cause it to be retained to higher temperatures.

Substantial efforts have been made toward improving the elevated-temperature properties of Fe_3Al aluminides by alloy additions of such elements as titanium, molybdenum, silicon, chromium, nickel, manganese, niobium, and tantalum (Ref 129, 130, 132–137). Among these alloying elements, titanium, molybdenum, and silicon are most effective in strengthening the aluminides through a solid-solution hardening effect (Ref 132). Figure 28 shows a substantial increase in the 650 °C (1200 °F) yield strength of a 30% Al aluminide brought about by alloying with molybdenum and titanium. In fact, the yield strength is tripled when the combined molybdenum and titanium content reaches 9%. The increase in strength has been related to an increase in the $DO_2 \rightarrow B2$ transition temperature and associated changes in the nature of the dislocations involved in the deformation processes. In terms of creep properties, alloying with molybdenum and additions of TiB_2 particles increases the stress for 100 h rupture life of a P/M aluminide from 28 to 193 MPa (4 to 28 ksi) at 650 °C (1200 °F) (Ref 129, 130). The solubility of niobium, tantalum, zirconium, and hafnium is low (<1%) in iron aluminides, and alloying with 1 or 2% of these elements substantially improves the room- and elevated-temperature strengths of iron aluminides through a precipitation-hardening effect (e.g., precipitation of $L2_1$ particles in Fe_3Al containing 2% Nb) (Ref 134, 137).

A recent alloy design of Fe_3Al showed that the ductility of the aluminide prepared by melting and casting and fabricated by hot rolling can be substantially improved by increasing the aluminum content from 25 to 28 or 30 at.% and by adding chromium at a level of 2 to 6% (Ref 135, 136). The increase in the aluminum concentration sharply decreases the yield strength of the aluminide. The beneficial effect of chromium may come from modifying the surface composition and reducing the water vapor and aluminum atom reaction, that is, reducing environmental embrittlement. The mechanical properties of the chromium-modified Fe_3Al alloys can be further improved by thermomechanical treatment and alloy additions of molybdenum and niobium (Ref 138). Some of these alloys show a tensile ductility of more than 15% at room temperature and a yield strength of close to 500 MPa (72.5 ksi) at 600 °C (1110 °F). These ductile Fe_3Al alloys are much stronger than austenitic and ferritic steels such as type 314 stainless steel and Fe-9Cr-1Mo steel. The refractory elements also substantially enhance the creep properties of the Fe_3Al alloys.

Table 6 lists several Fe_3Al alloys that were recently developed. Tensile and creep properties of wrought Fe_3Al alloys are compared with oxide dispersion strengthened (ODS) MA-956 alloy in Figures 29 and 30. The tensile properties of Fe_3Al alloys are similar to those of the ODS MA-956 alloy. However, the Y_2O_3 dispersion in MA-956

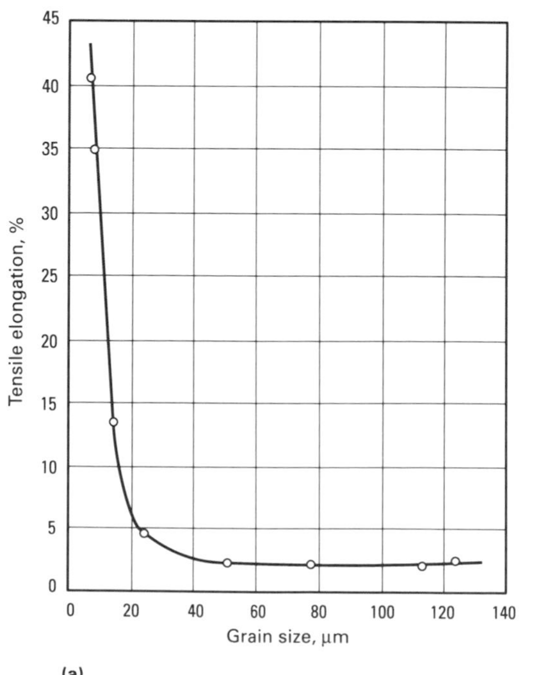

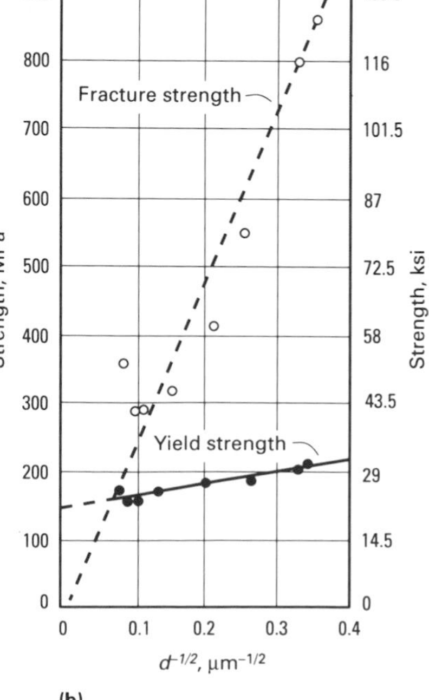

Fig. 20 Effect of grain size (*d*) on properties of NiAl at 673 K. (a) Tensile elongation. (b) Yield strength and fracture strength. Source: Ref 92

Table 5 Effect of selected test environments on room-temperature tensile properties of iron aluminides

Test environment (gas pressure)	Elongation, %	Yield strength, MPa	Yield strength, ksi	Ultimate tensile strength, MPa	Ultimate tensile strength, ksi
Fe_3Al (28% Al)(a)					
Air	4.1	387	56	559	81
Vacuum (~1 × 10^{-4} Pa)	12.8	387	56	851	123
Argon + 4% H_2 (6.7 × 10^4 Pa)	8.4	385	55.8	731	106
Oxygen (6.7 × 10^4 Pa)	12.0	392	56.8	867	126
Water vapor (1.3 × 10^3 Pa)	2.1	387	56	475	69
FeAl (36.5% Al)(a)					
Air	2.2	360	52.2	412	60
Vacuum (<1 × 10^{-4} Pa)	5.4	352	51	501	73
Argon + 4% H_2 (6.7 × 10^4 Pa)	6.2	379	55	579	84
Oxygen (6.7 × 10^4 Pa)	17.6	360	52.2	805	117
Water vapor (67 Pa)	2.4	368	53.4	430	62

(a) All specimens were annealed 1 h at 900 °C (1650 °F) + 2 h at 700 °C (1290 °F).

Table 6 Chemical compositions of selected Fe_3Al aluminides

Alloy (a)	Composition, wt%						
	Al	Cr	B	Zr	Nb	C	Fe
FAS	15.9	2.20	0.01	...	...	...	bal
FAL	15.9	5.5	0.01	0.15	...	...	bal
FA-129	15.9	5.5	...	...	1.0	0.05	bal

(a) Designations used by Oak Ridge National Laboratory, Oak Ridge, TN

Table 7 Chemical compositions of developmental FeAl (Fe-35.8 at.% Al) alloys

Alloy (a)	Composition, at.%						
	Cr	Nb	Ti	Mo	Zr	C	B
FA-350	...	...	...	...	0.05	...	0.24
FA-362	...	...	...	0.2	0.05	...	0.24
FA-372	...	...	...	0.2	0.05	...	...
FA-383	...	...	...	...	0.05	...	...
FA-384	2	...	...	0.2	0.05	...	...
FA-385	...	...	...	0.2	0.05	0.13	...
FA-386	...	...	...	0.2	0.05	0.24	...
FA-387	...	...	...	0.2	...	...	0.24
FA-388	...	...	...	0.2	...	0.25	...
FA-385M1	...	...	...	0.2	0.05	0.13	0.01
FA-385M2	...	...	...	0.2	0.05	0.13	0.021
FA-385M3	2	...	...	0.2	0.05	0.13	...
FA-385M4	...	0.5	...	0.2	0.05	0.13	...
FA-385M5	2	0.5	...	0.2	0.05	0.13	...
FA-385M6	2	0.5	...	0.2	0.05	0.25	...
FA-385M7	2	0.5	...	0.2	0.1	0.25	...
FA-385M8	2	0.5	0.05	0.2	0.05	0.13	...
FA-385M9	2	0.5	0.05	0.2	0.05	0.25	...
FA-385M10(b)	2	0.5	0.05	0.2	0.05	0.13	...
FA-385M11(c)	2	0.5	0.05	0.2	0.05	...	...

Designations used by Oak Ridge National Laboratory, Oak Ridge, TN. (b) Also contains 0.5% Ni, 0.3% Si, and 0.016% P (at.%). (c) Also contains 0.25% W (at.%). Source: Ref 141

significantly enhances its creep properties (see the article "Powder Metallurgy Superalloys" in this Volume). The oxide dispersion strengthening of Fe_3Al alloys is currently being evaluated as a means of improving the creep strength of iron aluminides (Ref 4).

Alloying Effect in FeAl Aluminides. FeAl aluminides containing 40% or more aluminum fail at room temperature by intergranular fracture with little tensile ductility (Ref 107, 108). Small additions of boron (0.05 to 0.2%) suppress grain-boundary fracture and allow a small increase in ductility (~3%) of Fe-40Al, but not of Fe-50Al (Ref 139). The beneficial effect of boron is not nearly as dramatic in FeAl as it is in Ni_3Al, but it is nevertheless significant. The ductility of boron-doped FeAl aluminides remains low because the alloys are still embrittled by the test environment (air). It has been found that boron-doped FeAl (40% Al) exhibits a high ductility (18%) when tested in dry oxygen to avoid environmental embrittlement (Ref 85).

Boron additions also increase the elevated-temperature strength of FeAl, especially in combination with niobium and zirconium. For example, the creep rate can be lowered by an order of magnitude at 825 °C (1520 °F) by the combination of 0.1% Zr and 0.2% B (Ref 140). Measurements of the activation energy for creep indicate that additions act by slowing diffusional processes rather than through precipitation reactions. Partial replacement of iron with nickel improves the creep properties of FeAl at high temperatures (Ref 87).

More recent alloy development studies have focused on a base alloy of Fe-36Al (at.%) alloyed with the elements listed in Table 7. The most effective elements for increasing high-temperature strength and room-temperature ductility of these FeAl alloys were small additions of molybdenum, zirconium, and boron in combination, with the synergistic effects being much more potent than the single element effects (Ref 141). While zirconium and boron additions were very important for improved room-temperature ductility, Mo+Zr+B additions produced the best tensile and creep-rupture strength at 600 °C (1110 °F) in an alloy designated FA-362 (Table 7). The FA-362 alloy also showed the highest room-temperature ductility in air (11.8%). Preoxidation at 700 °C (1290 °F) further increased the tensile ductility to 14.7% (Ref 141). Figure 31 compares the elevated-temperature yield strengths of various FeAl alloys listed in Table 7.

Weldability and Corrosion Resistance. Other properties of iron aluminides, including weldability (Ref 142) and corrosion resistance (Ref 106, 138), have been characterized to a limited extent. Fe_3Al is weldable with careful control of welding parameters and minor alloy additions. Additions of TiB_2 promote hot cracking and are detrimental to the weldability of Fe_3Al aluminides. Sound weldments have been achieved in Fe_3Al alloys using both electron beam and gas-tungsten arc welding processes.

The development of weldable FeAl alloys is also being pursued. One study involved testing various FeAl alloys for hot-cracking resistance. As shown in Fig. 32, micro-additions of boron (0.01 to 0.021 at.%) pushed the threshold hot-cracking stress to values of ≥200 MPa (29 ksi).

The iron aluminides are highly resistant to oxidation and sulfidation at elevated temperatures (Ref 103–106). This resistance stems from the ability of the aluminides to form highly protective Al_2O_3 scales. The oxidation resistance generally increases with increasing aluminum content; the major products are α-Al_2O_3 and trace amounts of iron oxides when the aluminides are oxidized at temperatures above 900 °C (1650 °F) (Ref 103). Cyclic oxidation of Fe-40Al alloyed with up to 1 at.% Hf, Zr, and B produced little degradation at temperatures up to 1000 °C (1830 °F) (Ref 104). Aluminide specimens tested at 700 and 870 °C (1290 and 1600 °F) showed no indication of attack in sulfidizing environments, except for the formation of a thin layer of oxides with a thickness in an interference color range (Ref 138). As shown in Fig. 33, the iron aluminide alloys exhibited corrosion rates lower than those of the best existing iron-base alloys (including coating material) by a couple of orders

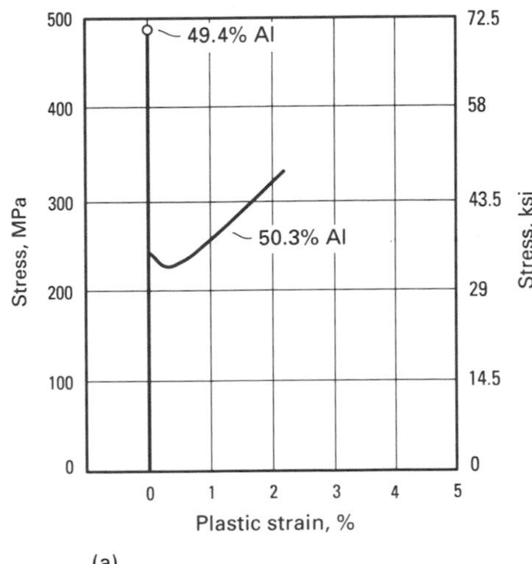

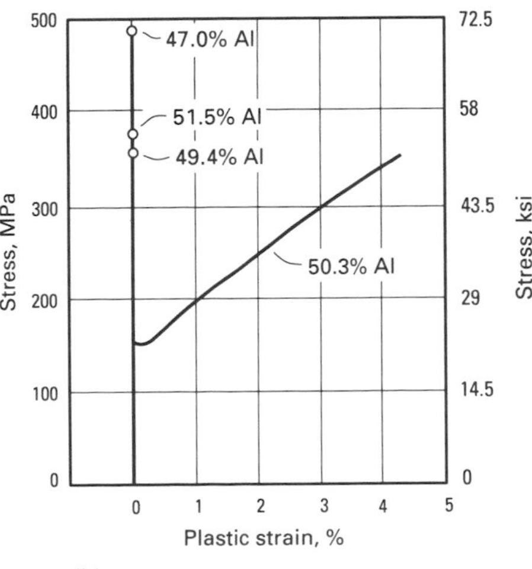

Fig. 21 Effect of stoichiometry on tensile properties of cast and extruded binary NiAl alloys. Nominal strain rate, 1.41×10^{-3}/s. (a) Tested at room temperature. (b) Tested at 473 K. Source: Ref 83

of magnitude when tested in a severe sulfidizing environment at 800 °C (1470 °F). In addition, the aluminides with more than 30% Al are very resistant to corrosion in molten nitrate salt environments at 650 °C (1200 °F).

Potential Applications. Iron aluminides were previously excluded from the realm of structural materials because of their brittleness at ambient temperatures and poor strength at elevated temperatures. Recent research and development activities have demonstrated that adequate engineering ductility (10 to 15%) can be achieved in the aluminides through the control of microstructure and alloy additions. Both tensile and creep strengths of the aluminides are substantially improved by alloying with refractory elements, which results in solution hardening and particle strengthening. The recently developed aluminide alloys are stronger than austenitic steels and ferritic low-alloy steels at ambient and elevated temperatures (Ref 138). Adequate ductility and

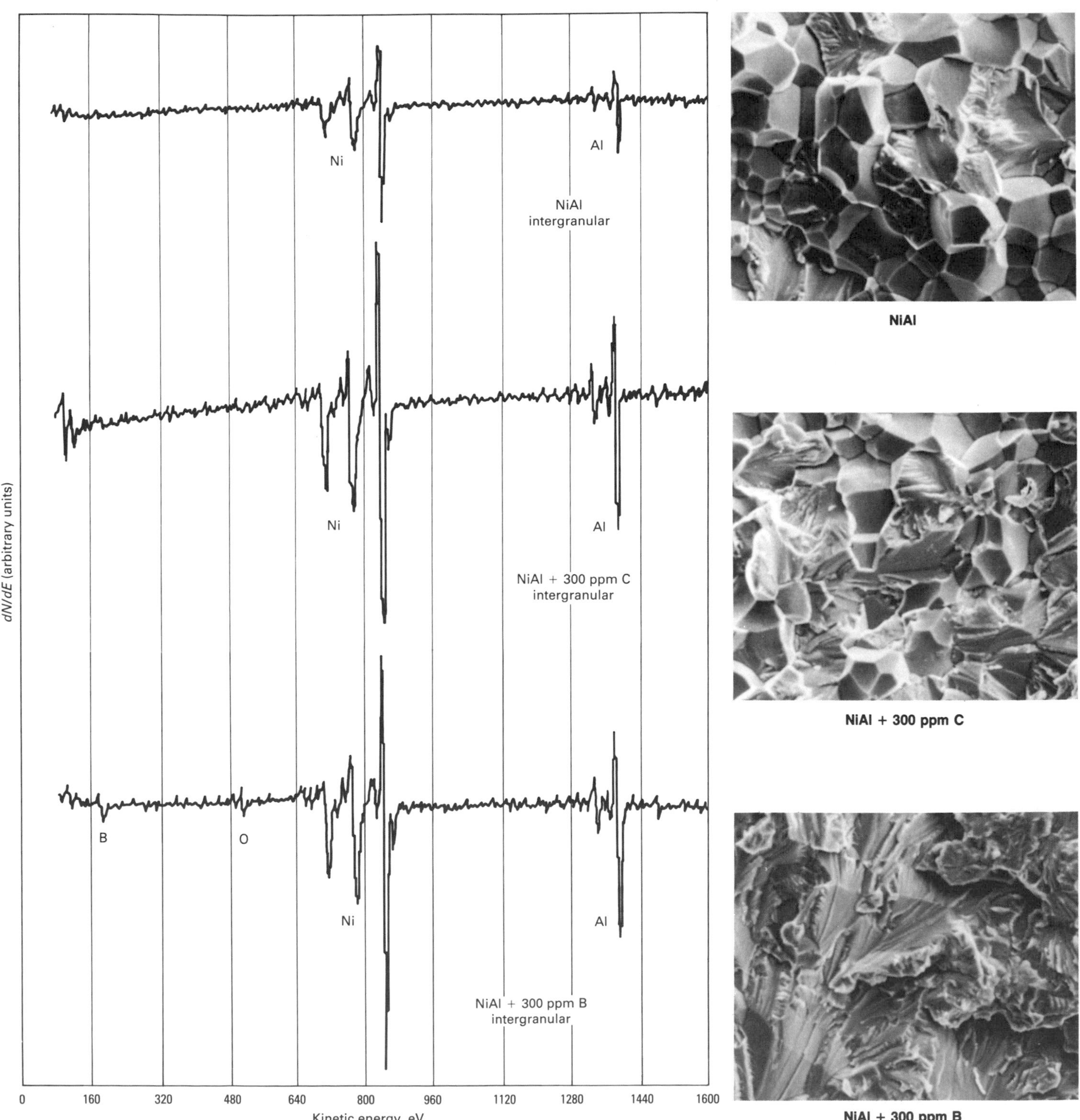

Fig. 22 Results of Auger electron analysis showing the effect of microalloying on the grain-boundary composition and the room-temperature fracture mode of NiAl. Top to bottom: unalloyed NiAl (mainly grain-boundary fracture), NiAl doped with 300 ppm C, and NiAl doped with 300 ppm B. All photomicrographs 400×. Source: Ref 94

strength combined with low cost, excellent oxidation and corrosion resistance, low density, and good fabricability make the aluminide alloys promising for structural use at temperatures up to 700 to 800 °C (1290 to 1470 °F). Potential applications include molten salt systems for chemical air separation, automotive exhaust systems, immersion heaters, heat exchangers, catalytic conversion vessels, chemical production systems, coal conversion systems, and so on. Several industrial companies have started to prepare large heats as a first step in the commercialization of iron aluminide alloys. Further research is required to develop a database (including information on tensile and creep properties, fracture toughness, low- and high-cycle fatigue, crack growth, and elastic modulus) on some promising aluminides for specific engineering applications.

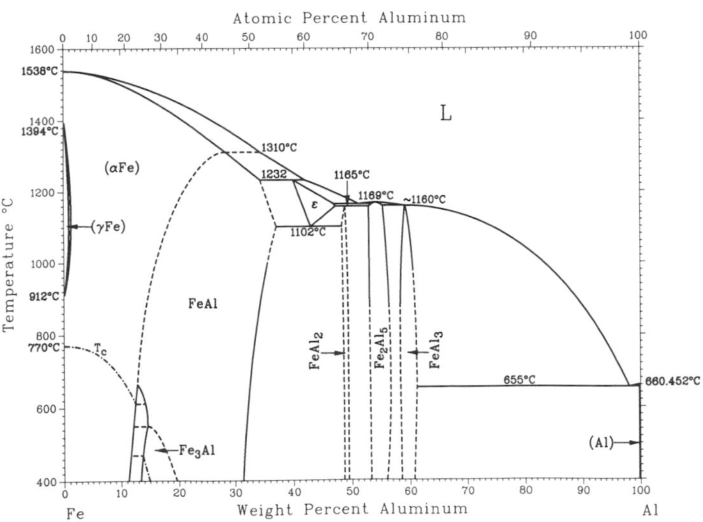

Fig. 23 The binary iron-aluminum phase diagram

Titanium Aluminides

Because of their low density, titanium aluminides based on Ti_3Al and TiAl are attractive candidates for applications in advanced aerospace engine and airframe components (Ref 20, 143–147). The characteristics of titanium aluminides are presented alongside those of other aluminides in Table 1; the creep behavior of titanium aluminides is compared with that of conventional titanium alloys in Fig. 34. Despite a lack of fracture resistance (low ductility, fracture toughness, and fatigue crack growth rate), the titanium aluminides Ti_3Al (α-2) and TiAl (γ) have great potential for enhanced performance. Properties of these aluminides are compared with those of conventional titanium alloys and superalloys in Table 8. Because they have slower diffusion rates than conventional titanium alloys, the titanium aluminides feature enhanced high-temperature properties such as strength retention, creep and stress rupture, and fatigue resistance (Ref 151).

Another negative feature of titanium aluminides, in addition to their low ductility at ambient temperatures, is their oxidation resistance, which is lower than desirable at elevated temperatures (Ref 152–154). The titanium aluminides are characterized by a strong tendency to form TiO_2, rather than the protective Al_2O_3, at high temperatures. Because of this tendency, a key factor in increasing the maximum-use temperatures of these aluminides is enhancing their oxidation resistance while maintaining adequate levels of creep and strength retention at elevated temperatures.

Alpha-2 Alloys

Crystal Structure and Deformation Behavior. Ti_3Al, which has an ordered DO_{19} structure, contains three linearly independent slip systems that account for dislocation motion on the basal {0001}, prism {1010}, and pyramidal {0221} planes (Fig. 35) (Ref 155, 156). Prism slip requires only a single dislocation without creating a near-neighbor anti-phase boundary, and additional slip requires movement of two disloca-

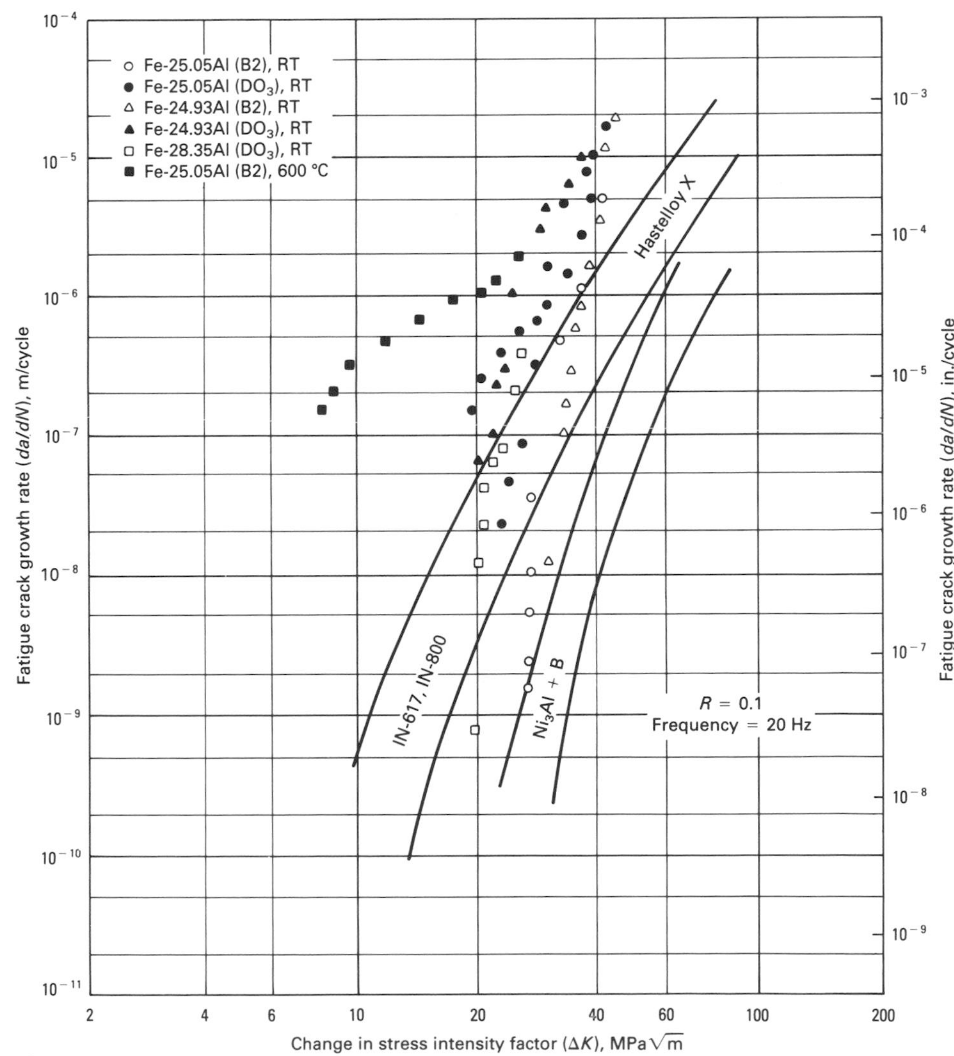

Fig. 24 Effects of aluminum content, crystal structure, and temperature on fatigue crack growth in Fe_3Al. Curves for nickel-base superalloys and Ni_3Al are shown for comparison. Stress ratio (*R*), 0.1; frequency, 20 Hz. RT, room temperature. Source: Ref 63

tions. In addition, two independent slip systems involving ($c + a$) slip occur to satisfy the Von Mises criterion for uniform deformation.

The semicommercial and experimental α-2 alloys developed are two phase (α-2 + β/*B*2), with contents of 23 to 25 at.% Al and 11 to 18 at.% Nb. Alloy compositions with current engineering significance are Ti-24Al-11Nb (Ref 157, 158), Ti-25Al-10Nb-3V-1Mo (Ref 147), Ti-25Al-17Nb-1Mo (Ref 159) and modified alloy compositions such as Ti-24.5Al-6Nb-6(Ta,Mo,Cr,V). Increasing the niobium content generally enhances most material properties, although excessive niobium can degrade creep performance. Niobium can be replaced by specific elements for improved strength (molybdenum, tantalum, or chromium), creep resistance (molybdenum), and oxidation resistance (tantalum, molybdenum). However, for full optimization of mechanical properties, control of the microstructure must be maintained, particularly for tensile, fatigue, and creep performance.

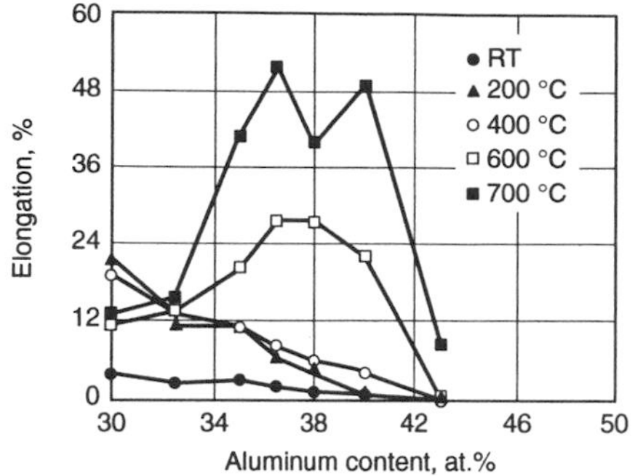

Fig. 25 A plot of tensile elongation as a function of aluminum concentration for FeAl alloys tested at temperatures up to 700 °C (1290 °F). RT, room temperature. Source: Ref 117

As shown on the titanium-aluminum phase diagram (Fig. 36), the α-2 (Ti_3Al) alloy has a wide range of compositional stability, with aluminum contents of 22 to 39 at.%. The compound is congruently disordered at a temperature of 1180 °C (2155 °F) and an aluminum content of 32 at.%. The stoichiometric composition, Ti-25Al, is stable up to about 1090 °C (1995 °F).

Material Processing. Microstructural features that can be varied by thermomechanical processing include primary α-2 grain size and volume fraction, secondary α-2 plate morphology and thickness, and the presence of secondary β grains. Beta processing generally results in elongated Widmanstätten α-2 in large primary β grains in a manner similar to that in conventional titanium alloys (Ref 143).

Up to 4 wt% H can be dissolved in titanium alloys at elevated temperatures. This hydrogen can then be used to improve processibility, and final mechanical properties are enhanced after its removal; removal of the hydrogen can be easily achieved by vacuum annealing (Ref 160, 161). This thermomechanical processing technique allows titanium aluminides to be processed at reduced temperatures (Ref 160–163) and results in a finer microstructure (Ref 155, 160, 161, 163).

Mechanical and Metallurgical Properties. Typical mechanical properties for a number of α-2 alloys are listed in Table 9. Production of two-phase alloys by alloying Ti_3Al with β-stabilizing elements results in up to a doubling of strength. Interface strengthening of the two-phase mixture appears to be predominantly responsible for the increased strength, but other strengthening factors, such as long-range order, solid solution, and texture effects, also contribute (Ref 164).

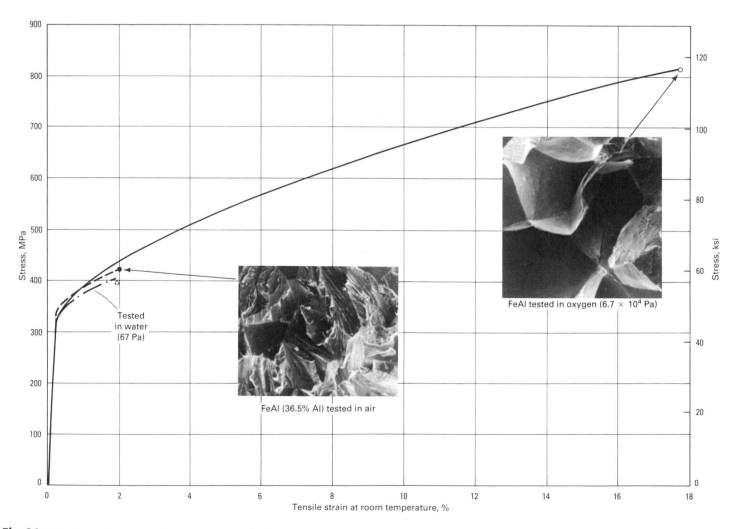

Fig. 26 Effect of test environment on the room-temperature ductility and fracture behavior of FeAl (36.5% Al)

A fine Widmanstätten microstructure or an aligned acicular α-2 grains exhibits better ductility than microstructures with a coarse Widmanstätten microstructure or an aligned acicular α-2 morphology. The fatigue properties of titanium alloys are strongly influenced by microstructure and work on conventional titanium alloys (Ref 143) suggests that high-ductility alloys perform best under low-cycle fatigue (LCF) conditions (Ref 165). The low ductility exhibited in material with Widmanstätten α plates in α + β alloys is responsible for low high-temperature LCF strength. Early data (Ref 166) suggest that fatigue crack growth rate is relatively insensitive to microstructure, although the coarse Widmanstätten microstructure exhibits the slowest fatigue crack growth rate at low stress intensities. Fracture toughness appears to depend on microstructures as well as alloy composition, but the precise relationship is yet to be defined. A detailed investigation into the effect of microstructure on creep behavior in Ti-25Al-10Nb-3V-1Mo has shown that the colony-type microstructure shows better creep resistance than other microstructures (Ref 167). Creep resistance of Ti-25-10-3-1 is raised by a factor of ten in the steady-state regime over that of conventional alloy Ti-1100 (Ti-6Al-3Sn-4Zr-0.4Mo-0.45Si) and two orders of magnitude over that of Ti-6Al-2Sn-4Zr-2Mo-0.1Si (Ref 167). However, 0.4% creep strain in Ti-25-10-3-1 is reached within 2 h.

Additions of silicon and zirconium appear to improve creep resistance (Ref 168), but the most significant improvement is attained by increasing the aluminum content to 25 at.% and limiting β-stabilizing elements to about 12 at.% (Ref 158, 169). However, the Ti-24.5Al-17Nb-1Mo alloy exhibits a rupture life superior to that of other α alloys (Ref 17).

Orthorhombic Alloys. At higher niobium levels, the α-2 phase evolves to a new ordered orthorhombic structure that is based on the composition Ti_2AlNb (O phase). This has been observed in titanium aluminides with compositions near Ti-(21-25)Al-(21-27)Nb (at.%). The crystal structures of the α-2 and ordered orthorhombic phases are compared in Fig. 37.

Although the orthorhombic alloys have a lower use temperature than do the γ aluminides described below (about 650 °C, or 1200 °F), they offer much higher absolute strengths. Room-temperature tensile strengths on the order of 1380 MPa (200 ksi) with close to 5% elongation have been reported. Woodfield and Lawless reported 0.2% yield strengths in the range of 590 to 690 MPa (86 to 100 ksi) at 700 °C (1290 °F) for Ti-21Al-25Nb (at.%) alloy (Ref 171). This material had a room-temperature yield strength of about 1070 MPa (155 ksi) with 3.5% elongation.

The ordered orthorhombic alloys having the best combination of tensile, creep, and fracture toughness properties are two-phase O+β alloys such as Ti-22Al-27Nb (at.%). Table 10 lists the elevated-temperature tensile properties of such alloys. Additional information on the processing and properties of Ti_2AlNb alloys can be found in Ref 172 to 177.

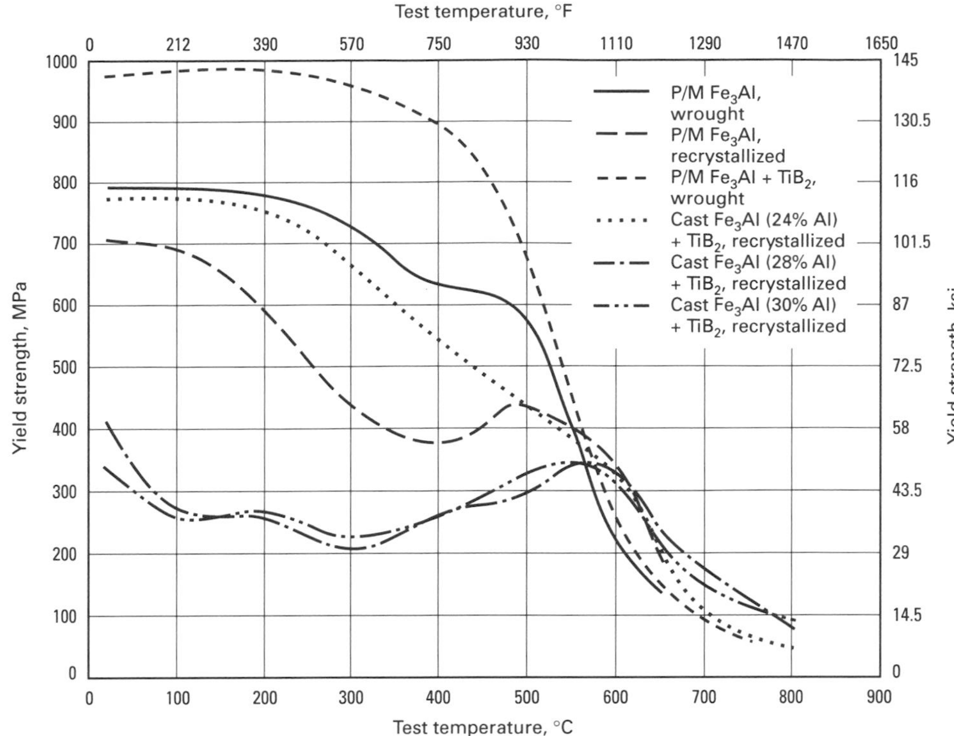

Fig. 27 Yield strength vs. test temperature for various Fe_3Al materials

Table 8 Properties of titanium aluminides, titanium-base conventional alloys, and nickel-base superalloys

Property	Conventional titanium alloys	Ti_3Al	TiAl	Nickel-base superalloys
Density, g/cm^3	4.5	4.1–4.7	3.7–3.9	8.3
Modulus, GPa (10^6 psi)	96–100 (14–14.5)	100–145 (14.5–21)	160–176 (23.2–25.5)	206 (30)
Yield strength, MPa (ksi)(a)	380–1150 (55–167)	700–990 (101–144)	400–650 (58–94)	...
Tensile strength, MPa (ksi)(a)	480–1200 (70–174)	800–1140 (116–165)	450–800 (65–116)	...
Creep limit, °C (°F)	600 (1110)	760 (1400)	1000 (1830)	1090 (1995)
Oxidation limit, °C (°F)	600 (1110)	650 (1200)	900 (1650)	1090 (1995)
Ductility at room temperature, %	20	2–10	1–4	3–5
Ductility at high temperature, %	High	10–20	10–60	10–20
Structure	hcp/bcc	$D0_{19}$	$L1_0$	fcc/$L1_2$

(a) At room temperature. Source: Ref 144–150

Table 9 Properties of α-2 Ti_3Al alloys with various microstructures

Alloy	Microstructure(a)	Yield strength, MPa	Yield strength, ksi	Ultimate tensile strength, MPa	Ultimate tensile strength, ksi	Elongation, %	Plane-strain fracture toughness (K_{Ic}), MPa$\sqrt{m}$	Plane-strain fracture toughness (K_{Ic}), ksi$\sqrt{in.}$	Creep rupture(b)
Ti-25Al	E	538	78	538	78	0.3	...	...	...
Ti-24Al-11Nb	W	787	114	824	119	0.7	...	...	44.7
	FW	761	110	967	140	4.8	...	...	...
Ti-24Al-14Nb	W	831	120	977	142	2.1	...	...	59.5
Ti-25Al-10Nb-3V-1Mo	W	825	119	1042	151	2.2	13.5	12.3	>360
	FW	823	119	950	138	0.8	...	...	...
	C+P	745	108	907	132	1.1	...	...	...
	W+P	759	110	963	140	2.6	...	...	...
	FW+P	942	137	1097	159	2.7	...	...	...
Ti-24.5Al-17Nb	W	952	138	1010	146	5.8	28.3	25.7	62
	W+P	705	102	940	136	10.0	...	...	...
Ti-25Al-17Nb-1Mo	FW	989	143	1133	164	3.4	20.9	19.0	476

(a) E, equiaxed α-2; W, Widmanstätten; FW, fine Widmanstätten; C, colony structure; P, primary α-2 grains. (b) Time to rupture, h, at 650 °C (1200 °F) and 380 MPa (55 ksi). Source: Ref 17

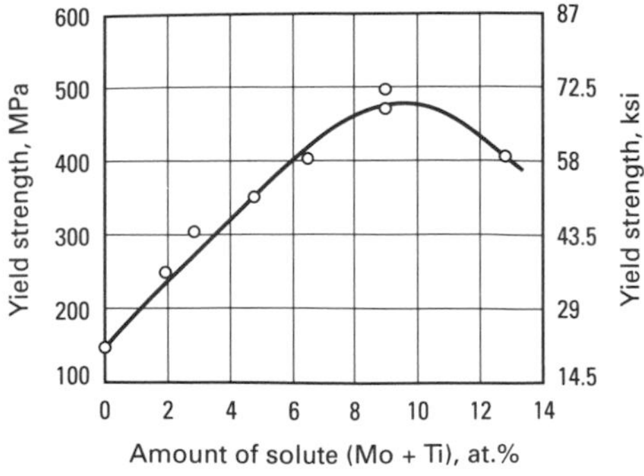

Fig. 28 Yield strength at 650 °C (1200 °F) as a function of the total solute content of molybdenum and titanium in FeAl (30 at.% Al). Source: Ref 132

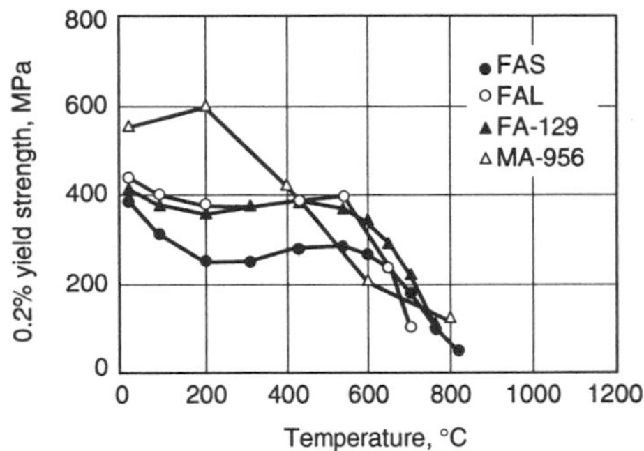

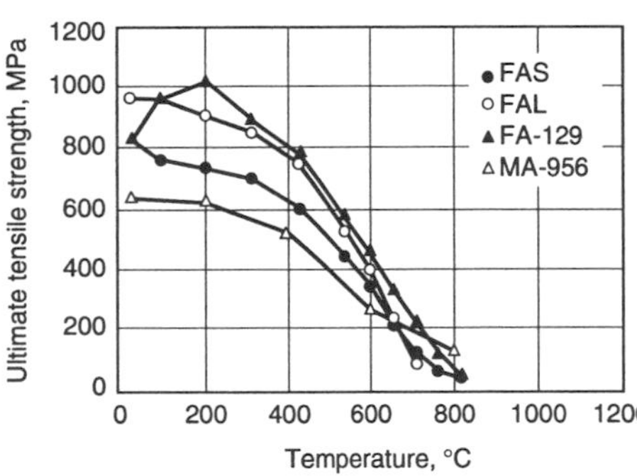

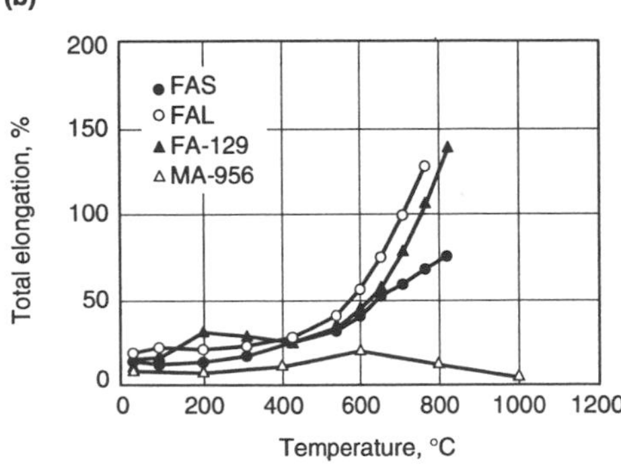

Fig. 29 Comparison of tensile properties of wrought Fe_3Al alloys with those of MA-956 alloy. (a) 0.2% yield strength. (b) Ultimate tensile strength. (c) Total elongation. See Table 6 for Fe_3Al alloy chemical compositions. Source: Ref 4

Gamma Alloys

Crystal Structure and Deformation Behavior. The γ-TiAl phase has an $L1_0$ ordered face-centered tetragonal structure (Ref 178–180), which has a wide range (49 to 66 at.% Al) of temperature-dependent stability (Fig. 36). At the equiatomic TiAl composition, the *c/a* ratio is 1.02; tetragonality increases up to *c/a* = 1.03 with increasing aluminum concentration (Ref 181–183). Within the compositional range specified at off-stoichiometric compositions, excess titanium or aluminum atoms occupy antisites without creating constitutional vacancies (Ref 184). The γ-TiAl phase apparently remains ordered up to its melting point of approximately 1450 °C (2640 °F).

The layered arrangement of titanium and aluminum atoms on successive (002) planes and the slight tetragonality of *c/a* = 1.02 (Fig. 38) gives rise to two types of dislocations with $\frac{1}{2}\langle 110\rangle$-type Burgers vectors on {111} in γ-TiAl: ordinary dislocations $\frac{1}{2}\langle 110\rangle$ and superdislocations $\langle 011\rangle = \frac{1}{2}\langle 011\rangle + \frac{1}{2}\langle 011\rangle$ that will leave the superlattice undisturbed (Fig. 38). Another superdislocation, with a Burgers vector of $\frac{1}{2}\langle 112\rangle$, has also been suggested (Ref 185, 186). The superdislocation core can dissociate further into other complex partial dislocations, which are energetically more favorable, involving planar defects such as stacking faults and antiphase boundaries (Ref 185–187). The slip systems and partial dislocations are shown in Fig. 38. The $\frac{1}{6}\langle 11\bar{2}\rangle$ on $\{111\}$ partials form twin dislocations, but $\frac{1}{6}\langle 211\rangle$ partials are forbidden as twinning dislocations in the $L1_0$ structure (Ref 188, 189).

In single-phase γ alloys containing 52 to 54 at.% Al, deformation at room temperature occurs by motion of both ordinary and superdislocations; however, the superdislocations [011] and [101] are largely immobile because segments of the trailing $\frac{1}{6}[112]$-type superpartials form faulted dipoles that must be extended as deformation progresses (Ref 185–187, 189, 190). Increasing temperature and decreasing aluminum content increase the $\frac{1}{2}\langle 110\rangle$ slip activity as the faulted dipoles disappear and twinning dominates (Ref 190, 191). In two-phase Ti-48Al, the deformation modes of primary γ grains are twinning with ⟨112⟩ twin dislocations and slip by $\frac{1}{2}[110]$-type dislocations.

The extremely low ductility values at ambient temperature and the increased ductility with increasing temperatures strongly influence the observed fracture mode. Tensile and fatigue specimens indicate that the predominant fracture modes are cleavage at low temperatures due to dislocation pileups and intergranular fracture at temperatures above the brittle-ductile transition (Ref 190–194).

The γ alloys of engineering importance contain approximately 45 to 48 at.% Al and 1 to 10 at.% M, with M being at least one of the following: vanadium, chromium, manganese, niobium, tantalum, and tungsten (Ref 146, 180, 195–199). These alloys can be divided into two categories:

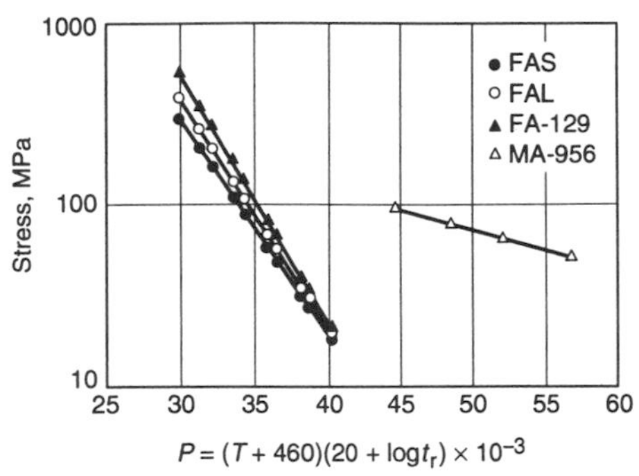

Fig. 30 Comparison of the creep-rupture strength of wrought Fe_3Al alloys with that of MA-956 alloy. Source: Ref 4

Table 10 Tensile properties of a two-phase (O + β) alloy (Ti-22Al-27Nb at.%)

Test temperature		Aging	Tensile yield strength		Ultimate tensile strength		Elongation,
°C	°F	treatment	MPa	ksi	MPa	ksi	%
22	72	None	1056	153	1152	167	3.4
		None	1028	149	1083	157	2.2
		540 °C (1000 °F), 100 h	1083	157	1166	169	3.3
		540 °C (1000 °F), 100 h	1090	158	1159	168	2.8
		650 °C (1200 °F), 100 h	1090	158	1145	166	2.6
		650 °C (1200 °F), 100 h	1076	156	1145	166	2.5
		760 °C (1400 °F), 100 h	987	143	1076	156	5.2
		760 °C (1400 °F), 100 h	966	140	1083	157	5.0
540	1000	None	849	123	1007	146	14.3
		None	856	124	1049	152	14.3
		540 °C (1000 °F), 100 h	876	127	1049	152	17.9
		540 °C (1000 °F), 100 h	890	129	1070	155	16.1
650	1200	None	794	115	938	136	14.3
		None	807	117	945	137	12.5
		650 °C (1200 °F), 100 h	794	115	938	136	10.7
		650 °C (1200 °F), 100 h	807	117	952	138	10.7
760	1400	None	559	81	787	114	10.7
		None	593	86	766	114	14.3
		760 °C (1400 °F), 100 h	462	67	649	94	21.4
		760 °C (1400 °F), 100 h	552	80	731	106	16.1

Source: Ref 172

single-phase (γ) alloys and two-phase (γ + α-2) materials (Ref 180). The (α-2 + γ)/γ phase boundary at 1000 °C (1830 °F) occurs at an aluminum content of approximately 49 at.%, depending on the type and level of solute M. Single-phase γ alloys contain third alloying elements such as niobium or tantalum that promote strengthening and further enhance oxidation resistance (Ref 200, 201). Third alloying elements in two-phase alloys can raise ductility (vanadium, chromium, and manganese) (Ref 146, 189, 197–199), increase oxidation resistance (niobium and tantalum) (Ref 200, 201), or enhance combined properties (Ref 180).

Material Processing. Gamma alloys are processed by conventional methods, including casting, ingot metallurgy, and P/M, and also by novel means. Figure 39 shows the major processing routes practiced to date for engineering gamma alloys. Important alloying/melting processes include induction skull melting, vacuum arc melting, and plasma melting.

Other methods under study include mechanical alloying, spray forming, shock reactive synthesis, physical vapor deposition, and hot pressing and rolling of elemental sheet into multilayer composite sheets.

The microstructure of the nominally γ alloys can be single-phase γ or, in slightly leaner compositions, two-phase γ + α-2. By appropriate thermomechanical processing, the morphology of the phases can be adjusted to produce either lamellar or equiaxed morphologies, or a mixture of the two (Ref 180).

The lamellar structure can lead to refinement of the microstructure, improved ductility (Ref 197, 199), and a decreased microstructure scale by recrystallizing the fine γ grains (Ref 203). Optimum ductility occurs at a content of about 10 vol% α-2; when the α-2 phase content exceeds 20 vol%, ductility can be degraded (Ref 198). This ductility behavior is consistent with the fact that α-2 becomes increasingly brittle with increasing aluminum content over 25 at.% (Ref 157). The α-2 plates contain approximately 35 at.% Al.

Control of the microstructure in single-phase γ alloys requires the optimization of grain size and morphology. In two-phase alloys, the volume ratio of lamellar to equiaxed gamma (LG/γ G) must also be controlled (Ref 180, 197, 198). A lamellar volume fraction of about 30% gives rise to the optimum combination of properties, with a desirable high-temperature creep resistance and acceptable levels of tensile strength and ductility (Ref 147). Heat treatment temperature and time strongly affect the LG/γ G volume ratio. Thermomechanical processing (TMP) refines the microstructure when processing is conducted in such a way that both the α and γ grains are

Table 11 Tensile properties and fracture toughness values of gamma titanium aluminides tested in air

Alloy designation and composition, at. %	Processing and microstructure	Temperature		Yield strength		Tensile strength		Elongation,	Fracture toughness, K_Q(a)	
		°C	°F	MPa	ksi	MPa	ksi	%	MPa √m	ksi √in.
48-1-(0.3C)/Ti-48Al-1V-0.3C-0.2O	Forging + HT/duplex	RT	RT	392	57	406	59	1.4	12.3	11.2
		437	819	...	...	...	...	...	22.8	20.7
		760	1400	320	46	470	68	10.8	...	...
	Casting/duplex	RT	RT	490	71	...	...	...	24.3	22.1
48-1(0.2C)/Ti-48Al-1V-0.2C-0.14O	Forging + HT/duplex + NL	RT	RT	480	70	530	77	1.5	...	...
		815	1500	360	52	450	65	...	...	...
48-2-2/Ti-48Al-2Cr-2Nb	Casting + HIP + HT/duplex	RT	RT	331	48	413	60	2.3	20–30	18–27
		760	1400	310	45	430	62	...	...	...
	Extrusion + HT/duplex	RT	RT	480	70	...	...	3.1	...	...
		760	1400	403	58	...	...	40	...	...
		870	1600	330	48	...	...	53	...	...
	Extrusion + HT/FL	RT	RT	454	66	...	...	0.5	...	...
		760	1400	405	59	...	...	3.0	...	...
		870	1600	350	51	...	...	19	...	...
	P/M extrusion + HT/NL	RT	RT	510	74	597	87	2.9	...	...
		700	1290	421	61	581	84	5.2	...	...
G1/Ti-47Al-1Cr-1V-2.6Nb	Forging + HT/duplex	RT	RT	480	70	548	79	2.3	12	10.8
		600	1110	383	56	507	74	3.1	16	14.6
		800	1470	324	47	492	71	55	...	...
	Forging + HT/FL	RT	RT	330	48	383	56	0.8	30–36	27–33
		800	1470	290	42	378	55	1.5	40–70	36–64
Sumitomo/Ti-45Al-1.6Mn	Reactive sintering/NL	RT	RT	465	67	566	82	1.4	...	...
		800	1470	370	54	540	78	14	...	...
ABB alloy/Ti-47Al-2W-0.5Si	Casting + HT/duplex	RT	RT	425	62	520	75	1.0	22	20
		760	1400	350	51	460	67	2.5	...	...
47XD/Ti-47Al-2Mn-2Nb-0.8TiB$_2$	Casting + HIP + HT/NL + TiB$_2$	RT	RT	402	58	482	70	1.5	15–16	13.6–14.6
		760	1400	344	50	458	66	...	...	...
45XD/Ti-45Al-2Mn-2Nb-0.8TiB$_2$	Casting + HIP + HT	RT	RT	570	83	695	101	1.5	15–19	13.6–17.3
		600	1110	440	64	650	94	...	...	...
		760	1400	415	60	510	74	19	...	...
GE alloy 204b/Ti-46.2Al-*x* Cr-*y* (Ta,Nb)	Casting + HIP + HT/NL	RT	RT	442	64	575	83	1.5	34.5	31.4
		760	1400	382	55.4	560	81	12.4	...	...
		840	1545	381	55.2	549	80	12.2	...	...
Ti-47Al-2Nb-2Cr-1Ta	Casting + HIP + HT/duplex	RT	RT	430	62	515	75	1.0	...	...
		800	1470	363	53	495	72	23.3	...	...
		870	1600	334	48	403	58	14.6	...	...
Ti-47Al-2Nb-1.75Cr	PM casting + HIP + HT	RT	RT	429	62	516	75	1.4	...	...
		760	1400	286	41	428	62	13.3	...	...
		815	1500	368	53	531	77	23.3	...	...
Alloy 7/Ti-46Al-4Nb-1W	Extrusion + HT/NL	RT	RT	648	94	717	104	1.6	...	...
		760	1400	517	75	692	100	...	...	...
Alloy K5/Ti-46.5Al-2Cr-3Nb-0.2W	Forging + HT/duplex	RT	RT	462	67	579	84	2.8	11	10
		800	1470	345	50	468	68	40	...	...
	Forging + HT/RFL	RT	RT	473	69	557	81	1.2	20–22	18–20
		800	1470	375	54	502	73	3.2	...	...
		870	1600	362	53	485	70	12.0	...	...

(a) K_q = the provisional K_{Ic} value. Abbreviations: HT, heat treated; HIP, hot-isostatically pressed; P/M, powder metallurgy; PM, permanent mold (casting); NL, nearly lamellar; FL, fully lamellar; RFL, refined fully lamellar; RT, room temperature. Source: Ref 202

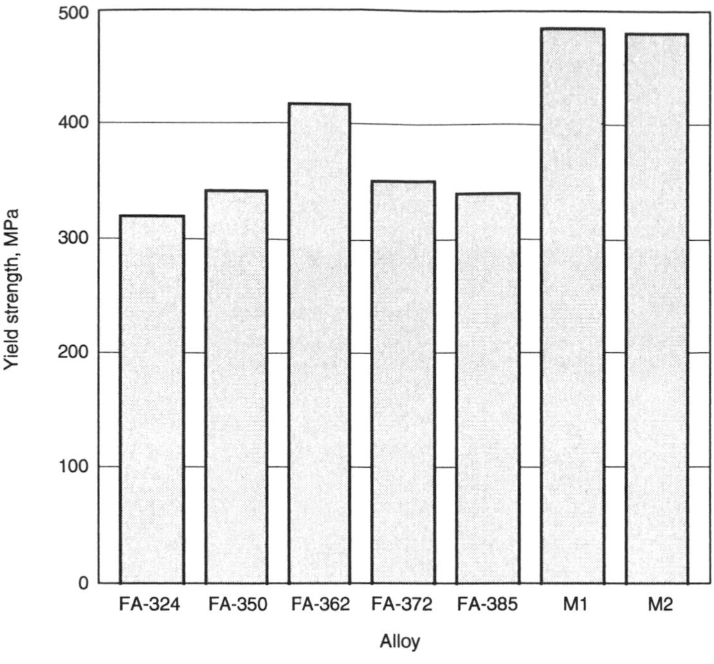

Fig. 31 Yield strength of several FeAl alloys tensile tested at 600 °C (1110 °F) in air. Specimens were punched from thin sheets, hot rolled at 850-900 °C (1560-1650 °F), and heat treated for 1 h at 700-800 °C (1290-1470 °F) prior to testing. Compositions cor-

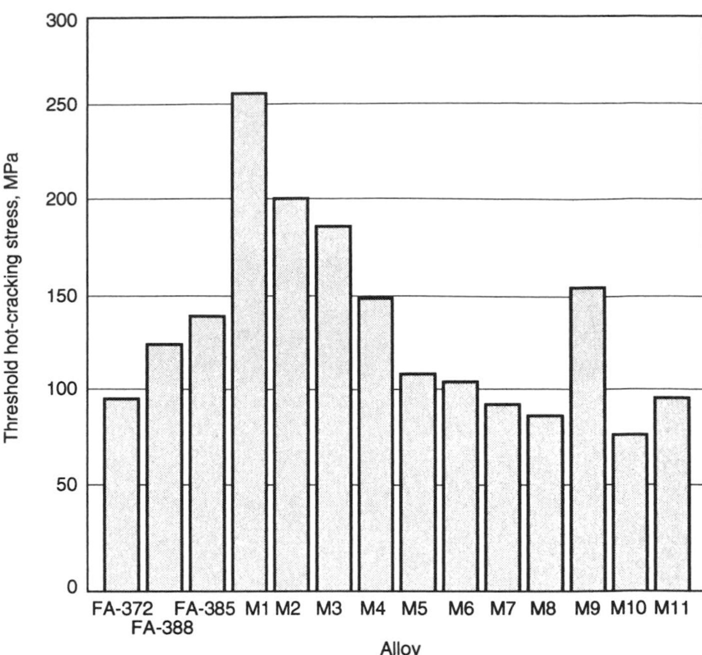

Fig. 32 Weldability of FeAl alloys as measured by the threshold hot-cracking stress for thin-sheet specimens using the Sigmajig test (see the article "Weldability Testing" in Volume 6 of the *ASM Handbook* for a description of this test). Compositions corresponding to those alloy designations can be found in Table 7.

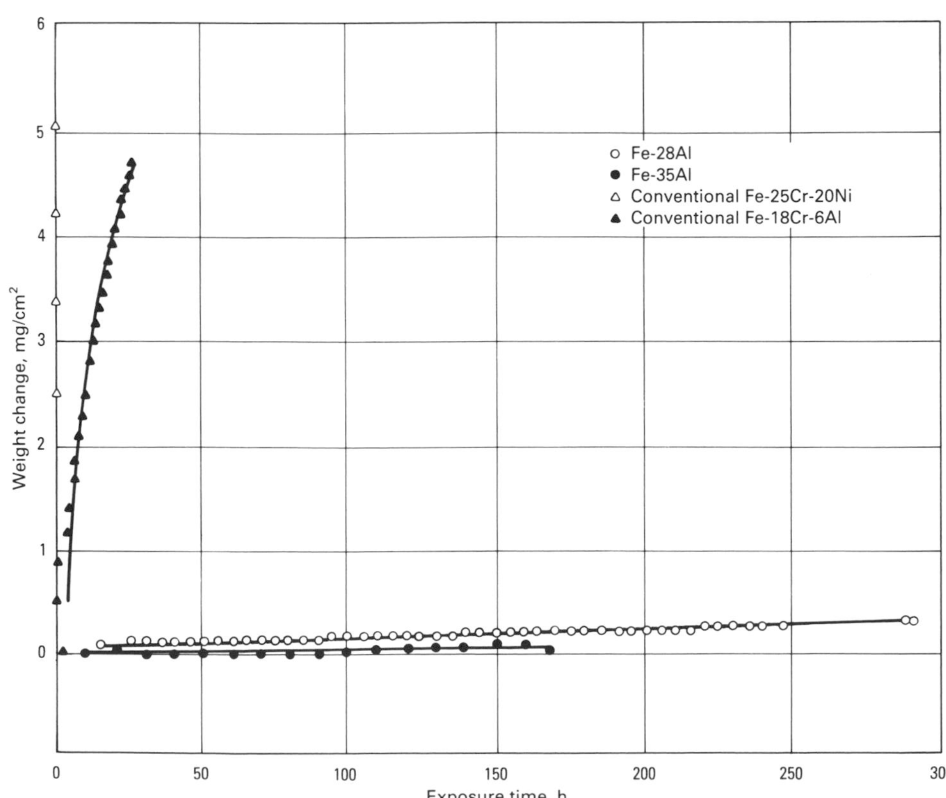

Fig. 33 Comparison of the corrosion behavior of iron aluminides with that of conventional iron-base alloys Fe-18Cr-6Al (the coating material) and Fe-25Cr-20Ni. All materials were exposed to a severe sulfidizing environment at 800 °C (1470 °F). Source: Ref 106

recrystallized in the ($\alpha + \gamma$) phase field. Grain morphology varies considerably depending on composition, solution treatment temperature and time, cooling rate, and stabilization temperature and time (Ref 180). Grain size decreases with reduced aluminum content and with additions of vanadium, manganese, and chromium (Ref 198, 199). The number of annealing twins in the γ phase increases as aluminum content decreases or when manganese or vanadium levels are increased (Ref 198). Chromium additions increase the volume fraction of the lamellar structure (Ref 199).

Mechanical and Metallurgical Properties. The strength and ductility of γ alloys are strongly dependent on alloy composition and TMP conditions (Ref 180). Figure 40 shows this variation in binary γ alloys after a number of TMP treatments. However, the Ti-52Al alloy demonstrates the lowest hardness value at room temperature, regardless of the TMP treatment (Ref 181, 204–207). At 1000 °C (1830 °F), however, strength tends to decrease gradually with increasing aluminum levels (Ref 180). Tensile strength and hardness vary in the same fashion with variations in aluminum content (Ref 180). Room-temperature tensile elongation is maximum at a composition of approximately Ti-48Al. Table 11 lists tensile properties and fracture toughness as functions of processing/microstructure and temperature.

Ternary alloys of composition Ti-48Al with approximately 1 to 3% V, Mn, or Cr exhibit enhanced ductility, but Ti-48Al alloys with approximately 1 to 3% Nb, Zr, Hf, Ta, or W show lower ductility than binary Ti-48Al (Ref 180). The brittle-ductile transition (BDT) occurs at 700

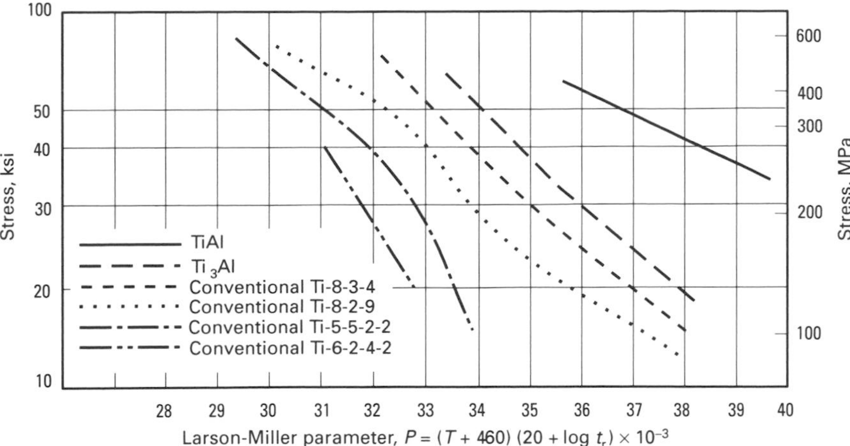

Fig. 34 Comparison of the creep behavior of conventional titanium alloys and titanium aluminide intermetallics. Source: Ref 148

°C (1290 °F) in Ti-56Al and at lower temperatures with decreasing aluminum levels. Increased room-temperature ductility generally results in a reduced BDT temperature. Above the BDT temperature, ductility increases rapidly with temperature, approaching 100% at 1000 °C (1830 °F) for the most ductile γ alloy compositions. The trend bands for variations in yield strength and tensile ductility with test temperature are shown in Fig. 41. The elastic moduli of γ alloys range from 160 to 176 GPa (23×10^6 to 25.5×10^6 psi) and decrease slowly with temperature (Ref 180).

Low-cycle fatigue experiments (Ref 147) suggest that fine grain sizes increase fatigue life at temperatures below 800 °C (1470 °F). Fatigue crack growth rates for γ alloys are more rapid than those for superalloys, even when density is normalized (Ref 196). Both fracture toughness and impact resistance are low at ambient temperatures, but fracture toughness increases with temperature; for example, the plane-strain fracture toughness (K_{Ic}) for Ti-48Al-1V-0.1C is 24 MPa$\sqrt{m}$ (21.8 ksi$\sqrt{in.}$) at room temperature (Ref 147). Fracture toughness is strongly dependent on the volume fraction of the lamellar phase. In a two-phase quaternary γ alloy, a fracture toughness of 12 MPa$\sqrt{m}$ (10.9 ksi$\sqrt{in.}$) is observed for a fine structure that is almost entirely γ; K_{Ic} is greater than 20 MPa$\sqrt{m}$ (18.2 ksi$\sqrt{in.}$) when a large volume fraction of lamellar grains are present (Ref 180). Creep properties of γ alloys, when normalized by density, are better than those of superalloys, but they are strongly influenced by alloy chemistry and TMP. Increased aluminum content and additions of tungsten (Ref 208) or carbon (Ref 147) increase creep resistance. Increasing the volume fraction of the lamellar structure enhances creep properties (Ref 147) but lowers ductility. The level of creep strain from elongation upon initial loading and primary creep is of concern because it can exceed projected design levels for maximum creep strain in the part. Figure 42 compares the effects of microstructure on steady-state or minimum creep rate at 800 °C (1470 °F) for various alloys.

Composites. Gamma titanium aluminides also serve as the matrix for titanium-matrix composites. Investment cast Ti-47Al-2Nb-2Mn (at.%) + 0.8 vol% TiB_2 and Ti-45Al-2Nb-2Mn (at.%) + 0.8 vol% TiB_2 XD (exothermic dispersion) alloys have been developed (Ref 209). These alloy form in situ titanium diborides that cause grain refinement and subsequent enhancement of mechanical and physical properties. The XD-type of in situ reinforcements offer better thermal stability than conventional metal matrix composites and the opportunity to introduce reinforcements in the micron scale range (at 0.8 vol%, the TiB_2 is simply serving as a grain refiner). The XD aluminides offer excellent fatigue properties, with endurance limits in the range of 95 to 115% of the yield strength (non-XD γ-alloys exhibit similar behavior). The high fatigue strength is attributed to the microstructural refinement mentioned above and a strain aging effect that occurs during loading (Ref 210). One major gas turbine engine manufacturer is studying XD γ-aluminide blades for the last stage low-pressure turbine in an aircraft engine that currently uses IN-713 blades. Service temperature is in the range of 650 to 700 °C (1200 to 1290 °F). XD titanium aluminide composites are also being used for applications such as missile fins (Ref 209). Ti-47Al-2V+7vol%TiB_2 has a higher strength than 17-4PH steel above 600 °C (1110 °F) and a higher modulus about 750 °C (1380 °F). These improvements extend the operating temperature of the missile wing, which led to a redesign utilizing the XD composite. Table 11 lists tensile properties of XD composites.

Oxidation Resistance. Although the γ class of titanium aluminides offers oxidation and interstitial (oxygen, nitrogen) embrittlement resistance superior to that of the α-2 and orthorhombic (Ti_2AlNb) classes of titanium aluminides, environmental durability is still a concern, especially at temperatures about 750 to 800 °C (1380 to 1470 °F) in air. This is because, as stated earlier in this article, all titanium aluminides are characterized by a strong tendency to form TiO_2, rather than the protective Al_2O_3 scale, at elevated temperature. To improve the oxidation resistance of γ aluminides, two areas are being explored: alloy development and the development of improved protective coatings.

Ternary and higher-order alloying additions can reduce the rate of oxidation of γ alloys (Ref 211). Of particular benefit are small (1 to 4%) ternary additions of tungsten, niobium, and tantalum. When combined with quaternary additions of 1 to 2% Cr or Mn, further improvement in oxidation resistance is gained. However, it is important to stress that these small alloying additions do not result in continuous Al_2O_3 scale formation. Rather, a complex intermixed Al_2O_3/TiO_2 scale is still formed, but the rate of growth of this scale is reduced. Additional information on the effects of alloying elements on the oxidation of γ aluminides can be found in the article "Design for Oxidation Resistance" in this Volume.

Further improvements in oxidation resistance can be obtained via protective coatings. Three general coating alloy approaches have been taken for protecting titanium aluminides: MCrAlY (M = Ni,Fe,Co), aluminizing, and silicides/ceramics

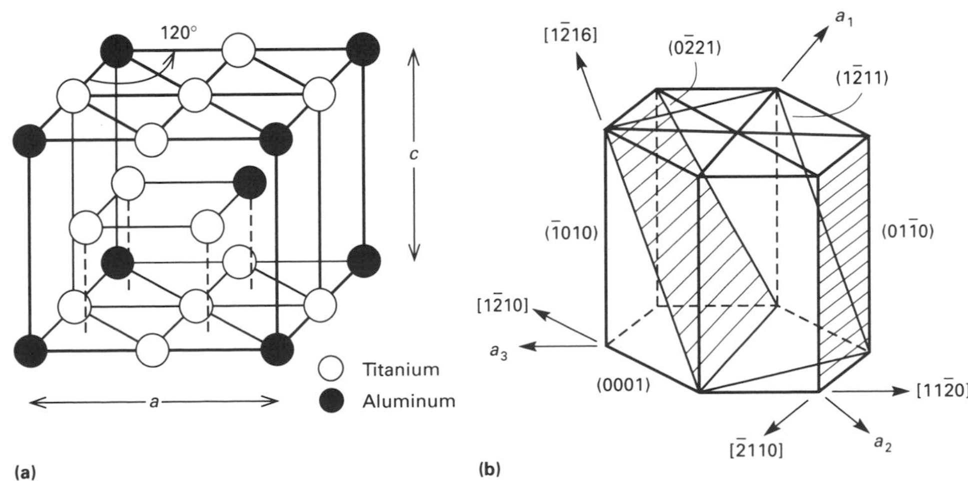

Fig. 35 Crystal structure of Ti_3Al. (a) $D0_{19}$ hexagonal superlattice structure of Ti_3Al with lattice constants of c = 0.420 nm and a = 0.577 nm. (b) Possible slip planes and slip vectors in the structure

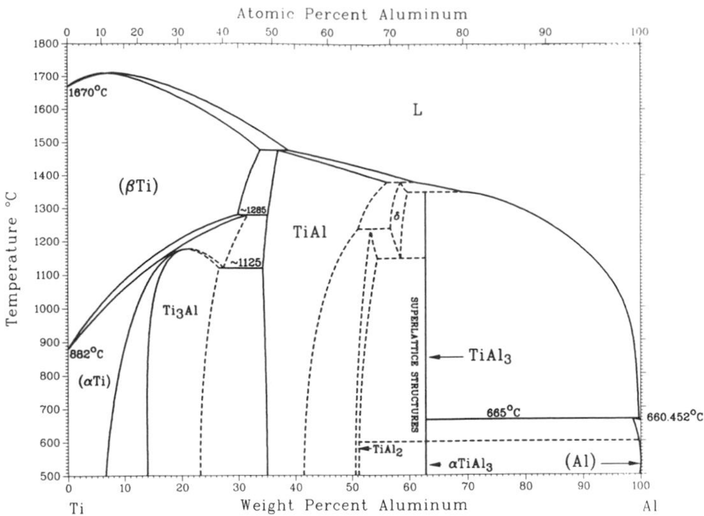

Fig. 36 The aluminum-titanium binary phase diagram.

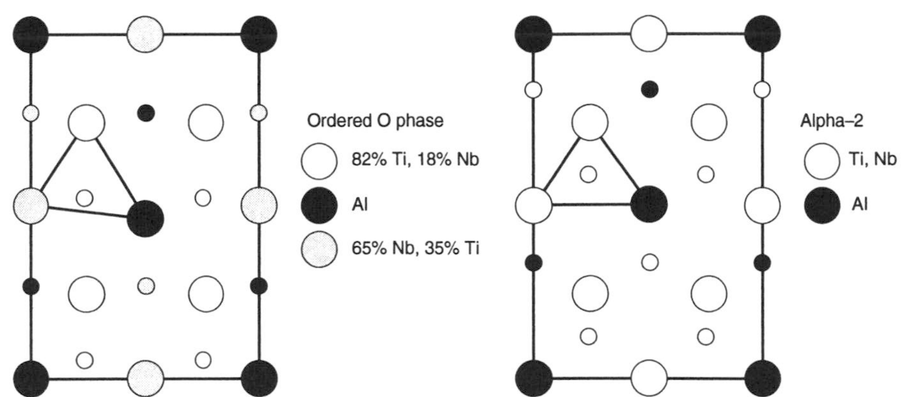

Fig. 37 Basal plane schematics of the ordered orthorhombic and α-2 phases. The large circles indicate atomic positions in the plane of the paper. Smaller circles represent atoms in planes above and below the plan of the paper. Source: Ref 170

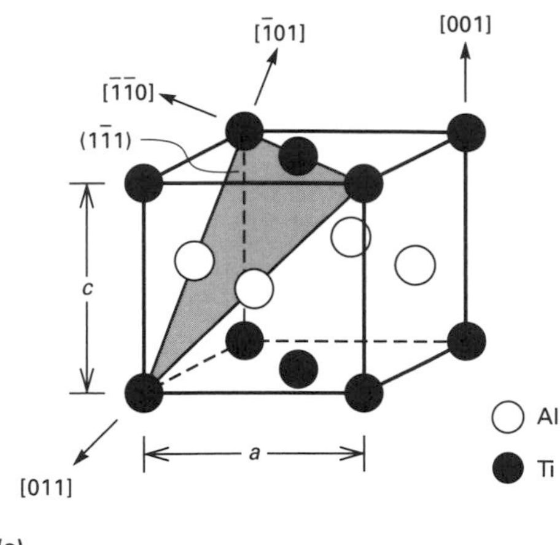

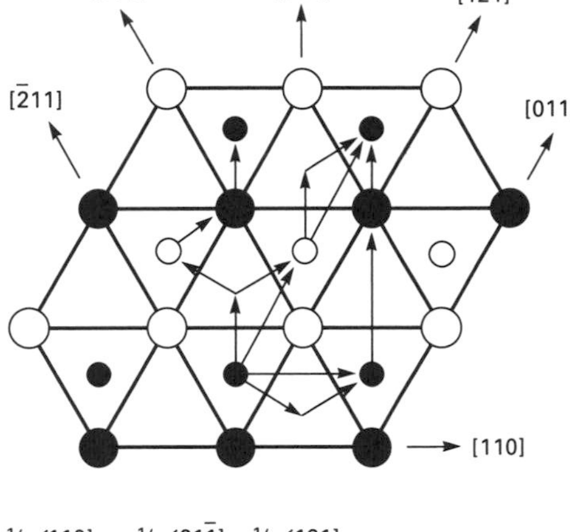

$\frac{1}{2}\langle 110] \rightarrow \frac{1}{6}\langle 21\bar{1}] + \frac{1}{6}\langle 121]$

$\langle 011] \rightarrow \frac{1}{6}\langle \bar{1}12] + \frac{1}{6}\langle 121] + \frac{1}{6}\langle \bar{1}12] + \frac{1}{6}\langle 121]$

$\rightarrow \frac{1}{6}\langle \bar{1}12] + \frac{1}{3}\langle \bar{1}12] + \frac{1}{2}\langle 110]$

$\frac{1}{2}\langle \bar{1}12] \rightarrow \frac{1}{6}\langle \bar{1}12] + \frac{1}{3}\langle \bar{1}12]$

$\rightarrow \frac{1}{6}\langle \bar{1}12] + \frac{1}{6}\langle \bar{2}11] + \frac{1}{6}\langle 121] + \frac{1}{6}\langle \bar{1}12]$

(b)

Fig. 38 Crystal structure of γ-TiAl alloys. (a) Ordered face-centered tetragonal (Ll_0) TiAl structure. Shaded area represents the ⟨111⟩ plane. (b) Slip dislocations on ⟨111⟩ plane, ordinary dislocations ½⟨110⟩, superdislocations ⟨011⟩ and ½⟨112⟩, and twin dislocations ⅙⟨112⟩ with possible dissociations. Source: Ref 181

(Ref 211). Protection of titanium aluminides under oxidizing conditions has been achieved with all three approaches; however, studies of such coatings on α-2 and orthorhombic-based titanium aluminides (monolithic and composite) report severe lifetime degradation under fatigue conditions. The fatigue life of coated material is often reduced to below that of uncoated material. Similar results are also expected for such coatings that are on γ titanium aluminides.

The degradation in the fatigue life of titanium aluminides by coatings results from three main factors: the formation of brittle coating/substrate reaction zones (chemical incompatibility), the brittleness of the coating alloy, and the differences in the coefficient of thermal expansion (CTE) between the coating and the substrate (Ref 211). MCrAlY coatings, which are successfully used to protect nickel-, iron-, and cobalt-base superalloys, are not chemically compatible with titanium aluminides and form brittle coating/substrate reaction zones at 800 °C. Aluminizing treatments result in the surface formation of the $TiAl_3$ and $TiAl_2$ phases, which are brittle and exhibit CTE mismatches with α-2 orthorhombic, and γ titanium aluminides. Silicide and ceramic coatings are also generally too brittle to survive fatigue conditions.

The ideal oxidation-resistant coating for γ alloys would be Ti-Al based for optimal chemical and mechanical compatibility with γ substrates, be capable of forming a continuous Al_2O_3 scale for protection from both oxidation and interstitial oxygen/nitrogen embrittlement, and possess reasonable mechanical properties to survive high-cycle fatigue (Ref 211). No ideal combination of these properties exists at present. However, reasonable compromises have been achieved with coating alloys based in the Ti-Al-Cr system. These coatings have been applied by spattering (Ti-44Al-28Cr on Ti-47Al-2Cr-2Ta), hot isostatic pressing (Ti-44Al-28Cr and Ti-50Al-20Cr coatings), and low-pressure plasma spraying. Figure 43 shows the results of interrupted weight gain oxidation data for Ti-48Al-2Cr-2Nb coated with Ti-51Al-12Cr. These tests showed that the coating successfully protected the substrate at 800 °C (1470 °F) and 1000 °C (1830 °F) in air.

Silicides

Silicides have been used as commercial materials because of their excellent oxidation and corrosion resistance in hostile environments. For example, $MoSi_2$ is currently used for heating elements at temperatures to 1800 °C (3270 °F). Ni_3Si is the major constituent of the commercial alloy Hastelloy D, a corrosion-resistant alloy with the unique ability to resist attack by sulfuric acid solutions.

Ni_3Si Alloys. The silicide Ni_3Si has an Ll_2 ordered crystal structure existing below the peritectic temperature of approximately 1035 °C (1900 °F). It is of commercial interest because of

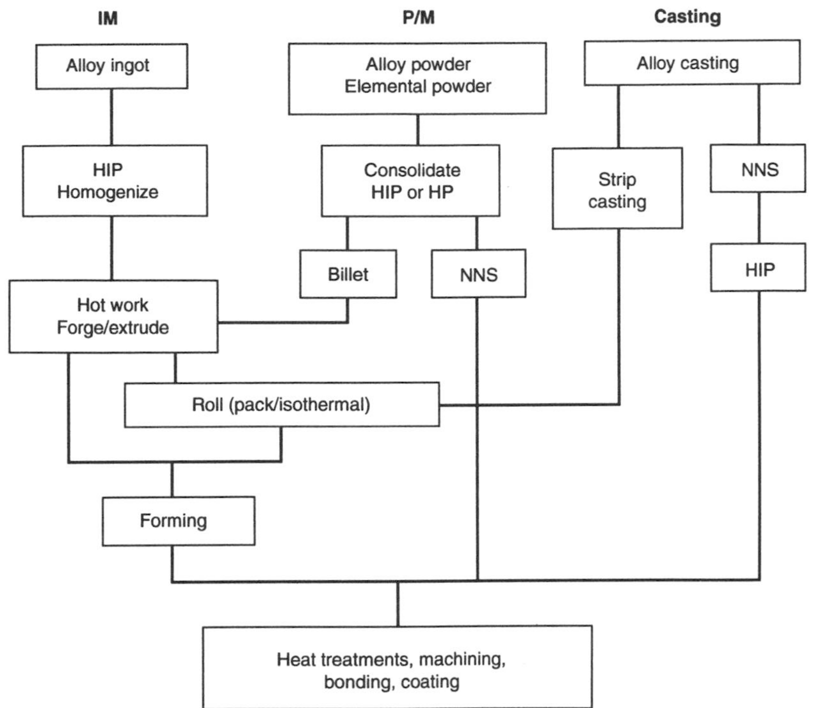

Fig. 39 Processing methods and routes practiced to date for engineering gamma titanium alloys. IM, ingot metallurgy; P/M, powder metallurgy; HIP, hot isostatic pressing, HP, hot pressing; NNS, near-net shape. Source: Ref 202

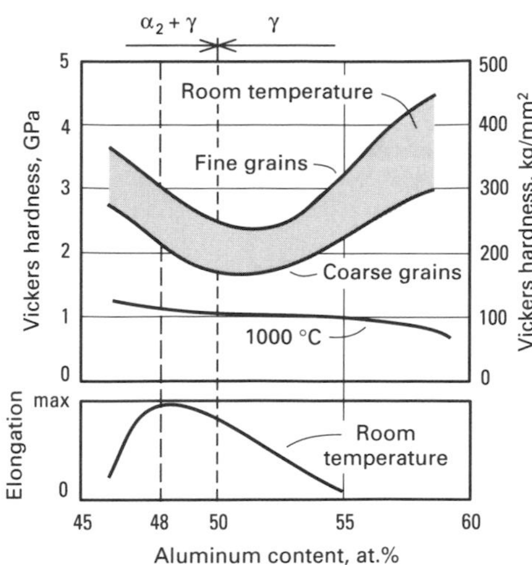

Fig. 40 Effect of aluminum content on room-temperature tensile elongation and hardness of binary γ titanium aluminide alloys. Hardness values at 1000 °C (1830 °F) are also shown. Note the single-phase γ region and the two phase ($\alpha_2 + \gamma$) region. Source: Ref 180

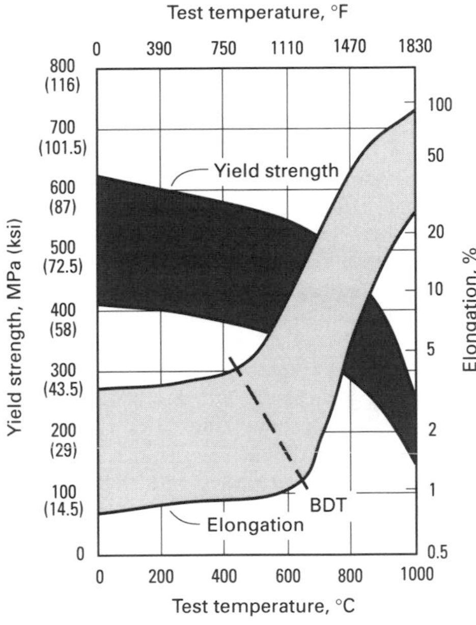

Fig. 41 Ranges of yield strength and tensile elongation as functions of test temperature for γ-TiAl alloys. BDT, brittle-ductile transition. Source: Ref 180

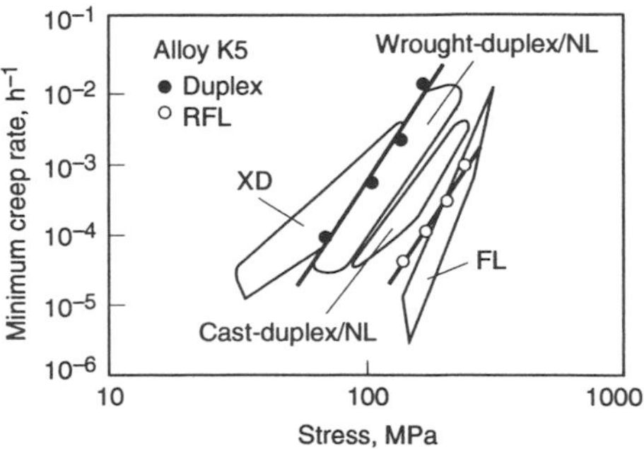

Fig. 42 Minimum creep rates against applied stress at 800 °C (1470 °F) in air for various gamma titanium alloys in different processing and/or microstructural conditions. XD, exothermic dispersion composites; NL, nearly lamellar; FL, fully lamellar. Source: Ref 202

its excellent corrosion resistance in acid environments, particularly sulfuric acid solutions. The engineering use of Ni_3Si is limited by its poor ductility at ambient temperatures and lack of fabricability at high temperatures (Ref 212–215). Because of these problems, the commercial Ni_3Si alloy, Hastelloy D, has to be used in the cast condition with relatively poor mechanical properties.

Grain boundaries in Ni_3Si, like those in Ni_3Al, are intrinsically brittle, as evidenced by Auger spectroscopic analyses (Ref 212–214). As shown in Fig. 44, the ductility of Ni_3Si can be effectively improved by reducing the silicon concentration below 20 at.% or by microalloying with boron, carbon, or beryllium. A dramatic improvement in room-temperature ductility was obtained by adding boron to Ni-18.9Si-3.2Ti (at.%). The increase in ductility from 3% for Ni-22.5Si to 30% for Ni-18.9Si-3.2Ti-0.1B is accomplished by a change in fracture mode from brittle grain-boundary fracture to ductile dimple fracture. Auger analyses showed that boron strongly segregates to grain boundaries, thereby suppressing intergranular fracture.

Just like Ni_3Al, ductile Ni_3Si alloys show a severe reduction in ductility at intermediate temperatures (400 to 800 °C, or 750 to 1470 °F) when tested in tension in air (Ref 215). The loss in ductility is due to dynamic embrittlement involving oxygen in air. The embrittlement can be reduced and the ductility improved by alloying with moderate amounts (4 to 6%) of chromium.

In the early 1970s, it was reported that additions of titanium greatly improved the as-cast properties of Ni_3Si alloys but reduced their hot fabricability (Ref 216, 217). Another systematic study (Ref 213, 215) of alloying effects for the development of wrought Ni_3Si alloys found that macroalloying with niobium and vanadium is as effective as that with titanium in improving the room-temperature ductility of Ni_3Si alloys. On the other hand, the alloying elements molybdenum, iron, and chromium are beneficial to hot fabricability. Some Ni_3Si alloys containing these three elements showed a superplastic behavior (approximate elongation of 500 to 600%) when tested at 1020 to 1100 °C (1870 to 2010 °F) in air. The strength of Ni_3Si at temperatures up to 700 °C (1290 °F) can be significantly improved by the addition of up to 1% Hf. Some alloys, such as Ni-18.9Si-3.2Cr-0.6Hf-0.15B (at.%) are, in fact, as strong as nickel-base superalloy IN-718 at room and intermediate temperatures. These ductile, strong Ni_3Si alloys have the potential to be used as structural materials for chemical and petrochemical applications.

Molybdenum disilicide ($MoSi_2$) is a line compound with a very high melting point (T_m = 2020 °C, or 3670 °F). It has an ordered tetragonal structure ($C11_b$) in which the unit cell contains three bcc lattices. $MoSi_2$ is attractive because of its high electrical and thermal conductivities and excellent oxidation resistance at high temperatures. The silicide is capable of forming a thin adhesive self-healing protective layer of silica glass on its surfaces when exposed to oxidizing atmospheres at temperatures up to 1800 °C (3270 °F). $MoSi_2$ has been used commercially for electrical heating elements in high-temperature furnaces under the trade name of Kanthal (Ref 218). The Kanthal materials, which contain roughly

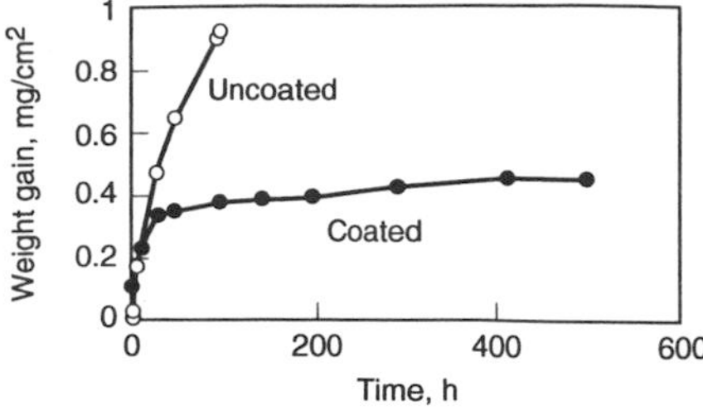

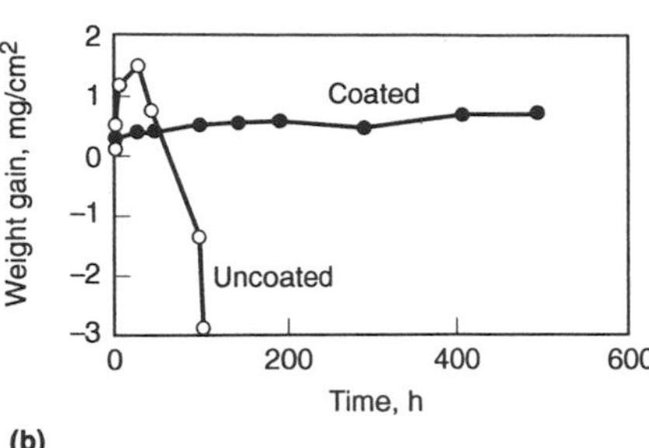

Fig. 43 Interrupted weight-gain oxidation data for Ti-51Al-12Cr coated by low-pressure plasma spraying and uncoated Ti-48Al-2Cr-2Nb at (a) 800 °C (1470 °F) and (b) 1000 °C (1830 °F) in air. Source: Ref 211

80% $MoSi_2$ and 20% glass ceramic compounds, are prepared by P/M processing. The ceramic component is added to improve the ductility of these materials at high temperatures and also to increase their electrical resistance. The lifetime of Kanthal heating elements can be as long as 5 years. However, $MoSi_2$ is prone to fast intergranular oxidation with severe material damage when exposed to air at 500 to 800 °C (930 to 1470 °F) (Ref 219). This problem can be alleviated by rapid heating to above 800 °C (1470 °F), or preheating to high temperatures for the formation of protective silica films.

Recently, considerable attention (Ref 219–221) has been given to the development of $MoSi_2$ as a high-temperature structural material for use at temperatures up to approximately 0.8 T_m (1550 °C, or 2820 °F). $MoSi_2$ is very brittle and hard (750 HV) at ambient temperatures. Figure 45 shows strength as a function of test temperature for $MoSi_2$. From 20 to 1250 °C (70 to 2280 °F), the yield strength is 320 MPa (46 ksi) and is independent of temperature; however, it decreases abruptly above 1300 °C (2370 °F). The silicide exhibits a BDT at about 925 °C (1700 °F), and above that temperature fracture strength increases sharply, as indicated by the line above the yield stress trend line in Fig. 45. Single crystals of $MoSi_2$ and WSi_2 have been successfully grown by floating-zone techniques. Studies of deformation in these crystals at 900 to 1500 °C (1650 to 2730 °F) revealed that slip occurs mainly on {110} and {013} plans (Ref 220).

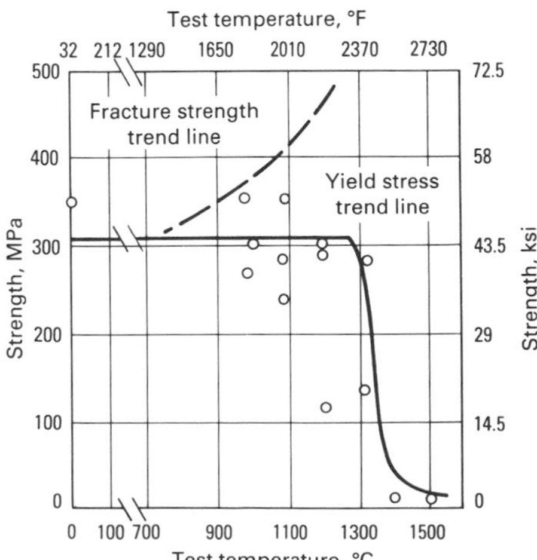

Fig. 45 Strength of $MoSi_2$ as a function of temperature. Data points were determined by various investigators and are compiled in Ref 221.

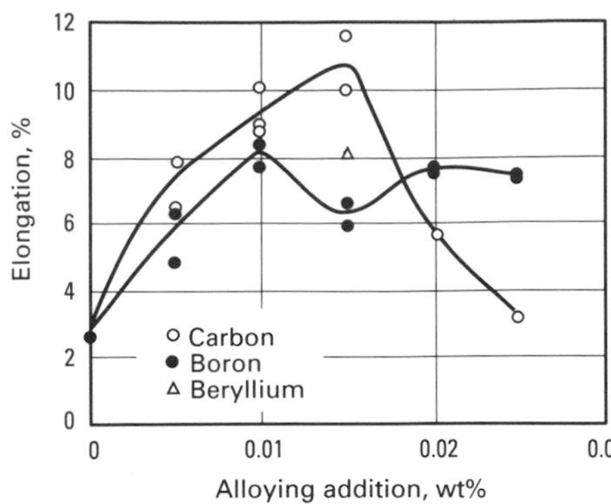

Fig. 44 Effect of microalloying with boron, carbon, and beryllium on the room-temperature ductility of Ni-22.5Si. Source: Ref 215

As shown in Fig. 45, $MoSi_2$ is not strong at high temperatures, particularly above 1300 °C (2370 °F). Consequently, efforts have been initiated to develop composite materials based on $MoSi_2$ (Ref 221, 222). Both ceramic fibers (or particulates) and ductile metal wires have been selected as reinforcement components. The ceramics, which include silicon carbide (SiC), titanium carbide (TiC), and zirconium diboride (ZrB_2), are compatible with $MoSi_2$, whereas the metal wires, which include niobium, tantalum, and tungsten, react with the silicide to form protective layers of Mo_5Si_3 that slow down interdiffusional processes (Ref 221). Some limited data have demonstrated pronounced improvements in high-temperature strength and room-temperature fracture toughness from these reinforcements. Further studies are required to optimize processing parameters.

ACKNOWLEDGEMENT

The information in this article is largely taken from C.T. Liu, J.O. Stiegler, and F.H. Froes, Ordered Intermetallics, *Properties and Selection: Nonferrous Alloys and Special-Purpose Materials,* Vol 2, *ASM Handbook* (formerly Vol 2, 10th ed., *Metals Handbook*), ASM International, 1990, p 911–942.

REFERENCES

1. C.T. Liu, E.H. Lee, and C.G. McKamey, *Scr. Metall.,* Vol 23, 1989, p 875
2. C.T. Liu, C.G. McKamey, and E.H. Lee, *Scr. Metall.,* Vol 24, 1990, p 385
3. E.A. Aitken, *Intermetallic Compounds,* J.H. Westbrook, Ed., Wiley, 1967, p 491–516
4. V.K. Sikka and S.C. Deevi, Intermetallics for Structural Applications, *Heat-Resistant Materials II,* ASM International, 1995, p 567–578
5. A.J. Bradley and A. Taylor, *Proc. R. Soc. (London) A,* Vol 136, 1932, p 210
6. A.J. Bradley and A. Taylor, *Proc. R. Soc. (London) A,* Vol 159, 1937, p 56
7. N. Ridley, *J. Inst. Met.,* Vol 94, 1966, p 255
8. M.J. Cooper, *Philos. Mag.,* Vol 8, 1963, p 805
9. S. Ohiai, Y. Oya, and T. Suzuki, *Acta Metall.,* Vol 32, 1984, p 289–298
10. *High-Temperature Ordered Intermetallic Alloys,* Materials Research Society Symposia Proceedings, Vol 39, C.C. Koch, C.T. Liu, and N.S. Stoloff, Ed., Materials Research Society, 1985
11. *High-Temperature Ordered Intermetallic Alloys II,* Materials Research Society Symposia Proceedings, Vol 81, N.S. Stoloff, C.C. Koch, C.T. Liu, and O. Izumi, Ed., Materials Research Society, 1987
12. *High-Temperature Ordered Intermetallic Alloys III,* Materials Research Society Symposia Proceedings, Vol 133, C.T. Liu, A.I. Taub, N.S. Stoloff, and C.C. Koch, Ed., Materials Research Society, 1989
13. *High-Temperature Ordered Intermetallic Alloys IV,* Materials Research Society Symposia Proceedings, Vol 213, L.A. Johnson, D.P. Pope, and J.O. Stiegler, Ed., Materials Research Society, 1991
14. *High-Temperature Ordered Intermetallic Alloys V,* Materials Research Society Symposia Proceedings, Vol 288, I. Baker, R. Darolia, J.D. Whittenberger, and M.H. Yoo, Ed., Materials Research Society, 1992
15. *High-Temperature Ordered Intermetallic Alloys VI,* Materials Research Society Symposia Proceedings, Vol 364, J. Horton, I. Baker, S. Hanada, R.D. Noebe, and D.S. Schwartz, Ed., Materials Research Society, 1994
16. C.T. Liu, R.N. Cahn, and S. Sauthoff, Ed., *Ordered Intermetallics—Physical Metallurgy and Mechanical Behavior,* Kluwer Academic Publishers, 1992
17. S.H. Whang, C.T. Liu, D.P. Pope, and J.O. Stiegler, Ed., *High-Temperature Aluminides and Intermetallics,* TMS-AIME, 1990
18. R. Darolia, J.J. Lewandowski, C.T. Liu, P.L. Martin, D.B. Miracle, and M.V. Nathal, Ed., *Structural Intermetallics,* TMS-AIME, 1993
19. T. Grobstein and J. Doychale, Ed., *Oxidation of High-Temperature Intermetallics,* TMS-AIME, 1989
20. Y.W. Kim and R.R. Boyer, Ed., *Microstructure/Property Relationships in Titanium Aluminides and Alloys,* TMS-AIME, 1991
21. C.T. Liu and C.C. Koch, *Trends in Critical Materials Requirements for Steels of the Future: Conservation and Substitution Technol-*

ogy for Chromium, NBSIR-83-2679-2, National Bureau of Standards, 1983

22. K. Aoki and O. Izumi, *Trans. Jpn. Inst. Met.,* Vol 19, 1978, p 203
23. E.M. Grala, in *Mechanical Properties of Intermetallic Compounds,* J.H. Westbrook, Ed., Wiley, 1960, p 358–404
24. C.T. Liu, C.L. White, and J.A. Horton, *Acta Metall.,* Vol 33, 1985, p 213–219
25. T. Takasugi, E.P. George, D.P. Pope, and O. Izumi, *Scr. Metall.,* Vol 19, 1985, p 551–556
26. T. Ogura, S. Hanada, T. Masumoto, and O. Izumi, *Metall. Trans. A,* Vol 16, 1985, p 441–443
27. A.I. Taub, S.C. Huang, and K.M. Chang, Stoichiometry Effects on the Strengthening and Ductilization of Ni_3Al by Boron Modification and Rapid Solidification, *Failure Mechanisms in High Performance Materials,* J.G. Early, T.R. Shives, and J.H. Smith, Ed., Cambridge University Press, 1985, p 57–65
28. C.L. White, R.A. Padgett, C.T. Liu, and S.M. Yalisove, *Scr. Metall.,* Vol 18, 1984, p 1417–1420
29. G.S. Painter and F.W. Averill, *Phys. Rev. Lett.,* Vol 58, 1987, p 234
30. S.P. Chen, A.F. Voter, and D.J. Srolovitz, *J. Phys.,* Vol 49, Oct 1988, p C5.157–C5.163
31. G.M. Bond, I.M. Robertson, and H.K. Birnbaum, *J. Mater. Res.,* Vol 2, 1987, p 436–440
32. E.M. Schulson, T.P. Weihs, D.V. Viens, and I. Baker, *Act. Metall.,* Vol 33, 1985, p 1587
33. P.S. Khadkikar, K. Vedula, and B.S. Shale, *Metall. Trans. A,* Vol 18, 1987, p 425
34. I. Baker, E.M. Schulson, and J.A. Horton, *Acta Metall.,* Vol 35, 1987, p 1533–1541
35. A.H. King and M.H. Yoo, in *High-Temperature Ordered Intermetallic Alloys II,* Materials Research Society Symposia Proceedings, Vol 81, N.S. Stoloff, C.C. Koch, C.T. Liu, and O. Izumi, Ed., Materials Research Society, 1987, p 99–104
36. I. Baker, E.M. Schulson, and J.R. Michael, *Philos. Mag.,* Vol B57, 1988, p 379
37. D.N. Sieloff, S.S. Brenner, and Hua Ming-Jian, in *High-Temperature Ordered Intermetallic Alloys III,* Materials Research Society Symposia Proceedings, Vol 133, C.T. Liu, A.I. Taub, N.S. Stoloff, and C.C. Koch, Materials Research Society, 1989, p 155–160
38. E.P. George, C.T. Liu, and R.A. Padgett, *Scr. Metall.,* Vol 23, 1989, p 979–982
39. R.A.D. Mackenzie and S.L. Sass, *Scr. Metall.,* Vol 22, 1988, p 1807
40. C.T. Liu, C.L. White, and E.H. Lee, *Scr. Metall.,* Vol 19, 1985, p 1247–1250
41. C.T. Liu and C.L. White, *Acta Metall.,* Vol 35, 1987, p 643
42. A.I. Taub, K.M. Chang, and C.T. Liu, *Scr. Metall.,* Vol 20, 1986, p 1613
43. C.A. Hippsley and J.H. DeVan, *Acta Metall.,* Vol 37, 1989, p 1485–1496
44. C.T. Liu and V.K. Sikka, *J. Met.,* Vol 38, 1986, p 19–21
45. C.T. Liu, in *Micon 86,* American Society for Testing and Materials, 1988, p 222–237
46. J.A. Horton, J.V. Cathcart, and C.T. Liu, *Oxid. Met.,* Vol 29, 1988, p 347–365
47. C.T. Liu and B.F. Oliver, *J. Mater. Res.,* Vol 4, 1989, p 294–299
48. C.T. Liu and C.L. White, in *High-Temperature Ordered Intermetallic Alloys,* Materials Research Society Symposia Proceedings, Vol 39, C.C. Koch, C.T. Liu, and N.S. Stoloff, Ed., Materials Research Society, 1985, p 365–380
49. B.H. Kear and H.G.F. Wilsdorf, *Trans. AIME,* Vol 224, 1962, p 383
50. B.H. Kear, *Acta Metall.,* Vol 12, 1964, p 555
51. M.H. Yoo, *Scr. Metall.,* Vol 20, 1986, p 915
52. M.H. Yoo, J.A. Horton, and C.T. Liu, *Acta Metall.,* Vol 36, 1988, p 2935
53. S. Ohiai, Y. Oya, and T. Suzuki, *Acta Metall.,* Vol 32, 1984, p 289–298
54. R.W. Guard and J.H. Westbrook, *Trans. AIME,* Vol 215, 1959, p 807–814
55. R.D. Rawlings and A. Staton-BeVan, *J. Mater. Sci.,* Vol 10, 1975, p 505–514
56. L.R. Curwick, Ph.D. dissertation, University of Minnesota, 1972
57. D.P. Pope and C.T. Liu, in *Superalloys, Supercomposites and Superceramics,* J.K. Tien and T. Caulfield, Ed., Academic Press, 1989, p 584–624
58. R.L. Fleischer, *Acta Metall.,* Vol 11, 1963, p 203
59. Y. Mishima, S. Ochiai, and T. Suzuki, *Acta Metall.,* Vol 33, 1985, p 1161
60. C.T. Liu, V.K. Sikka, J.A. Horton, and E.H. Lee, "Alloy Development and Mechanical Properties of Nickel Aluminides (Ni_3Al) Alloys," ORNL-6483, Oak Ridge National Laboratory, 1988
61. D.L. Anton, D.D. Pearson, and D.B. Snow, in *High-Temperature Ordered Intermetallic Alloys II,* Materials Research Society Symposia Proceedings, Vol 81, N.S. Stoloff, C.C. Koch, C.T. Liu, and O. Izumi, Ed., Materials Research Society, 1987, p 287–295
62. G.E. Fuchs, A.K. Kuruvilla, and N.S. Stoloff, private communication, 1989
63. N.S. Stoloff, G.E. Fuchs, A.K. Kuruvilla and S.J. Choe, in *Mechanical Properties of Intermetallic Compounds,* J.H. Westbrook, Ed., Wiley, 1959, p 247–260
64. G.M. Camus, D.J. Duquette, and N.S. Stoloff, in *High-Temperature Ordered Intermetallic Alloys III,* Materials Research Society Symposia Proceedings, Vol 133, C.T. Liu, A.I. Taub, N.S. Stoloff, and C.C. Koch, Ed., Materials Research Society, 1989, p 579–584
65. R.S. Bellows, E.A. Schwarkopf, and J.K. Tien, *Metall. Trans. A,* Vol 19, 1988, p 479–486
66. R.S. Bellows and J.K. Tien, *Scr. Metall.,* Vol 21, 1987, p 1659–1662
67. V.K. Sikka, Development of Nickel and Iron Aluminides and Their Applications, *Advances in High Temperature Structural Materials and Protective Coatings,* A.K. Koul, Ed., National Research Council of Canada, 1994, p 282–295
68. R.K. Williams, S.R. Graves, F.J. Weaver, and D.L. McElroy, Physical Properties of Ni_3Al Containing 24 and 25 Atomic Percent Aluminum, *Proc. Mater. Res. Soc.,* Vol 39, 1985, p 505–512
69. M.L. Santella, S.A. David, and C.L. White, in *High-Temperature Ordered Intermetallic Alloys,* Materials Research Society Symposia Proceedings, Vol 39, C.C. Koch, C.T. Liu, and N.S. Stoloff, Ed., Materials Research Society, 1985, p 495–503
70. S.A. David, W.A. Jemian, C.T. Liu, and J.A. Horton, *Weld. Res. Suppl.,* Jan 1985, p 22s-28s
71. M.L. Santella and S.A. David, *Weld. Res. Suppl.,* May 1986, p 129s–137s
72. G.L. Povirk, J.A. Horton, C.G. McKamey, T.N. Tiegs, and S.R. Nutt, *J. Mater. Sci.,* Vol 23, 1988, p 3945–3950
73. J.M. Yang, W.H. Kao, and C.T. Liu, *Metall. Trans. A,* Vol 20, 1989, p 2459–2469
74. G.E. Fuchs, in *High-Temperature Ordered Intermetallic Alloys III,* Materials Research Society Symposia Proceedings, Vol 133, C.T. Liu, A.I. Taub, N.S. Stoloff, and C.C. Koch, Ed., Materials Research Society, 1989, p 615–620
75. S. Nourbakhsh, F.L. Liang, and H. Margolin, *J. Phys. E. Sci. Instrum.,* Vol 21, 1988, p 898
76. S. Nourbakhsh, F.L. Liang, and H. Margolin, in *High-Temperature Ordered Intermetallic Alloys III,* Materials Research Society Symposia Proceedings, Vol 133, C.T. Liu, A.I. Taub, N.S. Stoloff, and C.C. Koch, Ed., Materials Research Society, 1989, p 459–464
77. J.L. Smialek, *Metall. Trans. A.,* Vol 9, 1978, p 309
78. J. Jedlinski and S. Miowic, *Mater. Sci. Eng.,* Vol 87, 1987, p 281
79. C.A. Barrett, *Oxid. Met.,* Vol 30, 1988, p 361
80. A. Ball and R.E. Smallman, *Acta Metall.,* Vol 14, 1966, p 1517
81. N.J. Zaluzec and H.L. Fraser, *Scr. Metall.,* Vol 8, 1974, p 1049
82. I. Baker and E.M. Schulson, *Metall. Trans. A.,* Vol 15, 1984, p 1129
83. K.H. Hahn and K. Vedula, *Scr. Metall.,* Vol 23, 1989, p 7
84. J.H. Westbrook, *J. Electrochem. Soc.,* Vol 103, 1956, p 54
85. C.T. Liu, Oak Ridge National Laboratory, unpublished research, 1989
86. K. Vedula, V. Pathare, I. Aslamidis, and R.H. Titran, in *High-Temperature Ordered Intermetallic Alloys,* Materials Research Society Symposia Proceedings, Vol 39, C.C. Koch, C.T. Liu, and N.S. Stoloff, Ed., Materials Research Society, 1985, p 411–421
87. I. Jung, M. Rudy, and G. Sauthoff, in *High-Temperature Ordered Intermetallic Alloys II,* Materials Research Society Symposia Proceedings, Vol 81, N.S. Stoloff, C.C. Koch, C.T. Liu, and O. Izumi, Ed., Materials Research Society, 1987, p 263–274
88. P.R. Strutt and B.H. Kear, in *High-Temperature Ordered Intermetallic Alloys,* Materials Research Society Symposia Proceedings, Vol 39, C.C. Koch, C.T. Liu, and N.S. Stoloff, Ed., Materials Research Society, 1985, p 279–292
89. J.D. Whittenberger, E. Arzt, and M.J. Luton, *J. Mater. Res.,* Vol 5, 1990, p 271

90. J.D. Whittenberger, in *Solid State Processing,* A.H. Claver, Ed., TMS-AIME, 1990, p 137
91. C.T. Liu et al., "Alloying Effects on Mechanical and Metallurgical Properties of NiAl," Martin Marietta Energy Systems, unpublished research
92. E.M. Schulson and D.R. Barker, *Scr. Metall.,* Vol 17, 1983, p 519
93. A.G. Rozner and R.J. Wasilewski, *J. Inst. Met.,* Vol 94, 1966, p 169
94. E.P. George and C.T. Liu, *J. Mater. Res.,* Vol 5, 1990, p 754
95. D.B. Miracle, S. Russell, and C.C. Law, in *High-Temperature Ordered Intermetallic Alloys III,* Materials Research Society Symposia Proceedings, Vol 133, C.T. Liu, A.I. Taub, N.S. Stoloff, and C.C. Koch, Ed., Materials Research Society, 1989, p 225–230
96. R. Darolia, D.F. Lahrman, R.D. Field, and A.J. Freeman, in *High-Temperature Ordered Intermetallic Alloys III,* Materials Research Society Symposia Proceedings, Vol 133, C.T. Liu, A.I. Taub, N.S. Stoloff, and C.C. Koch, Ed., Materials Research Society, 1989, p 113–118
97. S.M. Russell, C.C. Law, and M.J. Blackburn, in *High-Temperature Ordered Intermetallic Alloys III,* Materials Research Society Symposia Proceedings, Vol 133, C.T. Liu, A.I. Taub, N.S. Stoloff, and C.C. Koch, Ed., Materials Research Society, 1989, p 627–632
98. S. Guha, P. Munroe, and I. Baker, *Scr. Metall.,* Vol 23, 1989, p 897–900
99. A. Inoue, T. Masumoto, and H. Tomioka, *J. Mater. Sci.,* Vol 19, 1984, p 3097
100. S. Guha, P.R. Munroe, and I. Baker, in *High-Temperature Ordered Intermetallic Alloys III,* Materials Research Society Symposia Proceedings, Vol 133, C.T. Liu, A.I. Taub, N.S. Stoloff, and C.C. Koch, Ed., Materials Research Society, 1989, p 633–638
101. R.D. Field, D.D. Krueger, and S.C. Huang, in *High-Temperature Ordered Intermetallic Alloys III,* Materials Research Society Symposia Proceedings, Vol 133, C.T. Liu, A.I. Taub, N.S. Stoloff, and C.C. Koch, Ed., Materials Research Society, 1989, p 567–572
102. A.R.C. Westwood, *Metall. Trans. A.,* Vol 19, 1988, p 749
103. B. Schmidt, P. Nagpal, and I. Baker, in *High-Temperature Ordered Intermetallic Alloys III,* Materials Research Society Symposia Proceedings, Vol 133, C.T. Liu, A.I. Taub, N.S. Stoloff, and C.C. Koch, Ed., Materials Research Society, 1989, p 755–760
104. J.L. Smialek, J. Doychak, and D.J. Gaydosh, "Oxidation Behavior of FeAl + Hf,Zr,B," NASA TM-101402, NASA Lewis Research Center, 1988
105. C.G. McKamey et al., "Evaluation of Mechanical and Metallurgical Properties of Fe_3Al-Based Aluminides," ORNL/TM-10125, Oak Ridge National Laboratory, Sept 1986
106. J.H. DeVan, in *Oxidation of High-Temperature Intermetallics,* T. Grobstain and J. Doythak, Ed., TMS, 1989
107. I. Baker and D.J. Gaydosh, *Mater. Sci. Eng.,* Vol 96, 1987, p 147
108. M.G. Mendiratta, S.K. Ehlers, and D.K. Chatterjee, in *Rapid Solidification Processing: Principles and Technologies,* National Bureau of Standards, 1983, p 420
109. J.A. Horton, C.T. Liu, and C.C. Koch, in *High-Temperature Alloys: Theory and Design,* J.O. Stiegler, Ed., American Institute of Mining, Metallurgical and Petroleum Engineers, 1984, p 309–321
110. D.J. Gaydosh, S.L. Draper, and M.V. Nathal, *Metall. Trans. A,* Vol 20, 1989, p 1701
111. C.G. McKamey, J.A. Horton, and C.T. Liu, in *High-Temperature Ordered Intermetallic Alloys II,* Materials Research Society Symposia Proceedings, Vol 81, N.S. Stoloff, C.C. Koch, C.T. Liu, and O. Izumi, Ed., Materials Research Society, 1987, p 321–327
112. W.R. Kerr, *Metall. Trans. A,* Vol 17, 1986, p 2298
113. H. Inouye, in *High-Temperature Ordered Intermetallic Alloys,* Materials Research Society Symposia Proceedings, Vol 39, C.C. Koch, C.T. Liu, and N.S. Stoloff, Ed., Materials Research Society, 1985, p 255–261
114. G. Sainfort, P. Mouturat, P. Pepin, J. Petit, G. Cabane, and M. Salesse, *Mem. Étud. Sci. Rev. Métall.,* Vol 60, 1963, p 125
115. P. Morgnand, P. Mouturat, and G. Sainfort, *Acta Metall.,* Vol 16, 1968, p 807
116. S. Strothers and K. Vendula, in *Proceedings of the Powder Metallurgy Conference,* Vol 43, Metal Powder Industries Federation, 1987, p 597
117. C.T. Liu, V.K. Sikka, and C.G. McKamey, "Alloy Development of FeAl Aluminide Alloys for Structural Use in Corrosive Environments," ORNL Report, Martin Marietta Energy Systems, unpublished research
118. M.G. Mendiratta, H.K. Kim, and H.A. Lips, *Metall. Trans. A,* Vol 15, 1984, p 395
119. Y. Umakoshi and M. Yamaguchi, *Philos. Mag. A,* Vol 44, 1981, p 711
120. T. Takasugi and O. Izumi, *Acta Metall.,* Vol 34, 1986, p 607
121. N. Masahashi, T. Takasugi, and O. Izumi, *Metall Trans. A,* Vol 19, 1988, p 353
122. O. Izumi and T. Takasugi, *J. Mater. Res.,* Vol 3, 1988, p 426
123. T. Takasugi, N. Masahashi, and O. Izumi, *Scr. Metall.,* Vol 20, 1986, p 1317
124. N. Masahashi, T. Takasugi, and O. Izumi, *Acta Metall.,* Vol 36, 1988, p 1823–1836
125. T. Takasugi and O. Izumi, *Scr. Metall.,* Vol 19, 1985, p 903–907
126. A.K. Kuruvilla, S. Ashok, and N.S. Stoloff, in *Proc. Third International Congress on Hydrogen in Metals,* Vol 2, Pergamon Press, 1982, p 629
127. A.K. Kuruvilla and N.S. Stoloff, *Scr. Metall.,* Vol 19, 1985, p 83
128. G.M. Camus, N.S. Stoloff, and D.J. Duquette, *Acta Metall.,* Vol 37, 1989, p 1497–1501
129. R.G. Bordeau, "Development of Iron Aluminides," AFWAL-TR-87-4009, United Technologies Corp., Pratt and Whitney, 1987
130. M.G. Mendiratta, Tai-Il Mah, and S.K. Ehlers, "Mechanisms of Ductility and Fracture in Complex High-Temperature Materials," AFWAL-TR-85-4061, Materials Laboratory, U.S. Air Force Wright Aeronautic Laboratories, Airforce Systems Command, July 1985
131. J.O. Stiegler and C.T. Liu, in *Advances in Materials Science and Engineering,* R.W. Cahn, Ed., Pergamon Press, 1988, p 3–9
132. R.S. Diehm and D.E. Mikkola, in *High-Temperature and Ordered Intermediate Alloys II,* Materials Research Society Symposia Proceedings, Vol 81, N.S. Stoloff, C.C. Koch, C.T. Liu, and O. Izumi, Ed., Materials Research Society, 1987, p 329–334
133. C.G. McKamey and J.A. Horton, *Metall. Trans. A,* Vol 20, 1989, p 751–757
134. D.M. Dimiduk, M.G. Mendiratta, D. Banerjee, and H.A. Lipsitt, *Acta Metall.,* Vol 36, 1988, p 2947–2958
135. C.G. McKamey, J.A. Horton, and C.T. Liu, *Scr. Metall.,* Vol 22, 1988, p 1679
136. C.G. McKamey, J.A. Horton, and C.T. Liu, *J. Mater. Res.,* Vol 4, 1989, p 1156–1163
137. M.G. Mendiratta, S.K. Ehlers, D.M. Dimiduk, W.R. Kerr, S. Mazdiyasni, and H.R. Lipsitt, in *High-Temperature Ordered Intermetallic Alloys II,* Materials Research Society Symposia Proceedings, Vol 81, N.S. Stoloff, C.C. Koch, C.T. Liu, and O. Izumi, Ed., Materials Research Society, 1987, p 393–404
138. C.G. McKamey et al., "Development of Iron Aluminides for Gasification Systems," ORNL-TM-10793, Oak Ridge National Laboratory, July 1988
139. M.A. Crimp and K. Vedula, *J. Mater. Sci.,* Vol 78, 1986, p 193
140. K. Vedula and J.R. Stephens, in *High-Temperature Ordered Intermetallic Alloys II,* Materials Research Society Symposia Proceedings, Vol 81, N.S. Stoloff, C.C. Koch, C.T. Liu, and O. Izumi, Ed., Materials Research Society, 1987, p 381–391
141. P.J. Maziasz, C.T. Liu, and G.M. Goodwin, Overview of the Development of FeAl Intermetallic Alloys, *Heat Resistant Materials II,* ASM International, 1995, p 555–566
142. S.A. David et al., *Weld. J.,* Vol 68, 1989, p 372s–381s
143. F.H. Froes, D. Eylon, and H.B. Bomberger, Ed., *Titanium Technology: Present Status and Future Trends,* Titanium Development Association, 1985
144. P.J. Bania, An Advanced Alloy for Elevated Temperatures, *J. Met.,* Vol 40 (No. 3), 1988, p 20–22
145. H.H. Lipsitt, in *High-Temperature Ordered Intermetallic Alloys,* Materials Research Society Symposia Proceedings, Vol 39, C.C. Koch, C.T. Liu, and N.S. Stoloff, Ed., Materials Research Society, 1985, p 351–364
146. M.J. Blackburn and M.P. Smith, "Research to Conduct an Exploratory Experimental and Analytical Investigation of Alloys," Technical Report AFWAL-TR-80-4175, U.S. Air Force Wright Aeronautical Laboratories, 1980

147. M.J. Blackburn and M.P. Smith, "R&D on Composition and Processing of Titanium Aluminide Alloys for Turbine Engine," Technical Report AF-WAL-TR-82-4086, U.S. Air Force Wright Aeronautical Laboratories, 1982
148. F.H. Froes, *Mater. Edge,* No. 5, May 1988
149. Eli F. Bradley, "The Potential Structural Use of Aluminides in Jet Engines," paper presented at the Gorham Advanced Materials Institute Conference on Investment, Licensing and Strategic Partnering Opportunities, Emerging Technology, Applications, and Markets for Aluminides, Iron, Nickel and Titanium (Monterey, CA), Nov 1990
150. R.E. Schafrik, Dynamic Elastic Moduli of the Titanium Aluminides, *Metall. Trans. A,* Vol 8, 1977, p 1003–1006
151. H.A. Lipsitt, in *Advanced High Temperature Alloys: Processing and Properties,* S.S. Allen, R.M. Pellous, and R. Widmer, Ed., American Society for Metals, 1986
152. N.S. Choudhury, H.C. Graham, and J.W. Hinze, in *Properties of High Temperature Alloys With Emphasis on Environmental Effects,* Electrochemical Society, 1976, p 668–680
153. M. Khobaib and F.W. Vahldiek, in *Space Age Metals Technology,* Vol 2, F.H. Froes and R.A. Cull, Ed., Society for the Advancement of Material and Process Engineering, 1988, p 262–270
154. J. Subrahmanyam, Cyclic Oxidation of Aluminated Ti-14Al-24Nb Alloy, *J. Mater. Sci.,* Vol 23, 1988, p 1906–1910
155. W.J.S. Yang, Observations of Superdislocation Networks in Ti_3Al-Nb, *J. Mater. Sci. Lett.,* Vol 1, 1982, p 199–202
156. W.J.S. Yang, "C" Component Dislocations in Deformed Ti_3Al, *Metall. Trans. A,* Vol 13, 1982, p 324
157. M.J. Blackburn, D.L. Ruckle, and C.E. Bevau, "Research to Conduct an Exploratory Experimental and Analytical Investigation of Alloys," Technical Report AFML-TR-78-18, U.S. Air Force Materials Laboratory, 1978
158. M.J. Blackburn and M.P. Smith, "Research to Conduct an Exploratory Experimental and Analytical Investigation of Alloys," Technical Report AFML-TR-81-4046, U.S. Air Force Wright Aeronautical Laboratories, 1981
159. M.J. Blackburn and M.P. Smith, "Development of Improved Toughness Alloys Based on Titanium Aluminides," Interim Technical Report FR-19139, United Technologies, 1988
160. F.H. Froes and D. Eylon, *Hydrogen Effects on Materials Behavior,* A.W. Thompson and N.R. Moody, Ed., TMS, 1990
161. F.H. Froes, D. Eylon, and C. Suryanarayana, *J. Met.,* March 1990
162. W.H. Kao et al., in *Progress in Powder Metallurgy,* Vol 37, Metal Powder Industries Federation, 1982, p 289–301
163. C.H. Ward et al., in *Sixth World Conference on Titanium,* Part II, P. Lacombe, R. Tricot, and G. Beranger, Ed., Les Editions de Physique, 1989, p 1009–1014
164. C.H. Ward et al., in *Sixth World Conference on Titanium,* Part II, P. Lacombe, R. Tricot, and G. Beranger, Ed., Les Editions de Physique, 1989, p 1103–1108
165. R.W. Hertzberg, *Deformation and Fracture Mechanics of Engineering Materials,* 2nd ed., John Wiley and Sons, 1983
166. M.A. Stucke and H.A. Lipsitt, in *Titanium Rapid Solidification Technology,* F.H. Froes and D. Eylon, Ed., TMS, 1986, p 255–262
167. W. Cho, "Effect of Microstructure on Deformation and Creep Behavior of Ti-25Al-10Nb-3V-1Mo," Technical Report, U.S. Air Force Office of Scientific Research, Oct 1988
168. C.G. Rhodes, in *Sixth World Conference on Titanium,* Part I, P. Lacombe, R. Tricot, and G. Beranger, Ed., Les Editions de Physique, 1989, p 119–204
169. M.G. Mendiratta and H.A. Lipsitt, Steady-State Creep Behavior of Ti_3Al-Base Intermetallics, *J. Mater. Sci.,* Vol 15, 1980, p 2985–2990
170. B. Mossier et al., *CSR. Metall. Mater.,* Vol 24, 1990, p 2363–2368
171. A.P. Woodfield and B.H. Lawless, Orthorhombic Titanium Alloy Microstructure/Property Relationships, *Metallic Materials for Lightweight Applications,* M.G.H. Wells, E.B. Kula, and J.H. Beatty, Ed., Proceedings of the Sagamore Army Materials Research Conference
172. R.G. Rowe et al., *Mat. Res. Soc. Symp. Proc.,* Vol 213, 1991, p 703–708
173. R.G. Rowe, The Mechanical Properties of Titanium Aluminides near Ti-25Al-25Nb, *Microstructure/Property Relationships in Titanium Alloys and Titanium Aluminides,* Y. Kim and R.R. Boyer, Ed., TMS/AIME, 1990, p 387–398
174. R.G. Rowe, Recent Developments in Ti-Al-Nb Alloys, *High Temperature Aluminides and Intermetallics,* S.H. Whang, D.P. Pope, and J.O. Stigler, Ed., TMS/AIME, 1990, p 375–401
175. D. Banerjee, A New Ordered Orthorhombic Phase in a Ti_3Al-Nb Alloy, *Acta Metall.,* Vol 36, 1989, p 871–882
176. M.J. Kaufman, T.J. Broderick, C.H. Ward, and R.G. Rowe, Phase Relationships in the Ti_3Al + Nb System, *Sixth World Conference on Titanium,* P. Lacombe, R. Tricot, and G. Beranger, Ed., Les Editions de Physique, 1989, p 985–990
177. B. Moser, L.A. Bendersky, W.J. Boettinger, and R.G. Rowe, Neutron Powder Diffraction Study of the Orthorhombic Ti_2AlNb Phase, *Scr. Metall. Mater.,* Vol 24, 1990, p 2363–2368
178. H.R. Ogden et al., Constitution of Titanium-Aluminum Alloys, *Trans. AIME,* Vol 191, 1951, p 1150–1155
179. D. Clark, K.S. Kepson, and G.I. Lewis, A Study of the Titanium-Aluminum System up to 40 at.% Aluminum, *J. Inst. Met.,* Vol 91, 1962–1963, p 197
180. V.-W. Kim, Intermetallic Alloys Based on Gamma Titanium Aluminide, *J. Met.,* Vol 41 (No. 7), 1989, p 24–30
181. E.S. Bumps, H.D. Kessler, and M. Hansen, Titanium-Aluminum System, *Trans. AIME,* Vol 194, 1952, p 609–614
182. P. Duwez and J.L. Taylor, Crystal Structure of TiAl, *J. Met.,* 1952, p 70
183. S.C. Huang, E.L. Hall, and M.F.X. Gigliotti, in *High-Temperature Ordered Intermetallic Alloys II,* Materials Research Society Symposia Proceedings, Vol 81, N.S. Stoloff, C.C. Koch, C.T. Liu, and O. Izumi, Ed., Materials Research Society, 1987, p 481–486
184. R.P. Elliott and W. Rostoker, The Influence of Aluminum on the Occupation of Lattice Sites in the TiAl Phase, *Acta. Metall.,* Vol 2, 1954, p 884–885
185. G. Hug, A. Loiseau, and A. Lasalmonie, Nature and Dissociation of the Dislocations in TiAl Deformed at Room Temperature, *Philos. Mag. A,* Vol 54 (No. 1), 1986, p 47–65
186. T. Kawabata and O. Izumi, Dislocation Structures in TiAl Single Crystals Deformed at 77K, *Scr. Metall.,* Vol 21, 1987, p 433–434
187. G. Hug, A. Loiseau, and P. Veyssiere, Weak-Beam Observation of a Dissociation Transition in TiAl, *Philos. Mag. A,* Vol 57 (No. 3), 1988, p 499–523
188. D.W. Pashley, J.L. Robertson, and M.J. Stowell, The Deformation of Cu Au I, *Philos. Mag. A,* 8th series, Vol 19, 1969, p 83
189. D. Schechtman, M.J. Blackburn, and H.A. Lipsitt, The Plastic Deformation of TiAl, *Metall. Trans.,* Vol 5, 1974, p 1373
190. H.A. Lipsitt, D. Schechtman, and R.E. Schafrik, The Deformation and Fracture of TiAl at Elevated Temperatures, *Metall. Trans. A,* Vol 6, 1975, p 1991
191. E.L. Hall and S.-C. Huang, in *High-Temperature Ordered Intermetallic Alloys III,* Materials Research Society Symposia Proceedings, Vol 133, C.T. Liu, A.I. Taub, N.S. Stoloff, and C.C. Koch, Ed., Materials Research Society, 1989, p 693–698
192. T. Kawabata and O. Izumi, Dislocation Reactions and Fracture Mechanisms in TiAl Ll_0 Type Intermetallic Compound, *Scr. Metall.,* Vol 21, 1987, p 435–440
193. T. Kawabata et al., Bend Tests and Fracture Mechanisms of TiAl Single Crystals at 293-1083 K, *Acta Metall.,* Vol 36 (No. 4), 1988, p 963–975
194. S.M.L. Sastry and H.A. Lipsitt, Fatigue Deformation of TiAl Base Alloys, *Metall. Trans. A,* Vol 8, 1977, p 299
195. M.J. Blackburn and M.P. Smith, Titanium Alloys of the TiAl Type, U.S. Patent 4,294,615, 1981
196. M.J. Blackburn, J.T. Hill, and M.P. Smith, "R&D on Composition and Processing of Titanium Aluminide Alloys for Turbine Engines," Technical Report AFWAL-TR-84-4078, U.S. Air Force Wright Aeronautical Laboratories, 1984
197. S.-C. Huang and E.L. Hall, in *High-Temperature Ordered Intermetallic Alloys III,* Materials Research Society Symposia Proceedings, Vol 133, C.T. Liu, A.I. Taub, N.S. Stoloff, and C.C.

Koch, Ed., Materials Research Society, 1989, p 373–383
198. T. Tsujimoto and K. Hashimoto, in *High-Temperature Ordered Intermetallic Alloys III,* Materials Research Society Symposia Proceedings, Vol 133, C.T. Liu, A.I. Taub, N.S. Stoloff, and C.C. Koch, Ed., Materials Research Society, 1989, p 391–396
199. T. Kawabata, T. Tamura, and O. Izumi, in *High-Temperature Ordered Intermetallic Alloys III,* Materials Research Society Symposia Proceedings, Vol 133, C.T. Liu, A.I. Taub, N.S. Stoloff, and C.C. Koch, Ed., Materials Research Society, 1989, p 329–334
200. D.J. Maykuth, "Effects of Alloying Elements in Titanium," DMIC Report 136B, Battelle Memorial Institute, May 1961
201. I.A. Zelonkov and Y.N. Martynchik, Oxidation Resistance of Alloys of Compound TiAl with Niobium at 800 and 1000C, *Metallofiz., Nauk, Dumka,* Vol 42, 1972, p 63–66
202. Y-W. Kim, Ordered Intermetallic Alloys, Part III: Gamma Titanium Aluminides, *JOM,* July, 1994, p 30–39
203. C.R. Feng, D.J. Michel, and C.R. Crowe, in *High-Temperature Ordered Intermetallic Alloys III,* Materials Research Society Symposia Proceedings, Vol 133, C.T. Liu, A.I. Taub, N.S. Stoloff, and C.C. Koch, Ed., Materials Research Society, 1989, p 669–674
204. H.R. Ogden et al., Mechanical Properties of High Purity Ti-Al Alloys, *J. Met.,* Feb 1952
205. M.J. Blackburn and M.P. Smith, "The Understanding and Exploitation of Alloys Based on the Compound TiAl (Gamma Phase)," Technical Report AFML-TR-79-4056, U.S. Air Force Materials Laboratory, 1979
206. T. Tsujimoto et al., Structures and Properties of an Intermetallic Compound TiAl Based on Alloys Containing Silver, *Trans. Jpn. Inst. Met.,* Vol 27 (No. 5), 1986, p 341–350
207. S.-C. Huang, E.L. Hall, and M.F.X. Gigliotti, in *Sixth World Conference on Titanium,* Part II, P. Lacombe, R. Tricot, and G. Beranger, Ed., Les Editions de Physique, 1989, p 1109–1144
208. S.M. Barinov et al., Temperature Dependence of Strength and Ductility of the Decomposition of Titanium Aluminide, *Izv. Akad. Nauk SSSR,* Vol 5, 1983, p 170–174
209. K.S. Kumar, J.A.S. Green, D.E. Larsen, Jr., and L.D. Kramer, XD Titanium Aluminide Composites, *Adv. Mater. Proc.,* Vol 147 (No. 4), 1995, p 35–38
210. D.A. Wheeler, D.E. Larsen, Jr., and B. London, Elevated Temperature Behavior of Investment Cast Gamma and XD Gamma Titanium Aluminides, paper presented at the TMS Fall Meeting, Chicago, Nov 1992
211. M.P. Brady, W.J. Brindley, J.L. Smialek, and I.E. Locci, The Oxidation and Protection of Gamma Titanium Aluminides, *JOM,* Nov 1996, p 46–50
212. T. Takasugi, E.P. George, D.P. Pope, and O. Izumi, *Scr. Metall.,* Vol 19, 1985, p 551–556
213. W.C. Oliver and C.L. White, in *High-Temperature Ordered Intermetallic Alloys II,* Materials Research Society Symposia Proceedings, Vol 81, N.S. Stoloff, C.C. Koch, C.T. Liu, and O. Izumi, Ed., Materials Research Society, 1987, p 241–246
214. I. Baker, R.A. Padgett, and E.M. Schulson, *Scr. Metall.,* Vol 23, 1989, p 1969–1974
215. W.C. Oliver, in *High-Temperature Ordered Intermetallic Alloys III,* Materials Research Society Symposia Proceedings, Vol 133, C.T. Liu, A.I. Taub, N.S. Stoloff, and C.C. Koch, Ed., Materials Research Society, 1989, p 397–402
216. K.J. Williams, *J. Inst. Met.,* Vol 97, 1969, p 112
217. K.J. Williams, *J. Inst. Met.,* Vol 99, 1971, p 310
218. *Kanthal Super Handbook,* Kanthal Furnace Products, 1986
219. J. Schlichting, *High Temp.—High Press.,* Vol 10, 1978, p 241–269
220. Y. Ymakoshi, T. Hirano, T. Sakagami, and T. Yamane, *Scr. Metall.,* Vol 23, 1989, p 87–90
221. P. Meschter and D.S. Schwartz, *J. Met.,* Vol 41, 1989, p 52–55
222. J.-M. Yang, W. Kai, and S.M. Jeng, *Scr. Metall.,* Vol 23, 1989, p 1953–1958

Structural Ceramics

CERAMIC MATERIALS have long been used in a variety of applications that utilize their wear resistance, refractoriness, hardness, and high compression strength. Traditionally, they have not been used in tensile-loaded structures because they are brittle; that is, they experience catastrophic failure before permanent deformation. As Fig. 1 shows, traditional ceramics (pottery, floor tile, brick, etc.) exhibit fracture toughnesses that range from about only 0.7 to 3 $MPa\sqrt{m}$ (0.6 to 2.7 $ksi\sqrt{in.}$). These values fall far below those of polymers, composites (reinforced polymers), and metals used for engineering applications. However, recent advances in ceramic processing and microstructural control have resulted in a new family of advanced ceramics that are being used increasingly often for structural or load-bearing applications. These uses require materials that have high strength at room temperature and/or retain high strength at elevated temperatures, resist deformation (slow crack growth or creep), are damage tolerant, and resist corrosion and oxidation. Ceramics appropriate to these applications offer a significant weight savings over currently used metals (e.g., their density is about one-half that of steel). These applications include heat exchangers, automotive engine components such as turbocharger rotors and rotors and roller cam followers, power generation components, cutting tools, biomedical implants, and processing equipment used for fabricating a variety of polymer, metal, and ceramic parts. Table 1 lists some of the performance benefits that are attainable in typical applications.

The majority of high-performance structural ceramics under development for elevated-temperature applications are based on silicon nitride (Si_3N_4), silicon carbide (SiC), zirconia (ZrO_2), or alumina (Al_2O_3). Accordingly, the emphasis in this article is placed on these materials. The use of ceramic coatings to extend the high-temperature service lives of metallic materials or carbon-carbon composites components is discussed in the articles "Protective Coatings for Superalloys," "Titanium and Titanium Alloys," "Refractory Metals and Alloys," and "Carbon-Carbon Composites" in this Volume.

Silicon Nitride Ceramics

Silicon nitride has attracted much attention because of its unique combination of excellent high-temperature mechanical properties and resistance to oxidation and thermal shock. The intense interest in Si_3N_4 ceramics stems from the automotive industry, where use of ceramic components in engines would greatly improve operating efficiency. The automotive components of interest are turbocharger rotors, pistons, piston liners, and valves.

Processing. Silicon nitride may be produced by various methods to give a range of products, each one named after the fabrication route employed. The major types are:

- Reaction-bonded silicon nitride (RBSN)
- Hot-pressed silicon nitride (HPSN)
- Sintered (pressureless) silicon nitride (SSN)
- Sintered reaction-bonded silicon nitride (SRBSN)
- Hot isostatically pressed silicon nitride (HIPSN)

Dense Si_3N_4, in various forms, may be produced by sintering (with or without applied pressure) preformed Si_3N_4 with oxide additives that form liquid phases that allow more rapid diffusion of material and the attainment of near-theoretical density. This may be achieved by hot pressing, by pressureless sintering of both Si_3N_4 powder compacts and reaction-bonded Si_3N_4, or by hot isostatic pressing of either powder compacts or all of the previously sintered types. Dense Si_3N_4 products are used for elevated-temperature applications.

During firing, the following processes take place. Each particle of the original α-Si_3N_4 powder has around it a surface layer of silicon dioxide (SiO_2). An additive, such as magnesium oxide (MgO) or yttrium oxide (Y_2O_3), reacts with this silica and some of the nitride to form a liquid at the firing temperature, allowing sintering to take place in three stages:

1. *Rearrangement,* induced by the liquid phase
2. *Solution* of material from α into the liquid, followed by *diffusion* of silicon and nitrogen and then, when conditions permit, *precipitation* of the more thermodynamically stable β-silicon
3. *Coalescence,* where the liquid acts to eliminate final closed porosity and to produce a more rounded grain morphology

After cooling, the liquid either forms a glass, as with MgO additive, or crystallizes to form various oxynitride phases, as with Y_2O_3. The physical processes are shown schematically in Fig. 2. With little or no β present in the original powder, a certain supersaturation must be reached before nucleation occurs, and then growth of the fiber-like β grains takes place. The resulting morphology is rod shaped, and the resulting microstructure after cooling consists of interlocking β-Si_3N_4 grains with a residual secondary phase at the grain boundaries. More detailed information on the processing and forming of Si_3N_4 can be found in Ref 3.

Sialon is a generic term for a family of compositions produced by reacting Si_3N_4 with Al_2O_3 and aluminum nitride at high temperatures. In other words, a Sialon is a Si_3N_4-based ceramic in which some of the silicon has been replaced with aluminum and some of the nitrogen with oxygen. This results in a substituted solid solution, called a β'-Sialon, that has a compositional range of $Si_{6-x}Al_xO_xN_{8-x}$, based on the β-Si_6N_8 unit cell.

For a Sialon to be fabricated into a dense body, sintering aids must also be added. Rapid cooling from processing temperature produces a microstructure of β'-Sialon grains with a glassy intergranular phase. The properties of a Sialon depend

Table 1 Performance benefits attainable by using structural ceramics

Application	Benefit	Materials
Light-duty diesel (uncooled)	10–15% reduction in specific fuel consumption	ZrO_2, Si_3N_4, SiC, Al_2O_3, alumina titanate
Heavy-duty diesel (adiabatic)	22% reduction in specific fuel consumption	ZrO_2, Si_3N_4, SiC, Al_2O_3, alumina titanate
Light-duty automotive gas turbine	27% reduction in specific fuel consumption	Si_3N_4, SiC, lithium-aluminum-silicate, magnesium-aluminum-silicate
Recuperator for slot forging furnace	27% reduction in fuel consumption	SiC
Machining of grey cast iron	220% increase in productivity	Si_3N_4 (and Sialon)
Extrusion dies for brass	>220% increase in productivity	ZrO_2

Source: Ref 2

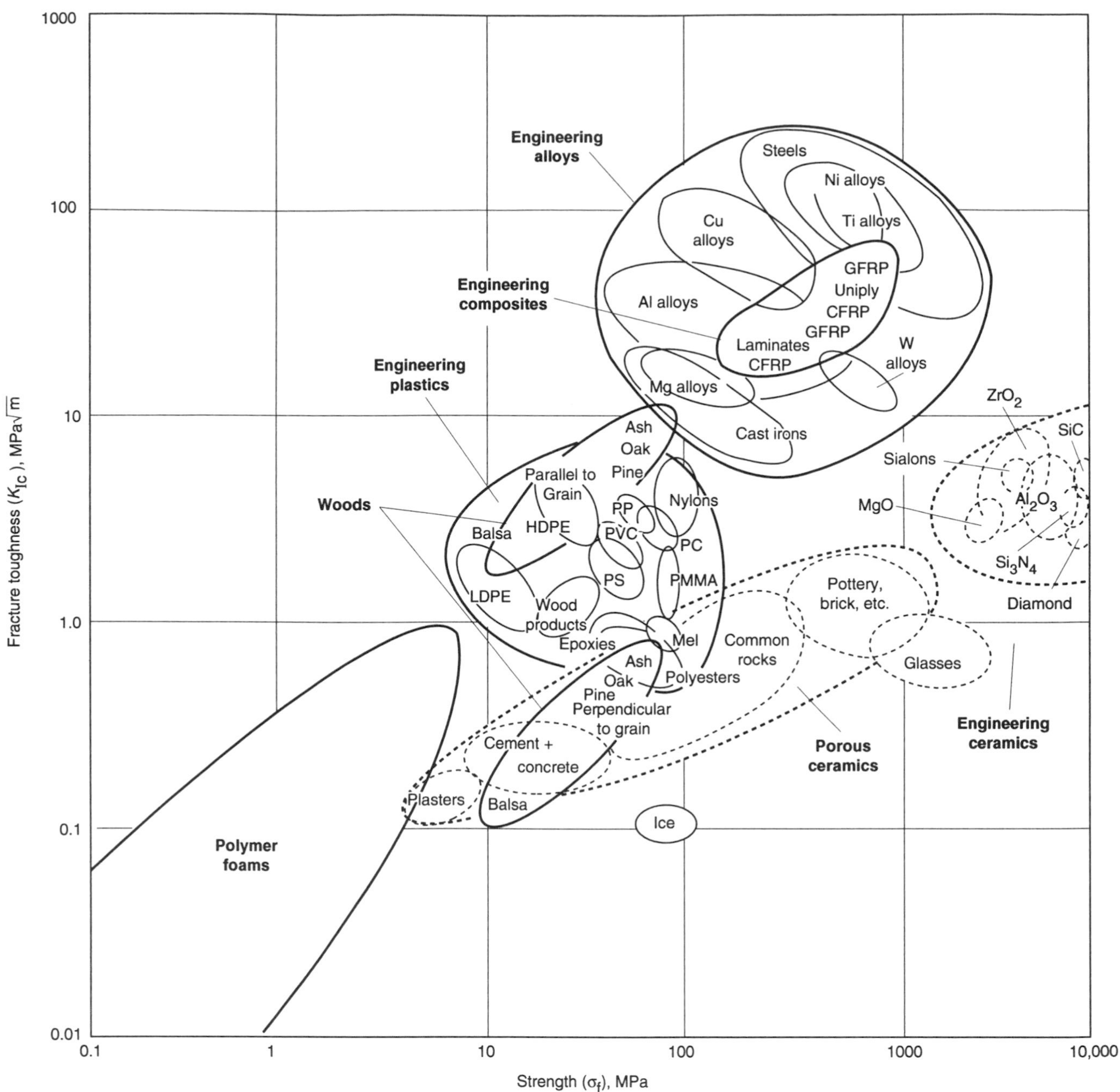

Fig. 1 Fracture toughness vs. strength for various engineered materials. The broken (dashed) property envelope lines for porous (traditional) ceramics and engineering (structural) ceramics indicate that the strength values are compressive, not tensile. Note that the fracture toughness values of recently developed structural ceramics match or exceed those of some metals. Source: Ref 1

Table 2 Thermal properties of Si_3N_4 ceramics

Property	Material type					
	RBSN	HPSN	SSN	SRBSN	HIP-SN	Sialon
Relative density, % theoretical	70–88	99–100	95–99	93–99	99–100	97–99
Coefficient of thermal expansion (α) from 25–1000 °C (77–1830 °F), 10^{-6}/°C	3.0	3.2–3.3	2.8–3.5	3.0–3.5	3.0–3.5	3.0–3.7
Thermal conductivity (λ) at:						
25 °C (77 °F), W/m · °C (Btu/ft · h · °F)	7–14 (4–8)	30–43 (17–25)	15–31 (8–18)	...	22 (12.27)	15–22 (8–13)
1000 °C (1830 °F), W/m · °C (Btu/ft · h · °F)	1.4–3 (0.8–1.7)	5–10 (3–6)	4–5 (2.3–3)	...	...	2.5 (1.4)
Specific heat (C_p) at:						
25 °C (77 °F), J/kg · °C (Btu/lb · °F)	700–1100 (0.17–0.26)	680–800 (0.16–0.19)	...	...	...	...
1000 °C (1830 °F), J/kg · °C (Btu/lb · °F)	1250 (0.30)	1200 (0.29)	...	...	...	...

Source: Ref 3

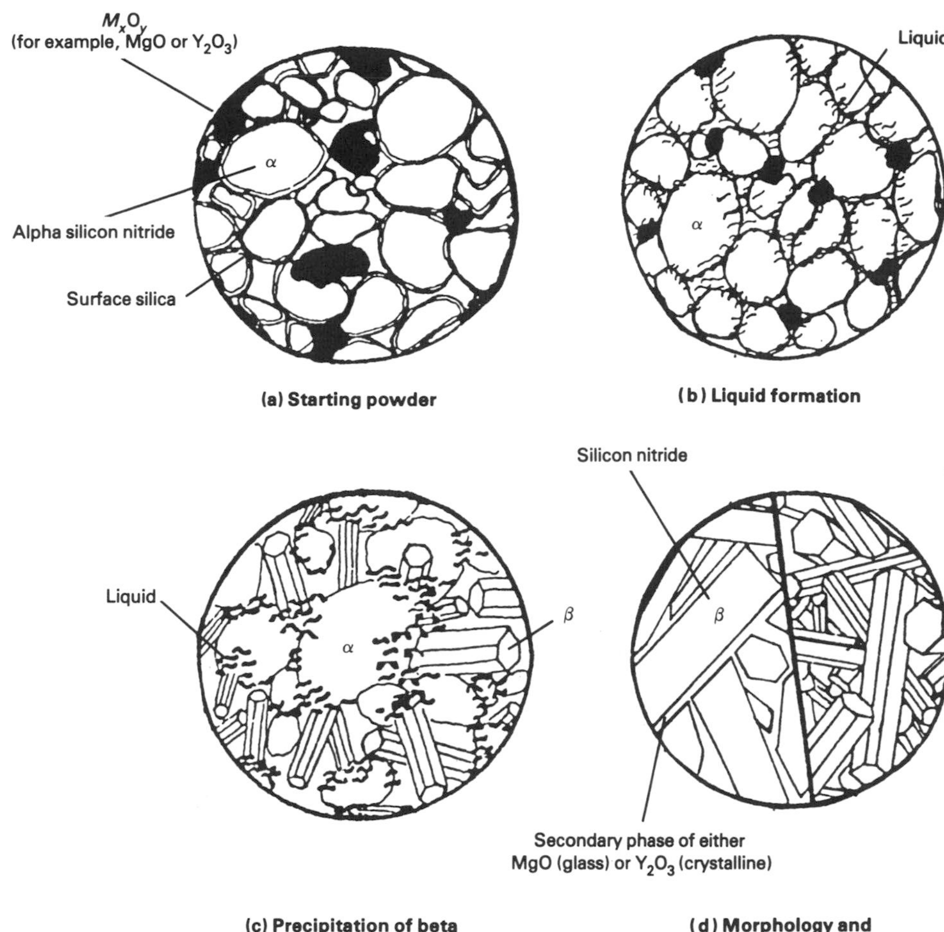

Fig. 2 Physical processes occurring during sintering of Si_3N_4 ceramics. Source: Ref 3

on the type and amount of sintering aid employed and the processing route followed during part fabrication.

The advantages of Sialons are their low coefficient of thermal expansion and good oxidation resistance. The potential applications are similar to those for Si_3N_4, namely automotive components, wear parts, and machine tool inserts.

A solid solution based on the α-Si_3N_4 structure has received less attention, but α′-Sialon shows promise commercially because it has greater hardness than β′-Sialon. In addition, it offers an improved product with less grain-boundary glass, as a result of the incorporation of the cation from the sintering additive (e.g., Y_2O_3) into a solid solution of $M_x(Si,Al)_{12}(O,N)_{16}$ based on the α-$Si_{12}Ni_{16}$ unit cell. Single-phase α-Sialons are currently available only in hot-pressed or hot isostatically pressed forms, but composites of β′ + α′ are produced by pressureless sintering.

Property/Processing/Structure Relationships. Key representative properties of Si_3N_4 in its various forms (including Sialons) are given in Tables 2 and 3. Wide ranges of property values are observed that depend not only on the processing route but also on the microstructural characteristics of the particular material type. In addition to the normal dependence on porosity levels, the properties are also crucially influenced by the amount and morphology of the β-form of Si_3N_4 and the amount, characteristics, and distribution of the grain-boundary phase.

The properties of Si_3N_4 ceramics must always be considered in relation to the processing route, the densification additives, and the resulting microstructure. The most desirable Si_3N_4 ceramics, in terms of strength and fracture toughness, consist of interlocking β grains with high aspect ratios, very little or no glass phase, and crystalline grain-boundary phases that are not oxidized at high temperatures. As shown schematically in Fig. 3, as a crack moves through Si_3N_4, it encounters a combination of fiberlike β and secondary grain-boundary material that tends to steer the crack along a tortuous path. Thus, more energy is required for propagation, resulting in an increase in toughness. A decreasing β-grain aspect ratio results in lower strength and lower fracture toughness.

Thus, there is an effective "microstructural toughening" in sintered Si_3N_4, and the concept of *microstructural engineering* has been applied to tailor the properties of Si_3N_4 ceramics to end applications. This involves manipulation chemistry of the Si_3N_4/additive system and heat treatment, both during and after sintering, to give the most beneficial secondary phases. Because regions of uncrystallized glass residues after sintering allow easy creep cavitation and increased creep strain, it is important to minimize their presence. Materials that are hot pressed or hot isostatically pressed require much lower volumes of liquid for densification, and in some cases they have been produced almost glass-free, but these processes are expensive.

The variation of strength with temperature for the different types of Si_3N_4 ceramics is shown in Fig. 4. Because of the presence of porosity, reaction-bonded Si_3N_4 has low flexural strength (150 to 350 MPa, or 22 to 50 ksi) at ambient temperature, but strength is retained up to very high temperatures (>1400 °C, or >2550 °F) because there is no glass phase.

Table 3 Mechanical properties of Si_3N_4 ceramics

Property	Material type							
	RBSN	HPSN	SSN	SRBSN	HIP-SN	HIP-RBSN	HIP-SSN	Sialon
Young's modulus (E), GPa (10^6 psi)	120–250 (17.4–36.2)	310–330 (44.9–47.8)	260–320 (37.7–46.4)	280–300 (40.6–43.5)	...	310–330 (44.9–47.8)	...	300 (43.5)
Poisson's ratio (ν)	0.20	0.27	0.25	0.23	0.23	0.27	...	0.23
Flexural strength (σ_f), MPa (ksi) at:								
25 °C (77 °F)	150–350 (21.7–50.7)	450–1000 (65.2–145)	600–1200 (86.9–173.8)	500–800 (72.5–115.9)	600–1200 (86.9–173.8)	500–800 (72.5–115.9)	600–1200 (86.9–173.8)	750–950 (108.7–137.7)
1350 °C (2460 °F)	140–340 (20.2–49.3)	250–450 (36.2–65.2)	340–550 (49.3–79.7)	350–450 (50.7–65.2)	350–550 (50.7–79.7)	250–450 (36.2–65.2)	300–520 (43.5–75.3)	300–550 (43.5–79.7)
Weibull modulus (m)	19–40	15–30	10–25	10–20	...	20–30	...	15
Fracture toughness (K_{Ic}), MPa$\sqrt{m}$ (ksi$\sqrt{in.}$)	1.5–2.8 (1.3–2.5)	4.2–7.0 (3.8–6.3)	5.0–8.5 (4.5–7.7)	5.0–5.5 (4.5–5.0)	4.2–7.0 (3.8–6.3)	2.0–5.8 (1.8–5.3)	4.0–8.0 (3.6–7.2)	6.0–8.0 (5.4–7.2)

Source: Ref 3

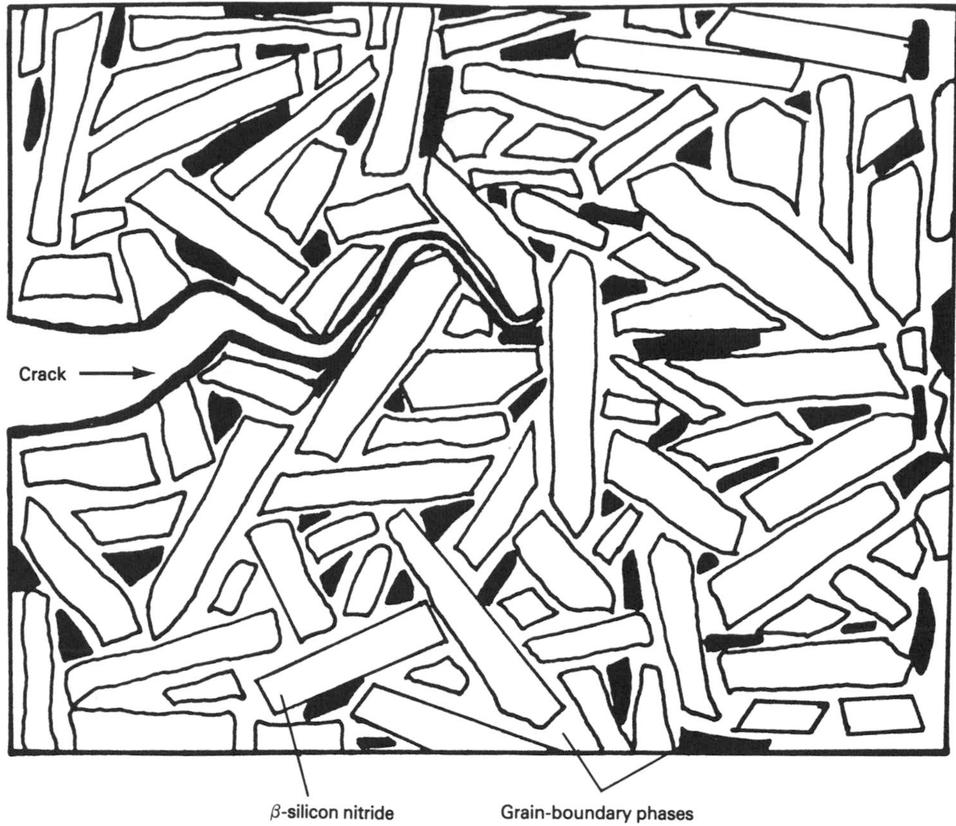

Fig. 3 Impeded crack path by microstructural toughening in dense Si_3N_4 ceramics with secondary grain-boundary phase. Source: Ref 3

With sintered Si_3N_4, much higher strengths are achieved (600 to 1200 MPa, or 87 to 174 ksi) at ambient temperature. At temperatures exceeding 1000 °C (1830 °F), however, the strength may decrease rapidly due to the softening of the intergranular glass. Strength is retained to much higher temperatures when the secondary phase is crystalline.

Sintered Sialons are of two types with different microstructural features: β′-Sialon grains plus glass and β′-Sialon grains plus crystalline yttrium-aluminum-garnet (YAG). The strength of Type 1 is similar to that of hot-pressed Si_3N_4 at ambient temperatures, but the strength decreases as the intergranular glass softens above 1000 °C (1830 °F). Type 2 Sialon has a lower ambient strength but retains a strength of 500 MPa (72.5 ksi) at temperatures of 1400 °C (2550 °F).

A characteristic property of Si_3N_4 ceramics containing intergranular glass phases is a transient rise in fracture toughness near the glass softening point. This is associated with the onset of subcritical crack growth (SCG) within a creep cavitation zone at the primary crack tip. The rise in fracture toughness is due to energy absorption by plastic deformation as a result of viscous flow of residual glass bridging crack surfaces. In type 2 Sialons containing the crystalline YAG phase, there is no evidence for a zone of SCG preceding rapid fracture, even at temperatures over 1450 °C (2640 °F) in an inert environment, yet this phenomenon results in degradation at much lower temperatures in type 1 Sialons and sintered Si_3N_4 ceramics containing glass.

The HPSN and HIP-SN grades, which have much lower levels of additives (hence less glass) and also zero porosity, generally have higher strengths than sintered Si_3N_4 or Sialons at higher temperatures. In applications such as metalcutting, it is important to retain strength and hardness to high temperatures. Sample hardness values are 750 HV (100 g load) for RBSN with density of 2.7 g/cm^3, 1600 to 1800 HV (100 g load) for HPSN materials, and 1350 to 1600 HV (100 g load) for sintered Si_3N_4. Hardness can be as high as 1800 HV in sintered Sialons containing both β′ and α′ phases, and the hardness differential due to the presence of α′ in these Sialon composites is retained up to high temperatures (Fig. 5). At 1000 °C (1830 °F), these materials are much harder than Al_2O_3 ceramics.

Stress-Rupture Behavior. Figure 6 shows stress-rupture behavior typical of hot-pressed, sintered, and reaction-bonded silicon nitrides, as well as a Sialon. The materials were all tested in air at 1200 °C (2190 °F). Although the reaction-bonded Si_3N_4 does not lose strength with time, its initial strength is significantly lower than the other silicon nitrides. Both the hot-pressed and sintered Si_3N_4 exhibit a large decrease in strength with time at temperature. The Syalon 201, a Sialon alloy with a crystallized grain boundary, has both very high strength and retention of strength for 1000 h under load at 1200 °C (2190 °F). Although the data shown in Fig. 6 indicate that most Si_3N_4 ceramics lose strength with time under load, all of these materials remain capable of carrying significant loads at 1200 °C (2190 °F) in air for times of at least 1000 h. No metallic alloy is capable of this. In fact, many alloys (aluminum, most copper, magnesium, and zinc) melt at temperatures below 1200 °C (2190 °F).

Creep behavior of Si_3N_4 ceramics is mainly controlled by grain-boundary sliding along the amorphous phase. In addition to the volume of glass, viscosity is an important consideration. Many sintered Si_3N_4 ceramics are densified with a combination of Y_2O_3 and Al_2O_3 additives. The β′-Sialons are also sintered with Y_2O_3. This system (Y-Si-Al-O-N) offers a combination of moderately low liquidus temperature with a comparatively high residual glass viscosity and glass transition temperature, particularly compared with glasses in the Mg-Si-O-N system. Thus, the Y-Si-Al-O-N system exhibits far superior high-temperature properties. Ceramics have been developed that have survived very long times to failure at temperatures well in excess of 1300 °C (2370 °F) and stresses above 400 MPa, or 58 ksi (Ref 5). This has been achieved by careful optimization of sintering and subsequent crystallization of the grain-boundary glass. This treatment suppresses creep cavitation and minimizes the creep rate at high applied stress levels.

Thermal shock resistance of Si_3N_4 ceramics varies widely and, as with the thermal and mechanical properties, depends on microstructural features in a complex way. Table 4 gives the critical temperature difference, ΔT_c, after water and oil quenching for the various types of Si_3N_4 ceramics. Unstable crack propagation was found to occur for all grades of Si_3N_4. With ΔT_c equated to the thermal shock resistance pa-

Table 4 Thermal shock resistance of Si_3N_4 ceramics

	Critical temperature change (ΔT_c) for:			
	Water quench		Oil quench	
Material type	Δ °C	Δ °F	Δ °C	Δ °F
RBSN	200–600	360–1080	750–1250	1350–2250
HPSN	400–800	720–1440	>1400	>2520
SSN	600–750	1080–1350	>1400	>2520
HIP-RBSN	800	1440	...	...
Sialon	300–550	540–990	...	...

Source: Ref 6

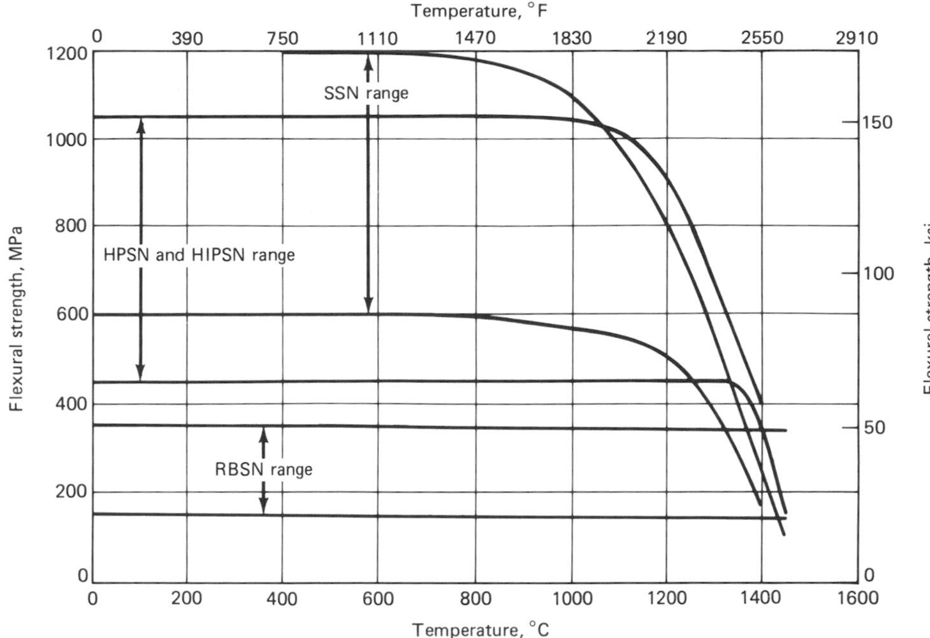

Fig. 4 Variations of flexural strength with temperature for various types of silicon nitride ceramics. Source: Ref 3

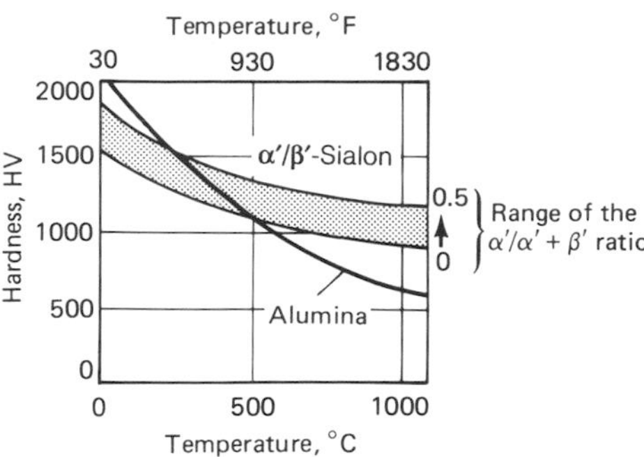

Fig. 5 Hot hardness behavior of β′/α′ Sialon composites. Source: Ref 4

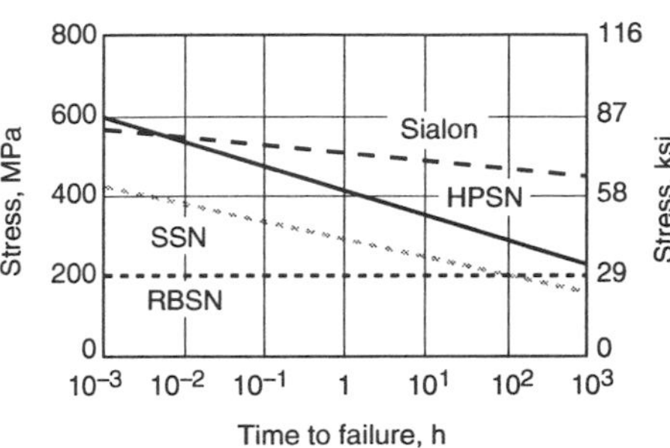

Fig. 6 Stress-rupture behavior of various Si_3N_4 ceramics. Source: Ref 2

rameter, *R*, the parameter for instantaneous temperature changes is:

$$R = \sigma_f(1 - \nu/\alpha E) \quad \text{(Eq 1)}$$

where σ_f is the fracture strength, ν is Poisson's ratio, α is the coefficient of thermal expansion, and *E* is Young's modulus.

For less severe quenching, a second parameter, $R' = R\lambda$, is defined. As the β content of Si_3N_4 ceramics increases, σ_f and the thermal conductivity, λ, also increase, and hence there is an increase in *R* and *R′*. Other effects on *R* and *R′* are as follows:

- An increase in grain size decreases σ_f but has little effect on λ or *E*; hence, *R* and *R′* also decrease with increasing grain size.
- An increase in either glass content or the extent of β′-Sialon solid solution formation has little effect on *R*. However, these increases result in a decrease in λ, so *R′* decreases.

Thermal Cycling Behavior. The retained fracture strength at ambient temperature, following thermal cycling from 1260 °C (2300 °F), is shown in Fig. 7 for various grades of RBSN and HPSN as a function of the number of cycles. Quenching was carried out at a maximum cooling rate of 80 °C/s (145 °F/s). Strength degradation occurs initially, and it is usually greater for RBSN than for HPSN grades. In spite of some loss of strength, Si_3N_4 retains a much higher strength after thermal cycling than other engineering ceramics (Ref 8, 9).

The interpretation of the thermal cycling behavior of the various types of Si_3N_4 is complex, because a superposition of various effects takes place. These effects include the induced thermal stress levels and plastic deformation (as well as oxidation) at high temperatures, which can result in crack healing and rounding of internal pores and flaws.

In all cases, the limiting temperature capability of Si_3N_4 ceramics appears to be around 1450 °C (2640 °F) in inert atmospheres. The added complication of using these ceramics in oxidizing atmospheres is a further limitation. Combined exposure with oxidation and high-temperature mechanical tests gives useful data on behavior (Ref 8, 9).

Oxidation of Si_3N_4 ceramics in air begins as low as 800 °C (1470 °F), and a thin protective layer of amorphous SiO_2 is formed on the surface of the Si_3N_4 according to:

$$Si_3N_4 + 3O_2 \rightarrow 3SiO_2 + 2N_2 \quad \text{(Eq 2)}$$

This simple reaction occurs in RBSN and dense Si_3N_4 at lower temperatures. However, oxidation of these materials at higher temperatures is much more complex. The rate of oxidation varies according to the amount and type of densification additive used. In the case of materials containing a grain-boundary glass phase, various investigations (Ref 10-12) have shown that the oxidation reaction is diffusion controlled, the rate being limited by outward diffusion of metallic impurity or dopant ions into the SiO_2 scale.

In the case of materials containing crystalline grain-boundary phases that are oxynitrides, the oxidation products may have substantially different specific volumes, giving rise to extensive cracking at the surface that exposes fresh surfaces to further attack. This is particularly a problem in hot-pressed Si_3N_4 doped with Y_2O_3 where the secondary phase is N-melilite (Ref 8); catastrophic oxidation in this case occurs at 900 to 1200 °C (1650 to 2200 °F). Knowledge of phase relationships and potential alternative grain-boundary phases in the various M-Si-O-N systems is thus important (Ref 13).

Si_3N_4-based composites employing SiC whisker or platelet reinforcement have been developed. The aim of this development effort was to produce a ceramic with improved toughness without the loss of high-temperature strength and other thermomechanical characteristics found in monolithic Si_3N_4.

During firing of the ceramic, the presence of the "inert" SiC whiskers inhibits shrinkage of the Si_3N_4 matrix, so pressureless sintering is more difficult than with monolithic types. To date, only hot pressing has been successful in producing fully dense Si_3N_4 composites.

On cooling from the sintering temperature, stresses arise as a result of thermal mismatch between whisker and matrix. The thermal expansion coefficient for SiC is 4.4×10^{-6}/K while that for Si_3N_4 is 3.2×10^{-6}/K. Thus, the whiskers will be in tension and the matrix will be in compres-

Table 5 Effect of whisker content on the properties of Si_3N_4-SiC_w composites

SiC whisker content, vol%	Toughness		Flexural strength	
	MPa$\sqrt{m}$	ksi$\sqrt{in.}$	MPa	ksi
0	5–7	4.6–6.4	400–650	60–95
10	6.5–9.5	5.9–8.6	400–500	60–75
30	7.5–10	6.8–9.1	350–450	50–65

Source: Ref 15

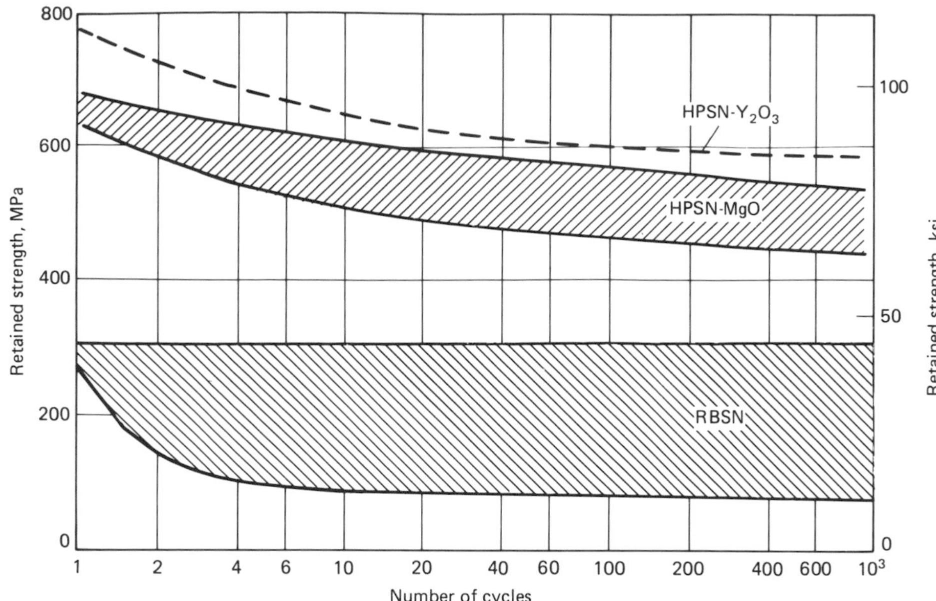

Fig. 7 Thermal cycling behavior of various Si_3N_4 ceramics, showing retained strength as a function of number of cycles from 1260 °C (2300 °F) to room temperature. Source: Ref 7

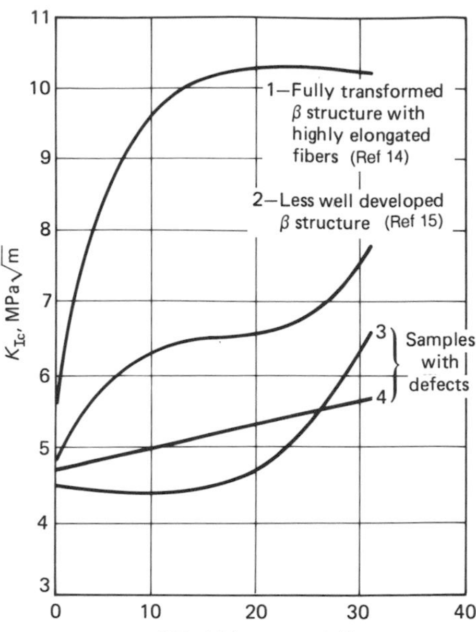

Fig. 8 Effect of SiC whiskers on the fracture toughness of different Si_3N_4 matrix composites

sion, and a higher stress will be required for matrix cracking (although an overall decrease in strength for the composite may also result). Radially, the SiC whiskers should shrink away from the matrix, resulting in a decrease in bond strength at the whisker/matrix interface and allowing toughening via crack deflection and whisker "pullout" (an example of whisker pullout is described below in the section on "Alumina Ceramics").

Figure 8 shows the effects of different amounts of SiC whiskers on the fracture toughness of Si_3N_4 that was hot pressed by different workers (Ref 14–16) under different conditions. The variations in sintering schedules led to differences in the composite matrix microstructure, which reflects the different results for fracture toughness. In case 1 (Ref 14) of Fig. 8, samples are fully dense, fully transformed β matrices with highly elongated fibrous grains and toughness values of 10 MPa$\sqrt{m}$ (9 ksi$\sqrt{in.}$). Case 2 has a less well-developed β grain microstructure (Ref 15). In cases 3 and 4, there is evidence that fabrication-related defects were the origins of fracture initiation, and composite toughness values are only fractionally better than those for monolithics.

Generally, results of mechanical testing reported for SiC whisker-reinforced Si_3N_4-based composites have shown a slight decrease in strength with whisker addition. However, simultaneous increases in fracture toughness and the modulus of rupture with whisker addition also have been reported (Ref 14). Table 5 shows the influence of whisker control on composite properties.

Whisker-reinforced Si_3N_4 composites are being considered for hot-section ceramic engine components and cutting tool inserts. Another key current application is can-making equipment and associated tooling. Of the ceramics examined for such applications, Si_3N_4-SiC_w offers the least affinity for aluminum metal pickup during two-piece can-making operations, compared with Al_2O_3 or ZrO_2 ceramics.

Silicon Carbide Ceramics

Processing. Silicon carbide can be produced with either a cubic (β) or hexagonal (α) crystal structure. The SiC ceramics include hot-pressed, direct-sintered, reaction-sintered, and chemically vapor deposited (CVD) materials. Typical properties of each class are presented in Table 6.

Hot-pressed SiC can be fabricated to essentially full density and high strength by using additions of boron and carbon or of Al_2O_3 to either α or β SiC starting powder (Ref 2). Hot pressing is typically accomplished at temperatures of 1900 to 2000 °C (3450 to 3630 °F) with pressures of 35 MPa (5 ksi).

In direct-sintered SiC, submicrometer SiC powder is compacted and sintered at temperatures in excess of 2000 °C (3600 °F), resulting in a high-purity product. Reaction-bonded SiC, on the other hand, is processed by forming a porous shape composed of SiC and carbon-powder particles. The shape is then infiltrated with silicon metal, which bonds the SiC particles. Both direct-sintered and reaction-sintered SiC ceramics have relatively low fracture toughness values, of the order of 3 to 4 MPa$\sqrt{m}$ (2.7 to 3.6 ksi$\sqrt{in.}$).

Chemical vapor infiltration (CVI) is an attractive process for fabricating fiber-reinforced composites. It is one of only a few processes capable of incorporating continuous SiC or alumino-sili-

Table 6 Typical properties of SiC ceramics

	Bend strength (4-point), MPa			Elastic modulus, E, GPa	Coefficient of thermal expansion, 10^{-6}/°C	Thermal conductivity, K, W/m · °C
	RT(a)	1000 °C	1375 °C			
Hot-pressed (Al_2O_3 additive)	655	585	520	449	4.5	35–85
Sintered (α phase)	310	310	310	407	4.8	50–100
Reaction-sintered (20% free Si by volume)	380	415	275	345	4.4	50–100
CVD	415	550	550	414	...	...

(a) Room temperature

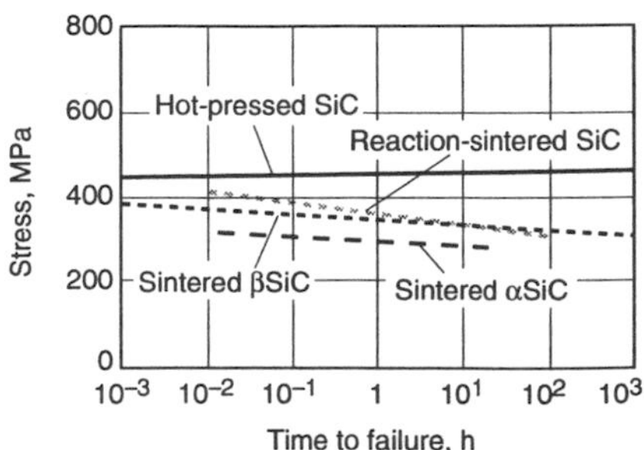

Fig. 9 Stress-rupture behavior of various SiC ceramics. Source: Ref 2

Table 7 Properties of zirconia and alumina ceramics

Material	Bulk density, g/cm³	Flexure strength MPa	Flexure strength ksi	Fracture toughness $MPa\sqrt{m}$	Fracture toughness $ksi\sqrt{in.}$	Hardness GPa	Hardness 10^6 psi	Elastic modulus GPa	Elastic modulus 10^6 psi
ZTA	4.1–4.3	600–700	87–101	5–8	4.6–7.3	15–16	2–2.3	330–360	48–52
Mg-PSZ	5.7–5.8	600–700	87–101	11–14	10–13	12	1.7	210	30
Y-TZP	6.1	900–1200	130–174	8–9	7.3–8.2	12	1.7	210	30
Al_2O_3–SiC_w	3.7–3.9	600–700	87–101	5–8	4.6–7.3	15–16	2–2.3	430–380	62–55

cate fibers in a ceramic matrix (often SiC) without chemically, thermally, or mechanically damaging the relatively fragile reinforcing fibers. The high strength (≈400 MPa, or ≈58 ksi) and exceptional fracture toughness (>20 $MPa\sqrt{m}$, or >18 $ksi\sqrt{in.}$) of these composites, combined with their refractoriness and resistance to erosion, corrosion, and wear, make them ideal candidates for numerous advanced high-temperature structural applications.

Chemically vapor deposited SiC is a fully dense material with no additives. This material is generally used as a coating for high-temperature oxidation resistance.

Stress-Rupture Behavior. Figure 9 illustrates the stress-rupture behavior typical of hot-pressed, sintered, and siliconized (reaction-sintered) SiC. These materials were tested in air at 1200 °C (2190 °F). There is less difference in stress-rupture behavior among SiC ceramics than is observed for Si_3N_4 ceramics (compare Fig. 6 and 9). The relative ranking of the strengths of the Si_3N_4 ceramics changes significantly with time under load, but the relative ranking of the strengths of the SiC ceramics remains fairly constant.

Oxidation Behavior. The SiC ceramics have oxidation resistances similar to those of Si_3N_4 ceramics, due to the protective silica layer that forms in oxidizing atmospheres. Quinn (Ref 17) found that after 360 h of exposure and 500 thermal cycles at 1370 °C (2500 °F) there was no degradation in the room-temperature strength of hot-pressed SiC. Siliconized SiC demonstrated a 32% reduction in strength.

Zirconia Ceramics

Pure ZrO_2 cannot be fabricated into a fully dense ceramic body using conventional processing techniques. The 3 to 5 vol% increase associated with the tetragonal-to-monoclinic phase transformation causes any pure ZrO_2 body to completely destruct upon cooling from the sintering temperature. Additives such as calcia (CaO), magnesia (MgO), yttria (Y_2O_3), or ceria (CeO_2) must be mixed with ZrO_2 to stabilize the material in either the tetragonal or cubic phase. Applications for cubic-stabilized ZrO_2 include various oxygen-sensor devices (cubic ZrO_2 has excellent ionic conductivity), induction heating elements for the production of optical fibers, resistance heating elements in new high-temperature oxidizing kilns, and inexpensive diamondlike gemstones.

Transformation-toughened zirconia (TTZ) is a generic term applied to stabilized ZrO_2 systems in which the tetragonal symmetry is retained as the primary phase. The most popular tetragonal-phase stabilizers are Y_2O_3, CaO, and MgO. The use of these additives results in two distinct microstructures.

MgO- and CaO-stabilized ZrO_2 consist of 0.1 to 0.25 μm cubic grains. Firing usually occurs within the single cubic-phase field, and phase assemblage is controlled during cooling. Interest in CaO-stabilized ZrO_2 has waned, but MgO-stabilized ZrO_2 (Mg-PSZ, where PSZ stands for partially stabilized zirconia) has enjoyed immense commercial success. Its combination of moderate to high strength (600 to 700 MPa, or 87 to 100 ksi), high fracture toughness (11 to 14 $MPa\sqrt{m}$, or 10 to 13 $ksi\sqrt{in.}$), and flaw tolerance makes it suitable for the most demanding structural ceramic applications. The elastic modulus is approximately 210 GPa (30 × 10^6 psi), and the hardness is approximately 12 to 13 GPa (1.7 to 1.9 × 10^6 psi). Among the applications for this material are extrusion nozzles in steel production, wire-drawing cap stands, foils for the paper-making industry, and compacting dies. Among the toughened or high-technology ceramic materials, MG-PSZ exhibits the best combination of mechanical properties and cost for room- and moderate-temperature structural applications. Properties of Mg-PSZ are listed in Table 7.

Yttria-stabilized ZrO_2 (Y-TZP, where TZP stands for tetragonal zirconia polycrystal) is a fine-grain, high-strength material with moderate to high fracture toughness. High-strength Y-TZP is manufactured by sintering at relatively low temperatures (1400 °C, or 2550 °F). Nearly 100% of the ZrO_2 is in the tetragonal symmetry, and the average grain size is approximately 0.6 to 0.8 μm. The tetragonal phase in this microstructure is very stable. Higher firing temperatures (1550 °C, or 2800 °F) result in a high-strength (1000 MPa, or 145 ksi), high-fracture-toughness (8.5 $MPa\sqrt{m}$, or 7.7 $ksi\sqrt{in.}$), fine-grain material with excellent wear resistance. The microstructure consists of a mixture of 1 to 2 μm tetragonal grains (90 to 95%) and 4 to 8 μm cubic grains (5

Fig. 10 Fracture surface of an Al_2O_3-SiC_w ceramic. Note hexagonal voids or holes due to whisker pullout upon fracture. 950×

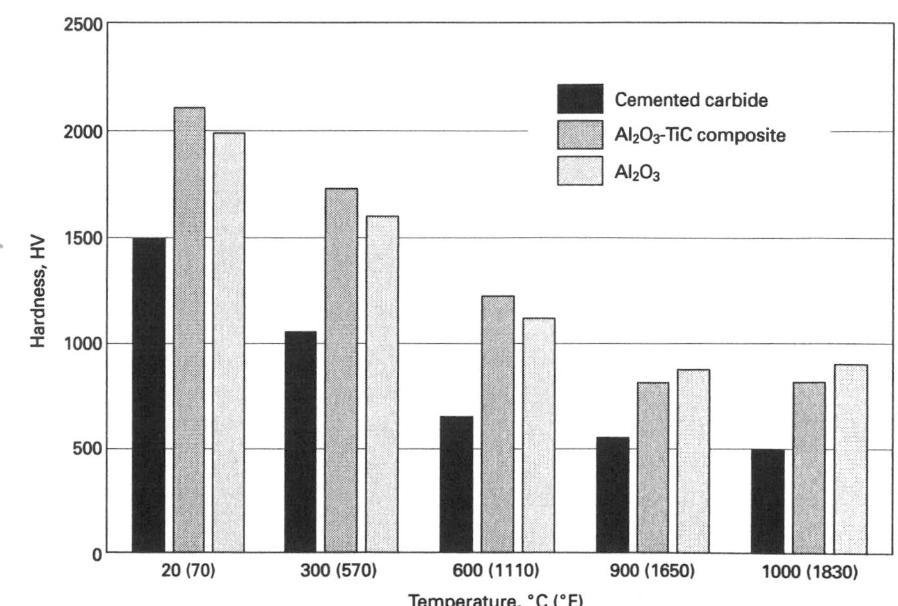

Fig. 11 Hot hardness comparison of Al_2O_3, Al_2O_3-TiC, and cemented carbide tool materials. Source: Ref 18

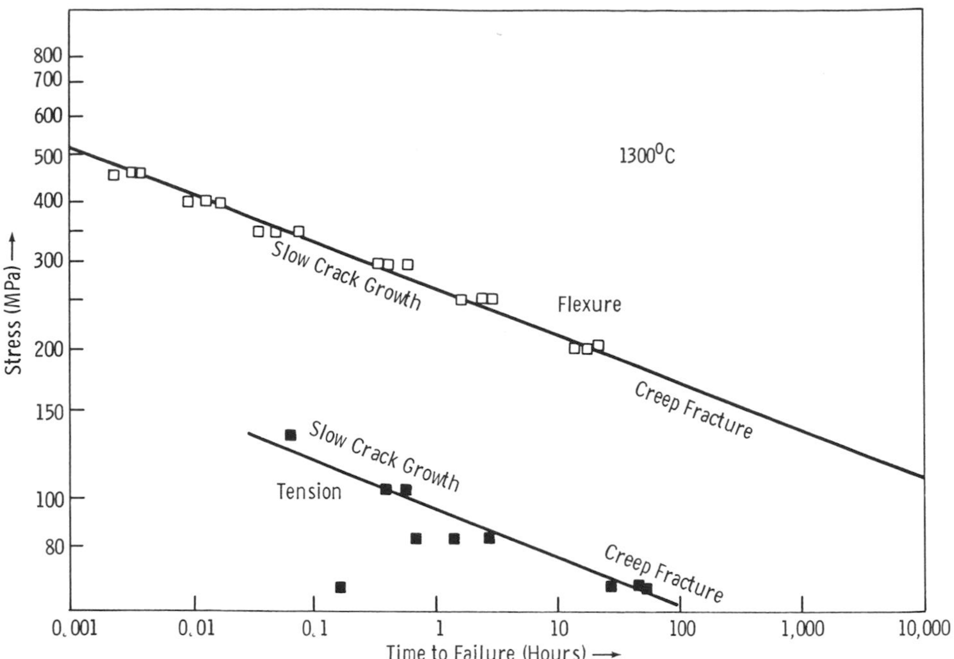

Fig. 12 Complementary flexure and tensile stress rupture data for Si_3N_4 at 1300 °C (2370 °F). Stress levels are different for the two methods. Source: Ref 23, 24

to 10%). The tetragonal phase in high-strength Y-TZP is more readily transformable because of the larger tetragonal grain size and the lower yttria content in the tetragonal phase, which result in a tougher material. Properties of Y-TZP are listed in Table 7.

One application for Y-TZP is ferrules for fiber optic assemblies. Material requirements include a very fine-grain microstructure, grain size control, dimensional control, excellent wear properties, and high strength. Fine-grain microstructure and good mechanical properties also make Y-TZP a candidate material for knife-edge applications, including scissors, slitter blades, knife blades, scalpels, and so forth. However, Y-TZP is more expensive than Mg-PSZ, has a lower fracture toughness, and is not nearly as flaw tolerant.

There are temperature limitations to TTZ materials. The mechanical strength of both Mg-PSZ and Y-TZP may start to deteriorate at temperatures as low as 500 °C (930 °F), and Y-TZP is susceptible to severe degradation between 200 and 300 °C (400 and 570 °F). The higher density of ZrO_2 ceramics also limits their use in some applications.

Alumina Ceramics

Pure aluminum oxide, Al_2O_3, has only one thermodynamically stable phase, the hexagonal α phase (corundum). Ceramics based on Al_2O_3 have been used for years. These traditional ceramics are not suitable for high-temperature structural applications, but toughened aluminas have been developed that have suitable properties to be considered for high-temperature structural applications.

Zirconia-toughened alumina (ZTA) is the generic term applied to Al_2O_3-ZrO_2 systems where Al_2O_3 is considered the primary or continuous (70 to 95%) phase. Zirconia particulate additions from 5 to 30% (either as pure ZrO_2 or stabilized ZrO_2) represent the second phase. The solubility of ZrO_2 in Al_2O_3 or Al_2O_3 in ZrO_2 is negligible. The ZrO_2 is present either in the tetragonal or monoclinic symmetry. ZTA is a material of interest primarily because it has a significantly higher strength and fracture toughness than Al_2O_3.

The microstructure and subsequent mechanical properties of ZTA can be tailored to specific applications. Higher ZrO_2 contents lead to increased fracture toughness and strength, with little reduction in hardness and elastic modulus, provided that most of the ZrO_2 can be retained in the tetragonal phase. Strengths up to 1050 MPa (152 ksi) and fracture toughness values as high as 7.5 MPa$\sqrt{m}$ (6.8 ksi$\sqrt{in.}$) have been measured (Table 7). Wear properties in some applications are better than those of Al_2O_3 ceramics because of mechanical property enhancement. These types of ZTA compositions have been used in some cutting tool applications.

Zirconia-toughened alumina has also been used in thermal shock applications where extensive use of monoclinic ZrO_2 might result in a severely microcracked ceramic body. The microstructure of ZTA allows thermal stresses to be distributed throughout a network of microcracks, where energy is expended opening and/or extending the microcracks and the bulk ceramic body is left intact.

Silicon carbide whisker-reinforced alumina (Al_2O_3-SiCw) was developed as a potential material for ceramic engine components. Incorporation of SiC whiskers (20 to 45 vol%) into an Al_2O_3 matrix with subsequent hot pressing results in a composite with significantly improved toughness. The whiskers, small fibers of single-crystal SiC about 0.5 to 1 μm in diameter and 10 to 125 μm long, have a higher thermal conductivity and a lower coefficient of thermal expansion than monolithic Al_2O_3. This improves thermal shock resistance.

The SiC whiskers in the Al_2O_3 matrix also produce a twofold increase in fracture toughness. As indicated in Table 7, fracture toughness as high as 8 MPa$\sqrt{m}$ (7.3 ksi$\sqrt{in.}$) is possible. The fracture toughness is enhanced by the occurrence of whisker pullout. A close examination of the fracture surface at high magnification reveals not only a clear indication of the whiskers, randomly dispersed throughout the matrix, but also obvious hexagonal voids where whiskers have actually been pulled out during the fracture process (Fig. 10). A large amount of energy is required to pull

Table 8 Selected ceramics used for heat exchanger applications

Material	Processing method	Manufacturer	Brand	Density, g/cm^3	Strength MPa	Strength ksi
SiC-Si	Extruded, slip cast	Norton	NC-430/CS101-K	3.07	160(a)	23(a)
SiC (100% α)	Extruded, slip cast	Carborundum	Hexaloy SA	3.10	320(b)	46(b)
					246(c)	35.7(c)
SiC	Extruded, slip cast	Norton	CS-101	2.7	91(a)	13(a)
					100–200(b)	14.5–29(b)
Si_3N_4-bonded SiC	Extruded	Norton	CX-589	2.51	84(a)	12.2(a)
SiC	Extruded, slip cast	Coors	SC-2	3.10	525(d)	76(d)
					175(c)	25(c)
SiC	Slip cast	Coors	RBSC-205	3.05	287(d)	41.5(d)
SiC	Chemical vapor infiltration	Amercom	...	2.7–3.2	300	43.5
Al_2O_3/ZrO_2	Sol-gel winding	B&W	...	3.7	350(c)	50(c)
SiC whisker-reinforced alumina	DIMOX(e)	Lanxide	...	...	460(c)	66.5(c)

(a) O-ring strength test. (b) Four-point bending flexural strength. (c) C-ring strength test. (d) Three-point bending flexural strength. (e) Proprietary process the whiskers out, and this greatly inhibits crack propagation.

Al_2O_3-SiC_w composites are principally used as cutting tool inserts. They have also been used as tooling for forming aluminum beverage cans.

Al_2O_3-TiC Composites. In the early 1970s, it was discovered that Al_2O_3 admixed with a refractory metal particulate (e.g., titanium carbide, TiC) could produce a ceramic with better hardness and fracture resistance than monolithic Al_2O_3. These hot-pressed or hot isostatically pressed composites consist of approximately 70% Al_2O_3 with 30% TiC particulate. Such composites are called black ceramics due to their color, which results from the presence of TiC. Cutting tool inserts are the primary application.

Dispersion of hard refractory particles increases the hardness of these composites at temperatures up to 800 °C (1470 °F) when compared to monolithic ceramics (Fig. 11). Simultaneously, the fracture toughness and bending strength are improved through the crack impediment, crack deflection, or crack branching that is caused by the dispersed hard particles. The higher hardness, in combination with the higher toughness, considerably increases the resistance to abrasive and erosive wear, and the lower thermal expansion and higher thermal conductivity improve thermal shock resistance over that of monolithic oxide ceramics. At temperatures exceeding 800 °C (1470 °F), however, the TiC particles oxidize and begin to lose their reinforcing properties, and the composite weakens.

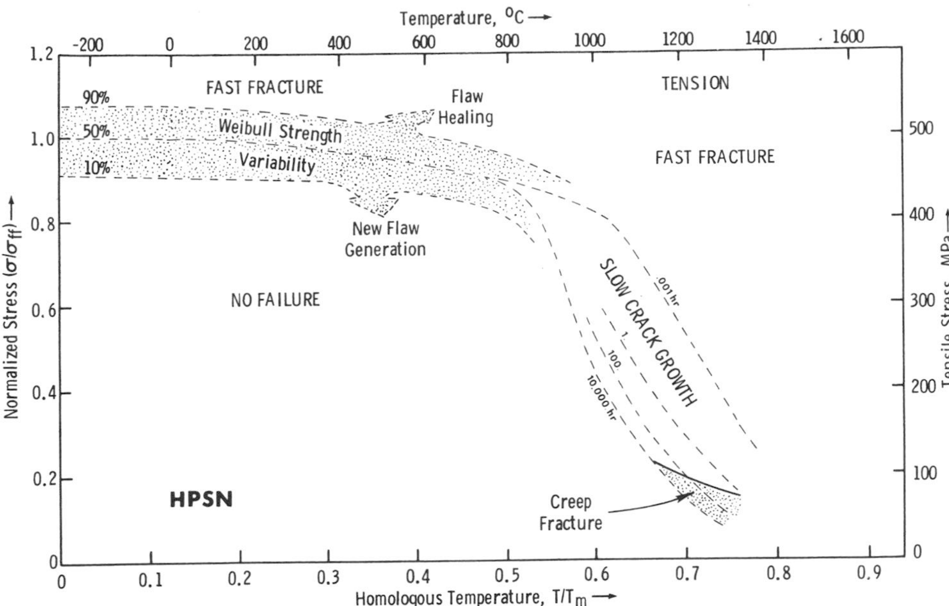

Fig. 13 Stress rupture data used to create fracture mechanism maps. Loci of constant failure time are labeled in hours. Source: Ref 23

Table 9 Ceramic corrosion in the presence of combustion products

Material	Corrosion depth data(a)		Projected corrosion	
	μm	mil	μm/y	mils/y
Cordierite	9.4	0.37	58.0	2.3
Mullite	0.85	0.034	8.2	0.32
Reaction-bonded SiC	0.20	0.0080	1.9	0.08
	0.16	0.0063	1.5	0.06
	0.18	0.0069	1.7	0.07
Sintered αSiC	0.37	0.014	3.5	0.14
Reaction-bonded Si_3N_4	0.90	0.035	8.6	0.34
Glass-enamel	1.4	0.054	13.3	0.52

(a) Data are based on a heat condensing application in which products were exposed to combustion products (H_2SO_4, HCl, and nitrous oxides) under cyclic conditions up to 290 °C (550 °F) for 900 h. Source: Ref 27

High-Temperature Strength Test Methods

Fast Fracture. The overwhelming majority of high-temperature strength tests have been done in four-point loading (flexural tests). Standards are being developed that are extensions of the low-temperature procedures. A variety of furnaces and environments can be used, typically up to 1600 °C (2910 °F) in air and up to 2000 °C (3630 °F) with some vacuum and inert gas systems. The test fixtures themselves must be dense ceramics, usually fairly pure forms of SiC, although occasionally Al_2O_3 fixtures are used at lower temperatures, and graphite fixtures are used in inert atmospheres.

The increasing use of the tension test for characterizing ceramics has been driven in large part by new programs to use ceramics at high temperatures in heat engines. As a result, most tensile test systems have been designed with high temperatures in mind. The gripping schemes must not only be elaborate enough to avoid stress concentrators and to align very precisely, but they must also be capable of being used in conjunction with furnaces. Most tension systems use cold grips with relatively long (for ceramics) specimens of 150 mm (6 in.). Such systems are described in Ref 19.

Creep and Stress Rupture. Direct tension tests of long duration are becoming more common, but most test systems are complicated and expensive. They typically are derivatives of the fast-fracture systems and make use of cold grips and long specimens. Most experiments are limited to 1000 h. An economical alternative test system with hot grips and a flat "dog bone" specimen configuration has been developed, and it is optimized for long-duration, low-stress creep experiments (Ref 20). A short, tapered specimen for similar experiments has been successfully used by Grathwohl (Ref 21). Strains must be measured with specialized extensometers because ceramic strains are extremely small, and resolutions of 1 μm (0.04 mil) must be recorded over the course of hours. The extensometers in use today are delicate mechanical units or lasers that either monitor distance between flags (specimen marks) or diffract when passed through a narrow slit between two flags. Reference 22 is an excellent compilation of several tension testing papers that discuss these methods in detail.

Most investigators have at some time resorted to using flexure testing, which is much less expensive and allows strains to be readily measured from the curvature in the specimen. In a sense, the bend specimen acts as a deflection magnifier,

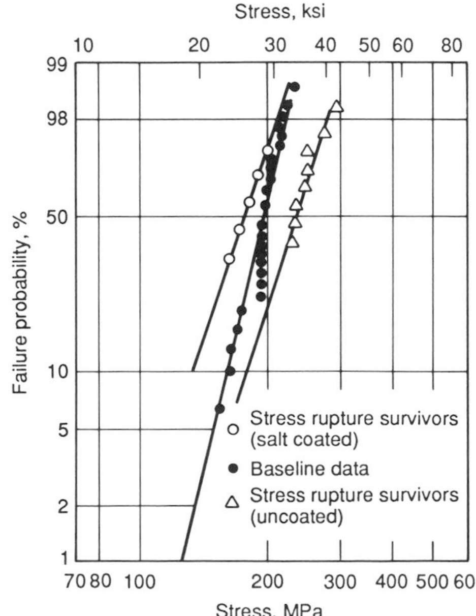

Fig. 14 Weibull plots for reaction-sintered SiC before and after 500 h stress rupture

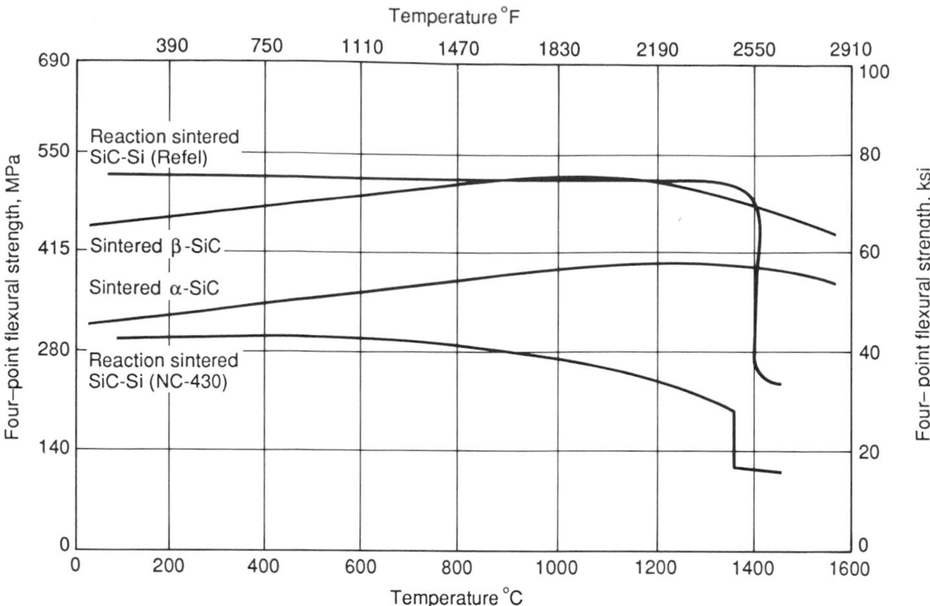

Fig. 15 Elevated-temperature flexural strength of SiC ceramics used for heat exchanger applications. Source: Ref 26

because the deflection associated with the integrated curvature is larger and easier to measure than the extension of a tension specimen. The drawback of the method is that there is a stress gradient in the specimen, which changes dramatically as the material creeps. Flexural creep testing is not a constant stress test. The strain is measured from the curvature, but this too must be adjusted for the proper constitutive equation.

In recent years, it has become evident that flexural creep data can be misleading or even erroneous, a consequence of the stress gradient, the relaxation of such gradient, and the complicated constitutive equations that apply to ceramics. Analytical attempts to deconvolute the tensile and compressive creep behavior are usually tainted or compromised by the assumptions that have to be made about the constitutive equations. It is far more rational to conduct direct tension or compression experiments for careful creep work. Flexure tests can be used for qualitative assessments of conditions for the onset of creep.

Stress-rupture data require extremely long-duration experiments. Some static fatigue phenomena occur in the absence of bulk creep deformation, and flexure testing may be eminently suitable in these cases. Failures out to 18,000 h have been reported (Ref 23), and data are readily accumulated, as shown in Fig. 12. Perhaps the best approach is to conduct complementary direct tension and flexure experiments, as discussed in Ref 23 and 24. It then becomes feasible to construct fracture mechanism maps, as shown in Fig. 13, patterned after those developed by Ashby et al. (Ref 25). In each case, fractography is essential for proper identification of the phenomena that cause failure.

Temperature-Resistant Applications

The high-temperature properties of advanced ceramics make them attractive for a variety of applications. For instance, the use of ceramic heat exchangers in waste heat recovery has received extensive attention over the last 10 to 15 years. Commercial products now include plate-fin and tubular recuperators. Two similar applications are process heat exchange and power generation heat exchange.

Engine components make use of both the temperature resistance and wear resistance of advanced ceramics. Ceramic materials and coatings are currently under development for application in conventional internal combustion engines, adiabatic diesel engines, and advanced gas turbines. The major advantages include reduced inertia, friction and fuel consumption, lower mass, and improved wear resistance. Sialon, SiC, and silicon nitrite are the primary material candidates in these engines.

Development of ceramics for aerospace applications has traditionally focused on coatings, used for their thermal insulating properties. However, considerable effort has been devoted in recent years to developing ceramic gas turbines for aerospace vehicles. Here the materials issues, especially reliability, are similar to those for the automotive gas turbine. Furthermore, the development of fiber-reinforced ceramic-based composites has raised the possibility of using these materials as structural components in advanced aerospace vehicles.

Heat Exchangers

Ceramic Materials. A variety of ceramic materials have been considered for use in heat exchange systems (Table 8). They have been used primarily in the critical zones in an effort to mini-

Table 10 Summary of ceramics applications for adiabatic engines

	Desired characteristics							
Adiabatic components	Low friction	Light weight	Insulation	Wear resistance	Heat resistance	Corrosion resistance	Expansion coefficient	High-technology ceramics
Piston	...	X	X	...	X	X	X	Si_3N_4, PSZ, TTA
Piston ring	...	...	...	X	X	...	...	SSN, PSZ, coating
Cylinder liner	X	...	...	X	X	X	X	Si_3N_4, PSZ, coating
Prechamber	...	...	X	...	X	X	...	PSZ, Si_3N_4
Valve	...	X	X	X	X	X	...	SSN, PSZ, composite
Valve seat insert	...	...	X	X	X	X	...	PSZ, SSN
Valve guides	X	...	X	...	X	X	...	PSZ, SSN, SiC
Exhaust/intake ports	...	...	X	...	X	X	...	ZrO_2, Si_3N_4, $TiO_2Al_2O_3$
Manifolds	...	...	X	...	X	X	...	ZrO_2, Si_3N_4, $TiO_2Al_2O_3$
Tappets	...	X	...	X	...	X	...	PSZ, SiC, Si_3N_4
Mechanical seals	X	...	...	X	...	X	...	SiC, Si_3N_4, PSZ
Turbocharger								
Turbine rotor	...	X	X	...	X	X	X	Si_3N_4, SiC
Turbine housing	...	...	X	...	X	X	X	LAS
Heat shield	...	...	X	...	X	X	X	ZrO_2, LAS
Ceramic bearings	X	X	...	X	X	X	X	SSN

LAS, lithium alumina silicate

mize the cost compared to that of metallic counterparts. Ceramics have been used for the main heat transfer surfaces (tubes, plates, etc.), seals, tubesheets, and insulation. Heat exchanger designs have ranged from all-ceramic to ceramic-metal hybrids with varying numbers of ceramic components.

Test Methods. Critical to the successful use of ceramic materials in heat exchangers has been the continuing development of design methodology and test methods. This has been a trend for all structural ceramics, but the performance of heat exchanger materials under potentially corrosive conditions has received substantial attention.

A variety of mechanical tests have been employed to characterize ceramic materials for heat exchangers. Strength tests have been run using C-ring or slit-ring tests (on tubes), flexural tests, and a lesser amount of burst or tensile testing. Compressive C-ring tests are simple tests that rely on tensile failure of outer surfaces where the most serious property degradation would occur. Several materials have been well characterized, such as those shown in Fig. 14 for which Weibull statistics have been plotted. Creep testing has also been performed on α-SiC and silicon-infiltrated SiC. Creep theories have been postulated, including void formation and coalescence, plastic flow of the free silicon, and grain-boundary sliding. A large data base of elevated-temperature flexural strength values has also been generated. The strength of SiC ceramics as a function of temperature is shown in Fig. 15.

Corrosion Tests. A number of corrosion tests have been run by universities, national laboratories, and private industry, using both simulated conditions and actual field exposure tests. The processes that have been targeted most often are steel reheating furnaces, various coal-fired processes such as gasifiers and combined cycles, and aluminum remelt furnaces. Typical of the industrial sector, the extent and chemistry of the corrosive species vary, depending on the application and control material performance. Corrosion testing is therefore critical to selecting the appropriate material for the specific application (Table 9).

Adiabatic Diesel Engines

Adiabatic diesel engines, as the name implies, are designed to reduce heat loss during engine operation. In practice, heat loss from an engine can be reduced by:

- Insulating the combustion chamber (cylinder liner, piston crown, and cylinder head), exhaust and intake ports, and the exhaust manifolds
- Eliminating the cooling system and its associated parasitic losses
- Utilizing exhaust heat by turbocompounding (which involves a turbine system geared to the crankshaft)

The simplest type of "adiabatic" engine may involve just insulation of the combustion chamber. In this case, material selection for hot engine components is not severely affected, because the insulated adiabatic engine still includes a cooling system. In a water-cooled diesel engine, for example, temperatures at the cylinder liner seldom exceed 200 °C (400 °F).

On the other hand, an uncooled adiabatic engine, without a constant, low-temperature sink on the cylinder liner, experiences an extremely high temperature on the cylinder liner wall. The piston and the piston rings also experience high temperatures (Fig. 16). The uncooled adiabatic engine also suffers higher compression work in the cycle, but due to the higher temperature and pressure after combustion, the expansion work is greater.

With cylinder wall temperatures exceeding 1000 °C (1830 °F), uncooled adiabatic engines have been considered a potentially useful application for emerging high-performance ceramics.

Table 11 Ceramic engine components under commercial production in Japan

Components	Production start	Engine	Manufacturer
Glow plug	1981	Isuzu	Kyocera
Swirl chamber	1983	Isuzu	Kyocera
Glow plug	1983	Mitsubishi	Kyocera
Intake heater	1983	Isuzu	Kyocera
Swirl chamber	1984	Toyota	Toyota
Rocker arm tip	1984	Mitsubishi	NGK
Glow plug	1985	Mazda	Kyocera
Glow plug	1985	Nissan	NTK
Turbocharger rotor	1985	Nissan	NTK
Rocker arm tip	1987	Nissan	NTK
Turbocharger rotor	1988	Isuzu	Kyocera
Link injector	1989	Cummins	Toshiba
Turbocharger rotor	1989	Toyota	Toyota
Turbocharger rotor	1990	Toyota	Kyocera
Rocker arm tip	1990	Mazda	Kyocera

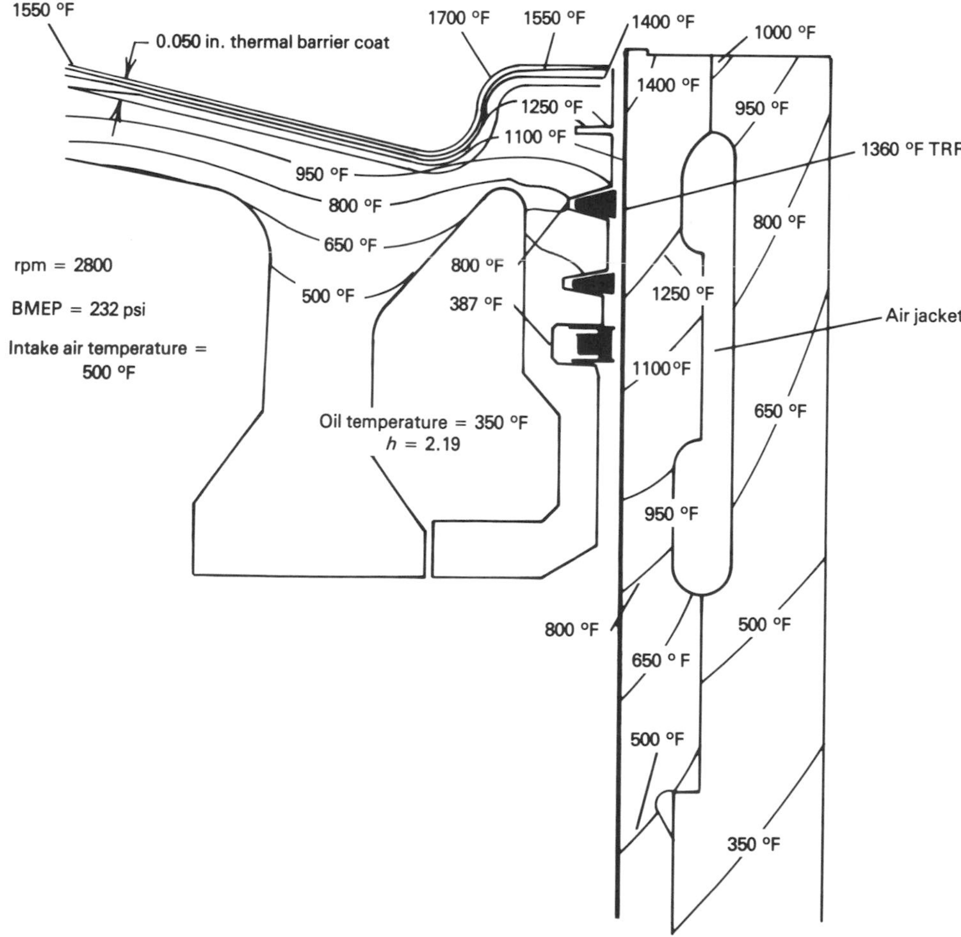

Fig. 16 Predicted cylinder and piston temperatures of an uncooled adiabatic diesel engine. TRR, top ring reversal; BMEP, brake mean effective pressure. Source: Ref 28

Table 12 Manufactured continuous-fiber-reinforced composites and their corresponding processes

Process	Composite(a)	Comments
Hot pressing	W-glass, Ni-glass, Mo-thoria, Mo-alumina, W-ceramic, stainless steel/alumina, C-glass, C/glass-ceramic, C-MgO, C-Al_2O_3, ZrO_2-MgO, ZrO_2-ZrO_2, SiC-glass, SiC/glass-ceramic, Al_2O_3-glass, C-Si_3N_4, Ta-Si_3N_4	Fibers and matrix powder are mixed together and hot pressed to produce low-porosity composites, with uncracked matrices, if thermal expansion coefficients are matched. Aligned continuous fiber composites can have very high strengths. Often combined with slurry infiltration of fibers or fiber preforms.
Cold pressing and sintering	C-glass, metal fiber/ceramic	Fibers and matrix are mixed, cold pressed, and sintered. Cracking can occur because of the large shrinkage of the matrix during sintering.
Devitrification	C/glass-ceramic, SiC/glass-ceramic	Fibers and glass powder are hot pressed at relatively low temperatures to give a reinforced glass. Further high-temperature heat treatment is used to devitrify the glass to a glass-ceramic.
Reaction bonding	Reinforced Si_3N_4	Fibers are incorporated into flame-sprayed silicon that is subsequently reaction sintered in nitrogen.
Slip casting	Ceramic fiber/fused silica	Ceramic fibers are incorporated into slips of finely divided fused silica, then fired. The increased porosity (due to the presence of fiber) usually results in degradation of properties.
Plasma spraying	Mo-Al_2O_3, W-Al_2O_3	Alumina powder is plasma sprayed. Processing is slow.
Chemical vapor infiltration and deposition	SiC fibers in SiC, C fibers in SiC	Fiber integrity can be preserved by lack of mechanical movement and relatively low process temperatures.
Directed metal oxidation (Lanxide Corp.)	SiC-Al_2O_3, SiC-SiC	Preform and parent metal alloy are heated until rapid oxidation occurs, and the oxidation reaction product becomes the matrix surrounding the reinforcing material in the preform.
Sol-gel (infiltration and sintering/hot pressing)	C-glass, mullite-mullite	Large shrinkage and matrix cracking can occur. Requires multiple infiltrations.
Polymer conversion (infiltration and pyrolysis)	C-C, C-SiC	Advantages are low shrinkage and easy fabrication.
Melt infiltration	SiC-$MoSi_2$, SiC-$CaSiO_3$, Si-SiC	Complex shapes can be fabricated, but high processing temperatures are required.

(a) Composites are given as fiber-matrix or fiber/matrix. Source: Ref 29

Table 13 Potential industrial applications for continuous-fiber-reinforced ceramic composites

Product area	Examples	Likely industrial market(s)
Advanced heat engines	Combustors, liners, wear parts, etc.	High-temperature gas turbines; possibly adiabatic diesels; promising market in gas turbine combustor retrofits
Heat recovery equipment	Air preheaters, recuperators	Indirect heating uses; energy-intensive industrial internals processes (e.g., aluminum remelters, steel reheaters, glass melters)
Burners and combustors	Radiant tube burners	Potentially any indirect-fired, high-temperature, and/or controlled-atmosphere heating/melting/heat-treating industrial application
Burners and combustors	Combustors	Low-nitrous-oxide clean fuel heating applications, including gas turbine combustors, industrial process heat
Burners and combustors	Low-temperature radiant combustors	Low-nitrous-oxide clean fuel heating applications, including small-scale (space heating) and large-scale (industrial process) applications
Process equipment	Reformers, reactors, hot isostatic pressing	Chemical process industry, petroleum refining
Waste incineration systems	Handling equipment, internals, cleanup	Conventional and advanced toxic/hazardous waste facilities, with or without energy recovery
Separation/filtration	Filters, substrates, centrifuges	Gas turbine, combined cycle, and other configurations; particulate traps for diesel exhausts, molten metal filters, sewage treatment

Source: Ref 30

Ceramic Materials. A summary of ceramic applications for adiabatic engines is shown in Table 10. There is also current interest in ceramic applications for other engine components, such as valve trains. For high-performance engines, lightweight Si_3N_4 is being seriously considered to reduce the valve train dynamics problems for greater output. The light weight and good hardness, strength, and tribological properties of Si_3N_4 may make it suitable for valve train components that will be reliable, durable, and have low friction and low dynamic stress. Light weight and good hardness also make Si_3N_4 suitable for other important applications, such as prechamber bowls, rocker arm tips, cams, tappets, turbochargers, and bearings.

In adiabatic engines, ceramics are most often used for their insulation characteristics and high-temperature properties. Insulation is used for pistons, cylinder heads, exhaust/intake ports, and cylinder liners.

Advanced Gas Turbines

The configuration and operating conditions of gas turbine engines provide the capability for greater fuel efficiency and lower gaseous and particulate emissions compared to conventional internal combustion (IC) engines (both spark ignition and diesel). Also, the gas turbine has multifuel capability and fewer moving parts than IC engines, and these attributes give it the potential to be more flexible, reliable, and cost-effective than IC engines. The use of ceramics in gas turbine engines enhances these benefits by increasing the allowable operating temperature. Projected fuel and cost savings range from 20 to 65% for both automotive and nonautomotive applications.

The demands placed on ceramics by the gas turbine engine environment are quite severe and can vary considerably, depending on the application. The ceramic gas turbine applications to be considered include missile, automotive, military, and commercial propulsion, and auxiliary power engines. Each of these applications has a different duty cycle, life requirement, operating temperature, and stress envelope (Fig. 17). The majority of the engines operate most of the time at power levels below maximum power. The specified duty cycle and engine life dictate the required strength, crack growth rate, creep rate, and environmental stability of the selected materials.

Most ceramic engine development programs have been carried out in the United States, Japan, and Europe (most notably Germany). Emphasis has been placed on monolithic Si_3N_4 and SiC. Ceramic components of interest include turbine rotors, stator vanes, combustion chamber linings, and transition duct linings. Japanese automotive companies have developed engine components that are already in production (Table 11).

Continuous-Fiber-Reinforced Composites

Continuous fiber reinforcement has the highest potential for improving stress-strain behavior and damage tolerance in structural ceramics, but it

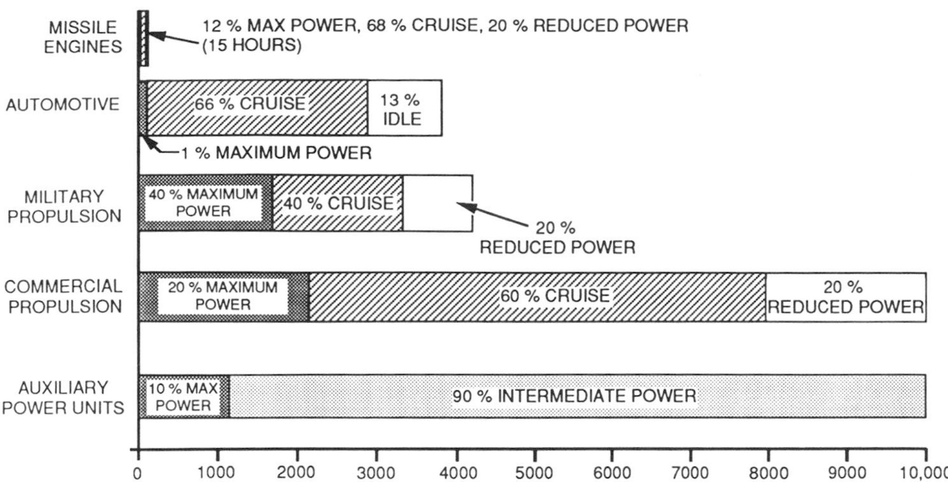

Fig. 17 Duty cycles and lifetime requirements for different types of gas turbine engines. Lifetime is given in hours.

also has the highest fabrication complexity and cost. Other problems in processing continuous-fiber-reinforced composites are the limited-temperature stability of the fibers and chemical attack of the fibers. The latter occurs because of reactions with the matrix material and/or with gaseous media of the sintering atmosphere.

Processing Methods. Table 12 shows the wide variety of fiber/matrix combinations that are under development, as well as their respective processing methods. Many of the densification and fabrication methods used for monolithic advanced ceramics can be adapted for ceramic-matrix composites. These include cold pressing and sintering, hot pressing, reaction bonding, and hot isostatic pressing. In addition, some of the conventional methods used in polymer-matrix composites can be applied to continuous-fiber-reinforced ceramics. Newer processing methods show considerable promise and can overcome some of the limitations of traditional methods, especially in terms of cost. Some involve infiltration of a preform in combination with conventional densification techniques. Others include chemical vapor deposition or infiltration, directed oxidation, self-propagating high-temperature synthesis, microwave processing, and in situ processing.

Overview of Applications. Applications can be divided between aerospace applications and all others. In the former, performance is the foremost consideration, while cost effectiveness dominates in the latter category.

In order to achieve high thrust-to-weight ratios, fast cruising speeds, high altitudes, and excellent flight performance in aerospace applications, materials must have high strength-to-density, high stiffness-to-density ratios, and excellent damage tolerance at high temperatures. Continuous-fiber-reinforced composites offer excellent mechanical properties that make them candidates for a variety of high-temperature aerospace applications, such as missiles, hypersonic radomes, hard armor, thermal protection systems, and turbine engines.

As indicated in Table 13, there are also a number of nonaerospace (industrial) applications for continuous-fiber-reinforced ceramics. Heat exchangers and recuperators made from ceramic materials can operate at higher temperatures and in more aggressive environments than can their metallic counterparts.

ACKNOWLEDGMENTS

The information in this article is largely taken from:

- J.R. Davis, Guide to Materials Selection, *Engineered Materials Handbook Desk Edition,* ASM International, 1995, p 106–154
- S. Hampshire, Engineering Properties of Nitrides, *Ceramics and Glasses,* Vol 4, *Engineered Materials Handbook,* ASM International, 1991, p 812-820
- G.D. Quinn, Strength and Proof Testing, *Ceramics and Glasses,* Vol 4, *Engineered Materials Handbook,* ASM International, 1991, p 585–598

REFERENCES

1. J.R. Davis, Guide to Materials Selection, *Engineered Materials Handbook Desk Edition,* ASM International, 1995, p 106–154
2. G.L. Leatherman and R. Nathan Katz, Structural Ceramics: Processing and Properties, *Superalloys, Supercomposites and Superceramics,* Academic Press, 1989, p 671–696
3. S. Hampshire, Engineering Properties of Nitrides, *Ceramics and Glasses,* Vol 4, *Engineered Materials Handbook,* ASM International, 1991, p 812–820
4. T. Ekstrom and N. Ingelstrom, Characterisation and Properties of Sialon Materials, *Non-Oxide Technical and Engineering Ceramics,* S. Hampshire, Ed., Elsevier-Applied Science, 1986, p 231–254
5. M.H. Lewis, S. Mason, and A. Szweda, Syalon Ceramic for Application at High Temperature and Stress, *Non-Oxide Technical and Engineering Ceramics,* S. Hampshire, Ed., Elsevier-Applied Science, 1986, p 175–190
6. G. Ziegler, Thermal Properties and Thermal Shock Resistance of Nitrogen Ceramics, *Progress in Nitrogen Ceramics,* F.L. Riley, Ed., Nijhoff, 1983, p 565–588
7. G. Ziegler, Thermal Cycling Behavior of Silicon Nitride, *Ceramic Components for Engines,* Proc. First Int. Symp., S. Somiya, E. Kanai, and K. Ando, Ed., KTK Scientific, 1983, p 232–248
8. G. Ziegler, J. Heinrich, and G. Wotting, Review: Relationships between Processing, Microstructure and Properties of Dense and Reaction-Bonded Silicon Nitride, *J. Mater. Sci.,* Vol 22, 1987, p 3041–3086
9. R.N. Katz and G.D. Quinn, Time-Temperature Effects in Nitride and Carbide Ceramics, *Progress in Nitrogen Ceramics,* F.L. Riley, Ed., Nijhoff, 1983, p 491–500
10. D. Cubicciotti, K.H. Lau, and R.L. Jones, Rate-Controlled Process in the Oxidation of Hot-Pressed Silicon Nitride, *J. Electrochem. Soc.,* Vol 124, 1977, p 1955–1956
11. M.H. Lewis and P. Barnard, Oxidation Mechanisms in Si-Al-O-N Ceramics, *J. Mater. Sci.,* Vol 15, 1980, p 443–448
12. M.J. Pomeroy and S. Hampshire, Oxidation Processes in Silicon Nitride Ceramics, *Mater. Sci. Eng.,* Vol A109, 1989, p 389–394
13. D.P. Thompson, Alternative Grain-Boundary Phases for Heat-Treated Si_3N_4 and β′-Sialon Ceramics, *Fabrication Technology,* R.W. Davidge and D.P. Thompson, Ed., *Br. Ceram. Proc.,* Vol 45, 1980, p 1–13
14. S.T. Buljan and V.K. Sarin, *Silicon Nitride Based Composites,* Vol 18, 1987, p 99–106
15. P.D. Shalek, J.J. Petrovic, G.F. Hurley, and F.D. Gac, Hot Pressed SiC Whisker/Si_3N_4 Matrix Composites, *Am. Ceram. Soc. Bull.,* Vol 65, 1986, p 351–356
16. A. Bellosi and G. De Portu, Hot-Pressed Si_3N_4-SiC Whisker Composites, *Mater. Sci. Eng.,* Vol A109, 1989, p 357–362
17. G.D. Quinn, R.N. Katz, and E.M. Lenoe, *Proc. 1977 DARPA/NAVSEA Ceramic Gas Turbine Review,* MCIC 78–36, March 1978
18. W.W. Gruss and K.M. Friedrich, Aluminum Oxide/Titanium Carbide Composite Cutting Tools, *Ceramic Cutting Tools,* E.D. Whitney, Ed., Noyes Publications, 1994, p 48–62
19. D. Lewis III, Tensile Testing of Ceramic and Ceramic-Matrix Composites, *Tensile Testing,* P. Han, Ed., ASM International, 1992, p 147–181
20. D.F. Carroll, S.M. Wiederhorn, and D.E. Roberts, Technique for Tensile Creep Testing of Ceramics, *J. Am. Ceram. Soc.,* Vol 72 (No. 9), 1989, p 1610–1614
21. G. Grathwohl, Current Testing Methods—A Critical Assessment, *Int. J. High Tech. Ceram.,* Vol 4, 1988, p 123–142
22. B.F. Dyson, R.D. Lohr, and R. Morrell, Ed., *Mechanical Testing of Engineering Ceramics at High Temperatures,* Elsevier, 1989
23. G.D. Quinn, Fracture Mechanism Maps for Advanced Structural Ceramics, *J. Mater. Sci.,* Vol 25, 1990, p 4361–4376

24. R. Govila, Uniaxial Tensile and Flexural Stress Rupture Strength of Hot-Pressed Si_3N_4, *J. Am. Ceram. Soc.,* Vol 65 (No. 1), 1982, p 15–21
25. C. Gandhi and M. Ashby, Fracture Mechanism Maps for Materials Which Cleave: F.C.C., B.C.C. and H.C.P. Metals and Ceramics, *Acta Metall.,* Vol 27, 1979, p 1565–1602
26. S.L. Richlan, A Survey of Ceramic Heat Exchanger Opportunities, *Ceramics in Heat Exchangers,* Vol 14, *Advances in Ceramics,* American Ceramic Society, 1985, p 3–14
27. I. Sekerciogiu, R. Razgaitus, and J. Lux, Evaluation of Ceramics for Condensing Heat Exchanger Applications, *Ceramics in Heat Exchangers,* Vol 14, *Advances in Ceramics,* American Ceramic Society, 1985, p 359–369
28. M.W. Woods, P.C. Glance, and E. Schwarz, "Advanced Insulated Titanium Piston for Adiabatic Engine," Paper 900623, Society of Automotive Engineers, 1990
29. D.C. Phillips, Fiber Reinforced Ceramics, *Fabrication of Composites,* Vol 4, *Handbook of Composite Materials,* A. Kelly and S.T. Mileiko, Ed., North Holland Publishing, 1983
30. K.K. Chawla, *Ceramic Matrix Composites,* Chapman & Hall, 1993, p 377-379

Carbon-Carbon Composites

CARBON-CARBON COMPOSITES began replacing fine-grained graphite as nose tips in rockets in the mid-1960s because they represented significant improvements in thermoshock behavior and erosion resistance. They are currently being introduced in fields that require their high specific strength and stiffness, in combination with their thermoshock resistance, chemical resistance, and fracture toughness, especially at high temperatures.

Carbon-carbon composites could be an ideal structural material if they did not experience severe oxidation at temperatures above 500 °C (930 °F). Intensive research is being conducted worldwide to develop protective coatings. The emphasis in this article is on coating technology for carbon-carbon composites.

Composite Manufacture

Carbon-carbon is a unique composite material in which a nonstructural carbonaceous matrix is reinforced by carbon fibers to create a heat-resistant structural material. The carbon fibers are generally employed as woven fabric in two-dimensional laminates, or as multidimensional preforms created by textile processing of multifilament tows (Ref 1, 2). Composite densification is typically accomplished using phenolic resins in a preforming process, followed by liquid impregnation, vapor phase carbon infiltration, or combinations of the two to achieve pore filling (Ref 3–10). High-temperature pyrolysis is used to convert liquid impregnants to carbon, whereas vapor phase infiltration is conducted in reduced-pressure reactors where flowing hydrocarbon gas infiltrates the composite and is thermally decomposed to deposit carbon. The result is a low-density (1.5 to 1.9 g/cm^3) composite that derives its mechanical performance from the carbon fiber reinforcement.

Figure 1 depicts the manufacturing route generally followed for producing carbon-carbon composites. Additional information on composite manufacture can be found in Volume 1, *Composites,* of the *Engineered Materials Handbook* published by ASM International.

Properties

Fibers. Commercial quantities of carbon fibers are derived from three feedstock or precursor sources: rayon, polyacrylonitride (PAN), and petroleum pitch. Each precursor category requires different processing techniques. There are several processes that are used by different manufacturers using the same type of precursor, but generally, conversion of the precursor to carbon fiber follows this sequence: stabilization, carbonization, graphitization (optional), surface treatments, application of sizings or finishes, and spooling.

Table 1 gives the names of some commercially available carbon fibers and their manufacturers. Data are provided to give an overview of the wide range of properties attainable with carbon fibers and to show that within each precursor group, the mechanical properties can be broadly tailored.

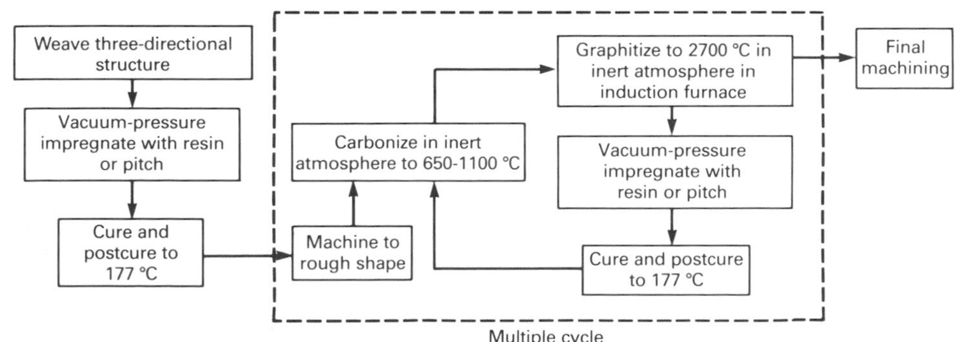

Fig. 1 Typical processing route for carbon-carbon composites. Source: Ref 11

Carbon fibers offer the highest modulus and highest strength of all reinforcing fibers (Fig. 2). The fibers do not suffer from stress corrosion or stress rupture failures at room temperature, as glass and organic polymer fibers do. At high temperatures, the strength and modulus are outstanding compared to those of other materials.

Carbon Fiber Structure. Carbon fibers can be used as reinforcements in many different architectures: random fibers; two-directional (2-D) fabrics in stacked, stitched, or pierced configurations; three-directional (3-D) geometries (cartesian or cylindrical coordinates) to increase the off-axis strength; or 3-D and multidirectional (4- to 11-direction) weaves to minimize the empty spaces between the rod (fiber) junctions.

The simplest type of multidirectional preform is based on a 3-D orthogonal construction, which is normally used to weave rectangular, block-type preforms. As shown in Fig. 3, this preform type consists of multiple yarn bundles located on cartesian coordinates. Each of the yarn bundles is straight in order to achieve the maximum structural capability of the fiber.

The type of multidirectional preform construction typically used for cylinders and other shapes of revolution (shown in Fig. 4) is a 3-D construction with yarns oriented on polar coordinates in the radial, axial, and circumferential directions. As with orthogonal block preforms, yarn type, spacing, and volume fraction can be varied in all three directions.

Matrix Precursor Impregnants. The two general categories of matrix precursors used for carbon-carbon densification are thermosetting resins, such as phenolics and furfurals, and pitches based on coal tar and petroleum. Properties such as viscosity, coke yield, density, microstructure, and degree of graphitization differ for these two impregnant categories. These properties are also influenced by the time-temperature-pressure relationships encountered during processing.

Composite Properties. Preform design, fiber type, matrix precursor, and processing all influence composite properties. There are no standard properties because the number of possible variations is almost limitless. The data in Table 2, presented as typical for 3-D orthogonal carbon-carbon composites, illustrate the following char-

Table 1 Typical mechanical property values of commercially available carbon fibers

Product name	Manufacturer	Precursor type(a)	Density, g/cm^3	Tensile strength, GPa	Tensile strength, 10^6 psi	Tensile modulus, GPa	Tensile modulus, 10^6 psi
AS-4	Hercules, Inc.	PAN	1.78	4.0	0.580	231	33.5
AS-6	Hercules, Inc.	PAN	1.82	4.5	0.652	245	35.5
IM-6	Hercules, Inc.	PAN	1.74	4.8	0.696	296	42.9
T300	Union Carbide/Toray	PAN	1.75	3.31	0.480	228	32.1
T500	Union Carbide/Toray	PAN	1.78	3.65	0.530	234	33.6
T700	Toray	PAN	1.80	4.48	0.650	248	36.0
T-40	Toray	PAN	1.74	4.50	0.652	296	42.9
Celion	Celanese/ToHo	PAN	1.77	3.55	0.515	234	34.0
Celion ST	Celanese/ToHo	PAN	1.78	4.34	0.630	234	39.0
XAS	Grafil/Hysol	PAN	1.84	3.45	0.500	234	34.0
HMS-4	Hercules, Inc.	PAN	1.78	3.10	0.450	338	49.0
PAN 50	Toray	PAN	1.81	2.41	0.355	393	57.0
HMS	Grafil/Hysol	PAN	1.91	1.52	0.220	341	49.4
G-50	Celanese/ToHo	PAN	1.78	2.48	0.360	359	52.0
GY-70	Celanese	PAN	1.96	1.52	0.220	483	70.0
P-55	Union Carbide	Pitch	2.0	1.73	0.250	379	55.0
P-75	Union Carbide	Pitch	2.0	2.07	0.300	517	75.0
P-100	Union Carbide	Pitch	2.15	2.24	0.325	724	100
HMG-50	Hitco/OCF	Rayon	1.9	2.07	0.300	345	50.0
Thornel 75	Union Carbide	Rayon	1.9	2.52	0.365	517	75.0

(a) PAN, polyacrylonitrile

Table 2 Typical properties of three-directional orthogonal carbon-carbon composites

Property	Direction Z	Direction X-Y
Density, g/cm^3	1.9	1.9
Tensile strength, MPa (ksi)		
at RT	310 (45)	103 (15)
at 1900 °K (2950 °F)	400 (58)	124 (18)
Tensile modulus, GPa (10^6 psi)		
at RT	152 (22)	62 (90)
at 1900 °K (2950 °F)	159 (23)	83 (120)
Compressive strength, MPa (ksi)		
at RT	159 (23)	117 (17)
at 1900 °K (2950 °F)	196 (28)	166 (24)
Compressive modulus, GPa (10^6 psi)		
at RT	131 (19)	69 (10)
at 1900 °K (2950 °F)	110 (16)	62 (90)
Thermal conductivity, W/m · K (Btu/ft · h · °F)		
at RT	246 (142)	149 (12)
at 1900 °K (2950 °F)	60 (5)	44 (4)
Coefficient of thermal expansion, 10^{-6}/K		
at RT	0 (0)	0 (0)
at 1900 °K (2950 °F)	3 (5)	4 (7)
at 3000 °K (4950 °F)	8 (14)	11 (20)

RT, room temperature. Source: Ref 14

acteristics, which are typical of most carbon-carbon composites:

- Low thermal expansion increases with temperature
- Strength increases with temperature
- Thermal conductivity decreases with temperature

Because carbon fibers tend to increase in strength with increasing temperature (Ref 15), carbon-carbon composites retain their tensile strength at extreme temperatures. Figure 5 compares the bend strength of a carbon-carbon composite at room temperature and 1600 °C (2190 °F).

Carbon-carbon composites provide unmatched specific stiffness and strength at temperatures from 1200 to 2200 °C (2192 to 3992 °F). At temperatures below 1000 °C (1832 °F), they exhibit specific strength equivalent to that of the most advanced superalloys (Ref 16). Figure 6 compares the elevated-temperature specific strength of carbon-carbon composites with that of other heat-resistant materials.

Oxidation of carbon-carbon composites can begin at temperatures as low as 400 °C (750 °F). The rate of oxidation depends on the perfection of the carbon structure and its purity. Highly disordered carbons, such as carbonized resins given low-temperature heat treatments, will oxidize at appreciable rates at 400 °C (750 °F). Highly graphitic structures, such as pitch-based carbon fibers, can be heated as high as 650 °C (1200 °F) before extensive oxidation occurs. At these low temperatures, carbons are very susceptible to catalytic oxidation by alkali metals, such as sodium, and by multivalent metals, such as iron and vanadium, at extremely low concentrations. Therefore, the oxidation rate often is determined by the initial purity of the carbon-carbon composite or by in-service contamination. Borates and particularly phosphates have been found to inhibit oxidation up to about 600 °C (1110 °F) (Ref 21). Oxidation at higher temperatures becomes more rapid, and by 1300 °C (2370 °F) it is completely limited by mass transport of oxygen to the surface and by transport of carbon monoxide and dioxide away from it. Oxidation protection at high temperatures is discussed in the following section of this article.

Protective Coatings

Coating technology for carbon-carbon has been driven primarily by the aerospace and defense industries, in applications where the composite is exposed to high-temperature oxidizing environments. Advanced applications include hot-section components for limited-life missile engines, exhaust components for fighter aircraft, hypersonic vehicle fuselage and wing components, and structures for space defense satellites (Ref 22). The most notable application of coated carbon-carbon is for the nose cap and wing leading edges of the Shuttle Orbiter vehicle (Ref 23–25). Over 40 successful missions have been flown, demonstrating the flight worthiness of coated carbon-carbon in reentry applications.

Fundamentals of Protecting Carbon-Carbon

Historical Development of Protecting Carbon Bodies. Many of the constituents and approaches for protecting carbon-carbon have grown from early research work aimed at protecting synthetic graphite bodies. Sixty years ago, a patent was issued to the National Carbon Company (Ref 26) for a coating method to render carbon articles oxidation resistant at high temperatures. Coating systems composed of an inner layer of SiC and outer glazes based on B_2O_3, P_2O_5, and SiO_2 were described. This work demonstrated the utility of glassy materials as coating constituents to enhance oxidation resistance. Work on JTA graphite for reentry applications (Ref 27–29) was particularly significant. In these

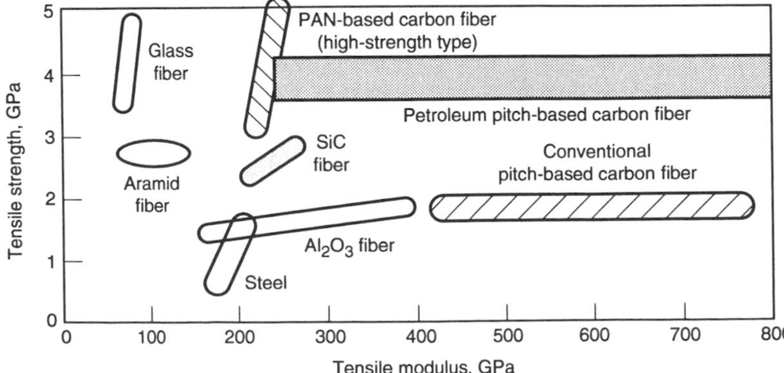

Fig. 2 Typical properties of various reinforcing fibers. PAN, polyacrylonitrile. Source: Ref 12

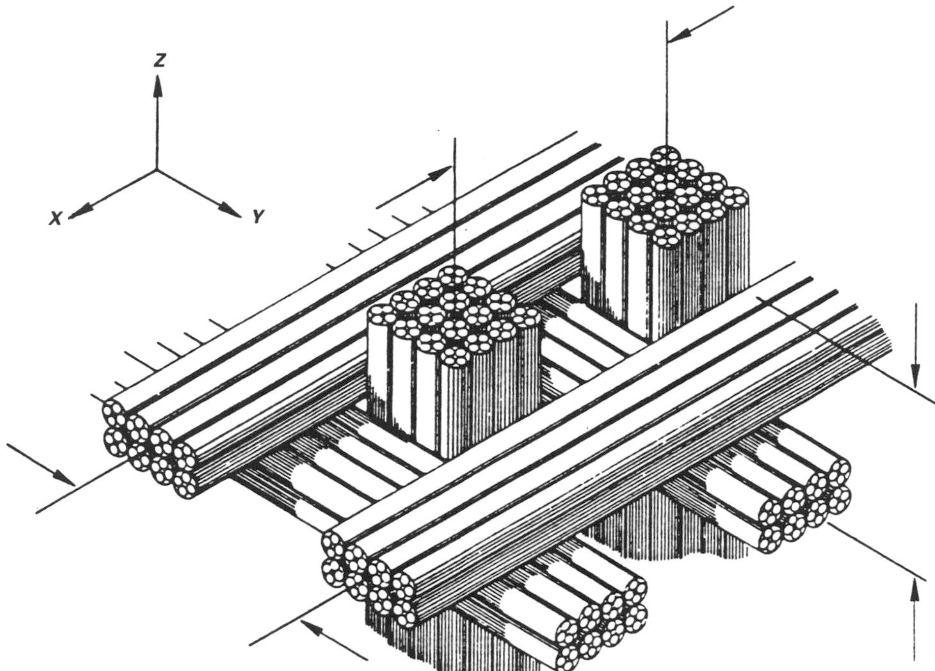

Fig. 3 Three-directional orthogonal preform construction. Source: Ref 13

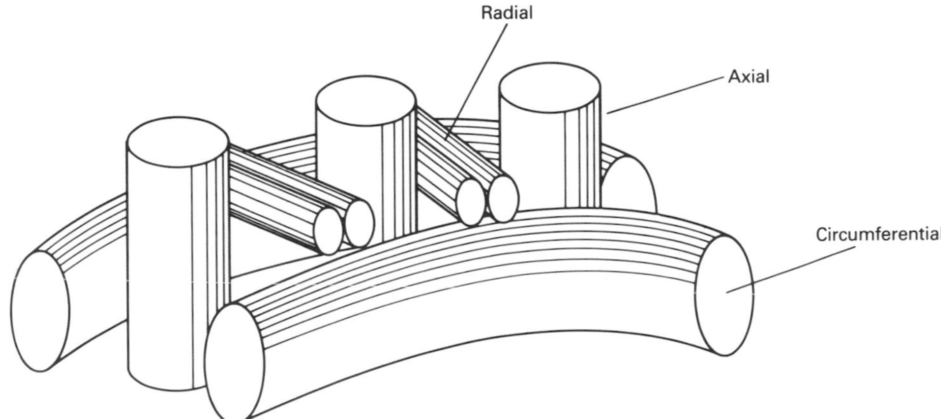

Fig. 4 Three-directional cylindrical preform construction. Source: Ref 13

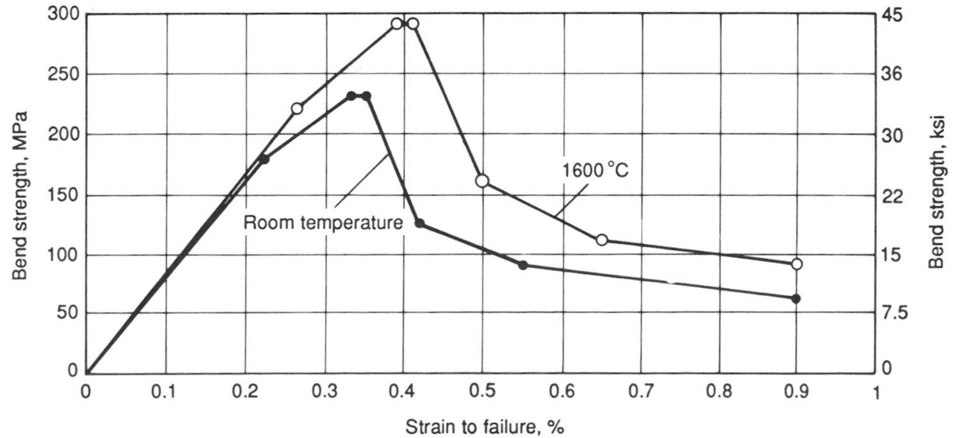

Fig. 5 Comparison of bend strength at room temperature and 1600 °C (2910 °F) for a two-directional weave carbon-carbon composite tested in an inert atmosphere

materials, refractory compounds containing boron, silicon, zirconium, and hafnium were used as additions to impart oxidation resistance. Oxidation resulted in formation of a borate glass coating that was protective for several hours up to 1700 °C (3092 °F). An important contribution of this work was enhancement of oxidation resistance through additions of boron to the body of carbon materials.

The investigation by Chown and coworkers of refractory carbides for protecting graphite (Ref 30) was an important early contribution to coating technology. Their results showed that chemical vapor deposition (CVD) of SiC could provide reliable protection for long periods of time at temperatures below 1700 °C (3092 °F). Experiments with a variety of refractory carbides and boride coatings formed by reaction sintering demonstrated that protection for short times up to 2200 °C (3992 °F) could be achieved with sintered ZrC and ZrB_2 coatings.

Coatings based on the use of iridium (Ref 31, 32) were investigated in the 1960s to protect graphite up to temperatures as high as 2100 °C (3812 °F). This concept relied on the very low carbon diffusivity and oxygen permeability of iridium, as a solid oxygen barrier. While some success was achieved, significant problems associated with volatile oxide formation, adherence, and the high thermal expansion of iridium limited the usefulness of this technology.

Carbon-carbon was thrust into the forefront as a high-temperature materials research topic by the requirement for a reusable, lightweight thermal protection system for the Shuttle Orbiter (Ref 33–36). The oxidation-protected carbon-carbon presently used for the shuttle nose cap and wing leading edge is based on a substrate containing low-strength, low-elastic-modulus rayon precursor fibers. The coating system is composed of a SiC coating, formed by conversion of the carbon-carbon in a pack process, and an outer silicate glaze coating filled with SiC powder (Ref 33–35). The process also involves sealing cracks in the coating system with silicon ethoxide (Ref 36). The SiC conversion layer is over 1 mm (0.025 in.) thick.

Throughout the 1980s, a significant level of research activity focused on the protection of high-performance carbon-carbon that used heat-stabilized polyacrylonitrile (PAN) or pitch-based fibers. These composites have higher strength, higher elastic moduli, and lower thermal expansion coefficients than the rayon-based materials. Applications have focused on using high-performance carbon-carbon in structural-weight-critical roles that require substrates and coatings of minimum thickness. For example, coating thicknesses normally are targeted to be 0.2 to 0.4 mm (0.005 to 0.010 in.). These physical factors, coupled with the significant increase in coating-substrate thermal mismatch stresses, have proved to be significant barriers to acceptable performance and broader use of carbon-carbon composites.

Carbon-Carbon Constituents and Microstructure. Applications requiring coatings typically use carbon fibers in laminated woven cloth or three-dimensional woven reinforcements. The fibers used are derived from rayon, PAN, or petroleum pitch and have a wide range of properties. For example, the elastic modulus along the fiber axis ranges from approximately 41.4 GPa (6×10^6 psi) for rayon fibers to 414 GPa (60×10^6 psi) for heat-stabilized PAN to 690 GPa (100×10^6 psi) for fibers. The axial fiber expansion coefficients become lower as the fiber modulus increases.

The characteristics of the matrix vary, depending on the method of densification. Generally, the matrix microstructure spans a range from being glasslike, with small, randomly oriented crystallites of turbostratic carbon, to having strongly oriented and highly graphitized large crystallites. Weak interfaces usually exist between the fibers and matrix, because strong covalent atomic bonding prevents the carbon constituents from sintering, even at very high temperatures. Because the mechanical properties of the matrix are substantially inferior to those of the fibers, the fibers generally control the mechanical performance and expansion characteristics of the composites. A rayon-fabric-reinforced laminated construction typically exhibits the following in-plane properties: a tensile strength of 51.7 MPa (7.5 ksi), a tensile elastic modulus of 13.8 GPa (2×10^6 psi), and a thermal expansion coefficient of 2.4×10^6/°C (1.3×10^{-6}/°F). Laminated constructions that have high-performance fibers exhibit the following typical in-plane properties: a tensile strength of 276 MPa (40 ksi), a tensile elastic modulus of 90 GPa (13×10^6 psi), and a thermal expansion coefficient of 1.4×10^{-6}/°C (0.8×10^{-6}/°F).

Matrix Inhibition. Carbon begins to oxidize at measurable rates at approximately 400 °C (750 °F). Carbon-carbon composites exhibit high internal surface areas due to the porous nature of the structure (typical levels of interconnected porosity are 10 to 15%). Adding inhibitor phases to the matrix has become an important part of an overall oxidation protection system, because inhibitors allow some control of oxidation that can occur through defects in coatings. Inhibitors can also prevent catastrophic oxidation failure due to coating separation at high temperatures.

Additions of boron, boron compounds, and phosphorus compounds have been effective in protecting carbon bodies (Ref 37–42) by true chemical inhibition and formation of internal and external glass layers that act as diffusion barriers. The practice of making boron additions to carbon-carbon for improved oxidation resistance was first disclosed in a 1978 patent (Ref 43). Since that time, many improvements and variations on this theme have been reported (Ref 44–53).

Internal chemical modifications can be made either by mixing the carbonaceous and nonoxide inhibitor powders and consolidating the constituents to form the carbon body, or by impregnating the porous body with liquids that contain the inhibitors, usually in oxide form. Boron and many nonoxide boron compounds are quite refractory, so the powder mixing and carbon processing route has often been used (Ref 40–42). In composite fabrication, submicron refractory compound additives are normally carried within impregnating resins and are dispersed through the fiber tows as well as between the fabric plies.

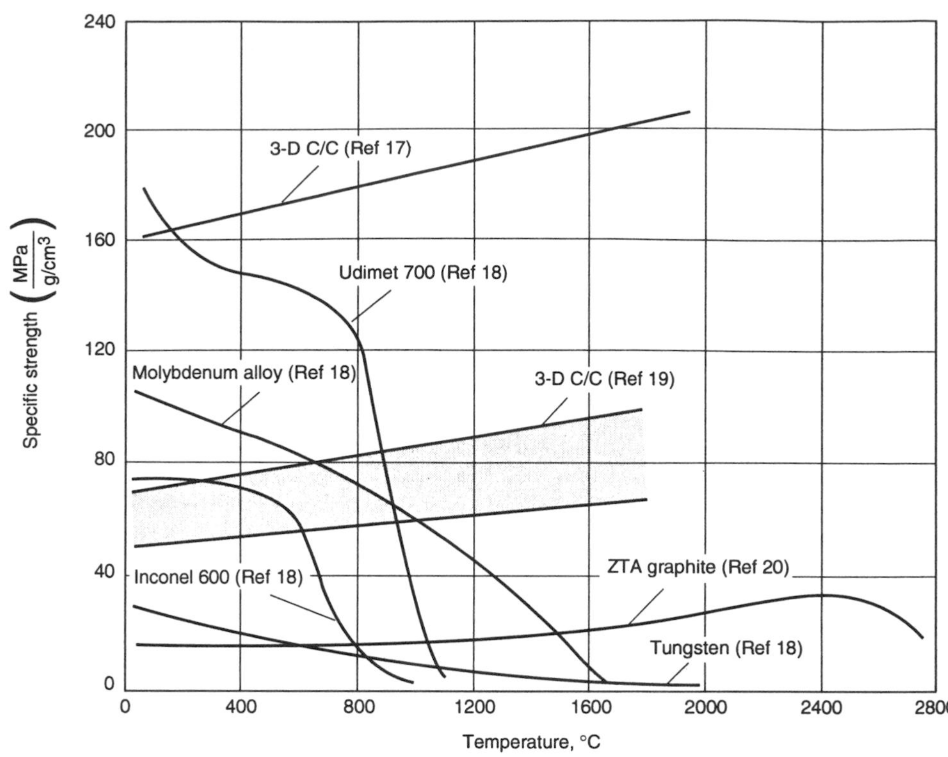

Fig. 6 Specific strength vs. temperature for various high-temperature materials. C/C, carbon-carbon composites; 3-D, three-directional fiber architecture

Coating Selection Principles. The most critical component of any coating architecture is the primary oxygen barrier. The oxygen barrier prevents oxygen ingress to the underlying composite by providing a physical permeation barrier and, in some cases, by gettering oxygen in the process. The critical parameters that guide the selection of the oxygen barrier are its oxidation characteristics, thermal expansion coefficient, and inherent oxygen permeability. A material that forms an adherent, low-permeability oxide scale is preferred as an oxygen barrier because it oxidizes slowly and has the potential to self-heal.

Figure 7 presents an Arrhenius plot of rate constants for oxidation of refractory materials that are typically considered for coating applications (Ref 54, 55). Scale growth as a function of time can be estimated from Fig. 7 using the relationship $x^2 = Kt$, where x is the scale thickness, K is the parabolic rate constant, and t is time in hours. Silicon-base ceramics exhibit substantially lower oxide growth kinetics than aluminum-, hafnium-, or zirconium-base ceramics. Time and temperature of service will dictate material selection and coating thickness. However, from the standpoint of forming thin protective scales in thermal cycles with peak temperatures in the range of 1400 to 1700 °C (2552 to 3092 °F), only Si_3N_4 and SiC exhibit sufficiently low

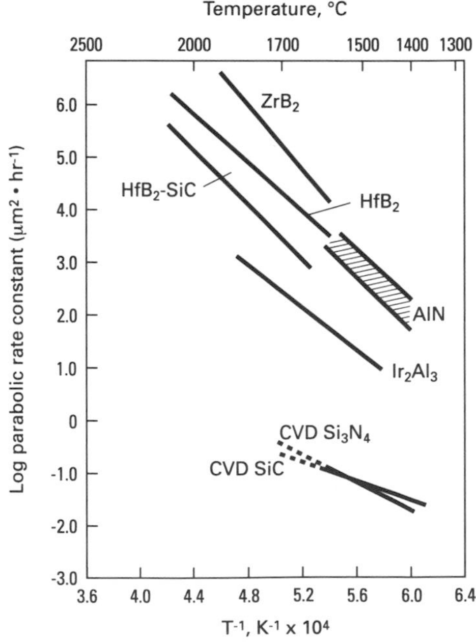

Fig. 7 Oxidation kinetics of refractory materials. CVD, chemical vapor deposition

rate constants for growth over extended time periods.

Figure 8 compares the thermal expansion behavior of refractory coating candidates with that measured for high-performance, fabric-reinforced carbon-carbon (Ref 16, 54, 55). The expansion of carbon-carbon in the in-plane direc-

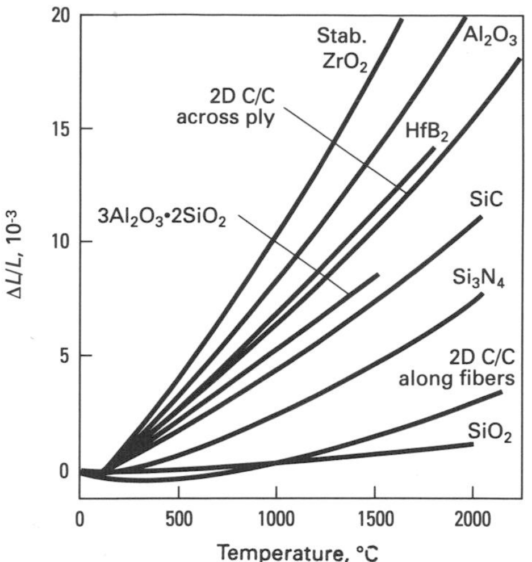

Fig. 8 Thermal expansion characteristics of ceramics and carbon-carbon laminates. C/C, carbon-carbon laminate; *L*, specimen length

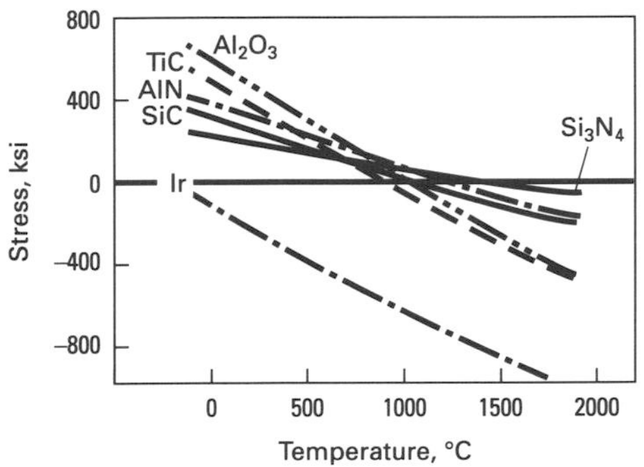

Fig. 9 Calculated thermal stresses for thin coatings on high-performance carbon-carbon laminates. Ratio of substrate thickness to coating thickness = 20.

tions is substantially lower than that of any of the refractory ceramics. This expansion difference, coupled with the high modulus of the refractory materials, results in significant thermal mismatch stresses when they are employed as coatings. An estimate of the thermal mismatch stresses when the coating thickness is small relative to the carbon-carbon substrate thickness can be calculated using the relationship

$$\sigma_c = \frac{E(\alpha_c - \alpha_s)\Delta T}{1 - \nu}$$

where σ_c is the stress in the coating, E is the coating elastic modulus, ν is the Poisson ratio of the coating, α_s is the thermal expansion coefficient of carbon-carbon, α_c is the thermal expansion coefficient of the coating, and ΔT is the difference between the deposition temperature and the selected temperature for stress calculation.

Table 3 summarizes the properties of refractory materials that have been used in deposition studies. Figure 9 presents the thermal stresses calculated as a function of temperature when these coatings were deposited onto high-performance two-dimensional carbon-carbon laminates. For the refractory ceramics, silicon nitride provides

Table 3 Properties of refractory materials deposited on carbon-carbon composites

Material	Deposition process	Deposition temperature, °C	Bulk properties: Modulus, psi × 10^6	Poisson's ratio	CTE (20–1900 °C), ppm/°C
SiC	CVD	1050	65	0.19	5.2
TiC	CVD	1000	65	0.19	9.5
Al_2O_3	CVD	1050	58	0.28	10.3
AlN	CVD	1250	50	0.3	6.1
Si_3N_4	CVD	1420	46	0.3	3.6
Ir	Sputtering	250	76	0.3	7.9
HfO_2	EB-PVD	1000	20	0.25	10.6

CTE, coefficient of thermal expansion; CVD, chemical vapor deposition; EB-PVD, electron-beam physical vapor deposition

the lowest thermal mismatch stresses of any of the ceramic coating candidates. These stresses are still high enough to cause cracking, however. Therefore, it is usually found that deposited ceramic coatings exhibit microcracking and that the crack pattern depends on the coating thickness and deposition temperature. Iridium metal deposited by electron-beam physical vapor deposition techniques can have low thermal mismatch stresses upon cooling. However, such a coating must then be able to withstand extremely high compressive stresses upon heating. In previously reported work (Ref 55), it has been shown that iridium-base coatings deposited by this technique onto high-performance carbon-carbon fail by compressive spalling at elevated temperatures.

Preferred Coating Approaches

Coating approaches are dictated by both application requirements and fundamental behavior. Generally speaking, SiC-based and Si_3N_4-based coatings have found broad use at temperatures below 1700 °C (3092 °F) because of minimum thermal mismatch stresses and low oxide-scale growth kinetics. In the higher temperature range (1700 to 2200 °C, or 3092 to 3992 °F), refractory carbides and borides have been used for short time periods. Coating deposition techniques that have been used include pack cementation, CVD, and slurry processes. Coating architectures are normally built using combinations of these techniques. In the following sections, typical coating architectures are discussed in accordance with the process used to deposit the primary oxygen barrier.

Pack Cementation. The coating system used on the Shuttle Orbiter vehicle is the preeminent example of the use of a pack process to create an oxidation protection system for carbon-carbon (Ref 25). In this process, the carbon-carbon part is packed in a retort with a dry pack mixture of alumina, silicon, and SiC. The retort is placed in a furnace, and under argon atmosphere a stepped time-temperature cycle is used to activate conversion of the carbon-carbon surface to SiC. Peak process temperature is approximately 1760 °C (3200 °F). This creates a porous SiC surface that is nominally 1.0 to 1.5 mm in thickness. Multiple impregnation and curing with an acid-activated tetraethoxysibcate liquid produces SiO_2 coating of the porous surfaces. A surface sealant consisting of a mixture of a commercial alkali silicate bonding liquid filled with SiC powder is then applied.

This system was designed to provide protection during multiple reentry cycles where surface temperatures of 1538 °C (2800 °F) are anticipated. The success of the shuttle missions and further testing (Ref 23–25) have proved this to be an effective approach for low-performance rayon-based composites. Attempts to use similar coatings modified with boron (Ref 56, 57) for other aerospace applications requiring high-performance carbon-carbon have met with only limited success.

Chemical Vapor Deposition. Attempts to expand carbon-carbon use to turbine engine hot-section and exhaust components fostered the need for protective coatings that could be applied as thin layers over the structural components without compromising mechanical performance. The coating architectures developed have been dependent on the application lifetime as well as on dynamic or static structural requirements. The CVD coatings are normally applied in multiple cycles to ensure even deposition rates over curved surfaces. A substrate pretreatment is normally used to enhance adherence. Silicon nitride overlay coatings have been shown to be effective for limited-life (<20 h) cycles where heating above the deposition temperature occurs rapidly and peak temperatures reach 1760 °C (3200 °F) (Ref 54, 55, 58). These coatings have employed a thin reaction layer of SiC (formed in a pack process, of the order of 5 μm) to serve as a reaction barrier and to enhance adherence. Si_3N_4 has been applied in thicknesses ranging from 125 to 250 μm (0.005 to 0.010 in.) in a multiple-step CVD process.

Other applications require that carbon-carbon withstand hundreds of hours of exposure to peak temperatures in the range of 1400 to 1500 °C (2552 to 2732 °F) and undergo thermal cycling to temperatures in the range of 600 to 1200 °C (1112 to 2192 °F). In these extended-life applications, a boron-rich inner layer is used to provide a source of glassy phase to seal microcracks in the outer coatings. Elemental boron, boron carbide, and combinations of boron compounds mixed with silicon or SiC are inner layer approaches. These layers are normally deposited in thicknesses in the range of 25 to 50 μm (0.001 to 0.002 in.) using CVD, conversion of the carbon surface, and slurry coating (Ref 59-61). Depositing SiC or Si_3N_4 by CVD is the preferred method to provide

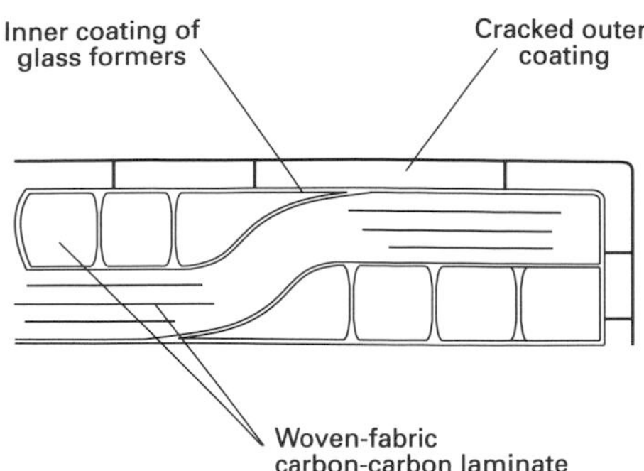

Fig. 10 Schematic of coating architecture used to protect carbon-carbon for extended life applications

hard, erosion-resistant surfaces that cover the boronated inner layers and inhibit vaporization of borate glass sealants (Ref 62, 63). Overlay thicknesses in the range of 200 to 300 μm (0.008 to 0.012 in.) are normally deposited in a multiple-step process. A typical coating architecture on an inhibited composite is shown schematically in Fig. 10.

Silicate glazes are frequently applied to fill the microcrack network that exists in SiC and Si_3N_4 coatings. Although the glaze is applied externally and is susceptible to vaporization and physical removal, it has been shown to improve cyclic oxidation lifetimes. Glaze overcoats are normally applied as aqueous sols incorporating boron and silicon that can be painted, sprayed, or dip coated. Typical processing involves air drying and firing above 1038 °C (1900 °F) in an argon atmosphere (Ref 62). The glaze can be periodically replenished.

Slurry coatings are produced by dispersing appropriate ceramic or metal powders in a liquid vehicle to make the slurry, applying the slurry as a paint to the component surface, evaporating or gelling the liquid to harden the coating, and then heating to a high temperature to stabilize and densify the coating. Slurries are applied by brushing, spraying, or dipping. The liquids can be water or volatile organics with organic binders in solution, inorganic or organic sols or solutions that form oxides, or thermosetting preceramic polymers or polymer solutions (Ref 64, 65).

Hardening produces a coating that is composed of the powder particles bound together and bonded to the substrate by the solid that is precipitated or condensed from the liquid. Heating to a high temperature decomposes the binder phase to form carbon or a ceramic. The shrinkage associated with binder decomposition and incomplete solid-state sintering of the powder particles resulted in a cracked, porous, and often weakly bonded coating unless a flowable and wetting liquid is formed by one of the constituents. This can be a glass, molten metal, or ceramic melt.

When coatings are meant to provide oxidation protection for graphite and carbon-carbon composite articles below 1000 °C (1832 °F), they often contain large amounts of boron in the form of elemental boron, B_4C, BN, metal borides, or B_2O_3 (Ref 66, 67). Employing B_2O_3 glass provides a wetting liquid at low temperatures on initial heating, and the nonoxides rapidly oxidize in use to produce the same result. Coatings composed mostly of refractory oxide particles, bound together with small amounts of borate glass, have shown utility at temperatures in the range of 1200 to 1500 °C (2192 to 2732 °F) in configurations where evaporation of the B_2O_3 is inhibited (Ref 46). In making such coatings the boron can be present in the powder constituents, the liquid vehicle, or both. Water or alcohol solutions of boric acid and liquid boron alkoxides are often used (Ref 46, 52). Solutions of preceramic boron polymers are also a possibility (Ref 68, 69).

Slurry coatings meant for higher temperatures, in which borates are replaced by glassy alkali silicates or aluminum phosphate, are prominent (Ref 36, 56). The use of siloxane fluids as preceramic polymers has also been disclosed (Ref 70). Converting the carbon surface to SiC is often recommended as a pretreatment to ensure the adherence of glassy borate, silicate, and phosphate coatings (Ref 36, 56).

The bonding and densification of slurry coatings with molten metal and melted ceramic phases has been used to produce protective layers with intermediate to very high temperature capabilities. Dense, very adherent coatings capable of extreme-temperature service can be made from paints containing fine refractory boride particles (Ref 30). The coatings are fully stabilized by heating the borides in contact with the carbon surface to temperatures over 2000 °C (3632 °F) in an inert environment to form a boride-carbon eutectic liquid. Coatings made by melting certain combinations of metal powders and reacting these with the carbon surface to form refractory carbides (Ref 71) are protective to 1800 °C (3272 °F). Recent work of this type (Ref 72) using a mixture of silicon, hafnium, and chromium powders reacted with the carbon surface at 1450 °C (2642 °F) has produced carbide coatings that provide excellent oxidation protection for short times at 1200 °C (2192 °F).

Practical Limitations of Coatings

As mentioned above, carbon-carbon composites can be used as structural materials up to at least 2200 °C (3992 °F). At the time of this writing, viable coating concepts to match this capability have not been consistently demonstrated, especially for times greater than a few hours. SiC and Si_3N_4 are limited thermodynamically to temperatures of approximately 1800 to 1815 °C (3272 to 3300 °F). At higher temperatures, the SiO_2 layers that form and protect these materials are disrupted by CO and N_2 interfacial pressures that become greater than 10^{-1} MPa (1 atm), causing the coatings to erode by uncontrolled oxidation (Ref 16, 55). Use of more refractory materials, such as HfC or HfB_2, is limited by the very rapid oxidation rates pointed out in Fig. 7. Rapid conversion of these films to high-expansion oxides leads to severe spallation in thermal cycles. Thus, above approximately 1760 °C (3200 °F), coating lifetimes are currently limited to a few hours.

For the range of applications where coating architectures incorporating borate sealant glasses are used, coating use temperatures are limited to approximately 1500 to 1550 °C (2732 to 2822 °F). When B_2O_3 contacts carbon at atmospheric pressure, the CO reaction product pressure will exceed 10^{-1} MPa (1 atm) at approximately 1575 °C (2867 °F). Borate glasses also cause dissolution of the protective SiO_2 scale that forms on SiC or Si_3N_4, leading to more rapid corrosion because of the high oxygen permeability of the mixed glass. Experience in test cycles with peak temperatures about 1400 °C (2552 °F) has shown that accelerated dissolution of coatings along microcrack boundaries eventually causes gross oxidation of boron-base inner layers, leading to massive dissolution of the silicon-base overlays.

Moisture sensitivity of borate glasses (Ref 73) can be a major limitation. Hydrolysis at low temperatures in moist air converts adherent B_2O_3-containing layers into loosely bonded boric acid particulate. Under long-term exposure, the sealant glasses that form beneath the hard overlays undergo moisture attack that leads to spallation. Subsequent heating cycles that rapidly release moisture can cause catastrophic failure. Finally, high-temperature exposure to moist environments makes borate glass susceptible to vaporization by the formation of HBO_2 (Ref 52).

Applications

The high cost of multidirectional carbon-carbon composites has restricted their use to aerospace and specialty applications. However, the development of cost-effective, automated, 3-D preform manufacturing techniques should lead to new applications. Some of the more important current and potential applications for carbon-carbon composites are summarized below.

Improved aircraft brakes became necessary as aircraft became heavier and faster. Aborted takeoffs that were terminated at maximum ground speed were uncertain with steel disk brakes. If successful in stopping the aircraft, the brakes were often destroyed because of warping or melting due to the intense heat generated during the stop. Carbon-carbon composites have a high melting point, are resistant to thermal shock, and have excellent friction and wear characteristics. They can be fabricated in shapes and sizes that are suitable for brake applications. Carbon-carbon brakes can provide superior stopping capability, survive several abortive stops, and last far longer than conventional aircraft brakes. Furthermore, this superior performance is accomplished with a significant weight savings (Ref 74–77). Such brakes are also being considered for high-speed train applications.

Disk brakes for aircraft are composed of a number of disks, half of which are keyed to the nonrotating brake mechanism. The other half rotate with the wheel to which they are keyed.

Braking is accomplished by forcing the disks together, at which time friction is converted to heat, which must be dissipated. This requires a material that is resistant to thermal shock, stable to very high temperatures, and has low thermal expansion and good thermal conductivity. In addition, the material should have a friction coefficient of about 0.3 to 0.5 for good stopping performance. Carbon-carbon composites have all of these important properties, along with a density of about 1.9 g/cm^3, which provides nearly four times the stopping power of copper or steel brakes. Some advanced aircraft require 820 to 1050 kJ/kg (350 to 450 Btu/lb) of the carbon-carbon brake to stop the aircraft. High-performance automobiles require only 300 to 520 kJ/kg (130 to 220 Btu/lb) (Ref 75). Carbon-carbon brakes with fibers in the plane of the flat surface have low wear and good frictional characteristics. Thus, carbon-carbon disks are made with layers of carbon fabric or with random fibers arrayed parallel to the braking surface.

Special modifications to the carbon-carbon brake system include the incorporation of refractory borides to avoid temporary reduction in the coefficient of friction after a period of idleness. The friction loss is probably caused by water absorption, and it can be overcome operationally by applying brake pressure several times before takeoff. The borides eliminate the need for this preconditioning. Metal caps on the teeth of the rotating disks are used on some systems to reduce wear on the teeth and to increase thermal conductivity.

Space Structures. Initiatives in the exploration and commercialization of space have highlighted the need for new high-performance composites for spacecraft. Many types of composites have applications in this area, including organic resin and metal matrix composites, but carbon-carbon composites have unique applications when the temperature of the structural components exceeds 1000 °C (1830 °F), even for short periods of time.

Carbon-carbon composites have several unique properties that make them attractive for high-temperature space applications. In some cases, they are the only material solution for specific design problems in spacecraft because they are inert in most space environments. Although they do suffer some oxidation from atomic oxygen in low earth orbit, carbon-carbon composites do not degrade or outgas in the high vacuum of deep space, as organic resin composites do. This property is attributed to the high manufacturing temperatures used to pyrolyze the organic precursors, such that carbon is the only remaining element.

Carbon-carbon composites are also attractive for space structures because of their relatively low density. Typical densities range from 1.4 g/cm^3 for flat panels and tubes to 2.0 g/cm^3 for composites reinforced in three directions with high-density fibers, such as pitch-based carbon. This low density (Ref 78) provides a significant weight advantage over such structural metals as aluminum, titanium, and niobium, as indicated below:

Material	Density, g/cm^3	Weight ratio to carbon-carbon materials
Carbon-carbon	1.4–2.0	1.0
Aluminum	2.7	1.6
Titanium	4.5	2.7
Niobium	8.4	4.9

Thermal expansion is a significant design consideration in space structures. The extent to which a space structure, such as a communications boom, is affected by solar radiation on one side and cold space on the other side is a key design parameter, especially as the sizes of such components increase. Carbon-carbon composites have very low expansion values over the range of –150 to 3000 °C (–240 to 5430 °F). The CTE at –150 °C (–240 °F) is typically 10×10^{-6}/K; at 3000 °C (5430 °F), it is 6×10^{-3}/K. In addition, there is a minimum in the thermal expansion curve between 150 and 300 °C (300 and 570 °F). This is of particular interest because in normal space systems, in which the temperature range is –100 to 100 °C (–150 to 212 °F), the thermal expansion of carbon-carbon composites being considered for spacecraft is nearly 0 (Ref 79).

Another property of interest in applying carbon-carbon composites in spacecraft is high thermal conductivity. It has been shown that the thermal conductivity of carbon fibers is a function of the degree of orientation of the graphite crystallites along the fiber axis. Therefore, carbon-carbon composites fabricated with carbon fibers with highly ordered crystallites, such as pitch-based carbon fibers, will have high thermal conductivity. The specific conductivity of carbon-carbon exceeds that of metals.

In many spacecraft applications, the mechanical properties of carbon-carbon composites become very important. Component designs that strive for the lowest possible weight result in very thin structural components that depend on materials with very high strength and stiffness. Carbon-carbon composites can meet these needs, especially when the temperature to which the component will be exposed exceeds 1000 °C (1830 °F). Carbon-carbon tubes have been made with a tensile strength greater than 400 MPa (60 ksi), which places it well above the category of advanced carbon-carbon composites (see Ref 80). In addition, the specific elastic modulus is three times that of steel.

The key to applying carbon-carbon composites to spacecraft is carbon fiber selection. For structural components that are required to be very thin and still have high strength and high modulus, the selection of carbon fibers with filament diameters of about 4 μm (160 μin.) is essential. Also important for spacecraft applications is the availability of high-modulus (and high-conductivity) pitch-based carbon fibers.

Other Applications. Bolts, screws, nuts, and washers are used where high-temperature and severe chemical conditions are present. Strength and stiffness at high temperatures guarantee high fastening stability. If graphite parts are screwed together, only low fastening moments are necessary. The system is self-fastening at the application temperature because of the anisotropic CTE. Parts are applied in the semiconductor industry, in furnace constructions, and other high-temperature equipment.

Interesting applications are as heaters, heat shields, and furnace walls in vacuum furnaces. Liners, plates, tubes, crucibles, sleeves, and other auxiliary aids are applied in the field of apparatus construction. The properties of carbon-carbon composites are also useful as tool segments, pressure plates, and resistance elements in hot sintering applications. In some industries, these composites are applied to replace asbestos.

Unidirectional carbon-carbon composites are being considered for engine valve applications because they have excellent mechanical strength (Ref 81). However, low interlaminar shear strength is a limitation for achieving the necessary mechanical performance in the valve head and the head-stem transition region. Careful modification of the design and layout of the composite in the valve body will be needed to overcome this limitation.

Carbon-carbon composites are biocompatible and can be tailored to be structurally compatible with bone for applications such as integral fixation of fractures (Ref 82). Multidirectional carbon-carbon composites are also being studied for potential use in hip joint replacements.

ACKNOWLEDGMENTS

The information in this article is largely taken from:

- J.R. Strife and J.E. Sheehan, Protective Coatings for Carbon-Carbon Composites, *Surface Engineering,* Vol 5, *ASM Handbook,* ASM International, 1994, p 887–891
- G.E. Ziegler and W. Hüttner, Engineering Properties of Carbon-Carbon and Ceramic-Matrix Composites, *Ceramics and Glasses,* Vol 4, *Engineered Materials Handbook,* ASM International, 1991, p 835–844
- H.D. Batha and C.R. Rowe, Structurally Reinforced Carbon-Carbon Composites, *Composites,* Vol 1, *Engineered Materials Handbook,* ASM International, 1987, p 922–924
- R.J. Diefendorf, Carbon/Graphite Fibers, *Composites,* Vol 1, *Engineered Materials Handbook,* ASM International, 1987, p 49–53
- R.J. Diefendorf, Continuous Carbon Fiber Reinforced Carbon Matrix Composites, *Composites,* Vol 1, *Engineered Materials Handbook,* ASM International, 1987, p 911–914
- N.W. Hansen, Carbon Fibers, *Composites,* Vol 1, *Engineered Materials Handbook,* ASM International, 1987, p 112–113
- L.E. McAllister, Multidirectionally Reinforced Carbon/Graphite Matrix Composites, *Composites,* Vol 1, *Engineered Materials Handbook,* ASM International, 1987, p 915–919

REFERENCES

1. L.E. McAllister and W.L. Lachman, *Handbook of Composites,* A. Kelly and S.T. Mileiko, Ed., North-Holland, 1983, p 111
2. F.K. Ko, *Carbon-Carbon Materials and Composites,* J.D. Buckley and D.D. Edie, Ed., Noyes, 1993, p 71
3. H.O. Pierson and M.L. Lieberman, *Carbon,* Vol 13, 1975, p 159
4. J.H. Cranmer, I.G. Plotzker, L.H. Peebles, and D.R. Uhlmann, *Carbon,* Vol 21, 1983, p 201
5. M.A. Forrest and H. Marsh, *J. Mater. Sci.,* Vol 18, 1983, p 973
6. G.S. Rellick, D.J. Chang, and R.J. Zaldivar, *J. Mater. Res.,* Vol 7, 1992, p 2798
7. N. Murdie, C.R Ju, J. Don, and M.A. Wright, *Carbon-Carbon Materials and Composites,* J.D. Buckley and D.D. Edie, Ed., Noyes, 1993, p 105
8. E. Fitzer, A. Gkogkidis, and M. Heine, *High Temp.-High Press.,* Vol 16, 1984, p 363
9. E. Fitzer and A. Gkogkisdis, *Petroleum-Derived Carbons,* J.D. Bacha, J.W. Newman, and J.L. White, Ed., American Chemical Society, 1986, p 346
10. R.L. Burns, *Carbon-Carbon Materials and Composites,* J.D. Buckley and D.D. Edie, Ed., Noyes, 1993, p 197
11. L.E. McAllister and A.R. Taverna, "The Development of High Strength Three Dimensionally Reinforced Graphite Composites," paper presented at the 73rd Annual Meeting, American Ceramic Society (Chicago), 1971
12. T. Kohno, A. Mutoh, Y. Kude, and Y. Sohda, Potential Qualities of Pitch-Based Carbon Fiber for High Temperature Composites, *Ceram. Eng. Sci. Proc.,* Vol 15 (No. 4), 1994, p 162–169
13. W.L. Lachman, J.A. Crawford, and L.E. McAllister, Multidirectionally Reinforced Carbon-Carbon Composites, *Proc. Internat. Conf. on Composite Materials,* B. Noton, R. Signorelli, K. Street, and L. Phillips, Ed., Metallurgical Society of AIME, 1978, p 1302–1319
14. A. Levine, "High Pressure Densified Carbon-Carbon Composites, Part II: Testing," Paper FC-21, presented at the 12th Biennial Conference on Carbon (Pittsburgh, PA), 1975
15. C.R. Rowe and D.L. Lowe, *Extended Abstracts of the 13th Biennial Conf. on Carbon,* American Carbon Society, July 1977, p 170
16. J.R. Strife and J.E. Sheehan, *Ceram. Bull.,* Vol 67 (No. 2), 1988, p 369
17. A. Levine, paper presented at the 12th Biennial Conference on Carbon (Pittsburgh, PA), 1975
18. J.G. Sessler and V. Weiss, Ed., *Aerospace Structural Metals Handbook,* Vol 11A, 4th ed., 1967
19. S.E. Hsu and C.I. Chen, The Processing and Properties of Some C/C Systems, *Superalloys, Supercomposites and Superceramics,* Academic Press, 1989, p 721–744
20. R.M. Hale and W.M. Fassel, Jr., WADD Tech. Report, Vol XIV, 1963, p 61–72
21. R.J. Diefendorf, Continuous Carbon Fiber Reinforced Carbon Matrix Composites, *Composites,* Vol 1, *Engineered Materials Handbook,* ASM International, 1987, p 911–914
22. L. Rubin, *Carbon-Carbon Materials and Composites,* J.D. Buckley and D.D. Edie, Ed., Noyes, 1993, p 267
23. H.G. Maahs, C.W. Ohlhorst, D.M. Barrett, P.O. Ransone, and J.W. Sawyer, *Materials Stability and Environmental Degradation,* MRS Symp. Proc., Vol 125, A. Barkatt, E.D. Verink, and L.R. Smith, Ed., Materials Research Society, 1988, p 15
24. R.C. Dickinson, *Materials Stability and Environmental Degradation,* MRS Symp. Proc., Vol 125, A. Barkatt, E.D. Verink, and L.R. Smith, Ed., Materials Research Society, 1988, p 3
25. D.M. Curry, E.H. Yuen, D.C. Chao, and C.N. Webster, *Damage and Oxidation Protection in High Temperature Composites,* Vol 1, G.K. Haritos and O.O. Ochoa, Ed., ASME, 1991, p 47
26. H.V. Johnson, "Oxidation Resisting Carbon Article," U.S. Patent 1,948,382, 20 Feb 1934
27. K.J. Zeitsch, *Modern Ceramics,* J.E. Hove and W.C. Riley, Ed., John Wiley, 1967, p 314
28. S.A. Bortz, *Ceramics in Severe Environments,* W.W. Kriegel and H. Palmour, Ed., Plenum, 1971, p 49
29. E.M. Goldstein, E.W. Carter, and S. Klutz, *Carbon,* Vol 4, 1966, p 273
30. J. Chown, R.F. Deacon, N. Singer, and A.E.S. White, *Special Ceramics,* P. Popper, Ed., Academic Press, 1963, p 81
31. J.M. Criscione, R.A. Mercuri, E.P. Schram, A.W. Smith, and H.F. Volk, "High Temperature Protective Coatings for Graphite," ML-TDR-64-173, Part H, Materials Laboratory, Wright-Patterson Air Force Base, Oct 1974
32. *High Temperature Oxidation Resistant Coatings,* National Academy of Sciences and Engineering, 1970, p 112
33. D.C. Rogers, D.M. Shuford, and J.I. Mueller, *Proc. Seventh National SAMPE Technical Conf.,* Society of Aerospace Material and Process Engineers, 1975, p 319
34. D.C. Rogers, R.O. Scott, and D.M. Shuford, *Proc. Eighth National SAMPE Technical Conf.,* Society of Aerospace Material and Process Engineers, 1976, p 308
35. Surface Seal for Carbon Parts, *NASA Technical Briefs,* Vol 6 (No. 2), MSC-18898, 1981
36. D.M. Shuford, "Enhancement Coating and Process for Carbonaceous Substrates," U.S. Patent 4,471,023, 11 Sept 1984
37. M.J. Lakewood and S.A. Taylor, "Oxidation-Resistant Graphite Article and Method," U.S. Patent 3,065,088, 20 Nov 1962
38. E.M. Goldstein, E.W. Carter, and S. Klutz, *Carbon,* Vol 4, 1966, p 273
39. W.E. Parker and J.F. Rakszawski, "Oxidation Resistant Carbonaceous Bodies and Method for Making," U.S. Patent 3,261,697, 19 July 1966
40. R.E. Woodley, *Carbon,* Vol 6, 1968, p 617
41. H.H. Strater, "Oxidation Resistant Carbon," U.S. Patent 3,510,347, 5 May 1970
42. K.J. Zeitsch, *Modern Ceramics,* J.E. Hove and W.C. Riley, Ed., John Wiley, 1967, p 314
43. L.C. Ehrenreich, "Reinforced Carbon and Graphite Articles," U.S. Patent 4,119,189, 10 Oct 1978
44. T. Vasilos, "Self-Healing Oxidation-Resistant Carbon Structure," U.S. Patent 4,599,256, 8 July 1986
45. P.E. Gray, "Oxidation Inhibited Carbon-Carbon Composites," U.S. Patent 4,795,677, 3 Jan 1989
46. D.W. McKee, *Carbon,* Vol 25, 1987, p 551
47. J.F. Rakszawski and W.E. Parker, *Carbon,* Vol 2, 1964, p 53
48. D.W. McKee, C.L. Spiro, and E.J. Lamby, *Carbon,* Vol 22, 1984, p 507
49. R.C. Shaffer, "Coating for Fibrous Carbon Materials in Boron Containing Composites," U.S. Patent 4,164,601, 14 Aug 1979
50. R.C. Shaffer and W.L. Tarasen, "Carbon Fabrics Sequentially Resin Coated with (1) A Metal-Containing Composition and (2) A Boron-Containing Composition Are Laminated and Carbonized," U.S. Patent 4,321,298, 23 March 1992
51. I. Jawed and D.C. Nagle, Oxidation Protection in Carbon-Carbon Composites, *Mat. Res. Bull.,* Vol 21, 1986, p 1391
52. D.W. McKee, *Carbon,* Vol 24, 1986, p 737
53. J.E. Sheehan and H.D. Batha, "C-C Composite Matrix Inhibition," paper presented at the 16th National Technical Conference, Society of Aerospace Material and Process Engineers, Oct 1984
54. J.R. Strife, *Damage and Oxidation Protection in High Temperature Composites,* G.K. Haritos and O.O. Ochoa, Ed., American Society of Mechanical Engineers, 1991, p 121
55. J.R. Strife, *Proc. Sixth Annual Conf. on Materials Technology,* M. Genisio, Ed., Southern Illinois University at Carbondale, 1990, p 166
56. D.M. Shuford, "Composition and Method for Forming a Protective Coating on Carbon-Carbon Substrates," U.S. Patent 4,465,777, 14 Aug 1984
57. T.E. Schmid, "Oxidation Resistant Carbon/Carbon Composites for Turbine Engine Aft Sections," AFWAL-TR-82-4159, Materials Laboratory, Wright-Patterson Air Force Base, Oct 1982
58. J.R. Strife, "Development of High Temperature Oxidation Protection for Carbon-Carbon Composites," NADC Report 91013-60, Naval Air Development Center, Warminster, PA, 1990
59. D.M. Shuford, "Composition and Method for Forming a Protective Coating on Carbon-Carbon Substrates," U.S. Patent 4,465,888, 14 Aug 1984
60. R.A. Holzl, "Self Protecting Carbon Bodies and Method for Making Same," U.S. Patent 4,515,860, 7 May 1985
61. D.A. Eitman, "Refractory Composite Articles," U.S. Patent 4,735,850, 5 April 1988
62. H. Dietrich, *Mater. Eng.,* Aug 1991, p 34
63. J.E. Sheehan, *Carbon-Carbon Materials and Composites,* J.D. Buckley and D.D. Edie, Ed., Noyes, 1993, p 2

64. C.W. Turner, Sol-Gel Process—Principles and Applications, *Ceram. Bull.,* Vol 70, 1991, p 1487
65. R.W. Rice, *Ceram. Bull.,* Vol 62, 1983, p 889
66. N.A. Hooton and N.E. Jannasch, "Coating for Protecting a Carbon Substrate in a Moist Oxidation Environment," U.S. Patent 3,914,508, 21 Oct 1975
67. G.R. Marin, "Oxidation Resistant Carbonaceous Bodies and Method of Producing Same," U.S. Patent 3,936,574, 3 Feb 1976
68. W.S. Coblenz, G.H. Wiseman, P.B. Davis, and R.W. Rice, Emergent Process Methods for High-Technology Ceramics, *Mater Sci. Res.,* Vol 17, 1984
69. L.G. Sneddon, K. Su, P.J. Fazen, A.T. Lynch, E.E. Remsen, and J.S. Beck, *Inorganic and Organometallic Oligomers and Polymers,* Kluwer Academic Publishers, 1991
70. M.S. Misra, "Coating for Graphite Electrodes," U.S. Patent 4,418,097, 29 Nov 1983
71. A.J. Valtschev and T. Nikolova, "Protecting Carbon Materials from Oxidation," U.S. Patent 3,348,929, 24 Oct 1967
72. H.S. Hu, A. Joshi, and J.S. Lee, *J. Vac. Sci.,* Vol A9, 1991, p 1535
73. P.B. Adams and D.L. Evans, *Mater. Sci. Res.,* Vol 12, 1978, p 525
74. J.P. Ruppe, Today and the Future in Aircraft Wheel and Brake Development, *Can. Aeronaut. Space J.,* Vol 27, 1981, p 212–216
75. L. Heraud and B. Broquere, Carbone pour le Freinage, Development in the Science and Technology of Composite Materials, *First European Conf. on Composite Materials and Exhibition,* A.R. Bunsell, P. Lamicq, and A. Massiah, Ed., Sept 1985, p 440–446
76. P. Turk, Carbon Brakes: The Competition Heats Up, *Interavia,* Sept 1984, p 980–982
77. T. Liu and E. Kartman, "Test Methods for the Evaluation of Carbon-Carbon Composite Friction Material," paper presented at Conference on Advances in High Performance Composite Technology (Clemson, SC), May 1986
78. R. Bacon and C.T. Moses, Carbon Fibers: From Light Bulbs to Outer Space," paper presented at the American Chemical Society Symposium, April 1986
79. R.A. Meyer, "Matrix Microstructure and Thermal Mechanical Property Behavior of Carbon-Carbon Composites," paper presented at the Baden-Baden Meeting, German Carbon Society, July 1986
80. A.J. Klein, Carbon-Carbon Composites, *Met. Prog.,* Vol 11, 1986, p 64–68
81. A. Kerkar, E. Kragness, and R. Rice, Mechanical Characterization of Unidirectional Carbon/Carbon Composites for Engine Valve Application, *Cer. Sci. Eng. Proc.,* Vol 13 (No. 9-10), 1992, p 770–787
82. E.W. Fitzer, L.M. Hüttner, L.M. Manocha, and D. Wolter, Carbon Fiber Reinforced Composites for Internal Bone-Plates, *Proceed. Fifth London Internat. Carbon and Graphite Conf.,* Vol I, Society of Chemical Industry, 1978, p 454–464

Special Topics

Assessment and Use of Creep-Rupture Data

USE OF CREEP-RUPTURE PROPERTIES to determine allowable stresses for service parts has evolved with experience, although guidelines for use differ among specifications. For temperatures in the creep range, the *ASME Boiler and Pressure Vessel Code* (Ref 1) uses the following criteria as a lower limit in determining creep values: 100% of the stress to produce a creep rate of 0.01%/1000 h, which is based on a conservative average of reported tests, as evaluated by an ASME committee (in assessing data, greater weight is given to those tests run for longer times); 67% of the average stress, or 80% of the minimum stress, required to produce rupture at the end of 100,000 h as determined from available extrapolated data.

For most commercial steels and alloys, the available raw data are obtained from many tests of durations ranging from a few hundred to a few thousand hours. Tests seldom run longer than 10,000 h, and durations of 100,000 h are rare. Measured secondary creep rates as low as 0.01%/1000 h are also unusual. Allowable stresses recommended in existing specifications usually are derived from extrapolations. Considerable scatter in test results may be observed, even for a given heat of an alloy, so interpolated creep and rupture strengths are not precisely known.

Although measurement and application of creep-rupture properties is often imprecise, general trends are evident. This article discusses methods for assessing creep-rupture properties (including nonclassical creep behavior), common interpolation and extrapolation procedures, and properties estimation based on insufficient data. Additional information on creep and creep-rupture can be found in the articles "Mechanical Properties at Elevated Temperatures" and "Design for Elevated-Temperature Applications" in this Volume.

Scatter of Creep-Rupture Test Data

Reliable creep-rupture property measurements require that the test specimens be representative of the material to be used in service, preferably in the product form and condition of intended service. Testing results may vary significantly with sampling procedures. For example, heavy sections may exhibit variations in strength level with depth after normalizing or quenching and tempering treatments. A common practice is to take samples midway between the center and the surface of the specimen.

Similar variations in strength can occur for fully annealed plate or bar that has been cold straightened. Re-annealing may be required before valid testing results can be obtained. For some materials (e.g., stainless steel castings several inches thick, or alloys subjected to elevated temperature after critical prior plastic deformation), the grains may become so coarse that a specimen cross section contains only one or several grains. Local creep rates may then vary considerably along the specimen gage length, according to the orientation of the individual large grains with respect to the loading direction.

Such effects caused cross sections of specimens from 127 mm (5 in.) thick castings of ACI CF-8M to distort from circular to oval, or even to nearly flat, as creep progressed in tests where rupture life at 595 to 870 °C (1100 to 1600 °F) ranged between a few hours and 40,000 h (Ref 2). In this example, time to rupture was not markedly affected by gage section diameter of 8.9 versus 12.7 mm (0.35 versus 0.5 in.) or by the nonuniform deformation of the gage section. Measured creep tended to be more variable with the smaller specimens.

For materials in which grain-boundary material and material in the body of the grains differ markedly in either composition or strength, the effect of number and orientation of grains in the cross section would be expected to result in greater scatter than was noted in the above example.

Sampling direction seldom affects creep and rupture properties for materials with uniform single-phase structures of small equiaxed grains. For such materials, specimen size also exhibits minimal influence, except for the greater relative importance of surface oxidation and similar effects in small specimens. Due to solidification and processing conditions, preferred orientation of secondary phases or grains can alter test results. In critical situations, the loading direction in the test specimen should parallel the major stress direction expected under service loading. For rolled materials, ASTM E 139 (Ref 3) recommends that specimens be taken in the direction of rolling, unless otherwise specified.

The presence of large discrete particles of lower ductility and strength (e.g., graphite flakes in gray cast iron, or large glassy inclusions in steel) may significantly lower the sound cross section of a small specimen, but have a lesser influence in a larger specimen. Specimen size may have the opposite effect if only a few scattered low-strength particles are present; consequently, a specimen small enough could be free of these large-scale areas of weakness. More subtle variations in local creep and rupture strengths arise at grain boundaries, precipitates, voids, composition gradients, and other regions of microscale nonuniformity.

Specimen Size. For many steels, the influence of specimen size is minimal compared to other variables. During testing of low-alloy steel (ASTM A193, grade B16) and a high-alloy steel (ASTM A453, grade 660) using five different specimen diameters ranging from 5 to 84 mm (0.2 to 3.3 in.), rupture strength appeared to be independent of size for unnotched specimens (Ref 4). Variation in rupture time with size was erratic for notched specimens, but the largest size had about three-quarters of the strength of the smallest geometrically similar specimen.

Scatter from Heterogeneities. Rupture properties were not uniform with position in the original bar, as well as among bars from the same heat that received the same treatment (Ref 4). In notched specimens sampled from the mid-radius location of the A193 alloy, fracture originated consistently in the area of the cross section nearest the outside of the original bar.

Three tests at 227.5 MPa/538 °C (33 ksi/1000 °F) produced rupture times from 10,330 to 12,000+ h (test incomplete) for specimens from one bar, whereas three specimens from the other two bars of the A193 steel lasted only 7198 to 7820 h under similar conditions. These six tests included four different specimen diameters, but scatter among bars significantly exceeded that due to test specimen size.

Heterogeneities in a low-alloy steel of commercial quality have been found to cause much of the typical 30% scatter encountered in determining creep rates under strict testing conditions (Ref 5). Scatter for rupture times is approximately half that magnitude with materials that are specially prepared to be uniform.

Temperature and Other Testing Variables. Even with the use of precise temperature controllers and high-quality pyrometric practice, care is required to hold the average indicated specimen temperature within usual specifications of ±2 °C (±3.6 °F) (Ref 3) with time. Variation from one location to another along the specimen gage length typically approaches this magnitude, even in furnaces with independent adjustment of power input within zones.

These variations in indicated temperature do not include errors in initial thermocouple and pyrometer calibrations, leadwire mismatch, or drift in thermocouple output with time. Actual temperature can differ from the reported value by at least 5 °C (9 °F) during some portion of a representative creep or rupture test.

For steel at 450 °C (842 °F), a temperature disparity of 5 °C (9 °F) corresponds approximately to 20% variation in rupture time (Ref 5). A 10% change in creep rate results from a load change of 1.5%, or a temperature change of 2.5 °C (4.5 °F) (Ref 4). An error of more than 1% in the applied load is unlikely under typical conditions.

Equipment for creep-rupture testing commonly includes features to promote uniform axial loading. However, bending loads may still occur. The increase in secondary creep rates and reduction in rupture life with eccentricity of loading is analyzed in Ref 6, which reports that extreme cases of bending introduced by threaded ends on cylindrical specimens reduce rupture life by as much as 60%. The largest effects occur for short or notched specimens and for materials with low ductility to rupture.

Eccentric loading exerts maximum influence on creep measurements in the early stages of a test. With initially bowed specimens, which are common when specimens are as-cast or are strip-type machined from thin material, early creep indications may be erroneous if the extensometer fails to average the strains accurately at opposite sides of the specimen.

Typical Data Scatter. Rolled bars from a single heat of 2.25Cr-1Mo steel were tested by 21 laboratories in eight countries (Ref 7). Following are the largest deviations in average results for one laboratory from the arithmetic mean of values from all laboratories for the stress to cause rupture in 1000, 3000, and 10,000 h:

Temperature, °C (°F)	Deviation, %
550 (1022)	+14.3 and −16.6
600 (1112)	+21.2 and −12.1

A group of 18 laboratories in seven countries cooperated in tests of Nimonic 105 alloy bars at 900 °C (1652 °F) (Ref 8). Mean rupture times adjusted to a common stress for individual laboratories (four tests each) ranged from a high of 1491.4 h to a low of 1090.2 h (15.5% deviation). Time to 0.5% total deformation varied from 75.4 to 182.3 h for 16 of these laboratories; deviation from the mean value thus increased to 41.5%.

Temperature control was found to be the most serious source of variation in rupture measurement. In particular, calibration drift of Chromel-Alumel thermocouples at this high test temperature was responsible for the long mean rupture times found by five of the laboratories. Not more than about 15% scatter generally can be attributed to testing variables, if a laboratory has followed standard procedures.

Multiple Heats and Product Forms. Scatter bands become much broader when data originate from tests on numerous heats, particularly when data include a broad range of product forms and sizes. Elevated-temperature properties for steel and superalloys are available (see, e.g., Ref 9–19). Typically, data for each grade of steel include plots of the stress for rupture in 1000, 10,000, or 100,000 h versus the test temperature. Where appropriate, distinctive shapes of data points identify different product forms or heat treatments.

When data at a given temperature are extensive, the reported range of derived stresses to produce rupture in a given time typically consists of a two-fold ratio of the highest to the lowest stress level. The corresponding spread of rupture times for a given test stress is on the order of 100-fold.

If a new set of test results does not fall within the broad scatter bands, careful review and confirmation is required before the new data are accepted as valid. Bias of new data tends toward the upper half of the scatter bands compiled for older data, because the increase of residual alloying elements from scrap recycling, higher nitrogen content, and improved alignment and temperature control in modern testers tend to raise indicated creep and rupture properties.

At a test temperature of 593 °C (1100 °F) and at a stress of 207 MPa (30 ksi), rupture life ranged from 84 to 2580 h and secondary creep rate ranged from 0.16 to 0.00077%/h for 20 heats of type 304 stainless steel in the re-annealed condition (Ref 20). Corresponding large variations were observed at all test temperatures and for tests on seven heats of type 316 stainless steel. Re-annealing of as-received material lowered time to rupture in some cases, but the degree of variation persisted in properties among heats.

Good linear correlation was obtained when the logarithm of rupture strength was plotted against ultimate tensile strength at the same temperature for various types of austenitic stainless steels in the annealed and cold-worked conditions at temperatures ranging from 538 to 816 °C (1000 to 1500 °F) and for test times approaching 10,000 h. Tensile strength, in turn, was reported to be essentially proportional to $(C + N)d^{-1/2}$, where C and N represent the weight percentage of carbon and nitrogen content, respectively, and d is average grain diameter.

Although long-time performance generally cannot be accurately predicted from short-time data, the location of a particular set of rupture data within a published scatter band should agree with the relative tensile strength of the material being tested in the range of tensile strengths spanned by all heats and product forms. Tests to measure low rates of secondary creep frequently are terminated after the creep rate appears to have become reasonably constant. If the test duration had been prolonged substantially, a continued slow decline in creep rate may have been observed.

A "false" minimum rate may occur in some tests before the classical secondary creep period has been reached. Observations of many comparative creep and rupture strengths for steels has established the following general relationships:

$$\frac{0.0001\%/\text{h creep strength}}{10{,}000 \text{ h rupture strength}} = 0.7 \text{ to } 0.8$$

$$\frac{0.00001\%/\text{h creep strength}}{100{,}000 \text{ h rupture strength}} = 0.5 \text{ to } 0.6$$

If a new set of test results differs from these patterns, verification is suggested before the new results are accepted.

Nonclassical Creep Behavior

The curve of creep deformation versus time traditionally displays three consecutive stages (Fig. 1). The longest period of substantially constant creep rate is preceded by a primary stage, during which the rate declines from an initial high value, and is followed by a tertiary stage of rising creep rate as rupture is approached (Fig. 2).

Although this classical pattern can be made to fit many materials and test conditions, the relative

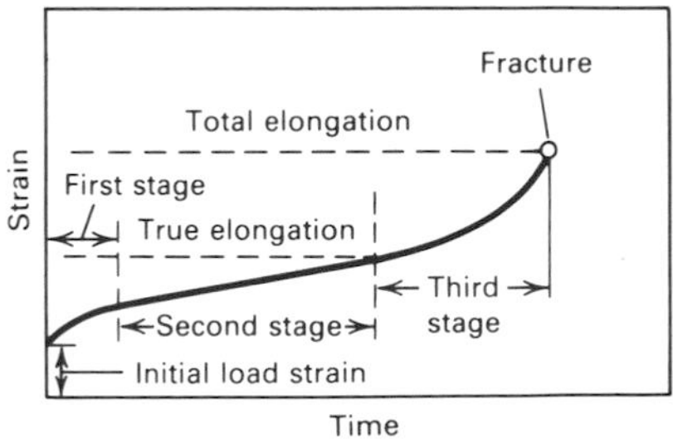

Fig. 1 Schematic tension-creep curve showing the three stages of creep

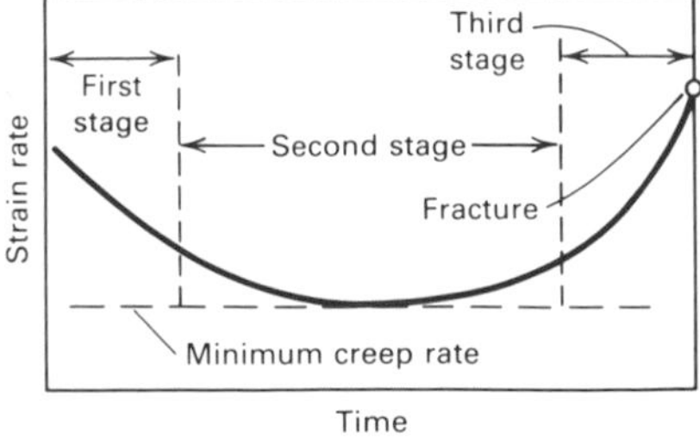

Fig. 2 Relationship of strain rate, or creep rate, and time during a constant-load creep test. The minimum creep rate is attained during second-stage creep.

duration of the three periods differs widely with materials and conditions. For example, in many superalloys and other materials in which a strengthening precipitate continues to age at creep temperatures, brief primary creep often shows transition to a long upward sweep of creep rate, with only a point of inflection for the secondary period.

Aging of normalized and tempered 0.5Cr-0.5Mo-0.25V steel during creep under 80 MPa (11.6 ksi) stress at 565 °C (1050 °F) has been reported to cause the creep curve to effectively exhibit only a continuously increasing creep rate to fracture (Ref 21). For twice the amount of stress, the creep curve in this case followed the classical trends. In other alloys, such as titanium alloys, with limited elongation before fracture, the tertiary stage may be brief and may show little increase in creep rate before rupture occurs.

A more obvious departure from classical behavior develops during the early portion of many tests when precise creep measurements are taken. When 34 ferritic steels were studied for as long as 100,000 h at temperatures ranging from 450 to 600 °C (842 to 1112 °F), step-form irregularities were observed, with an extended period of secondary creep preceded by a lower creep rate of shorter duration during primary creep (Ref 22).

Negative Creep. Because a variety of metallurgical processes can be involved and because the rates and direction of these processes can vary with time and temperature, departures from classical creep curves can take many forms and can be overlooked, unless accurate creep readings are taken at sufficiently close intervals, particularly during early stages of the test. For 2.25Cr-1Mo, Cr-Ni-Mo, and Cr-Mo-V steels, some tests have demonstrated an abrupt drop to negative creep rate (contraction) after a brief beginning period of positive primary creep. Once this contraction ceased, the remaining portion of the test displayed the classical succession of declining, steady, and then rising creep rates. Figure 3 gives an example for normalized and tempered 2.25Cr-1Mo steel tested at 275.8 MPa (40 ksi) at 482 °C (900 °F) (Ref 2).

In tests of boiler and pressure vessel materials, definite negative creep was noted in at least one test each at 482, 704, 816, and 871 °C (900, 1300, 1500, and 1600 °F) for cast CF8 austenitic stainless steel (Ref 2). Rupture times for these tests ranged from 1000 h to longer than 30,000 h. For some combinations of material lots and test temperatures, nearly all creep curves of these tests showed an early "false" minimum rate during part of the primary stage. Structural changes responsible for the measured contraction were undetermined.

Short-term negative creep also was observed in tests on quenched and tempered 2.25Cr-1Mo steel at 482 °C (900 °F) and at 482 and 538 °C (900 and 1000 °F) for the same steel in the normalized and tempered condition. Two steady-state creep stages for annealed 2.25Cr-1Mo steel have been reported (Ref 23), which were due to the interaction of molybdenum and carbon atoms with dislocations and the subsequent decrease in the number of these atoms as Mo_2C precipitated. A volume decrease associated with the precipitation process also could account for the observed creep curve trends.

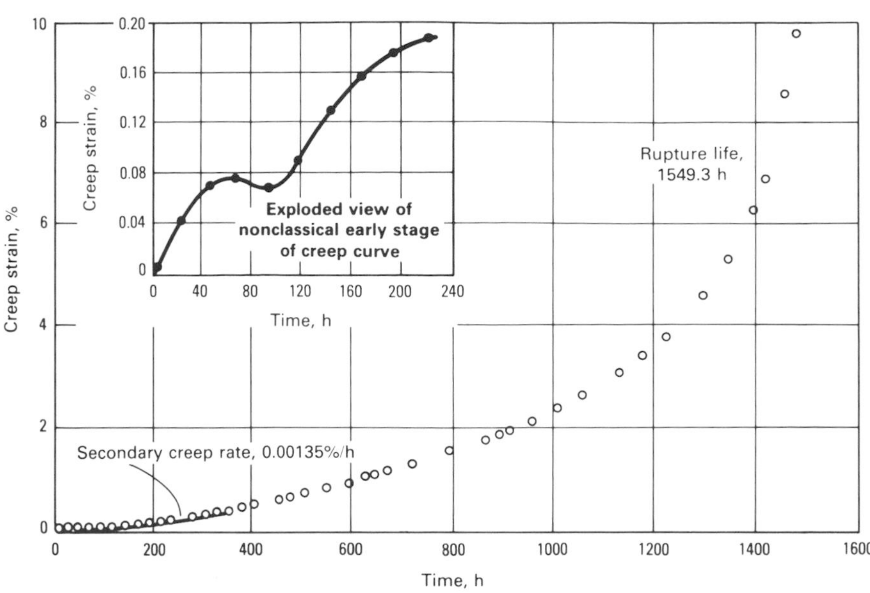

Fig. 3 Creep curve of 2.25Cr-1Mo steel with nonclassical early stage. Normalized and tempered to 607 MPa (88 ksi) tensile strength at room temperature. Tested at 482 °C (900 °F) at 275.8 MPa (40 ksi)

Interstitial diffusion of carbon and hydrogen into dislocations has been observed, and alloy strain-aging effects have been found to cause creep rate transitions noted for carbon steels and normalized 0.5% Mo steel (Ref 24). Negative creep in Nimonic 80A appears to be related to an ordering reaction in the Ni-Cr matrix and possible formation of Ni_3Cr (Ref 25).

Oxide and Nitride Strengthening. An entirely different source of variation from classical patterns occurs in creep tests at high temperature due to reaction with the air that forms the environment. Tests longer than 50 h with 80Ni-20Cr alloys at 816 and 982 °C (1500 and 1800 °F) showed a deceleration of creep after the normal tertiary stage was reached, resulting in a second period of steady-state creep and later another period of last-stage creep (Ref 26). Figure 4 illustrates a creep curve showing the effect of oxide strengthening.

This behavior, which prolonged rupture life and caused a slope decrease in curves of log stress versus log rupture life, was due to oxide and nitride formation on surfaces of the intercrystalline cracks that form extensively during tertiary creep. Observed interconnection of the bulk

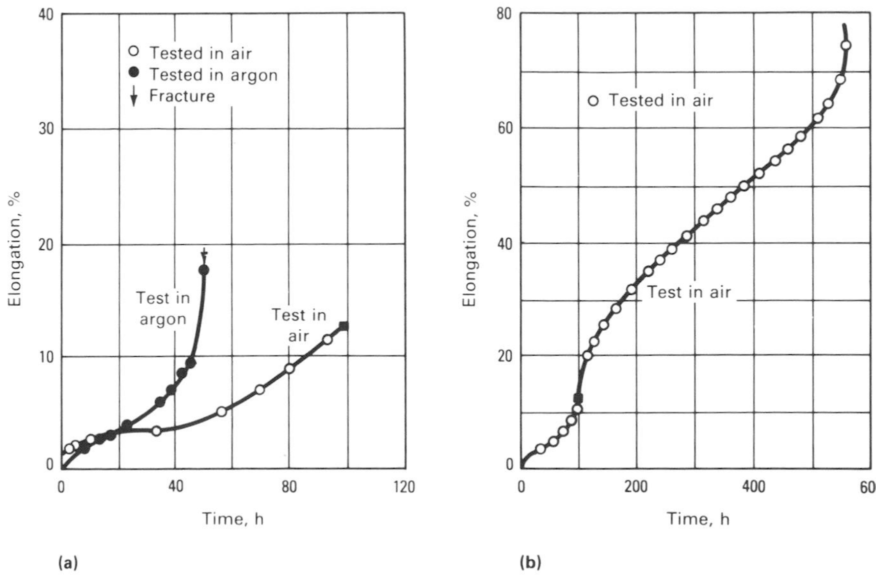

Fig. 4 Creep curves of alloy 2V tested at 980 °C (1800 °F) and 17.2 MPa (2500 psi). (a) Tests in argon and air for some duration. (b) Entire curve of specimen tested in air is shown. The square data point on both graphs represents the same point.

of these cracks added substantially to strengthening against creep deformation in the late stages of the tests.

This effect also was observed in 99.8% Ni tested at 816 °C (1500 °F) under 20.7 MPa (3000 psi) stress (Ref 27). Fracture after a prolonged time occurred in a lower-stressed section of the fillet. Fewer intergranular cracks in this region resulted in less oxidation strengthening than in the gage section.

Extrapolation and Interpolation Procedures

The determination of creep-rupture behavior under the conditions of intended service requires extrapolation and/or interpolation of raw data. No single method for determination of properties exists; however, a variety of techniques have evolved for data handling of most materials and applications of engineering interest. These techniques include graphical methods, time-temperature parameters, and methods used for estimations when data are sparse or hard to obtain.

Graphical Methods

Test results frequently are displayed as plots of log stress versus log rupture time and log stress versus log secondary creep rate, with a separate curve (isotherm) for each test temperature. For limited ranges of test variables, test points frequently fall in a straight line for each temperature. Nonlinearity of isotherms with broader ranges of test parameters has been treated variously, but common practice is to represent the data by two or more intersecting straight line segments. Figure 5 illustrates such treatment for an aluminum alloy.

Isotherms for lower temperatures characteristically display a flatter slope than those for higher temperatures. At a given temperature, when the test stresses drop below a given level that varies with alloy composition and metallurgical condition, the slope of the isotherm usually steepens. This steeper slope often approximates the slope for early times at the next higher test temperature.

Early investigators of engineering creep behavior introduced a "conservative" practice of using the slope from the next higher temperature when an isotherm had to be extended to longer times. Use of this method is limited to the specific temperatures of the test runs. Even under these conditions, extrapolations should be made only in the direction of longer times for the lower range of test temperatures.

The change in slope of log stress versus log time isotherms historically appeared to be associated with a gradual change in fracture mode, from transgranular at lower temperatures and higher stress to intergranular at relatively high temperatures and low stress. Therefore, the belief developed that once the slope of the longer time portion became established, further slope change would not occur. Experimental data available at that time provided no indication that these linear plots could not be extrapolated to long times with confidence. Subsequent long-time data demonstrate that such extrapolations may lead to erroneous results.

Upward Inflection of Log-Log Rupture Plots at Long Times. Review of 52 heats from 31 wrought and cast steels, each with test times longer than 50,000 h, indicated that some portion of the log stress versus log rupture time curves for all ferritic steels showed an increase in slope when tests were of sufficiently long duration (Ref 28). This upward inflection was pronounced depending on composition, heat treatment, and particularly test temperature.

A sharp inflection at one temperature (e.g., 500 °C, or 930 °F) was usually accompanied by a less distinct inflection covering a broader time range at a higher test temperature (e.g., 550 °C, or 1022 °F). Generally, these inflections shifted to shorter times and lower stresses with increasing test temperature. Existence of inflections appeared to be related to precipitation phenomena.

For the heat-treatable aluminum alloy 6061-T651, test stresses between about 20 and 50 MPa (2900 and 7250 psi) for temperatures ranging from 260 to 343 °C (500 to 650 °F) exhibited nearly the same slope on a plot of log stress versus log rupture time, which was steeper than for either higher or lower stress levels (Ref 29). The long-time rupture results obtained had been predicted (Ref 30) by separate graphical extrapolation of each of three regimes of rather constant slope (see Fig. 5).

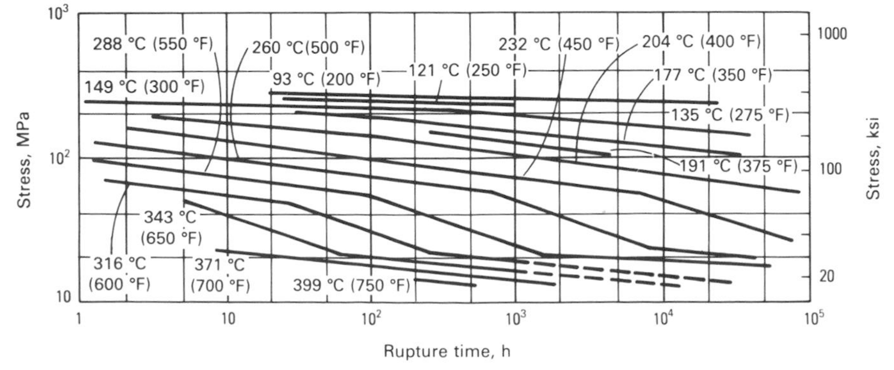

In this instance, the curves that were actually extended were lines for fixed stress levels (isostress lines) on plots of log rupture time versus temperature, or the reciprocal of absolute temperature. However, extrapolation could have been carried out on the usual log stress versus log rupture time plot by treating the data as a family of curves, with different portions of each curve falling into different slope regimes. Direct graphical extension of isostress lines appeared to provide better extrapolation of rupture data than other common methods (Ref 31).

Curves of log stress versus log rupture life for two chromium-molybdenum steels (ASTM A387, grades 22 and 11) typically show an increase followed by a decrease in steepness for tests at 538 to 566 °C (1000 to 1050 °F). Consequently, correct prediction of 100,000 h strengths requires that these changes in slope be incorporated into the analysis (Ref 32). This requirement applies to all evaluation methods. Unless the input data include results that encompass structural changes of the type expected under intended service conditions, accurate extrapolation cannot be expected.

Some metallurgists prefer a semilogarithmic plot of stress versus log rupture time. The sigmoidal shape of isotherms is thus more evident, but extrapolation difficulties remain. The double inflections (or sigmoidal shape) for rupture curves can be greatly accentuated when notched specimens are tested. In the intermediate stress regime, rupture life can actually decrease as the level of test stress is lowered.

Time-Temperature Parameters

Temperatures that are higher than those encountered in service have traditionally been used to shorten the time required to obtain creep-rupture results. One such approach incorporates time and temperature into an expression or parameter, such that a single master curve of stress or log stress can represent all data obtained for a given lot of material over a wide range of test conditions.

When the parameter calculated for a desired service time and temperature falls within the range of the master curve, the corresponding stress can be read directly from that curve. More than 30 parameters have been proposed; although not always developed the same way, several can be derived from the following:

$$P = \frac{(\log t/\sigma^Q) - \log t_A}{(T - T_A)^R} \qquad \text{(Eq 1)}$$

where t is rupture time in hours; σ is stress in psi; T is test temperature in °F; and T_A, $\log t_A$, Q, and R are constants determined from the experimental test data. Figure 6 illustrates geometric requirements for lines of constant stress for several parameter models on a plot of logarithm of time versus either temperature or its reciprocal. Of these, the Larson-Miller and Manson-Haferd parameters represent early developments in time-temperature parameters that retain considerable application.

Fig. 5 Logarithmic plot of stress vs. rupture life for aluminum alloy 6061-T651

The Larson-Miller Parameter. In 1953, Larson and Miller introduced the concept of time-temperature parameters to correlate and extrapolate creep-rupture data (Ref 33). For the Larson-Miller parameter, constant stress lines on a graph of log t versus $1/T_A$ converge to a point at $1/T_A = 0$ (Fig. 6a). At that point, log $t = C$ defines the optimum value of the constant, C, for the data involved.

In Fig. 7(a), an actual set of stress-rupture data (log stress versus log time to rupture) is shown for the nickel-base alloy Inconel 718. The data are then replotted in the form shown in Fig. 6(a) for the Larson-Miller parameter. This is accomplished by using constant-time intercepts, shown in Fig. 7(b) by the dashed lines, to arrive at the constant-time curves shown in Fig. 7(b).

Constant-stress curves are then plotted as shown in Fig. 7(c), using the intercepts shown by the dashed lines in Fig. 7(b). By extending the data in Fig. 7(c), a plausible set of convergent isostress lines meeting on the ordinate at a value of log $t = -25$ can be obtained. As illustrated in Fig. 6(a), the Larson-Miller equation for this set of data is:

$$P = f(\sigma) = T_A (\log t_R + 25) \qquad \text{(Eq 2)}$$

where T is temperature in °R and t_R is time to rupture in hours.

For each data point in the original set of stress, time, and temperature data, the proper value can be substituted in Eq 2 and plotted as shown in Fig. 7(d). This is the compact parameter form of graphical representation known as the "master curve." A common practice when input data are limited is to assume that $C = 20$, which has been found to be reasonably true for many materials.

Other Parameters. The Manson-Haferd parameter predicts that a constant stress plot of log t versus T yields a family of straight lines converging to a point $t(T_A, t_A)$, which defines the optimum constants for that particular data set (Fig. 6b). On these same coordinates, the Manson-Succop parameter requires that isostress lines be straight and parallel (Fig. 6d). These conflicting patterns and still different patterns for the additional parameters cannot occur simultaneously over the entire range of data from a given set.

A frequent finding is that different parameters provide best fit to different portions of the same data. For example, using data obtained from tests on a 1Cr-1Mo-0.25V steel, the Larson-Miller parameter gave the best extrapolation at 482 °C (900 °F), and the Manson-Haferd parameter was preferable at 538 °C (1000 °F). However, the Orr-Sherby-Dorn parameter gave the best fit at 593 °C (1100 °F) (Ref 34, 35).

Although numerous studies have considered the relative merits of these and other proposed parameters, no one parameter has emerged as universally superior to all others. Five representative parameters were compared in terms of correlating and extrapolating extensive sets of data on the creep and rupture properties of seven steels and superalloys (Cr-Mo, Cr-Mo-V, 12%Cr, A-286, Astroloy, René 41, and Inconel 718) (Ref 36). The difference in fit among parameter methods was found to be relatively small and inconsistent from one alloy to another. The largest source of variation in the fitted values for time to rupture, time to 1% creep, and minimum creep rate was the difference between alloys, regardless of the parameter used.

Results were marginal to poor when extrapolation was beyond the range of the fitted master curve. When prediction of long-time data was

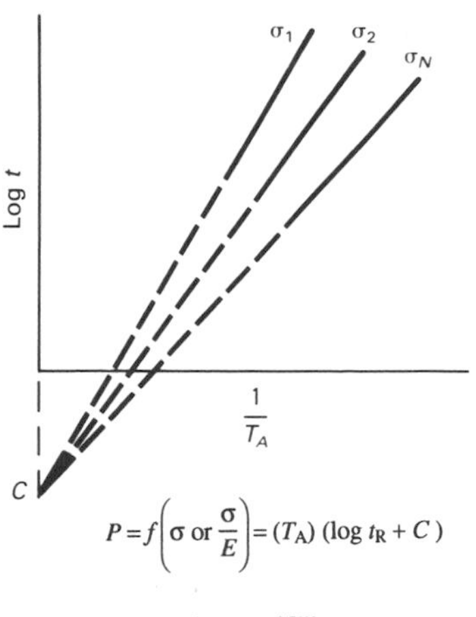

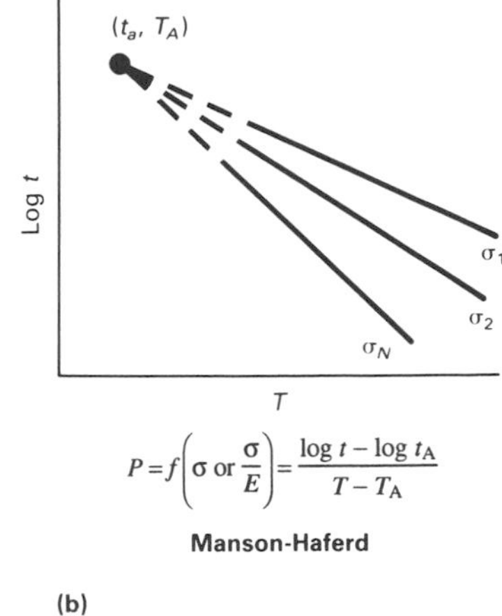

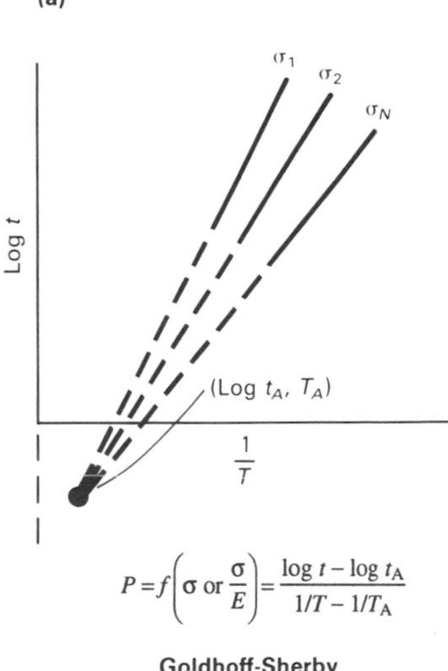

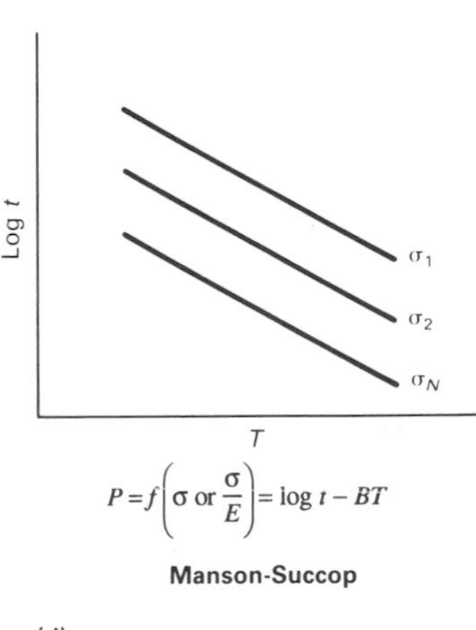

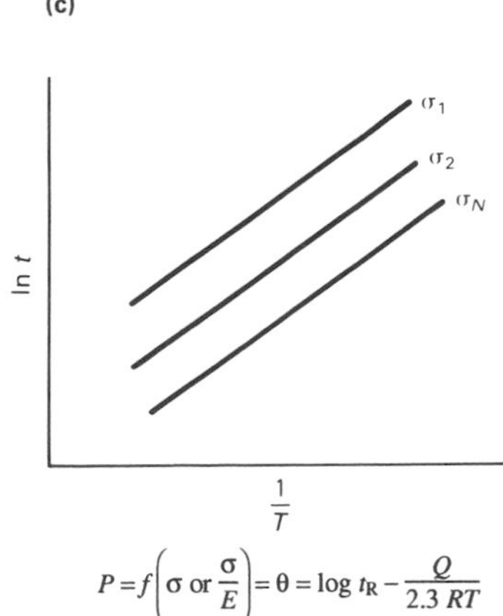

Fig. 6 Schematic representation of several time-temperature parameter models

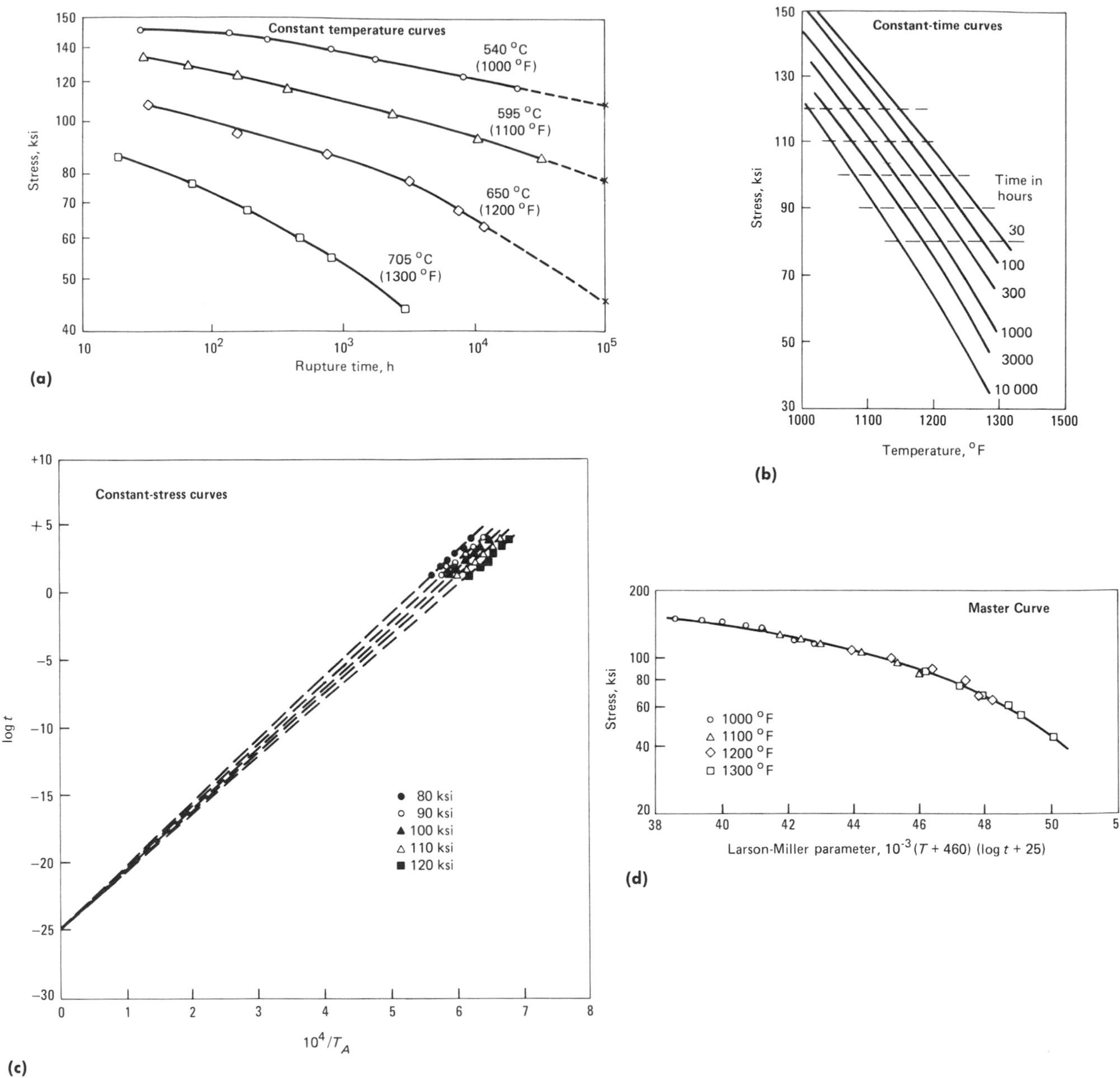

Fig. 7 Method of creating master Larson-Miller curve for Inconel 718 from experimental stress-rupture curves. See text for discussion of (a) through (d).

confined to the master curve derived using only short-time data, no single parametric method gave consistently superior results.

For critical evaluation of the comparative ability to predict known long-time rupture lives (11,000 to 64,000 h) for the ferritic steels, data up to 10,000 h were applied to establish fit to the parametric model. For the superalloys, only data up to 1000 h were used to predict known rupture times between 1200 and 33,000 h. Table 1 lists ranges for the ratio (predicted life/actual life) extracted from Ref 36 for 46 extrapolations, including one in which slight extension was required beyond the fit of each master curve.

Studies on 0.5Cr-0.5Mo-0.25V steel pipe (Ref 37) found the Manson-Haferd parameter significantly superior to predict known rupture times (8712 to 20,664 h) from data points of less than 6000 h duration than either the Larson-Miller or Orr-Sherby-Dorn parameters. The latter two parameters generally provide optimistic prediction of behavior. According to Ref 37, very short-time data should be eliminated from the analysis if predictions beyond 10,000 h are desired, because their inclusion distorts the correlation at long times. As with graphical methods, the accuracy of parametric extrapolations is related to the interval of test variables on which the prediction is based.

Minimum Commitment Method. Experimental data may deviate from the requirement imposed by the form of each parameter for linearity of isostress curves or for parallelism or convergence of families of such curves. The minimum commitment method starts with a time-temperature-stress relationship sufficiently general to include all commonly used parameters. The pattern of the data is not forced in advance; instead, the actual experimental data naturally lead to the most appropriate functional relationship for the particular material.

Manson applied a "station-function" approach to $f(\log t) + p(T) = g(\sigma)$; all parametric formulations can reduce to this equation. Each of the functions f, p, and g were represented by their discrete numerical magnitudes at specific values of the corresponding independent variable. Figure 8 illustrates treatment of Astroloy data given in Ref 36.

Temperature stations were arbitrarily chosen at T = 760, 816, 871, 927, and 982 °C (1400, 1500,

Table 1 Comparative extrapolation abilities

Parameter or method	Ratio: predicted life/actual life		
	Minimum	Average	Maximum
Larson-Miller	0.34	1.57	5.64
Manson-Haferd	0.44	1.51	6.30
Goldhoff-Sherby	0.36	1.64	8.85
Manson-Succop	0.39	1.53	4.96
Orr-Sherby-Dorn	0.11	1.09	4.01
Monkman-Grant	0.33	0.93	1.82

1600, 1700, and 1800 °F); the values of the p functions at these respective temperatures are designated p_1, p_2, p_3, p_4, and p_5. For times such that log t = 1.0, 1.25, 1.5, ..., 3.0, the respective corresponding f values are f_1, f_2, f_3, ..., f_8. The g values are designated g_1, g_2, g_3, ..., g_{11} for levels of log (T = 1.0, 1.1, 1.2, ..., 2.0, with σ given in ksi.

Consider the experimental point A (T = 927 °C, or 1700 °F; log t = 2.873; log σ = 1.322). At this point, p is directly p_4, but values of f and g must be interpolated. Higher-order interpolation can be easily accomplished, but simple linear interpolation was chosen for this illustration. For A between log t of f_8 = 2.75 and f_9 = 3.00:

$$f(2.873) = \frac{3.0 - 2.873}{3.0 - 2.75} f_8 + \frac{2.873 - 2.75}{3.0 - 2.75} f_9$$

$$= 0.508\, f_8 + 0.492\, f_9 \qquad \text{(Eq 3)}$$

In a similar manner, the relation for stress at this point becomes:

$$g(1.322) = \frac{1.4 - 1.322}{1.4 - 1.3} g_4 + \frac{1.322 - 1.3}{1.4 - 1.3} g_5$$

$$= 0.78\, g_4 + 0.2\, g_5 \qquad \text{(Eq 4)}$$

Introducing these results into the original general equation yields:

$$0.508 f_8 + 0.492 f_9 + p_4 = 0.78\, g_4 + 0.22\, g_5 \qquad \text{(Eq 5)}$$

A similar equation can be written for each experimental point. By choosing a sufficient number of stations, the number of equations available will exceed the number of unknowns, so that a least squares solution can be used to determine the unknowns. Once the equations have been solved, the function f(log t) may be extrapolated using graphical, polynomial, or recurrence relations.

Although a better fit to the data should result by the minimum commitment method than by forcing fit to an arbitrary parameter, extrapolations can still be imprecise. As with other methods, the degree of accuracy of predictions relies on having accurate and representative raw data from tests that reflect any structural changes expected to occur in the regime of the extrapolated conditions.

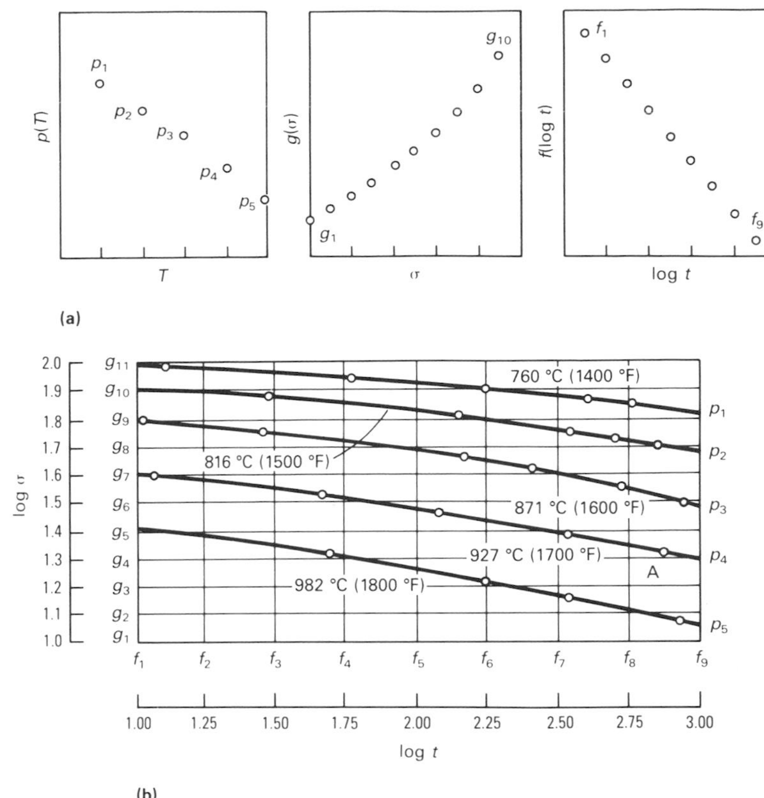

Fig. 8 Application of station function approach to Astroloy data. (a) Station-function representation of p (T), g (σ), and f (log t) at specific values of T, s, and log t. (b) Net point selections for solution. Source: Ref 38

A specialized form of the general equation developed in 1971 (Ref 39) contains a characteristic material constant A:

$$\log t + AP(T) \log t + P(T) = f(\log \sigma) \qquad \text{(Eq 6)}$$

The parameter A is a measure of structural stability, because the more unstable the material, the higher the negative value of A required to fit the data. On a plot of log stress versus log time, all isothermals converge at the extrapolated point log t = (1/A) (Ref 40). For the parallel isotherms of the Orr-Sherby-Dorn parameter, A = 0; Miller-Larson requires A = +0.5. With data fitting the Manson-Haferd parameter, negative A values result; the Astroloy data discussed above converge at about t = 6.36, resulting in A = –0.157.

Establishing an accurate individualized value of A for a specific material is difficult from a short-life database. One acceptable approach is to use universalized values of A = 0 for aluminum alloys and pure metals, and A = –0.05 for most steels and superalloys (Ref 40). Highly unstable materials like Astroloy, as well as carbon steels, require higher negative values of A; for these materials, A = –0.15 has been found to be adequate in most instances.

The minimum commitment method was initially developed for use with single-heat data. Application to multiple-heat analyses is discussed in Ref 41, including information on suitable computer programs.

Estimation of Required Properties Based on Insufficient Data

Complete independent evaluation of creep-rupture properties for a new lot of material, whether by graphical, parametric, or minimum commitment methods, requires numerous test points covering an extensive range of test variables. Frequently, the amount of data available is too limited for full treatment by usual procedures. Experimental difficulties often limit the ability to obtain accurate test results at conditions of interest, such as evaluation of creep-rupture properties near the low end of the temperature range in which time-dependent effects are significant.

Tests of short or moderate duration at these temperatures frequently require use of such high stress levels that the immediate high plastic strains at load application alter the nature of the material from that which exists during service under lower stresses. Testing at or near a stress of intended application often requires more time and/or expense than is feasible before the material is to be put into use. Approximate methods permit such difficulties to be treated in a generally satisfactory manner. Established correlations also permit estimation of some unmeasured properties from other types of available results.

The Monkman-Grant Relationship. Analysis of data for a variety of aluminum-, iron-, nickel-, titanium-, cobalt-, and copper-base alloys led Monkman and Grant to the following empirical relationship (Ref 42):

$$\log t_r + m \log (\text{mcr}) = C \quad \text{(Eq 7)}$$

where t_r is time to rupture, mcr is minimum creep rate, and m and C are constants that differ significantly among alloy groups, but exhibit nearly fixed values for a given heat of material, or for different lots within the same alloy group.

Equation 7 enables assessment of the reliability of each individual test by examining its fit within the scatter band for all tests. Once a minimum creep rate has been determined in a long test, rupture life can be estimated without running the test to failure. Although Monkman and Grant stated that this relationship was not intended for extrapolation, it can be used for that purpose, particularly when only low-stress tests are acceptable to prevent large initial plastic strains.

Table 1 includes the results obtained when the Monkman-Grant relationship is applied to data obtained on seven materials (Ref 36). The prediction of rupture life for these 46 extrapolations using this technique was more accurate than that provided by any of the five time-temperature parameters.

For additional materials (Ref 43) where good fit is obtained to a single linear plot on the coordinates of log time versus log secondary creep rate, extrapolation of a known secondary creep rate to the corresponding rupture life appears to be as good or better than by other extrapolation methods. One advantage of Eq 7 is that it can be applied successfully to as few as five or six data points, in contrast to the approximately 30 tests needed to establish the entire Manson-Haferd master curve (Ref 44). For the minimum commitment method, even more data points are usually required.

One advantage of this correlation, particularly with materials that exhibit structural instability under testing, is that the specimens used to determine the input data for secondary creep rates experience the same history of structural change that exists during the corresponding period of a test carried to rupture. Best predictions result by concentrating on tests encompassing a limited range of stresses and temperatures, thus discounting results obtained at considerably higher combined temperatures and stresses.

Reduced scatter was noted (Ref 45) for eight nonferrous alloys and two superalloys when the term log t_r in Eq 7 was replaced by log (t_r/ε_c), where ε_c is the total creep deformation at fracture. This trend was confirmed by Ref 46 in tests on a 2.25Cr-1Mo steel.

Although deformation-modified rupture time may improve correlation in some instances, other cases exist where use of the original relationship is sufficient or better. Data for 17 test points for 4% cold-worked type 304 stainless steel (Ref 47) exhibited a spread in creep elongation from 1.5 to 24%. Goodness of fit was identical (coefficient of determination r_2 = 0.86) for linear regression of the data treated by the original versus the modified log-log relationships.

Extrapolation is fast and direct when using the Monkman-Grant coordinates, but with the modified relationship, creep elongation at the given temperature and corresponding to the rupture time sought must first be estimated. This usually requires subjective extrapolation of only a few elongation values displaying wide scatter and with no evident single trend. Introduction of a creep elongation factor may have value when only correlation or interpolation of test results is desired, but it is not recommended for extrapolation.

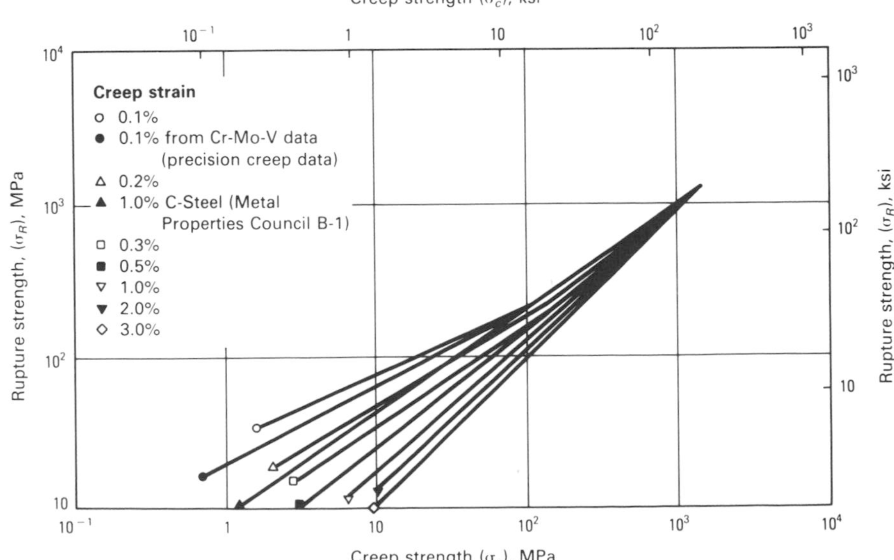

Fig. 9 Composite graph for the Gill-Goldhoff correlation. Source: Ref 52

One occasional problem in estimation of rupture life from creep data is uncertainty whether secondary creep has truly been established. Changes in creep rate with continuing test time are often sufficiently gradual and so close to the sensitivity of measurement that what appears to be a steady-rate condition may in fact still be a late portion of primary creep. Reference 48 illustrates successive apparent minimum creep rates of 2.05, 1.7, and 1.40%/10,000 h for respective test durations of 1000, 2000, and 5000 h.

A distinctive slope change in a plot of log creep rate versus time or log time often provides better assurance that the secondary creep period has been entered than study of the deformation-time curve itself. An equation expressing true strain in terms of elapsed time, secondary creep rate, and three constants deviates markedly from actual behavior during the early portion of primary creep, but a statistical analysis such as that detailed in Ref 49 may predict acceptable values of secondary creep rate from transient data.

For type 316 stainless steel tested at 704 to 830 °C (1300 to 1525 °F), the initial transient rate at t = 0 was found to be almost equal to 3.3 times the secondary creep rate in the same test (Ref 50). A significantly different magnitude (near 1000) for this ratio of initial and secondary creep rates was found in Ref 51 for a high-temperature alloy. A simple proportionality of this type and the more general analysis cited above are tempting alternatives for shortened test durations, but both suffer from the need for creep measurements that are more precise than those commonly obtained. Currently, neither method is capable of replacing long-time testing.

The Gill-Goldhoff Method. Many designs for elevated-temperature service require that deformation not exceed some maximum value; in these cases, creep strain rather than rupture life becomes the focus. Although published compilations and computer banks of data include rupture properties for most materials of engineering interest, corresponding information on the time-dependency of strain is frequently sparse or nonexistent. Many early studies did not include strain measurement during rupture tests. When creep data were obtained, accuracy was sometimes questionable due to inadequate control of temperature or low precision of strain measurements. Typically, the only listed creep data are minimum creep rates. Most of these results were obtained from tests that were terminated after a few thousand hours, or even less time, and true secondary creep rate may not have been established.

Studies by the Material Properties Council and similar groups attempt to report both the total strain on loading and the times to various levels of creep strain. Until such results are more universally available, estimates may still be required of the creep strain to be expected in given design situations.

Gill and Goldhoff (Ref 48, 52) found a log-log correlation between stress to cause rupture and stress for a given creep strain for the same time and temperature. Figure 9 shows their composite plot for aluminum-base alloys and stainless steels, including several superalloys.

To obtain this correlation, tests in which 0.1% creep occurred in less than 100 h were rejected to prevent intolerable data scatter. Despite this, the "universal" curves of Fig. 9 can be associated with fairly wide data scatter, particularly at low creep strains. Some deviations from the correlation were related to microstructural instabilities, which produce differing proportions of primary, secondary, and tertiary creep between alloys and for varying test conditions.

Despite occasional anomalies, the Gill-Goldhoff correlation meets some preliminary design needs, particularly if the technique is tailored to grades of alloys similar to those of immediate concern. In principle, this technique can also be used to predict rupture properties from early creep measurements from tests that are not continued to rupture. This use is limited by the short rupture times that are derived from tests terminated at creep strains of 1% or less. If these tests were continued to higher levels of creep, improved predictions of rupture could be obtained by determining the secondary creep rate and then applying the Monkman-Grant relationship.

Treatment of Isolated Test Points. Particularly at the start of a testing program, the need may arise to extract information from a single available test. The form of most parameters limits their use to situations in which multiple test results are available. The Larson-Miller parameter is an exception if the generalized constant $C = 20$ is used.

For the stress of the test, longer rupture times (within a factor of ten from the test duration) frequently can be estimated satisfactorily for temperatures below that of the test. If a master curve or graph of isothermals is available for another lot of like or similar alloy, a parallel curve passed through the coordinates of the test point for the new lot can serve as an approximate representation of expected behavior for limited ranges of variables from the test conditions.

Evaluating Creep Damage and Remaining Service Life

Specimens from service occasionally are tested to compare residual creep and rupture properties with the same or similar material that has not been used in service. Diverse evaluations may result, depending on the conditions selected for the tests and the criteria used to define damage. As shown below, a false prediction of drastic drop in rupture strength can result if tests are of rather short duration and if a simple direct ratio is taken of rupture life after service versus before service for similar test conditions.

Discrete Changes in Test Temperature or Load. When a creep-rupture test is interrupted by cooling and reheating at a moderate rate at constant load, and if the time under changing temperature is brief compared to the original test duration, the effect of the interruption on either the creep curve or rupture life usually cannot be readily detected, unless thermal gradients cause gross plastic deformation or spalling of surface layers.

Similar results can be expected when temperature and stress rise and fall in unison, as during startup and shutdown of a steam boiler. For the alternate situation, in which unloading occurs while the creep temperature is maintained, significant recovery of primary creep can ensue. Reapplying the load results in a period of primary creep.

Under step-wise alteration of load, temperature, or both during a test or service, performance frequently follows the "life-fraction rule" or "linear cumulative damage rule" (Ref 53), in which the percentage of total life consumed for each period of fixed temperature and stress is represented as:

$$\% \text{ total life} = \frac{\text{Actual time at the given conditions}}{\text{Rupture life at those conditions without alteration}} \times 100 \qquad \text{(Eq 8)}$$

Accuracy of this rule ranges from excellent to rather poor, with best results for multiple small excursions (Ref 54). Although solid-state reactions, which can reverse at different exposure temperatures, may introduce complications under some conditions, investigators have found the life-fraction rule to be more appropriate for steels under temperature changes than under stress changes (Ref 55).

Life-fraction summations at failure as low as 0.36 and as high as 2.43 have been reported (Ref 56). The spread was only from 0.75 to 1.50 for the same tests using damage fractions defined by $K(t/t_r) + (1 - K)(\varepsilon/\varepsilon_r)$, where t and ε are the time and strain under a period of fixed conditions, for which the rupture time and fracture strain are t_r and ε_r. K is a material constant ranging from zero to 1; the zero limit applies to materials that develop cracks early in life, and K approaches 1 for materials that exhibit no cracking until rupture is imminent. Typical values of K (Ref 56) are as follows:

Material and test temperature	K
Copper at 250 °C (482 °F)	0.3
A-286 alloy at 649 °C (1200 °F), solution treated at 1204 °C (2200 °F)	0.47
A-286 alloy at 649 °C (1200 °F), solution treated at 982 °C (1800 °F)	0.43
Inconel X-550 at 732 °C (1350 °F)	0.625

When data are insufficient for determination of K, acceptable results frequently can be obtained with an empirical rule (Ref 57), by which the life-fractions added are defined by $\sqrt{(t/t_r)\,(\varepsilon/\varepsilon_r)}$. With any of these cumulative damage rules, comparison usually is against rupture life from constant-load tests (i.e., with actual stress rising as creep reduces the cross section). When the same specimen undergoes load changes for different periods, respective stress levels have been based on the initial cross section. This corresponds to using the same original load if an interrupted test must be restarted. The actual stress at the time of test restart is $(\sigma_n)(A_o/A_c)$, where σ_n is the present nominal stress, A_o is the initial specimen cross section, and A_c is the specimen cross section after creep deformation up to the time of the test interruption.

When the specimen for a test in a later portion of creep has been machined from a part that has already undergone considerable reduction in cross section by prior creep, a corresponding area-modified stress must be employed for consistent interpretation of the results (Ref 58). The load applied to produce the desired nominal stress σ_n related to the virgin material is calculated to make the actual stress $\sigma_a = \sigma_n\ (A_o/A_c)$, where the latter term relates cross-sectional areas of the original part before and after creep. Use of such an area-modified stress has been reported to improve prediction of remaining life from post-service rupture tests (Ref 58, 59).

An approximation of remaining rupture life for components that have undergone prolonged service can be calculated by introducing best estimates for operating temperatures and stresses into the above damage rules. More exact evaluation, however, can be obtained by testing representative samples removed from service.

Measurement of Rupture Properties after Service. Direct measurement of the remaining life of a part at conditions near the service stress and temperature generally requires impractically lengthy testing times. Studies on material after creep service have involved increased temperature, stress, or both, with subsequent extrapolation of results to nominal service conditions.

Assessments made in this manner generally conclude that carbon, carbon-molybdenum, and chromium-molybdenum steels operated at or below allowable stresses (recommended by the *ASME Boiler and Pressure Vessel Code*) experienced negligible creep damage from service exposures up to 200,000 h (Ref 60). Significantly different, and presumably erroneous, conclusions can result from cursory studies or from poor selection of post-service tests.

Operating conditions for 1Cr-0.5Mo steel tubing removed from a superheater after 33,000 h of service (Ref 59) were sufficiently well known to permit true residual life to be determined by continuing samples to rupture under the same stress and temperature (66.26 MPa/557 °C, or 9610 psi/1035 °F). These tests indicated service to have accounted for 78% of the total life of the tubes.

In that study a series of accelerated tests at 66.26 MPa (9610 psi) stress, but with test temperatures ranging from 561 to 610 °C (1042 to 1130 °F), yielded service lives ranging from 6535 to 432 h. A plot of these results on coordinates of test temperature versus log test life was linear and extrapolated rather well to the total life established by the samples that continued to rupture at service stress and temperature. The simple life-fraction rule yielded values ranging from 0.93 to 1.05 (for the sum of test and service fractions when area-modified stresses were used) and from 0.96 to 1.11 (without the area correction).

Acceleration of tests by using stress levels ranging from 77.15 to 141.75 MPa (11.2 to 20.6 ksi) resulted in test durations between 2057 and 61 h at 557 °C (1035 °F). Departure from linearity of the plot of test stress versus log test life as the service stress was approached made extrapolation to expected total life at service conditions difficult. Deviation trends between mean ISO data and properties of this tubing before service may bias the life-fraction comparisons for increased temperature versus increased stress tests after service. However, the sum of test and service fractions were found to be only 0.83 for the highest test stress and were found to increase only marginally to 0.89 for the lowest accelerated stress (77.85 MPa, or 11.3 ksi) (Ref 60).

Life assessment based on extrapolation of temperature at the service stress is generally preferred to life assessment based on extrapolation of stress at the service temperature (Ref 61). However, many apparently successful evaluations of materials after creep service (Ref 62, 63) have used isothermal tests similar to those used to establish the original allowable stresses of the *ASME Boiler and Pressure Vessel Code* for creep temperatures.

A common finding for steels used to construct boilers and heated vessels has been that materials taken from service show lower rupture strengths than the same original material or other typical new materials made to the same specifications. However, the slope of the curve of log stress versus log rupture time is flatter than that for virgin samples, and each isotherm frequently approximates a single straight line. Extrapolation of that line characteristically predicts negligible change from original long-time strength values for specimens whose actual service stress was low enough to preclude extensive creep damage after prolonged operation.

Subjecting such essentially undamaged service specimens to a heat treatment similar to that originally given the component restores short-time rupture curves to approximately the level and form reported for the steel before service. Note that sigmoidal isotherms have been reported for these steels, with a final lower slope linear portion after time-dependent structural alterations produced a slope increase encompassing tests with intermediate durations to rupture (Ref 28).

Dependable application of these techniques to estimate remaining life requires that the loading direction for the final test correspond to the largest principal stress of service and that the specimen be representative of surface deterioration or other damage present in the part. Possible temperature and loading gradients in service must be kept in mind when selecting a sampling site and when applying test results to predictions of further serviceability. Despite these possible additional variables, published assessments of post-service rupture properties require only about the same number and duration of tests as conventional evaluation of any new lot of familiar material.

Creep-Rate Measurements after Service. An alternative testing approach has been to perform a creep test on the specimen from service and to compare the measured rate against creep for the same or like material without service. An ideal test would be conducted at the operating temperature and stress, but more extreme test conditions are common. Although simple in principle, this test can entail several difficulties.

If the specimen of previously crept material is heated to the test temperature before stressing, recovery effects are likely to produce a higher creep rate at the start of the test than the rate that existed when service was interrupted. Even if the test is brought to temperature with the load applied, an imprecise match of test and service stresses commonly leads to early nonsteady creep rates, even when a steady condition was previously established in service.

Reference creep rates, whether from tests on identical new material or not, usually are limited to reported extrapolated stresses for 0.0001%/h or 0.00001%/h secondary creep rate at several temperatures. To fix the creep rate to be expected for virgin material at the service stress and temperature requires nonlinear interpolation of a few tabulated data points.

The considerable scatter typical in creep measurement poses a particular problem, because under the specified stresses of most elevated-temperature designs, creep deformation obtained in a reasonable time can approach the sensitivity limit of most extensometer systems. A large increase in that sensitivity limit is impractical, because movements caused by thermal expansion due to unavoidable temperature fluctuations necessitate averaging of the resulting unsteady strain indications.

Even determination of test duration can be an important test decision. If the component in service was well within an extended period of secondary creep, a test conducted for about a thousand hours at near-service conditions should display a creep rate that is close to the latest period of service during the entire test. However, if secondary creep is not distinct and prolonged for the alloy under study, or if service exposure has reached a stage significantly earlier or later than the minimum rate, extended testing deviates increasingly from the last rate of service. Greatest concern arises when post-service test conditions are made more severe than those of service, either to shorten the study or to yield rates that are more convenient to measure.

Procedure modifications minimize these difficulties, but their use has not been documented. The "ideal" test does not require a virgin specimen or knowledge of its initial properties; it can be used with a single representative specimen and permits evaluation in no longer than a few months at conditions reasonably near those of service.

Assuming that prior operation typifies specified design, so that rupture life at service temperature T_S is 100,000 h or more, a suitable test procedure may be:

1. Load a cold post-service specimen to the estimated service stress (or, if this is unknown, to the applicable specified allowable stress).
2. Increase the specimen temperature to $T_S + 50$ °C ($T_S + 90$ °F) within 3 h. Immediately on reaching that temperature, gradually reduce furnace power to obtain a uniform specimen temperature of $T_S + 20$ °C ($T_S + 36$ °F) at the end of one additional hour.
3. Take creep measurements until a rate is established with a precision of two significant digits.
4. Increase the specimen temperature to $T_S + 50$ °C ($T_S + 90$ °F) (with the load maintained) and hold until about 0.5% additional creep strain occurs at that temperature.
5. Reduce the temperature to $T_S + 20$ °C ($T_S + 36$ °F) as in step 2 and obtain a second creep rate as in step 3.

This procedure minimizes anomalous recovery effects at the start of each creep-rate determination, permitting valid direct comparison of rates for the material as removed from service and after an additional exposure, which should account for another significant fraction of total rupture life.

A measured creep rate from step 5 that is lower than the rate determined in step 3 under identical conditions indicates that service has been so mild that steady secondary creep has not yet become fully established. The more common situation, in which the two rates are approximately the same or the second is faster than the first, would indicate that the component is within or past the secondary creep stage.

Creep and rupture tests appear to allow confident determination of remaining life of elevated-temperature materials; however, more rapid and less expensive methods of assessing creep damage continue to be sought.

Microstructural Evaluation of Creep Damage. Numerous studies beyond the realm of creep-rupture testing have sought immediate warning of creep damage, while monitoring its development, without the need for destructive sampling. Pending improvement and acceptance of such methods, it has been reported that remaining lifetime can be estimated from quantitative measurements of the microstructure (Ref 64). The procedure involves comparisons between a specimen taken from service and other specimens from accelerated tests at the service stress, but at increased temperatures. More detailed information on the effects of creep on microstructure can be found in the article "Mechanical Properties at Elevated Temperatures" in this Volume.

Planning a Test Program

The most important prerequisite of a reliable data analysis is the availability of "good" data. However, test data generally are not obtained with the specific goal of analysis. To ensure sound analysis of data, the design of experimental programs must be considered. Variables that significantly affect the data obtained are discussed below.

A simple method for planning a test program for creep and creep-rupture data is detailed in Ref 65. This method involves developing a test layout in matrix form by listing the test temperatures in

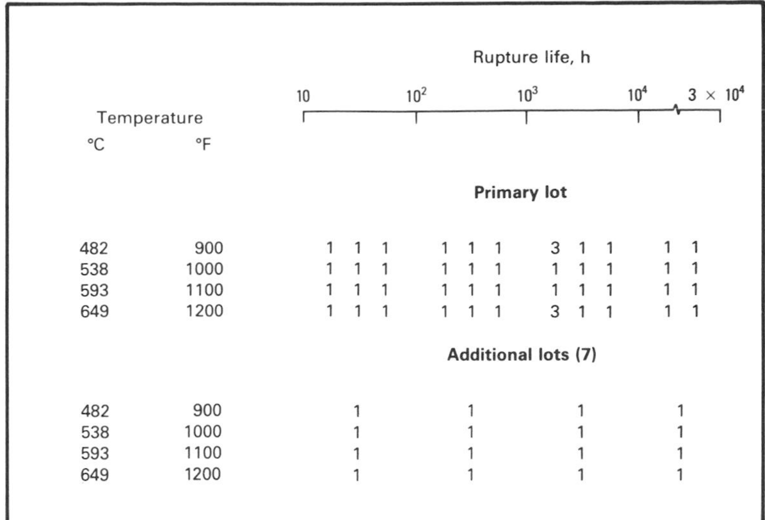

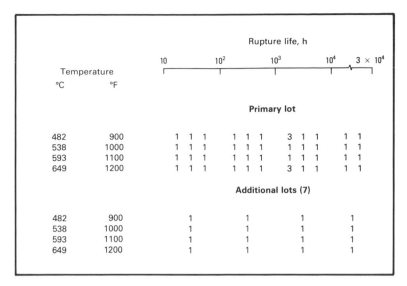

Fig. 10 Test matrix for creep testing

rows and tabulating the target test times in columns. Before the test matrix can be formed, however, the required temperature and life intervals must be determined.

Temperature. A range of temperature is usually required. For example, if information on material behavior is required over the range 482 to 649 °C (900 to 1200 °F), the intermediate temperatures must be chosen. This can be influenced by such factors as the expected magnitude of the temperature effect on properties, the likelihood that the different stress-property isothermals will be similar in shape and slope, and the anticipated occurrence of temperature-dependent instabilities or transformations.

In general, the test temperatures should be spaced so that at least one test stress level is common to each pair of adjacent isotherms. Temperature spacings of 55 °C (100 °F) are generally adequate for most applications. Rounded temperatures in degrees Fahrenheit should be used rather than Celsius, because existing data, if generated in the United States, are usually obtained at such temperatures. Thus, for direct comparison with previous data, Fahrenheit temperatures are suggested (e.g., testing at 900, 1000, 1100, and 1200 °F as opposed to 500, 550, 600, and 650 °C).

Life. Three tests, spaced about equally in log time, should be sufficient to characterize one log life cycle. Thus, between 100 and 1000 h, the target rupture lives would be 180, 320, and 560 h. Generally, times may be extrapolated by a factor of 2^n beyond the range of available data, where n is the number of log cycles spanned by the data.

Thus, if available data span times from 10 to 1000 h, they may be extrapolated to 4000 h with reasonable accuracy. If tests range from 10 to 30,000 h in duration, they may be extrapolated to approximately 300,000 h, which is a typical power plant lifetime. Tests with durations less than 10 h should not be factored into the calculation of n, however.

Stress. The above life estimates must, of course, be converted to stress before development of an actual test matrix. This requires some prior knowledge of the behavior of the material being tested. If such knowledge is not available, a series of scoping tests must be conducted to obtain stress estimates.

Lot-to-Lot Variations. Different lots of the same material, or batches of material from the same heat having a common processing and heat treatment history, can have significantly different creep properties. Test of several lots are required to yield the information needed for analysis of such data. Generally, data from at least eight lots, ideally from several different vendors, should be available to support analyses.

If possible, tests on the different lots should be conducted at common stresses and temperatures. At least one lot should be tested at least three times at a common stress and temperature for at least two temperatures to yield estimates of scatter. These repeated test conditions should result in test times of at least 1000 h.

A full test matrix need not be completed for every lot of material tested. However, at least one lot should be tested at all conditions in the matrix, and each lot should be tested at least once within each log cycle in time at each temperature.

To characterize completely a material over the range of 482 to 649 °C (900 to 1200 °F), for example, the following minimum test matrix should be completed:

1. Test one lot about 11 times (for times ranging from 10 to 30,000 h) at four temperatures (482, 538, 593, and 649 °C, or 900, 1000, 1100, and 1200 °F) (44 tests).
2. Repeat one of the above tests twice at 482 and 649 °C (900 and 1200 °F) (4 tests).
3. Test seven additional lots four times at each temperature (112 tests).

Thus, a minimum of about 160 tests is suggested to fully characterize a material. However, this characterization will apply only to lots representative of the population. This matrix is shown schematically in Fig. 10.

Selection of Rupture Stresses for Extensive Study

Research efforts by the American Petroleum Institute and the Materials Properties Council have involved testing of heavy sections (>300 mm, or >12 in., thick) of 2.25Cr-1Mo steel melted to restricted chemistry. The specified test matrix calls for rupture times of 100, 300, 1000, 3000, and 12,000 h each at temperatures of 454, 482, and 510 °C (850, 900, and 950 °F).

Conditions for Initial Tests. Room-temperature tensile tests for samples from four different thick plates averaged near 638 MPa (92.5 ksi). Published data for normalized and tempered 2.25Cr-1Mo steel with tensile strengths ranging from 620 to 655 MPa (90 to 95 ksi) suggested rupture strengths near the following levels:

Temperature, °C (°F)	Typical rupture strength from published data, MPa (ksi)		
	100 h	300 h	1000 h
510 (950)	317 (46)	290 (42)	262 (38)
480 (900)	358 (52)	331 (48)	303 (44)

The first test for each heat of steel was started at the highest temperature (510 °C, or 950 °F), with the 290 MPa (42 ksi) stress chosen to give a rupture time of about 300 h. If the composition and treatment for that heat result in higher or lower rupture strength than anticipated, the results are still acceptable as an approximation for the 1000 h point or the 100 h point.

Heat C produced 114.0 h rupture life at the 290 MPa (42 ksi) stress. Assuming the trend of the above table to be valid, the second specimen at 510 °C (950 °F) for this lot was started at 262 MPa (38 ksi), with goal life near 300 h, and a third specimen was started at the next lower temperature (480 °C, or 900 °F) at 331 MPa (48 ksi) to yield a goal life of 100 h. The latter test lasted 154.5 h.

In the initial test at 510 °C (950 °F) and 290 MPa (42 ksi), lower-strength heat L ruptured after only 40.4 h. The stress for the next test was therefore lowered to 248 MPa (36 ksi), and the first test at 480 °C (900 °F) used 290 MPa (42 ksi) for a goal life of 300 h.

Results from these six preliminary tests provided sufficient data for independent treatment by the Larson-Miller parameter with universalized $C = 20$ to set stresses for tests at all three temperatures to 1000 h goal life. As additional results accumulated, required stresses for the longest tests could be closely fixed by application of either refined parameter techniques or by graphical methods.

Allowance for Data Scatter. In an effort to minimize uncertainties from extrapolation procedures, undue emphasis may be placed on a test with the longest affordable duration. Normal data scatter still limits evaluation of expected performance from isolated points. Indeed, a more reliable prediction may be obtained when the same total test time is devoted to performing a greater number of tests with intermediate duration rather than only a few with extended life. An exception is when instabilities occur only with very long testing.

Some researchers study scatter by running multiple tests at one or more fixed combinations of stress and temperature. Again, a series of tests, each at different conditions, permits equally good statistical treatment to evaluate the scatter, while providing a broader indication of the material characteristics. Although emphasis has been placed on time to rupture, the extra expense of obtaining complete creep deformation data in all tests is often justified by the insight into the effects of structural changes.

ACKNOWLEDGMENTS

The information in this article is largely taken from:

- H.R. Voorhees, Assessment and Use of Creep-Rupture Properties, *Mechanical Testing,* Vol 8, *ASM Handbook* (formerly Vol 8, 9th ed., *Metals Handbook*), American Society for Metals, 1985, p 329–342
- M.K. Booker, Analysis of Creep and Creep-Rupture Data, *Mechanical Testing,* Vol 8, *ASM Handbook* (formerly Vol 8, 9th ed., *Metals Handbook*), American Society for Metals, 1985, p 685–694

REFERENCES

1. *ASME Boiler and Pressure Vessel Code,* Section 1, Rules for Construction of Power Boilers, Paragraph A-150, Basis for Estimating Stress Values, American Society of Mechanical Engineers, New York, 1983
2. Private business communication, Materials Technology Corporation data sheets submitted to the Metal Properties Council for tests under MPC Contract No. 174-1
3. "Standard Recommended Practice for Conducting Creep, Creep-Rupture, and Stress-Rupture Tests of Metallic Materials," E 139, *Annual Book of ASTM Standards,* Vol 03.01, American Society for Testing and Materials
4. A.K. Schmieder, Size Effect during Rupture Tests of Unnotched and Notched Specimens of Cr-Mo-V and Cr-Ni Steels, *Ductility and Toughness Considerations in Elevated Temperature Service,* American Society of Mechanical Engineers, 1978, p 31-48
5. B. Aronsson and A. Hede, Some Observations on the Reproducibility of Creep Rate Determinations, *High-Temperature Properties of Steel,* Iron and Steel Institute, London, 1967, p 41–45
6. D.R. Hayhurst, The Effects of Test Variables on Scatter in High-Temperature Tensile Creep-Rupture Data, *Int. J. Mech. Sci.,* Vol 16, 1974, p 829–841
7. W. Rottman, M. Krause, and K.J. Kremer, International Community Tests on Long-Term Behavior of 2-¼%Cr-1%Mo Steel, *High-Temperature Properties of Steel,* Iron and Steel Institute, London, 1967, p 23–29
8. D. Coutsouradis and D.K. Faurschou, Cooperative Creep Testing Program, AGARD Report 581, North Atlantic Treaty Organization, Advisory Group for Aerospace Research & Development, Neuilly-sur-Seine, France, 1971
9. *Report on the Elevated-Temperature Properties of Stainless Steels,* DS 5-S1, American Society for Testing and Materials, 1965
10. *An Evaluation of the Yield, Tensile, Creep, and Rupture Strengths of Wrought 304, 316, 321, and 347 Stainless Steels at Elevated Temperatures,* DS 5-S2, American Society for Testing and Materials, 1969
11. *Supplemental Report on the Elevated-Temperature Properties of Chromium-Molybdenum Steels,* DS 6-S1, American Society for Testing and Materials, 1966
12. *Supplemental Report on the Elevated-Temperature Properties of Chromium-Molybdenum Steels,* DS 6-S2, American Society for Testing and Materials, 1971
13. *Report on the Elevated-Temperature Properties of Selected Super-alloys,* DS 7-S1, American Society for Testing and Materials, 1968
14. *An Evaluation of the Elevated-Temperature Tensile and Creep-Rupture Properties of Wrought Carbon Steel,* DS 11-S1, American Society for Testing and Materials, 1969
15. *Evaluation of the Elevated-Temperature Tensile and Creep-Rupture Properties of C-Mo, Mn-Mo and Mn-Mo-Ni Steels,* DS 47, American Society for Testing and Materials, 1971
16. *Evaluation of the Elevated-Temperature Tensile and Creep-Rupture Properties of Steel,* DS 50, American Society for Testing and Materials, 1973
17. *Evaluation of the Elevated-Temperature Tensile and Creep-Rupture Properties of 3 to 9 Percent Cr-Mo Steels,* DS 58, American Society for Testing and Materials, 1975
18. *Evaluations of the Elevated-Temperature Tensile and Creep-Rupture Properties of 12 to 27 Percent Chromium Steels,* DS 59, American Society for Testing and Materials, 1980
19. *Compilation of Stress-Relaxation Data for Engineering Alloys,* DS 60, American Society for Testing and Materials, 1982
20. V.K. Sikka, H.E. McCoy, Jr., M.K. Booker, and C.R. Brinkman, "Heat-to-Heat Variation in Creep Properties of Types 304 and 316 Stainless Steels," Paper 75-PVP-26, American Society of Mechanical Engineers, 1975
21. K.R. Williams and B. Wilshire, Effects of Microstructural Instability on the Creep and Fracture Behavior of Ferritic Steels, *Mat. Sci. Eng.,* Vol 28, 1977, p 289–296
22. M. Wild, Analyse des Zeitstandverhaltens Warmfester Ferritischer Stahle bei Temperaturen von 450 bis 600 C, *Archiv. Eisenhütten-wes.,* Vol 34, Dec 1963, p 935–950
23. R.L. Klueh, Interaction Solid Solution Hardening in 2.25 Cr-1Mo Steel, *Mat. Sci. Eng.,* Vol 35, 1978, p 239–253
24. J. Glen, The Shape of Creep Curves, *Trans. ASME J. Basic Eng.,* Vol 85, Series D, No. 4, Dec 1963, p 595–600
25. J.P. Milan et al., "Negative Creep in Ni-Cr-Ti Alloys," paper presented at the Fourth Interamerican Conference on Materials Technology, Caracas, June-July 1975, p 102–106
26. R. Widmer and N.J. Grant, The Role of Atmosphere in the Creep-Rupture Behavior of 80Ni-20Cr Alloys, *Trans. ASME J. Basic Eng.,* Vol 82, Series D, No. 4, Dec 1960, p 882–886
27. P. Shahinian and M.R. Achter, Creep-Rupture of Nickel of Two Purities in Controlled Environments, *Proc. Joint Internat. Conf. on Creep,* Institute of Mechanical Engineers, London, 1963, p 7-49 to 7-57
28. J.H. Bennewitz, On the Shape of the Log Stress-Log Time Curve of Long Time Creep-Rupture Tests, *Proc. Joint Internat. Conf. on Creep,* Institute of Mechanical Engineers, London, 1963, p 5-81 to 5-92
29. W.C. Leslie, J.W. Jones, and H.R. Voorhees, Long-Time Creep-Rupture Tests of Aluminum Alloys, *J. Test. Eval.,* Vol 8 (No. 1), 1980, p 32–41
30. D.J. Wilson and H.R. Voorhees, Creep Rupture Testing of Aluminum Alloys to 100,000 Hours, *J. Mat.,* Vol 7 (No. 4), 1972, p 501–509
31. S.P. Agrawal, L.E. Byrnes, J.A. Yaker, and W.C. Leslie, Creep Rupture Testing of Aluminum Alloys: Metallographic Studies of Fractured Test Specimens, *J. Test. Eval.,* Vol 5 (No. 3), May 1977, p 161–173
32. D.J. Wilson, Extrapolation of Rupture Data for Type 304 (18Cr-10Ni), Grade 22 (2¼Cr-1Mo), and Grade 11 (1¼Cr-½Mo-¼Si) Steels, *Trans. ASME J. Eng. Mat. Technol.,* Jan 1974, p 22–33
33. F.R. Larson and J. Miller, A Time-Temperature Relationship for Rupture and Creep Stresses, *Trans. ASME,* Vol 74, 1952, p 765
34. R.M. Goldhoff, Comparison of Parameter Methods for Extrapolating High-Temperature Data, *Trans. ASME J. Basic Eng.,* Vol 81, Series D, No. 4, Dec 1959, p 629–644
35. R.L. Orr, O.D. Sherby, and J.E. Dorn, Correlations of Rupture Data for Metals at Elevated Temperatures, *Trans. ASM,* Vol 46, 1954, p 113–118
36. R.M. Goldhoff and G.J. Hahn, Correlation and Extrapolation of Creep-Rupture Data of Several Steels and Superalloys Using Time-Temperature Parameters, *Time-Temperature Parameters for Creep-Rupture Analysis,* American Society for Metals, 1968, p 199–245
37. W.M. Cummings and R.H. King, Extrapolation of Creep Strain and Rupture Properties of ½Cr-½Mo-¼V Pipe Steel, *Proc. Inst. Mech. Eng.,* Vol 185, 1970-71, p 285–299
38. S.S. Manson, Time-Temperature Parameters: A Re-evaluation and Some New Approaches, *Time-Temperature Parameters for Creep-Rupture Analysis,* American Society for Metals, 1968, p 1–113
39. S.S. Manson and C.R. Ensign, "A Specialized Model for Analysis of Creep Rupture Data by

the Minimum Commitment Station-Function Approach," NASA TM X-52999, 1971, p 1–14
40. S.S. Manson and C.R. Ensign, Interpolation and Extrapolation of Creep Rupture Data by the Minimum Commitment Method, Part 1: Focal-Point Convergence, *Characterization of Materials for Service at Elevated Temperature,* American Society of Mechanical Engineers, 1978, p 299–398
41. S.S. Manson and A. Muralidharen, Analysis of Creep Rupture Data for Five Multiheat Alloys by the Minimum Commitment Method Using Double Heat Centering Technique, *Progress in Analysis of Fatigue and Stress Rupture,* American Society of Mechanical Engineers, 1984, p 1–46
42. F.C. Monkman and N.J. Grant, An Empirical Relationship between Rupture Life and Minimum Creep Rate in Creep-Rupture Tests, *Proc. ASTM,* Vol 56, 1956, p 593–605
43. H.R. Voorhees, "Determination of Rupture Strength at Temperatures Near the Lower End of the Time-Dependent Range," findings reported to the Metal Properties Council, Inc., 1984
44. S.S. Manson and W.F. Brown, Jr., Discussion to a paper by F. Garofalo et al., *Trans. ASME,* Vol 78 (No. 7), Oct 1956, p 143
45. F. Dobes and K. Milicke, The Relation between Minimum Creep Rate and Time to Fracture, *Met. Sci.,* Vol 10, Nov 1976, p 382–384
46. D. Lonsdale and P.E.J. Flewitt, Relationship between Minimum Creep Rate and Time to Fracture for 2½%Cr-1%Mo Steel, *Met. Sci.,* May 1978, p 264–265
47. M. Gold, W.E. Leyda, and R.H. Zeisloft, The Effect of Varying Degree of Cold Work on the Stress-Rupture Properties of Type 304 Stainless Steel, *Trans. ASME J. Eng. Mat. Technol.,* Vol 97, Series H, No. 4, Oct 1975, p 305–312
48. R.F. Gill and R.M. Goldhoff, Analysis of Long-Time Creep Data for Determining Long-Term Strength, *Met. Eng. Quart.,* Vol 10 (No. 3), Aug 1970, p 30–39
49. P.L. Threadgill and B.L. Mordike, The Prediction of Creep Life from Transient Creep Data, *Z. Metallkd.,* Vol 68 (No. 4), 1977, p 266–269
50. F. Garofalo, C. Richmond, W.F. Domis, and F. von Gemmingen, Strain-Time, Rate-Stress and Rate-Temperature Relations during Large Deformations in Creep, *Proc. Joint Internat. Conf. on Creep,* Institution of Mechanical Engineers, London, 1963, p 1-31 to 1-39
51. P.L. Threadgill and B. Wilshire, Mechanisms of Transient and Steady-State Creep in a γ'-Hardened Austenitic Steel, *Creep Strength in Steel and High-Temperature Alloys,* The Metals Society, London, 1974, p 8–14
52. R.M. Goldhoff and R.F. Gill, A Method for Predicting Creep Data for Commercial Alloys on a Correlation between Creep Strength and Rupture Strength, *Trans. ASME J. Basic Eng.,* Vol 94, Series D, No. 1, March 1972, p 1–6
53. E.L. Robinson, Effect of Temperature Variation on the Creep Strength of Steels, *Trans. ASME,* Vol 60, 1938, p 253–259
54. P.N. Randall, Cumulative Damage in Creep-Rupture Tests of a Carbon Steel, *Trans. ASME J. Basic Eng.,* Vol 84, Series D, No. 2, June 1962, p 239–242
55. D.A. Woodford, Creep Damage and Remaining Life Concept, *Trans. ASME J. Eng. Mat. Technol.,* Vol 101 (No. 4), Dec 1979, p 311–316
56. M.M. Abo El Ata and I. Finnie, A Study of Creep Damage Rules, *Trans. ASME J. Basic Eng.,* Vol 94, Series D, No. 3, Sept 1972, p 533–543
57. J.W. Freeman and H.R. Voorhees, "Notch Sensitivity of Aircraft Structural and Engine Alloys, Part II: Further Studies with A-286 Alloys," Wright Air Development Center, Technical Report 57-58, Document 207,850, ASTIA, Jan 1959
58. R.V. Hart, Concept of Area-Modified Stress for Life-Fraction Summation during Creep, *Met. Technol.,* Sept 1977, p 447–448
59. R.V. Hart, Assessment of Remaining Creep Life Using Accelerated Stress-Rupture Tests, *Met. Technol.,* Jan 1976, p 1–7
60. J.W. Freeman and H.R. Voorhees, *Literature Survey on Creep Damage in Metals,* STP 391, American Society for Testing and Materials, 1965
61. B.J. Cane and K.R. Williams, Creep Damage Accumulation and Life Assessment of a ½Cr-½Mo-¼V Steel, in *Mechanical Behavior of Materials,* Vol 2, Pergamon Press, Oxford, 1979, p 255–264
62. J.J. Bodzin, J.W. Freeman, and I.A. Rohrig, Carbon and Carbon-Moly Steam Pipe after Long-Time Service, *Trans. ASME J. Basic Eng.,* Vol 88, Series D, No. 1, March 1966, p 14–20
63. T.M. Cullen, I.A. Rohrig, and J.W. Freeman, Creep-Rupture Properties of 1.25Cr-0.5Mo Steel after Service at 1000 Deg. F, *Trans. ASME J. Basic Eng.,* Vol 88, Series D, No. 3, Sept 1966, p 669–674
64. K.F. Hale, Creep-Failure Prediction from Observation of Microstructures in 2¼% Chromium-1% Molybdenum Steel, *Physical Metallurgy of Reactor Fuel Elements,* Central Electricity Generating Board, United Kingdom, 1975, p 193–201
65. R.C. Rice, Ed., "Reference Document for the Analysis of Creep and Stress Rupture Data in MIL-HDBK-5C," AFWAL-TR-81-4087, Air Force Wright Aeronautical Laboratories, Wright-Patterson Air Force Base, Dayton, OH, Sept 1981

Thermal and Thermomechanical Fatigue of Structural Alloys

STRUCTURAL ALLOYS are commonly subjected to a variety of thermal and thermomechanical loads. If the stresses in a component develop under thermal cycling without external loading, the term *thermal fatigue* (TF) or *thermal stress fatigue* is used. This process can be caused by steep temperature gradients in a component or across a section and can occur in a perfectly homogeneous isotropic material. For example, when the surface is heated it is constrained by the cooler material beneath the surface, and thus the surface undergoes compressive stresses. Upon cooling, the deformation is in the reverse direction, and tensile stresses could develop. Under heat/cool cycles, the surface will undergo TF damage. Examples of TF are encountered in railroad wheels subjected to brake-shoe action, which generates temperature gradients and, consequently, internal stresses (Ref 1, 2).

On the other hand, TF can develop even under conditions of uniform specimen temperature, instead caused by internal constraints such as different grain orientations at the microlevel or anisotropy of the thermal expansion coefficient of certain crystals (noncubic). Internal strains and stresses can be of sufficiently high magnitude to cause growth, distortion, and surface irregularities in the material (Ref 3). Consequently, thermal cycling results in damage and deterioration of the microstructure. This behavior has been observed in pure metals such as uranium, tin, and cadmium-base alloys and in duplex steels with ferritic/martensitic microstructures.

The term *thermomechanical fatigue* (TMF) describes fatigue under simultaneous changes in temperature and mechanical strain (Ref 4, 5). Mechanical strain is defined by subtracting the thermal strain from net strain, which should be uniform in a specimen. The mechanical strain arises from external constraints or externally applied loading. For example, if a specimen is held between two rigid walls and subjected to thermal cycling (and is not permitted to expand), it undergoes "external" compressive mechanical strain. Examples of TMF can be found in pressure vessels and piping; in the electric power industry, where structures experience pressure loadings and thermal transients with temperature gradients in the thickness direction; and in the aeronautical industry, where turbine blades and turbine disks undergo temperature gradients superimposed on stresses due to rotation. In the railroad application discussed earlier, when external loading due to rail/wheel contact is considered, then the material undergoes the more general case of TMF. The temperature rise on the surfaces of cylinders and pistons in automotive engines combined with applied cylinder pressures also represents TMF. Based on the mechanical strain range, the results of TF and TMF tests should correlate well.

A distinction must be drawn between isothermal high-temperature fatigue as cyclic straining under constant nominal temperature conditions versus TMF. As such, isothermal fatigue (IF) can be considered a special case of TMF. In most cases, the deformation and fatigue damage under TMF cannot be predicted based on IF information. Therefore, TMF experiments have been considered in studies of both stress-strain representation and damage evolution.

Sometimes the term *low-cycle thermal fatigue* or *low-cycle thermomechanical fatigue* is used. Low-cycle fatigue (LCF) can be identified two ways: (1) high-strain cycling where the inelastic strain range in the cycle exceeds the elastic strain range and (2) where the inelastic strains are of sufficient magnitude that they are spread uniformly over the microstructure. Fatigue damage at high temperatures develops as a result of this inelastic deformation where the strains are nonrecoverable. In low-cycle cases, the material suffers from damage in a finite (short) number of cycles. Thermomechanical fatigue is often a low-cycle fatigue issue. For example, in railroad wheels only severe braking applications—occurring infrequently over thousands of miles—contribute to damage, and fewer than 10,000 cycles take place during the lifetime of a wheel. Similarly, the largest thermal gradients and transients in jet engines develop during startup and shutdown. The total number of takeoffs and landings for an aircraft is fewer than 30,000 cycles over the lifetime of an aircraft. In the laboratory, investigations often are conducted under low-cycle conditions to complete the experiments in a reasonable period of time.

The inability to predict TMF damage from the IF database continues to challenge engineers and researchers. Thermomechanical fatigue encompasses several mechanisms in addition to "pure" fatigue damage, including high-temperature creep and oxidation, which directly contribute to damage. These mechanisms differ, depending on the strain-temperature history. They are different from those predicted by creep tests (with no reversals) and by stress-free (or constant-stress) oxidation tests. Microstructural degradation can occur under TMF in the form of (1) overaging, such as coarsening of precipitates or lamellae; (2) strain aging in the case of solute-hardened systems; (3) precipitation of second-phase particles; and (4) phase transformation within the temperature limits of the cycle. Also, variations in mechanical properties or thermal expansion coefficients between the matrix and strengthening particles (present in many alloys) result in local stresses and cracking. These mechanisms influence the deformation characteristics of the material, which inevitably couple with damage processes.

Because of the importance of TMF in real-world applications, considerable attention has been devoted to the problem in workshops and symposiums. Ever since the early 1950s and 1960s, TMF experiments have been reported by research groups in the United States, Europe, and Japan. A number of books, review articles, and symposia proceedings on the subject have been published (Ref 6-20). The advent of computer control and servohydraulic testing equipment has allowed simultaneous, accurate control of temperature and strain. Consequently, research in the field has flourished.

The database of TMF research, however, is small compared to the IF database. Experiments involving TMF remain difficult and expensive. The use of IF data to predict the performance of a material under TMF has been demonstrated to have drawbacks. The use of isothermal LCF and mechanical strain-range results at the maximum temperature end (or at a temperature with low IF resistance) may still be nonconservative. Attempts have been made to relate TMF crack growth to IF crack growth using linear elastic fracture mechanics (LEFM) concepts, but further refinements incorporating elastic-plastic fracture mechanics (EPFM) are needed. Many of the existing models do not account for the interaction of the mechanical strain with temperature. This interaction is rather complex and not well understood.

A distinction must be drawn between TMF and thermal shock (Ref 6, 21). Thermal shock involves a very rapid and sudden application of temperature (due to surface heating or internal heat generation), and the resulting stresses are often different from those produced under slow heating and cooling (i.e., quasi-static) conditions. Physical properties, such as specific heat and conductivity (which do not appear in low-strain-rate cases), appear explicitly in the thermal shock case. The rate of strain influences the material response and should be considered in damage due to thermal shock or in selection of materials for better thermal shock resistance.

Finally, if the body is subjected to thermal cycling conditions with superimposed net section loads, the component will undergo thermal ratcheting, which is the gradual accumulation of inelastic strains with cycles (Ref 22, 23). Failure due to thermal ratcheting involves both fatigue and ductile rupture mechanisms. The "two-bar structure" will be used to illustrate thermal ratcheting in a later section. Thermal ratcheting sometimes occurs unintentionally in thermomechanical tests when a region of the specimen is hotter than the surrounding regions, resulting in a bulge in the hot region.

This article provides an overview of the experimental methods in TF and TMF and presents experimental results on structural materials that have been considered in TF and TMF research. Life prediction models and constitutive equations suited for TF and TMF are also covered.

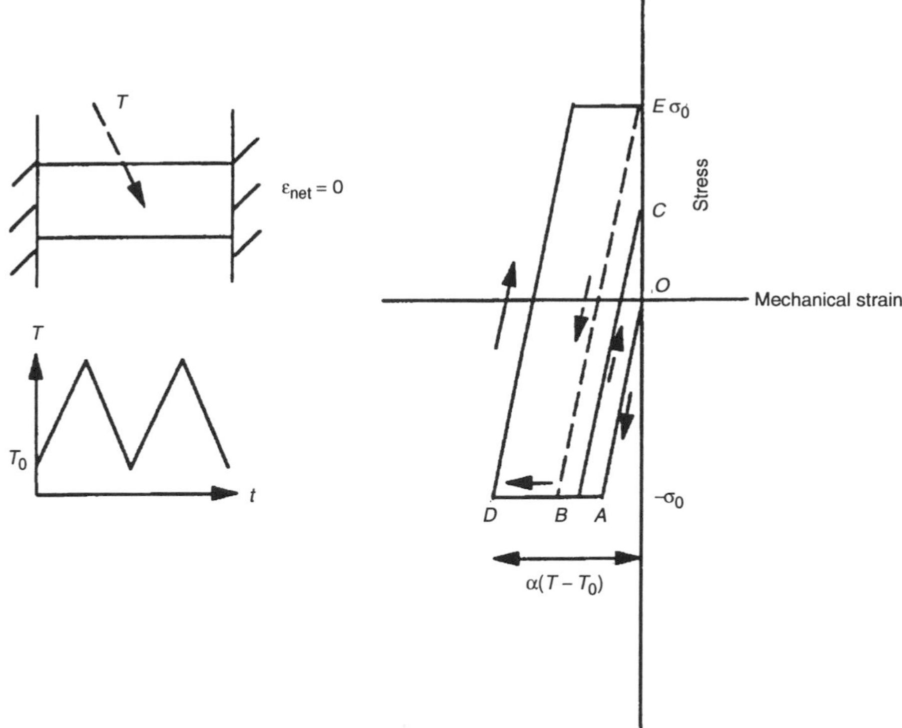

Fig. 1 Idealized stress-strain behavior under total constraint

Mechanical Strain and Thermal Strain

Free (unrestrained) thermal expansion and contraction produce no stresses. When the thermal expansion of a body is restrained upon uniform heating, thermal stresses develop. Consider the case where a bar is held between two rigid walls and subjected to thermal cycling. The length of the bar cannot change during heating and cooling. Let T_0 be the reference temperature at which the bar was placed under *total constraint*. The compatibility equation for this bar is given as:

$$\varepsilon_{net} = \varepsilon_{th} + \varepsilon_{mech} = \alpha\,(T - T_0) + \varepsilon_{mech}$$

In this case the net strain is zero, and all of the thermal strain is converted to mechanical strain. The thermal strain is defined as the product of coefficient of thermal expansion, α, and the temperature range $T - T_0$, where T is the current temperature. Then,

$$\varepsilon_{mech} = -\alpha\,(T - T_0)$$

Sometimes the total constraint case is identified as $\dot{\varepsilon}_{th}/\dot{\varepsilon}_{mech} = -1$. When this ratio is larger than −1, some free expansion and contraction occur, and the term *partial constraint* is used. If the $\dot{\varepsilon}_{th}/\dot{\varepsilon}_{mech}$ ratio is lower than −1, the condition is known as *overconstraint* (Ref 24). Therefore, the constraint influences the mechanical strain for a given thermal strain. Mechanical strain comprises elastic strain and inelastic strain (once yield stress is reached) and is the key parameter in TMF studies. The stress/mechanical-strain behavior shown in Fig. 1 is highly idealized; the material exhibits no hardening after yielding, the tension and compression strength are the same, and elastic modulus is independent of temperature. Upon heating, the bar is elastic and follows the stress-strain curve along *OA*. At *A*, the bar yields in compression, and upon further increase in temperature the mechanical strain on the bar increases along *AB*. The bar accumulates inelastic strain along *AB*. If the bar is cooled from *B*, it will deform in the reverse (i.e., tensile) direction. When the initial temperature is reached, the bar will return to zero mechanical strain, but a residual tensile stress will exist in the bar at point *C*. If the bar is again heated to the maximum temperature, the material will cycle between the stress point *B* and *C*. The bar is operating within the "shakedown" regime. It is unlikely that the bar will fail under these conditions because there is no plastic flow after the first reversal.

Next, consider the case when the thermal strain in the first heating portion of the cycle exceeded twice the elastic strain and a mechanical strain corresponding to point *D* is reached. Upon cooling back to the initial temperature, T_0, the bar will yield in tension and inelastic flow will occur until point *E* is reached. Upon reheating, the bar will deform in the reverse direction (dashed line) until it reaches point *D* in compression. A hysteresis loop develops as a result of this thermal cycle. Under alternate heat/cool cycles, forward and reverse yielding will occur every cycle, resulting in failure in a finite number of cycles.

The constrained bar model is conceptually easy to visualize, but in real structures the condition can be different from total constraint. This will be analyzed later in this article.

Experimental Techniques in TF and TMF

Table 1 summarizes different heating methods for TF and TMF. The advantages and disadvantages of each technique are listed, as are the materials examined. In early work, experimenters subjected specimens alternately to high and low temperatures with no external loading. One way to accomplish this procedure was by immersing the specimens in cold and hot fluidized beds, which can be operated up to 1150 °C (2100 °F) (Ref 49). Over the years, various wedge-shape specimen geometries have been used.

The crack growth can be observed and the data presented as a function of maximum temperature. If the results are to be compared to IF or TMF tests, the stresses and strains should be calculated (Ref 51, 55) with the finite-element method (FEM) or other numerical methods. The geometry and strain-temperature variations for the wedge specimen are shown in Fig. 2. Note that as the temperature increases, the strain-temperature variation is out of phase (OP). The minimum strain is reached within 10 s. At times beyond 10 s, the strain-temperature variation is in-phase (IP). Upon cooling, the reverse behavior was ob-

Table 1 Summary of TMF and TF test methods

Type of test	Heating method	Advantages	Disadvantages	Materials studied	Ref
TF	Immersion in hot and cold oil bath	Simplicity of the experiment	Transient stress strain could be present and should be calculated	Noncubic crystals, including tin, zinc, cadmium	3
TMF	Direct resistance	Rapid heating; allows space to mount the extensometer and pyrometer for crack growth measurements	Electric isolation of grips; local heating of crack tips	Conductive materials, stainless steel	25–31
TMF and TF	Induction (10–450 kHz, 5–40 kW capacity)	Rapid heating; complex specimen geometries permitted; inert environment testing using bellows	Experience with coil design required; electric noise in the strain signal due to high-frequency magnetic fields; high cost of unit	Aluminum, copper, steels, nickel-base superalloys	24, 32–45
TMF and TF	Quartz lamp (radiation)	Inexpensive; uniform temperature over different zones of the specimen	Shadow effects; slow cooling rates; enforced cooling needed	Nickel- and cobalt-base superalloys, metallic composites	46–48
TF	Fluidized bed	Good for screening TF resistance of materials	Stress-strain temperature transients must be calculated and surface oxidation removed	Nickel-base superalloys	49–52
TF	Burner heating; flame heating	Good for screening TF resistance of materials; surface hot corrosion damage representative of service	Stress-strain temperature transients must be calculated	Nickel-base superalloys, steels	53
TF	Thermal fatigue under bending	Under reversed bending one surface undergoes OP, the other undergoes IP	Stress-strain gradients must be calculated	Nimonic alloys	54
TF	Dynamometer (friction heating)	Very high temperatures on surface reached; representative of service	Oxides are wedged into cracks; friction characteristics change with time	0.5–0.7% C steels	1, 2

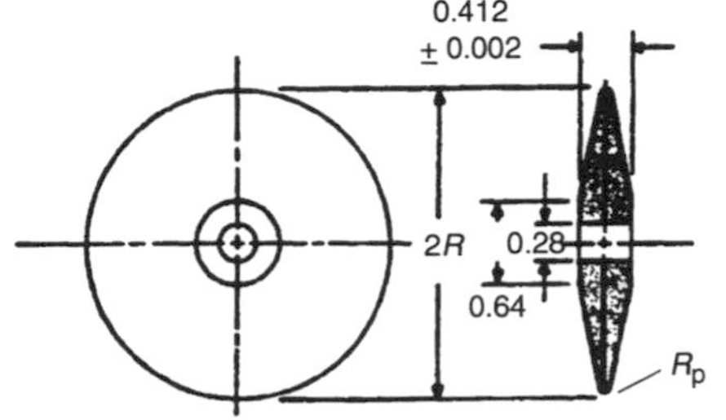

Part	$R_p \pm 0.001$	$2R \pm 0.01$
1	0.010	2.38
2	0.020	2.40
3	0.030	2.42
4	0.040	2.44

(a)

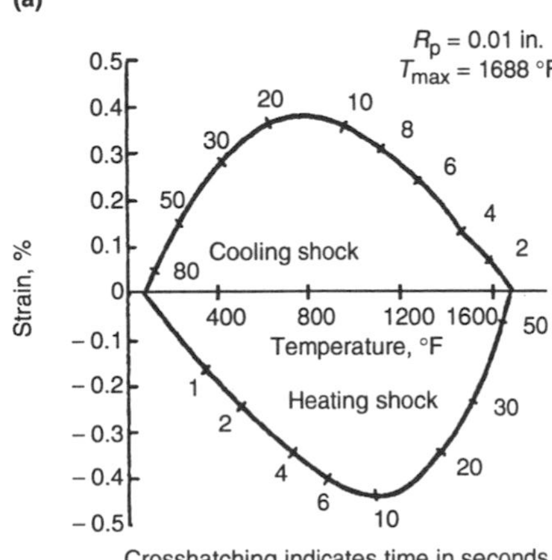

(b)

Fig. 2 (a) Wedge geometry for TF studies. Dimensions given in inches. (b) Strain-temperature variation in the fluidized-bed experiments. Source: Ref 13

served. Relative thermal fatigue resistance of many alloys can be classified with fluidized-bed experiments. This technique has proved to be of considerable value in examining the role of directional solidification, grain size, and γ' size and morphology in superalloys (Ref 51, 52, 56). The ε–T variation in Fig. 2(b) resembles the TMF diamond counterclockwise (DCCW) history that will be discussed later.

Instead of the fluidized-bed technique, burner heating and quartz lamp heating can be applied to the specimens. More recently, Remy and colleagues (Ref 47) used the quartz heating method to study the thermal fatigue behavior of nickel- and cobalt-base superalloys. Their specimen geometry was slightly different from the wedge used in early studies, but the principles of the method were the same. The use of quartz lamps is considerably more economical than other thermal fatigue heating methods. Simovich (Ref 34) developed a different specimen design: 5 cm diam disk (Fig. 3). This specimen was heated axisymmetrically using an induction heater with no external load, and a temperature gradient was developed in the radial direction. Cooling water was pumped through the large hole; the dark spot at the right marks a typical location for thermocouples. The maximum temperature considered was 650 °C (1200 °F) and the cycle time was approximately 60 s (controlled by induction heating). Using an axisymmetric model, Simovich calculated the circumferential stresses upon cooling and compared these results to experimental measurements. Under these conditions, cracks near 7 mm (0.3 in.) appeared in less than 2000 cycles in 0.7% C class steels.

All the experiments discussed thus far involved no external loading. Thermomechanical fatigue experiments with externally imposed strain were pioneered by Coffin (Ref 4, 25), who plotted the results versus plastic strain range. Both hollow and solid specimen designs were used. The hollow design allows more rapid heating and cooling. On the other hand, the solid specimen design lowers the possibility of buckling. Most IF experiments have been conducted on solid specimens; to obtain meaningful comparisons, such specimens should also be used in TMF studies.

Fig. 3 Disk specimen, showing radial cracks larger than 6.35 mm, used by Simovich (Ref 34) in thermal fatigue studies on steels. The dark spot at the right is used for temperature sensing.

Currently used techniques include resistance heating (Ref 4, 25–29), quartz lamp heating (Ref 46–48), and induction heating (Ref 24, 32–38). Induction heating is preferred, and the actual temperature gradient in the specimen should be known. Temperature measurements have been accomplished with spot-welded thermocouples, strapped-on thermocouples, or pyrometers. The temperature must be continuously monitored throughout the test. Infrared pyrometers are preferred in order to avoid potential failure originating from thermocouple beads or oxides formed at the thermocouple/specimen intersection. If thermocouples are chosen, a backup thermocouple is advised in case one should break off. A different temperature profile at different specimen locations could result in specimen barreling or instability effects. Depending on the thermal mass (i.e., grips) at the ends of the specimen and the "chimney" effect with induction or quartz heating, the coils or the lamp power in different zones of the specimen should be adjusted to avoid tem-

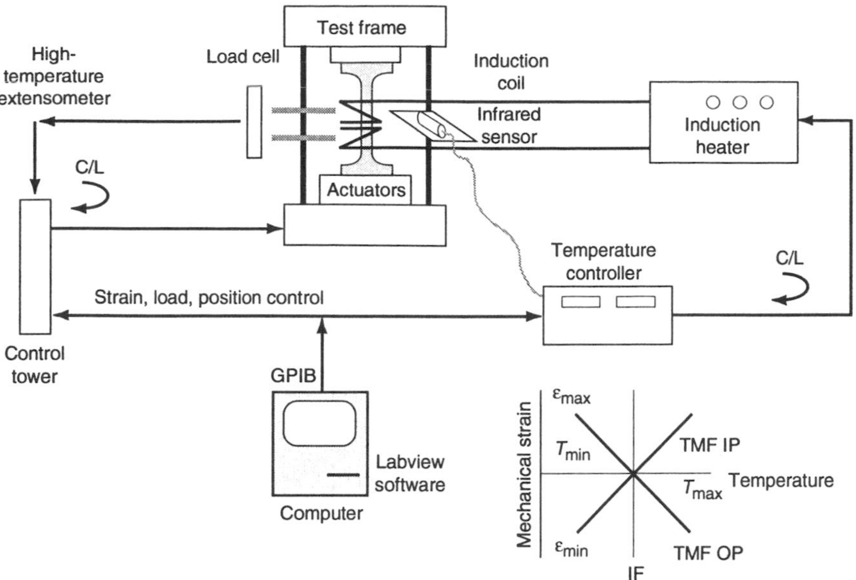

Fig. 4 Schematic of a modern TMF system

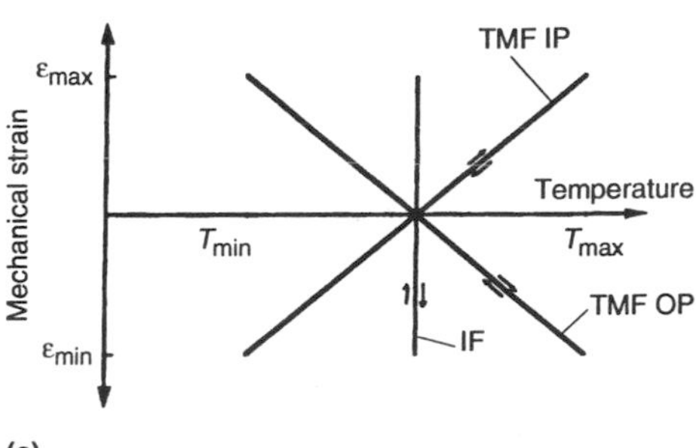

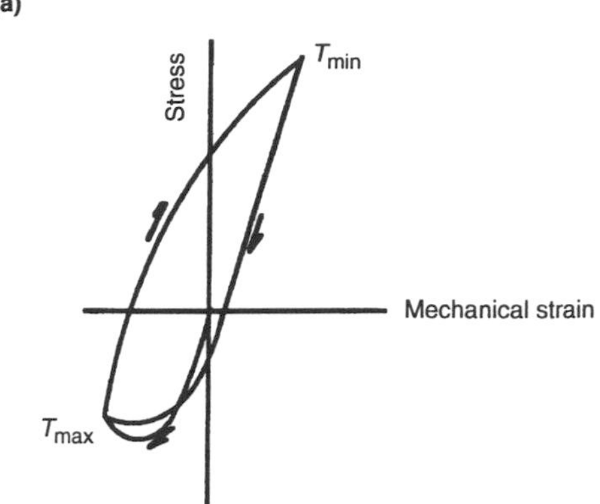

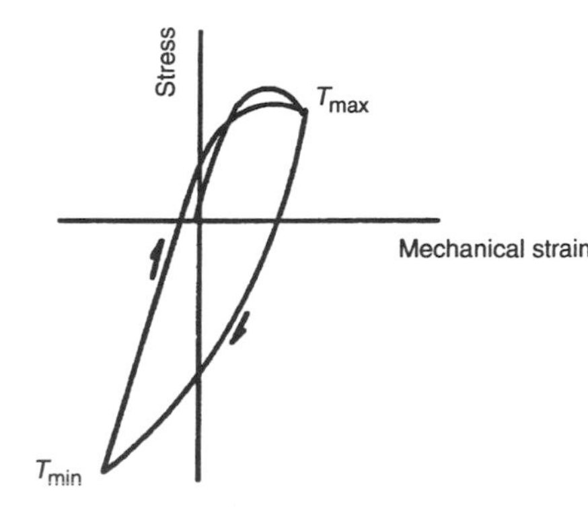

Fig. 5 (a) Mechanical strain/temperature variation in TMF OP, TMF IP, and IF. (b) TMF OP stress-strain response. (c) TMF IP stress-strain response

perature gradients more than 5 °C (9 °F). Coffin (Ref 57) has observed progressive thickening of the sample cross section at one region and progressive thinning at another region. Manson (Ref 6) has shown that if a local region of the specimen undergoes higher temperatures relative to the major length of the specimen, localized plastic strains and creep will occur in this region due to reduced yield stress. In some cases, when localized deformation as described above occurred, experimenters accounted for it in their analysis; interpretation of the results, however, is rather complex. Optimizing the dynamic rather than the static temperature profiles circumvents this problem and should be completed before a serious TMF research program is undertaken.

Quartz or alumina rod extensometers are used to control and measure the net strain during TMF experiments. Net strain is defined as the deflection divided by the initial gage length. Special attention should be paid in mounting the extensometer in the presence of an induction coil. The ends of the rods can be conical or chisel edged. At high temperatures, the spring load on the rods should be reduced to avoid penetration and notching of the specimen. In early studies, diametral strain measurements were made on hourglass specimens and converted to axial strain (Ref 4, 25, 28). The conversion requires Poisson's ratio and modulus of elasticity as a function of temperature and could have caused some errors in strain determination.

Thermal strain compensation is achieved by cycling the temperature at zero load before the test and determining the thermal strain as a function of temperature and time. Thermal strain can be defined using the coefficient of thermal expansion (CTE). Mechanical strain that produces stresses is defined by subtracting the thermal strain from the net strain. A good calibration and a good extensometry are required, because in TMF the mechanical strain range could be much lower than the thermal strain range.

Figure 4 shows a schematic of a modern TMF test system. The test machine is a digital-control servohydraulic test frame. There are two closed loops (C/L) in the control system. The control tower receives axial strain, position, and load signals from the test frame and sends them to a Macintosh computer fitted with a general-purpose instrumentation bus (GPIB) board. The computer, using Labview software, generates strain and temperature histories, which are transmitted to the temperature controller and to the control tower. Data collection is performed with the Labview software, and the results are displayed on the monitor during the experiment. A noncontact infrared pyrometer device has been used for temperature measurements. Specimens were heated using a high-frequency induction heater with a 15 kW capacity. The test system can perform TMF IP and OP tests, IF tests, and other complex strain-temperature variations.

TMF IP versus TMF OP

Mechanical strain/temperature waveform is classified according to the phase relation between mechanical strain and temperature. In-phase TMF means that peak strain coincides with maximum temperature; out-of-phase TMF means that peak strain coincides with minimum temperature. These two cases are shown in Fig. 5(a), along with the IF case. Generic hysteresis loops corresponding to the TMF OP and TMF IP cases are shown in Fig. 5(b) and (c), respectively. For a TMF cycle, the hysteresis loops are "unbalanced" in tension versus compression. In the TMF OP case, considerably more inelastic strains develop in compression relative to tension. The opposite behavior occurs in the TMF IP case. Some TMF experiments have been conducted under $R_\varepsilon = -1$ (i.e., completely reversed) conditions. Other TMF experiments have been conducted under $R_\varepsilon = -\infty$ (maximum mechanical strain is zero; see Fig. 1 [Ref 24]) and $R_\varepsilon = 0$ (minimum mechanical strain is zero [Ref 4]) conditions.

The inelastic strain is defined by subtracting the elastic strain from the mechanical strain:

$$\varepsilon_{in} = \varepsilon_{mech} - \frac{\sigma}{E[T(t)]} \quad \text{(Eq 1)}$$

For computational purposes, pairs of stress and temperature data points are needed. The variation in elastic modulus, $E[T(t)]$, as a function of temperature should be determined from isothermal experiments. A stress/inelastic strain hysteresis loop can be constructed using Eq 1. If there are hold periods during the TMF cycle, the equation will still be valid. The mechanical strain range, $\Delta\varepsilon_{mech}$, is shown in Fig. 6. The stress range in a TMF cycle is also shown for the OP case. The loop for the IP case is

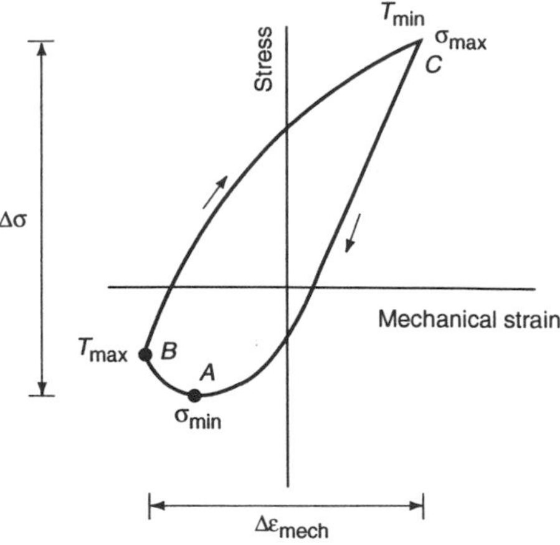

Fig. 6 Definitions of stress range and mechanical strain range in TMF

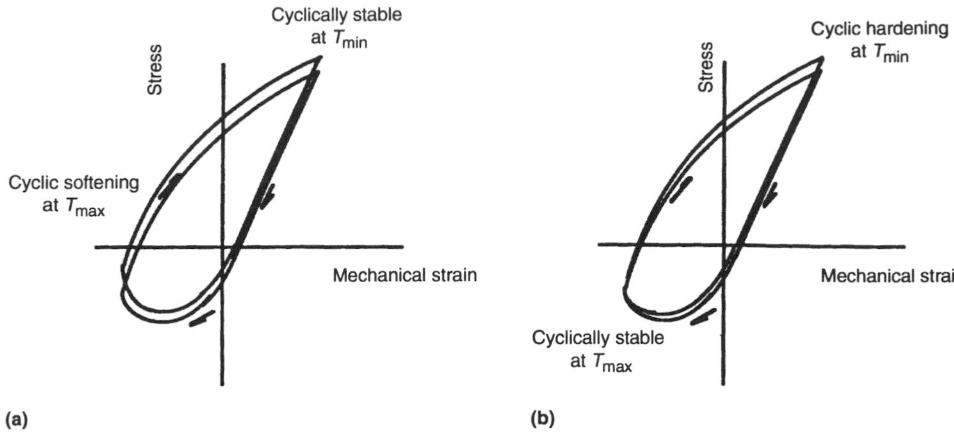

Fig. 7 Stress-strain response under cyclic softening (a) or cyclic hardening (b) conditions

similar, but reversed. Note that at the minimum strain (point B) the stress is not necessarily a minimum. Inelastic deformation with softening due to decrease in strength with increasing temperature is observed during AB. At B the maximum temperature is reached. Upon cooling, the behavior is elastic, followed by plastic deformation at the low-temperature end.

For engineering purposes, the inelastic strain range of a thermomechanical cycle can be determined to a first approximation by subtracting the elastic strains computed at the maximum and minimum strain levels. This gives:

$$\Delta\varepsilon_{in} \approx \Delta\varepsilon_{mech} - \left(\frac{|\sigma_B|}{E_B} + \frac{|\sigma_C|}{E_C}\right) \quad \text{(Eq 2)}$$

where E_C is the elastic modulus at the maximum strain and E_B is the elastic modulus corresponding to the minimum strain. Equation 2 slightly underestimates the inelastic strain range compared to the more exact equation. Note that the inelastic strain range includes the plastic strain, creep strain, and other strain components (e.g., transformation strain). Separation of plastic and creep strains in a TMF cycle is not straightforward. If needed, it can be done experimentally (Ref 58) by stress hold at selected points of the hysteresis loop or via constitutive models including plasticity and creep. Several constitutive models have been proposed for thermomechanical cyclic loadings and will be discussed later.

Just as in IF conditions, the TMF response of engineering materials involves cyclic hardening, cyclic softening, or cyclically stable behavior, depending on the microstructure, the maximum temperature level, and the phasing of strain and temperature. However, the behavior can be somewhat complex because of strain-temperature interaction. A material can harden, soften, or be cyclically stable at T_{max} of the cycle; likewise, at T_{min} the material can cyclically soften, harden, or be stable. Two possibilities are shown in Fig. 7. In Fig. 7(a), the material softens at T_{max} and remains cyclically stable at T_{min}. The material can cyclically soften at high temperature due to thermal recovery, causing coarsening of the microstructure, and in this case the hysteresis loops appear to "climb" in the tensile direction. Therefore, the tensile mean stress increases with increasing number of cycles. The microstructural coarsening could subsequently affect the strength at T_{min}, with the maximum stress in the cycle dropping with increasing number of cycles. Thereafter, the climbing of the hysteresis loops stops and the range of stress in the cycle decreases. This behavior has been documented in Ref 59. In the second example (Fig. 7b), stable behavior is observed at T_{max}, but the strength at T_{min} increases because of dynamic or static strain-aging effects. In this case, the hysteresis loops also climb in the tensile direction and, at the same time, overall stress range increases. Examples of this are discussed in Ref 60.

Other Strain-Temperature Variations in TF and TMF

Diamond (or baseball) TMF strain variation is obtained by changing the mechanical strain and temperature 90° or 270° out of phase. The diamond path should be specified as clockwise (DCW) or counterclockwise (DCCW), which could influence TMF life. The strain-temperature variation and the generic hysteresis loops for the DCCW case are shown in Fig. 8. In many structural alloys studied, the DCW and DCCW were not as damaging as TMF IP or TMF OP, because at the maximum temperature neither the strains nor the stresses were at a maximum. It is important to study the diamond TMF histories; they are encountered in many practical situations, such as turbine blades. Examples of more complex strain-temperature histories observed in service will be discussed in a later section.

A variation of the diamond history was proposed in the early 1970s (Ref 28). The term *bithermal fatigue* was coined in the mid-1980s by NASA researchers (Ref 30). In this case, the tensile portion of the loop is applied at one temperature, T_1, and the compressive portion of the

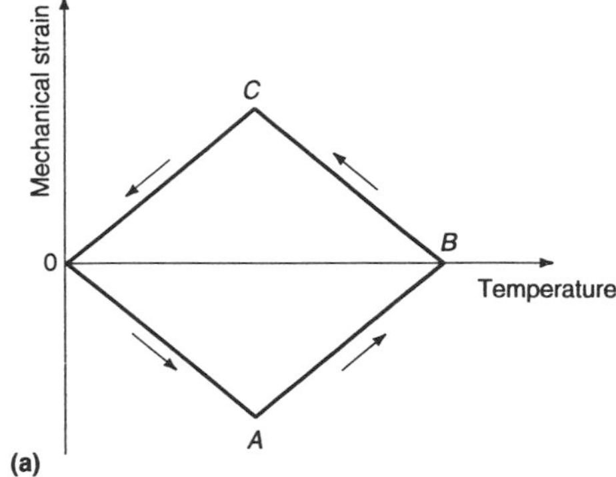

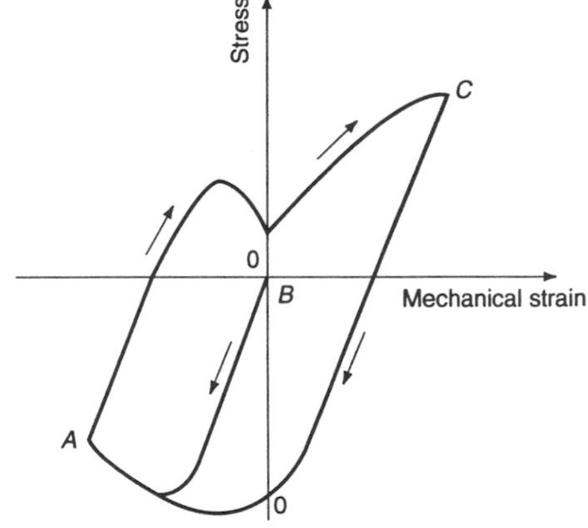

Fig. 8 Strain-temperature variation (a) and schematic of stress-strain response (b) for the DCCW case

loop is conducted at a different temperature, T_2. The temperature is changed, $T_1 \rightleftarrows T_2$, at zero stress. Advantages of this technique are that the tests can be conducted without the need for TMF computer software and the results more readily related to IF tests. If the thermal strains are large, however, the extensometer must have the range and the resolution to handle strain control at both temperature extremes. Also, some creep recovery due to internal stresses could occur during the zero stress temperature excursions.

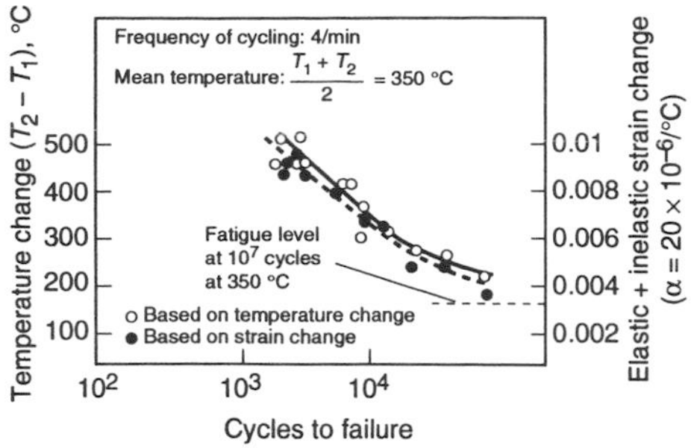

Fig. 9 Results of TMF experiments by Coffin (Ref 4) under total constraint for annealed type 347 stainless steel

TF and TMF of Carbon Steels, Low-Alloy Steels, and Stainless Steels

One of the early laboratory investigations of thermal fatigue in steels was conducted at the University of Illinois to further understand TMF in railroad wheels (Ref 1, 2). During the brake-shoe action on a railroad wheel, the rim is constrained by the surrounding cooler hub and the plate. Upon heating, circumferential compressive stresses develop; upon cooling, yielding in the tensile direction can occur. Under repeated brake applications, TF cracks can develop. This is simulated in the laboratory with a wheel dynamometer and a brake-shoe heating. Thermal cracks can grow to a size sufficient to exceed the fracture toughness of the material, resulting in catastrophic fracture.

Later experiments were conducted by Simovich (Ref 34), who subjected disk specimens to TF with induction heating. In this case, disks approximately 50 mm (2 in.) in diameter were heated and cooled with induction, generating a temperature gradient in the radial direction. The steel developed radial cracks that grew to sizes near 10 mm (0.4 in.). Simovich conducted a thermal analysis and stress analysis of the disk and correlated the fatigue results as a function of mechanical strain and stress range.

Well-controlled TMF experiments using direct resistance heating were conducted by Coffin at General Electric. The effect of prestrain on the fatigue life of type 347 stainless steel under thermal cycling has been established (Ref 4), as has the maximum temperature effect (up to 600 °C, or 1110 °F). The role of strain hold period in reducing TMF lives was established for periods of from 6 to 180 s. The influence of the hold period was explained as an increase in the inelastic strain range of the cycle. Coffin also examined the effect of thermal cycling on the subsequent stress-strain response and noted strain hardening of the material. Papers by Coffin (Ref 4) and Manson (Ref 5) were the first to propose a relationship between plastic strain range and life. This was later coined the low-cycle fatigue, or Coffin-Manson, equation:

$$N_f^{\alpha} \cdot \Delta\varepsilon_p = C$$

where N_f is the number of cycles to failure, $\Delta\varepsilon_p$ is the plastic strain range, and α and C are material constants.

Coffin's results on type 347 stainless steel are shown in Fig. 9. The specimens were subjected to total constraint ($\varepsilon_{net} = 0$). The mean temperature was maintained constant at 350 °C (660 °F), and the maximum temperature considered was as high as 650 °C (1200 °F). The horizontal axis (log scale) is the fatigue life, defined as fracture of the specimen. The vertical axis (linear coordinates) is given in terms of temperature range and mechanical strain. In the experiments of Coffin, these two quantities are not exactly equal because of deformation of supports and temperature distribution along the length of the tube. Therefore, the mechanical strain range is slightly lower than the thermal strain range in these experiments. The specimen is hot in tension and cold in compression and is undergoing TMF OP loading. Whether the specimen is clamped at the minimum temperature or the maximum temperature influences only the mean strain and has very little, if any, influence on fatigue life.

Coffin used resistance heating and developed a cam-and-lever mechanism to apply independent strains on the sample. This design has been duplicated in a number of subsequent TMF investigations in Japan, the United States, and the former Soviet Union. For example, research on constrained specimens was conducted on railroad wheel material, and hysteresis loops were established for carbon steels (0.4 to 0.7% C) at temperatures reaching 500 °C (930 °F) (Ref 39).

In Great Britain, the thermal fatigue resistance of carbon steels, alloy steels, and cast irons was investigated by Baron and Bloomfield (Ref 38) in the early 1960s. They used induction heating of an edge of a cold specimen. The strains were not measured or calculated for this case, but results have been displayed using T_{max} versus cycles to form a crack of finite size. The maximum temperature considered was 900 °C (1650 °F), where most steels transform to austenite, and martensite formed upon rapid cooling, resulting in rapid formation of cracks. At high temperatures austenitic stainless steels were found to be superior to other steels, while some of the nodular irons approached the TMF resistance of plain carbon steels.

Thermomechanical fatigue research on steels has attracted considerable attention in Japan. Kawamoto et al. (Ref 29) conducted TMF experiments on 0.7% C steels and 18-8 stainless steels. They confirmed that the hold periods reduced fatigue life and suggested that hold periods allow formation of metal carbides and oxides at grain boundaries. They found no considerable difference between IP and OP cycling when the results were compared based on mechanical strain range. They made the noteworthy observation that under TMF the lives were shorter than under IF, even when the IF test was conducted at the maximum temperature of the thermal cycle.

Taira and colleagues (Ref 35–37) have authored a number of key TMF papers covering a range of steels, including 1016 steel, chromium-molybdenum steels, and type 304 stainless steel. They used the mechanical strain range and plastic strain range to compare their data obtained under thermal cycling of 1016 steel in the temperature range of 100 to 600 °C (210 to 1110 °F). It is expected that considerable creep and oxidation effects are present in these steels at temperatures exceeding 500 °C (930 °F). Taira et al. (Ref 61, 62) also conducted thermal ratcheting tests under TMF OP conditions with stress control. Thermomechanical fatigue damage was predicted from creep-rupture data for these experiments.

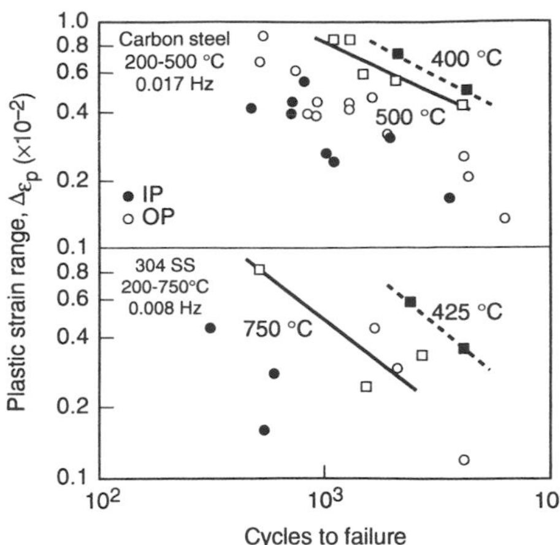

Fig. 10 Results of Fujino and Taira (Ref 63) on carbon steel and type 304 stainless steel, showing more damage in TMF IP relative to TMF OP

Fujino and Taira (Ref 63) demonstrated that for type 304 stainless steel (at 200 to 750 °C, or 390 to 1380 °F) the TMF IP lives were shorter than the TMF OP case by nearly a factor of four. Their results are shown in Fig. 10 for carbon steel and for type 304 stainless steel. Isothermal fatigue data at 425 and 750 °C (800 and 1380 °F) for type 304 and at 400 and 500 °C (750 and 930 °F) are given. The TMF OP lives were lower than IF lives in carbon steel, whereas for the stainless steel they were similar to the IF lives at maximum temperature of the cycle. The researchers made measurements of grain-boundary sliding and found evidence of it in TMF IP cases, but not in TMF OP or IF loadings. As the maximum temperature was lowered from 750 to 600 °C (1380 to 1110 °F) in TMF experiments (Ref 36), the TMF OP, TMF IP, and IF results at T_{max} of the cycle converged. It is clear that grain-boundary sliding due to unbalanced displacements at the microlevel becomes more pronounced as the maximum temperature in the cycle is increased.

Other studies in Japan were conducted by Udoguchi and Wada (Ref 31), who considered H46 martensitic stainless steel and type 347 stainless steel under TMF OP conditions with a maximum temperature of 700 °C (1290 °F) and 1040 steel with a maximum temperature of 400 °C (750 °F). Their results also confirmed that the TMF resistance is inferior to IF even when the results are compared at T_{max}. One of the most systematic investigations of TMF of steels was undertaken by Kuwabara and Nitta in the mid to

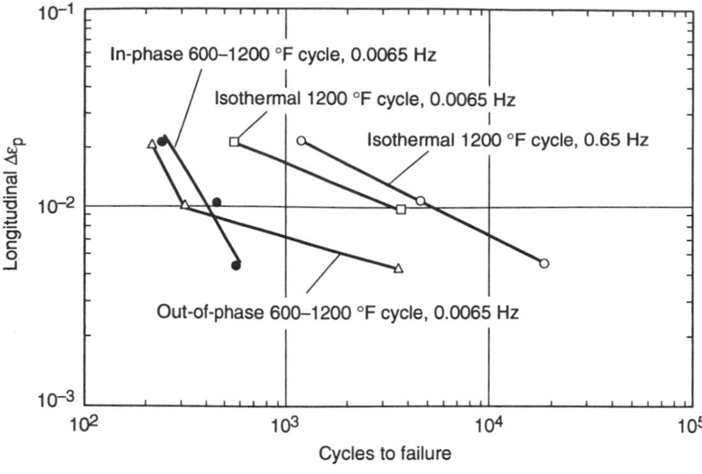

Fig. 11 Fatigue life results on type 304 stainless steel. Source: Ref 28

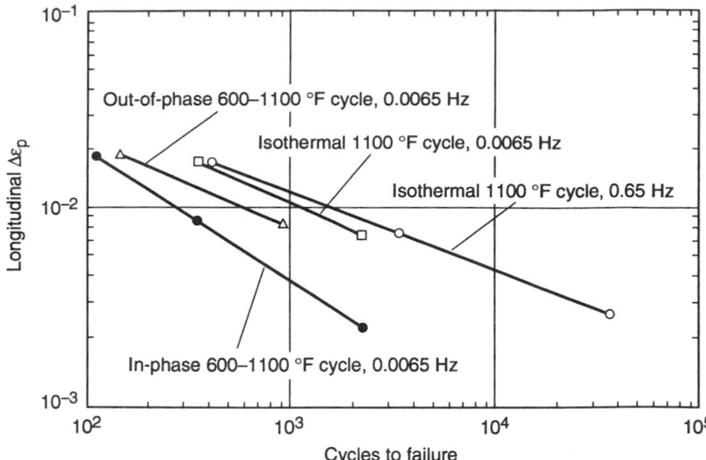

Fig. 12 Fatigue life results on A-286 precipitation-hardening stainless steel. Source: Ref 28

late 1970s (Ref 40, 41). They conducted TMF OP and TMF IP experiments on type 304 stainless steel under continuous cycling and also in the presence of tensile or compressive hold periods (300 to 600 °C, or 570 to 1110 °F). The TMF IP lives were shorter than TMF OP lives, but comparison with IF lives at 600 °C (1110 °F) showed that IF tests were more damaging. In another set of experiments on type 304, they showed that TMF IP and TMF OP were comparable when the maximum temperature was only 550 °C (1020 °F) (Ref 42); still, more intergranular cracking was observed for the IP case relative to the IF and OP cases. A hold period on these steels drastically reduced the TMF IP lives, but had little effect on the TMF OP behavior. The trends were somewhat reversed when chromium-molybdenum-vanadium steels were investigated; 1Cr-Mo-0.25V (Ref 43) (examined between 300 and 550 °C, or 570 and 1020 °F) and nickel-molybdenum-vanadium forged steel (Ref 44) exhibited shorter lives in TMF OP compared to TMF IP. This behavior is consistent with the propensity of these alloys to suffer from considerably higher oxidation damage relative to stainless steels. For both alloys, higher surface crack density was measured in the TMF OP case. At high strains TMF OP and TMF IP results converged. In early studies, Manson et al. (Ref 64) demonstrated the significant surface cracking due to oxidation effects in low-alloy steels; the TMF results were consistent with the shorter lives observed in PC (plasticity in tension reversed by creep in compression) type cycling relative to CP (creep in tension reversed by plasticity in compression) type cycling for this class of alloys.

Research at NASA (Ref 58) considered type 316 stainless steel subjected to temperature cycling in the range of 230 to 815 °C (445 to 1500 °F). Considerable creep strains were measured, and thermal recovery was present in these TMF experiments. Later, Halford and colleagues (Ref 30) conducted bithermal IP and bithermal OP tests on type 316, which showed good agreement with the earlier TMF OP and TMF IP tests. Their results showed that TMF IP was more damaging

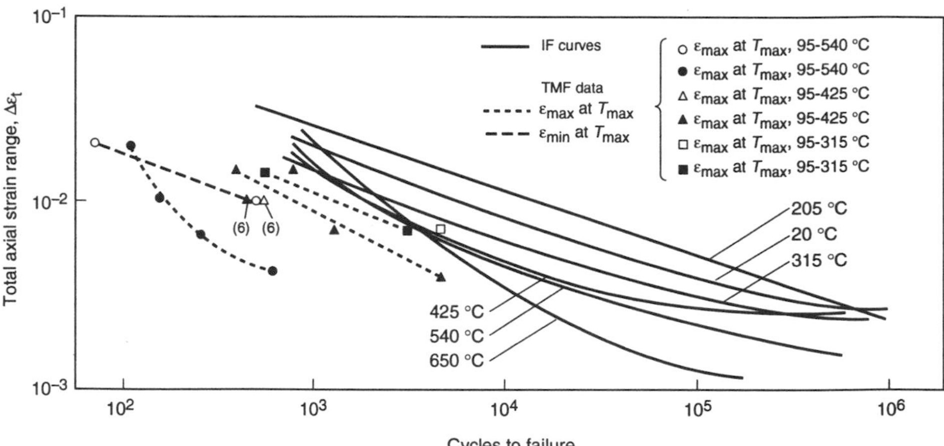

Fig. 13 Results of Jaske (Ref 66) on TMF of 1010 carbon steel. Note: (6) indicates a 6 min hold time at maximum temperature.

than TMF OP, a finding confirmed by Miller and Priest (Ref 19) on the same class of stainless steel. Sheffler (Ref 28) conducted bithermal TMF OP and TMF IP tests on type 304 stainless steel and demonstrated that both OP and IP lives were shorter than the IF data at maximum temperature of the cycle (Fig. 11). Similarly, for the A-286 alloy the bithermal IP lives were shortest compared to OP and IF results (Fig. 12). Sheffler made the TMF and IF comparisons for tests conducted at the same frequency and in ultrahigh vacuum. An interesting observation was that cavities formed due to unreversed grain-boundary displacements cannot be fully accommodated by intergranular sliding.

In Canada, Westwood (Ref 27, 65) conducted TMF tests on type 304 in the temperature range of 350 to 700 °C (660 to 1290 °F). The results showed good agreement between TMF IP and IF tests conducted at T_{max} of the thermal cycle, while the lives under TMF OP were longer. Although a larger difference between TMF IP and IF is expected at these temperatures, the lifetimes are generally consistent with previous data reported for similar materials.

Hysteresis response and life was studied by Jaske (Ref 66) on 1010 steel subjected to thermal cycling in the range of 95 to 540 °C (200 to 1000 °F). The cyclic hardening phenomenon was noted when the maximum temperature was below 425 °C (800 °F), possibly due to strain-aging effects. The TMF lives were significantly shorter than IF lives, even when the IF results from the maximum temperature were considered. The results of this study are shown in Fig. 13. The mechanical strain range versus life is plotted for TMF OP, TMF IP, and IF cases. There appears to be crossover in lives between TMF OP and TMF IP slightly below a strain range of 0.02. All the TMF data shown fall below the IF curves. This is consistent with the findings of Fujino and Taira (Ref 63). Similarly, Laub (Ref 67) studied 1010 steel used in heat exchangers and subjected specimens to total constraint TMF OP cycling where the mean temperature of the cycle was maintained constant. The most severe case examined was a mean temperature of 315 °C (600 °F) and a maximum temperature of 760 °C (1400 °F). Considerable oxidation of crack tips has been noted at temperatures exceeding 480 °C (900 °F).

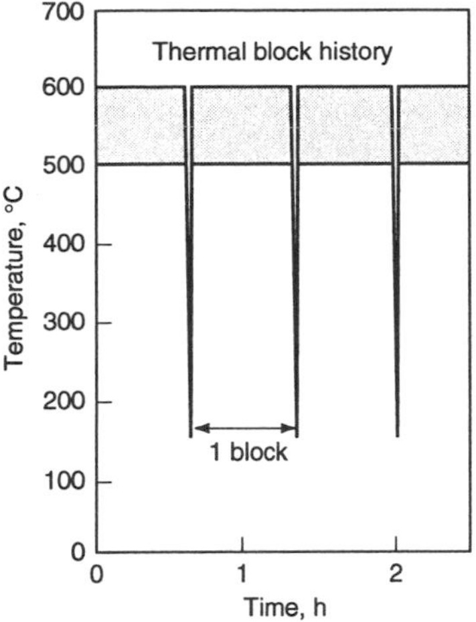

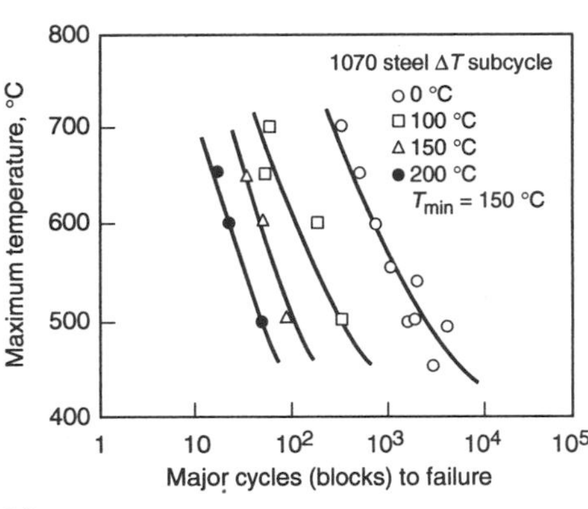

Fig. 14 (a) Temperature-time history under block loading. (b) Thermomechanical fatigue lives under TMF OP block loading. Source: Ref 59

Finally, in the work of Sehitoglu on 1070 steel, the stress-strain response was determined under total/partial and overconstraint TMF OP conditions with a minimum temperature of 150 °C (300 °F) and a maximum temperature of 700 °C (1290 °F). In later studies, Sehitoglu and his students investigated variable amplitude effects in TMF (Ref 59), environment effects in TMF (Ref 68, 69), phasing effects (TMF OP versus TMF IP) (Ref 68), strain-temperature changes conducive to strain aging (Ref 60), and notch effects and crack growth behavior under TMF (Ref 24, 60). A temperature-time history for a two-step TMF OP loading (total constraint) is shown in Fig. 14(a). One block includes one major cycle plus 100 minor (sub) cycles (Ref 33). In this case the major cycle underwent 150 ⇄ 600 °C (300 ⇄ 1110 °F) cycling under total constraint and the minor cycle experienced 500 ⇄ 600 °C (930 ⇄ 1110 °F). Because of considerable coarsening of the microstructure due to high-temperature exposure, the strength of the material at 150 °C (300 °F) is considerably lowered and the inelastic strain range of the cycle increases. The fatigue lives for these types of histories, where $\Delta T_{sub} = 0$, 100, 150, and 200 °C (30, 210, 300, and 390 °F), are shown in Fig. 14(b). This diagram indicates the dramatic deterioration of fatigue life in the presence of subcycles. For the same class of steels, Neu and Sehitoglu (Ref 68, 69) observed a typical crossover of the fatigue lives: At high strains IP tests were more damaging than OP, whereas the trend reversed at small strains. In tests conducted in a helium environment, the TMF IP experiments were more damaging than the TMF OP. The results of TMF OP and TMF IP experiments in air and in helium environment are shown in Fig. 15. The use of maximum-temperature IF data obtained for strain rates comparable to the TMF test predicted the trends, but a more sophisticated TMF life model has been proposed. Some of these studies will be discussed.

Mughrabi and his group in Germany (Ref 70) recently conducted both TMF IP and TMF OP tests on type 304L stainless steels. They observed a higher stress amplitude in TMF relative to IF when the TMF cycle coincides with temperatures near 450 °C (840 °F), where maximum dynamic strain aging occurs. Similar to the work of Sehitoglu et al. (Ref 33), they found that as the maximum temperature is increased, the maximum stress in the cycle occurs before the maximum temperature and maximum strain are reached. Thermomechanical fatigue IP tests revealed shorter lives when creep damage became more pronounced, whereas cavitation damage was not observed in the TMF OP case.

It is difficult to compare the results of one investigator to another, especially in TMF loading cases. This is because the TMF strain rate or frequency is dictated by the heating and cooling system, which is unique to the investigator. Even if the same heating method is used, there are no standards for TMF specimen geometry or for test control software, and the tests are often slowed down to ensure proper agreement between temperature and strain. Improved hardware and software would lead to greater reliability and consistency among different laboratories.

Environmental Effects. Coffin (Ref 71, 72) was the first to emphasize the significance of oxide damage in steels. At temperatures exceeding 500 °C (930 °F), an oxide layer forms on the surface of iron-base alloys. The iron oxides that form are brittle and facilitate crack advance into the substrate. This layer experiences a mechanical strain, which can result from one or a combination of the following: (1) strain from the applied mechanical loading in the material, (2) mismatch in the thermal expansion coefficients among the different stoichiometries of the oxides and substrate (Ref 33 and 73), (3) load due to the volume difference between the substrate and the various oxides (e.g., Fe_2O_3, Fe_3O_4, and FeO) (Ref 74, 75), and (4) other factors discussed in Ref 68. These mechanisms could affect the morphology of the surface oxide as well as the growing oxide-induced crack. Tensile oxide fracture facilitates crack initiation and crack growth, because the repeated oxide fracture can channel crack growth into the substrate. In TMF OP, the oxide forms near maximum temperature and upon cooling undergoes tension and fractures locally. Skelton (Ref 76) has shown that on 0.5Cr-Mo-V steels the crack growth rate in air is nearly an order of magnitude faster than in vacuum, with crack growth rates in steam environment falling between these two extremes.

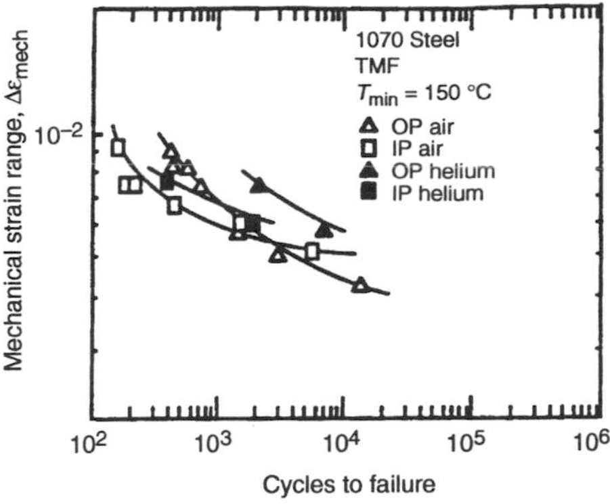

Fig. 15 Comparison of TMF OP and TMF IP experiments in air and in helium. Source: Ref 68, 69

One way to separate environmental damage from fatigue and creep damage is by performing tests in an inert or nonoxidizing atmosphere. Although a number of studies have been conducted on LCF under nonoxidizing environments (Ref 77–79), only two studies have been made on TMF of steels under an inert environment (Ref 28, 68). Sehitoglu and Neu devised a unique method of testing the specimen surrounded by bellows in which helium is trapped. The experiments were conducted under both TMF IP and TMF OP conditions. The increase in life relative to air results in the TMF OP case was nearly a factor of five, whereas in the TMF IP case the lives were not significantly influenced. The results are shown in Fig. 15.

Under conditions where creep mechanisms are dominant compared to environmental interaction effects, the fatigue life in air is about the same as in an inert atmosphere (Ref 78). However, when an environmental contribution exists, the fatigue life of smooth specimens is increased by a factor of 2 to 20 in a nonoxidizing atmosphere compared to tests performed in air (Ref 68, 76, 77, 79).

Strain Rate and Temperature Effects. Sehitoglu and his students conducted numerous investigations of 0.7% C steels (used in railroad wheels) and established the stress-strain behavior over the temperature range of 150 to 700 °C (300 to 1290 °F) (Ref 80–82). The effects of maximum temperature, strain aging, and thermal recovery due to spheroidization effects on stress-strain response have been identified. The effects of alloying were also examined. Early experiments have been reported on TMF behavior of carbon steels; unfortunately, the hysteresis loops have not been provided in these cases.

The influence of strain rate and temperature on life of type 304 stainless steel has been examined by Majumdar (Ref 83), who conducted experiments with a minimum temperature of 425 °C (800 °F) and a maximum temperature of 595 °C (1100 °F). They showed that at strain rates equal to or higher than 10^{-4}/s, the fractures are predominantly transgranular. For strain rates lower

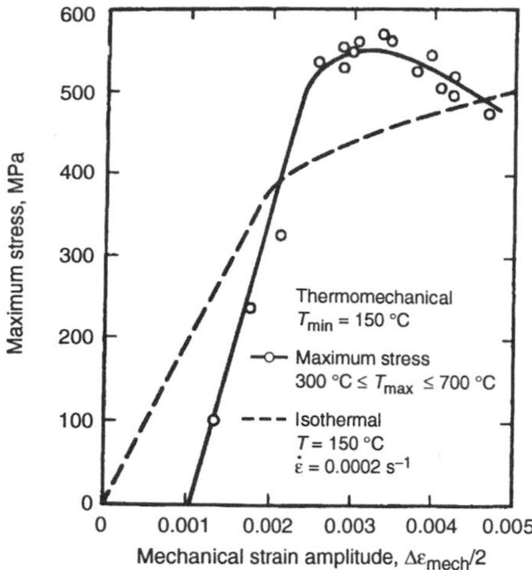

Fig. 16 Decrease in maximum stress in TMF OP case due to thermal recovery. Source: Ref 45

than 10^{-4}/s, however, the fractures become intergranular and the lives are shorter than the isothermal lives at maximum temperature. Majumdar also investigated the effect of hold time and demonstrated that hold periods reduce the cycles to failure.

The strain rates used in TMF studies vary in a narrow range. In his original study, Coffin (Ref 4) considered the influence of hold time and demonstrated that the cycles to failure decreased by a factor of three when the hold time increased from 6 to 180 s. Several high-strain-rate experiments were conducted by Taira and colleagues (Ref 61, 62) and Udoguchi and Wada (Ref 31) in their work on steels. Strain rates on the order of 10^{-4}/s were considered, which correspond to cycle times of 60 s. On the other extreme cycle, times near 30 min were considered (Ref 58). In most TMF research, the cycle time is on the order of 2 to 4 min, which corresponds to 5×10^{-5}/s. Direct resistance and induction heating methods can readily be used to produce strain rates on the order of 5×10^{-5}/s. Higher strain rates and the accompanying rapid temperature changes could produce temperature gradients in the specimen and make interpretation of the results difficult. Kuwabara and Nitta (Ref 84) examined the relationship between cycle time and TMF life in the range of 2 to 20 min. As the cycle time was increased, the fraction of intergranular cracks increased in the TMF IP case; however, the TMF OP results were not sensitive to strain rate. Commensurate with this finding is that TMF IP lives decreased with increasing cycle time while TMF OP lives remained constant.

Carbon steels undergo metallurgical changes in the form of coarsening of the pearlite lamellae and, ultimately, spheroidization at temperatures exceeding 400 °C (750 °F). Strain plays an important role in the spheroidization of pearlite. Deformation sets up subboundaries within the cementite, which are then rounded by diffusion driven by chemical potential gradients at the interface. This rounding of the interface edges leads to a complete band of ferrite separating the cementite. Many of these divisions occurring throughout the cementite break the lamellae up into segments, which then spheroidize. Strain-accelerated spheroidization can greatly reduce the time necessary to spheroidize a specimen at a given temperature.

In Fig. 16 the maximum stress in the TMF OP cycle is plotted versus the mechanical strain range in the cycle. Because the T_{min} was maintained constant in these experiments, higher strain ranges were achieved with higher maximum temperature. As the temperatures exceed the 500 °C (930 °F) value (or when the mechanical strain amplitude exceeds 0.003), the maximum stress decreases gradually with increasing mechanical strain. There are two main implications of this result: (1) the maximum stress cannot be used as a predictor of fatigue damage because for the same maximum stress there are two corresponding strain levels, one in the low-temperature and the other in the high-temperature regime (see discussion in Ref 33), and (2) the softening of the material at 150 °C (300 °F) means that the resistance of the material to deformation has decreased and the inelastic strain in the cycle has increased, producing enhanced damage.

Microstructural Changes in Steels. The mechanical properties below the transformation temperature (body-centered cubic, or bcc, phase) and above the transformation temperature (austenite face-centered cubic, or fcc, phase) are considerably different (Ref 85). Two series of creep tests on 0.7% C steels were performed under constant stress and temperature. Creep tests were conducted at temperatures of 400, 450, 500, and 550 °C (750, 840, 930, and 1020 °F) on 1070 steel well below the transformation temperature of 660 °C (1220 °F); the second series was conducted to investigate creep at temperatures above the austenitic transformation at temperatures of 700 and 800 °C (1290 and 1470 °F). It was found that both the transient and steady-state creep strain rates in the fcc phase were higher than the creep rates predicted with bcc phase properties by two orders of magnitude.

The second series of relaxation experiments (Ref 85) differed from the previous set in that, before each experiment, the specimen was heated to above its austenitic transformation temperature, to 925 °C (1700 °F), and was held at this temperature for 1 h. Then the specimen was rapidly cooled in air to the desired temperature of 500 or 550 °C (930 or 1020 °F), and the experiment proceeded as outlined above. Results indicate that the final stresses are similar for experiments with preheating and experiments without preheating to 925 °C (1700 °F), considering that preheat experiments indicate an initial stress about 50% lower than experiments without preheating.

The effect of phase transformations during thermal fatigue has been explored by Nortcott and Baron (Ref 86), who noted that repeated formation of austenite and martensite during the TF cycle generally leads to cracking of the material. Similarly, Sehitoglu (Ref 24) considered TMF experiments beyond 650 °C (1200 °F) on 1070 steels where the hysteresis loops recorded displayed the transformation effect. Thermal expansion characteristics are influenced by the nature of austenitic transformation. The mean value of α for 1070 steel below the transformation temperature is 8.34×10^{-6} 1/°F, whereas the mean value for α above the transformation temperature is 1.52×10^{-5} 1/°F. The coefficient of thermal expansion can be defined two ways: (1) tangent to the thermal strain-temperature curve or (2) as a secant modulus, the slope of the line connecting the thermal strain point to the origin. It is important to specify whether the CTE is a tangent or a secant value. When phase transformations occur, it is advisable to use a secant modulus; this avoids the problem of rapid changes in the tangent modulus upon phase transformation.

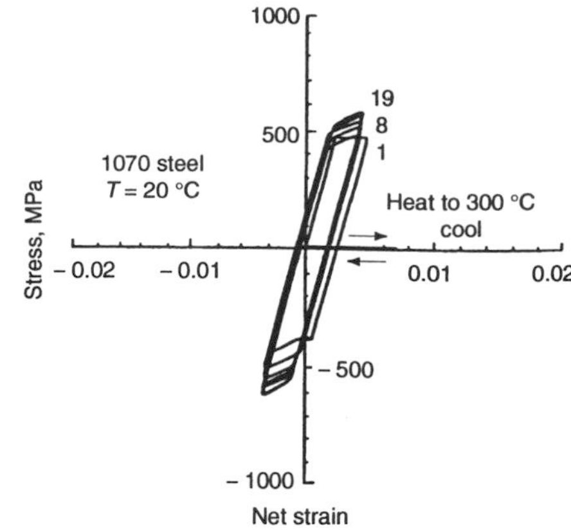

Fig. 17 Experimental σ–ε response under TMF strain-aging conditions. Source: Ref 80

Many steels undergo strain aging, which results in considerable hardening in a TMF test or a test that involves exposure of the material to temperatures below 400 °C (750 °F). Certain temperature-strain histories in solute-hardened materials produce strain aging, and thus strengthening, of the material. Thermomechanical fatigue studies under strain-aging conditions for steels and nickel-base superalloys have been discussed in Ref 20. Strengthening is caused by interstitial solute atoms, which anchor the dislocation motion. If the pinning of the dislocations occurs during deformation, the term *dynamic strain aging* is used (Ref 87). If the aging occurs under a constant load (after some plastic deformation), it is called *static strain aging* (Ref 87).

Static strain-aging experiments have been conducted on both 1020 and 1070 steels (Ref 80). The material was cycled at 20 °C (70 °F), but was exposed to the aging temperature time at zero stress every reversal. The experimental results are shown in Fig. 17. In these experiments, the deformation is at 20 °C (70 °F) but with intermittent exposure to 300 °C (570 °F) (up to 30 min) at zero load of the cycle. Increase in room-temperature strength as high as 30% has been measured after 40 reversals. The strain range of the hysteresis loops is $0.005 + \Delta\sigma/2E$, where $\Delta\sigma$ is the stress range and E is the elastic modulus at 20 °C (70 °F). Since this material is cyclically stable at

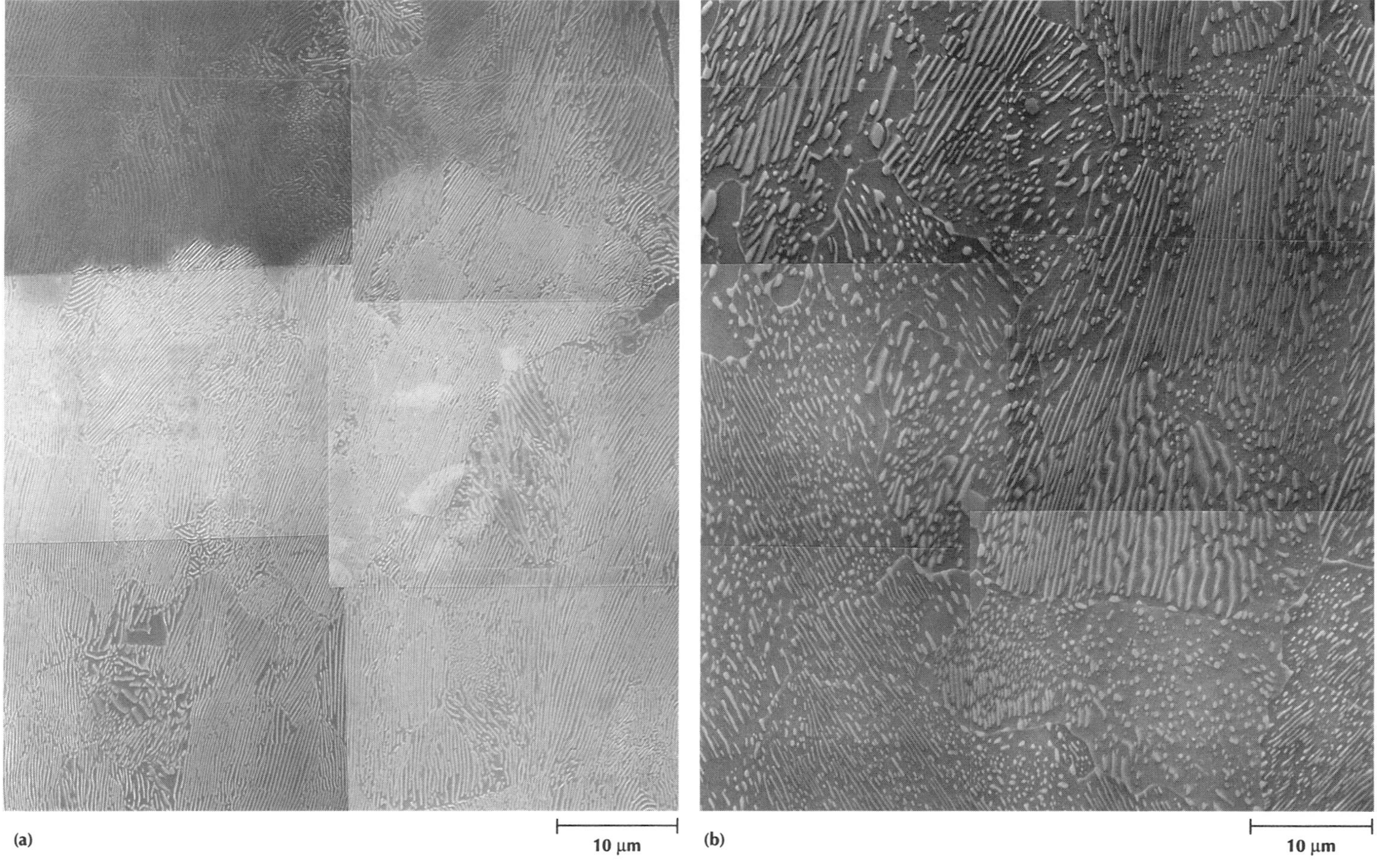

Fig. 18 Change in lamellae morphology upon exposure of steel at high temperature. See text for a discussion of these microstructures. Source: Ref 82

room temperature, the observed strengthening is attributed to strain aging. The specimen is cycled at a strain rate of 2×10^{-3}/s.

Thermal recovery effects have been observed in steels when temperatures exceed 500 °C (930 °F). In the case of pearlitic steels, the lamellae structure coarsens and, ultimately, a spheroidized material results. Examples of the microstructural change are illustrated in Fig. 18. Consequently, these changes alter the stress-strain response of the material. Figure 18(a) shows the mean lamellae thickness to be nearly 145 Å. As the microstructure becomes spheroidized (Fig. 18b), the mean spheroidite diameter is much larger than any of the cementite thicknesses. The coarsening process takes the form of an early breaking up of the lamellae, followed by spheroidite growth. This difference in size is apparent in Fig. 16. This phenomenon was documented in early studies (Ref 25, 34, 82). Table 2 summarizes the microstructural damage mechanisms identified by various experiments on steels.

Table 2 Summary of microstructural damage mechanisms in steels

Material	TMF IP	TMF OP	Ref
1070 steel	Crack growth at pearlite colony boundaries; ferrite-pearlite interfaces Internal oxygen attack of MnS particles; coarsening of pearlite lamellae; spheroidization Phase transformation, bcc-fcc; recrystallization	Strain-aging effect due to exposure at elevated temperature, followed by low temperature Formation and repeated fracture of oxides; internal oxygen attack of MnS particles Coarsening of pearlite lamellae; spheroidization	20, 24, 38, 45
Type 304 stainless steel	Strengthening due to strain aging; creep damage (grain-boundary triple points) in tensile stress part of the cycle Higher dislocation density compared to IF Mixture of dislocation arrangements compared to IF High density of intergranular cracks Higher grain-boundary sliding in tension relative to compression, resulting in ratcheting at the microlevel Grain-boundary residual stresses at low temperature relax at high temperatures, resulting in cavity nucleation	Strengthening due to strain aging; higher dislocation density compared to IF; mixture of dislocation arrangements compared to IF Lowest density of intergranular cracks compared to IF and IP	19, 28, 63, 70
1Cr-1Mo-0.25V steel	...	Higher density of crack formation in OP relative to IP and IF, possibly due to fracture of oxide scale	44, 76
Type 347 stainless steel	Creep damage at grain boundaries produced shortest lives for IP	...	31
1016 steel	Integral breadth of x-ray diffraction profiles as a measure of subgrain evolution during TMF	...	88

TMF of Aluminum Alloys

Only a handful of experiments has been reported on the elevated-temperature behavior of aluminum alloys. At temperatures exceeding 150 °C (300 °F), aluminum alloys undergo creep damage in the form of grain-boundary cavitation (Ref 89) and intergranular crack growth (Ref 90). Under TMF conditions with T_{mean} = 200 °C (390

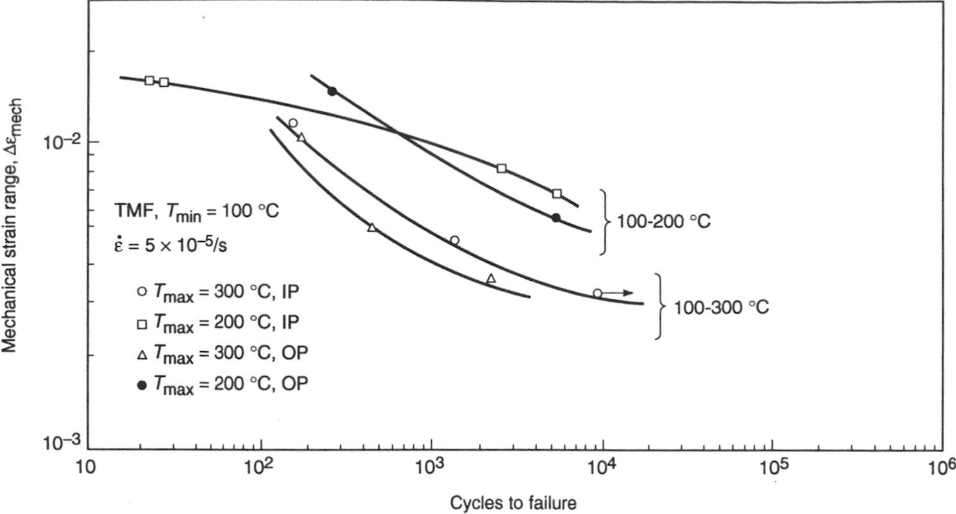

Fig. 19 Comparison of TMF OP and TMF IP lives for Al 2*xxx*-T4. Source: Ref 95

°F), creep damage is expected to occur under both OP and IP conditions. Extensive studies on the fatigue of aluminum (Ref 91-93) at room temperature revealed accelerated fatigue damage in air relative to a vacuum environment. At elevated temperatures the environment (oxidation) effect is expected to be more pronounced (Ref 94-96). Figure 19 compares TMF OP and TMF IP lives for aluminum alloys 2*xxx*-T4, a powder metallurgy material with minimal porosity level. In the experiments, $R_\varepsilon = -1$, the minimum temperature was 100 °C (210 °F), and the maximum temperatures were 200 and 300 °C (390 and 570 °F). A crossover in lives occurred for the 100 ⇄ 200 °C (210 ⇄ 390 °F) case, but there was no such crossover for the 100 ⇄ 300 °C (210 ⇄ 570 °F) case (Ref 90). Two other studies on TMF of aluminum alloys have been reported (Ref 97, 98). In Ref 97, a cast Al-Si-10Mg alloy was studied under total constraint TMF OP conditions; the mechanical strain increased proportionally with increasing maximum temperature. The minimum temperature was maintained constant at 50 °C (120 °F). The most severe case studied was under 50 ⇄ 350 °C (120 ⇄ 660 °F) conditions, and considerable cyclic softening was observed both at the low- and high-temperature ends of the cycle. In Ref 98, the cast alloys 319.0 and 356.0 were considered. This work studied the role of dendrite arm spacing, porosity level, composition, and heat treatment.

Relatively few studies have been conducted on the TMF aluminum alloys. The major issues are the following:

- Oxidation has an influence on fatigue damage both at room temperature and at elevated temperatures. A number of fundamental studies have been made of IF of aluminum at room temperature and a few vacuum tests at elevated temperatures (Ref 99), but there are no reported experiments under TMF loading. Based on the work of Bhat and Laird (Ref 99), considerable oxidation damage is present in polycrystalline aluminum alloys.
- If the maximum temperature exceeds the aging temperature, then considerable softening can be observed in TMF due to changes in the shape and size of the precipitates. The aging temperatures can vary from 150 to 200 °C (300 to 390 °F).
- Creep damage has been observed at temperatures exceeding 200 °C (390 °F) in TMF experiments. The creep damage is in the form of distributed cracks. When creep damage with diffuse cracks occurs, continuum damage mechanics concepts would be appropriate. In this case, the σ-ε behavior of a damaged material can be described by using effective stress and hydrostatic stress integrated over the cycle (Ref 100).

TMF of Nickel-Base High-Temperature Alloys

Much has been published on the high-temperature behavior of nickel-base superalloys—the development of which is ongoing, including the use of coating treatments. The major advantage of nickel-base superalloys over other metals is their useful TMF operating range, which extends to $T_{max}/T_m = 0.8$, where T_{max} is the maximum temperature in the cycle and T_m is the melting temperature in degrees Kelvin. On the other hand, when T_{max}/T_m values exceed 0.5, the TMF strength of steels is considerably lowered.

Depending on the test temperature, the superalloys may show either cyclic softening or cyclic hardening behavior. The hardening behavior is attributed to dislocation buildup at the precipitate/matrix interface. On the other hand, softening behavior is considered to be caused by precipitate shearing, increased dislocation climb facilitated by increased diffusion rates, and reduced dislocation densities caused by recovery processes (for a review, see Ref 101 and 102).

Similar to IF tests, TMF experiments exhibit a temperature and strain range dependence of cyclic stress response. Castelli et al. (Ref 103) observed cyclic hardening of Hastelloy X under OP cycling at $\Delta\varepsilon_m = 0.006$ over a temperature range of 600 to 800 °C (1110 to 1470 °F). When the temperature range was increased to 800 to 1000 °C (1470 to 1830 °F), cyclic softening was observed. Marchand et al. (Ref 104) tested B-1900+Hf under TMF OP and TMF IP at a temperature range of 400 ⇄ 925 °C (750 ⇄ 1700 °F). Cyclic stress-strain curves revealed cyclic hardening at low strain ranges and cyclic softening at high strain ranges when compared to cyclic stress-strain response at T_{max}. Sehitoglu and Boismier (Ref 105), working with polycrystalline Mar-M247 (500 ⇄ 870 °C, or 930 ⇄ 1600 °F), found gradual cyclic softening for most of the life at small strains, whereas cyclic hardening was observed at high strains. Stress-strain behavior has been found for René 80 (Ref 106), Mar-M247 (Ref 105), and Mar-M246 (Ref 107), CMSX-6 single crystals (Ref 108), and on Inconel 617 by Macherauch and colleagues (Ref 109). They studied Inconel 617 and reported significant hardening at T_{max} of 750 to 850 °C (1380 to 1560 °F). When the maximum temperature was higher than 950 °C (1740 °F), the response was stable.

A number of Japanese investigators have published results on TMF of superalloys. An extensive study of superalloys has been undertaken by Kuwabara et al. (Ref 42), who considered Inconel 718, Inconel 738LC, Inconel 939, Mar-M247, and René 80. For Inconel 718, the temperatures were 300 ⇄ 650 °C (570 ⇄ 1200 °F); for the other alloys, 300 ⇄ 900 °C (570 ⇄ 1650 °F). For Inconel 718, Inconel 939, and Mar-M247, shorter lives were demonstrated for the TMF IP case in the high strain range and a crossover in life at small strain levels. The Inconel 738LC and René 80 exhibited shorter lives for the TMF OP case relative to TMF IP. Taira et al. (Ref 36) considered Hastelloy X in the temperature ranges 300 ⇄ 900 °C and 300 ⇄ 750 °C (570 ⇄ 1650 °F and 570 ⇄ 1380 °F) and found that TMF IP lives are considerably shorter than the TMF OP case. These results are shown in Fig. 20.

In Great Britain, extensive thermal fatigue studies have been reported by Glenny and Taylor (Ref 110) on Nimonic and directionally solidifed nickel alloys. The duration of the thermal cycle (immersion times) and the maximum temperature effect (up to 920 °C, or 1690 °F) have been examined, and intergranular cracking has been noted. Tilly (Ref 111) used the tapered-disk geometry with the fluidized-bed technique under 20 ⇄ 920 °C (70 ⇄ 1690 °F) conditions. This author also conducted reverse bend tests with temperature cycling of 350 ⇄ 1000 °C (660 ⇄ 1830 °F) in air and in vacuum and showed an increase in fatigue life in vacuum relative to air of nearly a factor of two. The lives were lower than those predicted based on IF data at T_{max}.

Other TF experiments have been conducted by Woodford and Mowbray (Ref 50) on the nickel-base superalloys Inconel 738 and René 77 using the tapered-disk specimen. The hysteresis behav-

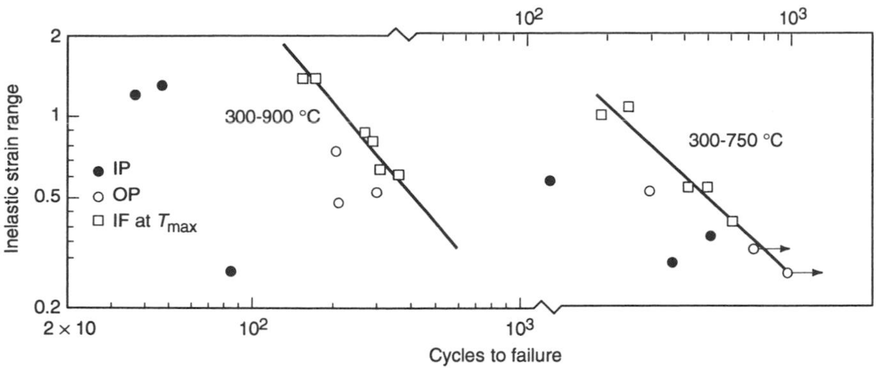

Fig. 20 Comparison of TMF OP and TMF IP lives for Hastelloy X. Source: Ref 36

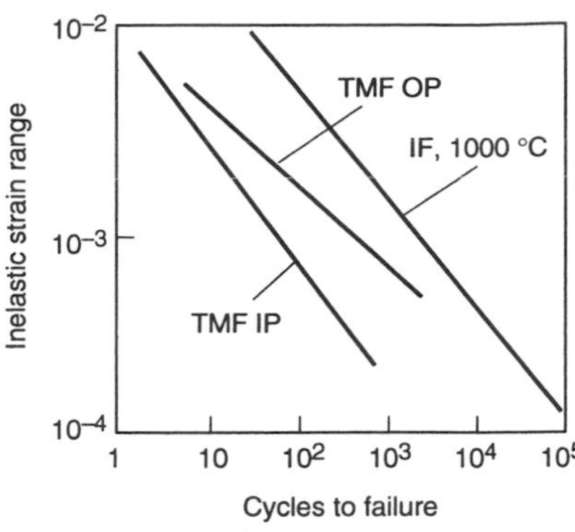

Fig. 22 Comparison of TMF OP and TMF IP lives for polycrystalline Mar-M200. Source: Ref 112

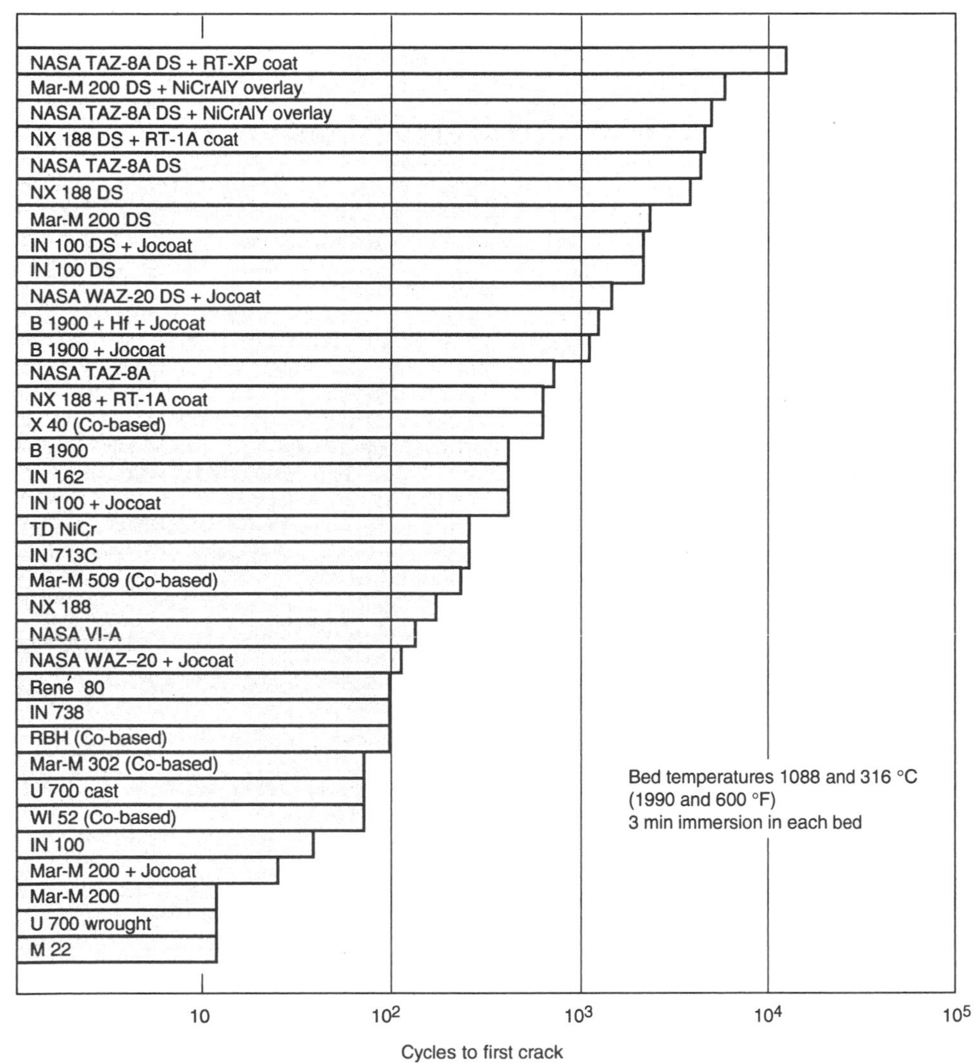

Fig. 21 Comparative thermal-stress fatigue resistances of coated and uncoated nickel- and cobalt-base superalloys. Protective coatings include: Jocoat, silicon-modified nickel-aluminide coating; RT-1A, chromium-aluminum duplex coating; RT-XP, aluminide-containing coating with a case depth of 70 μm (2.7 mils); NiCrAlY, electron-beam vapor-deposited coating about 135 μm (5.3 mils) thick. Source: Ref 51

ior was calculated at the disk periphery, and the temperature-strain phasing was similar to the DCW type. Crack length was monitored as a function of cycles. The temperature range was 22 ⇄ 920 °C (72 ⇄ 1690 °F) for both materials. These investigators made the important observation of γ′-depleted zones in the vicinity of crack tips. In these regions, aluminum is depleted and the γ′ structures break down. Bizon and Spera (Ref 51) considered 26 superalloys (both nickel- and cobalt-base) using the fluidized-bed technique. They noted the positive role of directional solidification (DS) and coatings on TF life (Fig. 21).

Most nickel-base superalloys exhibit a crossover of the TMF IP and TMF OP mechanical strain-life curves. In this case, TMF IP fatigue lives are shorter than TMF OP lives at high mechanical strain ranges, but are greater than TMF OP lives at low mechanical strain ranges. The crossover occurs at approximately $\Delta\varepsilon_m = 0.0045$. Kuwabara et al. (Ref 42) observed the crossover in life curves for Inconel 718 and Mar-M247 under temperature cycling of 300 ⇄ 650 °C and 300 ⇄ 900 °C (570 ⇄ 1200 °F and 570 ⇄ 1650 °F), respectively. Bill et al. (Ref 112) investigated Mar-M200 at mechanical strain ranges greater than 0.01 and over a temperature range of 495 ⇄ 1000 °C (925 ⇄ 1830 °F). These results are shown in Fig. 22.

Nelson et al. (Ref 113) studied B-1900+Hf at a temperature range of 540 ⇄ 870 °C (1000 ⇄ 1600 °F) and also observed a crossover corresponding to a mechanical strain range of 0.0045. Ramaswamy and Cook (Ref 114) conducted tests on Inconel 718 at $\Delta\varepsilon_m = 0.015$ and over temperature ranges of 345 ⇄ 565 °C and 345 ⇄ 650 °C (650 ⇄ 1050 °F and 650 ⇄ 1200 °F) and found TMF IP to be more damaging than TMF OP. However, they noted that René 80 (760 ⇄ 870 °C, or 1400 ⇄ 1600 °F) also showed a crossover in the TMF life curves. The TMF OP lives were shorter than the TMF IP cases (Ref 115); this has been attributed to high mean stresses in the TMF OP case. Gayda et al. (Ref 116) found that for coated PWA 1480 at inelastic strain ranges of less than 0.2%, the TMF OP lives are significantly longer than TMF IP, whereas in the high strain regime the TMF OP and TMF IP lives are comparable.

In the past, TMF life has been approximated by IF life at the maximum temperature of the TMF cycle using the same mechanical strain range. This appears to be applicable for TMF OP conditions. Experiments conducted by Bill et al. (Ref 112) on Mar-M200 revealed that IF life at T_{max} was slightly longer than TMF OP life (500 ⇄ 1000 °C, or 930 ⇄ 1830 °F). The IF lives may have been greater because the IF tests were conducted at a frequency 100 times greater than that

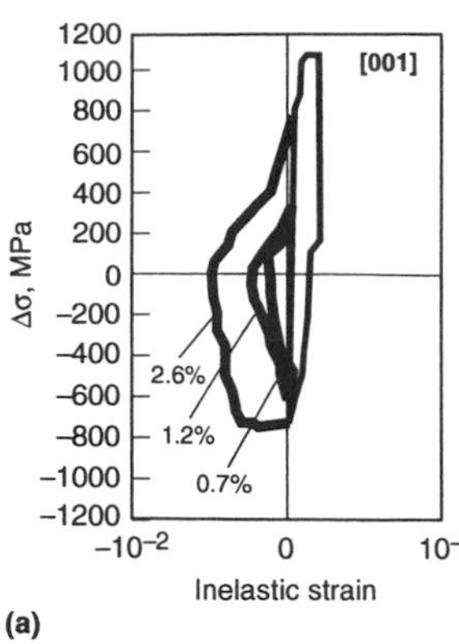

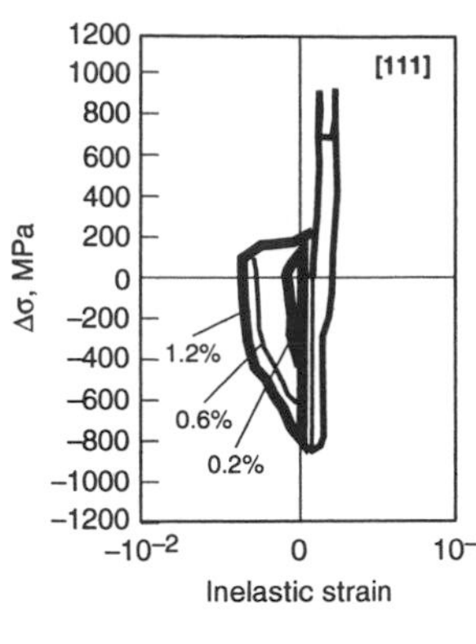

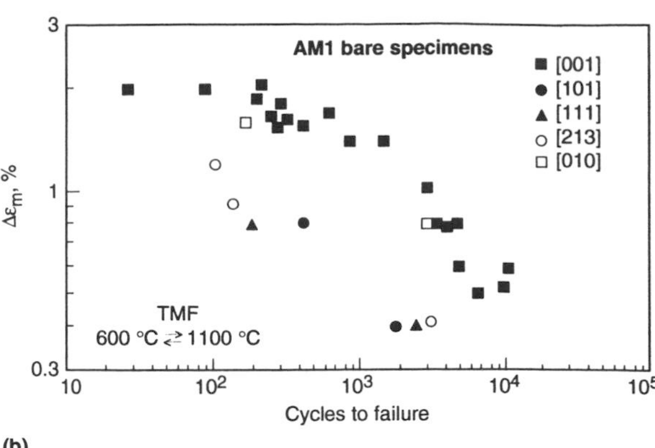

Fig. 23 Effect of crystallographic orientation on TMF behavior of AM1. Source: Ref 120

of the TMF tests. Nelson et al. (Ref 113) also found a correlation between IF life at T_{max} and OP life (540 to 870 °C, or 1000 to 1600 °F) for B-1900+Hf. Malpertu and Rémy (Ref 117) conducted TMF experiments on Inconel 100 utilizing a cycle similar to a counterclockwise 135° cycle over a temperature range of 600 to 1050 °C (1110 to 1920 °F). Initiation life was the same as the IF initiation life at T_{max}. There is a correlation between TMF OP life and IF life at T_{max}, but there are discrete differences in the damage mechanisms.

A few TMF studies have included nonproportional phasing of the mechanical strain and temperature. Nonproportional loading cycles are very important because they more closely approximate a service-induced strain-temperature history of an actual component. An example is the diamond-shape history, where the maximum and minimum mechanical strain occur at the median of the temperature range. Embley and Russell (Ref 115) conducted DCW history tests on Inconel 738 and found lives approaching two orders of magnitude longer than TMF OP and TMF IP lives over the same temperature range (425 to 870 °C, or 800 to 1600 °F). Nelson et al. (Ref 113) conducted TMF experiments on B-1900+Hf utilizing a counterclockwise elliptical strain-temperature cycle at 540 to 870 °C (1000 to 1600 °F). They discovered a fivefold increase in life over TMF OP lives.

Guedou and Honnorat (Ref 48) considered three alloys—Inconel 100, AM1, and DS 200—subjected to DCW and DCCW histories with a 650 ⇄ 1100 °C (1200 ⇄ 2010 °F) temperature range. The AM1 is a single-crystal alloy, DS-200 is directionally solidified, and Inconel 100 is polycrystalline. Based on mechanical strain range, AM1 exhibited the best properties. Isothermal fatigue data at T_{mean} was closest to the TMF mechanical strain life curve. Bernstein et al. (Ref 118) considered Inconel 738LC both in the coated and uncoated state under 425 ⇄ 870, 915, and 980 °C (800 ⇄ 1600, 1680, and 1800 °F) conditions. They found shorter lives for the coated material relative to the uncoated material. They discussed the turbine blade strain-temperatures extensively (in particular, the role of the startup and shutdown) and showed that the TMF OP cycle best describes the engine conditions.

Recently Halford et al. (Ref 30) have proposed the use of bithermal fatigue cycles as a simple alternative to TMF testing. Bithermal results have been interpreted with strain range partitioning (SRP) as a predominantly PC or CP type of loading. Bithermal experiments conducted on B-1900+Hf were directly related to TMF results by use of an appropriate damage rule. Other research on TMF OP of Hastelloy X has been reported by Kaufman and Halford (Ref 119) in the ranges 505 ⇄ 905 °C and 425 ⇄ 925 °C (940 ⇄1660 °F and 795 ⇄ 1695 °F).

Recent research on TMF in Europe centers around Remy et al. at École de Mines (Ref 117, 120), Guedou at SNECMA (Ref 48), Bressers and various colleagues at Petten (Ref 121), and Mughrabi and his students in Germany (Ref 70, 108). Remy and coworkers (Ref 120) studied the crystallographic orientation effect on the cyclic σ-ε behavior of AM1 superalloy in the temperature range of 600 ⇄ 1100 °C (1110 ⇄ 2010 °F). These results are summarized in Fig. 23(a) for the [001] and [111] orientations. The inelastic strain range in the cycle was found to be strongly orientation dependent, with [001] producing smaller inelastic strains than [111] (Fig. 23a). This is evident when the stress/inelastic-strain loops are compared for the case of mechanical strain range of 1.2%. The longest fatigue lives among five crystal orientations were found for the [001]-oriented specimens (Fig. 23b).

Bressers et al. (Ref 121) used a 135° OP cycle (i.e., DCCW) and studied the TMF behavior of SRR99 in the coated and uncoated condition. They considered both $R = 0$ and $R = -\infty$ cases and monitored the crack length as a function of cycles. The role of oxidation is emphasized in their model. Mughrabi and coworkers (Ref 108) studied CMSX-6 single-crystal superalloys of [001] orientation under 600 ⇄ 1100 °C (1110 ⇄ 2010 °F) conditions and documented the coarsening of precipitates during TMF. They confirmed that the mean stress in nickel-base superalloys play a considerable role. The life under TMF OP was considerably shorter than under TMF IP (five times), with DCW and DCCW cases between these two extremes. This study also confirms that inelastic strain range is not a good correlator of life when failure occurs in a finite number of cycles with very small $\Delta\varepsilon_{in}$ components. They noted that when the γ′ structure rafts, soft γ-matrix channels permit unconstrained dislocation motion. Also, during the high-temperature phase of the cycle dislocation climb and during the low-temperature end of the cycle, cutting of the particles has been observed.

Other work from Europe includes Marchionni et al. (Ref 122), who studied an oxide-dispersion (Y_2O_3)-strengthened Inconel alloy. The TMF OP and TMF IP results are similar and lie within the scatter of data. Macherauch and his group at Karlsruhe (Ref 109) have reported TMF OP and TMF IP experiments on Inconel 617, showing TMF IP damage to be more significant than TMF OP. The maximum temperature was in the range of 850 to 1050 °C (1560 to 1920 °F), while the minimum temperature was 600 °C (1110 °F).

Strain Rate and Frequency Effects. Strain rate can affect cyclic stress-strain response as well as fatigue life. In studies conducted on René 80 (Ref 123), Hastelloy X (Ref 103), Mar-M246 (Ref 49), and Mar-M200 (Ref 124), it has been reported that decreasing the frequency resulted in a decrease in the stress range; no change in the relative hardening and softening behavior has been observed. There is abundant information on the strain-rate effects under IF conditions, including decreasing frequency, lowering strain rates, or introducing hold times (Ref 118, 124–126). These effects are attributed to increased environmental and creep damage (Ref 126–128). Only in rare cases do the strain rate or hold times have no effect (Ref 112, 129) or does decreasing the strain rate or introducing hold periods increase fatigue life (Ref 124, 130, 131). This latter behavior can be explained based on a reduced creep component caused by reduced cyclic stresses when γ′ precipitate coarsening occurs.

There have been no systematic attempts to alter the strain rate (analogous to IF experiments), but some TMF experiments have introduced a hold period at maximum temperature. The effect of compressive hold periods on TMF of Inconel 738 has been established by Bernstein et al. (Ref 118), who reported shortened cycles to failure. In nickel-base superalloys, if the hold period in

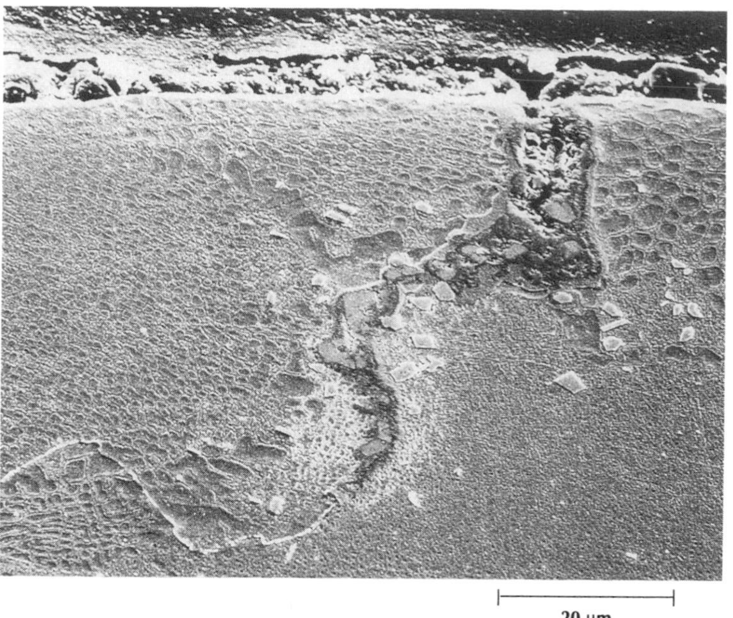

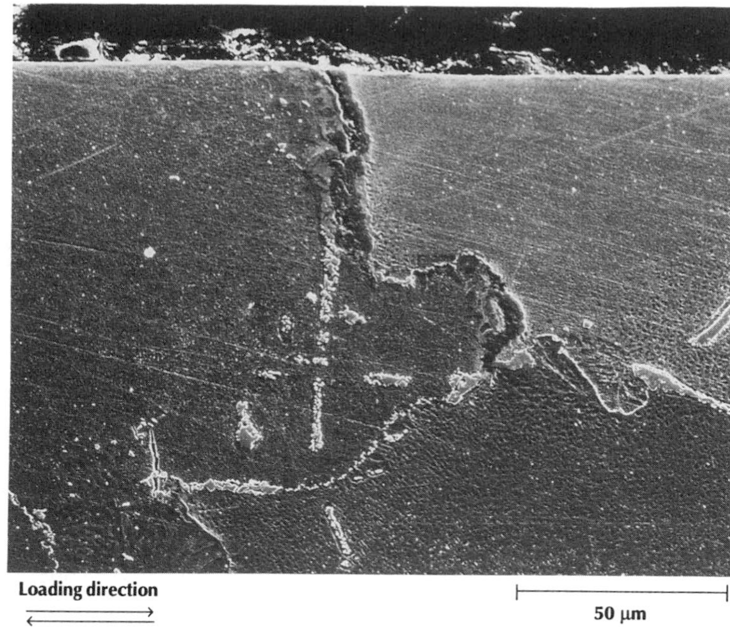

Fig. 24 Two views of grain-boundary oxidation and cracking in Mar-M247. Source: Ref 105

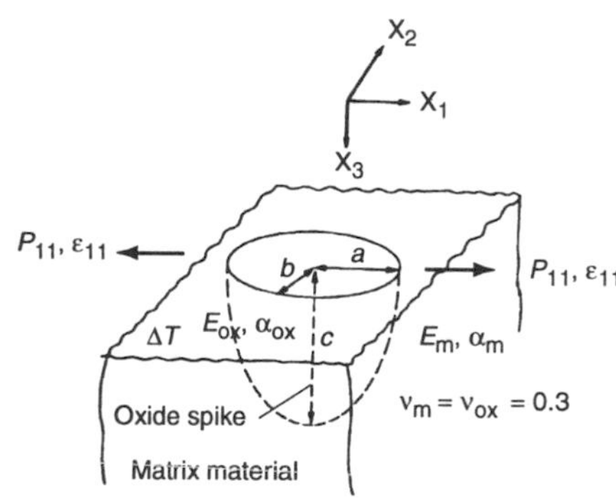

(a)

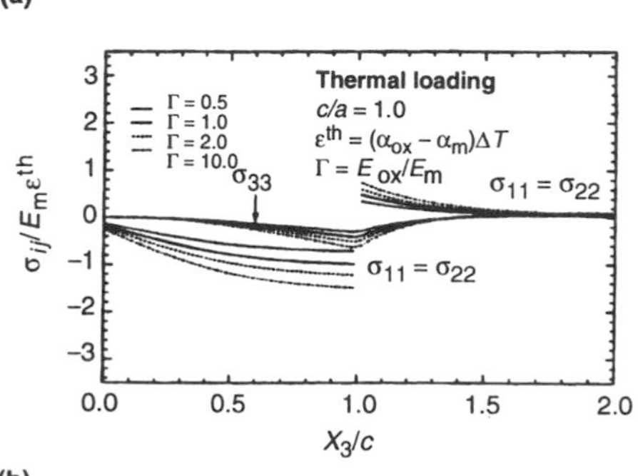

(b)

Fig. 25 (a) Geometry of an oxide spike (intrusion). (b) Oxide stresses as a function of α and E mismatch. Source: Ref 107

TMF OP results in stress relaxation in compression, high tensile mean stresses develop upon reversed loading—which is detrimental to fatigue life.

Environmental Effects. The effects of the environment on nickel-base superalloys at elevated temperatures are very complex. Environmental damage can affect both crack initiation and crack propagation and has a detrimental effect on fatigue life. Crack nucleation often originates from preferentially oxidized grain boundaries (Ref 126, 130, 132, 133). Grain boundaries are preferentially oxidized because they are paths of rapid diffusion and their composition may differ from that of the matrix (Ref 105, 107, 134, 135). An example of oxidation at grain boundaries and intergranular initiation in Mar-M247 subjected to TMF IP conditions is shown in Fig. 24. The experiment was conducted under TMF IP 500 ⇄ 870 °C (930 ⇄ 1600 °F) conditions.

Rémy et al. (Ref 136) oxidized precracked specimens and compared the crack growth with that of virgin specimens. These experiments revealed crack growth rates as much as three orders of magnitude higher than the virgin samples. They proposed a modified fracture mechanics approach to handle the crack growth under repeated oxide fracture; the oxidation constants were determined via integration over the cycle. The different TMF strain/temperature waveforms (600 ⇄ 1050 °C, or 1110 ⇄ 1920 °F) were predicted with their model.

Under elevated-temperature conditions, a protective oxide scale forms on the surface of the specimen, separating the substrate from the environment. However, spalling and cracking of the protective oxide scale occur due to stresses developed in the scale. The principal sources of stress in the oxide scale are the thermal stresses due to the difference between the thermal expansion coefficients of the oxide and the matrix. Although there is zero thermal stress at the oxide formation temperature, upon cooling by ΔT, a stress is generated in the oxide layer. Oxide spikes penetrate from the surface toward the inside of the substrate. The oxide spike morphology could form at the surface or at the coating/substrate interface upon failure of the coating. The problem of stress fields associated with oxide spikes has been studied by Kadioglu and Sehitoglu (Ref 107, 135). In their work, an oxide spike was modeled as a semispherical surface in homogeneity. The stress field in the vicinity of the oxide spike was calculated using a technique based on Eshelby's method. Then the calculated strain at the tip of the oxide spike was used in the life prediction model. Strains at the oxide tips increase considerably as the ratio of oxide elastic modulus to metal elastic modulus decreases. The stresses under different levels of thermal mismatch also were shown. Sample results are presented in Fig. 25. The geometry of the oxide intrusion (spike) is shown in Fig. 25(a). The variation of $\sigma_{ij}/(E_m\varepsilon^{th})$, which is the stress tensor normalized by the product of matrix modulus and thermal mismatch strain, as a function of distance measured from the surface is shown in Fig. 25(b). The term $\Gamma = E_{ox}/E_m$ is the oxide to matrix modulus ratio, and X_3/c represents the normalized distance normal to the free surface. The $X_3/c > 1$ represents the matrix region ahead of the oxide intrusion, while $X_3/c < 1$ represents the oxide. The critical parameter extracted from these studies is $\Delta\varepsilon_m^{ox}$, the mechanical strain range at the tip of the oxide. This result was used in the life prediction model described in Ref 135.

Oxidation characteristics of nickel-base superalloys can vary widely. The oxidation products formed vary with alloy composition, temperature, and time at temperature. A general oxidation characterization for nickel-base superalloys at 870 °C (1600 °F) can be drawn from Ref 137 and 138. Initially, a continuous film of Al_2O_3 forms. Diffusional mass transport of chromium through the Al_2O_3 layer allows the formation of an outer layer of Cr_2O_3. Eventually, spinels of $Ni(Cr,Al)_2O_4$ are formed. Some TiO_2 may also be formed. This sequence of events is a specific case. The oxides formed will vary from alloy to alloy and with variations in temperature and time of exposure.

Oxidized surfaces usually are associated with an adjacent zone of alloy depletion. This is char-

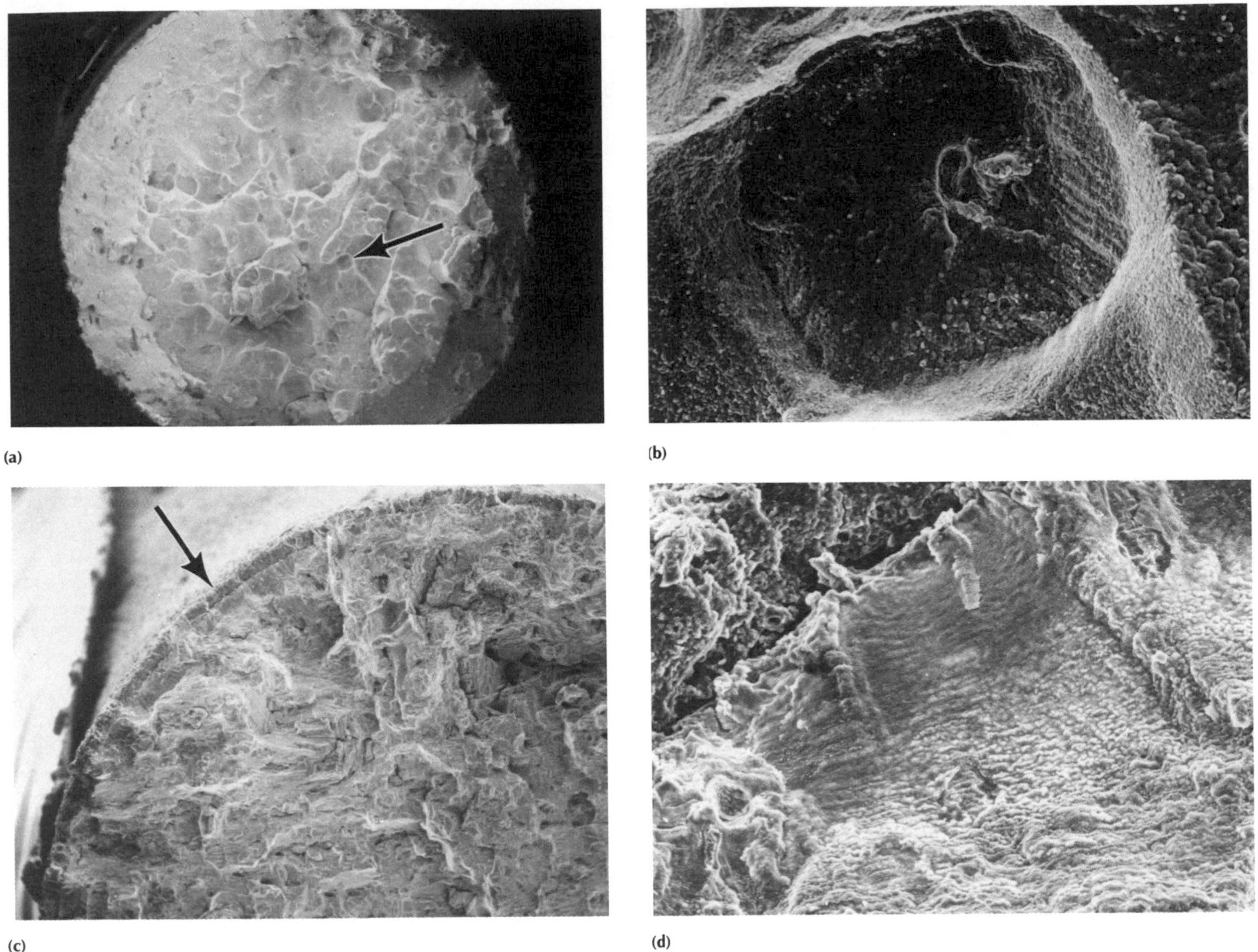

Fig. 26 Effect of thermal cycling on fatigue fracture of single-crystal PWA 1480 coated with NiCoCrAlY applied by the low-pressure plasma spray process. (a) Fracture surface of coated sample tested at 1050 °C (1920 °F) in isothermal low-cycle fatigue. The NiCoCrAlY coating was very ductile at the test temperature and did not crack. The well-protected superalloy failed via multiple internal cracking originating at microporosity (arrow). SEM, 12.5×. (b) High-magnification view of micropore at arrow in (a). SEM, 280×. (c) Fracture surface of coated sample cycled between 650 and 1050 °C (1200 and 1920 °F) in a thermomechanical fatigue test. The NiCoCrAlY coating was placed in tension during the low-temperature portion of the test, which caused it to crack in a few cycles. The nickel-base superalloy—now exposed to the atmosphere—failed at multiple surface locations (arrow) rather than at internal micropores, as in (a) and (b). SEM, 21×. (d) High-magnification view of fracture origin at arrow in (c). SEM, 260×. Courtesy of R.V. Miner and S.L. Draper, NASA Lewis Research Center

Oxidized surfaces usually are associated with an adjacent zone of alloy depletion. This is characterized by a zone depleted of γ' precipitates. Several studies have reported the existence of γ'-depleted zones (Ref 105, 107, 113, 126, 131–133). The γ'-depleted zone is caused by the loss of aluminum to the formation of oxides. This zone may also be depleted of solid-solution-strengthening elements such as chromium. The fatigue characteristics of such a layer may be markedly different in the initial cracking stages. Deformation bands develop in precipitate-free areas and lead to premature microcracking and fatigue failure (Ref 139). Due to oxidation of the crack tip, a region depleted of oxide-forming elements will be formed ahead of a fatigue crack. As a result, the crack will propagate into a region having changed mechanical properties. Crack growth in each cycle may be controlled by the size of the environmentally affected zone at the crack tip (Ref 126).

Steady-state formation of the oxide and alloy-depleted layers is governed by parabolic rate kinetics (Ref 128, 130, 138, 140). The rate of oxidation and alloy depletion is considered to be affected by the application of stress. It has been shown that oxidation and alloy depletion increase when stress is applied (Ref 108, 138). The effect of stress on environmental attack may vary with alloy composition and exposure conditions.

Coating Effects in Superalloys. Environmental degradation due to oxidation and corrosion may be prevented by using protective coatings. Various types of coatings have been used to reduce the deleterious effect of the environment (Ref 33, 45). Many of the coatings developed fulfill their protection role against oxidation or corrosion of the base material. Three main types of coatings have been used to protect superalloys: (1) diffusion coatings, (2) overlay coatings (MCrAlY, where M is nickel or cobalt), and (3) thermal barrier coatings (TBCs). More detailed information on these coating types can be found in the article "Protective Coatings for Superalloys" in this Volume.

The predominant oxide formed on the coating is Al_2O_3. The overlay coatings consist of Ni(Co,Fe),Cr, Al, and Y, and are called MCrAlY type. Finally, TBCs have been used to limit the heat flow into the base alloy. The materials most commonly considered as TBCs are general oxides such as ZrO_2 and Al_2O_3. The low thermal expansion coefficient of the ceramic coatings and the relatively high thermal expansion coefficient of the base alloy result in a large mismatch strain, which encourages the propagation of cracks in the coating. The presence of porosity, discontinuities, random microcracking, and a columnar structure with grain boundaries to the surface, known as coating segmentation, has also been observed.

Under TMF conditions, coatings undergo complex stress-strain changes and at the low-temperature end of the cycle could fracture. Figure 26 shows the results of thermal cycling on the fatigue fracture of a NiCoCrAlY-coated PWA 1480 single-crystal superalloy. Various researchers have found coatings to provide benefit, depending on the temperature (Ref 107, 128, 135, 141–144). In some cases, however, a reduction in the fatigue lives of some directionally solidified alloys (Ref 141) and other nickel-base superalloys (Ref 145, 146) has been noted. Goward (Ref 147) has investigated the TMF behavior of aluminide and CoCrAlY coatings. The tests were conducted under fully reverse condition by cycling the temperature between 425 and 925 °C (800 and 1695 °F). In these tests, the low-aluminum CoCrAlY coating exhibited a much higher resistance to crack initiation than the higher-aluminum CoCrAlY.

The effects of protective coatings on TF of superalloys have been studied by alternately immersing a variety of tapered disk and wedge-type specimens into hot and cold fluidized beds (Ref 148, 149). However, the strain/temperature cycle was not precisely known in these experiments. Thermomechanical fatigue tests in which the temperature and strain can be controlled separately have been used to investigate the effects of coatings on superalloys (Ref 150–152). Among the various forms of strain/temperature phase relations, the most damaging is TMF OP. In fact, it was reported that coated superalloys exhibited shorter lives under TMF OP than under TMF IP (Ref 150–151).

Under service conditions, additional strains on the coatings may arise due to thermal expansion mismatch, elastic moduli mismatch, diffusion between coating and substrate, phase transformation, or chemical reaction with the environment. These additional strains and stresses alter the crack initiation and propagation resistance of the materials, resulting in spallation of protective oxide scales and/or coating or early crack formation, which allows oxidation attack into the base alloy. To improve the fatigue performance of the coated components, these strains should be minimized by adjusting the mechanical and metallurgical properties of the coatings (provided that oxidation/corrosion resistance capability is maintained). Therefore, a life prediction methodology that will relate coating performance to the mechanical/physical properties of coating/substrate systems and environmental conditions is needed.

Coating cracking lives have been successfully correlated with total strain, which is the summation of the thermal expansion mismatch strain and the mechanical strain (Ref 152, 153). A fatigue crack growth model has also been proposed by Strangman (Ref 154). In this work, the penetration of a coating crack into the base metal has been analyzed using the fracture mechanics approach. In another approach (Ref 155), the life of a coated system was considered as the summation of the number of cycles to initiate a crack through the coating, the number of cycles for the coating crack to penetrate a small distance into the substrate, and the number of cycles to propagate the substrate to failure. In a recent study (Ref 151), the mechanical damages for the coating and the substrate have been calculated separately and then combined to produce an optimal prediction damage parameter. The predicted lives for coated superalloys were within a factor of two for TMF OP tests. The two-bar model representing the coating and the substrate has been used to determine the constitutive stress-strain loop for the coating under TMF conditions. Then, the number of cycles to initiate a crack in the coating has been estimated by the hysteretic energy method (Ref 156). With this approach, the fatigue life of overlay coatings was estimated within a factor of 2.5 in the case of TMF conditions (Ref 155). Although the two-bar model is one dimensional, it captures the first-order effects of the coatings on the behavior of base alloys. To study the effect of biaxiality requires nonlinear FEM due to the highly nonlinear behavior of the coating/substrate system (Ref 157).

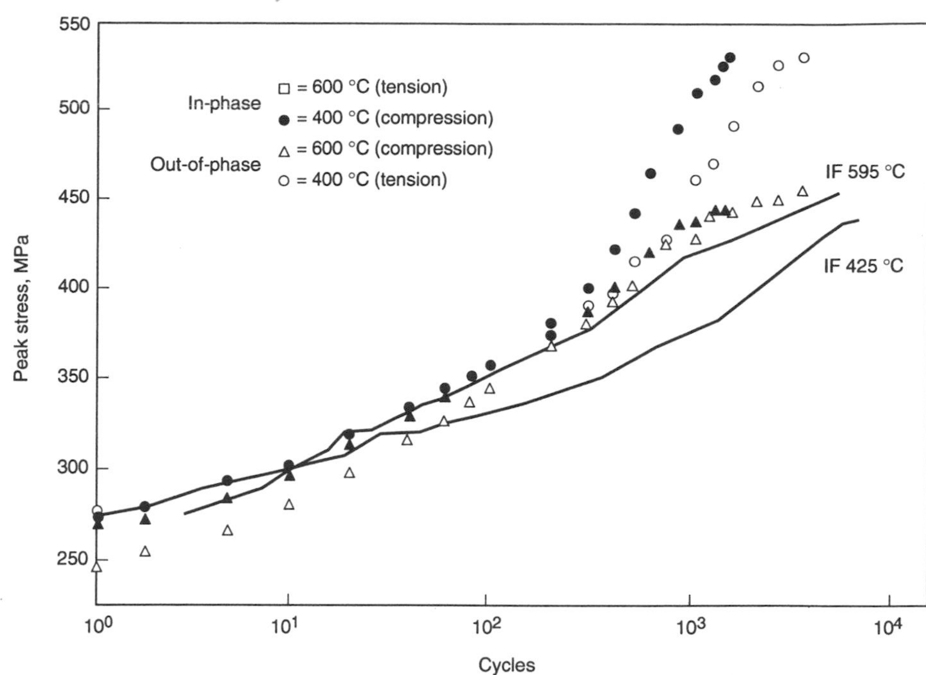

Fig. 27 Cyclic hardening for Hastelloy X under TMF conditions. Source: Ref 103

Swanson et al. (Ref 156) have conducted isothermal fatigue tests at 760, 925, and 1040 °C (1400, 1695, and 1905 °F) and TMF tests by cycling the temperature between 425 and 1040 °C (795 and 1905 °F) on PWA 286 (NiCoCrAlY+Si+Hf) overlay and PWA 273 (NiAl, outward diffusion) aluminide-coated single-crystal PWA 1480 alloy. Their tests used hollow tubes as test specimens. They found that, in many cases, coating cracks had progressed into the PWA 1480 alloy and directly caused failure. In some specimens, however, the coating cracks did not extend into the substrate, and failure was caused by a crack initiated from the uncoated inner surface. The coating cracks penetrated into the substrate in both OP and IF tests for specimens coated with PWA 273, but only in the OP tests for specimens with overlay coating. In this case, coating-initiated cracking was the dominant failure mode. It was difficult to draw a general conclusion about the effects of coatings on fatigue life from this work due to variation of specimen design and orientations, frequencies used, cycle type, and strain ranges applied.

Wright (Ref 128) has examined the oxidation-fatigue interactions in René N4. Isothermal tests were performed on uncoated, aluminide-coated, and preoxidized alloy at 1095 °C (2005 °F). The test results showed that although there were no differences in the fatigue life of coated and uncoated specimens in low-frequency tests ($f = \frac{1}{2}$ cycle/min), fatigue life increased significantly during high-frequency tests ($f = 20$ cycles/min). Glenny and Taylor (Ref 56), Bizon and Spera (Ref 51), and Woodford and Mowbray (Ref 50) have investigated the thermal fatigue characteristics of uncoated and coated superalloys using a variety of tapered-disk and wedge-type specimens. However, the available data from these studies are difficult to interpret due to the large variety of specimen shapes, thermal cycle shapes, and differences in the definition of failure criteria.

In summary, the following general rules apply for the TMF of nickel-base superalloys:

- For single crystals, the best TMF resistance has been obtained in the [001] direction. In this direction, the elastic modulus (and thus the stresses) developed is lower than in other directions; consequently, for a given mechanical strain range the plastic strain range is lowest among all possible directions. Directionally

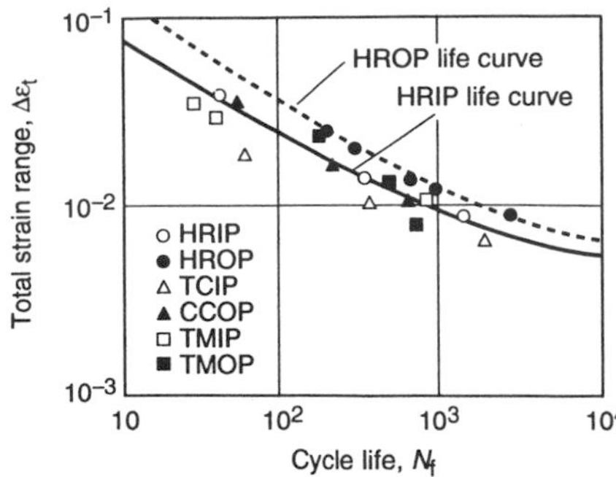

Fig. 28 Bithermal and thermomechanical fatigue data of Haynes 188 cycled between 316 and 760 °C (600 and 1400 °F). Test condition nomenclature: CCOP, compressive creep out-of-phase bithermal fatigue test; HRIP, high strain rate in-phase bithermal fatigue test; HROP, high strain rate out-of-phase bithermal fatigue test; TCIP, tensile creep in-phase bithermal fatigue test; TMIP, thermomechanical in-phase fatigue test; TMOP, thermomechanical out-of-phase fatigue test. Source: Ref 161

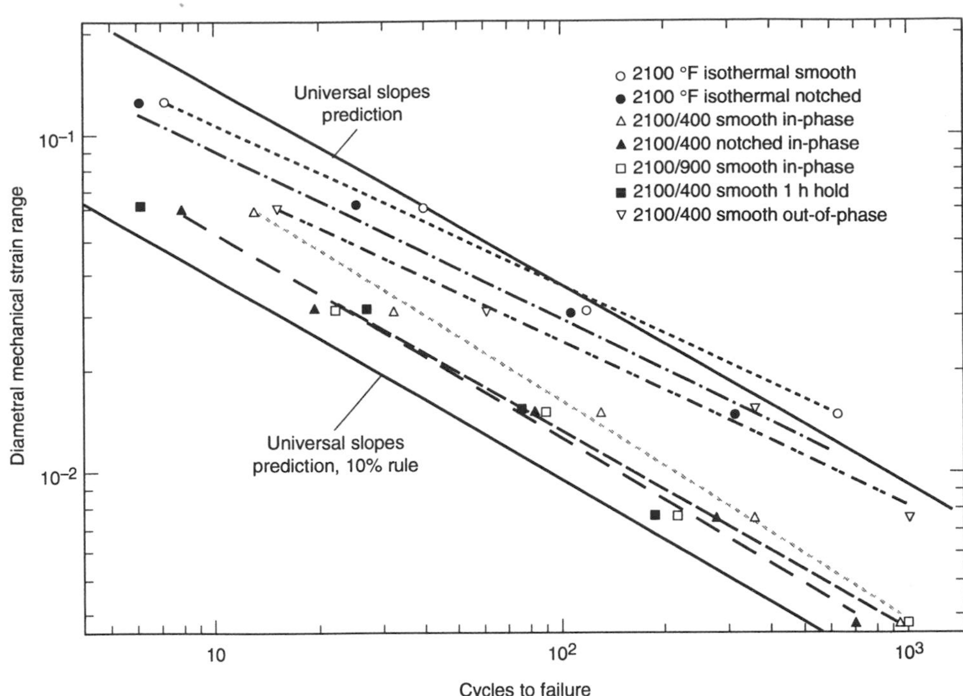

Fig. 29 Strain-controlled low-cycle fatigue results for Astar 811-C alloy (Ta-8W-1Re-0.7Hf-0.025C) tested in a hard vacuum ($<1 \times 10^{-7}$ torr) environment. Source: Ref 162

solidified alloys remove the grain boundaries transverse to the principal stress and also lower the elastic modulus in that direction relative to polycrystalline materials. For polycrystalline nickel-base superalloys, grain size and coatings influence TMF lives.

- Because of the unbalanced nature of inelastic deformation in TMF, mean stresses are sustained and do not relax. The mean stresses play a considerable role at finite lives, because the plastic strain range is smaller than the elastic strain range.
- Complex chemistries of oxides form with properties different from those of the substrate, resulting in internal stresses and oxide fracture that channels the crack into the material. Considerable depletion in the vicinity of oxides has been measured.
- At small strains and long lives, oxidation damage persists. Depending on stress and temperature, creep damage appears to be more significant at short lives.
- For the majority of nickel-base alloys at temperatures above 700 °C (1290 °F), TMF results display strain-rate sensitivity. Generally, as the strain rate is reduced or hold periods are introduced, the cycles to failure are lowered.
- For most nickel-base superalloys, TMF IP damage is larger than TMF OP damage at high strain amplitudes, whereas the trend is reversed at long lives. The diamond cycle often produces lives that fall between the TMF IP and TMF OP extremes.

Microstructural Changes. Under TMF conditions considerable changes in nickel-base superalloy microstructure have been known to occur, including changes in the size and morphology of γ' precipitates and the formation of dislocation networks around precipitates. For polycrystalline nickel-base superalloys exposed to temperatures above 800 °C (1470 °F), TMF OP loading results in transgranular propagation and TMF IP results in intergranular propagation.

Castelli et al. (Ref 103) and later Castelli and Ellis (Ref 32) observed cyclic hardening of Hastelloy X under TMF OP at $\Delta\varepsilon_m = 0.006$ over a temperature range of 600 to 800 °C (1110 to 1470 °F). In this alloy dynamic strain aging occurs in the region from 200 to 700 °C (390 to 1290 °F), and precipitation of chromium-rich precipitates also produces hardening. When the temperature range was increased to 800 to 1000 °C (1470 to 1830 °F), cyclic softening was observed. Since this is a solute-hardened superalloy, it undergoes considerable dynamic strain aging when the temperature is at 600 °C (1110 °F). Considerable precipitation of $M_{23}C_6$ carbides also occurs in the vicinity of dislocations, which coarsens at high temperatures and loses its effectiveness as the additional hardening mechanism. Examples of extensive hardening for Hastelloy X in TMF are shown in Fig. 27. Both TMF IP and TMF OP results are shown for 400 ⇄ 600 °C (750 ⇄ 1110 °F) conditions. The results are obtained at strain rates of 5×10^{-5}/s conditions. Strengthening increases by more than a factor of two over several thousand cycles. The IF results at 595 and 425 °C (1105 and 795 °F) are also shown for comparison.

Table 3 Reported damage mechanisms for TMF in-phase and TMF out-of-phase loadings of nickel-base superalloys

Material	TMF in-phase	TMF out-of-phase	Ref
Coating Ni-base superalloys	...	Fracture of coating upon cooling below its ductile brittle transition	158
Mar-M 247 (uncoated polycrystalline)	Intergranular crack initiation and growth	Repeated oxide damage rafting of the γ' structure different than IF or TMF IP	105
Mar-M247 (coated and uncoated)	Intergranular crack initiation and growth, internal crack initiation Rafting of the γ' structure different than IF or TMF OP	Fracture of coating upon cooling below its ductile brittle transition Rafting of the γ' structure different than IF or TMF IP	107, 135
Hastelloy X (solution strengthened Ni-base superalloy)	Strain rate dependent dynamic strain aging Precipitation hardening due to Cr-rich $M_{23}C_6$	Strain rate dependent dynamic strain aging Precipitation hardening due to $M_{23}C_6$	103
AM1 single crystals	...	Environment initiated damage in TMF differs from casting defect initiated damage in IF	48
IN 738 LC (coated and uncoated)	...	Lower lives for coated OP case because of coating fracture	118
CMSX-6 (single crystals)	Soft γ' matrix formation, cutting of particles at low temperatures, dislocation climb during high-temperature portion of the cycle	...	108

Thermal recovery processes in TMF have been documented by Sehitoglu and Boismier (Ref 105) and by Kadioglu and Sehitoglu (Ref 135) in the form of γ' coarsening and eventual rafting of the γ' microstructure.

Research on TF and TMF has described several microstructural changes that influence deformation (stress-strain) behavior as well as fatigue lifetime. Some, but not all, of the findings are listed in Table 3. The most important mechanisms are:

- Aging of the microstructure when exposed to high temperatures
- Repeated oxide rupture due to mismatch in mechanical and physical properties of matrix and oxide
- Strain aging in the case of solute-hardened materials at the elevated-temperature end, resulting in considerable hardening
- Enhanced grain-boundary damage due to unequal deformation during the cycle
- Carbide precipitation at grain boundaries at high temperature

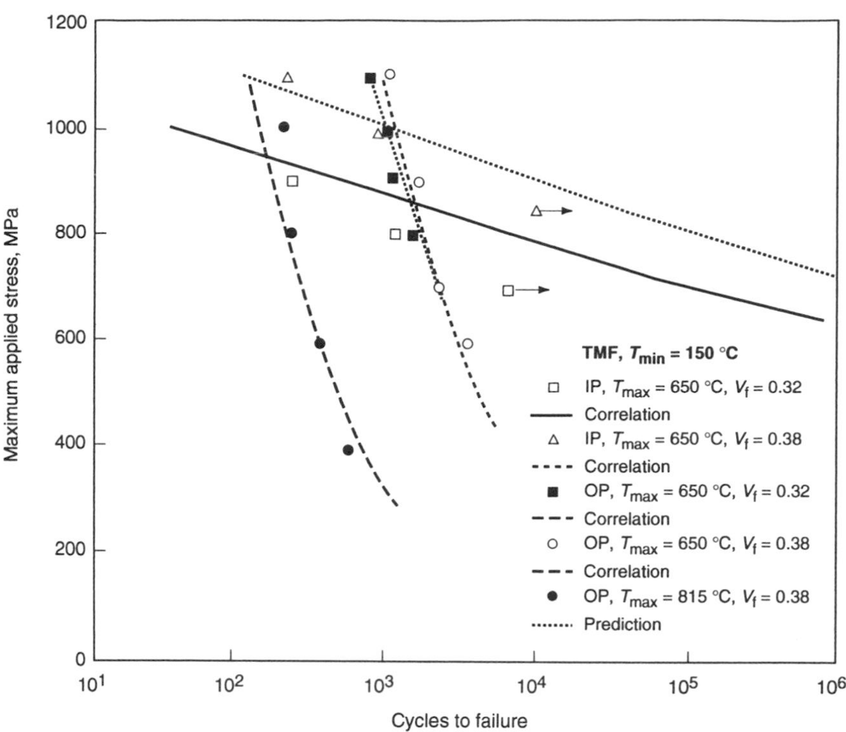

Fig. 30 TMF OP and TMF IP for SiC/titanium aluminide composite. Source: Ref 169

TMF of Other Structural Alloys

Several classes of advanced materials have been investigated under TMF loading conditions. Thermomechanical fatigue of titanium aluminide has been investigated by Wei et al. (Ref 159) under total constraint 25 ⇄ 750 °C, 25 ⇄ 900 °C (75 ⇄1380 °F, 75 ⇄ 1650 °F) conditions. The role of hydrogen and helium environment was investigated, and the lives were twice as long in helium environment relative to air. On a similar material (Ti_3Al) Mall et al. (Ref 160) studied TMF crack growth under OP and IP cases.

TMF data on cobalt-base superalloys have been published by Reuchet and Rémy (Ref 140) and Kalluri and Halford (Ref 161). The tests carried out by Kalluri and Halford (Ref 161) were conducted on Haynes 188 specimens fatigued under bithermal and TMF loading conditions between 316 and 760 °C (600 and 1400 °F). The bithermal and TMF fatigue data are plotted in terms of total strain range versus cyclic life in Fig. 28.

Early research by Sheffler and Doble (Ref 162) on TMF of tantalum alloys showed that the TMF IP lives are considerably shorter than the TMF OP case (Fig. 29). These materials have been investigated as an alternative to nickel-base superalloys, but their cost and/or performance characteristics have not been superior to nickel-base superalloys. Undoubtedly, they will find some specialty applications.

Thermomechanical fatigue is of considerable interest in the electronics industry. Failure mechanisms are currently being investigated in aluminum thin films in integrated circuits and in lead-tin solders undergoing temperature, current density, and mismatch in thermal expansion conditions. Recent TMF data on 63Sn-37Pb solder alloys (Ref 163) have been published. In these experiments the material is subjected to simultaneous shear loading and temperature cycling. It is important to note that if the specimen is subjected to shear and temperature, a set of material planes will experience TMF OP loading and the other orthogonal planes will undergo TMF IP loadings.

Starting in mid-1980, interest grew in metal-matrix composites (metal reinforced with ceramic particulates whiskers or fibers). These materials, although expensive, have been touted for lower thermal expansion coefficient, higher elastic modulus relative to matrix, and better high-temperature properties. Although some properties have improved with these new classes of "advanced" materials, the TMF resistance of these materials is not superior to that of the monolithic alloy. This is partly due to difficulties in the processing uniformity and detrimental residual stresses. Karayaka and Sehitoglu (Ref 164, 165), VanArsdell et al. (Ref 166, 167), and Sehitoglu (Ref 168) have authored a number of papers on TMF OP and IP of aluminum alloy 2024 reinforced with SiC particulates with volume fractions of SiC in the range 15 to 30%. Recent work (Ref 169, 170) focused on TMF of Timetal 21s (titanium alloy) reinforced with SiC (SCS-6) composites studied under TMF OP and TMF IP conditions (stress-control). Under TMF IP fiber damage was dominant and under TMF OP environment damage in the matrix or at interfaces was found to be most important. Sample results are shown in Fig. 30 for TMF OP and TMF IP loading conditions under stress control. The experimental techniques developed in Ref 46 were utilized. The results are plotted in an S_{max}-N_f format. Under TMF IP conditions, the lives are controlled by the fiber failure while for TMF OP case the damage mechanism is a combination of environmental and fatigue processes. It should be noted that at long lives the TMF OP is far more damaging. At high stresses there is crossover and TMF IP damage exceeds the TMF OP damage.

Multiaxial Effects in TF and TMF

The multiaxial effect is one of the least explored aspects of TF and TMF loadings. When a surface is heated the stresses are generally biaxial and this phenomenon has been discussed by Manson (Ref 6). Manson discusses thermal shock as well as slow heating cases where a two-dimensional stress state develops. Taira and Inoue (Ref 171) considered the biaxial stress fields in their TF analysis. In their experimental work they cooled a solid cylinder at one end and plotted their TF results using the von Mises equivalent strain range, showing good agreement with uniaxial tests on 0.16% steel. Recent research considers TMF under multiaxial loadings (Ref 172) where axial-torsional loading is applied simultaneously with temperature.

Crack Initiation and Crack Propagation

Crack Initiation in TF and TMF. Crack initiation within nickel-base superalloys can occur intergranularly at oxidized surface exposed grain boundaries or transgranularly. Transgranular initiation can be caused by heterogeneous planar slip which produces initiation along persistent slip bands at free surfaces (Ref 125, 139, 173). Transgranular initiation can also occur at pores, inclusions, and carbides (Ref 117, 139). Trans-

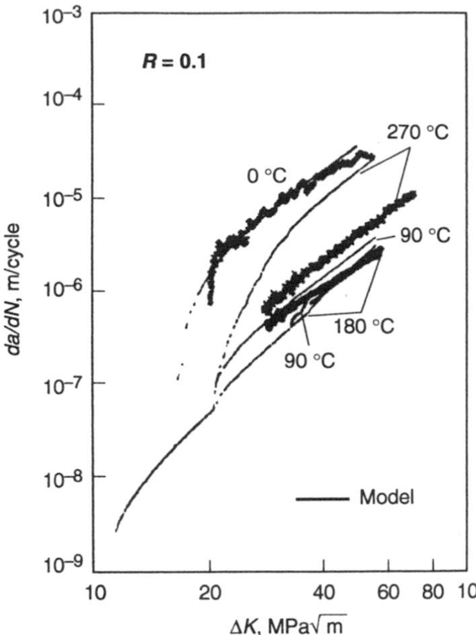

Fig. 31 Crack growth rates in TMF OP and TMF IP cases for Inconel 718. Source: Ref 177

granular crack initiation is more prominent at low temperatures and high frequencies. This is because the contributions from the creep and environmental components of damage are minimal. Gell and Leverant (Ref 126) indicated that one of the first observed effects of an increased creep component is a transition from transgranular to intergranular initiation. Runkle and Pelloux (Ref 174) found under IF conditions, as temperature is increased, initiation changes from transgranular to intergranular for Astroloy. A similar transition was observed for a decrease in strain rate by Nazmy (Ref 127) for IN 738 under IF loading at 900 °C (1650 °F).

At high temperatures crack initiation is predominantly intergranular (Ref 124, 130, 131, 174). This is attributed to increased damage contributions from creep and environmental attack. Environmental attack appears to be the more dominant of the two damage mechanisms. Intergranular crack initiation typically occurs at oxidized surface connected grain boundaries which are often accompanied by an adjacent zone of γ' depletion (Ref 50, 102, 118, 130, 131). Preferential grain boundary environmental attack occurs because of the easier path of oxygen diffusion. The decrease in fatigue life with increased temperature and decreased frequency can be partially attributed to the transition from transgranular to intergranular crack initiation. In general, intergranular crack initiation occurs at a faster rate than transgranular initiation (Ref 126). In the work of Kadioglu and Sehitoglu (Ref 107) the Mar-M246 exhibited intergranular crack initiation followed by a switch to transgranular crack growth.

Crack Propagation in TF and TMF. It is important to know the mode of crack propagation to develop physically meaningful fracture mechanics or micromechanical models to characterize the crack driving force. The interaction of creep damage and environmental attack and their effects on crack propagation under TMF conditions can be very complicated. In general, TMF OP produces transgranular crack propagation while intergranular fracture is observed for TMF IP case, within representative service temperature ranges (300 to 1000 °C, or 570 to 1830 °F)) at moderate strain rates (10^{-3} to 10^{-6}/s). TMF experiments conducted on B-1900+Hf at temperature ranges of 427 to 925 °C (800 to 1695 °F) (Ref 153) and 538 to 871 °C (1000 to 1600 °F) (Ref 113) revealed predominantly transgranular OP crack propagation and intergranular IP propagation. Kuwabara and Nitta (Ref 42) observed intergranular crack propagation for IN 718 (300 to 650 °C, or 570 to 1200 °F) and Mar-M247 (300 to 900 °C, or 570 to 1650 °F) under TMF IP loading. Milligan and Bill (Ref 124) performed TMF experiments on Mar-M200 (500 to 1000 °C, or 930 to 1830 °F) and found intergranular crack propagation and internal grain boundary cracking for TMF IP case, while TMF OP produced mixed transgranular and intergranular cracking. Ramaswamy and Cook (Ref 175) also observed transgranular cracking under TMF OP and intergranular cracking under TMF IP tests conducted on IN 718 at temperature ranges 343 to 565 °C and 343 to 649 °C (650 to 1050 °F and 650 to 1200 °F) and René 80 at 760 to 871 °C (1400 to 1600 °F). Intergranular crack propagation in TMF IP case is attributed to a tensile creep component that results in ratcheting at the microlevel resulting in weakening of the grain boundaries. In a study conducted on Mar-M247 similar results have been observed (Ref 105).

Rau et al. (Ref 176) performed TMF crack growth tests on Mar-M200 DS (427 to 927 °C, or 800 to 1700 °F) and B-1900+Hf (316 to 927 °C, or 600 to 1700 °F) under strain control. The Mar-M200 DS out-of-phase crack growth rates were greater than in-phase. The T_{max} IF crack growth rates for B-1900+Hf were the same as out-of-phase TMF crack growth rates. Leverant et al. (Ref 153) also tested B-1900+Hf (427 to 927 °C, or 800 to 1700 °F) under strain control. It was observed that out-of-phase and in-phase TMF crack propagation rates were the same. Heil et al. (Ref 177) conducted TMF crack growth tests on IN 718 under load control at a temperature range of 427 to 649 °C (800 to 1200 °F). They discovered that in-phase loading produced greater crack growth rates than out-of-phase loading. The subject of TMF crack growth rates needs to be investigated further. Figure 31 shows their result compared on the basis of the stress intensity range. These experiments were conducted under load control $R = 0.1$ conditions. The 90° refers to DCW and the 270° refers to DCCW load-temperature history. The 0° refers to TMF IP and 180° refers to TMF OP. They found that TMF IP is more damaging than TMF OP in 718 based on ΔK. Heil et al. forwarded a simple model incorporating crack growth from cycle and time-dependent damage. Their results are intuitively correct in view of the fact that the crack closure phenomenon will be minimum for the TMF IP case.

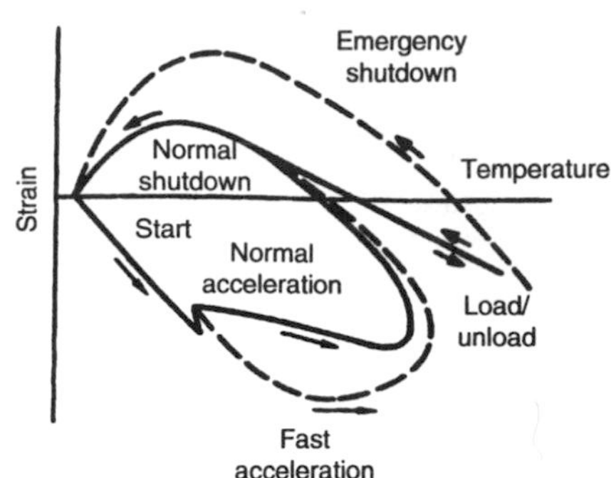

Fig. 32 Strain/temperature history on a turbine blade. Source: Ref 118

Okazaki and Koizumi (Ref 178, 179) in Japan used the J-integral to correlate crack growth rates under TMF IP conditions for low-alloy ferritic steel (Cr-Mo steel). Sehitoglu (Ref 24) correlated crack growth rates in TMF OP at temperatures of 150 ⇄ 400 °C and 150 ⇄ 600 °C (300 ⇄ 750 °F and 300 ⇄ 1110 °F) and IF results from room temperature using the range in crack opening displacement. Gemma et al. (Ref 180) considered crack growth in both DS Mar-M200 and B-1900 alloys under TMF OP conditions with 427 ⇄ 1038 °C (800 ⇄ 1900 °F) and showed that lower crack growth rates were obtained when the loading axis and the grain growth direction coincided confirming the benefit of DS alloys. In the United Kingdom, Skelton (Ref 181) studied crack growth in ferrous alloys (type 316, alloy 800, ½Cr-Mo-V) and used both the linear elastic and elastoplastic fracture mechanics parameters. The elastoplastic crack growth parameter was $C(\varepsilon^{in})^n a^Q$ where C, n, and Q are constants, a is crack length, and ε^{in} is inelastic strain. Skelton used this parameter for small cracks both for TMF and thermal shock conditions.

Recent research from Wright Patterson Air Force Base (Ref 160) has investigated TMF crack growth in titanium aluminide. Similar to IN 718 the TMF IP crack growth rates were higher than TMF OP. The LEFM parameter ΔK was used. The authors noted a distinct difference between superalloys and aluminides in that the crack tip blunts in aluminide and sustained load cracking do not occur.

Jordan and Meyers (Ref 182) conducted TMF OP, TMF IP experiments as well as a DCCW type cycle. Higher crack growth rates were noted for the TMF OP case relative to IF test at T_{max} for the case 427 ⇄ 871 °C (800 ⇄ 1600 °F). Several elastoplastic fracture mechanics parameters have been used including the strain-intensity range, stress-intensity range, and the range in J-integral.

Temperature-Strain Variations in Service

McKnight et al. (Ref 183) conducted finite-element analysis of turbine blades with a simplified mission cycle. The purpose was to analyze tip cracking due to thermomechanical fatigue. In their analysis of a hollow, air-cooled turbine

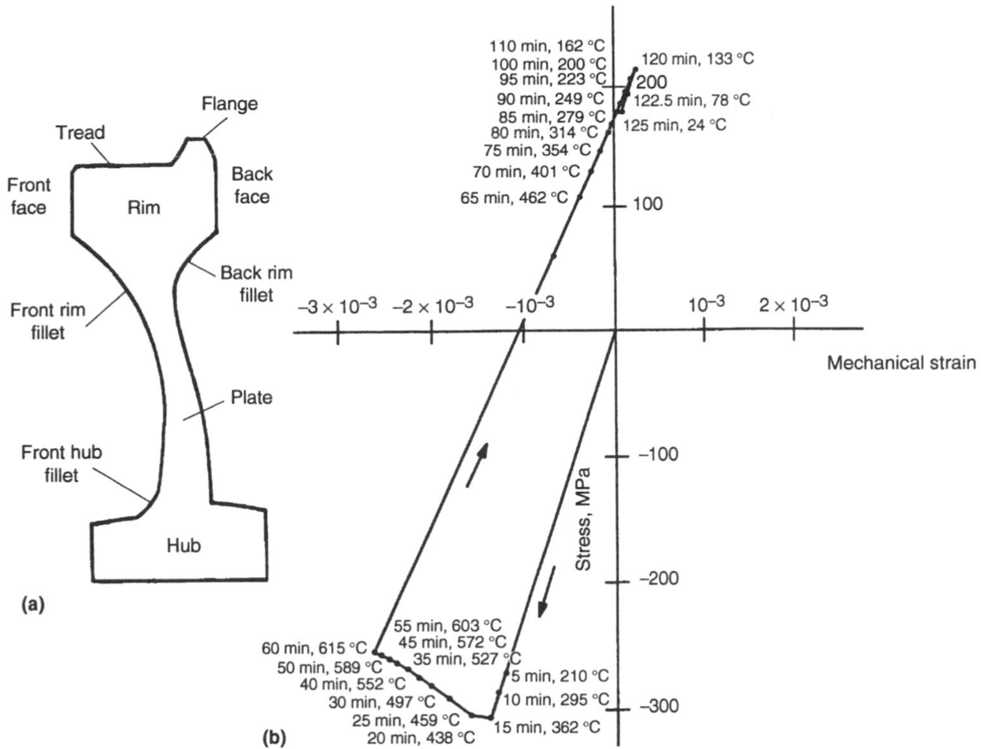

Fig. 33 Railroad wheel design (a) and stress/mechanical strain response at the tread surface (b). Source: Ref 184

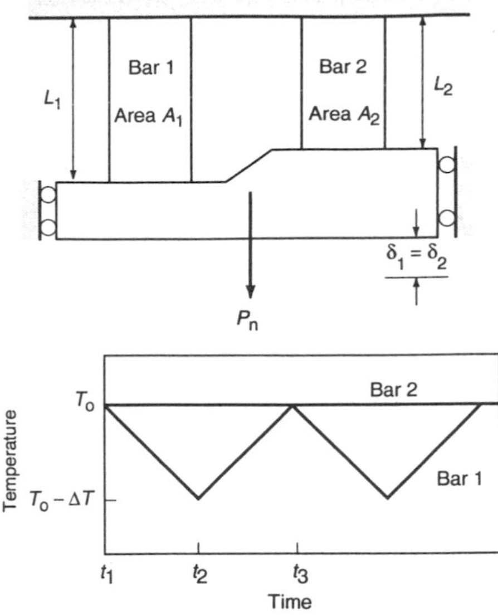

Fig. 34 Schematic of a two-bar structure and the temperature history on bars 1 and 2. Source: Ref 185

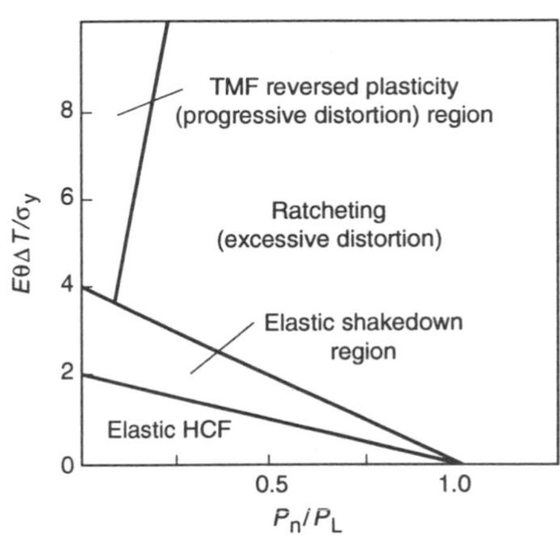

Fig. 35 Diagram showing the operating regimes of component behavior. Source: Ref 185

blade they did focus at a tip cap designated as a "squealer tip" just below the leading edge. They showed that at this critical location, the blade experiences a TMF OP type of cycle with a small amount of ratcheting in the compressive direction. The analysis indicated that the mechanical strain range (the principal strain component) in the critical location was near 0.31% after several cycles and the maximum temperature was as high as 1100 °C (2010 °F). Both the maximum and the minimum mechanical strains were negative. The strain temperature was predominantly a TMF OP cycle with an initial temperature of 300 °C (570 °F). Under these conditions the lifetime of René 80 simulated on a laboratory sample was 3000 cycles as established by Kaufman and Halford (Ref 119).

Embley and Russell (Ref 115) studied strain temperature phasings that are rather complex and representative of those experienced by turbine blades. Bernstein and colleagues (Ref 118) also studied "faithful" histories typical of leading edge of turbine blades. A schematic of this history is given in Fig. 32. The schematics in Embley and Russell (Ref 115) work and Bernstein's paper are similar. It was noted that during the startup the behavior is of TMF OP type. Higher strains in the compressive direction are achieved for "fast" acceleration relative to "normal" acceleration. The temperature does not reach a maximum at the acceleration peak; the temperature reaches a maximum during the steady-state operation (the load/unload portion). Upon normal shutdown the strain increases in the tensile direction as temperature is reduced. In the case of an emergency shutdown the tensile mechanical strains reached can be considerably higher.

Heating and cooling of the surface of a thick structure under the impingement of steam has been considered by Skelton (Ref 76). When the surface is exposed to the steam at 550 °C (1020 °F) it undergoes compression (OP behavior) because it is surrounded by the cooler material. With time, the temperature front moves into the structure, the temperature gradient reduces, and the surface unloads elastically at temperature. During steady-state conditions the strain remains constant and some relaxation of the stress can occur. When the steam is shut off, the surface wants to contract, but is again constrained by the warmer surrounding material. The surface then undergoes tension, and the maximum stress in the cycle is reached corresponding to the minimum temperature.

The current railroad wheel design has evolved as a result of TF experiments on actual wheels and also from thermomechanical elastoplastic analysis of stresses under braking conditions. The original wheel design had a straight plate region that produced higher stresses (due to the constraint) than the curved plate design. The curved plate design is shown in Fig. 33(a). Localized heating caused by friction occurs at the tread area. The results of a FEM analysis are given in Fig. 33(b). The circumferential stress/mechanical strain behavior under the brake-shoe (at the tread region of the wheel) is depicted in Fig. 33(b). The times and temperatures are also shown during the different stages of the heating process (Ref 184). The analysis was conducted for a 50HP application to the surface for a period of 1 h followed by 1 h cooling, then 5 min cool down to room temperature. The minimum stress develops 15 minutes into the 50HP application and upon subsequent rise in temperature the material softens. At the end of the heating period (60 min) the peak temperature at the surface has reached 615 °C (1139 °F). It was noted that the tensile stresses upon cooling are near 200 MPa (29 ksi).

Often in engineering applications an approximate measure of inelastic strains and stresses are needed without the execution of an elastoplastic FEM analysis. If the results of an elastic FEM analysis are available, the results of such an analysis can be used to estimate the inelastic strains. As pointed out earlier by Manson (Ref 6), the elastic strains calculated from FEM are assumed as total strains. Then, given the total strains, corresponding stresses and inelastic strains can be determined. This approach is referred to as "strain invariance principle" and has been successfully demonstrated for TF analysis of the wedge specimen (Ref 13).

Thermal Ratcheting

To understand thermal ratcheting, a two-bar structure shown schematically in Fig. 34 is considered. The two bars are in series and are subjected to a net section load, P_n, and one of the bars undergoes thermal cycling. In the present model, Bar 2 remains at a steady temperature of T_0 and Bar 1 undergoes thermal cycling such that the maximum temperature is T_0 and the minimum temperature is $T_0 - \Delta T$. Several regimes of material behaviors can be identified:

- Elastic (high-cycle fatigue)
- Elastic shakedown (high-cycle fatigue)
- Reversed plasticity (thermomechanical fatigue, low-cycle fatigue)
- Ratcheting regimes (shown as progressive and excessive distortion regions).

In the first mode, all strains are elastic. In the second mode, one of the two bars yields during the first half of the cycle followed by elastic response in both bars.

In the third mode, called plasticity, one of the bars yields plastically while the second bar remains elastic. TMF occurs in this regime. In the ratcheting case, one bar yields during the cooling half of the cycle while the second bar yields in the same direction during the heating portion of the cycle. This results in accumulation of strains in the tensile direction and ultimately failure due to a combined fatigue and ductility exhaustion mechanism. The equilibrium equation is

$$P_n = \sigma_1 A_1 + \sigma_2 A_2$$

and the compatibility equation ($\delta_1 = \delta_2$) is

$$\varepsilon_1 L_1 = \varepsilon_2 L_2$$

where ε_1 is the net strain on Bar 1 given as

$$\varepsilon_1 = \varepsilon_1^e + \varepsilon_1^{in} + \varepsilon_1^{th}$$

$$\varepsilon_2 = \varepsilon_2^e + \varepsilon_2^{in}$$

where ε_1^e, ε_1^{in}, ε_1^{th} represent elastic, inelastic, and thermal strain components.

A common way of representing the different deformation regimes is to plot the dimensionless parameter $E\theta\Delta T/\sigma_y$ versus the P_n/P_L where P_L is the limit load determined from the yield strength σ_y at T_0 and θ is the thermal expansion coefficient. It should be noted that similar results would be obtained by considering a pressure vessel subjected to a temperature gradient across the wall thickness and subjected to an internal pressure. The regime of "thermal ratcheting" and TMF are shown in Fig. 35. The experimental results obtained from various tests are also shown in the diagram. The letter "*P*" denotes plasticity, "*R*" denotes ratcheting, "*S*" stands for shakedown, and HCF stands for high-cycle fatigue. The stress-net strain response of a two-bar structure with $P_n/P_L = 0.5$ and $\Delta T = 150$ °C (270 °F) are shown in Fig. 36 for a 304 stainless steel. It should be noted that Bar 1 undergoes tension upon cooling. Therefore, the response depicted in Fig. 36 is termed "OP" (temperature is a minimum when stress is a maximum).

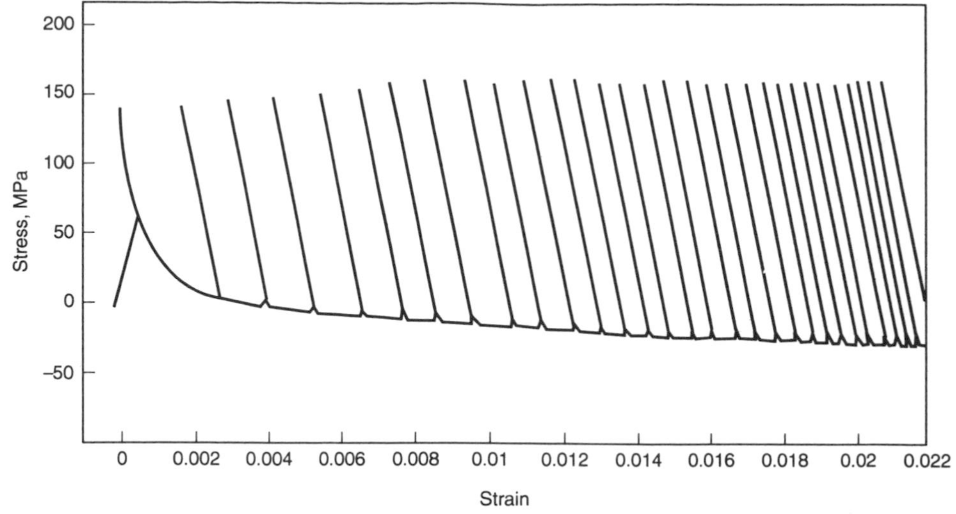

Fig. 36 Ratcheting of the two-bar structure for type 304 stainless steel. Source: Ref 185

These experiments were conducted by using two servohydraulic test systems with command signals from a microcomputer that enforces equilibrium and compatibility of the two-bar structure. Similar experiments have been reported by Swindeman and Robinson (Ref 186) using 2.25Cr-1Mo steel. In addition to thermal cycling they also considered compression hold period effects on ratcheting.

Thermal ratcheting experiments have been conducted under stress control cycling or constant stress with superimposed thermal cycling (Ref 61, 62). Taira (Ref 88) has proposed constitutive models for predicting the transient and steady-state stage of deformation under these conditions and forwarded creep rupture as the mechanism of failure under this type of cycling. Also, Russ et al. at Wright-Patternson Air Force Base (Ref 187) conducted stress control TMF experiments that resulted in thermal ratcheting in titanium aluminides.

Thermal Shock

Thermal shock occurs under rapid temperature change conditions and results in thermal stresses that cause fracture of the material. There are three main differences between thermal shock and thermal fatigue:

- Thermal shock is a single application of a temperature change, while thermal fatigue means repeated thermal cycling.
- Physical properties such as specific heat and thermal conductivity can influence the stresses developed; this is not the case with quasistatic thermal stress, but because the strain rates are high in thermal shock, the sensitivity of the material to strain rate is important.
- The stresses generated in thermal shock are high enough that fracture toughness level is important; it does not enter in thermal fatigue considerations.

In this section, only thermal shock behavior of structural metallic materials and not the more well studied thermal shock in ceramics will be considered. Thermal shock in metals can be encountered in hot-working manufacturing applications, such as the rapid temperature rise in the bore of a gun barrel, friction heating conditions such as in disk brakes, and rapid startups such as that in space shuttle main engines. The thrust chamber of the space shuttle has an inner liner with axial flow coolant channels constructed from copper or a copper-zirconium-silver alloy. During the firing cycle liquid hydrogen enters the channels to cool the chamber (or the hot-gas wall) and a severe temperature rise occurs within a second.

Flame heating has been used to simulate engine environment and involves very rapid heating of the surface (Ref 21). The Association of American Railroads used flame heating on the rim of a rotating wheel of 0.7% C steel which was then subjected to a water quench when the same region rotated through a water pool. In this case, because the surface temperatures were above the transformation temperature, thermal cracks developed upon quenching to room temperature. Baron (Ref 188) employed a tapered disk specimen similar to Glenny and Taylor (Ref 56) and using induction heating, reached heating of the order of 1 s. Rapid heating of a surface while the bulk remains cool can produce thermal shock conditions. For more details of thermal shock, the reader is referred to the text by Manson (Ref 6). More recently, Skelton (Ref 181) has studied crack growth under "thermal shock" conditions.

Another variation of thermal shock is called "thermal striping." In thermal striping fatigue the surface temperature fluctuations develop due to mixing of fluid streams that impinge on the surface of structures. A temperature gradient develops through the component wall, and the frequencies are generally much higher than in classical TF tests.

Life Prediction under TF and TMF

The question has been raised when correlating IF and TMF results as to what equivalent temperature should be used for meaningful comparisons. The maximum stress in a TMF experiment may not be at the maximum temperature and also the maximum strain may not coincide with the maximum temperature. Even though conventional IF data at the maximum temperature of a TMF cycle have been used to predict TMF lives, many researchers have shown that nonconservative predictions can still result (Ref 18, 40-44, 59, 60, 66-83) with this approach. It is essential to compare TMF and isothermal fatigue data at similar strain rates so that time-dependent damage mechanisms in both cases correspond.

Both environmental damage and creep damage on the material contribute to fatigue damage at elevated temperatures. A number of oxidation damage mechanisms have been proposed. These include (1) enhanced crack nucleation and crack growth by brittle surface oxide scale cracking (Ref 130, 189, 190), (2) grain-boundary oxidation that results in intergranular cracking (Ref 130, 134), and (3) preferential oxidation of second-phase particles (Ref 140). Oxidation and fatigue can interact, resulting in a much shorter life than when either of them act alone. TMF OP loading is more damaging whenever the rate of oxide fracture damage exceeds that of creep damage. Oxidation damage accelerates with increasing $\Delta\varepsilon$, T_{max}, or increasing thermal coefficient mismatch of metal and oxide. The low ductility of oxides promotes cracking. If repeated oxidation is very severe for a given alloy, shorter TMF OP lives compared with TMF IP lives can result over a wide range of life regimes. Low-carbon steels and low-alloy Cr-Mo steels exhibit shorter lives under TMF OP compared with TMF IP. These materials are known to be susceptible to oxidation damage. The TMF OP curve will also be lowered if the mean stress effects become significant. In the TMF OP case, tensile mean stresses develop that are conducive to minimal contact between crack surfaces and rapid crack growth rates.

Many mechanisms have been proposed to explain creep-induced damage and creep-fatigue interactions. These include:

- Coalescence of intergranular voids ahead of an advancing crack (Ref 191, 192)
- A greater crack tip plastic zone resulting from the summation of the plastic zones of voids ahead of a crack (Ref 193)
- Grain-boundary sliding initiating wedge-type cracks at grain boundaries (Ref 89) and at hard second-phase particles on the grain boundaries (Ref 194)
- Grain boundaries acting as weak paths for flow localization and crack growth (Ref 195)

Table 4 Summary of oxide failure mechanisms and models

Ref	Mechanisms of failure	Equation
197	Separation of an oxide film at a slip step	$\frac{\sigma_f}{E_{ox}} = [\varepsilon_f/0.38(1-K)]\,(h_c/a)^{1/4}\sin(w)$, Tensile failure $\frac{\tau_f}{E_{ox}} = [\varepsilon_f/(2(1+n))^{1/2}](h_c/a)^{1/2}\cos(w)$, Shear failure h = oxide thickness, E_{ox} = oxide modulus σ_f = fracture strength, τ_f = shear strength a = interfacial layer thickness, K = dimensionless parameter, w = geometry constant
198	Tensile failure due to α mismatch	$\sigma_f = \frac{E_{ox}\,\Delta T(\alpha_{ox}-\alpha_m)}{1+E_{ox/E_m}(h_c/h_m)}$ E = elastic modulus h_c = critical rupture thickness of oxide h_m = thickness of metal ΔT = uniform temperature change
199	Failure due to compressive hoop strain during growth	$h_c = \frac{d\,\varepsilon\,\rho_m(56+16x)}{112\,\rho_{ox}}$ FeO_x ρ_m = density of metal ρ_{ox} = density of oxide h_c = critical rupture thickness
200	Scale displacement (M) causing compressive stress on convex surfaces	metal loss $= H = \frac{12\gamma}{E_{ox}}\left[\frac{R^2}{M^2\phi}\right]^{1/3}$ R = radius of metal surface $m = \phi(1-\alpha)-(1-V)$ ϕ = oxide/metal ratio
200	Loss of scale integrity when oldest part of oxide sustains strains average stored energy	$S = \frac{1}{2}h\frac{E\varepsilon^2}{2\gamma}$ h = metal thickness lost ε = strain on oxide failure
201	Growth stresses	$\sigma_f = \frac{E_m(h_m)^2}{6rh_c}$ r = radius of curvature h_c = critical rupture thickness E_m = elastic modulus of metal h_m = thickness of base metal
202	Interfacial shear failure	$\tau_f = \frac{h_c E_{ox}\,\Delta T(\alpha_m-\alpha_{ox})}{L+[1+(E_m/E_{ox})(h_c/h_m)]}(1-\nu)$
203	Buckling of oxide	$\alpha\Delta T = \left[\frac{(h_c)^4}{a^4}+\frac{w}{E_{ox}h_c}\right]^{1/2}$ $\alpha\Delta T$ = thermal strain in oxide h_c = critical oxide rupture thickness a = crack length
68, 69	Fatigue failure of oxide	$h = \frac{B\,K_{peff}\,t^\beta}{\bar{h}_f}$ h = oxide intrusion depth $\bar{h}_f = \frac{\delta_0}{(\Delta\varepsilon_m)^2\phi\,\dot{\varepsilon}^a}$ B, β, δ_0, a = constants; K_{peff} = parabolic oxidation constant h_f = critical rupture thickness; $\Delta\varepsilon_m$ = applied mechanical strain range ϕ = phasing factor for thermomechanical loading
135	Fatigue failure of oxide	$\bar{h}_f = \frac{\delta_0}{(\Delta\varepsilon_m^{ox})^2\phi\,\dot{\varepsilon}^a}$ $\Delta\varepsilon_m^{ox}$ = local (oxide tip) mechanical strain range

- The modification of the crack tip strain fields in the absence of cavities (Ref 196)

It would be expected that a number of these creep mechanisms would operate under both isothermal and TMF conditions, depending on the alloy.

The models developed for TMF studies can broadly be divided into two categories: (1) the continuum-based models (using parameters such as strain range, stress range, product of maximum stress and strain range, and plastic work), and (2) physical damage equations (incorporating oxidation damage kinetics, microstructural observations, and creep damage mechanism). The major issue is how these equations perform at log lives where TMF data are scarce. Extrapolation of short-life data to long lives has limitations because the severity of the mechanisms will change.

Table 5 Summary of oxidation-fatigue laws

Ref	Experiments/mechanism	Material	Equation
72, 126	Isothermal in air and vacuum, surface and crack tip oxidation	A-286 steel	$\Delta\varepsilon = A(N_f)^b \nu^m$ A, b, m = constants N_f = cycles to failure $\Delta\varepsilon$ = strain range ν = cycle frequency
130	Isothermal in air, preoxidized samples, hold periods/surface and grain boundary oxidation	René 80	$\Delta\varepsilon_p = A\left[\frac{n}{1 + nt_h^{1/n}}\right]\exp\left(-\frac{Q}{RT}\right)(N_f)^b$ A, n, b, Q, R = constants T = temperature, t_h = hold period $\Delta\varepsilon_p$ = plastic strain range n = cycle frequency
134	Isothermal in air at different frequencies and hold periods, crack tip oxidation	IN 718, IN 750, Astroloy, 304 SS, ½CrMoV steel	$\frac{da}{dN} = \frac{D_{gb}}{n} f(K_{max})$ D_{gb} = oxygen diffusivity at grain boundary n = cycle frequency da/dN = fatigue crack growth rate K_{max} = maximum stress intensity
140	Isothermal in air and in vacuum, crack tip oxidation of matrix and carbides	Mar-M509	$\frac{da}{dN} = 0.51\,\Delta\varepsilon_p\left[\sec\frac{\pi}{2}\frac{s}{s_0} - 1\right]a + (1 - f_c)\,a_M\left(1 + \frac{\Delta\varepsilon_p}{2}\right)t_c + f_c a_c \exp(bs)t_c^{1/4}$ $\Delta\varepsilon_p$ = plastic strain range s, s_0 = applied stress, flow stress b, a_M, a_c = constants f_c = carbide fraction t_c = cycle period a = crack length
68, 69, 105, 107, 135	TMF in air and in helium, strain rates, oxide intrusions, repeated oxide rupture	0.7% C steel, Mar-M246, Mar-M247, Tiβ21s, Al2xxx-T4	$\frac{1}{N_f^{ox}} = \left[\frac{h_{cr}\delta_o}{B\phi^{ox}K_p^{eff}}\right]^{-1/\beta}\frac{2(\Delta\varepsilon_{mech})^{(2/\beta)+1}}{\dot{\varepsilon}^{1-(a/\beta)}}$ $h_{cr}, \delta_0, B, a, \beta$ = constants $\dot{\varepsilon}$ = strain rate ϕ^{ox} = phasing factor $\Delta\varepsilon_m$ = mechanical strain range, $1/N_{ox}$ = oxidation damage K_p^{eff} = parabolic oxidation constant
188	Isothermal in air with hold period, oxide rupture in tension	2¼Cr-1Mo steel	$A\Delta\varepsilon^m\left(1 - \frac{t_h}{t_c}\right) = N_f$ t_c = cycle period, t_h = hold period A, m = constants, $\Delta\varepsilon$ = strain range N_f = cycles to failure

Therefore, models based on physical mechanisms and not empirical fits to data are preferred when predictions in the long-life regime are considered. Finally, it should be noted that microstructural coarsening and other metallurgical instabilities are encountered in alloys at high temperatures, and this could result in a decrease in lives for both the TMF OP and TMF IP cases.

Oxidation Damage and Oxidation-Fatigue Laws. Specific modeling of oxide failure processes has been attempted by a number of investigators and are given in Ref 197 to 203. The range of mechanisms include separation of an oxide film at a slip step (Ref 197), failure due to mismatch of the coefficient of thermal expansion (Ref 198), failure due to growth stresses (Ref 199, 200, 201, 203), interfacial shear failure (Ref 202), failure due to oxide buckling (Ref 203), and fatigue failure of oxide (Ref 68, 69, 135). These models are listed in Table 4.

The majority of the models proposed characterize failure due to scale growth stresses or temperature change conditions. The effect of mechanical strains on oxidation is not well understood. Many of the models consider uniform oxide thickness. Experimental observations, however, indicate nonuniform thickness in the form of oxide intrusions (Ref 135).

A summary of oxidation-fatigue laws is given in Table 5 (Ref 1–4, 25–29, 34, 46–48). The models depicted in Table 5 specifically address fatigue plus oxidation damage. Despite the lack of quantitative understanding of oxidation fatigue, the models provide a qualitative description of the oxidation-fatigue process. The earliest oxidation-fatigue model is that proposed by Coffin (Ref 72, 123) and is referred to as the frequency modified life approach. Antolovich and coworkers (Ref 130) recognized the formation of oxide spikes at preferential grain boundaries in nickel alloys and proposed a strain-life relation incorporating oxidation kinetics. Liu and Oshida (Ref 134) modified fracture mechanics parameters and accounted for accelerated crack growth upon oxide film rupture at crack tips. Rémy and Reuchet's (Ref 140) model incorporates crack growth according to a Dugdale type of model modified to account for oxidation of metal and carbides. The model proposed by Sehitoglu and coworkers (Ref 68, 69, 105, 107, 135) incorporates an oxidation phasing factor and accelerated oxidation under mechanical straining conditions. This is based on measurement of oxide thicknesses and observations of oxide failure at the surface and at crack tips. The details of this model will be illustrated later.

Several observations should be made regarding these models. Because the experimental data that form their foundations have been obtained primarily under IF conditions, the use of these relations for thermomechanical loading is not recommended. The Neu-Sehitoglu model incorporates a varying oxidation severity depending on the strain-temperature history and can handle TMF cases.

Depending on the material, OP or IP loading condition, and whether or not tensile or compressive hold periods are encountered, oxidation damage could be significantly greater than creep damage. By performing experiments in controlled atmospheres, the mechanisms of oxide failure can be isolated and more easily interpreted.

Creep Damage and Creep-Fatigue Models. A summary of creep laws is given in Table 6. The strain range partitioning method (Ref 50, 62) separates the inelastic strain range into four generic components (pp = plastic-plastic, pc = plastic-creep, cp = creep-plastic, cc = creep-creep). Plastic-plastic stands for plasticity in tension reversed by plasticity in compression, plastic-creep stands for plasticity in tension reversed by creep in compression, and so forth. Recently, the method has been modified to handle time effects on the strain components, and a total strain range version has also been proposed. The application of the model to TMF has been outlined where stress-hold experiments conducted at various points around the stress-strain hysteresis loop are needed (Ref 50).

The time-cycle fraction rule (adopted as an ASME Code, Ref 204) involves linear summation of fatigue and creep damage, where the fatigue damage is expressed as cycle ratio and the creep damage is written as a time ratio. Creep damage is determined from stress-rupture diagrams, and fatigue damage is obtained from the strain-life equation. Because experimental results indicate that predictions based on the time-cycle fraction rule can be nonconservative, a modified time-cycle fraction rule has been proposed by Lemaitre and Plumtree (Ref 205) with applications involving cumulative damage. Recognizing that creep ductility is a function of strain rate, a ductility exhaustion approach has also been proposed that defines creep damage as strain rate to ductility ratio (Ref 206).

Table 6 Summary of creep-fatigue laws

Ref	Experiments/mechanism	Equation
58, 64	Isothermal: creep damage in tension versus compression	$\frac{1}{N_{pred}}=\frac{F_{pp}}{N_{pp}}+\frac{F_{pc}}{N_{pc}}+\frac{F_{cp}}{N_{pp}}+\frac{F_{cc}}{N_{cc}}$ N_{pred} = predicted life
204	Isothermal: time-cycle fraction rule, Miner-Robinson Rule	$D=\Sigma\frac{N}{N_f}+\Sigma\frac{t}{t_f}$ N_f = cycles to failure, t_f = time to failure
205	Isothermal: nonlinear damage	$D_f=1-(1-N/N_f)^{P+1}$ $D_c=1-(1-N/N_c)^{q+1}$
206	Isothermal	$D=\Sigma\frac{N}{N_f}+\Sigma\frac{\dot{\varepsilon}}{\varepsilon_f}$ $\dot{\varepsilon}$ = creep strain rate, ε_f = creep ductility p,q = constants, N_f = cycles to failure
207	Isothermal: void growth at crack tips	$\frac{da}{dt}=\alpha\left(\frac{K^2}{t}\right)^{\beta/n+1}$ β = void growth exponent, n = stress exponent K = stress intensity, t = time, da/dt = crack growth rate
83, 191	Isothermal: void growth in tension, void healing in compression	$\frac{dD_c}{dt}=\left(\frac{\varepsilon^m c}{\hat{\varepsilon}_c}\right)\lvert\dot{\varepsilon}^{in}\rvert$ $\frac{dD_f}{dt}=(1+\alpha D_c)(\dot{\varepsilon}_{f0})^{-(m+1)}\varepsilon^m\lvert\dot{\varepsilon}^{in}\rvert$ $\dot{\varepsilon}^{in}$ = creep strain rate, $\dot{\varepsilon}_{f0}$ = creep ductility $\alpha, m, m_c, \dot{\varepsilon}_{f0}, \hat{\varepsilon}_c$ = constants, D_c = creep damage, D_f = fatigue damage
208	Isothermal: decrease in crack closure load	$da/dN=C(U\Delta K)^m$ $U=U_0+(1-0.5/F(S_{max}/\sigma_0)\left\{1-\exp(-6\times10^3(S_{max}/\sigma_0)^m\frac{t_h\cdot\varepsilon_0}{(\sigma_c/\sigma_0)^m})\right\}^2$ U_0 = time independent effective stress ratio, σ_0 = yield strength, σ_c = creep strength, m = creep exponent S_{max} = maximum stress, t_h = hold period in tension
68, 69	Thermomechanical: intergranular, cracking	$D^{creep}=\phi^{creep}\int_0^{t_c}Ae^{(-\Delta H/RT)}\left(\frac{\alpha_1\bar{\sigma}+\alpha_2\sigma_H}{K}\right)^m dt$ (Terms defined in the text)
118	TMF life model	$N_f=C_0(\Delta\varepsilon)^{C_1}(t_h)^{C_2}\exp\left(\frac{C_3}{A}\right)$
209	Microcrack propagation in TMF	$\frac{da}{dN}=C_f(\Delta J)^{m_f}+C_c\hat{C}^{m_c}+C_0(\Delta J)^{m_0}(\Delta t)^{\Psi}$ (Summation of crack growth from fatigue, creep, and oxidation contributions)

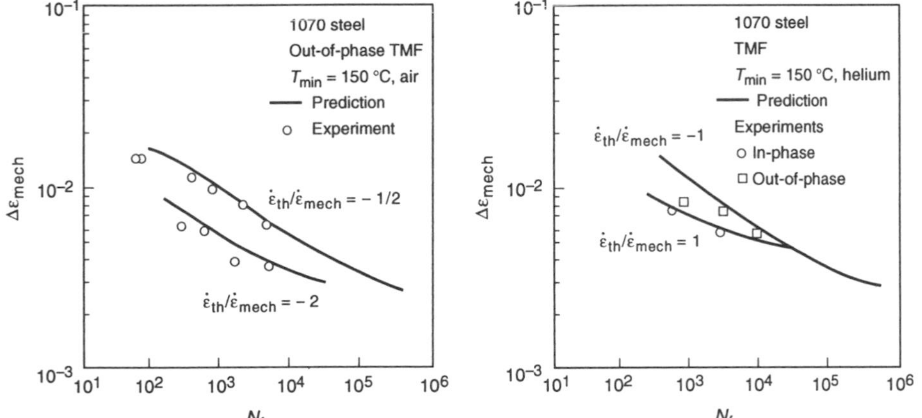

Fig. 37 Thermomechanical fatigue OP life prediction for steels under $\dot{\varepsilon}_{th}/\dot{\varepsilon}_{mech}=-1/2$ and $\dot{\varepsilon}_{th}/\dot{\varepsilon}_{mech}=-2$ conditions. Source: Ref 68, 69

Several researchers considered creep-fatigue to be a crack propagation-controlled problem where the creep mechanism is assumed to influence the fatigue crack growth or vice versa. Majumdar and Maiya (Ref 83) considered the influence of creep cavity growth ahead of a crack growing by fatigue in their damage-rate equations. In their model, sintering of cavities occurs in compression, effectively reversing the creep damage occurring in tension. The model is suitable for materials that exhibit copious cavitation. These damage-rate equations have recently been applied to thermomechanical fatigue loadings (Ref 191).

The model of Saxena and Bassani (Ref 207) accounts for cavity formation ahead of a crack tip and modified the crack tip stress fields; therefore, modified fracture mechanics parameters have been used to handle fatigue-creep crack growth. In the absence of cavities, the crack tip stress-strain fields are still modified; this is the basis for a modified crack closure model proposed by Sehitoglu and Sun (Ref 208). The model by Neu and Sehitoglu (Ref 68, 69) incorporates effective stress and hydrostatic stress components and a creep-phasing factor that depends on the thermomechanical history.

By performing tests in a helium environment, oxidation damage is eliminated leaving only the fatigue-creep damage. This allows the constants in the fatigue-creep damage term to be formulated directly.

In the Neu-Sehitoglu model (Ref 68, 69), to characterize the oxidation rate at different partial pressures of oxygen, the $(PO_2)^{1/q}$ term should be placed on the right-hand side of the growth law where q is a constant and PO_2 is the partial pressure of oxygen in the testing environment. In general, K_p will not be constant for a cycle that undergoes a varying temperature history. Therefore, an effective oxidation constant, K_p^{eff} is defined. The growth of voids and intergranular cracks occurs predominantly under tensile loading. Consequently, to take into account the load asymmetry, the creep damage term is a function of effective and hydrostatic stress components (see Ref 68, 69, and 105 for details). The constitutive model proposed by Slavik and Sehitoglu (Ref 80), which utilizes two state variables, was used in the simulations, but other constitutive models can also be used. Table 6 does not show some of the commonly used life prediction models including the $\sigma_{max}\Delta\varepsilon^{in.}/2$, a quantity related to tensile hysteresis energy, and the ΔW_p, plastic work or the total hysteresis energy parameters. As pointed out by Halford (Ref 18) the use of these parameters in TMF places rather unrealistic restrictions on the flow (σ–ε) and failure behavior of existing structural materials.

Prediction of TMF OP lives for 1070 steel are shown in Fig. 37. The minimum temperature is maintained at 150 °C (300 °F) in these experiments and the mechanical strain increases proportionally with maximum temperature with $\dot{\varepsilon}_{th}/\dot{\varepsilon}_{mech}=-1/2$ in the first case and $\dot{\varepsilon}_{th}/\dot{\varepsilon}_{mech}=-2$ in the second case. The predictions with the model are shown as solid lines. The $\dot{\varepsilon}_{th}/\dot{\varepsilon}_{mech}=-2$ is termed *partial constraint* and the case $\dot{\varepsilon}_{th}/\dot{\varepsilon}_{mech}=-1/2$ is termed *overconstraint* (Ref 24, 68, 69).

Constitutive Equations Suitable for TMF

Recent research on high-temperature deformation has produced a considerable number of constitutive models. Two classes of constitutive models suitable for thermomechanical loading have been reviewed by Slavik and Sehitoglu (Ref 81).

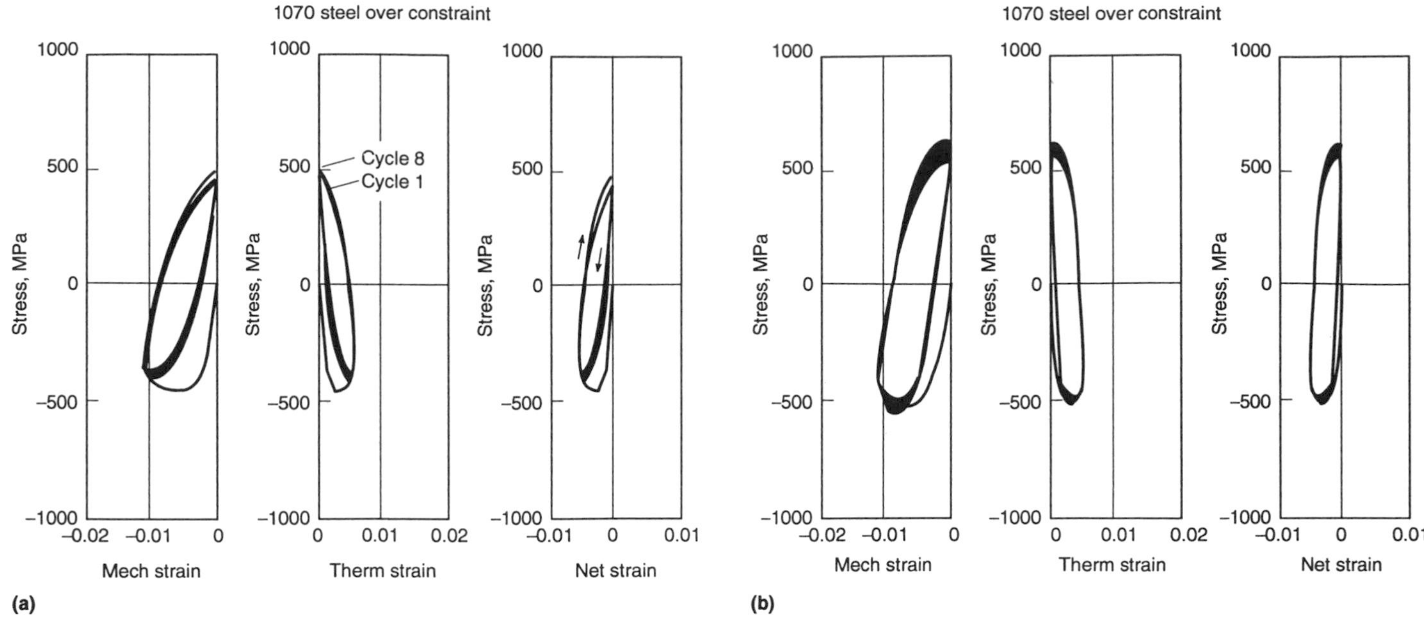

Fig. 38 Thermomechanical fatigue (out-of-phase) stress strain of 1070 steel. (a) Experimental. (b) Prediction using nonunified equations. Source: Ref 81

Table 7 Nonunified plasticity model

Elastic strain rate	$\dot{\varepsilon}^{e}_{ij} = ((1-\nu)\,\dot{\sigma}_{ij} - \nu\,\dot{\sigma}_{kk}\,\delta_{ij})/E - ((1-\nu)\sigma_{ij} - \nu\,\sigma_{kk}\,\delta_{ij})\dfrac{\partial E}{\partial T}\cdot\dfrac{\dot{T}}{E^2}$
Thermal strain rate	$\dot{\varepsilon}^{th}_{ij} = \theta\,\dot{T}\,\delta_{ij}$
Yield criteria	$f = \frac{1}{2}\left(S_{ij} - S^{c}_{ij}\right)\left(S_{ij} - S^{c}_{ij}\right) - k^2 = 0$
Plastic strain rate	$\dot{\varepsilon}^{p}_{ij} = H\dfrac{\partial f}{\partial S_{ij}}\left[\dfrac{\partial f}{\partial S_{kl}}\dot{S}_{kl} + \dfrac{\partial f}{\partial T}\dot{T}\right]$
Creep strain rate	$\dot{\varepsilon}^{c}_{ij} = B(\bar{\sigma})^{S-1}\,S_{ij}\exp\,(-\Delta H/RT)$
Plastic modulus	$H = A'\,J_2^{N/2}/(1 \pm \beta(\exp - p/p_0))$
Accumulated plastic strain	$p = \int_t \left(\frac{2}{3}\,\dot{\varepsilon}^{p}_{ij}\,\dot{\varepsilon}^{p}_{ij}\right)^{1/2} dt$

Source: Ref 81

Table 8 Two unified models used for TMF $\sigma - \varepsilon$ prediction

Ref	Flow rule	Back stress	Drag stress
80	$\dot{\varepsilon}^{in}_{ij} = \frac{3}{2} f(\bar{\sigma}/K)\dfrac{S_{ij} - S^{c}_{ij}}{\bar{\sigma}}$ $f\left(\dfrac{\bar{\sigma}}{K}\right) = \begin{cases} A\left(\dfrac{\bar{\sigma}}{K}\right)^{n} & \dfrac{\bar{\sigma}}{K} \le 1 \\ A\exp\left[\left(\dfrac{\bar{\sigma}}{K}\right)^{m} - 1\right] & \dfrac{\bar{\sigma}}{K} > 1 \end{cases}$	$\dot{S}^{c}_{ij} = \frac{2}{3} h_{\alpha}\dot{\varepsilon}^{in}_{ij} - r_{\alpha}S^{c}_{ij}$ $h_k = B(K_{sat} - K)\,\dot{\bar{\varepsilon}}^{in.}$ $r_k = C(K - K_{rec})$	$\dot{K} = h_k - r_k + \Theta\dot{T}$
211, 212	$\dot{\varepsilon}^{in}_{ij} = f\left(\dfrac{K^2}{J_2}\right)S_{ij}$ $f\left(\dfrac{K^2}{J_2}\right) = \left(\dfrac{D_0^2}{J_2}\exp\left[-\left(\dfrac{n+1}{n}\right)\left[\dfrac{K^2}{3J_2}\right]^{n}\right]\right)^{0.5}$		$\dot{K} = m(K_1 - K)\cdot\dot{W}^{in.} - A\left(\dfrac{K - K_i}{K_1^{*}}\right)^{r}\exp\left(\dfrac{-\Delta H}{RT}\right)$

Nonunified Creep Plasticity Models and Their Use in Thermal Loading. In the first class of models a time-dependent creep strain is added to the plastic strain resulting in the so-called nonunified models. The plastic strain can be described by the classical von Mises yield criteria and Prager or Ziegler rules. In the paper by Slavik and Sehitoglu (Ref 81), the Drucker-Palgen model (Ref 210) was used. The different strain-rate components and other important parameters are given in Table 7.

In Table 7 the first row gives the elastic strain rate of an isotopic material where the elastic modulus is a function of temperature. The second row is the thermal strain rate for an isotropic material. The yield criteria is the von Mises type where k is the yield stress in shear and the center of the yield surface is permitted to move in stress space. The equations for this translation are given in Ref 81. The plastic strain rate is normal to the yield surface and also changes with temperature as shown in Table 7. The creep strain rate is expressed as a function of equivalent stress and the creep rate is in the same direction as the deviatoric stress rate. The plastic modulus, the slope of the stress-plastic strain rate, varies with plastic strain history such that it exponentially approaches a steady-state value. The plastic strain history is described by the accumulated plastic strain which is a scalar positive quantity.

The advantages of this model are the following: (1) it can use the existing database of plastic and creep properties, (2) it conforms to existing FEM codes, and (3) it does not require special integration schemes. Experimental results and predictions of TMF OP behavior of 1070 steel with the above equations are given in Fig. 38(a) and (b) (Ref 81), respectively. In this case, 150 ⇄ 450 °C (300 ⇄ 840 °F) cycling under "overconstraint" ($\dot{\varepsilon}_{mech} = -2, \dot{\varepsilon}_{th}$) is shown. "Overconstraint" means that the mechanical strain is larger than the thermal strain because during heating the material is also subjected to a negative net strain as shown in Fig. 38(a). The prediction of the stress-strain response with the nonunified model is satisfactory (Fig. 38b).

Unified Creep Plasticity Models and Their Use in Thermal Loading. These models, termed *unified models*, have the potential to predict

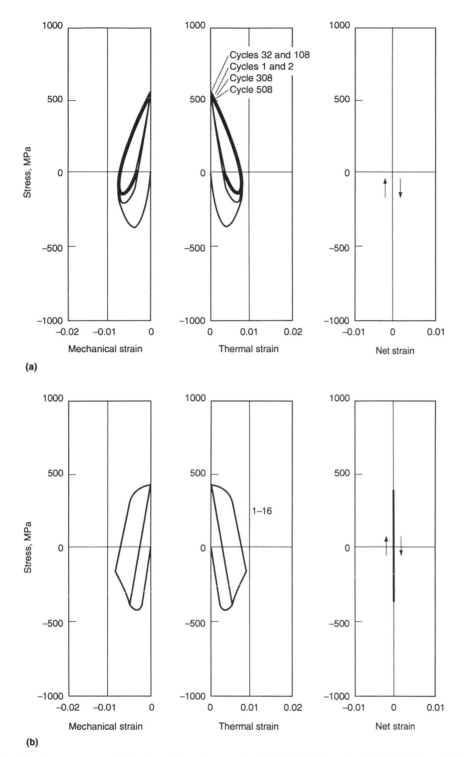

Fig. 39 Thermomechanical fatigue (out-of-phase) stress strain of 1070 steel. (a) Experimental. (b) Prediction using Bodner's model. Source: Ref 81

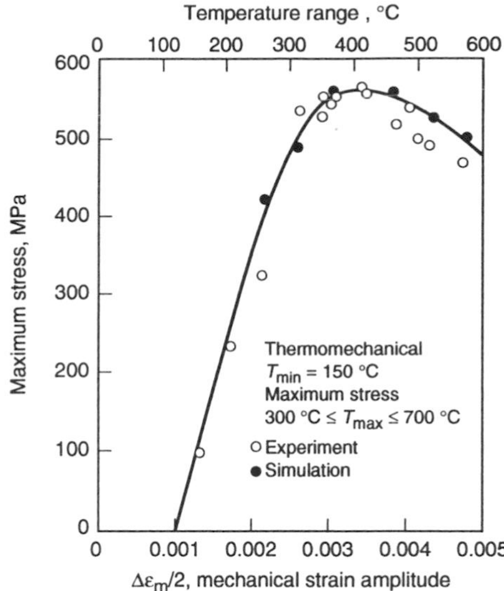

Fig. 40 Prediction of maximum stress in the TMF OP case in 1070 steel. Source: Ref 80

creep-plasticity interactions and strain rate effects more accurately than the nonunified models. A unified model that has been widely used was proposed by Bodner and Partom (Ref 211) and later modified for cyclic loading (Ref 212). The simulations of TMF OP with this model are given in Fig. 39 (Ref 81). Comparison of experimental results shown in Fig. 39(a) and predictions shown in Fig. 39(b) prove that the model is satisfactory. In Fig. 39(a) the stress-net strain response is shown under 150 ⇄ 600 °C (300 ⇄ 1110 °F) and total constraint ($\dot{\varepsilon}_{mech} = -\dot{\varepsilon}_{th}$) conditions. In this case, the net strain is zero and the mechanical strain is equal but opposite of thermal strain.

In the unified models, there is no yield surface assumption. Inelastic strain rates are permitted at small levels of effective stress which is usually the deviatoric stress-deviatoric backstress. The unified models are often composed of two state variables, the deviatoric back stress, S^c_{ij} and the drag stress, K. The back stress can be used to predict the Bauschinger effect in room-temperature loading and also the transient and steady-state creep response at high temperatures. The drag stress, K, accounts for cyclic hardening or softening, and the influence of plasticity on creep or vice versa. The strain rate sensitivity is determined by the flow rule. The general form of these unified relations is given in Table 8. We note that Bodner's original form did not include a back stress.

The first equation in Table 8 is the flow rule where $\dot{\varepsilon}^{in}_{ij}$ is the inelastic strain rate (which is the combination of plastic and creep strains), $\overline{\sigma}$ is the effective stress, S_{ij} is the deviatoric stress, S^c_{ij} is the deviatoric back stress, and K is the drag stress. The second equation describes the evolution of back (internal) stress where h_α is the hardening function for the back stress and r_α is the recovery function for the back stress. The third equation depicts the evolution of drag stress where h_k is the hardening function for the drag stress, r_k is the recovery function for the drag stress, and Θ is the thermally activated drag stress change term (defined as $\partial K_o/\partial T$). The term that distinguishes different models is the choice of the flow rule, $f(\overline{\sigma}/K)$ and the manner in which the hardening and recovery functions are determined. Different deformation mechanisms (plasticity, power law creep, diffusional creep) have been identified in deformation mechanism maps (Ref 47) but have not been explicitly considered. Sehitoglu and Slavik confirmed that at high strain rates the strain rate effective stress relation has the form of the exponential function (i.e., rate insensitive behavior) while at lower strain rates the relation between inelastic strain rate and effective stress is consistent with the power law creep relation. An experimentally based unified constitutive model had been proposed earlier by Sehitoglu and Slavik

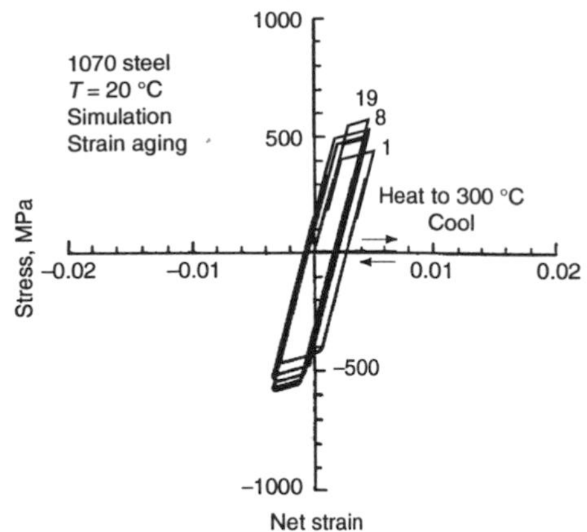

Fig. 41 Prediction of 1070 steel response under strain aging conditions. Source: Ref 80

Table 9 Constants for the unified model for selected materials

Material	E, MPa	θ, 1/°C	A, s^{-1}	n	m
1070 steel (Ref 80)	$202{,}250 - 31.0\,T$ $T \le 440$ °C $309{,}990 - 275.7\,T$ $T > 440$ °C	1.7×10^{-5}	$4.0 \times 10^{9} \exp[-25{,}300/(T + 273)]$	5.4	8.3
René 80 (Ref 215)	$192{,}170 - 60.7\,T$ $T < 871$ °C $310{,}990 - 197.1\,T$ $T \ge 871$ °C	1.6×10^{-5}(a)	2.33×10^{-10}, $T \le 650$ °C $8.14 \times 10^{15} \exp[-54{,}288/(T + 273)]$, $T > 650$ °C	8.06	5.69
Mar-M247 (Ref 105)	$253{,}900 - 107.8\,T$	1.6×10^{-5}	$1.33 \times 10^{23} \exp[-64{,}515/(T + 273)]$	11.6	17.5
Al2xxx-T4 (Ref 96, 164)	$72{,}750 - 50\,T$ $T < 150$ °C $82{,}000 - 90\,T$ $T \ge 150$ °C	3.0×10^{-5}	$9.8 \times 10^{11} \exp[-18{,}722/(T + 273)]$	4.6	10.1
Ti-β21S (Ref 214)	$114{,}000 - 42.3\,T$ $T \le 483$ °C $151{,}500 - 120\,T$ $T > 483$ °C	8.24×10^{-6} $+ 3.64 \times 10^{-9}\,T$	$7.7 \times 10^{5} \exp[-18{,}300(T + 273)]$	2.7	8.2

(a) Not determined from experiments

Table 10 Constants for the unified model for the drag stress term

Material	K_0, MPa	K_{sat}, MPa	B	C	K_{rec}, MPa
1070 steel	$262.7 - 0.04T$, $T \le 440$ °C $403.0 - 0.36T$, $T > 440$ °C	$256.0 + 1.4 \times 10^{-3} T^2$, $T \le 304$ °C $568.0 - 0.6T$, $T > 304$ °C	5.0	0, $T < 300$ °C $10^8 \exp[-20{,}000/(T + 273)]$, $T \ge 300$ °C	$548 - 0.62\,T$
René 80	384.0, $T \le 60$ °C 2.66×10^{-3} *E, $T > 760$ °C	NA	0	0	NA
Mar-M247	N/A	$886.1 - 0.376\,T$	NA(a)	0	NA
Al2xxx – T4	$226 - 0.15T$ $T < 150$ °C $256 - 0.28T$ $T \le 150$ °C	$620 - 1.66T$, $T < 150$ °C $420 - 0.30T$, $T \ge 150$ °C	5	$4.9 \times 10 - 10 + 4.0 \times 10^{-6} T - 3.2 \times 10^{-8} T^2 + 1.2 \times 10^{-10} T^3 - 2.3 \times 10^{-13} T^4 + 1.7 \times 10^{-16} T^5$	$260 - 0.8T$, $T < 150$ °C 20 $T \ge 150$ °C
Ti-β21S	$730 - 0.271T$ $T \le 483$ °C $970 - 0.768T$ $T > 483$ °C	$506.5 - 0.291T$	5(b)	0	NA

(a) $K = K_{sat}$ in this study. Constants were not determined for transient behavior. (b) Not determined from experiments

Table 11 Constants for the unified model for the back stress evolution term

Material	h_α	b	r_α
1070 steel	$b(17{,}200 - 20\bar{\alpha})$ $\alpha_{ij}\,\dot{\varepsilon}_{ij}^{in} \ge 0$ $b(17{,}200)$ $\alpha_{ij}\,\dot{\varepsilon}_{ij}^{in} < 0$	$0.89 + 2.58 \times 10^{-3} T + 1.53 \times 10^{-5} T^2 - 6.4 \times 10^{-8} T^3 + 5.15 \times 10^{-11} T^4$	$128.6 \exp[-19{,}460/(T + 273)]\,(\bar{\alpha})^{3.3}$
René 80	$(2.096 \times 10^{-5} + 1.50 \times 10^{-15}\,\lvert\bar{\alpha}\rvert^{b})^{-1}$	$5.02 + 2.25 \times 10^{-4} T$, $T \le 760$ °C $- 109.58 + 0.414T - 4.99 \times 10^{-4} T^2 + 2.005 \times 10^{-7} T^3$, $T > 760$ °C	0
Mar-M247	Not determined	Not determined	Not determined
Al2xxx-T4	$b(20{,}000 - 1000\bar{\alpha})$ $\alpha_{ij}\,\dot{\varepsilon}_{ij}^{in}$ $b(20{,}000)$ $\alpha_{ij}\,\dot{\varepsilon}_{ij}^{in} < 0$	$1.013 - 8.8 \times 10^{-4} T - 2.7 \times 10^{-6} T^2$	0
Ti-β21S	$b(14{,}500 - 422\,\bar{\alpha} + 3.0\,\bar{\alpha}^2)$ $\alpha_{ij}\,\dot{\varepsilon}_{ij}^{in} \ge 0$ $b(14{,}500)$ $\alpha_{ij}\,\dot{\varepsilon}_{ij}^{in} < 0$	$1 + 0.0125\,(T - 23) - 3.38 \times 10^{-5}\,(T - 23)^2 + 2.09 \times 10^{-8}\,(T - 23)^3$	0

(Ref 80, 81). A list of functions and corresponding experiments to determine these constants follow:

Function	Experimental determination
Flow rule, f	High and low strain rates, yield strength measurements in tensile or compressive monotonic tests
Hardening of back stress	High strain rate (room and high temperature)
Recovery of back stress	Low strain rate or creep (high temperature)
Hardening of drag stress	High strain rate cycling (room and high temperature)
Recovery of drag stress	Rest periods (high temperature)
Θ	Change of K_0 with temperature

The material constants for the Slavik-Sehitoglu model for different classes of materials are listed in Tables 9 to 11. These are 1070 steel (Ref 80), Mar-M247 (Ref 105), René 80 (Ref 213), Al2xxx-T4 (Ref 96, 164) and Ti-β21S (Ref 214). These tables are also explained in Ref 80 and summarized in Ref 215.

Predictions of material response utilizing this relation are given in Fig. 40 (Ref 80). The experimental results were shown earlier in Fig. 16. In these experiments (total constraint TMF OP), the minimum temperature was constant at 150 °C (300 °F). Because of thermal recovery effects, the drag stress decreases upon exposure to high temperatures and this results in loss of strength at the minimum temperature. Note that the model (solid rectangular points) predicts the decrease in maximum stress with increasing mechanical strain very accurately. Because of the thermal recovery effects, the maximum stress in the cycle decreases with increasing mechanical strain range. The agreement between experiment and prediction is excellent. It is possible to incorporate additional terms to predict such phenomena such as strain aging and dynamic recovery into the constitutive models. Other TMF and IF simulations with this unified model can be found in Ref 80, 96, 105, 164, and 213. Simulations for the case of strain aging for the 1070 steel have been demonstrated in Ref 80 and are further discussed below.

Certain temperature-strain histories in solute-hardened materials produce strain aging, hence strengthening of the material. Thermomechanical fatigue studies (under strain aging conditions for steels and nickel-base superalloys) have been discussed in Ref 60 and 215. Strengthening is caused by interstitial solute atoms which anchor the dislocation motion. If the pinning of the dislocations occurs during deformation, the term "dynamic strain aging" is used (Ref 87). If the aging occurs under a constant load (after some plastic deformation), it is called "static strain aging" (Ref 87).

Static strain aging experiments were conducted on 1070 steel. The material is cycled at 20 °C (70 °F) but is exposed to the aging temperature time at zero stress every reversal. A schematic of the strain-temperature history and the results for 1070 steel was shown earlier in Fig. 17. The predictions using the unified formulation are given in Fig. 41. The prediction of the strengthening under these conditions is very favorable. The details can be found in Ref 80 and 215.

Finally, it should be noted that other unified models have been tested under TMF loadings but not as extensively as the Slavik and Sehitoglu model. For example, Miller (Ref 216) proposed a unified model which he checked against experimental TMF IP data on PWA 663. His predictions were satisfactory. Walker (Ref 217) proposed a model for high-temperature/time-dependent loadings that was also applied to predict TMF behavior.

ACKNOWLEDGMENT

The information in this article is largely taken from H. Sehitoglu, Thermal and Thermomechanical Fatigue of Structural Alloys, *Fatigue and Fracture,* Vol 19, *ASM Handbook,* 1996, p 527–556.

REFERENCES

1. H.J. Schrader, The Friction of Railway Brake Shoes at High Speed and High Pressure, *Eng. Exp. Station Bull., Univ. Ill. Bull.*, Vol 35 (No. 72), May 1938
2. H.R. Wetenkamp, O.M. Sidebottom, and H.J. Schrader, The Effect of Brake Shoe Action on Thermal Cracking and on Failure of Wrought Steel Railway Car Wheels, *Eng. Exp. Station Bull., Univ. Ill. Bull.*, Vol 47 (No. 77), June 1950
3. W. Boas and R.W.K. Honeycombe, The Deformation of Tin-Base Bearing Alloys by Heating and Cooling, *Inst. Met. J.*, Vol 73, 1946–1947, p 33–444
4. L.F. Coffin, Jr., A Study of the Effects of Cyclic Thermal Stresses on a Ductile Metal, *Trans. ASME*, Vol 76 (No. 6), 1954, p 931–950
5. S. Manson, Behavior of Materials under Conditions of Thermal Stress, *Heat Transfer Symp., Univ. Mich. Eng. Res. Inst.*, Vol 27–38, 1953: see also NACA TN-2933, 1953
6. S.S. Manson, *Thermal Stress and Low-Cycle Fatigue*, McGraw-Hill, 1966
7. *Thermal and High-Strain Fatigue*, Monograph and Report Series No. 32, Institute of Metals, London, 1967
8. D.J. Littler, Ed., *Thermal Stresses and Thermal Fatigue*, Butterworths, London, 1971
9. R.P. Skelton, *Fatigue at High Temperature*, Applied Science Publishers, London, 1983
10. R.P. Skelton, *High Temperature Fatigue: Properties and Prediction*, Applied Science Publishers, London, 1983
11. A. Weronski and T. Hejwoski, *Thermal Fatigue of Metals*, Marcel Dekker, 1991
12. A.E. Carden, A.J. McEvily, and C.H. Wells, Ed., *Fatigue at Elevated Temperatures*, STP 520, ASTM, 1973
13. D.A. Spera and D.F. Mowbray, Ed., *Thermal Fatigue of Materials and Components*, STP 612, ASTM, 1976
14. H. Solomon, G. Halford, L. Kaisand, and B. Leis, Ed., *Low Cycle Fatigue,* STP 942, ASTM, 1988
15. H. Sehitoglu, Ed., *Thermo-Mechanical Fatigue Behavior of Materials*, STP 1186, ASTM, 1991
16. H. Sehitoglu and S.Y. Zamrik, Ed., *Thermal Stress, Material Deformation and Thermo-Mechanical Fatigue*, American Society of Mechanical Engineers, 1987
17. J. Bressers and L. Rémy, Ed., *Symp. Fatigue under Thermal and Mechanical Loading*, Kluwer Academic Publishers, May 1995
18. G. Halford, Low-Cycle Thermal Fatigue, TM 87225, National Aeronautics and Space Administration, 1986; see also Low-Cycle Thermal Fatigue, *Thermal Stress II*, R.B. Hetnarski, Ed., Elsevier, 1987
19. D.A. Miller and R.H. Priest, Material Response to Thermal-Mechanical Strain Cycling, *High Temperature Fatigue: Properties and Prediction*, R.P. Skelton, Ed., Applied Science Publishers, 1983, p 113–176
20. H. Sehitoglu, *Thermo-Mechanical Fatigue Life Prediction Methods*, STP 1122, ASTM, 1992, p 47–76
21. H.G. Baron, Thermal Shock and Thermal Fatigue, *Thermal Stress*, P.P. Benham and R.D. Hoyle, Ed., Pitman, London, 1964, p 182–206
22. D.R. Miller, Thermal-Stress Ratchet Mechanism in Pressure Vessels, *J. Basic Eng. (Trans. ASME)*, Vol 81, No. 2, 1959, p 190–196
23. J. Bree, Elastic-Plastic Behavior of Thin Tubes Subjected to Internal Pressure and Intermittent Heat Fluxes with Application to Fast-Nuclear-Reactor Fuel Elements, *J. Strain Anal.*, Vol 2 (No. 3), 1967, p 226–238
24. H. Sehitoglu, Constraint Effect in Thermo-Mechanical Fatigue, *J. Eng. Mater. Technol. (Trans. ASME)*, Vol 107, 1985, p 221–226
25. L.F. Coffin, Jr. and R.P. Wesley, Apparatus for Study of Effects of Cyclic Thermal Stresses on Ductile Metals, *Trans. ASME*, Vol 76 (No. 6), Aug 1954, p 923–930
26. A. Carden, Ed., *Thermal Fatigue Evaluation*, STP 465, ASTM, 1970, p 163–188
27. H.J. Westwood, High Temperature Fatigue of 304 Stainless Steel under Isothermal and Thermal Cycling Conditions, *Fracture 77: Advances in Research on the Strength and Fracture of Materials*, D.M.R. Taplin, Ed., Pergamon Press, 1978, p 755–765
28. K.D. Sheffler, Vacuum Thermal-Mechanical Fatigue Testing of Two Iron-Base High Temperature Alloys, *Thermal Fatigue of Materials and Components*, STP 612, D.A. Spera and D.F. Mowbray, Ed., 1976, p 214–226
29. M. Kawamoto, T. Tanaka, and H. Nakajima, Effect of Several Factors on Thermal Fatigue, *J. Mater.*, Vol 1 (No. 4), 1966, p 719–758
30. G. Halford, M.A. McGaw, R.C. Bill, and P. Fanti, Bithermal Fatigue: A Link between Isothermal and Thermomechanical Fatigue, *Low Cycle Fatigue*, STP 942, H. Solomon, G. Halford, L. Kaisand, and B. Leis, Ed., ASTM, 1988
31. T. Udoguchi and T. Wada, Thermal Effect on Low-Cycle Fatigue Strength of Steels, *Thermal Stresses and Thermal Fatigue,* D.J. Littler, Ed., Butterworths, London, 1971, p 109–123
32. M.G. Castelli and J.R. Ellis, Improved Techniques for Thermo-Mechanical Testing in Support of Deformation Modeling, *Thermo-Mechanical Fatigue Behavior of Materials*, STP 1186, H. Sehitoglu, Ed., ASTM, 1991, p 195–211
33. M. Karasek, H. Sehitoglu, and D. Slavik, Deformation and Damage under Thermal Loading, *Low Cycle Fatigue*, STP 942, H. Solomon, G. Halford, L. Kaisand, and B. Leis, Ed., ASTM, 1988, p 184–205
34. T.R. Simovich, "A Study of the Thermal Fatigue Characteristics of Several Plain Carbon Steels," M.S. thesis, University of Illinois, 1967
35. S. Taira, Relationship between Thermal Fatigue and Low-Cycle Fatigue at Elevated Temperature, *Fatigue at Elevated Temperatures*, STP 520, A.E. Carden, A.J. McEvily, and C.H. Wells, Ed., ASTM, 1973, p 80–101
36. S. Taira, M. Fujino, and R. Ohtani, Collaborative Study on Thermal Fatigue of Properties of High Temperature Alloys in Japan, *Fatigue Eng. Mater. Struct.*, Vol 1, 1979, p 495–508
37. S. Taira, M. Fujino, and S. Maruyama, Effects of Temperature and the Phase between Temperature and Strain on Crack Propagation in a Low Carbon Steel during Thermal Fatigue, *Mechanical Behavior of Materials*, Society of Materials Science, Kyoto, 1974, p 515–524
38. H.G. Baron and B.S. Bloomfield, Resistance to Thermal Stress Fatigue of Some Steels, Heat Resisting Alloys and Cast Irons, *J. Iron Steel Inst.*, Vol 197, 1961, p 223–232
39. D.R. Adolphson, "Stresses Developed in Uniaxially Restrained Railway Car Wheel Material When Subjected to Thermal Cycles," M.S. thesis, University of Illinois, 1957
40. K. Kuwabara and A. Nitta, Thermal-Mechanical Low Cycle Fatigue under Creep-Fatigue Interaction on Type 304 Stainless Steel, *Proc. ICM 3*, Vol 2, 1979, p 69–78
41. K. Kuwabara and A. Nitta, Effect of Strain Hold Time of High Temperature on Thermal Fatigue of Type 304 Stainless Steel, *ASME-MPC Symp. Creep-Fatigue Interaction*, American Society of Mechanical Engineers, 1976, p 161–177
42. K. Kuwabara, A. Nitta, and T. Kitamura, Thermal Mechanical Fatigue Life Prediction in High Temperature Component Materials for Power Plant, *ASME Int. Conf. Advances in Life Prediction*, D.A. Woodford and J.R. Whitehead, Ed., American Society of Mechanical Engineers, 1983, p 131–141
43. K. Kuwabara and A. Nitta, "Isothermal and Thermal Fatigue Strength of Cr-Mo-V Steel for Turbine Rotors," Report E277005, Central Research Institute of Electric Power Industry, Tokyo, 1977
44. A. Nitta, K. Kuwabara, and T. Kitamura, "Creep-Fatigue Damage in Power Plant Materials, Part 2: The Behavior of Elevated Tem-

perature Low Cycle Fatigue Crack Initiation and Propagation in Steam Turbine Rotor Steels," Report E282002, Central Research Institute of Electric Power Industry, Tokyo, 1982

45. H. Sehitoglu and M. Karasek, Observations of Material Behavior under Isothermal and Thermo-Mechanical Loading, *J. Eng. Mater. Technol. (Trans. ASME)*, Vol 108 (No. 2), 1986, p 192–198
46. G. Hartman, III, A Thermal Control System for Thermal Cycling, *J. Test. Eval.*, Vol 13 (No. 5), 1985, p 363–366
47. A. Koster, E. Chataigner, G. Cailletaud, and L. Rémy, Development of a Thermal Fatigue Facility to Simulate the Behavior of Superalloy Components, *Symp. Fatigue under Thermal and Mechanical Loading*, J. Bressers and L. Rémy, Ed., European Commission, Petten, May 1995
48. J. Guedou and Y. Honnorat, Thermo-Mechanical Fatigue of Turbo-Engine Blade Superalloys, *Thermo-Mechanical Fatigue Behavior of Materials*, STP 1186, H. Sehitoglu, Ed., ASTM, 1991, p 157–175
49. E. Glenny, J.E. Nortwood, S.W.K. Shaw, and T.A. Taylor, A Technique for Thermal-Shock and Thermal-Fatigue Testing, based on the Use of Fluidized Solids, *J. Inst. Met.*, Vol 87, 1958-59, p 294–302
50. D.A. Woodford and D.F. Mowbray, Effect of Material Characteristics and Test Variables on Thermal Fatigue of Cast Superalloys, *Mater. Sci. Eng.*, Vol 16, 1974, p 5–43
51. P.T. Bizon and D.A. Spera, Thermal-Stress Fatigue Behavior of Twenty-Six Superalloys, *Thermal Fatigue of Materials and Components*, STP 612, D.A. Spera and D.F. Mowbray, Ed., ASTM, 1976, p 106–122
52. M.A.H. Howes, Evaluation of Thermal Fatigue Resistance of Metals Using the Fluidized Bed Technique, *Fatigue at Elevated Temperatures*, STP 520, A.E. Carden, A.J. McEvily, and C.H. Wells, Ed., ASTM, 1973, p 242–254
53. F.K. Lampson, I.C. Isareff, Jr., and A.W.F. Green, Thermal Shock Testing under Stress of Certain High Temperature Alloys, *Proc. ASTM*, Vol 57, 1957, p 965–976
54. P.G. Forrest and K.B. Armstrong, The Thermal Fatigue Resistance of Nickel-Chromium Alloys, *Proc. Inst. Mech. Eng.*, 1963, p 3.1–3.7
55. D.F. Mowbray and J.E. McConnelee, Nonlinear Analysis of a Tapered Disk Specimen, *Thermal Fatigue of Materials and Components*, STP 612, D.A. Spera and D.F. Mowbray, Ed., ASTM, 1976, p 10–29
56. E. Glenny and T.A. Taylor, A Study of the Thermal Fatigue Behavior of Materials, *J. Inst. Met.*, Vol 88, 1959-60, p 449–461
57. L.F. Coffin, Instability Effects in Thermal Fatigue, *Thermal Fatigue of Materials and Components*, STP 612, D.A. Spera and D.F. Mowbray, Ed., ASTM, 1976, p 227–238
58. G.R. Halford and S.S. Manson, Life Prediction of Thermal-Mechanical Fatigue Using Strain Range Partitioning, *Thermal Fatigue of Materials and Components*, STP 612, D.A. Spera and D.F. Mowbray, Ed., ASTM, 1976, p 239–254
59. M. Karasek, H. Sehitoglu, and D. Slavik, Deformation and Damage under Thermal Loading, *Low Cycle Fatigue*, STP 942, H. Solomon, G. Halford, L. Kaisand, and B. Leis, Ed., ASTM, 1988, p 184–205
60. H. Sehitoglu, Crack Growth Studies under Selected Temperature-Strain Histories, *Eng. Fract. Mech.*, Vol 26 (No. 4), 1987, p 475–489
61. S. Taira and M. Ohnami, Fracture and Deformation of Metals Subjected to Thermal Cycling Combined with Mechanical Stress, *Joint Int. Conf. Creep*, Institution of Mechanical Engineers, 1963, p 57–62
62. S. Taira, M. Ohnami, and T. Kyogoku, Thermal Fatigue under Pulsating Thermal Stress Cycling, *Bull. Jpn. Soc. Mech. Eng.*, Vol 6, 1963, p 178–185
63. M. Fujino and S. Taira, Effect of Thermal Cycling on Low Cycle Fatigue Life of Steels and Grain Boundary Sliding Characteristics, *Proc. ICM 3*, Vol 2, 1979, p 49–58
64. S.S. Manson, G.R. Halford, and M.H. Hirschberg, Creep-Fatigue Analysis by Strain-Range Partitioning, *Design for Elevated Temperature Environment*, S.Y. Zamrik, Ed., American Society of Mechanical Engineers, 1971, p 12–28
65. J.J. Westwood and W.K. Lee, Creep-Fatigue Crack Initiation in 1/2 Cr-Mo-V Steel, *Proc. Int. Conf. Creep. Fract. Eng. Mater. Struct.*, Pineridge Press, 1981, p 517–530
66. C.E. Jaske, Thermal-Mechanical, Low-Cycle Fatigue of AISI 1010 Steel, *Thermal Fatigue of Materials and Components*, STP 612, D.A. Spera and D.F. Mowbray, Ed., ASTM, 1976, p 170–198
67. J.S. Laub, Some Thermal Fatigue Characteristics of Mild Steel for Heat Exchangers, *Thermal Fatigue of Materials and Components*, STP 612, D.A. Spera and D.F. Mowbray, Ed., ASTM, 1976, p 141–156
68. R. Neu and H. Sehitoglu, Thermo-Mechanical Fatigue Oxidation, Creep, Part I: Experiments, *Metall. Trans. A*, Vol 20A, 1989, p 1755–1767
69. R. Neu and H. Sehitoglu, Thermo-Mechanical Fatigue Oxidation, Creep, Part II: Life Prediction, *Metall. Trans. A*, Vol 20A, 1989, p 1769–1783
70. R. Zauter, F. Petry, H.-J. Christ, and H. Mughrabi, Thermo-Mechanical Fatigue of the Austenitic Stainless Steel AISI 304L, *Thermo-Mechanical Fatigue Behavior of Materials*, STP 1186, H. Sehitoglu, Ed., ASTM, 1993, p 70–90
71. L.F. Coffin, Jr., Cyclic Strain-Induced Oxidation of High Temperature Alloys, *Trans. ASM*, Vol 56, 1963, p 339–344
72. L.F. Coffin, Fr., The Effect of High Vacuum on the Low Cycle Fatigue Law, *Metall. Trans.*, Vol 3, July 1972, p 1777–1788
73. J.K. Tien and J.M. Davidson, Oxide Spallation Mechanisms, *Stress Effects and the Oxidation of Metals*, J.V. Cathcart, Ed., TMS-AIME, 1974, p 200–219
74. N.B. Pilling and R.E. Bedworth, The Oxidation of Metals at High Temperatures, *J. Inst. Met.*, Vol 29, 1923, p 529–582
75. M. Schutze, Deformation and Cracking Behavior of Protective Oxide Scales on Heat-Resistant Steels under Tensile Strain, *Oxid. Met.*, Vol 24 (No. 3/4), 1985, p 199–232
76. R.P. Skelton, Environmental Crack Growth in 0.5 Cr-Mo-V Steel During Isothermal High Strain Fatigue and Temperature Cycling, *Mater. Sci. Eng.*, Vol 35. 1978, p 287–298
77. E. Renner, H. Vehoff, H. Riedel, and P. Neumann, Creep Fatigue of Steels in Various Environments, *Low Cycle Fatigue and Elasto-Plastic Behavior of Materials*, 2nd Int. Conf. Low Cycle Fatigue and Elasto-Plastic Behavior of Materials (Munich), Elsevier, 1987, p 277–283
78. P.S. Maiya, Effects of Wave Shape and Ultrahigh Vacuum on Elevated Temperature Low Cycle Fatigue in Type 304 Stainless Steel, *Mater. Sci. Eng.*, Vol 47, 1981, p 13–21
79. C.R. Brinkman, High-Temperature Time-Dependent Fatigue Behavior of Several Engineering Structural Alloys, *Int. Met. Rev.*, Vol 30 (No. 5), 1985, p 235–258
80. D. Slavik and H. Sehitoglu, A Constitutive Model for High Temperature Loading, Part I: Experimentally Based Forms of the Equations; Part II: Comparison of Simulations with Experiments, *Thermal Stress, Material Deformation, and Thermo-Mechanical Fatigue*, H. Sehitoglu and S.Y. Zamrik, Ed., American Society of Mechanical Engineers, 1987, p 65–82
81. D. Slavik and H. Sehitoglu, Constitutive Models Suitable for Thermal Loading, *J. Eng. Mater. Technol. (Trans. ASME)*, Vol 108 (No. 4), 1986, p 303–312
82. H. Sehitoglu, Changes in State Variables at Elevated Temperatures, *J. Eng. Mater. Technol. (Trans. ASME)*, Vol 111, 1989, p 192–203
83. S. Majumdar, Thermo-Mechanical Fatigue of Type 304 Stainless Steel, *Thermal Stress, Material Deformation, and Thermo-Mechanical Fatigue*, H. Sehitoglu and S.Y. Zamrik, Ed., American Society of Mechanical Engineers, 1987, p 31–36
84. K. Kuwabara and A. Nitta, Thermal-Mechanical Low-Cycle Fatigue under Creep-Fatigue Interaction on Type 304 Stainless Steels, *Fatigue Fract. Eng. Mater. Struct.*, Vol 2, 1979, p 293–304
85. D. Mikrut, Elevated Temperature Time Dependent Behavior of 0.7% Carbon Steels, M.S. thesis, University of Illinois, 1989
86. L. Nortcott and H.G. Baron, The Craze Cracking of Metals, *J. Iron Steel Inst.*, 1956, p 385–408
87. J.D. Baird, Strain Aging of Steel—A Critical Review, *Iron Steel*, Vol 36, 1963, p 186–192, 368–374, 400–405
88. S. Taira, Relationship between Thermal Fatigue and Low Cycle Fatigue at Elevated Temperature, *Fatigue at Elevated Temperatures*, STP 520, A.E. Carden, A.J. McEvily, and C.H. Wells, Ed., ASTM, 1973, p 80–101
89. S. Baik and R. Raj, Wedge Type Creep Damage in Low Cycle Fatigue, *Metall. Trans. A*, Vol 13A, 1982, p 1207–1214

90. M. Karayaka and H. Sehitoglu, Thermomechanical Fatigue of Particulate Reinforced Aluminum 2*xxx*-T4, *Metall. Trans. A*, Vol 22A, 1991, p 697–707
91. D.A. Meyn, The Nature of Fatigue Crack Propagation in Air and in Vacuum for 2024 Aluminum, *Trans. ASM*, Vol 61, 1968, p 52–61
92. C.Q. Bowles and J. Schijve, Crack Tip Geometry for Fatigue Cracks Grown in Air and in Vacuum, *Fatigue Mechanisms: Advances in Quantitative Measurement of Physical Damage*, STP 811, ASTM, 1983, p 400–426
93. M.J. Hordon, Fatigue Behavior of Aluminum in Vacuum, *Acta Metall.*, Vol 14, 1966, p 1173–1178
94. M. Karayaka and H. Sehitoglu, Thermomechanical Fatigue of Al-SiCp Composites, *Proc. 4th Conf. Fatigue and Fracture Thresholds*, Vol 3, Materials and Components Engineering Publications, 1990, p 1693–1698
95. M. Karayaka and H. Sehitoglu, Thermomechanical Fatigue of Particulate Reinforced Aluminum 2*xxx*-T4, *Metall. Trans. A*, Vol 22A, 1991, p 697–707
96. M. Karayaka and H. Sehitoglu, Thermo-mechanical Deformation Modeling of Al2*xxx*-T4 Composites, *Acta Metall.*, Vol 41 (No. 1), 1993, p 175–189
97. B. Flaig, K.H. Lang, D. Lohe, and E. Macherauch, Thermal-Mechanical Fatigue of the Cast Aluminum Alloy GK-AlSi10Mgwa, *Symp. Fatigue under Thermal and Mechanical Loading*, J. Bressers and L. Rémy, Ed., European Commission, Petten, May 1995
98. R.B. Gundlach, B. Ross, A. Hetke, S. Valterra, and J.F. Mojica, Thermal Fatigue Resistance of Hypoeutectoid Aluminum Silica Casting Alloys, *AFS Trans.*, Vol 141, 1994, p 205–223
99. S. Bhat and C. Laird, Cyclic Stress-Strain Response and Damage Mechanisms at High Temperature, *Fatigue Mechanisms*, STP 675, Jeffrey Fong, Ed., ASTM, 1979, p 592–623
100. H. Sehitoglu and M. Karayaka, Prediction of Thermomechanical Fatigue Lives in Metal Matrix Composites, *Metall. Trans. A*, Vol 23A, 1992, p 2029–2038
101. R.V. Miner, Fatigue, *Superalloys II*, C.T. Sims, N.S. Stoloff, and W.C. Hagel, Ed., John Wiley & Sons, 1987, p 263–289
102. W.W. Milligan, E.S. Huron, and S.D. Antolovich, Deformation, Fatigue and Fracture Behavior of Two Cast Anisotropic Superalloys, *Fatigue '87*, Vol 3, 3rd Int. Conf. Fatigue and Fatigue Thresholds, Engineering Materials Advisory Services, 1987, p 1561–1591
103. M.G. Castelli, R.V. Miner, and D.N. Robinson, Thermo-Mechanical Deformation of a Dynamic Strain Aging Alloy, *Thermo-Mechanical Fatigue Behavior of Materials*, STP 1186, H. Sehitoglu, Ed., ASTM, 1991, p 106–125
104. N. Marchand, G.L. Espérance, and R.M. Pelloux, Thermal-Mechanical Cyclic Stress-Strain Responses of Cast B-1900+Hf, *Low Cycle Fatigue*, STP 942, H. Solomon, G. Halford, L. Kaisand, and B. Leis, Ed., ASTM, 1988, p 638–656
105. H. Sehitoglu and D.A. Boismier, Thermo-Mechanical Fatigue of Mar-M247, Part 1: Experiments, *J. Eng. Mater. Technol. (Trans. ASME)*, Vol 112, 1990, p 68–80; see also Thermo-Mechanical Fatigue of Mar-M247, Part 2: Life Prediction, *J. Eng. Mater. Technol. (Trans. ASME)*, Vol 112, 1990, p 80–90
106. T.S. Cook, K.S. Kim, and R.L. McKnight, Thermal Mechanical Fatigue of Cast René 80, *Low Cycle Fatigue*, STP 942, H. Solomon, G. Halford, L. Kaisand, and B. Leis, Ed., ASTM, 1988, p 692–708
107. Y. Kadioglu and H. Sehitoglu, Thermomechanical and Isothermal Fatigue Behavior of Bar and Coated Superalloys, *J. Eng. Mater. Technol. (Trans. ASME)*, Vol 118, 1996, p 94–102
108. S. Kraft, R. Zauter, and H. Mugrabi, Investigations on the High Temperature Low Cycle Thermomechanical Fatigue Behavior of the Monocrystalline Nickel Base Superalloys CMSX-6, *Symposium on Thermo-Mechanical Fatigue Behavior of Materials,* STP 1263, ASTM, 1996
109. Y. Pan, K. Lang, D. Lohe, and E. Macherauch, Cyclic Deformation and Precipitation Behavior of NiCr22Co12Mo9, *Phys. Status Solidi (a)*, Vol 138, 1993, p 133–145
110. E. Glenny and T.A. Taylor, A Study of the Thermal-Fatigue Behavior of Metals: The Effect of Test Conditions on Nickel-Base High-Temperature Alloys, *J. Inst. Met.*, Vol 88, 1959-60, p 449–461
111. G.P. Tilly, Laboratory Simulation of Thermal Fatigue Experienced by Gas Turbine Blading, *Thermal Stresses and Thermal Fatigue*, D.J. Littler, Ed., Butterworths, London, 1971, p 47–65
112. R.C. Bill, M.J. Verrilli, M.A. McGaw, and G.R. Halford, "Preliminary Study of Thermomechanical Fatigue of Polycrystalline MAR-M 200," TP-2280, National Aeronautics and Space Administration, Feb 1984
113. R.S. Nelson, J.F. Schoendorf, and L.S. Lin, "Creep Fatigue Life Prediction for Engine Hot Section Materials (Isotropic)—Interim Report," CR-179550, National Aeronautics and Space Administration, Dec 1986
114. V.G. Ramaswamy and T.S. Cook, "Cyclic Deformation and Thermomechanical Fatigue Model of Nickel Based Superalloys, Abstracts," presented at ASTM Workshop on Thermo-Mechanical Fatigue and Cyclic Deformation (Charleston, SC), ASTM, 1986
115. G.T. Embley and E.S. Russell, "Thermal-Mechanical Fatigue of Gas Turbine Bucket Alloys," presented at 1st Parsons Int. Turbine Conf. (Dublin), June 1984
116. J. Gayda, T.P. Gabb, and R.V. Miner, paper presented at NASA 4th TMF Workshop, National Aeronautics and Space Administration, 1987
117. J.L. Malpertu and L. Rémy, Thermomechanical Fatigue Behavior of a Superalloy, *Low Cycle Fatigue*, STP 942, H. Solomon, G. Halford, L. Kaisand, and B. Leis, Ed., ASTM, 1988, p 657–671
118. H. Bernstein, T.S. Grant, R.C. McClung, and J. Allen, Prediction of Thermal-Mechanical Fatigue Life for Gas Turbine Blades in Electric Power Generation, *Thermo-Mechanical Fatigue Behavior of Materials*, STP 1186, H. Sehitoglu, Ed., ASTM, 1991, p 212–238
119. A. Kaufman and G. Halford, "Engine Cyclic Durability by Analysis and Material Testing," TM-83557, National Aeronautics and Space Administration, 1984
120. E. Chataigner, E. Fleury, and L. Rémy, Thermo-Mechanical Fatigue Behavior of Coated and Bare Nickel-Base Superalloy Single Crystals, *Symp. Fatigue under Thermal and Mechanical Loading*, J. Bressers and L. Rémy, Ed., European Commission, Petten, May 1995
121. A. Arana, J.M. Martinez-Esnaola, and J. Bressers, Crack Propagation and Life Prediction in a Nickel Based Superalloy under TMF Conditions, *Symp. Fatigue under Thermal and Mechanical Loading*, J. Bressers and L. Rémy, Ed., European Commission, Petten, May 1995
122. M. Marchionni, D. Ranucci, and E. Picco, Influence of Cycle Shape and Specimen Geometry on TMF of an ODS Ni-Base Superalloy, *Symp. Fatigue under Thermal and Mechanical Loading*, J. Bressers and L. Rémy, Ed., European Commission, Petten, May 1995
123. L.F. Coffin, Jr., The Effect of Frequency on the Cyclic Strain and Fatigue Behavior of Cast René at 1600 °F, *Metall. Trans.*, Vol 5, May 1974, p 1053–1060
124. W.W. Milligan and R.C. Bill, "The Low Cycle Fatigue Behavior of Conventionally Cast MAR-M 200 at 1000 °C," TM-83769, National Aeronautics and Space Administration, Sept 1984
125. J. Gayda and R.V. Miner, Fatigue Crack Initiation and Propagation in Several Nickel-Based Superalloys at 650 °C, *Int. J. Fatigue*, Vol 5, July 1983, p 135–143
126. M. Gell and G.R. Leverant, Mechanisms of High Temperature Fatigue, *Fatigue at Elevated Temperatures*, ASTM 520, A.E. Carden, A.J. McEvily, and C.H. Wells, Ed., ASTM, 1973, p 37–67
127. M.Y. Nazmy, High Temperature Low Cycle Fatigue of IN 738 and Application of Strain Range Partitioning, *Metall. Trans. A*, Vol 14A, March 1983, p 449–461
128. P.K. Wright, Oxidation-Fatigue Interactions in a Single-Crystal Superalloy, *Low Cycle Fatigue*, STP 942, H. Solomon, G. Halford, L. Kaisand, and B. Leis, Ed., ASTM, 1988, p 558–575
129. L. Rémy, F. Rezai-Aria, R. Danzer, and W. Hoffelner, Evaluation of Life Prediction Methods in High Temperature Fatigue, *Low Cycle Fatigue*, STP 942, H. Solomon, G. Halford, L. Kaisand, and B. Leis, Ed., ASTM, 1988, p 1115–1132
130. S.D. Antolovich, R. Baur, and S. Liu, A Mechanistically Based Model for High Temperature LCF of Ni Base Superalloys, *Superal-*

loys 1980, American Society for Metals, 1980, p 605–613
131. S.D. Antolovich, S. Liu, and R. Baur, Low Cycle Fatigue Behavior of René 80 at Elevated Temperature, *Metall. Trans. A*, Vol 12A, March 1981, p 473–481
132. W.W. Milligan, E.S. Huron, and S.D. Antolovich, Deformation, Fatigue and Fracture Behavior of Two Cast Anisotropic Superalloys, *Fatigue '87*, Vol 3, 3rd Int. Conf. Fatigue and Fatigue Thresholds (Charlottesville, VA), 1987, p 1561–1591
133. C.J. McMahon and L.F. Coffin, Jr., Mechanisms of Damage and Fracture in High-Temperature, Low-Cycle Fatigue of a Cast Nickel-Based Superalloy, *Metall. Trans.*, Vol 1, Dec 1970, p 3443–3450
134. Y. Oshida and H.W. Liu, Grain Boundary Oxidation and an Analysis of the Effects of Oxidation on Fatigue Crack Nucleation Life, *Low Cycle Fatigue*, STP 942, H. Solomon, G. Halford, L. Kaisand, and B. Leis, Ed., ASTM, 1988, p 1199–1217
135. Y. Kadioglu and H. Sehitoglu, Modeling of Thermo-Mechanical Fatigue Damage in Coated Alloys, *Thermo-Mechanical Fatigue Behavior of Materials*, STP 1186, H. Sehitoglu, Ed., ASTM, 1991, p 17–34
136. L. Rémy, H. Bernard, J.L. Malpertu, and F. Rezai-Aria, Fatigue Life Prediction under Thermal-Mechanical Loading in a Nickel Base Superalloy, *Thermo-Mechanical Fatigue Behavior of Materials*, STP 1186, H. Sehitoglu, Ed., ASTM, 1991, p 3–16
137. S.T. Wlodek, The Oxidation of René 41 and Udimet 700, *Trans. Met. Soc. AIME*, Vol 230, Aug 1964, p 1078–1090
138. G.E. Wasielewski and R.A. Rapp, High-Temperature Oxidation, *The Superalloys*, C.T. Sims and W.C. Hagel, Ed., John Wiley & Sons, 1972, p 287–316
139. J.D. Varin, Microstructure and Properties of Superalloys, *The Superalloys*, C.T. Sims and W.C. Hagel, Ed., John Wiley & Sons, 1972, p 231–257
140. J. Reuchet and L. Rémy, Fatigue Oxidation Interaction in a Superalloy—Application to Life Prediction in High Temperature Low Cycle Fatigue, *Metall. Trans. A*, Vol 14A, Jan 1983, p 141–149
141. K. Schneider and H.W. Gruling, Mechanical Aspects of High Temperature Coatings, *Thin Film Solids*, Vol 107, 1983, p 395–416
142. R. Lane and N.M. Geyer, Superalloy Coatings for Gas Turbine Components, *J. Met.*, Vol 18, Feb 1966, p 186–191
143. G.F. Paskeit, D.H. Boone, and C.P. Sullivan, Effect of Aluminide Coating on the High-Cycle Fatigue Behavior of a Nickel Base High Temperature Alloy, *J. Inst. Met.*, Vol 100, Feb 1972, p 58–62
144. C.H. Wells and C.P. Sullivan, Low Cycle Fatigue of Udimet 700 at 1700 °F, *Trans. ASM*, Vol 61, March 1968, p 149–155
145. R.S. Bartocci, Behavior of High Temperature Coatings for Gas Turbines, *Hot Corrosion Problems Associated with Gas Turbines*, STP 421, ASTM, 1967, p 169–187
146. G. Liewelyn, Protection of Nickel Base Alloys against Sulfur Corrosion by Pack Aluminizing, *Hot Corrosion Problems Associated with Gas Turbines*, STP 421, ASTM, 1967, p 3–20
147. G.W. Goward, Current Research on Surface Protection of Superalloys for Gas Turbines, *J. Met.*, Oct 1970, p 31–39
148. D.H. Boone and C.P. Sullivan, *Fatigue at Elevated Temperatures*, STP 520, A.E. Carden, A.J. McEvily, and C.H. Wells, Ed., ASTM, 1972, p 401–415
149. A.J. Santhanam and C.G. Beck, *Thin Solid Films*, Vol 73, 1980, p 387–395
150. G.A. Swanson, I. Linask, D.M. Nissley, P.P. Norris, T.G. Meyer, and K.P. Walker, Report 179594, National Aeronautics and Space Administration, 1987
151. J.E. Heine, J.R. Warren, and B.A. Cowles, "Thermomechanical Fatigue of Coated Blade Materials," Final Report, Wright Research Development Center, 27 June 1989
152. K.R. Bain, The Effects of Coatings on the Thermomechanical Fatigue Life of a Single Crystal Turbine Blade Material, *AIAA/SAE/ASME/ASEE 21st Joint Propulsion Conf.*, 1985, p 1–6
153. G.R. Leverant, T.E. Strangman, and B.S. Langer, *3rd Int. Conf. Superalloys 1976*, Claitors, Vol 75, 1976, p 285
154. T.E. Strangman, "Thermal Fatigue of Oxidation Resistant Overlay Coatings for Superalloys," Ph.D. thesis, University of Connecticut, 1978: see also T.E. Strangman and S.W. Hopkins, Thermal Fatigue of Coated Superalloys, *Ceram. Bull.*, Vol 55 (No. 3), 1976, p 304–307
155. G.R. Halford, T.G. Meyer, R.S. Nelson, and D.M. Nissley, paper presented at 33rd Int. Gas Turbine and Aeroengine Congress (Amsterdam), ASME, June 1988
156. G.A. Swanson, I. Linask, D.M. Nissley, P.P. Norris, T.G. Meyer, and K.P. Walker, Report 179594, National Aeronautics and Space Administration, 1987
157. D.M. Nissley, Fatigue Damage Modeling for Coated Single Crystal Superalloys, *NASA Conf. Publ. 3003*, Vol 3, 1988, p 259–270
158. G.R. Leverant, T.E. Strangman, and B.S. Langer, Parameters Controlling the Thermal Fatigue Properties of Conventionally-Cast and Directionally-Solidified Turbine Alloys, *Superalloys: Metallurgy and Manufacture*, B.H. Kear, et al., Ed., Claitors, 1976, p 285–295
159. W. Wei, W. Dunfee, M. Gao, and R.P. Wei, The Effect of Environment on the Thermal Fatigue Behavior of Gamma Titanium Aluminide, *Symp. Fatigue under Thermal and Mechanical Loading*, J. Bressers and L. Rémy, Ed., European Commission, Petten, May 1995
160. S. Mall, T. Nicholas, J.J. Pernot, and D.G. Burgess, Crack Growth in a Titanium Aluminide Alloy under Thermal Mechanical Cycling, *Fatigue Fract. Eng. Mater. Struct.*, Vol 14 (No. 1), 1991, p 79–87
161. S. Kalluri and G. Halford, Damage Mechanisms in Bithermal and Thermo-Mechanical Fatigue of Haynes 188, *Thermo-Mechanical Fatigue Behavior of Materials*, STP 1186, H. Sehitoglu, Ed., ASTM, 1991, p 126–143
162. K.D. Sheffler and G.S. Doble, Thermal Fatigue Behavior of T-111 and Astar 811C in Ultrahigh Vacuum, *Fatigue at Elevated Temperatures*, STP 520, A.E. Carden, A.J. McEvily, and C.H. Wells, Ed., ASTM, 1973, p 482–489
163. P. Hacke, A Sprecher, and H. Conrad, Modeling of the Thermomechanical Fatigue of 63Sn-37Pb Alloy, *Thermo-Mechanical Fatigue Behavior of Materials*, STP 1186, H. Sehitoglu, Ed., ASTM, 1991, p 91–105
164. M. Karayaka and H. Sehitoglu, Thermomechanical Cyclic Deformation of Metal Matrix Composites: Internal Stress-Strain Fields, *ASTM Cyclic Deformation, Fracture and Nondestructive Evaluation of Advanced Materials*, ASTM, 1990
165. M. Karayaka and H. Sehitoglu, Thermomechanical Fatigue of Metal Matrix Composites, *Low Cycle Fatigue and Elasto-Plastic Behavior of Materials*, Vol 3, T.T. Rie, Ed., Elsevier, 1992, p 13–18
166. W. VanArsdell, "The Effect of Particle Size on the Thermo-mechanical Fatigue of Metal Matrix Composites," M.S. thesis, University of Illinois at Urbana, 1993
167. W. VanArsdell, H. Sehitoglu, and M. Mushiake, "The Effect of Particle Size on the Thermo-mechanical Fatigue of Metal Matrix Composites," *Fatigue '93*, EMAS, 1993
168. H. Sehitoglu, The Effect of Particle Size on the Thermo-Mechanical Fatigue Behavior of Metal Matrix Composites, *Symp. Fatigue under Thermal and Mechanical Loading*, J. Bressers and L. Rémy, Ed., Kluwer Academic Publishers, 1996
169. R. Neu, Thermo-Mechanical Fatigue Damage Mechanism Maps for Metal Matrix Composites, *2nd Symposium on Thermo-Mechanical Fatigue Behavior of Materials,* STP 1263, M. Verilli and M. Castelli, Ed., ASTM, 1996
170. R. Neu, A Mechanistic-Based Thermomechanical Fatigue Life Prediction Model for Metal Matrix Composites, *Fatigue Fract. Eng. Mater. Struct.*, Vol 16 (No. 8), 1993, p 811–828
171. S. Taira and T. Inoue, Thermal Fatigue under Multiaxial Thermal Stresses, *Thermal Stresses and Thermal Fatigue*, D.J. Littler, Ed., Butterworths, London, 1971, p 66–80
172. J. Meersman, J. Ziebs, H.-J. Kuhn, R. Sievert, J. Olscewski, and H. Frenz, The Stress-Strain Behavior of In 738LC under Thermo-mechanical Uni- and Multiaxial Fatigue Loading, *Symp. Fatigue under Thermal and Mechanical Loading*, J. Bressers and L. Rémy, Ed., European Commission, Petten, May 1995
173. M. Gell and D.J. Duquette, The Effects of Oxygen on Fatigue Fracture of Engineering Alloys, *Corrosion Fatigue: Chemistry, Mechanics and Microstructure*, NACE, 1971, p 366–378
174. J.C. Runkle and R.M. Pelloux, Micromechanisms of Low-Cycle Fatigue in Nickel-Based

Superalloys at Elevated Temperatures, *Fatigue Mechanisms*, STP 675, ASTM, 1979, p 501–527

175. V.G. Ramaswamy and T.S. Cook, "Cyclic Deformation and Thermomechanical Fatigue Model of Nickel Based Superalloys, Abstracts," presented at ASTM Workshop on Thermo-mechanical Fatigue and Cyclic Deformation (Charleston, SC), ASTM, 1986
176. C.A. Rau, Jr., A.E. Gemma, and G.R. Leverant, Thermal-Mechanical Fatigue Crack Propagation in Nickel- and Cobalt-Base Superalloys under Various Strain-Temperature Cycles, *Fatigue at Elevated Temperatures*, STP 520, A.E. Carden, A.J. McEvily, and C.H. Wells, Ed., ASTM, 1973, p 166–178
177. M.L. Heil, T. Nicholas, and G.K. Haritos, Crack Growth in Alloy 718 under Thermal-Mechanical Cycling, *Thermal Stress, Material Deformation, and Thermo-Mechanical Fatigue*, H. Sehitoglu and S.Y. Zamrik, Ed., American Society of Mechanical Engineers, 1987, p 23–29
178. M. Okazaki and T. Koizumi, Crack Propagation During Low Cycle Thermal-Mechanical and Isothermal Fatigue at Elevated Temperatures, *Metall. Trans. A*, Vol 14A (No. 8), Aug 1983, p 1641–1648
179. M. Okazaki and T. Koizumi, Effect of Strain Waveshape on Thermal-Mechanical Fatigue Crack Propagation in a Cast Low Alloy Steel, *J. Eng. Mater. Technol. (Trans. ASME)*, A.E. Carden, A.J. McEvily, and C.H. Wells, Ed., ASTM, Vol 81–87, 1983, p 1641–1648
180. A.E. Gemma, B.S. Langer, and G.R. Leverant, Thermo-Mechanical Fatigue Crack Propagation in an Anisotropic (Directionally Solidified) Nickel Base Superalloy, *Thermal Fatigue of Materials and Components*, STP 612, D.A. Spera and D.F. Mowbray, Ed., ASTM, p 199–213
181. R.P. Skelton, Crack Initiation and Growth in Simple Metal Components during Thermal Cycling, *Fatigue at High Temperature*, Applied Science Publishers, London, 1983, p 1–63
182. E.H. Jordan and G.J. Meyers, Fracture Mechanics Applied to Nonisothermal Fatigue Crack Growth, *Eng. Fract. Mech.*, Vol 23 (No. 2), 1986, p 345–358
183. R.L. McKnight, J.H. Laflen, and G.T. Spamer, "Turbine Blade Tip Durability Analysis," CR 165268, National Aeronautics and Space Administration, 1982
184. M.R. Johnson, private communication
185. D. Morrow, "Stress-Strain Response of a Two-Bar Structure Subject to Cyclic Thermal and Steady Net Section Loads," M.S. thesis, University of Illinois, 1982; see also E. Abrahamson, "Modeling the Behavior of Type 304 Stainless Steel with a Unified Creep-Plasticity Theory," Ph.D. thesis, University of Illinois, 1983
186. R.W. Swindeman and D.N. Robinson, Two-Bar Thermal Ratcheting of Annealed 2.25 Cr-1Mo Steel, *Thermal Stress, Material Deformation, and Thermo-Mechanical Fatigue*, H. Sehitoglu and S.Y. Zamrik, Ed., American Society of Mechanical Engineers, 1987, p 91–98
187. S.M. Russ, C.J. Boehlert, and D. Eylon, Out-of-Phase Thermomechanical Fatigue of Titanium Composite Matrices, *Mater. Sci. Eng.*, Vol A192/193, 1995, p 483–489
188. H.G. Baron, Thermal Shock and Thermal Fatigue, *Thermal Stress*, P.P. Benham and R.D. Hoyle, Ed., Pitman, London, 1964, p 182–206
189. K.D. Challenger, A.K. Miller, and C.R. Brinkman, An Explanation for the Effects of Hold Periods on the Elevated Temperature Fatigue Behavior of 2.25Cr-1Mo Steel, *J. Eng. Mater. Technol. (Trans. ASME)*, Vol 103, Jan 1981, p 7–14
190. J. Bressers, U. Schusser, and B. Ilschner, Environmental Effects on the Fatigue Behavior of Alloy 800H, *Low Cycle Fatigue and Elasto-Plastic Behavior of Materials*, 2nd Int. Conf. Low Cycle Fatigue and Elasto-Plastic Behaviour of Materials (Munich), Elsevier Applied Sciences, 1987, p 365–370
191. S. Majumdar and P.S. Maiya, A Mechanistic Model for Time-Dependent Fatigue, *J. Eng. Mater. Technol. (Trans. ASME)*, Vol 102, Jan 1980, p 159–167
192. B. Kirkwood and J.R. Weertman, Cavity Nucleation During Fatigue Crack Growth Caused by Linkage of Grain Boundary Cavities, *Micro and Macro Mechanics of Crack Growth*, K. Sadananda, B.B. Rath, and D.J. Michel, Ed., TMS-AIME, 1982, p 199–212
193. J. Wareing, Creep-Fatigue Interaction in Austenitic Stainless Steels, *Metall. Trans. A*, Vol 8A, May 1977, p 711–721
194. B.K. Min and R. Raj, Hold-Time Effects in High Temperature Fatigue, *Acta Metall.*, Vol 26, 1978, p 1007–1022
195. B. Tomkins, Fatigue: Mechanisms, *Creep and Fatigue in High Temperature Alloys*, J. Bressers, Ed., 1981, p 111–143
196. J.K. Tien, S.V. Nair, and V.C. Nardone, Creep-Fatigue Interaction in Structural Alloys, *Flow and Fracture at Elevated Temperatures*, R. Raj, Ed., American Society for Metals, 1985, p 179–213
197. J.C. Grosskreutz and M.B. McNeil, *J. Appl. Phys.*, Vol 40, 1969, p 355
198. D. Bruce and P. Hancock, Mechanical Properties and Adhesion of Surface Oxide Films on Iron and Nickel Measured during Growth, *J. Inst. Met.*, Vol 97, 1969, p 148–155
199. D. Bruce and P. Hancock, Influence of Specimen Geometry on the Growth and Mechanical Stability of Surface Oxides Formed on Iron and Steel in the Temperature Range 570 °C-800 °C, *J. Iron Steel Inst.*, Nov 1970, p 1021–1024
200. M.I. Manning, Geometrical Effects on Oxide Scale Integrity, *Corros. Sci.*, Vol 21 (No. 4), 1981, p 301–316
201. J. Stringer, Stress Generation and Relief in Growing Oxide Films, *Corros. Sci.*, Vol 10, 1970, p 513
202. J.K. Tien and J.M. Davidson, Oxide Spallation Mechanisms, *Stress Effects and the Oxidation of Metals*, J.V. Cathcart, Ed., TMS-AIME, 1974, p 200–219
203. C.H. Wells, P.S. Follansbee, and R.R. Dils, Mechanisms of Dynamic Degradation of Surface Oxides, *Stress Effects and the Oxidation of Metals*, J.V. Cathcart, Ed., TMS-AIME, 1974, p 220–244
204. ASME Boiler and Pressure Vessel Code, Case N-47-23, "Class 1 Components in Elevated Temperature Service," Section III, Division 1, American Society of Mechanical Engineers, 1986
205. J. Lemaitre and A. Plumtree, Application of Damage Concepts to Predict Creep-Fatigue Failures, *J. Eng. Mater. Technol. (Trans. ASME)*, Vol 101, 1979, p 284–292
206. R. Hales, A Method of Creep Damage Summation Based on Accumulated Strain for Assessment of Creep-Fatigue Endurance, *Fatigue Fract. Eng. Mater. Struct.*, Vol 6, 1983, p 121
207. A. Saxena and J.L. Bassani, Time-Dependent Fatigue Crack Growth Behavior at Elevated Temperature, *Fracture: Microstructure, Mechanisms and Mechanics*, J.M. Wells and J.D. Landes, Ed., TMS-AIME, 1984, p 357–383
208. H. Sehitoglu and W. Sun, The Significance of Crack Closure under High Temperature Fatigue Crack Growth with Hold Periods, *Eng. Fract. Mech.*, Vol 33, 1989, p 371–388
209. M. Miller, D.L. McDowell, R. Oehmke, and S. Antolovich, A Life Prediction Model for Thermomechanical Fatigue Based on Microcrack Propagation, *Thermo-Mechanical Fatigue Behavior of Materials*, STP 1186, H. Sehitoglu, Ed., ASTM, 1991, p 35–49
210. D. Drucker and L. Palgen, On the Stress-Strain Relations Suitable for Cyclic and Other Loading, *J. Appl. Mech. (Trans. ASME)*, Vol 48, 1981, p 479–485
211. S. Bodner and Y. Partom, Constitutive Equations for Elastic Viscoplastic Strain Hardening Materials, *J. Appl. Mech. (Trans. ASME)*, Vol 42, 1975, p 385–389
212. S. Bodner, I. Partom, and Y. Partom, Uniaxial Cyclic Loading of Elastic Viscoplastic Material, *J. Appl. Mech. (Trans. ASME)*, Vol 46, 1979, p 805–810
213. D. Slavik and T.S. Cook, A Unified Constitutive Model for Superalloys, *Int. J. Plast.*, Vol 6, 1990, p 651–664
214. R. Neu, private communication, 1995
215. H. Sehitoglu, "Thermomechanical Deformation of Engineering Alloys and Components—Experiments and Modeling," *Mechanical Behavior of Materials at High Temperature*, Kluwer, 1996, p 349–380
216. A.K. Miller, A Realistic Model for the Deformation Behavior of High Temperature Materials, *Fatigue at Elevated Temperatures*, STP 520, A.E. Carden, A.J. McEvily, and C.H. Wells, Ed., ASTM, 1973, p 613–624
217. K.P. Walker, "Research and Development Program for Nonlinear Structural Modeling with Advanced Time-Temperature Dependent Constitutive Relationships," CR-165533, National Aeronautics and Space Administration, 1981

Elevated-Temperature Crack Growth of Structural Alloys

HIGH-TEMPERATURE operating requirements for parts and equipment have drastically increased over the past 20 to 30 years, and businesses such as the utility, aircraft, and chemical industries are greatly dependent on the safe and efficient operation of such equipment. In the power-generation industry, for example, components operate for extended periods of time at temperatures between 0.3 to 0.5 of their absolute melting temperature and have design lives that are limited by creep. Defense and aerospace applications rely heavily on materials that maintain their integrity in the presence of combinations of high temperature and stress (Ref 1). Also, in other cases, as the original design lives are expiring, assessment of the remaining life of components currently in operation with the objective of extending the component service life has become an economic and safety consideration (Ref 2). Many facilities are now relying on retirement-for-cause philosophy to determine the end of service life for parts (Ref 3–5).

Despite the sophisticated methods of flaw detection that are available, defects and impurities are commonly present in all large components and can potentially escape detection. In the high-temperature regime, components fail by the accumulation of time-dependent creep strains at these defects, which with time evolve into cracks causing eventual failure. This results in financial losses in repair and downtime costs as well as possible loss of human life (Ref 6). This article focuses on the concepts for characterizing and predicting elevated-temperature crack growth in structural materials. Both creep and creep-fatigue crack growth will be considered, focusing mainly on test methods. For a discussion on the application of the data in life prediction, see the article "Creep-Fatigue Interaction" in this Volume and Ref 1, 5, and 7.

Creep and Creep-Fatigue

Failures that are attributed to creep result from either widespread or localized creep damage (Ref 8). If the part is subjected to uniform stress and temperatures, the damage is likely to be widespread and failure by creep rupture is apt to result. This is most commonly observed in thin section components such as steam pipes. Components with localized damage, which is a result of nonuniform stress and temperature distribution found most commonly in large structures, are more prone to fail as a result of creep crack propagation rather than stress rupture.

Service conditions experienced by components can also involve cyclic loading and unloading at elevated temperatures. Hence, in these situations, crack growth occurs not only under static loading (creep conditions), but creep-fatigue interactions play a major role in the initiation and growth of cracks. Components operating at high temperatures experience changes in conditions from beginning to end of each operating cycle, resulting in transient temperature gradients. Considering the case of steam turbines as an example, cracks in castings are typically located at the steam inlets of high-pressure and intermediate-pressure turbine sections because the local thermal stresses are higher. The primary cause of crack initiation and propagation in turbine casings is fatigue and creep-fatigue and occasionally brittle fracture due to high transient thermal stresses (Ref 4). Thermal stresses are responsible for fatigue and creep-fatigue crack growth (CFCG) in the lower-temperature regions, while creep contributes to crack growth in regions where temperature exceeds 425 °C (800 °F) (Ref 7).

Creep-Ductile Materials. Creep and creep-fatigue crack growth is a common occurrence in most engineering materials operating at high temperatures. Materials classified as creep-ductile have the ability to sustain significant amounts of crack growth prior to failure. Also, crack growth in these materials is accompanied by significant amounts of creep deformation at the crack tip. Therefore, a complete understanding of crack growth mechanics and damage mechanisms is required for accurate predictions of life of high-temperature components made from such materials. A typical flow diagram of this methodology is shown in Fig. 1. Examples of such materials include chromium-molybdenum steels, chromium-molybdenum-vanadium steels, and stainless steels.

This damage in creep-ductile materials at high temperatures is usually in the form of grain-boundary cavitation. It has been most commonly observed that this cavitation initiates at second-phase particles or defects on the grain boundaries (Ref 9). Nucleation and growth of these cavities lead to coalescence of these voids, eventual crack formation, and growth (Ref 10), which is the primary mechanism of creep crack growth. Failure due to creep-fatigue interaction can be described from two points of view (Ref 11): influence of cyclic loading on cavitation damage and influence of cavitation on cyclic initiation and propagation. These mechanisms are illustrated in Fig. 2, adapted from Ref 11. Three prominent mechanisms for fatigue crack growth at elevated temperatures, in the presence of hold times, are (1) alternating slip mechanism (crack-tip blunting mechanism), (2) fatigue crack growth caused by grain-boundary cavitation, and (3) influence of corrosive environment.

Creep-Brittle Materials. A second class of high-temperature structural materials is known as creep-brittle materials, which include high-temperature aluminum alloys, titanium alloys, nickel-base superalloys, intermetallics, and ceramic materials. The primary difference between this class of materials and the creep-ductile materials discussed previously is that creep crack growth in these materials is usually accompanied by small-scale creep deformation and by crack growth rates that are comparable to the rate at which creep deformation spreads in the cracked body. As discussed later in this article, this has a significant influence on the crack tip parameters that characterize crack growth rates. Because of these differences, the time-dependent fracture mechanics (TDFM) concepts described in the subsequent discussion will address these two types of material behavior separately.

Time-Dependent Fracture Mechanics

Stationary-Crack-Tip Parameters. As previously mentioned, crack growth in creep-ductile materials lags considerably behind the spreading of the creep zone. Therefore, a practical assumption of a stationary crack tip is made when searching for crack-tip parameters (although the limita-

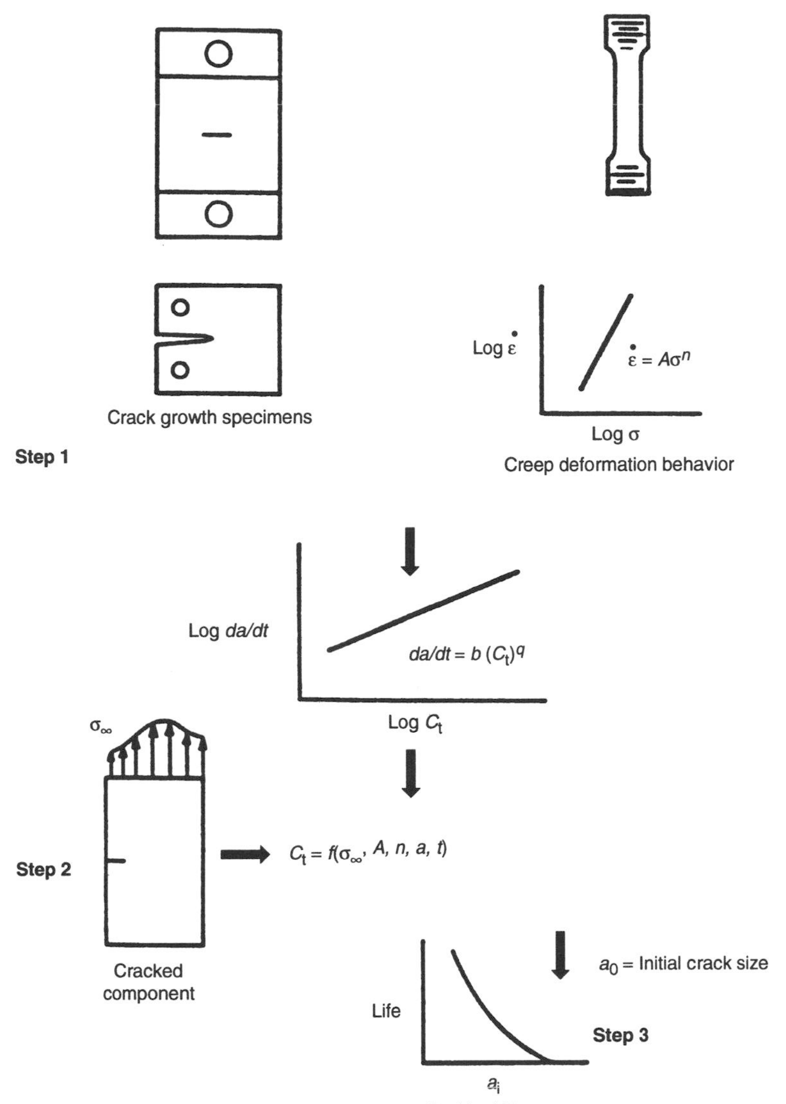

Fig. 1 Methodology for predicting crack propagation life using time-dependent fracture mechanics

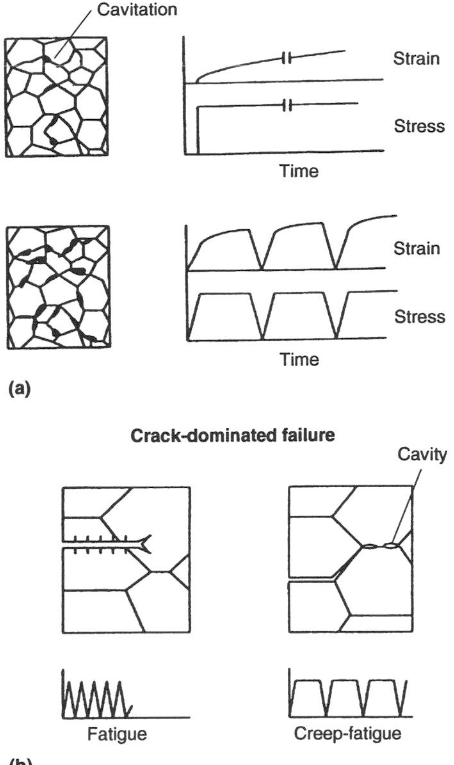

Fig. 2 Schematic representation of mechanistic aspects of creep-fatigue. (a) Effect of cycling on cavitation damage. (b) Effect of cavitation on cyclic crack growth. Source: Ref 11

tions and validity of this assumption in defining crack-tip parameters are discussed later in the section "Conditions with Growing Cracks.")

In order to describe the mechanics of creep and creep-fatigue crack growth, an understanding of the stress-strain-time response at the crack tip of a body subjected to a load in creep-temperature regime must be developed first. The stages of the evolving deformation zone ahead of a crack tip when a member is subjected to a load in the creep regime is shown in Fig. 3. The initial response of the body is elastic-plastic, and the crack-tip stress field is proportional to the stress-intensity factor, K, if the scale of plasticity is small compared with the crack size (Ref 12, 13). If the plastic zone is not small, the J-integral characterizes the instantaneous crack tip stresses and strains (Ref 14). With increasing time, creep deformation causes the relaxation of the stresses in the immediate vicinity of the crack tip, resulting in the formation of the creep zone, which continually increases in size with time. Because the parameters K and J are independent of time, they are not able to uniquely characterize the crack-tip stresses and strains within the creep zone. The parameters C^*, $C(t)$, and C_t have been developed to describe the evolution of time-dependent creep strains in the crack-tip region (Ref 15–18) and will be discussed later in this section. Within these creep regions, the crack-tip stress and strain fields resemble the Hutchinson-Rice-Rosengren (HRR) fields noted in elastic-plastic fracture mechanics (Ref 19, 20).

For a body undergoing creep, the uniaxial stress-strain-time response for a material that exhibits elastic, primary, secondary, and tertiary creep is given by:

$$\dot{\varepsilon} = \frac{\dot{\sigma}}{E} + A_1\,\varepsilon^{-p}\sigma^{n_1(1+p)} + A\sigma^n + A_3\sigma^{n_3}\,(\varepsilon - A\sigma^n t)^{p_3} \quad \text{(Eq 1)}$$

where ε and σ are the strain and stress, respectively, and $\dot{\varepsilon}$ and $\dot{\sigma}$ denote their time derivatives. The values of A, A_1, A_3, p, p_3, n, n_1, and n_3 are the creep regression constants derived from creep deformation data. The terms on the right-hand side of the equation represent the elastic, primary, secondary, and tertiary creep strain contributions, respectively. This equation is convenient for analyzing the creep deformation behavior of cracked bodies under creep loading conditions.

The crack-tip stress and strain behavior for a creeping body change with time, as a result of continuous evolution of the creep zone. The changes in the creep deformation zone follow the progression shown in Fig. 4. During the initial stage of small-scale creep, the creep zone is small compared with the crack size and the remaining ligament. Primary creep strains accumulate at a faster rate than the secondary creep strains; therefore, the primary creep strains initially dominate this region. Next, the primary creep zone continues to expand, and the secondary creep zone begins to evolve within the primary creep zone. Then the primary creep zone envelopes the entire remaining ligament, while the secondary creep zone continues to grow in size within the primary creep zone. Eventually, the secondary creep zone engulfs the entire remaining ligament. In heavily cavitating materials, the tertiary creep zone begins as a small zone near the crack tip, but can eventually cover the entire remaining ligament. In chromium-molybdenum steels, cavitation is usually limited to a small region near the crack tip, and the consideration of tertiary creep strains

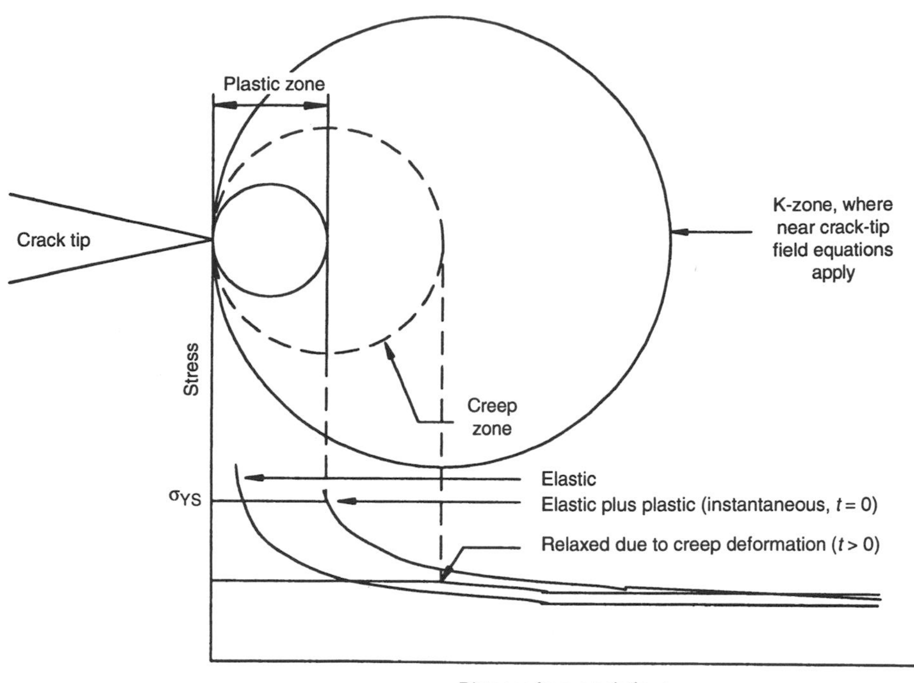

Fig. 3 Formation of deformation zones ahead of crack tip upon initial loading in the creep regime

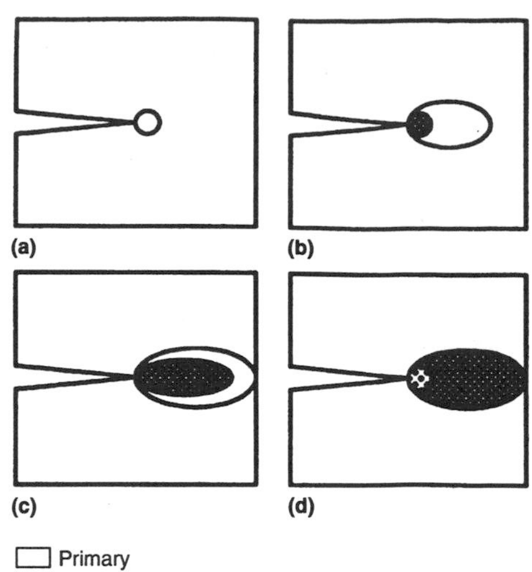

Fig. 4 Creep zone evolution. (a) Small scale primary creep conditions. (b) Secondary zone evolving within the primary zone. (c) Secondary zone becoming comparable in size with the extensive primary zone. (d) Extensive secondary zone enveloping entire ligament (steady-state conditions) and tertiary zone initiating

is not relevant in estimating the crack tip parameters.

Cracked Body Deforming under Steady-State Creep Conditions. When steady-state creep conditions dominate, as shown in Fig. 4(d), the relationship between stress and strain rate, Eq 1 simplifies to the so-called Norton relation:

$$\dot{\varepsilon} = A\,\sigma^n \tag{Eq 2}$$

For these conditions, the crack-tip parameter, C^*, was defined by Landes and Begley (Ref 15) and Nikbin, Webster, and Turner (Ref 21) by analogy to the J-integral. The C^*-integral is defined as follows:

$$C^* = \int_\Gamma W^*\,dy - T_i\left(\frac{\partial \dot{u}_i}{\partial x}\right)ds \tag{Eq 3}$$

where

$$W^* = \int_0^{\dot{\varepsilon}_{mn}} \sigma_{ij}\,d\dot{\varepsilon}_{ij} \tag{Eq 4}$$

is the strain energy rate density associated with the point stress and strain rate. In Eq 3, Γ is an arbitrary counterclockwise line contour starting at the lower crack surface and ending on the upper crack surface enclosing the crack tip and no other defect, T_i is the component of the traction vector in the direction of the outward normal, and ds is the increment in the contour path. Figure 5 illustrates this integral on a reference crack-tip coordinate system. A more detailed account of these notations can be found in Ref 22 to 24.

C^* is a path-independent integral whose value can be obtained by calculation of the integral along an arbitrary path, as mentioned before. The C^*-integral is also related to the energy rate or power difference between two identically loaded specimens having incrementally different crack lengths; therefore, C^* can be measured at the loading pins of the specimen and defined in that way by:

$$C^* = -\frac{1}{B}\frac{dU^*}{da} \tag{Eq 5}$$

where B is the specimen thickness and U^* is the steady-state energy rate (or stress-power) difference between two specimens in which the crack lengths differ by an incremental amount da, but are otherwise identical. The C^*-integral also describes the strength of the crack-tip stress and strain-rate singularities (Ref 25):

$$\sigma_{ij} = \left(\frac{C^*}{I_n A r}\right)^{1/(n+1)} \tilde{\sigma}_{ij}(\Theta) \tag{Eq 6}$$

$$\dot{\varepsilon}_{ij} = A\left(\frac{C^*}{I_n A r}\right)^{1/(n+1)} \tilde{\dot{\varepsilon}}_{ij}(\Theta) \tag{Eq 7}$$

where r is the distance from the crack tip, Θ is the angle from the plane of the crack, and $\tilde{\sigma}_{ij}(\Theta)$ and $\tilde{\dot{\varepsilon}}_{ij}(\Theta)$ are angular functions specified in Ref 26. A is the Norton law coefficient in the relation between stress and steady-state creep rate. I_n is a constant dependent on the steady-state creep exponent n, whose values may be found in tables (Ref 26). For most values of n of practical interest, I_n can be expressed approximately as 3 for plane-stress conditions and 4 for plane-strain conditions. Thus,

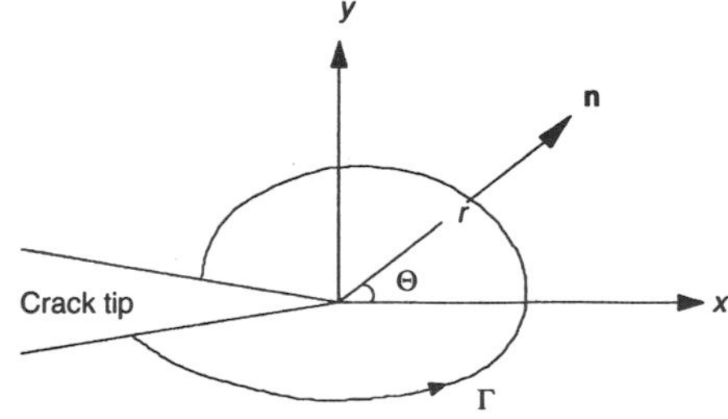

Fig. 5 Schematic of the contour integral in terms of crack-tip coordinate system used to define C^*. **n** is the unit normal vector.

C^* characterizes the strength of the crack-tip-stress singularity commonly known as the Hutchinson-Rosengren-Rice (HRR) singularity.

Using the load-line deflection rate, which can be measured directly from the specimen, the applied load and crack length, C^* can be determined by (Ref 27):

$$C^* = \frac{P\dot{V}_{SS}}{BW}\,\eta\left(\frac{a}{W}, n\right) \tag{Eq 8}$$

where W is the specimen width, B is the specimen thickness, $\dot{V}_{SS}$ is the steady-state load-line deflection rate, and η is a geometric function that is also dependent on the crack length to width ratio, a/W, and the secondary creep exponent, n. This method of determining C^* has been successfully used under laboratory conditions with test specimens, most commonly center crack panels and compact tension specimens. For the compact tension geometry, the value of η can be approximated as follows (Ref 2, 28–30):

$$\eta = \frac{n}{n+1}\left[\frac{2 + 0.522\left(1 - \frac{a}{W}\right)}{1 - \frac{a}{W}}\right] \quad \text{(Eq 9)}$$

where the term $n/n + 1$ on the right-hand side of the equation is strictly valid for secondary creep only and is replaced by $n_1/n_1 + 1$ when most of the test time is spent under primary creep conditions (Ref 31). The C^*-integral can also be determined from expressions analogous to the analytical expressions for estimating the fully plastic portion of the J-integral. This method is useful when the experimental values of the load-line deflection rates are not available and either plane-strain or plane-stress conditions prevail. When planar conditions are prevalent (assumption for most thick or very thin in-service members), C^* can be calculated as follows for compact tension specimens (Ref 12, 24, 32):

$$C^* = A\,(W - a)h_1\left[\frac{P}{\beta\zeta B(W - a)}\right]^{n+1} \quad \text{(Eq 10)}$$

where h_1 depends on a/W, n, and the state of stress (Ref 33), β equals either 1.455 or 1.071 for plane-strain or plane-stress conditions, respectively, and:

$$\zeta = \left[\left(\frac{2a}{W - a}\right)^2 + 2\left(\frac{2a}{W - a}\right) + 2\right]^{1/2} - \left[\left(\frac{2a}{W - a}\right) + 1\right] \quad \text{(Eq 11)}$$

Elastic plus Secondary Creep Conditions. As previously mentioned, the creep deformation zone changes with time and these zones evolve from small-scale to extensive creep conditions. For a cracked body to reach extensive secondary creep conditions, a characteristic transition time, t_T, has been proposed for when C^* becomes valid from the time of initial loading (Ref 32, 34):

$$t_T = \frac{K^2(1 - \nu^2)}{(n + 1)\,EC^*} \quad \text{(Eq 12)}$$

For times less than the calculated value of t_T, stress redistribution in the crack-tip region cannot be ignored. Thus Eq 2 must be modified to include the elastic term in addition to the power-law creep term. Under these circumstances, C^* is path-dependent and it no longer uniquely determines the crack-tip stress fields given by Eqs 6 and 7. The size of the secondary creep zone (r_c) can be determined by the relationship (Ref 32):

$$r_c\,(\theta, t, n) = \alpha K^2\,(EAt)^{2/(n-1)}\,\tilde{r}(\theta, n) \quad \text{(Eq 13)}$$

where n is the creep exponent and α is a function of the state of stress and n, and is given by:

$$\alpha = \frac{1}{2\pi}\left[\frac{(n + 1)I_n}{2\pi\,(1 - \nu^2)}\right]^{2/(n-1)}$$

and I_n is a dimensionless parameter related to the HRR stress field (Ref 26).

For several applications, the transition times may be large in comparison with the average operating time between start-up and shutdown for components. If operational shutdown is accepted as a part of normal operating mode, it is reasonable to infer that some components may never actually reach steady-state conditions, thereby spending their service life in the small-scale and transition creep regimes. Thus C^* is not applicable for characterizing creep crack growth in these components. Furthermore, primary creep behavior must be incorporated in the above analysis for it to be generally useful.

Even under small-scale creep, in the immediate vicinity of the crack tip, the creep strain rates exceed the elastic strain rates; therefore, selection of any integration path in the creep-dominated region will yield path-independence for the C^*-integral. The C^*-integral taken along a path near the crack tip has been called the $C(t)$ parameter and has been shown to characterize the amplitude of the crack-tip stress and strain fields (Ref 13, 35). For small-scale secondary creep (SSC) conditions, $C(t)$ can be approximated by the following equation (Ref 32, 34):

$$[C(t)]_{SSC} \cong \frac{(1 - \nu^2)\,K^2}{(n + 1)\,Et} \quad \text{(Eq 14)}$$

As the extensive creep conditions become prevalent, $C(t)$ becomes equal to C^*, and it also becomes completely path-independent. An interpolation formula for analytically estimating $C(t)$ during small-scale to extensive creep conditions for elastic-secondary creep conditions is given by (Ref 35):

$$C(t) \cong \frac{(1 - \nu^2)\,K^2}{(n + 1)\,Et} + C^* \quad \text{(Eq 15)}$$

Consideration of Primary Creep in Crack-Tip Parameters. Primary creep can be present in the small-scale as well as extensive creep conditions and can be of considerable importance in many elevated-temperature components such as chromium-molybdenum steels (Ref 24). Under extensive primary creep conditions, the second term in Eq 1 becomes dominant. Integrating this term and solving for ε results in $\varepsilon = \sigma^{n_1}[A_1 t(1 + p)]^{1/1+p}$. Because for a given material, A_1, p, and n_1 are constants, the accumulated strain is a function of stress and time. Furthermore, the strain and strain-rate dependence on stress and time is separable. Because of this property, the C^*-integral is path-independent for extensive primary creep; however, its value changes with time. Primary creep can also be included in the estimation of $C(t)$ under small-scale creep conditions. The transition time for the progression of small-scale primary creep conditions to evoke extensive primary creep conditions, t_1, is defined by (Ref 36):

$$t_1 = \frac{1}{n_1 + 1}\left(\frac{J}{C^*}\right)^{p+1} \quad \text{(Eq 16)}$$

where J is the J-integral (Ref 14). For $t > t_1$, extensive primary creep conditions prevail and as mentioned earlier, C^* is path-independent at a fixed time and thus defined as $C^*(t)$, whose value changes with time. It uniquely characterizes the instantaneous crack-tip stresses. $C^*(t)$-integral can be related to another path-independent integral, C_h^*, which is independent of time, by the following equation (Ref 36, 37):

$$C^*(t) = \frac{C_h^*}{(1 + p)t^{p/(1+p)}} \quad \text{(Eq 17)}$$

Thus, the time dependence of $C^*(t)$ can be separated from the crack-size and load-dependent parameter, C_h^*, which can be determined analytically much like J and C^*. For compact specimens (Ref 16):

$$C_h^* = [A_1\,(1 + p)]^{1/(1+p)}\,(W - a)h_1\left[\frac{P}{\beta\zeta B(W - a)}\right]^{n_1+1} \quad \text{(Eq 18)}$$

where A_1, p, and n_1 are the primary creep constants from Eq 1. With continuing evolution of the secondary creep zone within the primary creep zone, the elapsed time for the secondary zone to overtake the primary-zone boundary and engulf the remaining ligament is derived by (Ref 37):

$$t_2 = \left[\frac{(n + p + 1)\,C_h^*}{(n + 1)\,(1 + p)\,C^*}\right]^{(1+p)/p} \quad \text{(Eq 19)}$$

In a manner similar to the determination of the secondary creep zone size, the extent of the primary creep zone during small-scale creep can also be determined:

$$r_c(\theta,t,n) = \alpha'\,K^2\,[E(A_1 t)^{1/(1+p)}]^{2/(n_1-1)}\,\tilde{r}(\theta,n_1) \quad \text{(Eq 20)}$$

where α' is a function of the state of stress and the primary creep exponent n_1.

A condition commonly observed is one in which both primary and secondary creep strains occur simultaneously in the ligament. The $C^*(t)$-integral in this regime can be approximated by the following relationship (Ref 22):

$$C^*(t) \approx \frac{C_h^*}{(1 + p)t^{p/(1+p)}} + C_s^* \quad \text{(Eq 21)}$$

where C_s^* is the steady-state value of the C^*-integral. The $C(t)$-integral also characterizes the amplitude of the HRR fields under these conditions and a wide range expression for $C(t)$ is approximated by (Ref 16):

$$C(t) \approx [C(t)]_{SSC} + C^*(t) = [C(t)]_{SSC} + C^*\left[\left(\frac{t_2}{t}\right)^{p/(p+1)} + 1\right] \quad \text{(Eq 22)}$$

The parameter $C(t)$ is useful for characterizing the creep crack growth for the small-scale and steady-state regimes. However, one significant disadvantage of $C(t)$ is that it cannot be measured in the small-scale (transient) region and can only be calculated analytically. In the extensive creep regime, $C(t) = C^*$ so it can be measured from the load-line displacement readings directly from a test specimen as given earlier in Eq 8.

The C_t Parameter. Because $C(t)$ cannot be measured at the load-line under small-scale conditions, another parameter, C_t, has been proposed and shown to characterize creep-crack growth rates under a wide range of creep conditions (Ref 18, 38). The C_t parameter is defined as the instantaneous rate at which stress-power is dissipated and can be measured at the loading pins in the entire regime from small-scale to extensive creep. Thus by definition, C_t is equal to $C^*(t)$ and $C(t)$ in the extensive regime (Ref 24) and is given by (Ref 18):

$$C_t = \lim_{\Delta a \to 0} -\frac{\Delta U_t^*}{B\Delta a} = -\frac{1}{B}\frac{\partial U_t^*}{\partial a}\bigg|_{V_c} \quad \text{(Eq 23)}$$

where B is the specimen thickness, a is crack length, and ΔU_t^* is the instantaneous difference in the stress power between two cracked bodies that have incrementally differing crack lengths of Δa but are otherwise identical.

For small-scale creep conditions, the Irwin concept of effective crack length has been modified to define a stationary crack to accommodate the expression for C_t (Ref 18, 38):

$$a_{eff} = a + \chi \dot{r}_c \quad \text{(Eq 24)}$$

In this equation, χ is the scaling factor, which is approximately equal to ⅓ as determined by finite element analysis (Ref 39, 40), $\dot{r}_c$ is the creep zone size, a_{eff} is the effective crack length, and a is the physical crack length. This leads to an expression for C_t in the small-scale creep regime (Ref 18, 24) in which the load-line deflection can be directly measured during the test:

$$(C_t)_{SSC} = \frac{P(\dot{V}_c)_{SSC}}{BW}\frac{F'}{F} \quad \text{(Eq 25)}$$

where

$$F\left(\frac{a}{W}\right) = \frac{KB\sqrt{W}}{P}; \quad F'\left(\frac{a}{W}\right) = \frac{dF}{d\left(\frac{a}{W}\right)}$$

Analytically, the small-scale creep deflection rate can be determined by (Ref 38):

$$(\dot{V}_c)_{SSC} = \frac{2B(1-\nu^2)}{EP}K^2\chi\dot{r}_c \quad \text{(Eq 26)}$$

Substituting Eq 26 into Eq 25, an equation which directly relates $(C_t)_{ssc}$, K and $\dot{r}_c$ is determined:

$$(C_t)_{SSC} = 2(1-\nu^2)\,\chi\left(\frac{F'}{F}\right)\frac{K^2\dot{r}_c}{EW} \quad \text{(Eq 27)}$$

Using Eq 27, an analytical expression for a stationary crack can be derived for $(C_t)_{SSC}$ in which knowledge of load-line deflection is not needed, assuming that constants in the appropriate creep constitutive laws are available (Ref 38). For example, for elastic, secondary creep $(C_t)_{SSC}$ can be given by:

$$(C_t)_{SSC} = \frac{4\alpha\chi\tilde{r}_c(1-\nu^2)}{E(n-1)}(EA)^{2/(n-1)}\frac{K^4}{W}\frac{F'}{F}\left(\frac{1}{t}\right)^{(n-3)/(n-1)} \quad \text{(Eq 28)}$$

where α has been previously defined.

An expression for estimating C_t for a wide range of creep conditions from small-scale creep to extensive creep and also including primary creep has been derived that is very similar to the way in which $C(t)$ is derived (Ref 2, 18, 41):

$$C_t = (C_t)_{SSC} + C^*(t) \quad \text{(Eq 29)}$$

where the value of $(C_t)_{SSC}$ can be either experimentally determined from Eq 25 or analytically determined from Eq 27. If Eq 25 is used, the expression to measure C_t experimentally over the entire secondary creep range is given as (Ref 38):

$$C_t = \frac{P(\dot{V}_c - \dot{V}_{SS})}{BW}\left[\frac{F'}{F}\right] + C^* \quad \text{(Eq 30)}$$

where the load-line deflection rate in extensive creep conditions is subtracted from the total rate of deflection caused by creep.

Furthermore, a wide range expression for determining C_t in the presence of primary and secondary creep conditions has been determined in a similar manner as Eq 22 (Ref 41, 42):

$$C_t = \frac{P[\dot{V}_c - \dot{V}^*(t)]F'}{BW\;F} - C^*\left[\left(\frac{t_2}{t}\right)^{p/(1+p)} + 1\right] \quad \text{(Eq 31)}$$

where $\dot{V}_c^*(t)$ is the load-line deflection rate under extensive primary-secondary creep conditions.

Conditions with Growing Cracks. As previously stated, all the crack-tip parameters discussed thus far are based on the assumption of a stationary crack tip. Crack growth can significantly alter the crack-tip stress fields if the rate of crack growth is comparable to the rate at which creep deformation spreads at the crack tip. For a crack progressing with a velocity $\dot{a}$, the stress fields are dependent on the crack velocity by (Ref 43):

$$\sigma_{ij} = \alpha\left[\frac{\dot{a}}{AEr}\right]^{1/(n-1)}\tilde{\sigma}_{ij}(\Theta) \quad \text{(Eq 32)}$$

This stress field is alluded to as the Hui-Riedel (HR) field. By solving for r in Eq 6 and 32 and then setting them equal, it can be shown that for steady-state creep, stress distribution within the HR field is (Ref 44):

$$\sigma_{ij} = \left(\frac{EC^*}{\dot{a}}\right)^{1/2}\beta_1(\Theta) \quad \text{(Eq 33)}$$

It is intuitive from this equation that the extent of the HR field must be small in comparison to the extent of the HRR field for C^*(or C_t) to uniquely characterize the creep crack growth rate. By equating Eq 32 and 33, an estimation of the HR field size can be obtained. For materials in which the crack growth behavior is characterized by C^*(or C_t), the size of the HR zone has been estimated to be on the order of 0.01 Å which is negligible (Ref 43). This implies that C^* is a viable parameter even in the presence of growing cracks, provided the crack growth is slow in comparison to the rate of spread of creep deformation. This condition can be ensured by applying deflection-rate partitioning as shown below (Ref 27, 29):

$$\dot{V} = \dot{V}_e + \dot{V}_p + \dot{V}_c = \dot{a}\left[\left(\frac{\partial V_e}{\partial a}\right)_P + \left(\frac{\partial V_p}{\partial a}\right)_P\right] + \dot{V}_c \quad \text{(Eq 34)}$$

The term on the left side of Eq 34 is the total load-line deflection rate at constant load, while the first two terms on the right side are the deflection rates due to crack growth as a result of elastic and plastic compliance change, and the third term is due to creep deformation. The contribution of deflection due to creep is found by subtracting the deflection rates attributed to the change in elastic and plastic compliances from the total deflection rate:

$$\dot{V}_c = \dot{V} - \dot{a}\frac{B}{P}\left[\frac{2K^2}{E} + (m+1)J_p\right] \quad \text{(Eq 35)}$$

where J_p is the plastic portion of J and m is the plasticity exponent. Stationary crack tip parameters can only be used as long as the second term in the right-hand side contributes negligibly to $\dot{V}_c$. This condition is ensured by allowing only those data for which $\dot{V}_c/\dot{V} \geq 0.8$ (Ref 45).

Creep-Brittle Materials

In many situations, the crack growth rate is comparable to the rate of expansion of the creep zone, and the crack can no longer be assumed to be effectively stationary within an expanding creep field. These conditions are typical of creep-brittle materials, where the rate of creep strain accumulation at the crack tip is comparable to crack extension rates and where crack growth significantly perturbs the crack-tip stress field. Stated in another way, the HR field is no longer small in comparison to the extent of the HRR fields. Thus, the stationary-crack-tip parameters no longer characterize the crack-tip conditions and can no longer be expected to correlate uniquely with creep crack growth rate.

In creep-brittle materials, the rate of deflection caused by creep deformation represents only a small percentage of the total deflection rate; therefore, in the absence of significant plasticity, the rate of deflection caused by change in elastic compliance is comparable to the total deflection rate. Because the elastic contribution is analytically determined and the total deflection rate is experimentally measured, Eq 35 can sometimes erroneously yield negative creep deflection rates due to experimental error in the total deflection rate measurement. The negative creep deflection rates result in negative C_t values, which have no clear physical interpretation. Negative C_t values can also sometimes result if the creep zone size decreases (Ref 46), which is often the case in creep-brittle materials following incubation or toward the end of the test even in creep-ductile materials when stable crack growth sets in and the crack grows very rapidly. This aspect will be discussed in more detail in the next section.

Because Eq 35 is accurate only when the creep deflection rate dominates the total deflection rate, the equation lacks precision for creep contributions less than 80% of the total deflection (Ref 45). Therefore, the use of this equation and also the crack-tip parameters C_t and C^* is not suitable in creep-brittle materials.

Under special circumstances, time-independent fracture parameters, such as the J-integral (Ref 19, 20) or K, may correlate with the creep crack growth rate in creep-brittle materials. At elevated temperatures, some aluminum alloys have exhibited such creep-brittle behavior. For example, correlations between the creep crack growth rate and K have been established for aluminum alloys 2219-T851 (Ref 47, 48), 2519-T87 (Ref 49), and to a limited extent for 8009 (Ref 50). Similar correlations have also been demonstrated for other creep-brittle materials such as Ti-6242 (Ref 51) and for carbon-manganese steels at temperatures of 360 °C (680 °F) as discussed later in this article.

The precise conditions under which K or J characterize the crack growth behavior of creep-brittle materials are not yet well understood. However, the creep deformation resistance of creep-brittle materials is believed to be a significant factor. The accumulation of creep strain ahead of the crack tip is impeded in creep-brittle materials by microstructural features such as precipitates or dispersoids, and simultaneously decreasing rupture ductility increases the crack growth rate. As a result, the creep crack growth rate and the rate of creep strain accumulation in the crack-tip region are comparable. The movement of the crack perturbs the crack-tip stress fields, and it is no longer possible to represent the crack-tip fields using the Riedel-Rice formulation (Ref 32). In an idealized situation, one can imagine that the creep zone boundary and the crack tip in the plane of the crack move at equal speeds. Thus, the creep zone size at the crack tip remains constant, and using a coordinate system that moves with the crack tip as reference, the crack-tip stress field is also constant and completely determined by K. If it is assumed that the creep zone size remains small with respect to the pertinent length dimensions of the specimen and the shape also remains constant, creep crack growth is expected to correlate uniquely with K. In practice, all these conditions are most likely seldom met. Nevertheless, correlations between creep crack growth rates and K have been observed. However, the limitations of such correlations are not well understood. This remains an area of active research.

Incubation Period. When a cracked specimen is first loaded at elevated temperature, the material ahead of the crack tip is free from prior creep damage and, therefore, time-dependent crack growth does not begin instantly. Creep crack growth studies have shown that crack extension occurs following a specific time period, which has been termed incubation time. Incubation models based on ductility exhaustion and creep cavitation concepts have been developed (Ref 10, 52, 53); however, incubation time currently lacks a precise definition among researchers because it is unknown if the crack actually remains stationary during this period or if the crack grows at an indiscernible rate. Some researchers have defined the incubation period as the time required for the crack to grow through the initial creep zone size (Ref 54), while others have defined it based on a certain increase in the direct current potential drop output when utilizing the potential drop technique to monitor crack extension (Ref 55).

The incubation period is more pronounced in creep-brittle materials than in creep-ductile materials, and it can comprise nearly 90% of the total test time during creep crack growth testing of creep-brittle specimens. Thus, incubation period is vitally important when considering the creep crack growth characteristics of creep-brittle materials. The correlation between the creep crack growth rate and the stress-intensity factor displays a steep slope in creep-brittle materials, indicating that the crack growth rate rapidly increases for very small increases in K. The K-range, therefore, over which creep crack extension occurs prior to specimen fracture is small. Thus, once the crack begins to grow, the test material cannot sustain stable crack extension for a significant duration before failure occurs, a characteristic that limits the engineering application of the material. Some studies, however, indicate that the incubation period required prior to crack extension quickly increases for small decreases in the initial K-level (Ref 49). For some critical K-levels and below, the incubation time may be so large that the crack effectively remains stationary within the testing time frame. Total time to failure, therefore, must be considered to be a combination of the incubation period prior to crack initiation and the crack growth period following crack initiation. Evaluation of the test material for engineering applications cannot solely rely on the da/dt-K correlation, but must take into account the time required to initiate creep crack growth.

Crack-Tip Parameters for Creep-Fatigue Crack Growth

From previous sections, C_t appears to be the most appropriate crack-tip parameter for correlating creep crack growth over the regime from small-scale creep to extensive creep conditions for creep-ductile materials. The average value of C_t, $(C_t)_{avg}$, is used for characterizing the average crack growth rate, $(da/dt)_{avg}$, in creep-fatigue experiments (Ref 56–60).

The value of $(C_t)_{avg}$ is determined by two methods. The first method is suitable for specimens when load-line deflections during hold time are available, while the second is well suited for components when the load-line deflection must be calculated using the deformation laws for the material. Expressions for calculating $(C_t)_{avg}$ from laboratory experiments and in components have been developed (Ref 56, 61–63). The discussion here focuses on calculating $(C_t)_{avg}$ from experimental measurements because the primary emphasis in this article is on test methods for characterizing creep and creep-fatigue crack growth.

For materials whose deformation behavior is characterized by elastic, secondary creep, $(C_t)_{avg}$ can be determined as follows (Ref 63):

$$(C_t)_{avg} = \frac{\Delta P\, \Delta V_c}{B\, W\, t_h} \frac{F'}{F} - C^* \left(\frac{F'}{F} \frac{1}{\eta} - 1 \right) \qquad \text{(Eq 36)}$$

where ΔP is the applied load range, ΔV_c is the load-line deflection change due to creep during hold time t_h. Because the amount of crack extension during hold times is small, the change in total deflection during hold times is approximately equal to ΔV_c. The remaining variables have been previously defined. As previously noted for compact specimens, the ratio $(F'/F)/\eta \approx 1$, thus Eq 36 can be written as (Ref 57):

$$(C_t)_{avg} = \frac{\Delta P\, \Delta V_c}{B\, W\, t_h} \frac{F'}{F} \qquad \text{(Eq 37)}$$

It is also noted that the ΔV_c value in Eq 37 includes both primary and secondary creep contributions to the load-line deflection change. Thus Eq 37 also accounts for primary creep deformation in the estimation of $(C_t)_{avg}$. The $(da/dt)_{avg}$ is calculated as follows:

$$\left(\frac{da}{dt}\right)_{avg} = \frac{1}{t_h} \left[\left(\frac{da}{dN}\right) - \left(\frac{da}{dN}\right)_{cycle} \right] \qquad \text{(Eq 38)}$$

where $(da/dN)_{cycle}$ is the cyclic crack growth rate and has to be obtained from a continuous cycling

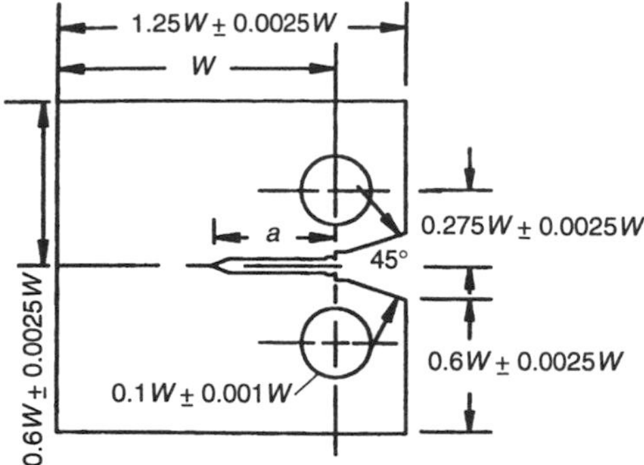

Fig. 6 Typical compact tension specimen dimensions (in inches) for a 1.00 in. thick (1T) specimen per ASTM E 1457. Specimen thickness (*B*) is related to width (*W*) as *B* = 0.5W ± 0.005 *W*.

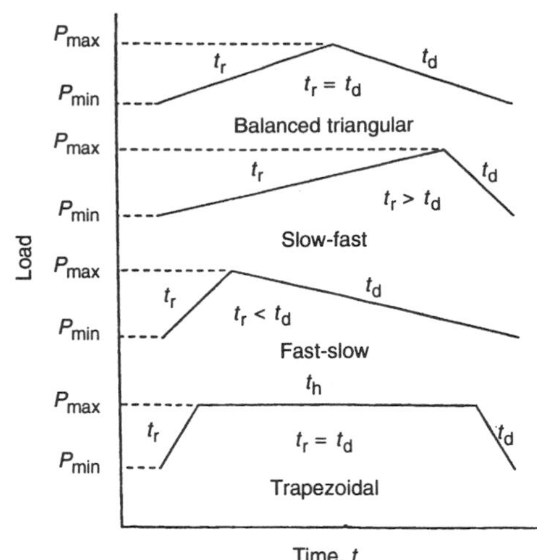

Fig. 7 Typical loading waveforms used during creep-fatigue crack growth testing. Source: Ref 66

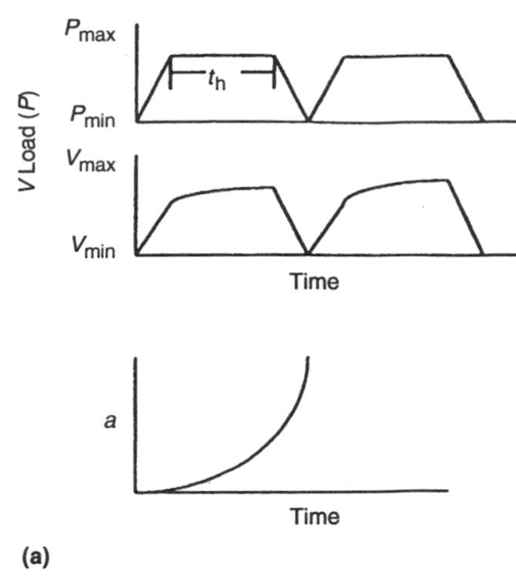

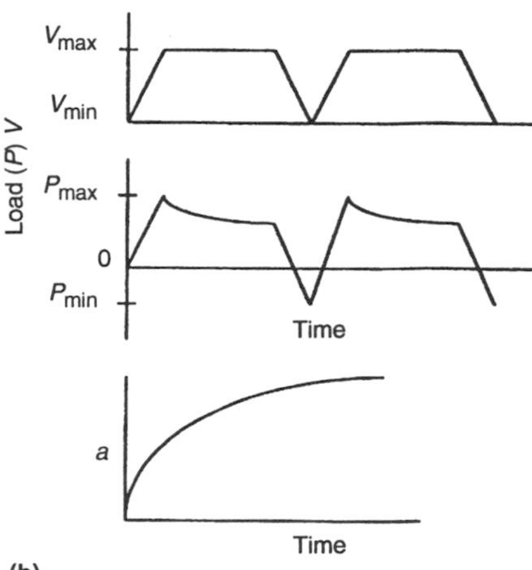

Fig. 8 Schematic comparison of load-controlled (a) and displacement-controlled (b) testing under trapezoidal loading. Source: Ref 66

fatigue crack growth rate test for which the rise and decay times are usually the same as the trapezoidal waveform used in the creep-fatigue crack growth tests. The overall crack growth rate, *da*/*dN*, is calculated from the crack length measurements during the creep-fatigue crack growth test.

Creep Crack Growth Testing

Creep crack growth tests are performed in accordance with ASTM E 1457, "Measurement of Creep Crack Growth Rates in Metals" (Ref 45) using the compact specimen geometry. The test procedure involves heating precracked specimens to the prescribed test temperature and applying a constant load until significant crack extension or specimen failure occurs. During the test, the crack length and load-line deflection are constantly monitored with time. The reduced test data compare the time rate of crack growth, *da*/*dt*, in terms of an elevated-temperature crack growth parameter described in the earlier section of this article.

Equipment

Specimen Configuration. The most widely used specimen configuration for creep and creep-fatigue crack growth testing is the compact-type (CT) specimen as shown in Fig. 6. Certain special dimensional requirements such as notch configuration and specimen width-to-thickness ratios can be found in ASTM specifications (Ref 45). Other configurations such as the center-cracked tensile (CCT) panel and single-edge notch (SEN) specimens have also been used; however, due to several reasons of convenience, the CT specimen geometry remains the most suitable geometry for creep and creep-fatigue crack growth testing. First, the transition time for extensive creep conditions to develop in CT specimens is longer than in CCT specimens for the same *K* and *a*/*W* for specimens of same width (*W*) (Ref 64). Because of the longer transition times in the CT specimens, the condition that $t_c/t_1 << 1$ during creep-fatigue testing, where t_c is the cycle time, is more easily satisfied. Another advantage of the CT specimen is that a clip-gage can be conveniently placed at the load-line to measure the deflection change. This is important for reliable and direct measurement of load-line deflection change, which is the measured quantity required to calculate the crack-tip parameters. Perhaps the most important advantage of CT specimens is that the magnitude of the applied load for the same applied value of *K* is significantly lower than for CCT specimens. Thus, smaller load capacity machines and smaller fixturing can be used for testing.

Test Machine. Creep crack growth testing should be performed in either deadweight or servomechanical test machines that are capable of maintaining a constant load over an extended period of time. During the duration of the testing, variations in loading must not exceed ±1.0% of the nominal value at any time. When using cantilever-type deadweight creep machines, it is important that the lever remain as close to horizontal as possible. The cantilever loading conditions are performed such that the loading ratios are 20 to 1 or 10 to 1, depending on the level length. Significant deviation of the cantilever from the level position results in significant variations in the load on the specimen. ASTM E 4 "Practices for Load Verification of Testing Machines" (Ref 65) details the required accuracy for the test equipment.

Creep-fatigue tests can be conducted under either load-controlled or displacement-controlled conditions. Figure 7 (Ref 66) shows various waveforms used for load-controlled testing in which the specimen is cycled between a fixed maximum and minimum load value. The displacement versus time and crack size versus time responses in these tests are shown schematically in Fig. 8(a). Displacement-controlled tests involve cycling between fixed displacements as shown schematically in Fig. 8(b) (Ref 66) along with the corresponding changes in the load-line displacement and the load versus time and crack size versus time responses.

The majority of the creep-fatigue testing performed to date has been under load-controlled conditions, largely because it is more convenient. Displacement-controlled testing requires specially designed grips that ensure that the specimen does not buckle or experience any torsional stresses during the compressive portion of the loading. However, this type of testing has some advantages over load-controlled testing, as discussed below.

In load-controlled tests, the net section stress ahead of the crack rises continuously as the crack grows. Thus, the stress-intensity factor continuously increases with crack growth while the size of the remaining ligament decreases. The scale of plasticity or creep in the specimen increases as the test progresses causing ratcheting in the specimen with accumulation of inelastic deflection after each cycle. In displacement-controlled tests, the applied load decreases with crack extension, as shown in Fig. 8(b), and the specimen deflection is forced to the minimum value at the end of each cycle. Therefore, ratcheting of displacement cannot occur (Ref 67), and data can be collected for greater crack extensions than in load-controlled tests. Load-controlled tests are suitable for low crack growth rates, and displacement amplitude tests are more suitable for higher crack growth rates (>4 × 10^{-6} mm/cycle) and longer hold time tests. Gladwin et al. (Ref 68) have noted that added complication due to static modes of fracture (e.g., tearing) may influence the creep-fatigue damage process in positive R load-controlled tests. Therefore, the apparent accelerations in crack growth rate associated with the hold time (creep damage) may be a result of, for example, stable tearing effects due to large deformations caused by ratcheting effects rather than the result of true creep-fatigue interaction (i.e., accelerated fatigue crack growth because of creep damaged material) (Ref 68). This problem is avoided when testing is performed under displacement range control.

Specimen heating is usually performed in electric resistance furnaces or convection laboratory ovens. Temperature control should be within ±2 °C (±3 °F) for tests performed at temperatures up to 1000 °C (1800 °F) and ±3 °C (±5 °F) for tests above 1000 °C (1800 °F). Specimens are usually tested in air at atmospheric pressure; however, inert atmosphere environments and vacuum conditions have been used and aid in the reduction of the effects of oxidation and other forms of corrosion. Thermocouples are used to monitor the specimen temperature. The thermocouples should be located in the uncracked ligament within a 2 to 5 mm (0.08 to 0.2 in.) region around the crack plane. Thermocouples should be in intimate contact with the specimen. Ceramic insulation is recommended for covering the individual lines to prevent shorting of the temperature circuit.

Fixturing. Pin-and-clevis assemblies are used to support test specimens in the load frames. This type of fixturing is used on both top and bottom specimen faces and allows in-plane rotation during specimen loading and during subsequent crack extension. Materials used for these fixtures must be creep resistant and able to withstand the loading and temperatures employed. The fixtures can be fabricated from grades 304 and 316 austenitic stainless steel, grade A286 precipitation-hardenable stainless steel, and nickel-base alloys Inconel 718 and Inconel X-750 in the annealed or solution-treated condition. After fabrication, hardenable parts should be heat treated so that they develop resistance to creep deformation.

Fatigue precracking is performed on creep test specimens to eliminate any effects of the machined notch and to provide a sharp crack for initial crack growth. The methodology for fatigue precracking is described extensively in ASTM E 399, "Test Method for Plane-Strain Fracture Toughness of Metallic Materials" (Ref 69). The precracking is to be carried out on the material in the same condition in which it is to be tested for creep crack growth behavior. The precracking can be performed between room temperature and the anticipated test temperature. The equipment used to precrack should allow symmetric load distribution in reference to the machined notch, and the maximum stress-intensity factor during the operation, K_{max}, should be controlled within ±5%. The procedure may be carried out at any frequency that allows accurate load application, and the specimen should be precracked to at least a length of 2.54 mm (0.100 in.).

The initial precracking is conducted at stress intensities high enough to allow crack initiation and growth out of the machined starter notch and with growth of the precrack, the load is decreased to avoid transient effects and to allow lower stress intensities for the creep test. The load values for precracking, P_f, are determined so as not to exceed the following value (Ref 45):

$$P_f = \frac{0.4B_N (W - a_0)^2 \sigma_{YS}}{(2W + a_0)} \quad \text{(Eq 41)}$$

where B_N is the corrected specimen thickness, W is the specimen width, and a_0 is the initial crack length measured from the load-line and σ_{YS} is the yield strength. During the final 0.64 mm (0.025 in.) of fatigue precrack extension, the maximum load should not be larger than P_f as determined from above or a load such that the ratio of the stress-intensity factor range to the Young's modulus ($\Delta K/E$) is equal to or less than 0.0025 mm$^{1/2}$ (0.0005 in.$^{1/2}$) (Ref 45), whichever is smaller. In doing so it is ensured that the final precrack loading does not exceed the initial load used during creep crack growth testing.

The crack length during precracking can be measured optically with a traveling microscope if the test is being performed at room temperature, and if the precracking is performed at elevated temperature, the electric potential drop technique can be used. Measurement of the fatigue precrack should be accurate within 0.1 mm (0.004 in.). Measurements should be made on both surfaces, and the value should not vary more than 1.25 mm (0.05 in.). If the surface cracks exceed this range, further extension is necessary until these criteria are met.

Required Measurements

Crack Length Measurement (Electric Potential Method). In monitoring the crack propagation during the elevated-temperature creep test, crack extensions must be resolved to at least 0.1 mm (0.004 in.). Because optical measurement techniques within an enclosed furnace are not feasible and the through-thickness crack fronts sometime differ significantly from observed surface lengths, crack length measurements are most commonly made during an elevated-temperature creep test using the electric potential drop technique. By applying a fixed electric current, any increase in crack length (a corresponding decrease in uncracked ligament) results in an increase in electric resistance, which is noted as an increase in output voltage across the output locations.

The current input and voltage output lead locations for typical CT and CCT specimens are shown in Fig. 9. These leads can either be welded onto the specimen or connected with screws. The choice of application method is dependent on material and test temperature. For softer materials tested at lower temperatures, threaded connections would be acceptable, but for harder materials and especially at higher temperatures, welding of leads is recommended. The leads should be

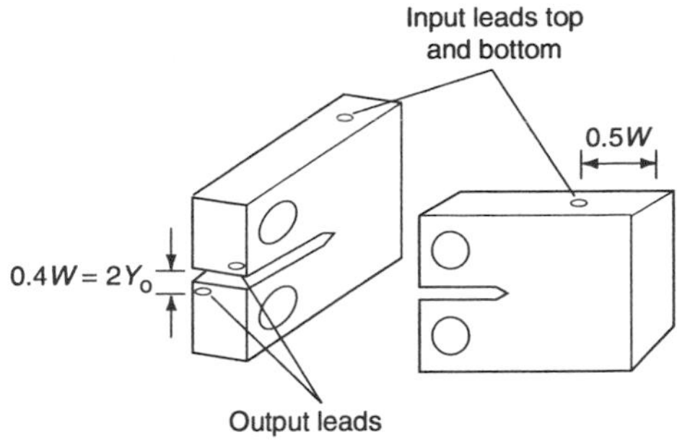

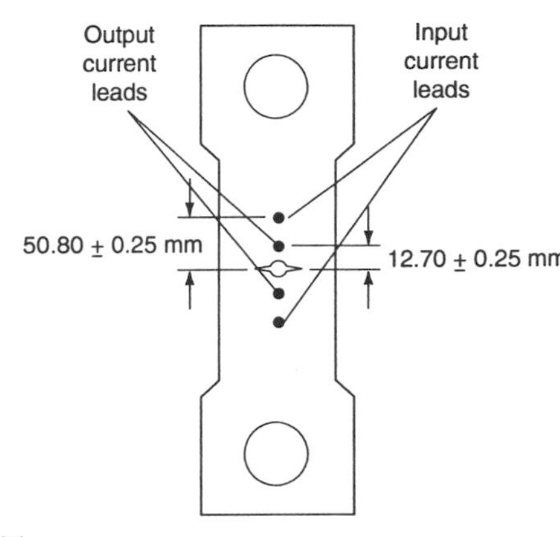

Fig. 9 Input current and voltage output lead locations for typical compact (a) and center-cracked tension specimens (b)

sufficiently long to allow current input devices and output voltage measuring instruments to be well away from the furnace to avoid excessive heating. The leads should also be approximately the same length and contain similar junctions to avoid excessive lead resistance, which contributes to the thermal voltage, V_{th} as described below. Use of 2 mm (0.08 in.) diameter stainless steel wires have been shown to work well because of superior oxidation resistance at elevated temperature; however, any oxidation-resistant material capable of carrying a current that is stable at the test temperature may be utilized as lead connectors. Nickel and copper wires have been successfully used as lead material for lower temperature tests.

When using the direct current electric potential method, the instantaneous voltage V and the initial voltage V_0 usually deviate from the indicated readings. This is due to the thermal voltage, V_{th}, which is caused by several factors, such as differences in the junction properties of the connectors used, differences in the resistance of the output leads, differing output lead lengths, and temperature differences in output leads themselves. Measurements should be taken of the V_{th} prior to the load application and at various times during the test. These measurements are made by turning off the current source and recording the output voltage. Before analyzing the crack length data, the values of V_{th} should be subtracted from the respective V and V_0 in order to determine the actual crack extension.

Knowing the corrected original voltage, V_0, and the corrected instantaneous change in voltage during crack extension, V, the crack length in a CT specimen can be computed by using the following closed form equation (Ref 55, 70):

$$a_i = \frac{2W}{\pi} \cos^{-1} \left\{ \frac{\cosh\left(\frac{\pi Y_0}{2W}\right)}{\cosh\left[\frac{V}{V_0} \cosh^{-1}\left(\frac{\cosh \frac{\pi Y_0}{2W}}{\cos \frac{\pi a_0}{2W}}\right)\right]} \right\} \quad \text{(Eq 42)}$$

where a_i is the instantaneous crack length, a_0 is the original crack length after precracking, Y_0 is the half separation distance between the voltage output leads, V_0 is the initial output voltage before load application, V is the instantaneous output voltage, and W is the specimen width.

Materials with high electrical conductivities can experience fluctuations in V_{th}. These fluctuations can be of the same magnitude as the voltage changes that accompany crack extension and could mask this information. Because of the potential variation in thermal voltage, the direct current electric potential method should not be the only nonvisual technique for crack length measurement. The use of more sophisticated electric potential setups, such as the reversing potential method, is recommended.

Crack lengths, both initial and final, are required to differ by no more than 5% across the specimen thickness. Maintaining a straight crack front is sometimes dependent on material and material thickness. Upon post test examination, thicker specimens have been noted to experience crack tunneling, or nonstraight crack extension. Crack tunneling (thumbnail-shaped crack fronts) is common in non-side-grooved (or parallel-sided) specimen configurations (Ref 8). This occurrence is a direct result of the conditions being closer to plane stress near the surfaces of the specimen and plane strain in the center, which results in higher crack growth rates near the specimen center. Side grooving of test specimens on the crack plane has been shown to greatly reduce this problem. Side grooves up to 25% in reduction are acceptable, but reductions of approximately 20% have been found to work well for many materials. The included angle of the grooves is typically less than 90° with a root radius less than or equal to 0.4 ± 0.2 mm (0.016 ± 0.008 in.). It is prudent to perform the side grooving after fatigue precracking because precracks are hard to see when located in the grooves.

Load-Line Deflection Measurements. Continuous displacement measurements are needed in the determination of the crack-tip parameters for the duration of the testing. These displacement readings should be taken directly from the load-line as much as possible. For CT specimens, the measuring device should be attached on the machined knife edges of the specimens. For CCT specimens, the deflection is to be measured on the load line at points that are ±35 mm (±1.40 in.) from the crack centerline. The measurement of the displacement can be directly measured by placing an elevated-temperature clip gage (either strain gages for temperatures up to approximately 150 °C (300 °F) or capacitance gages for higher temperatures) on the specimen and placing the entire assembly inside the furnace. If this type of device is not available, the displacements may be transferred outside the furnace with a rod-and-tube assembly that is connected to a displacement transducer—either a direct current displacement transducer (DCDT), linear variable displacement transducer (LVDT), or capacitance gage—outside the furnace. In these transfer-type displacement devices the transfer rod and tubes should be fabricated of material that experiences low thermal expansion and is thermally stable (Ref 45). The resolution of these deflection measurement devices should be a minimum of 0.01 mm (0.0004 in.) or less.

If measurements cannot be taken directly on the specimen load-line, deflection can be measured from the test machine crosshead movement with the use of dial gages and/or the displacement transducers noted above. Under the constant loading conditions, the crosshead/load-train deflection will be primarily due to the test specimen with the exception of the initial deflection, which will contain some elastic contributions.

With smaller-sized specimens, the reduced notch dimensions will not accommodate a clip-on load-line gage. Past research has used a modified rod-and-tube extensometer connected to an external DCDT, the arms of which were attached on the outer surface (top and bottom faces) of the specimens on the load-line just above the pin-holes (Ref 71). Clevises with deep throats can be used to accommodate this extensometry.

Data Acquisition. The measurements taken during the testing, electric potential voltage, load-line displacement, temperature, and cycles for creep-fatigue tests can be recorded continuously with the use of strip chart recorders, voltmeters, or digital data acquisition systems. The resolution of these acquisition systems should be at least one order of magnitude better than the measuring instrument. However, no matter which technique is used, it is important to remember the thermal voltage in the electric potential readout should be measured at least once every 24 h as described previously.

Typical Test Procedures

Test Setup. Because of the inherent scatter observed in most test situations, more than one test per condition is suggested so that data confidence intervals can be obtained. In creep and creep-fatigue crack growth testing, variables such as microstructural differences, load precision, environmental control, and, to a lesser degree, data processing contribute to scatter (Ref 45).

Prior to installation in the testing machine, the test specimen should be fitted with electric potential leads. The exposed surface of the potential leads that are on the interior of the furnace can be covered with ceramic insulators or other shielding to avoid direct contact with the furnace elements and other components (thermocouples, extensometry, etc.) inside the furnace. After securing of the test specimen in the clevises with clevis pins, a slight load not exceeding 10% of the intended test load may be applied to improve the axial stability of the load train. The extensometer is then placed on the load line of the specimen, and care is taken to ensure that the knife edges are securely in contact. The thermocouples are then placed in contact with the specimen on the crack plane in the uncracked ligament region. Figure 10 displays a close-up of a fully fitted 0.25T-CT test specimen prior to furnace heating. The furnace is then placed into position, sealed, and started. Specimen heating should be performed gradually to avoid overshooting the test temperature because the aforementioned temperature control limits also apply to specimen heating. The current in the electric potential system should also be on during the furnace heating because resistance heating of the specimen occurs by the applied current. Once the test temperature is achieved and stable, the specimen is allowed to "soak" for at least 1 h per 25 mm (1 in.) of specimen thickness at the test temperature prior to applying load. Once sufficient time at temperature has been achieved, a set of measurements are recorded in the no-load condition for the reference conditions. Then the load is carefully applied in order to avoid inertial loading. The choice of load or K

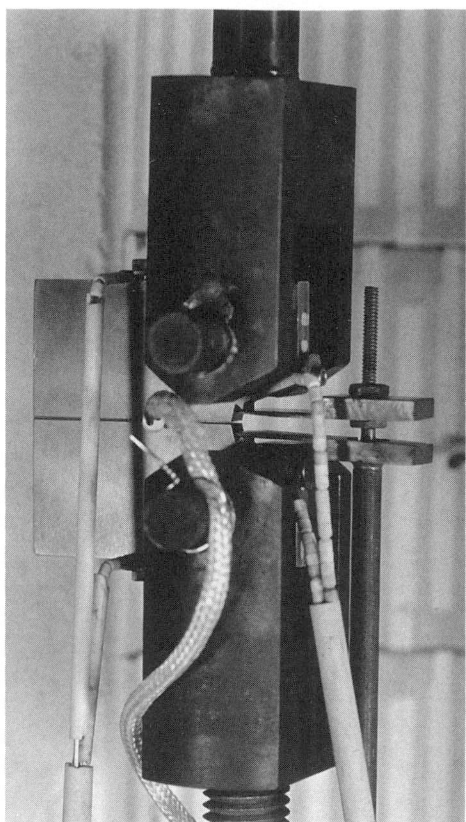

Fig. 10 Typical installed specimen ready for creep crack growth testing

level is dependent on the crack growth rates required during the test. Ideally, crack growth rates should be the same as those encountered by the material during service. The time of specimen loading should be as short as possible, and another set of measurements of electric potential and displacement are recorded immediately upon completion of loading as the initial loading condition (time = 0).

For creep-fatigue crack growth testing, additional consideration is given to specimen loading because cyclic tests are required. The hold time should be selected in conjunction with the K-level such that crack extensions of approximately 5 mm (0.2 in.) are obtained during the planned duration of the test. The hold time should also be selected such that the deflection that accumulates during the hold time is approximately three to five times the sensitivity of the displacement gage/amplifier system used. The loading waveshape should simulate the service loading conditions. As mentioned above, load-controlled testing can be performed under a variety of waveforms. For power-plant components and gas turbines used in airplanes, a trapezoidal waveform is a good approximation. The rise-decay and hold times should be representative of the relative times of fatigue and creep loading conditions in service.

Post Test Measurements. Once the test is completed, either due to specimen failure or by attainment of sufficient crack growth, the load is removed, the furnace is turned off, and the specimen is allowed to naturally cool and is then removed from the loading clevises. The original crack length (after precracking) and the final crack length (resulting from creep crack growth) are measured at nine equally spaced locations along the crack front. All the data are processed using computer programs that utilize either the secant method or seven-point polynomial method to calculate the deflection rates, dV/dt, crack growth rates, da/dt and the crack-tip parameters discussed previously. The details of these methods can be found in the ASTM E 1457 (Ref 45).

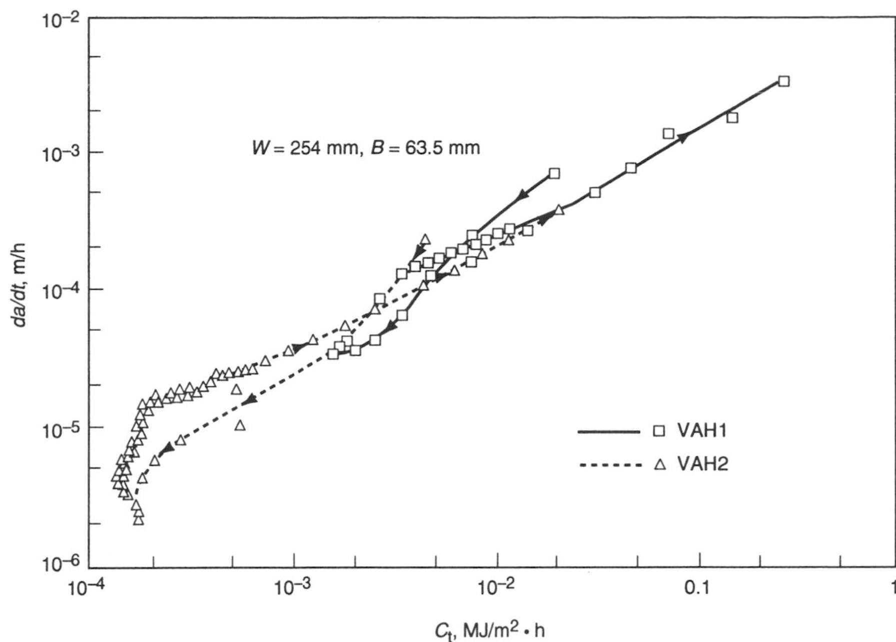

Fig. 11 Creep crack growth rate from specimens of 63.5 mm (2.5 in.) thickness in which crack growth occurred in small-scale, transition, and extensive creep regimes. Source: Ref 51, 72

Crack Growth Correlations

Creep Crack Growth. The experimental creep crack growth rate data for creep-ductile materials have been shown to correlate with the C^*-integral and the C_t parameter. As demonstrated in previous discussion, C_t and C^* are identical in the extensive creep regime. However, C_t is also valid in the small-scale and transition regimes where C^* is no longer path-independent and therefore does not uniquely characterize the crack-tip stress fields. Because C_t is more general than C^*, it has been chosen as the primary parameter for correlating creep crack growth in the discussion that follows. Figure 11 shows the creep crack growth rate of 1Cr-1Mo-¼V steel obtained from specimens that were 254 mm (10 in.) wide and had nominal thicknesses of 63.5 mm (2.5 in.) (Ref 72). These experiments were performed at the National Research Institute for Metals (NRIM) in Japan (Ref 72). In these tests, the initial crack growth rate occurred in the small-scale and the transition creep regions, while in the latter part of the test extensive creep conditions prevailed. The arrows mark the direction in which data were collected. Note that a high C_t value was obtained in the initial portion of the tests that progressively decreased to a minimum value and then subsequently increased. This was the pattern in both the tests identified by the specimen numbers VAH1 and VAH2. The authors note that the crack growth rates uniquely correlate with C_t during the increasing and decreasing portions of the tests, lending support to the theory that C_t can correlate crack growth rates over the wide range of conditions from small-scale to extensive creep.

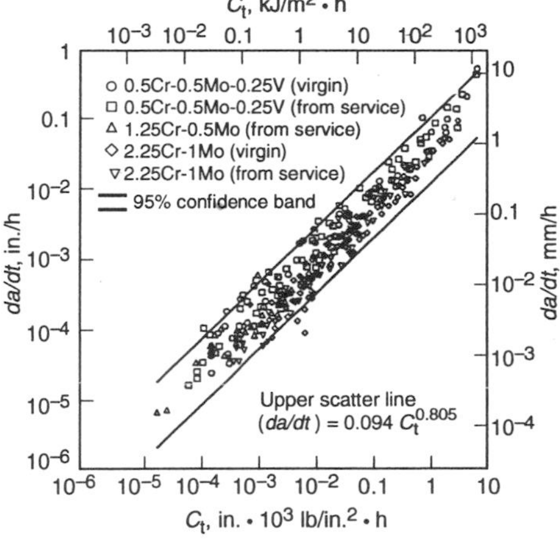

Fig. 12 Creep crack growth rate for chromium-molybdenum steels (tested at 1000 to 1022 °F) compiled from various laboratories. Source: Ref 7, 73

Figure 12 shows the creep crack growth rates for several chromium-molybdenum steels correlated with C_t (Ref 73). These data were consolidated from several experimental studies, essentially proving that consistency in the data can be obtained from one laboratory to another during creep crack growth testing. It is also observed that when da/dt is correlated with C_t (or C^* for that matter, in the case of extensive creep), a first-order normalization of temperature effects are obtained. Because C_t and C^* are obtained from the product of the load, P, and the creep displacement rate, $\dot{V}_c$, where one is applied and the other is a response dependent on temperature, it is expected that to some extent the correlation between da/dt and C_t is independent of temperature (Ref 73). However, when changing the temperature will result in fundamental changes in creep deformation and damage mechanisms, such normalization of temperature effects is not expected.

Figure 13 shows the correlation between da/dt and C^* for 304 stainless steel where compact as well as center crack tension geometries were used for obtaining the data (Ref 12). This demonstrates the geometry independence of such data. All tests in this study were in the extensive creep conditions, thus no distinction is made between C_t and C^*.

Figure 14 shows creep crack growth rate data from specimens of different sizes from the 1Cr-1Mo-¼V steel (Ref 37). It is noted that the data from the specimens 6.3 mm (0.25 in.) thick seem to lie at the lower end of the scatterband while the data from the specimens 64 mm (2.5 in.) thick seem to lie on the upper end of the scatterband. There seems to be a systematic effect of thickness indicating a state-of-stress effect. Therefore, in generating creep crack growth rate data, it is essential to give proper consideration to the

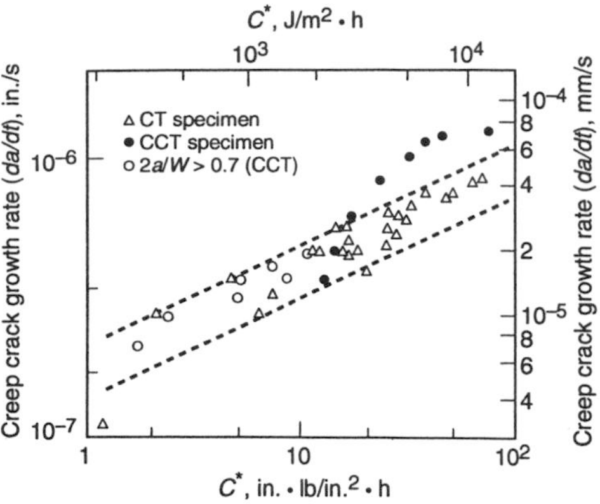

Fig. 13 Creep crack growth rate for 304 stainless steels at 595 °C (1100 °F) with differing specimen geometries. Source: Ref 12, 27

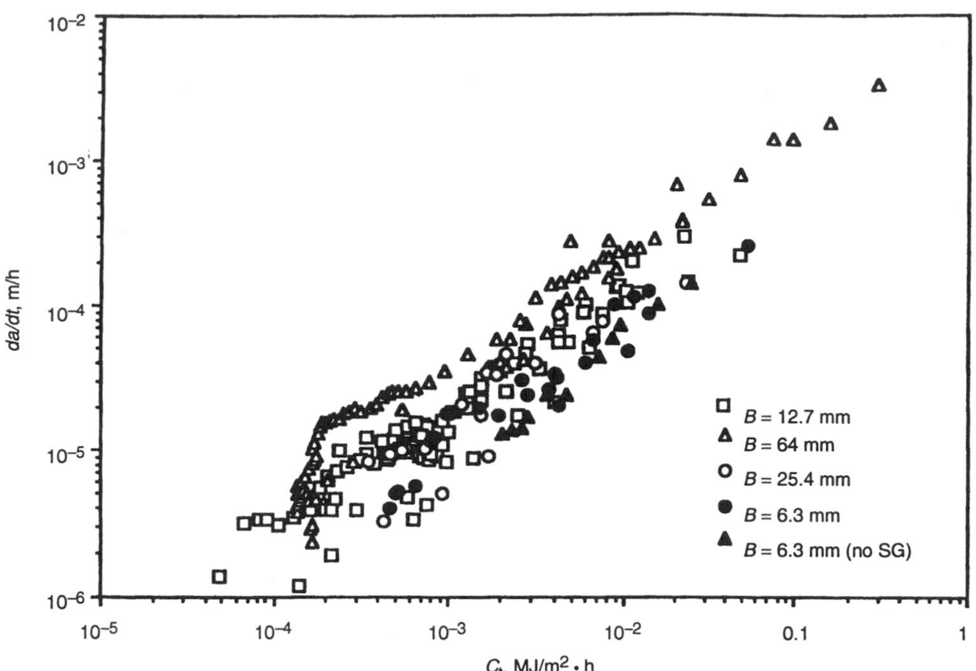

Fig. 14 Creep crack growth rate for 1Cr-1Mo-¼V steels with differing specimen sizes. SG, side grooved. Source: Ref 37, 51

thickness of the specimen, depending on the end use of the data.

Figure 15 shows the creep crack growth rate as a function of K for a highly cold-worked carbon-manganese steel tested at 360 °C (680 °F) (Ref 74). At this temperature, this material exhibits creep-brittle behavior. These correlations are very sensitive to the effects of temperature. Similar correlations have been shown for nickel-base alloys (Ref 75, 76), cold-worked 316 stainless steel (Ref 29), Ti-6242 alloy (Ref 51), and for 2519 Al alloy (Ref 49). Conditions for the unique correlation between da/dt and K are not well understood for creep-brittle materials, and this continues to be an area of considerable research.

Creep-Fatigue Crack Growth Correlations. Figure 16 (Ref 61) shows plots of da/dN versus ΔK for a 2.25Cr-1.0Mo steel at 595 °C (1100 °F). The regression line through the elevated-temperature fatigue test ($t_h = 0$) is used to get the cycle-dependent part in modeling the creep-fatigue data. The lack of correlation between da/dN and ΔK for the creep-fatigue tests is evident from the data scatter in this figure. Such lack of correlation has also been shown in Cr-Mo-V steel (Ref 77). An increase in the da/dN with increasing hold time for fixed ΔK has also been reported by Saxena and Bassani (Ref 78). This is due to the increasing contribution of time-dependent crack growth (Ref 79). Creep damage at the crack tip, influence of the environment, or microstructural changes such as formation of cavities that occur during loading at elevated temperatures could be responsible for this behavior (Ref 58).

The average time-dependent crack growth rates, $(da/dt)_{avg}$, are correlated with $(C_t)_{avg}$. Figure 17 (Ref 17) is a plot of $(da/dt)_{avg}$ versus $(C_t)_{avg}$ for 2.25Cr-1.0Mo steel, for various hold times at elevated temperatures. All data show a clear trend and fall into a narrow scatterband despite the range of hold times used. Similar trends have also been shown for 1.25Cr-0.5Mo steel as shown in Fig. 18 (Ref 60). This strongly indicates the usefulness of $(C_t)_{avg}$ in characterizing creep-fatigue rates. Furthermore, the creep crack growth (CCG) data for each of these materials has also been plotted on these graphs. However, da/dt has been correlated with C_t for the creep crack growth data. All the creep and creep-fatigue crack growth rate data show the same trend. This has the important implication that life prediction procedures for these materials would be considerably simplified because CCG data could be used to predict the life of components under creep-fatigue conditions and vice versa. In comparing a $(da/dt)_{avg}$ versus $(C_t)_{avg}$ relation of creep-fatigue with a da/dt versus C_t relation of CCG, it must be kept in mind that although C_t and $(C_t)_{avg}$ are equivalent parameters with the same physical interpretation, their exact values may differ slightly in the small-scale creep regime by a constant factor for a given material (Ref 60). The value of this constant ranges from 1 to 1.3 for different materials (Ref 60).

The time dependence of the life-prediction model is obtained by generating a regression line through all the data. The total fatigue crack growth rate per cycle is a linear summation of the cycle and time-dependent crack growth rates. Such an expression obtained for 2.25Cr-1.0Mo steel at 595 °C (1100 °F) under trapezoidal loading waveshapes is given in the following equation (Ref 57) for SI units (mm/h, MPa$\sqrt{m}$, and C_t in kJ/m² · h):

$$\frac{da}{dN} = 1.08 \times 10^{-6}\,\Delta K^{1.94} + 1.46 \times 10^{-2}\left[(C_t)_{avg}\right]^{0.722} t_h \quad \text{(SI)}$$

In English units (in./h, ksi$\sqrt{\text{in.}}$, and C_t in units 10^3 lb/in. · h) the relation is:

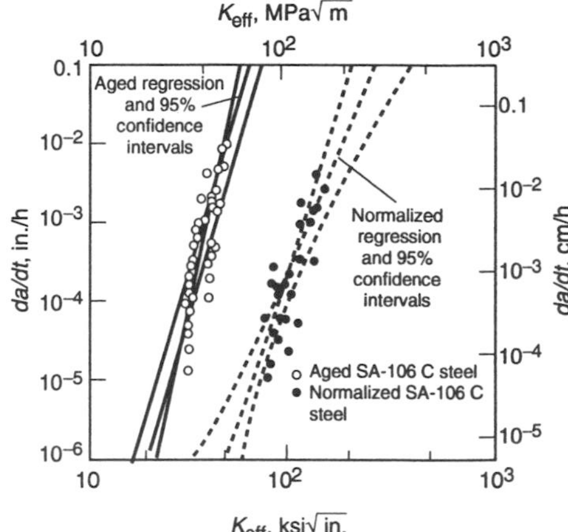

Fig. 15 Creep crack growth rate for a typical creep-brittle material. Source: Ref 74

$$\frac{da}{dN} = 5.11 \times 10^{-8}\,\Delta K^{1.94} + 2.40 \times 10^{-2}\left[(C_t)_{avg}\right]^{0.722} t_h \quad \text{(English)}$$

The first term in the equations above represents the cycle-dependent crack growth rate and the other term represents the time-dependent crack growth rate. These equations can be effectively used to predict the service life of high-temperature components made of 2.25Cr-1.0Mo steel under both creep and creep-fatigue crack growth conditions at 595 °C (1100 °F). An upper and lower scatterband can also be generated for the data in Fig. 17 and 18 for design purposes. This

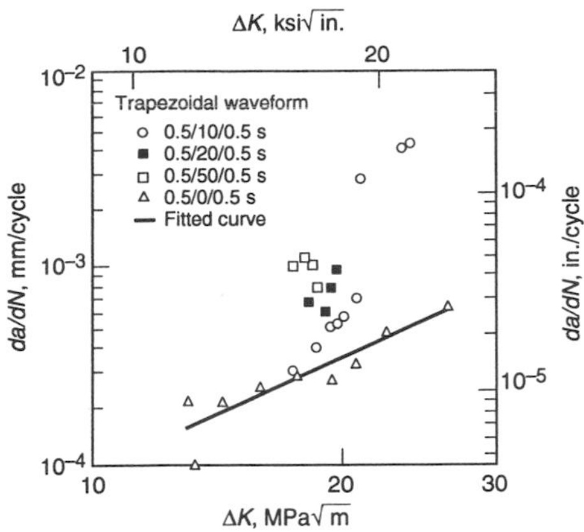

Fig. 16 da/dN versus ΔK for a 2¼Cr-1Mo steel at 595 °C (1100 °F) tested with and without hold times. Source: Ref 61

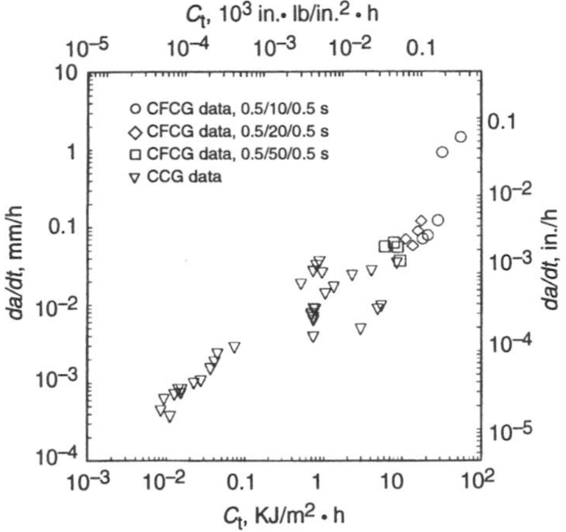

Fig. 17 Correlation of measured crack growth rates with the C_t calculated from experimental measurements (Ref 61) for 2.25Cr-1.0Mo steel at 595 °C (1100 °F). Note da/dt versus C_t plotted for the creep crack growth data and $(da/dt)_{avg}$ with $(C_t)_{avg}$ for the creep-fatigue data)

model has been established under the assumption that the crack growth during hold time is only due to creep deformation. Any other time-dependent effects like oxidation at the crack tip have not been considered (Ref 57). Neither have any synergistic effects due to any complicated interactions of the creep and fatigue mechanisms of crack growth during unloading/reloading been incorporated. However, with the assumption that the unloading/reloading times are much smaller compared with the hold times, their exclusion seems justified (Ref 57). If such effects were to be considered, depending on the material, an equation of the type presented above would be too simplistic in its description of the creep-fatigue behavior of a material. This remains a subject of future research.

ACKNOWLEDGMENT

The information in this article is largely taken from R.H. Norris, P.S. Grover, B.C. Hamilton, and A. Saxena, Elevated-Temperature Crack Growth, *Fatigue and Fracture,* Vol 19, *ASM Handbook,* 1996, p 507–519.

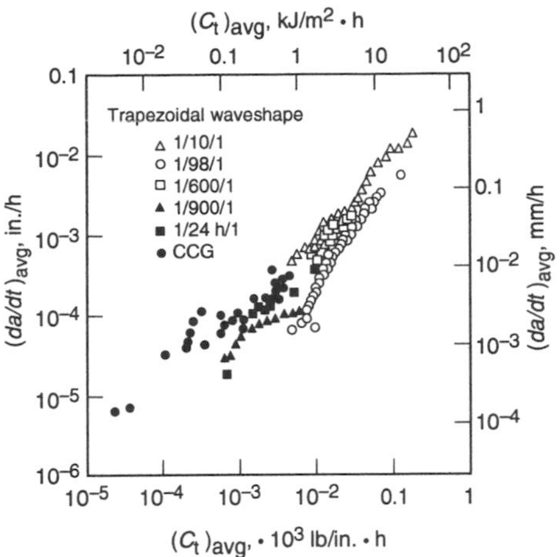

Fig. 18 Comparison between creep and creep-fatigue crack growth data in terms of the estimated $(C_t)_{avg}$ for 1.25Cr-0.5Mo steel at 540 °C (1000 °F). Source: Ref 59, 60

REFERENCES

1. J.A. Harris, Jr., D.L. Sims, and C.G. Annis, Jr., "Concept Definition: Retirement for Cause of F100 Rotor Components," AFWAL-Tr-80-4118, Air Force Wright Aeronautical Laboratories, Sept 1980
2. A. Saxena and P.K. Liaw, "Remaining Life Estimations of Boiler Pressure Parts—Crack Growth Studies," Final Report, CS 4688 per EPRI Contract RP 2253-7, 1986
3. Proc. EPRI Conf. on Life Extension and Assessment of Fossil Plants, June 1986, Electric Power Research Institute (Washington, D.C.)
4. W.A. Logsdon, P.K. Liaw, A. Saxena, and V. Hulina, *Eng. Fract. Mech.*, Vol 25, 1986, p 259
5. A. Saxena, P.K. Liaw, W.A. Logsdon, and V. Hulina, *Eng. Fract. Mech.*, Vol 25, 1986, p 290
6. The National Board of Boiler and Pressure Vessel Inspectors, *Natl. Board Bull.*, Vol 43 (No. 2), 1985
7. A. Saxena, "Life Assessment Methods and Codes," EPRI TR-103592, Electric Power Research Institute, Jan 1996
8. A. Saxena, "Recent Advances in Elevated Temperature Crack Growth and Models for Life Prediction," *Advances in Fracture Research: Proc. Seventh Int. Conf. on Fracture,* ICF-7, March 1989 (Houston, TX)
9. J.T. Staley, Jr., "Mechanisms of Creep Crack Growth in a Cu-1 wt.% Sb Alloy," M.S. thesis, Georgia Institute of Technology, March 1988
10. J.L. Bassani and V. Vitek, *Proc. Ninth National Congress of Applied Mechanics—Symposium on Non-Linear Fracture Mechanics,* L.B. Freund and C.F. Shih, Ed., ASME, 1982, p 127–133
11. R. Raj, *Flow and Fracture at Elevated Temperatures,* R. Raj, Ed., American Society for Metals, 1983, p 215–249
12. A. Saxena, *Fracture Mechanics—12,* STP 700, ASTM, 1980, p 131
13. J.L. Bassani and F.A. McClintock, Creep Relaxation of Stress around a Crack Tip, *Int. J. Solids Struct.*, Vol 7, 1981, p 479–492
14. J. Rice, A Path Independent Integral and the Approximate Analysis of Strain Concentration by Notches and Cracks, *J. Appl. Mech.*, Vol 35, 1986, p 379–386
15. J.D. Landes and J.A. Begley, A Fracture Mechanics Approach to Creep Crack Growth, *Mechanics of Crack Growth,* STP 590, ASTM, 1976, p 128–148
16. C.P. Leung, D.L. McDowell, and A. Saxena, "Influence of Primary Creep in the Estimation of C_t Parameter," Topical Report on Contract 2253-10, Electric Power Research Institute, 1988
17. H. Riedel and V. Detampel, Creep Crack Growth in Ductile, Creep Resistant Steels, *Int. J. Fracture,* Vol 24, 1987, p 239–262
18. A. Saxena, Creep Crack Growth under Non-Steady-State Conditions, *Fracture Mechanics—17,* STP 905, ASTM, 1986, p 185–201
19. J.W. Hutchinson, Singular Behavior at the End of a Tensile Crack in a Hardening Material, *J. Mech. Phys. Solids,* Vol 16, 1968, p 13–31
20. J.W. Rice and G.F. Rosengren, Plane Strain Deformation near Crack Tip in a Power Law Hardening Material, *J. Mech. Phys. Solids,* Vol 16, 1968, p 1–12
21. K.M. Nikbin, G.A. Webster, and C.E. Turner, Relevance of Nonlinear Fracture Mechanics to Creep Cracking, *Cracks and Fracture,* STP 601, ASTM, 1976, p 47–62
22. H. Riedel, *Fracture at High Temperatures,* Springer-Verlag, Berlin, 1987
23. M.F. Kanninen and C.H. Popelar, *Advanced Fracture Mechanics,* Oxford University Press, 1985, p 437
24. A. Saxena, Mechanics and Mechanisms of Creep Crack Growth, *Fracture Mechanics: Microstructures and Micromechanisms,* ASM International, 1989, p 283–334
25. N.L. Goldman and J.W. Hutchinson, Fully Plastic Crack Problems: The Center-Cracked Strip under Plane Strain, *Int. J. Solids Struct.*, Vol 11, 1975, p 575–591
26. C.F. Shih, "Table of Hutchinson-Rice-Rosengren Singular Field Quantities," Technical Report MRL E-147, Brown University, June 1983
27. A. Saxena and J.D. Landes, *Advances in Fracture Research,* ICF-6, Pergamon Press, 1984, p 3977–3988
28. D.J. Smith and G.A. Webster, *Elastic-Plastic Fracture,* Vol I, *Inelastic Crack Analysis,* STP 803, ASTM, 1983, p 654
29. A. Saxena, H.A. Ernst, and J.D. Landes, *Int. J. Fracture,* Vol 23, 1983, p 245–257
30. A. Saxena, T.T. Shih, and H.A. Ernst, *Fracture Mechanics—15,* STP 833, ASTM, 1984, p 516
31. A. Saxena, *Eng. Fract. Mech.*, Vol 40 (No. 4/5), 1991, p 721–736
32. H. Riedel and J.R. Rice, *Fracture Mechanics—12,* STP 700, ASTM, 1980, p 112

33. H.A. Ernst, *Fracture Mechanics—14*, Vol I, *Theory and Analysis*, STP 791, ASTM, 1983, p I-499
34. K. Ohji, K. Ogura, and S. Kubo, *Jpn. Soc. Mech. Eng.*, No. 790-13, 1979, p 18
35. R. Ehlers and H. Riedel, *Advances in Fracture Research*, ICF-5, Vol 2, Pergamon Press, 1981, p 691
36. H. Riedel, *J. Mech. Phys. Solids*, Vol 29, 1981, p 35
37. H. Riedel and V. Detampel, *Int. J. Fracture*, Vol 33, 1987, p 239
38. J. Bassani, D.E. Hawk, and A. Saxena, *Nonlinear Fracture Mechanics*, Vol I, *Time Dependent Fracture*, STP 995, ASTM, 1989, p 7
39. A. Saxena, *Mater. Sci. Eng. A*, Vol 108, 1988, p 125
40. C.P. Leung, D.L. McDowell, and A. Saxena, *Nonlinear Fracture Mechanics*, Vol I, *Time Dependent Fracture*, STP 995, ASTM, 1989, p 55
41. C.P. Leung and D.L. McDowell, *Int. J. Fracture*, Vol 46, 1990, p 81–104
42. C.P. Leung, Ph.D. dissertation, School of Mechanical Engineering, Georgia Institute of Technology, 1988
43. C.Y. Hui and H. Riedel, *Int. J. Fracture*, Vol 17, 1981, p 409–425
44. H. Riedel and W. Wagner, *Advances in Fracture Research*, ICF-5, Vol 2, Pergamon Press, 1985, p 683–688
45. "Standard Test Method for Measurement of Creep Crack Growth Rates in Metals," E 1457, ASTM, 1992
46. D.E. Hall, Ph.D. dissertation, School of Mechanical Engineering, Georgia Institute of Technology, 1995
47. P.L. Bensussan and R.M. Pelloux, Creep Crack Growth in 2219-T851 Aluminum Alloy: Applicability of Fracture Mechanics Concepts, *Advances in Fracture Research*, ICF-6, Vol 3, Pergamon Press, 1986, p 2167–2179
48. P.L. Bensussan, D.A. Jablonski, and R.M. Pelloux, *Metall. Trans. A*, Vol 15, 1984, p 107–120
49. B.C. Hamilton, M.S. thesis, School of Materials Science and Engineering, Georgia Institute of Technology, 1994
50. K.A. Jones, M.S. thesis, School of Materials Science and Engineering, Georgia Institute of Technology, 1993
51. B. Dogan, A. Saxena, and K.H. Schwalbe, *Mater. High Temp.*, Vol 10, 1992, p 138–143
52. P. Bensussan, *High Temperature Fracture Mechanisms and Mechanics: Proc. MECAMAT Int. Seminar on High Temperature Fracture Mechanisms and Mechanics,* P. Benussan and J. Mascavell, Ed., Mechanical Engineering Publications, Vol 3, 1990, p 1–17
53. P. Bensussan, G. Cailletaud, R. Pelloux, and A. Pineau, *The Mechanisms of Fracture*, V.S. Goel, Ed., American Society for Metals, 1986, p 587–595
54. T.S.P. Austin and G.A. Webster, *Fat. Fract. Eng. Mat. Struct.*, Vol 15 (No. 11), 1992, p 1081–1090
55. H.H. Johnson, *Mater. Res. Stand.*, Vol 5 (No. 9), 1965, p 442–445
56. P.S. Grover and A. Saxena, *Sãdhnã, Integrity Engineering Components*, Vol 20, Part I, 1995, p 53–85
57. P.S. Grover and A. Saxena, Characterization of Creep-Fatigue Crack Growth Behavior in 2¼Cr-1Mo Steel using $(C_t)_{avg}$, *Int. J. Fracture*, Vol 73, No. 4, 1995, p 273–286
58. A. Saxena, *JSME Int. J. Series A*, Vol 36 (No. 1), 1993, p 1–20
59. K.B. Yoon, Ph.D. dissertation, School of Mechanical Engineering, Georgia Institute of Technology, June 1990
60. K.B. Yoon, A. Saxena, and P.K. Liaw, *Int. J. Fracture*, Vol 59, 1993, p 95–114
61. P.S. Grover, M.S. thesis, School of Materials Science and Engineering, Georgia Institute of Technology, 1993
62. N. Adefris, Ph.D. dissertation, School of Materials Science and Engineering, Georgia Institute of Technology, 1993
63. K.B. Yoon, A. Saxena, and D.L. McDowell, *Fracture Mechanics—22*, STP 1131, ASTM, 1992, p 367–392
64. A. Saxena, Limits of Linear Elastic Fracture Mechanics in the Characterization of High-Temperature Fatigue Crack Growth, *Basic Questions in Fatigue: Vol II,* R. Wei and R. Gangloff, Ed., STP 924, ASTM, 1989, p 27–40
65. Practices of Load Verification of Testing Machines, E4-94, *Annual Book of Standards*, Vol 3.01, ASTM, 1994
66. P.S. Grover, Ph.D. dissertation, unpublished research, School of Materials Science and Engineering, Georgia Institute of Technology, 1995
67. A. Saxena, R.S. Williams, and T.T. Shih, *Fracture Mechanics—13*, STP 743, ASTM, 1981, p 86
68. D.N. Gladwin, D.J. Miller, and R.H. Priest, *Mater. Sci. Technol.*, Vol 5, Jan 1989, p 40–51
69. "Test Method for Plane-Strain Fracture Toughness of Metallic Materials," E 399, *Annual Book of ASTM Standards*, Vol 03.01, ASTM, 1994, p 680–714
70. K.H. Schwalbe and D.J. Hellman, *Test. Eval.*, Vol 9 (No. 3), 1981, p 218–221
71. R.H. Norris, Ph.D. dissertation, School of Materials Science and Engineering, Georgia Institute of Technology, 1994
72. A. Saxena, K. Yagi, and M. Tabuchi, *Fracture Mechanics: Vol 24*, STP 1207, ASTM, 1992, p 481–497
73. A. Saxena, J. Han, and K. Banerji, *J. Pressure Vessel Technol.*, Vol 110, 1988, p 137–146
74. Y. Gill, Ph.D. dissertation, School of Materials Science and Engineering, Georgia Institute of Technology, 1994
75. K. Sadananda and P. Shahinian, *Fracture Mechanics*, N. Perrone, et al., Ed., 1978, p 685–703
76. R.M. Pelloux and J.S. Huang, *Creep-Fatigue-Environment Interactions*, R.M. Pelloux and N.S. Stoloff, Ed., TMS-AIME, 1980, p 151–164
77. C.B. Harrison and G.N. Sandor, *Eng. Fract. Mech.*, Vol 3, 1971, p 403–420
78. A. Saxena and J.L. Bassani, *Fracture: Interactions of Microstructure, Mechanisms and Mechanics*, TMS-AIME, 1984, p 357–383
79. P.S. Grover and A. Saxena, *Structural Integrity: Experiments, Models and Applications,* ECF-10, K. Schwalbe and C. Bergin, Ed., Engineering Materials Advisory Services, 1994, p 1–21

Creep-Fatigue Interaction

Revised by Gary R. Halford, Senior Research Scientist, NASA–Lewis Research Center

CREEP-FATIGUE INTERACTION is a special phenomenon that can have a detrimental effect on the performance of metal parts or components operating at elevated temperatures. When temperatures are high enough, time-dependent creep strains as well as cyclic (i.e., fatigue) strains can be present; interpretation of the effect that one has on the other becomes extremely important. For example, it has been found that creep strains can seriously reduce fatigue life and/or that fatigue strains can seriously reduce creep life. It is the quantification of these effects and the application of this information to life prediction procedures that constitutes the primary objective in creep-fatigue interaction studies.

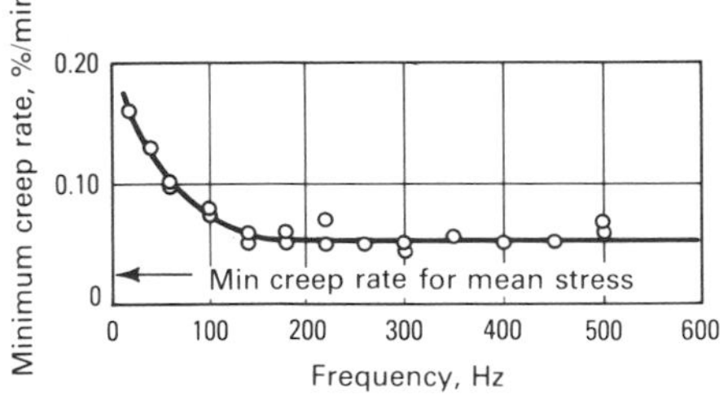

Fig. 1 Effect of frequency on minimum creep rate. Source: Ref 14

Creep-fatigue interaction is a problem that designers of elevated-temperature components must deal with to provide reliable service within the creep range. Creep and fatigue damage interactions may occur during either isothermal or nonisothermal (thermal or thermomechanical) cyclic loading at elevated temperatures. Historically, the effects of creep-fatigue interaction in metals and alloys have usually been studied under simpler, less expensive isothermal laboratory conditions for ease of interpretation of results. This course of action has been justifiable on the basis that service-induced loadings, although nonisothermal on a complete cycle basis, are nominally isothermal for extended periods of time during which creep or relaxation occurs. Hence, isothermal creep-fatigue studies are really a subset of the more general thermomechanical creep-fatigue interaction problem. The reader is referred to the article "Thermal and Thermomechanical Fatigue of Structural Alloys" in this Volume for further information on how creep interacts with nonisothermal fatigue.

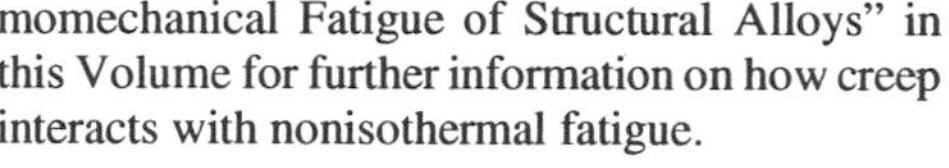

Historical Perspective

Creep-fatigue interaction has been investigated extensively since World War II. The earliest attempts (Ref 1–4) to evaluate combined creep and fatigue properties were made in Germany between 1936 and 1942. Immediately after the war, work was under way to develop methods for predicting creep-fatigue behavior (Ref 5, 6). Attempts were also made to apply basic material structural theories to the physical accumulation of damage in creep-fatigue (Ref 7–9).

Although creep-fatigue interfaces, environments, and properties had been discussed (Ref 10, 11), the first use of the term "creep-fatigue interaction" appears to have been in Ref 12. Furthermore, in an early review of the subject (Ref 13), it was noted that higher operating temperatures and stresses accentuated the problems of creep and fatigue.

According to Ref 12, "It is not satisfactory to assess the safety of the structure in terms of fatigue or creep alone since the presence of one phenomenon may lower the limit of the other." Several questions arise when the specifics of this statement are studied. The suggestion that the presence of one phenomenon *may* lower the limit implies that the limit may or *may not* be lowered. What determines how much the limit is lowered? What is meant by the limit and how is it quantified?

Most experimental studies of creep-fatigue behavior attempt to answer these questions and to develop a basic understanding of this behavior. Such understanding is closely linked to the development of life prediction methods. These methods outline the principles to be applied in predicting the expected operating life when a component is subjected to a creep influence of a certain magnitude in the presence of a fatigue influence of a certain magnitude.

In experiments (Ref 14) dealing with lead cable sheathing tested at 32 °C (90 °F), a mean stress of 5980 kPa (867 psi) was applied and a fatigue stress equivalent to 25% of the static value was

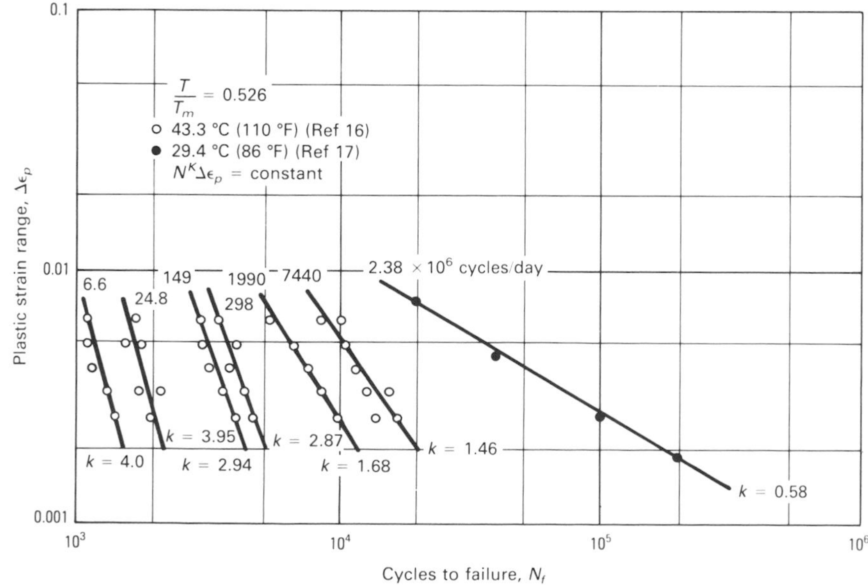

Fig. 2 Effect of frequency of cycling on low-cycle fatigue behavior of lead. Source: Ref 15

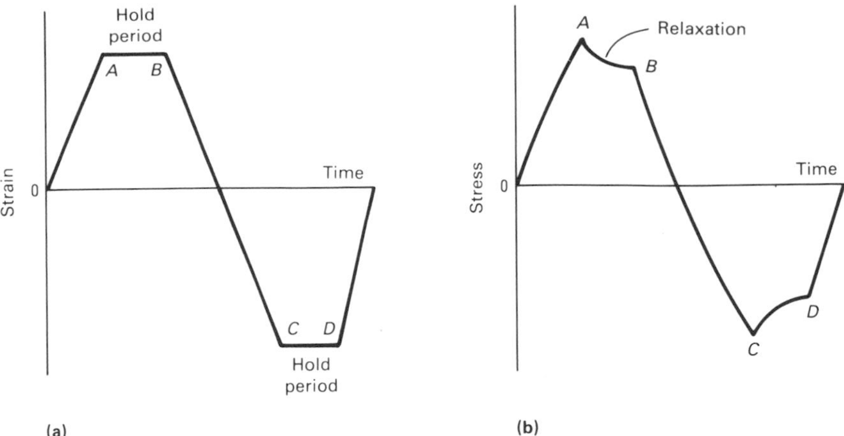

Fig. 3 Strain-controlled fatigue cycle with hold periods. (a) Imposed strain over time. (b) Stress response over time

superimposed (Fig. 1). In the range of frequency of the fatigue stress (from 10 to 200 Hz), the minimum creep rate associated with the mean stress was the greatest (0.17%/min) at the lowest frequency. At 150 Hz and above, the minimum creep rate stabilized at a value of 0.05%/min. A creep stress equal to the fatigue mean stress had a minimum creep rate of only 0.02%/min.

This example illustrates the presence of a fatigue or cyclic stress influence leading to a reduction of creep resistance measured in terms of an increase in the steady-state creep rate. Since the tests were load-controlled, a higher rate of creep in tension would lead to excessive tensile strains and a tensile necking instability in a shorter lifetime than would have been expected by neglecting the observed detrimental creep-fatigue interaction.

Another example of a combined creep-fatigue effect is discussed in Ref 15, which presents an evaluation of the effect of frequency on the strain-controlled fatigue behavior of lead tested at 43 °C (110 °F) (Ref 16) and at 29 °C (84 °F) (Ref 17). This effect is shown in Fig. 2, which plots plastic strain range ($\Delta\varepsilon_p$) versus cycles to failure (N_f). The test temperature is well into the creep range (above one-half the absolute melting temperature) for lead, and the presence of the creep influence is quite prominent. At a plastic strain range of 0.002, a decrease in frequency from 2.38×10^6 to 6.6 cycles/day led to a 100-fold decrease in fatigue life. The exponent k on the plastic strain-range expression in Fig. 2 increased from 0.58 to 4.0 over this range of frequencies. In this example, the presence of a creep influence led to a significant lowering of the limit for fatigue based on the fatigue life corresponding to a frequency of 2.38×10^6 cycles/day.

In these strain-controlled experiments, the failure mechanism was not tensile instability caused by excessive creep strain. Instead, the mode of cyclic crack initiation and growth was by intergranular fracturing (nominally associated with creep deformation mechanisms) rather than classical transgranular fatigue cracking mechanisms. This is in contrast to the previous example for load-controlled testing. Nevertheless, in either test mode, creep damage interacts with fatigue damage to reduce overall lifetime.

The primary emphasis of this article is placed on results obtained in cyclic strain-controlled testing. This condition best represents the boundary conditions imposed on material in localized regions of engineering structural components (notches, fillets, holes, etc.). This type of test yields extensive information about material response characteristics and provides the necessary data for developing the existing mathematical treatments of creep-fatigue interaction. Strain-controlled testing is discussed in Ref 18 and 19, and its development is detailed in Ref 20. Its use in providing a closer simulation of actual service conditions is described in Ref 19. Servocontrolled equipment capable of performing strain-controlled experiments became available in the 1960s. Since that time, computerized fatigue testing, data acquisition, data reduction, and plotting of results have become the norm.

Creep-Fatigue Effects

The introduction of hold periods (rather than simply reducing the continuous cycling frequency) into the fatigue cycle of high-temperature strain-controlled fatigue tests provided a new approach to the evaluation of creep-fatigue interaction. It also simulated many service conditions more closely and yielded material information that was more realistic for design purposes. Prior

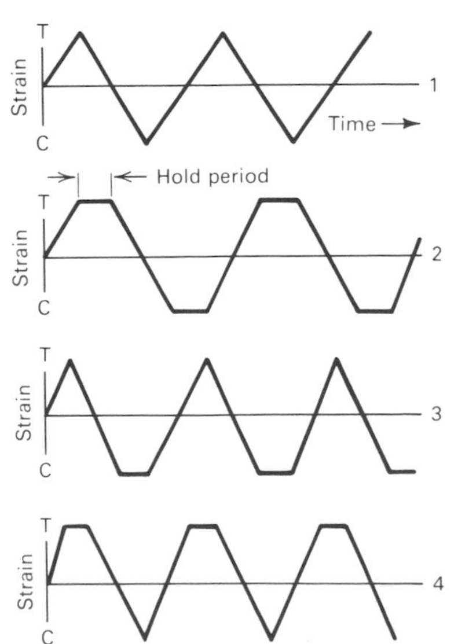

Fig. 4 Strain waveforms used in evaluating effects of hold time on low-cycle fatigue resistance. T, tension, C, compression. See text for an explanation of waveforms.

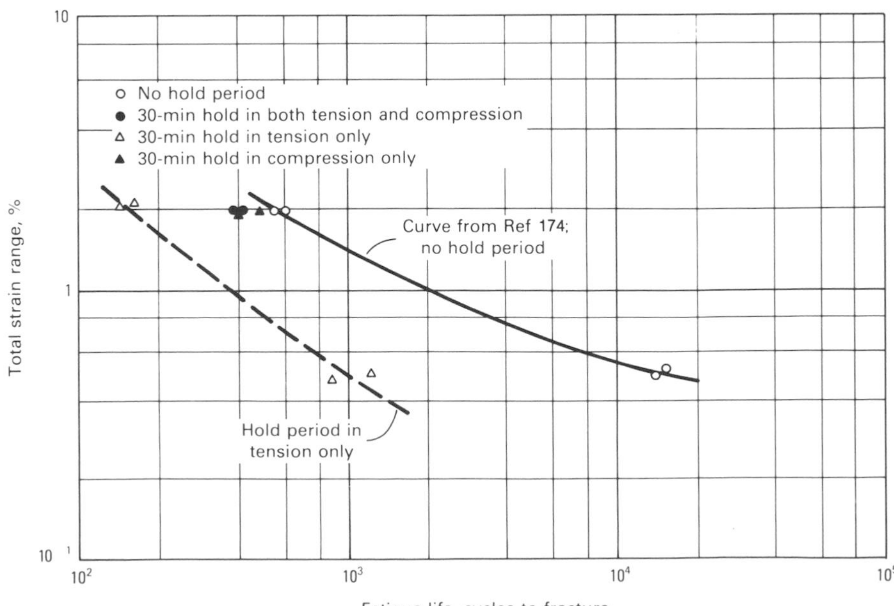

Fig. 5 Effect of hold period and strain waveform on the low-cycle fatigue resistance of AISI type 304 stainless steel. Tested in air at 650 °C (1200 °F) and at a strain rate of 4×10^{-3} s^{-1}

to 1964, the results of elevated-temperature, strain-controlled low-cycle fatigue tests without hold periods were considered to be relevant for design (Ref 18, 19, 21). Later, it was found that simple, uninterrupted low-cycle fatigue is seldom limiting in the design of pressure vessels operating at elevated temperatures (Ref 22, 23) and that the influence of repeated creep relaxation was an important consideration for hot pressure vessels (Ref 23).

Creep relaxation during hold periods in strain-controlled fatigue tests is illustrated in Fig. 3 for completely reversed strain cycling. In this case, the *R* ratio, or ratio of minimum strain to maximum strain, is equal to –1. The material is ramped at a constant strain rate along 0*A* (Fig. 3a); when point *A* is reached, the control system functions to maintain the strain value constant for the hold period of duration *AB*. At point *B*, the material is unloaded along *BC*, and a hold period of duration is introduced at the peak compressive strain level. Hold periods can be introduced at any strain level, but these usually occur near the peak strain levels in the cycle.

Stress relaxations during *AB* and *CD* are shown in Fig. 3(b). Stress relaxation is a form of creep that occurs during the hold periods at constant strain. To maintain the total strain constant, the stress is reduced to exchange elastic strain for creep (inelastic) strain.

Studies of several stainless steels using the waveforms shown in Fig. 4 have been reported (Ref 24). In tests of type 304 stainless steel at 650 °C (1200 °F), hold times in the tension portion of the cycle only (waveform 4 in Fig. 4) were found to be particularly detrimental and resulted in reductions in fatigue life much greater than those noted with waveforms 2 and 3. Figure 5 illustrates this behavior for a strain rate of 4×10^{-3} s^{-1} during the ramping portion of the strain cycle. On the basis of this total strain-range graph, a 30 min hold period in only the compression portion of the cycle leads to a slight reduction in fatigue resistance. A slightly greater reduction is observed when 30 min hold periods are used in both the tension and compression portions of the cycle. A significant reduction is observed for a 30 min hold period in only the tension portion of the cycle.

The data obtained in symmetrical-hold and compression-hold-only tests indicated a correlation with inelastic strain range. These results, presented in Fig. 6, indicate that on the basis of inelastic strain range the data obtained using waveforms 2 and 3 are essentially identical with those obtained in no-hold tests (waveform 1). In these hold-time tests, the inelastic strain range was based on the relaxed stress range, $\Delta\sigma_r$, at $N_f/2$.

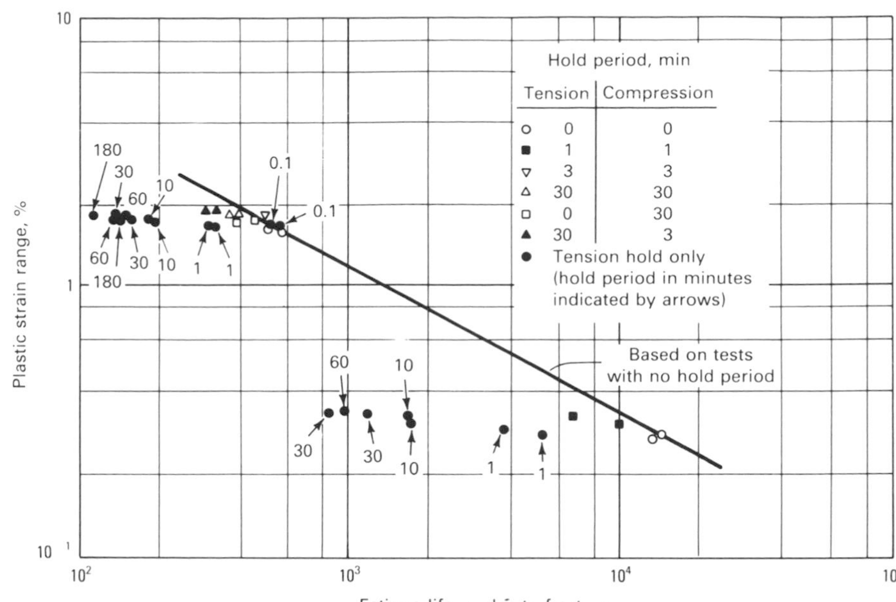

Fig. 6 Effect of hold period and strain waveform on the plastic-strain fatigue resistance of AISI type 304 stainless steel. Tested in air at 650 °C (1200 °F) and at a strain rate of 4×10^{-3} s^{-1}

These results suggest that the decreases in the low-cycle fatigue resistance noted in Fig. 5 (except for the tension-hold-only tests) are due only to increases in inelastic strain. Even though Fig. 5 is based on data obtained at certain constant values of total strain range, the inelastic strain ranges associated with all the data points at a given total strain range are not the same. The relaxation effects are somewhat different for each waveform, thus leading to variations in the inelastic strain-range component at a given value of the total strain range. When the data (except for tension-hold-only data) are compared at the same value of the inelastic strain range, a definite consistency is noted.

Data obtained in the tension-hold-only tests are inconsistent with the inelastic strain-range correlation. As shown in Fig. 6, these results describe a fatigue life that is much lower than would be predicted by the inelastic strain-range correlation. Apparently, holding only in the tension portion of the cycle leads to extensive material damage and drastically reduced fatigue life.

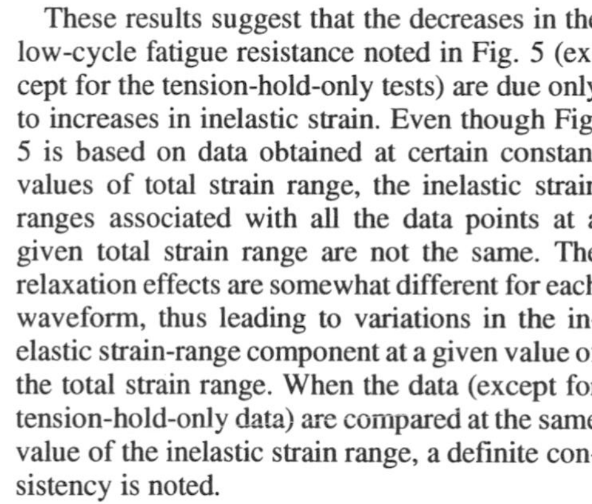

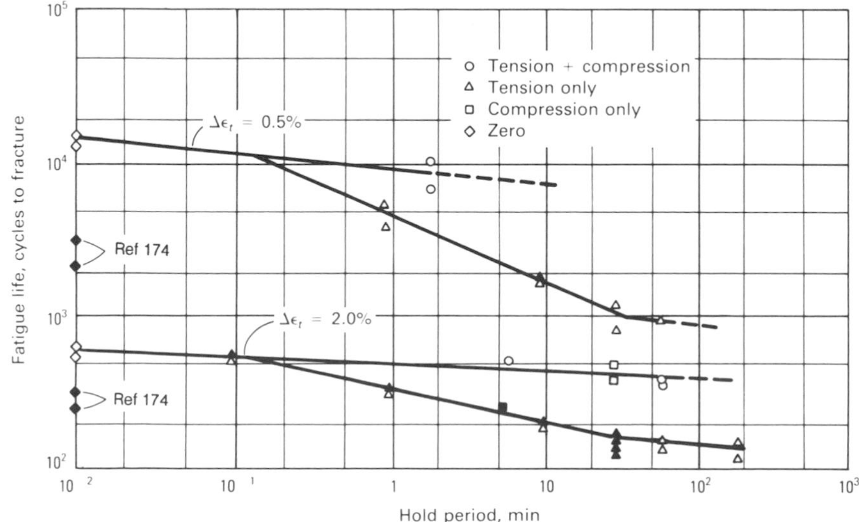

Fig. 7 Fatigue life versus hold-period time for AISI type 304 stainless steel tested in air at 650 °C (1200 °F) and at strain rates of 4×10^{-3} s^{-1} (open symbols) and 4×10^{-5} s^{-1} (closed symbols) using various strain waveforms

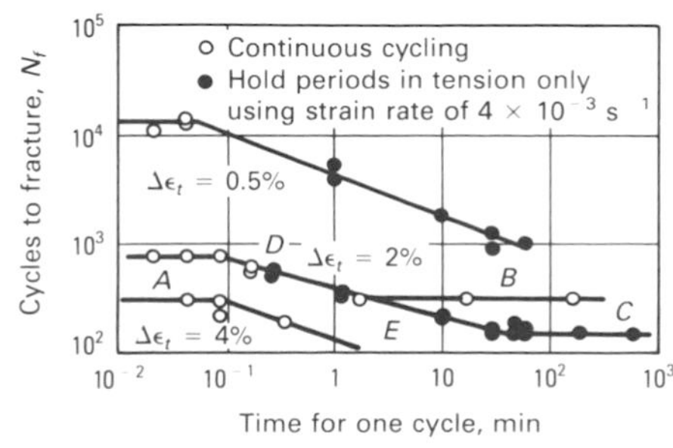

Fig. 8 Plot of N_f versus cycle time for AISI type 304 stainless steel tested in air at 650 °C (1200 °F). Two primed points refer to 30 min hold-period test using strain rate of 4×10^{-5} s^{-1}

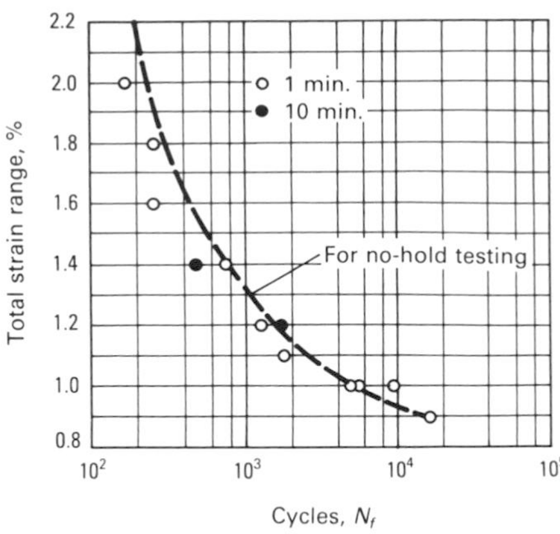

Fig. 9 Strain range versus N_f; hold periods in tension for René 95. Temperature, 650 °C (1200 °F); $R = -1$

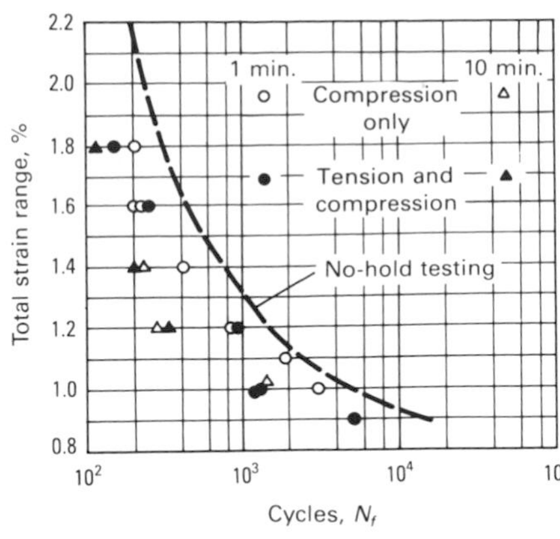

Fig. 10 Strain range versus N_f; hold period results for René 95. Temperature, 650 °C (1200 °F); $R = -1$

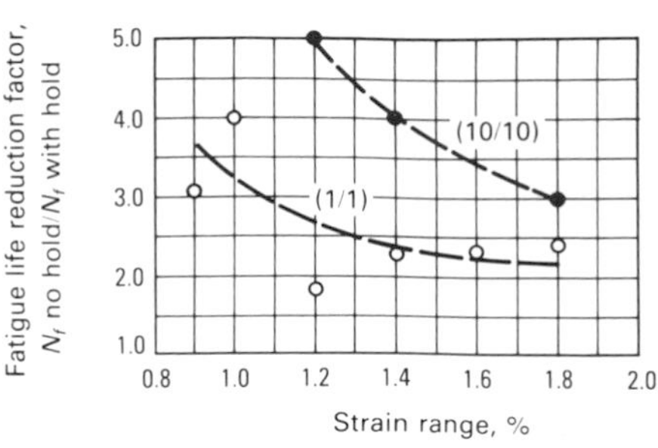

Fig. 11 Fatigue-life reduction factor versus strain range for René 95 for (1/1) and (10/10) tension-compression hold periods. Temperature, 650 °C (1200 °F); $R = -1$

Tension-hold-only tests have indicated that serious damage is encountered even when the hold period is on the order of 1 min. Table 1 summarizes the data obtained in the evaluation of this effect at 2.0% strain range. A saturation effect appears to be observed when the hold period approaches 30 min. These data also are presented in Fig. 7, in which the data obtained in the no-hold test have been arbitrarily plotted at a hold period of 10^{-2} min for reference.

For no-hold, compression-hold-only, and symmetrical-hold testing, a consistent behavior is indicated, as shown in Fig. 7. Data for tension-hold-only testing indicate deviations from this graph, and the direction of the deviation is toward reduced fatigue life. The hold period corresponding to the point of this deviation is about 10^{-1} min for both the 0.5 and 2.0% total strain range.

Some saturation is indicated at the 0.5% strain range to yield a behavior similar to that observed at the higher strain range. Only a few tests were performed at a strain rate of 4×10^{-5} s^{-1}. In tests at the 2.0% strain range, the tension-hold-only data at the slow strain rate were identical to the data in the saturation region of the higher strain rate. Because of the limited data available at the lower strain rate, no definite conclusions can be made regarding these observations.

Tests involving a 30 min hold period in tension plus a shorter hold period in compression (unsymmetrical holding) have shown that the detrimental effect of a hold period in tension can be significantly reduced by a short hold period in the compression portion of the cycle (Table 2). When the tension hold period is 30 min and a 3 min compression hold period is introduced, the fatigue life is within 80% of the fatigue life observed in the 30 min symmetrical-holding tests. Without this short hold period in compression, the fatigue life is reduced to about 40% of the 30 min symmetrical-holding fatigue life.

In this type of testing, the hold period in compression exerts a "healing" effect, or provides a mechanism that reduces the tendency for internal void formation. These results involving unsymmetrical holding appear to be the first data on the subject reported in the literature. They have served as a touchstone for the development of several of the numerous creep-fatigue life prediction methods to be discussed later in this article.

In summary, Fig. 8 illustrates fatigue behavior in terms of a plot of cycles to fracture versus the cycle period (Ref 25). Various regions are:

- *Region A:* N_f is independent of strain rate above about 8×10^{-3} s^{-1} (cycle time < 0.083 min).
- *Region B:* N_f is independent of strain rate below about 4×10^{-4} s^{-1} (cycle time > 1.67 min).
- *Region C:* N_f is independent of hold-period duration (for tension-hold period only) above about 50 min.
- *Region D:* N_f decreases with decreasing strain rate in the range from 8×10^{-3} s^{-1} to about 4×10^{-4} s^{-1}.
- *Region E:* N_f decreases with increasing hold-period length (tension-hold period only) in the range to about 50 min.

Region B is viewed as a saturation in frequency degradation for continuous cycling, while region C is a tension hold-time saturation.

These observations are based on the behavior of austenitic stainless steels and are representative of metals and alloys that suffer significant grain-boundary void formation and growth under creep loading conditions. Other classes of high-temperature alloys offer much greater resistance to grain-boundary creep damage of this type. Most notable are the nickel- and cobalt-base superalloys that have been metallurgically designed to resist grain-boundary creep. Consequently, the observed reductions in cyclic durability resistance of these alloys is generally the result of other time-dependent damage mechanisms, such as oxidation of exposed surfaces and interfaces. Additional time- and temperature-dependent mechanisms (metallurgical instabilities, phase transformations, recrystallization, grain

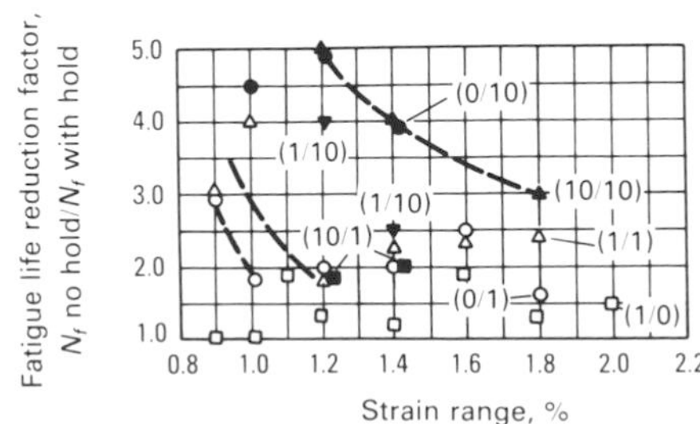

Fig. 12 Fatigue-life reduction factor versus strain range for René 95 for a range of tension-compression hold periods. Temperature, 650 °C (1200 °F); $R = -1$

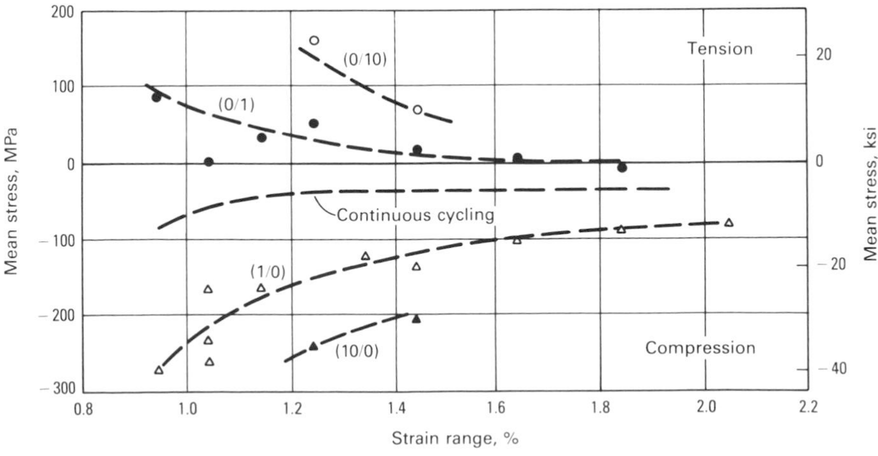

Fig. 13 Mean stress versus strain range for René 95. Temperature, 650 °C (1200 °F); $R = -1$

Table 1 Effect of hold-period length in tension-hold-only testing on fatigue resistance of AISI type 304 stainless steel

Tested in air at 650 °C (1200 °F) and a strain rate of 4×10^{-3} s^{-1} at a strain range of about 2.0%

Hold period, min	Cycles to failure, N_f	
	Test 1	Test 2
0	592	546
0.1	570	545
1.0	329	331
10.0	193	201
30.0	146	165
60.0	144	158
180.0	150	120

growth, etc.) can also contribute to losses in high-temperature fatigue resistance.

Studies of high-temperature fatigue have often included these other effects under the banner of "creep-fatigue" interaction. Because of the different active damaging mechanisms, the so-called creep-fatigue behavior of these alloys may differ appreciably from that of alloys that are grain-boundary creep void formers. For example, tensile hold periods may not result in the greatest degradation, nor will equal-duration hold periods in tension and compression necessarily produce a balanced, "healing" effect. A few specific examples are given in the following paragraphs of creep-fatigue behavior that deviates significantly from the classical behavior of the austenitic stainless steels.

A study of the effect of hold periods at peak strain on the fatigue life of René 95 indicated some interesting patterns (Ref 26). Waveforms with strain hold periods in tension, compression, and both tension and compression (balanced hold) were utilized. Although this testing effort focused primarily on hold-period durations of 1 min, some 10 min hold periods were used to identify the different effects of the two hold-period durations and the effect of the location of the hold period within the cycle.

The effect of a 1 min hold period in tension is shown in Fig. 9. Abbreviations of the form (1/10) are used to define the waveform imposed. The numerator represents the tension hold time in minutes, and the denominator represents compression hold time. Generally, the effect is negligible over the strain-range regime from 0.9 to 2.0%. A few data points at a strain range of 1.0% indicate a beneficial effect of the 1 min tension hold period. The two 10 min hold-period tests shown in Fig. 9 indicate essentially the same effect observed for the 1 min hold period. Consequently, for René 95 at 650 °C (1200 °F) ($R = -1$) and at strain ranges of 1.2 and 1.4%, hold-period durations of 1 and 10 min in tension have no detrimental effect on fatigue life.

Similar tests at 650 °C (1200 °F) with hold periods in compression led to results that were significantly different from those in Fig. 9. This behavior, shown in Fig. 10, indicates that 1 min hold periods exert a noticeable effect in reducing fatigue life below that observed in continuous cycling tests at the same strain range. The detrimental effect increases as the hold-period duration changes to 10 min. When a hold period is used in both tension and compression, the results for 1 min durations are about the same as for the 1 min hold in compression only. This same type of behavior also appears to follow for 10 min durations, although the reduction in fatigue life is somewhat greater than that observed with 1 min durations.

A more graphic summary of these hold-time effects is illustrated by plotting the fatigue-life reduction factor. This factor is simply the ratio of the fatigue life without a hold period to the fatigue life observed when a hold period is used at a given strain range. For certain waveforms, the fatigue-life reduction factor exhibited a definite trend toward higher values as the strain range was decreased. An example of this is shown in Fig. 11 for (1/1) and (10/10) types of tests.

A composite of all the hold-time results is shown in Fig. 12. The following trends are identifiable within reasonable scatter. One-minute hold periods in tension had a negligible effect on the fatigue life over the strain-range regime from 0.9 to 2.0%. One-minute hold periods in compression and in both tension and compression define a fatigue-life reduction factor of 2.0 for strain ranges from 1.8 to about 1.0%. Below 1.0%, there appears to be a sharp increase in the slope of this trend behavior, suggesting more detrimental effects in the lower strain-range regime. Because this lower strain-range regime is in the area of special design interest (i.e., fatigue-life values greater than 10,000 cycles), a more

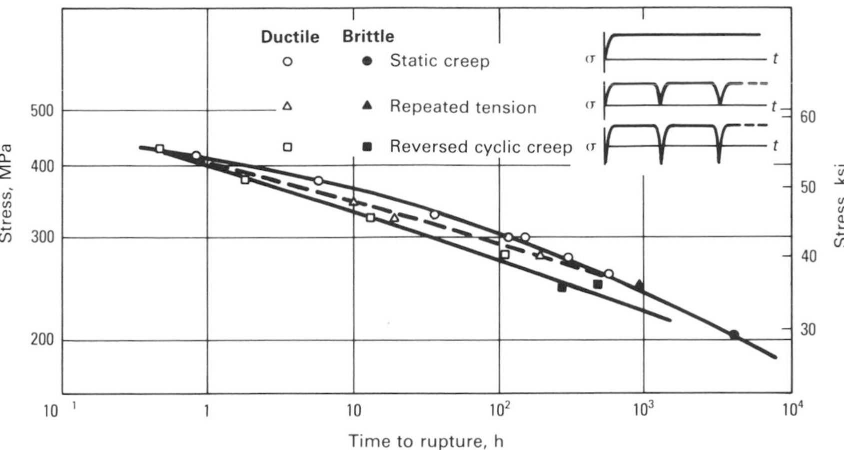

Fig. 14 Static, repeated tension, and reversed cyclic creep results for Cr-Mo-V steel. Source: Ref 27

Table 2 Test results of AISI type 304 stainless steel obtained using a 30 min hold period in tension plus a short hold period in compression

Tested in air at 650 °C (1200 °F) and a strain rate of 4×10^{-3} s^{-1}

Hold period tension, min	Hold period compression, min	Total strain range, %	Cycles to failure, N_f	
			Test 1	Test 2
0	0	1.98	592	546
30	30	1.98	380	416
30	0	2.08	146	...
30	0	2.02	...	165
30	3	1.98	308	...
30	3	2.00	...	336

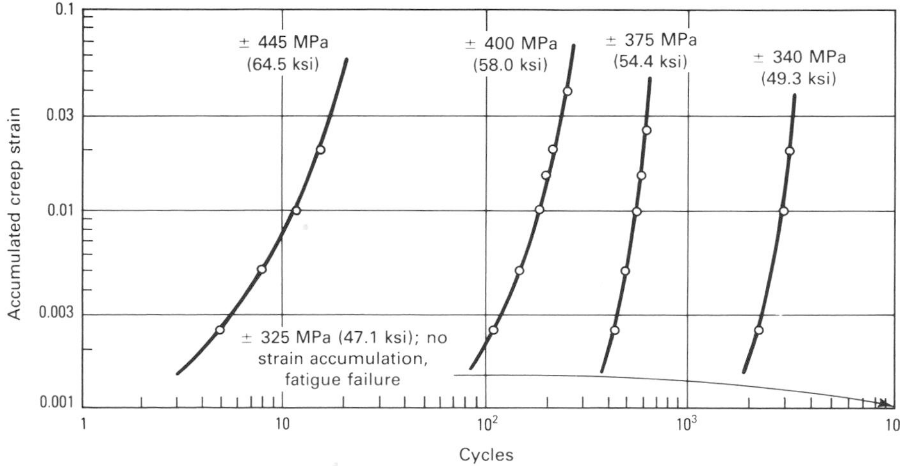

Fig. 15 Tensile strain accumulation in load-controlled continuous-cycling tests with zero mean load for Cr-Mo-V steel. Source: Ref 27

comprehensive study of hold-time effects in this regime appears warranted.

Ten-minute hold periods in compression and in both tension and compression define a trend behavior that begins at a fatigue-life reduction factor of about 5.0 in the strain-range regime near 1.0%. Further study of hold-time effect also is needed in the lower strain-range region.

The development of different mean stress values in the hold-time tests is an important consideration, and a comparison of such behavior is shown in Fig. 13. In continuous cycling tests at 20 cycles/min, the absolute value of the compressive stress component is always just slightly larger than the tensile stress component. This behavior is shown in Fig. 13; over much of the strain-range regime studied, a mean compressive stress of about 34.5 MPa (5 ksi) was exhibited.

In the region near a strain range of 1.0%, the compressive mean stress increased to a value near 70 MPa (10 ksi). This result can be compared with the behavior observed for various hold-period combinations, as shown in Fig. 13. For the 1/0 combination (1 min hold in tension only), the absolute magnitude of the mean compressive stress increases above the continuous cycling value and shows an increase to 275 MPa (40 ksi) compression in the low strain-range regime. This magnitude of the mean compressive stress is increased even more in the 10/0 combination. For the 0/1 combination (1 min hold period in compression only) the mean stress is close to zero in the high strain-range regime and gradually increases to about 70 MPa (10 ksi) tension at the lower strain ranges.

For 0/10, the mean tensile stress is even higher and appears to approach 275 MPa (40 ksi) in the strain-range regime near 1.0%. Data for the 1/1 and 10/10 combinations are not plotted, but are not too different from the continuous cycling behavior. In the case of the 10/1 combination, the mean stress behavior is similar to that exhibited by the 1/0 combination.

Results of studies of Cr-Mo-V steel tested at 565 °C (1050 °F) using static creep, repeated tension, and reversed cyclic creep are shown in Fig. 14 (Ref 27). In all these tests, a transition from ductile to brittle fracture occurred at a stress level near 260 MPa (38 ksi). Tensile strain measurements that were made during load-controlled continuous cycling tests at zero mean load revealed an accumulation of strain (Fig. 15) and creep-type failures, except at a stress level of 325 MPa (47.1 ksi), where no strain accumulation was noted and a fatigue failure was obtained.

Tests using prior fatigue exposure (10 and 50% of N_f) followed by a static creep test led to the results shown in Fig. 16. The creep-rupture life values were consistently shorter than the equivalent tests of virgin material. The cyclic straining in the fatigue cycles led to cyclic strain softening of the material and a substantial reduction in the subsequent creep-rupture life.

Titanium alloys exhibit unusual creep behavior in that they creep at relatively low temperatures. Room-temperature creep-fatigue interaction for Ti-6-2-1-1 (Ref 28) is also unusual. Prior amounts of creep strain (0.002 to 0.027) have been shown to be beneficial to subsequent fatigue life, and hold periods at peak tensile strain are actually beneficial to low-cycle fatigue life. Results of this nature would be difficult to predict based on experiences with austenitic stainless steels. In fact, one of the prominent creep-fatigue life prediction models—the time- and cycle-fraction model—considers only a degrading interaction between creep and fatigue.

Another example of unusual creep-fatigue behavior is that of the cast nickel-base superalloy, MAR-M 200 (Fig. 17a). Results of isothermal creep-fatigue tests at 925 °C (1700 °F) showed a nearly complete absence of creep-fatigue effects. Hold times were imposed under constant load, but the strain was limit-controlled (Ref 29, 30). Fractography revealed a consistent result. The dominant cracking was along interdendritic boundaries, neither classical transgranular fatigue nor intergranular creep cracking. The single cracking mechanism led to a single cyclic life for a given imposed strain range. Hence, no creep-fatigue effect was present. Again, for this example, classical creep-fatigue life prediction models (to be discussed later) would have predicted a significant interaction, whereas none was observed.

Isothermal creep-fatigue data for H-13 tool steel at 595 °C (1100 °F) (Ref 30) show a life-degrading interaction due to compressive creep hold periods (Fig. 17b). Tensile hold periods or balanced tension plus compression hold periods have negligible influence on cyclic lifetime. Specimens failed in a classical fatigue-related transgranular manner, with no evidence of creep-related grain-boundary void formation or intergranular cracking. Considerable oxidation was present on the free specimen surface and the newly formed crack faces.

Another interesting example of unusual creep-fatigue response is found for the cast nickel-base superalloy IN-792+Hf tested at three isotherms of 540, 760, and 925 °C (1000, 1400, and 1700 °F). Strain-controlled tests were conducted with imposed 2 or 5 min hold periods at peak tensile, compressive, or tensile and compressive strains. On a total strain range versus cyclic life basis, the differences in cyclic lives for a given temperature were negligible. When only the inelastic strain range was examined, the tensile hold and tensile plus compressive hold periods actually enhanced the observed cyclic lives by a factor of about 10

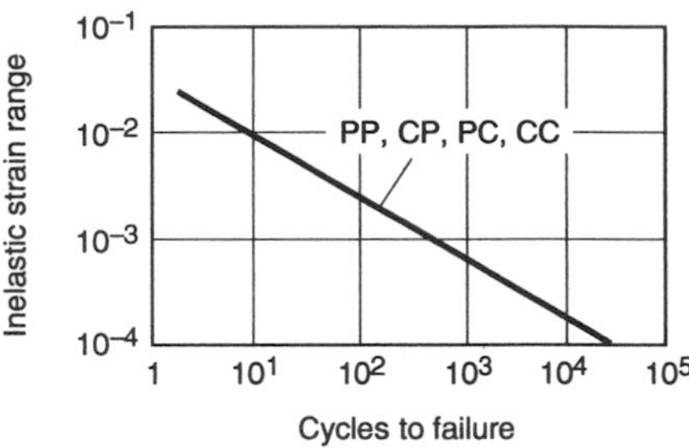

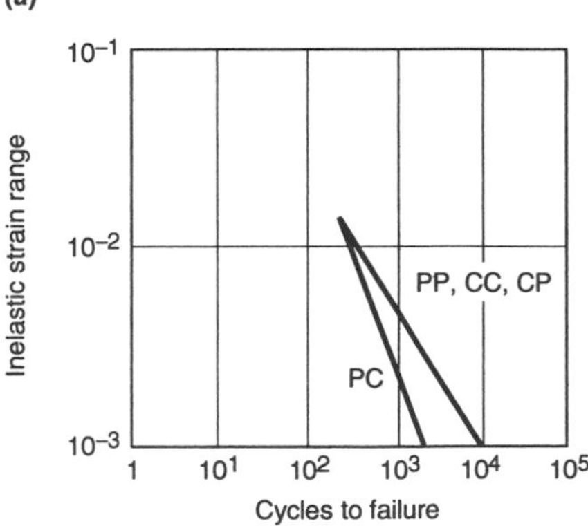

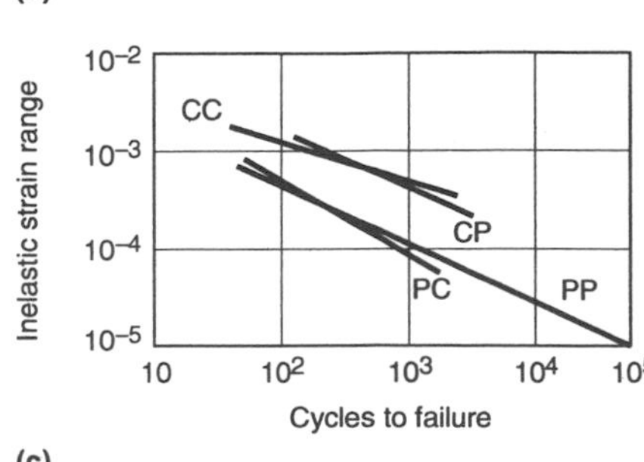

Fig. 17 Inelastic SRP life relationships. PP, no hold; CP, tensile hold; PC, compressive hold; CC, tensile + compressive hold. (a) MAR-M 200, polycrystalline, at 925 °C (1700 °F). (b) H-13 tool steel, at 595 °C (1100 °F). (c) IN 792 + Hf, at 540, 760, 925 °C (1000, 1400, 1700 °F). Source: Ref 30

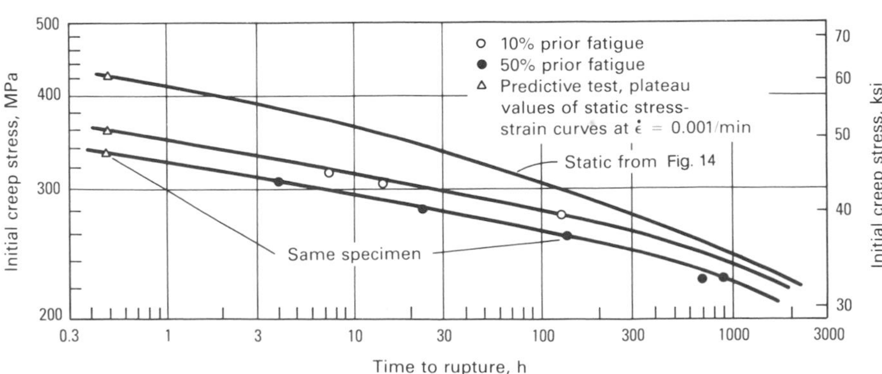

Fig. 16 Effect of prior fatigue on subsequent static creep rupture life for Cr-Mo-V steel. Source: Ref 27

compared to no hold period or only a compressive hold period (Fig. 17c, Ref 30). For this alloy, the creep ductility is twice the tensile ductility, reflecting the fact that the deformation mechanisms during creep were less damaging than for plastic deformation (Ref 31). Instead of concentrating strain locally, the tensile creep deformations were more homogeneously dispersed throughout the cast grain structure, thus delaying the initiation of cracks.

A common cracking mode for creep-fatigue interaction is intergranular cracking of grain boundaries perpendicular to the applied tensile stress. Elimination of transverse grain boundaries would exclude this failure mode and set the stage for prolonged high-temperature cyclic lives. Hence was born the original driving force for the directionally solidified (DS) and single-crystal (SC) nickel-base superalloys that are now used extensively in the hottest turbine stages of aeronautical gas turbine engines.

While this technical reasoning was perhaps responsible for the initial development work on nickel-base DS and SC turbine blade alloys, the features of these highly anisotropic alloys that have carried them to success is somewhat unrelated. Single-crystal (and DS) turbine blades are grown with a [001] cube axis in the radial direction. The elastic modulus along this cube axis is significantly lower (60 to 70%) than for a polycrystalline casting of the same alloy. At the same time, the coefficient of thermal expansion is isotropic and remains equal in value to its polycrystalline counterpart. This combination of thermal and mechanical properties gives rise to substantially lower thermal stresses in the SC blades even though the thermal strains are the same. Fortunately, the yield strength in a cube direction is not appreciably different from the polycrystalline yield strength, so the ratio of the induced thermal stress to the yield strength is lowered by the same ratio that the elastic modulus is lowered. Thermal fatigue lives of [001] single crystals is thus significantly greater than polycrystals.

An example of the creep-fatigue interaction for DS MAR-M 247LC (Ref 32) is given in Fig. 18 for samples with axes parallel to (0°), perpendicular to (90°), and inclined to (45°) the direction of solidification. The zero hold-time, strain-controlled tests (i.e., no creep) at 900 °C (1650 °F) apparently show the effect of orientation on the fatigue response that would be expected based on the previous discussion; 0° and 90° orientations are both along nominal [001] cube axis directions. However, the 45° oriented specimens would correspond to a direction of greater modulus of elasticity. For a nominally isotropic yield strength, a given strain range of, for example, 1.2% causes considerably more plasticity in 45° samples than in the [001] oriented samples. A higher plastic strain range would be expected to lower the fatigue life, in agreement with the results of Fig. 18. Only tensile hold-time creep-fatigue tests were performed. A 10 min tensile hold time caused cyclic life reductions of 10, 25, and 60% for the 0°, 90°, and 45° orientations, respectively.

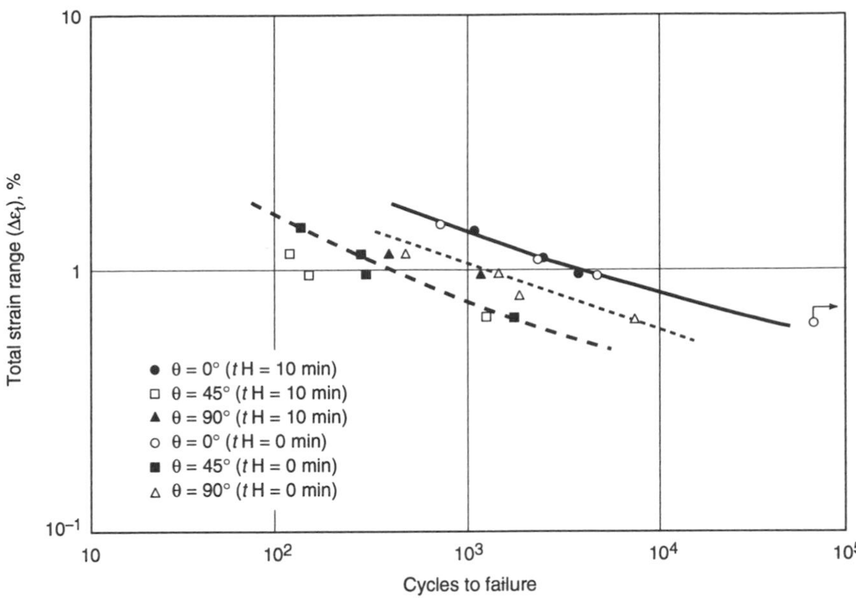

Fig. 18 Creep-fatigue behavior of DS nickel-base superalloy, MAR-M 247LC, at 900 °C (1650 °F). Angle θ denotes loading direction relative to [001] solidification direction. Source: Ref 32

Representative hysteresis loops are shown in Fig. 19 for these tests. Clearly, the hysteresis loop for the 45° orientation shows greater plasticity and a larger amount of stress relaxation (creep) than the 0° and 90° loops. The presence of a very few transverse grain boundaries for the 90° oriented specimen is likely responsible for its having a greater creep-fatigue effect than the 0° oriented specimen.

Most of the data reported for fatigue and creep-fatigue evaluations apply to testing in an air environment. However, it has been shown that the test environment can have a significant effect on fatigue and creep-fatigue results. Although this subject is beyond the scope of this article, environmental effects should be given proper consideration in any evaluation of creep-fatigue interac-

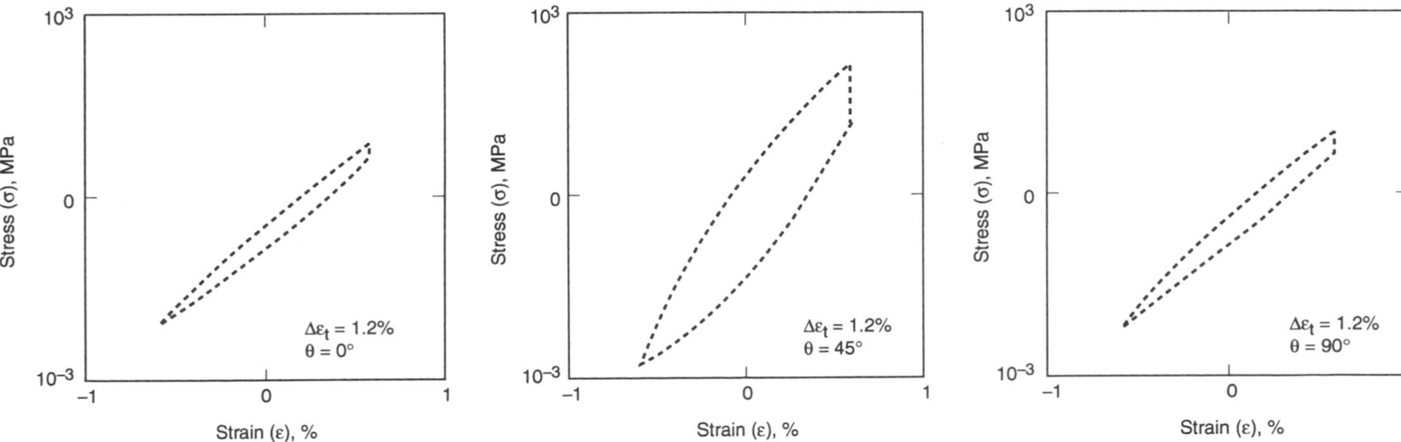

Fig. 19 Hysteresis loops for three loading directions relative to [001] growth direction of DS MAR-M 247LC at 900 °C (1650 °F); 10 min tensile strain hold-time tests. Source: Ref 32

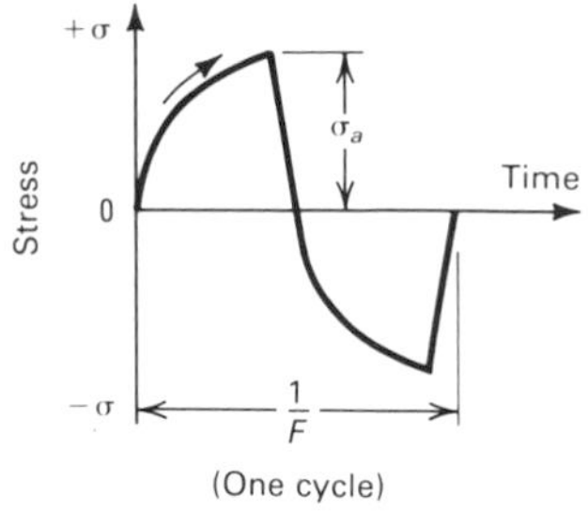

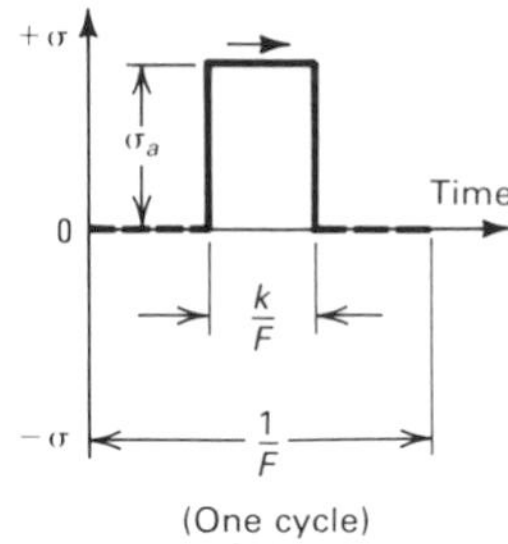

Fig. 20 Equivalent creep-rupture damage done by cyclic stress. k = effective fraction of each cycle spent at maximum tensile stress

tions. An excellent summary of the mechanisms of failure of oxides and their interaction with creep-fatigue, thermomechanical fatigue, and thermal fatigue is given in the article "Thermal and Thermomechanical Fatigue of Structural Alloys" in this Volume.

A vast quantity of creep-fatigue data for a wide variety of engineering metals and alloys has been generated since the 1960s. While no central repository exists for all these results, two nationally sponsored structural materials handbooks (Ref 33, 34) contain at least a small fraction of the publicly available information. If a material of interest does not have a broad base of industrial interest, however, it is unlikely that existing creep-fatigue data for that material will ever appear in these handbooks. Searching for creep-fatigue data on a specific material can become tedious. Fortunately, the advent of electronic search capabilities for seeking out publicly available technical articles and reports will make the task easier and certainly much faster.

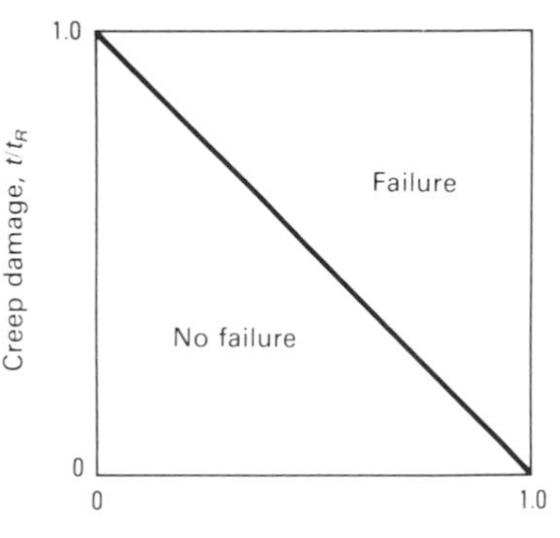

Fig. 21 Creep-fatigue interaction diagram and linear creep-fatigue interaction

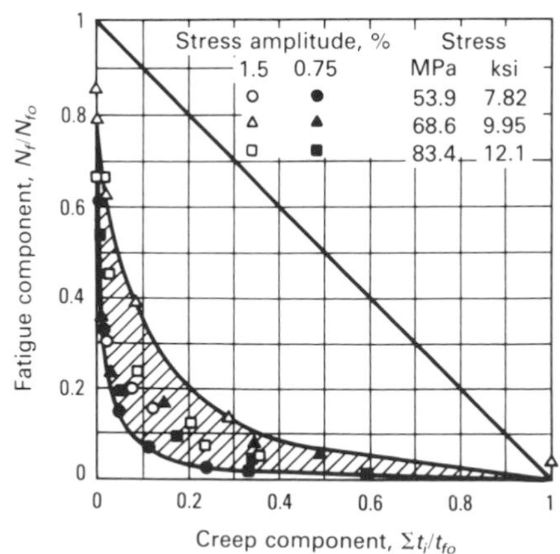

Fig. 22 Combined creep-fatigue data for austenitic stainless steel sheet tested in reversed bending at 700 °C (1290 °F). Source: Ref 36

Creep-Fatigue Prediction Techniques

Development of a mathematical formulation for life prediction is one of the most challenging aspects of creep-fatigue interaction. It is complicated by the fact that any proposed formulation must account for strain rate, relaxation at constant strain, creep at constant load, and the difference between tension and compression creep, and must be able to deal with the disparate creep-fatigue response of different classes of materials used in high-temperature applications.

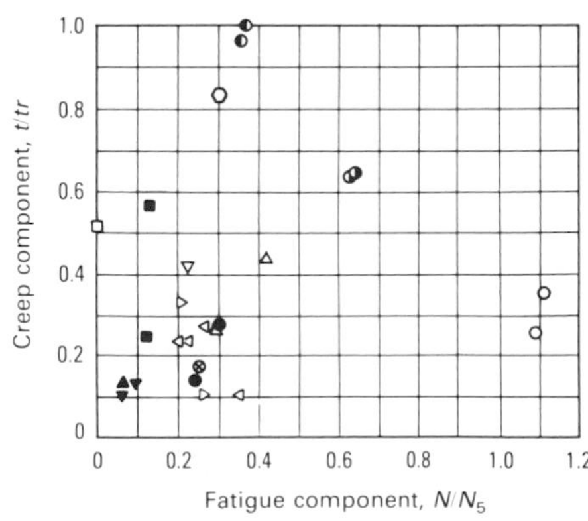

	Total strain range, %	Hold time, min		Total strain range, %	Hold time, min
□	0.25	10	◑	2.0	1
△	0.5	1	◐	2.0	10
■	0.5	10	◁	2.0	30
▼	0.5	30	▷	2.0	60
▲	0.5	60	●	2.0	180
▽	1.0	10	⊗	2.0	600
○	2.0	0.1	⬡	4.0	10

Fig. 23 Creep-fatigue interaction plot for hold-time data obtained in tests of type 304 stainless steel at 650 °C (1200 °F). Source: Ref 37

10% Rule. In an early attempt to provide quick estimates for fatigue life under combined creep and fatigue conditions (Ref 10), modifications to the Universal Slopes equation were considered (Ref 35). It was reasoned that the complexities introduced by high-temperature operation in the presence of a creep influence could not be accommodated in the Universal Slopes equation. Even when tensile properties determined at the temperature and strain rate of the fatigue test were used, the Universal Slopes equation led to fatigue-life estimates that were greater than those observed. It was concluded that the creep influence was introducing some intercrystalline cracking, which caused a portion of the crack initiation phase of the fatigue process to be bypassed. As a result, the "10% rule" was introduced; it proposed to take 10% of the Universal Slopes value and use it as the estimate for a creep-fatigue life. Generally, this approach yields conservative or lower-bound fatigue-life estimates. The 10% rule inherently excludes the possibility of directly accounting for the effects of frequency, hold time, mean load, and other influencing factors.

In an attempt to devise a semblance of sophistication that could permit extrapolation to very long hold periods per cycle, the addition of a term to define "time at temperature" was considered (Ref 10). Using the diagram in Fig. 20 to illustrate the stress pattern for completely reversed strain cycling ($R = -1.0$), it was assumed that the effect of the compression stress could be omitted and that the entire stress pattern could be replaced by one in which the stress is constant and equal to the maximum stress of the pattern it replaces.

Another part of this assumption was to choose the effective time period within which this peak stress is applied as being equal to k/F, where k is a fraction of the cycle time $1/F$ (F is frequency). In this approach, the "fatigue damage" effect was taken as the ratio of the number of cycles actually applied, N'_f, to the number that would be sustained in the absence of a creep effect (the latter being equal to the Universal Slopes value for N_f). In addition, the "creep damage" effect was taken to be the ratio of the time actually spent at stress, s_a, to the time required to cause rupture at this stress value. Then, assuming that failure occurs when:

$$\text{Creep damage} + \text{fatigue damage} = 1 \qquad \text{(Eq 1)}$$

$$\frac{t'}{t_R} + \frac{N'_f}{N_f} = 1 \qquad \text{(Eq 2)}$$

where:

$$t' = \frac{k}{F}(N'_f) \qquad \text{(Eq 3)}$$

This leads to:

$$N'_f = \frac{N_f}{1 + \dfrac{k}{AF}(N_f)^{(m+0.12)/m}} \qquad \text{(Eq 4)}$$

where A is obtained as a time intercept on the stress-rupture isotherm, and m is the slope of the stress-rupture isotherm on log-log coordinates. It was suggested (Ref 10) that the life be estimated by the 10% rule and by the application of Eq 4, and to then use the lower of the two values as a lower bound on fatigue life and a value of twice the lower-bound life as the average fatigue life.

The linear damage rule has been used extensively in the evaluation of creep-fatigue interaction. It is based on the simple relationship (similar to Eq 1) that fatigue damage can be expressed as a cycle fraction and that creep damage can be expressed as a time fraction. It is also assumed that these quantities can be added to represent damage accumulation. Failure occurs when this summation reaches a certain value. Expressed mathematically, this damage rule is:

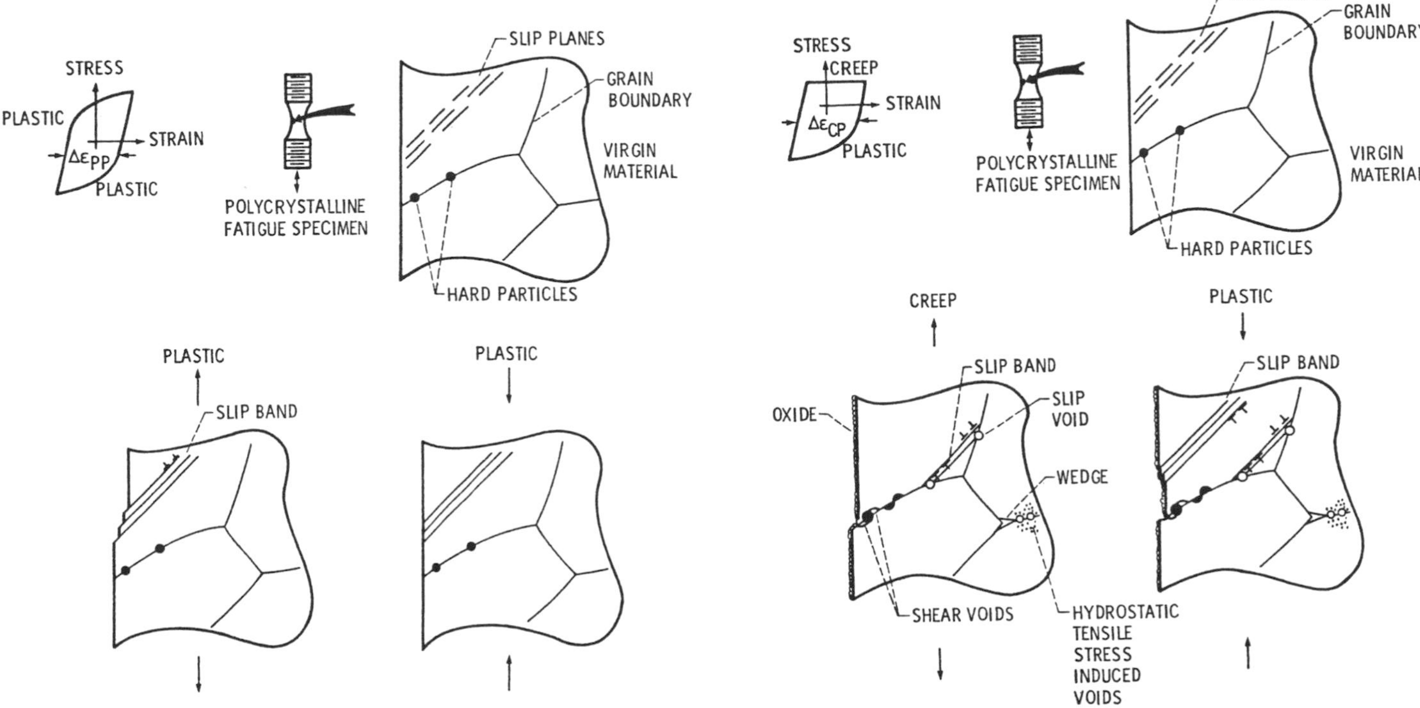

Fig. 24(a) Cyclic deformation and damage model for SRP, *PP* cycle. Source: Ref 30

Fig. 24(b) Cyclic deformation and damage model for SRP, *CP* cycle. Source: Ref 30

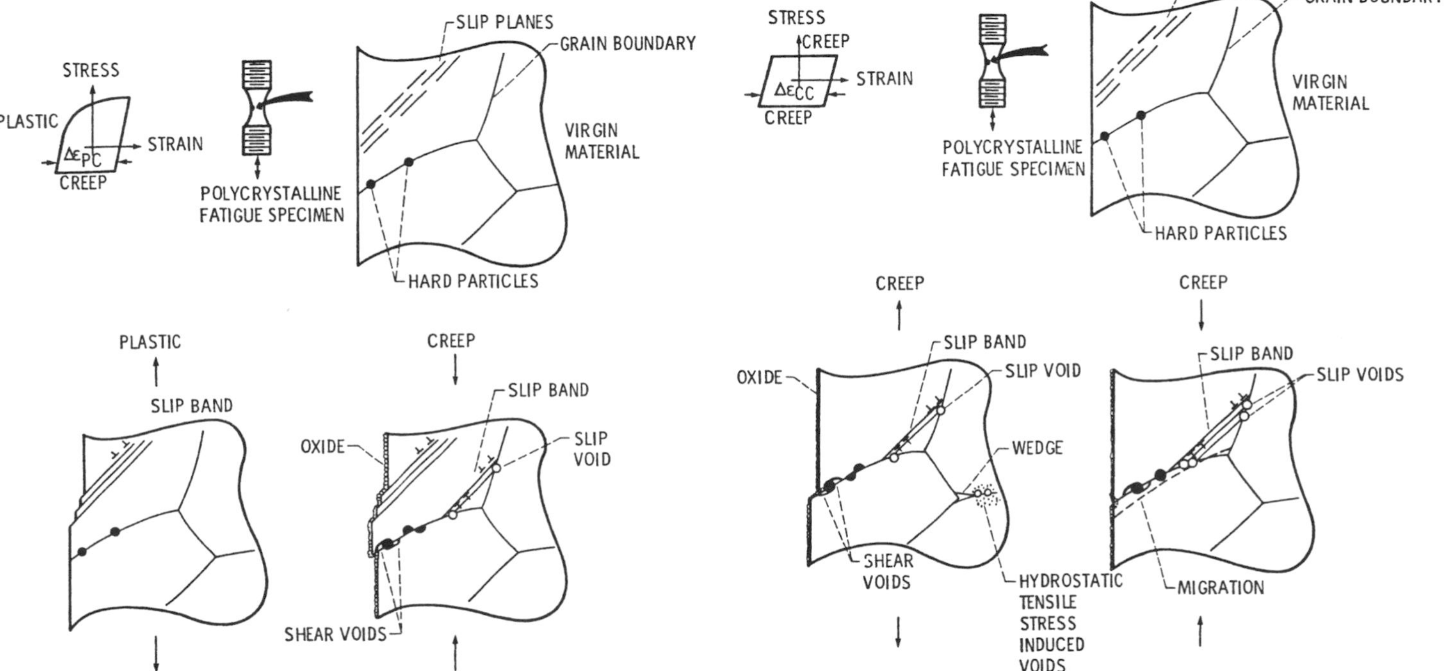

Fig. 24(c) Cyclic deformation and damage model for SRP, *PC* cycle. Source: Ref 30

Fig. 24(d) Cyclic deformation and damage model for SRP, *CC* cycle. Source: Ref 30

$$\sum_{n=1}^{f}\frac{N}{N_f}+\sum_{n=0}^{f}\frac{t}{t_R}=K \quad \text{(Eq 5)}$$

where N is the number of cycles of exposure at a given strain range; N_f is the cycles to failure at the given strain range; t is the time of exposure to a given stress-temperature combination; t_R is the time to rupture at the given stress-temperature combination; and f indicates that summation occurs over the entire exposure life, at which point the total damage accumulation is equal to K.

Usually, K is assumed to be unity, because the cycle ratio summation should be unity when no creep damage is present. This assumption has been proven to be false for multiple exposures, although K must be unity for a single strain-range exposure. Similarly, when no fatigue is present, the time ratio summation should be unity; this also has been proven false for multiple exposures.

Equation 5 differs from Eq 2 because t' is not used and N_f is a measured quantity rather than a calculated value based on the Universal Slopes equation. Furthermore, Eq 5 applies to the complete cycle because t' of Eq 2 is automatically accounted for due to the use of a measured N_f value. The time ratio term in Eq 5 applies when constant-stress or constant-strain hold periods are introduced. With constant-strain hold periods and for a hold-period duration of t_H, when relaxation occurs this creep damage would be calculated in integral form:

$$\sum_{n=1}^{N}\int_0^{t_H}\frac{dt}{t_R} \quad \text{(Eq 6)}$$

Creep-damage fractions (ratios) and fatigue-damage fractions (ratios) are usually presented graphically to form a creep-fatigue interaction diagram. In Fig. 21, the 45° diagonal has been drawn for $K = 1$ in Eq 5. The combinations of creep- and fatigue-damage fractions that define points in the region below the diagonal (for $K = 1$) represent safe operation. Points that fall on or above the diagonal correspond to failure conditions.

Generally, the creep-fatigue interaction diagram is not as shown in Fig. 21, because K usually is not equal to unity and the creep-fatigue interaction behavior usually is not linear. Several examples of creep-fatigue interaction diagrams are presented in Fig. 22 (Ref 36) and Fig. 23 (Ref 37). In Fig. 23, the fatigue-damage fraction is given in terms of the cycles to a 5.0% reduction in the peak cyclic stress. An example of a creep-fatigue damage summation for K greater than unity has been reported (Ref 38, 39) in creep-interspersion tests of annealed 2.25Cr-1Mo steel at 540 °C (1000 °F).

The frequency-modified fatigue equation has been proposed (Ref 40) to introduce a time-dependency term to properly account for behavior observed at high temperature. Based on Ref 16 and 41, fatigue resistance can be represented by:

$$\nu^k t_f = \text{constant} \quad \text{(Eq 7)}$$

where ν is the frequency, t_f is the time to failure, and k is a constant that depends only on temperature. The right side of Eq 7 has been expressed (Ref 40) in terms of the plastic strain range, $\Delta\varepsilon_p$, to yield:

$$\nu^k t_f = f(\Delta\varepsilon_p) \quad \text{(Eq 8)}$$

and

$$\nu^k t_f = \nu^k\left(\frac{N_f}{\nu}\right) = N_f\nu^{k-1} \quad \text{(Eq 9)}$$

and a definition of the frequency-modified fatigue life. Plotting this quantity versus $\Delta\varepsilon_p$ yields (Ref 40):

$$(N_f\nu^{k-1})^\beta\,\Delta\varepsilon_p = C_2 \quad \text{(Eq 10)}$$

where β and C_2 are constants. The modified Basquin relationship was then employed:

$$\Delta\varepsilon_e = \frac{\Delta\sigma}{E} = \frac{A'}{E}N_f^{-\beta'}\nu^{k'_1} \quad \text{(Eq 11)}$$

where $\Delta\varepsilon_e$ is the elastic strain range; $\Delta\sigma$ is the stress range; E is the elastic modulus; and A', β', and k'_1 are constants.

The total strain range is then obtained by adding the elastic and plastic strain-range terms of Eq 10 and 11 and then eliminating $\Delta\varepsilon_p$, to yield:

$$\Delta\varepsilon = \frac{AC_2^{\eta'}}{E}N_f^{-\beta n'}\nu^{k+(1-k)\beta n'} + C_2N_f^{-\beta}\nu^{(1-k)\beta} \quad \text{(Eq 12)}$$

Once the material constants are determined in supporting tests, Eq 12 can be used to estimate N_f values for given values of $\Delta\varepsilon$ and frequency. For hold-time tests, the frequency term is taken as the reciprocal of the cycle period (i.e., hold period plus ramp time).

Equation 12 does not account for wave shape and thus does not distinguish between different strain rates existing in loading and unloading or between compression holds, tension holds, and hold periods in tension and compression. Therefore, the concept of frequency separation was introduced (Ref 42). This involves a concept for separating the hysteresis loop for a cycle into two parts, with a time dependency associated with each part. Tension-going and compression-going frequencies are defined as ν_t and ν_c, respectively, where tension-going refers to the part of the loop in which the plastic strain rate is positive. Correspondingly, the compression-going part of the loop involves a negative plastic strain rate. Stress-range and plastic strain-range terms are also defined for use in the type of expression shown in Eq 12. These expressions involve the use of tension-going and compression-going strain ratios that can be converted into frequencies.

This approach has been applied (Ref 42) to data for many different alloys and has shown agreement between predicted and measured N_f values to within a factor of ±2 (i.e., the range of measured values is between the predicted value multiplied by 2 and the predicted value divided by 2).

Strain-range partitioning (SRP) focuses on the inelastic strain present in a cycle and the two directions of straining (Ref 43–45). These two directions are tension (positive inelastic strain rate) and compression (negative inelastic strain rate). In addition, two types of inelastic strain are considered: time dependent (creep) and time independent (plasticity).

By combining the two directions with the two types of strain, it is possible to achieve four types

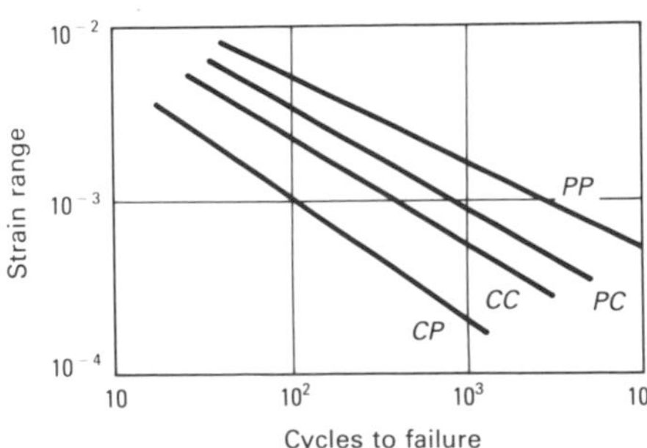

Fig. 25 Typical partitioned strain range/life relationships used to characterize material behavior in the creep-fatigue range. Source: Ref 45

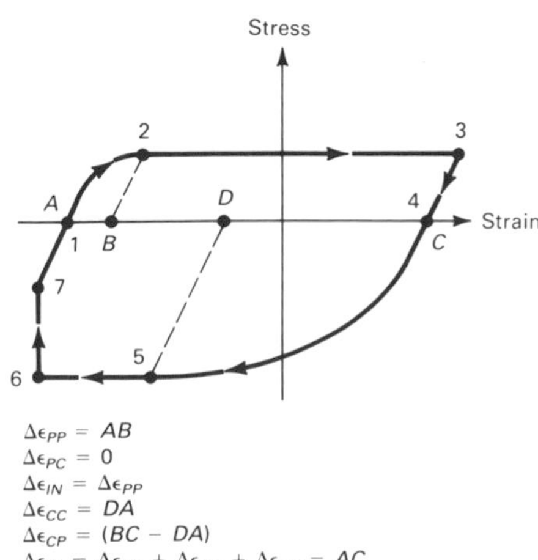

Fig. 26 Defining partitioned strain-range components of complex hysteresis loop. Source: Ref 45

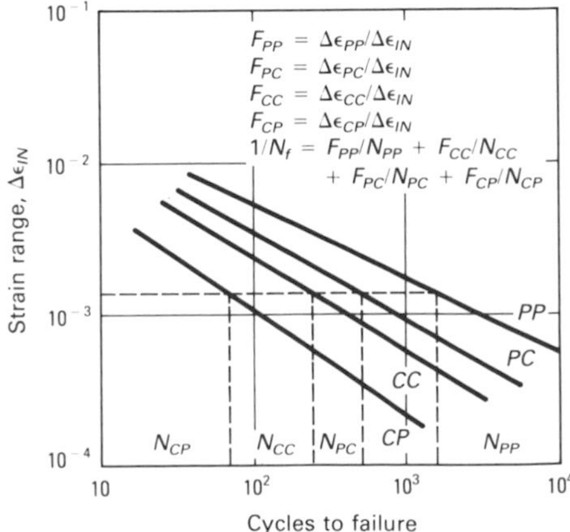

Fig. 27 Interaction damage rule. Source: Ref 45

of strain ranges that might be encountered in any conceivable hysteresis loop. These define the manner in which a tensile component of strain is balanced by a compressive component to close a hysteresis loop:

- Tensile plasticity reversed by compressive plasticity is designated a *PP* strain range and is represented by $\Delta\varepsilon_{PP}$.
- Tensile creep reversed by compression plasticity is designated a *CP* strain range and is represented by $\Delta\varepsilon_{CP}$.
- Tensile plasticity reversed by compressive creep is designated a *PC* strain range and is represented by $\Delta\varepsilon_{PC}$.
- Tensile creep reversed by compressive creep is designated a *CC* strain range and is represented by $\Delta\varepsilon_{CC}$.

The subscript notation for the strain ranges indicates the type of tensile strain first, followed by the type of compressive strain. Highly simplified micromechanistic models have been constructed (Ref 30) and are illustrated in Fig. 24(a) to 24(d) to demonstrate how creep and plastic strains might physically interact in a complete cycle of inelastic strain for each of the four distinct creep-plasticity cycles. Idealized hysteresis loops are included for each of the four cycles: *PP, CP, PC,* and *CC*.

Strain-range partitioning represents the premise that, in order to handle a complex high-temperature low-cycle fatigue problem, the inelastic strain range must first be divided or partitioned into its components. It was then proposed that each of the inelastic strain-range types be treated as individual modes for damage accumulation utilizing plots of log (inelastic strain range) versus log (N_f) to yield four straight lines, as shown in Fig. 25. These lines are established from results obtained in special tests that approximate the ideal hysteresis loops of Fig. 24(a) to 24(d). Once available, they are used in an SRP analysis of any type of cycle. This means of representing the creep-fatigue behavior of a material makes no prejudgments as to the extent of interaction between the two damage contributors, cyclic creep and plastic strains. While *PP*-type cyclic straining is usually the most benign, the SRP method does not require it to be so. An example of this type of behavior will be shown later in this section.

Figure 26 is a complex hysteresis loop composed of a series of loading sequences. Starting at point 1, load is applied rapidly to point 2, and then a constant stress is held until point 3 is reached. Elastic unloading then takes place to point 4, and compressive loading continues rapidly to point 5. The compressive stress is then held constant until point 6 is reached, when the strain is held constant and the stress is relaxed to point 7. The cycle is completed by unloading elastically to point 1, where the cycle started.

In this loop, the inelastic strain range is defined by the width of the loop *AC*. In going from *A* to *C* in the tension direction, a tensile plastic strain, *AB,* was accumulated along with a tensile creep strain, *BC*. In reversing the cycle, a compressive plastic strain, *CD,* was accumulated along with a compressive creep strain, *DA*. In this example, the tensile plastic strain *AB* was reversed by a portion of the compressive plastic strain *CD;* likewise, the entire compressive creep strain *DA* was reversed by only a portion of the tensile creep strain *BC*. Therefore, a *PP* strain range of magnitude *AB* was present, along with a *CC* strain range of magnitude *DA*. From the excess tensile creep strain and excess compressive plastic strain, a *CP* strain range of magnitude *BC-DA* or *CD-AB* (because $CD - AB = BC - DA$) is present.

In any hysteresis loop, it is possible to have a maximum of only three of the four types of strain ranges. It is not possible for the *PC* and *CP* strain ranges to be components of the same hysteresis loop.

For this model, the inelastic strain range, $\Delta\varepsilon_{IN} = AC$, is made up of three components: $\Delta\varepsilon_{PP} = AB$, $\Delta\varepsilon_{CC} = DA$, and $\Delta\varepsilon_{CP} = (BC - DA)$. In equation form, $\Delta\varepsilon_{IN} = \Delta\varepsilon_{PP} + \Delta\varepsilon_{CC} + \Delta\varepsilon_{CP}$.

Prediction of cyclic life is performed by using a modification of the linear damage rule. This has been termed the interaction damage rule (Ref 44) and is given by Eq 13:

$$\frac{F_{PP}}{N_{PP}} + \frac{F_{CC}}{N_{CC}} + \frac{F_{CP}}{N_{CP}} + \frac{F_{PC}}{N_{PC}} = \frac{1}{N_f} \quad \text{(Eq 13)}$$

where N_f is the predicted life, and the *F* terms are as defined in Fig. 27. For this example, $F_{PP} = \Delta\varepsilon_{PP}/\Delta\varepsilon_{IN} = AB/AC$, $F_{CC} = \Delta\varepsilon_{CC}/\Delta\varepsilon_{IN} = DA/AC$, and $F_{CP} = \Delta\varepsilon_{CP}/\Delta\varepsilon_{IN} = (BC - DA)/AC$. Note that $F_{PP} + F_{CC} + F_{CP} = 1$. For the inelastic strain range (in this case, $\Delta\varepsilon_{IN} = AC$), the values of cyclic lives N_{PP}, N_{CC}, and N_{CP} can be read directly from Fig. 27. It is not necessary to obtain a value of N_{PC} for this example, because it was determined that no such strain-range component exists ($\Delta\varepsilon_{PC} = 0$). The damage per cycle due to each of the components can be represented by F_{PP}/N_{PP}, F_{CC}/N_{CC}, and F_{CP}/N_{CP}. The total damage per cycle ($1/N_f$) is equal to the sum of the individual damage contributions.

This approach has been evaluated using data for numerous alloys of engineering significance (see, for example, Ref 30, 31, 45–57). The correlation between observed and model-calculated creep-fatigue lives is generally within a factor of ±2.

The SRP life prediction method has been addressed in more than 1000 publications since its inception in 1971. A substantial number of refinements have been incorporated into the life prediction method over time. The result is a broad-based life prediction system that addresses many of the practical aspects of high-temperature structural life prediction encountered in industrial applications. Included are means for dealing with upper and lower bounds on cyclic life, cumulative creep-fatigue damage analysis, mean stress effects, separation of the creep and plastic strain components of inelastic strain, multiaxial creep-fatigue, thermomechanical fatigue of monolithic materials and continuous-fiber-reinforced metal-matrix composites, long-term extrapolation, experimental and analytical partitioning of inelastic strains, strain ratcheting, low-strain (nominally elastic) creep-fatigue, ductility-normalized SRP estimation equations, and so on. Many of these features have been incorporated into a commercially available computer code (Ref 58).

One of the most significant of the refinements is the casting of the nominally inelastic strain-range approach into a total strain-range approach. Without this feature of the model, it could not be applied to most engineering design situations because knowledge, especially early on in a design, of the magnitude of the inelasticity involved is simply too inaccurate to be relied on. The following paragraphs briefly describe what is known as the total strain version of strain-range partitioning (TS-SRP) (Ref 59–61). This version has been

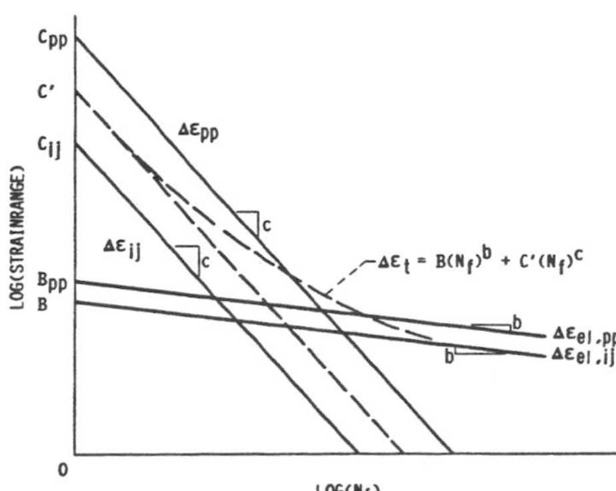

Fig. 28 Relation between total strain range and life for creep-fatigue cycles. Source: Ref 61

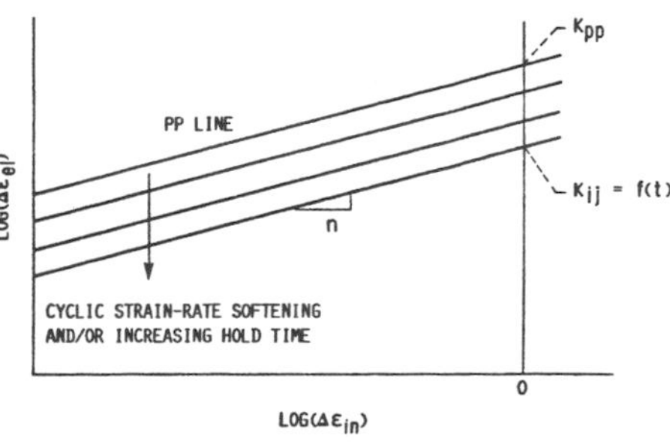

Fig. 29 Relations between inelastic and elastic strain ranges for creep-fatigue cycles. Cyclic strain-hardening coefficient K_{ij} is a function of hold time and strain-rate-hardening characteristics of an alloy. Source: Ref 61

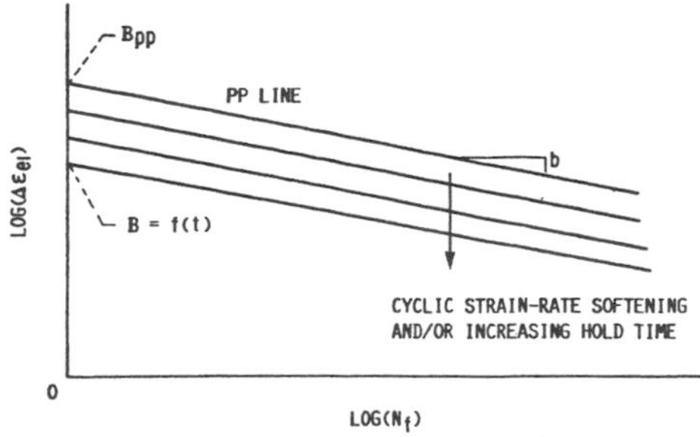

Fig. 30 Relation between elastic strain range and life for creep-fatigue cycles. Variation in elastic line intercept is a function of hold time and strain-rate-hardening characteristics of an alloy. Source: Ref 61

incorporated into a computer code (Ref 58) that is commercially available through COSMIC.

Expressing SRP in terms of total strain range does not change the fundamental premise of the approach. It remains based on the premise that cyclic life is a function of the inelastic strain range and the type and relative amounts of time-dependent strain (plasticity) and time-dependent strain (creep) present in a cyclic stress-strain hysteresis loop.

In formulating TS-SRP, the inelastic and elastic strain range versus life lines for isothermal creep-fatigue cycles are taken to be parallel to the corresponding failure lines for pure fatigue (*PP* cycles), as shown in Fig. 28. This is also the case for the frequency-modified approach.

Analysis shows that there is a relationship between failure behavior and flow behavior. Failure behavior is described by:

$$\Delta\varepsilon_{EL} = B(N_{f0})^b \quad \text{(Eq 14)}$$

$$\Delta\varepsilon_{IN} = C'(N_{f0})^c \quad \text{(Eq 15)}$$

where

$$C' = [\Sigma F_{ij}(C_{ij})^{1/c}]^c \quad \text{(Eq 16)}$$

and the subscripts *ij* denote the type of cycle (*PP, CC, CP,* or *PC*). Equation 16 is derived from the interaction damage rule (Ref 44) and the four generic SRP inelastic strain range versus cyclic life relations for a theoretical zero mean stress condition. Those life relationships are expressed as:

$$\Delta\varepsilon_{IN} = C_{ij}(N_{ij})^c \quad \text{(Eq 17)}$$

The intercepts (C_{ij}) may be expressed as functions of time if the material being characterized requires it. The interaction damage rule (Ref 44) is written as:

$$\Sigma\left(\frac{F_{ij}}{N_{ij}}\right) = \frac{1}{N_{f0}} \quad \text{(Eq 18)}$$

Equation 16 is obtained by Eq 17 to solve for N_{ij} and then substituting this term in Eq 18. Flow behavior is described by:

$$\Delta\varepsilon_{EL} = K_{ij}(\Delta\varepsilon_{ij})^n \quad \text{(Eq 19)}$$

where n = b/c. Assuming that the inelastic and elastic failure lines for creep-fatigue cycles are parallel to the corresponding failure lines for *PP* cycles, it follows that the strain-hardening exponent n in Eq 19 is a constant, as shown in Fig. 29. For isothermal conditions the strain-hardening coefficient K_{ij} is a function of temperature, total strain range, hold time, the manner in which creep strain is introduced into the cycle (stress hold, strain hold, etc.), and the strain rate-hardening characteristics of the alloy. For nonisothermal conditions K_{ij} is also a function of the maximum and minimum temperatures and the phase relation between strain and temperature.

The time-dependent behavior of the elastic line for creep cycles is shown in Fig. 30. Setting Eq 14 equal to Eq 19 and eliminating N_f by using Eq 15, the following equation relating flow and failure behavior is obtained:

$$B = K_{ij}(C')^n \quad \text{(Eq 20)}$$

The inelastic line intercepts C_{ij}, the exponent c, and the term C′ are considered to be failure terms. The strain fractions F_{ij}, the cyclic strain-hardening coefficients K_{ij}, and the strain-hardening exponent n are considered to be flow terms. Thus, the elastic line intercept B is determined by a combination of both flow and failure terms. We are now in a position to establish a total strain range versus life relation and thus predict life on a total strain basis. The total strain range is the sum of the elastic and inelastic strain ranges:

$$\Delta\varepsilon_t = \Delta\varepsilon_{EL} + \Delta\varepsilon_{IN} \quad \text{(Eq 21)}$$

From Eq 14 and 15, the following is obtained:

$$\Delta\varepsilon_t = B(N_{f0})^b + C'(N_{f0})^c \quad \text{(Eq 22)}$$

A schematic plot of Eq 22 is shown in Fig. 28. The predicted life, N_{f0}, represents the life for a zero mean stress condition. Correction for the presence of mean stress is necessary (Ref 61). An example application of TS-SRP is shown in Fig. 31(a) (Ref 60) for Inconel 718 at 650 °C (1200 °F). Life prediction by the original inelastic strain version of SRP is shown for comparison in Fig. 31(b). The TS-SRP approach has also been adapted to the life prediction of metal-matrix composites at elevated temperatures (Ref 62).

Other Creep-Fatigue Life Prediction Methods. Numerous creep-fatigue life prediction models have been proposed over the past half century. These are tabulated in the Appendix to this article.

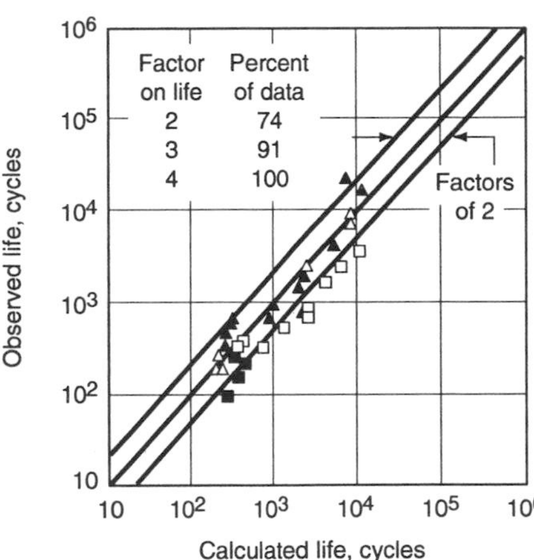

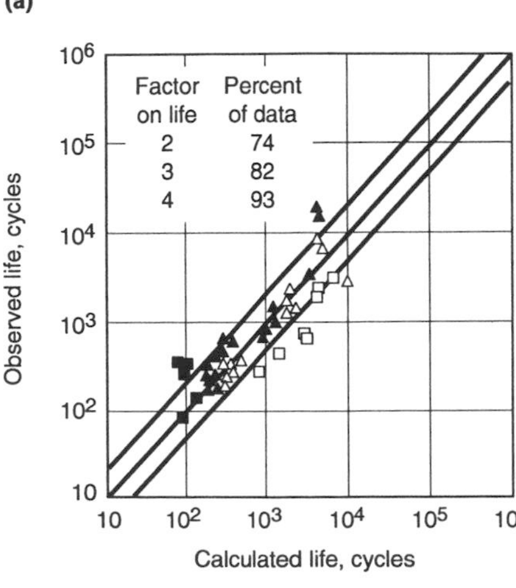

Fig. 31 Life predictions of creep-fatigue data for Inconel 718 at 650 °C (1200 °F). (a) Total strain-range approach. (b) Inelastic strain-range approach. Source: Ref 60

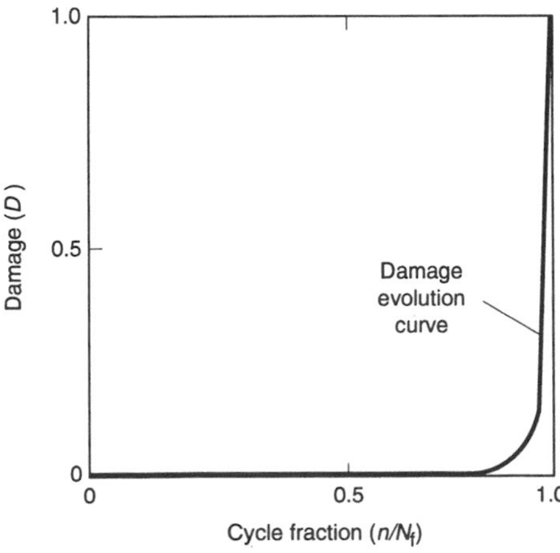

Fig. 32 Schematic of damage evolution during cyclic loading according to continuous damage mechanics

Continuum Damage Mechanics for Creep-Fatigue

The underlying concept of damage mechanics is that the rate of doing damage is related to the amount of damage present; that is, damage begets more damage. In its simplest representation, damage can be thought of as eroding the cross-sectional area that carries load. Kachanov (Ref 63) and Robotnov (Ref 64) used the concept to represent the monotonic creep-rupture process. As a coupon creeps at high temperature under a fixed tensile load, the cross-sectional area decreases, causing the effective stress on the material to increase and accelerating the creeping process. Part of the area decrease is due to Poisson contraction, but another part is the loss of internal load-bearing area due to the damage associated with internal void growth and cracking. Thus, damage, *D,* is the loss of effective load-bearing area:

$$\text{Effective stress} = \frac{\text{Nominal stress}}{1 - D} \quad \text{(Eq 23)}$$

Equation 23 gives rise to a highly nonlinear damage accumulation curve such as that shown schematically in Fig. 32.

Figure 33 (Ref 65) depicts the realm of utility of damage mechanics relative to microscopic and macroscopic scales of operation for a polycrystalline metal structure. Over the past three decades, research workers in France and elsewhere have developed a number of sophisticated models (Ref 66–71) for fatigue, creep-fatigue, creep-oxidation-fatigue, thermal fatigue, multiaxial fatigue, cumulative fatigue damage, and so on, whose origins stem from these basic notions. Recent developments (Ref 65) include application to the creep-fatigue life prediction of metal-matrix composite materials wherein internal damage initiation and propagation sites are well suited to the description of distributed damage. For this purpose, the linear life fraction rule discussed earlier is used to analytically combine the individual contributions from the evolution of creep and fatigue damage.

Appendix: Creep-Fatigue Life Prediction Models

Numerous creep-fatigue life prediction models have been proposed over the past half century. A 1991 survey (Ref 72) of the evolution of creep-fatigue life prediction models revealed the existence of more than 100 models. Each model was associated with one or more of the 14 categories listed in Table 3. These models have enjoyed some degree of success in dealing with a specific set of creep-fatigue data. However, most lack the generality that would enable them to achieve widespread acceptance in engineering applications. The list of models has been growing at an average rate of about five new or revised models per year.

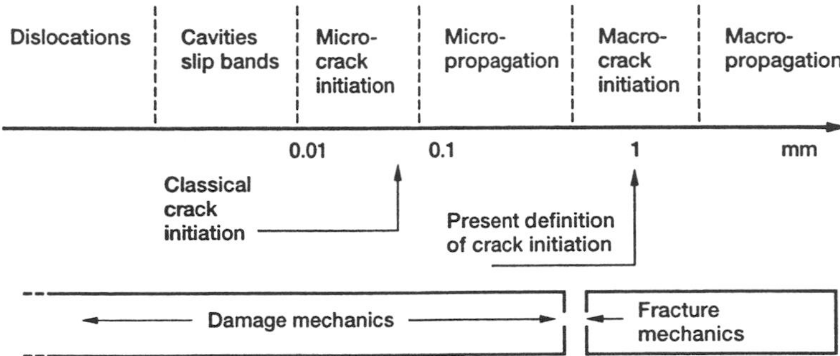

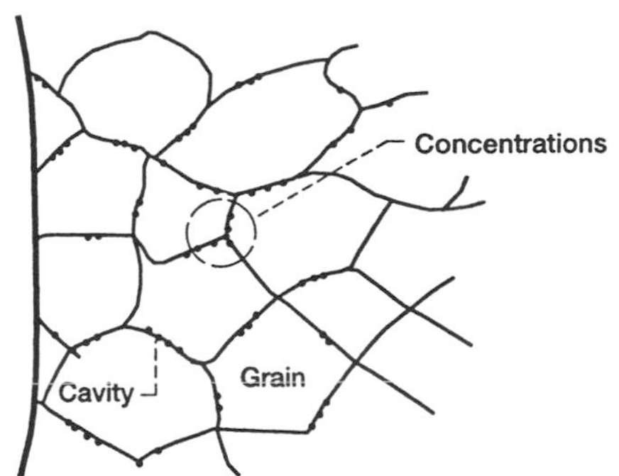

(a) Creep damage: coalescence of cavities and intergranular defects.

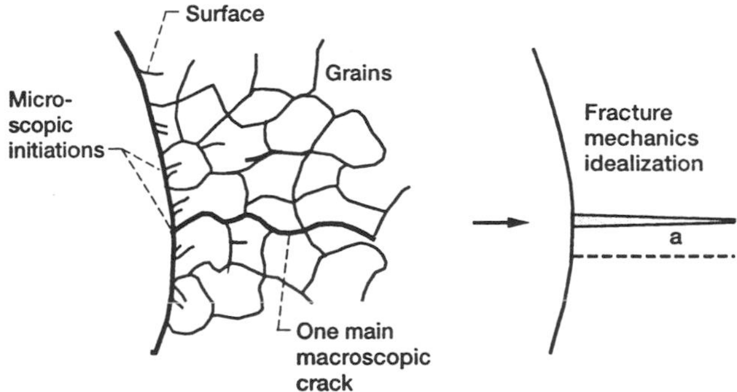

(b) Fatigue damage: nucleation of slip bands, microcracks and transgranular defects.

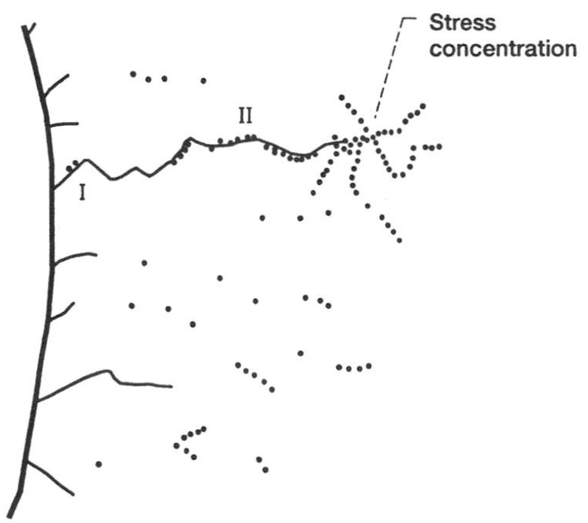

(c) Creep/fatigue damage: interaction of intergranular and transgranular defects.

Fig. 33 Schematic of different damage modes and the associated scale in a monolithic metal. Source: Ref 65

Table 3 Summary of creep-fatigue life prediction models

Type	Acronym	Title	Ref
A: Life or Damage Fraction Rules			
A	LCR	Linear Creep Rupture Damage Rule	73
A	LCF	Linear Creep Damage	74
A	TCF	Time + Cycle Fraction Rule	75
A	LFA	Life Fraction Approach	12
AD	TPR	Ten Percent Rule	76
ABD	LCD	Linear Creep Damage for Thermal Fatigue	77
ABD	MLF	Modified Life Fraction Rule	78
ABD	LFR	Life Fraction Rule	79
ABD	TCD	Turbine Component Design	80
ABD	ITC	Interactive Time-Cycle Fractions	36
ABD	RCF	Relaxation Creep Fatigue	81
ABE	CDC	Cumulative Damage under Creep	82
ABDE	PFC	Phenomenological Fatigue Creep	83
AE	FCD	Fatigue + Creep Damage Mechanisms	84
A	FNC	French Nuclear Code	85
ABDN	FCE	Fatigue-Creep-Environment Model	86, 87
B: Stress-Life Diagrams			
B	SRD	Stress Range Diagrams	88
B	SRD	Stress Range Diagrams	5
ABD	LCD	Linear Creep Damage for Thermal Fatigue	77
ABD	MLF	Modified Life Fraction Rule	78
ABD	LFR	Life Fraction Rule	79
ABD	TCD	Turbine Component Design	80
ABD	ITC	Interactive Time-Cycle Fractions	36
ABD	RCF	Relaxation Creep Fatigue	81
ABE	CDC	Cumulative Damage under Creep	82
ABDE	PFC	Phenomenological Fatigue Creep	83
ABDN	FCE	Fatigue-Creep-Environment Model	86, 87
C: Frequency Effect Equations			
C	FEE	Frequency Effect Equation	16
C	TTF	Time-to-Failure Mode	89
CD	MCS	Method of Characteristic Slopes	90
CD	FMF	Frequency-Modified Fatigue	40
C	HTE	Hold-Time Effects Mode	91
C	FSM	Frequency Separation Method	92
CF	THE	Tensile Hysteresis Energy Model	93, 94
CF	MTE	Modified Tensile Hysteretic Energy Model	95
C	EFE	Endochronic Frequency Equation	96
D: Strain-Range/Life Models			
D	PSF	Plastic Strain Fatigue Law	97
D	PSF	Plastic Strain Fatigue Law	98
D	TSR	Total Strain-Range/Life Model	99
D	MUS	Method of Universal Slopes	35
AD	TPR	Ten Percent Rule	76
D	TSR	Total Strain-Range/Life Model	100
ABD	LCD	Linear Creep Damage for Thermal Fatigue	77
D	PST	Plastic Strain-Range Time Mode	90
CD	FMF	Frequency-Modified Fatigue	40
CD	MCS	Method of Characteristic Slopes	101
DK	TCE	Thermal Cycling Equation	102
ABD	MLF	Modified Life Fraction Rule	78
ABD	LFR	Life Fraction Rule	79
ABD	TCD	Turbine Component Design	80
ABD	ITC	Interactive Time-Cycle Fractions	36
ABD	RCF	Relaxation Creep Fatigue	81
D	MSE	Modified Strain-Range Equation	103
D	TDE	Time-Dependent Exponent of Strain	104
ABDE	PFC	Phenomenological Fatigue Creep	83
DF	OCF	Overstress Concept of Creep-Fatigue	105
ABDN	FCE	Fatigue-Creep-Environment Model	86, 87
E: Damage Mechanics			
E	DMC	Damage Mechanics for Creep	63
E	CDM	Continuum Damage Model	66
E	CDM	Cyclic Damage Model	68
ABE	CDC	Cumulative Damage under Creep	82
AE	FCD	Fatigue + Creep Damage Mechanisms	84
ABDE	PFC	Phenomenological Fatigue Creep	83
F: Hysteresis Energy Models			
F	HEA	Hysteresis Energy Approach	100
CF	THE	Tensile Hysteresis Energy Model	93, 94
FI	PEM	Partitioned Energy Model	106
CF	MTE	Modified Tensile Hysteretic Energy Model	95
FH	CTE	Crack-Tip Energy Model	107, 108
F	CEG	Constant Enthalpy Gain Approach	109
FI	SEP	Strain Energy Partitioning	110
DF	OCF	Overstress Concept of Creep-Fatigue	105
F	CAB	Coated Anisotropic Blade Model	111
G: Ductility Exhaustion			
G	DEM	Ductility Exhaustion Model	112
G	LDE	Linear Creep Ductility Exhaustion	113
G: Ductility Exhaustion (continued)			
G	DET	Ductility Exhaustion for Thermal Fatigue	114
G	CPE	Creep-Plasticity Ductility Exhaustion	115
G	LPC	Linear Plastic/Creep Strain Exhaustion	116
G	CPE	Creep-Plastic Exhaustion	117
G	MDE	Modified Ductility Exhaustion Model	118
H: Crack Growth			
H	CGM	Crack Growth Model	119
H	HCG	High-Temperature Crack Growth	120
H	CGA	Crack Growth Approach	121
H	CPM	Crack Propagation Model	122
H	ETC	Elevated-Temperature Crack Propagation	123
H	CCT	Creep Crack-Tip Model	124
H	CCD	Cyclic Creep Damage Model	125
H	CIM	Crack Interaction Model	126
DH	MCI	Macrocrack Initiation, Fracture Mechanics	127
HN	LCT	Local Crack-Tip Model	128
FH	CTE	Crack-Tip Energy Model	107, 108
HKN	FAM	FATIGMOD	129
H	CFC	Creep-Fatigue Cracks Model	130
H	SCC	Short-Crack Creep Fatigue	131
HN	MMM	Microcracking, Creep-Fatigue-Environment	132
I: Strain-Range Partitioning			
I	SRP	Strain-Range Partitioning	43
Ia	IDR-SRP	Interaction Damage Rule	44
Ib	PDR-SRP	Product Damage Rule	31
Ic	ITF-SRP	Inelastic Thermal Fatigue, SRP	133
Id	DEX-SRP	Linear Ductility Exhaustion, SRP	134
Ie	PWA-SRP	Pratt & Whitney Combustor, SRP	135
If	MAF-SRP	Multiaxiality Factor, SRP	136
Ig	DNE-SRP	Ductility Normalized Equations, SRP	137
Ih	SRC-SRP	Strain-Range Conversion Principle, SRP	138, 139
Ii	MSE-SRP	Mean Stress Effects, SRP	53
Ij	CEP-SRP	Combustion Engineering Model, SRP	140
Ik	TSV-SRP	Total Strain Version of SRP	59
Il	DEP-SRP	Diesel Engine Piston, SRP	141
Im	NDR-SRP	Nonlinear Damage Rule, SRP	142
In	SDA-SRP	Statistical Data Analysis, SRP	143
Io	SRL-SRP	Statistically Refined Life, SRP	144
Ip	ETM-SRP	Exposure Time Modified, SRP	145
Iq	SSC-SRP	Steady-State Creep Rate, SRP	146, 147
Ir	TFM-SRP	Time-to-Failure Modified SRP	148
Is	TMF-SRP	Thermomechanical Fatigue, TSV-SRP	149
It	FMB-SRP	Fracture Mechanics Basis of SRP	150
FI	PEM	Partitioned Energy Model	106
FI	SEP	Strain Energy Partitioning	110
J: Time-to-Failure Model			
J	TCD	Time-Cycle Diagram	151
K: Macrophenomenological Models			
DK	TCE	Thermal Cycling Equation	102
K	PHF	Parametric High-Temperature Fatigue	152
K	SST	Stress-Strain-Temperature Empirical Model	153
HKN	FAM	FATIGMOD	129
K	ICC	Internal Cracking, Coated Single Crystals	154
L: Damage Rate Models			
L	DRM	Damage Rate Model	155
L	MDR	Modified Damage Rate Model	156
L	ILP	Incremental Life Prediction Law	157
L	TSM	Temperature and Strain Rate Model	158
M: Cyclic Damage Accumulation Models			
M	PDA	Preliminary Cyclic Damage Accumulation	159
M	CDA	Cyclic Damage Accumulation	160
N: Micromechanistic Models			
N	GBS	Grain Boundary Sliding Model	161
N	RVG	R-Void Growth Model	162
N	OCC	Oxide Cracking	163, 164
N	SOC	Stress-Oxidation Crack-Tip Model	165
HN	LCT	Local Crack-Tip Model	128
N	GBV	Grain-Boundary Void Model	166
N	IGW	Initiation and Growth of Wedge Cracks	167, 168
N	OFI	Oxidation Fatigue Interaction Model	169
N	ARN	Anelastic Recovery, ODS Alloys	170
HKN	FAM	FATIGMOD	129
N	CCC	Critical Cavity Criteria Model	171
N	CBS	Cohesive Boundary Strength Model	172
N	GBO	Grain-Boundary Oxidation Model	173
ABDN	FCE	Fatigue-Creep-Environment Model	86, 87
HN	MMM	Microcracking, Creep-Fatigue-Environment	132

Source: Ref 72

REFERENCES

1. A.H. Meleka, *Met. Rev.*, Vol 7 (No. 25), 1962, p 43
2. M. Hempel and H.E. Tillmanns, *Mitt. K.W. Inst. Eisenforsch.*, Vol 18, 1936, p 163
3. M. Hempel and F. Ardelt, *Mitt. K.W. Inst. Eisenforsch.*, Vol 21, 1939, p 115
4. M. Hempel and H. Krug, *Mitt. K.W. Inst. Eisenforsch.*, Vol 24, 1942, p 77
5. H.J. Tapsell, P.G. Forrest, and G.R. Tremain, Creep due to Fluctuating Stresses at Elevated Temperatures, *Engineering,* Vol 170, 1950, p 189
6. H.J. Tapsell, *Symp. High-Temperature Steels and Alloys for Gas Turbines,* Special Report No. 43, Iron and Steel Institute, London, 1952, p 169
7. A.J. Kennedy, *Proc. Int. Conf. Fatigue of Metals,* Institution of Mechanical Engineers, London, 1956, p 401
8. A.H. Meleka and A.V. Evershed, *J. Inst. Met.*, Vol 88, 1959–60, p 411
9. A.H. Meleka and G. B. Dunn, *Nature,* Vol 184, 1959, p 896
10. S.S. Manson, Interfaces between Fatigue, Creep, and Fracture, *Int. J. Fracture Mech.*, Vol 2 (No. 1), 1966, p 327
11. E.G. Ellison and E.M. Smith, *Fatigue at Elevated Temperatures,* STP 520, A.E. Carden, A.J. McEvily, and C.H. Wells, Ed., ASTM, 1972, p 575
12. D.S. Wood, The Effect of Creep on the High-Strain Fatigue Behavior of a Pressure Vessel Steel, *Weld. J. Res. Suppl.*, Vol 45, 1966, p 92s–96s
13. E.G. Ellison, *J. Mech. Eng. Sci.*, Vol 11 (No. 3), 1969, p 318–339
14. A.V. Evershed and G.B. Dunn, *Br. Iron Steel Res. Assoc. Rep.*, 1958, p 16
15. L.F. Coffin, Jr., Introduction to High-Temperature Low-Cycle Fatigue, *Exp. Mech.*, Vol 8, May 1968, p 218
16. J.F. Eckel, The Influence of Frequency on the Repeated Bending Life of Acid Lead, *Proc. ASTM,* Vol 51, 1951, p 745–756
17. G.R. Gohn and W.C. Ellis, The Fatigue Test as Applied to Lead Cable Sheath, *Proc. ASTM,* Vol 51, 1951, p 721
18. R.E. Peterson, Design Approaches for Low-Cycle Fatigue Problems in Power Apparatus, *10th Sagamore Conf.*, Syracuse University Research Institute, 1963
19. E. Krempl and B.M. Wundt, Hold-Time Effects in High-Temperature, Low-Cycle Fatigue: A Literature Survey and Interpretive Report, *STP 489,* ASTM, 1971
20. A.E. Carden, Fatigue at Elevated Temperatures: A Review of Test Methods, *Fatigue at Elevated Temperatures,* STP 520, A.E. Carden, A.J. McEvily, and C.H. Wells, Ed., ASTM, 1973, p 195–223
21. W.R. Berry and I. Johnson, Prevention of Cyclic Thermal-Stress Cracking in Steam Turbine Rotors, *J. Eng. Power,* Vol 86A, 1964, p 361–367
22. R.P. Kent, Some Aspects of Metallurgical Research and Development Applied to Large Steam Turbines, *Parsons J.*, Dec 1964
23. H.G. Edmunds, Repeated Cyclic Strains, *Proc. Inst. Mech. Eng.*, Vol 180 (Part 31), 1965–66, p 373–379
24. J.B. Conway, R.H. Stentz, and J.T. Berling, "Fatigue, Tensile, and Relaxation Behavior of Stainless Steels," TID-26135, U.S. Atomic Energy Commission, 1975
25. J.B. Conway, J.T. Berling, and R.H. Stentz, Strain Rate and Holdtime Saturation in Low-Cycle Fatigue: Design Parameter Plots, *Fatigue at Elevated Temperatures,* STP 520, A.E. Carden, A.J. McEvily, and C.H. Wells, Ed., ASTM, 1972, p 637–647
26. J.B. Conway and R.H. Stentz, "High Temperature, Low Cycle Fatigue Data for Three High Strength Nickel-Base Superalloys," Tech. Report AFWAL-TR-9-4077, Wright-Patterson Air Force Base, June 1980
27. E.G. Ellison and A.J.F. Paterson, Creep Fatigue Interactions in a 1Cr-Mo-V Steel, *Proc. Inst. Mech. Eng.*, Vol 190, 1976, p 321
28. H.P. Chu, B.A. McDonald, and O.P. Arora, Creep-Fatigue Interaction of Titanium Alloy Ti-6Al-2Cb-1Ta-0.8Mo at Room Temperature, *Titanium: Science and Technology,* Deutsche Gesellschaft für Metallkunde, Oberursel, West Germany, 1985, p 2395–2402
29. R.C. Bill, M.J. Verrilli, M.A. McGaw, and G.R. Halford, "A Preliminary Study of the Thermomechanical Fatigue of Polycrystalline MAR M-200," NASA TP-2280 (AVSCOM TR 83-C-6), National Aeronautics and Space Administration, 1984
30. S.S. Manson and G.R. Halford, Relation of Cyclic Loading Pattern to Microstructural Fracture in Creep-Fatigue, *Proc. 2nd Int. Conf. Fatigue and Fatigue Thresholds (Fatigue '84),* Vol 3, C.J. Beevers, Ed., Engineering Materials Advisory Services Ltd., Cradley Heath, Warley, U.K., 1984, p 1237–1255
31. C.G. Annis, M.C. VanWanderham, and R.M. Wallace, "Strainrange Partitioning Behavior of an Automotive Turbine Alloy," NASA CR-134974, National Aeronautics and Space Administration, 1976
32. T. Hasebe, M. Sakane, and M. Ohnami, Failure Life and Cyclic Constitutive Relation of Directionally Solidified MAR-M247 Superalloy in Creep-Fatigue, *Mechanical Behavior of Materials,* Vol 2, Pergamon Press, 1991, p 475–480
33. J.M. Holt, H. Mindlin, and C.Y. Ho, Ed., *Structural Alloys Handbook,* CINDAS/Purdue University, 1995
34. W.F. Brown, Jr., H. Mindlin, and C.Y. Ho, Ed., *Aerospace Structural Metals Handbook,* CINDAS/Purdue University, 1996
35. S.S. Manson, Fatigue: A Complex Subject—Some Simple Approximations, *Exp. Mech.*, Vol 5 (No. 7), 1965, p 193–226
36. R. Lagneborg and R. Attermo, The Effect of Combined Low-Cycle Fatigue and Creep on the Life of Austenitic Stainless Steels, *Metall. Trans.*, Vol 2, 1971, p 1821–1827
37. R.D. Campbell, Creep/Fatigue Interaction Correlation for 304 Stainless Steel Subjected to Strain-Controlled Cycling with Hold Times at Peak Strain, *J. Eng. Ind. (Trans. ASME),* Vol 93 (No. 4), 1971, p 88
38. R.M. Curran and B.M. Wundt, A Study of Low-Cycle Fatigue and Creep Interactions in Steels at Elevated Temperatures, *Reports of Current Work in Behavior of Materials at Elevated Temperatures,* A.O. Schaefer, Ed., Publ. G87, American Society of Mechanical Engineers, 1974, p 1–104
39. R.M. Curran and B.M. Wundt, Continuation of a Study of Low-Cycle Fatigue and Creep Interaction in Steels at Elevated Temperatures, *ASME-MPC Symp. Creep-Fatigue Interaction,* R.M. Curran, Ed., Publ. G00112 (MPC-3), American Society of Mechanical Engineers, 1976, p 203–282
40. L.F. Coffin, Jr., The Effect of Frequency on High Temperature, Low-Cycle Fatigue, *Proc. Air Force Conf. Fracture and Fatigue of Aircraft Structures,* AFDL-TR-70-144, 1970, p 301–312
41. A. Coles et al., The High Strain Fatigue Properties of Low-Alloy Creep Resisting Steels, *Proc. Int. Conf. Thermal and High-Strain Fatigue,* Monograph and Report Series No. 32, Metals and Metallurgy Trust, London, 1967
42. L.F. Coffin, Jr., et al., "Time-Dependent Fatigue of Structural Alloys: A General Assessment," ORNL-5073, Oak Ridge National Laboratory, 1975
43. S.S. Manson, G.R. Halford, and M.H. Hirschberg, Creep-Fatigue Analysis by Strain-Range Partitioning, *First Symp. Design for Elevated Temperature Environment,* American Society of Mechanical Engineers, 1971, p 12–28
44. S.S. Manson, The Challenge to Unify Treatment of High Temperature Fatigue—A Partisan Proposal Based on Strainrange Partitioning, *Fatigue at Elevated Temperatures,* STP 520, A.E. Carden, A.J. McEvily, and C.H. Wells, Ed., ASTM, 1973, p 744–775
45. M.H. Hirschberg and G.R. Halford, "Use of Strainrange Partitioning to Predict High-Temperature Low-Cycle Fatigue Life," NASA TN D-8072, National Aeronautics and Space Administration, 1976
46. K.D. Sheffler, "The Partitioned Strainrange Fatigue Behavior of Coated and Uncoated MAR-M-302 at 1000 °C (1832 °F) in Ultrahigh Vacuum," NASA CR-134626, National Aeronautics and Space Administration, 1974
47. K.D. Sheffler, Vacuum Thermal-Mechanical Fatigue Behavior of Two Iron-Base Alloys, *Thermal Fatigue of Materials and Components,* STP 612, D.A. Spera and D.F. Mowbray, Ed., ASTM, 1976, p 214–226
48. M.H. Hirschberg and G.R. Halford, "Use of Strainrange Partitioning to Predict High-Temperature Low-Cycle Fatigue Life," NASA TN D-8072, National Aeronautics and Space Administration, 1976
49. C.E. Jaske, R.C. Rice, R.D. Buchheit, D.B. Roach, and T.L. Porfilio, "Low-Cycle Fatigue

of Type 347 Stainless Steel and Hastelloy Alloy X in Hydrogen Gas and in Air at Elevated Temperatures," NASA CR-135022, National Aeronautics and Space Administration, 1976
50. J.F. Saltsman and G.R. Halford, Application of Strainrange Partitioning to the Prediction of Creep-Fatigue Lives of AISI Types 304 and 316 Stainless Steel, *Trans. ASME,* Vol 99 (No. 2), May 1977, p 264–271
51. G.R. Halford and A.J. Nachtigall, Strainrange Partitioning Behavior of the Nickel-Base Superalloys, René 80 and IN-100, *Proc. AGARD Conf. Characterization of Low Cycle High Temperature Fatigue by the Strainrange Partitioning Method,* AGARD CP-243, Advisory Group for Aerospace Research and Development, Neuilly-sur-Seine, France, 1978, p 2-1 to 2-14
52. J.F. Saltsman and G.R. Halford, Strainrange Partitioning Life Predictions of the Long Time Metal Properties Council Creep-Fatigue Tests, *Methods for Predicting Material Life in Fatigue,* American Society of Mechanical Engineers, 1979, p 101–132
53. G.R. Halford and A.J. Nachtigall, The Strainrange Partitioning Behavior of an Advanced Gas Turbine Disk Alloy, AF2-1DA, *J. Aircraft,* Vol 17 (No. 8), 1980, p 598–604
54. B.N. Leis, R. Rungta, and A.T. Hopper, "Creep Fatigue of Low-Cobalt Superalloys: Waspaloy, PM U 700, and Wrought U700," NASA CR-168260, National Aeronautics and Space Administration, 1983
55. G.R. Halford, L.R. Johnson, and J.A. Brown, High-Temperature LCF of Ni-201 and 304L SS, *Advanced Earth-to-Orbit Propulsion Technology 1986,* Vol 2, NASA CP-2437, National Aeronautics and Space Administration, 1986, p 172–204
56. G.R. Halford, T.G. Meyer, R.S. Nelson, D.M. Nissley, and G.A. Swanson, Fatigue Life Prediction Modeling for Turbine Hot Section Materials, *J. Eng. Gas Turbines Power (Trans. ASME),* Vol 111 (No. 2), April 1989, p 279–285
57. G.R. Halford, M.J. Verrilli, S. Kalluri, F.J. Ritzert, R.E. Duckert, and F.A. Holland, Thermomechanical and Bithermal Fatigue Behavior of Cast B1900 + Hf and Wrought Haynes 188, *Advances in Fatigue Lifetime Predictive Techniques,* STP 1122, M.R. Mitchell and R.W. Landgraf, Ed., ASTM, 1992, p 120–142
58. J.F. Saltsman, Computer Programs to Characterize Alloys and Predict Cyclic Life Using the Total Strain Version of Strainrange Partitioning, *Tutorial and Users Manual, Version 1.0,* NASA TM-4425, National Aeronautics and Space Administration, 1992
59. G.R. Halford and J.F. Saltsman, Strainrange Partitioning—A Total Strainrange Version, *Advances in Life Prediction Methods,* D.A. Woodford and J.R. Whitehead, Ed., American Society of Mechanical Engineers, 1983, p 17–26
60. J.F. Saltsman and G.R. Halford, An Update on the Total Strain Version of SRP, *Low Cycle Fatigue—Directions for the Future,* STP 942, H.D. Solomon, G.R. Halford, L.R. Kaisand, and B.N. Leis, Ed., ASTM, 1988, p 329–341
61. J.F. Saltsman and G.R. Halford, "Procedures for Characterizing an Alloy and Predicting Cyclic Life Using the Updated Total Strain Version of Strainrange Partitioning," NASA TM-4102, National Aeronautics and Space Administration, June 1989
62. G.R. Halford, B.A. Lerch, J.F. Saltsman, and V.K. Arya, Proposed Framework for TMF Life Prediction of MMCs, *Symp. Thermo-Mechanical Fatigue Behavior of Materials,* STP 1186, H. Sehitoglu, Ed., ASTM, 1993, p 176–194; see also NASA TM-3320, 1993
63. L.M. Kachanov, Time of the Rupture Process under Creep Conditions, *Izv. Akad. Nauk. SSR, Old. Tekh. Nauk.,* Vol 8, 1958, p 26–31
64. Y.N. Robotnov, Creep Rupture, *Proc. XII Int. Congress of Applied Mechanics,* Stanford-Springer, 1969
65. S. Kruch and S.M. Arnold, Creep Damage and Creep-Fatigue Damage Interaction Model for Unidirectional Metal Matrix Composites, *Symp. Applications of Continuum Damage Mechanics to Fatigue and Fracture,* STP 1315, ASTM, 1997
66. J. Lemaitre, J.L. Chaboche, and Y. Munakata, Method of Metal Characterization for Creep and Low Cycle Fatigue Prediction in Structures—Example of Udimet 700, *Proc. Symp. Mechanical Behavior of Materials,* Society of Materials Science, Kyoto, 1974, p 239–249
67. J. Lemaitre and J.-L. Chaboche, A Nonlinear Model of Creep-Fatigue Damage Cumulation and Interaction, *Mechanics of Visco-Plastic Media and Bodies,* J. Hult, Ed., Springer-Verlag, Berlin, 1975, p 291–301
68. A. Plumtree and J. Lemaitre, "Application of Damage Concepts to Predict Creep-Fatigue Failures," presented at *Joint ASME/CSME PVP Conf.,* Montreal, 1978
69. J.L. Chaboche, Continuous Damage Mechanics: A Tool to Describe Phenomena before Crack Initiation, *Nucl. Eng. Des.,* Vol 64, 1981, p 233–247
70. F. Gallerneau, D. Nouailhas, and J.-L. Chaboche, "A Fatigue Damage Model Including Interaction Effects with Oxidation and Creep Damage," Paper 1996-47, ONERA, Paris, 1996
71. F. Gallerneau, D. Nouailhas, and J.-L. Chaboche, "Fatigue Damage Behavior of a Coated Ni-Base Single Crystal Superalloy," Paper 1996-107, ONERA, Paris, 1996
72. G.R. Halford, Evolution of Creep-Fatigue Life Prediction Models, *Creep-Fatigue Interaction at High Temperature,* Vol 21, G.K. Haritos and O.O. Ochoa, Ed., American Society of Mechanical Engineers, 1991, p 43–57
73. E.L. Robinson, Effect of Temperature Variation on the Long-Time Rupture Strength of Steels, *Trans. ASME,* Vol 74 (No. 5), 1952, p 777–780
74. A. Berkovits, "Investigation of Three Analytical Hypotheses for Determining Material Creep Behavior under Varied Loads," NASA Technical Note D799, National Aeronautics and Space Administration, 1961
75. S. Taira, Lifetime of Structures Subjected to Varying Load and Temperature, *Colloquium on Creep in Structures,* N.J. Hoff, Ed., Springer-Verlag, Berlin, 1962, p 96–119
76. S.S. Manson and G.R. Halford, A Method of Estimating High-Temperature Low-Cycle Fatigue Behavior of Materials, *Proc. Int. Conf. Thermal and High-Strain Fatigue,* Metals and Metallurgy Trust, London, 1967, p 154–170
77. D.A. Spera, "A Linear Creep Damage Theory for Thermal Fatigue of Materials," Ph.D. dissertation, University of Wisconsin–Madison, 1968
78. S.S. Manson, G.R. Halford, and D.A. Spera, The Role of Creep in High Temperature Low-Cycle Fatigue, *Advances in Creep Design,* A.I. Smith and A.M. Nicolson, Ed., Halsted Press, 1971, p 229–249
79. Code Case 1331-5 (predecessor of current Code Case N-47), *Boiler and Pressure Vessel Code,* Section III, American Society of Mechanical Engineers, 1971
80. D.P. Timo, Designing Turbine Components for Low-Cycle Fatigue, *Proc. Int. Conf. Thermal Stresses and Thermal Fatigue,* D.J. Littler, Ed., Butterworths, London, 1971, p 453–469
81. P. Marshall and T.R. Cook, Prediction of Failure of Materials under Cyclic Loading, *Proc. Int. Conf. Thermal Stresses and Thermal Fatigue,* D.J. Littler, Ed., Butterworths, London, 1971, p 81–88
82. T. Bui-Quoc, An Engineering Approach for Cumulative Damage in Metals under Creep Loading, *J. Press. Vessel Technol. (Trans. ASME),* Vol 101, 1979, p 337–343
83. T. Bui-Quoc and A. Biron, A Phenomenological Approach for the Analysis of Combined Fatigue and Creep, *Nucl. Eng. Des.,* Vol 71, 1982, p 89–102
84. A. Plumtree and J. Lemaitre, in *Advances in Fracture Research,* D. Francois, Ed., Pergamon Press, 1982, p 2379
85. C. Heng, J.M. Grandemange, and A. Morel, RCC-M (Rules for Design and Construction of Nuclear Components), *Nuclear Engineering Design: Selected Papers from SMiRT-8 Seminar on Construction Codes and Engineering Mechanics* (26–27 Aug 1984, Paris), Vol 98 (No. 3), Jan 1985, p 265–277
86. R.W. Neu and H. Sehitoglu, Thermomechanical Fatigue, Oxidation, and Creep, Part I: Damage Mechanisms, *Metall. Trans. A,* Vol 20A, 1989, p 1755–1767
87. R.W. Neu and H. Sehitoglu, Thermomechanical Fatigue, Oxidation, and Creep, Part II: Life Prediction, *Metall. Trans. A,* Vol 20A, 1989, p 1769–1783
88. B.J. Lazan, Dynamic Creep and Rupture Properties under Tensile Fatigue Stress, *Proc. ASTM,* Vol 49, 1949, p 757
89. A. Coles and D. Skinner, Assessment of Thermal-Fatigue Resistance of High Temperature Alloys, *J. R. Aeronaut. Soc. London,* Vol 69, 1965, p 53–55
90. J.T. Berling and J.B. Conway, A Proposed Method for Predicting Low-Cycle Fatigue Behavior of 304 and 316 Stainless Steel, *Trans. Metall. Soc. AIME,* Vol 245, 1969, p 1137–1140

91. K. Wellinger and S. Sautter, Der Einfluss von Tempertur, Dehnungsgeschwindigkeit and Haltezeit auf das Zeitfestigkeitsverhalten von Stählen, *Arch. Eisenhüttenwes.*, Vol 44, 1973, p 47–55
92. L.F. Coffin, The Concept of Frequency Separation in Life Prediction for Time-Dependent Fatigue, *ASME-MPC Symp. Creep-Fatigue Interaction,* R.M. Curran, Ed., Publ. G00112 (MPC-3), American Society of Mechanical Engineers, 1976, p 349–363
93. W.J. Ostergren, A Damage Function and Associated Failure Equations for Predicting Hold Time and Frequency Effects in Elevated Temperature, Low Cycle Fatigue, *J. Test. Eval.,* Vol 4 (No. 5), 1976, p 327–339
94. W.J. Ostergren, Correlation of Hold Time Effects in Elevated Temperature Low Cycle Fatigue Using a Frequency Modified Damage Function, *ASME-MPC Symp. Creep-Fatigue Interaction,* R.M. Curran, Ed., Publ. G00112 (MPC-3), American Society of Mechanical Engineers, 1976, p 179–202
95. W.J. Ostergren and E. Krempl, A Uniaxial Damage Accumulation Law for Time-Varying Loading Including Creep-Fatigue Interaction, *J. Press. Vessel Technol. (Trans. ASME),* Vol 101, 1979, p 118–124
96. K.C. Valanis, On the Effect of Frequency on Fatigue Life, *Mechanics of Fatigue,* T. Mura, Ed., American Society of Mechanical Engineers, 1981, p 21–32
97. S.S. Manson, "Behavior of Materials under Conditions of Thermal Stress," NASA TN-2933, National Aeronautics and Space Administration, 1954
98. L.F. Coffin, A Study of the Effects of Cyclic Thermal Stresses on a Ductile Metal, *Trans. ASME,* Vol 76, 1954, p 931–950
99. S.S. Manson and M.H. Hirschberg, Fatigue Behavior in Strain Cycling in the Low- and Intermediate-Cycle Range, *Fatigue—An Interdisciplinary Approach,* J.J. Burke, N.L. Reed, and V. Weiss, Ed., Syracuse University Press, 1964, p 133–178
100. J. Morrow, "An Investigation of Plastic Strain Energy as a Criterion for Finite Fatigue Life," Report to Garrett Corp., 1960
101. J.T. Berling and J.B. Conway, A New Approach to the Prediction of Low-Cycle Fatigue Data, *Metall. Trans.,* Vol 1, 1970, p 805–809
102. T. Udoguchi and T. Wada, Thermal Effect on Low-Cycle Fatigue Strength of Steels, *Proc. Int. Conf. Thermal Stresses and Thermal Fatigue,* D.J. Littler, Ed., Butterworths, London, 1971, p 109–123
103. D. Sunamoto, T. Endo, and M. Fujihara, Hold Time Effects in High Temperature, Low Cycle Fatigue of Low Alloy Steels, *Creep and Fatigue in Elevated Temperature Applications,* Institution of Mechanical Engineers, London, 1974, C252
104. T. Udoguchi, Y. Asada, and I. Ichino, A Frequency Interpretation of Hold-Time Experiments on High Temperature Low-Cycle Fatigue of Steels for LMFBR, *Creep and Fatigue in Elevated Temperature Applications,* Institution of Mechanical Engineers, London, 1974, C211
105. M. Morishita, K. Taguchi, T. Asayama, A. Ishikawa, and Y. Asada, Application of the Overstress Concept for Creep-Fatigue Evaluation, *Low Cycle Fatigue—Directions for the Future,* STP 942, H.D. Solomon, G.R. Halford, L.R. Kaisand, and B.N. Leis, Ed., ASTM, 1988, p 487–499
106. B.N. Leis, An Energy-Based Fatigue and Creep-Fatigue Damage Parameter, *J. Press. Vessel Technol. (Trans. ASME),* Vol 99, 1977, p 524–533
107. V.M. Radhakrishnan, Damage Accumulation and Fracture Life in High-Temperature Low-Cycle Fatigue, *Low-Cycle Fatigue and Life Prediction,* STP 770, C. Amzallag, B.N. Leis, and P. Rabbe, Ed., ASTM, 1982, p 135–151
108. V.M. Radhakrishnan, Life Prediction in Time Dependent Fatigue, *Advances in Life Prediction Methods,* D.A. Woodford and J.R. Whitehead, Ed., American Society of Mechanical Engineers, 1983, p 143–150
109. P.W. Whaley, A Thermodynamic Approach to Material Fatigue, *Advances in Life Prediction Methods,* D.A. Woodford and J.R. Whitehead, Ed., American Society of Mechanical Engineers, 1983, p 41–50
110. J. He, Z. Duan, Y. Ning, and D. Zhao, Strain Energy Partitioning and Its Application to GH33A Nickel-Base Superalloy and 1Cr18Ni9Ti Stainless Steel, *Advances in Life Prediction Methods,* D.A. Woodford and J.R. Whitehead, Ed., American Society of Mechanical Engineers, 1983, p 27–32
111. D.M. Nissley, T.G. Meyer, and K.P. Walker, "Life Prediction and Constitutive Models for Engine Hot Section Anisotropic Materials Program," Final Report, NASA CR-189223, National Aeronautics and Space Administration, 1992
112. S.S. Manson, Thermal Stress in Design, Part 19: Cyclic Life of Ductile Materials, *Mach. Des.,* Vol 32, 1960, p 139–144
113. H.G. Edmunds and D.J. White, Observations of the Effect of Creep Relaxation on High-Strain Fatigue, *J. Mech. Eng. Sci.,* Vol 8 (No. 3), 1966, p 310–321
114. J.F. Polhemous, C.E. Spaeth, and W.H. Vogel, Ductility Exhaustion for Prediction of Thermal Fatigue and Creep Interaction, *Fatigue at Elevated Temperatures,* STP 520, A.E. Carden, A.J. McEvily, and C.H. Wells, Ed., ASTM, 1973, p 625–636
115. R.H. Priest and E.G. Ellison, A Combined Deformation Map-Ductility Exhaustion Approach to Creep-Fatigue Analysis, *Mater. Sci. Eng.,* Vol 49, 1981, p 7–17
116. D.A. Miller, C.D. Hamm, and J.L. Phillips, A Mechanistic Approach to the Prediction of Creep-Dominated Failure during Simulated Creep-Fatigue, *Mater. Sci. Eng.,* Vol 53, 1982, p 233–244
117. R.H. Priest, D.J. Beauchamp, and E.G. Ellison, Damage during Creep-Fatigue, *Advances in Life Prediction Methods,* D.A. Woodford and J.R. Whitehead, Ed., American Society of Mechanical Engineers, 1983, p 115–122
118. R. Hales, *Fatigue Eng. Mater. Struct.,* Vol 6, 1983, p 121
119. B. Tomkins, Fatigue Crack Propagation—An Analysis, *Philos. Mag.,* Vol 18, 1968, p 1041–1066
120. J. Wareing, B. Tomkins, and G. Sumner, Extent to Which Material Properties Control Fatigue Failure at Elevated Temperatures, *Fatigue at Elevated Temperatures,* STP 520, A.E. Carden, A.J. McEvily, and C.H. Wells, Ed., ASTM, 1973, p 123–138
121. H.D. Solomon, Frequency Modified Low Cycle Fatigue Crack Propagation, *Metall. Trans.,* Vol 4, 1973, p 341–347
122. A.E. Carden, Parametric Analysis of Fatigue Crack Growth, *Proc. Conf. Creep and Fatigue in Elevated Temperature Environment,* ASTM, 1974
123. B. Tomkins, The Development of Fatigue Crack Propagation Models for Engineering Applications at Elevated Temperatures, *J. Eng. Mater. Des. (Trans. ASME),* Vol 97, 1975, p 289–297
124. J. Wareing, *Metall. Trans. A,* Vol 8A, 1977, p 163
125. C.J. Franklin, in *High Temperature Alloys for Gas Turbines,* D. Coutsouradis, P. Felix, H. Fischmeister, L. Habraken, Y. Lindblom, and M.O. Speidel, Ed., Applied Science, London, 1978, p 513–547
126. J. Janson, Damage Model of Creep-Fatigue Interaction, *Eng. Fract. Mech.,* Vol 11, 1979, p 397–403
127. S. Taira, R. Ohtani, and T. Komatsu, Application of J-Integral to High-Temperature Crack Propagation, Part 1: Creep Crack Propagation, *J. Eng. Mater. Technol. (Trans. ASME),* Vol 101, 1979, p 154–161
128. A. Saxena, A Model for Predicting the Effect of Frequency on Fatigue Crack Growth Behavior at Elevated Temperature, *Fatigue Eng. Mater. Struct.,* Vol 3 (No. 3), 1981, p 247–255
129. K.S. Chan and A.K. Miller, FATIGMOD: A Unified Phenomenological Model for Predicting Fatigue Crack Initiation and Propagation, *ASME Int. Conf. Advances in Life Prediction Methods,* American Society of Mechanical Engineers, 1983, p 1–16
130. H. Riedel, Crack-Tip Stress Fields and Crack Growth under Creep-Fatigue Conditions, *Elastic-Plastic Fracture, 2nd Symp.,* Vol 1, *Inelastic Crack Analysis,* STP 803, C.F. Shih and J.P. Guda, Ed., ASTM, 1983, p 505–520
131. E. Renner, H. Vehoff, and P. Neumann, Prediction for Creep-Fatigue Based on the Growth of Short Cracks, *Fatigue Fract. Eng. Mater. Struct.,* Vol 12 (No. 6), 1989, p 569–584
132. D.L. McDowell and M.P. Miller, "Physically Based Microcrack Propagation Laws for Creep-Fatigue-Environment Interaction," presented at ASME Winter Annual Meeting (Atlanta), American Society of Mechanical Engineers, 1991
133. G.R. Halford and S.S. Manson, Life Prediction of Thermal-Mechanical Fatigue Using Strain-

range Partitioning, *Thermal Fatigue of Materials and Components,* STP 612, D.A. Spera and D.F. Mowbray, Ed., ASTM, 1976, p 239–254

134. S.S. Manson and G.R. Halford, Treatment of Multiaxial Creep-Fatigue by Strainrange Partitioning, *ASME-MPC Symp. Creep-Fatigue Interaction,* R.M. Curran, Ed., Publ. G00112 (MPC-3), American Society of Mechanical Engineers, 1976, p 299–322
135. W.H. Vogel, R.W. Soderquist, and B.C. Schlein, Application of Creep-LCF Cracking Model to Combustor Durability Prediction, *Fatigue Life Technology,* T.A. Cruse and J.P. Gallagher, Ed., American Society of Mechanical Engineers, 1977, p 22–31
136. S.S. Manson and G.R. Halford, Discussion of paper by J.J. Blass and S.Y. Zamrik, entitled "Multiaxial Low-Cycle Fatigue of Type 304 Stainless Steel," *J. Eng. Technol. (Trans. ASME),* Vol 99, 1977, p 283–286
137. G.R. Halford, J.F. Saltsman, and M.H. Hirschberg, Ductility Normalized-Strainrange Partitioning Life Relations for Creep-Fatigue Life Prediction, *Proc. Conf. Environmental Degradation of Engineering Materials,* Virginia Polytechnic Institute and State University, 1977, p 599–612
138. S.S. Manson, Some Useful Concepts for the Designer in Treating Cumulative Fatigue Damage at Elevated Temperatures, *Proc. 3rd Int. Conf. Mechanical Behavior of Materials,* Vol 1, K.J. Miller and R.F. Smith, Ed., Pergamon Press, 1979, p 13–45
139. S.S. Manson, The Strainrange Conversion Principle for Treating Cumulative Fatigue Damage in the Creep Range, *Random Fatigue Life Prediction,* Y.S. Shin and M.K. Au-Yang, Ed., *Pressure Vessels and Piping,* Vol 72, American Society of Mechanical Engineers, 1983, p 1–30
140. C.W. Lawton, Use of Low-Cycle Fatigue Data for Pressure Vessel Design, *Low-Cycle Fatigue and Life Prediction,* STP 770, C. Amzallag, B.N. Leis, and P. Rabbe, Ed., ASTM, 1982, p 585–599
141. O.T. Saugerud, Advances in Life Prediction of Thermally Loaded Diesel Engine Components, *Advances in Life Prediction Methods,* D.A. Woodford and J.R. Whitehead, Ed., American Society of Mechanical Engineers, 1983, p 229–240
142. W. Hoffelner, K.N. Melton, and C. Wüthrich, On Life Time Predictions with the Strain Range Partitioning Method, *Fatigue Eng. Mater. Struct.,* Vol 6, 1983, p 77–87
143. P.H. Wirsching and Y.T. Wu, "Reliability Considerations for the Total Strainrange Version of Strainrange Partitioning," NASA CR-174757, National Aeronautics and Space Administration, 1984
144. V. Bicego, A Nonstandard Technique for the Evaluation of the Basic Laws of the Strain Range Partitioning Method in Creep-Fatigue Life Prediction, *Proc. 2nd Int. Conf. Fatigue and Fatigue Thresholds (Fatigue '84),* Vol 3, C.J. Beevers, Ed., Engineering Materials Advisory Services Ltd., Cradley Heath, Warley, U.K., 1984, p 1257–1268
145. S. Kalluri and S.S. Manson, "Time Dependency of SRP Life Relationships," NASA CR-174946, NASA Grant NAG3-337, Case Western Reserve University, 1985
146. S. Kalluri, S.S. Manson, and G.R. Halford, Environmental Degradation of 316 Stainless Steel in High Temperature Low Cycle Fatigue, *3rd Int. Conf. Environmental Degradation of Engineering Materials,* R.P. McNitt and M.R. Louthan, Jr., Ed., Pennsylvania State University, 1987, p 503–519
147. S. Kalluri, S.S. Manson, and G.R. Halford, Exposure Time Considerations in High Temperature Low Cycle Fatigue, *Proc. 5th Int. Conf. Mechanical Behavior of Materials,* Vol 2, M.G. Yan, S.H. Zhang, and Z.M. Zheng, Ed., Pergamon Press, 1987, p 1029–1036
148. H.D. Solomon, Low-Frequency, High-Temperature Low Cycle Fatigue of 60Sn-40Pb Solder, *Low Cycle Fatigue—Directions for the Future,* STP 942, H.D. Solomon, G.R. Halford, L.R. Kaisand, and B.N. Leis, Ed., ASTM, 1988, p 342–370
149. J.F. Saltsman and G.R. Halford, "A Model for Life Prediction of Thermomechanical Fatigue Using the Total Strain Version of Strainrange Partitioning (SRP)—A Proposal," NASA TP-2779, National Aeronautics and Space Administration, 1988
150. T. Kitamura and G.R. Halford, "High Temperature Fracture Mechanics Basis for Strainrange Partitioning," NASA TM-4133, National Aeronautics and Space Administration, 1989
151. E.P. Esztergar and J.R. Ellis, Cumulative Fatigue Damage Concepts in Creep-Fatigue Life Predictions, *Proc. Int. Conf. Thermal Stresses and Thermal Fatigue,* D.J. Littler, Ed., Butterworths, London, 1969, p 128–156
152. E. Krempl, The Temperature Dependence of High-Strain Fatigue Life at Elevated Temperature in Parameter Representation, *Proc. Int. Conf. Thermal Stresses and Thermal Fatigue,* D.J. Littler, Ed., Butterworths, London, 1971, p 36–46
153. H. Bernstein, A Stress-Strain-Time Model (SST) for High Temperature Low Cycle Fatigue, *Methods for Predicting Material Life in Fatigue,* W.J. Ostergren and J.R. Whitehead, Ed., American Society of Mechanical Engineers, 1979, p 89–100
154. R.V. Miner, J. Gayda, and M.G. Hebsur, Creep-Fatigue Behavior of Ni-Co-Cr-Al-Y Coated PWA 1480 Superalloy Single Crystals, *Low Cycle Fatigue—Directions for the Future,* STP 942, H.D. Solomon, G.R. Halford, L.R. Kaisand, and B.N. Leis, Ed., ASTM, 1988, p 371–384
155. S. Majumdar and P.S. Maiya, A Damage Equation for Creep-Fatigue Interaction, *ASME-MPC Symp. Creep-Fatigue Interaction,* R.M. Curran, Ed., Publ. G001112 (MPC-3), American Society of Mechanical Engineers, 1976, p 323–336
156. S. Majumdar and P.S. Maiya, A Mechanistic Model for Time-Dependent Fatigue, *J. Eng. Mater. Technol. (Trans. ASME),* Vol 102, 1980, p 159–167
157. M. Satoh and E. Krempl, An Incremental Life Prediction Law for Creep-Fatigue Interaction, *Pressure Vessels and Piping,* Vol 60, American Society of Mechanical Engineers, 1982, p 71–79
158. A. Zhang, T. Bui-Quoc, and R. Gomuc, A Procedure for Low Cycle Fatigue Life Prediction for Various Temperatures and Strain Rates, *J. Eng. Mater. Technol. (Trans. ASME),* Vol 112 (No. 4), 1990, p 422–428
159. V. Moreno, D.M. Nissley, G.R. Halford, and J.F. Saltsman, "Application of Two Creep-Fatigue Life Prediction Models for the Prediction of Elevated Temperature Crack Initiation of a Nickel-Base Alloy," AIAA Preprint 85-1420, American Institute of Aeronautics and Astronautics, 1985
160. R.S. Nelson, "A High Temperature Fatigue Life Prediction Computer Code Based on the Cyclic Damage Accumulation (CDA) Model," Presented at ASME Winter Annual Meeting (Atlanta), American Society of Mechanical Engineers, 1991
161. D. McLean and A. Pineau, Grain-Boundary Sliding as a Correlating Concept for Fatigue Hold-Times, *Met. Sci.,* Vol 12, 1978, p 313–316
162. B.K. Min and R. Raj, Hold-Time Effects in High Temperature Fatigue, *Acta Metall.,* Vol 26, 1978, p 1007–1022
163. K.D. Challenger, A.K. Miller, and C.R. Brinkman, An Explanation for the Effects of Hold Periods on the Elevated Temperature Fatigue Behavior of 2Cr-1Mo Steel, *J. Eng. Mater. Technol. (Trans. ASME),* Vol 103, 1981, p 7–14
164. K.D. Challenger, A.K. Miller, and R.L. Langdon, *J. Mater. Energy Syst.,* Vol 3, 1981, p 51–61
165. S.D. Antolovich, S. Liu, and R. Baur, Low Cycle Fatigue Behavior of René 80 at Elevated Temperature, *Metall. Trans. A,* Vol 12A, 1981, p 473–481
166. J.R. Weertman, "A Study of the Role of Grain Boundary Cavitation in the Creep-Fatigue Interaction in High Temperature Fatigue," Final Technical Report, Northwestern University, 1982
167. S. Baik and R. Raj, Wedge Type Creep Damage in Low Cycle Fatigue, *Metall. Trans. A,* Vol 13A, 1982, p 1207–1214
168. S. Baik and R. Raj, Mechanisms of Creep-Fatigue Interaction, *Metall. Trans. A,* Vol 13A, 1982, p 1215–1221
169. J. Reuchet and L. Remy, Fatigue Oxidation Interaction in a Superalloy—Application to Life Prediction in High Temperature Low Cycle Fatigue, *Metall. Trans. A,* Vol 14A, 1983, p 141–149
170. V.C. Nardone, Doctoral dissertation, Columbia University, 1983
171. K.-T. Rie, R.-M. Schmidt, B. Ilschner, and S.W. Nam, A Model for Predicting Low Cycle

Fatigue Life under Creep-Fatigue Interaction, *Low Cycle Fatigue—Directions for the Future,* STP 942, H.D. Solomon, G.R. Halford, L.R. Kaisand, and B.N. Leis, Ed., ASTM, 1988, p 313–328

172. G.R. Romanoski, S.D. Antolovich, and R.M. Pelloux, A Model for Life Predictions of Nickel-Base Superalloys in High-Temperature Low Cycle Fatigue, *Low Cycle Fatigue—Directions for the Future,* STP 942, H.D. Solomon, G.R. Halford, L.R. Kaisand, and B.N. Leis, Ed., ASTM, 1988, p 456–467
173. Y. Oshida and H.W. Liu, Grain Boundary Oxidation and an Analysis of the Effects of Oxidation on Fatigue Crack Nucleation Life, *Low Cycle Fatigue—Directions for the Future,* STP 942, H.D. Solomon, G.R. Halford, L.R. Kaisand, and B.N. Leis, Ed., ASTM, 1988, p 1199–1217
174. J.T. Berling and T. Slot, Effect of Temperature and Strain Rate on Low Cycle Fatigue Resistance of AISI 304, 316 and 348 Stainless Steels, *Fatigue at High Temperature,* STP 459, ASTM, 1969, p 3–30; also USAEC Report GEMP-642, General Electric Co., 1968

SELECTED REFERENCES

- C. Amzallag, B.N. Leis, and P. Rabbe, Ed., *Low Cycle Fatigue and Life Prediction,* STP 770, ASTM, 1982
- J.T. Betting and J.B. Conway, A Proposed Method for Predicting the Low-Cycle Fatigue Behavior of 304 and 316 Stainless Steel, *Trans. Metall. Soc. AIME,* Vol 245, 1969, p 1137–1140
- G. Cailletaud, D. Nouailhas, J. Grattier, C. Levaillant, M. Mottot, J. Tortel, C. Escaravage, J. Heliot, and S. Kang, A Review of Creep Fatigue Life Prediction Methods: Identification and Extrapolation to Long Term and Low Strain Cyclic Loading, *Nucl. Eng. Des.,* Vol 83, 1984, p 267–292
- L.F. Coffin, Jr., "A Generalized Equation for Predicting High-Temperature Low-Cycle Fatigue, Including Hold Times," Report 69-C-401, General Electric Co., Research and Development Center, Dec 1969
- J.B. Conway, "Evaluation of Plastic Fatigue Properties of Heat-Resistant Alloys," USAEC Report GEMP-740, General Electric Co., 1969
- J.B. Conway, "Short-Term Tensile and Low-Cycle Fatigue Studies of Incoloy 800," USAEC Report GEMP-732, General Electric Co., 1969
- J.B. Conway and J.T. Berling, A New Correlation of Low-Cycle Fatigue Data Involving Hold Periods, *Metall. Trans.,* Vol 1 (No. 1), 1970, p 324
- J.B. Conway, J.T. Berling, and R.H. Stentz, A Brief Study of Cumulative Damage in Low-Cycle Fatigue Testing of AISI 304 Stainless Steel at 650 °C, *Metall. Trans.,* Vol 1, 1970, p 2034
- J.B. Conway, J.T. Berling, and R.H. Stentz, A Temperature Correlation of the Low-Cycle Fatigue Data for 304 Stainless Steel, *Metall. Trans.,* Vol 2 (No. 11), 1971, p 3247
- R.M. Curran and B.M. Wundt, A Program to Study Low-Cycle Fatigue and Creep Interaction in Steel at Elevated Temperatures, *2-1/4 Chrome 1 Molybdenum Steel in Pressure Vessels and Piping,* American Society of Mechanical Engineers, 1972, p 49
- J. Dubuc et al., Unified Theory of Cumulative Damage in Metal Fatigue, *Weld. Res. Counc. Bull.,* No. 162, 1971
- P.G. Forrest, The Fatigue Behavior of Mild Steels at Temperatures up to 500 °C, *J. Iron Steel Inst.,* Vol 200, 1962, p 452
- P.H. Frith, Fatigue Tests at Elevated Temperatures, *Symp. High Temperature Steels and Alloys for Gas Turbines,* Special Report No. 43, Iron and Steel Institute, London, 1951
- G.R. Halford, Cyclic Creep-Rupture Behavior of Three High-Temperature Alloys, *Metall. Trans.,* Vol 3 (No. 8), 1972, p 2247–2256
- G.R. Halford, J.F. Saltsman, and M.H. Hirschberg, Ductility Normalized-Strainrange Partitioning Life Relations for Creep-Fatigue Life Prediction, *Proc. Conf. Environmental Degradation of Engineering Materials,* Virginia Polytechnic Institute and State University, 1977, p 599–612
- S. Kalluri and G.R. Halford, Mechanisms of Failure in Non-Isothermal Fatigue of a Cobalt-Base Superalloy, *Conf. Advanced Earth-to-Orbit Propulsion Technology,* NASA CP-3283, National Aeronautics and Space Administration, 1994, p 304–313
- S. Kalluri, S.S. Manson, and G.R. Halford, Exposure Time Considerations in High Temperature Low Cycle Fatigue, *Proc. 5th Int. Conf. Mechanical Behavior of Materials,* Vol 2, M.G. Yan, S.H. Zhang, and Z.M. Zheng, Ed., Pergamon Press, 1987, p 1029–1036
- S. Kalluri, S.S. Manson, and G.R. Halford, Environmental Degradation of 316 Stainless Steel in High Temperature Low Cycle Fatigue, *3rd Int. Conf. Environmental Degradation of Engineering Materials,* R.P. McNitt and M.R. Louthan, Jr., Ed., Pennsylvania State University, 1987, p 503–519
- S. Kalluri, K.B.S. Rao, G.R. Halford, and M.A. McGaw, Deformation and Damage Mechanisms in Inconel 718 Superalloy, *Proc. Int. Symp. Superalloys 718, 625, 706, and Various Derivatives,* TMS-AIME, 1994, p 593–606
- A.J. Kennedy, *Processes of Creep and Fatigue in Metals,* John Wiley & Sons, 1962, p 155
- A.J. Kennedy, Interactions between Creep and Fatigue in Aluminum and in Certain of Its Alloys, *Int. Conf. Creep,* Institution of Mechanical Engineers, London, 1963, Session 3, p 17–24
- Y.S. Kim, M.J. Verrilli, and G.R. Halford, A Creep Cavity Growth Model for Creep-Fatigue Life Prediction of a Unidirectional W/Cu Composite, *Advances in Fatigue Lifetime Predictive Techniques,* Vol 2, STP 1211, M.R. Mitchell and R.W. Landgraf, Ed., ASTM, 1993, p 91–104
- M.M. Leven, The Interaction of Creep and Fatigue for a Rotor Steel, *Exp. Mech.,* Vol 13 (No. 9), 1973, p 353–372
- S.S. Manson and G.R. Halford, An Overview of High Temperature Material Fatigue: Aspects Covered by the 1973 International Conference on Creep and Fatigue, *Specialist Meeting on Low Cycle High Temperature Fatigue,* AGARD-CP-155, Advisory Group for Aerospace Research and Development, Neuilly-sur-Seine, France, April 1974, p 2-1 to 2-47
- S.S. Manson and G.R. Halford, Treatment of Multiaxial Creep-Fatigue by Strainrange Partitioning, *ASME-MPC Symp. Creep-Fatigue Interaction,* R.M. Curran, Ed., Publ. G00112 (MPC-3), American Society of Mechanical Engineers, 1976, p 299–322
- S.S. Manson and G.R. Halford, Discussion of paper by J.J. Blass and S.Y. Zamrik, entitled "Multiaxial Low-Cycle Fatigue of Type 304 Stainless Steel," *J. Eng. Mater. Technol. (Trans. ASME),* Vol 99, July 1977, p 283–286
- S.S. Manson, G.R. Halford, and A.J. Nachtigall, Separation of the Strain Components for Use in Strainrange Partitioning, *Advances in Design for Elevated Temperature Environment,* American Society of Mechanical Engineers, 1975, p 17–28
- R.L. McKnight, J.H. Laflen, and G.T. Spamer, "Turbine Blade Tip Durability Analysis," NASA CR-165268, National Aeronautics and Space Administration, 1981
- V. Moreno, "Combustor Liner Durability Analysis," NASA CR-165250, National Aeronautics and Space Administration, 1981
- K.B.S. Rao, H. Schuster, and G.R. Halford, Mechanisms of High Temperature Fatigue in Alloy 800H, *Metall. Mater. Trans. A,* Vol 27A, April 1996, p 851–861
- R.W. Swindeman, "The Strain Fatigue Properties of Inconel, Part II," ORNL-3250, Oak Ridge National Laboratory, 1962
- R.W. Swindeman, The Inter-Relation of Cyclic and Monotonic Creep Rupture, *Int. Conf. Creep,* Institution of Mechanical Engineers, London, 1963, Session 3, p 71–76
- M.T. Tong, P.A. Bartolotta, G.R. Halford, and A.D. Freed, Stirling Engine—Approach for Long-Term Durability Assessment, *Proc. 27th Intersociety Energy Conversion Engineering Conf.,* Vol 5, 1992, p 209–214
- D.J. White, Effect of Environment and Hold Time on the High Strain Fatigue Endurance of 1/2 Percent Molybdenum Steel, *Proc. Inst. Mech. Eng.,* Vol 184, 1969–70, p 223

Design for Elevated-Temperature Applications

APART FROM nineteenth-century steam boilers, machines and equipment for high-temperature operation have been developed principally in the present century. Energy conversion systems based on steam turbines, gas turbines, high-performance automobile engines, and jet engines provide the technological foundation for modern society. All of these machines have in common the use of metallic materials at temperatures where time-dependent deformation and fracture processes must be considered in their design. The single valued time-invariant strain associated with elastic or plastic design analysis in low-temperature applications is not applicable, nor is there in most situations a unique value of fracture toughness that may be used as a limiting condition for part failure. In addition to the phenomenological complexities of time-dependent behavior, there is now convincing evidence that the synergism associated with gaseous environmental interactions may have a major effect, in particular on high-temperature fracture.

This article reviews the basic mechanisms of elevated-temperature behavior and associated design considerations with emphasis on metals. Subsequently, the engineering analysis will be confined to presenting data in the form that a designer might use, with emphasis on design principles rather than detailed design analysis. Thus, multiaxial stresses, part analysis, and creep-fatigue interaction are not formally treated. However, remaining life assessment and the effect of nonsteady stresses are covered. A broader treatment of most of these aspects can be found in other articles that appear in this Volume, and in *Mechanical Testing,* Volume 8 of the *ASM Handbook.* Emphasis here is placed on developing an appreciation of the uses (and abuses) of creep and rupture testing, data presentation, data analysis, limitations of long-time tests, and alternative approaches to high-temperature design. The objective is to provide a solid foundation for design principles from a materials performance perspective.

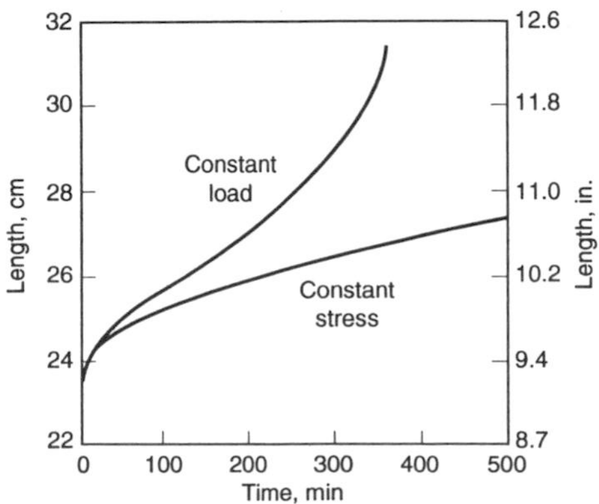

Fig. 1 Creep tests on lead wire. In both tests, initial lengths and initial loads were the same. Source: Ref 2

Historical Development of Creep Deformation Analysis

The phenomenon of time-dependent deformation was referred to as slow stretch by Philips (Ref 1) and as viscous flow by Andrade (Ref 2) at the beginning of this century and subsequently became known as creep. There were several seminal ideas in the Andrade work that have had a lasting impact on scientific studies and engineering dogma. The initial work was primarily on lead wires at room temperature (a high temperature relative to the melting point for lead) with some additional experiments on a 78.5% Sn 21.5% Pb alloy and copper. Andrade noted that after applying a fixed load the rate of extension initially decreased then became constant for a time, but finally increased and continued increasing until failure. He recognized that as the wire stretched, the load per unit area increased. Subsequently, he devised a scheme to compensate for this and maintain a constant stress on the wire. As a result of this, the extent of viscous flow, that is, extension linearly dependent on time, increased as shown in Fig. 1. Andrade also recognized that the length of wire being experimented on at any time is increasing and thus used the concept of true strain. He derived a formula to describe the observed deformation:

$$l = l_0(1 + \beta t^{1/3})e^{kt} \quad \text{(Eq 1)}$$

where l and l_0 are the current and initial specimen lengths, t is the time, and β and k are constants. The initial transient strain (later to be called primary creep) was referred to as beta creep and followed a time to the one-third law, the viscous region (later to be called steady-state creep) was proportional to time, and the accelerating strain region leading to fracture, which was not specifically treated by Andrade, later became known as tertiary creep. Much later, in a comprehensive study of creep in copper and aluminum, Wyatt (Ref 3) concluded that there are two types of transient creep in metals:

- At higher temperatures, beta creep predominates as in Andrade's experiments.
- At lower temperatures, the strain is proportional to log(time), and the flow is referred to as alpha creep.

From this early work, subsequent studies diverged into two investigative paths. The first sought understanding of creep deformation micromechanisms in pure metals and solid-solution alloys in relatively short-term tests, accepted the concept of steady-state creep (although testing was more often conducted at constant load rather than constant stress), and often assumed implicitly that viscous flow was history independent. This means that not only is there a steady creep rate associated with a given applied stress, but that this rate is obtained despite previous deformation at different stresses and temperatures. Although this might be a reasonable approximation for pure metals, it is manifestly wrong for most engineering alloys.

The second investigative path concentrated on generating long-time creep data on engineering materials. The testing was invariably at constant load, and data extracted included times for specific creep strains, minimum creep rates (although the term steady state was often used despite the fact that constant rates cannot be expected when the stress is changing), and time to failure (often referred to as rupture life). This latter measurement was of special significance because it became a basis for design against part failure, and later as a basis for estimating remaining life of operating components.

There thus emerged a framework for design against both creep deformation and fracture using a single testing procedure. It formed a basis for what might be called an uncracked body analysis and comprises the major part of this article. Analysis of cracked bodies involving fracture mechanics concepts as applied to creeping structures is not covered although some reference is made as appropriate. In particular, the importance of fatigue loading is emphasized in the article "Creep-Fatigue Interaction" in this Volume.

Until the last quarter century, virtually all creep and creep fracture studies were on metallic materials. However, as early as 1903, Philips (Ref 1) recognized that the phenomenon was not unique to the metallic bond and that materials with covalent and ionic bonds showed similar effects. In fact, creep of polymers is now of considerable importance in plastic automobile components and gas lines, and creep of ceramics is of interest in aerospace applications.

Basic Concepts of Elevated-Temperature Design

Time-dependent deformation and fracture of structural materials at elevated temperatures are among the most challenging engineering problems faced by materials engineers. In order to develop an improved design methodology for machines and equipment operating at high temperatures, several key concepts and their synergism must be understood. As described in this section, these include:

- Plastic instability at elevated temperatures
- Deformation mechanisms and strain components associated with creep processes
- Stress and temperature dependence
- Fracture at elevated temperatures
- Environmental effects

Design Phenomenology

The issues of interest from a design basis are the nature of primary creep, the validity of the concept of viscous steady-state creep, and the dependence of deformation on both temperature and stress. The simplest and most pervasive idea in creep of metals is an approach to an equilibrium microstructural and mechanical state. Thus a hardening associated with dislocation generation and interaction is countered by a dynamic microstructural recovery or softening. This process proceeds during primary creep and culminates in a steady-state situation. The idea was first presented by Bailey (Ref 4) and subsequently in the following mathematical form by Orowan (Ref 5):

$$d\sigma = \frac{\partial \sigma}{\partial \varepsilon} d\varepsilon + \frac{\partial \sigma}{\partial t} dt \qquad \text{(Eq 2)}$$

where $d\sigma$ represents the change in flow stress, $\partial\sigma/\partial\varepsilon$ represents the hardening that results from an increment of plastic strain $d\varepsilon$, and $\partial\sigma/\partial t$ represents the softening due to recovery in a time increment dt.

At constant stress (and temperature), the steady-state creep rate is given by:

$$\dot{\varepsilon} = -\frac{\partial\sigma/\partial t}{\partial\sigma/\partial\varepsilon} \qquad \text{(Eq 3)}$$

The numerator is frequently given the symbol r as the recovery rate associated with thermal softening, and the denominator is referred to as the strain-hardening coefficient, h. Although there is evidence that both hardening and softening processes occur during creep, and despite the fact that numerous studies have attempted to quantify Eq 2, it is in fact incorrect. As pointed out by McCartney (Ref 6), Eq 2 implies that an equation of state exists of the form:

$$\sigma = \sigma(\varepsilon, t) \qquad \text{(Eq 4)}$$

Equation 2 is the differential form of Eq 4, which assumes that the variables ε and t are independent. Since measured strain is in fact a function of elapsed time t, it follows that the partial derivatives have no meaning. Lloyd and McElroy (Ref 7) also concluded that the concept required a history-dependent term and the idea had serious deficiencies. They further concluded that the related concept of the applied stress being the sum of an internal stress opposing dislocation motion and an effective stress as the driving force for motion was inconsistent with real behavior. Their alternative theory draws on the observation of anelastic phenomena, which is considered in a subsequent section.

The concept of steady-state creep has been addressed rigorously in very few publications (Ref 8). From these limited studies, however, it can be stated that a constant creep rate cannot occur during the changing stress conditions of the common constant load test (or, if the creep rate appears constant, it cannot be steady state). Further, it can be said that most engineering alloys undergo purely time-dependent changes at temperature associated with an approach to thermodynamic equilibrium, such as precipitate coarsening. With the additional complication of strain-induced changes, it is unlikely that a steady state could be established. It is especially improbable that such a state could be history independent. Any search for a true steady state should, therefore, be limited to pure metals or solid-solution alloys and would require constant stress testing and true-strain plotting.

Plastic Instability

A major issue in the tensile creep test is the role of plastic instability in leading to tertiary creep. Understanding of the nature of plastic instability for time-dependent flow has depended on the theory of Hart (Ref 9). He showed that the condition for stable deformation is:

$$\gamma = m \geq 1 \qquad \text{(Eq 5)}$$

where m, which equals $[(\partial \ln \sigma)/(\partial \ln \dot{\varepsilon})]_\varepsilon$, is the strain-rate sensitivity, and γ, which equals $[(\partial \ln \sigma)/(\partial \varepsilon)]_{\dot{\varepsilon}}$, is a measure of the strain-hardening rate. For steady-state flow, γ is equal to 0. For constant stress tests, Burke and Nix (Ref 10) concluded that flow must be unstable when steady state is reached according to Hart's criterion but that macroscopic necking is insignificant and that the flow remains essentially homogeneous. They concluded that a true steady state does exist. Hart himself questioned the conclusions based on their analysis but did not rule out the possibility of a steady state for pure metals (Ref 8). In a very careful experimental analysis, Wray and Richmond (Ref 11) concluded that the concept of a family of steady states is valid. They advocated tests in which two of the basic parameters (stress, strain rate, and temperature) are held constant. However, they reported the intrusion of nonuniform deformation before the steady state was reached. They also pointed out the complexities associated with uncontrolled and often unmeasured loading paths, which produce different structures at the beginning of the constant stress or constant strain rate portions of the test. For constant stress tests in pure metals, although the concept of steady state (viscous flow in Andrade's terminology) is appealing, it appears not yet to have been rigorously demonstrated.

In constant load tests, steady-state behavior would of course result in an increasing creep rate after the minimum, as the true stress increases. As such, the test is inappropriate to evaluate the concept. However, it is by far the most common type of creep test and can be analyzed for instability (Ref 12). The condition for instability may be stated:

$$\ddot{A} \geq 0 \qquad \text{(Eq 6)}$$

where $\ddot{A}$ is the second derivative of specimen cross-sectional area with respect to time. This in turn leads to a point of instability expressed in terms of gage length:

$$\frac{\ddot{A}}{\dot{A}} = \frac{\ddot{L}}{\dot{L}} - \frac{2\dot{L}}{L} = 0 \qquad \text{(Eq 7)}$$

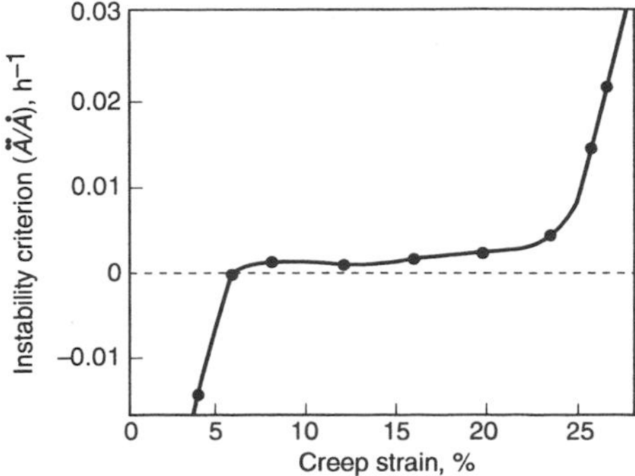

Fig. 2 Change in the parameter $\ddot{A}/\dot{A}$ with creep strain in nickel at 525 °C (980 °F) and 138 MPa (20 ksi). Source: Ref 12

This criterion is shown in Fig. 2 for constant load tests on nickel. The instability criterion is fulfilled at a strain very close to that of the minimum creep rate. However, the value of this criterion remains low up to 20 to 25% strain, at which separate measurements of specimen profiles indicate that macroscopic necking occurs. In this respect, the results are similar to constant stress results (Ref 10) in that although deformation is potentially unstable at the end of the primary stage, it is not grossly so.

Creep Processes

Creep behavior can be characterized either in terms of deformation mechanisms or in terms of strain constituents.

Deformation Mechanisms. Creep of metals is primarily a result of the motion of dislocations, but is distinct from time-independent behavior in that flow continues as obstacles, which may be dislocation tangles or precipitate particles, are progressively overcome. The rate-controlling step involves diffusion to allow climb of edge dislocations or cross slip of screw dislocations around obstacles. In steady-state theory, there is a balance between the hardening associated with this dislocation motion and interaction, and a dynamic recovery associated with the development of a dislocation substructure. Theory for such a process predicts a power-law dependence of creep rate on applied stress. For example, climb-controlled dislocation creep gives an exponent $n = 4$ in the following equation (Ref 13):

$$\dot{\varepsilon} = C\sigma^n \quad \text{(Eq 8)}$$

where C is a constant. Nothing in this or similar theories allows for history effects, and although the power function connection may be applicable, the value of n is not only invariably higher, but strongly history dependent in structural alloys (see the following section).

At very high homologous temperatures (T/T_m) and low stresses, creep may occur in both metals and ceramics by mass transport involving stress-directed flow of atoms from regions under compression to regions under tension. In this case, theory indicates that there is a stress dependence of unity and that the process is controlled either by bulk diffusion (Ref 14, 15) or by grain-boundary diffusion (Ref 16). These various processes of creep (dislocation controlled as well as diffusion controlled) may be represented on a deformation mechanism map to highlight regimes of stress and temperature where each mechanism, based on current theories, may be operating (Ref 17). However, such maps are only as good as the theories on which they are based and give no guidance on deformation path dependence.

Another important deformation process in metallic and ionic polycrystals at high temperature and low stresses is grain-boundary sliding (Ref 18). The resistance to sliding is determined by the mobility of grain-boundary dislocations and by the presence of hard particles at the boundary. This sliding leads to stress concentrations at grain junctions, which are important in nucleating cracks. In ductile materials, these stress concentrations may be relieved by creep and stress relaxation in the matrix or by grain-boundary migration (Ref 19).

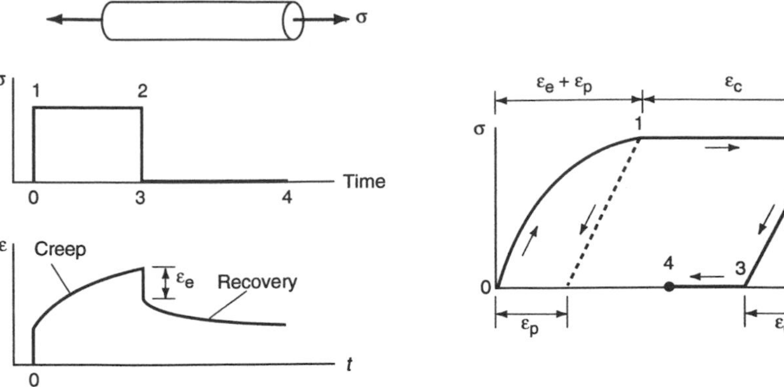

Fig. 3 Stress-time step applied to a material exhibiting strain response that includes time-independent elastic, time-independent plastic, time-dependent creep, and time-dependent anelastic (creep-recovery) components. Source: Ref 20

Strain Components. There are several different sources of strain at high temperature in response to an applied stress. The elastic strain is directly proportional to stress, and a modulus that is temperature dependent can be determined. For metallic materials and ceramics, although there is a strain-rate dependence of elastic modulus, it is small and often ignored. For polymers, by contrast, the elastic modulus is ill defined because of viscoelasticity.

Plastic strain for all materials may be treated as three separate constituents:

- Time-independent nonrecoverable, which may be thought of as an instantaneous deformation
- Time-dependent nonrecoverable, which may involve any or all of the micromechanisms described above
- Time-dependent recoverable

The first of these is unlikely to be significant in practical applications except in the region of stress concentrations since loading is normally well below the macroscopic yield stress. The second is the major source of creep in normal laboratory testing. The third constituent is not widely studied or analyzed, but may become very important at low stresses and under nonsteady conditions, that is, high-temperature service. It leads to what has been termed creep recovery and anelasticity.

At high temperatures, the application of a stress leads to creep deformation resulting from the motion of dislocations, mass transport by diffusion, or grain-boundary sliding. These processes in turn lead to a distribution of internal stresses that may relax on removal of the stress. This relaxation leads to a time-dependent contraction in addition to the elastic contraction and results in the phenomenon of creep recovery illustrated in Fig. 3. In polymers this phenomenon, which may account for nearly all the nonelastic strain, is termed viscoelastic recovery and is associated with the viscous sliding and unkinking of long molecular chains (Ref 21). In metals it is associated with the unbowing of pinned dislocations (Ref 7), rearrangement of dislocation networks (Ref 22), and local grain-boundary motion (Ref 23). In ceramics it appears to be primarily a grain-boundary phenomenon (Ref 24).

Whereas the importance of creep recovery is well recognized in polymer design, it has often been ignored in design of metallic and ceramic materials. A few extensive studies have been reported on metals (Ref 25–27) that have led to several broad conclusions:

- Creep-recovery strain increases linearly with stress for a fixed time at a given temperature, but is dependent on prestrain.
- The rate of creep recovery increases with increasing temperature.
- When the stress is low enough, essentially all transient creep is linear with stress and recoverable.
- Mathematically, the recovery may be described by a spectrum of spring dashpot combinations with a wide range of relaxation times.

Assuming that the measured recovery strain after unloading had made an equivalent contribution to forward creep (Ref 28), it was possible in these studies to separate the anelastic and plastic creep components as shown in Fig. 4. Because the anelastic component is linear with stress and the plastic component is a power function of stress (for the same time), at very low stresses the strain is entirely anelastic. This observation led to the definition of a plastic creep limit that was time dependent. For times up to 100 h in a low-alloy steel tested at 425 °C (800 °F), Lubahn (Ref 26) found this limit to be 140 MPa (20 ksi) (Ref 5); all creep below this stress was fully recoverable. In tests on a similar alloy at 538 °C (1000 °F), Goldhoff (Ref 27) found that the creep limit ranged from 150 MPa (22 ksi) for 1 h to zero at 5000 h. By plotting the ratio of anelastic to plastic strain for a fixed time (1000 h) as a function of stress (Fig. 5), Goldhoff (Ref 27) showed how the former became dominant at low stresses. Figure 5 also shows that a heat treatment that produces low ductility leads to higher ratios, suggesting a link between anelastic deformation and intergranular fracture, which was consistent with mi-

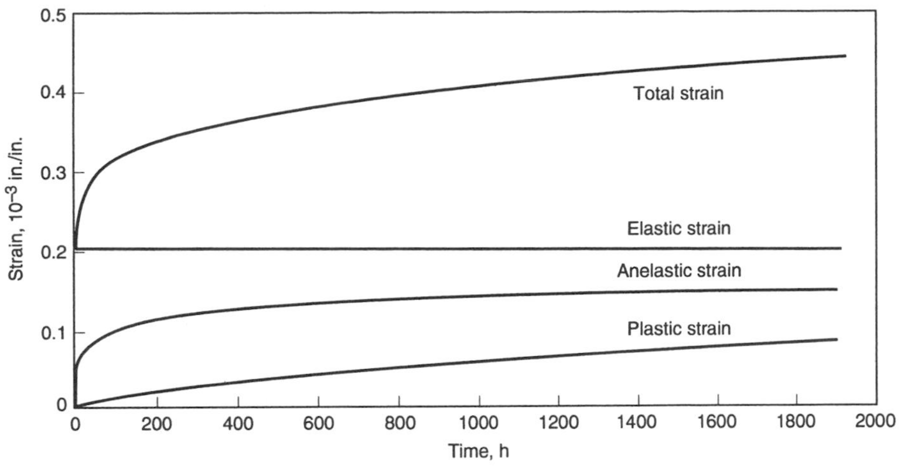

Fig. 4 The separation of strain components for a creep test on Cr-Mo-V steel at 538 °C (1000 °F) and 35 MPa (5 ksi). Source: Ref 27

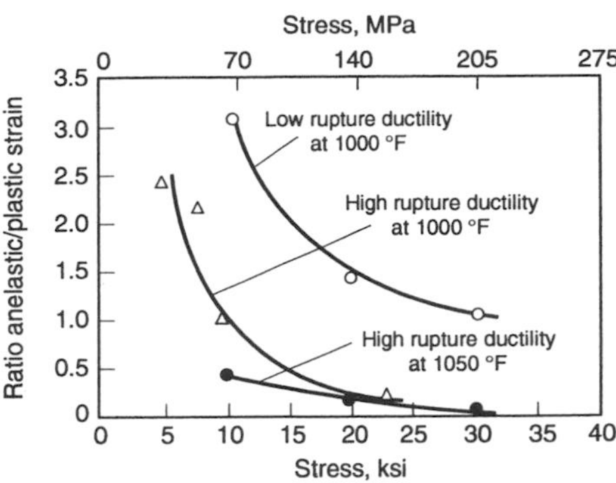

Fig. 5 Effect of ductility on recoverable creep strain for Cr-Mo-V steel after 1000 h creep exposure. Source: Ref 27

crostructural observations of fracture in this alloy.

There have been even fewer systematic studies of creep recovery in ceramics, but silicon carbide fibers have been shown to recover fully their creep strain between 1000 and 1400 °C (1830 and 2550 °F) (Ref 24). Additionally, provided an appropriate period was allowed for recovery after each stress cycle, tension-tension fatigue resulted in zero cumulative creep strain. This indicates the potential importance of anelastic phenomena in damage accumulation for nonsteady conditions. Very recent work on large specimens of silicon nitride have shown recovery of most of the accumulated strain after unloading from stress-relaxation tests (Fig. 6).

There are strong indications that anelastic phenomena should be included in design considerations. Anelastic contraction as well as extension can occur depending on whether the stress is decreased or increased, whereas plastic shortening never occurs. Although several authors have pointed out that, because of the linear stress dependence the analysis should be much simpler than for plastic creep analysis (Ref 7, 27), accurate measurements at the low stresses of interest for service applications are difficult. The possible link with fracture processes is also of great interest, but neither consideration has influenced design practice.

Stress and Temperature Dependence

The minimum creep rate in both constant load and constant stress tests is normally represented by a power function of stress (Eq 8), and the temperature by an Arrhenius expression including an activation energy term (Q) derived from chemical reaction rate theory (Ref 29):

$$\dot{\varepsilon} = S\sigma^n e^{-Q/RT} \quad \text{(Eq 9)}$$

where S, which is a constant, depends on structure. Although an exponential or hyperbolic sine stress function may provide a better fit in some cases, the power function has generally prevailed and has become strongly linked with mechanistic treatments. In pure metals, early studies indicated a stress exponent on the order of four and an activation energy close to that for self-diffusion (Ref 13, 29, 30). For engineering alloys, the stress exponents are generally higher and may not be constant (Ref 31), and the value of the activation energy may be much higher than that for the alloy matrix self-diffusion and may be sensitive to test temperature.

Because the basic formulation of Eq 9 is used to correlate much engineering data and is used in creep analysis of components, it is useful to examine critically some of the limitations in this analysis as they apply to engineering alloys. It was first shown by Lubahn (Ref 32) that, because of the rapidly decreasing creep rate in the primary stage, a strain-time plot of a portion of this stage always appears to show approximately constant rates at the longest times. This has led to many errors in the literature with false minimum creep rates. Some of these errors may lead to apparent n values close to one and consequent speculations about Newtonian viscous creep (Ref 33). Figure 7 shows results for minimum creep rates in a Cr-Mo-V steel in tests lasting up to 50,000 h. Also included are plots where time restrictions on the measurements were imposed to illustrate this potential for error. Nevertheless, the true minimum data points indicate n values ranging from 3.3 to 12.

As pointed out by Woodford (Ref 33), the curvature indicates that Eq 9 with S as a constant does not apply over the stress range, and it is meaningless to consider both n and S changing. In fact, the slope at any point has no clear physical significance because the structural state at the minimum creep rate is different for each test because of the different deformation history. To approximate a constant structure determination, creep rates have been measured under decreasing stress either by discrete stress drops (Ref 34) or during stress relaxation (Ref 35). The stress exponents measured from these data are much higher than those obtained from the minimum creep-rate data, but have clear physical significance because they relate to an approximately constant structure. An alternative approach for measuring stress dependence at close to constant structure is to monitor the creep rate and corresponding stress increase in constant load tests at strains beyond those corresponding to the minimum creep rates (Ref 33, 36). In this method $n = d \log \dot{\varepsilon}/d \log \sigma_0 (1 + \varepsilon)$ where σ_0 is the initial stress and ε the nominal strain. Results for the steel data are shown in Fig. 8 giving n values for individual

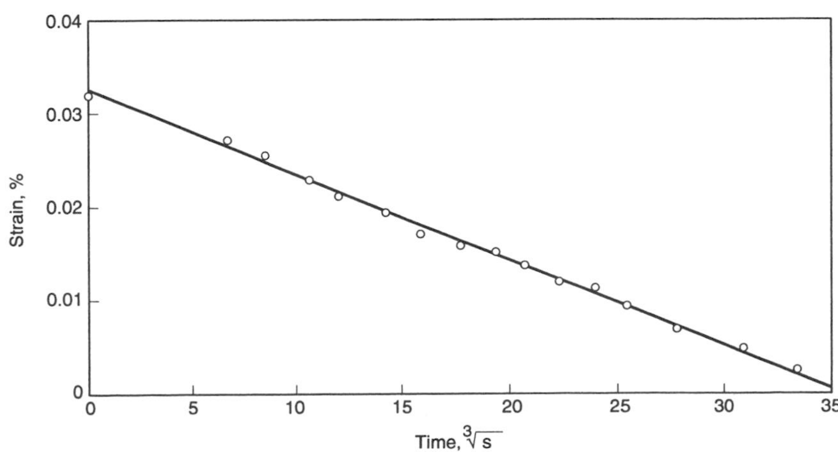

Fig. 6 Recovery of creep strain in silicon nitride at 1200 °C (2190 °F) after unloading from a stress-relaxation test started at 300 MPa (43.5 ksi), showing a time to the one-third dependence

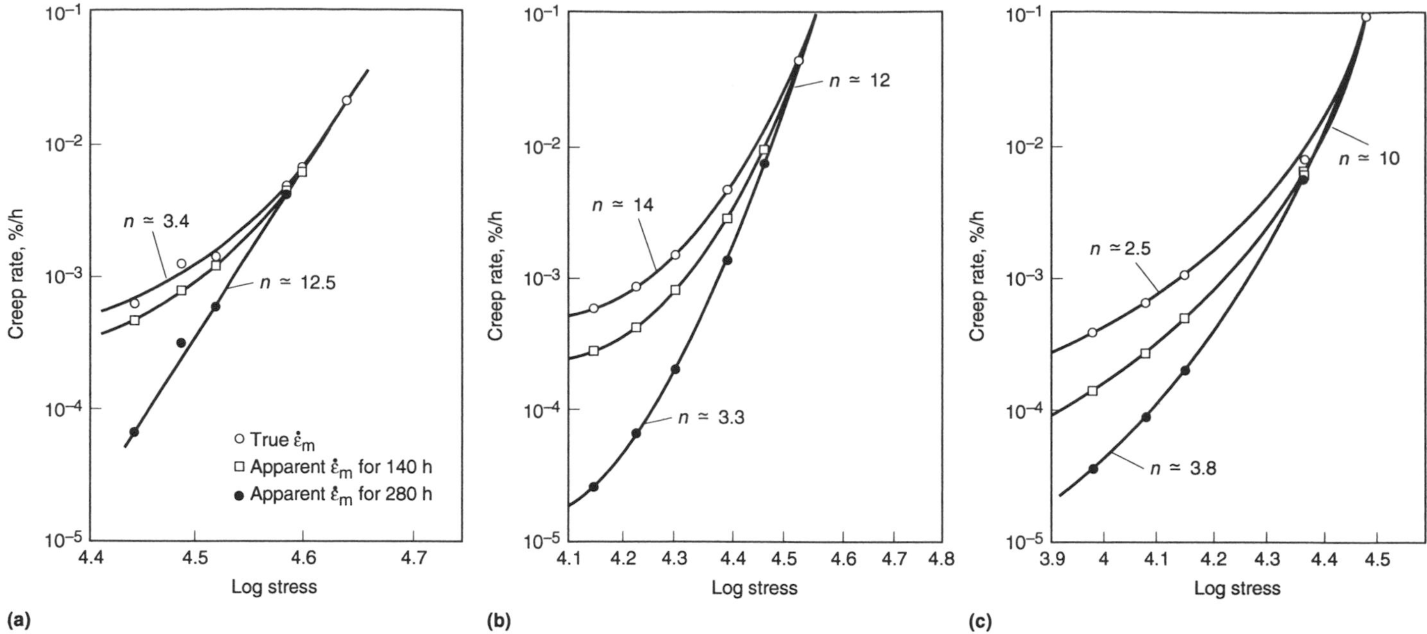

Fig. 7 Effect of test time restrictions on the apparent stress sensitivity of creep rate for a chromium-molybdenum steel at temperatures of (a) 510 °C (950 °F), (b) 565 °C (1050 °F), and (c) 593 °C (1100 °F). Source: Ref 33

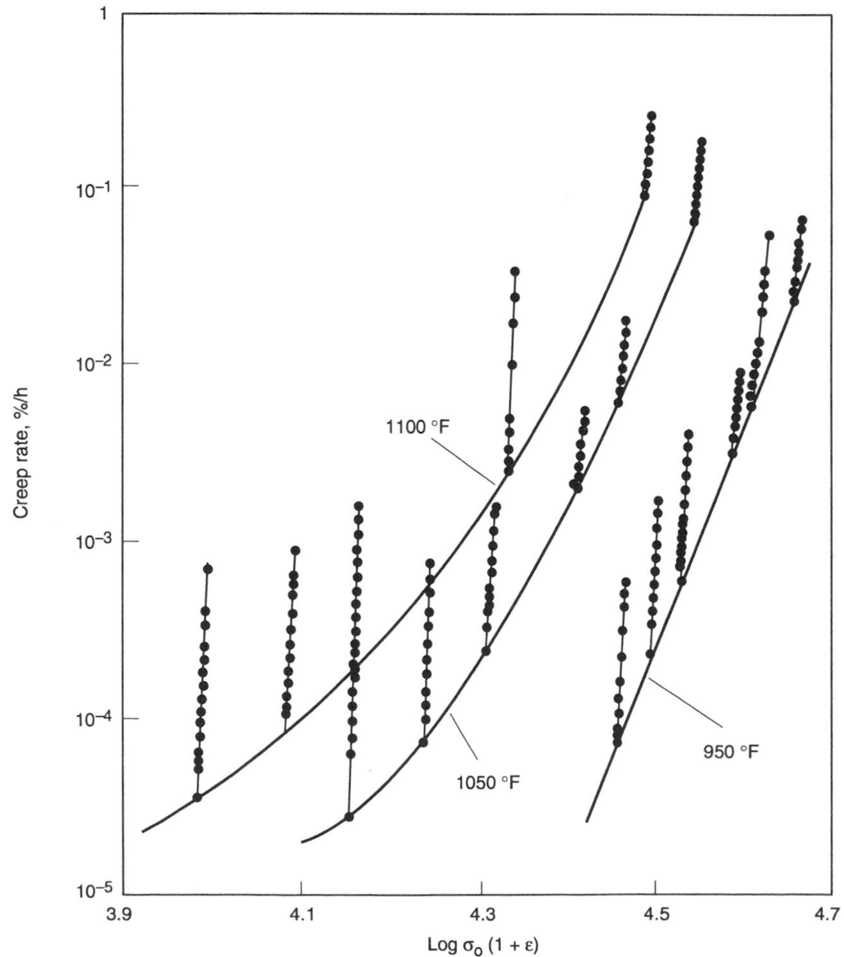

Fig. 8 Creep rate for a chromium-molybdenum steel as a function of the true stress showing that the stress sensitivity measured in a single test is different from that measured in separate tests. Source: Ref 33

tests between 30 and 100, which are much higher than values estimated from the slopes of the lines drawn through the minimum creep rates. It has been shown that, as in the stress-decrement measurements, the values of n may be related to a particular structural state. The reciprocal of these values gives a measure of strain-rate sensitivity and correlates well with elongation at fracture (Ref 33, 36, 37).

Although the representation of creep data in the engineering literature has been strongly influenced by the simple correlations reported for short-time tests for pure metals, it is clear that any physical significance is lost for most structural materials. The stress dependence of creep determined from the slope of a line drawn through minimum creep-rate data is expected to be quite different from that determined for a stress change on an individual specimen. The importance of deformation history is again apparent. Likewise, an exponential temperature dependence of minimum creep rates should be viewed as an empirical correlation. Temperature change experiments on a single specimen usually do not give the same activation energy, and because the structural state changes with temperature, a temperature change sequence effect on the apparent activation energy is also to be expected.

Fracture at Elevated Temperatures

As indicated previously, the constant load creep rupture test is the basis for design data for both creep strength (minimum creep rate or time to a specific creep strain) and failure (time to rupture). The various ways in which such data are presented, correlated, and extrapolated are addressed in subsequent sections. However, it is

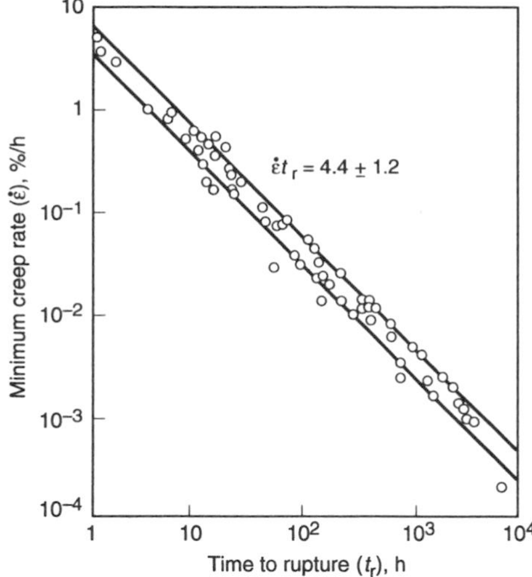

Fig. 9 Monkman-Grant relationship between minimum creep rate and time to rupture for a 2¼Cr-1Mo steel. Source: Ref 39

useful to note here that the well-known Monkman-Grant relationship (Ref 38) shown in Fig. 9 indicates that the time to rupture is reciprocally related to the minimum creep rate. This relationship is commonly observed in ductile materials and has been used to predict one property from the other. However, the true significance of the correlation is that the rupture life is principally a measure of creep strength rather than fracture resistance. This leads to a number of inconsistencies in design procedures that are discussed later in this article.

At this point, it is appropriate to consider the processes leading to fracture. Plastic instability in ductile materials has already been reviewed. This process may lead directly to fracture in pure metals and contribute significantly to fracture in engineering materials at moderately high stresses. However, of much greater concern are the processes leading to intergranular fracture with reduced ductility at low stresses and high temperatures. Here again, many of the basic studies have been conducted on pure metals and solid-solution alloys.

Crack Nucleation and Morphology. Two types of cracking have been identified: wedge-shaped cracks emanating from grain-boundary triple points and the formation of cavities or voids on grain-boundary facets often oriented perpendicular to the applied tensile stress (Ref 40). An example of creep cracks in nickel that appears to show both forms is given in Fig. 10. A fractographic study of creep cavities in tungsten concluded that the different crack morphologies actually reflected differences in growth rate. At low growth rates, surface diffusion allowed the cavities to reduce their surface tension by assuming nearly equiaxed polyhedral shapes. At higher growth rates, irregular two-dimensional cracks developed that on sectioning appeared as wedge cracks (Ref 42).

Although much work continues to model the nucleation and growth of these cracks and cavities (Ref 43), there are uncertainties in the mechanism of nucleation and in the identification of a failure criterion. For example, McLean has shown that a stress concentration up to 1000 is needed to nucleate a hole unless it is stabilized by internal pressure (Ref 44). As a consequence, the nucleation stage has been treated with less enthusiasm than has the modeling of growth. This issue may well be resolved on the basis of environmental interaction (see the section "Environmental Effects" in this article). Another major problem is the effect of temperature and stress on the extent of cracking at failure. Most theories assume that failure occurs at some critical cavity distribution or crack size. However, it has been shown that the extent of cavitation at failure or at any given fraction of the failure life is very sensitive to the test conditions (Ref 45, 46). Thus cavitation damage at failure at a high stress may be comparable to damage in the very early stage of a test at low stress. For stress-change experiments, there is therefore a loading sequence effect on rupture life, which is discussed later in this article, for engineering alloys.

Embrittlement Phenomena. As pointed out previously, rupture life is primarily a measure of creep strength; fracture resistance would be identified better with a separate measure that reflects the concern with embrittlement phenomena that may lead to component failure. Most engineering alloys lose ductility during high-temperature service. This has been shown to be a function of temperature and strain rate (Ref 47) so that there is a critical regime for maximum embrittlement. At a fixed strain rate, for example, ductility first decreases with increasing temperature. This is believed to be caused by grain boundaries playing an increasing role in the deformation process leading to the nucleation of intergranular cracks. At still higher temperatures, processes of recovery and relaxation at local stress concentrations lead to an improvement in ductility. Figure 11 is an example of a ductility contour map for a low-alloy steel based on measurements of reduction of area (RA) of long-term rupture tests (Ref 48). Maximum embrittlement occurred in a critical range of temperature and stress (or strain rate). This type of embrittlement generally coincides with a sensitivity to notches, which emphasizes its practical significance. For example, Fig. 12 shows that the ratio of notch strength to smooth strength for various test times passes through minima corresponding to ductility minima (Ref 49) based on reduction in area at failure in the notch. This tendency to develop so-called notch weakening at temperature is of great concern in selecting alloys and monitoring their service performance.

The embrittlement phenomena may be associated with precipitation in the alloy that interferes with the ability to accommodate stress concentrations at grain boundaries (Ref 48), with segrega-

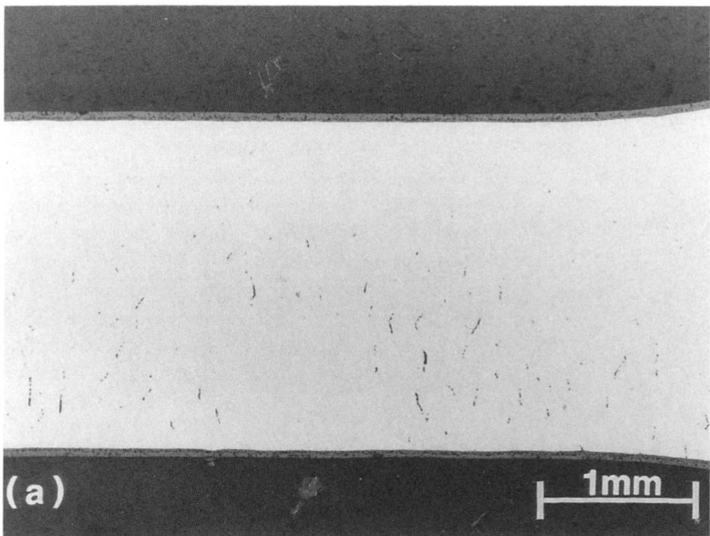

Fig. 10 Unetched microstructure of nickel samples after air testing at 15.8 MPa (2.3 ksi) and 800 °C (1470 °F). (a) Low-carbon Ni270 unloaded after 500 h slight cavitation. (b) Standard Ni270 after failure in 23 h. Source: Ref 41

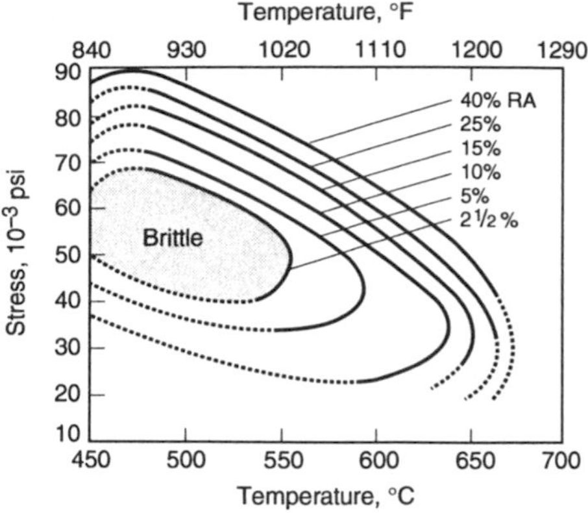

Fig. 11 Ductility contour map in stress/temperature space for a Cr-Mo-V steel. RA, reduction in area. Source: Ref 48

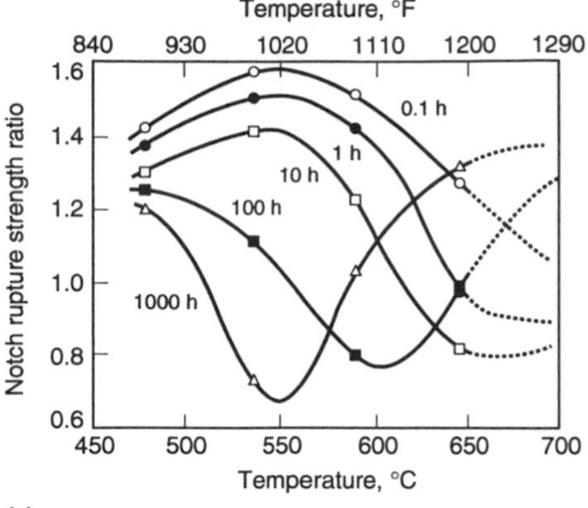

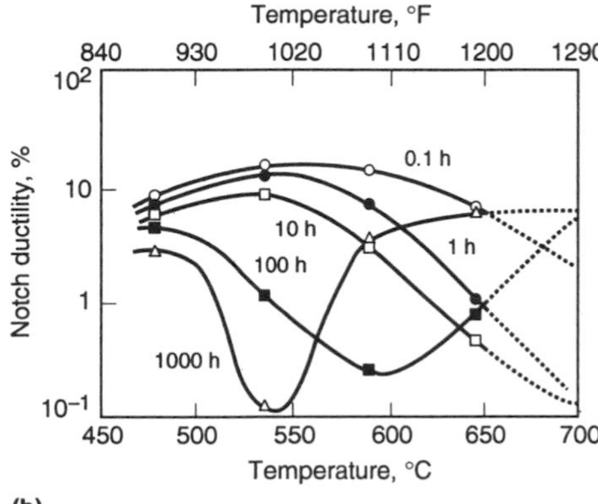

Fig. 12 Ratio of notched bar strength to smooth bar strength (a) and notch ductility (b) as a function of test temperature for various rupture times in a Cr-Mo-V steel. Source: Ref 49

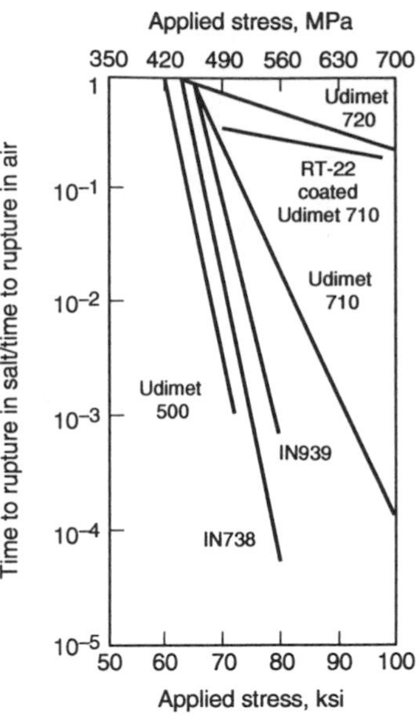

Fig. 13 Relative reductions in rupture life due to sulfate/chloride salt at 705 °C (1300 °F) for several superalloys. For RT-22 coated Udimet 710, rupture time in salt for coated alloy divided by time in air for uncoated alloy. Source: Ref 57

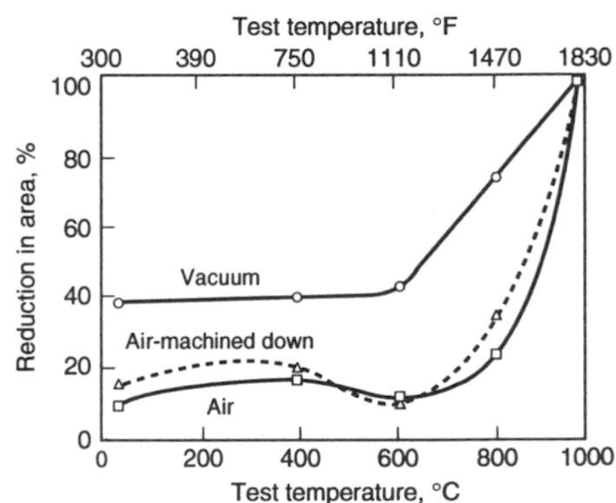

Fig. 14 Tensile ductility of IN903A after air and vacuum exposures at 1000 °C (1830 °F) for 100 h as a function of test temperature in vacuum tests. Embrittlement remained after reducing to half the initial diameter. Source: Ref 41

tion of embrittling species from the grain interior to the grain boundaries (Ref 50, 51), or with intergranular penetration of embrittling species from the environment (Ref 41).

Environmental Effects

It has long been known that test environment may affect creep-rupture behavior. Until quite recently, however, the work has been largely empirical with creep tests being conducted in various atmospheres and differences noted in creep rates and rupture lives (Ref 52). The effect on rupture life, in particular, was often less than a factor of ten in environments such as oxygen, hydrogen, nitrogen, carbon dioxide, and impure helium compared with vacuum. In many cases, it was not clear how inert the vacuum was, and little account was taken of specimen thickness. Often, effects on ductility were not reported, and there were very few studies of crack propagation.

A renewed interest developed in the 1970s as a result of observations of a dominant role played by the environment in high-temperature fatigue crack growth of superalloys (Ref 53). There were subsequent studies of sustained-load crack propagation that also showed very strong effects (Ref 54–56). In some cases at high stresses, the test environment was so severe, as in the case of sulfur, that profound changes were seen in smooth bar rupture life. Such an example is shown in Fig. 13, in which the time to rupture in common superalloys was reduced by several orders of magnitude in tests in a sulfate/chloride mixture at 705 °C (1300 °F) (Ref 57). These results were explained in terms of grain-boundary penetration of sulfur, which leads to rapid crack propagation. The coated specimen was far less susceptible, and the addition of a grain-boundary modifier (in this case, boron) in Udimet 720 gave an enormous improvement relative to Udimet 710.

Embrittling Effects of Oxygen. At about the same time that the ideas on environmental attack at an intergranular crack tip were being developed, it was also shown that short-term prior exposure in air at high temperature (greater than about 900 °C, or 1650 °F) could lead to profound embrittlement at intermediate temperatures (700 to 800 °C, or 1290 to 1470 °F) (Ref 41, 58–61). This was shown to be caused by intergranular diffusion of oxygen that penetrated on the order of millimeters in a few hours at 1000 °C (1830 °F). The embrittlement was monitored using measurements of tensile ductility at intermediate temperatures in iron-, nickel-, and cobalt-base alloys (Ref 41). An example of the results for the Fe-Ni-Co alloy, IN903A, is shown in Fig. 14, which also confirms the extent of damage penetration from tests in which the specimen diameter of 2.54 mm (0.1 in.) was reduced by half. Post-exposure tests on cast alloys showed that this embrittlement could also lead to a reduction in rupture life of several orders of magnitude. An example for alloy IN738 is shown in Fig. 15.

Using model alloys based on nickel, it was shown that oxygen in the elemental form in high-purity nickel did not embrittle; a chemical reaction was necessary (Ref 41). Three embrittling reactions were confirmed: a reaction with carbon to form carbon dioxide gas bubbles; a reaction with sulfides on grain boundaries to release sulfur, which does embrittle in the elemental form; and a reaction with oxide formers to form fine oxides that act to pin grain boundaries. These phenomena are believed to be the same processes that serve to embrittle the region ahead of a crack tip. Thus, oxygen attack may occur dynamically to account for the accelerated advance of a crack in air tests compared with inert environment tests, and it may occur during higher-temperature exposure with or without an applied stress to set up an embrittlement situation. Thermal fatigue in combustion turbines is a particularly challenging situation for oxygen attack since maximum strains develop at intermediate temperatures in the cycle, but holding may be at the maximum temperature (Ref 62).

Combined Effects of Oxygen and Carbon. Of special interest relative to the previous discussion of creep cavitation is the reaction between diffusing oxygen and carbon. In nickel, it was found that if this reaction were prevented, creep cavitation could not develop during creep tests.

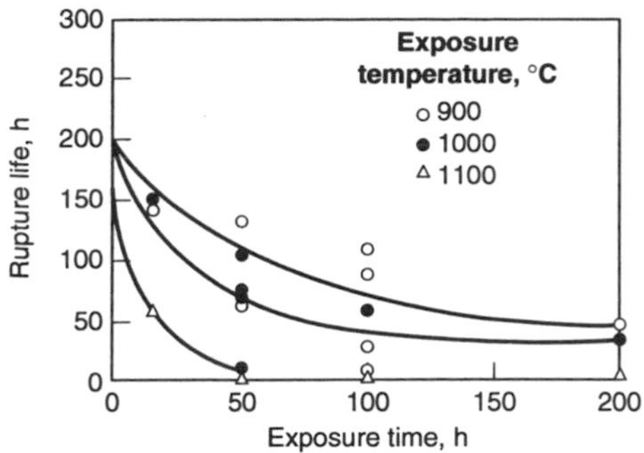

Fig. 15 Effect of exposure in air at various temperatures on stress-rupture life of IN738 at 800 °C (1470 °F) and 400 MPa (58 ksi). Source: Ref 59

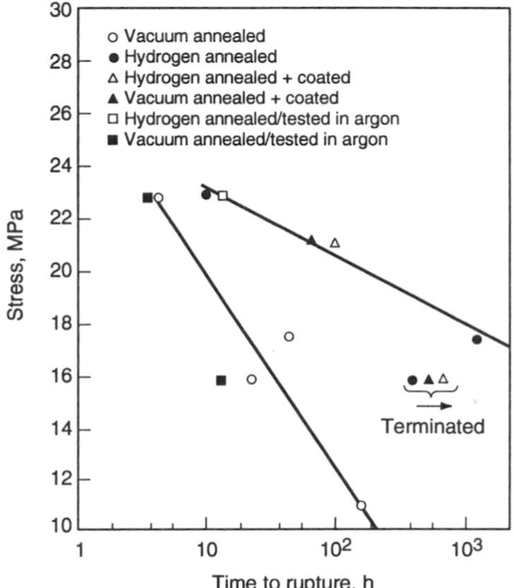

Fig. 16 Effect of environmental interaction on rupture life of Ni270 at 800 °C (1470 °F). Longer lives are obtained by preventing cavitation nucleation from carbon dioxide gas formation. This is achieved by decarburizing to eliminate carbon or by coating to prevent oxygen penetration. Source: Ref 63

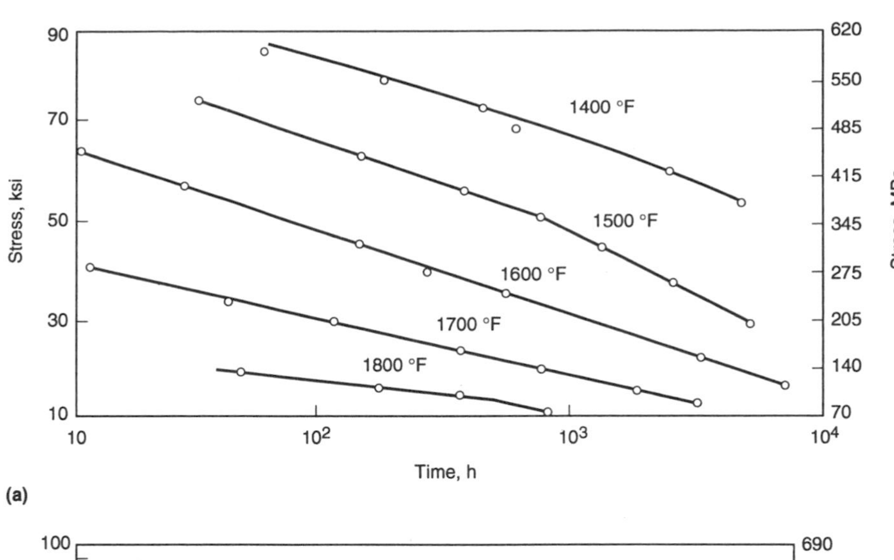

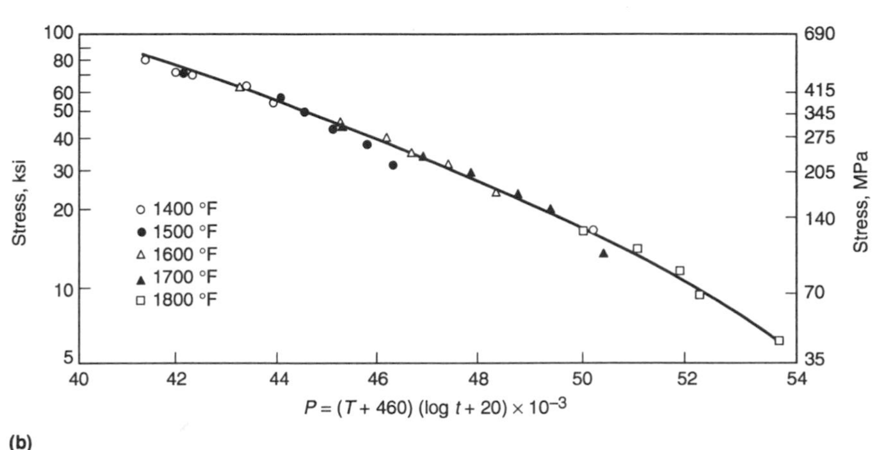

Fig. 17 Stress-rupture behavior of Astroloy. (a) Stress versus time curves. (b) Larson-Miller plot. Source: Ref 69

Prevention was achieved either by removing the carbon (decarburizing) or by applying an environmental protective diffusion or overlay coating (Ref 63). Air tests at 800 °C (1470 °F) and at various stresses showed an enormous increase in rupture life (Fig. 16) if the gas bubble formation did not occur. Since nickel has been used as an archetypical metal for the study of creep cavitation, the confirmation that the cavities are nucleated as gas bubbles now solves the problem of nucleation (Ref 44). The observation that even in superalloys gas bubbles are frequently nucleated at carbides, which may serve as cavity nuclei (Ref 59), and that cavitation during creep is often concentrated near the specimen surface (Ref 64), points to the likelihood that oxygen attack may be invariably associated with creep cavitation. The gas pressures developed in the bubbles appear to be quite adequate for nucleation (Ref 65).

Effect of Other Gaseous Elements. Hydrogen, chlorine, and sulfur may also embrittle as a result of penetration. Sulfur is particularly aggressive in that it diffuses more rapidly and embrittles more severely than does oxygen (Ref 66). It is also frequently found in coal gasification and oil-refining processes as well as industrial gas turbines operating on impure fuel.

Design Methodology

This section describes the basic presentation and analysis methods for creep rupture that are currently widely used. In addition to the application of these methods to materials selection and the setting of basic design rules, some consideration will be given to their application to remaining life assessment of operating components. The interaction with fatigue will not be included since some discussion of that complex topic is covered in the article "Creep-Fatigue Interaction" in this Volume. Also omitted from this discussion are multiaxial stress effects and sustained load (or creep) crack growth. However, it should be recognized that components in service normally operate under multiaxial stress systems, and detailed procedures are used for analysis that are based on effective stresses and strains. The considerable amount of work conducted in recent years on sustained-load crack growth, and the other topics alluded to, are reviewed in several recent texts (Ref 20, 39, 67) and in *Mechanical Testing,* Volume 8 of the *ASM Handbook.*

Creep Rupture Data Presentation. Laboratory creep tests are typically run between 100 and 10,000 h although a few are run for shorter times (for example, for acceptance tests), and occasionally some testing is conducted for longer times. Since most high-temperature components are expected to last ten years or more, service stresses are obviously lower than those used in the longest creep tests to generate data for most of the alloys used. Therefore, to provide data for creep rates and rupture lives that are appropriate for the setting of design stresses, it became necessary to develop methods for extrapolation. Over the years, a tremendous amount of effort has gone into optimizing methods of data extrapolation (Ref 68, 69).

One of the major considerations in such procedures must be statistical issues, such as the best estimate of the stress associated with a given median life or creep rate, the use of stress or time as the dependent variable in the data fitting, the treatment of variability among heats of the same alloy, and the analysis of data with run-outs. All of these issues have been treated with considerable rigor and shown to be important relative not

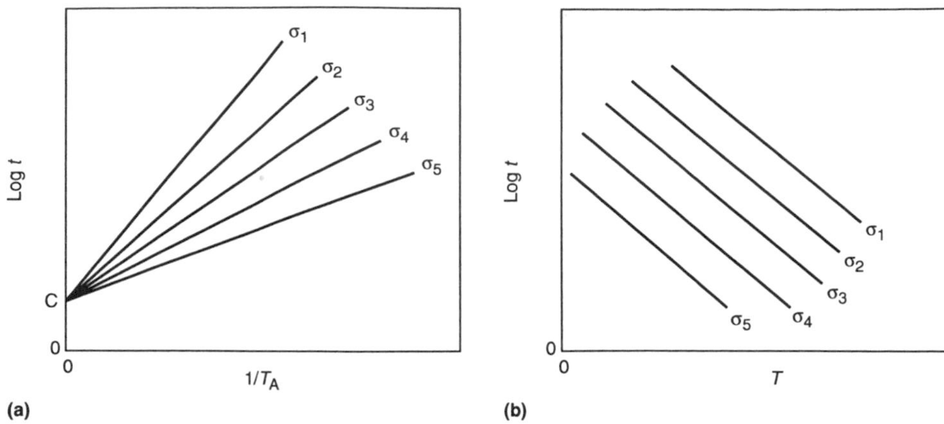

Fig. 18 Two common time-temperature parameters for rupture life. (a) Larson-Miller parameter. $f(\sigma) = T_A(\log t + C)$. (b) Manson-Succop parameter. $f(\sigma) = \log t - BT$. Source: Ref 69

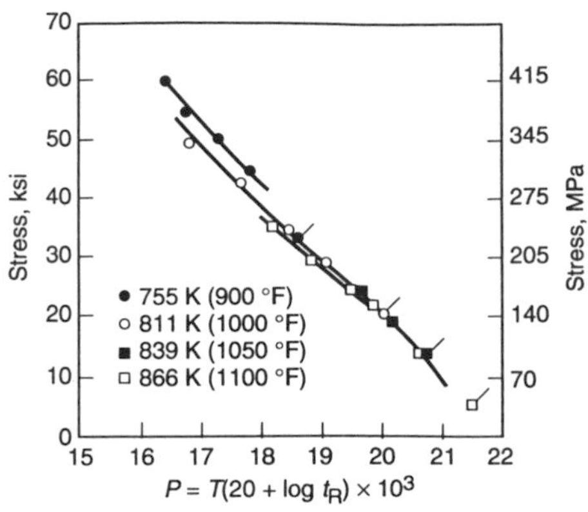

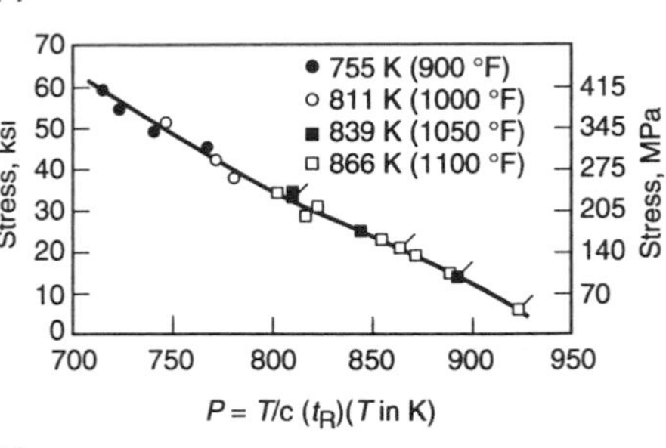

Fig. 20 Creep parameters for a Cr-Mo-V steel. (a) Larson-Miller plot using a constant of 20 showing segmenting of the data. (b) Same data using an optimized parameter based on the graphical optimization procedure (GOP) method. Source: Ref 76

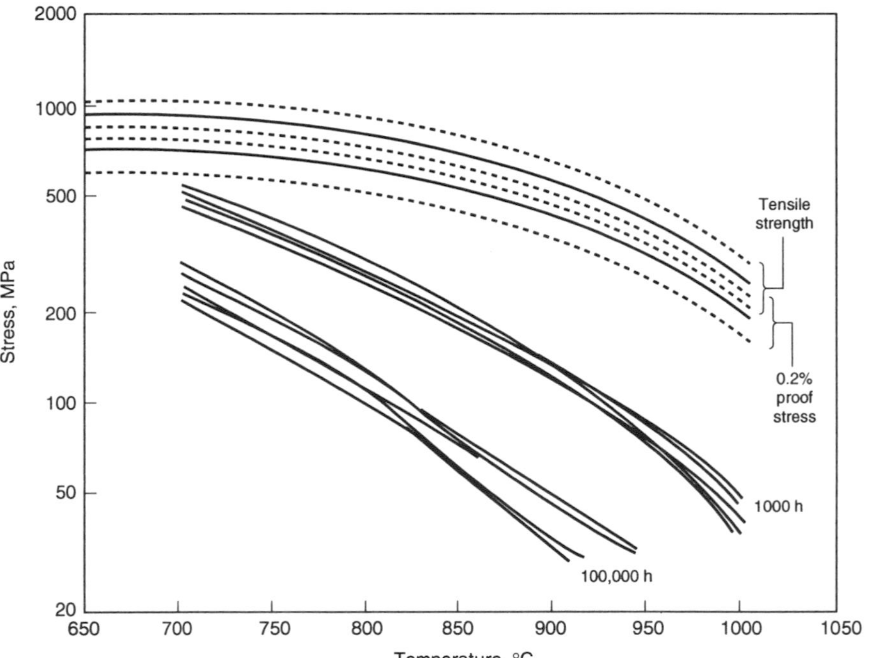

Fig. 19 Temperature dependence of 0.2% yield stress, tensile strength, and creep rupture strength at 1000 and 100,000 h for a nickel-base superalloy casting. Source: Ref 73

only to the proper interpretation of data, but to the proper design of experiments (Ref 69). In addition, there are different practices among testing laboratories that may have appreciable effects on results. These include specimen geometry, loading procedure, specimen alignment, furnace type, and temperature control.

Despite all these concerns regarding proper statistical treatment of data, a methodology has been developed based on time-temperature parameters that is now in widespread use. The approach may be used to achieve the following major design objectives:

- It allows the representation of creep rupture (or creep) data in a compact form, allowing interpolation of results that are not experimentally determined.
- It provides a simple basis for comparison and ranking of different alloys.
- Extrapolation to time ranges beyond those normally reached is straightforward.

Based on the Arrhenius rate equation and a previous tempering parameter (Ref 70), Larson and Miller (Ref 71) developed the most commonly used parameter:

$$P = f(\sigma) = T(\mathrm{C} + \log t_r) \qquad \text{(Eq 10)}$$

where σ is stress, T is absolute temperature, and t_r (or t_R) is time to rupture. In their original paper, the constant C was set equal to 20, which provided a good fit for a variety of alloys, and is still widely used today. A set of σ versus log t_r data for the alloy Astroloy and the parametric master curve derived from it is shown in Fig. 17.

Many other parametric forms have been developed (Ref 68, 69), the most common of which are described in the article "Assessment and Use of Creep-Rupture Data" in this Volume. The time-temperature parameter provides the means for interchanging time and temperature so that a long time may be computed from a short-time test at higher temperature. Figure 18 provides two parametric forms, the shape of which depend on whether isostress lines are parallel or converge on plots of log t versus $1/T$ (the Larson-Miller parameter in Fig. 18a) or T (the Manson-Succop parameter in Fig. 18b). The Manson-Succop parameter (Ref 72) has been used extensively by the Japanese National Research Institute for Metals in compiling data sheets on a wide range of alloys. For example, the data shown in Fig. 19 are for Ni-19Cr-18Co-4Mo-3Ti-3Al-B superalloy castings (Ref 73). The rupture data are statistically treated based on a Manson-Succop correlation, and the tensile data show 95% prediction intervals (dashed lines).

At least two approaches have been used to allow a given data set to determine objectively the best of the common parameters or to provide new parametric forms. The minimum commitment method (MCM) (Ref 74) and the graphical optimization procedure (GOP) (Ref 75) provide

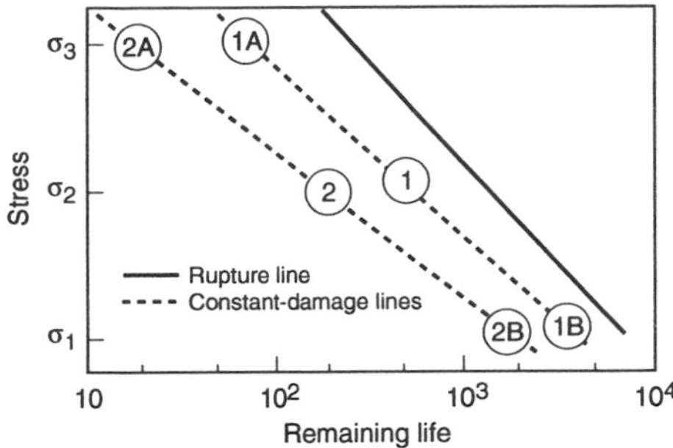

Fig. 21 Schematic plot illustrating construction of constant damage curves in terms of remaining life for different stresses. Source: Ref 83

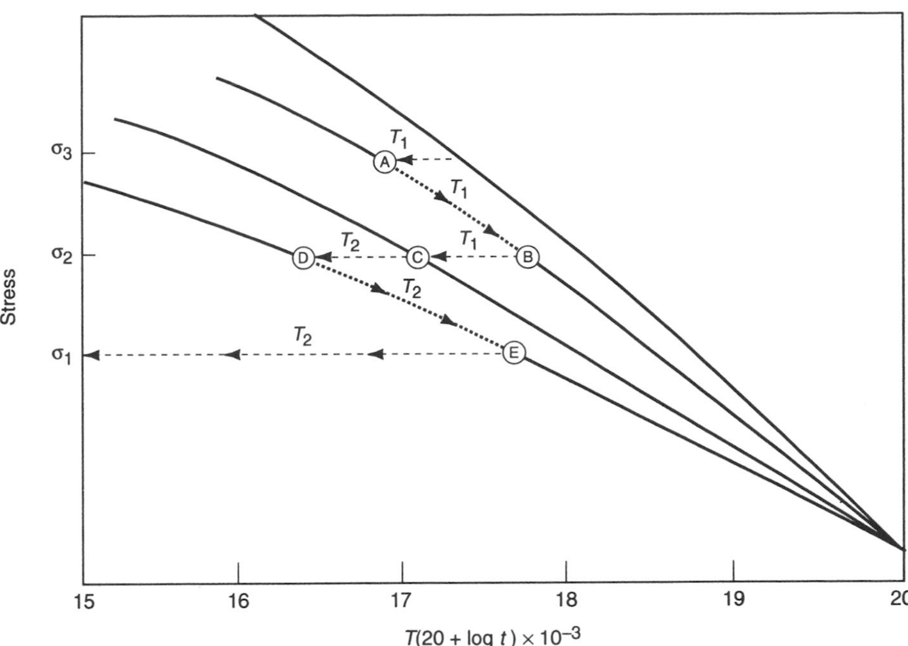

Fig. 22 Schematic plot illustrating construction of constant damage curves in terms of the Larson-Miller parameter. The sequence ABCDE is an example of a particular stress-temperature path. Source: Ref 83

optimal data presentation but lose the advantage of representing different alloys on the same parametric plot. The GOP method uses the fact that all parameters are in the form of one variable expressed as the product of functions of two others, for example, $t_r = H(\sigma)Q(T)$ or $T = H(\sigma)C(t_r)$. These functional forms are solved graphically to optimize the parameter. For example, Fig. 20(a) shows data on Cr-Mo-V steel plotted according to a Larson-Miller parameter. The curves at each temperature are clearly ordered, whereas Fig. 20(b) shows the same data using a parameter determined using the GOP method, which includes the optimized values of the $C(t_r)$ function. There is in this case no separation of the isothermal segments. The optimized value of the Larson-Miller "constant" actually varied from 28.5 at 10 h to 13.6 at 100,000 h (Ref 76) for this data set.

Design rules for high-temperature time-dependent deformation and fracture may be established based on formal codes or on proprietary manufacturers specifications. For example, an ASME code (Ref 77) is used for the design of fossil-fuel boilers and for pressure vessel and piping systems in the petroleum and chemical process industries. The allowable stresses are to be no higher than the lowest of:

- 100% of the stress to produce a creep rate of 0.01% in 1000 h
- 67% of the average stress to produce rupture in 100,000 h
- 80% of the minimum stress to produce rupture in 100,000 h

These stresses may be determined from parametric plots or from derived curves, such as those shown in Fig. 19. It is of interest to note that it is implicit in these rules that the rupture life provides a measure of creep strength. The connection between creep rate and rupture life may be made through the Monkman-Grant relationship (Ref 38) or through the Gill-Goldhoff correlation (Ref 78), which relates stress for rupture with stress for creep to a specific strain for a fixed time and temperature. Both of these methods are described in the article "Assessment and Use of Creep-Rupture Data" in this Volume.

With nickel-base superalloys, it has been found that surface cracks related to environmental attack may develop at strains as low as 0.5% (Ref 79). Since these cracks result in severe loss in fatigue life, this is an appropriate failure criterion rather than rupture life. Gas turbine blades may therefore be designed on the basis of time to 0.5% creep with a suitable safety factor on stress.

Damage Accumulation and Life Prediction

Engineering procedures for life management of operating components assume that the material is progressively degraded or damaged as creep strain increases and operating time accumulates. Damage may be in the form of precipitate changes that may result in softening (overaging) and reduced creep strength, or embrittlement and reduced resistance to fracture. The embrittlement may be due to segregation of harmful species, either from the interior or from the external environment, to interfaces, especially grain boundaries. Damage may also occur as a result of progressive intergranular cavitation and cracking, as previously described. Some of this damage may be reversible by suitable heat treatment or by hot isostatic pressing and may allow the possibility of component rejuvenation. However, for the purposes of component life management, which allows decisions to be made regarding part replacement, repair, or rejuvenation, there is a critical need to quantify the accumulation of damage as a function of operating conditions.

There are two basic approaches to using the concept of damage accumulation for life assessment:

- Based on a detailed knowledge of the operating conditions, including temperature and stress changes, the remaining life is estimated from the known original properties of the material of construction.
- Remaining life estimates are made using post-exposure measurements of microstructural changes, intergranular cavitation, or mechanical properties such as hardness, impact energy, or stress-rupture life.

Creep under Nonsteady Stress and Temperature. There are two common approaches to analyzing creep when the temperature and/or the stress change. The strain-hardening law assumes an equation of state of the form:

$$\dot{\varepsilon} = f_1(\sigma, T, \varepsilon_c) \quad \text{(Eq 11)}$$

which defines the state of the material in terms of creep strain (ε_c). The time-hardening law:

$$\dot{\varepsilon} = f_2(\sigma, T, t) \quad \text{(Eq 12)}$$

defines the state in terms of time. It should be clear that neither of these is physically rigorous. However, they provide analytical options, for example, when a stress change calls for a transfer to a new creep curve at the same strain or time, respectively. Although strain is intuitively a better compromise as a state variable, the time-hardening law is often easier to handle analytically for complex sequences of stress and temperature changes. Moreover, if the sequences involve multiple increases and decreases in stress and temperature, the two approaches may produce similar results. Modifications of the laws including normalizing the strain with the fracture strain, normalizing the time with the failure time, or using a mean function have also been used.

These same ideas have been extended to predict failure either in terms of strain fractions or time fractions. In this case, however, by far the

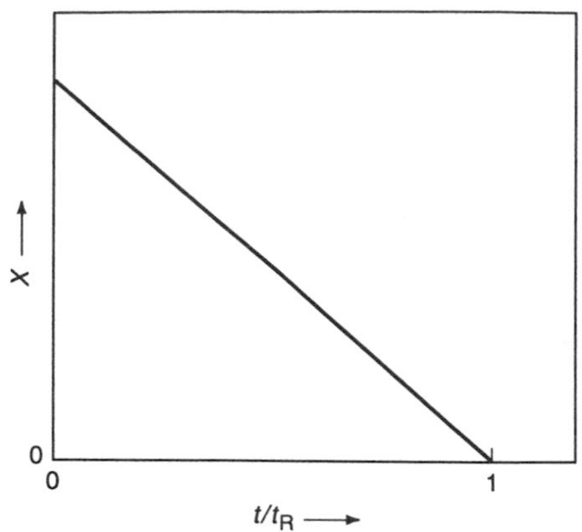

Fig. 23 Ideal behavior of a property suitable for damage monitoring as a function of life fraction

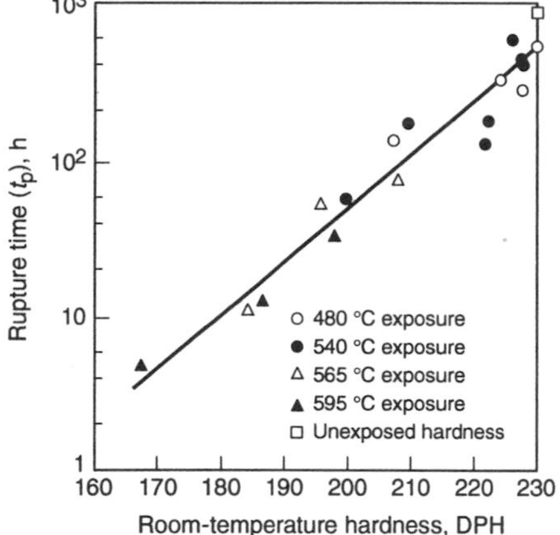

Fig. 24 Correlation between post-exposure rupture life for a Cr-Mo-V steel tested at 240 MPa (35 ksi) and 540 °C (1000 °F) and room-temperature hardness. DPH, diamond pyramid hardness. Source: Ref 85

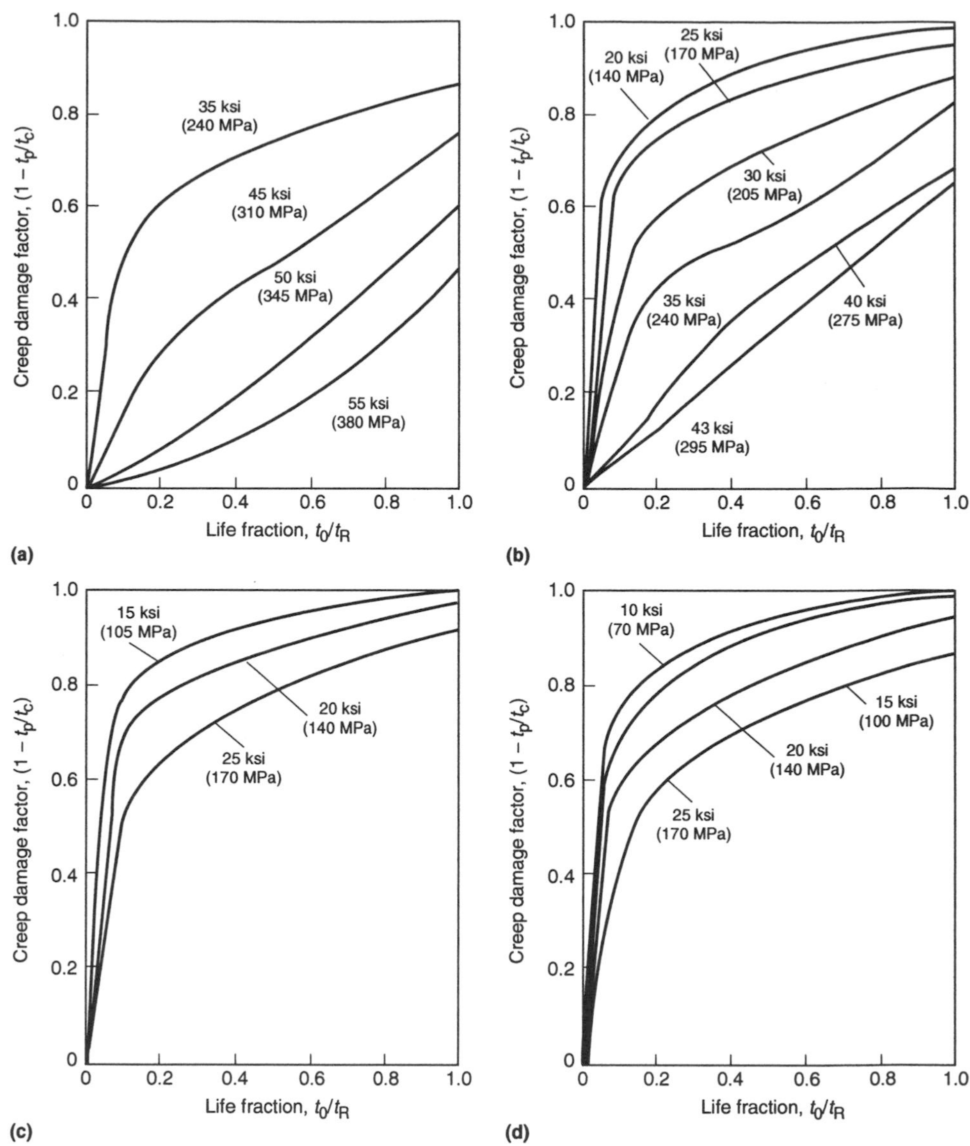

Fig. 25 Graphical description of used life as a function of life-fraction exposure for Cr-Mo-V steel at various stresses and temperatures. (a) At 480 °C (900 °F). (b) At 540 °C (1000 °F). (c) At 565 °C (1050 °F). (d) At 595 °C (1100 °F). Source: Ref 85

most extensively used in design analysis and life prediction has been the life-fraction approach. This concept was first introduced by Robinson (Ref 80, 81) as an accounting system for materials in which the rupture time is a power function of stress and the log rupture time is a linear function of temperature. The analytical solutions for such a material subjected to various temperature and stress cycles were given by Robinson (Ref 81), but the concept is now generally presented in the form:

$$\sum_{i=1}^{n} \frac{\Delta t_i}{t_{Ri}} = 1 \qquad \text{(Eq 13)}$$

where t is time spent at temperature T or stress σ, t_R is rupture time at temperature T or stress, and n is the number of temperature or stress changes.

Each fractional expenditure of life, or accumulated damage, is considered to be independent of all others. Failure is predicted when the sum of the fractions of life equals unity. This hypothesis remains a widely used tool for life assessment.

Considerable experimental work has attempted to confirm or refute the rule. However, when the results of multiple-stress or temperature-change experiments are analyzed in terms of the life-fraction rule, it is only possible to determine whether the rule is appropriate, but not why it fails. Only when damage is defined in terms of remaining life rather than used life is it possible to formulate the appropriate damage law (Ref 82). Figure 21 illustrates the procedure schematically for a stress-change experiment where the abscissa is the logarithm of the remaining life. The continuous line represents remaining life for the virgin material, that is, the stress-rupture data for the initial condition. Point 1 is the remaining life on two specimens exposed at stress σ_2. One of these specimens is then tested to failure at σ_3, and the measured remaining life is represented by point 1A. The stress on the second specimen is then reduced to σ_1, resulting in a measured remaining life indicated by 1B. The dashed line drawn through the points 1A, l, and 1B is then defined as a constant damage line in terms of remaining life. Clearly, any number of lines such as 2A, 2, and 2B may be constructed.

A necessary condition for the life-fraction rule to apply is that the constant damage lines should be displaced horizontally by a constant distance when remaining life is plotted on a logarithmic scale for both temperature and stress changes. For low-alloy steels, the life-fraction rule applies quite well for temperature changes but not for stress changes. For stress changes, the curves often converge at low stresses, leading to a loading sequence effect. For example, a stress increment leads to a life fraction at failure less than one and a stress decrement leads to a life fraction at failure greater than one. This behavior may be combined on a parametric plot (Fig. 22), where a sequence of both temperature and stress changes may be represented. For detailed discussion on the construction and use of the remaining life

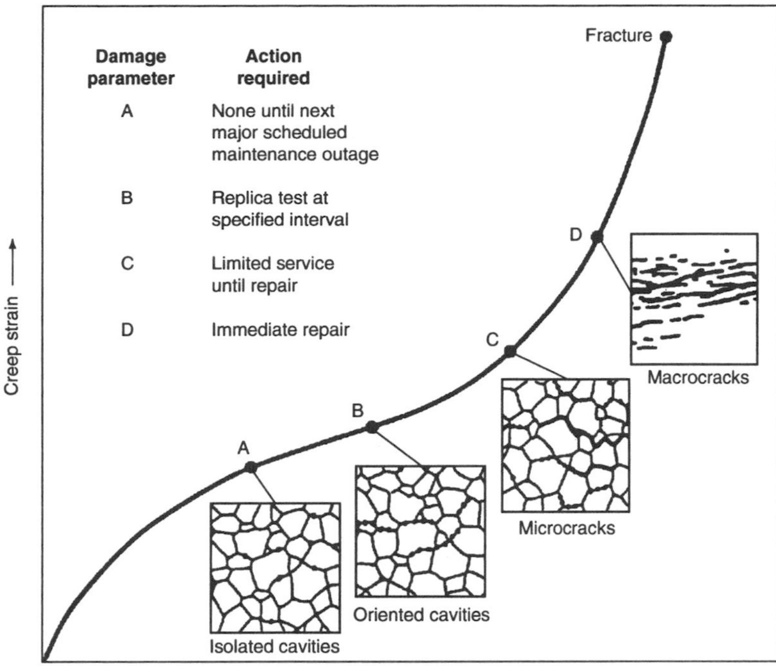

Fig. 26 Creep life assessment based on cavity classification in boiler steels. Source: Ref 87

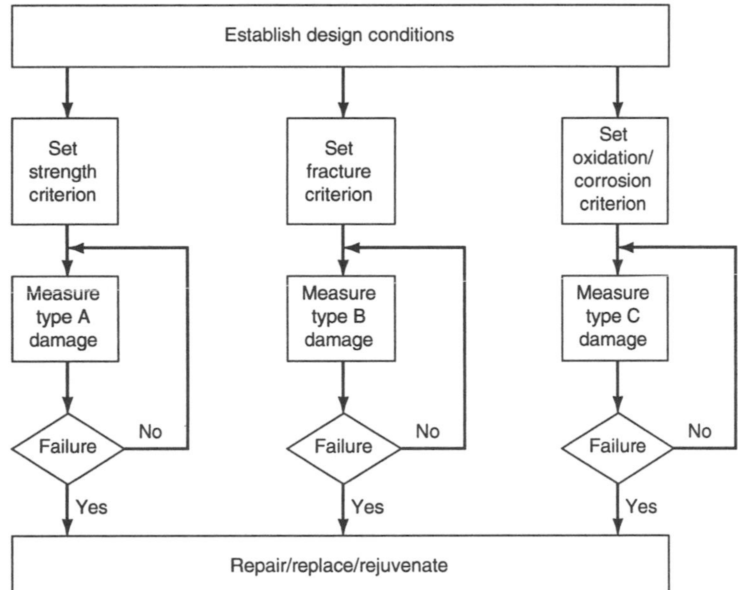

Fig. 27 The design-for-performance concept. Source: Ref 92

curves for nonsteady conditions and component life assessment, see Ref 82 to 84.

Post-Exposure Evaluation. Ideally, a microstructural characterization or mechanical property should change linearly with life fraction and independently of the test conditions, as shown in Fig. 23. In fact, the changes are invariably nonlinear and nonunique with life fraction. Both cavitation changes and creep strength changes are dependent on the test conditions (Ref 45, 46, 83). Thus, the amount of damage at failure or at a given life fraction is a strong function of the stress in particular.

A comprehensive study of a low-alloy steel in which miniature stress-rupture specimens were taken from large specimens, which had previously been exposed to various strains and times up to 60,000 h at four temperatures over a range of stresses, confirmed this complexity (Ref 85). The initial exposure led to progressive softening, which correlated well with the post-exposure rupture life (Fig. 24) in a standard test at 540 °C (1000 °F) and 240 MPa (35 ksi). Using parametric correlations with the original exposures, a series of damage curves were generated (Fig. 25) in which the ordinate is one minus the ratio of the post-exposure rupture life (t_p) divided by the mean rupture life of the unexposed material (t_c) under the same test conditions. This definition of creep damage increases to a limiting value of unity as t_p approaches zero.

It is readily apparent from these curves that (1) the damage accumulates in a nonlinear manner, (2) the damage at failure is dependent on the test conditions, and (3) the damage at a constant life fraction is strongly stress dependent but quite insensitive to temperature for a given stress. These points are consistent with the previous commentary on the limitations of the life-fraction rule and with the general applicability of time-temperature parameters.

Despite similar concerns with the use of microstructural observations of cavitation to predict remaining life, there have been extensive reported studies of such measurements (Ref 39, 86). The first detailed attempt to relate remaining life of power plant to cavitation observed on replicas was that of Neubauer and Wedel (Ref 87), who set up recommended procedures based on the observation of four levels of cavitation severity in steam pipes (Fig. 26). There were subsequent attempts to quantify the cavitation in terms of mechanistic concepts (Ref 88). Other studies have examined the cavitation in nickel-base alloys for gas turbines (Ref 89) and set up similar criteria for end of useful life.

The replica studies in particular have usually been limited to surface observations so that the possibility of environmental interaction is strong (see the section "Environmental Effects" in this article). Also, many of these studies have recently been criticized on the basis that the observation of cavities is very sensitive to preparation techniques (Ref 90).

Limitations and Alternative Design Approaches

In the increasingly competitive global market for manufactured components operating at high temperatures, there are three primary technical objectives:

- The development cycle for new materials and new designs must be reduced.
- The designs must be optimized for efficiency and performance.
- Procedures must be developed for component life management to allow timely repair or replacement.

Current Limitations. The primary design of most high-temperature components involves the application of some proportion of the stress for failure in a given time (usually in the range of 20,000 to 100,000 h) calculated from extrapolated constant load stress-rupture tests. It might

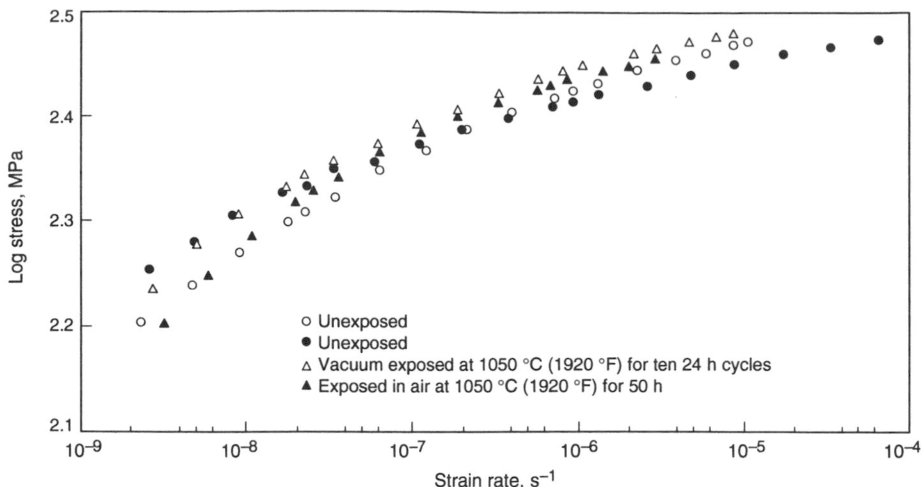

Fig. 28 Insensitivity of creep strength of IN738 to various thermal exposures as determined from stress versus creep-rate behavior calculated from stress-relaxation tests. Source: Ref 92

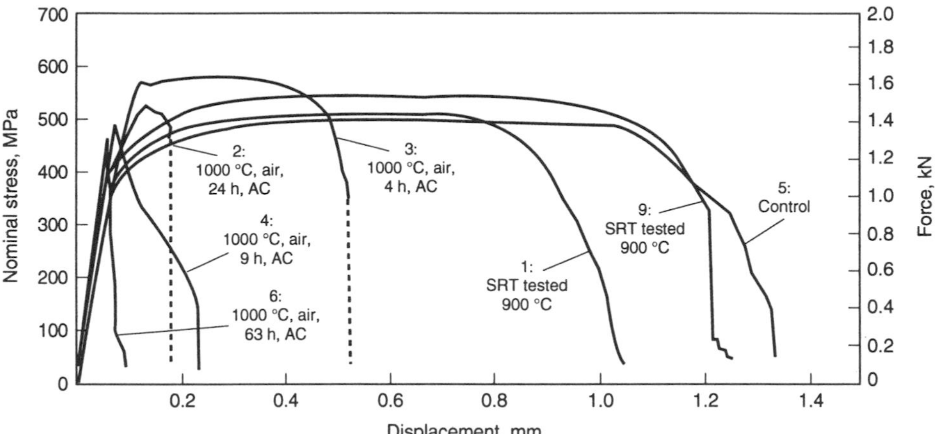

Fig. 29 Embrittlement of IN738 with increasing severity of exposure in air demonstrated in constant displacement rate (4 $\times 10^{-5}$ mm/s) tensile tests at 800 °C (1470 °F). SRT, stress-relaxation tests. AC, air cooled. Source: Ref 95

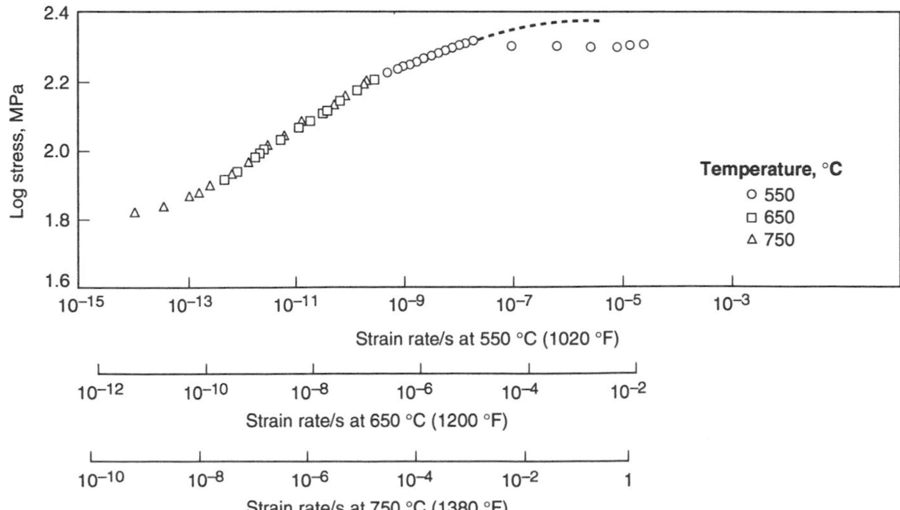

Fig. 30 Master stress versus creep-rate curve for Incoloy 800H between 550 and 750 °C (1020 and 1380 °F) achieved by horizontal translation of data generated on one specimen. Source: Ref 95

appear, therefore, that the most direct method of measuring remaining life after extended service would be to compute it based on the current remaining life of a sample taken from the component. Accordingly, methods are being developed to take miniature stress-rupture samples from components and assess remaining life based on extrapolation from such data. There are a number of reasons why this approach may be unsound:

- Component failure is often localized with little or no material degradation or damage remote from the failure.
- Cracks frequently initiate from the surface so that any post-exposure property measurement on material taken from the interior has limited value.
- Interactions with the operating gaseous environments, which may have profound effects on crack initiation and propagation, are generally ignored.
- The changing stress in a constant load test is not normally accounted for in summing life fractions to predict remaining life.
- The sources of scatter in experimental property measurement (for example, alignment, temperature control, precision of stress and strain measurement, specimen geometry) bear little connection to the sources of scatter in service.
- Time to rupture in an unnotched ductile alloy is principally a measure of deformation behavior rather than fracture resistance.

The failure to establish a clear separation of a strength requirement from a failure criterion (an appropriate analogy might be between yield strength and fracture toughness for many low-temperature components) leads to a paradox (Ref 91). This may be stated: When a component fails, the material of the component has a finite life, sometimes approaching the original design life, at the operating conditions of the component. It thus follows that a remaining life estimated from a sample taken from an operating component may bear no relationship to the actual component remaining life.

Low-temperature design is dominated by properties that uniquely characterize the mechanical state in terms of stiffness (modulus), strength (yield), and fracture resistance (K_c). If service-induced changes in state occur, for example, radiation hardening and embrittlement, these changes are monitored in terms of their effects on changes in the same short-time properties. Thus life management decisions are based on the same performance criteria as in the original design: there is no remaining life paradox in this case.

The basis for current methods of high-temperature design is different in that the objective is to incorporate time-dependent changes in the test methodology. The creep rates and rupture lives relate to the starting material only in terms of the specific deformation path being imposed, for example, a variable strain rate or a variable true stress test with an arbitrary interaction between creep deformation and fracture processes. This history is quite different from any real service history. To measure these properties, unlike in

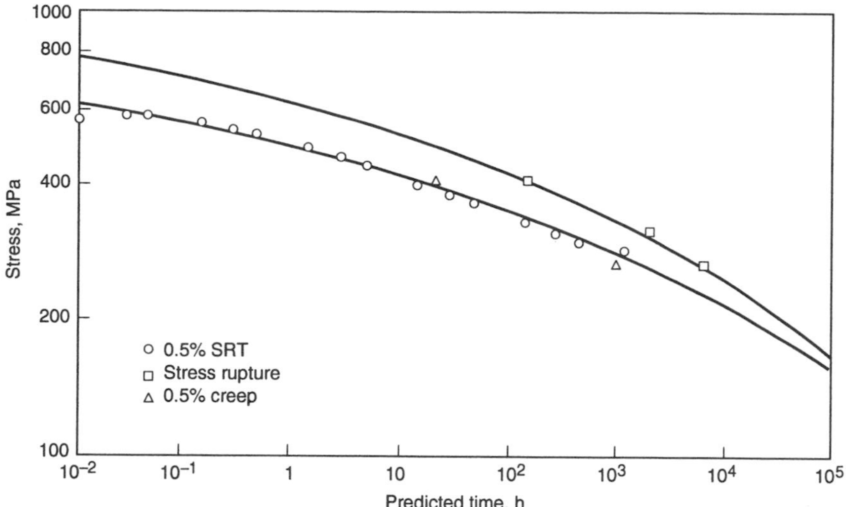

Fig. 31 Comparison of predicted time to 0.5% creep based on stress-relaxation measurements at 650 °C (1200 °F) for an austenitic iron-base alloy with measurements made on conventional creep tests. SRT, stress-relaxation tests

the low-temperature situation, the structural and mechanical states must be changed. As an example, the minimum creep rate at a given temperature and stress is a measure of the creep strength of the material in its current state; it relates to its initial state only for the particular deformation history. The designer may claim that long-time data are required for component creep strength evaluation, but in reality what is needed is low strain rate data. It will be shown that such data may be obtained with precision in short-time tests.

Simulating service complexity in material testing involves a hierarchy of increasing complexity and expense. Thus long-term creep tests are inadequate; nonsteady test conditions, complex stresses, cyclic stresses, environmental effects, and the synergism among them must be taken into account. Even with the most complex test plan, most service operations cannot be simulated.

Alternative Design Approaches and Tests. An alternative approach is to simplify the test methodology and develop tests to measure separately the high-temperature creep strength and fracture resistance, ideally to evaluate the current state in terms of these properties. The consequences of microstructural evolution, induced in service or in laboratory simulations, can then be assessed using the same short-time tests. Design is then based on minimum acceptable performance levels (Ref 92).

The concept is illustrated in Fig. 27 for three fundamental properties, including an environmental damage criterion, which could be critical in a number of high-temperature applications. Additional criteria could be added. However, it might not be necessary to include some important failure processes, such as high-cycle fatigue and thermal fatigue, because these in principle are either derivable from, or correlatable with, the other properties or are too complex and parameter sensitive to be assigned minimum performance values.

The creep strength criterion for high-temperature applications may be developed from a series of carefully conducted stress-relaxation tests (SRT). The stress versus time response is converted to a stress versus creep strain-rate response (Ref 92, 93). This is, in effect, a self-programmed variable stress creep test. Typically, a test lasting less than one day may cover five decades in creep rate. The accumulated nonelastic strain is usually less than 0.1%, so that several relaxation runs at different temperatures and from different stresses can be made on a single specimen with minimal change in the mechanical state. Thus an enormous amount of creep data can be generated in a short time.

A constant displacement rate (CDR) test (Ref 92, 94), in which a crack is enabled to initiate naturally at a grain boundary and propagate slowly under closed-loop displacement control, was developed as a separate basis to evaluate the fracture resistance. Both types of tests were first performed on large specimens to establish the broad principles of the methodology and then compared with data taken from miniature specimens machined from actual gas turbine blades (Ref 95). It was established that environmental interaction, specifically gas phase embrittlement (GPE) resulting from intergranular penetration of oxygen, could drastically reduce the fracture resistance with little or no effect on the creep strength (Ref 92, 95).

Figure 28 is an example of four relaxation runs at 900 °C (1650 °F) showing no appreciable effect of severe high-temperature exposure in air or vacuum on the stress versus creep-rate response for a cast nickel-base gas turbine blade made of alloy IN738. By contrast, Fig. 29 shows that there is a profound degradation in the CDR fracture resistance for high-temperature air exposures. This approach is clearly valuable for assessing performance and continued operation when embrittlement occurs with minimal effect on creep strength. It can also be used directly as a basis for creep design. For example, Fig. 30 shows how relaxation runs at different temperatures can be fitted by horizontal displacement of the strain-rate scales to allow stresses corresponding to very low creep rates to be estimated. In fact, this procedure allows design creep rates to be attained based on several one-day tests (Ref 92). It also allows comprehensive determination of the effect of prior exposure, either from laboratory simulations or component service, to be evaluated.

Recently, it has been shown how creep data obtained from stress-relaxation testing can be compared directly with long-time creep data in terms of time to specific creep strains. Since the two approaches use different deformation histories, neither of which is fundamentally preferred compared with component deformation, precise conformity should not be anticipated or required. However, for a number of alloys (Ref 95) and polymers (Ref 96), the agreement is surprisingly good. Figure 31 shows such a comparison for an iron-base superalloy.

Thus the approach offers accelerated creep strength evaluation, which may be used either directly or through calibration with long-time tests, and may be used to assess the creep strength of operating components. In particular, by recognizing high-temperature fracture as a property that is not uniquely connected to creep strength, the approach offers a basis to understand why some components fail very early in their design lives and why others appear undamaged after exceeding their design lives.

ACKNOWLEDGMENT

The information in this article is largely taken from D.A. Woodford, "Design for High-Temperature Applications" to be published in *Materials Selection and Design,* Volume 20, *ASM Handbook,* ASM International, 1997.

REFERENCES

1. F. Philips, The Slow Stretch in India Rubber, Glass and Metal Wire When Subjected to a Constant Pull, *Philos. Mag.,* Vol 9, 1905, p 513
2. E.N. da C. Andrade, The Viscous Flow in Metals and Allied Phenomena, *Proc. R. Soc.,* Vol A84, 1910, p 1–13
3. O.H. Wyatt, Transient Creep in Pure Metals, *Proc. Phys. Soc.,* Vol 66B, 1953, p 459–480
4. R.W. Bailey, *J. Inst. Met.,* Vol 35, 1926, p 27
5. E. Orowan, *J. West Scot. Iron and Steel Inst.,* Vol 54, 1946-1947, p 45
6. L.N. McCartney, No Time—Gentlemen Please, *Philos. Mag.,* Vol 33 (No. 4), 1976, p 689–695
7. G.J. Lloyd and R.J. McElroy, On the Anelastic Contribution to Creep, *Acta Metall.,* Vol 22, 1974, p 339–348
8. E.W. Hart, A Critical Examination of Steady State Creep, *Creep in Structures,* A.R.S. Ponter and D.R. Hayhurst, Ed., Springer-Verlag, 1981, p 90–102
9. E.W. Hart, Theory of the Tensile Test, *Acta Metall.,* Vol 15, 1967, p 351

10. M.A. Burke and W.D. Nix, *Acta Metall.*, Vol 23, 1975, p 793
11. P.J. Wray and O. Richmond, Experimental Approach to a Theory of Plasticity at Elevated Temperatures, *J. Appl. Physics,* Vol 39 (No. 12), 1968, p 5754–5761
12. D.A. Woodford, Creep Ductility and Dimensional Instability of Nickel at 500 and 525 °C, *Trans. ASM,* Vol 59, 1966, p 398–410
13. J. Weertman, *J. Appl. Phys.*, Vol 26, 1955, p 1213–1217
14. F.R.N. Nabarro, *Proc. Conf. Strength of Solids,* Physical Society, London, 1948, p 75
15. C. Herring, *J. Appl. Phys.*, Vol 21, 1950, p 437–445
16. R.L. Coble, *J. Appl. Phys.*, Vol 34, 1963, p 1679–1684
17. M.F. Ashby, *Acta Metall.,* Vol 20, 1572, p 887–897
18. R.L. Bell and T.G. Langdon, Grain Boundary Sliding, *Interfaces Conf.,* R.C. Gifkins, Ed., Butterworths, 1969, p 115–137
19. J.L. Walter and H.E. Cline, Grain Boundary Sliding, Migration, and Deformation in High-Purity Aluminum, *Trans. AIME,* Vol 242, 1968, p 1823–1830
20. N.E. Dowling, *Mechanical Behavior of Materials,* Prentice Hall, 1993
21. R.J. Crawford, *Plastics Engineering,* Pergamon Press, 1987, p 222
22. J.C. Gibeling and W.D. Nix, Observations of Anelastic Backflow Following Stress Reductions During Creep of Pure Metals, *Acta Metall.,* Vol 29, 1981, p 1769–1784
23. C.M. Zener, *Elasticity and Anelasticity of Metals,* University of Chicago, 1948
24. J.A. DiCarlo, *J. Mater. Sci.,* Vol 21, 1986, p 217–224
25. J. Henderson and J.D. Snedden, Creep Recovery of Aluminum Alloy DTD 2142, *Appl. Mater. Res.,* Vol 4 (No. 3), 1965, p 148–168
26. J.D. Lubahn, The Role of Anelasticity in Creep, Tension and Relaxation Behavior, *Trans. ASM,* Vol 45, 1953, p 787–838
27. R.M. Goldhoff, Creep Recovery in Heat Resistant Steels, *Advances in Creep Design,* A.I. Smith and A.M. Nicolson, Ed., John Wiley & Sons, 1971, p 81–109
28. T.S. Ke, Experimental Evidence of the Viscous Behavior of Grain Boundaries in Metals, *Phys. Rev.,* Vol 71, 1947, p 533
29. J.E. Dorn, Some Fundamental Experiments on High-Temperature Creep, *J. Mech. Phys. Solids,* Vol 3, 1954, p 85–116
30. O.D. Sherby, Factors Affecting the High Temperature Strength of Polycrystalline Solids, *Acta Metall.,* Vol 10, 1962, p 135–147
31. R. Viswanathan, The Effect of Stress and Temperature on the Creep and Rupture Behavior of a 1.25pct. Chromium-0.5 pct. Molybdenum Steel, *Metall. Trans. A,* Vol 8, 1977, p 877–993
32. J.D. Lubahn, "Creep of Metals," Symposium on Cold Working of Metals, American Society for Metals, 1949
33. D.A. Woodford, Measurement and Interpretation of the Stress Dependence of Creep at Low Stresses, *Mater. Sci. Eng.,* Vol 4, 1969, p 146–154
34. S.K. Mitra and D. McLean, Cold Work and Recovery in Creep at Ostensibly Constant Structure, *Met. Sci.,* Vol 1, 1967, p 192
35. E.W. Hart, "Phenomenological Theory: A Guide to Constitutive Relations and Fundamental Deformation Properties, *Constitutive Equations in Plasticity,* A.S. Argon, Ed., MIT Press, 1975
36. D.A. Woodford, Analysis of Creep Curves for a Magnesium-Zirconium Alloy, *J. Inst. Met.,* Vol 96, 1968, p 371–374
37. D.A. Woodford, Strain Rate Sensitivity as a Measure of Ductility, *Trans. ASM,* Vol 62, 1969, p 291–293
38. F.C. Monkman and N.J. Grant, *Proc. ASTM,* Vol 56, 1956, p 595
39. R. Viswanathan, *Damage Mechanisms and Life Assessment of High Temperature Components,* ASM International, 1989, p 82
40. F. Garofalo, *Fundamentals of Creep and Creep Rupture,* MacMillan, 1965
41. D.A. Woodford and R.H. Bricknell, Environmental Embrittlement of High Temperature Alloys by Oxygen, *Embrittlement of Engineering Alloys,* C.L. Briant and S.K. Banerji, Ed., Academic Press, 1983, p 157
42. J.O. Steigler, K. Farrell, B.T.M. Loh, and H.E. McCoy, Creep Cavitation in Tungsten, *Trans. ASM,* Vol 60, 1967, p 494
43. A.C.F. Cocks and M.F. Ashby, On Creep Fracture by Void Growth, *Prog. Mater. Sci.,* Vol 27, 1982, p 189–244
44. D. McLean, The Physics of High Temperature Creep in Metals, *Rept. Prog. Phys.,* Vol 29, 1966, p 1–33
45. D.A. Woodford, A Parametric Approach to Creep Damage, *Met. Sci. J.,* Vol 3, 1969, p 50–53
46. D.A. Woodford, "Density Changes During Creep in Nickel," *Met. Sci. J.,* Vol 3, 1969, p 234–240
47. F.N. Rhines and P.J. Wray, Investigation of the Intermediate Temperature Ductility Minimum in Metals, *Trans. ASM,* Vol 54, 1961, p 117–128
48. D.A. Woodford and R.M. Goldhoff, An Approach to the Understanding of Brittle Behavior of Steel at Elevated Temperatures, *Mater. Sci. Eng.,* Vol 5, 1970, p 303–324
49. W.E. Brown, M.H. Jones, and D.P. Newman, *Symp. on Strength and Ductility of Metals at Elevated Temperatures,* STP 128, ASTM, 1952, p 25
50. M.P. Seah, Grain Boundary Segregation, *Metal Phys.,* Vol 10, 1980, p 1043–1064
51. E.P. George, P.L. Li, and D.P. Pope, Creep Cavitation in Iron—Sulfides and Carbides as Nucleation Sites, *Acta Metall.,* Vol 35 (No. 10), 1987, p 2471–2486
52. R.H. Cook and R.P. Skelton, The Influence of Environment on High Temperature Mechanical Properties of Metals and Alloys, *Int. Met. Rev.,* Vol 19, 1974, p 199
53. L.F. Coffin, Fatigue at High Temperature, *Fatigue at Elevated Temperature,* A.E. Carden, A.J. McEvily, and C.H. Wells, Ed., STP 520, ASTM, 1972, p 5–36
54. K. Sadananda and P. Shahinian, The Effect of Environment on the Creep Crack Growth Behavior of Several Structural Alloys, *Mater. Sci. Eng.,* Vol 43, 1980, p 159–168
55. K.R. Bain and R.M. Pelloux, Effect of Environment on Creep Crack Growth in PM/HIP René 95, *Metall. Trans. A,* Vol 15, 1984, p 381–388
56. S. Floreen and R.H. Kane, Investigation of the Creep-Fatigue-Environment Interaction in a Nickel Base Superalloy, *Fatigue Eng. Mater. Struct.,* Vol 2, 1980, p 401
57. G.A. Whitlow, C.G. Beck, R. Viswanathan, and E.A. Crombie, The Effects of a Liquid Sulfate/Chloride Environment on Superalloys, *Metall. Trans. A,* Vol 15, 1984, p 23–28
58. W.H. Chang, *Proc. Conf. Superalloys—Processing,* Section V, MCIC-7210, AIME, 1972
59. D.A. Woodford, Environmental Damage of a Cast Nickel Base Superalloy, *Metall. Trans. A,* Vol 12, 1981, p 299–308
60. R.H. Bricknell and O.A. Woodford, The Embrittlement of Nickel Following High Temperature Air Exposure, *Metall. Trans. A,* Vol 12, 1981, p 425–433
61. M.C. Pandey, B.F. Dyson, and D.M.R. Taplin, Environmental, Stress-State and Section-Size Synergisms During Creep, *Proc. R. Soc. London A,* Vol 393, 1984, p 117–131
62. D.A. Woodford and D.F. Mowbray, Effect of Material Characteristics and Test Variables on Thermal Fatigue of Cast Superalloys, *Mater. Sci. Eng.,* Vol 16, 1974, p 5
63. R.H. Bricknell and D.A. Woodford, Cavitation in Nickel during Oxidation and Creep, *Int. Conf. on Creep and Fracture of Engineering Materials and Structures,* B. Wilshire and R.W. Evans, Ed., Pineridge Press, Inst. of Metals, 1991, p 249–262
64. E.C. Scaife and P.L. James, *Met. Sci. J.,* Vol 2, 1968, p 217
65. H. Reidel, Fracture at High Temperatures, Springer-Verlag, 1987
66. J.P. Beckman and D.A. Woodford, Gas Phase Embrittlement of Nickel by Sulfur, *Metall. Trans. A,* Vol 21, 1990, p 3049–3061
67. G.A. Webster and R.A. Ainsworth, *High Temperature Component Life Assessment,* Chapman and Hall, 1994
68. J.S. Conway, *Stress Rupture Parameters: Origin, Calculations and Use,* Gordon and Breach, 1969
69. R.M. Goldhoff, "Development of a Standard Methodology for the Correlation and Extrapolation of Elevated Temperature," EPRI FP-1062, Electric Power Research Institute, 1979
70. J.H. Holloman and L.C. Jaffe, Time-Temperature Relations in Tempering Steel, *Trans. AIME,* Vol 162, 1945, p 223–249
71. F.R. Larson and J. Miller, A Time-Temperature Relationship for Rupture and Creep Stresses, *Trans. ASME,* Vol 74, 1952, p 765–775
72. S.S. Manson and G. Succop, *Stress Rupture Properties of Inconel 700 and Correlation on the Basis of Several Time-Temperature Parameters,* ASTM, STP 174, 1956

73. NRIM Creep Data Sheet, No. 34B, National Research Institute for Metals, Tokyo, Japan, 1975
74. S.S. Manson and C.R. Ensign, "A Specialized Model for Analysis of Creep Rupture Data by the Minimum Commitment Method," Tech. Memo TMX 52999, National Aeronautics and Space Administration, 1971
75. D.A. Woodford, A Graphical Optimization Procedure for Time-Temperature Rupture Parameters, *Mater. Sci. Eng.,* Vol 15, 1974, p 169–175
76. D.A. Woodford, Perspectives in Creep and Stress Rupture, *Int. Conf. on Creep,* Tokyo, JSME, IMechE, ASME, ASTM, 1986, p 11–20
77. *ASME Boiler and Pressure Vessel Code,* Section 1, ASME
78. R.M. Goldhoff and R.F. Gill, A Method for Predicting Creep Data for Commercial Alloys on a Correlation between Creep Strength and Rupture Strength, *ASME J. Basic Eng.,* Vol 94, Series D, No. 1, 1972, p 1–6
79. W.L. Chambers, W.J. Ostergren, and J.H. Wood, Creep Failure Criteria for High Temperature Alloys, *J. Eng. Mater. Technol.,* Vol 101, 1979, p 374–379
80. E.L. Robinson, Effect of Temperature Variation on the Creep Strength of Steels, *Trans. ASME,* Vol 60, 1938, p 253–259
81. E.L. Robinson, Effect of Temperature Variation on the Long-Time Rupture Strength of Steels, *Trans. ASME,* Vol 74, 1952, p 777–781
82. D.A. Woodford, A Critical Assessment of the Life Fraction Rule for Creep-Rupture under Nonsteady Stress or Temperature, Paper No. 180.1, *Int. Conf. on Creep and Fatigue in Elevated Temperature Applications,* Inst. Mech. Eng., 1973-1974, p 1–6
83. D.A. Woodford, Creep Damage and the Remaining Life Concept, *J. Eng. Mater. Technol.,* Vol 101, 1979, p 311–316
84. R.V. Hart, Assessment of Remaining Creep Life Using Accelerated Stress Rupture Tests, *Met. Technol.,* Vol 13, 1976, p 1
85. R.M. Goldhoff and D.A. Woodford, *The Evaluation of Creep Damage in a Cr-Mo-V Steel,* STP 515, ASTM, 1972, p 89–106
86. R. Viswanathan and S.M. Gehl, Life-Assessment Technology for Power-Plant Components, *JOM,* Vol 44 (No. 2), 1992, p 34–42
87. B. Neubauer and U. Wedel, Restlife Estimation of Creeping Components by Means of Replicas, *Advances in Life Prediction Methods,* ASME, 1983, p 307–314
88. B.J. Cane and M.S. Shammas, "A Method for Remnant Life Estimation by Quantitative Assessment of Creep Cavitation on Plant," Report TPRD/U2645/N84 CEGB, Leatherhead Lab., 1984
89. S.A. Karllsson, C. Persson, and P.O. Persson, Metallographic Approach to Turbine Blade Lifetime Prediction, *Baltica III Conf. on Plant Condition and Life Management,* Helsinki, 1995, p 333–348
90. I. LeMay, T.L. da Silveira, and S.K.P. Cheung-Mak, Uncertainties in the Evaluation of High Temperature Damage in Power Stations and Petrochemical Plant, *Int. J. Press. Vess. Piping,* Vol 59, 1994, p 335–342
91. D.A. Woodford, The Remaining Life Paradox, *Int. Conf. on Fossil Power Plant Rehabilitation,* ASM International, 1989, p 149
92. D.A. Woodford, Test Methods for Accelerated Development, Design, and Life Assessment of High Temperature Materials, *Mater. Design,* Vol 14 (No. 4), 1993, p 231–242
93. E.W. Hart and H.D. Solomon, Stress Relaxation Testing of Aluminum, *Acta Metall.,* Vol 21, 1973, p 295
94. J.J. Pepe and D.C. Gonyea, Constant Displacement Rate Testing at Elevated Temperatures, *Int. Conf. on Fossil Power Plant Rehabilitation,* ASM International, 1989, p 39
95. D.A. Woodford, The Design for Performance Concept Applied to Life Management of Gas Turbine Blades, *Baltica III, Int. Conf. on Plant Condition and Life Management,* Helsinki, 1995, p 319–332
96. S.K. Reif, K.J. Amberge, and D.A. Woodford, Creep Design Analysis for a Thermoplastic from Stress Relaxation Measurements, *Mater. Design,* Vol 16 (No. 1), 1995, p 15–21

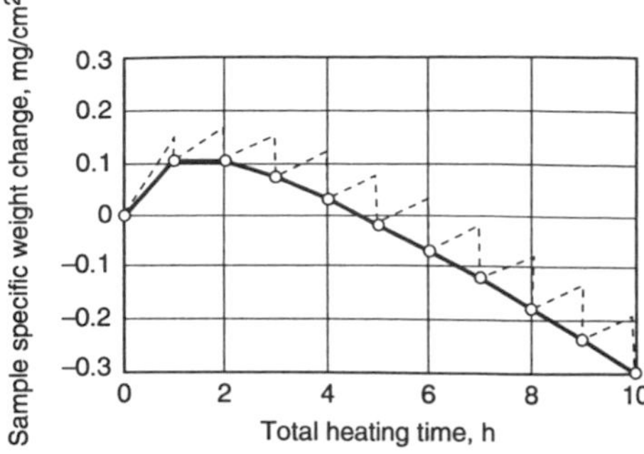

Fig. 3 Idealized cyclic oxidation weight change curve showing parabolic growth legs, punctuated by fractional spall events upon cooldowns. Source: Ref 2

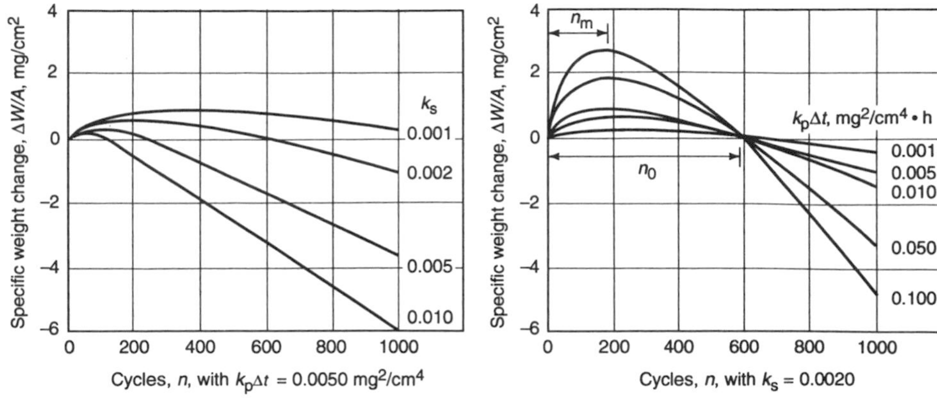

Fig. 4 Family of model cyclic oxidation curves for various parabolic rate constants, k_p, and spall constants, k_s. Source: Ref 5

zones) can also be readily distinguished. Scanning electron microscopy/energy dispersive spectroscopy (SEM/EDS) provides the next level of detail, with higher magnification and chemical analysis capability. This tool is especially useful in characterizing the morphology and distribution of features on the external surfaces. It is also used to determine the grain size of the oxide scales, the growth rate of which is often controlled by short-circuit grain-boundary diffusion. X-ray diffraction (XRD) of the external surface is widely used to determine the scale phases, and electron microprobe analysis of cross sections can accurately determine scale compositions and alloy-depletion profiles. More elaborate studies may employ transmission electron microscopy/scanning transmission electron microscopy (TEM/STEM), Auger spectroscopy, x-ray photoelectron spectroscopy, or chemical analyses of stripped corrosion products.

Alloy Design for Optimal Performance

The most general rule followed in alloy design for oxidation is that the alloys must form scales composed primarily of either Al_2O_3, Cr_2O_3, or SiO_2. For model iron-, cobalt-, or nickel-base binary alloys, this occurs at approximately 15 wt% Al, 20 wt% Cr, or 5 wt% Si. Unfortunately, these levels of aluminum, chromium, and silicon result in brittle alloys, and—in the case of silicon—very low melting temperatures. Silicon then is only used in small (1–3 wt%) additions to help stabilize Al_2O_3 or Cr_2O_3 scales. Many high-temperature alloys are, however, based on M-Cr systems having a good compromise of high-temperature mechanical properties and oxidation resistance. Small additions of manganese and silicon are widely used to stabilize Cr_2O_3 scales at concentrations below 20 wt% Cr. Similarly, chromium is widely used at 5 to 15 wt% levels to help stabilize the formation of Al_2O_3 scales at lower aluminum levels near 5 wt%.

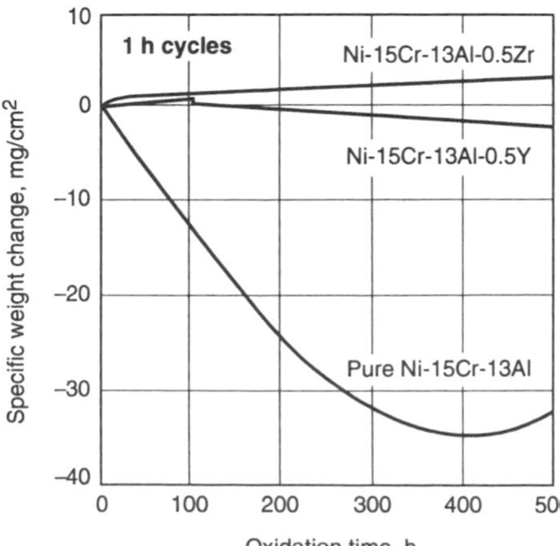

Fig. 5 The effect of a small addition of reactive elements (yttrium or zirconium) on scale adhesion and the 1100 °C (2010 °F) cyclic oxidation resistance of a model Ni-15Cr-13Al (wt%) coating alloy

The rationale for these secondary additions derives from the opposing trends of transient versus selective oxidation. Here, the initial oxidation process has been described as concurrent oxidation of all the major alloying elements present on the surface, even though one oxide is usually greatly favored from a thermodynamic standpoint. This proceeds until the initial rapid kinetics have transitioned to lower rates and the p_{O_2} at the oxide-metal interface has been lowered, both of which favor the formation of a healing layer of the most stable oxide. If silicon or chromium are added to a nickel-aluminum alloy, the contributions of slow growing SiO_2 or Cr_2O_3 nuclei during this transient period allow a complete transition to an Al_2O_3 inner healing layer at lower amounts of aluminum. The reduced aluminum level due to chromium additions can be seen in the Ni-Cr-Al ternary oxidation map of Fig. 2 that indicates which compositions form protective Al_2O_3 scales.

Another important consideration is that of scale adhesion during thermal cycling. The thermal expansion of oxides is much lower than that of most engineering metals such that large compressive stresses are built up in the scale upon cooling. As the scale thickens, these stresses eventually cause some form of scale fracture and spallation. For a strongly adherent scale, spallation often takes the form of microscopic segments in outer portions of the scale, thus relieving stresses without severe degradation. However, a weakly bonded scale can spall at the oxide/metal interface, exposing bare metal upon reheating. This results in more rapid consumption of the critical scale-forming elements, as indicated by Eq 3. Also, depletion zones are more likely to form, and subsequent oxidation products can now include more of the nonprotective transition scales, such as NiO.

The spalling process is of course directly reflected in the weight-change curve (Fig. 3), where each heating cycle is associated with a weight gain, and each cooling cycle produces a superimposed weight loss. For many alloys, the amount of scale spalled has been found to be a function of the (scale thickness)^2 times a spalling constant, Q_o. This process has been modeled in great detail, such that gravimetric curves can be predicted for various k_p growth rates and Q_o (or k_s) spall constants as shown in Fig. 4 (Ref 2–5).

Adherence is dramatically improved in alloys and coatings by the addition of about 0.01 to 0.1 wt% Y. An example of this effect is shown in Fig. 5. An explanation of the detailed mechanism is still evolving, but many critical experiments have indicated that interfacial segregation of sulfur is associated with scale debonding, presumably by weakening the interfacial bonds (Ref 6, 7). Conversely, the absence of sulfur segregation in desulfurized alloys is associated with good adhesion, even without the use of yttrium (Ref 8). When yttrium is added, it is believed to tie up the sulfur, present as a 1 to 10 ppm impurity in the bulk, by forming stable yttrium-sulfur sulfides or complexes. Other sulfur active elements, such as scandium, lanthanum, zirconium, and hafnium, have also shown this effect and are occasionally present in engineering alloys or coatings. The phenomenon is especially important for single-crystal superalloys, which have the potential of being very oxidation resistant, but are not easily manufactured with yttrium.

While the adhesion effect is prominent for Al_2O_3 and Cr_2O_3 scales, it has not been well documented for SiO_2 scales. Cr_2O_3 scales are also subject to fast outward growth when the alloy is not doped with reactive elements. Furthermore, all Cr_2O_3 scales are subject to degrada-

Table 1 Ranking of 34 superalloys by the 1100 °C (2010 °F), 200 h cyclic oxidation attack parameter, K_a, and primary composition makeup

Also shown are model K_a values calculated from Eq 5 (see text), final weight changes, ΔW/A, given in mg/cm^2, standard deviations, Σ, and number of duplicates, *n*.

Alloy	K_a (avg)	K_a (calc)	K_a (Σ)	ΔW/A (avg)	ΔW/A (Σ)	n	Composition, wt% (a)										½Cr + Al
							Cr	Al	Ta	Mo	W	Nb	Ti	Zr	Co	Other	
TRW-R	0.11	0.54	0.01	–0.96	0.16	2	8.0	5.3	6.0	3.0	4.0	0.3	0.8	0.1	8.0	1.0 Hf	9.3
B-1900	0.19	0.31	0.07	–1.40	0.54	7	8.0	6.0	4.3	6.0	0.1	0.1	1.0	0.1	10.0	...	10.0
NASA-VIA	0.33	0.35	0.10	–1.48	0.83	6	6.1	5.4	9.0	2.0	5.8	0.5	1.0	0.1	7.5	0.4 Hf, 0.5 Re	8.5
MAR-M-247	0.51	0.77	0.08	–4.22	1.17	5	8.2	5.5	3.0	0.6	10.0	0	1.0	0.1	10.0	1.50 Hf	9.6
TAZ-8A	0.61	0.52	0.43	0.56	2.93	11	6.0	6.0	8.0	4.0	4.0	2.5	0	1.0	0	...	9.0
IN-713 LC	0.72	0.94	...	–6.20	...	1	12.0	5.9	0	4.5	0	2.0	0.6	0.1	0	...	11.9
B-1900+Hf	0.72	0.33	0.89	–1.68	0.25	3	8.0	6.0	4.3	6.0	0.1	0.1	1.0	0.1	10.0	1.0 Hf	10.0
TRW-1800	0.73	0.87	...	–8.65	...	1	13.0	6.0	0	0	9.0	1.5	0.6	0.1	0.6	...	12.5
MAR-M-246	1.55	0.84	...	–24.44	...	1	11.0	5.0	2.0	0	0	0	1.5	0	11.0	0.1 Cu	10.5
René-125	3.02	2.06	2.46	–21.22	17.42	3	9.0	5.0	3.8	2.0	7.0	0	2.5	0.1	10.0	1.5 Hf	9.5
Astroloy	3.24	9.14	...	–30.25	...	1	15.0	4.4	0	5.3	4.0	0	3.5	0.1	15.0	...	11.9
NX-188	3.45	2.28	0.28	–48.76	13.71	2	0(b)	8.0	0	18.0(b)	0	0	0	0	0	...	8.0
René-120	6.85	8.86	...	–38.57	...	1	9.0	4.3	3.8	2.0	7.0	0	4.0	0.1	10.0	...	8.8
U-700	6.96	5.42	9.20	–55.21	66.80	27	15.0	4.3	0	4.5	0	0	3.5	0.1	18.5	...	11.8
Mar-M-200	8.21	14.35	5.26	–53.59	3.95	3	9.0	5.0	0	0	12.5	2.7	2.0	0.1	10.0	...	9.5
Mar-M-421	9.53	8.64	...	–74.11	...	1	15.8	4.3	0	2.0	3.8	2.0	1.8	0.1	9.5	...	12.2
Waspaloy	9.63	15.18	12.38	–79.29	113.00	4	19.5	1.3	0	4.3	0	0	3.0	0.1	13.5	...	11.1
Mar-M-200+Hf	17.31	16.18	3.94	–89.80	28.36	6	9.0	5.0	0	0	11.5	1.0	2.0	0.1	10.0	1.5 Hf	9.5
WAZ-20	21.15	15.09	9.40	–198.05	60.17	2	0	6.5	0	0	18.5(b)	0	0	0.2	0	...	6.5
IN-792	22.55	19.20	4.69	–166.26	17.64	7	12.7	3.2	3.9	2.0	3.9	0	4.2	0.1	9.0	0.8 Hf	9.6
Mar-M-509	25.43	25.67	4.11	–174.15	52.40	2	23.5	0	3.5	0	7.0	0	0.2	0.5	54.7	...	11.8
IN-718	28.57	29.09	...	–284.60	...	1	18.0	0.4	0	3.1	0	5.0	0.9	0	0	18.5 Fe	9.4
IN-625	28.72	11.28	...	–293.20	...	1	22.5	0.2	1.9	9.0	0	1.8	0.2	0	0	...	11.5
IN-738	29.33	19.60	9.56	–232.45	110.83	10	16.0	3.4	1.8	1.8	2.6	0.9	3.4	0.1	8.5	...	11.4
U-520	31.65	17.26	...	–172.80	...	1	19.0	2.0	0	6.0	1.0	0	3.0	0	12.0	...	11.5
U-720	32.34	19.29	...	–313.50	...	1	18.0	2.5	0	3.0	1.2	0	5.0	0	15.0	...	11.5
IN-939	32.58	30.14	...	–227.60	...	1	22.0	2.0	1.5	0	2.0	1.0	3.6	0.1	19.0	...	13.0
U-710	33.76	20.21	...	–270.20	...	1	18.0	2.5	0	3.0	1.5	0	5.0	0	15.0	...	11.5
X-40	35.57	24.46	...	–206.30	...	1	25.5	0	0	0	7.5	0	0	0	56.0	...	12.8
René-80	37.40	20.00	3.47	–330.35	135.84	2	14.0	3.0	0	4.0	0	0	5.0	0	9.5	...	10.0
R-150-SX	45.01	68.24	...	–596.40	...	1	5.0	5.5	6.0	1.0	5.0	0	0	0	12.0	3.0 Re, 2.2 V	8.0
W-152	45.29	54.96	42.86	–569.30	14.57	2	21.0	0	0	0	11.0	2.0	0	0	65.6	2.0 Fe	10.5
IN-100	46.06	24.31	70.06	–180.33	245.46	3	10.0	5.5	0	3.0	0	0	5.5	0.1	15.0	1.0 V	10.5
Mar-M-211	73.46	11.60	102.19	–269.76	360.82	2	9.0	5.0	0	2.5	5.0	2.7	2.0	0.1	10.0	...	9.5
						Range	**5–25.5**	**0–8**	**0–9**	**0–9**	**0–12**	**0–5**	**0–5.5**	**0–1**	...	...	**6.50–13**

(a) Ni = balance. (b) Rare exceptions out of the typical composition range. Source: Ref 18

tion by formation of a volatile CrO_3 gaseous species and are therefore less suitable for high-velocity applications. Many other phenomena may be encountered in the oxidation of high-temperature alloys, such as the preferential oxidation of carbide phases or grain boundaries. The fundamentals of these phenomena are not widely reported nor easily generalized. However, in practice some alloy-specific process may override otherwise protective behavior, such that performance cannot be projected without consulting test data of the alloy in question.

The preceding is a brief introduction to some general aspects of oxidation that apply to typical high-temperature alloys. They apply equally well to intermetallic compounds, aluminide and MCrAlY coatings, and bond coats for thermal barrier coatings. Excellent treatments of the fundamental aspects can be found in reference books by Kofstad (Ref 9) or Birks and Meier (Ref 10). More detailed information on a wide range of commercial alloys in a variety of environments (carburizing, nitriding, halogens, salts, liquid metals) can be found in Lai (Ref 11). Supplementary information on superalloy oxidation, corrosion, and coatings is present in Ref 1 and 12 to 16 and in the review of high-temperature oxidation of iron-, nickel-, and cobalt-base alloys in Ref 17.

Performance Characteristics of Superalloys

Cyclic Oxidation. The oxidation resistance of commercial superalloys can vary dramatically with alloy type, temperature, and specific environment. Thus, although many of the mechanisms referred to in the previous section are usually applicable, it is difficult to generalize the complex oxidation behavior inherent to these multielement systems. One can obtain a better appreciation of this fact by examining the behavior of a large number of commercial superalloys tested in one study (Ref 18). Here, 34 alloys were tested in cyclic oxidation in air at 1100 °C (2010 °F) for 200 h, often with multiple samples, and characterized by weight change and scale phases. Selected alloys were also tested at 1000 °C (1830 °F) for 500 h and at 1150 °C (2100 °F) for 100 h. The result is a database allowing alloy comparisons on a one-to-one basis as well as an illustration of broad, overall compositional effects.

The weight change curves were described by paralinear or linear kinetics, which had been previously correlated to metal loss by means of an attack parameter, K_a (Ref 4). This parameter defines the combined effect of parabolic-scale growth rates and spalling rates on the consumption of metal. It can therefore be used as an engineering design figure-of-merit related to the residual load-bearing cross section of a component.

The results are summarized in a ranking table (Table 1), which orders the alloys according to K_a, along with the final weight change, number of duplicates, and standard deviations. For the most part, weight change and K_a, track each other, to the extent that the 200 h weight loss at 1100 °C (2010 °F) equals 7.6 K_a, in mg/cm^2, from regression analysis. Exceptions occur when an unusually large spread in the data occur for duplicate samples. These spreads may bias the average performance parameters differently than weight change.

The compositions of the 34 alloys are also listed in Table 1. In most cases, multiple elemental changes occur with progression down the table, thus obscuring any simple compositional ef-

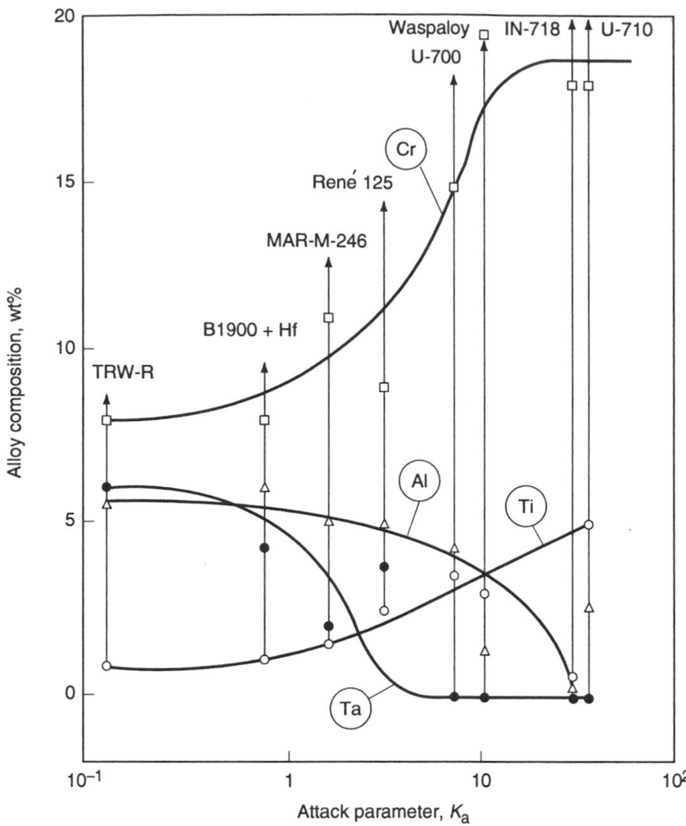

Fig. 6 Compositional trends of eight commercial superalloys with their 1100 °C (2010 °F) cyclic oxidation attack parameter. Source: Ref 18

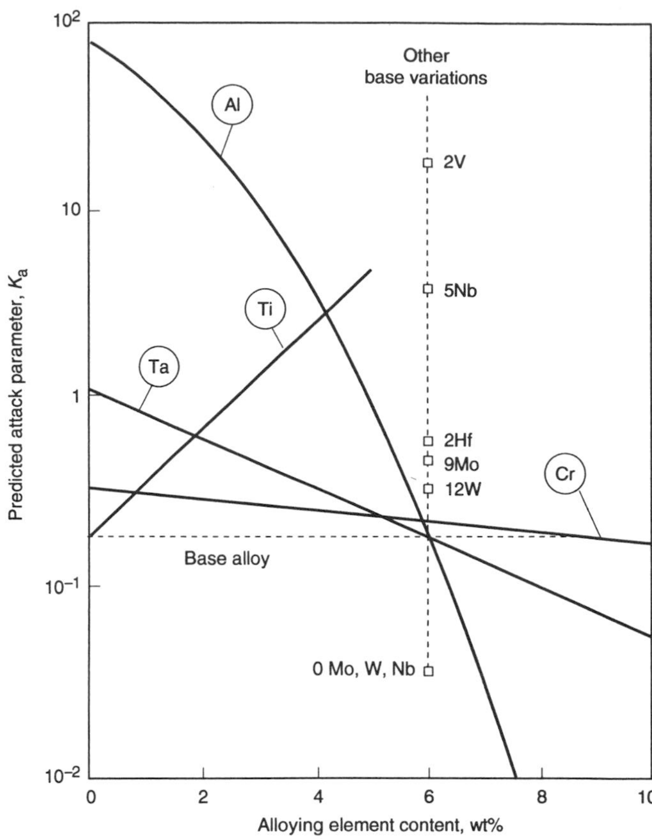

Fig. 7 Modeled (individual element) compositional effects on the 1100 °C (2010 °F) cyclic oxidation attack parameter for a Ni-8Cr-6Al-6Ta-4Mo-4W-1Nb base alloy

fects on performance. However, many broad general trends can be discerned. For example, fairly consistent compositional blocks have been drawn in. These show consistently high aluminum (>5%), high tantalum (>3%), and low titanium (<1%) for the top-performing alloys. Conversely, lower aluminum (<5%), lower tantalum (<2%), and higher titanium (3–5%) occur more commonly for the poorer-performing alloy groupings. The highest levels of chromium (20–25%) did not necessarily ensure good performance, but 5 to 15% Cr was present in all the best-performing alloys. The very high chromium alloys (20%) did not rank very highly because they also typically contained low aluminum (0–2%). In fact, it has been pointed out that chromium and aluminum contents are usually inversely correlated in commercial superalloys (Ref 1), such that:

$$\tfrac{1}{2}\mathrm{Cr} + \mathrm{Al} = 10 \qquad \text{(Eq 4)}$$

The value of this compositional parameter is also listed in Table 1 and can be seen to vary primarily between the narrow range of 8 to 13.

A graphical illustration of the points made above can be seen in Fig. 6. Here the chromium, aluminum, tantalum, and titanium contents of a select group of eight alloys is plotted against the 1100 °C (2010 °F) cyclic oxidation attack parameter. The broad trend of improved performance (low K_a) with higher aluminum, higher tantalum, and lower titanium contents is apparent. It also appears that good performance is associated with a low chromium content. However, because chromium is inversely correlated with aluminum content so strongly, a separable chromium effect cannot be claimed from this figure alone.

These eight alloys in Fig. 6 were selected from the master database in Table 1 because their chromium, aluminum, tantalum, and titanium contents all varied in a more or less monotonic fashion with increasing K_a. There were nonsystematic variations in other alloying elements, such as molybdenum, tungsten, and niobium, which are considered to have only secondary effects. Within reason, some compositional trends can be concluded from this simple two-dimensional plot. However, the complexity and interactions of systems containing ten or more alloying elements often obscures individual elemental effects and compositional dependencies. As a result, multiple linear regression was appropriately used to extract more precise compositional parameters affecting performance (Ref 18). An equation thus results that gives the best overall fit ($R^2 = 0.84$) and accounts for the simultaneous variations of 11 elements and temperature:

$$\log_{10} K_a = x_1(1/T_K) + x_2(\mathrm{Cr}/T_K) + x_3(\mathrm{Cr}) + x_4(\mathrm{CrW}) + x_5(\mathrm{Al}^2) + x_6(\mathrm{AlTa}) + x_7(\mathrm{AlMo}) + x_8(\mathrm{AlV}) + x_9(\mathrm{TaNb}) + x_{10}(\mathrm{TaTi}) + x_{11}(\mathrm{NbTi}) + x_{12}(\mathrm{NbHf}) + x_{13}(\mathrm{Ti}) + x_{14}(\mathrm{Re}) + x_{15} \qquad \text{(Eq 5)}$$

where:

$x_1 = -28{,}734$	$x_9 = 0.05346$
$x_2 = 924.75$	$x_{10} = 0.01932$
$x_3 = -0.71874$	$x_{11} = 0.08140$
$x_4 = 0.00373$	$x_{12} = 0.24155$
$x_5 = -0.05162$	$x_{13} = 0.08345$
$x_6 = -0.03008$	$x_{14} = 0.21293$
$x_7 = 0.01273$	$x_{15} = 22.756$
$x_8 = 0.16396$	

Equation 5 can be used to show the relative importance of various elemental changes to a hypothetical superalloy, where, for illustration purposes, only one element was varied at a time. For example, this technique allows the effects of aluminum and chromium to be decoupled. A base alloy was chosen near the composition of the best-performing group of alloys (i.e., Ni-8Cr-6Al-6Ta-4Mo-4W-1Nb-0Ti-0Hf-0V). The variations covered the compositional ranges typical for most of the superalloys in Table 1. Attack parameters predicted by Eq 5 for 1100 °C (2010 °F) oxidation are shown in Fig. 7. The performance of the base alloy is shown as the horizontal dashed line and its intersection with each of the elemental curves. It can be seen that aluminum additions are the most effective in reducing K_a. Tantalum and now chromium can be seen to be beneficial, in that order. Vanadium and titanium were the most deleterious alloying elements. Detrimental effects were also predicted for separately

Table 2(a) Ranking of iron-base ferritic alloys and austenitic stainless steels in a 10,000 h, 982 °C (1800 °F), interrupted oxidation test

See Tables 2(b), 2(c), and 2(d) for the rankings of cobalt- and nickel-base alloys that were also tested. A total of 68 commercial high-temperature alloys are ranked.

				Composition, wt%								
Alloy	Type	ΔW/A (max)	Rank	Ni	Fe	Cr	Al	Ti	Mo	Mn	Si	Other
RA-310	F	0.50	2	20.0	52.9	25.0	0	0	0	1.5	0.5	...
HOS-875	F	0.83	7	0	71.4	22.5	5.5	0	0	0	0.5	...
NASA-18T	F	1.25	9	0	76.5	18.0	2.0	0.5	0	0.4	1.3	1.3 Ta
TRW Valve	F	1.47	10	0.1	66.8	0	32.0	0	0.1	0.1	0.4	0.5 W
Thermenol	F	1.70	12	0	80.7	0	16.0	0	3.3	0	0	...
18SR	F	3.47	18	0.3	77.9	18.0	2.0	0.4	0	0.4	1	...
IN-800	A	−9.67	26	32.0	45.3	20.5	0.4	0.4	0	0.8	0.4	...
RA-330	A	−9.72	27	35.0	43.2	19.0	0	0	0	1.5	1.3	...
RA-26-1	F	−18.66	36	0.2	71.6	26.0	0	0.5	1.0	0.3	0.3	...
316SS	A	−51.31	41	12.0	65.4	17.0	0	0	2.5	2	1	...
T439SS	F	59.98	42	0.5	79.3	18.3	0.2	1.0	0	0.4	0.4	...
334SS	A	63.04	43	20.0	58.4	19.0	0.3	0.4	0	1	0.3	...
310SS	A	−92.36	46	20.5	50.8	25.0	0	0	0	2	1.5	...
321SS	A	112.18	51	10.5	69.0	17.0	0	0.4	0	2	1	...
430SS	F	−164.65	53	0	81.9	17.0	0	0	0	0	1	...
RA-309	A	−176.65	54	14.0	60.6	23.0	0	0	0	1.5	0.8	...
347SS	A	178.17	55	11.0	67.1	18.0	0	0	0	2	1	0.4 Nb, 0.4 Ta
409SS	F	184.59	57	0	88.5	11.0	0	0.5	0	0	0	...
309SS	A	−187.09	58	13.5	60.3	23.0	0	0	0	2	1	...
410SS	F	208.56	59	0	85.4	12.5	0	0	0	1	1	...
304SS	A	−227.65	60	10.0	67.9	19.0	0	0	0	2	1	...
Croloy 9	F	308.47	62	0	88.5	9.0	0	0	1.4	0.5	0.5	...
Croloy 5	F	310.63	63	0	93.7	5.0	0	0	0.6	0	0.5	...
Croloy 7	F	315.07	64	0	91.1	7.0	0	0	0.6	0.5	0.8	...
Multimet	A	−320.82	65	20.0	30.2	21.0	0	0	3.0	1.5	0.5	20 Co, 2.5 W, 0.5 Nb, 0.5 Ta

SS, stainless steel; A, austenitic; F, ferritic. Source: Ref 19

Table 2(b) Ranking of cobalt-base alloys in a 10,000 h, 982 °C (1800 °F), interrupted oxidation test

See Tables 2(a), 2(c), and 2(d) for the rankings of iron- and nickel-base alloys that were also tested. A total of 68 commercial high-temperature alloys are ranked.

			Composition, wt%									
Alloy	ΔW/A (max)	Rank	Ni	Co	Fe	Cr	Al	Ti	W	Nb	Ta	Other
X-40	−0.79	5	10.5	56.0	0	25.5	0	0	7.5	0	0	...
Belgian P-3	1.10	8	11.6	32.9	29.5	25.0	0	0	0	0	0	...
MAR-M-509	1.53	11	10.0	54.7	0	23.5	0	0.2	7.0	0	3.5	0.5 Zr
H-150	−4.41	20	3.0	47.6	20.0	28.0	0	0	0	0	0	...
HA-188	−4.87	21	22.0	39.4	1.5	22.0	0	0	14.0	0	0	...
Belgian S-57	−7.99	24	10.0	56.5	0	25.0	3.0	0	0	0	5.0	0.5 Y
L-605	−101.81	50	10.0	52.9	0	20.0	0	0	15.0	0	0	...
WI-52	−413.72	67	0	65.6	0	21.0	0	0	11.0	2.0	0	...

Source: Ref 19

adding the upper limit of the compositional range for niobium, hafnium, tungsten, or molybdenum. Alternatively, improved performance is predicted for removing the base levels of 4Mo, 4W, and 1Nb. No significant trend with cobalt content was found. It should be noted that although some of the modeled behavior exhibits $K_a < 0.1$, these low values were not experimentally observed in this database. The intent of this plot is to address broad trends rather than pinpoint the precise performance of alloys not included in the database.

A simplified equation was also obtained from a regression model that did not use any cross-term interactions. The resulting equation ($R^2 = 0.80$) thus extracts coefficients that apply only to individual elements:

$$\log_{10} K_a = \Sigma x_i(y_i) + x_{12} \quad \text{(Eq 6)}$$

where:

y_i =	x_i =
$1/T_K$	−17,305
Chromium	−0.08308
Aluminum	−0.33925
Tantalum	−0.15488
Titanium	0.26408
Molybdenum	0.04527
Niobium	0.24172
Hafnium	0.17782
Zirconium	0.37654
Rhenium	0.87295
Carbon	2.00000
x_{12} =	14.77172

Again, the relative benefits of aluminum, tantalum, and chromium and the detriment of titanium are evident. Negative effects of niobium, hafnium, zirconium, rhenium, and carbon are also indicated. It should be noted that of these only niobium and rhenium are typically added in excess of 1 wt%. Low hafnium additions (0.1 wt%) are generally considered beneficial to scale adhesion. However, "overdoping" with up to 2 wt% Hf to prevent grain-boundary cracking in columnar-grained superalloys can be detrimental to overall oxidation performance.

Finally, the 1100 and 1150 °C (2010 and 2100 °F) data, along with that from the few alloys tested at 1000 °C (1830 °F), was also incorporated by the model to obtain temperature effects. Thus, for the base alloy Eq 5 produced an Arrhenius relationship for the attack parameter, namely:

$$K_a = 6.5 \times 10^{14} \exp\{(-97{,}570 \text{ cal/mole})/RT\} \quad \text{(Eq 7)}$$

The apparent activation energy, 98 kcal/mole, is quite in line with those generally observed for diffusion in, and growth of nickel-, chromium-, and aluminum-base oxide scales (i.e., 50–125 kcal/mole).

Interrupted Long-Term Oxidation. Another large database was developed for a wider representation of high-temperature alloys (Ref 19). This study tested 68 commercial alloys, including iron-base ferritic and austenitic stainless steels, at 982 °C (1800 °F) for 10,000 h in air, cycled 10 times for 1000 h each. The results are summarized in Tables 2(a) through 2(d). These tables list the performance (final weight change) and major chemistry by 4 categories:

- Iron-base alloys (Table 2a)
- Cobalt-base alloys (Table 2b)

- Nickel-base sheet alloys (Table 2c)
- Nickel-base superalloys (Table 2d)

The best-performing group of alloys was the ferritic alloys with aluminum. These alloys typically contained from 2 to 32% Al and 0 to 18% Cr. They gained between just 1 to 4 mg/cm^2 and formed primarily protective Al_2O_3 scales, as expected from the composition (FeCrAl, FeAl, or Fe_3Al). Conversely, the ferritic alloys without aluminum (Croloys and ferritic stainless steels) oxidized the most as a group, gaining between 100 to 300 mg/cm^2 for all but one sample that lost 165 mg/cm^2. The major scales phases were Fe_2O_3 and occasionally Cr_2O_3 or $(Ni,Fe)Cr_2O_4$ spinel. The austenitic stainless steels performed midway between these groups, with weight changes of 19 to 228 mg/cm^2. One exception was alloy RA-310, which ranked among the better alloys with a gain of only 0.4 mg/cm^2. The major scale phases among this group were $(Ni,Fe)Cr_2O_4$ spinel and Cr_2O_3 or Fe_2O_3.

The cobalt-base alloys exhibited weight changes between a gain of 1 mg/cm^2 and a loss of 10 mg/cm^2 with two exceptions: L-605 samples lost 70 to 100 mg/cm^2 and WI-52 lost 414 mg/cm^2. All alloys exhibited MCr_2O_4 spinel and Cr_2O_3 as the primary scale phases, with no correlation with performance.

The nickel-base sheet alloys with aluminum were often among the best-performing alloys, with maximum weight changes less than 3 mg/cm^2.

The nickel-base alloys with iron exhibited a wide span of performance, with gains as low as 3 mg/cm^2 for IN-601 or losses ranging from 7 mg/cm^2 for Hastelloy X to 510 mg/cm^2 for Hastelloy N. The primary scale phases were usually $(Ni,Fe)Cr_2O_4$ and Cr_2O_3, except for the $NiMoO_4$ and MoO_4 observed on Hastelloy N.

The nickel-base alloys without iron also exhibited widely varying behavior, with weight gains ranging from 0.5 mg/cm^2 for U-700 to 178 mg/cm^2 for WAZ-20, and with weight losses ranging from 1.7 mg/cm^2 for Hastelloy S to −300 mg/cm^2 for IN-100. The scale phases were $Ni(Al,Cr)_2O_4$, NiO, Cr_2O_3, Al_2O_3, or $Cr(Ta,Mo,W)O_4$//$Ni(Ta,Mo,W)_2O_6$ tri-rutile phases, without an overall correlation with performance. However Al_2O_3-forming alloys did the best within this group.

The general picture obtained from this work is that Al_2O_3 formers and some Cr_2O_3 formers can be very oxidation resistant at 982 °C (1800 °F) for 10,000 h. The iron-base materials (ferritic and austenitic stainless steels) containing no aluminum were the least-protective classes of alloys, with a wide range of intermediate to excellent behavior exhibited by the nickel- and cobalt-base alloys.

Similar ranking behavior was observed for the same test conducted at 815 °C (1500 °F) for the majority of alloys (Ref 20). However, now a considerable number of chromia-formers performed almost as well as the alumina-formers. Especially notable improvements, on a relative basis, were observed for Hastelloy C-276, IN-600, IN-706, and type 304 stainless steel when tested at 815 °C (1500 °F) as compared to 982 °C (1800 °F).

Table 2(c) Ranking of nickel-base sheet alloys in a 10,000 h, 982 °C (1800 °F), interrupted oxidation test

See Tables 2(a), 2(b), and 2(d) for the rankings of iron- and cobalt-base alloys and nickel-base superalloys. A total of 68 commercial high-temperature alloys are ranked.

Alloy	$\Delta W/A$ (max)	Rank	Composition, wt %				
			Ni	Fe	Cr	Al	Other
DH-245, 0.74	4	74.1	0.7	20.0	3.5	...	
IN-702	0.80	6	80.5	0.5	15.4	3.1	0.4Ti
IN-601	2.71	16	60.3	14.1	23.0	1.4	...
Chromel A	−5.10	22	78.0	0.4	20.0	0	...
Chromel C	−11.65	29	59.3	24.0	15.0	0	...
DH-241	−12.07	30	76.3	0.3	20.0	0	...
Tophet 30	−12.21	31	68.6	0	30.0	0	1.4 Si
Chromel AA	−14.70	34	67.8	10.0	20.0	0	...
DH-242	−17.78	35	77.8	1.0	20.0	0	
Ni-40Cr	−21.90	38	60.0	0	40.0	0	...
Chromel P	75.70	44	89.4	0.2	10.0	0	...
Pure Ni-270	114.0	52	100.0	0	0	0	...

Source: Ref 19

Table 2(d) Ranking of nickel-base superalloys in a 10,000 h, 982 °C (1800 °F), interrupted oxidation test

See Tables 2(a), 2(b), and 2(c) for the rankings of iron- and cobalt-base alloys and nickel-base sheet alloys. A total of 68 commercial high-temperature alloys are ranked.

Alloy	$\Delta W/A$ (max)	Rank	Composition, wt %									
			Ni	Co	Fe	Cr	Al	Ti	Mo	W	Nb	Ta	Other
NASA-VIA	−0.08	1	62	7.5	0	6.1	5.4	1	2	5.8	0.5	9	0.4Hf, 0.5Re
U-700	0.63	3	54	18.5	0	15	4.3	3.5	4.5	0	0	0	...
Hastelloy S	−1.72	13	67.4	0	1	15.5	0.2	0	15.5	0	0	0	...
B-1900	−1.83	14	64.3	10	0	8	6	1	6	0.1	0.1	4.3	...
René-120	2.05	15	59.8	10	0	9	4.3	4	2	7	0	3.8	...
IN-617	−3.23	17	55.4	12.5	0	22	1	0	9	0	0	0	...
TAZ-8A	−4.11	19	68.4	0	0	6	6	0	4	4	2.5	8	1.0Zr
Hastelloy X	−7.00	23	46.3	1.5	18.5	22	0	0	9	0.6	0	0	...
IN-713LC	−8.67	25	74.8	0	0	12	5.9	0.6	4.5	0	2	0	...
IN-671	−11.02	28	51.6	0	0	48	0	0.4	0	0	0	0	...
RA-333	−12.35	32	45.2	3	18	25	0	0	3	3	0	0	...
MAR-M-200	−14.23	33	58.6	10	0	9	5	2	0	12.5	2.7	0	...
IN-706	−19.74	37	77.5	1	0	16	0.4	1.8	0	0	2.9	0	...
IN-600	−29.33	39	76.5	0	7.2	15.8	0	0	0	0	0	0	...
Hastelloy C-276	−34.36	40	55.4	2.5	5.5	15.5	0	0	16	3.8	0	0	0.35V
IN-750X	−80.82	45	73.6	0	6.8	15	0	2.5	0	0	0.9	0	...
IN-804	−93.81	47	42.6	0	25.4	29.5	0.3	0.4	0	0	0	0	...
Hastelloy G	−96.20	48	45.8	0	19.5	22	0	0	6.5	0.5	1	1	...
IN-738X	−96.51	49	61.4	8.5	0	16	3.4	3.4	1.8	2.6	0.9	1.8	...
WAZ-20	178.29	56	74.7	0	0	0	6.5	0	0	18.5	0	0	...
René-80	−239.97	61	64.3	9.5	0	14	3	5	4	0	0	0	...
IN-100	362.88	66	59.8	15	0	10	5.5	5.5	3	0	0	0	1.0V
Hastelloy N	−511.92	68	69.9	0.2	5	7	0	0	16.5	0	0	0	...

Source: Ref 19

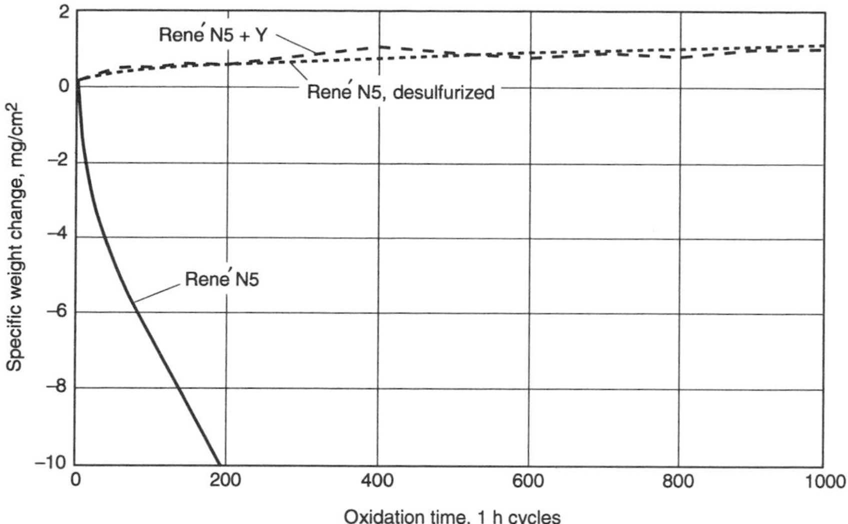

Fig. 8 The effect of desulfurization (1280 °C, or 2335 °F, 100 h hydrogen anneal) or yttrium additions (100–150 ppm) on the 1150 °C (2100 °F) cyclic oxidation behavior of René N5. Source: Ref 8

Performance Characteristics of Single-Crystal Superalloys

A great deal of current interest in the field of oxidation centers around advanced nickel-base single-crystal alloys. These alloys are used in the most demanding high-temperature applications of gas turbines. Some of the most common of these alloys are PWA 1480, 1484, and 1487 (Pratt and Whitney), René N4, N5, and N6 (General Electric), CMSX-4, -6, and -10 (Cannon-Muskegon), and SRR 99 and RR 2000 (Rolls-Royce). Most of these alloys fall within the following compositional range:

Element	wt%
Chromium	5–10
Aluminum	5–6
Tantalum	2–8
Molybdenum	1–3
Tungsten	4–6
Niobium	0–1
Rhenium	0–6
Titanium	0–5
Hafnium	0–0.3
Cobalt	5–10

These levels make possible, first and foremost, a continuous Al_2O_3 film, upon which excellent oxidation resistance depends. Many of the other elements also participate in scale formation as $Ni(Al,Cr)_2O_4$, $(Al,Cr)TaO_4$, and $NiTa_2O_6$ oxides. These phases, when located as an outer layer, may take part in a relatively innocuous transient growth process. Other nonprotective scales—(NiO, $NiTiO_3$, Cr_2O_3, Ta_2O_5, or $(Mo,W)O_3$—can form after an interval of cyclic oxidation if scale adhesion is lacking.

While this alloy class was not included in the 34 alloy database described earlier (see Table 1), Eq 5 can be used to predict attack parameters. For example, the 1100 °C (2010 °F) K_a for René N5 was calculated to be 0.22, which is comparable to the low value of 0.19 obtained for the base alloy in Fig. 7. Furthermore, the oxidation resistance of these alloys can be dramatically improved by maximizing Al_2O_3-scale adhesion. This is accomplished by the addition of reactive elements or by desulfurization by melt purification and/or hydrogen annealing (Ref 21–25). Examples of these effects can be seen in Fig. 8 to 10 for René N5, CMSX-4, and PWA 1480, respectively. Yttrium is difficult to homogeneously incorporate at high levels, but can be effective at just 15 ppm when the sulfur content is low (<2 ppm) (Ref 24). Alternatively, good adhesion is obtained without yttrium if the sulfur content is extremely low (<0.5 ppm).

Although some of the best oxidation resistance has been documented for single-crystal alloys, it is not due to the unique monocrystalline structure. Rather, the primary oxidation response stems from the high aluminum and tantalum contents, dopant levels of yttrium and hafnium, low sulfur, and the remainder of the compositional makeup. The absence of grain-boundary carbide strengtheners may also contribute to a more continuous Al_2O_3 scale.

The previous discussion has outlined some of the oxidation issues applying to bare (uncoated) alloys. High-temperature test data have been used for the purpose of discriminating compositional effects. It should be noted that many of the alloys will exhibit acceptable performance for intermediate- to low-temperature applications. But, for the most demanding applications, oxidation-resistant coatings combined with the best alloys become the most practical defense against aggressive environments.

Coating Concepts for Superalloys

Aluminides. The most widely used coating in gas turbines is the intermetallic compound NiAl. The monolithic material has been extensively studied and shown to form protective Al_2O_3 scales, even to the lower aluminum limits of the single-phase stoichiometry. However, in a cyclic mode, the hypostoichiometric compositions have been shown to exhibit degradation because the aluminum depletion from scale spallation can result in depletion zones with reduced oxidation resistance (Ref 26, 27). The use of reactive elements such as zirconium or yttrium have been shown to reduce interfacial spalling and produce some of the highest oxidation resistance ever measured for nickel-base material (Ref 28–30).

The coatings are generally produced by chemical vapor deposition (CVD) processes (aluminum halide gas or aluminum powder pack) by diffusional reaction with the superalloy substrates. Depending on the temperature and the aluminum activity of the pack, the coatings may be formed by outward nickel or inward aluminum diffusion (Ref 31). Inward diffusion coatings may require a higher temperature anneal to convert a very brittle Ni_2Al_3 phase into NiAl. Inward coatings also have a higher level of substrate elements present. At or below 1000 °C (1830 °F), very long oxidative lifetimes can be expected due to the low growth rate of Al_2O_3 scales and limited spalling. As temperature exceeds 1100 °C (1830 °F), interdiffusion with the substance becomes a major mode of aluminum loss, which then triggers initial coating degradation (Ref 27, 32). Once this dilution takes place, less-protective scales may form with excessive spalling and eventual failure. Interdiffusion is also responsible for potentially embrittling carbide or σ-Widmanstätten phases in the coating/substrate interdiffusion zone.

Dopant elements have been found to beneficially affect aluminide coating performance. Hafnium, zirconium and yttrium increase alumina scale adhesion. Silicon modifications can serve in this regard, but they are especially effective in stabilizing Al_2O_3-scale formation at very low aluminum concentrations. The most widespread aluminide modification is platinum. For example, a 20 μm layer of platinum is first electroplated onto the superalloy and then aluminized to form a $PtAl_2$-(Ni,Pt)Al two-phase mixture or layered structure. Platinum serves many purposes: it stabilizes an aluminum-rich surface phase for continued Al_2O_3 formation; it improves Al_2O_3-scale adhesion; and it improves the hot corrosion resistance of NiAl (Ref 33).

An example of a platinum-aluminide coating on a single-crystal superalloy, exposed in a high by-pass turbofan engine, is shown in Fig. 11. The outer layer is depleted NiAl, with particles of Ni_3Al forming at grain boundaries. The inner layer is an interdiffusion zone, enriched in alloying elements due to nickel dilution in forming the outer layer. Finally, a modified substrate structure is formed beneath this, with topologically closed-packed chromium-rich phases (σ).

Ni(Co)CrAlY. M-(10-30)Cr-(5-20)Al-(0.1-1)Y coatings are also widely used to solve the oxidation/corrosion problems of high-temperature turbine materials. They are generally applied by physical vapor deposition (PVD) or low-pressure (vacuum) plasma spraying (LPPS or VPS).

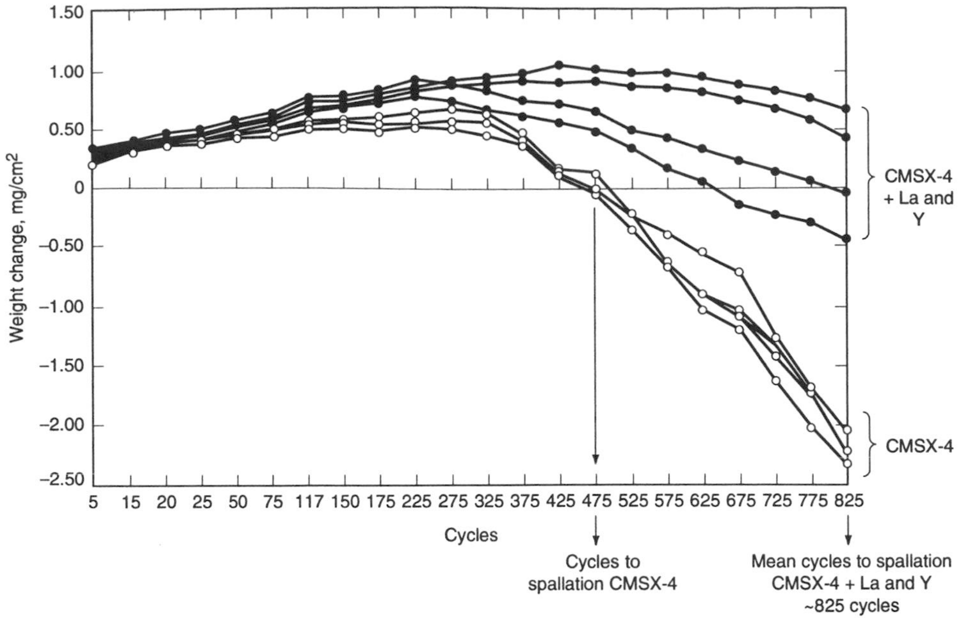

Fig. 9 The effect of 15 to 20 ppm of lanthanum and yttrium on the 1177 °C (2150 °F) cyclic oxidation behavior of low sulfur (<2 ppm) CMSX-4. Source: Ref 24

These coatings are also based on excellent Al_2O_3-forming capability (see Fig. 2) but at lower aluminum levels and with an increase in low-temperature ductility. The presence of chromium gives an added defense against sodium sulfate hot corrosion, for which Al_2O_3 formers are not as resistant as Cr_2O_3-formers. Because of the "overlay" coating process and the compositional similarity to the superalloy substrates, these coatings are much less susceptible to diffusional wearout than the aluminide conversion coatings. However, there is a reduced aluminum reservoir to sustain Al_2O_3-scale growth or reformation after spallation.

Thermal barrier coatings (TBCs) enable a large improvement in turbine durability or performance by reducing the operating temperature of the superalloy surface or by allowing hotter gas temperatures. The insulating layer is generally about 125 to 250 μm of 6 to 8% Y_2O_3-ZrO_2, generally referred to as yttria partially stabilized zirconia (YSZ). It is usually applied by plasma spraying (PS) or electron beam (EB)-PVD. A preliminary bond coat is necessary for optimal adhesion. For PS-YSZ, an LPPS NiCoCrAlY type of bond coat is conventionally used, whereas for EB-PVD coatings, a platinum aluminide diffusion coating is used.

The failure mechanism of the TBC in clean environments is a combination of bond-coat oxidation and thermal fatigue (Ref 34). As the Al_2O_3 scale thickens at the YSZ-bond coat interface, it increases the stresses on the adjacent zirconia. During cooldown, the higher-expansion substrate puts large compressive thermal stresses on the zirconia, encouraging it to buckle and pull the thermally grown Al_2O_3 scale off the bond coat. Continued cycling results in crack growth until a critical length is achieved and macrospalling of the ceramic coating occurs (failure).

The surface roughness of the plasma-sprayed system enhances the bond strength by the increased contact area and mechanical interlocking. However, this roughness also produces tensile stress concentrations at the tips of the bond coat asperities and thus initiates interfacial cracking. The porosity and weak, microcracked splat boundaries in the PS-YSZ accommodates much of this stress. In contrast, the EB-PVD coatings with aluminide bond coats have a much smoother interface and less stress concentration, albeit with less mechanical bonding. The thermal stress is accommodated by the columnar structure of the EB-PVD zirconia, which demonstrates a propensity to crack along column boundaries, rather than buckle and debond at the oxide/metal interface. An example of an EB-PVD coating applied to a platinum-aluminide-coated René N5 alloy is shown in Fig. 12. Both systems are expected to benefit by improved Al_2O_3-scale adhesion. This can be accomplished by low-sulfur alloys, low-sulfur coatings, and by reactive element dopants to both.

For more corrosive environments, such as shipboard marine-power turbines, low-melting $NaVO_3$ deposits may destabilize YSZ by leaching out the yttrium to form YVO_4. Destabilization allows a martensitic transformation to monoclinic zirconia accompanied by a large volume expansion and powdering of the coating. Scandia (+india)-stabilized zirconia has been found to be resistant to vanadate corrosion and may provide a benefit in the presence of sea salt and vanadium impurities in the fuel (Ref 35).

The state-of-the-art of processing, stability, service experience, thermal and stress modeling and design, and advanced concepts have been summarized in a recent NASA workshop on TBCs (Ref 36). In addition to aeroturbine components, the applications of this concept to diesel engine and land-based turbines has been described.

Chromizing. In the latter stages of the turbine (low-pressure turbine), the temperatures are reduced to below 900 °C (1650 °F). Thus, high-temperature oxidation is much less of a concern, but the conditions are optimal for Na_2SO_4 deposition and high-temperature corrosion (see the section on "Hot Corrosion" in this article). Chromizing is the CVD or pack deposition of elemental chromium, followed by a diffusion anneal to a nickel-chromium outer layer. This enrichment of chromium provides protection

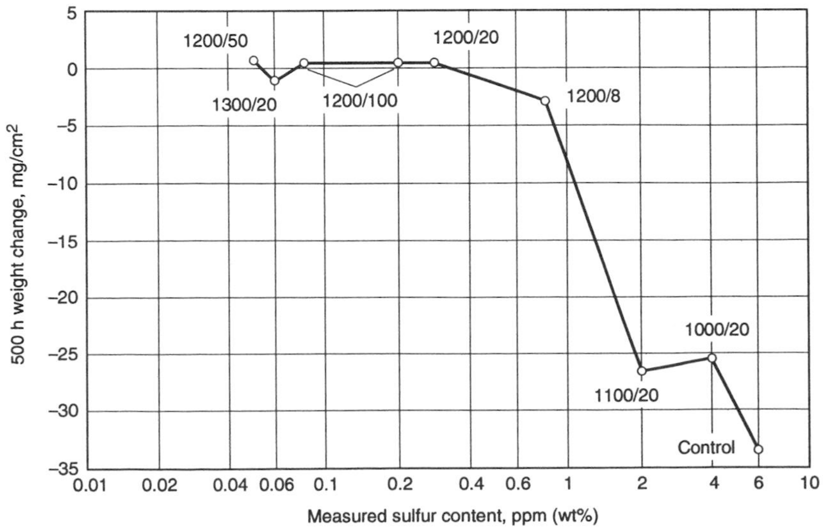

Fig. 10 The effect of sulfur content on the 500 h (cycles), 1100 °C (2010 °F) cyclic oxidation weight change of hydrogen-annealed PWA 1480. The values given inside the figure represent the annealing temperature/time in °C/h. Source: Ref 25

against the scale-fluxing attack that occurs under a molten sulfate deposit.

Performance Characteristics of Other High-Temperature Materials

Nickel and Titanium Aluminides. The oxidation resistance of NiAl has been actively studied since 1970. The fundamental mechanisms were described by Pettit, and regimes of protective behavior were mapped over the equilibrium diagram (Ref 37). Al_2O_3 was found to be stable over the entire temperature range for aluminum concentrations above about 30 at.%. Cyclic oxidation showed rapid spalling and depletion at 1100 °C (2010 °F) for compositions below about 40 at.% (Ref 27). Zirconium and other reactive element additions produced large improvements in scale adhesion, allowing protective behavior at 1200 °C (2190 °F) for 3000 h as shown in Fig. 13 (Ref 28, 29). Reactive element oxide dispersoids were also effective in this regard (Ref 30). Long-term durability is directly keyed to the large aluminum reservoirs in Ni-50Al-base alloys and the ability to supply aluminum to the surface with minimal depletion-zone formation until the bulk composition has dropped to 40 at.% (Ref 26). Impressive gains in low-temperature ductility and high-temperature strength were made for microalloyed single-crystal NiAl alloys (Ref 38, 39). However, widespread use as a free-standing structural material has not yet been realized.

Ni_3Al has been ductilized (by 0.1% B additions) to a much greater extent and has moderate oxidation resistance below 1000 °C (1830 °F) (Ref 40). However, an intermediate temperature embrittlement phenomenon is observed. Similarly, Fe_3Al and FeAl alloys have been engineered with moderate ductility (1–5%) and moderate oxidation resistance (Ref 41, 42). Some of the iron aluminide alloys have exhibited exceptional sulfidation resistance in recuperator environments from coal gasification plants and are also a low-cost alternative to nickel-base candidates.

Various titanium aluminides are being actively researched and considered for the latter stages of the compressor or turbine sections and have also found application in automotive valves and turbochargers. Ti-24Al-11Nb (at.%), Ti-22Al-24Nb (at.%), and Ti-48Al-2Cr-2Nb (at.%) (called the α_2-, orthorhombic-, and γ-phase alloys, respectively) have very attractive specific strength values because of their very low densities. The oxidation behavior has proved to be very complex, in that layered, intermixed $TiO_2 + Al_2O_3$ scales are the rule, with possible beneficial or detrimental effects of air versus oxygen environments (due to titanium nitride) (Ref 43–45). In general, the oxidation rate becomes excessive above 900 °C (1650 °F) for the α_2 and orthorhombic alloys. Furthermore, at 700 °C (1290 °F), surface hardening (embrittlement) of these two alloys has been observed due to oxygen interstitial diffusion (Ref 45). An effective coating would be very desirable; however, most attempts at aluminide, silicide, or MCrAlY coatings have only resulted in lowering fatigue lives.

The oxidation rate of γ-phase TiAl-base alloys is much more acceptable, and the embrittlement phenomena are much less prominent, requiring exposure above 800 °C (1470 °F) to achieve measurable hardening (Ref 45). However, TiAl alloys are inherently brittle, with maximum tensile ductilities on the order of just 1 to 2%. A predominance of Al_2O_3 in the scale on γ-phase alloys results in moderate oxidation resistance at 800 °C (1470 °F) as shown in Fig. 14. Considerable improvements are indicated by some experimental coatings that are based on the TiAl + Cr alloys (Fig. 15).

The oxidation behavior of nickel, iron, and titanium aluminides, as well as a number of other developmental intermetallic compounds, have been extensively reviewed. Recently published reviews can be found in Ref 46 to 49.

Refractory metals such as niobium, molybdenum, tantalum, tungsten, and rhenium are natural candidates for high-temperature applications because of their extremely high melting points (2250 to 3400 °C, or 4080 to 6150 °F). However, they exhibit some of the highest oxidation rates of structural metals to the extent that they are rarely used in oxygen-containing environments without coatings. Even with coatings, the potential for catastrophic failure generally limits their applications to less-critical components or short missions (e.g., rocket nozzles or exhaust nozzles of military turbines).

Coatings for refractory metals and their alloys are typically based on disilicides (MSi_2) that form protective SiO_2 scales. Some additives enable lower-melting silicates to flow and seal through-cracks that inevitably form because of thermal expansion mismatch stresses with the substrate. The coatings are generally applied by slurry spraying followed by vacuum sintering, where other additives are used to lower the processing fusion temperature of the coating. Typical components include molybdenum, niobium, titanium, iron, cobalt, chromium, and silicon (Ref 50).

While maximum lifetimes of hundreds of hours at 1500 °C (2730 °F) may be expected in an isothermal exposure for, say, a defect-free $MoSi_2$ coating, repeated cycling will trigger degradation by cracking. For example, bars of FS85 (Nb-28Ta-10W-1Zr) were slurry fusion coated with a commercial R512E coating (20Cr-20Fe-60Si) and tested in a 30 min cycle Mach 0.3 burner rig (Ref 51). This exposure produced coating breaches (defined by –5 mg/cm^2 weight

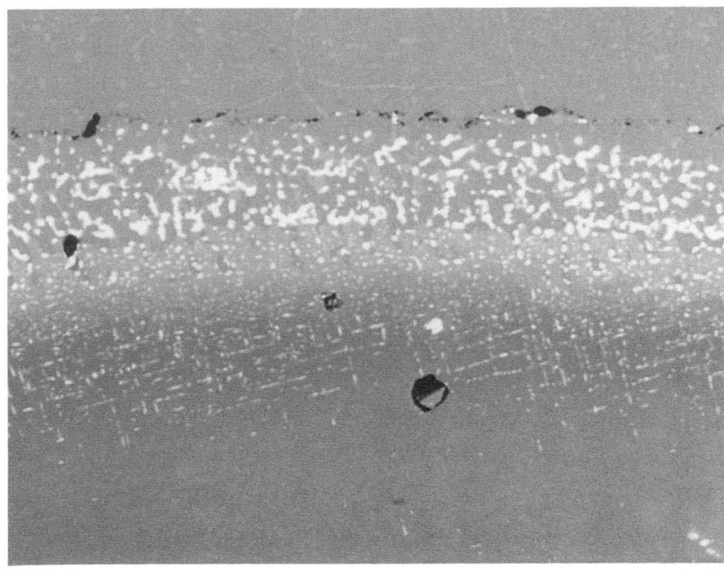

Fig. 11 Microstructure of a platinum-modified pack aluminide coating on a single-crystal superalloy after engine exposure. 600×

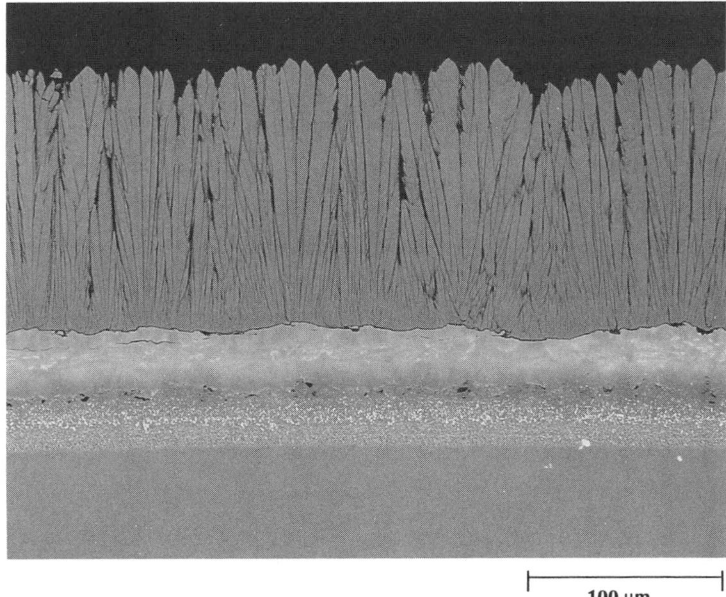

Fig. 12 Columnar structure of an EB-PVD 7% yttria-zirconia thermal barrier coating on a platinum-aluminide-coated single-crystal superalloy. 250×

losses) at 170, >200, 120, and 110 cycles for 982, 1093, 1260, and 1371 °C (1800, 2000, 2300, and 2500 °F) exposures, respectively. Except for the 1093 °C (2000 °F) exposure, all samples showed catastrophic oxidation pits or massive substrate attack at the specimen edges. Oxidation products were friable mixtures of Nb_2O_5, $FeNbO_4$, SiO_2, and amorphous silicates.

$MoSi_2$ has also been developed as a free-standing structural material. Its most common application is as ultrahigh temperature heating elements (SuperKanthal) for furnaces (up to 1700 °C, or 3090 °F). Its oxidation resistance is derived from extremely slow-growing SiO_2 scales (Fig. 1). However, at low temperatures (500 to 700 °C, or 930 to 1290 °F), $MoSi_2$ exhibits a classic "pest" catastrophic oxidation mechanism, where a sample literally turns to dust in a short amount of time. This mechanism has been shown by TEM techniques to be caused by stresses from a voluminous oxidation product of an inward-growing SiO_2 and MoO_x microfilament mixture (Ref 52, 53). Rapid inward diffusion of oxygen is allowed by the MoO_x microfilaments. In a defect-free and stress-free material, pest can be avoided or at least delayed. In practice, pest is prevented by compositing with silica-base glass, boron compounds, or silicon nitride (Ref 54). At high temperature, MoO_3 is very volatile and leaves the scale by vaporization, with little effect on the silica growth rate. More detailed information on high-temperature silicides can be found in Ref 55.

Ceramics also have very high melting (or decomposition) temperatures and are candidate materials for high-temperature applications. The primary corrosion and oxidation mechanisms and considerations have been reviewed in a number of publications by Jacobson (Ref 56, 57). From the standpoint of oxide ceramics, the conventional oxidation reaction is precluded. In actual combustion environments, however, there may be reactions with combustion products leading to volatile species and surface recession. For example, the reaction of water vapor with SiO_2 and some silicates leads to gaseous $Si(OH)_4$, with loss rates that could measurably decrease component thickness in high-temperature, high-velocity, high-pressure environments. Conversely, the reactivity of Al_2O_3 in a combustion environment is very limited, without any serious conversions to gaseous products, and zirconia reactions are virtually nonexistent.

The nonoxide ceramics, silicon carbide (SiC) and silicon nitride (Si_3N_4), are very refractory, oxidation-resistant materials that have been widely studied for high-temperature applications. In pure oxygen, these materials, like $MoSi_2$, are among the most oxidation-resistant (nonoxide) materials known. The low scale-growth rate is due to slow-growing SiO_2 scales, which may be amorphous below about 1300 °C (2370 °F) or for short oxidation times. At extremely high temperatures (1700 °C, or 3090 °F) the evolution of CO or N_2 reaction products may produce large bubbles that could disrupt the protective scales (Ref 58). In addition, at low p_{O_2} and high temperatures, a passive-to-active oxidation transition occurs in which the reaction product is now a nonprotective SiO gas. These silica-formers are also subject to the volatility issues in water vapor discussed above.

Other corrosive reactions with ceramics are not widely encountered. However, if alkali-containing deposits appear on silicon-base systems, a particularly damaging reaction may occur (Ref 56). For example, the catalyst for sea salt hot corrosion in turbine alloys, Na_2SO_4, has been shown to react with SiO_2 scales formed on SiC or Si_3N_4. A molten sodium-silicate product rapidly results, with bubble and pit formation in the sub-

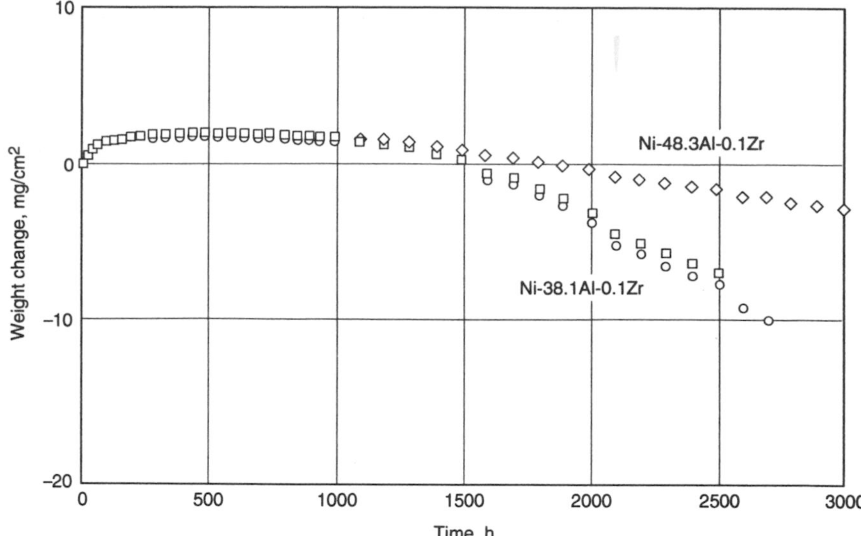

Fig. 13 Long-term protective 1200 °C (2190 °F) cyclic oxidation behavior of NiAl + 0.1 at.% Zr. Source: Ref 28

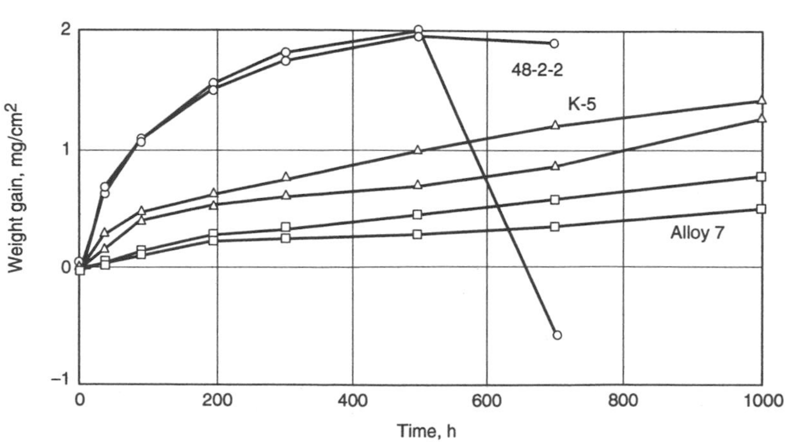

Fig. 14 Interrupted oxidation of structural γ-TiAl alloys at 800 °C (1470 °F) in air. Alloys tested include: Ti-48Al-2Cr-2Nb (48-2-2); Ti-46-5Al-3Nb-2Cr-0.2W (K-5); and Ti-46Al-5Nb-1W (Alloy 7). Source: Ref 45

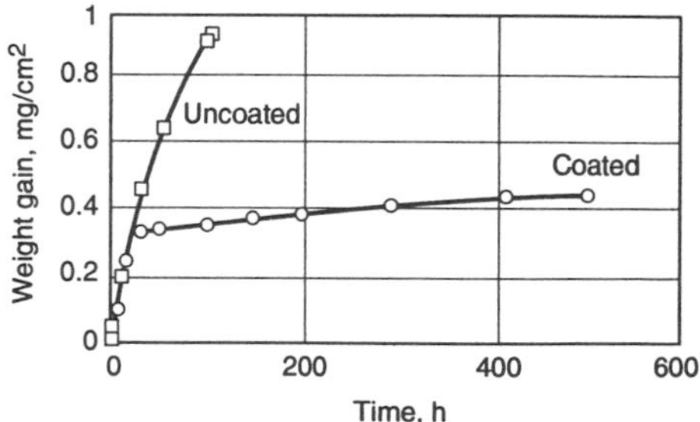

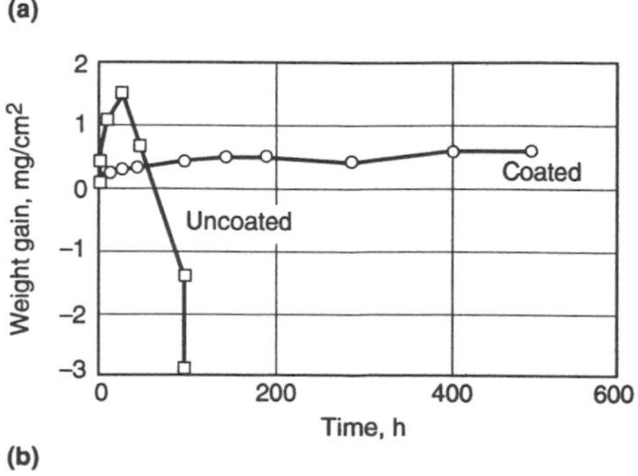

Fig. 15 The effect of a low-pressure plasma sprayed Ti-51Al-12Cr coating on the (a) 800 °C (1470 °F) and (b) 1000 °C (1830 °F) interrupted oxidation behavior of a Ti-48Al-2Cr-2Nb γ-TiAl alloy in air. Source: Ref 45

strate and fracture strength reductions of up to 50% (Ref 59, 60).

Degradation Due to Hot Corrosion and Particulate Ingestion

Most of the preceding discussions have evaluated alloy and coating performance on the basis of oxidation resistance and the appropriate mechanisms. However, there are other common environmental durability issues that in some cases supersede oxidation as the primary mode of attack of gas turbine components. As described below, these include hot corrosion of superalloys and airfoil degradation due to deposits resulting from ingested particles or sand.

Hot Corrosion

The subject of hot corrosion of superalloys is extremely complex and to a certain degree controversial. It entails the detailed chemical reactions involved with the effects of a molten Na_2SO_4 deposit on the oxidation process and the dissolution of oxide scales coupled with the formation of sulfides in the alloy. Na_2SO_4 forms from the reaction of ingested NaCl salt with the sulfur impurity in the fuel. Because deposition is not favored at high temperatures (>1000 °C, or 1830 °F), hot corrosion is generally limited to lower temperatures. Increased corrosion results from higher deposition rates (higher sodium and sulfur impurity levels, higher pressure, and lower temperature).

The propagation modes for hot corrosion are intimately related to the reactions between the molten deposits and the alloys. In particular, the deposits cause nonprotective reaction products to be formed. As will be briefly described below, the nonprotective reaction product is formed because of a "fluxing" action of the molten deposit. A more detailed review of the overall formalisms of hot corrosion is provided by Pettit and Giggins (Ref 14).

The thermodynamics of Na_2SO_4 dissociation define an SO_3 pressure and oxide ion (O^{2-}) activity product (equilibrium constant). This equilibrium is used to assess whether certain oxides will dissolve in the molten deposit. The term basic fluxing describes oxide dissolution when O^{2-} combines with the scale to form a soluble MO^{2-} radical. It is generally associated with temperatures >900 °C (1650 °F) and is so termed high-temperature hot corrosion (type I). Sulfide formation and alloy depletion ahead of the corrosion front is a typical characteristic.

The term acidic fluxing defines dissolution when a soluble metal M^{2+} ion and O^{2-} are produced from the scale. Acidic fluxing is generally associated with a temperature range of 650 to 800 °C (1200 to 1470 °F) and is thus termed low-temperature hot corrosion (type II). However, the presence of molybdenum, tungsten, or vanadium in the alloy can also induce these soluble M^{2+} ions to form, even at higher temperatures. Type II corrosion does not generally exhibit a depletion zone or sulfides ahead of the corrosion front.

For protection against hot corrosion, a high chromium content is the best alloying strategy under all conditions, unlike the recommendations outlined for simple oxidation. For basic fluxing, a sufficiently high aluminum content (10%) to ensure continuous Al_2O_3 formation can physically inhibit corrosion, but is not as chemically resistant as Cr_2O_3. Chromium also decreases the O^{2-} concentration and prevents NiO basic dissolution. Thus high chromium can protect both physically and chemically. For acidic fluxing, a high aluminum content is not effective in preventing corrosion. High molybdenum, tungsten, or vanadium contents exacerbate acidic corrosion, while chromium decreases the rate of acidic corrosion.

Hot Corrosion Data. A large database of alloy hot-corrosion resistance has been developed (Ref 61). Here 96 commercial and laboratory superalloys were exposed to a Mach 0.3 burner rig test, at 900 °C (1650 °F) for 300 h in 1-h cycles, using 0.5 ppm NaCl in the combustion air and jet A-1 fuel with 0.05 to 0.07 wt% S. The extent of corrosion was monitored by the percentage of metal cross section consumed, as determined by quantitative metallography.

The top 12 alloys exhibited less than 2.0% of consumed cross-sectional area. The only compositional distinction was that the chromium content varied from 12.6 to 17.9% and the aluminum content varied from 2.1 to 6.9%. Conversely, the 12 poorest-performing alloys contained 4.0 to 10.7% Cr and 4.9 to 6.5% Al. Thus, improved performance was definitely biased toward high-chromium alloys. High-aluminum/low-chromium alloys were not especially corrosion resistant. No distinction in the titanium, tungsten, molybdenum, tantalum, niobium, cobalt, or zirconium content was apparent in these two extreme groups.

Sulfur-Induced Hot Corrosion. Sulfidation is another corrosion mechanism that is of great concern in coal-fired power systems (boiler tubes, exhaust gas heat exchangers, etc.). Basically, nickel-, cobalt-, and iron-base alloys are not sulfidation resistant in high p_{S_2} or p_{SO_2} (low p_{O_2}) environments because diffusion is very rapid in the sulfide scales that form on these alloys. Nickel sulfide has a parabolic rate constant seven orders of magnitude higher than that of nickel oxide. These sulfides have a very defective (nonstoichiometric) lattice, allowing fast cation transport. Low-melting sulfides are common, which only intensifies the problem. The addition of chromium (>50 at.%) will decrease the nickel sulfidation rate by three orders of magnitude, but this rate is still a factor of 10^5 times that of Cr_2O_3 growth. A similar effect occurs with the addition of 20 wt% Al to Fe-20Cr, where the resulting improvement is vastly inferior to that produced in oxidation environments.

The corrosion products and mechanisms are much dependent on the partial pressures of the sulfur-containing gas, for example, H_2S/H_2, S_2/Ar, SO_2/O_2, as well as on temperature. Preoxidation has been found to delay the onset of sulfidation, especially with Al_2O_3 scales; however, the oxides of nickel, cobalt, iron, chromium, and aluminum have all been found to allow sulfur penetration. Thus, the design of materials for ultimate sulfidation resistance is lacking. Alloy design for immediate levels of sulfur-containing gases has been fruitful (Ref 40). More detailed information on sulfidation can be found in Ref 9 to 11.

Airfoil Degradation Caused by Particulate Deposits

In particularly dusty or sandy terrain (the Middle Eastern countries and the southwestern United States), substantial amounts of particulates can be ingested from airborne particles. These have a geological connection with the region and have been identified as silica sand, complex silicates, dolomite/calcium sulfates, and salts. The chemistries vary from pure SiO_2 sand to Ca-Al-Fe-Mg silicates, $(Ca,Mg)CO_3$, $CaSO_4$, and NaCl (Ref 62, 63). When passed through the combustor, these particles melt and are splat deposited on the surfaces of airfoils. The deposits have chemistries similar to the fines analyzed from sand samples. However, the carbon content is reduced because of gaseous CO_2 formation and the sulfur content is increased because of calcium and magnesium reactions with the sulfur impurity in the fuel which leads to sulfate deposits.

Although the calcium-magnesium-aluminum-silicate (CMAS) deposits are relatively refractory, they easily cover over cooling holes and decrease cooling effectiveness at the outer airfoil surfaces. The consequent temperature increase has an obvious structural debit, but it also increases the oxidation rate of the alloy. In extreme cases, burn-through was apparent in helicopter turbine airfoils. Furthermore, it poses the probability of accelerated release and diffusion of foreign elements from the deposit into the protective scales and alloy surfaces. Thus, sulfide formation or very low melting calcium-aluminum or magnesium-aluminum eutectics might result. For TBC-coated airfoils, these deposits have been found to infiltrate the porous zirconia and degrade resilience by filling beneficial microcracks and by increased high-temperature sintering (Ref 63). Any defense against this type of attack, however, is difficult to imagine from the standpoint of improved alloy/coating design. Filters for helicopter turbines and frequent inspections or washdowns would be more appropriate for known high-risk service routes or land-based turbine locations.

Limitations: Testing Techniques and Life Prediction

Isothermal thermogravimetric tests and cyclic furnace tests are relatively standard throughout the industry such that the measured k_p or weight change curves from different laboratories can be compared. Some ramifications occur because of variations in the length of the tests or the cycle duration. Burner rig testing becomes more problematic in that different rigs may operate at different velocities, may be pressurized, or may be operated at various levels of impurity dopants (salts) to the fuel. The most common Mach 0.3 atmospheric rig, with 1 h cycles is probably the closest standard oxidation rig test. However, to screen systems for corrosion resistance, individual laboratory preferences for salt level and cycle profile vary and depend on proprietary correlations with engine experience or models of atmospheric salt ingestion. Furnace corrosion tests are almost exclusively targeted toward basic understanding because of the high levels of salt deposits and unrealistically short incubation times. Thermal barrier coating performance is also difficult to faithfully simulate because the failure modes are stress related and the true stress state on a cooled airfoil is substantially different than that on a furnace coupon (no thermal gradient) or uncooled burner rig bar. Mechanical degradation of stressed high-temperature components (turbine airfoils) may also have an oxidation component, but the synergistic mechanisms are not as widely studied or recognized.

ACKNOWLEDGMENT

The information in this article is largely taken from J.L. Smialek, C.A. Barrett, and J.C. Schaeffer, "Design for Oxidation Resistance," *Materials Selection and Design,* Vol 20, *ASM Handbook,* ASM International, 1997

REFERENCES

1. J.L. Smialek and G.M. Meier, High-Temperature Oxidation, *Superalloys II,* C.T. Sims, N.S. Stoloff, and W.C. Hagel, Ed., John Wiley & Sons, 1987, p 293–323
2. C.E. Lowell, C.A. Barrett, R.W. Palmer, J.V. Auping, and H.B. Probst, COSP: A Computer Model of Cyclic Oxidation, *Oxid. Met.,* Vol 36 (No. 1/2), 1991, p 81–112
3. C.E. Lowell, J.L. Smialek, and C.A. Barrett, Cyclic Oxidation of Superalloys, *High Temperature Corrosion,* NACE-6, R.A. Rapp, Ed., National Association of Corrosion Engineers, 1983, p 219–261
4. C.A. Barrett and C.E. Lowell, Resistance of Ni-Cr-Al Alloys to Cyclic Oxidation at 1100 and 1200 °C, *Oxid. Met.* Vol 11 (No. 4), 1977, p 199–223
5. J.L. Smialek, Oxide Morphology and Spalling Model for NiAl, *Metall. Trans.,* Vol 9A, 1978, p 308
6. J.L. Smialek and R. Browning, Current Viewpoints on Oxide Adherence Mechanisms, *Electrochemical Society Symposium Proceedings on High Temperature Materials Chemistry III,* 1986, p 259–271
7. J.G. Smeggil, A.W. Funkenbusch, and N.S. Bornstein, *Metall. Trans.* A, Vol 17A, 1986, p 923–932
8. J.L. Smialek, D.T. Jayne, J.C. Schaeffer, and W.H. Murphy, Effects of Hydrogen Annealing, Sulfur Segregation and Diffusion on the Cyclic Oxidation Resistance of Superalloys: A Review, ICMC, *Thin Solid Films,* Vol 253, 1994, p 285–292
9. P. Kofstad, *High Temperature Corrosion,* Elsevier Applied Science, 1988
10. N. Birks and G.H. Meier, *Introduction of High Temperature Oxidation of Metals,* Arnold, 1983
11. G.Y. Lai, *High-Temperature Corrosion of Engineering Alloys,* ASM International, 1990
12. R. Wasielewski and R. Rapp, High Temperature Oxidation, *Superalloys,* C.T. Sims and W. Hagel, Ed., John Wiley & Sons, 1972
13. S.J. Grisaffe, Protective Coatings, *Superalloys,* C.T. Sims and W. Hagel, Ed., John Wiley & Sons, 1972
14. F.S. Pettit and C.S. Giggins, Hot Corrosion, *Superalloys II,* C.T. Sims, N.S. Stoloff, and W.C. Hagel, Ed., John Wiley & Sons, 1987, p 327–358
15. J.H. Wood and E. Goldman, Protective Coatings, *Superalloys II,* C.T. Sims, N.S. Stoloff, and W.C. Hagel, Ed., John Wiley & Sons, 1987, p 359–384
16. N. Birks, G.H. Meier, and F.S. Pettit, High Temperature Corrosion, *Superalloys, Supercomposites and Superceramics,* J.K. Tien and T. Caulfield, Ed., Academic Press, 1989, p 439–489
17. I.G. Wright, "Oxidation of Iron-, Nickel-, and Cobalt-Base Alloys," MCIC 72-07, Metals and Ceramics Information Center, June 1972
18. C.A. Barrett, "A Statistical Analysis of Elevated Temperature Gravimetric Cyclic Oxidation Data of Ni- and Co-Base Superalloys Based on an Oxidation Attack Parameter," TM-105934, NASA Lewis Research Center, Dec 1992
19. C.A. Barrett, "10,000-Hour Cyclic Oxidation Behavior at 982 °C (1800 °F) of 68 High-Temperature Co-, Fe-, and Ni-Base Alloys," TM107394, NASA Lewis Research Center, June 1987
20. C.A. Barrett, 10,000-Hour Cyclic Oxidation Behavior at 815 °C (1500 °F) of 33 High-Temperature Alloys, *Environmental Degradation of Engineering Materials,* College of Engineering, Virginia Tech, Blacksburg, VA, 10–12 Oct 1977, p 319–327
21. J.L. Smialek and B.K. Tubbs, Effect of Sulfur Removal on Scale Adhesion to PWA 1480, *Metall. Mater. Trans.,* Vol 26A, 1995, p 427–435
22. M.A. Smith, W.E. Frazier, and B.A. Pregger, Effect of Sulfur on the Cyclic Oxidation Behavior of a Single Crystalline Nickel-Based Superalloy, *Mater. Sci. Eng.,* Vol 203, p 388–398
23. M. Gobel, A. Rahmel, and M. Schutze, The Cyclic-Oxidation Behavior of Several Nickel-Base Single-Crystal Superalloys Without and With Coatings, *Oxid. Met.,* Vol 41 (No. 3/4), 1994, p 271–300
24. R.W. Broomfield, D.A. Ford, H.R. Bhangu, M.C. Thomas, D.J. Fraisier, P.S. Burkholder, K. Harris, G.L. Erikson, and J.B. Wahl, "Development and Turbine Engine Performance of Three Advanced Rhenium Containing Superalloys for Single Crystal and Directionally Solidified Blades and Vanes," Paper No. 97-GT-117, Presented at the ASME International Gas Turbine and Aeroengine Congress and Exhibition (Orlando, FL), June 1997
25. J.L. Smialek, "Oxidation Resistance and Critical Sulfur Content of Single Crystal Superalloys," Paper No. 96-GT-519, Presented at the ASME International Gas Turbine and Aeroengine Congress and Exhibition (Brussels, Belgium), June 1990
26. J.A. Nesbitt and E.J. Vinarcik, *Damage and Oxidation in High Temperature Composites,* G.K. Haritos and O.O. Ochoa, Ed., American Society of Mechanical Engineers, 1991, p 9–22
27. J.L. Smialek and C.E. Lowell, Effects of Diffusion on Aluminum Depletion and Degradation of NiAl Coatings, *J. Electrochem. Soc.,* Vol 121, 1974, p 80
28. C.A. Barrett, The Effect of 0.1 Atomic Percent Zirconium on the Cyclic Oxidation Behavior of β-NiAl for 3000 Hours at 1200 °C, *Oxidation of High-Temperature Intermetallics,* T. Grobstein and J. Doychak, Ed., The Minerals, Metals, & Materials Society, 1988, p 67–82
29. J. Doychak, J.L. Smialek, and C.A. Barrett, The Oxidation of Ni-Rich Ni-Al Intermetallics, *Oxidation of High-Temperature Intermetallics,* T. Grobstein and J. Doychak, Ed., The Minerals, Metals & Materials Society, 1988, p 41
30. B.A. Pint, The Oxidation Behavior of Oxide-Dispersed β-NiAl: I. Short-Term Performance at 1200 °C and II. Long-Term Performance at 1200 °C, Submitted to *Oxidation of Metals,* 1997
31. G.W. Goward and D.H. Boone, Mechanisms of Formation of Diffusion Aluminide Coatings on Nickel Base Superalloys, *Oxid. Met.,* Vol 3, 1971, p 475–495
32. M. Goebel, A. Rahmel, M. Schutze, M. Schorr, and W.T. Wu, Interdiffusion Between the Platinum-Modified Aluminide Coating RT22 and Nickel-Based Single-Crystal Superalloys at 1000 and 1200 °C, *Mater. High Temp.,* Vol 12 (No. 4), 1994
33. J.S. Smith and D.H. Boone, "Platinum Modified Aluminides—Present Status," Paper No. 90-GT-319, Presented at the ASME International Gas Turbine and Aeroengine Congress and Exhibition (Brussels, Belgium), June 1990
34. R.A. Miller and C.E. Lowell, Failure Mechanisms of Thermal Barrier Coatings Exposed to Elevated Temperatures, *Thin Solid Films,* Vol 67, 1984, p 517–521
35. R.L. Jones, Thermogravimetric Study of the 800 °C Reaction of Zirconia Stabilizing Oxides with SO_3-$NaVO_3$, *J. Electrochem. Soc.,* Vol 139 (No. 10), 1992, p 2794–2799
36. "First Thermal Barrier Coating Workshop," CP-3312, W.J. Brindley, Ed., NASA Lewis Research Center, Oct 1995
37. F.S. Pettit, Oxidation Mechanisms for Nickel-Aluminum Alloys at Temperatures Between 900 and 1300 °C, *Trans. TMS-AIME,* Vol 239, 1967, p 1296–1305
38. R. Darolia, NiAl Alloys for High-Temperature Structural Applications, *J. Met.,* Vol 43 (No. 3), 1991, p 44–49
39. R.D. Noebe, R.R. Bowman, and M.V. Nathal, "Review of the Physical and Mechanical Prop-

erties and Potential Applications of the B2 Compound NiAl," TM 105598, NASA Lewis Research Center, April 1992

40. J.H. DeVan and C.A. Hippsley, Oxidation of Ni_3Al Below 850 °C and Its Effect on Fracture Behavior, *Oxidation of High-Temperature Intermetallics,* T. Grobstein and J. Doychak, Ed., The Minerals, Metals, & Materials Society, 1988, p 31–40
41. J. H. DeVan, Oxidation Behavior of Fe_3Al and Derivative Alloys, *Oxidation of High-Temperature Intermetallics,* T. Grobstein and J. Doychak, Ed., The Minerals, Metals, & Materials Society, 1988, p 107–116
42. J.L. Corkum and W.W. Smeltzer, The Synergistic Effect of Aluminum and Silicon on the Oxidation Resistance of Iron Alloys, *Oxidation of High-Temperature Intermetallics,* T. Grobstein and J. Doychak, Ed., The Minerals, Metals, & Materials Society, 1988, p 97–106
43. A. Rahmel, W.J. Quadakkers, and M. Schutze, Fundamentals of TiAl Oxidation—A Critical Review, *Mater. Corros.,* Vol 46, 1995, p 271–285
44. G.H. Meier, D. Appalonia, R.A. Perkins, and K.T. Chiang, Oxidation of Ti-Base Alloys, *Oxidation of High-Temperature Intermetallics,* T. Grobstein and J. Doychak, Ed., The Minerals, Metals, & Materials Society, 1988, p 185–194
45. M.P. Brady, W.J. Brindley, J.L. Smialek, and I.E. Locci, The Oxidation and Protection of Gamma Titanium Aluminides, *JOM,* Vol 48 (No. 11), Nov 1996, p 46–50
46. *Oxidation of High-Temperature Intermetallics,* T. Grobstein and J. Doychak, Ed., The Minerals, Metals, and Materials Society, 1988
47. J. Doychak, Oxidation Behavior of High-Temperature Intermetallics, *Intermetallic Compounds,* J.H. Westbrook and R.L. Fleischer, Ed., John Wiley & Sons, 1994, p 977
48. G. Welsch, J.L. Smialek, J. Doychak, J. Waldman, and N.S. Jacobson, High Temperature Oxidation and Properties, *Oxidation and Corrosion of Intermetallic Alloys,* G. Welsh and P.D. Desai, Ed., Purdue Research Foundation, 1996, p 121–266
49. J.L. Smialek, J.A. Nesbitt, W.J. Brindley, M.P. Brady, J. Doychak, R.M. Dickerson, and D.R. Hull, Service Limitations for Oxidation Resistant Intermetallic Compounds, *Mater. Res. Soc. Symp.,* Vol 364, 1995, 1273–1284
50. C.M. Packer, Overview of Silicide Coatings for Refractory Metals, *Oxidation of High-Temperature Intermetallics,* T. Grobstein and J. Doychak, Ed., The Minerals, Metals, & Materials Society, 1988, p 235–244
51. J.L Smialek, M.D. Cuy, and D. Petrarca, unpublished research, NASA Lewis Research Center, 1992
52. J. Doychak, R.R. Dickerson, D. Hull, and M. Maloney, unpublished research, NASA Lewis Research Center, 1993
53. D. Berztiss, R.R. Cerchiara, E.A. Gulbransen, F.S. Pettit, and G.H. Meier, *Mater. Sci. Eng.,* Vol A155, 1992, p 164–181
54. M.G. Hebsur, Pest Resistant and Low CTE $MoSi_2$-Matrix for High Temperature Structural Applications, *Mater. Res. Soc. Symp. Proc.,* Vol 350, 1994, p 177–182
55. A.K. Vasudevan and J.J. Petrovic, High Temperature Structural Silicides, Proceedings of the First High Temperature Structural Silicides Workshop, *Mater. Sci. Eng.,* Vol 155, 1992
56. N.S. Jacobson, Corrosion of Silicon-Based Ceramics in Combustion Environments, *J. Am. Ceram. Soc.,* Vol 76 (No. 1) 1993, p 3–28
57. N.S. Jacobson, J.L. Smialek, and D.S. Fox, Molten Salt Corrosion of Ceramics, *Corrosion of Advanced Ceramics,* K.G. Nickel, Ed., Kluwer Academic Publishers, 1996, p 205–222
58. G.H. Schiroky, Oxidation Behavior of Chemically Vapor-Deposited Silicon Carbide, *Adv. Ceram. Mater.,* Vol 2 (No. 2), 1987, p 137–141
59. J.L. Smialek and N.S. Jacobson, Mechanisms of Strength Degradation for Hot Corrosion of α-SiC, *J. Am. Ceram. Soc.,* Vol 69, 1986, p 741–752
60. D.S. Fox and J.L. Smialek, Burner Rig Hot Corrosion of Silicon Carbide and Silicon Nitride, *J. Am. Ceram. Soc.,* Vol 73, 1990, p 303–311
61. C.A. Stearns, D.L. Deadmore, an C.A. Barrett, Effect of Alloy Composition on the Sodium-Sulfate Induced Hot Corrosion Attack of Cast Nickel-Base Superalloys at 900 °C, *Alternate Alloying for Environmental Resistance,* The Metallurgical Society, 1987, p 131–144
62. J.L. Smialek, F.A. Archer, and R.G. Garlick, Turbine Airfoil Degradation in the Persion Gulf War, *JOM,* Vol 46 (No. 12), Dec 1994, p 39–41
63. M.P. Borom, C.A. Johnson, and L.A. Peluso, "A Role of Environmental Deposits in Spallation of Thermal Barrier Coatings on Aeroengine and Land-Based Gas Turbine Hardware." Paper No. 96-GT-285, Presented at the ASME International Gas Turbine and Aeroengine Congress & Exhibition (Birmingham, UK), June 1996

Index

A

B

F

I

J

K

L

M

N

Q

R

S